FOR STUDENTS

► Adam Online Anatomy (AOA)
A 12-month, pre-paid subscription to this popular web-based product is included in your purchase of a new textbook. ***AOA***, an exciting new Internet tool allows you to explore the human body from skin to bone. ***AOA*** lets you click and dissect through the layers of the human body with over 20,000 detailed anatomical structures in fully detailed male and female bodies, including anterior, lateral, medial, and posterior views. You can activate and link to your subscription through the dedicated companion website for this text. Purchasing a new subscription to this product is available through *APCentral* at www.wiley.com/college/APCentral.

► A Photographic Atlas of the Human Body with Selected Cat, Sheep, and Cow Dissections (0-471-37487-3)
Packaged with every new copy of Tortora's ***Principles of Human Anatomy 9e,*** this atlas is designed to support your study and your laboratory experience. Organized by body system, the clearly labeled photographs provide a stunning visual reference to gross anatomy. Histological micrographs are also included. The atlas can be purchased separately from the textbook.

► Take Note! (0-471-41328-3)
This handy supplement allows you to organize your note taking and improve your understanding of anatomical structures. Following the sequence in the textbook, each left-handed page displays an unlabeled, black-and-white copy of every text figure. Students can fill in the labels during lecture or lab at the instructor's directions and take additional notes on the blank right-handed pages.

► Anatomy and Physiology: A Companion Coloring Book (0-471-39515-3)
This unique study tool is a breakthrough approach to learning and remembering the human body's anatomical structures and physiological processes. It features over 500 striking, original, illustrations that give students a clear and enduring understanding of anatomy and physiology.

FOR THE LABORATORY

► Laboratory Manual for Anatomy by William Radke (0-471-41413-1)
New to this edition is the perfect companion laboratory manual (publication in October of 2001). The manual is written in a clear and concise manner and contains many interactive activities including coloring and labeling of anatomical figures, critical thinking questions, and clinically oriented questions that promote problem solving. Lab exercises are designed to enhance students' ability to visualize anatomical structures and to encourage students to first apply information and to then critically evaluate it. Cat dissection exercises are included.

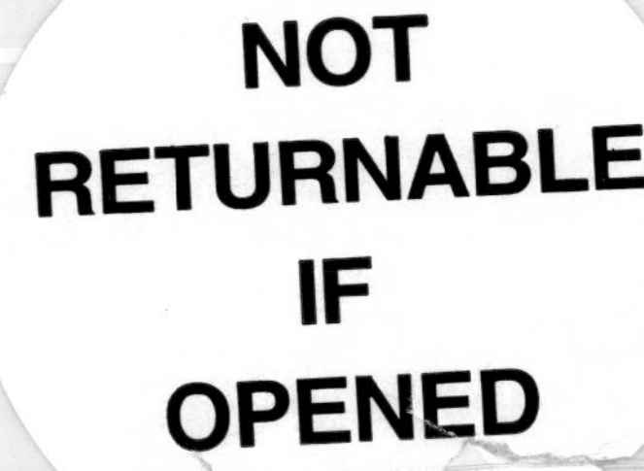
NOT
RETURNABLE
IF
OPENED

ABOUT THE AUTHOR

Jerry Tortora is Professor of Biology and former Biology Coordinator at Bergen Community College in Paramus, New Jersey, where he teaches human anatomy and physiology as well as microbiology. He received his bachelor's degree in biology from Fairleigh Dickinson University and his master's degree in science education from Montclair State College. He is a member of many professional organizations, such as the Human Anatomy and Physiology Society (HAPS), the American Society of Microbiology (ASM), American Association for the Advancement of Science (AAAS), National Education Association (NEA), and the Metropolitan Association of College and University Biologist (MACUB).

Above all, Jerry is devoted to his students and their aspirations. In recognition of this commitment, Jerry was the recipient of MACUB's 1992 President's Memorial Award. In 1996, he received a National Institute for Staff and Organizational Development (NISOD) excellent award from the University of Texas and was selected to represent Bergen Community College in a campaign to increase awareness of the contributions of community colleges to higher education.

Jerry is the author of several best-selling science textbooks and laboratory manuals, a calling that often requires an additional 40 hours per week beyond his teaching responsibilities. Nevertheless, he still makes time for four or five weekly aerobic workouts that include biking and running. He also enjoys attending college basketball and professional hockey games and performances at the Metropolitan Opera House.

To my children,

Lynne, Gerard, Kenneth, Anthony, and Andrew,

who make it all worthwhile.

G.J.T.

CONTRIBUTOR

Kevin Petti is a Professor teaching in the Department of Natural Sciences and is Chair of the Department of Health and Exercise Science at San Diego Miramar College. He teaches courses in human anatomy and physiology, human dissection, kinesiology, and health science. Kevin has also taught at the University of California San Diego, and the United States International University. Kevin is active in the Human Anatomy and Physiology Society (HAPS), serving on the Board of Directors and as HAPS membership chair. Having a keen interest in the history of science, Kevin has presented Man's Changing Image: The Historical Influence of Culture on Anatomy at several recent annual HAPS conferences.

Kevin has also authored an anatomy and physiology student study guide, and co-authored several peer reviewed original research articles relating to health, human behavior, and sport biomechanics. Kevin holds a bachelor's degree from Humboldt State University and a master's degree from San Diego State University. Presently, he is completing his doctorate at the University of San Diego.

Kevin is also devoted to staying physically active. He has completed more than ten triathlons and has technically rock climbed for years. Ascending the 2000' Northwest Face of Yosemite's Half Dome is his proudest achievement. Kevin is married and has two children.

PREFACE

P*rinciples of Human Anatomy*, ninth edition, is designed for introductory courses in human anatomy. The highly successful approach of previous editions–to provide students with an accurate, clearly written, and expertly illustrated presentation of the structure of the human body, to offer insights into the connections between structure and function, and to explore the practical and relevant applications of anatomical knowledge to everyday life and career development—has been retained. This new edition enhances and improves upon these strengths, while at the same time offering some new and innovative features to increase student motivation and success.

CHANGING IMAGES

From the ancient cultures of the East to the modern day culture of the Internet, the science and study of human anatomy has always been portrayed through artistic images that reflect the attitudes and knowledge of their time. These works of art often tell a dramatic story. While it may seem that the anatomy of man has changed little over the previous several centuries–a time frame that is miniscule in evolutionary terms–just how man views his structure is a much more dynamic matter. Predicated upon such issues as religion, culture, and historical era, representations of the human form from around the globe are quite varied. Indeed, comparing anatomical illustrations from 15th century Islam to 21st century computer generated images, is a lesson in the diversity of man's perception of himself.

In this ninth edition of *Principles of Human Anatomy*, these **Changing Images** of man and woman are celebrated and woven into the fabric of the text. Each chapter opens with a unique image and a thought-provoking question that is relevant to the chapter content. Later in each chapter, an expanded essay accompanies a larger version of the selected image. These essays, while tied to the chapter, are intended to also go a step further. Students will gain insight into the historical saga of anatomical studies, view images that reflect the evolution of the scientific method, and understand that the current body of anatomical knowledge is by contributions from the whole world. With visually beautiful images from Africa, Malaysia, Europe, and Japan, students are likely to "see themselves" and perhaps be more excited about their studies.

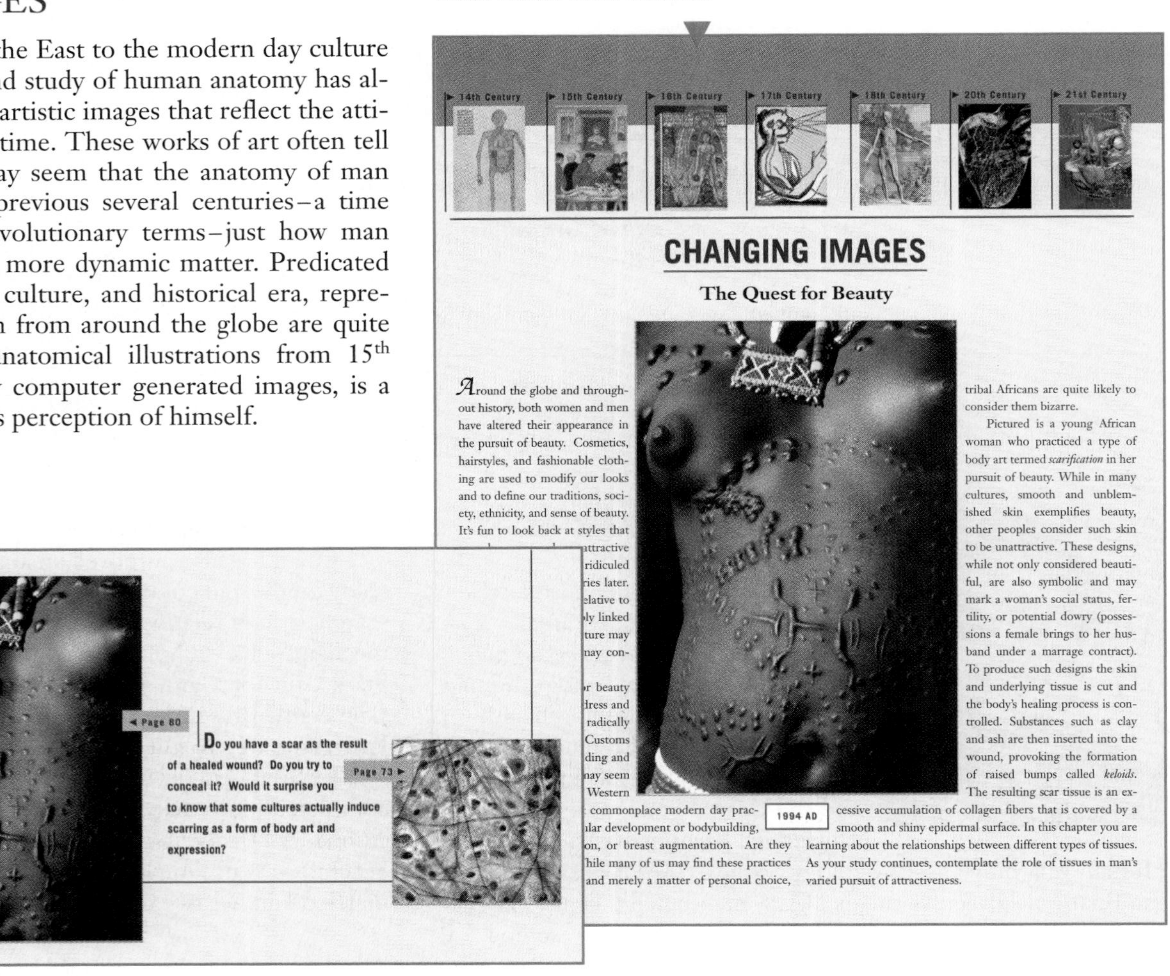
14th Century
15th Century
16th Century
17th Century
18th Century
20th Century
21st Century
CHANGING IMAGES
The Quest for Beauty
Around the globe and throughout history, both women and men have altered their appearance in the pursuit of beauty. Cosmetics, hairstyles, and fashionable clothing are used to modify our looks and to define our traditions, society, ethnicity, and sense of beauty. It's fun to look back at styles that
tribal Africans are quite likely to consider them bizarre.
Pictured is a young African woman who practiced a type of body art termed *scarification* in her pursuit of beauty. While in many cultures, smooth and unblemished skin exemplifies beauty, other peoples consider such skin to be unattractive. These designs, while not only considered beautiful, are also symbolic and may mark a woman's social status, fertility, or potential dowry (possessions a female brings to her husband under a marrage contract). To produce such designs the skin and underlying tissue is cut and the body's healing process is controlled. Substances such as clay and ash are then inserted into the wound, provoking the formation of raised bumps called *keloids*. The resulting scar tissue is an excessive accumulation of collagen fibers that is covered by a smooth and shiny epidermal surface. In this chapter you are learning about the relationships between different types of tissues. As your study continues, contemplate the role of tissues in man's varied pursuit of attractiveness.
1994 AD
◄ Page 80
Do you have a scar as the result of a healed wound? Do you try to conceal it? Would it surprise you to know that some cultures actually induce scarring as a form of body art and expression?
Page 73 ►

ORGANIZATION, SPECIAL TOPICS, AND CONTENT IMPROVEMENTS

Like most undergraduate anatomy courses, *Principles of Human Anatomy* follows a systems approach to the topic. The first two chapters introduce the nomenclature and conventions used to study human anatomy and review the fundamentals of cell biology, respectively. Following are chapters or groups of chapters covering the systems of the body. The text concludes with a chapter on developmental anatomy. Throughout the text, woven into the chapter content, special topics are considered.

Developmental Anatomy

Illustrated discussions of developmental anatomy are found near the conclusion of most body system chapters. Placing this coverage at the end of chapters enables students to master the concepts and anatomical terminology they need in order to learn about embryonic and fetal structures. A "fetus" icon highlights each of these sections.

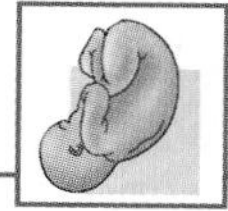

DEVELOPMENTAL ANATOMY OF THE SKELETAL SYSTEM

Objective

- Describe the development of the skeletal system and the limbs.

As noted earlier, both intramembranous and endochondral ossification begin when *mesenchymal cells,* connective tissue cells derived from **mesoderm,** migrate into the area where bone formation will occur. In some skeletal structures, mesenchymal cells develop into chondroblasts that form cartilage. In other skeletal structures, mesenchymal cells develop into osteoblasts that form *bone tissue* by intramembranous or endochondral ossification.

Aging

Anatomy is not static. As the body ages, its structure and related functions subtly change. Moreover, aging is a professionally relevant topic for the majority of this book's readers, who will go on to careers in health-related fields in which the median age of the client population is steadily advancing. For these reasons, many of the body system chapters explore age-related changes in anatomy and function.

AGING AND JOINTS

Objective

- Explain the effects of aging on joints.

The aging process usually results in decreased production of synovial fluid in joints. In addition, the articular cartilage becomes thinner with age, and ligaments shorten and lose some of their flexibility. The effects of aging on joints, which vary considerably from one person to another, are affected by genetic factors and by wear and tear. Although degenerative changes in joints may begin in individuals as young as 20 years of age, most do not occur until much later. In fact, by age 80, almost everyone develops some type of degeneration in the knees, elbows,

Exercise

Physical exercise can produce favorable changes in some anatomical structures, most notably those associated with the musculoskeletal and cardiovascular systems. This information has special relevance to readers embarking on careers in physical education, sports training, and dance. Key chapters include brief discussions of exercise, which are signaled by a distinctive "running shoe" icon.

EXERCISE AND THE HEART

Objective

- Explain the relationship between exercise and the heart.

No matter what a person's level of fitness, it can be improved at any age with regular exercise. Of the various types of exercise, some are more effective than others for improving the health of the cardiovascular system. **Aerobics,** any activity that works large body muscles for at least 20 minutes, elevates cardiac output and accelerates metabolic rate. Three to five such sessions a week are usually recommended for improving the health of the

Every chapter in the ninth edition of *Principles of Human Anatomy* incorporates a host of improvements to both the text and the art suggested by reviewers, educators, or students. Here are just some of the more noteworthy changes:

- **Chapter 2/Cells** This foundational chapter has been reorganized to present cellular structures in a more logical fashion. It has also been updated from front to back to reflect the latest thinking in molecular cell biology. Virtually every figure has been redrawn in a new, vibrant, three-dimensional style.
- **Chapter 3/Tissues** The much-imitated tissue tables in this chapter were redesigned to improve student comprehension of histology. Most photomicrographs are new and there is a new section on aging and tissues. There are new illustrations of the structural and functional types of multicellular exocrine glands.
- **Chapter 4/Integumentary System** New sections include coverage dealing with a comparison of eccrine and apocrine sweat glands, thin and thick skin, and aging and the integumentary system. Several new pieces of art enhance the pedagogy of this chapter.
- **Chapter 8/Joints** Reorganized, this chapter offers improved coverage and is further accentuated by clear and consistently stylized new range-of-motion photographs.
- **Chapter 10/The Muscular System** A new Exhibit and related art on the muscles of the larynx has been added.
- **Chapter 9/Muscle Tissue** Most of the art in this chapter has been redrawn.
- **Chapter 13/The Cardiovascular System: The Heart** Many new illustrations enhance the revised discussions of this vital organ. In particular, the details of cardiac muscle tissue have been carefully refined.
- **Chapter 14/The Cardiovascular System: Blood Vessels** This chapter's Exhibits have been revised to provide an even more consistent level of detail of arterial and venous circulation. Meticulously redrawn illustrations more clearly reveal the relevant blood vessel anatomy, and extended flow diagrams visually reinforce learning of relevant anatomy and blood vessels pathways.
- **Chapter 17/The Spinal Cord And Spinal Nerves** New illustrations for all Exhibits that deal with plexuses are included.
- **Chapter 20/Special Senses** Virtually all of the illustrations have been redrawn and most off the histology photos have been replaced.
- **Chapter 21/The Autonomic Nervous System** now has expanded coverage of ANS neurotransmitters and receptors and a new accompanying illustration.
- **Chapter 23/The Respiratory System** A new section on the relationship between exercise and the respiratory system has been added.
- **Chapter 24/The Digestive System** Includes many newly drawn illustrations.
- **Chapter 25/The Urinary System** Contains an expanded discussion of renal functions and many of the illustrations have been redrawn.
- **Chapter 26/The Reproductive System** Includes a new section on birth control methods and an accompanying Table has been added. In addition, there is a new illustration of meiosis.
- **Chapter 27/Developmental Anatomy** A new section on adjustments of the infant at birth related to the respiratory and cardiovascular system has been added.

ENHANCEMENTS TO THE ILLUSTRATION PROGRAM

New Design

A textbook with beautiful illustrations or photographs on most pages requires a carefully crafted and functional design. The new design for the ninth edition has been transformed to assist students in making the most of the text's many features and outstanding art. The larger trim size provides even more space for the already large and highly praised illustrations. Each page is carefully laid out to place related text, figures, and tables near one another, minimizing page-turning during the reading of a topic. New to this edition is the red print used to indicate the first mention of a figure or table. Not only is the reader alerted to refer to the figure or table, but the color print also serves as a place locator for easy return to the narrative.

Also new to the design of this edition are the distinctive icons incorporated throughout to signal special features and make them easy to find during review. These include the **key** with the Key Concept Statements; the **question mark** with the applicable questions that enhance every figure; the **stethoscope** indicating a clinical application within a chapter narrative; the fetus icon announcing the developmental anatomy section of a chapter; the **running shoe** highlighting content relevant to exercise, and the icons that indicate the distinctive types of **chapter-ending questions.**

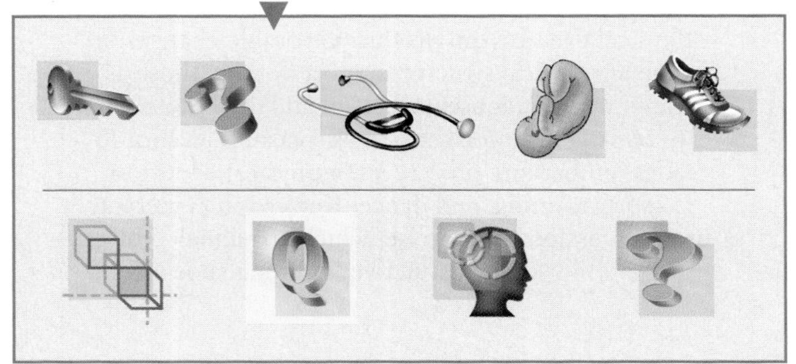

New Art

An outstanding illustration program has always been a signature feature of this text. Beautiful artwork, carefully chosen photographs and photomicrographs, and unique pedagogical enhancements all combine to make the visual appeal and usefulness of the illustration program in *Principles of Human Anatomy* distinctive. Continuing in this tradition, you will find exciting new three-dimensional illustrations gracing the pages of Chapter 2 (Cells), Chapter 4 (The Integumentary System), Chapter 5 (Bone Tissue), Chapter 9 (Muscle Tissue), Chapter 13 (The Cardiovascular System: The Heart), Chapter 14 (The Cardiovascular System: Blood Vessels), Chapter 20 (Special Senses), Chapter 24 (The Digestive System), and Chapter 25 (The Urinary System). Nearly every figure has been revised or improved in some way.

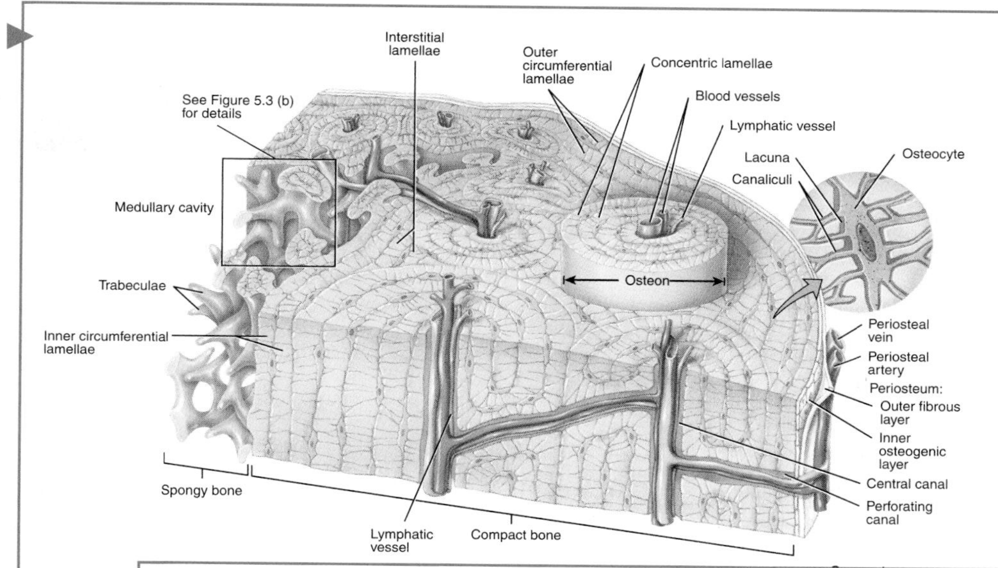

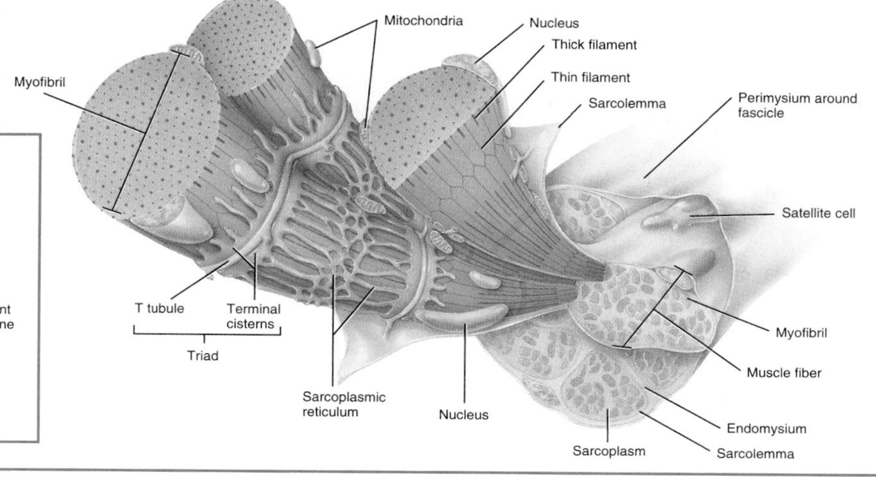

(b) Fenestrated capillary

New Histology-Based Art

As part of our plan of continuous improvement, many of the anatomical illustrations based on histological preparations have been replaced in this edition. See, for example, the beautiful rendered illustration of muscle tissue on page 232, the histology of the small intestine on page 741 or the new illustration of the interior of the eye on page 617.

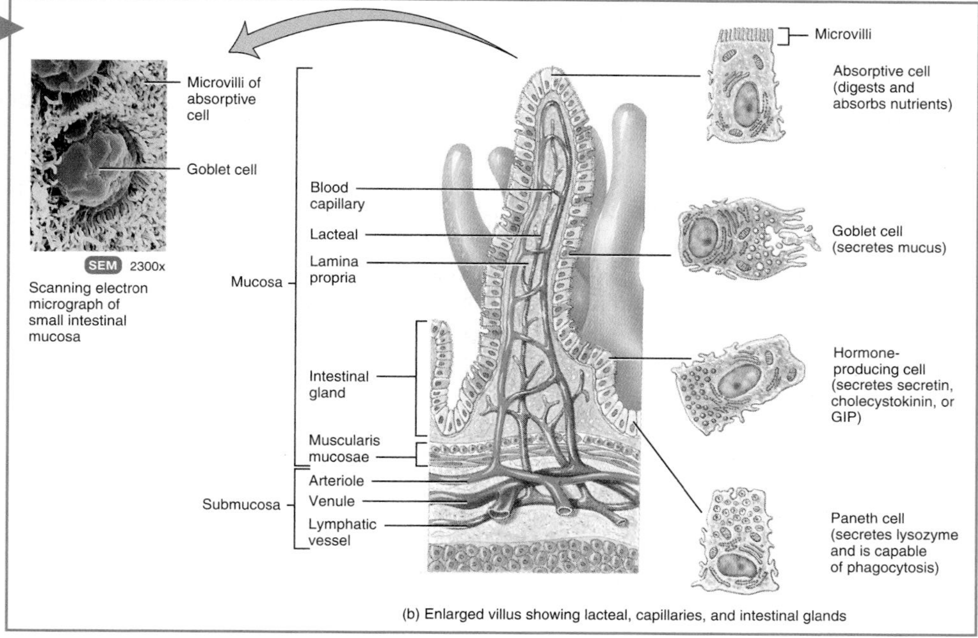

(b) Enlarged villus showing lacteal, capillaries, and intestinal glands

New Photomicrographs

Dr. Michael Ross of the University of Florida provided me with beautiful, customized photomicrographs of various tissues of the body. I have always considered Dr. Ross' photos among the best available and their inclusion in this edition greatly enhances the illustration program.

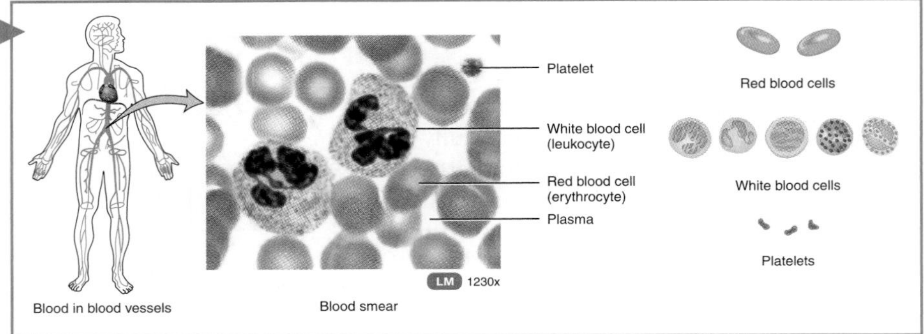

Cadaver Photos

An assortment of large, clear cadaver photos are included at strategic points in many chapters. What is more, many anatomy illustrations are keyed to the large cadaver photos included in my companion text, *A Photographic Atlas of the Human Body*. A copy of the *Atlas* is included with the purchase of every new text, or can be obtained separately from John Wiley & Sons, Inc.

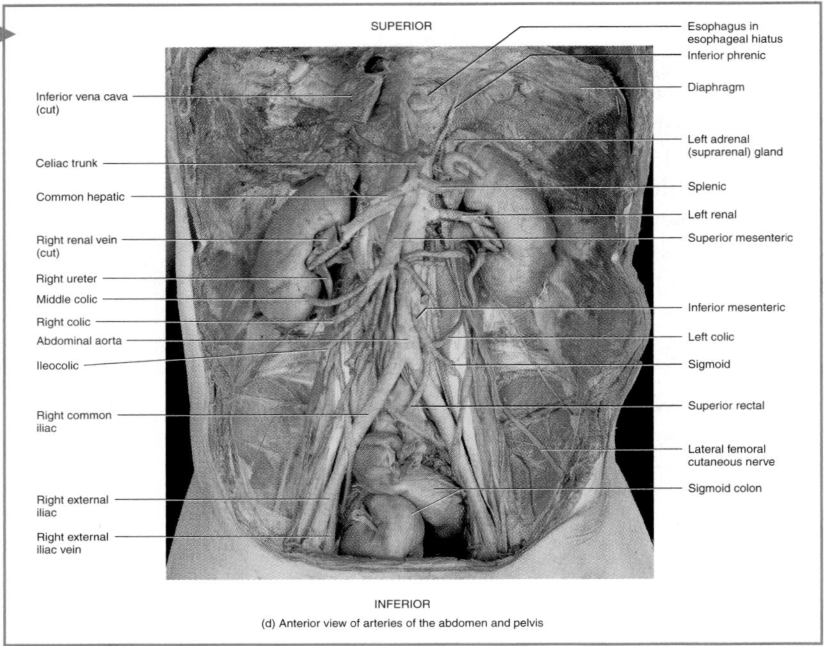

(d) Anterior view of arteries of the abdomen and pelvis

Orientation Diagrams

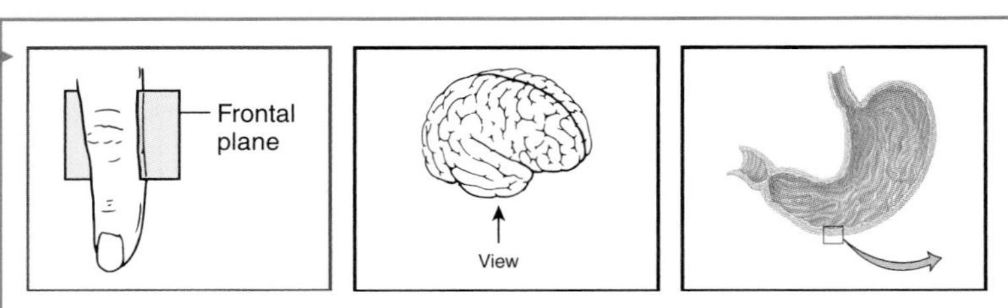

Students sometimes need help figuring out of the plane of view of anatomy illustrations—descriptions alone do not always suffice. An orientation diagram depicting and explaining the perspective of the view represented in the figure accompanies every major anatomy illustration. There are three types of diagrams: (1) planes used to indicate where certain sections are made when a part of the body is cut; (2) diagrams containing a directional arrow and the word "View" to indicate the direction from which the body part is viewed, e.g. superior, inferior, posterior, anterior; and (3) diagrams with arrows leading from or to them to direct attention to enlarged and detailed parts of illustrations.

Key Concept Statements

Included above every figure and denoted by the "key" icon, this feature summarizes an idea that is discussed in the text and demonstrated in the figure. A unique pedagogical feature to our text, these statements help students keep focused on the relevance of the figure to their understanding of specific content.

Revised Figure Questions

This highly applauded feature asks readers to synthesize verbal and visual information, think critically, or draw conclusions about what they see in a figure. Each Figure Question appears under its illustration and is highlighted in this edition by the distinctive "Question Mark" icon. Answers are located at the back of each chapter.

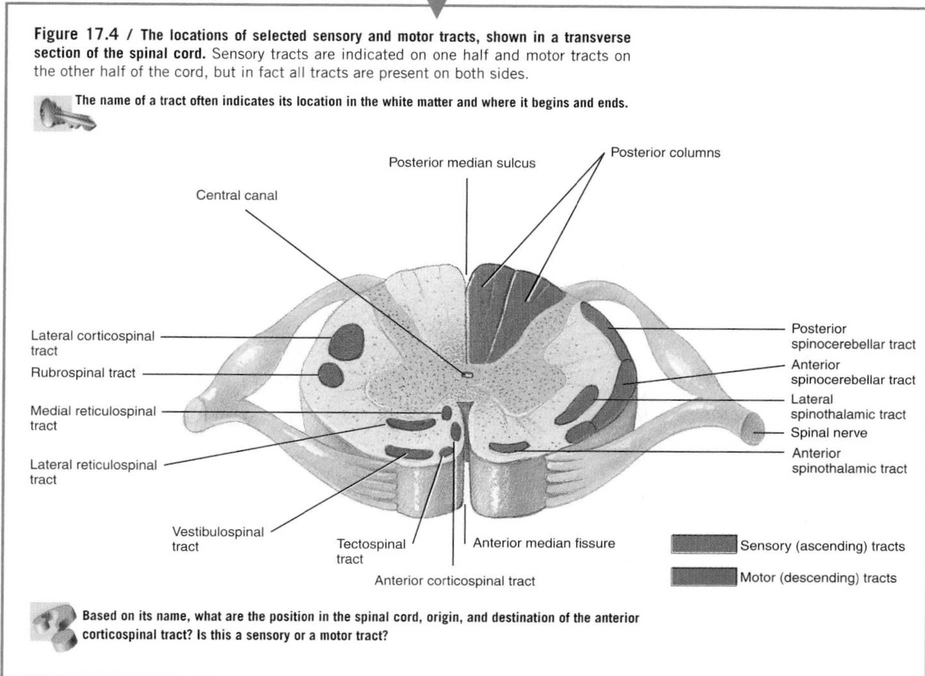

Figure 17.4 / The locations of selected sensory and motor tracts, shown in a transverse section of the spinal cord. Sensory tracts are indicated on one half and motor tracts on the other half of the cord, but in fact all tracts are present on both sides.

The name of a tract often indicates its location in the white matter and where it begins and ends.

Based on its name, what are the position in the spinal cord, origin, and destination of the anterior corticospinal tract? Is this a sensory or a motor tract?

Overviews of Functional Anatomy

This art-related feature succinctly lists the functions of the anatomical structure or system depicted in the adjacent figure. This juxtaposition of text and art further reinforces the connection between structure and function.

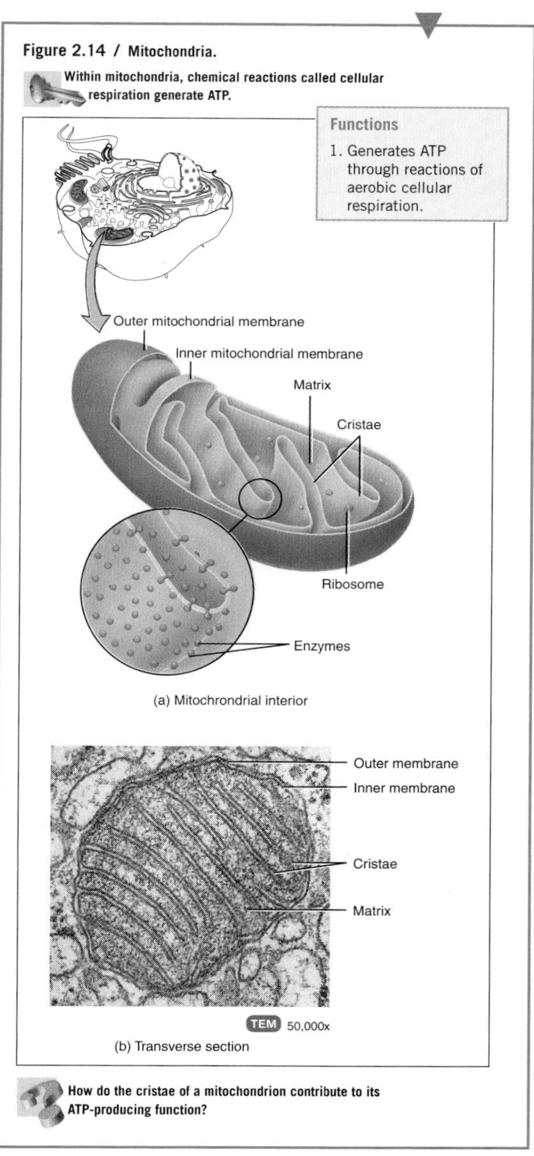

Figure 2.14 / Mitochondria.

Within mitochondria, chemical reactions called cellular respiration generate ATP.

How do the cristae of a mitochondrion contribute to its ATP-producing function?

ENHANCED CLINICAL AND PRACTICAL RELEVANCE

Principles of anatomy and functional anatomy sometimes are better understood by considering variations from normal. Moreover, most students are naturally curious about how the information in this book or in their anatomy course pertains to a potential career choice, or how it relates to them or family members. Most chapters in the text offer several opportunities for students to make connections between the normal and abnormal.

Clinical Applications

A perennial favorite among students, a variety of clinical perspectives—many proposed by students—are introduced within each chapter, directly following the discussions to which they relate. The distinctive stethoscope icon highlights these. Some applications are new to this edition, and all have been updated for this edition.

Torn Cartilage and Arthroscopy

The tearing of articular discs (menisci) in the knee, commonly called **torn cartilage,** occurs often among athletes. Such damaged cartilage will begin to wear and may precipitate arthritis unless surgically removed (meniscectomy). Surgical repair of the torn cartilage may be assisted by **arthroscopy** (ar-THROS-kō-pē; *-scopy* = observation). This minimally invasive procedure involves examination of the interior of a joint, usually the knee, with an arthroscope, a lighted pencil-thin instrument used for visualization. Arthroscopy is used to determine the nature and extent of damage following knee injury; to help remove torn cartilage and repair damaged cruciate ligaments in the knee; to help obtain tissue samples for analysis; to monitor the progression of disease and the effects of therapy; and to help perform surgery on other joints, such as the shoulder, elbow, ankle, and wrist.

Applications to Health

In the prior edition of this text, we packaged the Applications to Health discussions in a separate booklet with the text. Due to the feedback from many professors and students, we have updated, simplified, and returned this popular feature to the end of relevant chapters. These are concise discussions of major diseases, disorders, and medical conditions that illustrate departures from normal anatomy and function. They provide answers to many questions that students ask about medical disorders and diseases.

Key Medical Terms

Also found at the end of most chapters, vocabulary-building glossaries of selected medical terms and conditions have been fully updated for the ninth edition.

APPLICATIONS TO HEALTH

Osteoporosis

Osteoporosis (os′-tē-ō-pō-RŌ-sis; *por-* = passageway; *-osis* = condition) is literally a condition of porous bones (Figure 5.12). The basic problem is that bone resorption outpaces bone deposition. In large part this is due to depletion of calcium from the body—more calcium is lost in urine, feces, and sweat than is absorbed from the diet. Bone deposition also depends on certain hormones, such as estrogens. In response to low levels of calcium and estrogens, bone mass becomes so depleted that bones fracture, often spontaneously, under the mechanical stresses of everyday living. For example, a hip fracture might result from simply sitting down too quickly. In the United States, osteoporosis causes more than a million fractures a year, mainly in the hip, wrist, and vertebrae. Osteoporosis afflicts the entire skeletal system. In addition to fractures, osteoporosis causes shrinkage of vertebrae, height loss, hunched backs, and bone pain.

Thirty million people in the United States suffer from osteoporosis. The disorder primarily affects middle-aged and elderly people, 80% of them women. Older women suffer from osteoporosis more often than men for two reasons: Women's bones are less massive than men's bones, and production of estrogens in women declines dramatically at menopause, whereas production of the main androgen, testosterone, in older men wanes gradually and only slightly. Estrogen and testosterone increase osteoblast activity and synthesis of bone matrix. Besides gender, risk factors for developing osteoporosis include a family history of the disease, European or Asian ancestry, thin or small body build, an inactive lifestyle, cigarette smoking, a diet low in calcium and vitamin D, more than two alcoholic drinks a day, and the use of certain medications.

In postmenopausal women, treatment of osteoporosis may include estrogen replacement therapy (ERT; low doses of estrogens) or hormone replacement therapy (HRT; a combination of estrogens and progesterone, another sex steroid). Although such treatments help combat osteoporosis, they increase a woman's risk of breast cancer. The drug Raloxifene (Evista) mimics the beneficial effects of estrogens on bone without increasing the risk of breast cancer. A nonhormone drug called Alendronate (Fosamax) blocks resorption of bone by osteoclasts and is also used to treat osteoporosis.

Figure 5.12 / Comparison of spongy bone tissue from (a) a normal young adult and (b) a person with osteoporosis. Notice the weakened trabeculae in (b). Compact bone tissue is similarly affected by osteoporosis.

In osteoporosis, bone resorption outpaces bone formation, so bone mass decreases.

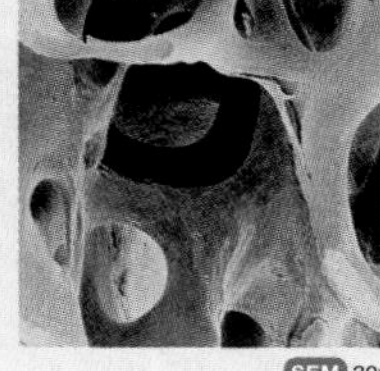

(a) Normal bone

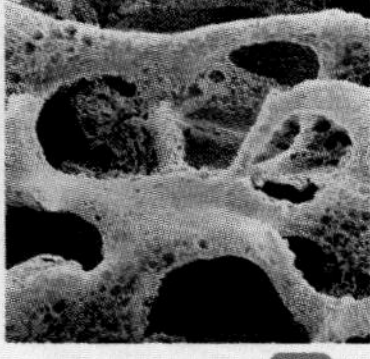

(b) Osteoporotic bone

If you wanted to develop a drug to lessen the effects of osteoporosis, would you look for a chemical that inhibits the activity of osteoblasts, or that of osteoclasts?

Perhaps more important than treatment is prevention. Adequate calcium intake and weight-bearing exercise in her early years may be more beneficial to a woman than drugs and calcium supplements when she is older.

Rickets and Osteomalacia

Rickets and **osteomalacia** (os′-tē-ō-ma-LĀ-she-a; *-malacia* = softness) are disorders in which bones fail to calcify. Although the organic matrix is still produced, calcium salts are not deposited, and the bones become "soft" or rubbery and easily deformed. Rickets affects the growing bones of children. Because new bone formed at the epiphyseal plates fails to ossify, bowed legs and deformities of the skull, rib cage, and pelvis are common results. In osteomalacia, sometimes called "adult rickets," new bone formed during remodeling fails to calcify. This disorder causes varying degrees of pain and tenderness in bones, especially in the hip and leg. Bone fractures also result from minor trauma.

KEY MEDICAL TERMS ASSOCIATED WITH BONE TISSUE

Osteoarthritis (os′-tē-ō-ar-THRĪ-tis; *arthr* = joint) The degeneration of articular cartilage such that the bony ends touch; the resulting friction of bone against bone worsens the condition. Usually associated with the elderly.

Osteogenic sarcoma (os′-tē-Ō-JEN-ik sar-KŌ-ma; *sarcoma* = connective tissue tumor) Bone cancer that primarily affects osteoblasts and occurs most often in teenagers during their growth spurt; the most common sites are the metaphyses of the thighbone (femur), shinbone (tibia), and arm bone (humerus). Metastases occur most often in lungs; treatment consists of multidrug chemotherapy and removal of the malignant growth, or amputation of the limb.

Osteomyelitis (os′-tē-ō-mī-el-Ī-tis) An infection of bone characterized by high fever, sweating, chills, pain, and nausea, pus formation, edema, and warmth over the affected bone and rigid overlying muscles. It is often caused by bacteria, usually *Staphylococcus aureus*. The bacteria may reach the bone from outside the body (through open fractures, penetrating wounds, or orthopedic surgical procedures); from other sites of infection in the body (abscessed teeth, burn infections, urinary tract infections, or upper respiratory infections) via the blood; and from adjacent soft tissue infections (as occurs in diabetes mellitus).

Osteopenia (os′-tē-ō-PĒ-nē-a; *penia* = poverty) Reduced bone mass due to a decrease in the rate of bone synthesis to a level insufficient to compensate for normal bone resorption; any decrease in bone mass below normal. An example is osteoporosis.

HALLMARK FEATURES

The ninth edition of *Principles of Human Anatomy* builds on the legacy of thoughtfully designed and class-tested pedagogical features that provide a complete learning system for students as they navigate their way through the text and course. Many—such as critical thinking questions and end of chapter quizzes—have been revised to reflect the enhancements to the text and art.

Contents at a Glance

A brief outline of the content to come opens each chapter of the text. For the ninth edition, this is paired with a preview of the Changing Image essay for the chapter as well as a question related to it and the chapter content designed to pique student interest.

Student Objectives

Student learning objectives are integrated into the body of the chapters, at the beginning of major sections of reading, where the act as "check points" for student reading.

Review Activities

Placed at strategic intervals within chapters, review questions give students the chance to validate their understanding as they read.

Helpful Exhibits and Tables

In the ninth edition, a new design improves the utility and readability of the helpful ***Tables*** included throughout the text.

In addition, ***Exhibits***, self-contained features developed to give students the extra help they need to learn the anatomy of complex body systems—most notably skeletal muscles, articulations, blood vessels, and nerves—have been redesigned. Each Exhibit consists of an objective, an overview, a tabular summary of the relevant anatomy, and an associated suite of illustrations or photographs. Some Exhibits contain a relevant clinical application as well. Students will find this clear and concise presentation to be the ideal study vehicle for organizing and learning these important systems.

Cross References

This new edition features cross-references that guide the reader to specific pages and figures. Most will help students relate new concepts to previously learned material. However, I acknowledge the really ambitious students by including some cross-references to material that has yet to be considered.

Pronunciations and Word Roots

Students—even the best—generally find it difficult at first to read and pronounce anatomical terms. As an educator in the largest metropolitan region of the United States, I teach and am sympathetic to the needs of the growing ranks of college students who speak English as a second language. Because of these reasons, the features included in *Principles of Human Anatomy* help students build a working and useful vocabulary are carefully scrutinized during development of each edition. New terms are highlighted in boldface type, and most highlighted terms in the text, tables, and exhibits are accompanied by easy-to-understand pronunciation guides. As a further aid, the most important highlighted terms include word roots, designed to provide an understanding of the meaning of new terms.

Study Outline

As always, readers will benefit from the popular end-of-chapter summary of major topics that is page referenced to the chapter discussions.

Self Quizzes

Quizzes at the end of every chapter include fill-in-the-blank, multiple choice, and matching questions. Many of the questions are new to this edition. These quizzes are meant not only for students to test their ability to memorize the facts presented in each chapter, but also to sharpen their critical thinking skills by applying the concepts and processes that are part of the way the human body is structured an how it functions. Answers to the Self-Quiz items are presented in an appendix at the end of the book.

Critical Thinking Applications

Critical Thinking Questions are provided at the end of each chapter. These questions indicated by a "thinking" head icon, are essay-style problems that encourage students to think about and apply the concepts they have studied in each chapter. The style of these questions ought to make students smile on occasion as well as think! Suggested answers to these questions appear in an appendix at the end of the book.

Glossary

A full glossary of terms with phonetic pronunciations appears at the end of the book. The basic building blocks of medical terminology—**Combining Forms, Word Roots, Prefixes, and Suffixes**—are listed inside the back cover.

ACKNOWLEDGEMENTS

I wish to especially thank several people for their helpful contributions to this edition. First, I want to express my thanks to Kevin Petti for the contributions he made to this edition by researching and writing the Changing Images Essays. It has been a rewarding experience integrating this new feature into the text, and Kevin's insights and knowledge have made it possible. Janice Meeking of Mt. Royal College contributed to the completion of this edition by updating and revising the Self-Quizzes at the end of each chapter. In the process she provided many new, challenging, and diverse test questions. Joan Barber of Delaware Technical & Community College revised or replaced the Critical Thinking Questions, also found at chapters' end. These questions not only challenge students to think critically, but they encourage them to do so through their spirited and often humorous tone. Happily, Dr. Michael Ross of the University of Florida agreed to provide new histology photographs especially to illustrate our work. I am grateful to all for their fine work in making this an even better edition for students to use.

The publication of the ninth edition of *Principles of Human Anatomy* is my first edition of this text with John Wiley and Sons, Inc. and it has been the occasion of many pleasant surprises for me. I have enjoyed the opportunity of collaborating with a fresh team of very enthusiastic, dedicated, and talented publishing professionals. Accordingly, I would like to recognize and thank the members of the Tortora team—a true team in every sense of the word.

My association with Wiley has reunited me with my editor Bonnie Roesch. Bonnie's multiplicity of talents was well known to me because I had the good fortune to work with her on previous editions of this book, as well as other textbooks. Under her leadership the books became even more successful best-sellers. Bonnie is one of the most professional, dedicated, talented, and knowledgeable editors in publishing. As expected, she has been the source of expert guidance, great ideas, and innovative improvements. Her reunion with me is one of the most gratifying aspects of this edition. I look forward to working with Bonnie for many years. Ellen Ford, my developmental editor, clearly understood the market for which the book is intended and directed her insightful comments, suggestions, and analysis in that direction. She shepherded the manuscript through its revision cycles and provided many perceptive suggestions, helping me to keep the book on target from beginning to end. The beautiful design for the book's interior is the brainchild of Karin Kincheloe, who also designed the book's cover. Moreover, Karin carefully laid out each page of the book to achieve the best possible arrangement of text, figures, and other elements. Both instructors and students will appreciate and benefit from the pedagogically effective and visually pleasing design elements that augment the changes made to this edition. Claudia Durell, my Art Coordinator, like Bonnie Roesch, has been with me for many years. I am grateful that Wiley has selected Claudia and has permitted me to maintain my collaboration with her. She remains a cornerstone of my projects. I owe a special thank you to Claudia for all her contributions. Her organizational skills, attention to detail, artistic ability, and understanding of our illustration preferences greatly enhance the visual appeal and style of the figures, surpassing our expectations. Kelly Tavares, Senior Production Editor, demonstrated her expertise and professionalism during each step of the production process. She coordinated all aspects of actually making and manufacturing the book; she also was "on press" as the book was being printed to ensure the highest possible quality. The many hours we spent on the telephone, including weekends, were worth the effort. Hillary Newman, Photo Researcher, provided me with all of the photos I requested and did it with efficiency, accuracy, and professionalism. Mary O'Sullivan, Assistant Editor, coordinated the development of the many supplements that support this text. Her efforts have made their imprint on this book and I am most appreciative of her input. Kelli Coaxum, Editorial Assistant, worked with Bonnie Roesch and helped her with all aspects of the project. She kept all the reviews and other information flowing smoothly to and from me. Her dedication and enthusiasm were obvious in all of her contacts with me. Thank you to everyone who helped during the revision and production of this book.

I am always extremely grateful to my colleagues who have reviewed the manuscript and offered numerous suggestions for improvement. The reviewers who have provided their time and expertise to maintain this book's accuracy are noted in the Reviewers list that follows:

Kay Azuma, Santa Monica College; Berit E. Blomquist, Salt Lake City Community College; Terri Borman, Pasadena City College; Alphonse R. Burdi, University of Michigan Medical School; Vincent A. Cobb, Northeastern State University; Ruth Heisler, University of Colorado-Boulder; Leslie Hendon, University of Alabama-Birmingham; John H. Langdon, University of Indianapolis; Steven Lewis, Penn Valley Community College; Malinda McMurray, Morehead State University; Scott Minton, Vanguard University; Izak Paul, Mt. Royal College; Dolores M. Schroeder, Indiana University; Mary Lynn Stephanou, Santa Monica College; Judith Tamburlin, SUNY Buffalo; Richard P. Tucker, University of California-Davis; Sara S. Tolsma, Northwestern College; Barbara Walton, University of Tennessee; Ophelia Weeks, Florida International University; Paul Wissmann, Santa Monica College.

As always, my family, colleagues, and friends have supported my writing activities in more ways than I could ever mention here. Their understanding and encouragement will always be appreciated and will never be taken for granted.

With this edition of *Principles of Human Anatomy* I have had another opportunity to work with Wiley throughout the entire publishing process. My journey has again united me with some former professionals and introduced me to some new ones. My association with the Wiley family continues to be most pleasant, satisfying, and rewarding. I have been given many reasons to "Celebrate the Difference" and I invite all of my readers to celebrate the difference as well. In addition, I would also like to invite all readers and users of the book to continue the tradition of sending comments and suggestions to me so that I can include them in the tenth edition.

Gerard J. Tortora
Department of Science and Health, S229
Bergen Community College
400 Paramus Road
Paramus, NJ 07652

NOTE TO STUDENTS

*Y*our book has a variety of special features that will make your time studying anatomy a more rewarding experience. These have been developed based on feedback from students—like you—who have used previous editions of the text. Below are some hints for using some of these helpful aids. A review of the preface will give you insight, both visually and in narrative, to all of the texts distinctive features.

Begin your study by anticipating what is to be learned from each chapter. A brief, page referenced overview, called **Contents at a Glance** begins each chapter. Then, as you begin each narrative section of the chapter, be sure to take note of the objectives at the beginning of the section to help you focus on what is important as you read it. At the end of the section, take time to try and answer the review questions placed there. If you can, then you are ready to move on to the next section. If you experience difficulty answering the questions you may want to re-read the section before continuing.

CONTENTS AT A GLANCE

BONE GROWTH

Objective

- Describe how bones grow in length and width.

During childhood, bones throughout the body grow in width by appositional growth, and long bones lengthen by the addition of bone material on the diaphyseal side of the epiphyseal plate. Bones stop growing in length at about age 25, although they may continue to thicken.

only detects abnormalities 3–6 months sooner than standard x-ray procedures, it exposes the patient to less radiation. ■

✓ Why is bone considered a connective tissue?
✓ Describe the four types of cells in bone tissue.
✓ What is the composition of the matrix of bone tissue?
✓ Distinguish between spongy and compact bone tissue in terms of its microscopic appearance, location, and function.

*S*tudying the figures (illustrations that include artwork and photographs) in this book is as important as reading the text. To get the most out of the visual parts of this book, use the tools we have added to the figures to help you understand the concepts being presented. Start by reading the **legend,** which explains what the figure is about. Next, study the **key concept statement,** which reveals a basic idea portrayed in the figure. Added to many figures you will also find an **orientation diagram** to help you understand the perspective from which you are viewing a particular piece of anatomical art. Finally, at the bottom of each figure you will find a **figure question.** If you try to answer these questions as you go along, they will serve as self-checks to help you understand the material. Often it will be possible to answer a questions by examining the figure itself. Other questions will encourage you to integrate the knowledge you've gained by carefully reading the text associated with the figure. Still other questions may prompt you to think critically about the topic at hand or predict a consequence in advance of its description in the text. You will find the answers to figure questions at the end of chapters.

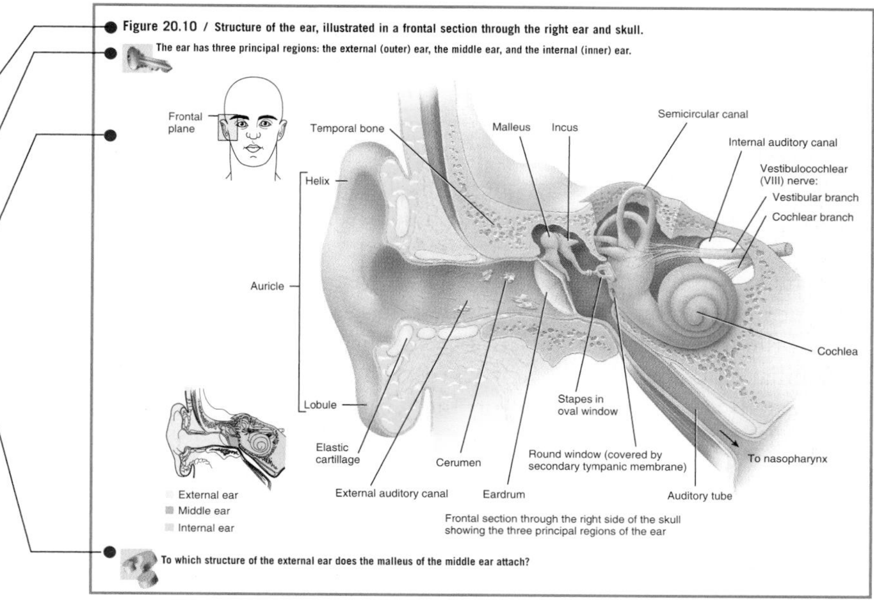

Figure 20.10 / Structure of the ear, illustrated in a frontal section through the right ear and skull.

The ear has three principal regions: the external (outer) ear, the middle ear, and the internal (inner) ear.

To which structure of the external ear does the malleus of the middle ear attach?

At the end of each chapter are other resources that you will find useful. The **Study Outline** is a concise statement of important topics discussed in the chapter. Page numbers are listed next to key concepts so you can easily refer to the specific passages in the text for clarification or amplification. The **Self-quiz Questions** are designed to help you evaluate your understanding of the chapter contents. **Critical Thinking Questions** are word problems that allow you to apply the concepts you have studied in the chapter to specific situations.

STUDY OUTLINE

Spinal Cord Anatomy (p. 516)

1. The spinal cord is protected by the vertebral column, meninges, cerebrospinal fluid, and denticulate ligaments.
2. The three meninges are coverings that run continuously around the spinal cord and brain; they are the dura mater, arachnoid, and pia mater.
3. The spinal cord begins as a continuation of the medulla oblongata and ends at about the second lumbar vertebra in an adult.
4. The spinal cord contains cervical and lumbar enlargements that serve as points of origin for nerves to the limbs.
5. The tapered inferior portion of the spinal cord is the conus

9. Parts of the spinal cord observed in transverse section are the gray commissure; central canal; anterior, posterior, and lateral gray horns; and anterior, posterior, and lateral white columns, which

SELF-QUIZ QUESTIONS

Choose the one best answer to the following questions:

1. Choose the true statement. (a) Tracts are located only in the spinal cord and not in the brain. (b) Tracts appear white because they consist of bundles of unmyelinated nerve fibers. (c) Ascending tracts are all motor tracts. (d) Ascending tracts are in the white matter; descending tracts are in the gray matter. (e) None of the above is true.

7. Lateral extensions of the pia mater that help to protect the spinal cord against displacement are called ________ .
8. The order of the meningeal layers from superficial to deep is ________ , ________ , ________ .
9. Individual nerve fibers are wrapped in a connective tissue covering known as ________ . Groups of nerve fibers are held in bundles by ________ . The entire nerve is wrapped with ________ .
10. Cell bodies of motor neurons leading to skeletal muscles are lo- ________ gray horns of the spinal cord, while cell urons supplying smooth muscle, cardiac muscle or

CRITICAL THINKING QUESTIONS

1. A high school senior dove headfirst into a murky pond. Unfortunately, he hit a submerged log with his head and is now paralyzed. Can you deduce the location of the inju likelihood of recovery from his injury?
HINT *He didn't just break bones.*
2. The spinal cord is covered in protective layers and vertebral column. How is it able to send and re from the periphery of the body?
HINT *How are you able to bend the vertebral column.*
3. Why doesn't the spinal cord "creep" up towards time you bend over? Why doesn't it get all twisted when you exercise?

4. Helga had spent many sleepless nights after moving in with her daughter, her daughter's three children, two dogs, and five cats.

ANSWERS TO FIGURE QUESTIONS

17.1 The superior boundary of the spinal dura mater is the foramen magnum of the occipital bone; the inferior boundary is the second sacral vertebra.

17.2 The cervical enlargement connects with sensory and motor nerves of the upper limbs.

17.3 In the spinal cord, a horn is an area of gray matter and a column is a region of white matter.

17.4 The anterior corticospinal tract is located on the anterior side of

17.8 Severing the spinal cord at level C2 causes respiratory arrest because it prevents descending nerve impulses from reaching the phrenic nerve, which stimulates contraction of the diaphragm.

17.9 The axillary, musculocutaneous, radial, median, and ulnar nerves are five important nerves that arise from the brachial plexus.

17.10 Injury to the median nerve affects sensations on the palm and fingers.

Throughout the text we have included **Pronunciations** and, sometimes, **Word Roots,** for many terms that may be new to you. These appear in parentheses immediately following the new words, and the pronunciations are repeated in the glossary at the back of the book. Look at the words carefully and say them out loud several times. Learning to pronounce a new word will help you remember it and make it a useful part of your medical vocabulary. Take a few minutes now to read the following pronunciation key, so it will be familiar as you encounter new words. The key is repeated at the beginning of the Glossary, page G-1.

Pronunciation Key

1. The most strongly accented syllable appears in capital letters, for example, bilateral (bī-LAT-er-al) and diagnosis (dī-ag-NŌ-sis).
2. If there is a secondary accent, it is noted by a prime (′), for example, physiology (fiz′-ē-OL-ō-jē). Any additional secondary accents are also noted by a prime, for example, decarboxylation (dē′-kar-bok′-si-LĀ-shun).
3. Vowels marked by a line above the letter are pronounced with the long sound as in the following common word:
   ā as in māke   ī as in īvy   ū as in neuron
   ē as in bē   ō as in pōle
4. Vowels not so marked are pronounced with the short sound as in the following words:
   a as in above   i as in sip   u as in but
   e as in bet   o as in not
5. One other phonetic symbol is used to indicate the following sound:
   oy as in oil

BRIEF TABLE OF CONTENTS

CONTENTS

1

AN INTRODUCTION TO THE HUMAN BODY

CONTENTS AT A GLANCE

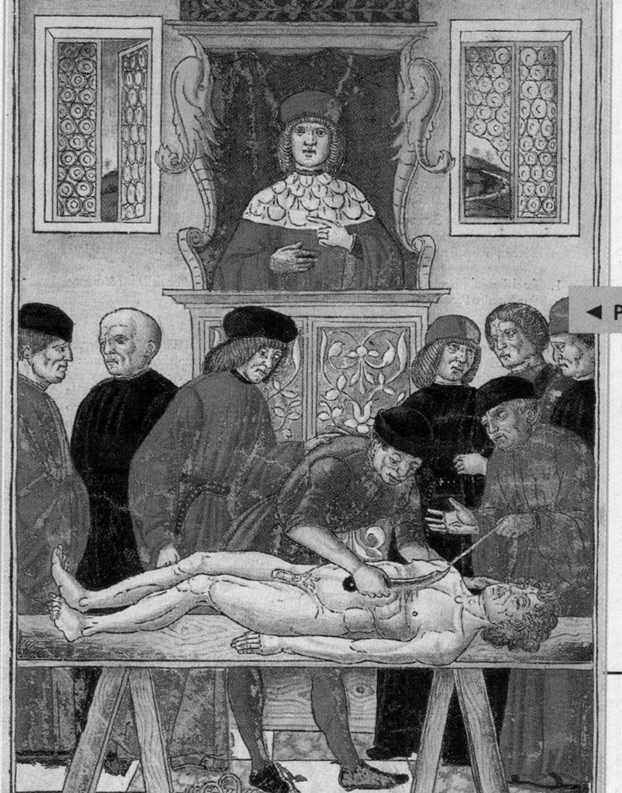

◀ Page 18

Can you picture yourself as a student in this classroom?

As you begin your study of the human body, think about how the study of human anatomy has developed and progressed. In addition to dissection, what other methods do we use today to visually explore our anatomy?

Page 21 ▶

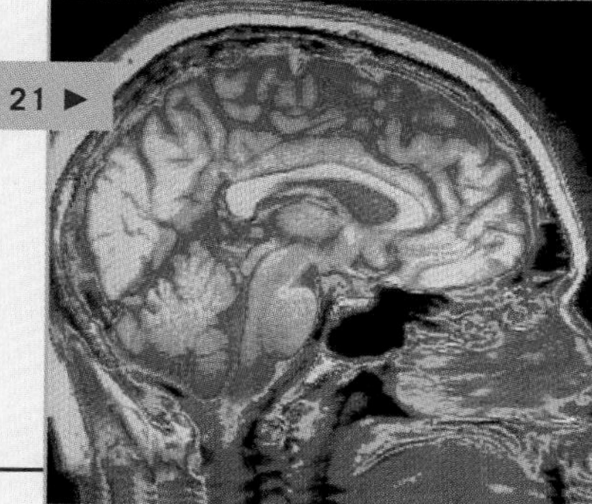

INTRODUCTION

You are about to begin a study of the human body in order to learn how it is organized and how it functions. This study involves many branches of science, each of which contributes to an understanding of how the body normally works. In order to understand what happens to the body when it is injured, diseased, or placed under stress, you must know how the body is organized and how its different parts normally work. Much of what you study in this chapter will help you visualize the body as anatomists do, and you will learn a basic anatomical vocabulary that will help you talk about the body in a way that is understood by professionals in various fields.

ANATOMY DEFINED

Objective

- Define anatomy and physiology, and name several subdisciplines of anatomy.

Anatomy (a-NAT-ō-mē; *ana* = upward; *tome* = to cut) is the study of *structure* and the relationships among structures. The anatomy of the human body can be studied at various levels of structural organization, ranging from microscopic to macroscopic. These levels and the different methods used to study them provide the basis for the subdisciplines of anatomy, several of which are described in Table 1.1.

Table 1.1 Selected Subdisciplines of Anatomy

Subdiscipline	Study of
Embryology (em′-brē-OL-ō-jē; *embry-* =embryo; *-ology* = study of)	Structures that emerge from the time of the fertilized egg through the eighth week in utero.
Developmental anatomy	Structures that emerge from the time of the fertilized egg to the adult form.
Cytology (sī-TOL-ō-jē; *cyt-* = cell)	Chemical and microscopic structure of cells.
Histology (hiss′-TOL-ō-jē; *hist-* = tissue)	Microscopic structure of tissues.
Surface anatomy	Anatomical landmarks on the surface of the body through visualization and palpation.
Gross anatomy	Structures that can be examined without using a microscope.
Systemic anatomy	Structure of specific systems of the body, such as the nervous or respiratory systems.
Regional anatomy	Specific regions of the body, such as the head or chest.
Radiographic anatomy (rā′-dē-ō-GRAF-ik; *radio-* = ray; *-graphic* = to write)	Body structures that can be visualized with x-rays.
Pathological anatomy (path′-ō-LOJ-i-kal; *path-* = disease)	Structural changes (from gross to microscopic) associated with disease.

Whereas anatomy deals with structures of the body, **physiology** (fiz′-ē-OL-ō-jē) deals with *functions* of body parts—that is, how they work. Because function cannot be completely separated from structure, you will learn about the human body by studying first its anatomy and then the necessary physiology. You will see how each structure of the body is designed to carry out a particular function and how the structure of a part often determines the functions it performs. For example, the hairs lining the nose filter air that you inhale. The bones of the skull are tightly joined to protect the brain. The bones of the fingers, by contrast, are more loosely joined to permit various movements. The external ear is shaped in such a way as to collect sound waves, which facilitates hearing. The lungs are filled with millions of air sacs that are so thin that they permit both the movement of oxygen into the blood for use by body cells and the movement of carbon dioxide out of the blood to be exhaled.

✓ Define anatomy. List and define the various subdivisions of anatomy. Define physiology.

✓ Give several examples of how structure and function are related.

Palpation, Auscultation, and Percussion

Three noninvasive techniques are commonly used by health-care professionals and students to assess certain aspects of body structure and function. In **palpation** (pal-PĀY-shun; *palpa-* = gently touching) the examiner feels body surfaces with the hands. An example is palpating a blood vessel (called an artery) to find the pulse and measure the heart rate. In **auscultation** (aus-cul-TĀY-shun; *ausculta-* = listening) the examiner listens to body sounds to evaluate the functioning of certain organs, often using a stethoscope to amplify the sounds. An example is auscultation of the lungs during breathing to check for crackling sounds associated with abnormal fluid accumulation in the lungs. In **percussion** (pur-KUSH-un; *percus-* = beat through) the examiner taps on the body surface with the fingertips and listens to the resulting echo. For example, percussion may reveal the abnormal presence of fluid in the lungs or air in the intestines. It is also used to reveal the size, consistency, and position of an underlying structure. ■

LEVELS OF BODY ORGANIZATION

Objectives

- Describe the levels of structural organization that make up the human body.
- Define the eleven systems of the human body, the organs present in each, and their general functions.

The levels of organization of a language—letters of the alphabet, words, sentences, paragraphs, and so on—provide a useful comparison to the levels of organization of the human body.

Your exploration of the human body will extend from some of the smallest body structures and their functions to the largest structure—an entire person. From the smallest to the largest size of their components, six levels of organization are relevant to understanding anatomy and physiology: the chemical, cellular, tissue, organ, system, and organismal levels of organization (Figure 1.1).

1 The *chemical level* includes **atoms,** the smallest units of matter that participate in chemical reactions, and **molecules,** two or more atoms joined together. Certain atoms, such as carbon (C), hydrogen (H), oxygen (O), nitrogen (N), phosphorus (P), and calcium (Ca), are essential for life. Familiar examples of molecules found in the body are deoxyribonucleic acid (DNA), the genetic material passed from one generation to the next; hemoglobin, a protein that carries oxygen in the blood; and glucose, commonly known as blood sugar.

2 Molecules, in turn, combine to form structures at the next level of organization—the *cellular level.* **Cells** are the basic structural and functional units of an organism and are the

Figure 1.1 / Levels of structural organization in the human body.

The levels of structural organization are chemical, cellular, tissue, organ, system, and organismal.

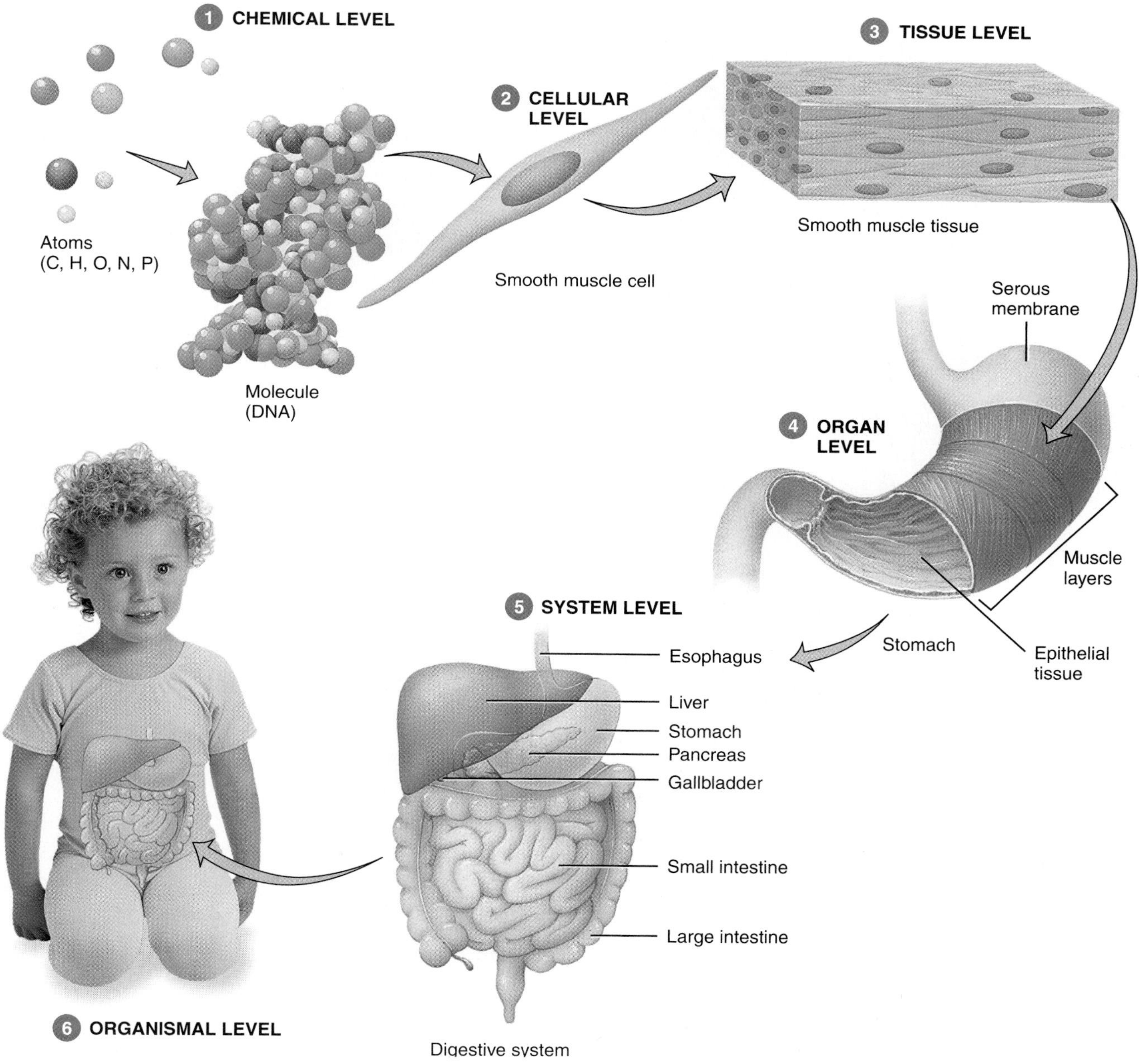

Which level of structural organization is composed of two or more different types of tissues that work together to perform a specific function?

smallest living units in the human body. Among the many kinds of cells in your body are muscle cells, nerve cells, and blood cells. Shown in Figure 1.1 is a smooth muscle cell, one of three different types of muscle cells in your body. The focus of Chapter 2 is the cellular level of organization.

3 The next level of structural organization is the *tissue level.* **Tissues** are groups of cells and the materials surrounding them that work together to perform a particular function. There are just four basic types of tissue in your body: *epithelial tissue*, *connective tissue*, *muscle tissue*, and *nervous tissue*, which are described in Chapter 3. Smooth muscle tissue consists of tightly packed smooth muscle cells.

4 When different kinds of tissues are joined together, they form the next level of organization, the *organ level.* **Organs** are structures that are composed of two or more different types of tissues; they have specific functions and usually have recognizable shapes. Examples of organs are the stomach, heart, liver, lungs, and brain. Figure 1.1 shows how several tissues make up the stomach. The outer covering is a *serous membrane*, a layer of epithelial tissue and connective tissue that protects the stomach and other organs and reduces friction when the stomach moves and rubs against other organs. Underneath are the *muscle tissue layers*, which contract to churn and mix food and push it on to the next

Table 1.2 Principal Systems of the Human Body: Representative Organs and Functions

Integumentary System

Components The skin and structures derived from it, such as hair, nails, and sweat and oil glands.

Functions Helps regulate body temperature; protects the body; eliminates some wastes; helps produce vitamin D; and detects sensations, such as pain, touch, hot, and cold.

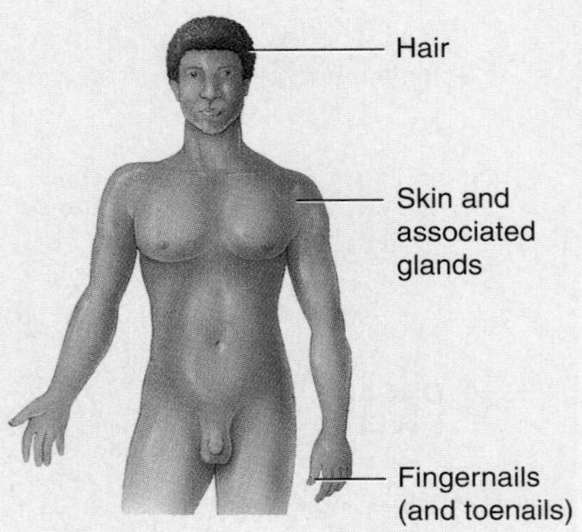

Skeletal System

Components All the bones and joints of the body and their associated cartilages.

Functions Supports and protects the body; assists in body movements; houses cells that give rise to blood cells; and stores minerals and lipids (fats).

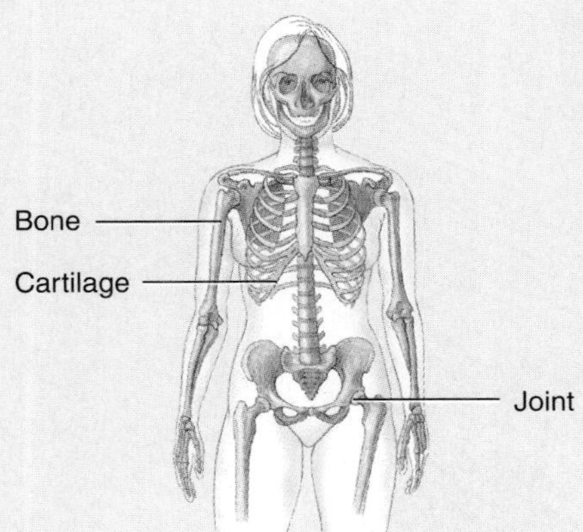

Muscular System

Components Refers specifically to skeletal muscle tissue, which is muscle usually attached to bones. Other muscle tissue types are smooth and cardiac.

Functions Powers movements of the body, such as walking and throwing a ball; stabilizes body positions (posture); and generates heat.

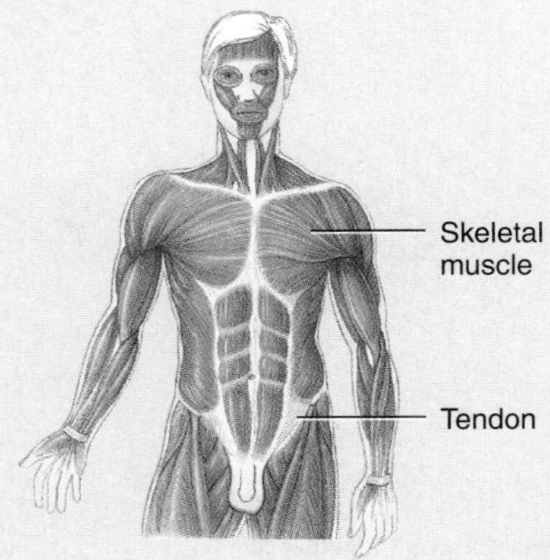

Cardiovascular System

Components Blood, heart, and blood vessels.

Functions Heart pumps blood through blood vessels; blood carries oxygen and nutrients to cells and carbon dioxide and wastes away from cells and helps regulate acid–base balance, temperature, and water content of body fluids; blood components help defend against disease and mend damaged blood vessels.

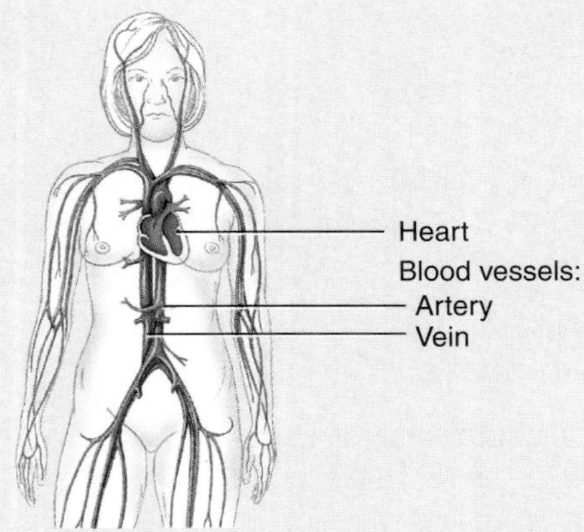

digestive organ, the small intestine. The innermost lining is an *epithelial tissue layer* that produces fluid and chemicals responsible for digestion in the stomach.

5 The next level of structural organization in the body is the *system level.* A **system** consists of related organs that have a common function. An example is the digestive system, which breaks down and absorbs food. Its organs include the mouth, salivary glands, pharynx (throat), esophagus (food tube), stomach, small intestine, large intestine, liver, gallbladder, and pancreas. Sometimes an organ is part of more than one system. The pancreas, for example, is part of both the digestive system and the hormone-producing endocrine system.

6 The largest organizational level is the *organismal level.* An **organism** is any living individual. All the parts of the human body functioning together constitute the total organism—one living person.

In the following chapters, you will study the anatomy and physiology of the major body systems. Table 1.2 on pages 4–6 introduces the components and functions of these systems in the order they are discussed in the book.

✓ Define the following terms: atom, molecule, cell, tissue, organ, system, and organism.

✓ Referring to Table 1.2, which body systems help eliminate wastes?

Lymphatic and Immune Systems

Components Lymph, lymphatic vessels, and structures or organs containing lymphatic tissue such as the spleen, thymus, lymph nodes, and tonsils. Lymphatic tissues contain large numbers of white blood cells called lymphocytes.

Functions Returns proteins and plasma (liquid portion of blood) to the cardiovascular system; transports triglycerides (fats) from the gastrointestinal tract to the cardiovascular system; serves as a site of maturation and proliferation of certain white blood cells; and helps protect against disease through the production of proteins called antibodies, as well as other responses.

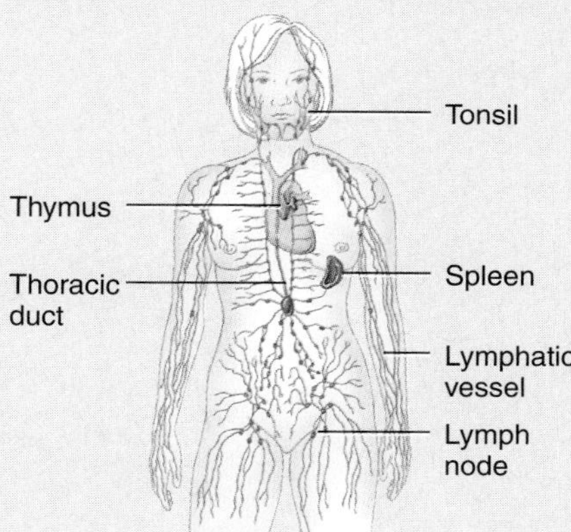

Nervous System

Components Brain, spinal cord, nerves, and special sense organs, such as the eyes and ears.

Functions Regulates body activities through action potentials (nerve impulses) stimulated by changes in the internal and external environments, interprets the changes, and responds to the changes by inducing muscular contractions or glandular secretions.

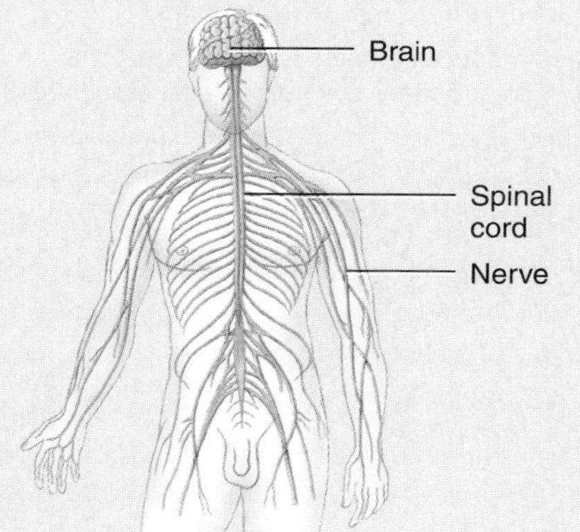

Endocrine System

Components All hormone-producing cells and glands such as the pituitary and thyroid glands and pancreas.

Functions Regulates body activities through hormones, chemicals transported in the blood to various target organs of the body.

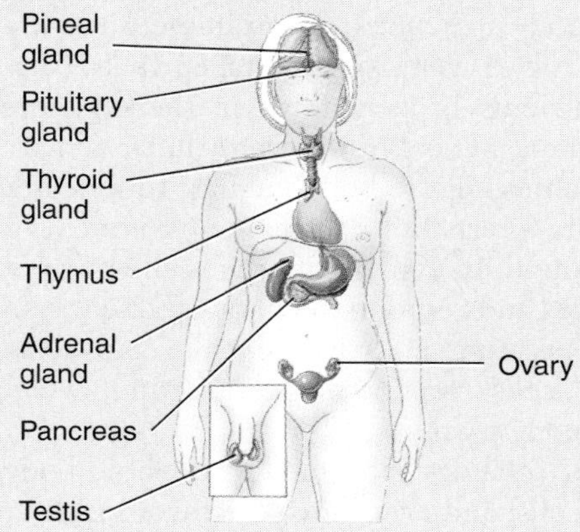

Respiratory System

Components Lungs and the airways leading into and out of them.

Functions Transfers oxygen from inhaled air to blood and carbon dioxide from blood to exhaled air; helps regulate acid–base balance of body fluids; air flowing out of lungs through vocal cords produces sounds.

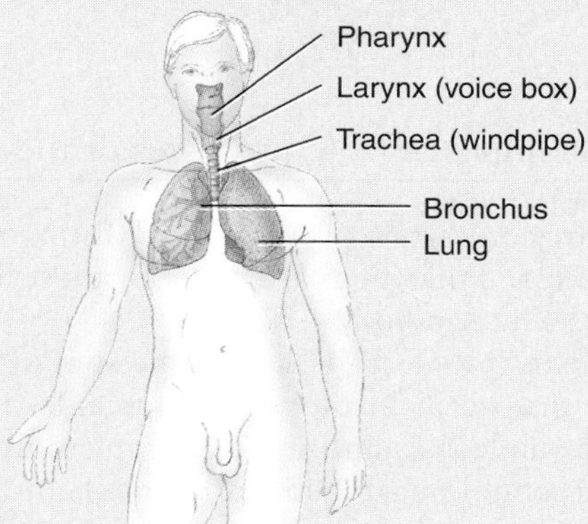

(continues)

Table 1.2 Principal Systems of the Human Body: Representative Organs and Functions (continued)

Digestive System

Components Organs of gastrointestinal tract, a long tube that includes the mouth, esophagus, stomach, intestines, and anus; also includes accessory organs that assist in digestive processes, such as the salivary glands, liver. gallbladder, and pancreas.

Functions Achieves physical and chemical breakdown of food; absorbs nutrients; eliminates solid wastes.

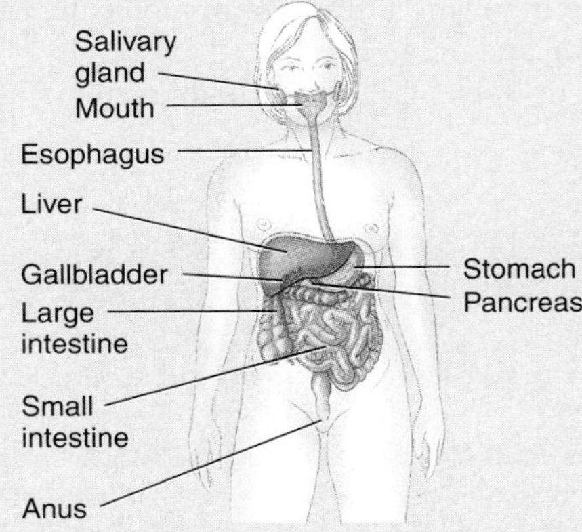

Urinary System

Components Kidneys, ureters, urinary bladder, and urethra that together produce, store, and eliminate urine.

Functions Regulates the volume and chemical composition of blood; eliminates metabolic wastes; regulates fluid and electrolyte balance; helps maintain the acid–base balance of body fluids and calcium balance of the body; and secretes a hormone that regulates red blood cell production.

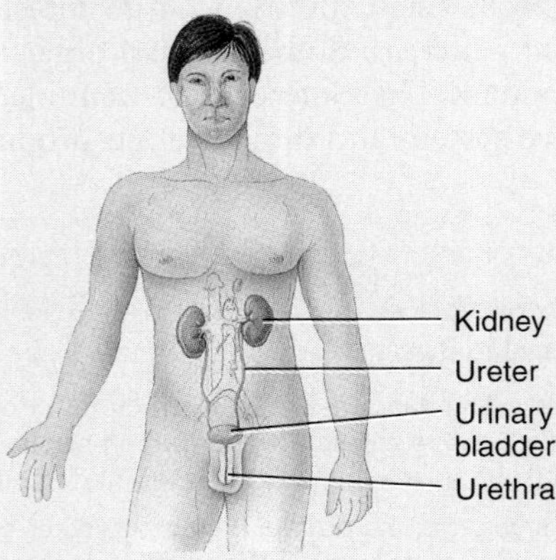

Reproductive Systems

Components Gonads (testes or ovaries) and associated organs: uterinetubes, uterus, and vagina in females and epididymis, ductus deferens, and penis in males.

Functions Gonads produce gametes (sperm or ova) that unite to form a new organism and release hormones that regulate reproduction and other body processes; associated organs transport and store gametes.

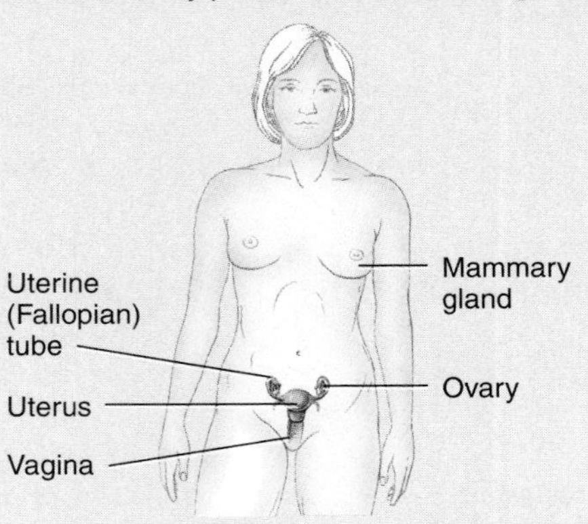

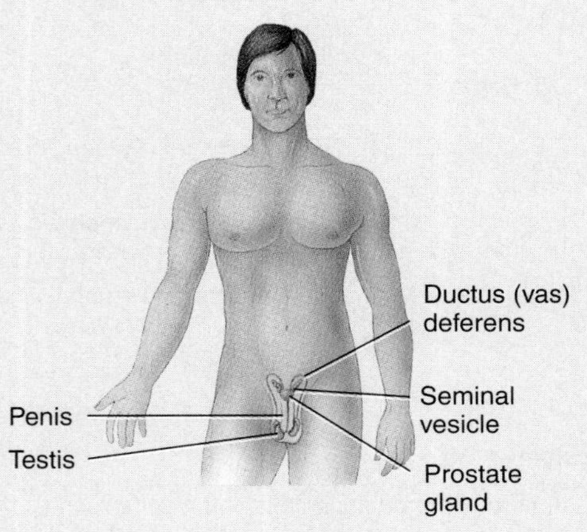

LIFE PROCESSES

Objective

- Define the important life processes of the human body.

Organisms carry on certain processes that distinguish them from nonliving things. Following are the six most important life processes of the human body:

Metabolism (me-TAB-ō-lizm) is the sum of all the chemical processes that occur in the body. It includes breaking down large, complex molecules into smaller, simpler ones and building the body's structural and functional components. For example, proteins in food are split into amino acids, the building blocks of proteins. Amino acids can then be used to build new proteins that make up body structures, for example, muscles and bones. Metabolism involves using oxygen taken in by the respiratory system and nutrients broken down in the digestive system to provide the chemical energy to power cellular activities.

Responsiveness is the body's ability to detect and respond to changes in its internal or external environment, for example, temperature, sound, or light. Different cells in the body detect different changes and respond in characteristic ways. Nerve cells respond by generating electrical signals, known as nerve impulses. Muscle cells respond by contracting, which generates force to move body parts.

Movement includes motion of the whole body, individual organs, single cells, and even tiny structures inside cells. For example, the coordinated action of several leg muscles moves your whole body from one place to another when you walk or run. After you eat a meal that contains fats, your gallbladder contracts and squirts bile into the gastrointestinal tract to aid in the digestion of fats. When a body tissue is damaged or infected, certain white blood cells move from the blood into the tissue to help clean up and repair the area. And inside individual cells, various sub-cellular structures move from one position to another to carry out their functions.

Growth is an increase in body size that results from an increase in the size of existing cells, the number of cells, or both. In addition, a tissue sometimes increases in size because the amount of material found between cells increases. In a growing bone, for example, mineral deposits accumulate around the bone cells, causing the bone to enlarge in length and width.

Each kind of cell in the body has a specialized structure and function. **Differentiation** is the process in which a cell develops from an unspecialized to a specialized state. Specialized cells differ in structure and function from the ancestor cells that gave rise to them. For example, red blood cells and several types of white blood cells differentiate from the same unspecialized ancestor cells in red bone marrow. Such ancestor cells, which can divide and give rise to progeny that undergo differentiation, are known as **stem cells.** Also, through differentiation, a fertilized egg develops into an embryo, and then into a fetus, an infant, a child, and finally an adult.

Reproduction refers either to the formation of new cells for growth, repair, or replacement, or to the production of a new individual. Some types of body cells, such as epithelial cells, reproduce continuously throughout a lifetime. Others, such as nerve and muscle cells, lose the ability to divide and proliferate and thus cannot be replaced if they are destroyed. Through the combining of sperm and ova, life continues from one generation to the next.

Although not all of these processes are occurring in cells throughout the body all of the time, when they cease to occur properly, the result is death of cells and then death of the human organism. Clinically, death in the human body is indicated by loss of the heartbeat, absence of spontaneous breathing, and loss of brain functions.

✓ Which life process in the human body sustains all the others?

BASIC ANATOMICAL TERMINOLOGY

Objectives

- Describe the orientation of the body in the anatomical position.
- Relate the common names to the corresponding anatomical descriptive terms for various regions of the human body.
- Define the anatomical planes and sections and the directional terms used to describe the human body.

Scientists and health-care professionals use a common language of special terms when referring to body structures and their functions. The language of anatomy and physiology has precisely defined meanings that allow us to communicate without using unneeded or ambiguous words. For example, is it correct to say, "The wrist is above the fingers?" This might be true if your arms are at your sides. But if you hold your hands up above your head, your fingers would be above your wrists. To prevent this kind of confusion, anatomists developed a standard anatomical position and a special vocabulary for relating body parts to one another.

Anatomical Position

In the study of anatomy, descriptions of any region or part of the human body assume that the body is in a specific stance called the **anatomical position.** In the anatomical position, the subject stands erect facing the observer, with the head level and the eyes facing directly forward. The feet are flat on the floor and directed forward, and the arms are at the sides with the palms turned forward (Figure 1.2). Once the body is in the anatomical position, it is easier to visualize and understand how it is organized into various regions.

In the anatomical position, the body is upright. Two terms describe a reclining body. If the body is lying face down, it is in the **prone** position. If the body is lying face up, it is in the **supine** position.

Regional Names

The human body is divided into several major regions that can be identified externally. The principal regions are the head, neck, trunk, upper limbs, and lower limbs (see Figure 1.2). The ***head*** consists of the skull and face. The *skull* encloses and protects the brain, while the *face* is the anterior (front) portion of the head that includes the eyes, nose, mouth, forehead, cheeks, and chin. The ***neck*** supports the head and attaches it to the trunk. The ***trunk*** consists of the chest, abdomen, and pelvis. Each ***upper limb*** *(extremity)* is attached to the trunk and consists of the shoulder, armpit, arm (portion of the limb from the shoulder to the elbow), forearm (portion of the limb from the elbow to the wrist), wrist, and hand. Each ***lower limb*** *(extremity)* is also attached to the trunk and consists of the buttock, thigh (portion of the limb from the buttock to the knee), leg (portion of the limb from the knee to the ankle), ankle, and foot. The *groin* is the area on the anterior surface of the body marked by a crease on each side, where the trunk attaches to the thighs.

Figure 1.2 shows the common names of major parts of the body. The corresponding anatomical descriptive form (adjective) for each part appears in parentheses next to the common name. For example, if you receive a tetanus shot in your *buttock*, it is a *gluteal* injection. Why does the descriptive form of a body part look different from its common name? The reason is that the descriptive form is based on a Greek or Latin word or "root" for the same part or area. The Latin word for armpit is *axilla* (ak-SIL-a), for example, and thus one of the nerves passing within the armpit is called the axillary nerve. You will learn more about the word roots of anatomical and physiological terms as you read this book.

Figure 1.2 / The anatomical position. The common names and corresponding anatomical terms (in parentheses) are indicated for specific body regions. For example, the head is the cephalic region.

In the anatomical position, the subject stands erect facing the observer with the head level and the eyes facing forward. The feet are flat on the floor and directed forward, and the arms are at the sides with the palms facing forward.

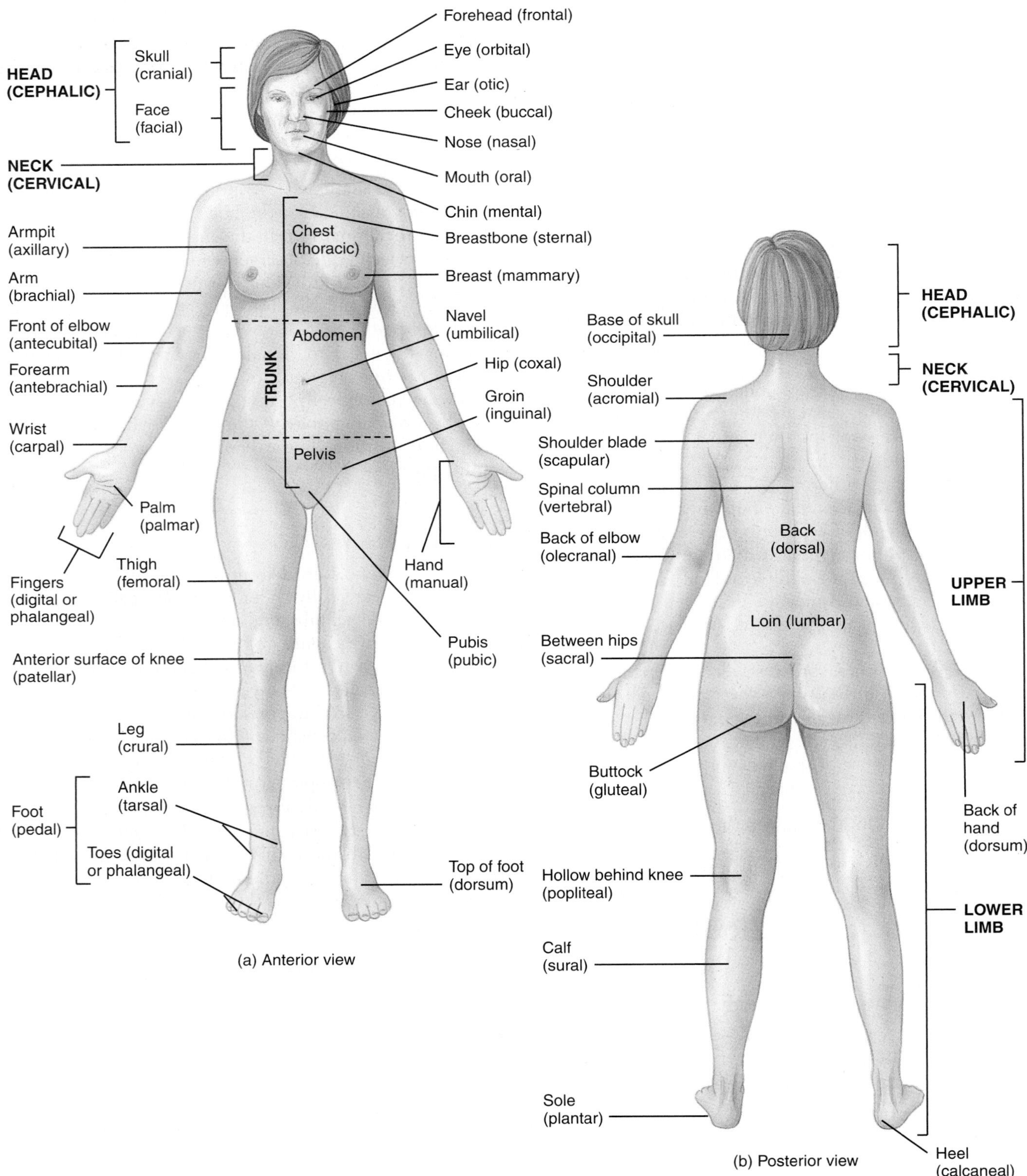

What is the usefulness of defining one standard anatomical position?

Planes and Sections

You will study parts of the body relative to planes—imaginary flat surfaces that pass through them (Figure 1.3). A **sagittal plane** (SAJ-i-tal; *sagitt-* = arrow) is a vertical plane that divides the body or organ into right and left sides. More specifically, when such a plane passes through the midline of the body or organ and divides it into *equal* right and left sides, it is called a **midsagittal plane,** or a **median plane.** If the sagittal plane does not pass through the midline but instead divides the body or organ into *unequal* right and left sides, it is called a **parasagittal plane** (*para-* = near). A **frontal,** or **coronal, plane** (kō-RŌ-nal; *corona* = crown) divides the body or organ into anterior (front) and posterior (back) portions. A **transverse plane** divides the body or organ into superior (upper) and inferior (lower) portions. A transverse plane may also be termed a cross-sectional or horizontal plane. Sagittal, frontal, and transverse planes are all at right angles to one another. An **oblique plane,** by contrast, passes through the body or organ at an angle between the transverse plane and either a sagittal or frontal plane.

When you study a body region, you often view it in **section,** meaning that you look at only one flat surface of the three-dimensional structure. It is important to know the plane of the section so you can understand the anatomical relationship of one part to another. Figure 1.4 indicates how three different sections—a *transverse section,* a *frontal section,* and a *midsagittal section*—provide different views of the brain.

Figure 1.3 / Planes through the human body.

Frontal, transverse, sagittal, and oblique planes divide the body in specific ways.

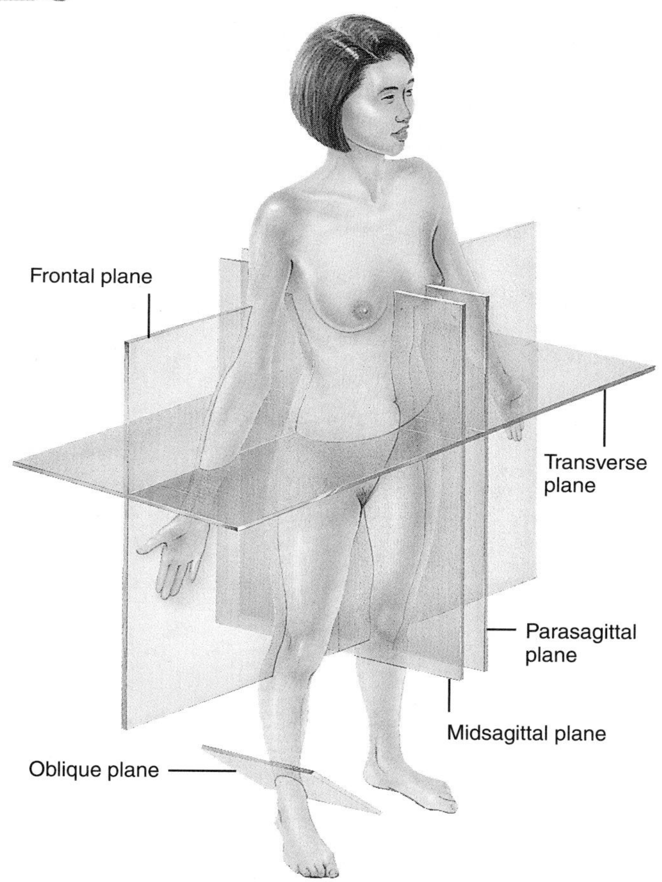

Which plane divides the heart into anterior and posterior portions?

Figure 1.4 / Planes and sections through different parts of the brain. The diagrams (left) show the planes, and the photographs (right) show the resulting sections. (**Note:** The arrows in the diagrams indicate the direction from which each section is viewed. This aid is used throughout the book to indicate viewing perspective.)

Planes divide the body in various ways to produce sections.

(a) Transverse plane — View — Transverse section

(b) Frontal plane — View — Frontal section

(c) Midsagittal plane — View — Midsagittal section

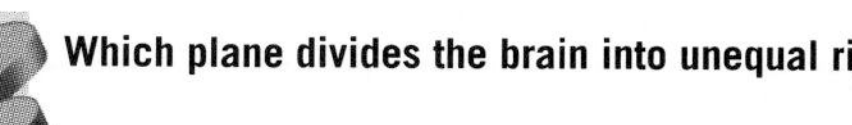
Which plane divides the brain into unequal right and left portions?

✓ Describe the anatomical position and explain why it is used.
✓ Locate each region on your own body, and then identify it by its common name and the corresponding anatomical descriptive form.
✓ What are the four principal types of planes that may be used to divide the body? Explain how each divides the body.
✓ Distinguish among a transverse, frontal, and midsagittal section.

Directional Terms

To locate various body structures, anatomists use specific **directional terms,** words that describe the position of one body part relative to another. Several directional terms can be grouped in pairs that have opposite meanings—for example, anterior (front) and posterior (back). Exhibit 1.1 presents the principal directional terms.

Exhibit 1.1 Directional Terms Used to Describe the Human Body (Figure 1.5)

► Most of the directional terms used to describe the human body can be grouped into pairs that have opposite meanings. For example, **superior** means toward the upper part of the body, whereas **inferior** means toward the lower part of the body. Moreover, it is important to understand that directional terms have relative meanings; they only make sense when used to describe the position of one structure relative to another. For example, your knee is superior to your ankle, even though both are located in the inferior half of the body. Study the directional terms in the following exhibit and the example of how each is used. As you read each example, look at Figure 1.5 to see the location of the structures mentioned.

Use each of the directional terms described in this exhibit in a sentence. (HINT: Refer to the structures in Figure 1.5.)

Directional Term	Definition	Example of Use
Superior (sū′-PEER-ē-or) (**cephalic** or **cranial**)	Toward the head, or the upper part of a structure.	The heart is superior to the liver.
Inferior (in′-FEER-ē-or) **(caudal)**	Away from the head, or the lower part of a structure.	The stomach is inferior to the lungs.
Anterior (an-TEER-ē-or) **(ventral)***	Nearer to or at the front of the body.	The sternum (breastbone) is anterior to the heart.
Posterior (pos-TEER-ē-or) **(dorsal)***	Nearer to or at the back of the body.	The esophagus (food tube) is posterior to the trachea (windpipe).
Medial (MĒ-dē-al)	Nearer to the midline† or midsagittal plane.	The ulna is on the medial side of the forearm.
Lateral (LAT-er-al)	Farther from the midline or midsagittal plane.	The lungs are lateral to the heart.
Intermediate (in′-ter-MĒ-dē-at)	Between two structures.	The transverse colon is intermediate between the ascending and descending colons.
Ipsilateral (ip-si-LAT-er-al)	On the same side of the body as another structure.	The gallbladder and ascending colon are ipsilateral.
Contralateral (CON-tra-lat-er-al)	On the opposite side of the body from another structure.	The ascending and descending colons are contralateral.
Proximal (PROK-si-mal)	Nearer to the attachment of a limb to the trunk; nearer to the point of origin.	The humerus is proximal to the radius.
Distal (DIS-tal)	Farther from the attachment of a limb to the trunk; farther from the point of origin.	The phalanges are distal to the carpals.
Superficial (sū′-per-FISH-al)	Toward or on the surface of the body.	The ribs are superficial to the lungs.
Deep (DĒP)	Away from the surface of the body.	The ribs are deep to the skin of the chest.

*Ventral refers to the belly side, whereas dorsal refers to the back side. In four-legged animals anterior = cephalic (toward the head), ventral = inferior, posterior = caudal (toward the tail), and dorsal = superior.

†The midline is an imaginary vertical line that divides the body into equal right and left sides.

Figure 1.5 / Directional terms.

Directional terms precisely locate various parts of the body relative to one another.

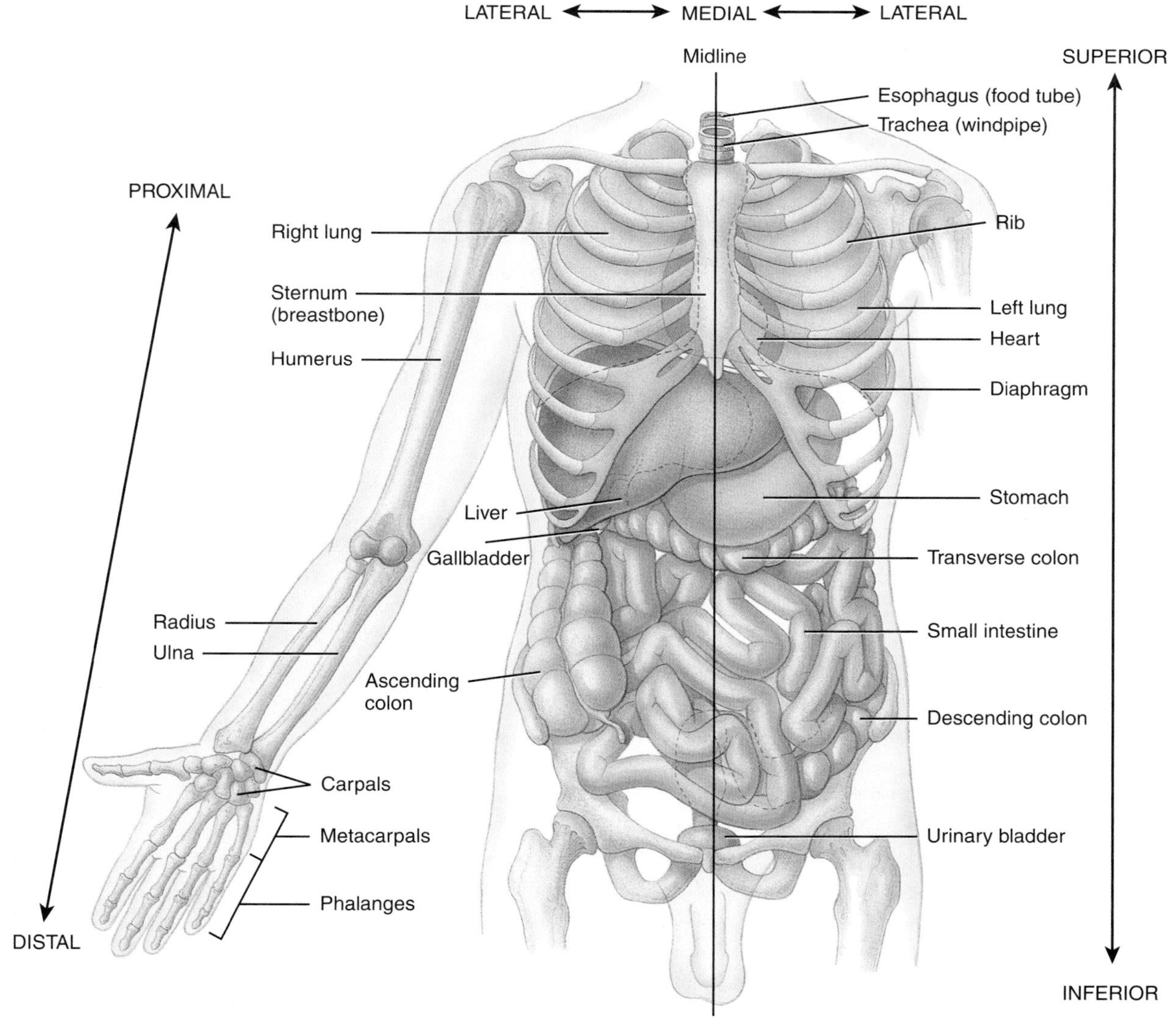

Anterior view of trunk and right upper limb

Is the radius proximal to the humerus? Is the esophagus anterior to the trachea? Are the ribs superficial to the lungs? Is the urinary bladder medial to the ascending colon? Is the sternum lateral to the descending colon?

Body Cavities

Objective

- Describe the principal body cavities, the organs they contain, and their associated linings.

Body cavities are spaces within the body that help protect, separate, and support internal organs. Bones, muscles, and ligaments separate the various body cavities from one another. The two principal cavities are the dorsal and ventral body cavities (Figure 1.6).

Dorsal Body Cavity

The **dorsal body cavity** is located near the dorsal (posterior) surface of the body and has two subdivisions, the cranial cavity and the vertebral canal. The **cranial cavity** is formed by the cranial bones and contains the brain. The **vertebral (spinal) canal** is formed by the bones of the vertebral column (backbone) and contains the spinal cord. Three layers of protective tissue, called **meninges** (me-NIN-jēz), line the dorsal body cavity.

Ventral Body Cavity

The other principal body cavity—the **ventral body cavity**—is located on the ventral (anterior) aspect of the body. The ventral body cavity also has two main subdivisions, the thoracic and the abdominopelvic cavities. The **diaphragm** (DĪ-a-fram; = partition or wall), the large dome-shaped muscle that powers lung expansion during breathing, forms the floor of the thoracic cavity and the roof of the abdominopelvic cavity. The organs inside the ventral body cavity are termed the **viscera** (VIS-er-a).

The superior portion of the ventral body cavity is the **thoracic cavity** (thor-AS-ik; *thorac-* = chest), or chest cavity (Figure 1.7). The thoracic cavity is encircled by the ribs, the muscles of the chest, the sternum (breastbone), and the thoracic portion of the vertebral column (backbone). Within the thoracic cavity are three smaller cavities: the **pericardial cavity** (per′-i-KAR-dē-al; *peri-* = around; *-cardial* = heart), a fluid-filled space that surrounds the heart, and two **pleural cavities** (PLŪR-al; *pleur-* = rib or side). Each pleural cavity surrounds one lung and contains a small amount of fluid. The central portion of the thoracic cavity is called the **mediastinum** (mē′-dē-as-TĪ-num; *me-*

Figure 1.6 / Body cavities. The dashed lines in (a) and (b) indicate the border between the abdominal and pelvic cavities. (See Tortora, *A Photographic Atlas of the Human Body,* Figures 6.6 and 11.13.)

The two principal body cavities are the dorsal and ventral body cavities.

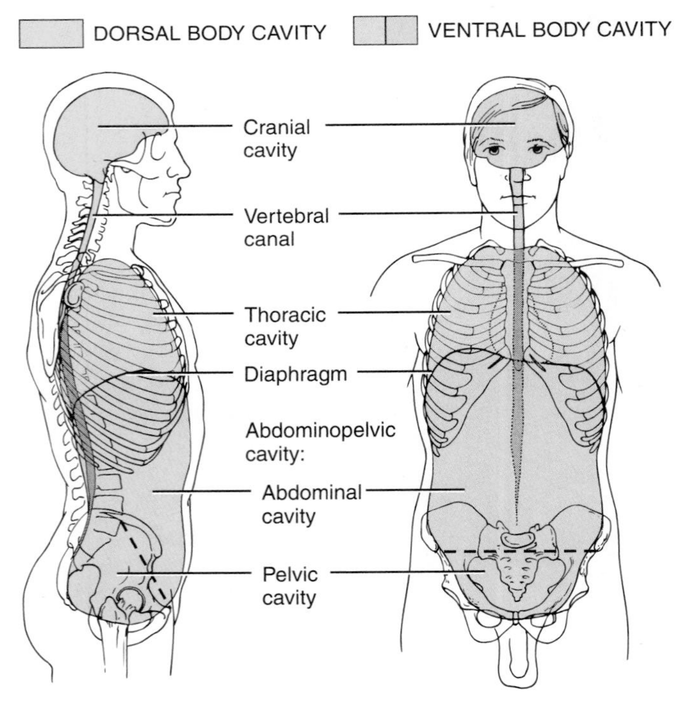

(a) Right lateral view (b) Anterior view

CAVITY	COMMENTS
DORSAL	
Cranial	Formed by cranial bones and contains brain and its coverings.
Vertebral	Formed by vertebral column and contains spinal cord and the beginnings of spinal nerves.
VENTRAL	
Thoracic	Chest cavity; separated from abdominal cavity by diaphragm.
Pleural (right and left)	Contain lungs.
Pericardial	Contains heart.
Mediastinum	Region between the lungs from the breastbone to backbone that contains heart, thymus, esophagus, trachea, bronchi, and many large blood and lymphatic vessels.
Abdominopelvic	Subdivided into abdominal and pelvic cavities.
Abdominal	Contains stomach, spleen, liver, gallbladder, small intestine, and most of large intestine.
Pelvic	Contains urinary bladder, portions of the large intestine, and internal female and male reproductive organs.

In which cavities are the following organs located: urinary bladder, stomach, heart, small intestine, lungs, internal female reproductive organs, thymus, spleen, liver? Use the following symbols for your response: T = thoracic cavity, A = abdominal cavity, or P = pelvic cavity.

Figure 1.7 / The thoracic cavity. The dashed lines in (a) and (c) indicate the borders of the mediastinum. Notice that the pericardial cavity surrounds the heart, and that the pleural cavities surround the lungs. **Note:** When transverse sections, such as those shown in (a) and (b), are viewed inferiorly (from below) the anterior aspect of the body appears on the top of the illustration and the left side of the body appears on the right side of the illustration. (See Tortora, *A Photographic Atlas of the Human Body,* Figure 6.6.)

The mediastinum is medial to the lungs; it extends from the sternum to the vertebral column and from the neck to the diaphragm.

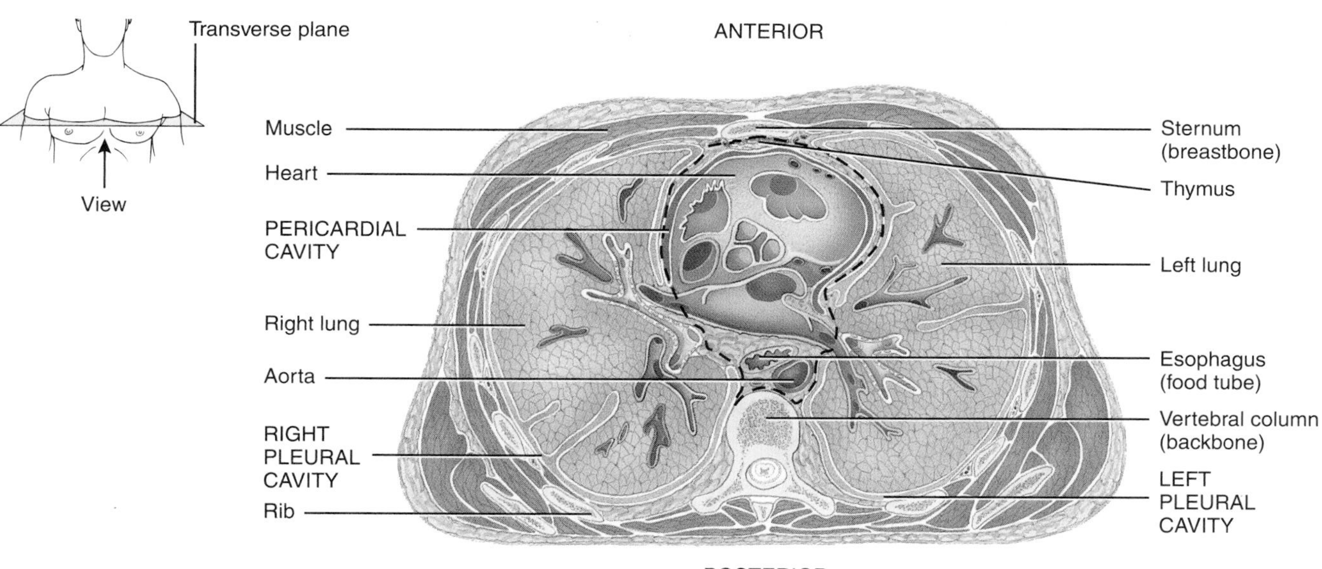

(a) Inferior view of transverse section of thoracic cavity

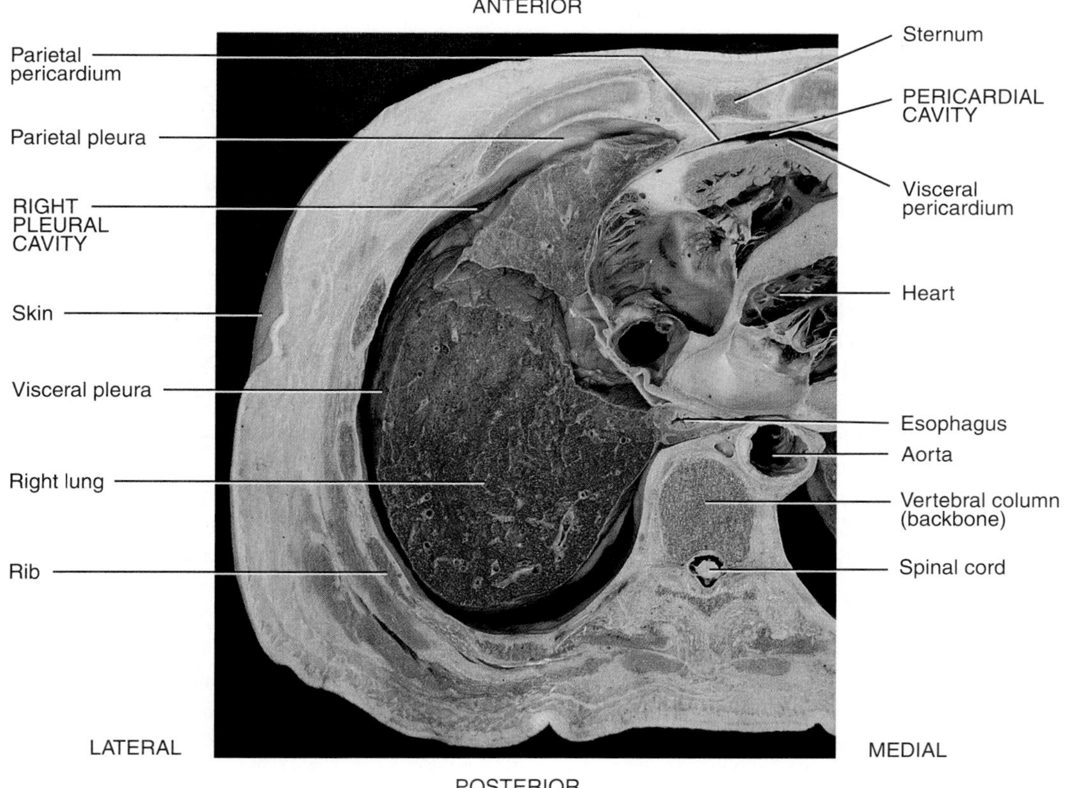

(b) Inferior view of transverse section of thoracic cavity

(Figure 1.7 continues)

Figure 1.7 / The thoracic cavity. (continued)

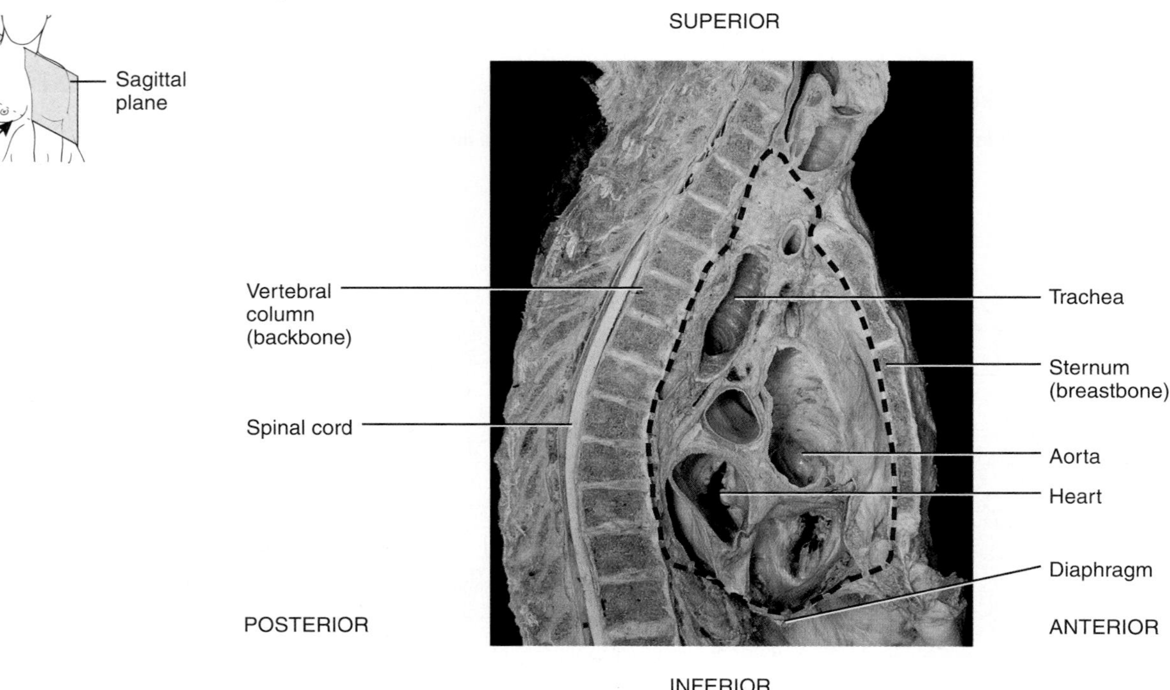

(c) Sagittal section of thoracic cavity

Which of the following structures are contained in the mediastinum: right lung, heart, esophagus, spinal cord, trachea, rib, thymus, left pleural cavity?

dia- = middle; *-stinum* = partition). It is located between the pleural cavities and extends from the sternum to the vertebral column, and from the neck to the diaphragm (Figure 1.7). The mediastinum contains all thoracic viscera except the lungs themselves. Among the structures in the mediastinum are the heart, esophagus, trachea, thymus, and several large blood vessels.

The inferior portion of the ventral body cavity is the **abdominopelvic cavity** (ab-dom′-i-nō-PEL-vik; Figure 1.8), which extends from the diaphragm to the groin and is encircled by the abdominal wall and the bones and muscles of the pelvis. As the name suggests, the abdominopelvic cavity is divided into two portions, even though no wall separates them (Figure 1.8). The superior portion, the **abdominal cavity** (*abdomin-* = belly), contains the stomach, spleen, liver, gallbladder, small intestine, and most of the large intestine. The inferior portion, the **pelvic cavity** (*pelv-* = basin), contains the urinary bladder, portions of the large intestine, and internal organs of the reproductive system.

Thoracic and Abdominal Cavity Membranes

A thin, slippery **serous membrane** covers the viscera within the thoracic and abdominal cavities and also lines the walls of the thorax and abdomen. The parts of a serous membrane are (1) the *parietal layer* (pa-RĪ-e-tal), which lines the walls of the cavities, and (2) the *visceral layer*, which covers and adheres to the viscera within the cavities. Serous fluid between the two layers reduces friction, allowing the viscera to slide somewhat during movements—for example, when the lungs expand and deflate when a person is breathing.

The serous membrane of the pleural cavities is called the **pleura** (PLŪ-ra) (see Figure 1.7b). The *visceral pleura* clings to the surface of the lungs, whereas the *parietal pleura* lines the chest wall. In between is the pleural cavity, filled with a small volume of serous fluid. The serous membrane of the pericardial cavity is the **pericardium** (see Figure 1.7b). The *visceral pericardium* covers the surface of the heart, whereas the *parietal pericardium* lines the chest wall. Between them is the pericardial cavity. The **peritoneum** (per-i-tō-NĒ-um) is the serous membrane of the abdominal cavity (see Figure 24.3a on page 719). The *visceral peritoneum* covers the abdominal viscera, whereas the *parietal peritoneum* lines the abdominal wall. Between them is the *peritoneal cavity*. Most abdominal organs are located in the peritoneal cavity. Some are located behind the parietal peritoneum—they are between it and the posterior abdominal wall. Such organs are said to be *retroperitoneal* (re′-trō-per-I-tō-NĒ-al; *retro* = behind). Examples include the kidneys, adrenal glands, pancreas, duodenum of the small intestine, ascending and descending

Figure 1.8 / The abdominopelvic cavity. The horizontal dashed black line approximates the point of separation of the abdominal and pelvic cavities. (See Tortora, *A Photographic Atlas of the Human Body,* Figure 12.2.)

The abdominopelvic cavity extends from the diaphragm to the groin.

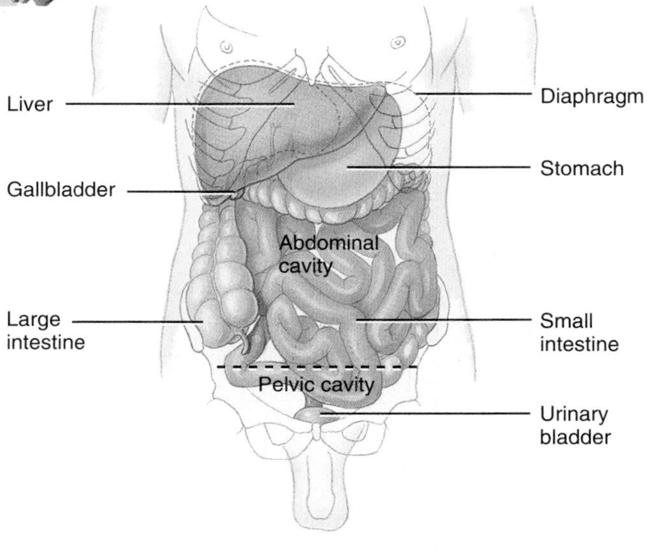

To which body systems do the organs shown here within the abdominal and pelvic cavities belong? (Hint: *Refer to Table 1.2 on pages 4–6*)

Table 1.3 Summary of Body Cavities and Their Membranes

Cavity	Comments
Dorsal	
Cranial	Formed by cranial bones and contains brain.
Vertebral	Formed by vertebral column and contains spinal cord.
Ventral	
Thoracic	Superior portion of ventral body cavity; contains pleural and pericardial cavities and the mediastinum.
Pleural	Each surrounds a lung; the serous membrane of the pleural cavities is the pleura.
Pericardial	Surrounds the heart; the serous membrane of the pericardial cavity is the pericardium.
Mediastinum	Central portion of thoracic cavity between the pleural cavities; extends from sternum to vertebral column and from neck to diaphragm; contains heart, thymus, esophagus, trachea, and several large blood vessels.
Abdominopelvic	Inferior portion of ventral body cavity; subdivided into abdominal and pelvic cavities.
Abdominal	Contains stomach, spleen, liver, gallbladder, small intestine, and most of large intestine; the serous membrane of the abdominal cavity is the peritoneum.
Pelvic	Contains urinary bladder, portions of large intestine, and internal organs of reproduction.

colons of the large intestine, and portions of the abdominal aorta and inferior vena cava.

A summary of body cavities and their membranes is presented in Table 1.3.

✓ What structures separate the various body cavities from one another?

✓ What is the mediastinum?

Abdominopelvic Regions and Quadrants

Objective

- Name and describe the abdominopelvic regions and the abdominopelvic quadrants.

To describe the location of the many abdominal and pelvic organs more easily, anatomists and clinicians use two methods of dividing the abdominopelvic cavity into smaller compartments. In the first method, two transverse and two vertical lines, aligned like a tick-tack-toe grid, partition this cavity into nine **abdominopelvic regions** (Figure 1.9a). The *subcostal* (top transverse) *line* is drawn just inferior to the rib cage, across the inferior portion of the stomach; the *transtubercular* (bottom transverse) *line* is drawn just inferior to the tops of the hip bones. The left and right *midclavicular* (two vertical) *lines* are drawn through the midpoints of the clavicles (collar bones), just medial to the nipples. The four lines divide the abdominopelvic cavity into a larger middle section and smaller left and right sections. The names of the nine abdominopelvic regions are **right hypochondriac, epigastric, left hypochondriac, right lumbar, umbilical, left lumbar, right inguinal (iliac), hypogastric (pubic),** and **left inguinal (iliac).** Note which organs and parts of organs are in the different regions by carefully examining Figure 1.9b–d and Table 1.4 on page 17. Figure 1.9b–d illustrates successively deeper views of the abdominopelvic contents in their respective regions. Although some parts of the body in these illustrations are most likely unfamiliar to you at this point, they will be discussed in detail in later chapters.

The second method is simpler and divides the abdominopelvic cavity into **quadrants** (KWOD-rantz; *quad-* = one-fourth), as shown in Figure 1.10 on page 18. In this method, a transverse plane and a midsagittal plane are passed through the **umbilicus** (um-bi-LĪ-kus; *umbilic-* = navel) or *belly button.* The names of the abdominopelvic quadrants are **right upper quadrant (RUQ), left upper quadrant (LUQ), right lower quadrant (RLQ),** and **left lower quadrant (LLQ).** Whereas the nine-region division is more widely used for anatomical studies, quadrants are more commonly used by clinicians for describing the site of abdominopelvic pain, tumor, or other abnormality.

Figure 1.9 / Abdominopelvic regions. (a) The nine regions. (b) The greater omentum has been removed. The greater omentum is a serous membrane that contains fatty tissue which covers some of the abdominal organs. (c) Some anterior organs have been removed, exposing the more posterior structures. (d) All anterior organs removed, fully revealing posterior structures. The internal reproductive organs in the pelvic cavity are shown in Figures 26.3 on page 784 and 26.13 on page 796.

The nine-region designation is used for anatomical studies.

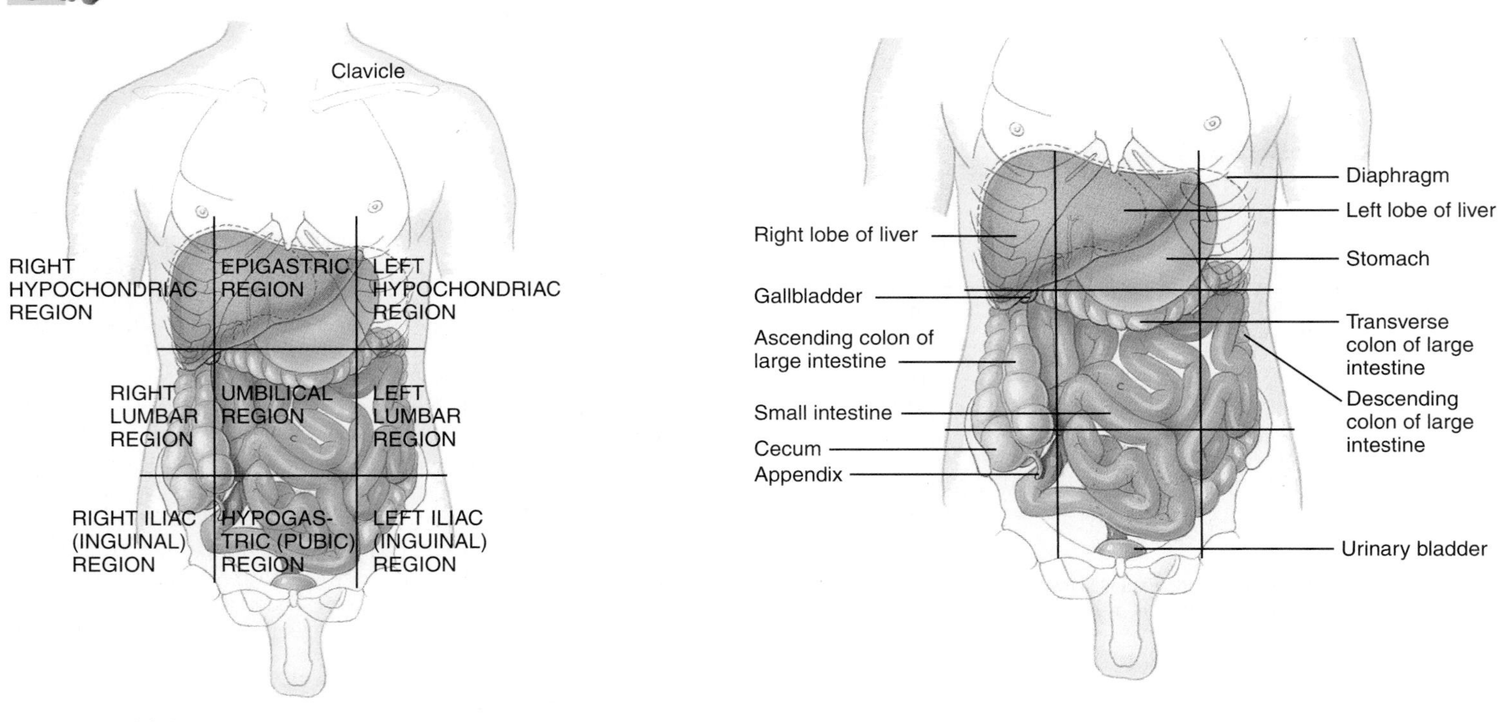

(a) Anterior view showing location of nine abdominopelvic regions

(b) Anterior superficial view

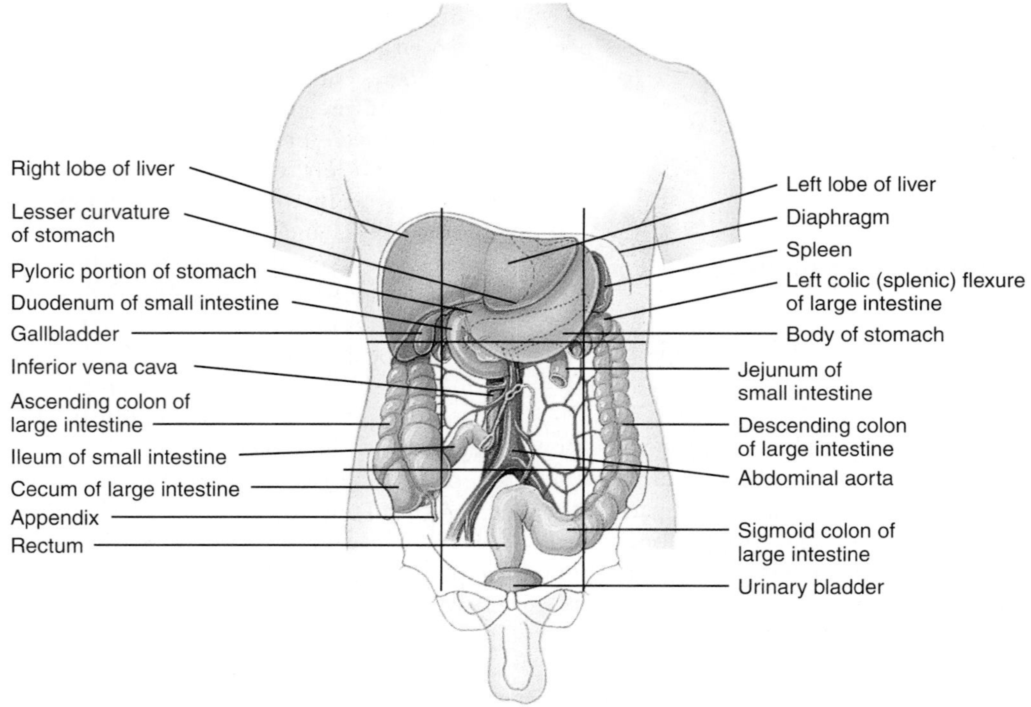

(c) Anterior deep view

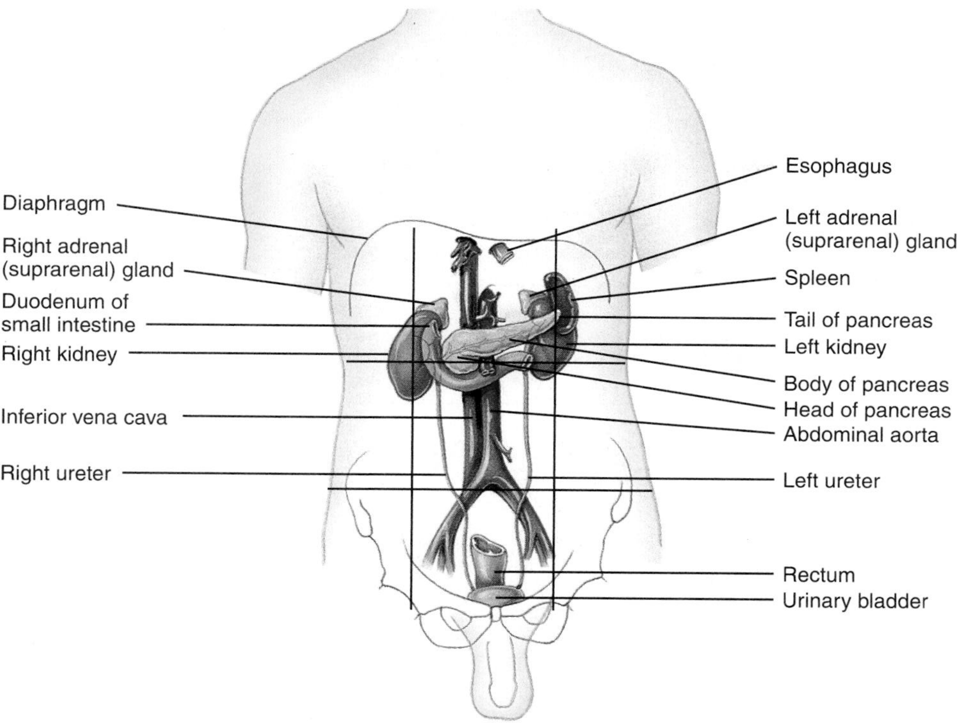

(d) Anterior deeper view

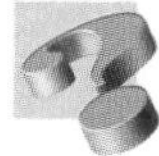

In which abdominopelvic region is each of the following found: left lobe of the liver, transverse colon, urinary bladder, spleen?

Table 1.4 Representative Structures Found in the Abdominopelvic Regions

Region	Representative Structures
Right hypochondriac (hī-pō-KON-drē-ak; *hypo* = under; *chondro* = cartilage)	Right lobe of liver, gallbladder, and upper superior third of right kidney.
Epigastric (ep-i-GAS-trik; *epi* = above; *gaster* = stomach)	Left lobe and medial part of right lobe of liver, pyloric portion and lesser curvature of stomach, superior and descending portions of duodenum, body and superior part of head of pancreas, and right and left adrenal (suprarenal) glands.
Left hypochondriac	Body and fundus of stomach, spleen, left colic (splenic) flexure, superior two-thirds of left kidney, and tail of pancreas.
Right lumbar (*lumbus* = loin)	Superior part of cecum, ascending colon, right colic (hepatic) flexure, inferior lateral portion of right kidney, and part of small intestine.
Umbilical (um-BIL-i-kul; *umbilikus* = navel)	Middle portion of tranverse colon, part of small intestine, and bifurcations (branching) of abdominal aorta and inferior vena cava.
Left lumbar	Descending colon, inferior third of left kidney, and part of small intestine.
Right iliac (inguinal) (IL-ē-ak; iliac refers to superior part of hip bone)	Lower end of cecum, appendix, and part of small intestine.
Hypogastric (pubic)	Urinary bladder when full, small intestine, and part of sigmoid colon.
Left iliac (inguinal)	Junction of descending and sigmoid parts of colon and part of small intestine.

▶ 14th Century

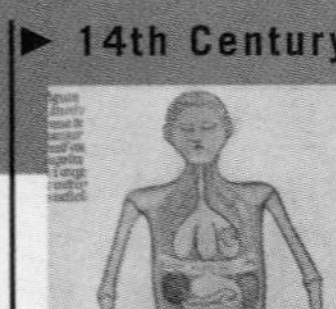

▶ 15th Century

▶ 16th Century

▶ 17th Century

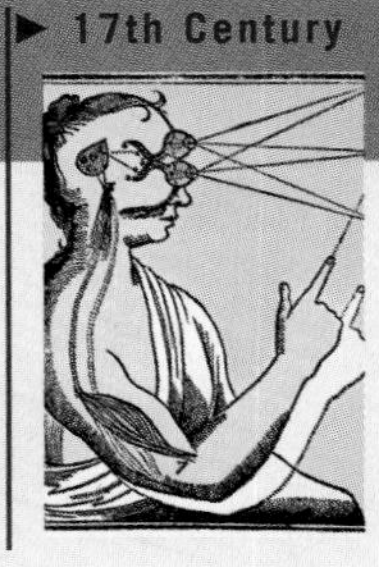

▶ 18th Century

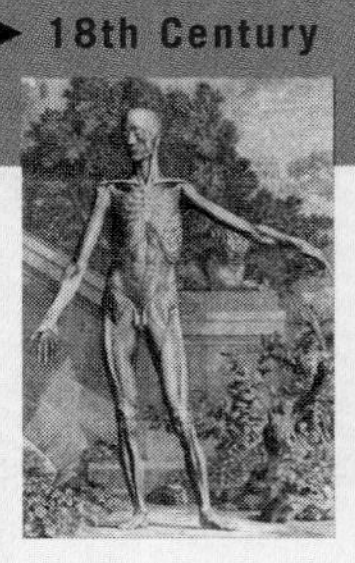

▶ 20th Century

▶ 21st Century

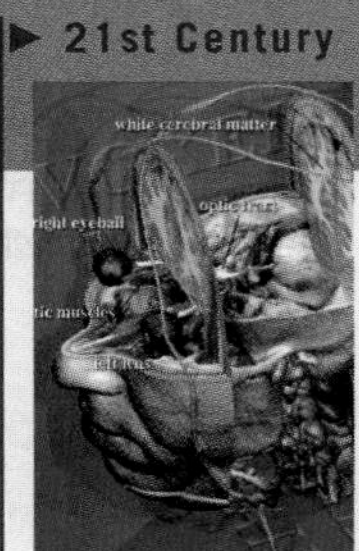

CHANGING IMAGES

Seeing is Believing. Or is It?

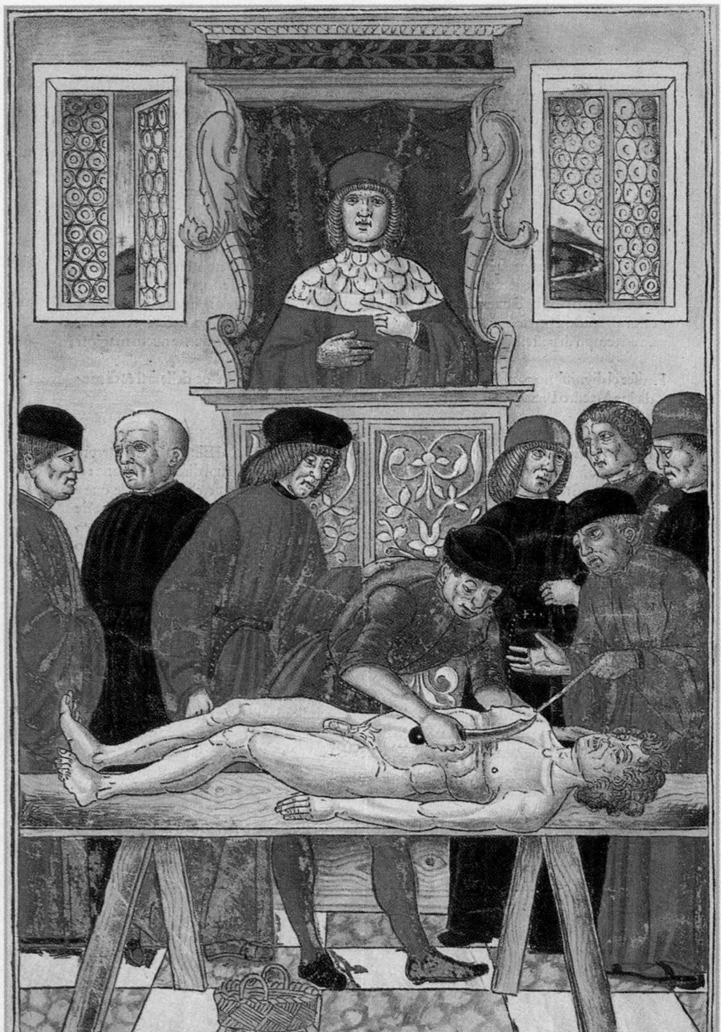

1493 AD

Human anatomy is perhaps the oldest biologic science. Its saga, however, is full of dogmatic teachings that often hindered its progress. The voluminous anatomical writings of Claudius Galen, a Greek living at the peak of the Roman Empire, were the foundation for an authoritarian attitude that persisted for nearly 1500 years. Since it was believed that all the anatomy that was needed to be known was in the writings of Galen, direct observation via dissection was infrequent throughout medieval Europe.

Pictured here is the frontispiece from one of Europe's most respected, Galen-based anatomical texts, John of Ketham's *Fasciculus Medicinae*, circa 1493. Here we see the *lector*, seated in his lofty *cathedra*, recite from Galenic texts. Professors and medical students stand about the cadaver while the short robed, lesser status *sector* performs the dissection. Despite glaring inconsistencies between Galen and what was directly observed, the lesson would be molded within the confines of his writings and rarely challenged.

Today, we too, gather around the cadaver while referencing texts like this one. Yet, other aspects of anatomical study have changed. Your professor does not condescend, but stands among you; students are encouraged to critically evaluate what they observe; and academic robes are replaced with modern dress. As you join the ranks of centuries of anatomy students, consider how far this old and proud science has come. As you progress through this text, you will be introduced to ideas that suggest where this discipline is headed and what excitement it holds for you.

Figure 1.10 / Quadrants of the abdominopelvic cavity. The two lines intersect at right angles at the umbilicus (navel).

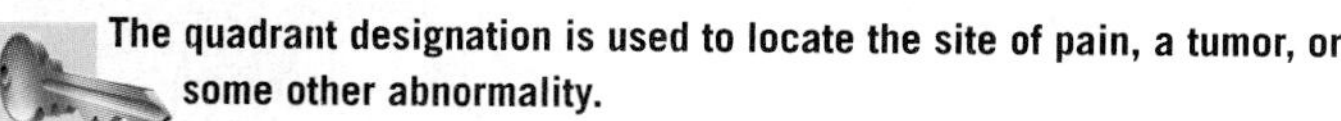
The quadrant designation is used to locate the site of pain, a tumor, or some other abnormality.

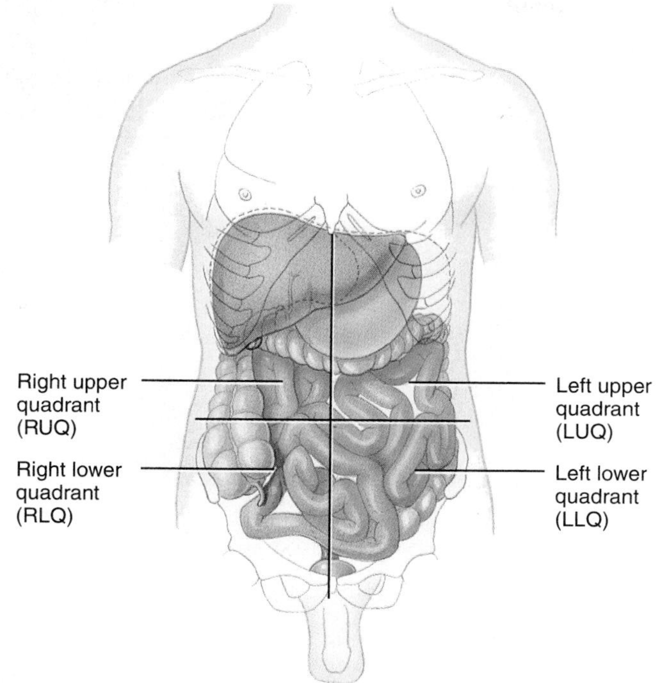

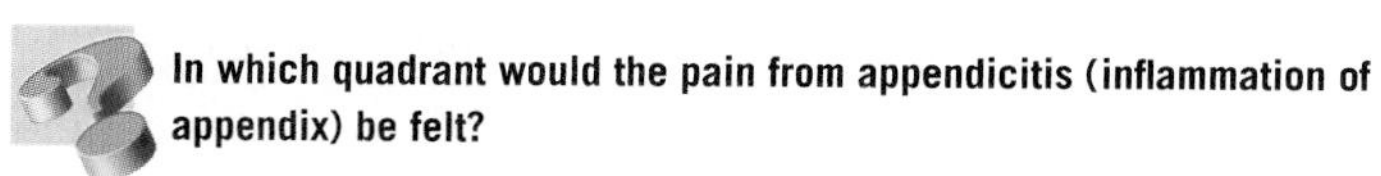
In which quadrant would the pain from appendicitis (inflammation of the appendix) be felt?

✓ Locate the abdominopelvic regions and the abdominopelvic quadrants on yourself, and list some of the organs found in each.

Autopsy

An **autopsy** (AW-top-sē; *auto-* = self; *-opsy* = to see) is a postmortem (after death) examination of the body and dissection of its internal organs to confirm or determine the cause of death. An autopsy can uncover the existence of diseases not detected during life, determine the extent of injuries, and explain how those injuries may have contributed to a person's death. It also may support the accuracy of diagnostic tests, establish the beneficial and adverse effects of drugs, reveal the effects of environmental influences on the body, provide more information about a disease, assist in the accumulation of statistical data, and educate health-care students. Moreover, an autopsy can reveal conditions that may affect offspring or siblings (such as congenital heart defects). An autopsy may be legally required, such as in the course of a criminal investigation, or to resolve disputes between beneficiaries and insurance companies about the cause of death.

A typical autopsy consists of three principal phases. The first phase is examination of the exterior of the body for the presence of wounds, scars, tumors, or other abnormalities. The second phase includes the dissection and gross examination of the major body organs, looking for pathological changes or evidence of violent destruction. The third phase of autopsy consists of microscopic examination of tissues to ascertain any pathology. Depending on the circumstances, techniques may also be used to detect and/or recover microbes or toxic chemicals and to determine the presence of foreign substances in the body. ■

MEDICAL IMAGING

Objective

- Describe the principles and importance of medical imaging procedures in the evaluation of organ functions and the diagnosis of disease.

Various kinds of **medical imaging** procedures allow visualization of structures inside our bodies and are increasingly helpful for precise diagnosis of a wide range of anatomical and physiological disorders. The grandparent of all medical imaging techniques is conventional radiography (x-rays), in medical use since the late 1940s. The newer imaging technologies not only contribute to diagnosis of disease, but they also are advancing our understanding of normal physiology. Table 1.5 describes some commonly used medical imaging techniques.

✓ Which forms of medical imaging use radiation? Which do not?

✓ Which imaging modality best reveals the physiology of a structure?

✓ Which imaging modality would best show a broken bone?

MEASURING THE HUMAN BODY

An important aspect of describing the body and understanding how it works is *measurement*—what the dimensions of an organ are, how much it weighs, how long it takes for a physiological event to occur. Such measurements also have clinical importance, for example, in determining how much of a given medication should be administered. As you will see, measurements involving time, weight, temperature, size, and volume are a routine part of your studies in a medical science program.

Whenever you come across a measurement in this text, the measurement will be given in metric units. The metric system is standardly used in the sciences. To help you compare the metric unit to a familiar unit, the approximate U.S. equivalent will also be given in parentheses directly after the metric unit. For example, you might be told that the length of a particular part of the body is 2.54 cm (1 in.).

To help you understand the correlation between the metric system and the U.S. system of measurement, there are three exhibits located in Appendix I: (1) metric units of length and some U.S. equivalents, (2) metric units of mass and some U.S. equivalents, and (3) metric units of volume and some U.S. equivalents.

Table 1.5 Common Medical Imaging Procedures

Radiography

Procedure: A single barrage of x-rays passes through the body, producing an image of interior structures on x-ray–sensitive film. The resulting two-dimensional image is a *radiograph* (RĀ-dē-ō-graf′), commonly called an *x-ray.*

Comments: Produces clear images of bony or dense structures, which appear bright, but poor images of soft tissues or organs, which appear hazy or dark.

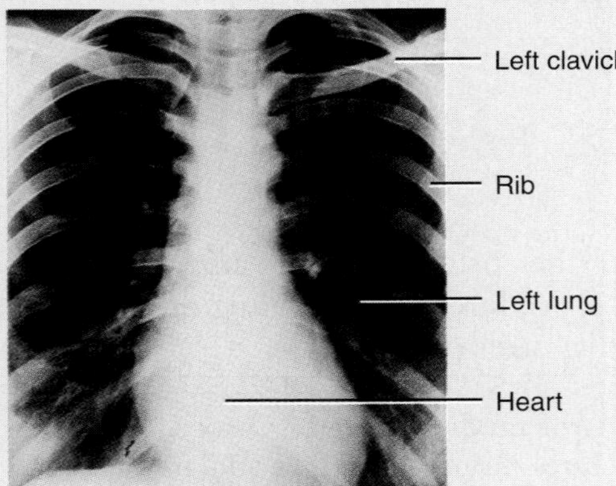

Anterior view of thorax

Digital Subtraction Angiography (DSA)

Procedure: A computer compares radiographs of a body region before and after a dye is injected into blood vessels. Tissues around the blood vessels are erased (digitally subtracted) from the second image. The result, shown on a monitor, is an unobstructed view of the blood vessels.

Comments: Used primarily to study blood vessels in the brain and heart.

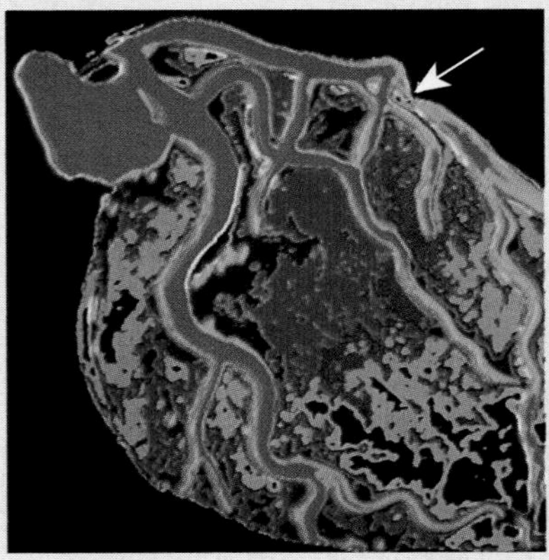

Blood vessels (red) surrounding heart (arrow indicates narrowed vessel)

Computed Tomography (CT)

[formerly called computerized axial tomography (CAT) scanning]

Procedure: Computer-assisted radiography in which an x-ray beam traces an arc at multiple angles around a section of the body. The resulting transverse section of the body, called a *CT scan,* is reproduced on a video monitor.

Comments: Visualizes soft tissues and organs with much more detail than conventional radiographs. Differing tissue densities show up as various shades of gray. Multiple scans can be assembled to build three-dimensional views of structures.

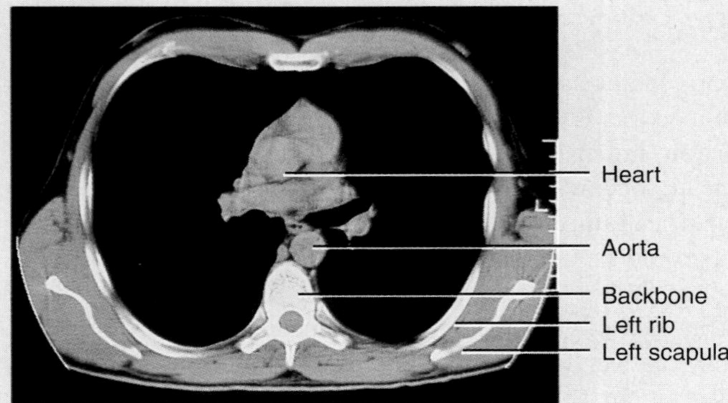

Inferior view of transverse section of the thorax

Sonography

Procedure: High-frequency sound waves produced by a handheld wand reflect off body tissues and are detected by the same instrument. The image, which may be still or moving, is called a *sonogram* (SŌ-nō-gram) and is reproduced on a video monitor.

Comments: Safe, noninvasive, painless, and uses no dyes. Most commonly used to visualize the fetus during pregnancy. Also used to observe the size, location, and actions of organs and blood flow through blood vessels.

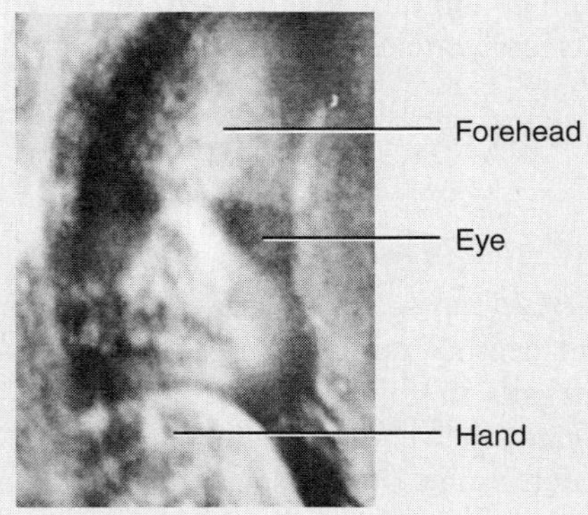

Courtesy of Andrew Joseph Tortora and Damaris Soler

Magnetic Resonance Imaging (MRI)

Procedure The body is exposed to a high-energy magnetic field, which causes protons (small positive particles within atoms, such as hydrogen) in body fluids and tissues to arrange themselves in relation to the field. Then a pulse of radiowaves "reads" these ion patterns, and a color-coded image is assembled on a video monitor. The resulting image is a two- or three-dimensional blueprint of cellular chemistry.

Comments Relatively safe, but can't be used on patients with metal in their bodies. Shows fine details for soft tissues but not for bones. Most useful for differentiating between normal and abnormal tissues. Used to detect tumors and artery-clogging fatty plaques, reveal brain abnormalities, and measure blood flow.

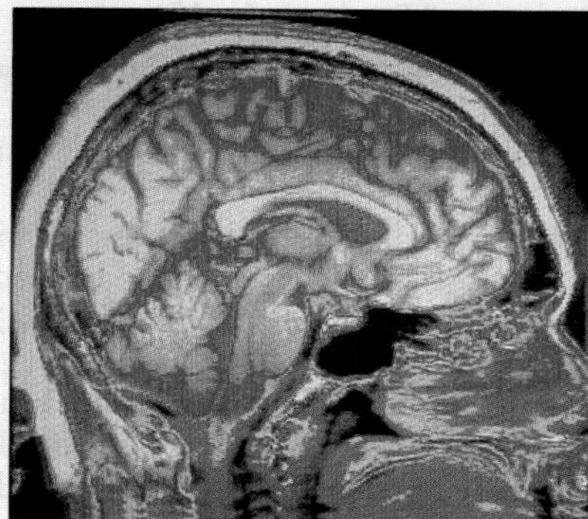

Sagittal section of brain

Positron Emission Tomography (PET)

Procedure A substance that emits positrons (positively charged particles) is injected into the body, where it is taken up by tissues. The collision of positrons with negatively charged electrons in body tissues produces gamma rays (similar to x-rays) that are detected by gamma cameras positioned around the subject. A computer receives signals from the gamma cameras and constructs a *PET scan* image, displayed in color on a video monitor. The PET scan shows where the injected substance is being used in the body.

Comments Used to study the physiology of body structures, such as metabolism in the brain or heart.

ANTERIOR

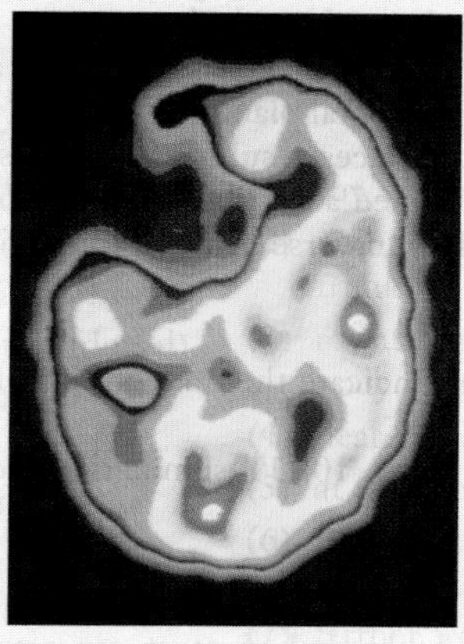

POSTERIOR

Transverse section showing blood flow through brain (darkened area at upper left indicates where a stroke has occurred)

STUDY OUTLINE

Anatomy Defined (p. 2)

1. Anatomy is the science of body structures and the relationships among structures; physiology is the science of body functions.
2. Some subdisciplines of anatomy are embryology, developmental anatomy, cytology, histology, surface anatomy, gross anatomy, systemic anatomy, regional anatomy, radiographic anatomy, and pathological anatomy (see Table 1.1 on page 2).

Levels of Body Organization (p. 2)

1. The human body consists of six levels of structural organization: chemical, cellular, tissue, organ, system, and organismal levels.
2. Cells are the basic structural and functional living units of an organism and the smallest living units in the human body.
3. Tissues are groups of cells and the materials surrounding them that work together to perform a particular function.
4. Organs are composed of two or more different types of tissues; they have specific functions and usually have recognizable shapes.
5. Systems consist of related organs that have a common function.
6. An organism is any living individual.
7. Table 1.2 on pages 4–6 introduces the eleven systems of the human organism: the integumentary, skeletal, muscular, nervous, endocrine, cardiovascular, lymphatic and immune, respiratory, digestive, urinary, and reproductive systems.

Life Processes (p. 6)

1. All organisms carry on certain processes that distinguish them from nonliving things.
2. The most important life processes of the human body are metabolism, responsiveness, movement, growth, differentiation, and reproduction.

Basic Anatomical Terminology (p. 7)

Anatomical Position (p. 7)

1. Descriptions of any region of the body assume the body is in the anatomical position, in which the subject stands erect facing the observer, with the head level and the eyes facing directly forward. The feet are flat on the floor and directed forward, and the arms are at the sides, with the palms turned forward.
2. A body lying face down is prone, whereas a body lying face up is supine.

ANSWERS TO FIGURE QUESTIONS

1.1 Organs have two or more different types of tissues that work together to perform a specific function.

1.2 Having one standard anatomical position allows directional terms to be clear, and any body part can be described in relation to any other part.

1.3 The frontal plane divides the heart into anterior and posterior portions.

1.4 The parasagittal plane divides the brain into unequal right and left portions.

1.5 No, No, Yes, Yes, No.

1.6 P, A, T, A, T, P, T, A, A.

1.7 Structures in the mediastinum include the heart, esophagus, trachea, and thymus.

1.8 The illustrated abdominal cavity organs all belong to the digestive system (liver, gallbladder, stomach, appendix, small intestine, and most of the large intestine). Illustrated pelvic cavity organs belong to the urinary system (the urinary bladder) and the digestive system (part of the large intestine).

1.9 The liver is mostly in the epigastric region; the transverse colon is in the umbilical region; the urinary bladder is in the hypogastric region; the spleen is in the left hypochondriac region.

1.10 The pain associated with appendicitis would be felt in the right lower quadrant (RLQ).

2

CELLS

CONTENTS AT A GLANCE

◄ Page 48

Do you believe that mice can arise from sweaty undergarments? Early biologists did.

Page 43 ►

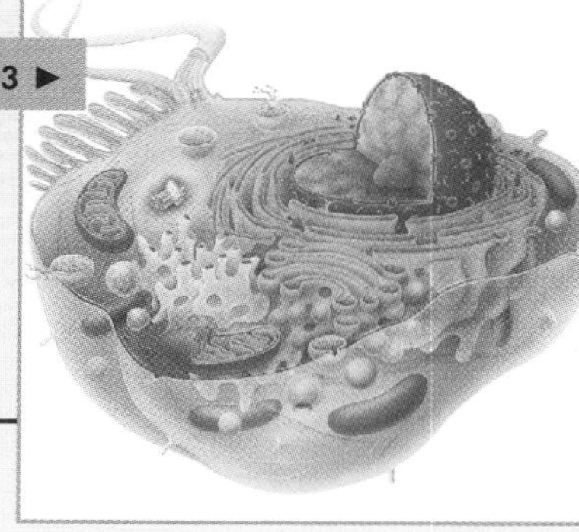

INTRODUCTION

Cells are the basic, living, structural and functional units of the body. Cells are composed of characteristic parts, whose coordinated functioning enables each cell type to fulfill a unique biochemical or structural role. **Cytology** (sī-TOL-ō-jē; *cyt-* = cell; *-ology* = study of) is the study of cellular structure, whereas **cell physiology** is the study of cellular function. As you study the various parts of a cell and their relationships to each other, you will learn that cell structure and function are interdependent and inseparable. Cells perform numerous chemical reactions to create life processes. One way that they do this is by compartmentalization, the isolation of specific kinds of chemical reactions within specialized structures inside the cell. The isolated reactions are coordinated with one another to maintain life in a cell, tissue, organ, system, and organism.

In very general terms, cells perform several basic kinds of work. For example, they regulate the inflow and outflow of materials to ensure that optimum conditions for life processes prevail inside the cell. Cells also use their genetic information (DNA) to ultimately guide the synthesis of most of their components and direct most of their chemical activities. Among these activities are the generation of ATP from the breakdown of nutrients, synthesis of molecules, transportation of molecules within and between cells, waste removal, and movement of cell parts or even entire cells.

A GENERALIZED CELL

Objective

- Name and describe the three principal parts of a cell.

Figure 2.1 is a cell that is a composite of many different cell types in the body. Most cells have many of the features shown in this diagram, but not necessarily all of them. For ease of study, we can divide a cell into three principal parts: the plasma membrane, cytoplasm, and nucleus.

- The **plasma membrane** forms the cell's sturdy outer surface, separating the cell's internal environment from the en-

Figure 2.1 / Generalized body cell.

The cell is the basic, living, structural and functional unit of the body.

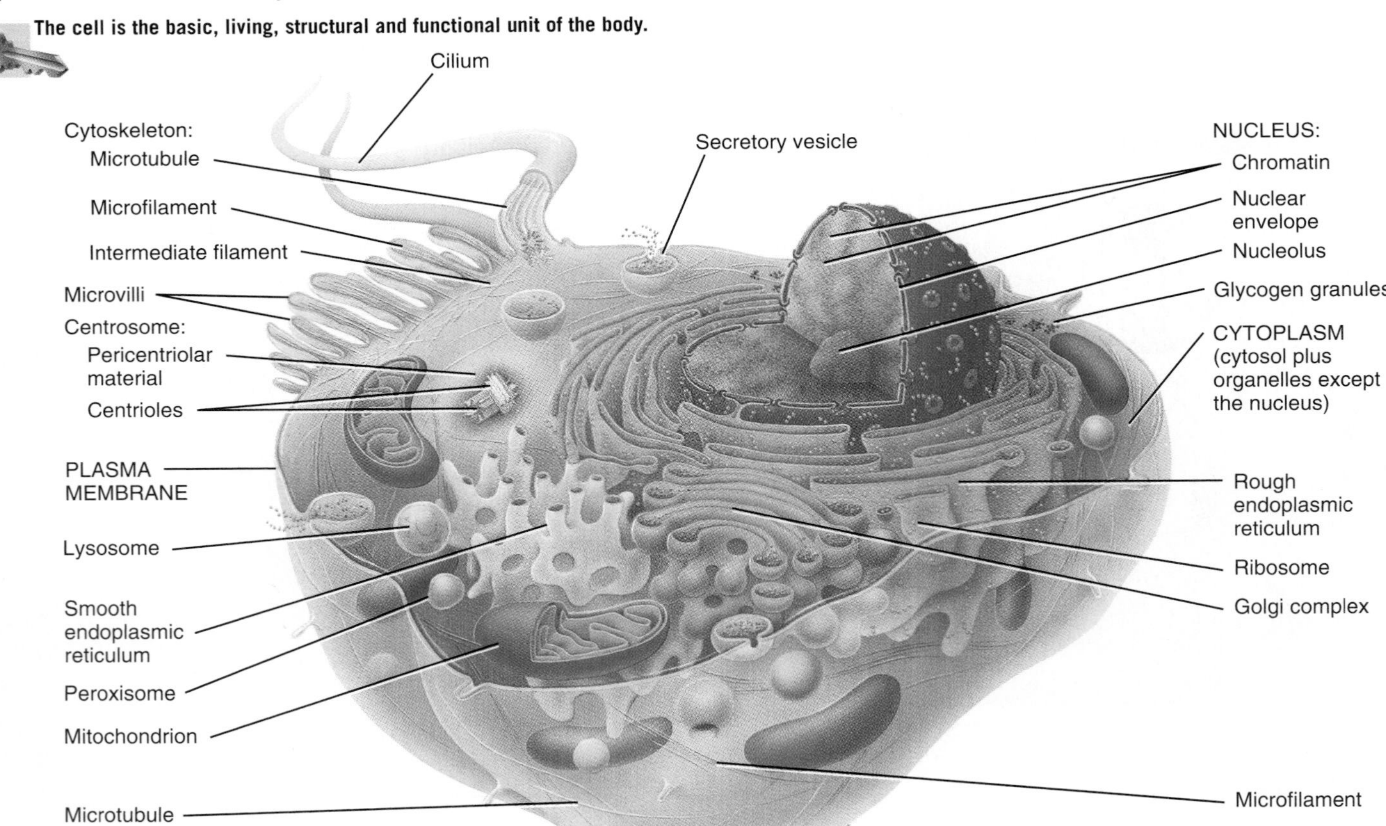

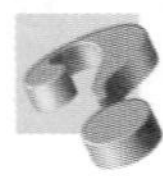

What are the three principal parts of a cell?

vironment outside the cell. The plasma membrane is a selective barrier that orchestrates the flow of materials into and out of a cell to establish and maintain the appropriate environment for normal cellular activities. It also plays a key role in communication both between cells and between cells and their external environment.

- The **cytoplasm** (SĪ-tō-plazm; *-plasm* = formed or molded) is all the cellular contents between the plasma membrane and the nucleus. This compartment has two components: the cytosol and organelles. **Cytosol** (SĪ-tō-sol) is the fluid portion of cytoplasm that consists mostly of water plus dissolved solutes and suspended particles. **Organelles** (or-ga-NELZ; = little organs) are highly organized subcellular structures, each having a characteristic shape and specific functions. Examples are the cytoskeleton, ribosomes, endoplasmic reticulum, Golgi complex, lysosomes, peroxisomes, and mitochondria.
- The **nucleus** (NŪ-klē-us; = nut kernel), although technically an organelle, is considered separately because of its numerous and diverse functions. Within the nucleus are chromosomes on which lie genes. Genes control cellular structure and most cellular activities.

THE PLASMA MEMBRANE

Objectives

- Describe the structure and functions of the plasma membrane.
- Describe the processes that transport substances across the plasma membrane.

The **plasma membrane** is a flexible yet sturdy barrier that surrounds and contains the cytoplasm of a cell. The *fluid mosaic model* describes its structure: The molecular arrangement of the plasma membrane resembles an ever-moving sea of lipids that contains a "mosaic" of many different proteins (Figure 2.2). The proteins may float freely like icebergs, be moored at specific locations, or be moved through the lipid sea. The lipids act as a barrier to the entry or exit of various substances, while some of the proteins in the plasma membrane act as "gatekeepers" that regulate the passage of other molecules and ions (charged particles) into and out of the cell. The plasma membranes of typical body cells are about a 50:50 mix by weight of proteins and lipids, but because membrane proteins are larger and more massive than membrane lipids, there are about 50 lipid molecules for each protein molecule.

Figure 2.2 / Structure of the plasma membrane.

The basic framework of the plasma membrane is the lipid bilayer.

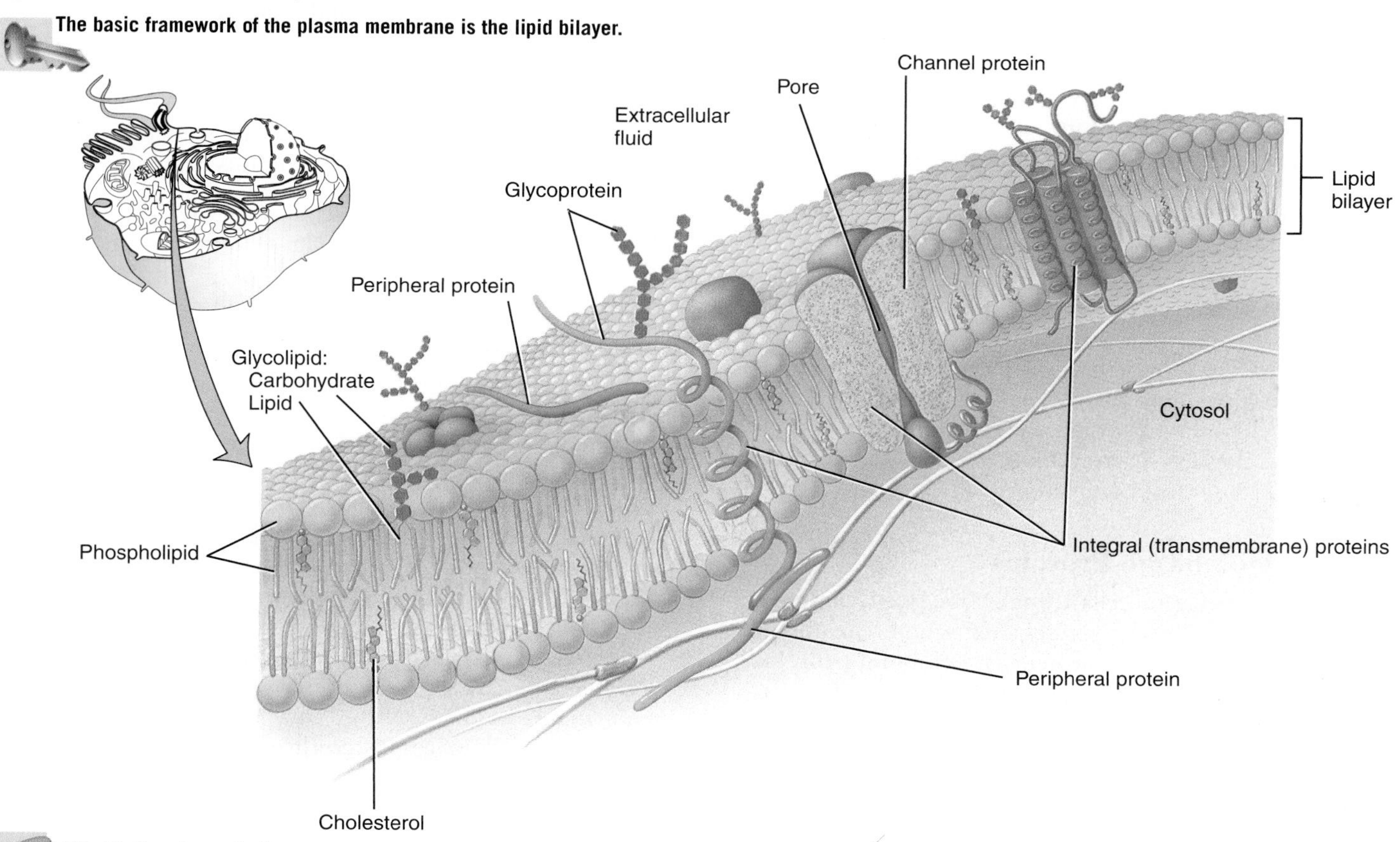

What is the glycocalyx?

Structure of the Membrane

The basic structural framework of the plasma membrane is the **lipid bilayer,** two back-to-back layers made up of three types of lipid molecules—phospholipids, cholesterol, and glycolipids (Figure 2.2). About 75% of the membrane lipids are *phospholipids*, lipids that contain phosphate groups. *Glycolipids*, which are lipids attached to carbohydrates (*glyco* = carbohydrate), account for about 5% of membrane lipids. The remaining 20% of plasma membrane lipids are *cholesterol* molecules, which are interspersed among the other lipids in both layers of the membrane.

There are several ways that membrane proteins may be associated with the lipid bilayer.

(1) Some stretch across the entire bilayer and project on both sides of the membrane. These are called *transmembrane proteins* (Figure 2.2). Many transmembrane proteins are *glycoproteins*, carbohydrates attached to proteins, that protrude into the extracellular fluid. As you will see shortly, different kinds of transmembrane proteins have different functions.

(2) Other membrane proteins are located at the outer surface of the membrane where they are attached to the lipid portion of the membrane (not illustrated). Both transmembrane proteins and lipid-linked proteins are together referred to as **integral proteins** because they can be separated from the membrane only by procedures that disrupt the lipid bilayer.

(3) Some proteins are attached to integral membrane proteins at the inner surface on the plasma membrane or the lipids at the outer surface of the membrane and are referred to as **peripheral proteins** (Figure 2.2). They are so named because they can be released from the membrane by relatively mild extraction procedures that leave the lipid bilayer intact.

The carbohydrate portions of glycolipids and glycoproteins form an extensive surface coat called the **glycocalyx** (gli′-kō-KĀL-iks), which has several important functions. It acts like a molecular "signature" that enables cells to recognize one another. For example, a white blood cell's ability to detect a "foreign" glycocalyx is one basis of the immune response that helps the body to destroy invading organisms. In addition, the glycocalyx enables cells to adhere to one another in some tissues, and it protects the cell from being digested by enzymes in the extracellular fluid. The glycocalyx also attracts a film of fluid to the surface of many cells, an action that makes red blood cells slippery as they flow through narrow blood vessels and protects cells that line the airways from drying out.

Functions of Membrane Proteins

The membranes of different cells feature remarkably different assortments of proteins, which largely determine membrane functions. Some transmembrane proteins are **channel proteins** that have a *pore*, or hole, through which a specific substance, such as potassium ions (K^+), can flow. Most channel proteins are *selective;* they allow only a single type of ion to pass through. Other transmembrane proteins act as transporters. **Transporters (carrier proteins)** bind a substance at one side of the membrane and move it to the other side of the membrane where it is released. The transport of substances is accomplished by a change in shape of the transporter. Integral proteins called **receptors** serve as cellular recognition sites. These proteins recognize and bind a specific molecule, such as the hormone insulin or a nutrient such as glucose, that is important for some cellular function. A specific molecule that binds to a receptor is called a *ligand* (LĪ-gand; *liga-* = tied) of that receptor. Some integral and peripheral proteins are **enzymes.** Membrane glycoproteins and glycolipids often are **cell-identity markers.** They may enable a cell to recognize other cells of the same kind during tissue formation or to recognize and respond to potentially dangerous foreign cells. The ABO blood type markers are one example of cell-identity markers. When you receive a blood transfusion, the blood type must be compatible with your own. Integral and peripheral proteins may serve as **linkers,** which anchor proteins in the plasma membranes of neighboring cells to one another or to filaments inside and outside the cell.

Membrane Permeability

A membrane is *permeable* to things that can pass through it and *impermeable* to those that cannot. Although plasma membranes are not completely permeable to any substance, they do permit some substances to pass more readily than others. This property of membranes is called **selective permeability.**

The lipid bilayer portion of the membrane is permeable to some molecules such as oxygen, carbon dioxide, and steroids, but is impermeable to ions and molecules such as glucose. It is also permeable to water. Transmembrane proteins that act as channels and transporters increase the plasma membrane's permeability to a variety of small- and medium-sized charged substances (including ions) that cannot cross the lipid bilayer. These proteins are very selective—each one helps only a specific molecule or ion to cross the membrane. Macromolecules, such as proteins, cannot pass through the plasma membrane except by vesicular transport (discussed later in this chapter).

✓ What is the composition of the lipid bilayer?
✓ Distinguish integral from peripheral proteins.
✓ What are the major functions of membrane proteins?

Transport Across the Plasma Membrane

Before discussing how materials move into and out of cells, we will describe the locations of the various fluids through which the substances move. Fluid within cells is called **intracellular** (*intra* = within, inside) **fluid (ICF)** or **cytosol.** Fluid outside body cells is called **extracellular** (*extra* = outside) **fluid (ECF)** and is found in several locations. (1) The ECF filling the microscopic spaces between the cells of tissues is called **interstitial** (in′-ter-STISH-al) **fluid** (*inter* = between), or *intercellular fluid.* (2) The ECF in blood vessels is termed **plasma** (Figure 2.3); in lymphatic vessels it is called **lymph.** Among the substances in extracellular fluid are gases, nutrients, and ions—all needed for the maintenance of life.

Extracellular fluid circulates through the blood and lymphatic vessels and into the spaces between the tissue cells. Thus, it is in constant motion throughout the body. Essentially, all body cells are surrounded by the same fluid environment. The movement of substances across a plasma membrane and across membranes within cells is essential to the life of the cell and the

Figure 2.3 / Body fluids. Intracellular fluid (ICF) is the fluid within cells. Extracellular fluid (ECF) is found outside cells, in blood vessels as plasma, lymphatic vessels as lymph, and between tissue cells as interstitial fluid.

Plasma membranes regulate fluid movements from one compartment to another.

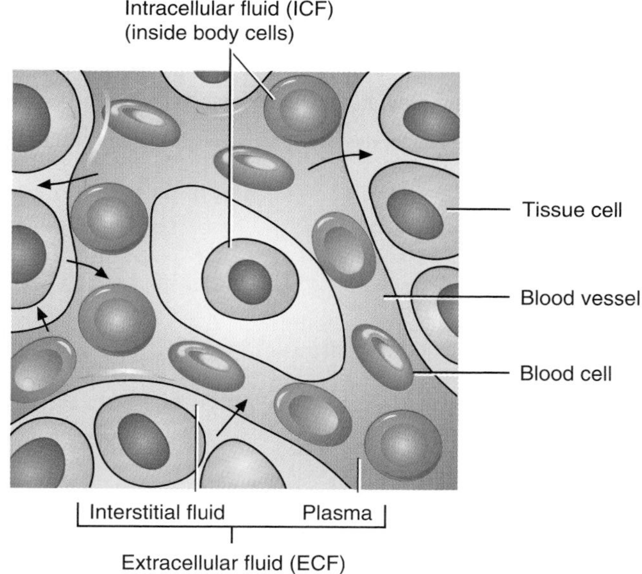

What is another name for intracellular fluid?

organism. Certain substances—oxygen, for example—must move into the cell to support life, whereas waste materials or harmful substances must be moved out. Plasma membranes regulate the movements of such materials.

The processes involved in the movement of substances across plasma membranes are classified as either passive or active. In *passive processes*, substances move across plasma membranes without the use of energy from the breakdown of ATP (adenosine triphosphate) by the cell. ATP is a molecule that stores energy for various cellular uses. The movement of substances in passive processes involves the *kinetic energy* (the energy of motion) of individual molecules or ions. The substances move on their own down a concentration gradient—that is, from an area where their concentration is higher to an area where their concentration is lower. The substances may also be forced across the plasma membrane by pressure from an area where the pressure is high to an area where it is low. In *active processes*, the cell uses some of its stored energy (from the breakdown of ATP) in moving the substance (usually ions) across the membrane since the substance typically moves against a concentration gradient—that is, from an area where its concentration is low to an area where its concentration is high. In addition, transporters in the membrane are involved in the movement of the ions.

Materials that cannot pass through the plasma membrane may enter or leave cells in vesicles. In **vesicular transport** (see Figures 2.4 and 2.5), tiny vesicles (spherical membranous sacs formed by budding off from an existing membrane) either detach from the plasma membrane while bringing materials into the cell or merge with the plasma membrane to release materials from the cell. Like active transport, vesicular transport requires the expenditure of energy by a cell. Large particles, such as whole bacteria and red blood cells, and macromolecules, such as polysaccharides and proteins, may enter and leave cells by vesicular transport.

Passive Processes

The passive processes considered here are simple diffusion, facilitated diffusion, osmosis, and filtration.

SIMPLE DIFFUSION **Simple diffusion** (*diffus* = spreading) is a *net* (greater) movement of molecules or ions from a region of their higher concentration to a region of their lower concentration—that is, the molecules move from an area where there are more of them to an area where there are fewer of them. The movement continues until the molecules are evenly distributed throughout the space. At this point, called equilibrium, molecules move in both directions at an *equal* rate; there is no net movement.

A good example of diffusion in the body is the movement of oxygen from the blood into the body cells and the movement of carbon dioxide from the cells back into the blood. This movement ensures that cells receive adequate amounts of oxygen and eliminate carbon dioxide as part of their normal metabolism.

FACILITATED DIFFUSION **Facilitated diffusion** is accomplished with the assistance of transmembrane proteins that function as transporters. In this process, some large molecules and molecules that are insoluble in lipids can still pass through the plasma membrane. Among these are various sugars, especially glucose. In facilitated diffusion, glucose binds to a transporter (carrier) protein on one side of the plasma membrane, the transporter undergoes a change in shape, and, as a result, glucose is released on the opposite side of the membrane. Thus, by altering its shape the transporter moves the glucose across the plasma membrane.

OSMOSIS **Osmosis** (oz-MŌ-sis) is the net movement of water molecules through a selectively permeable membrane from an area of higher water concentration to an area of lower water concentration. The water molecules pass through pores in integral proteins in the membrane and between neighboring phospholipid molecules, and the net movement continues until equilibrium is reached. Water moves between various compartments of the body by osmosis.

FILTRATION **Filtration** involves the movement of solvents such as water and dissolved substances such as sugar across a selectively permeable membrane by gravity or mechanical pressure, usually hydrostatic (water) pressure. In the body, the driving pressure is usually blood pressure generated by the pumping action of the heart. Most small- to medium-sized molecules, including nutrients, gases, ions, hormones, and vitamins, can be forced through cell membranes, but large proteins cannot. Filtration is a very important process in which water and nutrients in the blood are pushed into interstitial fluid for use by body cells. It is also the primary process in the initial stage of urine formation.

Active Process

The active process considered here is active transport.

Active Transport The process by which substances, usually ions, are transported across plasma membranes typically from an area of their lower concentration to an area of their higher concentration with the expenditure of energy by the cell is called **active transport.** In active transport the substance being moved makes contact with a specific site on a transporter (carrier) protein. Then the ATP splits, and the energy from the breakdown of ATP causes a change in the shape of the transporter protein that expels the substance on the opposite side of the membrane.

Active transport is vitally important in maintaining ion concentrations both in body cells and in extracellular fluids. For example, before a nerve cell can conduct a nerve impulse, the concentration of potassium ions (K^+) must be considerably higher inside the nerve cell than outside and the concentration of sodium ions (Na^+) must be higher outside than inside.

Vesicular Transport

A **vesicle,** as noted earlier, is a small, spherical, membranous sac formed by budding off from an existing membrane that transports substances from one structure to another within cells, takes in substances from extracellular fluid, and releases substances into extracellular fluid. In **endocytosis** (*endo-* = within), materials move into a cell in a vesicle formed from the plasma membrane. In **exocytosis** (*exo-* = out), materials move out of a cell by the fusion of vesicles with the plasma membrane. Both endocytosis and exocytosis require energy supplied by the breakdown of ATP.

Endocytosis Here we consider three types of endocytosis: receptor-mediated endocytosis, phagocytosis, and pinocytosis.

Receptor-mediated endocytosis is a highly selective type of endocytosis that enables a cell to take up particular ligands. A vesicle forms after a receptor protein in the plasma membrane recognizes and binds to a specific ligand in the extracellular fluid. By receptor-mediated endocytosis, cells take up transferrin (an iron-transporting protein in the blood), some vitamins, low density lipoproteins (LDLs), antibodies, certain hormones, and other specific macromolecules or particles. Receptor-mediated endocytosis occurs as follows (Figure 2.4):

1. *Binding.* On the extracellular side of the plasma membrane, a ligand binds to a specific receptor in the membrane to form a receptor-ligand complex. The receptors are integral membrane proteins that are concentrated in regions of the plasma membrane called *clathrin-coated pits,* because the membrane is lined on its cytoplasmic side by a peripheral protein called *clathrin.* The interaction between clathrin and the receptor-ligand complex causes the membrane to invaginate (fold inward).
2. *Vesicle formation.* The edges of the membrane around the clathrin-coated pit fuse and a portion of the membrane pinches off. The resulting vesicle, known as a *clathrin-coated vesicle,* contains the receptor-ligand complexes.
3. *Uncoating.* Almost immediately after it is formed, the clathrin-coated vesicle loses its clathrin coat to become an *uncoated vesicle.*
4. *Fusion with early endosome.* Several uncoated vesicles fuse with an *early endosome,* a vesicle that serves as a sorting compartment. Within an early endosome, ligands separate from their receptors. Ultimately, the contents of an early endosome are either recycled to the plasma membrane or transported to a lysosome, an organelle where degradation occurs.

Figure 2.4 / Receptor-mediated endocytosis.

Receptor-mediated endocytosis imports materials that are needed by cells.

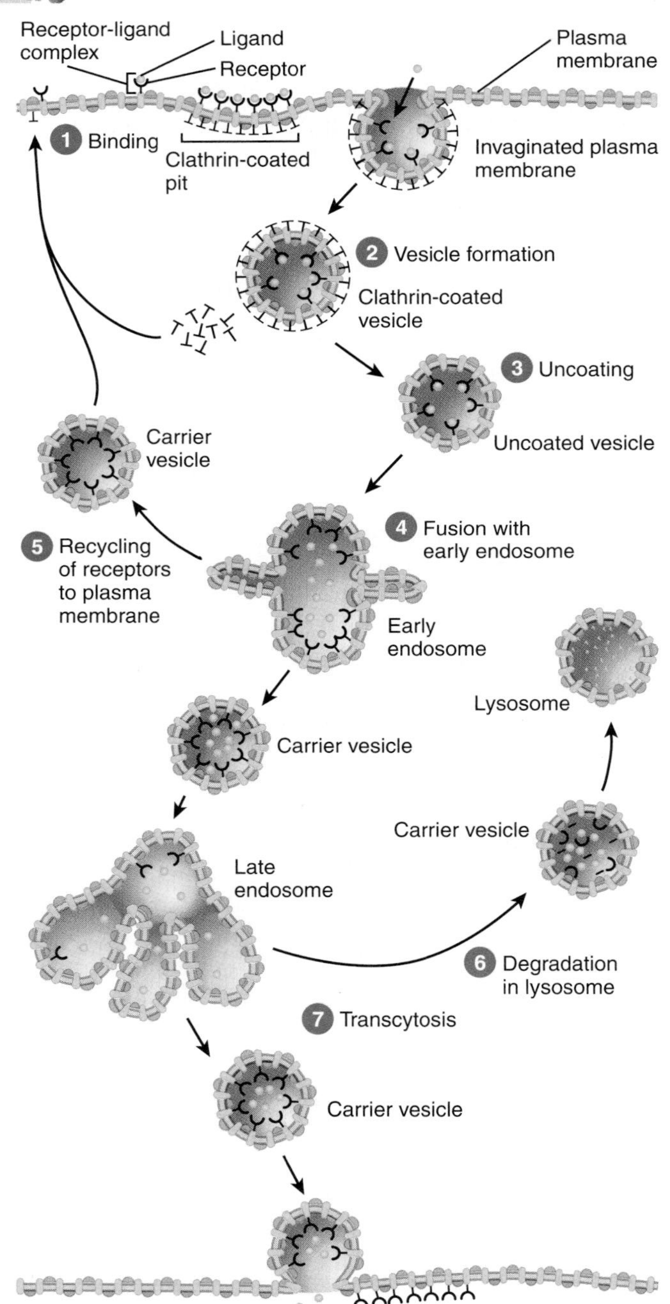

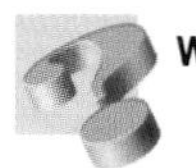
What are several examples of ligands?

5 *Recycling of receptors.* Many receptors within the early endosome enter a *carrier vesicle,* which buds from the early endosome and returns the receptors to the plasma membrane. Clathrin also is returned to the inner surface of the plasma membrane.

6 *Degradation in lysosomes.* Ligands (and their receptors) destined for degradation in lysosomes are transported from the early endosome, by one or more carrier vesicles, to a *late endosome.* The late endosome forms other carrier vesicles that fuse with lysosomes. Within lysosomes are digestive enzymes that break down a variety of large molecules such as proteins, polysaccharides, lipids, and nucleic acids.

7 *Transcytosis.* In some cases, ligands and receptors undergo **transcytosis,** a process whereby they are transported across the cell inside carrier vesicles that undergo exocytosis on the opposite side. As the carrier vesicles fuse with the plasma membrane, the vesicular contents are released into the extracellular fluid. Transcytosis occurs most often across the endothelial cells that line blood vessels and is a means for materials to move between blood plasma and interstitial fluid.

Viruses and Receptor-mediated Endocytosis

Although receptor-mediated endocytosis normally imports needed materials, some viruses take advantage of this mechanism to enter and infect body cells. For example, the human immunodeficiency virus (HIV), which causes acquired immunodeficiency syndrome (AIDS), enter cells by attaching to a receptor called CD4 in the plasma membrane. This receptor is present at the surface of some body cells, such as white blood cells called helper T cells. After binding to CD4, HIV enters the cell via receptor-mediated endocytosis. ■

Phagocytosis (fag′-ō-sī-TŌ-sis; *phago-* = to eat) is a form of endocytosis in which large solid particles, such as whole bacteria or viruses, are taken in by the cell (Figure 2.5).

Figure 2.5 / Phagocytosis.

Phagocytosis is a vital defense mechanism that helps protect the body from disease.

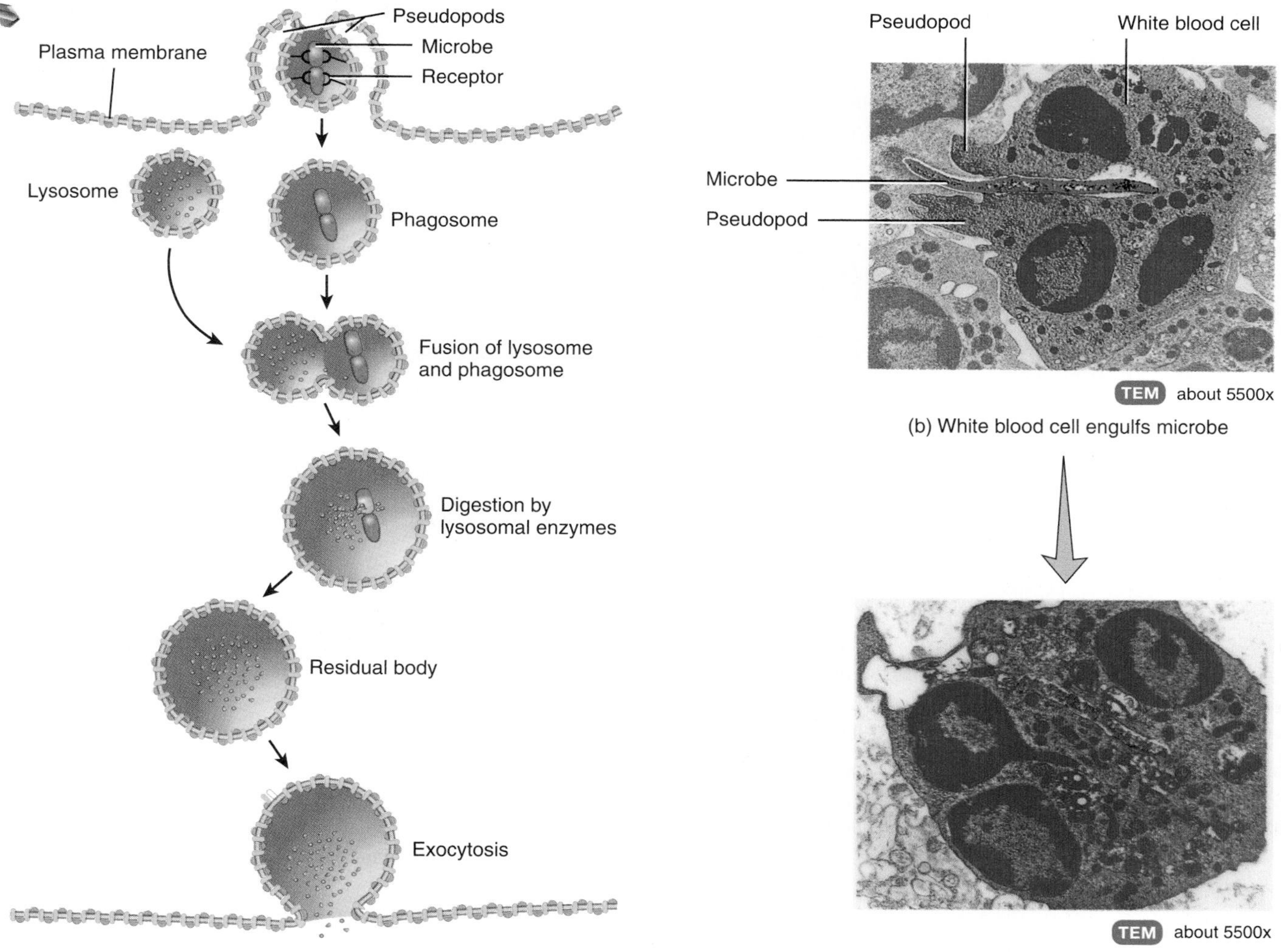

Pseudopods
Plasma membrane
Microbe
Receptor
Lysosome
Phagosome
Fusion of lysosome and phagosome
Digestion by lysosomal enzymes
Residual body
Exocytosis

(a) Diagram of the process

Pseudopod
White blood cell
Microbe
Pseudopod
TEM about 5500x

(b) White blood cell engulfs microbe

TEM about 5500x

(c) White blood cell destroys microbe

What triggers pseudopod formation?

Phagocytosis begins when the particle binds to a plasma membrane receptor. This causes the cell to extend projections of its plasma membrane and cytoplasm called **pseudopods** (SŪ-dō-pods; *pseudo-* = false; *-pods* = feet), which surround the particle outside the cell. The membranes of the pseudopods fuse to form a vesicle called a *phagosome*, which enters the cytoplasm. The phagosome fuses with one or more lysosomes, and the ingested material is broken down by lysosomal enzymes. In most cases, any undigested materials in the phagosome, now called a *residual body*, are released back into the extracellular fluid by exocytosis.

Phagocytosis occurs only in certain body cells, termed **phagocytes,** which are a class of cells that engulf and destroy bacteria and other foreign substances. Phagocytes include certain types of white blood cells and macrophages, which are pres-ent in most body tissues. The process of phagocytosis is a vital defense mechanism that helps protect the body from disease.

Pinocytosis (pi-nō-sī-TŌ-sis; *pino-* = to drink) is a form of endocytosis that involves the *nonselective* uptake of tiny droplets of extracellular fluid. No receptor proteins are involved; all solutes dissolved in the extracellular fluid are brought into the cell. In pinocytosis, the plasma membrane folds inward, forming a *pinocyte vesicle* containing a droplet of extracellular fluid. The pinocytic vesicle detaches, or "pinches off," from the plasma membrane and enters the cytosol. Within the cell, the pinocytic vesicle fuses with one or more lysosomes, where enzymes degrade the engulfed materials. As in phagocytosis, undigested materials accumulate in a residual body. In most cases the residual body fuses with the plasma membrane and expels its contents from the cell via exocytosis. Most body cells engage in pinocytosis.

EXOCYTOSIS As depicted in Figure 2.5, **exocytosis** involves the movement of materials out of a cell in vesicles that fuse with the plasma membrane. The ejected material may be either a waste product or a useful secretory product. All cells carry out exocytosis, but it is especially important in nerve cells, which release substances called *neurotransmitters*, and secretory cells, which secrete digestive enzymes or hormones. During exocytosis in secretory cells, membrane-enclosed vesicles called *secretory vesicles* form inside the cell, fuse with the plasma membrane, and release their contents into the extracellular fluid.

Portions of the plasma membrane lost through endocytosis are recovered or recycled by exocytosis. The balance between endocytosis and exocytosis keeps the surface area of the plasma membrane relatively constant. Membrane exchange is quite extensive in certain cells. Secretory cells of the pancreas, for example, recycle an amount of plasma membrane equal to the cell's entire surface area every 90 minutes.

Table 2.1 summarizes the processes by which materials are transported into and out of cells.

✓ Distinguish a passive process from an active process.
✓ Why are simple diffusion, facilitated diffusion, osmosis, filtration, and active transport important to the body?
✓ Define each type of vesicular transport. How is each important to the body?

Table 2.1 Processes by Which Substances Move Across Plasma Membranes

Process	Description
Passive Processes	Substances move down a concentration gradient from an area of higher to lower concentration or pressure; cell does not expend energy by the breakdown of ATP.
Simple Diffusion	Net movement of molecules or ions due to their kinetic energy from an area of higher to lower concentration until an equilibrium is reached.
Facilitated Diffusion	Movement of larger molecules across a selectively permeable membrane with the assistance of proteins in the membrane that serve as transporters (carriers).
Osmosis	Net movement of water molecules due to kinetic energy across a selectively permeable membrane from an area of higher to lower concentration of water until an equilibrium is reached.
Filtration	Movement of solvents (such as water) and solutes (such as glucose) across a selectively permeable membrane as a result of gravity or hydrostatic (water) pressure from an area of higher to lower pressure.
Active Process	Substances move against a concentration gradient from an area of lower to higher concentration; cell must expend energy released by the breakdown of ATP.
Active Transport	Movement of substances, usually ions, across a selectively permeable membrane from a region of lower to higher concentration by an interaction with transporters in the membrane.
Vesicular Transport	Movement of substances into or out of a cell in vesicles that bud from the plasma membrane; requires energy supplied by the breakdown of ATP.
Endocytosis	Movement of substances into a cell in vesicles.
Receptor-mediated endocytosis	Ligand-receptor complexes trigger infolding of a clathrin-coated pit that forms a vesicle containing ligands.
Phagocytosis	"Cell eating"; movement of a solid particle into a cell after pseudopods engulf it to form a phagosome.
Pinocytosis	"Cell drinking"; movement of extracellular fluid into a cell by infolding of plasma membrane, forming pinocytic vesicle.
Exocytosis	Movement of substances out of a cell in secretory vesicles that fuse with the plasma membrane and release their contents into the extracellular fluid.

CYTOPLASM

Objective

- Describe the structure and function of cytoplasm, cytosol, and organelles.

The cytoplasm includes the cytosol and a variety of organelles.

Cytosol

The **cytosol**, the fluid portion of the cytoplasm that surrounds organelles (see Figure 2.1), constitutes about 55% of total cell volume. Although it varies in composition and consistency from one part of a cell to another, cytosol is 75–90% water plus dissolved and suspended components. Among these are various ions, glucose, amino acids, fatty acids, proteins, lipids, ATP, and waste products. Also present are various organic molecules that aggregate into masses and are stored. These molecules may appear and disappear at various times in the life of a cell. Examples include *lipid droplets* that contain triglycerides and clusters of glycogen molecules called *glycogen granules* (see Figure 2.1).

The cytosol is the site of many chemical reactions required for a cell's existence. For example, enzymes in cytosol catalyze numerous chemical reactions. As a result of these reactions, energy is released and captured to drive cellular activities. In addition, these reactions provide the building blocks for maintaining cell structure, function, and growth.

Organelles

As noted previously, **organelles** are specialized structures that have characteristic shapes and that perform specific functions in cellular growth, maintenance, and reproduction. Despite the many chemical reactions occurring in a cell at the same time, there is little interference between one type of reaction and another because they occur in different organelles. Each type of organelle has its own set of enzymes that carry out specific reactions, and each is a functional compartment where specific physiological processes take place. Moreover, organelles often cooperate with each other to maintain homeostasis. The numbers and types of organelles vary in different cells, depending on the function of the cell. Although the nucleus is an organelle, it is discussed separately because of its special importance in directing the life of a cell.

The Cytoskeleton

The **cytoskeleton** is a network of several different kinds of protein filaments that extend throughout the cytosol (Figure 2.6).

Figure 2.6 / Cytoskeleton.

The cytoskeleton is a network of three kinds of protein filaments that extend throughout the cytoplasm: microfilaments, intermediate filaments, and microtubules.

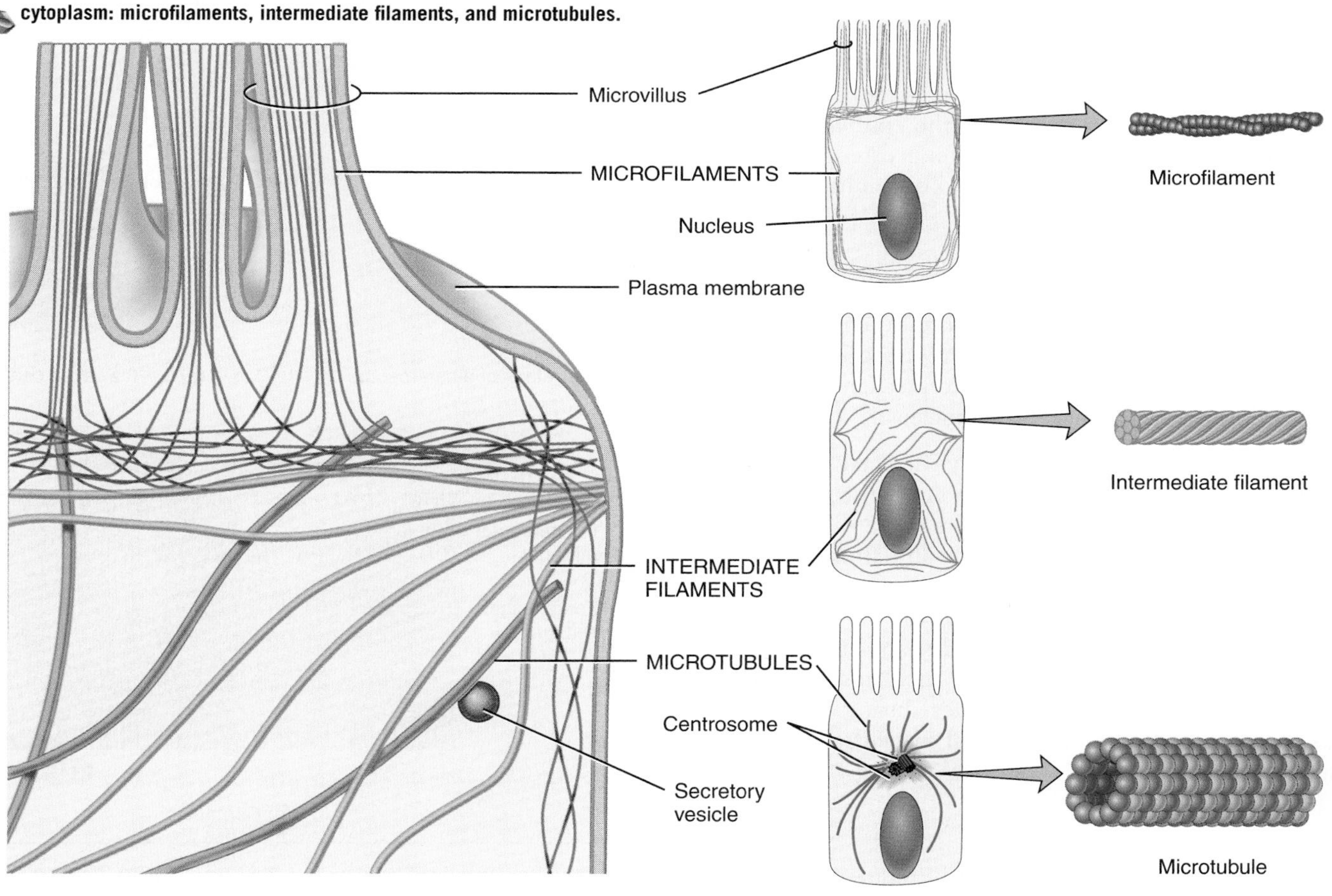

(a) Overview of cytoskeleton

(b) Distribution of cytoskeletal element (left) and detail of structure (right)

Which cytoskeletal component helps form the structure of centrioles, cilia, and flagella?

The cytoskeleton provides a structural framework for the cell, serving as a scaffold that helps to determine a cell's shape and organizes the cellular contents. The cytoskeleton is also responsible for cell movements, including the internal transport of organelles and some chemicals, the movement of chromosomes during cell division, and the movement of whole cells such as phagocytes. Although its name implies rigidity, the cytoskeleton is continually reorganizing as a cell moves and changes shape, as, for example, during cell division. The cytoskeleton is composed of three types of protein filaments. In the order of their increasing diameter, these structures are microfilaments, intermediate filaments, and microtubules.

MICROFILAMENTS These are the thinnest elements of the cytoskeleton. Microfilaments are concentrated at the periphery of a cell (Figure 2.6a and b) and are composed of the protein *actin* and have two general functions: movement and mechanical support. With respect to movement, microfilaments are involved in muscle contraction, cell division, and cell locomotion, such as occurs during the migration of embryonic cells during development, the invasion of tissues by white blood cells to fight infection, or the migration of skin cells during wound healing.

Microfilaments provide much of the mechanical support that is responsible for the basic strength and shapes of cells. They anchor the cytoskeleton to integral proteins in the plasma membrane. Microfilaments also provide mechanical support for cell extensions called **microvilli** (*micro-* = small; *-villi* = tufts of hair), which are nonmotile, microscopic fingerlike projections of the plasma membrane. Within a microvillus is a core of parallel microfilaments that supports it and attaches it to the cytoskeleton (Figure 2.6a and b). Microvilli are abundant on the surfaces of cells involved in absorption, such as the epithelial cells that line the small intestine. Some microfilaments extend beyond the plasma membrane and help cells attach to one another or to extracellular material.

INTERMEDIATE FILAMENTS As their name suggests, intermediate filaments are thicker than microfilaments but thinner than microtubules (Figure 2.6a and b). Several different proteins can compose intermediate filaments, which are exceptionally strong. They are found in parts of cells subject to mechanical stress and also help anchor organelles such as the nucleus.

MICROTUBULES The largest of the cytoskeletal components, microtubules are long, unbranched hollow tubes composed mainly of a protein called *tubulin*. An organelle called the centrosome (discussed shortly) serves as the initiation site for the assembly of microtubules. The microtubules grow outward from the centrosome toward the periphery of the cell (Figure 2.6a and b). Microtubules help determine cell shape and function in the intracellular transport of organelles, such as secretory vesicles, and the migration of chromosomes during cell division. They also participate in the movement of specialized cell projections, such as cilia and flagella.

Centrosome

The **centrosome,** located near the nucleus, consists of the pericentriolar material and centrioles (Figure 2.7a). The **pericentriolar material** is a region of the cytosol composed of a dense network of small protein fibers. This region is the organizing center for the mitotic spindle, which plays a critical role in cell division, and for microtubule formation in nondividing cells. Within the pericentriolar material is a pair of cylindrical structures called **centrioles,** each of which is composed of nine clusters of three microtubules (triplets) arranged in a circular pattern, an arrangement called a *9 + 0 array* (Figure 2.7b). The 9

Figure 2.7 / Centrosome.

The pericentriolar material of a centrosome organizes the mitotic spindle during cell division, whereas the centrioles help form or regenerate flagella and cilia.

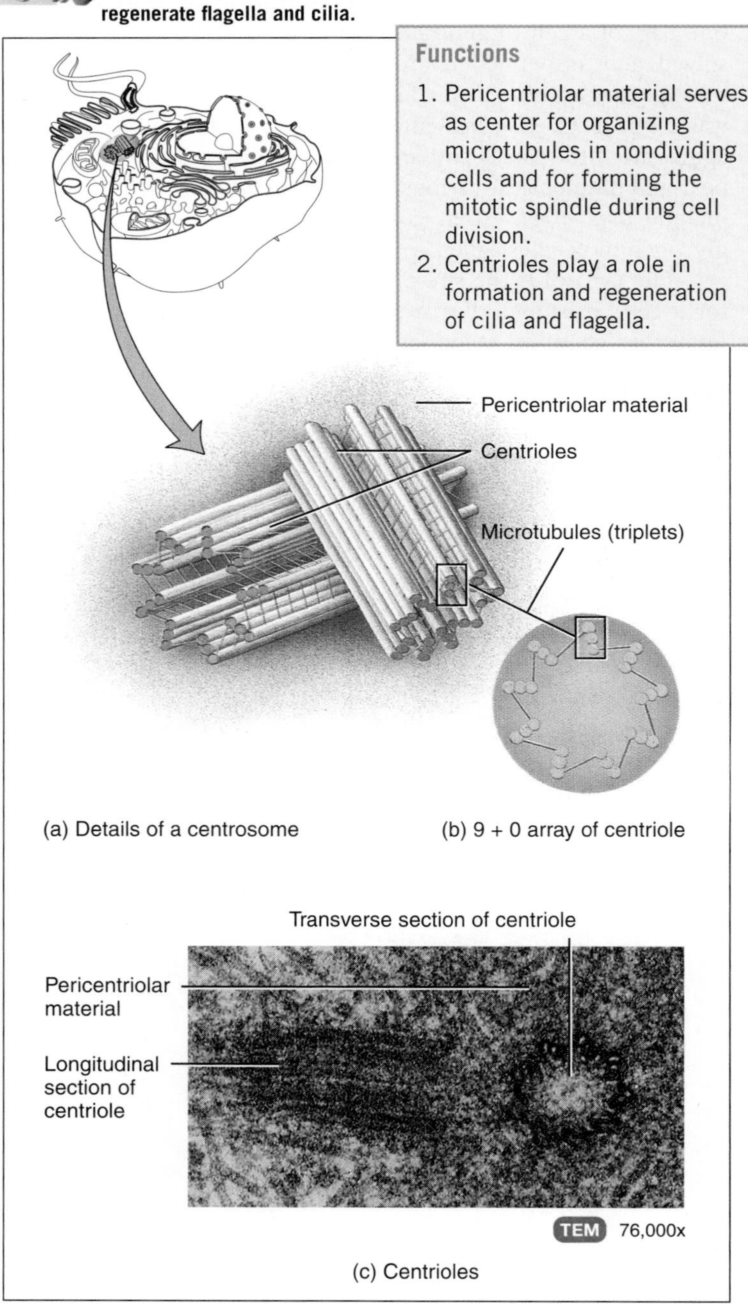
Functions
1. Pericentriolar material serves as center for organizing microtubules in nondividing cells and for forming the mitotic spindle during cell division.
2. Centrioles play a role in formation and regeneration of cilia and flagella.
Pericentriolar material
Centrioles
Microtubules (triplets)
Transverse section of centriole
Pericentriolar material
Longitudinal section of centriole
TEM 76,000x

(a) Details of a centrosome

(b) 9 + 0 array of centriole

(c) Centrioles

If you observed that a cell did not have a centrosome, what could you predict about its capacity for cell division?

refers to the nine clusters of microtubules, and the 0 refers to the absence of microtubules in the center. The long axis of one centriole is at a right angle to the long axis of the other (Figure 2.7c). Centrioles play a role in the formation and regeneration of cilia and flagella.

Cilia and Flagella

Microtubules are the dominant structural and functional components of cilia and flagella, both of which are motile projections of the cell surface. In the human body, cells that are firmly anchored in place use cilia to move fluids across their surfaces; motile cells, such as sperm cells, use flagella to propel themselves through a liquid medium.

Cilia (SIL-ē-a = eyelashes) are numerous, short, hairlike projections that extend from the surface of the cell (see Figure 2.1). Each cilium (the singular form) contains a core of microtubules surrounded by plasma membrane (Figure 2.8). The microtubules are arranged with one pair in the center surrounded by nine clusters of two microtubules (doublets), an arrangement called a *9 + 2 array.* Each cilium is anchored to a *basal body* just below the surface of the plasma membrane. A basal body is similar in structure to a centriole. In fact, basal bodies and centrioles are considered to be two different functional manifestations of the same structure. The function of a basal body is to initiate the assembly of cilia and flagella. The coordinated movement of numerous cilia on the surface of a cell ensures the steady movement of fluid over the cell's surface. Many cells of the respiratory tract, for example, have hundreds of cilia that help sweep foreign particles trapped in mucus away from the lungs. Their movement is paralyzed by nicotine in cigarette smoke. For this reason, smokers cough often to remove foreign particles from their airways. Cells that line the uterine (Fallopian) tubes also have cilia that sweep ova toward the uterus.

Flagella (fla-JEL-a; singular is *flagellum* = whip) are similar in structure to cilia but are much longer. Flagella usually move an entire cell. The only example of a flagellum in the human body is a sperm cell's tail, which propels the sperm toward its rendezvous with an ovum.

Ribosomes

Ribosomes (RĪ-bō-sōms) are sites of protein synthesis. These tiny organelles are packages of **ribosomal RNA (rRNA)** and many ribosomal proteins. Ribosomes are so named because of their high content of *ribo*nucleic acid. Structurally, a ribosome consists of two subunits, one about half the size of the other (Figure 2.9). The two subunits are made separately in the nucleolus, a spherical body inside the nucleus. Once produced, they exit the nucleus and join together in the cytosol.

Figure 2.8 / Cilia and flagella.

Whereas centrioles have a 9 + 0 array of microtubules, cilia and flagella have a 9 + 2 array.

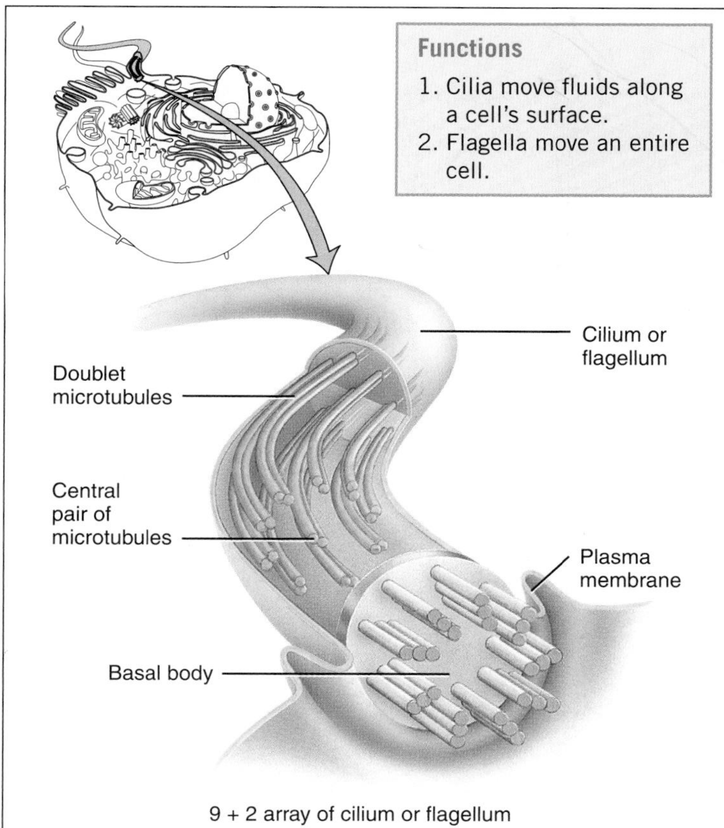

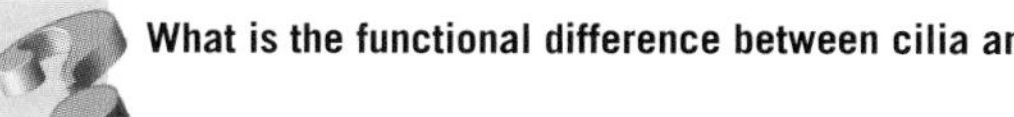
What is the functional difference between cilia and flagella?

Figure 2.9 / Ribosomes.

Ribosomes are the sites of protein synthesis.

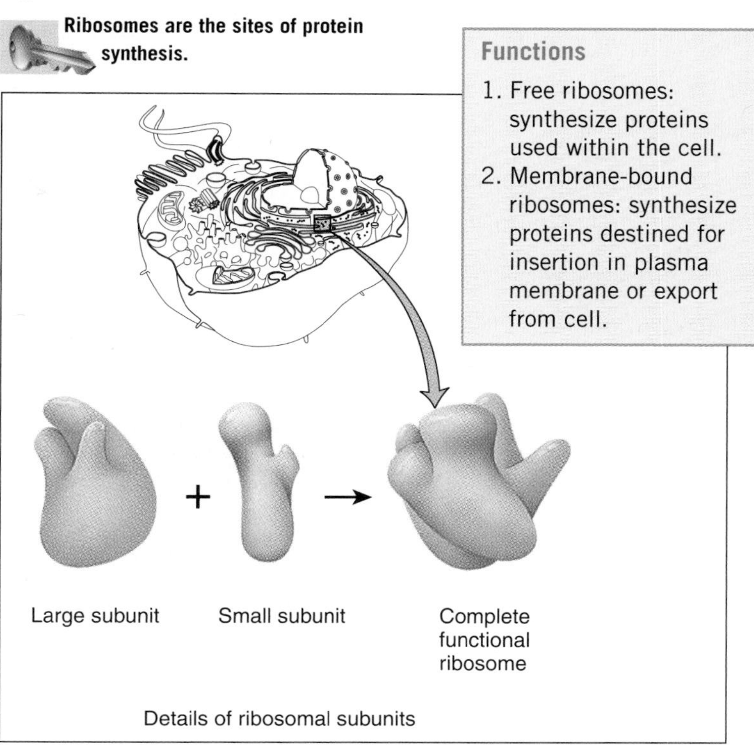

Functions
1. Free ribosomes: synthesize proteins used within the cell.
2. Membrane-bound ribosomes: synthesize proteins destined for insertion in plasma membrane or export from cell.
Large subunit
Small subunit
Complete functional ribosome
Details of ribosomal subunits

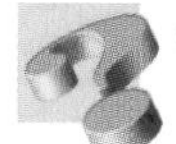
Where are subunits of ribosomes synthesized and assembled?

Some ribosomes, called *free ribosomes*, are unattached to any structure in the cytoplasm. Primarily, free ribosomes synthesize proteins used *inside* the cell. Other ribosomes, called *membrane-bound ribosomes*, attach to the nuclear membrane and to an extensively folded membrane called the endoplasmic reticulum (see Figure 2.10b). These ribosomes synthesize proteins destined for specific organelles, insertion in the plasma membrane, or for export from the cell. Ribosomes are also located within mitochondria, where they synthesize mitochondrial proteins. Sometimes 10–20 ribosomes occur in a stringlike arrangement called a *polyribosome*.

Endoplasmic Reticulum

The **endoplasmic reticulum** (en′-dō-PLAS-mik re-TIK-yū-lum; *-plasmic* = cytoplasm; *reticulum* = network), or **ER,** is a network of membranes that form flattened sacs or tubules called **cisternae** (sis-TER-nē; sistern = cavity) (Figure 2.10). The ER extends from the membrane around the nucleus (nuclear envelope), to which it is connected, throughout the cytoplasm. The ER is so extensive that it constitutes more than half of the membranous surfaces within the cytoplasm of most cells.

Cells contain two distinct forms of ER, which differ in structure and function. **Rough ER** is continuous with the nuclear membrane and usually is folded into a series of flattened sacs. The outer surface of rough ER is studded with ribosomes, the sites of protein synthesis. Proteins synthesized by ribosomes attached to rough ER enter cisternae within the ER for processing and sorting. In some cases, enzymes within the cisternae attach the proteins to carbohydrates to form glycoproteins. In other cases, enzymes attach the proteins to phospholipids, also synthesized by rough ER. These molecules may be incorporated into the membranes of organelles, or into the plasma membrane. Thus rough ER produces secretory proteins, membrane proteins, and many organellar proteins.

Figure 2.10 / Endoplasmic reticulum.

The endoplasmic reticulum is a network of membrane-enclosed sacs or tubules called cisternae that extend throughout the cytoplasm and connect to the nuclear envelope.

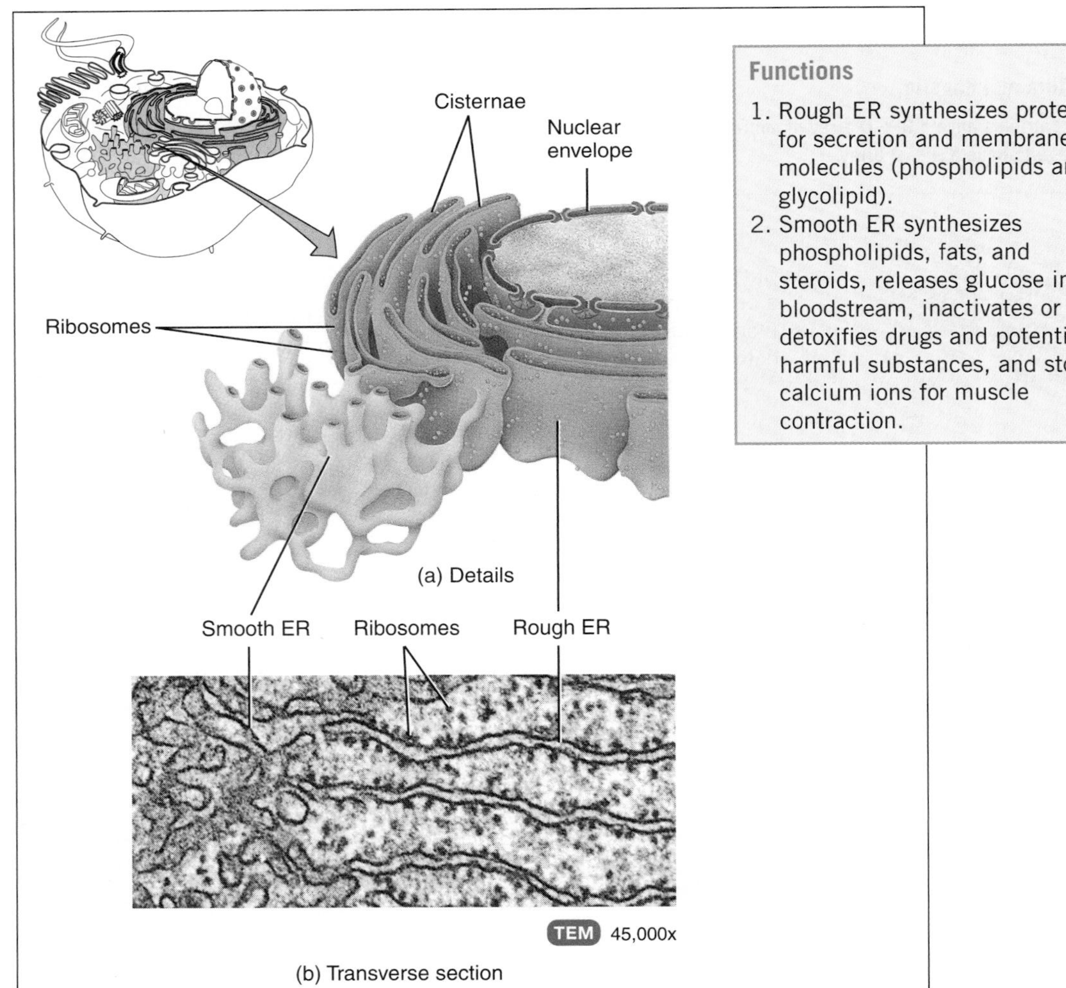

Functions

1. Rough ER synthesizes proteins for secretion and membrane molecules (phospholipids and glycolipid).
2. Smooth ER synthesizes phospholipids, fats, and steroids, releases glucose into bloodstream, inactivates or detoxifies drugs and potentially harmful substances, and stores calcium ions for muscle contraction.

What are the structural and functional differences between rough and smooth ER?

Smooth ER extends from the rough ER to form a network of membrane tubules (see Figure 2.10). Unlike rough ER, smooth ER does not have ribosomes on the outer surface of its membrane. However, smooth ER contains unique enzymes that make it functionally more diverse than rough ER. Although it does not synthesize proteins, smooth ER does synthesize phospholipids, as does rough ER. Smooth ER also synthesizes fats and steroids, such as estrogens and testosterone. In liver cells, enzymes of the smooth ER help release glucose into the bloodstream and inactivate or detoxify drugs and other potentially harmful substances, for example, alcohol. In muscle cells, calcium ions released from the sarcoplasmic reticulum, a form of smooth ER, trigger the contraction process.

Golgi Complex

Most of the proteins synthesized by ribosomes attached to rough ER are ultimately transported to other regions of the cell. The first step in the transport pathway is through an organelle called the **Golgi complex** (GOL-jē). It consists of 3–20 flattened, membranous sacs with bulging edges, called **cisternae,** that resemble a stack of pita bread (Figure 2.11). The cisternae are often curved, giving the Golgi complex a cuplike shape. Most cells have only one Golgi complex, although some may have several. The Golgi complex is more extensive in cells that secrete proteins into the extracellular fluid. This fact is a clue to the organelle's role in the cell.

The cisternae at the opposite ends of a Golgi complex differ from each other in size, shape, content, enzymatic activity, and vesicles present (described shortly). The convex **entry,** or ***cis* face,** is composed of cisternae that face the rough ER. The concave **exit,** or ***trans* face,** is composed of cisternae that face the plasma membrane. Cisternae between the entry and exit faces are called **medial cisternae.**

The entry face, medial cisternae, and exit face of the Golgi complex each contain different enzymes that permit them to modify, sort, and package proteins for transport to different destinations. For example, the entry face receives and modifies proteins produced by the rough ER. The medial cisternae add sugars to proteins and lipids to form glycoproteins and glycolipids, and they add proteins to lipids to form lipoproteins. The exit face modifies the molecules further and then sorts and packages them for transport to their destinations.

Figure 2.11 / Golgi complex.

The opposite faces of a Golgi complex differ in size, shape, content, enzymatic activity, and vesicles present.

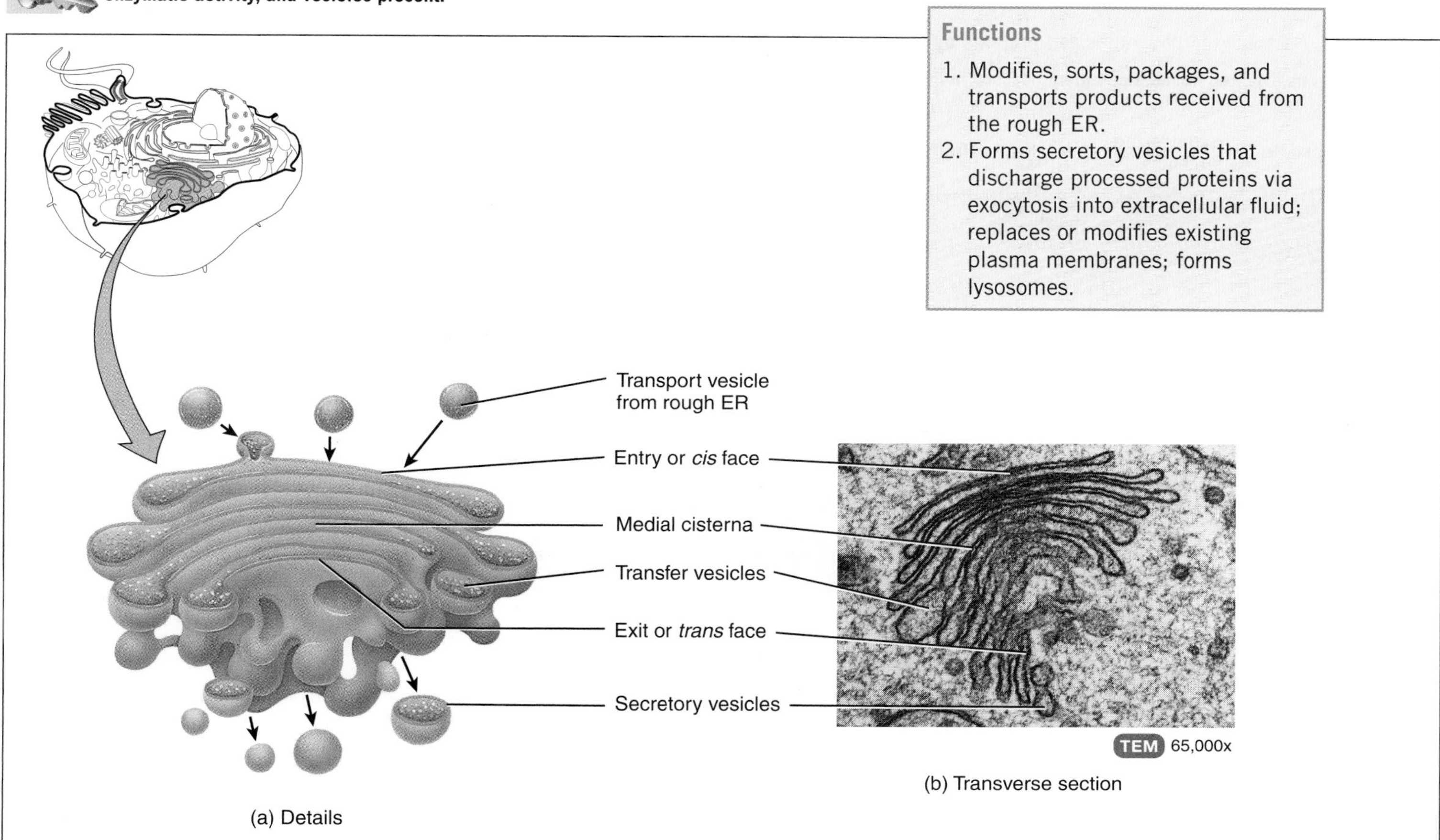

How do the entry and exit faces differ in function?

Proteins arriving at, passing through, and exiting the Golgi complex do so through a series of exchanges between vesicles (Figure 2.12):

1. Proteins synthesized by ribosomes on the rough ER are surrounded by a portion of the ER membrane, which eventually buds from the membrane surface to form a **transport vesicle.**
2. The transport vesicle moves toward the entry face of the Golgi complex.
3. The transport vesicle membrane fuses with the membrane at the entry face of the Golgi complex, releasing proteins into the cisternae.
4. The proteins are modified and move from the entry face into one or more medial cisternae via **transfer vesicles** that bud from the edges of the cisternae. Enzymes in the medial cisternae modify the proteins to form glycoproteins, glycolipids, and lipoproteins.
5. The products of the medial cisternae move via transfer vesicles into the interior of the exit face.
6. Within the exit face, the products are further modified, sorted, and packaged.
7. Some of the processed proteins leave the exit face in **secretory vesicles,** which deliver the proteins to the plasma membrane, where they are discharged by exocytosis into the extracellular fluid.
8. Other processed proteins leave the exit face in **membrane vesicles** that deliver their contents to the plasma membrane for incorporation into the membrane. In doing so, the Golgi complex adds new segments of plasma membrane as existing segments are lost, and it modifies the number and distribution of membrane molecules.
9. Finally, some processed proteins leave the exit face in **storage vesicles.** The major storage vesicles are lysosomes, whose structure and functions are discussed shortly.

Sometimes a disorder results from faulty routing of a molecule. Such is the case in **cystic fibrosis,** a deadly inherited disease that affects several body systems. The defective protein produced by the mutated cystic fibrosis gene fails to reach the plasma membrane, where it should be inserted to help pump chloride ions

Figure 2.12 / Packaging of synthesized proteins by the Golgi complex.

All proteins exported from the cell are processed in the Golgi complex.

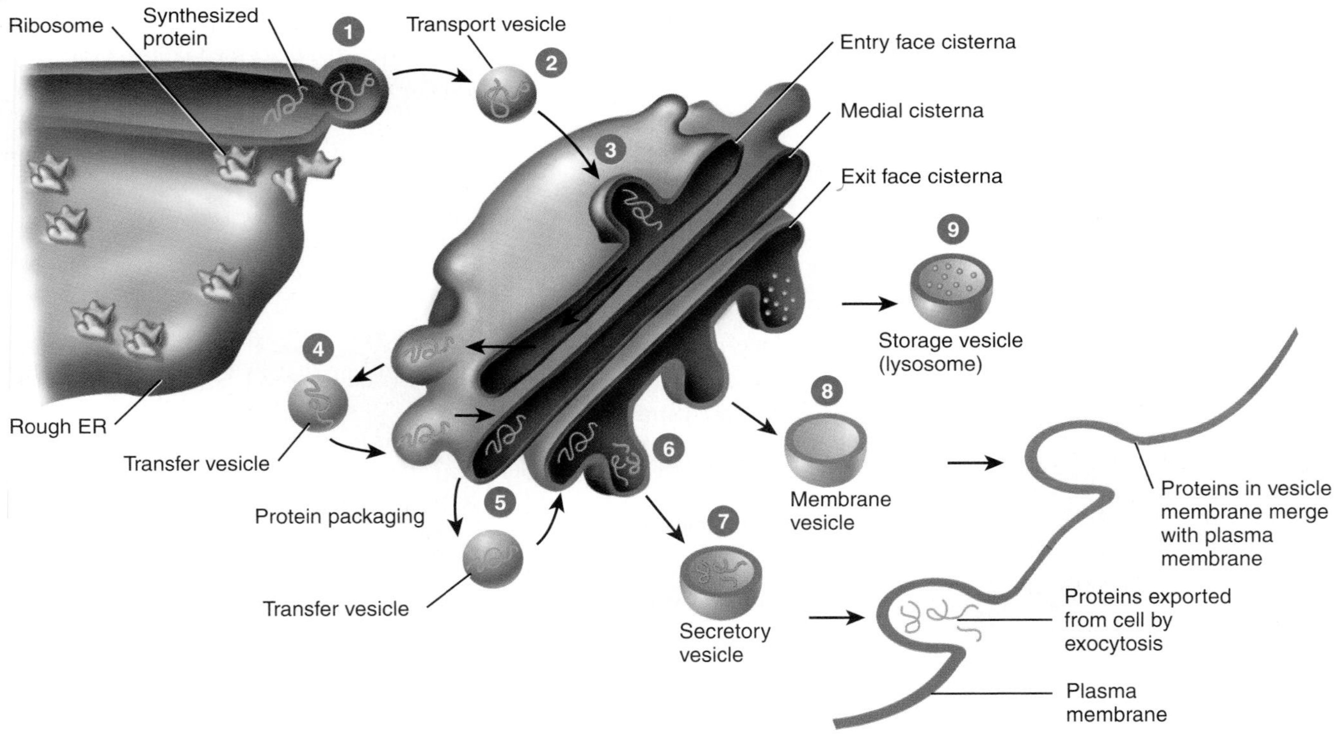

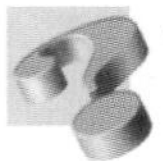

What are the three general destinations for proteins that leave the Golgi complex?

(Cl^-) out of certain cells. Evidently, the would-be pump protein becomes stuck in the endoplasmic reticulum or Golgi complex and never reaches its correct destination. The result is an imbalance in the transport of fluid and ions across the plasma membrane that causes the buildup of thick mucus outside certain types of cells. The accumulated mucus clogs the airways in the lungs, causing breathing difficulty, and prevents proper secretion of digestive enzymes by the pancreas, causing digestive disturbances. ■

Lysosomes

Lysosomes (LĪ-sō-sōms; *lyso-* = dissolving; -somes = bodies) are membrane-enclosed vesicles that form from the Golgi complex (Figure 2.13). Inside are as many as 40 kinds of powerful digestive enzymes that are capable of breaking down a wide variety of molecules. Lysosomes fuse with phagosomes and pinocytic vesicles and the lysosomal enzymes break down the contents of these structures. Proteins in the lysosomal membrane also allow the final products of digestion, such as sugars and amino acids, to be transported into the cytosol.

Lysosomal enzymes also recycle the cell's own structures. A lysosome can engulf another organelle, digest it, and return the digested components to the cytosol for reuse. In this way, old organelles are continually replaced. The process by which worn-out organelles are digested is called **autophagy** (aw-TOF-a-jē; *auto-* = self; *-phagy* = eating). A human liver cell, for example, recycles about half its cytoplasmic contents every week. In autophagy, the organelle to be digested is enclosed by a membrane derived from the ER to create a vesicle called an **autophagosome;** the vesicle then fuses with a lysosome. Lysosomal enzymes may also destroy the cell that contains them, a process known as **autolysis** (aw-TOL-i-sis). Autolysis occurs in some pathological conditions and also is responsible for the tissue deterioration that occurs immediately after death.

Although most of the digestive processes involving lysosomal enzymes occur with a cell, there are some instances in which the enzymes operate in extracellular digestion. One example is the release of lysosomal enzymes during fertilization. Enzymes from lysosomes in the head of a sperm cell help the sperm cell penetrate the surface of the ovum.

Figure 2.13 / Lysosomes.

Lysosomes arise in the Golgi complex and store several kinds of powerful digestive enzymes.

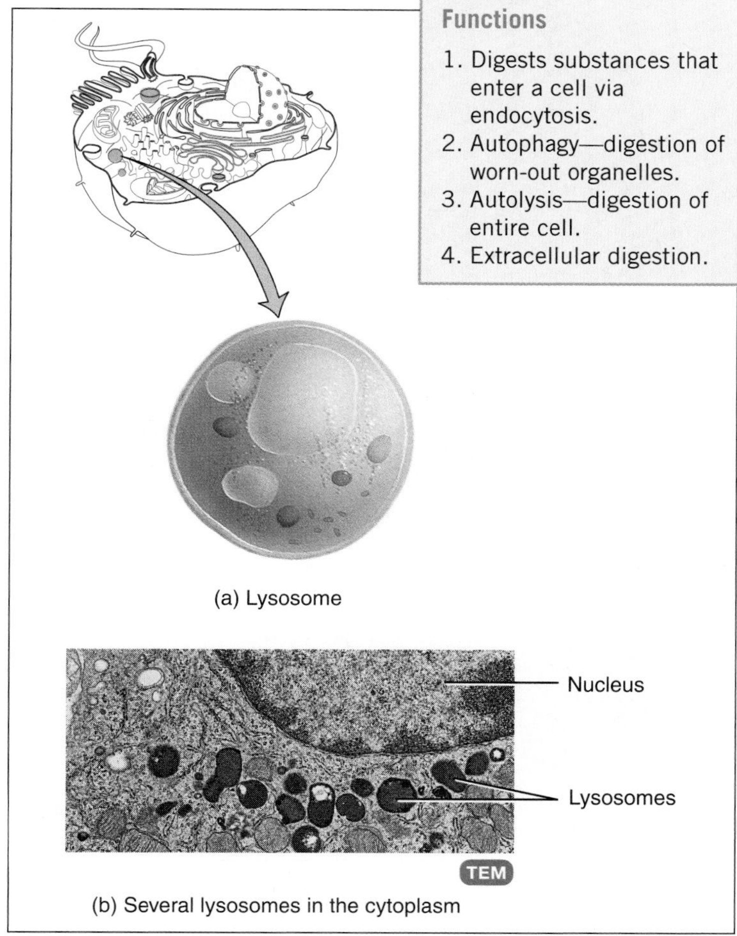

Functions
1. Digests substances that enter a cell via endocytosis.
2. Autophagy—digestion of worn-out organelles.
3. Autolysis—digestion of entire cell.
4. Extracellular digestion.
(a) Lysosome
Nucleus
Lysosomes
TEM
(b) Several lysosomes in the cytoplasm

What is the name of the process by which worn-out organelles are digested by lysosomes?

Some disorders are caused by faulty lysosomes. For example, **Tay-Sachs disease,** which most often affects children of Ashkenazi (eastern European Jewish) descent, is an inherited condition characterized by the absence of a single lysosomal enzyme. The missing enzyme normally breaks down a membrane glycolipid called ganglioside G_{M2} that is especially prevalent in nerve cells. As ganglioside G_{M2} accumulates, the nerve cells function less efficiently. Children with Tay-Sachs disease typically experience seizures and muscle rigidity. They gradually become blind, demented, and uncoordinated and usually die before the age of 5. The Tay-Sachs gene has been identified, and genetic testing now can detect whether an adult is a carrier of the defective gene. ■

Peroxisomes

Another group of organelles similar in structure to lysosomes, but smaller, are called **peroxisomes** (pe-ROKS-i-sōms; *peroxi* = peroxide; *somes* = bodies; see Figure 2.1). Although peroxisomes were once thought to form by budding off the ER, it now is generally agreed that they form by the division of preexisting peroxisomes.

Peroxisomes contain one or more enzymes that can oxidize (remove hydrogen atoms from) various organic substances. For example, substances such as amino acids and fatty acids are oxidized in peroxisomes as part of normal metabolism. In addition, enzymes in peroxisomes oxidize toxic substances, such as alcohol. A by-product of the oxidation reactions is hydrogen peroxide (H_2O_2), a potentially toxic compound. However, peroxisomes also contain an enzyme called *catalase*, which decomposes

H_2O_2. Because the generation and degradation of H_2O_2 occurs within the same organelle, peroxisomes protect other parts of the cell from the toxic effects of H_2O_2.

Mitochondria

Because of their function in generating ATP, **mitochondria** (mī-tō-KON-drē-a; *mito-* = thread; *-chondria* = granules) are the "powerhouses" of the cell. A cell may have as few as a hundred or as many as several thousand mitochondria (singular is mitochondrion). Physiologically active cells, such as those found in the muscles, liver, and kidneys, have a large number of mitochondria because they use ATP at a high rate. Mitochondria are usually located in a cell where the energy need is greatest, such as between the contractile proteins in muscle cells. A mitochondrion is bounded by two membranes that are similar in structure to the plasma membrane (Figure 2.14). The *outer mitochondrial membrane* is smooth, but the *inner mitochondrial membrane* is arranged in a series of folds called **cristae** (KRIS-tē; = ridges). The central fluid-filled cavity of a mitochondrion, enclosed by the inner membrane and cristae, is the **matrix.** The elaborate folds of the cristae provide an enormous surface area for the chemical reactions that are part of the aerobic phase of *cellular respiration.* These reactions produce most of a cell's ATP. Enzymes that catalyze cellular respiration are located in the matrix and on the cristae.

Like peroxisomes, mitochondria self-replicate, a process that occurs during times of increased cellular energy demand or before cell division. Each mitochondrion has multiple identical copies of a circular DNA molecule that contains 37 genes. The mitochondrial genes along with the genes in the cell's nucleus control the production of proteins that build mitochondrial components. Because ribosomes are also present in the mitochondrial matrix, some protein synthesis occurs inside mitochondria.

Although the nucleus of each somatic cell contains genes from both your mother and father, mitochondrial genes usually are inherited only from your mother. The head of a sperm (the part that penetrates and fertilizes an ovum) normally lacks most organelles, such as mitochondria, ribosomes, endoplasmic reticulum, and Golgi complexes.

✓ Distinguish between cytosol, cytoplasm, and organelles.
✓ What are the components and functions of the cytoskeleton?
✓ Briefly describe the structure and functions of centrosomes, cilia and flagella, ribosomes, endoplasmic reticulum, Golgi complex, lysosomes, peroxisomes, and mitochondria.

Figure 2.14 / Mitochondria.

Within mitochondria, chemical reactions called cellular respiration generate ATP.

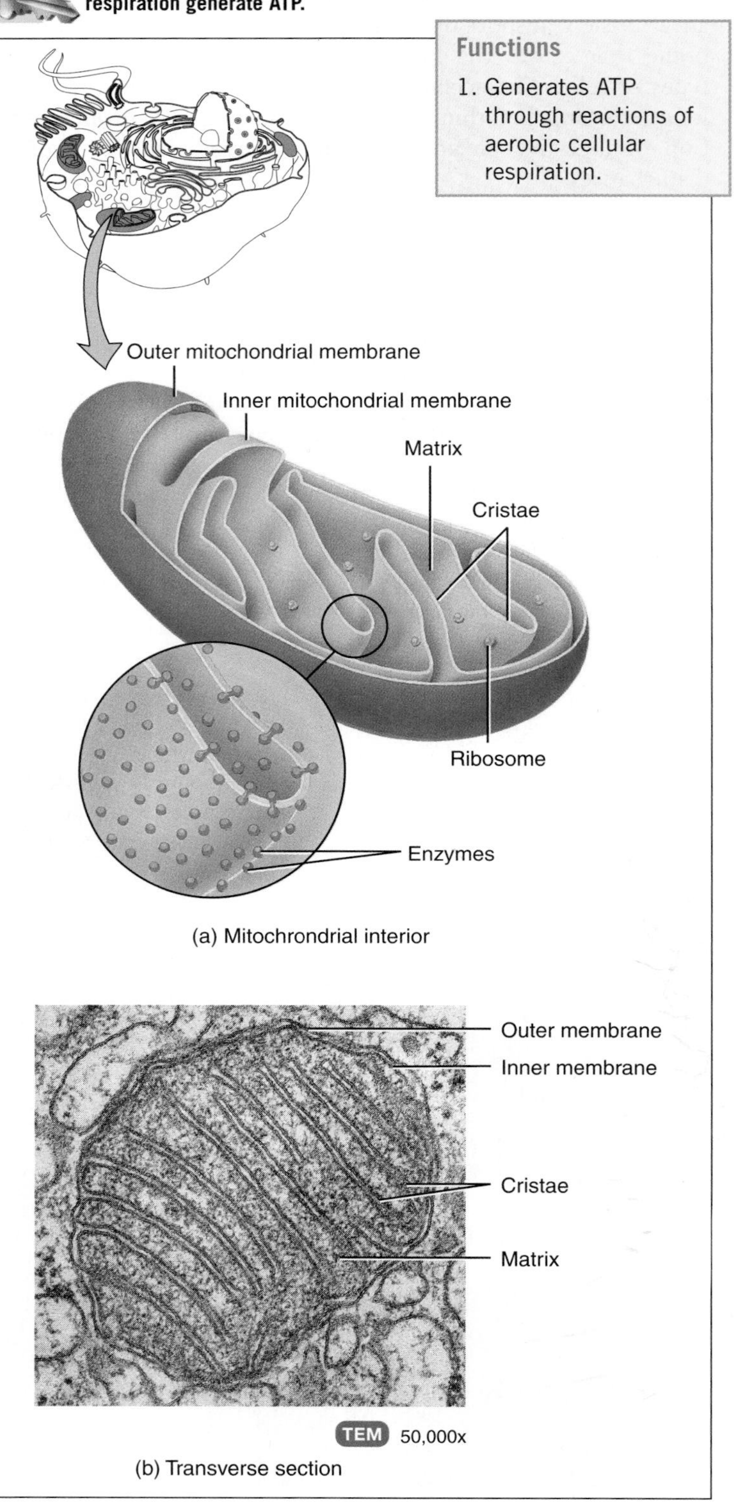

How do the cristae of a mitochondrion contribute to its ATP-producing function?

NUCLEUS

Objective

• Describe the structure and functions of the nucleus.

The **nucleus** is a spherical or oval-shaped structure that is usually the most prominent feature of a cell (Figure 2.15). Most

Figure 2.15 / Nucleus.

The nucleus contains most of a cell's genes, which are located on chromosomes.

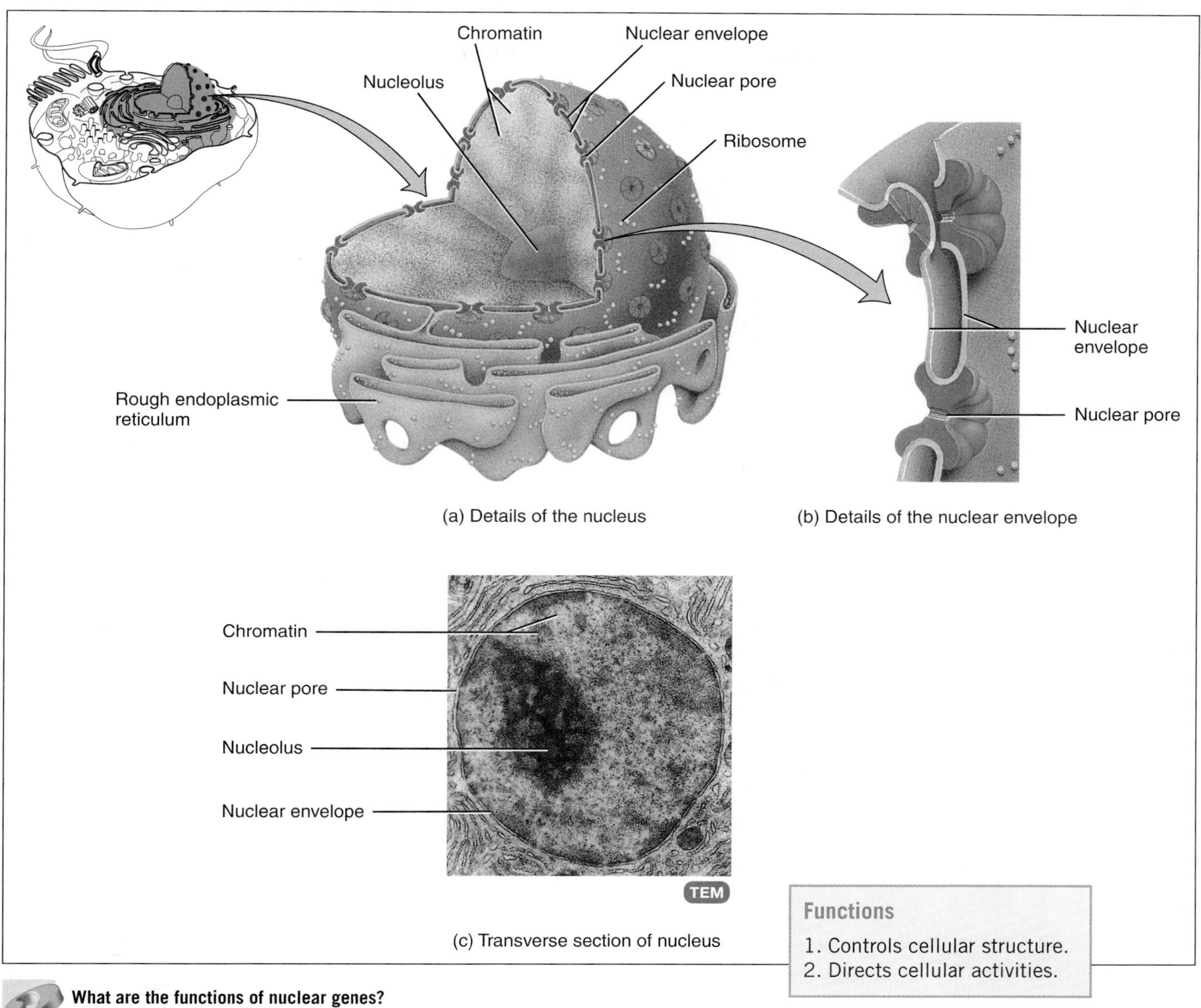

(a) Details of the nucleus

(b) Details of the nuclear envelope

(c) Transverse section of nucleus

Functions

1. Controls cellular structure.
2. Directs cellular activities.

What are the functions of nuclear genes?

body cells have a single nucleus, although some, such as mature red blood cells, have none. In contrast, skeletal muscle cells and a few other cells have several nuclei. A double membrane called the **nuclear envelope** separates the nucleus from the cytoplasm. Both layers of the nuclear envelope are lipid bilayers similar to the plasma membrane. The outer membrane of the nuclear envelope is continuous with rough ER and resembles it in structure. The nuclear envelope is perforated by numerous channels called **nuclear pores** (Figure 2.15c). Each pore consists of a circular arrangement of proteins that surrounds a large central channel. This channel is about 10 times larger than the pore of a channel protein in the plasma membrane.

Nuclear pores control the movement of substances between the nucleus and the cytoplasm. Small molecules and ions diffuse passively through open aqueous channels in the pores. Most large molecules, however, such as RNAs and proteins, cannot pass through the nuclear pores by diffusion. Instead, their passage involves an active transport process in which the molecules are recognized and selectively transported in one direction only. As large molecules are recognized and move through the regulated channels, the nuclear pores open up to accommodate them. This permits the selective transport of proteins from the cytosol into the nucleus and of RNAs from the nucleus into the cytosol.

One or more spherical bodies called **nucleoli** (nū′KLĒ-ō-lī) are present inside the nucleus. Nucleoli (singular is nucleolus) are clusters of protein, DNA, and RNA that are not enclosed by a membrane. They are the sites of the assembly of ribosomal subunits, which play a key role in protein synthesis. Nucleoli are quite prominent in cells that synthesize large amounts of protein, such as muscle and liver cells. Nucleoli disperse and disappear during cell division and reorganize once new cells are formed.

Within the nucleus are most of the cell's hereditary units, called **genes,** which control cellular structure and direct most cellular activities. Genes are arranged in single file along **chromosomes** (*chromo-* = colored). Human somatic (body) cells have 46 chromosomes, 23 inherited from each parent. Each chromosome is a long molecule of DNA that is coiled together with several proteins (Figure 2.16). In a cell that is not dividing, the 46 chromosomes appear as a diffuse, granular mass, which is called **chromatin.** Electron micrographs reveal that chromatin has a "beads-on-a-string" structure. Each "bead" is a **nucleosome** and consists of double-stranded DNA wrapped twice around a core of eight proteins called **histones,** which help organize the coiling and folding of DNA. The "string" between the "beads" is **linker DNA,** which holds adjacent nucleosomes together. Another histone promotes coiling of nucleosomes into a larger diameter **chromatin fiber,** which then folds into large loops. In cells that are not dividing, this is how DNA is packed. Just before cell division takes place, however, the DNA replicates (duplicates) and the loops condense even more, forming a pair of **sister chromatids.** Chromatids are easily seen as rod-shaped structures when viewed through a light microscope.

The main parts of a cell and their functions are summarized in Table 2.2.

Figure 2.16 / Packing of DNA into a chromosome in a dividing cell. When packing is complete, two identical DNAs and their histones form a pair of sister chromatids, held together by a centromere.

A chromosome is a highly coiled and folded DNA molecule that is combined with protein molecules.

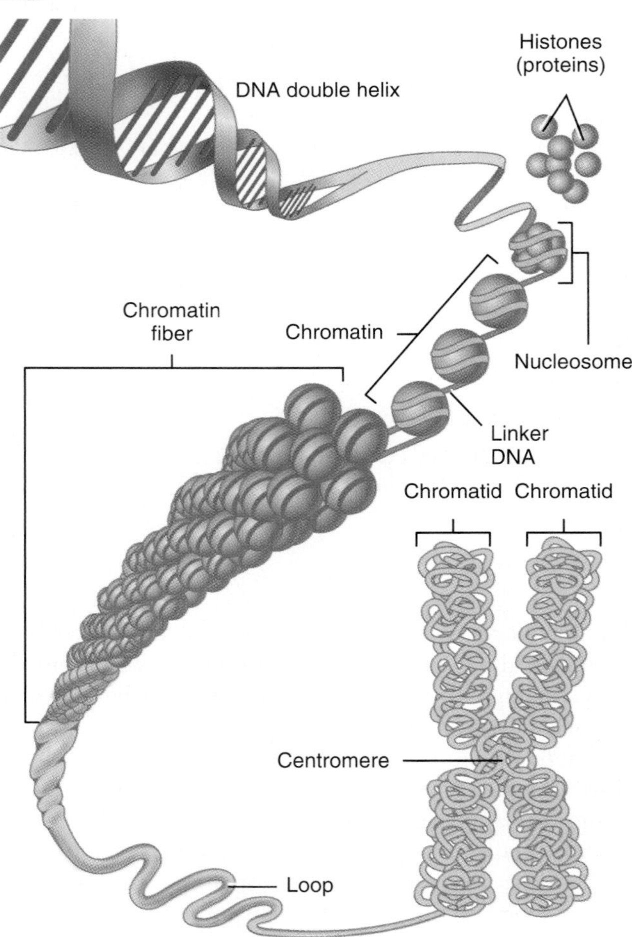

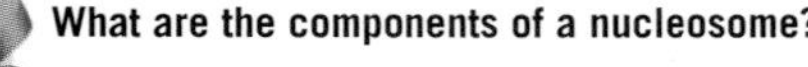
What are the components of a nucleosome?

SOMATIC CELL DIVISION

Objective

- Discuss the stages, events, and significance of somatic cell division.

Most cellular activities maintain the life of the cell on a day-to-day basis. However, as somatic cells become damaged, diseased, or worn out, they are replaced by **cell division,** the process whereby cells reproduce themselves. We distinguish between two kinds of cell division: somatic cell division and reproductive cell division.

In **somatic cell division,** a cell undergoes a nuclear division called **mitosis** and a cytoplasmic division called **cytokinesis** to produce two identical **daughter cells.** Each daughter cell has the same number and kind of chromosomes as the original cell. Somatic cell division replaces dead or injured cells and adds new ones for tissue growth.

Reproductive cell division is the mechanism that produces gametes—sperm and ova—the cells needed to form the next generation of sexually reproducing organisms. This process consists of a special two-step division called **meiosis,** in which the number of chromosomes in the nucleus is reduced by half. Meiosis is described in Chapter 28; here we focus on the division of somatic cells.

The Cell Cycle in Somatic Cells

The **cell cycle** is an orderly sequence of events by which a cell duplicates its contents and divides in two. Human cells, except for gametes, contain 23 pairs of chromosomes. The two chromosomes of each pair—one from the mother and one from the father—are called **homologous chromosomes.** They have similar genes, which are usually arranged in the same order. When a cell reproduces, it must replicate (duplicate) all its chromosomes so that its genes may be passed on to the next generation of cells. The cell cycle consists of two major periods: interphase,

Table 2.2 Cell Parts and Their Functions

Part	Structure	Functions
Plasma Membrane	Fluid-mosaic lipid bilayer consisting of phospholipids, cholesterol, and glycolipids studded with proteins; surrounds cytoplasm.	Protects cellular contents; makes contact with other cells; contains channels, transporters, receptors, enzymes, cell-identity markers, and linker proteins; mediates the entry and exit of substances.
Cytoplasm	Cellular contents between the plasma membrane and nucleus—cytosol and organelles. Cytosol is composed of water, solutes, suspended particles, lipid droplets, and glycogen granules. Organelles are specialized membranous or nonmembranous structures with characteristic shapes and specific functions.	
Cytoskeleton	Network of three kinds of protein filaments: microfilaments, intermediate filaments, and microtubules.	Maintains shape and general organization of cellular contents; responsible for cell movements.
Centrosome	Pericentriolar material plus paired centrioles (9 + 0 array of microtubules).	Pericentriolar material is organizing center for microtubules and mitotic spindle; centrioles form and regenerate cilia and flagella.
Cilia and Flagella	Motile cell surface projections composed of 9 + 2 array of microtubules and a basal body.	Cilia move fluids over a cell's surface; flagella move an entire cell.
Ribosome	Composed of two subunits containing ribosomal RNA and proteins; may be free in cytosol or attached to rough ER.	Protein synthesis.
Endoplasmic Reticulum (ER)	Membranous network of flattened sacs or tubules called cisternae. Rough ER is covered by ribosomes and is attached to nuclear membrane; smooth ER lacks ribosomes.	Rough ER synthesizes secretory proteins and membrane molecules (phospholipids and glycolipids); smooth ER synthesizes phospholipids, fats, and steroids, releases glucose into the bloodstream, inactivates or detoxifies drugs and potentially harmful substances, and stores calcium ions for muscle contraction.
Golgi Complex	3–20 flattened membranous sacs called cisternae; structurally and functionally divided into entry face, medial cisternae, and exit face.	Entry face accepts proteins from rough ER; medial cisternae form glycoproteins, glycolipids, and lipoproteins; exit face stores, packages, and exports medial cisternae products.
Lysosome	Vesicle formed from Golgi complex; contains digestive enzymes.	Fuses with and digests contents of late endosomes, pinocytic vesicles, and phagosomes; digests worn-out organelles (autophagy), entire cells (autolysis), and extracellular materials.
Peroxisome	Vesicle containing oxidative enzymes.	Detoxifies harmful substances.
Mitochondrion	Consists of outer and inner membranes, cristae, and matrix.	Site of aerobic cellular respiration reactions that produce most of a cell's ATP.
Nucleus	Consists of nuclear envelope with pores, nucleoli, and chromatin (or chromosomes).	Contains genes, which control cellular structure and most cellular activities.

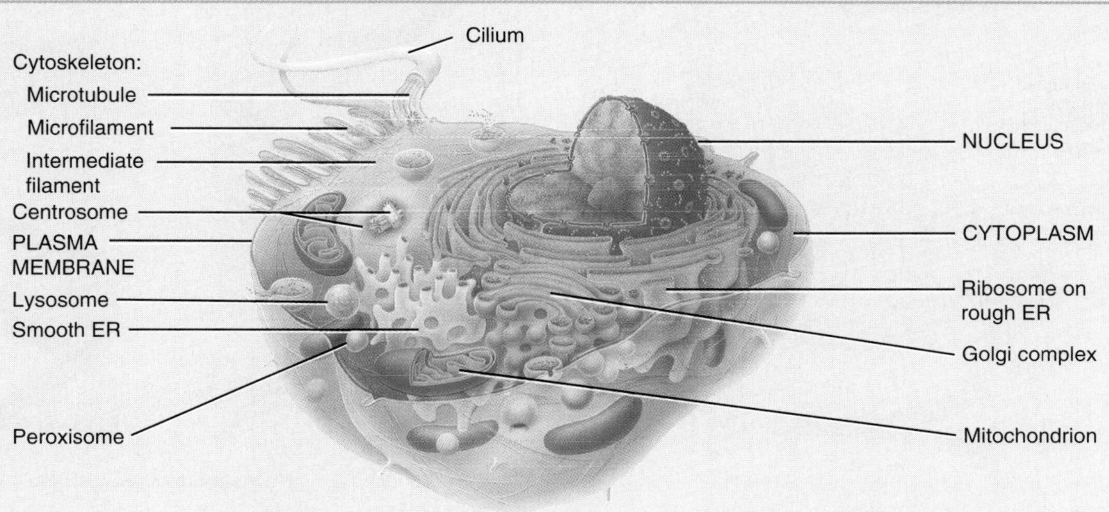

when a cell is not dividing, and the mitotic (M) phase, when a cell is dividing (Figure 2.17).

Interphase

During **interphase** the cell replicates its DNA. It also manufactures additional organelles and cytosolic components in anticipation of cell division. Interphase is a state of high metabolic activity, and during this time the cell does most of its growing.

Interphase consists of three phases: G_1, S, and G_2 (Figure 2.17). The S stands for synthesis of DNA. Because the G-phases are periods when there is no activity related to DNA duplication, they are thought of as gaps or interruptions in DNA duplication. The **G_1 phase** is the interval between the mitotic phase and the S phase. During G_1, the cell is metabolically active; it duplicates its organelles and cytosolic components, but not its DNA. Replication of centrosomes begins here but does not end until G_2; virtually all the cellular activities described in this chapter happen during G_1. For a typical body cell with a total cell cycle time of 24 hours, G_1 might last about 8–10 hours. The duration of this phase is quite variable, however, lasting from minutes, to hours, to years for different types of cells. The **S phase** is the interval between G_1 and G_2 and lasts about 6–8 hours. Its duration is also quite variable for different cell types. During the S phase, DNA replication occurs, ensuring that the two daughter cells formed from cell division will have identical genetic material. The **G_2 phase** is the interval between the S phase and the mitotic phase. It lasts about 4–6 hours. During G_2, cell growth continues, enzymes and other proteins are synthesized in preparation for cell division, and the replication of centrosomes is completed. Cells that remain in G_1 for a very long time, perhaps destined never to divide again, are said to be in the **G_0 state.** For example, most nerve cells are in this state. Once a cell enters the S phase, however, it is committed to go through cell division.

When DNA replicates during the S phase, its helical structure partially uncoils, and the two strands separate at the points where hydrogen bonds connect base pairs (Figure 2.18). Each exposed base then picks up a complementary base (with its associated sugar and phosphate group). This uncoiling and complementary base pairing continues until each of the two original

Figure 2.17 / The cell cycle for a typical body cell with a total cell cycle time of 24 hours. Not illustrated is cytokinesis, division of the cytoplasm, which occurs during late anaphase or early telophase of the mitotic phase.

In a complete cell cycle, a cell duplicates its contents and divides into two daughter cells.

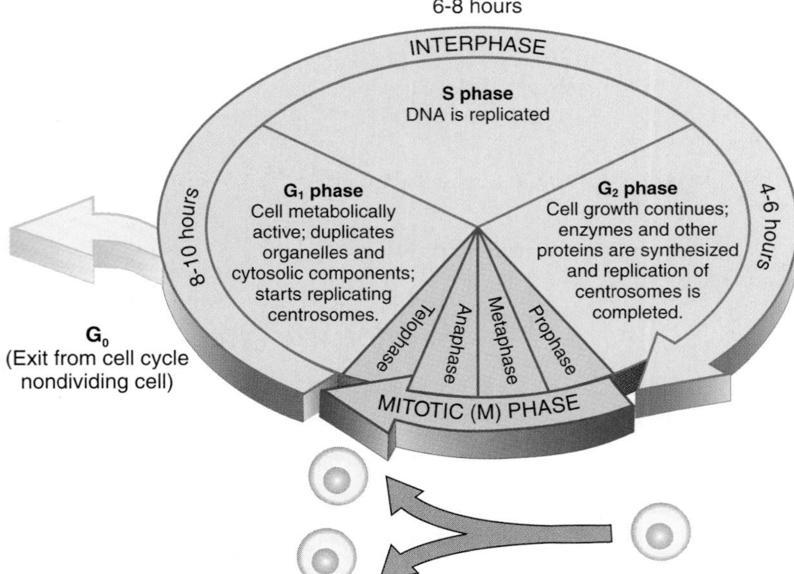

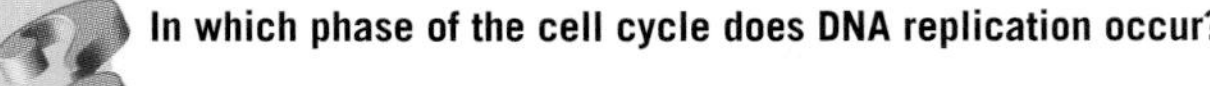
In which phase of the cell cycle does DNA replication occur?

Figure 2.18 / Replication of DNA. The two strands of the double helix separate by breaking the hydrogen bonds (shown as dotted lines) between nucleotides. New, complementary nucleotides attach at the proper sites, and a new strand of DNA is synthesized alongside each of the original strands. Arrows indicate hydrogen bonds forming again between pairs of bases.

Replication doubles the amount of DNA.

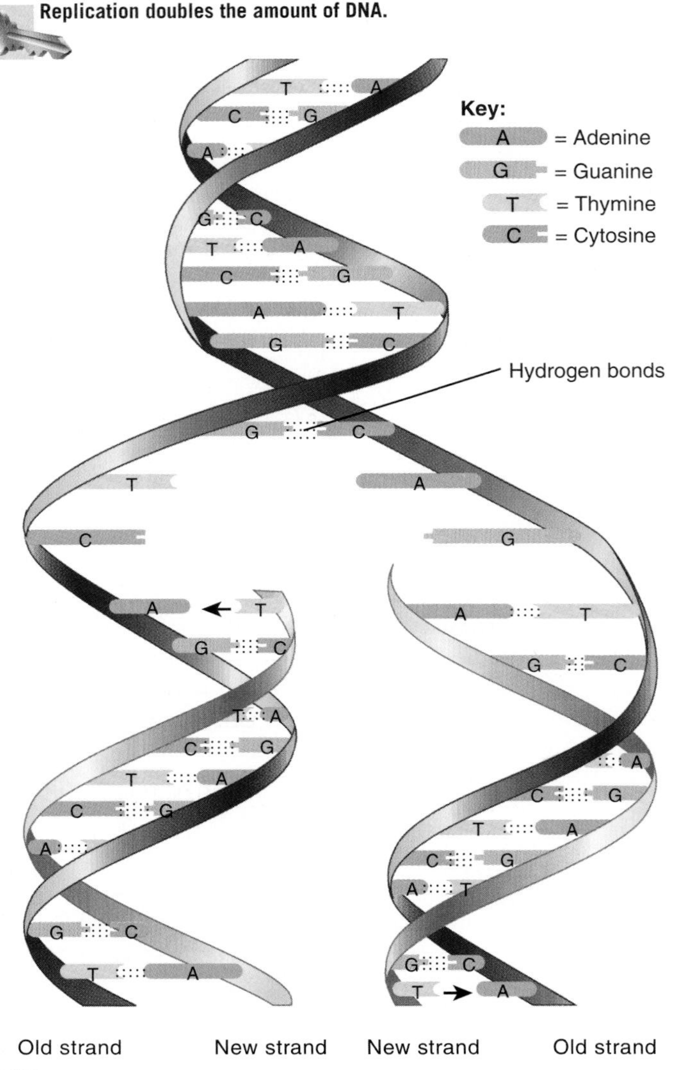

Why is it crucial that DNA replication occurs prior to cytokinesis in somatic cell division?

DNA strands is joined with a newly formed complementary DNA strand. The original DNA molecule has become two identical DNA molecules.

A microscopic view of a cell during interphase shows a clearly defined nuclear envelope, nucleolus, and chromatin (Figure 2.19a). The absence of visible chromosomes is another physical characteristic of interphase. Once a cell completes its activities during the G_1, S, and G_2 phases of interphase, the mitotic phase begins.

Mitotic Phase

The **mitotic (M) phase** of the cell cycle consists of nuclear division, or mitosis, and cytoplasmic division, or cytokinesis. The events that occur during this phase are plainly visible under a microscope because chromatin condenses into chromosomes.

NUCLEAR DIVISION: MITOSIS **Mitosis** is the distribution of the two sets of chromosomes, one set into each of two separate nuclei. The process results in the *exact* partitioning of genetic information. For convenience, biologists divide the process into four stages: prophase, metaphase, anaphase, and telophase. However, mitosis is a continuous process, one stage merges imperceptibly into the next.

- **Prophase.** During early prophase, the chromatin fibers condense and shorten (Figure 2.19b). The condensation process may prevent entangling of the long DNA strands as they move during mitosis. Because DNA replication took place during the S phase of interphase, each prophase chromosome contains a pair of identical double-stranded chromatids. Each chromatid pair is held together by a constricted region called a **centromere** that is required for the proper segregation of chromosomes. Attached to the outside of each centromere is a protein complex known as the **kinetochore** (ki-NET-ō-kor), whose function is described shortly.

  Later in prophase, the nucleolus disappears, the nuclear envelope breaks down, and each centrosome moves to an opposite pole (end) of the cell. As they do so, the pericentriolar material of the centrosomes starts to form the **mitotic spindle,** a football-shaped assembly of microtubules (Figure 2.19b). The lengthening of microtubules between centrosomes pushes the centrosomes to the poles of the cell so that the spindle extends from pole to pole. The spindle is responsible for the separation of chromatids to opposite poles of the cell.
- **Metaphase.** During metaphase, the microtubules align the centromeres of the chromatid pairs at the exact center of the mitotic spindle (Figure 2.19c). This midpoint region is called the **metaphase plate** or **equatorial plane region.**
- **Anaphase.** During anaphase, the centromeres split, separating sister chromatids. This enables one chromatid of each pair to move toward opposite poles of the cell (Figure 2.19d). Once separated, the sister chromatids are called **daughter chromosomes.** As the chromosomes are pulled by the microtubules during anaphase, they become V-shaped as the centromeres lead the way and seem to drag the trailing parts of the chromosomes toward the pole.
- **Telophase.** The final stage of mitosis, telophase, begins after chromosomal movement stops (Figure 2.19e). The identical sets of chromosomes now at opposite poles of the cell, uncoil and revert to threadlike chromatin. A nuclear envelope forms around each chromatin mass, nucleoli reappear in the daughter nuclei, and eventually the mitotic spindle breaks up.

CYTOPLASMIC DIVISION: CYTOKINESIS Division of a parent cell's cytoplasm and organelles is called **cytokinesis** (sī′-tō-ki-NĒ-sis; *cyto-* = cell; *-kinesis* = motion). This process begins in late anaphase or early telophase with formation of a **cleavage furrow,** a slight indentation of the plasma membrane, usually midway between the centrosomes, that extends around the center of the cell (see Figure 2.19d and e). The cleavage furrow is produced by the action of actin microfilaments that lie just inside the plasma membrane. The microfilaments form a *contractile ring* that pulls the plasma membrane progressively inward, constricting the center of the cell like a belt around the waist, and ultimately pinching it in two. The plane of the cleavage furrow is always perpendicular to the mitotic spindle, thereby ensuring that the two sets of chromosomes will be segregated into separate daughter cells. When cytokinesis is complete, interphase begins (Figure 2.19f).

Considering the cell cycle in its entirety, the sequence of events is

$$G_1 \text{ phase} \rightarrow \text{S phase} \rightarrow G_2 \text{ phase} \rightarrow \text{mitosis} \rightarrow \text{cytokinesis}$$

The events of the somatic cell cycle are summarized in Table 2.3.

Table 2.3 Events of the Somatic Cell Cycle

Phase	Activity
Interphase	Period between cell divisions; chromosomes not visible under light microscope.
G_1 phase	Metabolically active cell duplicates organelles and cytosolic components; centrosome replication begins.
S phase	DNA replicates.
G_2 phase	Cell growth, enzyme and protein synthesis continues; centrosome replication completed.
Mitotic Phase	Parent cell produces daughter cells with identical chromosomes; chromosomes visible under light microscope.
Mitosis	Nuclear division; distribution of two sets of chromosomes into separate nuclei.
Prophase	Chromatin fibers condense into paired chromatids; nucleolus and nuclear envelope disappear; centrosomes move to opposite poles of cell.
Metaphase	Centromeres of chromatid pairs line up at metaphase plate.
Anaphase	Centromeres divide; identical sets of chromosomes move to opposite poles of cell.
Telophase	Nuclear envelopes and nucleoli reappear; chromosomes resume chromatin form; mitotic spindle breaks up.
Cytokinesis	Cytoplasmic division; contractile ring forms cleavage furrow around center of cell, dividing cytoplasm into separate and equal portions.

Figure 2.19 / Cell division: mitosis and cytokinesis. Begin the sequence at (a) at the top of the figure and read clockwise until you complete the process. Although sister chromatids are identical, they are differentiated here by color (red and yellow) to more easily follow their movements.

In somatic cell division, a single diploid cell divides to produce two identical diploid daughter cells.

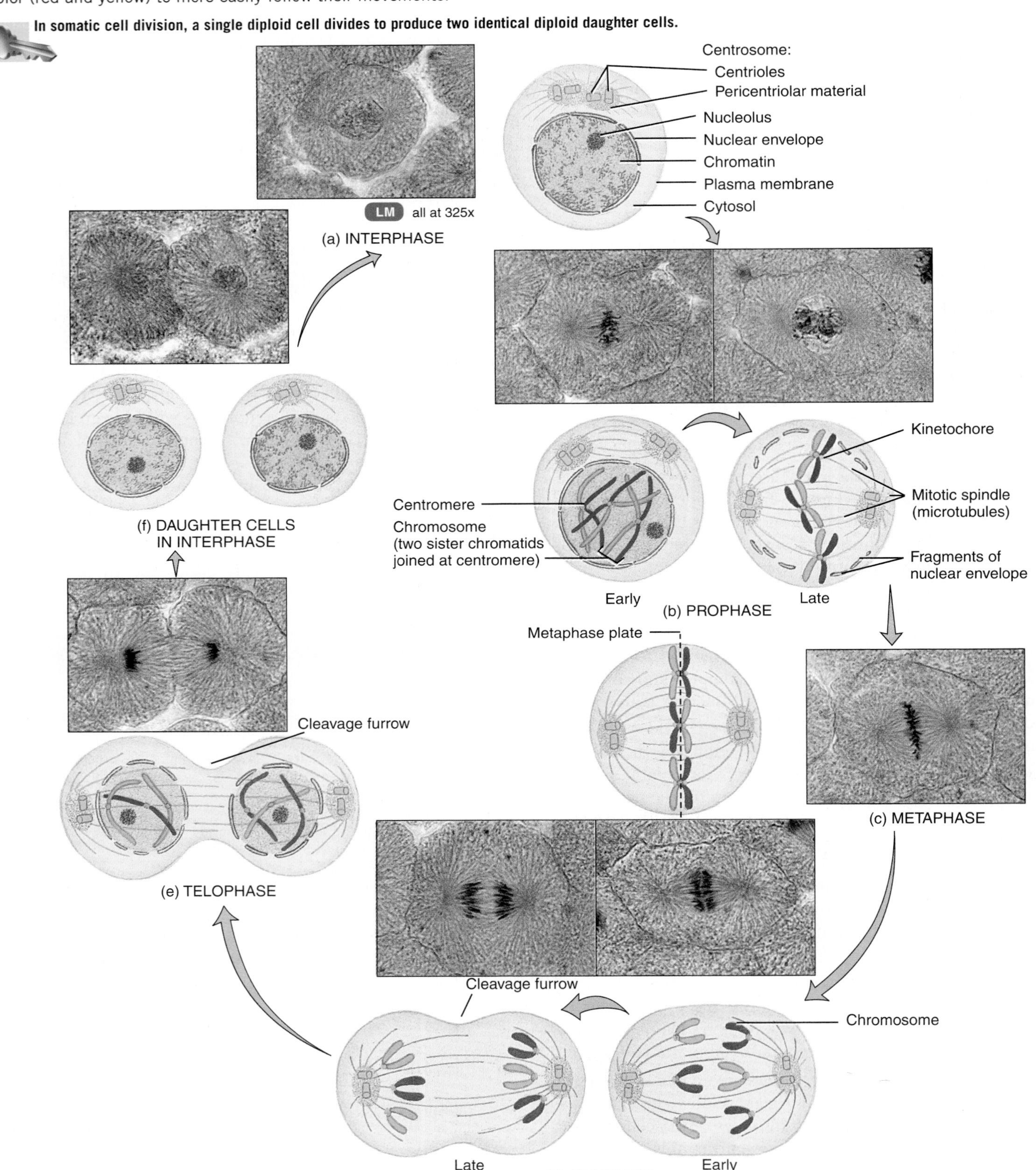

When does cytokinesis begin?

Control of Cell Destiny

Objective

- Discuss the signals that induce cell division, and describe apoptosis.

A cell has three possible destinies—to remain alive and functioning without dividing, to grow and divide, or to die. Homeostasis is maintained when there is a balance between cell proliferation and cell death. There are signals that tell a cell when to exist in the G_0 phase, when to divide, and when to die.

Throughout the lifetime of an organism, certain cells undergo an orderly, genetically programmed death, a process called **apoptosis** (ap-ō-TŌ-sis; = a falling off). In apoptosis, a triggering agent from either outside or inside the cell causes "cell-suicide" genes to produce enzymes that damage the cell in several ways, including disrupting its cytoskeleton and nucleus. As a result, the cell shrinks and pulls away from neighboring cells. The DNA within the nucleus fragments, and the cytoplasm shrinks, although the plasma membrane remains intact. Phagocytes in the vicinity then ingest the dying cell. Apoptosis, a normal type of cell death, contrasts with **necrosis** (ne-KRŌ-sis; = death), a pathological type of cell death that results from tissue injury. In necrosis, many adjacent cells swell, burst, and spill their cytoplasm into the interstitial fluid. The cellular debris usually stimulates an inflammatory response by the immune system that does not occur in apoptosis. Apoptosis is especially useful because it removes unneeded cells during development before birth. It continues to occur after birth to regulate the number of cells in a tissue and eliminate potentially dangerous cells such as cancer cells.

✓ Distinguish between the somatic and reproductive cell division. Why is each important?
✓ Define interphase. When does DNA replicate itself?
✓ Describe the principal events of each stage of the mitotic phase.
✓ How is cell destiny controlled?

AGING AND CELLS

Objective

- Explain the relationship of aging to cellular processes.

Aging is a normal process accompanied by a progressive alteration of the body's homeostatic adaptive responses. It produces observable changes in structure and function and increases vulnerability to environmental stress and disease. The specialized branch of medicine that deals with the medical problems and care of elderly persons is called **geriatrics** (jer′-ē-AT-riks; *ger-* = old age; *-iatrics* = medicine).

Although many millions of new cells normally are produced each minute, several kinds of cells in the body—heart muscle cells, skeletal muscle cells, and nerve cells—do not divide because they are arrested permanently in the G_0 phase. Experiments have shown that many other cell types have only a limited capability to divide. Normal cells grown outside the body divide only a certain number of times and then stop. These observations suggest that cessation of mitosis is a normal, genetically programmed event. According to this view, "aging genes" are part of the genetic blueprint at birth. These genes have an important function in normal cells but their activities slow over time. They bring about aging by slowing down or halting processes vital to life.

Glucose, the most abundant sugar in the body, plays a role in the aging process. Glucose is haphazardly added to proteins inside and outside cells, forming irreversible cross-links between adjacent protein molecules. With advancing age, more cross-links form, which contributes to the stiffening and loss of elasticity that occur in aging tissues.

Some free radicals are produced as part of normal cellular metabolism. Others are present in air pollution, radiation, and certain foods. Free radicals cause oxidative damage in lipids, proteins, nucleic acids by "stealing" electrons to accompany their unpaired electrons. Some effects are wrinkled skin, stiff joints, and hardened arteries. Although naturally occurring enzymes in peroxisomes and in the cytosol normally dispose of free radicals, the mechanism becomes less efficient with age. Certain dietary substances, such as vitamin E, vitamin C, beta-carotene, and selenium, are antioxidants that inhibit free radical formation.

Whereas some theories of aging explain the process at the cellular level, others concentrate on regulatory mechanisms operating within the entire organism. For example, the immune system may start to attack the body's own cells. This *autoimmune response* might be caused by changes in cell-identity markers at the surface of cells that cause antibodies to attach to and mark the cell for destruction. As changes in the proteins on the plasma membrane of cells increase, the autoimmune response intensifies, producing the well-known signs of aging.

The effects of aging on the various body systems are discussed in their respective chapters.

Progeria and Werner's Syndrome

Progeria (prō-JĒR-ē-a) is a noninherited disease characterized by normal development in the first year of life followed by rapid aging. The condition is expressed by dry and wrinkled skin, total baldness, and bird-like facial features. Death usually occurs around age 13.

Werner's syndrome is a rare, inherited disease that causes a rapid acceleration of aging, usually while the person is only in his or her twenties. It is characterized by wrinkling of the skin, graying of the hair and baldness, cataracts, muscular atrophy, and a tendency to develop diabetes mellitus, cancer, and cardiovascular disease. Most afflicted individuals die before age 50. Recently, the gene that causes Werner's syndrome has been identified. Researchers hope to use the information to gain insight into the mechanism of aging. ■

✓ What is aging?
✓ What are some of the cellular changes that occur with aging?

▶ 14th Century ▶ 15th Century

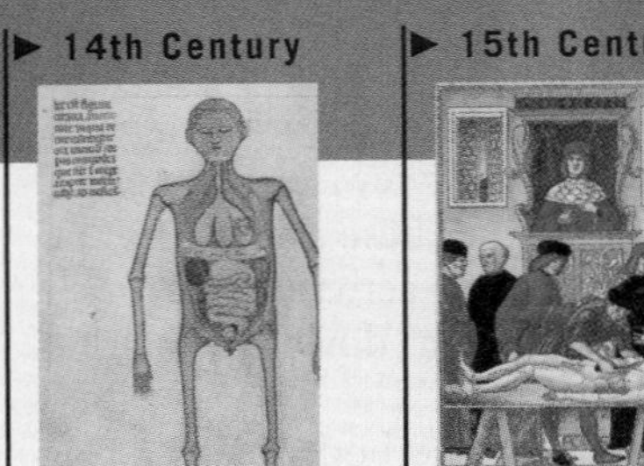

▶ 16th Century

▶ 17th Century

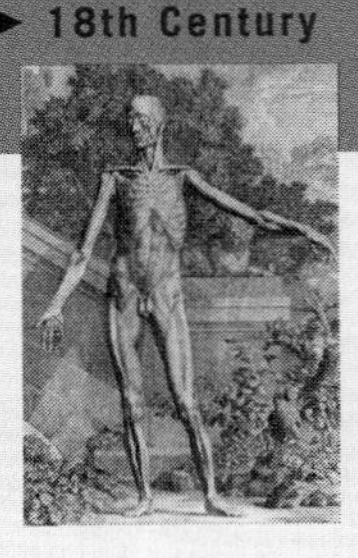

▶ 18th Century

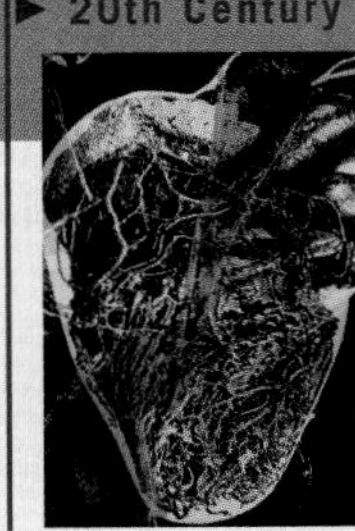

▶ 20th Century

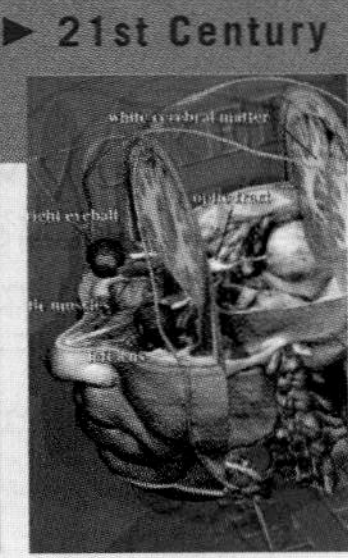

▶ 21st Century

CHANGING IMAGES

Hooked on Cells

Prior to the discovery that the basic structural unit of plants and animals is the cell, strange theories abounded concerning the basis and origins of life. Unraveling this mystery required the efforts of many scientists over many centuries. From Aristotle until the late nineteenth century, the prevailing theory was that life spontaneously arose from non-living substances. It may seem ludicrous today, but in the seventeenth century it was believed that placing sweaty undergarments in a jar with wheat husks would eventually produce mice. This theory of *spontaneous generation* or *abiogenesis* was prolific but eventually died a slow death. It was not until the invention of the microscope and the development of *cell theory* that abiogenesis was laid to rest.

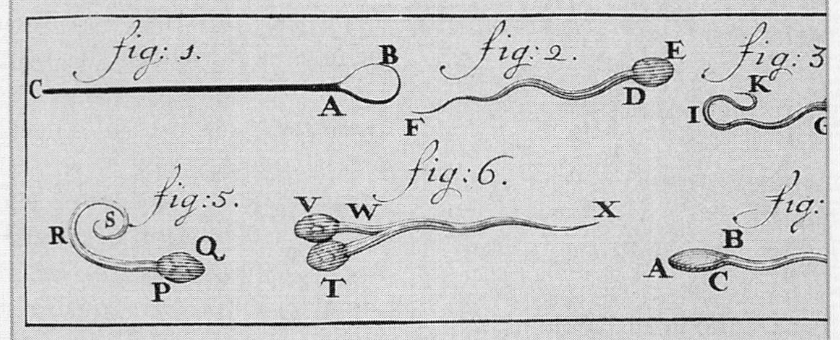

17th Century

The composite photo shown here depicts selected figures in this saga. Robert Hooke, one of the greatest experimental biologists of the seventeenth century, introduced the term *cell* while examining cork tissue (shown in the textbook on top) with the compound microscope and illumination system that he invented. The dead cellulose walls of cork cells that he observed reminded him of the cells resided in by monks. While it is Hooke who first identified plant cells, it is Antoni van Leeuwenhoek, pictured on the left, who was the first to identify human cells. Among his vast accomplishments is his discovery of the red blood cell and spermatozoa. His seventeenth century renderings of human spermatozoa are extremely accurate and pictured here as well in the illustration below. Another important figure, Theodore Schwann, pictured on the right, applied the developing plant cell theory to animal life. Schwann recognized the common features of cells to be a membrane, nucleus and cell body. As you will learn in Chapter 16, certain cells of the nervous system bear his name.

So as you sit here today and having read about cells, consider the sum total of knowledge that made this chapter possible. Also ponder that some of the theories we presently accept as truth may some day be considered amusing.

CELLULAR DIVERSITY

To this point, we have investigated the intracellular complementarity of a cell's structure and function. Most (but not all) cells have a full complement of organelles, each of which controls a specific set of physiologic processes within the cell. However, not all cells in the body look the same; nor do they perform identical functional roles in the body.

The body of an average human adult is composed of nearly 100 trillion cells, but all of these cells can be classified into about 200 different cell types. Cells vary considerably in size. High-powered microscopes are needed to see the smallest cells of the body. The largest cell, a single ovum, is barely visible to the unaided eye. The sizes of cells are measured in units called *micrometers*. One micrometer (μm) is equal to 1 one-millionth of a meter, or 10^{-6}m (1/25,000 of an inch). Whereas a red blood cell has a diameter of 8 μm, an ovum has a diameter of about 140 μm.

The shapes of cells also vary considerably (Figure 2.20). They may be round, oval, flat, cuboidal, columnar, elongated, star-shaped, cylindrical, or disc-shaped. A cell's shape is related to its function in the body. For example, a sperm cell has a long whiplike tail (flagellum) that it uses for locomotion. The disc-shape of a red blood cell gives it a large surface area that enhances its ability to pass oxygen to other cells. Some cells contain microvilli, which greatly increase their surface area. Microvilli are common in the epithelial cells that line the small intestine, where they enhance the absorption of digested food. Nerve cells have long extensions that permit them to transmit nerve impulses over great distances.

Cellular diversity is the stepping stone to more complex levels of biological organization such as tissues and organs.

Figure 2.20 / Diverse shapes and sizes of human cells. The relative difference in size between the smallest and largest cells is actually much greater than shown here.

The nearly 100 trillion cells in an average adult can be classified into about 200 different cell types.

Sperm cell

Smooth muscle cell

Nerve cell

Red blood cell

Epithelial cell

Why are sperm the only body cells that have a flagellum?

APPLICATIONS TO HEALTH

Cancer

Cancer is a group of diseases characterized by uncontrolled cell proliferation. When cells in a part of the body divide without control, the excess tissue that develops is called a **tumor** or **neoplasm** (NĒ-ō-plazm; *neo-* = new). The study of tumors is called **oncology** (on-KOL-ō-jē; *onco-* = swelling or mass). Tumors may be cancerous and sometimes fatal, or they may be harmless. A cancerous neoplasm is called a **malignant tumor** or **malignancy.** One property of some malignant tumors is their ability to undergo **metastasis** (me-TAS-ta-sis), the spread of cancerous cells to other parts of the body. A **benign tumor** is a noncancerous growth. Although benign tumors do not metastasize or invade, they may be surgically removed if they interfere with a normal body function or become disfiguring.

Types of Cancer

The name of the cancer is derived from the type of tissue in which it develops. Most human cancers are **carcinomas** (kar-sin-ŌM-az; *carcin-* = cancer; *-omas* = tumors), malignant tumors that arise from epithelial cells. **Melanomas** (mel-an-ŌM-az; *melan-* = black), for example, are cancerous growths of melanocytes, skin epithelial cells that produce the pigment melanin. **Sarcoma** (sar-KŌ-ma; *sarc-* = flesh) is a general term for any cancer arising from muscle cells or connective tissues. For example, **osteogenic sarcoma** (*osteo-* = bone; *-genic* = origin), the most frequent type of childhood cancer, destroys normal bone tissue. **Leukemia** (lū-KĒ-mē-a; *leuk-* = white; *-emia* = blood) is a cancer of blood-forming organs characterized by rapid growth of abnormal leukocytes (white blood cells). **Lymphoma** (lim-FŌ-ma) is a malignant disease of lymphatic tissue—for example, lymph nodes.

Growth and Spread of Cancer

Cells of malignant tumors duplicate rapidly and continuously. As malignant cells invade surrounding tissues, they trigger **angiogenesis,** the growth of new networks of blood vessels. Proteins that trigger angiogenesis in tumors are called **tumor angiogenesis factors (TAFs).** Tumor angiogenesis factors are balanced by inhibitors of angiogenesis. Thus the formation of new blood vessels can be accomplished either by over-production of TAFs or by the lack of naturally occurring angiogenesis inhibitors. As the cancer grows, it begins to compete with normal tissues for space and nutrients. Eventually, the normal tissue decreases in size and dies. Some malignant cells may detach from the initial (primary) tumor and invade a body cavity or enter the blood or lymph, then circulate to and invade other body tissues, establishing secondary tumors. Malignant cells resist the antitumor defenses of the body. The pain associated with cancer develops when the tumor presses on nerves or blocks a passageway in an organ so that secretions build up pressure.

Causes of Cancer

Several factors may trigger a normal cell to lose control and become cancerous. One cause is environmental agents: substances in the air we breathe, the water we drink, and the food we eat. A chemical agent or radiation that produces cancer is called a **carcinogen** (car-SIN-ō-jen). Carcinogens induce **mutations,** permanent structural changes in the DNA base sequence of a gene. The World Health Organization estimates that carcinogens may be associated with 60–90% of all human cancers. Examples of carcinogens are hydrocarbons found in cigarette tar, radon gas from the earth, and ultraviolet (UV) radiation in sunlight.

Viruses are a second cause of cancer. These agents are tiny packages of nucleic acids, either DNA or RNA, that are capable of infecting cells and converting them to virus-producers. The link between viruses and cancer is strongly established for a variety of animal cancers, and the link between viruses and human cancer is quite clear for some types of tumors, such as cervical cancer. The evidence suggests that chronic viral infections are associated with up to one-fifth of all cancers.

Intensive research efforts are now directed toward studying cancer-causing genes, or **oncogenes** (ON-kō-jēnz). These genes, when inappropriately activated, have the ability to transform a normal cell into a cancerous cell. Oncogenes develop when normal genes that regulate growth and development, called **proto-oncogenes,** are mutated or otherwise deregulated. These genes may undergo some change that either causes them to produce an abnormal product or disrupts their control so that they are expressed inappropriately, making their products in excessive amounts or at the wrong time. Scientists believe that some oncogenes cause excessive production of growth factors, chemicals that stimulate cell growth. Other oncogenes may cause changes in a surface receptor, causing it to send signals as though it were being activated by a growth factor. As a result, the growth pattern of the cell becomes abnormal.

Proto-oncogenes in every cell carry out normal cellular functions until a malignant change occurs. It appears that some proto-oncogenes are activated to oncogenes by mutations in which the DNA of the proto-oncogene is altered. Other proto-oncogenes are activated by a rearrangement of the chromosomes so that segments of DNA are exchanged. Rearrangement activates proto-oncogenes by placing them near genes that enhance their activity. Viruses are believed to cause cancer by inserting their own oncogenes or proto-oncogenes into the host cell's DNA.

Carcinogenesis: A Multistep Process

Carcinogenesis (kar′-si-nō-JEN-e-sis), the process by which cancer develops, is a multistep process in which as many as ten distinct mutations may have to accumulate in a cell before it becomes cancerous. The progression of genetic changes leading to cancer is best understood for colon (colorectal) cancer. Such cancers, as well as lung and breast cancer, take years or decades to develop. In colon cancer, the tumor begins as an area of increased cell proliferation that results from one mutation. This growth then progresses to abnormal, but noncancerous, growths called adenomas. When finally a mutation of *p53* (a tumor-suppressor gene) occurs, a carcinoma develops. **Tumor-suppressor genes** are genes found on chromosomes in normal cells that suppress unregulated cell growth. The fact that so many mutations are needed for a cancer to develop indicates that cell growth is normally controlled with many sets of checks and balances.

Treatment of Cancer

Many cancers are removed surgically. However, when cancer is widely distributed throughout the body or exists in organs such as the brain whose functioning would be greatly harmed by surgery, chemotherapy and radiation therapy may be used instead. Chemotherapy involves the administration of drugs that poison cancerous cells. Radiation therapy destroys the chromosomes of cancerous cells, thus preventing them from dividing. Because cancerous cells divide rapidly, they are more vulnerable to the destructive effects of chemotherapy and radiation therapy than are normal cells. Nevertheless, both chemotherapy and radiation therapy can kill or disrupt the function of some normal cells as well.

Treating cancer is difficult because it is not a single disease and because all the cells in a single tumor population rarely behave in the same way. Although most cancers are thought to derive from a single abnormal cell, by the time a tumor reaches a clinically detectable size, the cancer may contain a diverse population of abnormal cells. For example, some cancerous cells metastasize readily, and others do not; some are sensitive to chemotherapy drugs and some are drug resistant. Because of differences in drug resistance, a single chemotherapeutic agent may destroy susceptible cells but permit resistant cells to proliferate.

Another stumbling block encountered by blood-borne anticancer drugs is the physical barrier developed by solid tumors, such as those that arise in the breasts, lungs, colon, and other organs. The high-pressure areas deep within such tumors collapse blood vessels in the tumor. This makes it difficult, if not impossible, for blood-borne anticancer agents to penetrate the tumor.

KEY MEDICAL TERMS ASSOCIATED WITH CELLS

Note to the Student

Most chapters in this text are followed by a glossary of key medical terms that include both normal and pathological conditions. You should familiarize yourself with the terms because they will play an essential role in your medical vocabulary.

Some of these disorders, as well as disorders discussed in the text, are referred to as local or systemic. A *local disease* is one that affects one part or a limited area of the body. A *systemic disease* affects either the entire body or several parts.

The science that deals with why, when, and where diseases occur and how they are transmitted in a human community is known as **epidemiology** (ep′-i-dē-mē-OL-ō-jē; *epidemios* = prevalent; *logos* = study of). The science that deals with the effects and uses of drugs in the treatment of disease is called **pharmacology** (far′-ma-KOL-ō-jē; *pharmakon* = medicine; *logos* = study of).

Anaplasia (an′-a-PLĀ-zē-a; *an* = not; *plassein* = to shape) The loss of tissue differentiation and function that is characteristic of most malignancies.

Atrophy (AT-rō-fē; *a* = without; *-trophy* = nourishment) A decrease in the size of cells, with a subsequent decrease in the size of the affected tissue or organ; wasting away.

Dysplasia (dis-PLĀ-zē-a; *dys-* = abnormal; *plassein* = to shape) Alteration in the size, shape, and organization of cells due to chronic irritation or inflammation; may progress to neoplasia (tumor formation, usually malignant) or revert to normal if the irritation is removed.

Hyperplasia (hī-per-PLĀ-zē-a; *hyper-* = over) Increase in the number of cells of a tissue due to an increase in the frequency of cell division.

Hypertrophy (hī-PER-trō-fē) Increase in the size of cells without cell division.

Metaplasia (met′-a-PLĀ-zē-a; *meta-* = change) The transformation of one type of cell into another.

Progeny (PROJ-e-nē; *pro-* = forward; *-geny* = production) Offspring or descendants.

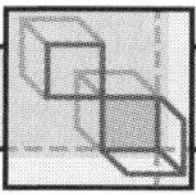

STUDY OUTLINE

Introduction (p. 26)

1. A cell is the basic, living, structural and functional unit of the body.
2. Cytology is the scientific study of cellular structure. Cell physiology is the study of cellular function.

A Generalized Cell (p. 26)

1. Figure 2.1 shows a cell that is a composite of many different cells in the body.
2. The principal parts of a cell are the plasma membrane; the cytoplasm, which consists of cytosol and organelles; and the nucleus.

The Plasma Membrane (p. 27)

Structure of the Membrane (p. 28)

1. The plasma membrane surrounds and contains the cytoplasm of a cell.
2. The membrane is composed of a 50:50 mix by weight of proteins and lipids.
3. According to the fluid mosaic model, the membrane is a mosaic of proteins floating like icebergs in a bilayer sea of lipids.
4. The lipid bilayer consists of two back-to-back layers of phospholipids, cholesterol, and glycolipids.
5. Integral proteins can be separated from the plasma membrane only by procedures that disrupt the lipid bilayer; peripheral proteins can be released from the membrane by relatively mild extraction procedures that leave the lipid bilayer intact.
6. Many integral proteins are glycoproteins that have sugar groups attached to the ends that face the extracellular fluid. Together with glycolipids, the glycoproteins form a glycocalyx on the extracellular surface of cells.

Function of Membrane Proteins (p. 28)

1. Membrane proteins have a variety of functions. Channel proteins and transporters are membrane proteins that help specific solutes across the membrane; receptors serve as cellular recognition sites; and linkers anchor proteins in the plasma membranes to filaments inside and outside the cell. Some membrane proteins are enzymes and others are cell-identity markers.
2. The membrane's selective permeability permits some substances to pass more readily than others. The lipid bilayer is permeable to water and molecules such as oxygen, carbon dioxide, and steroids. Channel proteins and transporters increase the plasma membrane's permeability to small- and medium-sized substances, including ions, that cannot cross the lipid bilayer.

Transport Across the Plasma Membrane (p. 28)

1. Passive processes depend on the concentration of substances and their kinetic energy.
2. Simple diffusion is the net movement of molecules or ions from an area of higher concentration to an area of lower concentration until an equilibrium is reached.
3. In facilitated diffusion, certain molecules, such as glucose, move through the membrane with the help of a transporter.
4. Osmosis is the movement of water through a selectively permeable membrane from an area of higher water concentration to an area of lower water concentration.
5. Filtration is the movement of water and dissolved substances across a membrane due to gravity or hydrostatic pressure.
6. Active processes depend on the use of ATP by the cell.
7. Active transport is the movement of a substance across a cell membrane from lower to higher concentration using energy derived from ATP.
8. Vesicular transport includes both endocytosis and exocytosis.
9. Receptor-mediated endocytosis is the selective uptake of large molecules and particles (ligands) that bind to specific receptors in membrane areas called clathrin-coated pits.

10. Phagocytosis is the ingestion of solid particles. It is an important process used by some white blood cells to destroy bacteria that enter the body.
11. Pinocytosis is the ingestion of extracellular fluid. In this process, the fluid becomes surrounded by a pinocytic vesicle.
12. Exocytosis involves movement of secretory or waste products out of a cell by fusion of vesicles with the plasma membrane.

Cytoplasm (p. 33)

1. Cytoplasm is all the cellular contents between the plasma membrane and the nucleus.
2. Cytoplasm consists of cytosol and organelles.

Cytosol (p. 33)

1. Cytosol is the fluid portion of cytoplasm.
2. It contains mostly water, plus ions, glucose, amino acids, fatty acids, proteins, lipids, ATP, and waste products.
3. Cytosol is the site of many chemical reactions required for a cell's existence.

Organelles (p. 33)

1. Organelles are specialized structures with characteristic shapes and specific functions.
2. Some organelles, such as the cytoskeleton, centrosome, cilia, flagella, and ribosomes, are nonmembranous (don't contain membranes).
3. Other organelles, such as the endoplasmic reticulum, Golgi complex, mitochondria, lysosomes, and peroxisomes, are membranous (contain one or more phospholipid membranes).
4. The cytoskeleton is a network of several kinds of protein filaments that extend throughout the cytoplasm.
5. Components of the cytoskeleton are microfilaments, intermediate filaments, and microtubules.
6. The cytoskeleton provides a structural framework for the cell and is responsible for cell movements.
7. The centrosome consists of pericentriolar material and centrioles.
8. The pericentriolar material organizes microtubules in nondividing cells and the miotic spindle in dividing cells.
9. Centrioles function in the formation and regeneration of cilia and flagella.
10. Cilia and flagella are motile projections of the cell surface.
11. Cilia move fluid along the cell surface.
12. Flagella move an entire cell.
13. Ribosomes, composed of ribosomal RNA and ribosomal proteins, consist of two subunits made in the nucleus.
14. Free ribosomes are not attached to any cytoplasmic structure; membrane-bound ribosomes are attached to endoplasmic reticulum.
15. Ribosomes are sites of protein synthesis.
16. Endoplasmic reticulum (ER) is a network of membranes that form flattened sacs or tubules called cisternae; it extends from the nuclear membrane throughout the cytoplasm.
17. Rough ER is covered by ribosomes, and its primary function is protein synthesis. It also forms glycoproteins and attaches proteins to phospholipids, which it also synthesizes.
18. Smooth ER lacks ribosomes. It synthesizes phospholipids, fats, and steroids; releases glucose from the liver into the bloodstream; inactivates or detoxifies drugs and other potentially harmful substances; and releases calcium ions that trigger contraction in muscle cells.
19. The Golgi complex consists of flattened sacs called cisternae that differ in size, shape, content, enzymatic activity, and vesicles present. Entry face cisternae face the rough ER, exit face cisternae face the plasma membrane, and medial cisternae are between the two.
20. The Golgi complex receives synthesized products from the rough ER and modifies, sorts, packages, and transports them within vesicles to different destinations.
21. Some processed proteins leave the cell in secretory vesicles, some are incorporated into the plasma membrane, and some enter lysosomes.
22. Lysosomes are membrane-enclosed vesicles that contain digestive enzymes.
23. Late endosomes, phagosomes, and pinocytic vesicles deliver materials to lysosomes for degradation.
24. Lysosomes function in digestion of worn-out organelles (autophagy), digestion of a host cell (autolysis), and extracellular digestion.
25. Peroxisomes are similar to lysosomes, but smaller.
26. Peroxisomes oxidize various organic substances such as amino acids, fatty acids, and toxic substances and, in the process, produce hydrogen peroxide.
27. The hydrogen peroxide produced from oxidation is degraded by an enzyme in peroxisomes called catalase.
28. Mitochondria consist of a smooth outer membrane, an inner membrane containing cristae, and a central cavity filled with a fluid called the matrix. Mitochondria are called the "powerhouses" of the cell because they produce ATP.

Nucleus (p. 40)

1. The nucleus consists of a double nuclear envelope; nuclear pores, which control the movement of substances between the nucleus and cytoplasm; nucleoli, which produce ribosomes; and genes arranged on chromosomes.
2. Genes control cellular structure and most cellular functions.

Somatic Cell Division (p. 42)

1. Cell division is the process by which cells reproduce themselves. It consists of nuclear division (mitosis or meiosis) and cytoplasmic division (cytokinesis).
2. Cell division that results in an increase in the number of body cells is called somatic cell division and involves a nuclear division called mitosis plus cytokinesis.
3. Cell division that results in the production of sperm and ova is called reproductive cell division and consists of a nuclear division called meiosis plus cytokinesis.

The Cell Cycle in Somatic Cells (p. 42)

1. The cell cycle is an orderly sequence of events in which a cell duplicates its contents and divides in two. It consists of interphase and a mitotic phase.
2. Before the mitotic phase, the DNA molecules, or chromosomes, replicate themselves so that identical chromosomes can be passed on to the next generation of cells.
3. A cell that is between divisions and is carrying on every life process except division is said to be in interphase, which consists of three phases: G_1, S, and G_2.
4. During the G_1 phase, the cell duplicates its organelles and cytosolic components; during the S phase, DNA replication occurs; during the G_2 phase, enzymes and other proteins are synthesized and centrosome replication is complete.
5. Mitosis is the replication of the cell chromosomes and the distribution of the two identical sets of chromosomes into separate and equal nuclei; mitosis consists of prophase, metaphase, anaphase, and telophase.

6. Cytokinesis usually begins in late anaphase and ends in telophase.
7. A cleavage furrow forms at the cell's metaphase plate and progresses inward, pinching in through the cell to form two separate portions of cytoplasm.

Control of Cell Destiny (p. 47)

1. A cell can either remain alive and functioning without dividing, grow and divide, or die.
2. Apoptosis is programmed cell death, a normal type of cell death. It first occurs during embryological development and continues for the lifetime of an organism.
3. Certain genes regulate both cell division and apoptosis. Abnormalities in these genes are associated with a wide variety of diseases and disorders.

Aging and Cells (p. 47)

1. Aging is a normal process accompanied by progressive alteration of the body's homeostatic adaptive responses.
2. Many theories of aging have been proposed, including genetically programmed cessation of cell division, the buildup of free radicals, and an intensified autoimmune response.
3. A cell's shape is related to its function.

Cellular Diversity (p. 49)

1. The almost 200 cell types in the body vary considerably in size and shape.
2. The sizes of cells are measured in micrometers. One micrometer equals 10^{-6}m (1/25,000 of an inch).
3. A cell's shape is related to its function.

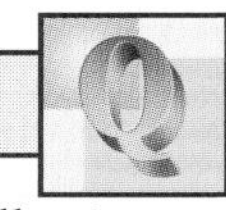

SELF-QUIZ QUESTIONS

Choose the one best answer to the following questions:

1. Which of the following is *not* part of the cytoplasm?
   a. ribosomes **b.** peroxisomes **c.** mitochondria
   d. nucleus **e.** endoplasmic reticulum
2. Respiratory gases move between the lungs and blood as a result of
   a. simple diffusion **b.** osmosis **c.** active transport
   d. phagocytosis **e.** pinocytosis
3. Which process does not belong with the others?
   a. phagocytosis **b.** osmosis **c.** simple diffusion
   d. facilitated diffusion **e.** filtration
4. The glycocalyx is a surface coat on cells that
   a. aids the movement of red blood cells through small blood vessels
   b. consists of carbohydrate portions of membrane glycolipids and glycoproteins
   c. serves as a recognition signal for other body cells
   d. facilitates the adherence of cells to each other in some tissues
   e. all of the above
5. The fluid located inside body cells is
   a. lymph **b.** intracellular fluid **c.** interstitial fluid
   d. intercellular fluid **e.** both c and d
6. As a result of somatic cell division, each daughter cell has
   a. half as many chromosomes as the parent cell
   b. twice as many chromosomes as the parent cell
   c. exactly the same number of chromosomes as the parent cell
   d. one-quarter as many chromosomes as the parent cell
   e. none of the above
7. The process by which worn-out cell organelles are digested is called
   a. autophagy **b.** hemolysis **c.** endocytosis
   d. autolysis **e.** exocytosis
8. Cytokinesis is the division of the cytoplasm of a cell. An equal division of the nucleus is a process called
   a. apoptosis **b.** mitosis **c.** meiosis **d.** metastasis
   e. pinocytosis

Complete the following.

9. The fluid portion of the cytoplasm is the ________ .
10. White blood cells exhibit the transport process known as ________ when they ingest bacteria.
11. Clathrin-coated pits participate in a membrane transport process called ________ .
12. The plasma membrane is composed of two main chemical components: a bilayer of ________ and protein.
13. DNA replication occurs during the ________ phase of the cell cycle.
14. A membrane that permits the passage of only certain materials is described as ________ .
15. The only example of a flagellum-bearing cell in the human is the ________ cell.

Are the following statements true or false?

16. Cilia exhibit an arrangement of microtubules known as a 9 + 2 array.
17. The cleavage furrow is a constriction of the cell membrane that occurs during cytokinesis due to the action of microfilaments.
18. The lipid bilayer of the cell membrane consists of cholesterol, phospholipids and glycolipids.
19. Match the following:

____	**(a)** directs cellular activities by means of genes located here	**(1)** centrosome
____	**(b)** sites of protein synthesis; may occur attached to ER or scattered freely in cytoplasm	**(2)** cilia
____	**(c)** system of membranous channels providing pathways for transport within cell and surface areas for chemical reactions; may be rough or smooth	**(3)** endoplasmic reticulum
____	**(d)** stacks of cisternae; involved in packaging and secretion of glycolipids, glycoproteins, and lipiproteins	**(4)** flagella
____	**(e)** may release enzymes that lead to autolysis of the cell	**(5)** lysosomes
____	**(f)** similar to lysosomes, but smaller; contain the enzyme catalase	**(6)** Golgi complex
		(7) microfilaments
		(8) mitochondria
		(9) nucleus
		(10) microtubules
		(11) ribosomes
		(12) peroxisomes

____ (g) cristae-containing structures, called "powerhouses of the cell" because ATP production occurs here

____ (h) part of cytoskeleton; peripherally located in cytoplasm; involved with cell migration and contraction

____ (i) part of cytoskeleton; give shape to cell; found in flagella, cilia, centrioles, and mitotic spindle

____ (j) helps organize mitotic spindle used in cell division

____ (k) long, hair-like struc-tures that help move an entire cell

____ (l) short, hair-like structures that move particles over cell surface

20. Match the following.

____ (a) chromosomes move toward opposite poles of the cell

____ (b) nuclear envelope disappears; chromatin thickens into distinct chromosomes; centrosomes move to opposite ends of cell and mitotic spindle forms

____ (c) the series of events is essentially the reverse of prophase; cytokinesis occurs

____ (d) chromatids line up on the metaphase plate

____ (e) the cell is not involved in cell division; it is in a phase that follows telophase

(1) anaphase
(2) interphase
(3) metaphase
(4) prophase
(5) telophase

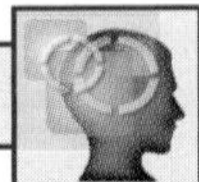

CRITICAL THINKING QUESTIONS

1. Tadpoles have gills and a fish-like tail but adult frogs have no tail at all. What explanation can you give for the disappearance of the tail?
   HINT *Human fetuses have frog-like webbed hands in utero but not at birth.*

2. You may inherit your brown eyes from either your mother or your father but some traits can only be passed down from mother to child. "maternal inheritance" is due to genetic material that is not located in the nucleus. Can you suggest an explanation?
   HINT *The sperm isn't even necessary in this type of inheritance.*

3. If the cell cycle in an actively growing region of the body lasted 24 hours, how many great-great-etc.-granddaughter cells would be produced from one parent cell at the end of one week.
   HINT *You'll have twice as many daughters on Monday as there were parent cell(s) on Sunday.*

4. A child was brought to the emergency room after eating rat poison containing arsenic. Arsenic kills rats by blocking the function of the mitochondria. What effect would the poison have on the child's body functions?
   HINT *The child may appear lethargic.*

5. Imagine that a new chemotherapy agent has been discovered that disrupts microtubules in cancerous cells but leaves normal cells unaffected. What effect would this agent have on the cancer cells?
   HINT *Think of microtubules as your cell's Lego building set.*

ANSWERS TO FIGURE QUESTIONS

2.1 Plasma membrane, cytoplasm, and nucleus.

2.2 The glycocalyx is a coat on the extracellular surface of the plasma membrane that is composed of the carbohydrate portions of membrane glycolipids and glycoproteins.

2.3 Cytosol

2.4 Cholesterol, iron, vitamins, and hormones are examples of ligands.

2.5 The binding of particles to a plasma membrane receptor triggers pseudopod formation.

2.6 Microtubules.

2.7 A cell without a centrosome probably would not be able to undergo cell division.

2.8 Cilia move fluids across cell surfaces, whereas flagella move an entire cell.

2.9 Large and small ribosomal subunits are synthesized in the nucleolus in the nucleus and then join together in the cytoplasm.

2.10 Rough ER has attached ribosomes, whereas smooth ER does not. Rough ER synthesizes proteins that will be exported from the cell; smooth ER is associated with lipid synthesis and other metabolic reactions.

2.11 The entry face receives and modifies proteins from rough ER while the exit face modifies, sorts, packages, and transports molecules.

2.12 Some proteins are discharged from the cell by exocytosis, some are incorporated into the plasma membrane, and some occupy storage vesicles that become lysosomes.

2.13 Digestion of worn-out organelles by lysosomes is called autophagy.

2.14 Mitochondrial cristae increase the surface area available for chemical reactions and contain the enzymes needed for ATP production.

2.15 They control cellular structure and direct most cellular activities.

2.16 A nucleosome is a double-stranded DNA wrapped twice around a core of eight histones (proteins).

2.17 Chromosomes replicate during the S phase.

2.18 DNA replication occurs before cytokinesis so that each of the new daughter cells will have a complete set of genes.

2.19 Cytokinesis usually starts in late anaphase or early telophase.

2.20 Sperm, which use the flagella for locomotion, are the only body cells required to move considerable distances.

3

TISSUES

CONTENTS AT A GLANCE

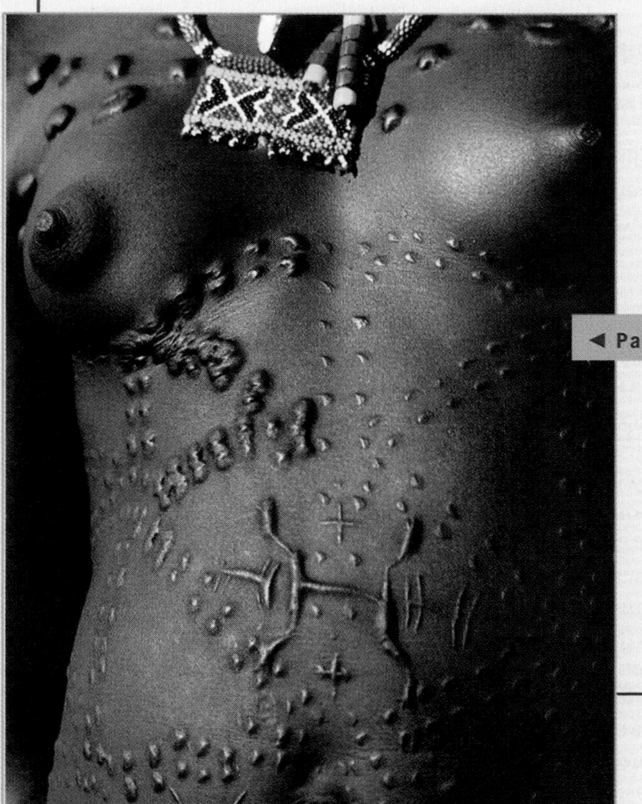

◄ Page 80

Do you have a scar as the result of a healed wound? Do you try to conceal it? Would it surprise you to know that some cultures actually induce scarring as a form of body art and expression?

Page 73 ►

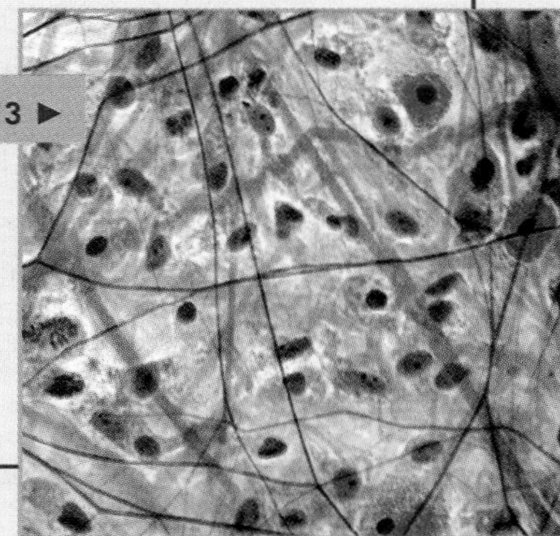

INTRODUCTION

A cell is a complex collection of compartments, each of which carries out a host of biochemical reactions that make life possible. However, a cell seldom functions as an isolated unit in the body. Instead, cells usually work together in groups called tissues. A **tissue** is a group of similar cells that usually has a common embryonic origin and functions together to carry out specialized activities. The structure and properties of a specific tissue are influenced by factors such as the nature of the extracellular material that surrounds the tissue cells and the connections between the cells that compose the tissue. Tissues may be hard, semisolid, or even liquid in their consistency, a range exemplified by bone, fat, and blood. In addition, tissues vary tremendously with respect to the kinds of cells present, how the cells are arranged, and the types of fibers present, if any. **Histology** (hiss′-TOL-ō-jē; *hist-* = tissue; *-ology* = study of) is the science that deals with the study of tissues. A **pathologist** (pa-THOL-ō-gist; *patho-* = disease) is a physician who specializes in laboratory studies of cells and tissues to help other physicians make accurate diagnoses. One of the principal functions of a pathologist is to examine tissues for any changes that might indicate disease.

TYPES OF TISSUES AND THEIR ORIGINS

Objective

- Name the four basic types of tissue that make up the human body, and state the characteristics of each.

Body tissues can be classified into four basic types according to their function and structure:

1. **Epithelial tissue** covers body surfaces and lines hollow organs, body cavities, and ducts. It also forms glands.
2. **Connective tissue** protects and supports the body and its organs. Various types of connective tissue bind organs together, store energy reserves as fat, and help provide immunity to disease-causing organisms.
3. **Muscle tissue** generates the physical force needed to make body structures move.
4. **Nervous tissue** detects changes in a variety of conditions inside and outside the body and responds by generating nerve impulses. The nervous tissue in the brain helps to maintain homeostasis.

Epithelial tissue and connective tissue, except for bone tissue and blood, are discussed in detail in this chapter. The general features of bone tissue and blood will be introduced here, but their detailed discussion is presented in Chapters 5 and 12, respectively. Similarly, the structure and function of muscle tissue and nervous tissue are examined in detail in Chapters 9 and 16, respectively.

All tissues of the body develop from three **primary germ layers,** which are called the **ectoderm, endoderm,** and **mesoderm,** the first tissues formed in a human embryo. Epithelial tissues develop from all three primary germ layers. All connective tissues and most muscle tissues derive from mesoderm. Nervous tissue develops from ectoderm. (Figure 27.5 on page 832 illustrates the primary germ layers and Table 27-1 on page 833 provides a list of structures derived from the primary germ layers.)

Normally, most cells within a tissue remain anchored to other cells, to basement membranes (described shortly), and to connective tissues. A few cells, such as phagocytes, move freely through the body, searching for invaders. Before birth, however, many cells migrate extensively as part of the growth and development process.

A **biopsy** (BĪ-op-sē; *bio-* = life, *-opsy* = to view) is the removal of a sample of living tissue for microscopic examination. This procedure is used to help diagnose many disorders, especially cancer, and to discover the cause of unexplained infections and inflammations. Both normal and potentially diseased tissues are removed for purposes of comparison. Once the tissue sample is removed, either surgically or through a needle and syringe, it may be preserved, stained to highlight special properties, or cut into thin sections for microscopic observation. Sometimes a biopsy is conducted while a patient is anesthetized during surgery to help a physician determine the most appropriate treatment. For example, if a biopsy of breast tissue reveals malignant cells, the surgeon can proceed with the most appropriate procedure immediately. ■

✓ Define a tissue.
✓ What are the four basic types of human tissue?

CELL JUNCTIONS

Objective

- Describe the structure and functions of the three main kinds of cell junctions.

As you will see, epithelial cells, and certain other body cells, are tightly joined to form a close functional unit. An important reason that tissue cells adhere to one another is that they contain specialized points of contact between their adjacent plasma membranes called **cell junctions.** In addition to providing cell-to-cell attachment and cell-to-extracellular material attachment, cell junctions also prevent the movement of molecules between some cells and provide channels for molecular communication between others.

There are three principal types of cell junctions: (1) tight junctions, (2) plaque-bearing junctions, and (3) gap junctions (Figure 3.1).

Figure 3.1 / Cell junctions.

Most epithelial cells and some muscle and nerve cells contain cell junctions.

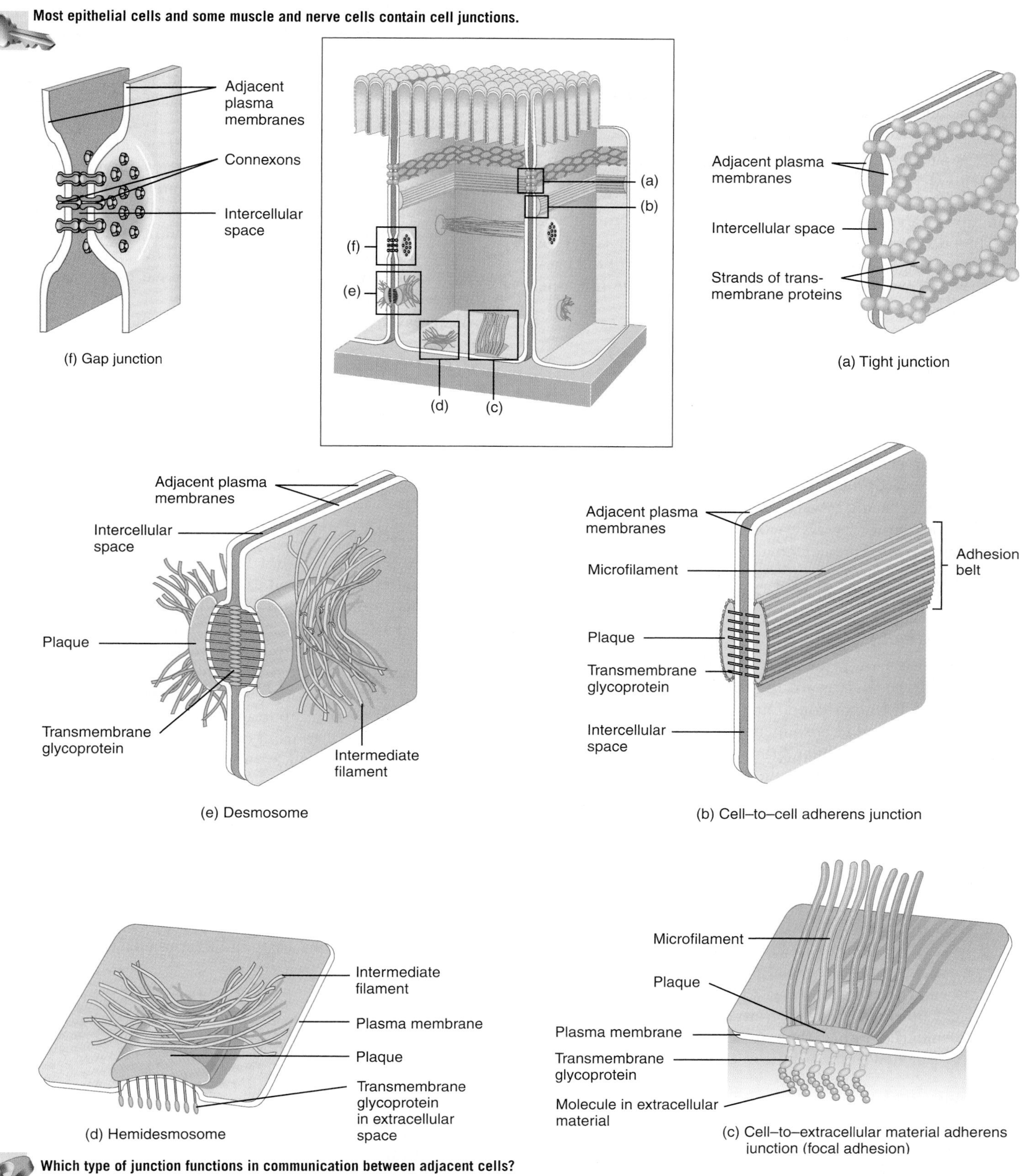

Which type of junction functions in communication between adjacent cells?

Tight Junctions

Cells that line the inner or outer surfaces of organs or body cavities are frequently connected by **tight junctions,** which completely encircle the cells (Figure 3.1a). In a tight junction, the outer surfaces of adjacent plasma membranes are fused together by their transmembrane proteins in a weblike pattern that enables the cells to firmly adhere to each other. This type of junction prevents the passage of molecules between cells. Tight junctions are common between epithelial cells such as those that line the stomach, intestines, and urinary bladder. In these organs, tight junctions prevent fluid and other substances from leaking into the blood or surrounding tissue, a situation that could be harmful. For example, a serious infection would result if the bacteria that are normally confined to the interior of the intestine escaped through the intestinal lining.

Plaque-Bearing Junctions

Plaque refers to a layer of dense proteins on the inside of the plasma membrane. The plaque surface facing away from the cytoplasm connects to a similar plaque surface in the plasma membrane on another cell or to extracellular material by means of transmembrane glycoproteins (integral membrane proteins). The plaque surface facing toward the cytoplasm is connected to the cytoskeleton, either by actin-containing microfilaments or keratin-containing intermediate filaments firmly attaching the cytoskeleton to the plasma membrane. There are two types of plaque-bearing junctions: adherens junctions and desmosomes.

Adherens Junctions

Adherens junctions are plaque-bearing cell junctions in which the plaque surface facing toward the cytoplasm is connected to the cytoskeleton by actin-containing microfilaments (Figure 3.1b). In a cell-to-cell adherens junction, the plasma membranes are separated by an intercellular space measuring about 20 nm*. Transmembrane glycoproteins embedded in the plaque of one cell cross this space and connect with the transmembrane proteins attached to the plaque of an adjacent cell to establish the adhesion between the two cells. In epithelial cells, these adherens junctions form extensive zones called *adhesion belts* that completely encircle the cell.

In addition to their function in attaching cells to each other, some adherens junctions bind cells to extracellular materials (Figure 3.1c). Such adherens junctions are called *focal adhesions* (*adhesion plaques*); they also contain plaque, actin-containing microfilaments, and transmembrane glycoproteins. The transmembrane glycoproteins, however, attach to molecules in the extracellular material rather than to identical transmembrane glycoproteins attached to the plaque of an adjacent cell.

Desmosomes

Desmosomes (DEZ-mō-sōms; *desmos* = bond; *soma* = body) are plaque-bearing cell junctions in which the plaque surface facing toward the cytoplasm is connected to the cytoskeleton by keratin-containing intermediate filaments (Figure 3.1e). This type of junction is especially numerous in cells that make up the epidermis (outermost layer of the skin) and the lining of the small intestine. Desmosomes are scattered over adjacent plasma membranes and form firm attachments between the cells. At a desmosome, the plasma membranes of adjacent cells are separated by an intercellular space measuring about 22–24 nm. As with adherens junctions, the plaques of adjoining cells are connected to each other by transmembrane glycoproteins. However, the intermediate filaments of a desmosome extend across the cytoplasm of the entire cell and thus attach to desmosomes on the opposite side of the cell. This arrangement contributes to the overall stability of the tissue by connecting the cytoskeletons of adjoining cells.

Whereas desmosomes connect adjacent cells, *hemidesmosomes* (*hemi* = half) connect cells to an extracellular material of some kind (Figure 3.1d). One such example is the basement membrane, an extracellular layer that attaches epithelial tissue to connective tissue and will be discussed in detail shortly. Hemidesmosomes look like half a desmosome, although the arrangement of intermediate filaments and the plaque proteins are different from those in desmosomes.

Gap Junctions

In a **gap junction** (Figure 3.1f), the outer layers of adjacent plasma membranes approach each other, leaving an intercellular gap of about 2 nm between them. The gap is bridged by membrane proteins called *connexons*, which form minute, fluid-filled tunnels. Ions and small molecules, such as glucose and amino acids, can pass directly from the cytoplasm of one cell into the cytoplasm of the next through the connexons.

Gap junctions are important because they allow the rapid spread of nerve impulses from one cell to the next in some parts of the nervous system and in the muscle tissue of the heart and gastrointestinal tract. In a developing embryo, chemical and electrical signals that regulate growth and differentiation may travel by way of gap junctions. Cancer cells lack gap junctions and therefore cannot communicate with each other. As a result, cell division of cancerous tissue is not coordinated and occurs in an uncontrolled manner.

✓ Define a cell junction.
✓ Which cell junctions are found in epithelial tissue?

EPITHELIAL TISSUE

Objective

- Describe the general features of epithelial tissue and the structure, location, and function of the different types of epithelium.

Epithelial tissue (ep-i-THĒ-lē-al), or **epithelium** (plural is *epithelia*) consists of cells arranged in continuous sheets, in either

*One nanometer (nm) = 0.000,000,001 m or 1×10^{-9} m.

single or multiple layers. The cells are densely packed and held tightly together by numerous cell junctions, so that there is little extracellular space between adjacent plasma membranes (Figure 3.2).

Epithelial cells have various surfaces that differ in structure and have specialized functions. The surfaces are referred to as **lateral surfaces,** which face neighboring cells; **apical surfaces,** which are exposed to the exterior of the body, a body cavity, or lumen (interior space) of an internal organ; and **basal surfaces,** which adhere to an extracellular material (Figure 3.2).

As you have already seen, lateral surfaces contain various cell junctions such as tight junctions, adherens junctions, desmosomes, and gap junctions, each with specialized functions. The apical surfaces contain modifications such as cilia, stereocilia (cilia of unusual length), and microvilli (described shortly). The attachment between the basal surfaces of epithelial cells and the underlying connective tissue is through a thin extracellular layer called the **basement membrane.** Hemidesmosomes in the basal surfaces of epithelial cells anchor the epithelium to the basement membrane. The basement membrane commonly consists of two layers: the basal lamina and the reticular lamina. The *basal lamina* (*lamina* = thin layer) is closer to the epithelial cells and is secreted by them. It contains primarily proteins, such as collagen fibers (described shortly), glycoproteins, and proteoglycans (also described shortly). The basal lamina functions as a selective filter that restricts the passage of large molecules from the underlying connective tissue to the epithelium. This is especially important in the kidneys. The basal lamina also forms a surface along which regenerating epithelial cells migrate in wound healing and assists in cell-to-cell interactions, such as the formation of nerve-cell junctions. The *reticular lamina* is closer to the connective tissue and contains reticular fibers produced by connective tissue cells called fibroblasts. The reticular lamina gives the basement membrane its strength. Fibers between the basal lamina and reticular lamina anchor the two layers together.

Epithelial tissue is **avascular** (*a-* = without; *-vascular* = vessel); that is, it lacks its own blood supply. The blood vessels that bring in nutrients and remove wastes are located in adjacent connective tissue. The exchange of these substances between connective tissue and epithelium occurs by diffusion. Although epithelial tissue is avascular, it has a nerve supply.

Because epithelial tissue forms boundaries between the body's organs or between the body and the external environment, it is repeatedly subject to physical breakdown and injury. But because it has a very high rate of cell division, epithelial tissue constantly renews and repairs itself by sloughing off dead or injured cells and replacing them with new ones. Epithelial tissue plays many different roles in the body, the most important of which include protection, filtration, secretion, absorption, and excretion. In addition, epithelial tissue combines with nervous tissue to form special organs for smell, hearing, vision, and touch.

Epithelial tissue may be divided into two types. **Covering and lining epithelium** forms the epidermis of the skin and the outer covering of some internal organs. It also forms the inner lining of blood vessels, ducts, and body cavities and the interior of the respiratory, digestive, urinary, and reproductive systems. **Glandular epithelium** constitutes the secreting portion of glands, such as the thyroid, adrenal, and sweat glands.

Figure 3.2 / Surfaces of epithelial cells and the structure and location of the basement membrane.

The basement membrane is found between epithelium and connective tissue.

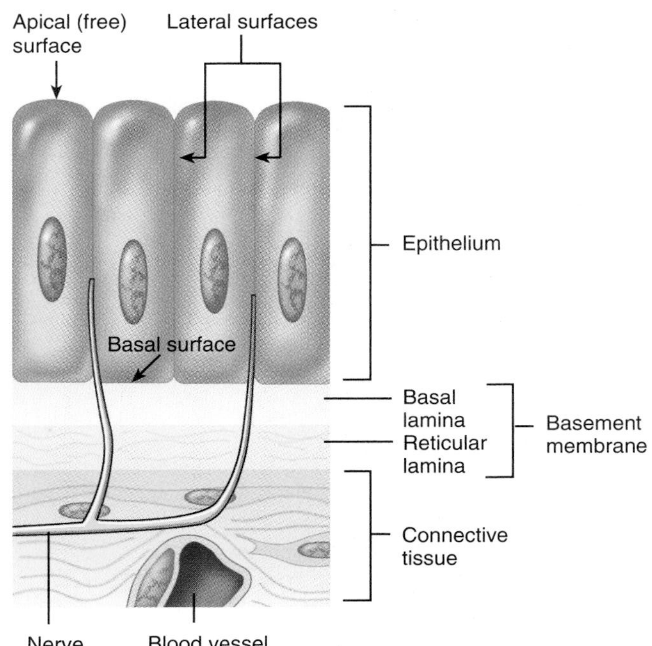

What are the functions of the basement membrane?

Covering and Lining Epithelium

The types of covering and lining epithelial tissue are classified according to two characteristics: the arrangement of cells into layers and the shapes of the cells.

1. Arrangement of layers. Covering and lining epithelium tends to be arranged in one or more layers depending on function:

   a. *Simple epithelium* is a single layer of cells that functions in diffusion, osmosis, filtration, *secretion* (the production and release of substances such as mucus, sweat, or enzymes), and *absorption* (the intake of fluids or other substances by cells).

   b. *Stratified epithelium* consists of two or more layers of cells that protect underlying tissues in locations where there is considerable wear and tear.

   c. *Pseudostratified epithelium* contains only a single layer of cells, but it appears to have multiple layers because cell nuclei lie at different levels and not all cells reach the apical surface. Cells that do extend to the apical surface are either ciliated or secrete mucus.

2. Cell shapes.
   a. *Squamous* cells (SKWĀ-mus = flat) are thin and arranged like floor tiles, which allows the rapid transport of substances through them.
   b. *Cuboidal* cells are as tall as they are wide and are shaped like cubes or hexagons. They function in either secretion or absorption.
   c. *Columnar* cells are much taller than they are wide, cylindrical, and protect underlying tissues. They may have cilia and may be specialized for secretion and absorption.
   d. *Transitional* cells change shape, from columnar to flat and back, as body parts stretch, expand, or move.

Combining the arrangements of layers and cell shapes provides the following classification scheme for covering and lining epithelium:

I. Simple epithelium
   A. Simple squamous epithelium
   B. Simple cuboidal epithelium
   C. Simple columnar epithelium (nonciliated and ciliated)
II. Stratified epithelium
   A. Stratified squamous epithelium (keratinized and nonkeratinized)*
   B. Stratified cuboidal epithelium*
   C. Stratified columnar epithelium*
   D. Transitional epithelium
III. Pseudostratified columnar epithelium (nonciliated and ciliated)

Each of these covering and lining epithelial tissues is described in the following sections and illustrated in Table 3.1. The tables throughout this chapter contain figures consisting of a photomicrograph, a corresponding diagram, and an inset that identifies a principal location of the tissue in the body. Along with the illustrations are descriptions, locations, and functions of the tissues.

Simple Epithelium

Simple squamous epithelium consists of a single layer of flat cells that resembles a tiled floor when viewed from the apical surface (Table 3.1A). The nucleus of each cell is oval or spheri-

*This classification is based on the shape of the cells at the apical surface.

Table 3.1 Epithelial Tissues

Covering and Lining Epithelium

A. Simple Squamous Epithelium

Description Single layer of flat cells; centrally located nucleus.

Location Lines heart, blood vessels, lymphatic vessels, air sacs of lungs, glomerular (Bowman's) capsule of kidneys, and inner surface of the tympanic membrane (eardrum); forms epithelial layer of serous membranes, such as the peritoneum.

Function Filtration, diffusion, osmosis, and secretion in serous membranes.

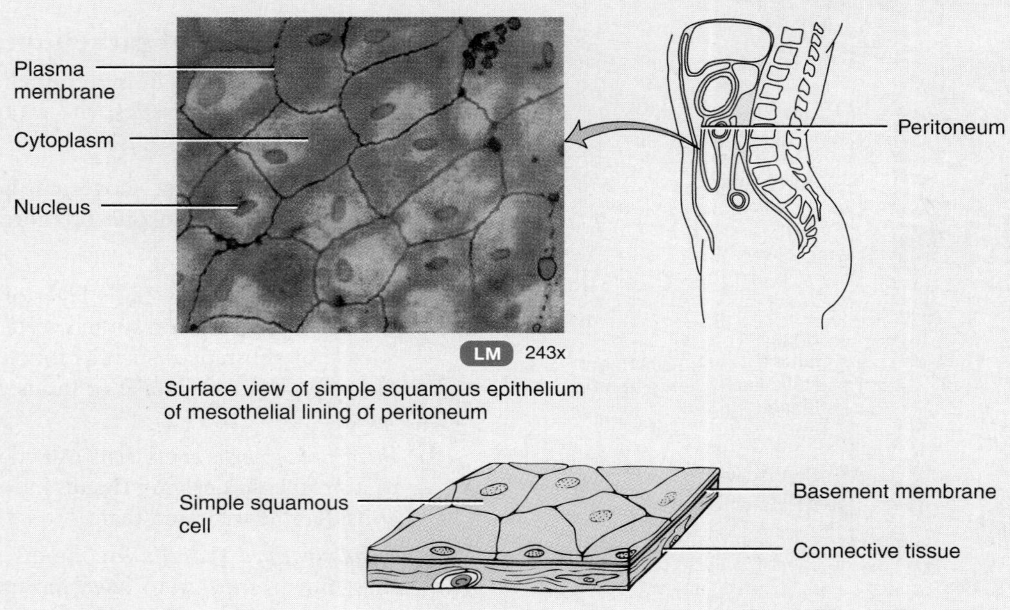

Surface view of simple squamous epithelium of mesothelial lining of peritoneum

Simple squamous epithelium

cal and centrally located. Simple squamous epithelium is found in parts of the body where filtration (kidneys) or diffusion (lungs) are priority processes. It is not found in body areas that are subject to wear and tear.

The simple squamous epithelium that lines the heart, blood vessels, and lymphatic vessels is known as **endothelium** (*endo-* = within; *-thelium* = covering); the type that forms the epithelial layer of serous membranes is called **mesothelium** (*meso-* = middle). Endothelium and mesothelium are both derived from the embryonic mesoderm.

Simple cuboidal epithelium has cells that are cuboidal in shape (Table 3.1B). This is obvious only when the tissue is sectioned and viewed from the side, as when a slice has been made through the epithelial tissue layer. Cell nuclei are usually round and centrally located. Simple cuboidal epithelium functions in secretion and absorption.

Simple columnar epithelium, when viewed from the side, has cells that appear rectangular with oval nuclei near the base of the cells. Simple columnar epithelium exists in two forms—nonciliated and ciliated.

Nonciliated simple columnar epithelium has microvilli and goblet cells (Table 3.1C). **Microvilli** are microscopic fingerlike cytoplasmic projections that increase the surface area of the plasma membrane (see Figure 2.1 on page 26). Their presence increases the rate of absorption by the cell. **Goblet cells** are modified columnar cells that secrete mucus, a slightly sticky fluid. Before it is released, mucus accumulates in the upper portion of the cell, causing that area to bulge out. The whole cell then resembles a goblet or wine glass. Secreted mucus serves as a lubricant for the linings of the digestive, respiratory, reproductive, and most of the urinary tracts. Mucus also helps to trap dust entering the respiratory tract, and it prevents destruction of the stomach lining by digestive enzymes.

Ciliated simple columnar epithelium (Table 3.1D) contains cells with cilia. In a few parts of the upper respiratory tract, ciliated columnar cells are interspersed with goblet cells. Mucus secreted by the goblet cells forms a film over the respiratory surface and traps inhaled foreign particles. The cilia wave in unison and move the mucus and any foreign particles toward the throat, where it can be coughed up and swallowed or spit out. Cilia also help to move an ovum through the uterine (Fallopian) tubes into the uterus.

Stratified Epithelium

In contrast to simple epithelium, stratified epithelium has at least two layers of cells. Thus, it is more durable and can better

Covering and Lining Epithelium

B. Simple Cuboidal Epithelium

Description Single layer of cube-shaped cells; centrally located nucleus.

Location Covers surface of ovary, lines anterior surface of capsule of the lens of the eye, forms the pigmented epithelium at the back of the eye, lines kidney tubules and smaller ducts of many glands, and makes up the secreting portion of some glands such as the thyroid gland and the ducts within some glands such as the pancreas.

Function Secretion and absorption.

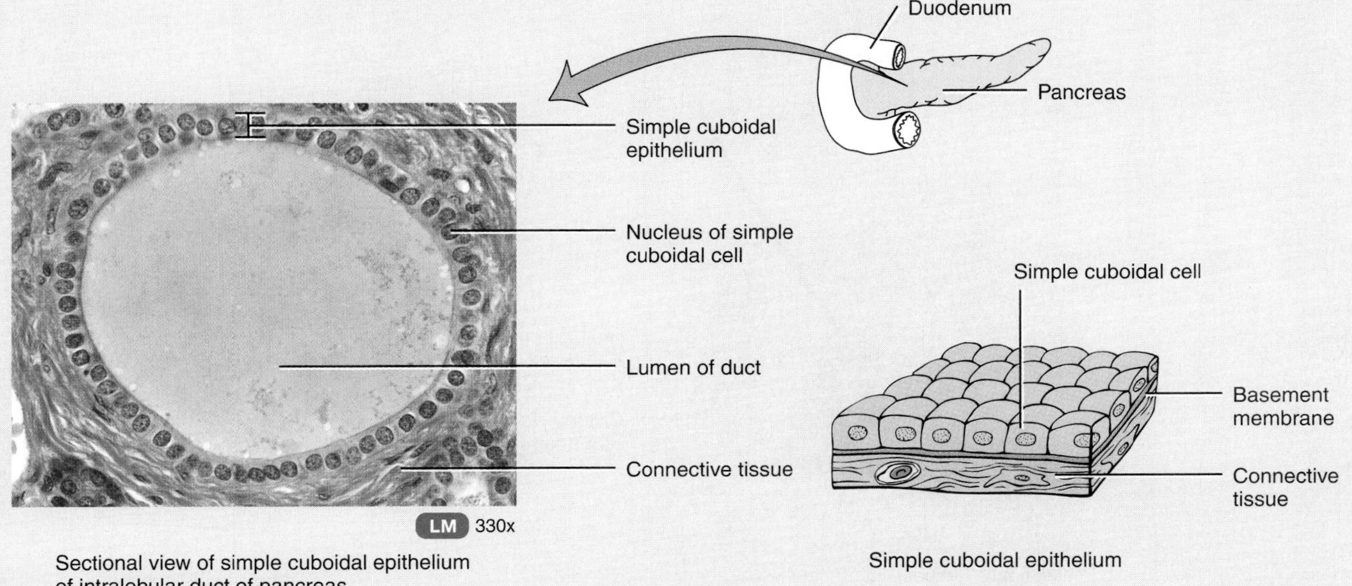

Sectional view of simple cuboidal epithelium of intralobular duct of pancreas

Simple cuboidal epithelium

(continues)

Table 3.1 Epithelial Tissues (continued)

Covering and Lining Epithelium

C. Nonciliated Simple Columnar Epithelium

Description Single layer of nonciliated rectangular cells; nucleus near base of cell; contains goblet cells and cells with microvilli in some locations.

Location Lines the gastrointestinal tract from the stomach to the anus, ducts of many glands, and gallbladder.

Function Secretion and absorption.

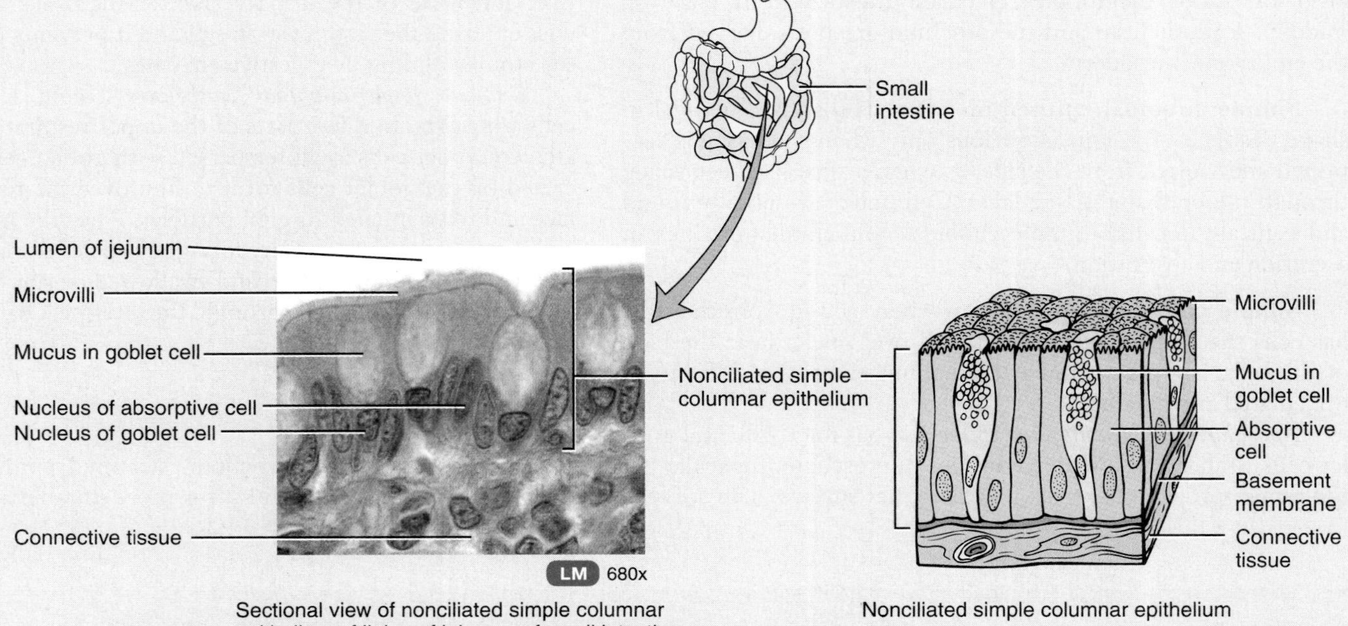

Sectional view of nonciliated simple columnar epithelium of lining of jejunum of small intestine

Nonciliated simple columnar epithelium

D. Ciliated Simple Columnar Epithelium

Description Single layer of ciliated rectangular cells; nucleus near base of cell; contains goblet cells in some locations.

Location Lines a few portions of upper respiratory tract, uterine (Fallopian) tubes, uterus, some paranasal sinuses, and central canal of spinal cord.

Function Moves mucus and other substances by ciliary action.

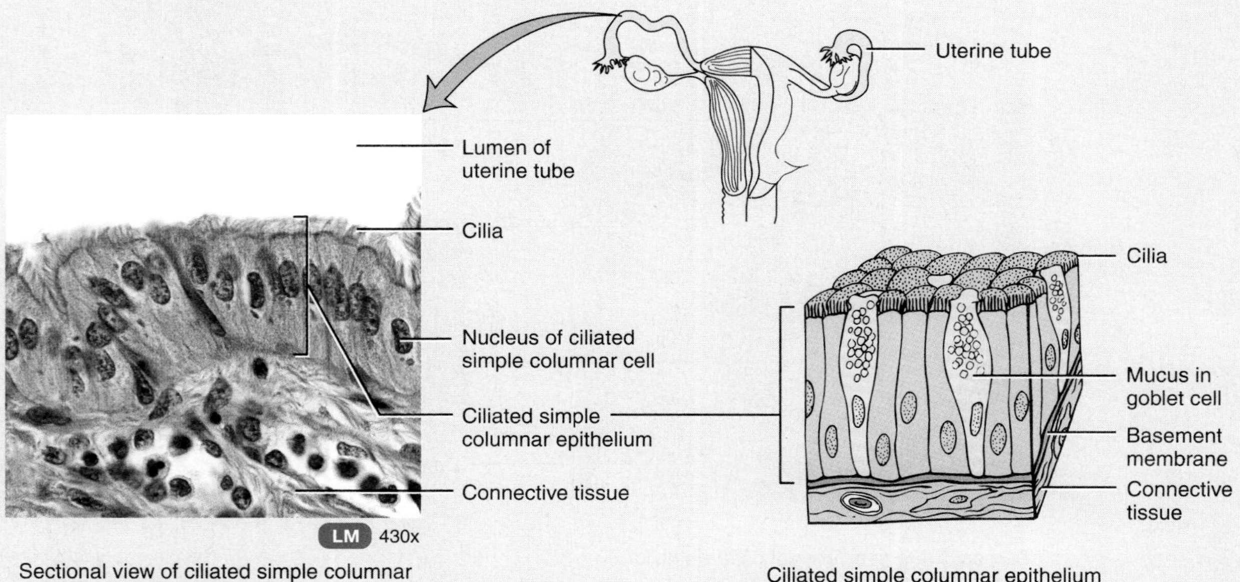

Sectional view of ciliated simple columnar epithelium of uterine tube

Ciliated simple columnar epithelium

Covering and Lining Epithelium

E. Stratified Squamous Epithelium

Description Several layers of cells; cuboidal to columnar shape in deep layers; squamous cells are at the apical surface layers; basal cells replace surface cells as they are lost.

Location Keratinized variety forms superficial layer of skin; nonkeratinized variety lines wet surfaces, such as lining of the mouth, esophagus, part of epiglottis, and vagina, and covers the tongue.

Function Protection.

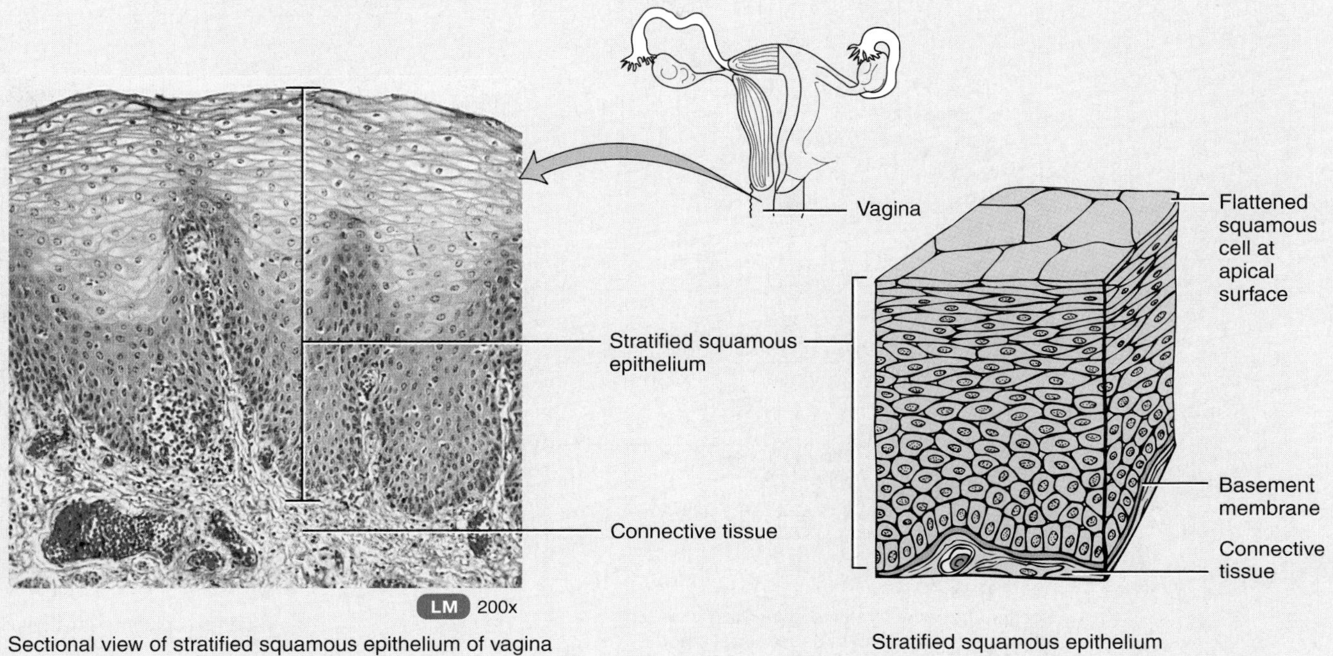

Sectional view of stratified squamous epithelium of vagina

Stratified squamous epithelium

F. Stratified Cuboidal Epithelium

Description Two or more layers of cells in which the cells in the apical layer are cube-shaped.

Location Ducts of adult sweat glands and esophageal glands and part of male urethra.

Function Protection and limited secretion and absorption.

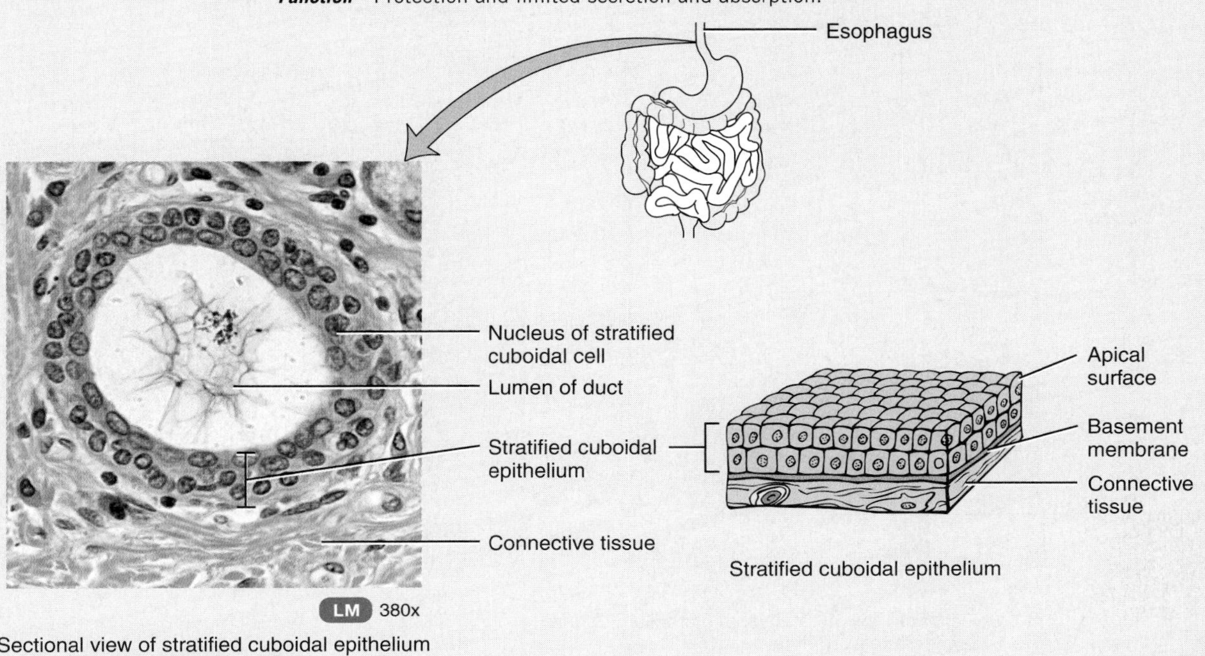

Sectional view of stratified cuboidal epithelium of the duct of esophageal gland

Stratified cuboidal epithelium

(continues)

Table 3.1 Epithelial Tissues (continued)

Covering and Lining Epithelium

G. Stratified Columnar Epithelium

Description Several layers of polyhedral cells; columnar cells are only in the apical layer.

Location Lines part of urethra, large excretory ducts of some glands such as esophageal glands, small areas in anal mucous membrane, and a part of the conjunctiva of the eye.

Function Protection and secretion.

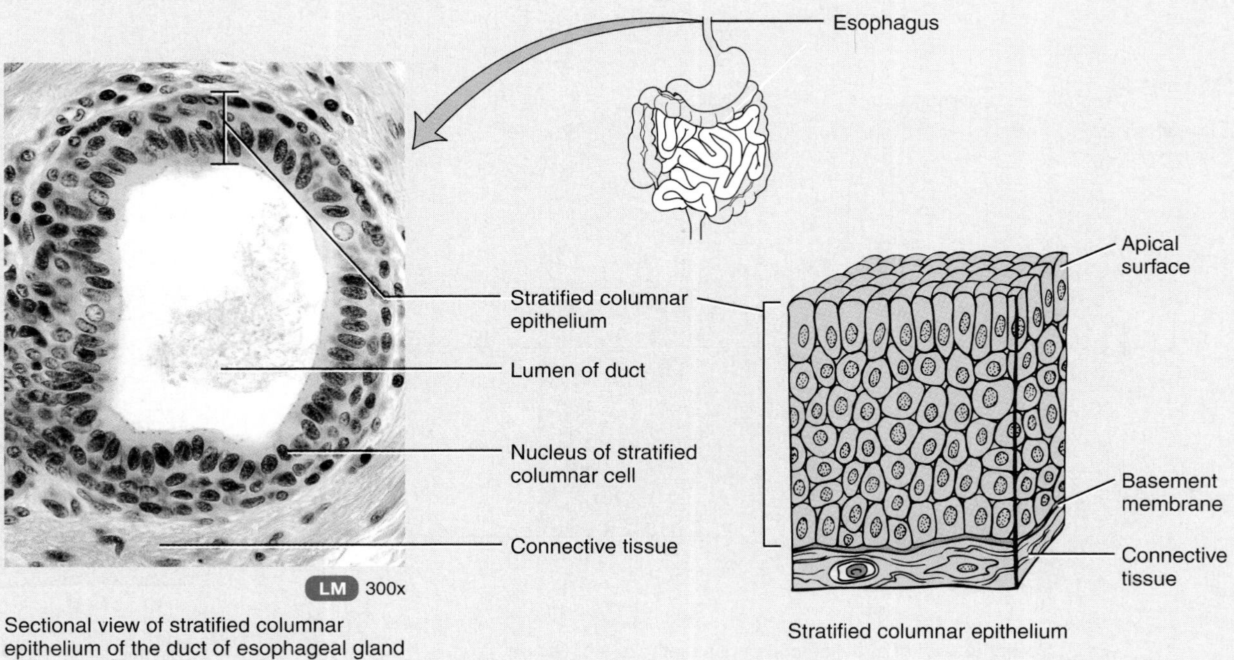

Sectional view of stratified columnar epithelium of the duct of esophageal gland

Stratified columnar epithelium

H. Transitional Epithelium

Description Appearance is variable (transitional); shape of cells at apical surface ranges from squamous (when stretched) to cuboidal (when relaxed).

Location Lines urinary bladder and portions of ureters and urethra.

Function Permits distention.

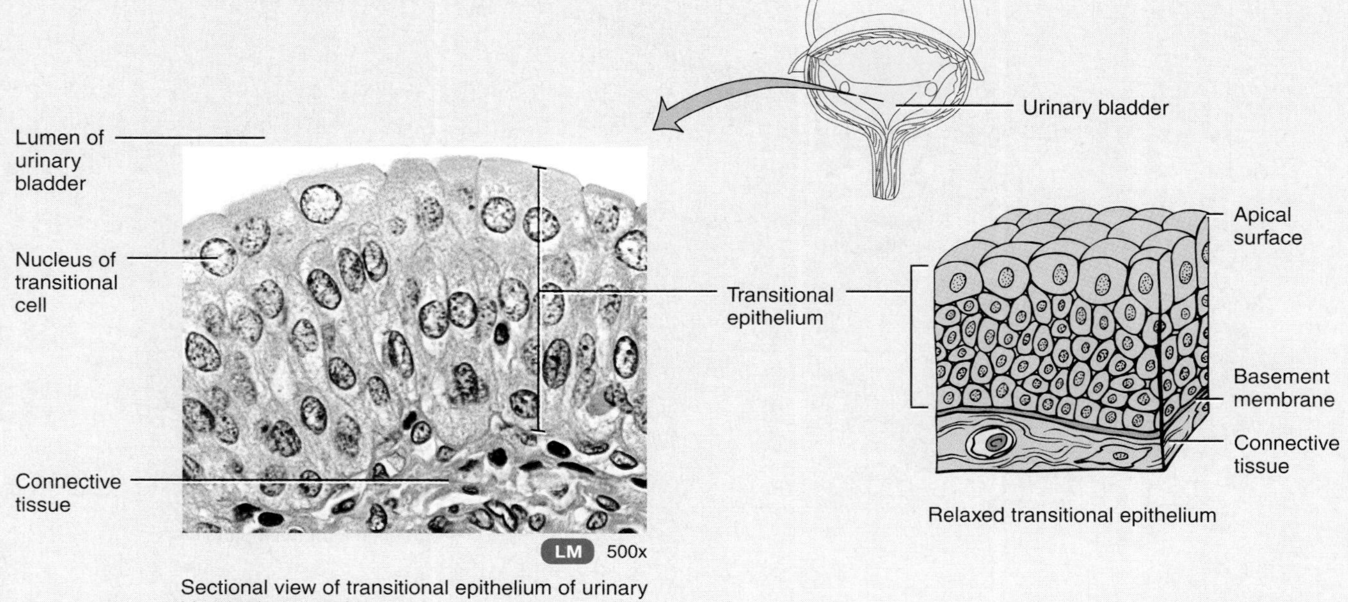

Sectional view of transitional epithelium of urinary bladder in relaxed state

Relaxed transitional epithelium

Covering and Lining Epithelium

I. Pseudostratified Columnar Epithelium

Description Not a true stratified tissue; nuclei of cells are at different levels; all cells are attached to basement membrane, but not all reach the apical surface.

Location Pseudostratified ciliated columnar epithelium lines the airways of most of upper respiratory tract; pseudostratified nonciliated columnar epithelium lines larger ducts of many glands, epididymis, and part of male urethra.

Function Secretion and movement of mucus by ciliary action.

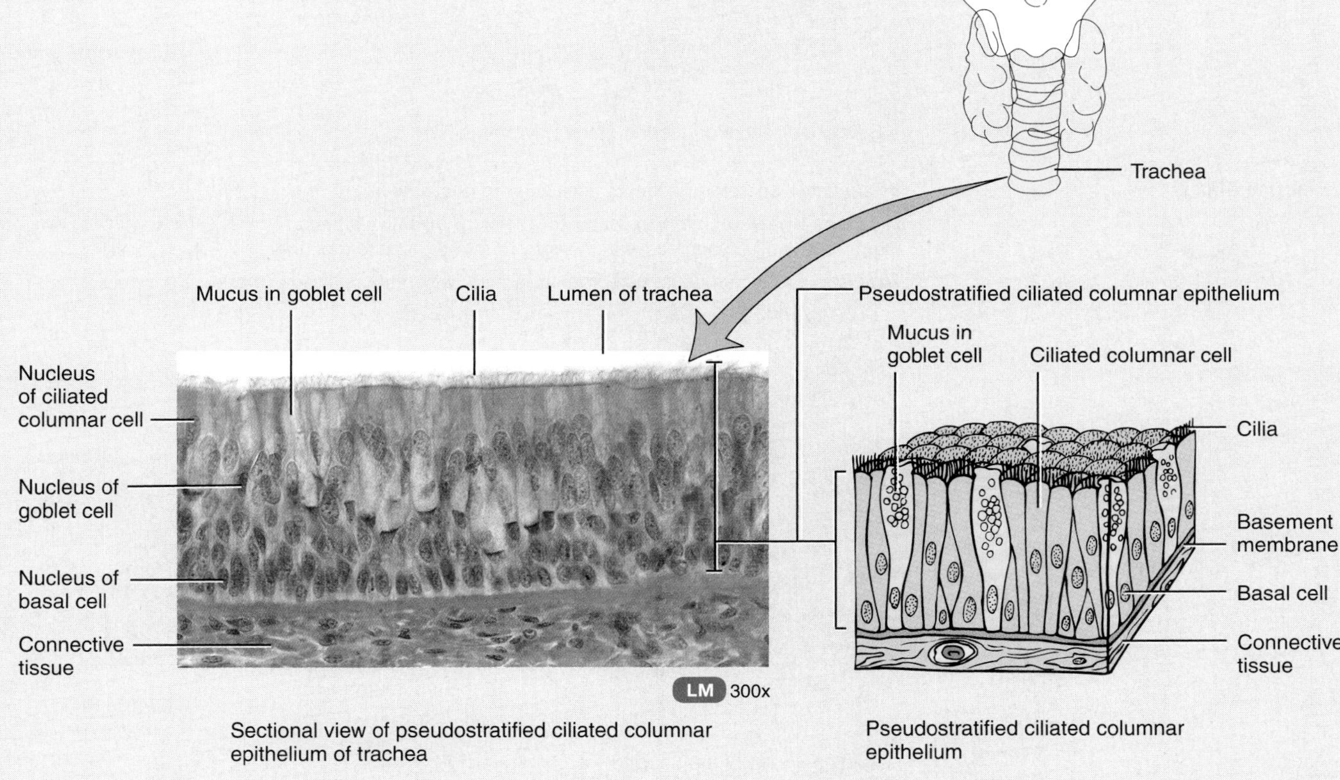

Sectional view of pseudostratified ciliated columnar epithelium of trachea

Pseudostratified ciliated columnar epithelium

(continues)

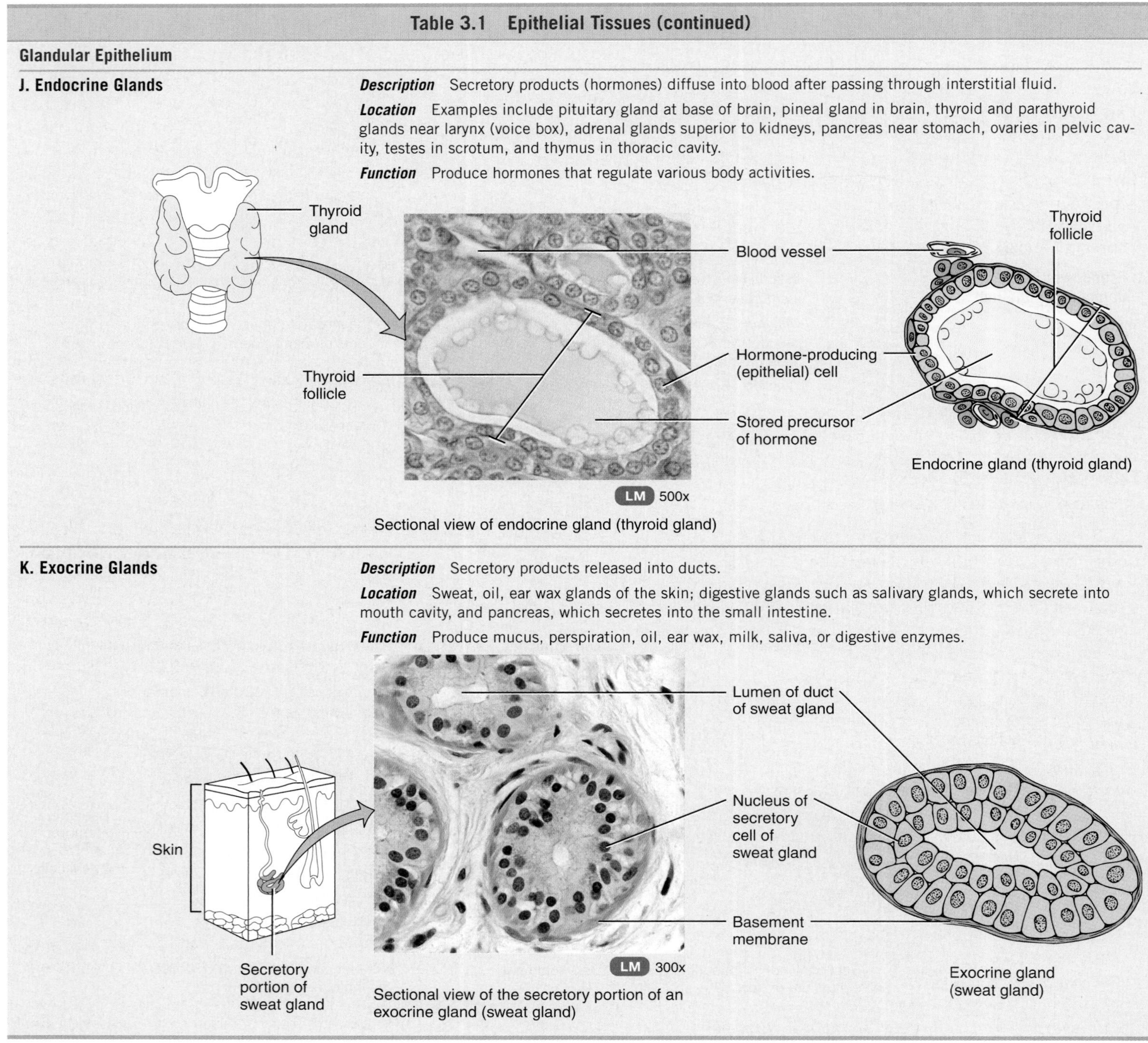

Table 3.1 Epithelial Tissues (continued)

Glandular Epithelium

J. Endocrine Glands

Description Secretory products (hormones) diffuse into blood after passing through interstitial fluid.

Location Examples include pituitary gland at base of brain, pineal gland in brain, thyroid and parathyroid glands near larynx (voice box), adrenal glands superior to kidneys, pancreas near stomach, ovaries in pelvic cavity, testes in scrotum, and thymus in thoracic cavity.

Function Produce hormones that regulate various body activities.

Sectional view of endocrine gland (thyroid gland)

K. Exocrine Glands

Description Secretory products released into ducts.

Location Sweat, oil, ear wax glands of the skin; digestive glands such as salivary glands, which secrete into mouth cavity, and pancreas, which secretes into the small intestine.

Function Produce mucus, perspiration, oil, ear wax, milk, saliva, or digestive enzymes.

Sectional view of the secretory portion of an exocrine gland (sweat gland)

protect underlying tissues. Some cells in stratified epithelia also produce secretions. The name of the specific kind of stratified epithelium depends on the shape of the cells at the apical surface.

Stratified squamous epithelium has flat cells in the superficial layers whereas in the deep layers, they vary in shape from cuboidal to columnar (Table 3.1E). The basal (deepest) cells continually undergo cell division. As new cells grow, the cells of the basal layer are pushed upward toward the apical surface. As they move farther from the deep layer and from their blood supply in the underlying connective tissue, they become dehydrated, shrunken, and harder. At the apical surface, the cells lose their cell junctions and are sloughed off, but they are replaced as new cells continually emerge from the basal cells.

Stratified squamous epithelium exists in both keratinized and nonkeratinized forms. In *keratinized stratified squamous epithelium*, a tough layer of keratin is deposited in the surface cells. **Keratin** is a protein that is resistant to friction and helps repel bacteria. *Nonkeratinized stratified squamous epithelium* does not contain keratin and remains moist.

A **Papanicolaou smear** (pa-pa-NI-kō-lō) or **Pap smear** involves collecting and microscopically examining epithelial cells that have sloughed off the surface of a tissue. A very common type of Pap smear involves examining the cells present in the secretions of the cervix and vagina. This type of Pap smear is performed mainly to detect early changes in the cells of the female reproductive system that may indicate cancer or a precancerous condition. Annual Pap smears are recommended for women over age 18 (earlier if sexually active) as part of a routine pelvic exam. ■

Stratified cuboidal epithelium is a fairly rare type of epithelium that sometimes consists of more than two layers of cells (Table 3.1F). Its function is mainly protective, but it also functions in limited secretion and absorption.

Stratified columnar epithelium, like stratified cuboidal epithelium, is also uncommon. Usually the basal layer or layers consist of shortened, irregularly polyhedral cells; only the apical cells are columnar (Table 3.1G). This type of epithelium functions in protection and secretion.

Transitional epithelium is limited to the urinary system and is variable in appearance, depending on whether the organ it lines is relaxed or distended (stretched). In its relaxed state (Table 3.1H), transitional epithelium looks similar to stratified cuboidal epithelium, except that the apical cells tend to be large and rounded. This allows the tissue to be distended without the outer cells separating from one another. As the cells are stretched, they become squamous in shape, giving the appearance of stratified squamous epithelium. Because of its elasticity, transitional epithelium lines hollow structures that are subjected to expansion from within, such as the urinary bladder. Its function is to help prevent rupture of these organs.

Pseudostratified Columnar Epithelium

The third category of covering and lining epithelium is called pseudostratified columnar epithelium (Table 3.1I). The nuclei of the cells are at various depths. Even though all the cells are attached to the basement membrane in a single layer, some cells do not extend to the apical surface. When viewed from the side, these features give the false impression of a multilayered tissue—thus the name *pseudo*stratified epithelium (*pseudo-* = false). In *pseudostratified ciliated columnar epithelium*, the cells that reach the apical surface either secrete mucus (goblet cells) or bear cilia that sweep away mucus and trapped foreign particles for eventual elimination from the body. *Pseudostratified nonciliated columnar epithelium* contains no cilia or goblet cells.

Glandular Epithelium

The function of glandular epithelium is secretion, which is accomplished by cells that often lie in clusters deep to the covering and lining epithelium. A **gland** may consist of one cell or a group of highly specialized epithelial cells that secrete substances into ducts, onto a surface, or into the blood. The production of such substances always requires active work by the cells and results in an expenditure of energy. Glands are classified as either endocrine or exocrine.

The secretions of **endocrine glands** (Table 3.1J) enter the extracellular fluid and then diffuse directly into the bloodstream without flowing through a duct. These secretions, called *hormones*, regulate many metabolic and physiological activities to maintain homeostasis. The pituitary, thyroid, and adrenal glands are examples of endocrine glands. Endocrine glands will be described in detail in Chapter 22.

Exocrine glands (*exo-* = outside; *-crine* = secretion; Table 3.1K) secrete their products into ducts (tubes) that empty at the surface of covering and lining epithelium or directly onto a free surface. The product of an exocrine gland may be released at the skin surface or into the lumen (interior space) of a hollow organ. The secretions of exocrine glands include mucus, sweat, oil, wax, and digestive enzymes. Examples of exocrine glands are sweat glands, which produce perspiration to help lower body temperature, and salivary glands, which secrete mucus and digestive enzymes.

Structural Classification of Exocrine Glands

Exocrine glands are classified into unicellular and multicellular types. **Unicellular glands** are single-celled. An important unicellular exocrine gland is the goblet cell. Although goblet cells do not contain ducts, they are classified as unicellular mucus secreting exocrine glands. Most glands are **multicellular glands,** composed of many cells that form a distinctive microscopic structure or macroscopic organ. Examples are sweat, oil, and salivary glands.

Multicellular glands are categorized according to two criteria: whether the ducts of the glands are branched or unbranched, and the shape of their secretory portions (Figure 3.3). If the duct of the gland does not branch, then it is a **simple gland;** if the duct branches, it is a **compound gland.** Glands with tubular secretory portions are **tubular glands;** those with flasklike secretory portions are **acinar glands** (AS-i-nar; *acin-* = berry). **Tubuloacinar glands** have both tubular and flasklike secretory portions.

Combinations of the degree of duct branching and the shape of the secretory portion are the criteria for the following structural classification scheme for multicellular exocrine glands:

I. **Simple gland:** single, nonbranched duct.
   A. **Simple tubular.** Secretory portion is straight and tubular. Example: large intestinal glands.
   B. **Simple branched tubular.** Secretory portion is branched and tubular. Example: gastric glands.
   C. **Simple coiled tubular.** Secretory portion is coiled. Example: sweat glands.
   D. **Simple acinar.** Secretory portion is flasklike. Example: glands of penile urethra.
   E. **Simple branched acinar.** Secretory portion is branched and flasklike. Example: sebaceous (oil) glands.

II. **Compound gland:** branched duct.
   A. **Compound tubular.** Secretory portion is tubular. Example: bulbourethral (Cowper's) glands.

Figure 3.3 / Multicellular exocrine glands. Blue represents the duct; gold represents the secretory portion.

Structural classification of multicellular exocrine glands is based on the branching pattern of the duct and the shape of the secreting portion.

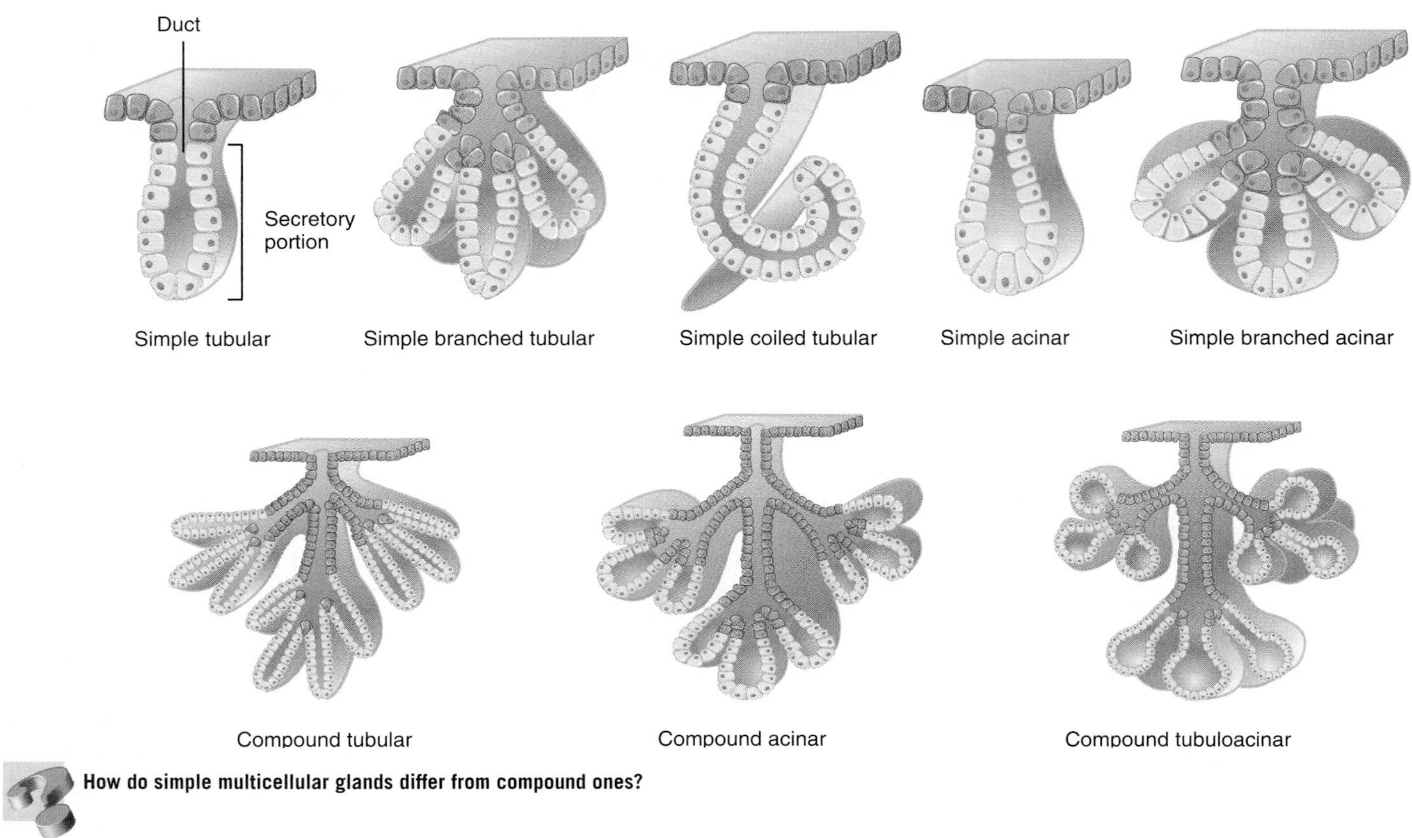

How do simple multicellular glands differ from compound ones?

B. **Compound acinar.** Secretory portion is flasklike. Example: mammary glands.
C. **Compound tubuloacinar.** Secretory portion is both tubular and flasklike. Example: acinar glands of the pancreas.

Functional Classification of Exocrine Glands

The functional classification of exocrine glands is based on whether the secretion alone is released from a cell or whether the secretion consists of whole or parts of glandular cells themselves. **Merocrine glands** (MER-ō-krin; *mero-* = a part) simply form the secretory product and release it from the cell. The secretory product is formed by ribosomes attached to rough ER; it is processed, sorted, and packaged by the Golgi complex, and released from the cell in secretory vesicles via exocytosis (Figure 3.4a). Most exocrine glands of the body produce merocrine secretions. Examples of merocrine glands are the salivary glands and pancreas. **Apocrine glands** (AP-ō-krin; *apo-* = from) accumulate their secretory product at the apical surface of the secreting cell. That portion of the cell pinches off from the rest of the cell to form the secretion (Figure 3.4b). The remaining part of the cell repairs itself and repeats the process. Electron micrographic studies have called into question whether humans have apocrine glands. What were once thought to be apocrine glands—for example, mammary glands—are probably merocrine glands. The cells of **holocrine glands** (HŌ-lō-krin; *holo-* = entire) accumulate a secretory product in their cytosol. As the secretory cell matures, it dies and becomes the secretory product (Figure 3.4c). The sloughed cell is replaced by a new cell. One example of a holocrine gland is a sebaceous (oil) gland of the skin.

✓ Describe the various layering arrangements and cell shapes of epithelium.
✓ What characteristics are common to all epithelial tissues?
✓ Describe how the structure of the following kinds of epithelium is related to their functions: simple squamous, simple cuboidal, simple columnar (nonciliated and ciliated), stratified squamous (keratinized and nonkeratinized), stratified cuboidal, stratified columnar, transitional, and pseudostratified columnar (ciliated and nonciliated).
✓ Define endothelium and mesothelium.
✓ What is a gland? Distinguish between endocrine glands and exocrine glands, and name and give examples of the three functional classes of exocrine glands.

Figure 3.4 / Functional classification of multicellular exocrine glands.

The functional classification of exocrine glands is based on whether a secretion is a product of a cell or consists of an entire or a partial glandular cell.

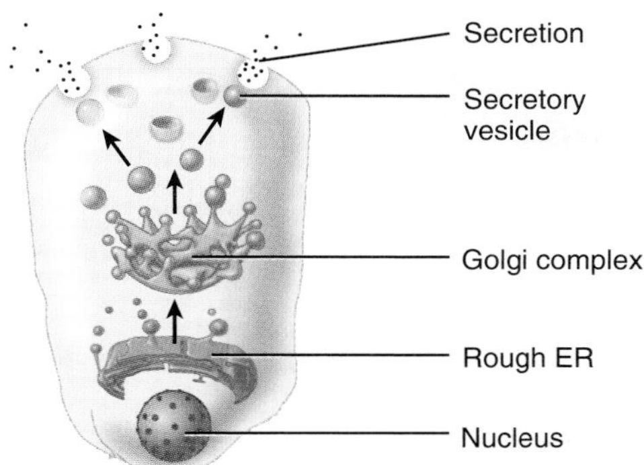

(a) Merocrine secretion

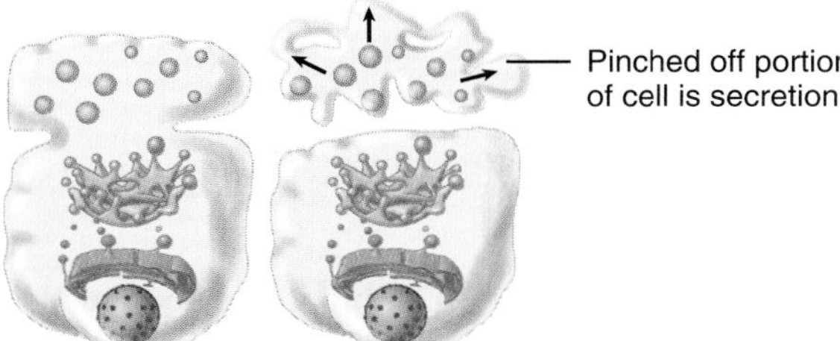

(b) Apocrine secretion

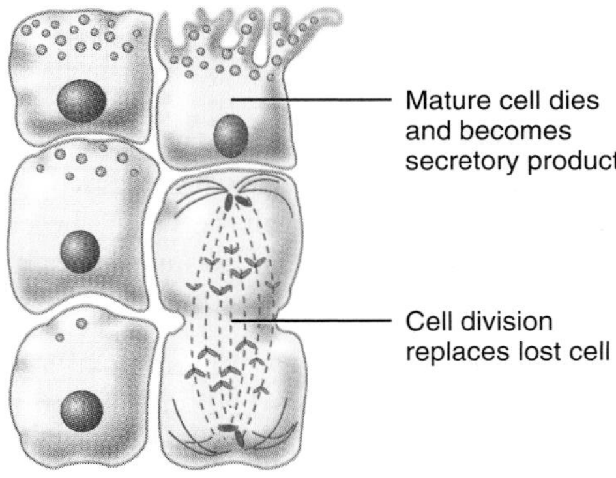

(c) Holocrine secretion

What class of glands are sebaceous (oil) glands? Salivary glands?

CONNECTIVE TISSUE

Objective

- Describe the general features of connective tissue and the structure, location, and function of the various types of connective tissues.

Connective tissue is one of the most abundant and widely distributed tissue in the body. It binds together, supports, and strengthens other body tissues; protects and insulates internal organs; compartmentalizes structures such as skeletal muscles; it is the major transport system within the body (blood is a fluid connective tissue); and it is the major site of stored energy reserves (adipose, or fat, tissue).

General Features of Connective Tissue

Connective tissue consists of two basic elements: cells and matrix. The **matrix** generally consists of fibers and a ground substance that occupies the space between the cells and the fibers. The matrix, which tends to prevent tissue cells from touching one another, may be fluid, semifluid, gelatinous, fibrous, or calcified. It is usually secreted by the connective tissue cells and determines the tissue's qualities. In blood, the matrix (which is not secreted by blood cells) is liquid. In cartilage, it is firm but pliable. In bone, it is hard and not pliable.

In contrast to epithelia, connective tissues do not usually occur on free surfaces, such as the covering or lining of internal organs, the lining of a body cavity, or the external surface of the body. However, joint cavities are lined by a type of connective tissue called areolar connective tissue. Also, unlike epithelia, connective tissues usually are highly vascular; that is, they have a rich blood supply. Exceptions include cartilage, which is avascular, and tendons, which have a scanty blood supply. Except for cartilage, connective tissues have a nerve supply.

Next we examine in detail the components of connective tissue.

Components of Connective Tissue

Connective Tissue Cells

The cells in connective tissue are derived from mesodermal embryonic cells called mesenchymal cells. Each major type of connective tissue contains an immature class of cells whose name ends in *-blast*, which means "to bud or sprout." These immature cells are called *fibroblasts* in loose and dense connective tissue, *chondroblasts* in cartilage, and *osteoblasts* in bone. Blast cells retain the capacity for cell division and secrete the matrix that is characteristic of the tissue. In cartilage and bone, once the matrix is produced, the fibroblasts differentiate into mature cells whose names end in *-cyte*, such as chondrocytes and osteocytes. Mature cells have a reduced capacity for cell division and matrix formation and are mostly involved in maintaining the matrix.

The types of cells present in various connective tissues depend on the type of tissue and include the following (Figure 3.5):

Figure 3.5 / Representative cells and fibers present in connective tissues.

Fibroblasts are usually the most numerous connective tissue cells.

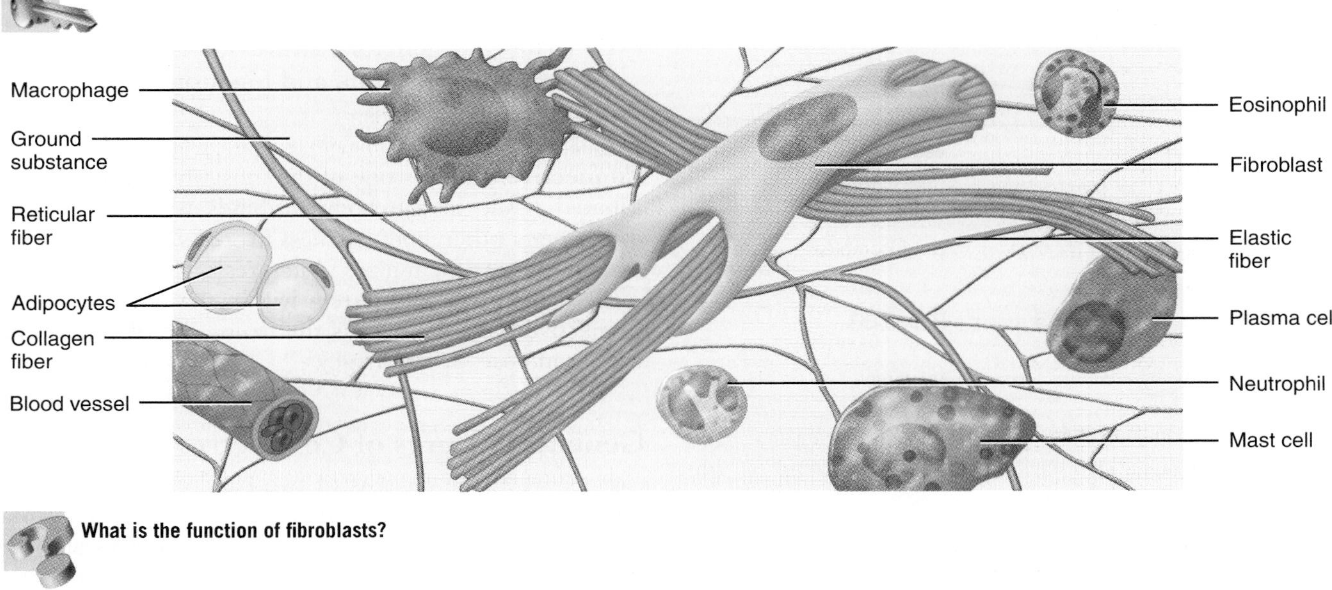

What is the function of fibroblasts?

1. **Fibroblasts** (FĪ-brō-blasts; *fibro-* = fibers) are large, flat, spindle-shaped cells with branching processes. They are present in all connective tissues, and they usually are the most numerous connective tissue cells. Fibroblasts migrate through the connective tissue, secreting the fibers and ground substance of the matrix.
2. **Macrophages** (MAK-rō-fā-jez; *macro-* = large; *-phages* = eaters), or *histiocytes,* develop from monocytes, a type of white blood cell. Macrophages are phagocytes that have an irregular shape with short branching projections and are capable of engulfing bacteria and cellular debris. Some are **fixed macrophages,** which means they reside in a particular tissue—for example, alveolar macrophages in the lungs or spleen macrophages in the spleen. Others are **wandering macrophages,** which roam the tissues and gather at sites of infection or inflammation.
3. **Plasma cells** are small and either round or irregular in shape. They develop from a type of white blood cell called a B lymphocyte. Plasma cells secrete *antibodies,* proteins that attack or neutralize foreign substances in the body. Thus, plasma cells are an important part of the body's immune system. Although they are found in many places in the body, most plasma cells reside in connective tissues, especially in the gastrointestinal tract and the mammary glands.
4. **Mast cells** are abundant alongside the blood vessels that supply connective tissue. They produce *histamine,* a chemical that dilates small blood vessels as part of the body's reaction to injury or infection.
5. **Adipocytes,** also called fat cells, are connective tissue cells that store triglycerides (fats). They are found deep to the skin and around organs such as the heart and kidneys.
6. **White blood cells** are not found in significant numbers in normal connective tissue. However, in response to certain conditions they migrate from blood into connective tissues, where their numbers increase significantly. For example, *neutrophils* increase in number at sites of infection, and *eosinophils* increase in number in response to allergic conditions and parasitic invasions.

Connective Tissue Matrix

Each type of connective tissue has unique properties, due to the accumulation of matrix materials between the cells. The matrix consists of a fluid, gel, or solid ground substance and protein fibers.

Ground substance is the component of a connective tissue between the cells and fibers. The ground substance supports cells, binds them together, and provides a medium through which substances are exchanged between the blood and cells. Until recently, it was thought to function mainly as an inert scaffolding that supports tissues. Now it is clear, however, that the ground substance plays an active role in how tissues develop, migrate, proliferate, and change shape, and in how they carry out their metabolic functions.

Ground substance contains water and an assortment of large molecules, many of which are complex combinations of polysaccharides and proteins. The carbohydrates are polysaccharides that include hyaluronic acid, chondroitin sulfate, dermatan sulfate, and keratin sulfate. Collectively, they are referred to as **glycosaminoglycans** (glī′-kōs-a-mē′-nō-GLĪ-kans) or **GAGs.** Except for hyaluronic acid, the GAGs are associated with proteins called **proteoglycans** (prō′-tē-ō-GLĪ-kans). The proteoglycans form a core protein that resembles the wire stem of a bottle brush and the GAGs project from the protein like the bristles of the brush. One of the most important properties of GAGs is that they trap water, making the ground substance more jelly-like. **Hyaluronic acid** (hī-a-lū-RON-ic) is a viscous, slippery substance that binds cells together, lubricates joints, and helps maintain the shape of the eyeballs. It also appears to play a role in helping phagocytes migrate through connective tissue during development and wound repair. White blood cells, sperm cells, and some bacteria produce *hyaluronidase*, an enzyme that breaks apart hyaluronic acid and causes the ground substance of connective tissue to become watery. The ability to produce hyaluronidase enables white blood cells to move through connective tissues to reach sites of infection and sperm cells to penetrate the ovum during fertilization. It also accounts for how bacteria spread through connective tissues. **Chondroitin sulfate** (kon-DROY-tin) is a jellylike substance that provides support and adhesiveness in cartilage, bone, skin, and blood vessels. The skin, tendons, blood vessels, and heart valves contain **dermatan sulfate,** whereas bone, cartilage, and the cornea of the eye contain **keratan sulfate.**

Also present in the ground substance are **adhesion proteins,** which are responsible for linking components of the ground substance to each other and to the surfaces of cells. The principal adhesion protein of connective tissue is *fibronectin,* which binds to both collagen fibers (discussed shortly) and ground substance and crosslinks them together. It also attaches cells to the ground substance.

Fibers in the matrix strengthen and support connective tissues. Three types of fibers are embedded in the matrix between the cells: collagen fibers, elastic fibers, and reticular fibers.

Collagen fibers (*colla* = glue), of which there are at least five different types, are very strong and resist pulling forces, but they are not stiff, which promotes tissue flexibility. These fibers often occur in bundles that lie parallel to one another (see Figure 3.5), an arrangement that affords great strength to the tissue. Chemically, collagen fibers consist of the protein *collagen.* This is the most abundant protein in your body, representing about 25% of the total protein. Collagen fibers are found in most types of connective tissues, especially bone, cartilage, tendons, and ligaments.

Elastic fibers, which are smaller in diameter than collagen fibers, branch and join together to form a network within a tissue. An elastic fiber consists of molecules of a protein called *elastin* surrounded by a glycoprotein named *fibrillin,* which strengthens and provides elasticity to elastic fibers. Because of their unique molecular structure, elastic fibers are strong but can be stretched up to 150% of their relaxed length without breaking. Equally important, elastic fibers have the ability to return to their original shape after being stretched, a property called *elasticity.* Elastic fibers are plentiful in skin, blood vessel walls, and lung tissue.

Reticular fibers (*reticul-* = net), which consist of *collagen* and a coating of glycoprotein, provide support in the walls of blood vessels and form branching networks around fat cells, nerve fibers, and skeletal and smooth muscle cells. Produced by fibroblasts, they are much thinner than collagen fibers. Like collagen fibers, reticular fibers provide support and strength and also form the **stroma** (= bed or covering) or supporting framework, of many soft organs, such as the spleen and lymph nodes. These fibers also form part of the basement membrane.

Marfan syndrome (MAR-fan) is an inherited disorder caused by a defective fibrillin gene. (Recall that fibrillin is a glycoprotein that strengthens and provides elasticity to elastic fibers). The result is abnormal development of elastic fibers. Tissues rich in elastic fibers are malformed or weakened. Structures affected most seriously are the covering layers of bones (periosteum), the ligament that suspends the lens of the eye, and the walls of the large arteries. People with Marfan syndrome tend to be tall and have disproportionately long arms, legs, fingers, and toes. The long, tapering fingers are called spider fingers (arachnodactyly). A common symptom is blurred vision caused by displacement of the lens of the eye. The most life-threatening complication of Marfan syndrome is weakening of the aorta (the main artery that emerges from the heart), which can suddenly burst. ■

Classification of Connective Tissue

Classifying connective tissues can be challenging because of the diversity of cells and matrix present, and the differences in their relative proportions. Thus, the grouping of connective tissues into categories is not always clear-cut. We will classify them as follows:

I. Embryonic connective tissue
  A. Mesenchyme
  B. Mucous connective tissue
II. Mature connective tissue
  A. Loose connective tissue
    1. Areolar connective tissue
    2. Adipose tissue
    3. Reticular connective tissue
  B. Dense connective tissue
    1. Dense regular connective tissue
    2. Dense irregular connective tissue
    3. Elastic connective tissue
  C. Cartilage
    1. Hyaline cartilage
    2. Fibrocartilage
    3. Elastic cartilage
  D. Bone tissue
  E. Blood tissue
  F. Lymph

Note that our classification scheme has two major subclasses of connective tissue: embryonic and mature. **Embryonic connective tissue** is present primarily in the *embryo,* the developing human from fertilization through the first two months of pregnancy, and in the *fetus,* the developing human from the third month of pregnancy to birth.

One example of embryonic connective tissue found almost exclusively in the embryo is **mesenchyme** (MEZ-en-kīm), the tissue from which all other connective tissues eventually arise (Table 3.2A). Mesenchyme is composed of irregularly shaped cells, a semifluid ground substance, and delicate reticular fibers. Another kind of embryonic tissue is **mucous connective tissue (Wharton's jelly),** found primarily in the umbilical cord of the fetus. Mucous connective tissue is a form of mesenchyme that

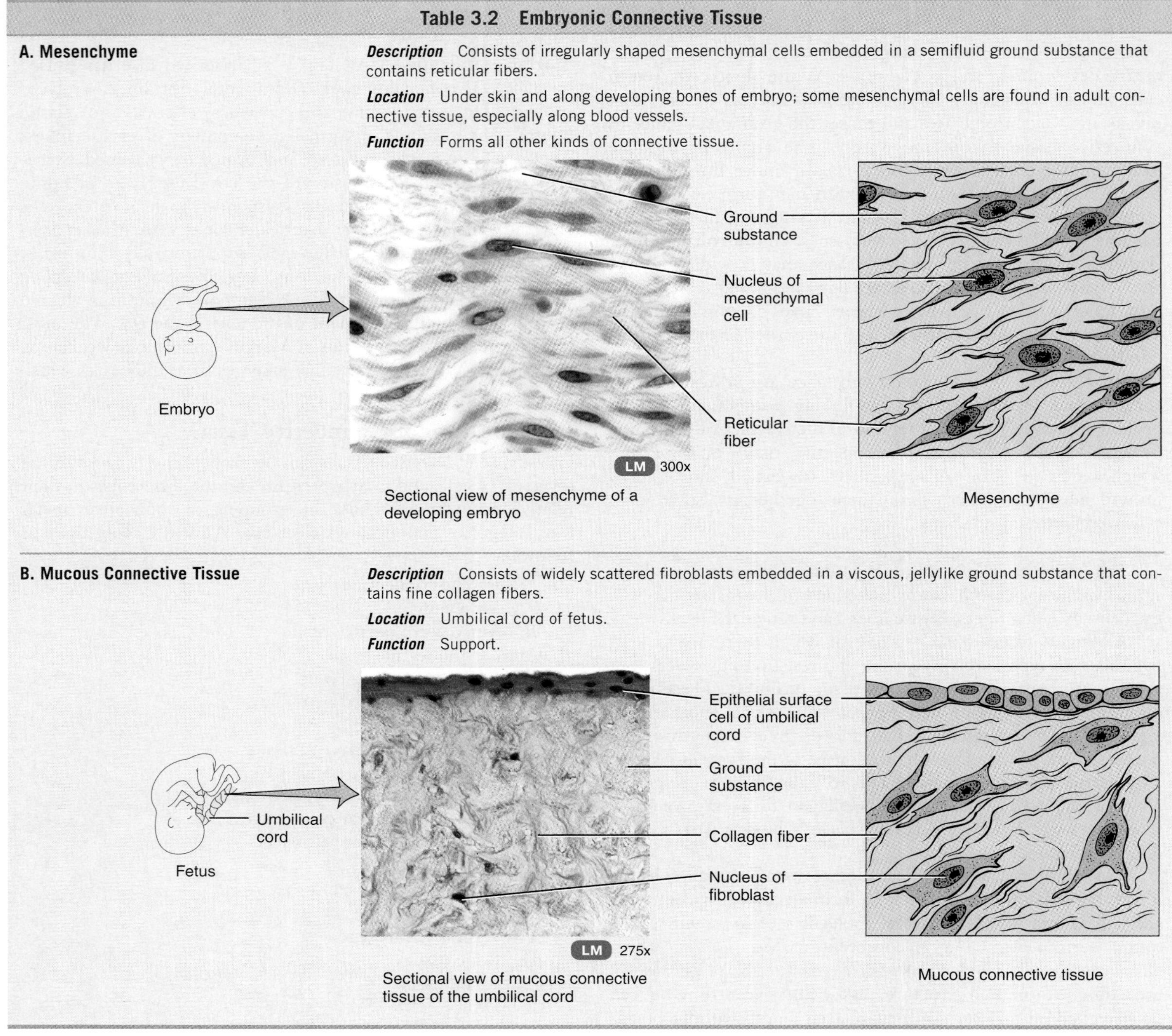

contains widely scattered fibroblasts, a more viscous jellylike ground substance, and collagen fibers (Table 3.2B).

Table 3.2 Embryonic Connective Tissue
A. Mesenchyme
Description Consists of irregularly shaped mesenchymal cells embedded in a semifluid ground substance that contains reticular fibers.
Location Under skin and along developing bones of embryo; some mesenchymal cells are found in adult connective tissue, especially along blood vessels.
Function Forms all other kinds of connective tissue.
Embryo
Ground substance
Nucleus of mesenchymal cell
Reticular fiber
LM 300x
Sectional view of mesenchyme of a developing embryo
Mesenchyme
B. Mucous Connective Tissue
Description Consists of widely scattered fibroblasts embedded in a viscous, jellylike ground substance that contains fine collagen fibers.
Location Umbilical cord of fetus.
Function Support.
Umbilical cord
Fetus
Epithelial surface cell of umbilical cord
Ground substance
Collagen fiber
Nucleus of fibroblast
LM 275x
Sectional view of mucous connective tissue of the umbilical cord
Mucous connective tissue

The second major subclass of connective tissue, **mature connective tissue,** is present in the newborn and has cells produced from mesenchyme. Mature connective tissue is of several types.

Types of Mature Connective Tissue

The six types of mature connective tissue are (1) loose connective tissue, (2) dense connective tissue, (3) cartilage, (4) bone tissue, (5) blood tissue, and (6) lymph.

Loose Connective Tissue

In **loose connective tissue** the fibers are loosely woven, and many cells are present. The types of loose connective tissue are areolar connective tissue, adipose tissue, and reticular connective tissue.

Areolar connective tissue (a-RĒ-ō-lar; *areol-* = a small space) is one of the most widely distributed connective tissues in the body. It contains several kinds of cells, including fibroblasts, macrophages, plasma cells, mast cells, adipocytes, and a few white blood cells (Table 3.3A). All three types of fibers—colla-

Table 3.3 Mature Connective Tissue

Loose Connective Tissue

A. Areolar Connective Tissue

Description Consists of fibers (collagen, elastic, and reticular) and several kinds of cells (fibroblasts, macrophages, plasma cells, adipocytes, and mast cells) embedded in a semifluid ground substance.

Location Subcutaneous layer deep to skin; papillary (superficial) region of dermis of skin; lamina propria of mucous membranes; and around blood vessels, nerves, and body organs.

Function Strength, elasticity, and support.

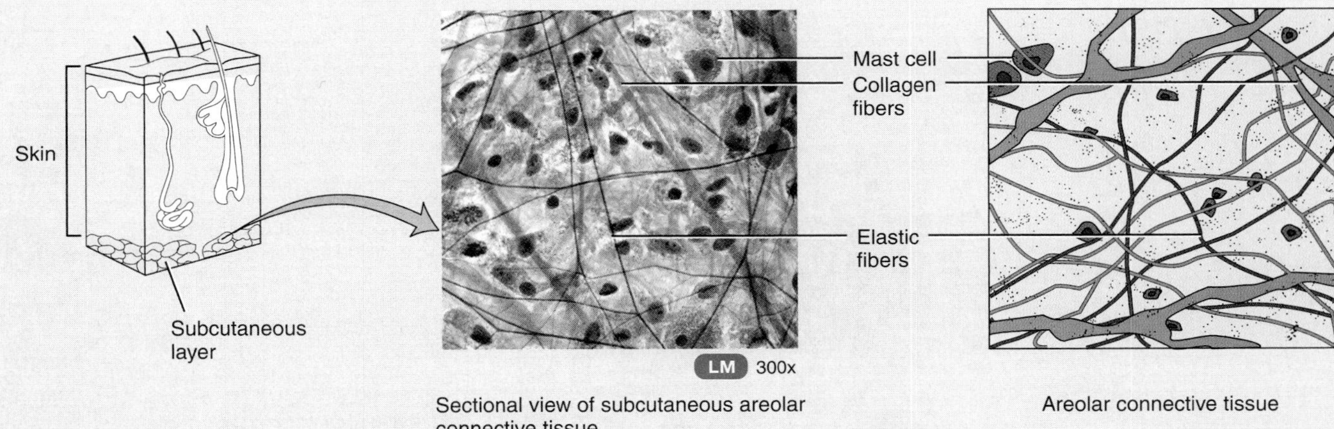

Sectional view of subcutaneous areolar connective tissue

Areolar connective tissue

B. Adipose Tissue

Description Consists of adipocytes, cells specialized to store triglycerides (fats) in a large central area of their cytoplasm; nuclei and cytoplasm are peripherally located.

Location Subcutaneous layer deep to skin, around heart and kidneys, yellow bone marrow of long bones, and padding around joints and behind eyeball in eye socket.

Function Reduces heat loss through skin, serves as an energy reserve, supports, and protects. In newborns, brown fat generates considerable heat that helps maintain proper body temperature.

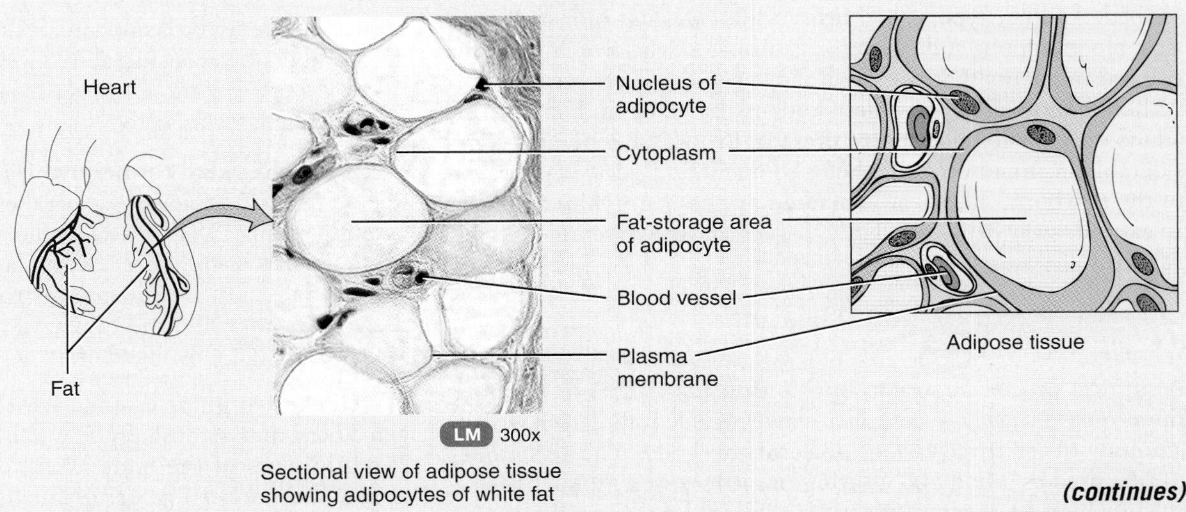

Sectional view of adipose tissue showing adipocytes of white fat

Adipose tissue

(continues)

gen, elastic, and reticular—are arranged randomly throughout the tissue. The ground substance contains hyaluronic acid, chondroitin sulfate, dermatan sulfate, and keratan sulfate. Combined with adipose tissue, areolar connective tissue forms the *subcutaneous layer*—the layer of tissue that attaches the skin to underlying tissues and organs.

Adipose tissue is a loose connective tissue in which the cells, called **adipocytes** (*adipo-* = fat), are specialized for storage of triglycerides (fats) (Table 3.3B). Adipocytes are derived from fibroblasts. Because the cell fills up with a single, large triglyceride droplet, the cytoplasm and nucleus are pushed to the periphery of the cell. Adipose tissue is found wherever there is areolar connective tissue. Adipose tissue is a good insulator and can therefore reduce heat loss through the skin. It is a major energy reserve and generally supports and protects various organs.

Table 3.3 Mature Connective Tissue (continued)

Loose Connective Tissue

C. Reticular Connective Tissue

Description Consists of a network of interlacing reticular fibers and reticular cells.

Location Stroma (supporting framework) of liver, spleen, lymph nodes; portion of red bone marrow that gives rise to blood cells; reticular lamina of the basement membrane; and around blood vessels and muscles.

Function Forms stroma of organs; binds together smooth muscle tissue cells.

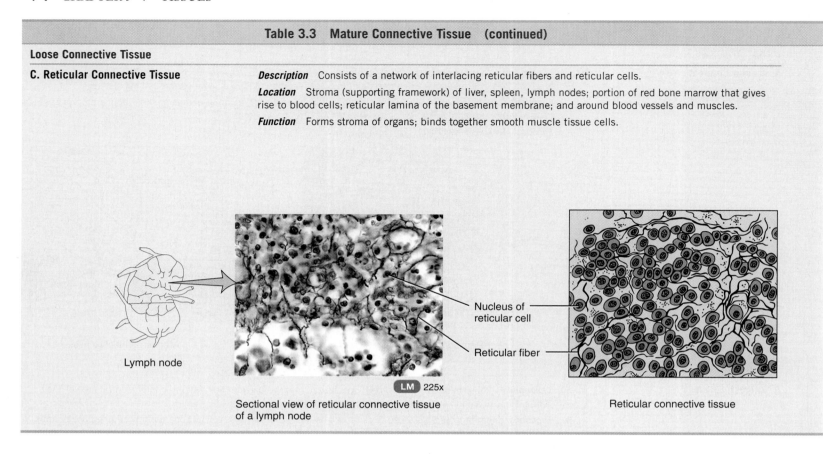

Sectional view of reticular connective tissue of a lymph node

Reticular connective tissue

Most of the fat in adults is **white fat,** the type just described. Another type, called **brown fat,** owes its color to a very rich blood supply and numerous mitochondria, which contain colored pigments that participate in aerobic cellular respiration. Although brown fat is widespread in the fetus and infant, in adults only small amounts are present. Brown fat generates considerable heat and probably helps to maintain body temperature in the newborn. The heat generated by the many mitochondria is carried away to other body tissues by the extensive blood supply.

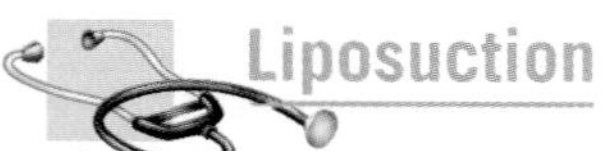

A surgical procedure, called **liposuction** (*lip-* = fat) or **suction lipectomy** (*-ectomy* = cut out), involves suctioning out small amounts of fat from various areas of the body. The technique can be used as a body-contouring procedure in a number of areas in which fat accumulates, such as the thighs, buttocks, arms, breasts, and abdomen. Postsurgical complications that may develop include fat emboli (clots), infection, fluid depletion, injury to internal structures, and severe postoperative pain. ■

Reticular connective tissue consists of fine interlacing reticular fibers and reticular cells, cells that are connected to each other and form a network (Table 3.3C). Reticular connective tissue forms the stroma (supporting framework) of certain organs and helps bind together smooth muscle cells.

Dense Connective Tissue

Dense connective tissue contains more numerous, thicker, and denser fibers but considerably fewer cells than loose connective tissue. There are three types: dense regular connective tissue, dense irregular connective tissue, and elastic connective tissue.

Dense regular connective tissue has bundles of collagen fibers *regularly* arranged in parallel patterns that confer great strength (Table 3.3D). The tissue structure withstands pulling along the axis of the fibers. Fibroblasts, which produce the fibers and ground substance, appear in rows between the fibers. The tissue is silvery white and tough, yet somewhat pliable. Examples are tendons and most ligaments.

Dense irregular connective tissue, which contains collagen fibers that are usually *irregularly* arranged (Table 3.3E), is found in parts of the body where pulling forces are exerted in various directions. The tissue usually occurs in sheets, such as in the dermis of the skin, which underlies the epidermis. Heart valves, the perichondrium (the membrane surrounding cartilage), and the periosteum (the membrane surrounding bone) are dense irregular connective tissues, although they have a fairly orderly arrangement of their collagen fibers.

Elastic connective tissue is characterized by freely branching elastic fibers that give the unstained tissue a yellowish color (Table 3.3F). Fibroblasts are present in the spaces between the fibers. Elastic connective tissue is quite strong and can recoil to

Dense Connective Tissue

D. Dense Regular Connective Tissue

Description Matrix looks shiny white; consists predominantly of collagen fibers arranged in bundles; fibroblasts present in rows between bundles.

Location Forms tendons (attach muscle to bone), most ligaments (attach bone to bone), and aponeuroses (sheetlike tendons that attach muscle to muscle or muscle to bone).

Function Provides strong attachment between various structures.

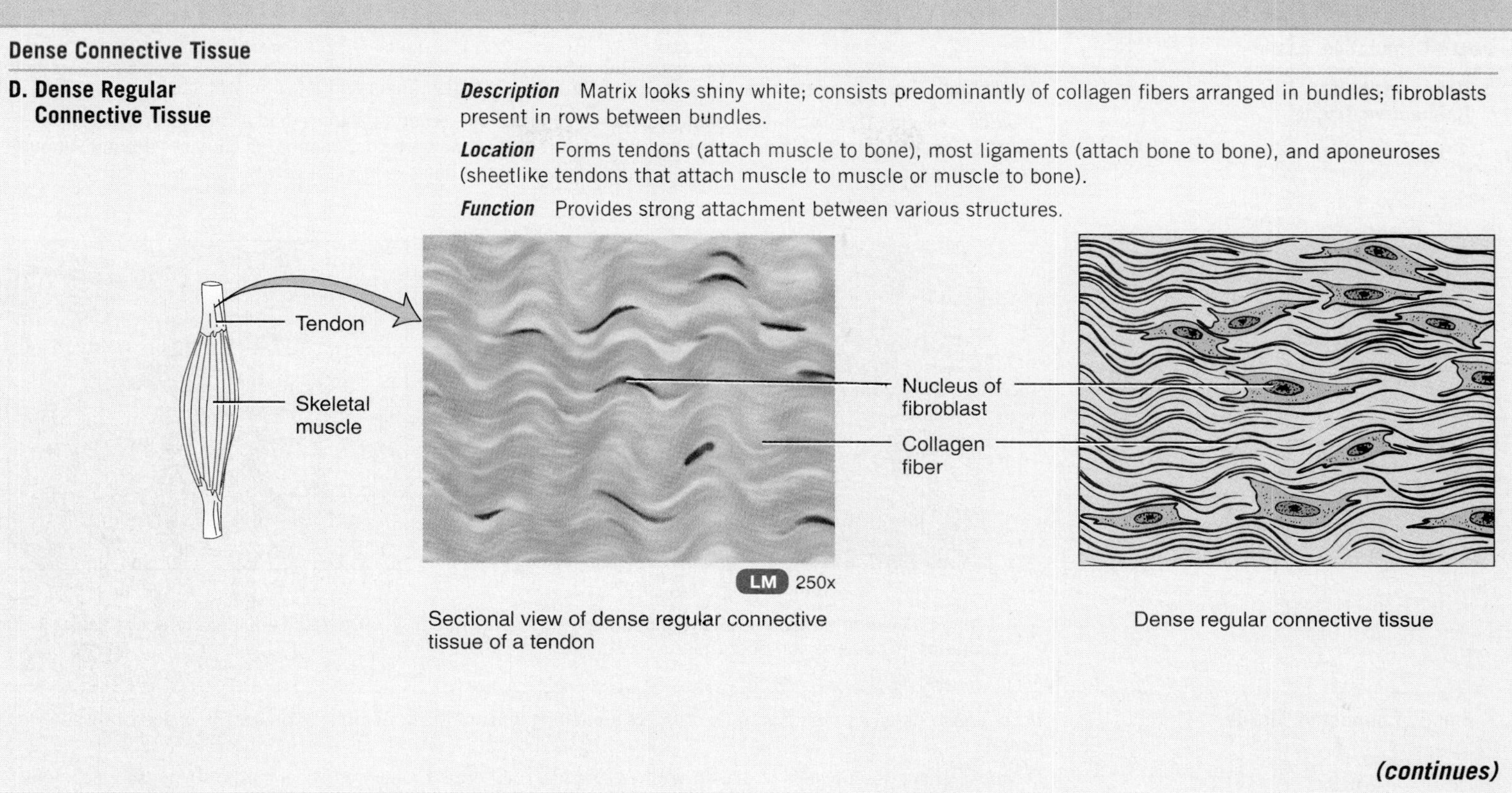

Sectional view of dense regular connective tissue of a tendon

Dense regular connective tissue

(continues)

its original shape after being stretched. Elasticity is important to the normal functioning of lung tissue, which recoils as you exhale, and elastic arteries, whose recoil between heartbeats helps maintain blood flow.

Cartilage

Cartilage consists of a dense network of collagen fibers and elastic fibers firmly embedded in chondroitin sulfate, a rubbery component of the ground substance. Cartilage can endure considerably more stress than loose and dense connective tissues. Whereas the strength of cartilage is due to its collagen fibers, its resilience (ability to assume its original shape after deformation) is due to chondroitin sulfate.

The cells of mature cartilage, called **chondrocytes** (KON-drō-sīts; *chondro-* = cartilage), occur singly or in groups within spaces called **lacunae** (la-KŪ-nē; = little lakes, the singular is lacuna) in the matrix. The surface of most cartilage is surrounded by a membrane of dense irregular connective tissue called the **perichondrium** (per′-i-KON-drē-um; *peri-* = around). Unlike other connective tissues, cartilage has no blood vessels or nerves, except in the perichondrium. The three kinds of cartilage are hyaline cartilage, fibrocartilage, and elastic cartilage.

Hyaline cartilage contains a resilient gel as its ground substance and appears in the body as a bluish-white, shiny substance. The fine collagen fibers are not visible with ordinary staining techniques, and prominent chondrocytes are found in lacunae (Table 3.3G). Most hyaline cartilage is surrounded by a perichondrium. The exceptions are the articular cartilage in joints and at the epiphyseal plates, the regions where bones lengthen as a person grows. Hyaline cartilage is the most abundant cartilage in the body. It affords flexibility and support and, at joints, reduces friction and absorbs shock. Hyaline cartilage is the weakest of the three types of cartilage.

Fibrocartilage combines strength and rigidity and is the strongest of the three types of cartilage. It has chondrocytes scattered among clearly visible bundles of collagen fibers within the matrix (Table 3.3H). Fibrocartilage lacks a perichondrium. One location for fibrocartilage is the discs between backbones.

Elastic cartilage provides strength and elasticity and maintains the shape of certain structures, such as the external ear. In this tissue, chondrocytes are located within a threadlike network of elastic fibers within the matrix (Table 3.3I). A perichondrium is present.

Bone Tissue

Cartilage, joints, and bone make up the skeletal system. Bones are organs composed of several different connective tissues, including **bone** or **osseous tissue** (OS-ē-us), the periosteum, red and yellow bone marrow, and the endosteum (a membrane that lines a space within bone that stores yellow bone marrow). Bone tissue is classified as either compact or spongy, depending on how the matrix and cells are organized.

Table 3.3 Mature Connective Tissue (continued)

Dense Connective Tissue

E. Dense Irregular Connective Tissue

Description Consists predominantly of randomly arranged collagen fibers and a few fibroblasts.

Location Fasciae (tissue beneath skin and around muscles and other organs), reticular (deeper) region of dermis of skin, periosteum of bone, perichondrium of cartilage, joint capsules, membrane capsules around various organs (kidneys, liver, testes, lymph nodes), pericardium of the heart, and heart valves.

Function Provides strength.

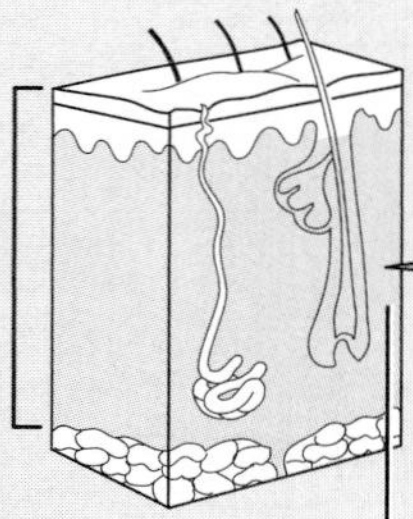

Skin

Dermis

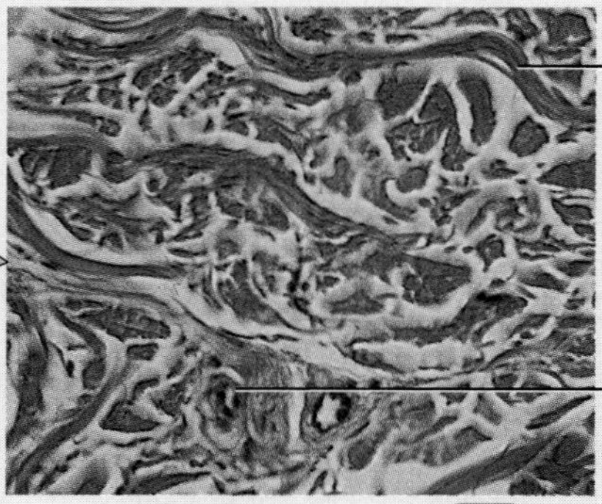

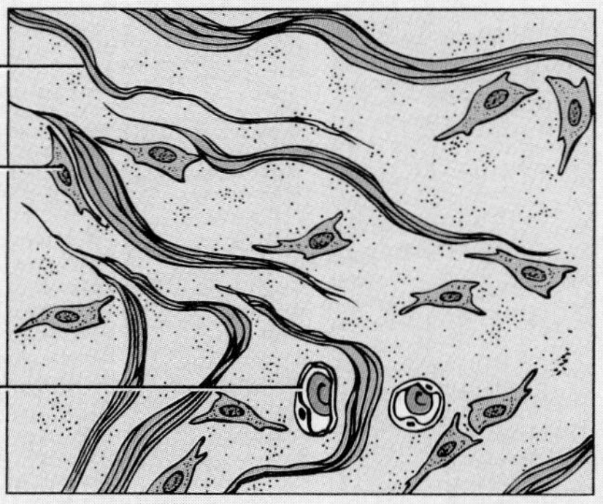

Collagen fiber

Fibroblast

Blood vessel

LM 275x

Sectional view of dense irregular connective tissue of reticular region of dermis

Dense irregular connective tissue

F. Elastic Connective Tissue

Description Consists predominantly of freely branching elastic fibers; fibroblasts present in spaces between fibers.

Location Lung tissue, walls of elastic arteries, trachea, bronchial tubes, true vocal cords, suspensory ligament of penis, and ligaments between vertebrae.

Function Allows stretching of various organs.

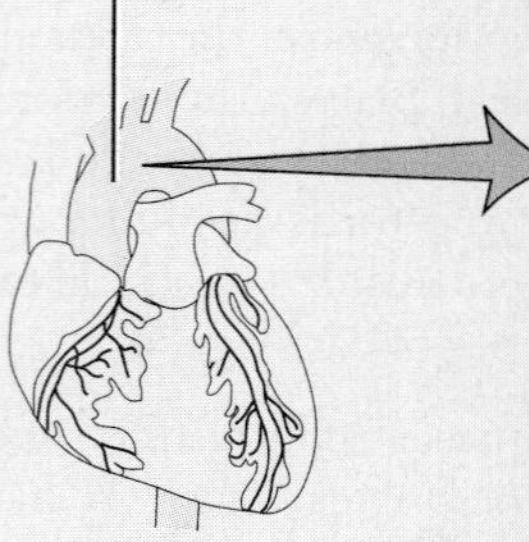

Aorta

Heart

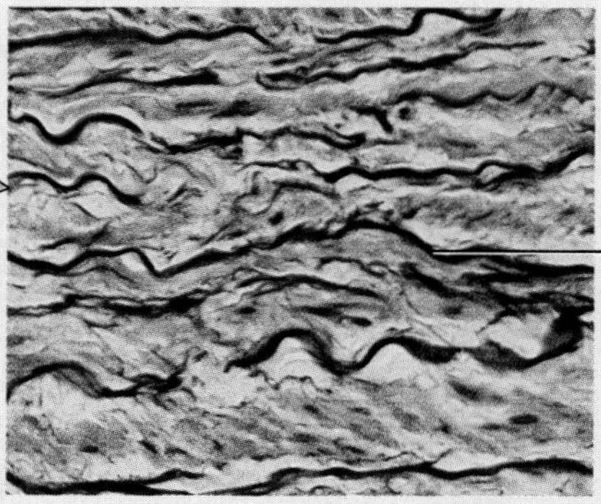

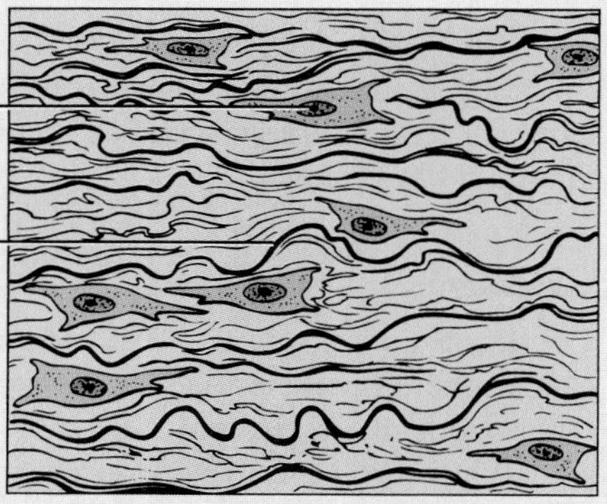

Nucleus of fibroblast

Elastic lamellae (sheets of elastic material)

LM 335x

Sectional view of elastic connective tissue of aorta

Elastic connective tissue

The basic unit of compact bone is an **osteon** or **Haversian system** (Table 3.3J). Each osteon is composed of four parts:

- The **lamellae** (la-MEL-lē; = little plates) are concentric rings of matrix that consist of mineral salts (mostly calcium and phosphates), which give bone its hardness, and collagen fibers, which give bone its strength.
- **Lacunae** (la-CŪ-nē) are small spaces between lamellae that contain mature bone cells called **osteocytes.**
- Projecting from the lacunae are **canaliculi** (CAN-a-lik-ū-lī; = little canals), networks of minute canals containing the processes of osteocytes. Canaliculi provide routes for nutrients to reach osteocytes and for wastes to leave them.
- A **central (Haversian) canal** contains blood vessels and nerves.

Spongy bone has no osteons; instead, it consists of columns of bone called **trabeculae** (tra-BEK-ū-lē; = little beams), which

Cartilage

G. Hyaline Cartilage

Description Consists of a bluish-white, shiny ground substance with fine collagen fibers; contains numerous chondrocytes; most abundant type of cartilage.

Location Ends of long bones, anterior ends of ribs, nose, parts of larynx, trachea, bronchi, bronchial tubes, and embryonic and fetal skeleton.

Function Provides smooth surfaces for movement at joints, as well as flexibility and support.

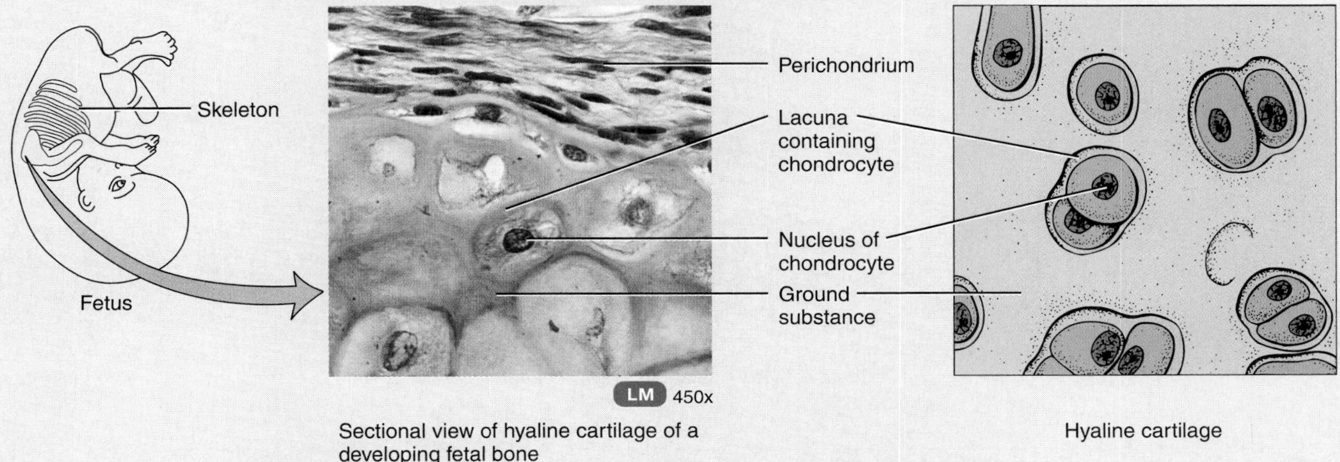

Sectional view of hyaline cartilage of a developing fetal bone

Hyaline cartilage

H. Fibrocartilage

Description Consists of chondrocytes scattered among bundles of collagen fibers within the matrix.

Location Pubic symphysis (point where hip bones join anteriorly), intervertebral discs (discs between vertebrae), menisci (cartilage pads) of knee, and portions of tendons that insert into cartilage.

Function Support and fusion.

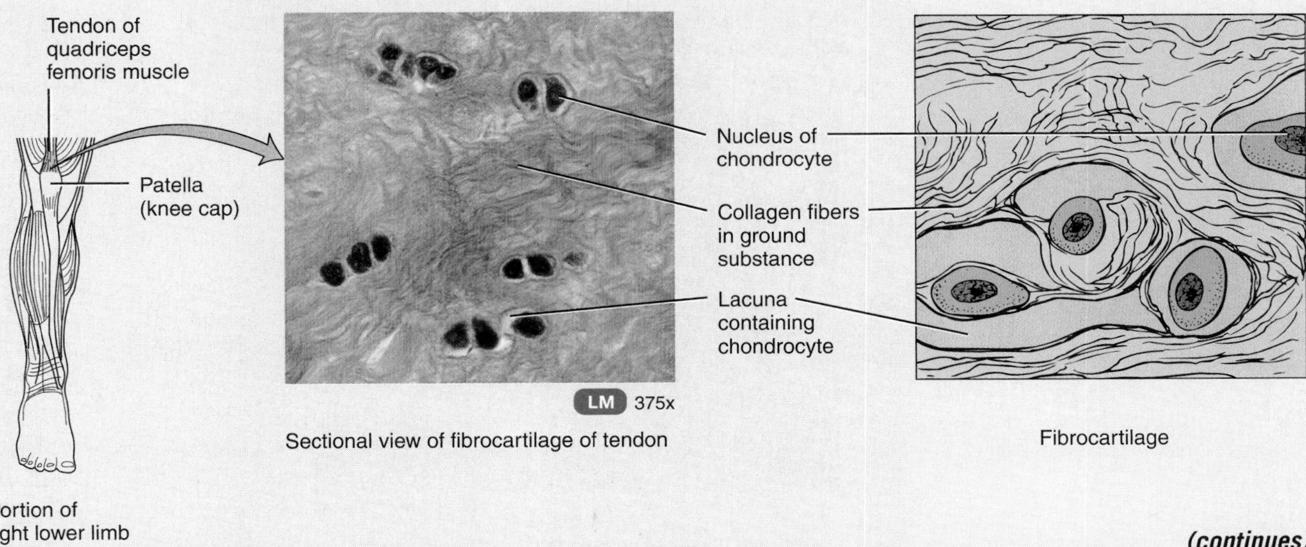

Sectional view of fibrocartilage of tendon

Fibrocartilage

(continues)

contain lamellae, osteocytes, lacunae, and canaliculi. Spaces between lamellae are filled with red bone marrow. Chapter 5 presents details of bone tissue histology.

The skeletal system has several functions. It supports soft tissues, protects delicate structures, and works with skeletal muscles to generate movement. Bone stores calcium and phosphorus; houses red bone marrow, which produces blood cells; and houses yellow bone marrow, which stores triglycerides as an energy source.

Blood Tissue

Blood tissue (or simply blood) is a connective tissue with a liquid matrix called **plasma,** a pale yellow fluid that consists mostly of water with a wide variety of dissolved substances—nutrients, wastes, enzymes, hormones, respiratory gases, and ions (Table 3.3K). Suspended in the plasma are formed elements: red blood cells (erythrocytes), white blood cells (leukocytes), and platelets. **Red blood cells** transport oxygen to body cells and remove carbon dioxide from them. **White blood cells** are involved in phago-

Table 3.3 Mature Connective Tissue (continued)

Cartilage

I. Elastic Cartilage

Description Consists of chondrocytes located in a threadlike network of elastic fibers within the matrix.

Location Auricle (part of external ear), auditory (Eustachian) tubes, and epiglottis.

Function Gives support and maintains shape.

Bone Tissue

J. Compact Bone

Description Compact bone tissue consists of osteons (Haversian systems) that contain lamellae, lacunae, osteocytes, canaliculi, and central (Haversian) canals. By contrast, spongy bone tissue (see Figure 5.3 on page 113) consists of thin columns called trabeculae; spaces between trabeculae are filled with red bone marrow.

Location Both compact and spongy bone tissue make up the various parts of bones of the body.

Function Support, protection, storage; houses blood-forming tissues; bones serve as levers that act together with muscle tissue to enable movement.

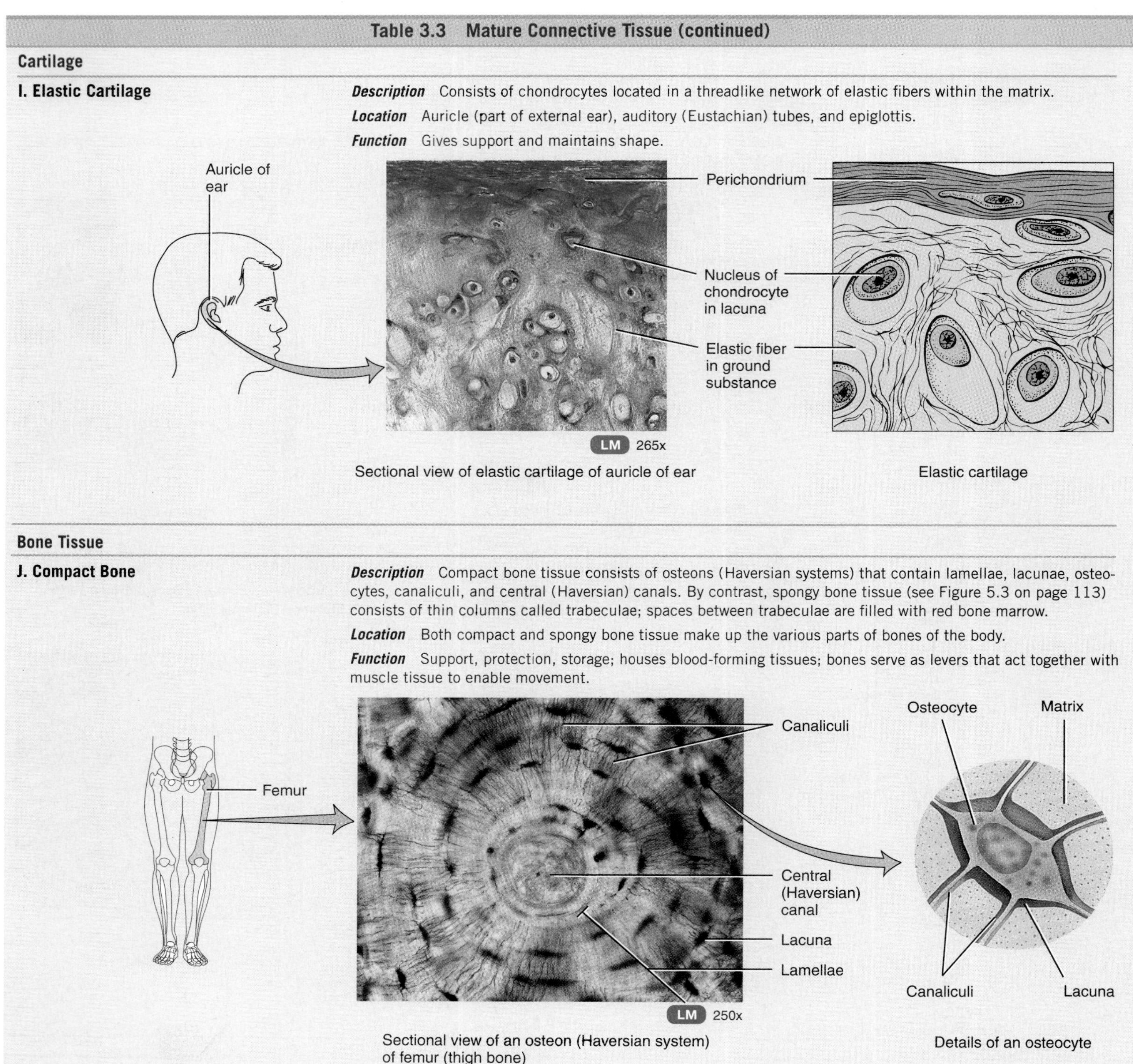

Sectional view of elastic cartilage of auricle of ear

Elastic cartilage

Sectional view of an osteon (Haversian system) of femur (thigh bone)

Details of an osteocyte

cytosis, immunity, and allergic reactions. **Platelets** participate in blood clotting. The details of blood are considered in Chapter 12.

Lymph

Lymph is interstitial fluid that flows in lymphatic vessels. It is a connective tissue that consists of a clear fluid similar to plasma but with much less protein. In the lymph are various cells and chemicals, whose composition varies from one part of the body to another. For example, lymph leaving lymph nodes includes many lymphocytes, a type of white blood cell, whereas lymph from the small intestine has a high content of newly absorbed dietary lipids. The details of lymph are considered in Chapter 15.

Blood Tissue

K. Blood

Description Consists of plasma and formed elements: red blood cells (erythrocytes), white blood cells (leukocytes), and platelets.

Location Within blood vessels (arteries, arterioles, capillaries, venules, and veins) and within the chambers of the heart.

Function Red blood cells transport oxygen and carbon dioxide; white blood cells carry on phagocytosis and are involved in allergic reactions and immune system responses; platelets are essential for the clotting of blood.

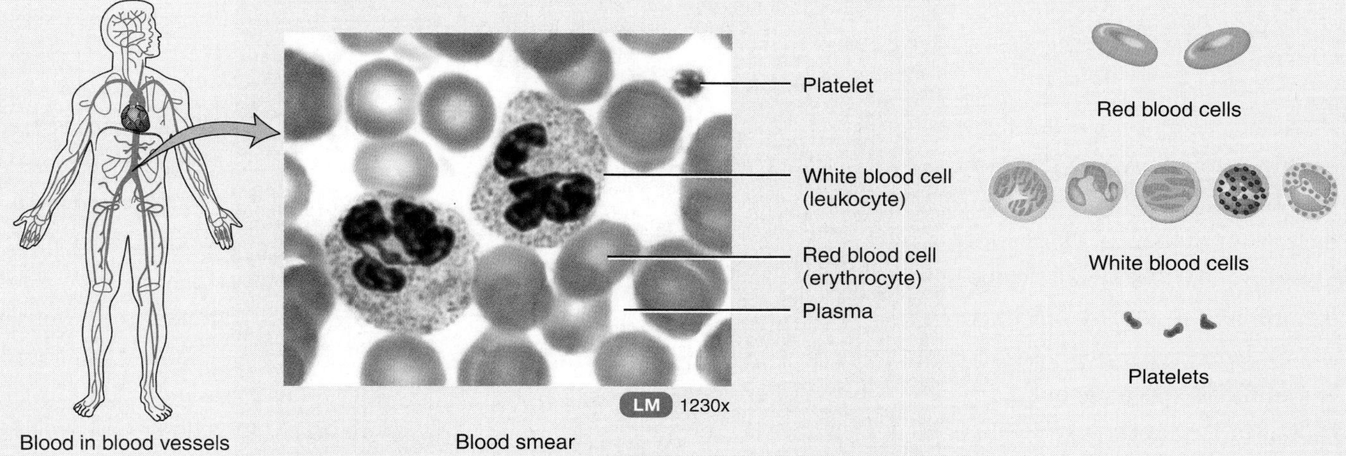

Blood in blood vessels

Blood smear

- ✓ List the ways in which connective tissue differs from epithelium.
- ✓ Describe the cells, ground substance, and fibers that make up connective tissue.
- ✓ How are connective tissues classified? List the various types.
- ✓ How are embryonic connective tissue and mature connective tissue distinguished?
- ✓ Describe how the structure of the following mature connective tissues are related to their functions: areolar connective tissue, adipose tissue, reticular connective tissue, dense regular connective tissue, dense irregular connective tissue, elastic connective tissue, hyaline cartilage, fibrocartilage, elastic cartilage, bone tissue, and blood tissue.

▶ 14th Century

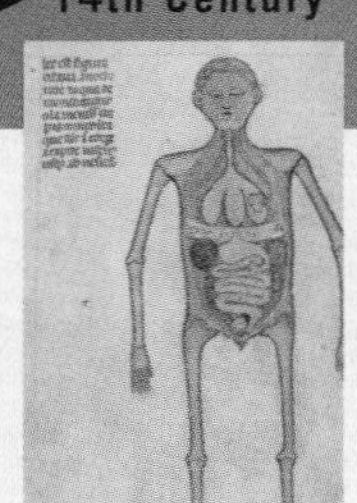

▶ 15th Century

▶ 16th Century

▶ 17th Century

▶ 18th Century

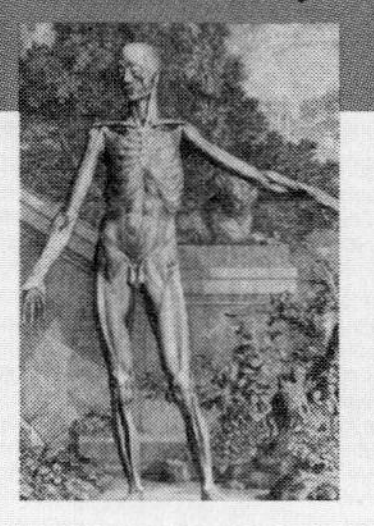

▶ 20th Century

▶ 21st Century

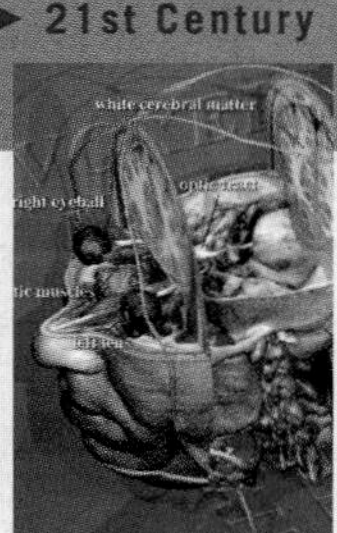

CHANGING IMAGES

The Quest for Beauty

Around the globe and throughout history, both women and men have altered their appearance in the pursuit of beauty. Cosmetics, hairstyles, and fashionable clothing are used to modify our looks and to define our traditions, society, ethnicity, and sense of beauty. It's fun to look back at styles that may have seemed so attractive at one time, only to be ridiculed decades, and even centuries later.

Beauty is certainly relative to era, but is also inextricably linked to culture. What one culture may find desirable, another may consider quite displeasing.

The global quest for beauty frequently goes beyond dress and includes practices that radically change one's anatomy. Customs such as Chinese foot binding and African head reshaping may seem radical or eccentric to Western culture. But think about commonplace modern day practices like excessive muscular development or bodybuilding, body piercing, liposuction, or breast augmentation. Are they really any less exotic? While many of us may find these practices not particularly unusual and merely a matter of personal choice, tribal Africans are quite likely to consider them bizarre.

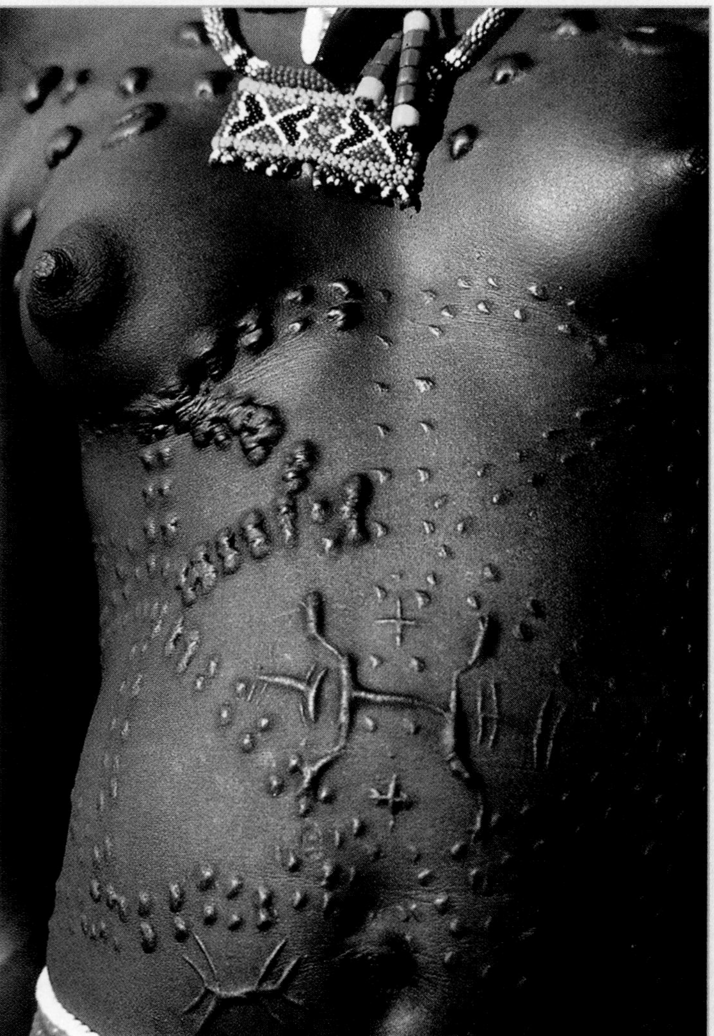

1994 AD

Pictured is a young African woman who practiced a type of body art termed *scarification* in her pursuit of beauty. While in many cultures, smooth and unblemished skin exemplifies beauty, other peoples consider such skin to be unattractive. These designs, while not only considered beautiful, are also symbolic and may mark a woman's social status, fertility, or potential dowry (possessions a female brings to her husband under a marrage contract). To produce such designs the skin and underlying tissue is cut and the body's healing process is controlled. Substances such as clay and ash are then inserted into the wound, provoking the formation of raised bumps called *keloids*. The resulting scar tissue is an excessive accumulation of collagen fibers that is covered by a smooth and shiny epidermal surface. In this chapter you are learning about the relationships between different types of tissues. As your study continues, contemplate the role of tissues in man's varied pursuit of attractiveness.

MEMBRANES

Objective

- Define a membrane, and describe the classification of membranes.

Membranes are flat sheets of pliable tissue that cover or line a part of the body. The combination of an epithelial layer and an underlying connective tissue layer constitutes an **epithelial membrane.** The principal epithelial membranes of the body are mucous membranes, serous membranes, and the cutaneous membrane, or skin. (Skin is discussed in detail in Chapter 4, and will not be discussed here.) Another kind of membrane, a **synovial membrane,** lines joints and contains connective tissue but no epithelium.

Epithelial Membranes

Mucous Membranes

A **mucous membrane,** or **mucosa,** lines a body cavity that opens directly to the exterior. Mucous membranes line the entire digestive, respiratory, and reproductive systems, and much of the urinary system. They consist of both a lining layer of epithelium and an underlying layer of connective tissue.

The epithelial layer of a mucous membrane is an important feature of the body's defense mechanisms because it is a barrier that microbes and other pathogens have difficulty penetrating. The cells are usually connected by tight junctions so materials cannot leak between them. Goblet cells and other cells of the epithelial layer secrete mucus, and this slippery fluid prevents the cavities from drying out. It also traps particles in the respiratory passageways and lubricates food as it moves through the gastrointestinal tract. In addition, the epithelial layer secretes some of the enzymes needed for digestion and is the site of food and fluid absorption in the gastrointestinal tract. The epithelia of mucous membranes vary greatly in different parts of the body. For example, the epithelium of the mucous membrane of the small intestine is nonciliated simple columnar (see Table 3.1C), whereas the epithelium of the large airways to the lungs is pseudostratified ciliated columnar (see Table 3.1I).

The connective tissue layer of a mucous membrane is areolar connective tissue and is called the **lamina propria** (LAM-i-na PRŌ-prē-a). The lamina propria is so named because it belongs to the mucous membrane (*propria* = one's own). The lamina propria supports the epithelium, binds it to the underlying structures, and allows some flexibility of the membrane. It also holds blood vessels in place and protects underlying muscles from abrasion or puncture. Oxygen and nutrients diffuse from the lamina propria to the epithelium covering it whereas carbon dioxide and wastes diffuse in the opposite direction.

Serous Membranes

A **serous membrane** (*serous* = watery), or **serosa,** lines a body cavity that does not open directly to the exterior, and it covers the organs that lie within the cavity. Serous membranes consist of areolar connective tissue covered by mesothelium (simple squamous epithelium) and are composed of two layers. The portion of a serous membrane attached to the cavity wall is called the **parietal layer** (pa-RĪ-e-tal; *paries* = wall); the portion that covers and attaches to the organs inside these cavities is the **visceral layer** (*viscer-* = body organ). The mesothelium of a serous membrane secretes **serous fluid,** a watery lubricating fluid that allows organs to glide easily over one another or to slide against the walls of cavities.

The serous membrane lining the thoracic cavity and covering the lungs is the **pleura.** Whereas the *parietal pleura* lines the thoracic cavity, the *visceral pleura* covers the lungs. The serous membrane lining the thoracic cavity and covering the heart is the **pericardium.** The *parietal pericardium* lines the thoracic cavity and the *visceral pericardium* covers the heart. The serous membrane lining the abdominal cavity and covering the abdominal organs is the **peritoneum.** The *parietal peritoneum* lines the abdominal cavity and the *visceral peritoneum* covers the abdominal organs.

Synovial Membranes

Synovial membranes (sin-Ō-vē-al) line the cavities of freely movable joints. (*Syn* = together, referring here to a place where bones come together.) Like serous membranes, synovial membranes line structures that do not open to the exterior. Unlike mucous, serous, and cutaneous membranes, they do not contain epithelium and are therefore not epithelial membranes. Synovial membranes are composed of areolar connective tissue with elastic fibers and varying numbers of adipocytes. Synovial membranes secrete **synovial fluid,** which lubricates and nourishes the cartilage covering the bones at movable joints; these membranes are called *articular synovial membranes.* Other synovial membranes line cushioning sacs, called *bursae*, and *tendon sheaths* in the hands and feet, thus easing the movement of muscle tendons.

✓ Define the following kinds of membranes: mucous, serous, cutaneous, and synovial.

✓ Where is each type of membrane located in the body? What are their functions?

MUSCLE TISSUE

Objective

- Describe the general features of muscle tissue, and contrast the structure, location, and mode of control of the three types of muscle tissue.

Muscle tissue consists of elongated fibers (cells) that are beautifully constructed to generate force and produce contraction. As a result of this characteristic, muscle tissue produces motion, maintains posture, and generates heat. Based on its location and structural and functional characteristics, muscle tissue is classified as: skeletal, cardiac, and smooth (Table 3.4).

Skeletal muscle tissue is named for its location—usually attached to the bones of the skeleton (Table 3.4A). It is also *striated*—the fibers contain alternating light and dark bands called *striations* that are perpendicular to the long axes of the fibers and

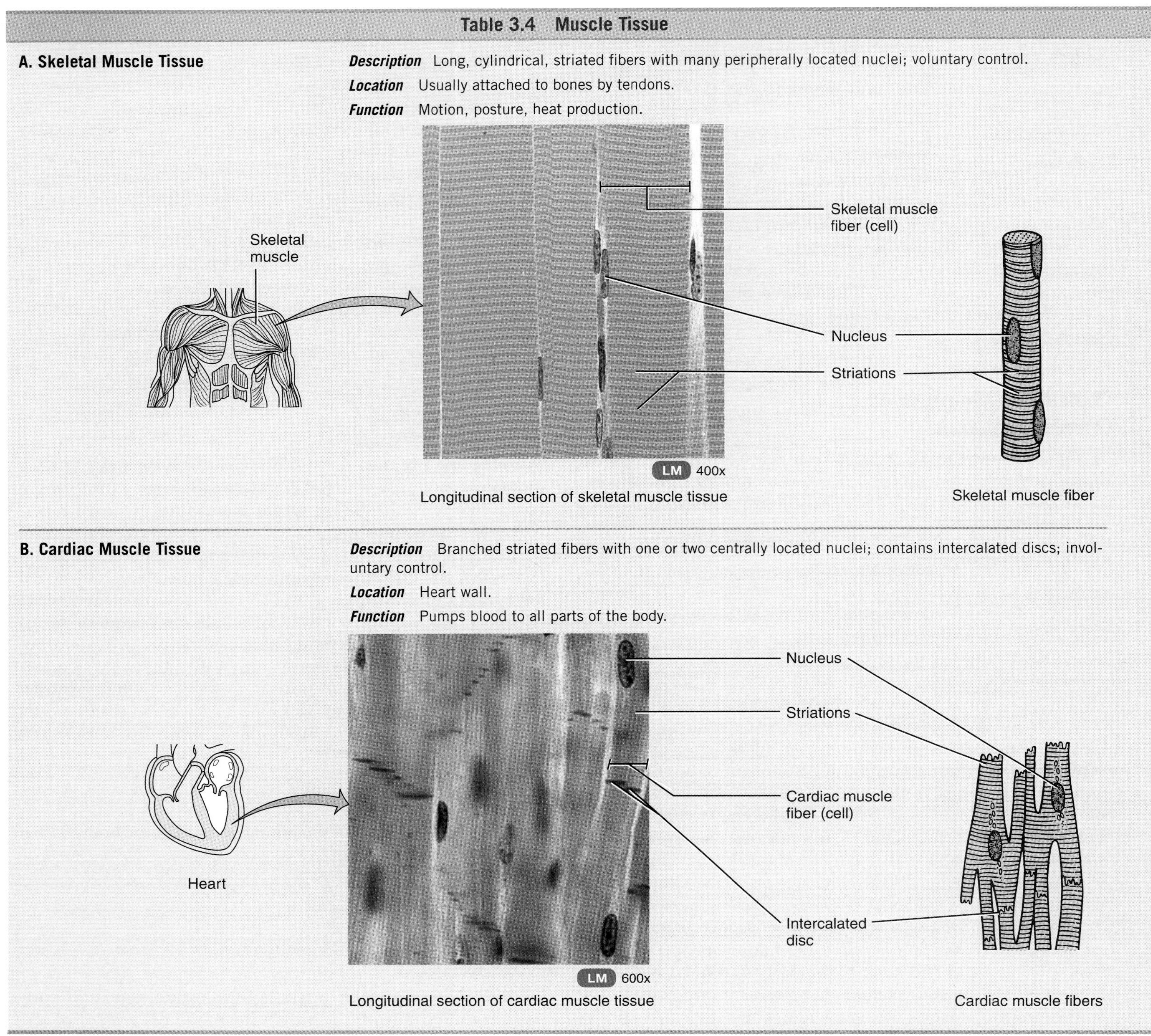
Table 3.4 Muscle Tissue

A. Skeletal Muscle Tissue

Description Long, cylindrical, striated fibers with many peripherally located nuclei; voluntary control.

Location Usually attached to bones by tendons.

Function Motion, posture, heat production.

Longitudinal section of skeletal muscle tissue

Skeletal muscle fiber

B. Cardiac Muscle Tissue

Description Branched striated fibers with one or two centrally located nuclei; contains intercalated discs; involuntary control.

Location Heart wall.

Function Pumps blood to all parts of the body.

Longitudinal section of cardiac muscle tissue

Cardiac muscle fibers

visible under a light microscope. Skeletal muscle is *voluntary* because it can be made to contract or relax by conscious control. A single skeletal muscle fiber is very long, roughly cylindrical in shape, and has many nuclei located at the periphery of the cell. Within a whole muscle, the individual fibers are parallel to each other.

Cardiac muscle tissue forms the bulk of the wall of the heart (Table 3.4B). Like skeletal muscle, it is striated. However, unlike skeletal muscle tissue, it is *involuntary;* its contraction is not consciously controlled. Cardiac muscle fibers are branched and usually have only one centrally located nucleus, although an occasional cell has two nuclei. Cardiac muscle fibers attach end to end to each other by transverse thickenings of the plasma membrane called **intercalated discs** (*intercalat-* = to insert between), which contain both desmosomes and gap junctions. Intercalated discs are unique to cardiac muscle. The desmosomes strengthen the tissue and hold the fibers together during their vigorous contractions. The gap junctions provide a route for quick conduction of muscle action potentials throughout the heart.

C. Smooth Muscle Tissue

Description Spindle-shaped, nonstriated fibers with one centrally located nucleus; involuntary control.

Location Walls of hollow internal structures such as blood vessels, airways to the lungs, stomach, intestines, gallbladder, urinary bladder, and uterus.

Function Motion (constriction of blood vessels and airways, propulsion of foods through gastrointestinal tract, contraction of urinary bladder and gallbladder).

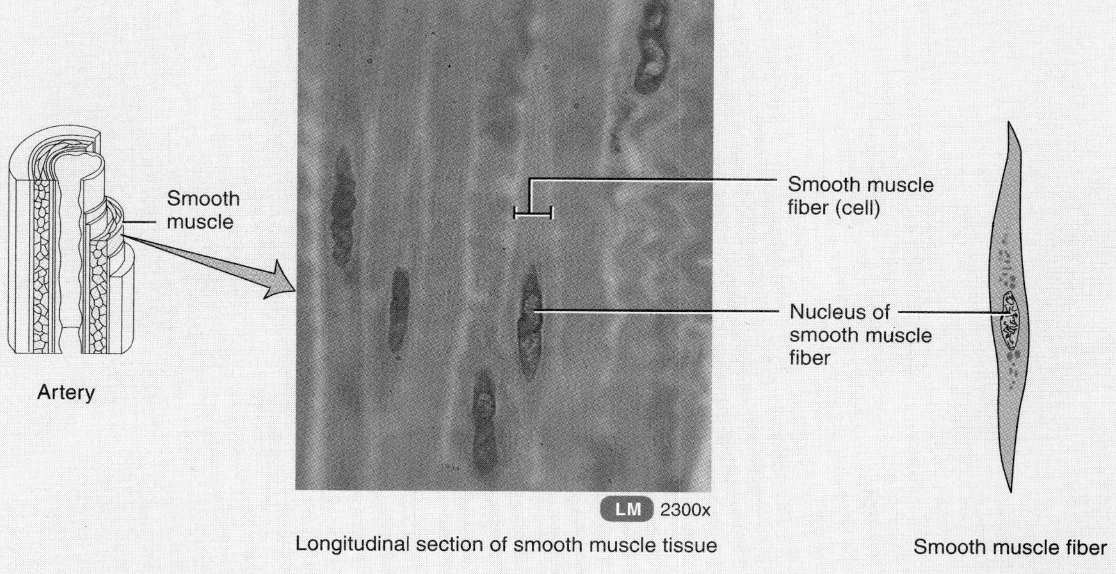

Longitudinal section of smooth muscle tissue

Smooth muscle fiber

Smooth muscle tissue is located in the walls of hollow internal structures such as blood vessels, airways to the lungs, the stomach, intestines, gallbladder, and urinary bladder (Table 3.4C). Its contraction helps constrict or narrow the lumen of blood vessels, physically break down and move food along the gastrointestinal tract, move fluids through the body, and eliminate wastes. Smooth muscle fibers are usually *involuntary*, and they are *nonstriated* (hence the term *smooth*). A smooth muscle fiber is small and spindle-shaped (thickest in the middle, with each end tapering) and contains a single, centrally located nucleus. In some smooth muscle tissue, such as in the wall of the intestines, many individual fibers are connected by gap junctions, which enable them to contract in synchrony. In other locations, such as the iris of the eye, smooth muscle fibers contract individually, like skeletal muscle fibers, because gap junctions are absent. Chapter 9 provides a detailed discussion of muscle tissue.

✓ Which muscles are striated and which are smooth?
✓ Which muscles have gap junctions?

Table 3.5 Nervous Tissue

Description Consists of neurons (nerve cells) and neuroglia. Neurons consist of a cell body and processes extending from the cell body, generally multiple dendrites and a single axon. Neuroglia do not generate or conduct nerve impulses but have other important functions.

Location Nervous system.

Function Exhibits sensitivity to various types of stimuli, converts stimuli into nerve impulses, and conducts nerve impulses to other neurons, muscle fibers, or glands.

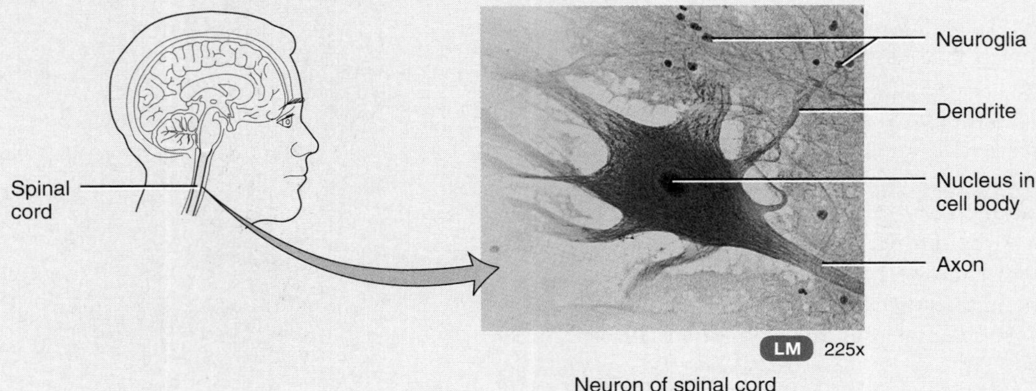

Neuron of spinal cord

NERVOUS TISSUE

Objective

- Describe the structure and functions of nervous tissue.

Despite the awesome complexity of the nervous system, it consists of only two principal kinds of cells: neurons and neuroglia. **Neurons,** or nerve cells, are sensitive to various stimuli. They convert stimuli into nerve impulses (action potentials) and conduct these impulses to other neurons, to muscle tissue, or to glands. Most neurons consist of a cell body and two kinds of cell processes—dendrites and axons (Table 3.5). The **cell body** contains the nucleus and other organelles. **Dendrites** (*dendr-* = tree) are tapering, highly branched, and usually short, cell processes. They are the major receiving or input portion of a neuron; they receive impulses from other neurons and conduct the impulses toward the cell body. The **axon** (*axo-* = axis) of a neuron is a single, thin, cylindrical process that may be very long. It is the output portion of a neuron, conducting nerve impulses away from the cell body and toward another neuron or to muscle fiber or glands.

Even though **neuroglia** (nū-RŌG-lē-a; *-glia* = glue) do not generate or conduct nerve impulses, these cells do have many important functions, such as supporting neurons (see Table 16.1 on page 504). The detailed structure and function of neurons and neuroglia are considered in Chapter 16.

In recent years a new technology called **tissue engineering** has emerged. Using a wide range of materials, scientists are learning to grow new tissues in the laboratory to replace damaged tissues in the body. Tissue engineers have already developed laboratory-grown versions of skin and cartilage. In the procedure, scaffolding beds of biodegradable synthetic materials or collagen are used as substrates that permit body cells such as skin cells or cartilage cells to be cultured. As the cells divide and assemble, the scaffolding degrades, and the new, permanent tissue is then implanted in the patient. Other structures being developed by tissue engineers include bones, tendons, heart valves, bone marrow, and intestines. Work is also underway to develop insulin-producing cells (pancreas) for diabetics, dopamine-producing cells (brain) for Parkinson's disease patients, and even entire livers and kidneys. ■

✓ Distinguish between the functions of neurons and neuroglia.

AGING AND TISSUES

Objective

- Explain the relationship of aging to tissues

Generally, tissues heal faster and leave less obvious scars in the young than in the aged. In fact, surgery performed on fetuses leaves no scars. The younger body is generally in a better nutritional state, its tissues have a better blood supply, and its cells have a higher metabolic rate. Thus, cells can synthesize needed materials and divide more quickly. The extracellular components of tissues also change with age. Collagen fibers, responsible for the strength of tendons, increase in number with aging. Changes in the collagen of arterial walls are as much responsible for their loss of extensibility as are the deposits associated with atherosclerosis, the deposition of fatty materials in arterial walls. Elastin, another extracellular component, is responsible for the elasticity of blood vessels and skin. It thickens, fragments, and acquires a greater affinity for calcium with age—changes that may also be associated with the development of atherosclerosis.

✓ What changes occur in tissues with aging?

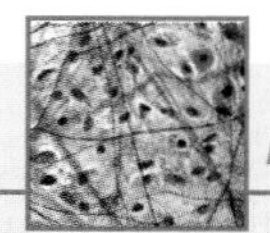

APPLICATIONS TO HEALTH

Systemic Lupus Erythematosus

Systemic lupus erythematosus (er-ē-the-ma-TŌ-sus), **SLE,** or simply lupus, is a chronic inflammatory disease of connective tissue occurring mostly in nonwhite women during their childbearing years. It is an autoimmune disease that can cause tissue damage in every body system. An **autoimmune disease** is one in which antibodies produced by the immune system fail to distinguish what is foreign from what is self and attack the body's own tissues. SLE, which can range from a mild condition in most patients to a rapidly fatal disease, is marked by periods of exacerbation and remission. The prevalence of SLE is about 1 in 2000 persons, with females more likely to be afflicted than males by a ratio of 8 : 1 or 9 : 1.

Although the cause of SLE is not known, genetic, environmental, and hormonal factors are implicated. The genetic component is suggested by studies of twins and family history. Environmental factors include viruses, bacteria, chemicals, drugs, exposure to excessive sunlight, and emotional stress. With regard to hormones, sex hormones, such as estrogens, may trigger SLE.

Signs and symptoms of SLE include painful joints, low-grade fever, fatigue, mouth ulcers, weight loss, enlarged lymph nodes and spleen, sensitivity to sunlight, rapid loss of large amounts of scalp hair, and anorexia. A distinguishing feature of lupus is an eruption across the bridge of the nose and cheeks called a "butterfly rash." Other skin lesions may occur, including blistering and ulceration. The erosive nature of some SLE skin lesions was thought to resemble the damage inflicted by the bite of a wolf—thus, the term *lupus* (= wolf). The most serious complications of the disease involve inflammation of the kidneys, liver, spleen, lungs, heart, brain, and gastrointestinal tract. Because there is no cure for SLE, treatment is supportive, including anti-inflammatory drugs, such as aspirin, and immunosuppressive drugs.

KEY MEDICAL TERMS ASSOCIATED WITH TISSUES

Atrophy (AT-rō-fē; *a-* = without; *-trophy* = nourishment) A decrease in the size of cells, with a subsequent decrease in the size of the affected tissue or organ.

Hypertrophy (hī-PER-trō-fē; *hyper-* = above or excessive) Increase in the size of a tissue because its cells enlarge without undergoing cell division.

Sjögren's syndrome An autoimmune disorder that causes inflammation and dysfunction of exocrine glands, especially the lacrimal (tear) glands and salivary glands. It is characterized by dry eyes and mouth, and salivary gland enlargement; systemic effects include arthritis, difficulty in swallowing, pancreatitis, pleuritis, and migraine; the disorder affects more females than males by a ratio of 9 : 1; individuals with the condition have a greatly increased risk of developing malignant lymphoma.

Tissue rejection An immune response of the body directed at foreign proteins in a transplanted tissue or organ; immunosuppressive drugs, such as cyclosporine, have largely overcome tissue rejection in heart-, kidney-, and liver-transplant patients.

Tissue transplantation The replacement of a diseased or injured tissue or organ. The most successful transplants involve use of a person's own tissues or those from an identical twin; less-successful transplants involve less-compatible tissues from a donor.

Xenotransplantation (zen′ō-trans-plan-TĀ-shun; *xeno-* = strange, foreign) The replacement of a diseased or injured tissue or organ with cells or tissues from an animal. Only a few cases of successful xenotransplantation exist to date.

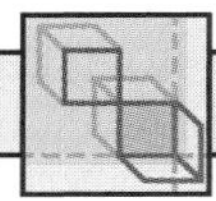

STUDY OUTLINE

Types of Tissues and Their Origins (p. 56)

1. A tissue is a group of similar cells that usually has a similar embryological origin and is specialized for a particular function.
2. The various tissues of the body are classified into four basic types: epithelial, connective, muscle, and nervous.
3. All tissues of the body develop from three primary germ layers, the first tissues that form in a human embryo: ectoderm, endoderm, and mesoderm.

Cell Junctions (p. 56)

1. Cell junctions are points of contact between plasma membranes of adjacent cells or between cells and extracellular material.
2. Types of cell junctions include (a) tight junctions, which prevent the passage of molecules between cells; (b) plaque-bearing junctions (cell-to-cell adherens junctions, cell-to-extracellular material junctions or focal adhesions, desmosomes, and hemidesmosomes, which attach cells to each other or to extracellular materials); and (c) gap junctions, which permit electrical or chemical signals to pass between cells.

Epithelial Tissue (p. 58)

1. The subtypes of epithelia include covering and lining epithelia and glandular epithelia.
2. An epithelium consists mostly of cells with little extracellular material between adjacent plasma membranes. The lateral, apical, and basal surfaces of epithelial cells are modified in various ways to carry out specific functions. Epithelium is arranged in sheets and attached to a basement membrane. Although it is avascular, it has a nerve supply. Epithelia are derived from all three primary germ layers and have a high capacity for renewal.

Covering and Lining Epithelium (p. 59)

1. Epithelial layers are simple (one layer), stratified (several layers), or pseudostratified (one layer that appears to be several); the cell

shapes may be squamous (flat), cuboidal (cubelike), columnar (rectangular), or transitional (variable).
2. Simple squamous epithelium consists of a single layer of flat cells (Table 3.1A). It is found in parts of the body where filtration or diffusion are priority processes. One type, endothelium, lines the heart and blood vessels. Another type, mesothelium, forms the serous membranes that both line the thoracic and abdominal cavities and cover the organs within them.
3. Simple cuboidal epithelium consists of a single layer of cube-shaped cells that function in secretion and absorption (Table 3.1B). It is found covering the ovaries, in the kidneys and eyes, and lining some glandular ducts.
4. Nonciliated simple columnar epithelium consists of a single layer of nonciliated rectangular cells (Table 3.1C). It lines most of the gastrointestinal tract. Specialized cells containing microvilli perform absorption. Goblet cells secrete mucus.
5. Ciliated simple columnar epithelium consists of a single layer of ciliated rectangular cells (Table 3.1D). It is found in a few portions of the upper respiratory tract, where it moves foreign particles trapped in mucus out of the respiratory tract.
6. Stratified squamous epithelium consists of several layers of cells; cells at the apical surface are flat (Table 3.1E). It is protective. A nonkeratinized variety lines the mouth; a keratinized variety forms the epidermis, the most superficial layer of the skin.
7. Stratified cuboidal epithelium consists of several layers of cells; cells at the apical surface are cube-shaped (Table 3.1F). It is found in adult sweat glands and a portion of the male urethra.
8. Stratified columnar epithelium consists of several layers of cells; cells at the apical surface are rectangular (Table 3.1G). It is found in a portion of the male urethra and large excretory ducts of some glands.
9. Transitional epithelium consists of several layers of cells whose appearance varies with the degree of stretching (Table 3.1H). It lines the urinary bladder.
10. Pseudostratified columnar epithelium has only one layer but gives the appearance of many (Table 3.1I). A ciliated variety contains goblet cells and lines most of the upper respiratory tract; a nonciliated variety has no goblet cells and lines ducts of many glands, the epididymis, and part of the male urethra.

Glandular Epithelium (p. 67)

1. A gland consists of one or a group of epithelial cells adapted for secretion.
2. Endocrine glands secrete hormones into the blood after passing through interstitial fluid (Table 3.1J).
3. Exocrine glands (mucous, sweat, oil, and digestive glands) secrete into ducts or directly onto a free surface (Table 3.1K).
4. Structural classification of exocrine glands includes unicellular and multicellular glands.
5. Functional classification of exocrine glands includes holocrine, apocrine, and merocrine glands.

Connective Tissue (p. 69)

1. Connective tissue is one of the most abundant tissues in the body.
2. Connective tissue consists of relatively few cells and an abundant matrix of ground substance and fibers. It does not usually occur on free surfaces, has a nerve supply (except for cartilage), and is highly vascular (except for cartilage, tendons, and ligaments).

Components of Connective Tissue (p. 69)

1. Cells in connective tissue are derived from mesenchymal cells.
2. Cell types include fibroblasts (secrete matrix), macrophages (perform phagocytosis), plasma cells (secrete antibodies), mast cells (produce histamine), adipocytes (store fat), and white blood cells (migrate from blood in response to infections).
3. The ground substance and fibers make up the matrix.
4. The ground substance supports and binds cells together, provides a medium for the exchange of materials, and is active in influencing cell functions.
5. Substances found in the ground substance include water and polysaccharides such as hyaluronic acid, chondroitin sulfate, dermatan sulfate, and keratan sulfate (glycosaminoglycans). Also present are proteoglycans and adhesion proteins.
6. The fibers in the matrix provide strength and support and are of three types: (a) collagen fibers (composed of collagen) are found in large amounts in bone, tendons, and ligaments; (b) elastic fibers (composed of elastin, fibrillin, and other glycoproteins) are found in skin, blood vessel walls, and lungs; and (c) reticular fibers (composed of collagen and glycoprotein) are found around fat cells, nerve fibers, and skeletal and smooth muscle cells.

Classification of Connective Tissue (p. 71)

1. The two major subclasses of connective tissue are embryonic connective tissue and mature connective tissue.
2. Mesenchyme is the embryonic connective tissue that forms all mature connective tissues (Table 3.2A).
3. Mucous connective tissue is the embryonic connective tissue that is found in the umbilical cord of the fetus, where it gives support (Table 3.2B).

Types of Mature Connective Tissue (p. 72)

1. Mature connective tissue differentiates from mesenchyme and is present in the newborn. It is subdivided into: loose connective tissue, dense connective tissue, cartilage, bone tissue, blood tissue, and lymph.
2. Loose connective tissue includes areolar connective tissue, adipose tissue, and reticular connective tissue.
3. Areolar connective tissue consists of the three types of fibers, several cells, and a semifluid ground substance (Table 3.3A). It is found in the subcutaneous layer, in mucous membranes, and around blood vessels, nerves, and body organs.
4. Adipose tissue consists of adipocytes, which store triglycerides (Table 3.3B). It is found in the subcutaneous layer, around organs, and in the yellow bone marrow of long bones.
5. Reticular connective tissue consists of reticular fibers and reticular cells and is found in the liver, spleen, and lymph nodes (Table 3.3C).
6. Dense connective tissue includes dense regular connective tissue, dense irregular connective tissue, and elastic connective tissue.
7. Dense regular connective tissue consists of parallel bundles of collagen fibers and fibroblasts (Table 3.3D). It forms tendons, most ligaments, and aponeuroses.
8. Dense irregular connective tissue usually has randomly arranged collagen fibers and a few fibroblasts (Table 3.3E). It is found in fasciae, the dermis of skin, and membrane capsules around organs.
9. Elastic connective tissue consists of branching elastic fibers and fibroblasts (Table 3.3F). It is found in the walls of large arteries and in the lungs, trachea, and bronchial tubes.
10. Cartilage contains chondrocytes and has a rubbery matrix (chondroitin sulfate) containing collagen and elastic fibers.
11. Hyaline cartilage is found in the embryonic skeleton, at the ends of bones, in the nose, and in respiratory structures (Table 3.3G). It is flexible, allows movement, and provides support.

12. Fibrocartilage is found in the pubic symphysis, intervertebral discs, and menisci (cartilage pads) of the knee joint (Table 3.3H).
13. Elastic cartilage maintains the shape of organs such as the epiglottis, auditory (Eustachian) tubes, and external ear (Table 3.3I).
14. Bone, or osseous tissue, consists of mineral salts and collagen fibers that contribute to its hardness, and cells called osteocytes that are located in lacunae (Table 3.3J). It supports, protects, functions in movement, stores minerals, and houses blood-forming tissue.
15. Blood tissue consists of plasma and red blood cells, white blood cells, and platelets (Table 3.3K). Its cells function to transport oxygen and carbon dioxide, carry on phagocytosis, participate in allergic reactions, provide immunity, and bring about blood clotting.
16. Lymph flows in lymphatic vessels and is a clear fluid whose composition varies from one part of the body to another.

Membranes (p. 81)

1. An epithelial membrane consists of an epithelial layer overlying a connective tissue layer. Examples are mucous, serous, and cutaneous membranes.
2. Mucous membranes line cavities that open to the exterior, such as the gastrointestinal tract.
3. Serous membranes line closed cavities (pleura, pericardium, peritoneum) and cover the organs in the cavities. These membranes consist of parietal and visceral layers.
4. The cutaneous membrane is the skin.
5. Synovial membranes line joint cavities, bursae, and tendon sheaths and consist of areolar connective tissue instead of epithelium.

Muscle Tissue (p. 81)

1. Muscle tissue consists of elongated fibers (cells) that are specialized for contraction. It provides motion, maintenance of posture, and heat production.
2. Skeletal muscle tissue is attached to bones and is striated; its action is voluntary (Table 3.4A).
3. Cardiac muscle tissue forms most of the heart wall and is striated; its action is involuntary (Table 3.4B).
4. Smooth muscle tissue is found in the walls of hollow internal structures (blood vessels and viscera) and is nonstriated; its action is involuntary (Table 3.4C).

Nervous Tissue (p. 84)

1. The nervous system is composed of neurons (nerve cells) and neuroglia (Table 3.5).
2. Neurons are sensitive to stimuli, convert stimuli into nerve impulses, and conduct nerve impulses.
3. Most neurons consist of a cell body and two types of processes called dendrites and axons.

Aging and Tissues (p. 84)

1. Tissues heal faster and leave less obvious scars in the young than in the aged; surgery performed on fetuses leaves no scars.
2. The extracellular components of tissues, such as collagen and elastic fibers, change with age.

SELF-QUIZ QUESTIONS

Choose the one best answer to the following questions:

1. A group of similar cells that has the same embryological origin and operates together to perform a specialized activity is called a(n) (a) organ, (b) tissue, (c) system, (d) organelle, (e) organism.
2. Which statement best describes covering and lining epithelium? (a) it is always arranged in a single layer of cells, (b) it contains large amounts of intercellular substance, (c) it has an abundant blood supply, (d) its free surface is exposed to the exterior of the body or to the interior of a hollow structure, (e) its cells are widely scattered.
3. A basement membrane is classified as a(n) (a) epithelial membrane, (b) plasma membrane, (c) mucous membrane, (d) synovial membrane, (e) none of the above.
4. An abdominal incision will pass through the skin, then subcutaneous tissue, then muscle, to reach which membrane lining the inside wall of the abdominal cavity? (a) visceral pericardium, (b) parietal pericardium, (c) visceral peritoneum, (d) parietal peritoneum, (e) parietal pleura.
5. Which tissue is characterized by the presence of cell bodies, dendrites, and axons? (a) muscle, (b) nervous, (c) vascular, (d) epithelial, (e) connective.
6. Which of the following structures forms a secretory product and discharges it by exocytosis? (a) holocrine gland, (b) connexon, (c) apocrine gland, (d) basal lamina, (e) merocrine gland.
7. Which statement best describes connective tissue? (a) usually contains a large amount of matrix, (b) always arranged in a single layer of cells, (c) primarily concerned with secretion, (d) cells are always very closely packed together, (e) is avascular.
8. Which of the following is involuntary and striated? (a) skeletal muscle tissue, (b) cardiac muscle tissue, (c) smooth muscle tissue, (d) none of the above, (e) both a and b.

Complete the following:

9. The muscle tissue that forms most of the wall of the heart is ________ muscle tissue.
10. A gland that secretes its product into a duct is referred to as a(n) ________ gland.
11. Bone tissue is known as ________ tissue. Compact bone consists of concentric rings called ________ .
12. Blood consists of a fluid called ________ containing three types of formed elements.
13. Hollow organs belonging to systems that do not open to the outside of the body are lined with a ________ membrane.
14. The connective tissue layer of a mucous membrane is called the ________ .

Are the following statements true or false?

15. A mucus-producing goblet cell is an example of a unicellular exocrine gland.
16. Tight junctions are common between epithelial cells lining the stomach or urinary bladder where they prevent fluid from passing between the lining cells.
17. Dense irregular connective tissue contains a parallel arrangement of bundles of collagen fibers and is located in tendons and ligaments.

18. Desmosomes connect lateral surfaces of adjacent cells to one another; hemidesmosomes connect the basal surface of a cell to a basement membrane.

19. Match the following:

____ **(a)** a mature bone cell; resides in a space called a lacuna
____ **(b)** large phagocytic cell derived from a monocyte; engulfs bacteria and cleans up debris; wandering or fixed
____ **(c)** derived from a fibroblast; specialized for triglyceride storage
____ **(d)** a red blood cell; functions in oxygen transport
____ **(e)** large, branched cell; secretes the matrix of connective tissues
____ **(f)** located along the walls of blood vessels; secrete histamine which dilates blood vessels during inflammation
____ **(g)** cartilage cell; resides in a space called a lacuna
____ **(h)** derived from a B lymphocyte; produces antibodies as part of an immune response

(1) fibroblast
(2) mast cell
(3) plasma cell
(4) osteocyte
(5) adipocyte
(6) chondrocyte
(7) macrophage
(8) erythrocyte

20. Match the following:

____ **(a)** lines inner surface of the stomach and intestine
____ **(b)** avascular connective tissue that occurs as three main types
____ **(c)** lines urinary bladder, ureters, and urethra
____ **(d)** forms fasciae and dermis of skin
____ **(e)** has cells that are specialized for triglyceride storage
____ **(f)** all cells attached to the basement membrane, but not all cells reach the surface of tissue
____ **(g)** subcutaneous tissue containing several kinds of cells and all three fiber types
____ **(h)** lines the mouth; present on the outer surface of the skin
____ **(i)** lines air sacs of the lungs; suited for diffusion of gases
____ **(j)** strong tissue found in the lungs; can recoil back to its original shape after being stretched

(1) dense irregular connective tissue
(2) adipose tissue
(3) simple columnar epithelium
(4) stratified squamous epithelium
(5) cartilage
(6) areolar connective tissue
(7) transitional epithelium
(8) pseudostratified columnar epithelium
(9) elastic connective tissue
(10) simple squamous epithelium

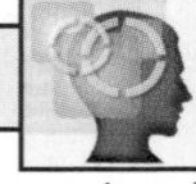

CRITICAL THINKING QUESTIONS

1. Noella experienced a burning pain while urinating and she could see blood in her urine. The doctor diagnosed an infection in her urinary bladder. Why did Noella have these symptoms?

HINT *Think about the urinary bladder's function to get a clue about its structure.*

2. You've gone out to eat at your favorite fast food joint: General Bellisimo's Fried Chicken. A health food zealot waves a chicken drumstick and declares "This is all fat!" Using your knowledge of tissues, defend the contents of the chicken leg.

HINT *Before it was a drumstick, it was a chicken's leg.*

3. Besides giving children something to do with their fingers, the mucous membrane of the nasal cavity has several important functions. Name some of these functions.

HINT *The nose is also part of the respiratory system.*

4. The neighborhood kids are walking around with common pins and sewing needles stuck through their fingertips. There is no visible bleeding. What type of tissue have they pierced? How do you know?

HINT *If you pierced all the way through your earlobe it would bleed.*

5. Imagine that you live 50 years in the future and you can custom-design a human to suit the environment. Your assignment is to customize the tissue makeup for life on a large planet with strong gravity, a cold climate, and a thin atmosphere. What adaptations would you incorporate into the structure of your custom human? Why?

HINT *Think of the adaptations in animals and people that live in the Artic versus those at the equator?*

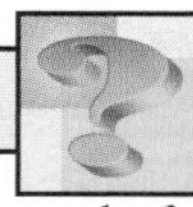

ANSWERS TO FIGURE QUESTIONS

3.1 Gap junctions allow the spread of electrical and chemical signals between adjacent cells.

3.2 The basement membrane provides physical support for epithelium, serves as a filter in the kidneys, and guides cell migration during development and tissue repair.

3.3 Simple multicellular exocrine glands have a nonbranched duct; compound multicellular exocrine glands have a branched duct.

3.4 Sebaceous (oil) glands are holocrine glands, and salivary glands are merocrine glands.

3.5 Fibroblasts secrete the fibers and ground substance of the matrix.

4

THE INTEGUMENTARY SYSTEM

CONTENTS AT A GLANCE

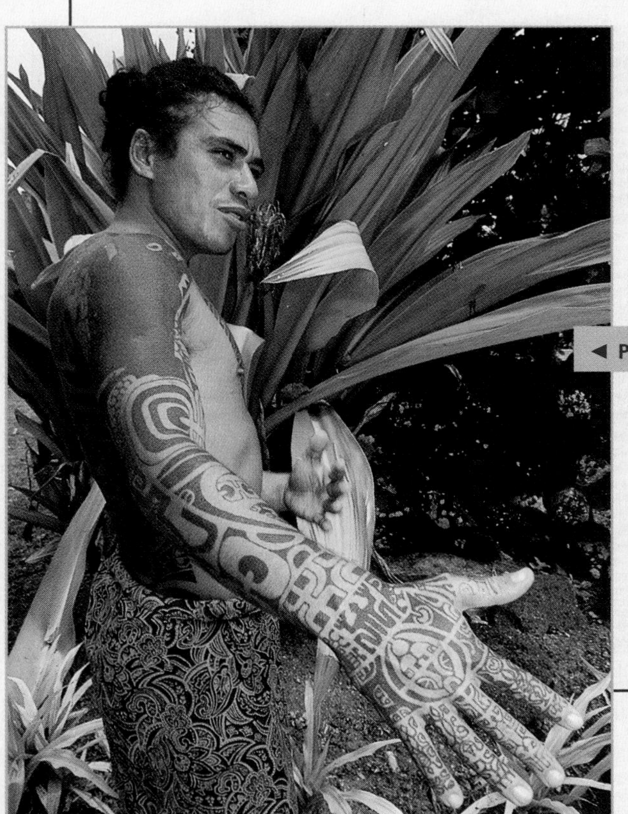

◀ Page 95

Has someone you know used the skin on their body as a canvas? Have you? What do you think about the role of tatoos as an expression of culture?

Page 92 ▶

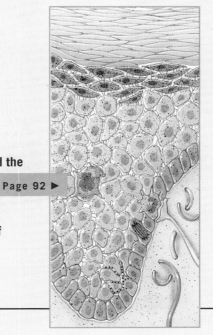

INTRODUCTION

Of all the body's organs, none is more easily inspected or more exposed to infection, disease, and injury than the skin. Because of its visibility, skin reflects our emotions and some aspects of normal physiology, as evidenced by frowning, blushing, and sweating. Changes in skin color may indicate disorders in the body; for example, a bluish skin color is one sign of heart failure. Abnormal skin eruptions or rashes such as chicken pox, cold sores, or measles may reveal systemic infections or diseases of internal organs. Other disorders may involve just the skin itself, such as warts, age spots, or pimples. The skin's location makes it vulnerable to damage from trauma, sunlight, microbes, and pollutants in the environment.

Skin and its accessory structures—hair and nails, along with various glands, muscles, and nerves—make up the **integumentary system** (in-teg-yū-MEN-tar-ē; (*inte-* = whole; *-gument* =

Figure 4.1 / Components of the integumentary system. The skin consists of a thin, superficial epidermis and a deep, thicker dermis. Deep to the skin is the subcutaneous layer, which attaches the dermis to underlying organs and tissues.

The integumentary system includes the skin and its accessory structures—hair, nails, and skin glands—along with associated muscles and nerves.

Functions

1. Body temperature regulation.
2. Reservoir for blood.
3. Protection from external environment.
4. Cutaneous sensations.
5. Excretion and absorption.
6. Vitamin D synthesis.

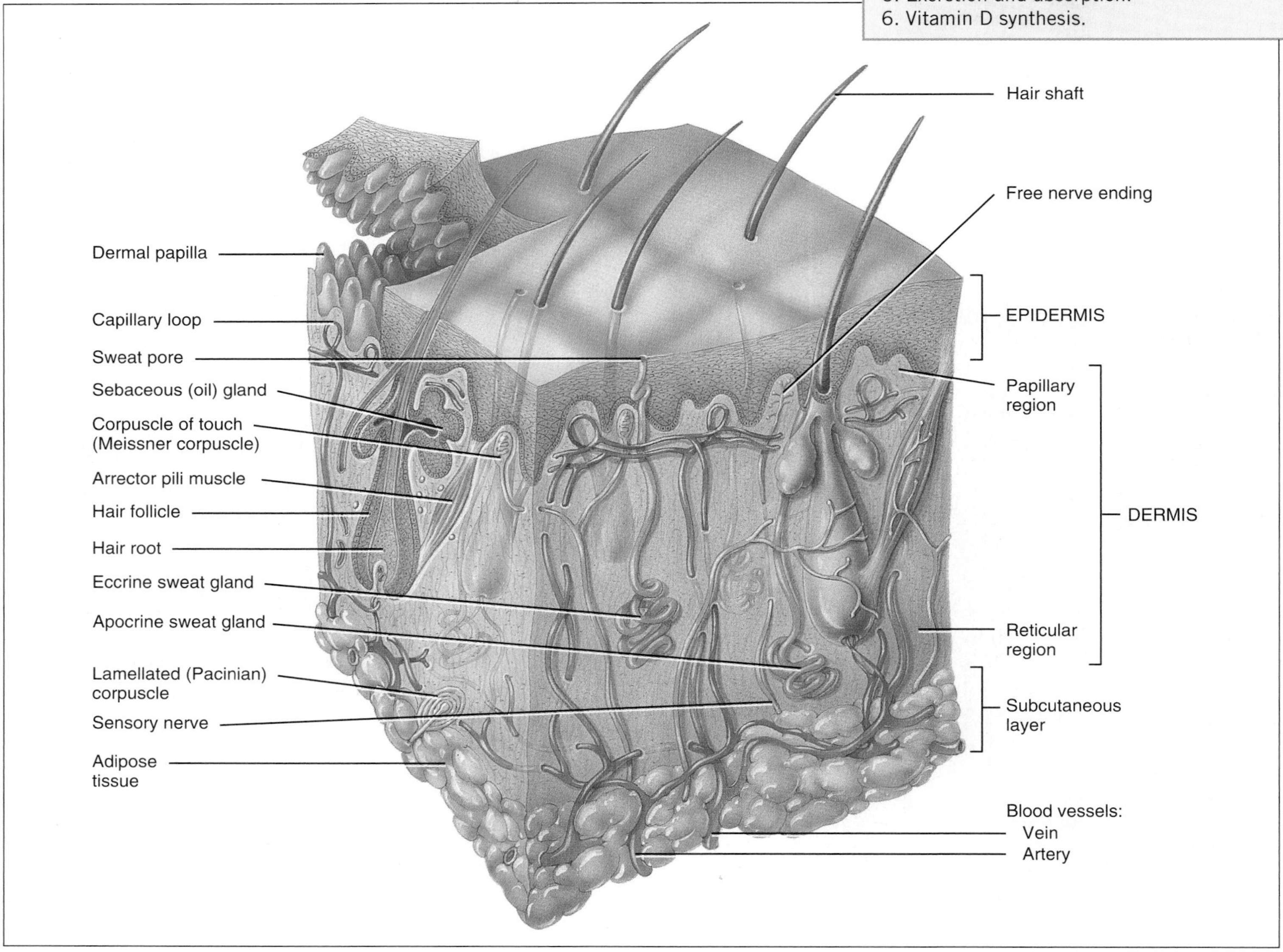

Dermal papilla
Capillary loop
Sweat pore
Sebaceous (oil) gland
Corpuscle of touch (Meissner corpuscle)
Arrector pili muscle
Hair follicle
Hair root
Eccrine sweat gland
Apocrine sweat gland
Lamellated (Pacinian) corpuscle
Sensory nerve
Adipose tissue
Hair shaft
Free nerve ending
EPIDERMIS
Papillary region
DERMIS
Reticular region
Subcutaneous layer
Blood vessels:
Vein
Artery

What kind of tissues make up the epidermis and the dermis?

body covering). The integumentary system guards the body's physical and biochemical integrity, maintains a constant body temperature, and provides sensory information about the surrounding environment. This chapter discusses the body's most visible system, one that often defines our very self-image.

STRUCTURE OF THE SKIN

Objective

- Describe the various layers of the epidermis and dermis, and the cells that compose them.

The **skin** consists of different tissues that are joined to perform specific functions. It is the largest organ of the body in surface area and weight. In adults, the skin covers an area of about 2 square meters (22 square feet) and weighs 4.5–5 kg (10–11 lb), about 16% of total body weight. It ranges in thickness from 0.5 mm (.02 in) on the eyelids to 4.0 mm (.16 in) on the heels. However, over most of the body it is 1–2 mm (.04–.08 in) thick. **Dermatology** (der′-ma-TOL-ō-jē; *dermat-* = skin; *-ology* = study of) is the branch of medicine that specializes in the diagnosis and treatment of skin disorders.

Structurally, the skin consists of two principal parts (Figure 4.1). The superficial, thinner portion, which is composed of *epithelial tissue,* is the **epidermis** (ep′-i-DERM-is; *epi-* = above). The deeper, thicker, *connective tissue* part is the **dermis.**

Deep to the dermis, and not part of the skin, is the **subcutaneous (subQ) layer,** or **hypodermis** (*hypo-* = below). This layer consists of areolar and adipose tissues. Fibers that extend from the dermis anchor the skin to the subcutaneous layer, which, in turn, attaches to underlying tissues and organs. The subcutaneous layer serves as a storage depot for fat and contains large blood vessels that supply the skin. This region (and sometimes the dermis) also contains nerve endings called **lamellated (Pacinian) corpuscles** (pa-SIN-ē-an) that are sensitive to pressure (Figure 4.1).

Epidermis

The **epidermis** is keratinized stratified squamous epithelium that contains four principal types of cells: the cell types include keratinocytes, melanocytes, Langerhans cells, and Merkel cells (Figure 4.2). About 90% of epidermal cells are **keratinocytes** (ker-a-TIN-ō-sīts; *keratino-* = hornlike; *-cytes* = cells) (Figure 4.2a) which are arranged in four or five layers. These cells produce **keratin,** a tough, fibrous protein that helps protect the skin and underlying tissues from heat, microbes, and chemicals. Keratinocytes also produce lamellar granules, which release a waterproofing sealant.

About 8% of the epidermal cells are **melanocytes** (MEL-a-nō-sīts; *melano-* = black), which develop from neural crest cells (ectoderm) of a developing embryo. Melanocytes (Figure 4.2b) produce the pigment melanin. Their long, slender projections extend between the keratinocytes and transfer melanin granules to them. **Melanin** is a brown-black pigment that contributes to skin color and absorbs damaging ultraviolet (UV) light. Once inside keratinocytes, the melanin granules cluster to form a protective veil over the nucleus on the side toward the skin surface. In this way they shield the nuclear DNA from UV light.

Figure 4.2 / Types of cells in the epidermis. Besides keratinocytes, the epidermis contains melanocytes, which produce the pigment melanin; Langerhans cells, which participate in immune responses; and Merkel cells, which function in the sensation of touch.

Most of the epidermis consists of keratinocytes, which produce the protein keratin (protects underlying tissues) and lamellar granules (contains a waterproof sealant).

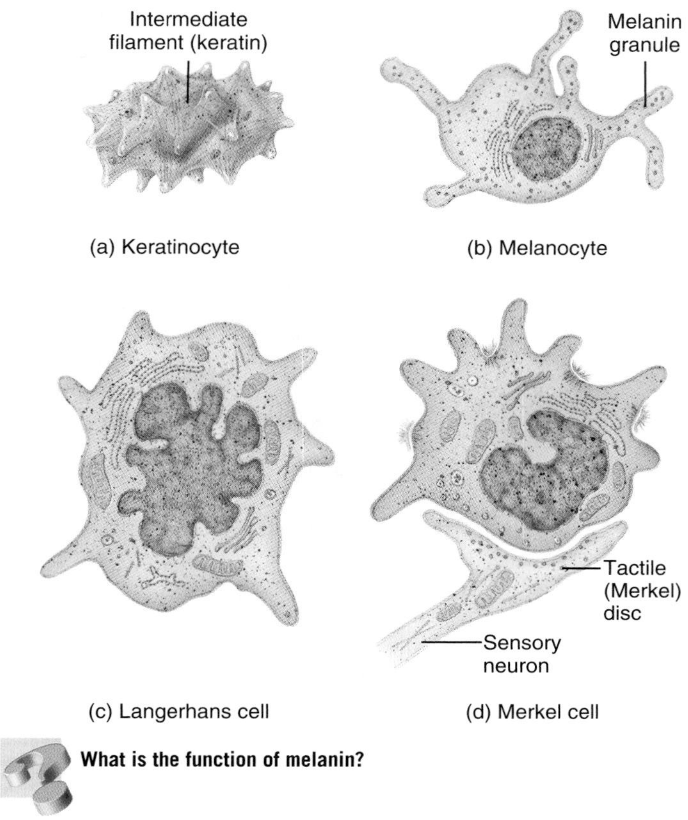

What is the function of melanin?

Langerhans cells (LANG-er-hans) arise from red bone marrow and migrate to the epidermis (Figure 4.2c), where they constitute a small proportion of the epidermal cells. They participate in immune responses mounted against microbes that invade the skin, and they are easily damaged by UV light.

Merkel cells are the least numerous of the epidermal cells. They are located in the deepest layer of the epidermis, where they contact the flattened process of a sensory neuron (nerve cell), a structure called a **tactile (Merkel) disc** (Figure 4.2d). Merkel cells and tactile discs function in the sensation of touch.

Several distinct layers of keratinocytes in various stages of development form the epidermis (Figure 4.3). In most regions of the body the epidermis has four strata or layers—stratum basale, stratum spinosum, stratum granulosum, and a thin stratum corneum. Where exposure to friction is greatest, as in the fingertips, palms, and soles, the epidermis has five layers—stra-

Figure 4.3 / Layers of the epidermis.

The epidermis consists of keratinized stratified squamous epithelium.

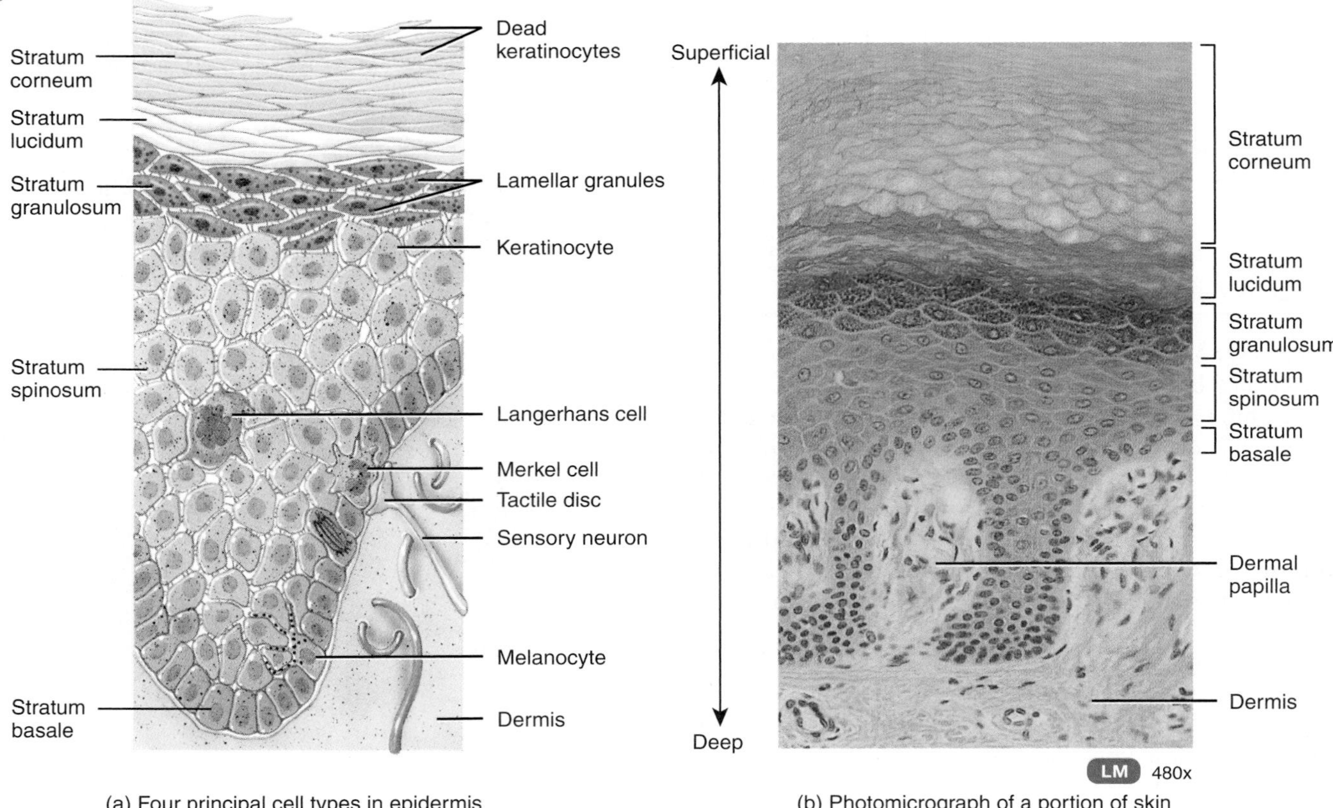

(a) Four principal cell types in epidermis

(b) Photomicrograph of a portion of skin

Which epidermal layer includes stem cells that continually undergo cell division?

tum basale, stratum spinosum, stratum granulosum, stratum lucidum, and a thick stratum corneum.

Stratum Basale

The deepest layer of the epidermis, the **stratum basale** (*basal-* = base), is composed of a single row of cuboidal or columnar keratinocytes, some of which are *stem cells* that undergo cell division to continually produce new keratinocytes. The nuclei of these keratinocytes are large, and their cytoplasm contains many ribosomes, a small Golgi complex, a few mitochondria, and some rough endoplasmic reticulum. The cytoskeleton within keratinocytes of the stratum basale includes scattered intermediate filaments called *tonofilaments*. Tonofilaments are composed of a protein that will later form keratin in the more superficial epidermal layers. Tonofilaments attach to desmosomes, which bind cells of the stratum basale to each other and to the cells of the adjacent stratum spinosum, and to hemidesmosomes, which bind the keratinocytes to the basement membrane between the epidermis and the dermis. Keratin also protects deeper layers from injury. Melanocytes, Langerhans cells, and Merkel cells with their associated tactile discs are scattered among the keratinocytes of the basal layer. The stratum basale is sometimes referred to as the **stratum germinativum** (jer′-mi-na-TĒ-vum; *germ-* = sprout) to indicate its role in forming new cells.

New skin cannot regenerate if an injury destroys the stratum basale and its stem cells. When 20 percent or more of the skin is badly damaged, there is serious fluid and electrolyte loss, infection from microbes, and inability to maintain normal body temperature. When skin wounds involve destruction of the stratum basale and its stem cells they require skin grafts in order to heal. A **skin graft** involves covering the wound with a patch of healthy skin taken from a donor site. To avoid tissue rejection, the transplanted skin is usually taken from the same individual *(autograft)* or an identical twin *(isograft)*. If skin damage is so extensive that an autograft would cause harm, a self-donation procedure called *autologous skin transplantation* (aw-TOL-ō-gus) may be used. In this procedure, performed most often for severely burned patients, small amounts of an individual's epidermis are removed, and the keratinocytes are cultured in the laboratory to produce thin sheets of skin. The new skin is transplanted back to the patient so that it covers the burn wound and generates a permanent skin. ■

Stratum Spinosum

Superficial to the stratum basale is the **stratum spinosum** (*spinos-* = thornlike), where 8–10 layers of polyhedral (many-sided) keratinocytes fit closely together. Cells in the more superficial portions of this layer become somewhat flattened. These keratinocytes have the same organelles as cells of the stratum basale, and some cells in this layer retain their ability to divide. When cells of the stratum spinosum are prepared for microscopic examination, they shrink and pull apart such that they appear to be covered with thornlike spines (Figure 4.3a), although they are rounded and larger in living tissue. At each spinelike projection, bundles of tonofilaments of the cytoskeleton, insert into desmosomes, which tightly join the cells to one another. This arrangement provides both strength and flexibility to the skin. Projections of both Langerhans cells and melanocytes also appear in this stratum.

Stratum Granulosum

At the middle of the epidermis, the **stratum granulosum** (*granulos-* = little grains) consists of three to five layers of flattened keratinocytes that are undergoing apoptosis. The nuclei and other organelles of these cells begin to degenerate, and tonofilaments become more apparent. A distinctive feature of cells in this layer is the presence of darkly staining granules of a protein called **keratohyalin** (ker′-a-tō-HĪ-a-lin), which converts the tonofilaments into keratin. Also present in the keratinocytes are membrane-enclosed **lamellar granules,** which release a lipid-rich secretion that fills the spaces between cells of the stratum granulosum and between more superficial cells of the epidermis. This secretion functions as a water-repellent sealant that retards loss of body fluids and entry of foreign materials. As their nuclei break down, the keratinocytes of the stratum granulosum can no longer carry on vital metabolic reactions, and they die. Thus, the stratum granulosum marks the transition between the deeper, metabolically active strata and the dead cells of the more superficial strata.

Stratum Lucidum

The **stratum lucidum** (*lucid-* = clear) is present only in the thick skin of the fingertips, palms, and soles. It consists of three to five layers of clear, flat, dead keratinocytes that contain large amounts of keratin and thickened plasma membranes.

Stratum Corneum

The **stratum corneum** (*corne-* = hard or hooflike) consists of 25–30 layers of dead, flat keratinocytes. These cells are continuously shed and replaced by cells from the deeper strata. The interior of the cells contains mostly keratin. Between the cells are lipids from lamellar granules. The stratum corneum thus protects against injury and microbes and serves as an effective water-repellent barrier. Constant exposure of skin to friction stimulates the formation of a *callus,* an abnormal thickening of the epidermis.

Keratinization and Growth of the Epidermis

Newly formed cells in the stratum basale are pushed slowly to the surface. As the cells move from one epidermal layer to the next, they accumulate more and more keratin a process called **keratinization** (ker-a-tin-i-ZĀ-shun). Then they undergo apoptosis—the nucleus fragments, other organelles disappear, and the cells die. Eventually the keratinized cells slough off and are replaced by underlying cells that, in turn, become keratinized. The whole process by which cells form in the stratum basale, rise to the surface, become keratinized, and slough off takes about four weeks in an average epidermis of 0.1 mm (.04 in) thickness. The rate of cell division in the stratum basale increases when the outer layers of the epidermis are stripped away, as occurs in abrasions and burns. The mechanisms that regulate this remarkable growth are not well understood, but hormonelike proteins such as **epidermal growth factor (EGF)** play a role.

Table 4.1 presents a summary of the distinctive features of the epidermal strata.

Psoriasis is a common and chronic skin disorder in which keratinocytes divide and move more quickly than normal from the stratum basale to the stratum corneum. They are shed prematurely in as little as 7–10 days. The immature keratinocytes make an abnormal keratin that forms flaky, silvery scales at the skin surface, most often on the knees, elbows, and scalp (dandruff). Effective treatments—various topical ointments and UV phototherapy—suppress cell division, decrease the rate of cell growth, or inhibit keratinization. ■

Dermis

The second, deeper part of the skin, the **dermis,** is composed mainly of connective tissue containing collagen and elastic

Table 4.1 Comparison of Epidermal Strata

Stratum	Description
Basale (germinativum)	Deepest layer, composed of a single row of cuboidal or columnar keratinocytes that contain scatters tonofilaments (intermediate filaments); stem cells undergo cell division to produce new keratinocytes; melanocytes, Langerhans cells, and Merkel cells associated with tactile discs are scattered among the keratinocytes.
Spinosum	Eight to ten rows of polyhedral keratinocytes with bundles of tonofilaments; includes projections of melanocytes and Langerhans cells.
Granulosum	Three to five rows of flattened keratinocytes in which organelles are beginning to degenerate; cells contain the protein keratohyalin, which converts tonofilaments into keratin, and lamellar granules, which release a lipid-rich, water-repellent secretion.
Lucidum	Present only in skin of fingertips, palms, and soles; consists of three to five rows of clear, flat, dead keratinocytes with large amounts of keratin.
Corneum	Twenty-five to 30 rows of dead, flat keratinocytes that contain mostly keratin.

fibers. The few cells present in the dermis include fibroblasts, macrophages, and some adipocytes. Blood vessels, nerves, glands, and hair follicles are embedded in dermal tissue. Based on its tissue structure, the dermis can be divided into a superficial papillary region and a deeper reticular region.

The **papillary region** is about one-fifth of the thickness of the total layer (see Figure 4.1). It consists of areolar connective tissue containing fine elastic fibers. Its surface area is greatly increased by small, fingerlike projections called **dermal papillae** (pa-PIL-ē; = nipples). These nipple-shaped structures indent the epidermis and some contain capillary loops. Other dermal papillae contain touch receptors called **corpuscles of touch,** or **Meissner corpuscles,** which contain sensitive nerve endings. Also present in the dermal papillae are **free nerve endings,** dendrites that lack any apparent structural specialization. Individual free nerve endings initiate signals that produce sensations of warmth, coolness, pain, tickling, and itching.

The deeper portion of the dermis is called the **reticular region** (*reticul-* = netlike). It consists of dense irregular connective tissue containing bundles of collagen and some coarse elastic fibers. The bundles of collagen fibers in the reticular region interlace in a netlike manner. Spaces between the fibers are occupied by a few adipose cells, hair follicles, nerves, sebaceous (oil) glands, and sudoriferous (sweat) glands.

The combination of collagen and elastic fibers in the reticular region provides the skin with strength, *extensibility* (ability to stretch), and *elasticity* (ability to return to its original shape after stretching). The extensibility of skin can readily be seen in pregnancy and obesity. Small tears that occur in the dermis caused by extreme stretching produce *striae* (STRĪ-ē; *striae* = streaks), or stretch marks, which are visible as red or silvery white streaks on the skin surface.

Figure 4.4 / Lines of cleavage.

Lines of cleavage in the skin indicate the predominant direction of underlying collagen fibers in the reticular region.

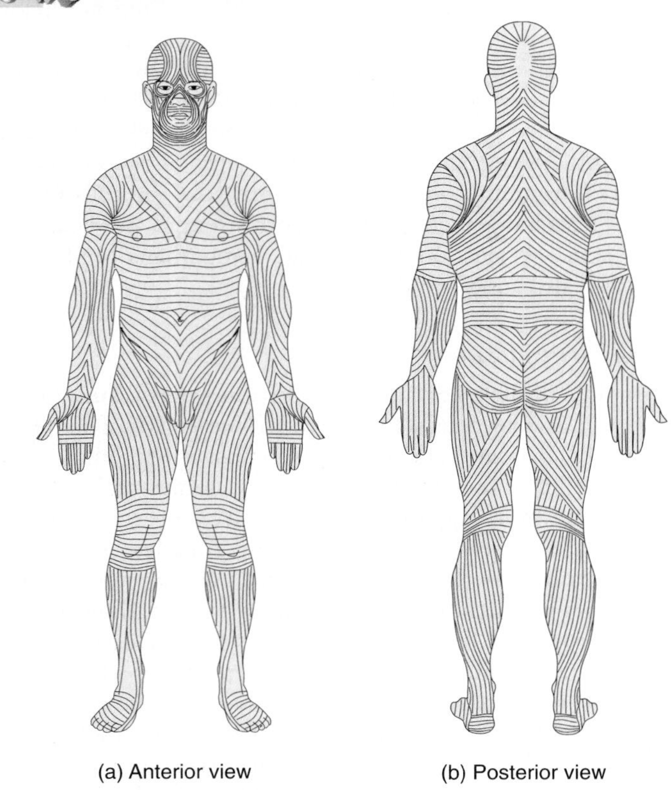

(a) Anterior view (b) Posterior view

Why are lines of cleavage clinically important?

Lines of Cleavage and Surgery

In certain regions of the body, collagen fibers tend to orient more in one direction than another. **Lines of cleavage (tension lines)** in the skin indicate the predominant direction of underlying collagen fibers (Figure 4.4). The lines are especially evident on the palmar surfaces of the fingers, where they are arranged parallel to the long axis of the digits. An incision running parallel to the collagen fibers will heal with only a fine scar. An incision made across the rows of fibers disrupts the collagen, and the wound tends to gape open and heal in a broad, thick scar. ■

The surfaces of the palms, fingers, soles, and toes are marked by series of ridges and grooves. They appear either as straight lines or as a pattern of loops and whorls, as on the tips of the digits. These **epidermal ridges** develop during the third and fourth fetal months as the epidermis conforms to the contours of the underlying dermal papillae of the papillary region. The ridges increase the surface area of the epidermis and thus function to increase the grip of the hand or foot by increasing friction. Because the ducts of sweat glands open on the tops of the epidermal ridges as sweat pores, the sweat and ridges form fingerprints (or footprints) when a smooth object is touched. The ridge pattern is genetically determined and is unique for each individual. Normally, the ridge pattern does not change during life, except to enlarge, and thus can serve as the basis for identification through fingerprints or footprints. A comparison of the structural features of the papillary and reticular regions of the dermis is presented in Table 4.2.

Table 4.2 Comparison of Papillary and Reticular Regions of the Dermis

Region	Description
Papillary	The superficial portion of the dermis (about one-fifth); consists of areolar connective tissue with elastic fibers; contains dermal papillae that house capillaries, corpuscles of touch, and free nerve endings.
Reticular	The deeper portion of the dermis (about four-fifths); consists of dense irregular connective tissue with bundles of collagen and some coarse elastic fibers. Spaces between fibers contain some adipose cells, hair follicles, nerves, sebaceous glands, and sudoriferous glands.

14th Century

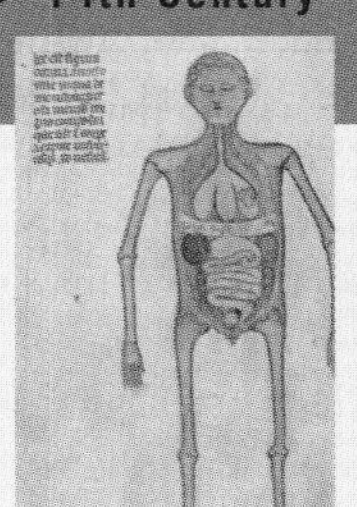

15th Century

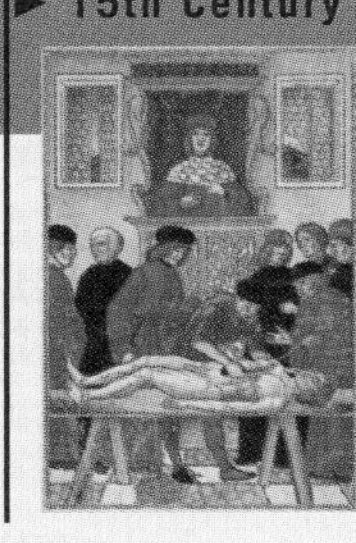

16th Century

17th Century

18th Century

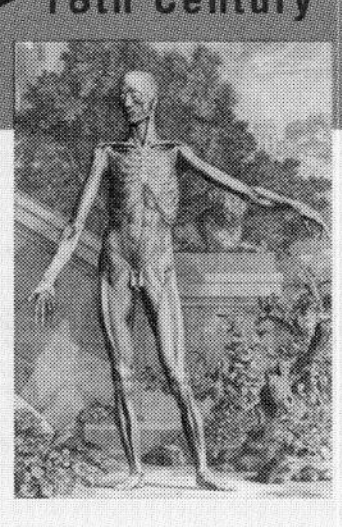

20th Century

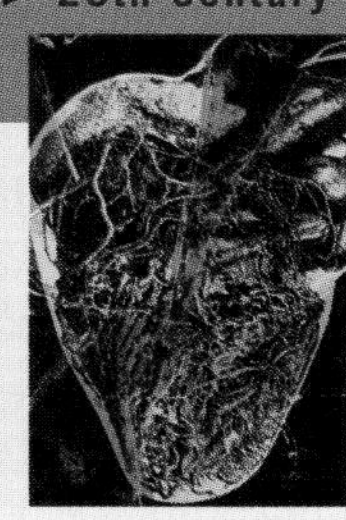

21st Century

CHANGING IMAGES

Living Art

1995 AD

Many researchers believe that tattooing was first practiced in Ancient Egypt. Archeological evidence suggests that body decoration by puncture tattoo can be traced back to Egypt between 4000 and 2000 B.C. From Egypt tattoo art traveled the globe only to disappear and then reappear several times throughout recorded history. While some cultures considered it merely ornamental, others believed it to be a divine gift and incorporated the practice into their religious rites and ceremonies. Today, tattooing in one form or another is practiced by nearly all the peoples of the globe.

In this image, a young Malaysian man bears intricate ornamental tattoos. His people have practiced tattooing for centuries. The images he chose to display on the living canvas of his skin may represent bravery, significant journeys, or may be merely artistic expressions. It is not uncommon to see these same designs woven into baskets by the local women. If he was tattooed in the ancestral method of Southeast Asia, traditional needles, or even thorns from a local bush, would be tapped with a hammer-like tool to puncture the designs into his flesh. Pigment composed of a variety of ingredients including soot, pig fat, and sugarcane juice would then be introduced into the wound.

As you continue this chapter on the integumentary system, consider ways in which the accessory structures of the system—hair and nails—are similarly used by cultures around the world to express beauty or beliefs.

The Structural Basis of Skin Color

Objective

- Explain the basis for skin color.

Melanin, carotene, and hemoglobin are three pigments that give skin a wide variety of colors. The amount of **melanin,** which is located mostly in the epidermis, causes the skin's color to vary from pale yellow to tan to black. There are actually two forms of melanin. *Eumelanin* is brownish-black and *pheomelanin* is a red to yellow pigment. The differences between these two forms of melanin is most apparent in the hair. Melanocytes are most plentiful in the mucous membranes, penis, nipples of the breasts and the area just around the nipples (areolae), face, and limbs. Because the *number* of melanocytes is about the same in all races, differences in skin color are due mainly to the *amount of pigment* the melanocytes produce and disperse to keratinocytes. In some people, melanin tends to accumulate in patches called *freckles.* As one grows older, *age (liver) spots* may develop. These are flat skin patches that look like freckles and range in color from light brown to black. Like freckles, age spots are accumulations of melanin.

Melanocytes synthesize melanin from the amino acid *tyrosine* in the presence of an enzyme called *tyrosinase.* Synthesis occurs in an organelle called a **melanosome.** Exposure to UV light increases the enzymatic activity within melanosomes and leads to increased melanin production. Both the amount and darkness of melanin increase, which gives the skin a tanned appearance and further protects the body against UV radiation. Thus, within limits, melanin serves a protective function. As you will see, however, repeatedly exposing the skin to UV light may cause skin cancer. A tan is lost when the melanin-containing keratinocytes are shed from the stratum corneum.

Carotene (KAR-ō-tēn; *carot-* = carrot), a yellow-orange pigment, is the precursor of vitamin A, which is used to synthesize pigments needed for vision. Carotene is found in the stratum corneum and fatty areas of the dermis and subcutaneous layer.

When little melanin or carotene are present, the epidermis appears translucent. Thus, the skin of white people appears pink to red, depending on the amount and oxygen content of the blood moving through capillaries in the dermis. The red color is due to **hemoglobin,** the oxygen-carrying pigment in red blood cells.

Albinism (AL-bin-izm; *albin-* = white) is the inherited inability of an individual to produce melanin. Most **albinos** (al-BĪ-nōs), people affected by albinism, have melanocytes that are unable to synthesize tyrosinase. Melanin is missing from their hair, eyes, and skin. In another condition, called **vitiligo** (vit-i-LĪ-gō), the partial or complete loss of melanocytes from patches of skin produces irregular white spots.

Skin Color as a Diagnostic Clue

The color of skin and mucous membranes can provide clues for diagnosing certain conditions. When blood is not picking up an adequate amount of oxygen in the lungs, such as in someone who has stopped breathing, the mucous membranes, nail beds, and skin appear bluish or **cyanotic** (sī-an-OT-ic; *cyan-* = blue). This occurs because hemoglobin that is depleted of oxygen looks deep, purplish blue. **Jaundice** (JON-dis; *jaund-* = yellow) is due to a buildup of the yellow pigment bilirubin in the blood. This condition gives a yellowish appearance to the whites of the eyes and the skin. Jaundice usually indicates liver disease. **Erythema** (er-e-THĒ-ma; *eryth-* = red), redness of the skin, is caused by engorgement of capillaries in the dermis with blood due to skin injury, exposure to heat, infection, inflammation, or allergic reactions. All skin color changes are observed most readily in people with lighter-colored skin and may be more difficult to discern in people with darker skin. ■

✓ What is the integumentary system?
✓ How does the subcutaneous layer relate to the skin?
✓ Compare the structure of the epidermis and dermis.
✓ List the distinctive features of the epidermal layers from deepest to most superficial.
✓ Compare the composition of the papillary and reticular regions of the dermis.
✓ How are epidermal ridges formed?
✓ Explain the factors that produce skin color.

ACCESSORY STRUCTURES OF THE SKIN

Objective

- Compare the structure, distribution, and functions of hair, skin glands, and nails.

Accessory structures of the skin—hair, skin glands, and nails—develop from the embryonic epidermis. They have a host of important functions: For example, hair and nails protect the body, and sweat glands help regulate body temperature.

Hair

Hairs, or *pili* (PI-lē), are present on most skin surfaces except the palms, palmar surfaces of the fingers, soles of the feet, and plantar surfaces of the toes. In adults, hair is usually most heavily distributed across the scalp, over the brows of the eyes, and around the external genitalia. Genetic and hormonal influences largely determine the thickness and pattern of distribution of hair.

Anatomy of a Hair

Each hair is composed of columns of dead, keratinized cells bonded together by extracellular proteins. The **shaft** is the superficial portion of the hair, most of which projects from the surface of the skin (Figure 4.5a).

The transverse-sectional shape of the shafts of hairs varies in relation to race. Straight hair, found in the Mongol races,

Figure 4.5 / Hair.

Hairs are growths of epidermis composed of dead, keratinized cells.

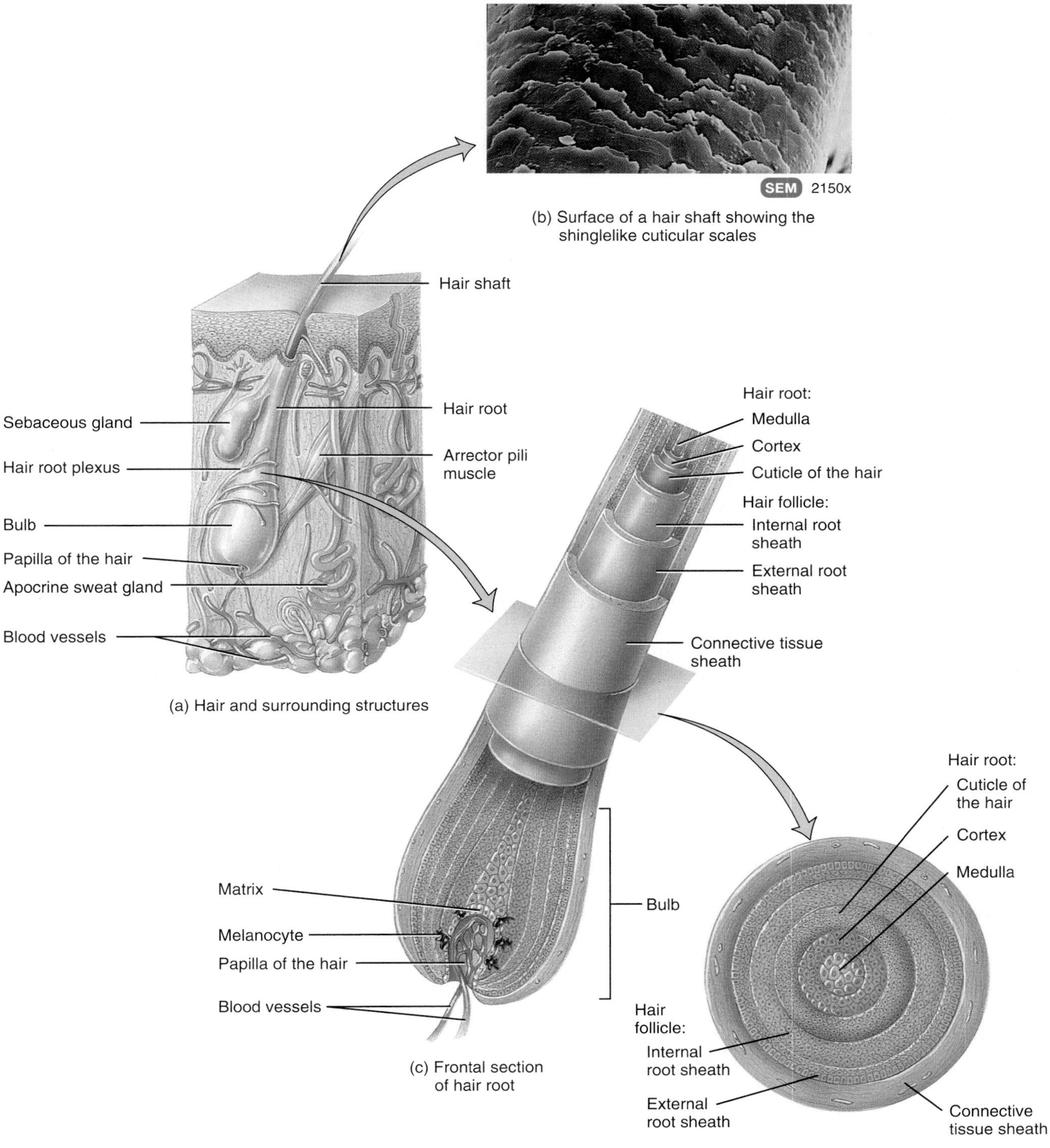

SEM 2150x

(b) Surface of a hair shaft showing the shinglelike cuticular scales

Hair shaft

Sebaceous gland

Hair root

Hair root plexus

Arrector pili muscle

Bulb

Papilla of the hair

Apocrine sweat gland

Blood vessels

(a) Hair and surrounding structures

Hair root:
Medulla
Cortex
Cuticle of the hair

Hair follicle:
Internal root sheath
External root sheath

Connective tissue sheath

Matrix

Melanocyte

Papilla of the hair

Blood vessels

Bulb

(c) Frontal section of hair root

Hair root:
Cuticle of the hair
Cortex
Medulla

Hair follicle:
Internal root sheath
External root sheath

Connective tissue sheath

(d) Transverse section of hair root

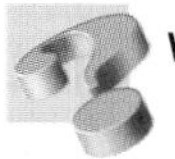

Why does it hurt when you pluck a hair out but not when you have a haircut?

Chinese, Eskimos, and Native Americans is round in transverse-section; the wavy hair found in Europeans is oval in transverse-section; and the curly hair of Blacks is kidney shaped in transverse-section. The **root** is the portion of the hair deep to the shaft that penetrates into the dermis, and sometimes into the subcutaneous layer. The shaft and root both consist of three concentric layers (Figure 4.5c, d). The inner *medulla* is composed of two or three rows of polyhedral cells containing pigment granules and air spaces. The middle *cortex* forms the major part of the shaft and consists of elongated cells that contain pigment granules in dark hair but mostly air in gray or white hair. The *cuticle* of the hair, the outermost layer, consists of a single layer of thin, flat cells that are the most heavily keratinized. Cuticle cells are arranged like shingles on the side of a house, with their free edges pointing toward the free end of the hair (Figure 4.5b).

Surrounding the root of the hair is the **hair follicle,** which is made up of an external root sheath and an internal root sheath (Figure 4.5c, d). The *external root sheath* is a downward continuation of the epidermis. Near the surface of the skin, it contains all the epidermal layers. At the base of the hair follicle, the external root sheath contains only the stratum basale. The *internal root sheath* is produced by the matrix (described shortly) and forms a cellular tubular sheath of epithelium between the external root sheath and the hair.

The base of each hair follicle is an onion-shaped structure, the **bulb** (Figure 4.5c). This structure houses a nipple-shaped indentation, the **papilla of the hair,** which contains areolar connective tissue and many blood vessels that nourish the growing hair follicle. The bulb also contains a germinal layer of cells called the **matrix.** Matrix cells arise from the stratum basale, the site of cell division. Hence, matrix cells are responsible for the growth of existing hairs, and they produce new hairs when old hairs are shed. This replacement process occurs within the same follicle. Matrix cells also give rise to the cells of the internal root sheath.

Hair Removal

A substance that removes superfluous hair is called a **depilatory.** It dissolves the protein in the hair shaft, turning it into a gelatinous mass that can be wiped away. Because the hair root is not affected, regrowth of the hair occurs. In **electrolysis,** the hair matrix is destroyed by an electric current so that the hair cannot regrow. ■

Sebaceous (oil) glands (discussed shortly) and a bundle of smooth muscle cells are also associated with hairs (Figure 4.5a). The smooth muscle is called **arrector pili** (a-REK-tor PI-lē; *arrect-* = to raise). It extends from the superficial dermis of the skin to the side of the hair follicle. In its normal position, hair emerges at an angle to the surface of the skin. Under physiologic or emotional stress, such as cold or fright, autonomic nerve endings stimulate the arrector pili muscles to contract, which pulls the hair shafts perpendicular to the skin surface. This action causes "goose bumps" or "gooseflesh" because the skin around the shaft forms slight elevations.

Surrounding each hair follicle are dendrites of neurons, called **hair root plexuses,** that are sensitive to touch (Figure 4.5a). The hair root plexuses generate nerve impulses if their hair shaft is moved.

Hair Growth

Each hair follicle goes through a growth cycle, which consists of a growth stage and a resting stage. During the **growth stage,** cells of the matrix differentiate, keratinize, and die. This process forms the root sheath and hair shaft. As new cells are added at the base of the hair root, the hair grows longer. In time, the growth of the hair stops and the **resting stage** begins. After the resting stage, a new growth cycle begins. The old hair root falls out or is pushed out of the hair follicle, and a new hair begins to grow in its place. Scalp hair grows for 2–6 years and rests for about 3 months. At any time, about 85% of scalp hairs are in the growth stage. Visible hair is dead, but until the hair is pushed out of its follicle by a new hair, portions of the root within the scalp are alive.

Normal hair loss in an adult scalp is about 70–100 hairs per day. Both the rate of growth and the replacement cycle can be altered by illness, diet, high fever, surgery, blood loss, severe emotional stress, and gender. Rapid weight-loss diets that severely restrict calories or protein increase hair loss. An increase in the rate of hair shedding can also occur with certain drugs, after radiation therapies for cancer, and for 3–4 months after childbirth.

Hair Color

The color of hair is due primarily to the amount and type of melanin in its keratinized cells. Melanin is synthesized by melanocytes scattered in the matrix of the bulb and passes into cells of the cortex and medulla (Figure 4.5c). Dark-colored hair contains mostly eumelanin (brownish-black), whereas blond and red hair contain mostly pheomelanin (yellow to red) in which there is iron and more sulfur. Graying hair occurs because of a progressive decline in tyrosinase, whereas white hair results from accumulation of air bubbles in the medullary shaft.

Functions of Hair

Although the protection it offers is limited, hair on the head guards the scalp from injury and the sun's rays. It also decreases heat loss from the scalp. Eyebrows and eyelashes protect the eyes from foreign particles, as does hair in the nostrils and in the external ear canal. Touch receptors associated with hair follicles (hair root plexuses) are activated whenever a hair is even slightly moved. Thus, hairs function in sensing light touch.

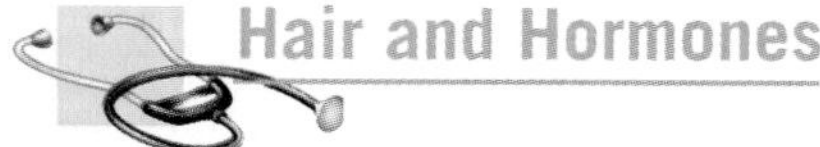

Hair and Hormones

At puberty, when their testes begin secreting significant quantities of androgens (male sex hormones), males develop the typical pattern of hair growth, including a beard and a hairy chest. In females, both the ovaries and the adrenal glands produce small quantities of androgens. Surprisingly, androgens also must be present for the most common form of baldness, **male-pattern baldness,** to occur. In genetically predisposed males, androgens somehow inhibit hair growth.

Occasionally, a tumor of one of these glands oversecretes androgens and causes **hirsutism** (hur-SŪ-tiz-um; *hirsutus* = shaggy). This is a condition of excessive hairiness of the upper lip, chin, chest, inner thighs, and abdomen in females or prepubertal males. ■

Skin Glands

Several kinds of exocrine glands are associated with the skin: sebaceous (oil) glands, sudoriferous (sweat) glands, ceruminous glands, and mammary glands. Mammary glands, which are specialized sudoriferous glands that secrete milk, are discussed in Chapter 26 with the female reproductive system.

Sebaceous Glands

Sebaceous glands (se-BĀ-shus; *sebace-* = greasy), or **oil glands,** are simple, branched acinar glands. With few exceptions, they are connected to hair follicles (see Figures 4.1 and 4.5a). The secreting portion of a sebaceous gland lies in the dermis and usually opens into the neck of the hair follicle. In other locations, such as the lips, glans penis, labia minora, and tarsal glands of the eyelids, sebaceous glands open directly onto the surface of the skin. Sebaceous glands, which vary in size and shape, are found in the skin over all regions of the body except the palms and soles. They are small in most areas of the trunk and limbs, but large in the skin of the breasts, face, neck, and upper chest.

Sebaceous glands secrete an oily substance called **sebum** (SĒ-bum), which is a mixture of fats, cholesterol, proteins, and inorganic salts. Sebum coats the surface of hairs and helps keep them from drying and becoming brittle. Sebum also prevents excessive evaporation of water from the skin, keeps the skin soft and pliable, and inhibits the growth of certain bacteria.

Acne is an inflammation of sebaceous glands that usually begins at puberty, when the sebaceous glands grow in size and increase their production of sebum. Although testosterone, a male sex hormone, appears to play the greatest role in stimulating sebaceous glands, other steroid hormones from the ovaries and adrenal glands stimulate sebaceous secretions in females. Acne occurs predominantly in sebaceous follicles that have been colonized by bacteria, some of which thrive in the lipid-rich sebum. The infection may cause a cyst or sac of connective tissue cells to form, which can destroy and displace epidermal cells. This condition, called **cystic acne,** can permanently scar the epidermis. ■

Sudoriferous Glands

There are three to four million **sweat glands,** or **sudoriferous glands** (sū′-dor-IF-er-us; *sudori-* = sweat; *-ferous* = bearing). The cells of sweat glands release their secretions by exocytosis and empty them onto the skin surface through pores or into hair follicles. Depending on their structure, location, and type of secretion, the glands are classified as eccrine and apocrine.

Eccrine sweat glands (*eccrine* = secreting outwardly) are simple, coiled tubular glands and are much more common than apocrine sweat glands. They are distributed throughout the skin, except for the margins of the lips, nail beds of the fingers and toes, glans penis, glans clitoris, labia minora, and eardrums. Eccrine sweat glands are most numerous in the skin of the forehead, palms, and soles; their density can be as high as 450 per square centimeter (3000 per square inch) in the palms. The secretory portion of eccrine sweat glands is located mostly in the deep dermis (sometimes in the upper subcutaneous layer). The excretory duct projects through the dermis and epidermis and ends as a pore at the surface of the epidermis (see Figure 4.1).

The sweat produced by eccrine sweat glands (about 600 mL per day) consists of water, ions (mostly Na^+ and Cl^-), urea, uric acid, ammonia, amino acids, glucose, and lactic acid. The main function of eccrine gland sweat is to help regulate body temperature through evaporation. As sweat evaporates, large quantities of heat energy leave the body surface. Sweat, or perspiration, usually occurs first on the forehead and scalp, extends to the face and the rest of the body, and occurs last on the palms and soles. Under conditions of emotional stress, however, the palms, soles, and axillae are the first surfaces to sweat. Eccrine sweat also plays a small role in eliminating wastes such as urea, uric acid, and ammonia. Sweat that evaporates from the skin before it is perceived as moisture is referred to as **insensible perspiration.** Sweat that is excreted in larger amounts and is perceived as moisture on the skin is called **sensible perspiration.**

Apocrine sweat glands are also simple, coiled tubular glands. They are found mainly in the skin of the axilla (armpit), groin, areolae (pigmented areas around the nipples) of the breasts, and bearded regions of the face in adult males. These glands were once thought to release their secretions in an apocrine manner—by pinching off a portion of the cell. We now know, however, that their secretion is released by exocytosis, which is characteristic of the release of secretions by merocrine glands (see page 68). Nevertheless, the term *apocrine* is still used. The secretory portion of these sweat glands is located mostly in the subcutaneous layer, and the excretory duct opens into hair follicles (see Figure 4.1). Their secretory product is slightly viscous compared to eccrine secretions and contains the same components as eccrine sweat plus lipids and proteins. In women, cells of apocrine sweat glands enlarge about the time of ovulation and shrink during menstruation. Whereas eccrine sweat glands start to function soon after birth, apocrine sweat glands do not begin to function until puberty. Apocrine sweat glands are stimulated during emotional stress and sexual excitement; these secretions are commonly known as a "cold sweat."

Table 4.3 presents a comparison of eccrine and apocrine sweat glands.

Ceruminous Glands

Modified sweat glands in the external ear, called **ceruminous glands** (se-RŪ-mi-nus; *cer-* = wax), produce a waxy secretion.

Table 4.3 Comparison of Eccrine and Apocrine Sweat Glands

Feature	Eccrine Sweat Glands	Appocrine Sweat Glands
Distribution	Throughout skin, except for margins of lips, nail beds, glans penis and clitoris, labia minora, and eardrums.	Skin of the axilla, groin, areolae, and bearded regions of the face.
Location of secretory portion	Mostly in deep dermis.	Mostly in subcutaneous layer.
Termination of excretory duct	Surface of epidermis.	Hair follicle.
Secretion	Less viscous; consists of water, ions (Na^+, Cl^-), urea, uric acid, ammonia, amino acids, glucose, lactic acid.	More viscous; consists of the same components as eccrine sweat glands plus lipids and proteins.
Functions	Regulation of body temperature and waste removal.	Stimulated during emotional stress and sexual excitement.
Onset of function	Soon after birth.	Puberty.

The secretory portions of ceruminous glands lie in the subcutaneous layer, deep to sebaceous glands. Their excretory ducts open either directly onto the surface of the external auditory canal (ear canal) or into ducts of sebaceous glands. The combined secretion of the ceruminous and sebaceous glands is called **cerumen,** or earwax. Cerumen, together with hairs in the external auditory canal, provides a sticky barrier that prevents the entrance of foreign bodies.

Impacted Cerumen

Some people produce an abnormally large amount of cerumen in the external auditory canal. The cerumen may accumulate until it becomes impacted (firmly wedged), which prevents sound waves from reaching the eardrum. The treatment for **impacted cerumen** is usually periodic ear irrigation with enzymes to dissolve the wax or removal of wax with a blunt instrument by trained medical personnel. The use of cotton-tipped swabs or sharp objects is not recommended for this purpose because they may push the cerumen farther into the external auditory canal and damage the eardrum. ■

Nails

Nails are plates of tightly packed, hard, keratinized epidermal cells. The cells form a clear, solid covering over the dorsal surfaces of the distal portions of the digits. Each nail consists of a nail body, a free edge, and a nail root (Figure 4.6). The **nail body** is the portion of the nail that is visible, the **free edge** is the part that may extend past the distal end of the digit, and the **nail root** is the portion that is buried in a fold of skin. Most of the nail body appears pink because of blood flowing through underlying capillaries. The free edge is white because there are no

Figure 4.6 / Nails. Shown is a fingernail.

Nail cells arise by transformation of superficial cells of the nail matrix into nail cells.

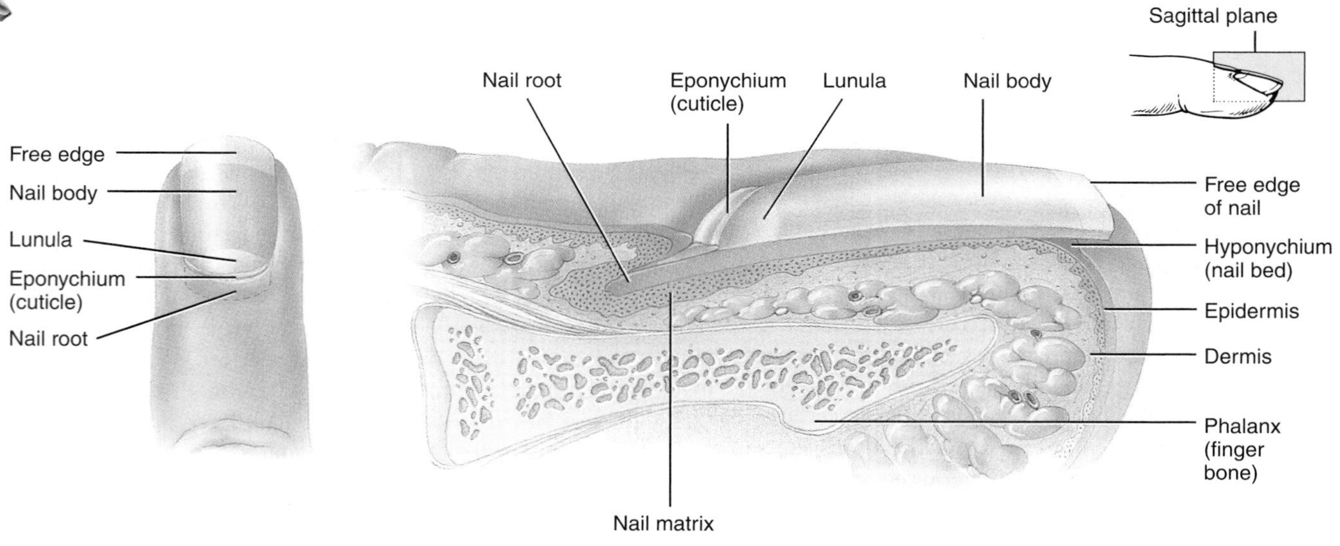

Free edge
Nail body
Lunula
Eponychium (cuticle)
Nail root

(a) Dorsal view

Sagittal plane
Nail root
Eponychium (cuticle)
Lunula
Nail body
Free edge of nail
Hyponychium (nail bed)
Epidermis
Dermis
Phalanx (finger bone)
Nail matrix

(b) Sagittal section showing internal detail

Why are nails so hard?

underlying capillaries. The whitish, crescent-shaped area of the proximal end of the nail body is called the **lunula** (LŪ-nyū-la; = little moon). It appears whitish because the vascular tissue underneath does not show through due to the thickened stratum basale in the area. Beneath the free edge is a thickened region of stratum corneum called the **hyponychium** (hī-′pō-NIK-ē-um; *hypo-* = below; *-onych* = nail), which secures the nail to the fingertip. The **eponychium** (ep′-ō-NIK-ē-um; *ep-* = above) or **cuticle** is a narrow band of epidermis that extends from and adheres to the margin (lateral border) of the nail wall. It occupies the proximal border of the nail and consists of stratum corneum.

The epithelium deep to the nail root is known as the **nail matrix.** The matrix cells divide mitotically to produce growth. Nail growth occurs by the transformation of superficial matrix cells into nail cells. In the process, the harder outer layer is pushed forward over the stratum basale. The growth rate of nails is determined by the rate at which matrix cells divide, which is influenced by factors such as a person's age, health, and nutritional status. Nail growth also varies according to the season, the time of day, and environmental temperature. The average growth in the length of fingernails is about 1 mm (.04 in.) per week. The growth rate is somewhat slower in toenails. The longer the digit, the faster the nail grows.

Functionally, nails help us grasp and manipulate small objects in various ways, provide protection against trauma to the ends of the digits, and allow us to scratch various parts of the body.

✓ Describe the structure of a hair. What produces "goose bumps"?
✓ Contrast the locations and functions of sebaceous (oil) glands, sudoriferous (sweat) glands, and ceruminous glands.
✓ Describe the principal parts of a nail.

TYPES OF SKIN

Objective

- Compare structural and functional differences in thin and thick skin.

Although the skin over the entire body is similar in structure, there are quite a few local variations related to thickness of the epidermis, strength, flexibility, degree of keratinization, distribution and type of hair, density and types of glands, pigmentation, vascularity (blood supply), and innervation (nerve supply). On the basis of certain structural and functional properties, we recognize two major types of skin: thin (hairy) and thick (hairless).

Thin skin covers all parts of the body except for the palms, palmar surfaces of the digits, and the soles. Its epidermis is thin, just 0.10–0.15 mm (.004–.006 in). A distinct stratum lucidum is lacking, and the strata spinosum and corneum are relatively thin. Thin skin has lower, broader, and fewer dermal papillae than thick skin and thus lacks epidermal ridges; it has hair follicles, arrector pili muscles, and sebaceous (oil) glands, but it has fewer sweat glands than thick skin. Finally, thin skin has a sparser distribution of sensory receptors than thick skin.

Table 4.4 Comparison of Thin and Thick Skin

Feature	Thin Skin	Thick Skin
Distribution	All parts of the body except palms, palmar surface of digits, and soles.	Palms, palmar surface of digits, and soles.
Epidermal thickness	0.10–0.15 mm (.004–.006 in).	0.6–4.5 mm (.024–.18 in).
Epidermal strata	Stratum lucidum essentially lacking; thinner strata spinosum and corneum.	Thick strata lucidum, spinosum, and corneum.
Epidermal ridges	Lacking due to poorly developed and fewer dermal papillae.	Present due to well-developed and more numerous dermal papillae.
Hair follicles and arrector pili muscles	Present.	Absent.
Sebaceous glands	Present.	Absent.
Sudoriferous glands	Fewer.	More numerous.
Sensory receptors	Sparser.	Denser.

Thick skin covers the palms, palmar surfaces of the digits, and soles. Its epidermis is relatively thick, 0.6–4.5 mm (.024–.18 in), and features a distinct stratum lucidum as well as thicker strata spinosum and corneum. The dermal papillae of thick skin are higher, narrower, and more numerous than those in thin skin, and thus thick skin has epidermal ridges. Thick skin lacks hair follicles, arrector pili muscles, and sebaceous glands, and has more sweat glands than thin skin. Also, sensory receptors are more densely clustered in thick skin.

Table 4.4 presents a summary of the features of thin and thick skin.

FUNCTIONS OF THE SKIN

Objective

- Describe how the skin contributes to body temperature regulation, protection, sensation, excretion and absorption, and synthesis of vitamin D.

Among the numerous functions of the integumentary system (mainly the skin) are the following:

1. **Regulation of body temperature.** In response to high environmental temperature or strenuous exercise, the evaporation of sweat from the skin surface helps lower an elevated body temperature to normal. In response to low environmental temperature, production of sweat is decreased, which helps conserve heat.
2. **Blood reservoir.** The dermis houses an extensive network of blood vessels that carry 8–10% of the total blood flow in

a resting adult. For this reason, the skin acts as a *blood reservoir.* During moderate exercise, the flow of blood through the skin increases, which increases the amount of heat radiated from the body. During very strenuous exercise, however, skin blood vessels constrict (narrow) somewhat, diverting more blood to contracting skeletal muscles and to the heart. Because of this shunting of blood away from the skin, however, less heat is lost from the skin, and body temperature tends to rise.

3. **Protection.** The skin covers the body and provides protection in various ways. Keratin in the skin protects underlying tissues from microbes, abrasion, heat, and chemicals and the tightly interlocked keratinocytes resist invasion by microbes. Lipids released by lamellar granules retard evaporation of water from the skin surface, thus protecting the body from dehydration; they also retard entry of water across the skin surface during showers and swims. The oily sebum from the sebaceous glands protects skin and hairs from drying out and contains bactericidal chemicals that kill surface bacteria. The pigment melanin provides some protection against the damaging effects of UV light. Protective functions that are immunological in nature are carried out by epidermal Langerhans cells, which alert the immune system to the presence of potentially harmful microbial invaders, and by macrophages in the dermis, which phagocytize bacteria and viruses that manage to penetrate the skin surface.
4. **Cutaneous sensations.** *Cutaneous sensations* are those that arise in the skin. These include tactile sensations—touch, pressure, vibration, and tickling—as well as thermal sensations such as warmth and coolness. Another cutaneous sensation, pain, usually is an indication of impending or actual tissue damage. Some of the wide variety of abundantly distributed nerve endings and receptors in the skin include the tactile discs of the epidermis, the corpuscles of touch in the dermis, and hair root plexuses around each hair follicle. Chapter 19 provides more details on the topic of cutaneous sensations.
5. **Excretion and absorption.** The skin plays minor roles in *excretion*, the elimination of substances from the body, and *absorption*, the passage of materials from the external environment into body cells. Besides removing water and heat (by evaporation), sweat also is the vehicle for excretion of small amounts of salts, carbon dioxide, and two organic products of protein breakdown—ammonia and urea. The absorption of water-soluble substances through the skin is negligible, but certain lipid-soluble materials do penetrate the skin. These include fat-soluble vitamins (A, D, E, and K) and oxygen and carbon dioxide gases. Toxic materials that can be absorbed through the skin include organic solvents such as acetone (in some nail polish removers) and carbon tetrachloride (dry-cleaning fluid); salts of heavy metals such as lead, mercury, and arsenic; and the toxins in poison ivy and poison oak.
6. **Synthesis of vitamin D.** What is commonly called vitamin D is actually a group of closely related compounds. Synthesis of vitamin D requires activation of a precursor molecule in the skin by UV rays in sunlight. Enzymes in the liver and kidneys then modify the activated molecule, finally producing *calcitriol*, the most active form of vitamin D. Calcitriol aids in the absorption of calcium in foods from the gastrointestinal tract into the blood.

Transdermal Drug Administration

Most drugs are either absorbed into the body through the digestive system or injected into subcutaneous tissue or muscle. An alternative route, **transdermal drug administration,** enables a drug contained within an adhesive skin patch to pass across the epidermis and into the blood vessels of the dermis. The drug is released at a controlled rate over one to several days. Because the major barrier to penetration of most drugs is the stratum corneum, transdermal absorption is most rapid in regions of the skin where this layer is thin, such as the scrotum, face, and scalp. A growing number of drugs are available for transdermal administration. These drugs include nitroglycerin, for prevention of angina pectoris (chest pain associated with heart disease); scopolamine, for motion sickness; estradiol, used for estrogen-replacement therapy during menopause; and nicotine, used to help people stop smoking. ■

✓ In what two ways does the skin help regulate body temperature?
✓ In what ways does the skin serve as a protective barrier?
✓ What sensations arise from stimulation of neurons in the skin?
✓ What types of molecules can penetrate the stratum corneum?

BLOOD SUPPLY OF THE INTEGUMENTARY SYSTEM

Objective

- Describe the blood supply of the integumentary system.

Although the epidermis is avascular, the dermis is well supplied with blood (see Figure 4.1). The arteries supplying the dermis are generally derived from branches of arteries supplying skeletal muscles in a particular region. Some arteries supply the skin directly. One plexus (network) of arteries, the **cutaneous plexus,** is located at the junction of the dermis and subcutaneous layer and sends branches that supply the sebaceous (oil) and sudoriferous (sweat) glands, the deep portions of hair follicles, and adipose tissue. The **papillary plexus,** formed at the level of the papillary region, sends branches that supply the capillary loops in the dermal papillae, sebaceous (oil) glands, and the superficial portion of hair follicles. The arterial plexuses are accompanied by venous plexuses that drain blood from the dermis into larger subcutaneous veins.

✓ How do the cutaneous and papillary plexus differ in distribution?

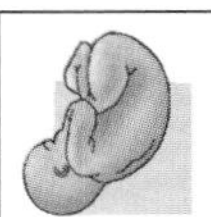

DEVELOPMENTAL ANATOMY OF THE INTEGUMENTARY SYSTEM

Objective

- Describe the development of the epidermis, its accessory structures, and the dermis.

At the end of many chapters throughout this book, we will discuss the developmental anatomy of body systems. Because the principal features of embryonic development are not described in detail until Chapter 27, it is necessary here to review a few terms and introduce others so that you can follow the development of organ systems.

As part of the earlier development of a fertilized egg, a portion of the developing embryo differentiates into three layers of tissue called **primary germ layers.** On the basis of position, the three primary germ layers are referred to as **ectoderm** (*ecto-* = outside), mesoderm (*meso-* = middle), and **endoderm** (*endo-* = within). They are the embryonic tissues from which all tissues and organs of the body eventually develop (see Table 27.1 on page 833).

The *epidermis* is derived from the **ectoderm.** At the beginning of the eighth week after fertilization the ectoderm consists of simple cuboidal epithelium. These cells become flattened and are known as the **periderm.** By the fourth month, all layers of the epidermis are formed, and each layer assumes its characteristic structure.

The *dermis* is derived from **mesodermal cells** in a zone beneath the ectoderm. There they undergo a process that changes them into the connective tissues that begin to form the dermis at about 11 weeks.

Nails are developed at about 10 weeks. Initially they consist of a thick layer of epithelium called the **primary nail field.** The nail itself is keratinized epithelium and grows distally from its base. It is not until the ninth month that the nails actually reach the tips of the digits.

As noted earlier *hair follicles* develop between the ninth and twelfth weeks as downgrowths of the stratum basale of the epidermis into the deeper dermis. The downgrowths soon differentiate into the bulb and the papilla of the hair, beginnings of the epithelial portions of sebaceous glands, and other structures associated with hair follicles. By the fifth or sixth month, the follicles produce **lanugo** (delicate fetal hair), first on the head and then on other parts of the body. The lanugo is usually shed prior to birth.

The epithelial (secretory) portions of *sebaceous glands* develop from the sides of the hair follicles at about 16 weeks and remain connected to the follicles.

The epithelial portions of *sudoriferous glands* are also derived from downgrowths of the stratum basale of the epidermis into the dermis. They appear at about 20 weeks on the palms and soles and a little later in other regions. The connective tissue and blood vessels associated with the glands develop from **mesoderm.**

✓ Which structures develop as downgrowths of the stratum basale?

AGING AND THE INTEGUMENTARY SYSTEM

Objective

- Describe the effects of aging on the integumentary system.

At about the sixth month of fetal development, secretions from sebaceous glands mix with sloughed off epidermal cells and hairs to form a fatty substance called **vernix caseosa** (VER-niks KĀ-sē-ō-sa; *vernix* = varnish, *caseosa* = cheese). This substance covers and protects the skin of the fetus from the constant exposure to the amniotic fluid in which it is bathed. In addition, the vernix caseosa facilitates the birth of the fetus because of its slippery nature. For most infants and children, relatively few problems are encountered with the skin as it ages. With the arrival of adolescence, however, some teens develop acne.

The pronounced effects of skin aging do not become noticeable until people reach their late forties. Most of the age-related changes occur in the dermis. Collagen fibers in the dermis begin to decrease in number, stiffen, break apart, and disorganize into a shapeless, matted tangle. Elastic fibers lose some of their elasticity, thicken into clumps, and fray, an effect that is greatly accelerated in the skin of smokers. Fibroblasts, which produce both collagen and elastic fibers, decrease in number. As a result, the skin forms the characteristic crevices and furrows known as wrinkles.

With further aging, Langerhans cells dwindle in number and macrophages become less-efficient phagocytes, thus decreasing the skin's immune responsiveness. Moreover, decreased size of sebaceous glands leads to dry and broken skin that is more susceptible to infection. Production of sweat diminishes, which probably contributes to the increased incidence of heat stroke in the elderly. There is a decrease in the number of functioning melanocytes, resulting in gray hair and atypical skin pigmentation. An increase in the size of some melanocytes produces pigmented blotching (liver spots). Walls of blood vessels in the dermis become thicker and less permeable, and subcutaneous fat is lost. Aged skin (especially the dermis) is thinner than young skin, and the migration of cells from the basal layer to the epidermal surface slows considerably. With the onset of old age, skin heals poorly and becomes more susceptible to pathological conditions such as skin cancer and pressure sores.

Growth of nails and hair slows during the second and third decades of life. The nails also may become more brittle with age, often due to dehydration or repeated use of cuticle remover or nail polish.

Photodamage

Although basking in the warmth of the sun may feel good, it is not a healthy practice. Both the longer wavelength ultraviolet A (UVA) rays and the shorter wavelength UVB rays cause **photodamage** of the skin. Light-skinned and dark-skinned individuals alike experience the effect of acute overexposure to UVB rays—a sunburn. Even if sunburn does not occur, the UVB rays can damage the DNA in epidermal cells, which can produce genetic mutations that cause skin cancer. As UVA rays penetrate to the dermis, they produce oxygen free radicals that disrupt collagen and elastic fibers in the extracellular matrix. This is the main reason for the severe wrinkling that occurs in people who spend a great deal of time in the sun without protection. ■

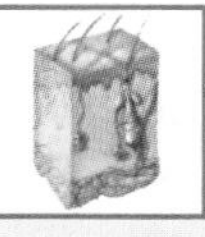

APPLICATIONS TO HEALTH

Skin Cancer

Excessive exposure to the sun has caused virtually all of the 1 million cases of **skin cancer** diagnosed in the United States in the year 2000. There are three common forms of skin cancer. **Basal cell carcinomas** account for about 78% of all skin cancers. The tumors arise from cells in the stratum basale of the epidermis and rarely metastasize. **Squamous cell carcinomas,** which account for about 20% of all skin cancers, arise from squamous cells of the epidermis, and they have a variable tendency to metastasize. Most arise from preexisting lesions of damaged tissue in sun-exposed skin. Basal and squamous cell carcinomas are together known as *non-melanoma skin cancer* and are 50% more common in males than in females. **Malignant melanomas** arise from melanocytes and account for about 2% of all skin cancers. They are the most prevalent life-threatening cancer in young women. The American Academy of Dermatology estimates that the lifetime risk of developing melanoma is currently 1 in 75, double the risk only 15 years ago. In part, this increase is due to depletion of the ozone layer, which absorbs some UV light high in the atmosphere. But the main reason for the increase is that more people are spending more time in the sun. Malignant melanomas metastasize rapidly and can kill a person within months of diagnosis.

The key to successful treatment of malignant melanoma is early detection. The early warning signs of malignant melanoma are identified by the acronym ABCD. *A* is for *asymmetry;* malignant melanomas tend to lack symmetry. *B* is for *border;* malignant melanomas have notched, indented, scalloped, or indistinct borders. *C* is for *color;* malignant melanomas have uneven coloration and may contain several colors. *D* is for *diameter;* ordinary moles tend to be smaller than 6 mm (0.25 in.), about the size of a pencil eraser. Once a malignant melanoma has the characteristics of A, B, and C, it is usually larger than 6 mm.

Among the risk factors for skin cancer are the following:

1. ***Skin type.*** Individuals with light-colored skin who never tan but always burn are at high risk.
2. ***Sun exposure.*** People who live in areas with many days of sunlight per year and at high altitudes (where ultraviolet light is more intense) have a higher risk of developing skin cancer. Likewise, people who engage in outdoor occupations and those who have suffered three or more severe sunburns have a higher risk.
3. ***Family history.*** Skin cancer rates are higher in some families than in others.
4. ***Age.*** Older people are more prone to skin cancer owing to longer total exposure to sunlight.
5. ***Immunological status.*** Individuals who are immunosuppressed have a higher incidence of skin cancer.

Burns

A **burn** is tissue damage caused by excessive heat, electricity, radioactivity, or corrosive chemicals that destroy (denature) the proteins in the skin cells. Burns destroy some of the skin's important contributions to homeostasis—protection against microbial invasion and desiccation, and thermoregulation.

Burns are graded according to their severity. A *first-degree burn* involves only the epidermis. It is characterized by mild pain and erythema (redness) but no blisters. Skin functions remain intact. The pain and damage caused by a first-degree burn may be lessened by immediately flushing it with cold water. Generally, a first-degree burn will heal in about 3–6 days and may be accompanied by flaking or peeling. One example of a first-degree burn is a mild sunburn.

A *second-degree burn* destroys a portion of the epidermis and possibly parts of the dermis. Some skin functions are lost. In a second-degree burn, redness, blister formation, edema, and pain result. (Blister formation is caused by separation of the epidermis from the dermis due to the accumulation of tissue fluid between the layers.) Associated structures, such as hair follicles, sebaceous glands, and sweat glands, usually are not injured. If there is no infection, second-degree burns heal without skin grafting in about 3–4 weeks, but scarring may result. First- and second-degree burns are collectively referred to as *partial-thickness burns.*

A *third-degree burn,* or *full-thickness burn,* destroys a portion of the epidermis, the underlying dermis, and associated structures. Most skin functions are lost. Such burns vary in appearance from marble-white to mahogany colored to charred, dry wounds. There is marked edema, and the burned region is numb because sensory nerve endings have been destroyed. Regeneration occurs slowly, and much granulation tissue forms before being covered by epithelium. Skin grafting may be required to promote healing and to minimize scarring.

The injury to the skin tissues directly in contact with the damaging agent is the *local effect* of a burn. Generally, however, the *systemic effects* of a major burn are a greater threat to life. The systemic effects of a burn may include (1) a large loss of water, plasma, and plasma proteins, which causes shock; (2) bacterial infection; (3) reduced circulation of blood; (4) decreased production of urine; and (5) diminished immune responses.

The seriousness of a burn is determined by its depth and extent of area involved, as well as the person's age and general health. According to the American Burn Association's classification of burn injury, a major burn includes third-degree burns over 10% of body surface area; or second-degree burns over 25% of body surface area; or any third-degree burns on the face, hands, feet, or perineum (per-i-NĒ-um, which includes the anal and urogenital regions). When the burn area exceeds 70%, more than half the victims die.

Pressure Sores

Pressure sores, also known as *decubitus ulcers* (dē-KYŪ-bi-tus), or *bedsores*, are caused by a constant deficiency of blood flow to tissues. Typically the affected tissue overlies a bony projection that has been subjected to prolonged pressure against an object such as a bed, cast, or splint. If the pressure is relieved in a few hours, redness occurs but no lasting tissue damage results. Blistering of the affected area may indicate superficial damage, whereas a reddish-blue discoloration may indicate deep tissue damage. Prolonged pressure results in tissue ulceration. Small breaks in the epidermis become infected, and the sensitive subcutaneous layer and deeper tissues are damaged. Eventually, the tissue dies. Pressure sores occur most often in patients who are bedridden. With proper care, pressure sores are preventable.

KEY MEDICAL TERMS ASSOCIATED WITH THE INTEGUMENTARY SYSTEM

Abrasion (a-BRĀ-shun; *ab* = away; *rasion* = scraped) An area where skin has been scraped away.

Alopecia (al′-ō-PĒ-shē-a) Partial or complete lack of hair; may result from aging, endocrine disorders, chemotherapy for cancer, genetic factors, or skin disease.

Athlete's foot A superficial fungus infection of the skin of the foot.

Cold sore A lesion, usually in oral mucous membrane, caused by Type 1 herpes simplex virus (HSV) transmitted by oral or respiratory routes. The virus remains dormant until triggered by factors such as ultraviolet light, hormonal changes, and emotional stress. Also called a *fever blister.*

Contact dermatitis (der-ma-TĪ-tis; *dermat-* = skin; *-itis* = inflammation of) Inflammation of the skin characterized by redness, itching, and swelling and caused by exposure of the skin to chemicals that bring about an allergic reaction, such as poison ivy toxin.

Contusion (kon-TŪ-shun; *contundere* = to bruise) Condition in which tissue deep to the skin is damaged, but the epidermis is not broken.

Corn A painful conical thickening of the stratum corneum of the epidermis found principally over toe joints and between the toes, often caused by friction or pressure. Corns may be hard or soft, depending on their location. Hard corns are usually found over toe joints, and soft corns are usually found between the fourth and fifth toes.

Cyst (SIST; *cyst* = sac containing fluid) A sac with a distinct connective tissue wall, containing a fluid or other material.

Hemangioma (he-man′-jē-Ō-ma; *hem-* = blood; *-angi-* = blood vessel; *-oma* = tumor) Localized tumor of the skin and subcutaneous layer that results from an abnormal increase in blood vessels. One type is a **portwine stain,** a flat, pink, red, or purple lesion present at birth, usually at the nape of the neck.

Hives Condition of the skin marked by reddened elevated patches that are often itchy. Most commonly caused by infections, physical trauma, medications, emotional stress, food additives, and certain food allergies. Also called *urticaria* (yūr-ti-KAR-ē-a).

Impetigo (im′-pe-TĪ-gō) Superficial skin infection caused by *Staphylococcus* bacteria; most common in children.

Intradermal (in-tra-DER-mal; *intra-* = within) Within the skin. Also called *intracutaneous.*

Keloid (KĒ-loid; *kelis* = tumor) An elevated, irregular darkened area of excess scar tissue caused by collagen formation during healing. It extends beyond the original injury and is tender and frequently painful. It occurs in the dermis and underlying subcutaneous tissue, usually after trauma, surgery, a burn, or severe acne; more common in people of African descent.

Keratosis (ker′-a-TŌ-sis; *kera-* = horn) Formation of a hardened growth of epidermal tissue, such as a *solar keratosis*, a premalignant lesion of the sun-exposed skin of the face and hands.

Laceration (las-er-Ā-shun; *lacer-* = torn) An irregular tear of the skin.

Nevus (NĒ-vus) A round, flat, or raised area of pigmented skin that may be present at birth or may develop later. Varies in color from yellow-brown to black. Also called a *mole* or *birthmark.*

Pruritus (prū-RĪ-tus; *pruri-* = to itch) Itching, one of the most common dermatological disorders. It may be caused by skin disorders (infections), systemic disorders (cancer, kidney failure), psychogenic factors (emotional stress), or allergic reactions.

Topical In reference to a medication, applied to the skin surface rather than ingested or injected.

Wart Mass produced by uncontrolled growth of epithelial skin cells; caused by a papilloma virus. Most warts are noncancerous.

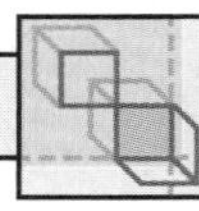

STUDY OUTLINE

Structure of the Skin (p. 91)

1. The integumentary system consists of the skin and its accessory structures—hair, nails, glands, muscles, and nerves.
2. The skin is the largest organ of the body in surface area and weight. The principal parts of the skin are the epidermis (superficial) and dermis (deep).
3. The subcutaneous layer (hypodermis) is deep to the dermis and not part of the skin. It anchors the dermis to underlying tissues and organs, and it contains lamellated (Pacinian) corpuscles.
4. The types of cells in the epidermis are keratinocytes, melanocytes, Langerhans cells, and Merkel cells.
5. The epidermal layers, from deep to superficial, are the stratum basale, stratum spinosum, stratum granulosum, stratum lucidum (in thick skin only), and stratum corneum. Stem cells in the stratum basale undergo continuous cell division, producing keratinocytes for the other layers.
6. The dermis consists of papillary and reticular regions. The papillary region is composed of areolar connective tissue containing fine elastic fibers, dermal papillae, and Meissner corpuscles. The reticular region is composed of dense irregular connective tissue containing interlaced collagen and coarse elastic fibers, adipose tissue, hair follicles, nerves, sebaceous (oil) glands, and ducts of sudoriferous (sweat) glands.
7. Epidermal ridges provide the basis for fingerprints and footprints.
8. The color of skin is due to melanin, carotene, and hemoglobin. The two forms of melanin are eumelanin (brownish-black) and pheomelanin (yellow to red).

Accessory Structures of the Skin (p. 96)

1. Accessory structures of the skin—hair, skin glands, and nails—develop from the embryonic epidermis.
2. A hair consists of a shaft, most of which is superficial to the surface, a root that penetrates the dermis and sometimes the subcutaneous layer, and a hair follicle.
3. Associated with each hair follicle is a sebaceous (oil) gland, an arrector pili muscle, and a hair root plexus.
4. New hairs develop from division of matrix cells in the bulb; hair replacement and growth occur in a cyclic pattern consisting of alternating growth and resting stages.
5. Hairs offer a limited amount of protection from the sun, heat loss, and entry of foreign particles into the eyes, nose, and ears. They also function in sensing light touch.
6. Sebaceous (oil) glands are usually connected to hair follicles; they are absent in the palms and soles. Sebaceous glands produce sebum, which moistens hairs and waterproofs the skin. Clogged sebaceous glands may produce acne.
7. There are two types of sudoriferous (sweat) glands: eccrine and apocrine. Eccrine sweat glands have an extensive distribution; their ducts terminate at pores at the surface of the epidermis. Apocrine sweat glands are limited to the skin of the axilla, groin, and areolae; their ducts open into hair follicles. They begin functioning at puberty and are stimulated during emotional stress and sexual excitement. Mammary glands are specialized sudoriferous glands that secrete milk.
8. Ceruminous glands are modified sudoriferous glands that secrete cerumen (ear wax). They are found in the external auditory canal (ear canal).
9. Nails are hard, keratinized epidermal cells over the dorsal surfaces of the distal portions of the digits.
10. The principal parts of a nail are the nail body, free edge, nail root, lunula, eponychium, and matrix. Cell division of the matrix cells produces new nails.

Types of Skin (p. 101)

1. Thin skin covers all parts of the body except for the palms, palmar surfaces of the digits, and the soles.
2. Thick skin covers the palms, palmar surfaces of the digits, and soles.

Functions of the Skin (p. 101)

1. Skin functions include body temperature regulation, protection, sensation, excretion and absorption, and synthesis of vitamin D.
2. The skin participates in thermoregulation by liberating sweat at its surface and functions as a blood reservoir.
3. The skin provides physical, chemical, and biological barriers that help protect the body.
4. Cutaneous sensations include touch, hot and cold, and pain.

Blood Supply of the Integumentary System (p. 102)

1. The epidermis is avascular.
2. The dermis is supplied by the cutaneous and papillary plexuses.

Developmental Anatomy of the Integumentary System (p. 103)

1. The epidermis develops from the embryonic ectoderm, and the accessory structures of the skin (hair, nails, and skin glands) are epidermal derivatives.
2. The dermis is derived from mesodermal cells.

Aging and the Integumentary System (p. 103)

1. Most effects of aging begin to occur when people reach their late forties.
2. Among the effects of aging are wrinkling, loss of subcutaneous fat, atrophy of sebaceous glands, and decrease in the number of melanocytes and Langerhans cells.

SELF-QUIZ QUESTIONS

Choose the one best answer to the following questions:

1. Which of the following statements about the function of skin is *not* true? (a) it helps regulate body temperature, (b) it protects against bacterial invasion and dehydration, (c) it absorbs water and salts, (d) it participates in the synthesis of vitamin D, (e) it detects stimuli related to temperature and pain.
2. The layer of the skin from which new epidermal cells are derived is the (a) stratum corneum, (b) stratum basale, (c) stratum lucidum, (d) stratum granulosum, (e) stratum spinosum
3. The substance that prevents excessive evaporation of water from the skin, keeps the skin soft and pliable, and inhibits the growth of bacteria is (a) sebum, (b) keratin, (c) cerumen, (d) sweat, (e) carotene.
4. The outermost layer of a hair is the (a) lunula, (b) cortex, (c) eponychium, (d) medulla, (e) none of the above.
5. The region of the dermis that is in direct contact with the epidermis is the (a) papillary region, (b) stratum corneum, (c) stratum basale, (d) hypodermis, (e) reticular region.
6. Which of the following statements are true?
   1. The lunula is located at the proximal end of the nail body.
   2. The free edge of the nail is white due to absence of underlying capillaries.
   3. Nails help us grasp and manipulate small objects.
   4. Nails protect the ends of the digits from trauma.
   5. The eponychium is located at the distal border of the nail.

   a. 1, 2, and 3 **b.** 1, 3, and 4 **c.** 1, 2, 3, and 4 **d.** 2, 3, and 4 **e.** 1, 2, and 5

Complete the following:

7. In the lining of the external auditory meatus are located ________ that secrete earwax called ________ .
8. The deeper portion of the skin, called the ________ , is made of ________ tissue.
9. Sweat glands are also known as ________ glands.

Are the following statements true or false?

10. Fat-soluble vitamins, oxygen, and carbon dioxide can all be absorbed through the skin.
11. A typical sunburn is an example of a first-degree burn.
12. The dermis receives its blood supply from two arterial networks or plexuses: the cutaneous plexus and the epidermal plexus.
13. The external root sheath is a downward continuation of the epidermis.
14. The dermis consists of two regions; the papillary region is superficial, and the reticular region is deep.
15. Racial differences in skin color are mainly due to the number of melanocytes in the epidermis.
16. Match the following:

____	**(a)** assist in immune responses; easily damaged by UV light	**(1)** melanocytes
____	**(b)** produce pigment that shields cell nuclei from UV light	**(2)** keratinocytes
____	**(c)** located in subcutaneous tissue; sensitive to pressure	**(3)** Merkel cells
____	**(d)** most abundant epidermal cells	**(4)** Langerhans cells
____	**(e)** touch receptor cells in the stratum basale layer	**(5)** Meissner corpuscles (corpuscles of touch)
____	**(f)** tactile receptors located in dermal papillae	**(6)** lamellated (Pacinian) corpuscles

17. Match the following:

____	**(a)** deepest layer, consisting of a single layer of cuboidal- to columnar-shaped keratinocytes	**(1)** spinosum
____	**(b)** eight to ten rows of polyhedral keratinocytes that exhibit spine-like projections	**(2)** corneum
____	**(c)** layer consisting of three to five rows of flattened, degenerating keratinocytes	**(3)** basale
____	**(d)** layer of flat, dead keratinocytes that are normally apparent only in thick skin of palms and soles	**(4)** lucidum
____	**(e)** most superficial layer of skin, consisting of 25 to 30 rows of flat, dead keratinocytes	**(5)** granulosum

CRITICAL THINKING QUESTIONS

1. Your 65-year-old aunt has spent every sunny day at the beach for as long as you can remember. Her skin looks a lot like a comfortable lounge chair—brown and wrinkled. Recently her dermatologist removed a suspicious growth from the skin of her face. What would you suspect is the problem and its likely cause?
   HINT *Tomatoes should be sun-dried but not people.*
2. Seung is fastidious about her personal hygiene. She's convinced that body secretions are unsanitary and wants to have all her exocrine glands removed. Is this wise?
   HINT *There are about 3 to 4 million sweat glands alone.*
3. Felicity is very concerned about her appearance. She heard that the skin should be protected against sun exposure and dryness to prevent wrinkles. Felicity's new plan for perfect skin is to only go out at night and to wear a waterproof covering over her entire body. Comment on Felicity's beauty plan.
   HINT *Think about the skin's functions, not its appearance.*
4. Your nephew has been learning about cells in science class and now refuses to take a bath. He asks "If all cells have a semipermeable membrane, and my skin is made out of cells, then won't I swell up and pop when I take a bath?" Explain this dilemma to your nephew before he starts attracting flies.
   HINT *The semipermeable membrane functions in a living cell.*
5. Little Wayne absolutely refused to wear anything on his feet except last year's cowboy boots until the soles of the boots wore clear through. When the shoe salesman was measuring his feet for new boots, he noticed that Wayne had several lumps on the back of his ankles, little toes, and soles of his feet. What are these lumps?
   HINT *Wayne's feet kept growing but the boots didn't.*

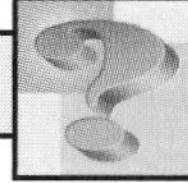

ANSWERS TO FIGURE QUESTIONS

4.1 The epidermis is epithelial tissue, whereas the dermis is connective.

4.2 Melanin protects the DNA of the nucleus of keratinocytes from damage from UV light.

4.3 The stratum basale is the layer of the epidermis that contains stem cells.

4.4 Surgical incisions parallel to lines of cleavage leave only fine scars.

4.5 Plucking a hair stimulates hair root plexuses in the dermis, some of which are sensitive to pain. Because the cells of a hair shaft are already dead and the hair shaft lacks nerves, cutting hair is not painful.

4.6 Nails are hard because they are composed of tightly packed, hard, keratinized epidermal cells.

5

BONE TISSUE

CONTENTS AT A GLANCE

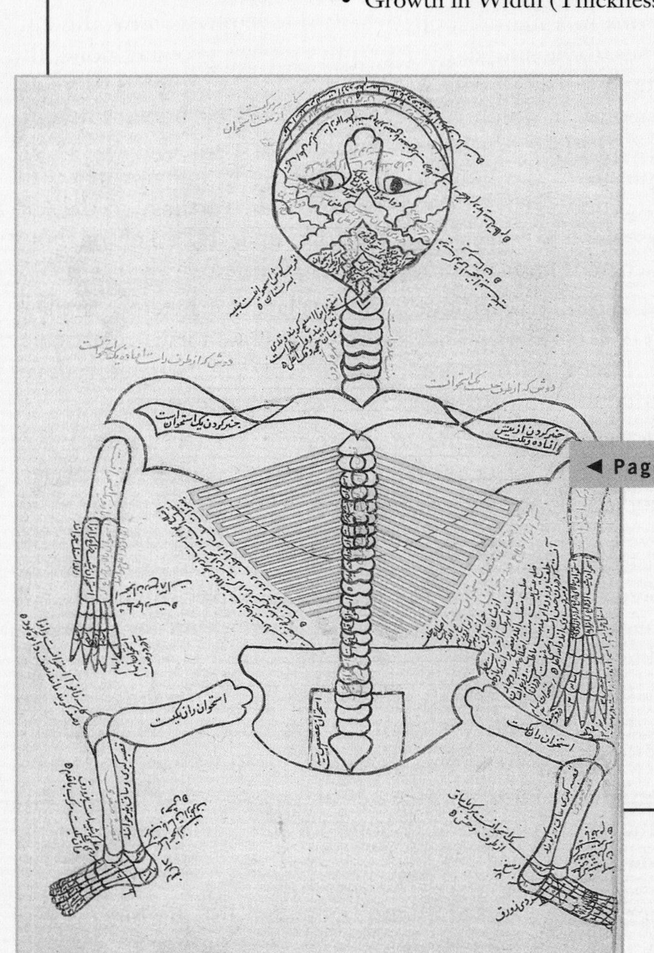

◄ Page 126

Can you guess from which culture and era this skeletal image originates?

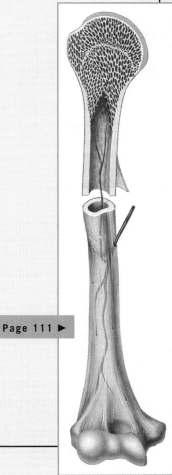

Page 111 ►

INTRODUCTION

Bone tissue is a complex and dynamic living tissue. It engages in a continuous process called remodeling—building new bone tissue and breaking down old bone tissue. A bone is made up of several different tissues working together: bone or osseous tissue, cartilage, dense connective tissues, epithelium, various blood-forming tissues, adipose tissue, and nervous tissue. For this reason, each individual bone is an organ. The entire framework of bones and their cartilages together constitute the **skeletal system.** This chapter will survey the various components of bones so you can understand how bones form, how they age, and how exercise affects the density and strength of bones. The study of bone structure and the treatment of bone disorders is called **osteology** (os-tē-OL-ō-jē; *oste-* = bone; *-ology* = study of).

FUNCTIONS OF THE SKELETAL SYSTEM

Objective

- Discuss the functions of the skeletal system.

Bone tissue and the skeletal system perform several basic functions:

1. **Support.** Bones serve as the structural framework for the body by supporting soft tissues and providing attachment points for the tendons of most skeletal muscles.
2. **Protection.** Bones protect many internal organs from injury. For example, cranial bones protect the brain, vertebrae protect the spinal cord, and the rib cage protects the heart and lungs.
3. **Assistance in movement.** When skeletal muscles contract, they act across joints to pull on bones to produce movement.
4. **Mineral storage and release.** Bone tissue stores several minerals, especially calcium and phosphorus, which contribute to the strength of the bone. Bone can release minerals into the bloodstream to maintain critical mineral balances and to distribute minerals to other organs.
5. **Blood cell production.** Within certain parts of bones a connective tissue called **red bone marrow** produces red blood cells, white blood cells, and platelets by a process called **hemopoiesis** (hēm-ō-poy-Ē-sis; *hemo-* = blood; *-poiesis* = to make). Red bone marrow, one of two types of bone marrow, consists of developing blood cells within a network of reticular fibers. Also present are adipocytes, macrophages, and fibroblasts.
6. **Triglyceride storage.** In the newborn, all bone marrow is red and is involved in hemopoiesis. However, with increasing age, blood cell production decreases, and most of the bone marrow changes from red to yellow. **Yellow bone marrow** consists primarily of adipocytes and a few scattered blood cells. Recall that adipocytes store triglycerides.

✓ What kinds of tissues make up the skeletal system?

✓ How do red and yellow bone marrow differ in composition and function?

ANATOMY OF A BONE

Objective

- Describe the parts of a long bone.

The structure of a bone may be analyzed by considering the parts of a long bone such as the humerus (the arm bone) or the femur (the thigh bone) (Figure 5.1). A long bone is one that has greater length than width. A typical long bone consists of the following parts:

1. The **diaphysis** (dī-AF-i-sis; *dia-* = through; *-physis* = growing) is the bone's shaft, or body—the long, cylindrical, main portion of the bone.
2. The **epiphyses** (e-PIF-i-sēz; *epi-* = over) are the distal and proximal ends of the bone. (The singular is epiphysis.)
3. The **metaphyses** (me-TAF-i-sēz; *meta-* = between) are the regions in a mature bone where the diaphysis joins the epiphyses. (The singular is metaphysis.) In a growing bone, the metaphyses are regions that include the epiphyseal plate, the point at which cartilage is replaced by bone. The **epiphyseal plate** is a layer of hyaline cartilage that allows the diaphysis of the bone to grow in length, but not in width. When bone growth in length stops, the cartilage in the epiphyseal plate is replaced by bone and the resulting bony structure is known as the **epiphyseal line.**
4. The **articular cartilage** is a thin layer of hyaline cartilage that covers each epiphysis where the bone forms an articulation (joint) with another bone. Articular cartilage reduces friction and absorbs shock at freely movable joints.
5. The **periosteum** (per′-ē-OS-tē-um; *peri-* = around) is a tough sheath of dense irregular connective tissue that surrounds the bone surface wherever it is not covered by articular cartilage. The periosteum contains bone-forming cells that enable bone to grow in width, but not in length. It also protects the bone, assists in fracture repair, helps nourish bone tissue, and serves as an attachment point for ligaments and tendons.
6. The **medullary cavity** (MED-yū-lar′-ē; *medulla-* = marrow, pith), or **marrow cavity,** is the space within the diaphysis that contains fatty yellow bone marrow in adults.
7. The **endosteum** (end-OS-tē-um; *endo-* = within) is a thin membrane that contains bone-forming cells and lines the medullary cavity.

✓ Diagram the parts of a long bone, and list the functions of each part.

Figure 5.1 / Parts of a long bone. The spongy bone tissue of the epiphysis and metaphysis contains red bone marrow, whereas the medullary cavity of the diaphysis contains yellow bone marrow (in adults).

A long bone is covered by articular cartilage at its proximal and distal epiphyses and by periosteum around the diaphysis.

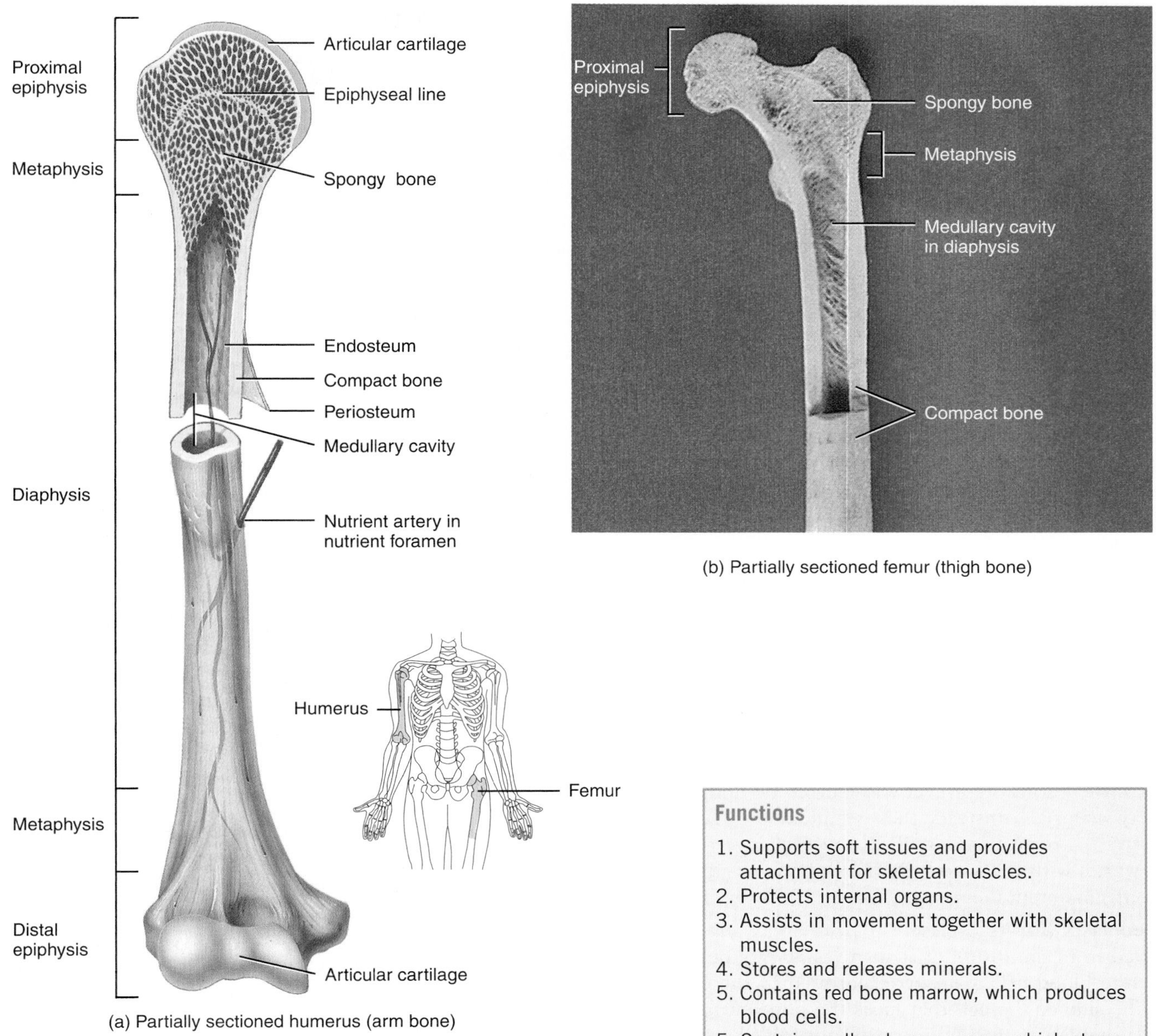

Proximal epiphysis
Metaphysis
Diaphysis
Metaphysis
Distal epiphysis
Articular cartilage
Epiphyseal line
Spongy bone
Endosteum
Compact bone
Periosteum
Medullary cavity
Nutrient artery in nutrient foramen
Humerus
Femur
Articular cartilage

(a) Partially sectioned humerus (arm bone)

Proximal epiphysis
Spongy bone
Metaphysis
Medullary cavity in diaphysis
Compact bone

(b) Partially sectioned femur (thigh bone)

Functions

1. Supports soft tissues and provides attachment for skeletal muscles.
2. Protects internal organs.
3. Assists in movement together with skeletal muscles.
4. Stores and releases minerals.
5. Contains red bone marrow, which produces blood cells.
5. Contains yellow bone marrow, which stores triglycerides (fats).

What is the functional significance of the periosteum?

HISTOLOGY OF BONE TISSUE

Objective

- Describe the histological features of bone tissue.

Like other connective tissues, **bone,** or **osseous tissue** (OS-ē-us), contains an abundant matrix of intercellular materials that surround widely separated cells. In bone, the matrix is about 25% water, 25% protein fibers, and 50% crystallized mineral salts. The four types of cells in bone tissue are: osteogenic cells, osteoblasts, osteocytes, and osteoclasts (Figure 5.2).

1. **Osteogenic cells** (os′-tē-ō-JEN-ik; *-genic* = producing) are unspecialized stem cells derived from mesenchyme, the tissue from which all connective tissues are formed. They are

Figure 5.2 / Types of cells in bone tissue.

Osteogenic cells undergo cell division and develop into osteoblasts, which secrete bone matrix.

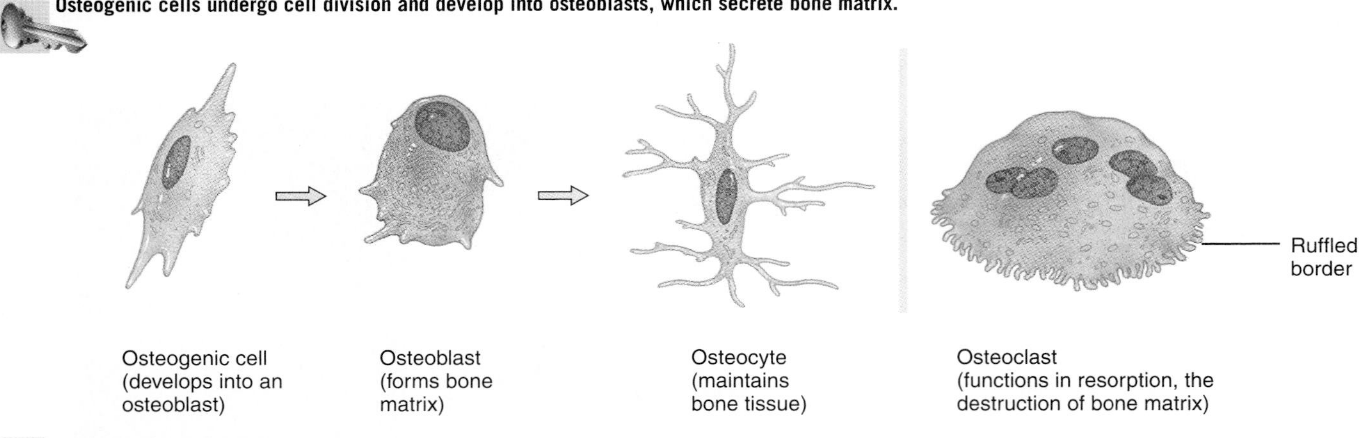

Why is bone resorption important?

the only bone cells to undergo cell division; the resulting daughter cells develop into osteoblasts. Osteogenic cells are found along the inner portion of the periosteum, in the endosteum, and in the canals within bone that contain blood vessels.

2. **Osteoblasts** (OS-tē-ō-blasts′; *-blasts* = buds or sprouts) are bone-building cells. They synthesize and secrete collagen fibers and other organic components needed to build the matrix of bone tissue, and they initiate calcification (described shortly). (Note: In bone or any other connective tissue, *blasts* secrete matrix.)
3. **Osteocytes** (OS-tē-ō-sīts′; *-cyte* = cell) are mature bone cells that are the principal cells of bone tissue. Osteocytes are derived from osteoblasts that have become entrapped in matrix secretions. However, osteocytes no longer secrete matrix materials. Instead, they maintain the daily cellular activities of bone tissue, such as the exchange of nutrients and wastes with the blood. (Note: In bone or any other tissue, *cytes* maintain the tissue.)
4. **Osteoclasts** (OS-tē-ō-clasts′; *-clast* = break) are huge cells derived from the fusion of as many as 50 monocytes, a type of white blood cell that is concentrated in the endosteum. On the side of the cell that faces the bone surface, the osteoclast's plasma membrane is deeply folded into a *ruffled border*. Here the cell releases powerful lysosomal enzymes and acids that digest the protein and mineral components of the underlying bone. This destruction of bone matrix is part of the normal development, growth, maintenance, and repair of bone.

The matrix of bone, unlike that of other connective tissues, contains abundant inorganic mineral salts, primarily *hydroxyapatite* (calcium phosphate) and some calcium carbonate. In addition, bone matrix includes small amounts of magnesium hydroxide, fluoride, and sulfate. As these mineral salts are deposited in the framework formed by the collagen fibers of the matrix, they crystallize, and the tissue hardens. This process of **calcification,** or **mineralization,** is initiated by osteoblasts.

Although a bone's *hardness* depends on crystallized inorganic mineral salts, its *flexibility* depends on its collagen fibers. Like reinforcing metal rods in concrete, collagen fibers and other organic molecules provide *tensile strength*, which is resistance to being stretched or torn apart. If the mineral salts in a bone are dissolved by soaking the bone in vinegar, for example, the bone becomes rubbery and flexible.

It was once thought that calcification simply occurred when enough mineral salts were present to form crystals. Now, however, we know that the process occurs only in the presence of collagen fibers. Mineral salts begin to crystallize in the microscopic spaces between collagen fibers. After the spaces are filled, mineral crystals accumulate around the collagen fibers. The combination of crystallized salts and collagen fibers is responsible for the hardness that is characteristic of bone.

Bone is not completely solid but has many small spaces between its hard components. Some spaces are channels for blood vessels that supply bone cells with nutrients. Other spaces are storage areas for red bone marrow. Depending on the size and distribution of the spaces, different regions of a bone are categorized as compact or spongy (see Figure 5.1). Overall, about 80% of the skeleton is compact bone and 20% is spongy bone.

Compact Bone Tissue

Compact bone tissue contains few spaces between its hard components. It forms the external layer of all bones and makes up the bulk of the diaphyses of long bones. Compact bone tissue provides protection and support and resists the stresses produced by weight and movement.

Compact bone tissue is arranged in units called **osteons** or **Haversian systems** (Figure 5.3a). Blood vessels, lymphatic vessels, and nerves from the periosteum penetrate the compact bone through transverse **perforating (Volkmann's) canals.** The vessels and nerves of the perforating canals connect with those of the medullary cavity, periosteum, and **central (Haversian) canals.** The central canals run longitudinally through the bone. Around the canals are **concentric lamellae**

Figure 5.3 / Histology of compact and spongy bone. (a) Sections through the diaphysis of a long bone, from the surrounding periosteum on the right, to compact bone in the middle, to spongy bone and the medullary cavity on the left. The inset at the upper right shows an osteocyte in a lacuna. (b and c) Details of spongy bone. A photomicrograph of compact bone tissue is provided in Table 3.3 on page 78 and a scanning electron micrograph of spongy bone tissue is provided in Figure 5.12.

Osteocytes lie in lacunae arranged in concentric circles around a central canal in compact bone, and in lacunae arranged irregularly in trabeculae of spongy bone.

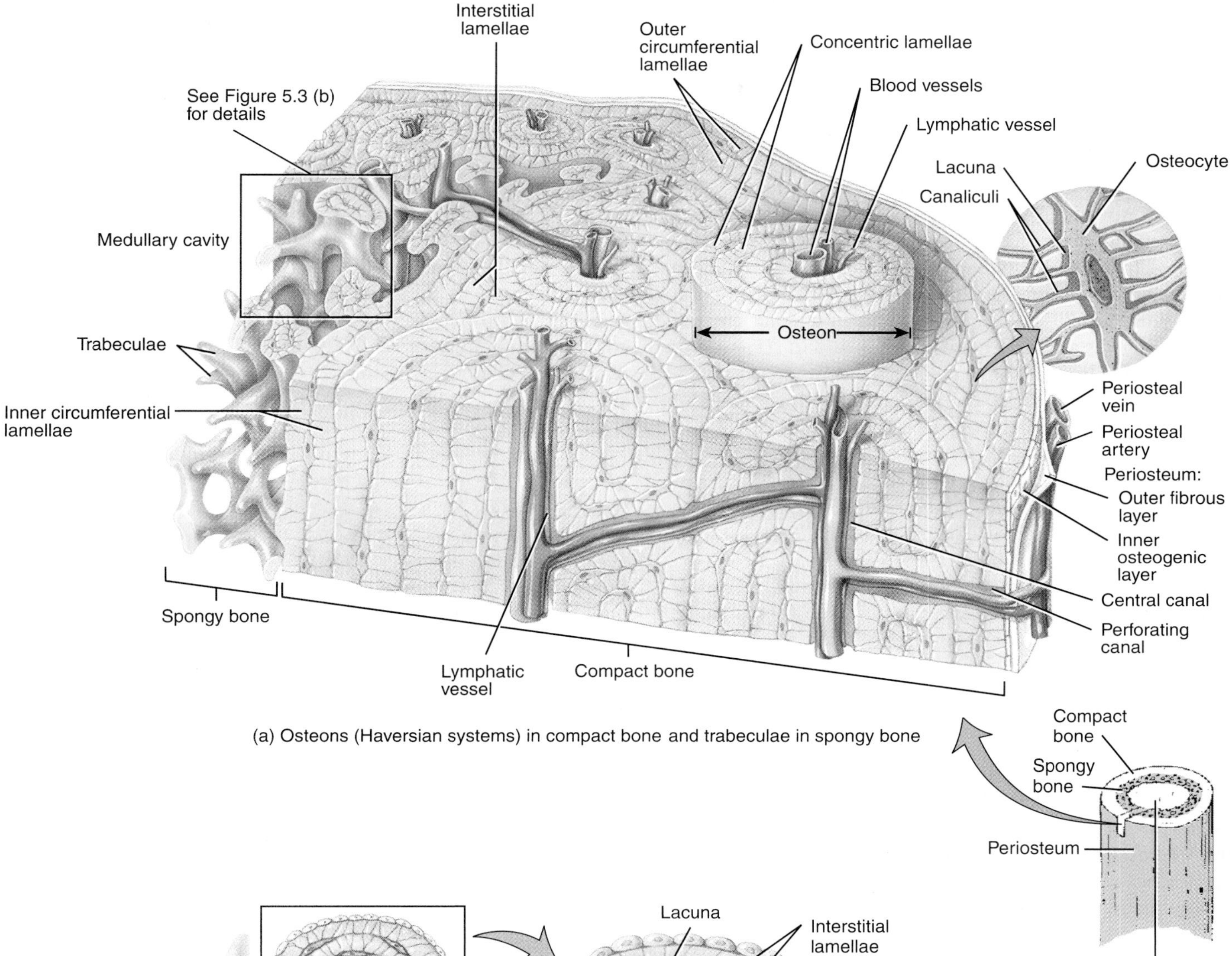

(a) Osteons (Haversian systems) in compact bone and trabeculae in spongy bone

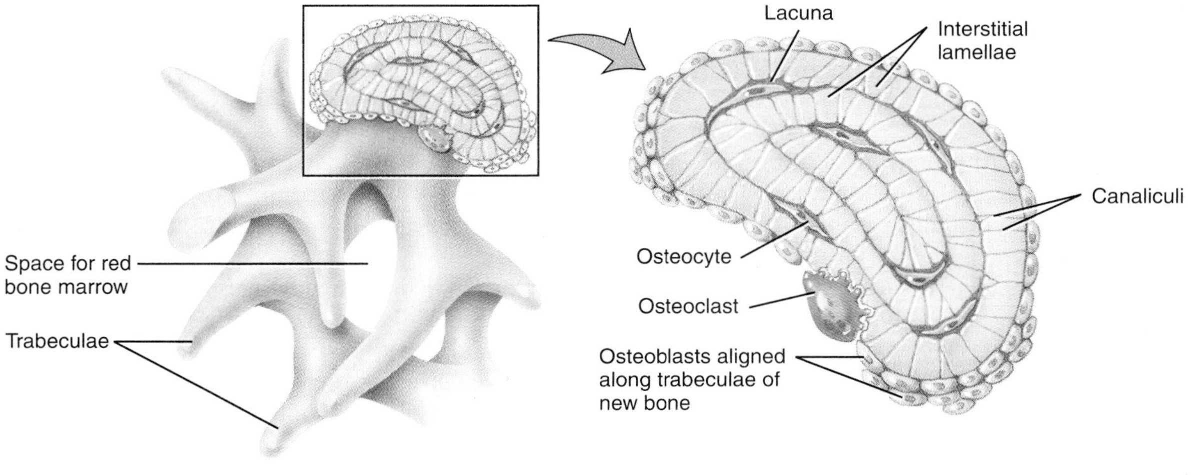

(b) Enlarged aspect of spongy bone trabeculae

(c) Details of a section of a trabecula

As people age, some central (Haversian) canals may become blocked. What effect would this have on the osteocytes?

(la-MEL-ē)—rings of hard, calcified matrix. Between the lamellae are small spaces called **lacunae** (la-KŪ-nē; = little lakes), which contain osteocytes. Radiating in all directions from the lacunae are tiny **canaliculi** (kan′-a-LIK-yū-lī; = small channels), which are filled with extracellular fluid. Inside the canaliculi are slender, fingerlike processes of osteocytes (see inset at top right of Figure 5.3a). The canaliculi connect lacunae with one another and with the central canals. This intricate branching network provides many routes for blood-borne nutrients and oxygen to diffuse through the fluid to the osteocytes and for wastes to diffuse back to the blood vessels.

Osteons in compact bone tissue are aligned in the same direction along lines of stress. In the shaft, for example, they are parallel to the long axis of the bone. As a result, the shaft of a long bone resists bending or fracturing even when considerable force is applied from either end. Compact bone tissue tends to be thickest in those parts of a bone where stresses are applied in relatively few directions. The lines of stress in a bone are not static. They change as a person learns to walk and in response to repeated strenuous physical activity, such as occurs when a person undertakes weight training. The lines of stress in a bone also can change as a result of fractures or physical deformity. Thus, the organization of osteons changes over time in response to the physical demands placed on the skeleton.

The areas between osteons contain **interstitial lamellae,** which also have lacunae with osteocytes and canaliculi. Interstitial lamellae are fragments of older osteons that have been partially destroyed during bone rebuilding or growth. Lamellae that encircle the bone just beneath the periosteum are called **outer circumferential lamellae;** those that encircle the medullary cavity are called **inner circumferential lamellae.**

Spongy Bone Tissue

In contrast to compact bone tissue, **spongy bone tissue** does not contain osteons (Figure 5.3b, c). It consists of lamellae that are arranged in an irregular lattice of thin columns of bone called **trabeculae** (tra-BEK-yū-lē; = little beams). The macroscopic spaces between the trabeculae of some bones are filled with red bone marrow, which produces blood cells. Within each trabecula (singular form) are osteocytes that lie in lacunae. Radiating from the lacunae are canaliculi. Osteocytes in the trabeculae receive nourishment directly from the blood circulating through the medullary cavities.

Spongy bone tissue makes up most of the bone tissue of short, flat, and irregularly shaped bones; most of the epiphyses of long bones; and a narrow rim around the medullary cavity of the diaphysis of long bones.

At first glance, the trabeculae of spongy bone tissue may appear to be much less well organized than the osteons of compact bone tissue. However, the trabeculae of spongy bone tissue are precisely oriented along lines of stress, a characteristic that helps bones resist stresses and transfer force without breaking. Spongy bone tissue tends to be located where bones are not heavily stressed or where stresses are applied from many directions.

Spongy bone tissue is different from compact bone tissue in two respects. First, spongy bone tissue is light, which reduces the overall weight of a bone so that it moves more readily when pulled by a skeletal muscle. Second, the trabeculae of spongy bone tissue support and protect the red bone marrow. Spongy bone in the hip bones, ribs, breastbone, backbones, and the ends of long bones is where red bone marrow is stored and hemopoiesis occurs in adults.

A **bone scan** is a diagnostic procedure that takes advantage of the fact that bone is living tissue. A small amount of a radioactive tracer compound that is readily absorbed by bone is injected intravenously. The degree of uptake of the tracer is related to the amount of blood flow to the bone. A scanning device measures the radiation emitted from the bones, and the information is translated into a photograph or diagram that can be read like an x-ray. Normal bone tissue is identified by a consistent gray color throughout because of its uniform uptake of the radioactive tracer. Darker or lighter areas, however, may indicate bone abnormalities. Darker areas, called "hot spots," are areas of increased metabolism that absorb more of the radioactive tracer. Hot spots may indicate bone cancer, abnormal healing of fractures, or abnormal bone growth. Lighter areas, called "cold spots," are areas of decreased metabolism that absorb less of the radioactive tracer. Cold spots may indicate problems such as degenerative bone disease, decalcified bone, fractures, bone infections, Paget's disease, and rheumatoid arthritis. A bone scan not only detects abnormalities 3–6 months sooner than standard x-ray procedures, it exposes the patient to less radiation. ■

✓ Why is bone considered a connective tissue?
✓ Describe the four types of cells in bone tissue.
✓ What is the composition of the matrix of bone tissue?
✓ Distinguish between spongy and compact bone tissue in terms of its microscopic appearance, location, and function.

BLOOD AND NERVE SUPPLY OF BONE

Objective

- Describe the blood and nerve supply of bone.

Bone is richly supplied with blood. Blood vessels, which are especially abundant in portions of bone containing red bone marrow, pass into bones from the periosteum. We will consider the blood supply to a long bone such as the mature tibia (shin bone) shown in Figure 5.4.

Periosteal arteries accompanied by nerves enter the diaphysis through numerous perforating (Volkmann's) canals and supply the periosteum and outer part of the compact bone (see Figure 5.3a). Near the center of the diaphysis, a large **nutrient artery** enters the compact bone through a hole called the **nutrient foramen** (Figure 5.4). The nutrient artery passes obliquely through the compact bone, and on entering the medullary cavity, divides into proximal and distal branches. These branches supply both the inner part of the diaphyseal compact bone tis-

Figure 5.4 / Blood supply of a mature long bone, the tibia (shinbone).

Bone is richly supplied with blood vessels.

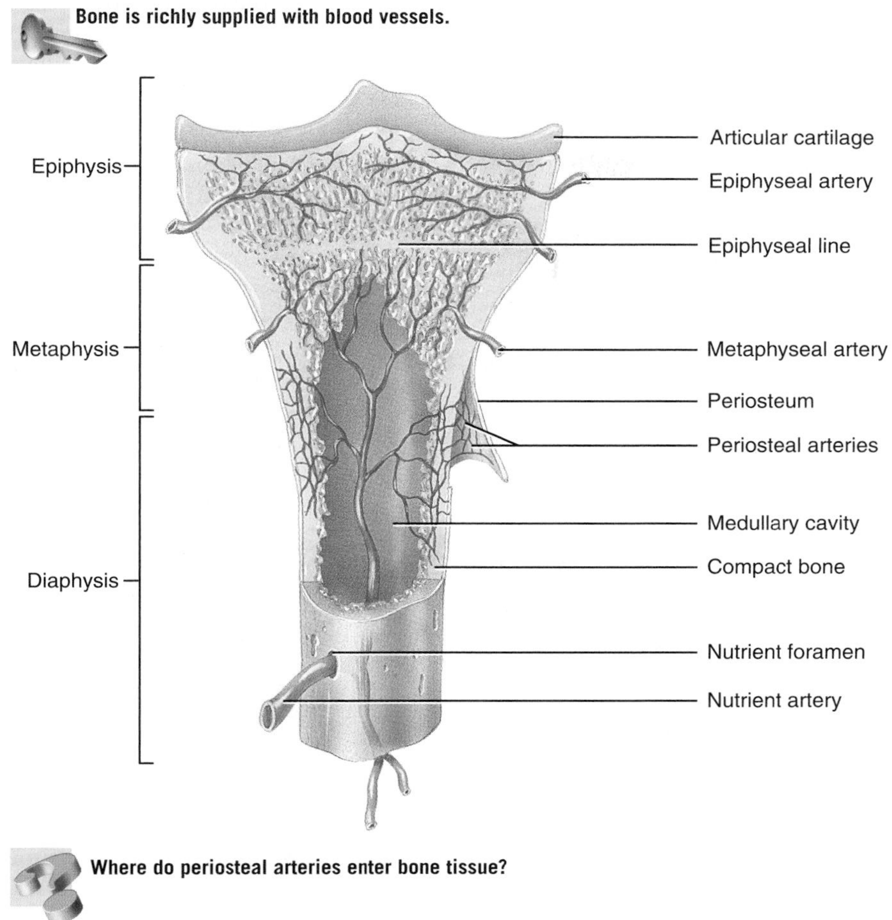

Where do periosteal arteries enter bone tissue?

sue and the spongy bone tissue and red marrow as far as the epiphyseal plates (or lines). Some bones, like the tibia, have only one nutrient artery; others like the femur (thigh bone) have several. The ends of long bones are supplied by the metaphyseal and epiphyseal arteries, which arise from arteries that supply the associated joint. The **metaphyseal arteries** enter the metaphyses of a long bone and, together with the nutrient artery, supply the red bone marrow and bone tissue of the metaphyses. The **epiphyseal arteries** enter the epiphyses of a long bone and supply the red bone marrow and bone tissue of the epiphyses.

Veins that carry blood away from long bones are evident in three places: (1) One or two **nutrient veins** accompany the nutrient artery in the diaphysis; (2) numerous **epiphyseal veins** and **metaphyseal veins** exit with their respective arteries in the epiphyses; and (3) many small **periosteal veins** exit with their respective arteries in the periosteum.

Nerves accompany the blood vessels that supply bones. The periosteum is rich in sensory nerves, some of which carry pain sensations. These nerves are especially sensitive to tearing or tension, which explains the severe pain resulting from a fracture or a bone tumor. For the same reason there is some pain associated with a bone marrow needle biopsy. In the procedure, a needle is inserted into the middle of the bone to withdraw a sample of red bone marrow to examine it microscopically for conditions such as leukemias, metastatic neoplasms, lymphoma, Hodgkin's disease, and aplastic anemia. As the needle penetrates the periosteum, pain is felt. Once it passes through, there is little pain.

✓ Explain the location and roles of the nutrient arteries, nutrient foramina, epiphyseal arteries, and periosteal arteries.

BONE FORMATION

Objective

- Describe the steps involved in intramembranous and endochondral ossification.

The process by which bone forms is called **ossification** (os′-i-fi-KĀ-shun; *ossi-* = bone; *-fication* = making) or **osteogenesis.** The "skeleton" of a human embryo is composed of fibrous connective tissue membranes formed by condensed embryonic connective tissue (mesenchyme) or pieces of hyaline cartilage that resemble the shape of bones. These embryonic tissues provide the template for subsequent ossification, which begins during the sixth or seventh week of embryonic development and follows one of two patterns. These two kinds of ossification do not lead to differences in the structure of mature bones; they are simply different methods for bone formation.

- **Intramembranous ossification** (in′-tra-MEM-bra-nus; *intra-* = within; *-membran* = membrane) is the formation of bone directly on or within fibrous connective tissue membranes formed by condensed mesenchymal cells. Such bones form *directly* from mesenchyme without first going through a cartilage stage.
- **Endochondral ossification** (en′-dō-KON-dral; *-chondr* = cartilage) is the formation of bone within hyaline cartilage. In this ossification process, mesenchymal cells are transformed into chondroblasts, which initially produce a hyaline cartilage "model" of the bone. Subsequently, osteoblasts gradually replace the cartilage with bone.

Intramembranous Ossification

Of the two processes of bone formation, intramembranous ossification involves the fewest steps. The flat bones of the skull and mandible (lower jawbone) are formed in this way. The process occurs as follows (Figure 5.5):

1. At the site of bone development, mesenchymal cells in fibrous connective tissue membranes condense (cluster) and differentiate, first into osteogenic cells and then into osteoblasts. The site of such a cluster is called a **center of ossification.** Osteoblasts secrete the organic bone matrix until they are completely surrounded by it.
2. The secretion of matrix stops. The osteoblasts become osteocytes, which lie in lacunae and extend their narrow cytoplasmic processes into canaliculi that radiate in all directions. Within a few days, calcium and other mineral salts are deposited, and the matrix hardens or calcifies.
3. The bone matrix develops into *trabeculae,* which fuse with one another to create spongy bone. Blood vessels grow into the spaces between the trabeculae and the mesenchyme along the surface of the newly formed bone. Connective tissue associated with the blood vessels in the trabeculae differentiates into red bone marrow.
4. On the outside of the bone, the mesenchyme condenses and develops into the periosteum. Eventually, most superficial layers of the spongy bone are replaced by a thin layer of compact bone, but spongy bone remains at the center. Much of the newly formed bone is remodeled (destroyed and reformed), a process that slowly transforms the bone into its adult size and shape.

Endochondral Ossification

The replacement of cartilage by bone is called endochondral ossification. Although most bones of the body are formed this way, the process is best observed in a long bone. It proceeds as follows (Figure 5.6 on page 118):

1. **Development of the cartilage model.** At the site of future bone formation, mesenchymal cells crowd together in the shape of the future bone. These cells differentiate into chondroblasts that produce a cartilage matrix; hence, the model consists of hyaline cartilage. In addition, a membrane called the **perichondrium** (per-i-KON-drē-um) develops around the cartilage model.
2. **Growth of the cartilage model.** Once chondroblasts become deeply buried in the cartilage matrix, they are called chondrocytes. The cartilage model grows in length by continual cell division of chondrocytes accompanied by further secretion of the cartilage matrix. This process results in an increase in length called **interstitial growth.** In contrast, growth of the cartilage in width is due mainly to the addition of more matrix to the periphery of the model by new chondroblasts that develop from the perichondrium. This growth pattern in which matrix is deposited on the cartilage surface is called **appositional growth** (a-pō-ZISH-a-nal).

   As the cartilage model continues to grow, chondrocytes in its midregion hypertrophy (increase in size), probably because they accumulate glycogen for ATP production and produce enzymes to catalyze additional chemical reactions. Some hypertrophied cells burst and release their contents, which increases the pH of the surrounding matrix. This change in pH triggers calcification. Other chondrocytes within the calcifying cartilage die because nutrients can no longer diffuse quickly enough through the matrix. As chondrocytes die, lacunae form and eventually merge into small cavities.
3. **Development of the primary ossification center.** Primary ossification proceeds *inward* from the external surface of the bone. A nutrient artery penetrates the perichondrium and the calcifying cartilage model through a nutrient foramen in the midregion of the cartilage model, stimulating osteogenic cells in the perichondrium to differentiate into osteoblasts. Beneath the perichondrium the osteoblasts secrete a thin shell of compact bone called the **periosteal bone collar.** Once the perichondrium starts to form bone, it is known as the **periosteum.** Near the middle of the model, periosteal capillaries extend into the disintegrating calcified cartilage. These vessels and the associated osteoblasts, osteoclasts, and red bone marrow cells are known as the **periosteal bud.** Upon growing into the cartilage model, the capillaries induce growth of a **primary ossification center,** a region where bone tissue will replace most of the cartilage. Osteoblasts then begin to deposit bone matrix over the remnants of calcified cartilage, forming spongy bone trabeculae. As the ossification center enlarges toward the ends of the bone, osteoclasts break down the newly formed spongy bone trabeculae. This activity leaves a cavity, the medullary (marrow) cavity, in the core of the model. The cavity then fills with red bone marrow.
4. **Development of secondary ossification centers.** The diaphysis (shaft), which was once a solid mass of hyaline cartilage, is replaced by compact bone that contains a medullary cavity filled with red bone marrow. When branches of the

Figure 5.5 / Intramembranous ossification. Illustrations 1 and 2 show a smaller field of vision at higher magnification than illustrations 3 and 4. Refer to this figure as you read the corresponding numbered paragraphs in the text.

Intramembranous ossification involves the formation of bone directly on or within fibrous connective tissue membranes formed by clusters of mesenchymal cells.

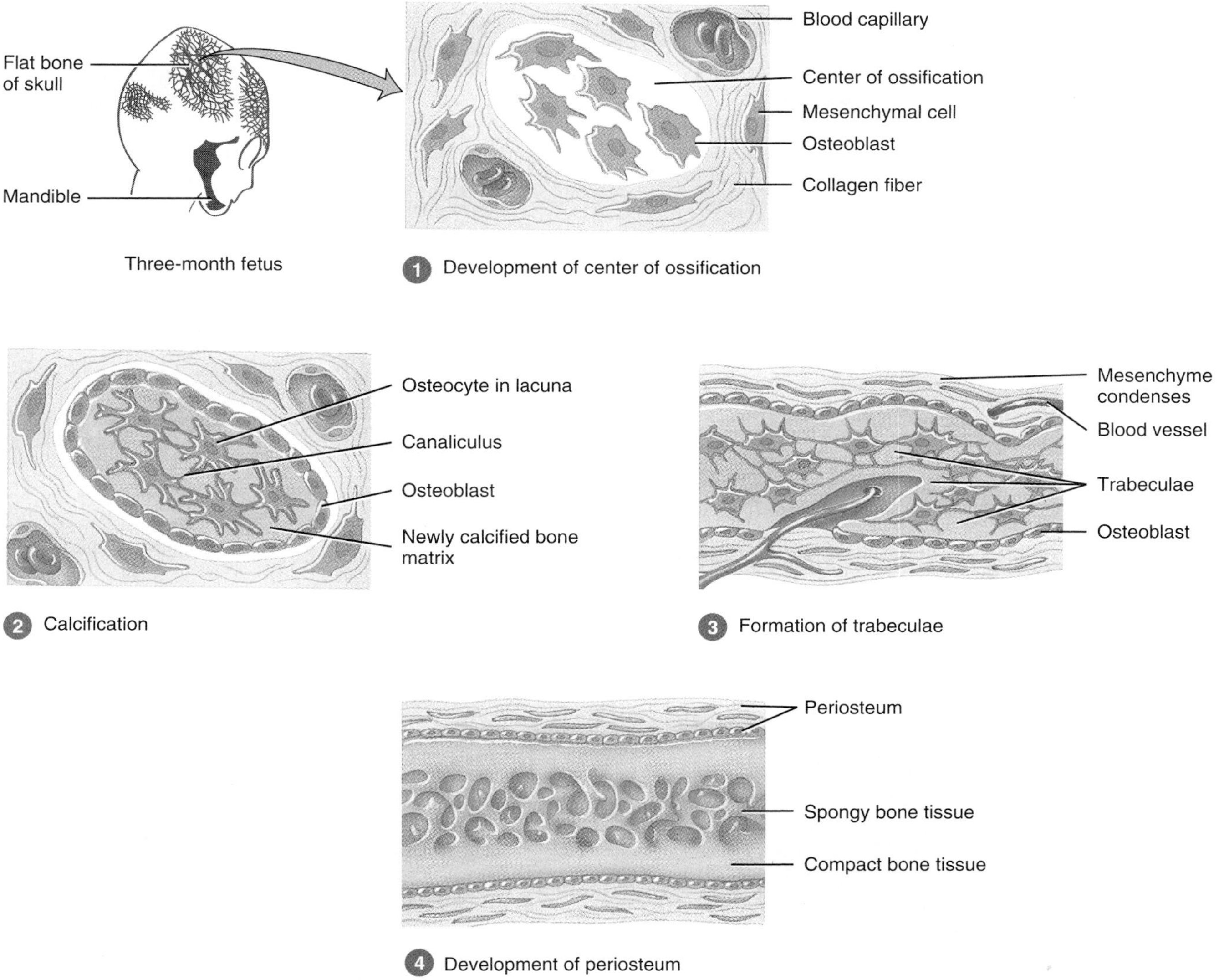

Flat bone of skull
Mandible
Three-month fetus

Blood capillary
Center of ossification
Mesenchymal cell
Osteoblast
Collagen fiber

1 Development of center of ossification

Osteocyte in lacuna
Canaliculus
Osteoblast
Newly calcified bone matrix

2 Calcification

Mesenchyme condenses
Blood vessel
Trabeculae
Osteoblast

3 Formation of trabeculae

Periosteum
Spongy bone tissue
Compact bone tissue

4 Development of periosteum

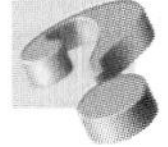

Which bones of the body develop by intramembranous ossification?

epiphyseal artery enter the epiphyses, **secondary ossification centers** develop, usually around the time of birth. Bone formation is similar to that in primary ossification centers. One difference, however, is that spongy bone remains in the interior of the epiphyses (no medullary cavities are formed there). Secondary ossification proceeds *outward* from the center of the epiphysis toward the outer surface of the bone.

5 **Formation of articular cartilage and epiphyseal plate.** The hyaline cartilage that covers the epiphyses becomes the articular cartilage. Prior to adulthood, hyaline cartilage remains between the diaphysis and epiphysis as the **epiphyseal plate,** which is responsible for the lengthwise growth of long bones.

✓ Outline the major events of intramembranous and endochondral ossification, and explain the main differences between the processes.

Figure 5.6 / Endochondral ossification.

During endochondral ossification, bone gradually replaces a cartilage model.

Perichondrium

Proximal epiphysis

Hyaline cartilage

Diaphysis

Uncalcified matrix

Calcified matrix

Periosteum

Distal epiphysis

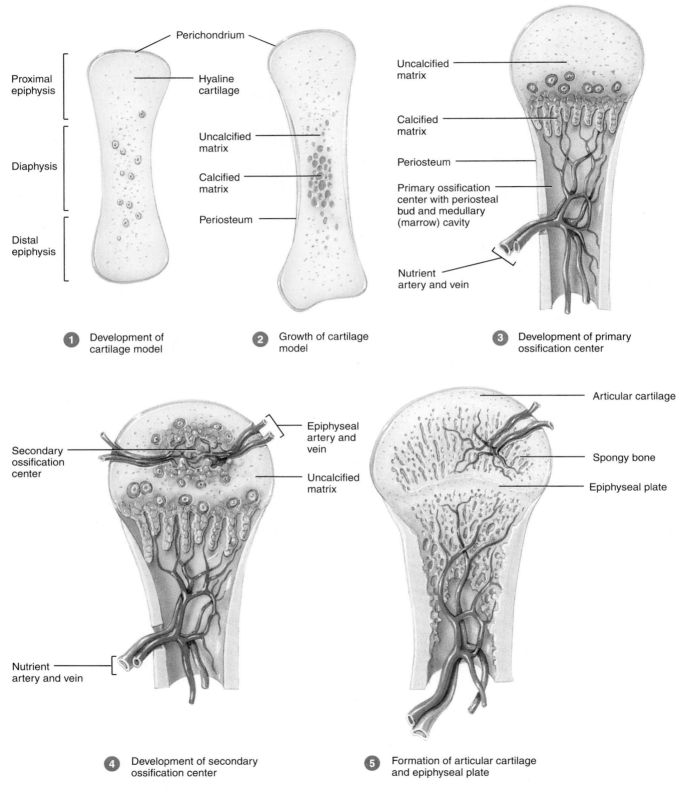

(1) Development of cartilage model

(2) Growth of cartilage model

Uncalcified matrix

Calcified matrix

Periosteum

Primary ossification center with periosteal bud and medullary (marrow) cavity

Nutrient artery and vein

(3) Development of primary ossification center

Secondary ossification center

Epiphyseal artery and vein

Uncalcified matrix

Nutrient artery and vein

(4) Development of secondary ossification center

Articular cartilage

Spongy bone

Epiphyseal plate

(5) Formation of articular cartilage and epiphyseal plate

If radiographs of an 18-year-old basketball star player show clear epiphyseal plates but no epiphyseal lines, is she likely to grow taller?

BONE GROWTH

Objective

- Describe how bones grow in length and width.

During childhood, bones throughout the body grow in width by appositional growth, and long bones lengthen by the addition of bone material on the diaphyseal side of the epiphyseal plate. Bones stop growing in length at about age 25, although they may continue to thicken.

Growth in Length

To understand how a bone grows in length, you need to know about the structure of the epiphyseal plate (Figure 5.7). The **epiphyseal plate** (ep′-i-FIZ-ē-al), a layer of hyaline cartilage in the metaphysis of a growing bone, consists of four zones (Figure 5.7b).

1. **Zone of resting cartilage.** This layer is nearest the epiphysis and consists of small, scattered chondrocytes. The cells do not function in bone growth (thus the term *resting*). Instead, they anchor the epiphyseal plate to the bone of the epiphysis.
2. **Zone of proliferating cartilage.** Slightly larger chondrocytes in this zone are arranged like stacks of coins. The chondrocytes divide to replace those that die at the diaphyseal side of the epiphyseal plate.
3. **Zone of hypertrophic cartilage.** In this layer, the chondrocytes are even larger and arranged in columns. The lengthening of the diaphysis is the result of cell divisions in the zone of proliferating cartilage and maturation of the cells in the zone of hypertrophic cartilage.
4. **Zone of calcified cartilage.** The final zone of the epiphyseal plate is only a few cells thick and consists mostly of dead chondrocytes because the matrix around them has calcified. The calcified cartilage is dissolved by osteoclasts, and the area is invaded by osteoblasts and capillaries from the diaphysis. The osteoblasts lay down bone matrix, replacing the calcified cartilage. As a result, the diaphyseal border of the epiphyseal plate is firmly cemented to the bone of the diaphysis.

The activity of the epiphyseal plate is the only way that the diaphysis can increase in length. As a bone grows, chondrocytes proliferate on the epiphyseal side of the plate. New chondrocytes cover older ones, which are then destroyed by calcification. Thus, the cartilage is replaced by bone on the diaphyseal side of the plate. In this way the thickness of the epiphyseal plate remains relatively constant, but the bone on the diaphyseal side increases in length.

Between the ages of 18 and 25, the epiphyseal plates close; that is, the epiphyseal cartilage cells stop dividing, and bone replaces all the cartilage. The epiphyseal plate fades, leaving a bony structure called the **epiphyseal line.** The appearance of the epiphyseal line signifies that the bone has stopped growing in length. The clavicle is the last bone to stop growing. If a bone fracture damages the epiphyseal plate, the fractured bone may be shorter than normal once adult stature is reached. This is because damage to cartilage, which is avascular, accelerates closure of the epiphyseal plate, and growth in length of the bone is inhibited.

Figure 5.7 / The epiphyseal plate is a layer of hyaline cartilage in the metaphysis of a growing bone. The epiphyseal plate appears as a dark band between whiter calcified areas in the radiograph.

The epiphyseal plate allows the diaphysis of a bone to increase in length.

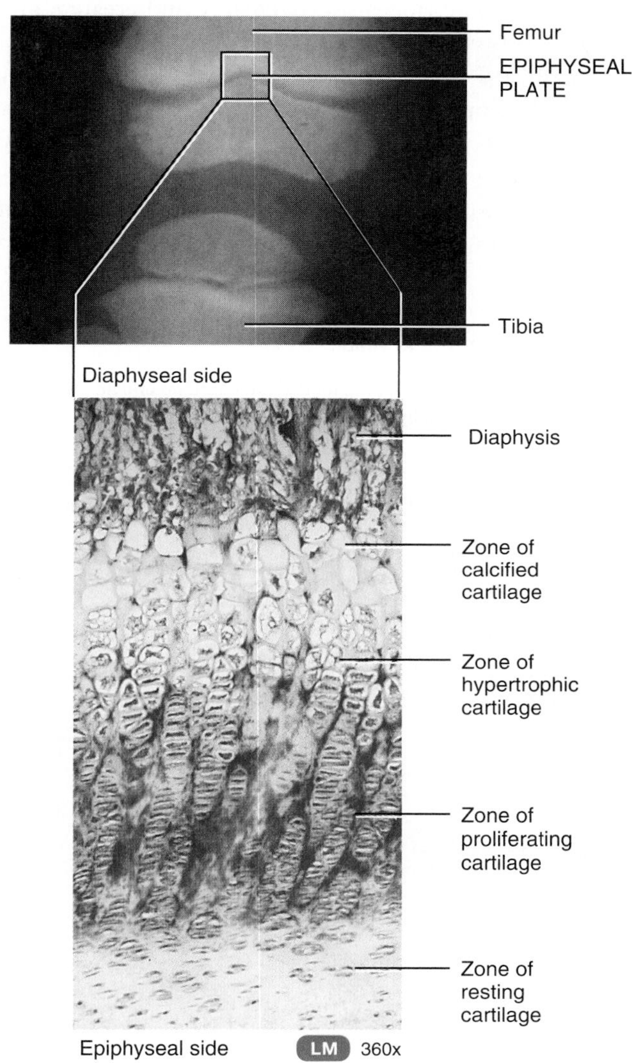

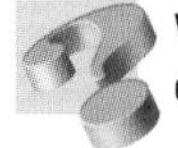

What activities of the epiphyseal plate account for the lengthwise growth of the diaphysis?

Growth in Width (Thickness)

Unlike cartilage, which can grow by both interstitial and appositional growth, bone can grow in width only by appositional growth. The process by which bone grows in width is as follows (Figure 5.8):

1. At the bone surface, periosteal cells differentiate into osteoblasts, which secret the collagen fibers and other organic molecules that form bone matrix. The osteoblasts become surrounded by matrix and develop into osteocytes. This process forms bone ridges on either side of a periosteal blood vessel. The ridges slowly enlarge and create a groove for the periosteal blood vessel.
2. Eventually, the ridges fold together and fuse, and the groove becomes a tunnel that encloses the blood vessel. The former periosteum now becomes the endosteum that lines the tunnel.
3. Bone deposition by osteoblasts from the endosteum forms new concentric lamellae. The formation of additional concentric lamellae proceeds inward toward the periosteal blood vessel. In this way, the tunnel fills in, and a new osteon is created.
4. As an osteon is forming, osteoblasts under the periosteum deposit new outer circumferential lamellae, thereby increasing the width of the bone. As additional periosteal blood vessels become enclosed as in Step 1, the growth process continues.

Recall that as new bone tissue is being deposited on the outer surface of bone, the bone tissue lining the medullary cavity is destroyed by osteoclasts in the endosteum. In this way the medullary cavity enlarges as the bone increases in width.

✓ Describe the zones of the epiphyseal plate and their functions.
✓ Explain how bone grows in length and width.
✓ What is the significance of the epiphyseal line?

Figure 5.8 / Bone growth in width.

As new bone tissue is deposited on the outer surface of bone by osteoblasts, the bone tissue lining the medullary cavity is destroyed by osteoclasts in the endosteum.

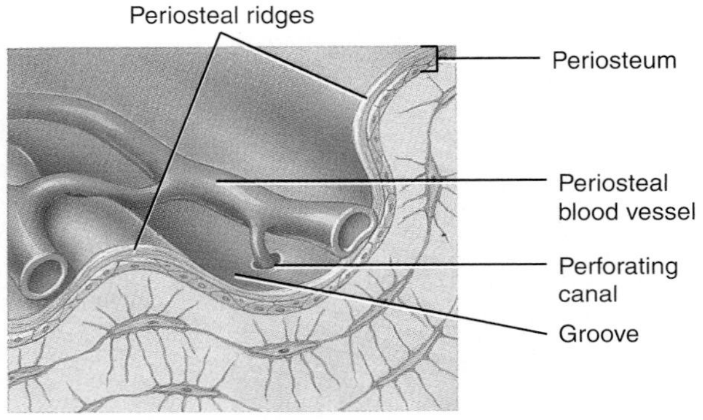

1 Ridges in periosteum create groove for periosteal blood vessel.

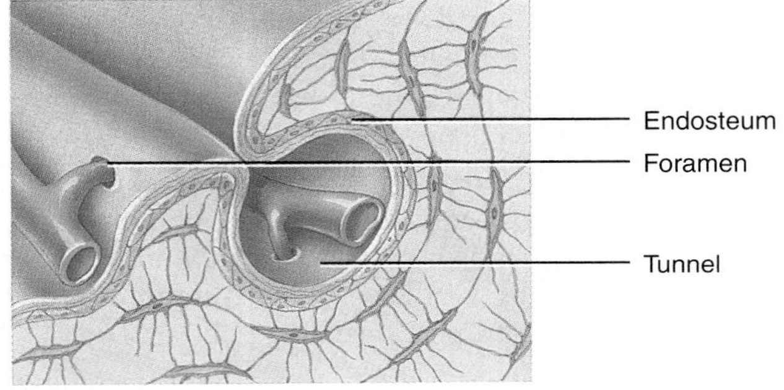

2 Periosteal ridges fuse, forming an endosteum-lined tunnel.

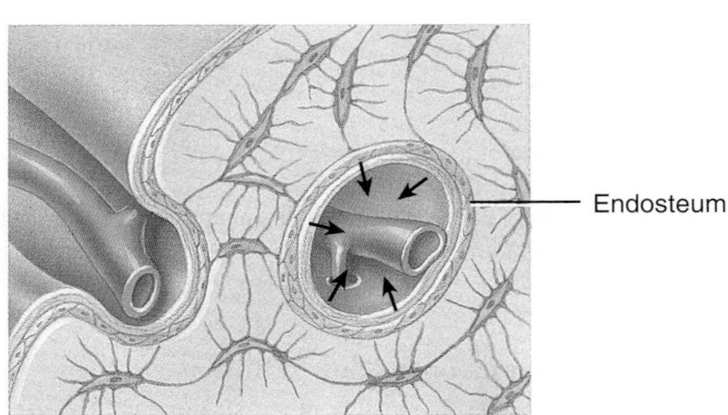

3 Osteoblasts in endosteum build new concentric lamellae inward toward center of tunnel, forming a new osteon.

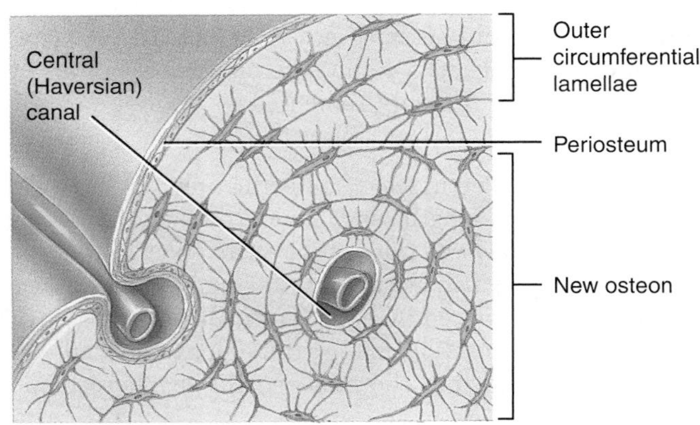

4 Bone grows outward as osteoblasts in periosteum build new outer circumferential lamellae. Osteon formation repeats as new periosteal ridges fold over blood vessels.

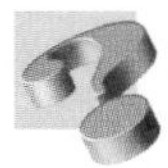

How does the medullary cavity enlarge during growth in width?

BONE REPLACEMENT

Objective

- Describe the processes involved in bone remodeling.

Bone, like skin, forms before birth but continually renews itself thereafter. **Remodeling** is the ongoing replacement of old bone tissue by new bone tissue. Bone constantly remodels and redistributes its matrix along lines of mechanical stress. Compact bone is formed from spongy bone. However, even after bones have reached their adult shapes and sizes, old bone is continually destroyed and new bone tissue is formed in its place. Remodeling also removes worn and injured bone, replacing it with new bone tissue. This process allows bone to serve as the body's reservoir for calcium. Many other tissues need calcium to perform their functions. Several hormones continually regulate exchanges of calcium between blood and bones.

Remodeling takes place at different rates in various body regions. The distal portion of the thighbone (femur) is replaced about every four months. By contrast, bone in certain areas of the shaft of the femur will not be completely replaced during an individual's life. Osteoclasts are responsible for bone resorption (destruction of matrix). A delicate balance exists between the actions of osteoclasts in removing minerals and collagen and of bone-making osteoblasts in depositing minerals and collagen. Should too much new tissue be formed, the bones become abnormally thick and heavy. If too much mineral material is deposited in the bone, the surplus may form thick bumps, or spurs, on the bone that interfere with movement at joints. A loss of too much calcium or tissue weakens the bones, and they may break, as occurs in osteoporosis, or they may become too flexible, as in rickets and osteomalacia. Abnormal acceleration of the remodeling process results in a condition called Paget's disease in which bone deformity results, especially in the bones of the limbs, pelvis, lower vertebrae, and skull. The newly formed bone becomes hard and brittle and fractures easily.

Normal bone growth in the young and bone replacement in the adult depend on the presence of (1) several minerals: calcium, phosphorus, fluoride, magnesium, iron, and manganese; (2) several vitamins: A, B_{12}, C, D, and K; and (3) several hormones: human growth hormone, sex hormones (estrogens and testosterone), thyroid hormones, calcitonin, parathyroid hormone, and insulin.

1. Define remodeling, and describe the roles of osteoblasts and osteoclasts in the process.

FRACTURE AND REPAIR OF BONE

Objective

- Describe the sequence of events in repair of a fracture.

A **fracture** is any break in a bone. Fractures are named according to their severity or the shape or position of the fracture line—or even the physician who first described them (Figure 5.9):

Figure 5.9 / Types of bone fractures.

A fracture is any break in a bone.

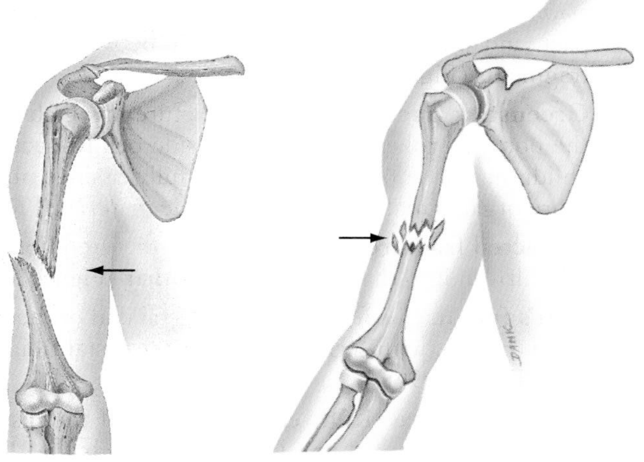

(a) Open fracture

(b) Comminuted fracture

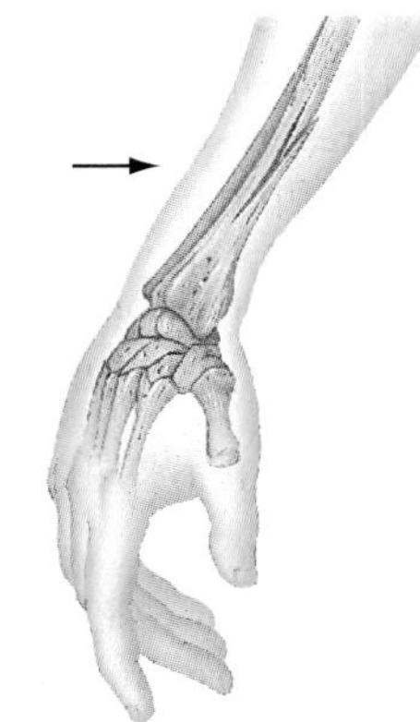

(c) Greenstick fracture

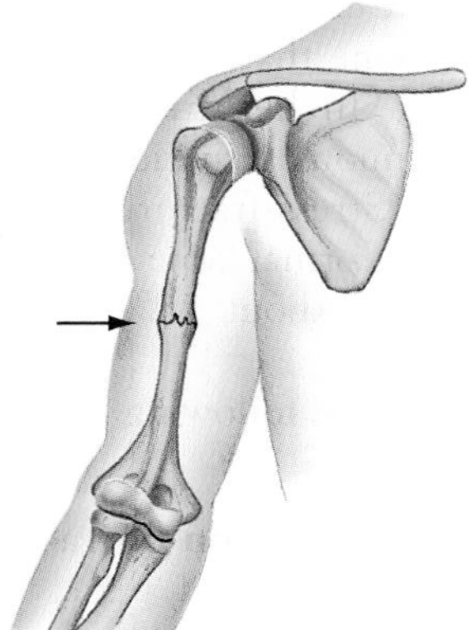

(d) Impacted fracture

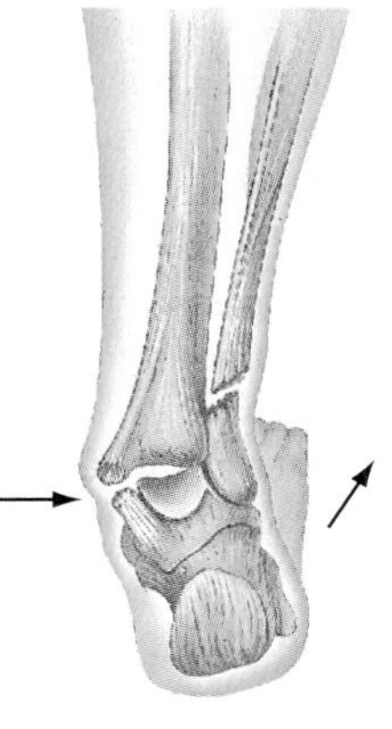

(e) Pott's fracture

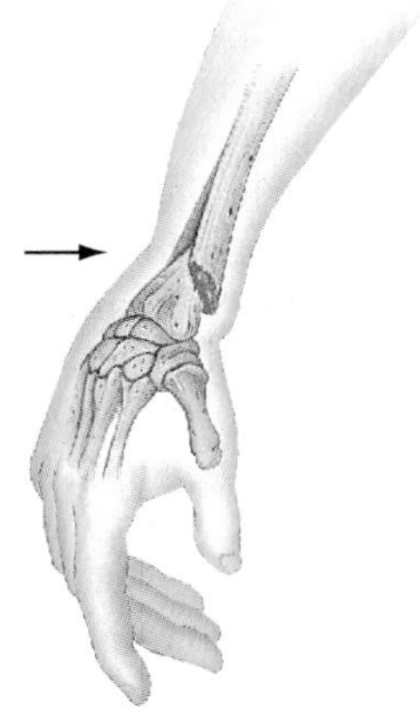

(f) Colles' fracture

What is a fibrocartilaginous callus?

EXERCISE AND BONE TISSUE

Objective

- Describe how exercise and mechanical stress affect bone tissue.

Within limits, bone has the ability to alter its strength in response to changes in mechanical stress. When placed under stress, bone tissue adapts by becoming stronger through increased deposition of mineral salts and production of collagen fibers. Another effect of stress is to increase the production of calcitonin, a hormone that inhibits bone resorption. Without mechanical stress, bone does not remodel normally because resorption outstrips bone formation. Removal of mechanical stress weakens bone through demineralization (loss of bone minerals) and decreased numbers of collagen fibers.

The main mechanical stresses on bone result from the contraction of skeletal muscles and the pull of gravity. If a person is bedridden or has a fractured bone in a cast, the strength of the unstressed bones diminishes. Astronauts subjected to the microgravity of space also lose bone mass. In both cases, bone loss can be dramatic—as much as 1% per week. Bones of athletes, which are repetitively and highly stressed, become notably thicker than those of nonathletes. Weight-bearing activities, such as walking or moderate weight lifting, help build and retain bone mass. In recent years it has become increasingly clear that adolescents and young adults should engage in regular weight-bearing exercise prior to the closure of the epiphyseal plates. This helps to build total bone mass prior to its inevitable reduction with aging. Even elderly people can strengthen their bones by engaging in weight-bearing exercise.

✓ Explain the types of mechanical stresses that may be used to strengthen bone tissue.

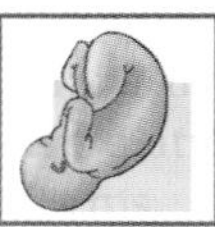

DEVELOPMENTAL ANATOMY OF THE SKELETAL SYSTEM

Objective

- Describe the development of the skeletal system and the limbs.

As noted earlier, both intramembranous and endochondral ossification begin when *mesenchymal cells*, connective tissue cells derived from **mesoderm,** migrate into the area where bone formation will occur. In some skeletal structures, mesenchymal cells develop into chondroblasts that form cartilage. In other skeletal structures, mesenchymal cells develop into osteoblasts that form *bone tissue* by intramembranous or endochondral ossification.

Discussion of the development of the skeletal system provides an excellent opportunity to trace the development of the limbs. The limbs make their appearance about the fifth week as small elevations called **limb buds** (Figure 5.11a) at the sides of the trunk. They consist of masses of general **mesoderm** covered by **ectoderm.** At this point, a mesenchymal skeleton exists in the limbs; some of the masses of mesoderm surrounding the developing bones will become the skeletal muscles of the limbs.

By the sixth week, the limb buds develop a constriction around the middle portion. The constriction produces distal segments of the upper buds called **hand plates** and distal segments of the lower buds called **foot plates** (Figure 5.11b). These plates represent the beginnings of the hands and feet, respectively. At this stage of limb development, a cartilaginous skeleton is present. By the seventh week (Figure 5.11c), the arm, forearm, and hand are evident in the upper limb bud, and the thigh, leg, and foot appear in the lower limb bud. Endochondral ossification has begun. By the eighth week (Figure 5.11d), as the shoulder, elbow, and wrist areas become apparent, the upper limb bud is appropriately called the upper limb, and the lower limb bud is now the lower limb.

The **notochord** is a flexible rod of mesoderm that defines the midline of the embryo and also gives it some rigidity. The notochord is situated where the future vertebral column will develop. As the vertebrae form, the notochord becomes surrounded by the developing vertebral bodies. The notochord eventually disappears, except for remnants that persist as the *nucleus pulposus* of the intervertebral discs (see Figure 6.17d on page 157).

✓ Describe when and how the limbs develop. ■

AGING AND BONE TISSUE

Objective

- Describe the effects of aging on bone tissue.

From birth through adolescence, more bone tissue is produced than is lost during bone remodeling. In young adults the rates of bone deposition and resorption are about the same. As the level of sex steroids diminishes during middle age, especially in women after menopause, a decrease in bone mass occurs because bone resorption outpaces bone deposition. In old age, loss of bone through resorption occurs more rapidly than bone gain. Because women's bones generally are smaller and less massive than men's bones to begin with, loss of bone mass in old age typically has a greater adverse effect in women.

There are two principal effects of aging on bone tissue: loss of bone mass and brittleness. The first effect results from the loss of calcium and other minerals from bone matrix (demineralization). This loss usually begins after age 30 in females, accelerates greatly around age 45 as levels of estrogens decrease, and continues until as much as 30% of the calcium in bones is lost by age 70. Once bone loss begins in females, about 8% of bone mass is lost every ten years. In males, calcium loss typically does not begin until after age 60, and about 3% of bone mass is lost every ten years. The loss of calcium from bones is one of the problems in osteoporosis (described shortly).

The second principal effect of aging on the skeletal system, brittleness, results from a lowered rate of protein synthesis,

Figure 5.11 / Features of a developing human embryo during weeks 5 through 8 of development. The liver and heart prominence are surface bulges resulting from the underlying growth of these organs. The embryo receives oxygen and nutrients from its mother through blood vessels in the umbilical cord.

After the limb buds develop, endochondral ossification of the limb bones begins during the seventh embryonic week.

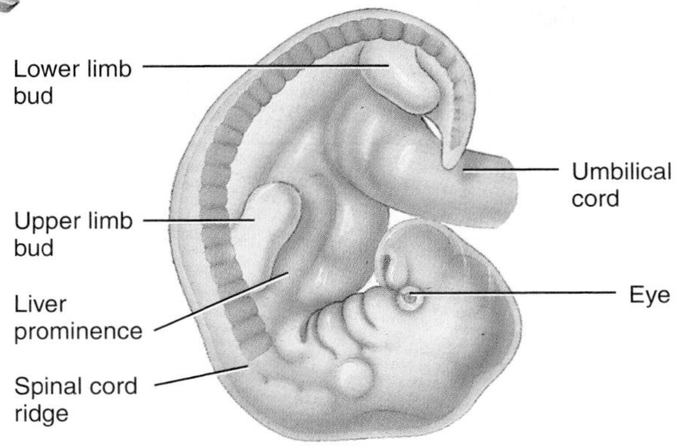

(a) Five-week embryo showing development of limb buds

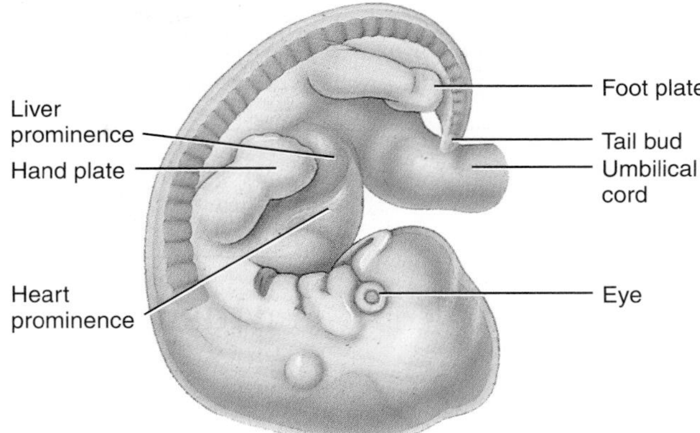

(b) Six-week embryo showing development of hand and foot plates

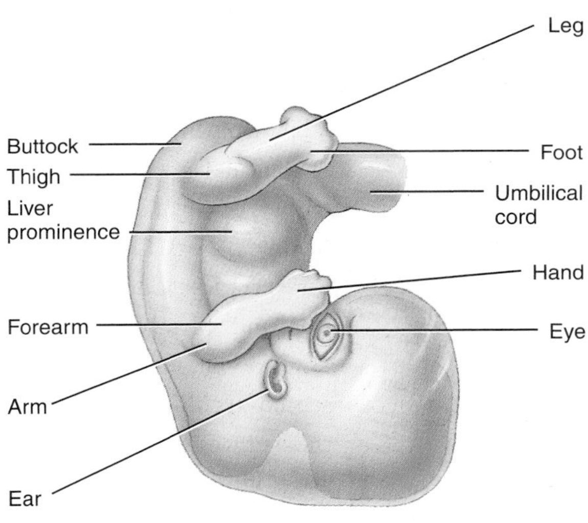

(c) Seven-week embryo showing development of arm, forearm, and hand in upper limb bud and thigh, leg, and foot in lower limb bud

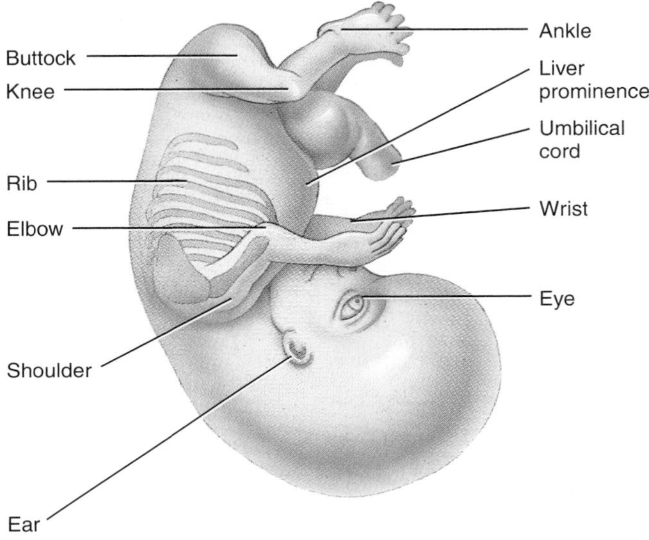

(d) Eight-week embryo in which limb buds have developed into upper and lower limbs

Which of the three basic embryonic tissues—ectoderm, mesoderm, and endoderm—gives rise to the skeletal system?

which diminishes the organic portion of bone matrix, mainly collagen fibers, that gives bone its tensile strength (resistance to being stretched or torn apart). As a consequence, inorganic minerals gradually constitute a greater proportion of the bone matrix. The loss of tensile strength causes the bones to become very brittle and susceptible to fracture. In some elderly people, collagen fiber synthesis slows, in part due to diminished production of human growth hormone. In addition to increasing the susceptibility to fractures, loss of bone mass also leads to deformity, pain, stiffness, loss of height, and loss of teeth.

✓ What is demineralization and how does it affect the functioning of bone?

✓ What changes occur in the organic portion of bone matrix with aging?

14th Century

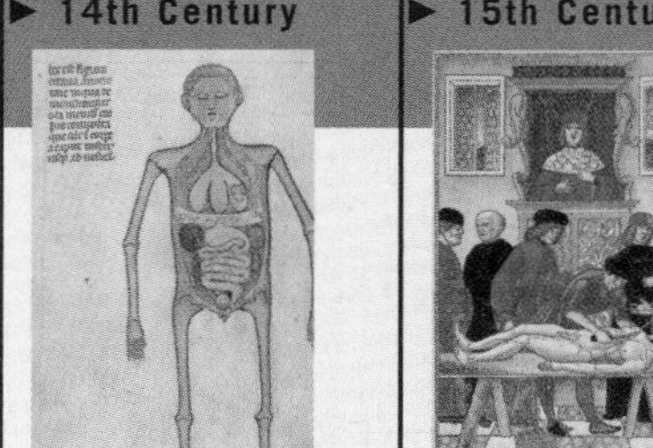

15th Century

16th Century

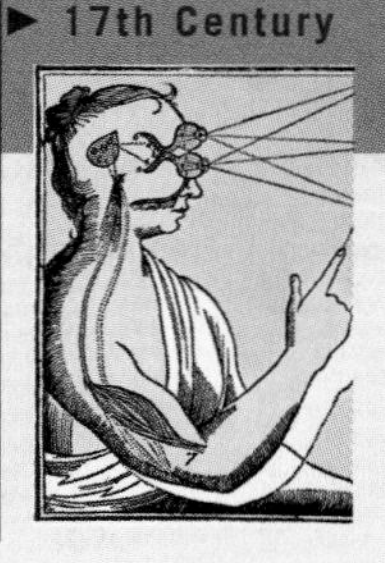

17th Century

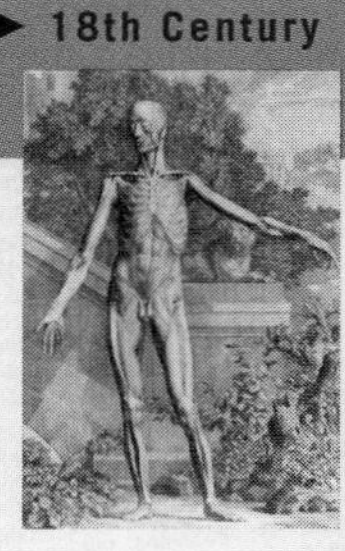

18th Century

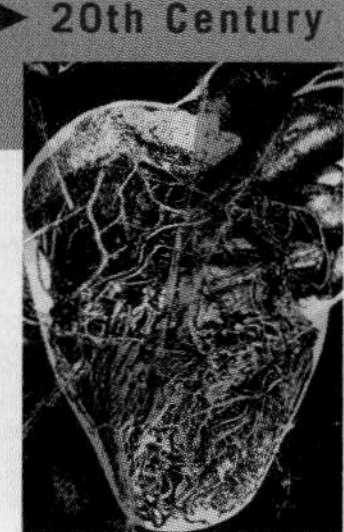

20th Century

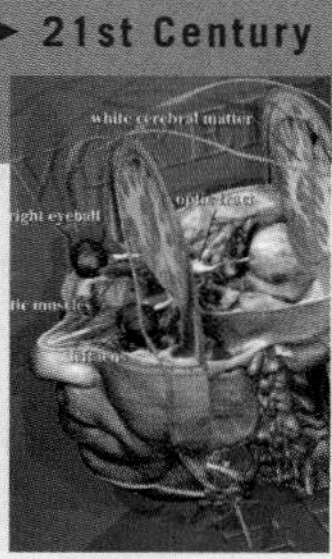

21st Century

CHANGING IMAGES

Muslim, Medieval, Modern

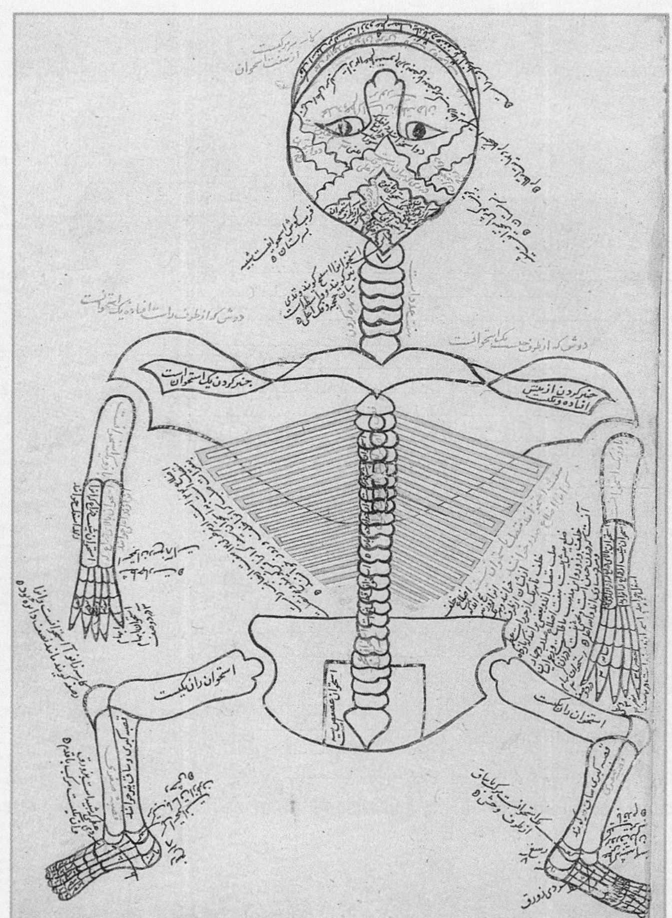

1493 AD

From the Iberian Peninsula to the Near East, an empire united under the faith of Islam rivaled that of Rome at its peak. Between the 7th and 9th century, the Mediterranean Basin and beyond was a colorful tapestry of Muslim culture that was woven with the threads of a great diversity of peoples and regions. As this empire grew, it tended to absorb the cultures of the subjugated. The writings of Greek scholars such as Galen, whom you may remember from Chapter 1, were translated into both Arabic and Persian. Muslim anatomists embraced these writings and utilized them for their own research. Indeed, the Medieval Latin versions of these Greek writings were translated from the libraries of the Muslim world. Had this not occurred, much of classic Greek science and philosophy would have been lost for posterity.

Pictured here is a skeletal figure from 15th century Islam. The "frog-like" anatomical position is typical for the era and was also employed throughout Europe in the Middle Ages. Examine this image for accuracy. Count the number of ribs and vertebrae. Are there too few, too many, or are they represented accurately? Can you discern the sutures of the skull? Did the artist intend exact anatomical accuracy, or are the bones and cranial sutures merely meant to be symbolic? To answer some of these questions you may have to look ahead to Chapter 6.

As you learn more about the rich cultural history of anatomy, one point to keep in mind is that it is based upon the contributions of *many* civilizations. As you enjoy viewing this skeletal image, examine it for not only its anatomical accuracy, but for how it represents the contributions of people who may be very different from yourself.

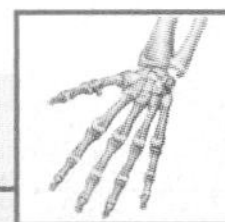

APPLICATIONS TO HEALTH

Osteoporosis

Osteoporosis (os′-tē-ō-pō-RŌ-sis; *por-* = passageway; *-osis* = condition) is literally a condition of porous bones (Figure 5.12). The basic problem is that bone resorption outpaces bone deposition. In large part this is due to depletion of calcium from the body—more calcium is lost in urine, feces, and sweat than is absorbed from the diet. Bone deposition also depends on certain hormones, such as estrogens. In response to low levels of calcium and estrogens, bone mass becomes so depleted that bones fracture, often spontaneously, under the mechanical stresses of everyday living. For example, a hip fracture might result from simply sitting down too quickly. In the United States, osteoporosis causes more than a million fractures a year, mainly in the hip, wrist, and vertebrae. Osteoporosis afflicts the entire skeletal system. In addition to fractures, osteoporosis causes shrinkage of vertebrae, height loss, hunched backs, and bone pain.

Thirty million people in the United States suffer from osteoporosis. The disorder primarily affects middle-aged and elderly people, 80% of them women. Older women suffer from osteoporosis more often than men for two reasons: Women's bones are less massive than men's bones, and production of estrogens in women declines dramatically at menopause, whereas production of the main androgen, testosterone, in older men wanes gradually and only slightly. Estrogen and testosterone increase osteoblast activity and synthesis of bone matrix. Besides gender, risk factors for developing osteoporosis include a family history of the disease, European or Asian ancestry, thin or small body build, an inactive lifestyle, cigarette smoking, a diet low in calcium and vitamin D, more than two alcoholic drinks a day, and the use of certain medications.

In postmenopausal women, treatment of osteoporosis may include estrogen replacement therapy (ERT; low doses of estrogens) or hormone replacement therapy (HRT; a combination of estrogens and progesterone, another sex steroid). Although such treatments help combat osteoporosis, they increase a woman's risk of breast cancer. The drug Raloxifene (Evista) mimics the beneficial effects of estrogens on bone without increasing the risk of breast cancer. A nonhormone drug called Alendronate (Fosamax) blocks resorption of bone by osteoclasts and is also used to treat osteoporosis.

Figure 5.12 / Comparison of spongy bone tissue from (a) a normal young adult and (b) a person with osteoporosis. Notice the weakened trabeculae in (b). Compact bone tissue is similarly affected by osteoporosis.

In osteoporosis, bone resorption outpaces bone formation, so bone mass decreases.

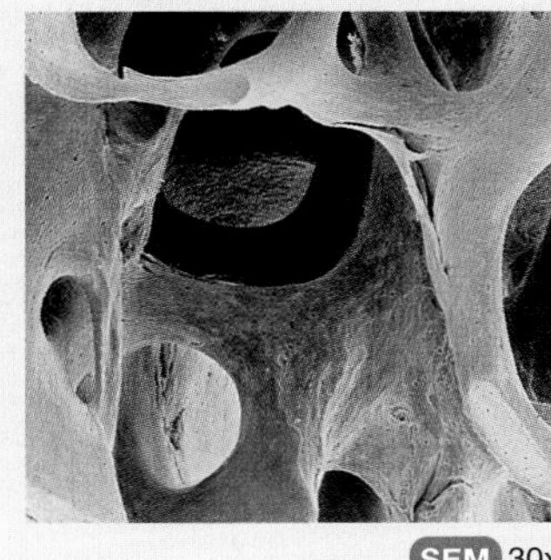

(a) Normal bone

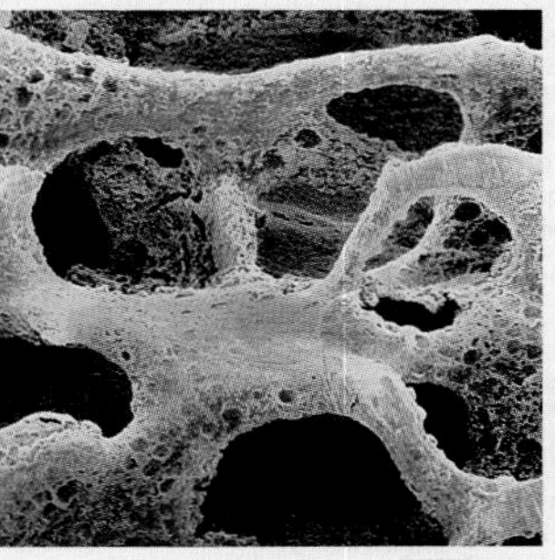

(b) Osteoporotic bone

If you wanted to develop a drug to lessen the effects of osteoporosis, would you look for a chemical that inhibits the activity of osteoblasts, or that of osteoclasts?

Perhaps more important than treatment is prevention. Adequate calcium intake and weight-bearing exercise in her early years may be more beneficial to a woman than drugs and calcium supplements when she is older.

Rickets and Osteomalacia

Rickets and **osteomalacia** (os′-tē-ō-ma-LĀ-she-a; *-malacia* = softness) are disorders in which bones fail to calcify. Although the organic matrix is still produced, calcium salts are not deposited, and the bones become "soft" or rubbery and easily deformed. Rickets affects the growing bones of children. Because new bone formed at the epiphyseal plates fails to ossify, bowed legs and deformities of the skull, rib cage, and pelvis are common results. In osteomalacia, sometimes called "adult rickets," new bone formed during remodeling fails to calcify. This disorder causes varying degrees of pain and tenderness in bones, especially in the hip and leg. Bone fractures also result from minor trauma.

KEY MEDICAL TERMS ASSOCIATED WITH BONE TISSUE

Osteoarthritis (os′-tē-ō-ar-THRĪ-tis; *arthr* = joint) The degeneration of articular cartilage such that the bony ends touch; the resulting friction of bone against bone worsens the condition. Usually associated with the elderly.

Osteogenic sarcoma (os′-tē-Ō-JEN-ik sar-KŌ-ma; *sarcoma* = connective tissue tumor) Bone cancer that primarily affects osteoblasts and occurs most often in teenagers during their growth spurt; the most common sites are the metaphyses of the thighbone (femur), shinbone (tibia), and arm bone (humerus). Metastases occur most often in lungs; treatment consists of multidrug chemotherapy and removal of the malignant growth, or amputation of the limb.

Osteomyelitis (os′-tē-ō-mī-el-Ī-tis) An infection of bone characterized by high fever, sweating, chills, pain, and nausea, pus formation, edema, and warmth over the affected bone and rigid overlying muscles. It is often caused by bacteria, usually *Staphylococcus aureus.* The bacteria may reach the bone from outside the body (through open fractures, penetrating wounds, or orthopedic surgical procedures); from other sites of infection in the body (abscessed teeth, burn infections, urinary tract infections, or upper respiratory infections) via the blood; and from adjacent soft tissue infections (as occurs in diabetes mellitus).

Osteopenia (os′-tē-ō-PĒ-nē-a; *penia* = poverty) Reduced bone mass due to a decrease in the rate of bone synthesis to a level insufficient to compensate for normal bone resorption; any decrease in bone mass below normal. An example is osteoporosis.

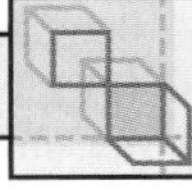

STUDY OUTLINE

Introduction (p. 110)

1. A bone is made up of several different tissues: bone, or osseous, tissue, cartilage, dense connective tissues, epithelium, various blood-forming tissues, adipose tissue, and nervous tissue.
2. The entire framework of bones and their cartilages constitute the skeletal system.

Functions of the Skeletal System (p. 110)

1. The skeletal system functions in support, protection, movement, mineral storage and release, blood cell production, and triglyceride storage.

Anatomy of a Bone (p. 110)

1. Parts of a typical long bone are the diaphysis (shaft), proximal and distal epiphyses (ends), metaphyses, articular cartilage, periosteum, medullary (marrow) cavity, and endosteum.

Histology of Bone Tissue (p. 111)

1. Bone tissue consists of widely separated cells surrounded by large amounts of matrix.
2. The four principal types of cells in bone tissue are osteogenic cells, osteoblasts, osteocytes, and osteoclasts.
3. The matrix of bone contains abundant mineral salts (mostly hydroxyapatite) and collagen fibers.
4. Compact bone tissue consists of osteons (Haversian systems) with little space between them.
5. Compact bone tissue lies over spongy bone tissue in the epiphyses and makes up most of the bone tissue of the diaphysis. Functionally, compact bone tissue protects, supports, and resists stress.
6. Spongy bone tissue does not contain osteons. It consists of trabeculae surrounding spaces filled with red bone marrow.
7. Spongy bone tissue forms most of the structure of short, flat, and irregular bones, and the epiphyses of long bones. Functionally, spongy bone tissue trabeculae offer resistance along lines of stress, support and protect red bone marrow, and make bones lighter for easier movement.

Blood and Nerve Supply of Bone (p. 114)

1. Long bones are supplied by periosteal, nutrient, and epiphyseal arteries; veins accompany the arteries.
2. Nerves accompany blood vessels in bone; the periosteum is rich in sensory neurons.

Bone Formation (p. 115)

1. Bone forms by a process called ossification (osteogenesis), which begins when mesenchymal cells are transformed into osteogenic cells, which undergo cell division and give rise to cells that differentiate into osteoblasts and osteoclasts.
2. Ossification begins during the sixth or seventh week of embryonic development. The two types of ossification, intramembranous and endochondral, involve the replacement of a preexisting connective tissue with bone.
3. Intramembranous ossification occurs within fibrous connective tissue membranes.
4. Endochondral ossification occurs within a hyaline cartilage model. The primary ossification center of a long bone is in the diaphysis. Cartilage degenerates, leaving cavities that merge to form the medullary cavity. Osteoblasts lay down bone. Next, ossification occurs in the epiphyses, where bone replaces cartilage, except for the epiphyseal plate.

Bone Growth (p. 119)

1. The epiphyseal plate consists of four zones: zones of resting cartilage, proliferating cartilage, hypertrophic cartilage, and calcified cartilage.
2. Because of the activity at the epiphyseal plate, the diaphysis of a bone increases in length.
3. Bone grows in width as a result of the addition of new bone tissue by periosteal osteoblasts around the outer surface of the bone (appositional growth).

Bone Replacement (p. 121)

1. Remodeling is the replacement of old bone tissue by new bone tissue.
2. Old bone tissue is constantly destroyed by osteoclasts, whereas new bone is constructed by osteoblasts.
3. Remodeling requires minerals (calcium, phosphorus, fluoride, magnesium, iron, and manganese), vitamins (A, B_{12}, C, D, and K), and hormones (human growth hormone, sex hormones, insulin, thyroid hormones, parathyroid hormone, and calcitonin).

Fracture and Repair of Bone (p. 121)

1. A fracture is any break in a bone.
2. Fracture repair involves formation of a fracture hematoma, of a fibrocartilaginous callus, of a bony callus, and bone remodeling.
3. Types of fractures include closed (simple), open (compound), comminuted, greenstick, impacted, stress, Pott's, and Colles'.

Exercise and Bone Tissue (p. 124)

1. Mechanical stress increases bone strength by increasing deposition of mineral salts and production of collagen fibers.
2. Removal of mechanical stress weakens bone through demineralization and collagen fiber reduction.

Developmental Anatomy of the Skeletal System (p. 124)

1. Bone forms from mesoderm by intramembranous or endochondral ossification.
2. Limbs develop from limb buds, which consist of mesoderm and ectoderm.

Aging and Bone Tissue (p. 124)

1. The principal effect of aging is a loss of calcium from bones, which may result in osteoporosis.
2. Another effect is a decreased production of matrix proteins (mostly collagen fibers), which makes bones more brittle and thus more susceptible to fracture.

SELF-QUIZ QUESTIONS

Choose the one best answer to the following questions:

1. Which of the following is *not* considered a function of bone tissue and the skeletal system? (a) attachment site of muscles, tendons, and ligaments, (b) stores calcium and phosphorus, (c) formation of blood cells, (d) support and protection of soft organs and tissues, (e) synthesis of vitamin D.

2. Consider the following four sets of vitamins and minerals. (1) vitamins B_{12} and K, (2) calcium and phosphorus, (3) vitamins A, C, and D, (4) copper and sodium. Which would you recommend in the diet of a child for the growth of bones? (a) all of the above, (b) 1, 2, and 4, (c) 2, 3, and 4, (d) 1, 2, and 3, (e) 2 and 3.

3. Which of the following is/are true? (1) Spongy bone tissue is found in the epiphyses of long bones and inside flat bones. (2) The microscopic structure of compact bone shows osteons. (3) Long bones consist entirely of compact bone, which is composed of osteons. (4) Spongy bone tissue consists of an irregular latticework of thin columns of bony material with many spaces.
   (a) 1 only, **(b)** 2 only, **(c)** 3 only, **(d)** 4 only, **(e)** 1, 2 and 4.

Complete the following:

4. The hardness of bone is due primarily to salts of ________ in the matrix.

5. Most bones of the body are formed from cartilage by the process of ________ ossification.

6. Typical of all connective tissues, bone consists mainly of ________ .

7. The region of the epiphyseal plate that is farthest from the diaphysis is the ________ .

8. The medullary cavity in the long bone of an adult contains _______ .

9. The cells responsible for the resorption (destruction) of bone tissue are ________ .

10. The process by which a bone grows in width is called ________ growth.

Are the following statements true or false?

11. The flat bones of the skull develop by intramembranous ossification.

12. Canaliculi are microscopic canals running longitudinally through bone.

13. Another name for the epiphysis is the shaft of the bone.

14. The epiphyseal plate appears earlier in life than the epiphyseal line.

15. A fracture hematoma is a mass of fibrocartilage that bridges the broken ends of the bones.

16. The organic matrix of bone tissue is secreted by osteoblasts.

17. Periosteal arteries enter the tissue of the diaphysis through perforating (Volkmann's) canals.

18. Match the following.

____ **(a)** thin layer of hyaline cartilage at end of long bone
____ **(b)** region of mature bone where diaphysis joins epiphysis
____ **(c)** covering over bone to which ligaments and tendons attach
____ **(d)** a layer of hyaline cartilage in the metaphysis of a growing bone
____ **(e)** lining of the medullary cavity; contains osteoprogenitor cells and osteoblasts

(1) articular cartilage
(2) endosteum
(3) fibrous periosteum
(4) metaphysis
(5) epiphyseal plate

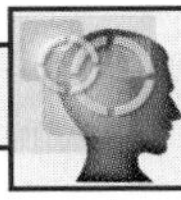

CRITICAL THINKING QUESTIONS

1. Lynda, a petite 55-year-old couch potato who smokes heavily, wants to lose 50 pounds before her next class reunion. Her diet consists mostly of diet soda and crackers. Explain the effects of her age and lifestyle on her bone composition.
   HINT *Bone is constantly remodeled.*

2. In a popular children's book, the bones in the hero's arm are magically removed by accident. (Don't worry—they grow back overnight.) What would be the results if your skeleton was to magically disappear?
   HINT *You've heard of jellyfish; how about jelly people?*

3. Aunt Edith is 95 years old today. She comments that she's been getting shorter every year and soon she'll fadeout all together. What's happening to Aunt Edith?
   HINT *Her mind is as sharp as ever; think about her bones.*

4. Astronaut John Glenn exercised everyday while in space, yet he and the other astronauts experienced weakness upon their return to earth. Why?
   HINT *It's harder to keep an object down in space than it is to pick it up.*

5. Chantal has never exercised but she finally started taking aerobics class three times a week. She's not sure if she'll stick with it so she wears her girlfriend's running suit and some old sneakers that she borrowed from her niece. She told her girlfriend, "I can deal with the sore muscles and sweating and stuff but my leg bones are killing me!" What Chantal's problem?
   HINT *You don't have to look good but you do need the right equipment.*

ANSWERS TO FIGURE QUESTIONS

5.1 The periosteum is essential for growth in bone width, bone repair, and bone nutrition. It also serves as a point of attachment for ligaments and tendons.

5.2 Bone resorption is necessary for the development, growth, maintenance, and repair of bone.

5.3 The central (Haversian) canals are the main blood supply to the osteocytes of an osteon (Haversian system), so their blockage would lead to death of the osteocytes.

5.4 Periosteal arteries enter bone tissue through perforating (Volkmann's) canals.

5.5 Flat bones of the skull and mandible (lower jawbone) develop by intramembranous ossification.

5.6 Yes, she probably will grow taller. Epiphyseal lines are indications of growth zones that have ceased to function. The absence of epiphyseal lines indicates that the bone is still lengthening.

5.7 The lengthwise growth of the diaphysis is caused by cell divisions in the zone of proliferating cartilage and maturation of the cells in the zone of hypertrophic cartilage.

5.8 The medullary cavity enlarges by activity of the osteoclasts in the endosteum.

5.9 A fibrocartilaginous callus is a mass of fibrocartilaginous repair tissue that bridges the ends of broken bones.

5.10 Healing of bone fractures can take months because calcium and phosphorus deposition is a slow process, and bone cells generally grow and reproduce slowly.

5.11 The skeletal system develops from mesoderm.

5.12 Inhibiting the activity of osteoclasts might lessen the effects of osteoporosis.

6

THE SKELETAL SYSTEM: THE AXIAL SKELETON

CONTENTS AT A GLANCE

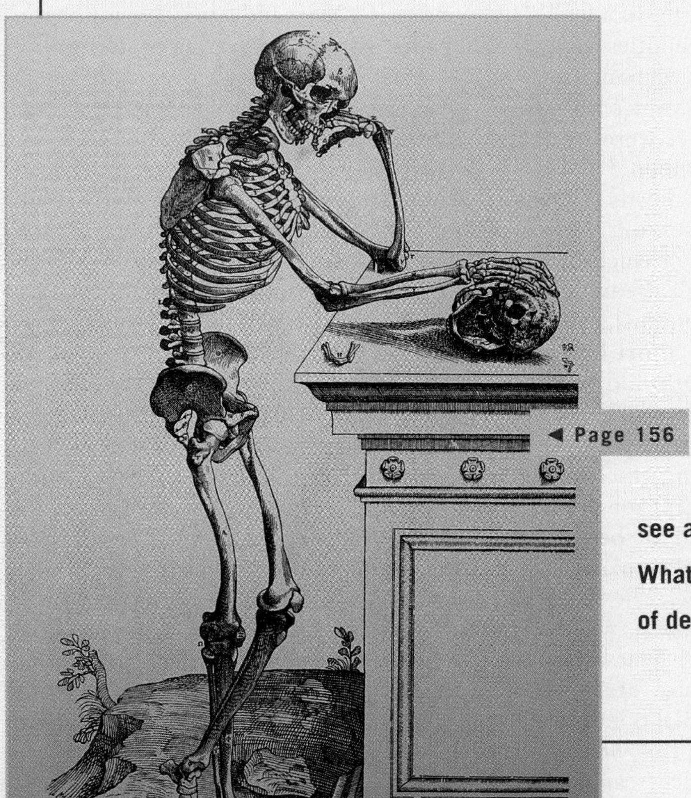

◄ Page 156

What do you think of when you see a picture of a skeleton? What words come to mind? Symbol of death or foundation of our bodies?

Page 140 ►

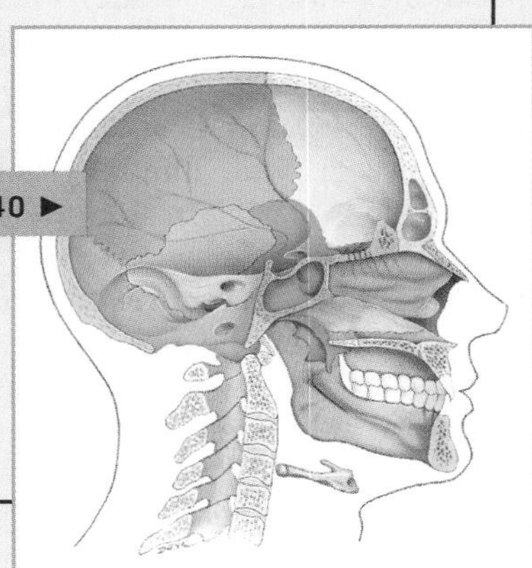

INTRODUCTION

Without bones, you would be unable to perform movements such as walking or grasping. The slightest blow to your head or chest could damage your brain or heart. It would even be impossible for you to chew food. Because the skeletal system forms the framework of the body, a familiarity with the names, shapes, and positions of individual bones will help you locate other organ systems. For example, the radial artery, the site where pulse is usually taken, is named for its proximity to the radius, the lateral bone of the forearm. The frontal lobe of the brain lies deep to the frontal (forehead) bone. The tibialis anterior muscle is located near the anterior surface of the tibia (shinbone). The ulnar nerve is named for its proximity to the ulna, the medial bone of the forearm. Parts of certain bones serve to locate structures within the skull and to outline the lungs and heart and abdominal and pelvic organs.

Movements such as throwing a ball, biking, and walking require an interaction between bones and muscles. To understand how muscles produce different movements, you need to learn where on bones the muscles attach and the types of joints acted on by the contracting muscles. The bones, muscles, and joints together form an integrated system called the **musculoskeletal system.**

DIVISIONS OF THE SKELETAL SYSTEM

Objective

- Describe how the skeleton is divided into axial and appendicular divisions.

The adult human skeleton consists of 206 named bones. The skeletons of infants and children have more than 206 bones because some of their bones, such as the hip bones and different backbones, fuse later in life. There are two principal divisions in the adult skeleton: the **axial skeleton** and the **appendicular skeleton.** Refer to Figure 6.1 to see how the two divisions join to form the complete skeleton. The bones of the axial skeleton are shown in blue. The longitudinal ***axis,*** or center, of the human body is a straight line that runs through the body's center of gravity. This imaginary line extends through the head and down to the space between the feet. The axial skeleton consists of the bones that lie around the axis: skull bones, auditory ossicles (ear bones), hyoid bone, ribs, breastbone, and bones of the backbone. Although the auditory ossicles are not considered part of the axial or appendicular skeleton, but rather as a separate group of bones, they are placed with the axial skeleton for convenience. The middle portion of each ear contains three auditory ossicles held together by a series of ligaments. The auditory ossicles vibrate in response to sound waves that strike the eardrum and have a key role in the mechanism of hearing. This is described in detail in Chapter 20.

The appendicular skeleton contains the bones of the **upper** and **lower limbs (extremities),** plus the bones called **girdles** that connect the limbs to the axial skeleton. Table 6.1 on page 134 presents the standard grouping of the 80 bones of the axial skeleton and the 126 bones of the appendicular skeleton.

We will organize our study of the skeletal system around the two divisions of the skeleton, with emphasis on how the many bones of the body are interrelated. In this chapter we focus on the axial skeleton, looking first at the skull and then at the bones of the vertebral column (backbone) and the chest. In Chapter 7 we explore the appendicular skeleton, examining in turn the bones of the pectoral (shoulder) girdle and upper limbs, and then the pelvic (hip) girdle and the lower limbs. But before we examine the skull we direct our attention to some rather general characteristics of bones.

✓ List the bones that make up the axial and appendicular divisions of the skeleton.

TYPES OF BONES

Objective

- Classify bones on the basis of shape and location.

Almost all the bones of the body can be classified into five principal types on the basis of shape: long, short, flat, irregular, and sesamoid (Figure 6.2 on page 134). **Long bones** have greater length than width and consist of a shaft and a variable number of extremities (ends). They are slightly curved for strength. A curved bone absorbs the stress of the body's weight at several different points so that the stress is evenly distributed. If such bones were straight, the weight of the body would be unevenly distributed and the bone would easily fracture. Long bones consist mostly of *compact bone tissue,* which is dense and has few spaces, but they also contain considerable amounts of *spongy bone tissue,* which has larger spaces. Long bones include those in the thigh (femur), leg (tibia and fibula), toes (phalanges), arm (humerus), forearm (ulna and radius), and fingers (phalanges).

Short bones are somewhat cube-shaped and nearly equal in length and width. They consist of spongy bone except at the surface, where there is a thin layer of compact bone. Examples of short bones are the wrist or carpal bones (except for the pisiform, which is a sesamoid bone) and the ankle or tarsal bones (except for the calcaneus, which is an irregular bone).

Flat bones are generally thin and composed of two nearly parallel plates of compact bone enclosing a layer of spongy bone. The layers of compact bone are called external and internal *tables,* whereas the spongy bone is referred to as *diploe* (DIP-lō-ē). Flat bones afford considerable protection and provide extensive areas for muscle attachment. Flat bones include the cranial bones, which protect the brain; the breastbone (sternum) and ribs, which protect organs in the thorax; and the shoulder blades (scapulae).

Irregular bones have complex shapes and cannot be grouped into any of the three categories just described. They also vary in the amount of spongy and compact bone present. Such bones include the backbones (vertebrae) and certain facial bones.

Figure 6.1 / Divisions of the skeletal system. The axial skeleton is indicated in blue, the appendicular skeleton in yellow. (Note the position of the hyoid bone in Figure 6.4a. See Tortora, *A Photographic Atlas of the Human Body,* Figure 3.1.)

The adult human skeleton consists of 206 bones grouped into axial and appendicular divisions.

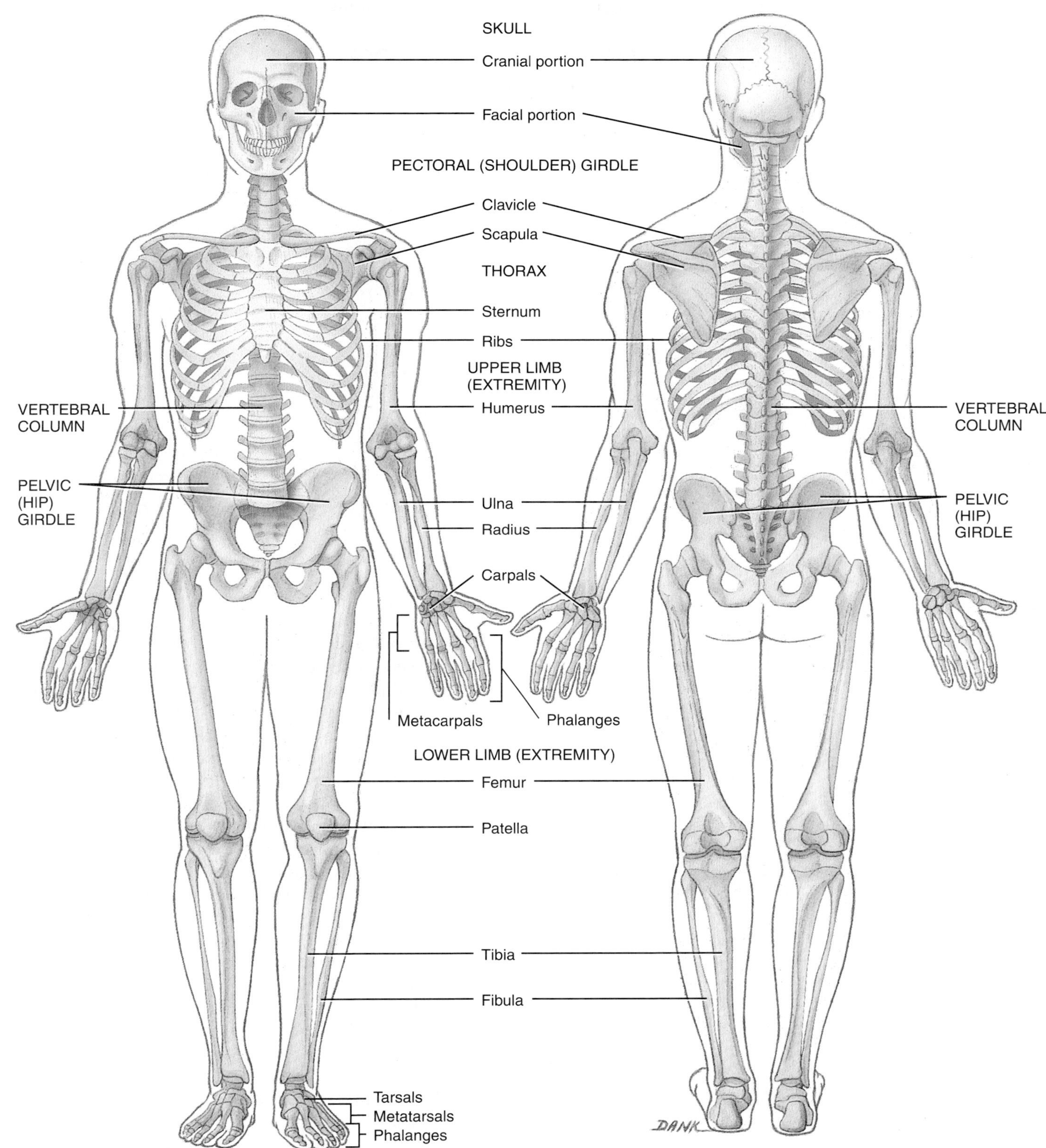

Which of the following structures are part of the axial skeleton, and which are part of the appendicular skeleton? Skull, clavicle, vertebral column, shoulder girdle, humerus, pelvic girdle, and femur.

Table 6.1 The Bones of the Adult Skeletal System

Division of the Skeleton	Structure	Number of Bones
Axial Skeleton		
	Skull	
	Cranium	8
	Face	14
	Hyoid	1
	Auditory ossicles	6
	Vertebral column	26
	Thorax	
	Sternum	1
	Ribs	24
		Subtotal = 80

Division of the Skeleton	Structure	Number of Bones
Appendicular Skeleton	**Pectoral (shoulder) girdles**	
	Clavicle	2
	Scapula	2
	Upper limbs (extremities)	
	Humerus	2
	Ulna	2
	Radius	2
	Carpals	16
	Metacarpals	10
	Phalanges	28
	Pelvic (hip) girdle	
	Hip, pelvic, or coxal bone	2
	Lower limbs (extremities)	
	Femur	2
	Fibula	2
	Tibia	2
	Patella	2
	Tarsals	14
	Metatarsals	10
	Phalanges	28
		Subtotal = 126
		Total = 206

Figure 6.2 / Types of bones based on shape. The bones are not drawn to scale.

The shapes of bones largely determine their functions.

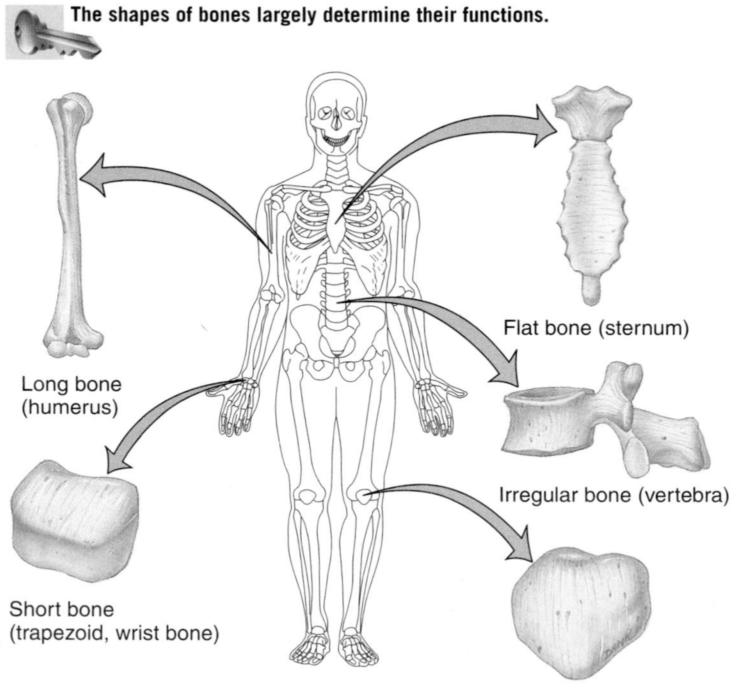

Which type of bone primarily protects and provides a large surface area for muscle attachment?

Sesamoid (meaning shaped like a sesame seed) **bones** develop in certain tendons where there is considerable friction, tension, and physical stress. They are not always completely ossified and measure only a few millimeters in diameter except for the two patellae (kneecaps), which are large sesamoid bones. The patellae are normally present in all individuals in the quadriceps femoris tendon. Otherwise, sesamoid bones vary in number from person to person. Functionally, sesamoid bones protect tendons from excessive wear and tear, and they often change the direction of pull of a tendon, which improves the mechanical advantage at a joint.

In the upper limbs, sesamoid bones normally occur only in the joints of the palmar surface of the hands. The two constant sesamoid bones are in the tendons of the adductor pollicis and flexor pollicis brevis muscles at the metacarpophalangeal joint of the thumb (see Figure 7.7). In the lower limbs, aside from the patellae, two constant sesamoid bones occur on the plantar surface of the foot in the tendons of the flexor hallucis brevis muscle at the metatarsophalangeal joint of the great (big) toe (see Figure 7.15).

An additional type of bone is not included in this classification by shape, but instead is classified by location. **Sutural** (SŪ-chur-al; *sutura* = seam) or **Wormian** (named after a Danish anatomist, O. Worm, 1588–1654) **bones** are small bones located within the sutures (joints) of certain cranial bones (see Figure 6.6). Their number varies greatly from person to person.

During active growth of the skeleton, red bone marrow is progressively replaced by yellow bone marrow in most of the long bones. In adults, red bone marrow is restricted to flat bones such as the ribs, sternum, and skull; irregular bones such as vertebrae and hip bones; long bones such as the proximal epiphyses of the femur and humerus; and some short bones.

✓ Give examples of long, short, flat, and irregular bones.

BONE SURFACE MARKINGS

Objective

- Describe the principal surface markings on bones and the functions of each.

Bones have characteristic **surface markings,** structural features adapted for specific functions. They are not present at birth but develop later in response to certain forces and are most prominent during adult life. In response to tension on a bone surface where tendons, ligaments, aponeuroses, and fasciae pull on the periosteum of bone, new bone is deposited, resulting in raised or roughened areas. Conversely, compression on a bone surface results in a depression. There are two major types of surface markings: (1) *depressions and openings,* which form joints or allow the passage of soft tissues (such as blood vessels and nerves), and (2) *processes,* which are projections or outgrowths that either help form joints or serve as attachment points for connective tissue (such as ligaments and tendons). Table 6.2 describes the various surface markings and provides examples of each.

Skeletal remains may persist for thousands of years after a person has died. They make it possible to trace patterns of disease and nutrition, evaluate the effects of certain social and economic changes, and deduce patterns of reproduction and mortality. Skeletal remains also may reveal an individual's sex, age, height, and race.

Many disorders can leave permanent effects on skeletal material. Three of the many common causes of bone pathologies are malnutrition, tumors, and infections. Each may cause specific changes in bone that permit diagnosis from skeletal remains. ■

✓ List and describe several bone surface markings, and give an example of each. Check your list against Table 6.2.

Table 6.2 Bone Surface Markings

Depressions and Openings: Sites allowing the passage of soft tissues (nerves, blood vessels, ligaments, tendons)

Marking	Description	Example
Fissure (FISH-ur)	Narrow slit between adjacent parts of bones through which blood vessels or nerves pass.	Superior orbital fissure of the sphenoid bone (Figure 6.13).
Foramen (fō-RĀ-men)	Opening (*foramen* = hole) through which blood vessels, nerves, or ligaments pass.	Optic foramen of the sphenoid bone (Figure 6.13).
Fossa (FOS-a)	Shallow depression (*fossa* = basin).	Coronoid fossa of the humerus (Figure 7.5a).
Sulcus (SUL-kus)	Furrow (*sulcus* = groove) along a bone surface that accommodates a blood vessel, nerve, or tendon.	Intertubercular sulcus of the humerus (Figure 7.5a).
Meatus (mē-Ā-tus)	Tubelike opening (*meatus* = passageway).	External auditory (acoustic) meatus of the temporal bone (Figure 6.4a).

Processes: Projections or outgrowths on bone that form joints or attachment points for connective tissue.

Marking	Description	Example
Processes that form joints:		
Condyle (KON-dīl)	Large, round protuberance (*condylus* = knuckle) at the end of a bone.	Lateral condyle of the femur (Figure 7.13b).
Facet	Smooth flat surface.	Superior articular facet of a vertebra (Figure 6.18c).
Head	Rounded articular projection supported on the neck (constricted portion) of a bone.	Head of the femur (Figure 7.13a).
Processes that form attachment points for connective tissue:		
Crest	Prominent ridge or elongated projection.	Iliac crest of the hip bone (Figure 7.10b).
Epicondyle	Projection above (*epi* = above) a condyle.	Medial epicondyle of the femur (Figure 7.13a).
Line	Long, narrow ridge or border (less prominent than a crest).	Linea aspera of the femur (Figure 7.13b).
Spinous process	Sharp, slender projection.	Spinous process of a vertebra (Figure 6.18a).
Trochanter (trō-KAN-ter)	Very large projection on the femur.	Greater trochanter of the femur (Figure 7.13a).
Tubercle (TŪ-ber-kul)	Small, rounded projection (*tube* = knob).	Greater tubercle of the humerus (Figure 7.5a).
Tuberosity	Large, rounded, usually roughened projection.	Ischial tuberosity of the hip bone (Figure 7.10b).

SKULL

Objective

- Name the cranial and facial bones and indicate the number of each.

The **skull,** which contains 22 bones, rests on the superior end of the vertebral column. It includes two sets of bones: cranial bones and facial bones (Table 6.3). The **cranial bones** (*crani-* = brain case) form the cranial cavity, which encloses and protects the brain. The eight cranial bones are the frontal bone, two parietal bones, two temporal bones, the occipital bone, the sphenoid bone, and the ethmoid bone. Fourteen **facial bones** form the face: two nasal bones, two maxillae (or maxillas), two zygomatic bones, the mandible, two lacrimal bones, two palatine bones, two inferior nasal conchae, and the vomer. Figures 6.3 through 6.8 illustrate these bones from different viewing directions.

Table 6.3 Summary of Bones of the Adult Skull

Cranial Bones	Facial Bones
Frontal (1)	Nasal (2)
Parietal (2)	Maxillae (2)
Temporal (2)	Zygomatic (2)
Occipital (1)	Mandible (1)
Spenoid (1)	Lacrimal (2)
Ethmoid (1)	Palatine (2)
	Inferior nasal conchae (2)
	Vomer (1)

The numbers in parentheses indicate how many of each bone are present.

Figure 6.3 / Skull.

The skull consists of cranial bones and facial bones.

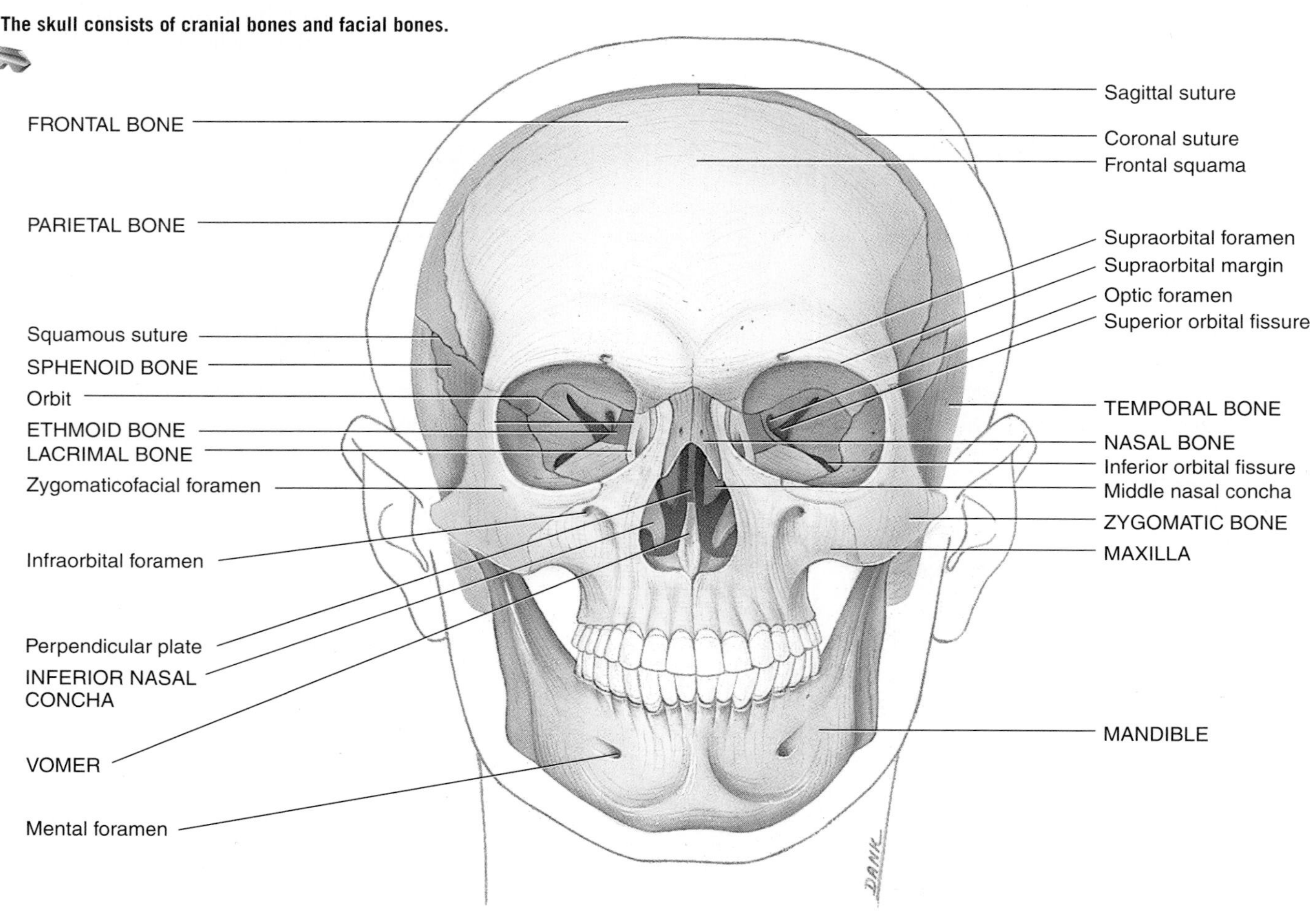

(a) Anterior view

General Features

Besides forming the large cranial cavity, the skull also forms several smaller cavities, including the nasal cavity and orbits (eye sockets), which open to the exterior. Certain skull bones also contain cavities that are lined with mucous membranes and are called paranasal sinuses. The sinuses open into the nasal cavity. Also within the skull are small cavities that house the structures involved in hearing and equilibrium.

Other than the auditory ossicles, which are involved in hearing and are located within the temporal bones, the mandible is the only movable bone of the skull. Most of the skull bones are held together by immovable joints called sutures, which are especially noticeable on the outer surface of the skull.

The skull has numerous surface markings, such as foramina and fissures through which blood vessels and nerves pass. You will learn the names of important skull bone surface markings when the various bones are described.

In addition to protecting the brain, the cranial bones also have other functions. Their inner surfaces attach to membranes (meninges) that stabilize the positions of the brain, blood vessels, and nerves. The outer surfaces of cranial bones provide large areas of attachment for muscles that move various parts of the head. The bones also provide attachment for some muscles that are involved in producing facial expressions. Besides forming the framework of the face, the facial bones protect and provide support for the entrances to the digestive and respiratory systems. Together, the cranial and facial bones protect and support the delicate special sense organs for vision, taste, smell, hearing, and equilibrium (balance).

Cranial Bones

Frontal Bone

The **frontal bone** forms the forehead (the anterior part of the cranium), the roofs of the *orbits* (eye sockets), and most of the anterior part of the cranial floor (Figure 6.3). Soon after birth the left and right sides of the frontal bone are united by the *metopic suture*, which usually disappears by age 6–8.

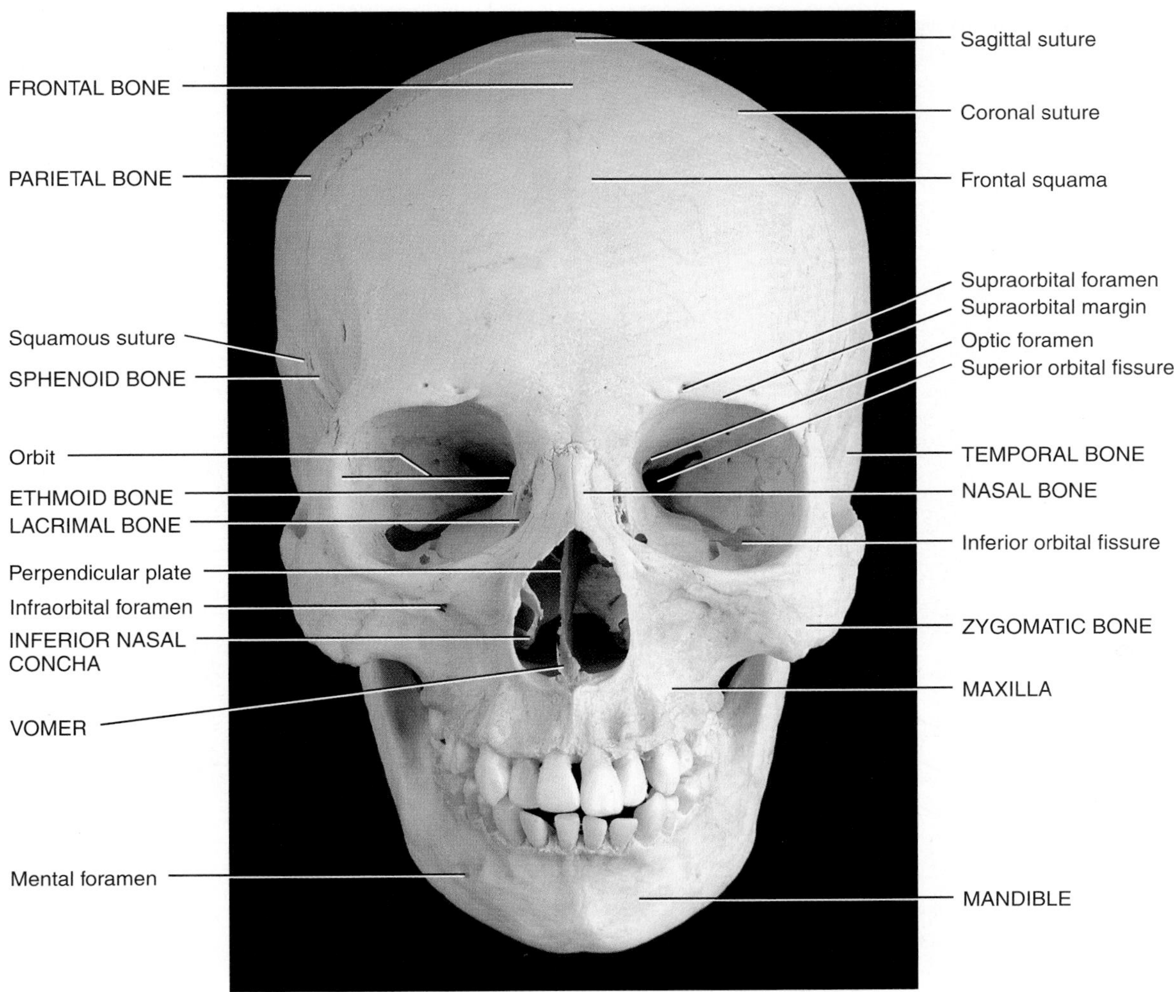

(b) Anterior view

Which of the bones shown here are cranial bones?

If you examine the anterior view of the skull in Figure 6.3, you will note the *frontal squama*, a scalelike plate of bone that forms the forehead. It gradually slopes inferiorly from the coronal suture on top of the skull, then angles abruptly and becomes almost vertical. Superior to the orbits the frontal bone thickens, forming the *supraorbital margin* (*supra-* = above; *-orbital* = wheel rut). From this margin the frontal bone extends posteriorly to form the roof of the orbit and part of the floor of the cranial cavity. Within the supraorbital margin, slightly medial to its midpoint, is a hole called the *supraorbital foramen* through which the supraorbital nerve and artery pass. The *frontal sinuses* lie deep to the frontal squama. Sinuses, or more technically, paranasal sinuses, are mucous-membrane–lined cavities in certain skull bones (described later). Among other functions, paranasal sinuses act as sound chambers that give the voice its resonance.

Black Eyes

Just superior to the supraorbital margin is a sharp ridge. A blow to the ridge often fractures the frontal bone or lacerates the skin over it, resulting in bleeding. Bruising of the skin over the ridge causes tissue fluid and blood to accumulate in the surrounding connective tissue. The resulting swelling and discoloration is called a **black eye**. ■

Parietal Bones

The two **parietal bones** (pa-RĪ-e-tal; *pariet-* = wall) form the greater portion of the sides and roof of the cranial cavity (Figure 6.4). The internal surfaces of the parietal bones contain many protrusions and depressions that accommodate the blood vessels supplying the dura mater, the superficial membrane (meninx) covering the brain. There are no foramina in the parietal bones.

Figure 6.4 / Skull. Although the hyoid bone is not part of the skull, it is included in the illustration for reference.

The zygomatic arch is formed by the zygomatic process of the temporal bone and the temporal process of the zygomatic bone.

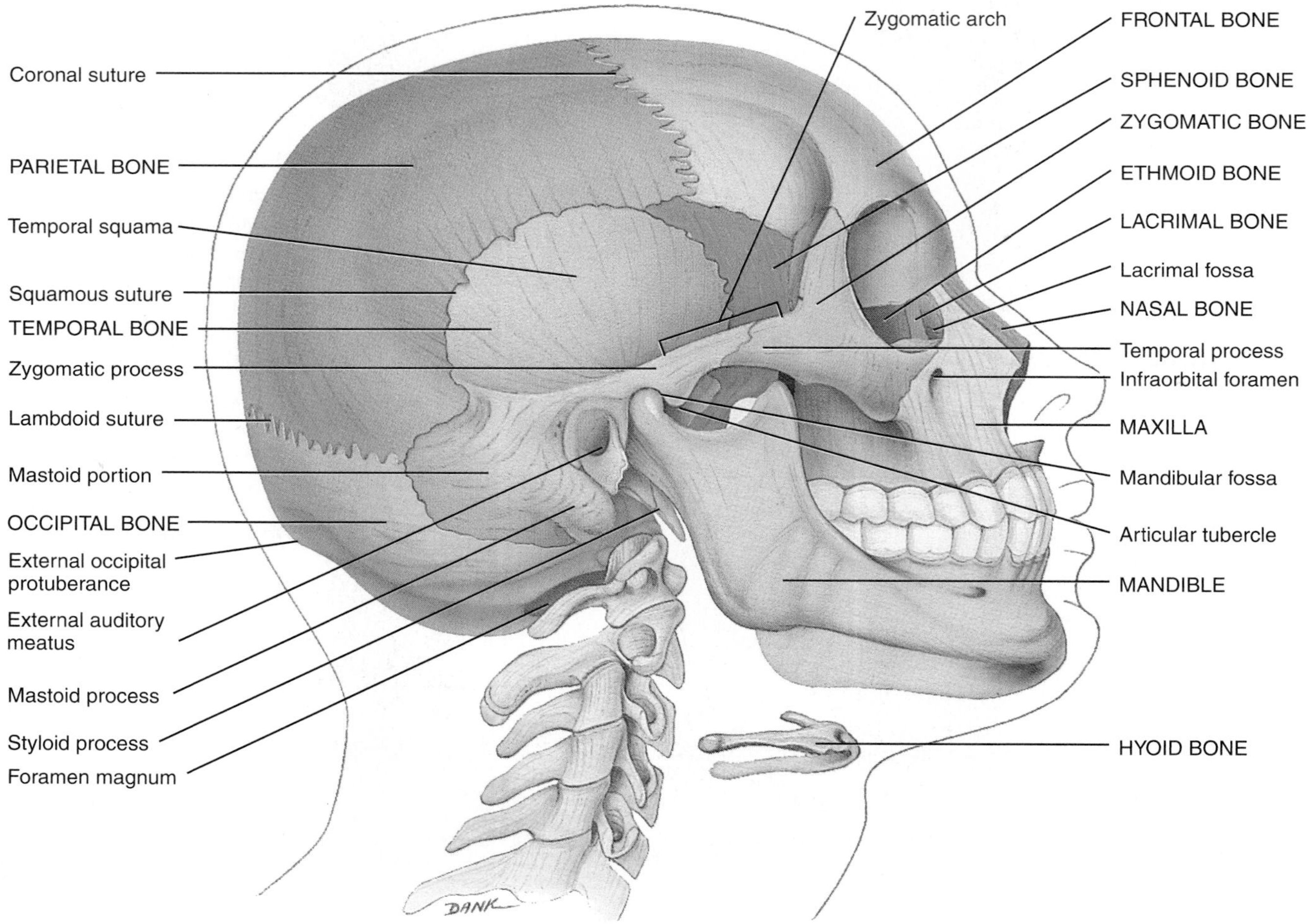

Coronal suture
PARIETAL BONE
Temporal squama
Squamous suture
TEMPORAL BONE
Zygomatic process
Lambdoid suture
Mastoid portion
OCCIPITAL BONE
External occipital protuberance
External auditory meatus
Mastoid process
Styloid process
Foramen magnum
Zygomatic arch
FRONTAL BONE
SPHENOID BONE
ZYGOMATIC BONE
ETHMOID BONE
LACRIMAL BONE
Lacrimal fossa
NASAL BONE
Temporal process
Infraorbital foramen
MAXILLA
Mandibular fossa
Articular tubercle
MANDIBLE
HYOID BONE

(a) Right lateral view

Temporal Bones

The two **temporal bones** (*tempor-* = temple) form the inferior lateral aspects of the cranium and part of the cranial floor. In the lateral view of the skull (Figure 6.4), note the *temporal squama*, the thin, flat portion of the temporal bone that forms the anterior and superior part of the temple. Projecting from the inferior portion of the temporal squama is the *zygomatic process*, which articulates (forms a joint) with the temporal process of the zygomatic (cheek) bone. Together, the zygomatic process of the temporal bone and the temporal process of the zygomatic bone form the *zygomatic arch*.

On the inferior posterior surface of the zygomatic process of the temporal bone is a socket called the *mandibular fossa*. Anterior to the mandibular fossa is a rounded elevation, the *articular tubercle* (Figure 6.4). The mandibular fossa and articular tubercle articulate with the mandible (lower jawbone) to form the *temporomandibular joint (TMJ)*.

Located posteriorly on the temporal bone is the *mastoid portion* (*mastid* = breast-shaped) (Figure 6.4). The mastoid portion is located posterior and inferior to the *external auditory meatus* (*meatus* = passageway), or ear canal, which directs sound waves into the ear. In the adult, this portion of the bone contains several *mastoid "air cells."* These tiny air-filled compartments are separated from the brain by thin bony partitions. **Mastoiditis,** (inflammation of the mastoid air cells), can spread to the middle ear and to the brain.

The *mastoid process* is a rounded projection of the mastoid portion of the temporal bone posterior to the external auditory meatus. It serves as a point of attachment for several neck muscles. The *internal auditory meatus* (Figure 6.5) is the opening through which the facial (VII) and vestibulocochlear (VIII) cranial nerves pass. The *styloid process* (*styl-* = stake or pole) projects inferiorly from the inferior surface of the temporal bone and serves as a point of attachment for muscles and ligaments of the tongue and neck (Figure 6.4). Between the styloid process

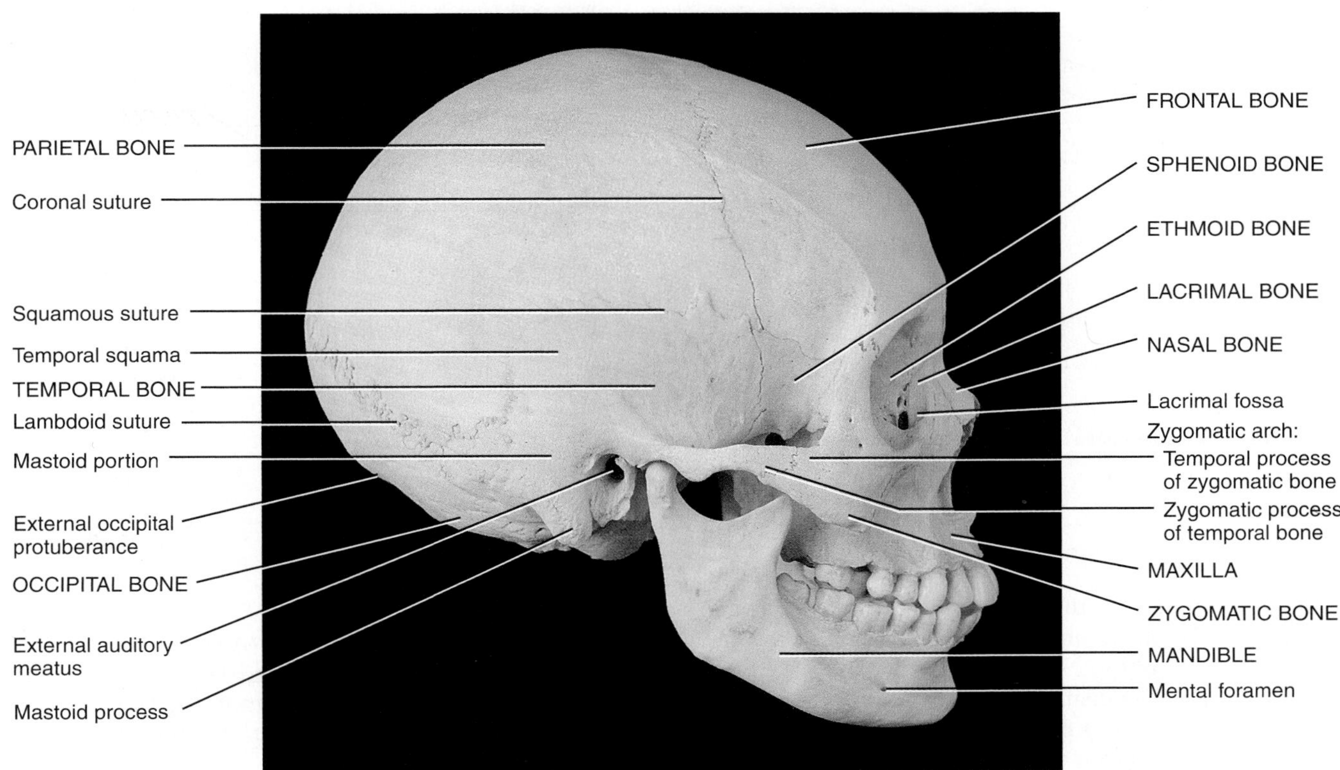

(b) Right lateral view

What are the major bones on either side of the squamous suture, the lambdoid suture, and the coronal suture?

Figure 6.5 / Skull.

The cranial bones are the frontal, parietal, temporal, occipital, sphenoid, and ethmoid bones. The facial bones are the nasal bone, maxillae, zygomatic bones, lacrimal bones, palatine bones, mandible, and vomer.

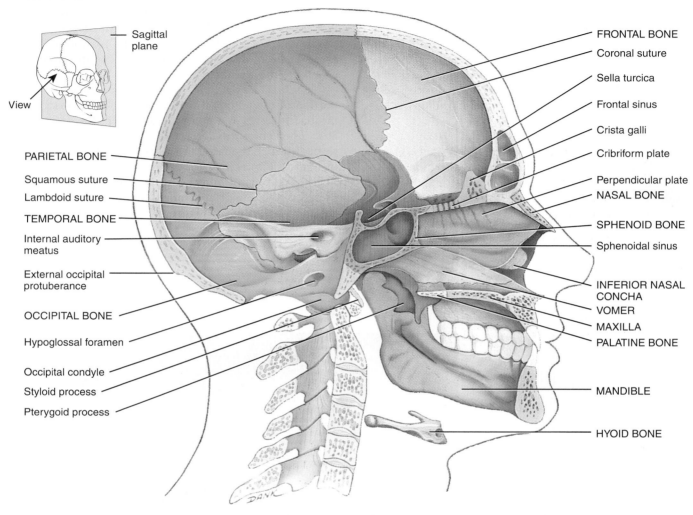

Medial view of sagittal section

With which bones does the temporal bone articulate?

and the mastoid process is the *stylomastoid foramen,* through which the facial (VII) nerve and stylomastoid artery pass (see Figure 6.7).

At the floor of the cranial cavity (see Figure 6.8a) is the *petrous portion* (*petrous* = rock) of the temporal bone. This portion is triangular and located at the base of the skull between the sphenoid and occipital bones. The petrous portion houses the internal ear and the middle ear, structures involved in hearing and equilibrium. It also contains the *carotid foramen,* through which the carotid artery passes (see Figure 6.7). Posterior to the carotid foramen and anterior to the occipital bone is the *jugular foramen,* a passageway for the jugular vein.

Occipital Bone

The **occipital bone** (ok-SIP-i-tal; *occipit-* = back of head) forms the posterior part and most of the base of the cranium (Figure 6.6; also see Figure 6.4). The *foramen magnum* (= large hole) is in the inferior part of the bone. Within this foramen, the medulla oblongata (inferior part of the brain) connects with the spinal cord. The vertebral and spinal arteries also pass through this foramen. The *occipital condyles* are oval processes with convex surfaces, one on either side of the foramen magnum (Figure 6.7 on pages 142–143). They articulate with depressions on the first cervical vertebra (atlas) to form the *atlanto-occipital joints.* Superior to each occipital condyle on the inferior surface of the

Figure 6.6 / Skull. The sutures are exaggerated for emphasis. (See Tortora, *A Photographic Atlas of the Human Body,* Figure 3.5.)

The occipital bone forms most of the posterior and inferior portions of the cranium.

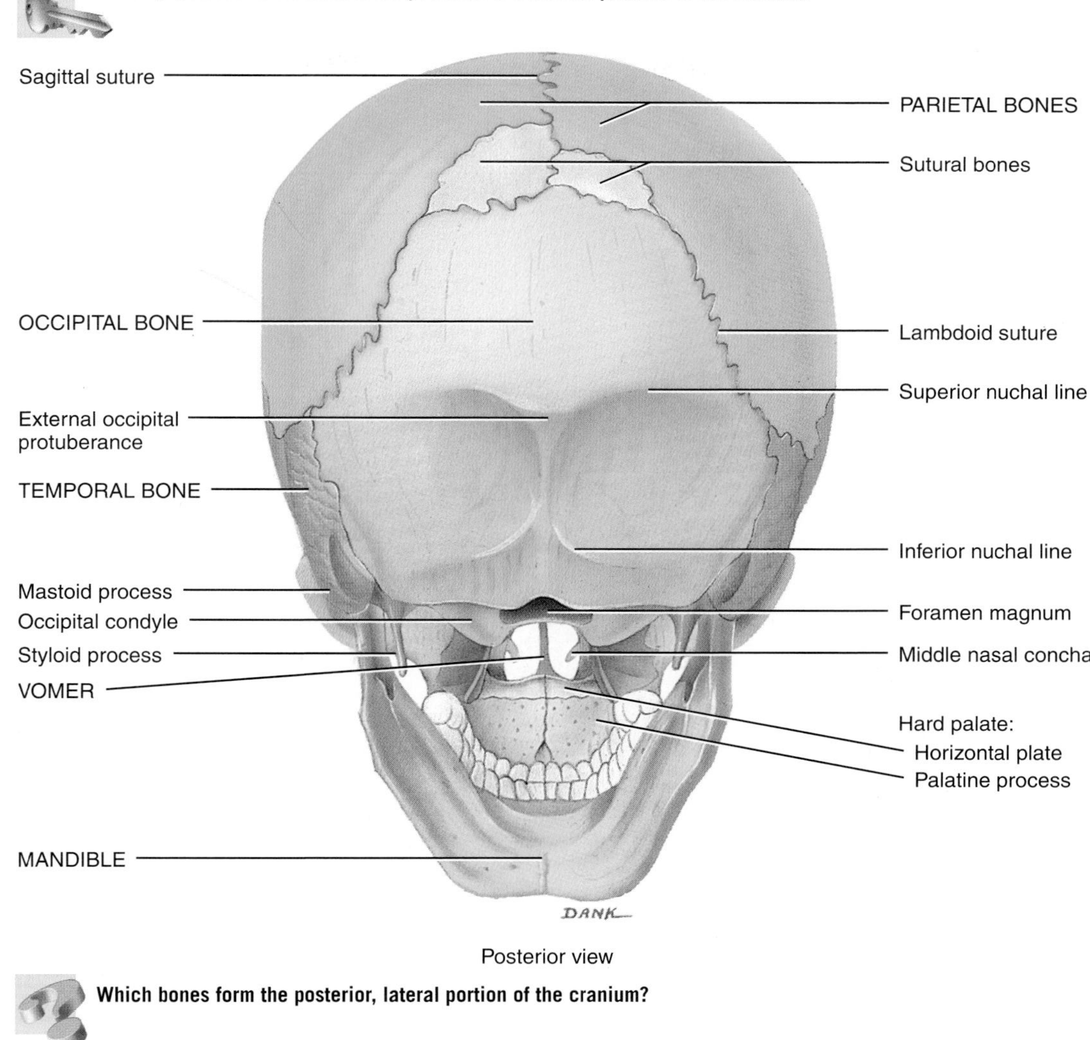

Posterior view

Which bones form the posterior, lateral portion of the cranium?

skull is the *hypoglossal foramen* (*hypo-* = under; *-glossal* = tongue), through which the hypoglossal (XII) nerve and a branch of the ascending pharyngeal artery pass (Figure 6.5).

The *external occipital protuberance* is a prominent midline projection on the posterior surface of the bone just superior to the foramen magnum. You may be able to feel this structure as a definite bump on the back of your head, just above your neck (Figure 6.4). A large fibrous, elastic ligament, the *ligamentum nuchae* (*nucha-* = nape of neck), which helps support the head, extends from the external occipital protuberance to the seventh cervical vertebra. Extending laterally from the protuberance are two curved lines, the *superior nuchal lines*, and below these are two *inferior nuchal lines*, which are areas of muscle attachment (Figure 6.7 on pages 142–143). It is possible to view the parts of the occipital bone, as well as surrounding structures, in the inferior view of the skull in Figure 6.7.

Sphenoid Bone

The **sphenoid bone** (SFĒ-noyd; = wedge-shaped) lies at the middle part of the base of the skull (Figure 6.7 and also Figure 6.8 on pages 144–145). This bone is called the keystone of the cranial floor because it articulates with all the other cranial bones, holding them together. Viewing the floor of the cranium superiorly (Figure 6.8a), note the sphenoid articulations: anteriorly with the frontal bone, laterally with the temporal bones, and posteriorly with the occipital bone. It lies posterior and slightly superior to the nasal cavity and forms part of the floor, sidewalls, and rear wall of the orbit (see Figure 6.13).

The shape of the sphenoid resembles a bat with outstretched wings (see Figure 6.8b). The *body* of the sphenoid is the cubelike medial portion between the ethmoid and occipital bones. It contains the *sphenoidal sinuses*, which drain into the nasal cavity (see Figure 6.11). On the superior surface of the

Figure 6.7 / Skull. The mandible (lower jawbone) has been removed.

The occipital condyles of the occipital bone articulate with the first cervical vertebra to form the atlanto-occipital joints.

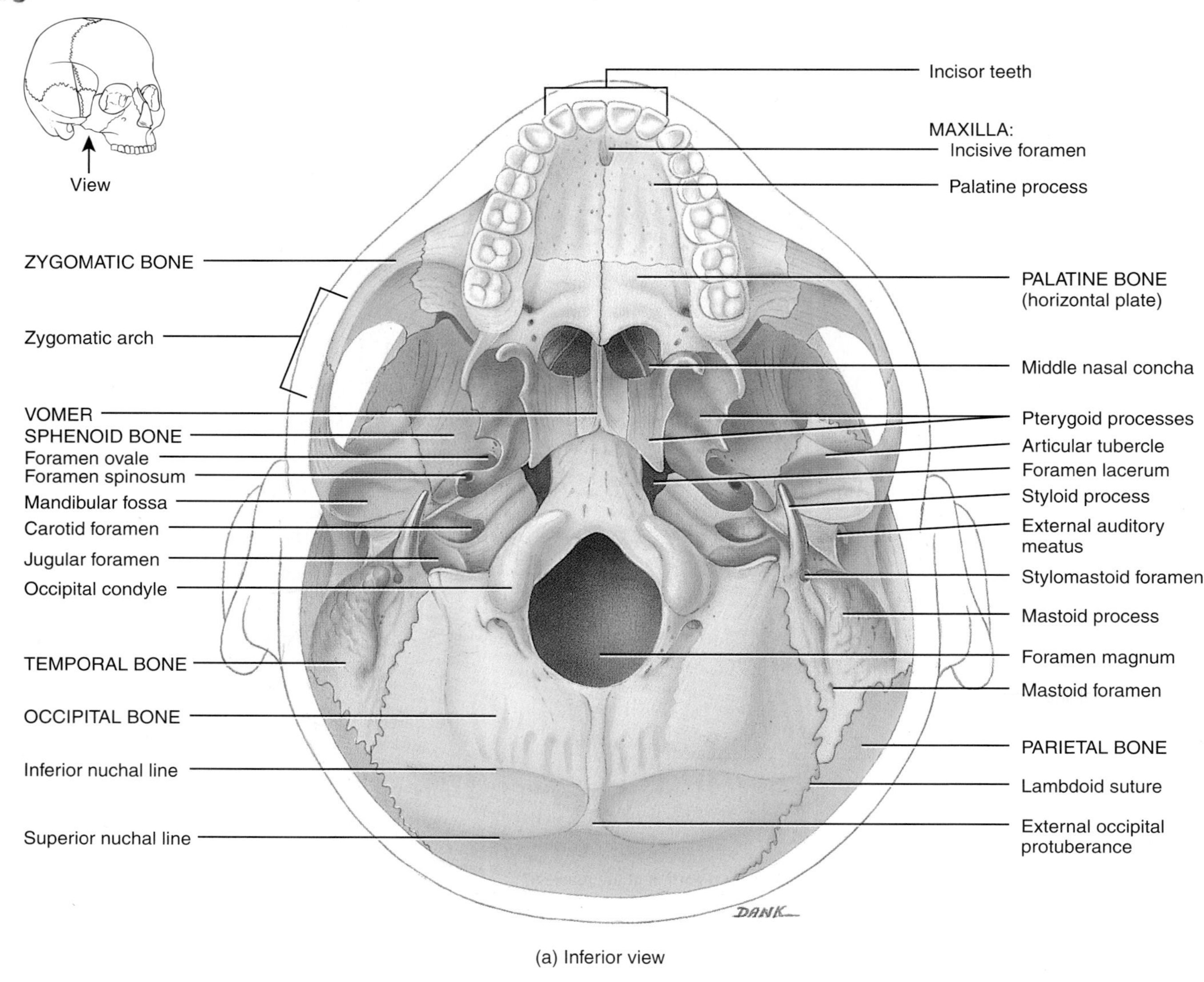

(a) Inferior view

body of the sphenoid is a depression called the *sella turcica* (SEL-a TUR-si-ka; *sella* = saddle; *turcica* = Turkish), which cradles the pituitary gland.

The *greater wings* of the sphenoid project laterally from the body, forming the anterolateral floor of the cranium. The greater wings also form part of the lateral wall of the skull just anterior to the temporal bone and can be viewed externally. The *lesser wings*, which are smaller than the greater wings, form a ridge of bone anterior and superior to the greater wings. They form part of the floor of the cranium and the posterior part of the orbit of the eye.

Between the body and lesser wing just anterior to the sella turcica is the *optic foramen* (*optic* = eye), through which the optic (II) nerve and ophthalmic artery pass. Lateral to the body between the greater and lesser wings is a somewhat triangular slit called the *superior orbital fissure*. This fissure may also be seen in the anterior view of the orbit in Figure 6.13.

In Figures 6.7 and 6.8b you can see the *pterygoid processes* (TER-i-goyd; = winglike) extending from the inferior part of the sphenoid bone. These structures project inferiorly from the points where the body and greater wings unite and form the lateral posterior region of the nasal cavity. Some of the muscles that move the mandible attach to the pterygoid processes. At the base of the lateral pterygoid process in the greater wing is the *foramen ovale* (= oval), an opening for the mandibular branch of the trigeminal (V) nerve. Another foramen, the *foramen spinosum* (= resembling a spine), lies at the posterior angle of the sphenoid and transmits the middle meningeal blood vessels. The

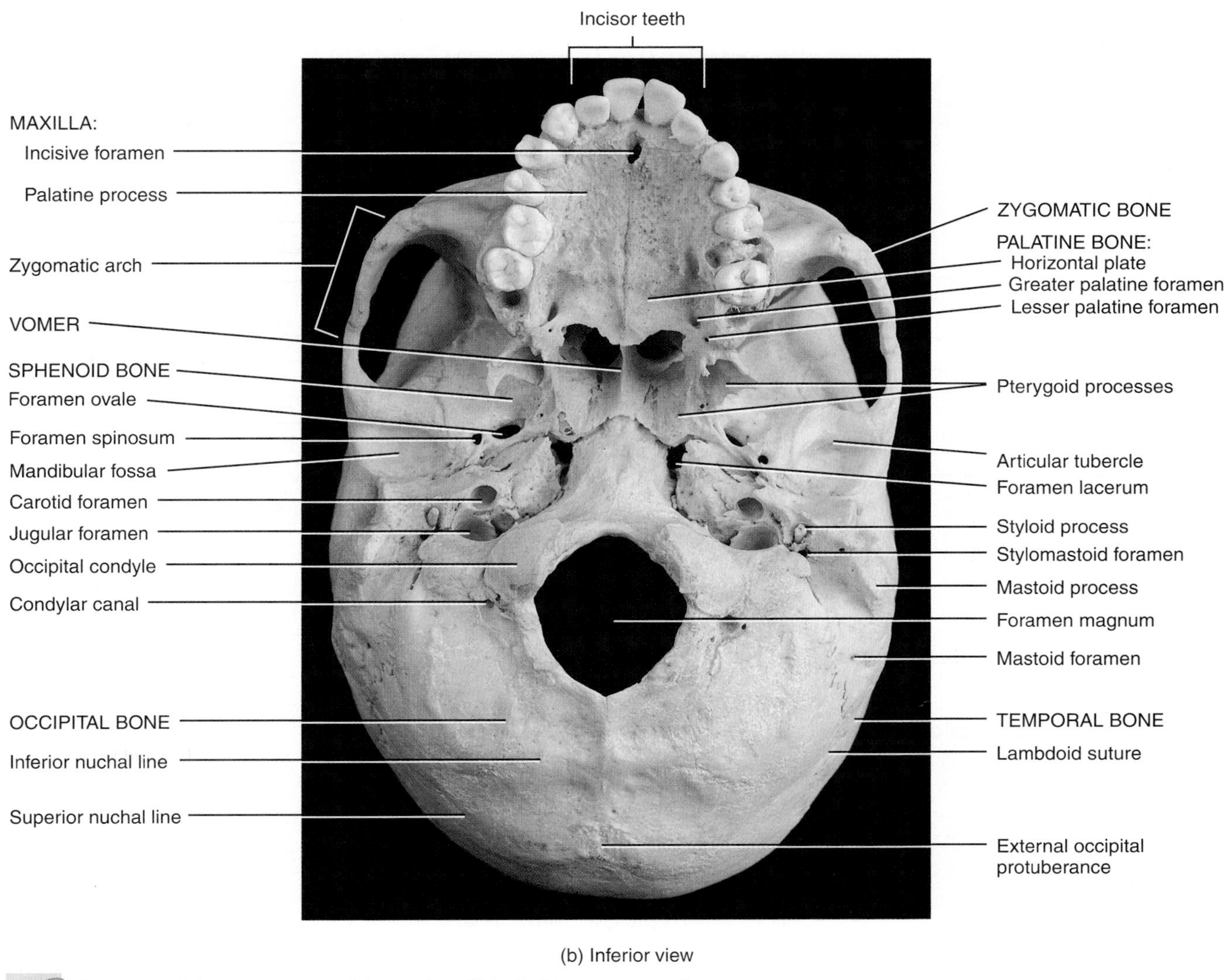

(b) Inferior view

What parts of the nervous system join together within the foramen magnum?

---

foramen lacerum (= lacerated) is bounded anteriorly by the sphenoid bone and medially by the sphenoid and occipital bones. The foramen is covered in part by a layer of fibrocartilage in living subjects. It transmits a branch of the ascending pharyngeal artery. Another foramen associated with the sphenoid bone is the *foramen rotundum* (= round) located at the junction of the anterior and medial parts of the sphenoid bone. Through it passes the maxillary branch of the trigeminal (V) nerve.

Ethmoid Bone

The **ethmoid bone** (ETH-moyd; = like a sieve) is a light, spongelike bone located on the midline in the anterior part of the cranial floor medial to the orbits (Figure 6.9 on pages 146–147). It is anterior to the sphenoid and posterior to the nasal bones. The ethmoid bone forms (1) part of the anterior portion of the cranial floor; (2) the medial wall of the orbits; (3) the superior portions of the nasal septum, a partition that divides the nasal cavity into right and left sides; and (4) most of the superior sidewalls of the nasal cavity. The ethmoid is a major superior supporting structure of the nasal cavity.

The *lateral masses* of the ethmoid bone compose most of the wall between the nasal cavity and the orbits. They contain 3 to 18 air spaces, or "cells." The ethmoidal cells together form the *ethmoidal sinuses* (see Figure 6.11). The *perpendicular plate* forms the superior portion of the nasal septum (see Figure 6.14). The *cribriform plate* (*cribri-* = sieve) lies in the anterior floor of the cranium and forms the roof of the nasal cavity. The cribriform plate contains the *olfactory foramina*

Figure 6.8 / Sphenoid bone. (See Tortora, *A Photographic Atlas of the Human Body,* Figures 3.8 and 3.9.)

The sphenoid bone is called the keystone of the cranial floor because it articulates with all other cranial bones, holding them together.

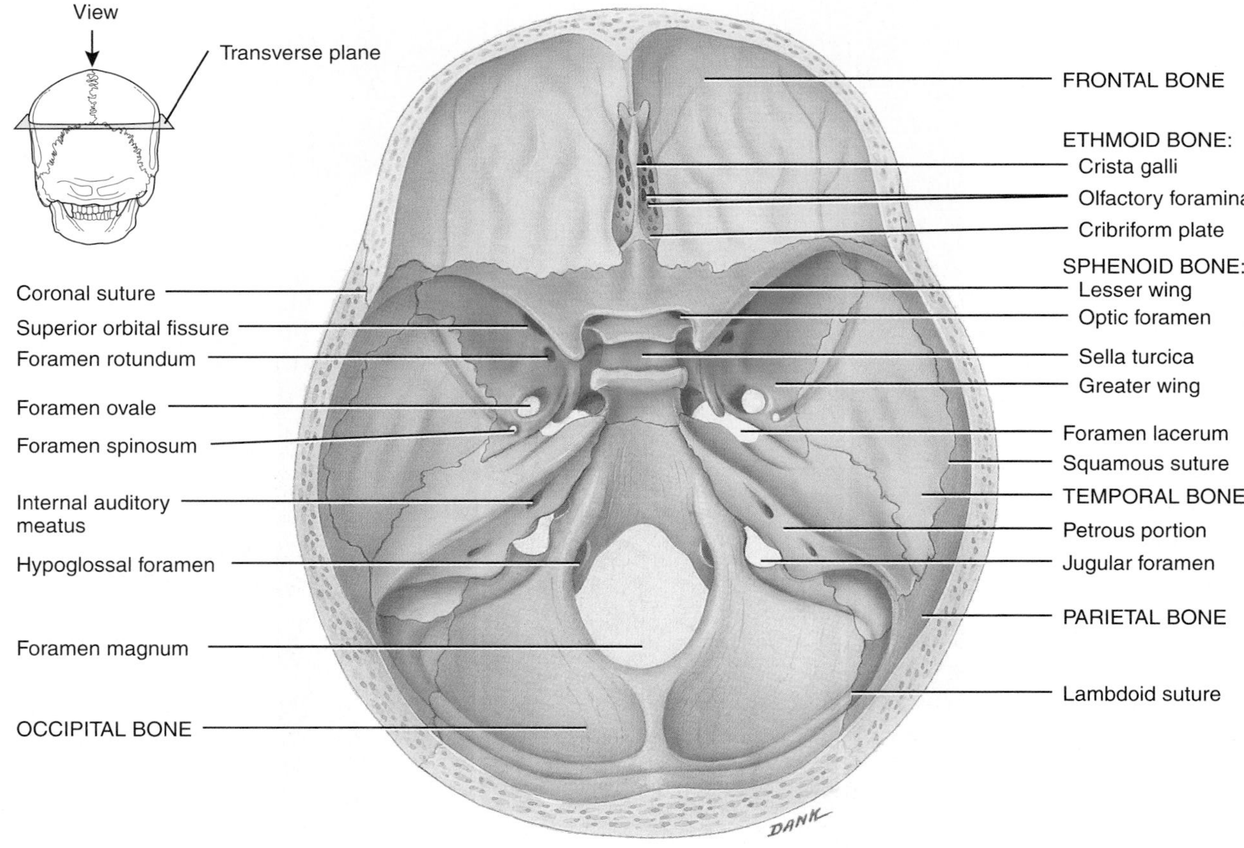

(a) Superior view of sphenoid bone in floor of cranium

(*olfact-* = smell), through which the olfactory (I) nerve passes. Projecting superiorly from the cribriform plate is a sharp triangular process called the *crista galli* (*crista* = crest; *galli* = cock). This structure serves as a point of attachment for the membranes that cover the brain.

The lateral masses of the ethmoid bone contain two thin, scroll-shaped projections lateral to the nasal septum. These are called the *superior nasal concha* (KONG-ka; = shell) or *turbinate* and the *middle nasal concha (turbinate).* A third pair of conchae (plural form), the inferior nasal conchae, are separate bones (discussed shortly). The conchae increase the vascular and mucous membrane surface area in the nasal cavities, which warms and moistens inhaled air before passing into the lungs. The conchae also cause inhaled air to swirl; the result is that many inhaled particles strike and become trapped in the mucus that lines the nasal passageways. The action of the conchae helps cleanse inhaled air before it passes into the rest of the respiratory tract. Air striking the mucous lining of the conchae is also warmed and moistened. The superior nasal conchae also participate in the sense of smell.

Facial Bones

The shape of the face changes dramatically during the first two years after birth. The brain and cranial bones expand, the teeth form and erupt, and the paranasal sinuses increase in size. Growth of the face ceases at about 16 years of age.

Nasal Bones

The paired **nasal bones** meet at the midline (Figure 6.3) and form part of the bridge of the nose. The major structural portion of the nose consists of cartilage.

Maxillae

The paired **maxillae** (mak-SIL-ē-; = jawbones) unite to form the upper jawbone and articulate directly with every bone of the face except the mandible, or lower jawbone (Figures 6.4 and 6.7). They form part of the floors of the orbits, part of the lateral walls and floor of the nasal cavity, and most of the hard palate. The hard palate is a bony partition formed by the pala-

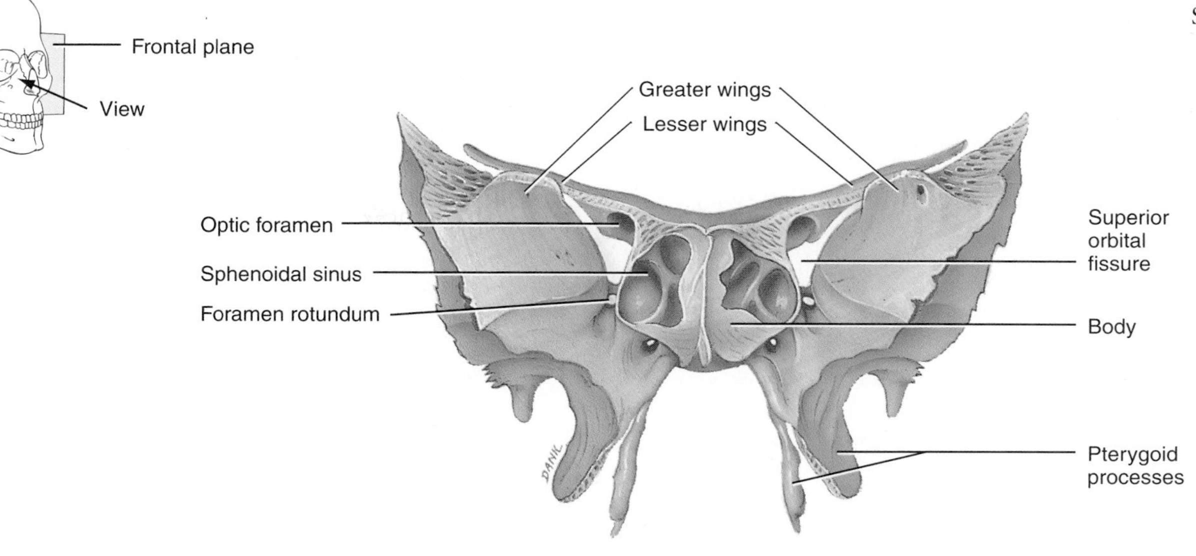

(b) Anterior view of sphenoid bone

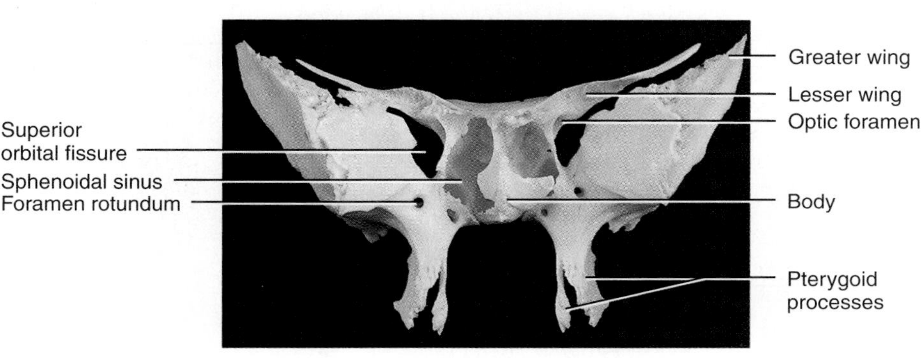

(c) Anterior view

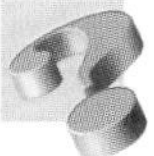

Name the bones that articulate with the sphenoid bone, starting at the crista galli of the ethmoid bone and going in a clockwise direction.

tine processes of the maxillae and horizontal plates of the palatine bones that forms the roof of the mouth.

Each maxilla (singular form) contains a large *maxillary sinus* that empties into the nasal cavity (see Figure 6.11). The *alveolar process* (al-VĒ-ō-lar; *alveol-* = small cavity) is an arch that contains the *alveoli* (sockets) for the maxillary (upper) teeth. The *palatine process* is a horizontal projection of the maxilla that forms the anterior three-quarters of the hard palate. The union and fusion of the maxillary bones is normally completed before birth.

The *infraorbital foramen* (*infra* = below; *orbital* = orbit), which can be seen in the anterior view of the skull in Figure 6.3, is an opening in the maxilla inferior to the orbit. Through it passes the infraorbital nerve and blood vessels and a branch of the maxillary division of the trigeminal (V) nerve. Another prominent foramen in the maxilla is the *incisive foramen* (= incisor teeth) just posterior to the incisor teeth (see Figure 6.7). It transmits branches of the greater palatine blood vessels and nasopalatine nerve. A final structure associated with the maxilla and sphenoid bone is the *inferior orbital fissure*, which is located between the greater wing of the sphenoid and the maxilla (see Figure 6.13).

Cleft Palate and Cleft Lip

Usually the palatine processes of the maxillary bones unite during weeks 10 to 12 of embryonic development. Failure to do so can result in one type of **cleft palate.** The condition may also involve incomplete fusion of the horizontal plates of the palatine bones (Figure 6.7). Another form of this condition, called **cleft lip,** involves a split in the upper lip. Cleft lip and cleft palate often occur together. Depending on the extent and position of the cleft, speech and swallowing may be affected. In addition, children with cleft palate tend to have many ear infections that can lead to hearing loss. Facial and oral surgeons recommend closure of cleft lip during the first few weeks following birth, and

Figure 6.9 / Ethmoid bone. (See Tortora, *A Photographic Atlas of the Human Body,* Figure 3.10.)

The ethmoid bone forms part of the anterior portion of the cranial floor, the medial wall of the orbits, the superior portions of the nasal septum, and most of the sidewalls of the nasal cavity.

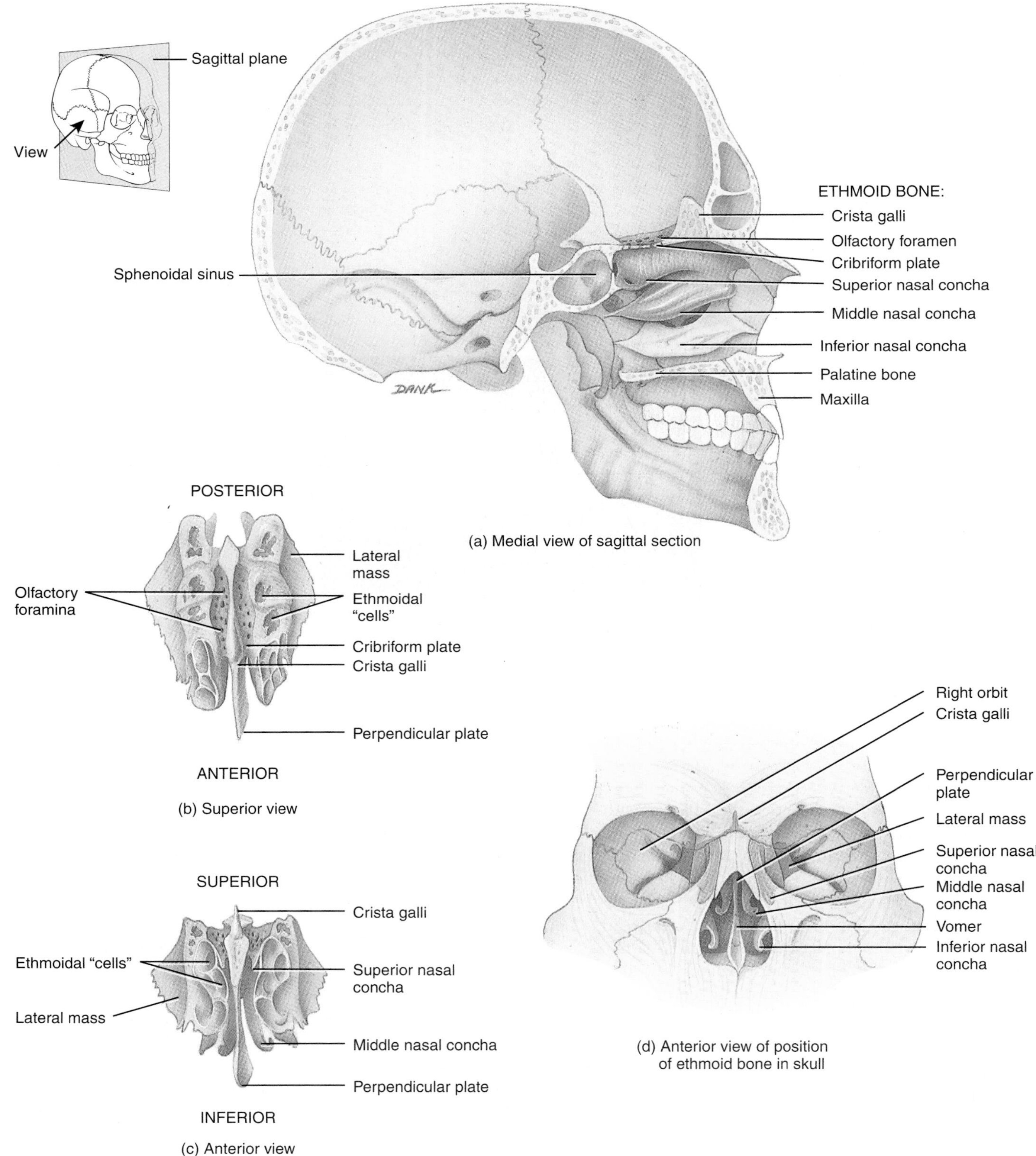

(a) Medial view of sagittal section

(b) Superior view

(c) Anterior view

(d) Anterior view of position of ethmoid bone in skull

(e) Frontal section through the head

What part of the ethmoid bone forms the superior part of the nasal septum? The medial walls of the orbits?

surgical results are excellent. Repair of cleft palate typically is done between 12 and 18 months of age, ideally before the child begins to talk. Speech therapy may be needed, because the palate is important for pronouncing consonants, and orthodontic therapy may be needed to align the teeth. Again, results are usually excellent. ■

Zygomatic Bones

The two **zygomatic bones** (*zygo-* = yokelike), commonly called cheekbones, form the prominences of the cheeks and part of the lateral wall and floor of each orbit (see Figure 6.13). They articulate with the frontal, maxilla, sphenoid, and temporal bones.

The *temporal process* of the zygomatic bone projects posteriorly and articulates with the zygomatic process of the temporal bone to form the *zygomatic arch* (Figure 6.4). A foramen associated with the zygomatic bone is the *zygomaticofacial foramen* near the center of the bone (Figure 6.3a). It transmits the zygomaticofacial nerve and vessels.

Lacrimal Bones

The paired **lacrimal bones** (LAK-ri-mal; *lacrim-* = teardrops) are thin and roughly resemble a fingernail in size and shape (Figures 6.3 and 6.4). These bones, the smallest bones of the face, are posterior and lateral to the nasal bones and form a part of the medial wall of each orbit. The lacrimal bones each contain a *lacrimal fossa*, a vertical tunnel formed with the maxilla, that houses the lacrimal sac, a structure that gathers tears and passes them into the nasal cavity (see Figure 6.13).

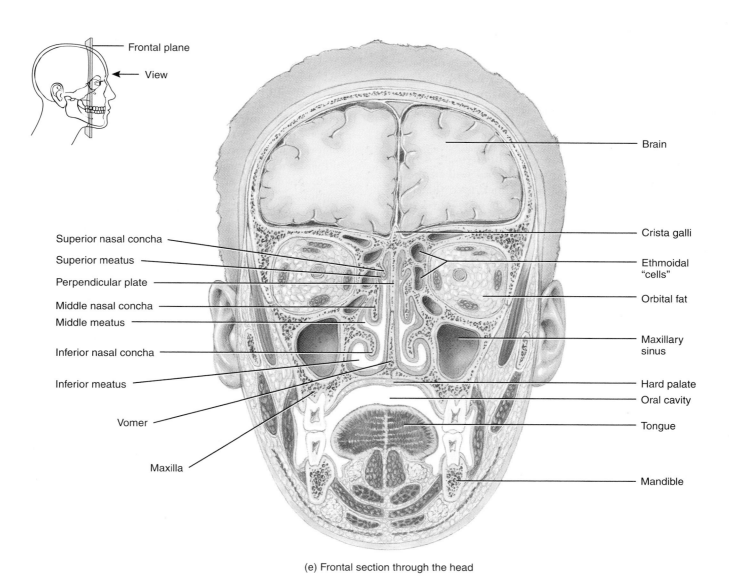

Palatine Bones

The two L-shaped **palatine bones** (PAL-a-tīn) form the posterior portion of the hard palate, part of the floor, and lateral wall of the nasal cavity, and a small portion of the floors of the orbits (Figure 6.7). The posterior portion of the hard palate, which separates the nasal cavity from the oral cavity, is formed by the *horizontal plates* of the palatine bones (Figures 6.6 and 6.7).

Inferior Nasal Conchae

The two **inferior nasal conchae** or **turbinates** are scroll-like bones that form a part of the inferior lateral wall of the nasal cavity and project into the nasal cavity inferior to the superior and middle nasal conchae of the ethmoid bone (Figures 6.3 and 6.9a). They serve the same function as the superior and middle nasal conchae of the ethmoid bone—that is, they help swirl and filter air before it passes into the lungs. Unlike the superior nasal conchae, the middle and inferior nasal conchae are not invloved in the sense of smell. The inferior nasal conchae are separate bones and not part of the ethmoid.

Vomer

The **vomer** (VŌ-mer; = plowshare) is a roughly triangular bone on the floor of the nasal cavity that articulates superiorly with the perpendicular plate of the ethmoid bone and inferiorly with both the maxillae and palatine bones along the midline (see Figures 6.3 and 6.7). It is one of the components of the nasal septum, the partition that divides the nasal cavity into right and left sides.

Mandible

The **mandible** (*mand-* = to chew), or lower jawbone, is the largest, strongest facial bone (Figure 6.10). It is the only movable skull bone (other than the auditory ossicles). In the lateral view, you can see that the mandible consists of a curved, horizontal portion, the *body*, and two perpendicular portions, the *rami* (RĀ-mī; = branches). The *angle* of the mandible is the area where each ramus (singular form) meets the body. Each ramus has a posterior *condylar process* (KON-di-lar) that articulates with the mandibular fossa and articular tubercle of the temporal bone (Figure 6.4) to form the **temporomandibular joint (TMJ).** It also has an anterior *coronoid process* (KOR-ō-noyd) to which the temporalis muscle attaches. The depression between the coronoid and condylar processes is called the *mandibular notch*. The *alveolar process* is an arch containing the *alveoli* (sockets) for the mandibular (lower) teeth.

The *mental foramen* (*ment-* = chin) is approximately inferior to the second premolar tooth. It is near this foramen that dentists reach the mental nerve when injecting anesthetics. Another foramen associated with the mandible is the *mandibular foramen* on the medial surface of each ramus, another site often used by dentists to inject anesthetics. The mandibular foramen is the beginning of the *mandibular canal*, which runs obliquely in the ramus and anteriorly to the body. Through the canal pass the inferior alveolar nerves and blood vessels, which are distributed to the mandibular teeth.

One problem associated with the temporomandibular joint (TMJ) is **TMJ syndrome.** It is characterized by dull pain around the ear, tenderness of the jaw muscles, a clicking or popping noise when opening or closing the mouth, limited or ab-

Figure 6.10 / Mandible.

The mandible is the largest and strongest facial bone.

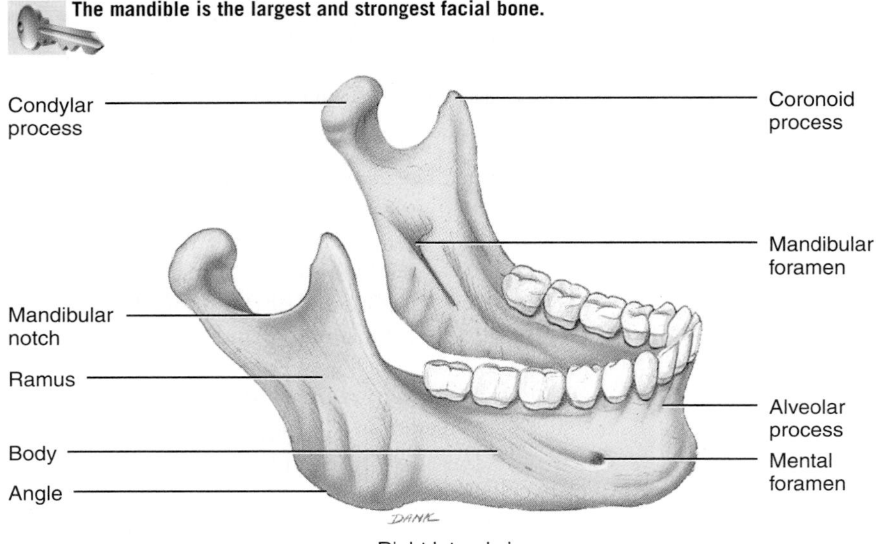

What is the distinctive functional feature of the mandible among all the skull bones?

normal opening of the mouth, headache, tooth sensitivity, and abnormal wearing of the teeth. TMJ syndrome can be caused by improperly aligned teeth, grinding or clenching the teeth, trauma to the head and neck, or arthritis. Treatment may involve applying moist heat or ice, eating a soft diet, taking pain relievers such as aspirin, muscle retraining, adjusting or reshaping the teeth, orthodontic treatment, or surgery. ■

SPECIAL FEATURES OF THE SKULL

Objective

- Describe the following special features of the skull: sutures, paranasal sinuses, fontanels, foramina, orbits, nasal septum, and cranial fossae.

Now that we have described the locations and parts of the cranial and facial bones, we will examine some of the special features of the skull, starting with sutures.

Sutures

A **suture** (SŪ-chur; = seam) is an immovable joint (in an adult) that is found only between skull bones and that holds most skull bones together. Sutures in the skulls of infants and children can be, and often are, movable. Several types of sutures can be distinguished depending on how the margins of the bones unite. In some sutures the margins of the bones are fairly smooth. In others, the margins overlap. In still others, the margins interlock in a jigsaw fashion. This latter arrangement provides added strength and decreases the chance of fracture.

The names of many sutures reflect the bones they unite. For example, the frontozygomatic suture is between the frontal bone and the zygomatic bone. Similarly, the sphenoparietal suture is between the sphenoid bone and the parietal bone. In other cases, however, the names of sutures are not so obvious. Of the many sutures found in the skull, we will identify only four prominent ones:

1. The **coronal suture** (kō-RŌ-nal; *coron-* = crown) unites the frontal bone and both parietal bones (Figure 6.4).
2. The **sagittal suture** (SAJ-i-tal; *sagitt-* = arrow) unites the two parietal bones on the superior midline of the skull (Figure 6.6). The sagittal suture is so named because in the infant, before the bones of the skull are firmly united, the suture and the fontanels (soft spots) associated with it somewhat resemble an arrow.
3. The **lambdoid suture** (LAM-doyd) unites both parietal bones and occipital bones. This suture is so named because of its resemblance to the Greek letter lambda (Λ), as can be seen in Figure 6.6. Sutural bones may be found within the sagittal and lambdoid sutures.
4. The **squamous sutures** (SKWĀ-mus; *squam-* = flat) unite the parietal and temporal bones on the lateral aspects of the skull (Figure 6.4).

Paranasal Sinuses

Even though they are not cranial or facial bones, this is an appropriate place to discuss the **paranasal sinuses** (*para-* = beside). These are paired cavities in certain cranial and facial bones near the nasal cavity and are clearly evident in a sagittal section of the skull (Figure 6.11). The paranasal sinuses are lined with mucous membranes that are continuous with the lining of the nasal cavity. Skull bones enclosing the paranasal sinuses are the frontal, sphenoid, ethmoid, and maxillary. Besides producing mucus, the paranasal sinuses serve as resonating chambers for sound as we speak or sing.

Secretions produced by the mucous membranes of the paranasal sinuses drain into the nasal cavity. An inflammation of the membranes due to an allergic reaction or infection is called **sinusitis.** If the membranes swell enough to block drainage into the nasal cavity, fluid pressure builds up in the paranasal sinuses, and a sinus headache results. Chronic sinusitis may also be caused by a severely deviated nasal septum or nasal polyps, growths that can be removed surgically. ■

Fontanels

The skeleton of a newly formed embryo consists of cartilage and fibrous connective tissue membrane structures shaped like bones. Gradually, ossification occurs—bone replaces the cartilage and fibrous connective tissue membranes. At birth, membrane-filled spaces called **fontanels** (fon-ta-NELZ; = little fountains) are found between cranial bones (Figure 6.12 on page 151). Commonly called "soft spots," fontanels are areas of fibrous connective tissue membranes that will eventually be replaced with bone by intramembranous ossification and become sutures. Functionally, the fontanels enable the fetal skull to modify its size and shape as it passes through the birth canal and permit rapid growth of the brain during infancy. The amount of closure in fontanels helps a physician to gauge the degree of brain development. In addition, the anterior fontanel serves as a landmark for withdrawal of blood for analysis from the superior sagittal sinus (a large vein on the midline surface of the brain). Although an infant may have many fontanels at birth, the form and location of several are fairly constant (Table 6.4 on page 151).

Foramina

Many **foramina** associated with the skull were mentioned in the descriptions of the cranial and facial bones they penetrate. As preparation for studying other systems of the body, especially the nervous and cardiovascular systems, these foramina and the structures passing through them are listed in Table 6.5 on page 152. For your convenience and for future reference, the foramina are listed alphabetically.

Figure 6.11 / Paranasal sinuses. (a and b) Location of the paranasal sinuses. (c) Openings of the paranasal sinuses into the nasal cavity. Portions of the conchae have been sectioned. (See Tortora, *A Photographic Atlas of the Human Body,* Figure 3.4.)

Paranasal sinuses are mucous-membrane–lined spaces in certain skull bones that connect to the nasal cavity.

Sagittal plane

View

Ethmoidal sinus

Frontal sinus

Ethmoidal "cell"

Sphenoidal sinus

Maxillary sinus

(a) Sagittal section

Frontal sinus

Ethmoidal sinus

Sphenoidal sinus

Maxillary sinus

(b) Radiograph of lateral view

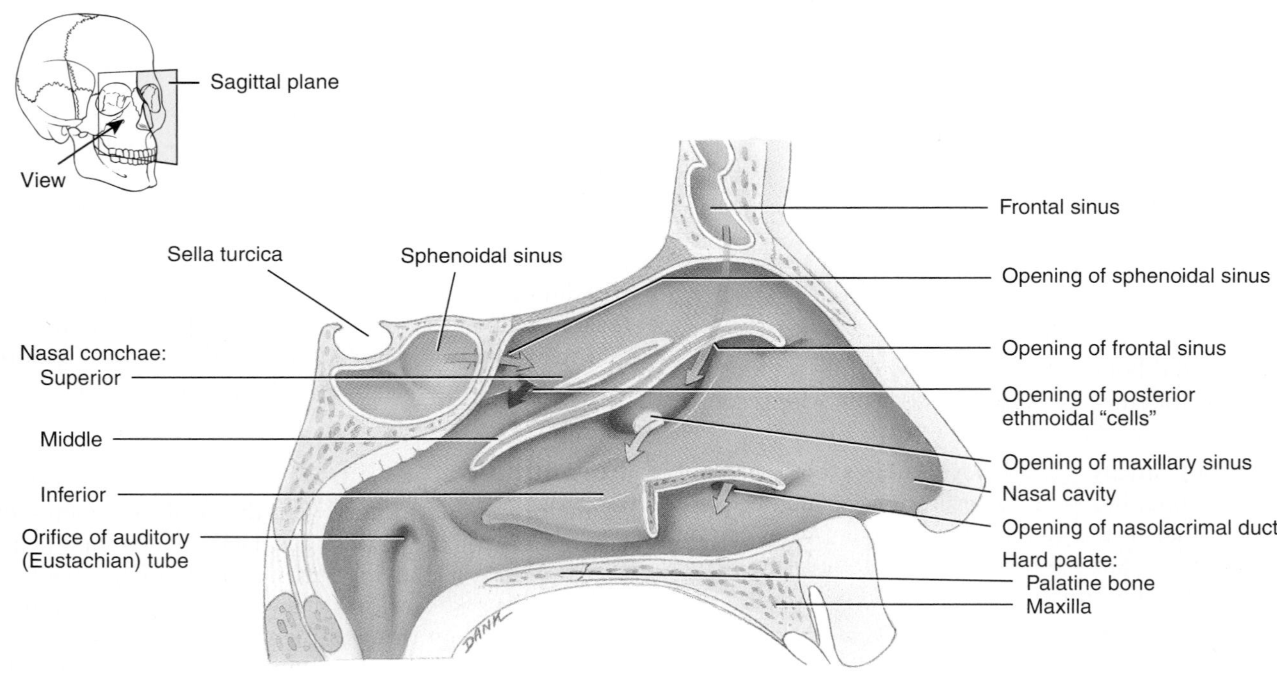

(c) Sagittal section

What are the functions of the paranasal sinuses?

Figure 6.12 / Fontanels at birth. (See Tortora, *A Photographic Atlas of the Human Body,* Figure 3.12.)

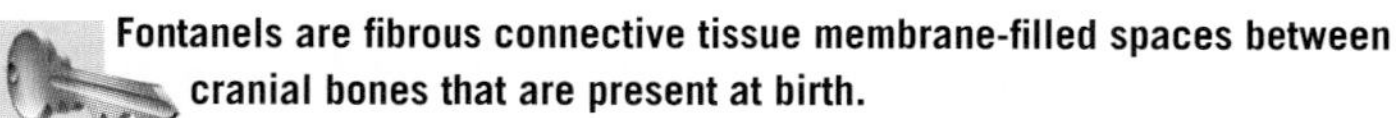

Fontanels are fibrous connective tissue membrane-filled spaces between cranial bones that are present at birth.

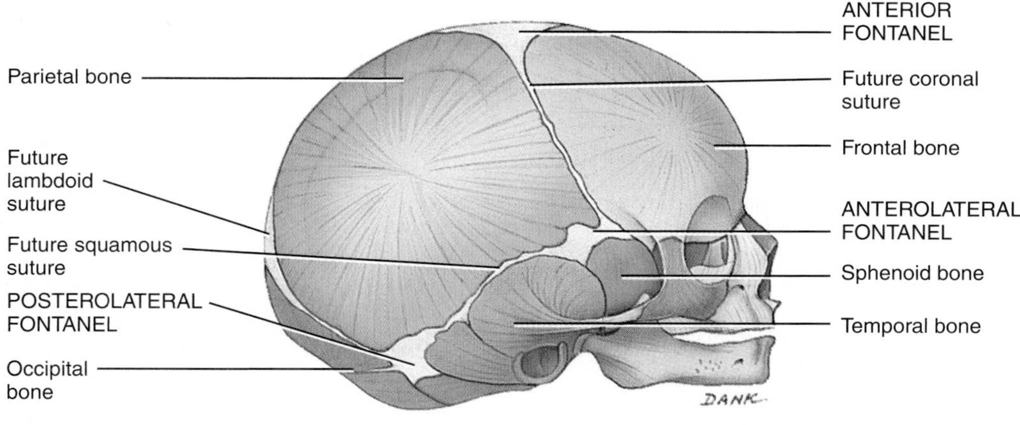

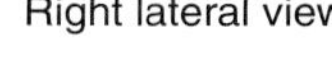

Right lateral view

Which fontanel is bordered by four different skull bones?

Table 6.4 Fontanels (see also Figure 6.12)

Fontanel	Location	Description
Anterior (frontal)	Between the two parietal bones and the frontal bone.	Roughly diamond-shaped, the largest of the six fontanels; usually closes 18–24 months after birth.
Posterior (occipital)	Between the two parietal bones and the occipital bone.	Diamond-shaped, considerably smaller than the anterior fontanel; generally closes about 2 months after birth.
Anterolateral (sphenoid)	One on each side of the skull between the frontal, parietal, temporal, and sphenoid bones.	Small and irregular in shape; normally close about 3 months after birth.
Posterolateral (mastoid)	One on each side of the skull between the parietal, occipital, and temporal bones.	Irregularly shaped; begin to close 1 or 2 months after birth, but closure is generally not complete until 12 months.

Table 6.5 Principal Foramina of the Skull

Foramen	Location	Structures Passing Through
Carotid (relating to carotid artery in neck)	Petrous portion of temporal bone (Figure 6.7a).	Internal carotid artery and sympathetic nerves for eyes.
Hypoglossal (*hypo* = under; *glossus* = tongue)	Superior to base of occipital condyles (Figure 6.8a).	Cranial nerve XII (hypoglossal) and branch of ascending pharyngeal artery.
Incisive (*incisive* = pertaining to incisor teeth)	Posterior to incisor teeth in maxilla (Figure 6.7a).	Branches of greater palatine blood vessels and nasopalatine nerve.
Infraorbital (*infra* = below)	Inferior to orbit in maxilla (Figure 6.13).	Infraorbital nerve and blood vessels and a branch of the maxillary division of cranial nerve V (trigeminal).
Jugular (*jugular* = throat)	Posterior to carotid canal between petrous portion of temporal bone and occipital bone (Figure 6.8a).	Internal jugular vein, cranial nerves IX (glossopharyngeal), X (vagus), and XI (accessory).
Lacerum (*lacerum* = lacerated)	Bounded anteriorly by sphenoid bone, posteriorly by petrous portion of temporal bone, and medially by the sphenoid bone and occipital bone (Figure 6.8a).	Branch of ascending pharyngeal artery in palatine bones.
Magnum (*magnum* = large)	Occipital bone (Figure 6.7a).	Medulla oblongata and its membranes (meninges), cranial nerve XI (accessory), and vertebral and spinal arteries.
Mandibular (*mand* = to chew)	Medial surface of ramus of mandible (Figure 6.10).	Inferior alveolar nerve and blood vessels.
Mastoid (breast-shaped)	Posterior border of mastoid process of temporal bone (Figure 6.7a).	Emissary vein to transverse sinus and branch of occipital artery to dura mater.
Mental (*ment-* = chin)	Inferior to second premolar tooth in mandible (Figure 6.10).	Mental nerve and vessels.
Olfactory (*olfact* = to smell)	Cribriform plate of ethmoid bone (Figure 6.8a).	Cranial nerve I (olfactory).
Optic (= eye)	Between superior and inferior portions of small wing of sphenoid bone (Figure 6.13).	Cranial nerve II (optic) and ophthalmic artery.
Ovale (*ovale* = oval)	Greater wing of sphenoid bone (Figure 6.8a).	Mandibular branch of cranial nerve V (trigeminal).
Rotundum (*rotundum* = round)	Junction of anterior and medial parts of sphenoid bone (Figure 6.8a).	Maxillary branch of cranial nerve V (trigeminal).
Spinosum (*spinosum* = resembling a spine)	Posterior angle of sphenoid bone (Figure 6.8a).	Middle meningeal blood vessels.
Stylomastoid (*stylo* = stake or pole)	Between styloid and mastoid processes of temporal bone (Figure 6.7a).	Cranial nerve VII (facial) and stylomastoid artery.
Supraorbital (*supra* = above)	Supraorbital margin of orbit in frontal bone (Figure 6.13).	Supraorbital nerve and artery.

Orbits

Each **orbit** (eye socket) is a pyramid-shaped structure that contains the eyeball and associated structures. It is formed by seven bones of the skull (Figure 6.13) and has four regions that converge posteriorly to form an *apex*:

1. The *roof* of the orbit consists of parts of the frontal and sphenoid bones.
2. The *lateral wall* of the orbit is formed by portions of the zygomatic and sphenoid bones. It is more posterior than the medial wall and accounts for the fact that injuries to the eyeball from the lateral side are more penetrating.
3. The *floor* of the orbit is formed by parts of the maxilla, zygomatic, and palatine bones.
4. The *medial wall* of the orbit is formed by portions of the maxilla, lacrimal, ethmoid, and sphenoid bones.

Associated with the orbit are five openings:

- Optic foramen at the junction of the roof and medial wall.
- Superior orbital fissure at the superior lateral angle of the apex.
- Inferior orbital fissure at the junction of the lateral wall and floor.
- Supraorbital foramen on the medial side of the supraorbital margin of the frontal bone.
- Lacrimal fossa in the lacrimal bone.

Figure 6.13 / Details of the orbit (eye socket). (See Tortora, *A Photographic Atlas of the Human Body,* Figure 3.11.)

The orbit is a pyramid-shaped structure that contains the eyeball and associated structures.

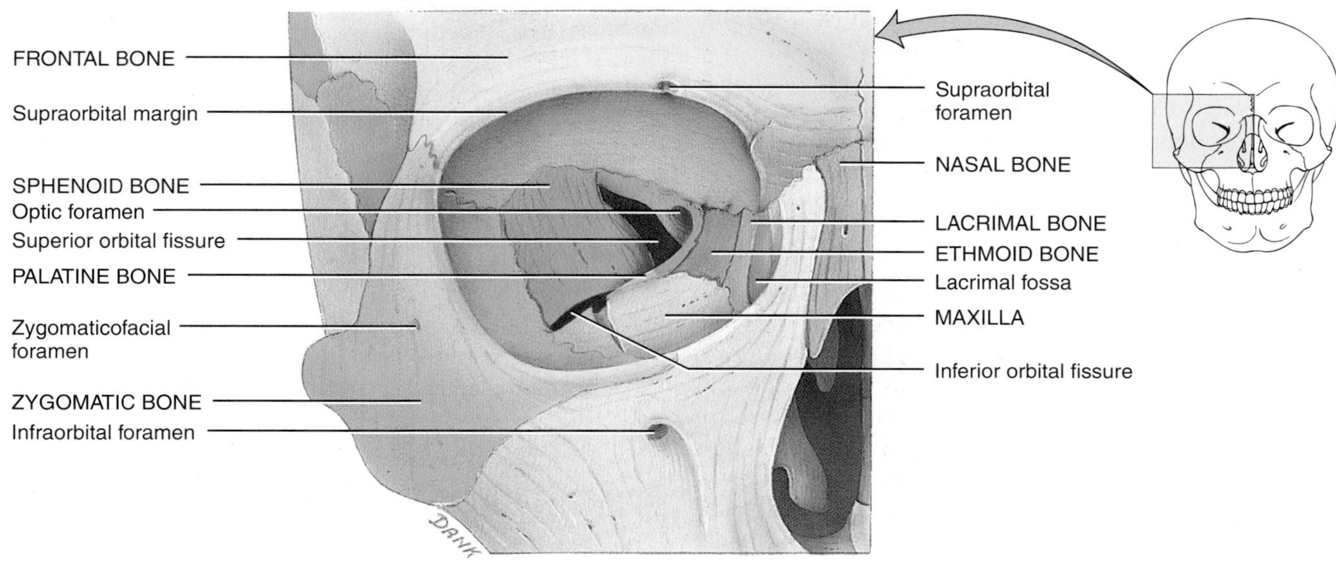

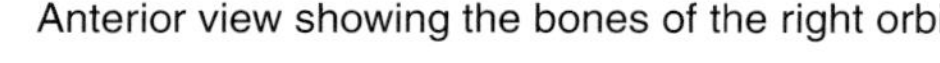
Anterior view showing the bones of the right orbit

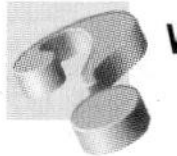
Which seven bones form the orbit?

Nasal Septum

The inside of the nose, called the nasal cavity, is divided into right and left sides by a vertical partition called the **nasal septum.** The septum is formed by the vomer, septal cartilage, and the perpendicular plate of the ethmoid bone (Figure 6.14). The anterior border of the vomer articulates with the septal cartilage, which is hyaline cartilage, to form the anterior portion of the septum. The superior border of the vomer articulates with the perpendicular plate of the ethmoid bone to form the remainder of the nasal septum.

Figure 6.14 / Nasal septum. (See Tortora, *A Photographic Atlas of the Human Body,* Figure 3.4.)

The structures that form the nasal septum are the perpendicular plate of the ethmoid bone, the vomer, and septal cartilage.

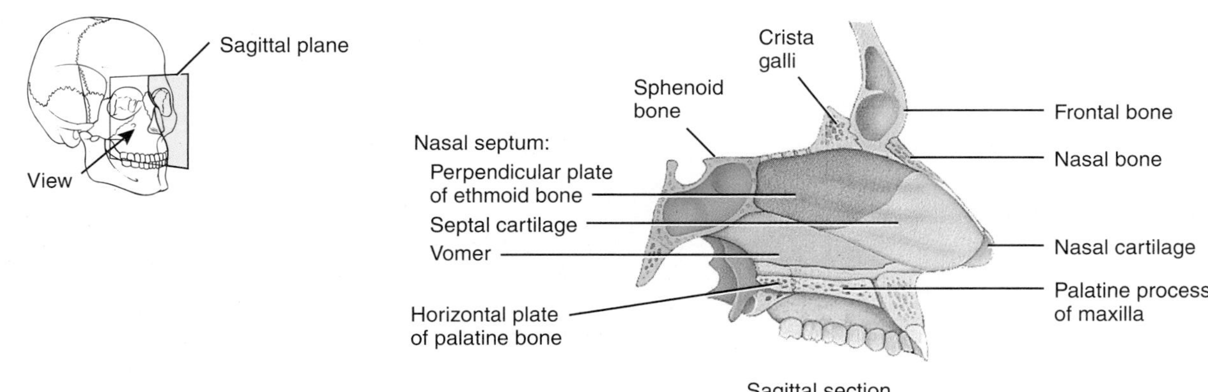

Sagittal section

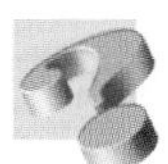
What is the function of the nasal septum?

A **deviated nasal septum** is one that is deflected laterally from the midline of the nose. The deviation usually occurs at the junction of the vomer with the septal cartilage. Septal deviations may occur as a result of a developmental abnormality or trauma. If the deviation is severe, it may entirely block the nasal passageway. Even a partial blockage may lead to infection. If inflammation occurs, it may cause nasal congestion, blockage of the paranasal sinus openings, chronic sinusitis, headache, and nosebleeds. The condition usually can be corrected surgically. ■

Cranial Fossae

The floor of the cranium contains three distinct levels, from anterior to posterior, called **cranial fossae** (Figure 6.15). The fossae contain depressions for the various brain convolutions, grooves for cranial blood vessels, and numerous foramina. From anterior to posterior, they are named the anterior cranial fossa, middle cranial fossa, and posterior cranial fossa. The highest level, the *anterior cranial fossa,* is formed largely by the portion of the frontal bone that constitutes the roof of the orbits and nasal cavity, the crista galli and cribriform plate of the ethmoid bone, and the lesser wings and part of the body of the sphenoid bone. This fossa houses the frontal lobes of the cerebral hemispheres of the brain. The rough surface of the frontal bone can lead to tearing of the frontal lobes of the cerebral hemispheres during head trauma. The *middle cranial fossa* is inferior and posterior to the anterior cranial fossa. It is shaped like a butterfly with a small median portion and two expanded lateral portions. The median portion is formed by part of the body of the sphenoid bone, and the lateral portions are formed by the greater wings of the sphenoid bone, temporal squama, and parietal bone. The middle cranial fossa cradles the temporal lobes of the cerebral hemispheres. The last fossa, at the most inferior level, is the *posterior cranial fossa,* the largest of the fossae. It is formed largely by the occipital bone and the petrous and mastoid portions of the temporal bone. It is a very deep fossa that accommodates the cerebellum, pons, and medulla oblongata of the brain.

- ✓ List the cranial and facial bones, and describe the general features of the skull.
- ✓ Define the following: suture, paranasal sinus, fontanel, foramen.
- ✓ What bones constitute the orbit?
- ✓ What structures make up the nasal septum?
- ✓ Name the cranial fossae from inferior to superior.

HYOID BONE

Objective

- Describe the relationship of the hyoid bone to the skull.

The single **hyoid bone** (= U-shaped) is a unique component of the axial skeleton because it does not articulate with any other

Figure 6.15 / Cranial fossae.

Cranial fossae are levels in the cranial floor that contain depressions for brain convolutions, grooves for blood vessels, and numerous foramina.

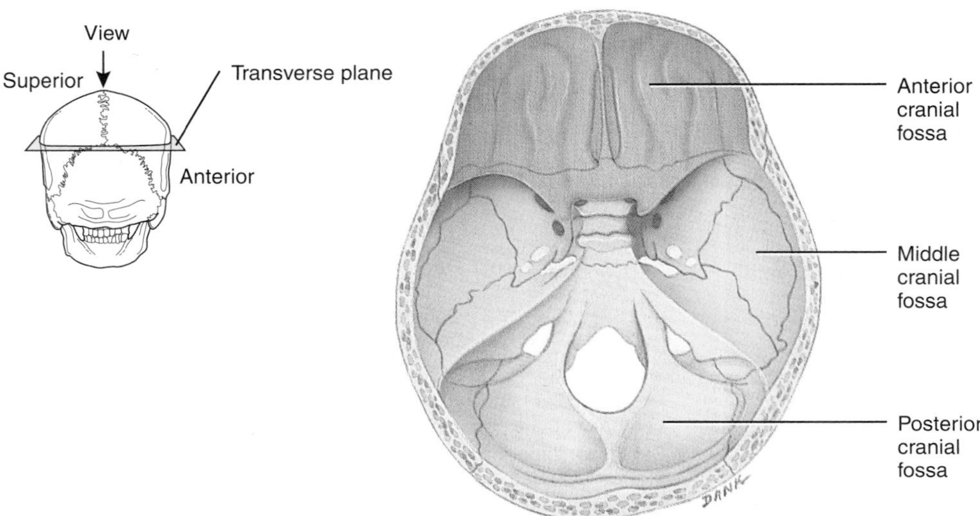

Which cranial fossae is the largest?

Figure 6.16 / Hyoid bone. (See Tortora, *A Photographic Atlas of the Human Body,* Figure 3.13.)

The hyoid bone supports the tongue, providing attachment sites for muscles of the tongue, neck, and pharynx.

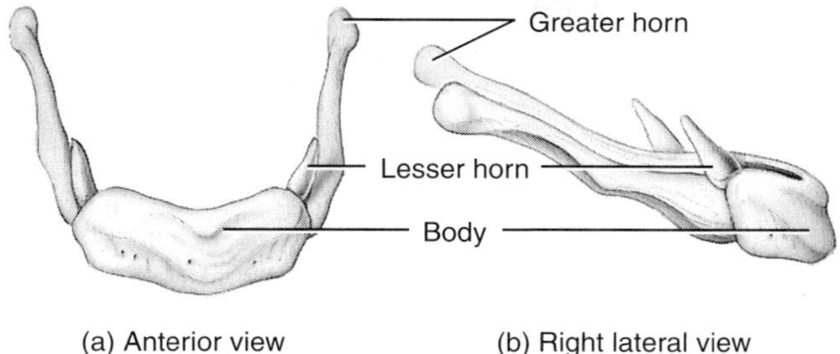

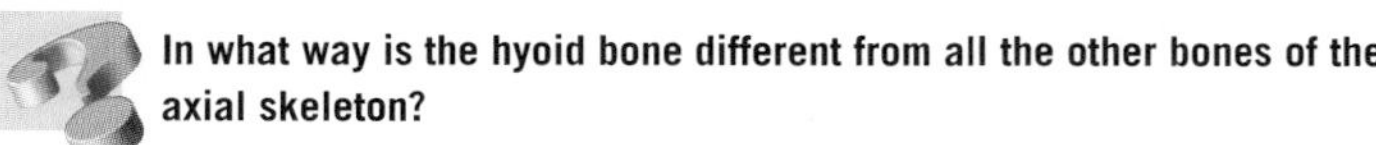
In what way is the hyoid bone different from all the other bones of the axial skeleton?

bone (Figure 6.4). Rather, it is suspended from the styloid processes of the temporal bones by ligaments and muscles. Located in the anterior neck between the mandible and larynx, the hyoid bone supports the tongue, providing attachment sites for some tongue muscles and for muscles of the neck and pharynx. The hyoid bone consists of a horizontal *body* and paired projections called the *lesser horns* and the *greater horns* (Figure 6.16). Muscles and ligaments attach to these paired projections.

The hyoid bone, as well as cartilages of the larynx and trachea, are often fractured during strangulation. As a result, they are carefully examined at autopsy when strangulation is suspected.

✓ What are the functions of the hyoid bone?

VERTEBRAL COLUMN

Objective

- Identify the regions and normal curves of the vertebral column.

The **vertebral column,** also called the *spine* or *backbone,* together with the sternum and ribs, forms the skeleton of the trunk of the body. The bone and connective tissue of the vertebral column form a strong, flexible rod that bends anteriorly, posteriorly, and laterally and rotates. It encloses and protects the spinal cord, supports the head, and serves as a point of attachment for the ribs, pelvic girdle, and muscles of the back.

The vertebral column makes up about two-fifths of the total height of the body and is composed of a series of bones called **vertebrae** (VER-te-brē; singular is *vertebra*). The length of the column is about 71 cm (28 in.) in an average adult male and about 61 cm (24 in.) in an average adult female. Between vertebrae are openings called **intervertebral foramina.** The thoracic and lumbar spinal nerves that connect the spinal cord to various parts of the body pass through these openings.

The adult vertebral column is divided into five regions that contain 26 bones distributed as follows (Figure 6.17a on page 157):

- The cervical region (*cervic-* = neck) contains seven cervical vertebrae in the neck.
- The thoracic region (*thorac-* = chest) contains 12 thoracic vertebrae that lie posterior to the thoracic cavity.
- The lumbar region (*lumb-* = loin) contains five lumbar vertebrae that support the lower back.
- The sacral region contains the *sacrum* (SĀ-krum; = sacred bone), one bone consisting of five fused sacral vertebrae.
- The coccygeal region (kok-SIJ-ē-al) contains one bone (sometimes two bones) called the *coccyx* (KOK-siks; = resembling the bill of a cuckoo), which consists of usually four fused coccygeal vertebrae.

Thus, before the sacral and coccygeal vertebrae fuse, the total number of vertebrae is 33. Whereas the cervical, thoracic, and lumbar vertebrae are movable, the sacrum and coccyx are not. We will discuss each of these regions in greater detail shortly.

Intervertebral Discs

Between the bodies of adjacent vertebrae from the second cervical vertebra to the sacrum are **intervertebral discs** (Figure 6.17d). Each disc has an outer fibrous ring consisting of fibrocartilage called the *annulus fibrosus* (*annulus* = ringlike) and an inner soft, pulpy, highly elastic substance called the *nucleus pulposus* (*pulposus* = pulplike). The discs form strong joints, permit various movements of the vertebral column, and absorb vertical shock. Under compression, they flatten, broaden, and bulge from their intervertebral spaces. Superior to the sacrum, the intervertebral discs constitute about one-fourth the length of the vertebral column.

Normal Curves

When viewed from the side, the vertebral column shows four **normal curves** (Figure 6.17b). The *cervical* and *lumbar curves* are anteriorly convex (bulging out), whereas the *thoracic* and *sacral curves* are anteriorly concave (cupping in). The curves of the vertebral column are important because they increase its strength, help maintain balance in the upright position, absorb shocks during walking and running, and help protect the column from fracture.

In the fetus, there is only a single anteriorly concave curve (Figure 6.17c). At approximately the third month after birth, when an infant begins to hold its head erect, the cervical curve develops. Later, when the child sits up, stands, and walks, the lumbar curve develops. The thoracic and sacral curves are called *primary curves* because they form first during fetal development. The cervical and lumbar curves are referred to as *secondary curves* because they begin to form later, several months after birth. All curves are fully developed by age 10. However, secondary curves may be progressively lost in elderly individuals. Later in the chapter we consider the causes of several types of abnormal curves—scoliosis, kyphosis, and lordosis (p. 169).

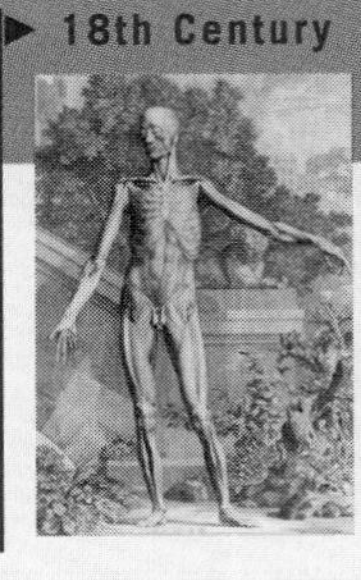

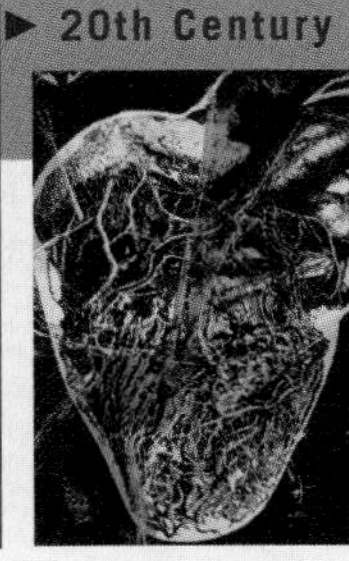

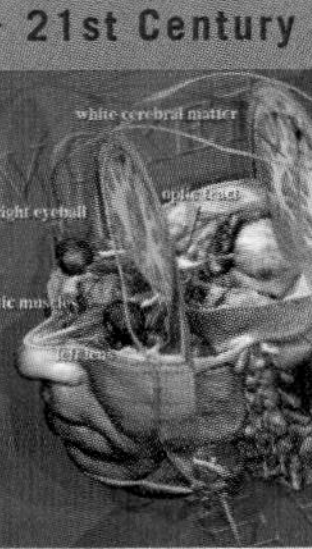

CHANGING IMAGES

A Framework for Life

Dry, lifeless and scattered, the human skeleton is often portrayed as a symbol of death. Not only is it an icon for Halloween, our language also contains phrases like *skeletal remains* and *skeletons in the closet* to reflect the cultural perception of our body's framework as being ghostly. This macabre view was also pervasive throughout Medieval Europe in anatomic illustrations.

The irony of this is that bone is comprised of skeletal tissue which is moist, responsive, and dynamic—the skeleton is indeed very much alive. The vitality of bone was celebrated famously by Renaissance anatomist Andreas Vesalius in his 1543 magnum opus *De Humani Corporis Fabrica* or simply translated: *The Fabric of the Human Body*. Indeed, Vesalius brought the skeleton to life. His work was richly illustrated with scores of detailed anatomic renderings. In Chapter 9 you will enjoy another of these images.

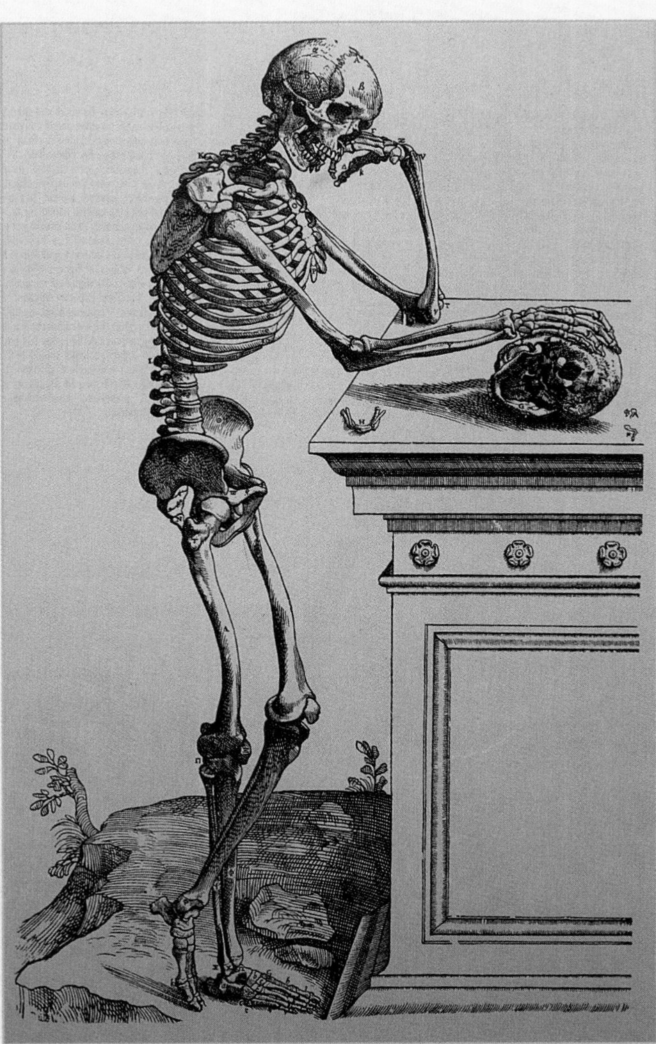

1543 AD

Pictured to the left is one of the more arresting figures from the osteologic series of the *Fabrica*. Vesalius appears to portray a contemplative pre-Shakespearean Hamlet who poses beside a crypt. Despite unparalleled artistic achievement, however, this image does exhibit anatomical errors. As a beginning student of anatomy, see if you can identify any of these inaccuracies. Be sure to examine closely the spinal curvature and pelvic tilt. Also, can you discern the bones that are placed atop the tomb on either side of the skull?

Vesalius believed that the skeletal system was perhaps the most important human system for an anatomy student to understand inasmuch as it is the foundation of the body. In that tradition, this text has placed skeletal system chapters near its beginning. As you continue to study the skeleton be sure to keep in mind its elegance and vitality.

Figure 6.17 / Vertebral column. The numbers in parentheses in (a) indicate the number of vertebrae in each region. In (d), the relative size of the disc has been enlarged for emphasis. A "window" has been cut in the annulus fibrosus so that the nucleus pulposus can be seen. (See Tortora, *A Photographic Atlas of the Human Body,* Figure 3.15.)

The adult vertebral column typically contains 26 vertebrae.

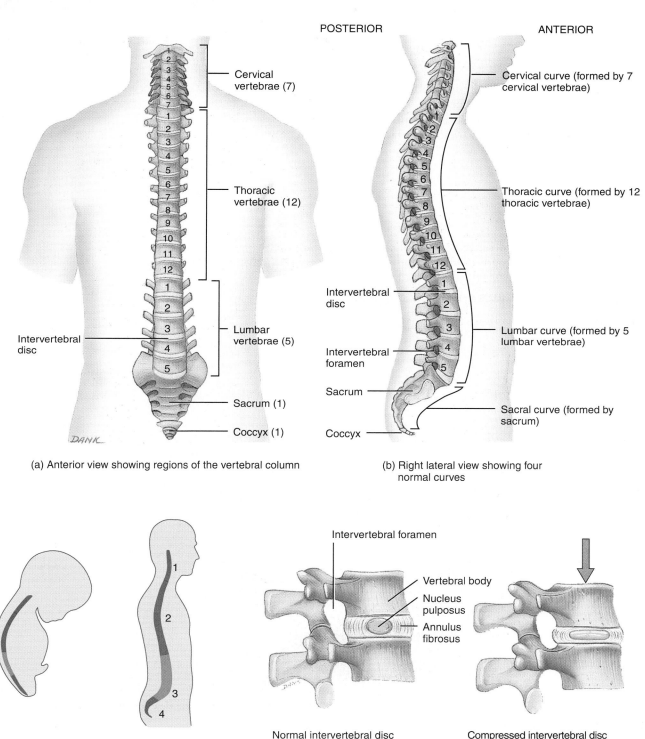

(a) Anterior view showing regions of the vertebral column

(b) Right lateral view showing four normal curves

Single curve in fetus

Four curves in adult

(c) Fetal and adult curves

Normal intervertebral disc

Compressed intervertebral disc in a weight-bearing situation

(d) Intervertebral disc

Which curves of the adult vertebral column are concave (relative to the anterior side of the body)?

Parts of a Typical Vertebra

Even though vertebrae in different regions of the spinal column vary in size, shape, and detail, they are similar enough that we can discuss the structures (and the functions) of a typical vertebra (Figure 6.18). Vertebrae typically consist of a body, a vertebral arch, and several processes.

Body

The *body* is the thick, disc-shaped anterior portion that is the weight-bearing part of a vertebra. Its superior and inferior surfaces are roughened for the attachment of cartilaginous intervertebral discs. The anterior and lateral surfaces contain nutrient foramina for blood vessels.

Vertebral Arch

The *vertebral arch* extends posteriorly from the body of the vertebra. It and the body of the vertebra surround the spinal cord. The vertebral arch is formed by two short, thick processes, the *pedicles* (PED-i-kuls; = little feet), which project posteriorly from the body to unite with the laminae. The *laminae* (LAM-i-nē; = thin layers) are the flat parts that join to form the posterior portion of the vertebral arch. The *vertebral foramen* lies between the vertebral arch and body and contains the spinal cord, fat, areolar connective tissue, and blood vessels. Collectively, the vertebral foramina of all vertebrae form the vertebral (spinal) canal, which is the inferior portion of the dorsal body cavity. The pedicles exhibit superior and inferior notches called *vertebral notches*. When the vertebral notches are stacked on top of

Figure 6.18 / Structure of a typical vertebra, as illustrated by a thoracic vertebra. In (b), only one spinal nerve has been included, and it has been extended beyond the intervertebral foramen for clarity. The sympathetic chain is part of the autonomic nervous system (see Figure 17.2).

A vertebra consists of a body, a vertebral arch, and several processes.

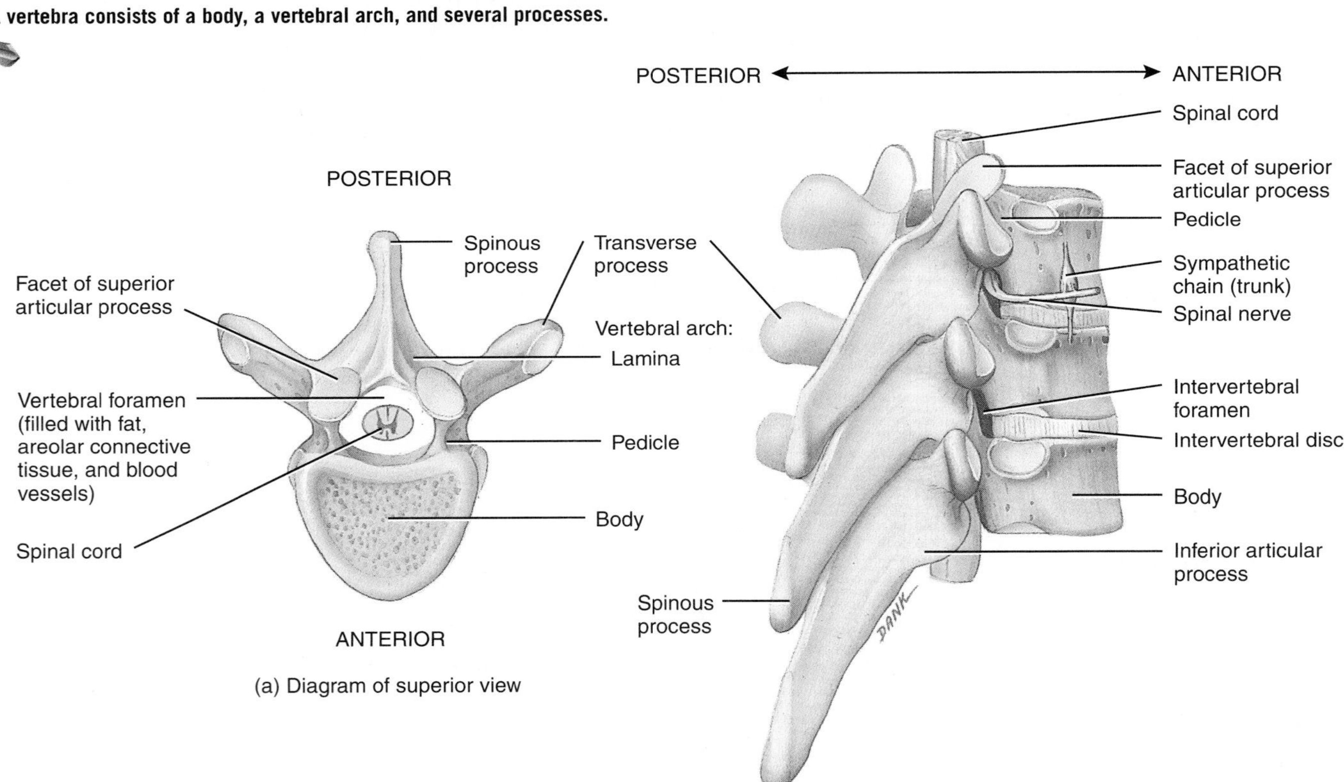

(a) Diagram of superior view

(b) Diagram of right posterolateral view of articulated vertebrae

one another, they form an opening (the intervertebral foramen) between adjoining vertebrae on each side of the column. Each opening permits the passage of a single spinal nerve (Figure 6.18b).

Processes

Seven *processes* arise from the vertebral arch. At the point where a lamina and pedicle join, a *transverse process* extends laterally on each side. A single *spinous process (spine)* projects posteriorly from the junction of the laminae. These three processes serve as points of attachment for muscles. The remaining four processes form joints with other vertebrae above or below. The two *superior articular processes* of a vertebra articulate with the two inferior articular processes of the vertebra immediately superior to them. The two *inferior articular processes* of a vertebra articulate with the two superior articular processes of the vertebra immediately inferior to them. The articulating surfaces of the articular processes are referred to as *facets* (= little faces). The articulations formed between the bodies and articular facets of successive vertebrae are called *intervertebral joints.*

Regions

We turn now to the five regions of the vertebral column, beginning superiorly and moving inferiorly. Note that vertebrae in each region are numbered in sequence, from superior to inferior.

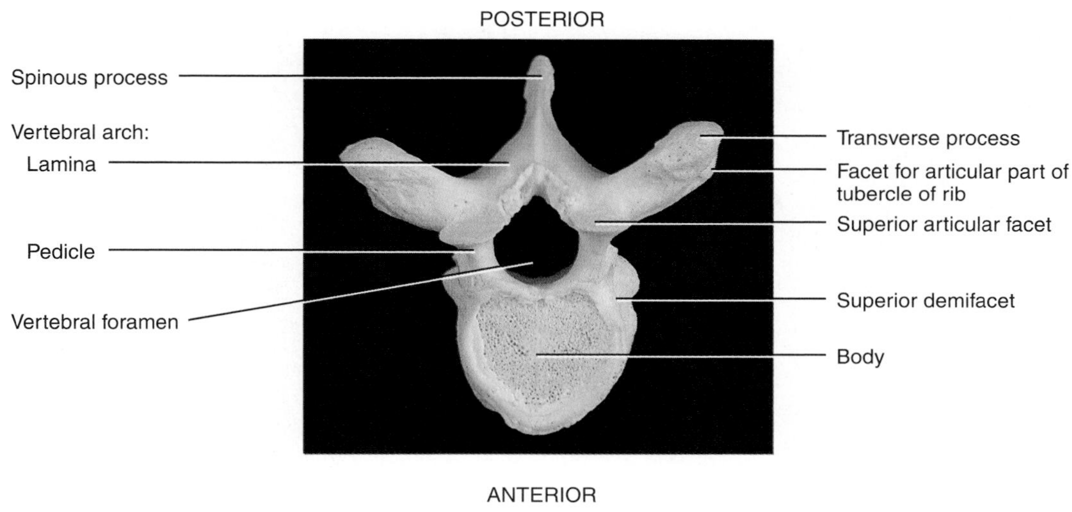

(c) Superior view

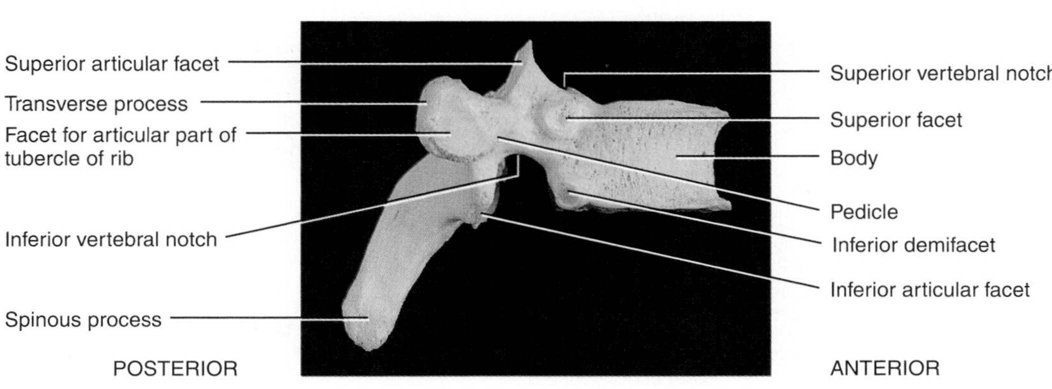

(d) Right lateral view

What are the functions of the vertebral and intervertebral foramina?

Cervical Region

The bodies of **cervical vertebrae** (C1–C7) are smaller than those of thoracic vertebrae (Figure 6.19a). The vertebral arches, however, are larger. All cervical vertebrae have three foramina: one vertebral foramen and two transverse foramina (Figure 6.19d). The vertebral foramina of cervical vertebrae are the largest in the spinal column because they house the cervical enlargement of the spinal cord. Each cervical transverse process contains a *transverse foramen* through which the vertebral artery and its accompanying vein and nerve fibers pass. The spinous processes of C2 through C6 are often *bifid*—that is, split into two parts (Figure 6.19a, d).

The first two cervical vertebrae differ considerably from the others. Like the mythological Atlas, who supported the world on his shoulders, the first cervical vertebra (C1), the **atlas,** supports the head (Figure 6.19a, b). The atlas is a ring of bone with *anterior* and *posterior arches* and large *lateral masses.* It lacks a

Figure 6.19 / Cervical vertebrae.

The cervical vertebrae are found in the neck region.

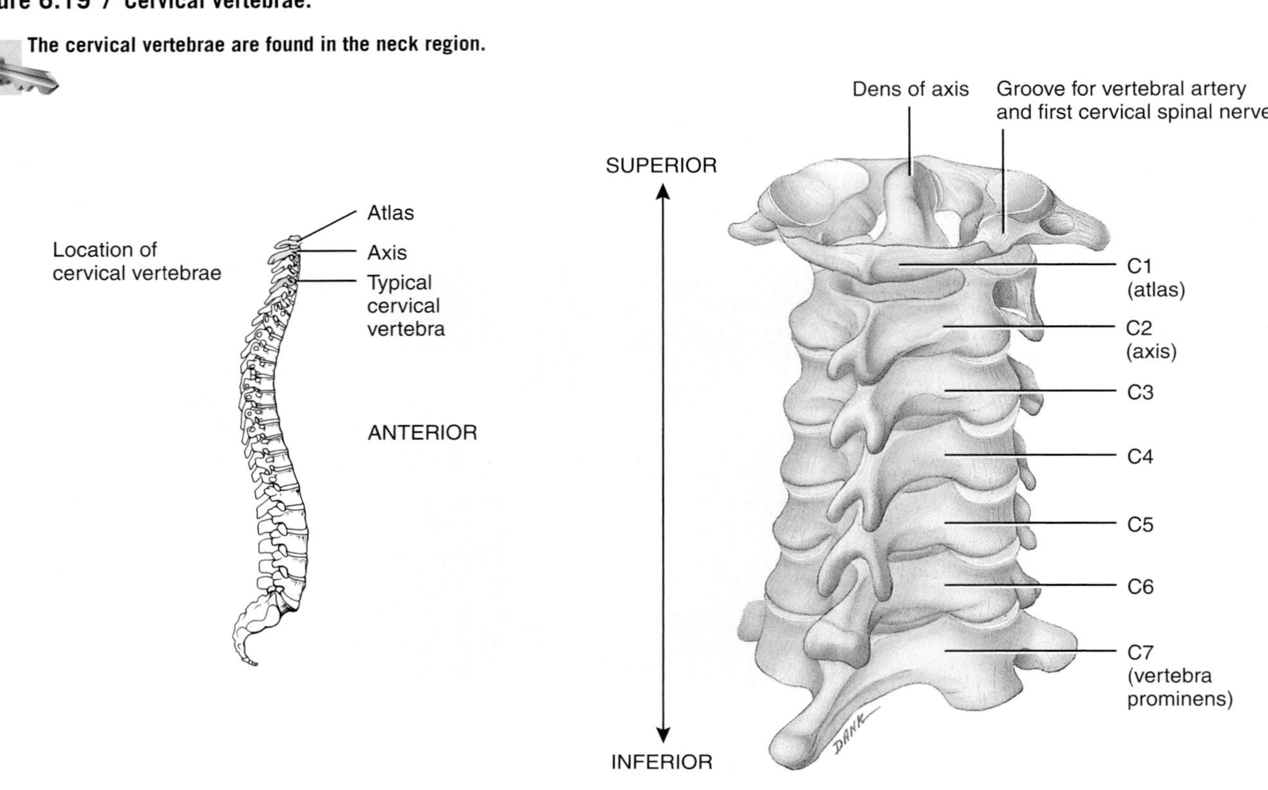

(a) Posterior view of articulated cervical vertebrae

body and a spinous process. The superior surfaces of the lateral masses, called *superior articular facets,* are concave and articulate with the occipital condyles of the occipital bone to form the *atlanto-occipital joints.* These articulations permit the movement seen when moving the head to signify yes. The inferior surfaces of the lateral masses, the *inferior articular facets,* articulate with the second cervical vertebra. The transverse processes and transverse foramina of the atlas are quite large.

The second cervical vertebra (C2), called the **axis** (Figure 6.19a, c), does have a body. A peglike process called the *dens* (= tooth) or *odontoid* process projects superiorly through the anterior portion of the vertebral foramen of the atlas. The dens makes a pivot on which the atlas and head rotate, as in moving the head to signify no. This arrangement permits side-to-side rotation of the head. The articulation formed between the anterior arch of the atlas and dens of the axis, and between their articular facets, is called the *atlanto-axial joint.* In some instances of trauma, the dens of the axis may be driven into the medulla ob-

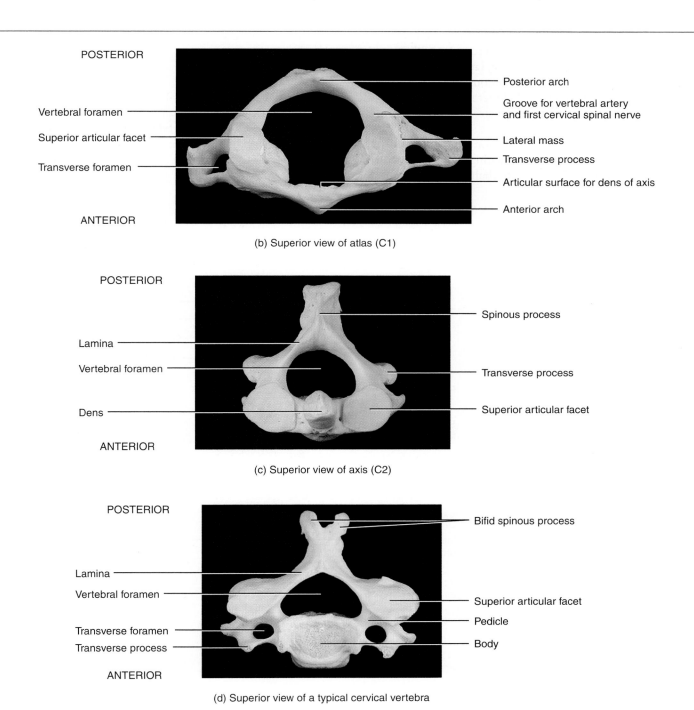

(b) Superior view of atlas (C1)

(c) Superior view of axis (C2)

(d) Superior view of a typical cervical vertebra

Which bones permit the movement of the head to signify no?

longata of the brain. When whiplash injuries result in death, this type of injury is the usual cause.

The third through sixth cervical vertebra (C3–C6), represented by the vertebra in Figure 6.19d, correspond to the structural pattern of the typical cervical vertebra previously described. The seventh cervical vertebra (C7), called the *vertebra prominens*, is somewhat different (Figure 6.19a). It has a single large spinous process that may be seen and felt at the base of the neck.

Thoracic Region

Thoracic vertebrae (T1–T12; Figure 6.20) are considerably larger and stronger than cervical vertebrae. In addition, the spinous processes on T1 and T2 are long, laterally flattened, and directed inferiorly. In contrast, the spinous processes on T11 and T12 are shorter, broader, and directed more posteriorly. Compared to cervical vertebrae, thoracic vertebrae also have longer and larger transverse processes.

The best distinguishing feature of thoracic vertebrae is that they articulate with the ribs. The articulating surfaces of the vertebrae are called *facets* and *demifacets* (half-facets). Except for T11 and T12, the transverse processes have facets for articulating with the *tubercles* of the ribs. The bodies of thoracic vertebrae also have facets or demifacets for articulation with the *heads* of the ribs. The articulations between the thoracic vertebrae and ribs are called *vertebrocostal joints.* As you can see in Figure 6.20a, T1 has a superior facet and an inferior demifacet, one on each side of the vertebral body. T2–T8 have a superior and inferior demifacet, one on each side of the vertebral body. T9 has a superior demifacet on each side of the vertebral body, and T10–T12 have a superior facet on each side of the vertebral body. Movements of the thoracic region are limited by thin intervertebral discs and by the attachment of the ribs to the sternum.

Lumbar Region

The **lumbar vertebrae** (L1–L5) are the largest and strongest in the vertebral column (Figure 6.21) because the amount of body weight supported by the vertebrae increases toward the inferior end of the backbone. Their various projections are short and thick. The superior articular processes are directed medially instead of superiorly, and the inferior articular processes are di-

Figure 6.20 / Thoracic vertebrae.

The thoracic vertebrae are found in the chest region and articulate with the ribs.

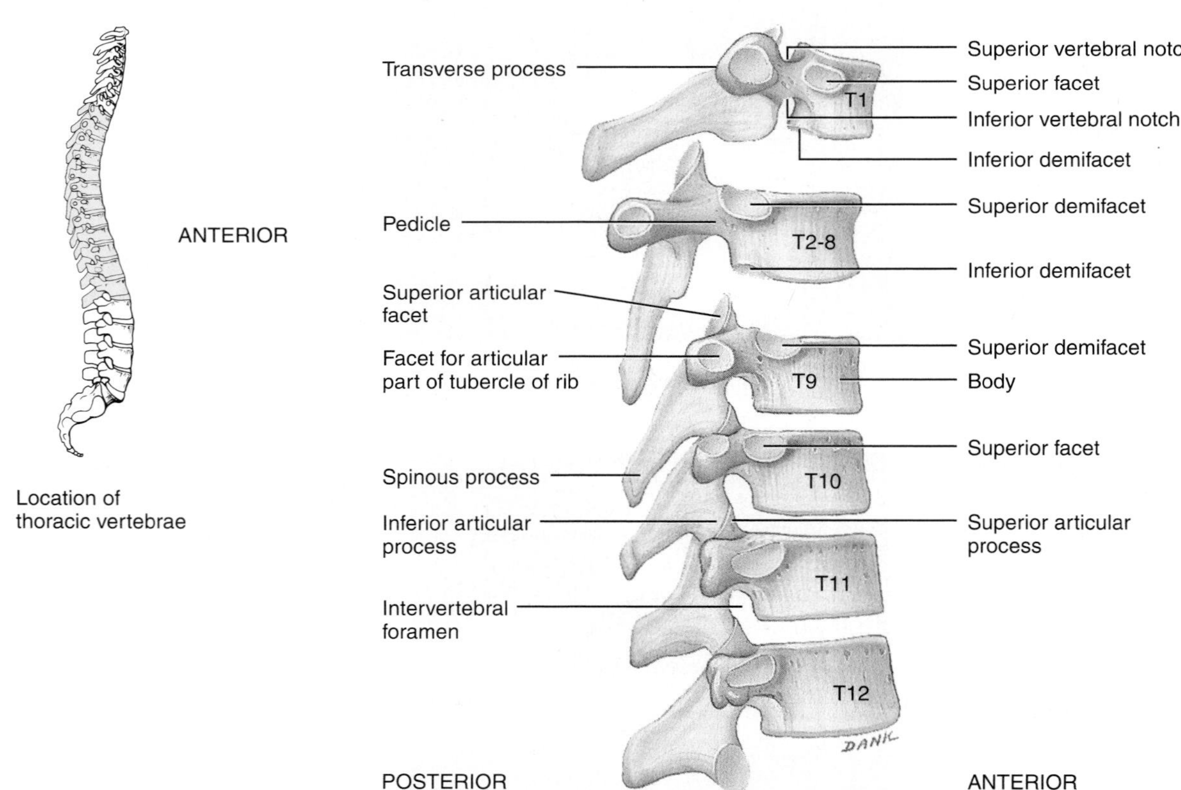

Right lateral view of several articulated thoracic vertebrae

Which parts of thoracic vertebrae articulate with the ribs?

Figure 6.21 / Lumbar vertebrae.

Lumbar vertebrae are found in the lower back.

ANTERIOR

Location of lumbar vertebrae

POSTERIOR

ANTERIOR

Superior articular process

Transverse process

Spinous process

Inferior articular facet

Intervertebral foramen

Intervertebral disc

Body

Superior vertebral notch

Inferior vertebral notch

DANK

(a) Diagram of right lateral view of articulated lumbar vertebrae

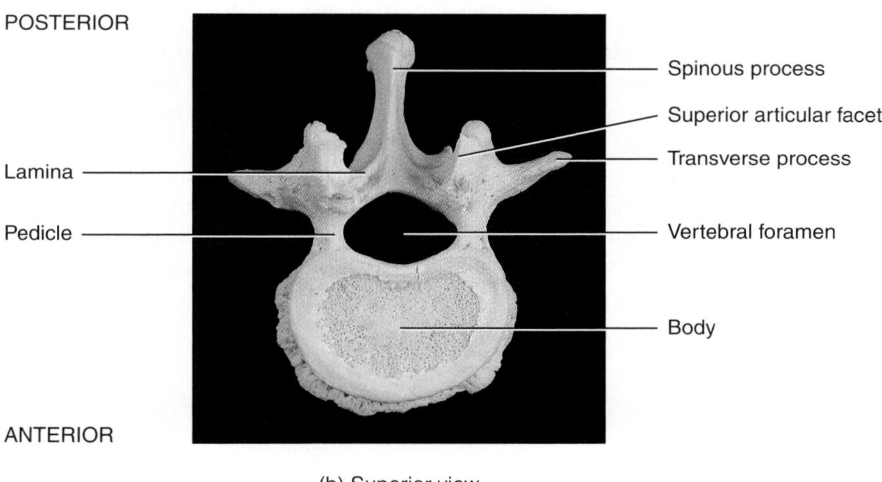

(b) Superior view

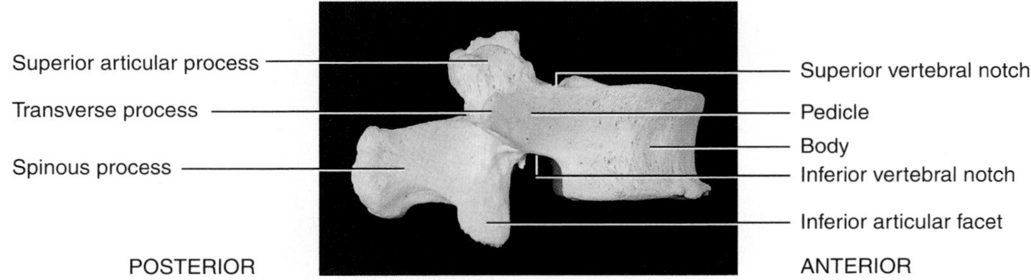

(c) Right lateral view

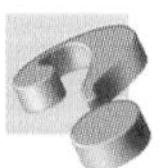

Why are the lumbar vertebrae the largest and strongest in the vertebral column?

rected laterally instead of inferiorly. The spinous processes are quadrilateral in shape, thick and broad, and project nearly straight posteriorly. The spinous processes are well-adapted for the attachment of the large back muscles.

A summary of the major structural differences among cervical, thoracic, and lumbar vertebrae is presented in Table 6.6.

Sacrum

The **sacrum** is a triangular bone formed by the union of five sacral vertebrae, indicated in Figure 6.22a as S1–S5. The sacral vertebrae begin to fuse in individuals between 16 and 18 years of age, a process usually completed by age 30. The sacrum serves as a strong foundation for the pelvic girdle. It is positioned at the posterior portion of the pelvic cavity medial to the two hip bones. The female sacrum is shorter, wider, and more curved between S2 and S3 than the male sacrum.

The concave anterior side of the sacrum faces the pelvic cavity. It is smooth and contains four *transverse lines (ridges)* that mark the joining of the sacral vertebral bodies (Figure 6.22a). At the ends of these lines are four pairs of *anterior sacral foramina*. The lateral portion of the superior surface contains a smooth surface called the *sacral ala* (= wing), which is formed by the fused transverse processes of the first sacral vertebra (S1).

The convex, posterior surface of the sacrum contains a *median sacral crest*, which is the fused spinous processes of the upper sacral vertebrae; a *lateral sacral crest*, which is the fused transverse processes of the sacral vertebrae; and four pairs of *posterior sacral foramina* (Figure 6.22b). These foramina communicate with the anterior sacral foramina through which nerves and blood vessels pass. The *sacral canal* is a continuation of the vertebral canal. The laminae of the fifth sacral vertebra, and sometimes the fourth, fail to meet. This leaves an inferior entrance to the vertebral canal called the *sacral hiatus* (hī-Ā-tus; = opening). On either side of the sacral hiatus are the *sacral cornua*, the inferior articular processes of the fifth sacral vertebra. They are connected by ligaments to the coccyx.

The narrow inferior portion of the sacrum is known as the *apex*. The broad superior portion of the sacrum is called the *base*. The anteriorly projecting border of the base, called the *sacral promontory* (PROM-on-tō′-rē), is one of the points used for measurements of the pelvis. On both lateral surfaces the sacrum has a large *auricular surface* (*auricular* = ear) that articulates with the ilium of each hip bone to form the *sacroiliac joint*. Posterior to the auricular surface is a roughened surface, the *sacral tuberosity*, that contains depressions for the attachment of ligaments. The sacral tuberosity is another surface of the sacrum that unites with the hip bones to form the sacroiliac joints. The *superior articular processes* of the sacrum articulate with the fifth lumbar vertebra, and the base of the sacrum articulates with the body of the fifth lumbar vertebra, to form the *lumbosacral joint*.

Coccyx

The **coccyx** is also triangular in shape and is usually formed by the fusion of four coccygeal vertebrae, indicated in Figure 6.22 as Co1–Co4. The coccygeal vertebrae fuse when a person is between 20 and 30 years of age. The dorsal surface of the body of the coccyx contains two long *coccygeal cornua* that are connected by ligaments to the sacral cornua. The coccygeal cornua are the pedicles and superior articular processes of the first coccygeal vertebra. On the lateral surfaces of the coccyx are a series of

Table 6.6 Comparison of Principal Structural Features of Cervical, Thoracic, and Lumbar Vertebrae

Characteristic	Cervical	Thoracic	Lumbar
Overall structure			
Size	Small	Larger	Largest
Foramina	One vertebral and two transverse	One vertebral	One vertebral
Spinous processes	Slender and often bifid (C2–C6)	Long and fairly thick (most project inferiorly)	Short and blunt (project posteriorly rather than inferiorly)
Transverse processes	Small	Fairly large	Large and blunt
Articular facets for ribs	No	Yes	No
Direction of articular facets			
Superior	Posterosuperior	Posterolateral	Medial
Inferior	Anterioinferior	Anteromedial	Lateral
Size of intervertebral discs	Thick relative to size of vertebral bodies	Thin relative to vertebral bodies	Massive

transverse processes, the first pair being the largest. The coccyx articulates superiorly with the apex of the sacrum. In females, the coccyx points inferiorly; in males, it points anteriorly (see Table 7.1).

Caudal Anesthesia

Anesthetic agents that act on the sacral and coccygeal nerves are sometimes injected through the sacral hiatus, a procedure called **caudal anesthesia** or **epidural block.** The procedure is used most often to relieve pain during labor and to provide anesthesia to the perineal area. Because the sacral hiatus is between the sacral cornua, the cornua are important bony landmarks for locating the hiatus. Anesthetic agents may also be injected through the posterior sacral foramina.

✓ What are the functions of the vertebral column?
✓ When do the secondary vertebral curves develop?
✓ What are the principal distinguishing characteristics of the bones of the various regions of the vertebral column?

THORAX

Objective

- Identify the bones of the thorax.

The term **thorax** refers to the entire chest. The skeletal portion of the thorax is a bony cage formed by the sternum, costal cartilages, ribs, and the bodies of the thoracic vertebrae (Figure 6.23). The thoracic cage is narrower at its superior end and broader at its inferior end and flattened from front to back. The thoracic cage encloses and protects the organs in the thoracic and superior abdominal cavities. It also provides support for the bones of the shoulder girdle and upper limbs.

Sternum

The **sternum,** or breastbone, is a flat, narrow bone measuring about 15 cm (6 in.) in length. It is located on the anterior midline of the thoracic wall. During thoracic surgery, it may be split midsagittally to allow surgeons access to structures in the thoracic cavity such as the thymus, heart, and great vessels of the heart.

Figure 6.22 / Sacrum and coccyx. (See Tortora, *A Photographic Atlas of the Human Body,* Figure 3.19.)

The sacrum is formed by the union of five sacral vertebrae, and the coccyx is formed by the union of usually four coccygeal vertebrae.

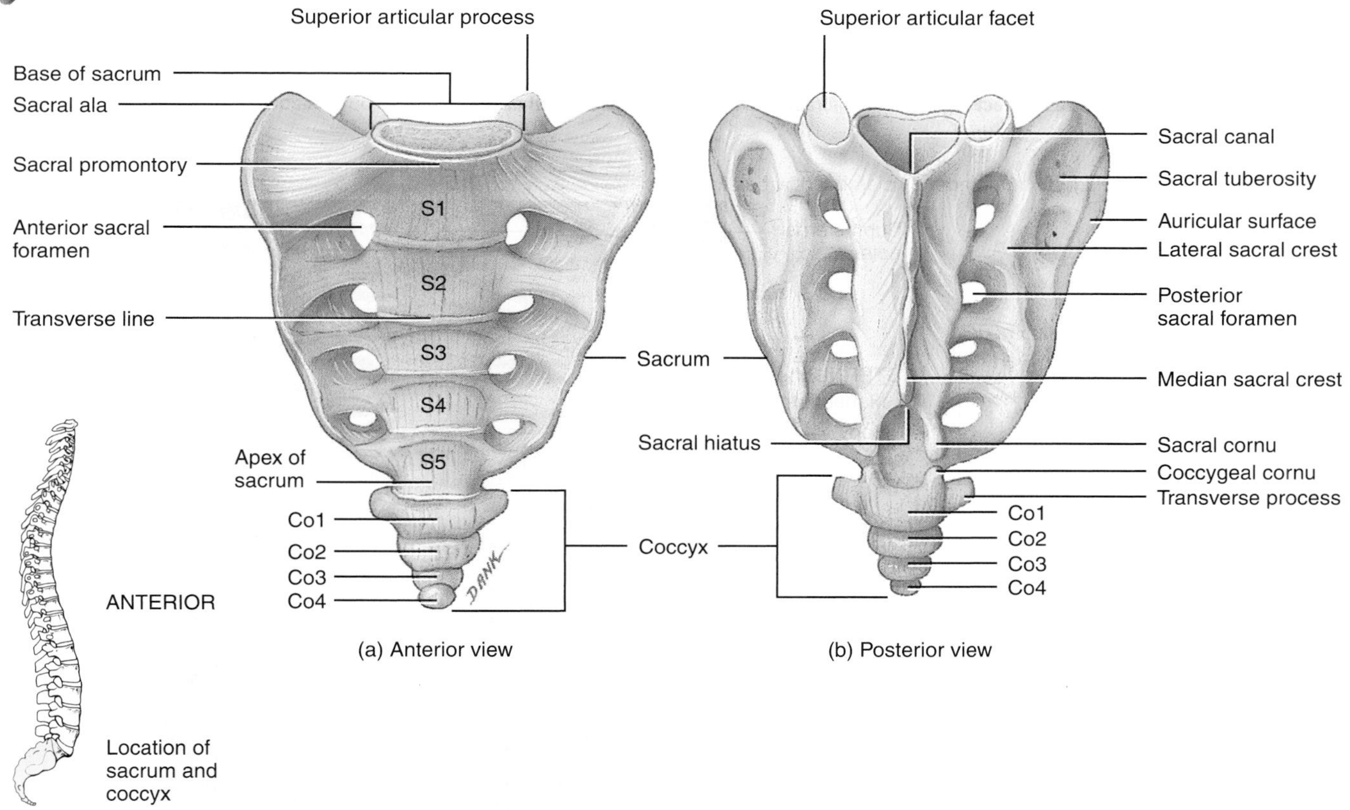

Superior articular process
Base of sacrum
Sacral ala
Sacral promontory
Anterior sacral foramen
Transverse line
S1
S2
S3
S4
S5
Apex of sacrum
Co1
Co2
Co3
Co4
ANTERIOR
Location of sacrum and coccyx
(a) Anterior view

Superior articular facet
Sacral canal
Sacral tuberosity
Auricular surface
Lateral sacral crest
Posterior sacral foramen
Sacrum
Median sacral crest
Sacral hiatus
Sacral cornu
Coccygeal cornu
Transverse process
Coccyx
Co1
Co2
Co3
Co4
(b) Posterior view

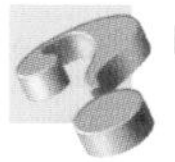

How many foramina pierce the sacrum, and what is their function?

In summary, the posterior portion of the rib is connected to a thoracic vertebra by its head and the articular part of a tubercle. The facet of the head fits into a facet on the body of one vertebra or into the demifacets of two adjoining vertebrae. The articular part of the tubercle articulates with the facet of the transverse process of the vertebra.

If you examine Figure 6.23, you will notice that the first rib is the shortest, broadest, and most sharply curved of the ribs. The first rib is an important landmark because of its close relationship to the nerves of the *brachial plexus*, which provides the entire nerve supply of the shoulder and upper limb, two major blood vessels, the subclavian artery and vein, and two skeletal muscles, the anterior and medial scalene muscles. The superior surface of the first rib has two shallow grooves, one for the subclavian vein and one for the subclavian artery and inferior trunk of the brachial plexus. The superior surface also serves as the point of attachment for the anterior and middle scalene muscles. The second rib is thinner, less curved, and considerably longer than the first. The tenth rib has a single articular facet on its head. The eleventh and twelfth ribs also have single articular facets on their heads, but no necks, tubercles, or costal angles.

Spaces between ribs, called *intercostal spaces*, are occupied by intercostal muscles, blood vessels, and nerves. Surgical access to the lungs or other structures in the thoracic cavity is commonly obtained through an intercostal space. Special rib retractors are used to create a wide separation between ribs. The costal cartilages are sufficiently elastic in younger individuals to permit considerable bending without breaking.

Structures passing between the thoracic cavity and the neck pass through an opening called the **superior thoracic aperture.** Among these structures are the trachea, esophagus, nerves, and blood vessels that supply and drain the head, neck, and upper limbs. The aperture is bordered by the first thoracic vertebra (posteriorly), the first pair of ribs and their cartilages, and the superior border of the manubrium of the sternum. Structures passing between the thoracic cavity and abdominal cavity pass through the **inferior thoracic aperture.** Through this large opening, which is closed by the diaphragm, pass structures such as the esophagus, nerves, and large blood vessels. This aperture is bordered by the twelfth thoracic vertebra (posteriorly), the eleventh and twelfth pair of ribs, the costal cartilages of ribs 7 through 10, and the joint between the body and xiphoid process of the sternum (anteriorly).

Rib Fractures

Rib fractures are the most common chest injuries, and they usually result from direct blows, most often from impact with a steering wheel, falls, and crushing injuries to the chest. Ribs tend to break at the point where the greatest force is applied, but they may also break at their weakest point—the site of greatest curvature, which is just anterior to the costal angle. In some cases, fractured ribs may puncture the heart, great vessels of the heart, lungs, trachea, bronchi, esophagus, spleen, liver, and kidneys. Rib fractures are usually quite painful. ■

✓ What bones form the skeleton of the thorax?
✓ What are the functions of the bones of the thorax?
✓ How are ribs classified on the basis of their attachment to the sternum?

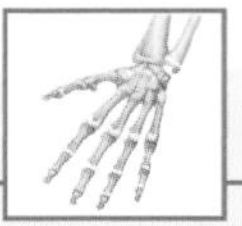

APPLICATIONS TO HEALTH

Herniated (Slipped) Disc

In their function as shock absorbers, intervertebral discs are constantly being compressed. If the anterior and posterior ligaments of the discs become injured or weakened, the pressure developed in the nucleus pulposus may be great enough to rupture the surrounding fibrocartilage (annulus fibrosus). If this occurs, the nucleus pulposus may herniate (protrude) posteriorly or into one of the adjacent vertebral bodies. This condition is called a **herniated (slipped) disc.** It occurs most often in the lumbar region because it bears much of the weight of the body, and it is the region of the most flexing and bending.

Most often the nucleus pulposus slips posteriorly toward the spinal cord and spinal nerves (Figure 6.25). This movement exerts pressure on the spinal nerves, causing acute pain. If the roots of the sciatic nerve, which passes from the spinal cord to the foot, are compressed, the pain radiates down the posterior thigh, through the calf, and occasionally into the foot. If pressure is exerted on the spinal cord itself, some of its neurons may be destroyed. A person with a herniated disc may undergo a

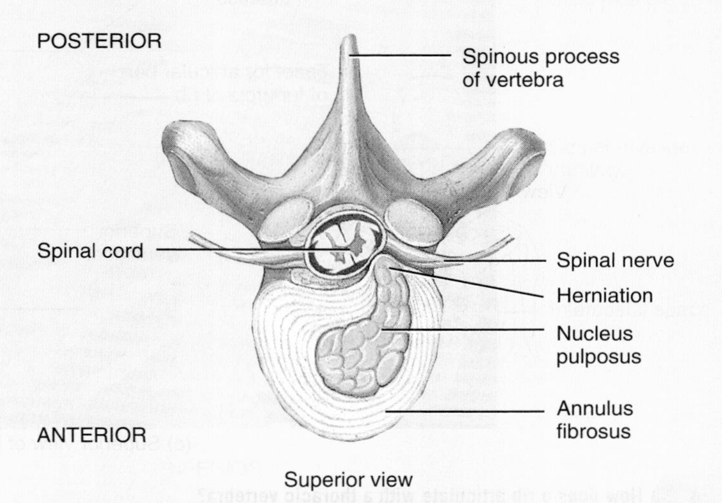

Figure 6.25 / Herniated (slipped) disc.

laminectomy, a procedure in which parts of the laminae of the vertebra and intervertebral disc are removed to relieve pressure on the nerves.

Abnormal Curves of the Vertebral Column

Various conditions may exaggerate the normal curves of the vertebral column, or the column may acquire a lateral bend, resulting in **abnormal curves** of the vertebral column.

Scoliosis (skō-lē-Ō-sis; *scolio-* = crooked) is a lateral bending of the vertebral column, usually in the thoracic region. This is the most common of the abnormal curves. It may result from congenitally (present at birth) malformed vertebrae, chronic sciatica, paralysis of muscles on one side of the vertebral column, poor posture, or one leg being shorter than the other.

Kyphosis (kī-FŌ-sis; *kyphos-* = hump) is an exaggeration of the thoracic curve of the vertebral column. In tuberculosis of the spine, vertebral bodies may partially collapse, causing an acute angular bending of the vertebral column. In the elderly, degeneration of the intervertebral discs leads to kyphosis. Kyphosis may also be caused by rickets and poor posture. It is also common in females with advanced osteoporosis. The term *round-shouldered* is an expression for mild kyphosis.

Lordosis (lor-DŌ-sis; *lord-* = bent backward), sometimes called *swayback,* is an exaggeration of the lumbar curve of the vertebral column. It may result from increased weight of the abdomen as in pregnancy or extreme obesity, poor posture, rickets, or tuberculosis of the spine.

Spina Bifida

Spina bifida (SPĪ-na BIF-i-da) is a congenital defect of the vertebral column in which laminae fail to unite at the midline. In serious cases, the membranes (meninges) around the spinal cord or the spinal cord itself protrude through the opening and produce perilous problems, such as partial or complete paralysis, partial or complete loss of urinary bladder control, and the absence of reflexes. An increased risk of spina bifida is associated with low levels of folic acid, one of the B vitamins, during pregnancy. Spina bifida may be diagnosed prenatally by a test of the mother's blood for a substance, called alphafetoprotein, produced by the fetus; by sonography; or by amniocentesis (withdrawal of amniotic fluid for analysis).

Fractures of the Vertebral Column

Fractures of the vertebral column often involve C1, C2, C4–T7, and T12–L2. Cervical or lumbar fractures usually result from a flexion–compression type of injury such as might be sustained in landing on the feet or buttocks after a fall or having a weight fall on the shoulders. Cervical vertebrae may be fractured or dislodged by a fall on the head with acute flexion of the neck, as might happen on diving into shallow water. Dislocation may result from the sudden forward-then-backward jerk ("whiplash") that may occur when an automobile crashes. Spinal cord or spinal nerve damage may occur as a result of fractures of the vertebral column.

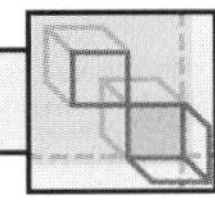

STUDY OUTLINE

Introduction (p. 132)

1. Bones protect soft body parts and make movement possible; they also serve as landmarks for locating parts of other body systems.
2. The musculoskeletal system is composed of the bones, joints, and muscles working together.

Divisions of the Skeletal System (p. 132)

1. The axial skeleton consists of bones arranged along the longitudinal axis. The parts of the axial skeleton are the skull, auditory ossicles (ear bones), hyoid bone, vertebral column, sternum, and ribs.
2. The appendicular skeleton consists of the bones of the girdles and the upper and lower limbs (extremities). The parts of the appendicular skeleton are the pectoral (shoulder) girdles, bones of the upper limbs, pelvic (hip) girdles, and bones of the lower limbs.

Types of Bones (p. 132)

1. On the basis of shape, bones are classified as long, short, flat, irregular, or sesamoid. Sesamoid bones develop in tendons or ligaments.
2. Sutural bones are found within the sutures of certain cranial bones.

Bone Surface Markings (p. 135)

1. Surface markings are structural features visible on the surfaces of bones.
2. Each marking—whether a depression, an opening, or a process—is structured for a specific function, such as joint formation, muscle attachment, or passage of nerves and blood vessels (see Table 6.2 on page 135).

Skull (p. 136)

1. The 22 bones of the skull include cranial bones and facial bones.
2. The eight cranial bones include the frontal, parietal (2), temporal (2), occipital, sphenoid, and ethmoid.
3. The 14 facial bones are the nasal (2), maxillae (2), zygomatic (2), lacrimal (2), palatine (2), inferior nasal conchae (2), vomer, and mandible.
4. Sutures are immovable joints that connect most bones of the skull. Examples are the coronal, sagittal, lambdoid, and squamous sutures.
5. Paranasal sinuses are cavities in bones of the skull that communicate with the nasal cavity. The sinuses are lined with mucous membranes. The cranial bones containing paranasal sinuses are the frontal, sphenoid, ethmoid, and maxillae.
6. Fontanels are fibrous, connective tissue membrane–filled spaces between the cranial bones of fetuses and infants. The major fontanels are the anterior, posterior, anterolaterals, and posterolaterals (see Table 6.4 on page 151). They fill in with bone and become sutures.
7. The foramina of the skull bones provide passages for nerves and blood vessels; see Table 6.5 on page 152.
8. Each of the orbits (eye sockets) is formed by seven bones of the skull.

9. The nasal septum consists of the vomer, perpendicular plate of the ethmoid, and septal cartilage. It divides the nasal cavity into left and right sides.
10. The three cranial fossae contain depressions for brain convolutions, grooves for cranial blood vessels, and numerous foramina.

Hyoid Bone (p. 154)

1. The hyoid bone is a U-shaped bone that does not articulate with any other bone.
2. It supports the tongue and provides attachment for some tongue muscles and for some muscles of the throat and neck.

Vertebral Column (p. 155)

1. The vertebral column, sternum, and ribs constitute the skeleton of the body's trunk.
2. The 26 bones of the adult vertebral column are the cervical vertebrae (7), the thoracic vertebrae (12), the lumbar vertebrae (5), the sacrum (5 fused vertebrae), and the coccyx (usually 4 fused vertebrae).
3. The vertebral column contains normal curves (cervical, thoracic, lumbar, and sacral) that give strength, support, and balance.
4. The vertebrae are similar in structure, each usually consisting of a body, vertebral arch, and seven processes. Vertebrae in the different regions of the column vary in size, shape, and detail.

Thorax (p. 165)

1. The thoracic skeleton consists of the sternum, ribs and costal cartilages, and thoracic vertebrae.
2. The thoracic cage protects vital organs in the chest area and upper abdomen.

SELF-QUIZ QUESTIONS

Choose the one best answer to the following questions:

1. A foramen is (a) a cavity within a bone, (b) a depression, (c) a hole for blood vessels and nerves, (d) a ridge, (e) a site for muscle attachment.
2. Which of the following structures is part of the frontal bone? (a) infraorbital foramen, (b) crista galli, (c) supraorbital foramen, (d) optic foramen, (e) zygomatic arch.
3. The second cervical vertebra (1) has a characteristic feature, the dens, (2) articulates with the first cervical vertebra in a pivot joint, (3) is called the vertebra prominens because of its long process.
   (a) 1 only, **(b)** 2 only, **(c)** 3 only, **(d)** both 1 and 2, **(e)** none of the above.
4. Which of the following bones is correctly matched with the process? (a) zygomatic bone-temporal process, (b) maxilla-pterygoid process, (c) vomer-mandibular process, (d) temporal-occipital process, (e) sphenoid-coronoid process.
5. All of the following articulate with the maxilla *except* the (a) nasal bone, (b) frontal bone, (c) palatine bone, (d) mandible, (e) zygomatic bone.
6. Which of the following is *not* a component of the orbit? (a) temporal, (b) ethmoid (c) zygomatic, (d) maxilla, (e) lacrimal.
7. The sella turcica (1) is a depression in the ethmoid bone, (2) is located in approximately the center of the floor of the cranium, (3) contains the pineal gland, (4) identifies the location of the auditory ossicles.
   (a) 1 only, **(b)** 2 only, **(c)** 3 only, **(d)** 4 only, **(e)** 1, 2, and 3.
8. Which of the following does *not* contain a paranasal sinus? (a) zygomatic, (b) sphenoid, (c) ethmoid, (d) frontal, (e) maxilla.
9. The skeleton of the thorax (a) is formed by 12 pairs of ribs and costal cartilages, the sternum, and the 12 thoracic vertebrae, (b) protects the internal chest organs, as well as the liver, (c) is narrower at its superior end, (d) aids in supporting the bones of the shoulder girdle, (e) is described by all of the above.
10. Ribs (a) always articulate with the vertebrae and the sternum, (b) always articulate with the sternum by means of cartilages, (c) may be classified as true, false, or floating, (d) may be considered part of the appendicular skeleton, (e) do not provide support for any part of the body.
11. Match the following:

____ **(a)** largest facial bone; it has an alveolar process		**(1)** zygomatic
____ **(b)** the cheekbones; they contribute to the structure of the lateral walls and floor of each orbit		**(2)** frontal
____ **(c)** the smallest facial bones; tear ducts pass through these		**(3)** nasal
____ **(d)** the bridge of the nose		**(4)** ethmoid
____ **(e)** contains openings for the carotid artery and jugular vein		**(5)** maxilla
____ **(f)** superior to the atlas; forms the base of the cranium		**(6)** temporal
____ **(g)** the name means "wall"; forms most of the roof and much of the side walls of the skull		**(7)** lacrimal
____ **(h)** the upper jaw; forms anterior portion of the hard palate		**(8)** occipital
____ **(i)** located in the roof and septum of the nasal cavity; contains superior and middle nasal conchae		**(9)** palatine
____ **(j)** the forehead; provides protection for the anterior portion of the brain		**(10)** parietal
____ **(k)** posterior portion of the hard palate		**(11)** vomer
____ **(l)** inferior part of the nasal septum		**(12)** mandible

Complete the following:

12. The first cervical vertebra is the ________ , and the second cervical vertebra is the ________ .
13. According to shape classification, phalanges are ________ bones, carpals are ________ bones, and ribs are ________ bones.
14. The sternum consists of three parts: ________ , ________ , and ________ .
15. Fibrous connective tissue membrane-filled spaces between the cranial bones of an infant are called ________ .

Are the following statements true or false?

16. The opening in the occipital bone where the medulla oblongata connects with the spinal cord is the foramen magnum.

17. The coronal suture is associated with the posterior fontanel of the fetal skull.

18. Sesamoid bones protect tendons from wear and tear and may improve the mechanical advantage at a joint by altering the direction of pull of a tendon.

19. The ramus of the mandible has an anterior condylar process and a posterior coronoid process.

20. The tubercle of a rib articulates with demifacets on the bodies of adjacent vertebrae.

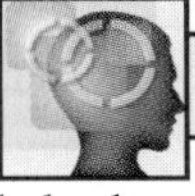

CRITICAL THINKING QUESTIONS

1. While investigating her new baby brother, 4 year-old girl Pattee finds a soft spot on the baby's skull and announces that the baby needs to go back because "it's not finished yet." Explain the presence of soft spots in the infant.
   HINT *An infant should be "soft in the head" but not an adult.*

2. Mike was caught sharing his exam answers with the student in the next seat. "Mike will do anything anyone asks," his friends said "he has no spine!" Can Mike really lack a spine? What is the correct anatomical term? Can you think of a reason for the common name "spine"?
   HINT *Look at the structure of the individual bones.*

3. The ad reads "New postureperfect mattress! Keeps spine perfectly straight—just like when you were born! A straight spine means a great sleep!" Would you buy a mattress from this company? Explain.
   HINT *Take a sideways view of the vertebral column.*

4. An "exploded skull" is a handy specimen for studying the anatomy of the cranium. The skull appears to be expanding outward after being blown apart, yet no bones are broken to prepare this specimen. How is this possible?
   HINT *You use this specimen to study the bones of the skull.*

5. John had complaints of severe headaches and a yellow-green nasal discharge. His physician ordered x-rays of his head. John was surprised to see several fuzzy-looking holes on the x-ray. Why does John have holes in his head?
   HINT *No insult intended to John—these holes are in his cranium.*

ANSWERS TO FIGURE QUESTIONS

6.1 Axial skeleton: skull, vertebral column. Appendicular skeleton: clavicle, shoulder girdle, humerus, pelvic girdle, femur.

6.2 Flat bones protect and provide a large surface area for muscle attachment.

6.3 The frontal, parietal, sphenoid, ethmoid, and temporal bones are cranial bones.

6.4 Squamous suture: parietal and temporal bones. Lambdoid suture: parietal and occipital bones. Coronal suture: parietal and frontal bones.

6.5 The temporal bone articulates with the parietal, sphenoid, zygomatic, and occipital bones.

6.6 The parietal bones form the posterior, lateral portion of the cranium.

6.7 The medulla oblongata of the brain connects with the spinal cord in the foramen magnum.

6.8 Crista galli of ethmoid bone, frontal, parietal, temporal, occipital, temporal, parietal, frontal, crista galli of ethmoid bone.

6.9 The perpendicular plate of the ethmoid bone forms the superior part of the nasal septum, and the lateral masses compose most of the medial walls of the orbits.

6.10 The mandible is the only movable skull bone, other than the auditory ossicles.

6.11 The paranasal sinuses produce mucus and serve as resonating chambers for vocalization.

6.12 The anterolateral fontanel is bordered by four different skull bones.

6.13 Bones forming the orbit are the frontal, sphenoid, zygomatic, maxilla, lacrimal, ethmoid, and palatine.

6.14 The nasal septum divides the nasal cavity into right and left sides.

6.15 The posterior cranial fossa is the largest.

6.16 The hyoid bone does not articulate with any other bone.

6.17 The thoracic and sacral curves are concave.

6.18 The vertebral foramina enclose the spinal cord, whereas the intervertebral foramina provide spaces for spinal nerves to exit the vertebral column.

6.19 The atlas moving on the axis permits movement of the head to signify "no."

6.20 The facets and demifacets on the bodies of the thoracic vertebrae articulate with the facets on the head of the ribs, and the facets on the transverse processes of these vertebrae articulate with the tubercles of the ribs.

6.21 The lumbar vertebrae are stout because the amount of body weight supported by vertebrae increases toward the inferior end of the vertebral column.

6.22 There are four pairs of sacral foramina, for a total of eight. Each anterior sacral foramen joins a posterior sacral foramen at the intervertebral foramen. Nerves and blood vessels pass through these tunnels in the bone.

6.23 True ribs (pairs 1–7), false ribs (pairs 8–12), and floating ribs (pairs 11 and 12).

6.24 The facet on the head of a rib fits into a facet on the body of a vertebra, and the articular part of the tubercle of a rib articulates with the facet of the transverse process of a vertebra.

7

THE SKELETAL SYSTEM: THE APPENDICULAR SKELETON

CONTENTS AT A GLANCE

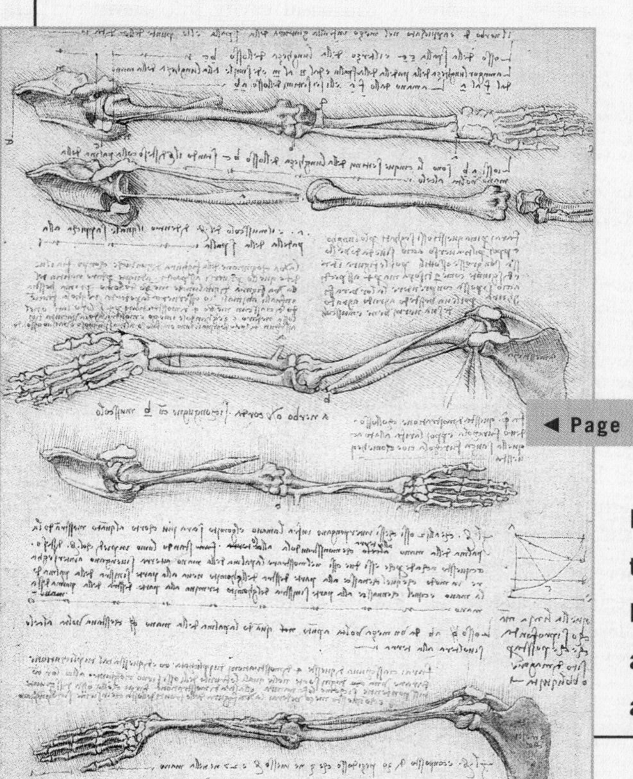

◄ Page 178

Most of us know Leonardo da Vinci for his fine art masterpieces such as the Mona Lisa and The Last Supper. Did you know that Leonardo was also an accomplished anatomist? Have you ever seen any of his anatomical renderings?

Page 181 ►

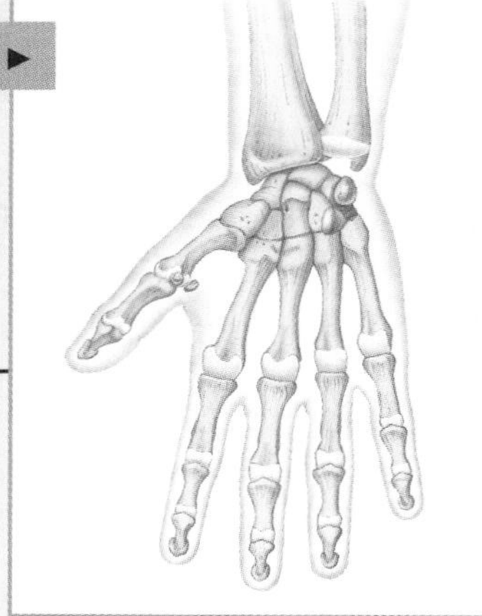

INTRODUCTION

As noted in Chapter 6, the two main divisions of the skeletal system are the axial skeleton and the appendicular skeleton. Whereas the general function of the axial skeleton is the protection of internal organs, the general function of the appendicular skeleton, the focus of this chapter, is primarily movement. The appendicular skeleton includes the bones that make up the upper and lower limbs as well as the bones, called girdles, that attach the limbs to the axial skeleton. As you progress through this chapter, you will see how the bones of the appendicular skeleton are connected with each other and with skeletal muscles, making possible a wide array of movements. They permit us to walk, write, use a computer, dance, swim, and play a musical instrument.

PECTORAL (SHOULDER) GIRDLE

Objective

- Identify the bones of the pectoral (shoulder) girdle and their principal markings.

The **pectoral** (PEK-tō-ral) or **shoulder girdles** attach the bones of the upper limbs to the axial skeleton (Figure 7.1). Each of the two pectoral girdles consists of a clavicle and a scapula. The *clavicle* is the anterior bone and articulates with the manubrium of the sternum at the *sternoclavicular joint*. The *scapula* is the posterior bone and articulates with the clavicle at the *acromioclavicular joint* and with the humerus at the *glenohumeral (shoulder) joint*. The pectoral girdles do not articulate with the vertebral column and are held in position instead by complex muscle attachments.

Figure 7.1 / Right pectoral (shoulder) girdle.

The clavicle is the anterior bone of the pectoral girdle and the scapula is the posterior bone.

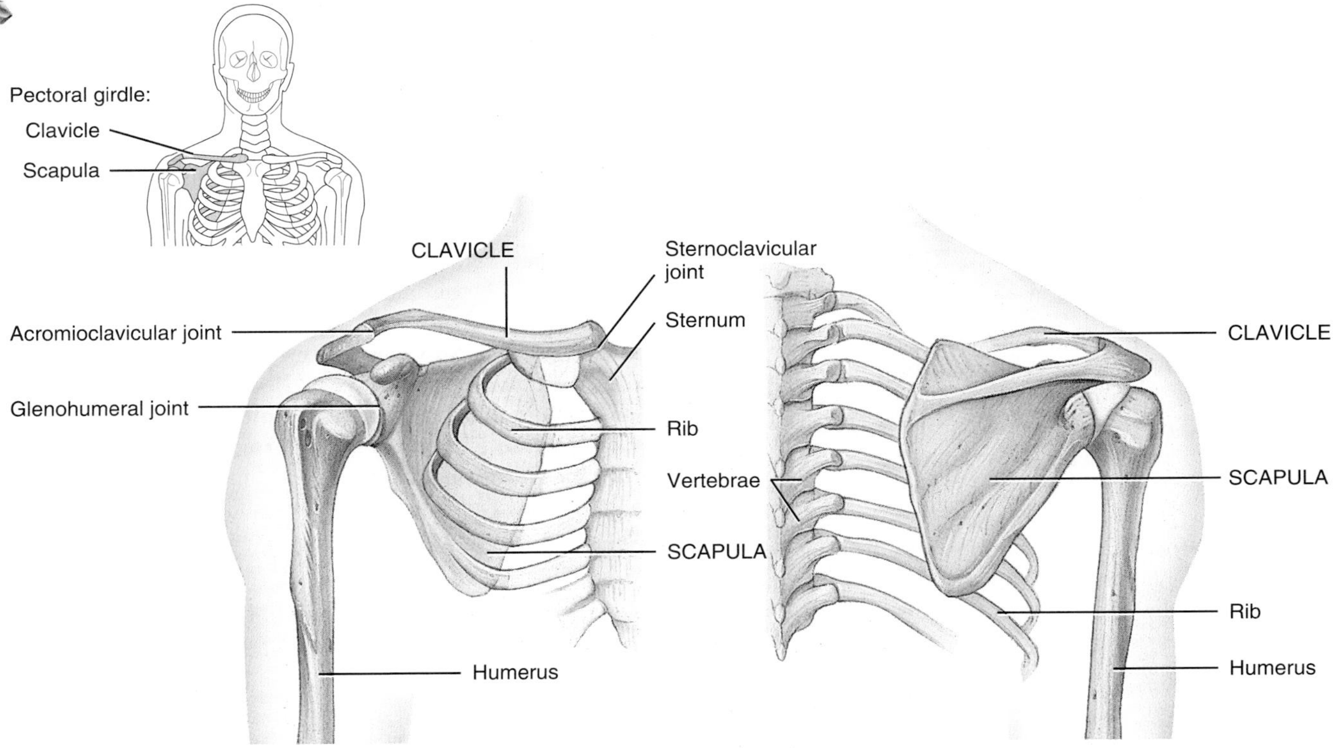

What is the function of the pectoral girdles?

Figure 7.2 / Right clavicle.

The clavicle articulates medially with the manubrium of the sternum and laterally with the acromion of the scapula.

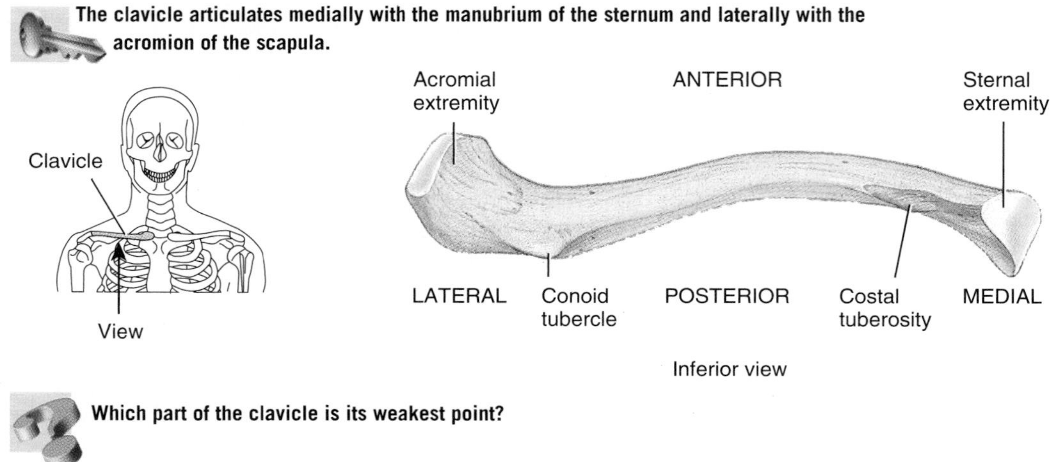

Which part of the clavicle is its weakest point?

Clavicle

Each slender, S-shaped **clavicle** (KLAV-i-kul = key), or *collarbone*, lies horizontally in the superior and anterior part of the thorax superior to the first rib (Figure 7.2). The medial half of the clavicle is convex anteriorly, whereas the lateral half is concave anteriorly. The medial end of the clavicle, the *sternal extremity*, is rounded and articulates with the manubrium of the sternum to form the *sternoclavicular joint.* The broad, flat, lateral end, the *acromial extremity* (a-KRŌ-mē-al), articulates with the acromion of the scapula. This joint is called the *acromioclavicular joint* (Figure 7.1). The *conoid tubercle* (KŌ-noyd = conelike) on the inferior surface of the lateral end of the bone is a point of attachment for the conoid ligament. The *costal tuberosity* on the inferior surface of the medial end is a point of attachment for the costoclavicular ligament.

The clavicle transmits mechanical force from the upper limb to the trunk. If the force transmitted to the clavicle is excessive, as in falling on one's outstretched arm, a **fractured clavicle** may result. The clavicle is one of the most frequently broken bones in the body. Because the middle of the clavicle is its weakest point, it is the most frequent fracture site. Since the clavicles provide attachment for several chest and shoulder muscles, they brace the upper limbs laterally. When a clavicle is fractured the shoulder is pulled medially and anteriorly and drops and the upper limb sags from loss of support. ■

Scapula

Each **scapula** (SCAP-yū-la), or *shoulder blade*, is a large, triangular, flat bone situated in the superior part of the posterior thorax between the levels of the second and seventh ribs (Figure 7.3). The medial borders of the scapulae (plural) lie about 5 cm (2 in.) from the vertebral column.

A sharp ridge, the *spine*, runs diagonally across the posterior surface of the flattened, triangular *body* of the scapula. The lateral end of the spine projects as a flattened, expanded process called the *acromion* (a-KRŌ-mē-on; *acrom-* = topmost), easily felt as the high point of the shoulder. Tailors measure the length of the upper limb from the acromion. The acromion articulates with the acromial extremity of the clavicle to form the *acromioclavicular joint.* Inferior to the acromion is a shallow depression, the *glenoid cavity*, that accepts the head of the humerus (arm bone) to form the *glenohumeral joint* (see Figure 7.1).

The thin edge of the bone near the vertebral column is the *medial (vertebral) border.* The thick edge closer to the arm is the *lateral (axillary) border.* The medial and lateral borders join at the *inferior angle.* The superior edge of the scapula, called the *superior border*, joins the vertebral border at the *superior angle.* The *scapular notch* is a prominent indentation along the superior border through which the suprascapular nerve passes.

At the lateral end of the superior border of the scapula is a projection of the anterior surface called the *coracoid process* (KOR-a-koyd = like a crow's beak), to which the tendons of muscles attach. Superior and inferior to the spine are two fossae: the *supraspinous fossa* (sū-pra-SPĪ-nus) and the *infraspinous fossa* (in-fra-SPĪ-nus), respectively. Both serve as surfaces of attachment for the tendons of shoulder muscles called the supraspinatus and infraspinatus muscles. On the anterior surface is a slightly hollowed-out area called the *subscapular fossa*, also a surface of attachment for the tendons of shoulder muscles.

✓ Which bones or parts of bones of the pectoral girdle form the sternoclavicular, acromioclavicular, and glenohumeral joints?

UPPER LIMB (EXTREMITY)

Objective

- Identify the bones of the upper limb and their principal markings.

Figure 7.3 / Right scapula (shoulder blade). (See Tortora, *A Photographic Atlas of the Human Body,* Figure 3.22.)

The glenoid cavity of the scapula articulates with the head of the humerus to form the shoulder joint.

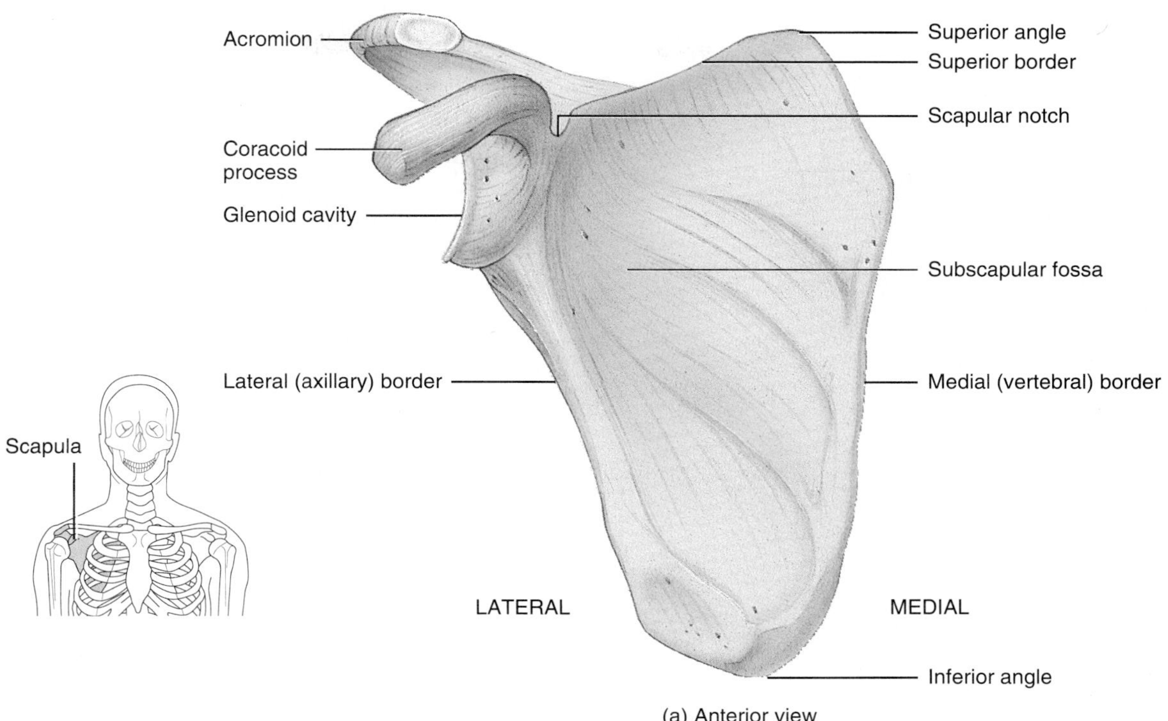

(a) Anterior view

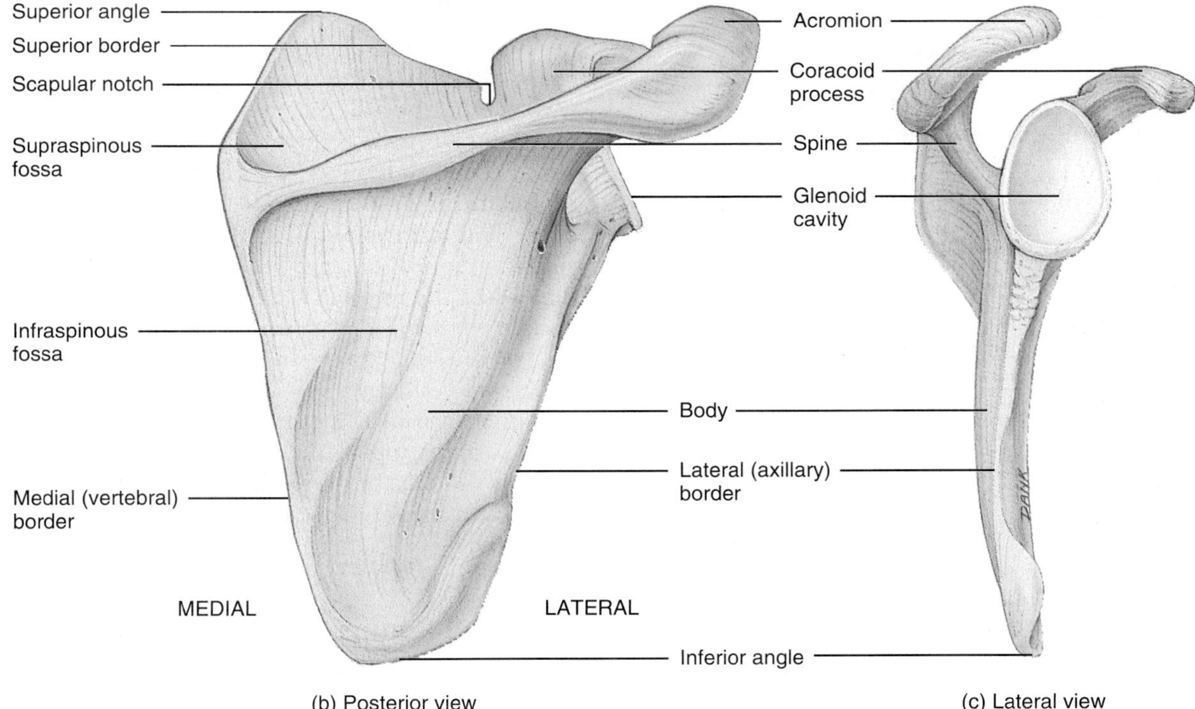

(b) Posterior view

(c) Lateral view

Which part of the scapula forms the high point of the shoulder?

Figure 7.4 / Right upper limb.

Each upper limb consists of a humerus, ulna, radius, carpals, metacarpals, and phalanges.

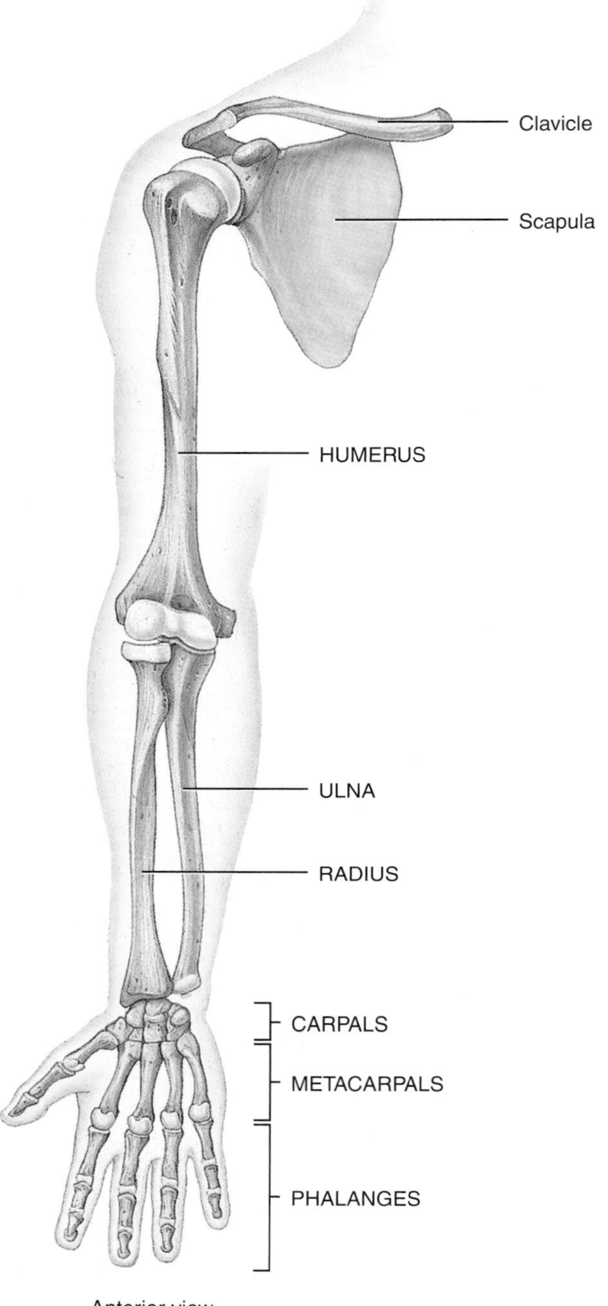

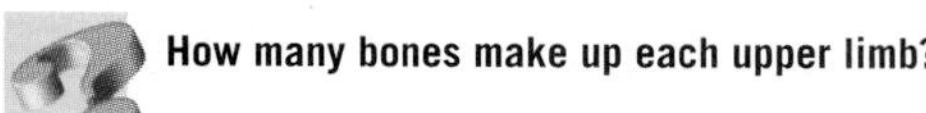
How many bones make up each upper limb?

Each **upper limb (extremity)** consists of 30 bones (Figure 7.4). These include the (1) humerus of the arm; (2) the ulna and radius of the forearm; and (3) the carpals of the carpus (wrist), the metacarpals of the metacarpus (palm), and the phalanges (bones of the digits) of the hand.

Humerus

The **humerus** (HŪ-mer-us), or arm bone, is the longest and largest bone of the upper limb (Figure 7.5). It articulates proximally with the scapula and distally at the elbow with both the ulna and the radius.

The proximal end of the humerus features a rounded *head* that articulates with the glenoid cavity of the scapula to form the *glenohumeral joint.* Distal to the head is the *anatomical neck*, the site of the epiphyseal line, which is visible as an oblique groove. The *greater tubercle* is a lateral projection distal to the anatomical neck. It is the most laterally palpable bony landmark of the shoulder region. The *lesser tubercle* projects anteriorly. Between both tubercles runs an *intertubercular sulcus.* The *surgical neck* is a constriction in the humerus just distal to the tubercles, where the head tapers to the shaft; it is so named because fractures often occur here.

The *body (shaft)* of the humerus is roughly cylindrical at its proximal end, but it gradually becomes triangular until it is flattened and broad at its distal end. Laterally, at the middle portion of the shaft, there is a roughened, V-shaped area called the *deltoid tuberosity.* This area serves as a point of attachment for the tendons of the deltoid muscle.

Several prominent features are evident at the distal end of the humerus. The *capitulum* (ka-PIT-yū-lum; *capit-* = head) is a rounded knob on the lateral aspect of the bone that articulates with the head of the radius. The *radial fossa* is an anterior depression that receives the head of the radius when the forearm is flexed (bent). The *trochlea* (TRŌK-lē-a), located medial to the capitulum, is a spool-shaped surface that articulates with the ulna. The *coronoid fossa* (KOR-o-noyd; = crown-shaped) is an anterior depression that receives the coronoid process of the ulna when the forearm is flexed. The *olecranon fossa* (ō-LEK-ra-non; = elbow) is a posterior depression that receives the olecranon of the ulna when the forearm is extended (straightened). The *medial epicondyle* and *lateral epicondyle* are rough projections on either side of the distal end to which the tendons of most muscles of the forearm are attached. The ulnar nerve lies on the posterior surface of the medial epicondyle and may easily be palpated by rolling a finger over the skin surface above the medial epicondyle.

Ulna and Radius

The **ulna** is located on the medial aspect (little-finger side) of the forearm and is longer than the radius (Figure 7.6 on page 179). It is connected to the radius by a broad, flat, fibrous connective tissue called the **interosseous membrane** (in-ter-OS-ē-us; *inter-* = between, *osse-* = bone). This membrane also provides a site of attachment for some tendons of deep skeletal muscles of the forearm. At the proximal end of the ulna (Figure

Figure 7.5 / Right humerus in relation to the scapula, ulna, and radius. (See Tortora, *A Photographic Atlas of the Human Body,* Figure 3.23.)

The humerus is the longest and largest bone of the upper limb.

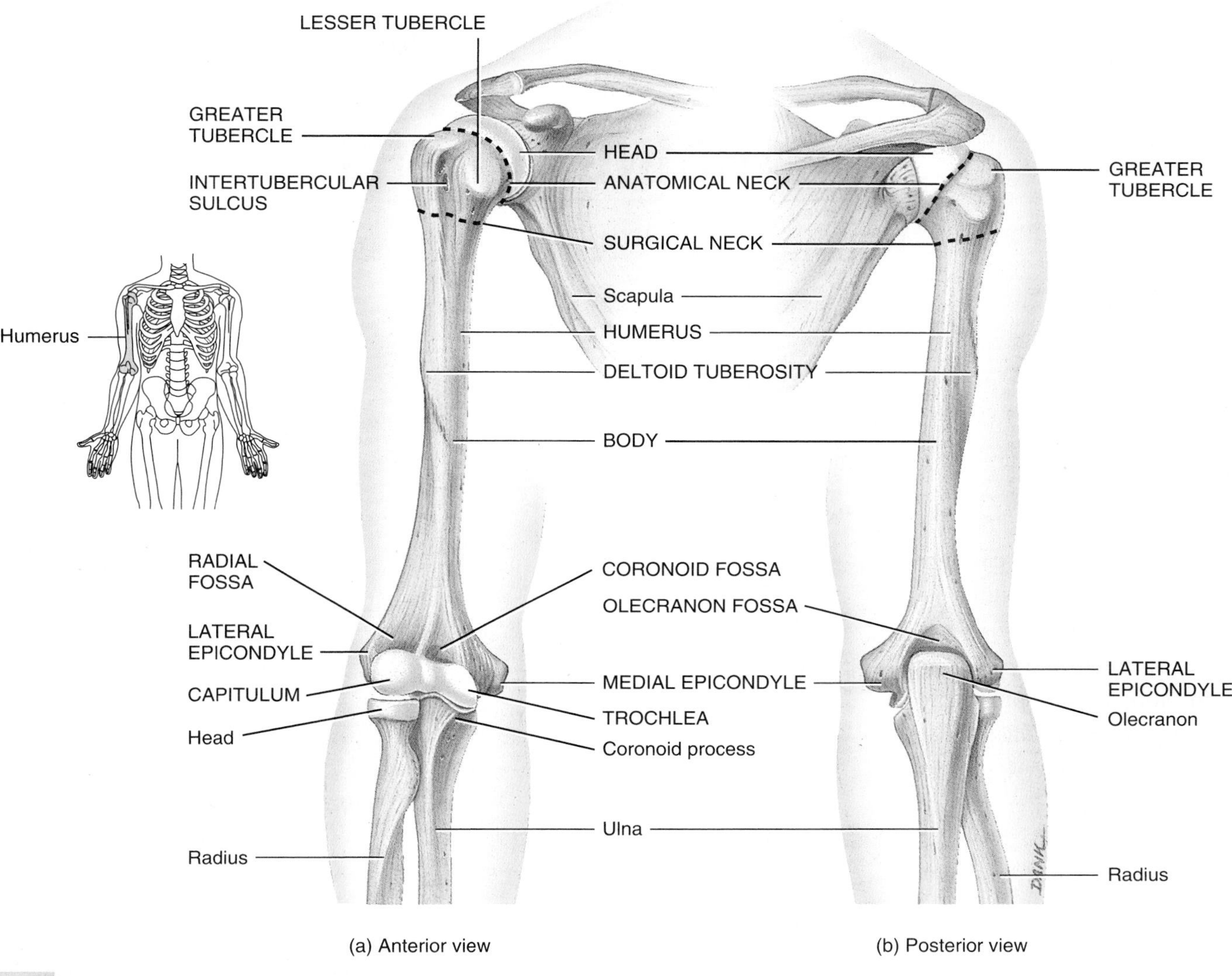

Which parts of the humerus articulate with the radius at the elbow? With the ulna at the elbow?

7.6b) is the *olecranon,* or *olecranon process,* which forms the prominence of the elbow. The *coronoid process* (Figure 7.6a) is an anterior projection that, together with the olecranon, receives the trochlea of the humerus. The *trochlear notch* is a large curved area between the olecranon and coronoid process that forms part of the elbow joint (Figure 7.7b). Just inferior to the coronoid process is the *ulnar tuberosity.* The distal end of the ulna consists of a *head* that is separated from the wrist by a fibrocartilage disc. A *styloid process* (*stylo-* = stake or pole) is on the posterior side of the distal end.

The **radius** is located on the lateral aspect (thumb side) of the forearm (Figure 7.6). The proximal end of the radius has a disc-shaped *head* that articulates with the capitulum of the humerus and the radial notch of the ulna. Inferior to the head is the constricted *neck.* A roughened area inferior to the neck on the medial side, called the *radial tuberosity,* is a point of attachment for the tendons of the biceps brachii muscle. The shaft of the radius widens distally to form a *styloid process* on the lateral side. Fracture of the distal end of the radius is the most common fracture in adults older than 50 years.

The *elbow joint* is where the ulna and radius articulate with the humerus. This occurs in two places: where the head of the

14th Century

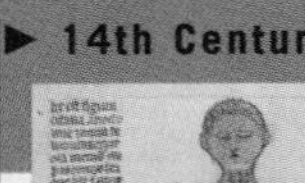

15th Century

16th Century

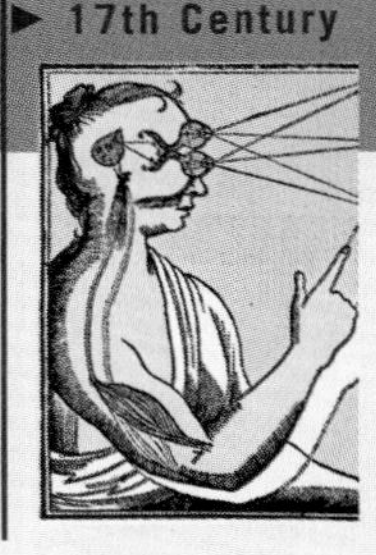

17th Century

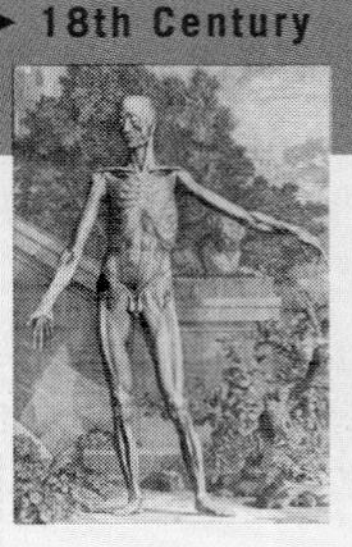

18th Century

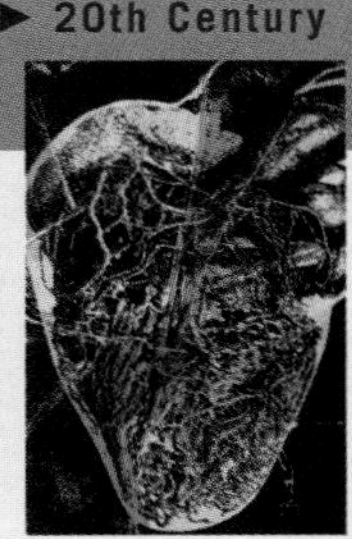

20th Century

21st Century

CHANGING IMAGES

Boning Up on the Skeletal System

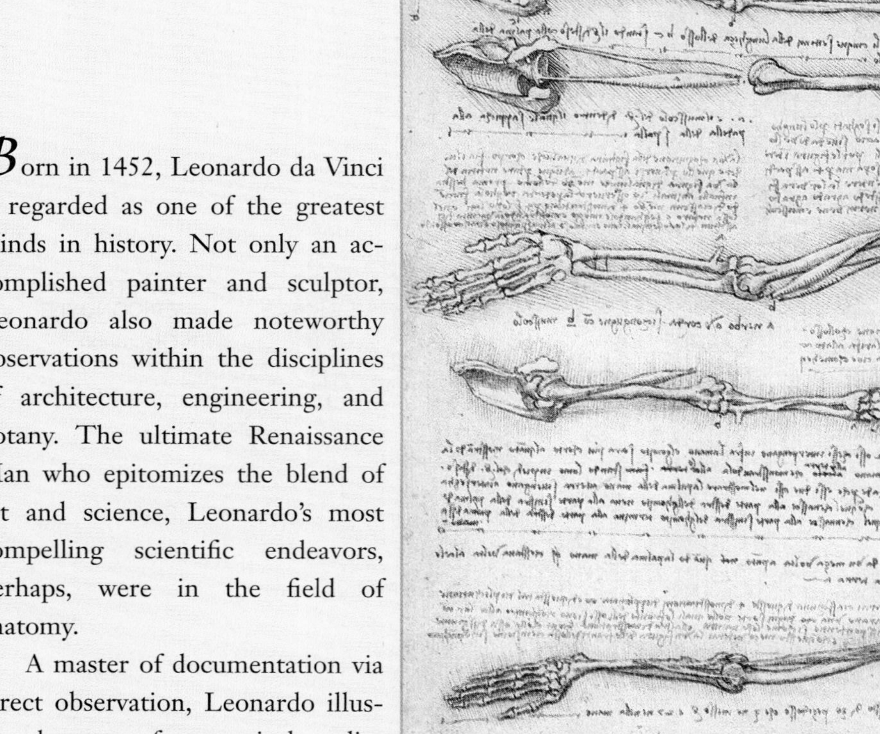

1510 AD

Born in 1452, Leonardo da Vinci is regarded as one of the greatest minds in history. Not only an accomplished painter and sculptor, Leonardo also made noteworthy observations within the disciplines of architecture, engineering, and botany. The ultimate Renaissance Man who epitomizes the blend of art and science, Leonardo's most compelling scientific endeavors, perhaps, were in the field of anatomy.

A master of documentation via direct observation, Leonardo illustrated scores of anatomical studies in his notebooks. Most likely these drawings were not intended for publication, but to improve Leonardo's artistic representations, as well as to satisfy his intellectual curiosity.

Pictured here is a drawing of the bones of the pectoral girdle and upper limb, circa 1510. Do you notice anything unusual about Leonardo's notes? He employed a remarkable *mirror writing* technique where Leonardo wrote from right to left, requiring the page to be held before a mirror to be read. Perhaps Leonardo was left-handed and was trying to avoid smeared ink, or maybe he wanted his ideas to be indecipherable to others. Either way, his ability to write in this manner is testament to his genius.

For all of his brilliance, however, Leonardo's anatomy was at times crude. So, as you read this chapter about the appendicular skeleton, refer to this image for its anatomical accuracy. Pay special attention to the posterior aspect of the scapula and the carpals of the hand. Can you identify any the muscles whose actions Leonardo is contemplating? Perhaps you will be able to after reading Chapter 10.

Figure 7.6 / Right ulna and radius in relation to the humerus and carpals. (See Tortora, *A Photographic Atlas of the Human Body,* Figure 3.24.)

In the forearm, the longer ulna is on the medial side, whereas the shorter radius is on the lateral side.

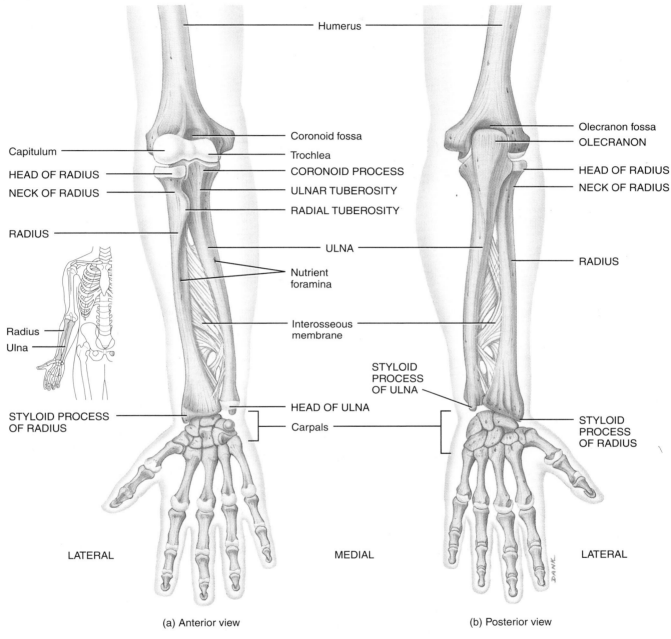

What part of the ulna is called the "elbow"?

radius articulates with the capitulum of the humerus (Figure 7.7a), and where the trochlea of the humerus is received by the trochlear notch of the ulna (Figure 7.7b).

In addition to being connected to each other by the interosseous membrane, the ulna and radius also articulate with one another proximally and distally. Proximally, the head of the radius articulates with the ulna's *radial notch,* a depression that is lateral and inferior to the trochlear notch (Figure 7.7b). This articulation is the *proximal radioulnar joint.* Distally, the head of the ulna articulates with the *ulnar notch* of the radius (Figure 7.7c). This articulation is the *distal radioulnar joint.* Finally, the distal end of the radius articulates with three bones of the wrist—the lunate, the scaphoid, and the triquetrum—to form the *radiocarpal (wrist) joint.*

Figure 7.7 / Articulations formed by the ulna and radius. (a) Elbow joint. (b) Joint surfaces at proximal end of the ulna. (c) Joint surfaces at distal ends of radius and ulna. The ulna and radius are also attached by the interosseous membrane.

The elbow joint is formed by the trochlear of the humerus, trochlear notch of the ulna, and the head of the radius.

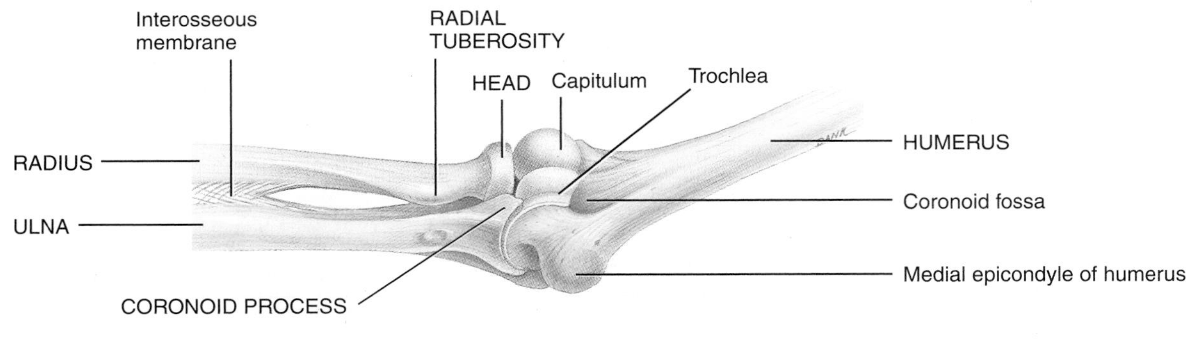

(a) Medial view in relation to humerus

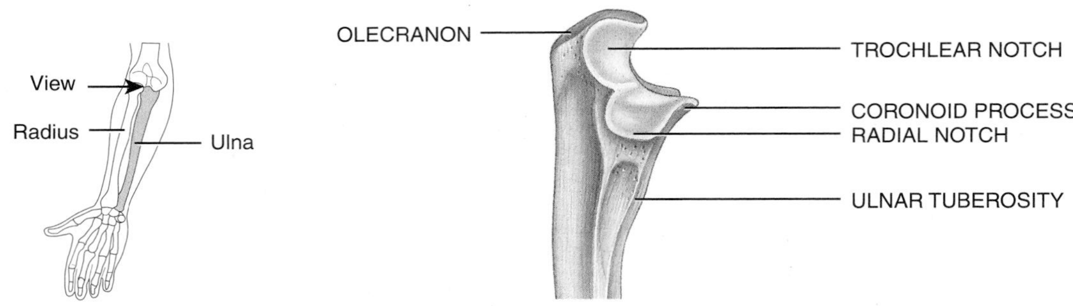

(b) Lateral view of proximal end of ulna

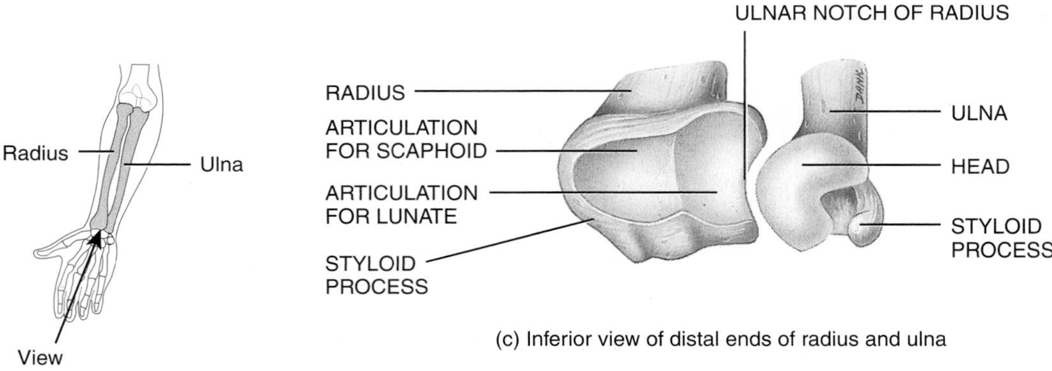

(c) Inferior view of distal ends of radius and ulna

How many joints are formed between the radius and ulna?

Carpals, Metacarpals, and Phalanges

The **carpus** (wrist) is the proximal region of the hand and consists of eight small bones, the **carpals,** joined to one another by ligaments (Figure 7.8). Articulations between carpal bones are called *intercarpal joints.* The carpals are arranged in two transverse rows of four bones each. Their names reflect their shapes. The carpals in the proximal row, from lateral to medial, are the **scaphoid** (SKAF-oid = boatlike), **lunate** (LŪ-nāt = moon-shaped), **triquetrum** (trī-KWĒ-trum = three-cornered), and **pisiform** (PĪ-si-form = pea-shaped). The carpals in the distal row, from lateral to medial, are the **trapezium** (tra-PĒ-zē-um = four-sided figure with no two sides parallel), **trapezoid** (TRAP-e-zoid = four-sided figure with two sides parallel), **capitate** (KAP-i-tāt = head-shaped) and **hamate** (HAM-āt = hooked).

The capitate is the largest carpal bone; its rounded projection, the head, articulates with the lunate. The hamate is named for a large hook-shaped projection on its anterior surface. In about 70% of carpal fractures, only the scaphoid is broken. This is because the force of a fall on an outstretched hand is transmitted from the capitate through the scaphoid to the radius.

Figure 7.8 / Right wrist and hand in relation to the ulna and radius.

The skeleton of the hand consists of the carpals in the proximal region of the hand, the metacarpals in the intermediate region of the hand, and phalanges in the distal region of the hand.

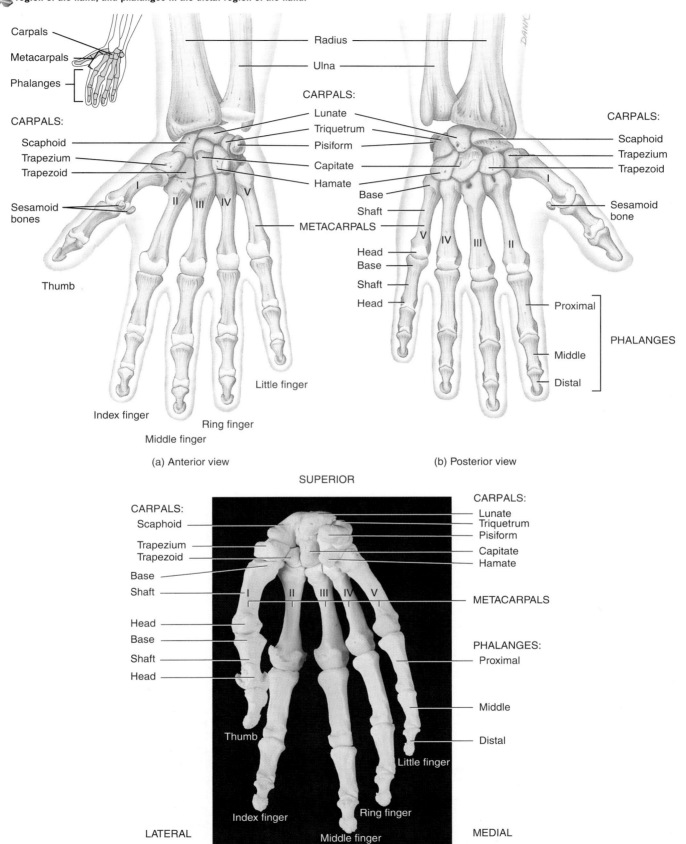

(a) Anterior view

(b) Posterior view

(c) Anterior view

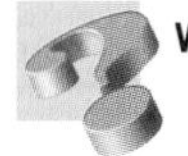

Which is the most frequently fractured wrist bone?

The concave space formed by the pisiform and hamate (on the ulnar side), and the scaphoid and trapezium (on the radial side), plus the *flexor retinaculum* (deep fascia) is the **carpal tunnel.** The long flexor tendons of the digits and thumb and the median nerve pass through the carpal tunnel. Narrowing of the carpal tunnel may give rise to a condition called carpal tunnel syndrome (described on page 312).

The **metacarpus** (*meta-* = beyond), or palm, is the intermediate region of the hand and consists of five bones called **metacarpals.** Each metacarpal bone consists of a proximal *base*, an intermediate *shaft*, and a distal *head* (Figure 7.8b). The metacarpal bones are numbered I to V (or 1–5), starting with the thumb, which is lateral. The bases articulate with the distal row of carpal bones to form the *carpometacarpal joints.* The heads articulate with the proximal phalanges to form the *metacarpophalangeal joints.* The heads of the metacarpals are commonly called "knuckles" and are readily visible in a clenched fist.

The **phalanges** (fa-LAN-jēz, *phalan-* = a battle line), or bones of the digits, make up the distal region of the hand. There are 14 phalanges in the five digits of each hand and, like the metacarpals, the digits are numbered I to V (or 1–5), beginning with the thumb, which is lateral. A single bone of a digit is referred to as a **phalanx** (FĀ-lanks). Each phalanx consists of a proximal *base*, an intermediate *shaft*, and a distal *head.* There are two phalanges in the thumb (*pollex*), and three phalanges in each of the other four digits. In order from the thumb, these other four digits are commonly referred to as the index finger, middle finger, ring finger, and little finger. The first row of phalanges, the *proximal row*, articulates with the metacarpal bones and second row of phalanges. The second row of phalanges, the *middle row*, articulates with the proximal row and the third row. The third row of phalanges, the *distal row*, articulates with the middle row. The thumb has no middle phalanx. Joints between phalanges are called *interphalangeal joints.*

✓ Name the bones that form the upper limb, from proximal to distal.
✓ Describe the joints of the upper limb.

PELVIC (HIP) GIRDLE

Objective

- Identify the bones of the pelvic girdle and their principal markings.

The **pelvic (hip) girdle** consists of the two **hip bones,** also called **coxal bones** (KOK-sal; *cox-* = hip) (Figure 7.9). The pelvic girdle accepts the bones of the lower limbs, connecting them to the axial skeleton. The hip bones are joined together anteriorly at a joint called the **pubic symphysis** (PYU̅-bik SIM-fi-sis). Posteriorly they unite with the sacrum at the *sacroiliac joints.* The complete ring composed of the hip bones, pubic symphysis, and sacrum forms a deep, basinlike structure called the **bony pelvis** (*pelv-* = basin). Functionally, the bony pelvis provides a strong and stable support for the vertebral column and pelvic organs.

Each of the two hip bones of a newborn consists of three bones separated by cartilage: a superior *ilium*, an inferior and anterior *pubis*, and an inferior and posterior *ischium*. By age 23, the three separate bones fuse together (Figure 7.10a). Although

Figure 7.9 / Bony pelvis. Shown here is the female bony pelvis. (See Tortora, *A Photographic Atlas of the Human Body,* Figure 3.27.)

The hip bones are united anteriorly at the pubic symphysis and posteriorly at the sacrum to form the bony pelvis.

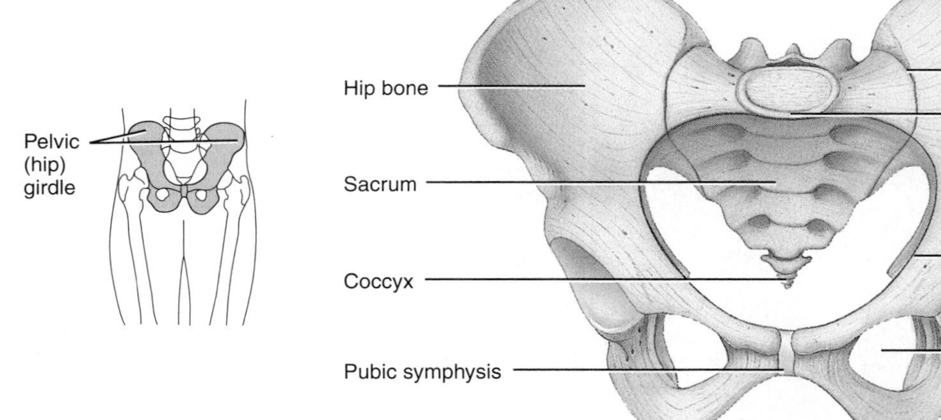

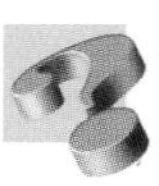
What are the functions of the bony pelvis?

Figure 7.10 / Right hip bone. The lines of fusion of the ilium, ischium, and pubis depicted in (a) are not always visible in an adult.

The acetabulum is the socket formed where the three parts of the hip bone converge.

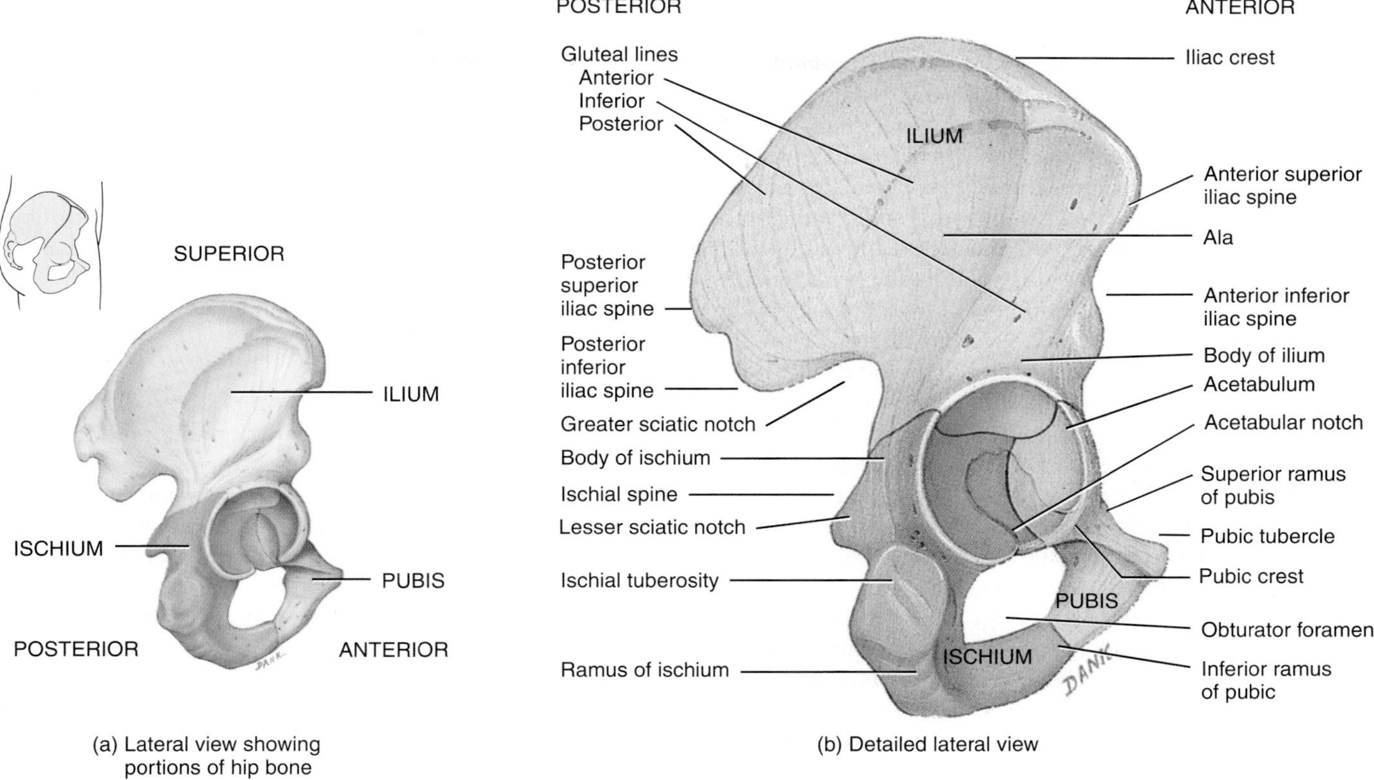

(a) Lateral view showing portions of hip bone

(b) Detailed lateral view

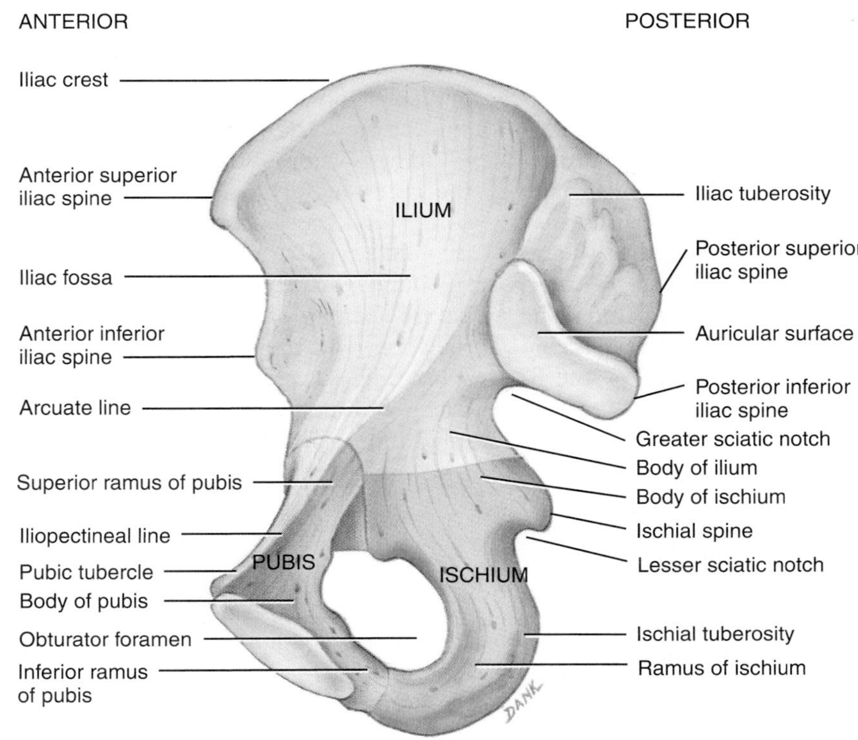

(c) Detailed medial view

Which part of the hip bone articulates with the femur? With the sacrum?

the hip bones function as single bones, anatomists commonly discuss them as though they still consisted of three bones.

Ilium

The **ilium** (= flank) is the largest of the three components of the hip bone (Figure 7.10b, c). The ilium is divided into a superior *ala* (= wing) and an inferior *body*, which enters into the formation of the *acetabulum*, the socket for the head of the femur. The superior border of the ilium, the *iliac crest*, ends anteriorly in a blunt *anterior superior iliac spine*. Below this spine is the *anterior inferior iliac spine*. Posteriorly, the iliac crest ends in a sharp *posterior superior iliac spine*. Below this spine is the *posterior inferior iliac spine*. The spines serve as points of attachment for the tendons of the muscles of the trunk, hip, and thighs. Below the posterior inferior iliac spine is the *greater sciatic notch* (sī-AT-ik), through which the sciatic nerve, the longest nerve in the body, passes. The medial surface of the ilium contains the *iliac fossa*, a concavity where the tendon of the iliacus muscle attaches. Posterior to this fossa are the *iliac tuberosity*, a point of attachment for the sacroiliac ligament, and the *auricular surface* (*auricular* = ear-shaped), which articulates with the sacrum to form the *sacroiliac joint* (Figure 7.9). Projecting anteriorly and inferiorly from the auricular surface is a ridge called the *arcuate line* (AR-kyū-āt; *arc-* = bow). The other conspicuous markings of the ilium are three arched lines on its lateral surface called the *posterior gluteal line* (*glut-* = buttock), the *anterior gluteal line*, and the *inferior gluteal line*. The tendons of the gluteal muscles attach to the ilium between these lines.

Ischium

The **ischium** (IS-kē-um; *ischi-* = hip) is the inferior, posterior portion of the hip bone (see Figure 7.10b, c). The ischium is composed of a superior *body* and an inferior *ramus* (*ram-* = branch), which joins the pubis. The ischium contains the prominent *ischial spine*, a *lesser sciatic notch* below the spine, and a rough and thickened *ischial tuberosity*. This prominent tuberosity may hurt someone's thigh when you sit on their lap. Together, the ramus and the pubis surround the *obturator foramen* (OB-tū-rā-ter; *obtur-* = closed up), the largest foramen in the skeleton. The foramen is so named because, even though blood vessels and nerves pass through it, it is nearly completely closed by the fibrous *obturator membrane*.

Pubis

The **pubis,** or **os pubis,** meaning pubic bone, is the anterior and inferior part of the hip bone (Figure 7.10b, c). The pubis consists of a *superior ramus*, an *inferior ramus*, and a *body* between the rami (plural) that contributes to the formation of the pubic symphysis. The anterior superior surface of the superior ramus is known as the *pubic crest*. At its lateral end is a projection, the *pubic tubercle*, which is the beginning of a raised line, the *iliopectineal line* (il-ē-ō-pek-TIN-ē-al). This line extends superiorly and laterally along the superior ramus to merge with the arcuate line of the ilium. These lines, as you will see shortly, are important landmarks for distinguishing the superior and inferior portions of the bony pelvis.

The *pubic symphysis* is the joint between the two hip bones (Figure 7.9). It consists of a disc of fibrocartilage. The pubic arch is formed by the convergence of the inferior rami of the two pubes (plural) (see Table 7.1). The *acetabulum* (as-e-TAB-yū-lum; = vinegar cup) is the deep fossa formed by the ilium, ischium, and pubis. It is the socket that accepts the rounded head of the femur. Together, the acetabulum and the femoral head form the *hip (coxal) joint*. On the inferior portion of the acetabulum is a deep indentation, the *acetabular notch*. It forms a foramen through which blood vessels and nerves pass, and it serves as a point of attachment for ligaments of the femur (for example, the ligament of the head of the femur).

True and False Pelves

The bony pelvis is divided into superior and inferior portions by a boundary called the *pelvic brim* (Figure 7.11a). You can trace the pelvic brim by following the landmarks around parts of the hip bones to form the outline of an oblique plane. Beginning posteriorly at the *sacral promontory* of the sacrum, trace laterally and inferiorly along the *arcuate lines* of the ilium. Continue inferiorly along the *iliopectineal lines* of the pubis. Finally, trace anteriorly to the superior portion of the *pubic symphysis*. Together, these points form an oblique plane that is higher in the back than in the front. The circumference of this plane is the pelvic brim.

The portion of the bony pelvis superior to the pelvic brim is the **false (greater) pelvis** (see Figure 7.11b). It is bordered by the lumbar vertebrae posteriorly, the upper portions of the hip bones laterally, and the abdominal wall anteriorly. The space enclosed by the false pelvis is part of the abdomen; it does not contain pelvic organs, except for the urinary bladder (when it is full) and the uterus during pregnancy.

The portion of the bony pelvis inferior to the pelvic brim is the **true (lesser) pelvis** (Figure 7.11b). It is bounded by the sacrum and coccyx posteriorly, inferior portions of the ilium and ischium laterally, and the pubic bones anteriorly. The true pelvis surrounds the pelvic cavity (see Figure 1.8 on page 15.). The superior opening of the true pelvis is the pelvic brim, also called the *pelvic inlet;* the inferior opening of the true pelvis is the *pelvic outlet*. The *pelvic axis* is an imaginary line that curves through the true pelvis and joins the central points of the planes of the pelvic inlet and outlet. During childbirth the pelvic axis is the route taken by the baby's head as it descends through the pelvis.

Pelvimetry is the measurement of the size of the inlet and outlet of the birth canal, which may be done by ultrasonography or physical examination. Measurement of the pelvic cavity in pregnant females is important because the fetus must pass through the narrower opening of the pelvis at birth. ■

✓ Distinguish between the true and false pelves.

Figure 7.11 / True and false pelves. Shown here is the female pelvis. For simplicity, the landmarks of the pelvic brim are shown only on the left side of the body, and the outline of the pelvic brim is shown only on the right side. The entire pelvic brim is shown in Figure 7.9.

The true and false pelves are separated by the pelvic brim.

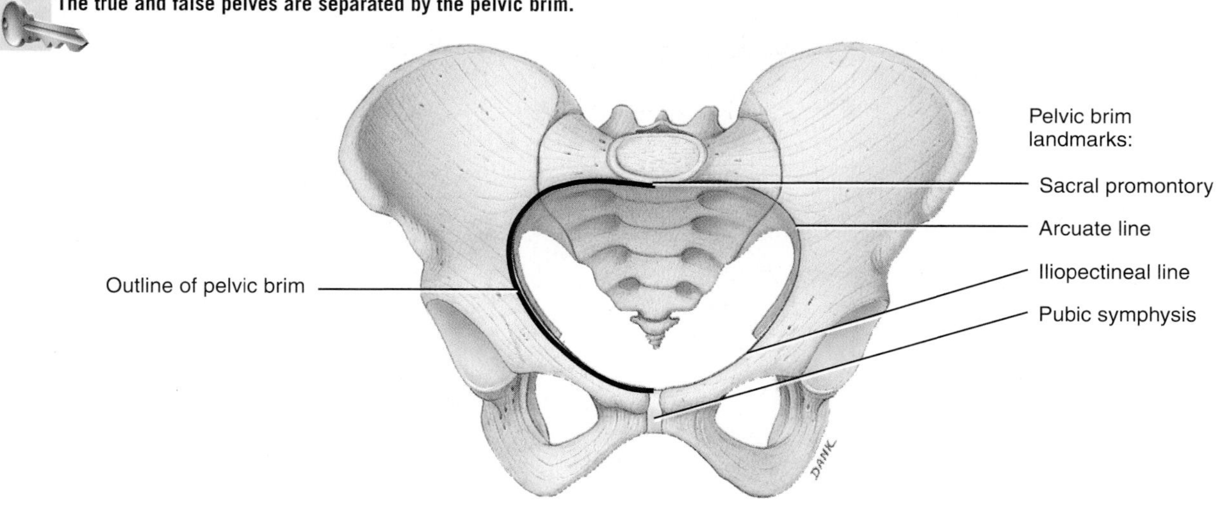

(a) Anterior view of borders of pelvic brim

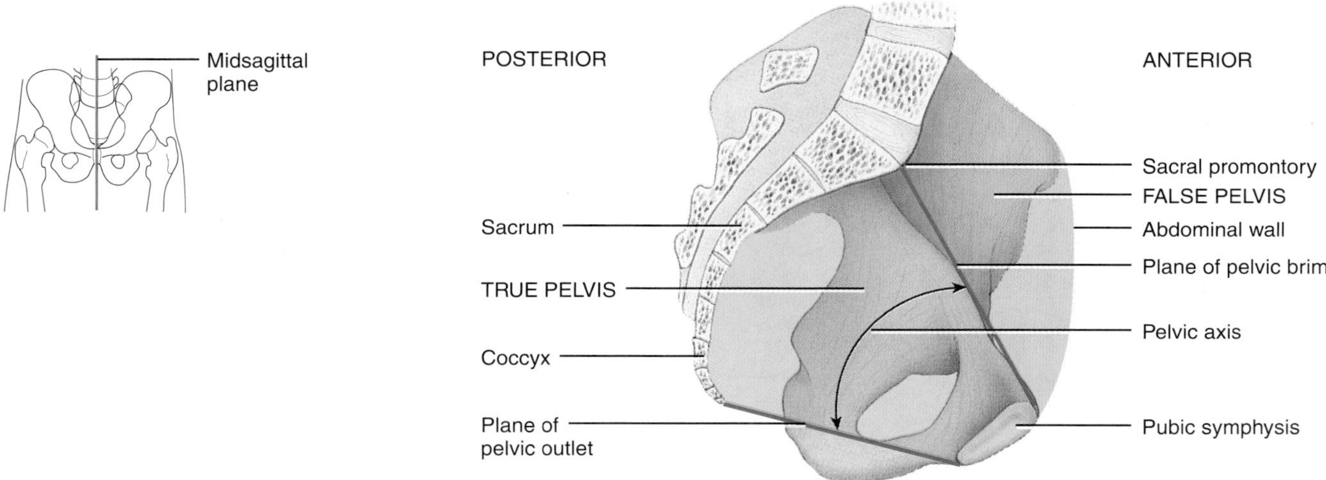

(b) Midsagittal section indicating locations of true and false pelves

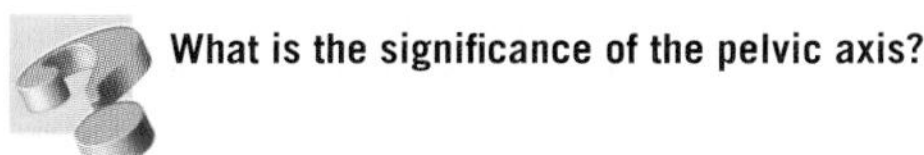

What is the significance of the pelvic axis?

COMPARISON OF FEMALE AND MALE PELVES

Objective

- Compare the principal structural differences between female and male pelves.

In the following discussion it is assumed that the male and female are comparable in age and physical stature. Generally, the bones of a male are larger and heavier than those of a female and have larger surface markings. Gender-related differences in the features of skeletal bones are readily apparent when comparing the female and male pelves. Most of the structural differences in the pelves are adaptations to the requirements of pregnancy and childbirth. The female's pelvis is wider and shallower than the male's. Consequently, there is more space in the true pelvis of the female, especially in the pelvic inlet and pelvic outlet, which accommodate the passage of the infant's head at birth. Other significant structural differences between pelves of females and males are listed and illustrated in Table 7.1.

✓ Why are structural differences between female and male pelves important?

Table 7.1 Comparison of Female and Male Pelves

Point of Comparison	Female	Male
General structure	Light and thin.	Heavy and thick.
False (greater) pelvis	Shallow.	Deep.
Pelvic brim (inlet)	Larger and more oval.	Smaller and heart-shaped.
Acetabulum	Small and faces anteriorly.	Large and faces laterally.
Obturator foramen	Oval	Round.
Pubic arch	Greater than 90° angle.	Less than 90° angle.

Pubic arch (greater than 90°)

Pubic arch (less than 90°)

Anterior views

Point of Comparison	Female	Male
Iliac crest	Less curved.	More curved.
Ilium	Less vertical.	More vertical.
Greater sciatic notch	Wide.	Narrow.
Coccyx	More movable and more curved anteriorly.	Less movable and less curved anteriorly.
Sacrum	Short, wide (see anterior view), and more curved anteriorly.	Long, narrow (see anterior view), and less curved anteriorly.

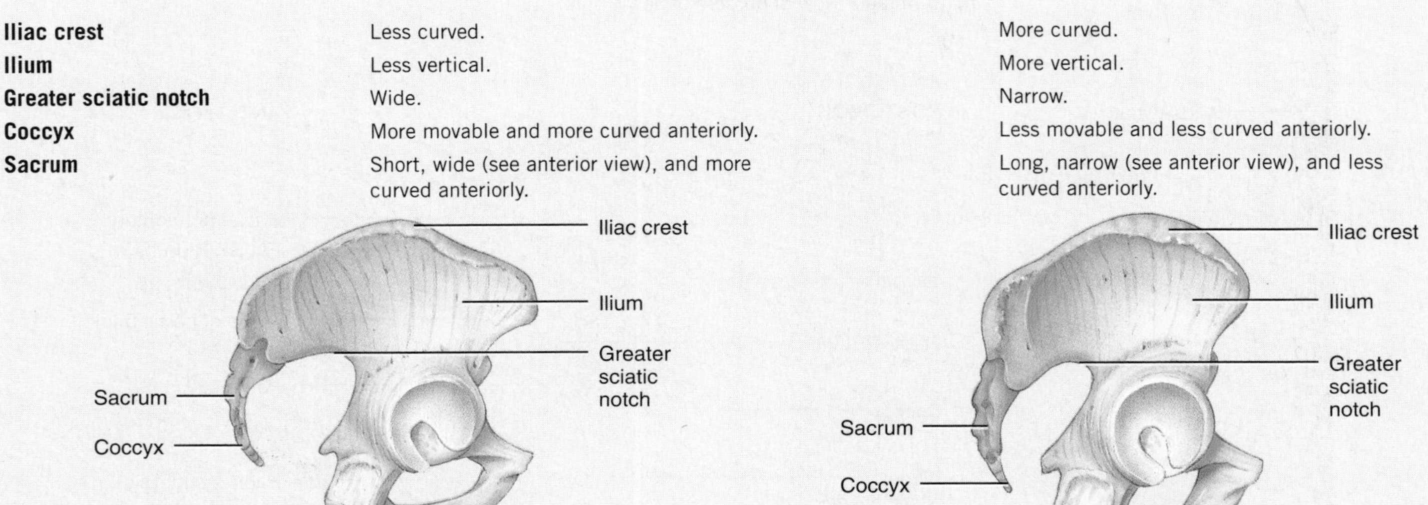

Right lateral views

Point of Comparison	Female	Male
Pelvic outlet	Wider.	Narrower.
Ischial tuberosity	Shorter, farther apart, and more laterally projecting.	Longer, closer together, and more medially projecting.

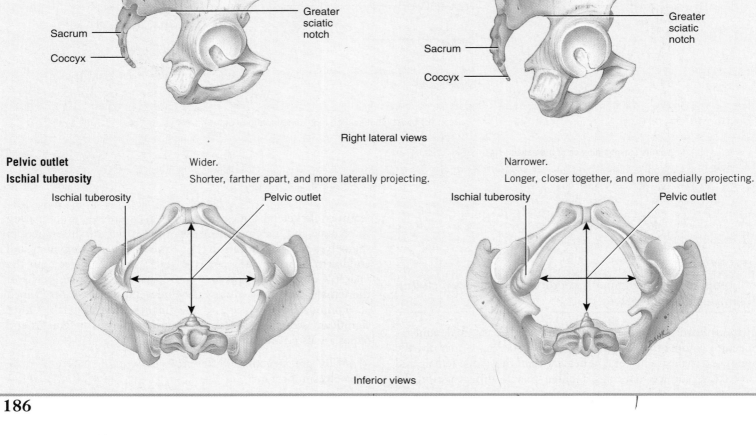

Inferior views

COMPARISON OF PECTORAL AND PELVIC GIRDLES

Objective

- Describe the structural differences between the pectoral and pelvic girdles.

Now that we've studied the structures of the pectoral and pelvic girdles, we can note some of their significant differences. The pectoral girdle does not directly articulate with the vertebral column, whereas the pelvic girdle directly articulates with the vertebral column via the sacroiliac joint. The sockets (glenoid fossae) for the upper limbs in the pectoral girdle are shallow and maximize movement, whereas the sockets (acetabula) for the lower limbs in the pelvic girdle are deep and allow less movement. Overall, the structure of the pectoral girdle offers more mobility than strength, whereas that of the pelvic girdle offers more strength than mobility.

✓ What is the overall difference between the pectoral and pelvic girdles?

LOWER LIMB (EXTREMITY)

Objective

- Identify the bones of the lower limb and their principal markings.

Each **lower limb (extremity)** is composed of 30 bones: (1) the femur of the thigh; (2) the patella (kneecap); (3) the tibia and fibula of the leg; and (4) the tarsals of the tarsus (ankle), the metatarsals of the metatarsus, and the phalanges (bones of the digits) of the foot (Figure 7.12).

Femur

The **femur**, or thighbone, is the longest, heaviest, and strongest bone in the body (Figure 7.13). Its proximal end articulates with the acetabulum of the hip bone, while its distal end articulates with the tibia and patella. The *body (shaft)* of the femur angles medially and, as a result, the knee joints are brought nearer to the midline. The angle of convergence is greater in females because the female pelvis is broader.

The proximal end of the femur consists of a rounded *head* that articulates with the acetabulum of the hip bone to form the *hip (coxal) joint.* The head contains a small centered depression (pit) called the *fovea capitis* (FŌ-vē-a CAP-i-tis). The ligament of the head of the femur connects the fovea capitis of the femur to the acetabulum of the hip bone. The *neck* of the femur is a constricted region distal to the head. The *greater trochanter* (trō-KAN-ter) and *lesser trochanter* are projections that serve as points of attachment for the tendons of some of the thigh and buttock muscles. The greater trochanter is the prominence felt and seen anterior to the hollow on the side of the hip. It is a landmark commonly used to locate the site for intramuscular injections

Figure 7.12 / Right lower limb.

Each lower limb consists of a femur, patella (kneecap), tibia, fibula, tarsals (ankle bones), metatarsals, and phalanges (bones of the digits).

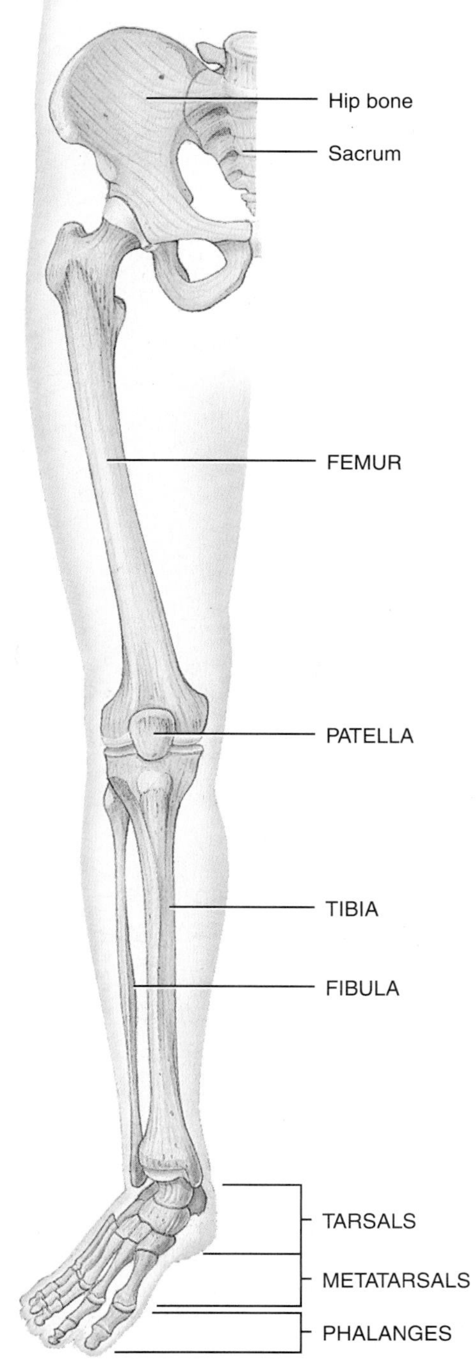

How many bones make up each lower limb?

Figure 7.13 / Right femur in relation to the hip bone, patella, tibia, and fibula. (See Tortora, *A Photographic Atlas of the Human Body,* Figure 3.28.)

The acetabulum of the hip bone and head of the femur articulate to form the hip joint.

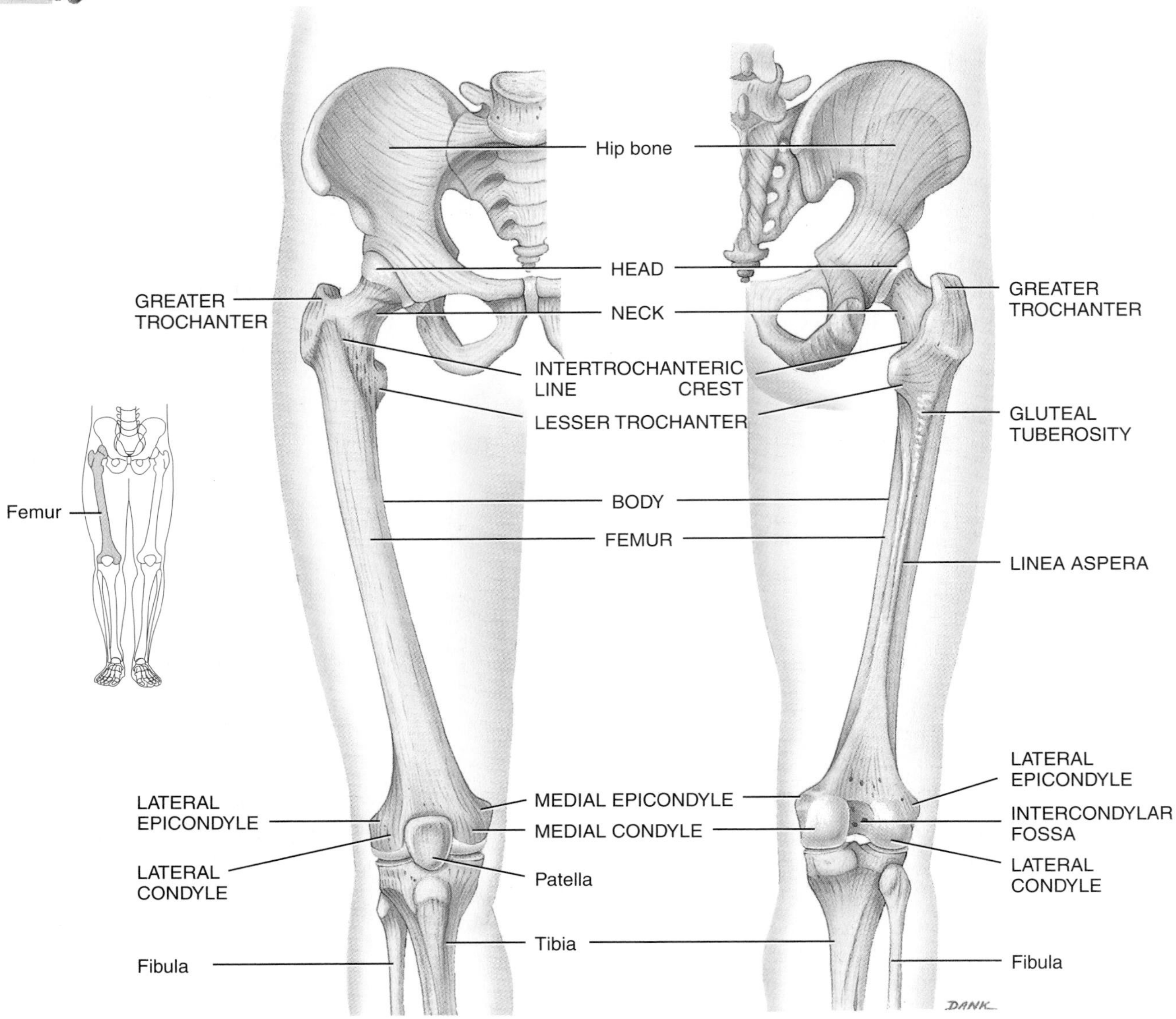

(a) Anterior view

(b) Posterior view

into the lateral surface of the thigh. The lesser trochanter is inferior and medial to the greater trochanter. Between the anterior surface of the trochanters is a narrow *intertrochanteric line* (Figure 7.13a); between the posterior surface of the trochanters is an *intertrochanteric crest* (Figure 7.13b).

Inferior to the intertrochanteric crest on the posterior surface of the body of the femur is a vertical ridge called the *gluteal tuberosity.* It blends into another vertical ridge called the *linea aspera* (LIN-ē-a AS-per-a; *asper-* = rough). Both ridges serve as attachment points for the tendons of several thigh muscles.

The distal end of the femur is expanded and includes the *medial condyle* and the *lateral condyle.* These articulate with the medial and lateral condyles of the tibia. Superior to the condyles are the *medial epicondyle* and the *lateral epicondyle.* A depressed area between the condyles on the posterior surface is called the *intercondylar fossa* (in-ter-KON-di-lar). The *patellar surface* is located between the condyles on the anterior surface.

Patella

The **patella** (= little dish), or kneecap, is a small, triangular bone located anterior to the knee joint (Figure 7.14). It is a sesamoid bone that develops in the tendon of the quadriceps femoris muscle. The broad superior end of the patella is called

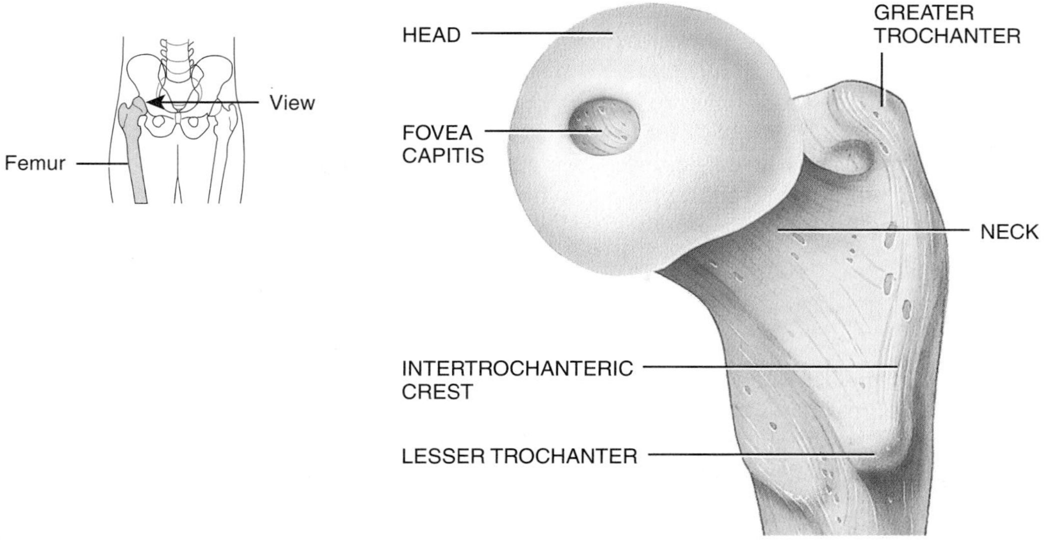

(c) Medial view of proximal end of femur

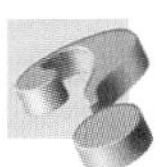

Why is the angle of convergence of the femurs greater in females than males?

the *base*. The pointed inferior end is the *apex*. The posterior surface contains two *articular facets*, one for the medial condyle of the femur and the other for the lateral condyle. The patellar ligament attaches the patella to the tibial tuberosity. The *patellofemoral joint*, between the posterior surface of the patella and the patellar surface of the femur, is the intermediate component of the *tibiofemoral (knee) joint*. The patella functions to increase the leverage of the tendon of the quadriceps femoris muscle, to maintain the position of the tendon when the knee is bent (flexed), and to protect the knee joint.

Patellofemoral Stress Syndrome

Patellofemoral stress syndrome ("runner's knee") is one of the most common problems in runners. During normal flexion and extension of the knee, the patella tracks (glides) up and down in the groove between the femoral condyles. In patellofemoral stress syndrome, normal tracking does not occur; instead, the patella also tracks laterally, and the increased pressure on the joint causes aching or tenderness around or under the patella.

Figure 7.14 / Right patella.

The patella articulates with the lateral and medial condyles of the femur.

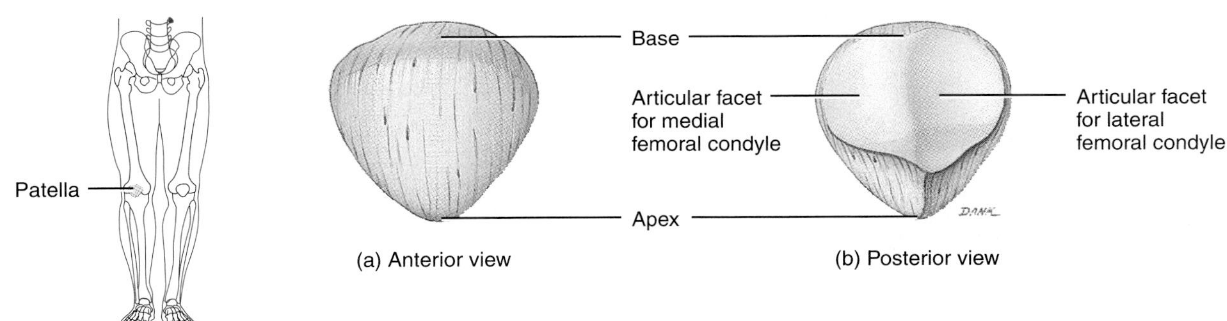

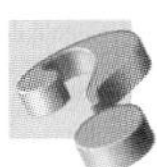

Because the patella develops in the tendon of a muscle, it is classified as which type of bone?

The pain typically occurs after a person has been sitting for a while, especially after exercise. It is worsened by squatting or walking down stairs. One cause of runner's knee is constantly walking, running, or jogging on the same side of the road. Because roads slope down on the sides, the knee that is closer to the center of the road endures greater mechanical stress because it does not fully extend during a stride. Other predisposing factors include having knock-knees, running on hills, and running long distances. ■

Tibia and Fibula

The **tibia,** or shin bone, is the larger, medial, weight-bearing bone of the leg (Figure 7.15). The tibia articulates at its proximal end with the femur and fibula and at its distal end with the fibula and the talus bone of the ankle. The tibia and fibula, like the ulna and radius, are connected by an interosseous membrane.

The proximal end of the tibia is expanded into a *lateral condyle* and a *medial condyle.* These articulate with the condyles of the femur to form the lateral and medial *tibiofemoral (knee) joints.* The inferior surface of the lateral condyle articulates with the head of the fibula. The slightly concave condyles are separated by an upward projection called the *intercondylar eminence* (Figure 7.15b). The *tibial tuberosity* on the anterior surface is a point of attachment for the patellar ligament. Inferior to and continuous with the tibial tuberosity is a sharp ridge that can be felt below the skin and is known as the *anterior border (crest)* or shin.

The medial surface of the distal end of the tibia forms the *medial malleolus* (mal-LĒ-ō-lus; = hammer). This structure articulates with the talus of the ankle and forms the prominence that can be felt on the medial surface of the ankle. The *fibular notch* (Figure 7.15c) articulates with the distal end of the fibula to form the *distal tibiofibular joint.*

The **fibula** is parallel and lateral to the tibia, but it is considerably smaller than the tibia. The *head* of the fibula, the proximal end, articulates with the inferior surface of the lateral condyle of the tibia below the level of the knee joint to form the *proximal tibiofibular joint.* The distal end has a projection called the *lateral malleolus* that articulates with the talus of the ankle.

Figure 7.15 / Right tibia and fibula in relation to the femur, patella, and talus. (See Tortora, *A Photographic Atlas of the Human Body,* Figure 3.30.)

The tibia articulates with the femur and fibula proximally, and with the fibula and talus distally.

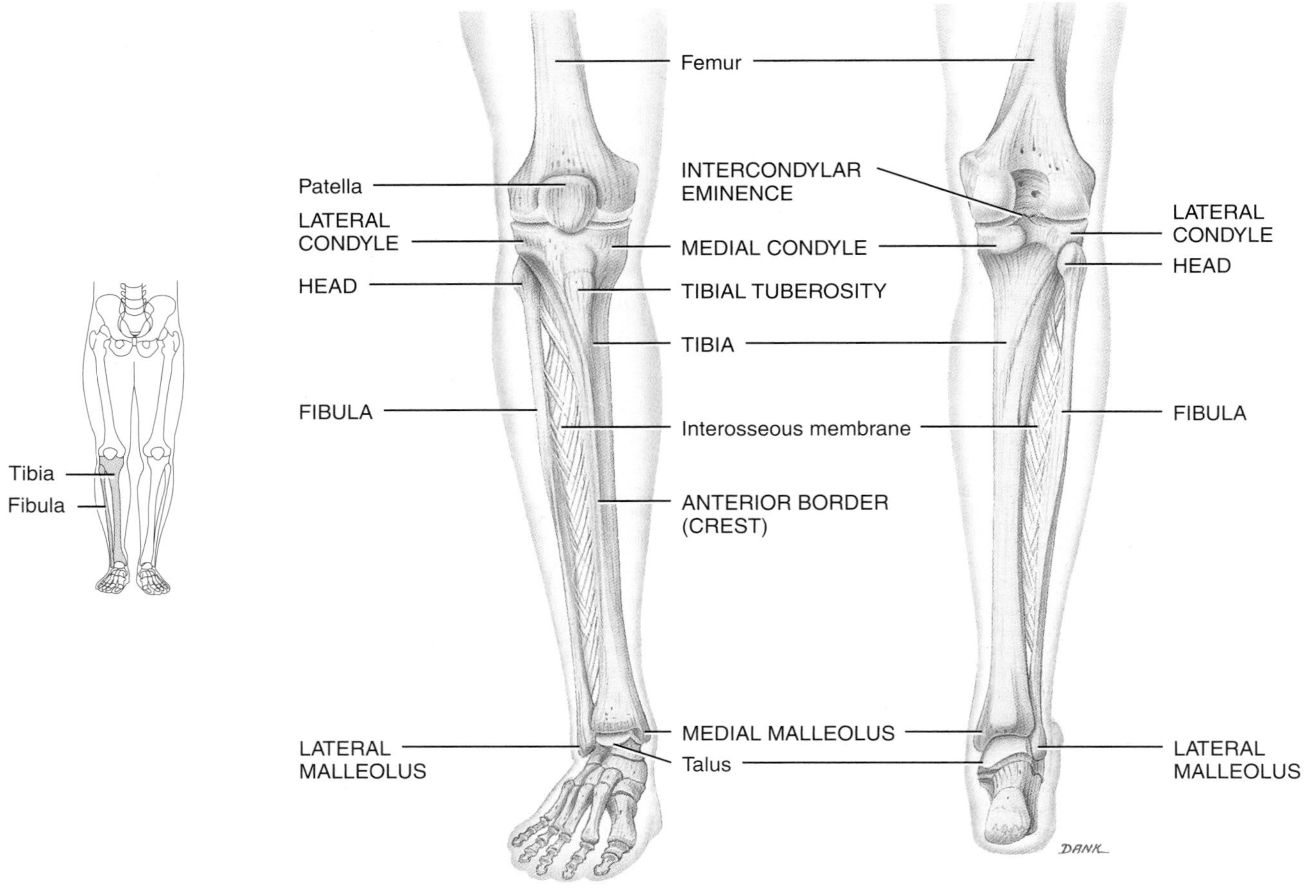

(a) Anterior view

(b) Posterior view

This forms the prominence on the lateral surface of the ankle. As noted, the fibula also articulates with the tibia at the fibular notch.

Tarsals, Metatarsals, and Phalanges

The **tarsus** (ankle) is the proximal region of the foot and consists of seven **tarsal bones** (Figure 7.16). They include the **talus** (TĀ-lus; = ankle bone) and **calcaneus** (kal-KĀ-nē-us; = heel), located in the posterior part of the foot. The calcaneus is the largest and strongest tarsal bone. The anterior tarsal bones are the **cuboid** (= cube-shaped), **navicular** (= like a little boat), and three **cuneiform bones** (= wedge-shaped) called the **first (medial), second (intermediate),** and **third (lateral) cuneiforms.** Joints between tarsal bones are called *intertarsal joints.* The talus, the most superior tarsal bone, is the only bone of the foot that articulates with the fibula and tibia. It articulates on one side with the medial malleolus of the tibia and on the other side with the lateral malleolus of the fibula. These articulations form the *talocrural (ankle) joint.* During walking, the talus transmits about half the weight of the body to the calcaneus. The remainder is transmitted to the other tarsal bones.

The **metatarsus** is the intermediate region of the foot and consists of five **metatarsal bones** numbered I to V (or 1–5) from the medial to lateral position (Figure 7.16). Like the metacarpals of the palm of the hand, each metatarsal consists of a proximal *base*, an intermediate *shaft*, and a distal *head.* The metatarsals articulate proximally with the first, second, and third cuneiform bones and with the cuboid to form the *tarsometatarsal joints.* Distally, they articulate with the proximal row of phalanges to form the *metatarsophalangeal joints.* The first metatarsal is thicker than the others because it bears more weight.

The **phalanges** comprise the distal component of the foot and resemble those of the hand both in number and arrangement. The toes are numbered I to V (or 1–5) beginning with the great toe, which is medial. Each also consists of a proximal *base*, an intermediate *shaft*, and a distal *head.* The great or big toe (*hallux*) has two large, heavy phalanges called proximal and distal phalanges. The other four toes each have three phalanges—proximal, middle, and distal. Joints between phalanges of the foot, like those of the hand, are called *interphalangeal joints.*

Arches of the Foot

The bones of the foot are arranged in two **arches** (Figure 7.17 on page 193). The arches enable the foot to support the weight of the body, provide an ideal distribution of body weight over the hard and soft tissues of the foot, and provide leverage when walking. The arches are not rigid; they yield as weight is applied and spring back when the weight is lifted, thus helping to absorb shocks. Usually, the arches are fully developed by the time children reach age 12 or 13.

The *longitudinal arch* has two parts, both of which consist of tarsal and metatarsal bones arranged to form an arch from the anterior to the posterior part of the foot. The *medial part* of the longitudinal arch originates at the calcaneus. It rises to the talus and descends through the navicular, the three cuneiforms, and the heads of the three medial metatarsals. The *lateral part* of the longitudinal arch also begins at the calcaneus. It rises at the cuboid and descends to the heads of the two lateral metatarsals.

The *transverse arch* is found between the medial and lateral aspects of the foot and is formed by the navicular, three cuneiforms, and the bases of the five metatarsals.

The bones composing the arches are held in position by ligaments and tendons. If these ligaments and tendons are weakened, the height of the medial longitudinal arch may decrease or "fall." The result is **flatfoot,** the causes of which include excessive weight, postural abnormalities, weakened supporting tissues, and genetic predisposition. A custom-designed arch support (orthotic) often is prescribed to treat flatfoot.

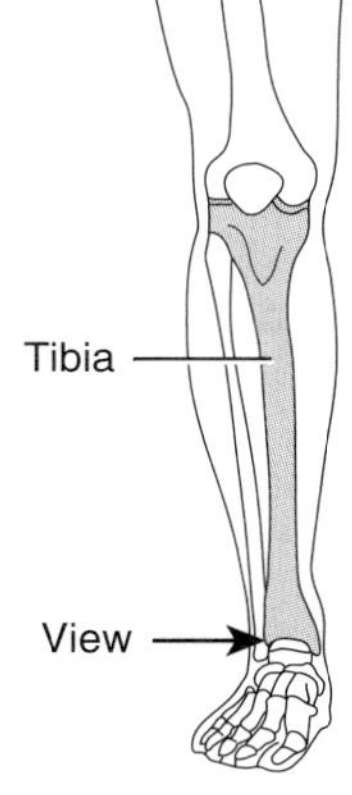

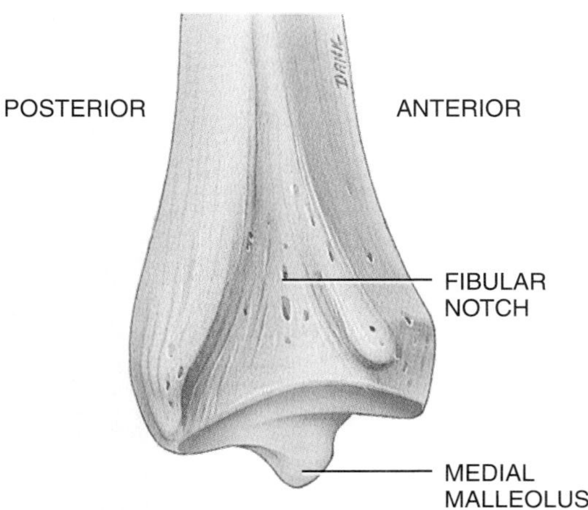

(c) Lateral view of distal end of tibia

Which leg bone bears the weight of the body?

Figure 7.16 / Right foot.

The skeleton of the foot consists of the tarsals in the proximal region of the foot, the metatarsals in the intermediate region of the foot, and the phalanges in the distal region of the foot.

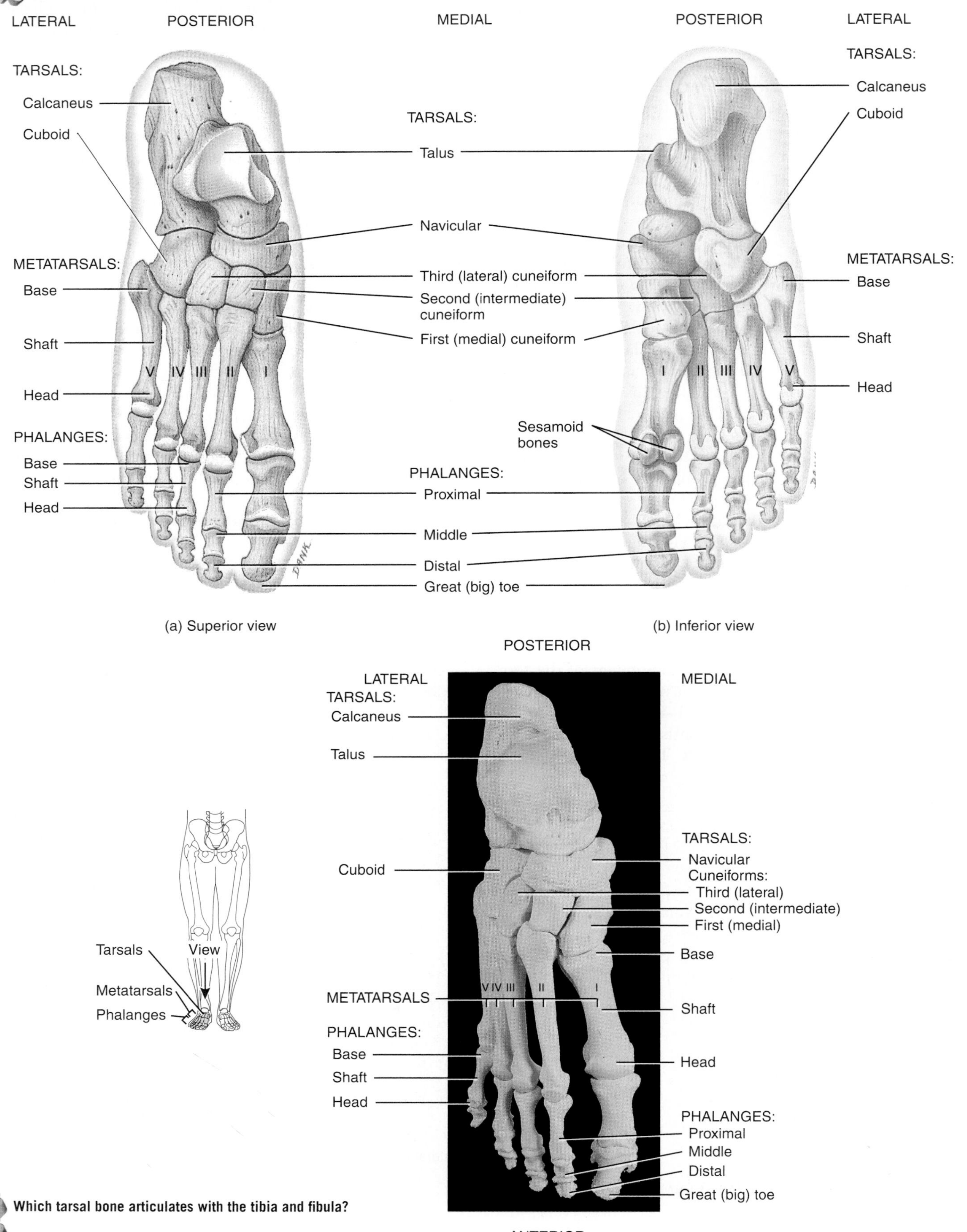

Which tarsal bone articulates with the tibia and fibula?

Figure 7.17 / Arches of the right foot.

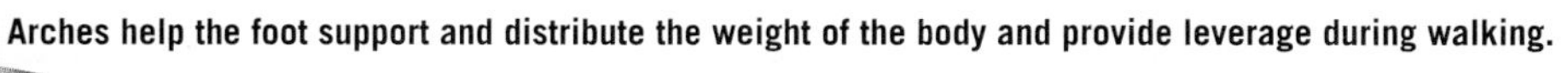

Arches help the foot support and distribute the weight of the body and provide leverage during walking.

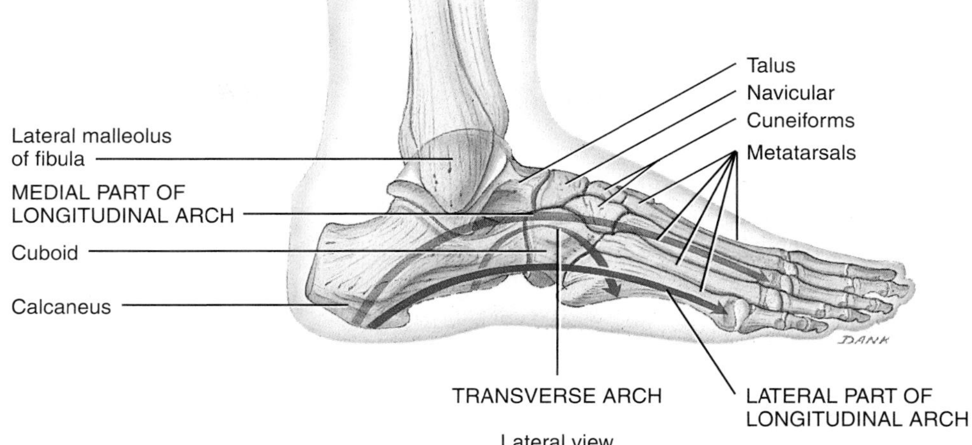

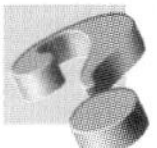

What structural feature of the arches allows them to absorb shocks?

Clawfoot is a condition in which the medial longitudinal arch is abnormally elevated. It is often caused by muscle deformities, such as may occur in diabetics whose neurological lesions lead to atrophy of muscles of the foot.

Clubfoot or **talipes equinovarus** (*-pes* = foot; *equino-* = horse) is an inherited deformity of the foot that occurs in 1 of every 1000 births. In this malformation, the foot is twisted inferiorly and medially, and the angle of the arch is increased. Treatment consists of manipulating the arch to a normal curvature by casts or adhesive tape, usually soon after birth. Corrective shoes or surgery may also be required. ■

✓ Name the bones that form the lower limb, from proximal to distal.
✓ Describe the joints of the lower limb.
✓ What are the functions of the arches of the foot?

APPLICATIONS TO HEALTH

Hip Fracture

Although any region of the hip girdle may fracture, the term **hip fracture** most commonly applies to a break in the bones associated with the hip joint—the head, neck, or trochanteric regions of the femur, or the bones that form the acetabulum. In the United States, 300,000–500,000 people sustain hip fractures each year. The incidence of hip fractures is increasing, due in part to longer life spans. Decreases in bone mass due to osteoporosis (which occurs more often in females) and an increased tendency to fall predispose elderly people to hip fractures.

Hip fractures often require surgical treatment, the goal of which is to repair and stabilize the fracture, increase mobility, and decrease pain. Sometimes the repair is accomplished by using surgical pins, screws, nails, and plates to secure the head of the femur. In severe hip fractures, the femoral head or the acetabulum of the hip bone may be replaced by prostheses (artificial devices). The procedure of replacing either the femoral head or the acetabulum is *hemiarthroplasty* (hem-ē-AR-thrō-plas-tē; *hemi-* = one half; *-arthro-* = joint; *-plasty* = molding). Replacement of both the femoral head and acetabulum is *total hip arthroplasty*. The acetabular prosthesis is made of plastic, whereas the femoral prosthesis is metal; both are designed to withstand a high degree of stress. The prostheses are attached to healthy portions of bone with acrylic cement and screws.

KEY MEDICAL TERMS ASSOCIATED WITH THE APPENDICULAR SKELETON

Genu valgum (JĒ-nū VAL-gum; *genu-* = knee; *valgum* = bent outward) A decreased space between the knees and increased space between the ankles due to a medial bending of the legs in relation to the thigh. Also called **knock-knee.**

Genu varum (JĒ-nū VAR-um; *varum* = bent toward the midline) An increased space between the knees due to a lateral bending of the legs relative to the thigh. Also called **bowleg.**

Hallux valgus (HAL-uks VAL-gus; *hallux* = great toe) Angulation of the great toe away from the midline of the body, typically caused by wearing tightly fitting shoes. Involves lateral deviation of the proximal phalanx of the great toe and medial displacement of metatarsal I. Also called **bunion.**

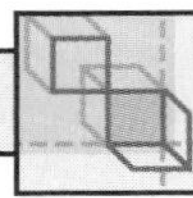

STUDY OUTLINE

Pectoral (Shoulder) Girdle (p. 173)

1. Each pectoral (shoulder) girdle consists of a clavicle and scapula.
2. Each pectoral girdle attaches an upper limb to the axial skeleton.

Upper Limb (p. 174)

1. Each of the two upper limbs contains 30 bones.
2. The bones of each upper limb include the humerus, the ulna, the radius, the carpals, the metacarpals, and the phalanges.

Pelvic (Hip) Girdle (p. 182)

1. The pelvic (hip) girdle consists of two hip bones.
2. Each hip bone consists of three fused bones: the ilium, pubis, and ischium.
3. The hip bones, sacrum, and pubic symphysis form the bony pelvis. It supports the vertebral column and pelvic viscera and attaches the lower limbs to the axial skeleton.
4. The true pelvis is separated from the false pelvis by the pelvic brim.

Comparison of Female and Male Pelves (p. 185)

1. Bones of males are generally larger and heavier than bones of females, with more prominent markings for muscle attachment.
2. The female pelvis is adapted for pregnancy and childbirth. Gender-related differences in pelvic structure are listed and illustrated in Table 7.1 on page 186.

Comparison of Pectoral and Pelvic Girdles (p. 187)

1. The pectoral girdle does not directly articulate with the vertebral column; the pelvic girdle does.
2. The glenoid fossae of the scapulae are shallow and maximize movement; the acetabula of the hip bones are deep and allow less movement.

Lower Limb (p. 187)

1. Each of the two lower limbs contains 30 bones.
2. The bones of each lower limb include the femur, the patella, the tibia, the fibula, the tarsals, the metatarsals, and the phalanges.
3. The bones of the foot are arranged in two arches, the longitudinal arch and the transverse arch, to provide support and leverage.

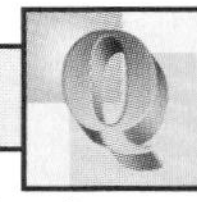

SELF-QUIZ QUESTIONS

Choose the one best answer to the following questions:

1. Which of the following are bones found in the axial division of the skeleton? (a) ribs, scapula, clavicle, (b) sphenoid, sternum, coccyx, (c) parietal, ulna, carpals, (d) lacrimal, sacrum, patella, (e) pubis, ischium, ilium.
2. On which bone are medial and lateral epicondyles located? (1) ulna, (2) femur, (3) radius, (4) humerus, (5) tibia.
   a. 1, 3, and 5, **b.** 2, 4, and 5, **c.** 1 and 2, **d.** 4 and 5, **e.** 2 and 4.
3. Which structures are on the posterior surface of the upper limb? (a) radial fossa and radial notch, (b) trochlea and capitulum, (c) coronoid process and coronoid fossa, (d) olecranon process and olecranon fossa, (e) lesser tubercle and intertubercular sulcus.
4. The bones of the pectoral girdle (a) articulate with the sternum anteriorly and the vertebrae posteriorly, (b) include both clavicles, both scapulae, and the manubrium of the sternum, (c) are considered to be part of the axial skeleton, (d) articulate with the head of the humerus, forming the shoulder joint, (e) are described by none of the above.
5. The anatomical name for the socket into which the humerus fits is the (a) acetabulum, (b) coronoid fossa, (c) glenoid cavity, (d) supraspinous fossa, (e) iliac fossa.
6. The radius (a) has a tuberosity on its distal end, (b) has an acromial process at its proximal end, (c) is longer and larger than the ulna, (d) has a styloid process on its distal end, (e) is the medial bone of the forearm.
7. The greater trochanter is a large bony prominence located (a) on the proximal part of the humerus, (b) on the proximal part of the femur, (c) near the tuberosity of the tibia, (d) on the ilium, (e) on the posterior surface of the scapula.
8. Which of the following is a part of the femur? (a) obturator foramen, (b) trochlear notch (c) patellar surface, (d) medial malleolus, (e) none of the above.
9. Which of the following statements regarding the male pelvis is *not* true? (a) The bones are heavier and thicker than in the female. (b) The male pelvis is narrow and deep. (c) The coccyx is less curved anteriorly. (d) The pelvic outlet is narrower than in the female. (e) none of the above.
10. Which of the following is *not* a carpal bone? (a) hamate, (b) cuboid, (c) pisiform, (d) trapezium, (e) scaphoid.

Complete the following.

11. The large depression on the anterior surface of the scapula is the ________ .
12. The anatomical name for the knuckle of your index finger is ________ .

13. The lesser trochanter is on the ________ surface of the ________ .

14. The portion of the pelvic above the pelvic brim is the ________ pelvis.

15. The acetabulum is part of the ________ bone.

Are the following statements true or false?

16. The metatarsals are proximal to the tarsals.

17. The phalanges are distal to the metatarsals and the tarsals.

18. The acromial extremity is located at the lateral end of the clavicle.

19. The capitulum is the lateral condyle of the radius.

20. Match the following bony landmarks and bones.

____	**(a)** fibular notch	**(1)** radius
____	**(b)** acromion	**(2)** femur
____	**(c)** coronoid process	**(3)** tibia
____	**(d)** lateral malleolus	**(4)** humerus
____	**(e)** deltoid tuberosity	**(5)** scapula
____	**(f)** radial tuberosity	**(6)** ischium
____	**(g)** anterior inferior iliac spine	**(7)** ulna
____	**(h)** linea aspera	**(8)** fibula
____	**(i)** ischial tuberosity	**(9)** ilium

CRITICAL THINKING QUESTIONS

1. The Local News reported that farmer Bob Ramsey caught his hand in a piece of machinery on Tuesday. He lost the lateral two fingers of his left hand. Science reporter Kent Clark, really a high school junior, reports that Farmer Ramsey has 3 remaining phalanges. Is Kent correct or is he anatomy-challenged?
   HINT *How many bones are in one finger?*

2. Rose had flat feet as a child and was told to take up ballet dancing to correct the condition. Now Rose has hallux valgus and trouble with her talocrural joint, but at least her flat feet are cured. Explain how Rose's problems relate to each other.
   HINT *Ballet dancers spend a lot of time on their toes.*

3. Since the hips and shoulders are located in the trunk of the body, why aren't they part of the axial skeleton like the rest of the bones composing the trunk?
   HINT *Have you ever seen a snake with shoulders?*

4. Amy was visiting her Grandmother Amelia in the hospital. Grandmother said she had a broken hip but when Amy peeked at the chart, it said she had a fractured femur. Explain the discrepancy.
   HINT *The area that people usually think of as the hip includes more than just the hip bone.*

5. Derrick, the high school gym teacher, jogs on the same banked high school track at the same time and in the same direction every day. Recently he has been experiencing an ache in one knee, especially after he sits and reads the newspaper following his jog. Nothing else in his routine has changed. What could be the problem with his knee?
   HINT *The answer is in his routine.*

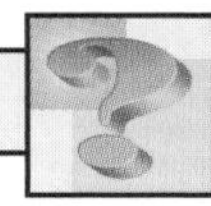

ANSWERS TO FIGURE QUESTIONS

7.1 The pectoral girdles attach the upper limbs to the axial skeleton.

7.2 The weakest part of the clavicle is its midregion.

7.3 The highest point of the shoulder is the acromion.

7.4 Each upper limb has 30 bones.

7.5 The radius articulates with the capitulum and radial fossa of the humerus. The ulna articulates with the trochlea, coronoid fossa, and olecranon fossa of the humerus.

7.6 The "elbow" part of the ulna is the olecranon.

7.7 The radius and ulna form the proximal and distal radioulnar joints. They are also connected by the interosseous membrane.

7.8 The most frequently fractured wrist bone is the scaphoid.

7.9 The bony pelvis attaches the lower limbs to the axial skeleton and supports the backbone and pelvic viscera.

7.10 The femur articulates with the acetabulum of the hip bone; the sacrum articulates with the auricular surface of the hip bone.

7.11 The pelvic axis is the course taken by a baby's head as it descends through the pelvis during childbirth.

7.12 Each lower limb has 30 bones.

7.13 The angle of convergence of the femurs is greater in females than males because the female pelvis is broader.

7.14 The patella is a sesamoid base—the largest one in the body.

7.15 The tibia is the weight-bearing bone of the leg.

7.16 The talus articulates with the tibia and the fibula.

7.17 The arches are not rigid; they yield when weight is applied and spring back when weight is lifted.

8

JOINTS

CONTENTS AT A GLANCE

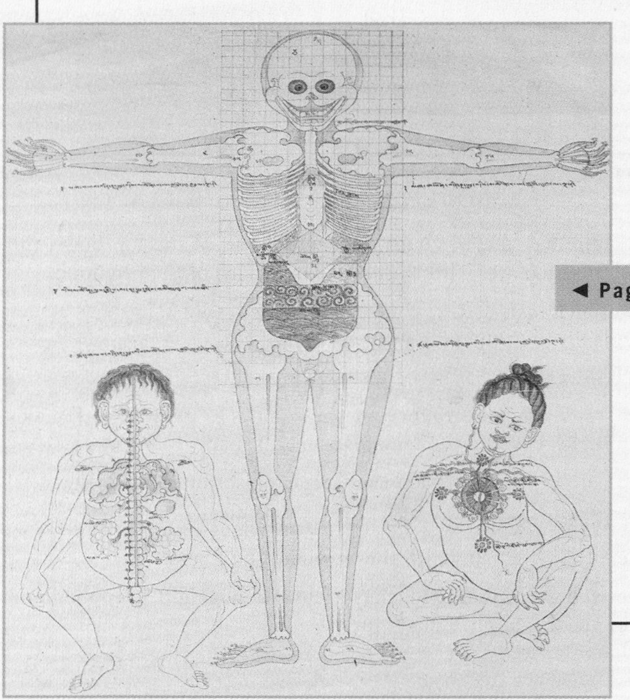

◄ Page 201

We often think of medicine as a scientific search for the truth. It's easy to forget that, like art or politics, medicine is also an expression of culture.

Page 219 ►

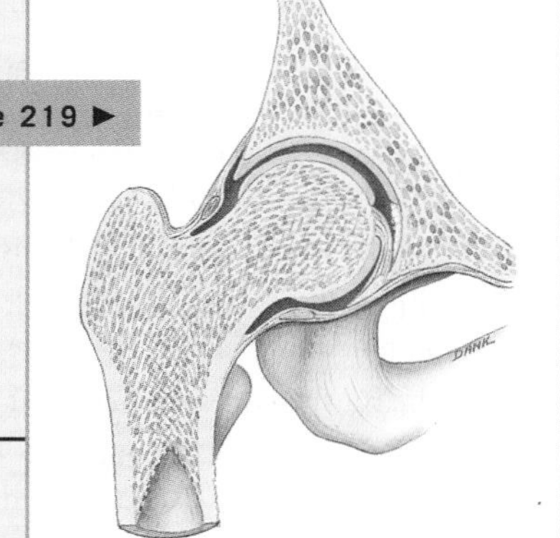

INTRODUCTION

Bones are too rigid to bend without being damaged. Fortunately, flexible connective tissues form joints that hold bones together while still permitting, in most cases, some degree of movement. A **joint,** also called an **articulation** or **arthrosis,** is a point of contact between two bones, between bone and cartilage, or between bone and teeth. When we say one bone *articulates* with another bone, we mean that the bones form a joint. Because most movements of the body occur at joints, you can appreciate their importance if you imagine how a cast over your knee joint makes walking difficult, or how a splint on a finger limits your ability to manipulate small objects. The scientific study of joints is termed **arthrology** (ar-THROL-ō-jē; *arthr-* = joint; *-ology* = study of). The study of motion of the human body is called **kinesiology** (ki-nē-sē′-OL-ō-jē; *kinesi-* = movement).

Some joints permit no movement, others permit slight movement, and still others afford fairly free movement. A joint's structure determines its combination of strength and flexibility. In general, the closer the fit at the point of contact, the stronger the joint. At tightly fitted joints, however, movement is restricted. The looser the fit, the greater the movement, but loosely fitted joints are prone to dislocation (displacement). Movement at joints is also determined by (1) the structure (shape) of articulating bones, (2) the flexibility (tension or tautness) of the connective tissue that binds the bones together, and (3) the position of associated ligaments, muscles, and tendons. Joint flexibility may also be affected by hormones. For example, relaxin, a hormone produced by the placenta and ovaries, relaxes the pubic symphysis and ligaments between the sacrum, hip bone, and coccyx toward the end of pregnancy, which assists in delivery.

JOINT CLASSIFICATIONS

Objective

- Describe the structural and functional classifications of joints.

Joints are classified according to their anatomical characteristics (structure) and the type of movement they permit (function).

The structural classification of joints is based on (1) the presence or absence of a space between the articulating bones, called a synovial cavity, and (2) the type of connective tissue that binds the bones together. Structurally, joints are classified as one of the following types:

- **Fibrous joints:** The bones are held together by fibrous connective tissue that is rich in collagen fibers and there is no synovial cavity.
- **Cartilaginous** (kar-ti-LAJ-i-nus) **joints:** The bones are held together by cartilage, and there is no synovial cavity.
- **Synovial joints:** The bones forming the joint have a synovial cavity and are united by the dense irregular connective tissue of an articular capsule, and often by accessory ligaments.

The functional classification of joints relates to the degree of movement they permit. Functionally, joints are classified as one of the following types:

- **Synarthrosis** (sin′-ar-THRŌ-sis; *syn-* = together): An immovable joint. The plural is *synarthroses.*
- **Amphiarthrosis** (am′-fē-ar-THRŌ-sis; *amphi-* = on both sides): A slightly movable joint. The plural is *amphiarthroses.*
- **Diarthrosis** (dī-ar-THRŌ-sis; = movable joint): A freely movable joint. The plural is *diarthroses.* All diarthroses are synovial joints. They have a variety of shapes and permit several different types of movements.

The following sections present the joints of the body according to their structural classification. As we examine the structure of each type of joint, we will also explore its functional attributes.

FIBROUS JOINTS

Objective

- Describe the structure and functions of the three types of fibrous joints.

Fibrous joints lack a synovial cavity, and the articulating bones are held very closely together by fibrous connective tissue. These joints, which permit little or no movement, include sutures, syndesmoses, and gomphoses.

Sutures

A **suture** (SŪ-chur; *sutur-* = seam) is a fibrous joint composed of a thin layer of dense fibrous connective tissue that unites only bones of the skull. An example is the coronal suture between the parietal and frontal bones (Figure 8.1a). The irregular, interlocking edges of sutures give them added strength and decrease their chance of fracturing. Because a suture is immovable, it is classified functionally as a synarthrosis.

Some sutures, although present during childhood, are replaced by bone in the adult. Such a suture is called a **synostosis** (sin′-os-TŌ-sis; *os-* = bone), or bony joint—a joint in which there is a complete fusion of bone across the suture line. An example is the metopic suture between the left and right sides of the frontal bone that begins to fuse during infancy. A synostosis is also classified functionally as a synarthrosis.

Syndesmoses

A **syndesmosis** (sin′-dez-MŌ-sis; *syndesmo-* = band or ligament) is a fibrous joint in which there is a greater distance between the articulating bones and more fibrous connective tissue

Figure 8.1 / Fibrous joints.

At a fibrous joint the bones are held together by fibrous connective tissue.

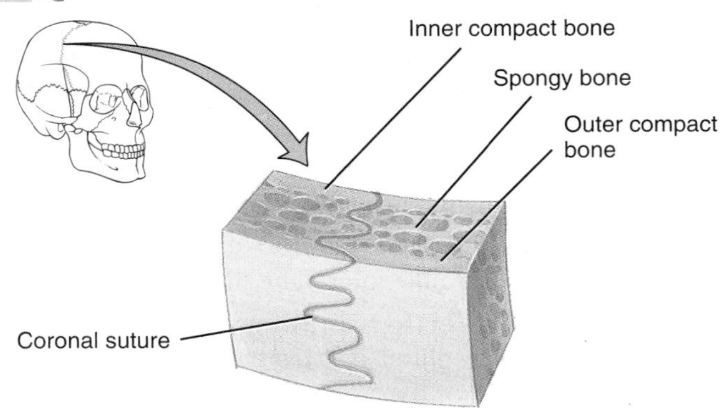

(a) Suture between skull bones

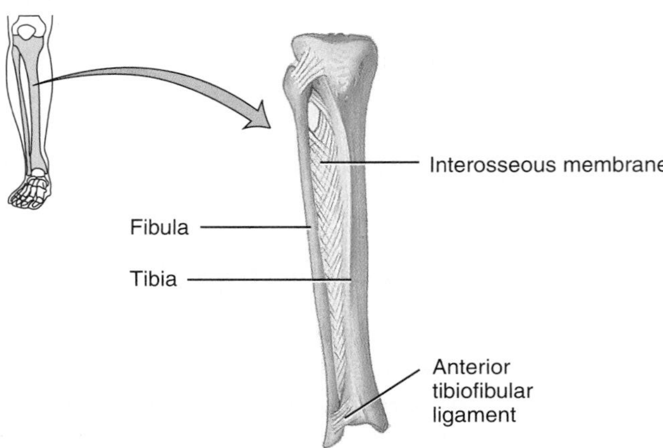

(b) Syndesmoses between tibia and fibula

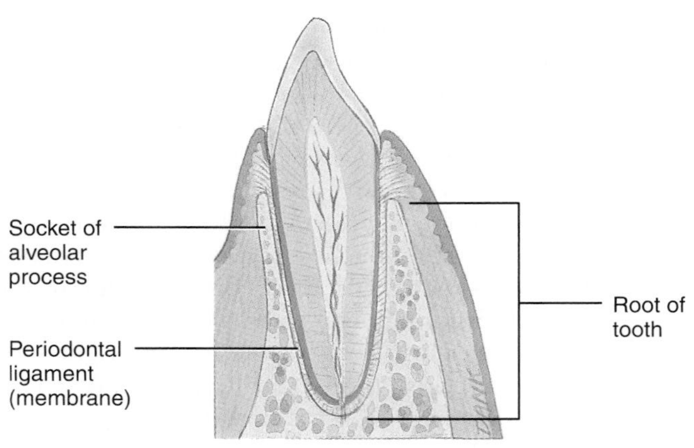

(c) Gomphosis between tooth and socket of alveolar process

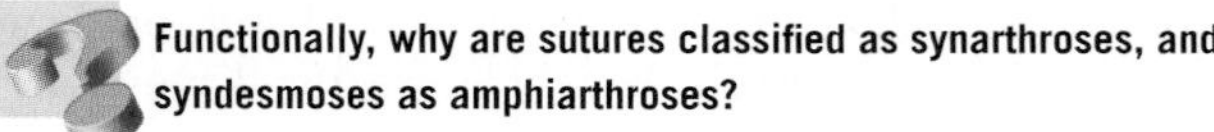
Functionally, why are sutures classified as synarthroses, and syndesmoses as amphiarthroses?

than in a suture. The fibrous connective tissue is arranged either as a bundle (ligament) or as a sheet (interosseous membrane) (Figure 8.1b). One example of a syndesmosis is the distal tibiofibular joint, where the anterior tibiofibular ligament connects the tibia and fibula. Another example is the interosseous membrane between the parallel borders of the tibia and fibula. Because it permits slight movement, a syndesmosis is classified functionally as an amphiarthrosis.

Gomphoses

A **gomphosis** (gom-FŌ-sis; *gompho-* = a bolt or nail) is a type of fibrous joint in which a cone-shaped peg fits into a socket. The only examples of gomphoses are the articulations between the roots of the teeth and their sockets (alveoli) in the maxillae and mandible (Figure 8.1c). The dense fibrous connective tissue between a tooth and its socket is the periodontal ligament (membrane). A gomphosis is classified functionally as a synarthrosis, an immovable joint.

CARTILAGINOUS JOINTS

Objective

- Describe the structure and functions of the two types of cartilaginous joints.

Like a fibrous joint, a **cartilaginous joint** lacks a synovial cavity and allows little or no movement. Here the articulating bones are tightly connected by either fibrocartilage or hyaline cartilage. The two types of cartilaginous joints are synchondroses and symphyses.

Synchondroses

A **synchondrosis** (sin′-kon-DRŌ-sis; *chondro-* = cartilage) is a cartilaginous joint in which the connecting material is hyaline cartilage. An example of a synchondrosis is the epiphyseal plate that connects the epiphysis and diaphysis of a growing bone (Figure 8.2a). A photomicrograph of the epiphysis is shown in Figure 5.7 on page 120. Functionally, a synchondrosis is a synarthrosis. When bone elongation ceases, bone replaces the hyaline cartilage, and the synchondrosis becomes a synostosis, a bony joint. Another example of a synchondrosis is the joint between the first rib and the manubrium of the sternum, which also ossifies during adult life and becomes an immovable synostosis.

Symphyses

A **symphysis** (SIM-fi-sis; = growing together) is a cartilaginous joint in which the ends of the articulating bones are covered with hyaline cartilage, but the bones are connected by a broad, flat disc of fibrocartilage. All symphyses occur in the midline of the body. The pubic symphysis between the anterior surfaces of the hip bones is one example of a symphysis (Figure 8.2b). This type of joint is also found at the junction of the manubrium and body of the sternum (see Figure 6.23 on page 166) and at the in-

Figure 8.2 / Cartilaginous joints.

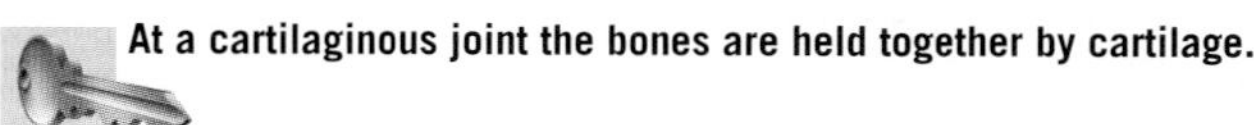

At a cartilaginous joint the bones are held together by cartilage.

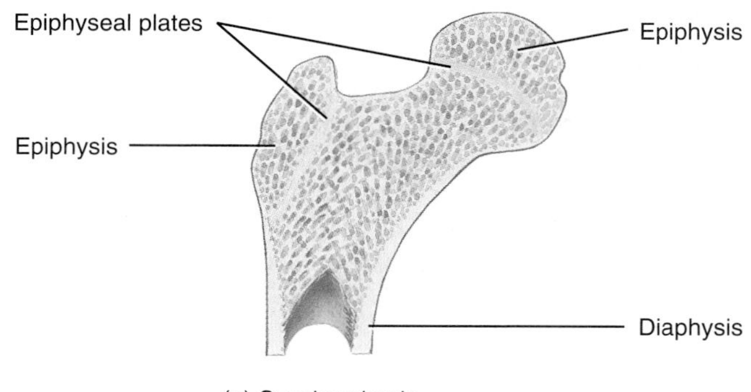

(a) Synchondrosis

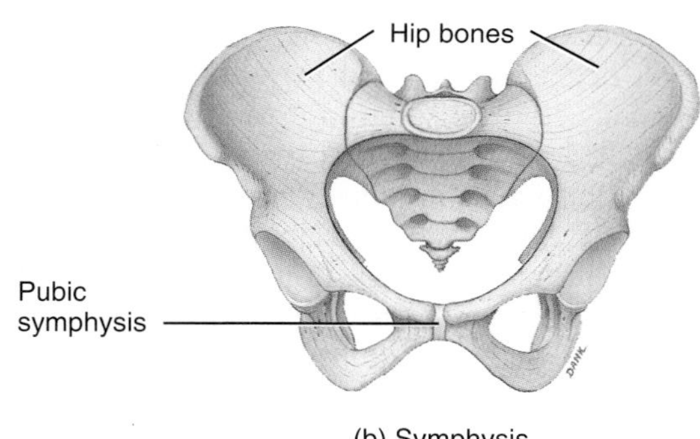

(b) Symphysis

What is the structural difference between a synchondrosis and a symphysis?

tervertebral joints between the bodies of vertebrae (see Figure 6.21a on page 163). A portion of the intervertebral disc is fibrocartilage. A symphysis is an amphiarthrosis, a slightly movable joint.

✓ List the types of fibrous and cartilaginous joints. Name a joint of each type in the body, and name the type of tissue that holds the articulating bones together.

✓ Which of the fibrous and cartilaginous joints are classified as amphiarthroses?

SYNOVIAL JOINTS

Objective

- Describe the structure of synovial joints.

Synovial joints (si-NŌ-vē-al) have a space called a **synovial (joint) cavity** between the articulating bones (Figure 8.3). The structure of these joints allows the bones to move freely. Hence, all synovial joints are classified functionally as diarthroses.

Structure of Synovial Joints

The bones at a synovial joint are covered by **articular cartilage,** which is hyaline cartilage. The cartilage provides a smooth, slippery surface for the articulating bones, but it does not bind them together. Articular cartilage reduces friction between bones of the joint during movement and helps to absorb shock.

Articular Capsule

A sleevelike **articular capsule** surrounds a synovial joint, encloses the synovial cavity, and unites the articulating bones. The articular capsule is composed of two layers, an outer fibrous capsule and an inner synovial membrane (Figure 8.3). The outer layer, the **fibrous capsule,** usually consists of dense, irregular connective tissue that attaches to the periosteum of the articulating bones. The flexibility of the fibrous capsule permits considerable movement at a joint while its great tensile strength (resistance to stretching) helps prevent the bones from dislocating. The fibers of some fibrous capsules are arranged in parallel bundles that are highly adapted for resisting strains. Such fiber bundles are called **ligaments** (*liga-* = bound or tied) and are often designated by individual names. The strength of ligaments is one of the principal mechanical factors that hold bones close together in a synovial joint. The inner layer of the articular capsule, the **synovial membrane,** is composed of areolar connective tissue with elastic fibers. At many synovial joints, the synovial membrane includes accumulations of adipose tissue called **articular fat pads.** An example is the infrapatellar fat pad in the knee (see Figure 8.15c).

Synovial Fluid

The synovial membrane secretes **synovial fluid** (*ov-* = egg), which forms a thin film over the surfaces within the articular capsule. This viscous, clear or pale yellow fluid was named for its similarity in appearance and consistency to uncooked egg white or albumin. Synovial fluid consists of hyaluronic acid secreted by fibroblast-like cells in the synovial membrane and interstitial fluid filtered from blood plasma. Its several functions include reducing friction by lubricating the joint and supplying nutrients to and removing metabolic wastes from the chondrocytes within articular cartilage. (Recall that cartilage is an avascular tissue.) Synovial fluid also contains phagocytic cells that remove microbes and the debris that results from normal wear and tear in the joint. When a synovial joint is immobile for a time, the fluid becomes quite viscous (gel-like), but as joint movement increases, the fluid becomes less viscous. One of the benefits of warming up before exercise is that it stimulates the production and secretion of synovial fluid.

Accessory Ligaments and Articular Discs

Many synovial joints also contain **accessory ligaments** called extracapsular ligaments and intracapsular ligaments. *Extracapsular ligaments* lie outside the articular capsule. Examples are the fibular and tibial collateral ligaments of the knee joint (see Figure 8.15d). *Intracapsular ligaments* occur within the articular capsule but are excluded from the synovial cavity by folds of the synovial membrane. Examples are the anterior and posterior cruciate ligaments of the knee joint (see Figure 8.15d).

Inside some synovial joints, such as the knee, are pads of fibrocartilage that lie between the articular surfaces of the bones and are attached to the fibrous capsule. These pads are called **articular discs** or **menisci** (men-IS-ī or men-IS-kī; the singular is *meniscus*). Figure 8.15d depicts the lateral and medial menisci in the knee joint. The discs usually subdivide the synovial cavity into two separate spaces. This separation can allow separate movements to occur in each space. As you will see later, separate movements occur in the respective compartments of the temporomandibular joint (TMJ). By modifying the shape of the joint surfaces of the articulating bones, articular discs allow two bones of different shapes to fit more tightly. Articular discs also help to maintain the stability of the joint and direct the flow of synovial fluid to the areas of greatest friction.

Figure 8.3 / Structure of a typical synovial joint. Note the two layers of the articular capsule—the fibrous capsule and the synovial membrane. Synovial fluid fills the joint cavity between the synovial membrane and the articular cartilage.

The distinguishing feature of a synovial joint is the synovial cavity between the articulating bones.

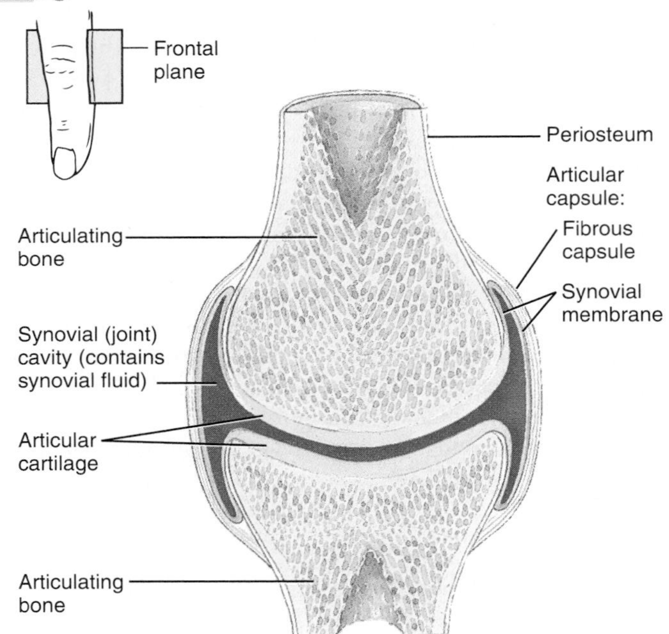

(a) Frontal section

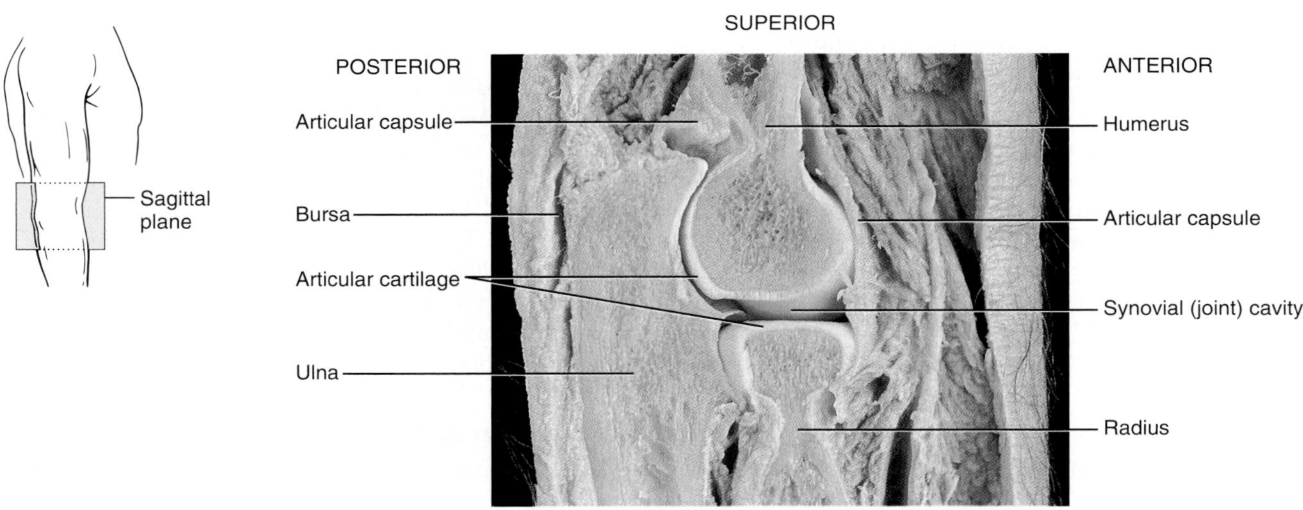

(b) Sagittal section of right elbow joint

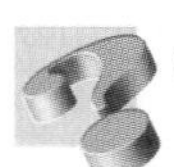

What is the functional classification of synovial joints?

Torn Cartilage and Arthroscopy

The tearing of articular discs (menisci) in the knee, commonly called **torn cartilage,** occurs often among athletes. Such damaged cartilage will begin to wear and may precipitate arthritis unless surgically removed (meniscectomy). Surgical repair of the torn cartilage may be assisted by **arthroscopy** (ar-THROS-kō-pē; *-scopy* = observation). This minimally invasive procedure involves examination of the interior of a joint, usually the knee, with an arthroscope, a lighted pencil-thin instrument used for visualization. Arthroscopy is used to determine the nature and extent of damage following knee injury; to help remove torn cartilage and repair damaged cruciate ligaments in the knee; to help obtain tissue samples for analysis; to monitor the progression of disease and the effects of therapy; and to help perform surgery on other joints, such as the shoulder, elbow, ankle, and wrist.

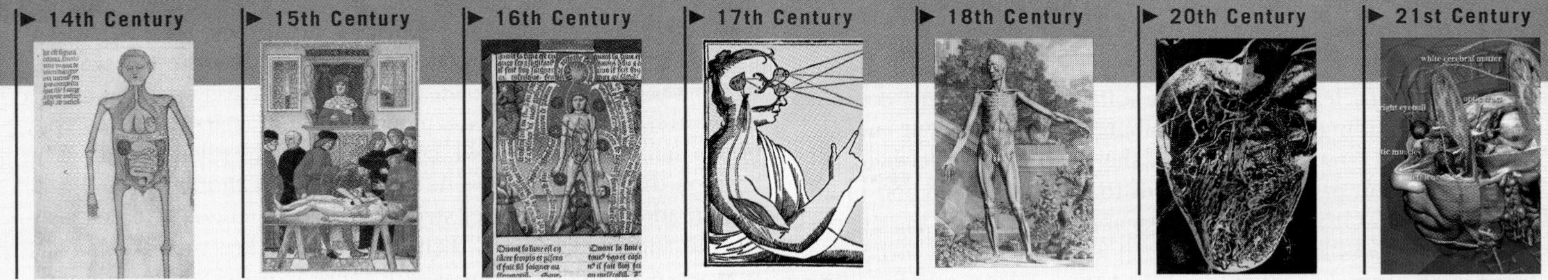

CHANGING IMAGES

Healing—As Art, Science, and Philosophy

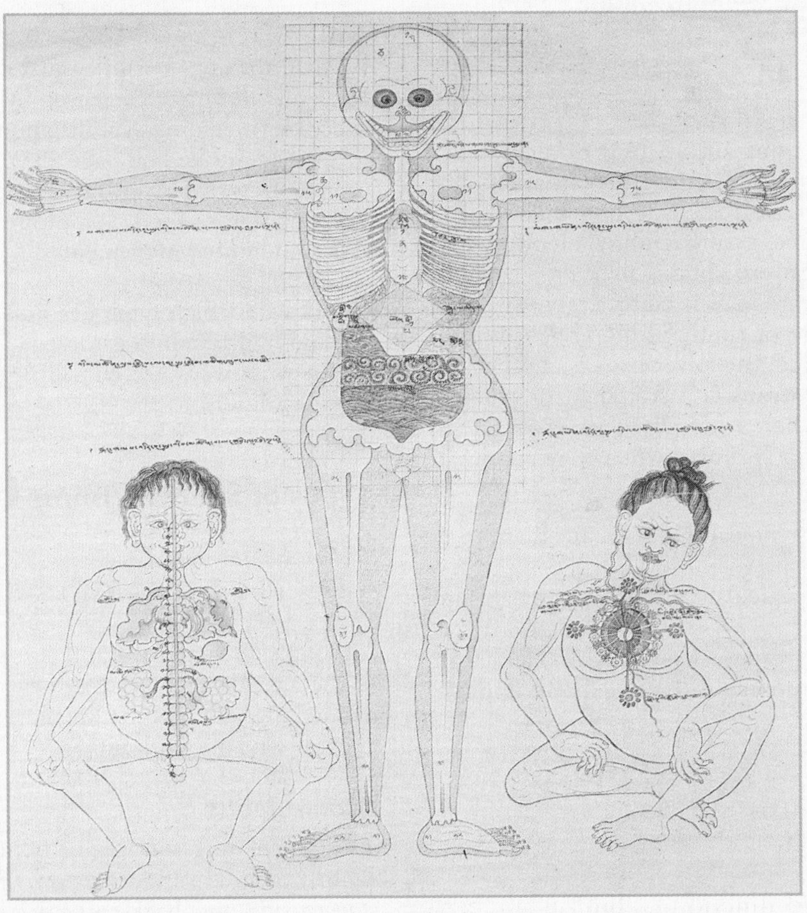

17th Century

Although not well known in the West, the Tibetan science of healing is considered alongside the great medical traditions of the world including the ancient Greek and Chinese. Shown here is an image from the book *Tantra of Secret Instructions on the Eight Branches, the Essence of the Elixir of Immortality*. Also known as the ancient *Four Tantras*, this text is considered the fundamental treatise on Tibetan medicine and contains 77 *Thangkas*,—amazingly detailed and artistically beautiful illustrations. A *Thangka* is often an elaborate wall displayed art form that employs painting, embroidery, and decorative woodwork. The *Thangkas* of the *Four Tantras* are devoted to everything from diet and the source of disease, to thoughts concerning the cure of demonic possession. Based partially upon 10th century medical texts, the *Four Tantras* was authored in 17th century Tibet.

The illustration selected here displays an anterior view of human anatomy. The skeletal system is highlighted, as are the contents of the abdominopelvic cavity. Inset are images of the skull and vertebral column, as are Sanskrit notations. Across the top, iconography represents Tibetan medical and spiritual philosophy.

As you continue to read this chapter devoted to articulations, look closely at the joints in this image. Examine the acromioclavicular, glenohumeral and hip joints. How can you explain such anatomical errors? Could it be that Buddhism, like other world religions, forbids dissection? Or, as was discussed concerning the Islamic skeletal rendering in Chapter 5, is the anatomy meant to be merely figurative? Either way, the blend of art and anatomy in this image is breathtaking.

Nerve and Blood Supply

The nerves that supply a joint are the same as those that supply the skeletal muscles that move the joint. Synovial joints contain many nerve endings that are distributed to the articular capsule and associated ligaments. Some of the nerve endings convey information about pain from the joint to the spinal cord and brain for processing. Other nerve endings are responsive to the degree of movement and stretch at a joint. This information, too, is relayed to the spinal cord and brain, which may respond by sending impulses through different nerves to the muscles to adjust body movements.

Arteries in the vicinity of a synovial joint send out numerous branches that penetrate the ligaments and articular capsule to deliver oxygen and nutrients. Veins remove carbon dioxide and wastes from the joints. The arterial branches from several different arteries typically join together around a joint before penetrating the articular capsule. The articulating portions of a synovial joint receive nourishment from synovial fluid, whereas all other joint tissues are supplied by blood capillaries.

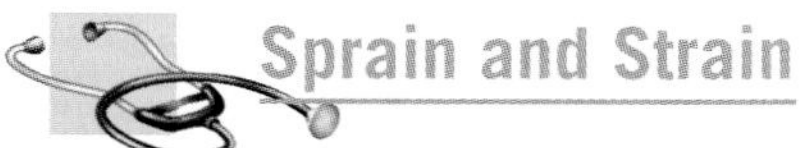

Sprain and Strain

A **sprain** is the forcible wrenching or twisting of a joint that stretches or tears its ligaments but does not dislocate the bones. It occurs when the ligaments are stressed beyond their normal capacity. Sprains also may damage surrounding blood vessels, muscles, tendons, or nerves. Severe sprains may be so painful that the joint cannot be moved. There is considerable swelling, which results from hemorrhage of ruptured blood vessels. The ankle joint is most often sprained; the lower back is another frequent location for sprains. A **strain** is a stretched or partially torn muscle. It often occurs when a muscle contracts suddenly and powerfully—for example, in sprinters when they accelerate too quickly. ■

- ✓ How does the structure of synovial joints permit them to be diarthroses?
- ✓ What are the functions of articular cartilage, synovial fluid, and articular discs?
- ✓ What types of sensations are perceived at joints, and from what sources do joint components receive their nourishment?

Bursae and Tendon Sheaths

Objective

- Describe the structure and function of bursae and tendon sheaths.

The various movements of the body create friction between moving parts. Saclike structures called **bursae** (BER-sē = purses) are strategically situated to alleviate friction in some joints, such as the shoulder and knee joints (see Figures 8.12 and 8.15c). Bursae (singular is *bursa*) are not strictly parts of synovial joints, but they do resemble joint capsules because their walls consist of connective tissue lined by a synovial membrane. They are also filled with a fluid similar to synovial fluid. Bursae are located between the skin and bone in places where skin rubs over bone. They are also found between tendons and bones, muscles and bones, and ligaments and bones. The fluid-filled bursal sacs cushion the movement of one body part over another.

In addition to bursae, structures called tendon sheaths also reduce friction at joints. **Tendon sheaths** are tubelike bursae that wrap around tendons where there is considerable friction. This occurs where tendons pass through synovial cavities, such as the tendon of the biceps brachii muscle at the shoulder joint (see Figure 8.12c). Other areas of considerable friction where tendon sheaths are found are at the wrist and ankle, where many tendons come together in a confined space (see Figure 10.24 on page 332), and in the fingers and toes, where there is a great deal of movement (see Figure 10.20 on page 314).

Bursitis

An acute or chronic inflammation of a bursa is called **bursitis.** The condition may be caused by trauma, by an acute or chronic infection (including syphilis and tuberculosis), or by rheumatoid arthritis (described on page 225). Repeated, excessive exertion of a joint often results in bursitis, with local inflammation and the accumulation of fluid. Symptoms include pain, swelling, tenderness, and limited movement. ■

- ✓ Explain which types of joints are nonaxial, monaxial, biaxial, and multiaxial.
- ✓ In what ways are bursae similar to joint capsules? How do they differ?

Types of Synovial Joints

Objective

- Describe the six subtypes of synovial joints.

Although all synovial joints are similar in structure, the shapes of the articulating surfaces vary. Accordingly, synovial joints are divided into six subtypes: planar, hinge, pivot, condyloid, saddle, and ball-and-socket joints.

Planar Joints

The articulating surfaces of bones in a **planar joint** are flat or slightly curved (Figure 8.4a). Planar joints primarily permit side-to-side and back-and-forth gliding movements (described shortly). These joints are said to be *nonaxial* because the motion they allow does not occur around an axis or plane (line). However, x-ray films made during wrist and ankle movements reveal some rotation of the small carpal and tarsal bones in addition to their predominant gliding movements. Some examples of planar joints are the intercarpal joints (between carpal bones at the wrist); intertarsal joints (between tarsal bones at the ankle); sternoclavicular joints (between the manubrium of the ster-

Figure 8.4 / Subtypes of synovial joints. For each subtype, a drawing of the actual joint and a simplified diagram are shown.

Synovial joints are classified into subtypes on the basis of the shapes of the articulating bone surfaces.

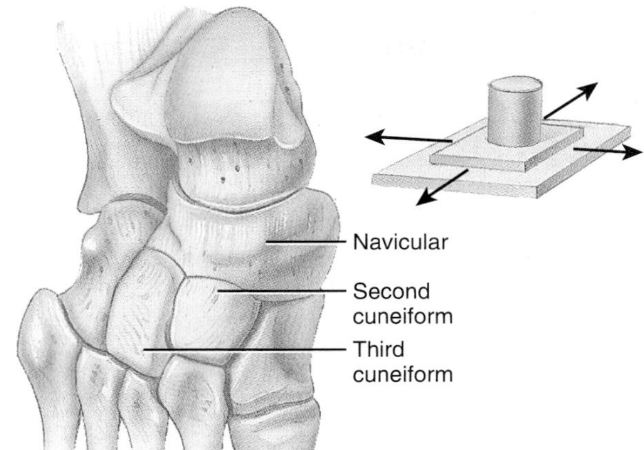

(a) Planar joint between the navicular and second and third cuneiforms of the tarsus in the foot

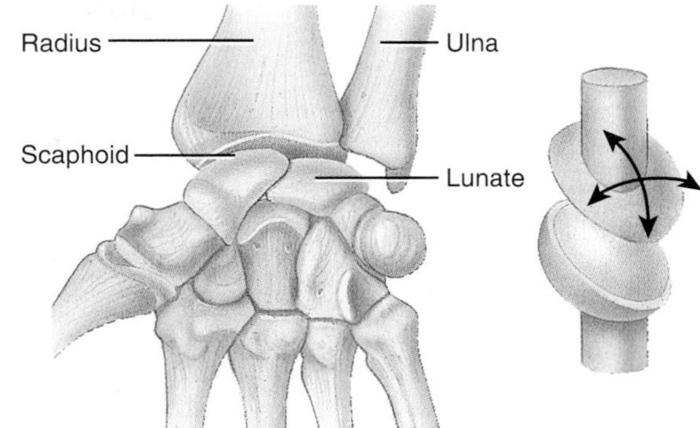

(d) Condyloid joint between radius and scaphoid and lunate bones of the carpus (wrist)

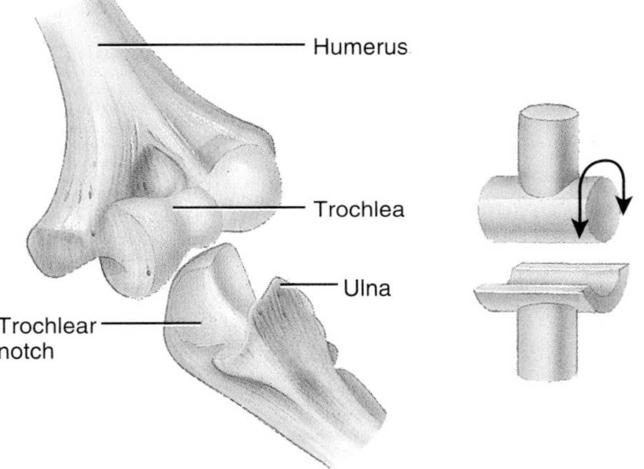

(b) Hinge joint between trochlea of humerus and trochlear notch of ulna at the elbow

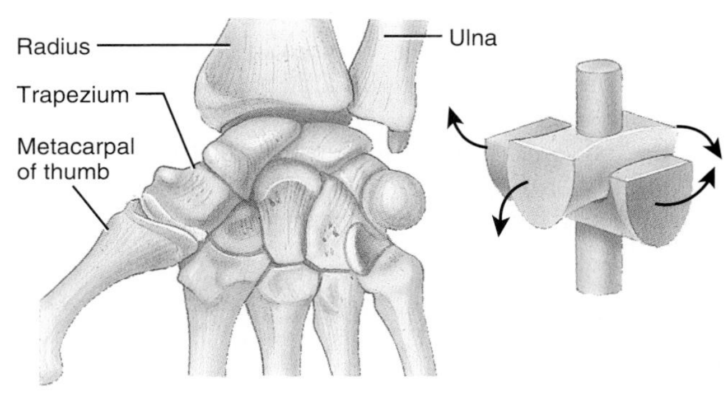

(e) Saddle joint between trapezium of carpus (wrist) and metacarpal of thumb

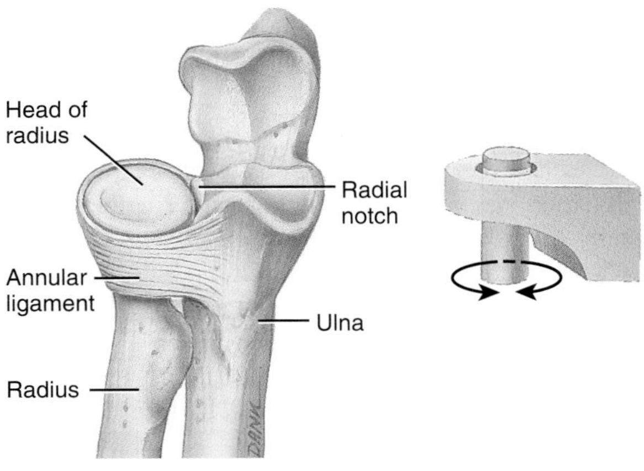

(c) Pivot joint between head of radius and radial notch of ulna

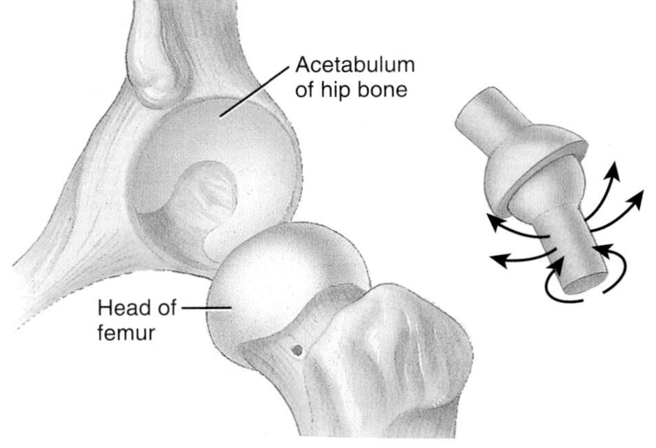

(f) Ball-and-socket joint between head of the femur and acetabulum of the hip bone

Which of the joints shown here are biaxial?

num and the clavicle); acromioclavicular joints (between the acromion of the scapula and the clavicle); sternocostal joints (between the sternum and ends of the costal cartilages at the tips of the second through seventh pairs of ribs); and vertebrocostal joints (between the heads and tubercles of ribs and transverse processes of thoracic vertebrae).

Hinge Joints

In a **hinge joint,** the convex surface of one bone fits into the concave surface of another bone (Figure 8.4b). As the name implies, hinge joints produce an angular, opening-and-closing motion like that of a hinged door. In most joint movements, one bone remains in a fixed position while the other moves around an axis. If a joint permits movement in only one axis, it is said to be *monaxial* (mon-AKS-ē-al) or *uniaxial.* Hinge joints are said to be uniaxial, or monaxial, because they typically allow motion around a single axis. Examples of hinge joints are the knee, elbow, ankle, and interphalangeal joints.

Pivot Joints

In a **pivot joint,** the rounded or pointed surface of one bone articulates with a ring formed partly by another bone and partly by a ligament (Figure 8.4c). A pivot joint is monaxial because it allows rotation around its own longitudinal axis only. Examples of pivot joints are the atlantoaxial joint, in which the atlas rotates around the axis and permits the head to turn from side to side as in signifying "no" (see Figure 8.9a), and the radioulnar joints that enable the palms to turn anteriorly and posteriorly (see Figure 8.10h).

Condyloid Joints

In a **condyloid joint** (KON-di-loyd; *condyl-* = knuckle) or *ellipsoidal* joint, a convex, oval projection of one bone fits into an oval depression in another bone (Figure 8.4d). Examples are the wrist and metacarpophalangeal joints for the second through fifth digits. A condyloid joint is *biaxial* (bī-AKS-ē-al) because it permits movement around two axes. Notice that you can move your index finger both up and down and from side to side.

Saddle Joints

In a **saddle joint,** the articular surface of one bone is saddle-shaped, and the articular surface of the other bone fits into the "saddle" as a sitting rider would sit (Figure 8.4e). A saddle joint is a modified condyloid joint in which the movement is somewhat freer. Saddle joints are *biaxial,* producing side-to-side and up-and-down movements. An example of a saddle joint is the carpometacarpal joint between the trapezium of the carpus and metacarpal of the thumb.

Ball-and-Socket Joints

A ball-and-socket joint consists of the ball-like surface of one bone fitting into a cuplike depression of another bone (Figure 8.4f). Such joints are *multiaxial (polyaxial)* because they permit movement around three axes plus all directions in between. Examples of functional ball-and-socket joints are the shoulder and hip joints. At the shoulder joint the head of the humerus fits into the glenoid cavity of the scapula. At the hip joint the head of the femur fits into the acetabulum of the hip bone.

Table 8.1 summarizes the structural and functional categories of joints.

TYPES OF MOVEMENTS AT SYNOVIAL JOINTS

Objective

- Describe the types of movements that can occur at synovial joints.

Anatomists, physical therapists, and kinesiologists use specific terminology to designate movements that can occur at synovial joints. These precise terms may indicate the form of motion, the direction of movement, or the relationship of one body part to another during movement. Movements at synovial joints are grouped into four main categories: (1) gliding, (2) angular movements, (3) rotation, and (4) special movements. This last category includes movements that occur only at certain joints.

Gliding

Gliding is a simple movement in which relatively flat bone surfaces move back and forth and from side to side with respect to one another (Figure 8.5). There is no significant alteration of the angle between the bones. Gliding movements are limited in range due to the structure of the articular capsule and associated ligaments and bones. Gliding occurs at planar joints.

Angular Movements

In **angular movements,** there is an increase or a decrease in the angle between articulating bones. The principal angular movements are flexion, extension, lateral extension, hyperextension, abduction, adduction, and circumduction. These movements are discussed with respect to the body in the anatomical position.

Flexion, Extension, Lateral Flexion, and Hyperextension

Flexion and extension are opposite movements. In **flexion** (FLEK-shun; *flex-* = to bend) there is a decrease in the angle between articulating bones; in **extension** (eks-TEN-shun; *exten-* = to stretch out) there is an increase in the angle between articulating bones, often to restore a part of the body to the anatomical position after it has been flexed (Figure 8.6 on page 206). Both movements usually occur in the sagittal plane. Hinge, pivot, condyloid, saddle, and ball-and-socket joints all permit flexion and extension.

All of the following are examples of flexion:

- bending the head toward the chest at the atlanto-occipital joint between the atlas (the first vertebra) and the occipital bone of the skull, and at the cervical intervertebral joints between the cervical vertebrae (Figure 8.6a)
- bending the trunk forward at the intervertebral joints

Table 8.1 Summary of Structural Categories and Functional Classification of Joints

Structural Category	Description	Functional Classification	Example
Fibrous	Articulating bones held together by fibrous connective tissue; no synovial cavity.		
Suture	Articulating bones united by a thin layer of dense fibrous connective tissue; found between bones of the skull. With age, some sutures are replaced by a synostosis, in which the bones fuse across the former suture.	Synarthrosis (immovable).	Frontal suture.
Syndesmosis	Articulating bones united by dense fibrous connective tissue, either a ligament or an interosseous membrane.	Amphiarthrosis (slightly movable).	Distal tibiofibular joint.
Gomphosis	Articulating bones united by periodontal ligament; cone-shaped peg fits into a socket.	Synarthrosis.	At roots of teeth in sockets of maxillae and mandible.
Cartilaginous	Articulating bones united by cartilage; no synovial cavity.		
Synchondrosis	Connecting material is hyaline cartilage; becomes a synostosis when bone elongation ceases.	Synarthrosis.	Epiphyseal plate at joint between the diaphysis and epiphysis of a long bone.
Symphysis	Connecting material is a broad, flat disc of fibrocartilage.	Amphiarthrosis.	Intervertebral joints (between adjacent vertebral bodies), pubic symphysis, and junction of manubrium with body of sternum.
Synovial	Characterized by a synovial cavity, articular cartilage, and an articular capsule; may contain accessory ligaments, articular discs, and bursae.		
Planar	Articulated surfaces are flat or slightly curved.	Nonaxial diarthrosis (freely movable); gliding motion.	Intercarpal, intertarsal, sternocostal (between sternum and the 2nd–7th pairs of ribs), and vertebrocostal joints.
Hinge	Convex surface fits into a concave surface.	Monaxial diarthrosis; angular motion.	Elbow, ankle, and interphalangeal joints.
Pivot	Rounded or pointed surface fits into a ring formed partly by bone and partly by a ligament.	Monaxial diarthrosis; rotation.	Atlantoaxial and proximal radioulnar joints.
Condyloid	Oval-shaped projection fits into an oval-shaped depression.	Biaxial diarthrosis; angular motion.	Radiocarpal and most metacarpophalangeal joints.
Saddle	Articular surface of one bone is saddle-shaped, and the articular surface of the other bone "sits" in the saddle.	Biaxial diarthrosis; angular motion.	Carpometacarpal joint between trapezium and first metacarpal.
Ball-and-socket	Ball-like surface fits into a cuplike depression.	Multiaxial diarthrosis; angular motion and rotation.	Shoulder and hip joints.

Figure 8.5 / Gliding movements at synovial joints.

Gliding motion consists of side-to-side and back-and-forth movements.

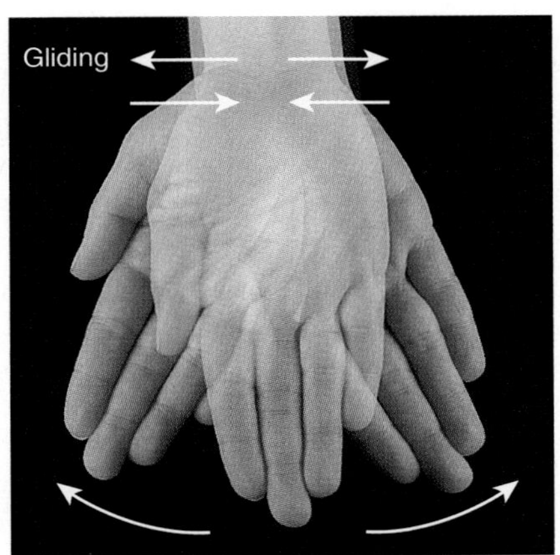

Intercarpal joints

Which types of synovial joints exhibit gliding movements?

Figure 8.6 / Angular movements at synovial joints—flexion, extension, hyperextension, and lateral flexion.

In angular movements, there is an increase or decrease in the angle between articulating bones.

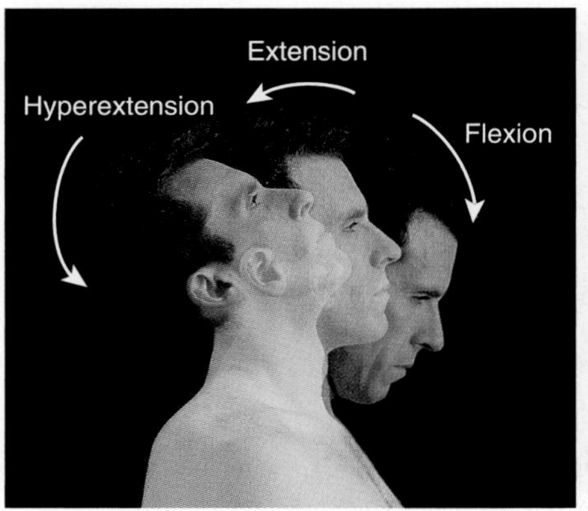

(a) Atlanto-occipital and cervical intervertebral joints

(b) Shoulder joint

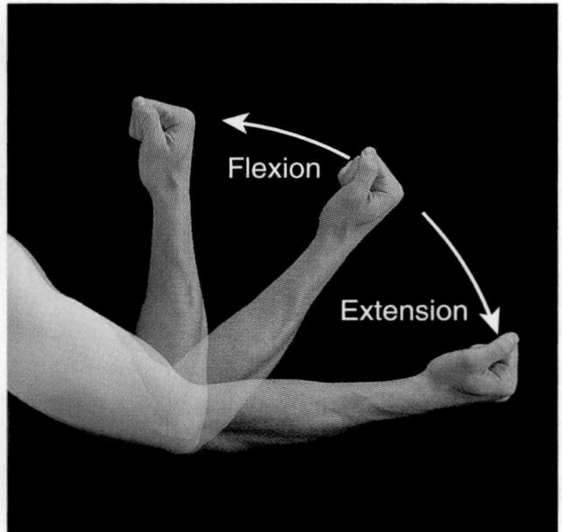

(c) Elbow joint

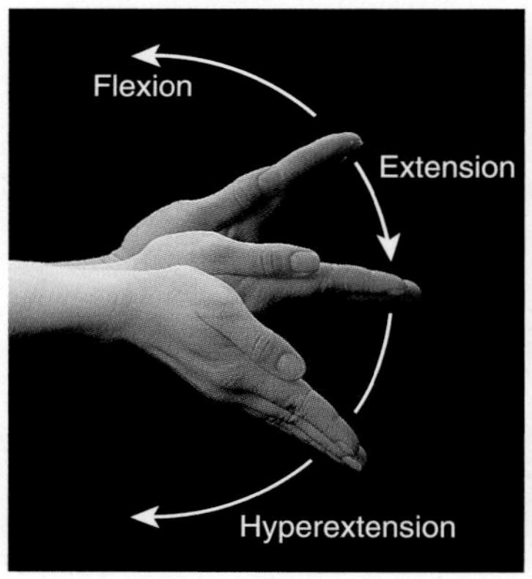

(d) Wrist joint

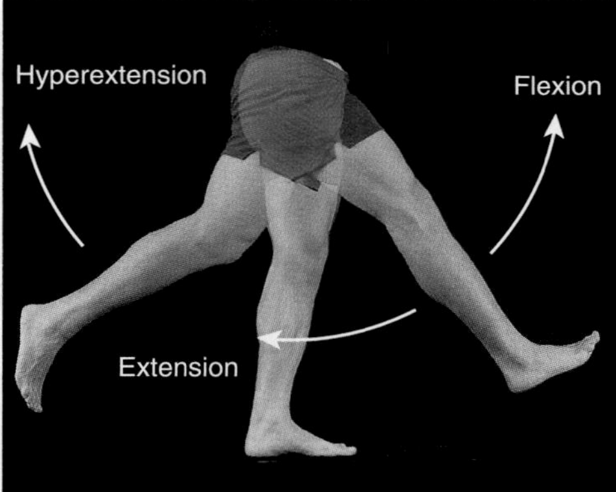

(e) Hip joint

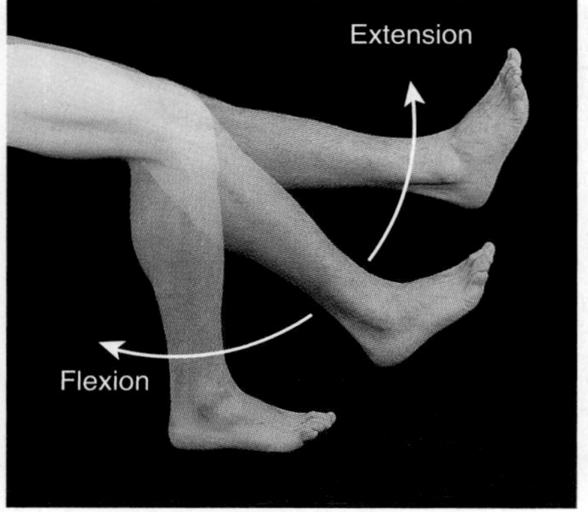

(f) Knee joint

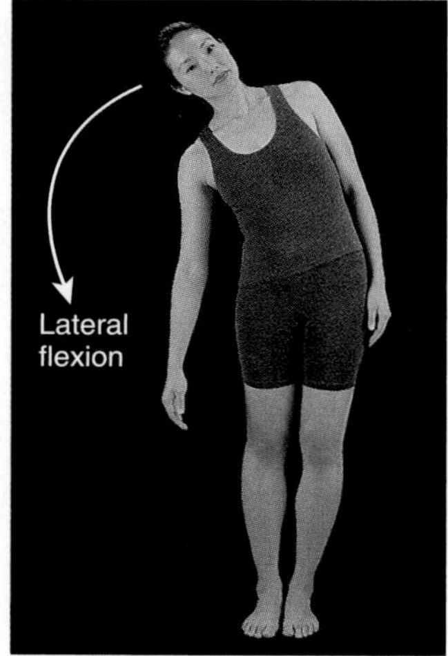

(g) Intervertebral joints

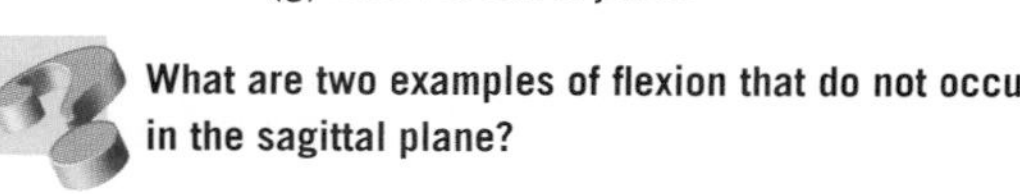

What are two examples of flexion that do not occur in the sagittal plane?

- moving the humerus forward at the shoulder joint, as in swinging the arms forward while walking (Figure 8.6b)
- moving the forearm toward the arm at the elbow joint between the humerus, ulna, and radius (Figure 8.6c)
- moving the palm toward the forearm at the wrist or radiocarpal joint between the radius and carpals (Figure 8.6d)
- bending the digits of the hand or feet at the interphalangeal joints between phalanges
- moving the femur forward at the hip joint between the femur and hip bone, as in walking (Figure 8.6e)
- moving the leg toward the thigh at the knee or tibiofemoral joint between the tibia, femur, and patella, as occurs when bending the knee (Figure 8.6f)

Although flexion and extension usually occur in the sagittal plane, there are a few exceptions. For example, flexion of the thumb involves movement of the thumb medially across the palm at the carpometacarpal joint between the trapezium and metacarpal of the thumb, as in touching the thumb to the oppo-

Figure 8.7 / Angular movements at synovial joints—abduction and adduction.

Abduction and adduction usually occur in the frontal plane.

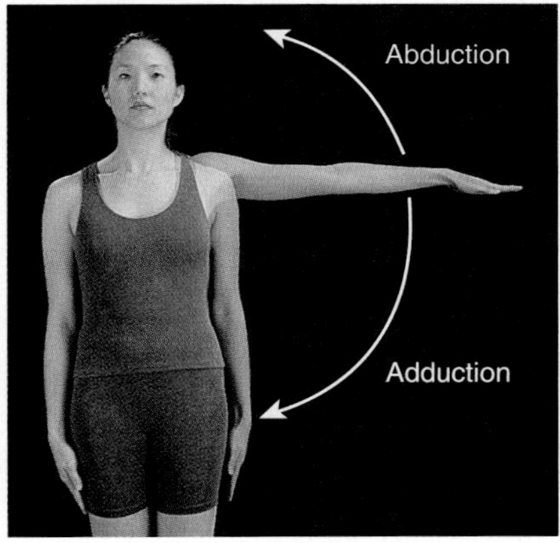

(a) Shoulder joint

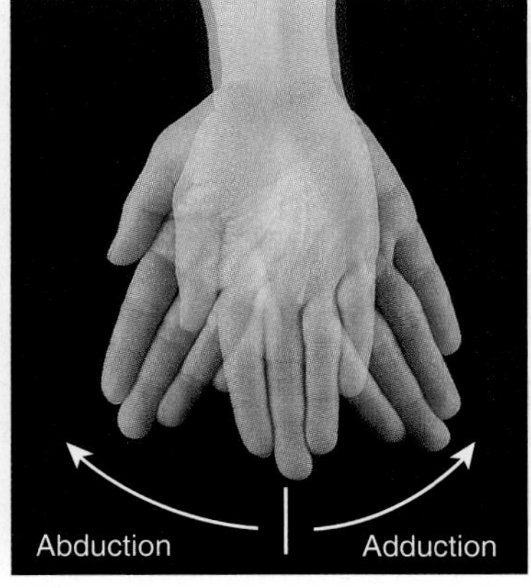

(b) Wrist joint

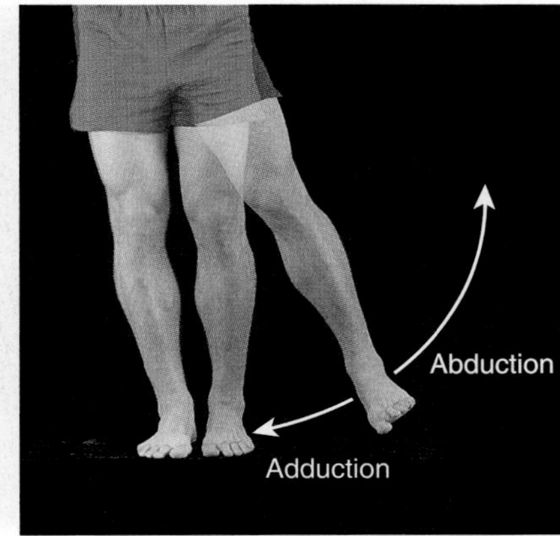

(c) Hip joint

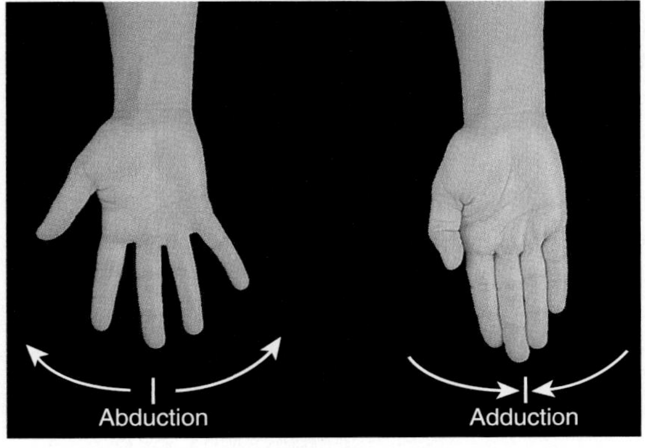

(d) Metacarpophalangeal joints of the fingers (not the thumb)

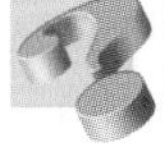

In what way is considering adduction as "adding your limb to your trunk" an effective learning device?

site side of the palm. Another example is movement of the trunk sideways to the right or left at the waist. This movement, which occurs in the frontal plane and involves the intervertebral joints, is called **lateral flexion** (Figure 8.6g).

Continuation of extension beyond the anatomical position is called **hyperextension** (*hyper-* = beyond or excessive). Examples of hyperextension include:

- bending the head backward at the atlanto-occipital and cervical intervertebral joints (Figure 8.6a)
- bending the trunk backward at the intervertebral joints
- moving the humerus backward at the shoulder joint, as in swinging the arms backward while walking (Figure 8.6b)
- moving the palm backward at the wrist joint (Figure 8.6d)
- moving the femur backward at the hip joint, as in walking (Figure 8.6e)

Hyperextension of hinge joints, such as the elbow, interphalangeal, and knee joints, is usually prevented by factors such as the arrangement of ligaments and the anatomical alignment of the bones.

Abduction, Adduction, and Circumduction

Abduction (ab-DUK-shun; *ab-* = away; *-duct* = to lead) is the movement of a bone away from the midline, whereas **adduction** (ad-DUK-shun; *ad-* = toward) is the movement of a bone toward the midline. Both movements usually occur in the frontal plane. Condyloid, saddle, and ball-and-socket joints permit abduction and adduction. Examples of abduction include moving the humerus laterally at the shoulder joint, moving the palm laterally at the wrist joint, and moving the femur laterally at the hip joint (Figure 8.7a–c). The movement that returns each of these body parts to the anatomical position is adduction (Figure 8.7a–c).

Figure 8.8 / Angular movements at synovial joints—circumduction.

Circumduction is the movement of the distal end of a body part in a circle.

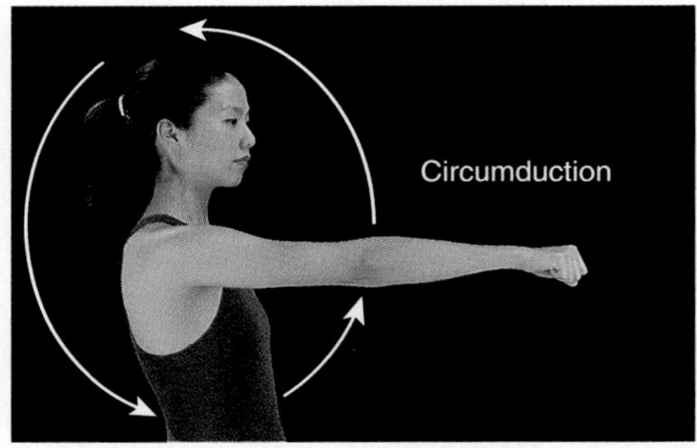

(a) Shoulder joint

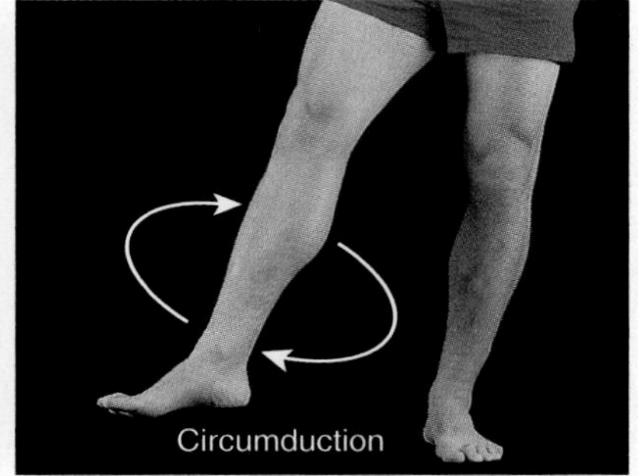

(b) Hip joint

Which movements in continuous sequence produce circumduction?

With respect to the digits, the midline of the body is not used as a point of reference for abduction and adduction. In abduction of the fingers (but not the thumb), an imaginary line is drawn through the longitudinal axis of the middle (longest) finger, and the fingers move away (spread out) from the middle finger (Figure 8.7d). In abduction of the thumb, the thumb moves away from the palm in the sagittal plane (see Figure 10.20f on page 315). Abduction of the toes is relative to an imaginary line drawn through the second toe. Adduction of the fingers and toes involves returning them to the anatomical position. Adduction of the thumb moves the thumb toward the palm in the sagittal plane (see Figure 10.20f).

Circumduction (ser-kum-DUK-shun; *circ-* = circle) is movement of the distal end of a body part in a circle (Figure 8.8). Circumduction occurs as a result of a continuous sequence of flexion, abduction, extension, and adduction. Examples of circumduction are moving the humerus in a circle at the shoulder joint (Figure 8.8a), moving the hand in a circle at the wrist joint, moving the thumb in a circle at the carpometacarpal joint, moving the fingers in a circle at the metacarpophalangeal joints (between the metacarpals and phalanges), and moving the femur in a circle at the hip joint (Figure 8.8b). Both the shoulder and hip joints permit circumduction, but flexion, abduction, extension, and adduction are more limited in the hip joints than the shoulder joints due to the tension on certain ligaments and muscles (see Exhibits 8.2 and 8.4).

Rotation

In **rotation** (rō-TĀ-shun; *rota-* = revolve), a bone revolves around its own longitudinal axis. Pivot and ball-and-socket joints permit rotation. One example is turning the head from side to side at the atlanto-axial joint, as in signifying "no" (Figure 8.9a). Another is turning the trunk from side to side at the interverte-

Figure 8.9 / Rotation at synovial joints.

In rotation, a bone revolves around its own longitudinal axis.

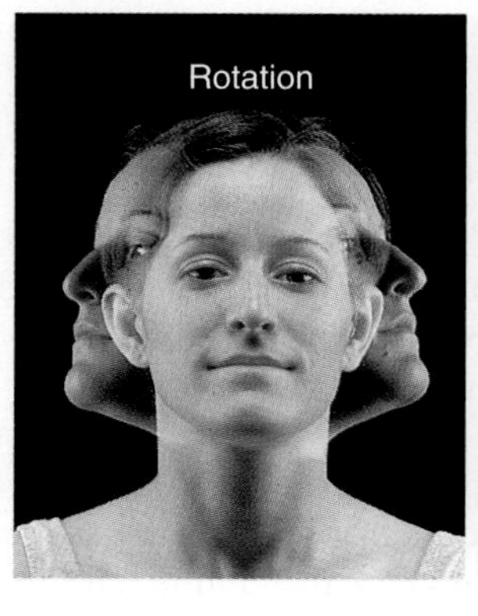

(a) Atlanto-axial joint

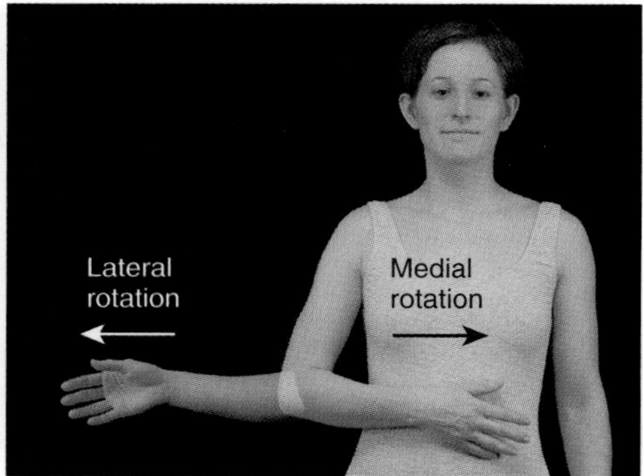

(b) Shoulder joint

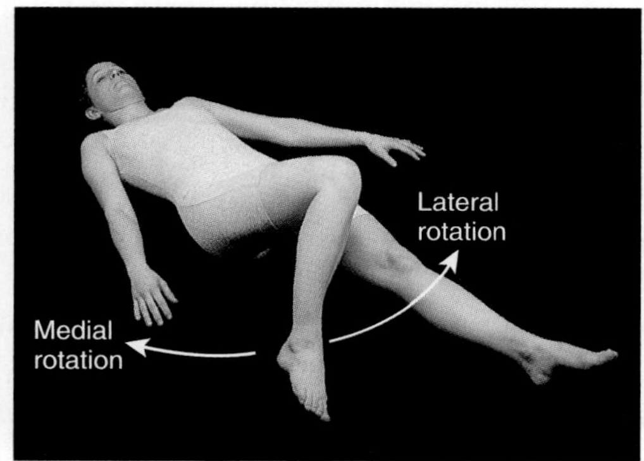

(c) Hip joint

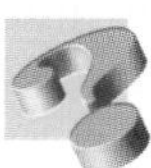

How do medial and lateral rotation differ?

bral joints while keeping the hips and lower limbs in the anatomical position. In the limbs, rotation is defined relative to the midline, and specific qualifying terms are used. If the anterior surface of a bone of the limb is turned toward the midline, the movement is called *medial (internal) rotation.* You can medially rotate the humerus at the shoulder joint as follows: Starting in the anatomical position, flex your elbow and then draw your palm across the chest (Figure 8.9b). Medial rotation of the forearm at the radioulnar joints (between the radius and ulna) involves turning the palm medially from the anatomical position (see Figure 8.10h). You can medially rotate the femur at the hip joint as follows: Lie on your back, bend your knee, and then move your leg and foot laterally from the midline. Although you are moving your leg and foot laterally, the femur is rotating medially (Figure 8.9c). Medial rotation of the leg at the knee joint can be produced by sitting on a chair, bending your knee, raising your lower limb off the floor, and turning your toes medially. If the anterior surface of the bone of a limb is turned away from the midline, the movement is called *lateral (external) rotation* (see Figure 8.9b, c).

Special Movements

As noted previously, **special movements** occur only at certain joints. They include elevation, depression, protraction, retraction, inversion, eversion, dorsiflexion, plantar flexion, supination, pronation, and opposition (Figure 8.10):

- **Elevation** (el-e-VĀ-shun; = to lift up) in an upward movement of a part of the body, such as closing the mouth at the temporomandibular joint (between the mandible and temporal bone) to elevate the mandible (Figure 8.10a) or shrugging the shoulders at the acromioclavicular joint to elevate the scapula.
- **Depression** (dē-PRESH-un; = to press down) is a downward movement of a part of the body, such as opening the mouth to depress the mandible (Figure 8.10b) or returning shrugged shoulders to the anatomical position to depress the scapula.
- **Protraction** (prō-TRAK-shun; = to draw forth) is a movement of a part of the body anteriorly in the transverse plane. You can protract your mandible at the temporomandibular joint by thrusting it outward (Figure 8.10c) or protract your clavicles at the acromioclavicular and sternoclavicular joints by crossing your arms.
- **Retraction** (rē-TRAK-shun; = to draw back) is a movement of a protracted part of the body back to the anatomical position (Figure 8.10d).
- **Inversion** (in-VER-zhun; = to turn inward) is movement of the soles medially at the intertarsal joints (between the tarsals) (Figure 8.10e).
- **Eversion** (ē-VER-zhun; = to turn outward) is a movement of the soles laterally at the intertarsal joints (Figure 8.10f).

Figure 8.10 / Special movements at synovial joints.

Special movements occur only at certain synovial joints.

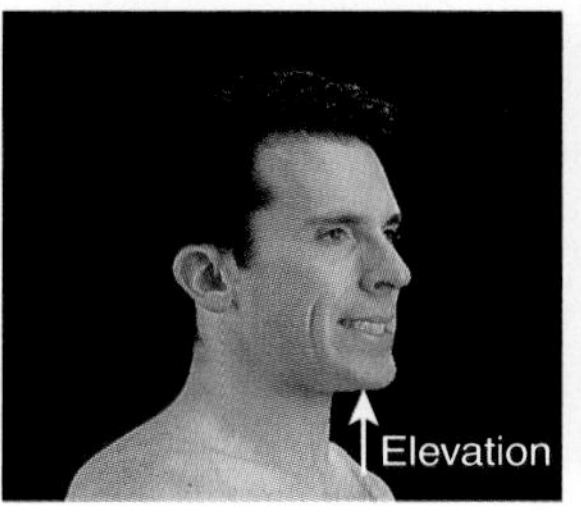

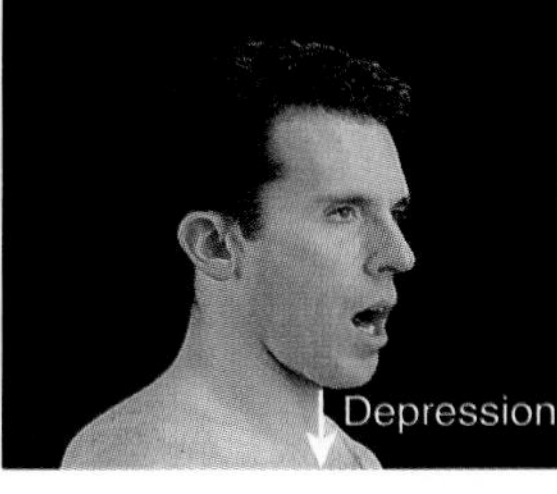

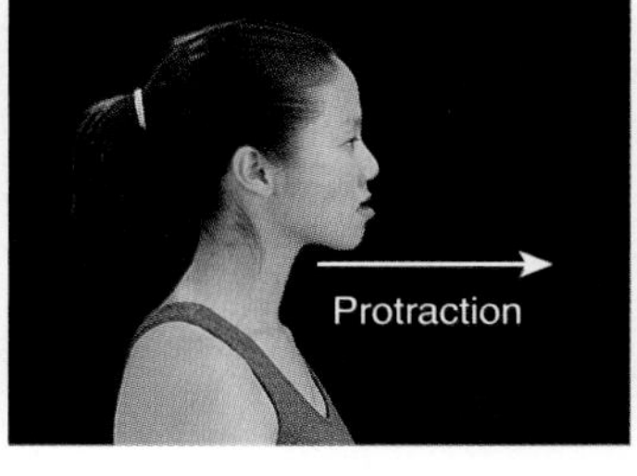

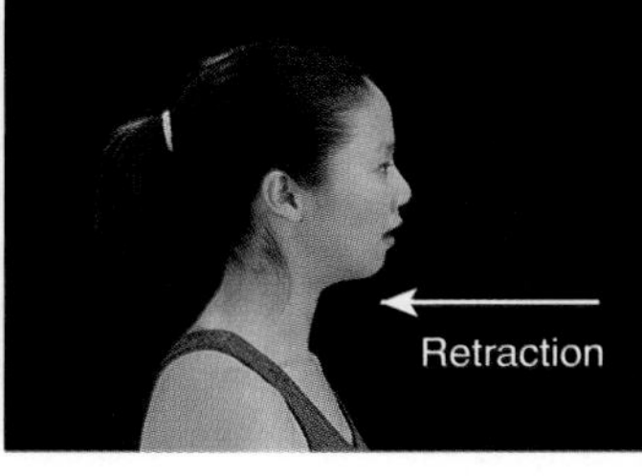

(a) Temporomandibular joint (b) (c) Temporomandibular joint (d)

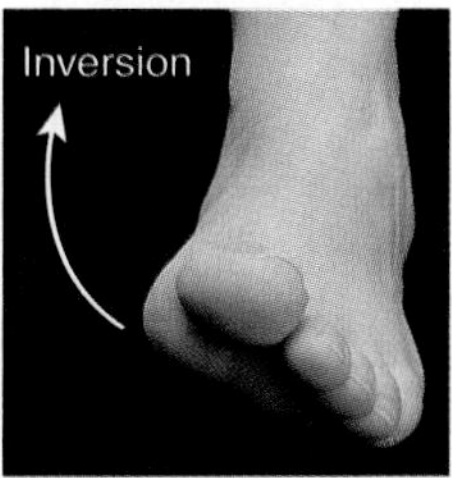

Eversion

Dorsiflexion

Plantar flexion

(e) Intertarsal joints (f) (g) Ankle joint

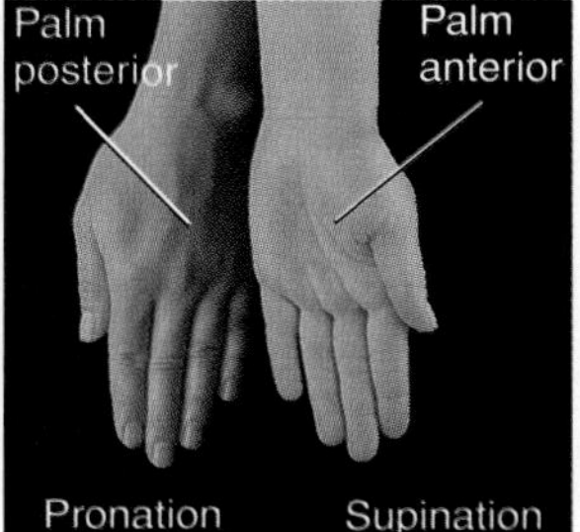

(h) Radioulnar joint

What movement of the shoulder girdle is involved in bringing the arms forward until the elbows touch?

- **Dorsiflexion** (dor-si-FLEK-shun) refers to bending of the foot at the ankle or talocrural joint (between the tibia, fibula, and talus) in the direction of the dorsum (superior surface) (Figure 8.10g). Dorsiflexion occurs when you stand on your heels.
- **Plantar flexion** involves bending of the foot at the ankle joint in the direction of the plantar or inferior surface (see Figure 8.10g), as when standing on your toes. Whereas dorsiflexion is true flexion, plantar flexion is true extension.
- **Supination** (sū-pi-NĀ-shun) is a movement of the forearm at the proximal and distal radioulnar joints in which the palm is turned anteriorly or superiorly (Figure 8.10h). This position of the palms is one of the defining features of the anatomical position.
- **Pronation** (prō-NĀ-shun) is a movement of the forearm at the proximal and distal radioulnar joints in which the distal end of the radius crosses over the distal end of the ulna and the palm is turned posteriorly or inferiorly (see Figure 8.10h).
- **Opposition** (op-ō-ZISH-un) is the movement of the thumb at the carpometacarpal joint (between the trapezium and metacarpal of the thumb) in which the thumb moves across the palm to touch the tips of the fingers on the same hand (see Figure 10.20f on page 315). This is the distinctive digital movement that gives humans and other primates the ability to grasp and manipulate objects very precisely.

A summary of the movements that occur at synovial joints is presented in Table 8.2.

✓ On yourself or with a partner, demonstrate each movement listed in Table 8.2.

SELECTED JOINTS OF THE BODY

In this section of the text we examine in detail four selected joints of the body in a series of exhibits. Each exhibit considers a specific synovial joint and contains (1) a definition—a description of the type of joint and the bones that form the joint; (2) the anatomical components—a description of the major connecting ligaments, articular disc, articular capsule, and other distinguishing features of the joint; and (3) the joint's possible movements. Each exhibit also refers you to a figure that illustrates the joint. The joints described are the temporomandibular joint (TMJ), shoulder (humeroscapular or glenohumeral) joint, elbow joint, hip (coxal) joint, and knee (tibiofemoral) joint.

In previous chapters we discussed the major bones and their markings. In this chapter we have examined how joints are classified according to their structure and function, and we have introduced the movements that occur at joints. Please refer to Table 8.3, which will help you integrate the information you learned about bones with the information about joint classifications and movements. This table lists some of the major joints of

Table 8.2 Summary of Movements at Synovial Joints

Movement	Description
Gliding	Movement of relatively flat bone surfaces back and forth and from side to side over one another; little change in the angle between bones.
Angular	Increase or decrease in the angle between bones.
Flexion	Decrease in the angle between articulating bones, usually in the sagittal plane.
Lateral flexion	Movement of the spine in the frontal plane.
Extension	Increase in the angle between articulating bones, usually in the sagittal plane.
Hyperextension	Extension beyond the anatomical position.
Abduction	Movement of a bone away from the midline, usually in the frontal plane.
Adduction	Movement of a bone toward the midline, usually in the frontal plane.
Circumduction	Flexion, abduction, extension, and adduction in succession, in which the distal end of a body part moves in a circle.
Rotation	Movement of a bone around its longitudinal axis; in the limbs, it may be medial (toward midline) or lateral (away from midline).
Special	Occurs at specific joints.
Elevation	Superior movement of a body part.
Depression	Inferior movement of a body part.
Protraction	Anterior movement of a body part in the transverse plane.
Retraction	Posterior movement of a body part in the transverse plane.
Inversion	Medial movement of the soles so that they face each other.
Eversion	Lateral movement of the soles so that they face away from each other.
Dorsiflexion	Bending the foot in the direction of the dorsum (superior surface).
Plantar flexion	Bending the foot in the direction of the plantar surface (sole).
Supination	Movement of the forearm that turns the palm anteriorly or superiorly.
Pronation	Movement of the forearm that turns the palm posteriorly or inferiorly.
Opposition	Movement of the thumb across the palm to touch fingertips on the same hand.

Table 8.3 Selected Joints According to Articular Components, Structural and Functional Classification, and Movements

Joint	Articular Components	Classification	Movements
Axial Skeleton			
Suture	Between skull bones.	*Structural:* fibrous. *Functional:* synarthrosis.	None.
Atlanto-occipital	Between superior articular facets of atlas and occipital condyles of occipital bone.	*Structural:* synovial (condyloid). *Functional:* diarthrosis.	Flexion and extension of head and slight lateral flexion of head to either side.
Atlanto-axial	(1) Between dens of axis and anterior arch of atlas and (2) between lateral masses of atlas and axis.	*Structural:* synovial (pivot) between dens and anterior arch, and synovial (planar) between lateral masses. *Functional:* diarthrosis.	Rotation of head.
Intervertebral	(1) Between vertebral bodies and (2) between vertebral arches.	*Structural:* cartilaginous (symphysis) between vertebral bodies, and synovial (planar) between vertebral arches. *Functional:* amphiarthrosis between vertebral bodies, and diarthrosis between vertebral arches.	Flexion, extension, lateral flexion, and rotation of vertebral column.
Vertebrocostal	(1) Between facets of heads of ribs and facets of bodies of adjacent thoracic vertebrae and intervertebral discs between them and (2) between articular part of tubercles of ribs and facets of transverse processes of thoracic vertebrae.	*Structural:* synovial (planar). *Functional:* diarthrosis.	Slight gliding.
Sternocostal	Between sternum and first seven pairs of ribs.	*Structural:* cartilaginous (synchondrosis) between sternum and first pair of ribs, and synovial (planar) between sternum and second through seventh pairs of ribs. *Functional:* synarthrosis between sternum and first pair of ribs, and diarthrosis between sternum and second through seventh pairs of ribs.	None between sternum and first pair of ribs; slight gliding between sternum and second through seventh pairs of ribs.
Lumbosacral	(1) Between body of fifth lumbar vertebra and base of sacrum and (2) between inferior articular facets of fifth lumbar vertebra and superior articular facets of first vertebra of sacrum.	*Structural:* cartilaginous (symphysis) between body and base, and synovial (planar) between articular facets. *Functional:* amphiarthrosis between body and base, and diarthrosis between articular facets.	Flexion, extension, and lateral flexion of vertebral column.
Pectoral Girdles and Upper Limbs			
Sternoclavicular	Between sternal end of clavicle, manubrium of sternum, and first costal cartilage.	*Structural:* synovial (planar and pivot). *Functional:* diarthrosis.	Gliding, with limited movements in nearly every direction.
Acromioclavicular	Between acromion of scapula and acromial end of clavicle.	*Structural:* synovial (planar). *Functional:* diarthrosis.	Gliding and rotation of scapula on clavicle.
Radioulnar	Proximal radioulnar joint between head of radius and radial notch of ulna; distal radioulnar joint between ulnar notch of radius and head of ulna.	*Structural:* synovial (pivot). *Functional:* diarthrosis.	Rotation of forearm.
Wrist (radiocarpal)	Between distal end of radius and scaphoid, lunate, and triquetrum of carpus.	*Structural:* synovial (condyloid). *Functional:* diarthrosis.	Flexion, extension, abduction, adduction, circumduction, and slight hyperextension of wrist.
Intercarpal	Between proximal row of carpal bones, distal row of carpal bones, and between both rows of carpal bones (midcarpal joints).	*Structural:* synovial (planar), except for hamate, scaphoid, and lunate (midcarpal) joint, which is synovial (saddle). *Functional:* diarthrosis.	Gliding plus flexion and abduction at midcarpal joints.

(continues)

Table 8.3 Selected Joints According to Articular Components, Structural and Functional Classification, and Movements (continued)

Joint	Articular Components	Classification	Movements
Pectoral Girdles and Upper Limbs			
Carpometacarpal	Carpometacarpal joint of thumb between trapezium of carpus and first metacarpal; carpometacarpal joints of remaining digits formed between carpus and second through fifth metacarpals.	*Structural:* synovial (saddle) at thumb and synovial (planar) at remaining digits. *Functional:* diarthrosis.	Flexion, extension, abduction, adduction, and circumduction at base of first metacarpal and gliding at remaining digits.
Metacarpophalangeal and metatarsophalangeal	Between heads of metacarpals (or metatarsals) and bases of proximal phalanges.	*Structural:* synovial (condyloid). *Functional:* diarthrosis.	Flexion, extension, abduction, adduction, and circumduction of phalanges.
Interphalangeal	Between heads of phalanges and bases of more distal phalanges.	*Structural:* synovial (hinge). *Functional:* diarthrosis.	Flexion and extension of phalanges.
Pelvic Girdle and Lower Limbs			
Sacroiliac	Between auricular surfaces of sacrum and ilia of hip bones.	*Structural:* synovial (planar). *Functional:* diarthrosis.	Slight gliding (even more so during pregnancy).
Pubic symphysis	Between anterior surfaces of hip bones.	*Structural:* cartilaginous (symphysis). *Functional:* amphiarthrosis.	Slight movements (even more so during pregnancy).
Tibiofibular	Proximal tibiofibular joint between lateral condyle of tibia and head of fibula; distal tibiofibular joint between distal end of fibula and fibular notch of tibia.	*Structural:* synovial (planar) at proximal joint, and fibrous (syndesmosis) at distal joint. *Functional:* diarthrosis at proximal joint, and amphiarthrosis at distal joint.	Slight gliding at proximal joint, and slight rotation of fibula during dorsiflexion of foot.
Ankle (talocrural)	(1) Between distal end of tibia and its medial malleolus and talus and (2) between lateral malleolus of fibula and talus.	*Structural:* synovial (hinge). *Functional:* diarthrosis.	Dorsiflexion and plantar flexion of foot.
Intertarsal	Subtalar joint between talus and calcaneus of tarsus; talocalcaneonavicular joint between talus and calcaneus and navicular of tarsus; calcaneocuboid joint between calcaneus and cuboid of tarsus.	*Structural:* synovial (planar) at subtalar and calcaneocuboid joints, and synovial at talocalcaneonavicular joint. *Functional:* diarthrosis.	Inversion, eversion, abduction, and adduction of foot.
Tarsometatarsal	Between three cuneiforms of tarsus and bases of five metatarsal bones.	*Structural:* synovial (planar). *Functional:* diarthrosis.	Slight gliding.

the body according to their articular components (the bones that enter into their formation), their structural and functional classification, and the type(s) of movement that occurs at each joint. Because the temporomandibular, shoulder, elbow, hip, and knee joints are described in Exhibits 8.1 through 8.5, they are not included in Table 8.3.

Exhibit 8.1 Temporomandibular Joint or TMJ (Figure 8.11)

Objective

► Describe the anatomical components of the temporomandibular joint and explain the movements that can occur.

Definition

► Combined hinge and gliding joint formed by the condylar process of the mandible and the mandibular fossa and articular tubercle of the temporal bone. The two temporomandibular joints are the only movable joints between skull bones; all other skull joints are sutures and therefore immovable.

Anatomical Components

1. *Articular disc (meniscus).* Fibrocartilage disc that separates the joint cavity into a superior and inferior compartment, each with a synovial membrane.
2. *Articular capsule.* Thin, fairly loose envelope around the circumference of the joint.
3. *Lateral ligament.* Two short bands on the lateral surface of the articular capsule that extend inferiorly and posteriorly from the inferior border and tubercle of the zygomatic process of the temporal bone to the lateral and posterior aspect of the neck of the mandible. The lateral ligament is covered by the parotid gland and helps prevent displacement of the mandible.
4. *Sphenomandibular ligament.* Thin band that extends inferiorly and anteriorly from the spine of the sphenoid bone to the ramus of the mandible.
5. *Stylomandibular ligament.* Thickened band of deep cervical fascia that extends from the styloid process of the temporal bone to the inferior and posterior border of the ramus of the mandible. The ligament separates the parotid from the submandibular gland.

Movements

► Only the mandible moves because the maxilla is firmly anchored to other bones by sutures. Accordingly, the mandible may function in depression (jaw opening) and elevation (jaw closing), which occurs in the inferior compartment and protraction, retraction, lateral displacement, and slight rotation, which occur in the superior compartment (see Figure 8.10a–d).

Dislocated Mandible

A **dislocation,** or *luxation* (luks-Ā-shun; *luxation* = dislocation) is the displacement of a bone from a joint with tearing of ligaments, tendons, and articular capsules. It is usually caused by a blow or fall, although unusual physical effort may be a factor. For example, during yawning or taking a large bite, the condylar processes of the mandible may pass anterior to the articular tubercles.This results in a **dislocated mandible** (anterior displacement). When the mandible is displaced in this manner, the mouth remains wide open and the person is unable to close it. This may be corrected by pressing the thumbs downward on the lower molar teeth and pushing the mandible backward. Other causes of a dislocated mandible include a lateral blow to the chin when the mouth is open and fractures of the mandible.

Why is the temporomandibular joint unique among joints of the skull?

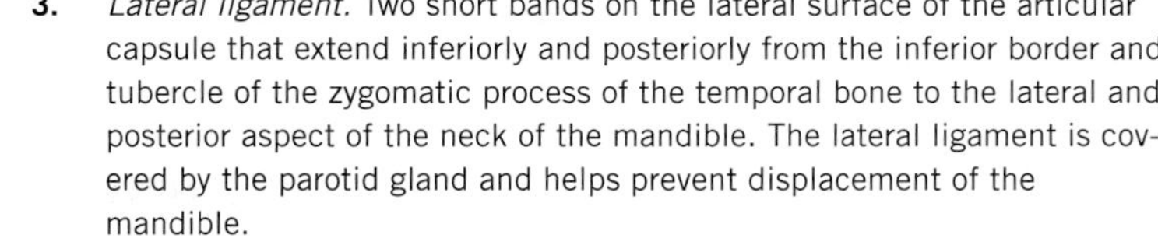

Figure 8.11 / Right temporomandibular joint (TMJ).

The TMJ is the only movable joint between skull bones.

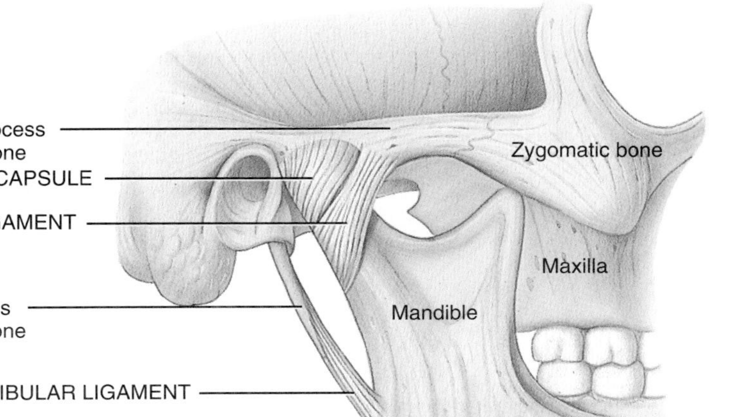

(a) Right lateral view

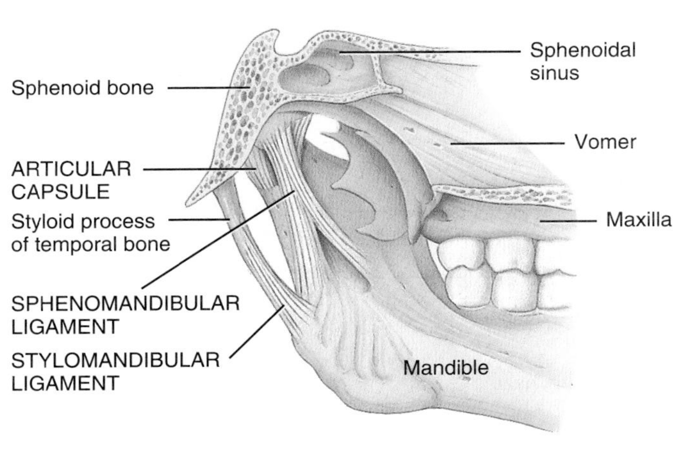

(b) Medial view

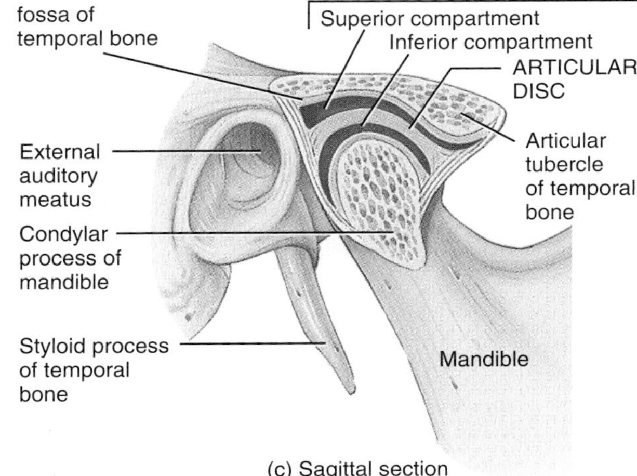

(c) Sagittal section

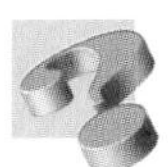

Which ligament prevents displacement of the mandible?

Exhibit 8.2 Shoulder (Humeroscapular or Glenohumeral) Joint (Figure 8.12)

Objective

► Describe the anatomical components of the shoulder joint and the movements that can occur.

Definition

► Ball-and-socket joint formed by the head of the humerus and the glenoid cavity of the scapula.

Anatomical Components

1. *Articular capsule.* Thin, loose sac that completely envelops the joint. It extends from the glenoid cavity to the anatomical neck of the humerus. The inferior part of the capsule is its weakest area.
2. *Coracohumeral ligament.* Strong, broad ligament that strengthens the superior part of the articular capsule and extends from the coracoid process of the scapula to the greater tubercle of the humerus.
3. *Glenohumeral ligaments.* Three thickenings of the articular capsule over the anterior surface of the joint. They extend from the glenoid cavity to the lesser tubercle and anatomical neck of the humerus. These ligaments are often indistinct or absent and provide only minimal strength.
4. *Transverse humeral ligament.* Narrow sheet extending from the greater tubercle to the lesser tubercle of the humerus.
5. *Glenoid labrum.* Narrow rim of fibrocartilage around the edge of the glenoid cavity. It slightly deepens and enlarges the glenoid cavity.
6. *Four bursae* are associated with the shoulder joint. They are the *subscapular bursa, subdeltoid bursa, subacromial bursa,* and *subcoracoid bursa.*

Movements

► Produces flexion, extension, abduction, adduction, medial rotation, lateral rotation, and circumduction of the arm (see Figures 8.6–8.9).

► The shoulder joint has more freedom of movement than any other joint of the body. This freedom results from the looseness of the articular capsule and shallowness of the glenoid cavity in relation to the large size of the head of the humerus.

► Although the ligaments of the shoulder joint strengthen it to some extent, most of the strength results from the muscles that surround the joint, especially the *rotator cuff muscles.* These muscles (supraspinatus, infraspinatus, teres minor, and subscapularis) join the scapula to the humerus (see also Figure 10.17 on pages 300–301). The tendons of the muscles, together called the **rotator cuff,** encircle the joint (except for the inferior portion) and fuse with the articular capsule. The rotator cuff muscles work as a group to hold the head of the humerus in the glenoid cavity.

Dislocated and Separated Shoulder

The joint most commonly dislocated in adults is the shoulder joint because its socket is quite shallow. Usually in a **dislocated shoulder,** the

Figure 8.12 / Right shoulder (humeroscapular or glenohumeral) joint. (See Tortora, *A Photographic Atlas of the Human Body,* Figure 4.1.)

Most of the stability of the shoulder joints results from the arrangement of the rotator cuff muscles.

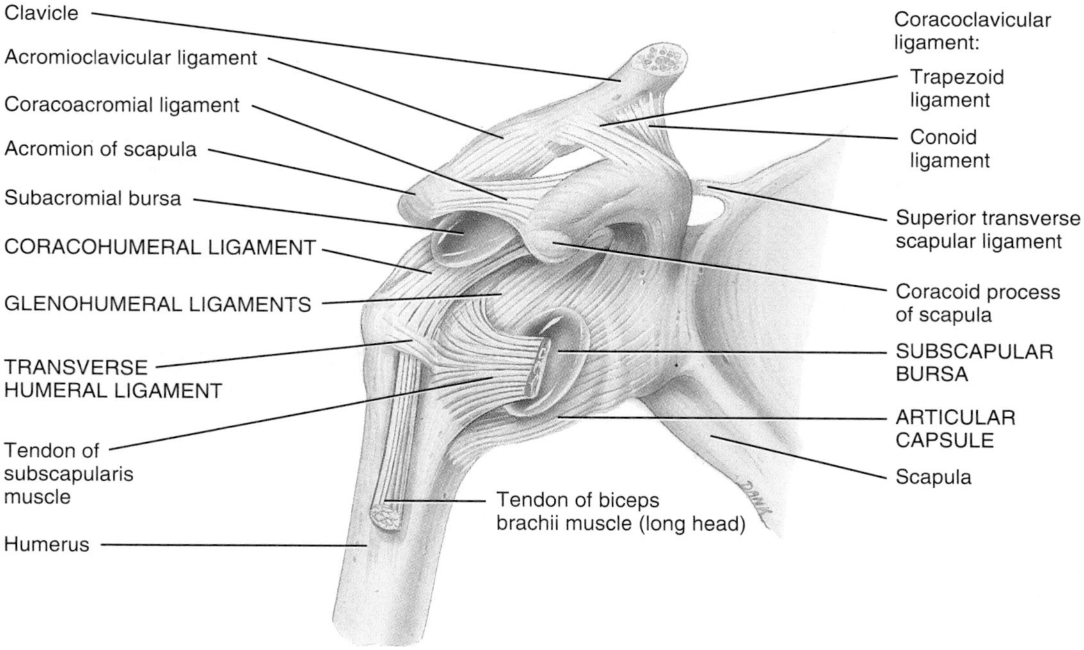

(a) Anterior view

head of the humerus becomes displaced inferiorly, where the articular capsule is least protected. Dislocations of the mandible, elbow, fingers, knee, or hip are less common.

A **separated shoulder** refers to an injury of the acromioclavicular joint, a joint formed by the acromion of the scapula and the acromial end of the clavicle. This condition is usually the result of forceful trauma to the joint, as when the shoulder strikes the ground.

✓ Which tendons at the shoulder joint of a baseball pitcher are most likely to be torn due to excessive circumduction?

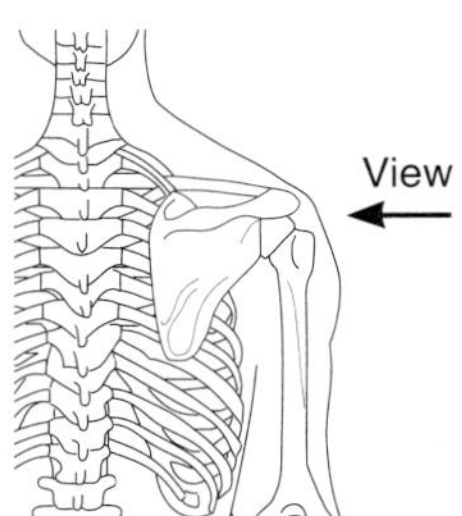

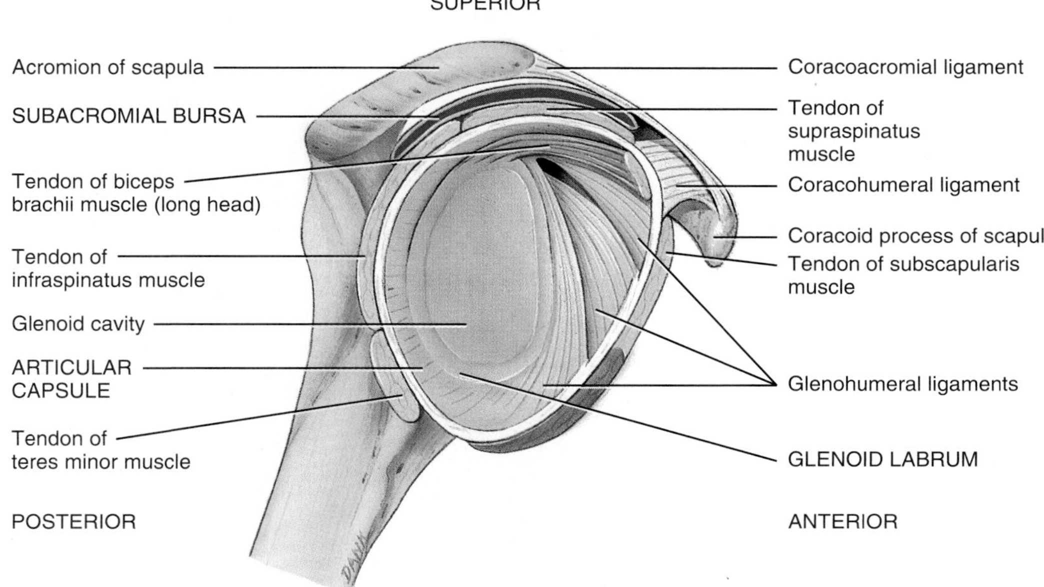

(b) Lateral view (opened)

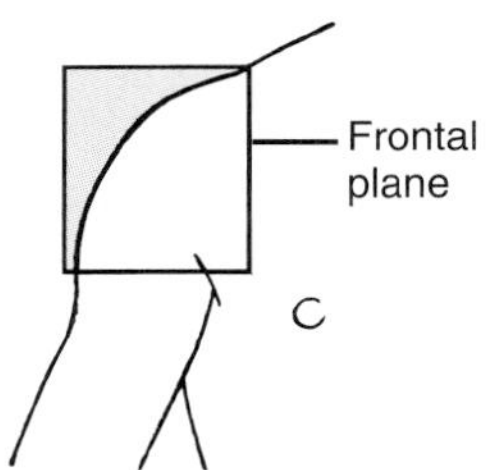

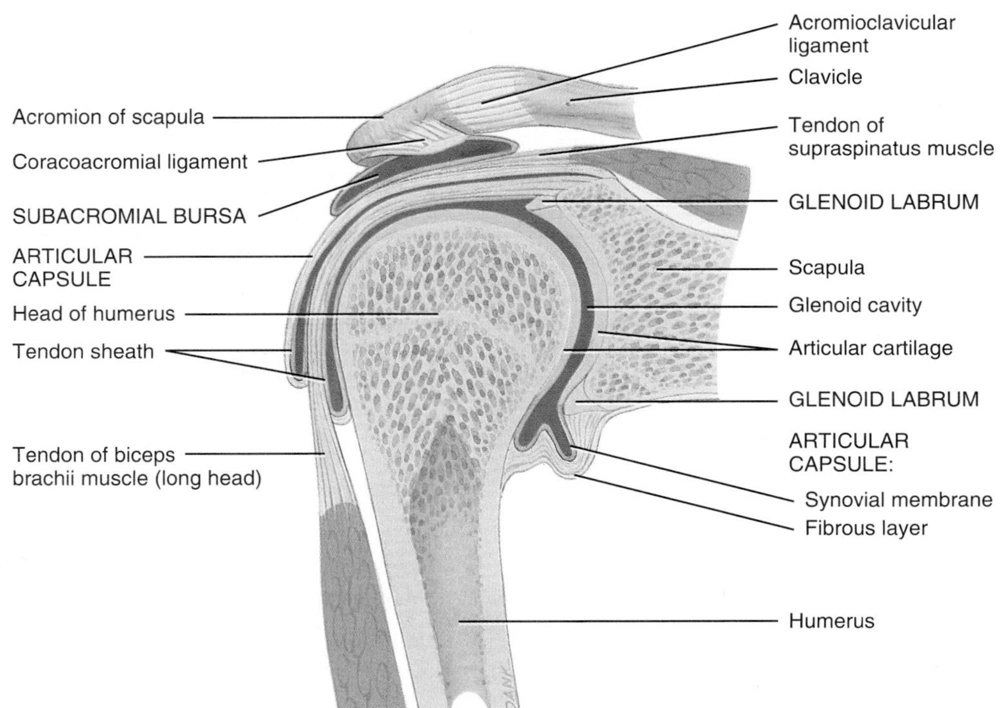

(c) Frontal section

(continues)

Exhibit 8.2 Shoulder (Humeroscapular or Glenohumeral) Joint (Figure 8.12 continued)

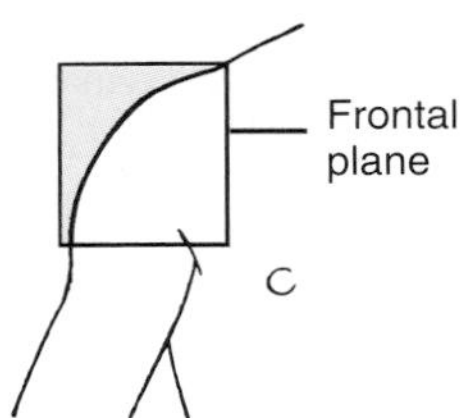

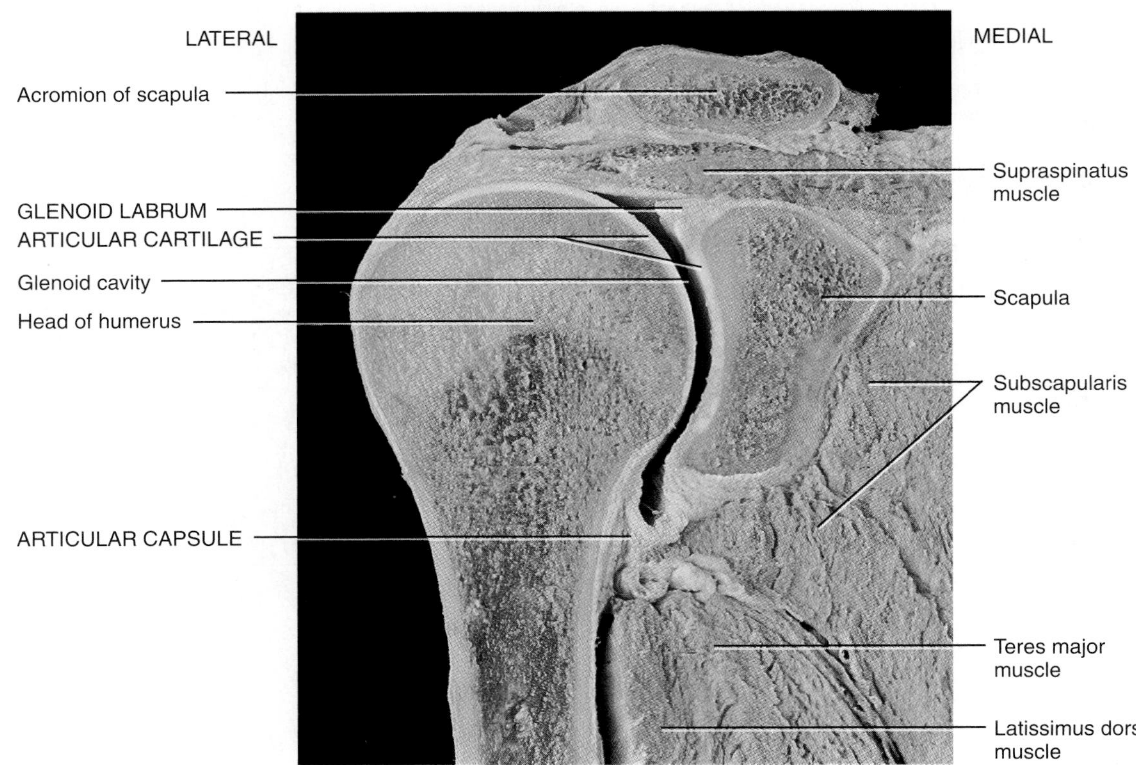

(d) Frontal section

Why does the shoulder joint have more freedom of movement than any other joint of the body?

Exhibit 8.3 Elbow Joint (Figure 8.13)

Objective

▶ Describe the anatomical components of the elbow joint and the movements that can occur.

Definition

▶ Hinge joint formed by the trochlea and capitulum of the humerus, the trochlear notch of the ulna, and the head of the radius.

Anatomical Components

1. *Articular capsule.* The anterior part covers the anterior part of the joint, from the radial and coronoid fossae of the humerus to the coronoid process of the ulna and the annular ligament of the radius. The posterior part extends from the capitulum, olecranon fossa, and lateral epicondyle of the humerus to the annular ligament of the radius, the olecranon of the ulna, and the ulna posterior to the radial notch.
2. *Ulnar collateral ligament.* Thick, triangular ligament that extends from the medial epicondyle of the humerus to the coronoid process and olecranon of the ulna.
3. *Radial collateral ligament.* Strong, triangular ligament that extends from the lateral epicondyle of the humerus to the annular ligament of the radius and the radial notch of the ulna.

Movements

▶ Flexion and extension of the forearm (see Figure 8.6c).

Exhibit 8.3 Elbow Joint (Figure 8.13 continued)

Tennis Elbow, Little-League Elbow, and Dislocation of the Radial Head

Tennis elbow most commonly refers to pain at or near the lateral epicondyle of the humerus, usually caused by an improperly executed backhand. The extensor muscles strain or sprain, resulting in pain.

Little-league elbow typically develops as a result of a heavy pitching schedule and/or a schedule that involves throwing curve balls, especially among youngsters. In this disorder, the elbow may enlarge, fragment, or separate.

When a strong pull is applied to the forearm while it is extended and supinated, the head of the radius may slide past or rupture the radial annular ligament, a ligament that is attached to the radial notch of the ulna and forms a collar around the head of the radius at the proximal radioulnar joint. Such a **dislocation of the radial head** is the most common upper limb dislocation in children and may occur when swinging a child around with outstretched arms.

✓ At the elbow joint, which ligaments connect (a) the humerus and the ulna, and (b) the humerus and the radius?

Figure 8.13 / Right elbow joint. (See Tortora, *A Photographic Atlas of the Human Body,* Figure 4.3.) See also Figure 8.3b in the textbook.

The elbow joint is formed by parts of three bones: humerus, ulna, and radius.

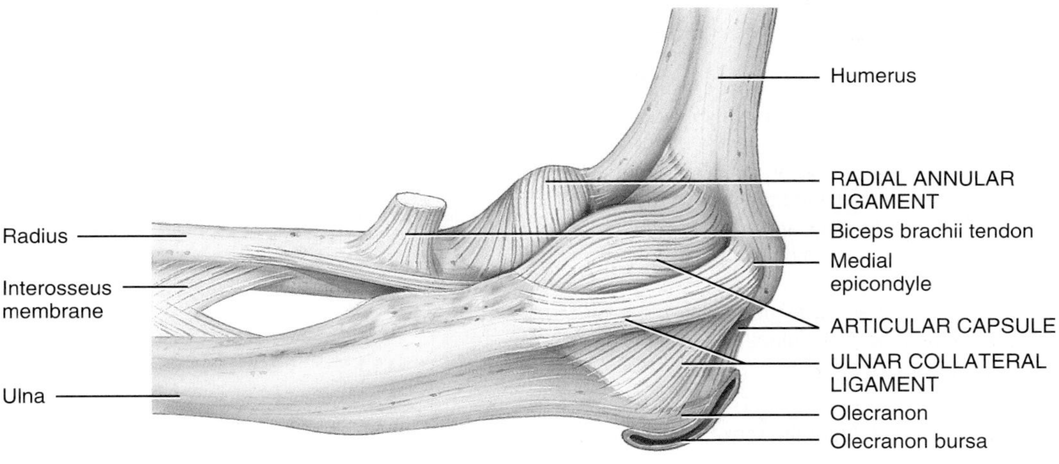

(a) Medial aspect

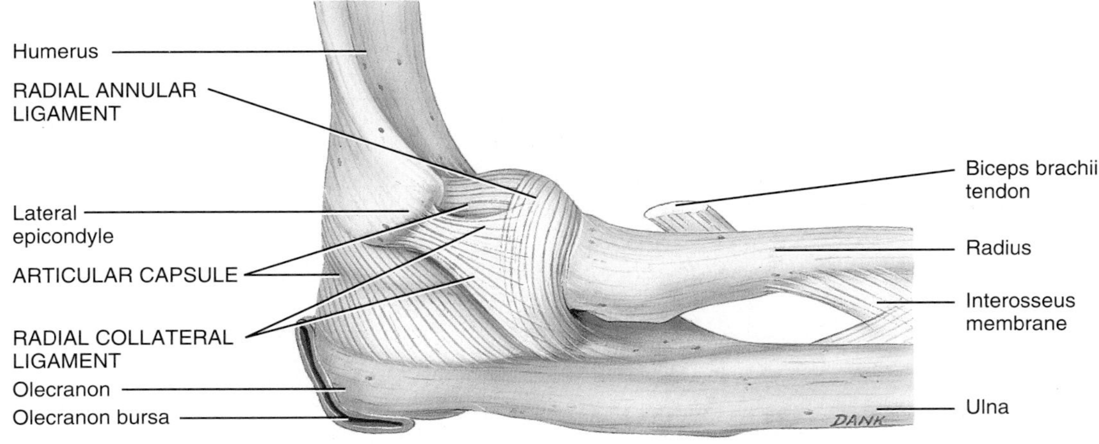

(b) Lateral aspect

Which movements are possible at a hinge joint?

Exhibit 8.4 Hip (Coxal) Joint (Figure 8.14)

Objective

► Describe the anatomical components of the hip joint and the movements that can occur.

Definition

► Ball-and-socket joint formed by the head of the femur and the acetabulum of the hip bone.

Anatomical Components

1. *Articular capsule.* Very dense and strong capsule that extends from the rim of the acetabulum to the neck of the femur. One of the strongest structures of the body, the capsule consists of circular and longitudinal fibers. The circular fibers, called the *zona orbicularis,* form a collar around the neck of the femur. The longitudinal fibers are reinforced by accessory ligaments known as the iliofemoral ligament, pubofemoral ligament, and ischiofemoral ligament.
2. *Iliofemoral ligament.* Thickened portion of the articular capsule that extends from the anterior inferior iliac spine of the hip bone to the intertrochanteric line of the femur.
3. *Pubofemoral ligament.* Thickened portion of the articular capsule that extends from the pubic part of the rim of the acetabulum to the neck of the femur.
4. *Ischiofemoral ligament.* Thickened portion of the articular capsule that extends from the ischial wall of the acetabulum to the neck of the femur.
5. *Ligament of the head of the femur.* Flat, triangular band that extends from the fossa of the acetabulum to the fovea capitis of the head of the femur.
6. *Acetabular labrum.* Fibrocartilage rim attached to the margin of the acetabulum that enhances the depth of the acetabulum. Because the diameter of the acetabular labrum is smaller than that of the head of the femur, dislocation of the femur is rare.
7. *Transverse ligament of the acetabulum.* Strong ligament that crosses over the acetabular notch. It supports part of the acetabular labrum and is connected with the ligament of the head of the femur and the articular capsule.

Movements

► Flexion, extension, abduction, adduction, circumduction, medial rotation, and lateral rotation of thigh (see Figures 8.6–8.9).

► The extreme stability of the hip joint is related to the very strong articular capsule and its accessory ligaments, the manner in which the femur fits into the acetabulum, and the muscles surrounding the joint. Although the shoulder and hip joints are both ball-and-socket joints, the movements at the hip joints do not have as wide a range of motion. Flexion is limited by the anterior surface of the thigh coming into contact with the anterior abdominal wall when the knee is flexed and by tension of the hamstring muscles when the knee is extended. Extension is limited by tension of the iliofemoral, pubo-femoral, and ischiofemoral ligaments. Abduction is limited by the tension of the pubofemoral ligament, and adduction is limited by contact with the opposite limb and tension in the ligament of the head of the femur. Medial rotation is limited by the tension in the ischiofemoral ligament, and lateral rotation is limited by tension in the iliofemoral and pubofemoral ligaments.

✓ What factors limit the degree of flexion and abduction at the hip joint?

Figure 8.14 / Right hip (coxal) joint.

The articular capsule of the hip joint is one of the strongest structures in the body.

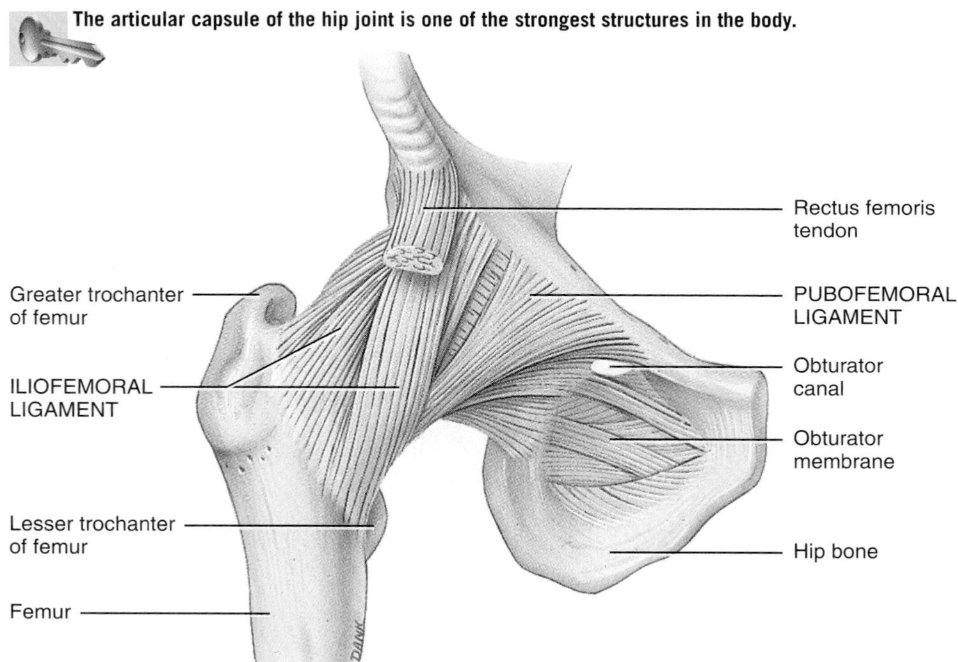

(a) Anterior view

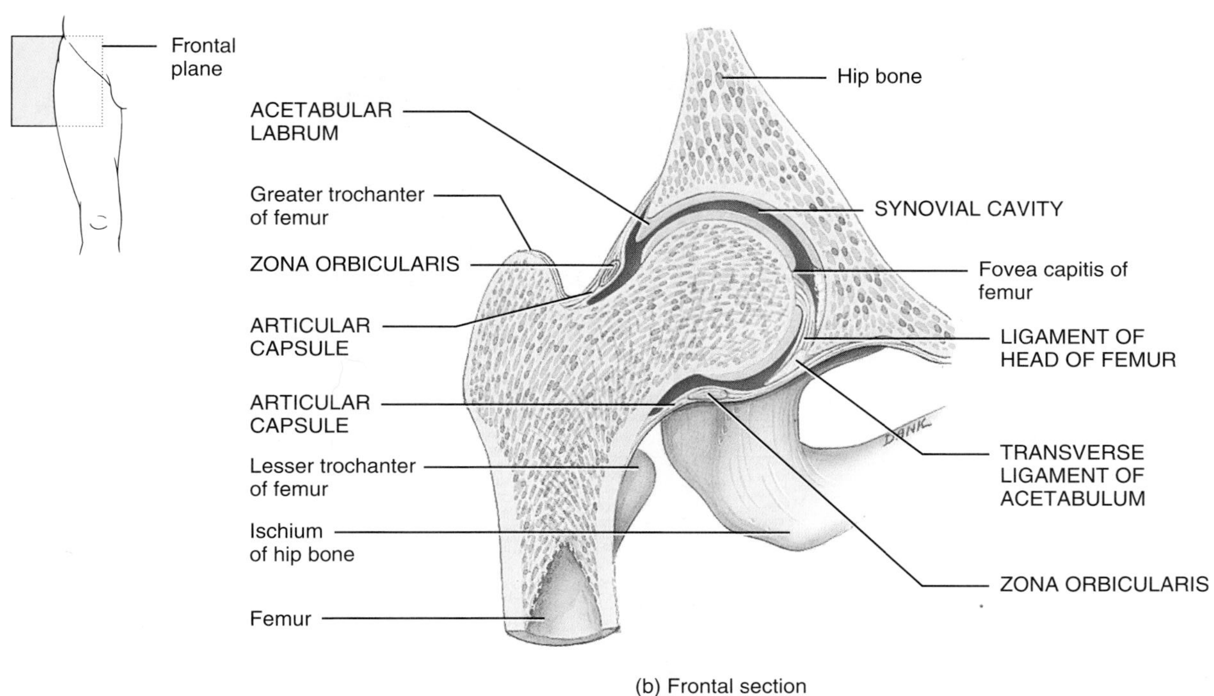

(b) Frontal section

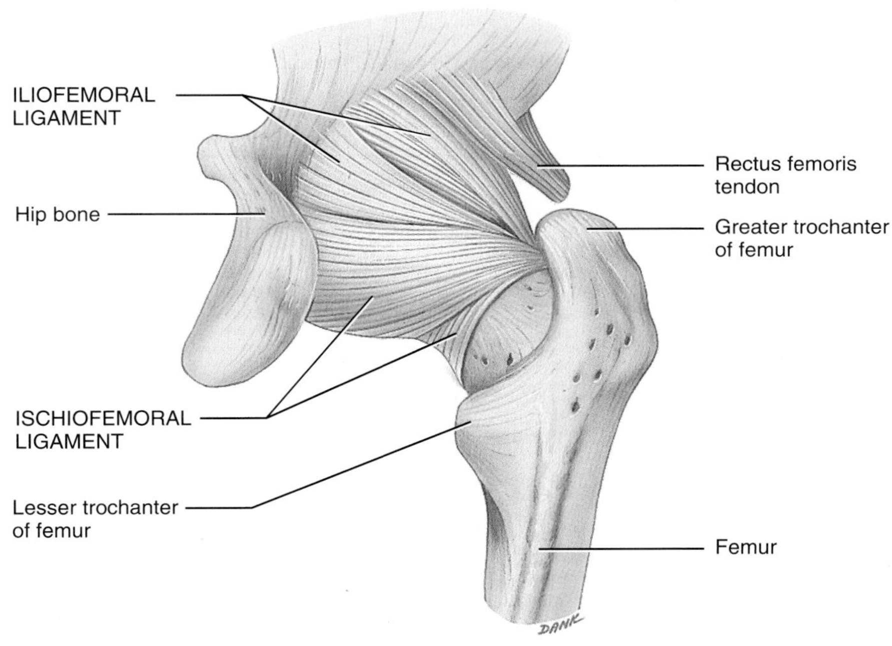

(c) Posterior view

(continues)

Exhibit 8.4 Hip (Coxal) Joint (Figure 8.14 continued)

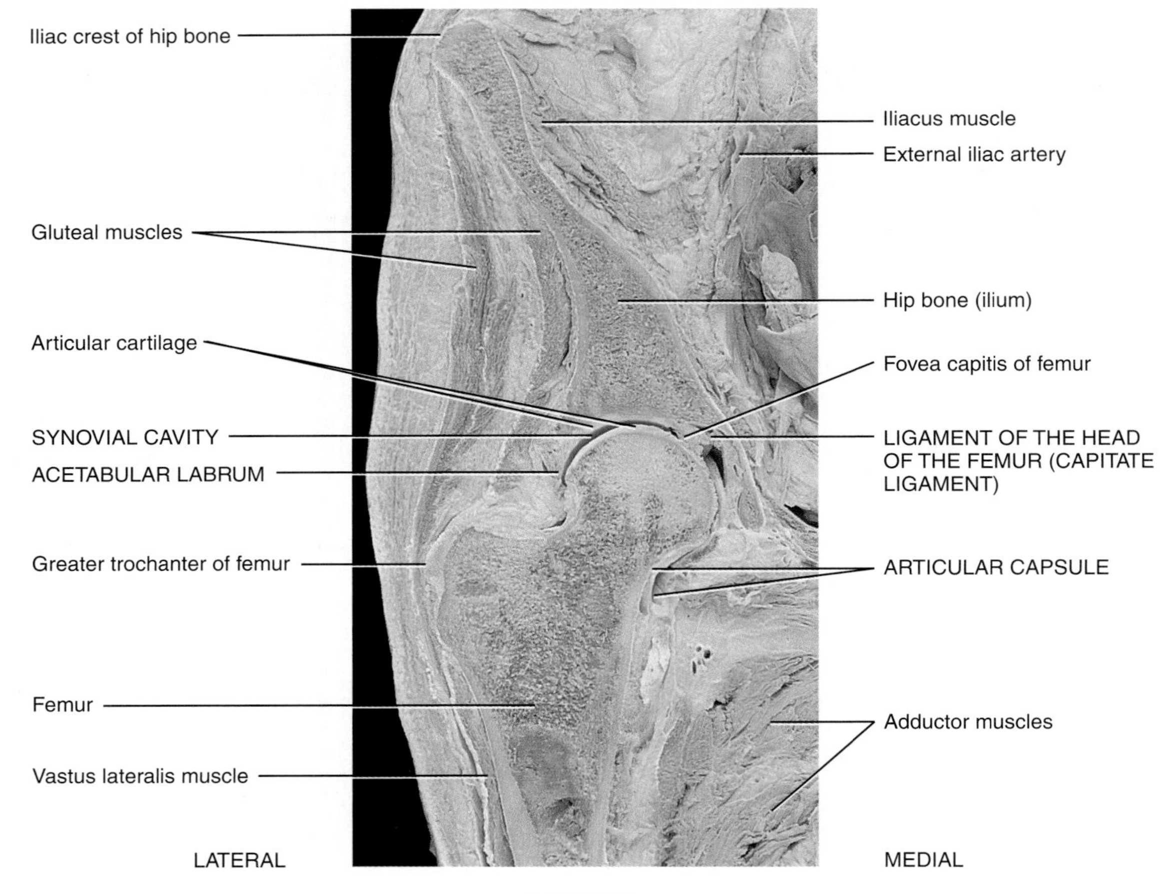

(d) Frontal section

Which ligaments limit the degree of extension that is possible at the hip joint?

Exhibit 8.5 Knee (Tibiofemoral) Joint (Figure 8.15)

Objective

► Describe the anatomical components of the knee joint and explain the movements that can occur.

Definition

► The largest and most complex joint of the body, actually consisting of three joints within a single synovial cavity: (1) an intermediate patellofemoral joint between the patella and the patellar surface of the femur, which is a plantar joint; (2) a lateral tibiofemoral joint between the lateral condyle of the femur, lateral meniscus, and lateral condyle of the tibia, which is a modified hinge joint; and (3) a medial tibiofemoral joint between the medial condyle of the femur, medial meniscus, and medial condyle of the tibia, which is also a modified hinge joint.

Anatomical Components

1. *Articular capsule.* No complete, independent capsule unites the bones. The ligamentous sheath surrounding the joint consists mostly of muscle tendons or expansions of them. There are, however, some capsular fibers connecting the articulating bones.
2. *Medial and lateral patellar retinacula.* Fused tendons of insertion of the quadriceps femoris muscle and the fascia lata (deep fascia of thigh) that strengthen the anterior surface of the joint.
3. *Patellar ligament.* Continuation of the common tendon of insertion of the quadriceps femoris muscle that extends from the patella to the tibial tuberosity. This ligament also strengthens the anterior surface of the joint. The posterior surface of the ligament is separated from the synovial membrane of the joint by an infrapatellar fat pad.
4. *Oblique popliteal ligament.* Broad, flat ligament that extends from the intercondylar fossa of the femur to the head of the tibia. The tendon of the semimembranosus muscle is superficial to the ligament and passes from the medial condyle of the tibia to the lateral condyle of the femur. The ligament and tendon strengthen the posterior surface of the joint.
5. *Arcuate popliteal ligament.* Extends from the lateral condyle of the femur to the styloid process of the head of the fibula. It strengthens the lower lateral part of the posterior surface of the joint.
6. *Tibial collateral ligament.* Broad, flat ligament on the medial surface of the joint that extends from the medial condyle of the femur to the medial condyle of the tibia. The ligament is crossed by tendons of the sartorius, gracilis, and semitendinosus muscles, all of which strengthen the medial aspect of the joint. Because the tibial collateral ligament is firmly attached to the medial meniscus, tearing of the ligament frequently results in tearing of the meniscus and damage to the anterior cruciate ligament, described under 8a in next column.
7. *Fibular collateral ligament.* Strong, rounded ligament on the lateral surface of the joint that extends from the lateral condyle of the femur to the lateral side of the head of the fibula. It strengthens the lateral aspect of the joint. The ligament is covered by the tendon of the biceps femoris muscle. The tendon of the popliteal muscle is deep to the ligament.
8. *Intracapsular ligaments.* Ligaments within the capsule that connect the tibia and femur. The anterior and posterior cruciate (*cruciatus* = a cross) ligaments are named on the basis of their origins relative to the intercondylar area of the tibia. Following their originations, they cross on the way to their destinations on the femur.
   a. *Anterior cruciate ligament (ACL).* Extends posteriorly and laterally from a point *anterior* to the intercondylar area of the tibia to the posterior part of the medial surface of the lateral condyle of the femur. This ligament is stretched or torn in about 70% of all serious knee injuries. The ACL limits hyperextension of the knee and prevents the anterior sliding of the tibia on the femur.
   b. *Posterior cruciate ligament (PCL).* Extends anteriorly and medially from a depression on the *posterior* intercondylar area of the tibia and lateral meniscus to the anterior part of the medial surface of the medial condyle of the femur. The PCL prevents the posterior sliding of the tibia (and anterior sliding of the femur) when the knee is flexed. This is very important when walking down a steep incline or stairs.
9. *Articular discs (menisci).* Two fibrocartilage discs between the tibial and femoral condyles that help compensate for the irregular shapes of the bones and circulate synovial fluid.
   a. *Medial meniscus.* Semicircular piece of fibrocartilage (C-shaped). Its anterior end is attached to the anterior intercondylar fossa of the tibia, anterior to the anterior cruciate ligament. Its posterior end is attached to the posterior intercondylar fossa of the tibia between the attachments of the posterior cruciate ligament and lateral meniscus.
   b. *Lateral meniscus.* Nearly circular piece of fibrocartilage (approaches an incomplete O in shape). Its anterior end is attached anterior to the intercondylar eminence of the tibia, and lateral and posterior to the anterior cruciate ligament. Its posterior end is attached posterior to the intercondylar eminence of the tibia, and anterior to the posterior end of the medial meniscus. The medial and lateral menisci are connected to each other by the *transverse ligament* and to the margins of the head of the tibia by the *coronary ligaments* (not illustrated).
10. The more important *bursae* of the knee include the following:
   a. *Prepatellar bursa* between the patella and skin.
   b. *Intrapatellar bursa* between superior part of tibia and patellar ligament.
   c. *Suprapatellar bursa* between inferior part of femur and deep surface of quadriceps femoris muscle.

Movements

► Flexion, extension, slight medial rotation and slight lateral rotation of leg in flexed position (see Figures 8.6f and 8.9c).

Knee Injuries

The knee joint is the joint most vulnerable to damage because of the stresses to which it is subjected and because there is no interlocking of the articulating bones; reinforcement is strictly by ligaments and tendons. The most common type of **knee injury** in football is rupture of the tibial (medial) collateral ligaments, often associated with tearing of the anterior cruciate ligament and medial meniscus (torn cartilage). Usually, a hard blow to the lateral side of the knee while the foot is fixed on the ground causes the injury. When a knee is examined for such an injury, the three Cs are kept in mind: collateral ligament, cruciate ligament, and cartilage, the "unhappy triad."

A **swollen knee** may occur immediately or hours after an injury. Immediate swelling is due to escape of blood from damaged blood vessels adjacent to areas involving rupture of the anterior cruciate ligament, damage to synovial membranes, torn menisci, fractures, or collateral ligament sprains. Delayed swelling is due to excessive production of synovial fluid as a result of irritation of the synovial membrane, a condition commonly referred to as "water on the knee."

A **dislocated knee** refers to the displacement of the tibia relative to the femur. Accordingly, such dislocations are classified as anterior, posterior, medial, lateral, or rotatory. The most common type is anterior dislocation, resulting from hyperextension of the knee. A frequent consequence of a dislocated knee is damage to the popliteal artery.

✓ What are the opposing functions of the anterior and posterior cruciate ligaments?

(continues)

Exhibit 8.5 Knee (Tibiofemoral) Joint (Figure 8.15 continued)

Figure 8.15 / Right knee (tibiofemoral) joint. (See Tortora, *A Photographic Atlas of the Human Body,* Figure 4.7.)

The knee joint is the largest and most complex joint in the body.

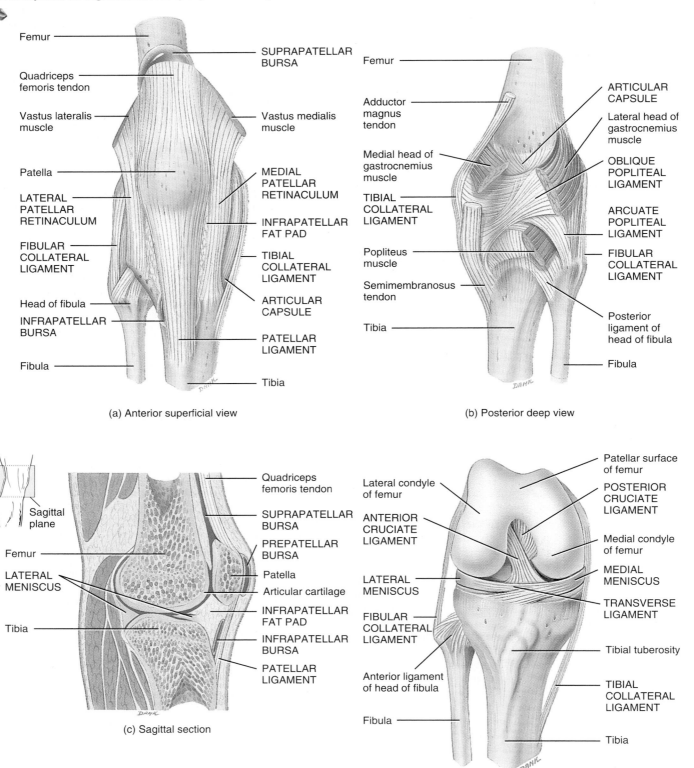

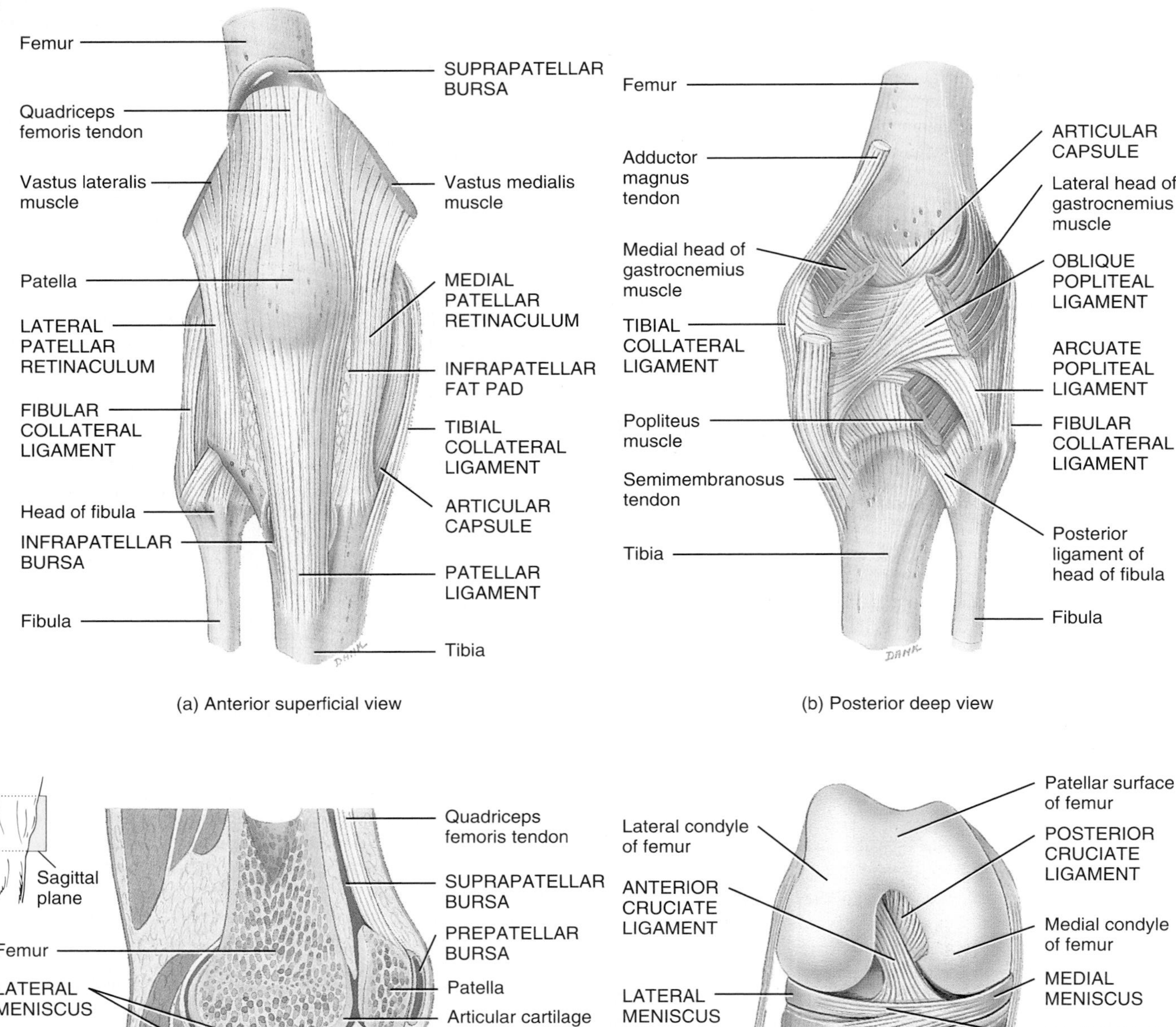

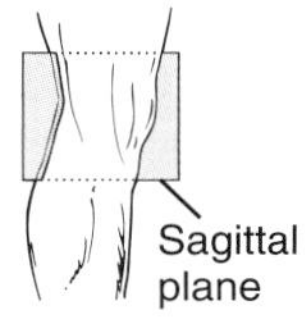

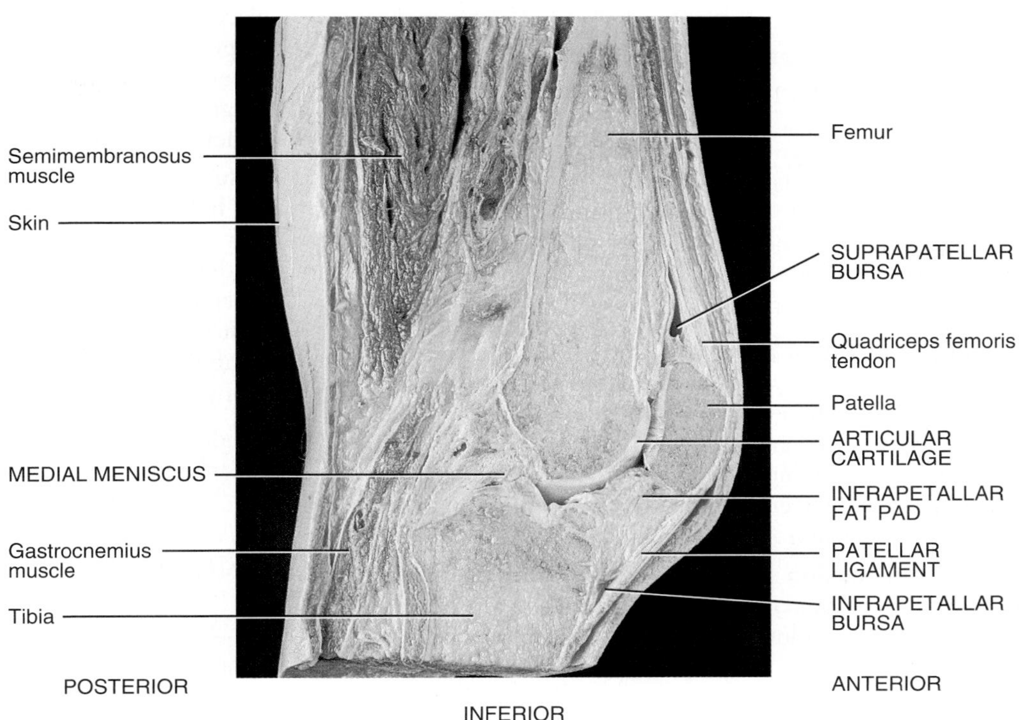

(e) Sagittal section

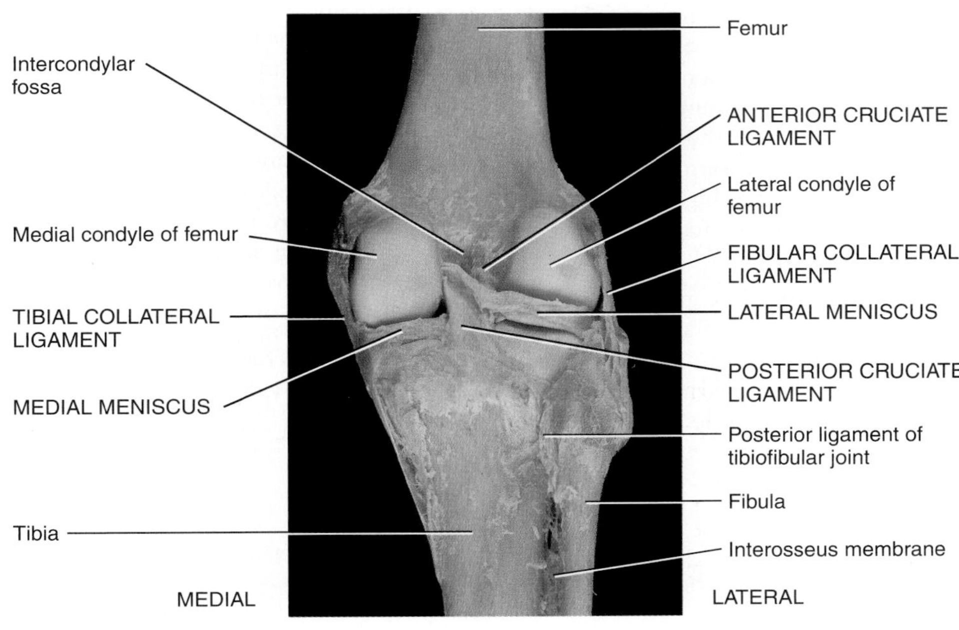

(f) Posterior view

What movement occurs at the knee joint when the quadriceps femoris muscle contracts?

FACTORS AFFECTING CONTACT AND RANGE OF MOTION AT SYNOVIAL JOINTS

Objective

- Describe six factors that influence the type of movement and range of motion possible at a synovial joint.

The articular surfaces of synovial joints contact one another and determine the type and range of motion that is possible. **Range of motion (ROM)** refers to the range, measured in degrees of a circle, through which the bones of a joint can be moved. The following factors contribute to keeping the articular surfaces in contact and affect range of motion:

1. **Structure or shape of the articulating bones.** The structure or shape of the articulating bones determines how closely they can fit together. The articular surfaces of some bones interlock with one another. This spatial relationship is very obvious at the hip joint, where the head of the femur articulates with the acetabulum of the hip bone. An interlocking fit allows rotational movement.
2. **Strength and tension (tautness) of the joint ligaments.** The different components of a fibrous capsule are tense or taut only when the joint is in certain positions. Tense ligaments not only restrict the range of motion but also direct the movement of the articulating bones with respect to each other. In the knee joint, for example, the anterior cruciate ligament is taut and the posterior cruciate ligament is loose when the knee is straightened, and the reverse occurs when the knee is bent.
3. **Arrangement and tension of the muscles.** Muscle tension reinforces the restraint placed on a joint by its ligaments, and thus restricts movement. A good example of the effect of muscle tension on a joint is seen at the hip joint. When the thigh is raised with the knee straight, the movement is restricted by the tension of the hamstring muscles on the posterior surface of the thigh. But if the knee is bent, the tension on the hamstring muscles is lessened, and the thigh can be raised farther.
4. **Apposition of soft parts.** The point at which one body surface contacts another may limit mobility. For example, if you bend your arm at the elbow, it can move no farther after the anterior surface of the forearm meets with and presses against the biceps brachii muscle of the arm. Joint movement may also be restricted by the presence of adipose tissue.
5. **Hormones.** Joint flexibility may also be affected by hormones. For example, relaxin, a hormone produced by the placenta and ovaries, increases the flexibility of the fibrocartilage of the pubic symphysis and loosens the ligaments between the sacrum, hip bone, and coccyx toward the end of pregnancy. These changes permit expansion of the pelvic outlet, which assists in delivery of the baby.
6. **Disuse.** Movement at a joint may be restricted if a joint has not been used for an extended period. For example, if an elbow joint is immobilized by a cast, only a limited range of motion at the joint may be possible for a time after the cast is removed. Restricted movement due to disuse may result from decreased amounts of synovial fluid, from diminished flexibility of ligaments and tendons, and from muscular atrophy, a reduction in size or wasting of a muscle.

✓ Besides the hip joint, which joints have bones that interlock and allow rotational movement?

AGING AND JOINTS

Objective

- Explain the effects of aging on joints.

The aging process usually results in decreased production of synovial fluid in joints. In addition, the articular cartilage becomes thinner with age, and ligaments shorten and lose some of their flexibility. The effects of aging on joints, which vary considerably from one person to another, are affected by genetic factors and by wear and tear. Although degenerative changes in joints may begin in individuals as young as 20 years of age, most do not occur until much later. In fact, by age 80, almost everyone develops some type of degeneration in the knees, elbows, hips, and shoulders. Males also commonly develop degenerative changes in the vertebral column, resulting in a hunched-over posture and pressure on nerve roots. As you will see in the following section, one type of arthritis, called osteoarthritis, is at least partially age-related. Nearly everyone over age 70 has evidence of some osteoarthritic changes.

✓ Which joints show evidence of degeneration in nearly all individuals as aging progresses?

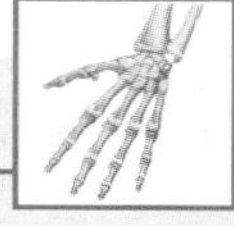

APPLICATIONS TO HEALTH

Rheumatism and Arthritis

Rheumatism (RŪ-ma-tizm) is any of a variety of painful disorders of the supporting structures of the body—bones, ligaments, tendons, or muscles. **Arthritis** is a form of rheumatism in which the joints have become inflamed. Inflammation, pain, and stiffness may also involve adjacent muscles. Arthritis afflicts about 40 million people in the United States.

Three important classes of arthritis are (1) diffuse connective tissue diseases, such as rheumatoid arthritis; (2) degenerative joint disease, such as osteoarthritis; and (3) metabolic and endocrine diseases with associated arthritis, such as gouty arthritis.

Rheumatoid Arthritis

Rheumatoid arthritis (RA) is an autoimmune disease in which the immune system of the body attacks its own tissues—in this case, its own cartilage and joint linings. RA is characterized by inflammation of the joint, which causes swelling, pain, and loss of function. Usually, this form of arthritis occurs bilaterally: If one wrist is affected, the other is also likely to be affected, although usually not to the same degree.

The primary symptom of rheumatoid arthritis is inflammation of the synovial membrane. If untreated, the membrane thickens, and synovial fluid accumulates. The resulting pressure causes pain and tenderness. The membrane then produces an abnormal granulation tissue, called *pannus*, that adheres to the surface of the articular cartilage and sometimes erodes the cartilage completely. When the cartilage is destroyed, fibrous tissue joins the exposed bone ends. The fibrous tissue ossifies and fuses the joint so that it becomes immovable—the ultimate crippling effect of rheumatoid arthritis. The growth of the granulation tissue causes the distortion of the fingers that characterizes hands of RA sufferers.

Osteoarthritis

Osteoarthritis (OA) (os′-tē-ō-ar-THRĪ-tis) is a degenerative joint disease that apparently results from a combination of aging, irritation of the joints, and wear and abrasion. It is commonly known as "wear-and-tear" arthritis and is the leading cause of disability in older individuals.

Osteoarthritis is a progressive disorder of synovial joints, particularly weight-bearing joints. It is characterized by the deterioration of articular cartilage and by the formation of new bone in the subchondral areas and at the margins of the joint. The cartilage slowly degenerates, and as the bone ends become exposed, spurs (small bumps) of new osseous tissue are deposited on them. These spurs decrease the space of the joint cavity and restrict joint movement. Unlike rheumatoid arthritis, osteoarthritis affects mainly the articular cartilage, although the synovial membrane often becomes inflamed late in the disease. A major distinction between osteoarthritis and rheumatoid arthritis is that osteoarthritis strikes the larger joints (knees, hips) first, whereas rheumatoid arthritis first strikes smaller joints.

Gouty Arthritis

Uric acid (a substance that gives urine its name) is a waste product of nucleic acid (DNA and RNA) metabolism. A person who suffers from **gout** (GOWT) either produces excessive amounts of uric acid or is not able to excrete as much as normal. The result is a buildup of uric acid in the blood. This excess acid reacts with sodium to form a salt called sodium urate. Crystals of this salt accumulate in soft tissues such as the kidneys and in the cartilage of the ears and joints.

In **gouty arthritis,** sodium urate crystals are deposited in the soft tissues of the joints. The crystals irritate and erode the cartilage, causing inflammation, swelling, and acute pain. Eventually, the crystals destroy all joint tissues. If the disorder is untreated, the ends of the articulating bones fuse, and the joint becomes immovable.

Lyme Disease

In 1975 a cluster of disease cases that were first diagnosed as rheumatoid arthritis were reported near the town of Lyme, Connecticut. Today this disease is known as **Lyme disease,** and it is now found in most states across the country. It is caused by a spiral-shaped bacterium (*Borrelia burgdorferi*) that is transmitted to humans mainly by deer ticks (*Ixodes dammini*) that are so small their bites often go unnoticed.

Within a few weeks of the tick bite, a rash may appear at the site. The rash often resembles a bull's eye, but there are many variations, and some people never develop a rash. Other symptoms include joint stiffness, fever and chills, headache, stiff neck, nausea, and low back pain. In advanced stages of the disease, arthritis is the principal complication. It usually involves the larger joints such as the knee, ankle, hip, elbow, or wrist. Lyme disease responds well to antibiotics, especially if they are given in the early stages. However, some of the symptoms may linger for years.

Ankylosing Spondylitis

Ankylosing spondylitis (ang′-ki-LŌ-sing spon′-di-LĪ-tis; *ankyle* = stiff; *spondyl* = vertebra) is an inflammatory disease that affects joints between vertebrae (intervertebral) and between the sacrum and hip bone (sacroiliac joint). The cause is unknown. The disease is more common in males and has its onset between the ages of 20 and 40 years. It is characterized by pain and stiffness in the hips and lower back that progress upward along the backbone. Inflammation can lead to *ankylosis* (severe or complete loss of movement at a joint) and *kyphosis* (hunchback). Treatment consists of anti-inflammatory drugs, heat, massage, and supervised exercise.

KEY MEDICAL TERMS ASSOCIATED WITH JOINTS

Arthralgia (ar-THRAL-jē-a; *arthr-* = joint; *-algia* = pain) Pain in a joint.

Bursectomy (bur-SEK-tō-mē; *-ectomy* = removal of) Removal of a bursa.

Chondritis (kon-DRĪ-tis; *chondr-* = cartilage) Inflammation of cartilage.

Dislocation (dis′-lō-KĀ-shun; *dis-* = apart) The displacement of a bone from a joint, with resultant tearing of ligaments, tendons, and articular capsules; usually caused by a blow or fall. Also termed **luxation** (luks-Ā-shun; *luxatio* = dislocation).

Subluxation (sub-luks-Ā-shun) A partial or incomplete dislocation.

Synovitis (sin′-ō-VĪ-tis) Inflammation of a synovial membrane in a joint.

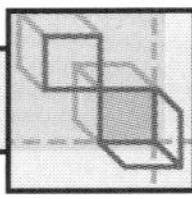

STUDY OUTLINE

Introduction (p. 197)

1. A joint (articulation or arthrosis) is a point of contact between two bones, between bone and cartilage, or between bone and teeth.
2. A joint's structure may permit no movement, slight movement, or free movement.

Joint Classifications (p. 197)

1. Structural classification is based on the presence or absence of a synovial cavity and the type of connecting tissue. Structurally, joints are classified as fibrous, cartilaginous, or synovial.
2. Functional classification of joints is based on the degree of movement permitted. Joints may be synarthroses (immovable), amphiarthroses (slightly movable), or diarthroses (freely movable).

Fibrous Joints (p. 197)

1. Bones held by fibrous connective tissue are fibrous joints.
2. These joints include immovable sutures (found between skull bones), slightly movable syndesmoses (such as the distal tibiofibular joint), and immovable gomphoses (roots of teeth in the sockets in the mandible and maxilla).

Cartilaginous Joints (p. 198)

1. Bones held together by cartilage are cartilaginous joints.
2. These joints include immovable synchondroses united by hyaline cartilage (epiphyseal plates between diaphyses and epiphyses) and slightly movable symphyses united by fibrocartilage (pubic symphysis).

Synovial Joints (p. 199)

1. Synovial joints contain a space between bones called the synovial cavity. All synovial joints are diarthroses.
2. Other characteristics of synovial joints are the presence of articular cartilage and an articular capsule, made up of a fibrous capsule and a synovial membrane.
3. The synovial membrane secretes synovial fluid, which forms a thin, viscous film over the surfaces within the articular capsule.
4. Many synovial joints also contain accessory ligaments (extracapsular and intracapsular) and articular discs (menisci).
5. Synovial joints contain an extensive nerve and blood supply. The nerves convey information about pain, joint movements, and the degree of stretch at a joint. Blood vessels penetrate the articular capsule and ligaments.
6. Bursae are saclike structures, similar in structure to joint capsules, that alleviate friction in joints such as the shoulder and knee joints.
7. Tendon sheaths are tubelike bursae that wrap around tendons where there is considerable friction.
8. Subtypes of synovial joints are planar, hinge, pivot, condyloid, saddle, and ball-and-socket.
9. Planar joint: articulating surfaces are flat; bone glides back and forth and side to side (nonaxial); found between carpals and tarsals.
10. Hinge joint: convex surface of one bone fits into concave surface of another; motion is angular around one axis (monaxial); examples are the elbow, knee, and ankle joints.
11. Pivot joint: a round or pointed surface of one bone fits into a ring formed by another bone and a ligament; movement is rotation (monaxial); examples are the atlantoaxial and radioulnar joints.
12. Condyloid joint: an oval projection of one bone fits into an oval cavity of another; motion is angular around two axes (biaxial); examples are the wrist joint and metacarpophalangeal joints for the second through fifth digits.
13. Saddle joint: articular surface of one bone is shaped like a saddle and the other bone fits into the "saddle" like a sitting rider; motion is angular around two axes (biaxial); an example is the carpometacarpal joint between the trapezium and the metacarpal of the thumb.
14. Ball-and-socket joint: ball-shaped surface of one bone fits into cuplike depression of another; motion is angular and rotational around three axes and all direction in between (multiaxial); examples are the shoulder and hip joints.
15. Table 8.1 on page 205 summarizes the structural and functional categories of joints.

Types of Movements at Synovial Joints (p. 204)

1. In a gliding movement, the nearly flat surfaces of bones move back and forth and from side to side. Gliding movements occur at planar joints.
2. In angular movements, a change in the angle between bones occurs. Examples are flexion-extension, lateral flexion, hyperextension, abduction-adduction. Circumduction refers to flexion, abduction, extension, and adduction in succession. Angular movements occur at hinge, condyloid, saddle, and ball-and-socket joints.
3. In rotation, a bone moves around its own longitudinal axis. Rotation can occur at pivot and ball-and-socket joints.
4. Special movements occur at specific synovial joints. Examples are elevation-depression, protraction-retraction, inversion-eversion, dorsiflexion-plantar flexion, supination-pronation, and opposition.
5. Table 8.2 on page 210 summarizes the various types of movements at synovial joints.

Selected Joints of the Body (p. 210)

1. A summary of selected joints of the body, including articular components, structural and functional classifications, and movements, is presented in Table 8.3 on pages 211–212.
2. The temporomandibular joint (TMJ) is between the condyle of the mandible and mandibular fossa and articular tubercle of the temporal bone (Exhibit 8.1, page 213).
3. The shoulder (humeroscapular or glenohumeral) joint is between the head of the humerus and glenoid cavity of the scapula (Exhibit 8.2 on page 214).
4. The elbow joint is between the trochlea of the humerus, the trochlear notch of the ulna, and the head of the radius (Exhibit 8.3 on page 216).
5. The hip (coxal) joint is between the head of the femur and acetabulum of the hip bone (Exhibit 8.4 on page 218).
6. The knee (tibiofemoral) joint is between the patella and patellar surface of femur; lateral condyle of femur, lateral meniscus, and lateral condyle of tibia; and medial condyle of femur, medial meniscus, and medial condyle of tibia (Exhibit 8.5 on page 221).

Factors Affecting Contact and Range of Motion at Synovial Joints (p. 224)

1. The ways that articular surfaces of synovial joints contact one another determines the type of movement possible.
2. Factors that contribute to keeping the surfaces in contact and affect range of motion are structure or shape of the articulating bones, strength and tension of the joint ligaments, arrangement and tension of the muscles, apposition of soft parts, hormones, and disuse.

Aging and Joints (p. 224)

1. With aging, a decrease in synovial fluid, thinning of articular cartilage, and decreased flexibility of ligaments occur.
2. Most individuals experience some degeneration in the knees, elbows, hips, and shoulders due to the aging process.

SELF-QUIZ QUESTIONS

Choose the one best answer to the following questions:

1. Choose the pair of terms that is most closely associated or matched. (a) wrist joint-pronation, (b) intertarsal joints-inversion, (c) elbow joint-hyperextension, (d) ankle joint-eversion, (e) interphalangeal joint-circumduction.
2. Which of the following statements about synovial joints is (are) true? (1) They are enclosed in a synovial-lined capsule and have ligaments uniting the bones of the joint. (2) They constitute relatively few of the joints of the body. (3) They always permit flexion, extension, abduction, and adduction.
   (a) 1 only **(b)** 2 only **(c)** 3 only **(d)** a, b, and c **(e)** none of the above.
3. Which of the following structures is *not* associated with the knee joint? (a) glenoid labrum, (b) patellar ligament, (c) infrapatellar bursa, (d) cruciate ligaments, (e) tibial collateral ligament.
4. Which of the following statements about an articulation is (are) correct? (1) It is a point of contact between bones. (2) It is a point of contact between bone and cartilage. (3) It is a point of contact between bone and muscle. (4) It is a point of contact between bones and teeth.
   (a) 1 only **(b)** 1 and 2 **(c)** 1, 2, and 3 **(d)** 1, 2, and 4 **(e)** 1, 2, 3, and 4.
5. Choose the pair of terms that is most closely associated or matched. (a) wrist joint-radioulnar joint, (b) shoulder joint-acetabulum, (c) elbow joint-radioulnar joint, (d) cartilaginous joint-joint cavity, (e) knee joint-lateral and medial menisci.
6. Which of the following statements is (are) true about the joint between the epiphysis and the diaphysis of a long bone? (1) It is one example of a synchondrosis, (2) It is a synarthrosis, (3) It is temporary and becomes a synostosis.
   (a) 1 only **(b)** 2 only **(c)** 3 only **(d)** 1 and 3 **(e)** 1, 2, and 3.
7. The synovial membrane (1) secretes synovial fluid into a joint cavity, (2) lines the fibrous capsule of movable joints, (3) is an epithelial membrane.
   (a) 1 only **(b)** 2 only **(c)** 3 only **(d)** 1 and 2 **(e)** 1, 2, and 3.

Complete the following:

8. A slightly movable joint is classified as ________ .
9. The largest, most complex joint of the body is the ________ joint.
10. The functional classification of the temporomandibular joint is ________ .
11. The ends of bones at synovial joints are covered with a protective layer of ________ .
12. Sac-like structures that reduce friction between body parts at joints are called ________ .

Are the following statements true or false?

13. Sutures, syndesmoses, and symphyses are kinds of fibrous joints.
14. A shoulder dislocation is an injury to the acromioclavicular joint.
15. A tendon sheath is a tube-like bursa found at the wrist and ankle joints.
16. All fibrous joints are synarthrotic and all cartilaginous joints are amphiarthrotic.
17. The nerves that supply a joint are different from those that supply the skeletal muscles that move that joint.
18. Match the following.

____ **(a)** distal tibiofibular joint
____ **(b)** pubic symphysis
____ **(c)** coronal suture
____ **(d)** tooth in alveolar socket
____ **(e)** atlanto-axial joint
____ **(f)** intercarpal joints
____ **(g)** elbow joint
____ **(h)** carpometacarpal joint of thumb
____ **(i)** epiphyseal plate

(1) synovial, saddle
(2) gomphosis
(3) syndesmosis
(4) synovial, pivot
(5) cartilaginous (fibrocartilage)
(6) cartilaginous (hyaline cartilage)
(7) fibrous joint, synarthrosis
(8) synovial, gliding
(9) synovial, hinge

19. Match the following.

____ **(a)** monoaxial joint; only rotation possible
____ **(b)** sternoclavicular and intercarpal joints
____ **(c)** shoulder and hip joints
____ **(d)** monaxial joint; only flexion and extension possible
____ **(e)** biaxial joint

(1) ball-and-socket
(2) condyloid
(3) gliding
(4) hinge
(5) pivot

20. Match the following.

____ **(a)** movement of a body part anteriorly, horizontally to the ground
____ **(b)** horizontal movement of an anteriorly projected body part back into the anatomical position
____ **(c)** movement of the soles medially at the intertarsal joints
____ **(d)** movement of the soles laterally
____ **(e)** action that occurs when you stand on your heels
____ **(f)** position of foot when heel is on the floor and rest of foot is raised
____ **(g)** movement of the forearm to turn the palm anteriorly
____ **(h)** movement of the forearm to turn the palm posteriorly

(1) pronation
(2) plantar flexion
(3) eversion
(4) retraction
(5) inversion
(6) protraction
(7) dorsiflexion
(8) supination

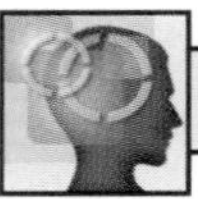

CRITICAL THINKING QUESTIONS

1. Burt and Al have been golf partners for 50 years. Burt's golf game improved by 5 points this spring and he credits the hip replacement he had last year. Al's knee has been bothering him for years but when he asked his orthopedist about a new knee joint he was told "it's not that simple". Al told Burt "one joint's like any other joint" and wants a second opinion. What will the new doctor say?
   HINT *Al knows more about golf than he does about joints.*

2. After your second Human Anatomy exam, you dropped to one knee, raised your arm over your head with one hand clenched into a fist, pumped your arm up and down, bent your head back, looked straight up, and yelled "YES!" Use the proper terms to describe the movements at the various joints.
   HINT *Think about how the position of each joint varies from anatomical position.*

3. Lars was just getting the hang of bodysurfing during here first trip to the shore when he got caught in the break of a wave. While he was being rolled in the surf, he felt his shoulder "pop". When Lars finally got back to his towel he was out of breath, in pain and his arm was hanging at an odd angle. What's the prognosis for the rest of Lars' vacation at the shore?
   HINT *The more range a joint has, the looser the fit.*

4. Taylor's workouts on the exercise machines resulted in the back muscles on her right side being stronger than the muscles on the left. She complains of strange popping noises in her back and hip and occasional pain along her back from her shoulders to her buttocks. What is causing Taylor's problems?
   HINT *Taylor's back feels better when her trainer "Pops" it back into alignment.*

5. In a new medical technique, a sample of articular cartilage is taken from a healthy joint and used to grow a new supply of cartilage cells that are then injected back into a damaged joint. Russ, an ex-rugby player, wants a sample of the cartilage in his ear used to grow new cells to repair the torn cartilage in his knee. "This way I get a new knee and I can fill the hole in my ear with a big stud." Will this work?
   HINT *Russ may need to look for another spot for that stud.*

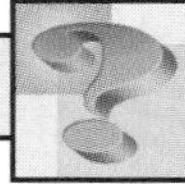

ANSWERS TO FIGURE QUESTIONS

8.1 Functionally, sutures are classified as synarthroses because they are immovable; syndesmoses are classified as amphiarthroses because they are slightly movable.

8.2 The structural difference between a synchondrosis and a symphysis is the type of cartilage that holds the joint together: hyaline cartilage in a synchondrosis and fibrocartilage in a symphysis.

8.3 Functionally, synovial joints are diarthroses, freely movable joints.

8.4 Condyloid and saddle joints are biaxial joints.

8.5 Gliding movements occur at planar joints.

8.6 Two examples of flexion that do not occur in the sagittal plane are flexion of the thumb and lateral flexion of the trunk.

8.7 When you adduct your arm or leg, you bring it closer to the midline of the body, thus "adding" it to the trunk.

8.8 Circumduction involves flexion, abduction, extension, and adduction in continuous sequence.

8.9 The anterior surface of a bone or limb rotates toward the midline in medial rotation, and away from the midline in lateral rotation.

8.10 Bringing the arms forward until the elbows touch is an example of protraction.

8.11 The lateral ligament prevents displacement of the mandible.

8.12 The shoulder joint is so freely movable because of the looseness of its articular capsule and the shallowness of the glenoid cavity in relation to the size of the head of the humerus.

8.13 A hinge joint permits flexion and extension.

8.14 Tension in three ligaments—iliofemoral, pubofemoral, and ischiofemoral—limit the degree of extension at the hip joint.

8.15 Contraction of the quadriceps causes extension at the knee joint.

9

MUSCLE TISSUE

CONTENTS AT A GLANCE

◄ Page 237

From the Changing Images you have seen so far, can you recognize whose work this may be? What is the intent of this image? Is it merely a chart displaying muscles that are to be memorized? Or is it a reference plate for painters and sculptors?

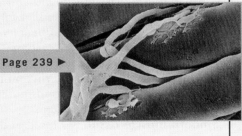

Page 239 ►

INTRODUCTION

Although bones provide leverage and form the framework of the body, they cannot move body parts by themselves. Motion results from the alternating contraction and relaxation of muscles, which constitute 40–50% of total body weight. Your muscular strength reflects the prime function of muscle—changing chemical energy into mechanical energy to generate force, perform work, and produce movement. In addition, muscle tissue stabilizes the body's position, regulates organ volume, generates heat, and propels fluids and food through various body systems. The scientific study of muscles is known as **myology** (mī-OL-ō-jē; *my-* = muscle; *-ology* = study of).

OVERVIEW OF MUSCLE TISSUE

Objective

- Correlate the three types of muscle tissue with their functions and special properties.

Types of Muscle Tissue

There are three types of muscle tissue: skeletal, cardiac, and smooth muscle (see the comparison in Table 3.4 on pages 82–83). Whereas the three types of muscle tissue share some properties, they differ from one another in their microscopic anatomy, location, and control by the nervous and endocrine systems.

Skeletal muscle tissue is so-named because the function of most skeletal muscles is to move bones of the skeleton. (A few skeletal muscles attach to structures other than bone such as the skin or even other skeletal muscles.) Skeletal muscle tissue is termed *striated* because alternating light and dark bands (*striations*) are visible when the tissue is examined under a microscope (see Figure 9.4). Skeletal muscle tissue works primarily in a *voluntary* manner because its activity can be consciously controlled.

Cardiac muscle tissue is found only in the heart, where it forms most of the heart wall. Like skeletal muscle, cardiac muscle is *striated,* but its action is *involuntary*—its alternating contraction and relaxation cannot be consciously influenced. The heart beats because it has a pacemaker that initiates each contraction; this built-in or intrinsic rhythm is called *autorhythmicity.* Several hormones and neurotransmitters adjust heart rate by speeding or slowing the pacemaker.

Smooth muscle tissue is located in the walls of hollow internal structures, such as blood vessels, airways, and most organs in the abdominopelvic cavity. It is also attached to hair follicles in the skin. Under a microscope, this tissue looks *nonstriated* or *smooth.* The action of smooth muscle is usually *involuntary,* and some smooth muscle tissue also has autorhythmicity. Both cardiac muscle and smooth muscle are regulated by neurons that are part of the autonomic (involuntary) division of the nervous system and by hormones released by endocrine glands.

Functions of Muscle Tissue

Through sustained contraction or alternating contraction and relaxation, muscle tissue has four key functions: producing body movements, stabilizing body positions, storing and moving substances within the body, and generating heat.

1. **Producing body movements.** Total body movements such as walking and running, and localized movements such as grasping a pencil or nodding the head, rely on the integrated functioning of bones, joints, and skeletal muscles.
2. **Stabilizing body positions.** Skeletal muscle contractions stabilize joints and help maintain body positions, such as standing or sitting. Postural muscles contract continuously when a person is awake; for example, sustained contractions in neck muscles hold the head upright.
3. **Storing and moving substances within the body.** Sustained contractions of ringlike bands of smooth muscles called *sphincters* may prevent outflow of the contents of a hollow organ. Temporary storage of food in the stomach or urine in the urinary bladder is possible because smooth muscle sphincters close off the outlets of these organs. Cardiac muscle contractions pump blood through the body's blood vessels. Contraction and relaxation of smooth muscle in the walls of blood vessels help adjust their diameter and thus regulate the rate of blood flow. Smooth muscle contractions also move food and substances such as bile and enzymes through the gastrointestinal tract, push gametes (sperm and oocytes) through the reproductive systems, and propel urine through the urinary system. Skeletal muscle contractions promote the flow of lymph and aid the return of blood to the heart.
4. **Producing heat.** As muscle tissue contracts, it also produces heat. Much of the heat released by muscle is used to maintain normal body temperature. Involuntary contractions of skeletal muscle, known as shivering, can increase the rate of heat production several fold.

Properties of Muscle Tissue

Muscle tissue has four special properties that enable it to function and contribute to homeostasis:

1. **Electrical excitability,** a property of both muscle cells and neurons, is the ability to respond to certain stimuli by producing electrical signals—for example, *action potentials (impulses).* The action potentials propagate (travel) along the plasma membrane due to the presence of specific ion channels. For muscle cells, the stimuli that trigger action potentials may be electrical signals arising in the muscle tissue itself, such as occurs in the heart's pacemaker, or chemical stimuli, such as neurotransmitters released by neurons, hormones distributed by the blood, or even local changes in pH.
2. **Contractility** is the ability of muscle tissue to contract forcefully when stimulated by an action potential. When muscle contracts, it generates tension (force of contraction)

while pulling on its attachment points. In an **isometric contraction** (*iso-* = equal; *-metric* = measure or length), the muscle develops tension but does not shorten. An example is holding a book in an outstretched hand. If the tension generated is great enough to overcome the resistance of the object to being moved, the muscle shortens and movement occurs. In an **isotonic contraction** (*-tonic* = tension), the tension developed by the muscle remains almost constant while the muscle shortens. An example is lifting a book off a table.

3. **Extensibility** is the ability of muscle to stretch without being damaged. Extensibility allows a muscle to contract forcefully even if it is already stretched. Normally, smooth muscle is subject to the greatest amount of stretching. For example, each time the stomach fills with food, the muscle in the wall is stretched. Cardiac muscle also is stretched each time the heart fills with blood.
4. **Elasticity** is the ability of muscle tissue to return to its original length and shape after contraction or extension.

Skeletal muscle is the focus of much of this chapter. Cardiac muscle and smooth muscle are described briefly here and are discussed in more detail in Chapter 13 (the heart), in Chapter 21 (the autonomic nervous system), and the various organs containing smooth muscle.

✔ What features distinguish the three types of muscle tissue?
✔ Summarize the functions of muscle tissue.
✔ Describe the properties of muscle tissue.

SKELETAL MUSCLE TISSUE

Objectives

- Explain the relation of connective tissue components, blood vessels, and nerves to skeletal muscles.
- Describe the microscopic anatomy of a skeletal muscle fiber.
- Explain how a skeletal muscle fiber contracts and relaxes.
- Why is muscle tone important?

Each skeletal muscle is a separate organ composed of hundreds to thousands of cells. The cells are called **fibers** because of their elongated shape. Connective tissues surround muscle fibers and whole muscles, and blood vessels and nerves penetrate into muscle (Figure 9.1). To understand how skeletal muscle contraction can generate tension, one first needs to understand its gross and microscopic anatomy.

Connective Tissue Components

Connective tissue surrounds and protects muscle tissue. A **fascia** (FASH-ē-a; = bandage) is a sheet or broad band of fibrous connective tissue that supports and surrounds organs of the body. The **superficial fascia** (or **subcutaneous layer**) separates muscle from skin (see Figure 10.23 on page 329). It is composed of areolar connective tissue and adipose tissue and provides a pathway for nerves, blood vessels, and lymphatic vesssels to enter and exit muscles. The adipose tissue of superficial fascia stores most of the body's triglycerides, serves as an insulating layer that reduces heat loss, and protects muscles from physical trauma. **Deep fascia** is dense, irregular connective tissue that lines the body wall and limbs and holds muscles with similar functions together (see Figure 10.23 on page 329). Deep fascia allows free movement of muscles, carries nerves and blood and lymphatic vessels, and fills spaces between muscles.

Three layers of connective tissue extend from the deep fascia to further protect and strengthen skeletal muscle (Figure 9.1). The outermost layer, which encircles the whole muscle, is the **epimysium** (ep-i-MĪZ-ē-um; *epi-* = upon). **Perimysium** (per-i-MĪZ-ē-um; *peri-* = around) surrounds groups of 10 to 100 or more individual muscle fibers, separating them into bundles called **fascicles** (FAS-i-kuls; = little bundles). Many fascicles are large enough to be seen with the naked eye. They give a cut of meat its characteristic "grain," and if you tear a piece of meat, it rips apart along the fascicles. Both epimysium and perimysium are dense irregular connective tissue. Penetrating the interior of each fascicle and separating individual muscle fibers from one another is **endomysium** (en′-dō-MĪZ-ē-um; *endo-* = within), a thin sheath of areolar connective tissue.

Deep fascia, epimysium, perimysium, and endomysium are all continuous with, and contribute collagen fibers to, the connective tissue that attaches skeletal muscle to other structures, such as bone or another muscle. All three connective tissue layers may extend beyond the muscle fibers to form a **tendon**—a cord of dense regular connective tissue that attaches a muscle to the periosteum of a bone. An example is the calcaneal (Achilles) tendon of the gastrocnemius (calf) muscle, which attaches the muscle to the calcaneus (see Figure 10.24c on page 333). When the connective tissue elements extend as a broad, flat layer, the tendon is called an **aponeurosis** (*apo-* = from; *neur-* = a sinew). An example of an aponeurosis is the galea aponeurotica on top of the skull between the frontalis and occipitalis muscles (see Figure 10.4c on page 266).

Certain tendons, especially those of the wrist and ankle, are enclosed by tubes of fibrous connective tissue called **tendon (synovial) sheaths,** which are similar in structure to bursae. The inner layer of a tendon sheath, the *visceral layer*, is attached to the surface of the tendon. The outer layer is known as the *parietal layer* and is attached to bone (see Figure 10.20a on page 314). Between the layers is a cavity that contains a film of synovial fluid. Tendon sheaths reduce friction as tendons slide back and forth.

A **ganglion (synovial) cyst** is a swelling, usually on the dorsum of the wrist and hand, that appears periodically, especially following activities that involve repetitive hand motions. The cyst contains fluid within fibrous connective tissue and is an extension of a tendon sheath that encloses a long extensor tendon in

Figure 9.1 / Organization of skeletal muscle and its connective tissue coverings.

A skeletal muscle consists of individual muscle fibers (cells) bundled into fascicles and surrounded by three connective tissue layers that are extensions of the deep fascia.

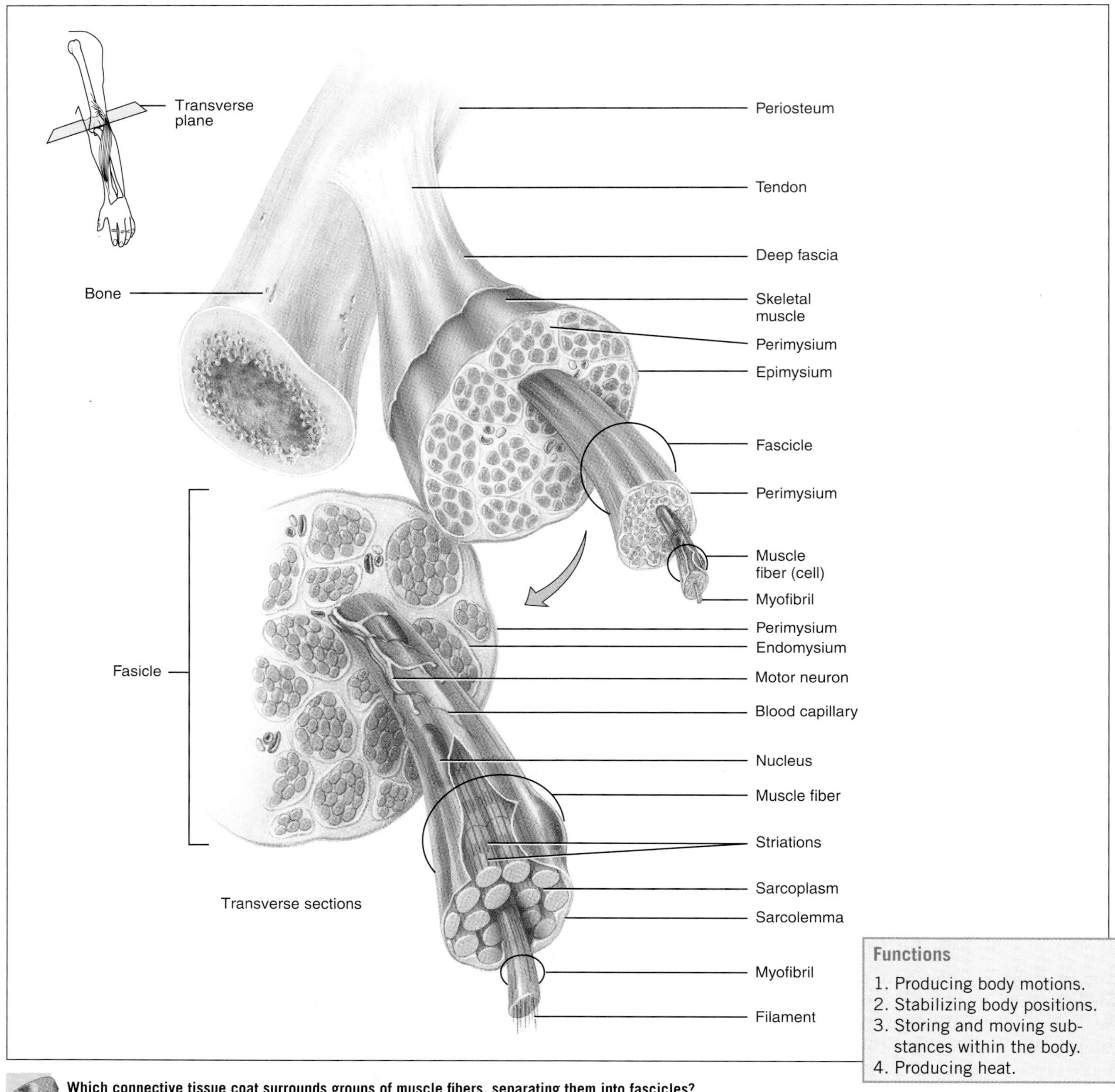

Functions

1. Producing body motions.
2. Stabilizing body positions.
3. Storing and moving substances within the body.
4. Producing heat.

Which connective tissue coat surrounds groups of muscle fibers, separating them into fascicles?

the wrist. Although they are usually not painful, ganglion cysts do cause some discomfort when they are inadvertently hit or when the wrist is flexed or extended. They result from an accumulation of fluid that has leaked from a tendon sheath or joint. As a rule they are harmless, generally requiring no treatment and having little effect on activities. ■

Microscopic Anatomy of a Skeletal Muscle Fiber

During embryonic development, each skeletal muscle fiber arises from the fusion of a hundred or more small mesodermal cells called *myoblasts*. Hence, each mature skeletal muscle fiber has a hundred or more nuclei. Once fusion has occurred, the muscle fiber loses its ability to undergo cell division. Thus, the number of skeletal muscle fibers is set before birth, and most of these fibers last a lifetime. The dramatic muscle growth that occurs after birth is achieved mainly by enlargement of existing fibers. A few myoblasts do persist in mature skeletal muscle as *satellite cells*, which retain the capacity to fuse with one another or with damaged muscle fibers to regenerate functional muscle fibers (Figure 9.2). Mature muscle fibers lie parallel to one another and range from 10 to 100 μm* in diameter. Although a typical length is 100 mm (4 in.), some are up to 30 cm (12 in.) long.

Sarcolemma, T Tubules, and Sarcoplasm

The multiple nuclei of a skeletal muscle fiber are located just beneath the **sarcolemma** (*sarc-* = flesh; *-lemma* = sheath), the fiber's plasma membrane (Figure 9.2). Thousands of tiny invaginations of the sarcolemma, called **T (transverse) tubules**, tunnel from the surface in toward the center of each muscle fiber. T tubules are open to the outside of the fiber and thus are filled with extracellular fluid. Muscle action potentials propagate

*One micrometer (μm) is 10^{-6} meter (1/25,000 in.).

Figure 9.2 / Microscopic organization of skeletal muscle. The sarcolemma of the muscle fiber encloses sarcoplasm and myofibrils and sarcoplasmic reticulum (SR) wraps around each myofibril. Thousands of T tubules filled with extracellular fluid invaginate from the sarcolemma toward the center of the muscle fiber. A T tubule and the two terminal cisterns of the SR on either side of it form a triad. A photomicrograph of skeletal muscle tissue is shown in Table 3.4A on page 82.

The contractile elements of muscle fibers are the myofibrils, which contain overlapping thick and thin filaments.

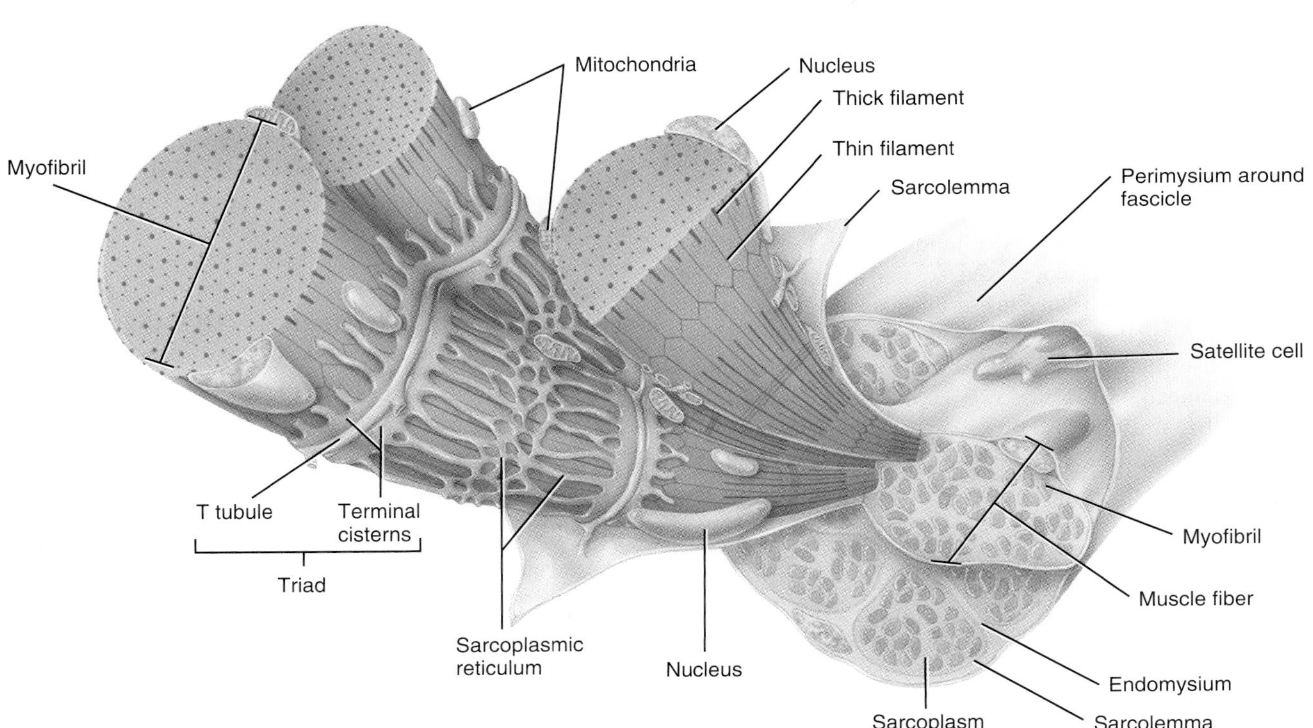

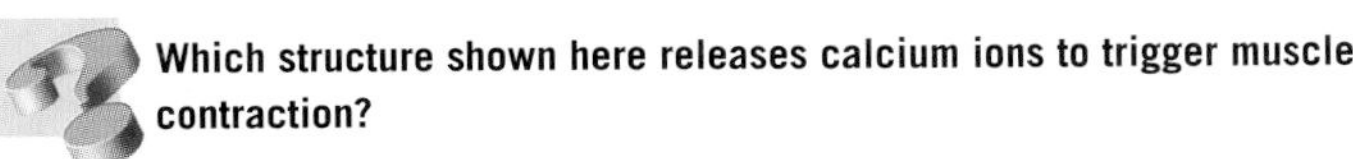

Which structure shown here releases calcium ions to trigger muscle contraction?

along the sarcolemma and through the T tubules, quickly spreading throughout the muscle fiber. This arrangement ensures that all the superficial and deep parts of the muscle fiber become excited by an action potential virtually simultaneously.

Within the sarcolemma is the **sarcoplasm,** the cytoplasm of a muscle fiber. Sarcoplasm includes a substantial amount of glycogen, which can be split into glucose that is used for ATP synthesis. In addition, the sarcoplasm contains **myoglobin** (mī-ō-GLŌB-in), a red, oxygen-binding protein, found only in muscle fibers. The oxygen is needed for ATP production within mitochondria, which lie in rows throughout the muscle fiber, strategically close to the muscle proteins that use ATP during contraction.

Myofibrils and Sarcoplasmic Reticulum

At high magnification the sarcoplasm appears stuffed with little threads. These small structures are the contractile elements of skeletal muscle, the **myofibrils** (Figure 9.2). Myofibrils are about 2 μm in diameter and extend the entire length of the muscle fiber. Their prominent striations make the whole muscle fiber look striated.

A fluid-filled system of membranous sacs called the **sarcoplasmic reticulum (SR)** encircles each myofibril (Figure 9.2). This elaborate system of sacs is similar to smooth endoplasmic reticulum in nonmuscle cells. Dilated end sacs called **terminal cisterns** (5 reservoirs) butt against the T tubule from both sides. A transverse tubule and the two terminal cisterns on either side of it form a **triad** (*tri-* = three). In a relaxed muscle fiber, the sarcoplasmic reticulum stores calcium ions (Ca^{2+}). Release of Ca^{2+} from the terminal cisterns of the sarcoplasmic reticulum triggers muscle contraction.

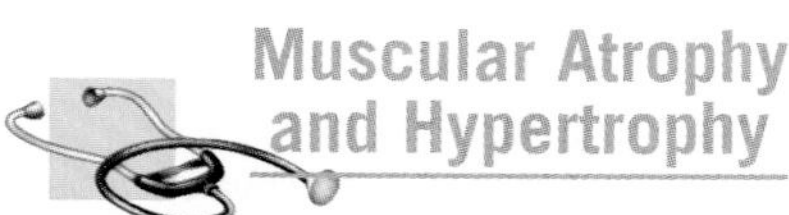

Muscular Atrophy and Hypertrophy

Muscular atrophy (A-trō-fē; *a-* = without, *-trophy* = nourishment) is a wasting away of muscles. Individual muscle fibers decrease in size as a result of progressive loss of myofibrils. The atrophy that occurs if muscles are not used is termed *disuse atrophy.* Bedridden individuals and people with casts experience disuse atrophy because the flow of nerve impulses to inactive muscle is greatly reduced. If the nerve supply to a muscle is disrupted or cut, the muscle undergoes *denervation atrophy.* Over a period of from 6 months to 2 years, the muscle shrinks to about one-fourth its original size, and the muscle fibers are replaced by fibrous connective tissue. The transition to connective tissue, when complete, cannot be reversed.

Muscular hypertrophy (hī-PER-trō-fē; *hyper-* = above or excessive) is an increase in the diameter of muscle fibers owing to the production of more myofibrils, mitochondria, sarcoplasmic reticulum, and so forth. It results from very forceful, repetitive muscular activity, such as strength training. Because hypertrophied muscles contain more myofibrils, they are capable of more forceful contractions. ■

Filaments and the Sarcomere

Within myofibrils are two types of even smaller structures called **filaments,** which are only 1–2 μm long. The diameter of the *thin filaments* is about 8 nm,* and that of the *thick filaments* is about 16 nm (Figure 9.2). The filaments inside a myofibril do not extend the entire length of a muscle fiber; instead, they are arranged in compartments called **sarcomeres** (*-mere* = part), which are the basic functional units of a myofibril (Figure 9.3a). Narrow, plate-shaped regions of dense material called **Z discs** separate one sarcomere from the next.

The thick and thin filaments overlap one another to a greater or lesser extent, depending on whether the muscle is contracted, relaxed, or stretched. The pattern of their overlap forms a variety of zones and bands (Figure 9.3b) and creates the striations that can be seen both in single myofibrils and in whole muscle fibers. The darker middle portion of the sarcomere is the **A band,** which extends the entire length of the thick filaments (Figure 9.3b). Toward each end of the A band is a **zone of overlap,** where the thick and thin filaments lie side by side. The **I band** is a lighter, less dense area that contains thin filaments but no thick filaments (Figure 9.3b). A Z disc passes through the center of each I band. A narrow **H zone** in the center of each A band contains thick but not thin filaments. Supporting proteins that hold the thick filaments together at the center of the H zone form the **M line,** so-named because it is at the *middle* of the sarcomere. Figure 9.4 shows the relations of the zones, bands, and lines as seen in a transmission electron micrograph.

Exercise-Induced Muscle Damage

Comparison of electron micrographs of muscle tissue taken from athletes before and after intense exercise reveal considerable exercise-induced muscle damage, including torn sarcolemmas in some muscle fibers, damaged myofibrils, and disrupted Z discs. Microscopic muscle damage after exercise is also indicated by increases in blood levels of proteins, such as myoglobin and the enzyme creatine kinase, that are normally confined within muscle fibers. From 12 to 48 hours after a period of strenuous exercise, skeletal muscles often become sore. Such **delayed onset muscle soreness (DOMS)** is accompanied by stiffness, tenderness, and swelling. Although the causes of DOMS are not completely understood, microscopic muscle damage appears to be a major factor. ■

Muscle Proteins

Myofibrils are built from three kinds of proteins: (1) contractile proteins, which generate force during contraction; (2) regulatory proteins, which help switch the contraction process on and off; and (3) structural proteins, which keep the thick and thin filaments in the proper alignment, give the myofibril elasticity and extensibility, and link the myofibrils to the sarcolemma and extracellular matrix.

*One nanometer (nm) is 10^{-9} meter (0.001 μm).

Figure 9.3 / The arrangement of filaments within a sarcomere. A sarcomere extends from one Z disc to the next.

Myofibrils contain two types of filaments: thick filaments and thin filaments.

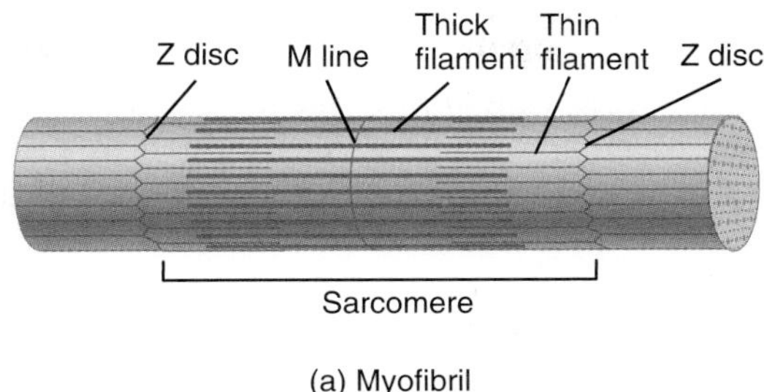

(a) Myofibril

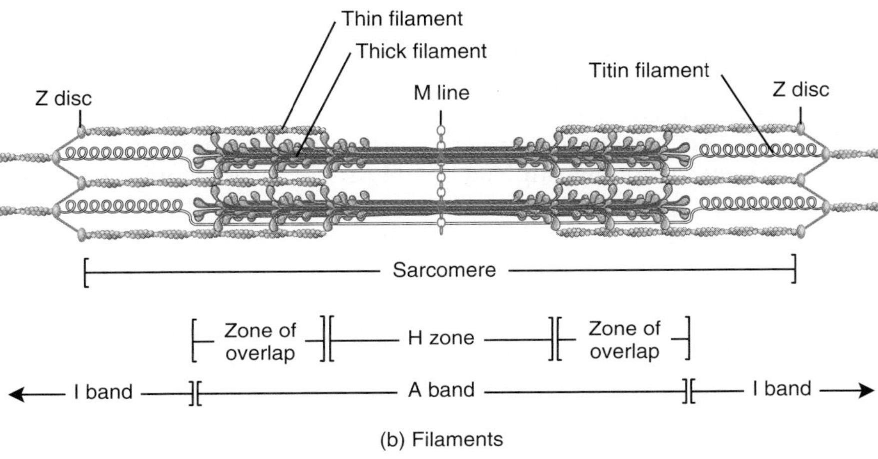

(b) Filaments

Among the following, which is smallest: muscle fiber, thick filament, or myofibril? Which is largest?

Figure 9.4 / Transmission electron micrograph showing the characteristic zones and bands of a sarcomere.

The striations of skeletal muscle are alternating darker A bands and lighter I bands.

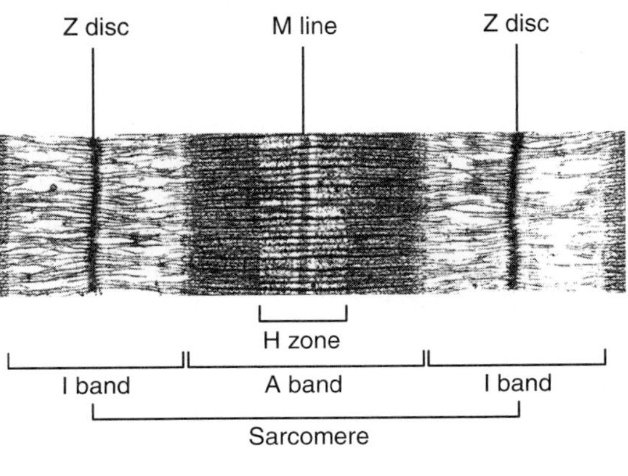

What is the ratio between thick and thin filaments in skeletal muscle?

The two *contractile proteins* in muscle are myosin and actin, which are the main components of thick and thin filaments, respectively. **Myosin** functions as a *motor protein* in all three types of muscle tissue. Motor proteins push or pull their cargo to achieve movement by converting the chemical energy in ATP to mechanical energy of motion or force production. About 300 molecules of myosin form a single thick filament. Each myosin molecule is shaped like two golf clubs twisted together (Figure 9.5a). The *myosin tail* (golf club handles) points toward the M line in the center of the sarcomere. Tails of neighboring myosin molecules lie parallel to one another, forming the shaft of the thick filament. The two projections of each myosin molecule (golf club heads) are called *myosin heads* or *crossbridges*. The heads project outward from the shaft in a spiraling fashion, each extending toward one of the six thin filaments that surround the thick filament.

Thin filaments extend from anchoring points within the Z discs (see Figure 9.3b). Their main component is the protein **actin.** Individual actin molecules join to form an actin filament that is twisted into a helix (Figure 9.5b). On each actin molecule is a *myosin-binding site*, where a myosin head can attach. Smaller amounts of two *regulatory proteins*—**tropomyosin** and **troponin**—are also part of the thin filament. In relaxed muscle, myosin is blocked from binding to actin because tropomyosin covers the *myosin-binding site* on actin. The tropomyosin strand, in turn, is held in place by troponin.

Besides contractile and regulatory proteins, muscle contains about a dozen *structural proteins* that contribute to the alignment, stability, elasticity, and extensibility of myofibrils. One such protein is *titin* (*titan* = gigantic), the third most plentiful protein in skeletal muscle (after actin and myosin). This molecule's name reflects its huge size; with a molecular weight of about 3 million daltons, titin is 50 times larger than an average protein. Each titin molecule spans half a sarcomere, from a Z disc to an M line, a distance of 1–1.2 μm in relaxed muscle. Titin anchors a thick filament to both a Z disc and the M line, thereby helping to stabilize the position of the thick filament. The portion of the titin molecule that extends from the Z disc to the beginning of the thick filament is very elastic. Because it can stretch to at least four times its resting length and then spring back unharmed, titin accounts for much of the elasticity and extensibility of myofibrils. Titin probably helps the sarcomere return to its resting length after a muscle has contracted or been stretched. Another structural protein is *dystrophin*, a cytoskeletal protein that links thin filaments of the sarcomere to integral membrane proteins of the sarcolemma. In turn, the membrane proteins attach to proteins in the connective tissue matrix that surrounds muscle fibers. Hence, dystrophin and its associated proteins are thought to reinforce the sarcolemma and help transmit to the tendons the tension generated by the sarcomeres.

Figure 9.5 / Structure of thick and thin filaments. (a) The thick filament contains about 300 myosin molecules, one of which is pictured. The myosin tails form the shaft of the thick filament, whereas the myosin heads project outward toward the surrounding thin filaments. (b) Thin filaments contain actin, troponin, and tropomyosin.

Contractile proteins (myosin and actin) generate force during contraction, whereas regulatory proteins (troponin and tropomyosin) help switch contraction on and off.

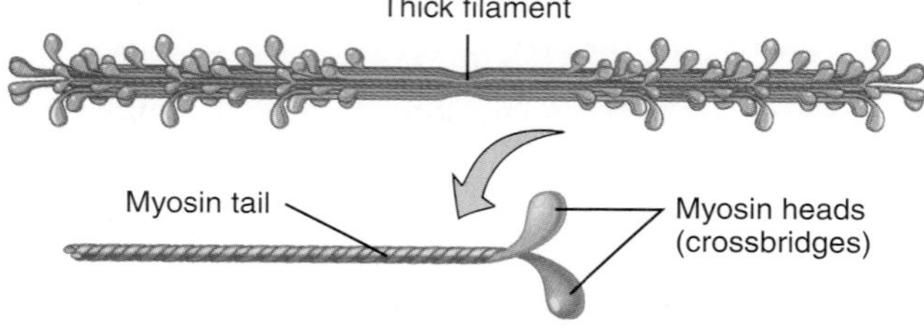

(a) One thick filament and a myosin molecule

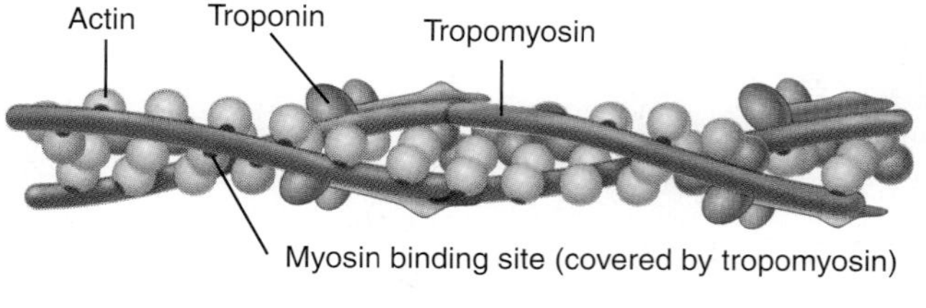

(b) Portion of a thin filament

Which proteins connect to the Z disc? Which proteins are present in the A band? In the I band?

Nerve and Blood Supply

Skeletal muscles are well supplied with nerves and blood vessels. Generally, an artery and one or two veins accompany each nerve that penetrates a skeletal muscle. The neurons (nerve cells) that stimulate skeletal muscle fibers to contract are called **motor neurons.** A motor neuron has a threadlike extension, called an **axon,** from the brain or spinal cord to a group of skeletal muscle fibers. A muscle fiber contracts in response to one or more action potentials propagating along its sarcolemma and through its T tubule system. Muscle action potentials arise at the **neuromuscular junction (NMJ),** the synapse between a motor neuron and a skeletal muscle fiber (Figure 9.6a). A **synapse** is a region where communication occurs between two neurons, or between a neuron and a target cell—for example, between a motor neuron and a muscle fiber. At most synapses a small gap, called the **synaptic cleft,** separates the two cells. Because the cells do not physically touch, the action potential from one cell cannot "jump the gap" to directly excite the next cell. Instead, the first cell communicates with the second indirectly, by releasing a chemical called a **neurotransmitter.**

At a neuromuscular junction, the motor neuron axon terminal divides into a cluster of **synaptic end bulbs** (Figure 9.6a). Suspended in the cytosol within each bulb are hundreds of membrane-enclosed sacs called **synaptic vesicles.** Inside each synaptic vesicle are thousands of molecules of **acetylcholine**

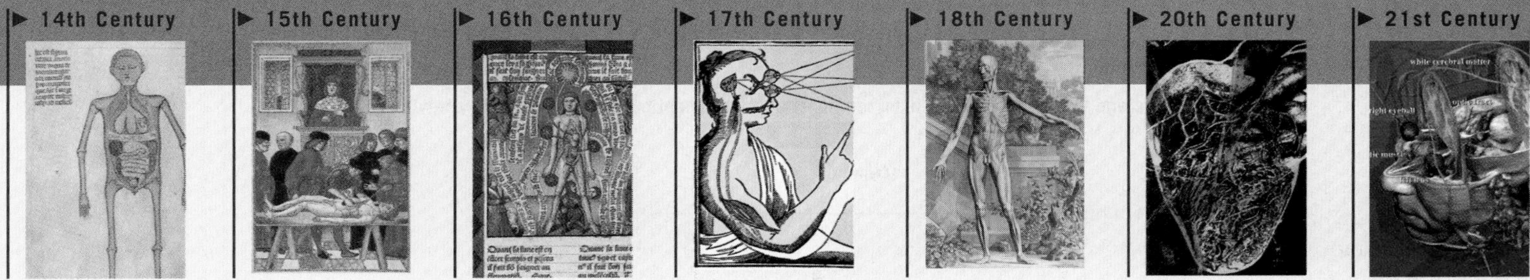

CHANGING IMAGES

The Artist Within Us

In Chapter 6 you were introduced to the work of Andreas Vesalius, the 16th century physician and anatomist whose treasure, *De Humani Corporis Fabrica*, is regarded as one of the greatest literary blends of art and science. The *Fabrica* marks the beginning of direct observation as the foundation of modern science. This shift signaled the beginning-of-the-end for Galen, who you may remember from Chapter 1, as the unquestionable authority in matters of anatomy and medicine.

The image shown here is commonly referred to as the *Fabrica's Second Muscular Figure.* While his predecessors rendered anatomical studies that appeared so very lifeless, (see Chapter 24) Vesalius toiled to bring anatomy to life. Indeed, the figure here is dynamically posed, as if in the midst of an elegant countryside dance. Clearly, this plate, which is one in a series of 14, is intended to maximally display the musculature for his medical students. But Vesalius had more than just science in mind. He writes his purpose also to be "... to display a total view of the scheme of muscles such as only painters and sculptors are wont to consider ..."

Vesalius was concerned with the anatomist and the artist; both of whom embraced the *Fabrica.* So much so was he also interested in the *Fabrica's* artistic composition, it was rediscovered within the last hundred years, that placing the plates of his muscle series side by side reveals a continuous landscape. Illustrated are the hills near the University of Padua, which was the center of 16th century artistic and scientific renaissance, and where Vesalius studied and taught anatomy.

As you further your anatomical studies, consider the other disciplines being addressed. Do you now appreciate the human form as a work of art? Are you now better able to illustrate the body with greater accuracy? Is there a connection between art and science? Perhaps anatomy should be studied not only for more than just scientific pursuits, but for artistic reasons as well.

1543 AD

Figure 9.6 / Structure of the neuromuscular junction (NMJ), the synapse between a motor neuron and a skeletal muscle fiber.

Synaptic end bulbs at the tips of axon terminals contain synaptic vesicles filled with acetylcholine.

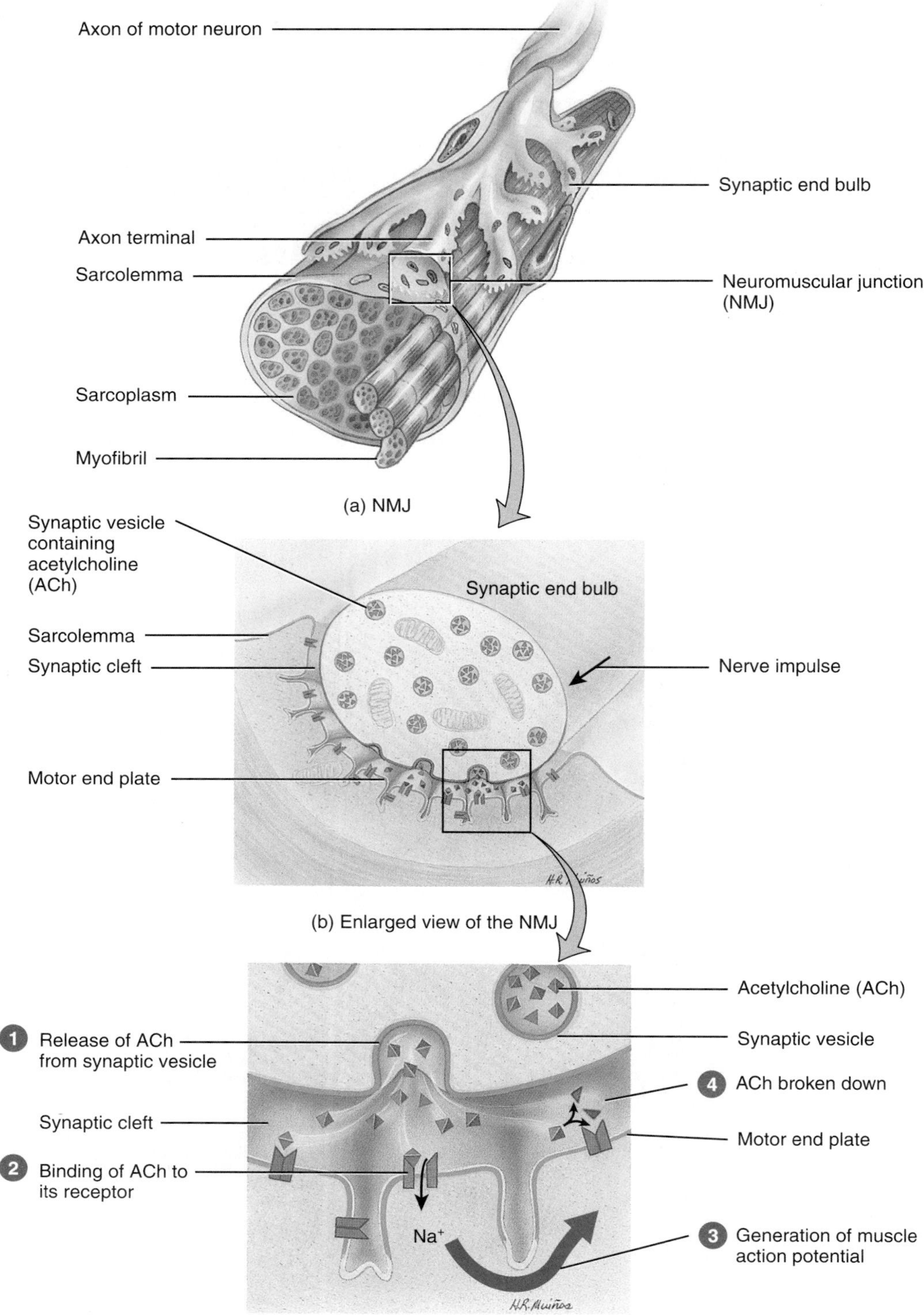

(c) Binding of acetylcholine to ACh receptors in the motor end plate

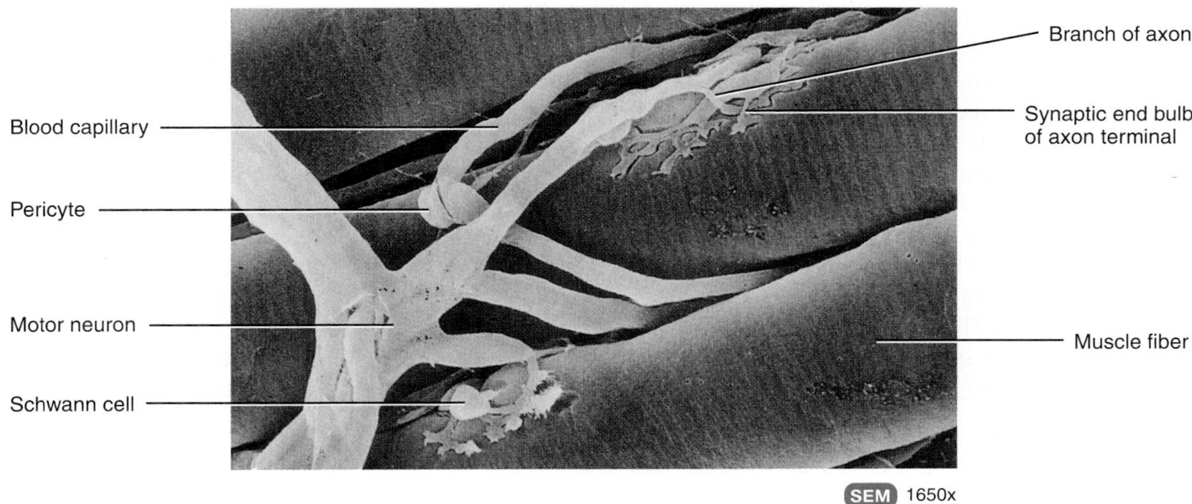

(d) Motor neuron and muscle fibers of a motor unit

What portion of the sarcolemma contains acetylcholine receptors?

(as′-ē-til-KŌ-lēn), abbreviated **ACh,** the neurotransmitter released at the NMJ. The region of the sarcolemma that is adjacent to the synaptic end bulbs is called the **motor end plate.** It contains 30–40 million *acetylcholine receptors*, which are integral transmembrane proteins that bind specifically to ACh. A neuromuscular junction thus includes all the synaptic end bulbs on one side of the synaptic cleft, plus the motor end plate of the muscle fiber on the other side.

Even though each skeletal muscle fiber has only a single neuromuscular junction, the axon of a motor neuron branches out and forms neuromuscular junctions with many different muscle fibers. A motor neuron plus all the skeletal muscle fibers it stimulates is called a **motor unit.** A single motor neuron makes contact with an average of 150 muscle fibers and all muscle fibers in one motor unit contract in unison. Typically, the muscle fibers of a motor unit are dispersed throughout a muscle rather than clustered together. Muscles that control precise movements consist of many small motor units. For example, muscles of the larynx (voice box) that control voice production have as few as two or three muscle fibers per motor unit, and muscles controlling eye movements may have 10–20 muscle fibers per motor unit. In contrast, some motor units in skeletal muscles responsible for large-scale and powerful movements, such as the biceps brachii muscle in the arm and the gastrocnemius muscle in the leg, may have 2000–3000 muscle fibers each. Remember, all the muscle fibers of a motor unit contract and relax together. Accordingly, the total strength of a contraction depends, in part, on how large the motor units are and how many motor units are activated at the same time.

Microscopic blood vessels called capillaries are plentiful in muscle tissue; each muscle fiber is in close contact with one or more capillaries (see Figure 9.1). Capillary blood brings oxygen and nutrients to the muscle fibers and removes heat and the waste products of muscle metabolism. Especially during contraction, a muscle fiber synthesizes and uses considerable ATP (adenosine triphosphate); these reactions require oxygen, glucose, fatty acids, and other substances that are supplied in the blood.

Contraction and Relaxation of Skeletal Muscle Fibers

Arrival of the nerve impulse (nerve action potential) at the synaptic end bulbs causes synaptic vesicles to fuse with the motor neuron's plasma membrane and liberate ACh. The ACh diffuses across the synaptic cleft between the motor neuron and the motor end plate (Figure 9.6c). Binding of ACh to its receptor in the sarcolemma allows small cations, most importantly Na^+, to flow across the membrane. The inflow of Na^+ makes the inside of the muscle fiber more positively charged than the outside, which changes the membrane potential and triggers a muscle action potential. The muscle action potential then propagates along the sarcolemma through the T tubule system and to the sarcoplasmic reticulum (SR), where it causes the release of calcium ions (Ca^{2+}) into the cytosol of the muscle fiber. In the presence of Ca^{2+} and ATP, the skeletal muscle shortens because the thick and thin filaments slide past each other **(sliding filament mechanism).** What happens is that the myosin heads attach to and "walk" along the thin filaments at both ends of a sarcomere, progressively pulling the thin filaments toward the M line (Figure 9.7). As a result, the thin filaments slide inward and meet at the center of a sarcomere. They may even move so far inward that their ends overlap (Figure 9.7c). As the thin filaments slide inward, the Z discs come closer together, and the sarcomere shortens. However, the lengths of the individual thick and thin filaments do not change. Shortening of the sarcomeres causes shortening of the whole muscle fiber, and ultimately shortening of the entire muscle.

The effect of ACh binding lasts only briefly because the neurotransmitter is rapidly broken down by an enzyme called **acetylcholinesterase (AChE),** which is attached to collagen fibers in the extracellular matrix of the synaptic cleft. AChE breaks down ACh into products that cannot activate the ACh receptor. If another nerve impulse releases more acetylcholine, ACh binds to its receptor, Na^+ flows inside the sarcolemma, and another muscle action potential is triggered. When action potentials cease in the motor neuron, ACh release stops and AChE rapidly breaks down the ACh already present in the synaptic cleft. This ends the generation of muscle action potentials, and the Ca^{2+} moves back into the sarcoplasmic reticulum membrane.

Because skeletal muscle fibers often are very long cells, the NMJ usually is located near the midpoint of the fiber. Muscle action potentials arise at the NMJ and then propagate toward both ends of the fiber. This arrangement permits nearly simultaneous activation (and thus contraction) of all parts of the fiber.

Figure 9.7 / Sliding filament mechanism of muscle contraction, as it occurs in two adjacent sarcomeres.

During muscle contractions, thin filaments move toward the M line of each sarcomere.

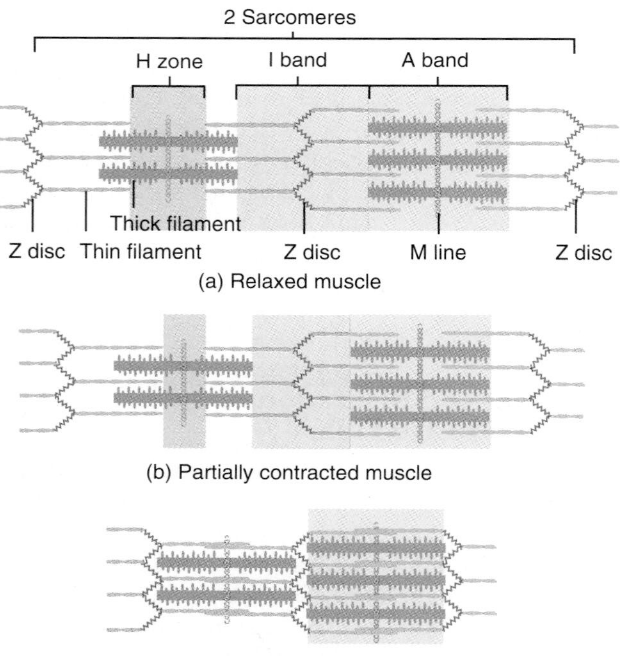

What happens to the I band and H zone during contraction? Do the lengths of the thick and thin filaments change?

After death, cellular membranes start to become leaky. Calcium ions leak out of the sarcoplasmic reticulum into the cytosol and allow myosin heads to bind to actin. ATP synthesis has ceased, however, so the crossbridges cannot detach from actin. The resulting condition, in which muscles are in a state of rigidity (cannot contract or stretch), is called **rigor mortis** (rigidity of death). Rigor mortis begins 3–4 hours after death and lasts about 24 hours; then it disappears as proteolytic enzymes from lysosomes digest the crossbridges.

Muscle Tone

In a skeletal muscle, a small number of motor units are involuntarily activated to produce a sustained contraction of their muscle fibers while the majority of the motor units are not activated and their muscle fibers remain relaxed. This process gives rise to **muscle tone** (*tonos* = tension), which keeps muscles firm. In order to sustain muscle tone, small groups of motor units become active and then become inactive, in a constantly shifting pattern. Although muscle tone keeps skeletal muscles firm, it does not result in a contraction strong enough to produce movement. For example, when the muscles in the back of the neck are in normal tonic contraction, they keep the head upright and prevent it from slumping forward on the chest, but they do not generate enough force to draw the head backward into hyperextension. Muscle tone is important in smooth muscle such as that found in the gastrointestinal tract, where the walls of the digestive organs maintain a steady pressure on their contents. The tone of smooth muscles surrounding the walls of blood vessels plays a crucial role in maintaining a steady blood pressure.

Hypotonia (*hypo* = below) refers to decreased or lost muscle tone. Such muscles are said to be **flaccid** (FLAK-sid or FLAS-sid). Flaccid muscles are loose and appear flattened rather than rounded; the affected limbs are hyperextended. Certain disorders of the nervous system and electrolyte disturbances (especially sodium and calcium and to a lesser extent magnesium) may result in *flaccid paralysis*, which is characterized by loss of muscle tone, loss or reduction of tendon reflexes, and atrophy (wasting away) and degeneration of muscles.

Hypertonia (*hyper* = above) refers to increased muscle tone and is expressed in two ways: spasticity or rigidity. **Spasticity** is characterized by increased muscle tone (stiffness) associated with an increase in tendon reflexes and pathological reflexes (such as the Babinski sign, in which the great toe extends with or without fanning of the other toes in response to stroking the outer margin of the sole). Certain disorders of the nervous system and electrolyte disturbances such as those previously noted may result in *spastic paralysis*, partial paralysis in which the muscles exhibit spasticity. **Rigidity** refers to increased muscle tone, although reflexes are not affected. ■

✔ Describe the types of fascia that cover skeletal muscles.
✔ Describe the components of a sarcomere. How do thin and thick filaments differ structurally?
✔ What proteins comprise skeletal muscle tissue and what are their functions?
✔ Describe the nerve supply to a skeletal muscle fiber.
✔ Why is a rich blood supply so important to muscle contraction?

TYPES OF SKELETAL MUSCLE FIBERS

Objective

- Compare the structure and function of the three types of skeletal muscle fibers.

Skeletal muscle fibers are not all alike in either composition or function. For example, muscle fibers vary in their content of myoglobin, the red protein that binds oxygen in muscle fibers. Skeletal muscle fibers that have a high myoglobin content are called *red muscle fibers*, while those that have a low myoglobin content are called *white muscle fibers*. Red muscle fibers also contain more mitochondria for ATP production and are supplied by more blood capillaries than are white muscle fibers.

Skeletal muscle fibers contract and relax with different velocities. A fiber is categorized as either slow or fast depending on how rapidly the ATPase in its myosin heads hydrolyzes ATP. In addition, we have seen that skeletal muscle fibers vary in the metabolic reactions they use to generate ATP and in how quickly they fatigue. Based on these structural and functional characteristics, skeletal muscle fibers are classified into three main types: (1) slow fibers, (2) fast fibers, and (3) intermediate fibers.

1. **Slow fibers.** These fibers, which are the smallest in diameter, contain large amounts of myoglobin and many blood capillaries so that they thus are dark red. Slow fibers have the fewest myofibrils among the three types. Because slow fibers generate ATP by aerobic (oxygen-requiring) reactions, they have large numbers of mitochondria. They use ATP at a slow rate and, as a result, have a slow contraction velocity. However, they are capable of sustained contractions. These fibers are very resistant to fatigue and are adapted for maintaining posture and endurance-type activities, such as running a marathon.
2. **Fast fibers.** These fibers, which are largest in diameter, contain more myofibrils than slow fibers and thus can generate more powerful contractions. They have a low myoglobin content, relatively few blood capillaries, and appear white in color. Fast fibers generate ATP by anaerobic (oxygen-free) reactions and thus have relatively few mitochondria. Anaerobic reactions cannot supply these fibers continuously with sufficient ATP, and they fatigue quickly. Fast fibers contain large numbers of glycogen granules, which store glucose to provide the fuel for ATP generation. Owing to their large size and their ability to use ATP at a high rate, fast fibers contract strongly and rapidly. They are adapted for intense movements of short duration, such as weight lifting or throwing a ball.
3. **Intermediate fibers.** These fibers are intermediate in diameter between slow and fast fibers. Like slow fibers, they contain large amounts of myoglobin, many blood capillaries, and thus are dark red. Also like slow fibers, they generate ATP by aerobic reactions and have many mitochondria. However, because they use ATP at a rapid rate like fast fibers, their contraction velocity is fast. Intermediate fibers are fatigue-resistant, but not quite as much as slow fibers. They are adapted for activities such as walking and sprinting.

Most skeletal muscles of the body are a mixture of all three types of fibers, but the proportions vary, depending on the usual action of the muscle. For example, postural muscles of the neck, back, and legs have a high proportion of slow fibers. Muscles of the shoulders and arms are not constantly active but are used intermittently, usually for short periods of time, to produce large amounts of tension such as in lifting and throwing. These muscles have a high proportion of fast fibers. Leg muscles not only support the body but are also used for walking and running. Such muscles have large numbers of both slow and intermediate fibers.

Even though most skeletal muscles are a mixture of all three types of skeletal muscle fibers, the skeletal muscle fibers of any one motor unit are all the same type. But the different motor units in a muscle may be used in various ways, depending on need. For example, if only a weak contraction is needed to perform a task, only slow motor units are activated. If a stronger contraction is needed, the motor units of fast fibers are recruited. And if a maximal contraction is required, motor units of intermediate fibers are also called into action. Activation of various motor units is determined in the brain and spinal cord.

Table 9.1 summarizes the characteristics of the three types of skeletal muscle fibers.

Table 9.1 Comparison of Structural and Functional Characteristics of Skeletal Muscle Fibers

Characteristic	Slow Fibers	Intermediate Fibers	Fast Fibers
Fiber diameter	Smallest.	Intermediate.	Largest.
Myoglobin content	Large amount.	Large amount.	Small amount.
Mitochondria	Many.	Many.	Few.
Capillaries	Many.	Many.	Few.
Color	Red.	Red.	White (pale).
Capacity to generate ATP	High capacity by aerobic (oxygen-requiring) reactions.	Intermediate capacity by aerobic (oxygen-requiring) reactions.	Low capacity by anaerobic (nonoxygen-requiring) reactions.
Rate of ATP use	Slow.	Fast.	Fast.
Contraction velocity	Slow.	Fast.	Fast.
Resistance to fatigue	High.	High, but not as resistant as slow fibers.	Low.
Location where fibers are abundant	Postural muscles such as those of the neck.	Leg muscles.	Arm muscles.
Primary functions of fibers	Maintaining posture and endurance-type activities.	Walking, sprinting.	Intense movements of short duration.

Slow fiber
Fast fiber
Intermediate fiber
LM 440x

Three types of skeletal muscle fibers in transverse section

EXERCISE AND SKELETAL MUSCLE TISSUE

The relative percentages of fast and slow fibers in each muscle are genetically determined, and the relative proportions of each help account for individual differences in physical performance. For example, people with a higher proportion of fast fibers would do very well in activities that require periods of intense activity, such as weight lifting. People with higher percentages of slow fibers are somewhat better able to perform activities that require endurance, such as running a marathon.

Although the total number of skeletal muscle fibers usually does not change, the characteristics of those present can be altered to some extent. Various types of exercises can induce changes in the fibers in a skeletal muscle. Endurance-type exercises cause a gradual transformation of some fast fibers into intermediate fibers. The transformed muscle fibers show slight increases in diameter, number of mitochondria, blood supply, and strength. Endurance (aerobic) exercises result in cardiovascular and respiratory changes that cause skeletal muscles to receive better supplies of oxygen and nutrients but do not increase muscle mass. On the other hand, exercises that require great strength for short periods of time produce an increase in the size and strength of fast fibers. The increase in size is due to increased synthesis of thin and thick filaments. The overall result is muscle enlargement.

The process by which muscles increase in size with training is not clear. Some studies show increases in the size of individual muscle fibers, whereas others claim increases in the overall number of muscle fibers. However, the number of muscle fibers is not thought to increase significantly after birth. During childhood, the increase in the *size* of muscle fibers appears to be at least partially under the control of human growth hormone, which is produced by the anterior pituitary gland. In males, a further increase in the size of muscle fibers is due to the hormone testosterone, produced by the testes.

Anabolic Steroids

The illegal use of **anabolic steroids** by athletes has received widespread attention. These steroid hormones, similar to testosterone, are taken to increase muscle size and thus strength during athletic contests. The large doses needed to produce an effect, however, have damaging, sometimes even devastating side effects, including liver cancer, kidney damage, increased risk of heart disease, stunted growth, wide mood swings, and increased irritability and aggression. Additionally, females may experience atrophy of the breasts and uterus, menstrual irregularities, sterility, facial hair growth, and deepening of the voice; males may experience diminished testosterone secretion, atrophy of the testes, and baldness. ■

✔ What is the basis for classifying skeletal muscle fibers into three types?

CARDIAC MUSCLE TISSUE

Objective

- Describe the main structural and functional characteristics of cardiac muscle tissue.

The principal tissue in the heart wall is **cardiac muscle tissue.** Although it is striated like skeletal muscle, its activity cannot be controlled voluntarily. Also, certain cardiac muscle fibers (as well as some smooth muscle fibers and nerve cells in the brain and spinal cord) display **autorhythmicity,** the ability to repeatedly generate spontaneous action potentials. In the heart, these action potentials cause alternating contraction and relaxation of the heart muscle fibers.

Cardiac muscle fibers are shorter in length, larger in diameter, and not as circular in transverse section as skeletal muscle fibers. They also exhibit branching, which gives an individual fiber a Y-shaped appearance (Figure 9.8a). The endomysium (areolar connective tissue) and numerous capillaries occupy the spaces between cardiac muscle fibers. A typical cardiac muscle fiber is 50–100 μm long and has a diameter of about 14 μm. Usually there is only one centrally located nucleus, although an occasional cell may have two nuclei. The sarcolemma of cardiac muscle fibers is similar to that of skeletal muscle, but the sarcoplasm is more abundant and the mitochondria are larger and more numerous. Cardiac muscle fibers have the same arrangement of actin and myosin and the same bands, zones, and Z discs as skeletal muscle fibers (Figure 9.8b). The transverse tubules of mammalian cardiac muscle are wider but less abundant than those of skeletal muscle: there is only one transverse tubule per sarcomere, located at the Z disc. Also, the sarcoplasmic reticulum of cardiac muscle fibers is scanty compared with the SR of skeletal muscle fibers. As a result, cardiac muscle has a limited intracellular reserve of Ca^{2+}. During contraction, a substantial amount of Ca^{2+} enters cardiac muscle fibers from extracellular fluid.

Although cardiac muscle fibers branch and interconnect with each other, they form two separate networks. The muscular walls and the partition of the superior chambers of the heart (atria) compose one network. The muscular walls and the partition of the inferior chambers of the heart (ventricles) compose the other network. The ends of each fiber in a network connect to its neighbors by irregular transverse thickenings of the sarcolemma called **intercalated** (in-TER-ka-lāted; *intercalare* = to insert between) **discs.** The discs contain **desmosomes,** which hold the fibers together, and **gap junctions,** which allow muscle action potentials to spread from one muscle fiber to another. As a consequence, when a single fiber of either network is stimulated, all the other fibers in the network become stimulated as well. Thus each network contracts as a functional unit. When the fibers of the atria contract as a unit, blood moves into the ventricles. Then, when the ventricular fibers contract as a unit, they pump blood out of the heart into arteries.

Under normal resting conditions, cardiac muscle tissue contracts and relaxes about 75 times a minute. This continuous, rhythmic activity is a major functional difference between cardiac and skeletal muscle tissue.

Another difference is the source of stimulation. Skeletal muscle tissue contracts only when stimulated by acetylcholine released by an action potential in a motor neuron. In contrast, cardiac muscle tissue can contract without extrinsic (outside) nervous or hormonal stimulation. Its source of stimulation is a conducting network of specialized cardiac muscle fibers within the heart. Stimulation from the body's nervous system merely causes the conducting fibers to increase or decrease their rate of discharge. Also, cardiac muscle tissue remains contracted 10 to 15 times longer than skeletal muscle tissue. Cardiac muscle tissue also has a long *refractory period* (period of lost excitability), lasting several tenths of a second, that allows time for the heart chambers to relax and fill with blood between beats. The long refractory period permits the heart rate to increase significantly but prevents the heart from undergoing *tetanus* (sustained contraction). If heart muscle could undergo tetanus, blood flow would cease.

✔ How do cardiac and skeletal muscle tissue differ structurally and functionally?

SMOOTH MUSCLE TISSUE

Objective

- Describe the main structural and functional characteristics of smooth muscle tissue.

Like cardiac muscle tissue, **smooth muscle tissue** is usually activated involuntarily. Of the two types of smooth muscle tissue, the more common type is **visceral (single-unit) smooth muscle tissue)** (Figure 9.9a on page 245). It is found in wraparound sheets that form part of the walls of small arteries and veins and of hollow viscera such as the stomach, intestines, uterus, and urinary bladder. Like cardiac muscle, visceral smooth muscle is autorhythmic. Because the fibers connect to one another by gap junctions, muscle action potentials spread throughout the network. When a neurotransmitter, hormone, or autorhythmic signal stimulates one fiber, the muscle action potential spreads to neighboring fibers, which then contract in unison, as a single unit.

The second kind of smooth muscle tissue is **multiunit smooth muscle tissue** (Figure 9.9b). It consists of individual fibers, each with its own motor neuron terminals and with few gap junctions between neighboring fibers. Whereas stimulation of one visceral muscle fiber causes contraction of many adjacent

Figure 9.8 / Histology of cardiac muscle. A photomicrograph of cardiac muscle tissue is shown in Table 3.4B on page 82.

Muscle fibers of the atria form one functional network, whereas muscle fibers of the ventricles form a second functional network.

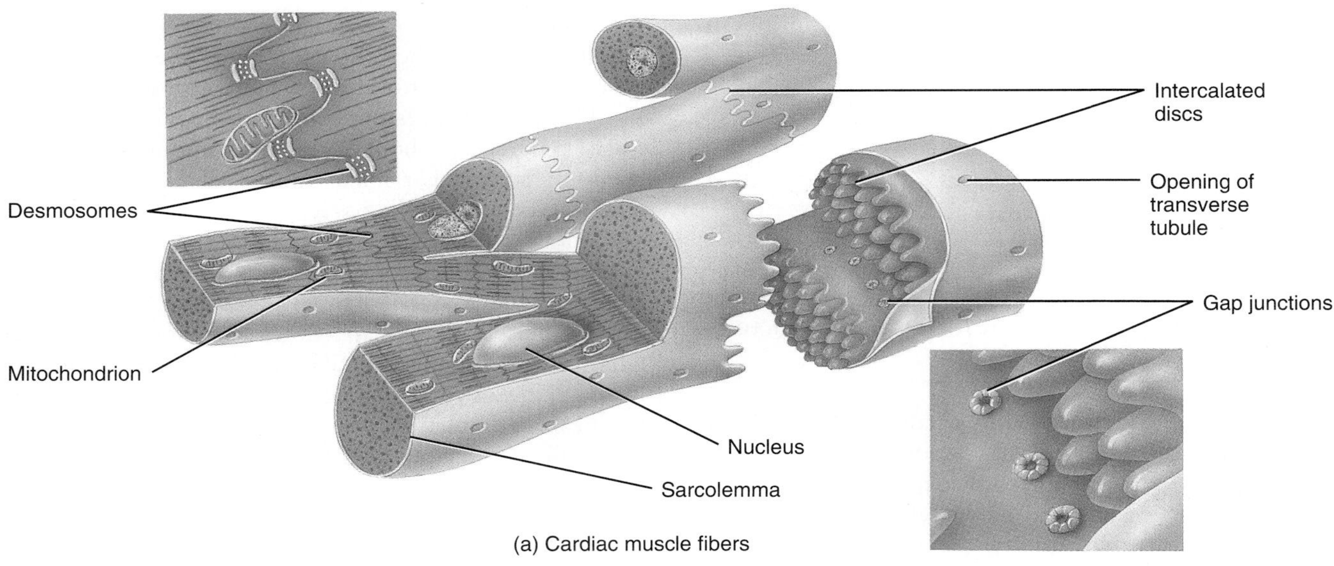

(a) Cardiac muscle fibers

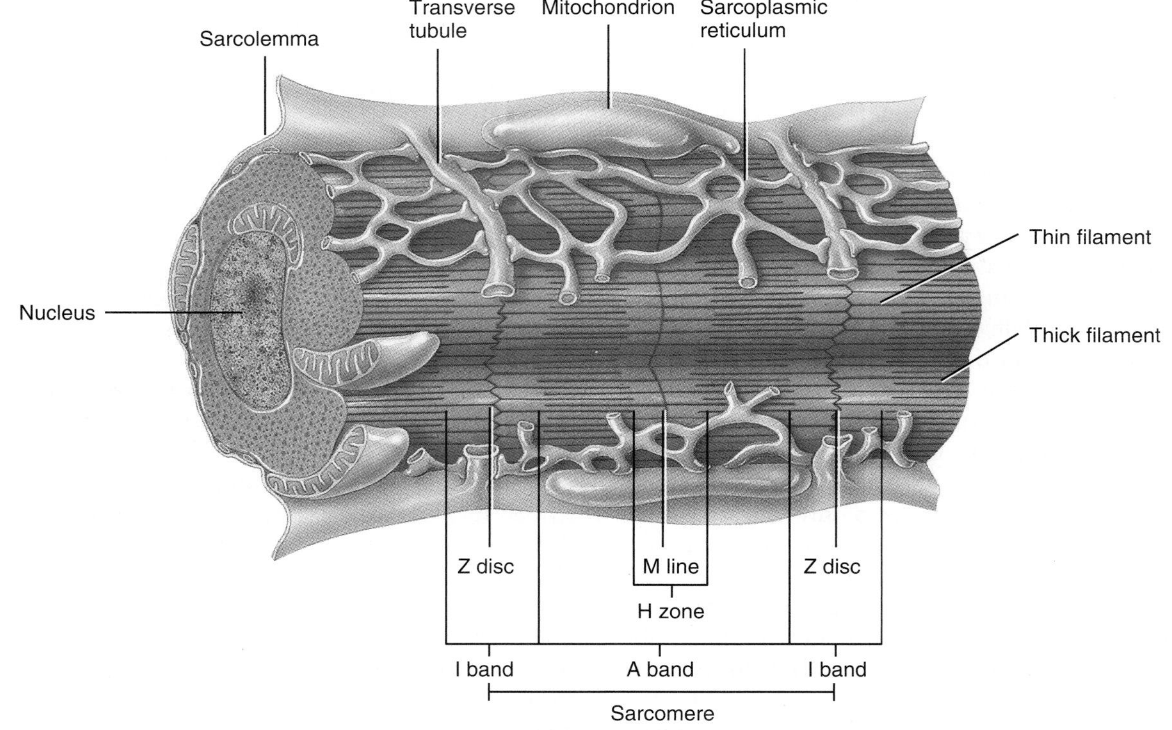

(b) Cardiac myofibrils based on an electron micrograph

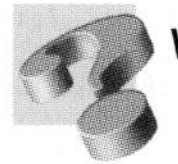

What are the functions of intercalated discs in cardiac muscle fibers?

Figure 9.9 / Histology of smooth muscle tissue. In (a), one autonomic motor neuron synapses with several visceral smooth muscle fibers, and action potentials spread to neighboring fibers through gap junctions. In (b), three autonomic motor neurons synapse with individual multiunit smooth muscle fibers. Stimulation of one multiunit fiber causes contraction of that fiber only. A photomicrograph of smooth muscle tissue is shown in Table 3.4C on page 83.

Smooth muscle fibers have thick and thin filaments but no transverse tubules and scanty sarcoplasmic reticulum.

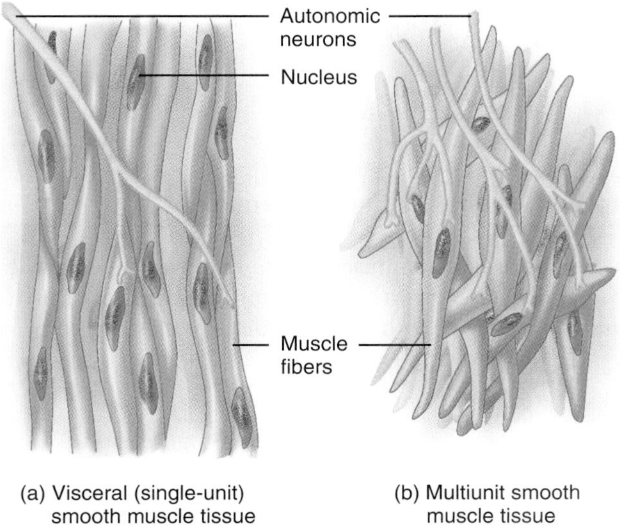

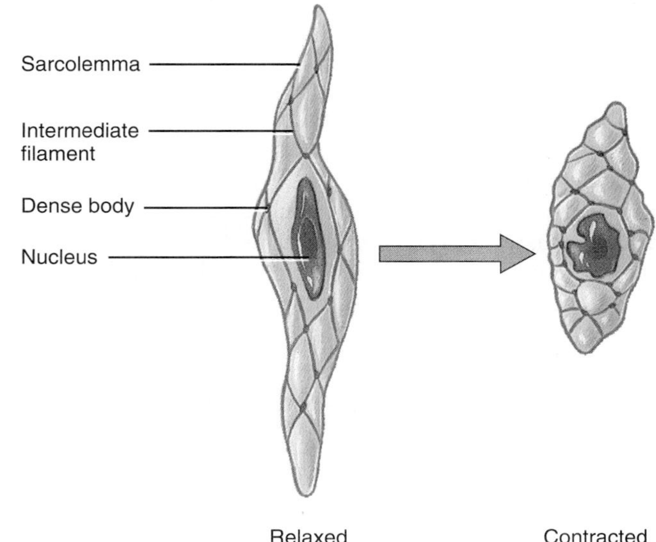

Which type of smooth muscle is more like cardiac muscle than skeletal muscle, with respect to both its structure and function?

fibers, stimulation of one multiunit fiber causes contraction of that fiber only. Multiunit smooth muscle tissue is found in the walls of large arteries, in airways to the lungs, in the arrector pili muscles that attach to hair follicles, in the muscles of the iris that adjust pupil diameter, and in the ciliary body that adjusts focus of the lens in the eye.

Endomysium surrounds smooth muscle fibers, which are considerably smaller than skeletal muscle fibers. A single relaxed smooth muscle fiber is 30–200 μm long, thickest in the middle (3–8 μm), and tapered at each end (Figure 9.9c). Within each fiber is a single, oval, centrally located nucleus. The sarcoplasm of smooth muscle fibers contain both *thick filaments* and *thin filaments*, in ratios between about 1 : 10 and 1 : 15, but they are not arranged in orderly sarcomeres as in striated muscle. Smooth muscle fibers also contain *intermediate filaments.* Because the various filaments have no regular pattern of overlap, smooth muscle fibers do not exhibit striations. This is the reason for the name *smooth.* Smooth muscle fibers also lack transverse tubules and have only scanty sarcoplasmic reticulum for storage of Ca^{2+}.

In smooth muscle fibers, intermediate filaments attach to structures called **dense bodies,** which are functionally similar to Z discs in striated muscle fibers. Some dense bodies are dispersed throughout the sarcoplasm; others are attached to the sarcolemma. Bundles of intermediate filaments stretch from one dense body to another (Figure 9.9c). During contraction, the sliding filament mechanism involving thick and thin filaments generates tension that is transmitted to intermediate filaments. These, in turn, pull on the dense bodies attached to the sarcolemma, causing a shortening of the muscle fiber. This shortening produces a bubblelike expansion of the sarcolemma (Figure 9.9c). When a smooth muscle fiber contracts, it turns like a corkscrew and rotates in the opposite direction as it relaxes.

Although the principles of contraction are similar in all three types of muscle tissue, smooth muscle tissue exhibits some important physiological differences. Compared with contraction in a skeletal muscle fiber, contraction in a smooth muscle fiber starts more slowly and lasts much longer. Moreover, smooth muscle can both shorten and stretch to a greater extent than other muscle types.

An increase in the concentration of Ca^{2+} in smooth muscle cytosol initiates contraction, just as in striated muscle. Sarcoplasmic reticulum (the reservoir for Ca^{2+} in striated muscle) is scanty in smooth muscle. Calcium ions flow into smooth muscle cytosol from both the extracellular fluid and sarcoplasmic reticulum, but because there are no transverse tubules in smooth muscle fibers, it takes longer for Ca^{2+} to reach the filaments in the center of the fiber and trigger the contractile process. This accounts, in part, for the slow onset and prolonged contraction of smooth muscle.

Calcium ions also move out of the muscle fiber slowly, which delays relaxation. The prolonged presence of Ca^{2+} in the cytosol provides for **smooth muscle tone,** a state of continued partial contraction. Smooth muscle tissue can thus sustain long-term tone, which is important in the gastrointestinal tract, where the walls maintain a steady pressure on the contents of the tract, and in the walls of blood vessels called arterioles, which maintain a steady pressure on blood.

Most smooth muscle fibers contract or relax in response to action potentials from the autonomic nervous system. In addition, many smooth muscle fibers contract or relax in response to stretching, hormones, or local factors such as changes in pH, oxygen and carbon dioxide levels, temperature, and ion concentrations.

Unlike striated muscle fibers, smooth muscle fibers can stretch considerably and still maintain their contractile function. When smooth muscle fibers are stretched, they initially contract, developing increased tension. Within a minute or so, the tension decreases. This phenomenon, termed the **stress–relaxation response,** allows smooth muscle to undergo great changes in length while still retaining the ability to contract effectively. Thus even though smooth muscle in the walls of blood vessels and hollow organs such as the stomach, intestines, and urinary bladder can stretch, the pressure on the contents within them changes very little. After the organ empties, however, the smooth muscle in the wall rebounds, and the wall retains its firmness.

✔ How do visceral and multiunit smooth muscle differ?
✔ Make a table comparing the properties of skeletal and smooth muscle.

REGENERATION OF MUSCLE TISSUE

Objective

- Explain how muscle fibers regenerate.

Because mature skeletal muscle fibers have lost the ability to undergo cell division, growth of skeletal muscle after birth is due mainly to hypertrophy, the enlargement of existing cells, rather

Table 9.2 Summary of the Principal Features of the Three Types of Muscle Tissue

Characteristic	Skeletal Muscle	Cardiac Muscle	Smooth Muscle
Microscopic appearance and features	Long cylindrical fiber with many peripherally located nuclei; striated.	Branched cylindrical fiber with one centrally located nucleus; intercalated discs join neighboring fibers; striated.	Spindle-shaped fiber with one centrally positioned nucleus; not striated.
Location	Attached primarily to bones by tendons.	Heart.	Walls of hollow viscera, airways, blood vessels, iris and ciliary body of eye, arrector pili of hair follicles.
Fiber diameter	Very large (10–100 μm).	Large (14 μm).	Small (3–8 μm).
Connective tissue components	Endomysium, perimysium, and epimysium.	Endomysium.	Endomysium.
Fiber length	100 μm to 30 cm.	50–100 μm.	30–200 μm.
Contractile proteins organized into sarcomeres	Yes.	Yes.	No.
Sarcoplasmic reticulum	Abundant.	Some.	Scanty.
Transverse tubules present?	Yes, aligned with each A-I band junction.	Yes, aligned with each Z disc.	No.
Junctions between fibers	None.	Intercalcated discs containing gap junctions and desmosomes.	Gap junctions in visceral smooth muscle; none in multiunit smooth muscle.
Autorhythmicity	No.	Yes.	Yes, in visceral smooth muscle.
Source of Ca^{2+} for contraction	Sarcoplasmic reticulum.	Sarcoplasmic reticulum and extracellular fluid.	Sarcoplasmic reticulum and extracellular fluid.
Speed of contraction	Fast.	Moderate.	Slow.
Nervous control	Voluntary.	Involuntary.	Involuntary.
Capacity for regeneration	Limited, via satellite cells.	None.	Considerable, via pericytes (compared with other muscle tissues, but limited compared with epithelium).

than to hyperplasia, an increase in the number of fibers. Satellite cells divide slowly and fuse with existing fibers to assist both in muscle growth and in repair of damaged fibers. In response to muscle injury or a disease that causes muscle degeneration, additional cells derived from red bone marrow migrate into the muscle and participate in regeneration of the damaged fibers. The number of new skeletal muscle fibers formed in these ways, however, is not sufficient to compensate for significant skeletal muscle damage or degeneration. In such cases, skeletal muscle tissue undergoes **fibrosis,** the replacement of muscle fibers by fibrous scar tissue. For this reason, skeletal muscle tissue has only limited powers of regeneration.

Cardiac muscle does not have cells comparable to the satellite cells of skeletal muscle. Damaged cardiac muscle fibers are not repaired or replaced, and healing occurs by fibrosis. Cardiac muscle fibers can undergo hypertrophy, however, in response to increased workload. For this reason, many athletes have enlarged hearts.

Smooth muscle tissue, like skeletal and cardiac muscle tissue, can undergo hypertrophy. In addition, certain smooth muscle fibers, such as those in the uterus, retain their capacity for division and thus can grow by hyperplasia. Also, new smooth muscle fibers can arise from cells called *pericytes,* stem cells found in association with blood capillaries and small veins (see Figure 9.6d). Smooth muscle fibers can also proliferate in certain pathological conditions, such as occur in the development of atherosclerosis. Compared with the other two types of muscle tissue, smooth muscle tissue has considerably greater powers of regeneration. Such powers are still limited when compared with other tissues, such as epithelium.

Table 9.2 on page 246 summarizes the principal characteristics of the three types of muscle tissue.

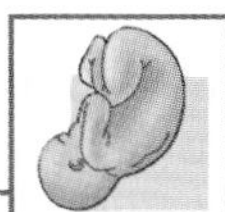

DEVELOPMENTAL ANATOMY OF THE MUSCULAR SYSTEM

Objective

- Describe the development of the muscular system.

In this brief discussion of the development of the human muscular system, we concentrate mostly on skeletal muscles. Except for the muscles of the iris of the eyes and the arrector pili muscles attached to hairs, all muscles of the body are derived from **mesoderm.** As the mesoderm develops, a portion of it becomes arranged in dense columns on either side of the developing nervous system. These columns of mesoderm undergo segmentation into a series of blocks of cells called **somites** (Figure 9.10a). The first pair of somites appears on the 20th day of embryonic development. By the 30th day, 44 pairs of somites have formed.

With the exception of the skeletal muscles of the head and limbs, *skeletal muscles* develop from the **mesoderm of somites.** Because there are very few somites in the head region of the embryo, most of the skeletal muscles there develop from the **general mesoderm** in the head region. The skeletal muscles of the limbs develop from masses of general mesoderm around developing bones in embryonic limb buds (origins of future limbs; see Figure 5.11a on page 126).

The cells of a somite are differentiated into three regions: (1) **myotome,** which forms some of the skeletal muscles; (2) **dermatome,** which forms the connective tissues, including the dermis; and (3) **sclerotome,** which gives rise to the vertebrae (Figure 9.10b).

Cardiac muscle develops from **mesodermal cells** that migrate to and envelop the developing heart while it is still in the form of primitive heart tubes (see in Figure 13.13 on page 407).

Smooth muscle develops from **mesodermal cells** that migrate to and envelop the developing gastrointestinal tract and viscera.

✔ From which embryonic tissue do most muscles develop?

Figure 9.10 / Location and structure of somites, key structures in the development of the muscular system.

Most muscles are derived from mesoderm.

(a) Dorsal aspect of an embryo showing somites

(b) Transverse section of a somite

Which portion of a somite differentiates into skeletal muscle?

AGING AND MUSCLE TISSUE

Objective

- Explain how aging affects skeletal muscle.

Beginning at about 30 years of age, humans undergo a slow progressive loss of skeletal muscle mass that is replaced largely by fibrous connective tissue and adipose tissue. In part, this decline is due to increasing inactivity. Accompanying the loss of muscle mass is a decrease in maximal strength and a slowing of muscle reflexes. In some muscles, a selective loss of muscle fibers of a given type may occur. With aging, the relative number of slow fibers appears to increase. This could be due either to atrophy of the other fiber types or their conversion into slow fibers. Whether this is an effect of aging itself or mainly reflects the more limited physical activity of older people is still an unresolved question. Nevertheless, endurance and strength training programs are effective in older people and can slow or even reverse the age-associated decline in muscular performance.

✔ Why do skeletal muscle fibers have only limited powers of regeneration?

✔ Why does muscle strength decrease with aging?

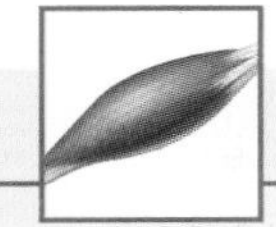

APPLICATIONS TO HEALTH

Muscle function may be abnormal due to disease or damage of any of the components of a motor unit: somatic motor neurons, neuromuscular junctions, or muscle fibers. The term **neuromuscular disease** encompasses problems at all three sites, whereas the term **myopathy** (mī-OP-a-thē; *-pathy* = disease) signifies a disease or disorder of the skeletal muscle tissue itself.

Myasthenia Gravis

Myasthenia gravis (mī-as-THĒ-nē-a GRAV-is) is an autoimmune disease that causes chronic, progressive damage of the neuromuscular junction. In people with myasthenia gravis the immune system inappropriately produces antibodies that bind to and block some ACh receptors, thereby decreasing the number of functional ACh receptors at the motor end plates of skeletal muscles (see Figure 9.6c). Because 75% of patients with myasthenia gravis have hyperplasia or tumors of the thymus, it is possible that thymic abnormalities may cause the disorder. As the disease progresses, more ACh receptors are affected. Muscles become increasingly weaker, fatigue more easily, and may eventually cease to function.

Myasthenia gravis occurs in about 1 in 10,000 people and is more common in women, who typically are ages 20–40 at onset, than in men, who usually are ages 50–60 at onset. The muscles of the face and neck are most often affected. Initial symptoms include a weakness of the eye muscles, which may produce double vision, and difficulty in swallowing. Later, the individual has difficulty chewing and talking. Eventually the muscles of the limbs may become involved. Death may result from paralysis of the respiratory muscles, but often the disorder does not progress to this stage.

Anticholinesterase drugs such as pyridostigmine (Mestinon) or neostigmine are the first line of treatment for myasthenia gravis. They act as inhibitors of acetylcholinesterase, the enzyme that breaks down ACh. Thus, they raise the level of ACh that is available to bind with still-functional receptors. More recently, steroid drugs, such as prednisone, have been used with success to reduce antibody levels. Another treatment is plasmapheresis, a procedure that removes the antibodies from the blood. Often, surgical removal of the thymus (thymectomy) is helpful.

Muscular Dystrophy

Muscular dystrophy is a group of inherited muscle-destroying diseases that cause progressive degeneration of skeletal muscle fibers. The most common form of muscular dystrophy is *Duchenne muscular dystrophy* (*DMD;* dū-SHĀN). Because the mutated gene is on the X chromosome, and because males have only one copy of it, DMD strikes boys almost exclusively. Worldwide, about 1 in every 3500 male babies—21,000 in all—are born with DMD each year. The disorder usually becomes apparent between the ages of 2 and 5, when parents notice the child falls often and has difficulty running, jumping, and hopping. By age 12 most boys with DMD are unable to walk. Respiratory or cardiac failure usually causes death between the ages of 20 and 30.

In DMD, the gene that codes for dystrophin is mutated, and little or no dystrophin is present. Without the reinforcing effect of dystrophin, the sarcolemma easily tears during muscle contraction. Because their plasma membrane is damaged, muscle fibers slowly rupture and die. The dystrophin gene was discovered in 1987, and by 1990 the first attempts were made to treat DMD patients with gene therapy. The muscles of three boys with DMD were injected with myoblasts bearing functional dystrophin genes, but only a few muscle fibers gained the ability to produce dystrophin. Similar clinical trials with additional patients have also failed. An alternate approach to the problem is to find a way to induce muscle fibers to produce a protein called utrophin, which is structurally similar to dystrophin. Experiments with dystrophin-deficient mice suggest this approach may work.

Abnormal Contractions of Skeletal Muscle

One kind of abnormal muscular contraction is a **spasm,** a sudden involuntary contraction of a single muscle in a large group of muscles. A painful spasmodic contraction is known as a

cramp. A **tic** is a spasmodic twitching made involuntarily by muscles that are ordinarily under voluntary control. Twitching of the eyelid and facial muscles are examples of tics. A **tremor** is a rhythmic, involuntary, purposeless contraction that produces a quivering or shaking movement. A **fasciculation** is an involuntary, brief twitch of an entire motor unit that is visible under the skin; it occurs irregularly and is not associated with movement of the affected muscle. Fasciculations may be seen in multiple sclerosis (see page 512) or in amyotrophic lateral sclerosis (Lou Gehrig's disease). A **fibrillation** is a spontaneous contraction of a single muscle fiber that is not visible under the skin but can be recorded by electromyography. Fibrillations may signal destruction of motor neurons.

KEY MEDICAL TERMS ASSOCIATED WITH MUSCLE TISSUE

Charley horse Tearing of a muscle as a result of forceful impact, accompanied by bleeding and severe pain. It often occurs in contact sports and typically affects the quadriceps femoris muscle on the anterior surface of the thigh. The condition is treated by ice immediately after the injury, rest, a supportive wrap, and elevation of the limb.

Hypertonia (*hyper-* = above; *tonia* = tension) Increased muscle tone, either rigidity (muscle stiffness) or spasticity (muscle stiffness associated with an increase in tendon reflexes).

Hypotonia (*hypo-* = below or deficient) Decreased or lost muscle tone, usually due to damage to somatic motor neurons.

Myalgia (mī-AL-jē-a; *my-* = muscle; *-algia* = painful condition) Pain in or associated with muscles.

Myoma (mī-Ō-ma; *-oma* = tumor) A tumor consisting of muscle tissue.

Myomalacia (mī′-ō-ma-LĀ-shē-a; *-malacia* = soft) Softening of a muscle.

Myopathy (mī-OP-a-thē; *pathos* = disease) Any disease of muscle tissue.

Myositis (mī′-ō-SĪ-tis; *-itis* = inflammation of) Inflammation of muscle fibers.

Myotonia (mī′-ō-TŌ-nē-a) Increased muscular excitability and contractility, with decreased power of relaxation; tonic spasm of the muscle.

Volkmann's contracture (FŌLK-manz kon-TRAK-tur; *contra-* = against) Permanent shortening (contracture) of a muscle due to replacement of destroyed muscle fibers by fibrous connective tissue, which lacks extensibility. Destruction of muscle fibers may occur from interference with circulation caused by a tight bandage, a piece of elastic, or a cast.

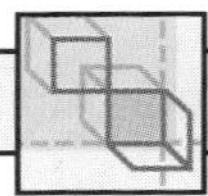

STUDY OUTLINE

Introduction (p. 230)

1. Muscles constitute 40–50% of total body weight.
2. The prime function of muscle is changing chemical energy into mechanical energy to perform work.

Overview of Muscle Tissue (p. 230)

1. The three types of muscle tissue are skeletal, cardiac, and smooth. Skeletal muscle tissue is primarily attached to bones; it is striated and under voluntary control. Cardiac muscle tissue forms the wall of the heart; it is striated and involuntary—not under voluntary control. Smooth muscle tissue is located primarily in internal organs; it is nonstriated (smooth) and involuntary.
2. Through contraction and relaxation, muscle tissue performs four important functions: producing body movements, stabilizing body positions, storing and moving substances within the body, and producing heat.
3. Four special properties of muscle tissues are electrical excitability, the property of responding to stimuli by producing action potentials; contractility, the ability to generate tension to do work; extensibility, the ability to be extended (stretched); and elasticity, the ability to return to original shape after contraction or extension.
4. An isotonic contraction occurs when a muscle shortens and moves a constant load; tension remains almost constant. An isometric contraction occurs without shortening of the muscle; tension increases greatly.

Skeletal Muscle Tissue (p. 231)

1. Connective tissues surrounding skeletal muscles are the epimysium, covering the entire muscle; perimysium, covering fascicles; and the endomysium, covering muscle fibers. Superficial fascia separates muscle from skin.
2. Tendons and aponeuroses are extensions of connective tissue beyond muscle fibers that attach the muscle to bone or to other muscle.
3. Each skeletal muscle fiber has 100 or more nuclei because it arises from fusion of many myoblasts. Satellite cells are myoblasts that persist after birth. The sarcolemma is a muscle fiber's plasma membrane; it surrounds the sarcoplasm. T tubules are invaginations of the sarcolemma.
4. Each fiber contains myofibrils, the contractile elements of skeletal muscle. Sarcoplasmic reticulum surrounds each myofibril. Within a myofibril are thin and thick filaments arranged in compartments called sarcomeres.
5. The overlapping of thick and thin filaments produces striations; darker A bands alternate with lighter I bands.
6. Myofibrils are built from three types of proteins: contractile, regulatory, and structural. The contractile proteins are myosin (thick filament) and actin (thin filament). Regulatory proteins are tropomyosin and troponin (both are part of the thin filament). Structural proteins include titin (links Z disc to M line and stabilizes thick filament).

7. Projecting myosin heads (crossbridges) contain actin-binding and ATP-binding sites and are the motor proteins that power muscle contraction.
8. Skeletal muscles are well supplied with nerves and blood vessels. Generally, an artery and one or two veins accompany each nerve that penetrates a skeletal muscle. Blood capillaries bring in oxygen and nutrients and remove heat and waste products of muscle metabolism.
9. Motor neurons provide the nerve impulses that stimulate skeletal muscle to contract.
10. The neuromuscular junction (NMJ) is the synapse between a somatic motor neuron and a skeletal muscle fiber. The NMJ includes the axon terminals and synaptic end bulbs of a motor neuron, plus the adjacent motor end plate of the muscle fiber sarcolemma.
11. When an action potential reaches the end bulbs of a somatic motor neuron, it triggers exocytosis of the synaptic vesicles. ACh diffuses across the synaptic cleft and binds to ACh receptors, initiating a muscle action potential. Acetylcholinesterase then quickly destroys ACh.
12. Muscle contraction occurs because myosin heads attach to and "walk" along the thin filaments at both ends of a sarcomere, progressively pulling the thin filaments toward the center of a sarcomere. As the thin filaments slide inward, the Z discs come closer together, and the sarcomere shortens.

Types of Skeletal Muscle Fibers (p. 241)

1. On the basis of their structure and function, skeletal muscle fibers are classified as slow, fast, and intermediate fibers.
2. Most skeletal muscles contain a mixture of all three fiber types, but their proportions vary with the typical action of the muscle.
3. Table 9.1 on page 242 summarizes the three types of fibers.

Exercise and Skeletal Muscle Tissue (p. 242)

1. Various types of exercises can induce the transformation of one skeletal muscle fiber type into another.
2. The increase in size of fast skeletal muscle fiber is due to increased synthesis of thin and thick filaments.

Cardiac Muscle Tissue (p. 243)

1. This muscle tissue is found only in the heart. It is striated and involuntary.
2. The fibers are branching cylinders and usually contain a single centrally located nucleus.
3. Compared to skeletal muscle tissue, cardiac muscle tissue has more sarcoplasm, more mitochondria, less well-developed sarcoplasmic reticulum, and wider transverse tubules located at Z discs rather than at A–I band junctions. Filaments are not arranged in discrete myofibrils.
4. Cardiac muscle fibers branch and are connected end to end via desmosomes.
5. Intercalated discs provide strength and aid in conduction of muscle action potentials by way of gap junctions located in the discs.
6. Unlike skeletal muscle tissue, cardiac muscle tissue contracts and relaxes rapidly, continuously, and rhythmically.
7. Cardiac muscle tissue can contract without extrinsic stimulation and can remain contracted longer than skeletal muscle tissue.

Smooth Muscle Tissue (p. 243)

1. Smooth muscle tissue is nonstriated and involuntary.
2. Smooth muscle fibers contain intermediate filaments and dense bodies that function as Z discs.
3. Visceral (single-unit) smooth muscle is found in the walls of viscera and small blood vessels. The fibers are arranged in a network.
4. Multiunit smooth muscle is found in large blood vessels, arrector pili muscles, and the iris of the eye. The fibers operate independently rather than in unison.
5. The duration of contraction and relaxation of smooth muscle is longer than in skeletal muscle.
6. Smooth muscle fibers contract in response to nerve impulses, hormones, and local factors.
7. Smooth muscle fibers can stretch considerably without developing tension.

Regeneration of Muscle Tissue (p. 246)

1. Skeletal muscle fibers cannot divide and have limited powers of regeneration; cardiac muscle fibers can neither divide nor regenerate; smooth muscle fibers have limited capacity for division and regeneration.
2. Table 9.2 on page 246 summarizes the principal characteristics of the three types of muscle tissue.

Developmental Anatomy of the Muscular System (p. 247)

1. With few exceptions, muscles develop from mesoderm.
2. Skeletal muscles of the head and limbs develop from general mesoderm; the remainder of the skeletal muscles develop from the mesoderm of somites.

Aging and Muscle Tissue (p. 248)

1. Beginning at about 30 years of age, humans undergo a slowly progressive loss of skeletal muscle, which is replaced by fibrous connective tissue and fat.
2. Aging also results in a decrease in muscle strength and in diminished muscle reflexes.

SELF-QUIZ QUESTIONS

Choose the one best answer to these questions.

1. The specialized region of the sarcolemma at the neuromuscular junction is called the (a) synapse, (b) synaptic cleft, (c) motor end plate, (d) cisterna, (e) motor unit
2. The ability of muscle tissue to respond to a stimulus by producing action potentials is referred to as (a) contractility, (b) excitability, (c) elasticity, (d) extensibility, (e) conductivity.
3. Which of the following is a function of muscle tissue? (a) motion, (b) regulation of organ volume, (c) maintenance of posture, (d) heat production, (e) all of the above.
4. Which of the following pairs is *not* appropriately matched? (a) slow fibers-small diameter, (b) fast fibers-low myoglobin content, (c) slow fibers-few blood capillaries, (d) fast fibers-large diameter, (e) slow fibers-high myoglobin content.

5. Choose the *false* statement about cardiac muscle. (a) Cardiac muscle has a relatively long refractory period. (b) Cardiac muscle usually has one centrally located nucleus per fiber. (c) Cardiac muscle cells have less sarcoplasmic reticulum than skeletal muscle, and therefore store limited amounts of calcium. (d) Cardiac muscle fibers are separated at their ends by intercalated discs. (e) Cardiac muscle fibers are spindle-shaped with no striations.

6. A motor unit is defined as a (a) nerve and a muscle, (b) single neuron and a single muscle fiber, (c) neuron and all the muscle fibers it stimulates, (d) single muscle fiber and the nerves that innervate it, (e) muscle and the motor and sensory nerves that innervate it.

7. Arrange the following from largest to smallest. (1) myofibril, (2) filament, (3) muscle fiber, (4) fascicle

(a) 1, 3, 2, 4 **(b)** 3, 4 ,2, 1 **(c)** 3, 1, 4, 2 **(d)** 4, 3, 1, 2 **(e)** 4, 2, 3, 1

Complete the following.

8. In a relaxed skeletal muscle fiber, the ________ stores calcium ions.

9. Red skeletal muscle fibers have a high ________ content and also contain numerous ________ for ATP production.

10. The ________ of muscle fibers allow for rapid spread of an action potential into the interior of the muscle cell.

11. Each muscle cell (also known as a ________) is enclosed by a cell membrane called the ________ and contains cytoplasm called ________.

12. To describe a sarcomere during contraction, complete each statement with one of the following: L (lengthens), S (shortens), or U (is unchanged in length). (a) The sarcomere ________. (b) Each thick myofilament (A band) ________. (c) Each thin myofilament ________. (d) The I band ________. (e) The H zone ________.

13. The muscle tissues that exhibit autorhythmicity are ________ and ________.

Are the following statements true or false?

14. The two contractile proteins in muscle fibers are actin and myosin.

15. In adult life, growth of skeletal muscle is due to an increase in the number of muscle cells rather than an increase in the size of existing cells.

16. Elasticity is the property of muscle tissue that allows it to be stretched.

17. Stimulation of one visceral (single-unit) smooth muscle fiber causes contraction of several adjacent fibers; stimulation of a multiunit fiber causes contraction of that fiber only.

18. Intermediate skeletal muscle fibers are adapted for walking, and running.

19. Match the following.

____ **(a)** involuntary muscle found in blood vessels and intestine
____ **(b)** involuntary striated muscle
____ **(c)** striated voluntary muscle attached to bones
____ **(d)** spindle-shaped cells
____ **(e)** intercalated discs
____ **(f)** no transverse tubules
____ **(g)** several nuclei, peripherally arranged

(1) cardiac
(2) skeletal
(3) smooth

20. Match the following.

____ **(a)** encircles individual muscles, fascicles, and fibers
____ **(b)** cord of dense regular connective tissue that most often attaches muscle to the periosteum of bone
____ **(c)** tube of fibrous connective tissue; similar in structure to bursae; reduce friction
____ **(d)** a broad, flat layer of dense connective tissue; connects skeletal muscle to other structures
____ **(e)** lines body wall and limbs; separates muscles into functional groups
____ **(f)** under the skin; provides insulation

(1) tendon sheath
(2) epimysium
(3) deep fascia
(4) superficial fascia
(5) tendon
(6) aponeurosis

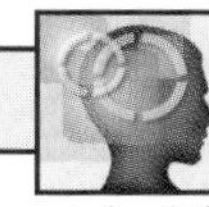

CRITICAL THINKING QUESTIONS

1. A marathon runner and a weight-lifter were discussing muscle types while working out. Both were convinced that their own muscle types were the best. How does the leg muscle of the runner compare to the arm of the lifter?

HINT *The speed of the muscle fiber is just as important as the speed of the runner.*

2. Bill tore some ligaments in his knee while skiing. He was in a toe-to-thigh cast for 6 weeks. When the cast was removed, the newly healed leg was noticeably thinner than the uncasted leg. What happened to his leg?

HINT *If you don't use it, you lose it.*

3. The newspaper reported several cases of botulism poisoning following a fund-raiser potluck dinner for the local clinic. The cause appeared to be three-bean salad "flavored" with the bacterium *Colstridium botulinum*. The C. *botulinum* produces a toxin that blocks the release of acetylcholine from the motor neuron. What would be the affect of botulism poisoning on muscle function.

HINT *How does the neuron's action potential "jump" to the muscle?*

4. Research is underway to grow new cardiac muscle cells for ailing hearts. Skeletal muscle transplants have been tried, but they don't work as well as cardiac muscle. Both skeletal and cardiac muscle are striated so why does cardiac muscle support rhythmic contraction while skeletal muscle does not?

HINT *Tetanus is OK for a skeletal muscle but not for the heart.*

5. The detective in the murder mystery was called to the scene of the crime. She declared that the unfortunate demise of the murder victim must have happened less than 24 hours ago due to the rigid appearance of the body. Explain her pronouncement?

HINT *ATP synthesis will stop after death.*

ANSWERS TO FIGURE QUESTIONS

9.1 Perimysium bundles groups of muscle fibers into fascicles.

9.2 The sarcoplasmic reticulum releases calcium ions to trigger muscle contraction.

9.3 Size, from smallest to largest: thick filament, myofibril, muscle fiber.

9.4 There are two thin filaments for each thick filament in skeletal muscle.

9.5 Actin and titin connect to the Z disc. A bands contain myosin, actin, troponin, tropomyosin, and titin; I bands contain actin, troponin, tropomyosin, and titin.

9.6 The portion of the sarcolemma that contains acetylcholine receptors is the motor end plate.

9.7 The I bands and H zones disappear. The lengths of the filaments do not change.

9.8 The intercalated discs contain desmosomes that hold the cardiac muscle fibers together and gap junctions that enable action potentials to spread from one muscle fiber to another.

9.9 Visceral smooth muscle and cardiac muscle are similar in that both contain gap junctions, which allow action potentials to spread from one cell to its neighbors.

9.10 As its name suggests, the myotome of a somite differentiates into skeletal muscle.

10

THE MUSCULAR SYSTEM

CONTENTS AT A GLANCE

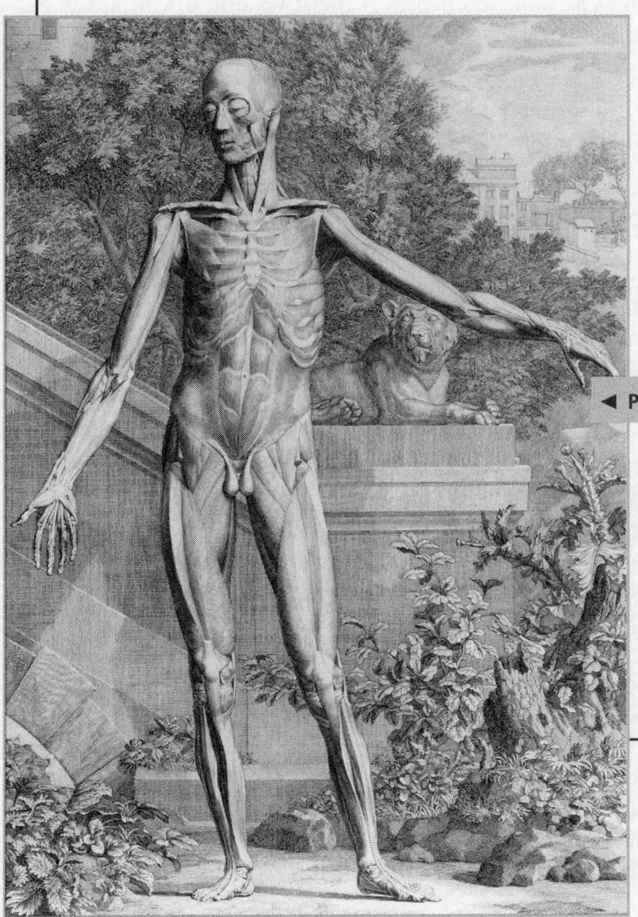

◄ Page 262

Have you ever pulled a hamstring or experienced a shin splint as a result of an athletic activity? Understanding the relation of muscles to movement can explain how these injuries occur.

Page 260 ►

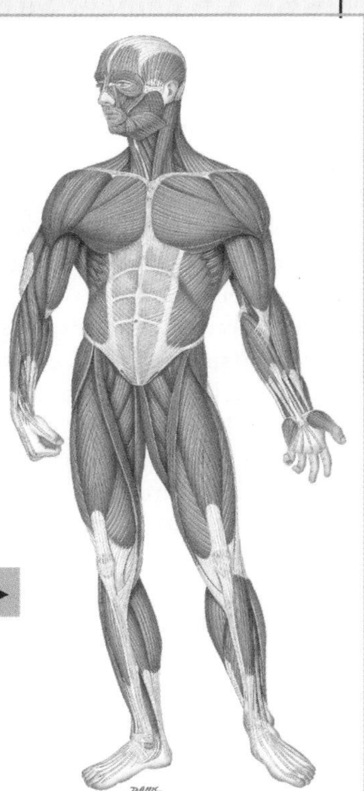

INTRODUCTION

Although there are three types of muscle tissue in the body (skeletal, cardiac, and smooth), the term **muscular system,** refers to the voluntary *skeletal* muscle system: the skeletal muscle tissue and connective tissues that make up individual muscle organs, such as the biceps brachii muscle.

To understand how movement occurs at specific joints, you must be able to identify the body's major skeletal muscle groups. Moreover, it is important to know the attachment sites of each muscle, as well as its innervation (nerve supply). Developing a working knowledge of these key points of skeletal muscle anatomy will enable you to analyze how movements happen normally. This knowledge is exceptionally important for a host of allied health and physical rehabilitation professionals who work with people whose normal patterns of movement and physical mobility have been disrupted by circumstances such as physical trauma, surgery, or muscular paralysis. Accordingly, in this chapter we will identify the principal skeletal muscle groups and consider how each muscle group produces voluntary movement.

HOW SKELETAL MUSCLES PRODUCE MOVEMENTS

Muscle Attachment Sites: Origin and Insertion

Objective

- Describe the relationship between bones and skeletal muscles in producing body movements.

As you will see later, not all skeletal muscles produce movements. For example, some skeletal muscles stabilize bones so that another skeletal muscle can execute movements more efficiently. Skeletal muscles that do produce movements do so by exerting force on tendons, which in turn pull on bones or other structures (such as skin). Most muscles cross at least one joint and are usually attached to articulating bones that form the joint (Figure 10.1a).

When a skeletal muscle contracts, it draws one of the articulating bones toward the other. The two articulating bones usually do not move equally in response to contraction. One bone remains stationary or near its original position, either because other muscles stabilize that bone by contracting and pulling it in the opposite direction or because its structure makes it less movable. Ordinarily, the attachment of a muscle's tendon to the stationary bone is called the **origin;** the attachment of the muscle's other tendon to the movable bone is called the **insertion.** A good analogy is a spring on a door. In this example, the part of the spring attached to the frame is the origin; the part attached to the door represents the insertion. A useful rule of thumb is that the origin is usually proximal and the insertion distal, especially in the limbs and the insertion is usually pulled toward the origin.

The fleshy portion of the muscle between the tendons is called the **belly** (*gaster*). Keep in mind that muscles that move a body part often do not cover the moving part. Figure 10.1b shows that although one of the functions of the biceps brachii muscle is to move the forearm, the belly of the muscle lies over the humerus, not the forearm. You will also see that muscles that cross two joints, such as the rectus femoris and sartorius, have more complex actions than muscles that cross only one joint.

In the limbs, skeletal muscles and their associated blood vessels and nerves are grouped into functional groups called compartments. In the upper limbs, for example, flexor compartment muscles are on the anterior surface whereas extensor compartment muscles are on the posterior surface.

Tenosynovitis (ten′-ō-sin-ō-VĪ-tis) is an inflammation of the tendons, tendon sheaths, and synovial membranes surrounding certain joints. The tendons most often affected are at the wrists, shoulders, elbows (resulting in a condition called tennis elbow), finger joints (resulting in a condition called trigger finger), ankles, and feet. The affected sheaths sometimes become visibly swollen because of fluid accumulation. Tenderness and pain are frequently associated with movement of the body part. The condition often follows trauma, strain, or excessive exercise. Tenosynovitis of the dorsum of the foot may be caused by tying shoelaces too tightly. Also, gymnasts are prone to developing the condition as a result of chronic, repetitive, and maximum hyperextension at the wrists. ■

✔ Using the terms origin, insertion, and belly in your discussion, describe how skeletal muscles produce body movements by pulling on bones.

Lever Systems and Leverage

Objective

- Define a lever and a fulcrum, and compare the three types of levers on the basis of location of the fulcrum, effort, and resistance.

In producing movements, bones act as levers, and joints function as the fulcrums of these levers. A **lever** may be defined as a rigid rod that moves around some fixed point called a **fulcrum,** symbolized by F. A lever is acted on at two different points by two different forces: the **effort** (E), which causes movement, and the **resistance** R, or *load,* which opposes movement. The effort is the force exerted by muscular contraction, whereas the resistance is typically the weight of the body part that is moved. Motion occurs when the effort applied to the bone at the insertion exceeds the resistance (load). Consider the biceps brachii flexing the forearm at the elbow as an object is lifted (Figure 10.1b). When the forearm is raised, the elbow is the fulcrum. The weight of the forearm plus the weight of the object in the hand is the resistance. The force of contraction of the biceps brachii pulling the forearm up is the effort.

Figure 10.1 / Relationship of skeletal muscles to bones. (a) Muscles are attached to bones by tendons known as the origin and insertion. (b) Skeletal muscles produce movements by pulling on bones. Bones serve as levers, and joints act as fulcrums for the levers. Here the lever-fulcrum principle is illustrated by the movement of the forearm. Note where the resistance and effort are applied in this example.

In the limbs, the origin of a muscle is usually proximal and the insertion is usually distal.

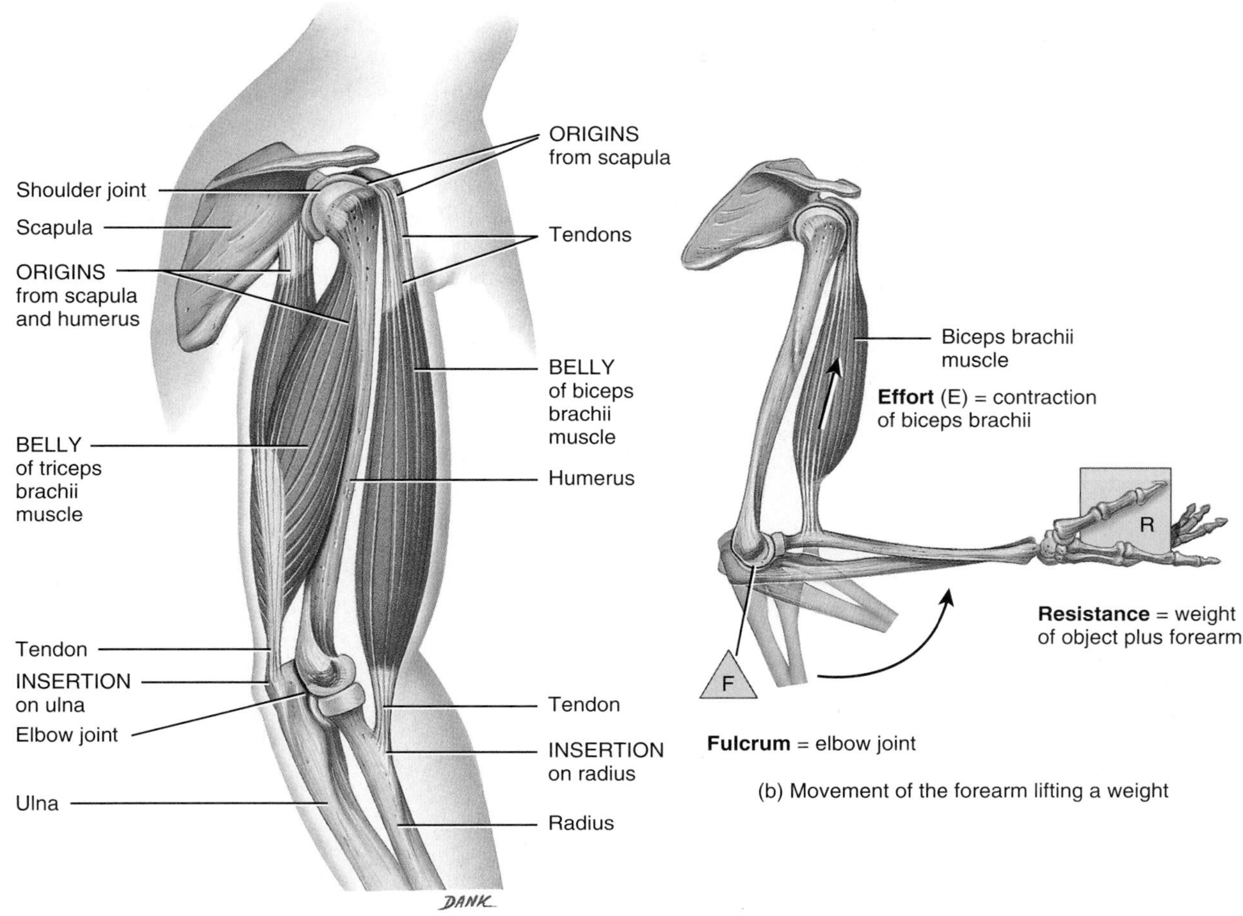

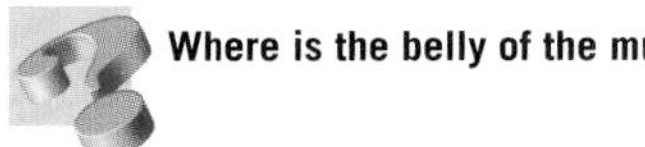

Where is the belly of the muscle that extends the forearm located?

Levers produce trade-offs between effort and the speed and range of motion. In one situation, a lever operates at a *mechanical advantage*—has **leverage**—when a less forceful effort can move a more forceful resistance. Here the tradeoff is that the effort must move farther (must have a longer range of motion) and faster than the resistance. Recall that range of motion refers to the range, measured in degrees of a circle, through which the bones of a joint can be moved. The lever formed by the mandible at the temporomandibular joints (fulcrums) and the effort provided by contraction of the jaw muscles produce a powerful mechanical advantage that crushes food. In another situation, a lever operates at a *mechanical disadvantage* when a more forceful effort moves a less forceful resistance. In this case the trade-off is that the effort moves a shorter distance and less quickly than the resistance. The lever formed by the humerus at the shoulder joint (fulcrum) and the effort provided by the back and shoulder muscles produces a mechanical "disadvantage" that enables a big-league pitcher to hurl a baseball at nearly 100 miles per hour!

The positions of the effort, resistance, and fulcrum on the lever determine whether the lever operates at a mechanical advantage or disadvantage. When the resistance is close to the fulcrum and the effort is applied farther away, the lever operates at a mechanical advantage. When you chew food, the resistance (the food) is positioned close to the fulcrums (your temporomandibular joints) while your jaw muscles exert effort farther out from the joints. On the other hand, when the effort is applied close to the fulcrum and the resistance is applied farther away, the lever operates at a mechanical disadvantage. When a pitcher throws a baseball, the back and shoulder muscles apply intense effort very close to the fulcrum (the shoulder joint) while the lighter resistance (the ball) is propelled at the far end of the lever (the arm bone).

Levers are categorized into three types according to the positions of the fulcrum, the effort, and the resistance:

1. In **first-class levers** the fulcrum is between the effort and the resistance (Figure 10.2a). (Think E*F*R.) Scissors and seesaws are examples of first-class levers. A first-class lever can produce either a mechanical advantage or disadvantage depending on whether the effort or the resistance is closer to the fulcrum. As we've seen in the preceding examples, if the effort is farther from the fulcrum than the resistance, a strong resistance can be moved, but not very far or fast. If the effort is closer to the fulcrum than the resistance, only a weaker resistance can be moved, but it moves far and fast.

   There are few first-class levers in the body. One is the lever formed by the head resting on the vertebral column (Figure 10.2a). When the head is raised, the weight of the anterior portion of the skull is the resistance. The atlanto-occipital joint between the atlas and the occipital bone forms the fulcrum. The contraction of the posterior neck muscles provides the effort.
2. In **second-class levers** the resistance is between the fulcrum and the effort (Figure 10.2b). (Think F*R*E.) These levers operate like a wheelbarrow. They always produce a mechanical advantage because the resistance is always closer to the fulcrum than the effort. This arrangement sacrifices speed and range of motion for force. Most authorities believe that there are no second-class levers in the body.

Figure 10.2 / Types of levers.

Levers are divided into three types based on the placement of the fulcrum, effort, and resistance.

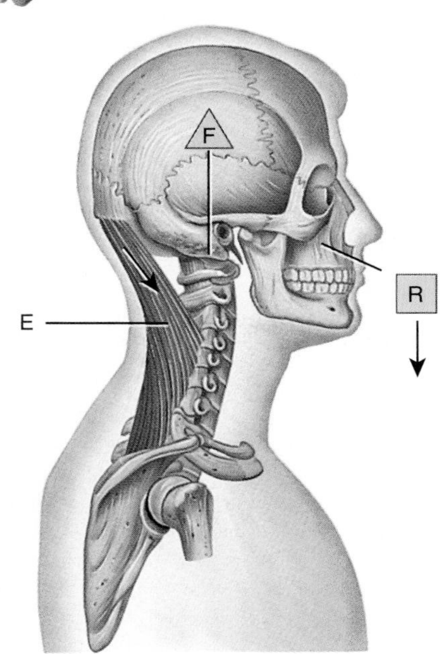

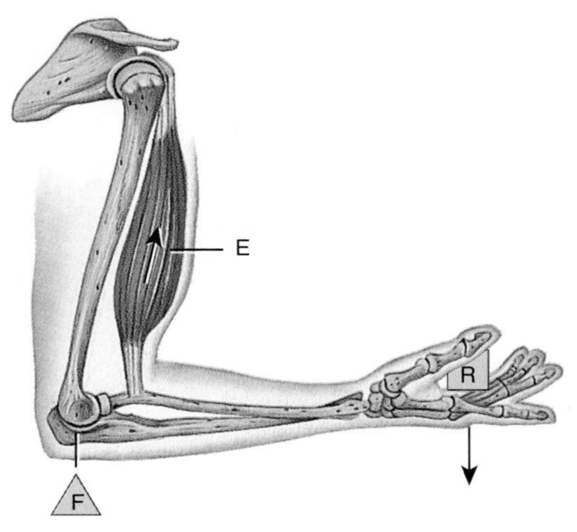

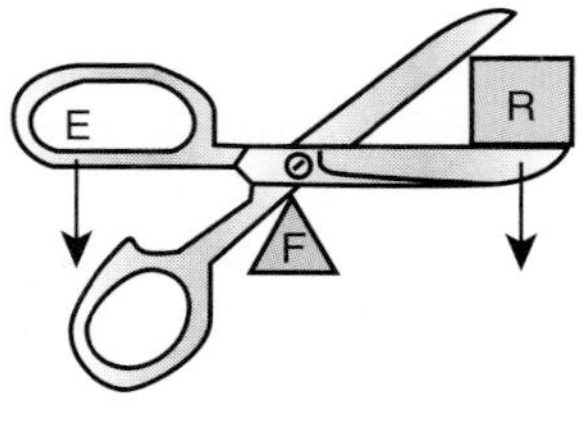

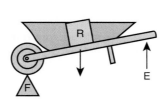

(a) First-class lever

(b) Second-class lever

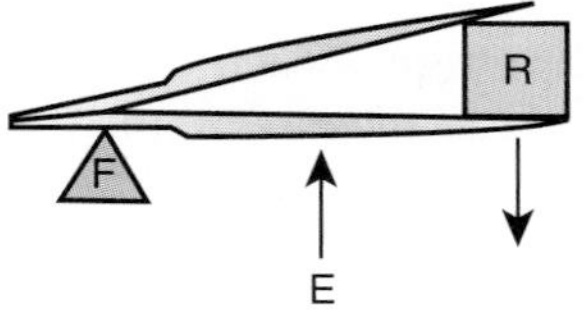

(c) Third-class lever

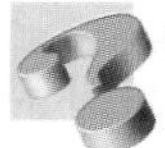

Which type of lever produces the most force?

3. In **third-class levers** the effort is between the fulcrum and the resistance (Figure 10.2c). (Think F*E*R.) These levers operate like a pair of forceps and are the most common levers in the body. Third-class levers always produce a mechanical disadvantage because the effort is always closer to the fulcrum than the resistance. In the body, this arrangement favors speed and range of motion over force. The elbow joint, the bones of the arm and forearm, and the biceps brachii muscle make up a third-class lever (Figure 10.2c). As we have seen, in flexing the forearm at the elbow, the weight of the hand and forearm is the resistance, the elbow joint is the fulcrum, and the contraction of the biceps brachii muscle provides the effort. Another example of the action of a third-class lever is adduction of the thigh, in which the thigh is the resistance, the hip joint is the fulcrum, and the contraction of the adductor muscles is the effort.

Effects of Fascicle Arrangement

Objective

- Identify the various arrangements of muscle fibers in a skeletal muscle, and relate the arrangements to strength of contraction and range of motion.

Recall from Chapter 9 that skeletal muscle fibers (cells) are arranged within the muscle in bundles called **fascicles.** The muscle fibers are arranged in a parallel fashion within each bundle, but the arrangement of the fascicles with respect to the tendons may take one of five characteristic patterns: parallel, fusiform (cigar-shaped), circular, triangular, or pennate (shaped like a feather) (Table 10.1).

Fascicular arrangement affects a muscle's power and range of motion. When a muscle fiber contracts, it shortens to a length about 70% of its resting length. Thus, the longer the

Table 10.1 Arrangement of Fascicles

Parallel	Fusiform
Fascicles parallel to longitudinal axis of muscle; terminate at either end in flat tendons.	Fascicles nearly parallel to longitudinal axis of muscle; terminate in flat tendons; muscle tapers toward tendons, where diameter is less than at belly.
Example: Stylohyoid muscle (see Figure 10.8)	*Example:* Digastric muscle (see Figure 10.8)
Circular	**Triangular**
Fascicles in concentric circular arrangements form sphincter muscles that enclose an orifice (opening).	Fascicles spread over broad area converge at thick central tendon; gives muscle a triangular appearance.
Example: Orbicularis oculi muscle (see Figure 10.4)	*Example:* Pectoralis major muscle (see Figure 10.3a)

Pennate

Short fascicles in relation to total muscle length; tendon extends nearly entire length of muscle.

Unipennate	Bipennate	Multipennate
Fascicles are arranged on only one side of tendon.	Fascicles are arranged on both sides of centrally positioned tendon.	Fascicles attach obliquely from many directions to several tendons.
Example: Extensor digitorum longus muscle (see Figure 10.24b)	*Example:* Rectus femoris muscle (see Figure 10.22a)	*Example:* Deltoid muscle (see Figure 10.12a)

fibers in a muscle, the greater the range of motion it can produce. By contrast, the strength of a muscle depends on the total number of fibers it contains, because a short fiber can contract as forcefully as a long one. Because a given muscle can contain either a small number of long fibers or a large number of short fibers, fascicular arrangement represents a compromise between power and range of motion. Pennate muscles, for example, have a large number of fascicles distributed over their tendons, giving them greater power but a smaller range of motion. Parallel muscles, in contrast, have comparatively few fascicles that extend the length of the muscle; thus, they have a greater range of motion but less power.

Intramuscular Injections

An **intramuscular (IM) injection** penetrates the skin and subcutaneous tissue to enter the muscle itself. Intramuscular injections are preferred when prompt absorption is desired, when larger doses than can be given subcutaneously are indicated, or when the drug is too irritating to give subcutaneously. The common sites for intramuscular injections include the gluteus medius muscle of the buttock (see Figure 10.3b), lateral side of the thigh in the midportion of the vastus lateralis muscle (see Figure 10.3a), and the deltoid muscle of the shoulder (see Figure 10.3b). Muscles in these areas, especially the gluteal muscles in the buttock, are fairly thick, and absorption is promoted by the extensive blood supply to such large muscles. To avoid injury, intramuscular injections are given deep within the muscle and away from major nerves and blood vessels. ■

Coordination within Muscle Groups

Objective

- Explain how the prime mover, antagonist, synergist, and fixator in a muscle group work together to produce movements.

Most movements are the result of several skeletal muscles acting as a group rather than individually. Most skeletal muscles are arranged in opposing (antagonistic) pairs at joints—that is, flexors-extensors, abductors-adductors, and so on. Within opposing pairs, one muscle, called the **prime mover** or **agonist** (= leader), contracts to cause a desired action while the other muscle, the **antagonist** (*anti-* = against), stretches and yields to the effects of the prime mover. In the process of flexing the forearm at the elbow, for instance, the biceps brachii is the prime mover, and the triceps brachii is the antagonist (see Figure 10.1). The antagonist and prime mover are usually located on opposite sides of the bone or joint, as is the case in this example.

With an opposing pair of muscles, the roles of the prime mover and antagonist can be exchanged. For example, in the process of extending the forearm at the elbow, the triceps brachii serves as the prime mover, and the biceps brachii functions as the antagonist; their roles are reversed. Note that if the prime mover and antagonist contract simultaneously with equal force, there will be no movement.

There are many examples of a prime mover crossing several joints before it reaches the joint at which its primary action occurs. The biceps brachii, for example, spans both the shoulder and elbow joints, and its primary action is on the forearm. In order to prevent unwanted movements at intermediate joints or to otherwise aid the movement of the prime mover, muscles called **synergists** (SIN-er-gists; *syn* = together; *ergon* = work) contract and stabilize the intermediate joints. As an example, muscles that flex the fingers (prime movers) cross the intercarpal and radiocarpal joints (intermediate joints). If movement at these intermediate joints was unrestrained, you would not be able to flex your fingers without flexing the wrist at the same time. Synergistic contraction of the wrist extensors stabilizes the wrist joint and prevents it from moving (unwanted movement) while the flexor muscles of the fingers contract to bring about efficient flexion of the fingers (primary action). Synergists are usually located very close to the prime mover.

Some muscles in a group also act as **fixators,** which stabilize the origin of the prime mover so that the prime mover can act more efficiently. Fixators steady the proximal end of a limb while movements occur at the distal end. For example, the scapula in the pectoral (shoulder) girdle is a freely movable bone that serves as the origin for several muscles that move the arm. However, when the arm muscles contract, the scapula must be held steady. This is accomplished by fixator muscles that hold the scapula firmly against the back of the chest. In abduction of the arm, the deltoid muscle serves as the prime mover, whereas fixators (pectoralis minor, trapezius, subclavius, serratus anterior muscles, and others) hold the scapula firmly (see Figure 10.16). The scapula serves as the attachment site for the origin of the deltoid muscle, whereas the insertion of the muscle pulls on the humerus to abduct the arm. Under different conditions—that is, for different movements—many muscles may act, at various times, as prime movers, antagonists, synergists, or fixators.

✔ Describe the three types of levers, and give an example of each type found in the body.
✔ Describe the various arrangements of fascicles.
✔ Define the roles of the prime mover (agonist), antagonist, synergist, and fixator in producing movements.

HOW SKELETAL MUSCLES ARE NAMED

Objective

- Explain seven characteristics used in naming skeletal muscles.

The names of most of the nearly 700 skeletal muscles are based on several characteristics. Most skeletal muscles are named on the basis of combinations of these characteristics. Learning the terms that refer to these characteristics will help you remember the names of muscles. Although many characteristics may be reflected in a name, the most important include the direction in which the muscle fibers run; the size, shape, action, number of origins, and location of the muscle; and the sites of origin and insertion of the muscle. Study Table 10.2 to become familiar with the terms used in muscle names.

✔ Select ten muscles presented in Figure 10.3 and identify the bases of their names. (HINT: *Use the prefix, suffix, and root of each muscle's name as a guide.*)

Table 10.2 Characteristics Used to Name Muscles

Name	Meaning	Example	Figure
DIRECTION: Orientation of muscle fibers relative to the body's midline.			
Rectus	parallel to midline	Rectus abdominis	10.12b
Transverse	perpendicular to midline	Transversus abdominis	10.12b
Oblique	diagonal to midline	External oblique	10.12a
SIZE: Relative size of the muscle.			
Maximus	largest	Gluteus maximus	10.22c
Minimus	smallest	Gluteus minimus	10.22c
Longus	longest	Adductor longus	10.22a
Brevis	shortest	Peroneus brevis	10.24b
Latissimus	widest	Latissimus dorsi	10.17b
Longissimus	longest	Longissimus capitis	10.21a
Magnus	large	Adductor magnus	10.22a
Major	larger	Pectoralis major	10.17a
Minor	smaller	Pectoralis minor	10.16a
Vastus	great	Vastus lateralis	10.22a
SHAPE: Relative shape of the muscle.			
Deltoid	triangular	Deltoid	10.12b
Trapezius	trapezoid	Trapezius	10.3b
Serratus	saw-toothed	Serratus anterior	10.16b
Rhomboideus	diamond-shaped	Rhomboideus major	10.17c
Orbicularis	circular	Orbicularis oculi	10.4a
Pectinate	comblike	Pectineus	10.22a
Piriformis	pear-shaped	Piriformis	10.22c
Platys	flat	Platysma	10.4a
Quadratus	square	Quadratus femoris	10.22c
Gracilis	slender	Gracilis	10.22a
ACTION: Principal action of the muscle.			
Flexor	decreases joint angle	Flexor carpi radialis	10.19a
Extensor	increases joint angle	Extensor carpi ulnaris	10.19c
Abductor	moves bone away from midline	Abductor pollicis longus	10.19c
Adductor	moves bone closer to midline	Adductor longus	10.22a
Levator	produces superior movement	Levator scapulae	10.16a
Depressor	produces inferior movement	Depressor labii inferioris	10.4b
Supinator	turns palm superiorly or anteriorly	Supinator	10.19b
Pronator	turns palm inferiorly or posteriorly	Pronator teres	10.19a
Sphincter	decreases size of opening	External anal sphincter	10.14
Tensor	makes a body part rigid	Tensor fasciae latae	10.22a
Rotator	moves bone around longitudinal axis	Obturator externus	10.22b
NUMBER OF ORIGINS: Number of tendons of origin.			
Biceps	two origins	Biceps brachii	10.18a
Triceps	three origins	Triceps brachii	10.18b
Quadriceps	four origins	Quadriceps femoris	10.22a
LOCATION: Structure near which a muscle is found. *Example:* Frontalis, a muscle near the frontal bone (Figure 10.4a).			
ORIGIN AND INSERTION: Sites where muscle originates and inserts. *Example:* Sternocleidomastoid, originating on the sternum and clavicle inserting on the mastoid process of the temporal bone (Figure 10.3a).			

Figure 10.3 / Principal superficial skeletal muscles.

Most movements require several skeletal muscles acting in groups rather than individually.

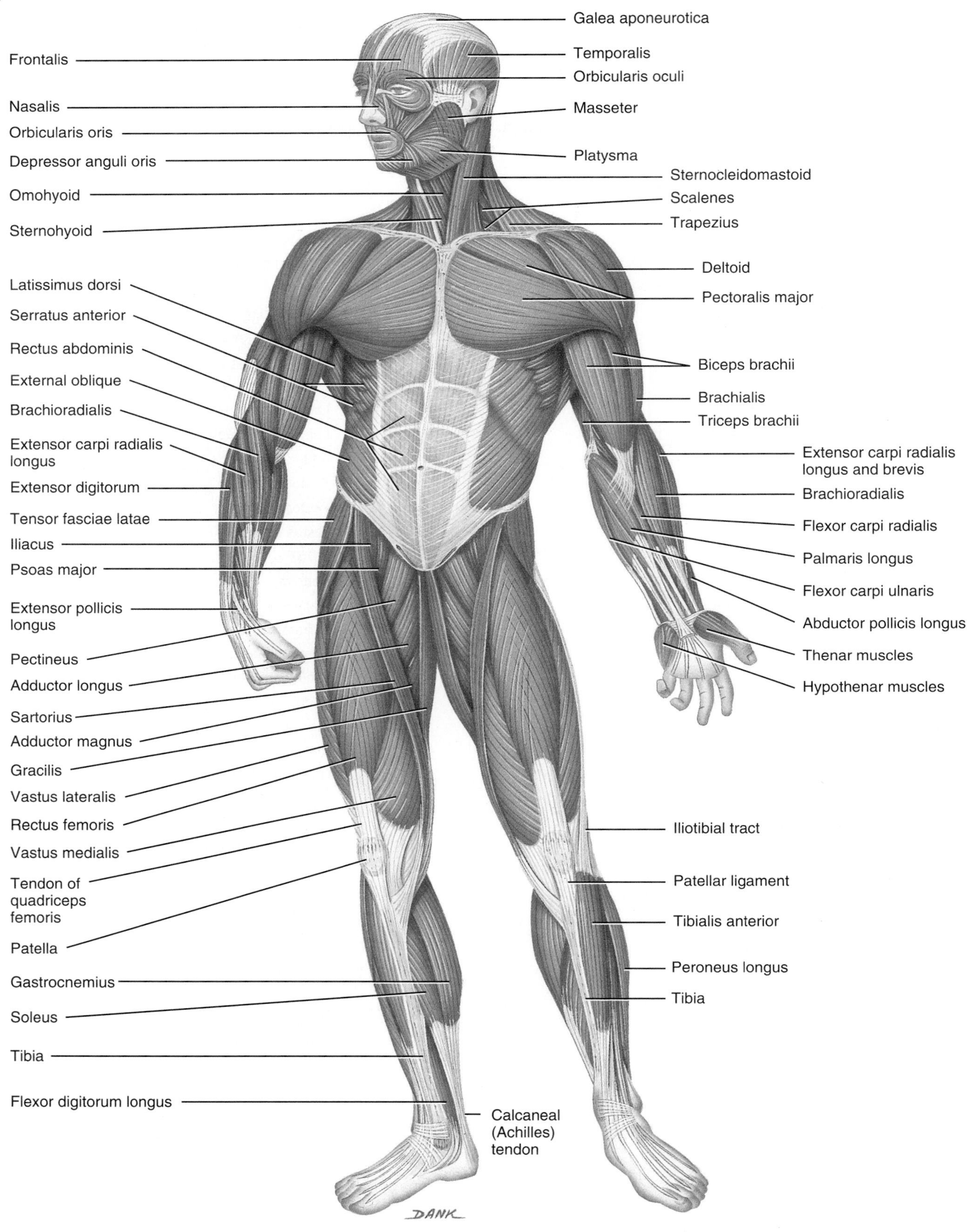

(a) Anterior view

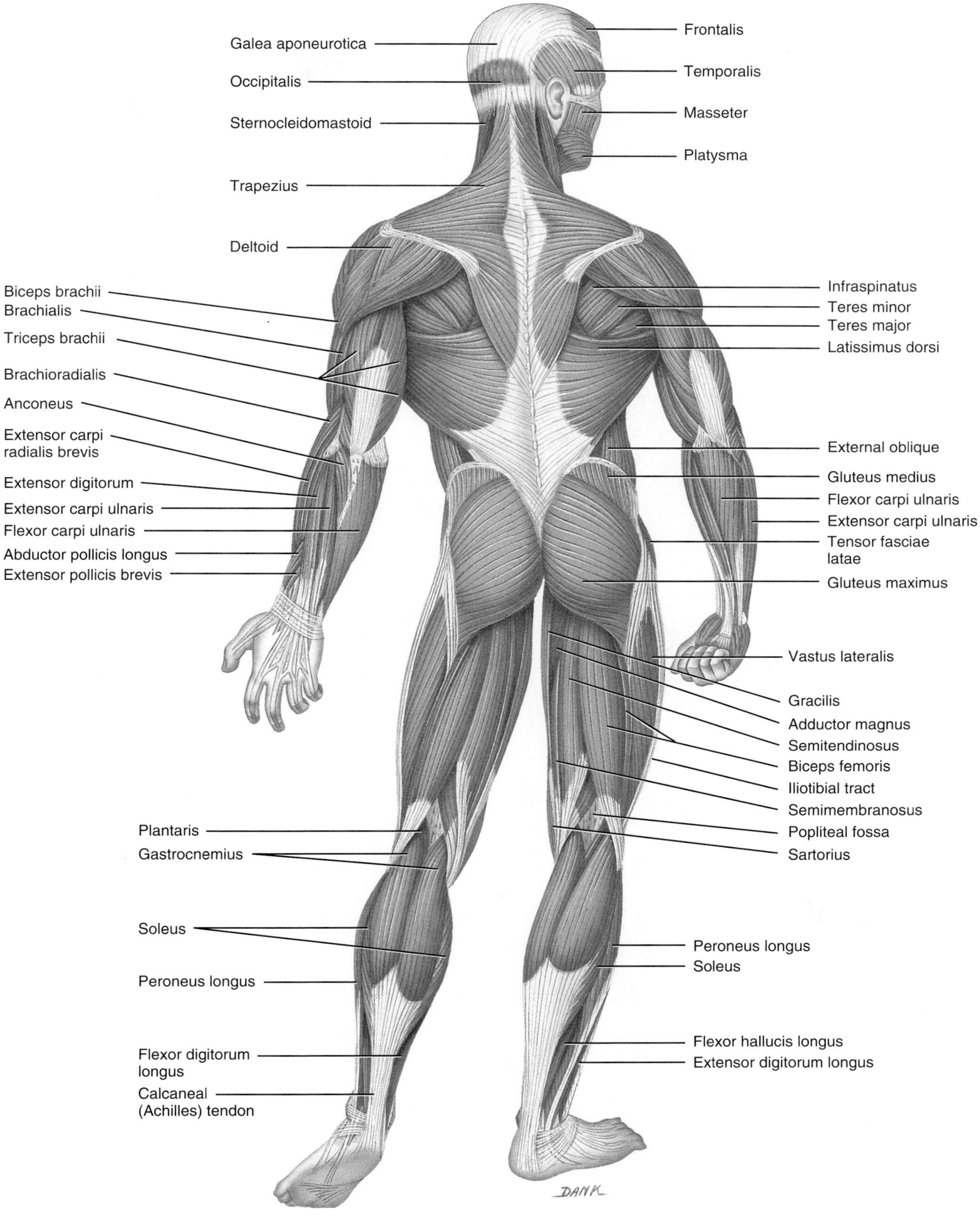

Galea aponeurotica
Occipitalis
Sternocleidomastoid
Trapezius
Deltoid
Frontalis
Temporalis
Masseter
Platysma
Biceps brachii
Brachialis
Triceps brachii
Brachioradialis
Anconeus
Extensor carpi radialis brevis
Extensor digitorum
Extensor carpi ulnaris
Flexor carpi ulnaris
Abductor pollicis longus
Extensor pollicis brevis
Infraspinatus
Teres minor
Teres major
Latissimus dorsi
External oblique
Gluteus medius
Flexor carpi ulnaris
Extensor carpi ulnaris
Tensor fasciae latae
Gluteus maximus
Vastus lateralis
Gracilis
Adductor magnus
Semitendinosus
Biceps femoris
Iliotibial tract
Semimembranosus
Popliteal fossa
Sartorius
Plantaris
Gastrocnemius
Soleus
Peroneus longus
Flexor digitorum longus
Calcaneal (Achilles) tendon
Peroneus longus
Soleus
Flexor hallucis longus
Extensor digitorum longus
DANK

(b) Posterior view

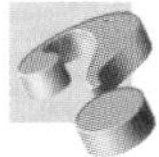

Give an example of a muscle named for each of the following characteristics: direction of fibers, shape, action, size, origin and insertion, location, and number of tendons of origin.

▶ 14th Century

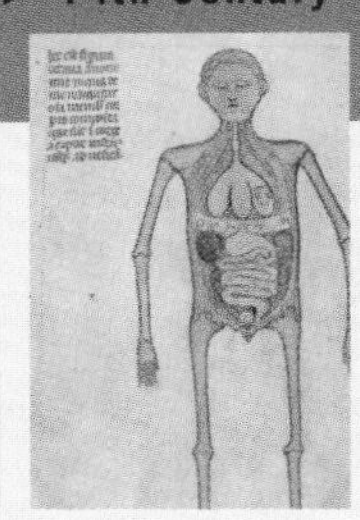

▶ 15th Century

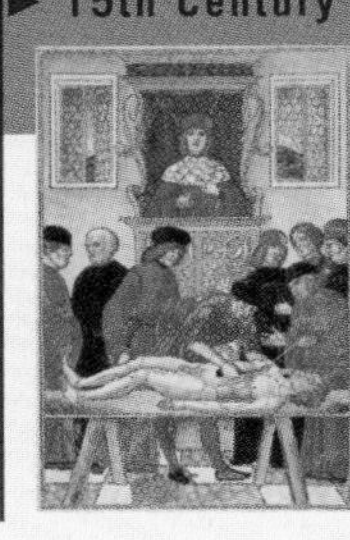

▶ 16th Century

▶ 17th Century

▶ 18th Century

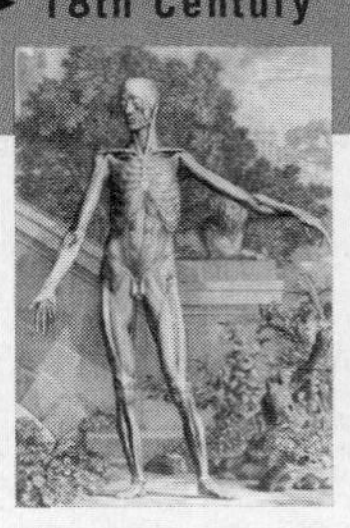

▶ 20th Century

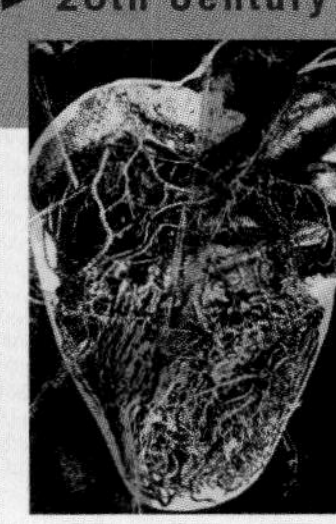

▶ 21st Century

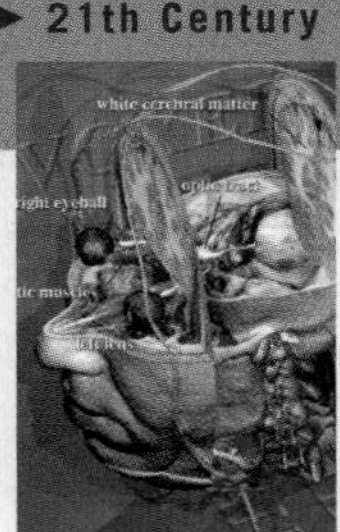

CHANGING IMAGES

Put Some Muscle Into It!

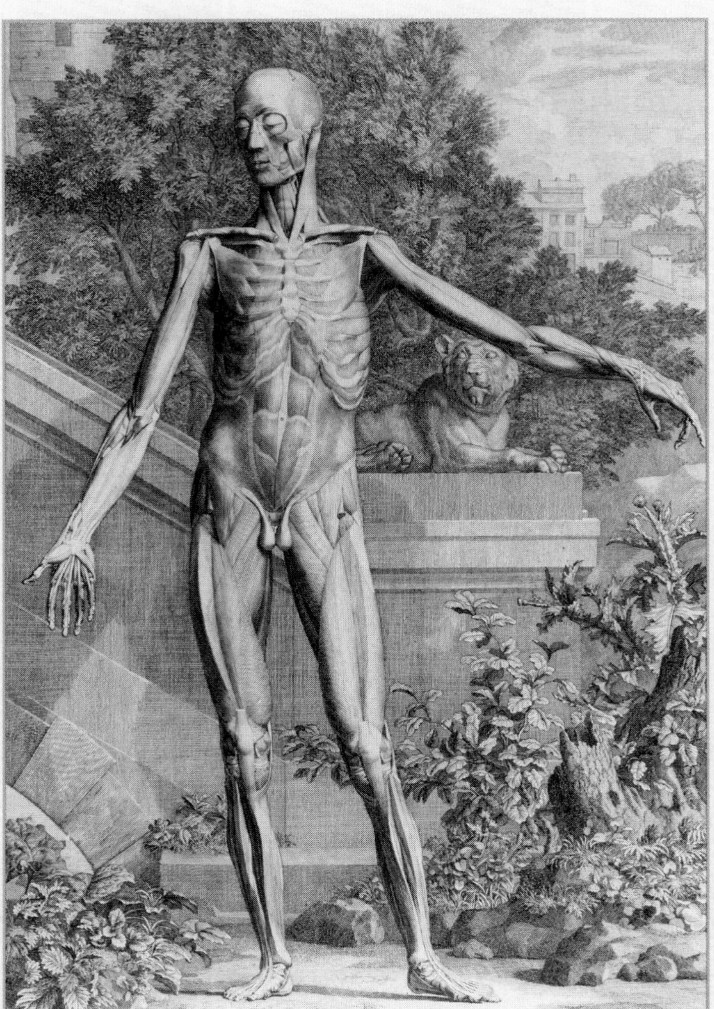

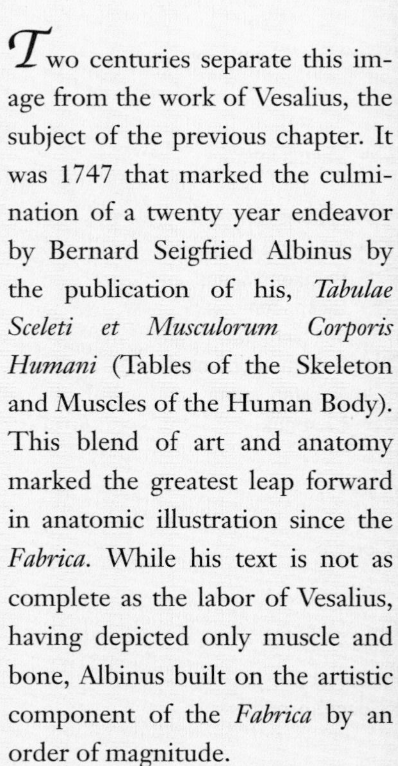

Two centuries separate this image from the work of Vesalius, the subject of the previous chapter. It was 1747 that marked the culmination of a twenty year endeavor by Bernard Seigfried Albinus by the publication of his, *Tabulae Sceleti et Musculorum Corporis Humani* (Tables of the Skeleton and Muscles of the Human Body). This blend of art and anatomy marked the greatest leap forward in anatomic illustration since the *Fabrica*. While his text is not as complete as the labor of Vesalius, having depicted only muscle and bone, Albinus built on the artistic component of the *Fabrica* by an order of magnitude.

Employing engraved copper plates, as opposed to the woodcuts of the *Fabrica*, Albinus in collaboration with his master artist-engraver Jan Wandelaar, fashioned skeletal and muscular renderings with elaborate landscapes that are unparalleled. Albinus was so concerned with artistic presentation, that he refused to clutter the plates with reference labels. Instead, the facing page of each image was that of an outline of the same figure. It was here that he placed the identifying legend. Even the pose is elegant; note how the left calcaneus bears no weight.

The image depicted here is referred to as *The Second Order of Muscles, Front View*. After you have learned the skeletal muscles in this chapter, see if you can identify each of the muscles represented. Are you able to distinguish which muscles have been removed? Pay careful attention to the chest, abdomen, and anterior thigh. As you perform this examination, bear in mind the place this image holds in the history of anatomic illustration.

1747 AD

PRINCIPAL SKELETAL MUSCLES

Exhibits 10.1 through 10.20 will assist you in learning the names of the principal skeletal muscles in various regions of the body. The muscles in the exhibits are divided into groups according to the part of the body on which they act. As you study groups of muscles in the exhibits, refer to Figure 10.3 to see how each group is related to the others.

The exhibits contain the following information:

- *Objectives.* These statements describe the principal outcomes that can be expected after studying the exhibit.
- *Overview.* This information provides a general orientation to the muscles under consideration and emphasizes how the muscles are organized within various regions. The discussion also highlights any distinguishing or interesting features about the muscles.
- *Muscle names.* The derivations indicate how the muscles are named. Once you have mastered the naming of the muscles, their actions will have more meaning.
- *Origins, insertions, and actions.* For each muscle, you are also given its origin and insertion (points of attachment to bones or other structures) and actions, the principal movements that occur when the muscle contracts.
- *Innervation.* This section indicates the nerve supply for each muscle. In general, muscles in the head region are served by cranial nerves, which arise from the lower portions of the brain. Muscles in the rest of the body are served by spinal nerves, which arise from the spinal cord within the vertebral column. Cranial nerves are designated by both a name and a Roman numeral in parentheses—for example, the facial (VII) cranial nerve. Spinal nerves are numbered in groups according to the portion of the spinal cord from which they arise: C = cervical (neck region), T = thoracic (chest region), L = lumbar (lower back region), and S = sacral (buttocks region). An example is T1, the first thoracic spinal nerve.
- *"Relating Muscles to Movements."* These exercises will help you organize the muscles in a body region according to the actions they share in common.
- *Figures.* The figures in the exhibits present superficial and deep, anterior and posterior, or medial and lateral views to show each muscle's position as clearly as possible. *The muscle names in all capital letters are specifically referred to in the table portion of the exhibit.* Many of the reference structures (such as blood vessels and nerves) will be discussed more fully in subsequent chapters.

The exhibits and accompanying figures that follow describe the principal skeletal muscles:

Exhibit 10.1 Muscles of Facial Expression (Figure 10.4)

Objective

► Describe the origin, insertion, action, and innervation of the muscles of facial expression.

The muscles of facial expression provide humans with the ability to express a wide variety of emotions. The muscles themselves lie within the layers of superficial fascia. They usually originate in the fascia or bones of the skull and insert into the skin. Because of their insertions, the muscles of facial expression move the skin rather than a joint when they contract.

Among the noteworthy muscles in this group are those surrounding the orifices (openings) of the head such as the eyes, nose, and mouth. These muscles function as *sphincters* (SFINGK-ters), which close the orifices, and *dilators,* which open the orifices. For example, the **orbicularis oculi** muscle closes the eye, whereas the **levator palpebrae superioris** muscle opens the eye. As another example, the **orbicularis oris** muscle closes the mouth, whereas several other muscles (**zygomaticus major, levator labii superioris, depressor labii inferioris, mentalis,** and **risorius**) radiate out from the lips and open the mouth. The **occipitofrontalis** is another unusual muscle in this group because it is made up of two parts: a posterior part called the **occipitalis,** which is superficial to the occipital bone, and an anterior part called the **frontalis,** which is superficial to the frontal bone. The two muscular portions are held together by a strong aponeurosis (sheetlike tendon), the **galea aponeurotica** (GĀ-lē-a ap-ō-nū-RŌ-ti-ka; *galea* = helmet) or *epicranial aponeurosis,* which covers the superior and lateral surfaces of the skull. The **buccinator** muscle forms the major muscular portion of the cheek. It functions in whistling, blowing, and sucking and also assists in chewing. The duct of the parotid gland (salivary gland) pierces the buccinator muscle to reach the oral cavity. The buccinator muscle is so named because it compresses the cheeks (*bucca* = cheek) during blowing—for example, when a musician plays a wind instrument such as a trumpet.

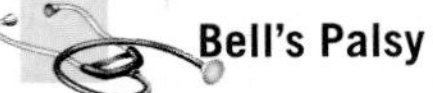

Bell's Palsy

Bell's palsy, also known as **fascial paralysis,** is a unilateral paralysis of the muscles of facial expression as a result of damage or disease of the facial (VII) cranial nerve. Although the cause is unknown, inflammation of the facial nerve and a relationship to the herpes simplex virus have been suggested. The paralysis causes the entire side of the face to droop in severe cases, and the person cannot wrinkle the forehead, close the eye, or pucker the lips on the affected side. Difficulty in swallowing and drooling also occur. Eighty percent of patients recover completely within a few weeks to a few months. For others, paralysis is permanent.

Relating Muscles to Movements

Arrange the muscles in this exhibit into two groups: (1) those that act on the mouth and (2) those that act on the eyes.

✔ What muscles would you use to do the following: show surprise, express sadness, bare your upper teeth, pucker your lips, squint, blow up a balloon?

Innervation

All of the muscles of facial expression are innervated by the facial (VII) cranial nerve, except for the levator palpebrae superioris muscle, which is innervated by the oculomotor (III) cranial nerve.

Muscle	Origin	Insertion	Action
Occipitofrontalis (ok-sip′-i-tō-fron-TAL-is)			
Frontalis (fron-TA-lis; *frontalis* = in front)	Galea aponeurotica.	Skin superior to supraorbital margin.	Draws scalp anteriorly, raises eyebrows, and wrinkles skin of forehead horizontally.
Occipitalis (ok-sip′-i-TA-lis; *occipito* = base of skull)	Occipital bone and mastoid process of temporal bone.	Galea aponeurotica.	Draws scalp posteriorly.
Orbicularis oris (or-bi′-kyū-LAR-is OR-is; *orb* = circular; *or* = mouth)	Muscle fibers surrounding opening of mouth.	Skin at corner of mouth.	Closes and protrudes lips, compresses lips against teeth, and shapes lips during speech.
Zygomaticus major (zī-gō-MA-ti-kus; *zygomatic* = cheek bone; *major* = greater)	Zygomatic bone.	Skin at angle of mouth and orbicularis oris.	Draws angle of mouth superiorly and laterally, as in smiling or laughing.
Zygomaticus minor (*minor* = lesser)	Zygomatic bone.	Upper lip.	Elevates upper lip, exposing maxillary teeth.
Levator labii superioris (le-VĀ-tor LĀ-bē-ī sū-per′-ē-OR-is; *levator* = raises or elevates; *labii* = lip; *superioris* = upper)	Superior to infraorbital foramen of maxilla.	Skin at angle of mouth and orbicularis oris.	Elevates upper lip.
Depressor labii inferioris (de-PRE-sor LĀ-bē-ī in-fer′-ē-OR-is; *depressor* = depresses or lowers; *inferioris* = lower)	Mandible.	Skin of lower lip.	Depresses (lowers) lower lip.
Depressor anguli oris (*angul* = angle or corner)	Mandible.	Angle of mouth.	Draws angle of mouth laterally and inferiorly, as in opening mouth.
Buccinator (BUK-si-nā′-tor; *bucca* = cheek)	Alveolar processes of maxilla and mandible and pterygomandibular raphe (fibrous band extending from the pterygoid process to the mandible).	Orbicularis oris.	Presses cheeks against teeth and lips, as in whistling, blowing, and sucking; draws corner of mouth laterally; and assists in mastication (chewing) by keeping food between the teeth (and not between teeth and cheeks).

Muscle	Origin	Insertion	Action
Mentalis (men-TA-lis; *mentum* = chin)	Mandible	Skin of chin.	Elevates and protrudes lower lip and pulls skin of chin up, as in pouting.
Platysma (pla-TIZ-ma; *platy* = flat, broad)	Fascia over deltoid and pectoralis major muscles.	Mandible, muscles around angle of mouth, and skin of lower face.	Draws outer part of lower lip inferiorly and posteriorly as in pouting; depresses mandible.
Risorious (ri-ZOR-ē-us; *risor* = laughter)	Fascia over parotid (salivary) gland.	Skin at angle of mouth.	Draws angle of mouth laterally, as in tenseness.
Orbicularis oculi (or-bi′-kyū-LAR-is OK-yū-lī; *oculus* = eye)	Medial wall of orbit.	Circular path around orbit.	Closes eye.
Corrugator supercilii (KOR-a-gā′-tor sū-per-SI-lē-ī; *corrugo* = wrinkle; *supercilium* = eyebrow)	Medial end of superciliary arch of frontal bone.	Skin of eyebrow.	Draws eyebrow inferiorly and wrinkles skin of forehead vertically as in frowning.
Levator palpebrae superioris PAL-pe-brē sū-per′-ē-OR-is; *palpebrae* = eyelids) (see also Figure 10.5a)	Roof of orbit (lesser wing of sphenoid bone).	Skin of upper eyelid.	Elevates upper eyelid (opens eye).

Figure 10.4 / Muscles of facial expression. (See Tortora, *Photographic Atlas of the Human Body,* Figures 5.2 through 5.4.)

When they contract, muscles of facial expression move the skin rather than a joint.

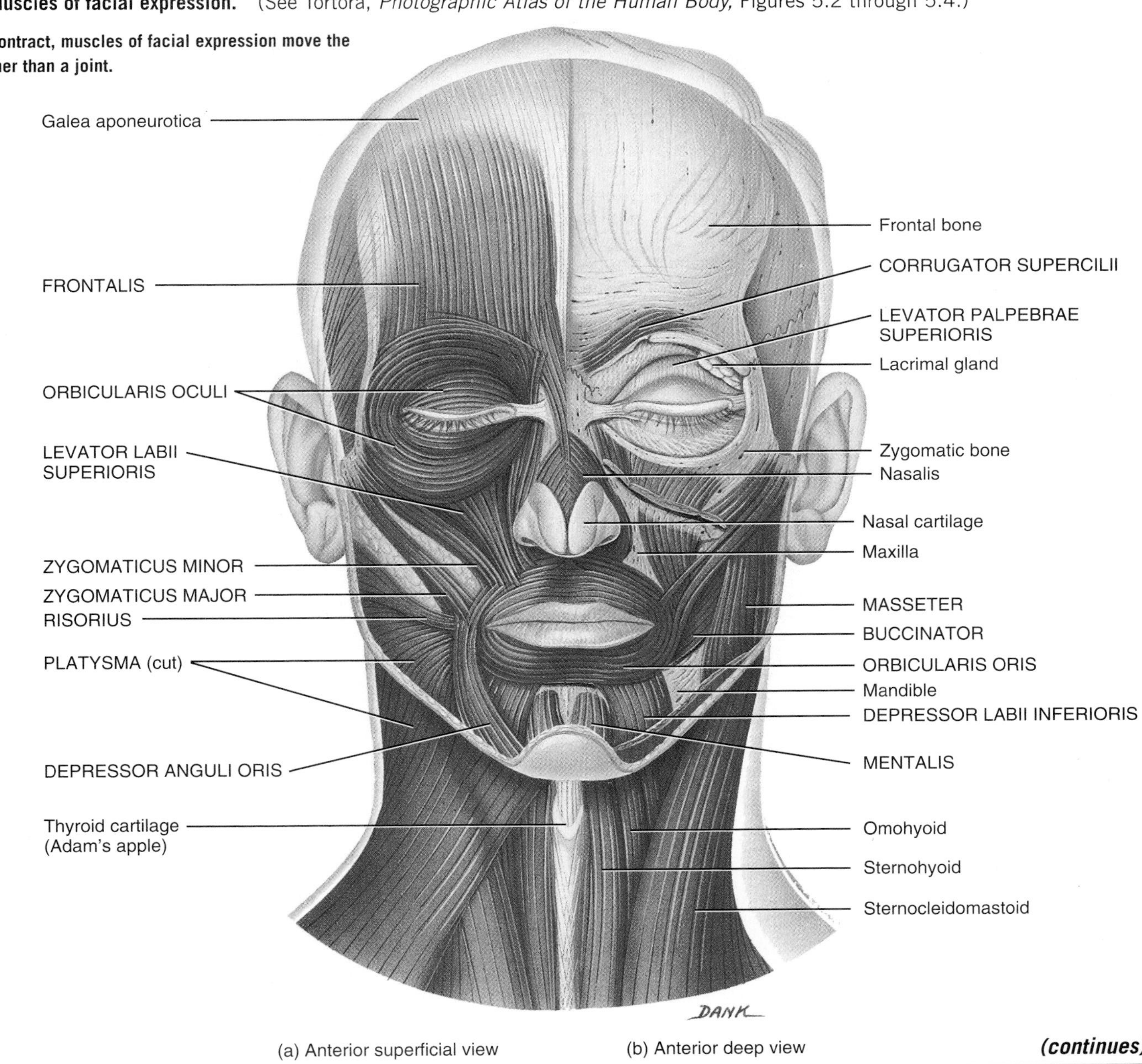

(a) Anterior superficial view (b) Anterior deep view

(continues)

Exhibit 10.1 Muscles of Facial Expression (continued)

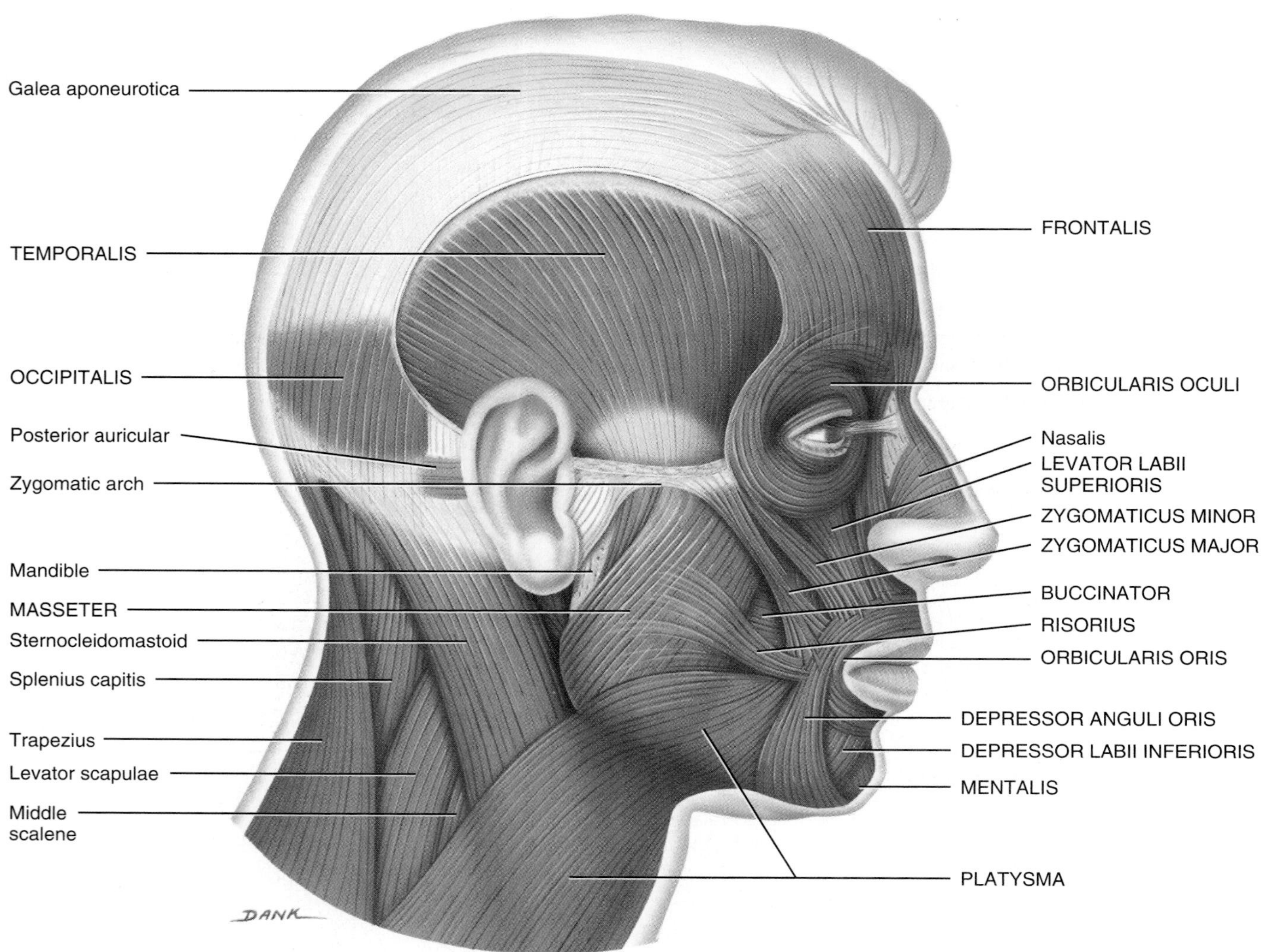

(c) Right lateral superficial view

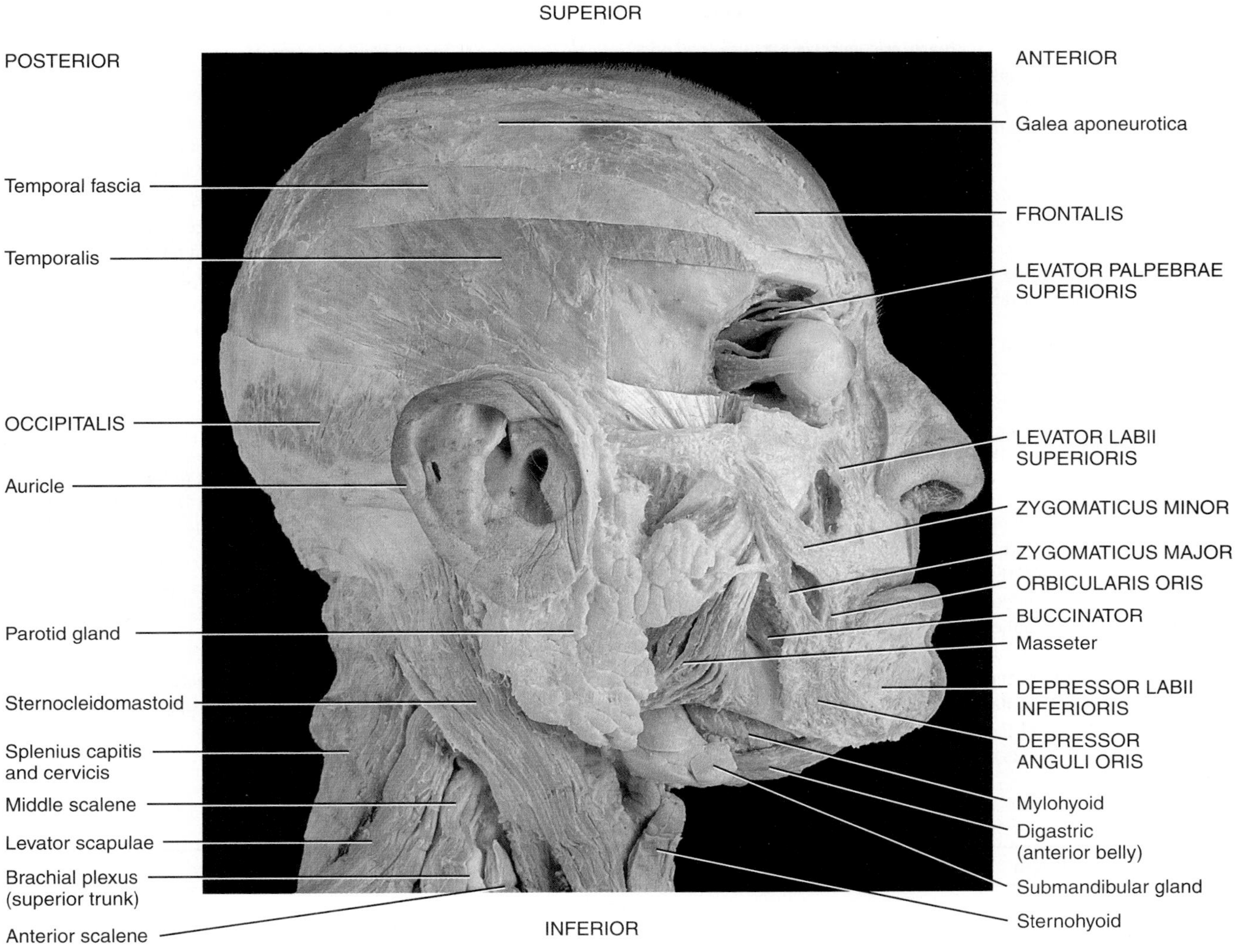

SUPERIOR

POSTERIOR

ANTERIOR

Galea aponeurotica

Temporal fascia

FRONTALIS

Temporalis

LEVATOR PALPEBRAE SUPERIORIS

OCCIPITALIS

LEVATOR LABII SUPERIORIS

Auricle

ZYGOMATICUS MINOR

ZYGOMATICUS MAJOR

ORBICULARIS ORIS

BUCCINATOR

Parotid gland

Masseter

DEPRESSOR LABII INFERIORIS

Sternocleidomastoid

DEPRESSOR ANGULI ORIS

Splenius capitis and cervicis

Mylohyoid

Middle scalene

Digastric (anterior belly)

Levator scapulae

Brachial plexus (superior trunk)

Submandibular gland

Sternohyoid

Anterior scalene

INFERIOR

(d) Right lateral superficial view

Which muscles of facial expression cause frowning, smiling, pouting, and squinting?

Exhibit 10.2 Muscles That Move the Eyeballs—Extrinsic Muscles (Figure 10.5)

Objective

- Describe the origin, insertion, action, and innervation of the extrinsic muscles of the eyeballs.

Muscles that move the eyeballs are called **extrinsic muscles** because they originate outside the eyeballs (in the orbit) and are inserted on the outer surface of the sclera ("white of the eye"). The extrinsic muscles of the eyeballs are among the fastest contracting and most precisely controlled skeletal muscles in the body.

Movements of the eyeballs are controlled by three pairs of extrinsic muscles: (1) superior and inferior recti, (2) lateral and medial recti, and (3) superior and inferior oblique. The four recti muscles (superior, inferior, lateral, and medial) arise from a tendinous ring in the orbit and insert into the sclera of the eye. The **superior** and **inferior recti** lie in the same vertical plane, and the **medial** and **lateral recti** in the same horizontal plane. The actions of the recti muscles can be deduced from their insertions on the sclera. The superior and inferior recti move the eyeballs superiorly and inferiorly, respectively; the lateral and medial recti move the eyeballs laterally and medially, respectively. It should be noted that neither the superior nor the inferior rectus muscle pulls directly parallel to the long axis of the eyeballs, and as a result both muscles also move the eyeballs medially.

It is not as easy to deduce the actions of the oblique muscles (superior and inferior) because of their paths through the orbits. For example, the **superior oblique** muscle originates posteriorly near the tendinous ring and then passes anteriorly and ends in a round tendon, which runs through a pulleylike loop called the *trochlea* (*trochlea* = pulley) in the anterior and medial part of the roof of the orbit. The tendon then turns and inserts on the posterolateral aspect of the eyeballs. Accordingly, the superior oblique muscle moves the eyeballs inferiorly and laterally. The **inferior oblique** muscle originates on the maxilla at the anteromedial aspect of the floor of the orbit. It then passes posteriorly and laterally and inserts on the posterolateral aspect of the eyeballs. Because of this arrangement, the inferior oblique muscle moves the eyeballs superiorly and laterally.

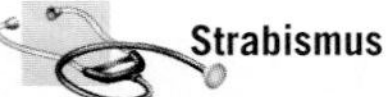

Strabismus

Strabismus is a condition in which the two eyes are not properly aligned. This can be hereditary or it can be due to birth injuries, poor attachments of the muscles, or localized disease. In strabismus, the two eyes do not work as a team. Each eye sends an image to a different area of the brain. This confuses the brain, which usually ignores the messages sent by one of the eyes. As a result, the ignored eye becomes weaker, hence "lazy eye" or *amblyopia* develops. *External strabismus* results when a lesion on the oculomotor (III) cranial nerve, which controls the superior, inferior, and medial recti, and inferior oblique muscles of the eye, causes the eyeball to move laterally when at rest, and results in an inability to move the eyeball medially and inferiorly. A lesion in the abducens (VI) cranial nerve, which innervates the lateral rectus muscle, results in *internal strabismus,* a condition where the eyeball moves medially when at rest and cannot move laterally. ■

Relating Muscles to Movements

Arrange the muscles in this exhibit according to their actions on the eyeballs: (1) elevation, (2) depression, (3) abduction, (4) adduction, (5) medial rotation, and (6) lateral rotation. The same muscle may be mentioned more than once.

✔ Which muscles contract and relax in each eye as you gaze to your left without moving your head?

Innervation

The superior rectus, inferior rectus, medial rectus, and inferior oblique muscles are innervated by the oculomotor (III) cranial nerve; the lateral rectus is innervated by the abducens (VI) cranial nerve; and the superior oblique is innervated by the trochlear (IV) cranial nerve.

Muscle	Origin	Insertion	Action
Superior rectus	Common tendinous ring (attached to orbit around optic foramen).	Superior and central part of eyeball.	Moves eyeball superiorly (elevation) and medially (adduction), and rotates it medially.
Inferior rectus	Same as above.	Inferior and central part of eyeball.	Moves eyeball inferiorly (depression) and medially (adduction), and rotates it laterally.
Lateral rectus	Same as above.	Lateral side of eyeball.	Moves eyeball laterally (abduction).
Medial rectus	Same as above.	Medial side of eyeball.	Moves eyeball medially (adduction).
Superior oblique	Sphenoid bone, superior and medial to the tendinous ring in the orbit.	Eyeball between superior and lateral recti. The muscle inserts into the superior and lateral surfaces of the eyeball via a tendon that passes through the trochlea.	Moves eyeball inferiorly (depression) and laterally (abduction), and rotates it medially.
Inferior oblique	Maxilla in floor of orbit.	Eyeball between inferior and lateral recti.	Moves eyeball superiorly (elevation) and laterally (abduction), and rotates it laterally.

Figure 10.5 / Extrinsic muscles of the eyeball. (See Tortora, *A Photographic Atlas of the Human Body,* Figure 5.4.)

The extrinsic muscles of the eyeball are among the fastest contracting and most precisely controlled skeletal muscles in the body.

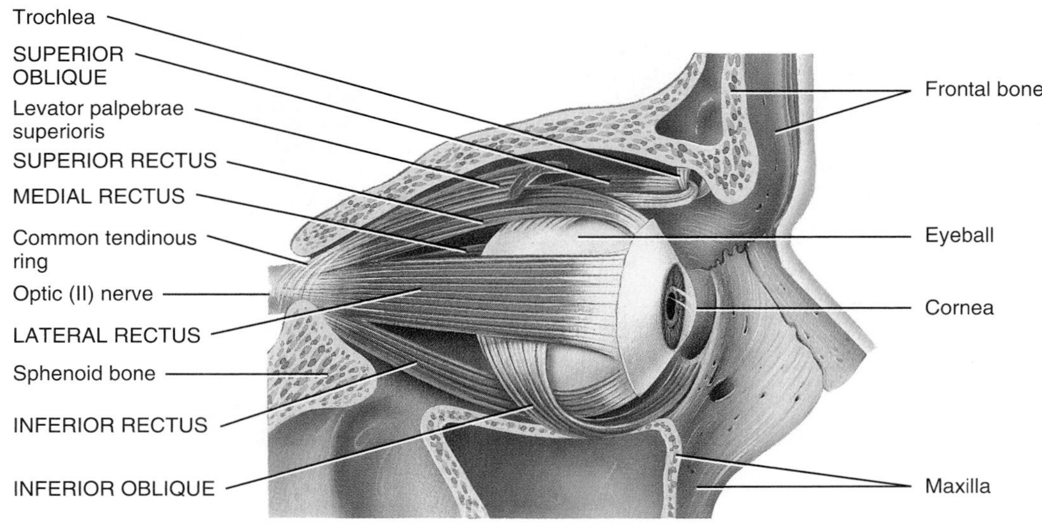

(a) Lateral view of right eyeball

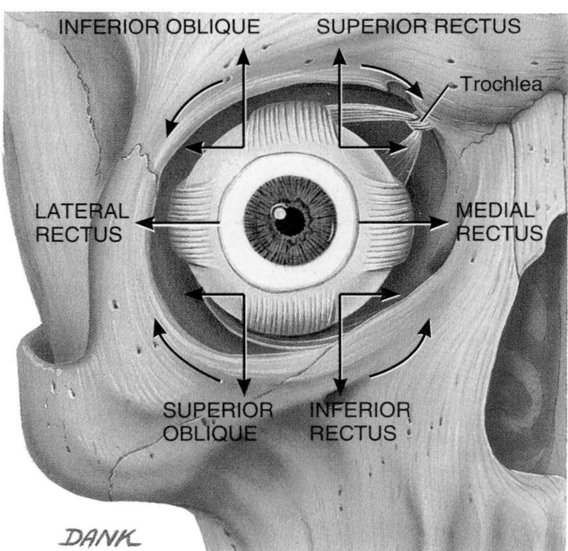

(b) Movements of right eyeball in response to contraction of extrinsic muscles

(continues)

Exhibit 10.2 Muscles That Move the Eyeballs—Extrinsic Muscles (Figure 10.5)

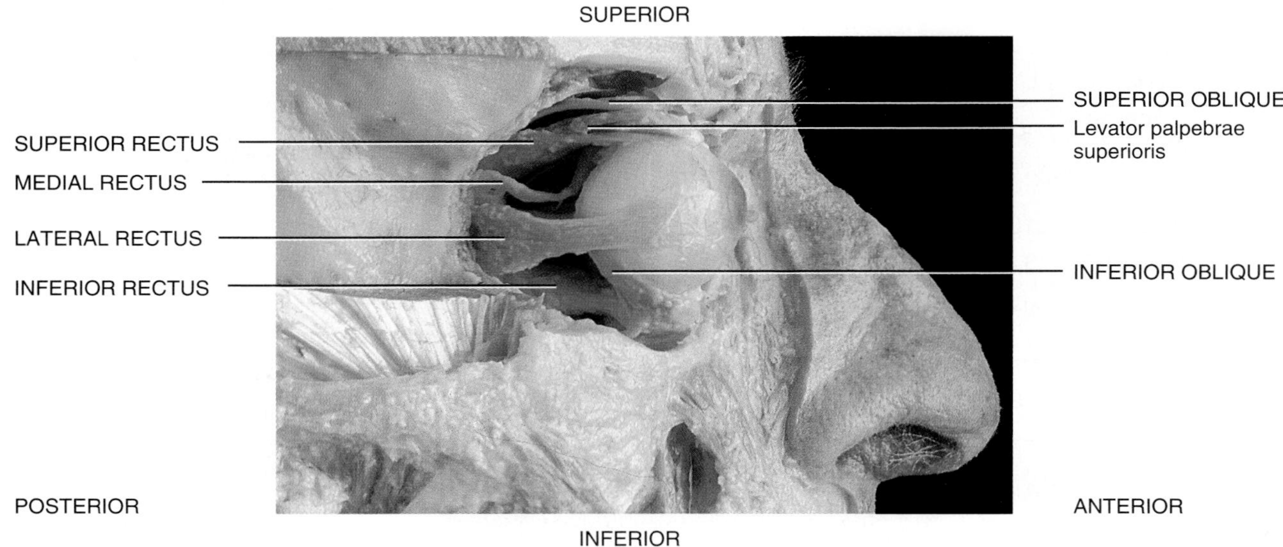

(c) Lateral view of right eyeball

Why does the inferior oblique muscle move the eyeball superiorly and laterally?

Exhibit 10.3 Muscles That Move the Mandible (Lower Jaw) (Figure 10.6)

Objective

► Describe the origin, insertion, action, and innervation of the muscles that move the mandible.

The muscles that move the mandible (lower jaw) at the temporomandibular joint (TMJ) are known as the muscles of mastication because they are involved in chewing (mastication). Of the four pairs of muscles involved in mastication, three are powerful closers of the jaw and account for the strength of the bite: **masseter, temporalis,** and **medial pterygoid.** Of these, the masseter is the strongest. The medial and **lateral pterygoid** muscles assist in mastication by moving the mandible from side to side to help grind food. These muscles protrude the mandible.

In 1996, researchers at the University of Maryland reported that they had identified a new muscle in the skull, tentatively named the *sphenomandibularis muscle.* It extends from the lateral surface of the sphenoid bone to the medial aspect of the condylar process and ramus of the mandible. The muscle is believed to be either a fifth muscle of mastication or a previously unidentified component of an already identified muscle (temporalis or medial pterygoid).

Relating Muscles to Movements

Arrange the muscles in this exhibit according to their actions on the mandible: (1) elevation, (2) depression, (3) retraction, (4) protraction, and (5) side-to-side movement. The same muscle may be mentioned more than once.

✔ What would happen if you lost tone in the masseter and temporalis muscles?

Innervation

All muscles are innervated by the mandibular division of the trigeminal (V) cranial nerve, except the sphenomandibularis muscle, which is innervated by the maxillary branch of the trigeminal (V) cranial nerve.

Muscle	Origin	Insertion	Action
Masseter (MA-se-ter; *maseter* = chewer) (see Figure 10.4c)	Maxilla and zygomatic arch.	Angle and ramus of mandible.	Elevates mandible, as in closing mouth, and retracts (draws back) mandible.
Temporalis (tem′-por-A-lis; *tempora* = temples)	Temporal bone.	Coronoid process and ramus of mandible.	Elevates and retracts mandible.
Medial pterygoid (TER-i-goid; *medial* = closer to midline; *pterygoid* = like a wing)	Medial surface of lateral portion of pterygoid process of sphenoid bone; maxilla.	Angle and ramus of mandible.	Elevates and protracts (protrudes) mandible and moves mandible from side to side.
Lateral pterygoid (TER-i-goid; *lateral* = farther from midline)	Greater wing and lateral surface of lateral portion of pterygoid process of sphenoid bone.	Condyle of mandible; temporomandibular joint (TMJ).	Protracts mandible, depresses mandible as in opening mouth, and moves mandible from side to side.

(continues)

Exhibit 10.3 Muscles That Move the Mandible (Lower Jaw) (Figure 10.6) continued

Figure 10.6 / Muscles that move the mandible (lower jaw).

The muscles that move the mandible are also known as muscles of mastication.

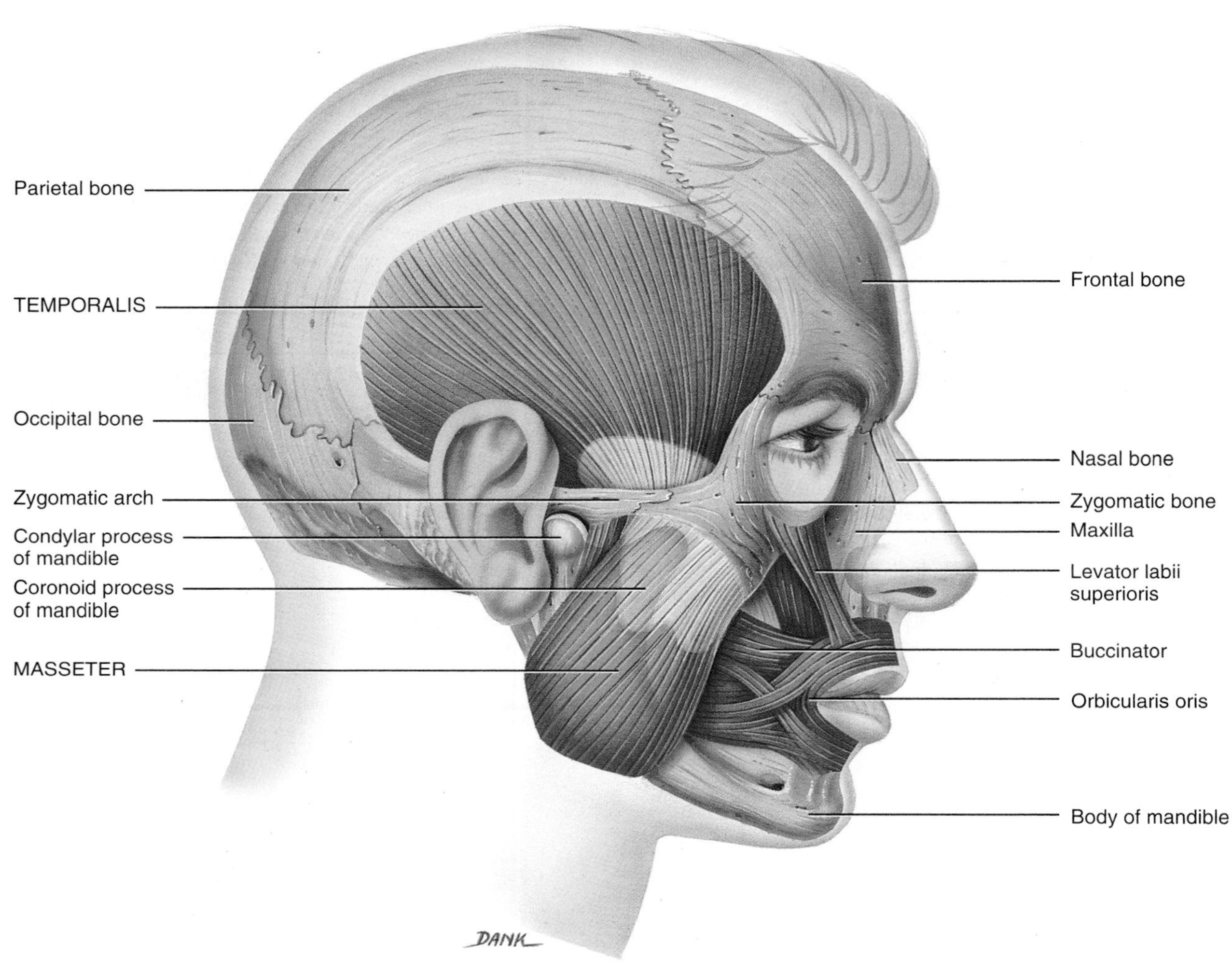

(a) Right lateral superficial view

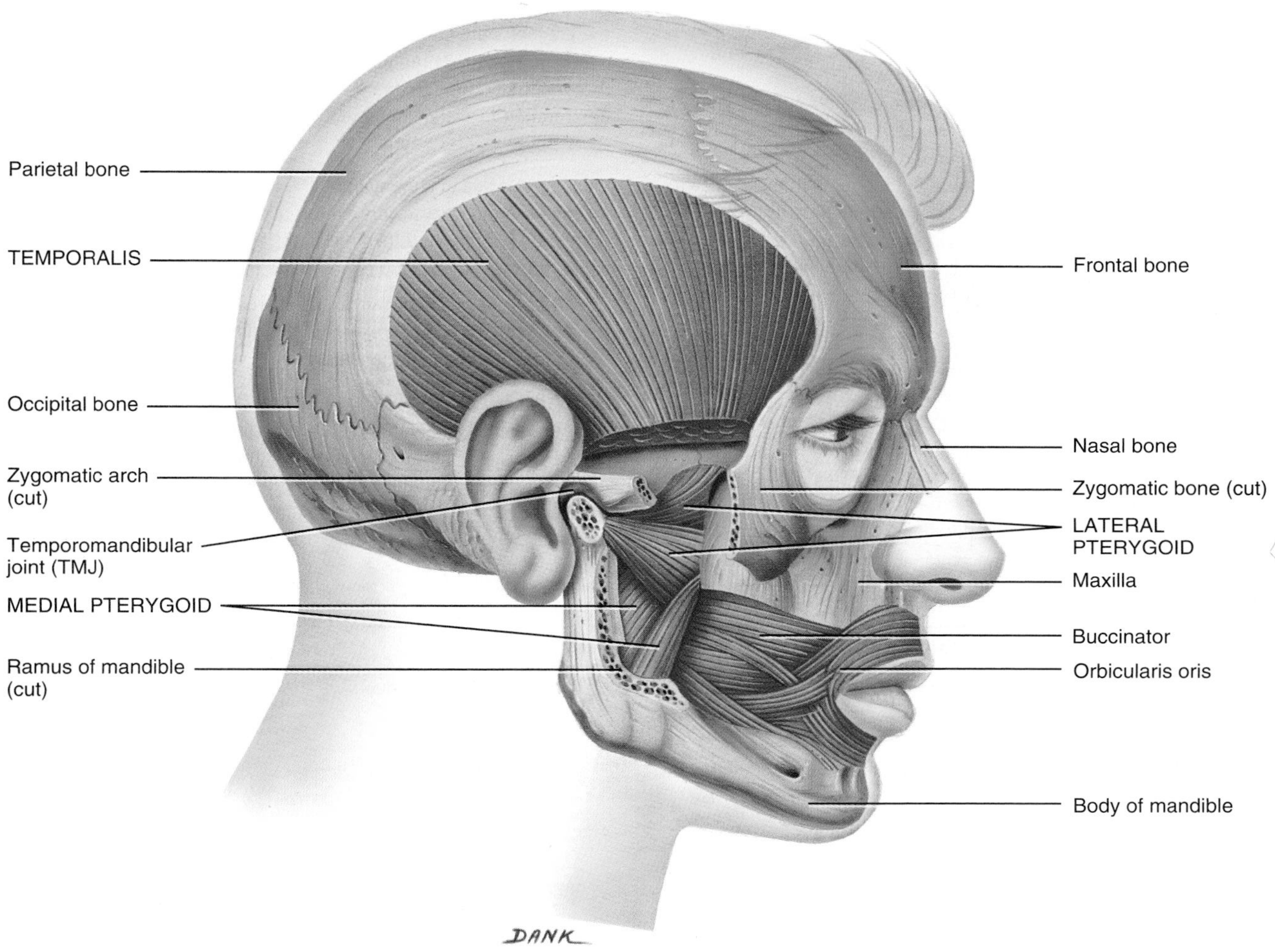

Parietal bone
TEMPORALIS
Occipital bone
Zygomatic arch (cut)
Temporomandibular joint (TMJ)
MEDIAL PTERYGOID
Ramus of mandible (cut)
Frontal bone
Nasal bone
Zygomatic bone (cut)
LATERAL PTERYGOID
Maxilla
Buccinator
Orbicularis oris
Body of mandible
DANK

(b) Right lateral deep view

Which is the strongest muscle of mastication?

Exhibit 10.4 Muscles That Move the Tongue—Extrinsic Muscles (Figure 10.7)

Objective

► Describe the origin, insertion, action, and innervation of the extrinsic muscles of the tongue.

The tongue is a highly mobile structure that is vital to digestive functions such as mastication, perception of taste, and deglutition (swallowing). It is also important in speech. The tongue's mobility is greatly aided by its suspension from the mandible, styloid process of the temporal bone, and hyoid bone.

The tongue is divided into lateral halves by a median fibrous septum. The septum extends throughout the length of the tongue and is attached inferiorly to the hyoid bone. Muscles of the tongue are of two principal types: extrinsic and intrinsic. **Extrinsic muscles** originate outside the tongue and insert into it. They move the entire tongue in various directions, such as anteriorly, posteriorly, and laterally. **Intrinsic muscles** originate and insert within the tongue. These muscles alter the shape of the tongue rather than moving the entire tongue. The extrinsic and intrinsic muscles of the tongue insert into both lateral halves of the tongue.

The names of the extrinsic muscles of the tongue all end in *glossus,* meaning tongue. The actions of the muscles are obvious, considering the positions of the mandible, styloid process, hyoid bone, and soft palate, which serve as origins for these muscles. For example, the **genioglossus** (originates on the mandible) pulls the tongue downward and forward, the **styloglossus** (originates on the styloid process) pulls the tongue upward and backward, the **hyoglossus** (originates on the hyoid bone) pulls the tongue downward and flattens it, and the **palatoglossus** (originates on the soft palate) raises the back portion of the tongue.

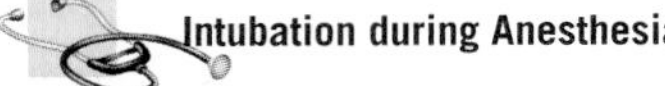

Intubation during Anesthesia

When general anesthesia is administered during surgery, a total relaxation of the genioglossus muscle results. This causes the tongue to fall posteriorly, which may obstruct the airway to the lungs. To avoid this, the mandible is either manually thrust forward and held in place, or a tube is inserted from the lips through the laryngopharynx (inferior portion of the throat) into the trachea (endotracheal intubation). ■

Relating Muscles to Movements

Arrange the muscles in this exhibit according to the following actions on the tongue: (1) depression, (2) elevation, (3) protraction, and (4) retraction. The same muscle may be mentioned more than once.

✔ When your physician says, "Open your mouth, stick out your tongue, and say ahh," so she can examine the inside of your mouth for possible signs of infection, which muscles do you contract?

Innervation

All muscles are innervated by the hypoglossal (XII) cranial nerve, except the palatoglossus, which is innervated by the pharyngeal plexus, which contains axons from both the vagus (X) and the accessory (XI) cranial nerves.

Muscle	Origin	Insertion	Action
Genioglossus (jē′-nē-ō-GLOS-us; *geneion* = chin; *glossus* = tongue)	Mandible.	Undersurface of tongue and hyoid bone.	Depresses tongue and thrusts it anteriorly (protraction).
Styloglossus (stī′-lō-GLOS-us; *stylo* = stake or pole; styloid process of temporal bone)	Styloid process of temporal bone.	Side and undersurface of tongue.	Elevates tongue and draws it posteriorly (retraction).
Palatoglossus (pal′-a-tō-GLOS-us; *palato* = palate)	Anterior surface of soft palate.	Side of tongue.	Elevates posterior portion of tongue and draws soft palate down on tongue.
Hyoglossus (hī′-ō-GLOS-us)	Greater horn and body of hyoid bone.	Side of tongue.	Depresses tongue and draws down its sides.

Figure 10.7 / Muscles that move the tongue.

The extrinsic and intrinsic muscles of the tongue are arranged in both lateral halves of the tongue.

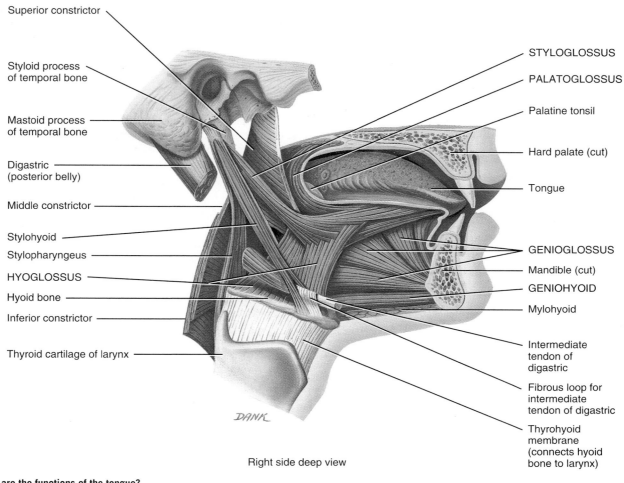

Superior constrictor
Styloid process of temporal bone
Mastoid process of temporal bone
Digastric (posterior belly)
Middle constrictor
Stylohyoid
Stylopharyngeus
HYOGLOSSUS
Hyoid bone
Inferior constrictor
Thyroid cartilage of larynx
STYLOGLOSSUS
PALATOGLOSSUS
Palatine tonsil
Hard palate (cut)
Tongue
GENIOGLOSSUS
Mandible (cut)
GENIOHYOID
Mylohyoid
Intermediate tendon of digastric
Fibrous loop for intermediate tendon of digastric
Thyrohyoid membrane (connects hyoid bone to larynx)
DANK

Right side deep view

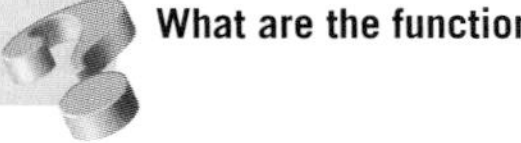

What are the functions of the tongue?

Exhibit 10.5 Muscles of the Floor of the Oral Cavity (Mouth) (Figure 10.8)

Objective

► Describe the origin, insertion, action, and innervation of the muscles of the floor of the oral cavity.

Two groups of muscles are associated with the anterior aspect of the neck: (1) the **suprahyoid muscles,** so called because they are located superior to the hyoid bone, and (2) the **infrahyoid muscles,** so named because they lie inferior to the hyoid bone. Acting with the infrahyoid muscles, the suprahyoid muscles fix the hyoid bone, thus the bone serves as a firm base upon which the tongue can move. In this exhibit we will consider the suprahyoid muscles, which are associated with the floor of the oral cavity; Exhibit 10.7 describes the infrahyoid muscles.

As a group, the suprahyoid muscles elevate the hyoid bone, the floor of the oral cavity, and the tongue during swallowing. As its name suggests, the **digastric** muscle has two bellies, anterior and posterior, united by an intermediate tendon that is held in position by a fibrous loop (see Figure 10.7). This muscle elevates the hyoid bone and larynx (voice box) during swallowing and speech and also depresses the mandible. The **stylohyoid** muscle elevates and draws the hyoid bone posteriorly, thus elongating the floor of the oral cavity during swallowing. The **mylohyoid** muscle elevates the hyoid bone and helps press the tongue against the roof of the oral cavity during swallowing to move food from the oral cavity into the throat. The **geniohyoid** muscle elevates and draws the hyoid bone anteriorly to shorten the floor of the oral cavity and to widen the throat to receive food that is being swallowed. It also depresses the mandible.

Relating Muscles to Movements

Arrange the muscles in this exhibit according to the following actions on the hyoid bone: (1) elevating it, (2) drawing it anteriorly, and (3) drawing it posteriorly. The same muscle may be mentioned more than once.

✔ Which tongue, facial, and mandibular muscles do you use when chewing food?

Innervation

The anterior belly of the digastric and the mylohyoid are innervated by the mandibular division of the trigeminal (V) cranial nerve; the posterior belly of the digastric and the stylohyoid are innervated by the facial (VII) cranial nerve; and the geniohyoid is innervated by the first cervical spinal nerve.

Muscle	Origin	Insertion	Action
Digastric (dī′-GAS-trik; *di* = two; *gaster* = belly)	Anterior belly from inner side of inferior border of mandible; posterior belly from mastoid process of temporal bone.	Body of hyoid bone via an intermediate tendon.	Elevates hyoid bone and depresses mandible, as in opening the mouth.
Stylohyoid (stī′-lō-HĪ-oid; *stylo* = stake or pole, styloid process of temporal bone; *hyoedes* = U-shaped, pertaining to hyoid bone)	Styloid process of temporal bone.	Body of hyoid bone.	Elevates hyoid bone and draws it posteriorly.
Mylohyoid (mī′-lō-HĪ-oid) (*myle* = mill)	Inner surface of mandible.	Body of hyoid bone.	Elevates hyoid bone and floor of mouth and depresses mandible.
Geniohyoid (jē′-nē-ō-HĪ-oid; *geneion* = chin) (see also Figure 10.7)	Inner surface of mandible.	Body of hyoid bone.	Elevates hyoid bone, draws hyoid bone and tongue anteriorly, and depresses mandible.

Figure 10.8 / Muscles of the floor of the oral cavity and front of the neck.

The suprahyoid muscles elevate the hyoid bone, the floor of the oral cavity, and the tongue during swallowing.

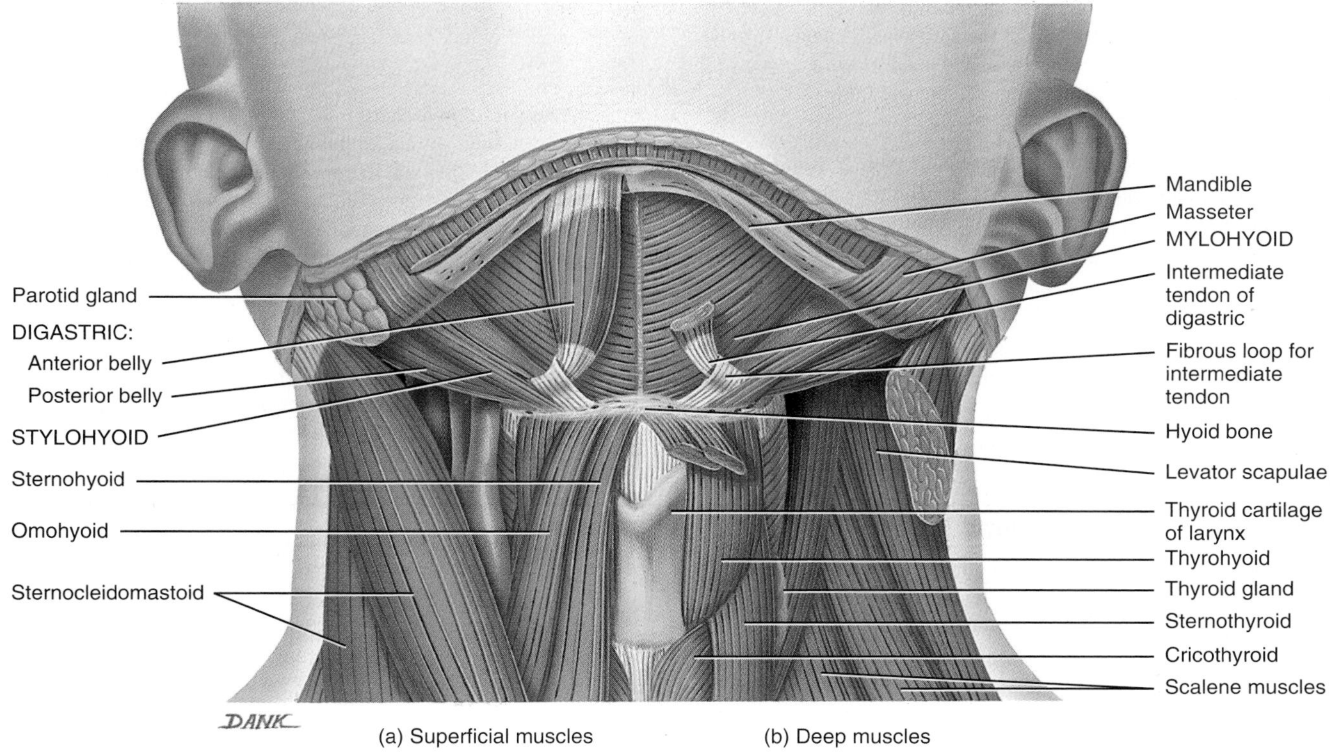

(a) Superficial muscles (b) Deep muscles

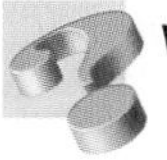

What is the combined action of the suprahyoid and infrahyoid muscles?

Exhibit 10.6 Muscles That Move the Head (Figure 10.9)

Objective

► Describe the origin, insertion, action, and innervation of the muscles that move the head.

The head is attached to the vertebral column at the atlantooccipital joint formed by the atlas and occipital bone. Balance and movement of the head on the vertebral column involves the action of several neck muscles. For example, contraction of the two **sternocleidomastoid** muscles together (bilaterally) flexes the cervical portion of the vertebral column and extends the head. Acting singly (unilaterally), the sternocleidomastoid muscle laterally flexes and rotates the head. Bilateral contraction of the **semispinalis capitis, splenius capitis,** and **longissimus capitis** muscles extends the head. However, when these same muscles contract unilaterally, their actions are quite different, involving primarily rotation of the head.

The sternocleidomastoid muscle is an important landmark that divides the neck into two major triangles: anterior and posterior. The triangles are important because of the structures that lie within their boundaries.

The **anterior triangle** is bordered superiorly by the mandible, inferiorly by the sternum, medially by the cervical midline, and laterally by the anterior border of the sternocleidomastoid muscle. The anterior triangle is subdivided into an unpaired submental triangle and three paired triangles: submandibular, carotid, and muscular. The anterior triangle contains submental, submandibular, and deep cervical lymph nodes; the submandibular salivary gland and a portion of the parotid salivary gland; the facial artery and vein; carotid arteries and internal jugular vein; and the glossopharyngeal (IX), vagus (X), accessory (XI), and hypoglossal (XII) cranial nerves.

The **posterior triangle** is bordered inferiorly by the clavicle, anteriorly by the posterior border of the sternocleidomastoid muscle, and posteriorly by the anterior border of the trapezius muscle. The posterior triangle is subdivided into two triangles, occipital and supraclavicular, by the inferior belly of the omohyoid muscle. The posterior triangle contains part of the subclavian artery, external jugular vein, cervical lymph nodes, brachial plexus, and accessory (XI) cranial nerve.

Relating Muscles to Movements

Arrange the muscles in this exhibit according to the following actions on the head: (1) flexion, (2) lateral flexion, (3) extension, (4) rotation to side opposite contracting muscle, and (5) rotation to same side as contracting muscle. The same muscle may be mentioned more than once.

✔ What muscles do you use when you move your head to signify "yes" and "no"?

Innervation

The sternocleidomastoid is innervated by the accessory (XI) cranial nerve; the capitis muscles are all innervated by cervical spinal nerves.

Muscle	Origin	Insertion	Action
Sternocleidomastoid (ster′-nō-klī′-dō-MAS-toid; *sternum* = breastbone; *cleido* = clavicle; *mastoid* = mastoid process of temporal bone)	Sternum and clavicle.	Mastoid process of temporal bone.	Acting together (bilaterally), flex cervical portion of vertebral column, extend head, and elevate sternum during forced inspiration; acting singly (unilaterally), laterally flex and rotate head to side opposite contracting muscle.
Semispinalis capitis (se′-mē-spi-NA-lis KAP-i-tis; *semi* = half; *spine* = spinous process; *caput* = head) (see Figure 10.21a)	Transverse processes of first six or seven thoracic vertebrae and seventh cervical vertebra, and articular processes of fourth, fifth, and sixth cervical vertebrae.	Occipital bone between superior and inferior nuchal lines.	Acting together, extend head; acting singly, rotate head to side opposite contracting muscle.
Splenius capitis (SPLĒ-nē-us KAP-i-tis; *splenion* = bandage) (see Figure 10.21a)	Ligamentum nuchae and spinous processes of seventh cervical vertebra and first three or four thoracic vertebrae.	Occipital bone and mastoid process of temporal bone.	Acting together, extend head; acting singly, laterally flex and rotate head to same side as contracting muscle.
Longissimus capitis (lon-JIS-i-mus KAP-i-tis; *longissimus* = longest) (see Figure 10.21a)	Transverse processes of upper four thoracic vertebrae and articular processes of last four cervical vertebrae.	Mastoid process of temporal bone.	Acting together, extend head; acting singly, laterally flex and rotate head to same side as contracting muscle.

Figure 10.9 / Triangles of the neck.

The sternocleidomastoid muscle divides the neck into two principal triangles: anterior and posterior.

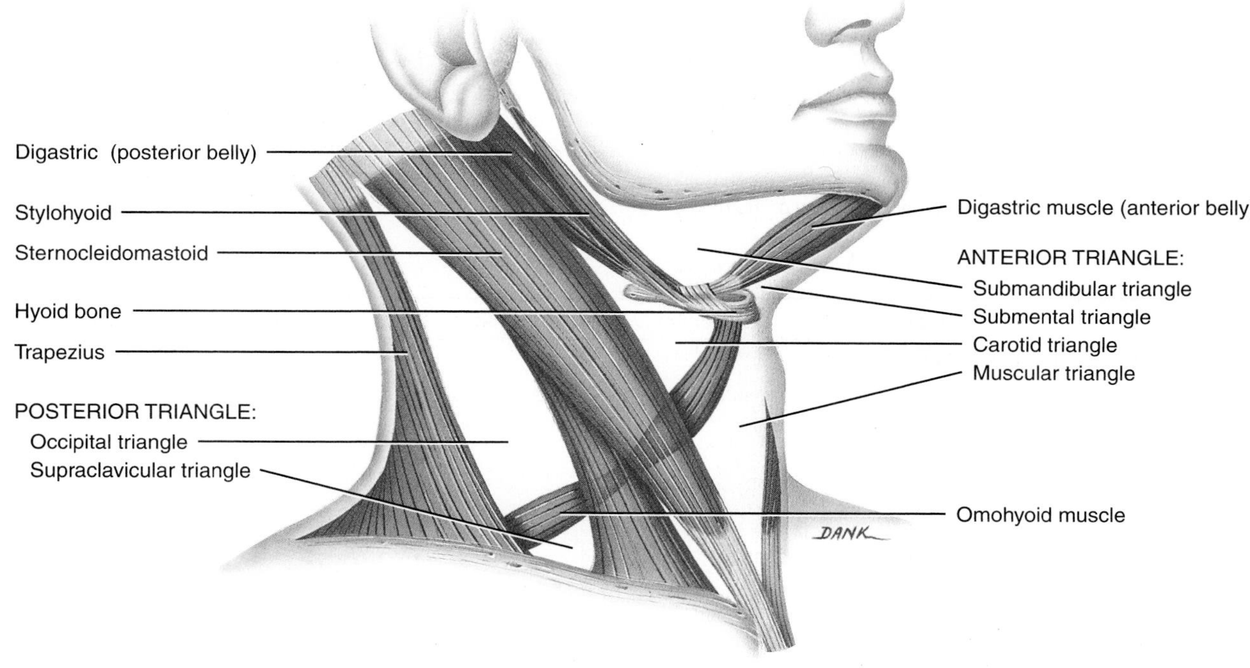

Right side

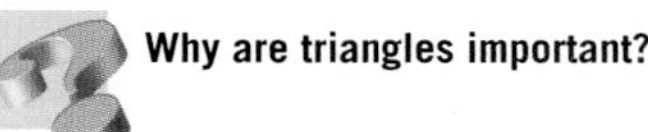

Why are triangles important?

Exhibit 10.7 Muscles of the Larynx (Voice Box) (Figure 10.10)

The muscles of the larynx (voice box), like those of the eyeballs and tongue, are grouped into **extrinsic** and **intrinsic muscles.** The extrinsic muscles of the larynx, which are associated with the anterior aspect of the neck, are called **infrahyoid muscles** because they lie inferior to the hyoid bone. (These muscles are sometimes called "strap" muscles because of their ribbonlike appearance.) Most of the extrinsic muscles depress the hyoid bone and larynx during swallowing and speech.

The **omohyoid** muscle, like the digastric muscle, is composed of two bellies connected by an intermediate tendon. In this case, however, the two bellies are referred to as *superior* and *inferior,* rather than anterior and posterior. Together, the omohyoid, **sternohyoid,** and **thyrohyoid** muscles depress the hyoid bone. In addition, the **sternothyroid** muscle depresses the thyroid cartilage (Adam's apple) of the larynx, whereas the thyrohyoid muscle elevates the thyroid cartilage. These actions are necessary during the production of low and high tones, respectively, during phonation.

Whereas the extrinsic muscles move the larynx as a whole, the intrinsic muscles only move parts of the larynx. Based upon their actions, the intrinsic muscles may be grouped into three functional sets according to their actions. The first set includes the **cricothyroid** and **thyroarytenoid** muscles, which regulate the tension of the vocal folds (true vocal cords). The second set varies the size of the *rima glottidis* (space between the vocal folds). These muscles include the **lateral cricoarytenoid,** which brings the vocal folds together (adduction), thus closing the rima glottidis, and the **posterior cricoarytenoid,** which moves the vocal folds apart (abduction), thus opening the rima glottidis. The **transverse arytenoid** closes the posterior portion of the rima glottidis. The last intrinsic muscle functions as a sphincter to control the size of the inlet of the larynx, which is the opening anteriorly from the pharynx (throat) into the larynx. This muscle is the **oblique arytenoid.**

Relating Muscles to Movements

After you have studied the muscles in this exhibit, arrange them according to the following actions: (1) depression of the thyroid cartilage; (2) elevation of the vocal cords; (3) increasing tension on the vocal cords; (4) moving the vocal cords apart, and (5) moving the vocal cords together. The same muscle may be mentioned more than once.

✔ How are the intrinsic muscles of the larynx grouped functionally?

Innervation

The omohyoid, sternohyoid, and sternothyroid are innervated by branches of the ansa cervicalis (C1–C3); the thyrohyoid is innervated by branches of the ansa cervicalis (C1–C2) and the descending hypoglossal (XII) nerve; and the intrinsic muscles are innervated by the recurrent laryngeal branch of the vagus (X) nerve.

Muscle	Origin	Insertion	Action
Extrinsic Muscles of the Larynx (Infrahyoid Muscles)			
Omohyoid (ō-mō-HĪ-oid; *omo* = relationship to the shoulder; *hyoedes* = U-shaped, pertaining to hyoid bone)	Superior border of scapula and superior transverse ligament.	Body of hyoid bone.	Depresses hyoid bone.
Sternohyoid (ster′-nō-HĪ-oid; *sterno* = sternum)	Medial end of clavicle and manubrium of sternum.	Body of hyoid bone.	Depresses hyoid bone.
Sternothyroid (ster′-nō-THĪ-roid; *thyro* = thyroid gland)	Manubrium of sternum.	Thyroid cartilage of larynx.	Depresses thyroid cartilage of larynx.
Thyrohyoid (thī′-rō-HĪ-oid)	Thyroid cartilage of larynx.	Greater horn of hyoid bone.	Elevates thyroid cartilage and depresses hyoid bone.
Intrinsic Muscles of the Larynx			
Cricothyroid (kri-kō-THĪ-roid, *crico* = cricoid cartilage of larynx)	Anterior and lateral portion of cricoid cartilage of larynx.	Anterior border of thyroid cartilage of larynx and posterior part of inferior border of thyroid cartilage of larynx.	Elongates and places tension on vocal folds.
Thyroarytenoid (thī′-rō-ar′-i-TĒ-noid)	Inferior portion of thyroid cartilage of larynx and middle of cricothyroid ligament.	Base and anterior surface of arytenoid cartilage of larynx.	Shortens and relaxes vocal folds.
Lateral cricoarytenoid (kri′-kō-ar′-i-TĒ-noid)	Superior border of cricoid cartilage of larynx.	Anterior surface of arytenoid cartilage of larynx.	Brings vocal folds together (adduction), thus closing the rima glottidis.
Posterior cricoarytenoid (kri′-kō-ar′-i-TĒ-noid; *arytaina* = shaped like a jug)	Posterior surface of cricoid cartilage of larynx.	Posterior surface of arytenoid cartilage of larynx.	Moves the vocal folds apart (abduction), thus opening the rima glottidis.
Transverse and **oblique arytenoid** (ar′-i-TĒ-noid)	Posterior surface and lateral border of one arytenoid cartilage of larynx.	Corresponding parts of opposite arytenoid cartilage of larynx.	The transverse arytenoid closes the posterior portion of the rima glottidis; the oblique arytenoid regulates the size of the inlet of the larynx.

Figure 10.10 / Muscles of the larynx (voice box). (See Tortora, *A Photographic Atlas of the Human Body,* Figure 5.6.)

Intrinsic muscles of the larynx adjust the tension of the vocal folds and open or close the rima glottidis.

Anterior superficial view (a) Anterior deep view

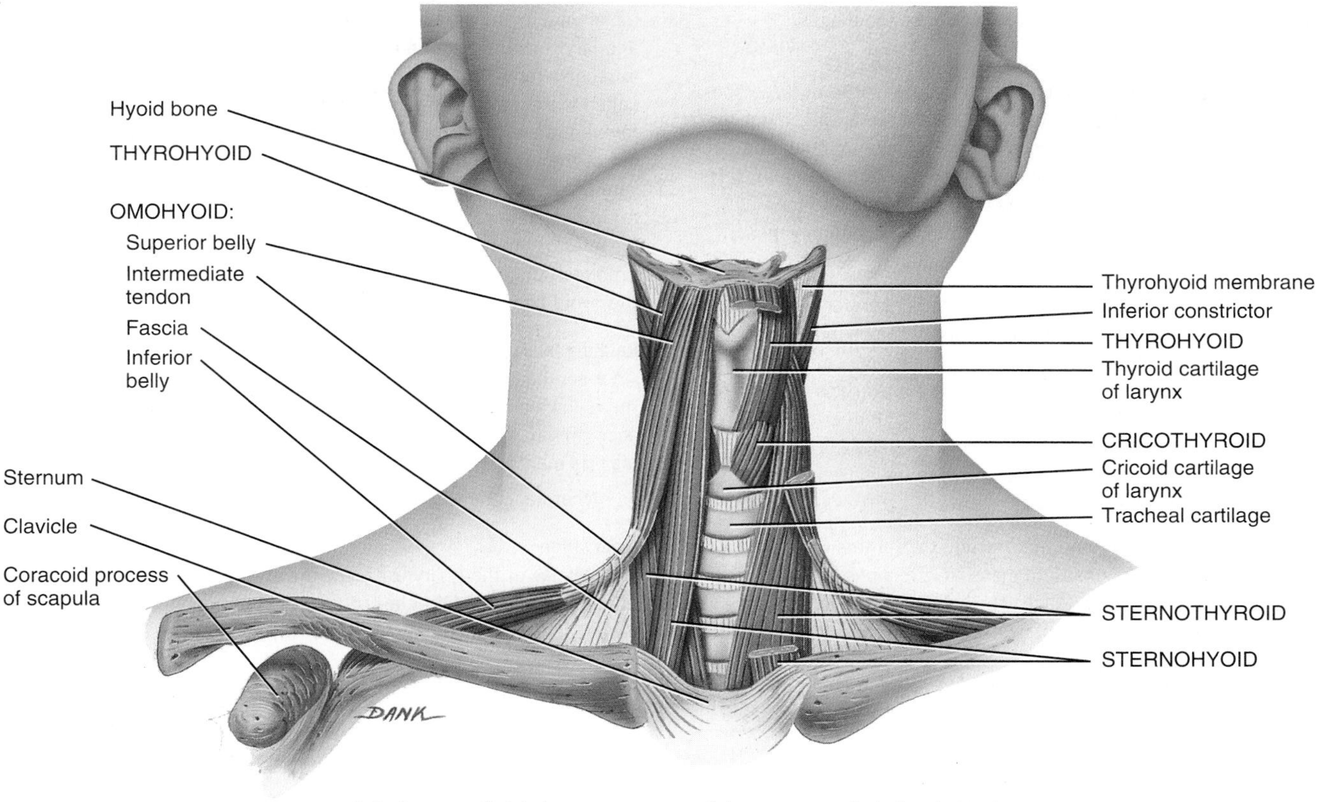

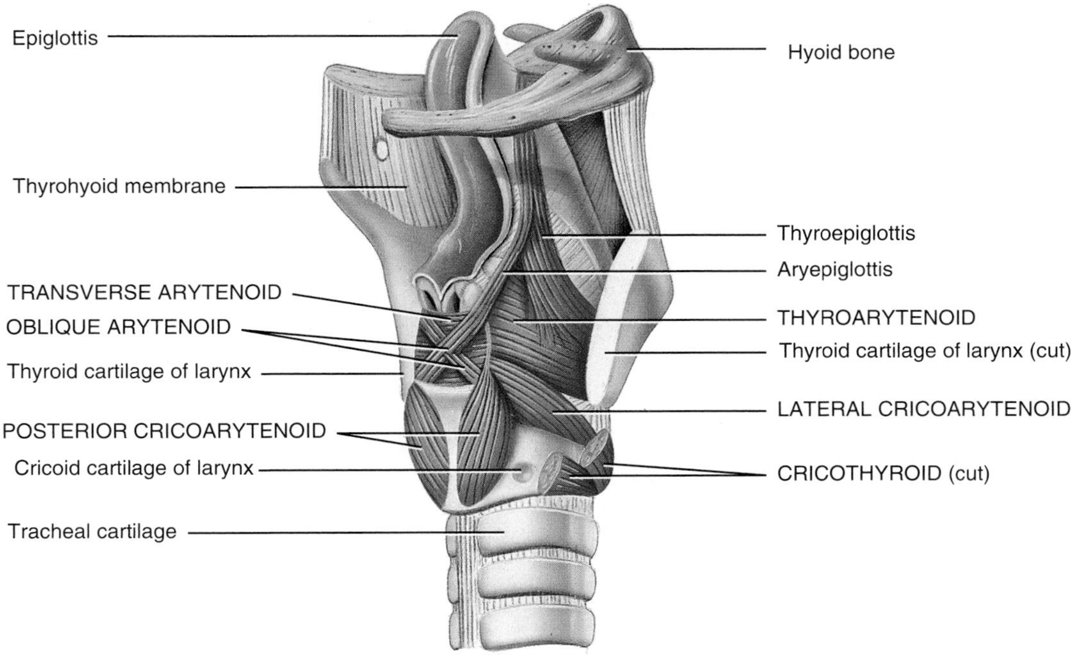

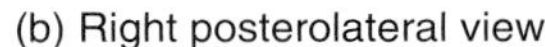

(b) Right posterolateral view

Elevation and depression of the thyroid cartilage is necessary for what aspect of phonation?

Exhibit 10.8 Muscles of the Pharynx (Throat) (Figure 10.11)

The pharynx (throat) is a somewhat funnel-shaped tube posterior to the nasal and oral cavities. It is a common chamber for the respiratory and digestive systems, opening into the anterior larynx and posterior esophagus.

The muscles of the pharynx are arranged in two layers, an outer circular layer and an inner longitudinal layer. The **circular layer** is composed of three constrictor muscles, each overlapping the muscle above it, an arrangement that resembles stacked flowerpots. Their names are the inferior, middle, and superior constrictor muscles. The **longitudinal layer** is composed of three muscles that descend from the styloid process of the temporal bone, auditory (Eustachian) tube, and soft palate. Their names are the stylopharyngeus, salpingopharyngeus, and palatopharyngeus, respectively.

The **inferior constrictor** muscle is the thickest of the constrictor muscles. Its inferior fibers are continuous with the musculature of the esophagus, whereas its superior fibers overlap the middle constrictor. The **middle constrictor** is fan-shaped and smaller than the inferior constrictor, and it overlaps the superior constrictor. The **superior constrictor** is quadrilateral and thinner than the other constrictors. As a group, the constrictor muscles constrict the pharynx during swallowing. The sequential contraction of these muscles moves food and drink from the mouth into the esophagus.

The **stylopharyngeus** muscle is a long, thin muscle that enters the pharynx between the middle and superior constrictors and inserts along with the palatopharyngeus on the thyroid cartilage. The **salpingopharyngeus** muscle is a thin muscle that descends in the lateral wall of the pharynx and also inserts along with the palatopharyngeus muscle. The **palatopharyngeus** muscle also descends in the lateral wall of the pharynx and inserts along with the stylopharyngeus. All three muscles of the longitudinal layer elevate the pharynx and larynx during swallowing and speaking. Elevation of the pharynx widens it to receive food and liquids, and elevation of the larynx causes a structure called the epiglottis to close over the rima glottidis (space between the vocal folds) and seal the respiratory passageway. Food and drink are further kept out of the respiratory tract by the suprahyoid muscles of the larynx, which elevate the hyoid bone, and thus the larynx, which is attached to the hyoid bone. Additionally, the respiratory passageway is sealed by the action of the intrinsic muscles of the larynx, which bring the vocal folds together to close off the rima glottidis. After swallowing, the infrahyoid muscles of the larynx depress the hyoid bone and larynx.

Relating Muscles to Movements

After you have studied the muscles in this exhibit, arrange them according to the following actions on the pharynx: (1) constriction and (2) elevation, and according to the following action on the larynx: elevation. The same muscle may be mentioned more than once.

✔ What happens when the pharynx and larynx are elevated?

Innervation

All muscles of the pharynx except the stylopharyngeus are innervated by the pharyngeal plexus, branches of the vagus (X) nerve. The stylopharyngeus is innervated by the glossopharyngeal (IX) nerve.

Muscle	Origin	Insertion	Action
Circular Layer			
Inferior constrictor (*inferior* = below; *constrictor* = decreases diameter of a lumen)	Cricoid and thyroid cartilages of larynx.	Posterior median raphe (slender band of collagen fibers) of pharynx.	Constricts inferior portion of pharynx to propel food and drink into esophagus.
Middle constrictor	Greater and lesser horns of hyoid bone and stylohyoid ligament.	Posterior median raphe of pharynx.	Constricts middle portion of pharynx to propel food and drink into esophagus.
Superior constrictor (*superior* = above)	Pterygoid process of sphenoid, pterygomandibular raphe, and medial surface of mandible.	Posterior median raphe of pharynx.	Constricts superior portion of pharynx to propel food and drink into esophagus.
Longitudinal Layer			
Stylopharyngeus (stī′-lō-far-IN-jē-us; *stylo* = stake or pole; styloid process of temporal bone; *pharyngo* = pharynx) (see also Figure 10.7)	Styloid process of temporal bone.	Thyroid cartilage with the palatopharyngeus.	Elevates larynx and pharynx.
Salpingopharyngeus (sal-pin′-gō-far-IN-jē-us; *salping* = pertaining to the auditory or uterine tube)	Inferior portion of auditory (Eustachian) tube.	Thyroid cartilage with the palatopharyngeus.	Elevates larynx and pharynx and opens orifice of auditory (Eustachian) tube.
Palatopharyngeus (pal′-a-tō-far-IN-jē-us; *palato* = palate)	Soft palate.	Thyroid cartilage with the stylopharyngeus.	Elevates larynx and pharynx and helps close nasopharynx during swallowing.

Figure 10.11 / Muscles of the pharynx.

The muscles of the pharynx assist in swallowing and in closing off the respiratory passageway.

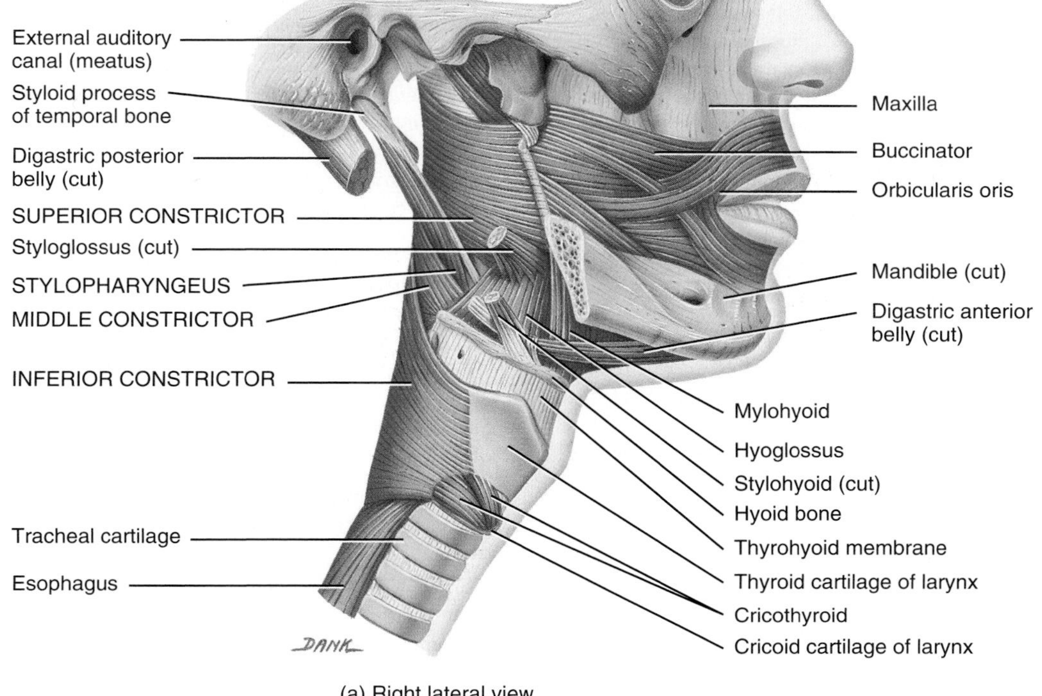

(a) Right lateral view

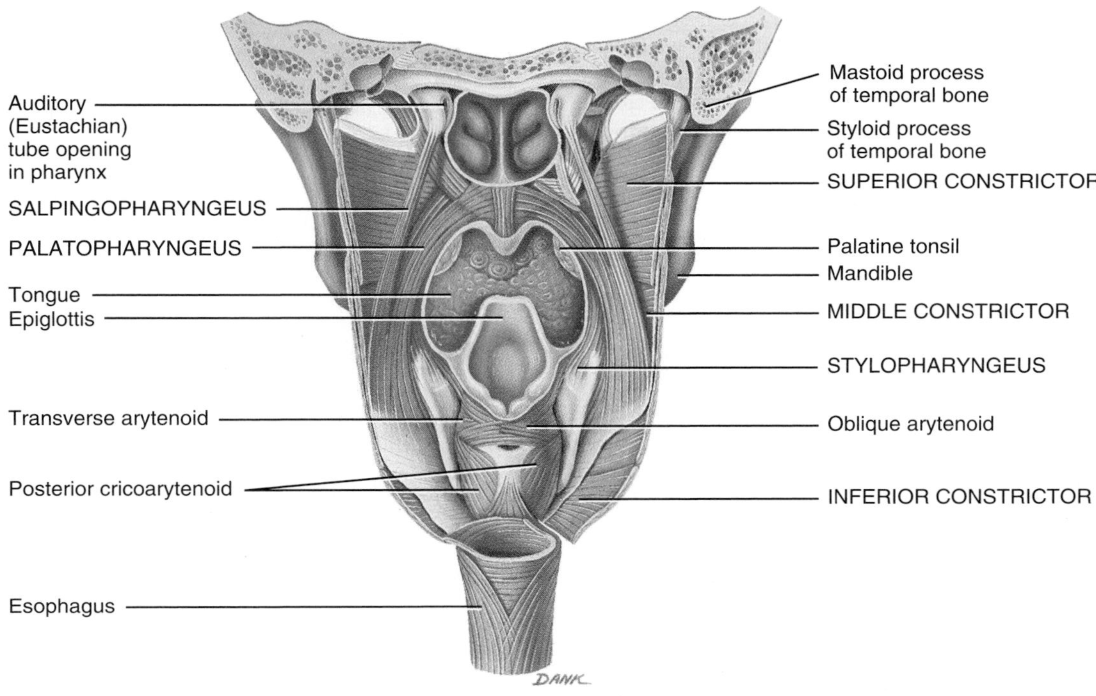

(b) Posterior view

What are the antagonistic roles of the longitudinal muscles of the pharynx and the infrahyoid muscles?

Exhibit 10.9 Muscles That Act on the Abdominal Wall (Figure 10.12)

Objective

▶ Describe the origin, insertion, action, and innervation of the muscles that act on the abdominal wall.

The anterolateral abdominal wall is composed of skin, fascia, and four pairs of muscles: the external oblique, internal oblique, transversus abdominis, and rectus abdominis. The first three muscles are flat muscles arranged from superficial to deep. The **external oblique** is the superficial flat muscle with its fibers directed inferiorly and medially. The **internal oblique** is the intermediate flat muscle with its fibers directed at right angles to those of the external oblique. The **transversus abdominis** is the deepest of the flat muscles, with most of its fibers directed horizontally around the abdominal wall. Together, the external oblique, internal oblique, and transversus abdominus form three layers of muscle around the abdomen. The muscle fibers of each layer run cross-directionally to one another, a structural arrangement that affords considerable protection to the abdominal viscera, especially when the muscles have good tone.

The **rectus abdominis** muscle is a long, straplike muscle that extends the entire length of the anterior abdominal wall, from the pubic crest and pubic symphysis to the cartilages of ribs 5–7 and the xiphoid process of the sternum. The anterior surface of the muscle is interrupted by three transverse fibrous bands of tissue called **tendinous intersections,** believed to be remnants of septa that separated myotomes during embryological development.

As a group, the muscles of the anterolateral abdominal wall help contain and protect the abdominal viscera; flex, laterally flex, and rotate the vertebral column at the intervertebral joints; compress the abdomen during forced expiration; and produce the force required for defecation, urination, and childbirth.

The aponeuroses of the external oblique, internal oblique, and transversus abdominis muscles form the **rectus sheaths,** which enclose the rectus abdominis muscles. The sheaths meet at the midline to form the **linea alba** (= white line), a tough, fibrous band that extends from the xiphoid process of the sternum to the pubic symphysis. In the latter stages of pregnancy, the linea alba stretches to increase the distance between the rectus abdominis muscles. The inferior free border of the external oblique aponeurosis, plus some collagen fibers, forms the **inguinal ligament,** which runs from the anterior superior iliac spine to the pubic tubercle (see also Figure 10.22a). Just superior to the medial end of the inguinal ligament is a triangular slit in the aponeurosis referred to as the **superficial inguinal ring,** the outer opening of the **inguinal canal** (see Figure 26.4 on page 786). The canal contains the spermatic cord and ilioinguinal nerve in males, and the round ligament of the uterus and ilioinguinal nerve in females.

The posterior abdominal wall is formed by the lumbar vertebrae, parts of the ilia of the hip bones, psoas major and iliacus muscles (described in Exhibit 10.19), and quadratus lumborum muscle. Whereas the anterolateral abdominal wall can contract and distend, the posterior abdominal wall is bulky and stable by comparison.

Inguinal Hernia

The inguinal region is a weak area in the abdominal wall. It is often the site of an **inguinal hernia,** a rupture or separation of a portion of the inguinal area of the abdominal wall resulting in the protrusion of a part of the small intestine. Hernia is much more common in males than in females because the inguinal canals in males are larger to accommodate the spermatic cord and ilioinguinal nerve and represent especially weak points in the abdominal wall. ■

Relating Muscles to Movements

Arrange the muscles in this exhibit according to the following actions on the vertebral column: (1) flexion, (2) lateral flexion, (3) extension, and (4) rotation. The same muscle may be mentioned more than once.

✔ Which muscles do you contract when you "suck in your tummy," thereby compressing the anterior abdominal wall?

Innervation

The rectus abdominis is innervated by branches of thoracic nerves T7–T12; the external oblique is innervated by branches of thoracic nerves T7–T12 and the iliohypogastric nerve; the internal oblique and transversus abdominis are innervated by branches of thoracic nerves T8–T12 and the iliohypogastric and ilioinguinal nerves; the quadratus lumborum is innervated by branches of thoracic nerve T12 and lumbar nerves L1–L3 or L1–L4.

Muscle	Origin	Insertion	Action
Rectus abdominis (REK-tus ab-DOM-in-is; *rectus* = fibers parallel to midline; *abdomino* = abdomen)	Pubic crest and pubic symphysis.	Cartilage of fifth to seventh ribs and xiphoid process.	Flexes vertebral column, especially lumbar portion, and compresses abdomen to aid in defecation, urination, forced expiration, and childbirth.
External oblique (ō-BLĒK; *external* = closer to surface; *oblique* = fibers diagonal to midline)	Inferior eight ribs.	Iliac crest and linea alba.	Acting together (bilaterally), compress abdomen and flex vertebral column; acting singly (unilaterally), laterally flex vertebral column, especially lumbar portion, and rotate vertebral column.
Internal oblique (ō-BLĒK; *internal* = farther from surface)	Iliac crest, inguinal ligament, and thoracolumbar fascia.	Cartilage of last three or four ribs and linea alba.	Acting together, compress abdomen and flex vertebral column; acting singly, laterally flex vertebral column, especially lumbar portion, and rotate vertebral column.
Transversus abdominis (tranz-VER-sus ab-DOM-in-is; *transverse* = fibers perpendicular to midline)	Iliac crest, inguinal ligament, lumbar fascia, and cartilages of inferior six ribs.	Xiphoid process, linea alba, and pubis.	Compresses abdomen.
Quadratus lumborum (kwod-RĀ-tus lum-BOR-um; *quad* = four; *lumbo* = lumbar region) (see Figure 10.13)	Iliac crest and iliolumbar ligament.	Inferior border of twelfth rib and transverse processes of first four lumbar vertebrae.	Acting together, pull twelfth ribs inferiorly during forced expiration, fix twelfth ribs to prevent their elevation during deep inspiration, and help extend lumbar portion of vertebral column; acting singly, laterally flex vertebral column, especially lumbar portion.

(continues)

Exhibit 10.9 Muscles That Act on the Abdominal Wall (Figure 10.12) continued

Figure 10.12 / Muscles of the male anterolateral abdominal wall.

The anterolateral abdominal muscles protect the abdominal viscera, move the vertebral column, and assist in forced expiration, defecation, urination, and childbirth.

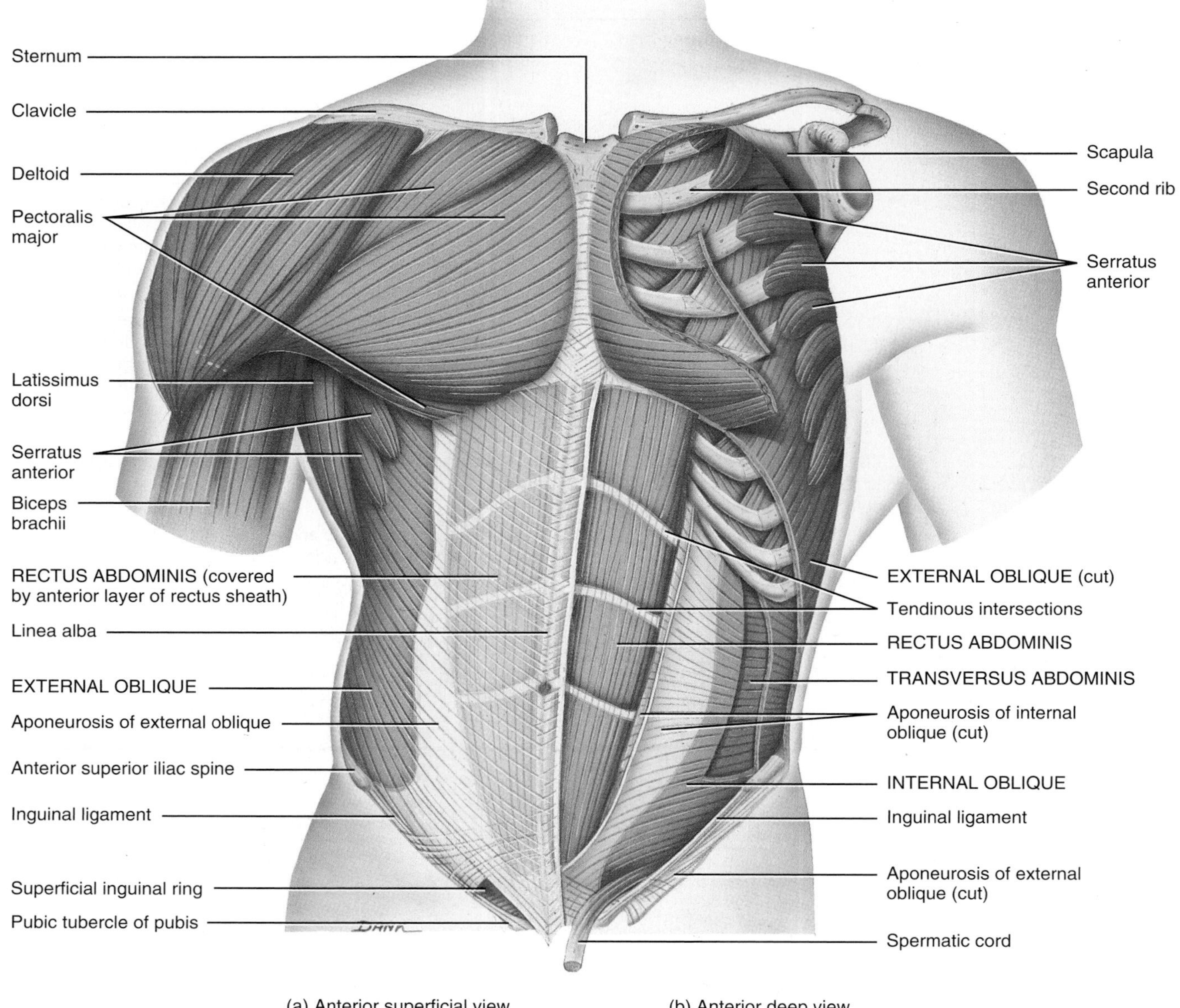

(a) Anterior superficial view

(b) Anterior deep view

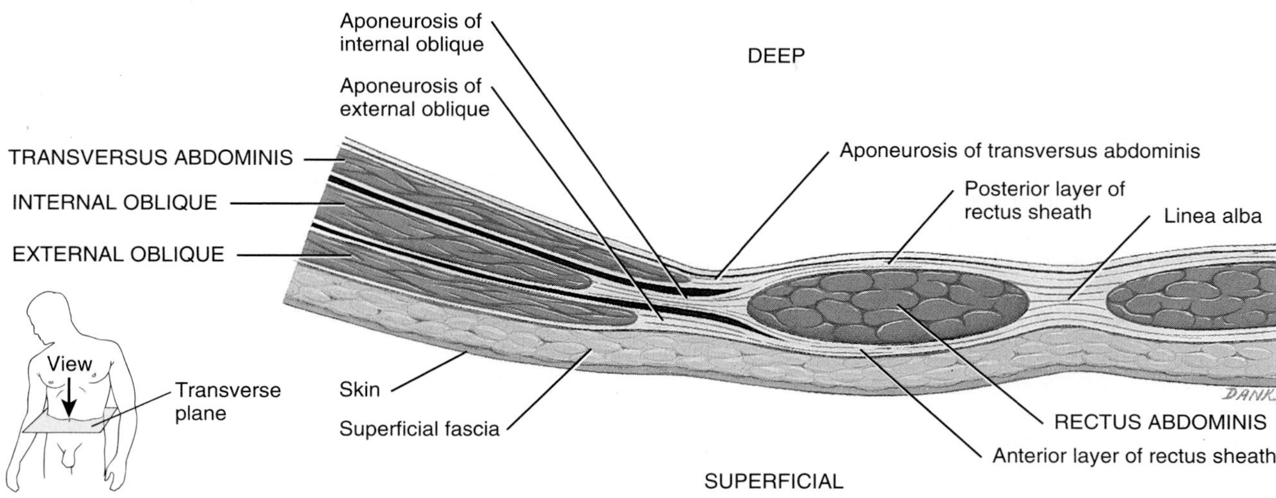

(c) Transverse section of anterior abdominal wall superior to umbilicus (navel)

SUPERIOR

Serratus anterior
Latissimus dorsi
EXTERNAL OBLIQUE
INTERNAL OBLIQUE
TRANSVERSUS ABDOMINIS
Tendinous intersection
Linea alba
RECTUS ABDOMINIS
POSTERIOR
ANTERIOR

INFERIOR

(d) Right anterolateral view

Which abdominal muscle aids in urination?

Exhibit 10.10 Muscles Used in Breathing (Figure 10.13)

Objective

▶ Describe the origin, insertion, action, and innervation of the muscles used in breathing.

The muscles described here alter the size of the thoracic cavity so that breathing (ventilation) can occur. Inspiration (breathing in) occurs when the thoracic cavity increases in size, and expiration (breathing out) occurs when the thoracic cavity decreases in size.

The **diaphragm** is the most important muscle in breathing. It is a dome-shaped musculotendinous partition that separates the thoracic and abdominal cavities. The diaphragm is composed of a peripheral muscular portion and a central portion called the central tendon. The **central tendon** is a strong aponeurosis that serves as the tendon of insertion for all the peripheral muscular fibers of the diaphragm. It fuses with the inferior surface of the fibrous pericardium (external covering of the heart) and the parietal pleurae (external coverings of the lungs).

Movements of the diaphragm also help to return to the heart venous blood passing through the abdomen. Together with the anterolateral abdominal muscles, the diaphragm helps to increase intraabdominal pressure to evacuate the pelvic contents during defecation, urination, and childbirth. This mechanism is further assisted when you take a deep breath and close the rima glottidis (the space between vocal folds). The trapped air in the respiratory system prevents the diaphragm from elevating. The increase in intraabdominal pressure as just described also helps support the vertebral column and prevents flexion during weight lifting. This greatly assists the back muscles in lifting a heavy weight.

The diaphragm has three major openings through which various structures pass between the thorax and abdomen. These structures include the aorta, along with the thoracic duct and azygos vein, which pass through the **aortic hiatus;** the esophagus with accompanying vagus (X) cranial nerves, which pass through the **esophageal hiatus;** and the inferior vena cava, which passes through the **foramen for the vena cava.** In a condition called a hiatus hernia, the stomach protrudes superiorly through the esophageal hiatus.

The other muscles involved in breathing are the **intercostal** muscles, which occupy the intercostal spaces, the spaces between ribs. These muscles are arranged in three layers. The 11 **external intercostal** muscles occupy the superficial layer, and their fibers run obliquely inferiorly and anteriorly from the rib above to the rib below. They elevate the ribs during inspiration to help expand the thoracic cavity. The 11 **internal intercostal** muscles occupy the intermediate layer of the intercostal spaces. The fibers of these muscles run obliquely inferiorly and posteriorly from the inferior border of the rib above to the superior border of the rib below. They draw adjacent ribs together during forced expiration to help decrease the size of the thoracic cavity. The deepest muscle layer is made up of the **transversus thoracis** muscles. These poorly developed muscles (not described in the exhibit) run in the same direction as the internal intercostals, and they may have the same role.

Relating Muscles to Movements

Arrange the muscles in this exhibit according to the following actions on the size of the thorax: (1) increase in vertical length, (2) increase in lateral and anteroposterior dimensions, and (3) decrease in lateral and anteroposterior dimensions.

✔ What are the names of the three openings in the diaphragm, and which structures pass through each?

Innervation

The diaphragm is innervated by the phrenic nerve, which contains axons from cervical spinal nerves C3–C5; the external and internal intercostals are innervated by thoracic spinal nerves T2–T12.

Muscle	Origin	Insertion	Action
Diaphragm (DĪ-a-fram; *dia* = across; *phragma* = wall)	Xiphoid process of the sternum, costal cartilages of inferior six ribs, and lumbar vertebrae.	Central tendon.	Forms floor of thoracic cavity; pulls central tendon inferiorly during inspiration, and, as dome of diaphragm flattens, increases vertical length of thorax.
External intercostals (in′-ter-KOS-tals; *external* = closer to surface; *inter* = between; *costa* = rib)	Inferior border of rib above.	Superior border of rib below.	Elevate ribs during inspiration and thus increase lateral and anteroposterior dimensions of thorax.
Internal intercostals (in′-ter-KOS-tals; *internal* = farther from surface)	Superior border of rib below.	Inferior border of rib above.	Draw adjacent ribs together during forced expiration and thus decrease lateral and anteroposterior dimensions of thorax.

Figure 10.13 / Muscles used in breathing, as seen in a male.

Openings in the diaphragm permit the passage of the aorta, esophagus, and inferior vena cava.

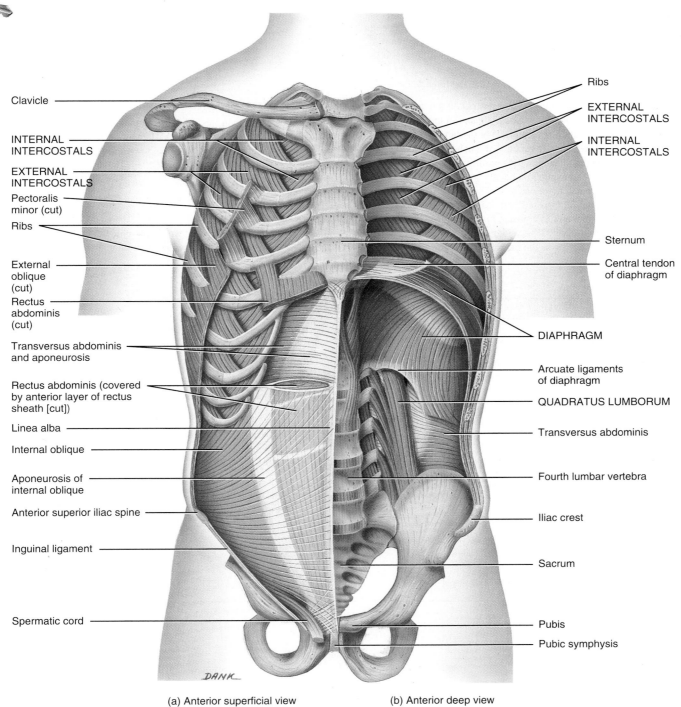

(a) Anterior superficial view

(b) Anterior deep view

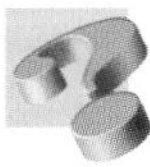

Which muscle associated with breathing is innervated by the phrenic nerve?

Exhibit 10.11 Muscles of the Pelvic Floor (Figure 10.14)

Objective

► Describe the origin, insertion, action, and innervation of the muscles of the pelvic floor.

The muscles of the pelvic floor are the levator ani and coccygeus. Together with the fascia covering their internal and external surfaces, these muscles are referred to as the **pelvic diaphragm,** which stretches from the pubis anteriorly to the coccyx posteriorly, and from one lateral wall of the pelvis to the other. This arrangement gives the pelvic diaphragm the appearance of a funnel suspended from its attachments. The pelvic diaphragm is pierced by the anal canal and urethra in both sexes, and also by the vagina in females.

The **levator ani** muscle is divisible into two muscles called the **pubococcygeus** and **iliococcygeus.** These muscles in the female are shown in Figure 10.14 and in the male in Figure 10.15. The levator ani is the largest and most important muscle of the pelvic floor. It supports the pelvic viscera and resists the inferior thrust that accompanies increases in intraabdominal pressure during functions such as forced expiration, coughing, vomiting, urination, and defecation. The muscle also functions as a sphincter at the anorectal junction, urethra, and vagina. During childbirth, the levator ani muscle supports the head of the fetus, and the muscle may be injured during a difficult childbirth or traumatized during an episiotomy (a cut made with surgical scissors to prevent or direct tearing of the perineum during the birth of a baby). This may cause urinary stress incontinence in which there is a leakage of urine whenever intraabdominal pressure is increased—for example, during coughing. In addition to assisting the levator ani, the **coccygeus** muscle pulls the coccyx anteriorly after it has been pushed posteriorly following defecation or childbirth. In order to treat urinary stress incontinence, it may be necessary to strengthen and tighten the muscles that support the pelvic viscera. This is accomplished by *Kegel exercises,* the alternate contraction and relaxation of muscles of the pelvic floor.

Relating Muscles to Movements

Arrange the muscles in this exhibit according to the following actions on the pelvic viscera: supporting and maintaining their position, and resisting an increase in intraabdominal pressure; and according to the following action on the anus, urethra, and vagina: constriction. The same muscle may be mentioned more than once.

✔ Which muscles are strengthened by Kegel exercises?

Innervation

The pubococcygeus and iliococcygeus are innervated by sacral spinal nerves S2–S4; the coccygeus is innervated by sacral spinal nerves S4–S5.

Muscle	Origin	Insertion	Action
Levator ani (le-VĀ-tor Ā-nē; *levator* = raises; *ani* = anus)	This muscle is divisible into two parts, the pubococcygeus muscle and the iliococcygeus muscle.		
Pubococcygeus (pyū′-bō-kok-SIJ-ē-us; *pubo* = pubis; *coccygeus* = coccyx)	Pubis.	Coccyx, urethra, anal canal, central tendon of perineum, and anococcygeal raphe (narrow fibrous band that extends from anus to coccyx).	Supports and maintains position of pelvic viscera; resists increase in intraabdominal pressure during forced expiration, coughing, vomiting, urination, and defecation; constricts anus, urethra, and vagina; and supports fetal head during childbirth.
Iliococcygeus (il′-ē-ō-kok-SIJ-ē-us; *ilio* = ilium)	Ischial spine.	Coccyx.	As above.
Coccygeus (kok-SIJ-ē-us)	Ischial spine.	Lower sacrum and upper coccyx.	Supports and maintains position of pelvic viscera; resists increase in intraabdominal pressure during forced expiration, coughing, vomiting, urination, and defecation; and pulls coccyx anteriorly following defecation or childbirth.

Figure 10.14 / Muscles of the pelvic floor, as seen in the female perineum.

The pelvic diaphragm supports the pelvic viscera.

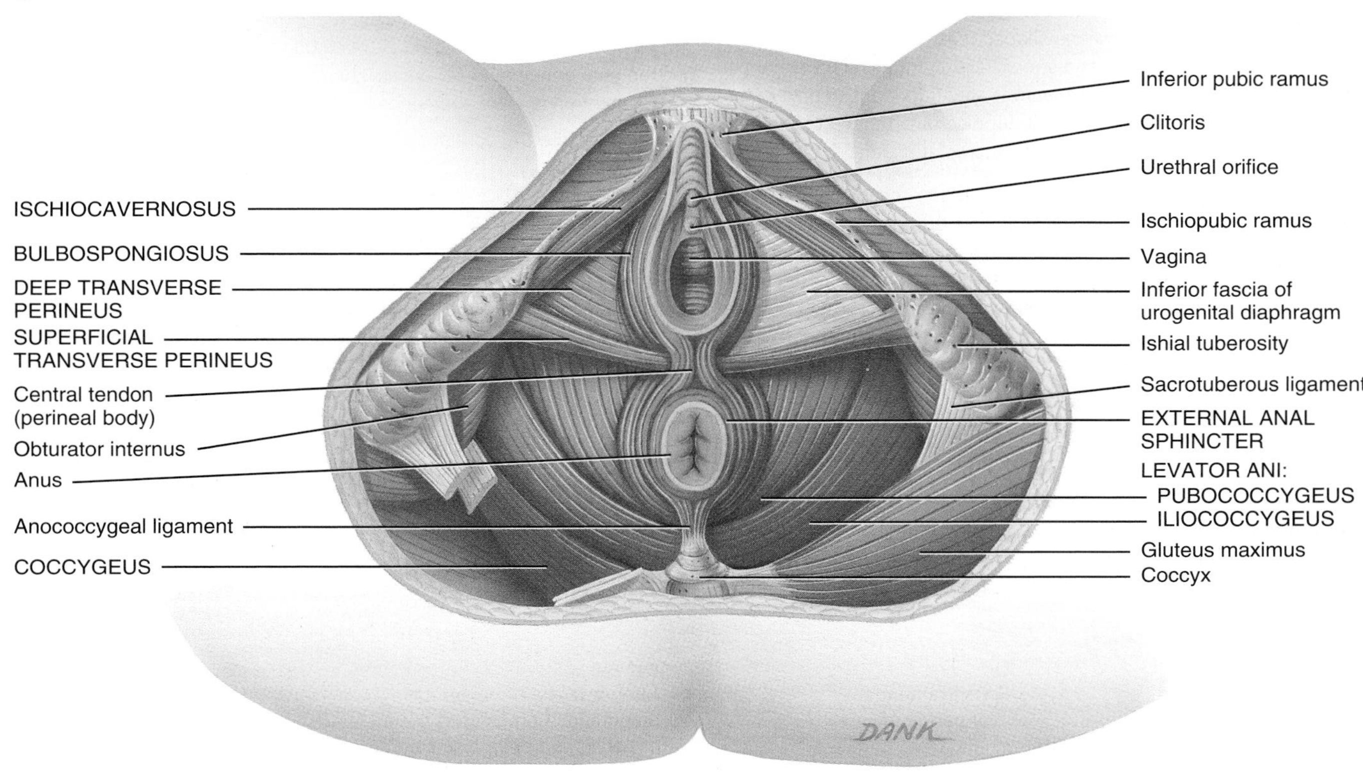

Inferior superficial view

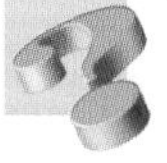

What are the borders of the pelvic diaphragm?

Exhibit 10.12 Muscles of the Perineum (Figures 10.14 and 10.15)

Objective

▶ Describe the origin, insertion, action, and innervation of the muscles of the perineum.

The **perineum,** the region of the trunk inferior to the pelvic diaphragm, is a diamond-shaped area that extends from the pubic symphysis anteriorly, to the coccyx posteriorly, and to the ischial tuberosities laterally. The female and the male perineums may be compared in Figures 10.14 and 10.15, respectively. A transverse line drawn between the ischial tuberosities divides the perineum into an anterior **urogenital triangle** that contains the external genitals and a posterior **anal triangle** that contains the anus (see Figure 26.20 on page 804). In the center of the perineum is a wedge-shaped mass of fibrous tissue called the **central tendon (perineal body).** It is a strong tendon into which several perineal muscles insert. Clinically, the perineum is very important to physicians who care for women during pregnancy and treat disorders related to the female genital tract, urogenital organs, and the anorectal region.

The muscles of the perineum are arranged in two layers: **superficial** and **deep.** The muscles of the superficial layer are the superficial transverse perineus, bulbospongiosus, and ischiocavernosus. The **superficial transverse perineus** muscle is a narrow muscle that passes more or less transversely across the perineum anterior to the anus. It joins with the external sphincter posteriorly. The **bulbospongiosus** muscle assists in urination and erection of the penis in males and in decreasing the vaginal orifice and erection of the clitoris in females. The **ischiocavernosus** muscle assists in maintaining erection of the penis and clitoris. The deep muscles of the perineum are the **deep transverse perineus** muscle and **external urethral sphincter** (see Figure 25.10 on page 773). The deep transverse perineus, external urethral sphincter, and their fascia are known as the **urogenital diaphragm.** The muscles of this diaphragm assist in urination and ejaculation in males and urination in females. The **external anal sphincter** closely adheres to the skin around the margin of the anus and assists in defecation.

Relating Muscles to Movements

Arrange the muscles in this exhibit according to the following actions: (1) expulsion of urine and semen, (2) erection of the clitoris and penis, (3) closing the anal orifice, and (4) constricting the vaginal orifice. The same muscle may be mentioned more than once.

✔ Describe the borders and contents of the urogenital triangle and the anal triangle.

Innervation

Muscles of the perineum are innervated by the pudendal nerve of the sacral plexus (shown in Exhibit 17.4 on page 523). More specifically, all muscles are innervated by the perineal branch of the pudendal nerve, except for the external anal sphincter, which is innervated by sacral spinal nerve S4 and the inferior rectal branch of the pudendal nerve.

Muscle	Origin	Insertion	Action
Superficial Perineal Muscles			
Superficial transverse perineus (per-i-NĒ-us; *superficial* = closer to surface; *transverse* = across; *perineus* = perineum)	Ischial tuberosity.	Central tendon of perineum.	Helps stabilize central tendon of perineum.
Bulbospongiosus (bul′-bō-spon′-jē-Ō-sus; *bulbus* = bulb; *spongia* = sponge)	Central tendon of perineum.	Inferior fascia of urogenital diaphragm, corpus spongiosum of penis, and deep fascia on dorsum of penis in male; pubic arch and root and dorsum of clitoris in female.	Helps expel urine during urination, helps propel semen along urethra, assists in erection of the penis in male; constricts vaginal orifice and assists in erection of clitoris in female.
Ischiocavernosus (is′-kē-ō-ka′-ver-NŌ-sus; *ischion* = hip)	Ischial tuberosity and ischial and pubic rami.	Corpus cavernosum of penis in male and clitoris in female.	Maintains erection of penis in male and clitoris in female.
Deep Perineal Muscles			
Deep transverse perineus (per-i-NĒ-us; *deep* = farther from surface)	Ischial rami.	Central tendon of perineum.	Helps expel last drops of urine and semen in male and urine in female.
External urethral sphincter (yū-RĒ-thral SFINGK-ter)	Ischial and pubic rami.	Median raphe in male and vaginal wall in female.	Helps expel last drops of urine and semen in male and urine in female.
External anal sphincter (Ā-nal)	Anococcygeal ligament.	Central tendon of perineum.	Keeps anal canal and anus closed.

Figure 10.15 / Muscles of the male perineum.

The urogenital diaphragm assists in urination in females and males, ejaculation in males, and helps strengthen the pelvic floor.

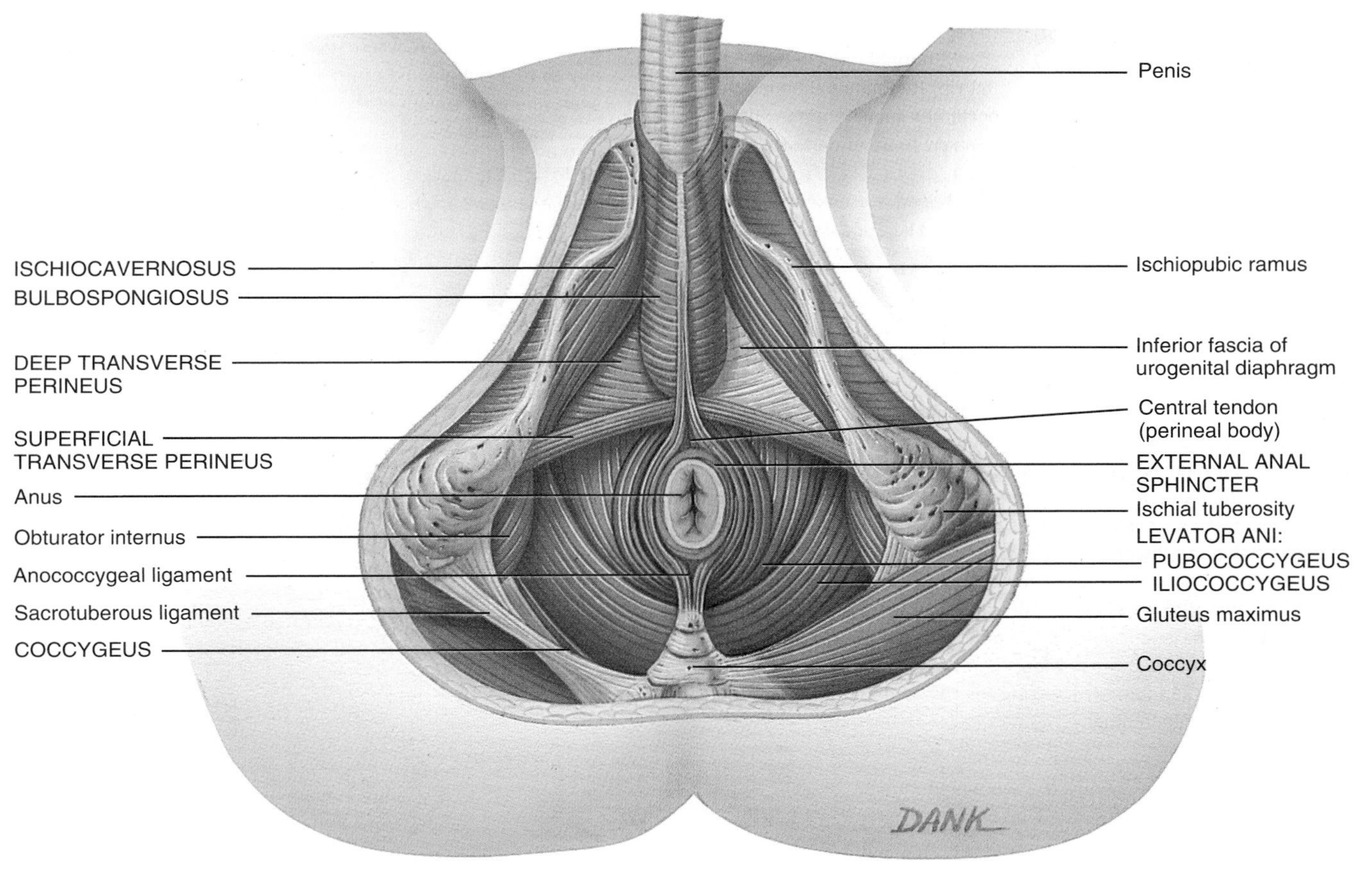

Inferior superficial view

What are the borders of the perineum?

Exhibit 10.13 Muscles That Move the Pectoral (Shoulder) Girdle (Figure 10.16)

Objective

▶ Describe the origin, insertion, action, and innervation of the muscles that move the pectoral girdle.

The principal action of the muscles that move the pectoral girdle is to fix the scapula so it can function as a stable point of origin for most of the muscles that move the humerus. Because scapular movements usually accompany humeral movements in the same direction, the muscles also move the scapula to increase the range of motion of the humerus. For example, it would not be possible to abduct the humerus past the horizontal if the scapula did not move with the humerus. During abduction, the scapula follows the humerus by rotating upward.

Muscles that move the pectoral girdle can be classified into two groups based on their location in the thorax: **anterior** and **posterior thoracic muscles.** The anterior thoracic muscles are the subclavius, pectoralis minor, and serratus anterior (Figure 10.16a). The **subclavius** is a small, cylindrical muscle under the clavicle that extends from the clavicle to the first rib. It steadies the clavicle during movements of the pectoral girdle. The **pectoralis minor** is a thin, flat, triangular muscle that is deep to the pectoralis major. In addition to its role in movements of the scapula, the pectoralis minor muscle assists in forced inspiration. The **serratus anterior** is a large, flat, fan-shaped muscle between the ribs and scapula. It is named because of the saw-toothed appearance of its origins on the ribs.

The posterior thoracic muscles are the trapezius, levator scapulae, rhomboideus major, and rhomboideus minor (Figure 10.16b). The **trapezius** is a large, flat, triangular sheet of muscle extending from the skull and vertebral column medially to the pectoral girdle laterally. It is the most superficial back muscle and covers the posterior neck region and superior portion of the trunk. The two trapezius muscles form a trapezoid (diamond-shaped quadrangle)—hence its name. The **levator scapulae** is a narrow, elongated muscle in the posterior portion of the neck. It is deep to the sternocleidomastoid and trapezius muscles. As its name suggests, one of its actions is to elevate the scapula. The **rhomboideus major** and **rhomboideus minor** lie deep to the trapezius and are not always distinct from each other. They appear as parallel bands that pass inferolaterally from the vertebrae to the scapula. They are named on the basis of their shape—that is, a rhomboid (an oblique parallelogram). The rhomboideus major is about two times wider than the rhomboideus minor. Both muscles are used when forcibly lowering the raised upper limbs, as in driving a stake with a sledgehammer.

To understand the actions of muscles that move the scapula, it is first helpful to describe the various movements of the scapula:

- **Elevation.** Superior movement of the scapula, such as shrugging the shoulders or lifting a weight over the head.
- **Depression.** Inferior movement of the scapula, as in doing a "pull-up."
- **Abduction (protraction).** Movement of the scapula laterally and anteriorly, as in doing a "push-up" or punching.
- **Adduction (retraction).** Movement of the scapula medially and posteriorly, as in pulling the oars in a rowboat.
- **Upward rotation.** Movement of the inferior angle of the scapula laterally so that the glenoid cavity is moved upward. This movement is required to abduct the humerus past the horizontal.
- **Downward rotation.** Movement of the inferior angle of the scapula medially so that the glenoid cavity is moved downward. This movement is seen when the weight of the body is supported on the hands by a gymnast on parallel bars.

Relating Muscles to Movements

Arrange the muscles in this exhibit according to the following actions on the scapula: (1) depression, (2) elevation, (3) abduction, (4) adduction, (5) upward rotation, and (6) downward rotation. The same muscle may be mentioned more than once.

✔ What muscles in this exhibit are used to raise your shoulders, lower your shoulders, join your hands behind your back, and join your hands in front of your chest?

Innervation

Muscles that move the shoulder are innervated mainly by nerves that emerge from the cervical and brachial plexuses (shown in Exhibits 17.1 and 17.2 on pages 528–533). More specifically, the subclavius is innervated by the subclavian nerve; the pectoralis minor is innervated by the medial pectoral nerve; the serratus anterior is innervated by the long thoracic nerve; the trapezius is innervated by the accessory (XI) cranial nerve and cervical spinal nerves C3–C5; the levator scapulae is innervated by the dorsal scapular nerve and cervical spinal nerves C3–C5; and the rhomboideus major and rhomboideus minor are innervated by the dorsal scapular nerve.

Muscle	Origin	Insertion	Action
Anterior Thoracic Muscles			
Subclavius (sub-KLĀ-vē-us; *sub* = under; *clavius* = clavicle)	First rib.	Clavicle.	Depresses and moves clavicle anteriorly and helps stabilize pectoral girdle.
Pectoralis minor (pek′-tor-A-lis; *pectus* = breast, chest, thorax; *minor* = lesser)	Second through fifth, third through fifth, or second through fourth ribs.	Coracoid process of scapula.	Abducts scapula and rotates it downward; elevates third through fifth ribs during forced inspiration when scapula is fixed.
Serratus anterior (ser-Ā-tus; *serratus* = saw-toothed; *anterior* = before)	Superior eight or nine ribs.	Vertebral border and inferior angle of scapula.	Abducts scapula and rotates it upward; elevates ribs when scapula is fixed; known as "boxer's muscle."
Posterior Thoracic Muscles			
Trapezius (tra-PĒ-zē-us; *trapezoides* = trapezoid-shaped)	Superior nuchal line of occipital bone, ligamentum nuchae, and spines of seventh cervical and all thoracic vertebrae.	Clavicle and acromion and spine of scapula.	Superior fibers elevate scapula and can help extend head; middle fibers adduct scapula; inferior fibers depress scapula; superior and inferior fibers together rotate scapula upward; stabilizes scapula.
Levator scapulae (le-VĀ-tor SKA-pyū-lē; *levare* = to raise; *scapulae* = scapula)	Superior four or five cervical vertebrae.	Superior vertebral border of scapula.	Elevates scapula and rotates it downward.
Rhomboideus major (rom-BOID-ē-us; *rhomboides* = rhomboid or diamond-shaped) (see Figure 10.17c)	Spines of second to fifth thoracic vertebrae.	Vertebral border of scapula inferior to spine.	Elevates and adducts scapula and rotates it downward; stabilizes scapula.
Rhomboideus minor (rom-BOID-ē-us) (see Figure 10.17c)	Spines of seventh cervical and first thoracic vertebrae.	Vertebral border of scapula superior to spine.	Elevates and adducts scapula and rotates it downward; stabilizes scapula.

(continues)

Exhibit 10.13 Muscles That Move the Pectoral (Shoulder) Girdle (Figure 10.16) continued

Figure 10.16 / Muscles that move the pectoral (shoulder) girdle. (See Tortora, *A Photographic Atlas of the Human Body,* Figure 5.8.)

Muscles that move the pectoral girdle originate on the axial skeleton and insert on the clavicle or scapula.

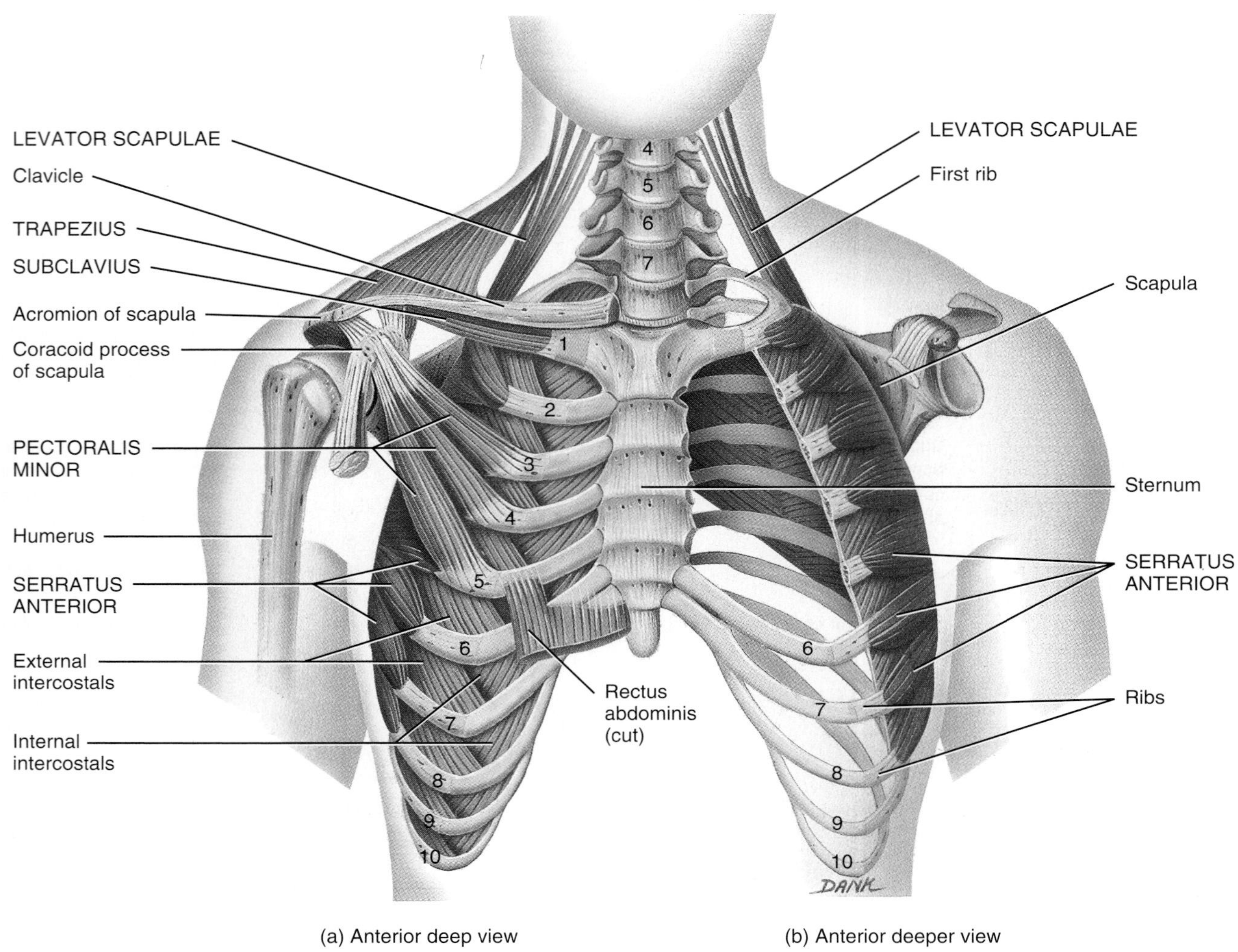

(a) Anterior deep view

(b) Anterior deeper view

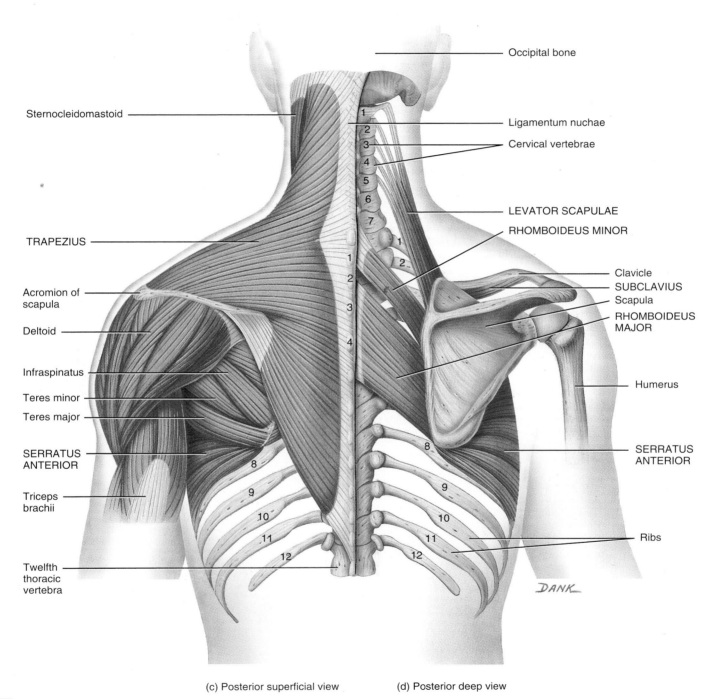

Occipital bone
Sternocleidomastoid
Ligamentum nuchae
Cervical vertebrae
LEVATOR SCAPULAE
RHOMBOIDEUS MINOR
TRAPEZIUS
Clavicle
SUBCLAVIUS
Scapula
Acromion of scapula
RHOMBOIDEUS MAJOR
Deltoid
Infraspinatus
Humerus
Teres minor
Teres major
SERRATUS ANTERIOR
SERRATUS ANTERIOR
Triceps brachii
Ribs
Twelfth thoracic vertebra
DANK

(c) Posterior superficial view (d) Posterior deep view

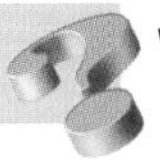

What is the main action of the muscles that move the pectoral girdle?

Exhibit 10.14 Muscles That Move the Humerus (Arm) (Figure 10.17)

Objective

► Describe the origin, insertion, action, and innervation of the muscles that move the humerus.

Of the nine muscles that cross the shoulder joint, only two of them (pectoralis major and latissimus dorsi) do not originate on the scapula. These two muscles are thus called **axial muscles,** because they originate on the axial skeleton. The remaining seven muscles are called the **scapular muscles.**

Of the two axial muscles that move the humerus, the **pectoralis major** is a large, thick, fan-shaped muscle that covers the superior part of the thorax. It has two origins: a smaller clavicular head and a larger sternocostal head. The **latissimus dorsi** is a broad, triangular muscle located on the inferior part of the back. It is commonly called the "swimmer's muscle" because its many actions are used while swimming.

Among the scapular muscles, the **deltoid** is a thick, powerful shoulder muscle that covers the shoulder joint and forms the rounded contour of the shoulder. This muscle is a frequent site of intramuscular injections. As you study the deltoid, note that its fibers originate from three different points and that each group of fibers moves the humerus differently. The **subscapularis** is a large triangular muscle that fills the subscapular fossa of the scapula and forms part of the posterior wall of the axilla. The **supraspinatus** is a rounded muscle, named for its location in the supraspinous fossa of the scapula. It lies deep to the trapezius. The **infraspinatus** is a triangular muscle, also named for its location in the infraspinous fossa of the scapula. The **teres major** is a thick, flattened muscle inferior to the teres minor that also helps form part of the posterior wall of the axilla. The **teres minor** is a cylindrical, elongated muscle, often inseparable from the infraspinatus, which lies along its superior border. The **coracobrachialis** is an elongated, narrow muscle in the arm that is pierced by the musculocutaneous nerve.

The strength and stability of the shoulder joint are not provided by the shape of the articulating bones or its ligaments. Instead, four deep muscles of the shoulder—subscapularis, supraspinatus, infraspinatus, and teres minor—strengthen and stabilize the shoulder joint. These muscles join the scapula to the humerus. Their flat tendons fuse together to form a nearly complete circle around the shoulder joint, like the cuff on a shirt sleeve. This arrangement is referred to as the **rotator (musculotendinous) cuff.** The supraspinatus muscle is especially predisposed to wear and tear because of its location between the head of the humerus and acromion of the scapula, which compress its tendon during shoulder movements.

Impingement Syndrome

One of the most common causes of shoulder pain and dysfunction in athletes is known as **impingement syndrome.** The repetitive overhead motion that is common in baseball, overhead racquet sports, spiking a volleyball, and swimming puts these athletes at risk for developing this syndrome. Continual pinching of the supraspinatus tendon causes it to become inflamed and results in pain. If the condition persists, the tendon may degenerate near the attachment to the humerus and ultimately may tear away from the bone (rotator cuff injury). ■

Relating Muscles to Movements

Arrange the muscles in this exhibit according to the following actions on the humerus at the shoulder joint: (1) flexion, (2) extension, (3) abduction, (4) adduction, (5) medial rotation, and (6) lateral rotation. The same muscle may be mentioned more than once.

✔ Why are two muscles that cross the shoulder joint called axial muscles, and the seven other called scapular muscles?

Innervation

Muscles that move the arm are innervated by nerves that emerge from the brachial plexus (shown in Exhibit 17.2 on pages 530–533). More specifically, the pectoralis major is innervated by the medial and lateral pectoral nerves; the latissimus dorsi is innervated by the thoracodorsal nerve; the deltoid and teres minor are innervated by the axillary nerve; the subscapularis is innervated by the upper and lower subscapular nerves; the supraspinatus and infraspinatus are innervated by the suprascapular nerve; the teres major is innervated by the lower subscapular nerve; and the coracobrachialis is innervated by the musculocutaneous nerve.

Muscle	Origin	Insertion	Action
Axial Muscles That Move the Humerus			
Pectoralis major (pek'-tor-A-lis) (see also Figure 10.12a)	Clavicle (clavicular head); sternum and costal cartilages of second to sixth ribs and sometimes first to seventh ribs (sternocostal head).	Greater tubercle and intertubercular sulcus of humerus.	As a whole, adducts and medially rotates arm at shoulder joint; clavicular head alone flexes arm, and sternocostal head alone extends arm at shoulder joint.
Latissimus dorsi (la-TIS-i-mus DOR-sī; *latissimus* = widest; *dorsum* = back)	Spines of inferior six thoracic vertebrae, lumbar vertebrae, crests of sacrum and ilium, inferior four ribs.	Intertubercular sulcus of humerus.	Extends, adducts, and medially rotates arm at shoulder joint; draws arm inferiorly and posteriorly.
Scapular Muscles That Move the Humerus			
Deltoid (DEL-toyd; *deltoides* = triangular) (see also Figure 10.12a)	Acromial extremity of clavicle (anterior fibers), acromion of scapula (lateral fibers), and spine of scapula (posterior fibers).	Deltoid tuberosity of humerus.	Lateral fibers abduct arm at shoulder joint; anterior fibers flex and medially rotate arm at shoulder joint; posterior fibers extend and laterally rotate arm at shoulder joint.
Subscapularis (sub-scap'-yū-LA-ris; *sub* = below; *scapularis* = scapula)	Subscapular fossa of scapula.	Lesser tubercle of humerus.	Medially rotates arm at shoulder joint.
Supraspinatus (sū'-pra-spi-NĀ-tus; *supra* = above; *spinatus* = spine of scapula)	Supraspinous fossa of scapula.	Greater tubercle of humerus.	Assists deltoid muscle in abducting arm at shoulder joint.
Infraspinatus (in'-fra-spi-NĀ-tus; *infra* = below)	Infraspinous fossa of scapula.	Greater tubercle of humerus.	Laterally rotates and adducts arm at shoulder joint.
Teres major (TE-rēz; *teres* = long and round)	Inferior angle of scapula.	Intertubercular sulcus of humerus.	Extends arm at shoulder joint and assists in adduction and medial rotation of arm at shoulder joint.
Teres minor (TE-rēz)	Inferior lateral border of scapula.	Greater tubercle of humerus.	Laterally rotates, extends, and adducts arm at shoulder joint.
Coracobrachialis (kor'-a-kō-BRĀ-kē-a'-lis; *coraco* = coracoid process)	Coracoid process of scapula.	Middle of medial surface of shaft of humerus.	Flexes and adducts arm at shoulder joint.

(continues)

Exhibit 10.14 Muscles That Move the Humerus (Arm) (Figure 10.17) continued

Figure 10.17 / Muscles that move the humerus (arm). (See Tortora, *A Photographic Atlas of the Human Body,* Figure 5.10.)

The strength and stability of the shoulder joint are provided by the tendons that form the rotator cuff.

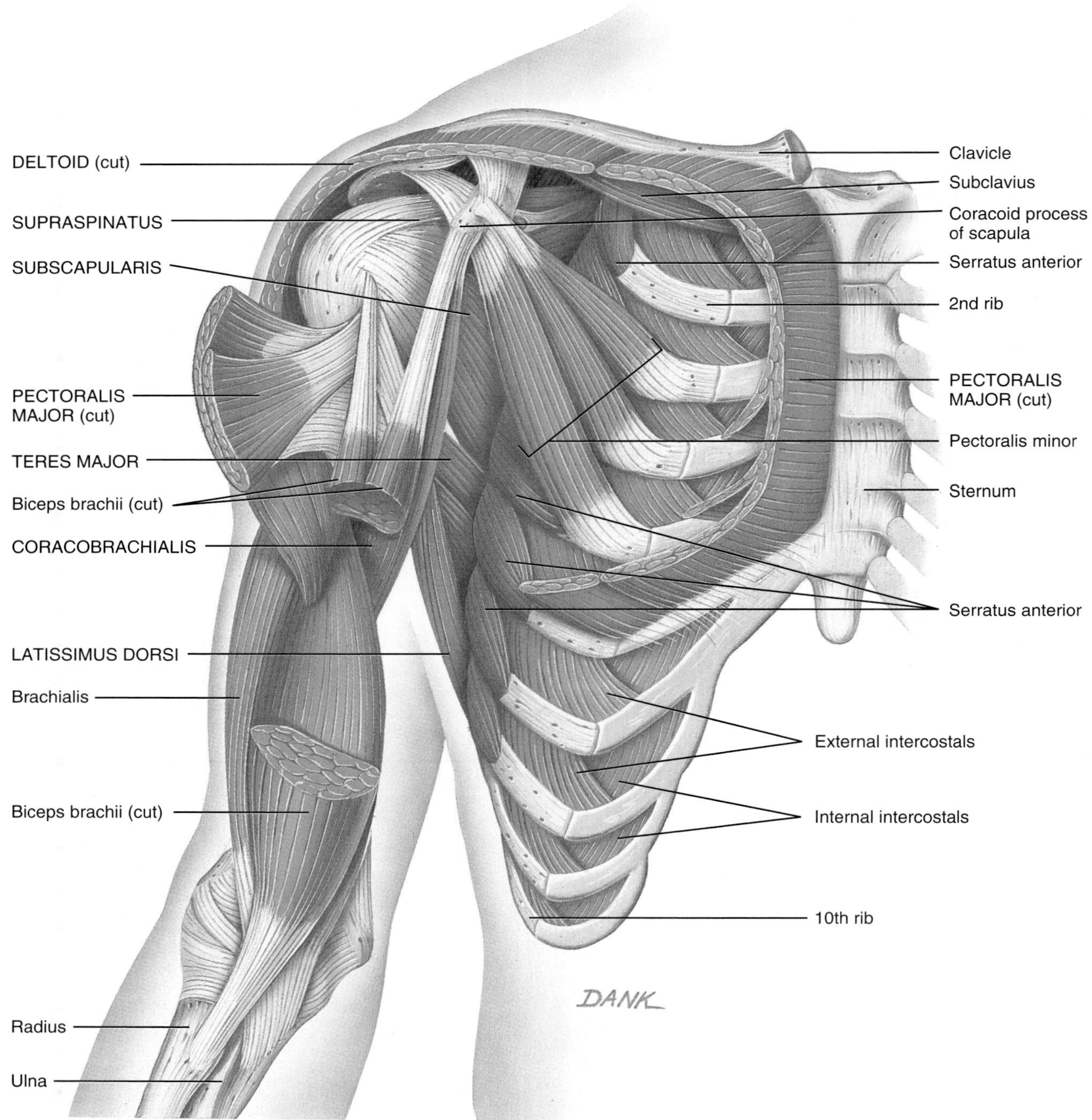

(a) Anterior deep view (the intact pectoralis major muscle is shown in figure 10.12a)

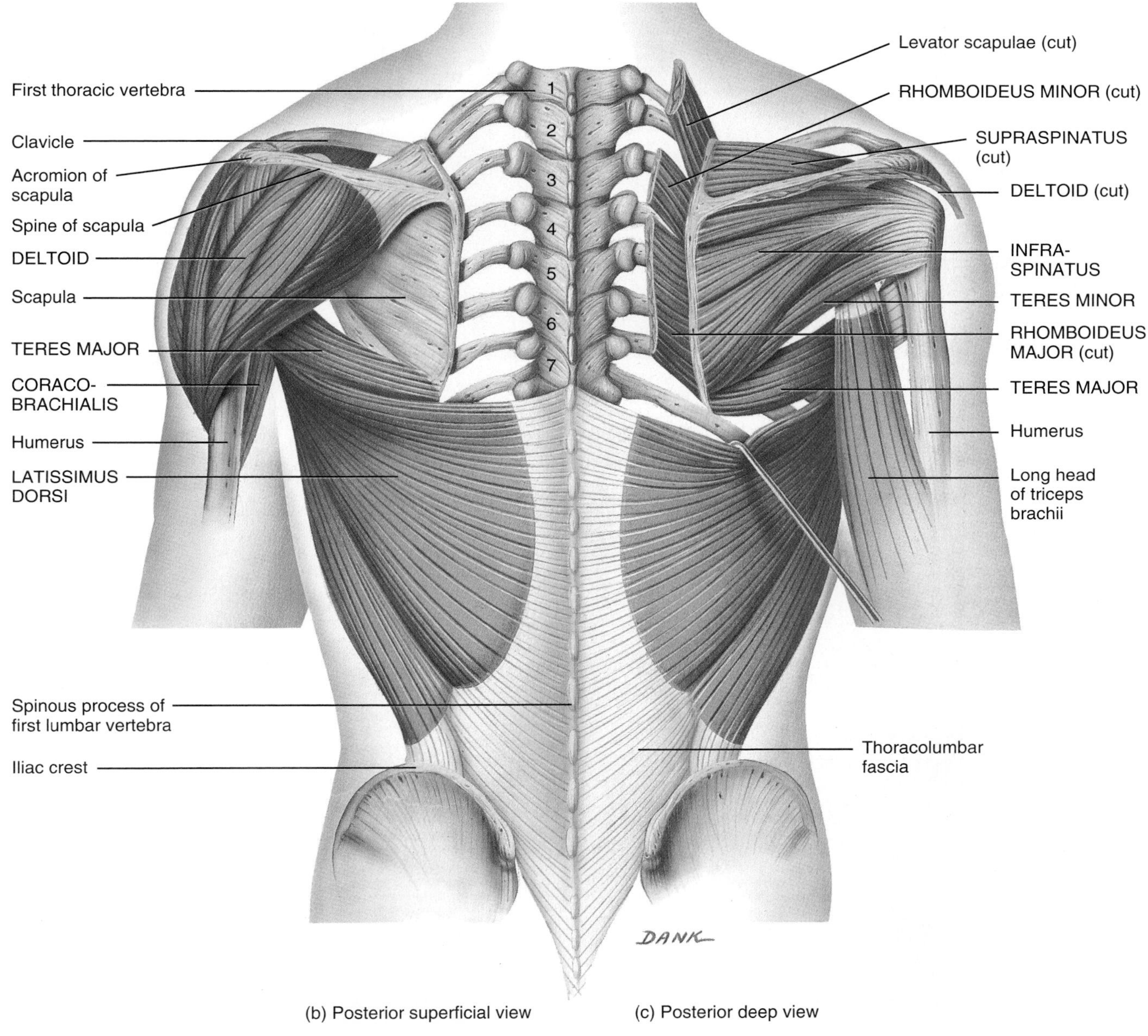

(b) Posterior superficial view (c) Posterior deep view

(continues)

Exhibit 10.14 Muscles That Move the Humerus (Arm) (Figure 10.17) continued

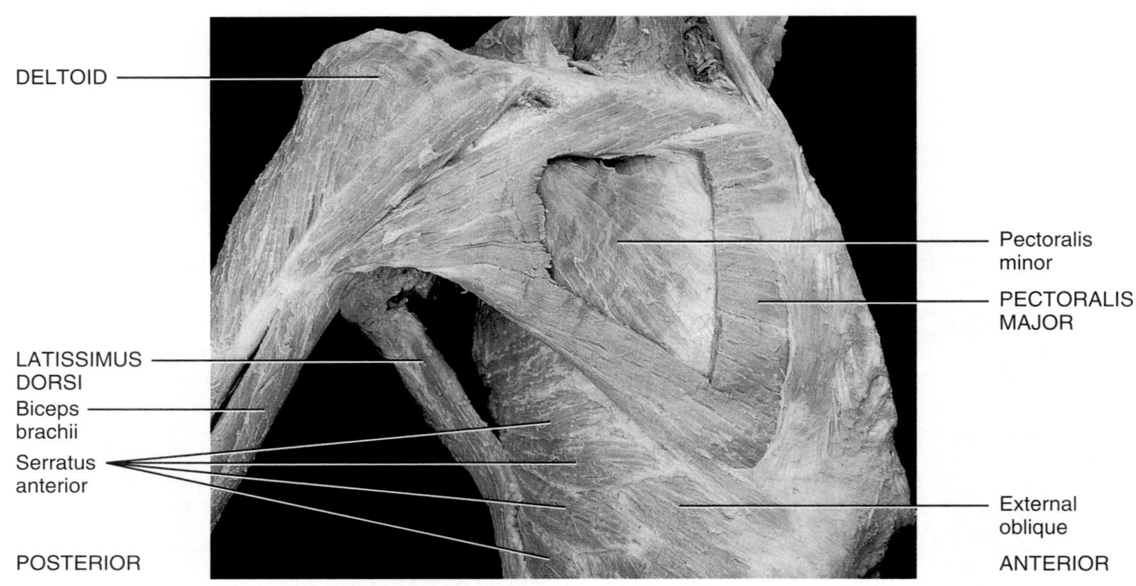

(d) Right lateral superficial view

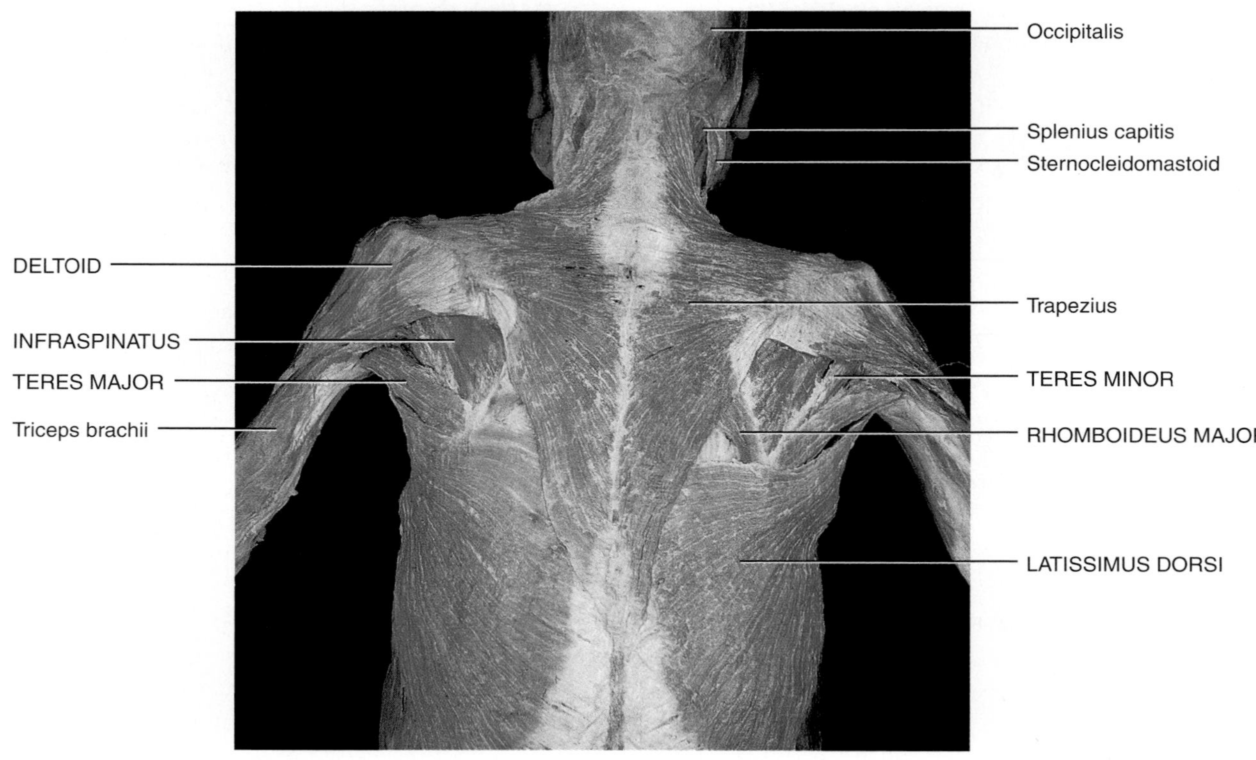

(e) Posterior view

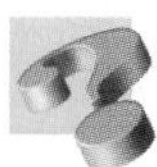

Which tendons make up the rotator cuff?

Exhibit 10.15 Muscles That Move the Radius and Ulna (Forearm) (Figure 10.18)

Objective

► Describe the origin, insertion, action, and innervation of the muscles that move the radius and ulna.

Most of the muscles that move the radius and ulna (forearm) are involved in flexion and extension at the elbow, which is a hinge joint. The biceps brachii, brachialis, and brachioradialis muscles are the flexor muscles. The extensor muscles are the triceps brachii and the anconeus.

The **biceps brachii** is the large muscle located on the anterior surface of the arm. As indicated by its name, it has two heads of origin (long and short), both from the scapula. The muscle spans both the shoulder and elbow joints. In addition to its role in flexing the forearm at the elbow joint, it also supinates the forearm at the radioulnar joints and flexes the arm at the shoulder joint. The **brachialis** is deep to the biceps brachii muscle. It is the most powerful flexor of the forearm at the elbow joint. For this reason, it is called the "workhorse" of the elbow flexors. The **brachioradialis** flexes the forearm at the elbow joint, especially when a quick movement is required or when a weight is lifted slowly during flexion of the forearm.

The **triceps brachii** is the large muscle located on the posterior surface of the arm. It is the more powerful of the extensors of the forearm at the elbow joint. As its name implies, it has three heads of origin, one from the scapula (long head) and two from the humerus (lateral and medial heads). The long head crosses the shoulder joint; the other heads do not. The **anconeus** is a small muscle located on the lateral part of the posterior aspect of the elbow that assists the triceps brachii in extending the forearm at the elbow joint.

Some muscles that move the radius and ulna are involved in pronation and supination at the radioulnar joints. The pronators, as suggested by their names, are the **pronator teres** and **pronator quadratus** muscles. The supinator of the forearm is the aptly named **supinator** muscle. You use the powerful action of the supinator when you twist a corkscrew or turn a screw with a screwdriver.

As noted earlier, in the limbs, functionally related skeletal muscles and their associated blood vessels and nerves are grouped together by fascia into regions called **compartments.** In the arm, the biceps brachii, brachialis, and coracobrachialis muscles constitute the *flexor compartment;* the triceps brachii muscle forms the *extensor compartment.*

Relating Muscles to Movements

Arrange the muscles in this exhibit according to the following actions on the elbow joint: (1) flexion and (2) extension; the following actions on the forearm at the radioulnar joints: (1) supination and (2) pronation; and the following actions on the humerus at the shoulder joint: (1) flexion and (2) extension. The same muscle may be mentioned more than once.

✔ Which muscles are in the anterior and posterior compartments of the arm?

Innervation

Muscles that move the radius and ulna are innervated by nerves derived from the brachial plexus (shown in Exhibit 17.2 on pages 530–533). More specifically, the biceps brachii is innervated by the musculocutaneous nerve; the brachialis is innervated by the musculocutaneous and radial nerves; the brachioradialis, triceps brachii, and anconeus are innervated by the radial nerve; the pronator teres and pronator quadratus are innervated by the median nerve; and the supinator is innervated by the deep radial nerve.

Muscle	Origin	Insertion	Action
Forearm Flexors			
Biceps brachii (BĪ-ceps BRĀ-kē-ī; *biceps* = two heads of origin; *brachion* = arm)	*Long head* originates from tubercle above glenoid cavity of scapula (supraglenoid tubercle); *short head* originates from coracoid process of scapula.	Radial tuberosity and bicipital aponeurosis.*	Flexes forearm at elbow joint, supinates forearm at radioulnar joints, and flexes arm at shoulder joint.
Brachialis (brā-kē-A-lis)	Distal, anterior surface of humerus.	Ulnar tuberosity and coronoid process of ulna.	Flexes forearm at elbow joint.
Brachioradialis (bra′-kē-ō-rā-dē-A-lis; *radialis* = radius) (see Figure 10.19a)	Lateral border of distal end of humerus.	Superior to styloid process of radius.	Flexes forearm at elbow joint and supinates and pronates forearm at radioulnar joints to neutral position.
Forearm Extensors			
Triceps brachii (TRĪ-ceps BRĀ-kē-ī; *triceps* = three heads of origin)	*Long head* originates from a projection inferior to glenoid cavity (infraglenoid tubercle) of scapula; *lateral head* originates from lateral and posterior surface of humerus superior to radial groove; *medial head* originates from entire posterior surface of humerus inferior to a groove for the radial nerve.	Olecranon of ulna.	Extends forearm at elbow joint and extends arm at shoulder joint.

(continues)

*The bicipital aponeurosis is a broad aponeurosis from the tendon of insertion of the biceps brachii muscle that descends medially across the brachial artery and fuses with deep fascia over the forearm flexor muscles.

Exhibit 10.15 Muscles That Move the Radius and Ulna (Forearm) (Figure 10.18) (continued)

Muscle	Origin	Insertion	Action
Anconeus (an-KŌ-nē-us; *anconeal* = pertaining to elbow) (see Figure 10.19c)	Lateral epicondyle of humerus.	Olecranon and superior portion of shaft of ulna.	Extends forearm at elbow joint.
Forearm Pronators			
Pronator teres (PRŌ-nā-tor TE-rēz; *pronation* = turning palm downward or posteriorly) (see Figure 10.19a)	Medial epicondyle of humerus and coronoid process of ulna.	Midlateral surface of radius.	Pronates forearm at radioulnar joints and weakly flexes forearm at elbow joint.
Pronator quadratus (PRŌ-nā-tor kwod-RĀ-tus; *quadratus* = squared, four-sided) (see Figure 10.19a)	Distal portion of shaft of ulna.	Distal portion of shaft of radius.	Pronates forearm at radioulnar joints.
Forearm Supinator			
Supinator (SŪ-pi-nā-tor; *supination* = turning palm upward or anteriorly) (see Figure 10.19a)	Lateral epicondyle of humerus and ridge near radial notch of ulna (supinator crest).	Lateral surface of proximal one-third of radius.	Supinates forearm at radioulnar joints.

Figure 10.18 / Muscles that move the radius and ulna (forearm).

Whereas the anterior arm muscles flex the forearm, the posterior arm muscles extend it.

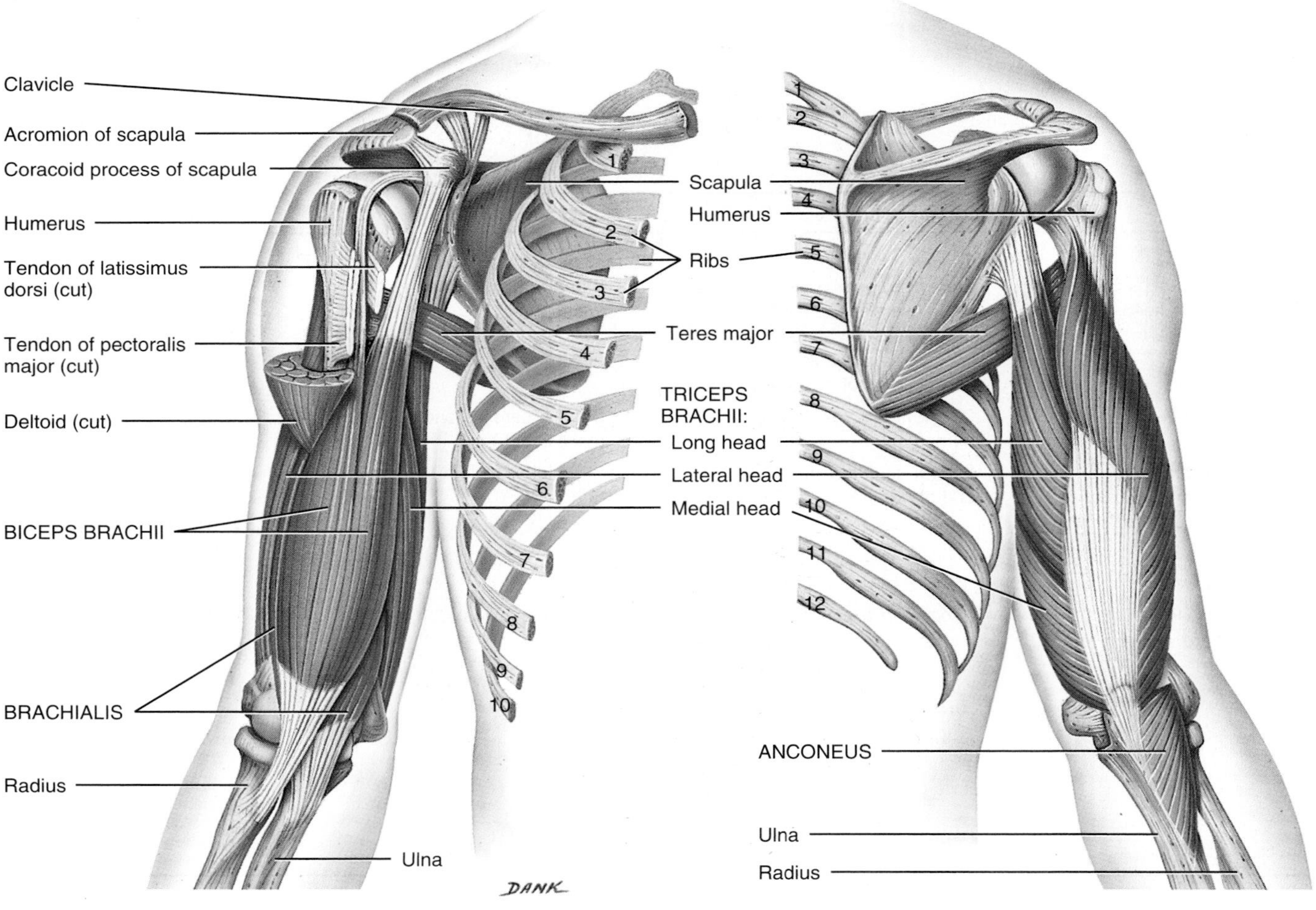

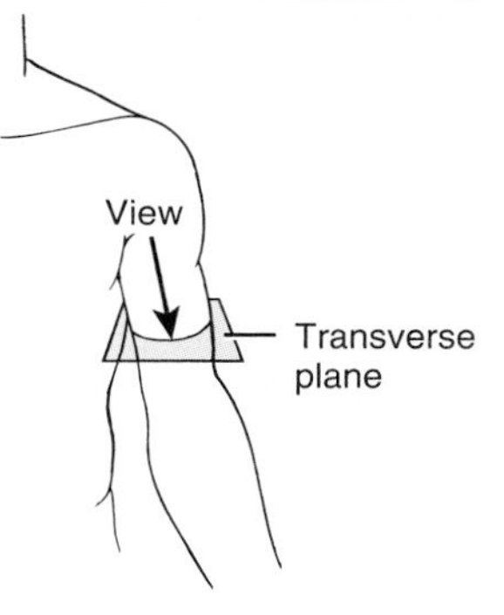

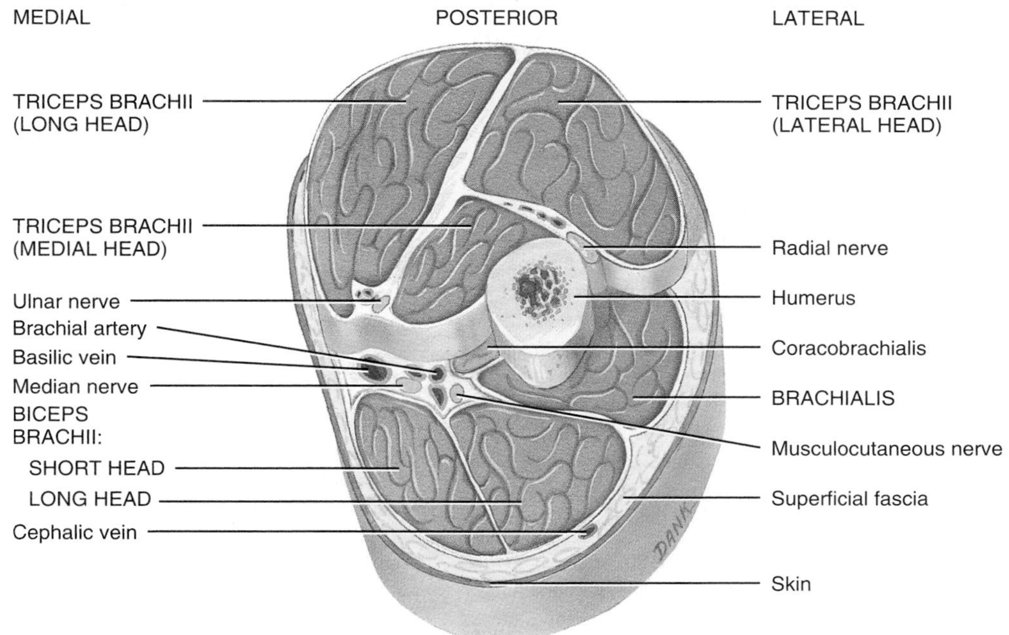

(c) Superior view of transverse section of arm

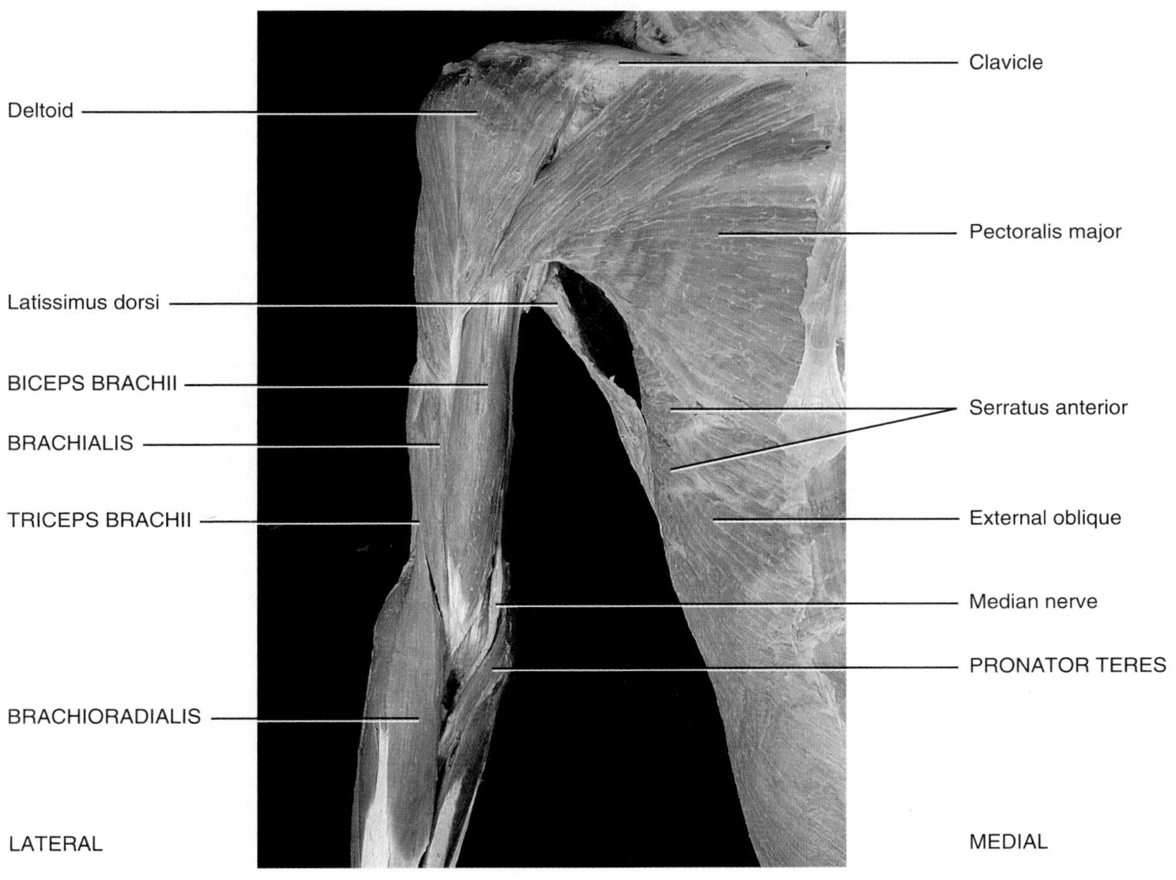

(d) Anterior superficial view

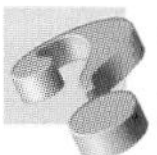

Which muscles are the most powerful flexor and the most powerful extensor of the forearm?

Exhibit 10.16 Muscles That Move the Wrist, Hand, and Digits (Figure 10.19)

Objective

► Describe the origin, insertion, action, and innervation of the muscles that move the wrist, hand, and digits.

Muscles that move the wrist, hand, and digits are many and varied. Those in this group that act on the digits are known as **extrinsic muscles** because they originate *outside* the hand and insert within it. As you will see, the names for the muscles that move the wrist, hand, and digits give some indication of their origin, insertion, or action. On the basis of location and function, the muscles are divided into two groups: (1) anterior compartment muscles and (2) posterior compartment muscles. The **anterior compartment** muscles originate on the humerus; typically insert on the carpals, metacarpals, and phalanges; and function as flexors. The bellies of these muscles form the bulk of the forearm. The **posterior compartment** muscles originate on the humerus; insert on the metacarpals and phalanges; and function as extensors. Within each compartment, the muscles are grouped as superficial and deep.

The **superficial anterior compartment** muscles (Figure 10.19a) are arranged in the following order from lateral to medial: **flexor carpi radialis, palmaris longus** (absent in about 10% of the population), and **flexor carpi ulnaris** (the ulnar nerve and artery are just lateral to the tendon of this muscle at the wrist). The **flexor digitorum superficialis** muscle is actually deep to the other three muscles and is the largest superficial muscle in the forearm.

The **deep anterior compartment** muscles (Figure 10.19b) are arranged in the following order from lateral to medial: **flexor pollicis longus** (the only flexor of the distal phalanx of the thumb) and **flexor digitorum profundus** (ends in four tendons that insert into the distal phalanges of the fingers).

The **superficial posterior compartment** muscles (Figure 10.19c) are arranged in the following order from lateral to medial: **extensor carpi radialis longus, extensor carpi radialis brevis, extensor digitorum** (occupies most of the posterior surface of the forearm and divides into four tendons that insert into the middle and distal phalanges of the fingers), **extensor digiti minimi** (a slender muscle generally connected to the extensor digitorum), and the **extensor carpi ulnaris.**

The **deep posterior compartment** muscles (Figure 10.19d) are arranged in the following order from lateral to medial: **abductor pollicis longus, extensor pollicis brevis, extensor pollicis longus,** and **extensor indicis.**

The tendons of the muscles of the forearm that attach to the wrist or continue into the hand, along with blood vessels and nerves, are held close to bones by strong fascial structures. The tendons are also surrounded by tendon sheaths. At the wrist, the deep fascia is thickened into fibrous bands called **retinacula** (*retinere* = retain). The **flexor retinaculum** (*transverse carpal ligament*) is located over the palmar surface of the carpal bones. Through it pass the long flexor tendons of the digits and wrist and the median nerve. The **extensor retinaculum** (*dorsal carpal ligament*) is located over the dorsal surface of the carpal bones. Through it pass the extensor tendons of the wrist and digits.

Relating Muscles to Movements

Arrange the muscles in this exhibit according to the following actions on the wrist joint: (1) flexion, (2) extension, (3) abduction, and (4) adduction; the following actions on the fingers at the metacarpophalangeal joints: (1) flexion and (2) extension; the following actions on the fingers at the interphalangeal joints: (1) flexion and (2) extension; the following actions on the thumb at the carpometacarpal, metacarpophalangeal, and interphalangeal joints: (1) extension and (2) abduction; and the following action on the thumb at the interphalangeal joint: flexion. The same muscle may be mentioned more than once.

✔ Which muscles and actions of the wrist, hand, and digits are used when writing?

Innervation

Muscles that move the wrist, hand, and digits are innervated by nerves derived from the brachial plexus (shown in Exhibit 17.2 on pages 530–533). More specifically, the flexor carpi radialis, palmaris longus, flexor digitorum superficialis, and flexor pollicis longus are innervated by the median nerve; the flexor carpi ulnaris is innervated by the ulnar nerve; the flexor digitorum profundus is innervated by the median and ulnar nerves; the extensor carpi radialis longus, extensor carpi radialis brevis, and extensor digitorum are innervated by the radial nerve; and the extensor digiti minimi, extensor carpi ulnaris, and all of the deep muscles of the posterior compartment are innervated by the deep radial nerve.

Muscle	Origin	Insertion	Action
Superficial Anterior Compartment (Flexors)			
Flexor carpi radialis (FLEK-sor KAR-pē-rā'-dē-A-lis; *flexor* = decreases angle at joint; *carpus* = wrist; *radialis* = radius)	Medial epicondyle of humerus.	Second and third metacarpals.	Flexes and abducts hand (radial deviation) at wrist joint.
Palmaris longus (pal-MA-ris LON-gus; *palma* = palm; *longus* = long)	Medial epicondyle of humerus.	Flexor retinaculum and palmar aponeurosis (deep fascia in center of palm).	Weakly flexes hand at wrist joint.
Flexor carpi ulnaris (FLEK-sor KAR-pē ul-NAR-is; *ulnaris* = ulna)	Medial epicondyle of humerus and superior posterior border of ulna.	Pisiform, hamate, and base of fifth metacarpal.	Flexes and adducts hand (ulnar deviation) at wrist joint.
Flexor digitorum superficialis (FLEK-sor di'-ji-TOR-um sū'-per-fish'-ē-A-lis; *digit* = finger or toe; *superficialis* = closer to surface)	Medial epicondyle of humerus, coronoid process of ulna, and a ridge along lateral margin of anterior surface (anterior oblique line) of radius.	Middle phalanx of each finger.*	Flexes middle phalanx of each finger at proximal interphalangeal joint, proximal phalanx of each finger at metacarpophalangeal joint, and hand at wrist joint.

*Reminder: The thumb or pollex is the first digit and has two phalanges: proximal and distal. The remaining digits, the fingers, are numbered 2–5, and each has three phalanges: proximal, middle, and distal.

Muscle	Origin	Insertion	Action
Deep Anterior Compartment (Flexors)			
Flexor pollicis longus (FLEK-sor POL-li-sis LON-gus; *pollex* = thumb)	Anterior surface of radius and interosseous membrane (sheet of fibrous tissue that holds shafts of ulna and radius together).	Base of distal phalanx of thumb.	Flexes distal phalanx of thumb at interphalangeal joint.
Flexor digitorum profundus (FLEK-sor di′-ji-TOR-um prō-FUN-dus; *profundus* = deep)	Anterior medial surface of body of ulna.	Base of distal phalanx of each finger.	Flexes distal and middle phalanges of each finger at interphalangeal joints, proximal phalanx of each finger at metacarpophalangeal joint, and hand at wrist joint.
Superficial Posterior Compartment (Extensors)			
Extensor carpi radialis longus (eks-TEN-sor KAR-pē rā′-dē-A-lis LON-gus; *extensor* = increases angle at joint)	Lateral supracondylar ridge of humerus.	Second metacarpal.	Extends and abducts hand at wrist joint.
Extensor carpi radialis brevis (eks-TEN-sor KAR-pē rā′-dē-A-lis BREV-is; *brevis* = short)	Lateral epicondyle of humerus.	Third metacarpal.	Extends and abducts hand at wrist joint.
Superficial Posterior Compartment (Extensors)			
Extensor digitorum (eks-TEN-sor di′-ji-TOR-um)	Lateral epicondyle of humerus.	Distal and middle phalanges of each finger.	Extends distal and middle phalanges of each finger at interphalangeal joints, proximal phalanx of each finger at metacarpophalangeal joint, and hand at wrist joint.
Extensor digiti minimi (eks-TEN-sor DIJ-i-tē MIN-i-mē; *minimi* = little finger)	Lateral epicondyle of humerus.	Tendon of extensor digitorum on fifth phalanx.	Extends proximal phalanx of little finger at metacarpophalangeal joint and hand at wrist joint.
Extensor carpi ulnaris (eks-TEN-sor KAR-pē ul-NAR-is)	Lateral epicondyle of humerus and posterior border of ulna.	Fifth metacarpal.	Extends and adducts hand at wrist joint.
Deep Posterior Compartment (Extensors)			
Abductor pollicis longus (ab-DUK-tor POL-li-sis LON-gus; *abductor* = moves part away from midline)	Posterior surface of middle of radius and ulna and interosseous membrane.	First metacarpal.	Abducts and extends thumb at carpometacarpal joint and abducts hand at wrist joint.
Extensor pollicis brevis (eks-TEN-sor POL-li-sis BREV-is)	Posterior surface of middle of radius and interosseous membrane.	Base of proximal phalanx of thumb.	Extends proximal phalanx of thumb at metacarpophalangeal joint, first metacarpal of thumb at carpometacarpal joint, and hand at wrist joint.
Extensor pollicis longus (eks-TEN-sor POL-li-sis LON-gus)	Posterior surface of middle of ulna and interosseous membrane.	Base of distal phalanx of thumb.	Extends distal phalanx of thumb at interphalangeal joint, first metacarpal of thumb at carpometacarpal joint, and abducts hand at wrist joint.
Extensor indicis (eks-TEN-sor IN-di-kis; *indicis* = index)	Posterior surface of ulna.	Tendon of extensor digitorum of index finger.	Extends distal and middle phalanges of index finger at interphalangeal joints, proximal phalanx of index finger at metacarpophalangeal joint, and hand at wrist joint.

(continues)

Figure 10.19 / Muscles that move the wrist, hand, and digits. (See Tortora, *A Photographic Atlas of the Human Body,* Figures 5.12 and 5.13.)

The anterior compartment muscles function as flexors, and the posterior compartment muscles function as extensors.

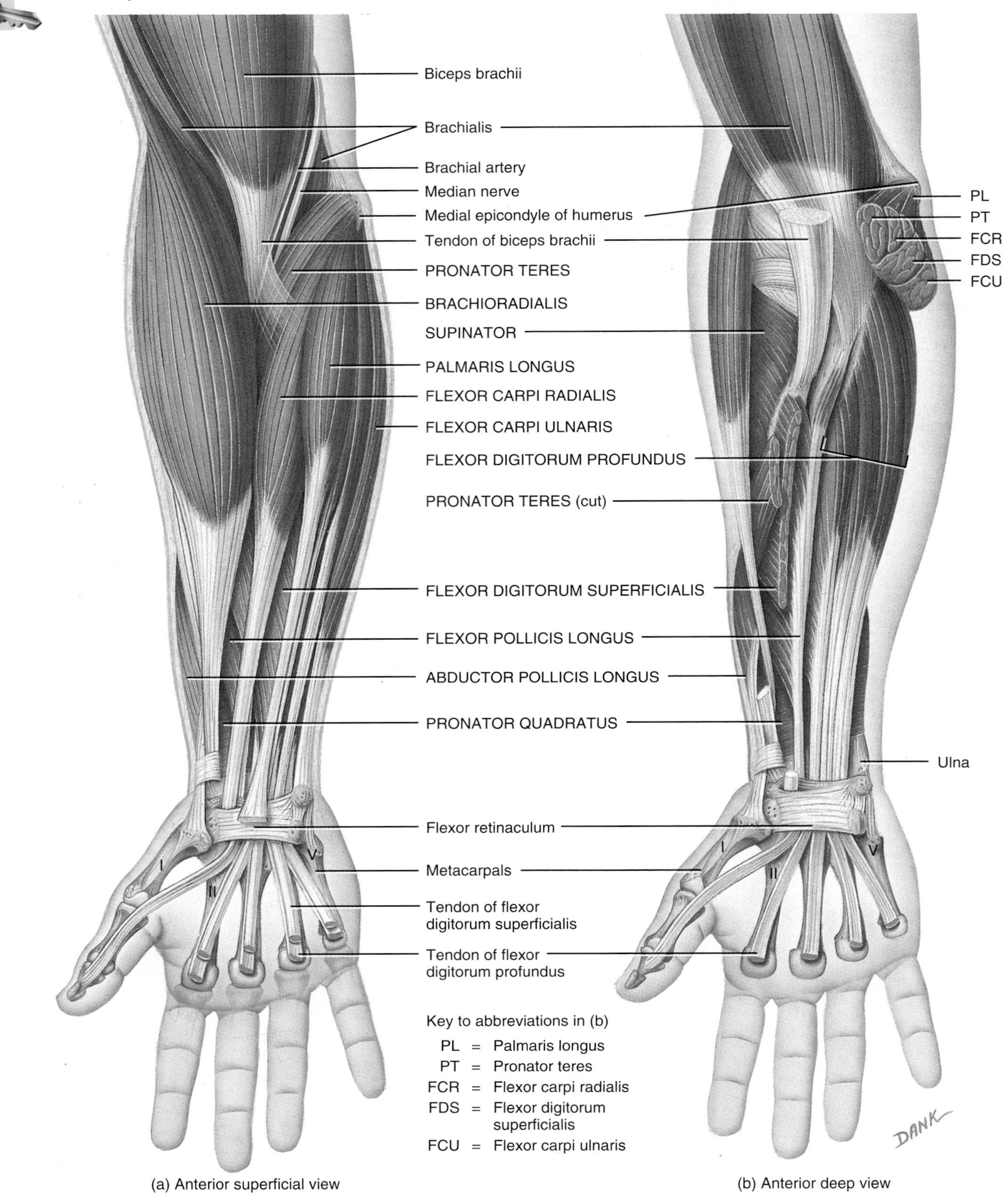

(a) Anterior superficial view

(b) Anterior deep view

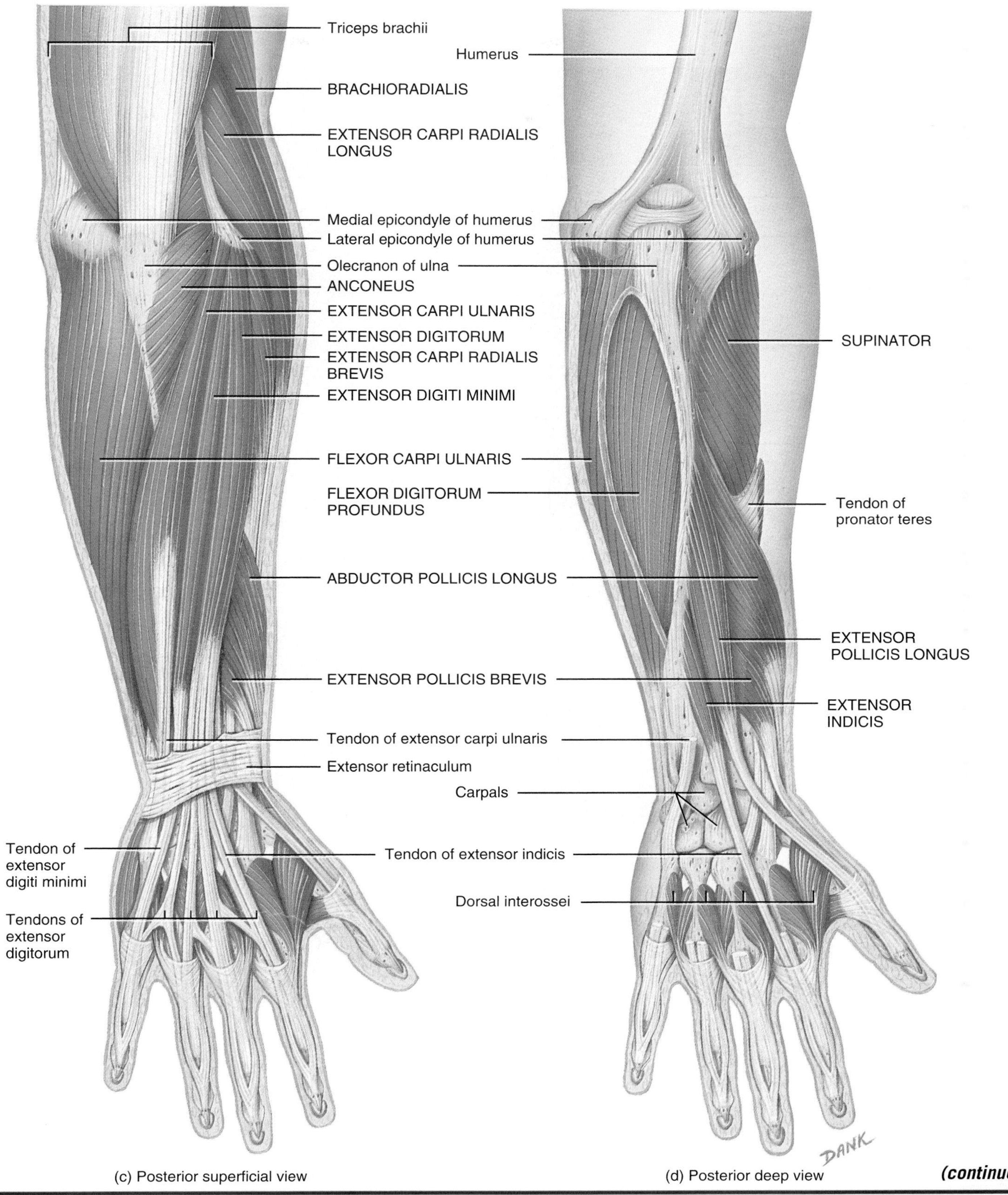

(c) Posterior superficial view

(d) Posterior deep view

(continues)

Exhibit 10.16 Muscles That Move the Wrist, Hand, and Digits (Figure 10.19) continued

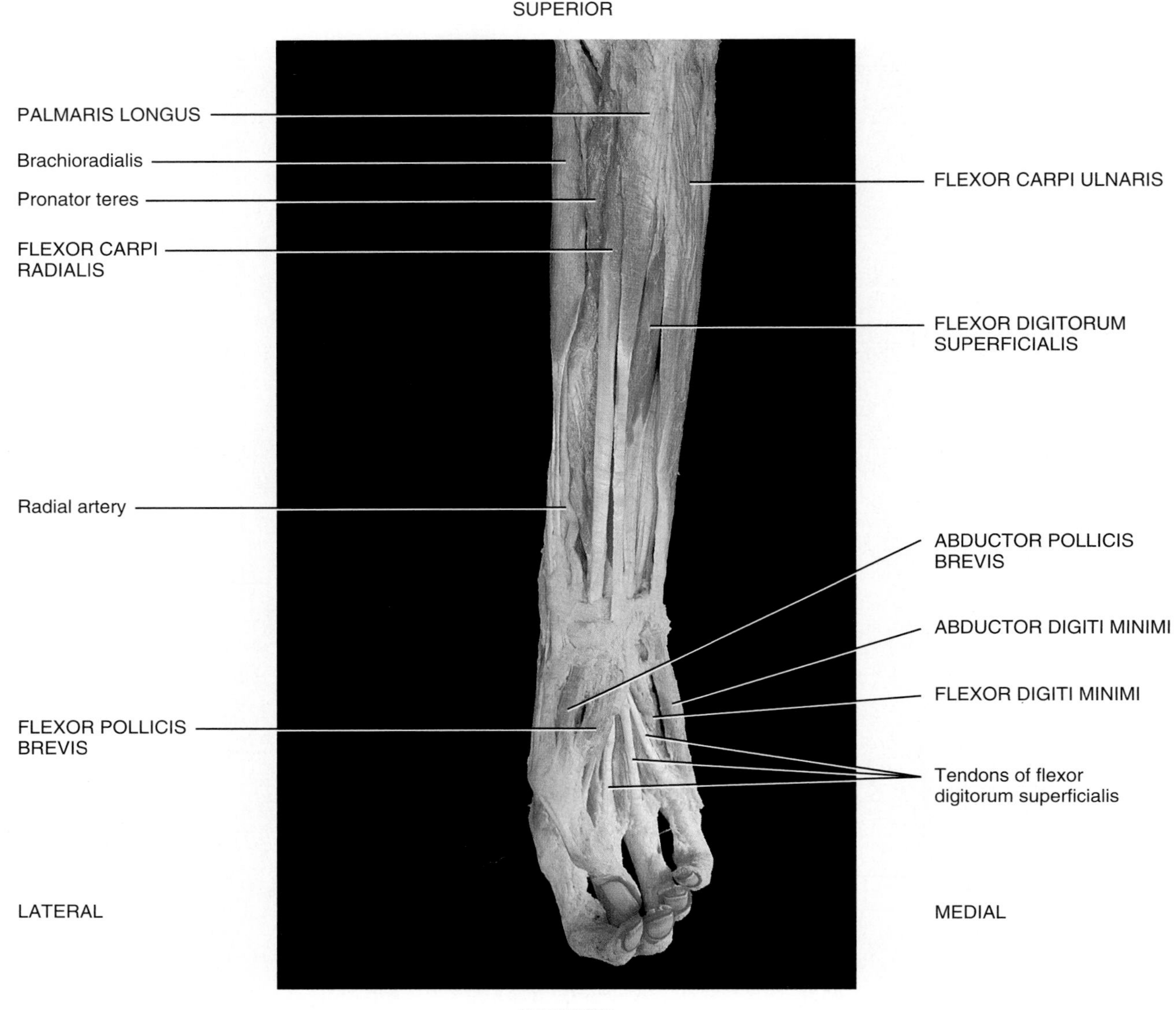

(e) Anterior superficial view

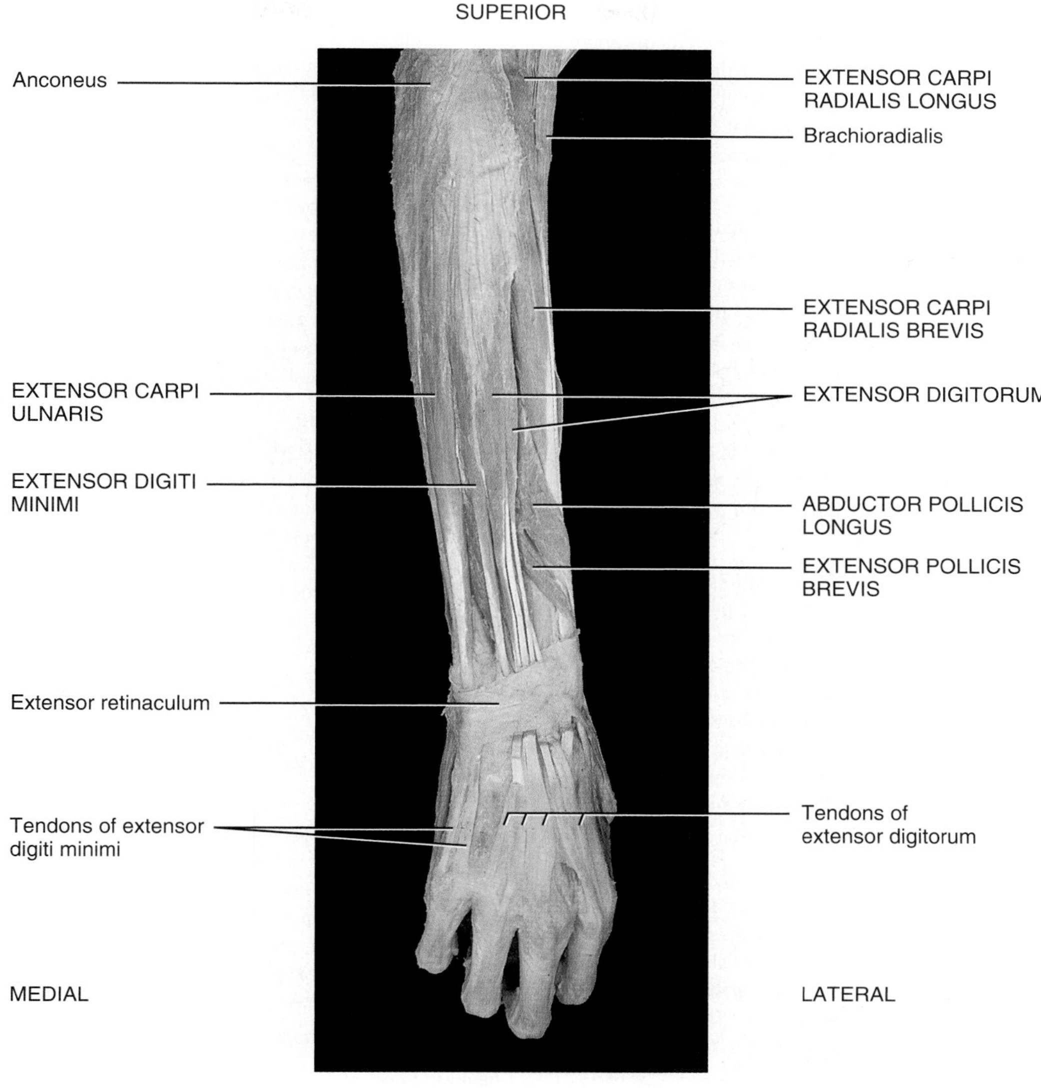

(f) Posterior superficial view

What structures pass through the flexor retinaculum?

Exhibit 10.17 Intrinsic Muscles of the Hand (Figure 10.20)

Objective

► Describe the origin, insertion, action, and innervation of the intrinsic muscles of the hand.

Several of the muscles discussed in Exhibit 10.16 move the digits in various ways and are known as extrinsic muscles. They produce the powerful but crude movements of the digits. The **intrinsic muscles** in the palm produce the weak but intricate and precise movements of the digits that characterize the human hand. The muscles in this group are so named because their origins and insertions are *within* the hand.

The intrinsic muscles of the hand are divided into three groups: (1) **thenar,** (2) **hypothenar,** and (3) **intermediate.** The four thenar muscles act on the thumb and form the **thenar eminence,** the lateral rounded contour on the palm that is also called the ball of the thumb. The thenar muscles include the abductor pollicis brevis, opponens pollicis, flexor pollicis brevis, and adductor pollicis. The **abductor pollicis brevis** is a thin, short, relatively broad superficial muscle on the lateral side of the thenar eminence. The **oppenens pollicus** is a small, triangular muscle that is deep to the abductor pollicis brevis muscle. The **flexor pollicis brevis** is a short, wide muscle that is medial to the abductor pollicis brevis muscle. The **adductor pollicis** is fan-shaped and has two heads (oblique and transverse) separated by a gap through which the radial artery passes.

The three hypothenar muscles act on the little finger and form the **hypothenar eminence,** the medial rounded contour on the palm that is also called the ball of the little finger. The hypothenar muscles are the abductor digiti minimi, flexor digiti minimi brevis, and opponens digiti minimi. The **abductor digiti minimi** is a short, wide muscle and is the most superficial of the hypothenar muscles. It is a powerful muscle that plays an important role in grasping an object with outspread fingers. The **flexor digiti minimi brevis** muscle is also short and wide and is lateral to the abductor digiti minimi muscle. The **opponens digiti minimi** muscle is triangular and deep to the other two hypothenar muscles.

The 12 intermediate (midpalmar) muscles act on all the digits except the thumb. The intermediate muscles include the lumbricals, palmar interossei, and dorsal interossei. The **lumbricals,** as their name indicates, are worm-shaped. They originate from and insert into the tendons of other muscles (flexor digitorum profundus and extensor digitorum). The **palmar interossei** are the smaller and most superficial of the interossei muscles. The **dorsal interossei** are the deep interossei muscles. Both sets of interossei muscles are located between the metacarpals and are important in abduction, adduction, flexion, and extension of the fingers, and in movements in skilled activities such as writing, typing, and playing a piano.

The functional importance of the hand is readily apparent when one considers that certain hand injuries can result in permanent disability. Most of the dexterity of the hand depends on movements of the thumb. The general activities of the hand are free motion, power grip (forcible movement of the fingers and thumb against the palm, as in squeezing), precision handling (a change in position of a handled object that requires exact control of finger and thumb positions, as in winding a watch or threading a needle), and pinch (compression between the thumb and index finger or between the thumb and first two fingers).

Movements of the thumb are very important in the precise activities of the hand, and they are defined in different planes from comparable movements of other digits because the thumb is positioned at a right angle to the other digits. The five principal movements of the thumb are illustrated in Figure 10.20f and include *flexion* (movement of the thumb medially across the palm), *extension* (movement of the thumb laterally away from the palm), *abduction* (movement of the thumb in an anteroposterior plane away from the palm), *adduction* (movement of the thumb in an anteroposterior plane toward the palm), and *opposition* (movement of the thumb across the palm so that the tip of the thumb meets the tip of a finger). Opposition is the single most distinctive digital movement that gives humans and other primates the ability to grasp and manipulate objects precisely.

Carpal Tunnel Syndrome

The **carpal tunnel** is a narrow passageway formed anteriorly by the flexor retinaculum and posteriorly by the carpal bones. Through this tunnel pass the median nerve, the most superficial structure, and the long flexor tendons for the digits. Structures within the carpal tunnel, especially the median nerve, are vulnerable to compression, and the resulting condition is called **carpal tunnel syndrome.** Compression of the median nerve leads to sensory changes over the lateral side of the hand and muscle weakness in the thenar eminence. This results in pain, numbness, and tingling of the fingers. The condition may be caused by inflammation of the digital tendon sheaths, fluid retention, excessive exercise, infection, trauma, and repetitive activities that involve flexion of the wrist, such as keyboarding, cutting hair, and playing a piano. ■

Relating Muscles to Movements

Arrange the muscles in this exhibit according to the following actions on the thumb at the carpometacarpal and metaphalangeal joints: (1) abduction, (2) adduction, (3) flexion, and (4) opposition; and the following actions on the fingers at the metacarpophalangeal and interphalangeal joints: (1) abduction, (2) adduction, (3) flexion, and (4) extension. The same muscle may be mentioned more than once.

✔ Compare the actions of the extrinsic and intrinsic muscles of the hand.

Innervation

Intrinsic muscles of the hand are innervated by nerves derived from the brachial plexus (shown in Exhibit 17.2 on pages 530–533). More specifically, the abductor pollicis brevis and opponens pollicis are innervated by the median nerve. The adductor pollicis, abductor digiti minimi, flexor digiti minimi brevis, opponens digiti minimi, dorsal interossei, and palmar interossei are innervated by the ulnar nerve. The flexor pollicis brevis and lumbricals are innervated by the median and ulnar nerves.

Muscle	Origin	Insertion	Action
Thenar			
Abductor pollicis brevis (*abductor* = moves part away from middle; *pollex* = thumb; *brevis* = short)	Flexor retinaculum, scaphoid, and trapezium.	Lateral side of proximal phalanx of thumb.	Abducts thumb at carpometacarpal joint.
Opponens pollicis (*oppenens* = opposes)	Flexor retinaculum and trapezium.	Lateral side of first metacarpal (thumb).	Moves thumb across palm to meet little finger (opposition) at the carpometacarpal joint.
Flexor pollicis brevis (*flexor* = decreases angle at joint)	Flexor retinaculum, trapezium, capitate, and trapezoid.	Lateral side of proximal phalanx of thumb.	Flexes thumb at carpometacarpal and metacarpophalangeal joints.
Adductor pollicis (*adductor* = moves part toward midline)	Oblique head: capitate and second and third metacarpals; transverse head: third metacarpal.	Medial side of proximal phalanx of thumb by a tendon containing a sesamoid bone.	Adducts thumb at carpometacarpal and metacarpophalangeal joints.
Hypothenar			
Abductor digiti minimi (*digit* = finger or toe; *minimi* = little)	Pisiform and tendon of flexor carpi ulnaris.	Medial side of proximal phalanx of little finger.	Abducts and flexes little finger at metacarpophalangeal joint.
Flexor digiti minimi brevis	Flexor retinaculum and hamate.	Medial side of proximal phalanx of little finger.	Flexes little finger at carpometacarpal and metacarpophalangeal joints.
Opponens digiti minimi	Flexor retinaculum and hamate.	Medial side of fifth metacarpal (little finger).	Moves little finger across palm to meet thumb (opposition) at the carpometacarpal joint.
Intermediate (Midpalmar)			
Lumbricals (LUM-bri-kals; *lumbricus* = earthworm) (four muscles)	Lateral sides of tendons and flexor digitorum profundus of each finger.	Lateral sides of tendons of extensor digitorum on proximal phalanx of each finger.	Flex each finger at metacarpophalangeal joints and extend each finger at interphalangeal joints.
Palmar interossei (in′-ter-OS-ē-ī (*palmar* = palm; *inter* = between; *ossei* = bones) (four muscles)	Sides of shafts of metacarpals of all digits (except the middle one)	Sides of bases of proximal phalanges of all digits (except the middle one).	Adduct each finger at metacarpophalangeal joints; flex each finger at metacarpophalangeal joints.
Dorsal interossei (in′-ter-OS-ē-ī; *dorsal* = back surface) (four muscles)	Adjacent sides of metacarpals.	Proximal phalanx of each finger.	Abduct fingers 2–4 at metacarpophalangeal joints; flex fingers 2–4 at metacarpophalangeal joints; and extend each finger at interphalangeal joints.

(continues)

Exhibit 10.17 Intrinsic Muscles of the Hand (Figure 10.20) continued

Figure 10.20 / Intrinsic muscles of the hand. (See Tortora, *A Photographic Atlas of the Human Body,* Figures 5.14 and 5.15.)

The intrinsic muscles of the hand produce the intricate and precise movements of the digits that characterize the human hand.

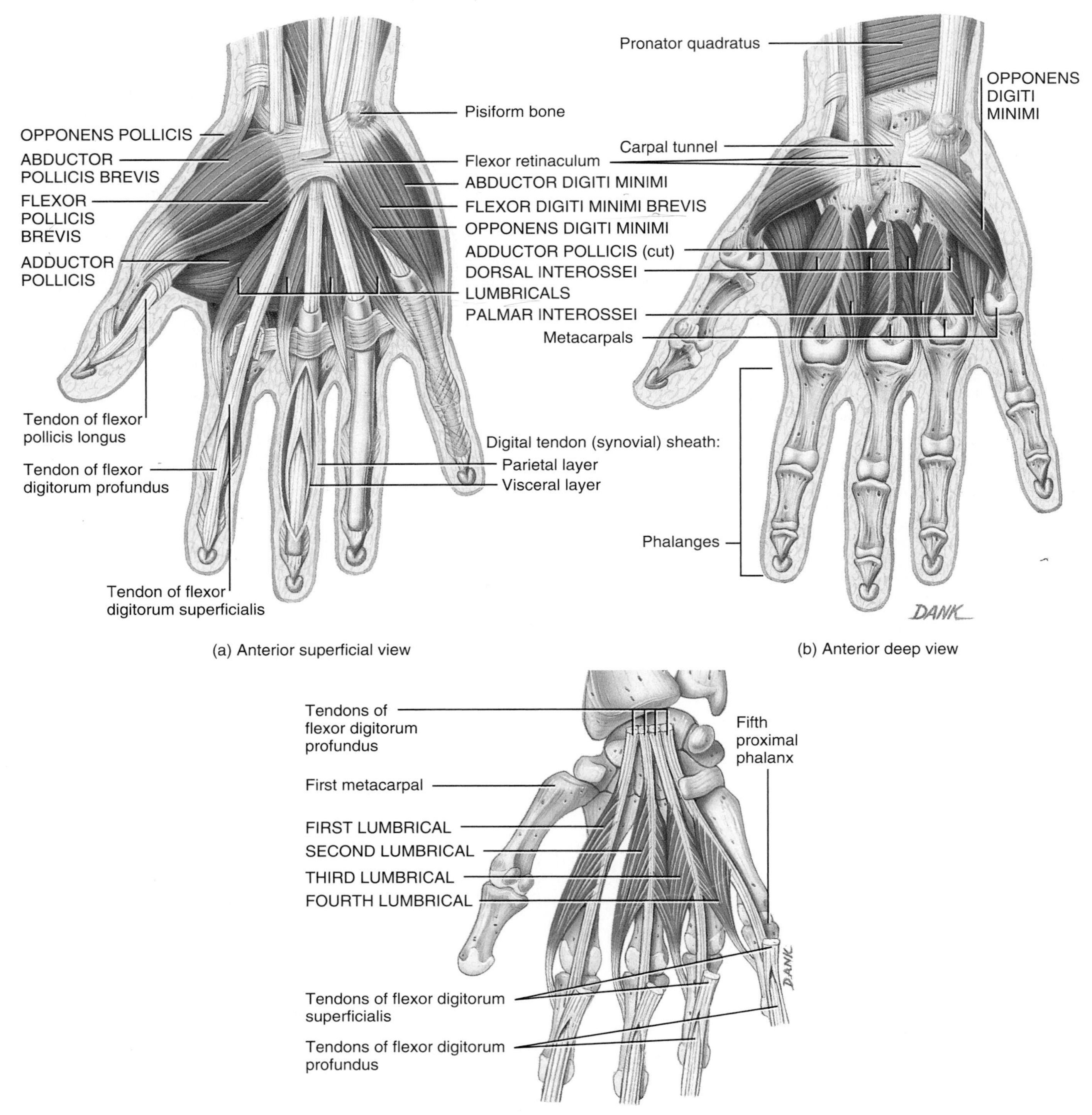

(a) Anterior superficial view

(b) Anterior deep view

(c) Anterior deep view showing lumbricals

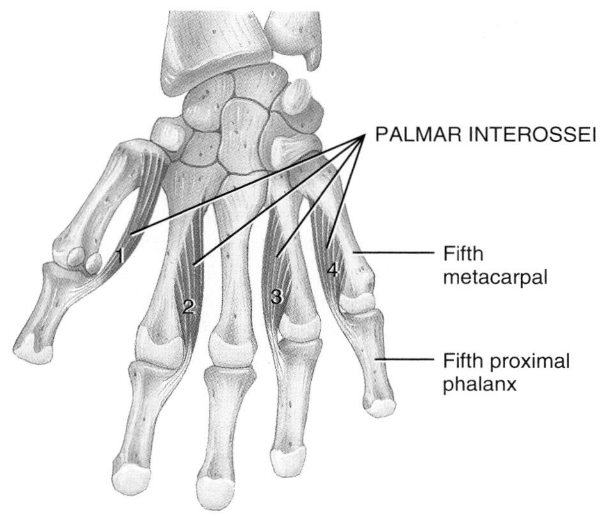

(d) Anterior deep view of palmar interossei

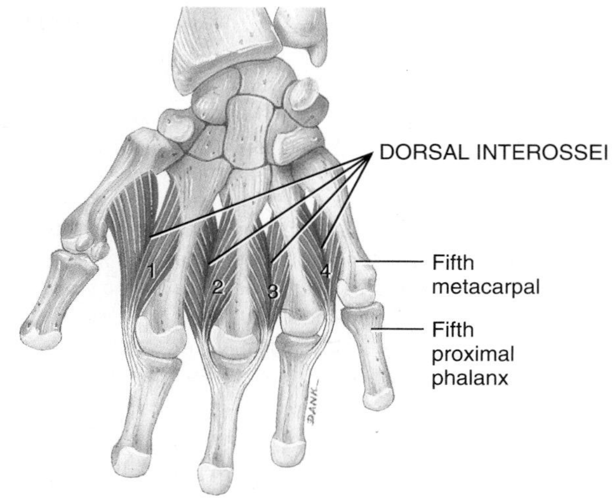

(e) Anterior deep view of dorsal interossei

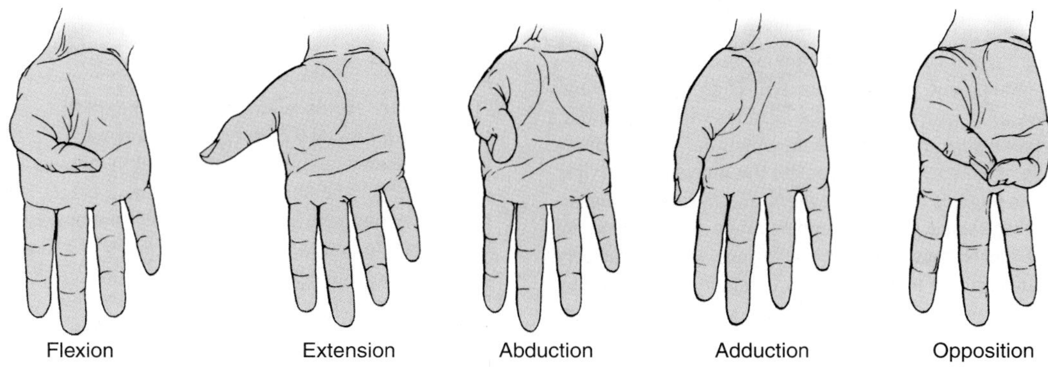

(f) Movements of the thumb

Muscles of the thenar eminence act on which digit?

Exhibit 10.18 Muscles That Move the Vertebral Column (Backbone) (Figure 10.21)

Objective

► Describe the origin, insertion, action, and innervation of the muscles that move the vertebral column.

The muscles that move the vertebral column (backbone) are quite complex because they have multiple origins and insertions and there is considerable overlap among them. One way to group the muscles is on the basis of the general direction of the muscle bundles and their approximate lengths. For example, the splenius muscles arise from the midline and extend laterally and superiorly to their insertions (Figure 10.21a). The erector spinae (sacrospinalis) muscle arises from either the midline or more laterally but usually runs almost longitudinally, with neither a significant lateral nor medial direction as it is traced superiorly. The transversospinalis muscles arise laterally but extend toward the midline as they are traced superiorly. Deep to these three muscle groups are small segmental muscles that extend between spinous processes or transverse processes of vertebrae. Because the scalene muscles also assist in moving the vertebral column, they are included in this exhibit. Note in Exhibit 10.9 that the rectus abdominis, external oblique, internal oblique, and quadratus lumborum muscles also play a role in moving the vertebral column.

The bandage-like **splenius** muscles are attached to the sides and back of the neck. The two muscles in this group are named on the basis of their superior attachments (insertions): **splenius capitis** (head region) and **splenius cervicis** (cervical region). They extend the head and laterally flex and rotate the head.

The **erector spinae (sacrospinalis)** is the largest muscle mass of the back, forming a prominent bulge on either side of the vertebral column. It is the chief extensor of the vertebral column. It is also important in controlling flexion, lateral flexion, and rotation of the vertebral column and in maintaining the lumbar curve, because the main mass of the muscle is in the lumbar region. It consists of three groups: iliocostalis (laterally placed), longissimus (intermediately placed), and spinalis (medially placed). These groups, in turn, consist of a series of overlapping muscles, and the muscles within the groups are named according to the regions of the body with which they are associated. The **iliocostalis group** consists of three muscles: the **iliocostalis cervicis** (cervical region), **iliocostalis thoracis** (thoracic region), and **iliocostalis lumborum** (lumbar region). The **longissimus group** resembles a herring bone and consists of three muscles: the **longissimus capitis** (head region), **longissimus cervicis** (cervical region), and **longissimus thoracic** (thoracic region). The **spinalis group** also consists of three muscles: the **spinalis capitis, spinalis cervicis,** and **spinalis thoracis.**

The **transversospinalis** muscles are named because their fibers run from the transverse processes to the spinous processes of the vertebrae. The semispinalis muscles in this group are also named according to the region of the body with which they are associated: **semispinalis capitis** (head region), **semispinalis cervicis** (cervical region), and **semispinalis thoracic** (thoracic region). These muscles extend the vertebral column and rotate the head. The **multifidus** muscle in this group, as its name implies, is split into several bundles. It extends and laterally flexes the vertebral column and rotates the head. The **rotatores** muscles of this group are short and are found along the entire length of the vertebral column. They extend and rotate the vertebral column.

Within the **segmental** muscle group (Figure 10.21b) are the **interspinales** and **intertransversarii** muscles, which unite the spinous and transverse processes of consecutive vertebrae. They function primarily in stabilizing the vertebral column during its movements.

Within the **scalene** group (Figure 10.21c), the **anterior scalene** muscle is anterior to the middle scalene muscle, the **middle scalene** muscle is intermediate in placement and is the longest and largest of the scalene muscles, and the **posterior scalene** muscle is posterior to the middle scalene muscle and is the smallest of the scalene muscles. These muscles flex, laterally flex, and rotate the head and assist in deep inspiration.

Relating Muscles to Movements

Arrange the muscles in this exhibit according to the following actions on the head at the atlantooccipital and intervertebral joints: (1) flexion, (2) extension, (3) lateral flexion, (4) rotation to same side as contracting muscle, and (5) rotation to opposite side as contracting muscle; the following actions on the vertebral column at the intervertebral joints: (1) flexion, (2) extension, (3) lateral flexion, (4) rotation, and (5) stabilization; and the following action on the ribs: elevation during deep inspiration. The same muscle may be mentioned more than once.

✔ What are the four major groups of muscles that move the vertebral column?

Innervation

The splenius capitis is innervated by the middle cervical nerves; the splenius cervicis is innervated by the inferior cervical nerves; the iliocostalis cervicis and semispinalis capitis are innervated by the cervical and thoracic nerves; the iliocostalis thoracis is innervated by thoracic nerves; the iliocostalis lumborum is innervated by lumbar nerves; the longissimus capitis is innervated by the middle and inferior cervical nerves; the longissimus cervicis, longissimus thoracic, all spinalis muscles, multifidus, rotatores, and interspinales are innervated by cervical, thoracic, and lumbar nerves; the semispinalis cervicis and semispinalis thoracis are innervated by cervical and thoracic nerves; the intertransversarii are innervated by spinal nerves; the anterior scalene is innervated by cervical nerves C5–C6; the middle scalene is innervated by cervical nerves C3–C8; and the posterior scalene is innervated by cervical nerves C6–C8.

Muscle	Origin	Insertion	Action
Splenius (SPLĒ-nē-us)			
Splenius capitis (KAP-i-tis; *splenium* = bandage; *caput* = head)	Ligamentum nuchae and spinous processes of seventh cervical vertebra and first three or four thoracic vertebrae.	Occipital bone and mastoid process of temporal bone.	Acting together (bilaterally), extend head; acting singly (unilaterally), laterally flex and rotate head to same side as contracting muscle.
Splenius cervicis (SER-vi-kis; *cervix* = neck)	Spinous processes of third through sixth thoracic vertebrae.	Transverse processes of first two or four cervical vertebrae.	Acting together, extend head; acting singly, laterally flex and rotate head to same side as contracting muscle.
Erector Spinae (e-REK-tor SPI-nē) Consists of iliocostalis, longissimus, and spinalis muscles.			
Iliocostalis Group (Lateral)			
Iliocostalis cervicis (il′-ē-ō-kos-TAL-is SER-vi-kis; *ilium* flank; *costa* = rib)	Superior six ribs.	Transverse processes of fourth to sixth cervical vertebrae.	Acting together, muscles of each region (cervical, thoracic, and lumbar) extend and maintain erect posture of vertebral column of their respective regions; acting singly, laterally flex vertebral column of their respective regions.
Iliocostalis thoracis (il′-ē-ō-kos-TAL-is thō-RA-kis; *thorax* = chest)	Inferior six ribs.	Superior six ribs.	
Iliocostalis lumborum (il′-ē-ō-kos-TAL-is lum-BOR-um)	Iliac crest.	Inferior six ribs.	
Longissimus Group (Intermediate)			
Longissimus capitis (lon-JIS-i-mus KAP-i-tis; *longissimus* = longest)	Transverse processes of superior four thoracic vertebrae and articular processes of inferior four cervical vertebrae.	Mastoid process of temporal bone.	Acting together, both longissimus capitis muscles extend head; acting singly, rotate head to same side as contracting muscle. Acting together, longissimus cervicis and both longissimus thoracis muscles extend vertebral column of their respective regions; acting singly, laterally flex vertebral column of their respective regions.
Longissimus cervicis (lon-JIS-i-mus SER-vi-kis)	Transverse processes of fourth and fifth thoracic vertebrae.	Transverse processes of second to sixth cervical vertebrae.	
Longissimus thoracis (lon-JIS-i-mus thō-RA-kis)	Transverse processes of lumbar vertebrae.	Transverse processes of all thoracic and superior lumbar vertebrae and ninth and tenth ribs.	
Spinalis Group (Medial)			
Spinalis capitis (spi-NA-lis KAP-i-tis; *spinalis* = vertebral column)	Arises with semispinalis capitis.	Occipital bone.	Acting together, muscles of each region (cervical, thoracic, and lumbar) extend vertebral column of their respective regions.
Spinalis cervicis (spi-NA-lis SER-vi-kis)	Ligamentum nuchae and spinous process of seventh cervical vertebra.	Spinous process of axis.	
Spinalis thoracis (spi-NA-lis thō-RA-kis)	Spinous processes of superior lumbar and inferior thoracic vertebrae.	Spinous processes of superior thoracic vertebrae.	
Transversospinalis (trans-ver′-sō-spi-NA-lis)			
Semispinalis capitis (sem′-ē-spi-NA-lis KAP-i-tis; *semi* = partially or one-half)	Transverse processes of first six or seven thoracic vertebrae and seventh cervical vertebra, and articular processes of fourth, fifth, and sixth cervical vertebrae.	Occipital bone.	Acting together, extend head; acting singly, rotate head to side opposite contracting muscle.
Semispinalis cervicis (sem′-ē-spi-NA-lis SER-vi-kis)	Transverse processes of superior five or six thoracic vertebrae.	Spinous processes of first to fifth cervical vertebrae.	Acting together, the semispinalis cervicis and semispinalis thoracis muscles extend the vertebral column of their respective regions; acting singly, rotate head to side opposite contracting muscle.
Semispinalis thoracis (sem′-ē-spi-NA-lis thō-RA-kis)	Transverse processes of sixth to tenth thoracic vertebrae.	Spinous processes of superior four thoracic and last two cervical vertebrae.	

(continues)

Muscle	Origin	Insertion	Action
Transversospinalis (trans-ver′-sō-spi-NA-lis)			
Multifidus (mul-TIF-i-dus; *multi* = many; *findere* = to split)	Sacrum, ilium, transverse processes of lumbar, thoracic, and inferior four cervical vertebrae.	Spinous process of a more superior vertebra.	Acting together, extend vertebral column; acting singly, laterally flex vertebral column and rotate head to side opposite contracting muscle.
Rotatores (rō′-ta-TŌ-rez; *rotare* = to turn)	Transverse processes of all vertebrae.	Spinous process of vertebra superior to the one of origin.	Acting together, extend vertebral column; acting singly, rotate vertebral column to side opposite contracting muscle.
Segmental (seg-MEN-tal)			
Interspinales (in-ter-SPĪ-nāl-ēz; *inter* = between)	Superior surface of all spinous processes.	Inferior surface of spinous process of vertebra superior to the one of origin.	Acting together, extend vertebral column; acting singly, stabilize vertebral column during movement.
Intertransversarii (in′-ter-trans-vers-AR-ē-ī; *inter* = between)	Transverse processes of all vertebrae.	Transverse process of vertebra superior to the one of origin.	Acting together, extend vertebral column; acting singly, laterally flex vertebral column and stabilize it during movements.
Scalene (SKĀ-lēn)			
Anterior scalene (SKĀ-lēn; *anterior* = front; *skalenos* = uneven)	Transverse processes of third through sixth cervical vertebrae.	First rib.	Acting together, both anterior scalene and middle scalene muscles flex head and elevate first ribs during deep inspiration; acting singly, laterally flex head and rotate head to side opposite contracting muscle.
Middle scalene (SKĀ-lēn)	Transverse processes of inferior six cervical vertebrae.	First rib.	
Posterior scalene (SKĀ-lēn)	Transverse processes of fourth through sixth cervical vertebrae.	Second rib.	Acting together, flex head and elevate second ribs during deep inspiration; acting singly, laterally flex head and rotate head to side opposite contracting muscle.

Figure 10.21 / Muscles that move the vertebral column (backbone).

The erector spinae group (iliocostalis, longissimus, and spinalis muscles) is the largest muscular mass of the body and is the chief extensor of the vertebral column.

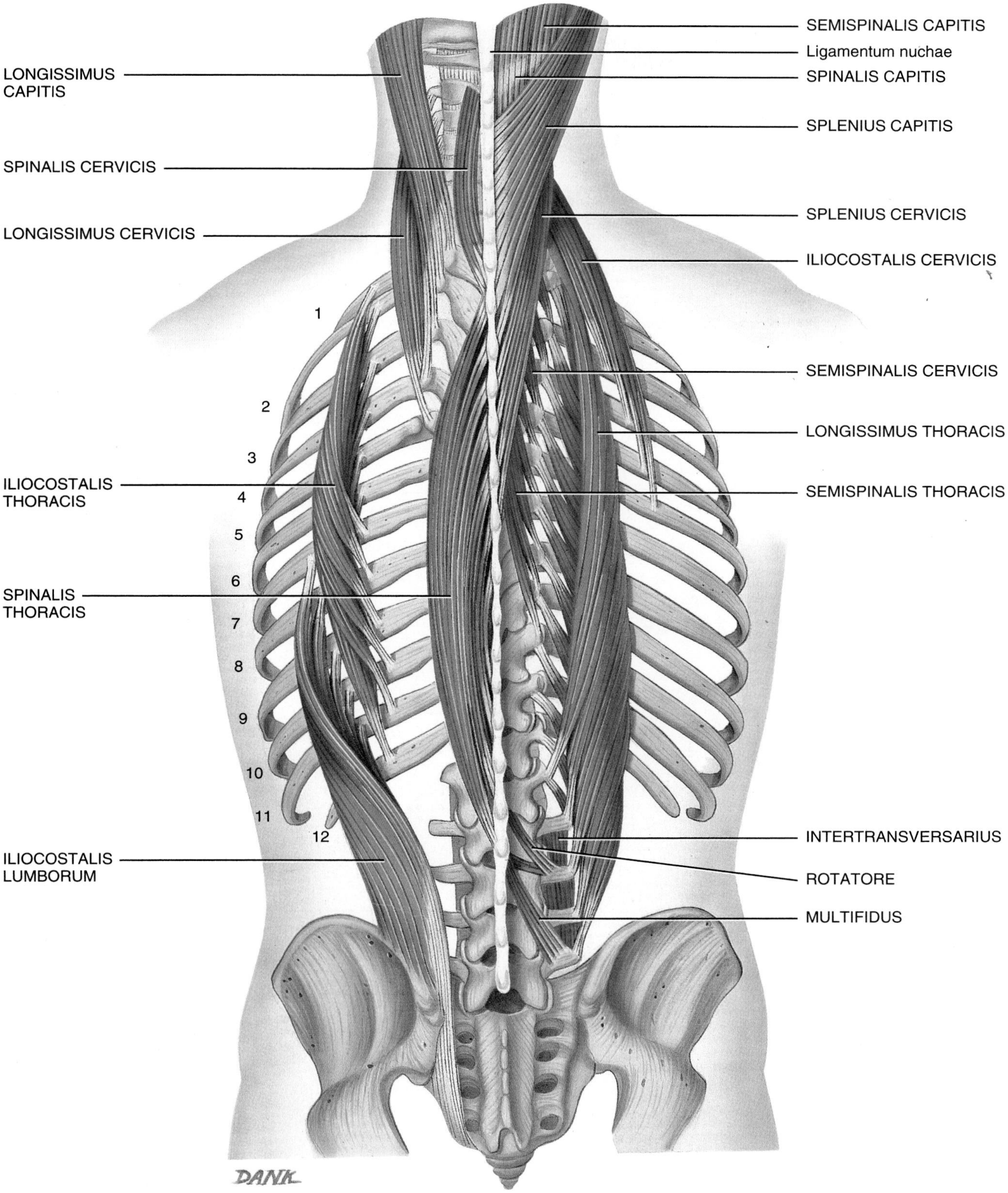

LONGISSIMUS CAPITIS
SPINALIS CERVICIS
LONGISSIMUS CERVICIS
ILIOCOSTALIS THORACIS
SPINALIS THORACIS
ILIOCOSTALIS LUMBORUM
SEMISPINALIS CAPITIS
Ligamentum nuchae
SPINALIS CAPITIS
SPLENIUS CAPITIS
SPLENIUS CERVICIS
ILIOCOSTALIS CERVICIS
SEMISPINALIS CERVICIS
LONGISSIMUS THORACIS
SEMISPINALIS THORACIS
INTERTRANSVERSARIUS
ROTATORE
MULTIFIDUS

(a) Posterior view

(continues)

Exhibit 10.18 Muscles That Move the Vertebral Column (Backbone) (Figure 10.21) continued

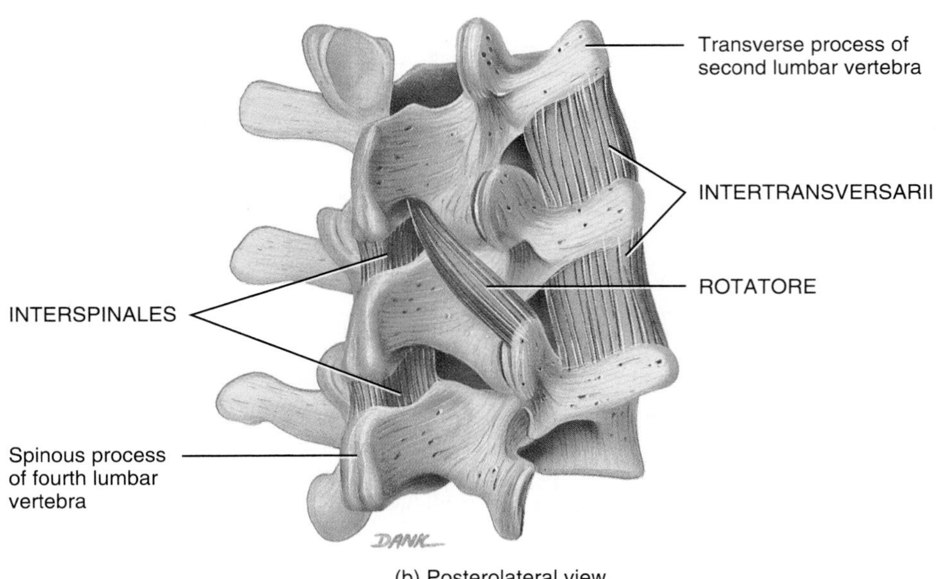

(b) Posterolateral view

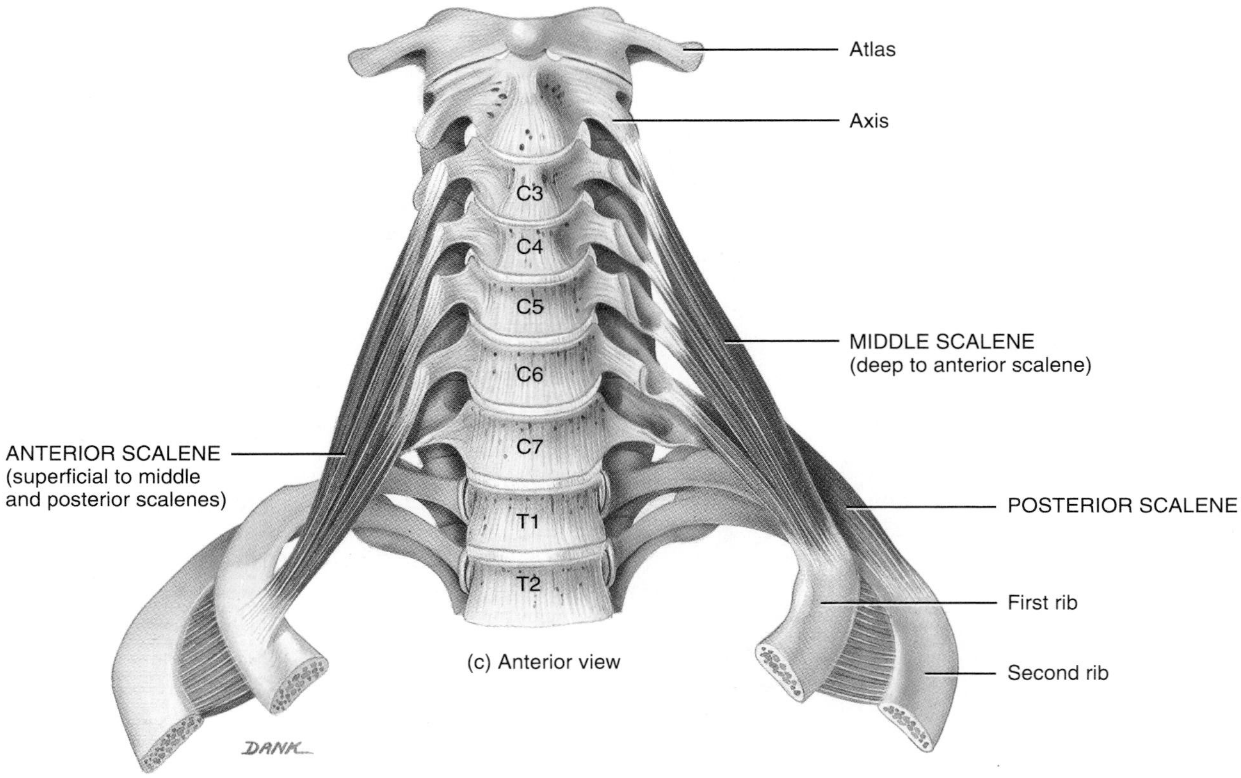

(c) Anterior view

Which muscles originate at the midline and extend laterally and upward to their insertion?

Exhibit 10.19 Muscles That Move the Femur (Thigh) (Figure 10.22)

Objective

▶ Describe the origin, insertion, action, and innervation of the muscles of the femur.

As you will see, muscles of the lower limbs are larger and more powerful than those of the upper limbs because lower limb muscles function in stability, locomotion, and maintenance of posture. Upper limb muscles are characterized by versatility of movement. In addition, muscles of the lower limbs often cross two joints and act equally on both.

The majority of muscles that move the femur originate on the pelvic girdle and insert on the femur. The **psoas major** and **iliacus** muscles have a common insertion and are together referred to as the **iliopsoas** (il′-ē-ō-SŌ-as) muscle because they share a common insertion (lesser trochanter of femur). There are three gluteal muscles: gluteus maximus, gluteus medius, and gluteus minimus. The **gluteus maximus** is the largest and heaviest of the three muscles and is one of the largest muscles in the body. It is the chief extensor of the femur. The **gluteus medius** is mostly deep to the gluteus maximus and is a powerful abductor of the femur at the hip joint. It is a common site for an intramuscular injection. The **gluteus minimus** is the smallest of the gluteal muscles and lies deep to the gluteus medius.

The **tensor fasciae latae** muscle is located on the lateral surface of the thigh. A layer of deep fascia composed of dense connective tissue encircles the entire thigh and is referred to as the **fascia lata.** It is well-developed laterally where, together with the tendons of the tensor fasciae and gluteus maximus muscles, it forms a structure called the **iliotibial tract.** The tract inserts into the lateral condyle of the tibia.

The **piriformis, obturator internus, obturator externus, superior gemellus, inferior gemellus,** and **quadratus femoris** muscles are all deep to the gluteus maximus muscle and function as lateral rotators of the femur at the hip joint.

Three muscles on the medial aspect of the thigh are the **adductor longus, adductor brevis,** and **adductor magnus.** They originate on the hip bone and insert on the femur. All three muscles adduct, flex, and medially rotate the femur at the hip joint. The **pectineus** muscle also adducts and flexes the femur at the hip joint.

Technically, the adductor muscles and pectineus muscles are components of the medial compartment of the thigh and could also be included in Exhibit 10.20. However, they are included here because they act on the femur.

Pulled Groin

Certain athletic and other activities may cause a **pulled groin,** a strain, stretching, or tearing of the distal attachments of the medial muscles of the thigh, iliopsoas, and/or adductor muscles. This generally results from activities that involve quick sprints, as might occur during soccer, tennis, football, and running. ■

Relating Muscles to Movements

Arrange the muscles in this exhibit according to the following actions on the thigh at the hip joint: (1) flexion, (2) extension, (3) abduction, (4) adduction, (5) medial rotation, and (6) lateral rotation. The same muscle may be mentioned more than once.

✔ What structures form the iliotibial tract?

Innervation

Muscles that move the femur are innervated by nerves derived from the lumbar and sacral plexuses (shown in Exhibits 17.3 and 17.4 on pages 534–538). More specifically, the psoas major is innervated by lumbar nerves L2–L3; the iliacus and pectineus are innervated by the femoral nerve; the gluteus maximus is innervated by the inferior gluteal nerve; the gluteus medius and minimus and tensor fasciae latae are innervated by the superior gluteal nerve; the piriformis is innervated by sacral nerves S1 or S2, mainly S1; the obturator internus and superior gemellus are innervated by the nerve to obturator internus; the obturator externus, adductor longus, and adductor brevis are innervated by the obturator nerve; the inferior gemellus and quadratus femoris are innervated by the nerve to quadratus femoris; and the adductor magnus is innervated by the obturator and sciatic nerves.

Muscle	Origin	Insertion	Action
Psoas major (SŌ-as; *psoa* = muscle of loin)	Transverse processes and bodies of lumbar vertebrae.	With iliacus into lesser trochanter of femur.	Both psoas major and iliacus muscles acting together flex thigh at hip joint, rotate thigh laterally, and flex trunk on the hip as in sitting up from the supine position.
Iliacus (il′-ē-AK-us; *iliac* = ilium)	Iliac fossa.	With psoas major into lesser trochanter of femur.	
Gluteus maximus (GLŪ-tē-us MAK-si-mus; *glutos* = buttock; *maximus* = largest)	Iliac crest, sacrum, coccyx, and aponeurosis of sacrospinalis.	Iliotibial tract of fascia lata and lateral part of linea aspera under greater trochanter (gluteal tuberosity) of femur.	Extends thigh at hip joint and laterally rotates thigh.
Gluteus medius (GLŪ-tē-us MĒ-dē-us; *media* = middle)	Ilium.	Greater trochanter of femur.	Abducts thigh at hip joint and medially rotates thigh.
Gluteus minimus (GLŪ-tē-us MIN-i-mus; *minimus* = smallest)	Ilium.	Greater trochanter of femur.	Abducts thigh at hip joint and medially rotates thigh.
Tensor fasciae latae (TEN-sor FA-shē-ē LĀ-tē; *tensor* = makes tense; *fascia* = band; *latus* = wide)	Iliac crest.	Tibia by way of the iliotibial tract.	Flexes and abducts thigh at hip joint.

(continues)

Exhibit 10.19 Muscles that Move the Femur (Thigh) (Figure 10.22) continued

Muscle	Origin	Insertion	Action
Piriformis (pir-i-FOR-mis; *pirum* = pear; *forma* = shape)	Anterior sacrum.	Superior border of greater trochanter of femur.	Laterally rotates and abducts thigh at hip joint.
Obturator internus (OB-tū-rā′-tor in-TER-nus; *obturator* = obturator foramen; *internus* = inside)	Inner surface of obturator foramen, pubis, and ischium.	Greater trochanter of femur.	Laterally rotates and abducts thigh at hip joint.
Obturator externus (OB-tū-rā′-tor ex-TER-nus; *externus* = outside)	Outer surface of obturator membrane.	Deep depression inferior to greater trochanter (trochanteric fossa) of femur.	Laterally rotates and abducts thigh at hip joint.
Superior gemellus (jem-EL-lus; *superior* = above; *gemellus* = twins)	Ischial spine.	Greater trochanter of femur.	Laterally rotates and abducts thigh at hip joint.
Inferior gemellus (jem-EL-lus; *inferior* = below)	Ischial tuberosity.	Greater trochanter of femur.	Laterally rotates and abducts thigh at hip joint.
Quadratus femoris (kwod-RĀ-tus FEM-or-is; *quad* = four; *femoris* = femur)	Ischial tuberosity.	Elevation superior to midportion of intertrochanteric crest (quadrate tubercle) on posterior femur.	Laterally rotates and stabilizes hip joint.
Adductor longus (LONG-us; *adductor* = moves part closer to midline; *longus* = long)	Pubic crest and pubic symphysis.	Linea aspera of femur.	Adducts and flexes thigh at hip joint and medially rotates thigh.
Adductor brevis (BREV-is; *brevis* = short)	Inferior ramus of pubis.	Superior half of linea aspera of femur.	Adducts and flexes thigh at hip joint and medially rotates thigh.
Adductor magnus (MAG-nus; *magnus* = large)	Inferior ramus of pubis and ischium to ischial tuberosity.	Linea aspera of femur.	Adducts thigh at hip joint and medially rotates thigh; anterior part flexes thigh at hip joint, and posterior part extends thigh at hip joint.
Pectineus (pek-TIN-ē-us; *pecten* = comb-shaped)	Superior ramus of pubis.	Pectineal line of femur, between lesser trochanter and linea aspera.	Flexes and adducts thigh at hip joint.

Figure 10.22 / Muscles that move the femur (thigh).

Most muscles that move the femur originate on the pelvic (hip) girdle and insert on the femur.

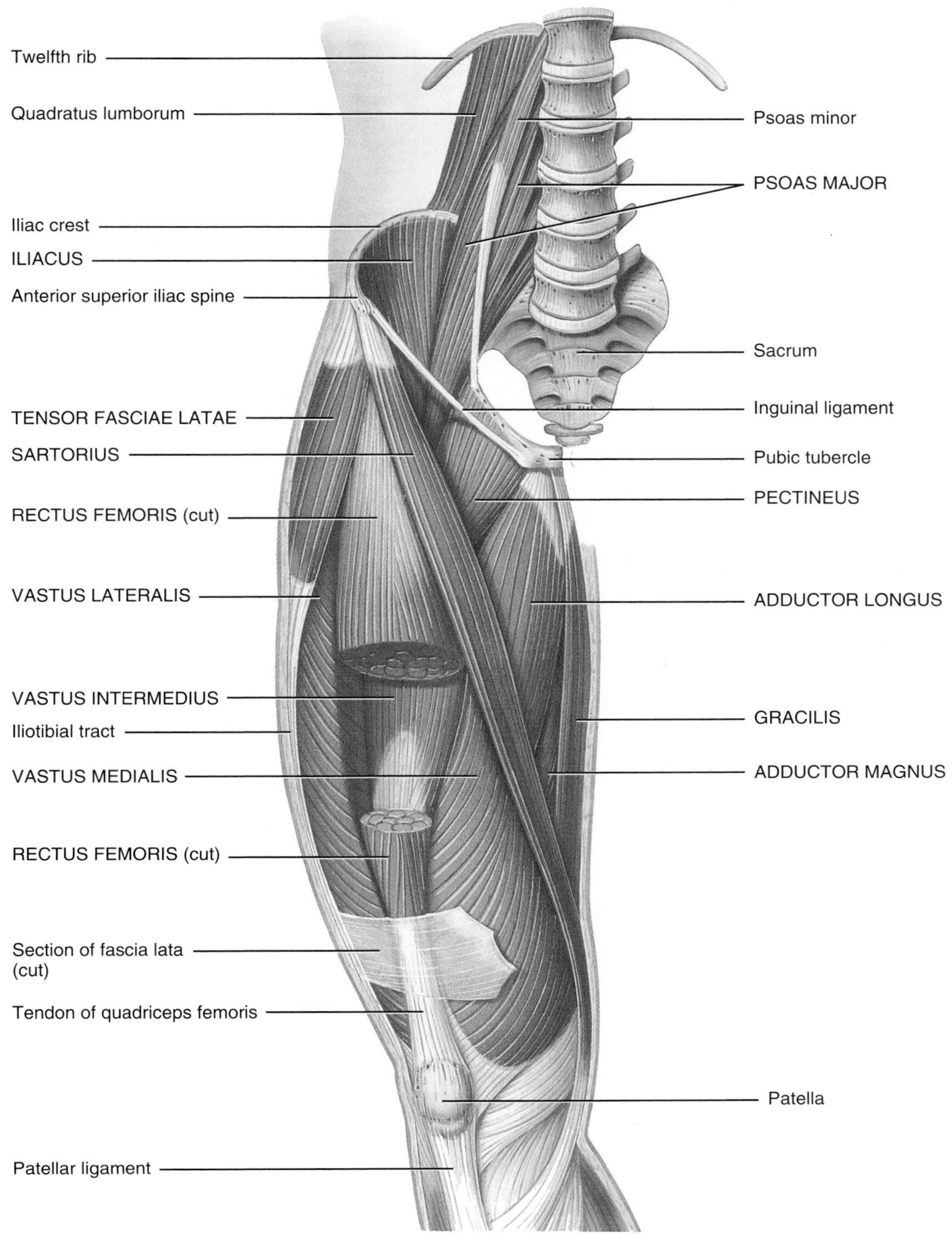

(a) Anterior superficial view

(continues)

Exhibit 10.19 Muscles that Move the Femur (Thigh) (Figure 10.22) continued

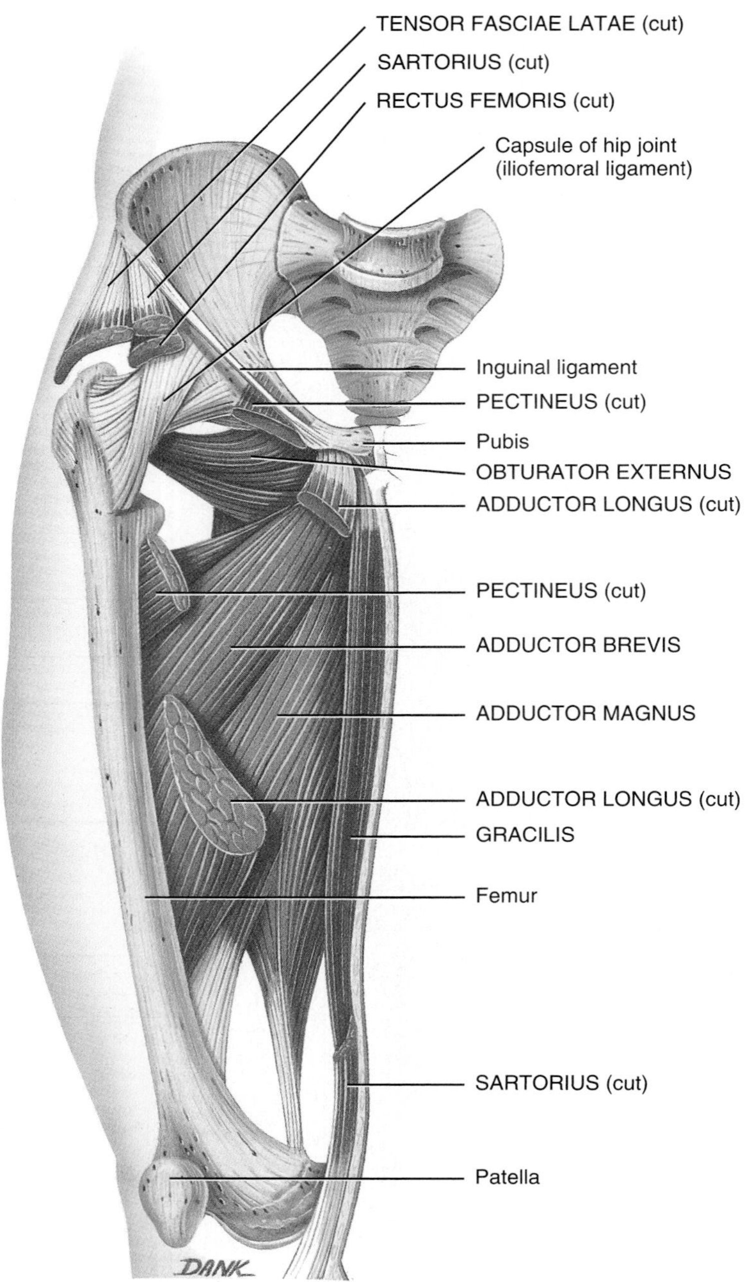

(b) Anterior deep view (femur rotated laterally)

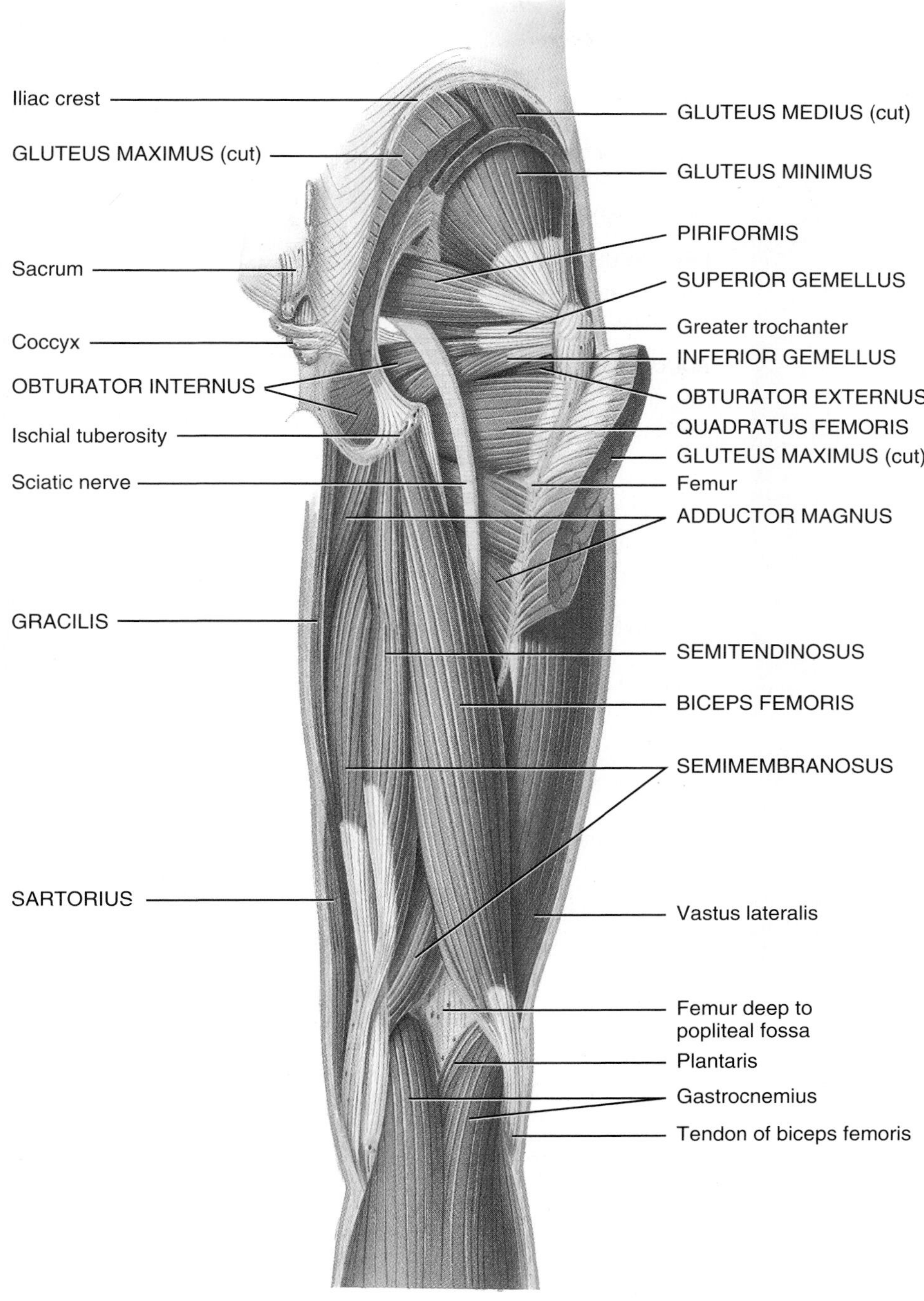

(c) Posterior superficial view

(continues)

Exhibit 10.19 Muscles that Move the Femur (Thigh) (Figure 10.22) continued

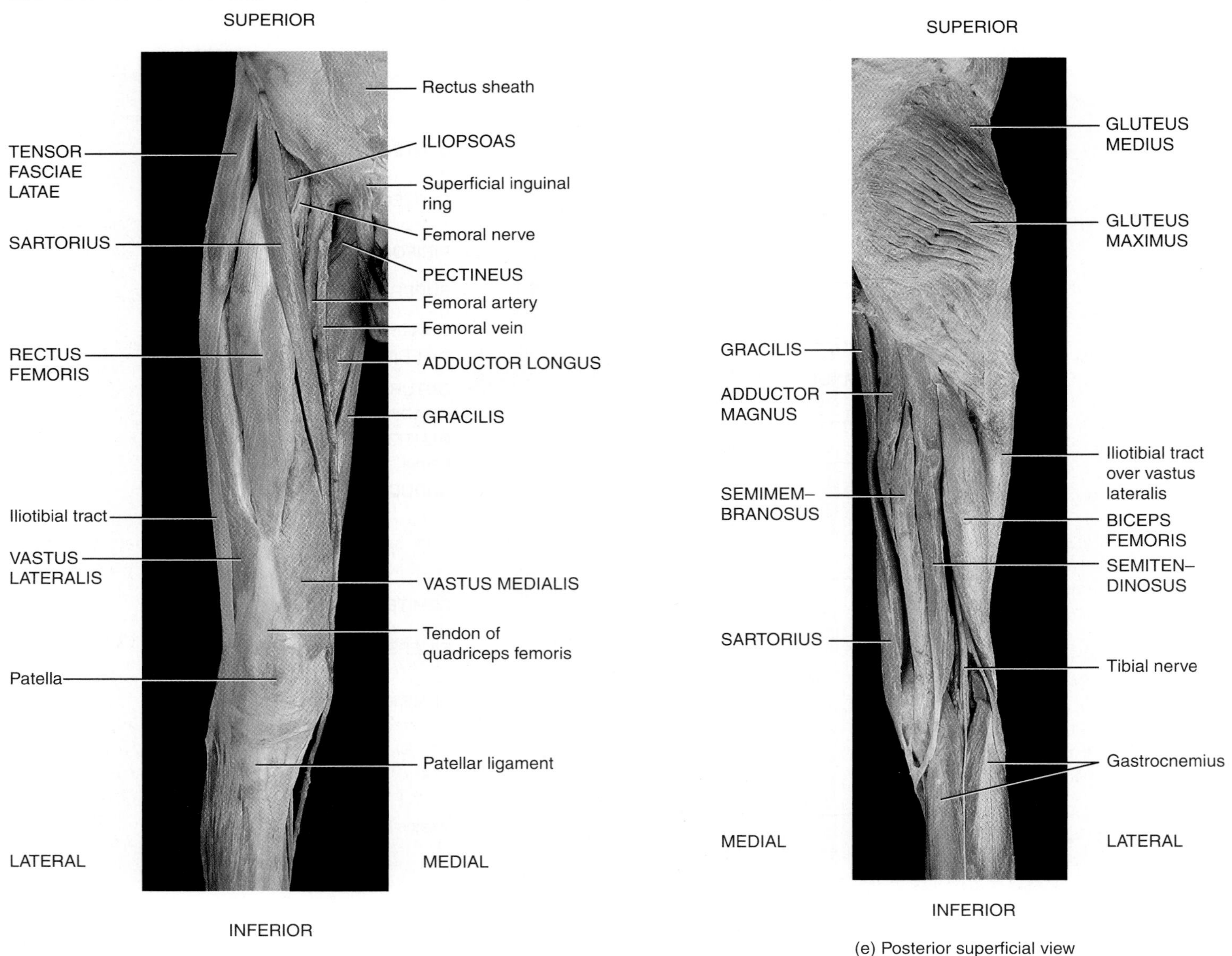

(d) Anterior superficial view

(e) Posterior superficial view

What are the principal differences between the muscles of the upper and lower limbs?

Exhibit 10.20 Muscles That Act on the Femur (Thigh) and Tibia and Fibula (Leg) (Figure 10.23)

Objective

► Describe the origin, insertion, action, and innervation of the muscles that act on the femur and tibia and fibula.

The muscles that act on the femur (thigh) and tibia and fibula (leg) are separated by deep fascia into medial, anterior, and posterior compartments. The **medial (adductor) compartment** is so named because its muscles adduct the femur at the hip joint. (See the adductor magnus, adductor longus, adductor brevis, and pectineus, which are components of the medial compartment, in Exhibit 10.19.) The **gracilis,** the other muscle in the medial compartment, not only adducts the thigh, but also flexes the leg at the knee joint. For this reason, it is discussed in this exhibit. The gracilis is a long, straplike muscle that lies on the medial aspect of the thigh and knee.

The **anterior (extensor) compartment** is so designated because its muscles extend the leg (and also flex the thigh). This compartment is composed of the quadriceps femoris and sartorius muscles. The **quadriceps femoris** muscle is the biggest muscle in the body, covering most of the anterior surface and medial and lateral aspects of the thigh. The muscle is actually a composite muscle, usually described as four separate muscles: (1) **rectus femoris,** on the anterior aspect of the thigh; (2) **vastus lateralis,** on the lateral aspect of the thigh; (3) **vastus medialis,** on the medial aspect of the thigh; and (4) **vastus intermedius,** located deep to the rectus femoris between the vastus lateralis and vastus medialis. The common tendon for the four muscles is known as the **quadriceps tendon,** which inserts into the patella. The tendon continues inferior to the patella as the **patellar ligament,** which attaches to the tibial tuberosity. The quadriceps femoris muscle is the great extensor muscle of the leg. The **sartorius** is a long, narrow muscle that forms a band across the thigh from the ilium of the hip bone to the medial side of the tibia. The various movements it produces help effect the cross-legged sitting position in which the heel of one limb is placed on the knee of the opposite limb. It is known as the tailor's muscle because tailors frequently assume this cross-legged sitting position. (Because the major action of the sartorius muscle is to move the thigh rather than the leg, it could also have been included in Exhibit 10.19.)

The **posterior (flexor) compartment** is so named because its muscles flex the leg (and also extend the thigh). This compartment is composed of three muscles collectively called the **hamstrings:** (1) **biceps femoris,** (2) **semitendinosus,** and (3) **semimembranosus.** The hamstrings are so named because their tendons are long and stringlike in the popliteal area, and from an old practice of butchers in which they hung hams for smoking by these long tendons.

Because the hamstrings span two joints (hip and knee), they are both extensors of the thigh and flexors of the leg. The **popliteal fossa** is a diamond-shaped space on the posterior aspect of the knee bordered laterally by the tendons of the biceps femoris muscle and medially by the tendons of the semitendinosus and semimembranosus muscles.

Pulled Hamstrings

A strain or partial tear of the proximal hamstring muscles is referred to as **pulled hamstrings** or **hamstring strains.** They are common sports injuries in individuals who run very hard and/or are required to perform quick starts and stops. Sometimes the violent muscular exertion required to perform a feat tears off part of the tendinous origins of the hamstrings, especially the biceps femoris, from the ischial tuberosity. This is usually accompanied by a contusion (bruising), tearing of some of the muscle fibers, and rupture of blood vessels, producing a hematoma (collection of blood) and pain. Adequate training with good balance between the quadriceps femoris and hamstrings and stretching exercises before running or competing are important in preventing this injury. ■

Relating Muscles to Movements

Arrange the muscles in this exhibit according to the following actions on the thigh at the hip joint: (1) abduction, (2) adduction, (3) lateral rotation, (4) flexion, and (5) extension; and according to the following actions on the leg at the knee joint: (1) flexion and (2) extension. The same muscle may be mentioned more than once.

✔ Organize the muscles that act on the femur (thigh) and tibia and fibula (leg) into medial, anterior, and posterior compartments.

Innervation

Muscles that act on the femur and tibia and fibula are innervated by the obturator and femoral nerves derived from the lumbar plexus (shown in Exhibit 17.3 on pages 534–536) and the sciatic nerve derived from the sacral plexus (shown in Exhibit 17.4 on pages 537–538). More specifically, the gracilis is innervated by the obturator nerve; the quadriceps femoris and sartorius are innervated by the femoral nerve; the biceps femoris is innervated by the tibial and common peroneal nerves from the sciatic nerve; and the semimembranosus and semitendinosus are innervated by the tibial nerve from the sciatic nerve.

(continues)

Exhibit 10.20 Muscles That Act on the Femur (Thigh) and Tibia and Fibula (Leg) (Figure 10.23) continued

Muscle	Origin	Insertion	Action
Medial (adductor) compartment			
Adductor magnus (MAG-nus)	See Exhibit 10.19.		
Adductor longus (LONG-us)			
Adductor brevis (BREV-is)			
Pectineus (pek-TIN-ē-us)			
Gracilis (GRAS-i-lis; *gracilis* = slender)	Pubic symphysis and pubic arch.	Medial surface of body of tibia.	Adducts thigh at hip joint, medially rotates thigh, and flexes leg at knee joint.
Anterior (extensor) compartment			
Quadriceps femoris (KWOD-ri-ceps FEM-or-is; *quadriceps* = four heads of origin; *femoris* = femur)			
Rectus femoris (REK-tus FEM-or-is; *rectus* = fibers parallel to midline)	Anterior inferior iliac spine.	Patella via quadriceps tendon and then tibial tuberosity via patellar ligament.	All four heads extend leg at knee joint; rectus femoris muscle acting alone also flexes thigh at hip joint.
Vastus lateralis (VAS-tus lat′-er-A-lis; *vastus* = large; *lateralis* = lateral)	Greater trochanter and linea aspera of femur.		
Vastus medialis (VAS-tus mē′-dē-A-lis; *medialis* = medial)	Linea aspera of femur.		
Vastus intermedius (VAS-tus in′-ter-MĒ-dē-us; *intermedius* = middle)	Anterior and lateral surfaces of body of femur.		
Sartorius (sar-TOR-ē-us; *sartor* = tailor; longest muscle in body)	Anterior superior iliac spine.	Medial surface of body of tibia.	Flexes leg at knee joint; flexes, abducts, and laterally rotates thigh at hip joint.
Posterior (flexor) compartment			
Hamstrings A collective designation for three separate muscles.			
Biceps femoris (BĪ-ceps FEM-or-is; *biceps* = two heads of origin)	Long head arises from ischial tuberosity; short head arises from linea aspera of femur.	Head of fibula and lateral condyle of tibia.	Flexes leg at knee joint and extends thigh at hip joint.
Semitendinosus (sem′-ē-TEN-di-nō-sus; *semi* = half; *tendo* = tendon)	Ischial tuberosity.	Proximal part of medial surface of shaft of tibia.	Flexes leg at knee joint and extends thigh at hip joint.
Semimembranosus (sem′-ē-MEM-bra-nō-sus; *membran* = membrane)	Ischial tuberosity.	Medial condyle of tibia.	Flexes leg at knee joint and extends thigh at hip joint.

Figure 10.23 / Muscles that act on the femur (thigh) and tibia and fibula (leg).

Muscles that act on the leg originate in the hip and thigh and are separated into compartments by deep fascia.

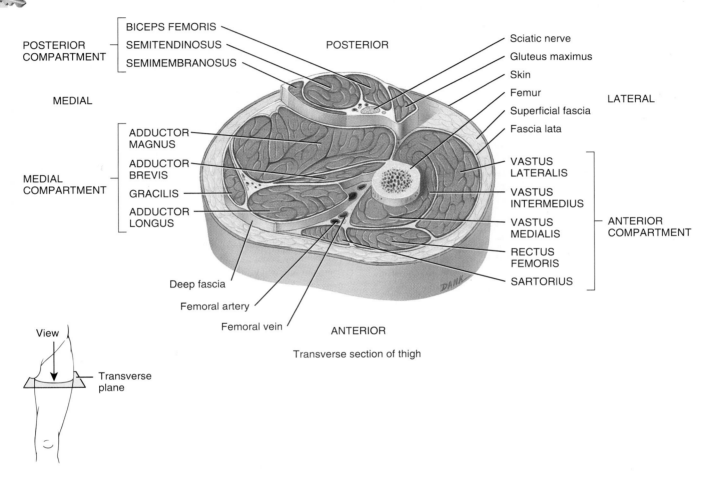

POSTERIOR COMPARTMENT
BICEPS FEMORIS
SEMITENDINOSUS
SEMIMEMBRANOSUS
POSTERIOR
Sciatic nerve
Gluteus maximus
Skin
Femur
Superficial fascia
Fascia lata
MEDIAL
LATERAL
MEDIAL COMPARTMENT
ADDUCTOR MAGNUS
ADDUCTOR BREVIS
GRACILIS
ADDUCTOR LONGUS
ANTERIOR COMPARTMENT
VASTUS LATERALIS
VASTUS INTERMEDIUS
VASTUS MEDIALIS
RECTUS FEMORIS
SARTORIUS
Deep fascia
Femoral artery
Femoral vein
ANTERIOR
View
Transverse plane

Transverse section of thigh

Which muscles constitute the quadriceps femoris and hamstring muscles?

Exhibit 10.21 Muscles That Move the Foot and Toes (Figure 10.24)

Objective

► Describe the origin, insertion, action, and innervation of the muscles that move the foot and toes.

Muscles that move the foot and toes are located in the leg. The musculature of the leg, like that of the thigh, is divided by deep fascia into three compartments: anterior, lateral, and posterior. The **anterior compartment** consists of muscles that dorsiflex the foot. In a situation analogous to the wrist, the tendons of the muscles of the anterior compartment are held firmly to the ankle by thickenings of deep fascia called the **superior extensor retinaculum** (*transverse ligament of the ankle*) and **inferior extensor retinaculum** (*cruciate ligament of the ankle*).

Within the anterior compartment, the **tibialis anterior** is a long, thick muscle against the lateral surface of the tibia, where it is easy to palpate. The **extensor hallucis longus** is a thin muscle between and partly deep to the **tibialis anterior** and **extensor digitorum longus** muscles. This latter muscle is featherlike and lies lateral to the tibialis anterior muscle, where it can easily be palpated. The **peroneus tertius** muscle is actually part of the extensor digitorum longus, with which it shares a common origin.

The **lateral (peroneal) compartment** contains two muscles that plantar flex and evert the foot: **peroneus longus** and **peroneus brevis.**

The **posterior compartment** consists of muscles that are divisible into superficial and deep groups. The superficial muscles share a common tendon of insertion, the **calcaneal (Achilles) tendon,** the strongest tendon of the body, which inserts into the calcaneal bone of the ankle. The superficial and most of the deep muscles plantar flex the foot at the ankle joint. The superficial muscles of the posterior compartment are the gastrocnemius, soleus, and plantaris—the so-called calf muscles. The large size of these muscles is directly related to our upright stance, a characteristic of humans. The **gastrocnemius** is the most superficial muscle and forms the prominence of the calf. The **soleus** is broad, flat, and lies deep to the gastrocnemius. It derives its name from its resemblance to a flat fish (sole). The **plantaris** is a small muscle that may be absent; sometimes there are two of them in each leg. It runs obliquely between the gastrocnemius and soleus muscles.

The deep muscles of the posterior compartment are the popliteus, tibialis posterior, flexor digitorum longus, and flexor hallucis longus. The **popliteus** is a triangular muscle that forms the floor of the popliteal fossa. The **tibialis posterior** is the deepest muscle in the posterior compartment. It lies between the flexor digitorum longus and flexor hallucis longus muscles. The **flexor digitorum longus** is smaller than the **flexor hallucis longus,** even though the former flexes four toes, whereas the latter flexes only the great toe at the interphalangeal joint.

Shinsplint Syndrome

Shinsplint syndrome, or simply **shinsplints,** refers to pain or soreness along the tibia, specifically the medial, distal two-thirds. It may be caused by tendinitis of the anterior compartment muscles, especially the tibialis anterior muscle, inflammation of the periosteum (periostitis) around the tibia, or stress fractures of the tibia. The tendinitis usually occurs when poorly conditioned runners run on hard or banked surfaces with poorly supportive running shoes. The condition may also occur as a result of vigorous activity of the legs following a period of relative inactivity. The muscles in the anterior compartment (mainly the tibialis anterior) can be strengthened to balance the stronger posterior compartment muscles. ■

Relating Muscles to Movements

Arrange the muscles in this exhibit according to the following actions on the foot at the ankle joint: (1) dorsiflexion and (2) plantar flexion; according to the following actions on the foot at the intertarsal joints: (1) inversion and (2) eversion; and according to the following actions on the toes at the metatarsophalangeal and intertarsal joints: (1) flexion and (2) extension. The same muscle may be mentioned more than once.

✔ What is the superior extensor retinaculum? Inferior extensor retinaculum?

Innervation

Muscles that move the foot and toes are innervated by nerves derived from the sacral plexus (shown in Exhibit 17.4 on pages 537–538). More specifically, muscles of the anterior compartment are innervated by the deep peroneal nerve; muscles of the lateral compartment are innervated by the superficial peroneal nerve; and muscles of the posterior compartment are innervated by the tibial nerve.

Muscle	Origin	Insertion	Action
Anterior Compartment			
Tibialis anterior (tib′-ē-A-lis; *tibialis* = tibia; *anterior* = front)	Lateral condyle and body of tibia and interosseous membrane (sheet of fibrous tissue that holds shafts of tibia and fibula together).	First metatarsal and first (medial) cuneiform.	Dorsiflexes foot at ankle joint and inverts foot at intertarsal joints.
Extensor hallucis longus (HAL-yū-sis LON-gus; *extensor* = increases angle at joint; *hallucis* = hallux or great toe; *longus* = long)	Anterior surface of fibula and interosseous membrane.	Distal phalanx of great toe.	Dorsiflexes foot at ankle joint and extends proximal phalanx of great toe at metatarsophalangeal joint.
Extensor digitorum longus (di′-ji-TOR-um LON-gus)	Lateral condyle of tibia, anterior surface of fibula, and interosseous membrane.	Middle and distal phalanges of toes 2–5.*	Dorsiflexes foot at ankle joint and extends distal and middle phalanges of each toe at interphalangeal joints and proximal phalanx of each toe at metatarsophalangeal joint.
Peroneus tertius (per′-ō-NĒ-us TER-shus; *perone* = fibula; *tertius* = third)	Distal third of fibula and interosseous membrane.	Base of fifth metatarsal.	Dorsiflexes foot at ankle joint and everts foot at intertarsal joints.
Lateral (Peroneal) Compartment			
Peroneus longus (per′-ō-NĒ-us LON-gus)	Head and body of fibula and lateral condyle of tibia.	First metatarsal and first cuneiform.	Plantar flexes foot at ankle joint and everts foot at intertarsal joints.
Peroneus brevis (per′-ō-NĒ-us BREV-is; *brevis* = short)	Body of fibula.	Base of fifth metatarsal.	Plantar flexes foot at ankle joint and everts foot at intertarsal joints.
Superficial Posterior Compartment			
Gastrocnemius (gas′-trok-NĒ-mē-us; *gaster* = belly; *kneme* = leg)	Lateral and medial condyles of femur and capsule of knee.	Calcaneus by way of calcaneal (Achilles) tendon.	Plantar flexes foot at ankle joint and flexes leg at knee joint.
Soleus (SŌ-lē-us; *soleus* = sole of foot)	Head of fibula and medial border of tibia.	Calcaneus by way of calcaneal (Achilles) tendon.	Plantar flexes foot at ankle joint.
Plantaris (plan-TA-ris; *plantar* = sole of foot)	Femur superior to lateral condyle.	Calcaneus by way of calcaneal (Achilles) tendon.	Plantar flexes foot at ankle joint and flexes leg at knee joint.
Deep Posterior Compartment			
Popliteus (pop-LIT-ē-us; *poples* = posterior surface of knee)	Lateral condyle of femur.	Proximal tibia.	Flexes leg at knee joint and medially rotates tibia to unlock the extended knee.
Tibialis (tib′-ē-A-lis) **posterior** (*posterior* = back)	Tibia, fibula, and interosseous membrane.	Second, third, and fourth metatarsals; navicular; all three cuneiforms; and cuboid.	Plantar flexes foot at ankle joint and inverts foot at intertarsal joints.
Flexor digitorum longus (di′-ji-TOR-um LON-gus; *digitorum* = finger or toe)	Posterior surface of tibia.	Distal phalanges of toes 2–5.	Plantar flexes foot at ankle joint; flexes distal and middle phalanges of each toe at interphalangeal joints and proximal phalanx of each toe at metatarsophalangeal joint.
Flexor hallucis longus (HAL-yū-sis LON-gus; *flexor* = decreases angle at joint)	Inferior two-thirds of fibula.	Distal phalanx of great toe.	Plantar flexes foot at ankle joint; flexes distal phalanx of great toe at interphalangeal joint and proximal phalanx of great toe at metatarsophalangeal joint.

*Reminder: The great toe or hallux is the first toe and has two phalanges: proximal and distal. The remaining toes are numbered 2–5, and each has three phalanges: proximal, middle, and distal.

(continues)

Exhibit 10.21 Muscles That Move the Foot and Toes (Figure 10.24) continued

Figure 10.24 / Muscles that move the foot and toes.

The superficial muscles of the posterior compartment share a common tendon of insertion, the calcaneal (Achilles) tendon, that inserts into the calcaneal bone of the ankle.

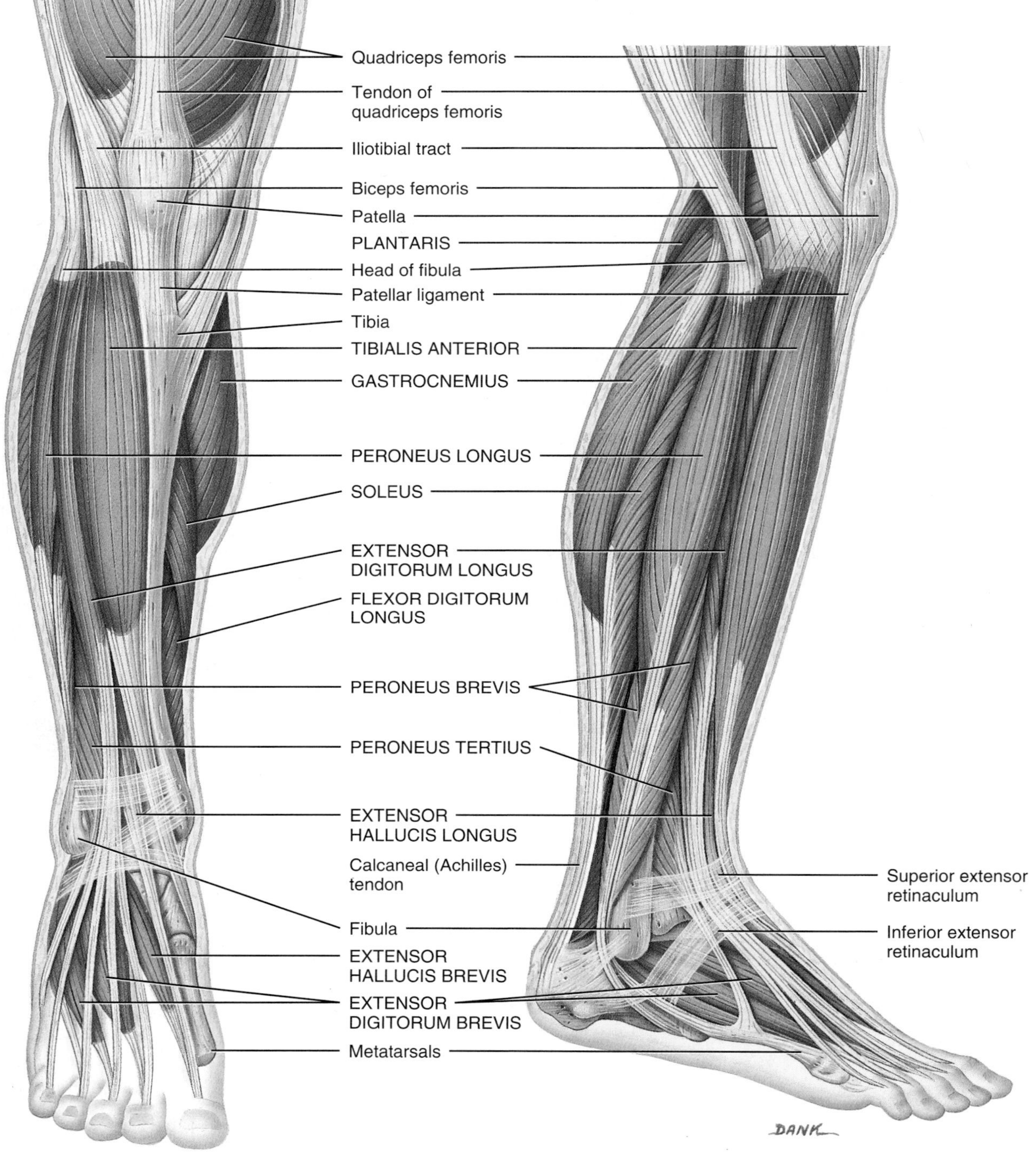

(a) Anterior superficial view

(b) Right lateral superficial view

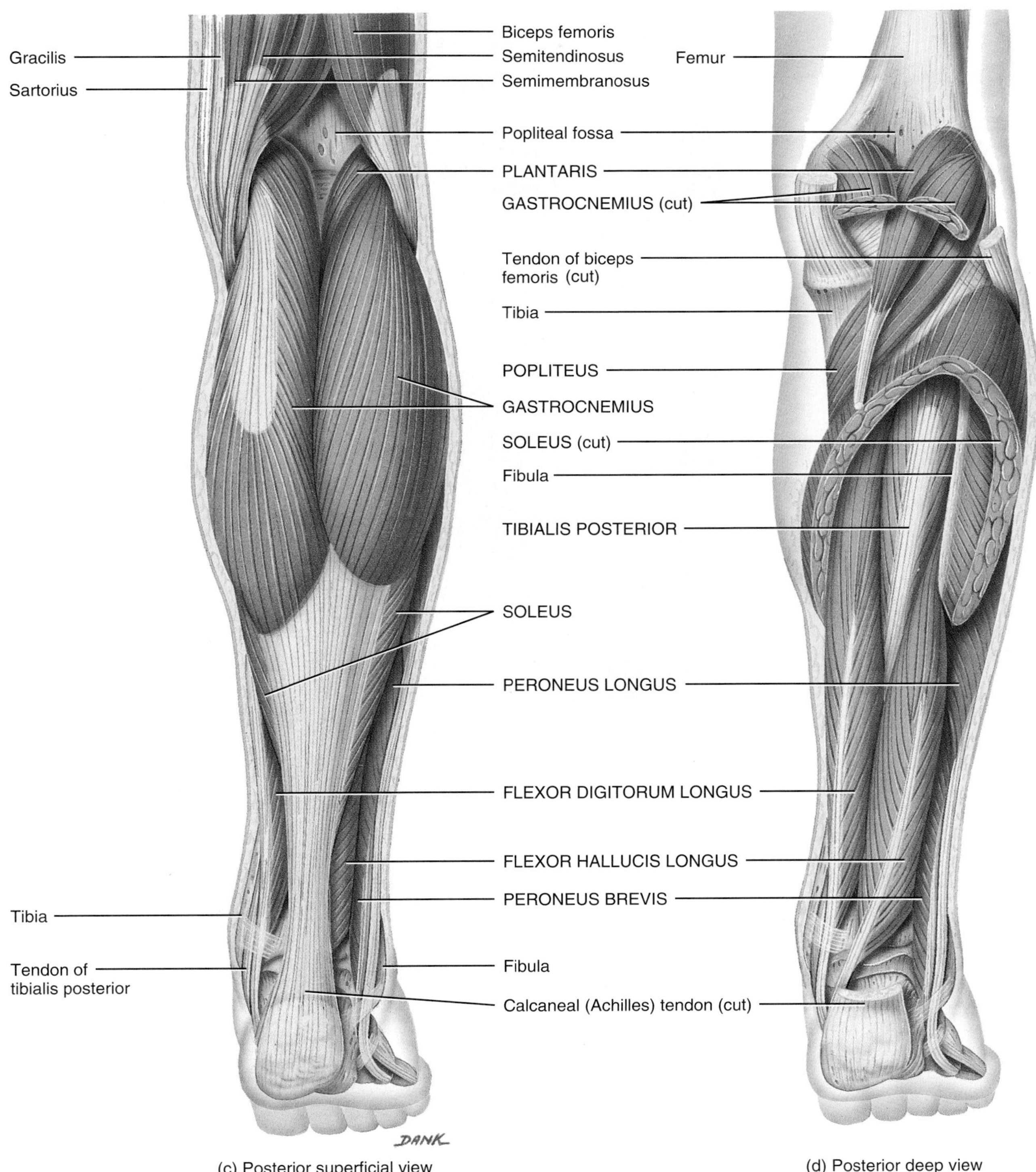

(c) Posterior superficial view

(d) Posterior deep view

(continues)

Exhibit 10.21 Muscles That Move the Foot and Toes (Figure 10.24) continued

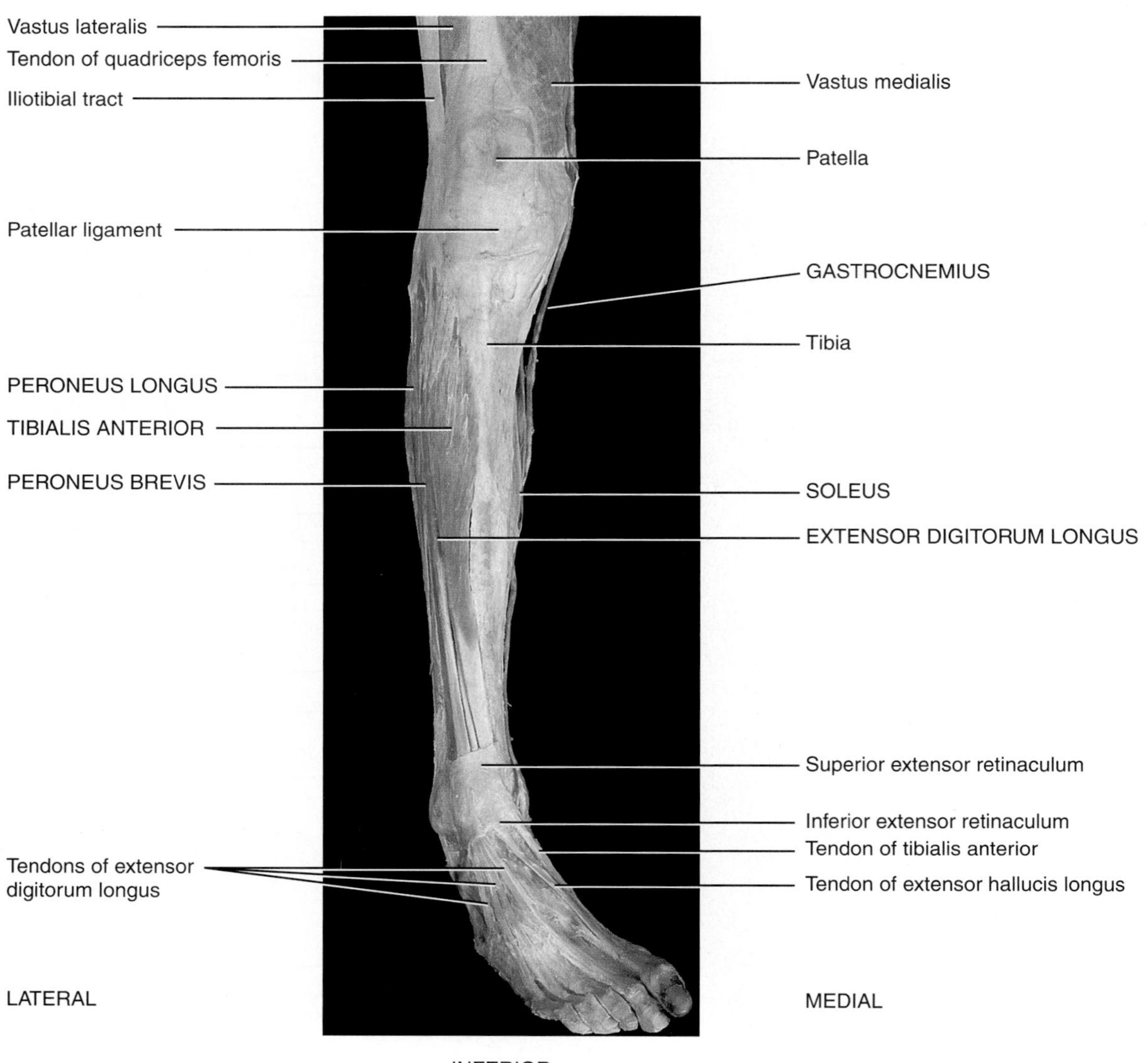

(e) Anterior superficial view

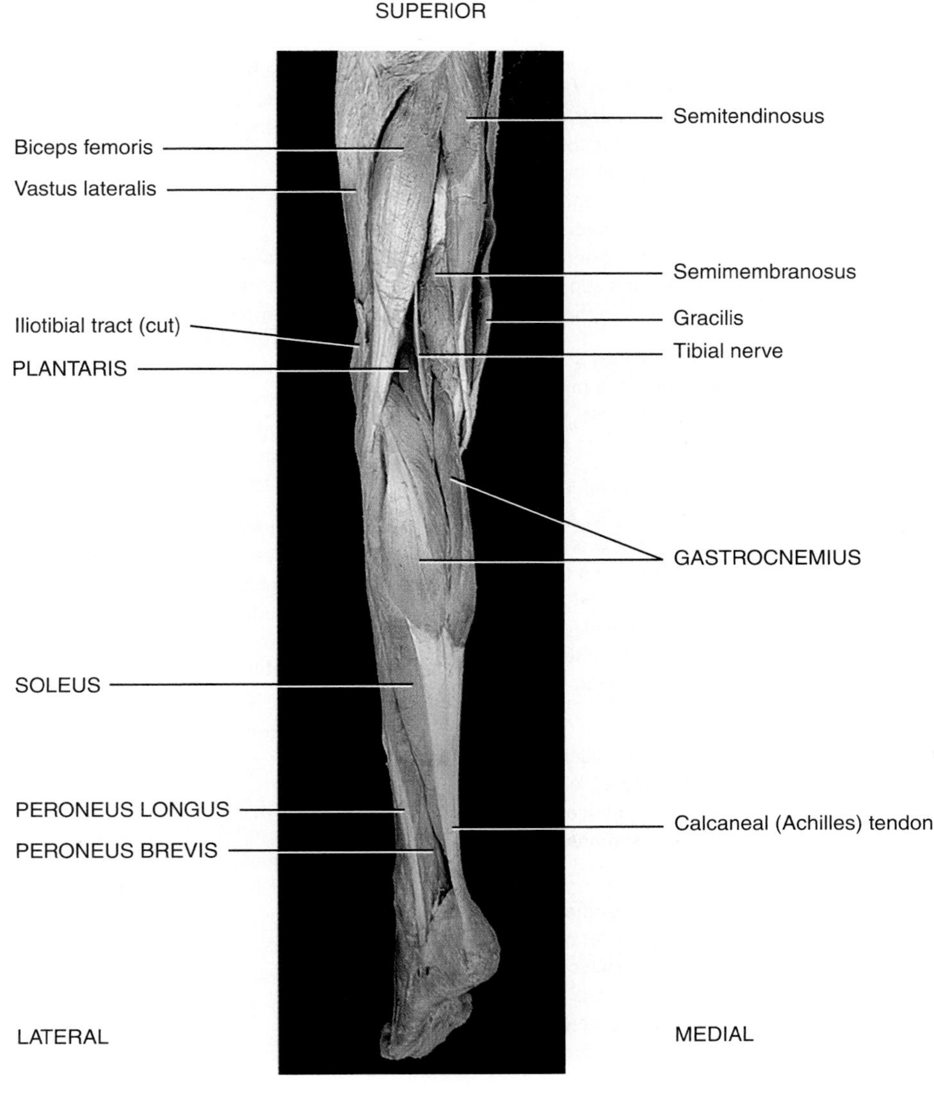

(f) Posterior superficial view

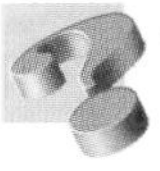

What structures firmly hold the tendons of the anterior compartment muscles to the ankle?

Exhibit 10.22 Intrinsic Muscles of the Foot (Figure 10.25)

Objective

► Describe the origin, insertion, action, and innervation of the intrinsic muscles of the foot.

Several of the muscles discussed in Exhibit 10.21 move the toes in various ways and are known as extrinsic muscles. The muscles in this exhibit are called **intrinsic muscles** because they originate and insert *within* the foot. The intrinsic muscles of the foot are, for the most part, comparable to those in the hand. Whereas the muscles of the hand are specialized for precise and intricate movements, those of the foot are limited to support and locomotion. The deep fascia of the foot forms the **plantar aponeurosis (fascia)** that extends from the calcaneus bone to the phalanges of the toes. The aponeurosis supports the longitudinal arch of the foot and encloses the flexor tendons of the foot.

The intrinsic muscles of the foot are divided into two groups: **dorsal** and **plantar.** There is only one dorsal muscle, the **extensor digitorum brevis,** a four-part muscle deep to the tendons of the extensor digitorum longus muscle, which extends toes 2–5 at the metatarsophalangeal joints.

The plantar muscles are arranged in four layers, the more superficial layer being referred to as the **first layer.** The muscles in the first layer are the **abductor hallucis,** which lies along the medial border of the sole, is comparable to the abductor pollicis brevis in the hand, and abducts the great toe at the metatarsophalangeal joint; the **flexor digitorum brevis,** which lies in the middle of the sole and flexes toes 2–5 at the interphalangeal and metatarsophalangeal joints; and the **abductor digiti minimi,** which lies along the lateral border of the sole, is comparable to the same muscle in the hand, and abducts the small toe.

The **second layer** consists of the **quadratus plantae,** a rectangular muscle that arises by two heads and flexes toes 2–5 at the metatarsophalangeal joints, and the **lumbricals,** four small muscles that are similar to the lumbricals in the hands. They flex the proximal phalanges and extend the distal phalanges of toes 2–5.

The **third layer** is composed of the **flexor hallucis brevis,** which lies adjacent to the plantar surface of the metatarsal of the great toe, is comparable to the same muscle in the hand, and flexes the great toe; the **adductor hallucis,** which has an oblique and transverse head like the adductor pollicis in the hand and adducts the great toe; and the **flexor digiti minimi brevis,** which lies superficial to the metatarsal of the small toe, is comparable to the same muscle in the hand, and flexes the small toe.

The **fourth layer** is the deepest and consists of the **dorsal interossei,** four muscles that adduct toes 2–4, flex the proximal phalanges, and extend the distal phalanges, and the three **plantar interossei,** which adduct toes 3–5, flex the proximal phalanges, and extend the distal phalanges. The interossei of the feet are similar to those of the hand, except that their actions are relative to the midline of the second digit rather than to the third as in the hand.

Plantar Fasciitis

Plantar fasciitis (fas-ē-Ī-tis) or **painful heel syndrome** is an inflammatory reaction due to chronic irritation of the plantar aponeurosis (fascia) at its origin on the calcaneus (heel bone). The aponeurosis becomes less elastic with age. This condition is also related to weight-bearing activities (walking, jogging, lifting heavy objects), improperly constructed or fitted shoes, excess weight (puts pressure on the feet), and poor biomechanics (flat feet, high arches, and abnormalities in gait may cause uneven distribution of weight on the feet). Plantar fasciitis is the most common cause of heel pain in runners and arises in response to the repeated impact of running. ■

Relating Muscles to Movements

Arrange the muscles in this exhibit according to the following actions on the great toe at the metatarsophalangeal joint: (1) flexion, (2) extension, (3) abduction, and (4) adduction; and according to the following actions on toes 2–5 at the metatarsophalangeal and interphalangeal joints: (1) flexion, (2) extension, (3) abduction, and (4) adduction. The same muscle may be mentioned more than once.

✔ How do the intrinsic muscles of the hand and foot differ in function?

Innervation

Intrinsic muscles of the foot are innervated by nerves derived from the sacral plexus (shown in Exhibit 17.4 on pages 537–538). More specifically, the extensor digitorum brevis is innervated by the deep peroneal nerve; the abductor hallucis and flexor digitorum brevis are innervated by the medial plantar nerve; the abductor digiti minimi, quadratus plantae, adductor hallucis, flexor digiti minimi brevis, and dorsal and plantar interossei are innervated by the lateral plantar nerve; and the lumbricals are innervated by the medial and lateral plantar nerves.

Muscle	Origin	Insertion	Action
Dorsal			
Extensor digitorum brevis (*extensor* = increases angle at joint; *digit* = finger or toe; *brevis* = short) (see Figure 10.24a, b)	Calcaneus and inferior extensor retinaculum.	Tendons of extensor digitorum longus on toes 2–4 and proximal phalanx of great toe.*	Extensor hallucis brevis extends great toe at metatarsophalangeal joint and extensor digitorum brevis extends toes 2–4 at interphalangeal joints.
Plantar			
First Layer (most superficial)			
Abductor hallucis (*abductor* = moves part away from midline; *hallucis* = hallux or great toe)	Calcaneus, plantar aponeurosis, and flexor retinaculum.	Medial side of proximal phalanx of great toe with the tendon of the flexor hallucis brevis.	Abducts and flexes great toe at metatarsophalangeal joint.
Flexor digitorum brevis (*flexor* = decreases angle at joint)	Calcaneus and plantar aponeurosis.	Sides of middle phalanx of toes 2–5.	Flexes toes 2–5 at proximal interphalangeal and metatarsophalangeal joints.
Abductor digiti minimi (*minimi* = little)	Calcaneus and plantar aponeurosis.	Lateral side of proximal phalanx of small toe with the tendon of the flexor digiti minimi brevis.	Abducts and flexes small toe at metatarsophalangeal joint.
Second Layer			
Quadratus plantae (*quad* = four; *planta* = sole)	Calcaneus.	Tendon of flexor digitorum longus.	Assists flexor digitorum longus to flex toes 2–5 at interphalangeal and metatarsophalangeal joints.
Lumbricals (LUM-bri-kals; *lumbricus* = earthworm)	Tendons of flexor digitorum longus.	Tendons of extensor digitorum longus on proximal phalanges of toes 2–5.	Extend toes 2–5 at interphalangeal joints and flex toes 2–5 at metatarsophalangeal joints.
Third Layer			
Flexor hallucis brevis	Cuboid and third (lateral) cuneiform.	Medial and lateral sides of proximal phalanx of great toe via a tendon containing a sesamoid bone.	Flexes great toe at metatarsophalangeal joint.
Adductor hallucis	Metatarsals 2–4, ligaments of 3–5 metatarsophalangeal joints, and tendon of peroneus longus.	Lateral side of proximal phalanx of great toe.	Adducts and flexes great toe at metatarsophalangeal joint.
Flexor digiti minimi brevis	Metatarsal 5 and tendon of peroneus longus.	Lateral side of proximal phalanx of small toe.	Flexes small toe at metatarsophalangeal joint.
Fourth Layer (deepest)			
Dorsal interossei (not illustrated)	Adjacent side of metatarsals.	Proximal phalanges: both sides of toe 2 and lateral side of toes 3 and 4.	Abduct and flex toes 2–4 at metatarsophalangeal joints and extend toes at interphalangeal joints.
Plantar interossei	Metatarsals 3–5.	Medial side of proximal phalanges of toes 3–5.	Adduct and flex proximal metatarsophalangeal joints and extend toes at interphalangeal joints.

*The tendon that inserts into the proximal phalanx of the great toe, together with its belly, is frequently described as a separate muscle, the extensor hallucis brevis.

(continues)

Exhibit 10.22 Intrinsic Muscles of the Foot (Figure 10.25) continued

Figure 10.25 / Intrinsic muscles of the foot.

Whereas the muscles of the hand are specialized for precise and intricate movements, those of the foot are limited to support and movement.

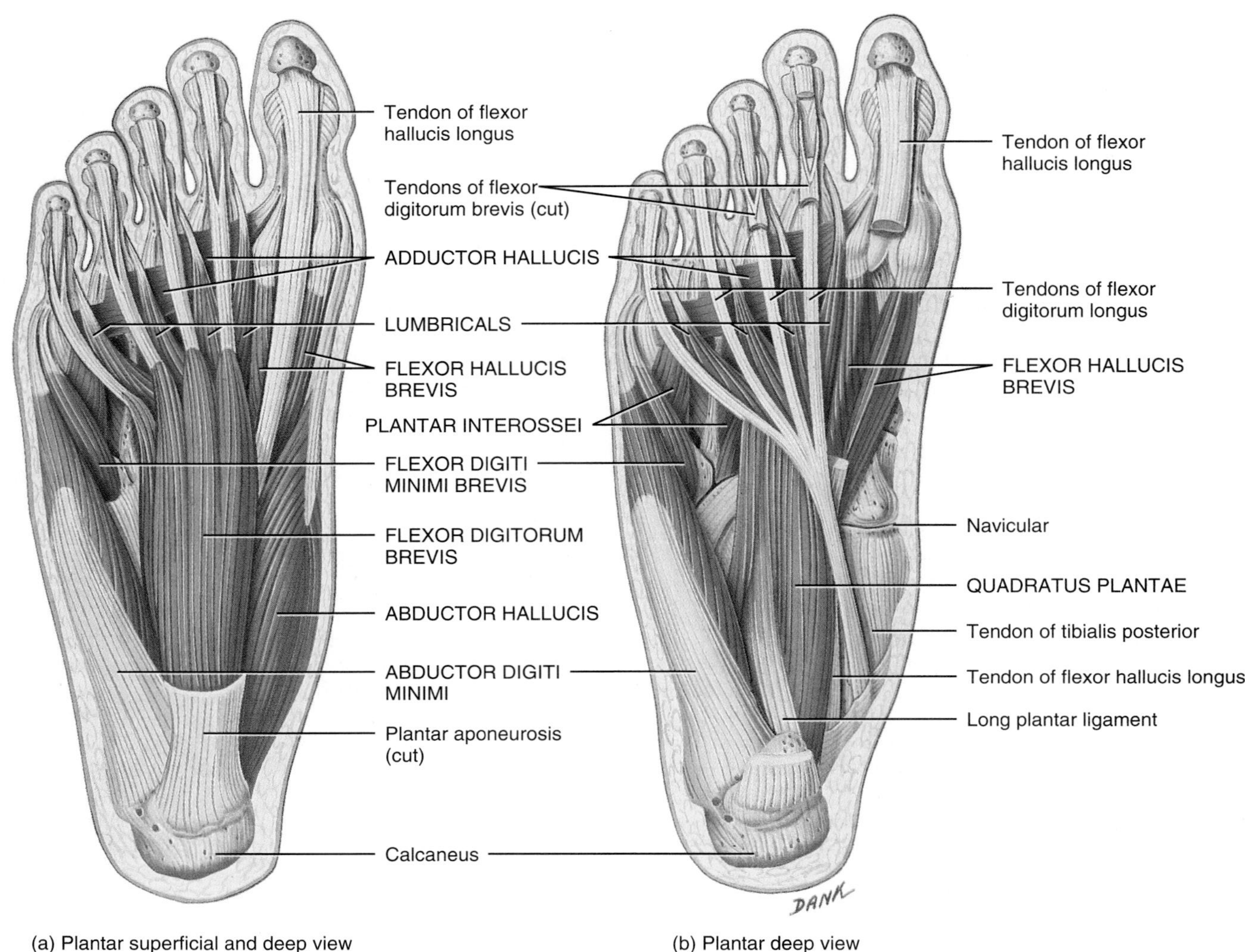

(a) Plantar superficial and deep view

(b) Plantar deep view

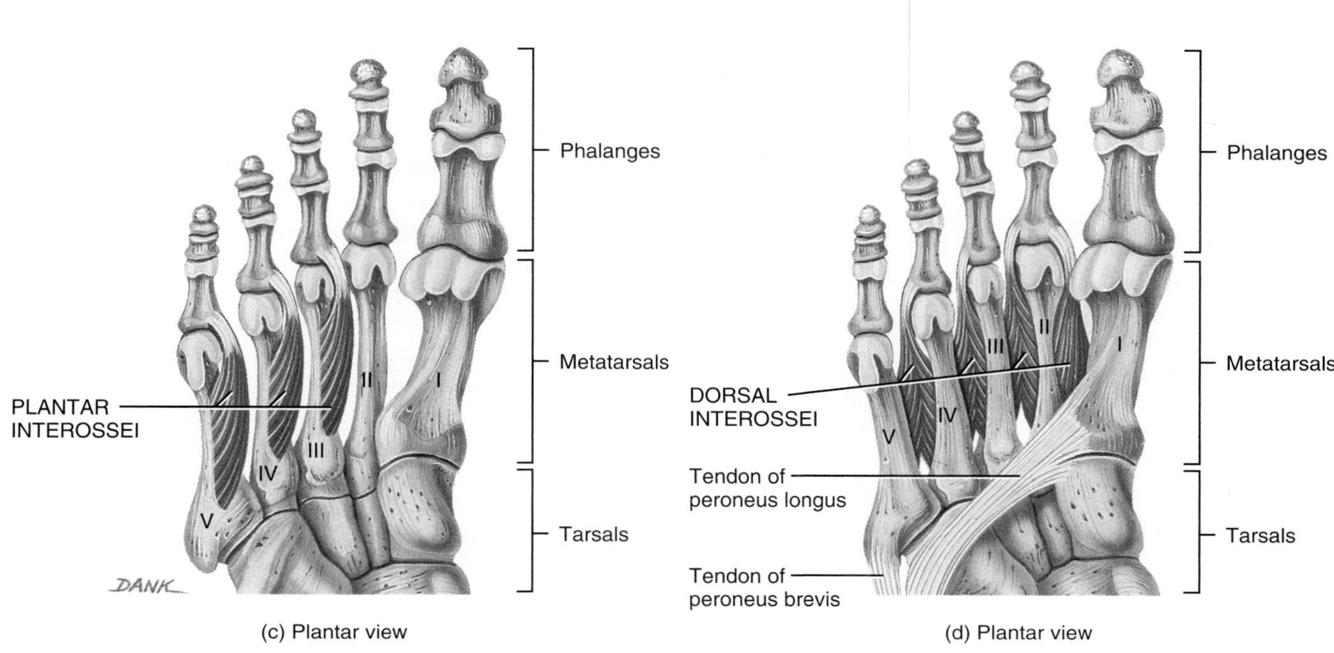

(c) Plantar view

(d) Plantar view

What structure supports the longitudinal arch and encloses the flexor tendons of the foot?

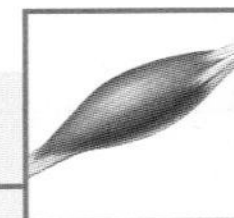

APPLICATIONS TO HEALTH

Running Injuries

It is estimated that nearly 70% of those who jog or run will sustain some type of running-related injury. Even though most such injuries are minor, some are quite serious; moreover, untreated or inappropriately treated minor injuries may become chronic. Among runners, common sites of injury include the ankle, knee, calcaneal (Achilles) tendon, hip, groin, foot, and back. Of these, the knee often is the most severely injured area.

Running injuries are frequently related to faulty training techniques. This may involve improper (or lack of) warm-up routines, running too much, or running too soon following an injury. Or it might involve extended running on hard and/or uneven surfaces. Poorly constructed or worn-out running shoes can also contribute to injury. Any biomechanical problem (such as a fallen arch) aggravated by running can also contribute to injuries.

Most sports injuries should be treated initially with RICE therapy, which stands for Rest, Ice, Compression, and Elevation. Immediately apply ice, and rest and elevate the injured part. Then apply an elastic bandage, if possible, to compress the injured tissue. Continue using RICE for 2–3 days, and resist the temptation to apply heat, which may worsen the swelling. Follow-up treatment may include alternating moist heat and ice massage to enhance blood flow in the injured area. Sometimes, nonsteroidal anti-inflammatory drugs (NSAIDs) or local injections of corticosteroids are needed. During the recovery period, it is important to keep active using an alternate fitness program that does not worsen the original injury. This activity should be determined in consultation with a physician. Finally, careful exercise is needed to rehabilitate the injured area itself.

Compartment Syndrome

In several exhibits in this chapter we noted that skeletal muscles in the limbs are organized into units called compartments. This compartmental arrangement causes problems for some individuals under certain conditions. In a disorder called **compartment syndrome,** some external or internal pressure constricts the structures within a compartment, resulting in damaged blood vessels and subsequent reduction of the blood supply (ischemia) to the structures within the compartment. Common causes of compartment syndrome include crushing and penetrating injuries, contusion (damage to subcutaneous tissues without the skin being broken), muscle strain (overstretching of a muscle), or an improperly fitted cast. As a result of hemorrhage, tissue injury, and edema (buildup of interstitial fluid), the pressure increase in the compartment can have serious consequences. Because the fasciae that enclose the compartments are very strong, accumulated blood and interstitial fluid cannot escape, and the increased pressure can literally choke off the blood flow and deprive nearby muscles and nerves of oxygen. One treatment option is *fasciotomy* (fash-ē-OT-ō-me), a surgical procedure in which muscle fascia is cut to relieve the pressure. Without intervention, nerves suffer damage, and muscles develop scar tissue that results in permanent shortening of the muscles, a condition called *contracture.*

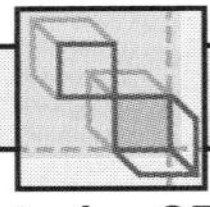

STUDY OUTLINE

How Skeletal Muscles Produce Movements (p. 254)

1. Skeletal muscles that produce movements do so by pulling on bones.
2. The attachment to the stationary bone is the origin; the attachment to the movable bone is the insertion.
3. Bones serve as levers, and joints serve as fulcrums. The lever is acted on by two different forces: resistance and effort.
4. Levers are categorized into three types—first-class, second-class, and third-class (most common)—according to the positions of the fulcrum, the effort, and the resistance on the lever.
5. Fascicular arrangements include parallel, fusiform, circular, triangular, and pennate. Fascicular arrangement affects a muscle's power and range of motion.
6. The prime mover produces the desired action; the antagonist produces an opposite action. The synergist assists the prime mover by reducing unnecessary movement. The fixator stabilizes the origin of the prime mover so that it can act more efficiently.

How Skeletal Muscles Are Named (p. 258)

1. Skeletal muscles are named on the basis of distinctive criteria: direction of muscle fibers; size, shape, action, number of origins (or heads), and location of the muscle; and sites of origin and insertion of the muscle.
2. Most skeletal muscles are named on the basis of combinations of these criteria.

Principal Skeletal Muscles (p. 263)

1. Muscles of facial expression move the skin rather than a joint when they contract, and they permit us to express a wide variety of emotions.
2. The extrinsic muscles that move the eyeballs are among the fastest contracting and most precisely controlled skeletal muscles in the body. They permit us to elevate, depress, abduct, adduct, and medially and laterally rotate the eyeballs.
3. Muscles that move the mandible (lower jaw) are also known as the muscles of mastication because they are involved in chewing.
4. The extrinsic muscles that move the tongue are important in chewing, swallowing, and speech.
5. Muscles of the floor of the oral cavity (mouth) are called suprahyoid muscles because they are located above the hyoid bone. They elevate the hyoid bone, oral cavity, and tongue during swallowing.

6. Muscles that move the head alter the position of the head and help balance the head on the vertebral column.
7. Most muscles of the larynx (voice box) depress the hyoid bone and larynx during swallowing and speech and move the internal portion of the larynx during speech.
8. Muscles of the pharynx are arranged into a circular layer, which functions in swallowing, and a longitudinal layer, which functions in swallowing and speaking.
9. Muscles that act on the abdominal wall help contain and protect the abdominal viscera, move the vertebral column, compress the abdomen, and produce the force required for defecation, urination, vomiting, and childbirth.
10. Muscles used in breathing alter the size of the thoracic cavity so that ventilation can occur and assist in venous return of blood to the heart.
11. Muscles of the pelvic floor support the pelvic viscera, resist the thrust that accompanies increases in intraabdominal pressure, and function as sphincters at the anorectal junction, urethra, and vagina.
12. Muscles of the perineum assist in urination, erection of the penis and clitoris, ejaculation, and defecation.
13. Muscles that move the pectoral (shoulder) girdle stabilize the scapula so it can function as a stable point of origin for most of the muscles that move the humerus.
14. Muscles that move the humerus (arm) originate for the most part on the scapula (scapular muscles); the remaining muscles originate on the axial skeleton (axial muscles).
15. Muscles that move the radius and ulna (forearm) are involved in flexion and extension at the elbow joint and are organized into flexor and extensor compartments.
16. Muscles that move the wrist, hand, and digits are many and varied, and those muscles that act on the digits are called extrinsic muscles.
17. The intrinsic muscles of the hand are important in skilled activities and provide humans with the ability to grasp precisely and manipulate objects.
18. Muscles that move the vertebral column are quite complex because they have multiple origins and insertions and because there is considerable overlap among them.
19. Muscles that move the femur (thigh) originate for the most part on the pelvic girdle and insert on the femur, and these muscles are larger and more powerful than comparable muscles in the upper limb.
20. Muscles that move the femur (thigh) and tibia and fibula (leg) are separated into medial (adductor), anterior (extensor), and posterior (flexor) compartments.
21. Muscles that move the foot and toes are divided into anterior, lateral, and posterior compartments.
22. Intrinsic muscles of the foot, unlike those of the hand, are limited to the functions of support and locomotion.

SELF-QUIZ QUESTIONS

Choose the one best answer to the following questions:

1. Which of these statements is false when you hyperextend your head as if to look at the sky? (a) The weight of the face and jaw serves as the R (resistance). (b) The posterior neck muscles provide the E (effort). (c) The F (fulcrum) is the atlanto-occipital joint. (d) This is an example of a second-class lever. (e) All statements are false.

2. Which generalization concerning movement by skeletal muscles is *not* true? (a) Muscles produce movements by pulling on bones. (b) During contractions, the two articulating bones move equally. (c) The tendon attachment to the stationary bone is the origin. (d) Bones serve as levers and joints as fulcrums of the levers. (e) In the limbs, the insertion is usually distal.

3. Which muscles are contracted during expiration? (a) diaphragm only, (b) diaphragm and internal intercostals, (c) internal intercostals only, (d) diaphragm and external intercostals, (e) none of the above.

4. Choose the muscle(s) that depress(es) the mandible. (a) lateral pterygoid, (b) digastric, (c) medial pterygoid, (d) all of the above, (e) both a and b.

5. The action of the pectoralis major muscle is to (a) abduct the arm and rotate the arm laterally, (b) flex, adduct, and rotate that arm medially, (c) adduct the arm and rotate the arm laterally, (d) abduct the arm and rotate the arm medially, (e) abduct and raise the arm.

6. The muscles that form the anterior compartment of the forearm (1) originate on the medial epicondyle of the humerus or portions of the radius and ulna (2) originate on the lateral epicondyle of the humerus or portions of the radius and ulna, (3) flex the forearm at the elbow, (4) flex the wrist and fingers, (5) extend the forearm at the elbow, (6) extend the wrist and fingers.
   a. 1, 3 and 4, **b.** 2, 5 and 6 **c.** 1 and 6 **d.** 2 and 4, **e.** 1 and 4

7. Muscles of which group share a common origin on the ischium and act to extend the thigh and flex the leg? (a) gluteal muscles, (b) quadriceps muscles, (c) adductor muscles, (d) hamstring muscles, (e) peroneal muscles

Complete the following:

8. A muscle that contracts to cause the desired action is called the ________ . Muscles that assist or cooperate with the muscle that causes the desired action are known as ________ .

9. Fascicular arrangement may take one of five characteristic patterns: ________ , ________ , ________ , ________ , or ________ .

10. Name the skeletal muscle that: (a) depresses the mandible, (b) closes the eye, (c) assists in mastication by pressing the cheeks against the teeth so that food is kept between the teeth, (d) elevates eyebrows and wrinkles the forehead to show surprise.

11. The ________ muscle, which assists in speech and swallowing, has two bellies united by an intermediate tendon held by a fibrous loop to the hyoid.

12. The sternocleidomastoid and longissimus capitis muscles insert on the ________ of the temporal bone.

13. The ________ tendon is the tendon of insertion for the diaphragm.

14. Three posterior thoracic muscles that adduct the scapula are ________ , ________ , and ________ .

15. Whereas the ________ muscles act on the thumb, the ________ , muscles act on the little finger.

16. The largest muscle mass of the back, the ________ , is the chief extensor of the vertebral column and helps to maintain the ________ curve.

17. The pectineus, psoas major, adductor longus, and iliacus muscles have one action in common: ________ of the thigh.

18. Quadriceps femoris is the main muscle mass on the ________ surface of the thigh. The name of this muscle mass indicates it has ________ heads of origin. All converge to insert on the ________ bone by means of the patellar ligament. All therefore act to ________ the leg.

19. The three muscles that utilize the calcaneal tendon to insert on the calcaneus are ________ , ________ , and ________ .

Are the following statements true or false?

20. The flexor retinaculum is located on the anterior surface of the wrist and serves to hold the tendons, nerves and blood vessels close to the bones of the wrist.

21. The superior oblique muscle originates from the orbit, passes through the trochlea and then re-inserts on the orbit.

22. The muscles that plantarflex the foot at the ankle joint are located in the anterior compartment of the leg.

23. Match each of these muscles with the clue(s) given by the parts of its name.

____ **(a)** transversus abdominis
____ **(b)** gluteus minimus
____ **(c)** biceps brachii
____ **(d)** sternocleidomastoid
____ **(e)** adductor magnus
____ **(f)** tibialis anterior
____ **(g)** rhomboideus major

(1) action
(2) direction of fibers
(3) location
(4) number of tendons of origin
(5) sites of origin and/or insertion
(6) size or shape

24. Match the following.

____ **(a)** FER
____ **(b)** FRE
____ **(c)** EFR
____ **(d)** for example, the elbow joint, the bones of the forearm, and the biceps brachii
____ **(e)** for example, the atlanto-occipital joint, the weight of the anterior skull, and the muscles of the posterior neck
____ **(f)** operates like a wheelbarrow
____ **(g)** the most common lever system in the body

(1) first-class lever
(2) second-class lever
(3) third-class lever

25. Match the following muscles with the appropriate nerves.

____ **(a)** most of the muscles of facial expression
____ **(b)** muscles that move the mandible
____ **(c)** most muscles that move the tongue
____ **(d)** most muscles of the pharynx
____ **(e)** muscles that move the head
____ **(f)** muscles of the antero-lateral abdominal wall
____ **(g)** diaphragm
____ **(h)** quadriceps femoris muscles
____ **(i)** muscles of the anterior compartment of the leg
____ **(j)** muscles of the lateral compartment of the leg
____ **(k)** muscles of the posterior compartment of the leg

(1) hypoglossal (XII) nerve
(2) thoracic nerves 7-12
(3) facial (VII) nerve
(4) tibial nerve
(5) femoral nerve
(6) trigeminal (V) nerve, mandibular division
(7) deep peroneal nerve
(8) accessory (XI) and cervical spinal nerves
(9) superficial peroneal nerve
(10) phrenic nerve
(11) vagus (X) nerve

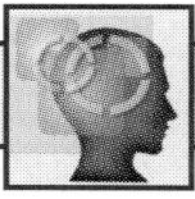

CRITICAL THINKING QUESTIONS

1. Three-year-old Ming likes to form her tongue into a cylinder shape and use it as a straw when she drinks her milk. Name the muscles that Ming uses to protrude her lips and tongue and to suck up the milk.
   HINT *"Glossus" means tongue.*
2. Baby Eddie weighed 13 lbs 4 oz at birth and he's been gaining ever since. His mom has noticed a sharp pain between her shoulder blades when she picks up Eddie to put him in his high chair (again). Which muscle may Eddie's mother have strained?
   HINT *She feels like she's going to pull her arms and shoulders right off her back if Eddie gets any bigger.*
3. A group of dental students were working out while singing "we might, we might, we might improve our bite!" Which muscles work to elevate and depress the jaw?
   HINT *Chewing gum gives these muscles a good work out.*
4. When asked to name their favorite muscle, most women in the class named a muscle they enjoyed viewing on men; most men in the class named a muscle they liked looking at in the mirror. They didn't name the same muscle. The women's favorite is one of the largest muscles in the body and extends the thigh. The men's choice has two origins and flexes the forearm. Identify the muscles.
   HINT *The men can't see the women's favorite when they admire themselves in the mirror.*
5. Baby Eddie's mother had a difficult time delivering the baby. Since the birth, she has been complaining of stress incontinence. What may be the cause of her problem?
   HINT *You need a pretty large passageway for 13 lb 4 oz baby.*

ANSWERS TO FIGURE QUESTIONS

10.1 The belly of the muscle that extends the forearm, the triceps brachii, is located posterior to the humerus.

10.2 Second-class levers produce the most force.

10.3 Possible correct responses: direction of fibers: external oblique; shape: deltoid; action: extensor digitorum; size: gluteus maximus; origin and insertion: sternocleidomastoid; location: tibialis anterior; number of tendons of origin: biceps brachii.

10.4 Frowning: frontalis; smiling: zygomaticus major; pouting: platysma; squinting: orbicularis oculi.

10.5 The inferior oblique muscle moves the eyeball superiorly and laterally because it originates at the anteromedial aspect of the floor of the orbit and inserts on the posterolateral aspect of the eyeball.

10.6 The masseter is the strongest of the chewing muscles.

10.7 Functions of the tongue include chewing, perception of taste, swallowing, and speech.

10.8 The suprahyoid and infrahyoid muscles fix the hyoid bone to assist in tongue movements.

10.9 Triangles are important because of the structures that lie within their boundaries.

10.10 Producing high and low tones.

10.11 The longitudinal muscles elevate the larynx, whereas the infrahyoid muscles depress the larynx.

10.12 The rectus abdominis muscle aids in urination.

10.13 The diaphragm is innervated by the phrenic nerve.

10.14 The borders of the pelvic diaphragm are the pubic symphysis anteriorly, the coccyx posteriorly, and the walls of the pelvis laterally.

10.15 The borders of the perineum are the pubic symphysis anteriorly, the coccyx posteriorly, and the ischial tuberosities laterally.

10.16 The main action of the muscles that move the pectoral girdle is to stabilize the scapula to assist in movements of the humerus.

10.17 The rotator cuff consists of the flat tendons of the subscapularis, supraspinatus, infraspinatus, and teres minor muscles that form a nearly complete circle around the shoulder joint.

10.18 Most powerful forearm flexor: brachialis; most powerful forearm extensor: triceps brachii.

10.19 Flexor tendons of the digits and wrist and the median nerve pass through the flexor retinaculum.

10.20 Muscles of the thenar eminence act on the thumb.

10.21 The splenius muscles extend laterally and upward from the ligamentum nuchae at the midline.

10.22 Upper limb muscles exhibit diversity of movement; lower limb muscles function in stability, locomotion, and maintenance of posture. Moreover, lower limb muscles usually cross two joints and act equally on both.

10.23 Quadriceps femoris: rectus femoris, vastus lateralis, vastus medialis, and vastus intermedius; hamstrings: biceps femoris, semitendinosus, and semimembranosus.

10.24 The superior and inferior extensor retinacula firmly hold the tendons of the anterior compartment muscles to the ankle.

10.25 The plantar aponeurosis supports the longitudinal arch and encloses the flexor tendons of the foot.

11

SURFACE ANATOMY

CONTENTS AT A GLANCE

◄ Page 354

You have now studied the skeletal and muscular systems. With that knowledge, can you identify any surface anatomy landmarks in this painting?

Page 359 ►

INTRODUCTION

In Chapter 1 we introduced several branches of anatomy and noted their relationship to how we understand the body's structure. In this chapter we will take a closer look at surface anatomy. Recall that **surface anatomy** is the study of the anatomical landmarks on the exterior of the body. A knowledge of surface anatomy will not only help you to identify structures on the body's exterior, but it will also assist you in locating the positions of various internal structures.

The study of surface anatomy involves two related, although distinct, activities: visualization and palpation. **Visualization** involves looking in a very selective and purposeful manner at a specific part of the body. **Palapation** (pal-PĀ-shun) means using the sense of touch to determine the location of an internal part of the body through the skin. Like visualization, palpation is performed selectively and purposefully, and it supplements information already gained through other methods, including visualization. Because of the varying depth of different structures from the surface of the body and due to differences in the thickness of the subcutaneous layer over different parts of the body between females (thicker) and males (thinner), palpations may range from light, to moderate, to deep.

Knowledge of surface anatomy has many applications, both anatomically and clinically. From an anatomical viewpoint, a knowledge of surface anatomy can provide valuable information regarding the location of structures such as bones, muscles, blood vessels, nerves, lymph nodes, and internal organs. Clinically, a knowledge of surface anatomy is the basis for conducting a physical examination and performing certain diagnostic tests. Health care professionals use their understanding of surface anatomy to learn where to take the pulse, measure blood pressure, draw blood, stop bleeding, insert needles and tubes, make surgical incisions, reduce (straighten) fractured bones, and listen to sounds made by the heart, lungs, and intestines. Clinicians also draw on a knowledge of surface anatomy to assess the status of lymph nodes and to identify the presence of tumors or other unusual masses in the body.

Recall that there are five principal regions of the body: (1) head, (2) neck, (3) trunk, (4) upper limbs, and (5) lower limbs. Using these regions as our guide, we will first study the surface anatomy of the head and then discuss the other regions, concluding with the lower limbs. Because living bodies are best suited for studying surface anatomy, you will find it helpful to visualize and/or palpate the various structures described on your own body as you study each region. Following a brief introduction to a region of the body, you will be presented with a list of the prominent structures to locate. This directed approach will help to organize your learning efforts. Labeled photographs illustrate most of the structures listed in each region.

SURFACE ANATOMY OF THE HEAD

Objective

- Describe the surface anatomy features of the head.

The **head** (*cephalic region*, or *caput*) contains the brain and sense organs (eyes, ears, nose, and tongue) and is divided into the cranium and face. The **cranium** (*skull*, or *brain case*) is that portion of the head that surrounds and protects the brain; the **face** is the anterior portion of the head.

Regions of the Cranium and Face

The cranium and face are divided into several regions, which are described here and illustrated in Figure 11.1.

- **Frontal region:** forms the front of the skull and includes the frontal bone.
- **Parietal region:** forms the crown of the skull and includes the parietal bones.
- **Temporal region:** forms the sides of the skull and includes the temporal bones.
- **Occipital region:** forms the base of the skull and includes the occipital bone.
- **Orbital (ocular) region:** includes the eyeballs, eyebrows, and eyelids.
- **Infraorbital region:** the region inferior to the orbit.
- **Zygomatic region:** the region inferolateral to the orbit that includes the zygomatic bones.
- **Nasal region:** region of the nose.
- **Oral region:** region of the mouth.
- **Mental region:** anterior portion of the mandible.
- **Buccal region:** region of the cheek.
- **Auricular region:** region of the external ear.

In addition to locating the various regions of the head, it is also possible to visualize and/or palpate certain structures within each region. We will examine various bone and muscle structures of the head and then look at specific areas such as the eyes, ears, and nose.

Many of the bony landmarks of the head and muscles of facial expression are labeled in (Figure 11.2 on page 347). As you read about each feature, refer to the figure, and try to visualize and/or palpate each feature on your own head or on the head of a study partner.

Figure 11.1 / Principal regions of the cranium and face.

The head consists of an anterior face and a cranium that surrounds the brain.

(a) Anterior view

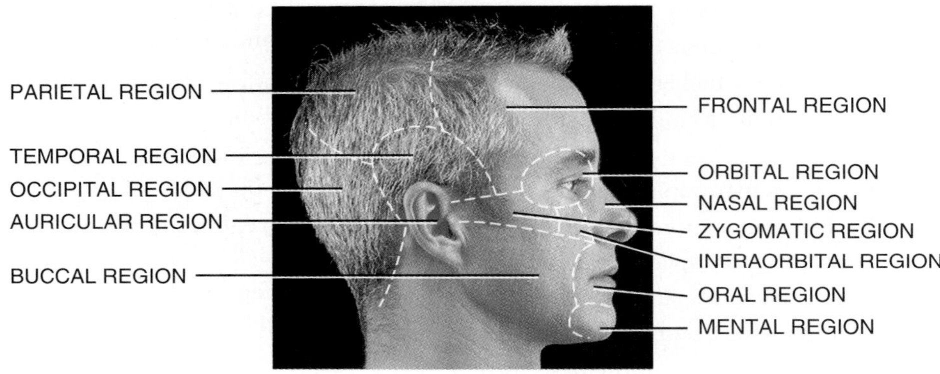

(b) Right lateral view

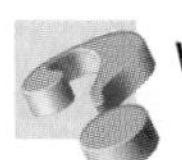
Which regions are named for bones that are deep to them?

Bony Landmarks of the Head

Several bony structures of the skull can be detected through palpation, including the following:

- **Sagittal suture.** Move the fingers from side to side over the superior aspect of the scalp in order to palpate this suture.
- **Coronal** and **lambdoid sutures.** Use the same side-to-side technique to palpate these structures located on the frontal and occipital regions of the skull.
- **External occipital protuberance.** This is the most prominent bony landmark on the occipital region of the skull.
- **Orbit.** Deep to the eyebrow, the superior aspect of the orbit, the **supraorbital margin** of the frontal bone, can be palpated. Actually, the entire circumference of the orbit can be palpated.
- **Nasal bones.** These can be palpated in the nasal region between the orbits on either side of the midline.
- **Mandible.** The **ramus** (vertical portion), **body** (horizontal portion), and **angle** (area where the ramus meets the body) of the mandible can be palpated easily at the mental and buccal regions of the head.
- **Zygomatic arch.** This can be palpated in the zygomatic region.

Muscles of Facial Expression

Several muscles of facial expression can be palpated while they are contracting:

- **Frontalis** and **occipitalis.** By raising and lowering the eyebrows, it is possible to palpate the two muscles in the frontal

Figure 11.2 / Surface anatomy of the head.

Several muscles of facial expression can be palpated while they are contracting.

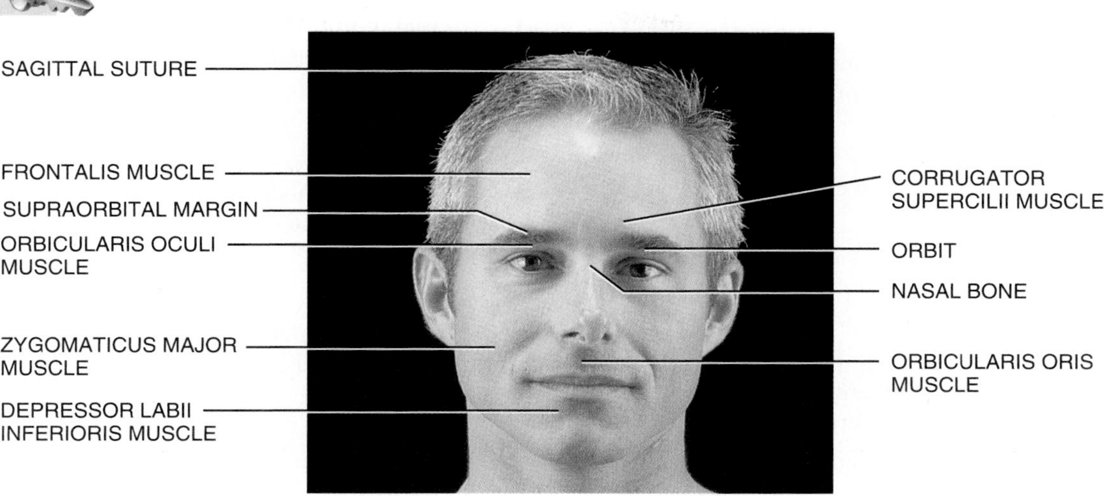

(a) Anterior view

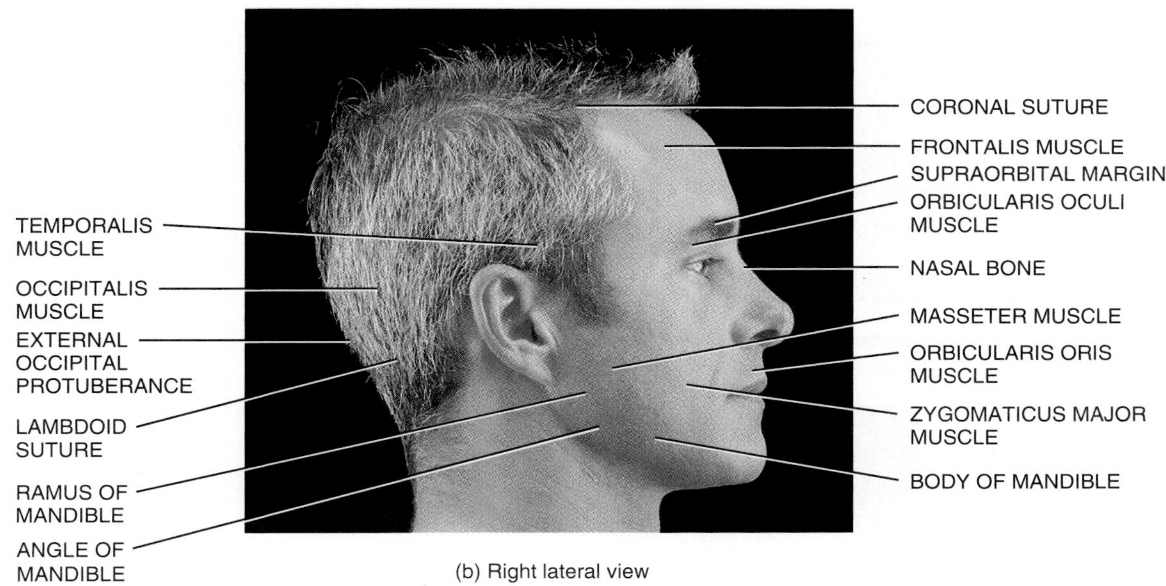

(b) Right lateral view

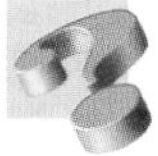

When palpating the sagittal suture, what sheetlike tendon is also palpated deep to the skin?

and occipital regions as they alternately contract and the scalp moves forward and backward.

- **Orbicularis oculi.** By closing the eyes and placing the fingers on the eyelids, the muscle can be palpated in the orbital region by tightly squeezing the eyes shut.
- **Corrugator supercilii.** By frowning, this muscle can be felt above the nose near the medial end of the eyebrow.
- **Zygomaticus major.** By smiling, this muscle can be palpated between the corner of the mouth and zygomatic (cheek) bone.
- **Depressor labii inferioris.** This muscle can be palpated in the mental region between the lower lip and chin when moving the lower lip inferiorly to expose the lower teeth.
- **Orbicularis oris.** By closing the lips tightly, this muscle can be palpated in the oral region around the margin of the lips.
- **Temporalis.** By alternately clenching the teeth and then opening the mouth, it is possible to palpate this muscle in the temporal region just superior to the zygomatic arch.

- **Masseter.** Again, by alternately clenching the teeth and then opening the mouth, this muscle can be palpated just superior and anterior to the angle of the mandible.

✔ What is surface anatomy? Why are visualization and palpation important in learning surface anatomy?
✔ What are some of the applications of a knowledge of surface anatomy?
✔ What are the divisions of the head? Describe the regions of the head and name one structure in each.

Surface Features of the Eyes

Now we will examine the surface features of the eyes, illustrated in Figure 11.3.

- **Iris** (= rainbow). Circular pigmented muscular structure behind the cornea (transparent covering over the iris).
- **Pupil.** Opening in the center of the iris through which light travels.
- **Sclera** (*skleros* = hard). "White" of the eye; a coat of fibrous tissue that covers the entire eyeball except for the cornea.
- **Conjunctiva.** Membrane that covers the exposed surface of the eyeball and lines the eyelids.
- **Eyelids (palpebrae).** Folds of skin and muscle lined on their inner aspect by conjunctiva.
- **Palpebral fissure.** Space between the eyelids when they are open; when the eye is closed, it lies just inferior to the level of the pupil.
- **Medial commissure.** Medial (near the nose) site of the union of the upper and lower eyelids.
- **Lateral commissure.** Lateral (away from the nose) site of the union of the upper and lower eyelids.
- **Lacrimal** (*lacrim* = tears) **caruncle.** Fleshy, yellowish projection of the medial commissure that contains modified sudoriferous (sweat) and sebaceous (oil) glands.
- **Eyelashes.** Hairs on the margin of the eyelids, usually arranged in two or three rows.
- **Eyebrows (supercilia).** Several rows of hairs superior to the upper eyelids.

Surface Features of the Ears

Next, locate the surface features of the ears in Figure 11.4.

- **Auricle.** Shell-shaped portion of the external ear; also called the *pinna.* It funnels sound waves into the external auditory canal. Just posterior to it, the **mastoid process** of the temporal bone can be palpated.
- **Tragus** (*tragos* = goat; hair that grows on the tragus resembles the beard of a goat). Cartilaginous projection anterior to the external auditory canal. Just anterior to the tragus and

Figure 11.3 / Surface anatomy of the right eye.

The eyeball is protected by the eyebrow, eyelashes, and eyelids.

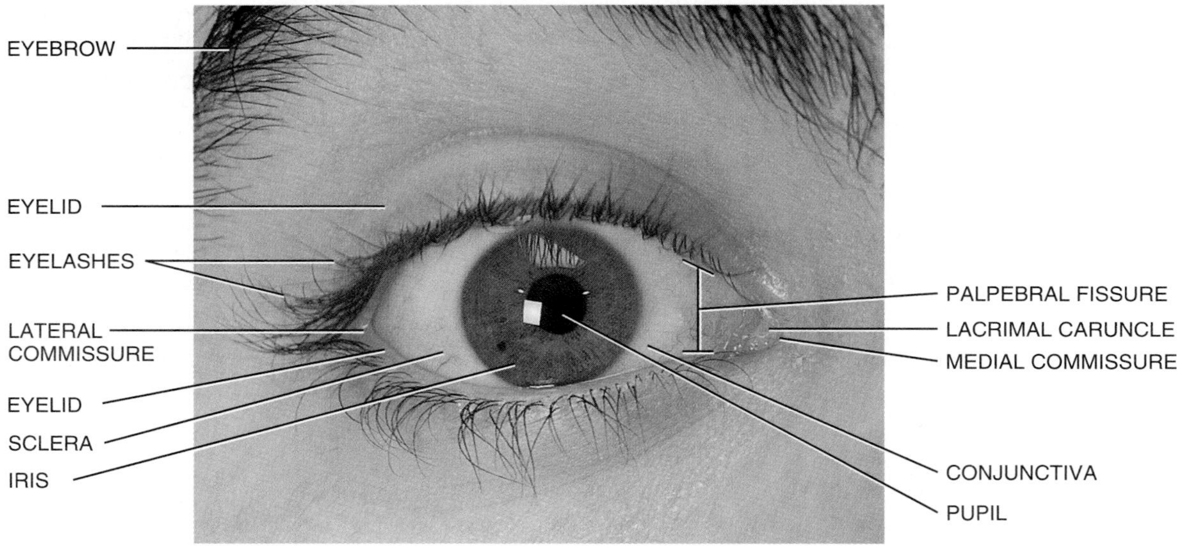

Anterior view

Through which part of the eye does light enter?

Figure 11.4 / Surface anatomy of the right ear.

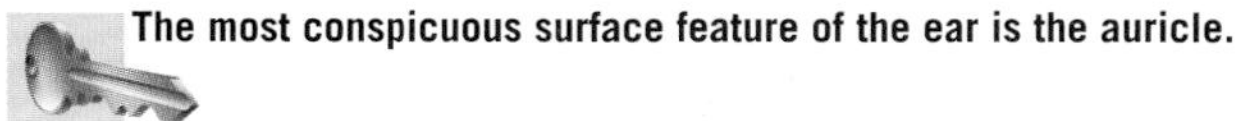

The most conspicuous surface feature of the ear is the auricle.

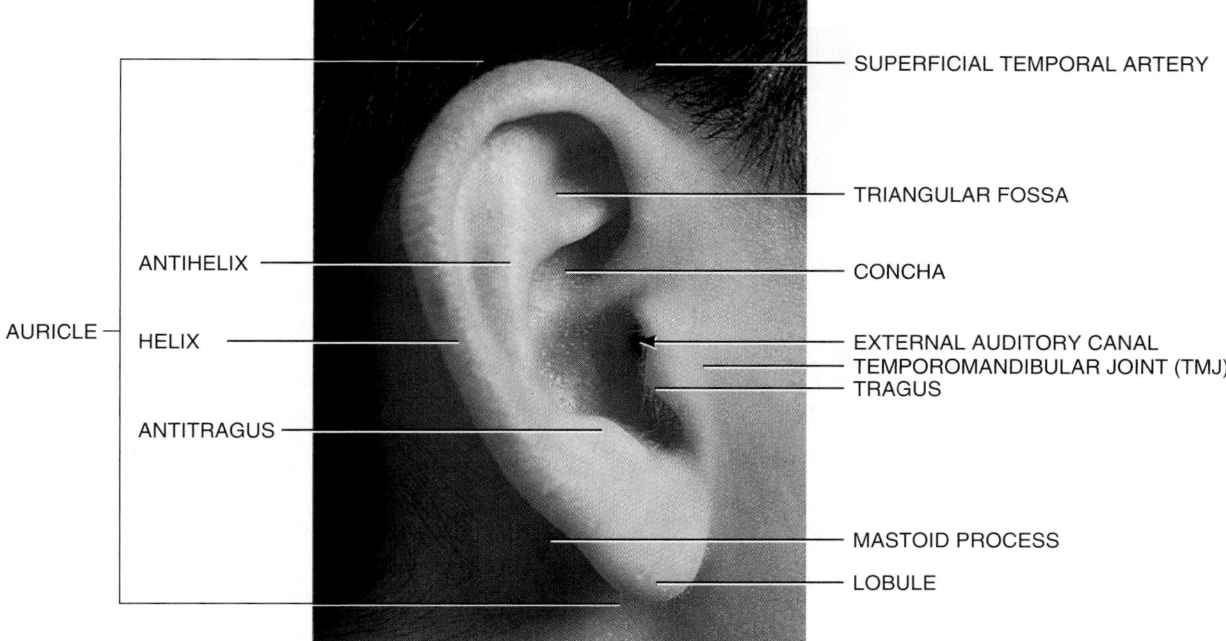

Right lateral view

Which part of the temporal bone can be palpated just posterior to the auricle?

between it and the neck of the mandible you can palpate the **superficial temporal artery** and feel the pulse in the vessel. Inferior to this point, directly anterior to the tragus, it is possible to feel the **temporomandibular joint (TMJ).** As you open and close your jaw, you can feel the mandibular condyle moving.

- **Antitragus.** Cartilaginous projection opposite the tragus.
- **Concha.** Hollow of the auricle.
- **Helix.** Superior and posterior free margin of the auricle.
- **Antihelix.** Semicircular ridge superior and posterior to the tragus.
- **Triangular fossa.** Depression in the superior portion of the antihelix.
- **Lobule.** Inferior portion of the auricle; it does not contain cartilage.
- **External auditory canal (meatus).** Canal about 3 cm (1 in.) long extending from the external ear to the eardrum (tympanic membrane). It contains ceruminous glands that secrete cerumen (earwax). The **condylar process** of the mandible can be palpated by placing your little finger in the canal and opening and closing your mouth.

Surface Features of the Nose and Mouth

To complete the discussion of the surface anatomy of the head, locate the following features of the nose and mouth (Figure 11.5):

- **Root.** Superior attachment of the nose at the forehead, between the eyes.
- **Apex.** Tip of the nose.
- **Dorsum nasi.** Rounded anterior border connecting the root and apex; in profile, it may be straight, convex, concave, or wavy.
- **External naris.** External opening into the nose.
- **Ala.** Convex flared portion of the inferior lateral surfaces of the nose.
- **Bridge.** Superior portion of the dorsum nasi, superficial to the nasal bones.
- **Philtrum** (FIL-trum; = depression on upper lip). The vertical groove on the upper lip that extends along the midline to the inferior portion of the nose.
- **Lips (labia).** Superior and inferior fleshy borders of the oral cavity.

Figure 11.5 / Surface anatomy of the nose and lips.

The external nares permit air to move into and out of the nose during breathing.

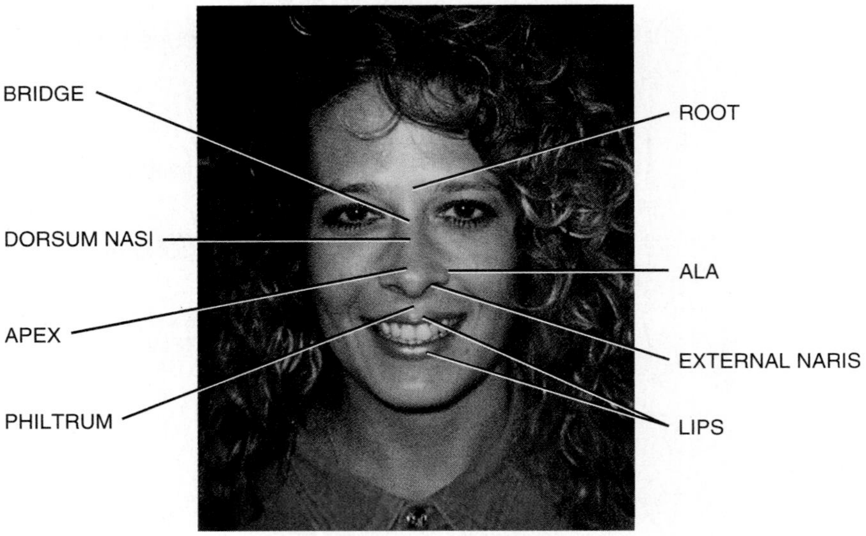

Anterior view

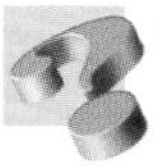

What is the name of the anterior border of the nose that connects the root and apex?

- ✔ List and define five surface features of the eyes, ears, and nose.
- ✔ How would you locate the superficial temporal artery, temporomandibular joint, zygomatic arch, mastoid process, and condylar process of the mandible?

SURFACE ANATOMY OF THE NECK

Objective

- Describe the surface anatomy features of the neck.

The **neck** *(cervical)* connects the head to the trunk and is divided into an *anterior cervical region*, two *lateral cervical regions*, and a *posterior cervical region (nucha)*.

The following are the major surface features of the neck, most of which are illustrated in Figure 11.6:

- **Thyroid cartilage (Adam's apple).** The largest of the cartilages that compose the larynx (voice box). It is the most prominent structure in the midline of the anterior cervical region.
- **Hyoid bone.** Located just superior to the thyroid cartilage. It is the first structure palpated in the midline inferior to the chin.
- **Cricoid cartilage.** An inferior laryngeal cartilage located just inferior to the thyroid cartilage, it attaches the larynx to the trachea (windpipe). After you pass your finger over the cricoid cartilage moving inferiorly, your fingertip sinks in. The cricoid cartilage is used as a landmark for performing a tracheostomy, which is described on page 694.
- **Thyroid gland.** Two-lobed gland with one lobe on either side of the trachea.
- **Sternocleidomastoid muscle.** Forms the major portion of the lateral aspect of the neck. If you rotate your head to either side, you can palpate the muscle from its origin on the sternum and clavicle to its insertion on the mastoid process of the temporal bone.
- **Common carotid artery.** Lies just deep to the sternocleidomastoid muscle along its anterior border.
- **Internal jugular vein.** Located lateral to the common carotid artery.
- **Subclavian artery.** Located just lateral to the inferior portion of the sternocleidomastoid muscle. Pressure on this artery can stop bleeding in the upper limb because this vessel supplies blood to the entire limb.
- **External carotid artery.** Located superior to the larynx, just anterior to the sternocleidomastoid muscle, this artery is the site of the carotid (neck) pulse.
- **External jugular vein.** Located superficial to the sternocleidomastoid muscle, this vessel is readily seen if you are angry or your collar is too tight.

Figure 11.6 / Surface anatomy of the neck.

The anatomical subdivisions of the neck are the anterior cervical region, lateral cervical regions, and posterior cervical region.

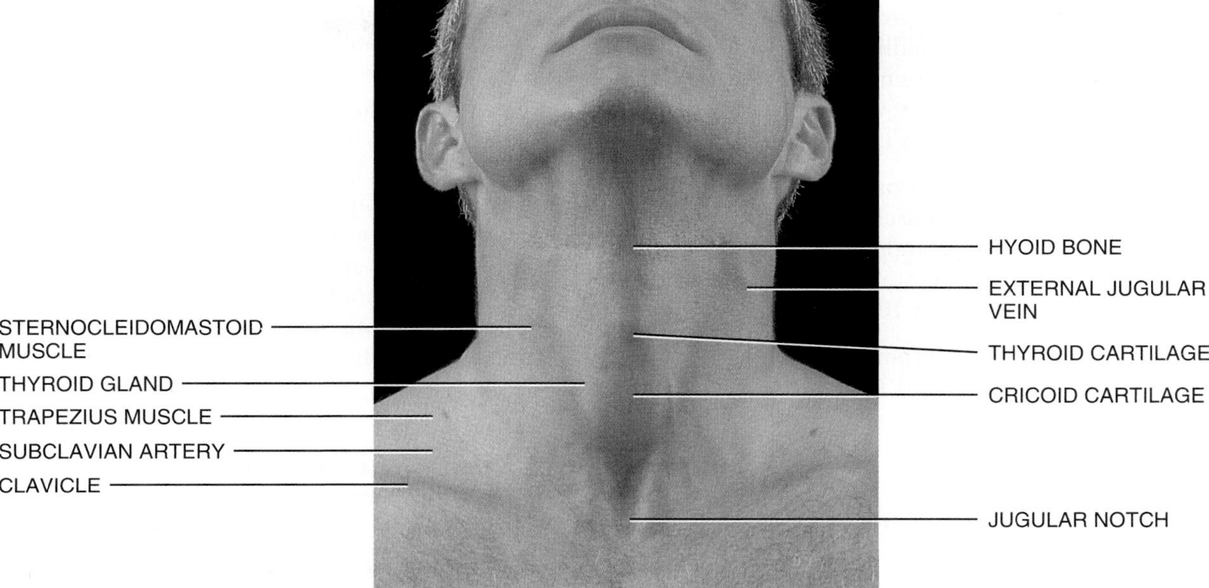

(a) Anterior view of surface anatomy features

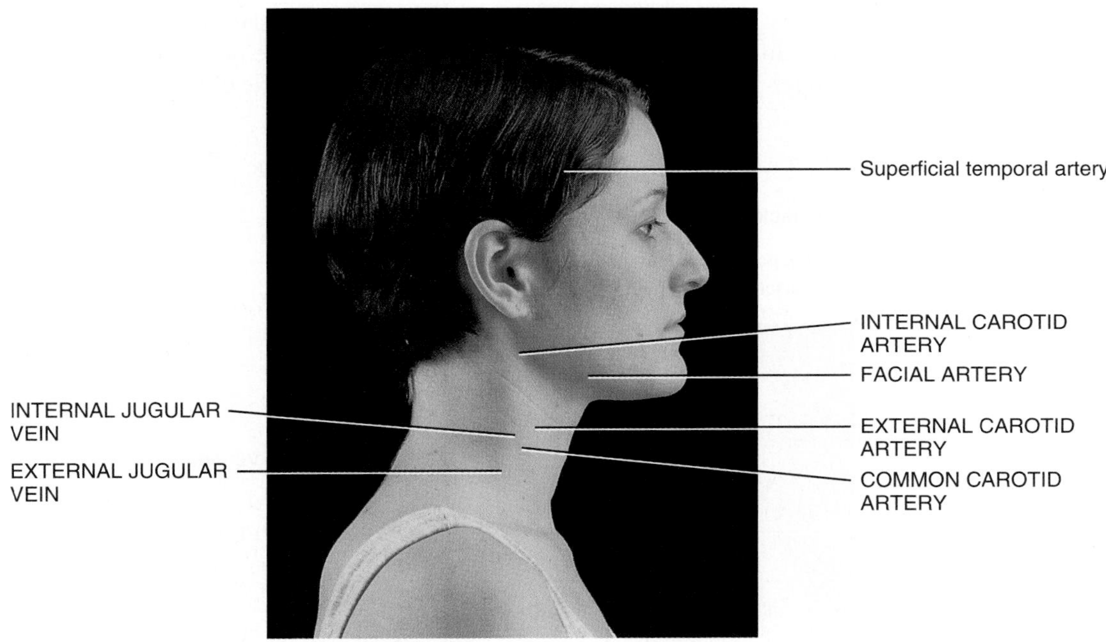

(b) Right lateral view

Which muscle divides the neck into anterior and posterior triangles?

- **Trapezius muscle.** Extends inferiorly and laterally from the base of the skull and occupies a portion of the lateral cervical region. A "stiff neck" is frequently associated with inflammation of this muscle.
- **Vertebral spines.** The spinous processes of the cervical vertebrae may be felt along the midline of the posterior region of the neck. Especially prominent at the base of the neck is the spinous process of the seventh cervical vertebra, called the vertebra prominens (see Figure 11.7).

The sternocleidomastoid muscle is not only a landmark for several arteries and veins, it is also the muscle that divides the neck into two major triangles: anterior and posterior (see Figure 10.9). The triangles are important because of the structures that lie within their boundaries (see Exhibit 10.6 on page 278).

✔ What is the neck? Why is it important?
✔ How would you palpate the hyoid bone, cricoid cartilage, thyroid gland, common carotid artery, external carotid artery, and external jugular vein?

SURFACE ANATOMY OF THE TRUNK

Objective

- Describe the surface anatomy of the various features of the trunk—back, chest, abdomen, and pelvis.

The **trunk** is divided into the back, chest, and abdomen and pelvis. We will consider each region separately.

Surface Features of the Back

Among the prominent features of the **back,** or *dorsum*, are several superficial bones and muscles (Figure 11.7).

- **Vertebral spines.** The spinous processes of vertebrae, especially the thoracic and lumbar vertebrae, are quite prominent when the vertebral column is flexed.
- **Scapulae.** These easily identifiable surface landmarks on the back lie between ribs 2–7. In fact, it is also possible to palpate some ribs on the back. Depending on how lean a person is, it might be possible to palpate various parts of the scapula, such as the **vertebral border, axillary border, inferior angle, spine,** and **acromion.** The **spinous process of T3** is at about the same level as the spine of the scapula, and the **spinous process of T7** is about opposite the inferior angle of the scapula.
- **Latissimus dorsi muscle.** A broad, flat, triangular muscle of the lumbar region that extends superiorly to the axilla.
- **Erector spinae (sacrospinalis) muscle.** Located on either side of the vertebral column between the twelfth ribs and iliac crests.
- **Infraspinatus muscle.** Located inferior to the spine of the scapula.
- **Trapezius muscle.** Extends from the cervical and thoracic vertebrae to the spine and acromion of the scapula and the lateral end of the clavicle. It also occupies a portion of the lateral region of the neck and forms the posterior border of the posterior triangle of the neck.

Figure 11.7 / Surface anatomy of the back.

The posterior boundary of the axilla, the posterior axillary fold, is formed mainly by the latissimus dorsi and teres major muscles.

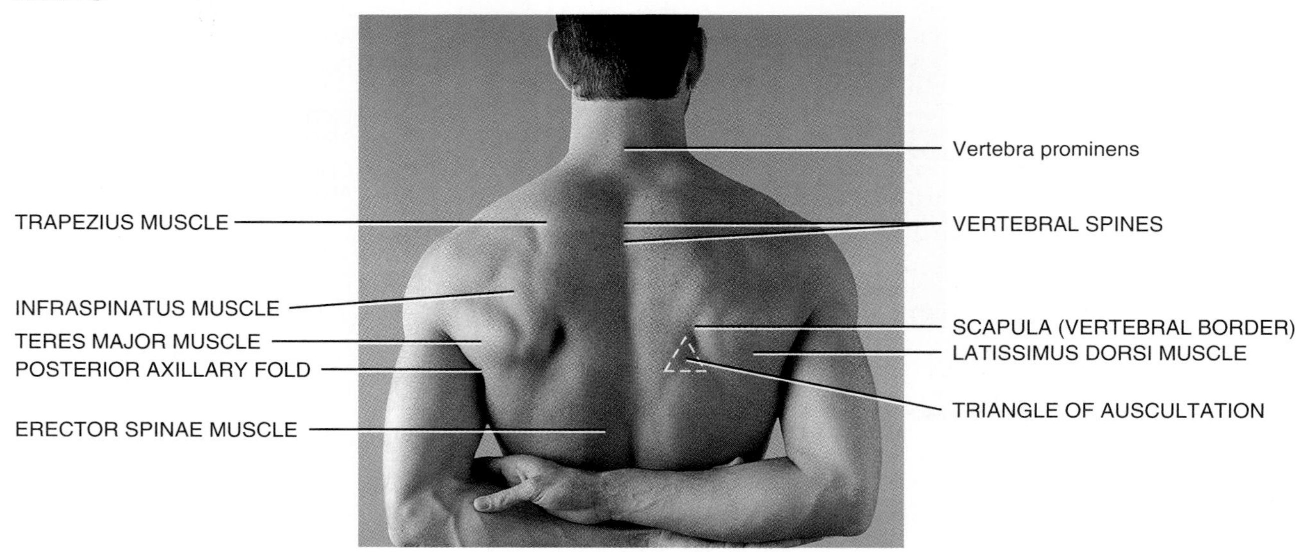

Posterior view

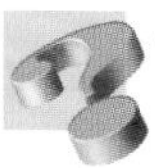
What is the clinical significance of the triangle of auscultation?

- **Teres major muscle.** Located inferior to the infraspinatus muscle; together with the latissimus dorsi muscle forms the inferior border of the posterior axillary fold (posterior wall of the axilla).
- **Posterior axillary fold.** Formed by the latissimus dorsi and teres major muscles, it can be palpated between the fingers and thumb at the posterior aspect of the axilla; forms the posterior wall of the axilla.
- **Triangle of auscultation** (aw-skul-TĀ-shun; *ausculto* = listening). A triangular region of the back just medial to the inferior part of the scapula, where the rib cage is not covered by superficial muscles. It is bounded by the latissimus dorsi and trapezius muscles and vertebral border of the scapula.

Triangle of Auscultation and Lung Sounds

The triangle of auscultation is a landmark of clinical significance because in this area respiratory sounds can be heard clearly through a stethoscope pressed against the skin. If a patient folds the arms across the chest and bends forward, the lung sounds can be heard clearly in the intercostal space between ribs 6 and 7.

✔ What are the divisions of the trunk?
✔ Describe the location of the scapulae relative to the ribs and vertebrae.
✔ What is the triangle of auscultation? Why is it important clinically?

Surface Features of the Chest

The surface anatomy of the **chest,** or *thorax,* includes several distinguishing surface landmarks (Figure 11.8), as well as some surface markings used to identify the location of the heart and lungs within the chest.

- **Clavicle.** Visible at the junction of the neck and thorax. Inferior to the clavicle, especially where it articulates with the manubrium (superior portion) of the sternum, the **first rib** can be palpated. In a depression superior to the medial end of the clavicle, just lateral to the sternocleidomastoid muscle, the **trunks of the brachial plexus** can be palpated (see Figure 17.9a on page 531).
- **Suprasternal (jugular) notch of sternum.** A depression on the superior border of the manubrium of the sternum between the medial ends of the clavicles; the **trachea** (windpipe) can be palpated posterior to the notch.
- **Manubrium of sternum.** Superior portion of the sternum at the same levels as the bodies of the third and fourth thoracic vertebrae and anterior to the arch of the aorta.
- **Sternal angle of sternum.** Formed at the junction between the manubrium and body of the sternum, located about 4 cm (1.5 in.) inferior to the suprasternal notch. It is palpable deep to the skin and locates the costal cartilage of the second rib. It is the most reliable landmark of the chest and the starting point from which ribs are counted. At or inferior to the sternal angle and slightly to the right, the trachea divides into right and left primary bronchi.
- **Body of sternum.** Midportion of the sternum anterior to the heart and the vertebral bodies of T5–T8.
- **Xiphoid process of sternum.** Inferior portion of the sternum, medial to the seventh costal cartilages. The joint between the xiphoid process and body of the sternum is called the **xiphisternal joint.** The heart lies on the diaphragm deep to this joint.
- **Costal margin.** Inferior edges of the costal cartilages of ribs 7 through 10. The first costal cartilage lies inferior to

Figure 11.8 / Surface anatomy of the chest.

Thoracic viscera are protected by the sternum, ribs, and thoracic vertebrae, which form the skeleton of the thorax.

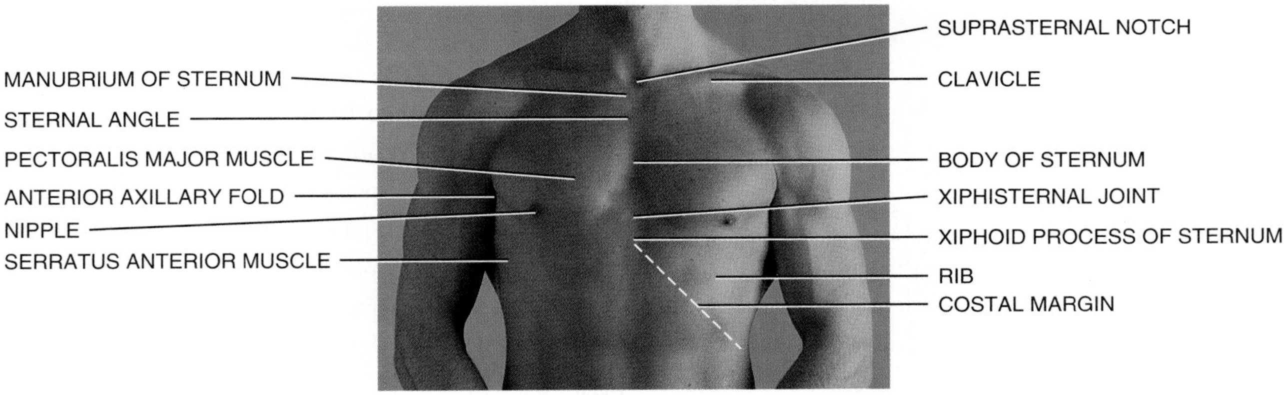

Anterior view of surface features of the chest

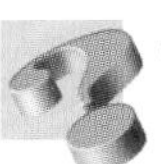

What is the most reliable landmark of the chest?

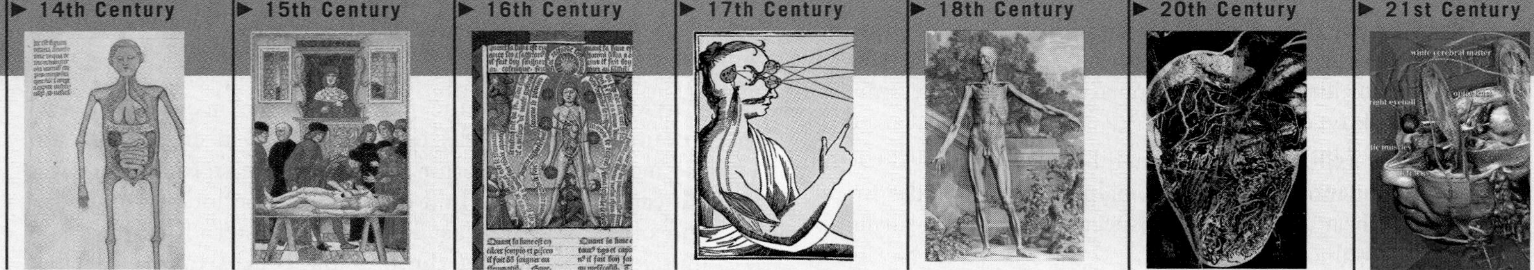

CHANGING IMAGES

Looking Up

1512 AD

High atop scaffolding, arching his back and bending his neck, Michelangelo Buonarroti toiled between 1508 and 1512 to create what is considered by many to be the world's greatest fresco. Michelangelo, a master of the High Renaissance, painted nine scenes from the Old Testament on the ceiling of the Sistine Chapel in the Vatican. These scenes depict scores of dramatically posed nudes. The anatomical accuracy and correct positioning of the figures demonstrate Michelangelo's exceptional command of human anatomy.

While painting, and perhaps more so while sculpting, which was his preferred craft, Michelangelo was required to call upon an understanding of human anatomy that could only have been realized by extensive direct observation. Indeed, evidence suggests that he studied anatomy while in his teens, and may have even collaborated on an anatomical text in his advanced years.

Pictured here is a fragment of the scene *The Creation of Adam* from the book of Genesis. Closely examine this image. Are you able to discern any bony landmarks of the pectoral girdle? How many muscles can you identify? Pay careful attention to the chest, abdominal wall, and upper limb. Is it possible for Michelangelo to have achieved such anatomical exactness without studying anatomy?

As you continue your study of surface anatomy, in order to enhance your clinical skills, consider how such knowledge can strengthen artistic abilities as well. This is especially true for those of you with an artistic inclination. Your future renderings may now appear more real. Consider this: Should anatomy courses be required for art students today, much like it was during the Renaissance?

the medial end of the clavicle; the seventh costal cartilage is the most inferior costal cartilage to articulate directly with the sternum; the tenth costal cartilage forms the most inferior part of the costal margin when viewed anteriorly. At the superior end of the costal margin is the xiphisternal joint.

- **Serratus anterior muscle.** Inferior and lateral to the pectoralis major muscle.
- **Ribs.** Twelve pairs of ribs help to form the bony cage of the thoracic cavity. Depending on body leanness, they may or may not be visible. One site for listening to the heartbeat in adults is the left fifth intercostal space, just medial to the left midclavicular line.
- **Mammary glands.** Accessory organs of the female reproductive system located inside the breasts. These glands overlie the pectoralis major muscle (two-thirds) and serratus anterior muscle (one-third). After puberty, the mammary glands enlarge to their hemispherical shape, and in young adult females they extend from the second through the sixth ribs and from the lateral margin of the sternum to the midaxillary line. The **midaxillary line** (see Figure 23.9e on page 698) is a vertical line that extends downward on the lateral thoracic wall from the center of the axilla.
- **Nipples.** Superficial to the fourth intercostal space or fifth rib, about 10 cm (4 in.) from the midline in males and most females. The position of the nipples in females is variable depending on the size and pendulousness of the breasts. The right dome of the diaphragm is just inferior to the right nipple, the left dome is about 2–3 cm (1 in.) inferior to the left nipple, and the central tendon is at the level of the xiphisternal joint.
- **Anterior axillary fold.** Formed by the lateral border of the pectoralis major muscle; can be palpated between the fingers and thumb; forms the anterior wall of the axilla.
- **Pectoralis major muscle.** Principal upper chest muscle; in the male, the inferior border of the muscle forms a curved line leading to the anterior wall of the axilla and serves as a guide to the fifth rib. In the female, the inferior border is mostly covered by the breast.

Although structures within the chest are almost totally hidden by the sternum, ribs, and thoracic vertebrae, it is important to indicate the surface markings of the heart and lungs. The projection (shape of an organ on the surface of the body) of the heart on the anterior surface of the chest is indicated by four points as follows (see Figure 13.1c on page 389): The *inferior left point* is the apex (inferior, pointed end) of the heart, which projects downward and to the left and can be palpated in the fifth intercostal space, about 9 cm (3.5 in.) to the left of the midline. The *inferior right point* is at the lower border of the costal cartilage of the right sixth rib, about 3 cm to the right of the midline. The *superior right point* is located at the superior border of the costal cartilage of the right third rib, about 3 cm to the right of the midline. The *superior left point* is located at the inferior border of the costal cartilage of the left second rib, about 3 cm to the left of the midline. If you connect the four points, you can determine the heart's location and size (about the size of a person's closed fist).

The lungs almost totally fill the thorax. The *apex* (superior, pointed end) of the lungs lies just superior to the medial third of the clavicles and is the only area that can be palpated (see Figure 23.9a on page 698). The anterior, lateral, and posterior surfaces of the lungs lie against the ribs. The *base* (inferior, broad end) of the lungs is concave and fits over the convex area of the diaphragm. It extends from the sixth costal cartilage anteriorly to the spinous process of the tenth thoracic vertebra posteriorly. (The base of the right lung is usually slightly higher than that of the left due to the position of the liver below it.) The covering around the lungs is called the *pleura.* At the base of each lung, the pleura extends about 5 cm below the base from the sixth costal cartilage anteriorly to the twelfth rib posteriorly. Thus, the lungs do not completely fill the pleural cavity in this area. A needle is typically inserted into this space in the pleural cavity to withdraw fluid from the chest.

✔ Describe the location of the sternum relative to the ribs and vertebrae.
✔ What is the significance of the xiphisternal joint as a landmark?
✔ Where would you palpate the first rib and brachial plexus?
✔ Why is the midaxillary line an important anatomical landmark?
✔ How would you indicate the surface markings of the heart?
✔ Describe the position of the lungs in the thorax.
✔ Why do the lungs not completely fill the pleural cavity? Why is this important clinically?

Surface Features of the Abdomen and Pelvis

Following are some of the prominent surface anatomy features of the **abdomen** and **pelvis** (Figure 11.9):

- **Umbilicus.** Also called the *navel;* it marks the site of attachment of the umbilical cord to the fetus. It is level with the intervertebral disc between the bodies of vertebrae L3 and L4. The **abdominal aorta,** which branches into the right and left common iliac arteries anterior to the body of vertebra L4, can easily be palpated through the upper part of the anterior abdominal wall just to the left of the midline. The **inferior vena cava** lies to the right of the abdominal aorta and is wider; it arises anterior to the body of vertebra L5.
- **External oblique muscle.** Located inferior to the serratus anterior muscle. The aponeurosis of the muscle on its inferior border is the **inguinal ligament,** a structure along which hernias frequently occur.
- **Rectus abdominis muscles.** Located just lateral to the midline of the abdomen. They can be seen by raising the shoulders while in the supine position without using the arms.
- **Linea alba.** Flat, tendinous raphe forming a furrow along the midline between the rectus abdominis muscles. The fur-

row extends from the xiphoid process to the pubic symphysis. It is broad above the umbilicus and narrow below it. The linea alba is a frequently selected site for abdominal surgery because an incision through it severs no muscles and only a few blood vessels and nerves.

- **Tendinous intersection.** A fibrous band that runs transversely or obliquely across the rectus abdominis muscles. One intersection is at the level of the umbilicus, one is at the level of the xiphoid process, and one is midway between. The rectus abdominis muscles thus contain four prominent bulges and three tendinous intersections.
- **Linea semilunaris.** The lateral edge of the rectus abdominis muscle that crosses the costal margin at the top of the ninth costal cartilage.
- **McBurney's point.** Located two-thirds of the way down an imaginary line drawn between the umbilicus and anterior superior iliac spine.

McBurney's Point and the Appendix

McBurney's point is an important landmark related to the appendix. When the appendix must be removed in an operation called an *appendectomy*, an oblique incision is made through McBurney's point. Moreover, pressure of the finger on McBurney's point produces pain in acute *appendicitis*, inflammation of the appendix. ■

- **Iliac crest.** Superior margin of the ilium of the hip bone. It forms the outline of the superior border of the buttock. When you rest your hands on your hips, they rest on the iliac crests. A horizontal line drawn across the highest point of each iliac crest is called the **supracristal line,** which intersects the spinous process of the fourth lumbar vertebra. This vertebra is a landmark for performing a spinal tap (see page 520).
- **Anterior superior iliac spine.** The anterior end of the iliac crest that lies at the upper lateral end of the fold of the groin.
- **Posterior superior iliac spine.** The posterior end of the iliac crest, indicated by a dimple in the skin that coincides with the middle of the sacroiliac joint, where the hip bone attaches to the sacrum.
- **Pubic tubercle.** Projection on the superior border of the pubis of the hip bone. Attached to it is the medial end of the inguinal ligament. The lateral end of the ligament is attached to the anterior superior iliac spine. The **inguinal ligament** is the inferior free edge of the aponeurosis of the external oblique muscle that forms the **inguinal canal.** The spermatic cord in males and the round ligament of the uterus in females pass through the inguinal canal.
- **Pubic symphysis.** Anterior joint of the hip bones; palpated as a firm resistance in the midline at the inferior portion of the anterior abdominal wall.
- **Mons pubis.** An elevation of adipose tissue covered by skin and pubic hair that is anterior to the pubic symphysis.
- **Sacrum.** The fused spinous processes of the sacrum, called the median sacral crest, can be palpated beneath the skin in the superior of the **gluteal cleft,** a depression along the midline that separates the buttocks (and is part of the discussion on the buttocks on page 363).
- **Coccyx.** The inferior surface of the tip of the coccyx can be palpated in the gluteal cleft, about 2.5 cm (1 in.) posterior to the anus.

Chapter 1 discussed the division of the abdomen and pelvis into nine regions (see Figure 1.9 on page 16). The nine-region

Figure 11.9 / Surface anatomy of the abdomen and pelvis.

The linea alba is a frequent site for an abdominal incision because cutting through it severs no muscles and few blood vessels and nerves.

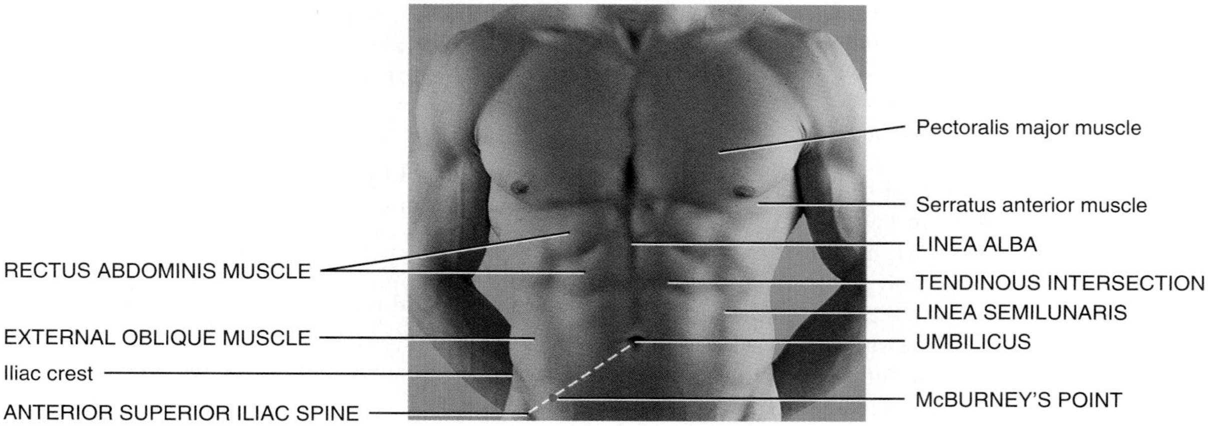

(a) Anterior view of abdomen

designation subdivided the abdomen and pelvis by drawing two vertical lines (left and right clavicular), an upper horizontal line (subcostal), and a lower horizontal line (transtubercular). Take time to examine Figure 1.9 carefully, so that you can see which organs or parts of organs lie within each region.

In Chapter 10, as part of your study of skeletal muscles, you were introduced to the **perineum,** the diamond-shaped region medial to the thigh and buttocks. It is bounded by the pubic symphysis anteriorly, the ischial tuberosities laterally, and the coccyx posteriorly. A transverse line drawn between the ischial tuberosities divides the perineum into an anterior *urogenital triangle* that contains the external genitals, and a posterior *anal triangle* that contains the anus. Details concerning the perineum are provided in Chapter 26. At this point keep in mind that the musculature of the perineum constitutes the floor of the pelvic cavity.

✔ Why are the linea alba, linea semilunaris, McBurney's point, and supracristal line important landmarks?
✔ What is the inguinal ligament, and why is it important?
✔ How would you palpate the sacrum and coccyx?
✔ What is the perineum?

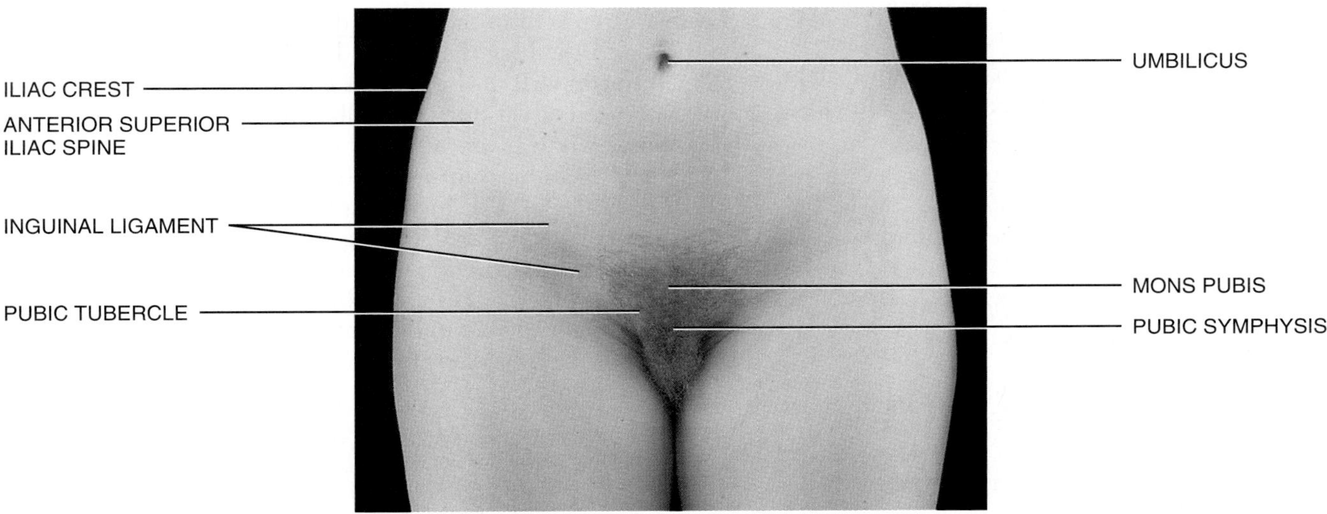

(b) Anterior view of pelvis

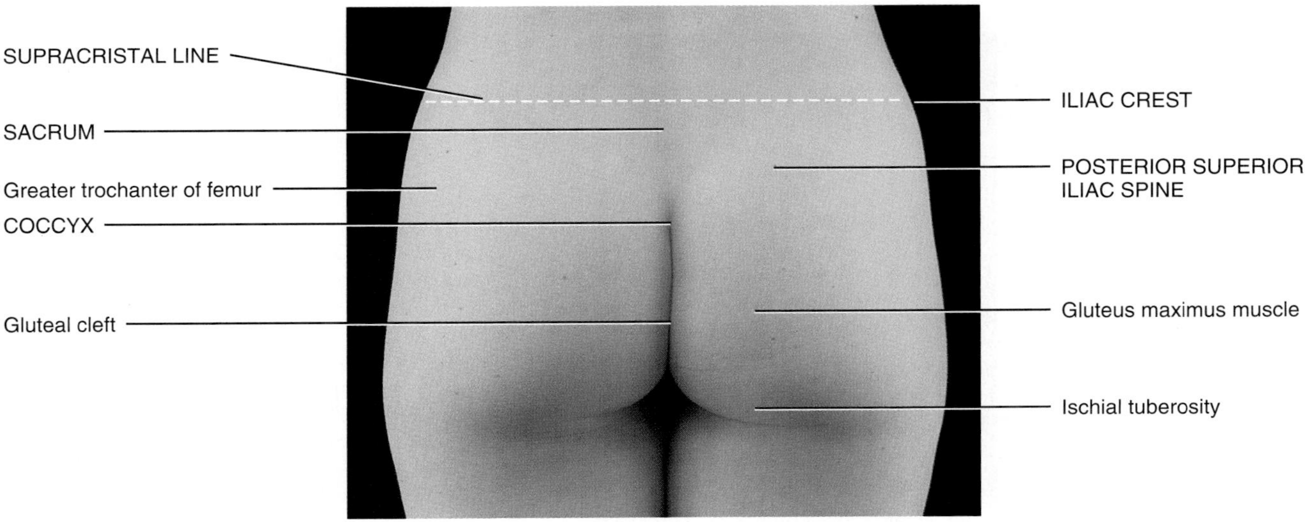

(c) Posterior view of pelvis

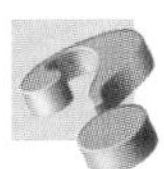

What is the clinical significance of McBurney's point?

SURFACE ANATOMY OF THE UPPER LIMB (EXTREMITY)

Objective

- Describe the surface anatomy of the various regions of the upper limb—armpit, shoulder, arm, elbow, forearm, wrist, and hand.

The **upper limb** consists of the shoulder, armpit, arm, elbow, forearm, wrist, and hand.

Surface Features of the Shoulder

The **shoulder,** or *acromial region,* is located at the lateral aspect of the clavicle, where the clavicle joins the scapula and the scapula joins the humerus. The region presents several conspicuous surface features (Figure 11.10).

- **Acromioclavicular joint.** A slight elevation at the lateral end of the clavicle. It is the joint between the acromion of the scapula and the clavicle.
- **Acromion.** The expanded lateral end of the spine of the scapula that forms the top of the shoulder. It can be palpated about 2.5 cm (1 in.) distal to the acromioclavicular joint.
- **Humerus.** The **greater tubercle** of the humerus may be palpated on the superior aspect of the shoulder. It is the most laterally palpable bony structure.
- **Deltoid muscle.** Triangular muscle that forms the rounded prominence of the shoulder.

Shoulder Intramuscular Injection

The *deltoid muscle* is a frequent site for an intramuscular injection. To avoid injury to major blood vessels and nerves, the injection is given in the midportion of the muscle about 2–3 finger-breadths inferior to the acromion of the scapula and lateral to the axilla. ■

Surface Features of the Armpit

The **armpit,** or *axilla,* is a pyramid-shaped area at the junction of the arm and the chest that enables blood vessels and nerves to pass between the neck and the upper limbs (Figure 11.11a). The **apex** of the axilla is surrounded by the clavicle, scapula, and first rib. The **base** of the axilla is formed by the concave skin and fascia that extends from the arm to the chest wall. It contains hair, and deep to the base the axillary lymph nodes can be palpated. The **anterior wall** of the axilla is composed mainly of the pectoralis major muscle (anterior axillary fold; see also Figure 11.8). The **posterior wall** of the axilla is formed mainly by the teres major and latissimus dorsi muscles (posterior axillary fold; see also Figure 11.7). The **medial wall** of the axilla is formed by ribs 1–4 and their corresponding intercostal muscles, plus the overlying serratus anterior muscle. Finally, the **lateral wall** of the axilla is formed by the coracobrachialis muscle and the superior portion of the shaft of the humerus. Passing through the axilla are the axillary artery and vein, branches of the brachial plexus, and axillary lymph nodes.

✔ What are the major divisions of the upper limb?

Figure 11.10 / Surface anatomy of the shoulder.

The deltoid muscle gives the shoulder its rounded prominence.

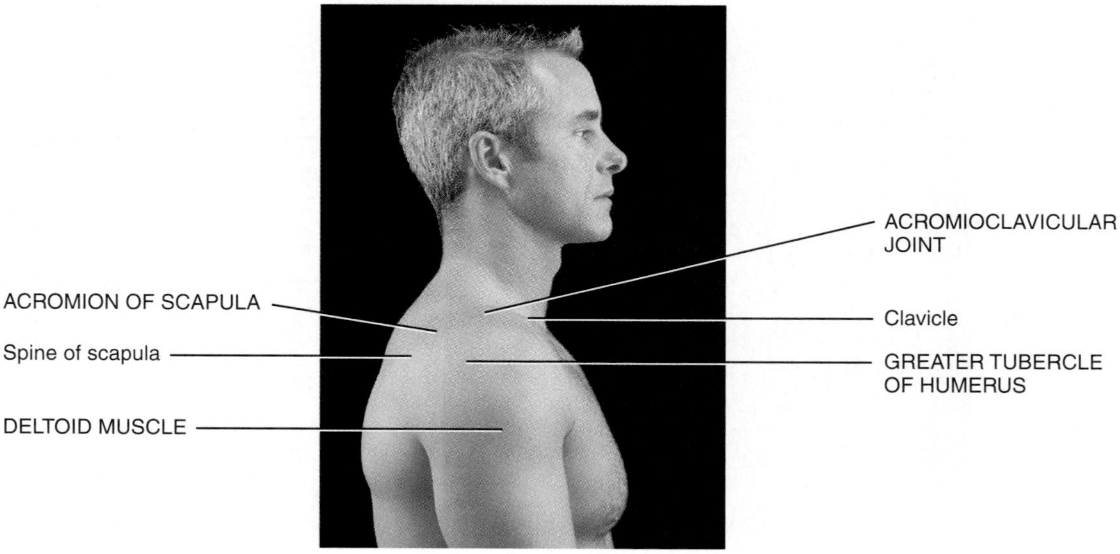

ACROMIOCLAVICULAR JOINT
ACROMION OF SCAPULA
Clavicle
Spine of scapula
GREATER TUBERCLE OF HUMERUS
DELTOID MUSCLE

Right lateral view

Which structure forms the top of the shoulder?

- ✔ What three bones are present at the shoulder? How would you palpate each?
- ✔ What is the clinical significance of the deltoid muscle?
- ✔ Define the axilla. What are its borders? Why is it important?

Figure 11.11 / Surface anatomy of the arm and elbow.

The biceps brachii and triceps brachii muscles form the bulk of the musculature of the arm.

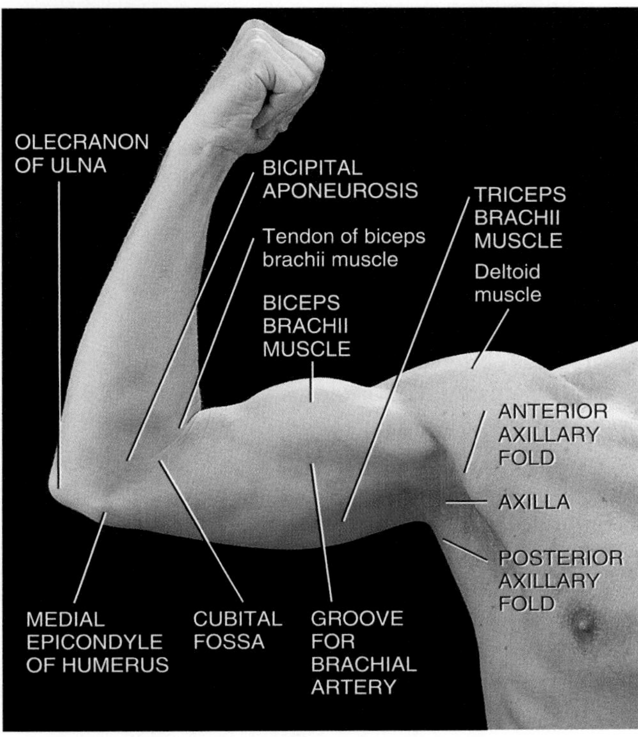

(a) Medial view of arm

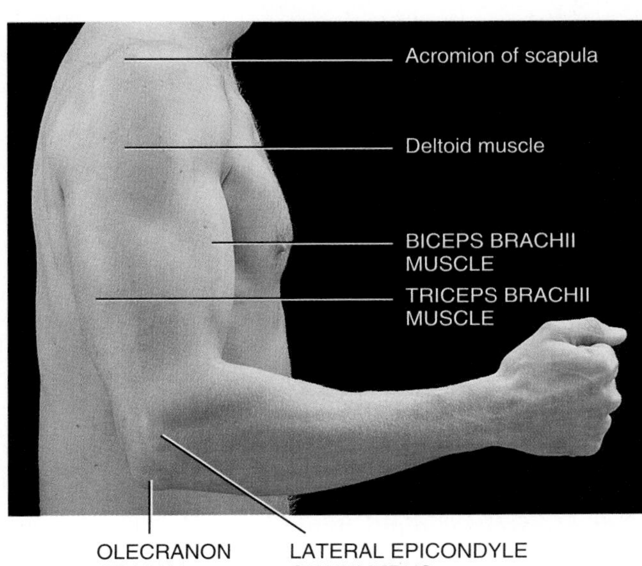

(b) Right lateral view of arm

Surface Features of the Arm and Elbow

The **arm,** or *brachium,* is the region between the shoulder and elbow. The **elbow,** or *cubitus,* is the region where the arm and forearm join. The arm and elbow present several surface anatomy features (Figure 11.11).

- **Humerus.** This may be palpated along its entire length, especially at the elbow.
- **Biceps brachii muscle.** Forms the bulk of the anterior surface of the arm. On the medial side of the muscle is a groove that contains the **brachial artery.**
- **Triceps brachii muscle.** Forms the bulk of the posterior surface of the arm.
- **Medial epicondyle.** Medial projection of the humerus near the elbow.
- **Lateral epicondyle.** Lateral projection of the humerus near the elbow.
- **Olecranon.** Projection of the proximal end of the ulna between and slightly superior to the epicondyles when the forearm is extended; it forms the elbow.
- **Ulnar nerve.** Can be palpated in a groove posterior to the medial epicondyle. The "funny bone" is the region where the ulnar nerve rests against the medial epicondyle. Hitting the nerve at this point produces a sharp pain along the medial side of the forearm.
- **Cubital fossa.** Triangular space in the anterior region of the elbow bounded proximally by an imaginary line between the humeral epicondyles, laterally by the medial border of the brachioradialis muscle, and medially by the lateral border of the pronator teres muscle; contains the tendon of the

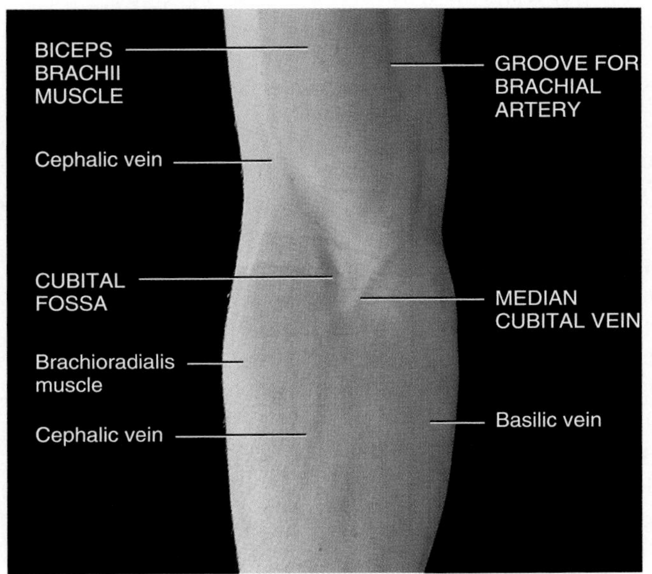

(c) Anterior view of cubital fossa

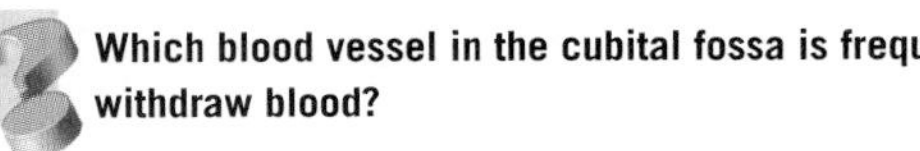

Which blood vessel in the cubital fossa is frequently used to withdraw blood?

biceps brachii muscle, the median cubital vein, brachial artery and its terminal branches (radial and ulnar arteries), and parts of the median and radial nerves.

- **Median cubital vein.** Crosses the cubital fossa obliquely and connects the lateral cephalic with the medial basilic veins of the arm. The median cubital vein is frequently used to withdraw blood from a vein for diagnostic purposes or to introduce substances into blood, such as medications, contrast media for radiographic procedures, nutrients, and blood cells and/or plasma for transfusion.
- **Brachial artery.** Continuation of the axillary artery that passes posterior to the coracobrachialis muscle and then medial to the biceps brachii muscle. It enters the middle of the cubital fossa and passes deep to the bicipital aponeurosis, which separates it from the median cubital vein.
- **Bicipital aponeurosis.** An aponeurotic band that inserts the biceps brachii muscle into the deep fascia in the medial aspect of the forearm. It can be felt when the muscle contracts.

Blood pressure is usually measured in the brachial artery, when the cuff of a *sphygmomanometer* is wrapped around the arm and a stethoscope is placed over the brachial artery in the cubital fossa. Pulse can also be detected in the artery in the cubital fossa. In cases of severe hemorrhage in the forearm and hand, pressure may be applied to the brachial artery to help stop the bleeding.

✔ In which arteries of the upper limb can a pulse be detected?
✔ What is the "funny bone"? Why is it so designated?
✔ Define the borders and significance of the cubital fossa.
✔ What is the clinical importance of the median cubital vein?
✔ What artery is normally used to take blood pressure?

Surface Features of the Forearm and Wrist

The **forearm,** or *antebrachium,* is the region between the elbow and wrist. The **wrist,** or carpus, is between the forearm and palm. Following are some prominent surface anatomy features of the forearm and wrist (Figure 11.12).

- **Ulna.** The medial bone of the forearm. It can be palpated along its entire length from the olecranon (see Figure 11.11a, b) to the **styloid process,** a projection on the distal end of the bone at the medial side of the wrist. The **head of the ulna** is a conspicuous enlargement just proximal to the styloid process.
- **Radius.** When the forearm is rotated, the distal half of the radius can be palpated; the proximal half is covered by muscles. The **styloid process** of the radius is a projection on the distal end of the bone at the lateral side of the wrist.
- **Muscles.** Because of their close proximity, it is difficult to identify individual muscles of the forearm. Instead, it is much easier to identify tendons as they approach the wrist and then trace them proximally to the following muscles:

  Brachioradialis muscle. Located at the superior and lateral aspect of the forearm.

  Flexor carpi radialis muscle. The tendon of this muscle is on the lateral side of the forearm about 1 cm medial to the styloid process of the radius.

  Palmaris longus muscle. The tendon of this muscle is medial to the flexor carpi radialis tendon and can be seen if the wrist is slightly flexed and the base of the thumb and little finger are drawn together.

  Flexor digitorum superficialis muscle. The tendon of this muscle is medial to the palmaris longus tendon and can be palpated by flexing the fingers at the metacarpophalangeal and proximal interphalangeal joints.

  Flexor carpi ulnaris muscle. The tendon of this muscle is on the medial aspect of the forearm.

- **Radial artery.** Located on the lateral aspect of the wrist between the flexor carpi radialis tendon and styloid process of the radius. It is frequently used to take a pulse.
- **Pisiform bone.** Medial bone of the proximal row of carpals that can be palpated as a projection distal and anterior to the styloid process of the ulna.
- **"Anatomical snuffbox."** A triangular depression between tendons of extensor pollicis brevis and extensor pollicis longus muscles. It derives its name from a habit of previous centuries in which a person would take a pinch of snuff (powdered tobacco or scented powders) and place it in the depression before sniffing it into the nose. The styloid process of the radius, the base of the first metacarpal, trapezium, scaphoid, and radial artery can all be palpated in the depression.
- **Wrist creases.** Three more or less constant lines on the anterior aspect of the wrist (named *proximal*, *middle*, and *distal*) where the skin is firmly attached to underlying deep fascia.

Surface Features of the Hand

The **hand,** or *manus,* is the region from the wrist to the termination of the upper limb; it has several conspicuous surface features (Figure 11.13 on page 362).

- **Knuckles.** Commonly refers to the dorsal aspect of the distal ends of metacarpals II–V (or 2–5), but also includes the dorsal aspects of the metacarpophalangeal and interphalangeal joints.
- **Dorsal venous arch.** Superficial veins on the dorsum of the hand that drain blood into the cephalic vein. It can be displayed by compressing the blood vessels at the wrist for a few moments as the hand is opened and closed.
- **Tendon of extensor digiti minimi muscle.** This can be seen on the dorsum of the hand in line with the phalanx of the little finger.

Figure 11.12 / Surface anatomy of the forearm and wrist.

Muscles of the forearm are most easily identified by locating their tendons near the wrist and tracing them proximally.

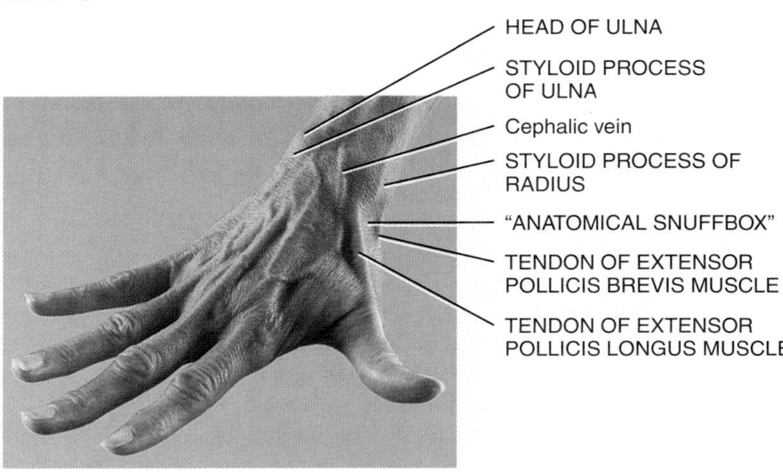

(a) Dorsum of wrist

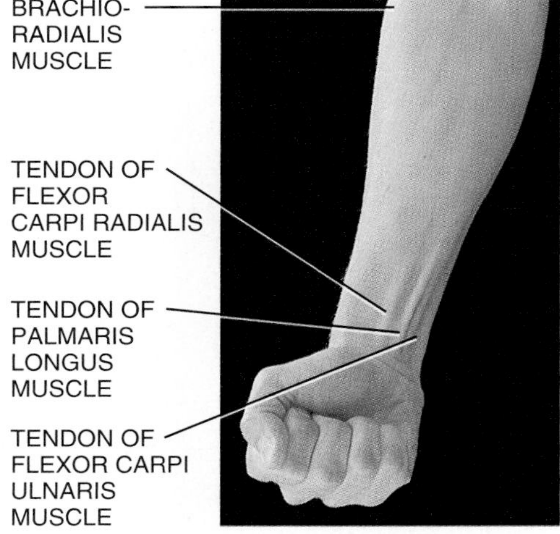

(b) Anterior aspect of forearm and wrist

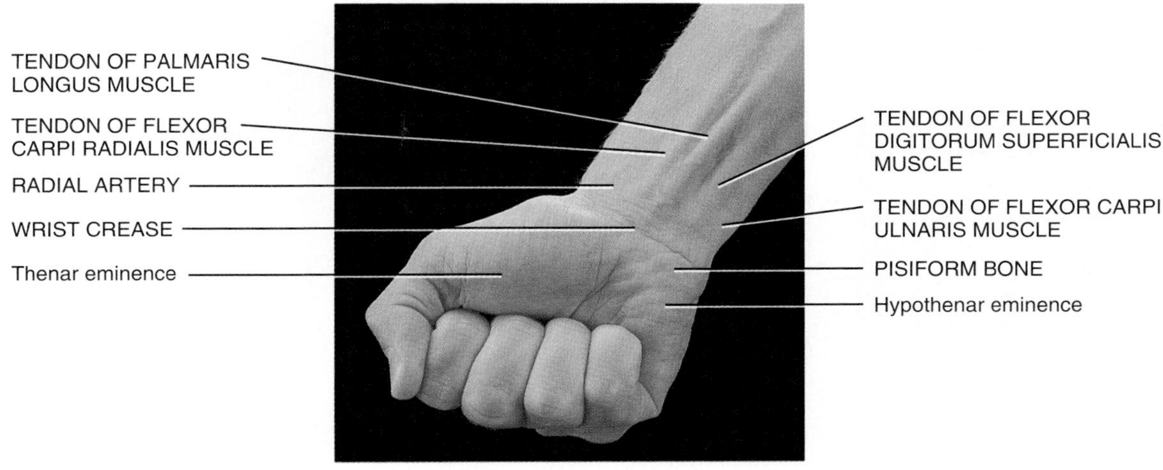

(c) Anterior aspect of wrist

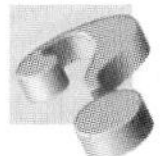

Which blood vessel is frequently used to take a pulse?

- **Tendons of extensor digitorum muscle.** These can be seen on the dorsum of the hand in line with the phalanges of the ring, middle, and index fingers.
- **Tendon of the extensor pollicis brevis muscle.** This tendon (described previously) is in line with the phalanx of the thumb (see Figure 11.12a).
- **Thenar eminence.** Larger, rounded contour on the lateral aspect of the palm formed by muscles that move the thumb. Also called the ball of the thumb.
- **Hypothenar eminence.** Smaller, rounded contour on the medial aspect of the palm formed by muscles that move the little finger. Also called the ball of the little finger.
- **Palmar flexion creases.** Skin creases on the palm.
- **Digital flexion creases.** Skin creases on the anterior surface of the fingers.

✔ How would you palpate the ulna, styloid process of the ulna, head of the ulna, and styloid process of the radius?

✔ Explain the easiest way to identify the muscles of the forearm.

✔ What is the "anatomical snuffbox"?

✔ What are the knuckles? Why are the thenar and hypothenar eminences important?

Figure 11.13 / Surface anatomy of the hand.

Several tendons on the dorsum of the hand can be identified by their alignment with the phalanges of the digits.

KNUCKLES
THENAR EMINENCE
HYPOTHENAR EMINENCE

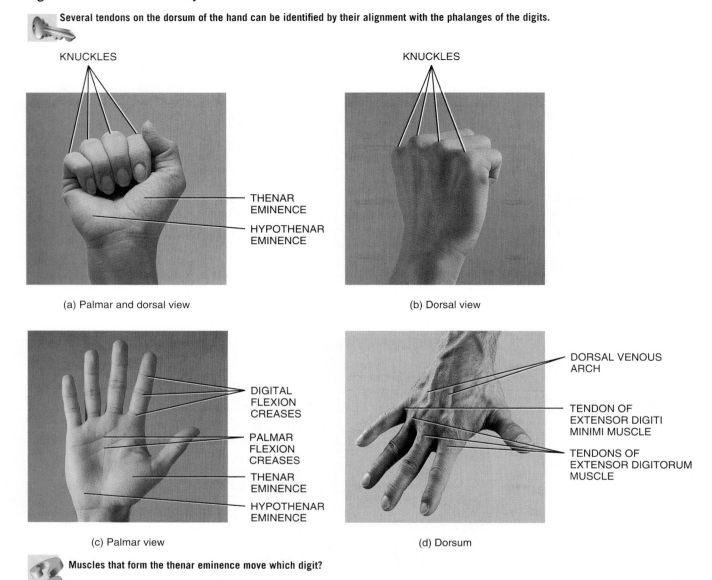

(a) Palmar and dorsal view

KNUCKLES

(b) Dorsal view

DIGITAL FLEXION CREASES
PALMAR FLEXION CREASES
THENAR EMINENCE
HYPOTHENAR EMINENCE

(c) Palmar view

DORSAL VENOUS ARCH
TENDON OF EXTENSOR DIGITI MINIMI MUSCLE
TENDONS OF EXTENSOR DIGITORUM MUSCLE

(d) Dorsum

Muscles that form the thenar eminence move which digit?

SURFACE ANATOMY OF THE LOWER LIMB (EXTREMITY)

Objective

- Describe the surface features of the various regions of the lower limb—buttock, thigh, knee, leg, ankle, and foot.

The **lower limb** consists of the buttock, thigh, knee, leg, ankle, and foot.

Surface Features of the Buttock

The **buttock,** or *gluteal region,* is formed mainly by the gluteus maximus muscle. The outline of the superior border of the buttock is formed by the iliac crests (see Figure 11.9c). Following are some surface features of the buttock (Figure 11.14).

- **Gluteus maximus muscle.** Forms the major portion of the prominence of the buttocks. The sciatic nerve is deep to this muscle.
- **Gluteus medius muscle.** Superior and lateral to the gluteus maximus muscle.

Gluteal Intramuscular Injection

Another common site for an intramuscular injection is the *gluteus medius muscle.* In order to give this injection, the buttock is divided into quadrants and the upper outer quadrant is used as an injection site. The iliac crest serves as the landmark for this quadrant. This site is chosen because the gluteus medius muscle in this area is quite thick, and there is less chance of injury to the sciatic nerve or major blood vessels.

Figure 11.14 / Surface anatomy of the buttocks.

The gluteus medius muscle is a frequent site for an intramuscular injection.

Iliac crest
GREATER TROCHANTER
GLUTEUS MEDIUS MUSCLE
GLUTEUS MAXIMUS MUSCLE
GLUTEAL (NATAL) CLEFT
ISCHIAL TUBEROSITY
GLUTEAL FOLD

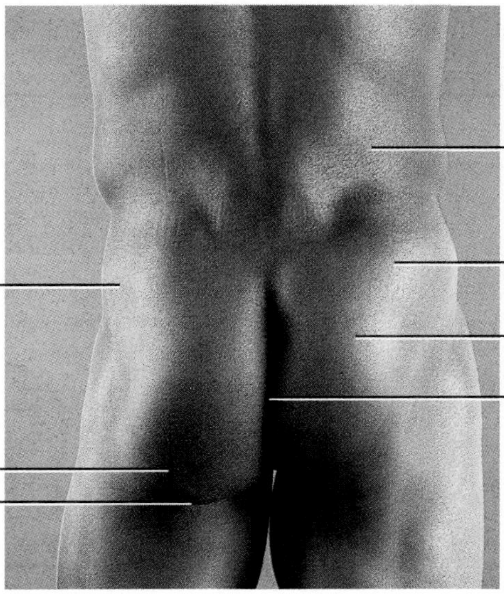

Posterior view

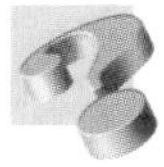

Which muscle forms the bulk of the buttocks?

- **Gluteal (natal) cleft.** Depression along the midline that separates the left and right buttocks.
- **Gluteal fold.** Inferior limit of the buttock that roughly corresponds to the inferior margin of the gluteus maximus muscle.
- **Ischial tuberosity.** Just superior to the medial side of the gluteal fold, the tuberosity bears the weight of the body when seated.
- **Greater trochanter.** A projection of the proximal end of the femur on the lateral side of the thigh. It is about 20 cm (8 in.) inferior to the highest point of the iliac crest.

✔ What are the divisions of the lower limb?
✔ Define the buttock. What is the gluteal cleft? Gluteal fold?
✔ Why is the greater trochanter of the femur important?

Surface Features of the Thigh and Knee

The **thigh,** or *femoral region,* is the region from the hip to the knee. The **knee,** or *genu,* is the region where the thigh and leg join. Several muscles are clearly visible in the thigh. Following are several surface features of the thigh and knee (Figure 11.15).

- **Sartorius muscle.** Superficial anterior muscle that can be traced from the lateral aspect of the thigh to the medial aspect of the knee.
- **Quadriceps femoris muscle.** Three of the four components of the muscle can be seen: **rectus femoris** at the midpoint of the anterior aspect of the thigh; **vastus medialis** at the anteromedial aspect of the thigh; and **vastus lateralis** at the anterolateral aspect of the thigh. The fourth component, the vastus intermedius, is deep to the rectus femoris.

Thigh Intramuscular Injection

The *vastus lateralis muscle* of the quadriceps femoris group is another site that may be used for an intramuscular injection. This site is located at a point midway between the greater trochanter (see Figure 11.14) of the femur and the patella (knee cap). Injection in this area reduces the chance of injury to major blood vessels and nerves. ■

- **Adductor longus muscle.** Located at the superior aspect of the medial thigh. It is the most superficial of the three adductor muscles (adductor magnus, adductor brevis, and adductor longus).
- **Femoral triangle.** A large space formed by the inguinal ligament superiorly, the sartorius muscle laterally, and the adductor longus muscle medially. The triangle contains the femoral artery, vein, and nerve; inguinal lymph nodes; and the terminal portion of the great saphenous vein. The triangle is an important arterial pressure point in cases of severe hemorrhage of the lower limb. Hernias frequently occur in this area.
- **Hamstring muscles.** Superficial, posterior thigh muscles located below the gluteal folds. They are the **biceps femoris,** which lies more laterally as it passes inferiorly to the knee, and the **semitendinosus** and **semimembranosus,** which lie medially as they pass inferiorly to the knee. The

Figure 11.15 / Surface anatomy of the thigh and knee.

The quadriceps femoris and hamstrings form the bulk of the musculature of the thigh.

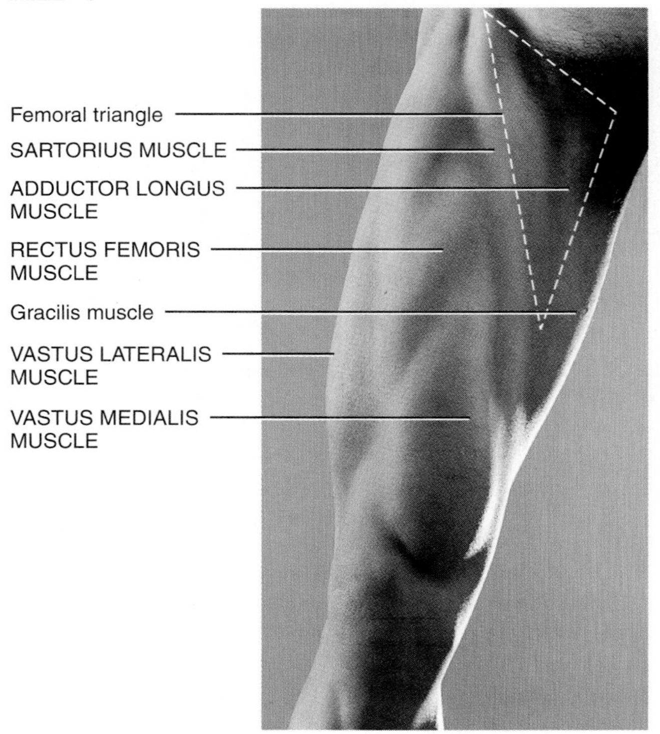

(a) Anterior view of thigh

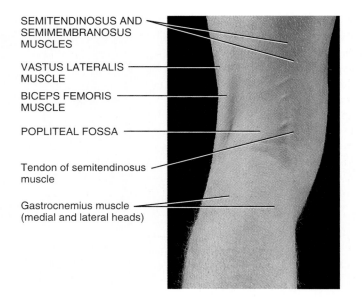

SEMITENDINOSUS AND SEMIMEMBRANOSUS MUSCLES
VASTUS LATERALIS MUSCLE
BICEPS FEMORIS MUSCLE
POPLITEAL FOSSA
Tendon of semitendinosus muscle
Gastrocnemius muscle (medial and lateral heads)

(b) Posterior view of popliteal fossa

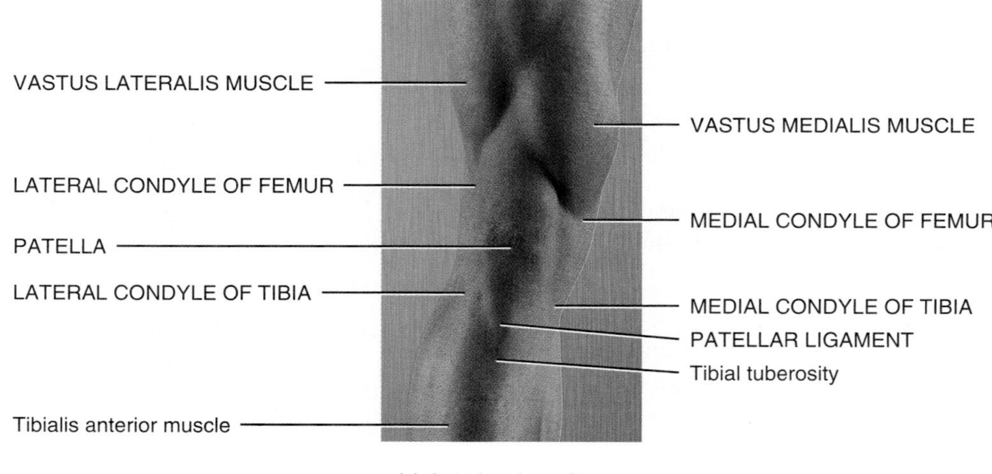

(c) Anterior view of knee

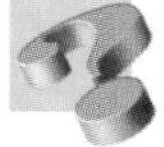

Which muscles form the borders of the popliteal fossa?

tendons of the hamstring muscles can be palpated laterally and medially on the posterior aspect of the knee.

The knee presents several distinguishing surface features.

- **Patella.** Kneecap. A large sesamoid bone located within the tendon of the quadriceps femoris muscle on the anterior surface of the knee along the midline.
- **Patellar ligament.** Continuation of the quadriceps femoris tendon inferior to the patella. Infrapatellar fat pads are found on both sides of it.
- **Medial condyle of femur.** Medial projection on the distal end of the femur.
- **Medial condyle of tibia.** Medial projection on the proximal end of the tibia.

- **Lateral condyle of femur.** Lateral projection on the distal end of the femur.
- **Lateral condyle of tibia.** Lateral projection on the proximal end of the tibia. All four condyles can be palpated just inferior to the patella on either side of the patellar ligament.
- **Popliteal fossa.** A diamond-shaped area on the posterior aspect of the knee that is clearly visible when the knee is flexed. The fossa is bordered superolaterally by the biceps femoris muscle, superomedially by the semimembranosus and semitendinosus muscles, and inferolaterally and inferomedially by the lateral and medial heads of the gastrocnemius muscle, respectively. The **head of the fibula** can easily be palpated on the lateral side of the popliteal fossa. The fossa also contains the popliteal artery and vein. It is sometimes possible to detect a pulse in the popliteal artery.

Surface Features of the Leg, Ankle, and Foot

The **leg**, or *crus*, is the region between the knee and ankle. The **ankle**, or *tarsus*, is between the leg and foot and the **foot** is the region from the ankle to the termination of the lower limb. Following are several surface anatomy features of the leg, ankle, and foot (Figure 11.16).

- **Tibial tuberosity.** Bony prominence on the superior, anterior surface of the tibia into which the patellar ligament inserts.
- **Tibialis anterior muscle.** Lies against the lateral surface of the tibia, where it is easy to palpate, particularly when the foot is dorsiflexed. The distal tendon of the muscle can be traced to its insertion on the medial cuneiform and base of the first metatarsal.
- **Tibia.** The medial surface and anterior border (shin) of the tibia are subcutaneous and can be palpated throughout the length of the bone.
- **Peroneus longus muscle.** A superficial lateral muscle that overlies the fibula.
- **Gastrocnemius muscle.** Forms the bulk of the midportion and superior portion of the posterior aspect of the leg. The medial and lateral heads can be clearly seen in a person standing on tiptoe.
- **Soleus muscle.** Located deep to the gastrocnemius muscle; it and the gastrocnemius together are referred to as the calf muscles.
- **Calcaneal (Achilles) tendon.** Prominent tendon of the gastrocnemius and soleus muscles on the posterior aspect of the ankle; inserts into the **calcaneus** (heel) bone of the foot.
- **Lateral malleolus of fibula.** Projection of the distal end of the fibula that forms the lateral prominence of the ankle. The head of the fibula, at the proximal end of the bone, lies at the same level as the tibial tuberosity.
- **Medial malleolus of tibia.** Projection of the distal end of the tibia that forms the medial prominence of the ankle.

Figure 11.16 / Surface anatomy of the leg, ankle, and foot.

The calcaneal tendon is a common tendon for the gastrocnemius and soleus (calf muscles) and inserts into the calcaneus (heel bone).

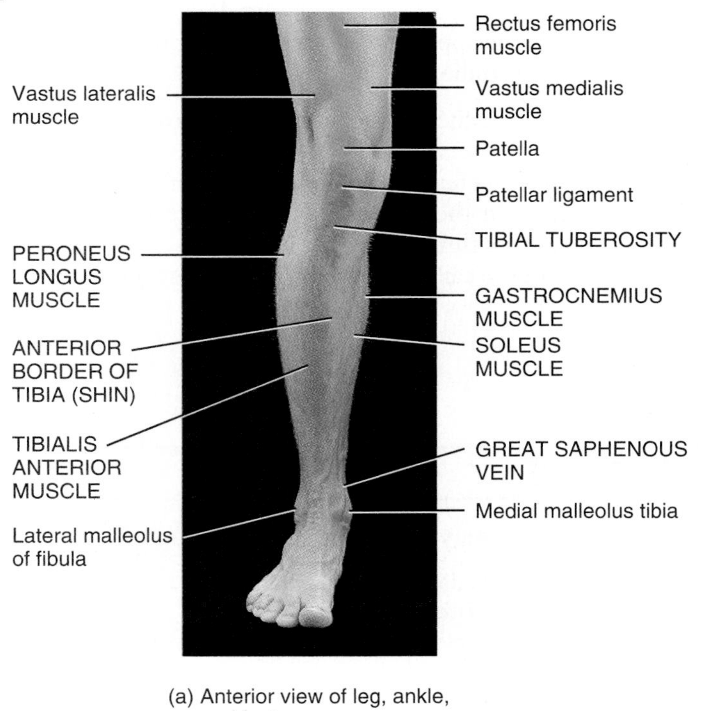

(a) Anterior view of leg, ankle, and foot

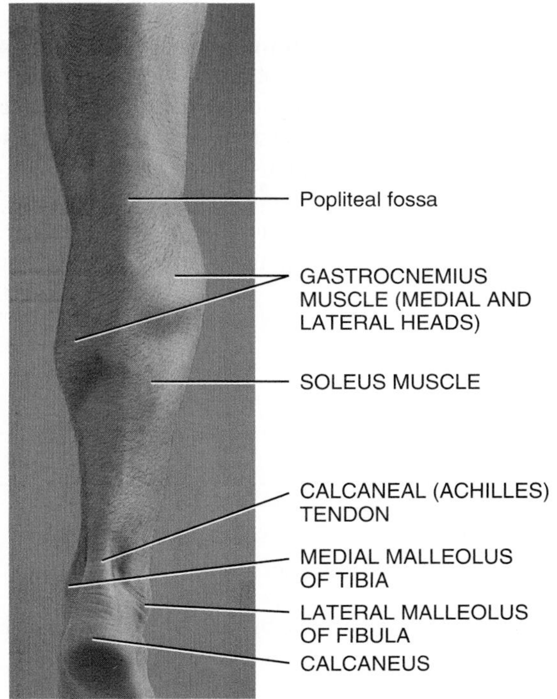

(b) Posterior view of leg and ankle

(continues)

Figure 11.16 / Surface anatomy of the leg, ankle, and foot. (continued)

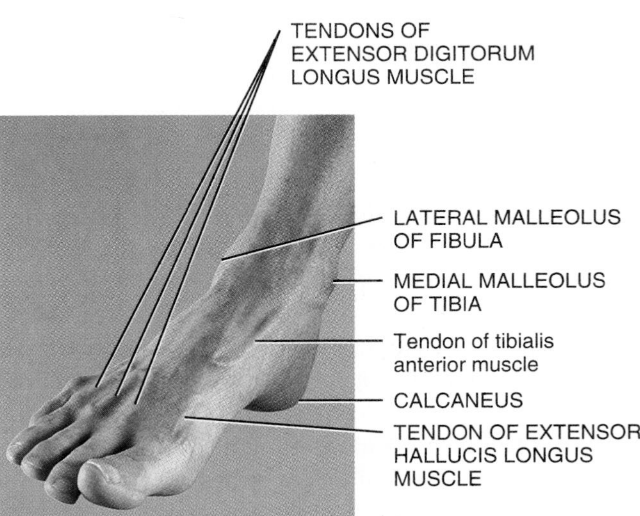

(c) Dorsum of foot

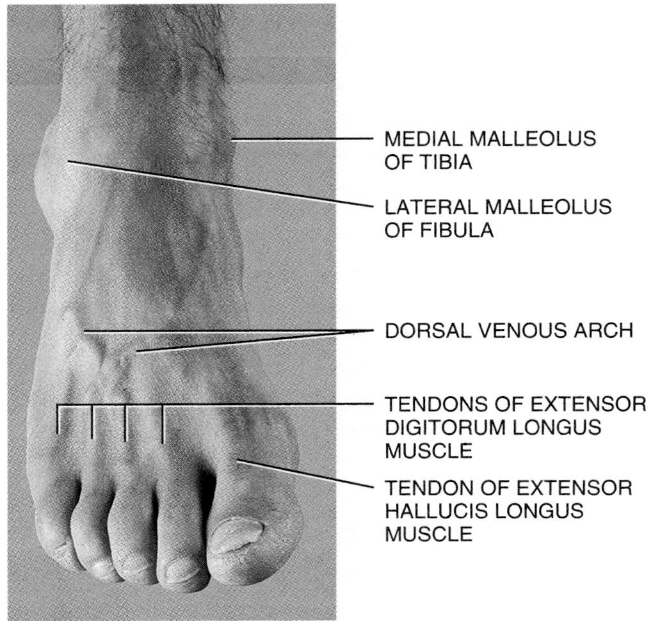

(d) Dorsum of foot

Which artery is just lateral to the tendon of the extensor digitorum muscle and is a site where a pulse may be detected?

- **Dorsal venous arch.** Superficial veins on the dorsum of the foot that unite to form the small and great saphenous veins. The great saphenous vein is the longest vein of the body.
- **Tendons of extensor digitorum longus muscle.** Visible in line with phalanges II–V (or 2–5).
- **Tendon of extensor hallucis longus muscle.** Visible in line with phalanx I (great toe). Pulsations in the dorsalis pedis artery may be felt in most people just lateral to this tendon where the blood vessel passes over the navicular and cuneiform bones of the tarsus.

Now that you have studied some of the surface features of the principal regions of the body, it is hoped you can appreciate their importance in locating the positions of many internal structures, especially bones, joints, and muscles. In later chapters you will learn to relate surface features to various blood vessels, lymph nodes, nerves, and organs in the chest, abdomen, and pelvis. Once you have visualized and palpated as many of these structures as possible on your own body, you can likely better appreciate the application of surface anatomy to your anatomical and clinical studies.

✔ What are the major anterior and posterior muscles of the thigh?
✔ Define the femoral triangle. Why is it important?
✔ What are the borders of the popliteal fossa? What structures pass through it?
✔ Name the calf muscles. Where do they insert?

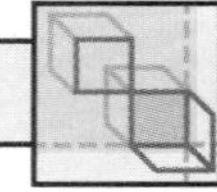

STUDY OUTLINE

Introduction (p. 345)

1. Surface anatomy is the study of anatomical landmarks on the exterior of the body.
2. Surface features may be noted by visualization and/or palpation.
3. Knowledge of surface anatomy has many anatomical and clinical applications.
4. The regions of the body include the head, neck, trunk, upper limbs, and lower limbs.

Surface Anatomy of the Head (p. 345)

1. The head is divided into a cranium and face, each of which is divided into distinct regions.
2. Many muscles of facial expression and skull bones are easily palpated.
3. The eyes, ears, and nose present numerous readily identifiable surface features.

Surface Anatomy of the Neck (p. 350)

1. The neck connects the head to the trunk.
2. It contains several important arteries and veins.
3. The neck is divisible into several triangles by specific muscles and bones.

Surface Anatomy of the Trunk (p. 352)

1. The trunk is divided into the back, chest, abdomen, and pelvis.
2. The back presents several muscles and superficial bones that serve as landmarks.
3. The triangle of auscultation is a region of the back not covered by superficial muscles where respiratory sounds can be heard clearly.
4. The skeleton of the chest protects the organs within and provides several surface landmarks for identifying the position of the heart and lungs.
5. The abdomen and pelvis contain several important landmarks, such as the linea alba, linea semilunaris, McBurney's point, and supracristal line.
6. The perineum forms the floor of the pelvis.

Surface Anatomy of the Upper Limb (Extremity) (p. 358)

1. The upper limb consists of the shoulder, armpit, arm, elbow, forearm, wrist, and hand.
2. The prominence of the shoulder is formed by the deltoid muscle, a frequent site of an intramuscular injection.
3. The axilla is a pyramid-shaped area at the junction of the arm and chest that contains blood vessels and nerves that pass between the neck and upper limbs.
4. The arm contains several blood vessels used for withdrawing blood, introducing fluids, taking the pulse, and measuring blood pressure.
5. Several muscles of the forearm are more easily identifiable by their tendons. The radial artery is a principal blood vessel for detecting a pulse.
6. The hand contains the origin of the cephalic vein and extensor tendons on the dorsum, and the thenar and hypothenar eminences on the palm.

Surface Anatomy of the Lower Limb (Extremity) (p. 362)

1. The lower limb consists of the buttock, thigh, knee, leg, ankle, and foot.
2. The buttock is formed mainly by the gluteus maximus muscle.
3. The major muscles of the thigh are the anterior quadriceps femoris and the posterior hamstrings. An important landmark in the thigh is the femoral triangle.
4. The popliteal fossa is a diamond-shaped area on the posterior aspect of the knee.
5. The gastrocnemius muscle forms the bulk of the midportion of the posterior aspect of the leg and together with the soleus muscle forms the calf of the leg.
6. The foot contains the origin of the great saphenous vein, the largest vein in the body.

SELF-QUIZ QUESTIONS

Choose the one best answer to these questions.

1. Which of the following is true of the sternal angle? (a) It is palpated to locate the costal cartilage of the second rib. (b) It is the junction between the manubrium and body of the sternum. (c) It is external to the point at which the trachea branches into the right and left primary bronchi. (d) a and b are correct (e) a, b, and c are correct.
2. The face is subdivided into several regions. Which of the following does *not* belong? (a) temporal, (b) nasal, (c) mental, (d) oral, (e) auricular.
3. The superior left point of the ________ and the apex of the ________ can be palpated just superior to the medial third of the clavicle. (a) heart, left lung (b) heart, heart (c) left lung, heart (d) left lung, left lung

Complete the following.

4. The head is divided into the ________ region and the ________ region.
5. The ________ muscle is a useful landmark for locating the carotid artery.
6. The ________ artery is commonly used to measure blood pressure.
7. Use the directional terms (anterior, posterior, superior, inferior, proximal, distal, medial, lateral) to complete the following. (a) The mental region is ________ to the oral region. (b) The orbital region is ________ to the frontal region. (c) The antebrachium is ________ to the brachium. (d) The sternal angle is ________ to the xiphoid process. (e) The tragus of the ear is ________ to the external auditory canal. (f) The iris of the eye is ________ to the medial commissure. (g) The lobule is the ________ portion of the auricle.
8. The inguinal ligament, sartorius muscle, and the adductor longus muscle border the ________ triangle.
9. Match the following terms with their descriptions.

____ (a) prominent posterior projection in the midline at the base of the neck		(1) thyroid cartilage
____ (b) most prominent structure in the midline of the anterior cervical region		(2) iliac crest
____ (c) superior border of the buttock		(3) ischial tuberosity
____ (d) forms the tip of the shoulder		(4) acromion
____ (e) located deep to the dimple that coincides with the middle of the sacroiliac joint		(5) vertebra prominens
____ (f) bears most of the weight of a person when sitting		(6) calcaneus
____ (g) attachment point for calcaneal (Achilles) tendon		(7) styloid process
____ (h) protuberance at the distal end of the ulna		(8) posterior superior iliac spine

Are the following statements true or false?

10. The ulnar nerve lies in a groove posterior to the medial epicondyle.

11. All of the following surface features are located on the lower limbs: tibial tuberosity, calcaneal (Achilles) tendon, lateral malleolus.

12. The trapezius muscle forms the posterior border of the posterior triangle of the lateral cervical region.

13. McBurney's point is located at the half way point along a line between the umbilicus and the anterior superior iliac spine.

14. A landmark for performing a tracheostomy is the cricoid cartilage.

15. The latissimus dorsi, trapezius, and sacrum form the triangle of auscultation.

16. A vein of the upper limb often used for intravenous therapy is located in the popliteal fossa.

17. Match the following common and anatomical terms.

____	(a) skull	(1) supercilia
____	(b) eyebrows	(2) palpebrae
____	(c) cheek	(3) auricular region
____	(d) hand	(4) buccal region
____	(e) ear	(5) cranium
____	(f) eyelids	(6) gluteal region
____	(g) buttock	(7) manus

18. Match the following.

____	(a) frequent site for intramuscular injections in the upper limb	(1) palmaris longus
____	(b) a "stiff neck" results from an inflammation of this muscle	(2) triceps brachii
____	(c) divides neck into anterior and posterior (lateral) triangles	(3) trapezius
____	(d) contributes to the anterior axillary fold	(4) deltoid
____	(e) occupies most of the posterior surface of the arm	(5) vastus lateralis
____	(f) tendon is visible on the anterior surface of the wrist when making a fist	(6) extensor pollicus longus
____	(g) tendon forms one border of the "anatomical snuffbox"	(7) sternocleidomastoid
____	(h) frequently used as a site for insulin injection	(8) pectoralis major

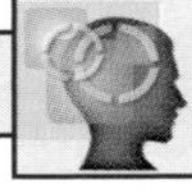

CRITICAL THINKING QUESTIONS

1. Fred, the football fanatic, ran into the goal post and landed on his back. He told the coach that he felt all right but when he reached behind his head, he felt a bump at the base of his neck. The team physician examined Fred and told him that the bump was nothing to worry about; that it had always been there. What is the bump?
HINT *Feel your neck.*

2. Great Aunt Edith heard you're bringing your new boyfriend, the one with an anatomically correct heart tattooed on his bicep (he can even make it beat) and "I break for ambulances" bumper sticker, to Thanksgiving dinner. She told your mother that she's having "palpations." Your new boyfriend, a first year medical student, said she probably meant "palpitations." Is there a difference?
HINT *Do you palpate or palpitate someone's pulse?*

3. An ad for a new telephoto camera appeared in a medical journal. It states "you'll see the pores on your neighbor's ala, the sweat on his philtrum, the hairs in his external nares, the studs in his helix, the ring in his supercilia and the color of his sclera." Is this a great camera? Explain.
HINT *This gives new meaning to "close up."*

4. Juan donated a pint of blood at the Red Cross blood drive. After donating, he can hold the gauze pad in place without using his hand. Use correct terminology to describe where Juan got stuck.
HINT *The phlebotomist prepared the commonly used site to insert the needle.*

5. While the new mom and dad were diapering the baby, the mom said that the baby had the cutest little dimples in her cheeks. The dad agreed that their baby was the cutest one in the world but said he didn't remember seeing any dimples on the baby's face. Help dad out with a quick anatomy lesson.
HINT *She's talking about the OTHER cheeks.*

ANSWERS TO FIGURE QUESTIONS

11.1 Frontal, parietal, temporal, occipital, zygomatic, and nasal.

11.2 Galea aponeurotica.

11.3 Pupil.

11.4 Mastoid process.

11.5 Dorsum nasi.

11.6 Sternocleidomastoid.

11.7 With the aid of a stethoscope, respiratory sounds can be heard clearly in the triangle.

11.8 Sternal angle.

11.9 Pain produced by pressure of the finger on McBurney's point indicates acute appendicitis; the incision for an appendectomy is made through it.

11.10 Acromion of the scapula.

11.11 Median cubital vein.

11.12 Radial artery.

11.13 The thumb.

11.14 Gluteus maximus.

11.15 The fossa is bordered superolaterally by the biceps femoris, superomedially by the semimembranosus and semitendinosus, and inferolaterally and inferomedially by the lateral and medial heads of the gastrocnemius.

11.16 Dorsalis pedis artery.

12

THE CARDIOVASCULAR SYSTEM: BLOOD

CONTENTS AT A GLANCE

◀ Page 377

Can you guess what is being practiced in this illustration? Can you speculate as to the culture of the people in this image?

Page 373 ▶

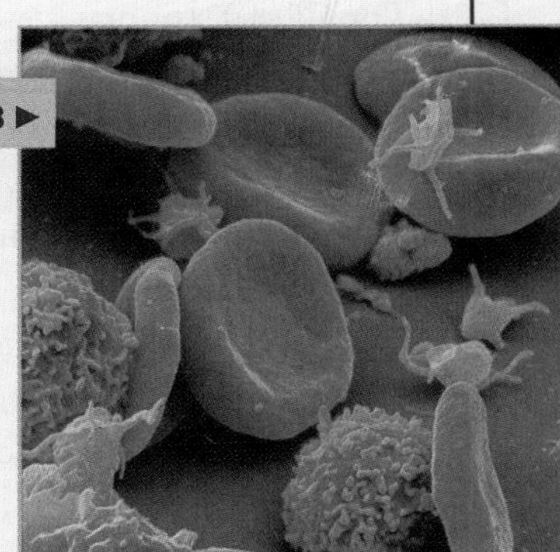

INTRODUCTION

The **cardiovascular system** (*cardio* = heart; *vascular* = blood or blood vessels) consists of the blood, heart, and blood vessels. The focus of this chapter is blood, while the next two chapters cover the heart and blood vessels, respectively.

Most cells of a multicellular organism cannot move around to obtain oxygen and nutrients and get rid of carbon dioxide and other wastes. Instead, these needs are met by two fluids: blood and interstitial fluid. **Blood** is a connective tissue composed of a liquid portion called plasma (the matrix) and a cellular portion consisting of various cells and cell fragments. **Interstitial fluid** is the fluid that bathes body cells. Oxygen brought into the lungs and nutrients brought into the gastrointestinal tract are transported by the blood to the cells of the body. The oxygen and nutrients diffuse from the blood into the interstitial fluid and then into body cells. Carbon dioxide and other wastes move in the reverse direction from the body cells into the interstitial fluid and then into the blood. Blood then transports the wastes to various organs—the lungs, kidneys, skin, and digestive system—for elimination from the body. As you will see in this chapter, blood not only transports various substances throughout the body, it also helps regulate several life processes and affords protection against disease.

In order for blood to reach all the cells, it must be moved throughout the body. The *heart* is the pump that circulates the blood. Blood vessels convey blood from the heart to body cells and from body cells back to the heart. The largest blood vessels, which take blood away from the heart, are called *arteries.* Arteries branch into smaller vessels called *arterioles.* As an arteriole enters a tissue, it divides into numerous microscopic vessels called *capillaries.* Substances exchanged between the blood and interstitial fluid pass through the thin walls of capillaries. Before leaving a tissue, capillaries unite to form small vessels called *venules,* which in turn merge to form progressively larger vessels called *veins.* Veins convey blood back to the heart.

For all of its similarities in origin, composition, and functions, blood is as unique from one person to another as are skin, bone, and hair. Health care professionals routinely examine and analyze its differences through various blood tests when attempting to determine the cause of different diseases. Yet, despite these differences, blood is the most easily and widely shared of human tissues, saving many thousands of lives every year through blood transfusions.

The branch of science concerned with the study of blood, blood-forming tissues, and the disorders associated with them is **hematology** (hēm-a-TOL-ō-jē; *hem-* or *hemat-* = blood; *-ology* = study of).

FUNCTIONS OF BLOOD

Objective

- List and describe the functions of blood.

Blood, the only liquid connective tissue, has three general functions:

1. **Transportation.** Blood transports oxygen from the lungs to the cells of the body and carbon dioxide from body cells to the lungs. It also carries nutrients from the gastrointestinal tract to body cells, heat and waste products away from body cells, and hormones from endocrine glands to other body cells.
2. **Regulation.** Blood helps regulate pH through buffers. It helps regulate body temperature through the capacity of the water in plasma to absorb and give off heat and through its variable rate of flow through the skin, where excess heat can be lost from the blood to the environment. Blood osmotic pressure also influences the water content of cells, mainly through interactions involving dissolved ions and proteins.
3. **Protection.** Blood can clot, which protects against its excessive loss from the cardiovascular system after an injury. In addition, white blood cells protect against disease by carrying on phagocytosis and producing proteins called antibodies. Blood also contains proteins called interferons and complement that help protect against disease.

✔ What substances does blood transport?

PHYSICAL CHARACTERISTICS OF BLOOD

Objective

- List the principal physical characteristics of blood.

Blood is denser and more viscous (sticky) than water, which is part of the reason it flows more slowly than water. The temperature of blood is about 38°C (100.4°F), which is slightly higher than normal body temperature, and it has a slightly alkaline pH ranging from 7.35 to 7.45. Blood constitutes about 8% of the total body weight. The blood volume is 5–6 liters (1.5 gal) in an average-sized adult male and 4–5 liters (1.2 gal) in an average-sized adult female.

Blood samples for laboratory testing may be obtained in several ways. The most frequently used procedure is **venipuncture,** withdrawal of blood from a vein using a hypodermic needle and syringe. A tourniquet is wrapped around the arm above the venipuncture site to cause the blood to accumulate in the vein. This increased blood volume causes the vein to stand out. Opening and closing the fist further causes it to stand out, making the venipuncture more successful. Another method of drawing blood is through a **finger** or **heel stick.** This is frequently used by diabetic patients who monitor their daily blood sugar as well as for drawing blood from infants and children. A common site for venipuncture is the median cubital vein anterior to the elbow (see Figure 14.13b on page 451).

✔ How much does the blood of a 150 pound person weigh?

COMPONENTS OF BLOOD

Objective

- Describe the principal components of blood.

Whole blood is composed of blood plasma, a watery liquid that contains dissolved substances, and formed elements, which are cells and cell fragments. If a sample of blood is centrifuged (spun) in a small glass tube, the cells sink to the bottom of the tube while the lighter-weight plasma forms a layer on top (Figure 12.1a). Blood is about 45% formed elements and about 55% plasma. Normally, more than 99% of the formed elements are red-colored red blood cells (RBCs). Pale or colorless white blood cells (WBCs) and platelets occupy less than 1% of total blood volume. They form a very thin layer, called the *buffy coat,* between the packed RBCs and plasma in centrifuged blood. Figure 12.1b shows the composition of blood plasma and the numbers of the various types of formed elements in blood.

Blood Plasma

When the formed elements are removed from blood, the straw-colored liquid called **blood plasma** is left. Plasma is about 91.5% water and 8.5% solutes, most of which (7% by weight) are proteins. Some of the proteins in plasma are also found elsewhere in the body, but those confined to blood are called **plasma proteins.** Hepatocytes (liver cells) synthesize most of the plasma proteins, which include the **albumins** (54% of plasma proteins), **globulins** (38%), and **fibrinogen** (7%). Certain blood cells develop into cells that produce gamma globulins, an important type of globulin. These plasma proteins are also called **antibodies** or **immunoglobulins** because they are produced during certain immune responses. Foreign invaders such as bacteria and viruses stimulate production of antibodies. An antibody binds specifically to the foreign substance, called an **antigen,** that provoked its production; thus, there are millions of different types of antibodies. When the antigen and antibody come together, they form an **antigen–antibody complex,** which disables the invading antigen.

Besides proteins, other solutes in plasma include electrolytes, nutrients, regulatory substances such as enzymes and hormones, gases, and waste products such as urea, uric acid, creatinine, ammonia, and bilirubin.

Table 12.1 on page 373 describes the chemical composition of blood plasma.

Formed Elements

The **formed elements** of the blood include **red blood cells (RBCs), white blood cells (WBCs),** and **platelets** (Figure 12.2 on page 373). Although RBCs and WBCs are living cells, platelets are cell fragments. Unlike RBCs and platelets, which perform limited roles, WBCs have a number of more general functions. There are several distinct types of WBCs—neutrophils, lymphocytes, monocytes, eosinophils, and basophils—each having a unique microscopic appearance. The roles of each type of WBC are discussed later in this chapter.

The percentage of total blood volume occupied by RBCs is called the **hematocrit** (he-MAT-ō-krit). For example, a hematocrit of 40 means that 40% of the volume of blood is composed of RBCs. The normal range of hematocrit for adult females is 38–46% (average = 42); for adult males it is 40–54% (average = 47). The hormone testosterone, which is present in much higher concentration in males than in females, stimulates synthesis of a hormone called erythropoietin by the kidneys. This hormone, which stimulates production of RBCs, contributes to higher hematocrits in males. Lower values in women during their reproductive years may be due to excessive loss of blood during menstruation. A significant drop in hematocrit indicates anemia, a lower than normal number of RBCs. In *polycythemia* the percentage of RBCs is abnormally high, and the hematocrit may be 65% or higher. Polycythemia may be caused by conditions such as an unregulated increase in RBC production, tissue hypoxia, dehydration, and blood doping by athletes.

In recent years, some athletes have been tempted to try **blood doping** in efforts to enhance their performance. In this procedure, blood cells are removed from the body, stored for a month or so, and then reinjected a few days before an athletic event. Because delivery of oxygen to muscle is a limiting factor in muscular feats, and given that red blood cells carry oxygen, increasing the oxygen-carrying capacity of the blood was expected to increase muscular performance. Indeed, blood doping can improve athletic performance in endurance events. The practice is dangerous, however, because it increases the workload of the heart. The increased number of RBCs raises the viscosity of the blood, making the blood more difficult for the heart to pump. Moreover, the procedure is banned by the International Olympics Committee.

✔ Distinguish between blood plasma and formed elements.
✔ What are the major constituents of blood plasma? What does each do?

FORMATION OF BLOOD CELLS

Objective

- Explain the origin of blood cells.

Although some lymphocytes have a lifetime measured in years, most formed elements of the blood are continually dying and being replaced within hours, days, or weeks. Various mechanisms regulate the total number of RBCs and platelets in circulation, and the numbers normally remain steady. The abundance

Figure 12.1 / Components of blood in a normal adult.

Blood is a connective tissue that consists of plasma (liquid) plus formed elements (red blood cells, white blood cells, and platelets).

Functions of Blood

1. Transport of oxygen, carbon dioxide, nutrients, hormones, heat, and wastes.
2. Regulation of pH, body temperature, and water content of cells.
3. Protection against blood loss through clotting, and against disease through phagocytic white blood cells and antibodies.

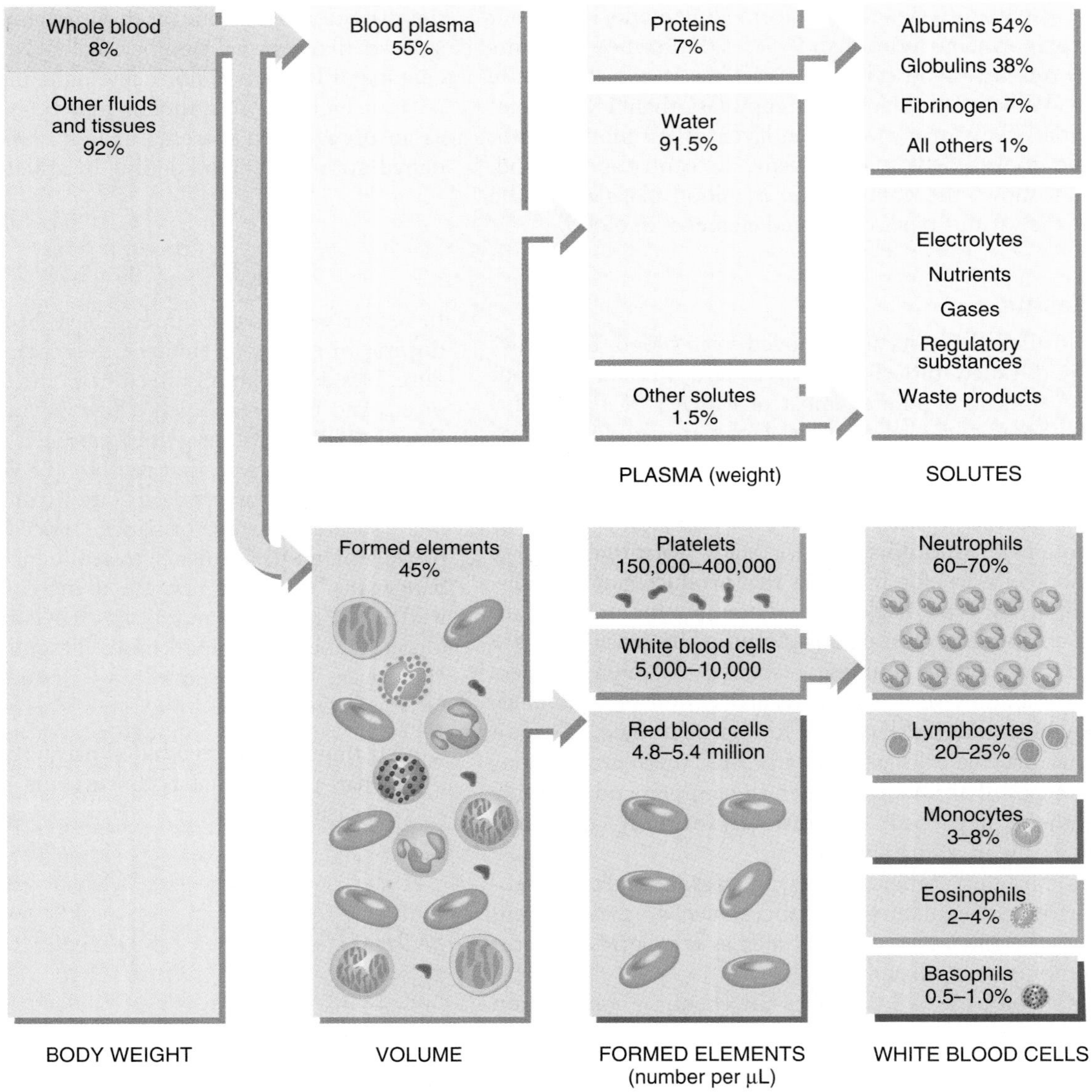

Plasma (55%)

Buffy coat, composed of white blood cells and platelets

Red blood cells (45%)

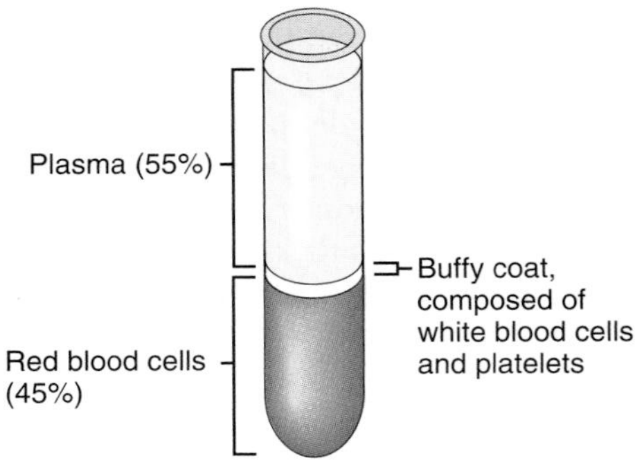

(a) Appearance of centrifuged blood

BODY WEIGHT

Whole blood 8%

Other fluids and tissues 92%

VOLUME

Blood plasma 55%

Formed elements 45%

PLASMA (weight)

Proteins 7%

Water 91.5%

Other solutes 1.5%

SOLUTES

Albumins 54%

Globulins 38%

Fibrinogen 7%

All others 1%

Electrolytes

Nutrients

Gases

Regulatory substances

Waste products

FORMED ELEMENTS (number per µL)

Platelets 150,000–400,000

White blood cells 5,000–10,000

Red blood cells 4.8–5.4 million

WHITE BLOOD CELLS

Neutrophils 60–70%

Lymphocytes 20–25%

Monocytes 3–8%

Eosinophils 2–4%

Basophils 0.5–1.0%

(b) Components of blood

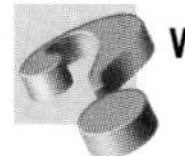

What is the approximate volume of blood in your body?

Table 12.1 Substances in Blood Plasma

Constituent	Description
Water (91.5%)	Liquid portion of blood. Acts as solvent and suspending medium for components of blood; absorbs, transports, and releases heat.
Proteins (7.0%)	Exert osmotic pressure, which helps maintain water balance between blood and tissues and regulate blood volume.
Albumins	Smallest and most numerous plasma proteins; produced by liver. Function as transport proteins for several steroid hormones and for fatty acids.
Globulins	Protein group to which antibodies (immunoglobulins) belong. Produced by liver and by plasma cells, which develop from B lymphocytes. Antibodies help attack viruses and bacteria. Alpha and beta globulins transport iron, lipids, and fat-soluble vitamins.
Fibrinogen	Produced by liver. Plays essential role in blood clotting.
Other Solutes (1.5%)	
Electrolytes	Inorganic salts. Positively charged ions (cations) include Na^+, K^+, Ca^{2+}, Mg^{2+}; negatively charged ions (anions) include Cl^-, HPO_4^{2-}, SO_4^{2-}, and HCO_3^-. Help maintain osmotic pressure and play essential roles in the various functions of cells.
Nutrients	Products of digestion pass into blood for distribution to all body cells. Include amino acids (from proteins), glucose (from carbohydrates), fatty acids and glycerol (from triglycerides), vitamins, and minerals.
Gases	Oxygen (O_2), carbon dioxide (CO_2), and nitrogen (N_2). Whereas more O_2 is associated with hemoglobin inside red blood cells, more CO_2 is dissolved in plasma. N_2 has no known function in the body.
Regulatory substances	Enzymes, produced by body cells, catalyze chemical reactions. Hormones, produced by endocrine glands, regulate growth and development in the body.
Wastes	Most are breakdown products of protein metabolism and are carried by blood to organs of excretion. Include urea, uric acid, creatine, creatinine, bilirubin, and ammonia.

of the different types of WBCs, however, varies in response to challenges by invading pathogens and other foreign antigens.

The process by which the formed elements of blood develop is called **hemopoiesis** (hē-mō-poy-Ē-sis; *-poiesis* = making) or *hematopoiesis*. About 0.05–0.1% of red bone marrow cells are **pluripotent stem cells** (Figure 12.3), which are derived from mesenchyme. These cells replenish themselves, proliferate, and differentiate into cells that give rise to *all* the formed elements of the blood.

Originating from the pluripotent stem cells are two other types of stem cells, called *myeloid stem cells* and *lymphoid stem cells.* They also have the ability to replenish themselves and to differentiate into other formed elements of blood. Myeloid stem cells begin their development in red bone marrow and give rise to red blood cells, platelets, monocytes, neutrophils, eosinophils, and basophils. Lymphoid stem cells begin their development in red bone marrow but complete it in lymphatic tissues; they give rise to lymphocytes. Although the various stem cells have distinctive cell identity markers in their plasma membranes, they cannot be distinguished histologically and resemble lymphocytes.

During hemopoiesis, the myeloid stem cells differentiate into **progenitor cells**—cells that no longer are capable of replenishing themselves and are committed to giving rise to more specific elements of blood. Some progenitor cells are referred to as *colony-forming units (CFUs).* After the CFU designation is an abbreviation that designates the mature elements in blood they will produce: CFU-E ultimately produces red blood cells (erythrocytes), CFU-Meg ultimately produces platelets, and CFU-GM ultimately produces neutrophils and monocytes. Progenitor cells, like stem cells, resemble lymphocytes and cannot be distinguished by their microscopic appearance alone. Other myeloid stem cells develop directly into cells called precursor cells (de-

Figure 12.2 / Scanning electron micrograph of the formed elements of blood.

The formed elements of blood are red blood cells (RBCs), white blood cells (WBCs), and platelets.

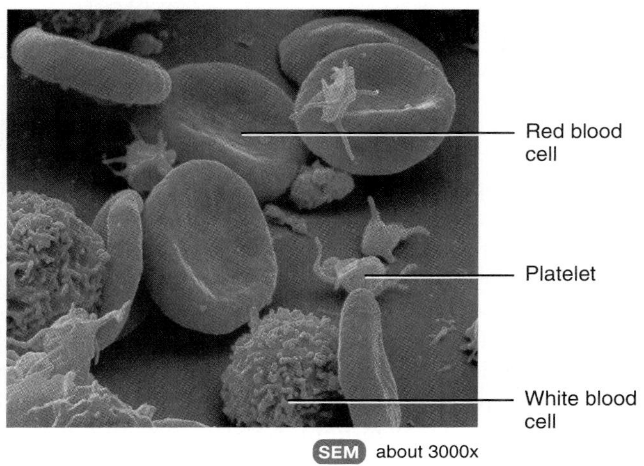

Which formed elements of the blood are cell fragments?

Figure 12.3 / Origin, development, and structure of blood cells. Some of the generations of some cell lines have been omitted.

Blood cell production is called hemopoiesis and occurs only in red bone marrow after birth.

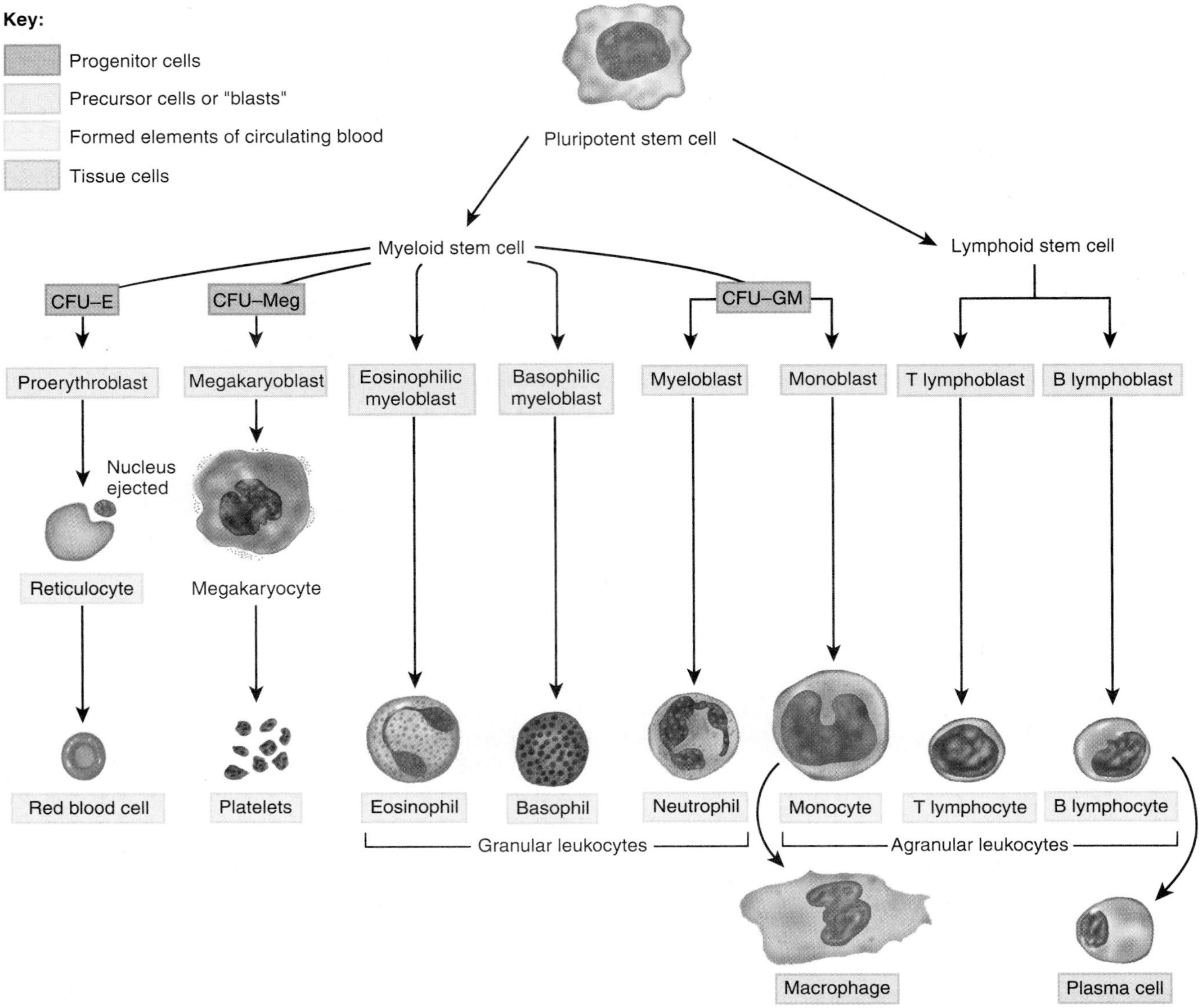

Key:
Progenitor cells
Precursor cells or "blasts"
Formed elements of circulating blood
Tissue cells

Pluripotent stem cell
Myeloid stem cell
Lymphoid stem cell
CFU–E
CFU–Meg
CFU–GM
Proerythroblast
Megakaryoblast
Eosinophilic myeloblast
Basophilic myeloblast
Myeloblast
Monoblast
T lymphoblast
B lymphoblast
Nucleus ejected
Reticulocyte
Megakaryocyte
Red blood cell
Platelets
Eosinophil
Basophil
Neutrophil
Monocyte
T lymphocyte
B lymphocyte
Granular leukocytes
Agranular leukocytes
Macrophage
Plasma cell

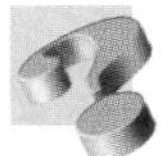

From which connective tissue do pluripotent stem cells develop?

scribed next). Lymphoid stem cells differentiate into pre-B cells and prothymocytes, which ultimately develop into B lymphocytes and T lymphocytes, respectively.

In the next generation, the cells are called **precursor cells,** or **blasts.** Over several cell divisions, they develop into the actual formed elements of blood. For example, monoblasts develop into monocytes, eosinophilic myeloblasts develop into eosinophils, and so on. Precursor cells exhibit recognizable histological characteristics.

During embryonic and fetal life, the yolk sac, liver, spleen, thymus, lymph nodes, and red bone marrow all participate at various times in producing the formed elements. After birth, however, hemopoiesis takes place *in red bone marrow only*, although small numbers of stem cells circulate in the bloodstream. Red bone marrow is found in the epiphyses (ends) of long bones such as the humerus and femur; in flat bones such as the sternum, ribs, and cranial bones; in vertebrae; and in the pelvis. With the exception of lymphocytes, formed elements do not divide once they leave the bone marrow.

Several hormones called **hemopoietic growth factors** regulate the differentiation and proliferation of particular progenitor cells. **Erythropoietin** (e-rith′-rō-POY-e-tin) or **EPO,** produced mainly by the kidneys, increases the numbers of red blood cell precursors. **Thrombopoietin** (throm′-bō-POY-e-tin) or **TPO** is produced by the liver and stimulates formation of platelets (thrombocytes). Several different cytokines regulate development of different blood cell types. **Cytokines** are produced by red bone marrow cells, leukocytes, macrophages, fibroblasts, and endothelial cells. They stimulate proliferation of progenitor cells in red bone marrow and regulate the activities of cells involved in nonspecific defenses (such as phagocytes) and immune responses (such as B cells and T cells). Two important families of cytokines that stimulate white blood cell formation are **colony-stimulating factors (CSFs)** and **interleukins.**

Medical Uses of Hemopoietic Growth Factors

Hemopoietic growth factors made available through recombinant DNA technology hold tremendous potential for medical uses when a person's natural ability to form blood cells is diminished or defective. Recombinant erythropoietin (EPO) is very effective in treating the diminished red blood cell production that accompanies end-stage kidney disease. Granulocyte–macrophage colony-stimulating factor (GM-CSF) and granulocyte CSF (G-CSF) are given to stimulate white blood cell formation in cancer patients who are receiving chemotherapy, which tends to kill their red bone marrow cells as well as the cancer cells. Thrombopoietin shows great promise for preventing platelet depletion during chemotherapy. CSFs and thrombopoietin also improve the outcome of patients who receive bone marrow transplants. ■

- ✔ Briefly explain the origin of blood cells.
- ✔ Describe the hemopoietic growth factors that regulate differentiation and proliferation of particular progenitor cells.

RED BLOOD CELLS

Objective

- Describe the structure, functions, life cycle, and production of red blood cells.

Red blood cells (RBCs) or **erythrocytes** (e-RITH-rō-sīts; *erythro-* = red; *-cyte* = cell) contain the oxygen-carrying protein **hemoglobin,** which is a pigment that gives whole blood its red color. A healthy adult male has about 5.4 million red blood cells per microliter (μL) of blood*, and a healthy adult female has about 4.8 million. (One drop of blood is about 50 μL.) To maintain normal quantities of RBCs, new mature cells must enter the circulation at the astonishing rate of at least 2 million per second, a pace that balances the equally high rate of RBC destruction.

* 1 μL = 1 mm^3 = 10^{-6} liter

RBC Anatomy

RBCs are biconcave discs with a diameter of 7–8 μm (Figure 12.4a). Mature red blood cells have a simple structure. Their plasma membrane is both strong and flexible, which allows them to deform without rupturing as they squeeze through narrow capillaries. As you will see later, certain glycolipids in the plasma membrane of RBCs are antigens that account for the various blood groups such as the ABO and Rh groups. RBCs lack a nucleus and other organelles and can neither reproduce nor carry on extensive metabolic activities. The cytosol of RBCs contains dissolved hemoglobin molecules, which were synthesized before loss of the nucleus during RBC production; hemoglobin constitutes about 33% of the cell's weight.

RBC Functions

Red blood cells are highly specialized for their oxygen transport function. Each one contains about 280 million hemoglobin molecules. Because circulating RBCs have no nucleus, all their internal space is available for oxygen transport. Moreover, RBCs lack mitochondria and generate ATP anaerobically (without oxygen); as a result, they do not consume any of the oxygen they transport. Even the shape of an RBC facilitates its function. A biconcave disc has a much greater surface area for its volume than, say, a sphere or a cube. This shape provides a large surface area for the diffusion of gas molecules into and out of the RBC.

A hemoglobin molecule consists of a protein called **globin,** composed of four polypeptide chains (two alpha and two beta chains), plus four nonprotein pigments called **hemes** (Figure 12.4b). Each heme is associated with one polypeptide chain and

Figure 12.4 / The shapes of a red blood cell (RBC) and a hemoglobin molecule. In (b), note that each of the four polypeptide chains of a hemoglobin molecule (blue) has one heme group (gold), which contains an iron ion (Fe^{2+}), shown in red.

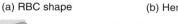

The iron portion of a heme group binds oxygen for transport by hemoglobin.

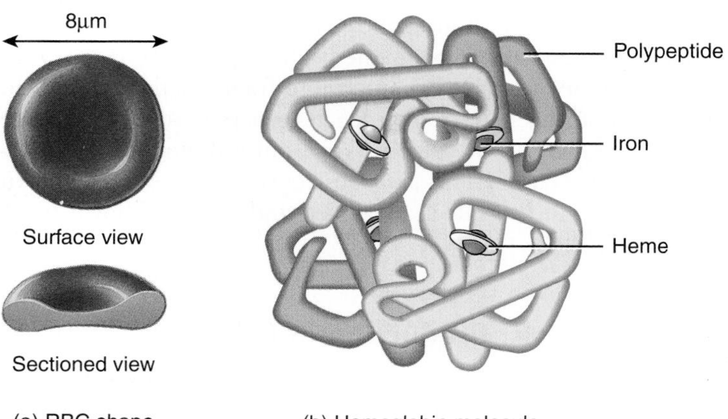

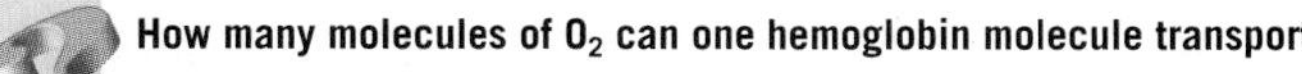

contains an iron ion (Fe^{2+}) that can combine reversibly with one oxygen molecule. The oxygen picked up in the lungs is transported to other tissues of the body bound to the iron of the heme group. In the tissues, the iron–oxygen reaction reverses. Hemoglobin releases oxygen, which diffuses first into the interstitial fluid and then into cells.

Hemoglobin also transports about 23% of the total carbon dioxide, a waste product of metabolism. Blood flowing through tissue capillaries picks up carbon dioxide, some of which combines with amino acids in the globin portion of hemoglobin. As blood flows through the lungs, the carbon dioxide is released from hemoglobin and then exhaled.

In addition to its key role in transporting oxygen and carbon dioxide, hemoglobin also assumes a function in blood pressure regulation, a recent and surprising discovery. The iron ions in the heme portion of hemoglobin have a strong affinity for *nitric oxide (NO)*, a gas produced by the endothelial cells that line blood vessels. Hemoglobin that has already circulated through the body and given up its oxygen to tissues passes into the lungs, where it releases both carbon dioxide to be exhaled and NO. Then the hemoglobin picks up a fresh supply of oxygen and a different form of NO called *super nitric oxide (SNO)*, probably produced by lung cells. In body tissues, hemoglobin releases oxygen and SNO and picks up carbon dioxide. The released SNO causes *vasodilation*, an increase in blood vessel diameter that occurs when the smooth muscle in the vessel wall relaxes. At the tissue level, hemoglobin also picks up excess NO, which tends to cause *vasoconstriction*, a decrease in blood vessel diameter that occurs when the smooth muscle in the vessel wall contracts. Vasodilation decreases blood pressure, whereas vasoconstriction increases it. By ferrying NO and SNO throughout the body, hemoglobin helps regulate blood pressure by adjusting the amount of NO or SNO to which blood vessels are exposed. This newly appreciated role of hemoglobin may influence development of drugs to treat hypertension (high blood pressure).

RBC Life Cycle

Red blood cells live only about 120 days because of the wear and tear their plasma membranes undergo as they squeeze through blood capillaries. Without a nucleus and other organelles, RBCs cannot synthesize new components to replace damaged ones. The plasma membranes become more fragile with age, and the cells are more likely to burst, especially as they squeeze through narrow channels in the spleen. Worn-out red blood cells are removed from circulation and destroyed by fixed phagocytic macrophages in the spleen and liver, and the breakdown products are recycled.

Erythropoiesis: Production of RBCs

Erythropoiesis (e-rith′-rō-poy-Ē-sis), the production of RBCs, starts in the red bone marrow with a precursor cell called a proerythroblast (see Figure 12.3). The proerythroblast divides several times, producing cells that begin to synthesize hemoglobin. Ultimately, a cell near the end of the developmental sequence ejects its nucleus and becomes a **reticulocyte** (re-TIK-yū-lō-sīt). Loss of the nucleus causes the center of the cell to indent, producing a distinctive biconcave shape. Reticulocytes, which are about 34% hemoglobin and retain some mitochondria, ribosomes, and endoplasmic reticulum, pass from red bone marrow into the bloodstream by squeezing between the endothelial cells of blood capillaries. Reticulocytes usually develop into erythrocytes, or mature red blood cells, within 1–2 days after their release from red bone marrow.

Normally, erythropoiesis and red blood cell destruction proceed at roughly the same pace. If the oxygen-carrying capacity of the blood falls because erythropoiesis is not keeping up with RBC destruction, RBC production is increased. Cellular oxygen deficiency, called **hypoxia** (hī-POKS-ē-a), may occur if not enough oxygen enters the blood. For example, the reduced oxygen content of air at high altitudes results in a reduced level of oxygen in the blood. Oxygen delivery may also fall due to anemia, which has many causes: lack of iron, lack of certain amino acids, and lack of vitamin B_{12} are but a few (see page 381). Circulatory problems that reduce blood flow to tissues may also reduce oxygen delivery. Whatever the cause, hypoxia stimulates the kidneys to step up the release of erythropoietin. This hormone circulates through the blood to the red bone marrow, where it speeds the development of proerythroblasts into reticulocytes. Premature newborns often exhibit anemia, due in part to inadequate production of erythropoietin. During the first weeks after birth, the liver, not the kidneys, produces most EPO. Because the liver is less sensitive than the kidneys to hypoxia, newborns have a smaller EPO response to anemia than do adults.

Reticulocyte Count

The rate of erythropoiesis is measured by a **reticulocyte count.** Normally, a little less than 1% of the oldest RBCs are replaced by newcomer reticulocytes on any given day. It then takes 1–2 days for the reticulocytes to lose the last vestiges of endoplasmic reticulum and become mature RBCs. Thus, reticulocytes account for about 0.5–1.5% of all RBCs in a normal blood sample. A low "retic" count in a person who is anemic might indicate an inability of the red bone marrow to respond to erythropoietin, perhaps because of a nutritional deficiency or leukemia. A high "retic" count might indicate a good red bone marrow response to previous blood loss or to iron therapy in someone who had been iron deficient. ■

- ✔ Describe the size, microscopic appearance, and functions of RBCs.
- ✔ Define erythropoiesis. Relate erythropoiesis to hematocrit. What factors accelerate and slow erythropoiesis?

Blood Group Systems

More than 100 kinds of antigens have been detected on the surface of red blood cells. These antigens are genetically determined and are referred to as **isoantigens** or **agglutinogens**

14th Century

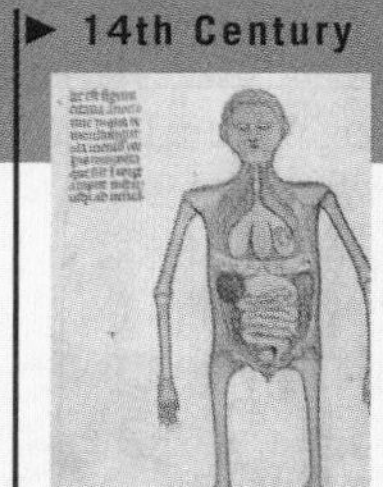

15th Century

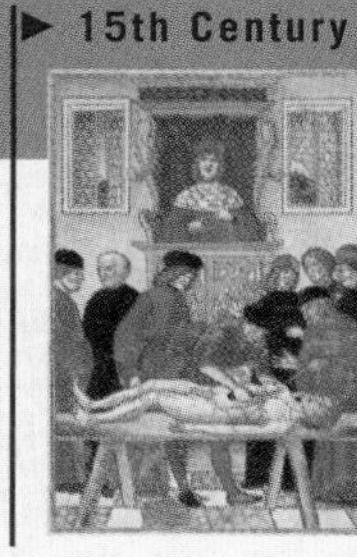

16th Century

17th Century

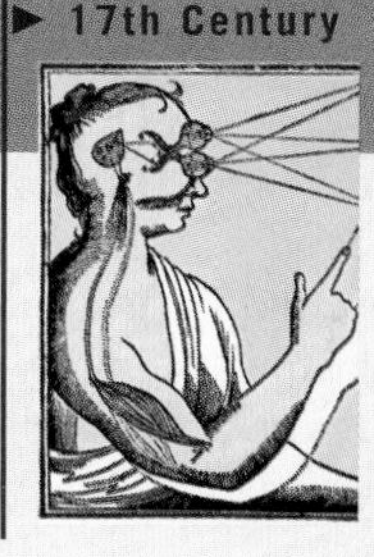

18th Century

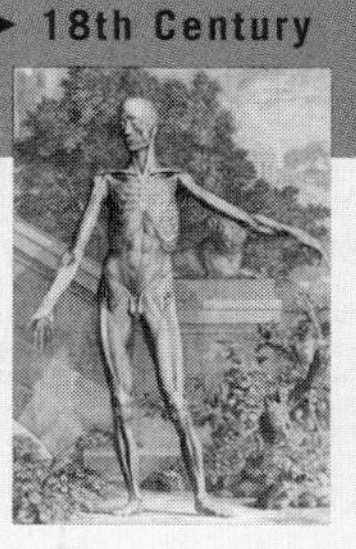

20th Century

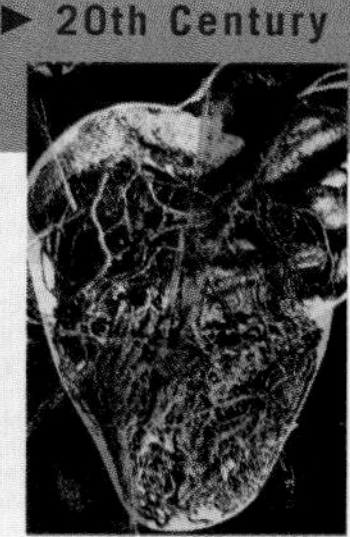

21st Century

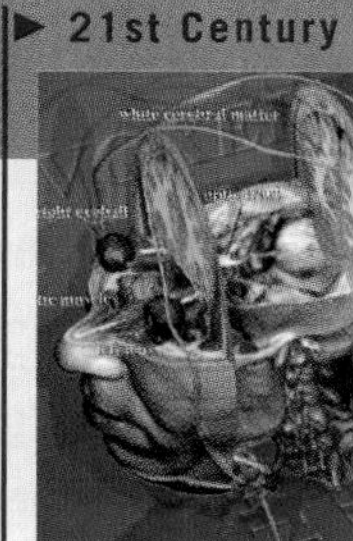

CHANGING IMAGES

Bloodletting

1699 AD

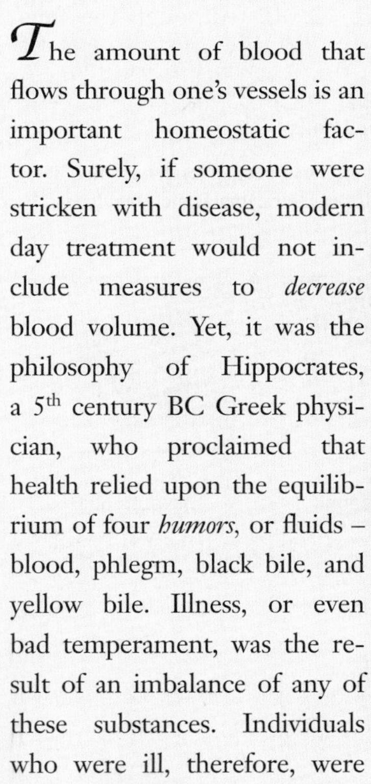

The amount of blood that flows through one's vessels is an important homeostatic factor. Surely, if someone were stricken with disease, modern day treatment would not include measures to *decrease* blood volume. Yet, it was the philosophy of Hippocrates, a 5th century BC Greek physician, who proclaimed that health relied upon the equilibrium of four *humors*, or fluids – blood, phlegm, black bile, and yellow bile. Illness, or even bad temperament, was the result of an imbalance of any of these substances. Individuals who were ill, therefore, were frequently treated by being drained of blood – a procedure termed *bloodletting*. Hippocrates' *Humoral Theory*, as well as the procedure of bloodletting, was followed well into the 19th century and was embraced nearly worldwide.

Pictured here is an illustration titled, *The Indians' Manner of Bloodletting* from Lionel Wafer's 1699 book, *A New Voyage and Description of the Isthmus of America*. Wafer's 17th century adventure to present day eastern Panama demonstrates that bloodletting was also practiced by the native peoples of Central America. Here we see the wife of a tribal chief being treated for an unknown illness. Wafer writes: "The patient is seated on a stone in the river, and one with a small bow shoots little arrows into the naked body."

How is it that such a dangerous practice could survive for over two thousand years? As you read about blood in this chapter, can you explain the perils of this procedure? What other human systems could be affected? Consider issues of hydration, urine formation, blood pressure, and immunity.

(ag′-lū-TIN-ō-jens). Many of these antigens appear in characteristic patterns, a fact that enables scientists or health care professionals to identify a person's blood as belonging to one or more blood grouping systems; at least 14 are currently recognized. Each system is characterized by the presence or absence of specific antigens on the surface of a red blood cell's plasma membrane. The two major blood groups distinguished on the basis of these antigens are called the ABO and Rh blood groups.

The *ABO blood grouping system* is based on two antigens, symbolized as *A* and *B*. Individuals whose erythrocytes manufacture only antigen *A* are said to have blood type A. Those who manufacture only antigen *B* are type B. Individuals who manufacture both *A* and *B* are type AB. Those who manufacture neither are type O.

The *Rh blood grouping system* is so named because it was first worked out using the blood of the *Rh*esus monkey. Individuals whose erythrocytes have the Rh antigens (D antigens) are designated *Rh*$^+$. Those who lack Rh antigens are designated *Rh*$^-$.

As just noted, the presence or absence of certain antigens on red blood cells is the basis for classifying blood into several different groups. Such information is very important when a transfusion is given. A transfusion is the transfer of whole blood or blood components (red blood cells or plasma, for example) into the bloodstream. A transfusion may be given to treat low blood volume, anemia, or a low platelet count. However, in an incompatible blood transfusion, one person's red blood cell antigens may be recognized as foreign when transfused into someone with a different blood type. In this case, the transfused cells undergo hemolysis (burst), releasing hemoglobin into the plasma, and the liberated hemoglobin may cause kidney damage.

WHITE BLOOD CELLS

Objective

- Describe the structure, functions, and production of white blood cells.

WBC Anatomy and Types

Unlike red blood cells, white blood cells, or **leukocytes** (LŪ-kō-sīts; *leuko-* = white), have a nucleus and do not contain hemoglobin (Figure 12.5). WBCs are classified as either granular or agranular, depending on whether they contain conspicuous cytoplasmic vesicles (originally called granules) that are made visible by staining. *Granular leukocytes* include neutrophils, eosinophils, and basophils; *agranular leukocytes* include lymphocytes and monocytes. As shown in Figure 12.3, monocytes and granular leukocytes develop from myeloid stem cells. In contrast, lymphocytes develop from lymphoid stem cells.

Granular Leukocytes

After staining, each of the three types of granular leukocytes display conspicuous granules with distinctive coloration that can be recognized under a light microscope. The large, uniform-sized granules of an **eosinophil** (ē-ō-SIN-ō-fil) are *eosinophilic* (= eosin-loving)—they stain red-orange with acidic dyes (Figure 12.5a). The granules usually do not cover or obscure the nucleus, which most often has two or three lobes connected by either a thin strand or a thick strand. The round, variable-sized granules of a **basophil** (BĀ-sō-fil) are *basophilic* (= basic loving)—they stain blue-purple with basic dyes (Figure 12.5b). The granules commonly obscure the nucleus, which has two lobes. The granules of a **neutrophil** (NŪ-trō-fil) are smaller, evenly distributed, and pale lilac in color (Figure 12.5c); the nucleus has two to five lobes, connected by very thin strands of chromatin. As the cells age, the number of nuclear lobes increases. Because older neutrophils thus have several differently shaped nuclear lobes, they are often called *polymorphonuclear leukocytes (PMNs)*, polymorphs, or "polys." Younger neutrophils are often called *bands* because their nucleus is more rod-shaped.

Agranular Leukocytes

Even though so-called agranular leukocytes possess cytoplasmic granules, the granules are not visible under a light microscope because of their small size and poor staining qualities.

The nucleus of a **lymphocyte** (LIM-fō-sīt) stains dark and is round or slightly indented. The cytoplasm stains sky blue and forms a rim around the nucleus. The larger the cell, the more cytoplasm is visible. Lymphocytes are classified as large or small based on cell diameter: 6–9 μm in small lymphocytes and 10–14 μm in large lymphocytes (Figure 12.5d). (Although the func-

Figure 12.5 / Structure of white blood cells.

White blood cells are distinguished from one another by the shape of their nuclei and the staining properties of their cytoplasmic granules.

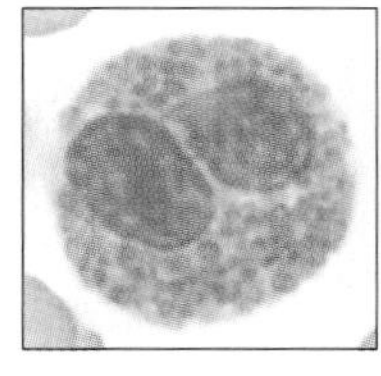

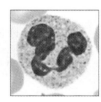

(a) Eosinophil

(b) Basophil

(c) Neutrophil

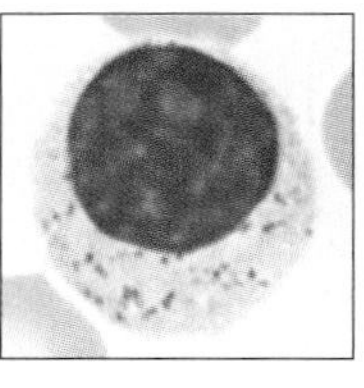

(d) Small lymphocyte

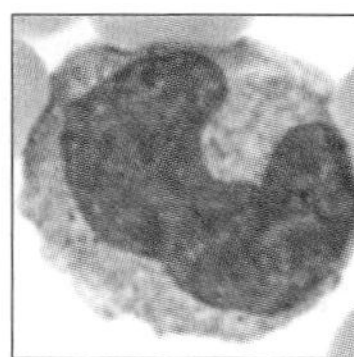

(e) Monocyte

All LM 1420x

Which WBCs are called granular leukocytes? Why?

tional significance of the size difference between small and large lymphocytes is unclear, the distinction is still useful clinically because an increase in the number of large lymphocytes has diagnostic significance in acute viral infections and in some immunodeficiency diseases.)

Monocytes (MON-ō-sīts) are 12–20 μm in diameter (Figure 12.5e). The nucleus of a monocyte is usually kidney-shaped or horseshoe-shaped, and the cytoplasm is blue-gray and has a foamy appearance. This color and appearance are due to very fine *azurophilic granules* (az′-yū-rō-FIL-ik; *azur* = blue; *philos* = loving), which are lysosomes. The blood is merely a conduit for monocytes, which migrate from the blood into the tissues, where they enlarge and differentiate into **macrophages** (= large eaters). Some become **fixed macrophages,** which means they reside in a particular tissue; examples are alveolar macrophages in the lungs, macrophages in the spleen, or stellate reticuloendothelial (Kupffer) cells in the liver. Others become **wandering macrophages,** which roam the tissues and gather at sites of infection or inflammation.

White blood cells and other nucleated body cells have proteins, called *major histocompatibility (MHC) antigens,* protruding from their plasma membranes into the extracellular fluid. These cell identity markers are unique for each person (except identical twins). Although RBCs possess blood group antigens, they lack the MHC antigens.

WBC Functions

In a healthy body, some WBCs, especially lymphocytes, can live for several months or years, but most live only a few days. During a period of infection, phagocytic WBCs may live only a few hours. WBCs are far less numerous than red blood cells, about 5000–10,000 cells per μL of blood. RBCs therefore outnumber white blood cells by about 700 : 1. **Leukocytosis** (lū′-kō-sī-TŌ-sis), an increase in the number of WBCs (above 10,000/μL), is a normal, protective response to stresses such as invading microbes, strenuous exercise, anesthesia, and surgery. An abnormally low level of white blood cells (below 5000/μL) is termed **leukopenia** (lū-kō-PĒ-nē-a). It is never beneficial and may be caused by radiation, shock, and certain chemotherapeutic agents.

The skin and mucous membranes of the body are continuously exposed to microbes and their toxins. Some of these microbes can invade deeper tissues to cause disease. Once pathogens enter the body, the general function of white blood cells is to combat them by phagocytosis or immune responses. To accomplish these tasks, many WBCs leave the bloodstream and collect at points of pathogen invasion or inflammation. Once granulocytes and monocytes leave the bloodstream to fight injury or infection, they never return to it. Lymphocytes, on the other hand, continually recirculate—from blood to interstitial spaces of tissues to lymphatic fluid and back to blood. Only 2% of the total lymphocyte population is circulating in the blood at any given time; the rest are in lymphatic fluid and organs such as skin, lungs, lymph nodes, and spleen.

WBCs leave the bloodstream by a process termed **emigration** (em′-i-GRĀ-shun; *e-* = out; *migra-* = wander) in which they roll along the endothelium, stick to it, and then squeeze between endothelial cells. The precise signals that stimulate emigration through a particular blood vessel vary for the different types of WBCs.

Neutrophils and macrophages are active in **phagocytosis;** they can ingest bacteria and dispose of dead matter. Several different chemicals released by microbes and inflamed tissues attract phagocytes, a phenomenon called **chemotaxis.**

Among WBCs, neutrophils respond most quickly to tissue destruction by bacteria. After engulfing a pathogen during phagocytosis, a neutrophil unleashes several destructive chemicals to destroy the ingested pathogen. These chemicals include the enzyme **lysozyme,** which destroys certain bacteria, and **strong oxidants,** such as the superoxide anion (O_2^-), hydrogen peroxide (H_2O_2), and the hypochlorite anion (OCl^-), which is similar to household bleach. Neutrophils also contain **defensins,** proteins that exhibit a broad range of antibiotic activity against bacteria and fungi. Within a neutrophil, granules containing defensins merge with phagosomes containing microbes. Defensins form peptide "spears" that poke holes in microbe membranes; the resulting loss of cellular contents kills the invader.

Monocytes take longer to reach a site of infection than do neutrophils, but they arrive in large numbers and destroy more microbes. Upon arrival they enlarge and differentiate into wandering macrophages, which clean up cellular debris and microbes by phagocytosis following an infection.

Eosinophils leave the capillaries and enter tissue fluid. They are believed to release enzymes, such as histaminase, that combat the effects of histamine and other mediators of inflammation in allergic reactions. Eosinophils also phagocytize antigen–antibody complexes and are effective against certain parasitic worms. A high eosinophil count often indicates an allergic condition or a parasitic infection.

Basophils are also involved in inflammatory and allergic reactions. They liberate heparin, histamine, and serotonin substances that intensify the inflammatory reaction and are involved in hypersensitivity (allergic) reactions.

The major types of lymphocytes are B cells, T cells, and natural killer cells, which are the major combatants in immune responses (described in detail in Chapter 15). B cells are particularly effective in destroying bacteria and inactivating their toxins. T cells attack viruses, fungi, transplanted cells, cancer cells, and some bacteria. Immune responses carried out by B cells and T cells help combat infection and provide protection against some diseases. T cells also are responsible for transfusion reactions, allergies, and rejection of transplanted organs. Natural killer cells attack a wide variety of infectious microbes and certain spontaneously arising tumor cells.

An increase in the number of circulating WBCs usually indicates inflammation or infection. A physician may order a **differential white blood cell count** to detect infection or inflammation, determine the effects of possible poisoning by chemicals or drugs, monitor blood disorders (for example, leukemia) and effects of chemotherapy, or detect allergic reactions and parasitic infections. Because each type of white blood cell plays a different role, determining the *percentage* of each type in the blood assists in diagnosing the condition. Table 12.2 lists the significance of both elevated and depressed WBC counts.

Table 12.2 Significance of Elevated and Depressed White Blood Cell Counts

WBC Type	High Count May Indicate	Low Count May Indicate
Neutrophils	Bacterial infection, burns, stress, inflammation.	Radiation exposure, drug toxicity, vitamin B_{12} deficiency, systemic lupus erythematosus (SLE).
Lymphocytes	Viral infections, some leukemias.	Prolonged illness, immunosuppression, treatment with cortisol.
Monocytes	Viral or fungal infections, tuberculosis, some leukemias, other chronic diseases.	Bone marrow depression, treatment with cortisol.
Eosinophils	Allergic reactions, parasitic infections, autoimmune diseases.	Drug toxicity, stress.
Basophils	Allergic reactions, leukemias, cancers, hypothyroidism.	Pregnancy, ovulation, stress, hyperthyroidism.

Table 12.3 Summary of the Formed Elements in Blood

Name and Appearance	Number	Characteristics*	Functions
Red blood cells (RBCs) or erythrocytes	4.8 million/μL in females. 5.4 million/μL in males.	7–8 μm diameter, biconcave discs, without a nucleus; live for about 120 days.	Hemoglobin within RBCs transports most of the oxygen and part of the carbon dioxide in the blood. It also transports NO and SNO, which help adjust blood pressure.
White blood cells (WBCs) or leukocytes	5000–10,000/μL.	Most live for a few hours to a few days.†	Combat pathogens and other foreign substances that enter the body.
Granular leukocytes			
Neutrophils	60–70% of all WBCs.	10–12 μm diameter; nucleus has 2–5 lobes connected by thin strands of chromatin; cytoplasm has very fine, pale lilac granules.	Phagocytosis. Destruction of bacteria with lysozyme, defensins, and strong oxidants, such as superoxide anion, hydrogen peroxide, and hypochlorite anion.
Eosinophils	2–4% of all WBCs.	10–12 μm diameter; nucleus has 2 or 3 lobes; large, red-orange granules fill the cytoplasm.	Combat the effects of histamine in allergic reactions, phagocytize antigen–antibody complexes, and destroy certain parasitic worms.
Basophils	0.5–1% of all WBCs.	8–10 μm diameter; nucleus has 2 lobes; large cytoplasmic granules appear deep blue-purple.	Liberate heparin, histamine, and serotonin in allergic reactions that intensify the overall inflammatory response.
Agranular leukocytes			
Lymphocytes (T cells, B cells, and natural killer cells)	20–25% of all WBCs.	Small lymphocytes are 6–9 μm in diameter; large lymphocytes are 10–14 μm in diameter; nucleus is round or slightly indented; cytoplasm forms a rim around the nucleus that looks sky blue; the larger the cell, the more cytoplasm is visible.	Mediate immune responses, including antigen–antibody reactions. B cells develop into plasma cells, which secrete antibodies. T cells attack invading viruses, cancer cells, and transplanted tissue cells, Natural killer cells attack a wide variety of infectious microbes and certain spontaneously arising tumorcells.
Monocytes	3–8% of all WBCs.	12–20 μm diameter; nucleus is kidney-shaped or horseshoe-shaped; cytoplasm is blue-gray and has foamy appearance.	Phagocytosis (after transforming into fixed or wandering macrophages).
Platelets (thrombocytes)	150,000–400,000/μL.	2–4 μm diameter cell fragments that live for 5–9 days; contain many granules but no nucleus.	Form platelet plug in hemostasis; release chemicals that promote vascular spasm and blood clotting.

*Colors are those seen when using Wright's stain.
†Some lymphocytes, called T and B memory cells, can live for many years once they are established.

A **bone marrow transplant** is the intravenous transfer of red bone marrow from a healthy donor to a recipient, with the goal of establishing normal hemopoiesis and consequently normal blood cell counts in the recipient. In patients with cancer or certain genetic diseases, the defective red bone marrow first must be destroyed by high doses of chemotherapy and whole body radiation. The donor marrow must be very closely matched to that of the recipient, or rejection occurs. When a transplant is successful, stem cells from the donor marrow reseed and grow in the recipient's red bone marrow cavities. Bone marrow transplants have been used to treat aplastic anemia, certain types of leukemia, severe combined immunodeficiency disease (SCID), Hodgkin's disease, non-Hodgkin's lymphoma, multiple myeloma, thalassemia, sickle-cell disease, breast cancer, ovarian cancer, testicular cancer, and hemolytic anemia. ■

- ✔ Explain the importance of emigration, chemotaxis, and phagocytosis in fighting bacterial invaders.
- ✔ Distinguish between leukocytosis and leukopenia.
- ✔ What is a differential white blood cell count?
- ✔ What functions are performed by B cells, T cells, and natural killer cells?

PLATELETS

Objective

- Describe the structure, function, and origin of platelets.

Besides the immature cell types that develop into erythrocytes and leukocytes, hemopoietic stem cells also differentiate into cells that produce platelets. Under the influence of the hormone **thrombopoietin,** myeloid stem cells develop into megakaryocyte-colony-forming cells that, in turn, develop into precursor cells called megakaryoblasts (see Figure 12.3 on page 380). Megakaryoblasts transform into megakaryocytes, huge cells that splinter into 2000–3000 fragments. Each fragment, enclosed by a piece of the cell membrane, is a **platelet,** or **thrombocyte.** Platelets break off from the megakaryocytes in red bone marrow and then enter the blood circulation. Between 150,000 and 400,000 platelets are present in each μL of blood. They are disc-shaped, 2–4 μm in diameter, and exhibit many granules but no nucleus. Platelets help stop blood loss from damaged blood vessels by forming a platelet plug. Their granules also contain chemicals that, once released, promote blood clotting. Platelets have a short life span, normally just 5–9 days. Aged and dead platelets are removed from the circulation by fixed macrophages in the spleen and liver.

Table 12.3 on page 380 summarizes the formed elements in blood.

A **complete blood count (CBC)** is a valuable test that screens for anemia and various infections. Usually included are counts of RBCs, WBCs, and platelets per μL of whole blood; hematocrit; and differential white blood cell count. The amount of hemoglobin in grams per mL of blood also is determined. Normal hemoglobin ranges are: infants, 14–20 g/100 mL of blood; adult females, 12–16 g/100 mL of blood; and adult males, 13.5–18 g/100 mL of blood.

- ✔ Compare RBCs, WBCs, and platelets with respect to size, number per μL, and life span.

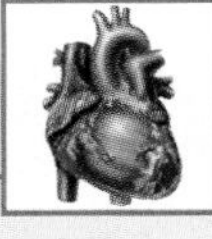

APPLICATIONS TO HEALTH

Anemia

Anemia is a condition in which the oxygen-carrying capacity of blood is reduced. Many kinds of anemia exist; all are characterized by reduced numbers of RBCs or a decreased amount of hemoglobin in the blood. The person feels fatigued and is intolerant of cold, both of which are related to lack of oxygen needed for ATP and heat production. Also, the skin appears pale, due to the low content of red-colored hemoglobin circulating in skin blood vessels. Among the most important types of anemia are the following:

- *Iron-deficiency anemia,* the most prevalent kind of anemia, is caused by inadequate absorption of iron, excessive loss of iron, increased iron requirement, or insufficient intake of iron. Women are at greater risk for iron-deficiency anemia due to menstrual blood losses and increased iron demands of the growing fetus during pregnancy. Gastrointestinal losses such as occurs with malignancy or ulceration also contribute to this type of anemia.
- *Pernicious anemia* is caused by insufficient hemopoiesis resulting from an inability of the stomach to produce intrinsic factor, which is needed for absorption of vitamin B_{12} in the small intestine.
- *Hemorrhagic anemia* is due to an excessive loss of RBCs through bleeding resulting from large wounds, stomach ulcers, or especially heavy menstruation.
- In *hemolytic anemia,* RBC plasma membranes rupture prematurely, and their hemoglobin pours into the plasma. The condition may result from inherited defects such as abnormal red blood cell enzymes, or from outside agents such as parasites, toxins, or antibodies from incompatible transfused blood.
- *Thalassemia* (thal′-a-SĒ-mē-a) is a group of hereditary he-

molytic anemias associated with deficient synthesis of hemoglobin. The RBCs are small (microcytic), pale (hypochromic), and short-lived. Thalassemia occurs primarily in populations from countries bordering the Mediterranean Sea.

- *Aplastic anemia* results from destruction of the red bone marrow. Toxins, gamma radiation, and certain medications that inhibit enzymes needed for hemopoiesis are causes.

Sickle-Cell Disease

The RBCs of a person with **sickle-cell disease (SCD)** contain Hb-S, an abnormal kind of hemoglobin. When Hb-S gives up oxygen to the interstitial fluid, it forms long, stiff, rodlike structures that bend the erythrocyte into a sickle shape (Figure 12.6). The sickled cells rupture easily. Even though erythropoiesis is stimulated by the loss of the cells, it cannot keep pace with hemolysis; hemolytic anemia is the result. Prolonged oxygen reduction may eventually cause extensive tissue damage.

Sickle-cell disease is inherited. People with two sickle-cell genes have severe anemia, whereas those with only one defective gene may have minor problems. Sickle-cell genes are found primarily among populations, or descendants of populations, that live in the malaria belt around the world, including parts of Mediterranean Europe, sub-Saharan Africa, and tropical Asia. Some 1–2% of African Americans have SCD. The gene responsible for the tendency of the RBCs to sickle also alters the permeability of the plasma membranes of sickled cells, causing potassium ions to leak out. Low levels of potassium kill the malarial parasites that infect sickled cells. Because of this effect, a person with one normal gene and one sickle-cell gene has a high resistance to malaria. The possession of a single sickle-cell gene thus confers a survival advantage.

Treatment consists of administration of analgesics to relieve pain, fluid therapy to maintain hydration, oxygen to reduce the stimulus for the crisis, antibiotics to counter infections, and blood transfusions. People who suffer sickle-cell disease have normal fetal hemoglobin (Hb-F), a slightly different form of hemoglobin that predominates at birth and normally is present in small amounts thereafter. In some patients with sickle-cell disease, a drug called hydroxyurea promotes transcription of the normal Hb-F gene, elevates the level of Hb-F, and reduces the chance that the RBCs will sickle. Unfortunately, this drug also has toxic effects on the bone marrow; thus, its safety for long-term use is questionable.

Figure 12.6 / Red blood cells from a patient with sickle-cell disease.

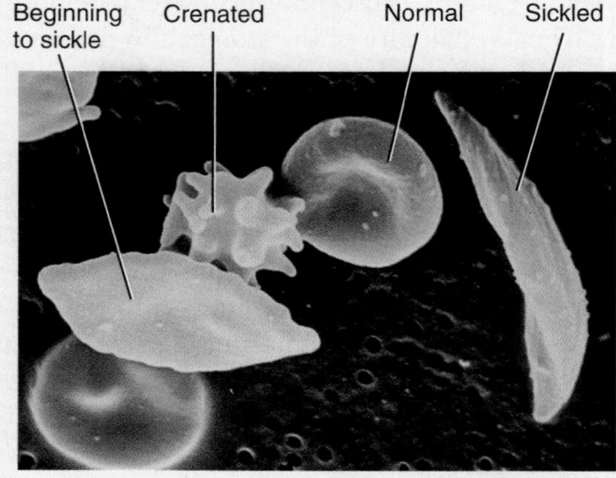

Hemophilia

Hemophilia (*-philia* = loving) is an inherited deficiency of clotting in which bleeding may occur spontaneously or after only minor trauma. Different types of hemophilia are due to deficiencies of different blood clotting factors and exhibit varying degrees of severity, ranging from mild to severe bleeding tendencies. The most common type (classic hemophilia) is hemophilia A, in which factor VIII is absent. People with hemophilia B lack factor IX. Hemophilia A and B occur primarily among males because these are sex-linked recessive disorders, whereas hemophilia C affects both males and females. Hemophilia C is caused by a lack of factor XI (which in turn activates factor IX) and is much less severe than hemophilia A or B because an alternate activator of factor IX is present—namely, factor VII. Hemophilia is characterized by spontaneous or traumatic subcutaneous and intramuscular hemorrhaging, nosebleeds, blood in the urine, and hemorrhages in joints that produce pain and tissue damage. Treatment involves transfusions of fresh plasma or concentrates of the deficient clotting factor to relieve the tendency to bleed.

Disseminated Intravascular Clotting

Disseminated intravascular clotting (DIC) is a disorder of hemostasis characterized by simultaneous and unregulated blood clotting and hemorrhage throughout the body. DIC is associated primarily with endothelial damage, tissue damage, and direct activation of clotting factor X, all of which activate the blood clotting sequence. Common causes of DIC are infections, hypoxia, low blood flow rates, trauma, tumors, hypotension, and hemolysis. DIC may be severe and life threatening. The clots decrease blood flow and eventually cause ischemia, infarction, and necrosis, resulting in multisystem organ dysfunction. A peculiar feature of DIC is that the patient often begins to bleed despite forming clots. This occurs because so many clotting factors are removed by the widespread clotting that too few remain to allow normal clotting of the remaining blood. Thus, DIC presents the paradoxical presence of blood clotting and bleeding at the same time.

Leukemia

Acute leukemia is a malignant disease of blood-forming tissues characterized by uncontrolled production and accumulation of immature leukocytes. In **chronic leukemia,** mature leukocytes accumulate in the bloodstream because they do not die at the end of their normal life span. The *human T cell leukemia-lymphoma virus-1 (HTLV-1)* is strongly associated with some types of leukemia. The abnormal accumulation of immature leukocytes may be reduced by treatment with x-rays and antileukemic drugs. In some cases, a red bone marrow transplant can cure the leukemia.

KEY MEDICAL TERMS ASSOCIATED WITH BLOOD

Acute normovolemic hemodilution (nor-mō-vō-LĒ-mik hē-mō-di-LŪ-shun) Removal of blood immediately before surgery and its replacement with a cell-free solution to maintain sufficient blood volume for adequate circulation. At the end of surgery, once bleeding has been controlled, the collected blood is returned to the body.

Autologous preoperative transfusion (aw-TOL-o-gus trans-FYŪ-zhun; *auto-* = self) Donating one's own blood; can be done up to 6 weeks before elective surgery. Also called *predonation.*

Blood bank A facility that collects and stores a supply of blood for future use by the donor or others. Because blood banks have additional and diverse functions (immunohematology reference work, continuing medical education, bone and tissue storage, and clinical consultation), they are more appropriately referred to as **centers of transfusion medicine.**

Cyanosis (sī-a-NŌ-sis; *cyano-* = blue) Slightly bluish/dark-purple skin discoloration, most easily seen in the nail beds and mucous membranes, due to an increased quantity of reduced hemoglobin (hemoglobin not combined with oxygen) in systemic blood.

Embolus (EM-bō-lus = plug) A blood clot, bubble of air or fat from broken bones, mass of bacteria, or other foreign material transported by the blood.

Gamma globulin (GLOB-yū-lin) Solution of immunoglobulins from blood consisting of antibodies that react with specific pathogens, such as viruses. It is prepared by injecting the specific virus into animals, removing blood from the animals after antibodies have accumulated, isolating the antibodies, and injecting them into a human to provide short-term immunity.

Hemochromatosis (hē-mō-krō-ma-TŌ-sis; *chroma* = color) Disorder of iron metabolism characterized by excessive absorption of ingested iron and excess deposits of iron in tissues (especially the liver, heart, pituitary gland, gonads, and pancreas) that result in bronze discoloration of the skin, cirrhosis, diabetes mellitus, and bone and joint abnormalities.

Hemorrhage (HEM-or-ij; *rhegnynai* = bursting forth) Loss of a large amount of blood; can be either internal (from blood vessels into tissues) or external (from blood vessels directly to the surface of the body).

Jaundice (*jaund-* = yellow) An abnormal yellowish discoloration of the sclerae of the eyes, skin, and mucous membranes due to excess bilirubin (yellow-orange pigment) in the blood. The three main categories of jaundice are *prehepatic jaundice,* due to excess production of bilirubin; *hepatic jaundice,* due to abnormal bilirubin processing by the liver caused by congenital liver disease, cirrhosis (scar tissue formation) of the liver, or hepatitis (liver inflammation); and *extrahepatic jaundice,* due to blockage of bile drainage by gallstones or cancer of the bowel or pancreas.

Multiple myeloma (mī-e-LŌ-ma) Malignant disorder of plasma cells in red bone marrow; symptoms (pain, osteoporosis, hypercalcemia, thrombocytopenia, kidney damage) are caused by the growing tumor cell mass or antibodies produced by malignant cells.

Phlebotomist (fle-BOT-ō-mist; *phlebo-* = vein; *-tom* = cut) A technician who specializes in withdrawing blood.

Septicemia (sep′-ti-SĒ-mē-a; *septic-* = decay; *-emia* = condition of blood) Toxins or disease-causing bacteria in the blood. Also called "blood poisoning."

Thrombocytopenia (throm′-bō-sī′-tō-PĒ-nē-a; *-penia* = poverty) Very low platelet count that results in a tendency to bleed from capillaries.

Thrombus (THROM-bus = clot) A clot in the cardiovascular system formed in an unbroken blood vessel (usually a vein) from constituents in blood called clotting factors; the clot consists of a network of insoluble fibrin threads in which the formed elements of blood are trapped.

Transfusion (trans-FYŪ-shun) Transfer of whole blood, blood components (red blood cells only or plasma only), or red bone marrow directly into the bloodstream.

Venesection (vē′-ne-SEK-shun; *ven-* = vein) Opening of a vein for withdrawal of blood. Although **phlebotomy** (fle-BŌT-ō-me) is a synonym for venesection, in clinical practice phlebotomy refers to therapeutic bloodletting, such as the removal of some blood to lower its viscosity in a patient with polycythemia.

Whole blood Blood containing all formed elements, plasma, and plasma solutes in natural concentrations.

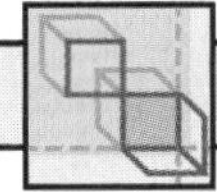

STUDY OUTLINE

Introduction (p. 370)

1. The cardiovascular system consists of the blood, heart, and blood vessels.
2. Blood is a connective tissue composed of plasma (liquid portion) and formed elements (cells and cell fragments).

Functions of Blood (p. 370)

1. Blood transports oxygen, carbon dioxide, nutrients, wastes, and hormones.
2. It helps regulate pH, body temperature, and water content of cells.
3. It provides protection through clotting and by combating toxins and microbes, a function of certain phagocytic white blood cells or specialized plasma proteins.

Physical Characteristics of Blood (p. 370)

1. Physical characteristics of blood include a viscosity greater than that of water; a temperature of 38°C (100.4°F); and a pH of 7.35–7.45.
2. Blood constitutes about 8% of body weight, and its volume is 4–6 liters in adults.

Components of Blood (p. 371)

1. Blood consists of about 55% plasma and about 45% formed elements.
2. The hematocrit is the percentage of total blood volume occupied by red blood cells.
3. Plasma consists of 91.5% water and 8.5% solutes.
4. Principal solutes include proteins (albumins, globulins, fibrinogen), nutrients, vitamins, hormones, respiratory gases, electrolytes, and waste products.
5. The formed elements in blood include red blood cells (erythrocytes), white blood cells (leukocytes), and platelets.

Formation of Blood Cells (p. 371)

1. Hemopoiesis is the formation of blood cells from hemopoietic stem cells in red bone marrow.
2. Myeloid stem cells form RBCs, platelets, granulocytes, and monocytes. Lymphoid stem cells give rise to lymphocytes.
3. Several hemopoietic growth factors stimulate differentiation and proliferation of the various blood cells.

Red Blood Cells (p. 375)

1. Mature RBCs are biconcave discs that lack nuclei and contain hemoglobin.
2. The function of the hemoglobin in red blood cells is to transport oxygen and some carbon dioxide.
3. RBCs live about 120 days. A healthy male has about 5.4 million RBCs/μL of blood; a healthy female, about 4.8 million/μL.
4. After phagocytosis of aged RBCs by macrophages, hemoglobin is recycled.
5. RBC formation, called erythropoiesis, occurs in adult red bone marrow of certain bones. It is stimulated by hypoxia, which stimulates release of erythropoietin by the kidneys.
6. A reticulocyte count is a diagnostic test that indicates the rate of erythropoiesis.

White Blood Cells (p. 378)

1. WBCs are nucleated cells. The two principal types are granulocytes (neutrophils, eosinophils, and basophils) and agranulocytes (lymphocytes and monocytes).
2. The general function of WBCs is to combat inflammation and infection. Neutrophils and macrophages (which develop from monocytes) act by phagocytosis.
3. Eosinophils combat the effects of histamine in allergic reactions, phagocytize antigen–antibody complexes, and combat parasitic worms; basophils liberate heparin, histamine, and serotonin in allergic reactions that intensify the inflammatory response.
4. B lymphocytes, in response to the presence of foreign substances called antigens, differentiate into plasma cells that produce antibodies. Antibodies attach to the antigens and render them harmless. This antigen–antibody response combats infection and provides immunity. T lymphocytes destroy foreign invaders directly. Natural killer cells attack infectious microbes and tumor cells.
5. Except for lymphocytes, which may live for years, WBCs usually live for only a few hours or a few days. Normal blood contains 5000–10,000 WBCs/μL.

Platelets (p. 381)

1. Platelets (thrombocytes) are disc-shaped structures without nuclei.
2. They are fragments derived from megakaryocytes and are involved in clotting.
3. Normal blood contains 150,000–400,000 platelets/μL.

SELF-QUIZ QUESTIONS

Choose the one best answer to these questions.

1. Which of the following is *not* a site of hemopoiesis in the body? (a) sternum, (b) liver, (c) vertebrae, (d) head of humerus, (e) parietal bones.
2. Which of the following statements is false? (a) Blood is thicker than water. (b) Blood normally has a pH of 6.5–7.0. (c) The adult male body normally contains about 5–6 liters of blood. (d) Normal blood temperature is 38 degrees C. (e) Blood constitutes about 8 % of total body weight.
3. The normal red blood cell count in healthy adult males is (a) 5.4 million RBCs/μL, (b) 2 million RBCs/μL, (c) 0.5 million RBCs/μL, (d) 250,000 RBCs/μL, (e) 8,000 RBCs/μL.
4. Erythrocytes (a) lack nuclei, (b) are disc-shaped, (c) bear ABO and Rh antigens on their cell membranes, (d) all of the above, (e) none of the above.
5. Which of the following is *not* a function of blood? (a) protection against fluid loss, (b) transportation of gases and nutrients, (c) regulation of body temperature, (d) protection against microbes and toxins, (e) synthesis of plasma proteins.
6. Pluripotent stem cells: (a) are derived from mesenchyme in bone marrow. (b) give rise to red blood cells. (c) give rise to white blood cells. (d) give rise to platelets. (e) all of the above.
7. Which of the following statements about neutrophils is *false*? (a) They are actively phagocytic. (b) They are the most abundant type of leukocyte. (c) Their number decreases during most infections. (d) They combat bacteria with lysozyme, oxidants, and defensins. (e) They contain MHC antigens on their cell membranes.
8. Agranular leukocytes develop from which precursor cells? (a) megakaryoblasts, (b) lymphoblasts, (c) monoblasts, (d) All of the above are correct answers. (e) Only b and c are correct answers.
9. Place the following cells in the correct order for the production of red blood cells: (1) proerythroblasts, (2) CFU-E cells, (3) myeloid stem cells, (4) erythroblast, (5) pluripotent stem cell, (6) reticulocyte

   a. 5, 3, 2, 1, 4, 6 **b.** 3, 2, 5, 6, 1, 4 **c.** 4, 5, 3, 6, 1, 2
   d. 5, 2, 1, 3, 4, 6 **e.** 2, 3, 5, 1, 4, 6
10. Lymphocytes may develop into (a) B cells, (b) T cells (c) natural killer cells, (d) helper cells, (e) all of the above.

11. Platelets function in the (a) transport of carbon dioxide, (b) production of vitamin K, (c) utilization of calcium and phosphorus, (d) destruction of bacteria, (e) clotting of blood.

12. Place the following cells in the correct order for the process of neutrophil production: (1) myeloblasts, (2) pluripotent stem cell, (3) CFU-GM cells, (4) myeloid stem cell

a. 2, 3, 4, 1 **b.** 2, 4, 3, 1 **c.** 4, 2, 1, 3 **d.** 3, 2, 4, 1
e. 3, 1, 2, 4

Complete the following.

13. A test to determine the percentage of each type of white blood cell is known as a ________ count.

14. Blood is a connective tissue that consists of about ________ % extracellular material and about ________ % formed elements.

15. The ability of white blood cells to crawl through capillaries and reach an injured or infected body tissue is ________ .

16. The process of red blood cell formation is called ________ .

17. The three types of proteins known as plasma proteins are ________ , ________ , and ________ .

18. Neutrophils produce chemical weapons that they use to destroy ingested pathogens. These chemicals include strong oxidants such as superoxide anion and hydrogen peroxide; the enzyme ________ that destroys certain bacteria, and proteins called ________ that destroy the cell membranes of bacteria and fungi.

19. Platelets are formed when large cells called ________ break apart into fragments.

Are the following statements true or false?

20. Nucleated cells (including white blood cells) have surface MHC antigens.

21. The pigment hemoglobin is responsible for the color of blood, the transport of respiratory gases, and is also involved in blood pressure regulation through its role in transport of nitric oxide (NO) and super nitric oxide (SNO).

22. The three kinds of granular leukocytes are neutrophils, monocytes, and eosinophils.

23. Match the following terms with their description:

____ **(a)** constitute the largest percentage (about 60%-70%) of leukocytes
____ **(b)** make up 20% - 25 % of leukocytes
____ **(c)** are involved in immunity; form plasma cells for antibody production
____ **(d)** are involved in allergic response; release serotonin, heparin, and histamine
____ **(e)** are involved in allergic response; release antihistamines
____ **(f)** develop into wandering macrophages that clean up sites of infection
____ **(g)** are important in phagocytosis (two answers)
____ **(h)** contains large reddish granules in the cytoplasm
____ **(i)** contains large deep blue or purple granules in cytoplasm
____ **(j)** cytoplasm is blue-gray and foamy in appearance

(1) basophils
(2) eosinophils
(3) lymphocytes
(4) monocytes
(5) neutrophils

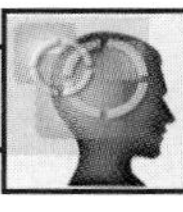

CRITICAL THINKING QUESTIONS

1. A professional bicyclist finished second in a grueling cross country bike race. Although he appeared to be in great physical condition, he suffered a heart attack a few hours after the race. The press speculated that his condition was caused by blood doping. Explain.
HINT *Blood "dope"-ing doesn't refer to his intelligence in this case, but it ought to.*

2. Maddy has been sniffling and sneezing her way through Human Anatomy class all semester. While checking a smear of her own blood, she noticed that her blood had a lot more of the bluish-black granular cells than her lab partner's blood. Maddy's eyes are watery and itchy so she asked her instructor to confirm her finding. What are these cells and why does Maddy have a higher number of them than usual?
HINT *Maddy used to take antihistamines to relieve her symptoms but she quit because they made her sleepy.*

3. While looking at slides of their own blood under the microscope, Tony teased Renee "my blood's redder than your blood. My blood's redder than yours!"(sung to the tune of Oscar Meyer wiener.) In a way, he's right. How does blood differ in males versus females?
HINT *What makes blood red?*

4. Imagine that a new cream was advertised "Heat those icy feet! New "NO-cold" cream increases your body's NO when rubbed on the skin of your feet." Would this treatment help cold feet everywhere? Explain.
HINT *NO comes in two forms.*

5. Raoul plans on having some elective surgery performed in an about a month, after the semester is over. His doctor suggested that he donate some of his own blood now. Raoul's not sure about this; he figures he's going to need all the blood he has for the surgery. Why should he give up some of his blood now?
HINT *An ounce (or a pint) of prevention is worth a pound of cure.*

ANSWERS TO FIGURE QUESTIONS

12.1 Blood volume is about 6 liters in males and 4–5 liters in females, representing about 8% of body weight.

12.2 Platelets are cell fragments.

12.3 Pluripotent stem cells develop from mesenchyme.

12.4 One hemoglobin molecule can transport four O_2 molecules—one bound to each heme group.

12.5 Neutrophils, eosinophils, and basophils are called granulocytes because all have cytoplasmic granules that are visible through a light microscope when stained.

13

THE CARDIOVASCULAR SYSTEM: THE HEART

CONTENTS AT A GLANCE

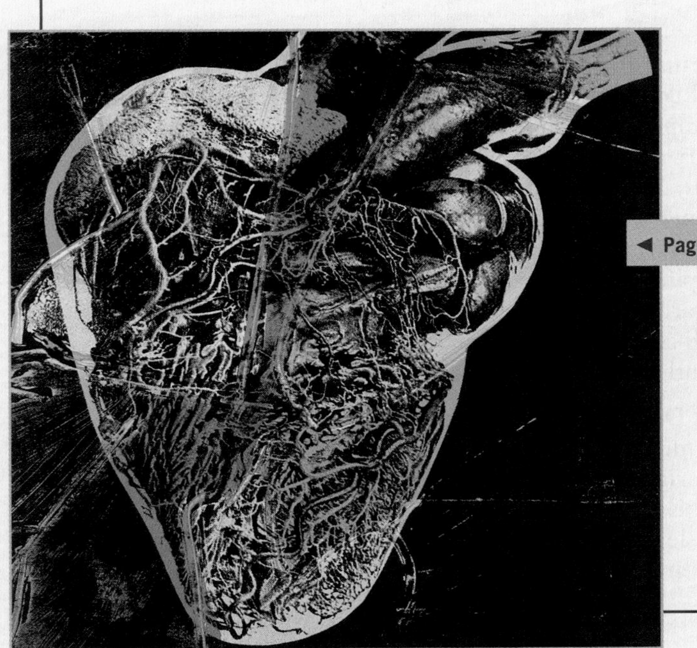

◄ Page 391

Can you recognize whose work this is? Is this a modern or historical composition?

Page 393 ►

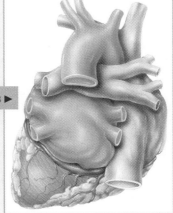

INTRODUCTION

The cardiovascular system consists of the blood, heart, and blood vessels. In the previous chapter we examined the composition and functions of blood. For blood to reach body cells and exchange materials with them, it must be constantly propelled through blood vessels. The heart is the pump that circulates the blood through an estimated 100,000 km (60,000 mi) of blood vessels. Even while you are sleeping, your heart pumps 30 times its own weight each minute, about 5 liters (5.3 qt) to the lungs and the same volume to the rest of the body. At this rate, the heart pumps more than 14,000 liters (3,600 gal) of blood in a day, or 10 million liters (2.6 million gal) in a year. You don't spend all your time sleeping, however, and your heart pumps even more vigorously when you are active. Thus, the actual blood volume the heart pumps in a single day is much larger. The study of the normal heart and the diseases associated with it is **cardiology** (kar-dē-OL-ō-jē; *cardio-* = heart; *-ology* = study of). This chapter explores the design of the heart and the unique properties that permit it to pump for a lifetime without rest.

LOCATION AND SURFACE PROJECTION OF THE HEART

Objective

- Describe the location of the heart, and trace its outline on the surface of the chest.

For all its might, the cone-shaped heart is relatively small, roughly the same size as a closed fist—about 12 cm (5 in.) long, 9 cm (3.5 in.) wide at its broadest point, and 6 cm (2.5 in.) thick. Its mass averages 250 g (8 oz) in adult females and 300 g (10 oz) in adult males. The heart rests on the diaphragm, near the midline of the thoracic cavity in the **mediastinum** (mē-dē-a-STĪ-num), a mass of tissue that extends from the sternum to the vertebral column and between the coverings (pleurae) of the lungs (Figure 13.1a, b).

About two-thirds of the mass of the heart lies to the left of the body's midline. The position of the heart in the mediastinum is more readily appreciated by examining its ends, surfaces, and borders (Figure 13.1b). Visualize the heart as a cone lying on its side. The pointed end of the heart is the **apex,** which is directed anteriorly, inferiorly, and to the left. The broad portion of the heart opposite the apex is the **base,** which is directed posteriorly, superiorly, and to the right. In addition to the apex and base, the heart has several surfaces and borders (margins) that are useful in determining its surface projection (described shortly). The **anterior surface** is deep to the sternum and ribs. The **inferior surface** is the portion of the heart that rests mostly on the diaphragm and is found between the apex and right border. The **right border** faces the right lung and extends from the inferior surface to the base; the **left border,** also called the pulmonary border, faces the left lung and extends from the base to the apex.

Determining an organ's **surface projection** means outlining its dimensions with respect to landmarks on the surface of the body. This practice is useful when conducting diagnostic procedures (for example, a lumbar puncture), auscultation (for example, listening to heart and lung sounds), and anatomical studies. We can project the heart on the anterior surface of the chest by locating the following landmarks (Figure 13.1c): The **superior right point** is located at the superior border of the third right costal cartilage, about 3 cm (1 in.) to the right of the midline. The **superior left point** is located at the inferior border of the second left costal cartilage, about 3 cm to the left of the midline. A line connecting these two points corresponds to the base of the heart. The **inferior left point** is located at the apex of the heart in the fifth left intercostal space, about 9 cm (3.5 in.) to the left of the midline. A line connecting the superior and inferior left points corresponds to the left border of the heart. The **inferior right point** is located at the superior border of the sixth right costal cartilage, about 3 cm to the right of the midline. A line connecting the inferior left and right points corresponds to the inferior surface of the heart, and a line connecting the inferior and superior right points corresponds to the right border of the heart. When all four points are connected, they form an outline that roughly reveals the size and shape of the heart.

Cardiopulmonary Resuscitation

Because the heart lies between two rigid structures—the vertebral column and the sternum (Figure 13.1a)—external pressure (compression) on the chest can be used to force blood out of the heart and into the circulation. In cases in which the heart suddenly stops beating, **cardiopulmonary resuscitation (CPR)**—properly applied cardiac compressions, performed in conjunction with artificial ventilation of the lungs—saves lives by keeping oxygenated blood circulating until the heart can be restarted. Some evidence suggests that the outcome after CPR with cardiac compression alone is similar to that after mouth-to-mouth ventilation of the lungs. ■

- ✔ Describe the position of the heart in the mediastinum by defining its apex, base, anterior and posterior surfaces, and right and left borders.
- ✔ Explain the location of the superior right point, superior left point, inferior left point, and inferior right point. Why are these points significant?

Figure 13.1 / Position of the heart and associated structures in the mediastinum (dashed outline), and the points of the heart that correspond to its surface projection. (See Tortora, *A Photographic Atlas of the Human Body,* Figures 6.5 and 6.6.)

The heart is located in the mediastinum; two-thirds of its mass is to the left of the midline.

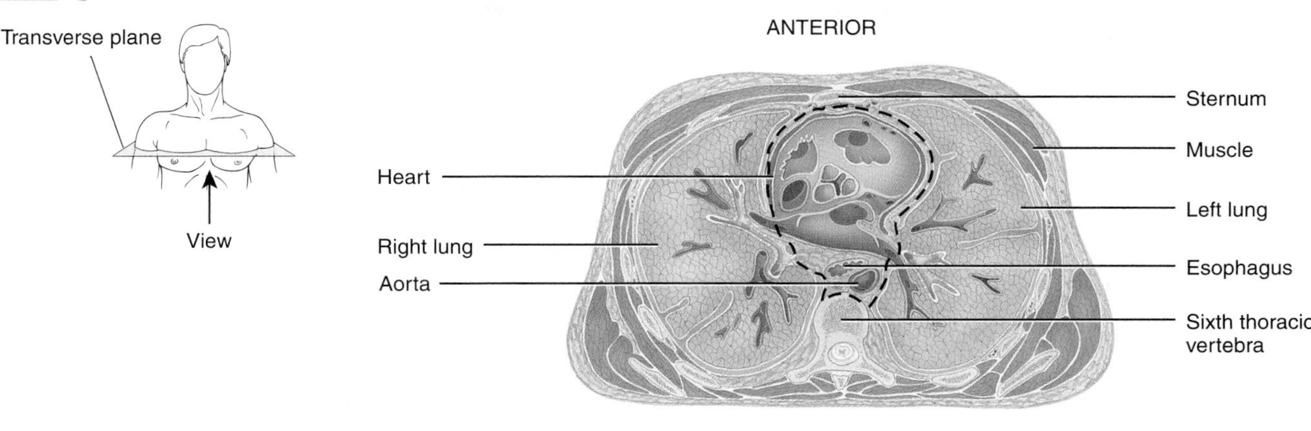

(a) Inferior view of transverse section of thoracic cavity showing the heart in the mediastinum

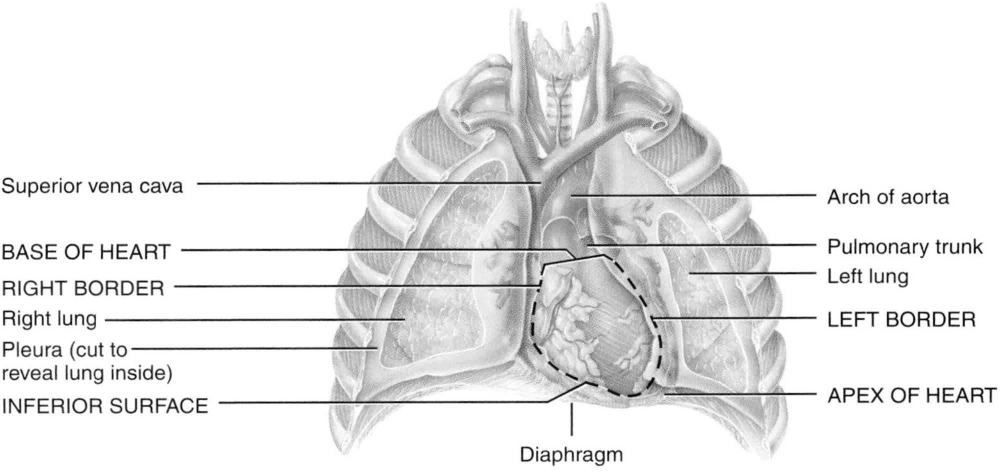

(b) Anterior view of the heart in the mediastinum

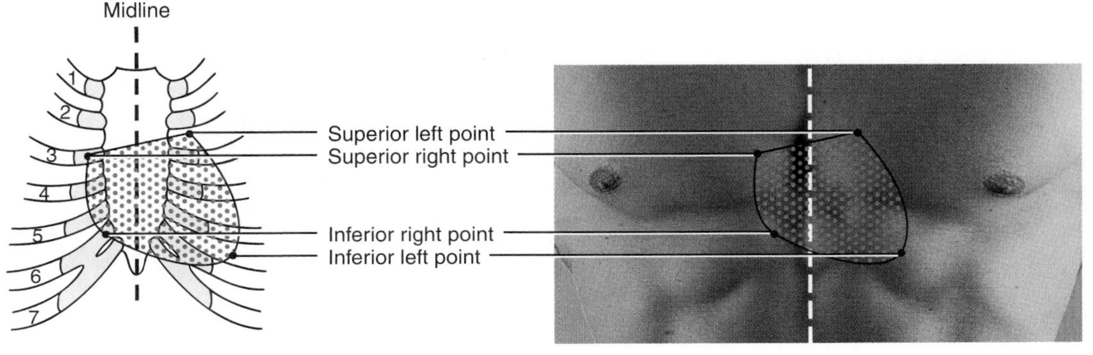

(c) Surface projection of the heart

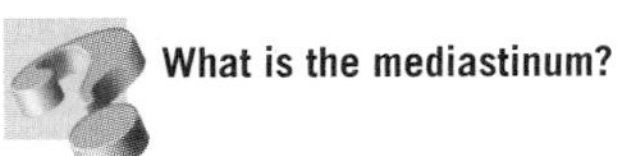
What is the mediastinum?

STRUCTURE AND FUNCTION OF THE HEART

Objectives

- Describe the structure of the pericardium and the heart wall.
- Discuss the external and internal anatomy of the chambers of the heart.
- Describe the structure and function of the valves of the heart.

Pericardium

The membrane that surrounds and protects the heart is the **pericardium** (*peri-* = around). It confines the heart to its position in the mediastinum, while allowing sufficient freedom of movement for vigorous and rapid contraction. The pericardium consists of two principal portions: the fibrous pericardium and the serous pericardium (Figure 13.2a, b). The superficial **fibrous pericardium** is a tough, inelastic, dense irregular connective tissue. It resembles a bag that rests on and attaches to the diaphragm; its open end is fused to the connective tissues of the blood vessels entering and leaving the heart. The fibrous pericardium prevents overstretching of the heart, provides protection, and anchors the heart in the mediastinum.

The deeper **serous pericardium** is a thinner, more delicate membrane that forms a double layer around the heart (Figure 13.2a, b). The outer **parietal layer** of the serous pericardium is fused to the fibrous pericardium. The inner **visceral layer** of the serous pericardium, also called the **epicardium** (*epi-* = on top of), adheres tightly to the surface of the heart. Between the parietal and visceral layers of the serous pericardium is a thin film of serous fluid. This fluid, known as **pericardial fluid,** is a slippery secretion of the pericardial cells that reduces friction between the membranes as the heart moves. The space that contains the few milliliters of pericardial fluid is called the **pericardial cavity.**

Figure 13.2 / Pericardium and heart wall.

The pericardium is a triple-layered sac that surrounds and protects the heart.

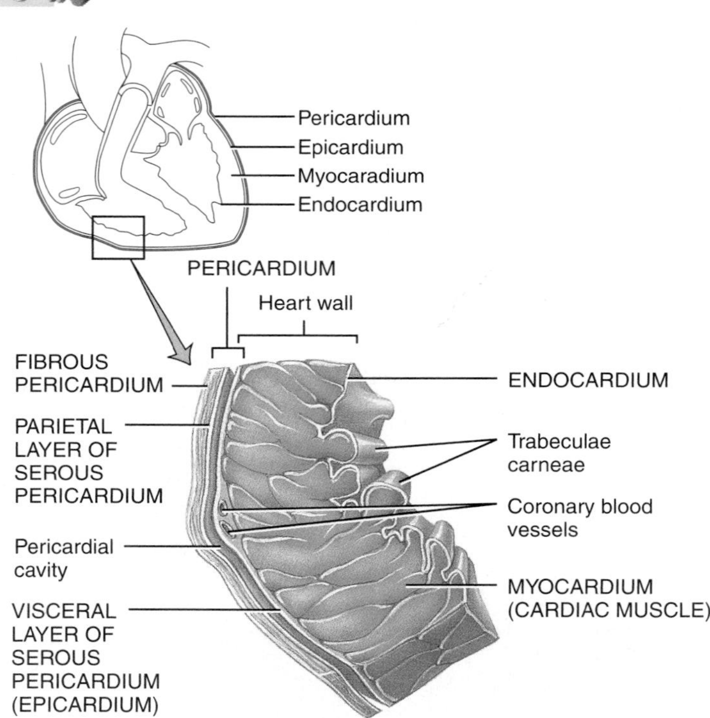

(a) Portion of pericardium and right ventricular heart wall showing the divisions of the pericardium and layers of the heart wall

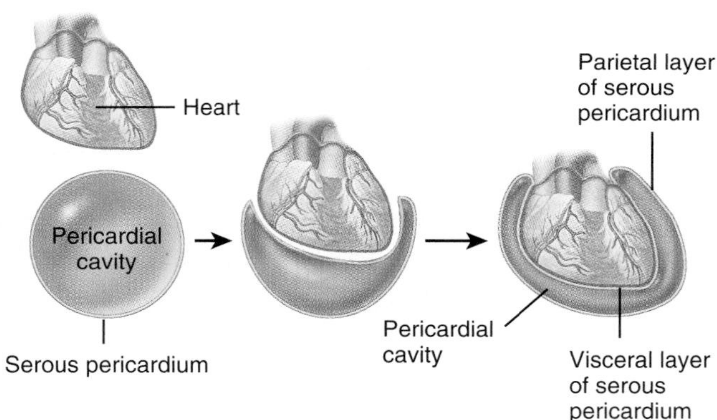

(b) Simplified relationship of the serous pericardium to the heart

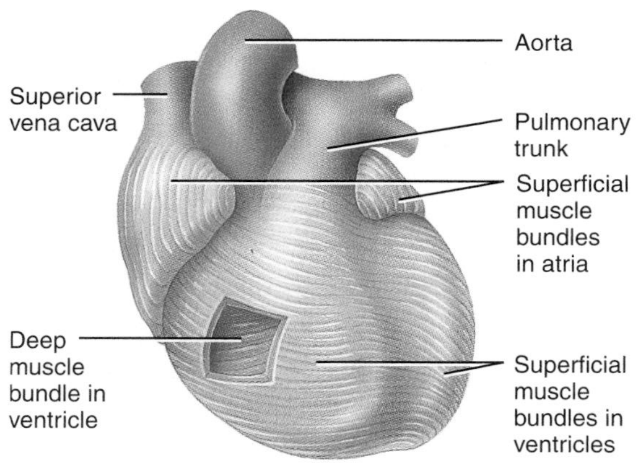

(c) Cardiac muscle bundles of the myocardium

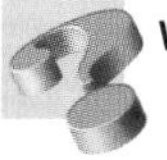

Which layer is both a part of the pericardium and a part of the heart wall?

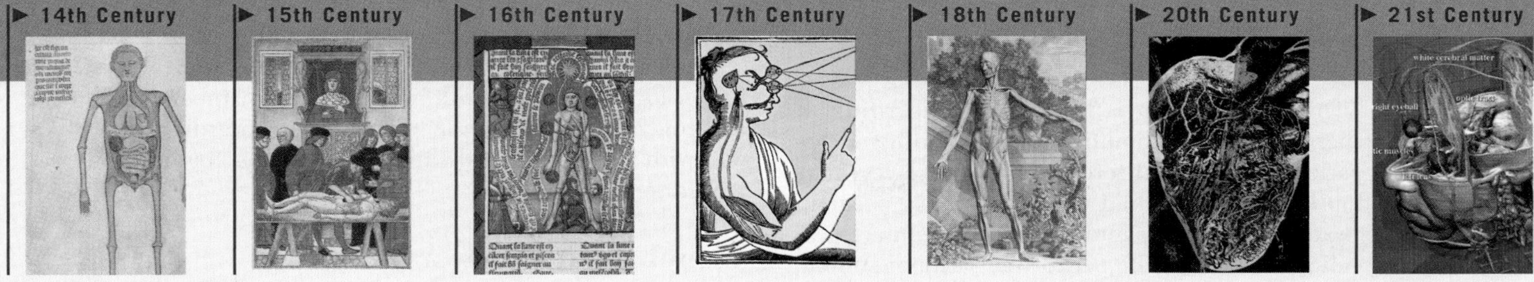

CHANGING IMAGES

You Gotta Have Heart

In these *Changing Images*, the story of anatomy as art has focused on centuries old pieces with scientific objectives. Anatomy pioneers such as Vesalius (Chapters 5 and 9) and Harvey (Chapter 14) commissioned artists to illustrate their medically based texts. Leonardo Da Vinci and Michelangelo (Chapters 7 and 11) painted and sculpted for much more than science, as they were both artists and anatomists.

But we should not think of this blend of two very different disciplines as something solely from the past; or consider it a lost art. There are examples of 20th century contemporary artists employing anatomy as a subject. Avant-garde exhibitions of photography, sculpture, video, and live performance art have provided new perspectives on how we see ourselves.

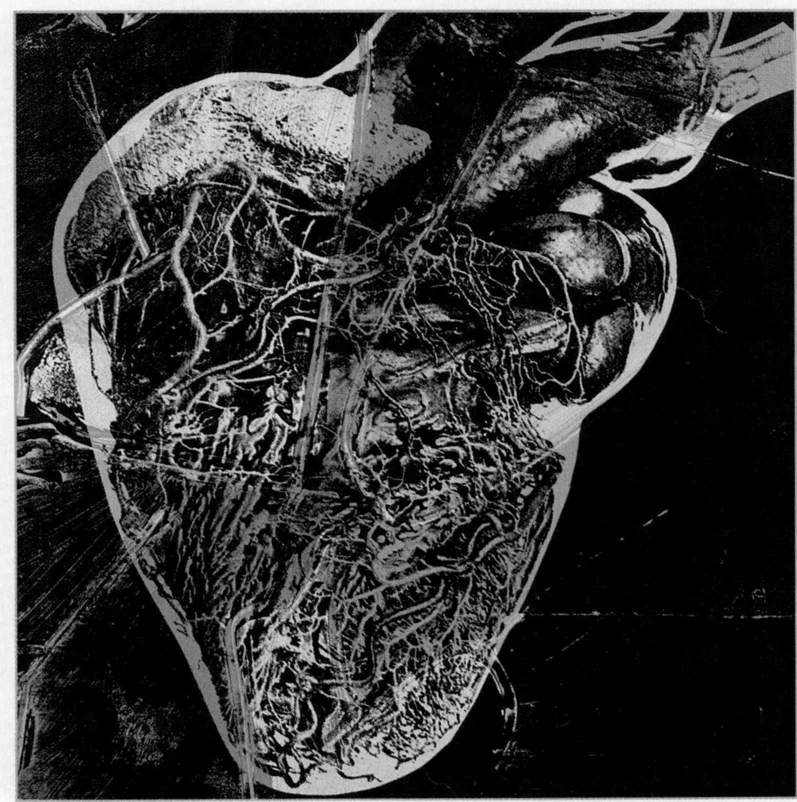

1979 AD

Andy Warhol, pioneer of *Pop Art*, incorporated ordinary objects as well as the human form into his work. From the early 1960's to the late 1980's his paintings embraced both familiarity and celebrity - from soup cans to Marilyn Monroe portraits. Warhol's methods were innovative and controversial. He used repeated, silk-screened imagery - a technique that was both celebrated and criticized in his lifetime. Like many of the artists discussed in this text, Warhol's work is likely to be characteristic of its time. Pictured here is a Warhol heart rendering, a testament to anatomy's potential for the modern artist. Do you find this piece to be artistic? Is anatomy art? Compare it to the illustrations throughout this chapter. Is it an accurate rendering of the heart? Can you tell whether it is a human heart?

Pericarditis and Cardiac Tamponade

Inflammation of the pericardium is known as **pericarditis.** If production of pericardial fluid diminishes, painful rubbing together of the parietal and visceral serous pericardial layers may result. A buildup of pericardial fluid (which may also occur in pericarditis) or extensive bleeding into the pericardium are life-threatening conditions. Because the pericardium cannot stretch, the buildup of fluid or blood compresses the heart. This compression, known as **cardiac tamponade** (tam′-pon-ĀD), can stop the beating of the heart.

Layers of the Heart Wall

The wall of the heart consists of three layers (Figure 13.2a): the epicardium (external layer), the myocardium (middle layer), and the endocardium (inner layer). The outermost **epicardium,** also called the *visceral layer of the serous pericardium*, is the thin, transparent outer layer of the wall. It is composed of mesothelium and delicate connective tissue that imparts a smooth, slippery texture to the outermost surface of the heart. The middle **myocardium** (*myo-* = muscle), which is cardiac muscle tissue, makes up the bulk of the heart and is responsible for its pumping action. Although it is striated like skeletal muscle, cardiac muscle is involuntary like smooth muscle. The cardiac muscle fibers swirl diagonally around the heart in interlacing bundles (Figure 13.2c). The innermost **endocardium** (*endo-* = within) is a thin layer of endothelium overlying a thin layer of connective tissue. It provides a smooth lining for the chambers of the heart and covers the valves of the heart. The endocardium is continuous with the endothelial lining of the large blood vessels attached to the heart.

Chambers of the Heart

The heart contains four chambers. The two upper chambers are the **atria** (= entry halls or chambers), and the two lower chambers are the **ventricles** (= little bellies). On the anterior surface of each atrium is a wrinkled pouchlike structure called an **auricle** (OR-i-kul; *auri-* = ear), so named because of its resemblance to a dog's ear (Figure 13.3). Each auricle slightly increases the capacity of an atrium so that it can hold a greater volume of blood. Also on the surface of the heart are a series of grooves, called **sulci** (SUL-sē), that contain coronary blood vessels and a variable amount of fat. Each sulcus (SUL-kus) marks the external boundary between two chambers of the heart. The deep **coronary sulcus** (*coron-* = resembling a crown) encircles most of the heart and marks the boundary between the superior atria and inferior ventricles. The **anterior interventricular sulcus** is a shallow groove on the anterior surface of the heart that marks the boundary between the right and left ventricles. This sulcus continues around to the posterior surface of the heart as the **posterior interventricular sulcus,** which marks the boundary

Figure 13.3 / Structure of the heart: surface features.

Sulci are grooves that contain blood vessels and fat and mark the boundaries between the various chambers.

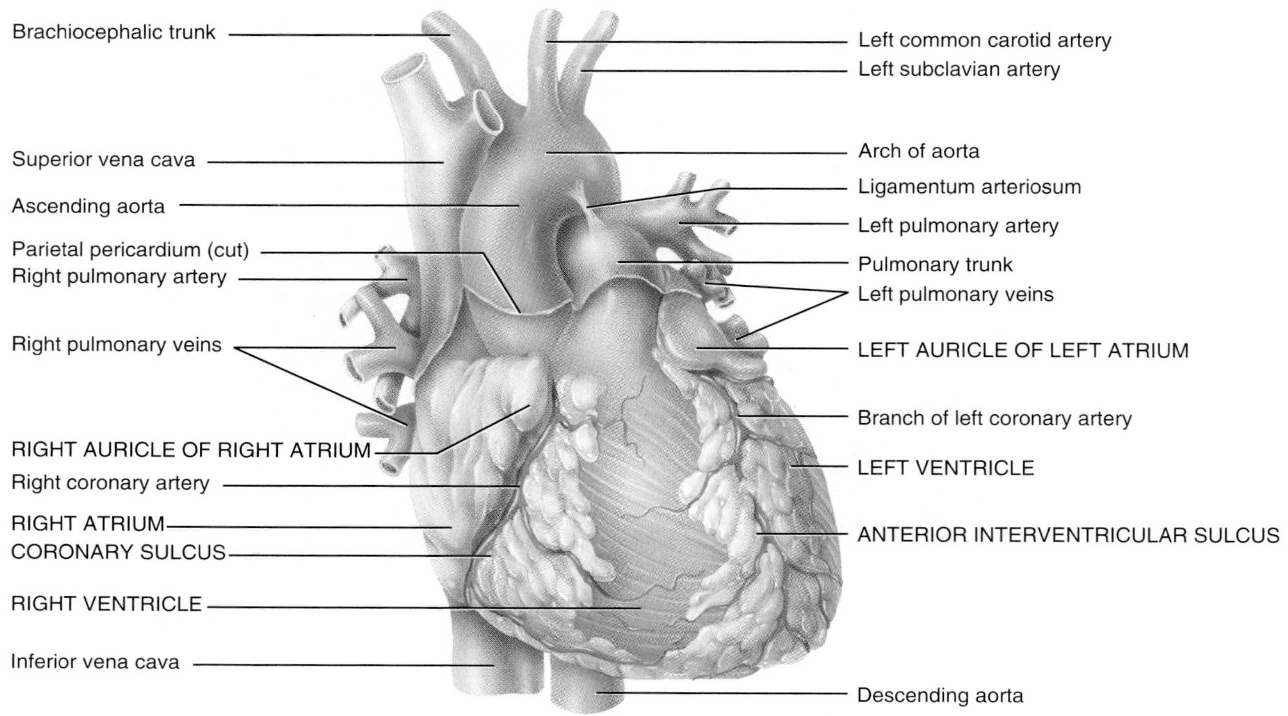

(a) Anterior external view showing surface features

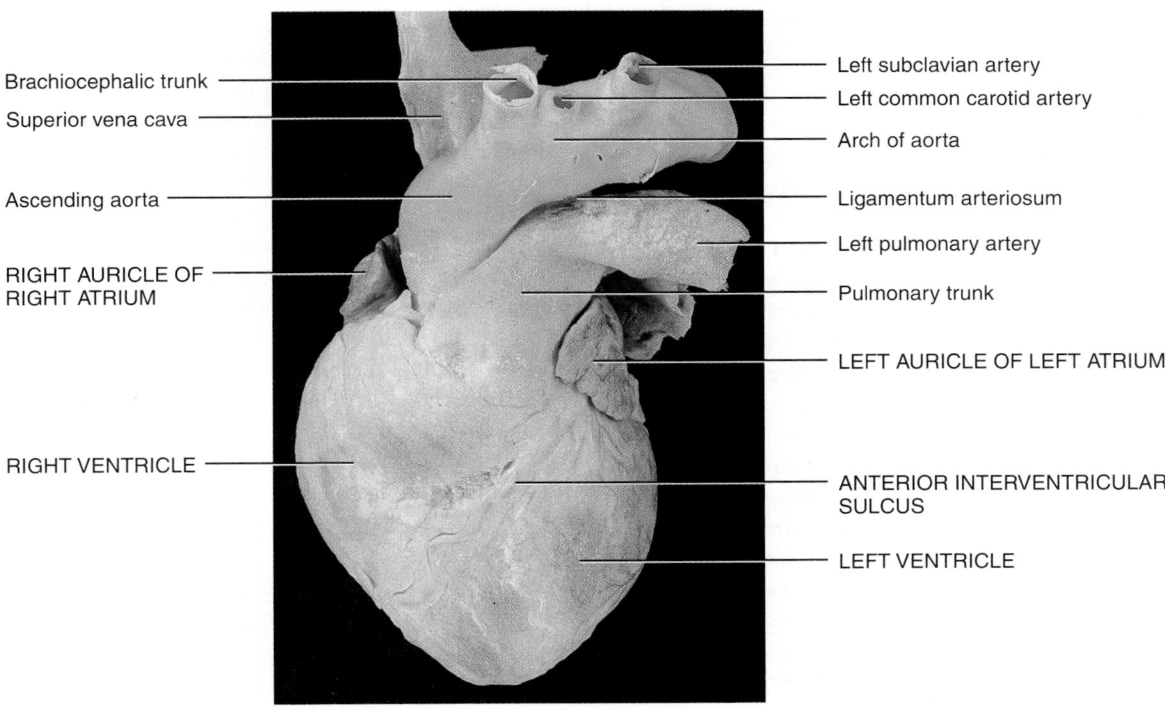

(b) Anterior external view showing surface features

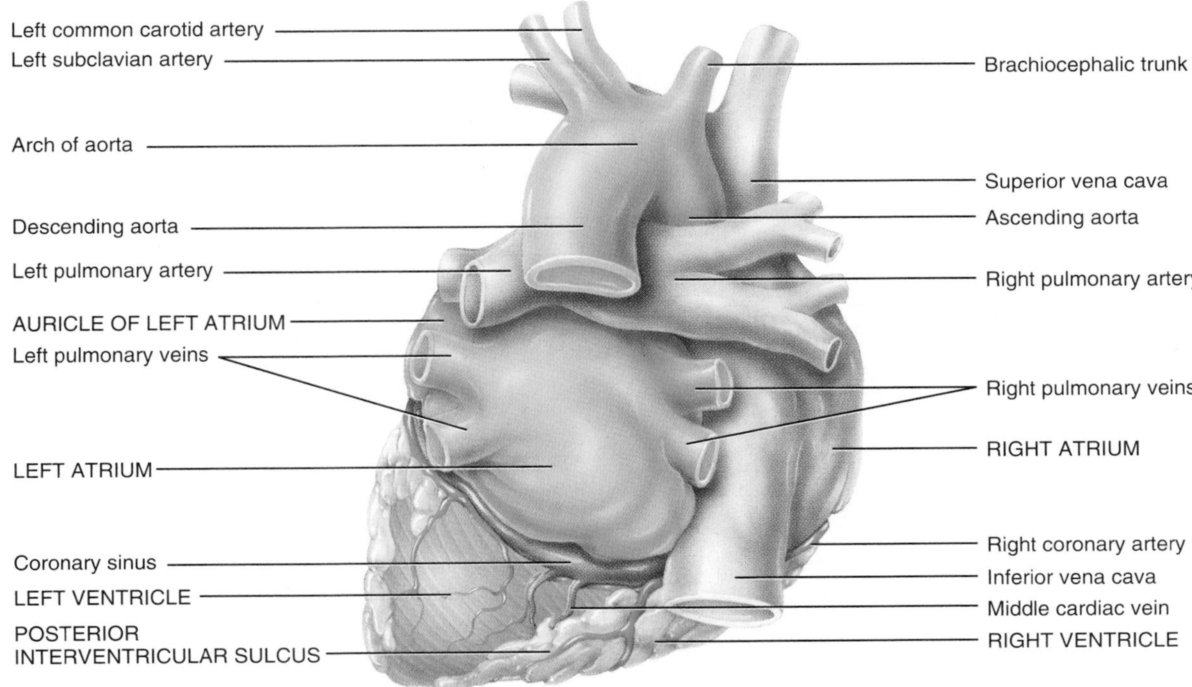

(c) Posterior external view showing surface features

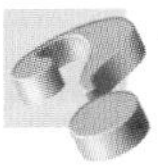 **The coronary sulcus forms a boundary between which chambers of the heart?**

between the ventricles on the posterior aspect of the heart (Figure 13.3c).

Right Atrium

The **right atrium** forms the right border of the heart (see Figure 13.1b). It receives blood from three veins: *superior vena cava, inferior vena cava,* and *coronary sinus* (Figure 13.4a). The anterior and posterior walls within the right atrium differ considerably. Whereas the posterior wall is smooth, the anterior wall is rough due to the presence of muscular ridges called **pectinate muscles** (*pectin* = comb), which also extend into the auricle (Figure 13.4b). Between the right atrium and left atrium is a thin partition called the **interatrial septum** (*inter-* = between; *septum* = a dividing wall or partition). A prominent feature of this septum is an oval depression called the **fossa ovalis,** which is the remnant of the foramen ovale, an opening in the interatrial septum of the fetal heart that normally closes soon after birth (see Figure 14.18 on page 466). Blood passes from the right atrium into the right ventricle through a valve called the **tricuspid valve** (tri-KUS-pid; *tri-* = three; *cuspid* = point) because it consists of three leaflets or cusps (Figure 13.4a). The valves of the heart are composed of dense connective tissue covered by endocardium.

Right Ventricle

The **right ventricle** forms most of the anterior surface of the heart. The inside of the right ventricle contains a series of ridges formed by raised bundles of cardiac muscle fibers called **trabeculae carneae** (tra-BEK-yū-lē KAR-nē-ē; *trabeculae* = little beams; *carneae* = fleshy). Some of the trabeculae carneae convey part of the conduction system of the heart (described on page 401). The cusps of the tricuspid valve are connected to tendon-like cords, the **chordae tendineae** (KOR-dē ten-DIN-ē-ē; *chord-* = cord; *tend-* = tendon), which, in turn, are connected to cone-shaped trabeculae carneae called **papillary muscles** (*papill-* = nipple). The right ventricle is separated from the left ventricle by a partition called the **interventricular septum.** Blood passes from the right ventricle through the **pulmonary valve** into a large artery called the *pulmonary trunk*, which divides into right and left *pulmonary arteries.*

Left Atrium

The **left atrium** forms most of the base of the heart (Figure 13.1b). It receives blood from the lungs through four *pulmonary veins.* Like the right atrium, the inside of the left atrium has a smooth posterior wall. Because pectinate muscles are confined to the auricle of the left atrium, the anterior wall of the left

Figure 13.4 / Structure of the heart: internal anatomy.

The thickness of the four chambers varies according to their functions.

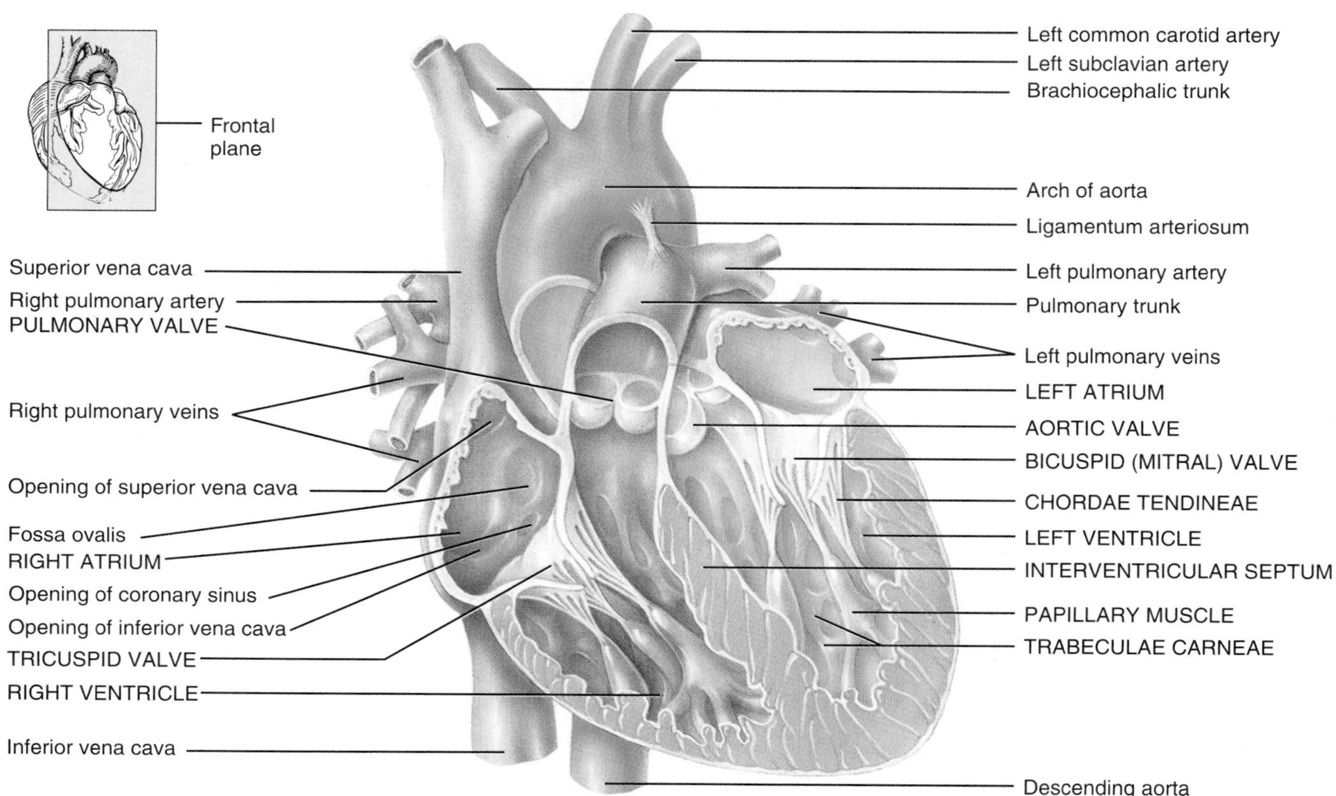

(a) Anterior view of frontal section showing internal anatomy

atrium is smooth. Blood passes from the left atrium into the left ventricle through the **bicuspid (mitral) valve** (*bi-* = two), which has two cusps. The term mitral refers to its resemblance to a bishop's miter (hat), which is two-sided.

Left Ventricle

The **left ventricle** forms the apex of the heart (Figure 13.1b). Like the right ventricle, the left ventricle contains trabeculae carneae and has chordae tendinae that anchor the cusps of the bicuspid valve to papillary muscles. Blood passes from the left ventricle through the **aortic valve** into the largest artery of the body, the *ascending aorta* (*aorte* = to suspend, because the aorta once was believed to lift up the heart). Some of the blood in the aorta flows into the *coronary arteries*, which branch from the ascending aorta and carry blood to the heart wall; the remainder of the blood passes into the *arch of the aorta* and *descending aorta* (*thoracic aorta* and *abdominal aorta*). Branches of the arch of the aorta and descending aorta carry blood throughout the body.

During fetal life, a temporary blood vessel, called the *ductus arteriosus*, shunts blood from the pulmonary trunk into the aorta, so that only a small amount of blood enters the nonfunctioning fetal lungs (see Figure 14.18 on page 466). The ductus arteriosus normally closes shortly after birth, leaving a remnant known as the **ligamentum arteriosum,** which connects the arch of the aorta and pulmonary trunk (Figure 13.4a).

Myocardial Thickness and Function

The thickness of the myocardium of the four chambers varies according to function. The atria are thin-walled because they deliver blood into the adjacent ventricles; since the ventricles

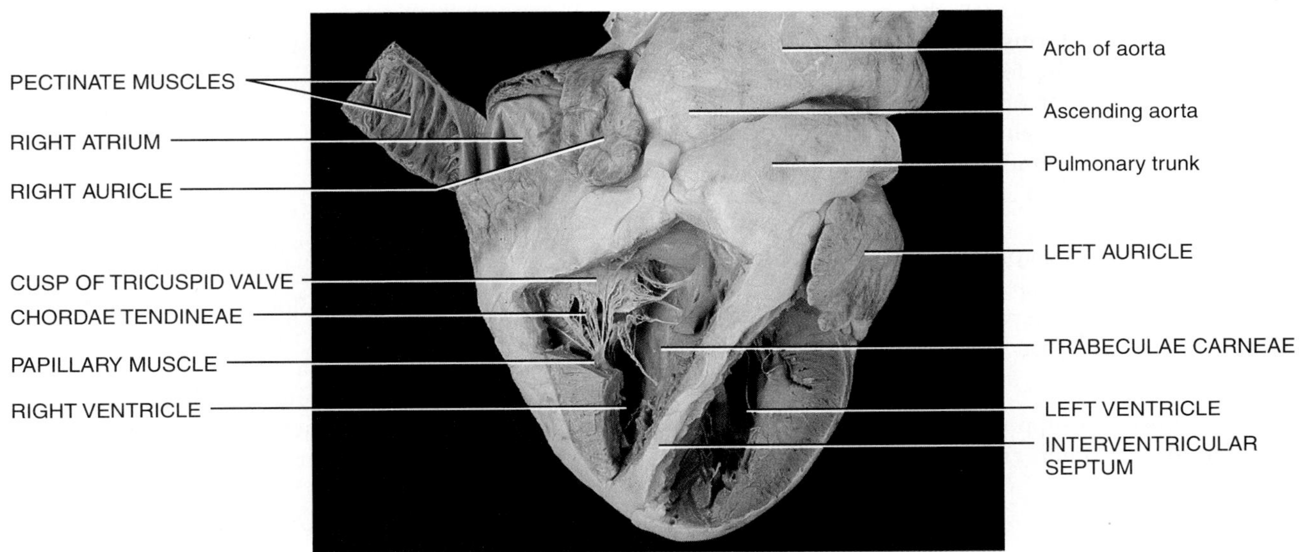

(b) Partially sectioned heart in anterior view showing internal anatomy

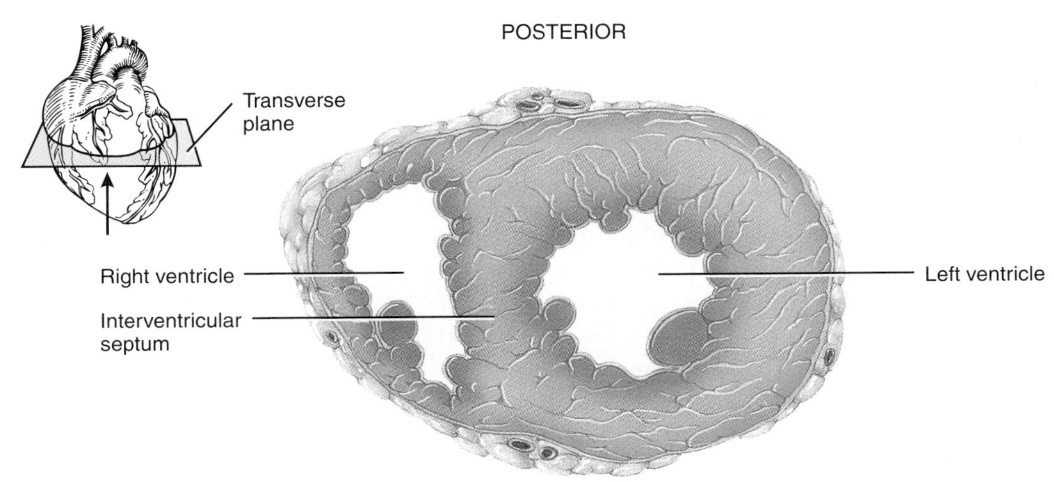

(c) Inferior view of transverse section showing differences in thickness of ventricular walls

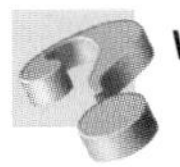

Which chamber has the thickest wall?

pump blood greater distances, their walls are thicker (Figure 13.4a). Although the right and left ventricles act as two separate pumps that simultaneously eject equal volumes of blood, the right side has a much smaller workload. It pumps blood a short distance to the lungs and there is low resistance to blood flow. The left ventricle pumps blood great distances to all other parts of the body and the resistance to blood flow is higher. Thus, the left ventricle works harder than the right ventricle to maintain the same rate of blood flow. The anatomy of the two ventricles confirms this functional difference: The muscular wall of the left ventricle is considerably thicker than that of the right ventricle (Figure 13.4c). Note also that the perimeter of the lumen (space) of the left ventricle is circular, whereas that of the right ventricle is crescent-shaped.

Fibrous Skeleton of the Heart

In addition to cardiac muscle tissue, the heart wall contains dense connective tissue that forms the **fibrous skeleton of the heart.** The skeleton forms the foundation to which the heart valves attach, serves as a point of insertion for cardiac muscle bundles (see Figure 13.2b), prevents overstretching of the valves as blood passes through them, and acts as an electrical insulator that prevents the direct spread of action potentials from the atria to the ventricles.

Essentially, the fibrous skeleton consists of dense connective tissue rings that surround the valves of the heart, fuse with one another, and merge with the interventricular septum. The components of the fibrous skeleton of the heart are as follows (Figure 13.5):

1. Four **fibrous rings** that support the four valves of the heart and are fused to each other: **right atrioventricular fibrous ring, left atrioventricular fibrous ring, pulmonary fibrous ring,** and **aortic fibrous ring.**
2. **Right fibrous trigone** (TRĪ-gōn), a relatively large triangular mass formed by the fusion of the fibrous connective tissue of the left atrioventricular, aortic, and right atrioventricular fibrous rings.
3. **Left fibrous trigone,** a smaller mass formed by the fusion of the fibrous connective tissue of the left atrioventricular and aortic fibrous rings.
4. **Conus tendon,** a mass formed by the fusion of the fibrous connective tissue of the pulmonary and aortic fibrous rings.

Heart Valves

As each chamber of the heart contracts, it pushes a volume of blood into a ventricle or out of the heart into an artery. Valves open and close in response to pressure changes as the heart contracts and relaxes. Each of the four valves helps to ensure one-way flow of blood by opening to let blood through and closing to prevent its backflow.

Atrioventricular Valves

Because they are located between an atrium and a ventricle, the tricuspid and bicuspid valves are termed **atrioventricular (AV) valves.** When an AV valve is open, the pointed ends of the cusps project into the ventricle. Blood moves from the atria into the ventricles through open AV valves when atrial pressure is higher than ventricular pressure (Figure 13.6a, c). At this time, the papillary muscles are relaxed, and the chordae tendineae are slack. When the ventricles contract, the pressure of the blood drives the cusps upward until their edges meet and close the opening

Figure 13.5 / Fibrous skeleton of the heart.

The fibrous skeleton provides a base for the attachment of heart valves, prevents overstretching of the valves, serves as a point of insertion for cardiac muscle bundles, and prevents the direct spread of action potentials from the atria to the ventricles.

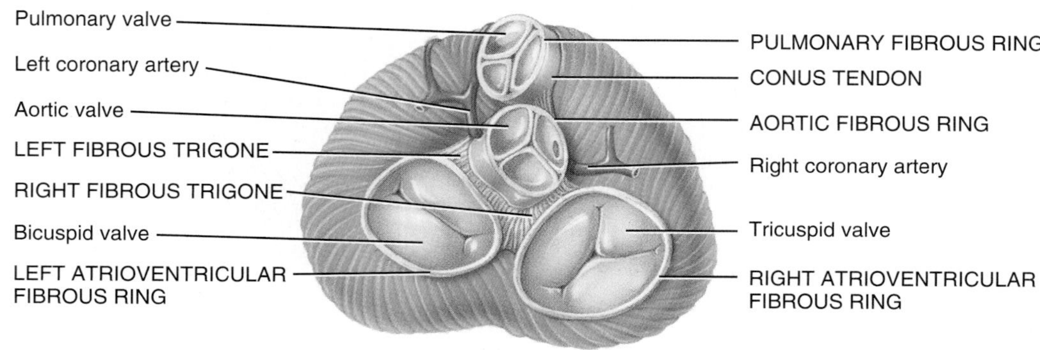

Superior view (the atria have been removed)

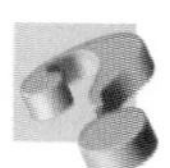

Which component of the fibrous skeleton supports the heart valves?

Figure 13.6 / Valves of the heart.

Heart valves prevent the backflow of blood.

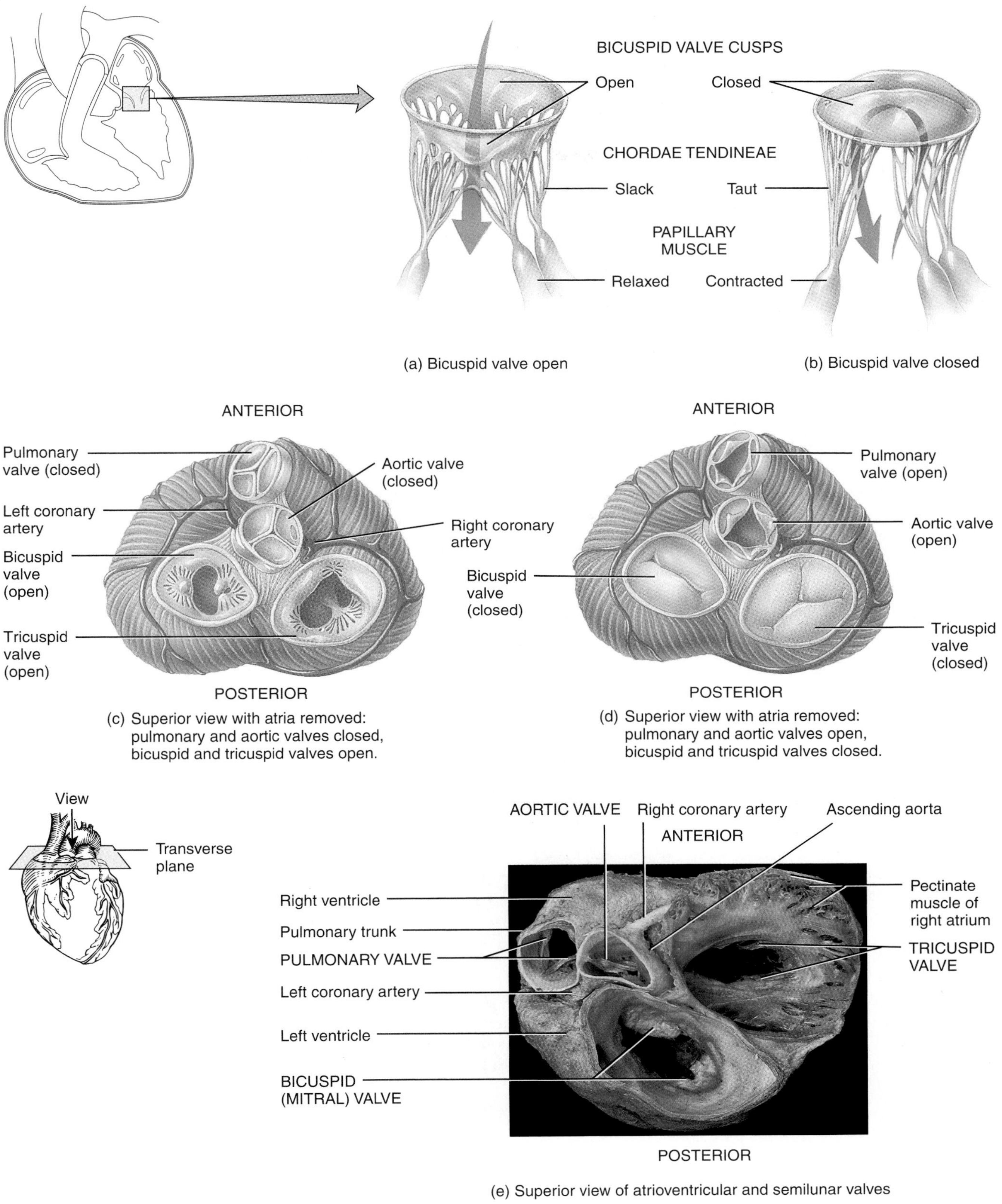

BICUSPID VALVE CUSPS
Open
Closed
CHORDAE TENDINEAE
Slack
Taut
PAPILLARY MUSCLE
Relaxed
Contracted

(a) Bicuspid valve open

(b) Bicuspid valve closed

ANTERIOR
Pulmonary valve (closed)
Aortic valve (closed)
Left coronary artery
Right coronary artery
Bicuspid valve (open)
Tricuspid valve (open)
POSTERIOR

(c) Superior view with atria removed: pulmonary and aortic valves closed, bicuspid and tricuspid valves open.

ANTERIOR
Pulmonary valve (open)
Aortic valve (open)
Bicuspid valve (closed)
Tricuspid valve (closed)
POSTERIOR

(d) Superior view with atria removed: pulmonary and aortic valves open, bicuspid and tricuspid valves closed.

View
Transverse plane

AORTIC VALVE
Right coronary artery
Ascending aorta
ANTERIOR
Right ventricle
Pulmonary trunk
PULMONARY VALVE
Left coronary artery
Left ventricle
BICUSPID (MITRAL) VALVE
Pectinate muscle of right atrium
TRICUSPID VALVE
POSTERIOR

(e) Superior view of atrioventricular and semilunar valves

How do papillary muscles prevent AV valve cusps from everting or swinging upward into the atria?

(Figure 13.6b, d). At the same time, the papillary muscles are also contracting, which pulls on and tightens the chordae tendineae, preventing the valve cusps from everting (opening in the opposite direction into the atria due to the high ventricular pressure). If the AV valves or chordae tendineae are damaged, blood may regurgitate (flow back) into the atria when the ventricles contract.

Semilunar Valves

The two **semilunar (SL) valves** (*semi* = half; *lunar* = moon-shaped), the pulmonary and aortic valves, allow ejection of blood from the heart into arteries, but prevent backflow of blood into the ventricles. Both SL valves consist of three crescent-shaped cusps (Figure 13.6c), each attached to the artery wall by its convex outer margin. The free borders of the cusps curve outward and project into the lumen of the artery. When the ventricles contract, pressure builds up within them. The semilunar valves open when pressure in the ventricles exceeds the pressure in the arteries, permitting ejection of blood from the ventricles into the pulmonary trunk and aorta (Figure 13.6d). As the ventricles relax, blood starts to flow back toward the heart. This backflowing blood fills the valve cusps, which tightly closes the semilunar valves (Figure 13.6c).

Rheumatic Fever

Certain infectious diseases can damage or destroy the heart valves. One example is **rheumatic fever,** an acute systemic inflammatory disease. It usually occurs after a streptococcal infection of the throat and can affect many of the body's connective tissues. The streptococcal bacteria trigger an immune response in which antibodies produced to destroy the bacteria attack and inflame the connective tissues in joints, heart valves, and other organs. Even though the entire heart wall may be weakened, rheumatic fever most often damages the bicuspid (mitral) and

Figure 13.7 / Systemic and pulmonary circulations. Throughout this book, blood vessels that carry oxygenated blood are colored red, whereas those that carry deoxygenated blood are colored blue.

The left side of the heart pumps freshly oxygenated blood into the systemic circulation to all tissues of the body except the air sacs (alveoli) of the lungs; the right side of the heart pumps deoxygenated blood into the pulmonary circulation to the air sacs (alveoli) of the lungs.

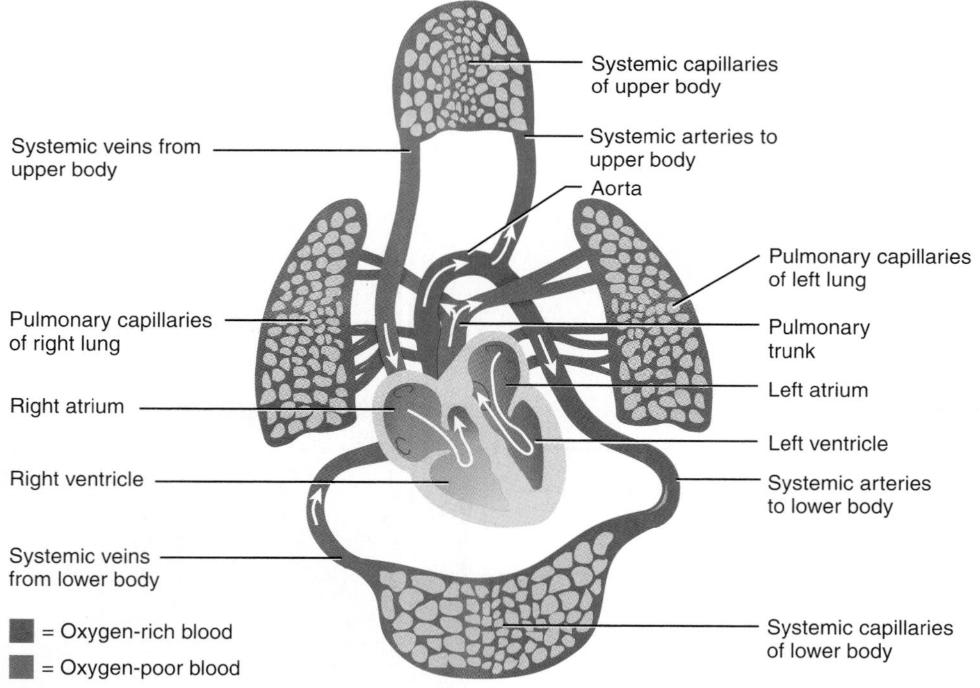

(a) Systemic and pulmonary circulations

aortic valves, which then may fail to open and close properly or may constantly leak. The damage to valves produced by rheumatic fever is permanent.

- ✔ Define the following external features of the heart: auricle, coronary sulcus, anterior interventricular sulcus, and posterior interventricular sulcus.
- ✔ Describe the characteristic internal features of each chamber of the heart.
- ✔ For each chamber of the heart, list the blood vessels that deliver blood to it or receive ejected blood, and name the valve that blood passes through on its way to the next heart chamber or blood vessel.
- ✔ Describe the relationship between wall thickness and function for each heart chamber.
- ✔ How does the fibrous skeleton of the heart assist the operation of heart valves?
- ✔ What causes the heart valves to open and to close?

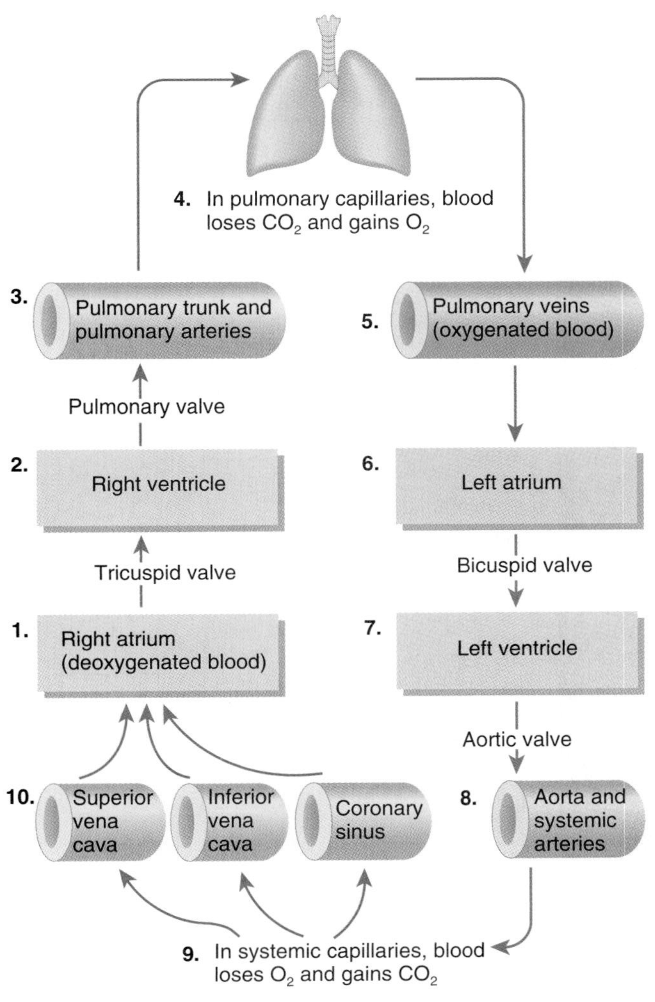

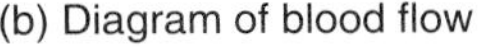
(b) Diagram of blood flow

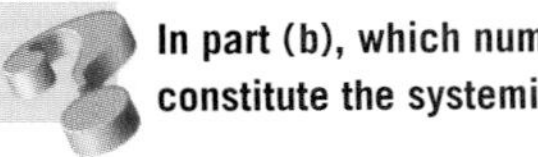
In part (b), which numbers constitute the pulmonary circulation? Which constitute the systemic circulation?

CIRCULATION OF BLOOD

Objectives

- Describe the flow of blood through the chambers of the heart and through the systemic and pulmonary circulations.
- Discuss the coronary circulation.

Systemic and Pulmonary Circulations

With each beat, the heart pumps blood into two circuits—the **systemic circulation** and the **pulmonary circulation** (*pulmon-* = lung). As you will see later, the two circuits are arranged in series so that the output of one becomes the input of the other. The left side of the heart, which receives bright red freshly oxygenated (oxygen-rich) blood from the lungs, is the pump for the systemic circulation. The left ventricle ejects blood into the *aorta*, which branches into (Figure 13.7a on page 398) progressively smaller *systemic arteries* that carry the blood to all organs throughout the body—except for the air sacs (alveoli) of the lungs, which are supplied by the pulmonary circulation. In systemic tissues, arteries give rise to smaller-diameter *arterioles*, which finally lead into extensive beds of *systemic capillaries*. Exchange of nutrients and gases occurs across the thin capillary walls: In the tissues, blood unloads O_2 (oxygen) and picks up CO_2 (carbon dioxide). In most cases, blood flows through only one capillary and then enters a *systemic venule*. Venules carry deoxygenated (oxygen-poor) blood away from tissues and merge to form larger *systemic veins*, and ultimately the blood flows back to the right atrium.

The right side of the heart is the pump for the pulmonary circulation; it receives all the dark red deoxygenated blood returning from the systemic circulation. Blood ejected from the right ventricle flows into the *pulmonary trunk*, which branches into *pulmonary arteries* that carry blood to the right and left lungs. In pulmonary capillaries, blood unloads CO_2, which is exhaled, and picks up O_2. The freshly oxygenated blood then flows into pulmonary veins and returns to the left atrium. Figure 13.7b reviews the route of blood flow through the chambers and valves of the heart and the pulmonary and systemic circulations.

Coronary Circulation

Nutrients could not possibly diffuse from blood in the chambers of the heart through all the layers of cells that make up the heart tissue. For this reason, the wall of the heart has its own blood vessels. The flow of blood through the many vessels that pierce the myocardium is called the **coronary (cardiac) circulation.** The arteries of the heart encircle it like a crown encircles the head. While it is contracting, the heart receives little oxygenated blood by way of the **coronary arteries,** which branch from the ascending aorta (Figure 13.8a). When the heart relaxes, however, the high pressure of blood in the aorta propels blood through the coronary arteries, into capillaries, and then into **coronary veins** (Figure 13.8b).

Coronary Arteries

Two coronary arteries, the right and left coronary arteries, branch from the ascending aorta and supply oxygenated blood

Figure 13.8 / Coronary (cardiac) circulation. These views of the heart from the anterior aspect in (a) and (b) are drawn as if the heart were transparent to reveal blood vessels on the posterior aspect. (See Tortora, *A Photographic Atlas of the Human Body,* Figure 6.9.)

The right and left coronary arteries deliver blood to the heart; the coronary veins drain blood from the heart into the coronary sinus.

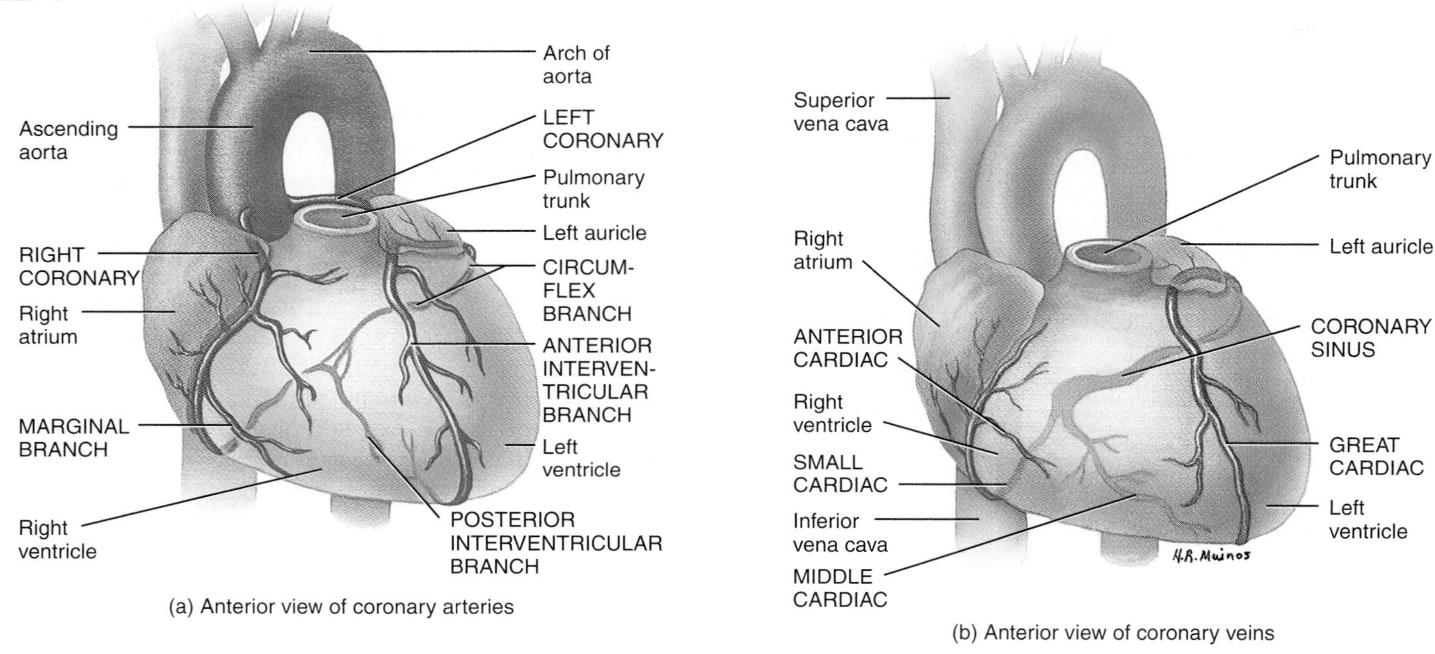

(a) Anterior view of coronary arteries

(b) Anterior view of coronary veins

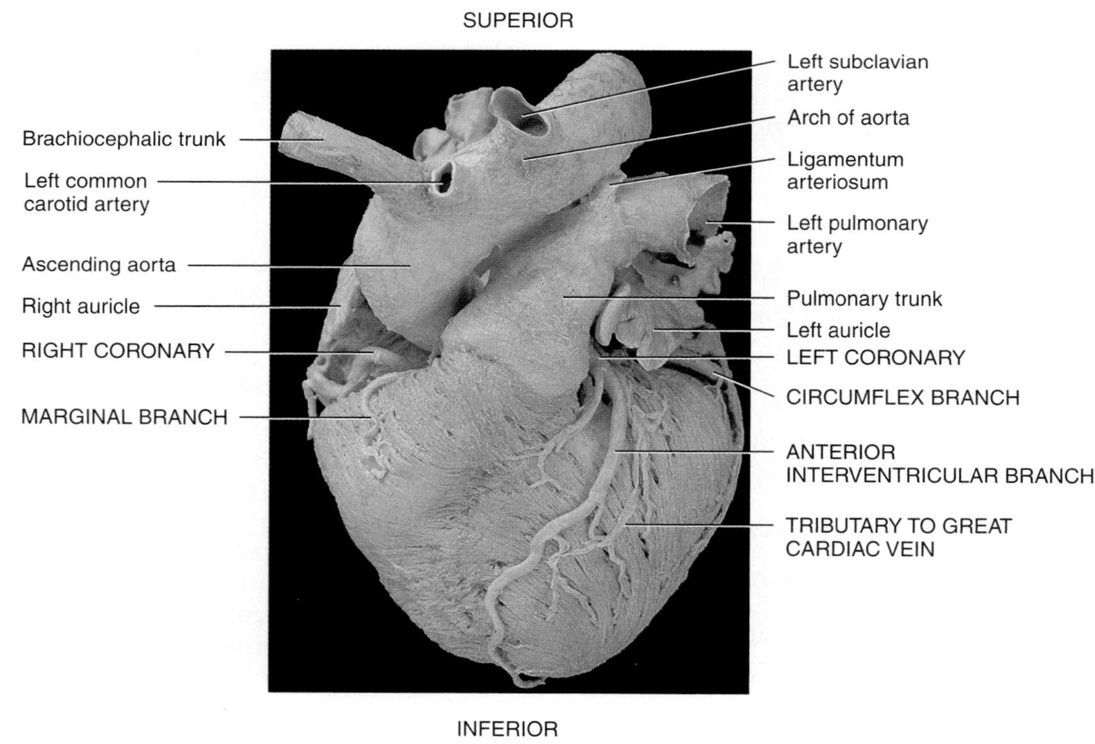

(c) Anterior view

Which coronary blood vessel delivers oxygenated blood to the left atrium and left ventricle?

to the myocardium (Figure 13.8a). The **left coronary artery** passes inferior to the left auricle and divides into the anterior interventricular and circumflex branches. The **anterior interventricular branch** or **left anterior descending (LAD) artery** is in the anterior interventricular sulcus and supplies oxygenated blood to the walls of both ventricles. The **circumflex branch** lies in the coronary sulcus and distributes oxygenated blood to the walls of the left ventricle and left atrium.

The **right coronary artery** supplies small branches (atrial branches) to the right atrium. It continues inferior to the right auricle and divides into the posterior interventricular and marginal branches. The **posterior interventricular branch** follows the posterior interventricular sulcus and supplies the walls of the two ventricles with oxygenated blood. The **marginal branch** in the coronary sulcus transports oxygenated blood to the myocardium of the right ventricle.

Most parts of the body receive blood from branches of more than one artery, and where two or more arteries supply the same region, they usually connect. These connections, called **anastomoses** (a-nas′-tō-MŌ-sēs), provide alternate routes for blood to reach a particular organ or tissue. The myocardium contains many anastomoses that connect branches of a given coronary artery or extend between branches of different coronary arteries. They provide detours for arterial blood if a main route becomes obstructed. Thus, the heart muscle may receive sufficient oxygen even if one of its coronary arteries is partially blocked.

Coronary Veins

After blood passes through the arteries of the coronary circulation, where it delivers oxygen and nutrients to the heart muscle, it passes into capillaries, where it collects carbon dioxide and wastes, and then into veins. The deoxygenated blood then drains into a large vascular sinus on the posterior surface of the heart, called the **coronary sinus** (Figure 13.8b), which empties into the right atrium. A vascular sinus is a thin-walled vein that has no smooth muscle to alter its diameter. The principal tributaries carrying blood into the coronary sinus are the **great cardiac vein,** which drains the anterior aspect of the heart, and the **middle cardiac vein,** which drains the posterior aspect of the heart.

- ✔ In correct sequence, list the heart chambers, heart valves, and blood vessels encountered by a drop of blood as it flows out of the right atrium until it reaches the aorta.
- ✔ Which arteries deliver oxygenated blood to the myocardium of the left and right ventricles?

CARDIAC MUSCLE AND THE CARDIAC CONDUCTION SYSTEM

Objectives

- Describe the structural and functional characteristics of cardiac muscle tissue.
- Explain the structural and functional features of the conduction system of the heart.
- Explain the meaning of an electrocardiogram and its diagnostic importance.

Histology of Cardiac Muscle

Compared to skeletal muscle fibers, cardiac muscle fibers are shorter in length, larger in diameter, and not as circular in transverse section (Figure 13.9). They also exhibit branching, which gives an individual fiber a Y-shaped appearance (see Table 3.4 on page 82). A typical cardiac muscle fiber is 50–100 μm long and has a diameter of about 14 μm. Usually there is only one centrally located nucleus, although an occasional cell may have two nuclei. The sarcolemma of cardiac muscle fibers is similar to that of skeletal muscle, but the sarcoplasm is more abundant and the mitochondria are larger and more numerous. Cardiac muscle fibers have the same arrangement of actin and myosin, and the same bands, zones, and Z discs, as skeletal muscle fibers. The transverse tubules of cardiac muscle are wider but less abundant than those of skeletal muscle; there is only one transverse tubule per sarcomere, located at the Z disc. Also, the sarcoplasmic reticulum of cardiac muscle fibers is scanty compared with the SR of skeletal muscle fibers. As a result, cardiac muscle has a limited intracellular reserve of Ca^{2+}. During contraction, a substantial amount of Ca^{2+} enters cardiac muscle fibers from extracellular fluid.

Although cardiac muscle fibers branch and interconnect with each other, they form two separate functional networks. The muscular walls and partition of the atria compose one network, whereas the muscular walls and partition of the ventricles compose the other network. The ends of each fiber in a network connect to its neighbors by irregular transverse thickening of the sarcolemma called **intercalated discs** (in-TER-ka-lāt-ed; *intercalat-* = to insert between). The discs contain **desmosomes,** which hold the fibers together, and **gap junctions,** which allow action potentials to spread from one muscle fiber to another. As a consequence, when a single fiber of either network is stimulated, all the other fibers in the network become stimulated as well. Thus, each network contracts as a functional unit. When the fibers of the atria contract as a unit, blood moves into the ventricles; when the ventricular fibers contract as a unit, they eject blood out of the heart into arteries.

The Conduction System

During embryonic development, about 1% of the cardiac muscle fibers become *autorhythmic cells,* that is, cells that repeatedly and rhythmically generate action potentials. Autorhythmic cells act as a **pacemaker,** setting the rhythm for the contraction of the entire heart, and they form the **conduction system,** the route that delivers action potentials throughout the heart muscle. The conduction system assures that cardiac chambers are stimulated to contract in a coordinated manner, which makes the heart an effective pump. Cardiac action potentials propagate through the following components of the conduction system (Figure 13.10 on page 403):

Figure 13.9 / Histology of cardiac muscle.

Muscle fibers of the atria form one functional network, and muscle fibers of the ventricles form a second functional network.

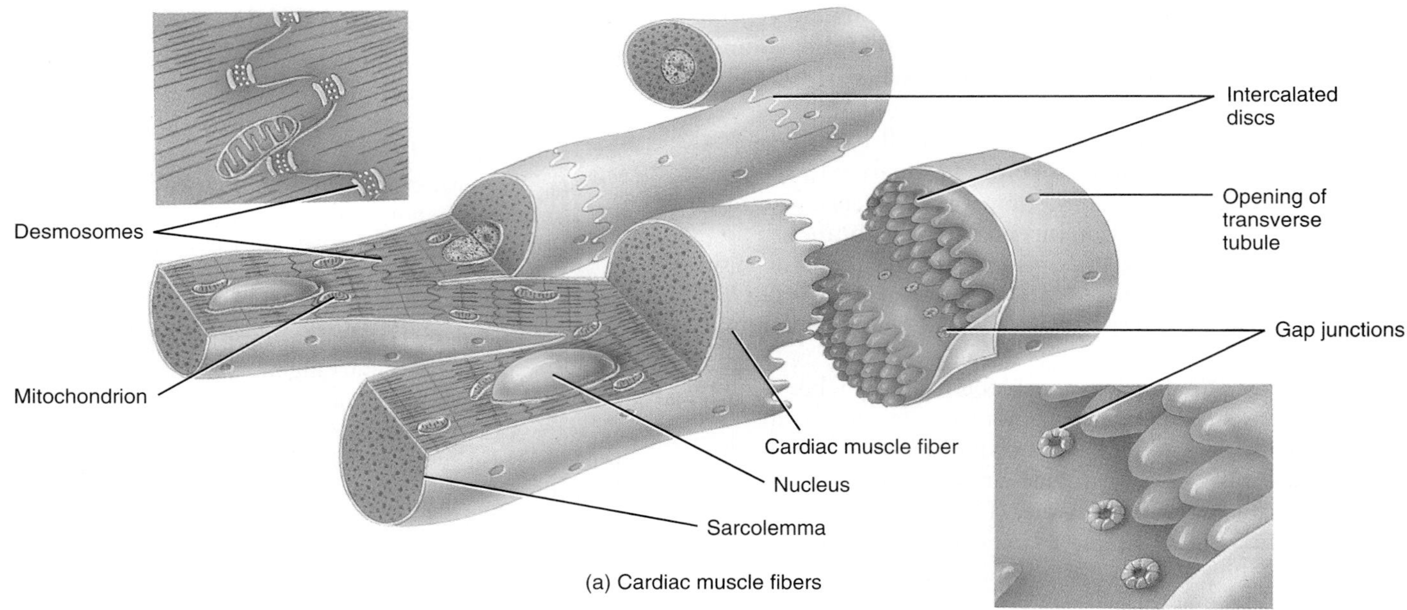

(a) Cardiac muscle fibers

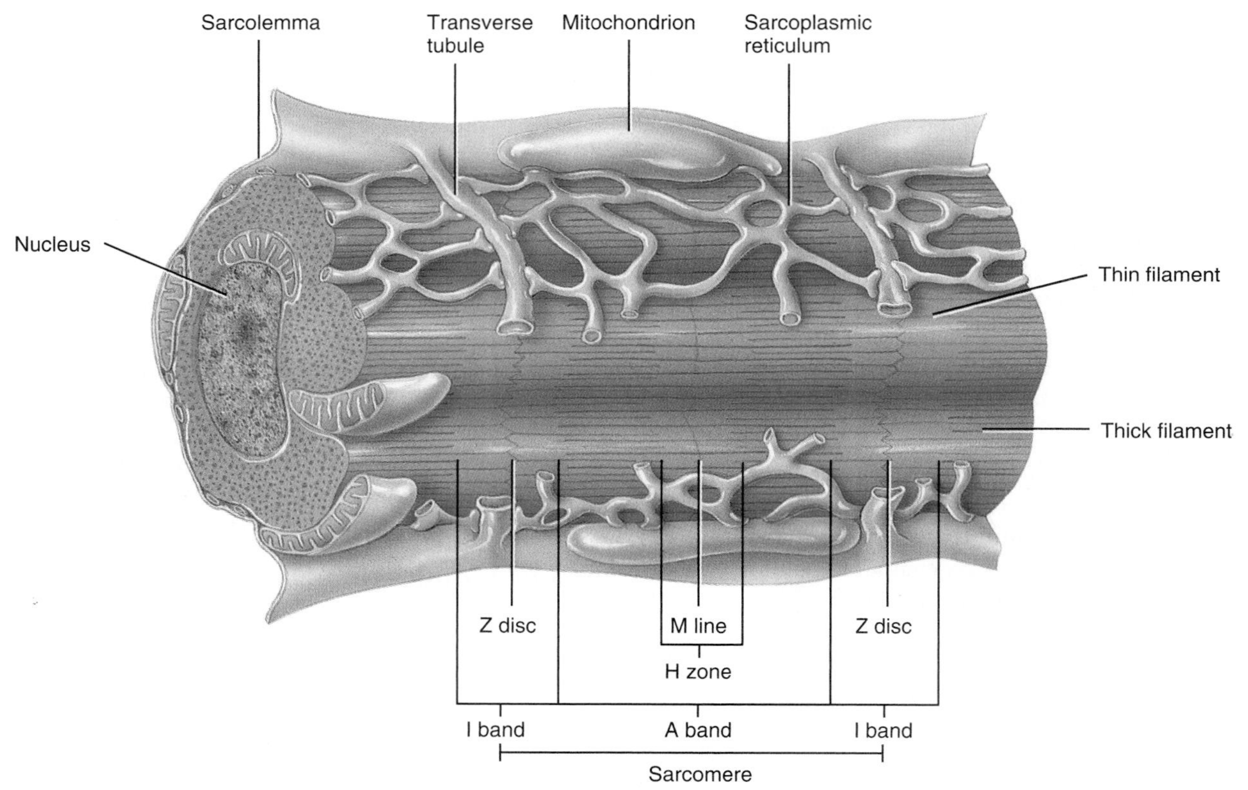

(b) Cardiac myofibrils based on an electron micrograph

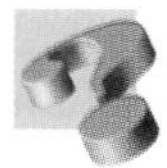

What are the functions of intercalated discs in cardiac muscle fibers?

Figure 13.10 / The conduction system of the heart and a normal electrocardiogram. The route of action potentials through the numbered components of the conduction system is described in the text.

The conduction system ensures that cardiac chambers contract in a coordinated manner.

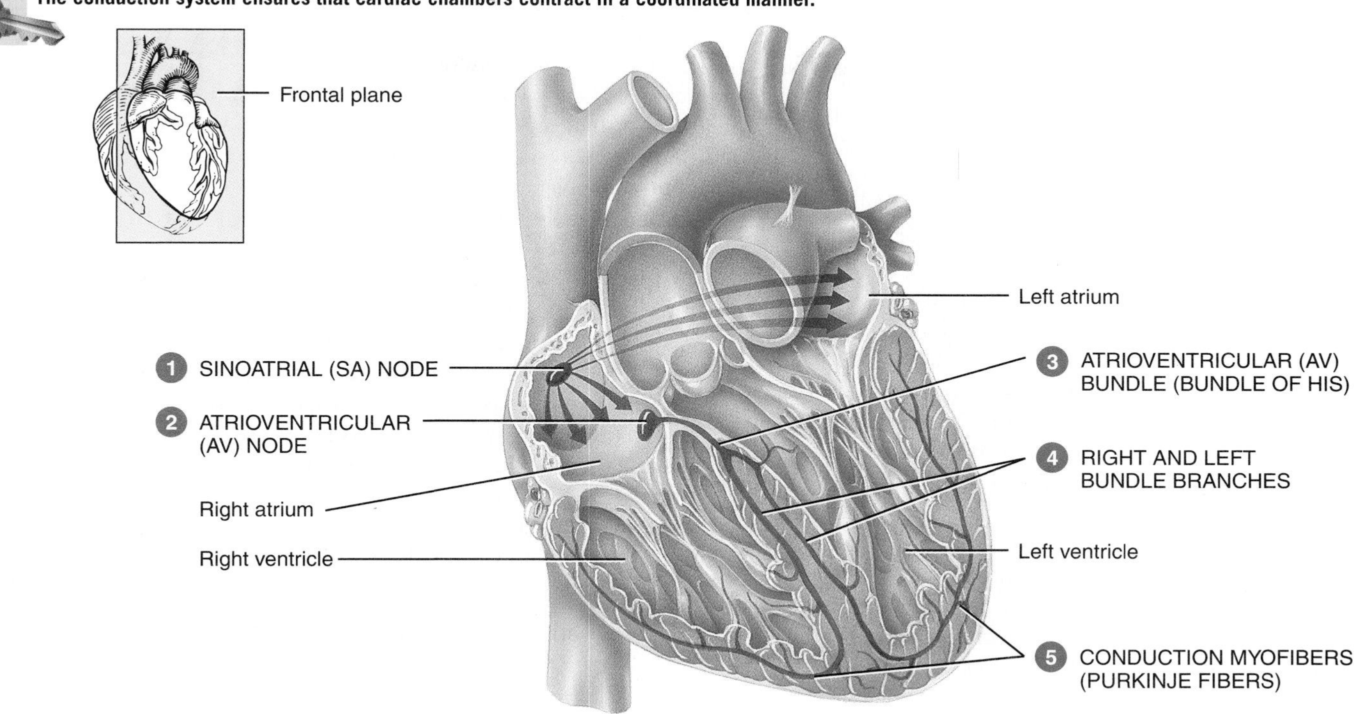

(a) Anterior view of frontal section

1. Normally, cardiac excitation begins in the **sinoatrial (SA) node,** located in the right atrial wall just inferior to the opening of the superior vena cava. Each action potential from the SA node propagates throughout both atria via gap junctions in the intercalated discs of atrial fibers. With the arrival of the action potential, the atria contract.
2. By propagating along atrial muscle fibers, the action potential reaches the **atrioventricular (AV) node,** located in the septum between the two atria, just anterior to the opening of the coronary sinus. At the AV node, the action potential slows considerably, providing time for the atria to empty their blood into the ventricles.
3. From the AV node, the action potential enters the **atrioventricular (AV) bundle** (also known as the **bundle of His**), the only site where action potentials can conduct from the atria to the ventricles. (Elsewhere, the fibrous skeleton of the heart electrically insulates the atria from the ventricles.)
4. After conducting along the AV bundle, the action potential then enters both the **right** and **left bundle branches** that course through the interventricular septum toward the apex of the heart.
5. Finally, the large-diameter **conduction myofibers (Purkinje fibers)** rapidly conduct the action potential, first to the apex of the ventricular myocardium and then upward to the remainder of the ventricular myocardium, pushing the blood upward to the semilunar valves. About 0.20 sec (200 milliseconds) after the atria contract, the ventricles contract.

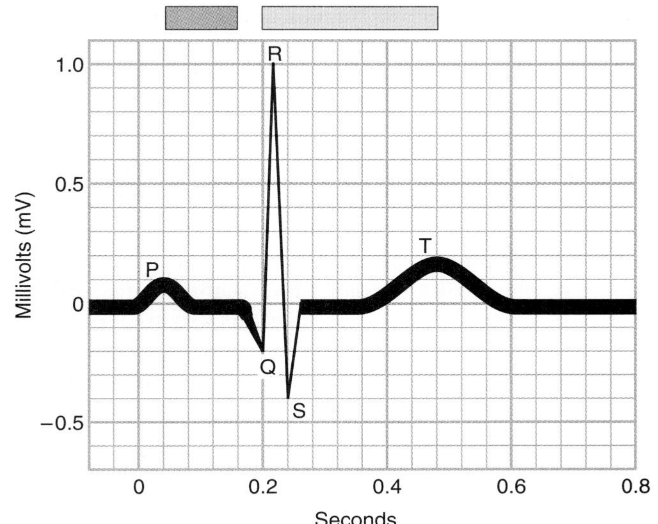

(b) Waves associated with a normal electrocardiogram of a single heartbeat

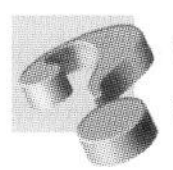

Which component of the conduction system provides the only electrical connection between the atria and the ventricles?

The SA node initiates action potentials 90 to 100 times per minute, faster than any other region of the conducting system. Thus, the SA node sets the rhythm for contraction of the heart—it is the *pacemaker* of the heart. Various hormones and neurotransmitters can speed or slow pacing of the heart by SA node fibers. In a person at rest, for example, acetylcholine released by the parasympathetic division of the ANS typically slows SA node pacing to about 75 action potentials per minute, causing 75 heartbeats per minute. If the SA node becomes diseased or damaged, the slower AV node fibers can become the pacemaker. With pacing by the AV node, however, heart rate is slower, only 40 to 50 beats/min. If the activity of both nodes is suppressed, the heartbeat may still be maintained by the AV bundle, a bundle branch, or conduction myofibers. These fibers generate action potentials very slowly, about 20 to 40 times per minute. At such a low heart rate, blood flow to the brain is inadequate. When this condition occurs, normal heart rhythm can be restored and maintained by surgically implanting an **artificial pacemaker,** a device that sends out small electrical currents to stimulate the heart to maintain adequate cardiac output. Many of the newer pacemakers, called activity-adjusted pacemakers, automatically speed up the heartbeat during exercise.

Sometimes, a site other than the SA node develops abnormal self-excitability and becomes the pacemaker. Such a site is called an *ectopic pacemaker* (ek-TOP-ik; *ectop-* = displaced). An ectopic pacemaker may operate only occasionally, producing an irregular heartbeat, or it may pace the heart for some period of time. Triggers of ectopic activity include caffeine and nicotine, electrolyte imbalances, hypoxia, and toxic reactions to drugs such as digitalis.

Transmission of action potentials through the conduction system generates an electric current that can be detected on the body's surface. A recording of the electrical changes that accompany the heartbeat is called an **electrocardiogram** (e-lek′-trō-KAR-dē-ō-gram), which is abbreviated as either *ECG* or *EKG.* The action potentials are graphed as a series of up-and-down waves during an ECG. Three clearly recognizable waves normally accompany each cardiac cycle (Figure 13.10b). The first, called the **P wave,** is the spread of an action potential from the SA node through the two atria. A fraction of a second after the P wave begins, the atria contract. The second wave, called the **QRS wave,** is the spread of the action potential through the ventricles. Shortly after the QRS wave begins, the ventricles contract. The third wave, the **T wave,** indicates ventricular relaxation. There is no wave to show atrial relaxation because the stronger QRS wave masks this event.

Variations in the size and duration of deflection waves of an ECG are useful in diagnosing abnormal cardiac rhythms and conduction patterns and in following the course of recovery from a heart attack. It can also detect the presence of a living fetus.

Sometimes it is necessary to evaluate the heart's response to the stress of physical exercise. Such a test is called a **stress electrocardiogram,** or **stress test.** Although narrowed coronary arteries may carry adequate oxygenated blood while a person is at rest, during strenuous exercise they will be unable to meet the heart's increased need for oxygen, creating changes that can be noted on an electrocardiogram.

Arrhythmias

Arrhythmia (a-RITH-mē-a) or *dysrhythmia* is a general term referring to an irregularity in heart rhythm resulting from a defect in the conduction system of the heart. Arrhythmias are caused by such factors as caffeine, nicotine, alcohol, and certain other drugs; anxiety; hyperthyroidism; potassium deficiency; and certain heart diseases. One serious arrhythmia is *heart block,* in which propagation of action potentials along the conduction system is either slowed or blocked. The most common site of blockage is the atrioventricular node, a condition called *atrioventricular (AV) block.* In *atrial flutter,* the atria contract abnormally rapidly, about 300 times per minute. *Atrial fibrillation* is asynchronous contraction of atrial fibers (about 400–600 times per minute), whereas *ventricular fibrillation* is asynchronous contractions of ventricular contractile fibers. In both cases, the fibrillating chambers fail to pump blood because some muscle fibers are contracting while others are relaxing. In a strong heart, atrial fibrillation reduces the pumping effectiveness of the heart by only 20–30%, which is compatible with life. In ventricular fibrillation, however, death occurs quickly because blood is not being ejected from the ventricles. ■

- ✔ How do cardiac muscle fibers differ structurally and functionally from skeletal muscle fibers?
- ✔ What are autorhythmic cells?
- ✔ Trace an action potential through the conduction system of the heart.
- ✔ What is an electrocardiogram? What is its diagnostic significance?

CARDIAC CYCLE (HEARTBEAT)

Objective

- Describe the phases associated with a cardiac cycle.

The **cardiac cycle** comprises all the events associated with one heartbeat. In a normal cardiac cycle, the two atria contract while the two ventricles relax. Then while the two ventricles contract, the two atria relax. **Systole** (SIS-tō-lē) refers to the phase of contraction of a chamber of the heart; **diastole** (dī′-AS-tō-lē) is the phase of relaxation. For the purposes of our discussion, we will divide the cardiac cycle into the following phases illustrated in Figure 13.11:

1. **Relaxation period.** At the end of a cardiac cycle when the ventricles start to relax, all four chambers are in diastole. This is the beginning of the *relaxation period.* As the ventricles relax, pressure within the chambers drops, and blood starts to flow from the pulmonary trunk and aorta back toward the ventricles. As this blood becomes trapped in the semilunar cusps, however, the semilunar valves close. As the ventricles continue to relax, the space inside expands, and the pressure falls. When ventricular pressure drops below atrial pressure, the AV valves open and ventricular filling begins.

Figure 13.11 / The cardiac cycle (heartbeat).

A cardiac cycle comprises all the events associated with a single heartbeat.

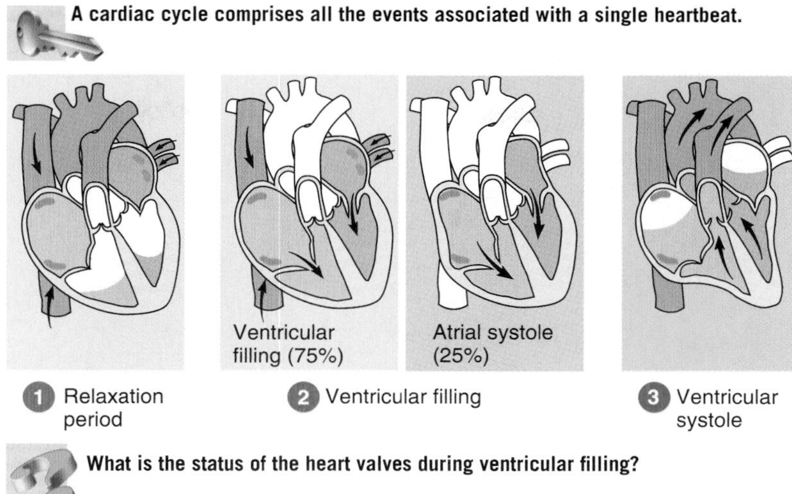

What is the status of the heart valves during ventricular filling?

2 **Ventricular filling.** The major part of ventricular filling (75%) occurs just after the AV valves open. This occurs *without atrial systole.* Atrial contraction accounts for the remaining 25% of the blood that fills the ventricles. Throughout the period of ventricular filling, the AV valves are open and the semilunar valves are closed.

3 **Ventricular systole (contraction).** Ventricular contraction pushes blood up against the AV valves, forcing them shut. For a very brief period, all four valves are closed again. As ventricular contraction continues, pressure inside the chambers rises sharply. When left ventricular pressure rises above the pressure in the arteries, both semilunar valves open, and ejection of blood from the heart begins. This lasts until the ventricles start to relax. Then, the semilunar valves close and another relaxation period begins.

✔ What is a cardiac cycle?
✔ Describe the major events of the phases of a cardiac cycle.

HEART SOUNDS

Objective

- Describe the source of the two heart sounds normally heard with a stethoscope.

The act of listening to sounds within the body is called **auscultation** (aws-kul-TĀ-shun; *ausculta-* = listening), and it is usually done with a stethoscope. The sound of the heartbeat comes primarily from blood turbulence caused by the closing of the heart valves. During each cardiac cycle, four **heart sounds** are generated, but in a normal heart only the first and second heart sounds (S1 and S2) are loud enough to be heard by listening through a stethoscope.

The first sound (S1), which can be described as a **lubb** sound, is louder and a bit longer than the second sound. The lubb is the sound created by blood turbulence associated with closure of the AV valves soon after ventricular systole begins. The second sound (S2), which is shorter and not as loud as the first, can be described as a **dupp** sound. S2 is caused by blood turbulence associated with closure of the semilunar valves at the beginning of ventricular diastole. Although these heart sounds are due to blood turbulence associated with the closure of valves, they are best heard at the surface of the chest in locations that differ slightly from the actual locations of the valves (Figure 13.12). The sound associated with closure of the aortic valve (first part of S2) is best heard near the superior right point, the sound associated with closure of the pulmonary valve (second part of S2) near the superior left point, the sound associated with closure of the bicuspid valve (first part of S1) near the inferior left point, and the sound associated with closure of the tricuspid valve (second part of S1) near the inferior right point.

Heart sounds provide valuable information about the mechanical operation of the heart. A **heart murmur** is an abnormal sound consisting of a rushing or gurgling noise that is heard before, between, or after the normal heart sounds, or that may mask the normal heart sounds. Although some heart murmurs are "innocent," meaning they are not associated with a significant heart problem, most often a murmur indicates a valve disorder. ■

Among the valvular abnormalities that may contribute to murmurs are **mitral stenosis** (narrowing of the mitral valve by scar formation or a congenital defect), **mitral insufficiency** (backflow or regurgitation of blood from the left ventricle into the left atrium due to a damaged mitral valve or ruptured chordae tendineae), **aortic stenosis** (narrowing of the aortic valve), and **aortic insufficiency** (backflow of blood from the aorta into the left ventricle). Another cause of a heart murmur is **mitral valve prolapse (MVP),** an inherited disorder in which one or

Figure 13.12 / Location of valves (purple) and auscultation sites (red) for heart sounds.

Listening to sounds within the body is called auscultation; it is usually done with a stethoscope.

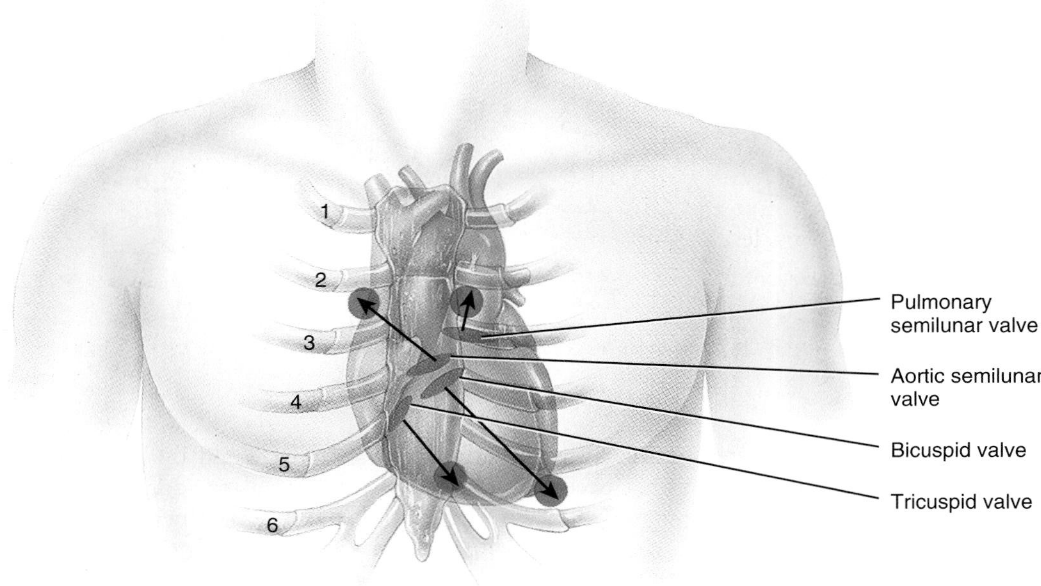

Which heart sound is related to blood turbulence associated with closure of the AV valves?

both cusps of the mitral valve protrude into the left atrium during ventricular contraction. Although a small volume of blood may flow back into the left atrium during ventricular contraction and result in a murmur, mitral valve prolapse does not always pose a serious threat. MVP occurs in 10–15% of the population, more often in females (65%).

✔ Where are the two heart sounds best heard on the surface of the chest?

EXERCISE AND THE HEART

Objective

- Explain the relationship between exercise and the heart.

No matter what a person's level of fitness, it can be improved at any age with regular exercise. Of the various types of exercise, some are more effective than others for improving the health of the cardiovascular system. **Aerobics,** any activity that works large body muscles for at least 20 minutes, elevates cardiac output and accelerates metabolic rate. Three to five such sessions a week are usually recommended for improving the health of the cardiovascular system. Brisk walking, running, bicycling, cross-country skiing, and swimming are examples of aerobic activities.

Sustained exercise increases the oxygen demand of the muscles. Whether the demand is met depends primarily on the adequacy of cardiac output and proper functioning of the respiratory system. After several weeks of training, a healthy person increases maximal cardiac output, thereby increasing the maximal rate of oxygen delivery to the tissues. Oxygen delivery also rises because hemoglobin level increases and skeletal muscles develop more capillary networks in response to long-term training.

During strenuous activity, a well-trained athlete can achieve a cardiac output double that of a sedentary person, in part because training causes hypertrophy (enlargement) of the heart. Even though the heart of a well-trained athlete is larger, *resting* cardiac output is about the same as in a healthy untrained person, because stroke volume is increased while heart rate is decreased. The resting heart rate of a trained athlete often is only 40–60 beats per minute (resting bradycardia). Regular exercise also helps to reduce blood pressure, anxiety, and depression; control weight; and increase the body's ability to dissolve blood clots by increasing fibrinolytic activity.

✔ What are some of the cardiovascular benefits of regular exercise?

DEVELOPMENTAL ANATOMY OF THE HEART

Objective

- Describe the development of the heart.

The *heart*, a derivative of **mesoderm,** begins to develop before the end of the third week of gestation. It begins its development in the ventral region of the embryo inferior to the foregut (see Figure 24.21 on pages 749–750). The first step in its development is the formation of a pair of tubes, the **endothelial (endocardial) tubes,** which develop from mesodermal cells (Figure 13.13). These tubes then unite to form a common tube, referred to as the **primitive heart tube.** Next, the primitive heart tube develops into five distinct regions: (1) **truncus arteriosus,** (2) **bulbus cordis,** (3) **ventricle,** (4) **atrium,** and (5) **sinus venosus.** Because the bulbus cordis and ventricle grow most rapidly, and because the heart enlarges more rapidly than its superior and inferior attachments, the heart first assumes a U-shape and then an S-shape. The flexures of the heart reorient the regions so that the atrium and sinus venosus eventually come to lie superior to the bulbus cordis, ventricle, and truncus arteriosus. Contractions of the primitive heart begin by day 22; they originate in the sinus venosus and force blood through the tubular heart.

At about the seventh week of development, a partition called the **interatrial septum** forms in the atrial region. This septum divides the atrial region into a *right atrium* and *left atrium.* There is an opening in the partition called the **foramen ovale,** which normally closes at birth and later forms a depression called the *fossa ovalis.* An **interventricular septum** also develops; it partitions the ventricular region into a *right ventricle* and *left ventricle.* The bulbus cordis and truncus arteriosus divide into two vessels, the *aorta* (arising from the left ventricle) and the *pulmonary trunk* (arising from the right ventricle). The great veins of the heart, the *superior vena cava* and the *inferior vena cava,* develop from the venous end of the primitive heart tube.

✔ What structures develop from the bulbus cordis and truncus arteriosus? ■

Figure 13.13 / Development of the heart. Arrows within the structures indicate the direction of blood flow.

The heart begins its development from mesoderm before the end of the third week of gestation.

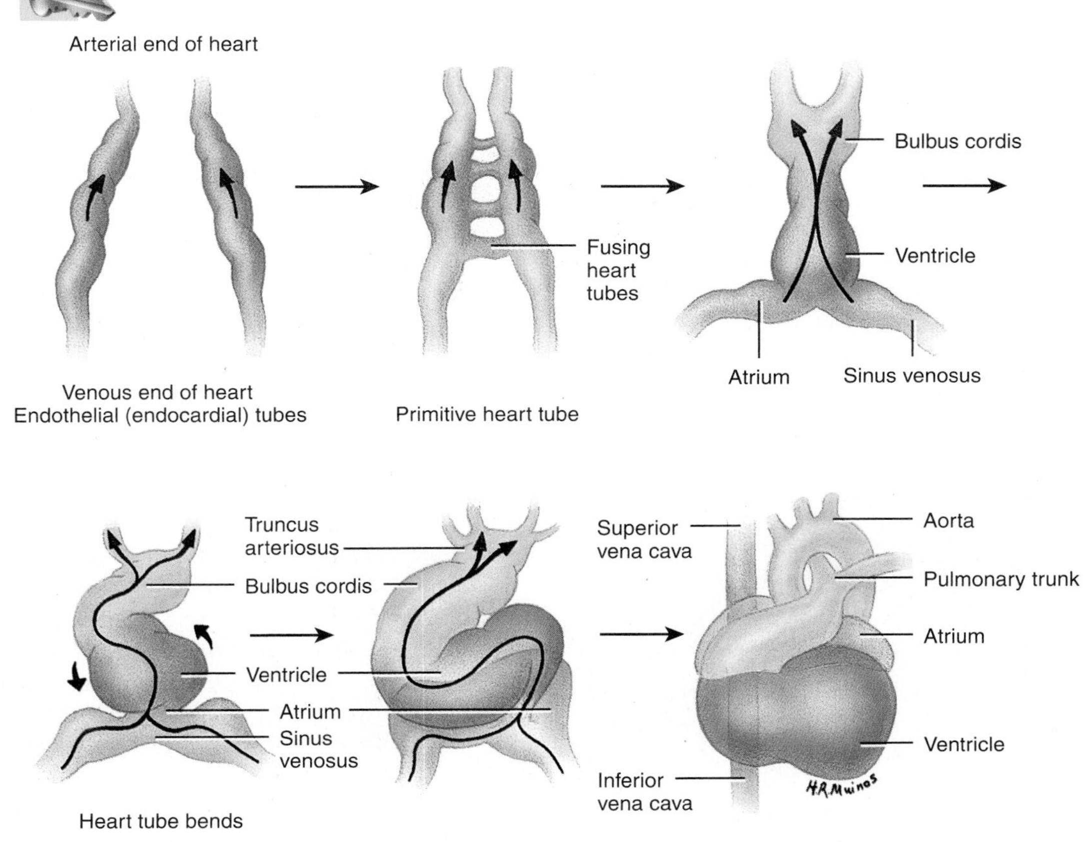

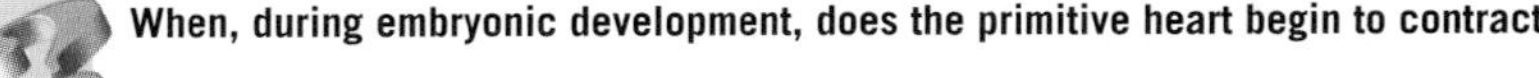
When, during embryonic development, does the primitive heart begin to contract?

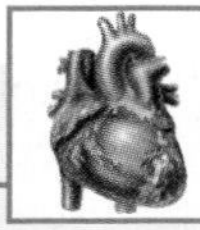

APPLICATIONS TO HEALTH

Coronary Artery Disease

Coronary artery disease (CAD) is a serious medical problem that annually affects about 7 million people and causes nearly three quarters of a million deaths in the United States. CAD is defined as the effects of the accumulation of atherosclerotic plaques (described shortly) in coronary arteries that lead to a reduction in blood flow to the myocardium. Some individuals have no signs or symptoms; others experience angina pectoris (chest pain), and still others suffer a heart attack.

Risk Factors for CAD

People who possess combinations of certain risk factors are more likely to develop CAD. *Risk factors* are characteristics, symptoms, or signs present in a disease-free person that are statistically associated with a greater chance of developing a disease. Some risk factors in CAD are modifiable; that is, they can be altered by changing diet and other habits or can be controlled by taking medications. Among these risk factors for CAD are high blood cholesterol level, high blood pressure, cigarette smoking, obesity, diabetes, "type A" personality, and sedentary lifestyle. Other risk factors are unmodifiable—that is, beyond our control—including genetic predisposition (family history of CAD at an early age), age, and gender. (For example, even though adult males are more likely than adult females to develop CAD, after age 70 the risks are roughly equal.)

Development of Atherosclerotic Plaques

Although the following discussion of atherosclerosis applies to coronary arteries, the process can also occur in arteries outside the heart. Thickening of the walls of arteries and loss of elasticity are the main characteristics of a group of diseases called **arteriosclerosis** (ar-tē-rē-ō-skle-RŌ-sis; *sclero-* = hardening). One form of arteriosclerosis is **atherosclerosis** (ath-er-ō-skle-RŌ-sis), a progressive disease characterized by the formation in the walls of large and medium-sized arteries of lesions called **atherosclerotic plaques** (Figure 13.14). Atherosclerosis is initiated by one or more unknown factors that cause damage to the endothelial lining of the arterial wall. Factors that may initiate the process include high circulating LDL levels, cytomegalovirus (a common herpes virus), prolonged high blood pressure, carbon monoxide in cigarette smoke, and high glucose levels in diabetes mellitus. Atherosclerosis is thought to begin when one of these factors injures the endothelium of an artery, promoting the aggregation of platelets and also attracting phagocytes.

At the injury site, cholesterol and triglycerides collect in the inner layer of the arterial wall. As part of the inflammatory process, macrophages also arrive at the site. Contact with platelets, lipids, and other components of blood stimulates smooth muscle cells and collagen fibers in the arterial wall to proliferate abnormally. In response to this proliferation and to the buildup of lipids, an atherosclerotic plaque develops, progressively obstructing blood flow as it enlarges. An additional danger is that a plaque may provide a roughened surface that attracts platelets, initiating clot formation, further obstructing blood flow. Moreover, a thrombus or piece of a thrombus may dislodge to become an embolus and obstruct blood flow in other vessels.

Diagnosis of CAD

Cardiac catheterization (kath′-e-ter-i-ZĀ-shun) is an invasive procedure used to visualize the heart's coronary arteries, chambers, valves, and great vessels. It may also be used to measure pressure in the heart and blood vessels; to assess function, cardiac output, and diastolic properties of the left ventricle; to measure the flow of blood through the heart and blood vessels, the oxygen content of blood, and the status of heart valves and conduction system; and to identify the exact location of septal and valvular defects. The basic procedure involves inserting a long, flexible, radiopaque **catheter** (plastic tube) into a peripheral vein (for right heart catheterization) or a peripheral artery (for left

Figure 13.14 / Photomicrographs of a transverse section of (a) a normal artery and (b) one partially obstructed by an atherosclerotic plaque.

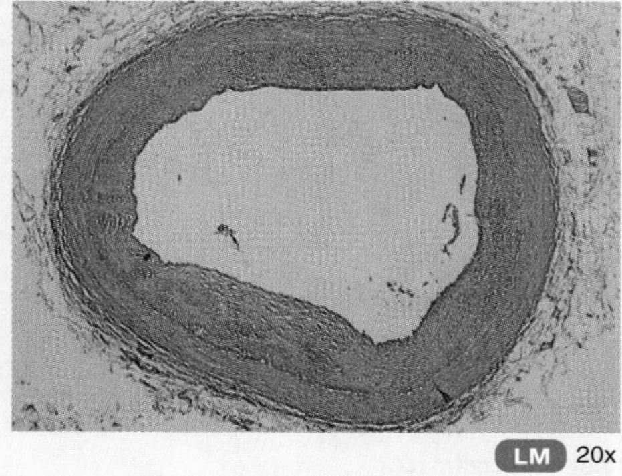

(a) Normal artery

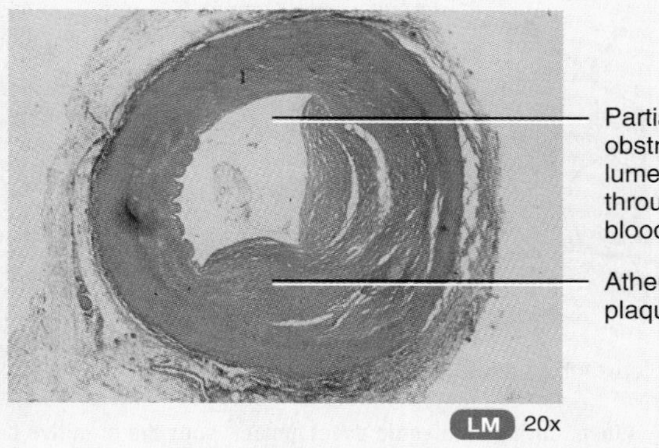

(b) Obstructed artery

heart catheterization) and guiding it under fluoroscopy (x-ray observation).

Cardiac angiography (an′-jē-OG-ra-fē) is another invasive procedure in which a cardiac catheter is used to inject a radiopaque contrast medium into blood vessels or heart chambers. The procedure may be used to visualize coronary arteries, the aorta, pulmonary blood vessels, and the ventricles to assess structural abnormalities in blood vessels (such as atherosclerotic plaques and emboli), ventricular volume, wall thickness, and wall motion. Angiography can also be used to inject clot-dissolving drugs, such as streptokinase or tissue plasminogen activator (t-PA), into a coronary artery to dissolve an obstructing thrombus.

Treatment of CAD

Treatment options for CAD include drugs (nitroglycerine, beta blockers, and cholesterol-lowering and clot-dissolving agents) and various surgical and nonsurgical procedures designed to increase the blood supply to the heart.

Coronary artery bypass grafting (CABG) is a surgical procedure in which a blood vessel from another part of the body is attached ("grafted") to a coronary artery so as to bypass an area of blockage. A piece of the grafted blood vessel is sutured between the aorta and the unblocked portion of the coronary artery (Figure 13.15a).

A nonsurgical procedure used to treat CAD is termed **percutaneous transluminal coronary angioplasty (PTCA)** (*percutaneous* = through the skin; *trans-* = across; *lumen* = an opening or channel in a tube; *angio-* = blood vessel; *-plasty* = to mold or to shape). In this procedure, a balloon catheter (plastic tube) is inserted into an artery of an arm or leg and gently guided into a coronary artery (Figure 13.15b). Then, while dye is released, angiograms (x-rays of blood vessels) are taken to locate the plaques. Next, the catheter is advanced to the point of obstruction, and a balloonlike device is inflated with air to squash the plaque against the blood vessel wall. Because about 30–50% of PTCA-opened arteries fail due to restenosis (renarrowing) within six months after the procedure is done, a special device called a stent may be inserted via a catheter. A *stent* is a stainless steel device, resembling a spring coil, that is permanently placed in an artery to keep the artery patent (open), permitting blood to circulate (Figure 13.5c). Restenosis may be due to damage from the procedure itself, for PTCA may damage the coronary artery wall, leading to platelet activation, proliferation of smooth muscle fibers, and plaque formation.

Myocardial Ischemia and Infarction

Partial obstruction of blood flow in the coronary arteries may cause **myocardial ischemia** (is-KĒ-mē-a; *ische-* = to obstruct; *-emia* = in the blood). Usually, ischemia causes **hypoxia** (reduced oxygen supply), which may weaken cells without killing them. **Angina pectoris** (an-JĪ-na, or AN-ji-na, PEK-to-ris), literally meaning "strangled chest," is a severe pain that usually accompanies myocardial ischemia. Typically, sufferers describe it as a tightness or squeezing sensation, as though the chest were in a vise. Angina pectoris often occurs during exertion, when the heart demands more oxygen, and then disappears with rest. The

Figure 13.15 / Three procedures for reestablishing blood flow in occluded coronary arteries.

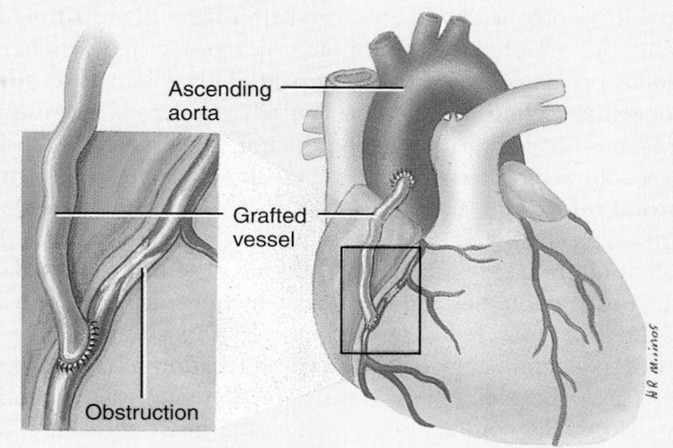

(a) Coronary artery bypass grafting (CABG)

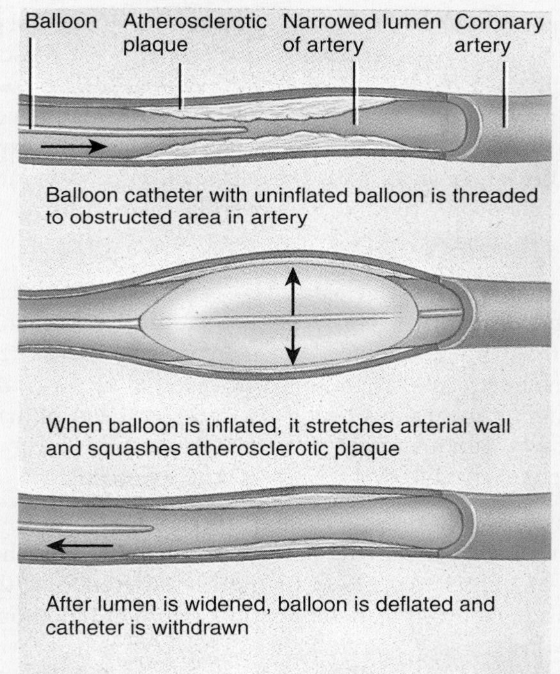

(b) Percutaneous transluminal coronary angioplasty (PTCA)

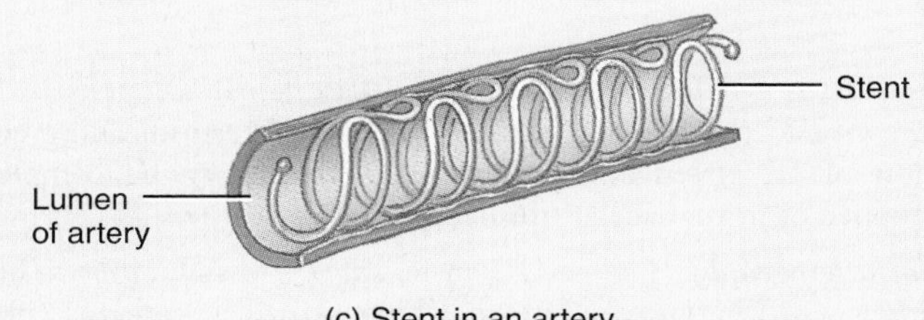

(c) Stent in an artery

pain associated with angina pectoris is often referred to the neck, chin, or down the left arm to the elbow. In some people, especially people with diabetes who suffer from neuropathy (disease of the peripheral or autonomic nervous system), ischemic episodes occur without producing pain. This is known as **silent myocardial ischemia** and is particularly dangerous because the person has no forewarning of an impending heart attack.

A complete obstruction to blood flow may result in a **myocardial infarction** (in-FARK-shun), or **MI,** commonly called a heart attack. *Infarction* means the death of an area of tissue because of interrupted blood supply. Because the heart tissue distal to the obstruction dies and is replaced by noncontractile scar tissue, the heart muscle loses at least some of its strength. The aftereffects depend partly on the size and location of the infarcted (dead) area. Besides killing normal heart tissue, the infarction may disrupt the conduction system of the heart and cause sudden death by triggering ventricular fibrillation. Treatment for a myocardial infarction may involve injection of a thrombolytic (clot-dissolving) agent such as streptokinase or t-PA, plus heparin (an anticoagulant), or performing coronary angioplasty or coronary artery bypass grafting. Fortunately, heart muscle can remain alive in a resting person if it receives as little as 10–15% of its normal blood supply, but such a person may have little ability to engage in physical activity. Furthermore, the extensive anastomoses of coronary blood vessels are a significant factor in helping some people to survive myocardial infarctions.

Congenital Heart Defects

A defect that is present at birth, and usually before, is called a **congenital defect.** Many such defects are not serious and may go unnoticed for a lifetime, and others heal themselves, but some are life-threatening and must be repaired by surgical techniques ranging from simple suturing to replacement of malfunctioning parts with synthetic materials. Among the several congenital defects that affect the heart are the following:

- **Coarctation** (kō′-ark-TĀ-shun) **of the aorta.** In this condition, a segment of the aorta is too narrow, and thus the flow of oxygenated blood to the body is reduced, the left ventricle is forced to pump harder, and high blood pressure develops.
- **Patent ductus arteriosus.** In some babies, the ductus arteriosus, a temporary blood vessel between the aorta and the pulmonary trunk, remains open rather than closing shortly after birth. As a result, aortic blood flows into the lower-pressure pulmonary trunk, thus increasing the pulmonary trunk blood pressure and overworking both ventricles.
- **Septal defect.** This defect is an opening in the septum that separates the interior of the heart into left and right sides. In an **interatrial septal defect** the fetal foramen ovale between the two atria fails to close after birth. An **interventricular septal defect** is caused by incomplete closure of the interventricular septum, which permits oxygenated blood to flow directly from the left ventricle into the right ventricle, where it mixes with deoxygenated blood.
- **Tetralogy of Fallot** (tet-RAL-ō-jē of fal-Ō). This condition is actually a combination of four defects: an interventricular septal defect, an aorta that emerges from both ventricles instead of from the left ventricle only, a stenosed pulmonary semilunar valve, and an enlarged right ventricle. Deoxygenated blood from the right ventricle enters the left ventricle through the interventricular septum, mixes with oxygenated blood, and is pumped into the systemic circulation. Moreover, because the aorta emerges from the right ventricle and the pulmonary trunk is stenosed, very little blood reaches the pulmonary circulation. This causes cyanosis, the bluish discoloration most easily seen in nail beds and mucous membranes when the level of deoxygenated hemoglobin is high. For this reason, tetralogy of Fallot is one of the conditions that cause a "blue baby."

Congestive Heart Failure

In **congestive heart failure (CHF),** the heart is a failing pump. Causes of CHF include coronary artery disease, congenital defects, long-term high blood pressure, myocardial infarctions (regions of dead heart tissue due to a previous heart attack), and valve disorders. As the pump becomes less effective, more blood remains in the ventricles at the end of each cycle. Often, one side of the heart starts to fail before the other. If the left ventricle fails first, it can't pump out all the blood it receives. As a result, blood backs up in the lungs. The result is *pulmonary edema*, fluid accumulation in the lungs that can suffocate an untreated person. If the right ventricle fails first, blood backs up in the systemic vessels. In this case, the resulting *peripheral edema* is usually most noticeable in the feet and ankles.

KEY MEDICAL TERMS ASSOCIATED WITH THE HEART

Cardiac arrest (KAR-dē-ak a-REST) A clinical term meaning cessation of an effective heartbeat. The heart may be completely stopped or in ventricular fibrillation.

Cardiac angiography (an′-jē-OG-ra-fē; *angio* = vessel; *cardio* = heart; *graph* = writing) Procedure in which a cardiac catheter (plastic tube) is used to inject a radiopaque contrast medium into blood vessels or heart chambers. The procedure may be used to visualize blood vessels and the ventricles to assess structural abnormalities. Angiography can also be used to inject clot-dissolving drugs, such as streptokinase or tissue plasminogen activator (t-PA), into a coronary artery to dissolve an obstructing thrombus.

Cardiac catheterization (kath′-e-ter-i-ZĀ-shun) Procedure used to visualize the heart's coronary arteries, chambers, valves, and great vessels. It may also be used to measure pressure in the heart and blood vessels; to assess cardiac output; and to measure the flow of blood through the heart and blood vessels, the oxygen content of blood, and the status of heart valves and conduction system. The basic procedure involves inserting a catheter (plastic tube) into a peripheral vein (for right heart catheterization) or artery (for left heart catheterization) and guiding it under fluoroscopy (x-ray observation).

Cardiomegaly (kar′-dē-ō-MEG-a-lē; *mega* = large) Heart enlargement.

Cor pulmonale (CP) (KOR pul-mōn-ALE; *cor-* = heart; *pulmon-* = lung) A term referring to right ventricular hypertrophy from disorders that bring about hypertension (high blood pressure) in the pulmonary circulation.

Cardiopulmonary resuscitation (kar′-dē-ō-PUL-mō-ner-ē re-sus′-i-TĀ-shun) *(CPR)* The artificial establishment of normal or near-normal respiration and circulation. The **A, B, C's** of cardiopulmonary resuscitation are **Airway, Breathing,** and **Circulation,** meaning the rescuer must establish an airway, provide artificial ventilation if breathing has stopped, and reestablish circulation if there is inadequate cardiac action. The procedure should be performed in that order (A, B, C).

Incompetent valve (in-KOM-pe-tent) Any valve that does not close properly and thus permits a backflow of blood; also called *valvular insufficiency*.

Palpitation (pal′-pi-TĀ-shun) A fluttering of the heart or an abnormal rate or rhythm of the heart.

Paroxysmal tachycardia (par′-ok-SIZ-mal tak′-e-KAR-dē-a) A period of rapid heartbeats that begins and ends suddenly.

Sudden cardiac death The unexpected cessation of circulation and breathing due to an underlying heart disease such as ischemia, myocardial infarction, or a disturbance in cardiac rhythm.

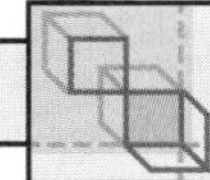

STUDY OUTLINE

Location and Surface Projection of the Heart (p. 388)

1. The heart is located in the mediastinum; about two-thirds of its mass is to the left of the midline.
2. The heart is basically a cone lying on its side; it consists of an apex, a base, anterior and inferior surfaces, and right and left borders.
3. Four points are used to project the heart's location to the surface of the chest.

Structure and Function of the Heart (p. 390)

1. The pericardium is the membrane that surrounds and protects the heart; it consists of an outer fibrous layer and an inner serous pericardium, which is composed of a parietal and a visceral layer.
2. Between the parietal and visceral layers of the serous pericardium is the pericardial cavity, a potential space filled with a few milliliters of pericardial fluid that reduces friction between the two membranes.
3. The wall of the heart has three layers: epicardium (visceral layer of the serous pericardium), myocardium, and endocardium.
4. The epicardium consists of mesothelium and connective tissue, the myocardium is composed of cardiac muscle tissue, and the endocardium consists of endothelium and connective tissue.
5. The heart chambers include two superior chambers, the right and left atria, and two inferior chambers, the right and left ventricles.
6. External features of the heart include the auricles (flaps on each atrium that increase their volume), the coronary sulcus between the atria and ventricles, and the anterior and posterior sulci between the ventricles on the anterior and posterior surfaces of the heart, respectively.
7. The right atrium receives blood from the superior vena cava, inferior vena cava, and coronary sinus. It is separated from the left atrium by the interatrial septum, which contains the fossa ovalis. Blood exits the right atrium through the tricuspid valve.
8. The right ventricle receives blood from the right atrium. It is separated from the left ventricle by the interventricular septum and pumps blood to the lungs through the pulmonary valve and pulmonary trunk.
9. Oxygenated blood enters the left atrium from the pulmonary veins and exits through the bicuspid (mitral) valve.
10. The left ventricle pumps oxygenated blood into the systemic circulation through the aortic valve and aorta.
11. The thickness of the myocardium of the four chambers varies according to the chamber's function. The left ventricle has the thickest wall because of its high workload.
12. The fibrous skeleton of the heart is dense connective tissue that surrounds and supports the valves of the heart.
13. Heart valves prevent backflow of blood within the heart.
14. The atrioventricular (AV) valves, which lie between atria and ventricles, are the tricuspid valve on the right side of the heart and the bicuspid (mitral) valve on the left. The chordae tendineae and papillary muscles stabilize the flaps of the AV valves and stop blood from backing into the atria.
15. Each of the two arteries that leaves the heart has a semilunar valve (aortic and pulmonary).

Circulation of Blood (p. 399)

1. The left side of the heart is the pump for the systemic circulation, the circulation of blood throughout the body except for the air sacs of the lungs. The left ventricle ejects blood into the aorta, and blood then flows into systemic arteries, arterioles, capillaries, venules, and veins, which carry it back to the right atrium.
2. The right side of the heart is the pump for pulmonary circulation, the circulation of blood through the lungs. The right ventricle ejects blood into the pulmonary trunk, and blood then flows into pulmonary arteries, pulmonary capillaries, and pulmonary veins, which carry it back to the left atrium.
3. The flow of blood through the heart is called the coronary (cardiac) circulation.
4. The principal arteries of the coronary circulation are the left and right coronary arteries; the principal veins are the cardiac vein and the coronary sinus.

Cardiac Muscle and the Cardiac Conduction System (p. 401)

1. Cardiac muscle fibers usually contain a single centrally located nucleus. Compared to skeletal muscle fibers, cardiac muscle fibers have more sarcoplasm, more mitochondria, less well-developed sarcoplasmic reticulum, and wider transverse tubules, which are located at Z discs rather than at A–I band junctions.

2. Cardiac muscle fibers branch and are connected via end-to-end intercalated discs, which provide strength and aid in conduction of muscle action potentials by way of gap junctions located in the discs.
3. Autorhythmic cells form the conduction system; these are cardiac muscle fibers that spontaneously generate action potentials.
4. Components of the conduction system are the sinoatrial (SA) node (pacemaker), atrioventricular (AV) node, atrioventricular (AV) bundle (bundle of His), bundle branches, and conduction myofibers (Purkinje fibers).
5. The record of electrical changes during the course of cardiac cycles is called an electrocardiogram (ECG).
6. A normal ECG consists of a P wave (atrial depolarization), a QRS complex (onset of ventricular depolarization), and a T wave (ventricular repolarization).

Cardiac Cycle (Heartbeat) (p. 404)

1. A cardiac cycle consists of the systole (contraction) and diastole (relaxation) of both atria, plus the systole and diastole of both ventricles.
2. The phases of the cardiac cycle are (a) relaxation period, (b) ventricular filling, and (c) ventricular systole.

Heart Sounds (p. 405)

1. S1, the first heart sound (lubb), is caused by blood turbulence associated with the closing of the atrioventricular valves.
2. S2, the second sound (dupp), is caused by blood turbulence associated with the closing of semilunar valves.

Exercise and the Heart (p. 406)

1. Sustained exercise increases oxygen demand on muscles.
2. Among the benefits of aerobic exercise are increased cardiac output, decreased blood pressure, weight control, and increased fibrinolytic activity.

Developmental Anatomy of the Heart (p. 407)

1. The heart develops from mesoderm.
2. The endothelial tubes develop into the four-chambered heart and great vessels of the heart.

SELF-QUIZ QUESTIONS

Choose the one best answer to these questions.

1. Which of the following vessels carries oxygenated blood into the walls of the atria? (a) right coronary artery, (b) pulmonary artery (c) marginal branch, (d) aorta, (e) anterior interventricular branch (left anterior descending artery).
2. Which of the following statements is true of the right atrium? (1) It contains the pacemaker. (2) It receives blood from the superior and inferior vena cavae. (3) It receives blood directly from the lungs. (4) It empties into the aorta.
   a. 1 only **b.** 2 only **c.** 3 only **d.** 4 only **e.** 1 and 2.
3. Which of the following is the correct route of blood through the heart from the systemic circulation to the pulmonary circulation and back to the systemic circulation?
   a. right atrium, tricuspid valve, right ventricle, pulmonary semilunar valve, left atrium, mitral valve, left ventricle, aortic semilunar valve **b.** left atrium, triscuspid valve, left ventricle, pulmonary semilunar valve, right atrium, mitral valve, right ventricle, aortic semilunar valve **c.** left atrium, pulmonary semilunar valve, right atrium, tricuspid valve, left ventricle, aortic semilunar valve, right ventricle, mitral valve **d.** left ventricle, mitral valve, left atrium, pulmonary semilunar valve, right ventricle, tricuspid valve, right atrium, aortic semilunar valve **e.** right atrium, mitral valve, right ventricle, pulmonary semilunar valve, left atrium, tricuspid valve, left ventricle, aortic semilunar valve.
4. Which of the following vessels carries oxygenated blood? (a) superior vena cava, (b) coronary sinus, (c) inferior vena cava, (d) pulmonary veins, (e) none of the above.
5. Which of the following vessels is *not* associated with the right side of the heart? (a) superior vena cava, (b) pulmonary trunk, (c) inferior vena cava, (d) aorta, (e) coronary sinus.
6. Which of the following sequences correctly represents the conduction of an impulse through the heart? (a) SA node, AV node, AV bundle, bundle branches, (b) SA node, AV bundle, AV node, bundle branches, (c) AV node, SA node, AV bundle, bundle branches, (d) SA node, bundle branches, AV node, AV bundle, (e) AV node, AB bundle, SA node, bundle branches.

Complete the following.

7. Two types of junctions found in intercalated discs are ________ and ________ .
8. The first two vessels to branch from the ascending aorta are ________ and ________ .
9. The first heart sound is created by turbulence of blood at the closing of the ________ valves.
10. The pacemaker of the heart is the ________ of the conduction system.
11. During the period of ventricular filling, the AV valves are in the ________ position, and the semilunar valves are in the ________ position.
12. The heart is derived from ________-derm. The heart begins to develop during the ________ month. Its initial formation consists of two endothelial tubes that unite to form the ________ tube.
13. The heart chamber with the thickest wall is the ________ .
14. The ________ point of the heart is located in the fifth left intercostal space, about 9 cm to the left of the midline.

Are the following statements true or false?

15. Trabeculae carneae are ridges located on the inner surface of the right ventricle.
16. The right side of the heart is the pump for the pulmonary circulation.

17. A semilunar valve consists of two half-moon shaped cusps.

18. The fibrous skeleton of the heart is a layer of dense connective tissue that anchors the heart to the surrounding structures of the mediastinum.

19. Most of the ventricular filling occurs during atrial systole.

20. The conduction myofibers (Purkinje fibers) distribute action potentials first to the apex of the ventricles, then superiorly to the remainder of the ventricles.

21. Chordae tendineae attach AV valve cusps to pectinate muscles.

22. The sound of the bicuspid valves is best heard near the inferior right point.

23. Match the following.

____ (a) also called the mitral valve
____ (b) prevents backflow of blood into right atrium
____ (c) prevents backflow of blood into right ventricle
____ (d) prevents backflow of blood into left ventricle
____ (e) located between the right atrium and right ventricle
____ (f) located between the left atrium and left ventricle

(1) aortic semilunar valve
(2) pulmonary semilunar
(3) bicuspid valve
(4) tricuspid valve

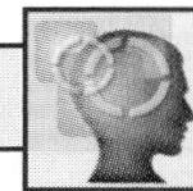

CRITICAL THINKING QUESTIONS

1. Why don't the AV valves flip open backwards when the ventricles contract?
HINT *Why doesn't a parachute turn inside out?*

2. Mr. Williams was diagnosed with blockages in his anterior interventricular branch or left anterior descending (LAD) and circumflex branch of the left coronary artery. What regions of the heart may be affected by these blockages?
HINT *Use the initials to remind you of the vessel location.*

3. The heart is constantly beating which means it's constantly moving. Why doesn't the heart beat its way out of position? Why doesn't the muscle contraction pull the fibers apart?
HINT *To keep a boat in one place, you use an anchor.*

4. Brittany and Sergio were arguing about why the heart is described as a double pump. Brittany said it's because of the two ventricles. Sergio said it's because of the division between the upper and lower chambers. Relate the heart's structure and function to the double pump.
HINT *You can look at this more than one way.*

5. Why doesn't the action potential travel from the atria through the gap junctions to the ventricles?
HINT *What other tissue is present in the heart besides cardiac muscle?*

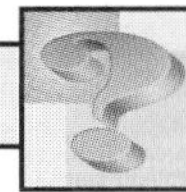

ANSWERS TO FIGURE QUESTIONS

13.1 The mediastinum is the mass of tissue that extends from the sternum to the vertebral column between the pleurae of the lungs.

13.2 The visceral layer of the serous pericardium (epicardium) is both a part of the pericardium and a part of the heart wall.

13.3 The coronary sulcus forms a boundary between the atria and ventricles.

13.4 The left ventricle has the thickest wall.

13.5 The four fibrous rings.

13.6 The papillary muscles contract, which pulls on the chordae tendineae and prevents valve cusps from everting.

13.7 Numbers 2 (right ventricle) through 6 depict the pulmonary circulation, whereas numbers 7 (left ventricle) through 10 and 1 (right atrium) depict the systemic circulation.

13.8 The circumflex artery delivers oxygenated blood to the left atrium and left ventricle.

13.9 The intercalated discs hold the cardiac muscle fibers together and enable action potentials to propagate from one muscle fiber to another.

13.10 The only electrical connection between the atria and the ventricles is the atrioventricular bundle.

13.11 The atrioventricular valves are open and the semilunar valves are closed.

13.12 The first sound (S1), or lubb, is associated with the closure of AV valves.

13.13 The heart begins to contract by the 22nd day of gestation.

14

THE CARDIOVASCULAR SYSTEM: BLOOD VESSELS

CONTENTS AT A GLANCE

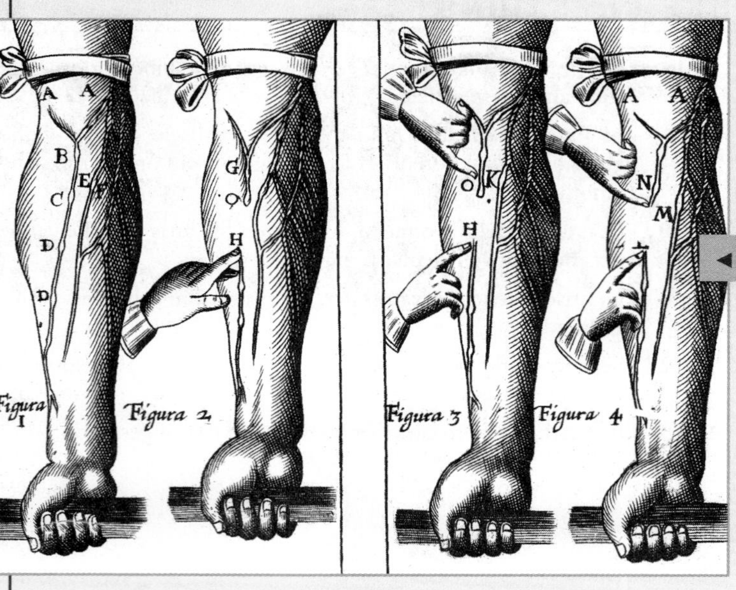

◄ Page 422

Can you determine what is being demonstrated in this image?

Page 451 ►

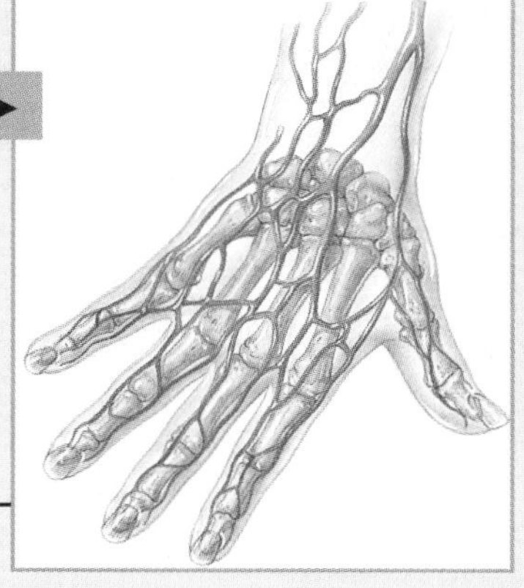

INTRODUCTION

Blood vessels form a system of tubes that carries blood away from the heart, transports it to the tissues of the body, and then returns it to the heart. This chapter focuses on the structure and functions of the various types of blood vessels and on the vessels that constitute the major circulatory routes.

ANATOMY OF BLOOD VESSELS

Objective

- Contrast the structure and function of arteries, arterioles, capillaries, venules, and veins.

Arteries (*ar-* = air; *ter-* = to carry) are vessels that carry blood from the heart to other organs. Large, elastic arteries leave the heart and divide into medium-sized, muscular arteries that branch out into the various regions of the body. Medium-sized arteries then divide into small arteries, which, in turn, divide into still smaller arteries called **arterioles** (ar-TER-ē-ōls). As the arterioles enter a tissue, they branch into countless microscopic vessels called **capillaries** (KAP-i-lar′-ēs). Substances are exchanged between the blood and body tissues through the thin walls of capillaries. Before leaving the tissue, groups of capillaries unite to form small veins called **venules** (VEN-yūls), which merge to form progressively larger blood vessels called veins. **Veins** (VĀNZ) then convey blood from the tissues back to the heart. Because blood vessels require oxygen (O_2) and nutrients just like other tissues of the body, larger blood vessels are served by their own blood vessels, called **vasa vasorum** (literally, vasculature of vessels), located within their walls.

Arteries

In ancient times, **arteries** were found empty at death and thus were thought to contain only air. The wall of an artery has three coats, or tunics: (1) tunica interna, (2) tunica media, and (3) tunica externa (Figure 14.1). The innermost coat, the **tunica interna (intima),** is composed of a lining of simple squamous epithelium called *endothelium*, a *basement membrane*, and a layer of elastic tissue called the *internal elastic lamina*. The endothelium is a continuous layer of cells that line the inner surface of the entire cardiovascular system (the heart and all blood vessels). Normally, the only tissue that blood comes in contact with is endothelium. The tunica interna is closest to the **lumen,** the hollow center through which blood flows. The middle coat, or **tunica media,** is usually the thickest layer; it consists of elastic fibers and smooth muscle fibers (cells), which are arranged in rings around the lumen. Due to their plentiful elastic fibers, arteries normally have high *compliance*, which means that their walls easily stretch or expand without tearing in response to a small increase in pressure. The outer coat, the **tunica externa,** is composed principally of elastic and collagen fibers. In muscular arteries (described shortly), an *external elastic lamina* composed of elastic tissue separates the tunica externa from the tunica media.

Sympathetic fibers of the autonomic nervous system innervate vascular smooth muscle. An increase in sympathetic stimulation typically stimulates the smooth muscle to contract, squeezing the vessel wall and narrowing the lumen. Such a decrease in the diameter of the lumen of a blood vessel is called **vasoconstriction.** In contrast, when sympathetic stimulation decreases, or in the presence of certain chemicals (such as nitric oxide, K^+, H^+, and lactic acid), smooth muscle fibers relax. The resulting increase in lumen diameter is called **vasodilation.** Additionally, when an artery or arteriole is damaged, its smooth muscle contracts, producing vascular spasm of the vessel. Such a vasospasm limits blood flow through the damaged vessel and helps reduce blood loss if the vessel is small.

Elastic Arteries

The largest-diameter arteries, termed **elastic arteries** because the tunica media contains a high proportion of elastic fibers, have walls that are relatively thin in proportion to their overall diameter. Elastic arteries perform an important function: They help propel blood onward while the ventricles are relaxing. As blood is ejected from the heart into elastic arteries, their highly elastic walls stretch, accommodating the surge of blood. By stretching, the elastic fibers momentarily store mechanical energy, functioning as a **pressure reservoir.** Then, the elastic fibers recoil and convert stored (potential) energy in the vessel into kinetic energy of the blood. Thus, blood continues to move through the arteries even while the ventricles are relaxed. Because they conduct blood from the heart to medium-sized, more-muscular arteries, elastic arteries also are called *conducting arteries*. The aorta and the brachiocephalic, common carotid, subclavian, vertebral, pulmonary, and common iliac arteries are elastic arteries (see Figure 14.6).

Muscular Arteries

Medium-sized arteries are called **muscular arteries** because their tunica media contains more smooth muscle and fewer elastic fibers than elastic arteries. Thus, muscular arteries are capable of greater vasoconstriction and vasodilation to adjust the rate of blood flow. The large amount of smooth muscle makes the walls of muscular arteries relatively thick. Muscular arteries also are called *distributing arteries* because they distribute blood to various parts of the body. Examples include the brachial artery in the arm and radial artery in the forearm (see Figure 14.6).

Arterioles

An **arteriole** (= small artery) is a very small (almost microscopic) artery that delivers blood to capillaries (Figure 14.2 on page 417). Arterioles near the arteries from which they branch have a tunica interna like that of arteries, a tunica media composed of smooth muscle and very few elastic fibers, and a tunica externa composed mostly of elastic and collagen fibers. In the smallest-diameter arterioles, which are closest to capillaries, the tunics consist of little more than a ring of endothelial cells surrounded by a few scattered smooth muscle fibers.

Arterioles play a key role in regulating blood flow from arteries into capillaries. When the smooth muscle of arterioles contracts, causing vasoconstriction, blood flow into capillaries

Figure 14.1 / Comparative structure of blood vessels. The relative size of the capillary in (c) is enlarged.

Arteries carry blood from the heart to tissues; veins carry blood from tissues to the heart.

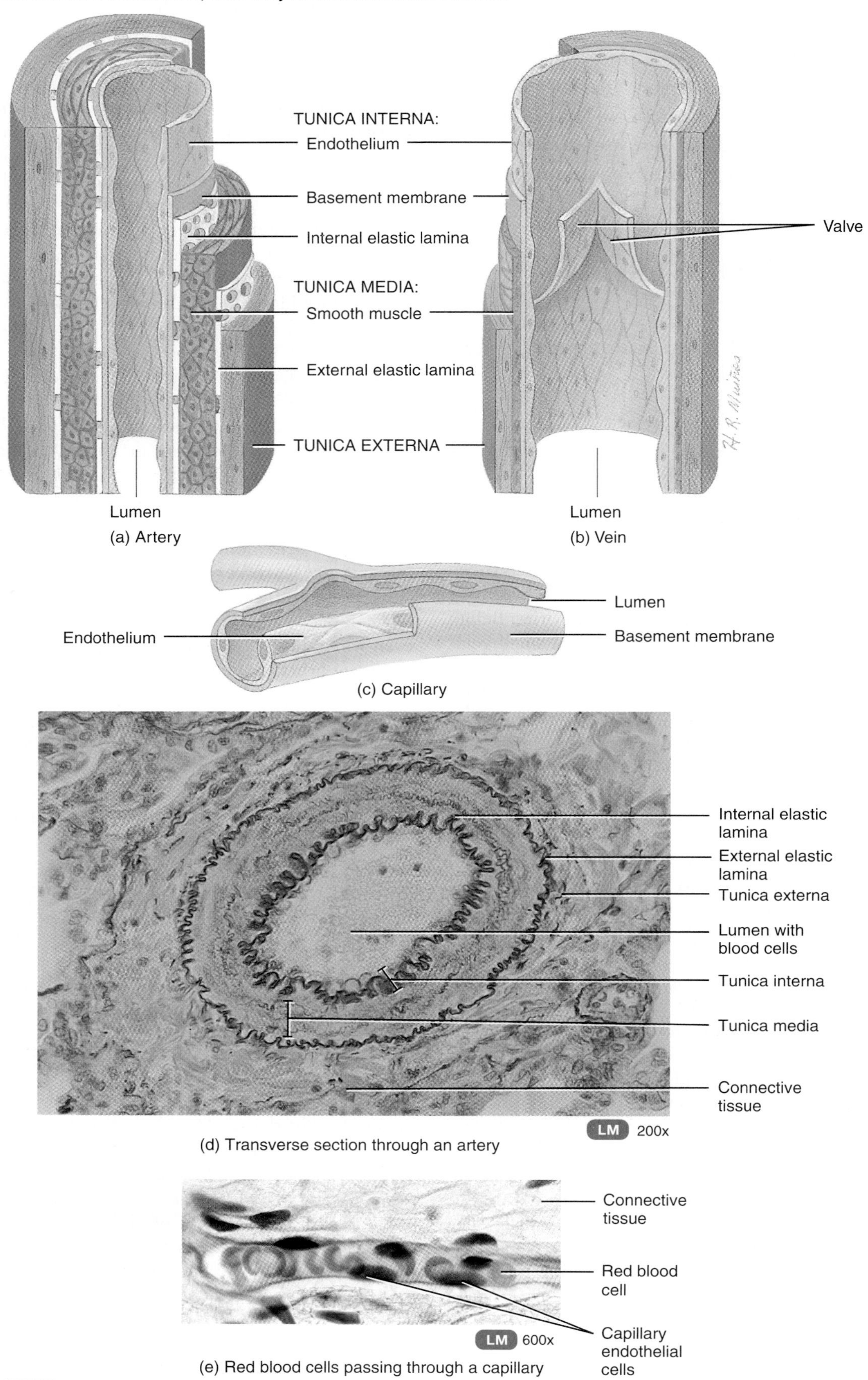

(a) Artery

(b) Vein

(c) Capillary

(d) Transverse section through an artery

(e) Red blood cells passing through a capillary

Which vessel—the femoral artery or the femoral vein—has a thicker wall? Which has a wider lumen?

Figure 14.2 / Arteriole, capillaries, and venule.

Arterioles regulate blood flow into capillaries, where nutrients, gases, and wastes are exchanged between the blood and interstitial fluid.

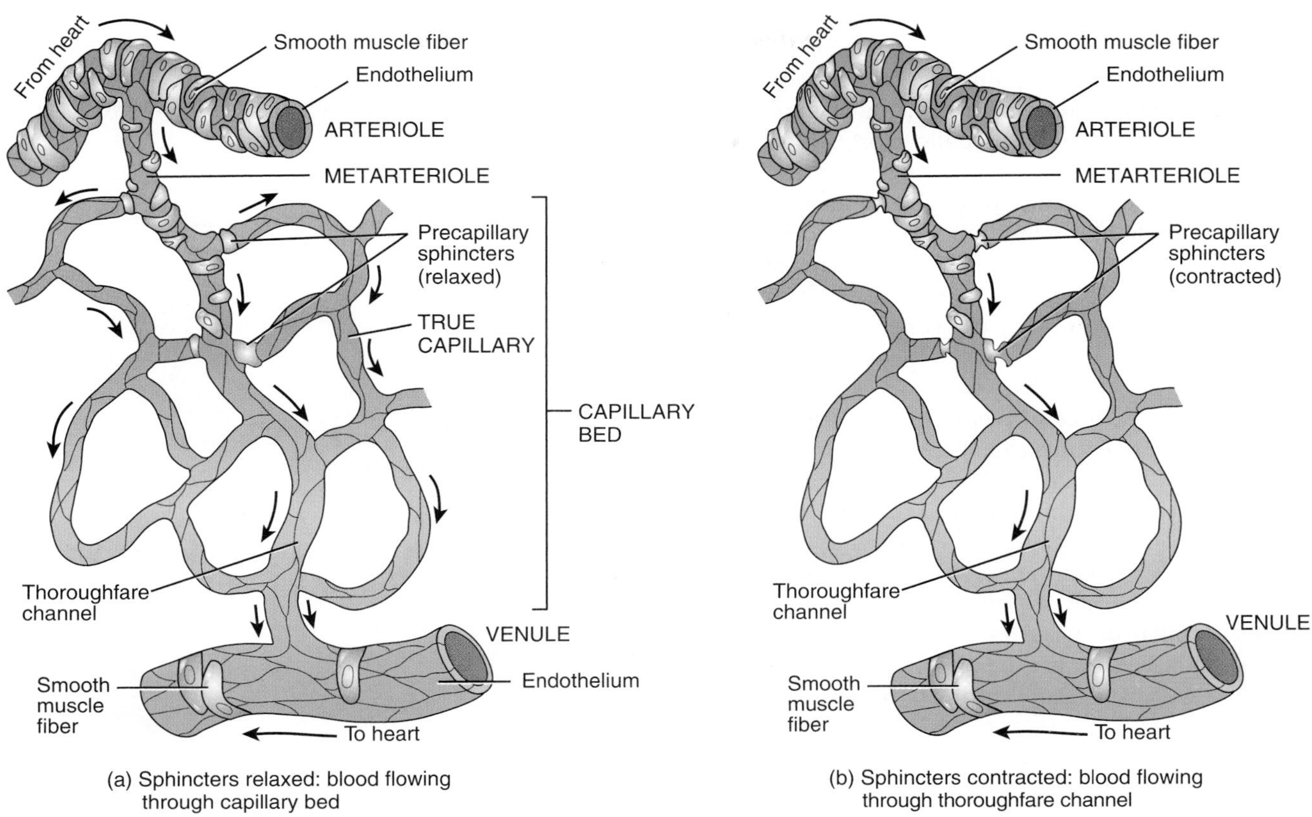

(a) Sphincters relaxed: blood flowing through capillary bed

(b) Sphincters contracted: blood flowing through thoroughfare channel

Why do metabolically active tissues have extensive capillary networks?

decreases; when the smooth muscle relaxes, arterioles vasodilate and blood flow into capillaries increases. A change in diameter of arterioles can also significantly affect blood pressure. A major function of arterioles is to control resistance, the opposition to blood flow principally due to friction between blood and the inner walls of blood vessels. They do so by changing their diameters and thus control blood pressure and blood flow to a tissue. For this reason, arterioles are known as *resistance vessels.*

Capillaries

Capillaries (*capillar-* = hairlike) are microscopic vessels that usually connect arterioles and venules (Figure 14.2). The flow of blood from arterioles to venules through capillaries is called the **microcirculation.** Capillaries are found near almost every cell in the body, but their distribution varies with the metabolic activity of the tissue they serve. Body tissues with high metabolic requirements, such as muscles and the liver, kidneys, and nervous system, use more O_2 and nutrients and thus have extensive capillary networks. Tissues with lower metabolic requirements, such as tendons and ligaments, contain fewer capillaries. A few tissues—all covering and lining epithelia, the cornea and lens of the eye, and cartilage—lack capillaries.

The primary function of capillaries is to permit the exchange of nutrients and wastes between the blood and tissue cells through the interstitial fluid and for this reason blood capillaries are known as *exchange vessels.* The structure of capillaries is admirably suited to this purpose: Because capillary walls are composed of only a single layer of epithelial cells (endothelium) and a basement membrane and have no tunica media or tunica externa (see Figure 14.1c), a substance in the blood must pass through just one cell layer to reach the interstitial fluid and tissue cells. Exchange of materials occurs only through the walls of capillaries and the beginning of venules; the walls of arteries, arterioles, most venules, and veins present too thick a barrier. Capillaries form extensive branching networks that increase the surface area available for rapid exchange of materials. In most tissues, blood flows through only a small portion of the capillary network when metabolic needs are low. However, when a tissue is active, such as contracting muscle, the entire capillary network fills with blood.

A **metarteriole** (*met-* = beyond) is a vessel that emerges from an arteriole and supplies a group of 10–100 capillaries that constitute a **capillary bed** (Figure 14.2). The proximal portion of a metarteriole is surrounded by scattered smooth muscle fibers whose contraction and relaxation help regulate blood flow through the capillary bed. The distal portion of a metarteriole, which empties into a venule, has no smooth muscle fibers and is called a **thoroughfare channel.** Blood flowing through a thoroughfare channel bypasses the capillary bed.

True capillaries emerge from arterioles or metarterioles. At their sites of origin, a ring of smooth muscle fibers called a **precapillary sphincter** controls the flow of blood into a true capillary. When the precapillary sphincters are relaxed (open), blood flows into the capillary bed (Figure 14.2a); when precapillary sphincters contract (close or partially close), blood flow through the capillary bed ceases or decreases (Figure 14.2b). Typically, blood flows intermittently through a capillary bed due to alternating contraction and relaxation of the smooth muscle of metarterioles and the precapillary sphincters. This intermittent contraction and relaxation, which may occur 5–10 times per minute, is called **vasomotion.** In part, vasomotion is due to chemicals, such as nitric oxide. At any given time, blood flows through only about 25% of a capillary bed.

The body contains different types of capillaries (Figure 14.3). Many capillaries are **continuous capillaries,** in which the plasma membranes of endothelial cells form a continuous tube that is interrupted only by **intercellular clefts,** which are gaps between neighboring endothelial cells (Figure 14.3a). Continuous capillaries are found in skeletal and smooth muscle, connective tissues, and the lungs. Other capillaries of the body are **fenestrated capillaries** (*fenestr-* = window), which differ from continuous capillaries in that the plasma membranes of the endothelial cells have many **fenestrations,** or pores, ranging from 70 to 100 nm in diameter (Figure 14.3b). Fenestrated capillaries are found in the kidneys, villi of the small intestine, choroid plexuses of the ventricles in the brain, ciliary processes of the eyes, and endocrine glands.

Sinusoids are wider and more winding than other capillaries. Their endothelial cells may have unusually large fenestrations. In addition to an incomplete or absent basement membrane (Figure 14.3c), sinusoids have very large intercellular clefts that allow proteins and in some cases even blood cells to pass from a tissue into the bloodstream. For example, newly formed blood cells enter the bloodstream through the sinusoids of red bone marrow. In addition, sinusoids contain specialized lining cells that are adapted to the function of the tissue. Sinusoids in the liver, for example, contain phagocytic cells that remove bacteria and other debris from the blood. Besides the liver and red bone marrow, sinusoids are also present in the spleen, anterior pituitary gland, and parathyroid glands.

Materials that are exchanged between blood and interstitial fluid can move across capillary walls in four basic ways: through intercellular clefts, through fenestrations, within pinocytic vesicles that undergo transcytosis, and by simple diffusion through the plasma membranes of the endothelial cells.

Venules

When several capillaries unite, they form small veins called **venules** (= little vein). Venules collect blood from capillaries and drain into veins. The smallest venules, those closest to the capillaries, consist of a tunica interna of endothelium and a tunica media that has only a few scattered smooth muscle fibers and fibroblasts (see Figure 14.2). Like capillaries, the walls of the smallest venules are very porous and through which many phagocytic white blood cells emigrate from the bloodstream into an inflamed or infected tissue. As venules become larger and converge to form veins, they contain the tunica externa characteristic of veins.

Figure 14.3 / Types of capillaries (shown in transverse sections).

Capillaries are microscopic blood vessels that connect arterioles and venules.

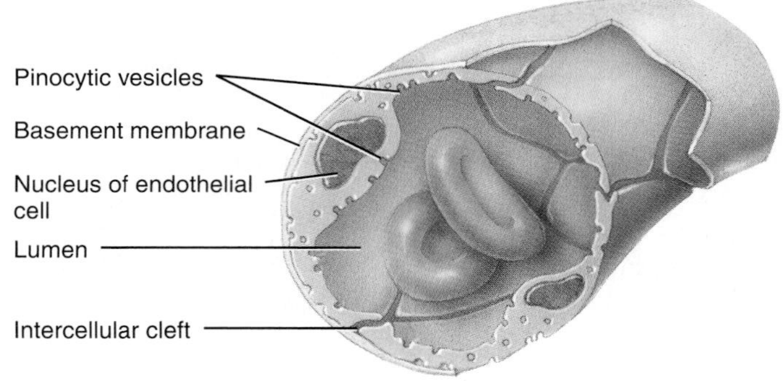

(a) Continuous capillary formed by endothelial cells

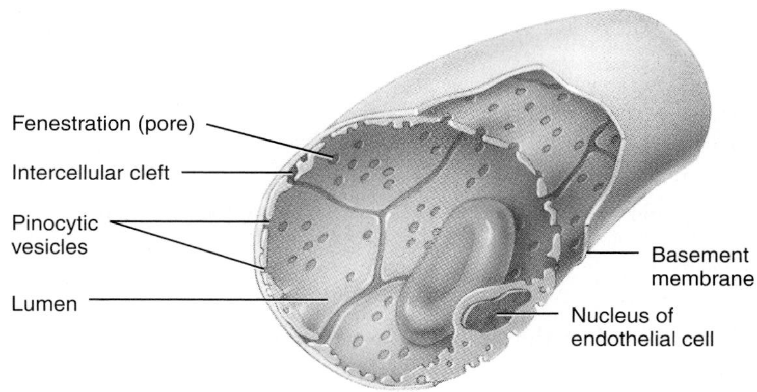

(b) Fenestrated capillary

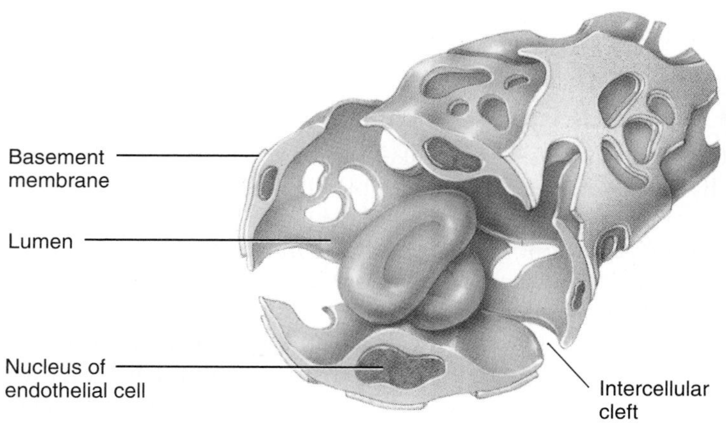

(c) Sinusoid

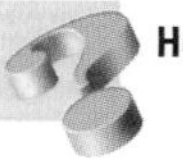

How do materials cross capillary walls?

Veins

Although **veins** are composed of essentially the same three coats as arteries, the relative thicknesses of the layers are different. The tunica interna of veins is thinner than that of arteries; the tunica media of veins is much thinner than in arteries, with relatively little smooth muscle and elastic fibers. The tunica externa of veins is the thickest layer and consists of collagen and elastic fibers; the tunica externa of the inferior vena cava also contains longitudinal fibers of smooth muscle. Veins lack the external or internal elastic laminae found in arteries (see Figure 14.1b). Because of these differences, veins are still distensible enough to adapt to variations in the volume and pressure of blood passing through them, although they are not designed to withstand high pressure. Furthermore, the lumen of a vein is larger than that of a comparable artery, and veins frequently appear collapsed (flattened) when sectioned.

The pumping action of the heart is a major factor in moving venous blood back to the heart. The contraction of skeletal muscles in the lower limbs also helps boost venous return to the heart. Valves in veins, described shortly, also assume a key function in returning venous blood to the heart. The average blood pressure in veins is considerably lower than in arteries. The difference in pressure can be noticed when blood flows from a cut vessel. Blood leaves a cut vein in an even, slow flow but spurts rapidly from a cut artery. Most of the structural differences between arteries and veins reflect this pressure difference. For example, the walls of veins are not as strong as those of arteries. Many veins feature **valves** (see Figure 14.1b), which are needed because venous blood pressure is so low. When you stand, the pressure pushing blood up the veins in your lower limbs is barely enough to overcome the force of gravity pulling it back down.

Each valve is composed of two or more thin folds of tunica interna that form flaplike cusps; the cusps project into the lumen of the veins pointing toward the heart (Figure 14.4). Veins pass between groups of skeletal muscles, and when these muscles contract venous pressure is increased; the valve opens as the cusps are pushed against the wall of the vein by the blood as it flows through the valve toward the heart. When the muscles relax, blood would tend to move back toward the feet, but is stopped from doing so as the cusps come together and close the lumen of the vein. As a result, blood is prevented from moving away from the heart. In this way, valves prevent backflow of blood and aid in moving blood in one direction only—toward the heart.

A **vascular (venous) sinus** is a vein with a thin endothelial wall that has no smooth muscle to alter its diameter. In a vascular sinus, the surrounding dense connective tissue replaces the tunica media and tunica externa in providing support. For example, dural venous sinuses, which are supported by the dura mater, convey deoxygenated blood from the brain to the heart. Another example of a vascular sinus is the coronary sinus of the heart.

Varicose Veins

Leaky venous valves can cause veins to become dilated and twisted in appearance, a condition called **varicose veins** (*varic-* = a swollen vein). The condition may occur in the veins of al-

Figure 14.4 / Role of skeletal muscle contractions and venous valves in returning blood to the heart. (a) When skeletal muscles contract, the proximal valve opens, and blood is forced toward the heart. (b) Sections of a venous valve.

Venous return depends on the pumping action of the heart, skeletal muscle contractions, and valves in veins.

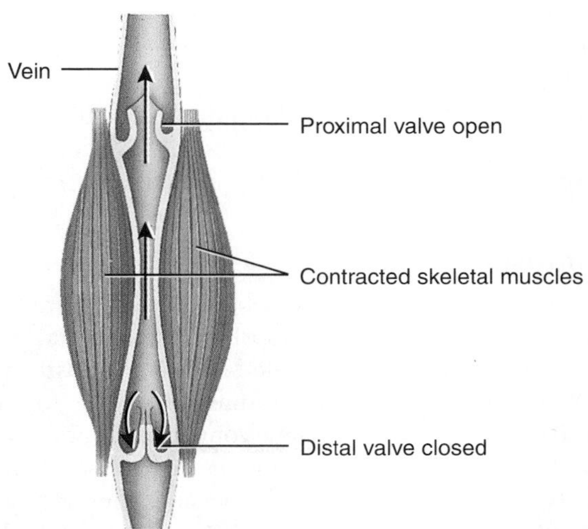

(a) Diagram of contracted skeletal muscles

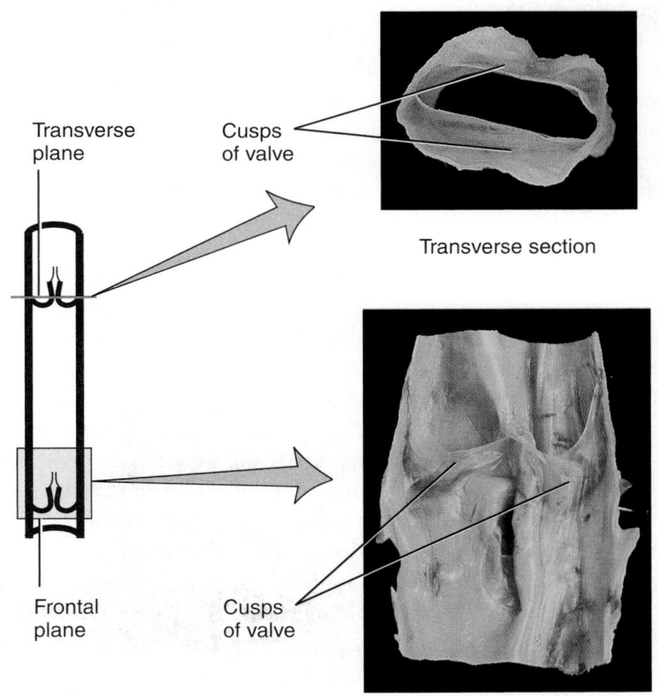

(b) Photograph of a valve in a vein

Why are valves more important in arm veins and leg veins than in neck veins?

most any body part, but it is most common in the esophagus and in superficial veins of the lower limbs. The valvular defect may be congenital or may result from mechanical stress (prolonged standing or pregnancy) or aging. The leaking venous valves allow the backflow of blood, which causes pooling of blood. This, in turn, creates pressure that distends the vein and allows fluid to leak into surrounding tissue. As a result, the affected vein and the tissue around it may become inflamed and painfully tender. Veins close to the surface of the legs, especially the saphenous vein, are highly susceptible to varicosities, whereas deeper veins are not as vulnerable because surrounding skeletal muscles prevent their walls from stretching excessively. Varicosed veins in the anal canal are referred to as hemorrhoids. ■

Anastomoses

Most tissues of the body receive blood from more than one artery. The union of the branches of two or more arteries supplying the same body region is called an **anastomosis** (a-nas-tō-MŌ-sis; = connecting; plural is **anastomoses**). Anastomoses between arteries provide alternate routes for blood to reach a tissue or organ. If blood flow stops momentarily when normal movements compress a vessel, or if a vessel is blocked by disease, injury, or surgery, then circulation to a part of the body is not necessarily stopped. The alternate route of blood flow to a body part through an anastomosis is known as **collateral circulation.** Anastomoses may also occur between veins and between arterioles and venules. Arteries that do not anastomose are known as **end arteries.** Obstruction of an end artery interrupts the blood supply to a whole segment of an organ, producing necrosis (death) of that segment. Alternate blood routes may also be provided by nonanastomosing vessels that supply the same region of the body.

Blood Distribution

The largest portion of your blood volume at rest—about 60%—is in systemic veins and venules. Systemic capillaries hold only about 5% of the blood volume, and arteries and arterioles about 15%. Because systemic veins and venules contain a large percentage of the blood volume, they function as **blood reservoirs** from which blood can be diverted quickly if the need arises. For example, when there is increased muscular activity, the cardiovascular center in the brain stem sends more sympathetic impulses to veins. The result is *venoconstriction*, constriction of veins, which reduces the volume of blood in reservoirs and allows a greater blood volume to flow to skeletal muscles, where it is needed most. A similar mechanism operates in cases of hemorrhage, when blood volume and pressure decrease; in this case, venoconstriction helps counteract the drop in blood pressure. Among the principal blood reservoirs are the veins of the abdominal organs (especially the liver and spleen) and the veins of the skin.

✓ Discuss the importance of elastic fibers and smooth muscle in the tunica media of arteries.
✓ Distinguish between elastic and muscular arteries in terms of location, histology, and function.
✓ Describe the structural features of capillaries that allow exchange of materials between blood and body cells.
✓ What are the main structural and functional differences between arteries and veins?
✓ Describe the relationship between anastomoses and collateral circulation.

CIRCULATORY ROUTES

Arteries, arterioles, capillaries, venules, and veins are organized into routes that deliver blood throughout the body. We can now look at the basic routes the blood takes as it is transported through its vessels.

Figure 14.5 shows the **circulatory routes** for blood flow. The routes are parallel; that is, in most cases a portion of the cardiac output flows separately to each tissue of the body. Thus each organ receives its own supply of freshly oxygenated blood. The two basic postnatal (after birth) routes for blood flow are the systemic and pulmonary circulations. The **systemic circulation** includes all the arteries and arterioles that carry oxygenated blood from the left ventricle to systemic capillaries plus the veins and venules that carry deoxygenated blood returning to the right atrium after flowing through body organs. Blood leaving the aorta and flowing through the systemic arteries is a bright red color. As it moves through capillaries, it loses some of its oxygen and picks up carbon dioxide, so that blood in systemic veins is a dark red color.

Some subdivisions of the systemic circulation are the **coronary (cardiac) circulation** (see Figure 13.10), which supplies the myocardium of the heart; **cerebral circulation,** which supplies the brain (see Figure 14.8c); and the **hepatic portal circulation,** which extends from the gastrointestinal tract to the liver (see Figure 14.16). The nutrient arteries to the lungs, such as the bronchial arteries, also are part of the systemic circulation.

When blood returns to the heart from the systemic route, it is pumped out of the right ventricle through the **pulmonary circulation** to the lungs (see Figure 14.17). In capillaries of the air sacs (alveoli) of the lungs, the blood loses some of its carbon dioxide and takes on oxygen. Bright red again, it returns to the left atrium of the heart and reenters the systemic circulation as it is pumped out by the left ventricle.

Another major route—the **fetal circulation**—exists only in the fetus and contains special structures that allow the developing fetus to exchange materials with its mother (see Figure 14.18).

Systemic Circulation

The systemic circulation carries oxygen and nutrients to body tissues and removes carbon dioxide and other wastes and heat from the tissues. All systemic arteries branch from the aorta. Deoxygenated blood returns to the heart through the systemic veins. All the veins of the systemic circulation drain into the **superior vena cava, inferior vena cava,** or **coronary sinus,** which in turn empty into the right atrium.

The principal arteries and veins of the systemic circulation are described and illustrated in Exhibits 14.1–14.12 and Figures 14.6–14.15 to assist you in learning their names. The blood vessels are organized in the exhibits according to regions of the body. Figure 14.6a shows an overview of the major arteries, and

Figure 14.5 / Generalized view of circulatory routes. Heavy black arrows indicate the systemic circulation (detailed in Exhibits 14.1–14.12), thin black arrows the pulmonary circulation (detailed in Figure 14.17), and red arrows the hepatic portal circulation (detailed in Figure 14.16). Refer to Figure 13.8 on page 400 for details of the coronary circulation, and to Figure 14.18 for details of the fetal circulation.

Blood vessels are organized into various routes that deliver blood to tissues of the body.

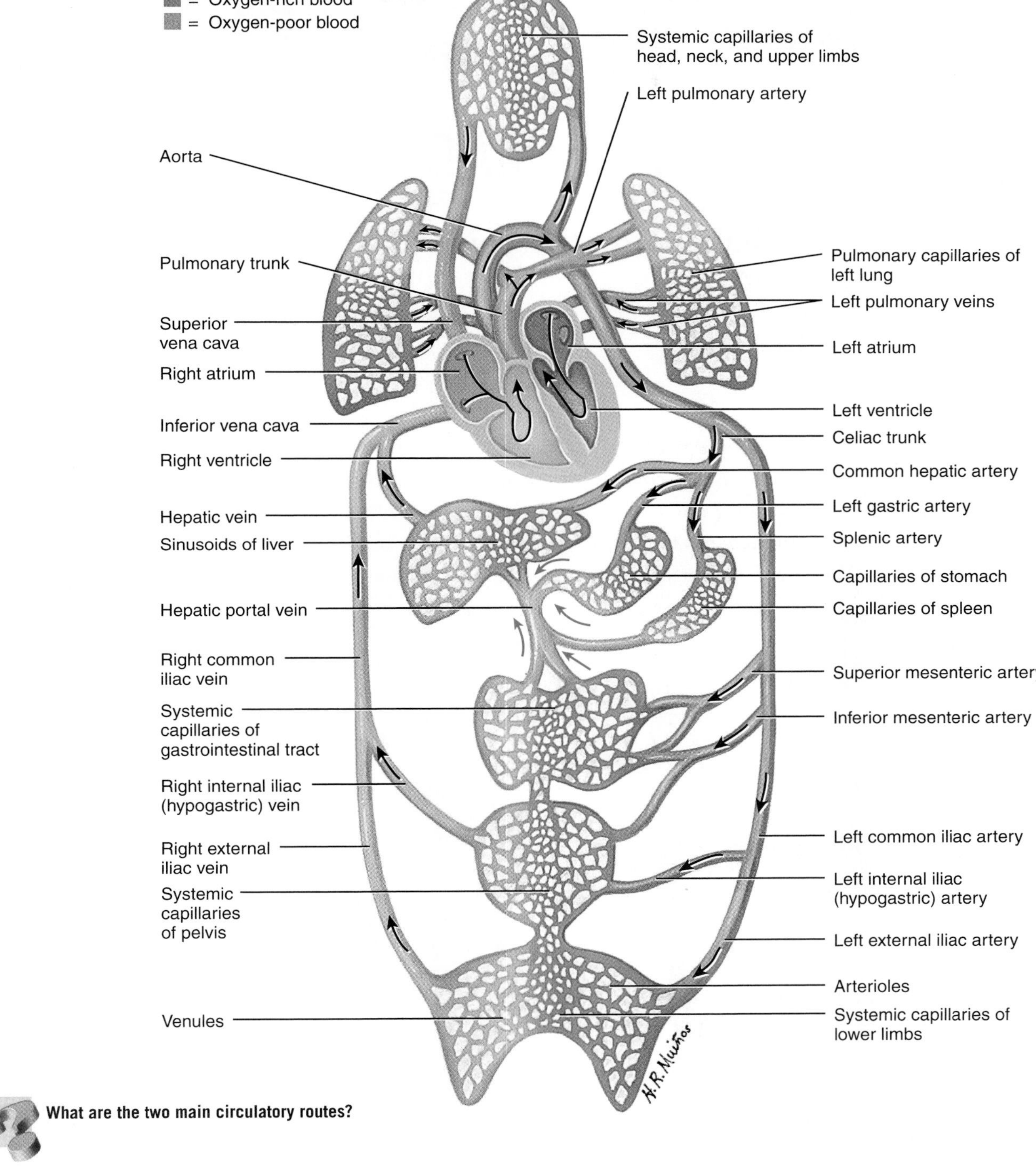

What are the two main circulatory routes?

Figure 14.11 shows an overview of the major veins. As you study the various blood vessels in the exhibits, refer to these two figures to see the relationships of the blood vessels under consideration to other regions of the body.

Each of the exhibits contains the following information:

- **An overview.** This information provides a general orientation to the blood vessels under consideration, with emphasis on how the blood vessels are organized into various regions and on distinguishing and/or interesting features about the blood vessels.
- **Blood vessel names.** Students often have difficulty with the pronunciations and meanings of blood vessels' names. To learn

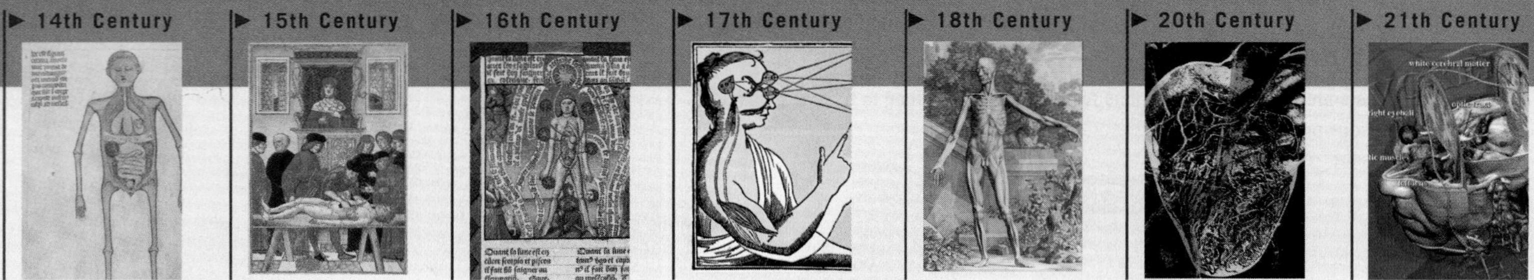

CHANGING IMAGES

Challenging the Spirits

Blood is converted from food and infused with the natural spirit by the liver. It ebbs and flows in veins only, and is consumed as fuel by the body. Systemic arteries pulse not with blood, but pneuma, inhaled air that travels from the lungs to the left ventricle. Here it is warmed by the heart and converted to the vital spirit. Once this vital spirit travels to the brain, the animal spirit arises, and is then dispersed through the nerves. These spirits respectively govern human life, growth and development, sensation and movement. This is the pneumatic theory.

Based upon the writings of Hippocrates, a 5th century B.C. Greek, and refined by Galen in the 2nd century A.D., this model of the circulatory system survived until the 17th century. It required the brilliance of William Harvey, an Englishmen, who in 1628 published his treatise, *Anatomical Essay of the Motion of the Heart and Blood in Animals*, to crumble centuries of unchallenged thought. Laying the foundation for the modern scientific method, Harvey demonstrated, through direct observation and quantification, that the heart is a central pump that propels the blood in a circular course. This notion may seem rudimentary today, but in the 17th century it was heretical to confront Galen. Though unlike Vesalius who was vilified until his death, Harvey lived to be celebrated for his genius.

Pictured here is an image from Harvey's landmark work. Demonstrated is the function of venous valves in controlling the direction of blood flow. Harvey used such observations to illustrate that indeed venous circulation does not ebb and flow but courses toward the heart.

1628 AD

In this chapter you are studying blood vessels of the cardiovascular system and their routes in the body. As you do so, appreciate that these circulatory principles were hard fought for acceptance. Also consider that future scientific advancement requires critical evaluation of current beliefs.

Exhibit 14.1 The Aorta and Its Branches (Figure 14.6)

Objective

▶ Identify the four principal divisions of the aorta and locate the major arterial branches arising from each division.

The **aorta** (*aortae* = to lift up) is the largest artery of the body, with a diameter of 2–3 cm (about 1 in.). Its four principal divisions are the ascending aorta, arch of the aorta, thoracic aorta, and abdominal aorta. The portion of the aorta that emerges from the left ventricle posterior to the pulmonary trunk is the **ascending aorta.** The beginning of the aorta contains the aortic semilunar valve (see Figure 13.4a on page 394). The ascending aorta gives off two coronary artery branches that supply the myocardium of the heart. Then it turns to the left, forming the **arch of the aorta,** which descends and ends at the level of the intervertebral disc between the fourth and fifth thoracic vertebrae. As the aorta continues to descend, it lies close to the vertebral bodies, passes through the diaphragm, and divides at the level of the fourth lumbar vertebra into two **common iliac arteries,** which carry blood to the lower limbs. The section of the aorta between the arch of the aorta and the diaphragm is called the **thoracic aorta;** the section between the diaphragm and the common iliac arteries is the **abdominal aorta.** Each division of the aorta gives off arteries that branch into distributing arteries that lead to organs. Within the organs, the arteries divide into arterioles and then into capillaries that service the systemic tissues (all tissues except the alveoli of the lungs).

✓ What general regions are supplied by each of the four principal divisions of the aorta?

Division and Branches	Region Supplied
Ascending aorta	
Right and left coronary arteries	Heart
Arch of the aorta	
Brachiocephalic trunk (brā′-kē-ō-se-FAL-ik)	
Right common carotid artery (ka-ROT-id)	Right side of head and neck
Right subclavian artery (sub-KLĀ-vē-an)	Right upper limb
Left common carotid artery	Left side of head and neck
Left subclavian artery	Left upper limb
Thoracic aorta (*thorac* = chest)	
Intercostal arteries (in′-ter-KOS-tal)	Intercostal and chest muscles and plurae
Superior phrenic arteries (FREN-ik)	Posterior and superior surfaces of diaphragm
Bronchial arteries (BRONG-kē-al)	Bronchi of lungs
Esophageal arteries (e-sof′-a-JĒ-al)	Esophagus
Abdominal aorta	
Inferior phrenic arteries (FREN-ik)	Inferior surface of diaphragm
Celiac trunk (SĒ-lē-ak)	
Common hepatic artery (he-PAT-ik)	Liver
Left gastric artery (GAS-trik)	Stomach and esophagus
Splenic artery (SPLĒN-ik)	Spleen, pancreas, and stomach
Superior mesenteric artery (MES-en-ter′-ik)	Small intestine, cecum, ascending and transverse colons, and pancreas
Suprarenal arteries (sū-pra-RĒ-nal)	Adrenal (suprarenal) glands
Renal arteries (RĒ-nal)	Kidneys
Gonadal arteries (gō-NAD-al)	
Testicular arteries (tes-TIK-yū-lar)	Testes (male)
Ovarian arteries (ō-VAR-ē-an)	Ovaries (female)
Inferior mesenteric artery	Transverse, descending, and sigmoid colons; rectum
Common iliac arteries (IL-ē-ak)	
External iliac arteries	Lower limbs
Internal iliac (hypogastric) arteries	Uterus (female), prostate gland (male), muscles of buttocks, and urinary bladder

(continues)

them more easily, study the phonetic pronunciations and word derivations that indicate how blood vessels get their names.

- **Region supplied or drained.** For each artery listed, there is a description of the parts of the body that receive blood from the vessel. For each vein listed, there is a description of the parts of the body that are drained by the vessel.
- **Illustrations and photographs.** The figures that accompany the exhibits contain several elements. There is always a drawing of the blood vessels under consideration. In many cases, flow diagrams are provided to indicate the patterns of blood distribution or drainage. Cadaver photographs provide more realistic views of the blood vessels.

(Text continues p. 462)

Exhibit 14.1 The Aorta and Its Branches (Figure 14.6) (continued)

Figure 14.6 / Aorta and its principal branches.

All systemic arteries branch from the aorta.

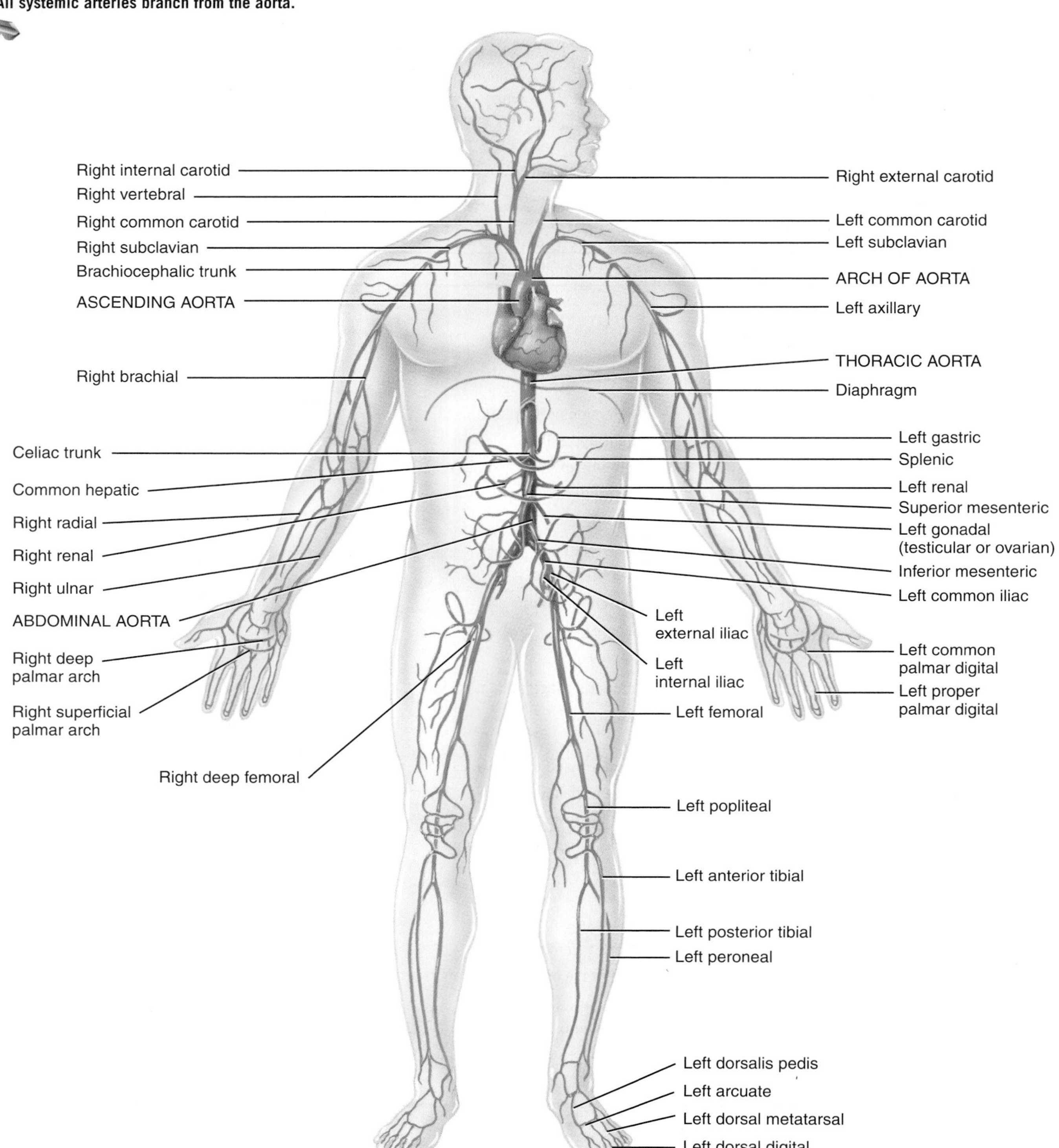

(a) Overall anterior view of the principal branches of the aorta

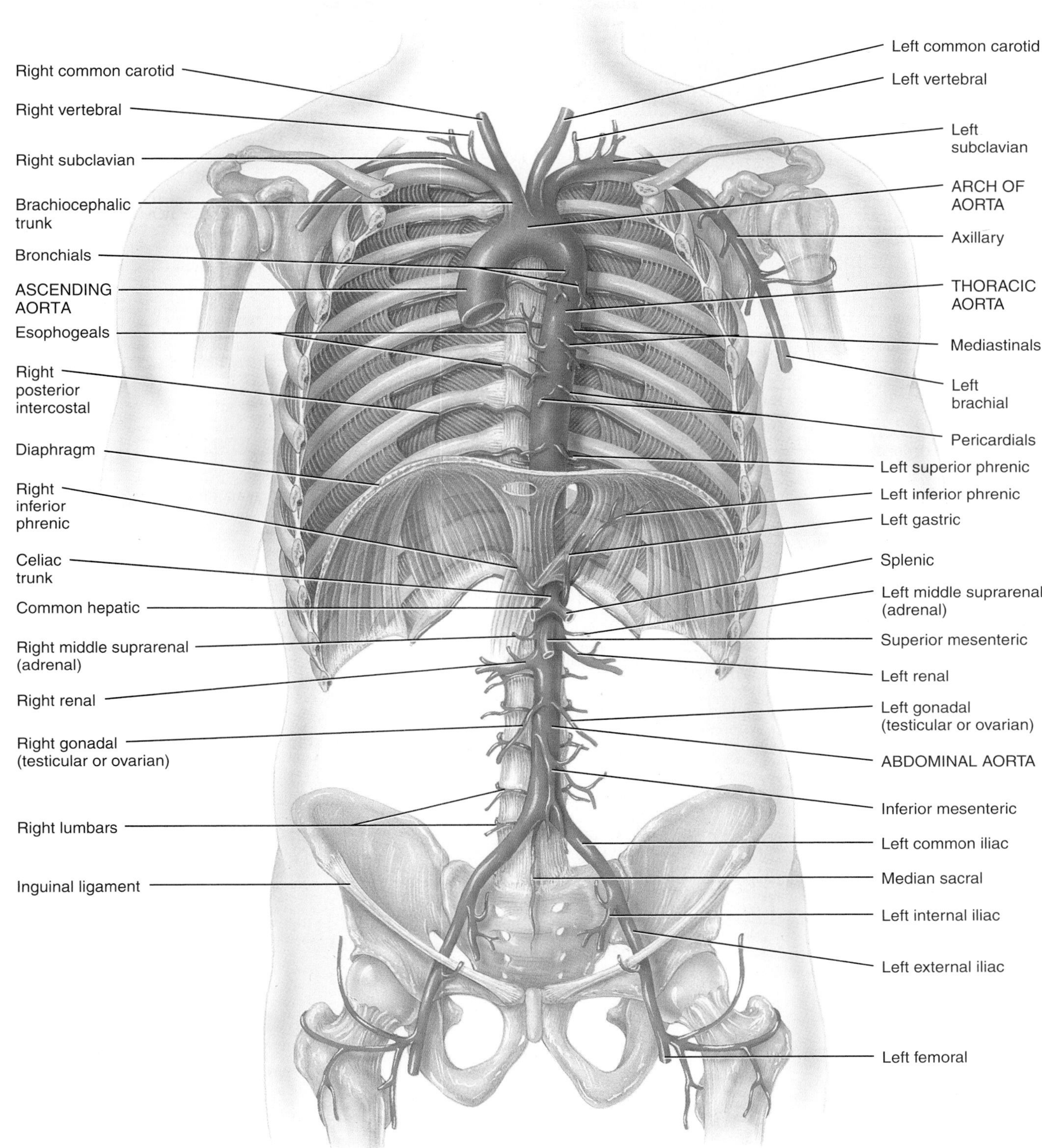

(b) Detailed anterior view of the principal branches of the aorta

(continues)

SUPERIOR

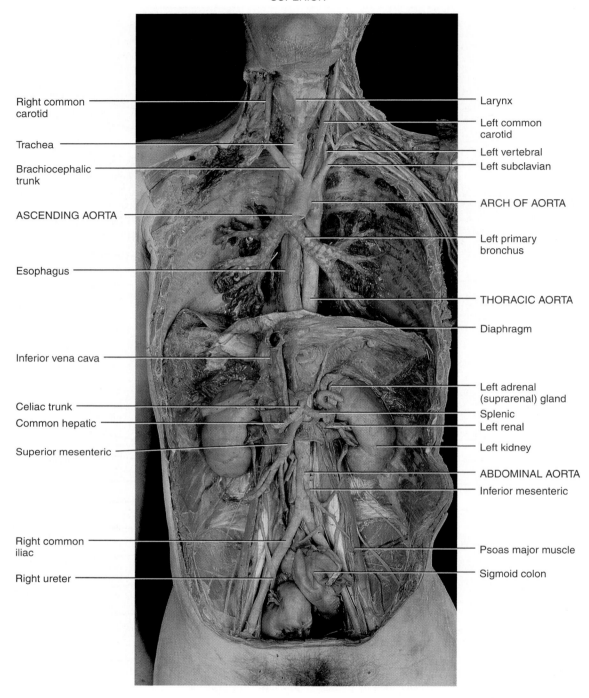

Right common carotid

Trachea

Brachiocephalic trunk

ASCENDING AORTA

Esophagus

Inferior vena cava

Celiac trunk

Common hepatic

Superior mesenteric

Right common iliac

Right ureter

Larynx

Left common carotid

Left vertebral

Left subclavian

ARCH OF AORTA

Left primary bronchus

THORACIC AORTA

Diaphragm

Left adrenal (suprarenal) gland

Splenic

Left renal

Left kidney

ABDOMINAL AORTA

Inferior mesenteric

Psoas major muscle

Sigmoid colon

INFERIOR

(c) Anterior view of the principal branches of the aorta

What are the four principal subdivisions of the aorta?

Exhibit 14.2 Ascending Aorta (Figure 14.7)

Objective

▶ Identify the two primary arterial branches of the ascending aorta.

The **ascending aorta** is about 5 cm (2 in.) in length and begins at the aortic semilunar valve. It is directed superiorly, slightly anteriorly, and to the right; it ends at the level of the sternal angle, where it becomes the arch of the aorta. The beginning of the ascending aorta is posterior to the pulmonary trunk and right auricle; the right pulmonary artery is posterior to it. At its origin, the ascending aorta contains three dilations called aortic sinuses. Two of these, the right and left sinuses, give rise to the right and left coronary arteries, respectively. The ascending aorta is more susceptible to aneurysms than succeeding parts of the aorta because of the higher pressure of blood from the left ventricle during ventricular systole (contraction).

The right and left **coronary arteries** (*coron-* = crown) arise from the ascending aorta just superior to the aortic semilunar valve. They form a crownlike ring around the heart, giving off branches to the atrial and ventricular myocardium. The **posterior interventricular branch** (in-ter-ven-TRIK-yū-lar; *inter-* = between) of the right coronary artery supplies both ventricles, and the **marginal branch** supplies the right ventricle. The **anterior interventricular (left anterior descending) branch** of the left coronary artery supplies both ventricles, and the **circumflex branch** (SER-kum-flex; *circum-* = around; *-flex* = to bend) supplies the left atrium and left ventricle.

✓ Which branches of the coronary arteries supply the left ventricle? Why does the left ventricle have such an extensive arterial blood supply?

Figure 14.7 / Ascending aorta and its branches. (See Tortora, *A Photographic Atlas of the Human Body,* Figures 6.8 and 6.9.)

The ascending aorta is the first division of the aorta.

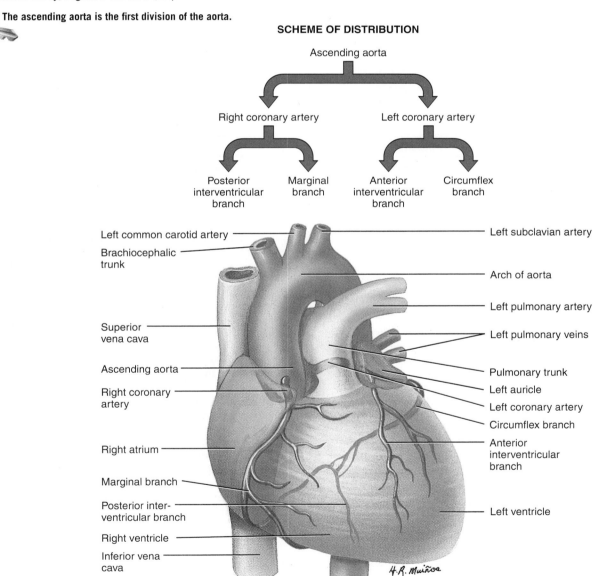

SCHEME OF DISTRIBUTION

Ascending aorta

Right coronary artery — Posterior interventricular branch; Marginal branch

Left coronary artery — Anterior interventricular branch; Circumflex branch

Left common carotid artery
Brachiocephalic trunk
Superior vena cava
Ascending aorta
Right coronary artery
Right atrium
Marginal branch
Posterior interventricular branch
Right ventricle
Inferior vena cava
Left subclavian artery
Arch of aorta
Left pulmonary artery
Left pulmonary veins
Pulmonary trunk
Left auricle
Left coronary artery
Circumflex branch
Anterior interventricular branch
Left ventricle
H.R. Muñoz

Anterior view of ascending aorta and its branches

Which arteries arise from the ascending aorta?

Exhibit 14.3 The Arch of the Aorta (Figure 14.8)

Objective

► Identify the three principal arteries that branch from the arch of the aorta.

The **arch of the aorta** is 4–5 cm (almost 2 in.) in length and is the continuation of the ascending aorta. It emerges from the pericardium posterior to the sternum at the level of the sternal angle. Initially, the arch is directed superiorly, posteriorly, and to the left, and then inferiorly on the left side of the body of the fourth thoracic vertebra. The arch of the aorta ends at the level of the intervertebral disc between the fourth and fifth thoracic vertebrae, where it becomes the thoracic aorta. The thymus lies anterior to the arch of the aorta, whereas the trachea lies posterior to it.

Interconnecting the arch of the aorta and the pulmonary trunk is a structure called the **ligamentum arteriosum.** It is a remnant of a fetal blood vessel called the ductus arteriosus (see Figure 14.18). Within the arch of the aorta are receptors (nerve cells) called baroreceptors that are sensitive to changes in blood pressure. Situated near the arch of the aorta are other receptors, referred to as **aortic bodies,** that are sensitive to changes in blood concentrations of chemicals such as O_2, CO_2, and H^+.

Three major arteries branch from the superior aspect of the arch of the aorta: the brachiocephalic trunk, the left common carotid, and the left subclavian. The first and largest branch is the **brachiocephalic trunk** (brā′-kē-ō-se-FAL-ik; *brachio-* = arm; *-cephalic* = head). It extends superiorly, bending slightly to the right, and divides at the right sternoclavicular joint to form the right subclavian artery and right common carotid artery. The second branch is the **left common carotid artery** (ka-ROT-id), which divides into basically the same branches with the same names as the right common carotid artery. The third branch is the **left subclavian artery** (sub-KLĀ-vē-an), which distributes blood to the left vertebral artery and vessels of the left upper limb. Arteries branching from the left subclavian artery are similar in distribution and name to those branching from the right subclavian artery. The accompanying table focuses on the principal arteries originating from the brachiocephalic trunk.

✓ What general regions are supplied by the arteries that arise from the arch of the aorta?

Branch	Description and Region Supplied
Brachiocephalic trunk	The **brachiocephalic trunk** divides to form the right subclavian artery and right common carotid artery (Figure 14.8a). There is no left brachiocephalic trunk.
Right subclavian artery (sub-KLĀ-vē-an)	The **right subclavian artery** extends from the brachiocephalic trunk to the first rib and then passes into the armpit (axilla). The general distribution of the artery is to the brain and spinal cord, neck, shoulder, thoracic viscera and wall, and scapular muscles.
Axillary artery (AK-si-ler-ē; = armpit)	Continuation of the right subclavian artery into the axilla is called the **axillary artery** (Figure 14.8a). (Note that the right subclavian artery, which passes deep to the clavicle, is a good example of the practice of giving the same vessel different names as it passes through different regions.) Its general distribution is the shoulder, thoracic and scapular muscles, and humerus.
Brachial artery (BRĀ-kē-al; = arm)	The **brachial artery** is the continuation of the axillary artery into the arm (Figure 14.8a,d). The brachial artery provides the main blood supply to the arm and is superficial and palpable along its course. It begins at the tendon of the teres major muscle and ends just distal to the bend of the elbow. At first, the brachial artery is medial to the humerus, but as it descends it gradually curves laterally and passes through the cubital fossa, a triangular depression anterior to the elbow where you can easily detect the pulse of the brachial artery and listen to the various sounds when taking a person's blood pressure. Just distal to the bend in the elbow, the brachial artery divides into the radial artery and ulnar artery.
Radial artery (RĀ-dē-al; = radius)	The **radial artery** is the smaller branch and is a direct continuation of the brachial artery (Figure 14.8a,d). It passes along the lateral (radial) aspect of the forearm and then through the wrist and hand, supplying these structures with blood. At the wrist, the radial artery comes into contact with the distal end of the radius, where it is covered only by fascia and skin. Because of its superficial location at this point, it is a common site for measuring radial pulse.
Ulnar artery (UL-nar; = ulna)	The **ulnar artery,** the larger branch of the brachial artery, passes along the medial (ulnar) aspect of the forearm and then into the wrist and hand, supplying these structures with blood (Figure 14.8a,d). In the palm, branches of the radial and ulnar arteries anastomose to form the superficial palmar arch and the deep palmar arch.

Branch	Description and Region Supplied
Superficial palmar arch (*palma* = palm)	The **superficial palmar arch** is formed mainly by the ulnar artery, with a contribution from a branch of the radial artery (Figure 14.8a,d). The arch is superficial to the long flexor tendons of the fingers and extends across the palm at the bases of the metacarpals. It gives rise to **common palmar digital arteries,** which supply the palm. Each of these divides into a pair of **proper palmar digital arteries,** which supply the fingers.
Deep palmar arch	The **deep palmar arch** is formed mainly by the radial artery, with a contribution from a branch of the ulnar artery. The arch is deep to the long flexor tendons of the fingers and extends across the palm, just distal to the bases of the metacarpals. Arising from the deep palmar arch are **palmar metacarpal arteries,** which supply the palm and anastomose with the common palmar digital arteries of the superficial palmar arch.
Vertebral artery (VER-te-bral)	Before passing into the axilla, the right subclavian artery gives off a major branch to the brain called the **right vertebral artery** (Figure 14.8b). The right vertebral artery passes through the foramina of the transverse processes of the sixth through first cervical vertebrae and enters the skull through the foramen magnum to reach the inferior surface of the brain. Here it unites with the left vertebral artery to form the **basilar artery** (BAS-i-lar). The vertebral artery supplies the posterior portion of the brain with blood. The basilar artery passes along the midline of the anterior aspect of the brain stem and supplies the cerebellum and pons of the brain and the internal ear.
Right common carotid artery	The **right common carotid artery** begins at the bifurcation of the brachiocephalic trunk, posterior to the right sternoclavicular joint, and passes superiorly in the neck to supply structures in the head (Figure 14.8b). At the superior border of the larynx (voice box), it divides into the right external and right internal carotid arteries.
External carotid artery	The **external carotid artery** begins at the superior border of the larynx and terminates near the temporomandibular joint in the substance of the parotid gland, where it divides into two branches: the superficial temporal and maxillary arteries. The carotid pulse can be detected in the external carotid artery just anterior to the sternocleidomastoid muscle at the superior border of the larynx. The general distribution of the external carotid artery is to structures *externai* to the skull.
Internal carotid artery	The internal carotid artery has no branches in the neck and supplies structures *internal* to the skull. It enters the cranial cavity through the carotid foramen in the temporal bone. At the proximal portion of the internal carotid artery there is a slight dilation called the **carotid sinus,** which contains receptors (nerve cells) called **baroreceptors** that monitor changes in blood pressure. Near the baroreceptors is a small mass of tissue at the bifurcation of the common carotid artery called the **carotid body.** It contains receptors that monitor changes in concentrations of chemicals in blood such as O_2, CO_2, and H^+. The internal carotid artery supplies blood to the eyeball and other orbital structures, ear, most of the cerebrum of the brain, pituitary gland, and external nose. The terminal branches of the internal carotid artery are the **anterior cerebral artery,** which supplies most of the medial surface of the cerebrum, and the **middle cerebral artery,** which supplies most of the lateral surface of the cerebrum (Figure 14.8c). Inside the cranium, anastomoses of the left and right internal carotid arteries along with the basilar artery form an arrangement of blood vessels at the base of the brain near the sella turcica called the **cerebral arterial circle (circle of Willis).** From this circle (Figure 14.8c) arise arteries supplying most of the brain. Essentially, the cerebral arterial circle is formed by the union of the **anterior cerebral arteries** (branches of internal carotids) and **posterior cerebral arteries** (branches of basilar artery). The posterior cerebral arteries are connected with the internal carotid arteries by the **posterior communicating arteries**. The anterior cerebral arteries are connected by the **anterior communicating arteries.** The **internal carotid arteries** are also considered part of the cerebral arterial circle. The functions of the cerebral arterial circle are to equalize blood pressure to the brain and provide alternate routes for blood flow to the brain, should the arteries become damaged.
Left common carotid artery	See description in the overview on page 428.
Left subclavian artery	See description in the overview on page 428.

(continues)

Exhibit 14.3 The Arch of the Aorta (Figure 14.8) (continued)

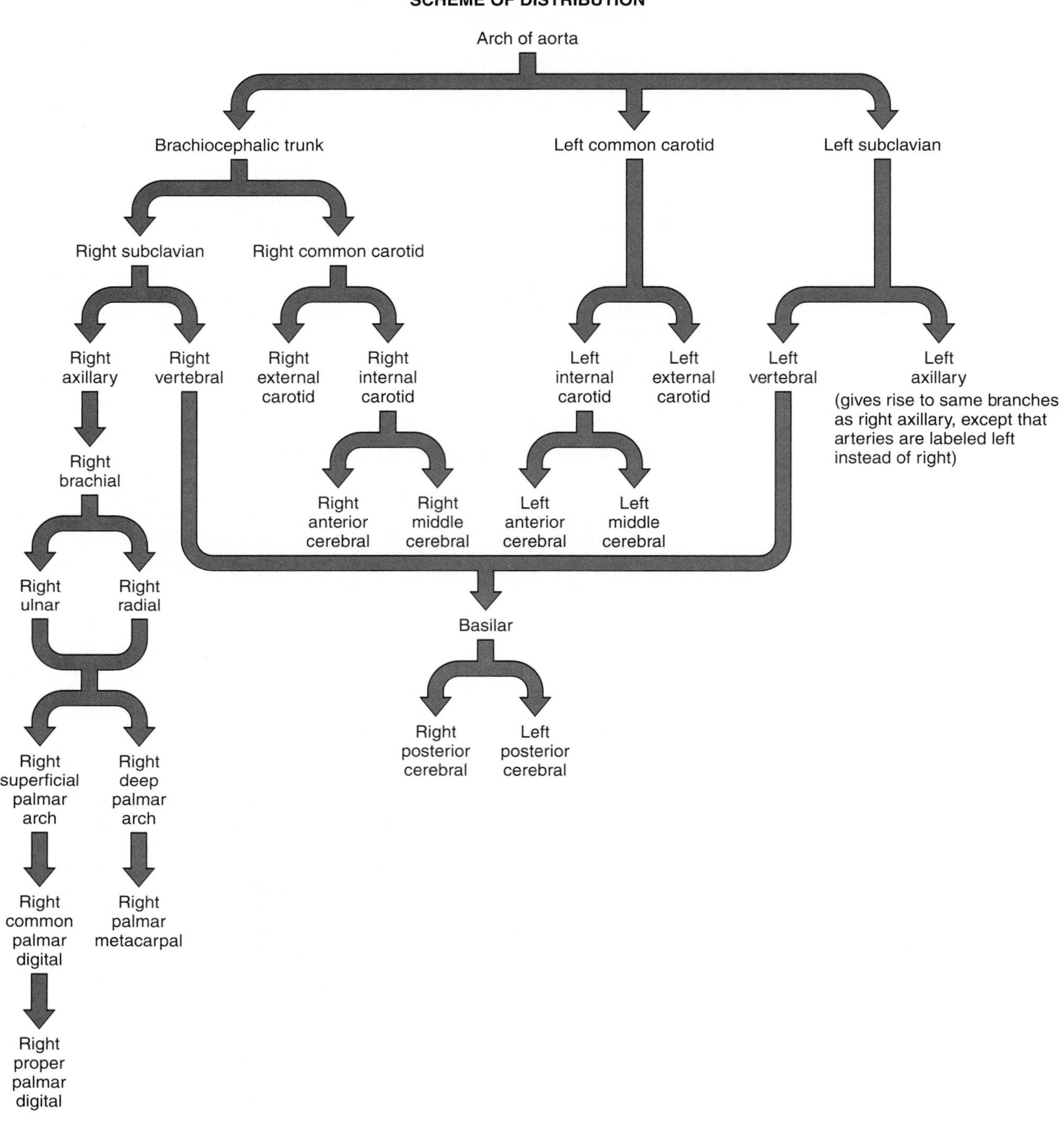

Figure 14.8 / Arch of the aorta and its branches. Note in (c) the arteries that constitute the cerebral arterial circle (circle of Willis). (See Tortora, *A Photographic Atlas of the Human Body,* Figures 6.11 and 6.12.)

The arch of the aorta ends at the level of the intervertebral disc between the fourth and fifth thoracic vertebrae.

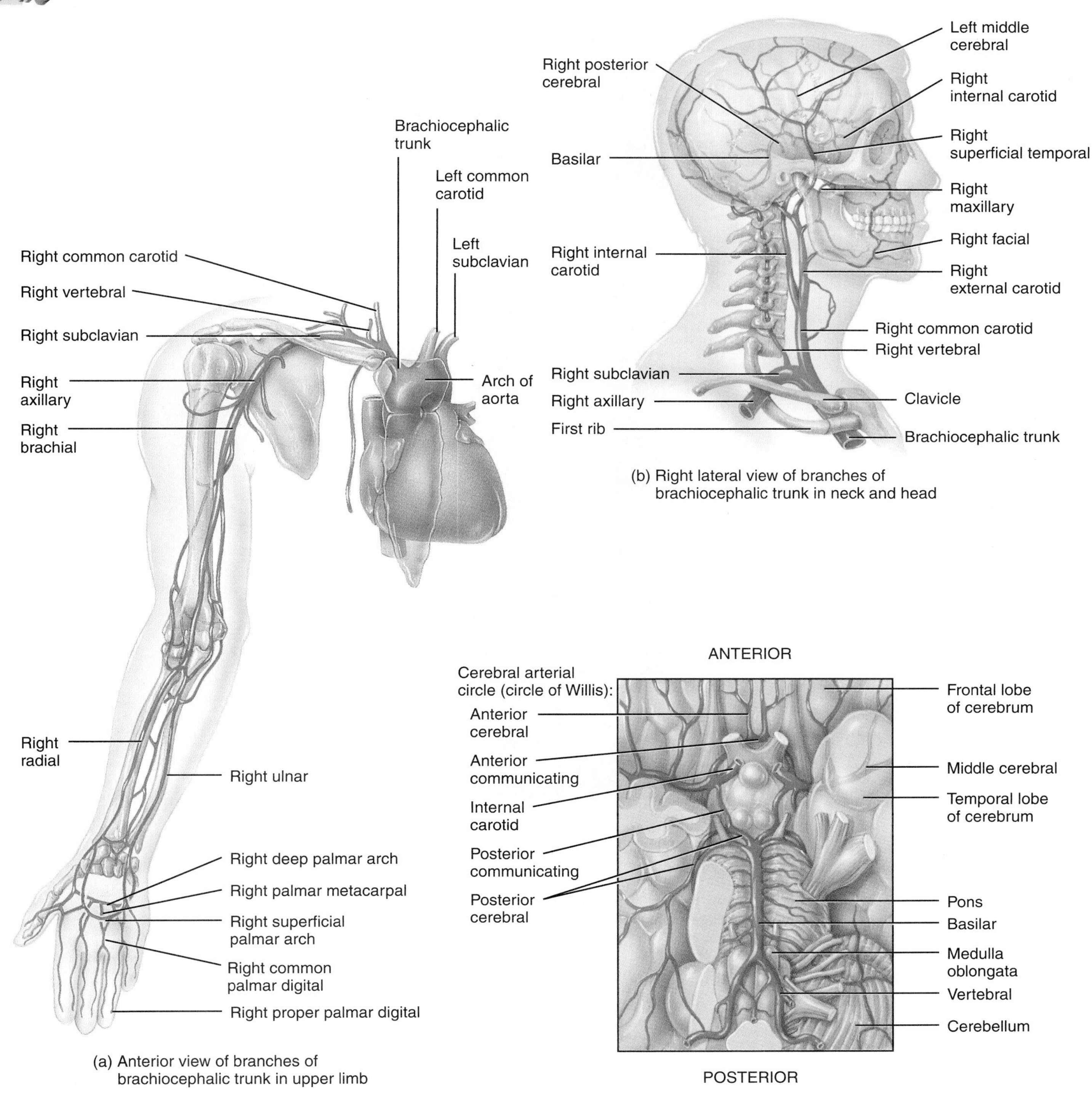

(a) Anterior view of branches of brachiocephalic trunk in upper limb

(b) Right lateral view of branches of brachiocephalic trunk in neck and head

(c) Inferior view of base of brain showing cerebral arterial circle

(continues)

Exhibit 14.3 The Arch of the Aorta (Figure 14.8) (continued)

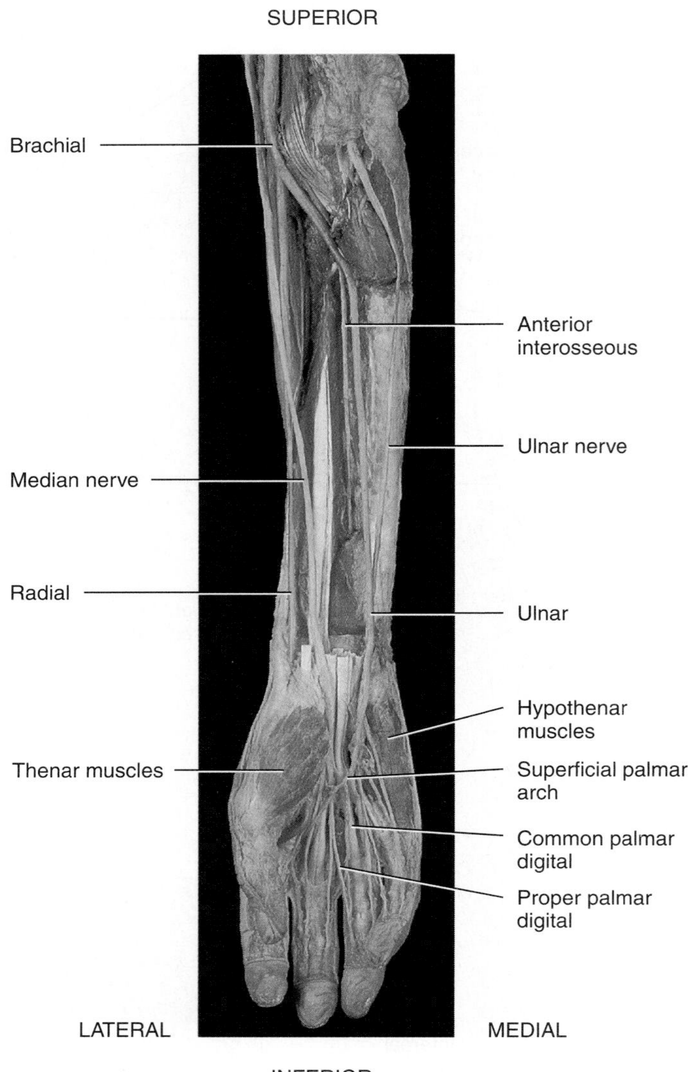

(d) Anterior view of arteries of upper limb

What are the three major branches of the arch of the aorta, in order of their origination?

Exhibit 14.4 Thoracic Aorta (see Figure 14.6b)

Objective

► Identify the visceral and parietal branches of the thoracic aorta.

The **thoracic aorta** is about 20 cm (8 in.) long and is a continuation of the arch of the aorta. It begins at the level of the intervertebral disc between the fourth and fifth thoracic vertebrae, where it lies to the left of the vertebral column. As it descends, it moves closer to the midline and ends at an opening in the diaphragm (aortic hiatus) anterior to the vertebral column at the level of the intervertebral disc between the twelfth thoracic and first lumbar vertebrae.

Along its course the thoracic aorta sends off numerous small arteries, **visceral branches** to viscera and **parietal branches** to body wall structures.

✓ What general regions are supplied by the visceral and parietal branches of the thoracic aorta?

Branch	Description and Region Supplied
Visceral	
Pericardial arteries (per′-i-KAR-dē-al); *peri* = around; *cardia* = heart)	Two or three minute **pericardial arteries** supply blood to the pericardium.
Bronchial arteries (BRONG-kē-al; = windpipe)	One right and two left **bronchial arteries** supply the bronchial tubes, pleurae, bronchial lymph nodes, and esophagus. (Whereas the right bronchial artery arises from the third posterior intercostal artery, the two left bronchial arteries arise from the thoracic aorta).
Esophageal arteries (e-sof′-a-JĒ-al; *oisein* = to carry; *phage* = food)	Four or five **esophageal arteries** supply the esophagus.
Mediastinal arteries (mē′-dē-as-TĪ-nal)	Numerous small **mediastinal arteries** supply blood to structures in the mediastinum.
Parietal	
Posterior intercostal arteries (in′-ter-KOS-tal; *inter* = between; *costa* = rib)	Nine pairs of **posterior intercostal arteries** supply the intercostal, pectoralis major and minor, and serratus anterior muscles; overlying subcutaneous tissue and skin; mammary glands; and vertebrae, meninges, and spinal cord.
Subcostal arteries (sub-KOS-tal; *sub* = under)	The left and right **subcostal arteries** have a distribution similar to that of the posterior intercostals.
Superior phrenic arteries (FREN-ik; = diaphragm)	Small **superior phrenic arteries** supply the superior and posterior surfaces of the diaphragm.

SCHEME OF DISTRIBUTION

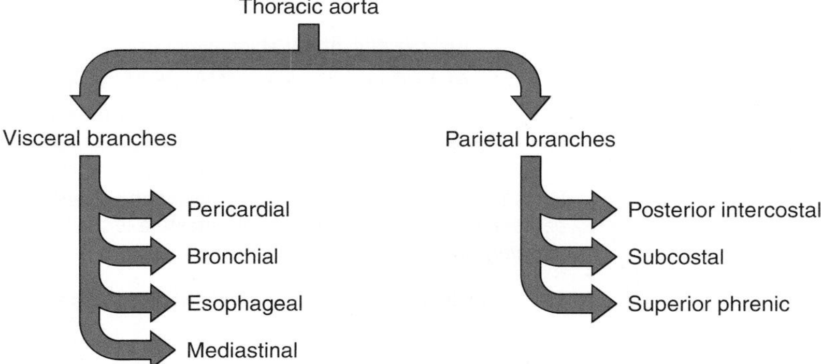

Exhibit 14.5 Abdominal Aorta (Figure 14.9)

Objective

▶ Identify the visceral and parietal branches of the abdominal aorta.

The **abdominal aorta** is the continuation of the thoracic aorta (see Figure 14.6). It begins at the aortic hiatus in the diaphragm and ends at about the level of the fourth lumbar vertebra, where it divides into the right and left common iliac arteries. The abdominal aorta lies anterior to the vertebral column.

As with the thoracic aorta, the abdominal aorta gives off visceral and parietal branches. The unpaired visceral branches arise from the anterior surface of the aorta and include the **celiac trunk** and the **superior mesenteric** and **inferior mesenteric arteries** (see Figure 14.6b). The paired visceral branches arise from the lateral surfaces of the aorta and include the **suprarenal, renal,** and **gonadal arteries.** The unpaired parietal branch is the **median sacral artery.** The paired parietal branches arise from the posterolateral surfaces of the aorta and include the **inferior phrenic** and **lumbar arteries.**

✓ Name the paired visceral and parietal branches and the unpaired visceral and parietal branches of the abdominal aorta, and indicate the general regions they supply.

Branch	Description and Region Supplied
Unpaired Visceral Branches	
Celiac trunk (SĒ-lē-ak)	The **celiac trunk (artery)** is the first visceral branch from the aorta inferior to the diaphragm, at about the level of the twelfth thoracic vertebra (Figure 14.9a). Almost immediately, the celiac trunk divides into three branches—the left gastric, splenic, and common hepatic arteries (Figure 14.9a). **1.** The **left gastric artery** (GAS-trik; = stomach) is the smallest of the three branches. It passes superiorly to the left toward the esophagus and then turns to follow the lesser curvature of the stomach. It supplies the stomach and esophagus. **2.** The **splenic artery** (SPLĒN-ik = spleen) is the largest branch of the celiac trunk. It arises from the left side of the celiac trunk distal to the left gastric artery, and passes horizontally to the left along the pancreas. Before reaching the spleen, it gives rise to three arteries: • **Pancreatic artery** (pan-krē-AT-ik), which supplies the pancreas. • **Left gastroepiploic artery** (gas′-trō-ep′-i-PLŌ-ik; *epiplo-* = omentum), which supplies the stomach and greater omentum. • **Short gastric artery,** which supplies the stomach. **3.** The **common hepatic artery** (he-PAT-ik; = liver) is intermediate in size between the left gastric and splenic arteries. Unlike the other two branches of the celiac trunk, the common hepatic artery arises from the right side; it gives rise to three arteries: • **Proper hepatic artery,** which supplies the liver, gallbladder, and stomach. • **Right gastric artery,** which supplies the stomach. • **Gastroduodenal artery** (gas′-trō-dū′-ō-DĒ-nal), which supplies the stomach, duodenum of the small intestine, pancreas, and greater omentum.
Superior mesenteric (MES-en-ter′-ik; *meso* = middle; *enteric* = intestines)	The **superior mesenteric artery** (Figure 14.9b) arises from the anterior surface of the abdominal aorta about 1 cm inferior to the celiac trunk at the level of the first lumbar vertebra. It extends inferiorly and anteriorly and between the layers of mesentery, which is a portion of the peritoneum that attaches the small intestine to the posterior abdominal wall. It anastomoses extensively and has five branches: **1.** The **inferior pancreaticoduodenal artery** (pan′-krē-at′-i-kō-dū′-ō-DĒ-nal) supplies the pancreas and duodenum. **2.** The **jejunal** (je-JŪ-nal) and **ileal arteries** (IL-ē-al) supply the jejunum and ileum of the small intestine, respectively. **3.** The **ileocolic artery** (il′ē-ō-KŌL-ik) supplies the ileum and ascending colon of the large intestine. **4.** The **right colic artery** (KŌL-ik) supplies the ascending colon. **5.** The **middle colic artery** supplies the transverse colon of the large intestine.

Branch	Description and Region Supplied
Inferior mesenteric	The **inferior mesenteric artery** (Figure 14.9c) arises from the anterior aspect of the abdominal aorta at the level of the third lumbar vertebra and then passes inferiorly to the left of the aorta. It anastomoses extensively and has three branches: **1.** The **left colic artery** supplies the transverse colon and descending colon of the large intestine. **2.** The **sigmoid arteries** (SIG-moyd) supply the descending colon and sigmoid colon of the large intestine. **3.** The **superior rectal artery** (REK-tal) supplies the rectum of the large intestine.
Paired Visceral Branches	
Suprarenal arteries (sū′-pra-RĒ-nal; *supra* = above; *ren* = kidney)	Although there are three pairs of **suprarenal (adrenal) arteries** that supply the adrenal (suprarenal) glands (superior, middle, and inferior), only the middle pair originates directly from the abdominal aorta (see Figure 14.6b). The middle suprarenal arteries arise at the level of the first lumbar vertebra at or superior to the renal arteries. The superior suprarenal arteries arise from the inferior phrenic artery, and the inferior suprarenal arteries originate from the renal arteries.
Renal arteries (RĒ-nal; = kidney)	The right and left **renal arteries** usually arise from the lateral aspects of the abdominal aorta at the superior border of the second lumbar vertebra, about 1 cm inferior to the superior mesenteric artery (see Figure 14.6b). The right renal vein, which is longer than the left, arises slightly lower than the left and passes posterior to the right renal vein and inferior vena cava. The left renal artery is posterior to the left renal vein and is crossed by the inferior mesenteric vein. The renal arteries carry blood to the kidneys, adrenal (suprarenal) glands, and ureters. Their distribution within the kidneys is discussed in Chapter 25.
Gonadal (gō-NAD-al; *gon* = **seed) [testicular** (test-TIK-yū-lar) **or ovarian arteries** (ō-VA-rē-an)**]**	The **gonadal arteries** arise from the abdominal aorta at the level of the second lumbar vertebra just inferior to the renal arteries (see Figure 14.6b). In males, the gonadal arteries are specifically referred to as the **testicular arteries.** They pass through the inguinal canal and supply the testes, epididymis, and ureters. In females, the gonadal arteries are called the **ovarian arteries.** They are much shorter than the testicular arteries and supply the ovaries, uterine (Fallopian) tubes, and ureters.
Unpaired Parietal Branch	
Median sacral artery (SĀ-kral; = sacrum)	The **median sacral artery** arises from the posterior surface of the abdominal aorta about 1 cm superior to the bifurcation of the aorta into the right and left common iliac arteries (see Figure 14.6b). The median sacral artery supplies the sacrum and coccyx.
Paired Parietal Branches	
Inferior phrenic arteries (FREN-ik; = diaphragm)	The **inferior phrenic arteries** are the first paired branches of the abdominal aorta, immediately superior to the origin of the celiac trunk (see Figure 14.6b). (They may also arise from the renal arteries.) The inferior phrenic arteries are distributed to the inferior surface of the diaphragm and adrenal (suprarenal) glands.
Lumbar arteries (LUM-bar; = loin)	The four pairs of **lumbar arteries** arise from the posterolateral surface of the abdominal aorta (see Figure 14.6b). They supply the lumbar vertebrae, spinal cord and its meninges, and the muscles and skin of the lumbar region of the back.

(continues)

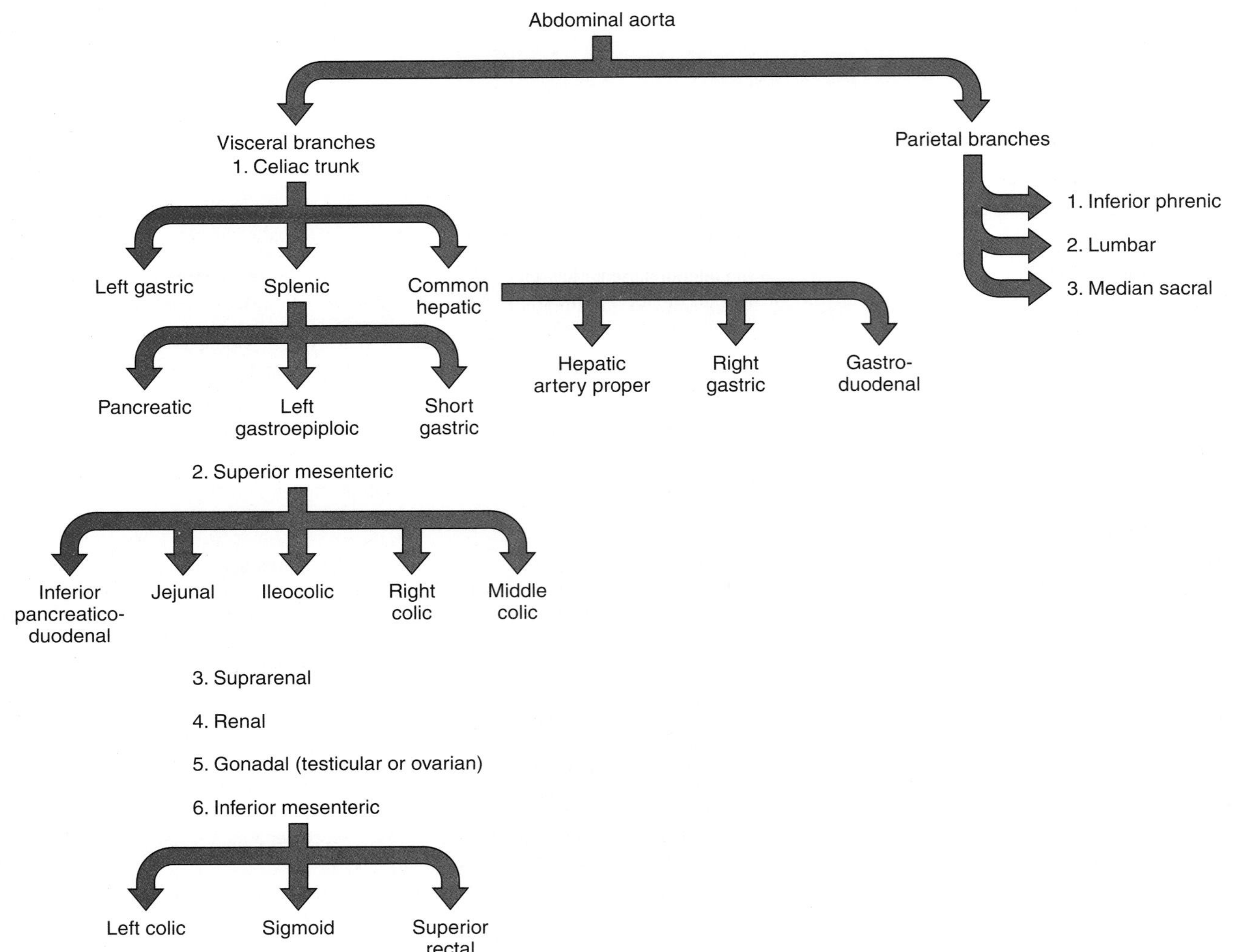
SCHEME OF DISTRIBUTION
Abdominal aorta
Visceral branches
1. Celiac trunk
Left gastric
Splenic
Common hepatic
Pancreatic
Left gastroepiploic
Short gastric
Hepatic artery proper
Right gastric
Gastro-duodenal
2. Superior mesenteric
Inferior pancreatico-duodenal
Jejunal
Ileocolic
Right colic
Middle colic
3. Suprarenal
4. Renal
5. Gonadal (testicular or ovarian)
6. Inferior mesenteric
Left colic
Sigmoid
Superior rectal
Parietal branches
1. Inferior phrenic
2. Lumbar
3. Median sacral

Figure 14.9 / Abdominal aorta and its principal branches. (See Tortora, *A Photographic Atlas of the Human Body,* Figure 6.14.)

The abdominal aorta is the continuation of the thoracic aorta.

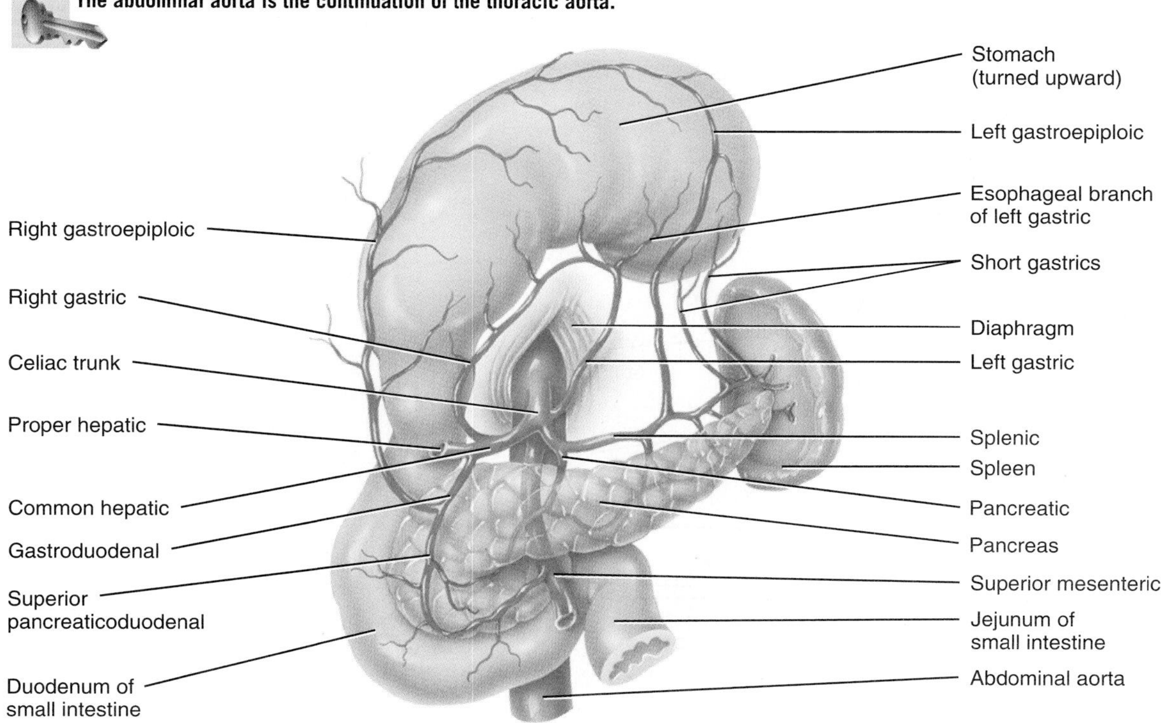

(a) Anterior view of celiac trunk and its branches

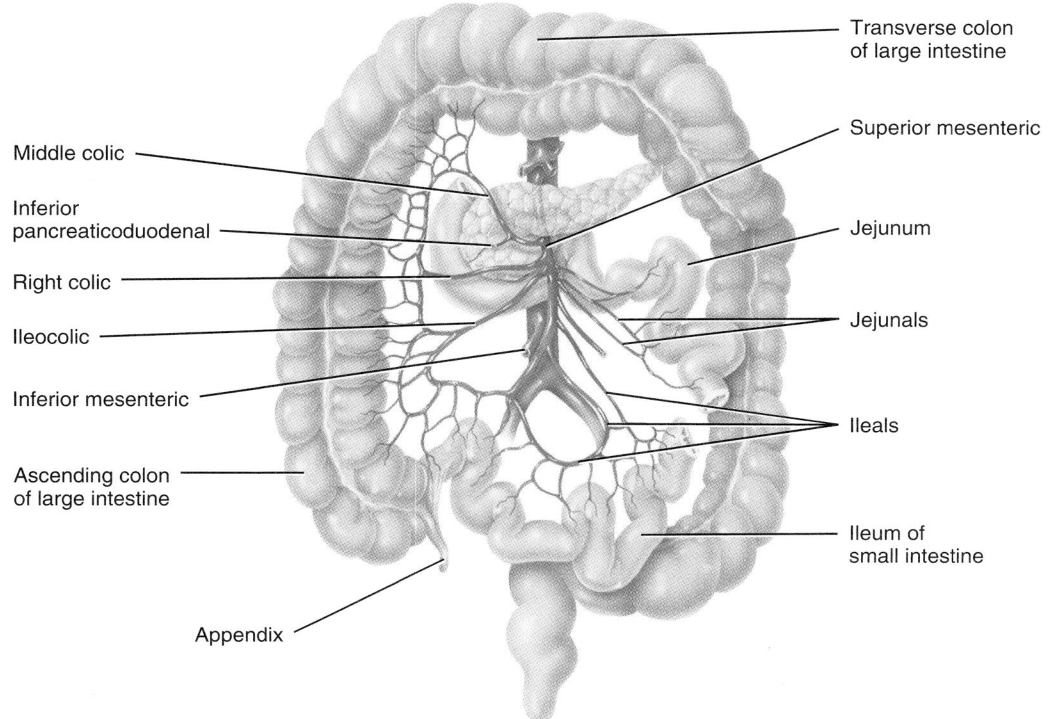

(b) Anterior view of superior mesenteric artery and its branches

(continues)

Exhibit 14.5 Abdominal Aorta (Figure 14.9) (continued)

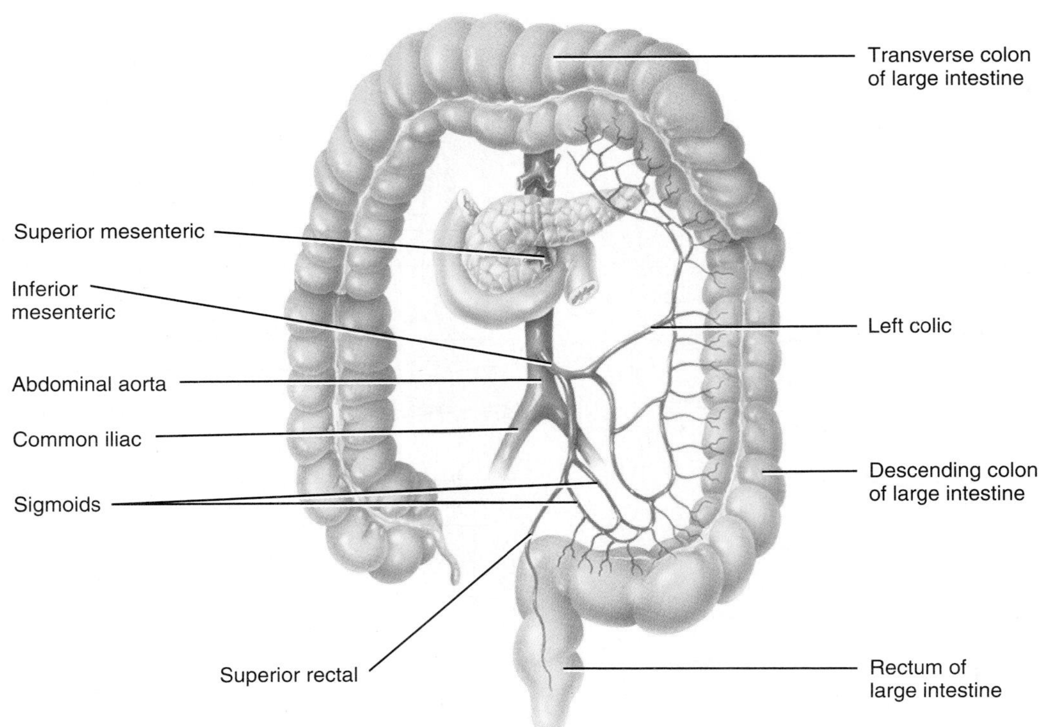

(c) Anterior view of inferior mesenteric artery and its branches

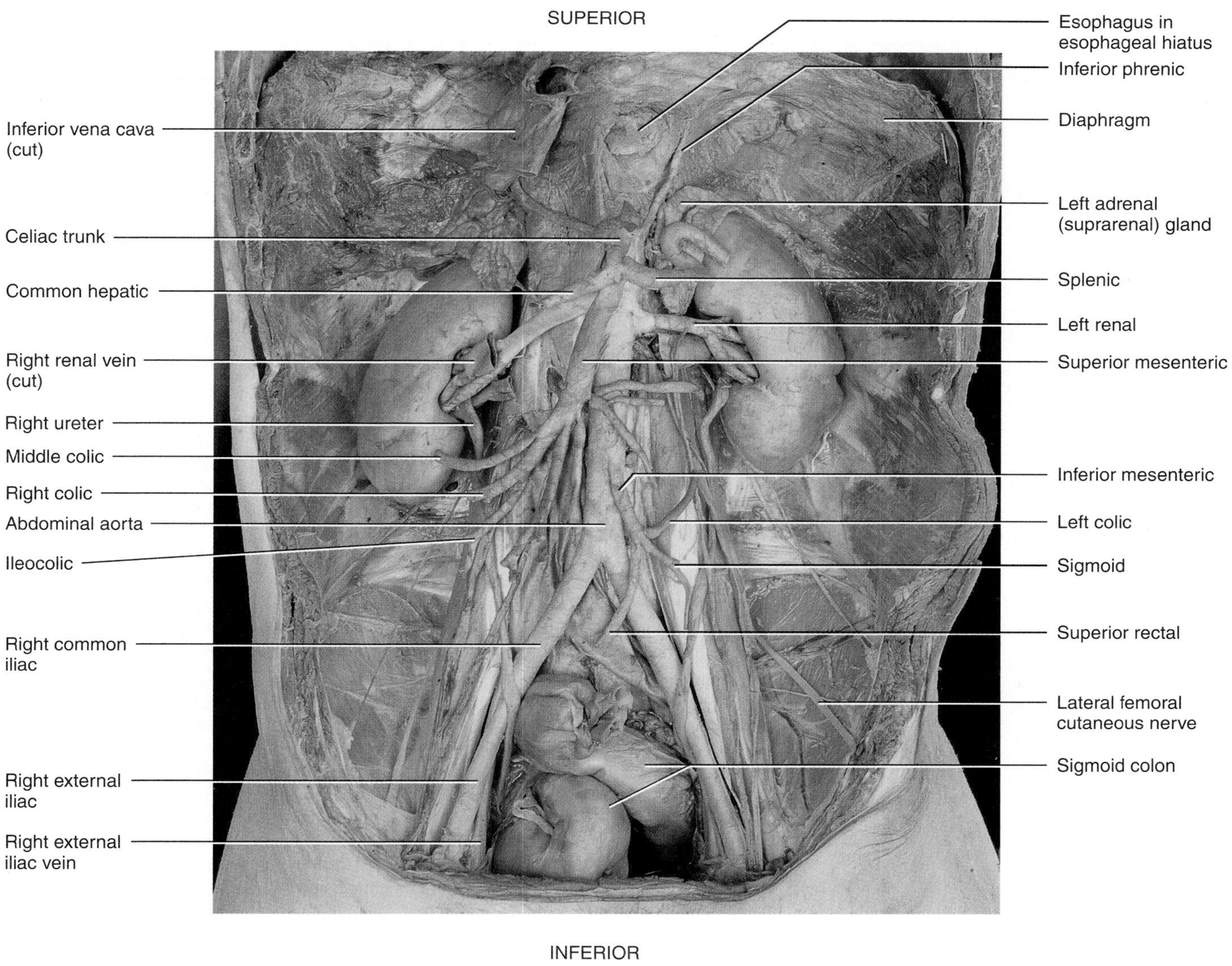

SUPERIOR

Esophagus in esophageal hiatus
Inferior phrenic
Diaphragm
Inferior vena cava (cut)
Left adrenal (suprarenal) gland
Celiac trunk
Splenic
Common hepatic
Left renal
Right renal vein (cut)
Superior mesenteric
Right ureter
Middle colic
Inferior mesenteric
Right colic
Abdominal aorta
Left colic
Ileocolic
Sigmoid
Right common iliac
Superior rectal
Lateral femoral cutaneous nerve
Sigmoid colon
Right external iliac
Right external iliac vein

INFERIOR

(d) Anterior view of arteries of the abdomen and pelvis

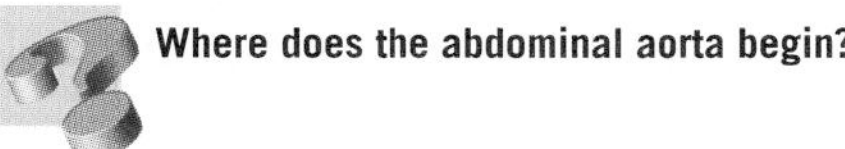

Where does the abdominal aorta begin?

Exhibit 14.6 Arteries of the Pelvis and Lower Limbs (Figure 14.10)

Objective

► Identify the two major branches of the common iliac arteries.

The abdominal aorta ends by dividing into the right and left **common iliac arteries.** These, in turn, divide into the **internal** and **external iliac arteries.** In sequence, the external iliacs become the **femoral arteries** in the thighs, the **popliteal arteries** posterior to the knee, and the **anterior** and **posterior tibial arteries** in the legs.

✓ What general regions are supplied by the internal and external iliac arteries?

Branch	Description and Region Supplied
Common iliac arteries (IL-ē-ak; = ilium)	At about the level of the fourth lumbar vertebra, the abdominal aorta divides into the right and left **common iliac arteries,** the terminal branches of the abdominal aorta. Each passes inferiorly about 5 cm (2 in.) and gives rise to two branches: the internal iliac and external iliac arteries. The general distribution of the common iliac arteries is to the pelvis, external genitals, and lower limbs.
Internal iliac arteries	The **internal iliac (hypogastric) arteries** are the primary arteries of the pelvis. They begin at the bifurcation of the common iliac arteries anterior to the sacroiliac joint at the level of the lumbosacral intervertebral disc. They pass posteromedially as they descend in the pelvis and divide into anterior and posterior divisions. The general distribution of the internal iliac arteries is to the pelvis, buttocks, external genitals, and thigh.
External iliac arteries	The **external iliac arteries** are larger than the internal iliac arteries. Like the internal iliac arteries, they begin at the bifurcation of the common iliac arteries. They descend along the medial border of the psoas major muscles following the pelvic brim, pass posterior to the midportion of the inguinal ligaments, and become the femoral arteries. The general distribution of the external iliac arteries is to the lower limbs. Specifically, branches of the external iliac arteries supply the muscles of the anterior abdominal wall, the cremasteric muscle in males and the round ligament of the uterus in females, and the lower limbs.
Femoral arteries (FEM-o-ral; = thigh)	The **femoral arteries** descend along the anteromedial aspects of the thighs to the junction of the middle and lower third of the thighs. Here they pass through an opening in the tendon of the adductor magnus muscle, where they emerge posterior to the femurs as the popliteal arteries. A pulse may be felt in the femoral artery just inferior to the inguinal ligament. The general distribution of the femoral arteries is to the lower abdominal wall, groin, external genitals, and muscles of the thigh.
Popliteal arteries (pop′-li-TĒ-al; = posterior surface of the knee)	The **popliteal arteries** are the continuation of the femoral arteries through the popliteal fossa (space behind the knee). They descend to the inferior border of the popliteus muscles, where they divide into the anterior and posterior tibial arteries. A pulse may be detected in the popliteal arteries. In addition to supplying the adductor magnus and hamstring muscles and the skin on the posterior aspect of the legs, branches of the popliteal arteries also supply the gastrocnemius, soleus, and plantaris muscles of the calf, knee joint, femur, patella, and fibula.
Anterior tibial arteries (TIB-ē-al; = shin bone)	The **anterior tibial arteries** descend from the bifurcation of the popliteal arteries. They are smaller than the posterior tibial arteries. The anterior tibial arteries descend through the anterior muscular compartment of the leg. They pass through the interosseous membrane that connects the tibia and fibula, lateral to the tibia. The anterior tibial arteries supply the knee joints, anterior compartment muscles of the legs, skin over the anterior aspects of the legs, and ankle joints. At the ankles, the anterior tibial arteries become the **dorsalis pedis arteries** (PED-is; *ped* = foot), also arteries from which a pulse may be detected. The dorsalis pedis arteries supply the muscles, skin, and joints on the dorsal aspects of the feet. On the dorsum of the feet, the dorsalis pedis arteries give off a transverse branch at the first (medial) cuneiform bone. These arteries are the **arcuate arteries** (*arcuat-* = bowed), which run laterally over the bases of the metatarsals. From the arcuate arteries branch **dorsal metatarsal arteries,** which supply the feet. The dorsal metatarsal arteries terminate by dividing into the **dorsal digital arteries,** which supply the toes.
Posterior tibial arteries	The **posterior tibial arteries,** the direct continuations of the popliteal arteries, descend from the bifurcation of the popliteal arteries. They pass down the posterior muscular compartment of the legs posterior to the medial malleolus of the tibia. They terminate by dividing into the medial and lateral plantar arteries. Their general distribution is to the muscles, bones, and joints of the leg and foot. Major branches of the posterior tibial arteries are the **peroneal arteries** (per′-o-NĒ-al; *perone* = fibula). They supply the peroneal, soleus, tibialis posterior, and flexor hallucis muscles; fibula; tarsus; and lateral aspect of the heel. The bifurcation of the posterior tibial arteries into the medial and lateral plantar arteries occurs deep to the flexor retinaculum on the medial side of the feet. The **medial plantar arteries** (PLAN-tar = sole of foot) supply the abductor hallucis and flexor digitorum brevis muscles and the toes. The **lateral plantar arteries** unite with a branch of the dorsalis pedis arteries to form the **plantar arch.** The arch begins at the base of the fifth metatarsal and extends medially across the metacarpals. As the arch crosses the foot, it gives off **plantar metatarsal arteries,** which supply the feet. These terminate by dividing into **plantar digital arteries,** which supply the toes.

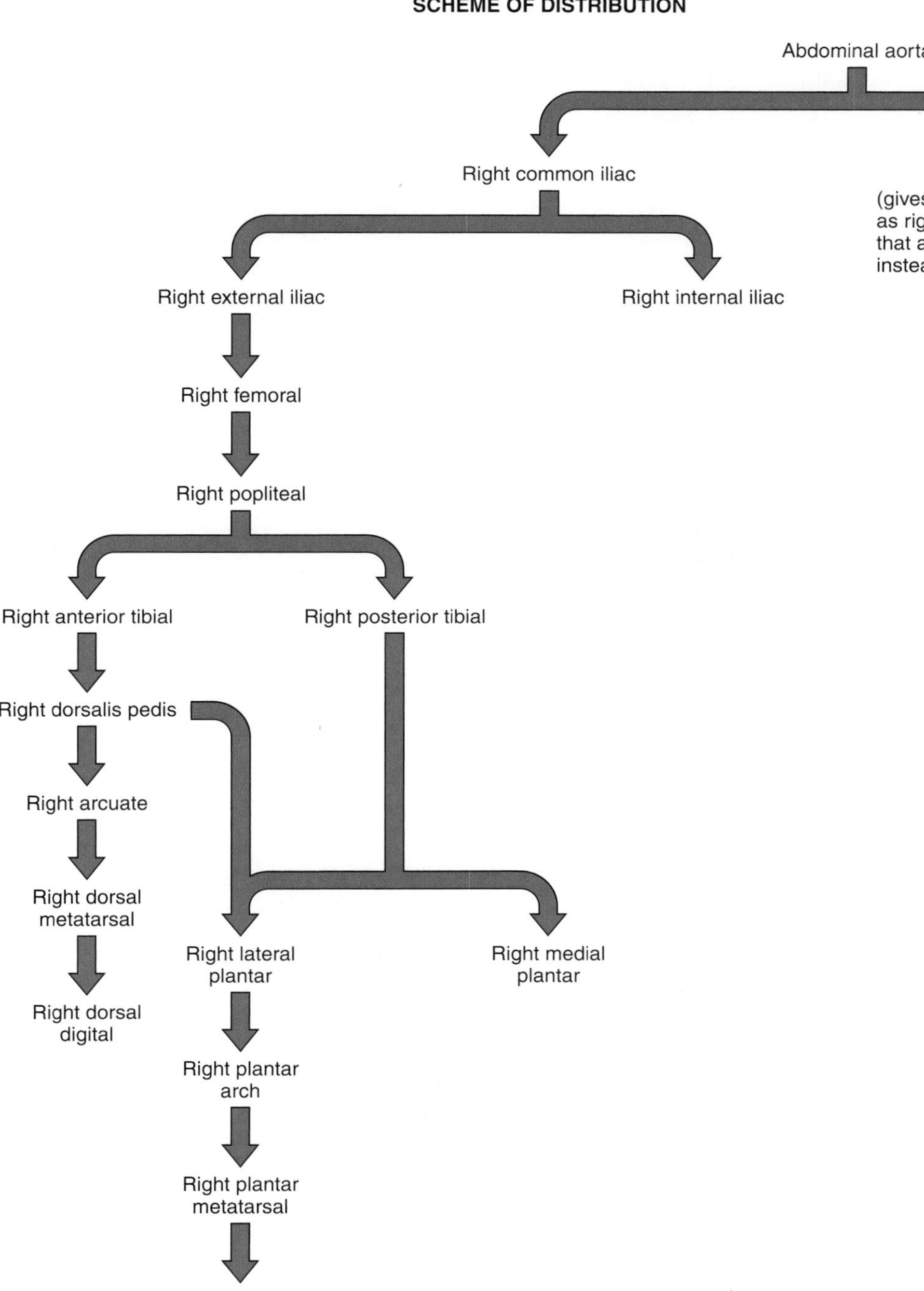
SCHEME OF DISTRIBUTION
Abdominal aorta
Right common iliac
Left common iliac
(gives rise to same branches as right common iliac, except that arteries are labeled left instead of right)
Right external iliac
Right internal iliac
Right femoral
Right popliteal
Right anterior tibial
Right posterior tibial
Right dorsalis pedis
Right arcuate
Right dorsal metatarsal
Right dorsal digital
Right lateral plantar
Right medial plantar
Right plantar arch
Right plantar metatarsal
Right plantar digital

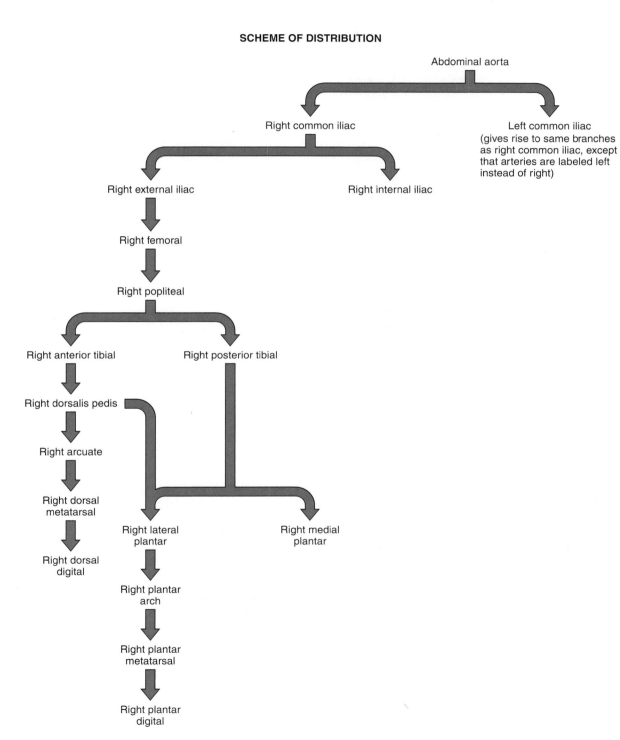

(continues)

Exhibit 14.6 Arteries of the Pelvis and Lower Limbs (Figure 14.10) (continued)

Figure 14.10 / Arteries of the pelvis and right lower limb.

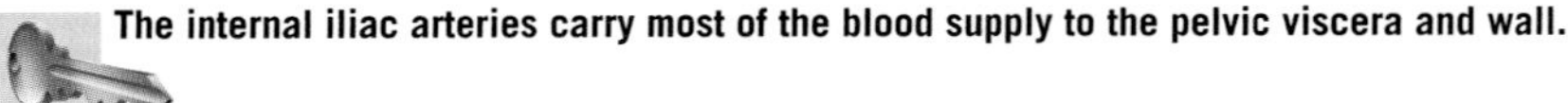

The internal iliac arteries carry most of the blood supply to the pelvic viscera and wall.

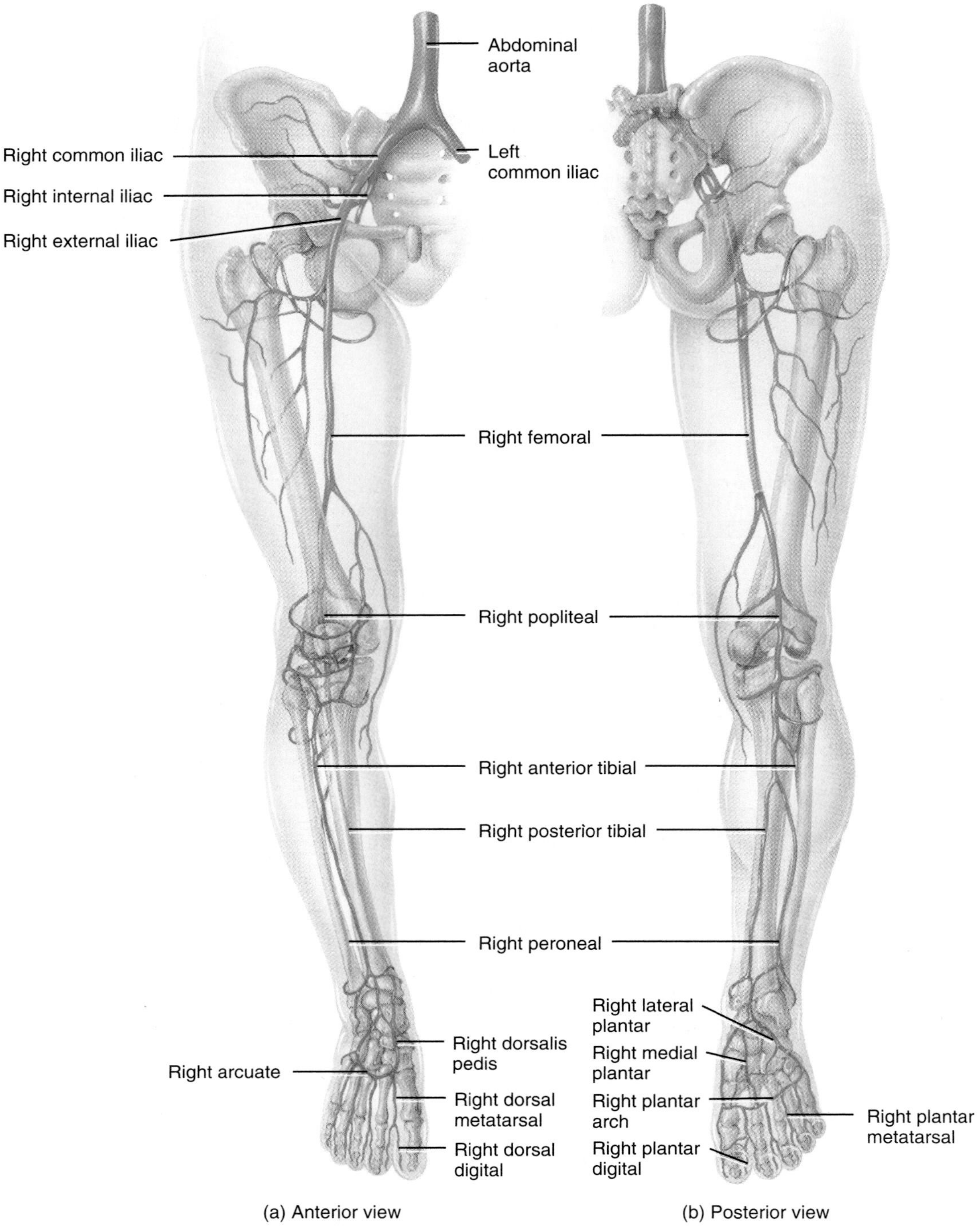

SUPERIOR

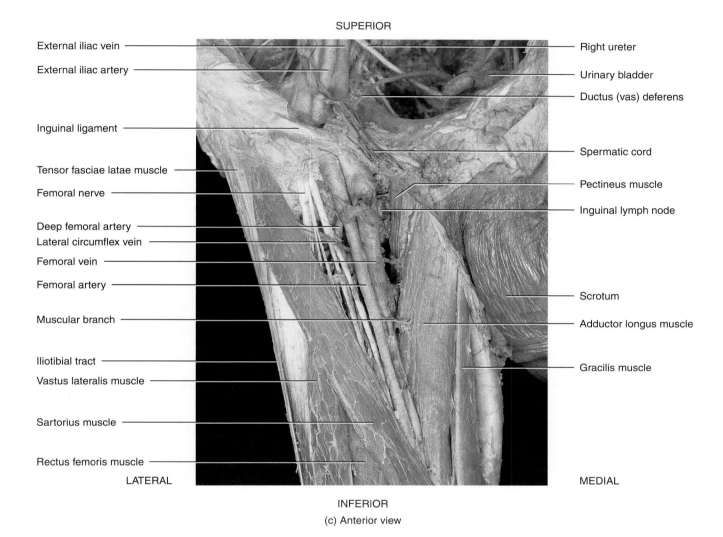

External iliac vein
External iliac artery
Inguinal ligament
Tensor fasciae latae muscle
Femoral nerve
Deep femoral artery
Lateral circumflex vein
Femoral vein
Femoral artery
Muscular branch
Iliotibial tract
Vastus lateralis muscle
Sartorius muscle
Rectus femoris muscle

Right ureter
Urinary bladder
Ductus (vas) deferens
Spermatic cord
Pectineus muscle
Inguinal lymph node
Scrotum
Adductor longus muscle
Gracilis muscle

LATERAL

MEDIAL

INFERIOR

(c) Anterior view

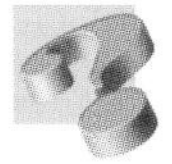

At what point does the abdominal aorta divide into the common iliac arteries?

Exhibit 14.7 Veins of the Systemic Circulation (Figure 14.11)

Objective

▶ Identify the three systemic veins that return deoxygenated blood to the heart.

Whereas arteries distribute blood to various parts of the body, veins drain blood away from them. For the most part, arteries are deep, whereas veins may be superficial or deep. Superficial veins are located just beneath the skin and can easily be seen. Because there are no large superficial arteries, the names of superficial veins do not correspond to those of arteries. Superficial veins are clinically important as sites for withdrawing blood or giving injections. Deep veins generally travel alongside arteries and usually bear the same name. Arteries usually follow definite pathways; veins are more difficult to follow because they connect in irregular networks in which many tributaries merge to form a large vein. Although only one systemic artery, the aorta, takes oxygenated blood away from the heart (left ventricle), three systemic veins, the **coronary sinus, superior vena cava,** and **inferior vena cava,** deliver deoxygenated blood to the heart (right atrium). The coronary sinus receives blood from the cardiac veins; the superior vena cava receives blood from other veins superior to the diaphragm, except the air sacs (alveoli) of the lungs; the inferior vena cava receives blood from veins inferior to the diaphragm.

✓ What are the three tributaries of the coronary sinus?

Vein	Description and Region Supplied
Coronary sinus (KOR-ō-nar-ē; *corona* = crown)	The **coronary sinus** is the main vein of the heart; it receives almost all venous blood from the myocardium. It is located in the coronary sulcus (see Figure 13.3c) and opens into the right atrium between the orifice of the inferior vena cava and the tricuspid valve. It is a wide venous channel into which three veins drain. It receives the **great cardiac vein** (in the anterior interventricular sulcus) into its left end, and the **middle cardiac vein** (in the posterior interventricular sulcus) and the **small cardiac vein** into its right end. Several **anterior cardiac veins** drain directly into the right atrium.
Superior vena cava (SVC) (VĒ-na CĀ-va; *vena* = vein; *cava* = cavelike)	The **SVC** is about 7.5 cm. (3 in.) long and 2 cm (1 in.) in diameter and empties its blood into the superior part of the right atrium. It begins posterior to the right first costal cartilage by the union of the right and left brachiocephalic veins and ends at the level of the right third costal cartilage, where it enters the right atrium. The SVC drains the head, neck, chest, and upper limbs.
Inferior vena cava (IVC)	The **IVC** is the largest vein in the body, about 3.5 cm (1½ in.) in diameter. It begins anterior to the fifth lumbar vertebra by the union of the common iliac veins, ascends behind the peritoneum to the right of the midline, pierces the costal tendon of the diaphragm at the level of the eighth thoracic vertebra, and enters the inferior part of the right atrium. The IVC drains the abdomen, pelvis, and lower limbs. The inferior vena cava is commonly compressed during the later stages of pregnancy by the enlarging uterus, producing edema of the ankles and feet and temporary varicose veins.

Figure 14.11 / Principal veins.

Deoxygenated blood returns to the heart via the superior and inferior venae cavae and the coronary sinus.
(See Tortora, *A Photographic Atlas of the Human Body,* Figure 6.18.)

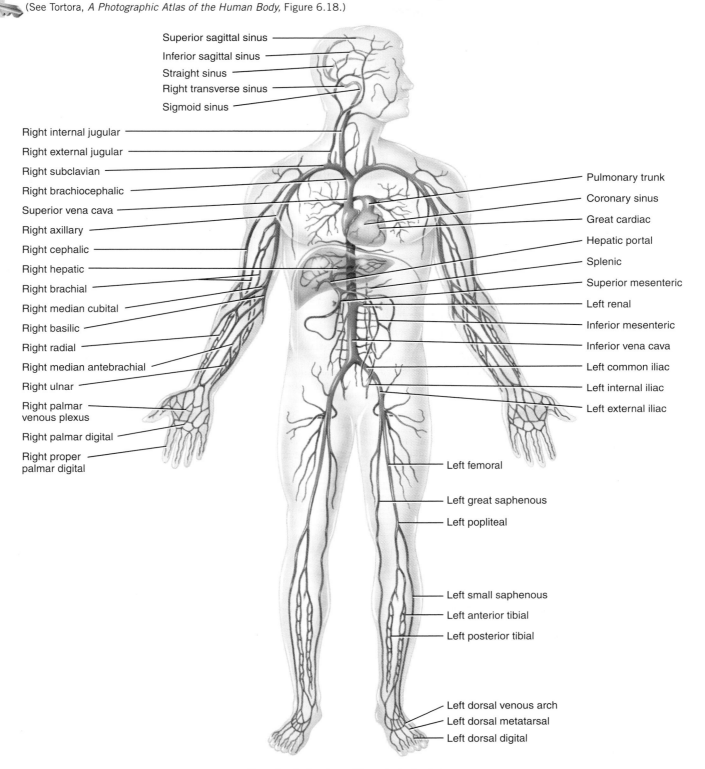

Superior sagittal sinus
Inferior sagittal sinus
Straight sinus
Right transverse sinus
Sigmoid sinus
Right internal jugular
Right external jugular
Right subclavian
Right brachiocephalic
Superior vena cava
Right axillary
Right cephalic
Right hepatic
Right brachial
Right median cubital
Right basilic
Right radial
Right median antebrachial
Right ulnar
Right palmar venous plexus
Right palmar digital
Right proper palmar digital
Pulmonary trunk
Coronary sinus
Great cardiac
Hepatic portal
Splenic
Superior mesenteric
Left renal
Inferior mesenteric
Inferior vena cava
Left common iliac
Left internal iliac
Left external iliac
Left femoral
Left great saphenous
Left popliteal
Left small saphenous
Left anterior tibial
Left posterior tibial
Left dorsal venous arch
Left dorsal metatarsal
Left dorsal digital

Overall anterior view of the principal veins

Which general regions of the body are drained by the superior vena cava and the inferior vena cava?

Exhibit 14.8 Veins of the Head and Neck (Figure 14.12)

Objective

► Identify the three major veins that drain blood from the head.

Most blood draining from the head passes into three pairs of veins: the **internal jugular, external jugular,** and **vertebral veins.** Within the brain, all veins drain into dural venous sinuses and then into the internal jugular veins. **Dural venous sinuses** are endothelial-lined venous channels between layers of the cranial dura mater.

✓ Which general areas are drained by the internal jugular, external jugular, and vertebral veins?

Vein	Description and Region Drained
Internal jugular veins (JUG-yū-lar; *juglar* = throat)	The flow of blood from the dural venous sinuses into the internal jugular veins is as follows (Figure 14.12): The **superior sagittal sinus** (SAJ-i-tal; *sagitallis* = straight) begins at the frontal bone, where it receives a vein from the nasal cavity, and passes posteriorly to the occipital bone. Along its course, it receives blood from the superior, medial, and lateral aspects of the cerebral hemispheres, meninges, and cranial bones. The superior sagittal sinus usually turns to the right and drains into the right transverse sinus. The **inferior sagittal sinus** is much smaller than the superior sagittal sinus; it begins posterior to the attachment of the falx cerebri and receives the great cerebral vein to become the straight sinus. The great cerebral vein drains the deeper parts of the brain. Along its course the inferior sagittal sinus also receives tributaries from the superior and medial aspects of the cerebral hemispheres. The **straight sinus** is formed by the union of the inferior sagittal sinus and the great cerebral vein. The straight sinus also receives blood from the cerebellum and usually drains into the left transverse sinus. The **transverse (lateral) sinuses** begin near the occipital bone, pass laterally and anteriorly, and become the sigmoid sinuses near the temporal bone. The transverse sinuses receive blood from the cerebrum, cerebellum, and cranial bones. The **sigmoid sinuses** (SIG-moid = S-shaped) are located along the temporal bone. They pass through the jugular foramina, where they terminate in the internal jugular veins. The sigmoid sinuses drain the transverse sinuses. **The cavernous sinuses** (KAV-er-nus = cavelike) are located on either side of the sphenoid bone. They receive blood from the ophthalmic veins from the orbits and from the cerebral veins from the cerebral hemispheres. They ultimately empty into the transverse sinuses and internal jugular veins. The cavernous sinuses are unique because they have nerves and a major blood vessel passing through them on their way to the orbit and face. The oculomotor (III), trochlear (IV), and ophthalmic and maxillary divisions of the trigeminal (V) nerve, as well as the internal carotid arteries, pass through the cavernous sinuses. The right and left **internal jugular veins** pass inferiorly on either side of the neck lateral to the internal carotid and common carotid arteries. They then unite with the subclavian veins posterior to the clavicles at the sternoclavicular joints to form the right and left **brachiocephalic veins** (brā-kē-ō-se-FAL-ik; *bracium* = arm; *cephalic* = head). From here blood flows into the superior vena cava. The general structures drained by the internal jugular veins are the brain (through the dural venous sinuses), face, and neck.
External jugular veins	The right and left **external jugular veins** begin in the parotid glands near the angle of the mandible. They are superficial veins that descend through the neck across the sternocleidomastoid muscles. They terminate at a point opposite the middle of the clavicle, where they empty into the subclavian veins. The general structures drained by the external jugular veins are external to the cranium, such as the scalp and superficial and deep regions of the face. In cases of heart failure, the venous pressure in the right atrium may rise. In such patients the pressure in the column of blood in the external jugular vein rises so that, even with the patient at rest and sitting in a chair, the external jugular vein will be visibly distended. Temporary distention of the vein is often seen in healthy adults when the intrathoracic pressure is raised during coughing and physical exertion.
Vertebral veins (VER-te-bral; *vertebra* = vertebrae)	The right and left **vertebral veins** originate inferior to the occipital condyles. They descend through successive transverse foramina of the first six cervical vertebrae and emerge from the foramina of the sixth cervical vertebra to enter the brachiocephalic veins in the root of the neck. The vertebral veins drain deep structures in the neck such as the cervical vertebrae, cervical spinal cord, and some neck muscles.

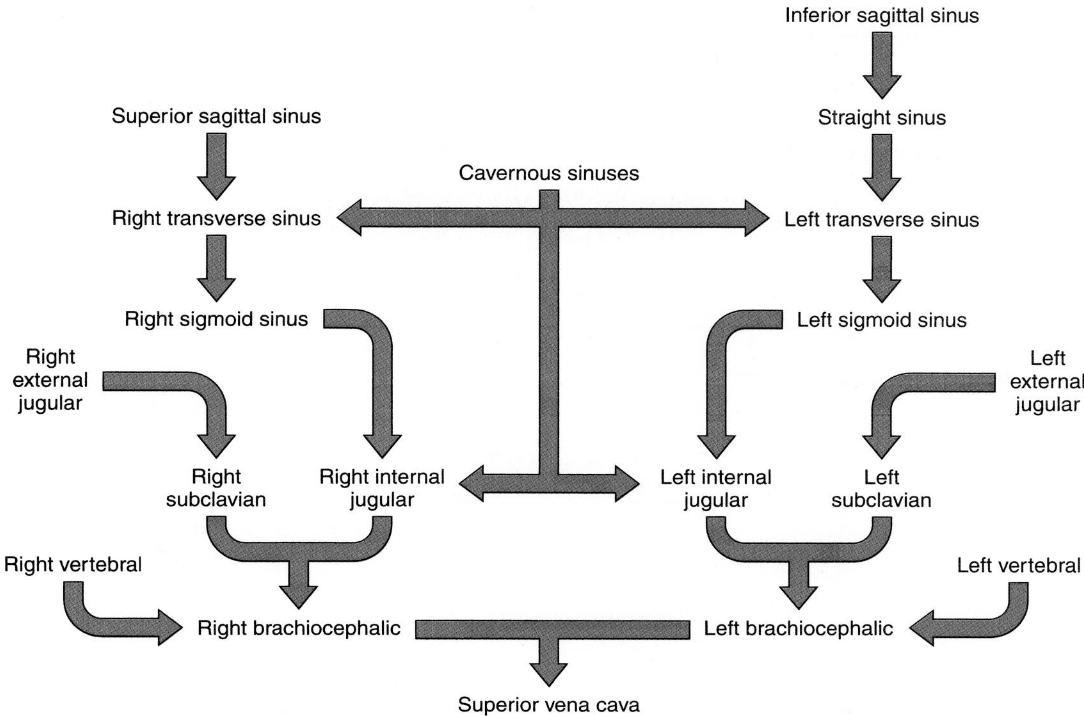
SCHEME OF DRAINAGE
Inferior sagittal sinus
Superior sagittal sinus
Straight sinus
Cavernous sinuses
Right transverse sinus
Left transverse sinus
Right sigmoid sinus
Left sigmoid sinus
Right external jugular
Left external jugular
Right subclavian
Right internal jugular
Left internal jugular
Left subclavian
Right vertebral
Left vertebral
Right brachiocephalic
Left brachiocephalic
Superior vena cava

(continues)

Exhibit 14.8 Veins of the Head and Neck (Figure 14.12) (continued)

Figure 14.12 / Principal veins of the head and neck.

Blood draining from the head passes into the internal jugular, external jugular, and vertebral veins.

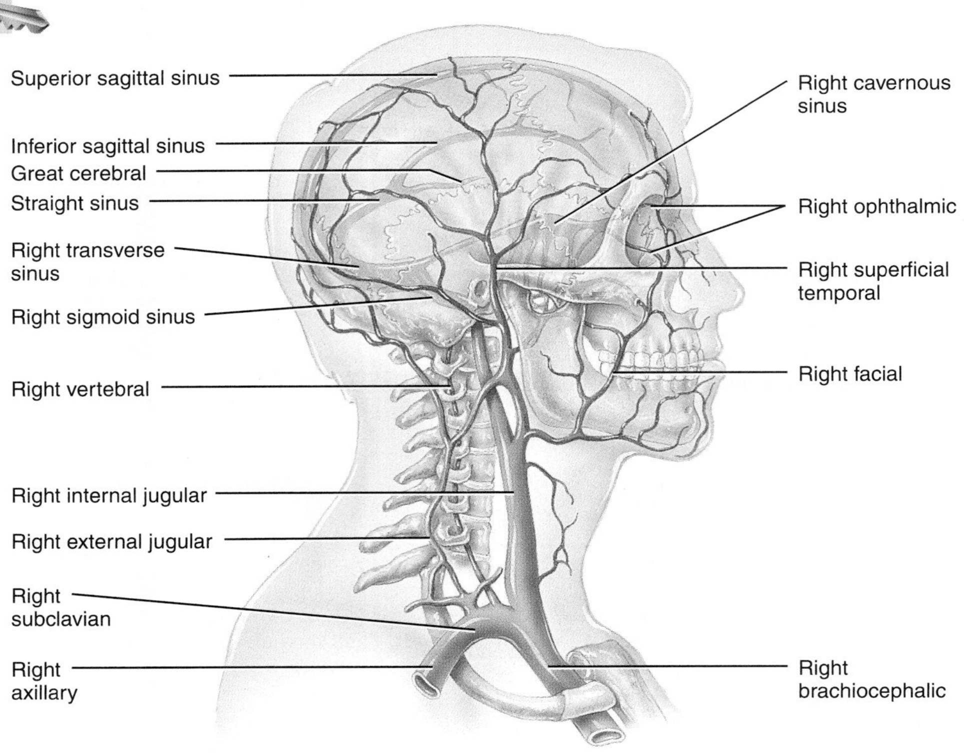

Superior sagittal sinus
Inferior sagittal sinus
Great cerebral
Straight sinus
Right transverse sinus
Right sigmoid sinus
Right vertebral
Right internal jugular
Right external jugular
Right subclavian
Right axillary
Right cavernous sinus
Right ophthalmic
Right superficial temporal
Right facial
Right brachiocephalic

(a) Right lateral view

SUPERIOR

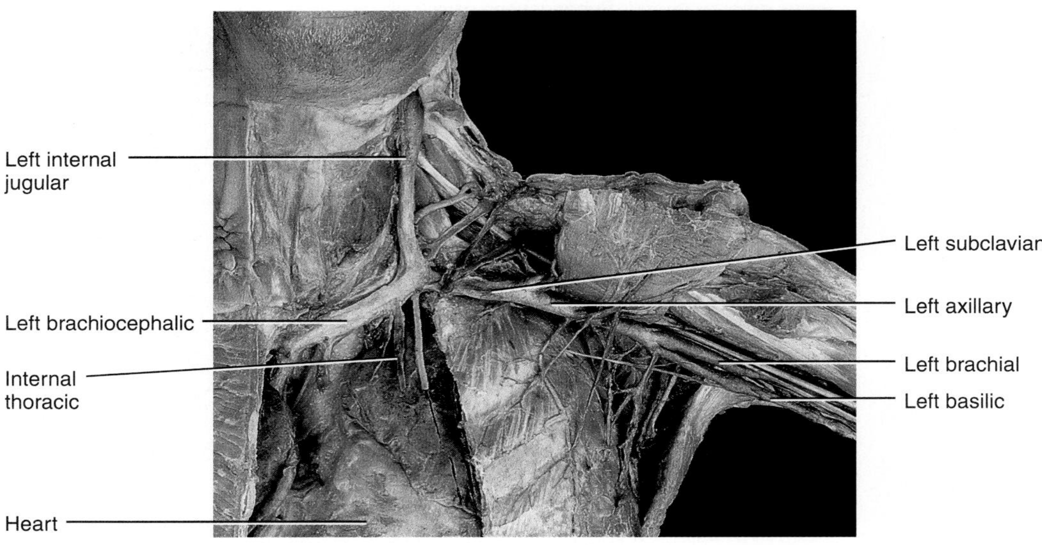

INFERIOR

(b) Anterior view

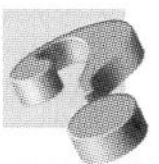

Into which veins in the neck does all venous blood in the brain drain?

Exhibit 14.9 Veins of the Upper Limbs (Figure 14.13)

Objective

- Identify the principal veins that drain the upper limbs.

Blood from the upper limbs is returned to the heart by both superficial and deep veins. Both sets of veins have valves, which are more numerous in the deep veins. **Superficial veins** are located just deep to the skin and are often visible. They anastomose extensively with one another and with deep veins, and they do not accompany arteries. Superficial veins are larger than deep veins and return most of the blood from the upper limbs. **Deep veins** are located deep in the body. They usually accompany arteries and have the same names as the corresponding arteries.

✓ Where do the cephalic, basilic, median antebrachial, radial, and ulnar veins originate?

Vein	Description and Region Drained
Superficial	
Cephalic veins (se-FAL-ik; *cephalic* = head)	The principal superficial veins that drain the upper limbs are the cephalic and basilic veins. They originate in the hand and convey blood from the smaller superficial veins into the axillary veins. The **cephalic veins** begin on the lateral aspect of the **dorsal venous arches** (VĒ-nus), networks of veins on the dorsum of the hands formed by the **dorsal metacarpal veins** (Figure 14.13a). These veins, in turn, drain the **dorsal digital veins,** which pass along the sides of the fingers. Following their formation from the dorsal venous arches, the cephalic veins arch around the radial side of the forearms to the anterior surface and ascend through the entire limbs along the anterolateral surface. The cephalic veins end where they join the axillary veins, just inferior to the clavicles. **Accessory cephalic veins** originate either from a venous plexus on the dorsum of the forearms or from the medial aspects of the dorsal venous arches and unite with the cephalic veins just inferior to the elbow. The cephalic veins drain blood from the lateral aspect of the upper limbs.
Basilic veins (ba-SIL-ik; *basilikos* = royal)	The **basilic veins** begin on the medial aspects of the dorsal venous arches and ascend along the posteromedial surface of the forearm and anteromedial surface of the arm (Figure 14.13b). They drain blood from the medial aspects of the upper limbs. Anterior to the elbow, the basilic veins are connected to the cephalic veins by the **median cubital veins** (*cubitus* = elbow), which drain the forearm. If veins must be punctured for an injection, transfusion, or removal of a blood sample, the medial cubital veins are preferred. After receiving the median cubital veins, the basilic veins continue ascending until they reach the middle of the arm. There they penetrate the tissues deeply and run alongside the brachial arteries until they join the brachial veins. As the basilic and brachial veins merge in the axillary area, they form the axillary veins.
Median antebrachial veins (an′-tē-BRĀ-kē-al; *ante* = before, in front of; *brachium* = arm)	The **median antebrachial veins (median veins of the forearm)** begin in the **palmar venous plexuses,** networks of veins in the palms. The plexuses drain the **palmar digital veins** in the fingers. The median antebrachial veins ascend anteriorly in the forearms to join the basilic or median cubital veins, sometimes both. They drain the palms and forearms.
Deep	
Radial veins (RĀ-dē-al = radius)	The paired radial veins begin at the **deep palmar venous arches** (Figure 14.13c). These arches drain the **palmar metacarpal veins** in the palms. The radial veins drain the lateral aspects of the forearms and pass alongside the radial arteries. Just inferior to the elbow joint, the radial veins unite with the ulnar veins to form the brachial veins.
Ulnar veins (UL-nar = ulna)	The paired **ulnar veins,** which are larger than the radial veins, begin at the **superficial palmar venous arches.** These arches drain the **common palmar digital veins** and the **proper palmar digital veins** in the fingers. The ulnar veins drain the medial aspect of the forearms, pass alongside the ulnar arteries, and join with the radial veins to form the brachial veins.
Brachial veins (BRĀ-kē-al; *brachium* = arm)	The paired **brachial veins** accompany the brachial arteries. They drain the forearms, elbow joints, arms, and humerus. They pass superiorly and join with the basilic veins to form the axillary veins.
Axillary veins (AK-si-ler′-ē; *axilla* = armpit)	The **axillary veins** ascend to the outer borders of the first ribs, where they become the subclavian veins. The axillary veins receive tributaries that correspond to the branches of the axillary arteries. The axillary veins drain the arms, axillas, and superolateral chest wall.
Subclavian veins (sub-KLĀ-vē-an; *sub* = under; *clavicula* = clavicle)	The **subclavian veins** are continuations of the axillary veins that terminate at the sternal end of the clavicles, where they unite with the internal jugular veins to form the brachiocephalic veins. The subclavian veins drain the arms, neck, and thoracic wall. The thoracic duct of the lymphatic system delivers lymph into the left subclavian vein at the junction with the internal jugular. The right lymphatic duct delivers lymph into the right subclavian vein at the corresponding junction (see Figure 15.3 on page 477).

(continues)

Exhibit 14.9 Veins of the Upper Limbs (Figure 14.13) (continued)

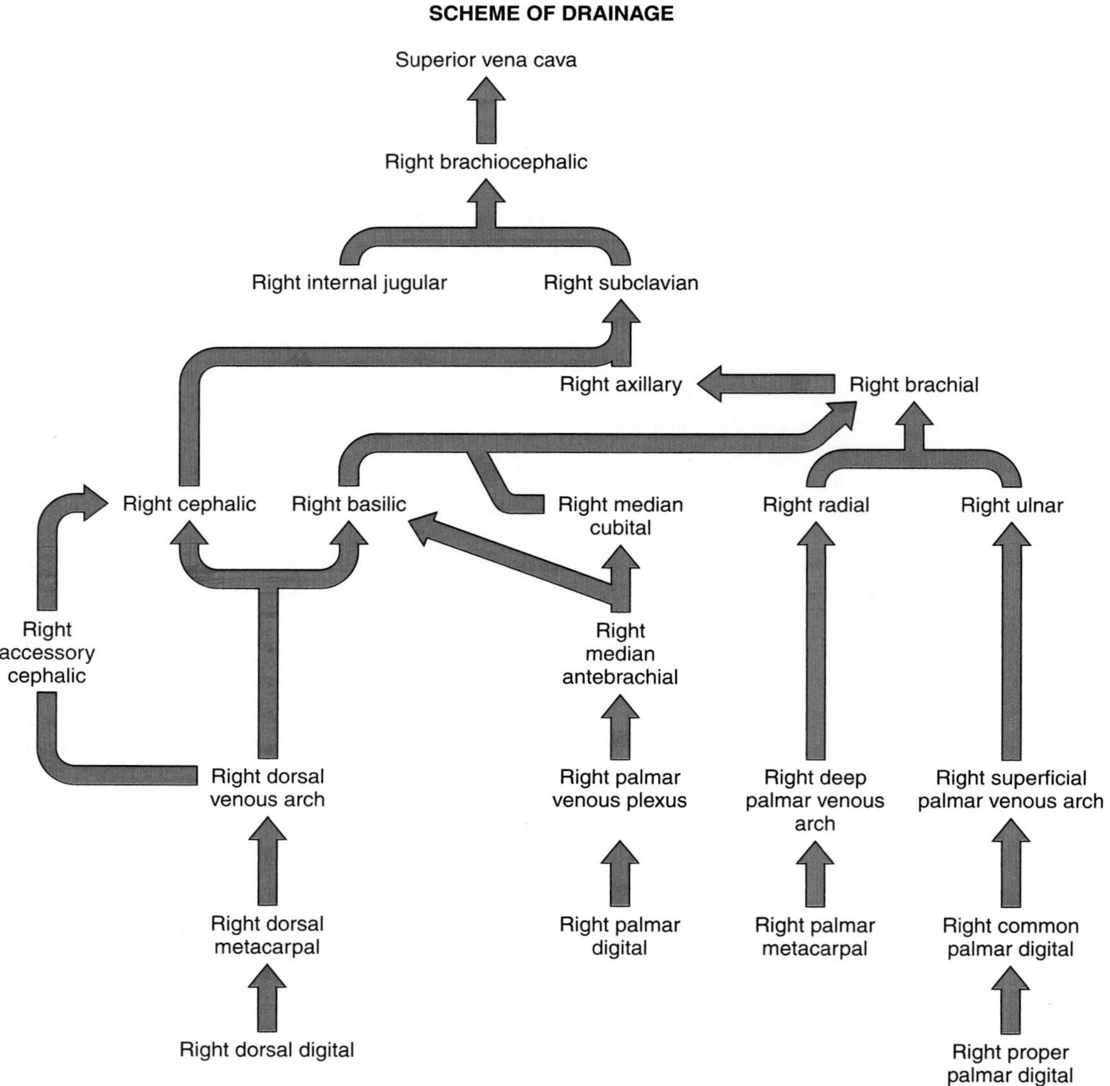
SCHEME OF DRAINAGE
Superior vena cava
Right brachiocephalic
Right internal jugular
Right subclavian
Right axillary
Right brachial
Right cephalic
Right basilic
Right median cubital
Right radial
Right ulnar
Right accessory cephalic
Right median antebrachial
Right dorsal venous arch
Right palmar venous plexus
Right deep palmar venous arch
Right superficial palmar venous arch
Right dorsal metacarpal
Right palmar digital
Right palmar metacarpal
Right common palmar digital
Right dorsal digital
Right proper palmar digital

Figure 14.13 / Principal veins of the right upper limb. (See Tortora, *A Photographic Atlas of the Human Body,* Figure 6.20.)

Deep veins usually accompany arteries that have similar names.

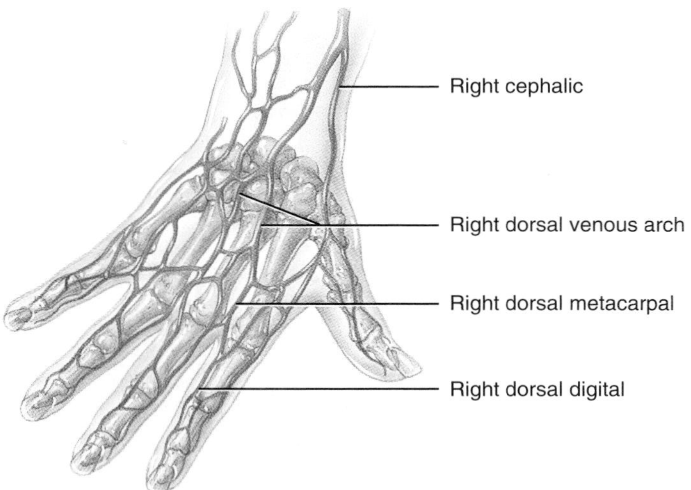

(a) Posterior view of superficial veins of the hand

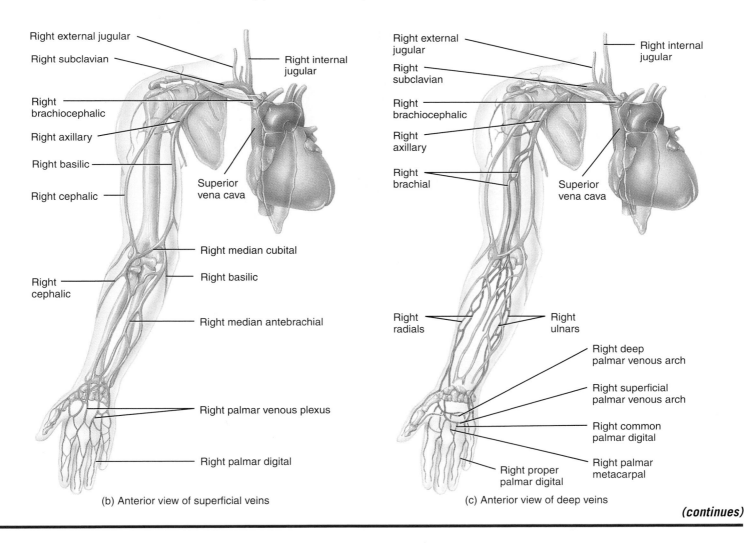

Right external jugular
Right subclavian
Right internal jugular
Right brachiocephalic
Right axillary
Right basilic
Superior vena cava
Right cephalic
Right median cubital
Right basilic
Right cephalic
Right median antebrachial
Right palmar venous plexus
Right palmar digital

(b) Anterior view of superficial veins

Right external jugular
Right internal jugular
Right subclavian
Right brachiocephalic
Right axillary
Right brachial
Superior vena cava
Right radials
Right ulnars
Right deep palmar venous arch
Right superficial palmar venous arch
Right common palmar digital
Right palmar metacarpal
Right proper palmar digital

(c) Anterior view of deep veins

(continues)

Exhibit 14.9 Veins of the Upper Limbs (Figure 14.13) (continued)

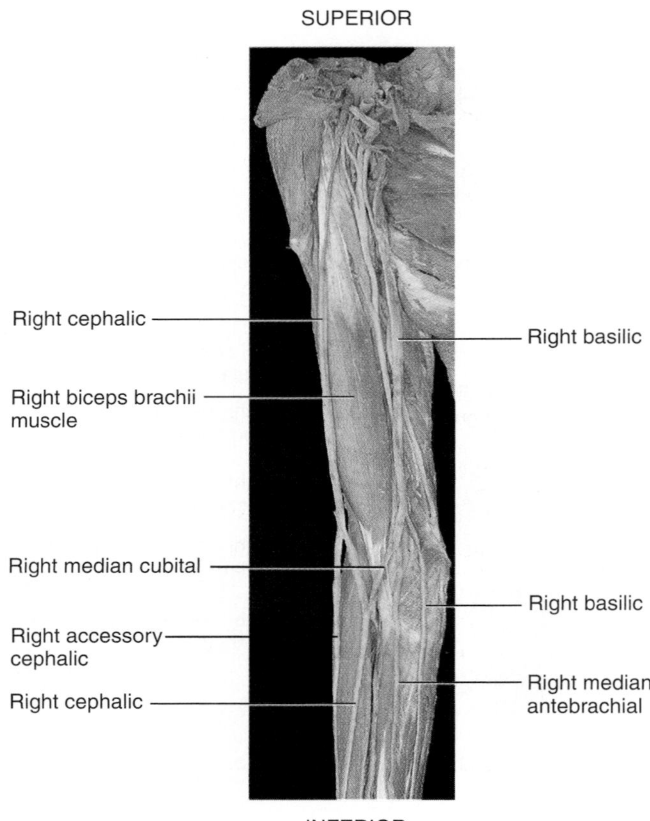

(d) Anterior view of superficial veins of arm and forearm

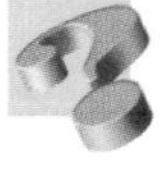

From which vein in the upper limb is a blood sample often taken?

Exhibit 14.10 Veins of the Thorax (Figure 14.14)

Objective

- ▶ Identify the components of the azygos system of veins.

Although the brachiocephalic veins drain some portions of the thorax, most thoracic structures are drained by the **azygos system,** a network of veins that runs on either side of the vertebral column. The system consists of three veins—the **azygous, hemiazygos,** and **accessory hemiazygos veins**. These veins show considerable variation in origin, course, tributaries, anastomoses, and termination. Ultimately they empty into the superior vena cava.

✓ What is the importance of the azygos system relative to the inferior vena cava?

Vein	Description and Region Drained
Brachiocephalic vein (brā′-kē-ō-se-FAL-ik; *brachium* = arm; *cephalic* = head)	The right and left **brachiocephalic veins,** formed by the union of the subclavian and internal jugular veins, drain blood from the head, neck, upper limbs, mammary glands, and superior thorax. The brachiocephalic veins unite to form the superior vena cava. Because the superior vena cava is to the right of the body's midline, the left brachiocephalic vein is longer than the right. The right brachiocephalic vein is anterior and to the right of the brachiocephalic trunk. The left brachiocephalic vein is anterior to the brachiocephalic trunk, the left common carotid and left subclavian arteries, the trachea, and the left vagus (X) and phrenic nerves.
Azygos system (az-Ī-gos = unpaired)	The **azygos system,** besides collecting blood from the thorax and abdominal wall, may serve as a bypass for the inferior vena cava that drains blood from the lower body. Several small veins directly link the azygos system with the inferior vena cava. Large veins that drain the lower limbs and abdomen conduct blood into the azygos system. If the inferior vena cava or hepatic portal vein becomes obstructed, the azygos system can return blood from the lower body to the superior vena cava.
Azygos vein	The **azygos vein** lies anterior to the vertebral column, slightly to the right of the midline. It usually begins at the junction of the right ascending lumbar and right subcostal veins near the diaphragm. At the level of the fourth thoracic vertebra, it arches over the root of the right lung to end in the superior vena cava. Generally, the azygos vein drains the right side of the thoracic wall, thoracic viscera, and abdominal wall. Specifically, the azygos vein receives blood from most of the **right posterior intercostal, hemiazygos, accessory hemiazygos, esophageal, mediastinal, pericardial,** and **bronchial veins.**
Hemiazygos vein (HEM-ē-az-ī-gos; *hemi* = **half**)	The **hemiazygos vein** is anterior to the vertebral column and slightly to the left of the midline. It frequently begins at the junction of the left ascending lumbar and left subcostal veins. It terminates by joining the azygos vein at about the level of the ninth thoracic vertebra. Generally, the hemiazygos vein drains the left side of the thoracic wall, thoracic viscera, and abdominal wall. Specifically, the hemiazygos vein receives blood from the ninth through eleventh **left posterior intercostal, esophageal, mediastinal,** and sometimes the **accessory hemiazygos veins.**
Accessory hemiazygos vein	The **accessory hemiazygos vein** is also anterior to the vertebral column and to the left of the midline. It begins at the fourth or fifth intercostal space and descends from the fifth to the eighth thoracic vertebra or ends in the hemiazygos vein. It terminates by joining the azygos vein at about the level of the eighth thoracic vertebra. The accessory hemiazygos vein drains the left side of the thoracic wall. It receives blood from the fourth through eighth **left posterior intercostal veins** (the first through third left posterior intercostal veins open into the left brachiocephalic vein), **left bronchial,** and **mediastinal veins.**

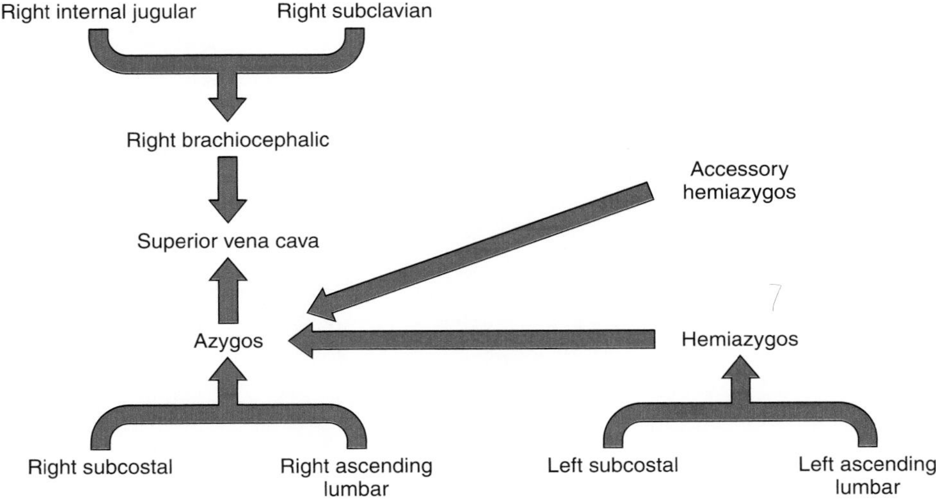

(continues)

Exhibit 14.10 Veins of the Thorax (Figure 14.14) (continued)

Figure 14.14 / Principal veins of the thorax, abdomen, and pelvis.

Most thoracic structures are drained by the azygos system of veins.

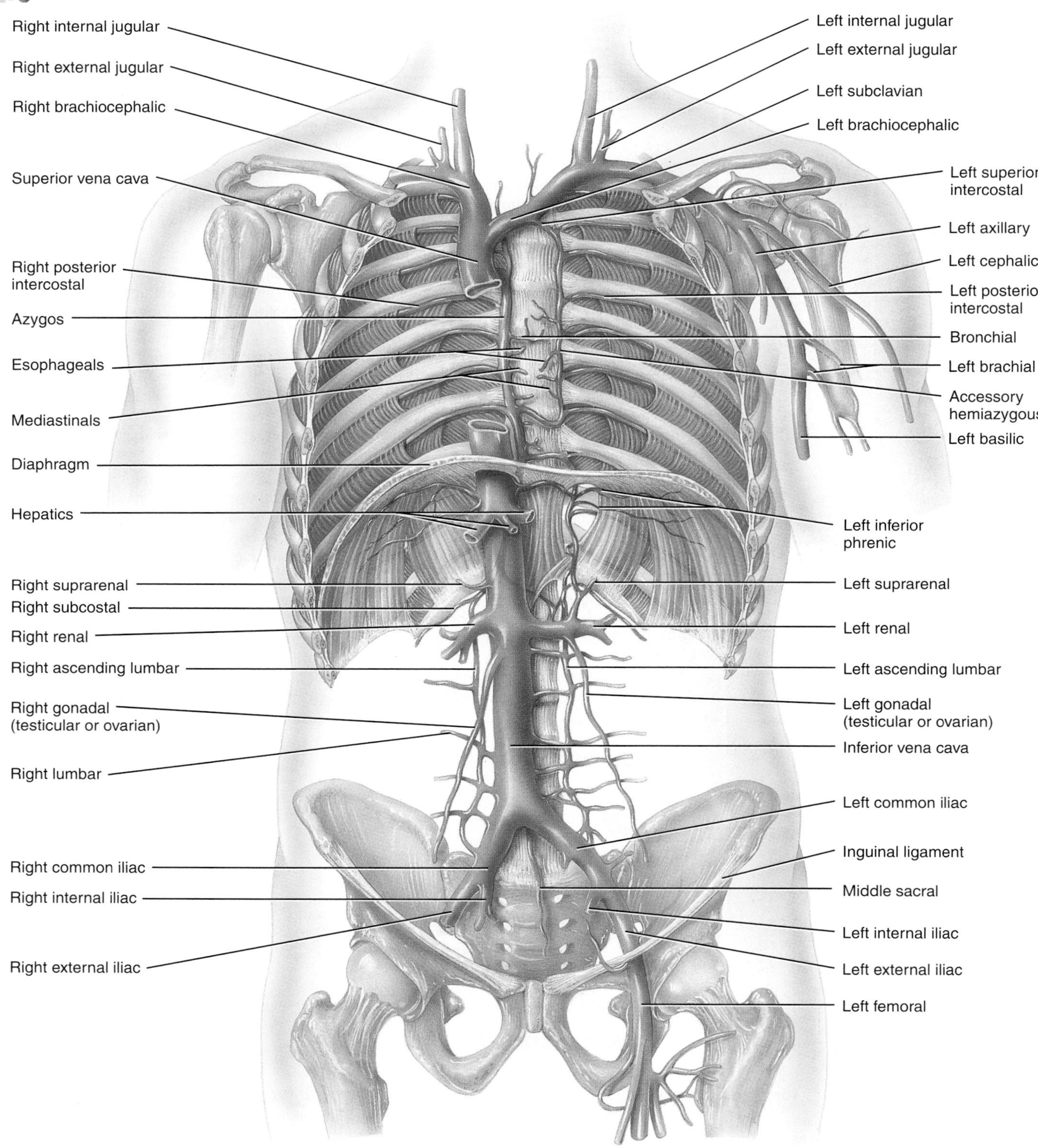

(a) Anterior view

SUPERIOR

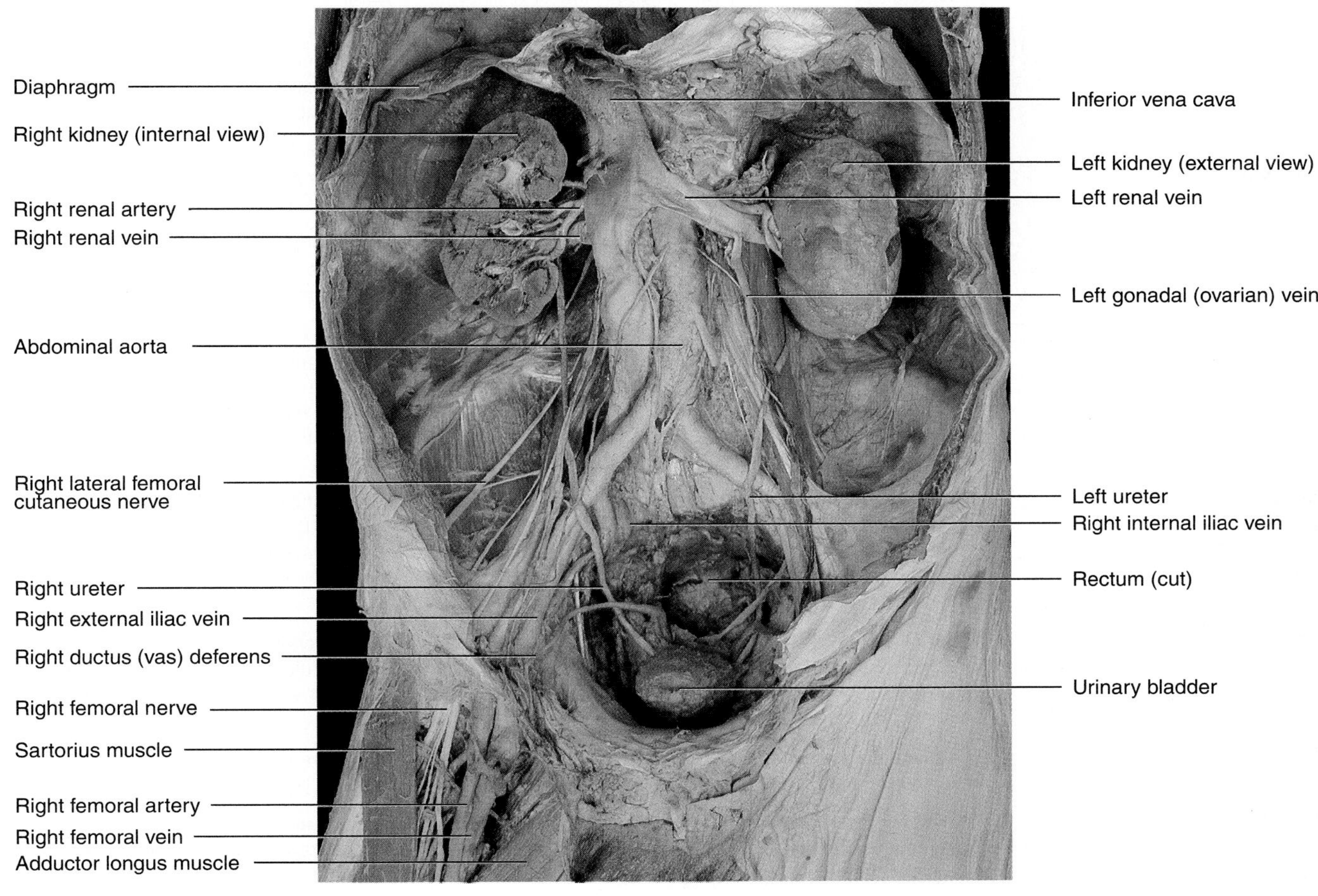

INFERIOR

(b) Anterior view

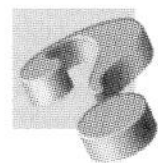

Which vein returns blood from the abdominopelvic viscera to the heart?

Exhibit 14.11 Veins of the Abdomen and Pelvis (Figure 14.14)

Objective

► Identify the principal veins that drain the abdomen and pelvis.

Blood from the abdominal and pelvic viscera and abdominal wall returns to the heart via the **inferior vena cava.** Many small veins enter the inferior vena cava. Most carry return flow from parietal branches of the abdominal aorta, and their names correspond to the names of the arteries.

The inferior vena cava does not receive veins directly from the gastrointestinal tract, spleen, pancreas, and gallbladder. These organs pass their blood into a common vein, the **hepatic portal vein,** which delivers the blood to the liver. The hepatic portal vein is formed by the union of the superior mesenteric and splenic veins (see Figure 14.16). This special flow of venous blood is called the **hepatic portal circulation,** which is described shortly. After passing through the liver for processing, blood drains into the hepatic veins, which empty into the inferior vena cava.

✓ What structures are drained by the lumbar, gonadal, renal, suprarenal, inferior phrenic, and hepatic veins?

Vein	Description and Region Drained
Inferior vena cava (VĒ-na CĀ-va; *vena* = vein; *cava* = cavelike)	The **inferior vena cava** is formed by the union of two common iliac veins that drain the lower limbs, pelvis, and abdomen. The inferior vena cava extends superiorly through the abdomen and thorax to the right atrium.
Common iliac veins (IL-ē-ak; *iliac* = pertaining to the ilium)	The **common iliac veins** are formed by the union of the internal and external iliac veins anterior to the sacroiliac joint and represent the distal continuation of the inferior vena cava at their bifurcation. The right common iliac vein is much shorter than the left and is also more vertical. Generally, the common iliac veins drain the pelvis, external genitals, and lower limbs.
Internal iliac veins	The **internal (hypogastric) iliac veins** begin near the superior portion of the greater sciatic notch and run medial to their corresponding arteries. Generally, the veins drain the thigh, buttocks, external genitals, and pelvis.
External iliac veins	The **external iliac veins** are companions of the internal iliac arteries and begin at the inguinal ligaments as continuations of the femoral veins. They end anterior to the sacroiliac joint where they join with the internal iliac veins to form the common iliac veins. The external iliac veins drain the lower limbs, cremasteric muscle in males, and the abdominal wall.
Lumbar veins (LUM-bar; *lumbar* = loin)	A series of parallel **lumbar veins,** usually four on each side, drain blood from both sides of the posterior abdominal wall, vertebral canal, spinal cord, and meninges. The lumbar veins run horizontally with the lumbar arteries. They connect at right angles with the right and left **ascending lumbar veins,** which form the origin of the corresponding azygos or hemiazygos vein. The lumbar veins drain some blood into the ascending lumbars and then run to the inferior vena cava, where they release the remainder of the flow.
Gonadal veins (gō-NAD-al; *gono* = seed) **[testicular** (test-TIK-yū-lar) or **ovarian** (ō-VAR-ē-an)]	The **gonadal veins** ascend with the gonadal arteries along the posterior abdominal wall. In the male, the gonadal veins are called the testicular veins. The **testicular veins** drain the testes (the left testicular vein empties into the left renal vein, and the right testicular vein drains into the inferior vena cava). In the female, the gonadal veins are called the ovarian veins. The **ovarian veins** drain the ovaries (the left ovarian vein empties into the left renal vein, and the right ovarian vein drains into the inferior vena cava).
Renal veins (RĒ-nal; *renal* = kidney)	The **renal veins** are large and pass anterior to the renal arteries. The left renal vein is longer than the right renal vein and passes anterior to the abdominal aorta. It receives the left testicular (or ovarian), left inferior phrenic, and usually left suprarenal veins. The right renal vein empties into the inferior vena cava posterior to the duodenum. The renal veins drain the kidneys.
Suprarenal veins (sū′-pra-RĒ-nal; *supra* = above)	The **suprarenal veins** drain the adrenal (suprarenal) glands (the left suprarenal vein empties into the left renal vein, and the right suprarenal vein empties into the inferior vena cava).
Inferior phrenic veins (FREN-ik; *phrenic* = diaphragm)	The **inferior phrenic veins** drain the diaphragm (the left inferior phrenic vein usually sends one tributary to the left suprarenal vein, which empties into the left renal vein, and another tributary that empties into the inferior vena cava; the right inferior phrenic vein empties into the inferior vena cava).
Hepatic veins (he-PAT-ik; *hepatic* = liver)	The **hepatic veins** drain the liver.

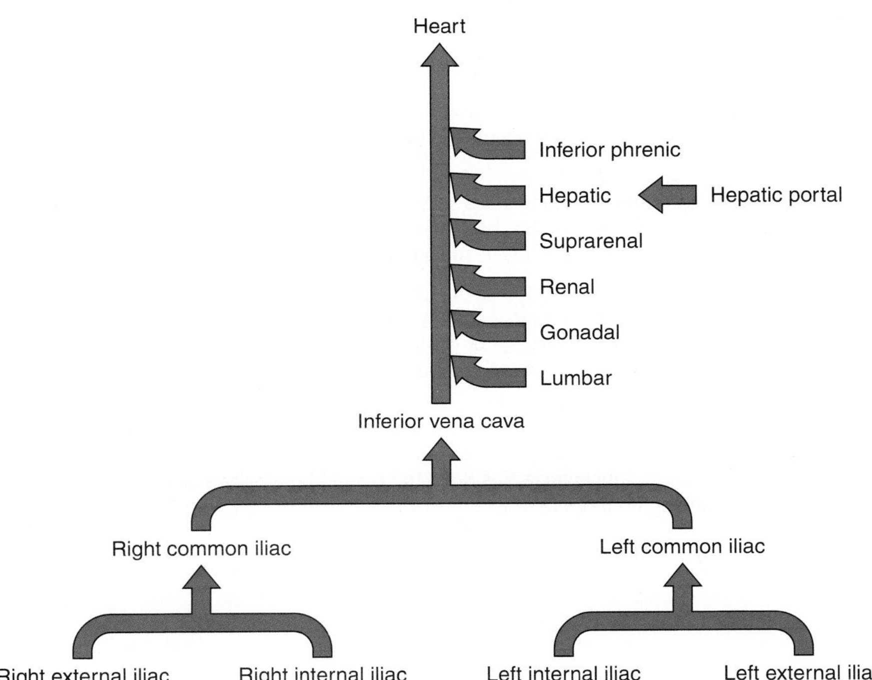
SCHEME OF DRAINAGE
Heart
Inferior phrenic
Hepatic
Hepatic portal
Suprarenal
Renal
Gonadal
Lumbar
Inferior vena cava
Right common iliac
Left common iliac
Right external iliac
Right internal iliac
Left internal iliac
Left external iliac

Exhibit 14.12 Veins of the Lower Limbs (Figure 14.15)

Objective

► Identify the principal superficial and deep veins that drain the lower limbs.

As with the upper limbs, blood from the lower limbs is drained by both **superficial** and **deep veins.** The superficial veins often anastomose with each other and with deep veins along their length. Deep veins, for the most part, have the same names as corresponding arteries. All veins of the lower limbs have valves, which are more numerous than in veins of the upper limbs.

✓ Why are the great saphenous veins clinically important?

Vein	Description and Region Drained
Superficial Veins	
Great saphenous veins (sa-FĒ-nus; *saphenes* = clearly visible)	The **great (long) saphenous veins,** the longest veins in the body, ascend from the foot to the groin in the subcutaneous layer. They begin at the medial end of the dorsal venous arches of the foot. The **dorsal venous arches** (VĒ-nus) are networks of veins on the dorsum of the foot formed by the **dorsal digital veins,** which collect blood from the toes, and then unite in pairs to form the **dorsal metatarsal veins,** which parallel the metatarsals. As the dorsal metatarsal veins approach the foot, they combine to form the dorsal venous arches. The great saphenous veins pass anterior to the medial malleolus of the tibia and then superiorly along the medial aspect of the leg and thigh just deep to the skin. They receive tributaries from superficial tissues and connect with the deep veins as well. They empty into the femoral veins at the groin. Generally, the great saphenous veins drain mainly the medial side of the leg and thigh, the groin, external genitals, and abdominal wall. Along their length, the great saphenous veins have from 10 to 20 valves, with more located in the leg than the thigh. These veins are more likely to be subject to varicosities than other veins in the lower limbs because they must support a long column of blood and are not well supported by skeletal muscles. The great saphenous veins are often used for prolonged administration of intravenous fluids. This is particularly important in very young children and in patients of any age who are in shock and whose veins are collapsed. The great saphenous veins are also often used as a source of vascular grafts, especially for coronary bypass surgery. In the procedure, the vein is removed and then reversed so that the valves do not obstruct the flow of blood.
Small saphenous veins	The **small (short) saphenous veins** begin at the lateral aspect of the dorsal venous arches of the foot. They pass posterior to the lateral malleolus of the fibula and ascend deep to the skin along the posterior aspect of the leg. They empty into the popliteal veins in the popliteal fossa, posterior to the knee. Along their length, the small saphenous veins have from 9 to 12 valves. The small saphenous veins drain the foot and posterior aspect of the leg. They may communicate with the great saphenous veins in the proximal thigh.

Vein	Description and Region Drained
Deep Veins	
Posterior tibial veins (TIB-ē-al)	The **plantar digital veins** on the plantar surfaces of the toes unite to form the **plantar metatarsal veins,** which parallel the metatarsals. They unite to form the **deep plantar venous arches.** From each arch emerges the **medial** and **lateral plantar veins.** The paired **posterior tibial veins,** which sometimes unite to form a single vessel, are formed by the medial and lateral plantar veins, posterior to the medial malleolus of the tibia. They accompany the posterior tibial artery through the leg. They ascend deep to the muscles in the posterior aspect of the leg and drain the foot and posterior compartment muscles. About two-thirds of the way up the leg, the posterior tibial veins drain blood from the **peroneal veins** (per′-ō-NĒ-al; *perone* = fibula), which drain the lateral and posterior leg muscles. The posterior tibial veins unite with the anterior tibial veins just inferior to the popliteal fossa to form the popliteal veins.
Anterior tibial veins	The paired **anterior tibial veins** arise in the dorsal venous arch and accompany the anterior tibial artery. They ascend in the interosseous membrane between the tibia and fibula and unite with the posterior tibial veins to form the popliteal vein. The anterior tibial veins drain the ankle joint, knee joint, tibiofibular joint, and anterior portion of the leg.
Popliteal veins (pop′-li-TĒ-al; *popliteus* = hollow behind knee)	The **popliteal veins** are formed by the union of the anterior and posterior tibial veins. The popliteal veins also receive blood from the small saphenous veins and tributaries that correspond to branches of the popliteal artery. The popliteal veins drain the knee joint and the skin, muscles, and bones of portions of the calf and thigh around the knee joint.
Femoral veins (FEM-o-ral)	The **femoral veins** accompany the femoral arteries and are the continuations of the popliteal veins just superior to the knee. The femoral veins extend up the posterior surface of the thighs and drain the muscles of the thighs, femurs, external genitals, and superficial lymph nodes. The largest tributaries of the femoral veins are the **deep femoral veins.** Just before penetrating the abdominal wall, the femoral veins receive the deep femoral veins and the great saphenous veins. The veins formed from this union penetrate the body wall and enter the pelvic cavity. Here they are known as the **external iliac veins.**

(continues)

Exhibit 14.12 Veins of the Lower Limbs (Figure 14.15) (continued)

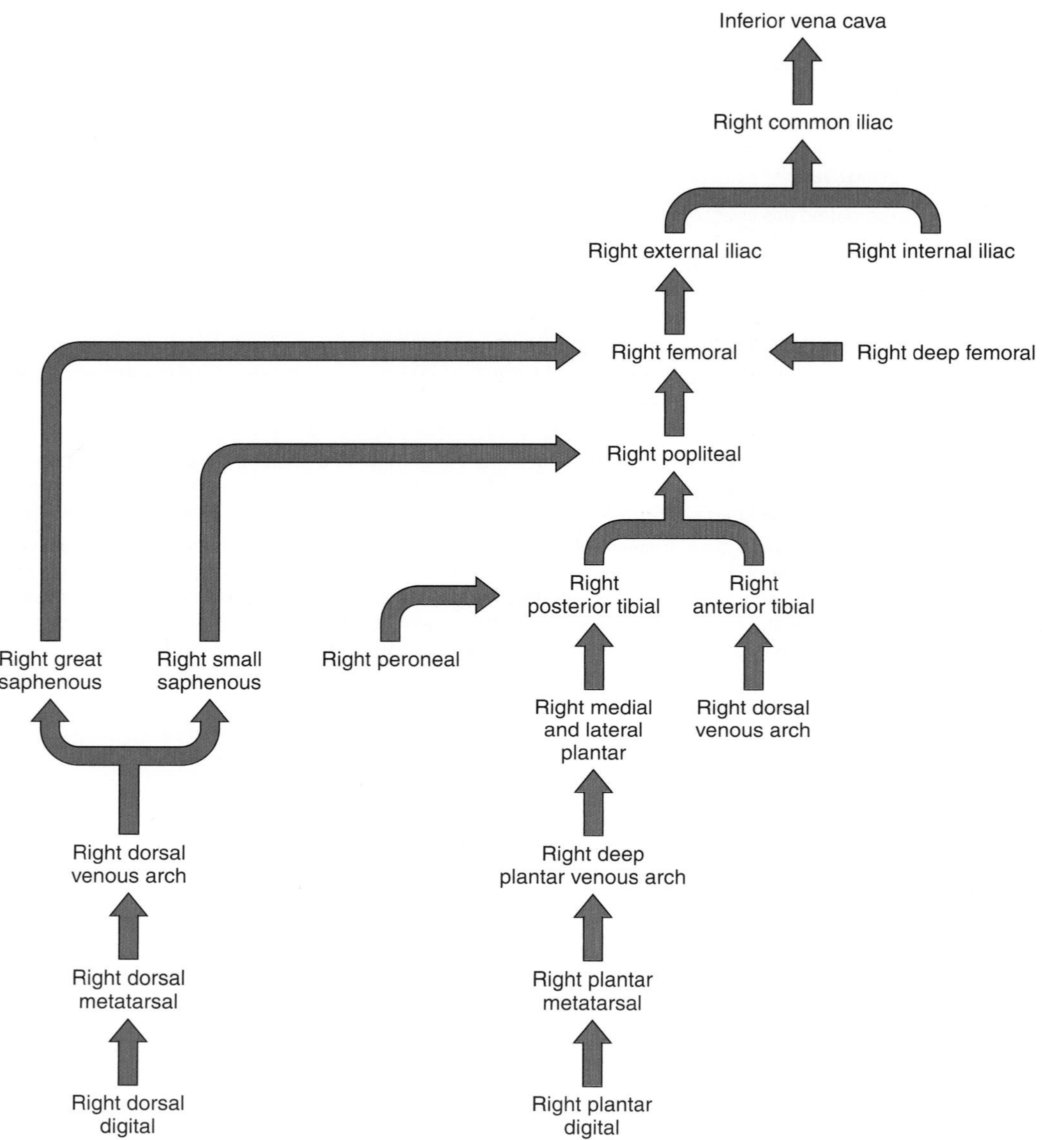

Figure 14.15 / Principal veins of the pelvis and lower limbs. (See Tortora, *A Photographic Atlas of the Human Body,* Figure 6.22.)

Deep veins usually bear the names of their companion arteries.

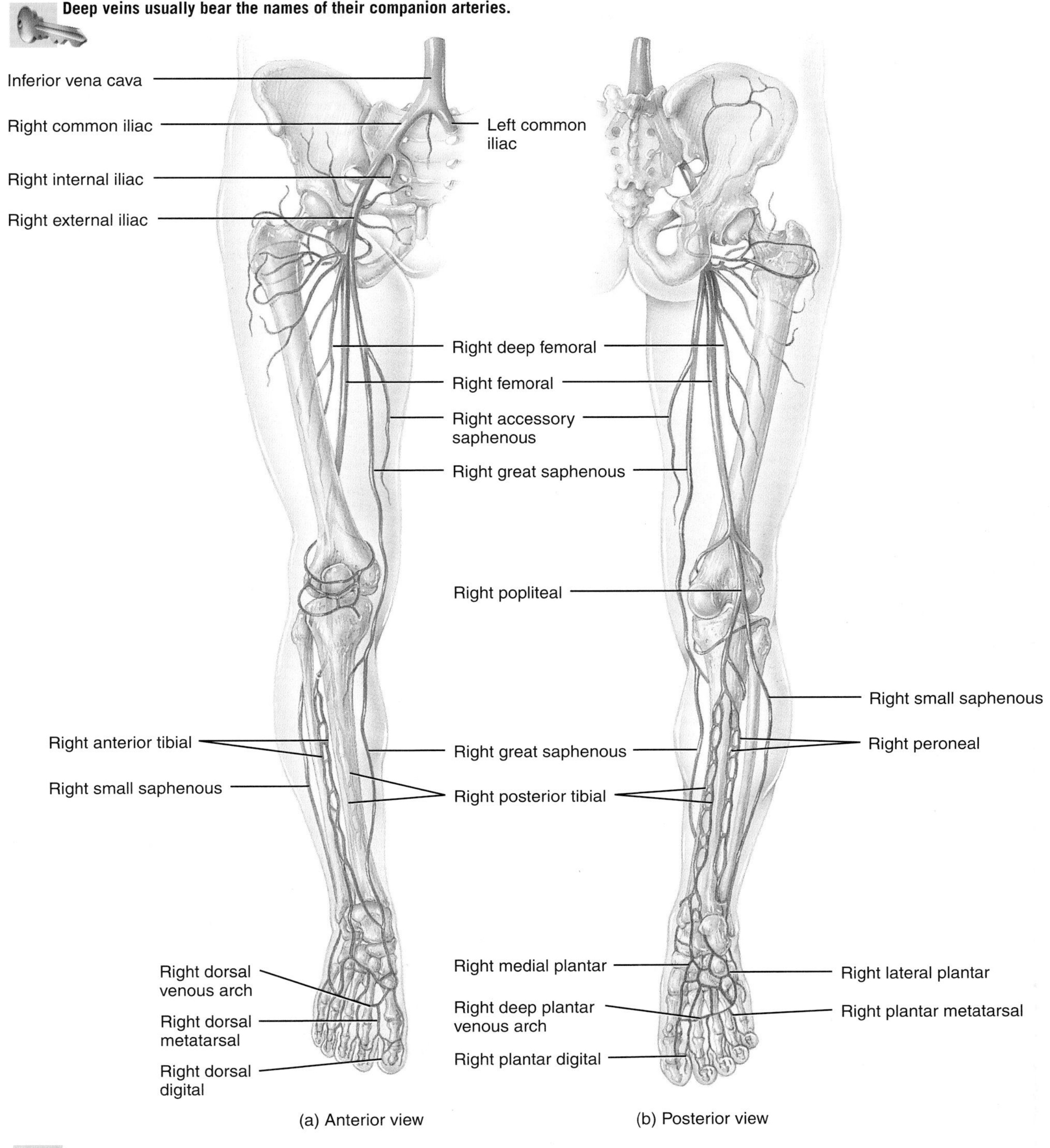

Which veins of the lower limb are superficial?

The Hepatic Portal Circulation

Objective

- Identify the blood vessels and route of the hepatic portal circulation.

We noted earlier that the two principal circulatory routes are the systemic and pulmonary circulations. Before looking at the pulmonary circulation, we will examine the hepatic portal circulation, a subdivision of the systemic circulation. The **hepatic portal circulation** (*hepat-* = liver) detours venous blood from the gastrointestinal organs and spleen through the liver before it returns to the heart (Figure 14.16). A *portal system* carries blood between two capillary networks, from one location in the body to another, without passing through the heart. In this case, blood flows from capillaries of the gastrointestinal tract to sinu-

Figure 14.16 / Hepatic portal circulation. A schematic diagram of blood flow through the liver, including arterial circulation, is shown in (b); deoxygenated blood is indicated in blue, oxygenated blood in red.

The hepatic portal circulation delivers venous blood from the organs of the gastrointestinal tract and spleen to the liver.

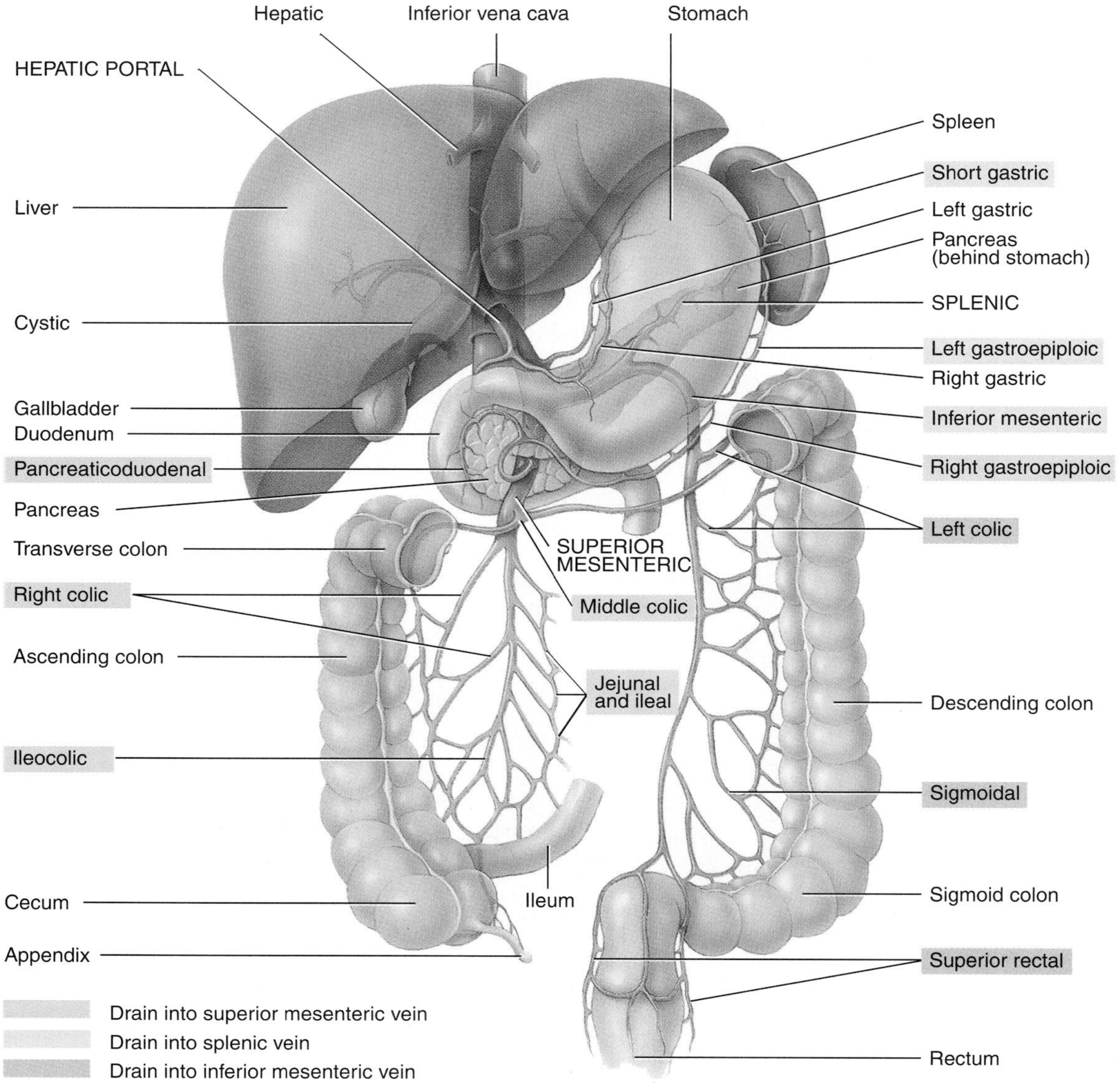

(a) Anterior view of veins draining into the hepatic portal vein

soids of the liver. After a meal, hepatic portal blood is rich in substances absorbed from the gastrointestinal tract. The liver stores some of them and modifies others before they pass into the general circulation. For example, the liver converts glucose into glycogen for storage, thereby helping to maintain homeostasis of blood glucose during a fast; the liver also detoxifies harmful substances such as alcohol that have been absorbed from the gastrointestinal tract and destroys bacteria by phagocytosis.

The **hepatic portal vein** is formed by the union of the superior mesenteric and splenic veins. The **superior mesenteric vein** drains blood from the small intestine and portions of the large intestine, stomach, and pancreas through the *jejunal, ileal, ileocolic, right colic, middle colic, pancreaticoduodenal,* and *right gastroepiploic veins.* The **splenic vein** drains blood from the stomach, pancreas, and portions of the large intestine through the *short gastric, left gastroepiploic, pancreatic,* and *inferior mesenteric veins.* The inferior mesenteric vein, which passes into the splenic vein, drains portions of the large intestine through the superior *rectal, sigmoidal,* and *left colic veins.* The *right* and *left gastric veins,* which open directly into the hepatic portal vein, drain the stomach. The *cystic vein,* which also opens into the hepatic portal vein, drains the gallbladder.

At the same time that the liver receives nutrient-rich deoxygenated blood via the hepatic portal system, it also receives oxygenated blood via the proper hepatic artery, a branch of the celiac trunk. Ultimately, all blood leaves the liver through the **hepatic veins,** which drain into the inferior vena cava.

The Pulmonary Circulation

Objective

- Identify the blood vessels and route of the pulmonary circulation.

The **pulmonary circulation** (*pulmo-* = lung) carries deoxygenated blood from the right ventricle to the air sacs (alveoli) within the lungs and returns oxygenated blood from the air sacs to the left atrium (Figure 14.17). The **pulmonary trunk** emerges from the right ventricle and passes superiorly, posteriorly, and to the left. It then divides into two branches: the **right pulmonary artery** to the right lung and the **left pulmonary artery** to the left lung. After birth, the pulmonary arteries are the only arteries that carry deoxygenated blood. On entering the lungs, the branches divide and subdivide until finally they form capillaries around the air sacs within the lungs. CO_2 passes from

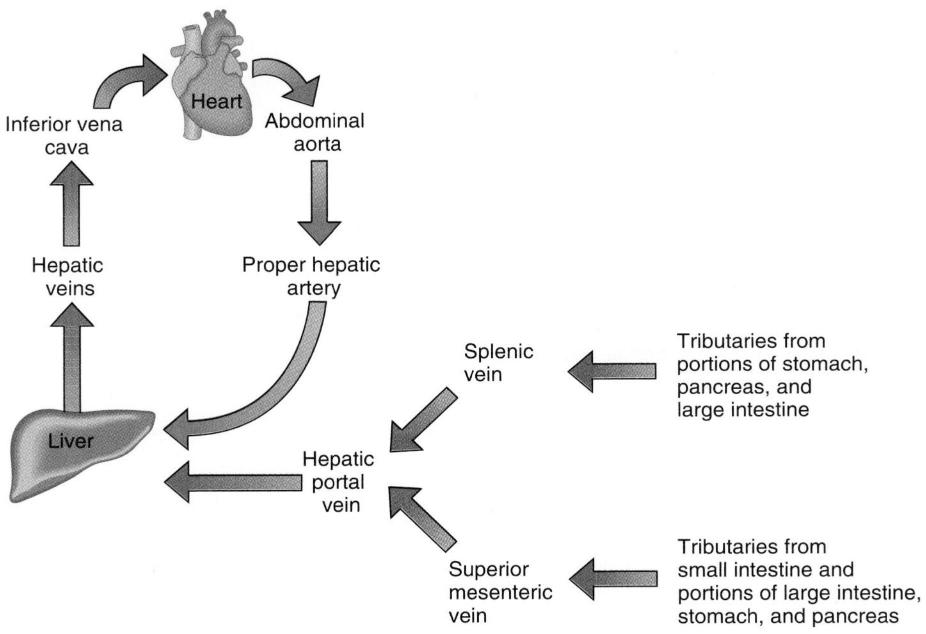

(b) Scheme of principal blood vessels of hepatic portal circulation and arterial supply and venous drainage of liver

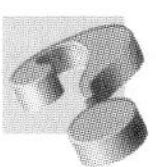

Which veins carry blood away from the liver?

Figure 14.17 / Pulmonary circulation.

The pulmonary circulation brings deoxygenated blood from the right ventricle to the lungs and returns oxygenated blood from the lungs to the left atrium.

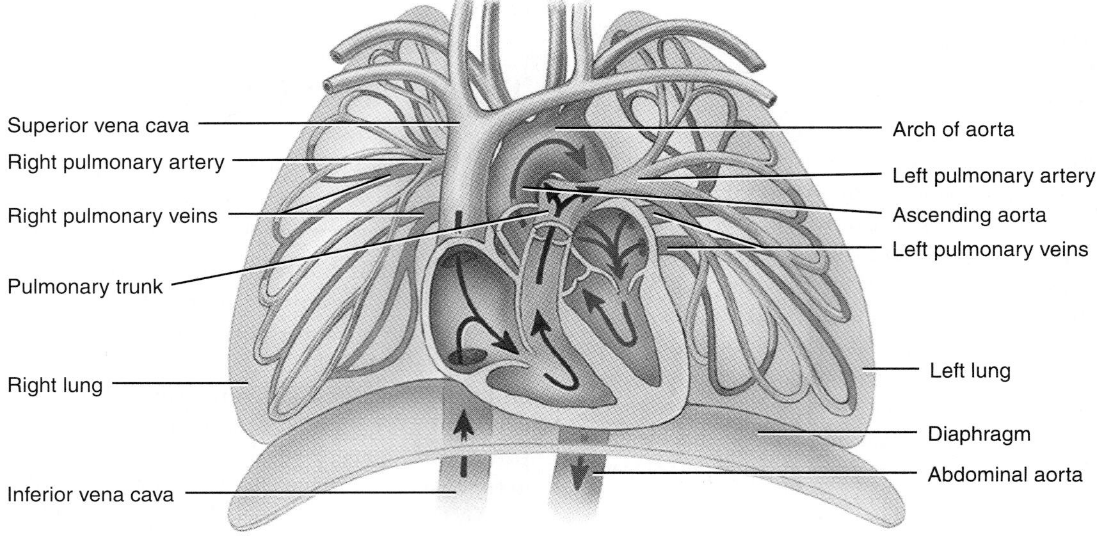

(a) Anterior view

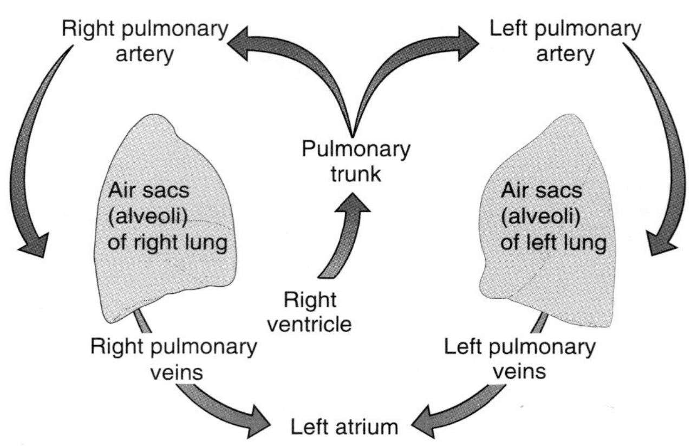

(b) Scheme of pulmonary circulation

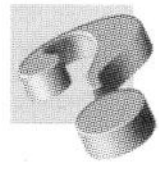

After birth, which are the only arteries that carry deoxygenated blood?

the blood into the air sacs and is exhaled. Inhaled O_2 passes from the air within the lungs into the blood. The pulmonary capillaries unite to form venules and eventually **pulmonary veins,** which exit the lungs and transport the oxygenated blood to the left atrium. Two left and two right pulmonary veins enter the left atrium. After birth, the pulmonary veins are the only veins that carry oxygenated blood. Contractions of the left ventricle then eject the oxygenated blood into the systemic circulation.

The pulmonary and systemic circulations are different in two important ways. First, blood in the pulmonary circulation need not be pumped as far as blood in the systemic circulation. Second, compared to systemic arteries, pulmonary arteries have larger diameters, thinner walls, and less elastic tissue; as a result, the resistance to pulmonary blood flow is very low, which means that less pressure is needed to move blood through the lungs. The peak systolic pressure in the right ventricle is about 20% of that in the left ventricle.

Because resistance in the pulmonary circulation is low, normal *pulmonary* capillary hydrostatic pressure, the principal force that moves fluid out of capillaries into interstitial fluid, is only 10 mm Hg. This compares to the average *systemic* capillary pressure of about 25 mm Hg. The relatively low capillary hydrostatic pressure tends to prevent pulmonary edema. However, if capillary blood pressure in the lungs increases (due to increased left atrial pressure as may occur in mitral valve stenosis) or capillary permeability increases (as may occur from bacterial toxins), edema may develop. Pulmonary edema reduces the rate of diffusion of oxygen and carbon dioxide and thus slows the exchange of respiratory gases in the lungs.

The Fetal Circulation

Objective

- Identify the blood vessels and route of the fetal circulation.

The circulatory system of a fetus, called the **fetal circulation,** differs from the postnatal (after birth) circulation because the lungs, kidneys, and gastrointestinal organs do not begin to function until birth. The fetus obtains its O_2 and nutrients by diffusion from the maternal blood and eliminates its CO_2 and wastes by diffusion into the maternal blood.

The exchange of materials between fetal and maternal circulations occurs through the **placenta** (pla-SEN-ta), which forms inside the mother's uterus and attaches to the umbilicus (navel) of the fetus by the **umbilical cord** (um-BIL-i-kal). The placenta communicates with the mother's cardiovascular system through many small blood vessels that emerge from the uterine wall. The umbilical cord contains blood vessels that branch into capillaries in the placenta. Wastes from the fetal blood diffuse out of the capillaries, into spaces containing maternal blood (intervillous spaces) in the placenta, and finally into the mother's uterine veins (see Figure 27.8 on page 836). Nutrients travel the opposite route—from the maternal blood vessels to the intervillous spaces to the fetal capillaries. Normally, there is no direct mixing of maternal and fetal blood because all exchanges occur by diffusion through capillary walls.

Blood passes from the fetus to the placenta via two **umbilical arteries** (Figure 14.18). These branches of the internal iliac (hypogastric) arteries are within the umbilical cord. At the placenta, fetal blood picks up O_2 and nutrients and eliminates CO_2 and wastes. The oxygenated blood returns from the placenta via a single **umbilical vein.** This vein ascends to the liver of the fetus, where it divides into two branches. Whereas some blood flows through the branch that joins the hepatic portal vein and enters the liver, most of the blood flows into the second branch, the **ductus venosus** (DUK-tus ve-NŌ-sus), which drains into the inferior vena cava.

Deoxygenated blood returning from the inferior regions mingles with oxygenated blood from the ductus venosus in the inferior vena cava. This mixed blood then enters the right atrium. Deoxygenated blood returning from the superior regions of the fetus enters the superior vena cava and passes into the right atrium.

Most of the fetal blood does not pass from the right ventricle to the lungs, as it does in postnatal circulation, because an opening called the **foramen ovale** (fō-RĀ-men ō-VAL-ē) exists in the septum between the right and left atria. About one third of the blood passes through the foramen ovale directly into the systemic circulation. The blood that does pass into the right ventricle is pumped into the pulmonary trunk, but little of this blood reaches the nonfunctioning fetal lungs. Instead, most is sent through the **ductus arteriosus** (ar-tē-rē-Ō-sus), a vessel that connects the pulmonary trunk with the aorta, so that most blood bypasses the fetal lungs. The blood in the aorta is carried to all fetal tissues through the systemic circulation. When the common iliac arteries branch into the external and internal iliacs, part of the blood flows into the internal iliacs, into the umbilical arteries, and back to the placenta for another exchange of materials. The only fetal vessel that carries fully oxygenated blood is the umbilical vein.

After birth, when pulmonary (lung), renal, and digestive functions begin, the following vascular changes occur (Figure 14.18b).

1. When the umbilical cord is tied off, blood no longer flows through the umbilical arteries, they fill with connective tissue, and the distal portions of the umbilical arteries become fibrous cords called the **medial umbilical ligaments.** Although the arteries are closed functionally only a few minutes after birth, complete obliteration of the lumens may take 2–3 months.
2. The umbilical vein collapses but remains as the **ligamentum teres (round ligament),** a structure that attaches the umbilicus to the liver.
3. The ductus venosus collapses but remains as the **ligamentum venosum,** a fibrous cord on the inferior surface of the liver.
4. The placenta is expelled as the **"afterbirth."**
5. The foramen ovale normally closes shortly after birth to become the **fossa ovalis,** a depression in the interatrial sep-

Figure 14.18 / Fetal circulation and changes at birth. The boxes between parts (a) and (b) describe the fate of certain fetal structures once postnatal circulation is established.

The lungs and gastrointestinal organs do not begin to function until birth.

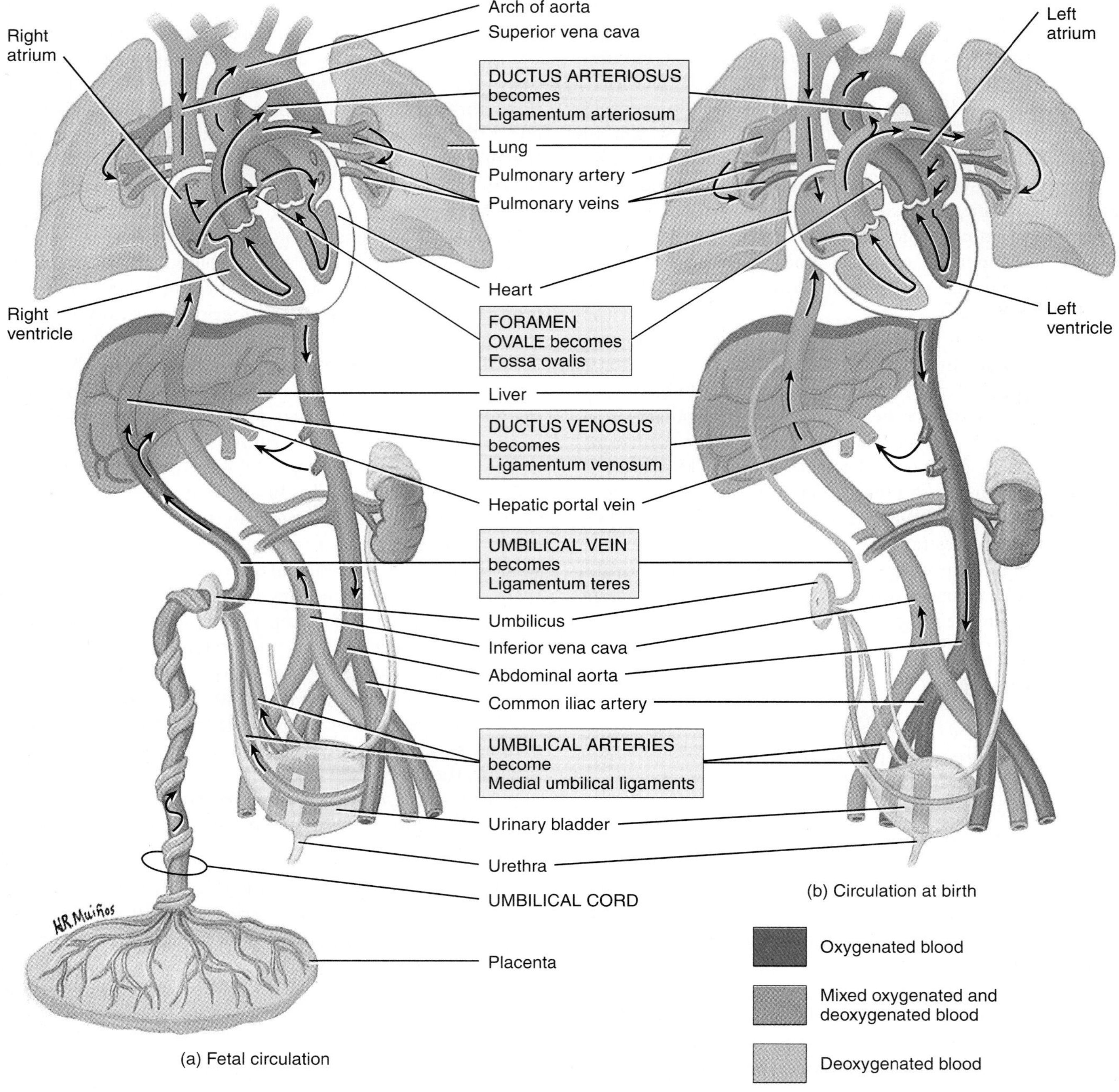

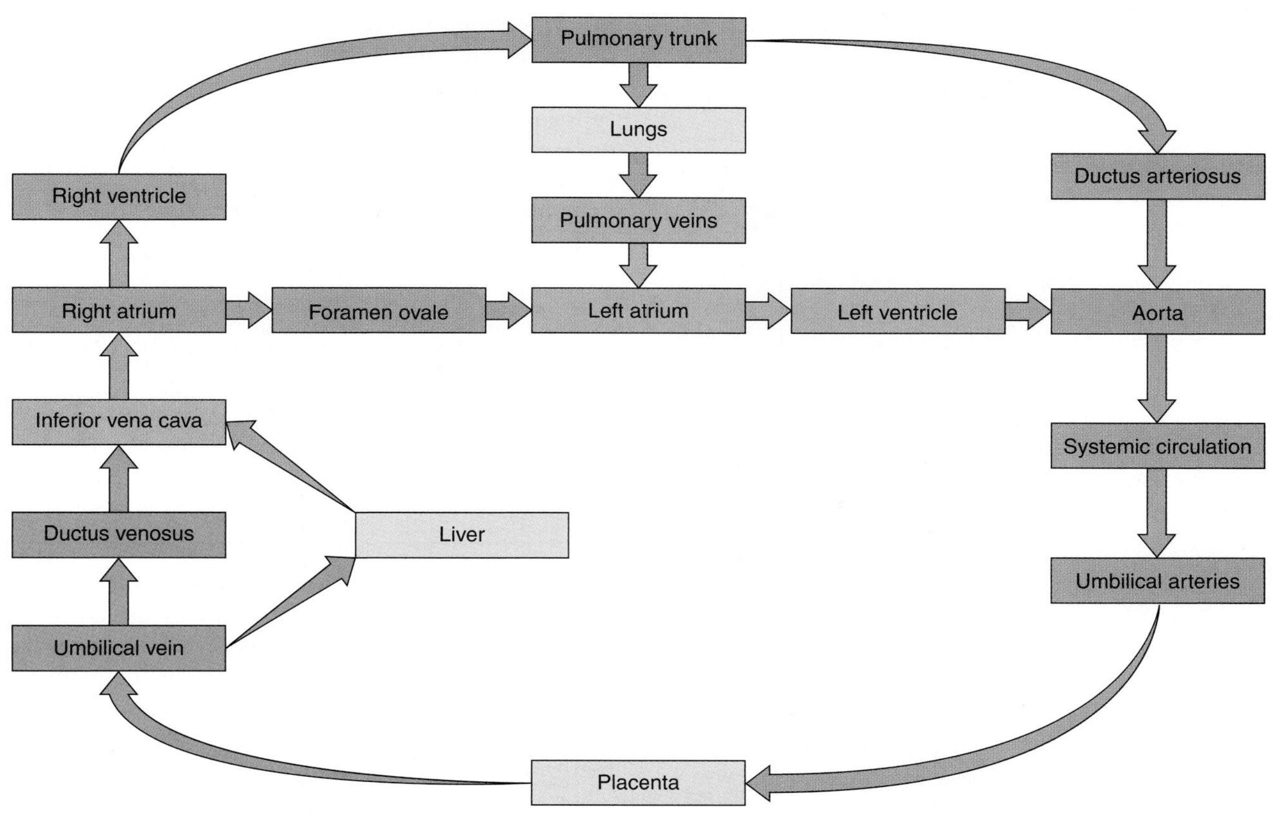

(c) Scheme of fetal circulation

Through which structure does the exchange of materials between mother and fetus occur?

tum. When an infant takes its first breath, the lungs expand and blood flow to the lungs increases. Blood returning from the lungs to the heart increases pressure in the left atrium. This closes the foramen ovale by pushing the valve that guards it against the interatrial septum. Permanent closure occurs in about a year.

6. The ductus arteriosus closes by vasoconstriction almost immediately after birth and becomes the **ligamentum arteriosum.** Complete anatomical obliteration of the lumen takes 1–3 months.

✓ Prepare a diagram to show the hepatic portal circulation. Why is this route important?

✓ Prepare a diagram to show the route of the pulmonary circulation.

✓ Discuss the anatomy and physiology of the fetal circulation. Indicate the function of the umbilical arteries, umbilical vein, ductus venosus, foramen ovale, and ductus arteriosus.

DEVELOPMENTAL ANATOMY OF BLOOD VESSELS AND BLOOD

Objective

- Describe the development of blood vessels and blood.

The human yolk sac has little yolk to nourish the developing embryo. Blood and blood vessel formation start as early as 15–16 days in the **mesoderm** of the yolk sac, chorion, and body stalk.

Blood vessels develop from isolated masses and cords of mesenchyme in the mesoderm called **blood islands** (Figure 14.19). Spaces soon appear in the islands and become the lumens of the blood vessels. Some of the mesenchymal cells immediately around the spaces give rise to the *endothelial lining of the blood vessels.* Mesenchyme around the endothelium forms the *tunics* (interna, media, externa) of the larger blood vessels. Growth and

Figure 14.19 / Development of blood vessels and blood cells from blood islands.

Blood vessel development begins in the embryo on about day 15 or 16.

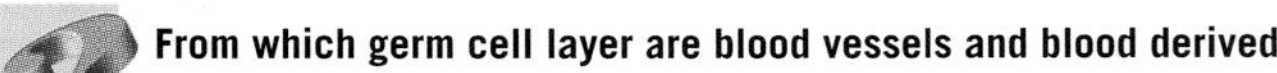

From which germ cell layer are blood vessels and blood derived?

fusion of blood islands form an extensive network of blood vessels throughout the embryo.

Blood plasma and *blood cells* are produced by the endothelial cells and appear in the blood vessels of the yolk sac and allantois quite early. Blood formation in the embryo itself begins at about the second month in the liver and spleen, a little later in red bone marrow, and much later in lymph nodes.

✓ Compare the origin of blood vessels and blood.

AGING AND THE CARDIOVASCULAR SYSTEM

Objective

- Explain the effects of aging on the cardiovascular system.

General changes in the cardiovascular system associated with aging include decreased compliance of the aorta, reduction in cardiac muscle fiber size, progressive loss of cardiac muscular strength, reduced cardiac output, a decline in maximum heart rate, and an increase in systolic blood pressure. Total blood cholesterol tends to increase with age, as does low-density lipoprotein (LDL); high-density lipoprotein (HDL) tends to decrease. There is an increase in the incidence of coronary artery disease (CAD), the major cause of heart disease and death in older Americans. Congestive heart failure, a set of symptoms associated with impaired pumping of the heart, is also prevalent in older individuals. Changes in blood vessels that serve brain tissue—for example, atherosclerosis—reduce nourishment to the brain and result in the malfunction or death of brain cells. By age 80, cerebral blood flow is 20% less and renal blood flow is 50% less than in the same person at age 30.

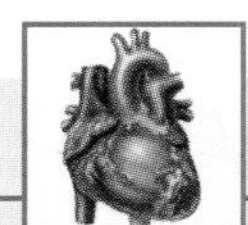

APPLICATIONS TO HEALTH

Hypertension

Hypertension, or persistently high blood pressure, is defined as systolic blood pressure of 140 mm Hg or greater and diastolic blood pressure of 90 mm Hg or greater. Recall that a blood pressure of 120/80 is normal and desirable in a healthy adult. In industrialized societies, hypertension is the most common disorder affecting the heart and blood vessels; it is a major cause of heart failure, kidney disease, and stroke. The classification system adopted in 1997 ranks blood pressure values for adults as follows:

Optimal	Systolic less than 120; diastolic less than 80
Normal	Systolic less than 130; diastolic less than 85
High-normal	Systolic 130–139; diastolic 85–89
Hypertension	Systolic 140 or greater; diastolic 90 or greater
Stage 1	Systolic 140–159; diastolic 90–99
Stage 2	Systolic 160–179; diastolic 100–109
Stage 3	Systolic 180 or greater; diastolic 110 or greater

Amazingly, 90–95% of all cases of hypertension are classified as **primary hypertension,** which is a persistently elevated blood pressure that cannot be attributed to any identifiable cause. The remaining 5–10% of cases are **secondary hypertension,** which has an identifiable underlying cause, such as kidney disease.

Damaging Effects of Untreated Hypertension

Hypertension is a major risk factor for the number one (heart disease) and number three (stroke) causes of death in the United States. High blood pressure is known as the "silent killer" because it can cause considerable damage to the blood vessels, heart, brain, and kidneys before it causes pain or other noticeable symptoms. In blood vessels, hypertension causes thickening of the tunica media, accelerates development of atherosclerosis and coronary artery disease (CAD), and increases systemic vascular resistance. In the heart, hypertension forces the ventricles to work harder to eject blood. The response to an increased workload is hypertrophy of the myocardium, especially in the wall of the left ventricle. If the coronary blood flow cannot meet the increased demand for oxygen, angina pectoris or even myocardial infarction may result. When myocardial hypertrophy can no longer compensate for the increased afterload, the left ventricle becomes dilated and weakened. Because arteries in the brain are usually less protected by surrounding tissues than are the major arteries in other parts of the body, prolonged hypertension can eventually cause them to rupture, and a stroke occurs due to a brain hemorrhage. Hypertension also damages kidney arterioles, causing them to thicken, which narrows the lumen; because the blood supply to the kidneys is thereby reduced, the kidneys secrete more renin, which elevates the blood pressure even more.

Lifestyle Changes to Reduce Hypertension

Although several categories of drugs (described next) can reduce elevated blood pressure, the following lifestyle changes are also effective in managing hypertension.

- *Lose weight.* This is the best treatment for high blood pressure short of using drugs. Loss of even a few pounds helps reduce blood pressure in overweight hypertensive individuals.
- *Limit alcohol intake.* Drinking in moderation may lower the risk for coronary heart disease, mainly among males over 45 and females over 55. Moderation is defined as no more than one 12-oz beer per day for females and no more than two 12-oz beers per day for males.
- *Exercise.* Becoming more physically fit by engaging in moderate activity (such as brisk walking) several times a week for 30–45 minutes can lower systolic blood pressure by about 10 mm Hg.
- *Reduce intake of sodium (salt).* Roughly half the people with hypertension are "salt sensitive." For them, a high-salt diet appears to promote hypertension, and a low-salt diet can lower their blood pressure.
- *Maintain recommended dietary intake of potassium, calcium, and magnesium.* Higher levels of potassium, calcium, and magnesium in the diet are associated with a lower risk of hypertension.
- *Don't smoke.* Smoking has devastating effects on the heart and can augment the damaging effects of high blood pressure by promoting vasoconstriction.
- *Manage stress.* Various meditation and biofeedback techniques help some people reduce high blood pressure. These methods may work by decreasing the daily release of epinephrine and norepinephrine by the adrenal medulla.

Drug Treatment of Hypertension

Drugs having several different mechanisms of action are effective in lowering blood pressure. Many people are successfully treated with *diuretics*, agents that decrease blood pressure by decreasing blood volume because they increase elimination of water and salt in the urine. *ACE (angiotensin converting enzyme) inhibitors* block formation of angiotensin II and thereby promote vasodilation and decrease the liberation of aldosterone. *Beta blockers* reduce blood pressure by inhibiting the secretion of renin and by decreasing heart rate and contractility. *Vasodilators* relax the smooth muscle in arterial walls, causing vasodilation and lowering blood pressure by lowering systemic vascular resistance. An important category of vasodilators are the *calcium channel blockers*, which slow the inflow of Ca^{2+} into vascular smooth muscle cells. They reduce the heart's work load by slowing Ca^{2+} entry into myocardial fibers, thereby decreasing the force of myocardial contraction.

KEY MEDICAL TERMS ASSOCIATED WITH BLOOD VESSELS

Aneurysm (AN-yū-rizm) A thin, weakened section of the wall of an artery or a vein that bulges outward, forming a balloonlike sac. Common causes are atherosclerosis, syphilis, congenital blood vessel defects, and trauma. If untreated, the aneurysm enlarges and the blood vessel wall becomes so thin that it bursts. The result is massive hemorrhage with shock, severe pain, stroke, or death.

Angiogenesis (an′-jē-ō-JEN-e-sis) Formation of new blood vessels.

Aortography (ā′-or-TOG-ra-fē) X-ray examination of the aorta and its main branches after injection of a radiopaque dye.

Arteritis (ar′-te-RĪ-tis; *-itis* = inflammation of) Inflammation of an artery, probably due to an autoimmune response.

Carotid endarterectomy (ka-ROT-id end′-ar-ter-EK-tō-mē) The removal of atherosclerotic plaque from the carotid artery to restore greater blood flow to the brain.

Claudication (klaw′-di-KĀ-shun) Pain and lameness or limping caused by defective circulation of the blood in the vessels of the limbs.

Deep venous thrombosis The presence of a thrombus (blood clot) in a deep vein of the lower limbs. It may lead to (1) pulmonary embolism, if the thrombus dislodges and then lodges within the pulmonary arterial blood flow, and (2) postphlebitic syndrome, which consists of edema, pain, and skin changes due to destruction of venous valves.

Hypotension (hī-pō-TEN-shun) Low blood pressure; most commonly used to describe an acute drop in blood pressure, as occurs during excessive blood loss.

Normotensive (nor′-mō-TEN-siv) Characterized by normal blood pressure.

Occlusion (ō-KLŪ-shun) The closure or obstruction of the lumen of a structure such as a blood vessel. An example is an atherosclerotic plaque in an artery.

Orthostatic hypotension (or′-thō-STAT-ik; *ortho-* = straight; *-static* = causing to stand) An excessive lowering of systemic blood pressure when a person assumes an erect or semierect posture; it is usually a sign of a disease. May be caused by excessive fluid loss, certain drugs, and cardiovascular or neurogenic factors. Also called **postural hypotension.**

Phlebitis (fle-BĪ-tis; *phleb-* = vein) Inflammation of a vein, often in a leg.

Raynaud's disease (rā-NŌZ) A vascular disorder, primarily of females, characterized by bilateral attacks of ischemia, usually of the fingers and toes, in which the skin becomes pale and exhibits burning and pain. It is brought on by cold temperatures or emotional stimuli.

Thrombectomy (throm-BEK-tō-mē; *thrombo-* = clot) An operation to remove a blood clot from a blood vessel.

Thrombophlebitis (throm′-bō-fle-BĪ-tis) Inflammation of a vein involving clot formation. Superficial thrombophlebitis occurs in veins under the skin, especially in the calf.

White coat (office) hypertension A stress-induced syndrome found in patients who have elevated blood pressure when being examined by health-care personnel, but otherwise have normal blood pressure.

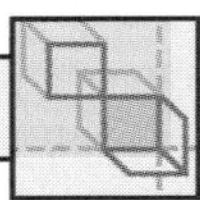

STUDY OUTLINE

Anatomy of Blood Vessels (p. 415)

1. Arteries carry blood away from the heart. The wall of an artery consists of a tunica interna, a tunica media (which maintains elasticity and contractility), and a tunica externa.
2. Large arteries are termed elastic (conducting) arteries, and medium-sized arteries are called muscular (distributing) arteries.
3. Many arteries anastomose: The distal ends of two or more vessels unite. An alternate blood route from an anastomosis is called collateral circulation. Arteries that do not anastomose are called end arteries.
4. Arterioles are small arteries that deliver blood to capillaries.
5. Through constriction and dilation, arterioles assume a key role in regulating blood flow from arteries into capillaries and in altering arterial blood pressure.
6. Capillaries are microscopic blood vessels through which materials are exchanged between blood and tissue cells; some capillaries are continuous, whereas others are fenestrated.
7. Capillaries branch to form an extensive network throughout a tissue. This network increases the surface area available for the exchange of materials between the blood and the body tissue; it also allows rapid exchange of large quantities of materials.
8. Precapillary sphincters regulate blood flow through capillaries.
9. Microscopic blood vessels in the liver are called sinusoids.
10. Venules are small vessels that form from the merging capillaries; venules merge to form veins.
11. Veins consist of the same three tunics as arteries but have a thinner tunica interna and media. The lumen of a vein is also larger than that of a comparable artery.
12. Veins contain valves to prevent backflow of blood.
13. Weak valves can lead to varicose veins.
14. Vascular (venous) sinuses are veins with very thin walls.
15. Systemic veins are collectively called blood reservoirs because they hold a large volume of blood. If the need arises, this blood can be shifted into other blood vessels through vasoconstriction of veins.
16. The principal blood reservoirs are the veins of the abdominal organs (liver and spleen) and skin.

Circulatory Routes (p. 420)

1. The two basic postnatal circulatory routes are the systemic and pulmonary circulations.
2. Among the subdivisions of the systemic circulation are the coronary (cardiac) and the hepatic portal circulations.
3. Fetal circulation exists only in the fetus.
4. The systemic circulation carries oxygenated blood from the left ventricle through the aorta to all parts of the body (including some lung tissue, but *not* the air sacs of the lungs) and returns the deoxygenated blood to the right atrium.
5. The aorta is divided into the ascending aorta, the arch of the aorta, and the descending aorta. Each section gives off arteries that branch to supply the whole body.
6. Blood returns to the heart through the systemic veins. All veins of the systemic circulation drain into the superior or inferior venae cavae or the coronary sinus; these, in turn, empty into the right atrium.
7. The principal blood vessels of the systemic circulation may be reviewed in Exhibits 14.1–14.12.
8. The hepatic portal circulation detours venous blood from the gastrointestinal organs and spleen and directs it into the hepatic portal vein of the liver before it is returned to the heart. It enables the liver to utilize nutrients and detoxify harmful substances in the blood.
9. The pulmonary circulation takes deoxygenated blood from the right ventricle to the alveoli within the lungs and returns oxygenated blood from the alveoli to the left atrium. It allows blood to be oxygenated for systemic circulation.
10. The fetal circulation involves the exchange of materials between fetus and mother.
11. The fetus derives O_2 and nutrients and eliminates CO_2 and wastes through the maternal blood supply via the placenta.
12. At birth, when pulmonary (lung), digestive, and liver functions begin, the special structures of fetal circulation are no longer needed.

Developmental Anatomy of Blood Vessels and Blood (p. 467)

1. Blood vessels develop from isolated masses of mesenchyme in mesoderm called blood islands.
2. Blood is produced by the endothelium of blood vessels.

Aging and the Cardiovascular System (p. 468)

1. General changes associated with aging include reduced elasticity of blood vessels, reduction in cardiac muscle size, reduced cardiac output, and increased systolic blood pressure.
2. The incidence of coronary artery disease (CAD), congestive heart failure (CHF), and atherosclerosis increases with age.

SELF-QUIZ QUESTIONS

Choose the one best answer to the following questions.

1. Which statement best describes arteries? (a) All carry oxygenated blood to the heart. (b) All contain valves to prevent the backflow of blood. (c) All carry blood away from the heart. (d) Only large arteries are lined with endothelium. (e) All branch from the descending aorta.

2. Which statement is true of veins? (a) Their tunica interna is thicker than in arteries. (b) Their tunica externa is thicker than in arteries. (c) Most veins in the limbs have valves. (d) They always carry deoxygenated blood. (e) All ultimately empty into the inferior vena cava.

3. Put the following vessels in the correct order to trace the route of a drop of blood moving from the right side of the heart to the left side of the heart. (1) pulmonary vein, (2) pulmonary artery, (3) arterioles, (4) venules, (5) capillaries.
   a. 1,4,5,3,2 **b.** 1,3,5,4,2 **c.** 2,4,5,3,1 **d.** 2,5,4,1,3 **e.** 2,3,5,4,1

4. Which of the following is a muscular (distributing) artery? (a) radial, (b) subclavian, (c) axillary, (d) common iliac, (e) aorta.

5. In fetal circulation, blood bypasses pulmonary circulation by passing through the (a) ductus venosus, (b) ductus arteriosus, (c) fossa ovalis, (d) all of the above, (e) b and c.

6. Which of the following arteries carry blood into the cerebral arterial circle (Circle of Willis)? (1) internal carotid, (2) external carotid, (3) posterior cerebral, (4) anterior cerebral, (5) basilar.
   a. 3 and 4 **b.** 1 and 5 **c.** 1, 2, and 5 **d.** 2, 3, and 4 **e.** 1, 4, and 5

7. Blood is supplied to the pelvic viscera by way of the (a) inferior vena cava, (b) superior mesenteric artery, (c) external iliac artery, (d) internal iliac artery, (e) femoral artery.

8. Put the following vessels in the correct order to trace the route of a drop of blood moving from the small intestine to the heart. (1) hepatic portal vein, (2) hepatic vein, (3) inferior vena cava, (4) superior mesenteric vein, (5) small vessels within the liver.
   a. 4,1,5,2,3 **b.** 3,1,2,5,4 **c.** 1,4,5,3,2 **d.** 4,2,1,5,3 **e.** 4,3,2,5,1

Complete the following.

9. The three branches of the celiac trunk are the ________, ________, and ________.

10. The three vessels that return deoxygenated blood to the right atrium are ________, ________, and ________.

11. Hypertension, a major cause of stroke, kidney disease and heart failure, is defined as systolic blood pressure of ________ mm Hg or greater and diastolic blood pressure of ________ mm Hg or greater.

12. The two common iliac veins unite to form the ________.

13. Blood from gastrointestinal organs and the spleen is transported to the liver via the ________ vein.

14. The union of the distal branches of two or more arteries supplying the same body region is called a (an) ________.

15. Most of the smooth muscle of arteries is in the tunica ________.

16. A decrease in the size of the lumen of a blood vessel due to contraction of smooth muscle is called ________.

17. Blood from all the dural venous sinuses eventually drains into the ________ veins, which descend the neck.

18. The ________ veins are the only veins that carry oxygenated blood after birth.

Are the following statements true or false?

19. Match the following.

____	**(a)** passes through cervical transverse foramina	**(1)** anterior tibial artery
____	**(b)** continues as left axillary artery	**(2)** esophageal artery
____	**(c)** gives rise to the left common carotid artery	**(3)** cephalic vein
____	**(d)** is a branch of the thoracic aorta	**(4)** sigmoid sinus
____	**(e)** abdominal vessel that branches to supply blood to the pancreas, small intestine and part of the large intestine	**(5)** coronary sinus
____	**(f)** becomes the dorsalis pedis artery at the ankle	**(6)** posterior tibial vein
____	**(g)** receives venous blood from the myocardium	**(7)** left subclavian artery
____	**(h)** is located in the arm	**(8)** arch of aorta
____	**(i)** is the longest vein in the body	**(9)** great saphenous vein
____	**(j)** is completely located in the leg	**(10)** vertebral artery
____	**(k)** is completely located in the head	**(11)** superior mesenteric artery

20. Arteries in the pulmonary circulation are larger in diameter and have thinner, less elastic walls than arteries in the systemic circulation.

21. The external jugular veins join with the internal jugular veins to form the superior vena cava.

22. In order for blood to flow from the left brachial vein to the right brachial artery, it must pass through the heart and capillaries in the lungs.

23. The median cubital veins connect the basilic and cephalic veins anterior to the elbow.

24. Match the following.

____ **(a)** They carry blood away from the microcirculation.

____ **(b)** These vessels have no tunica media or tunica externa and are known as exchange vessels.

____ **(c)** They act momentarily as a pressure reservoir as their elastic walls stretch to accommodate the surge of blood as it is ejected from the heart.

____ **(d)** They have thin porous walls that permit emigration of white blood cells.

____ **(e)** They are known as resistance vessels because their diameters change, thus altering blood pressure and blood flow.

____ **(f)** They have relatively thick walls due to the amount of smooth muscle in the tunica media.

____ **(g)** They are the largest diameter arteries, but they have relatively thin walls in proportion to their diameter.

____ **(h)** An example is the brachial artery.

____ **(i)** An example is the common carotid artery.

____ **(j)** They are known as distributing arteries because they distribute blood to various parts of the body.

(1) muscular arteries
(2) elastic arteries
(3) capillaries
(4) arterioles
(5) venules

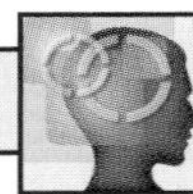

CRITICAL THINKING QUESTIONS

1. Which structures present in the fetal circulation are absent in the adult circulatory system? Why do these changes occur?
HINT *A mother may want to do everything for her baby but she can't breathe for him after he's born.*

2. Mike spent the week before the exam lying on a beach studying the back of his eyelids. "What was that question about sinuses doing on the circulatory test?" he complained. "We did that back with bones!" See if you can enlighten Mike.
HINT *In bone, sinuses are filled with air; these aren't.*

3. You've read about varicose veins. Why aren't there varicose arteries?
HINT *What do veins have that arteries don't?*

4. When asked to describe the function of the superior vena cava, Hua answered that it drained the arms, head, and heart muscle. "Not quite; try again," commented the instructor. Hua is stumped. Can you help?
HINT *Hua's answer included one organ too many and left out one region.*

5. Geena is an engineering major. She signed up for human anatomy because "How hard can it be? We learned the parts of the body in kindergarten!" Now Geena is confused. "What's all this stuff about arteries leading into a circle? That's a terrible traffic pattern!" Explain these arteries to Geena.
HINT *Try telling Geena that banks aren't the only users of "collateral."*

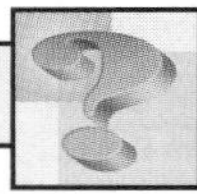

ANSWERS TO FIGURE QUESTIONS

14.1 The femoral artery has the thicker wall; the femoral vein has the wider lumen.

14.2 Metabolically active tissues use O_2 and produce wastes more rapidly than inactive tissues.

14.3 Materials cross capillary walls through intercellular clefts and fenestrations, via transcytosis in pinocytic vesicles, and through the plasma membranes of endothelial cells.

14.4 When you are standing, gravity tends to cause pooling of blood in the veins of the limbs. The valves prevent backflow as the blood proceeds toward the right atrium after each heartbeat. When you are erect, gravity aids the flow of blood in neck veins back toward the heart.

14.5 The two main circulatory routes are the systemic and pulmonary circulations.

14.6 The four principal subdivisions of the aorta are the ascending aorta, arch of the aorta, thoracic aorta, and abdominal aorta.

14.7 The left and right coronary arteries arise from the ascending aorta.

14.8 Branches of the arch of the aorta are the brachiocephalic trunk, left common carotid artery, and left subclavian artery.

14.9 The abdominal aorta begins at the aortic hiatus in the diaphragm.

14.10 The abdominal aorta divides into the common iliac arteries at about the level of L4.

14.11 The superior vena cava drains regions above the diaphragm, and the inferior vena cava drains regions below the diaphragm.

14.12 All venous blood in the brain drains into the internal jugular veins.

14.13 The median cubital vein is often used for withdrawing blood.

14.14 The inferior vena cava returns blood from abdominopelvic viscera to the heart.

14.15 Superficial veins of the lower limbs are the dorsal venous arch and the great saphenous and small saphenous veins.

14.16 The hepatic veins carry blood away from the liver.

14.17 The pulmonary arteries carry deoxygenated blood.

14.18 Exchange of materials between mother and fetus occurs across the placenta.

14.19 Blood vessels and blood are derived from mesoderm.

15

THE LYMPHATIC SYSTEM

CONTENTS AT A GLANCE

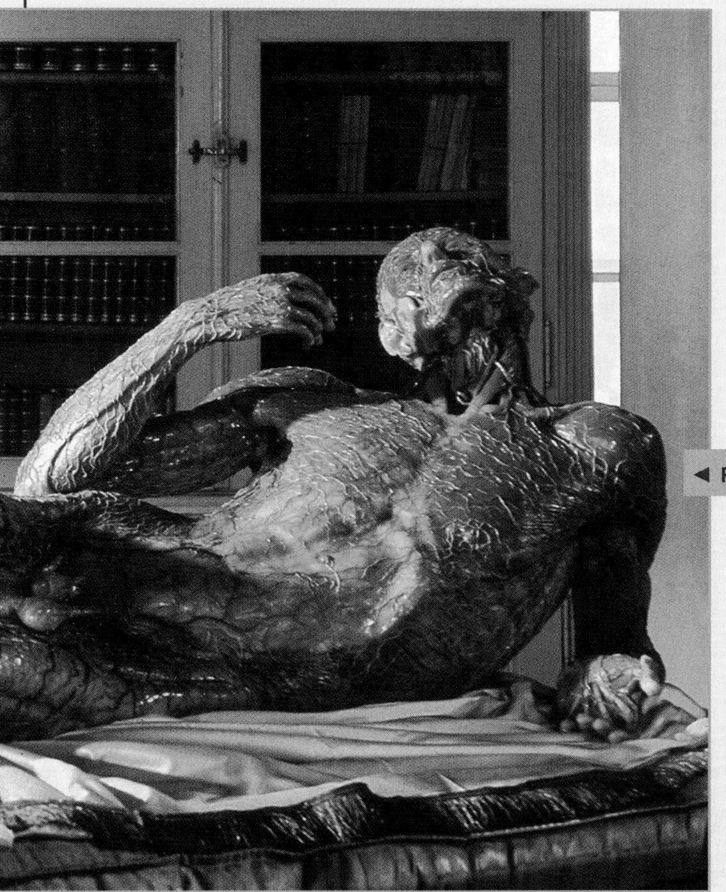

◄ Page 479

If you were asked to construct an anatomical model, which methods and materials would you use? From what material is this model constructed? Which human structures are being demonstrated?

Page 481 ►

INTRODUCTION

The ability of the body to ward off disease is called **resistance;** conversely, vulnerability or lack of resistance is termed **susceptibility.** Resistance to disease is of two types: nonspecific and specific. **Nonspecific resistance** to disease includes defense mechanisms that provide immediate but general protection against invasion by a wide range of **pathogens,** which are disease-producing microbes such as bacteria, viruses, and parasites. The first line of defense in nonspecific resistance is provided by mechanical and chemical barriers of the skin and mucous membranes; the acidity of the stomach contents, for example, kills many bacteria ingested in food. **Specific resistance,** or **immunity,** develops more slowly and involves activation of specific lymphocytes that combat a particular pathogen or other foreign substance. The body system responsible for immunity is the lymphatic system.

THE LYMPHATIC SYSTEM

Objectives

- Describe the general components of the lymphatic system, and list its functions.
- Describe the organization of lymphatic vessels.
- Describe the formation and flow of lymph.
- List and describe the primary and secondary lymphatic organs and tissues.

The **lymphatic system** (lim-FAT-ik) consists of a fluid called lymph flowing within lymphatic vessels, several structures and organs that contain lymphatic tissue, and red bone marrow, which houses stem cells that develop into lymphocytes (Figure 15.1). As you will see shortly, most components of blood plasma filter through blood capillary walls to form interstitial fluid. When interstitial fluid drains into lymphatic vessels, it is called lymph. Therefore, the composition of interstitial fluid and lymph are basically the same; the major difference between the two is location. Whereas interstitial fluid is found between cells, lymph is located within lymphatic vessels and lymphatic tissue. After fluid passes from interstitial spaces into lymphatic vessels, it is called **lymph** (LIMF; = clear fluid). Lymphatic tissue is a specialized type of connective tissue that contains large numbers of lymphocytes. Recall from Chapter 12 that lymphocytes are agranular white blood cells. The two major types of lymphocytes are T cells and B cells.

Functions of the Lymphatic System

The lymphatic system has three primary functions:

1. **Draining excess interstitial fluid.** Lymphatic vessels drain excess interstitial fluid from tissue spaces and return it to the blood.
2. **Transporting dietary lipids.** Lymphatic vessels transport the lipids and lipid-soluble vitamins (A, D, E, and K) absorbed by the gastrointestinal tract to the blood.
3. **Facilitating immune responses.** Lymphatic tissue initiates highly specific responses directed against particular microbes or abnormal cells. Lymphocytes, aided by macrophages, recognize foreign cells, microbes, toxins, and cancer cells and respond to them in two basic ways: Lymphocytes called T cells destroy the intruders by causing them to rupture or by releasing cytotoxic (cell-killing) substances; lymphocytes called B cells differentiate into plasma cells, which secrete antibodies—proteins that combine with and cause destruction of specific foreign cells or substances. In carrying out specific immune responses, the lymphatic system concentrates foreign substances in certain lymphatic organs, circulates lymphocytes through the organs to make contact with the foreign substances, and destroys the foreign substances and eliminates them from the body.

Lymphatic Vessels and Lymph Circulation

Lymphatic vessels begin as **lymphatic capillaries,** tiny tubes closed at one end and located in the spaces between cells (Figure 15.2 on page 476). Just as blood capillaries converge to form venules and veins, lymphatic capillaries unite to form larger **lymphatic vessels** (see Figure 15.1), which resemble veins in structure but have thinner walls and more valves. At various intervals along the lymphatic vessels, lymph flows through lymphatic tissue structures called **lymph nodes.** In the skin, lymphatic vessels lie in subcutaneous tissue and generally follow veins; lymphatic vessels of the viscera generally follow arteries, forming plexuses (networks) around them.

Lymphatic Capillaries

Lymphatic capillaries are slightly larger in diameter than blood capillaries and are found throughout the body, except in avascular tissues (such as cartilage, the epidermis, and the cornea of the eye), the central nervous system, portions of the spleen, and red bone marrow. The unique structure of lymphatic capillaries permits interstitial fluid to flow into but not out of them. The margins of endothelial cells that make up the wall of a lymphatic capillary overlap; when pressure is greater in the interstitial fluid than in lymph, the cells separate slightly, like a one-way valve opening, and fluid enters the lymphatic capillary (Figure 15.2b). When pressure is greater inside the lymphatic capillary, the cells adhere more closely, so lymph cannot flow back into interstitial fluid. *Anchoring filaments,* which lie at right angles to the lymphatic capillary, attach lymphatic endothelial cells to surrounding tissues. When excess interstitial fluid accumulates and causes tissue swelling, the anchoring filaments are pulled, making the openings between cells even larger so that more fluid can flow into the lymphatic capillary.

In the small intestine, specialized lymphatic capillaries called **lacteals** (LAK-tē-als; *lact-* = milky) carry dietary lipids into lymphatic vessels and ultimately into the blood. The presence of these lipids causes the lymph draining the small intestine

Figure 15.1 / Components of the lymphatic system.

The lymphatic system consists of lymph, lymphatic vessels, lymphatic tissues, and red bone marrow.

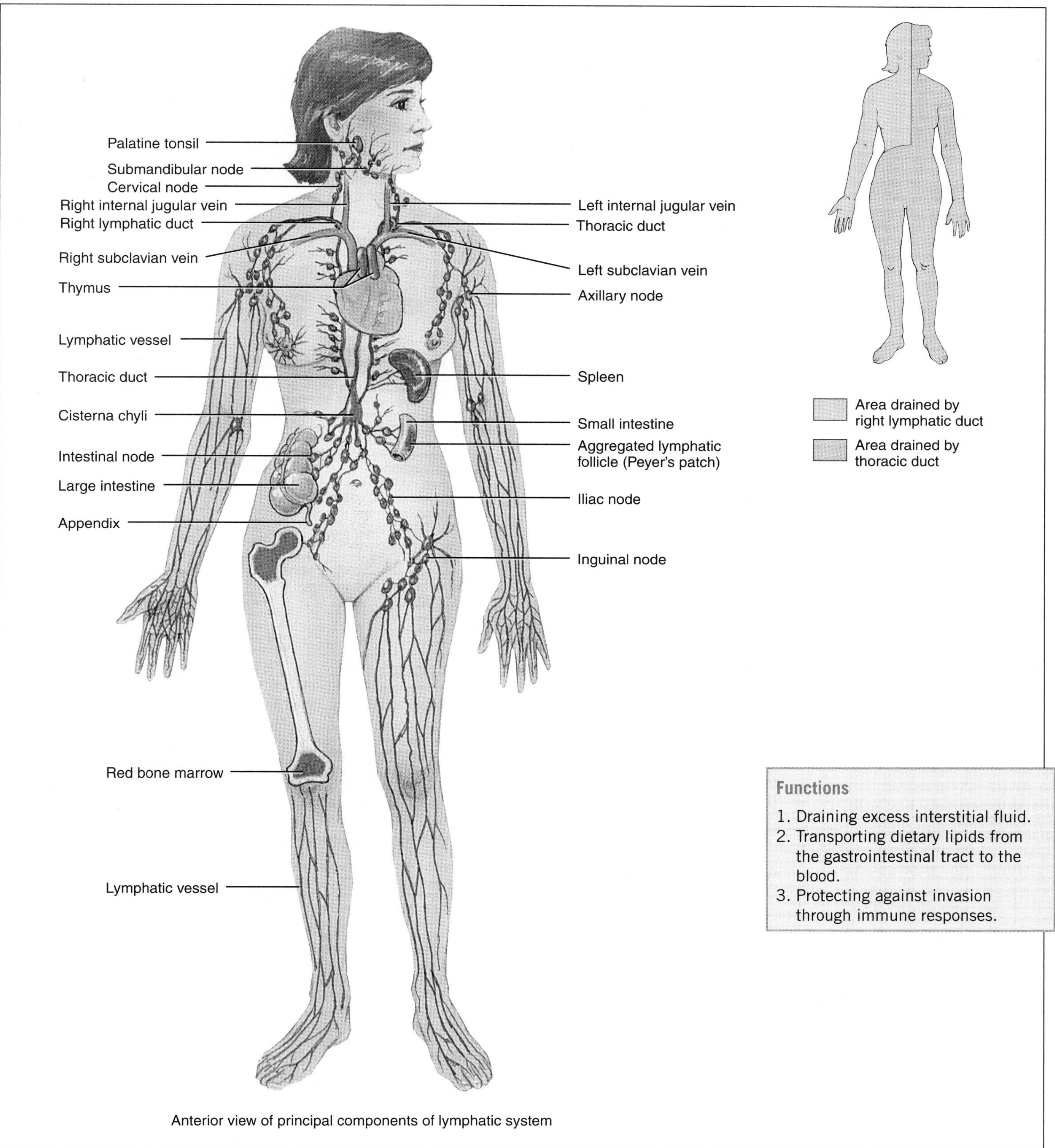

Anterior view of principal components of lymphatic system

What tissue contains stem cells that develop into lymphocytes?

Figure 15.2 / Lymphatic capillaries.

Lymphatic capillaries are found throughout the body except in avascular tissues, the central nervous system, portions of the spleen, and red bone marrow.

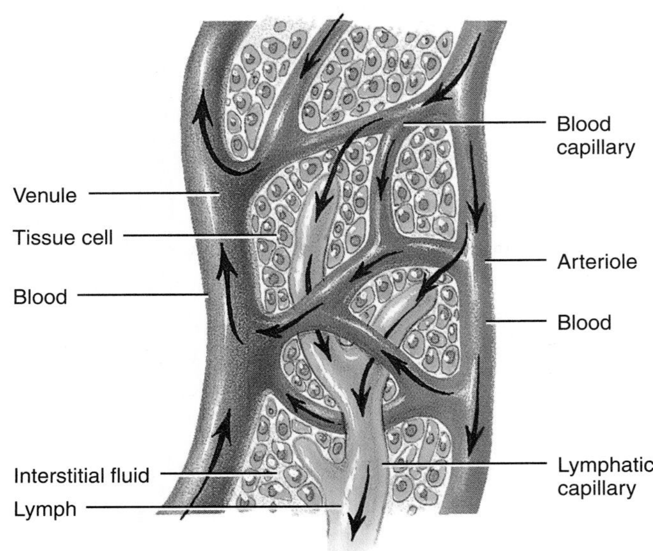

(a) Relationship of lymphatic capillaries to tissue cells and blood capillaries

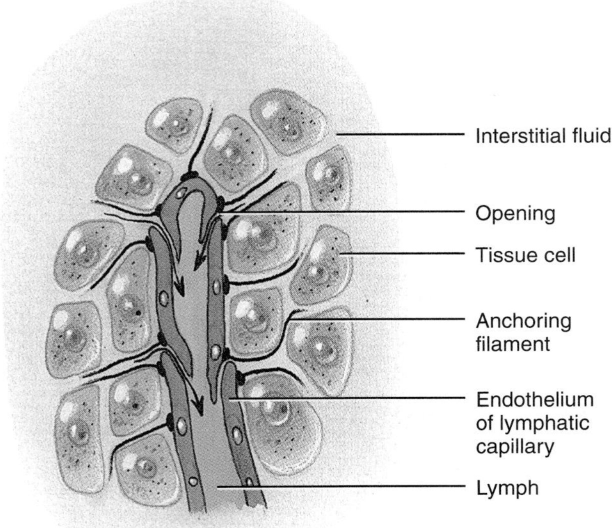

(b) Details of a lymphatic capillary

Is lymph most similar to blood plasma or to interstitial fluid? Why?

to appear creamy white; such lymph is referred to as **chyle** (KĪL; = juice). Elsewhere, lymph is a clear, pale yellow fluid.

Lymph Trunks and Ducts

Lymph passes from lymphatic capillaries into lymphatic vessels and then through lymph nodes. Lymphatic vessels exiting lymph nodes pass lymph either toward another node within the same group or on to another group of nodes. From the most proximal group of each chain of nodes, the exiting vessels unite to form **lymph trunks.** The principal trunks are the **lumbar, intestinal, bronchomediastinal, subclavian,** and **jugular trunks** (Figure 15.3). The principal trunks pass their lymph into two main channels, the thoracic duct and the right lymphatic duct. Lymph passes from these ducts into venous blood.

The **thoracic (left lymphatic) duct** is about 38–45 cm (15–18 in.) long and begins as a dilation called the **cisterna chyli** (sis-TER-na KĪ-lē; *cisterna* = cavity or reservoir) anterior to the second lumbar vertebra. The thoracic duct, the main collecting duct of the lymphatic system, receives lymph from the left side of the head, neck, and chest, the left upper limb, and the entire body inferior to the ribs. The thoracic duct drains lymph into venous blood via the **left subclavian vein.**

The cisterna chyli receives lymph from the right and left lumbar trunks and from the intestinal trunk. The lumbar trunks drain lymph from the lower limbs, the wall and viscera of the pelvis, the kidneys, the adrenal glands, and the deep lymphatic vessels that drain lymph from most of the abdominal wall. The intestinal trunk drains lymph from the stomach, intestines, pancreas, spleen, and part of the liver.

In the neck, the thoracic duct also receives lymph from the left jugular, left subclavian, and left bronchomediastinal trunks. The left jugular trunk drains lymph from the left side of the head and neck, and the left subclavian trunk drains lymph from the left upper limb. The left bronchomediastinal trunk drains lymph from the left side of the deeper parts of the anterior thoracic wall, the superior part of the anterior abdominal wall, the anterior part of the diaphragm, the left lung, and the left side of the heart.

The **right lymphatic duct** (Figure 15.3) is about 1.25 cm (0.5 in.) long and drains lymph from the upper right side of the body into venous blood via the **right subclavian vein.** Three lymphatic trunks drain into the right lymphatic duct: the right jugular trunk, which drains the right side of the head and neck; the right subclavian trunk, which drains the right upper limb; and the right bronchomediastinal trunk, which drains the right side of the thorax, the right lung, the right side of the heart, and part of the liver.

Formation and Flow of Lymph

Most components of blood plasma freely filter through the capillary walls to form interstitial fluid. Some of this material is reabsorbed. However, more fluid filters out of blood capillaries than returns to them by reabsorption. The excess filtered fluid—about 3 liters per day—drains into lymphatic vessels and be-

Figure 15.3 / Routes for drainage of lymph from lymph trunks into the thoracic and right lymphatic ducts.

All lymph returns to the bloodstream through the thoracic (left) lymphatic duct and right lymphatic duct.

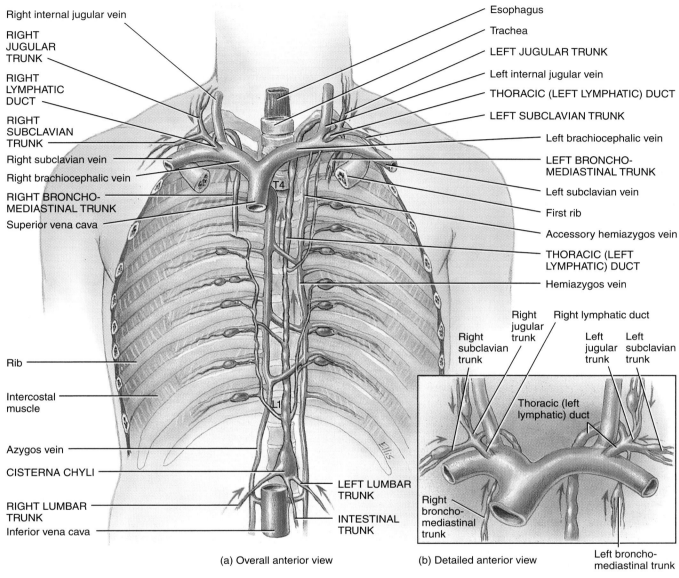

(a) Overall anterior view

(b) Detailed anterior view

Which lymphatic vessels empty into the cisterna chyli, and which duct receives lymph from the cisterna chyli?

comes lymph. Because most plasma proteins are too large to leave blood vessels, interstitial fluid contains only small amounts of protein. Proteins that do leave plasma, however, cannot return to the blood directly by diffusion because the concentration gradient (high level of proteins inside blood capillaries, low level outside) prevents such movement. The proteins, however, can move readily through the more permeable lymphatic capillaries into the lymph. Thus, an important function of lymphatic vessels is to return lost plasma proteins to the bloodstream.

Ultimately, lymph drains into venous blood through the right lymphatic duct and the thoracic duct at the junction of the internal jugular and subclavian veins (Figure 15.3). Thus, the sequence of fluid flow is blood capillaries (blood) $\rightarrow$ interstitial spaces (interstitial fluid) $\rightarrow$ lymphatic capillaries (lymph) $\rightarrow$ lymphatic vessels (lymph) $\rightarrow$ lymphatic ducts (lymph) $\rightarrow$ subclavian veins (blood). This sequence, as well as the relationship of the lymphatic and cardiovascular systems, is illustrated in Figure 15.4.

Figure 15.4 / Schematic diagram showing the relationship of the lymphatic system to the cardiovascular system.

The sequence of fluid flow is blood capillaries (blood) → interstitial spaces (interstitial fluid) → lymphatic capillaries (lymph) → lymphatic vessels (lymph) → lymphatic ducts (lymph) → subclavian veins (blood).

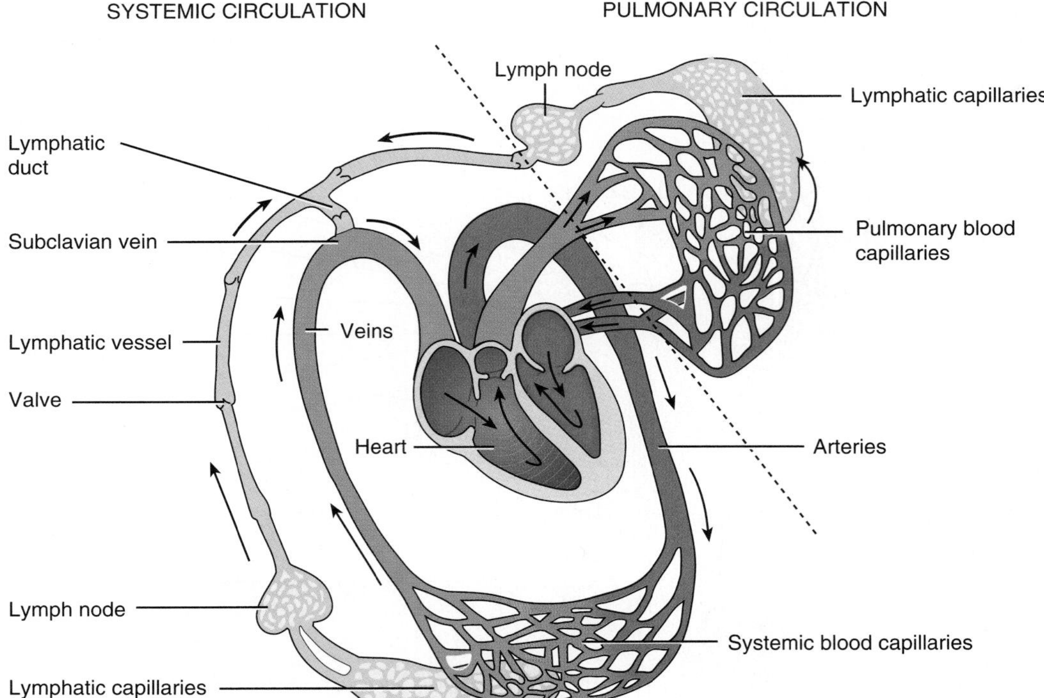

Arrows show direction of flow of lymph and blood

Does inhalation promote or hinder the flow of lymph?

The skeletal muscle and respiratory pumps promote the flow of lymph from tissue spaces to the large lymphatic ducts to the subclavian veins. Muscle contractions compress lymphatic vessels and force lymph toward the subclavian veins in a kind of milking action. Lymphatic vessels contain one-way valves, similar to those found in veins, that prevent backflow of lymph. During respirations, the pressure present in the lymphatic system changes. With each inhalation, lymph flows from the abdominal region, where the pressure is higher, toward the thoracic region, where it is lower. Likewise, the fall in abdominal pressure during exhalation promotes the flow of lymph from more distal vessels into the abdominal vessels. In addition, when a lymphatic vessel distends, the smooth muscle in its wall contracts, which helps move lymph from one segment of the vessel to the next.

Edema (*edemas* = swelling), an excessive accumulation of interstitial fluid in tissue spaces, may be caused by an obstruction, such as an infected lymph node or a blockage in a lymphatic vessel. It may also be due to a pregnant uterus pressing on abdominal lymphatic vessels. Another cause is excessive lymph formation and increased permeability of blood capillary walls such as occurs during inflammation. A rise in capillary blood pressure, in which interstitial fluid is formed faster than it is passed into lymphatic vessels, also may result in edema. ■

Lymphatic Organs and Tissues

The organs and tissues of the lymphatic system, which are widely distributed throughout the body, are classified into two groups based on their functions. **Primary lymphatic organs** provide the appropriate environment for stem cells to divide and mature into B cells and T cells, the lymphocytes that carry out immune responses. The primary lymphatic organs are the **red bone marrow** (in flat bones and the epiphyses of long bones of adults) and the **thymus.** Pluripotent stem cells in red bone marrow give rise to mature B cells and to pre-T cells that migrate to the thymus, where they mature. The **secondary lymphatic organs** and **tissues,** which are the sites where most immune responses occur, include **lymph nodes,** the **spleen,** and **lymphatic nodules.** The thymus, lymph nodes, and spleen are considered organs because each is surrounded by a connective tissue capsule; lymphatic nodules, in contrast, are not organs because they lack a capsule.

14th Century

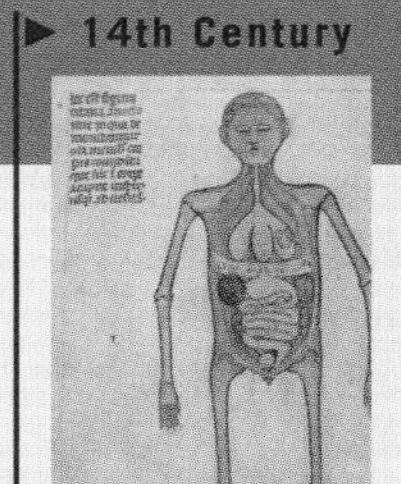

15th Century

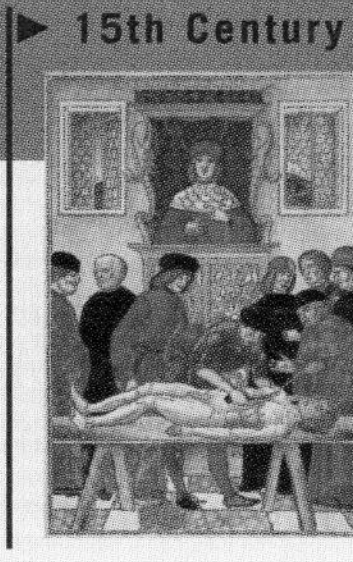

16th Century

17th Century

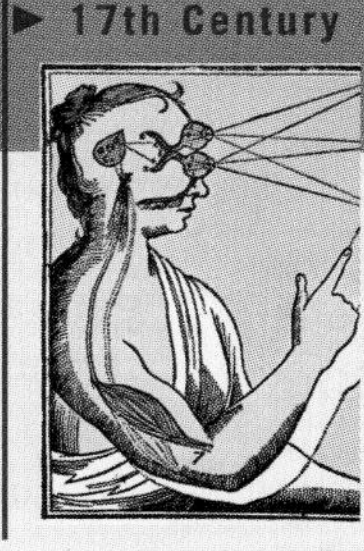

18th Century

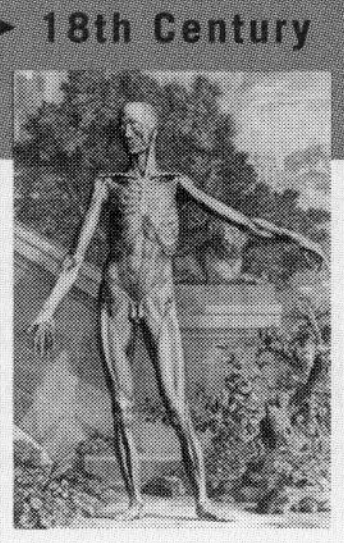

20th Century

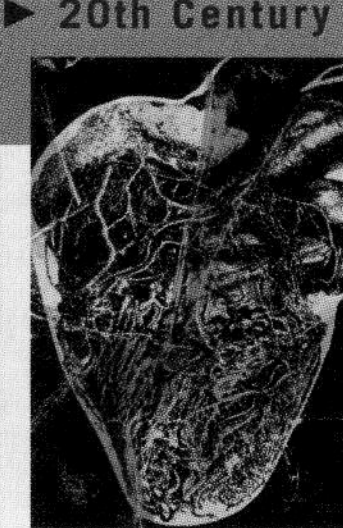

21st Century

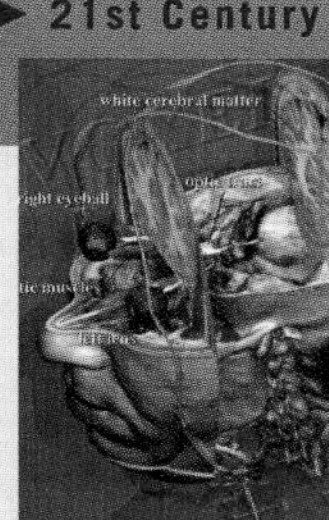

CHANGING IMAGES

Know Thyself

Thus far in these Changing Images essays, we have examined artistic representation of anatomy through a variety of mediums—from medieval drawings and Renaissance paintings, to Malaysian tattooing and Tibetan Thangkas. The subject of this discussion now focuses on one of the more extraordinary methods of anatomical inquiry.

The figure shown here, circa late 18th century, is part of the remarkable collection of anatomical wax models from the La Specola Museum of the University of Florence, Italy. Anatomists of this era were venturing beyond the flat, two-dimensional representations of drawings, paintings and woodcuts to create accurate, three-dimensional models. By dissecting cadavers to expose regions of interest, anatomists and artisans collaborated to make clay models and plaster casts of organs. A mixture of waxes, resins, and dyes (exact composition unknown) was poured into these molds. This process continued until a complete human statue of incredible detail was fashioned. Archival evidence suggests that, due to the inability to preserve cadavers, it took nearly 200 specimens to produce a single, whole body figure. Since they were to be used in teaching situations, many of the 26 whole body wax models at the La Specola Museum are designed to be dismantled and reassembled.

Beginning in the 17th century, anatomists had discovered the lymphatic system's role in the absorption of lipids, and the transportation of fluids along an elaborate network of vessels and regional lymph nodes. The anatomy of the lymphatic vessels and superficial veins is dramatically demonstrated here. Are you able to identify any anatomical landmarks? Look for the superficial inguinal lymph nodes (see page 487). Can you identify any the veins of the brachial and axillary regions?

This wax model summarizes 18th century anatomical knowledge of the lymphatic system. Perhaps more so it symbolizes the Socratic admonition to "know thyself."

1771-1850 AD

Thymus

The **thymus** usually has two lobes and is located in the mediastinum, posterior to the sternum (Figure 15.5a). An enveloping layer of connective tissue holds the two **thymic lobes** closely together, but a connective tissue **capsule** encloses each lobe separately. Extensions of the capsule, called **trabeculae** (tra-BEK-yū-lē; = little beams) penetrate inward and divide the lobes into **lobules** (Figure 15.5b).

Each lobule consists of a deeply staining outer **cortex** and a lighter staining central **medulla.** The cortex is composed of tightly packed lymphocytes, epithelial cells called **reticular epithelial cells** that surround clusters of lymphocytes, and macrophages. The medulla consists mostly of reticular epithelial cells and more widely scattered lymphocytes. Although only some of their functions are known, the reticular epithelial cells produce thymic hormones, which are thought to aid in the maturation of T cells. In addition, the medulla contains characteristic **thymic (Hassall's) corpuscles,** concentric layers of flattened reticular epithelial cells filled with keratohyalin granules and keratin (Figure 15.5c). Their functions are unknown.

The thymus is large in infants, having a mass of about 70 g (about 2.5 oz). After puberty, adipose and areolar connective tissue begin to replace the thymic tissue. By the time a person reaches maturity, the thymus has atrophied considerably, and in old age it may weigh only 3 g. Before the thymus atrophies, it has populated the lymph nodes, spleen, and other lymphatic organs and tissues with T cells. The T cells continue to proliferate throughout an individual's lifetime. Although the thymus is not a site of immune responses, it populates other structures where immune responses do occur.

Lymph Nodes

The approximately 600 bean-shaped organs located along lymphatic vessels are called **lymph nodes.** They are scattered throughout the body, both superficially and deep, and usually in groups (see Figure 15.1). Lymph nodes are heavily concentrated near the mammary glands and in the axillae and groin. Later in this chapter, the principal groups of lymph nodes will be presented in a series of Exhibits.

Lymph nodes are 1–25 mm (0.04–1 in.) long and are covered by a **capsule** of dense connective tissue that extends into the node (Figure 15.6). The capsular extensions, called **trabeculae,** divide the node into compartments, provide support, and provide a route for blood vessels into the interior of a node.

Figure 15.5 / Thymus. (See Tortora, *A Photographic Atlas of the Human Body,* Figure 7.2a.)

The bilobed thymus is largest at puberty and then atrophies with age.

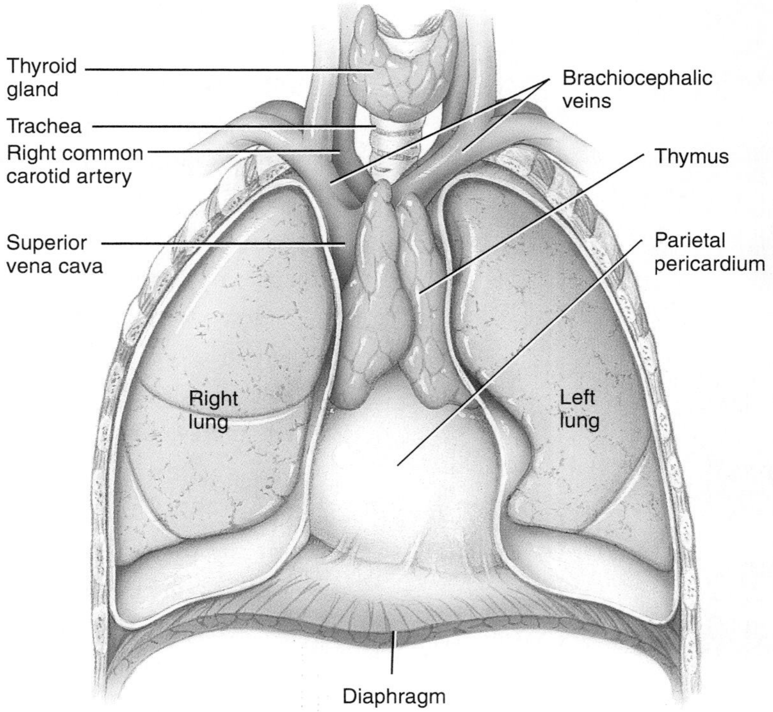

(a) Thymus of adolescent

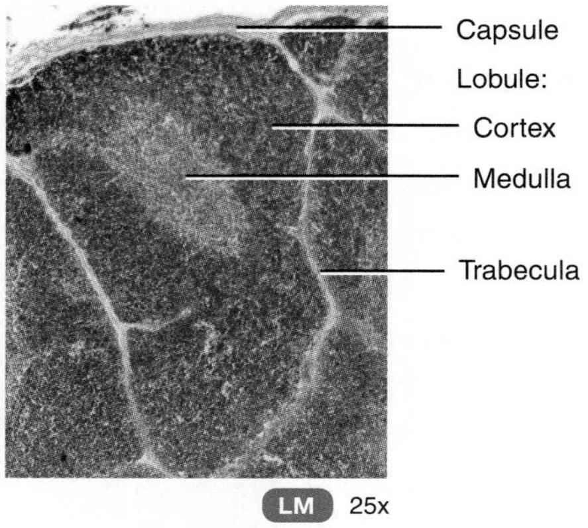

(b) Thymic lobules

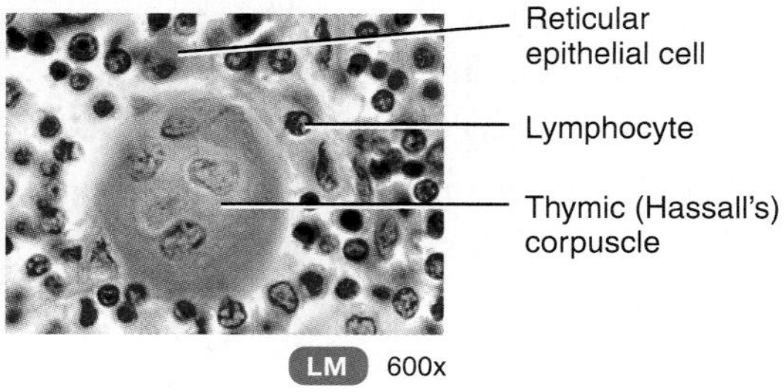

(c) Thymic corpuscle

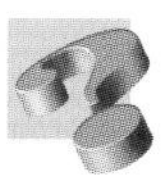

Which lymphocytes mature in the thymus?

Figure 15.6 / Structure of a lymph node. Arrows indicate the direction of lymph flow through a lymph node.

Lymph nodes are present throughout the body, usually clustered in groups.

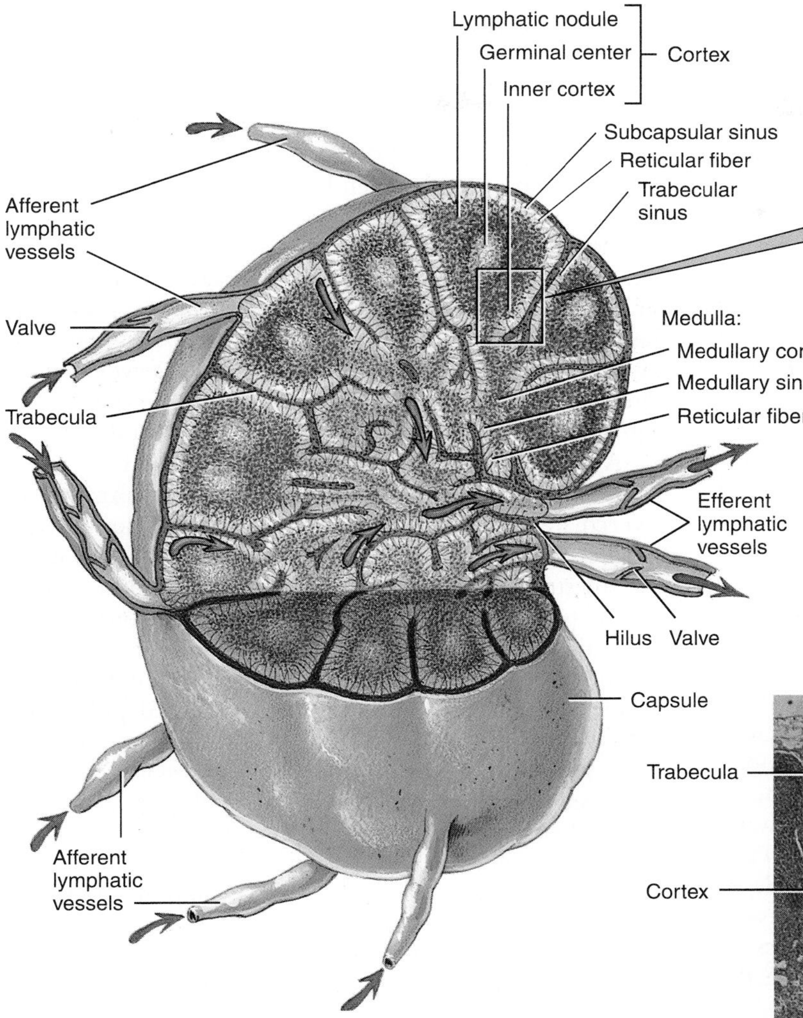

Lymphatic nodule
Germinal center
Inner cortex
Cortex
Subcapsular sinus
Reticular fiber
Trabecular sinus
Afferent lymphatic vessels
Valve
Trabecula
Medulla:
Medullary cord
Medullary sinus
Reticular fiber
Efferent lymphatic vessels
Hilus
Valve
Capsule
Afferent lymphatic vessels

(a) Partially sectioned lymph node

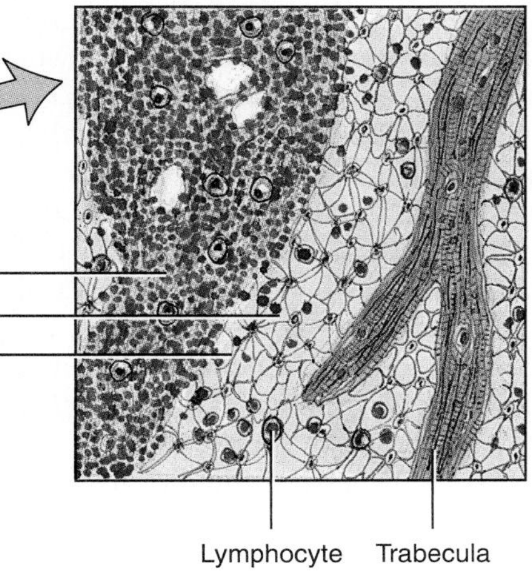

Lymphocyte
Trabecula

Details of a lymph node

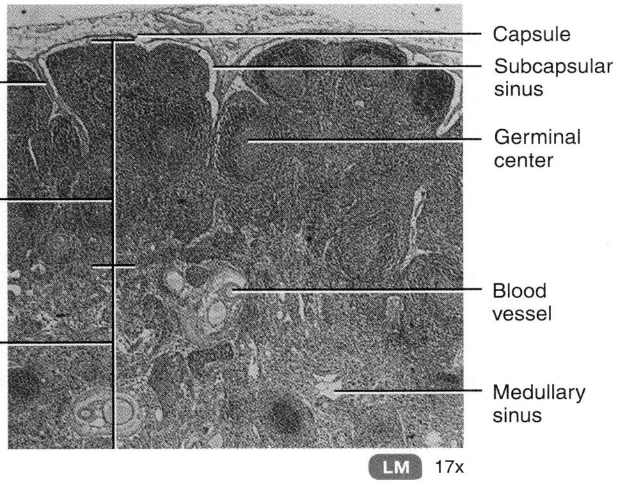

Capsule
Subcapsular sinus
Trabecula
Germinal center
Cortex
Blood vessel
Medulla
Medullary sinus
LM 17x

(b) Portion of a lymph node

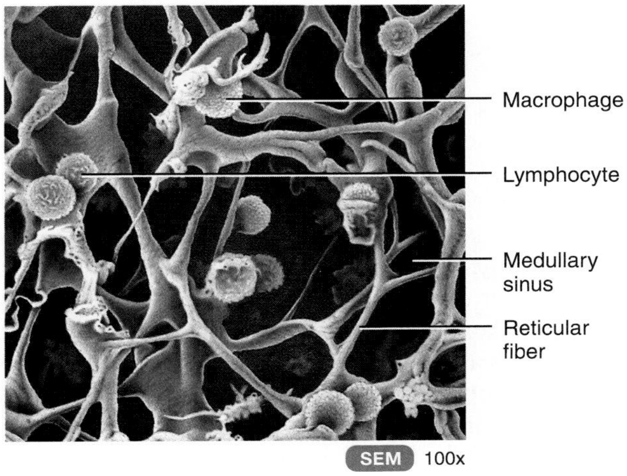

Macrophage
Lymphocyte
Medullary sinus
Reticular fiber
SEM 100x

(c) Portion of a lymph node

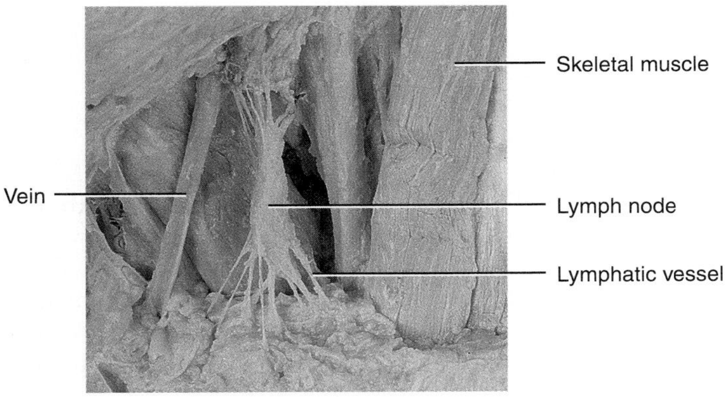

Skeletal muscle
Vein
Lymph node
Lymphatic vessel

(d) Anterior view of a lymph node

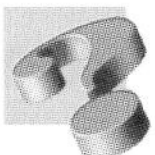

What happens to foreign substances in the lymph when they enter a lymph node?

Internal to the capsule is a supporting network of reticular fibers and fibroblasts. The capsule, trabeculae, reticular fibers, and fibroblasts constitute the *stroma*, or framework, of a lymph node. The *parenchyma* of a lymph node is specialized into two regions: a superficial cortex and a deep medulla. The cortex of a lymph node is divided into outer and inner regions. The **outer cortex** contains lymphatic nodules of B cells. Lighter-staining areas in the nodules, called **germinal centers,** are where B cells proliferate into antibody-secreting plasma cells. The reticular cells of the germinal centers, called **dendritic cells,** serve as antigen-presenting cells that help initiate immune responses. The germinal centers also contain macrophages. Deep to the outer cortex is the **inner cortex,** which contains T cells. The **medulla** of a lymph node contains B cells and plasma cells in tightly packed strands called **medullary cords.**

Lymph flows through a node in one direction only. It enters through **afferent lymphatic vessels** (*afferent* = to carry toward), which penetrate the convex surface of the node at several points. The afferent vessels contain valves that open toward the center of the node, such that the lymph is directed *inward.* Within the node, lymph enters **sinuses,** which are a series of irregular channels that contain branching reticular fibers, lymphocytes, and macrophages. From the afferent lymphatic vessels, lymph flows into the **subcapsular sinus** immediately beneath the capsule, then into **trabecular sinuses,** which extend through the cortex and run parallel to the trabeculae. From there, lymph flows into **medullary sinuses,** which extend through the medulla. The medullary sinuses drain into one or two **efferent lymphatic vessels** (*efferent* = to carry away), which are wider than afferent vessels and fewer in number. They contain valves that open away from the center of the node to convey lymph *out* of the node. Efferent lymphatic vessels emerge from one side of the lymph node at a slight depression called a **hilus** (HĪ-lus). Blood vessels also enter and leave the node at the hilus.

Only lymph nodes filter lymph. As lymph enters one end of a node, foreign substances are trapped by the reticular fibers within the sinuses of the node. Macrophages then destroy some foreign substances by phagocytosis while lymphocytes destroy others by a variety of immune responses. Filtered lymph then leaves the other end of the node. Plasma cells and T cells that have proliferated within a lymph node can also leave in lymph and circulate to other parts of the body.

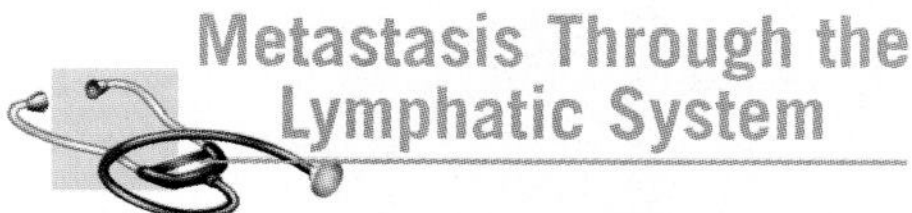

Metastasis Through the Lymphatic System

Metastasis (me-TAS-ta-sis; *meta-* = beyond; *stasis* = to stand) is the spread of disease from one organ to another not directly connected to it. It is a characteristic of malignant tumors. Cancer cells may travel via the blood or lymphatic system and establish new tumors where they lodge. When metastasis occurs via the lymphatic system, secondary tumor sites can be predicted according to the direction of lymph flow from the primary tumor site. Cancerous lymph nodes feel enlarged, firm, nontender, and fixed to underlying structures. By contrast, most lymph nodes that are enlarged due to an infection are not firm, are moveable, and are very tender. ■

Spleen

The oval **spleen** is the largest single mass of lymphatic tissue in the body, measuring about 12 cm (5 in.) in length (Figure 15.7a). It is located in the left hypochondriac region between the stomach and diaphragm. The superior surface of the spleen is smooth and convex and conforms to the concave surface of the diaphragm. Neighboring organs make indentations in the visceral surface of the spleen—the gastric impression (stomach), the renal impression (left kidney), and the colic impression (left flexure of colon). Like lymph nodes, the spleen has a hilus. Through the hilus of the spleen pass the splenic artery, splenic vein, and efferent lymphatic vessels.

A capsule of dense connective tissue surrounds the spleen. Trabeculae extend inward from the capsule, which, in turn, is covered by a serous membrane, the visceral peritoneum. The capsule plus trabeculae, reticular fibers, and fibroblasts constitute the stroma of the spleen; the parenchyma of the spleen consists of two different kinds of tissue called white pulp and red pulp (Figure 15.7b). **White pulp** is lymphatic tissue, mostly lymphocytes and macrophages, arranged around branches of the splenic artery called central arteries. The **red pulp** consists of **venous sinuses** filled with blood and cords of splenic tissue called **splenic (Billroth's) cords.** Splenic cords consist of red blood cells, macrophages, lymphocytes, plasma cells, and granulocytes. Veins are closely associated with the red pulp.

Blood flowing into the spleen through the splenic artery enters the central arteries of the white pulp. Within the white pulp, B cells and T cells carry out immune functions while spleen macrophages destroy blood-borne pathogens by phagocytosis. Within the red pulp, the spleen performs three functions related to blood cells: (1) removal by macrophages of worn out or defective blood cells and platelets; (2) storage of platelets, perhaps up to one-third of the body's supply; and (3) production of blood cells (hemopoiesis) during fetal life.

Ruptured Spleen

The spleen is the organ most often damaged in cases of abdominal trauma. Severe blows over the inferior left chest or superior abdomen may fracture the protecting ribs. Such a crushing injury may rupture the spleen, which causes severe intraperitoneal hemorrhage and shock. Prompt removal of the spleen, called a **splenectomy,** is needed to prevent the patient from bleeding to death. Other structures, particularly red bone marrow and the liver, can take over functions normally carried out by the spleen. The spleen's absence can place the patient at higher risk for sepsis, a blood infection resulting from the loss of the filtering and phagocytic functions of the spleen. Patients with a history of a splenectomy take prophylactic antibiotics before any invasive procedures to reduce the risk of sepsis. ■

Figure 15.7 / Structure of the spleen.

The spleen is the largest single mass of lymphatic tissue in the body.

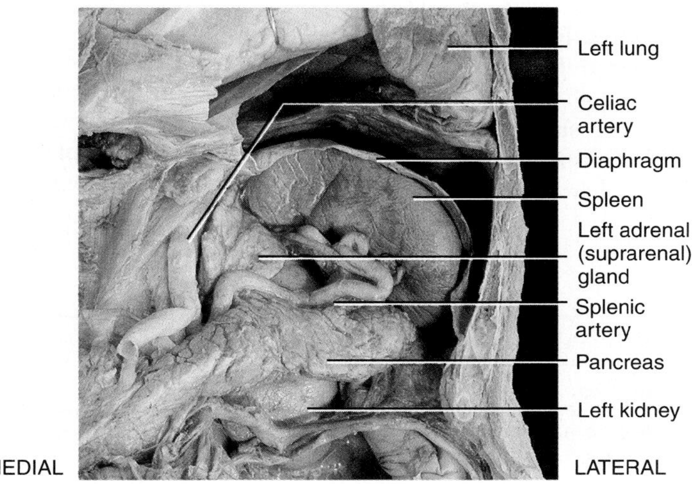

(a) Anterior view of a portion of the abdominal cavity

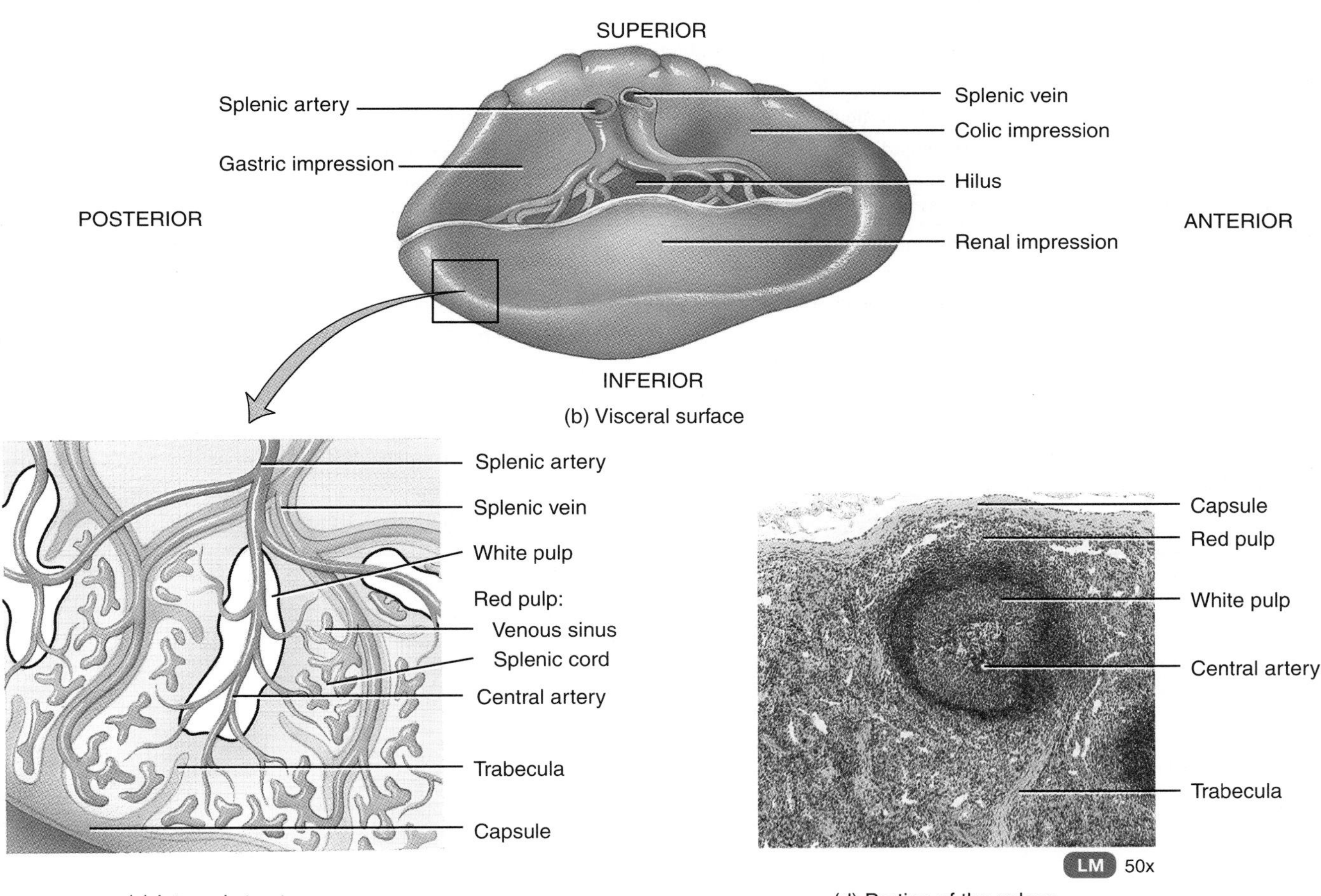

(b) Visceral surface

(c) Internal structure

(d) Portion of the spleen

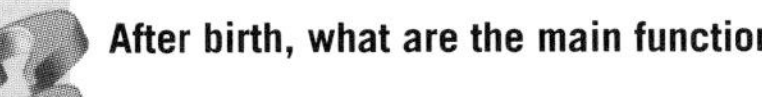

After birth, what are the main functions of the spleen?

Lymphatic Nodules

Lymphatic nodules are oval-shaped concentrations of lymphatic tissue that are not surrounded by a capsule. Because they are scattered throughout the lamina propria (connective tissue) of mucous membranes lining the gastrointestinal, urinary, and reproductive tracts and the respiratory airways, lymphatic nodules are also referred to as **mucosa-associated lymphoid tissue (MALT).**

Although many lymphatic nodules are small and solitary, some occur in multiple large aggregations in specific parts of the body. Among these are the tonsils in the pharyngeal region and the aggregated lymphatic follicles (Peyer's patches) in the ileum of the small intestine. Aggregations of lymphatic nodules also occur in the appendix. Usually there are five **tonsils** that form a ring at the junction of the oral cavity and oropharynx and at the junction of the nasal cavity and nasopharynx (see Figure 23.2b on page 687). Thus, the tonsils are strategically positioned to participate in immune responses against foreign substances that are inhaled or ingested. The single **pharyngeal tonsil** (fa-RIN-jē-al) or **adenoid** is embedded in the posterior wall of the nasopharynx. The two **palatine tonsils** (PAL-a-tīn) lie at the posterior region of the oral cavity, one on either side; these are the

Exhibit 15.1 Principal Lymph Nodes of the Head and Neck (Figure 15.8)

Lymph Nodes of the Head	Location	Drainage
Occipital nodes	Near trapezius and semispinalis capitis muscles.	Occipital portion of scalp and upper neck.
Retroauricular nodes	Posterior to ear.	Skin of ear and posterior parietal region of scalp.
Preauricular nodes	Anterior to ear.	Auricle of ear and temporal region of scalp.
Parotid nodes	Embedded in and inferior to parotid gland.	Root of nose, eyelids, anterior temporal region, external auditory meatus, tympanic cavity, nasopharynx, and posterior portions of nasal cavity.
Facial nodes	Consist of three groups: infraorbital, buccal, and mandibular.	
Infraorbital nodes	Inferior to the orbit.	Eyelids and conjunctiva.
Buccal nodes	At angle of mouth.	Skin and mucous membrane of nose and cheek.
Mandibular nodes	Over mandible.	Skin and mucous membrane of nose and cheek.

Lymph Nodes of the Neck	Location	Drainage
Submandibular nodes	Along inferior border of mandible.	Chin, lips, nose, nasal cavity, cheeks, gums, inferior surface of palate, and anterior portion of tongue.
Submental nodes	Between diagastric muscles.	Chin, lower lip, cheeks, tip of tongue, and floor of mouth.
Superficial cervical nodes	Along external jugular vein.	Inferior part of ear and parotid region.
Deep cervical nodes	Largest group of nodes in neck, consisting of numerous large nodes forming a chain extending from base of skull to root of neck. They are arbitrarily divided into superior deep cervical nodes and inferior deep cervical nodes.	
Superior deep cervical nodes	Deep to sternocleidomastoid muscle.	Posterior head and neck, auricle, tongue, larynx, esophagus, thyroid gland, nasopharynx, nasal cavity, palate, and tonsils.
Inferior deep cervical nodes	Near subclavian vein.	Posterior scalp and neck, superficial pectoral region, and part of arm.

tonsils commonly removed in a tonsillectomy. The paired **lingual tonsils** (LIN-gwal), located at the base of the tongue, may also require removal during a tonsillectomy.

✓ How are interstitial fluid and lymph similar, and how do they differ?
✓ How do lymphatic vessels differ in structure from veins?
✓ Construct a diagram that shows the route of lymph circulation.
✓ What is the role of the thymus in immunity?
✓ What functions do lymph nodes serve?
✓ Describe the functions of the spleen and tonsils.

PRINCIPAL GROUPS OF LYMPH NODES

Objective

- Identify the locations and drainage regions of the principal groups of lymph nodes.

Exhibits 15.1–15.5 (pages 484–491) provide a listing of the principal groups of lymph nodes, by region, and the general areas that they drain.

Figure 15.8 / Principal lymph nodes of the head and neck.

The facial lymph nodes include the infraorbital, buccal, and mandibular lymph nodes.

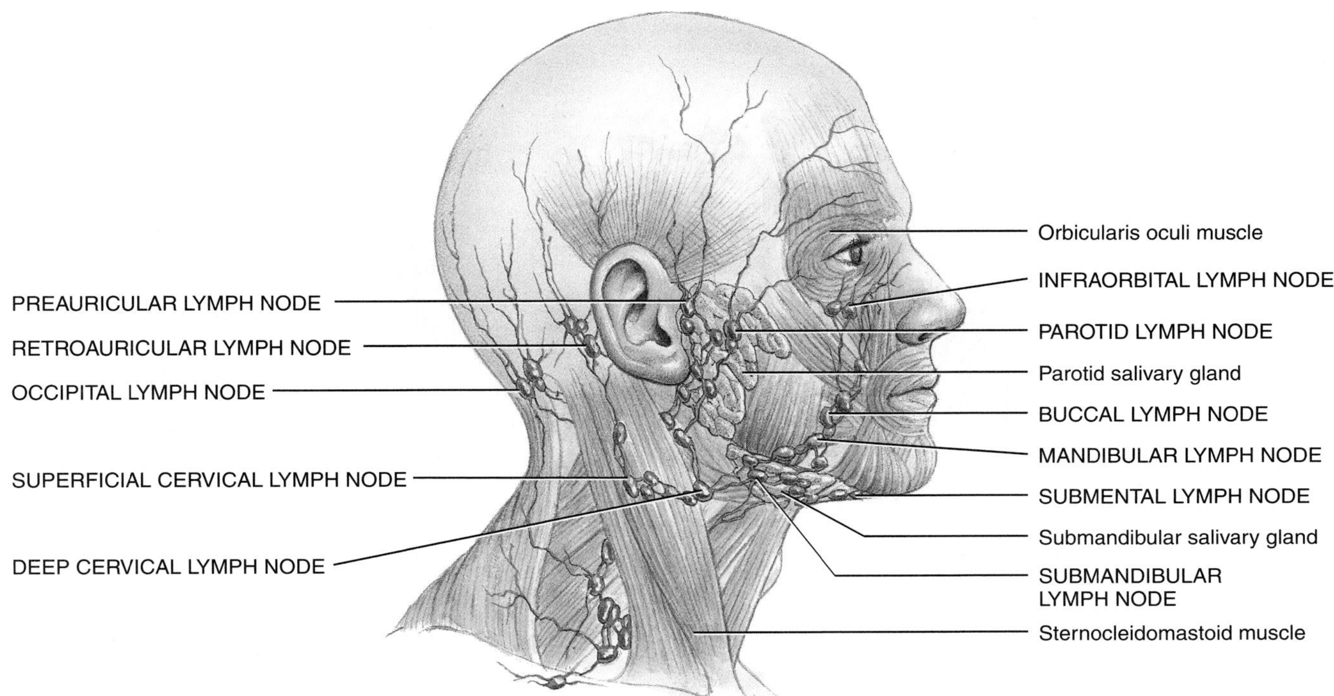

Lateral view of lymph nodes of the head and neck

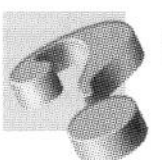

Which is the largest group of lymph nodes in the neck?

Exhibit 15.2 Principal Lymph Nodes of the Upper Limbs (Figure 15.9)

Lymph Nodes	Location	Drainage
Supratrochlear nodes	Superior to medial epicondyle of humerus.	Medial fingers, palm, and forearm.
Deltopectoral nodes	Inferior to clavicle.	Lymphatic vessels on radial side of upper limb.
Axillary nodes	Most deep lymph nodes of the upper limbs are in the axilla and are called the axillary nodes. They are large.	
Lateral nodes	Medial and posterior aspects of axillary artery.	Most of entire upper limb.*
Pectoral (anterior) nodes	Along inferior border of the pectoralis minor muscle.	Skin and muscles of anterior and lateral thoracic walls and central and lateral portions of mammary gland.
Subscapular (posterior) nodes	Along subscapular artery.	Skin and muscles of posterior part of neck and thoracic wall.
Central (intermediate) nodes	Base of axilla embedded in adipose tissue.	Lateral, pectoral (anterior), and subscapular (posterior) nodes.
Subclavicular (apical) nodes	Posterior and superior to pectoralis minor muscle.	Deltopectoral nodes.

*Because infection or malignancy of the upper limbs may cause tenderness and swelling in the axilla, the axillary nodes, especially the lateral group, are clinically important in that they filter lymph from much of the upper limb.

Figure 15.9 / Principal lymph nodes of the upper limbs. In (b) the direction of drainage is indicated by arrows.

Most of the lymph drainage of the breast is to the pectoral group of axillary lymph nodes.

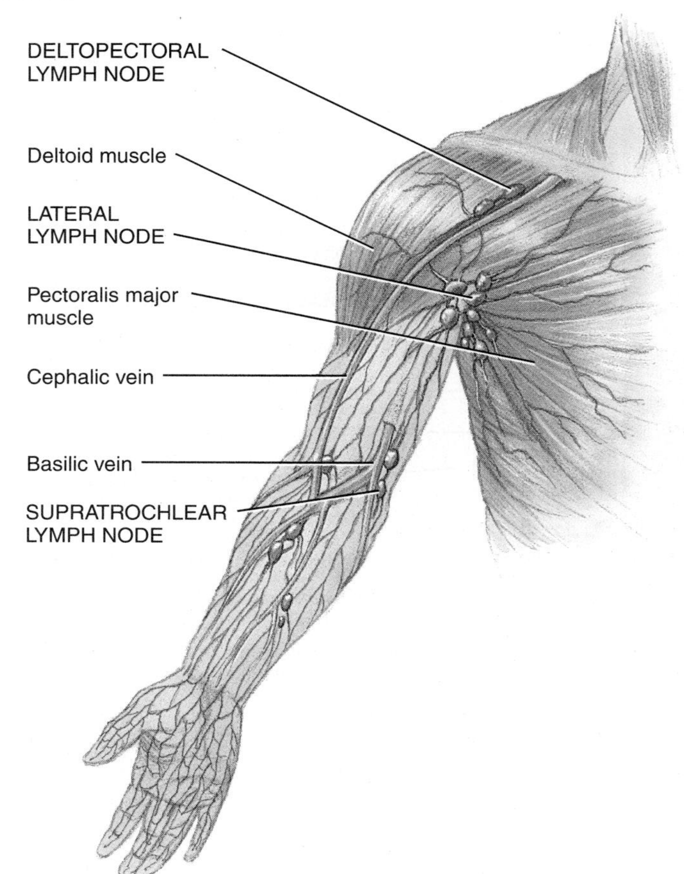

(a) Anterior view of lymph nodes of the upper limb

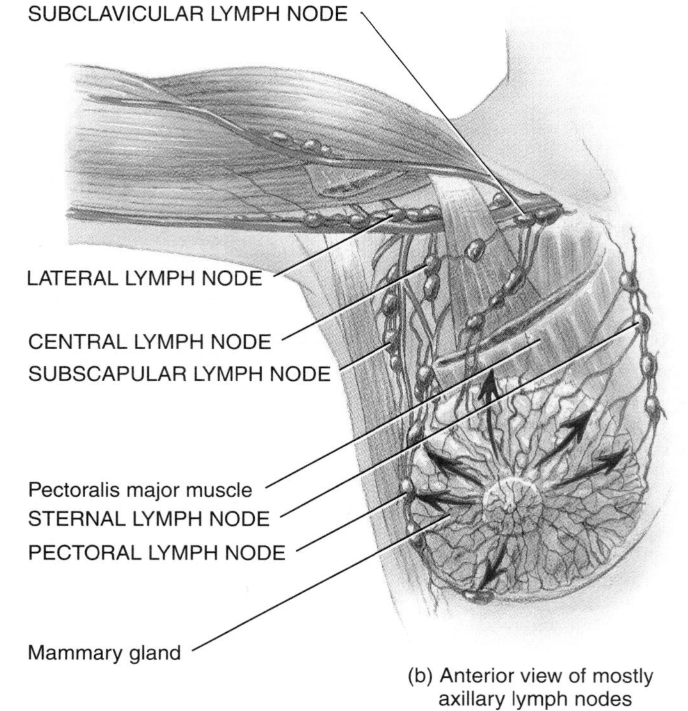

(b) Anterior view of mostly axillary lymph nodes

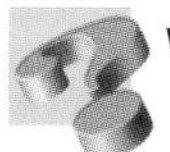

Which lymph nodes drain most of the upper limb?

Exhibit 15.3 Principal Lymph Nodes of the Lower Limbs (Figure 15.10)

Lymph Nodes	Location	Drainage
Popliteal nodes	In adipose tissue in popliteal fossa.	Knee and portions of leg and foot, especially heel.
Superficial inguinal nodes	Parallel to saphenous vein.	Anterior and lateral abdominal wall to level of umbilicus, gluteal region, external genitals, perineal region, and entire superficial lymphatics of lower limb.
Deep inguinal nodes	Medial to femoral vein.	Deep lymphatics of lower limb, penis, and clitoris.

Figure 15.10 / Principal lymph nodes of the lower limbs.

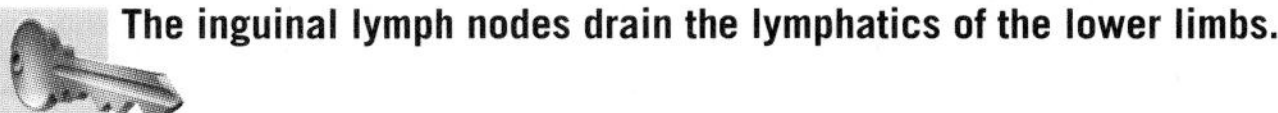

The inguinal lymph nodes drain the lymphatics of the lower limbs.

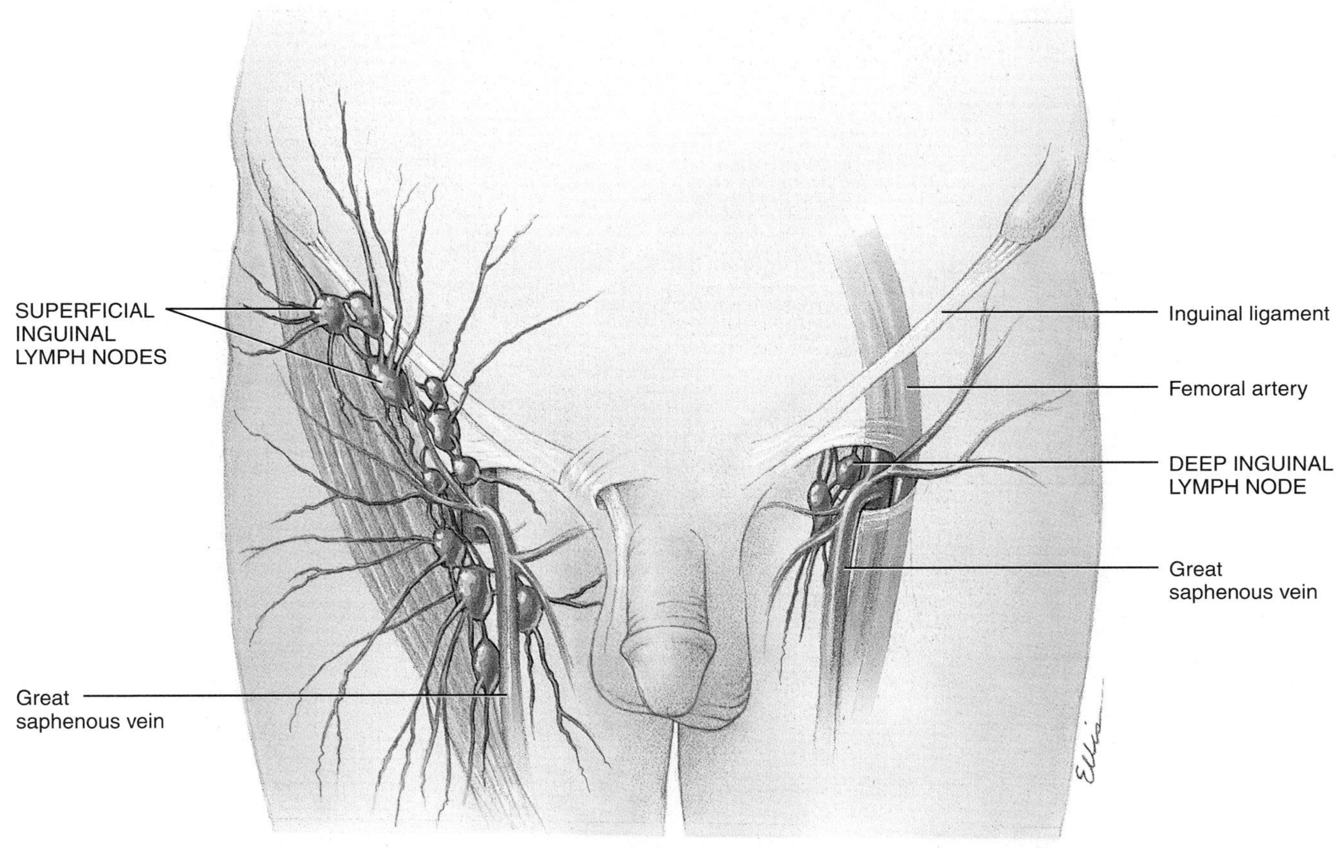

Anterior view of inguinal lymph nodes

Which lymph nodes are parallel to the saphenous vein?

Exhibit 15.4 Principal Lymph Nodes of the Abdomen and Pelvis (Figure 15.11)

Parietal Lymph Nodes	Location	Drainage
Parietal nodes are located retroperitoneally (behind the parietal peritoneum) and in close association with larger blood vessels.		
External iliac nodes	Along external iliac vessels.	Deep lymphatics of abdominal wall inferior to umbilicus, adductor region of thigh, urinary bladder, prostate gland, ductus (vas) deferens, seminal vesicles, prostatic and membranous urethra, uterine (Fallopian) tubes, uterus, and vagina.
Common iliac nodes	Along course of common iliac vessels.	Pelvic viscera.
Internal iliac nodes	Near internal iliac artery.	Pelvic viscera, perineum, gluteal region, and posterior surface of thigh.
Sacral nodes	In hollow of sacrum.	Rectum, prostate gland, and posterior pelvic wall.
Lumbar nodes	From aortic bifurcation to diaphragm; arranged around aorta and designated as *right lateral aortic nodes, left lateral aortic nodes, preaortic nodes,* and *retroaorticnodes.*	Efferents from testes, ovaries, uterine (Fallopian) tubes, uterus, kidneys, adrenal (suprarenal) glands, abdominal surface of diaphragm, and lateral abdominal wall.

(continues)

Figure 15.11 / Principal lymph nodes of the abdomen and pelvis.

The parietal lymph nodes are retroperitoneal and in close association with larger blood vessels.

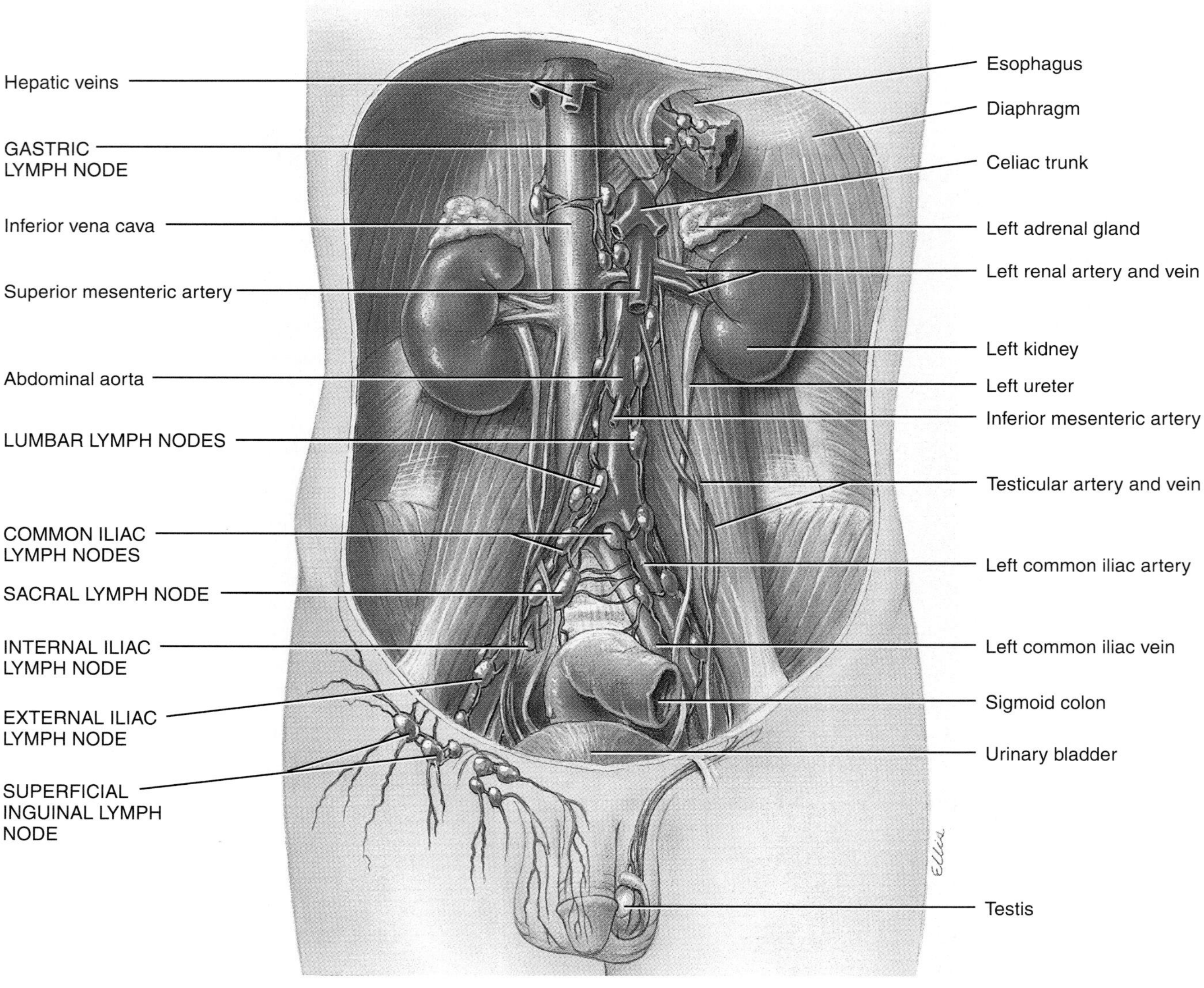

(a) Anterior view of abdominal and pelvic lymph nodes

Visceral Lymph Nodes	Location	Drainage
Visceral nodes are found in association with visceral arteries.		
Celiac nodes	Consist of three groups: gastric, hepatic, and pancreaticosplenic.	
Gastric nodes	Along lesser curvature of stomach.	Lesser curvature of stomach; inferior, anterior, and posterior aspects of stomach; esophagus.
Hepatic nodes	Along hepatic artery.	Stomach, duodenum, liver, gallbladder, and pancreas.
Pancreaticosplenic nodes	Along splenic artery.	Stomach, spleen, and pancreas.
Superior mesenteric nodes	Consist of three groups: mesenteric, ileocolic, and transverse mesocolic.	
Mesenteric nodes	Along superior mesenteric artery.	Jejunum and all parts of ileum, except for terminal portion.
Ileocolic nodes	Along ileocolic artery.	Terminal portion of ileum, appendix, cecum, and ascending colon.
Transverse mesocolic nodes	Between layers of transverse mesocolon.	Descending iliac and sigmoid parts of colon.
Inferior mesenteric nodes	Near left colic, sigmoid, and superior rectal arteries.	Descending, iliac, and sigmoid parts of colon; superior part of rectum; superior anal canal.

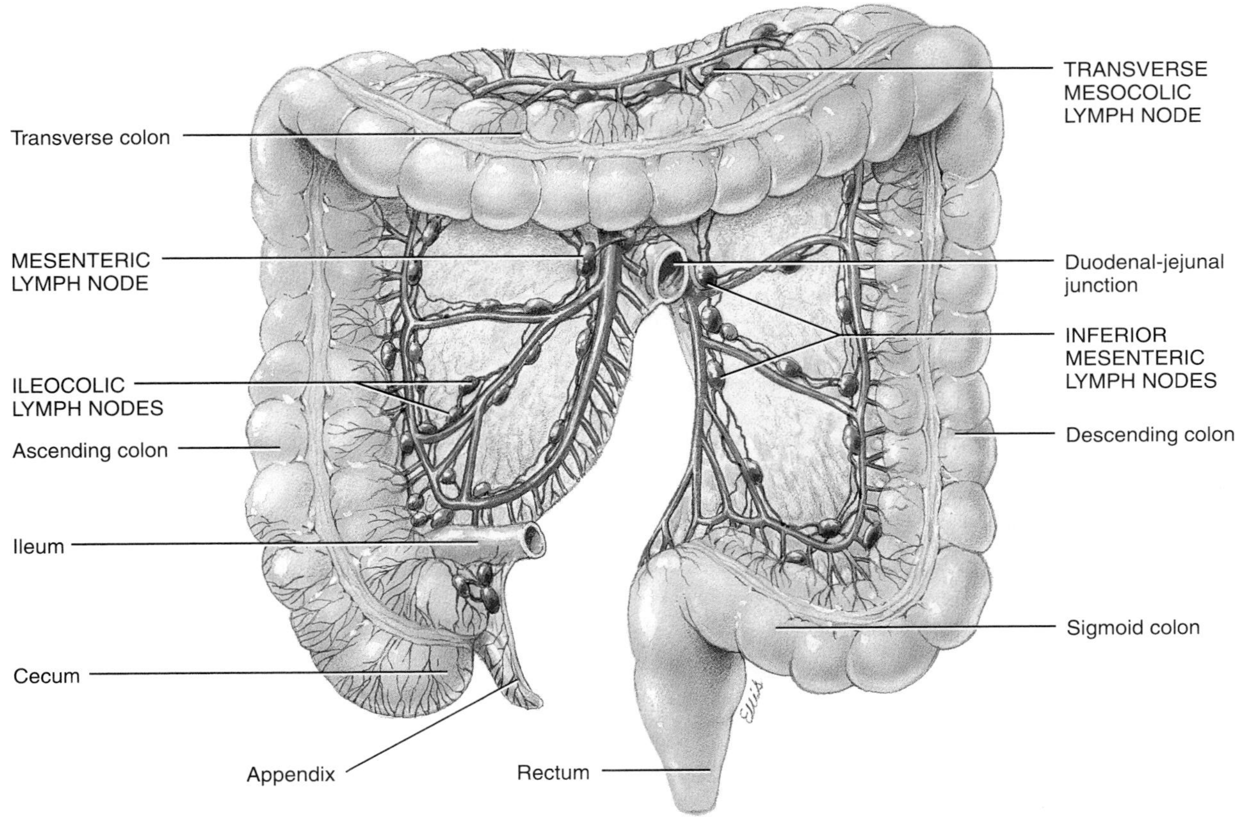

(b) Anterior view of superior and inferior mesenteric lymph nodes

What are the three groups of celiac lymph nodes?

Exhibit 15.5 Principal Lymph Nodes of the Thorax (Figure 15.12)

Parietal Lymph Nodes	Location	Drainage
Parietal nodes drain the wall of the thorax.		
Sternal (parasternal) nodes	Alongside internal thoracic artery.	Central and lateral parts of mammary gland, deeper structures of anterior abdominal wall superior to umbilicus, diaphragmatic surface of liver, and deeper parts of anterior portion of thoracic wall.
Intercostal nodes	Near heads of ribs at posterior parts of intercostal spaces.	Posterolateral aspect of thoracic wall.
Phrenic (diaphragmatic) nodes	On thoracic aspect of diaphragm and divisible into three sets called anterior phrenic, middle phrenic, and posterior phrenic.	
Anterior phrenic nodes	Posterior to base of xiphoid process.	Convex surface of liver, diaphragm, and anterior abdominal wall.
Middle phrenic nodes	Close to phrenic nerves where they pierce diaphragm.	Medial part of diaphragm and convex surface of liver.
Posterior phrenic nodes	Posterior surface of diaphragm near aorta.	Posterior part of diaphragm.

Visceral Lymph Nodes	Location	Drainage
Visceral nodes drain the viscera in the thorax.		
Anterior mediastinal nodes	Anterior part of superior mediastinum anterior to arch of aorta.	Thymus and pericardium.
Posterior mediastinal nodes	Posterior to pericardium.	Esophagus, posterior aspect of the pericardium, diaphragm, and convex surface of liver.
Tracheobronchial nodes	Are divided into five groups: tracheal, superior and inferior tracheobronchial, bronchopulmonary, and pulmonary nodes.	
Tracheal nodes	Either side of trachea.	Trachea and upper esophagus.
Superior tracheobronchial nodes	Between trachea and bronchi.	Trachea and bronchi.
Inferior tracheobronchial nodes	Between bronchi.	Trachea and bronchi.
Bronchopulmonary nodes	In hilus of each lung.	Lungs and bronchi.
Pulmonary nodes	Within lungs on larger bronchial tube branches.	Lungs and bronchi.

More than 75% of the lymph drainage of the breast is to the pectoral (anterior) group of axillary lymph nodes. In breast cancer, it is possible that cancer cells leave the breast and lodge in the pectoral nodes. From here, metastasis may develop in other axillary nodes. Most of the remaining lymphatic drainage is to the sternal (parasternal) lymph nodes (see Figures 15.9b and 15.12). ■

DEVELOPMENTAL ANATOMY OF THE LYMPHATIC SYSTEM

Objective

- Describe the development of the lymphatic system.

The lymphatic system begins to develop by the end of the fifth week of embryonic life. *Lymphatic vessels* develop from **lymph sacs** that arise from developing veins, which are derived from **mesoderm.**

Figure 15.12 / Principal lymph nodes of the thorax.

The parietal lymph nodes drain the thoracic wall, while the visceral lymph nodes drain the viscera of the thorax.

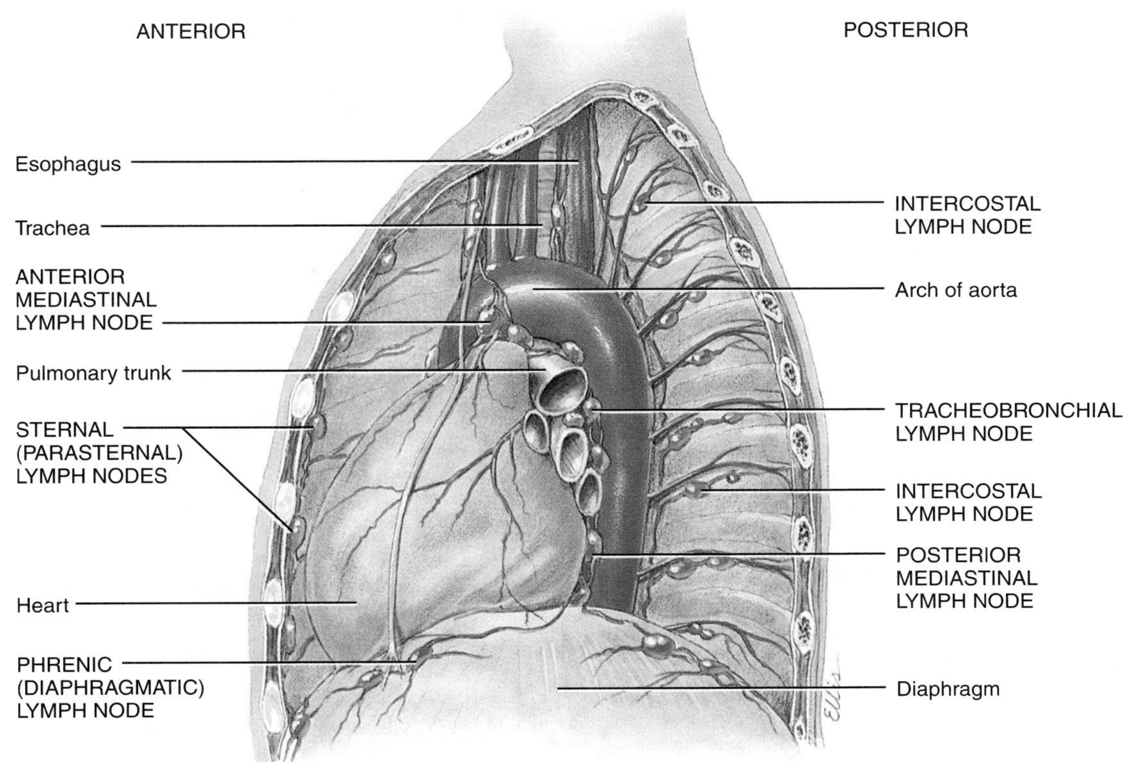

View from left side showing thoracic lymph nodes

Which general group of lymph nodes drains mostly the diaphragm?

The first lymph sacs to appear are the paired **jugular lymph sacs** at the junction of the internal jugular and subclavian veins (Figure 15.13). From the jugular lymph sacs, capillary plexuses spread to the thorax, upper limbs, neck, and head. Some of the plexuses enlarge and form lymphatic vessels in their respective regions. Each jugular lymph sac retains at least one connection with its jugular vein, the left one developing into the superior portion of the thoracic duct (left lymphatic duct).

The next lymph sac to appear is the unpaired **retroperitoneal lymph sac** at the root of the mesentery of the intestine. It develops from the primitive vena cava and mesonephric (primitive kidney) veins. Capillary plexuses and lymphatic vessels spread from the retroperitoneal lymph sac to the abdominal viscera and diaphragm. The sac establishes connections with the cisterna chyli but loses its connections with neighboring veins.

At about the time the retroperitoneal lymph sac is developing, another lymph sac, the **cisterna chyli,** develops inferior to the diaphragm on the posterior abdominal wall. It gives rise to the inferior portion of the *thoracic duct* and the *cisterna chyli* of the thoracic duct. Like the retroperitoneal lymph sac, the cisterna chyli also loses its connections with surrounding veins.

Figure 15.13 / Development of the lymphatic system.

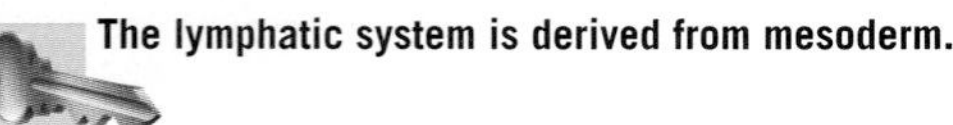
The lymphatic system is derived from mesoderm.

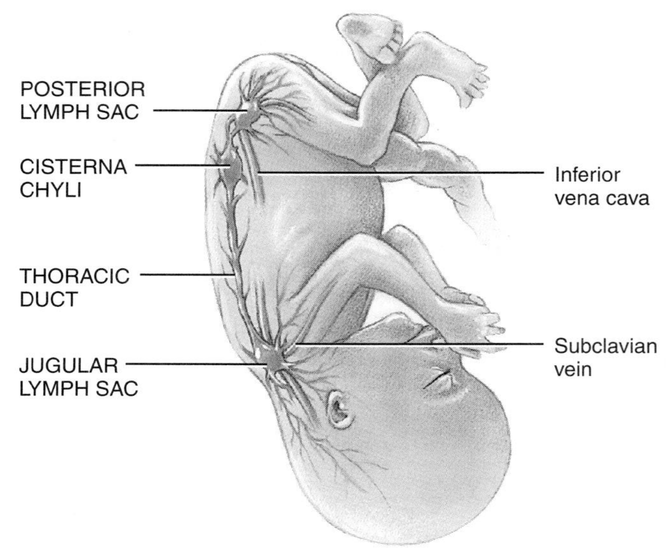

When does the lymphatic system begin to develop?

The last of the lymph sacs, the paired **posterior lymph sacs,** develop from the iliac veins. The posterior lymph sacs produce capillary plexuses and lymphatic vessels of the abdominal wall, pelvic region, and lower limbs. The posterior lymph sacs join the cisterna chyli and lose their connections with adjacent veins.

With the exception of the anterior part of the sac from which the cisterna chyli develops, all lymph sacs become invaded by **mesenchymal cells** and are converted into groups of *lymph nodes.*

The *spleen* develops from **mesenchymal cells** between layers of the dorsal mesentery of the stomach. The *thymus* arises as an outgrowth of the **third pharyngeal pouch** (see Figure 22.8 on page 676).

✓ Name the four lymph sacs from which lymphatic vessels develop.

AGING AND THE LYMPHATIC SYSTEM

Objective

- Describe the effects of aging on the lymphatic system.

With advancing age, elderly individuals become more susceptible to all types of infections and malignancies. Their response to vaccines is decreased, and they tend to produce more autoantibodies (antibodies against their body's own molecules). In addition, the immune system exhibits lowered levels of function. For example, T cells become less responsive to antigens, and fewer T cells respond to infections. This may result from age-related atrophy of the thymus or decreased production of thymic hormones. Because the T cell population decreases with age, B cells are also less responsive. Consequently, antibody levels do not increase as rapidly in response to a challenge by an antigen, resulting in increased susceptibility to various infections. It is for this key reason that elderly individuals are encouraged to get influenza (flu) vaccinations each year.

✓ What changes occur in the lymphatic system with advancing age?

APPLICATIONS TO HEALTH

AIDS: Acquired Immunodeficiency Syndrome

Acquired immunodeficiency syndrome (AIDS) is a condition in which a person experiences a telltale assortment of infections as a result of the progressive destruction of immune system cells by the **human immunodeficiency virus (HIV).** As we will see, AIDS represents the end stage of infection by HIV. A person infected with HIV may be symptom-free for many years, even while the virus is actively attacking the immune system. HIV infection is serious and usually fatal because it takes control of and destroys the very cells that the body deploys to attack the virus.

Epidemiology

Because HIV is present in the blood and some body fluids, it is most effectively transmitted by actions or practices that involve the exchange of blood or body fluids between people. Within most populations, HIV is transmitted in semen or vaginal fluid during unprotected (without a condom) anal, vaginal, or oral sex. The likelihood of transmission of HIV from an infected person to an uninfected person varies significantly, depending on the type of exposure or contact involved. The risk of becoming infected with HIV through unprotected oral sex is lower than that of unprotected anal or vaginal sex. HIV is also transmitted by direct blood-to-blood contact, such as occurs among intravenous drug users who share hypodermic needles. People at high risk are the sexual partners of HIV-infected individuals; those at lesser risk include health-care professionals who may be accidentally stuck by HIV-contaminated hypodermic needles. In addition, HIV may be transmitted from a mother to her fetus or suckling infant. In the United States and Europe prior to 1985, HIV was unknowingly spread by the transfusion of blood and blood products containing the virus. Effective HIV screening of blood instituted after 1985 has largely eliminated this mode of HIV transmission in the United States and other developed nations. By contrast, in sub-Saharan Africa, where two-thirds of those infected with HIV live, 25% of transfused blood is not screened for HIV.

Most people in economically advantaged, industrial nations who are diagnosed as having AIDS are either homosexual men who have engaged in unprotected anal intercourse or intravenous drug users. The rate of new HIV infections in the United States paints a different and disturbing picture: The greatest increases in new HIV infections are among people of color, women, and teenagers. In developing nations, HIV is largely transmitted during unprotected heterosexual intercourse, but the problem is compounded by a contaminated blood supply. Of the 40 million people infected with HIV worldwide in the year 2000, about half were women and one-quarter were children.

HIV is a very fragile virus; it cannot survive for long outside the human body. The virus is not transmitted by insect bites. It is important to understand that one cannot become infected by casual physical contact with an HIV-infected person, such as by hugging or sharing household items. The virus can be eliminated from personal care items and medical equipment by exposing them to heat (135°F for 10 minutes) or by cleaning them with common disinfectants such as hydrogen peroxide, rubbing alcohol, Lysol, household bleach, or germicidal cleansers such as Betadine or Hibiclens. Standard dishwashing and clothes washing also kills HIV.

The epidemiology of HIV also suggests ways the disease can be prevented. The chance of transmitting or of being infected by HIV during vaginal or anal intercourse can be greatly minimized—although not entirely eliminated—by the use of latex condoms. Public health programs aimed at encouraging intravenous drug users not to share needles have proved effective at checking the increase in new HIV infections among this population. Also, the prophylactic administration of certain drugs such as AZT (discussed shortly) to pregnant HIV-infected women has proved remarkably effective in minimizing the transmission of the virus to their newborn babies.

Pathogenesis of HIV Infection

HIV consists of a single strand of genetic material surrounded by a protective protein coat. To replicate, a virus must enter a host cell, where it uses the cell's enzymes, ribosomes, and nutrients to reproduce copies of its genetic information. Unlike most viruses, however, HIV is a form of **retrovirus,** a virus whose genetic information is carried in RNA instead of DNA. Once inside the host cell, a viral enzyme called **reverse transcriptase** reads the viral RNA strand and makes a DNA copy.

The coat that surrounds HIV's RNA and reverse transcriptase is composed of many molecules of a protein termed P24. In addition, the coat is wrapped by an envelope composed of a lipid bilayer penetrated by a distinctive assortment of glycoproteins (Figure 15.14). One of these glycoproteins, GP120, functions as the docking protein that binds to CD4 molecules on T cells, macrophages, and dendritic cells. In addition, HIV must simultaneously attach to a coreceptor in the host cell's plasma membrane—a molecule designated CCR5 in antigen-presenting cells and CXCR4 in T cells—before it can gain entry into the cell. Another glycoprotein, GP41, helps the viral lipid bilayer fuse with the host cell's lipid bilayer.

The binding of HIV's docking proteins with the receptors and coreceptors in the host cell's plasma membrane causes receptor-mediated endocytosis, which brings the virus into the cell's cytoplasm. Once inside the cell, HIV sheds its protein coat. The reverse transcriptase enzyme makes a DNA copy of the viral RNA, and the viral DNA copy becomes integrated into the cell's DNA. Thus, the viral DNA is duplicated along with the host cell's DNA during normal cell division. In addition, under circumstances that are still not well understood, the viral DNA can cause the infected cell to begin producing millions of copies of viral RNA and to assemble new protein coats for each copy. The new HIV copies bud off from the cell's plasma membrane and circulate in the blood to infect other cells.

HIV mainly damages T4 (helper T) cells, and it does so in various ways. Over 100 billion viral copies may be synthesized each day. The viruses bud so rapidly from an infected cell's plasma membrane that cell lysis eventually occurs. In addition, the body's defenses attack the infected cells, killing them as well as the viruses they harbor.

Figure 15.14 / Human immunodeficiency virus (HIV), the causative agent of AIDS. The core contains RNA and reverse transcriptase, plus several other enzymes. The protein coat (capsid) around the core consists of a protein called P24. The envelope consists of a lipid bilayer studded with glycoproteins (GP120 and GP41) that play a vital role when HIV binds to and enters certain target cells.

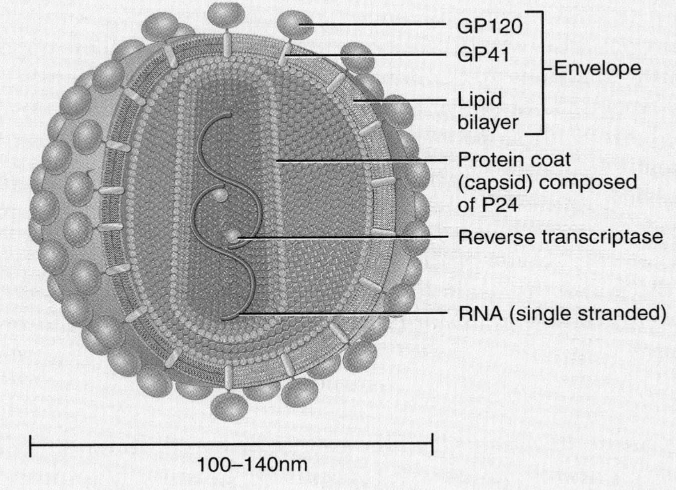

As we saw in this chapter, T4 cells are instrumental in orchestrating the actions of the immune system. In most HIV-infected individuals, the body is able to replace HIV-infected T4 cells at about the same rate that they are destroyed. This process continues for many years while the body's ability to replace T4 cells is slowly exhausted; the number of T4 cells in circulation progressively declines, at an estimated rate of about 20 million cells per day.

Signs, Symptoms, and Diagnosis of HIV Infection

Immediately following infection with HIV, most people experience a brief flu-like illness. Common signs and symptoms are fever, fatigue, rash, headache, joint pain, sore throat, and swollen lymph nodes. In addition, about 50% of infected people have night sweats. After three to four weeks, plasma cells begin secreting antibodies to components of the HIV protein coat. These antibodies are detectable in blood plasma and form the basis for some of the screening tests for HIV. When people test "HIV-positive," this usually means they have antibodies to HIV in their bloodstream. If an acute HIV infection is suspected but the antibody test is negative, laboratory tests based on detection of HIV's RNA or P24 coat protein in blood plasma can confirm the presence of HIV.

Progression to AIDS

After a period of 2–10 years, the virus destroys enough T4 cells that most infected people begin to experience symptoms of immunodeficiency. HIV-infected people commonly have enlarged lymph nodes and experience persistent fatigue, involuntary weight loss, night sweats, skin rashes, diarrhea, and various lesions of the mouth and gums. In addition, the virus may begin to infect neurons in the brain, affecting the person's memory and producing visual disturbances.

As the immune system slowly collapses, an HIV-infected person becomes susceptible to a host of opportunistic infections, diseases caused by microorganisms that are normally held in check but now proliferate because of the defective immune system. AIDS is diagnosed when the T4 cell count drops below 200 cells per nanoliter (= cubic millimeter) of blood or when opportunistic infections arise, whichever occurs first. Typically, it is these opportunistic infections that eventually cause the death of the person.

About 5% of individuals infected with HIV have not developed AIDS; these people are called long-term nonprogressors. Their T4 count is stable, and they are symptom free. Their survival may be due to infection by a weak strain of HIV or the presence of potent natural killer cells and strong antibodies against HIV. Some of these individuals have mutations in both of the genes that code for the coreceptor CCR5, which likely prevents certain strains of HIV from entering their antigen-presenting cells.

Treatment of HIV Infection

At present, infection with HIV cannot be cured, and despite intensive research, no effective vaccine is yet available to provide immunity against HIV. However, two categories of drugs have proved successful in extending the life of many HIV-infected individuals. The first category, *reverse transcriptase inhibitors*, interferes with the action of the reverse transcriptase enzyme that the virus uses to convert its RNA into a DNA copy. Among the drugs in this category are azidovudine (AZT), didanosine (ddI), dideoxycytidine (ddC), and stavudine (d4T). The second and more recently discovered category is the *protease inhibitors*. These drugs interfere with the action of protease, a viral enzyme that cuts proteins into pieces that are assembled into the coat of newly produced HIV particles. Drugs in this category include nelfinavir, saquinavir, ritonaxir, and indinavir.

Beginning in late 1995, researchers discovered that most HIV-infected individuals who receive *triple therapy*—a combination of two differently acting reverse transcriptase inhibitors and one protease inhibitor—experience a drastic reduction in viral load (the number of copies of HIV RNA in a milliliter of plasma) and an increase in the number of T4 cells in whole blood. Not only does triple therapy delay the progression of HIV infection to AIDS, but many individuals with AIDS have seen the remission or disappearance of opportunistic infections and an apparent return to health. Unfortunately, triple therapy is very costly (exceeding $10,000 per year), the dosing schedule is grueling, and not all people can tolerate the harsh side effects of these drugs. It is currently thought that people who find the drugs helpful must keep taking them for a long time, perhaps for life.

Allergic Reactions

A person who is overly reactive to a substance that is tolerated by most other people is said to be **hypersensitive (allergic).** Whenever an allergic reaction occurs, some tissue injury results. The antigens that induce an allergic reaction are called **allergens.** Common allergens include certain foods (milk, peanuts, shellfish, eggs), antibiotics (penicillin, tetracycline), vaccines (pertussis, typhoid), venoms (honeybee, wasp, snake), cosmetics, chemicals in plants such as poison ivy, pollens, dust, molds, iodine-containing dyes used in certain x-ray procedures, and even microbes.

Type I (anaphylactic) reactions are the most common hypersensitivity reactions and occur within a few minutes after a person sensitized to an allergen is reexposed to it. **Anaphylaxis** (an′-a-fi-LAK-sis) results from the interaction of allergens with certain antibodies on the surface of mast cells and basophils. In response to certain allergens, some people produce these antibodies that bind to the surface of mast cells and basophils. The next time the same allergen enters the body, it attaches to the antibodies already present. In response, the mast cells and basophils release histamine, prostaglandins, leukotrienes, and kinin. Collectively, these mediators cause vasodilation, increased blood capillary permeability, increased smooth muscle contraction in the airways of the lungs, and increased mucus secretion. As a result, a person may experience inflammatory responses, difficulty in breathing through the constricted airways, and a runny nose from excess mucus secretion. In **anaphylactic**

shock, which may occur in a susceptible individual who has just received a triggering drug or been stung by a wasp, wheezing and shortness of breath as airways constrict are usually accompanied by shock due to vasodilation and fluid loss from blood. This life-threatening emergency is usually treated by injecting epinephrine to dilate the airways and strengthen the heartbeat.

Lymphomas

Lymphomas (lim-FŌ-mas; *lympha* = clear water; *oma* = tumor) are cancers of the lymphatic system, especially the lymph nodes. Most have no known cause. The two principal types of lymphomas are Hodgkin's disease and non-Hodgkin's lymphoma.

Hodgkin's disease (HD) is characterized by a painless, non-tender enlargement of one or more lymph nodes, most commonly in the neck, chest, and axilla. If the disease has metastasized from these sites, fever, night sweats, weight loss, and bone pain also occur. HD primarily affects individuals between ages 15 and 35 and those over 60, and it is more common in males. If diagnosed early, HD has a 90–95% cure rate. Treatment consists of radiation therapy, chemotherapy, and bone marrow transplantation.

Non-Hodgkin's lymphoma (NHL), which is more common than HD, occurs in all age groups, the incidence increasing with age to a maximum between ages 45 and 70. NHL may start the same way as HD but may also include an enlarged spleen, anemia, and general malaise. Up to half of all individuals with NHL are cured or survive for a lengthy period. Treatment options include radiation therapy, chemotherapy, and bone marrow transplantation.

Infectious Mononucleosis

Infectious mononucleosis (IM) is a contagious disease caused by the *Epstein-Barr virus (EBV).* It occurs mainly in children and young adults, and more often in females than in males by a 3:1 ratio. The virus most commonly enters the body through intimate oral contact such as kissing; it then multiplies in lymphatic tissues and spreads into the blood, where it infects and multiplies in B lymphocytes, the primary host cells. As a result of this infection, the B cells become enlarged and abnormal in appearance such that they resemble monocytes, the primary reason for the term *mononucleosis.* Signs and symptoms include an elevated white blood cell count with an abnormally high percentage of lymphocytes, fatigue, headache, dizziness, sore throat, enlarged and tender lymph nodes, and fever. There is no cure for infectious mononucleosis, but the disease usually runs its course in a few weeks.

KEY MEDICAL TERMS ASSOCIATED WITH THE LYMPHATIC SYSTEM

Adenitis (ad′-e-NĪ-tis; *aden-* = gland; *-itis* = inflammation of) Enlarged, tender, and inflamed lymph nodes resulting from an infection.

Allograft (AL-ō-graft; *allo-* = other) A transplant between genetically distinct individuals of the same species. Skin transplants from other people and blood transfusions are allografts.

Autograft (AW-tō-graft; *auto-* = self) A transplant in which one's own tissue is grafted to another part of the body (such as skin grafts for burn treatment or plastic surgery).

Autoimmune disease (aw-tō-i-MYŪN) A disease in which the immune system fails to recognize self-antigens and attacks the person's own cells. Examples are rheumatoid arthritis (RA), systemic lupus erythematosus (SLE), rheumatic fever, hemolytic and pernicious anemias, Addison's disease, Graves' disease, insulin-dependent diabetes mellitus, myasthenia gravis, multiple sclerosis (MS), and ulcerative colitis. Also called **autoimmunity.**

Gamma globulin (GLOB-yū-lin) Suspension of immunoglobulins from blood consisting of antibodies that react with a specific pathogen. It is prepared by injecting the pathogen into animals, removing blood from the animals after antibodies have been produced, isolating the antibodies, and injecting them into a human to provide short-term immunity.

Hypersplenism (hī′-per-SPLĒN-izm; *hyper* = over) Abnormal splenic activity due to splenic enlargement and associated with an increased rate of destruction of normal blood cells.

Lymphadenopathy (lim-fad′-e-NOP-a-thē; *lymph-* = clear fluid; *-pathy* = disease) Enlarged, sometimes tender lymph glands.

Lymphedema (lim′-fe-DĒ-ma; *edema* = swelling) Accumulation of lymph producing subcutaneous tissue swelling.

Lymphomas (lim-FŌ-mas; *-omas* = tumors) Cancers of the lymphatic system, especially the lymph nodes. The two principal types of lymphomas are Hodgkin's disease and non-Hodgkin's lymphoma.

Splenomegaly (splē′-nō-MEG-a-lē; *mega-* = large) Enlarged spleen.

Systemic lupus erythematosus (er-e′-thēm-a-TŌ-sus), ***SLE,*** or ***lupus*** (*lupus* = wolf) An autoimmune, noncontagious, inflammatory disease of connective tissue, occurring mostly in young women. In SLE, damage to blood vessel walls result in the release of chemicals that mediate inflammation. Symptoms of SLE include joint pain, slight fever, fatigue, oral ulcers, weight loss, enlarged lymph nodes and spleen, photosensitivity, rapid loss of large amounts of scalp hair, and sometimes an eruption across the bridge of the nose and cheeks called a "butterfly rash."

Tonsillectomy (ton′-si-LEK-tō-mē; *-ectomy* = excision) Removal of a tonsil.

Xenograft (ZEN-ō-graft; *xeno-* = strange or foreign) A transplant between animals of different species. Xenografts from porcine (pig) or bovine (cow) tissue may be used in a human as a physiological dressing for severe burns.

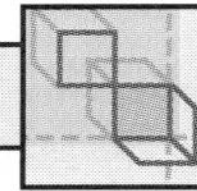

STUDY OUTLINE

Introduction (p. 474)

1. The ability to ward off disease is called resistance. Lack of resistance is called susceptibility.
2. Nonspecific resistance refers to a wide variety of body responses against a wide range of pathogens; specific resistance or immunity involves activation of specific lymphocytes to combat a particular foreign substance.

The Lymphatic System (p. 474)

1. The lymphatic system carries out immune responses and consists of lymph, lymphatic vessels, and structures and organs that contain lymphatic tissue (specialized reticular tissue containing many lymphocytes).
2. The lymphatic system drains interstitial fluid, transports dietary lipids, and protects against pathogens through immune responses.
3. Lymphatic vessels begin as lymph capillaries with one closed end in tissue spaces between cells.
4. Interstitial fluid drains into lymphatic capillaries, thus forming lymph.
5. Lymph capillaries merge to form larger vessels, called lymphatic vessels, which convey lymph into and out of structures called lymph nodes.
6. The route of lymph flow is from lymph capillaries to lymphatic vessels to lymph trunks to the thoracic duct or right lymphatic duct to the subclavian veins.
7. Lymph flows as a result of skeletal muscle contractions and respiratory movements. It is also aided by valves in lymphatic vessels.
8. The primary lymphatic organs are red bone marrow and the thymus. Secondary lymphatic organs are lymph nodes, spleen, and lymphatic nodules.
9. The thymus lies between the sternum and the large blood vessels above the heart. It is the site of T cell maturation.
10. Lymph nodes are encapsulated, oval structures located along lymphatic vessels.
11. Lymph enters nodes through afferent lymphatic vessels, is filtered, and exits through efferent lymphatic vessels.
12. Lymph nodes are the site of proliferation of plasma cells and T cells.
13. The spleen is the largest single mass of lymphatic tissue in the body. It is a site of B cell proliferation into plasma cells and phagocytosis of bacteria and worn-out red blood cells.
14. Lymphatic nodules are scattered throughout the mucosa of the gastrointestinal, respiratory, urinary, and reproductive tracts. This lymphatic tissue is termed mucosa-associated lymphoid tissue (MALT).

Principal Groups of Lymph Nodes (p. 485)

1. Lymph nodes are scattered throughout the body in superficial and deep groups.
2. The principal groups of lymph nodes are found in the head and neck, upper limbs, lower limbs, abdomen and pelvis, and thorax.

Developmental Anatomy of the Lymphatic System (p. 490)

1. Lymphatic vessels develop from lymph sacs, which arise from developing veins. Thus, they are derived from mesoderm.
2. Lymph nodes develop from lymph sacs that become invaded by mesenchymal cells.

Aging and the Lymphatic System (p. 492)

1. With advancing age, individuals become more susceptible to infections and malignancies, respond less well to vaccines, and produce more autoantibodies.
2. Immune responses also diminish with age.

SELF-QUIZ QUESTIONS

Choose the one best answer to the following questions.

1. Put the following items in the correct order for the pathway of lymph from the stomach to the blood. (1) thoracic duct, (2) lymphatic vessels and lymph nodes, (3) lymph capillaries, (4) interstitial fluid, (5) left subclavian vein.
   a. 2,3,4,5,1, **b.** 4,3,2,1,5, **c.** 1,5,3,2,4, **d.** 4,2,1,5,3 **e.** 3,2,4,1,5.
2. Which of the following is *not* a major site of lymphatic tissue? (a) tonsils, (b) thymus, (c) kidneys, (d) spleen, (e) lymph nodes
3. Which of the following is an important function of the lymphatic system? (a) controlling body temperature by evaporation of sweat, (b) manufacturing all white blood cells, (c) transporting fluids out to, and back from, the body tissues, (d) returning fluid and proteins to the cardiovascular system, (e) all of the above.
4. The spleen (a) serves as storage site for blood platelets, (b) is an organ in which phagocytosis of aged erythrocytes takes place, (c) is an organ in which phagocytosis of aged or damaged platelets occurs, (d) is a site of blood formation in the fetus, (e) all of the above.
5. A major difference between the spleen and lymph nodes is that (a) lymph nodes have afferent and efferent lymphatic vessels; the spleen has only efferent lymphatic vessels (b) the spleen is located in the abdomen; lymph nodes are not, (c) lymph nodes are not enclosed in a capsule; whereas the spleen is, (d) lymph nodes have afferent and efferent lymphatic vessels; the spleen has neither, (e) lymph flows through sinuses in the spleen, but not through sinuses in the lymph nodes.
6. Which of the following statements about HIV transmission is true? (a) Standard dishwashing and clothes washing kills HIV. (b) The transmission of HIV requires the transfer of, or direct contact with, infected body fluids. (c) Organ transplants and artificial insemination are routes for HIV transmission. (d) HIV may be present in semen, blood, and breast milk. (e) All of the above.

7. Both the thoracic duct and the right lymphatic duct empty directly into the (a) axillary lymph nodes, (b) superior vena cava, (c) cisterna chyli, (d) subclavian arteries, (e) subclavian veins.

8. Which of the following statements is true about the flow of lymph? (1) It is aided by a pressure gradient difference—toward the thoracic region. (2) It is maintained primarily by the milking action of muscles. (3) It is possible because of valves in lymphatic vessels.

   a. 1 only, **b.** 2 only, **c.** 3 only **d.** all of the above, **e.** 2 and 3.

Complete the following.

9. The lymphatic organ known as the ________ does not function as an important site for immune responses; it is the site of maturation of T cells that populate other lymphatic organs.

10. Lymphatic vessels are similar in structure to ________ but have thinner walls and more valves.

11. The ________ and the ________ are the primary lymphatic organs because they produce the lymphocytes for the immune system.

12. The thoracic duct starts in the lumbar region as a dilation known as the ________ .

13. The lymphatic system derives from the ________ -derm.

14. Lymphatic nodules in lamina propria of mucous membranes are collectively referred to as ________ .

Are the following statements true or false?

15. The group of tonsils usually consists of one palatine tonsil, 2 pharyngeal tonsils, and three lingual tonsils

16. Lymph is conveyed into a node at several points by afferent lymphatic vessels.

17. The main difference between lymph and interstitial fluid is location.

18. Most of the lymph nodes that drain viscera bear the same names as arteries that supply these viscera.

19. Lymphatic capillaries are slightly smaller in diameter and less porous than systemic capillaries.

20. Match the following lymph nodes with their descriptions.

____	**(a)** facial nodes	**(1)** drain the knee, leg, and heel
____	**(b)** axillary nodes	**(2)** three groups of nodes that drain the stomach, liver, pancreas, and spleen
____	**(c)** popliteal nodes	**(3)** largest group of nodes in the neck
____	**(d)** common iliac nodes	**(4)** three groups of nodes that drain the mucous membranes of the cheeks, eyelids, and nose
____	**(e)** celiac nodes	**(5)** anterior to sacrum; drain the rectum, prostate gland, and pelvic wall
____	**(f)** phrenic nodes	**(6)** three groups of nodes that drain parts of the liver, diaphragm, and abdominal wall
____	**(g)** sacral nodes	**(7)** deep nodes; drain the upper limbs, skin and muscles of the thorax, and part of the neck
____	**(h)** deep cervical nodes	**(8)** arranged along blood vessels of the same name; drain the pelvic viscera

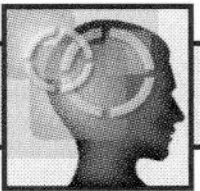

CRITICAL THINKING QUESTIONS

1. Jamal got a splinter in his right heel while he was playing beach volleyball. He neglected to clean it properly and it became infected. The first aid station warned him to have the wound taken care of before the infection spreads to his blood. How can an infection travel from his foot to his cardiovascular system?
   HINT *The lymphatic and cardiovascular systems can be thought of together as the circulatory system.*

2. Four-year old Kelsey had a history of repeated throat infections, averaging 10 per year. She breathed loudly through her mouth and even snored. Following the pediatrician's recommendation, she had an operation to remove the troublesome organs. When she returned to preschool, she told the other kids about all the ice cream she got to eat following her "tonsil-X-me." Where are these organs located and what is their function?
   HINT *Kelsey needed all that ice cream "'cause my throat hurt something awful!"*

3. Jason was hit by a drunk driver and may have a ruptured spleen. His mother is worried that he's suffered a fatal injury. Can you live without a spleen? What happens if you need to have your spleen removed?
   HINT *Some organs are more essential than others.*

4. With all the talk about AIDS in the news, your younger brother is worried about catching it from a germy bathroom door knob. Should he be concerned? Explain the cause of AIDS.
   HINT *Antibiotics won't cure AIDS so it's not a bacterial infection.*

5. An infection with a tropical parasite may block a lymphatic vessel. What would be the effect of blockage of the left subclavian trunk?
   HINT *If the pipe is blocked in your sink, it won't drain.*

ANSWERS TO FIGURE QUESTIONS

15.1 Red bone marrow contains stem cells that develop into lymphocytes.

15.2 Lymph is more similar to interstitial fluid than to plasma because the protein content of lymph is low.

15.3 The left and right lumbar trunks and the intestinal trunk empty into the cisterna chyli, which then drains into the thoracic duct.

15.4 Inhalation promotes the movement of lymph from abdominal lymphatic vessels toward the thoracic region.

15.5 T cells mature in the thymus.

15.6 Foreign substances that enter a lymph node may be phagocytized by macrophages or attacked by lymphocytes that mount immune responses.

15.7 White pulp of the spleen functions in immunity; red pulp of the spleen performs functions related to blood cells.

15.8 Deep cervical lymph nodes.

15.9 Lateral lymph nodes.

15.10 Superficial inguinal lymph nodes.

15.11 Gastric, hepatic, and pancreaticosplenic nodes.

15.12 Phrenic lymph nodes.

15.13 By the end of the fifth week of gestation.

16

NERVOUS TISSUE

CONTENTS AT A GLANCE

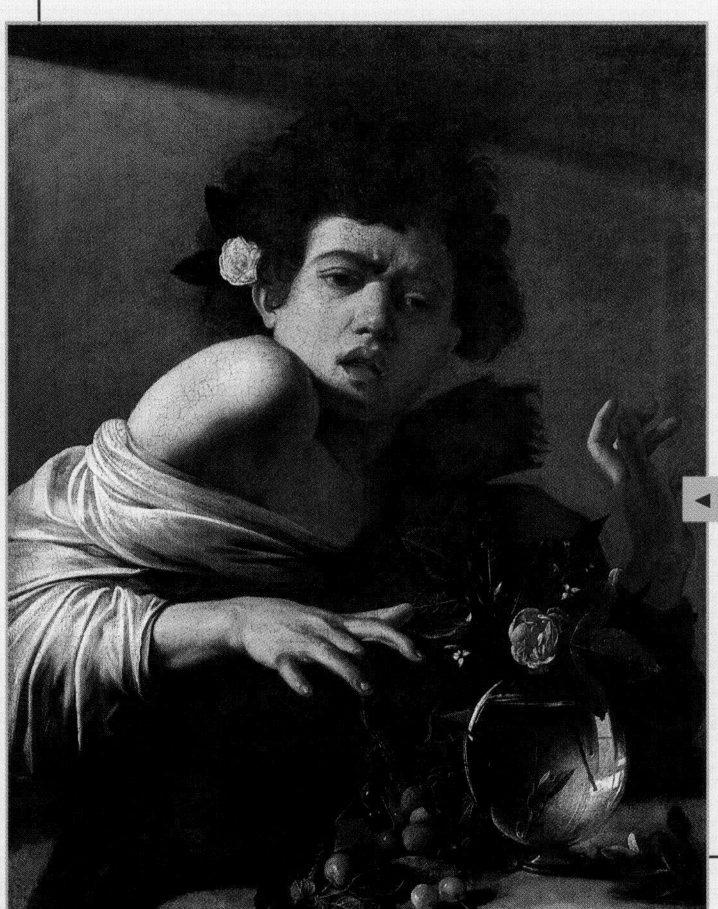

◀ Page 502

A lizard's jaws clamp down on a boy's finger. The pain causes him to flinch. Which kinds of tissues allow for this process to occur?

Page 509 ▶

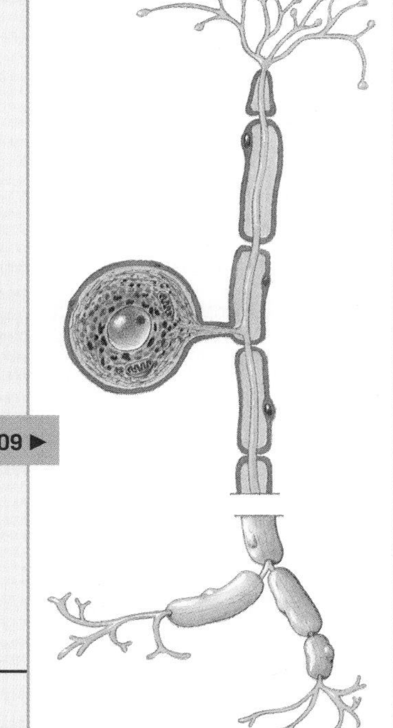

INTRODUCTION

The body has two control centers, one for rapid response—the nervous system—and one that makes slower but no less important adjustments for maintaining homeostasis—the endocrine system. Because the nervous system is quite complex, we will consider its parts in several related chapters. This chapter focuses on nervous tissue. We will discuss the basic structure and functions of nerve cells and supporting cells called neuroglia. In chapters that follow, we will examine the structure and functions of the spinal cord and spinal nerves and the brain and cranial nerves. Then we will discuss the general senses and the motor and sensory pathways to understand how nerve impulses pass from receptors to the spinal cord and brain or from the spinal cord and brain to muscles to glands. Our journey continues with a discussion of the special senses: smell, taste, vision, hearing, and equilibrium. We conclude our study of the nervous system by examining the autonomic nervous system, the part of the nervous system that operates without conscious control.

The branch of medical science that deals with the normal functioning and disorders of the nervous system is **neurology** (nū-ROL-ō-jē; *neur-* = nerve or nervous system; *-ology* = study of).

OVERVIEW OF THE NERVOUS SYSTEM

Objectives

- List the structures and describe the basic functions of the nervous system.
- Describe the organization of the nervous system.

Structure and Functions of the Nervous System

The **nervous system** is a complex, highly organized network of billions of neurons (NŪ-rons) and even more neuroglia. The structures that make up the nervous system include the brain, cranial nerves and their branches, the spinal cord, spinal nerves and their branches, ganglia, enteric plexuses, and sensory receptors (Figure 16.1). The **brain** is housed within the skull and con-

Figure 16.1 / Major structures of the nervous system.

The nervous system includes the brain, cranial nerves, spinal cord, spinal nerves, ganglia, enteric plexuses, and sensory receptors.

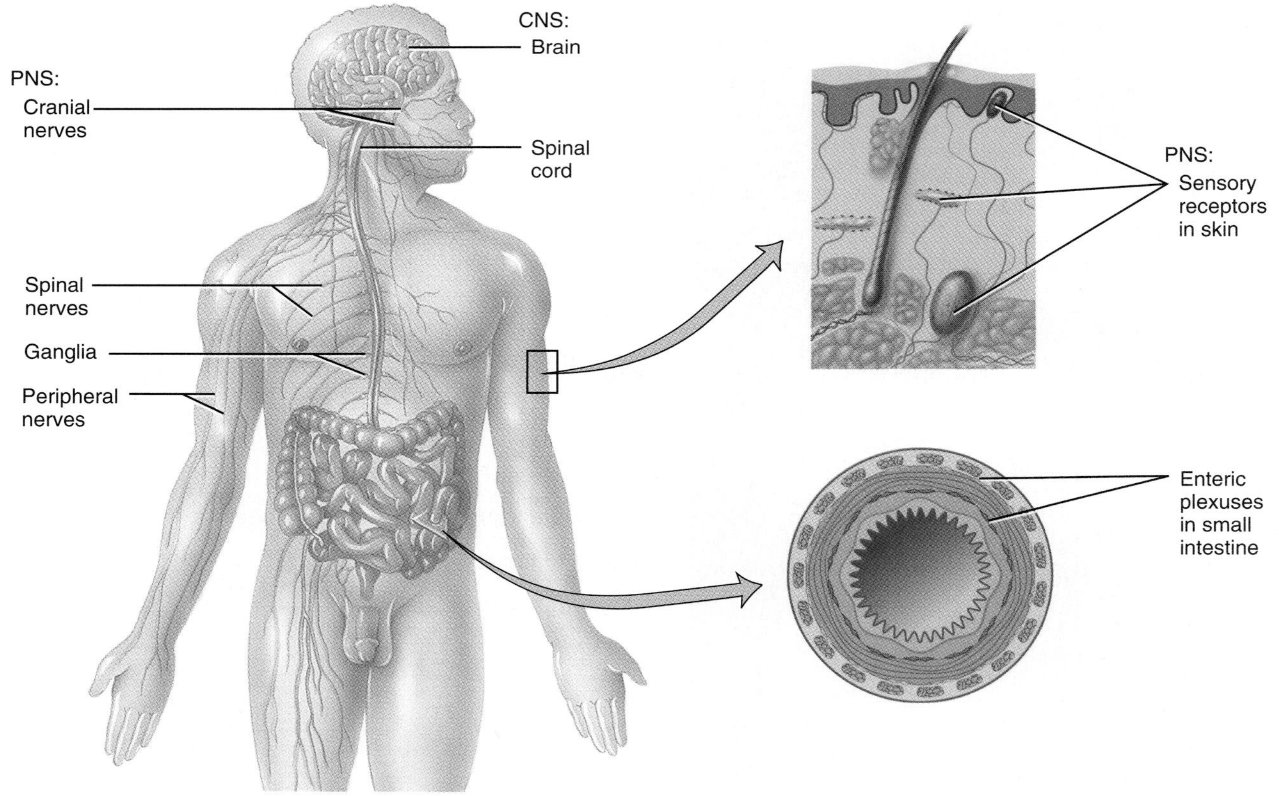

What is the total number of cranial and spinal nerves in the human body?

tains about 100 billion neurons. Twelve pairs (right and left) of **cranial nerves,** numbered I through XII, emerge from the base of the brain. Each nerve follows a defined path and serves a specific region of the body. For example, the right cranial nerve I carries signals for the sense of smell from the right side of the nose to the brain.

The **spinal cord**, which is continuous with the brain, contains about 100 million neurons. The spinal cord is encircled by the bones of the vertebral column, and passes through the foramen magnum of the skull to the brain. Thirty-one pairs of **spinal nerves** emerge from the spinal cord, each serving a specific region on the right or left side of the body. In the walls of organs of the gastrointestinal tract are extensive networks of neurons, called **enteric plexuses,** that help regulate the digestive system. **Sensory receptors** are either parts of neurons or separate, specialized cells that monitor changes in the internal or external environment.

The nervous system has three basic functions:

- **Sensory function. Sensory receptors** *detect* internal stimuli, such as an increase in blood acidity, and external stimuli, such as a raindrop landing on your arm. The neurons that carry sensory information into the brain and spinal cord are **sensory,** or **afferent, neurons.**
- **Integrative function.** The nervous system processes, or *integrates,* sensory information by analyzing and storing some of it and by making decisions regarding appropriate responses. The neurons that serve this function, **interneurons** *(association neurons),* make up the vast majority of neurons in the body.
- **Motor function.** The nervous system's motor function involves *responding* to integration decisions. The neurons that serve this function are **motor,** or **efferent, neurons,** which carry information out of the brain and spinal cord. The cells and organs innervated by motor neurons are termed **effectors;** examples are muscle fibers and glandular cells.

Organization of the Nervous System

The nervous system is composed of two subsystems: the **central nervous system (CNS)** and the **peripheral** (pe-RIF-er-al) **nervous system (PNS).** The CNS consists of the brain and spinal cord, which integrate and correlate many different kinds of incoming sensory information. The CNS is also the source of thoughts, emotions, and memories. Most nerve impulses that stimulate muscles to contract and glands to secrete originate in the CNS. The PNS includes all nervous tissue outside the CNS: cranial nerves and their branches, spinal nerves and their branches, ganglia, and sensory receptors.

The PNS may be subdivided into a **somatic nervous system (SNS)** (*somat-* = body), an **autonomic nervous system (ANS)** (*auto-* = self; *-nomic* = law), and an **enteric nervous system (ENS)** (*enter-* = intestines) (Figure 16.2). The SNS consists of (1) sensory neurons that convey information from somatic and special sensory receptors primarily in the head, body wall, and limbs to the CNS, and (2) motor neurons from the CNS that conduct impulses to *skeletal muscles* only. Because these motor responses can be consciously controlled, the action of this part of the PNS is *voluntary.*

Figure 16.2 / Organization of the nervous system. Subdivisions of the PNS are the somatic nervous system (SNS), the autonomic nervous system (ANS), and the enteric nervous system (ENS).

The two main subdivisions of the nervous system are (1) the central nervous system (CNS), consisting of the brain and spinal cord, and (2) the peripheral nervous system (PNS), consisting of all nervous tissue outside the CNS.

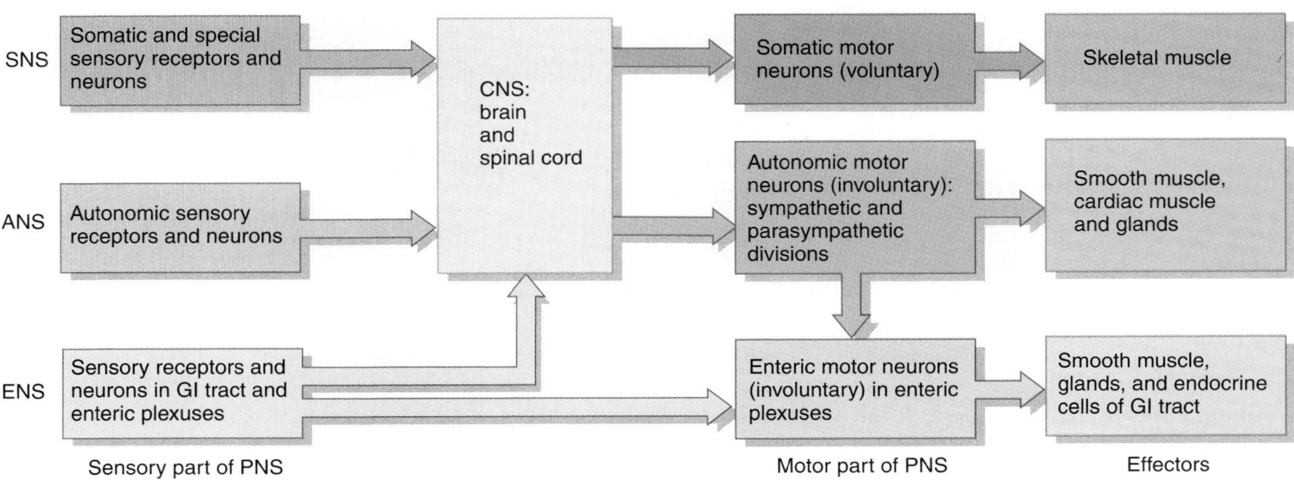

What terms are given to neurons that carry input to the CNS? That carry output from the CNS?

14th Century

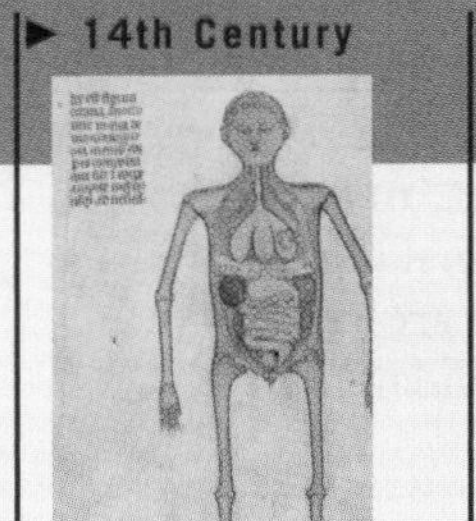

15th Century

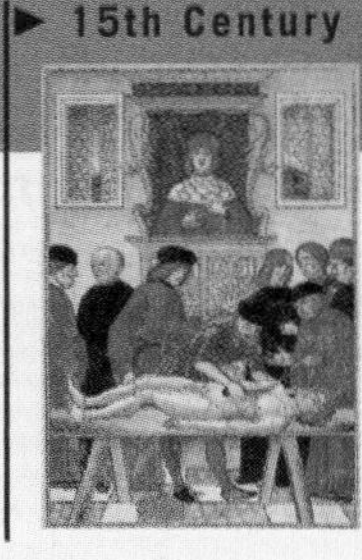

16th Century

17th Century

18th Century

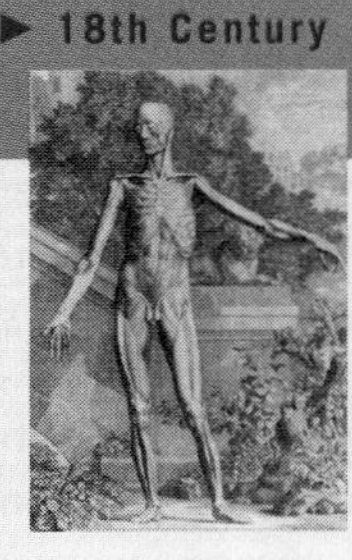

20th Century

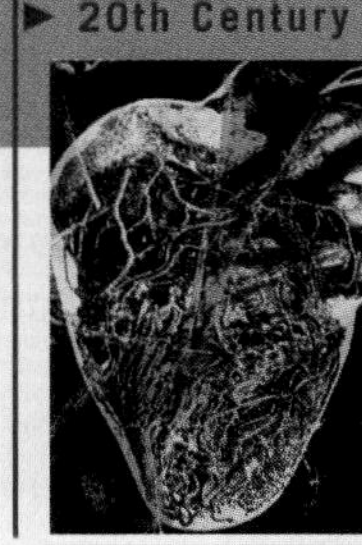

21st Century

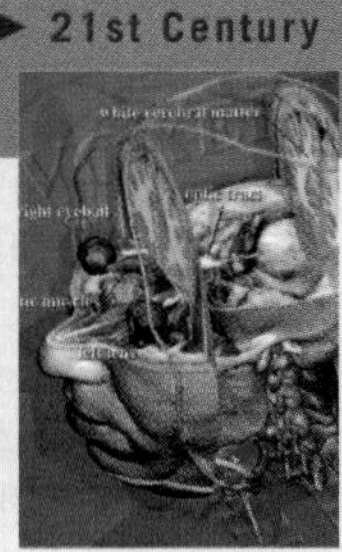

CHANGING IMAGES

Response to Change

The ability of humans to respond quickly to stimuli diminishes threats in both the internal and external environments. Drastic changes in blood pressure are alleviated or a hand on a hot stove is quickly removed. Without the ability to immediately react to an ever-changing environment, homeostasis would be constantly disrupted, and life itself would be impossible.

This ability is achieved through a network of connections that wires our bodies with the ability to sense and respond to various stimuli. But at the most basic level, what is it that allows for these functions? The answer lies in the tissues that comprise the nervous system - the most complex system of our body.

Pictured here is a 1594 painting, *Boy Bitten by a Lizard*, by Italian Baroque painter Michelangelo Merisi Caravaggio. This image dramatically demonstrates the functions of nervous tissue. Can you sense the feeling of pain? Notice the hands quickly withdrawing. Examine the facial muscles contracting in a painful grimace. As you read this chapter about nervous tissue and learn of the various types of neurons and neuroglia, think about how important the nervous system is to you in your daily activities.

1594 AD

The ANS consists of (1) sensory neurons that convey information from autonomic sensory receptors, located primarily in the viscera, to the CNS, and (2) motor neurons from the CNS that conduct nerve impulses to *smooth muscle*, *cardiac muscle*, and *glands*. As its motor responses are not normally under conscious control, the action of the ANS is *involuntary*. The motor portion of the ANS consists of two branches, the **sympathetic division** and the **parasympathetic division.** With a few exceptions, effectors are innervated by both, and usually the two divisions have opposing actions. For example, sympathetic neurons increase heart rate, whereas parasympathetic neurons decrease it.

The ENS is the "brain of the gut," and its operation is involuntary. Once considered part of the ANS, the ENS consists of approximately 100 million neurons in enteric plexuses that extend the entire length of the gastrointestinal (GI) tract. Many of the neurons of the enteric plexuses function independently of the ANS and CNS to some extent, although they also communicate with the CNS via sympathetic and parasympathetic neurons. Sensory neurons of the ENS monitor chemical changes within the GI tract and the stretching of its walls. Enteric motor neurons govern contraction of GI tract smooth muscle, secretions of the GI tract organs such as acid secretion by the stomach, and activity of GI tract endocrine cells.

✓ Discuss the three basic functions of the nervous system.
✓ What are the functional differences between typical sensory and motor neurons? Define interneuron.
✓ Construct a table to compare the components and operation of the SNS, ANS, and ENS.
✓ Relate the terms voluntary and involuntary to the various subdivisions of the nervous system.

HISTOLOGY OF NERVOUS TISSUE

Objectives

- Contrast the histological characteristics and the functions of neuroglia and neurons.
- Distinguish between gray matter and white matter.

Nervous tissue consists of only two principal kinds of cells: neurons and neuroglia. Neurons are responsible for most special functions attributed to the nervous system: sensing, thinking, remembering, controlling muscle activity, and regulating glandular secretions. Neuroglia support, nourish, and protect the neurons and maintain homeostasis in the interstitial fluid that bathes neurons.

Neuroglia

Neuroglia (nū-RŌG-lē-a; *-glia* = glue) or **glia** constitute about half the volume of the CNS. Their name derives from early microscopists' idea that they were the "glue" that held nervous tissue together. We now know that neuroglia are not merely passive bystanders but rather active participants in the operation of nervous tissue. Generally, neuroglia are smaller than neurons and 5–50 times more numerous. In contrast to neurons, glia do not generate or propagate nerve impulses, and they can multiply and divide in the mature nervous system. In cases of injury or disease, neuroglia multiply to fill in the spaces formerly occupied by neurons. Brain tumors derived from glia, called **gliomas,** tend to be highly malignant and grow rapidly. Of the six types of neuroglia, four—astrocytes, oligodendrocytes, microglia, and ependymal cells—are found only in the CNS. The remaining two types—Schwann cells and satellite cells—are present in the PNS. The structure and functions of neuroglia are summarized in Table 16.1.

Neurons

Neurons have the ability to respond to stimuli and convert them into nerve impulses. Very simply, a **nerve impulse (nerve action potential)** is a sudden change in electrochemical force that propagates itself along the surface of the membrane of a neuron. Among other things, a nerve impulse is sustained by the movement of sodium, potassium, and other ions between interstitial fluid and the inside of a neuron. For a nerve impulse to begin, a stimulus of adequate strength must be applied to the neuron. A **stimulus** is any change in the environment of sufficient strength to initiate a nerve impulse. The ability of a neuron to respond to a stimulus and convert it into a nerve impulse is known as **excitability.**

Some neurons are tiny and relay signals over a short distance (less than 1 mm) within the CNS. Others are the longest cells in your body. Motor neurons that cause muscles to wiggle your toes, for example, extend from the lumbar region of your spinal cord (just above waist level) to your foot. Some sensory neurons are even longer. Those that allow you to feel the position of your wiggling toes stretch all the way from your foot to the lower portion of your brain. Nerve impulses travel these great distances at speeds ranging from 0.5 to 130 meters per second (1 to 280 mi/hr).

The functional contact between two neurons or between a neuron and an effector (muscle or gland) cell is called a **synapse.** The synapse between a motor neuron and a muscle fiber is called a **neuromuscular junction** (see Figure 9.6 on pages 238–239), whereas the synapse between a neuron and glandular cells is called a **neuroglandular junction.**

Parts of a Neuron

Most neurons have three parts: (1) a cell body, (2) dendrites, and (3) an axon (Figure 16.3 on page 505). The **cell body** contains a nucleus surrounded by cytoplasm that includes typical organelles such as lysosomes, mitochondria, and a Golgi complex. Many neurons also contain *lipofuscin*, a pigment that occurs as clumps of yellowish brown granules in the cytoplasm. Lipofuscin is probably a product of neuronal lysosomes that accumulates as the neuron ages, but it does not seem to harm the neuron. The cell body also contains prominent clusters of rough endoplasmic reticulum, termed *Nissl bodies*. Newly synthesized proteins produced by Nissl bodies are used to replace cellular

Table 16.1 Neuroglia

Type	Appearance	Functions
Central Nervous System		
Astrocytes (AS-trō-sīts; *astro-* = star; *-cyte* = cell)	Star-shaped, with many processes. 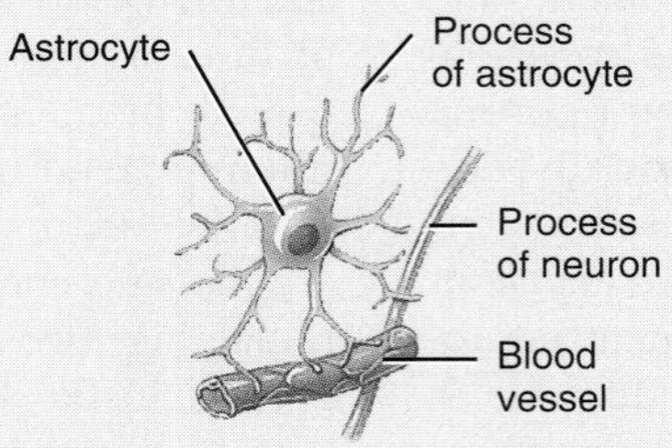	Help maintain appropriate chemical environment for generation of neuron action potentials; provide nutrients to neurons; take up excess neurotransmitters; participate in the metabolism of neurotransmitters; maintain proper balance of Ca^{2+} and K^{+}; assist with migration of neurons during brain development; help form the blood-brain barrier.
Oligodendrocytes (OL-i-gō-den′-drō-sīts; *oligo-* = few; *dendro-* = tree)	Smaller than astrocytes, with fewer processes; round or oval cell body. 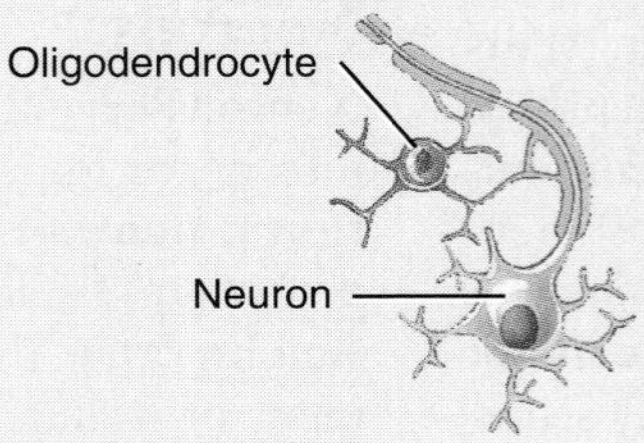	Form supporting network around CNS neurons; produce myelin sheath around several adjacent axons of CNS neurons.
Microglia (mī-KROG-lē-a; *micro-* = small)	Small cells with few processes; derived from mesodermal cells that also give rise to monocytes and macrophages. 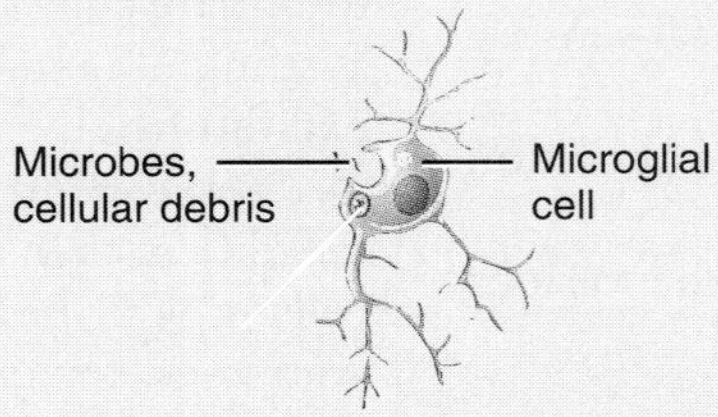	Protect CNS cells from disease by engulfing invading microbes; clear away debris of dead cells; migrate to areas of injured nerve tissue.
Ependymal cells (ep-EN-di-mal; *epen-* = above; *dym-* = garment)	Epithelial cells arranged in a single layer; range in shape from cuboidal to columnar; many are ciliated. 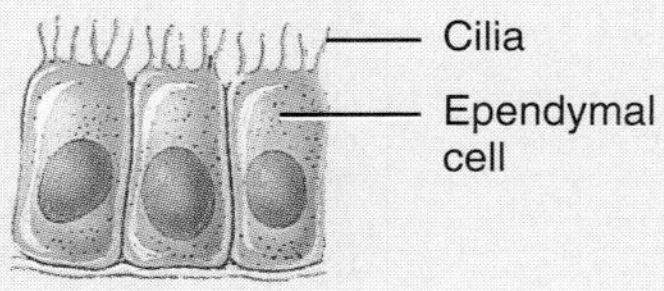	Line ventricles of the brain (spaces filled with cerebrospinal fluid) and central canal of the spinal cord; form cerebrospinal fluid and assist in its circulation.
Peripheral Nervous System		
Schwann cells (SCHVON) or neurolemmocytes	Flattened cells that encircle PNS axons. 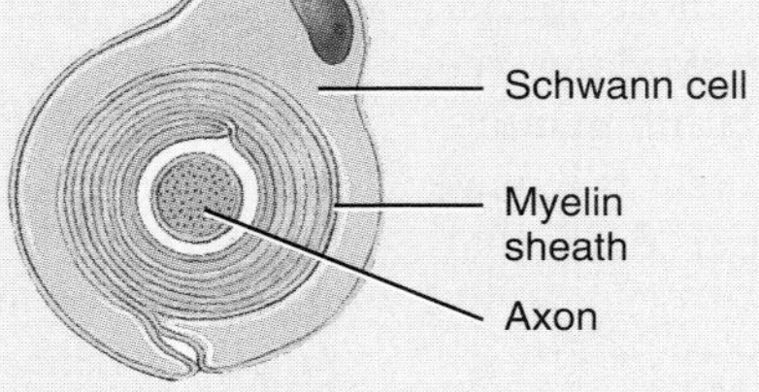	Each cell produces part of the myelin sheath around a single axon of a PNS neuron; participate in regeneration of PNS axons.
Satellite cells (SAT-i-līt)	Flattened cells arranged around the cell bodies of neurons in ganglia. 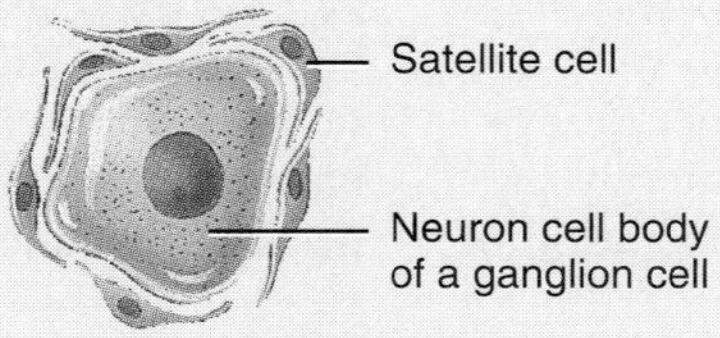	Support neurons in PNS ganglia.

Figure 16.3 / Structure of a typical neuron. Arrows indicate the direction of information flow: dendrites → cell body → axon → axon terminals. The break in the figure indicates that the axon actually is longer than shown.

The basic parts of a neuron are dendrites, a cell body, and an axon.

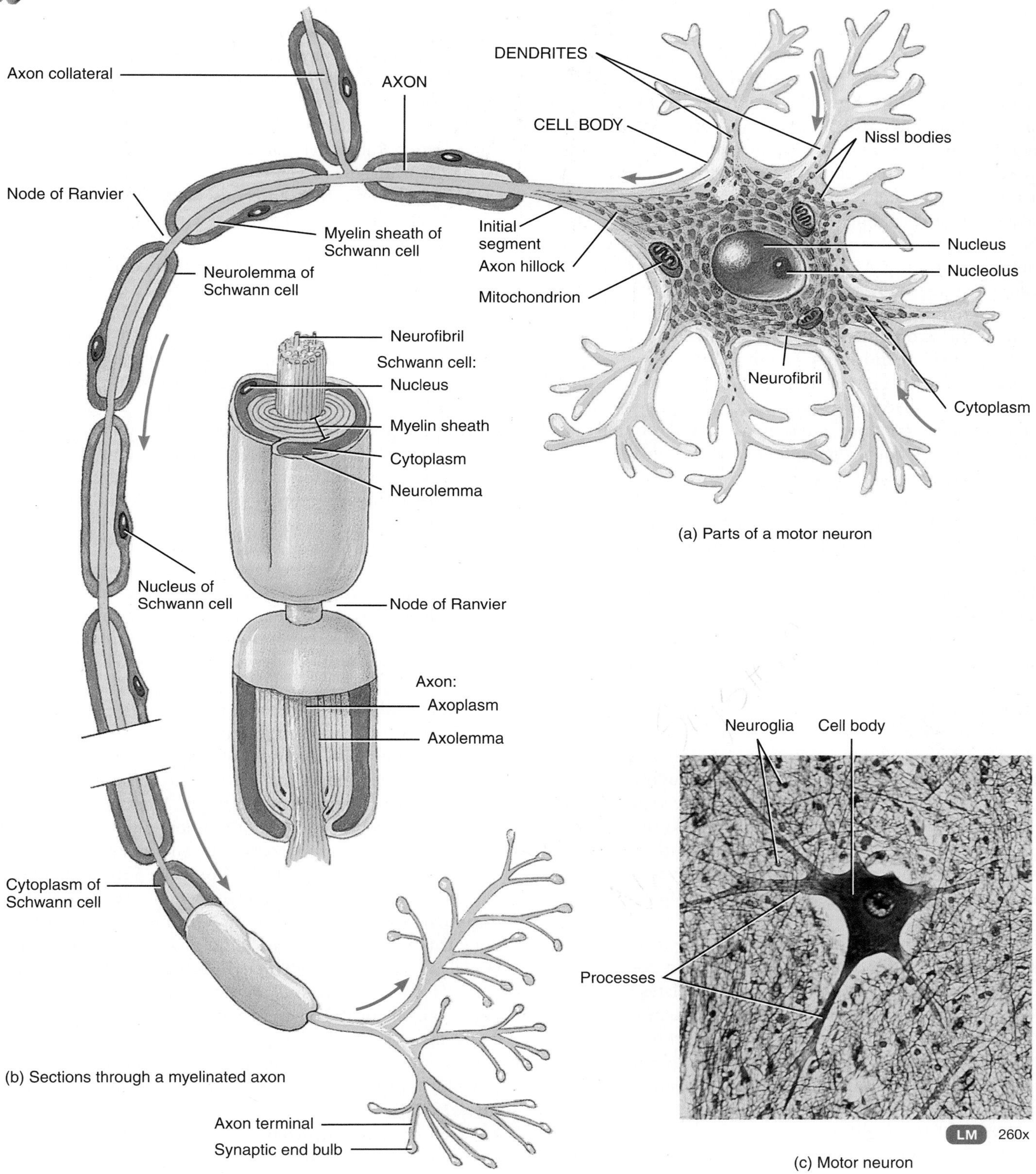

DENDRITES
CELL BODY
Nissl bodies
Initial segment
Axon hillock
Mitochondrion
Nucleus
Nucleolus
Neurofibril
Cytoplasm

(a) Parts of a motor neuron

Axon collateral
AXON
Node of Ranvier
Myelin sheath of Schwann cell
Neurolemma of Schwann cell
Neurofibril
Schwann cell:
Nucleus
Myelin sheath
Cytoplasm
Neurolemma
Nucleus of Schwann cell
Node of Ranvier
Axon:
Axoplasm
Axolemma
Cytoplasm of Schwann cell
Axon terminal
Synaptic end bulb

(b) Sections through a myelinated axon

Neuroglia
Cell body
Processes
LM 260x

(c) Motor neuron

What roles do the dendrites, cell body, and axon play in communication of signals?

components, as material for growth of neurons, and to regenerate damaged axons in the PNS. The cytoskeleton includes both *neurofibrils*, composed of bundles of intermediate filaments that provide the cell shape and support, and *microtubules*, which assist in moving materials between the cell body and axon.

Two kinds of processes or extensions emerge from the cell body of a neuron: multiple dendrites and a single axon. *Nerve fiber* is a general term for any neuronal process (dendrite or axon). **Dendrites** (= little trees) are the receiving or input portions of a neuron. They usually are short, tapering, and highly branched. In many neurons the dendrites form a tree-shaped array of processes extending from the cell body. Dendrites usually are not myelinated. Their cytoplasm contains Nissl bodies, mitochondria, and other organelles.

The second type of process, the **axon** (= axis) propagates nerve impulses toward another neuron, a muscle fiber, or a gland cell. Each neuron has only one axon, a long, thin, cylindrical projection that often joins the cell body at a cone-shaped elevation called the **axon hillock** (= small hill). The first part of the axon is called the **initial segment.** In most neurons, impulses arise at the junction of the axon hillock and the initial segment, which is called the **trigger zone,** and then travel along the axon. An axon contains mitochondria, microtubules, and neurofibrils. Because rough endoplasmic reticulum is not present, protein synthesis does not occur in the axon. The cytoplasm, called **axoplasm,** is surrounded by a plasma membrane known as the **axolemma** (*lemma* = sheath or husk). Along the length of an axon, side branches called **axon collaterals** may branch off, typically at a right angle to the axon. The axon and its collaterals end by dividing into many fine processes called **axon terminals.**

The site of communication between two neurons or between a neuron and an effector cell is called a **synapse.** The tips of some axon terminals swell into bulb-shaped structures called **synaptic end bulbs,** whereas others exhibit a string of swollen bumps called **varicosities.** Both synaptic end bulbs and varicosities contain many minute membrane-enclosed sacs called **synaptic vesicles** that store a chemical **neurotransmitter.** Neurons were long thought to liberate just one type of neurotransmitter at all synaptic end bulbs. We now know that many neurons contain two or even three neurotransmitters. The neurotransmitter molecules released from synaptic vesicles then influence the activity of other neurons, muscle fibers, or gland cells.

The axons of most mammalian neurons are surrounded by a **myelin sheath,** a multilayered lipid and protein covering produced by neuroglia. The sheath electrically insulates the axon of a neuron and increases the speed of nerve impulse conduction. Axons with such a covering are said to be *myelinated*, whereas those without it are *unmyelinated.* However, electron micrographs reveal that even unmyelinated axons are surrounded by a thin coat of neuroglial plasma membrane.

Two types of neuroglia produce myelin sheaths: Schwann cells (in the PNS) and oligodendrocytes (in the CNS). In the PNS, **Schwann cells** begin to form myelin sheaths around axons during fetal development. Each Schwann cell wraps about 1 millimeter (1 mm = 0.04 in.) of a single axon's length by spiraling many times around the axon (Figure 16.4b). Eventually, multiple layers of glial plasma membrane surround the axon, with the Schwann cell's cytoplasm and nucleus forming the outermost layer. The inner portion, consisting of up to 100 layers of Schwann cell membrane, is the myelin sheath. The outer nucleated cytoplasmic layer of the Schwann cell, which encloses the myelin sheath, is the **neurolemma (sheath of Schwann).** A neurolemma is found only around axons in the PNS. When an axon is injured, the neurolemma aids regeneration by forming a regeneration tube that guides and stimulates regrowth of the axon. Gaps in the myelin sheath, called **nodes of Ranvier** (RON-vē-ā), appear at intervals along the axon (see Figure 16.3). Each Schwann cell wraps one axon segment between two nodes.

In the CNS, an **oligodendrocyte** myelinates parts of many axons in somewhat the same manner that a Schwann cell myelinates part of a single PNS axon (see Table 16.1). The oligodendrocyte puts forth an average of 15 broad, flat processes that spiral around CNS axons, forming a myelin sheath. A neurolemma is not present, however, because the oligodendrocyte cell body and nucleus do not envelop the axon. Nodes of Ranvier are present, but they are fewer in number. Axons in the CNS display little regrowth after injury. This is thought to be due, in part, to the absence of a neurolemma, and in part to an inhibitory influence exerted by the oligodendrocytes on axon regrowth.

The amount of myelin increases from birth to maturity, and its presence greatly increases the speed of nerve impulse conduction. An infant's responses to stimuli are neither as rapid nor as coordinated as those of an older child or an adult, in part because myelination is still in progress during infancy. Certain diseases, such as multiple sclerosis (see page 512) destroy myelin sheaths.

Nerve fiber is a general term for any neuronal process (dendrite or axon). Most often, it refers to an axon and its sheaths. A **nerve** is a bundle of many nerve fibers that course along the same path in the PNS, such as the ulnar nerve in the arm or the sciatic nerve in the thigh. Most nerves include bundles of both sensory and motor fibers and are surrounded by three connective tissue coats, the endoneurium, perineurium, and epineurium (see Figure 17.6 on page 526). Individual axons, whether myelinated or unmyelinated, are wrapped in **endoneurium** (en′-dō-NŪ-rē-um). Groups of axons with their endoneurium are arranged in bundles called **fascicles,** and each fascicle is wrapped in **perineurium** (per′-i-NŪ-rē-um). The superficial covering around the entire nerve is the **epineurium** (ep′-i-NŪ-rē-um). You may recall from Chapter 10 that the connective tissue coverings of skeletal muscles—endomysium, perimysium, and epimysium—are similar in organization to those of nerves. This organization is common to excitable tissues.

Cell bodies of neurons in the PNS generally are clustered together to form **ganglia** (GANG-lē-a; *ganglion* = knot). A **tract** is a bundle of nerve fibers, without connective tissue elements, in the CNS. Tracts may interconnect different regions of

Figure 16.4 / Myelinated and unmyelinated axons.

Axons of most mammalian neurons are surrounded by a myelin sheath produced by Schwann cells in the PNS and by oligodendrocytes in the CNS.

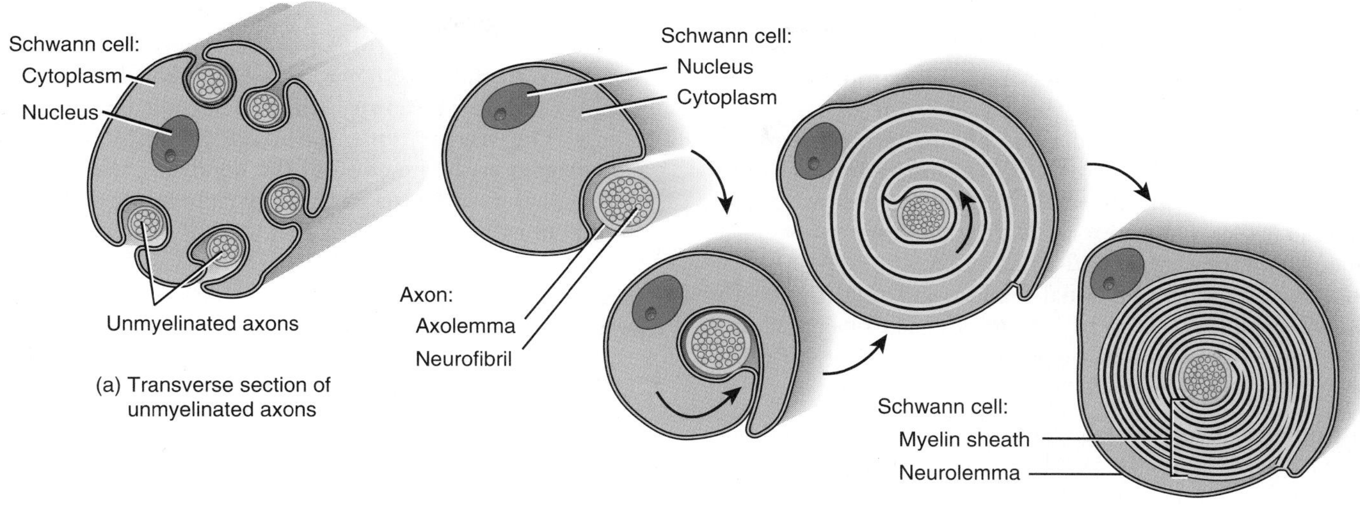

(a) Transverse section of unmyelinated axons

(b) Transverse sections of stages in the formation of a myelin sheath

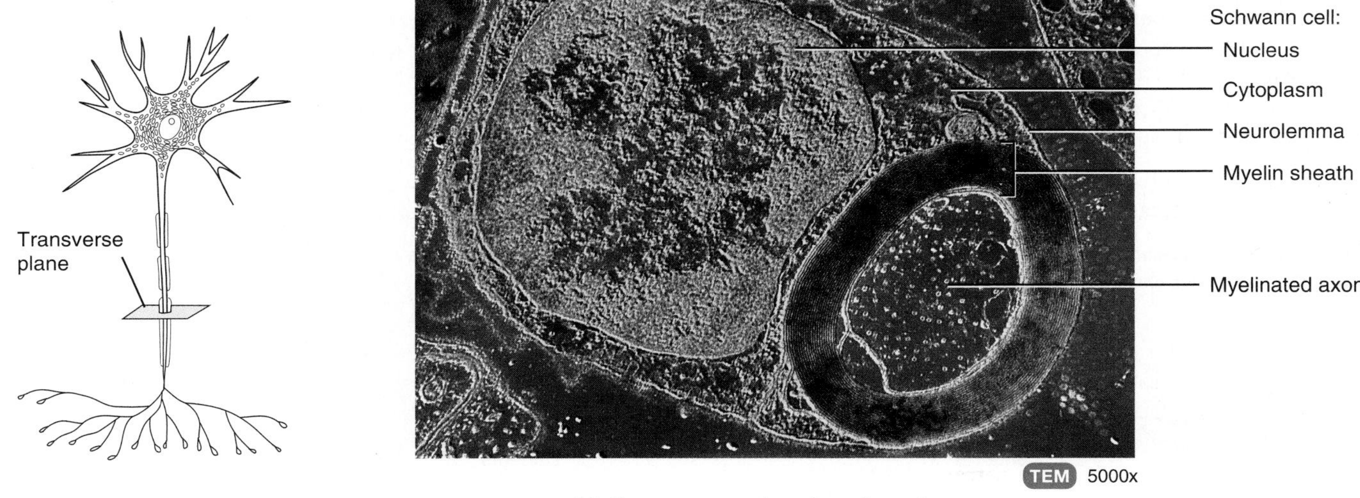

(c) Transverse section of myelinated axon

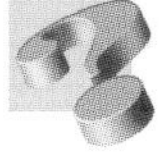

What is the functional advantage of myelination?

the brain or extend long distances up or down the spinal cord and connect with special regions of the brain. A **nucleus** is a collection of nerve cell bodies in the CNS.

Structural Diversity in Neurons

Neurons display great diversity in size and shape. For example, their cell bodies range in diameter from 5 micrometers (μm) (smaller than a red blood cell) up to 135 μm (barely large enough to see with the unaided eye). The pattern of dendritic branching is varied and distinctive for neurons in different parts of the nervous system. A few small neurons lack an axon, and many others have very short axons, but the longest neurons have axons that extend for a meter (3.2 ft) or more.

Both structural and functional features are used to classify the various neurons in the body. *Structurally*, neurons are classified according to the number of processes extending from the cell body (Figure 16.5).

1. **Multipolar neurons** usually have several dendrites and one axon (see also Figure 16.3). Most neurons in the brain and spinal cord are of this type.
2. **Bipolar neurons** have one main dendrite and one axon; they are found in the retina of the eye, in the inner ear, and in the olfactory area of the brain.
3. **Unipolar neurons** are sensory neurons that originate in the embryo as bipolar neurons. During development, the axon and dendrite fuse into a single process that divides into two branches a short distance from the cell body. Both branches have the characteristic structure and function of an axon: They are long, cylindrical processes that may be myelinated, and they propagate action potentials. However, the axon branch that extends into the periphery has unmyelinated dendrites at its distal tip, whereas the axon branch that extends into the CNS ends in synaptic end bulbs. The dendrites monitor a sensory stimulus such as touch or stretching. The trigger zone for nerve impulses in a unipolar neuron is at the junction of the dendrites and axon (Figure 16.5c). The impulses then propagate toward the synaptic end bulbs.

Functionally, neurons are classified according to the direction in which they transmit nerve impulses.

1. **Sensory (afferent) neurons** (Figure 16.6) transmit nerve impulses from receptors in the skin, sense organs, muscles, joints, and viscera *toward* the brain and spinal cord.
2. **Motor (efferent) neurons** convey nerve impulses *from* the brain and spinal cord to effectors, which may be either muscles or glands.
3. **Interneurons (association neurons)** are all other neurons that are not specifically sensory or motor neurons. Most neurons in the body, perhaps 90%, are interneurons of thousands of different types. Interneurons often are named for the histologist who first described them; examples are

Figure 16.5 / Structural classification of neurons. Breaks indicate that axons are longer than shown.

A multipolar neuron has many processes extending from the cell body, whereas a bipolar neuron has two and a unipolar neuron has one.

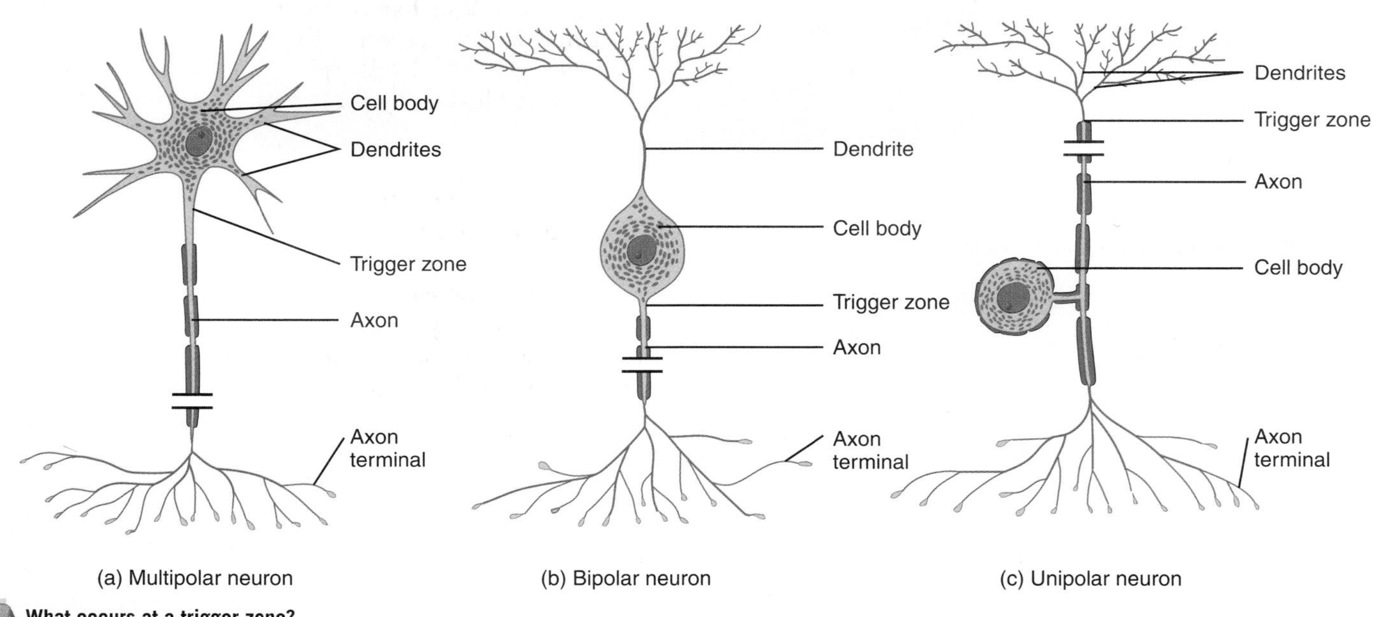

What occurs at a trigger zone?

Figure 16.6 / Structure of a typical sensory (afferent) neuron. Arrows indicate the direction of information flow. The break indicates that the process is actually longer than shown.

Sensory neurons transmit nerve impulses from receptors to the central nervous system.

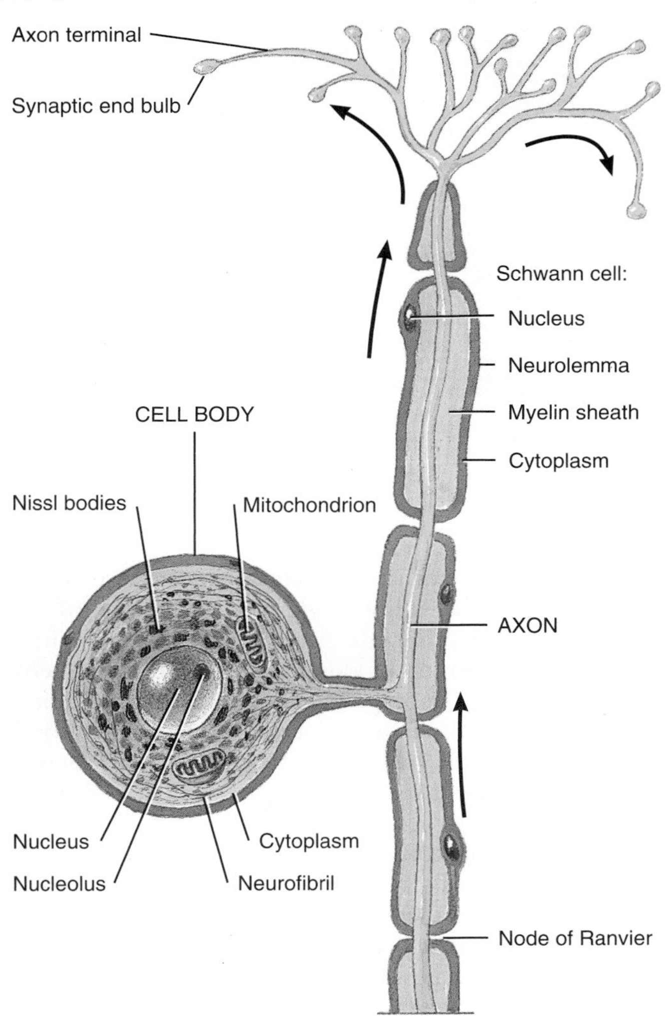

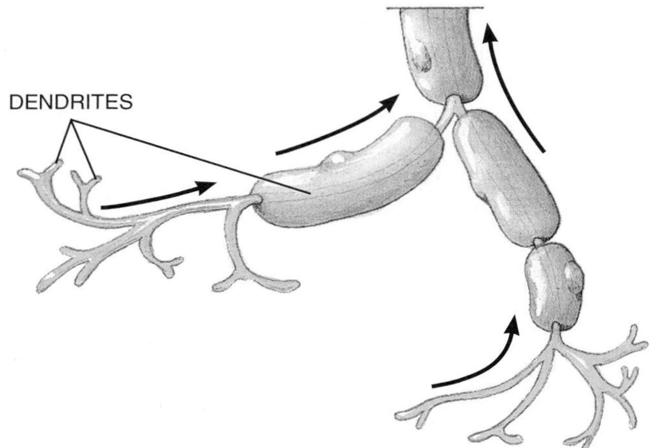

Where are the receptors that transmit nerve impulses to the sensory neurons located?

Figure 16.7 / Two examples of interneurons. Arrows indicate the direction of information flow.

Interneurons carry nerve impulses from one neuron to another.

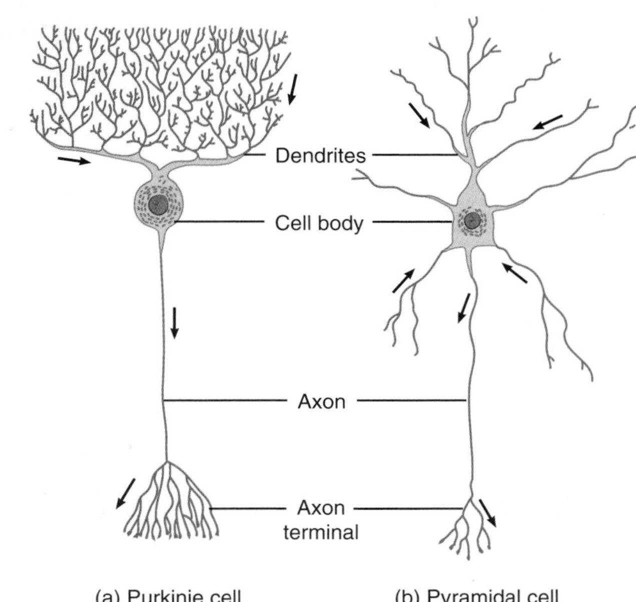

Do the dendrites of interneurons receive input or provide output signals?

Purkinje cells (pur-KIN-jē) in the cerebellum (Figure 16.7a) and **Renshaw cells** in the spinal cord. Others are named for some aspect of their shape or appearance. For example, **pyramidal cells** (pi-RAM-i-dal), found in the brain, have a cell body shaped like a pyramid (Figure 16.7b).

Gray and White Matter

In a freshly dissected section of the brain or spinal cord, some regions look white and glistening, whereas others appear gray (Figure 16.8). The **white matter** is aggregations of myelinated processes from many neurons. The whitish color of myelin gives white matter its name. The **gray matter** of the nervous system contains neuronal cell bodies, dendrites, unmyelinated axons, axon terminals, and neuroglia. It looks grayish rather than white because there is little or no myelin in these areas. Blood vessels are present in both white and gray matter.

In the spinal cord, the white matter surrounds an inner core of gray matter shaped like a butterfly or the letter H; in the brain, a thin shell of gray matter covers the surface of the largest portions of the brain, the cerebrum and cerebellum (Figure 16.8). Many nuclei of gray matter also lie deep within the brain.

Figure 16.8 / Distribution of gray and white matter in the spinal cord and brain.

White matter consists of myelinated processes of many neurons. Gray matter consists of neuron cell bodies, dendrites, axon terminals, bundles of unmyelinated axons, and neuroglia.

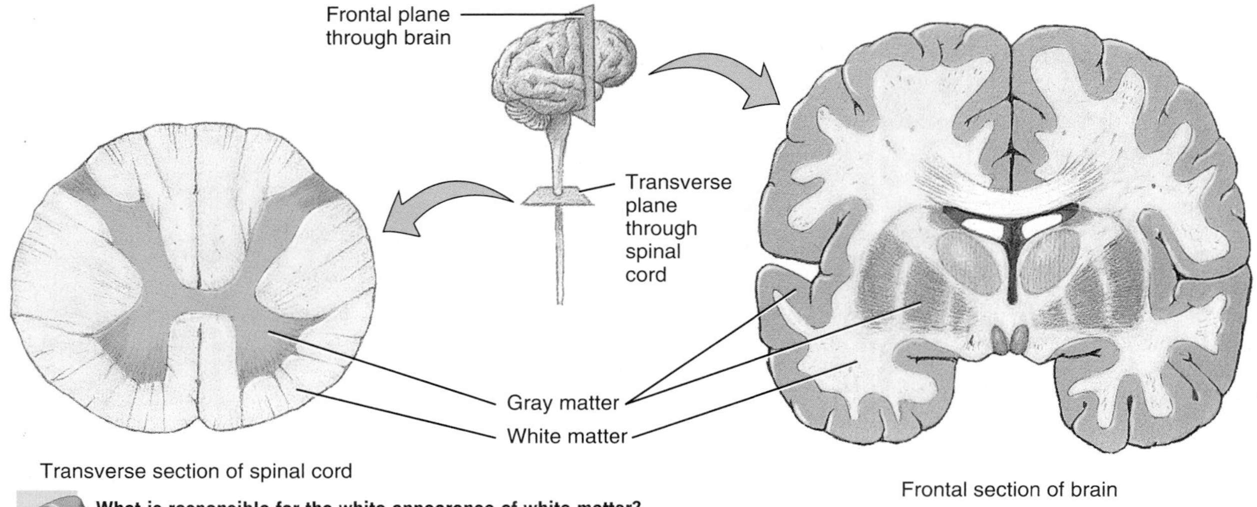

What is responsible for the white appearance of white matter?

When used to describe nervous tissue, a **nucleus** is a cluster of neuronal cell bodies within the CNS. The arrangements of gray and white matter in the spinal cord and brain are described more extensively in Chapters 17 and 18, respectively.

- ✓ Describe the parts of a neuron and the functions of each.
- ✓ Give several examples of the structural diversity of neurons.
- ✓ Define neurolemma. Why is it important?
- ✓ With reference to the nervous system, what is a nucleus?

NEURONAL CIRCUITS

Objective

- Identify the various types of neuronal circuits in the nervous system.

The CNS contains billions of neurons organized into complicated networks called **neuronal circuits** over which nerve impulses are conducted. In a **simple series circuit,** a presynaptic neuron stimulates only a single neuron. The second neuron then stimulates another, and so on. Most neuronal circuits, however, are more complex.

A single presynaptic neuron may synapse with several postsynaptic neurons. Such an arrangement, called **divergence,** permits one presynaptic neuron to influence several postsynaptic neurons (or several muscle fibers or gland cells) at the same time. In a **diverging circuit,** the nerve impulse from a single presynaptic neuron causes the stimulation of increasing numbers of cells along the circuit (Figure 16.9a). For example, a small number of neurons in the brain that govern a particular body movement stimulate a much larger number of neurons in the spinal cord. Sensory signals also feed into diverging circuits and are often relayed to several regions of the brain.

In another arrangement, called **convergence,** several presynaptic neurons synapse with a single postsynaptic neuron. This arrangement permits more effective stimulation or inhibition of the postsynaptic neuron. In one type of **converging circuit** (Figure 16.9b), the postsynaptic neuron receives nerve impulses from several different sources. For example, a single motor neuron that synapses with skeletal muscle fibers at neuromuscular junctions receives input from several pathways that originate in different brain regions.

Some circuits are constructed so that once the presynaptic cell is stimulated, it will cause the postsynaptic cell to transmit a series of nerve impulses. One such circuit is called a **reverberating (oscillatory) circuit** (Figure 16.9c). In this pattern, the incoming impulse stimulates the first neuron, which stimulates the second, which stimulates the third, and so on. Branches from later neurons synapse with earlier ones, however, sending impulses back through the circuit again and again. The output signal may last from a few seconds to many hours, depending on the number of synapses and the arrangement of neurons in the circuit. Inhibitory neurons may turn off a reverberating circuit after a period of time. Among the body responses thought to be the result of output signals from reverberating circuits are breathing, coordinated muscular activities, waking up, sleeping (when reverberation stops), and short-term memory.

A fourth type of circuit is the **parallel after-discharge circuit** (Figure 16.9d). In this circuit, a single presynaptic cell stimulates a group of neurons, each of which synapses with a common postsynaptic cell. It is thought that parallel after-discharge circuits may be involved in precise activities such as mathematical calculations.

- ✓ What is a neuronal circuit?
- ✓ What are the functions of diverging, converging, reverberating, and parallel after-discharge circuits?

Figure 16.9 / Examples of neuronal circuits.

Neuronal circuits are groups of neurons arranged in patterns over which nerve impulses are conducted.

Input
Outputs
(a) Diverging circuit

Input
Input
Input
Output
(b) Converging circuit

Input
Output
(c) Reverberating circuit

Input
Output
(d) Parallel after-discharge circuit

A motor neuron in the spinal cord typically receives input from neurons that originate in several different regions of the brain. Is this an example of convergence or divergence?

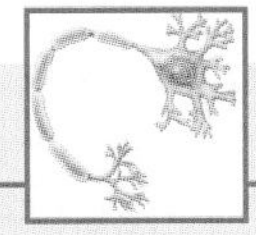

APPLICATIONS TO HEALTH

Damage and Repair of Peripheral Neurons

Mammalian neurons have very limited powers of **regeneration,** the capability to replicate or repair themselves. In the PNS, damage to dendrites and myelinated axons may be repaired if the cell body remains intact and if the Schwann cells that produce myelination remain active. In the CNS, little or no repair of damage to neurons occurs; even when the cell body remains intact, a severed axon cannot be repaired or regrown.

Neurogenesis—the birth of new neurons from undifferentiated stem cells—occurs regularly in some animals. For example, new neurons appear and disappear every year in some songbirds. Until recently, the dogma in humans and other primates was "no new neurons" in the adult brain. Then, in 1992, Canadian researchers published their unexpected finding that **epidermal growth factor (EGF)** stimulated cells taken from the brains of adult mice to proliferate into both neurons and astrocytes. Previously, EGF was known to trigger mitosis in a variety of nonneuronal cells and to promote wound healing and tissue regeneration. In 1998 scientists discovered that significant numbers of new neurons do arise in the adult human hippocampus, an area of the brain that is crucial for learning.

The nearly complete lack of neurogenesis in other regions of the brain and spinal cord, however, seems to result from two factors: (1) inhibitory influences from neuroglia, particularly oligodendrocytes, and (2) absence of growth-stimulating cues that were present during fetal development. Axons in the CNS are myelinated by oligodendrocytes, which do not form neurolemmas (sheaths of Schwann). In addition, CNS myelin is one of the factors inhibiting regeneration of neurons. Perhaps this is the same mechanism that stops axonal growth once a target region has been reached during development. Also, after axonal damage, nearby astrocytes proliferate rapidly, forming a type of scar tissue and constituting a physical barrier to regeneration. Thus, injury of the brain or spinal cord usually is permanent. Nevertheless, ongoing research seeks ways to stimulate dormant stem cells to replace neurons lost through damage or disease and to improve the environment to allow existing axons to bridge the gap of the injury, especially in spinal cord injuries. Also, tissue-cultured neurons might be useful for transplantation purposes.

Axons and dendrites that are associated with a neurolemma may undergo repair if the cell body is intact, if the Schwann cells are functional, and if scar tissue formation does not occur too rapidly. Most nerves in the PNS consist of processes that are covered with a neurolemma. A person who injures axons of a nerve in an upper limb, for example, has a good chance of regaining nerve function.

When there is damage to an axon, changes usually occur both in the cell body of the affected neuron and in the portion of the axon distal to the site of injury. Changes also may occur in the portion of the axon proximal to the site of injury.

About 24–48 hours after injury to a process of a normal central or peripheral neuron, the Nissl bodies break up into fine granular masses. This alteration is called **chromatolysis** (krō′-ma-TOL-i-sis; *chromato-* = color; *-lysis* = dissolution). It begins between the axon hillock and nucleus and then spreads throughout the cell body. As a result of chromatolysis, the cell body swells, reaching a maximum size between 10 and 20 days after injury.

By the third to fifth day, the part of the process distal to the damaged region becomes slightly swollen and then breaks up into fragments; the myelin sheath also deteriorates. Degeneration of the distal portion of the neuronal process and myelin sheath is called **Wallerian degeneration.** Following degeneration, macrophages phagocytize the debris.

The changes in the proximal portion of the axon, called **retrograde degeneration,** are similar to those that occur during Wallerian degeneration. The main difference in retrograde degeneration is that the changes extend only to the first node of Ranvier.

Following chromatolysis, signs of recovery in the cell body become evident. Synthesis of RNA and protein accelerates, which favors rebuilding or **regeneration** of the axon. Recovery often takes several months. Even though the neuronal process and myelin sheath degenerate, the neurolemma remains. The Schwann cells on either side of the injured site multiply by mitosis, grow toward each other, and may form a **regeneration tube** across the injured area. The tube is composed of neurolemma and endoneurium and guides growth of new processes from the proximal area across the injured area into the distal area previously occupied by the original axon. New axons cannot grow if the gap at the site of injury is too large or if the gap becomes filled with collagen fibers.

During the first few days following damage, buds of regenerating axons begin to invade the tube formed by the Schwann cells. Axons from the proximal area grow at a rate of about 1.5 mm (0.06 in.) per day across the area of damage, find their way into the distal regeneration tubes, and grow toward the distally located receptors and effectors. Thus, some sensory and motor connections are reestablished and some functions restored. In time, the Schwann cells form a new myelin sheath.

Multiple Sclerosis

Multiple sclerosis (MS) is a progressive destruction of myelin sheaths of neurons in the CNS that afflicts about half a million people in the United States, and 2 million people worldwide. It usually appears in people between the ages of 20 and 40, affecting females twice as often as males. Like rheumatoid arthritis, MS is an autoimmune disease—the body's own immune system spearheads the attack. The condition's name describes the anatomical pathology; myelin sheaths deteriorate to *scleroses,* which are hardened scars or plaques, in *multiple* regions. Magnetic resonance imaging (MRI) studies reveal numerous plaques in the white matter of the brain and spinal cord. The destruction of myelin sheaths slows and then short-circuits conduction of nerve impulses.

The most common form of MS is called relapsing-remitting MS. Usually, the first symptoms, including a feeling of heaviness or weakness in the muscles, abnormal sensations, or double vision, occur in early adult life. An attack is followed by a period of remission during which the symptoms temporarily disappear. Sometime later, a new series of plaques develop, and the person suffers a second attack. One attack follows another over the years, usually every year or two. The result is a progressive loss of function interspersed with remission periods, during which symptoms abate. Although the cause of MS is unclear, both genetic susceptibility and exposure to some environmental factor (perhaps a herpesvirus) appear to contribute. Since 1993, many patients with relapsing-remitting MS have been treated with injections of interferon beta, which lengthens the time between relapses, decreases the severity of relapses, and slows formation of new lesions in some cases. Unfortunately, not all MS patients can tolerate interferon beta, and therapy becomes less effective as the disease progresses.

Epilepsy

The second most common neurological disorder after stroke (rupture or blockage of a brain blood vessel) is **epilepsy,** which afflicts about 1% of the world's population. Epilepsy is characterized by short, recurrent, periodic attacks of motor, sensory, or psychological malfunction. The attacks, called *epileptic seizures,* are initiated by abnormal, synchronous electrical discharges from millions of neurons in the brain, perhaps resulting from abnormal reverberating circuits. The discharges stimulate many of the neurons to send nerve impulses over their conduction pathways; as a result, lights, noise, or smells may be sensed when the eyes, ears, and nose have not been stimulated. Moreover, the skeletal muscles of a person undergoing an attack may contract involuntarily. *Partial seizures* begin in a small focus on one side the brain and produce milder symptoms, whereas *generalized seizures* involve larger areas on both sides of the brain and loss of consciousness.

Epilepsy has many causes, including brain damage at birth (the most common cause); metabolic disturbances (hypoglycemia, hypocalcemia, uremia, hypoxia); infections (encephalitis or meningitis); toxins (alcohol, tranquilizers, hallucinogens); vascular disturbances (hemorrhage, hypotension); head injuries; and tumors and abscesses of the brain. However, most epileptic seizures are idiopathic—they have no demonstrable cause. Epilepsy almost never affects intelligence.

Epileptic seizures often can be eliminated or alleviated by antiepileptic drugs, such as phenytoin, carbamazepine, and valproate sodium. An implantable device that stimulates the vagus nerve (cranial nerve X) also has produced dramatic results in reducing seizures in some patients whose epilepsy was not well-controlled by drugs.

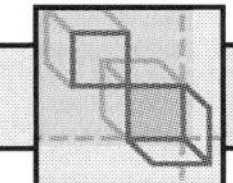

STUDY OUTLINE

Overview of the Nervous System (p. 500)

1. Structures that make up the nervous system are the brain, 12 pairs of cranial nerves and their branches, the spinal cord, 31 pairs of spinal nerves and their branches, ganglia, enteric plexuses, and sensory receptors.
2. The nervous system helps maintain homeostasis and integrates all body activities by sensing changes (sensory function), interpreting them (integrative function), and reacting to them (motor function).
3. On the basis of their function, neurons are either sensory neurons, interneurons, or motor neurons.
4. The central nervous system (CNS) consists of the brain and spinal cord; the peripheral nervous system (PNS) consists of all nervous tissue outside the CNS—cranial and spinal nerves, ganglia, and sensory receptors.
5. Sensory (afferent) neurons provide input to the CNS; motor (efferent) neurons carry output from the CNS to effectors.
6. The PNS is subdivided into the somatic nervous system (SNS), autonomic nervous system (ANS), and enteric nervous system (ENS).
7. The SNS consists of (1) neurons that conduct impulses from somatic and special sense receptors to the CNS and (2) motor neurons from the CNS to skeletal muscles.
8. The ANS contains (1) sensory neurons from visceral organs and (2) motor neurons that convey impulses from the CNS to smooth muscle tissue, cardiac muscle tissue, and glands.
9. The ENS consists of neurons in two enteric plexuses that extend the length of the gastrointestinal (GI) tract and function independently of the ANS and CNS to some extent. The ENS monitors sensory changes and controls operation of the GI tract.

Histology of Nervous Tissue (p. 503)

1. Nervous tissue consists of two principal kinds of cells: neurons (nerve cells) and neuroglia.
2. Neuroglia support, nurture, and protect the neurons and maintain homeostasis in the interstitial fluid that bathes neurons. Neuroglia include astrocytes, oligodendrocytes, microglia, ependymal cells, Schwann cells, and satellite cells (summarized in Table 16.1 on page 504).
3. Two types of neuroglia produce myelin sheaths: oligodendrocytes myelinate axons in the CNS, and Schwann cells myelinate axons in the PNS.
4. Neurons are responsible for most special functions attributed to the nervous system: sensing, thinking, remembering, controlling muscle activity, and regulating glandular secretions. Most neurons consist of many dendrites, which are the main receiving or input region; a cell body that includes typical cellular organelles; and a single axon that propagates nerve impulses toward another neuron, a muscle fiber, or a gland cell.
5. Synapses are the sites of functional connections between two neurons or a neuron and an effector (muscle or gland). Axon terminals contain synaptic vesicles filled with neurotransmitter molecules.
6. On the basis of their structure, neurons are either multipolar, bipolar, or unipolar.
7. On the basis of function, neurons are sensory (afferent) neurons, which conduct impulses from receptors to the CNS: motor (efferent) neurons, which conduct impulses to effectors; or interneurons (association neurons).
8. White matter consists of aggregations of myelinated processes, whereas gray matter contains neuron cell bodies, dendrites, and axon terminals or bundles of unmyelinated axons and neuroglia.
9. In the spinal cord, gray matter forms an H-shaped inner core that is surrounded by white matter. In the brain, a thin, superficial shell of gray matter covers the cerebral and cerebellar hemispheres.

Neuronal Circuits (p. 510)

1. Neurons in the central nervous system are organized into networks called circuits.
2. Circuits include simple series, diverging, converging, reverberating (oscillatory), and parallel after-discharge circuits.

SELF-QUIZ QUESTIONS

Choose the one best answer to the following questions.

1. Which of the following statements is true? (a) A neuron usually has many axons that are connected to other neurons. (b) Most unipolar neurons are found in the central nervous system. (c) Multipolar neurons possess numerous dendrites and only one axon. (d) Sensory neurons are usually bipolar. (e) All of the above are true.
2. The myelin sheath of peripheral nervous system neurons is produced by (a) Schwann cells, (b) astrocytes, (c) oligodendrocytes, (d) microglia, (e) ependymal cells.
3. The term nerve fiber usually refers to (a) a synaptic end bulb, (b) an axon terminal, (c) an axon and its sheaths, (d) a cell body, (e) a bundle of axons.
4. The point of contact between a nerve fiber and a muscle is called the (a) axon hillock, (b) neurofibral node (c) trigger zone (d) neuromuscular junction, (e) ganglion.
5. Which of the following functions does *not* pertain to any of the various types of specialized cells called neuroglia? (a) regulation of entry of substances into the brain, (b) production of a myelin sheath around central system neurons, (c) engulfment and destruction of microbes, (d) assistance in the production and circulation of cerebrospinal fluid, (e) replacement of damaged neurons with new ones.

Complete the following.

6. The structures of the peripheral nervous system that connect to the central nervous system (CNS) are the ________ nerves and the ________ .
7. The subdivisions of the peripheral nervous system are the ________ , ________ , and ________ .
8. A tract is a bundle of nerve fibers located in the ________ nervous system.
9. A nerve is made up of groups of axons, and is surrounded by a layer of connective tissue called ________ .
10. The part of a neuron that conducts nerve impulses toward the cell body is the ________ .
11. Sacs in the axon terminals that store neurotransmitters are called ________ .

12. In a ________ circuit, a single postsynaptic neuron receives stimulation or inhibition from several presynaptic neurons.

Are the following statements true or false?

13. Match the following.

____	**(a)** connects an axon to the cell body	**(1)** axon
____	**(b)** yellowish pigment that increases with age; appears to be a by-product of lysosomes	**(2)** axon terminal
____	**(c)** plasma membrane of the axon	**(3)** axon hillock
____	**(d)** long, thin fibrils composed of bundles of intermediate filaments	**(4)** chromatophilic substance (Nissl bodies)
____	**(e)** orderly arrangement of rough endoplasmic reticulum; site of protein syntheses	**(5)** dendrite
____	**(f)** conducts nerve impulses toward cell body	**(6)** lipofuscin
____	**(g)** conducts nerve impulses away from cell body; has synaptic end bulbs that contain neurotransmitters	**(7)** axolemma
____	**(h)** fine filaments that are branching ends of axons and axon collaterals	**(8)** neurofibrils

14. Match the following.

____	**(a)** association neurons	**(1)** sensory neurons
____	**(b)** unipolar neurons	**(2)** motor neurons
____	**(c)** afferent neurons	**(3)** interneurons
____	**(d)** efferent neurons	
____	**(e)** most central nervous system neurons are of this type	
____	**(f)** these neurons conduct impulses away from the central nervous system	
____	**(g)** these neurons conduct impulses toward the central nervous system	

15. A unipolar neuron originates as a bipolar neuron, the axon and dendrite of which fuse close to the cell body.

16. The two principal divisions of the peripheral nervous system are the brain and the spinal cord.

17. Motor responses of the autonomic nervous system are under subconscious control.

18. White matter refers to aggregations of myelinated processes from many axons.

19. A nucleus is a mass of nerve cell bodies and dendrites inside the central nervous system.

20. A neuron with several dendrites and one axon is classified as a sensory neuron.

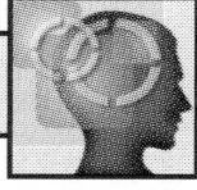

CRITICAL THINKING QUESTIONS

1. Varudhini was reviewing her anatomy notes when she heard a catchy advertising jingle on the radio. Hours later, on her way to class, the jingle was still running through her mind. Varudhini was suddenly reminded of a neuronal circuit from her anatomy notes. Name the type of circuit and explain its uses in the body.
   HINT *If the temperature of your oven oscillates, it goes up and down, up and down, and so on.*

2. Varudhini was looking for a parking spot at the mall. She kept going around and around the parking lot waiting for a close parking space. "Now I know what a neuroglial cell feels like," she muttered. (Varudhini really needs to lay off the books for awhile.) Name Varudhini's neuroglial cell and explain its function.
   HINT *She had the same thought when she woke up with her blanket wrapped tightly around her.*

3. When asked to define "gray matter", Jen answered that it's white matter from a really old person. How would you define gray matter?
   HINT *The difference between gray and white has to due with its structure, not with age.*

4. The buzzing of the alarm clock woke Mohamed. He yawned, stretched, and started to salivate as he smelled the coffee brewing. He could feel his stomach rumble. List the divisions of the nervous system involved in each of these actions.
   HINT *Remember Pavlov's dogs? Salivation is automatic.*

5. Samantha has multiple sclerosis. In some ways, her own movements remind her of those of Sam, her new grandson. Compare the effects of Samantha's MS to Sam's developing nervous system.
   HINT *The neurons can't do it all alone.*

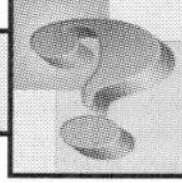

ANSWERS TO FIGURE QUESTIONS

16.1 The total number of cranial and spinal nerves in your body is $(12 \times 2) + (31 \times 2) = 86$.

16.2 Sensory or afferent neurons carry input to the CNS; motor or efferent neurons carry output from the CNS.

16.3 Dendrites of motor neurons or interneurons receive input signals and dendrites of sensory neurons generate input signals; the cell body also receives input signals; the axon conducts nerve impulses (action potentials) and transmits the message to another neuron or effector cell by releasing a neurotransmitter at its synaptic end bulbs.

16.4 Myelination increases the speed of nerve impulse conduction.

16.5 Nerve impulses arise at the trigger zone.

16.6 The skin, sense organs, muscles, joints, and viscera.

16.7 Dendrites receive input signals.

16.8 Myelin makes white matter look shiny and white.

16.9 A motor neuron receiving input from several other neurons is an example of convergence.

17

THE SPINAL CORD AND THE SPINAL NERVES

CONTENTS AT A GLANCE

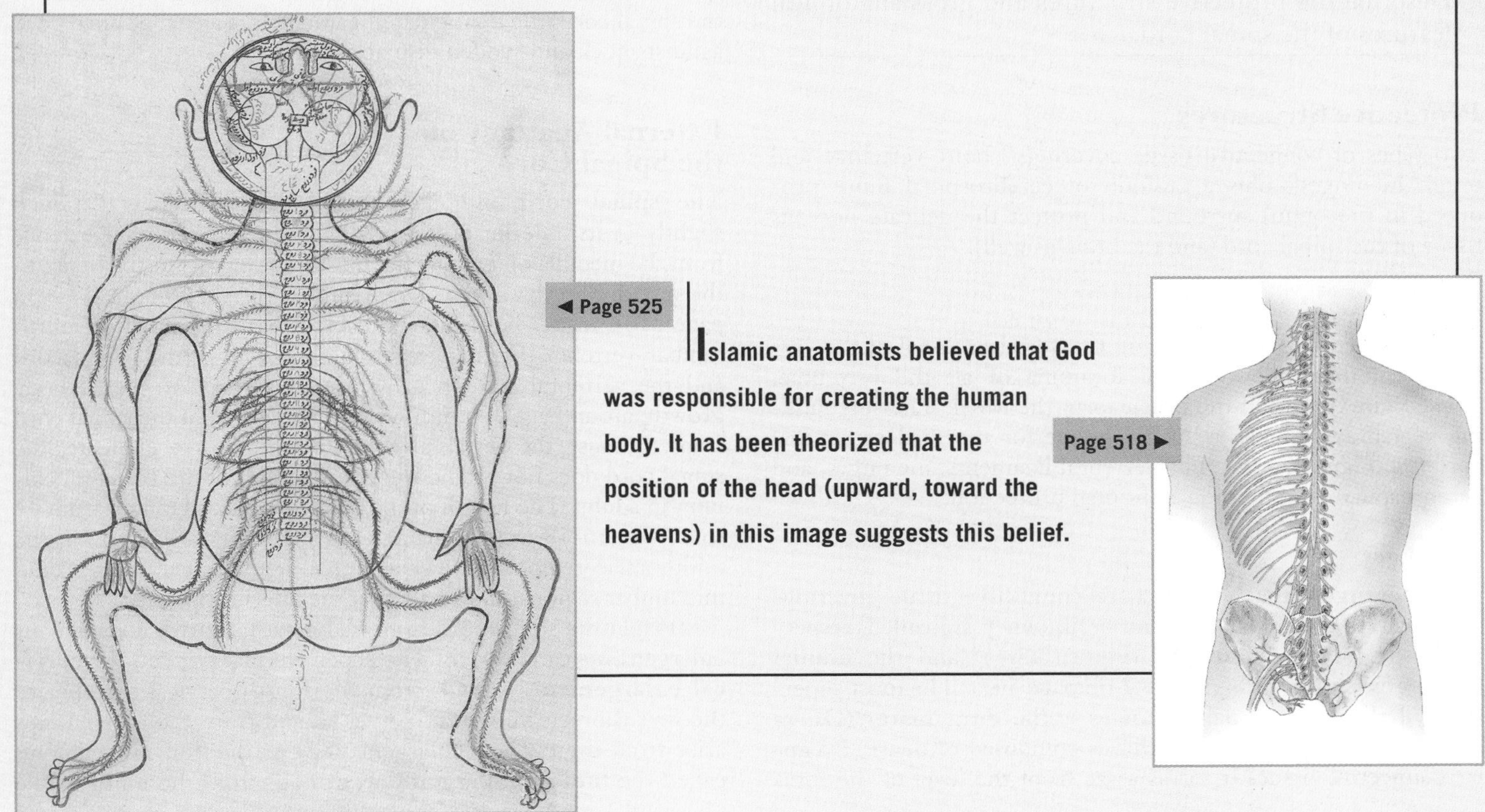

Islamic anatomists believed that God was responsible for creating the human body. It has been theorized that the position of the head (upward, toward the heavens) in this image suggests this belief.

INTRODUCTION

Together, the spinal cord and spinal nerves contain neuronal circuits that mediate some of your quickest reactions to environmental changes. If you pick up something hot, for example, the grasping muscles may relax and you may drop it even before the sensation of extreme heat or pain reaches your conscious perception. This is an example of a spinal cord reflex—a quick, automatic response to certain kinds of stimuli that involves neurons only in the spinal nerves and spinal cord. Besides processing reflexes, the spinal cord also is the site for integrating excitation or inhibition of neurons that arises locally or is triggered by nerve impulses from the peripheral nervous system (the periphery) and the brain. Moreover, the spinal cord is the highway traveled by sensory nerve impulses headed for the brain, and by motor nerve impulses destined for spinal nerves. Keep in mind that the spinal cord is continuous with the brain and that together they constitute the central nervous system (CNS).

SPINAL CORD ANATOMY

Objective

- Describe the protective structures and gross anatomical features of the spinal cord.

Protective Structures

Two types of connective tissue coverings—bony vertebrae and tough meninges—plus a cushion of cerebrospinal fluid (produced in the brain) surround and protect the delicate nervous tissue of the spinal cord (and the brain as well).

Vertebral Column

The spinal cord is located within the vertebral canal of the vertebral column. The vertebral foramina of all the vertebrae, stacked one on top of the other, form the canal. The surrounding vertebrae provide a sturdy shelter for the enclosed spinal cord (see Figure 17.1b). The vertebral ligaments, meninges, and cerebrospinal fluid provide additional protection.

Meninges

The **meninges** (me-NIN-jēz) are connective tissue coverings that encircle the spinal cord and brain. They are called, respectively, the **spinal meninges** (Figure 17.1a) and the **cranial meninges** (shown in Figure 18.2 on page 547). The most superficial of the three spinal meninges is the **dura mater** (DŪ-ra MĀ-ter; = tough mother), which is composed of dense, irregular connective tissue. It forms a sac from the level of the foramen magnum in the occipital bone, where it is continuous with the dura mater of the brain, to the second sacral vertebra, where it is close-ended. The spinal cord is also protected by a cushion of fat and connective tissue located in the **epidural space,** a space between the dura mater and the wall of the vertebral canal (Figure 17.1b).

The middle **meninx** (MĒ-ninks, singular form of meninges) is an avascular covering called the **arachnoid** (a-RAK-noyd; *arachn-* = spider; *-oid* = similar to) because of its spider's web arrangement of delicate collagen fibers and some elastic fibers. It is deep to the dura mater and is also continuous with the arachnoid of the brain. Between the dura mater and the arachnoid is a thin **subdural space,** which contains interstitial fluid.

The innermost meninx is the **pia mater** (PĒ-a MĀ-ter; *pia* = delicate), a thin, transparent connective tissue layer that adheres to the surface of the spinal cord and brain. It consists of interlacing bundles of collagen fibers and some fine elastic fibers, and it contains many blood vessels that supply oxygen and nutrients to the spinal cord. Between the arachnoid and the pia mater is the **subarachnoid space,** which contains cerebrospinal fluid. Inflammation of the meninges is known as *meningitis.*

All three spinal meninges cover the spinal nerves up to the point of exit from the spinal column through the intervertebral foramina. Triangular-shaped membranous extensions of the pia mater suspend the spinal cord in the middle of its dural sheath. These extensions, called **denticulate ligaments** (den-TIK-yū-lāt; = small tooth), are thickenings of the pia mater. They project laterally and fuse with the arachnoid and inner surface of the dura mater between the anterior and posterior nerve roots of spinal nerves on either side. Extending all along the length of the spinal cord, the denticulate ligaments protect the spinal cord against shock and sudden displacement.

External Anatomy of the Spinal Cord

The **spinal cord,** although roughly cylindrical, is flattened slightly in its anterior-posterior dimension. In adults, it extends from the medulla oblongata, the most inferior part of the brain, to the superior border of the second lumbar vertebra (Figure 17.2 on page 518). In newborn infants, it extends to the third or fourth lumbar vertebra. During early childhood, both the spinal cord and the vertebral column grow longer as part of overall body growth; around age 4 or 5, however, elongation of the spinal cord stops. Because the vertebral column continues to elongate, the spinal cord does not extend the entire length of the vertebral column in adults. The length of the adult spinal cord ranges from 42 to 45 cm (16–18 in.). Its diameter is about 2 cm (0.75 in.) in the midthoracic region, somewhat larger in the lower cervical and midlumbar regions, and smallest at the inferior tip.

When the spinal cord is viewed externally, two conspicuous enlargements can be seen. The superior enlargement, the **cervical enlargement,** extends from the fourth cervical vertebra to the first thoracic vertebra. Nerves to and from the upper limbs arise from the cervical enlargement. The inferior enlargement, called the **lumbar enlargement,** extends from the ninth to the

Figure 17.1 / Gross anatomy of the spinal cord. The spinal meninges are evident in both views.

Meninges are connective tissue coverings that surround the spinal cord and brain.

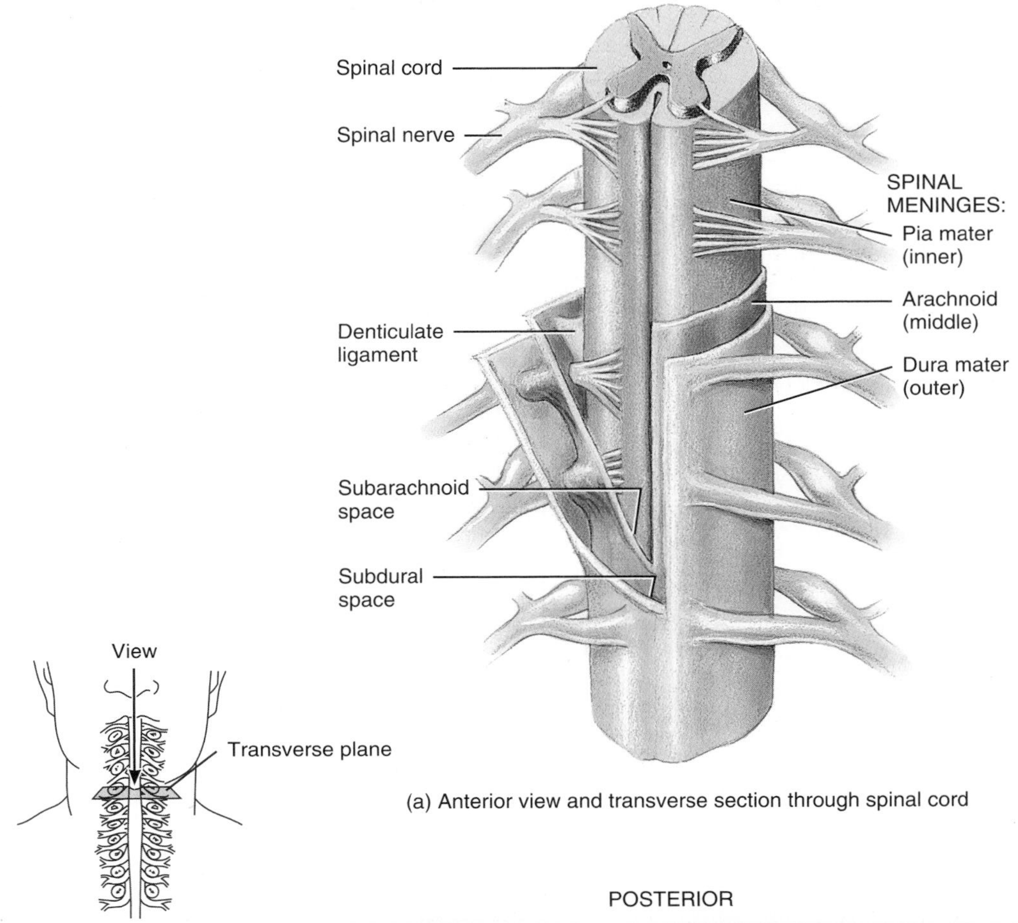

Spinal cord
Spinal nerve
SPINAL MENINGES:
Pia mater (inner)
Arachnoid (middle)
Dura mater (outer)
Denticulate ligament
Subarachnoid space
Subdural space
View
Transverse plane

(a) Anterior view and transverse section through spinal cord

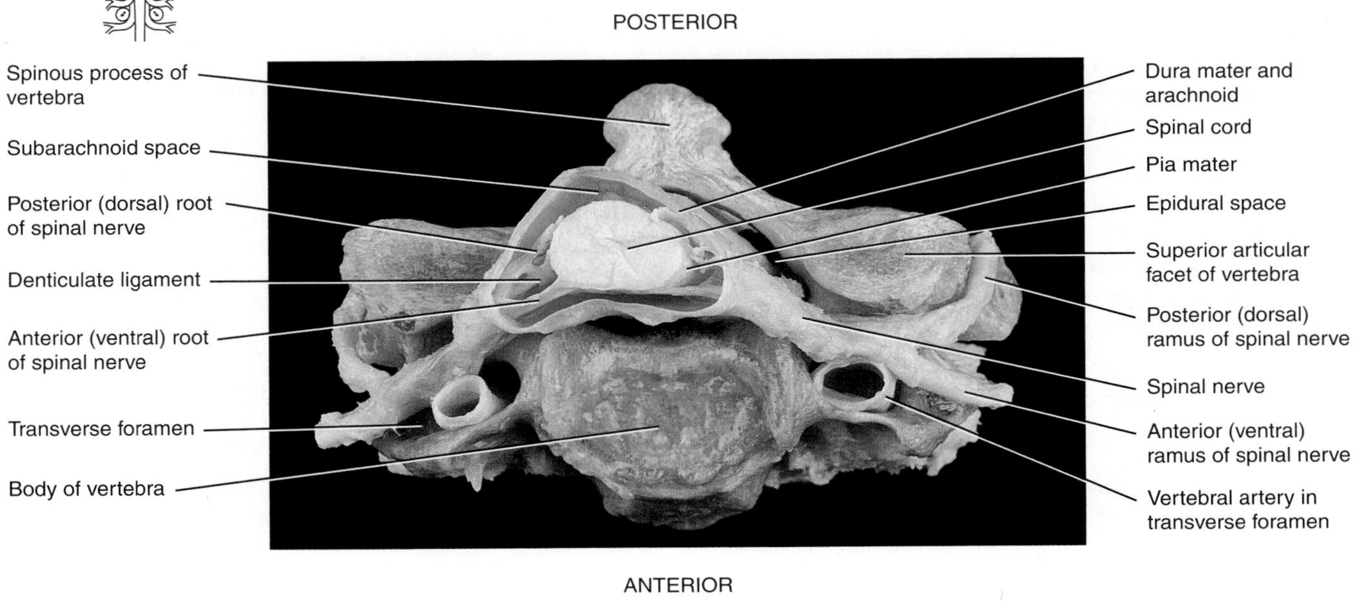

POSTERIOR
Spinous process of vertebra
Subarachnoid space
Posterior (dorsal) root of spinal nerve
Denticulate ligament
Anterior (ventral) root of spinal nerve
Transverse foramen
Body of vertebra
Dura mater and arachnoid
Spinal cord
Pia mater
Epidural space
Superior articular facet of vertebra
Posterior (dorsal) ramus of spinal nerve
Spinal nerve
Anterior (ventral) ramus of spinal nerve
Vertebral artery in transverse foramen
ANTERIOR

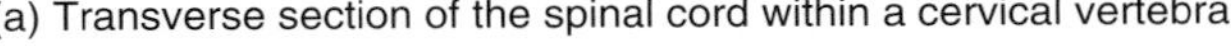

(a) Transverse section of the spinal cord within a cervical vertebra

What are the superior and inferior boundaries of the spinal dura mater?

Figure 17.2 / External anatomy of the spinal cord and the spinal nerves.

The spinal cord extends from the medulla oblongata of the brain to the superior border of the second lumbar vertebra.

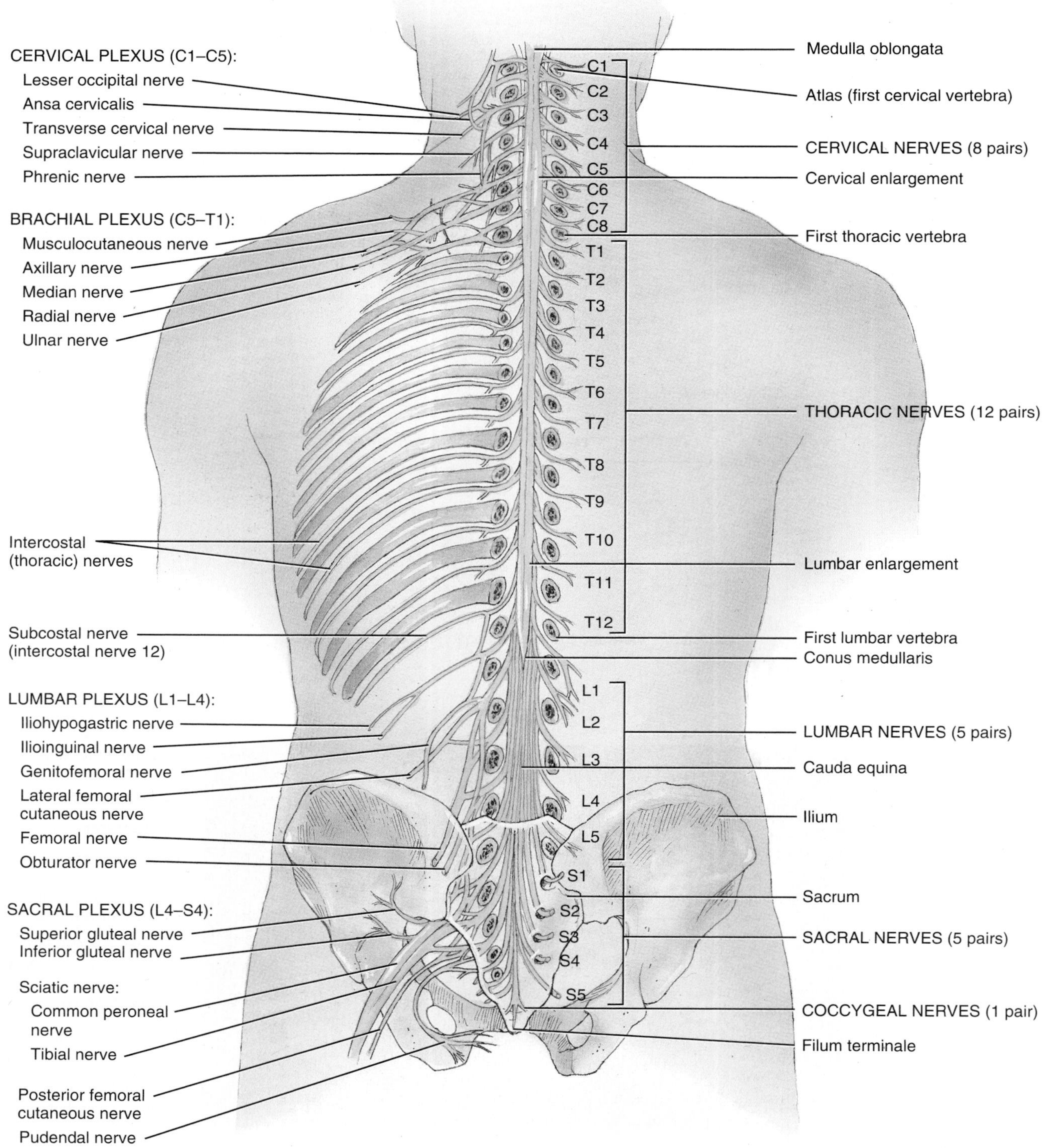

(a) Posterior view of entire spinal cord and portions of spinal nerves

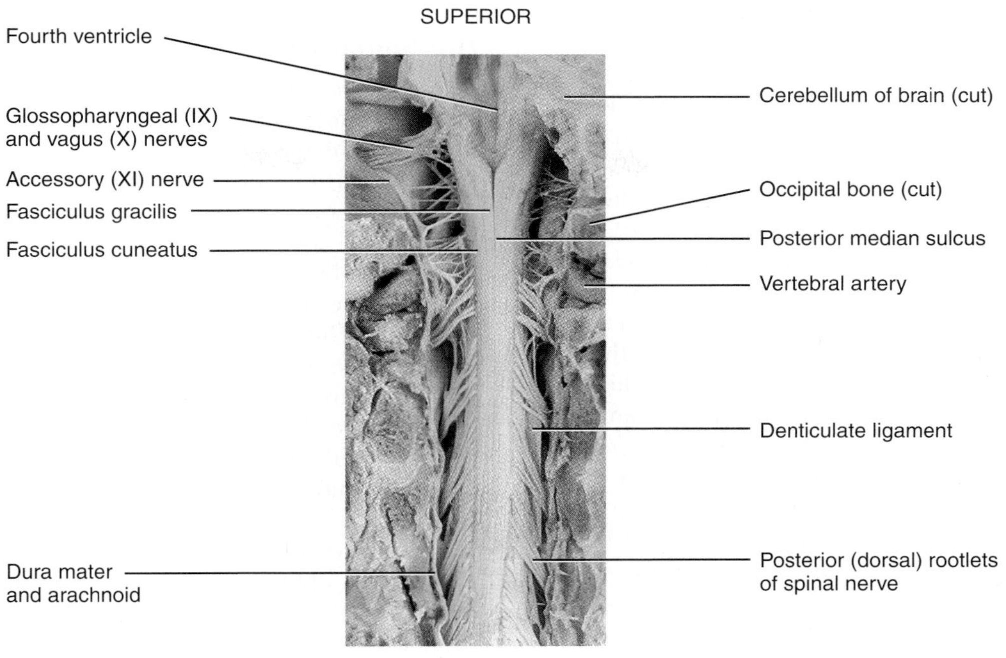

(b) Posterior view of cervical region of spinal cord

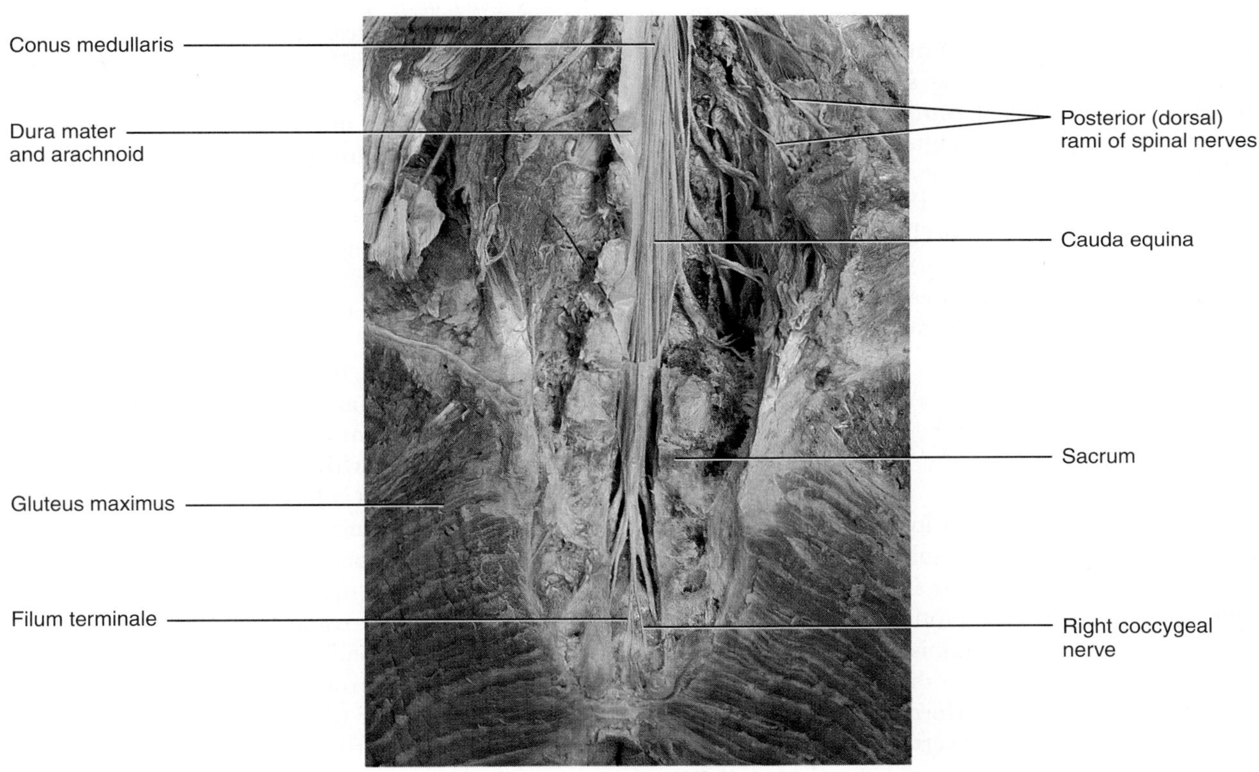

(c) Posterior view of inferior portion of spinal cord

What portion of the spinal cord connects with sensory and motor nerves of the upper limbs?

twelfth thoracic vertebra. Nerves to and from the lower limbs arise from the lumbar enlargement.

Inferior to the lumbar enlargement, the spinal cord tapers to a conical portion known as the **conus medullaris** (KŌ-nus med-yū-LAR-is; *conus* = cone), which ends at the level of the intervertebral disc between the first and second lumbar vertebrae in adults. Arising from the conus medullaris is the **filum terminale** (FĪ-lum ter-mi-NAL-ē; = terminal filament), an extension of the pia mater that extends inferiorly and anchors the spinal cord to the coccyx.

Nerves that arise from the inferior part of the spinal cord do not leave the vertebral column at the same level as they exit from the cord. The roots (points of attachment to the spinal cord) of these nerves angle inferiorly in the vertebral canal from the end of the spinal cord like wisps of hair. Appropriately, the roots of these nerves are collectively named the **cauda equina** (KAW-da ē-KWĪ-na), meaning "horse's tail."

Spinal cord organization appears to be segmented because the 31 pairs of spinal nerves emerge at regular intervals (Figure 17.2). Indeed, each pair of spinal nerves is said to arise from a *spinal segment.* Within the spinal cord, however, there is no obvious segmentation dividing up the white matter or the gray matter. The naming of spinal nerves and spinal segments is based on their location. There are eight pairs of *cervical nerves* (represented as C1–C8), twelve pairs of *thoracic nerves* (T1–T12), five pairs of *lumbar nerves* (L1–L5), five pairs of *sacral nerves* (S1–S5), and one pair of coccygeal nerves.

Spinal nerves are the paths of communication between the spinal cord and the nerves innervating specific regions of the body. Two bundles of axons, called **roots,** connect each spinal nerve to a segment of the cord (see Figure 17.3a). The **posterior** or **dorsal root** contains only sensory fibers, which conduct nerve impulses from the periphery into the central nervous system. Each posterior root also has a swelling, the **posterior** or **dorsal root ganglion,** which contains the cell bodies of sensory neurons. The **anterior** or **ventral root** contains axons of motor neurons, which conduct impulses from the CNS to effector organs and cells.

In a **spinal tap (lumbar puncture),** a local anesthetic is given and a long needle is inserted into the subarachnoid space. The procedure is used to withdraw cerebrospinal fluid (CSF) for diagnostic purposes; to introduce antibiotics, contrast media for myelography, or anesthetics; to administer chemotherapy; to measure CSF pressure; and to evaluate the effects of treatment. In adults, a spinal tap is normally performed between the third and fourth or fourth and fifth lumbar vertebrae. Because this region is inferior to the lowest portion of the spinal cord, it provides relatively safe access. (A line drawn across the highest points of the iliac crests, called the *supracristal line* (see Figure 11.9c on page 357), passes through the spinous process of the fourth lumbar vertebra.) ■

Internal Anatomy of the Spinal Cord

Two grooves penetrate the white matter of the spinal cord and divide it into right and left sides (Figure 17.3). The **anterior median fissure** is a deep, wide groove on the anterior (ventral) side, whereas the **posterior median sulcus** is a shallower, narrow groove on the posterior (dorsal) surface. The gray matter of the spinal cord is shaped like the letter H or a butterfly and is surrounded by white matter. The gray matter consists primarily of cell bodies of neurons, neuroglia, unmyelinated axons, and dendrites of interneurons and motor neurons. The white matter consists of bundles of myelinated and unmyelinated axons of sensory neurons, interneurons, and motor neurons. The **gray commissure** (KOM-mi-shur) forms the crossbar of the H. In the center of the gray commissure is a small space called the **central canal,** which extends the entire length of the spinal cord. At its superior end, the central canal is continuous with the fourth ventricle (a space that contains cerebrospinal fluid) in the medulla oblongata of the brain. Anterior to the gray commissure is the **anterior (ventral) white commissure,** which connects the white matter of the right and left sides of the spinal cord.

In the gray matter of the spinal cord and brain, clusters of neuronal cell bodies form functional groups called nuclei. *Sensory nuclei* receive input from receptors via sensory neurons, whereas *motor nuclei* provide output to effector tissues via motor neurons. The gray matter on each side of the spinal cord is subdivided into regions called **horns.** The **anterior (ventral) gray horns** contain cell bodies of somatic motor neurons and motor nuclei, which provide nerve impulses for contraction of skeletal muscles. The **posterior (dorsal) gray horns** contain somatic and autonomic sensory nuclei. Between the anterior and posterior gray horns are the **lateral gray horns,** which are present only in the thoracic, upper lumbar, and sacral segments of the cord. The lateral horns contain cell bodies of autonomic motor neurons that regulate activity of smooth muscle, cardiac muscle, and glands.

The white matter, like the gray matter, is organized into regions. The anterior and posterior gray horns divide the white matter on each side into three broad areas called **columns:** (1) **anterior (ventral) white columns,** (2) **posterior (dorsal) white columns,** and (3) **lateral white columns.** Each column, in turn, contains distinct bundles of nerve axons having a common origin or destination and carrying similar information. These bundles, which may extend long distances up or down the spinal cord, are called **tracts. Sensory (ascending) tracts** consist of axons that conduct nerve impulses toward the brain. Tracts consisting of axons that carry nerve impulses down the cord are called **motor (descending) tracts.** Sensory and motor tracts of the spinal cord are continuous with sensory and motor tracts in the brain.

The various spinal cord segments vary in size, shape, relative amounts of gray and white matter, and distribution and shape of gray matter. These features are summarized in Table 17.1 on page 522.

Figure 17.3 / Internal anatomy of the spinal cord: the organization of gray matter and white matter. For simplicity, dendrites are not shown in this and several other illustrations of transverse sections of the spinal cord. Blue and red arrows in (a) indicate the direction of nerve impulse propagation.

In the spinal cord, white matter surrounds the gray matter.

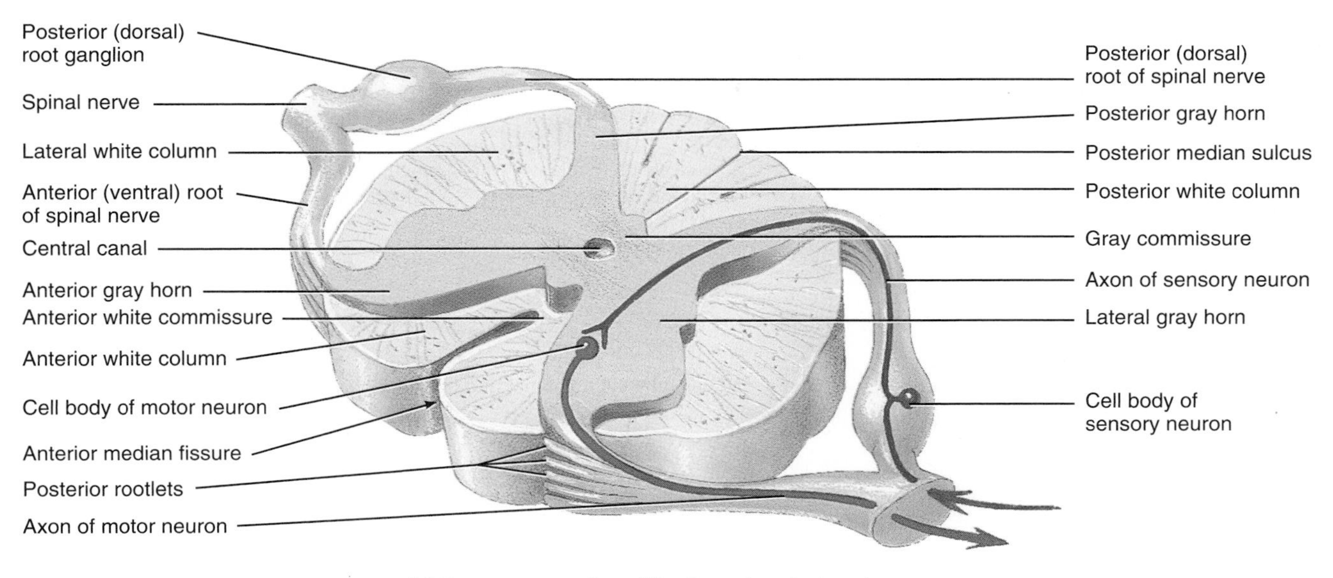

(a) Transverse section of the thoracic spinal cord

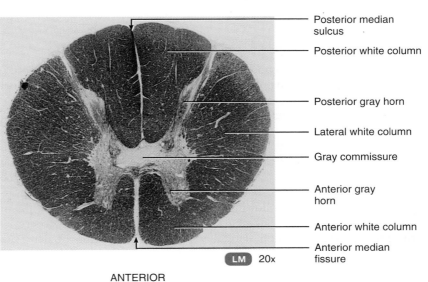

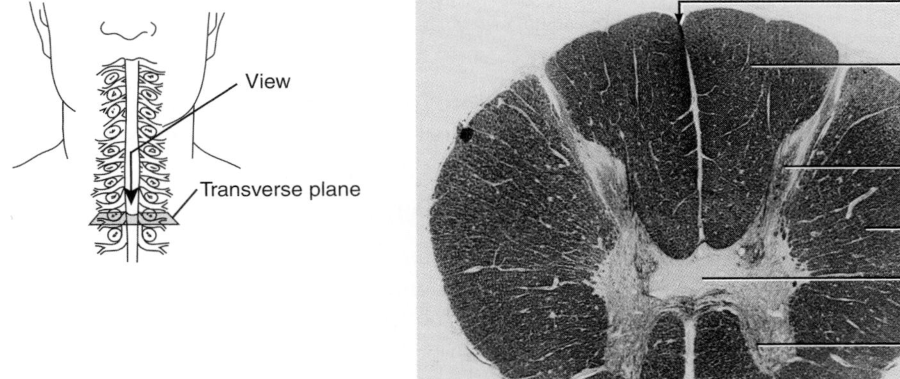

(b) Transverse section of the thoracic spinal cord

Functions

1. White matter tracts propragate sensory inpulses from the periphery to the brain and motor impulses from the brain to the periphery.
2. Gray matter receives and integrates incoming and outgoing information.

What is the difference between a horn and a column in the spinal cord?

Table 17.1 Comparison of Various Spinal Cord Segments

Segment	Distinguishing Characteristics
Cervical (Segment C1) (Segment C8)	Relatively large diameter, relatively large amounts of white matter, oval in shape; in upper cervical segments (C1–C6), posterior gray horn is large, but anterior gray horn is relatively small; in lower cervical segments (C6 and below), posterior gray horns are enlarged and anterior gray horns are well developed.
Thoracic (Segment T2)	Small diameter is due to relatively small amounts of gray matter; except for first thoracic segment, anterior and posterior gray horns are relatively small; a small lateral gray horn is present.
Lumbar (Segment L4)	Nearly circular; very large anterior and posterior gray horns; relatively less white matter than cervical segments.
Sacral (Segment S3)	Relatively small, but with relatively large amounts of gray matter; relatively small amounts of white matter; anterior and posterior gray horns are large and thick.
Coccygeal	Resemble lower sacral spinal segments, but much smaller.

- ✓ Explain the locations and compositions of the spinal meninges. Describe the locations of the epidural, subdural, and subarachnoid spaces.
- ✓ Describe the location of the spinal cord. What are the cervical and lumbar enlargements?
- ✓ Define conus medullaris, filum terminale, and cauda equina. What is a spinal segment? How is the spinal cord partially divided into right and left sides?
- ✓ Based on your knowledge of the structure of the spinal cord in transverse section, define the following: gray commissure, central canal, anterior gray horn, lateral gray horn, posterior gray horn, anterior white column, lateral white column, posterior white column, ascending tract, and descending tract.

SPINAL CORD FUNCTIONS

Objectives

- Describe the functions of the principal sensory and motor tracts of the spinal cord.
- Describe the functional components of a reflex arc and how reflexes contribute to homeostasis.

The spinal cord has two principal functions in maintaining homeostasis: it conducts nerve impulses and integrates information.

Sensory and Motor Tracts

The white matter tracts in the spinal cord are highways for nerve impulse conduction. Along these tracts, sensory impulses flow from the periphery to the brain, and motor impulses flow from the brain to the periphery.

Often, the name of a tract indicates its position in the white matter, where it begins and ends, and, by extension, the direction of nerve impulse propagation. For example, the anterior spinothalamic tract is located in the *anterior* white column; it begins in the *spinal cord* and ends in the *thalamus* (a region of the brain). Notice that the location of the neuronal cell bodies and dendrites comes first in the name, and the location of the axon terminals comes last. This regularity in naming allows you to determine the direction of information flow along any tract named according to this convention. Thus, because the anterior spinothalamic tract conveys nerve impulses from the cord toward the brain, it is a sensory (ascending) tract. Figure 17.4 highlights the principal sensory and motor tracts in the spinal cord. These tracts are described in detail in Chapter 19 and summarized in Table 19.1 on page 592.

Figure 17.4 / The locations of selected sensory and motor tracts, shown in a transverse section of the spinal cord. Sensory tracts are indicated on one half and motor tracts on the other half of the cord, but in fact all tracts are present on both sides.

The name of a tract often indicates its location in the white matter and where it begins and ends.

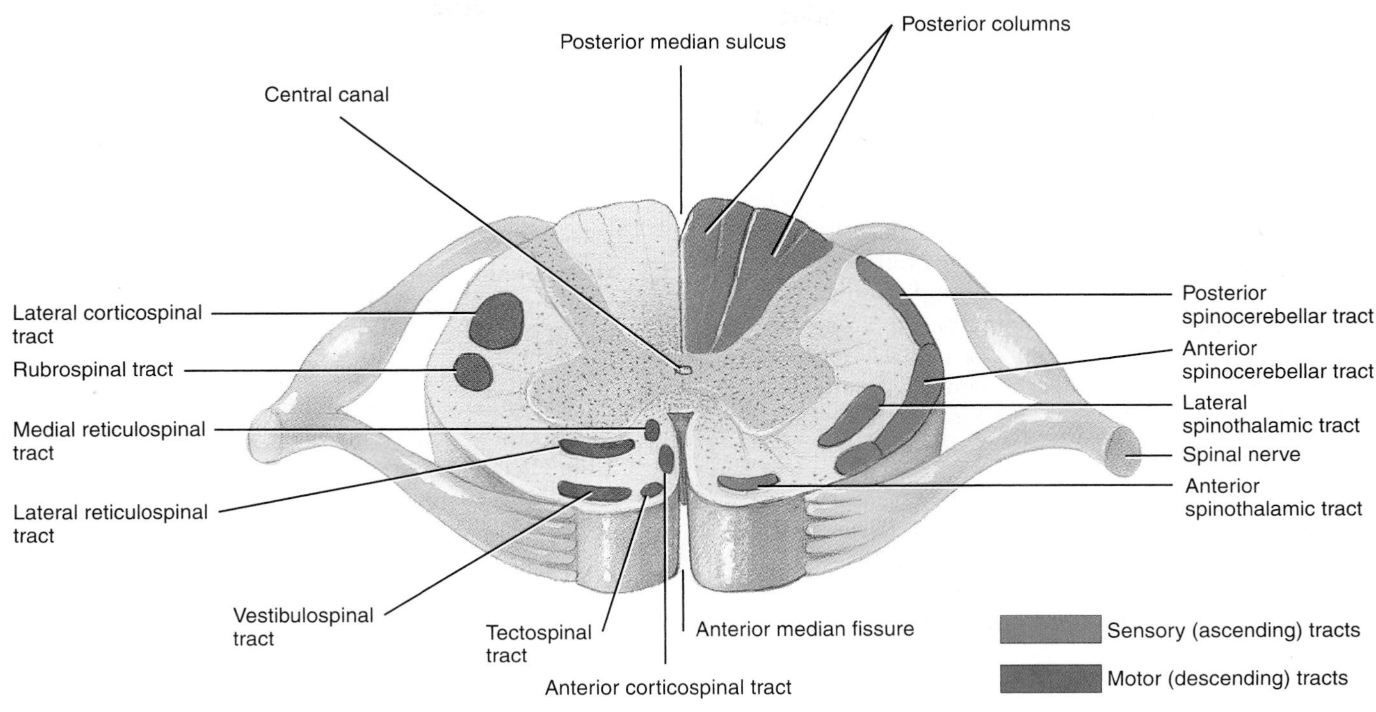

Based on its name, what are the position in the spinal cord, origin, and destination of the anterior corticospinal tract? Is this a sensory or a motor tract?

Sensory information from receptors travels up the spinal cord to the brain along two main routes on each side of the cord: the spinothalamic tracts and the posterior columns. The two **spinothalamic tracts** convey impulses for sensing pain, temperature, deep pressure, and a crude, poorly localized sense of touch. The **posterior columns** carry nerve impulses for sensing (1) proprioception, awareness of the movements of muscles, tendons, and joints; (2) fine touch, the ability to feel exactly what part of the body is touched; (3) two-point discrimination, the ability to distinguish the touching of two different points on the skin, even though they are close together; (4) pressure; and (5) vibration.

The sensory systems keep the CNS informed of changes in the external and internal environments. Responses to this information are brought about by motor systems, which enable us to move about and change our physical relationship to the world around us. As sensory information is conveyed to the CNS, it becomes part of a large pool of sensory input. Each piece of incoming information is integrated with all the other information arriving from activated sensory neurons.

Through the activity of interneurons, the integration process occurs not just in one region of the CNS, but in many regions. Integration occurs within the spinal cord and several regions of the brain. As a result, motor responses to make a muscle contract or a gland secrete can be initiated at several levels. Most regulation of involuntary activities of smooth muscle, cardiac muscle, and glands by the autonomic nervous system originates in the brain stem (which is continuous with the spinal cord) and in a nearby brain region called the hypothalamus. More information is given in Chapter 18.

The cerebral cortex (superficial gray matter of the cerebrum) plays a major role in controlling precise, voluntary muscular movements, whereas other brain regions provide important integration for regulation of automatic movements, such as arm swinging during walking. Motor output to skeletal muscles travels down the spinal cord in two types of descending pathways: direct and indirect. **Direct pathways** (the lateral corticospinal, anterior corticospinal, and corticobulbar tracts) convey nerve impulses that originate in the cerebral cortex and are destined to cause precise, *voluntary* movements of skeletal muscles. **Indirect pathways** (rubrospinal, tectospinal, and vestibulospinal tracts) convey nerve impulses from the brain stem and other parts of the brain that are destined to program automatic movements, help coordinate body movements with visual stimuli, maintain skeletal muscle tone and posture, and play a major role in equilibrium by regulating muscle tone in response to movements of the head.

Reflexes

The second way the spinal cord promotes homeostasis is by serving as an integrating center for **spinal reflexes.** Integration takes place in the spinal gray matter. **Reflexes** are fast, predictable, automatic responses to changes in the environment. Some reflexes, which occur through the brain stem rather than the spinal cord, involve cranial nerves and are called **cranial reflexes.** You are probably most aware of **somatic reflexes,** which involve contraction of skeletal muscles. Equally important, however, are the **autonomic (visceral) reflexes,** which generally are not consciously perceived. They involve responses of smooth muscle, cardiac muscle, and glands. Body functions such as heart rate, digestion, urination, and defecation are controlled by the autonomic nervous system through autonomic reflexes.

Nerve impulses propagating into, through, and out of the CNS follow specific pathways, depending on the kind of information, its origin, and its destination. The pathway followed by nerve impulses that produce a reflex response is a **reflex arc.** A reflex arc includes the following five functional components (Figure 17.5):

1. **Sensory receptor.** The distal end of a sensory neuron (dendrite) or an associated sensory structure serves as a sensory receptor. It responds to a specific **stimulus**—a change in the internal or external environment—by triggering one or more nerve impulses in the sensory neuron.
2. **Sensory neuron.** The nerve impulses propagate along the axon of the sensory neuron to the axon terminals, which are located in the gray matter of the spinal cord or brain stem.
3. **Integrating center.** One or more regions of gray matter within the CNS act as an integrating center. In the simplest type of reflex, the integrating center is a single synapse between a sensory neuron and a motor neuron. A reflex pathway having only one synapse in the CNS is termed a **monosynaptic reflex arc.** More often the integrating center consists of one or more interneurons, which may relay the impulse to other interneurons as well as to a motor neuron. A **polysynaptic reflex arc** involves more than two types of neurons and more than one CNS synapse.
4. **Motor neuron.** Impulses triggered by the integrating center propagate out of the CNS along a motor neuron to the part of the body that will respond.
5. **Effector.** The part of the body that responds to the motor nerve impulse, such as a muscle or gland, is the effector. If the effector is skeletal muscle, the reflex is a **somatic reflex.** If the effector is smooth muscle, cardiac muscle, or a gland, the reflex is an **autonomic (visceral) reflex.**

Because reflexes are normally so predictable, they provide useful information about the health of the nervous system and can greatly aid diagnosis of disease. If a reflex is absent or abnor-

Figure 17.5 / General components of a reflex arc. The arrows show the direction of nerve impulse propagation.

Reflexes are fast, predictable, automatic responses to changes in the environment.

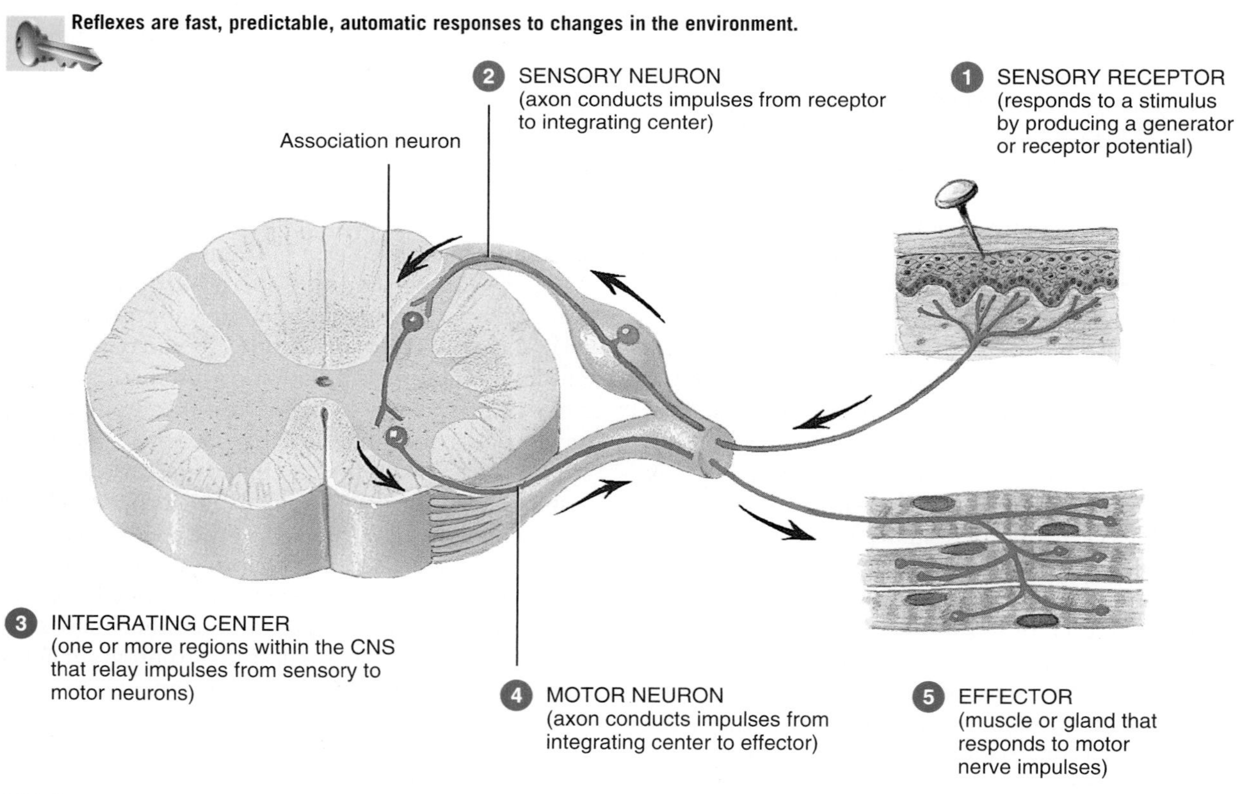

2 SENSORY NEURON
(axon conducts impulses from receptor to integrating center)
1 SENSORY RECEPTOR
(responds to a stimulus by producing a generator or receptor potential)
Association neuron
3 INTEGRATING CENTER
(one or more regions within the CNS that relay impulses from sensory to motor neurons)
4 MOTOR NEURON
(axon conducts impulses from integrating center to effector)
5 EFFECTOR
(muscle or gland that responds to motor nerve impulses)

Which reflexes involve skeletal muscle contraction?

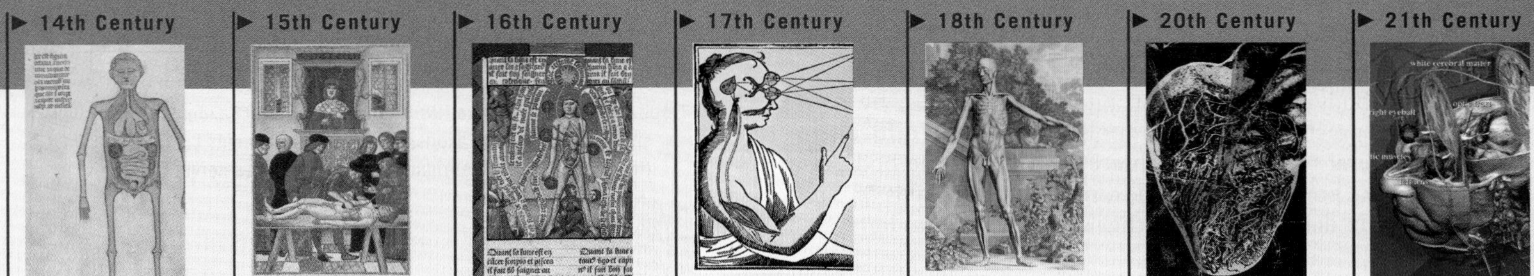

CHANGING IMAGES

A Higher Authority?

As in Europe, anatomical knowledge in medieval Islam was based on the writings of Galen. The authority of this second century Greek physician rarely would be questioned in these cultures and throughout the centuries. A unique aspect of the Islamic anatomist, however, was his *teleological* perspective, that is, an ultimate purpose explains all natural phenomena. The focus of Islamic anatomical texts, therefore, often centered on the wisdom of God in the design of man, more so than exact description. Anatomy was viewed in a sacred light, rather than as a precise science.

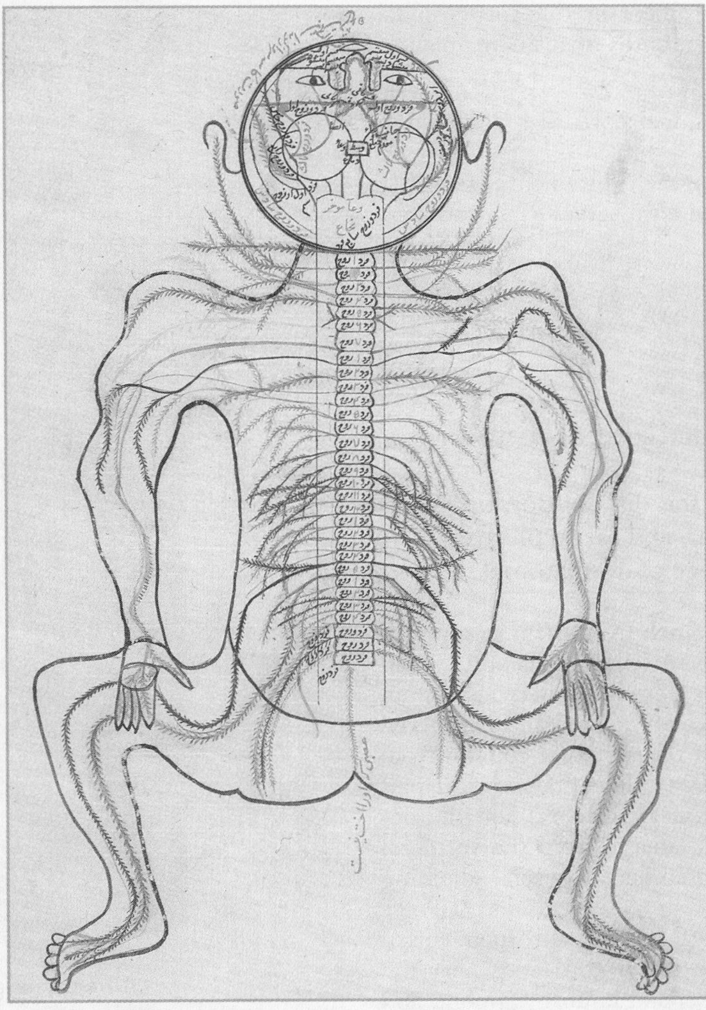

Pictured here is an illustration of the nerves from a 14th century Persian manuscript that is considered to be the oldest surviving Islamic culture writing in anatomical studies. This image is from *Tashrih-i- badan-i insan or The Anatomy of the Human Body* by Mansur ibn Ilyas. *Mansur's Anatomy*, as it is often called, describes what was then believed to be the five systems of the body: bones, muscles, nerves, veins, and arteries. The view in the illustration is from a posterior perspective, with the head hyperextended, and the mouth facing upward. The spinal nerves can be seen exiting between the vertebrae. Can you see the relation of the spinal nerves depicted here to those in Figure 17.2a? It also seems that the artist is representing the ventricles (spaces filled with cerebrospinal fluid) of the brain, a concept you will study in the next chapter.

Scientists traditionally explain natural phenomena by direct observation, data collection, and controlled experiments that generate reproducible results. The laws of nature are thus described. Yet, consistent with the view of the Islamic anatomist of centuries ago, have you ever considered the role of God in the natural world? Are nature's laws divine in their origin? Is the universe random in its existence? Science has difficulty answering such questions. Yet, it seems self -evident that there is no conflict between science and theology to consider a divine influence in human anatomy.

14th Century

mal, the damage may be anywhere along a particular pathway. Somatic reflexes generally can be tested simply by tapping or stroking the body's surface. Most autonomic reflexes, by contrast, are not practical diagnostic tools because it is difficult to stimulate visceral receptors, which are deep inside the body. An exception is the pupillary light reflex, in which the pupils of both eyes decrease in diameter when either eye is exposed to light.

Plantar Flexion Reflex and Babinski Sign

Several reflexes have clinical significance. Among the most important is the **plantar flexion reflex,** elicited by gently stroking the lateral outer margin of the sole. The normal response is a curling under of all the toes. Damage to descending motor pathways, such as the corticospinal tract, alters this reflex such that stroking the sole causes the great toe to extend, with or without fanning of the other toes, a response called the **Babinski sign.** In children under 18 months of age, the Babinski sign is normal due to incomplete myelination of fibers in the corticospinal tract. The normal response after 18 months of age is the plantar flexion reflex. ■

✓ Describe the function of the spinal cord as a conduction pathway.

✓ What is a reflex? List and define the components of a reflex arc, and describe how the spinal cord serves as an integrating center for reflexes.

SPINAL NERVES

Objectives

- Describe the components, connective tissue coverings, and branches of a spinal nerve.
- Define a plexus, and identify the distribution of nerves of the cervical, brachial, lumbar, and sacral plexuses.
- Describe the clinical significance of dermatomes.

Spinal nerves and the nerves that branch from them to serve all parts of the body connect the CNS to sensory receptors, muscles, and glands and are part of the peripheral nervous system (PNS). The 31 pairs of spinal nerves are named and numbered according to the region and level of the vertebral column from which they emerge (see Figure 17.2a). The first cervical pair emerges between the atlas (first cervical vertebra) and the occipital bone; all other spinal nerves emerge from the vertebral column through intervertebral foramina between adjoining vertebrae.

Not all spinal cord segments are in line with their corresponding vertebrae. Recall that the spinal cord ends near the level of the superior border of the second lumbar vertebra, and that the roots of the lower lumbar, sacral, and coccygeal nerves descend at an angle to reach their respective foramina before emerging from the vertebral column. This arrangement constitutes the cauda equina (see Figure 17.2a).

Figure 17.6 / Organization and connective tissue coverings of a spinal nerve.

Three layers of connective tissue wrappings protect axons: endoneurium surrounds individual axons, perineurium surrounds bundles of axons (fascicles), and epineurium surrounds an entire nerve.

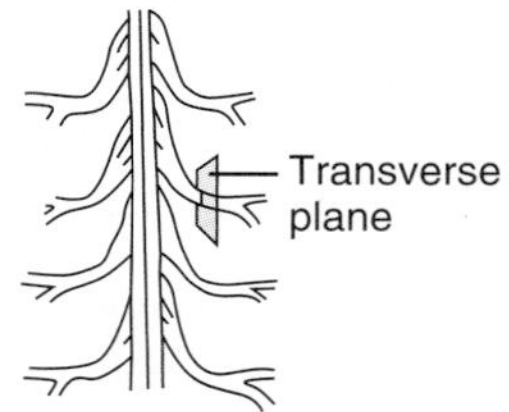

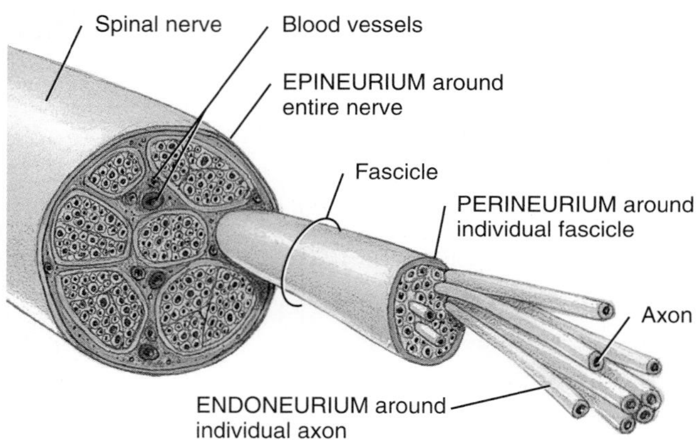

(a) Transverse sections showing the coverings of a spinal nerve

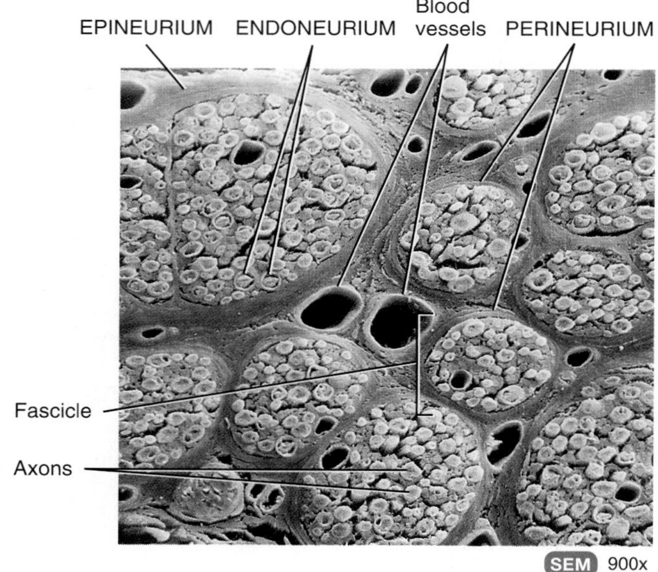

(b) Transverse section of 12 nerve fascicles

From Richard Kessel and Randy Kardon, *Tissues and Organs: A Text-Altas of Scanning Electron Microscopy,* W. H. Freeman and Company. Reprinted by permission.

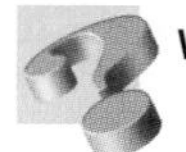

Why are all spinal nerves classified as mixed nerves?

As noted earlier, a typical **spinal nerve** has two connections to the cord: a posterior root and an anterior root (see Figure 17.3a). The posterior and anterior roots unite to form a spinal nerve at the intervertebral foramen. Because the posterior root contains sensory axons and the anterior root contains motor axons, a spinal nerve is a **mixed nerve.** The posterior root contains a ganglion in which cell bodies of sensory neurons are located.

Connective Tissue Coverings of Spinal Nerves

Each cranial nerve and spinal nerve contains layers of protective connective tissue coverings (Figure 17.6 on page 526). Individual axons, whether myelinated or unmyelinated, are wrapped in **endoneurium** (en′-dō-NŪ-rē-um; *endo-* = within or inner). Groups of axons with their endoneurium are arranged in bundles called **fascicles,** each of which is wrapped in **perineurium** (per′-i-NŪ-rē-um; *peri-* = around). The superficial covering over the entire nerve is the **epineurium** (ep′-i-NŪ-rē-um; *epi-* = over). The dura mater of the spinal meninges fuses with the epineurium as the nerve passes through the intervertebral foramen. Note the presence of many blood vessels, which nourish nerves, within the perineurium and epineurium (Figure 17.6b). You may recall from Chapter 10 that the connective tissue coverings of skeletal muscles—endomysium, perimysium, and epimysium—are similar in organization to those of nerves.

Distribution of Spinal Nerves

Branches

A short distance after passing through its intervertebral foramen, a spinal nerve divides into several branches (Figure 17.7). These branches are known as **rami** (RĀ-mī = branches). The **posterior (dorsal) ramus** (RĀ-mus; singular form) serves the deep muscles and skin of the dorsal surface of the trunk. The **anterior (ventral) ramus** serves the muscles and structures of the upper and lower limbs and the skin of the lateral and ventral surfaces of the trunk. In addition to posterior and anterior rami, spinal nerves also give off a **meningeal branch,** which reenters the vertebral canal through the intervertebral foramen and supplies the vertebrae, vertebral ligaments, blood vessels of the spinal cord, and meninges. Other branches of a spinal nerve are

Figure 17.7 / Branches of a typical spinal nerve, shown in transverse section through the thoracic portion of the spinal cord. (See also Figure 17.1b.)

The branches of a spinal nerve are the posterior ramus, the anterior ramus, the meningeal branch, and the rami communicantes.

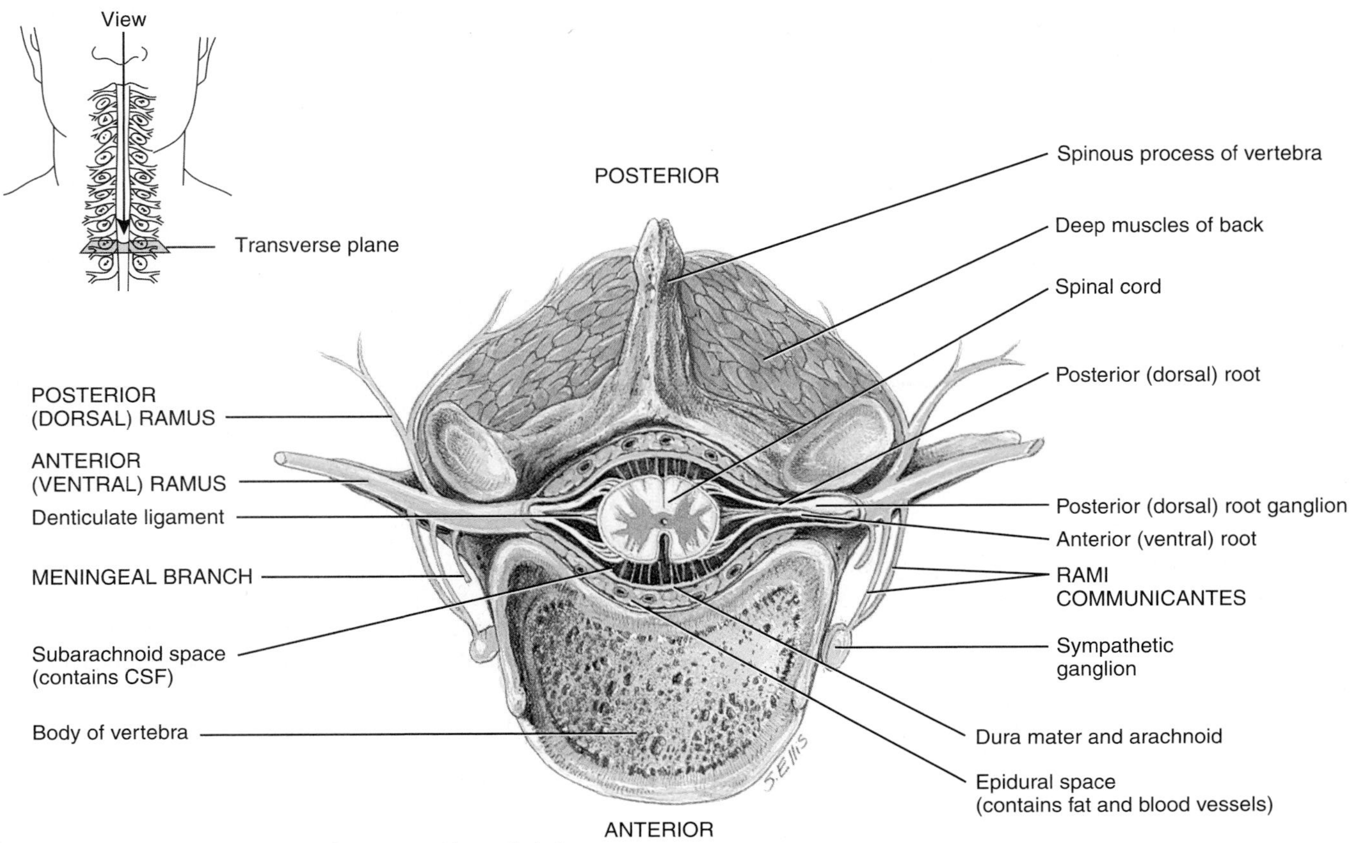

Which spinal nerve branches serve the upper and lower limbs?

Exhibit 17.1 Cervical Plexus (Figure 17.8)

Objective

- ▶ Describe the origin, distribution, and effects of damage to the cervical plexus.

The **cervical plexus** (SER-vi-kul) is formed by the anterior (ventral) rami of the first four cervical nerves (C1–C4), with contributions from C5. There is one on each side of the neck alongside the first four cervical vertebrae. The roots of the plexus indicated in Figure 17.8 are the anterior rami.

The cervical plexus supplies the skin and muscles of the head, neck, and superior part of the shoulders and chest. The phrenic nerves arise from the cervical plexuses and supply motor fibers to the diaphragm. Branches of the cervical plexus also run parallel to the accessory (XI) and hypoglossal (XII) cranial nerves.

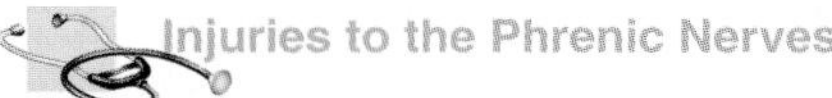

Injuries to the Phrenic Nerves

Complete severing of the spinal cord above the origin of the phrenic nerves (C3, C4, and C5) causes respiratory arrest. Breathing stops because the phrenic nerves no longer send impulses to the diaphragm.

✓ What is the origin of the cervical plexus?

Nerve	Origin	Distribution
Superficial (Sensory) Branches		
Lesser occipital	C2	Skin of scalp posterior and superior to ear.
Great auricular (aw-RIK-yū-lar)	C2–C3	Skin anterior, inferior, and over ear, and over parotid glands.
Transverse cervical	C2–C3	Skin over anterior aspect of neck.
Supraclavicular	C3–C4	Skin over superior portion of chest and shoulder.
Deep (Largely Motor) Branches		
Ansa cervicalis (AN-sa ser-vi-KAL-is)		This nerve divides into superior and inferior roots.
Superior root	C1	Infrahyoid and geniohyoid muscles of neck.
Inferior root	C2–C3	Infrahyoid muscles of neck.
Phrenic (FREN-ik)	C3–C5	Diaphragm.
Segmental branches	C1–C5	Prevertebral (deep) muscles of neck, levator scapulae, and middle scalene muscles.

the **rami communicantes** (kō-myū-ni-KAN-tēz), components of the autonomic nervous system whose structure and function are discussed in Chapter 21.

Plexuses

The anterior rami of spinal nerves, except for thoracic nerves T2–T12, do not go directly to the body structures they supply. Instead, they form networks on both the left and right sides of the body by joining with various numbers of nerve fibers from anterior rami of adjacent nerves. Such a network is called a **plexus** (= braid). The principal plexuses are the **cervical plexus, brachial plexus, lumbar plexus,** and **sacral plexus.** Refer to Figure 17.2a to see their relationships to one another. Emerging from the plexuses are nerves bearing names that are often descriptive of the general regions they serve or the course they take. Each of the nerves, in turn, may have several branches named for the specific structures they innervate.

Figure 17.8 / Cervical plexus in anterior view. (See Tortora, *A Photographic Atlas of the Human Body,* Figure 8.7.)

The cervical plexus supplies the skin and muscles of the head, neck, superior portion of the shoulders and chest, and diaphragm.

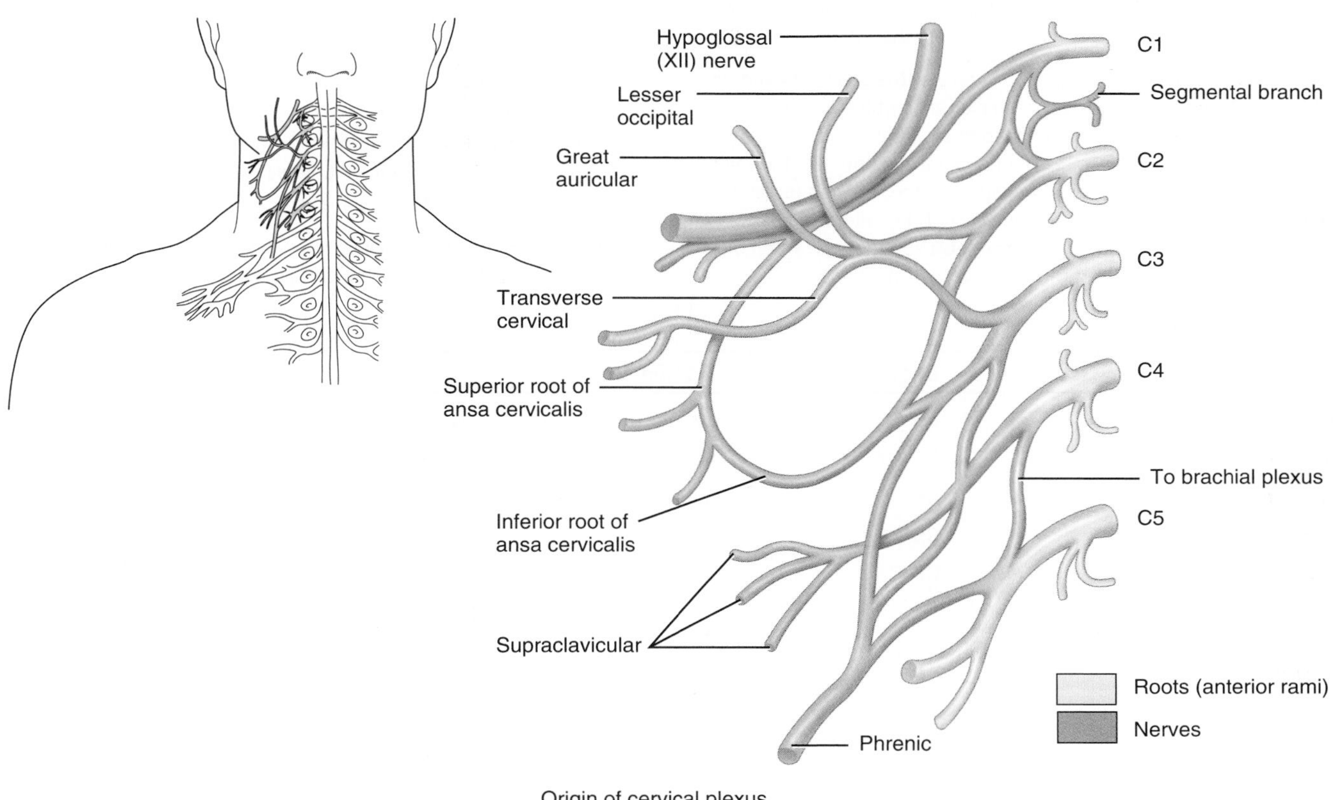

Origin of cervical plexus

Why does complete severing of the spinal cord at level C2 cause respiratory arrest?

Exhibits 17.1–17.4 summarize the principal plexuses. The anterior rami of spinal nerves T2–T12 are called intercostal nerves and will be discussed next.

Intercostal Nerves

The anterior rami of spinal nerves T2–T12 do not enter into the formation of plexuses and are known as **intercostal,** or **thoracic, nerves.** These nerves directly innervate the structures they supply. After leaving its intervertebral foramen, the anterior ramus of nerve T2 innervates the intercostal muscles of the second intercostal space and supplies the skin of the axilla and posteromedial aspect of the arm. Nerves T3–T6 pass in the costal grooves of the ribs and then to the intercostal muscles and skin of the anterior and lateral chest wall. Nerves T7–T12 supply the intercostal muscles and abdominal muscles, and the overlying skin. The posterior rami of the intercostal nerves supply the deep back muscles and skin of the posterior aspect of the thorax.

(Text continues p. 538)

Exhibit 17.2 Brachial Plexus (Figures 17.9 and 17.10)

Objective

- Describe the origin, distribution, and effects of damage to the brachial plexus.

The anterior (ventral) rami of spinal nerves C5–C8 and T1 form the **brachial plexus** (BRĀ-kē-al), which extends inferiorly and laterally on either side of the last four cervical and first thoracic vertebrae (Figure 17.9a). It passes superior to the first rib posterior to the clavicle and then enters the axilla.

The brachial plexus provides the entire nerve supply of the shoulders and upper limbs (Figure 17.9b). Five important nerves arise from the brachial plexus. (1) The **axillary nerve** supplies the deltoid and teres minor muscles. (2) The **musculocutaneous nerve** supplies the flexors of the arm. (3) The **radial nerve** supplies the muscles on the posterior aspect of the arm and forearm. (4) The **median nerve** supplies most of the muscles of the anterior forearm and some of the muscles of the hand. (5) The **ulnar nerve** supplies the anteromedial muscles of the forearm and most of the muscles of the hand.

Injuries to Nerves Emerging from the Brachial Plexus

Injury to the superior roots of the brachial plexus (C5–C6) may result from forceful pulling away of the head from the shoulder, as might occur from a heavy fall on the shoulder or during childbirth in which the infant's head is excessively stretched. The presentation of this injury is characterized by an upper limb in which the shoulder is adducted, the arm is medially rotated, the elbow is extended, the forearm is pronated, and the wrist is flexed (Figure 17.10a). This condition is called **Erb-Duchene palsy** or **waiter's tip position.** There is loss of sensation along the lateral side of the arm.

Radial (and axillary) **nerve injury** can be caused by improperly administered intramuscular injections into the deltoid muscle. The radial nerve may also be

Nerve	Origin	Distribution
Dorsal scapular (SKAP-yū-lar)	C5	Levator scapulae, rhomboideus major, and rhomboideus minor muscles.
Long thoracic (thor-RAS-ik)	C5–C7	Serratus anterior muscle.
Nerve to subclavius (sub-KLĀ-ve-us)	C5–C6	Subclavius muscle.
Suprascapular	C5–C6	Supraspinatus and infraspinatus muscles.
Musculocutaneous (mus′-kyū-lo-kyū-TĀN-ē-us)	C5–C7	Coracobrachialis, biceps brachii, and brachialis muscles.
Lateral pectoral (PEK-to-ral)	C5–C7	Pectoralis major muscle.
Upper subscapular	C5–C6	Subscapularis muscle.
Thoracodorsal (tho-RA-kō-dor-sal)	C6–C8	Latissimus dorsi muscle.
Lower subscapular	C5–C6	Subscapularis and teres major muscles.
Axillary (AK-si-lar-ē) or **circumflex** (SER-kum-fleks)	C5–C6	Deltoid and teres minor muscles; skin over deltoid and superior posterior aspect of arm.
Median	C5–T1	Flexors of forearm, except flexor carpi ulnaris and some muscles of the hand (lateral palm); skin of lateral two-thirds of palm of hand and fingers.
Radial	C5–C8, T1	Triceps brachii and other extensor muscles of arm and extensor muscles of forearm; skin of posterior arm and forearm, lateral two-thirds of dorsum of hand, and fingers over proximal and middle phalanges.
Medial pectoral	C8–T1	Pectoralis major and pectoralis minor muscles.
Medial brachial cutaneous (BRĀ-kē-al kyū′-TĀ-nē-us)	C8–T1	Skin of medial and posterior aspects of distal third of arm.
Medial antebrachial cutaneous (an′-tē-BRĀ-kē-el)	C8–T1	Skin of medial and posterior aspects of forearm.
Ulnar	C8–T1	Flexor carpi ulnaris, flexor digitorum profundus, and most muscles of the hand; skin of medial side of hand, little finger, and medial half of ring finger.

injured when a cast is applied too tightly around the mid-humerus. Radial nerve injury is indicated by **wrist drop,** the inability to extend the wrist and fingers (Figure 17.10b). Sensory loss is minimal due to the overlap of sensory innervation by adjacent nerves.

Median nerve injury is indicated by numbness, tingling, and pain in the palm and fingers. There is also inability to pronate the forearm and flex the proximal interphalangeal joints of all digits and the distal interphalangeal joints of the second and third digits (Figure 17.10c). In addition, wrist flexion is weak and is accompanied by adduction, and thumb movements are weak.

Ulnar nerve injury is indicated by an inability to abduct or adduct the fingers, atrophy of the interosseus muscles of the hand, hyperextension of the metacarpophalangeal joints, and flexion of the interphalangeal joints, a condition called **claw-hand** (Figure 17.10d). There is loss of sensation over the little finger.

Long thoracic nerve injury results in paralysis of the serratus anterior muscle. The medial border of the scapula protrudes, giving it the appearance of a wing. When the arm is raised, the vertebral border and inferior angle of the scapula pull away from the thoracic wall and protrude outward, a condition called **winged scapula** (Figure 17.10e). The arm cannot be abducted beyond the horizontal position.

✓ Injury of which nerve could cause paralysis of the serratus anterior muscle?

Figure 17.9 / Brachial plexus in anterior view. (See Tortora, *A Photographic Atlas of the Human Body,* Figure 8.9.)

The brachial plexus supplies the shoulders and upper limbs.

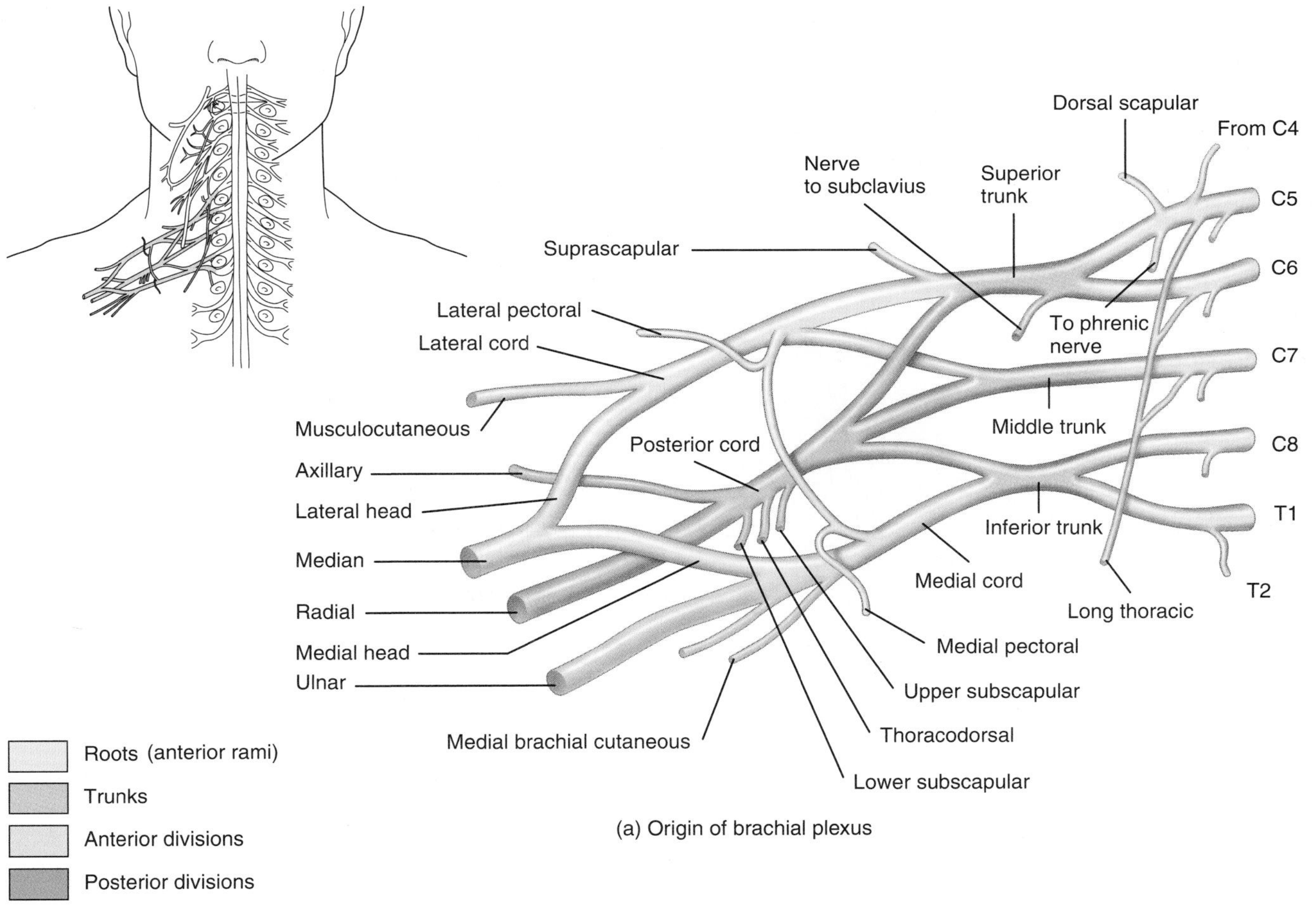

(a) Origin of brachial plexus

(continues)

Exhibit 17.2 Brachial Plexus (Figures 17.9 and 17.10) (continued)

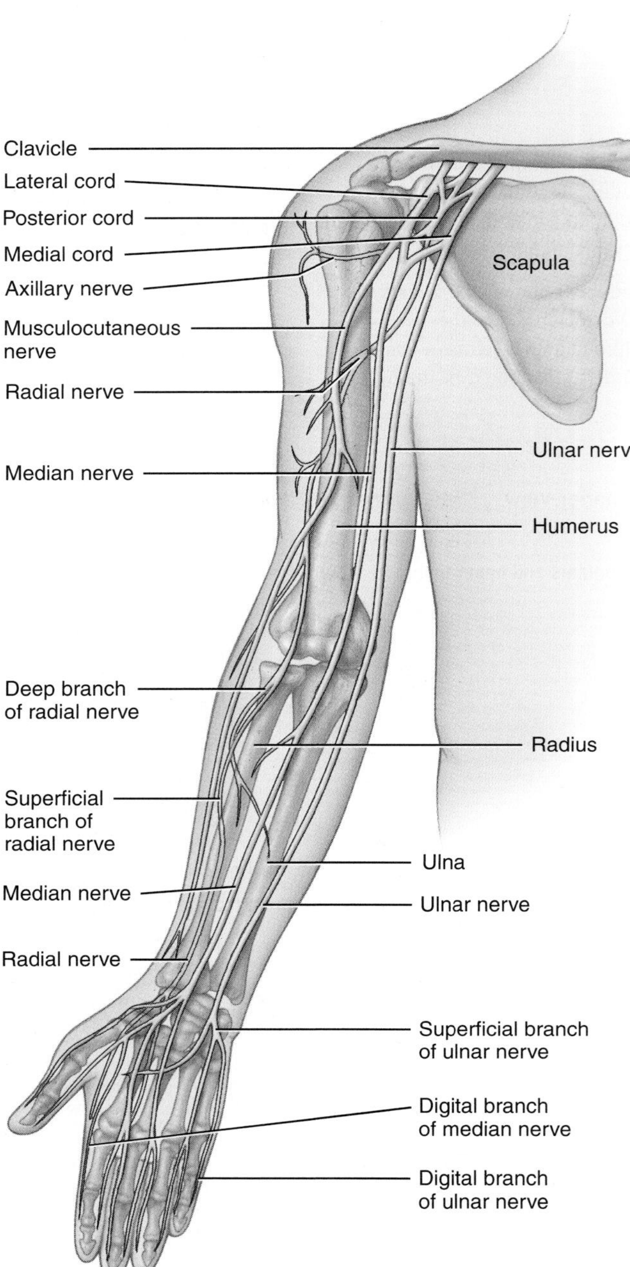

(b) Distribution of nerves from the brachial plexus

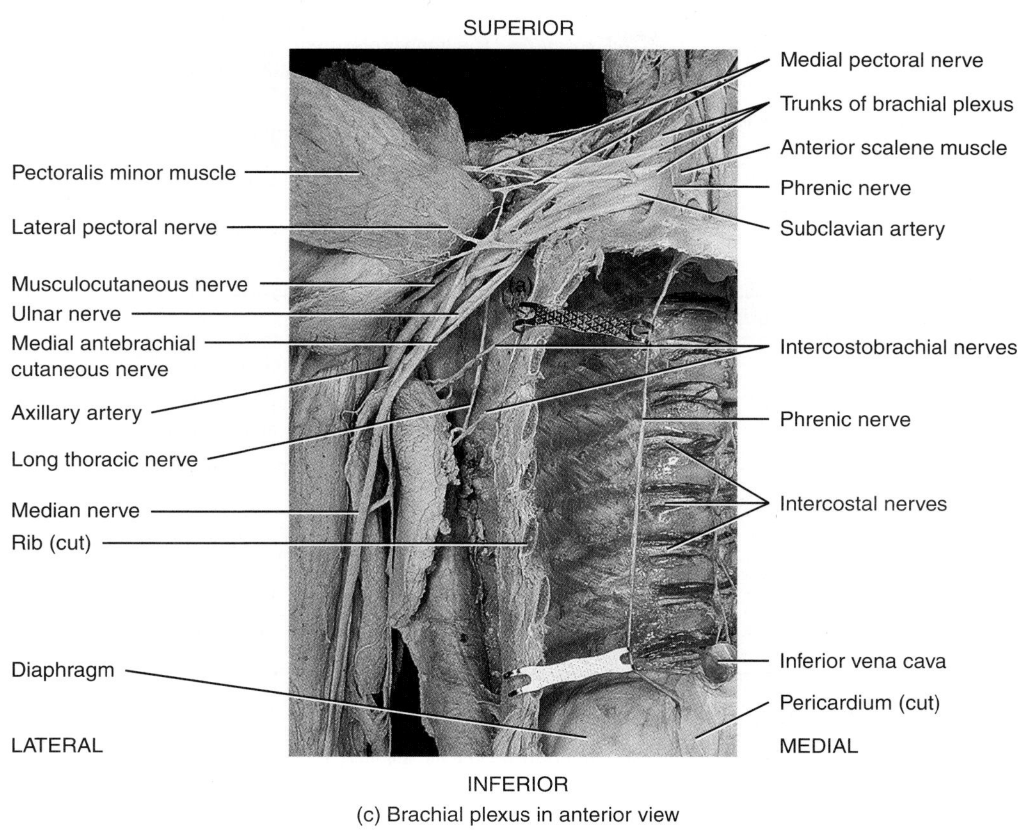

(c) Brachial plexus in anterior view

What five important nerves arise from the brachial plexus?

Figure 17.10 / Injuries to the brachial plexus.

Injuries to the brachial plexus affect the sensations and movements of the upper limbs.

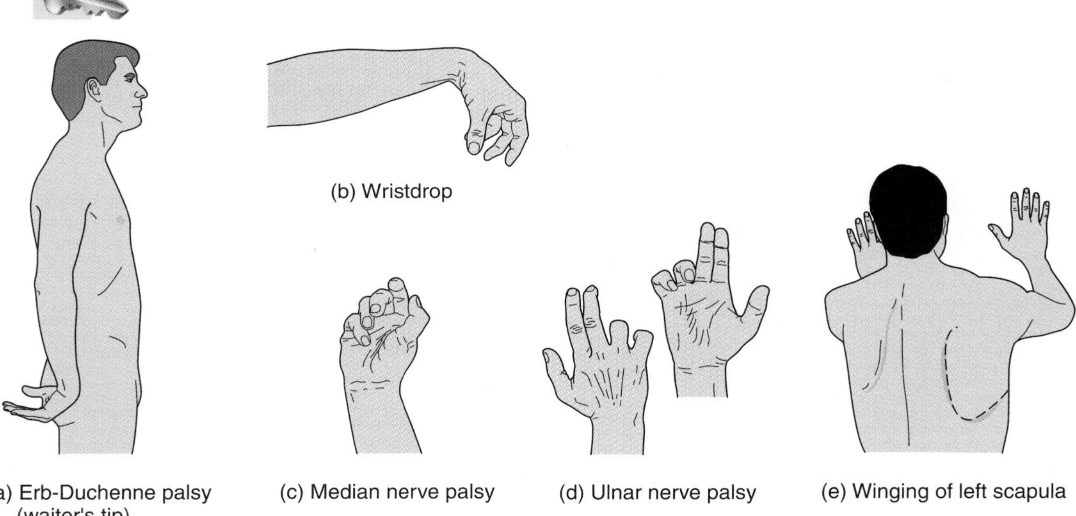

Injury to which nerve of the brachial plexus affects sensations on the palm and fingers?

Exhibit 17.3 Lumbar Plexus (Figure 17.11)

Objective

- Describe the origin, distribution, and effects of damage to the lumbar plexus.

Anterior (ventral) rami of spinal nerves L1–L4 form the **lumbar** (LUM-bar) **plexus.** Unlike the brachial plexus, there is no intricate intermingling of fibers in the lumbar plexus. On either side of the first four lumbar vertebrae, the lumbar plexus passes obliquely outward, posterior to the psoas major muscle and anterior to the quadratus lumborum muscle. It then gives rise to its peripheral nerves.

The lumbar plexus supplies the anterolateral abdominal wall, external genitals, and part of the lower limbs.

The largest nerve arising from the lumbar plexus is the femoral nerve. **Femoral nerve injury,** as in stab or gunshot wounds, is indicated by an inability to extend the leg and by loss of sensation in the skin over the anteromedial aspect of the thigh.

Nerve	Origin	Distribution
Iliohypogastric (iL′-ē-ō-hī-pō-GAS-trik)	L1	Muscles of anterolateral abdominal wall; skin of inferior abdomen and buttock.
Ilioinguinal (iL′-ē-ō-IN-gwi-nal)	L1	Muscles of anterolateral abdominal wall; skin of superior medial aspect of thigh, root of penis and scrotum in male, and labia majora and mons pubis in female.
Genitofemoral (jen′-i-tō-FEM-or-al)	L1–L2	Cremaster muscle; skin over middle anterior surface of thigh, scrotum in male, and labia majora in female.
Lateral femoral cutaneous	L2–L3	Skin over lateral, anterior, and posterior aspects of thigh.
Femoral	L2–L4	Flexor muscles of thigh and extensor muscles of leg; skin over anterior and medial aspect of thigh and medial side of leg and foot.
Obturator (OB-tū-rā-tor)	L2–L4	Adductor muscles of leg; skin over medial aspect of thigh.

Figure 17.11 / Lumbar plexus in anterior view.

The lumbar plexus supplies the anterolateral abdominal wall, external genitals, and part of the lower limbs.

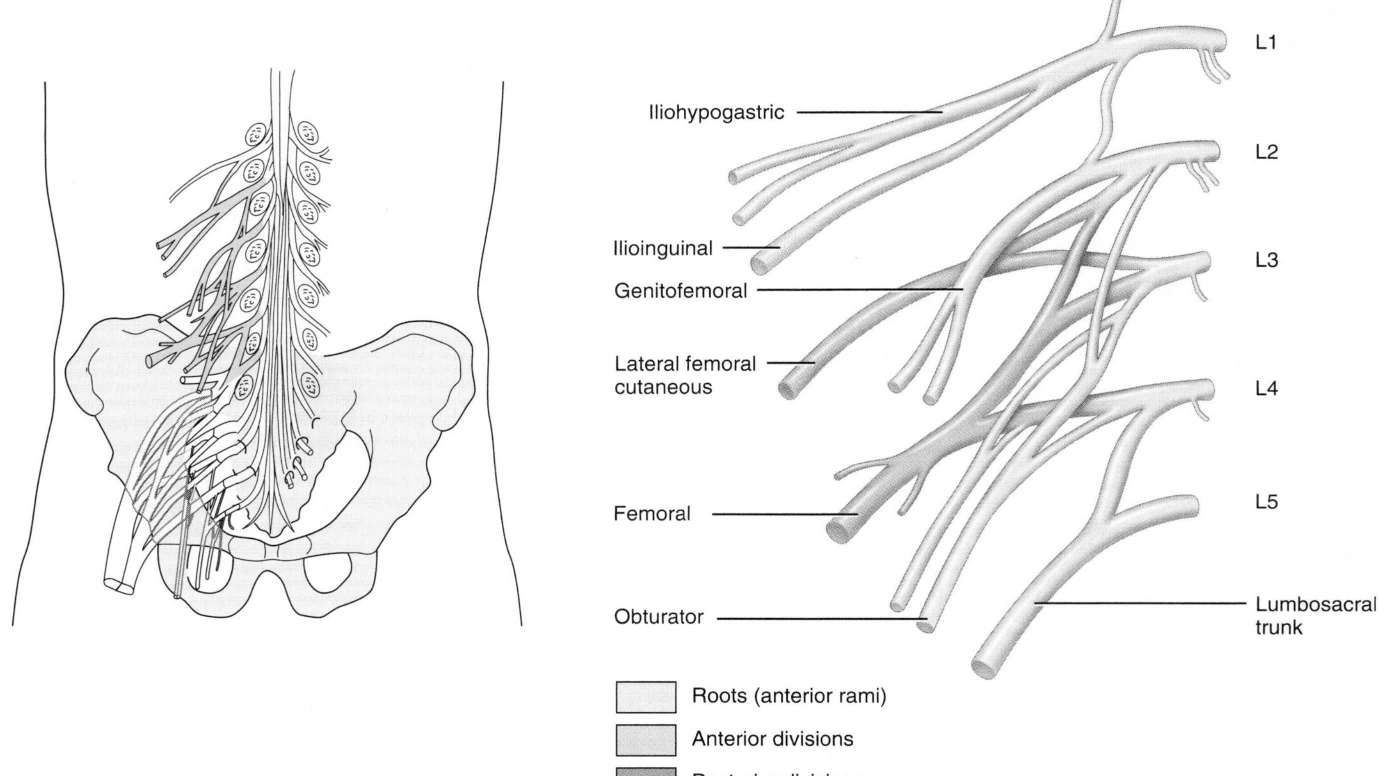

(a) Origin of lumbar plexus

Obturator nerve injury, a common complication of childbirth, results in paralysis of the adductor muscles of the leg and loss of sensation over the medial aspect of the thigh.

✓ Injury of which nerve could cause loss of sensation in the buttocks?

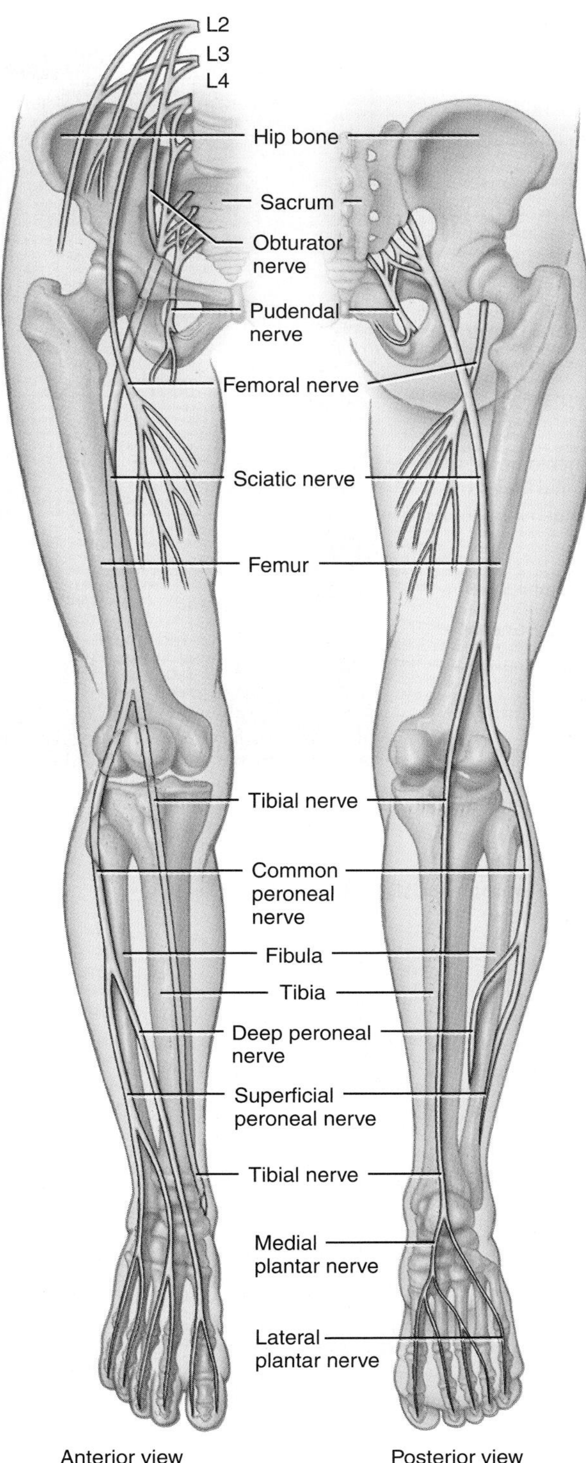

(b) Distribution of nerves from the lumbar and sacral plexuses

(continues)

Exhibit 17.3 Lumbar Plexus (Figure 17.11) (continued)

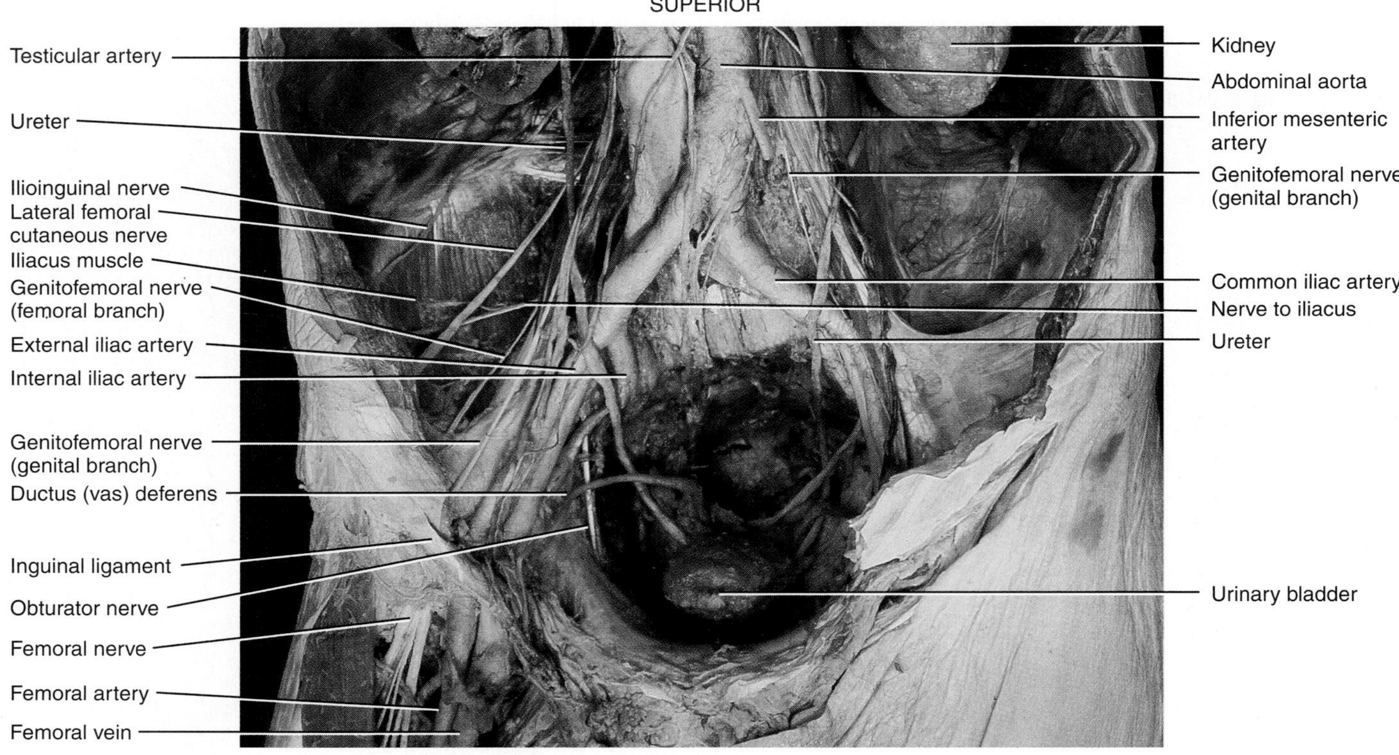

(c) Anterior view

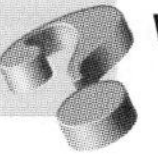

What are the signs of femoral nerve injury?

Exhibit 17.4 Sacral Plexus (Figure 17.12)

Objective

- Describe the origin, distribution, and effects of damage to the sacral plexus.

The anterior (ventral) rami of spinal nerves L4–L5 and S1–S4 form the **sacral plexus** (SĀ-kral). It is situated largely anterior to the sacrum. The sacral plexus supplies the buttocks, perineum, and lower limbs. The largest nerve in the body—the sciatic nerve—arises from the sacral plexus.

Injury to the sciatic nerve and its branches results in **sciatica,** pain that may extend from the buttock down the posterior and lateral aspect of the leg and the lateral aspect of the foot. The sciatic nerve may be injured because of a herniated (slipped) disc, dislocated hip, osteoarthritis of the lumbosacral spine, pressure from the uterus during pregnancy, or an improperly administered gluteal intramuscular injection.

In the majority of sciatic nerve injuries, the common peroneal portion is usually affected, frequently from fractures of the fibula or by pressure from casts or splints. Damage to the common peroneal nerve causes the foot to be plantar flexed, a condition called **footdrop,** and inverted, a condition called **equinovarus.** There is also loss of function along the anterolateral aspects of the leg and dorsum of the foot and toes. Injury to the tibial portion of the sciatic nerve results in dorsiflexion of the foot plus eversion, a condition called **calcaneovalgus.** Loss of sensation on the sole also occurs.

✓ Injury of which nerve causes footdrop?

Nerve	Origin	Distribution
Superior gluteal (GLŪ-tē-al)	L4–L5 and S1	Gluteus minimus and gluteus medius muscles and tensor fasciae latae.
Inferior gluteal	L5–S2	Gluteus maximus muscle.
Nerve to piriformis (pir-i-FORM-is)	S1–S2	Piriformis muscle.
Nerve to quadratus femoris (quod-RĀ-tus FEM-or-is) **and inferior gemellus** (jem-EL-us)	L4–L5 and S1	Quadratus femoris and inferior gemellus muscles.
Nerve to obturator internus (OB-tū-rā′-tor in-TER-nus) **and superior gemellus**	L5–S2	Obturator internus and superior gemellus muscles.
Perforating cutaneous (kyū′-TĀ-ne-us)	S2–S3	Skin over inferior medial aspect of buttock.
Posterior femoral cutaneous	S1–S3	Skin over anal region, inferior lateral aspect of buttock, superior posterior aspect of thigh, superior part of calf, scrotum in male, and labia majora in female.
Sciatic (sī-AT-ik)	L4–S3	Actually two nerves—tibial and common peroneal—bound together by common sheath of connective tissue. It splits into its two divisions, usually at the knee. (See below for distributions). As sciatic nerve descends through the thigh, it sends branches to hamstring muscles and adductor magnus.
Tibial (TIB-ē-al)	L4–S3	Gastrocnemius, plantaris, soleus, popliteus, tibialis posterior, flexor digitorum longus, and flexor hallucis longus muscles. Branches of tibial nerve in foot are medial plantar nerve and lateral plantar nerve.
Medial plantar (PLAN-tar)		Abductor hallucis, flexor digitorum brevis, and flexor hallucis brevis muscles; skin over medial two-thirds of plantar surface of foot.
Lateral plantar		Remaining muscles of foot not supplied by medial plantar nerve; skin over lateral third of plantar surface of foot.
Common peroneal (per′-ō-NĒ-al)	L4–S2	Divides into a superficial peroneal and a deep peroneal branch.
Superficial peroneal		Peroneus longus and peroneus brevis muscles; skin over distal third of anterior aspect of leg and dorsum of foot.
Deep peroneal		Tibialis anterior, extensor hallucis longus, peroneus tertius, and extensor digitorum longus and brevis muscles; skin on adjacent sides of great and second toes.
Pudendal (pyū-DEN-dal)	S2–S4	Muscles of perineum; skin of penis and scrotum in male and clitoris, labia majora, labia minora, and vagina in female.

(continues)

Exhibit 17.4 Sacral Plexus (Figure 17.12) (continued)

Figure 17.12 / Sacral plexus in anterior view. The distribution of the nerves of the sacral plexus is shown in Figure 17.11b. (See Tortora, *A Photographic Atlas of the Human Body,* Figure 8.11.)

The sacral plexus supplies the buttocks, perineum, and lower limbs.

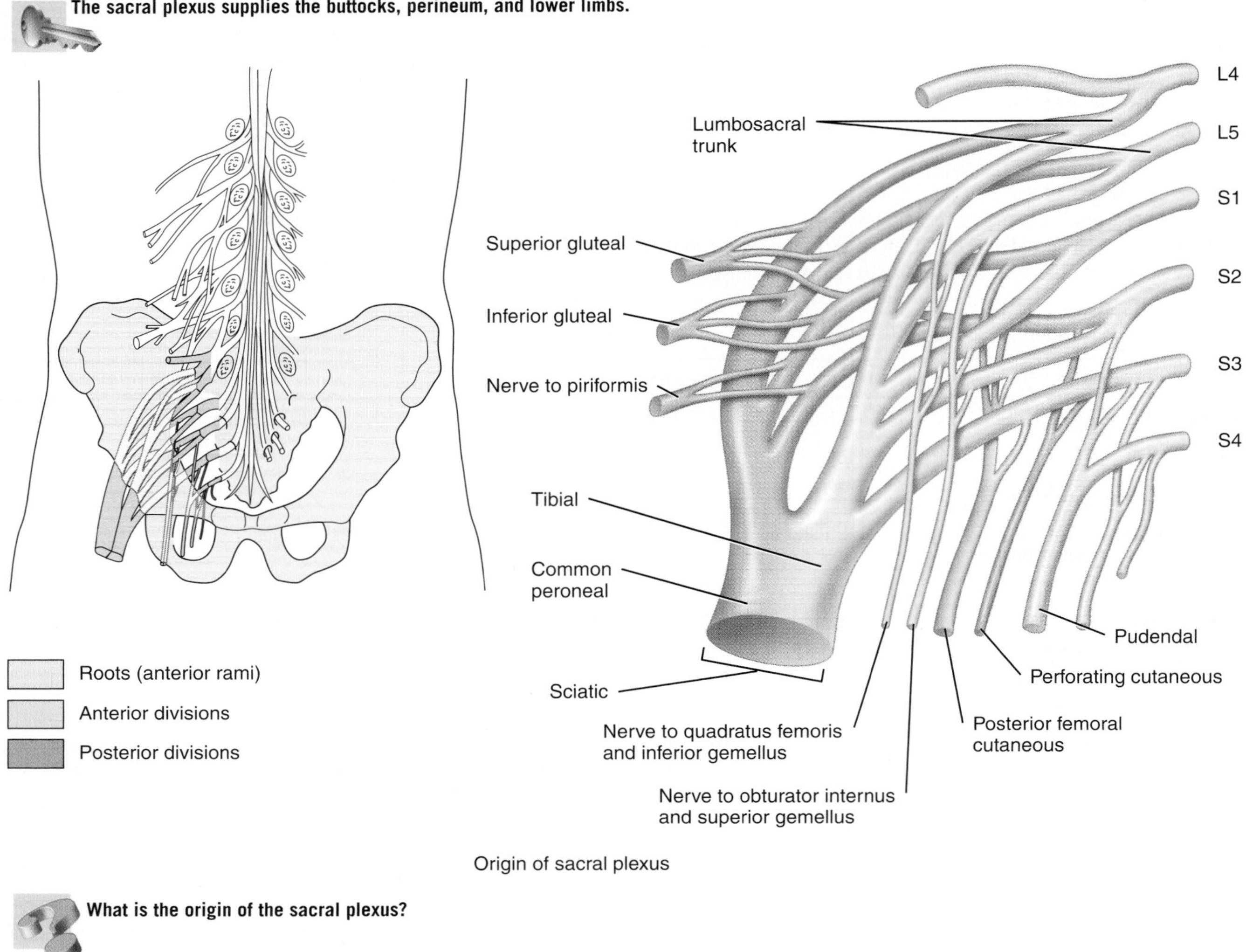

Origin of sacral plexus

What is the origin of the sacral plexus?

Dermatomes

The skin over the entire body is supplied by somatic sensory neurons that carry nerve impulses from the skin into the spinal cord and brain stem. Likewise, the underlying skeletal muscles are innervated by somatic motor neurons that carry impulses out of the spinal cord. Each spinal nerve contains sensory neurons that serve a specific, predictable segment of the body. Most of the skin of the face and scalp is served by cranial nerve V (the trigeminal nerve). The area of the skin that provides sensory input to one pair of spinal nerves or to cranial nerve V (for the face) is called a **dermatome** (*derma-* = skin; *-tome* = thin segment) (Figure 17.13). The nerve supply in adjacent dermatomes

Figure 17.13 / Distribution of dermatomes.

A dermatome is an area of skin that provides sensory input via the posterior roots of one pair of spinal nerves or via cranial nerve V (trigeminal nerve).

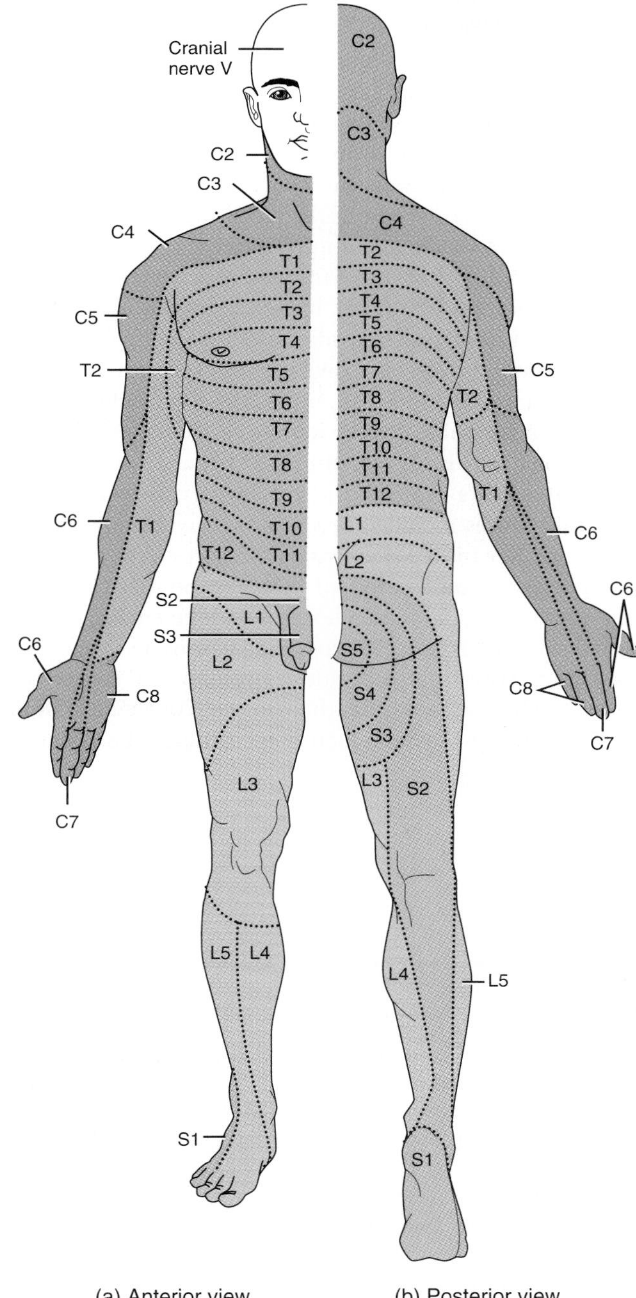

(a) Anterior view (b) Posterior view

Which is the only spinal nerve that does not have a corresponding dermatome?

overlaps somewhat. In regions where the overlap is considerable, little loss of sensation may result if only one of the nerves supplying the dermatome is damaged.

Knowing which spinal cord segments supply each dermatome makes it possible to locate damaged regions of the spinal cord. If the skin in a particular region is stimulated but the sensation is not perceived, the nerves supplying that dermatome are probably damaged. Information about the innervation patterns of spinal nerves can also be used therapeutically. Cutting roots or infusing local anesthetics can block pain either permanently or transiently. But because dermatomes overlap, deliberate production of a region of complete anesthesia may require that at least three adjacent spinal nerves be cut or blocked by an anesthetic drug.

Spinal Cord Transection and Muscle Function

An injury that entirely severs the spinal cord is said to cause complete **transection** (tran-SEK-shun; *trans-* = across; *-sectus* = to cut). After transection, a person will have permanent loss of sensations in dermatomes below the injury because ascending nerve impulses cannot propagate past the transection to reach the brain. At the same time, voluntary muscle contractions will be lost below the transection because nerve impulses descending from the brain also cannot pass. The extent of paralysis of skeletal muscles depends on the level of injury. The following list outlines which muscle functions may be retained at progressively lower levels of spinal cord transection.

- C1–C3: no function maintained from the neck down; ventilator needed for breathing
- C4–C5: diaphragm, which allows breathing
- C6–C7: some arm and chest muscles, which allows feeding, some dressing, and propelling wheelchair
- T1–T3: intact arm function
- T4–T9: control of trunk above the umbilicus
- T10–L1: most thigh muscles, which allows walking with long leg braces
- L1–L2: most leg muscles, which allows walking with short leg braces ■

✓ How are spinal nerves named and numbered? Why are all spinal nerves classified as mixed nerves?
✓ Describe how a spinal nerve is connected with the spinal cord.
✓ Describe the branches and innervations of a typical spinal nerve.
✓ Describe the principal plexuses and the regions they supply.

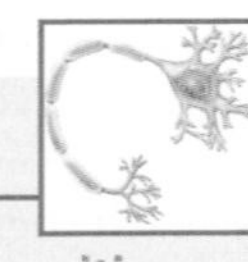

APPLICATIONS TO HEALTH

Neuritis

Inflammation of one or several nerves is termed **neuritis** (*neur-* = nerve; *-itis* = inflammation). It may result from irritation to the nerve produced by direct blows, bone fractures, contusions, or penetrating injuries. Additional causes include infections, vitamin deficiency (usually thiamine), and poisons such as carbon monoxide, carbon tetrachloride, heavy metals, and some drugs.

Shingles

Shingles is an acute infection of the peripheral nervous system caused by herpes zoster (HER-pēz ZOS-ter), the virus that also causes chickenpox. After a person recovers from chickenpox, the virus retreats to a posterior root ganglion. If the virus is reactivated, the immune system usually prevents it from spreading. From time to time, however, the reactivated virus overcomes a weakened immune system, leaves the ganglion, and travels down sensory neurons of the skin. The result is pain, discoloration of the skin, and a characteristic line of skin blisters. The line of blisters marks the distribution (dermatome) of the particular cutaneous sensory nerve belonging to the infected root ganglion.

Poliomyelitis

Poliomyelitis, or simply **polio,** is caused by a virus called poliovirus. The onset of the disease is marked by fever, severe headache, a stiff neck and back, deep muscle pain and weakness, and loss of certain somatic reflexes. In its most serious form, the virus produces paralysis by destroying cell bodies of motor neurons, specifically those in the anterior horns of the spinal cord and in the nuclei of the cranial nerves. Polio can cause death from respiratory or heart failure if the virus invades the brain cells of the vital medullary centers that control breathing and heart functions. Even though the availability of polio vaccines has virtually eradicated polio in the United States, outbreaks of polio continue throughout the world. Due to international travel, polio could easily be reintroduced into North America if individuals are not vaccinated appropriately.

Several decades after suffering a severe attack of polio and following their recovery from it, some individuals develop a condition called **post-polio syndrome.** This neurological disorder is characterized by progressive muscle weakness, extreme fatigue, loss of function, and pain, especially in muscles and joints. Post-polio syndrome seems to involve a slow degeneration of motor neurons that innervate muscle fibers. Triggering factors appear to be a fall, a minor accident, surgery, or prolonged bed rest. Possible causes include overuse of surviving motor neurons over time, smaller motor neurons as a result of the initial infection by the virus, reactivation of dormant polio virus particles, immune-mediated responses, hormone deficiencies, and environmental toxins. Treatment consists of muscle-strengthening exercises, administration of pyridostigminine to enhance the action of acetylcholine in stimulating muscle contraction, and administration of nerve growth factors to stimulate both nerve and muscle growth.

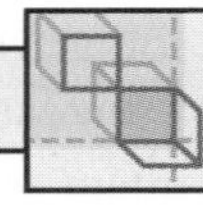

STUDY OUTLINE

Spinal Cord Anatomy (p. 516)

1. The spinal cord is protected by the vertebral column, meninges, cerebrospinal fluid, and denticulate ligaments.
2. The three meninges are coverings that run continuously around the spinal cord and brain; they are the dura mater, arachnoid, and pia mater.
3. The spinal cord begins as a continuation of the medulla oblongata and ends at about the second lumbar vertebra in an adult.
4. The spinal cord contains cervical and lumbar enlargements that serve as points of origin for nerves to the limbs.
5. The tapered inferior portion of the spinal cord is the conus medullaris, from which arise the filum terminale and cauda equina.
6. Spinal nerves connect to each segment of the spinal cord by two roots: The posterior or dorsal root contains sensory nerve fibers, and the anterior or ventral root contains motor neuron axons.
7. The spinal cord is partially divided into right and left sides by the anterior median fissure and the posterior median sulcus.
8. The gray matter in the spinal cord is divided into horns, and the white matter into columns. In the center of the spinal cord is the central canal, which runs the length of the spinal cord.
9. Parts of the spinal cord observed in transverse section are the gray commissure; central canal; anterior, posterior, and lateral gray horns; and anterior, posterior, and lateral white columns, which contain ascending and descending tracts. Each part has specific functions.
10. The spinal cord conveys sensory and motor information by way of ascending and descending tracts, respectively.

Spinal Cord Functions (p. 522)

1. A major function of the spinal cord is to propagate nerve impulses from the periphery to the brain (sensory tracts) and to conduct motor impulses from the brain to the periphery (motor tracts).
2. Sensory information travels along two main routes in the white matter of the spinal cord: the posterior columns and the spinothalamic tracts.
3. In the white matter of the spinal cord, motor information travels along direct pathways and indirect pathways.
4. A second major function of the spinal cord is to serve as an integrating center for spinal reflexes. This integration occurs in the gray matter.

5. A reflex is a fast, predictable, automatic response to changes in the environment that helps maintain homeostasis.
6. Reflexes may be spinal or cranial and somatic or autonomic (visceral).
7. A reflex arc is the simplest type of pathway that connects sensory input to motor output.
8. The components of a reflex arc are sensory receptor, sensory neuron, integrating center, motor neuron, and effector.

Spinal Nerves (p. 526)

1. The 31 pairs of spinal nerves are named and numbered according to the region and level of the spinal cord from which they emerge.
2. There are 8 pairs of cervical, 12 pairs of thoracic, 5 pairs of lumbar, 5 pairs of sacral, and 1 pair of coccygeal nerves.
3. Spinal nerves typically are connected with the spinal cord by a posterior root and an anterior root. All spinal nerves contain both sensory and motor axons (are mixed nerves).
4. Three connective tissue coverings associated with spinal nerves are the endoneurium, perineurium, and epineurium.
5. Branches of a spinal nerve include the posterior ramus, anterior ramus, meningeal branch, and rami communicantes.
6. The anterior rami of spinal nerves, except for T2–T12, form networks of nerves called plexuses.
7. Emerging from the plexuses are nerves bearing names that typically describe the general regions they supply or the route they follow.
8. Nerves of the cervical plexus supply the skin and muscles of the head, neck, and upper part of the shoulders; they connect with some cranial nerves and innervate the diaphragm.
9. Nerves of the brachial plexus supply the upper limbs and several neck and shoulder muscles.
10. Nerves of the lumbar plexus supply the anterolateral abdominal wall, external genitals, and part of the lower limbs.
11. Nerves of the sacral plexus supply the buttocks, perineum, and part of the lower limbs.
12. Anterior rami of nerves T2–T12 do not form plexuses and are called intercostal (thoracic) nerves. They are distributed directly to the structures they supply in intercostal spaces.
13. Sensory neurons within spinal nerves and cranial nerve V (trigeminal) serve specific, constant segments of the skin called dermatomes.
14. Knowledge of dermatomes helps a physician determine which segment of the spinal cord or which spinal nerve is damaged.

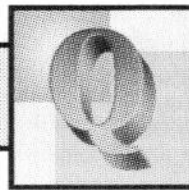

SELF-QUIZ QUESTIONS

Choose the one best answer to the following questions:

1. Choose the true statement. (a) Tracts are located only in the spinal cord and not in the brain. (b) Tracts appear white because they consist of bundles of unmyelinated nerve fibers. (c) Ascending tracts are all motor tracts. (d) Ascending tracts are in the white matter; descending tracts are in the gray matter. (e) None of the above is true.
2. Which of the following statements is true of a reflex arc? (1) It always includes at least a sensory and a motor neuron. (2) It always has its integrating center in the brain or the spinal cord. (3) It always terminates in muscle or gland.
   (a) 1 only, **(b)** 2 only, **(c)** 3 only, **(d)** all of the above, **(e)** none of the above.
3. The posterior (dorsal) root ganglion contains (a) cell bodies of motor neurons, (b) cell bodies of sensory neurons, (c) cranial nerve axons, (d) synapses, (e) all of the above.
4. Which sequence best represents the course of a nerve impulse over a reflex arc? (a) receptor, integrating center, sensory neuron, motor neuron, effector, (b) effector, sensory neuron, integrating center, motor neuron, receptor, (c) receptor, sensory neuron, integrating center, motor neuron, effector, (d) receptor, motor neuron, integrating center, sensory neuron, effector, (e) effector, integrating center, sensory neuron, motor neuron, receptor.
5. The spinothalamic tracts convey sensory information to the brain regarding (a) pain, (b) two-point discrimination, (c) temperature, (d) proprioception, (e) both a and c.

Complete the following.

6. The spinal cord extends from the ________ of the brain to the level of the ________ of the vertebral column.
7. Lateral extensions of the pia mater that help to protect the spinal cord against displacement are called ________ .
8. The order of the meningeal layers from superficial to deep is ________ , ________ , ________ .
9. Individual nerve fibers are wrapped in a connective tissue covering known as ________ . Groups of nerve fibers are held in bundles by ________ . The entire nerve is wrapped with ________ .
10. Cell bodies of motor neurons leading to skeletal muscles are located in the ________ gray horns of the spinal cord, while cell bodies of neurons supplying smooth muscle, cardiac muscle or glands lie in the ________ gray horns of the spinal cord.
11. There are ________ pairs of spinal nerves that emerge from the spinal cord, named and numbered as follows:
    ________ pair(s) of cervical nerves, ________ pair(s) of thoracic nerves, ________ pair(s) of lumbar nerves, ________ pair(s) of sacral nerves, ________ pair(s) of coccygeal nerves.

Are the following statements true or false?

12. Cell bodies of motor neurons are located in the posterior (dorsal) root ganglia.
13. The anterior median fissure is the deeper, wider longitudinal groove on the external surface of the spinal cord.
14. The tapering inferior end of the spinal cord at the level of the first two lumbar vertebrae is called the conus medullaris.
15. The main functions of the spinal cord are that it serves as a reflex center and it is the site where sensations are felt.
16. A tract is a bundle of nerve fibers inside the central nervous system.
17. Networks called plexuses are formed by the anterior (ventral) rami of spinal nerves, except for T2-T12.

18. Match the following:

____ **(a)** provides the entire nerve supply for the shoulder and upper limbs

____ **(b)** gives rise to the phrenic nerve that supplies the diaphragm

____ **(c)** gives rise to the median, radial, and axillary nerves

____ **(d)** not a plexus at all, but rather segmentally arranged nerves

____ **(e)** supplies fibers to the scalp, neck, and part of the shoulder and chest

____ **(f)** supplies fibers to the femoral nerve that innervates the flexor muscles of the thigh and the extensor muscles of the leg

____ **(g)** gives rise to the largest nerve in the body, the sciatic nerve, that supplies the posterior aspect of the thigh and the leg

(1) brachial plexus
(2) cervical plexus
(3) sacral plexus
(4) lumbar plexus
(5) intercostal nerves

19. Match the following:

____ **(a)** any region of spinal cord from which one pair of spinal nerves arises

____ **(b)** non-nervous inferior extension of pia mater; anchors spinal cord in place

____ **(c)** "horses tail"; extension of spinal nerve roots in lumbar and sacral regions within subarachnoid space

____ **(d)** gives rise to nerves that serve the upper limbs

____ **(e)** gives rise to nerves that serve the lower limbs

(1) cauda equina
(2) cervical enlargement
(3) lumbar enlargement
(4) filum terminale
(5) spinal segment

20. Match the following:

____ **(a)** the joining together of the anterior rami of adjacent nerves

____ **(b)** spinal nerve branches that serve the deep muscles and skin of the posterior surface of the trunk

____ **(c)** spinal nerve branches that serve the muscles and structures of the upper and lower limbs and the lateral and ventral trunk

____ **(d)** area of the spinal cord from which nerves to the upper limbs arise

____ **(e)** area of the spinal cord from which nerves to the lower limbs arise

____ **(f)** contains motor neuron axons and conducts impulses from the spinal cord to the periphery

____ **(g)** contains sensory nerve fibers and conducts impulses from the periphery into the spinal cord

(1) cervical enlargement
(2) lumbar enlargement
(3) posterior root
(4) anterior root
(5) posterior ramus
(6) anterior ramus
(7) plexus

CRITICAL THINKING QUESTIONS

1. A high school senior dove headfirst into a murky pond. Unfortunately, he hit a submerged log with his head and is now paralyzed. Can you deduce the location of the injury. What is the likelihood of recovery from his injury?
HINT *He didn't just break bones.*

2. The spinal cord is covered in protective layers and enclosed by the vertebral column. How is it able to send and receive messages from the periphery of the body?
HINT *How are you able to bend the vertebral column.*

3. Why doesn't the spinal cord "creep" up towards the head every time you bend over? Why doesn't it get all twisted out of position when you exercise?
HINT *How would you prevent a boat form floating away from a dock?*

4. Helga had spent many sleepless nights after moving in with her daughter, her daughter's three children, two dogs, and five cats. This morning she woke up with burning pain on her left side and a line of blisters extending along the waist band of her pajama bottoms on the left side. Helga groaned, "What is this-flea bites?" The pets are not to blame. What is causing Helga's condition?
HINT *Ignore the dogs and cats—think chicken.*

5. After examining Helga's blisters, the doctor tapped her knees and stroked the bottom of her feet. The toes on her right foot fanned out, the great toe extending straight out, while the toes on her left foot curled under. Helga is not pleased by her asymmetrical feet. Explain the problem to Helga.
HINT *Her one year old grandson has the same response as Helga's right foot, but with both his feet.*

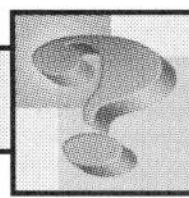

ANSWERS TO FIGURE QUESTIONS

17.1 The superior boundary of the spinal dura mater is the foramen magnum of the occipital bone; the inferior boundary is the second sacral vertebra.

17.2 The cervical enlargement connects with sensory and motor nerves of the upper limbs.

17.3 In the spinal cord, a horn is an area of gray matter and a column is a region of white matter.

17.4 The anterior corticospinal tract is located on the anterior side of the spinal cord, originates in the cortex of the cerebrum, and ends in the spinal cord. It contains descending fibers and thus is a motor tract.

17.5 Somatic reflexes.

17.6 A spinal nerve is mixed (has sensory and motor components) because it is formed by the union of the posterior root containing sensory axons and the anterior root containing motor axons.

17.7 The anterior rami serve the upper and lower limbs.

17.8 Severing the spinal cord at level C2 causes respiratory arrest because it prevents descending nerve impulses from reaching the phrenic nerve, which stimulates contraction of the diaphragm.

17.9 The axillary, musculocutaneous, radial, median, and ulnar nerves are five important nerves that arise from the brachial plexus.

17.10 Injury to the median nerve affects sensations on the palm and fingers.

17.11 Signs of femoral nerve injury include inability to extend the leg and loss of sensation in the skin over the anterolateral aspect of the thigh.

17.12 The origin of the sacral plexus is the anterior rami of spinal nerves L4–L5 and S1–S4.

17.13 The only spinal nerve without a corresponding dermatome is C1.

18

THE BRAIN AND THE CRANIAL NERVES

CONTENTS AT A GLANCE

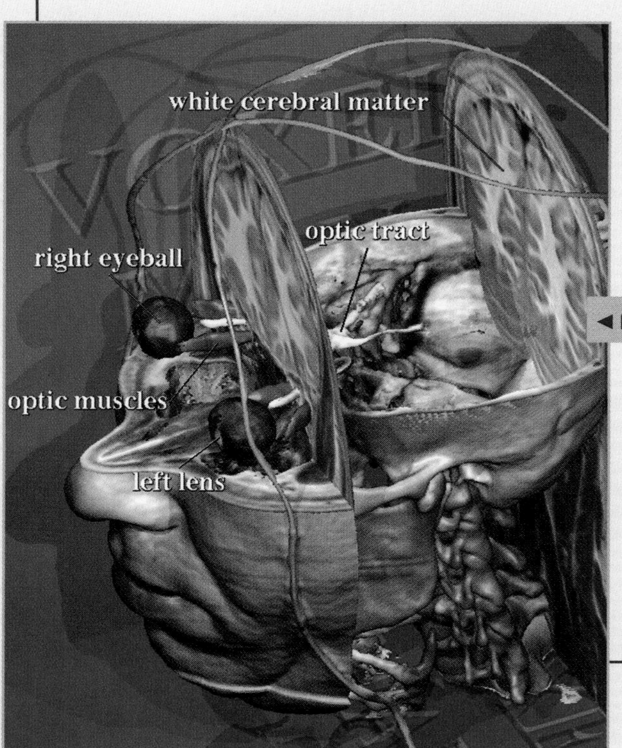

◄ Page 564

What is your first reaction when looking at this image? Can you guess how such an image was generated? Is this a painting, a photograph, or a virtual cadaver?

Page 566 ►

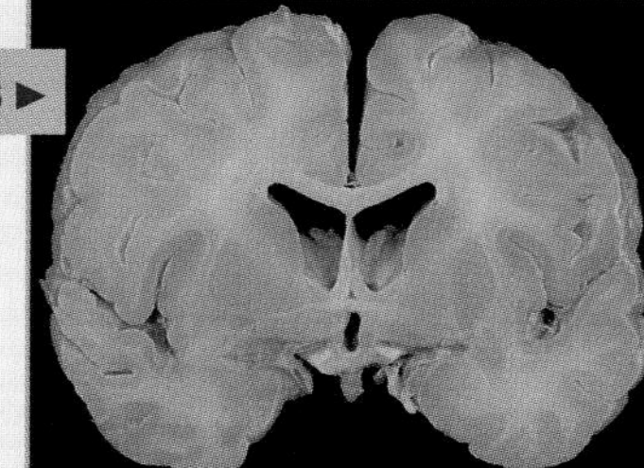

INTRODUCTION

Solving an equation, feeling hungry, breathing—each of these processes is mediated by different regions of the **brain,** that portion of the central nervous system contained within the cranium. The adult brain is made up of about 100 billion neurons and 1000 billion neuroglia; it is one of the largest organs in the body, with a mass of about 1300 g (almost 3 lb.). The brain is the center for registering sensations, correlating them with one another and with stored information, making decisions, and taking actions. It also is the center for the intellect, emotions, behavior, and memory. But the brain encompasses yet a larger domain: It directs our behavior toward others. With ideas that excite, artistry that dazzles, or rhetoric that mesmerizes, one person's thoughts and actions may influence and shape the lives of many others. As you will see, different regions of the brain are specialized for different functions, and many parts of the brain work together to accomplish a particular function. This chapter explores the principal parts of the brain, how the brain is protected and nourished, and how it is related to both the spinal cord and the 12 pairs of cranial nerves.

OVERVIEW OF BRAIN ORGANIZATION AND BLOOD SUPPLY

Objectives

- Identify the principal parts of the brain.
- Describe how the brain is protected.

Principal Parts of the Brain

The brain consists of four principal parts: brain stem, cerebellum, diencephalon, and cerebrum (Figure 18.1). The **brain**

Figure 18.1 / The brain. The infundibulum and pituitary gland are discussed with the endocrine system in Chapter 22. (See Tortora, *A Photographic Atlas of the Human Body,* Figure 8.13.)

The four principal parts of the brain are the brain stem, cerebellum, diencephalon, and cerebrum.

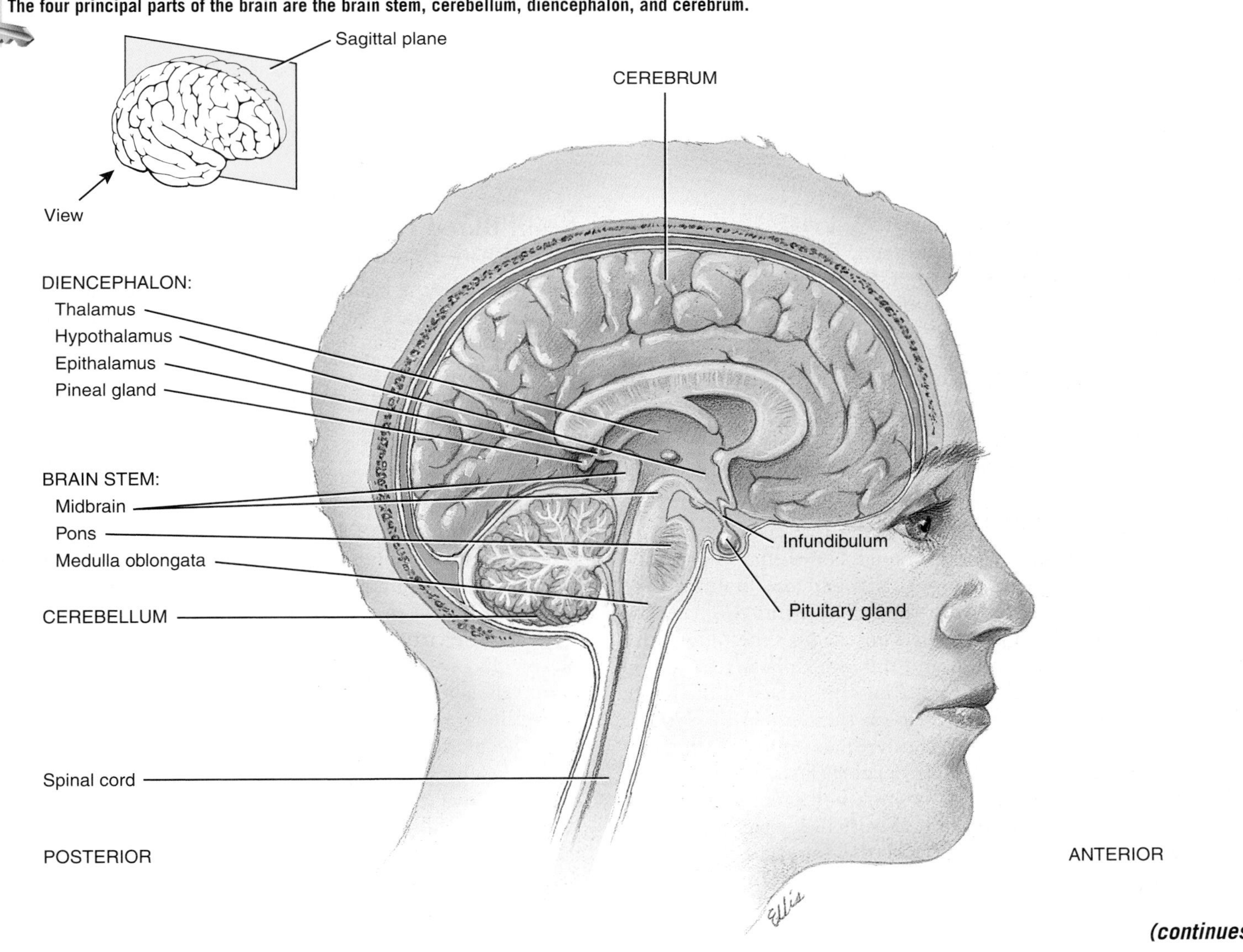

(a) Sagittal section, medial view

(continues)

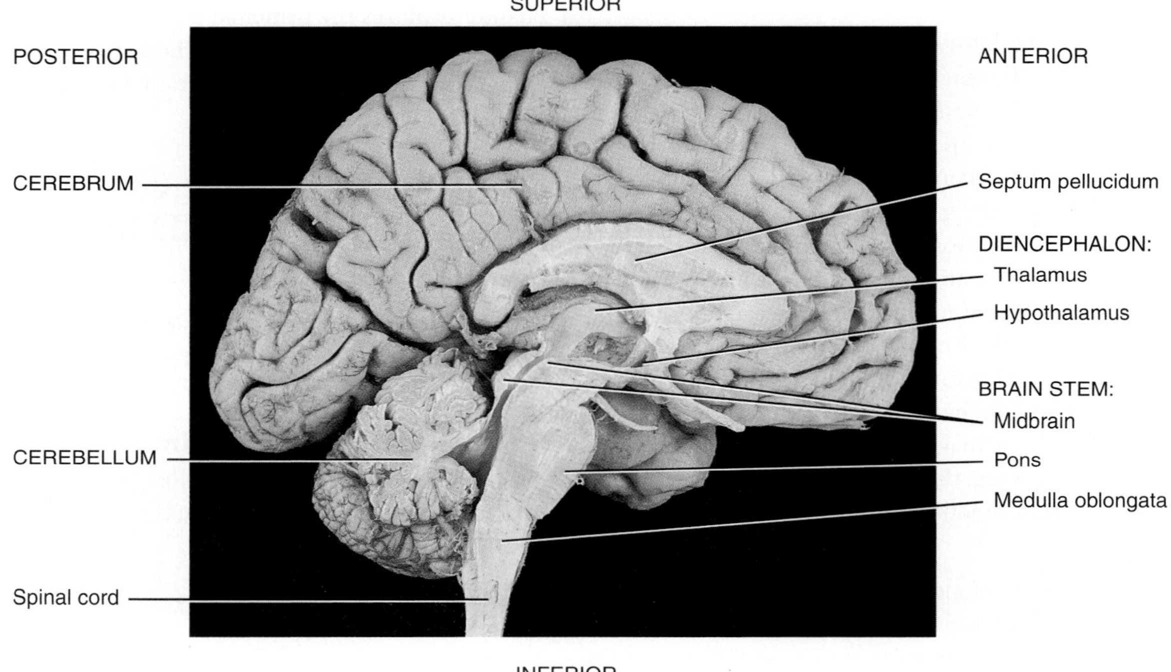

(b) Sagittal section, medial view

Which part of the brain is the largest?

stem is continuous with the spinal cord and consists of the medulla oblongata, pons, and midbrain. Posterior to the brain stem is the **cerebellum** (ser′-e-BEL-um; = little brain). Superior to the brain stem is the **diencephalon** (dī′-en-SEF-a-lon; *di-* = through; *-encephalon* = brain) consisting primarily of the thalamus and hypothalamus and including the epithalamus and the subthalamus. The **cerebrum** (se-RĒ-brum; = brain) spreads over the diencephalon like a mushroom cap and occupies most of the cranium.

Protective Coverings of the Brain

The brain is protected by the cranial bones and the cranial meninges. The **cranial meninges** surround the brain (Figure 18.2). They are continuous with the spinal meninges, have the same basic structure, and bear the same names: the outer **dura mater,** the middle **arachnoid,** and the inner **pia mater.** Blood vessels that enter brain tissue pass along the surface of the brain, and as they penetrate inward, they are surrounded by a loose-fitting layer of pia mater. Three extensions of the dura mater separate parts of the brain: the **falx cerebri** (FALKS CER-e-brē; *falx* = sickle-shaped) separates the two hemispheres (sides) of the cerebrum; the **falx cerebelli** (cer-e-BEL-ī) separates the two hemispheres of the cerebellum; and the **tentorium cerebelli** (ten-TŌ-rē-um; = tent) separates the cerebrum from the cerebellum.

Brain Blood Flow and the Blood–Brain Barrier

Blood flows to the brain mainly via the internal carotid and vertebral arteries (see Figure 14.8b on page 431); the veins that return blood from the head to the heart are shown in Figure 14.12a (page 448).

In an adult, the brain represents only 2% of total body weight, but consumes about 20% of the oxygen and glucose used at rest. Neurons synthesize ATP almost exclusively from glucose via reactions that use oxygen. When activity of neurons and neuroglia increases in a region of the brain, blood flow to that area also increases. Even a brief slowing of brain blood flow may cause unconsciousness. Typically, an interruption in blood flow for 1 or 2 minutes impairs neuronal function, and total deprivation of oxygen for about 4 minutes causes permanent injury. Because virtually no glucose is stored in the brain, the supply of glucose also must be continuous. If blood entering the brain has a low level of glucose, mental confusion, dizziness, convulsions, and loss of consciousness may occur.

The **blood–brain barrier (BBB)** protects brain cells from harmful substances and pathogens by preventing passage of many substances from the blood into brain tissue. Tight junctions seal together the endothelial cells of brain capillaries, which also are surrounded by a continuous basement membrane. Pressed up against the capillaries are the processes of large numbers of astrocytes (one type of neuroglia), which are

Figure 18.2 / The protective coverings of the brain.

Cranial bones and the cranial meninges protect the brain.

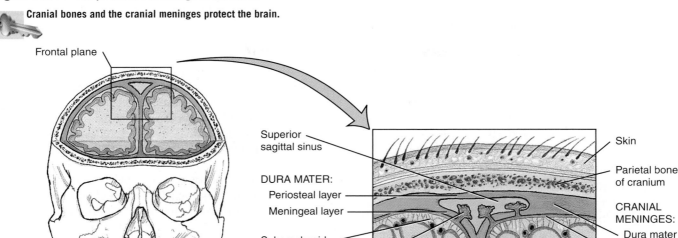

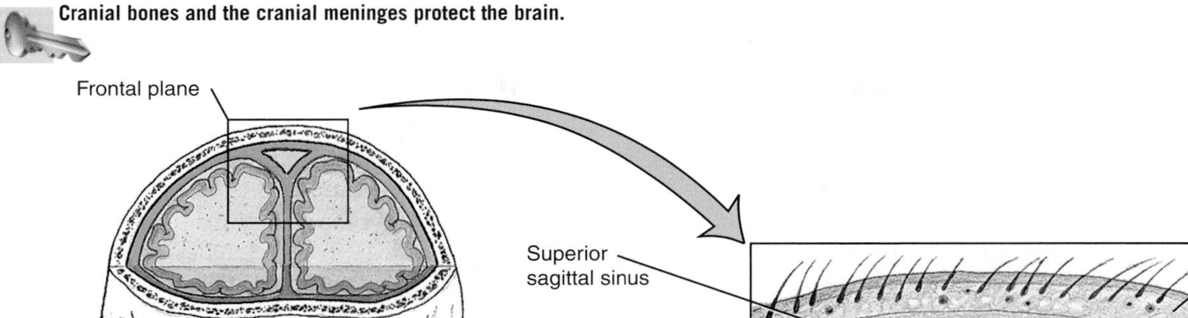

Frontal section through skull showing the cranial meninges

What are the three layers of the cranial meninges, from superficial to deep?

thought to selectively pass some substances from the blood but inhibit the passage of others. A few water-soluble substances (for example, glucose) cross the BBB by active transport; other substances, such as creatinine, urea, and most ions, cross the BBB very slowly. Still other substances—proteins and most antibiotic drugs—do not pass at all from the blood into brain tissue. However, the BBB does not prevent passage of lipid-soluble substances, such as oxygen, carbon dioxide, alcohol, and most anesthetic agents, into brain tissue.

Breaching the Blood–Brain Barrier

We have seen how the BBB prevents the passage into brain tissue of potentially harmful substances. But another consequence of the BBB's efficient protection is that it also prevents the passage of certain drugs that could be therapeutic for brain cancer or other CNS disorders. Researchers are finding ways to move drugs past the BBB. In one method, the drug is injected in a concentrated sugar solution. The high osmotic pressure of the sugar solution causes the endothelial cells of the capillaries to shrink, which opens gaps between their tight junctions. As a result, the drug can enter the brain tissue. ■

✓ Compare the sizes and locations of the cerebrum and cerebellum.
✓ Describe the locations of the cranial meninges.
✓ Explain the blood supply to the brain and the blood–brain barrier.

CEREBROSPINAL FLUID PRODUCTION AND CIRCULATION IN VENTRICLES

Objective

- Explain the formation and circulation of cerebrospinal fluid.

Cerebrospinal fluid (CSF) is a clear, colorless liquid that protects the brain and spinal cord against chemical and physical injuries and carries oxygen, glucose, and other needed substances from the blood to neurons and neuroglia. This fluid circulates continuously through the cavities within the brain and spinal cord and around the brain and spinal cord in the subarachnoid space (between the arachnoid and pia mater).

Figure 18.3 shows the four CSF-filled cavities within the brain, which are called **ventricles** (VEN-tri-kuls; = little cavities). A **lateral ventricle** is located in each hemisphere of the cerebrum. Anteriorly, the lateral ventricles are separated by a thin membrane, the **septum pellucidum** (SEP-tum pe-LŪ-si-dum; *pellucid* = transparent). The **third ventricle** is a narrow cavity along the midline superior to the hypothalamus and between the right and left halves of the thalamus. The **fourth ventricle** lies between the brain stem and the cerebellum.

The total volume of CSF is 80–150 mL (3–5 oz) in an average adult. CSF contains glucose, proteins, lactic acid, urea, and ions (Na^+, K^+, Ca^{2+}, Mg^{2+}, Cl^- and HCO_3^-); it also contains some white blood cells. The CSF contributes to homeostasis in three main ways:

1. **Mechanical protection.** CSF serves as a shock-absorbing medium that protects the delicate tissue of the brain and spinal cord from jolts that would otherwise cause them to hit the bony walls of the cranial and vertebral cavities. The fluid also buoys the brain so that it "floats" in the cranial cavity.
2. **Chemical protection.** CSF provides an optimal chemical environment for accurate neuronal signaling. Even slight changes in the ionic composition of CSF within the brain can seriously disrupt production of action potentials.
3. **Circulation.** CSF is a medium for exchange of nutrients and waste products between the blood and nervous tissue.

The sites of CSF production are the **choroid plexuses** (KŌ-royd; = membranelike), which are networks of capillaries in the walls of the ventricles. The capillaries are covered by ependymal cells that form cerebrospinal fluid from blood plasma by filtration and secretion. Because the ependymal cells are joined by tight junctions (shown in Figure 3.1 on page 57), materials entering CSF from choroid capillaries cannot leak between these cells; instead, they must pass through the ependymal cells. This **blood–cerebrospinal fluid barrier** permits certain substances to enter the CSF but excludes others, protecting the brain and spinal cord from potentially harmful blood-borne substances.

The CSF formed in the choroid plexuses of each lateral ventricle flows into the third ventricle through a pair of narrow, oval openings, the **interventricular foramina (foramina of Monro)** (Figure 18.4a). More CSF is added by the choroid plexus in the roof of the third ventricle. The fluid then flows through the **cerebral aqueduct (aqueduct of Sylvius)**, which passes through the midbrain, into the fourth ventricle. The choroid plexus of the fourth ventricle contributes more fluid. CSF enters the subarachnoid space through three openings in the roof of the fourth ventricle: a **median aperture (of Magendie)** and the paired **lateral apertures (of Luschka)**, one on each side. CSF then circulates in the central canal of the spinal cord and in the subarachnoid space around the surface of

Figure 18.3 / Locations of ventricles within a "transparent" brain. One interventricular foramen on each side connects a lateral ventricle to the third ventricle, and the cerebral aqueduct connects the third ventricle to the fourth ventricle.

Ventricles are cerebrospinal fluid-filled cavities within the brain.

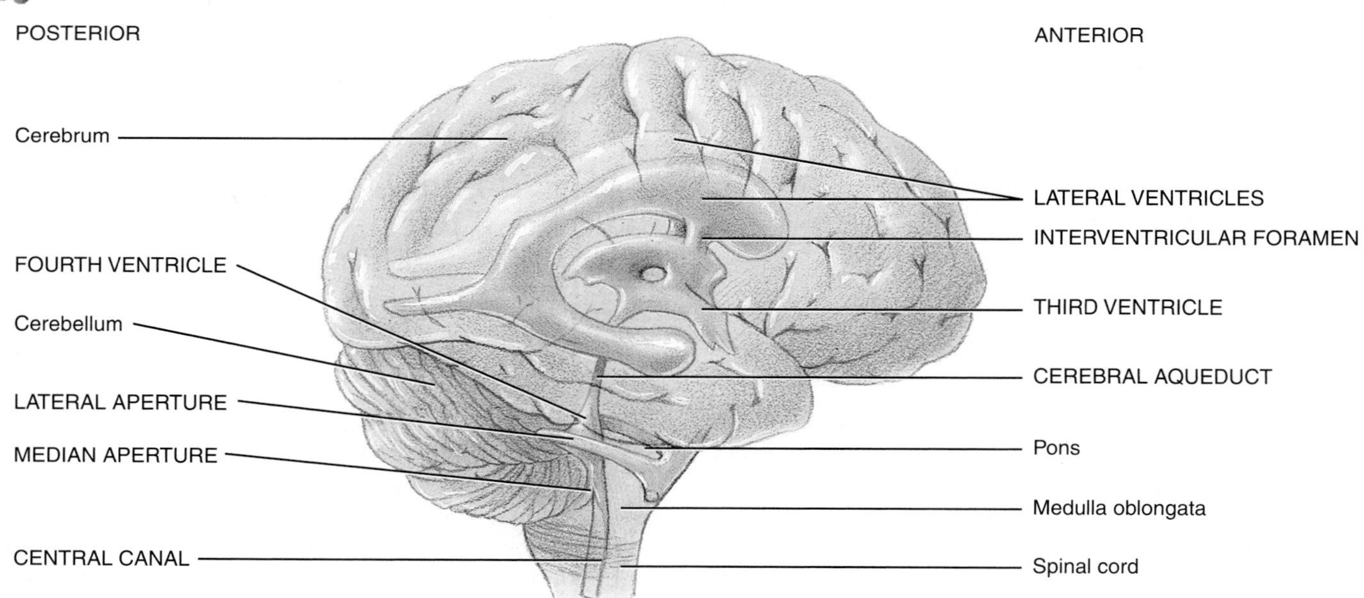

Which brain region is anterior to the fourth ventricle? Which is posterior to it?

Figure 18.4 / Pathways of circulating cerebrospinal fluid. (See Tortora, *A Photographic Atlas of the Human Body,* Figure 8.15.)

CSF is formed by ependymal cells that cover the choroid plexuses of the ventricles.

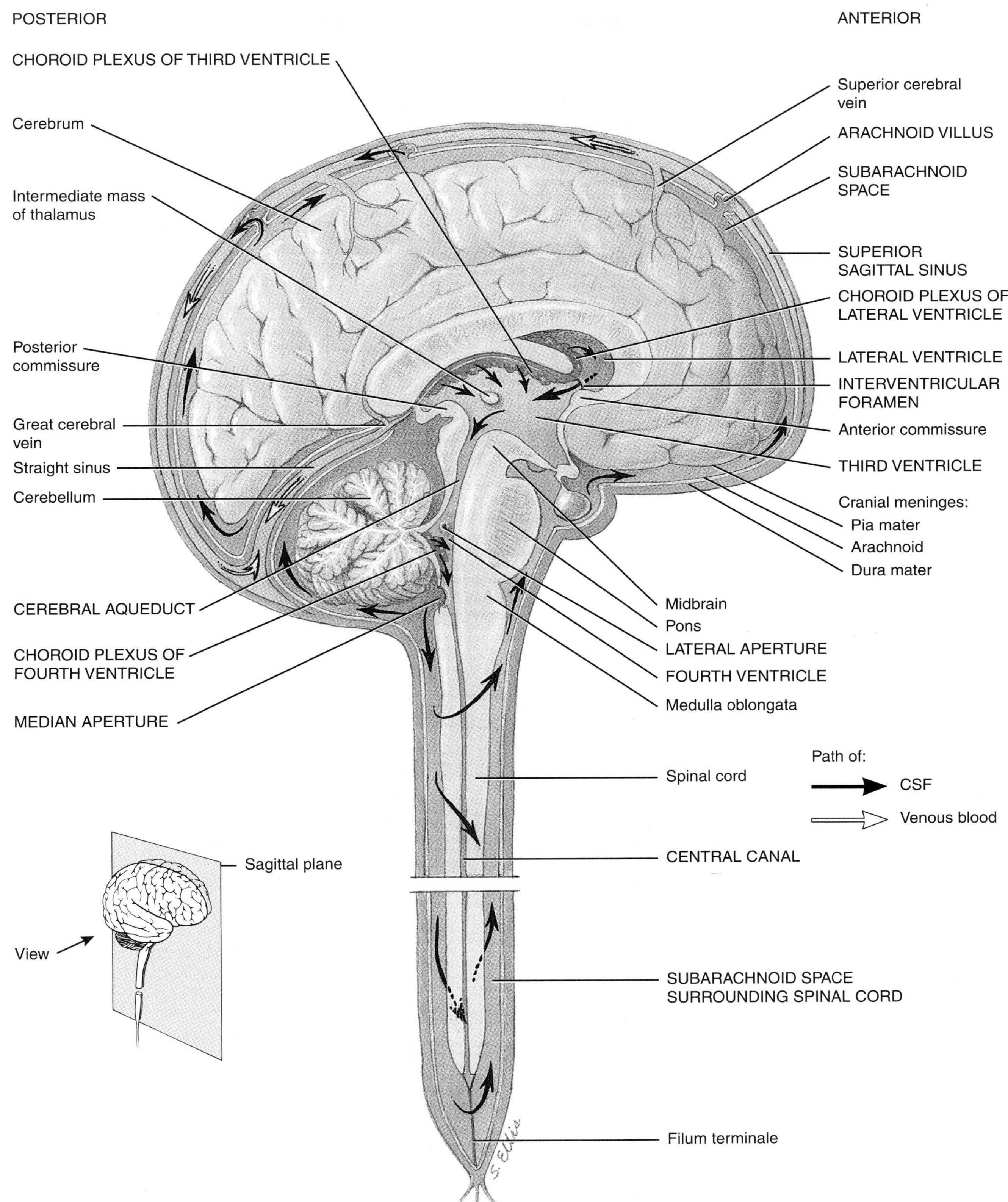

(a) Sagittal section of brain and spinal cord

(continues)

Figure 18.4 (continued)

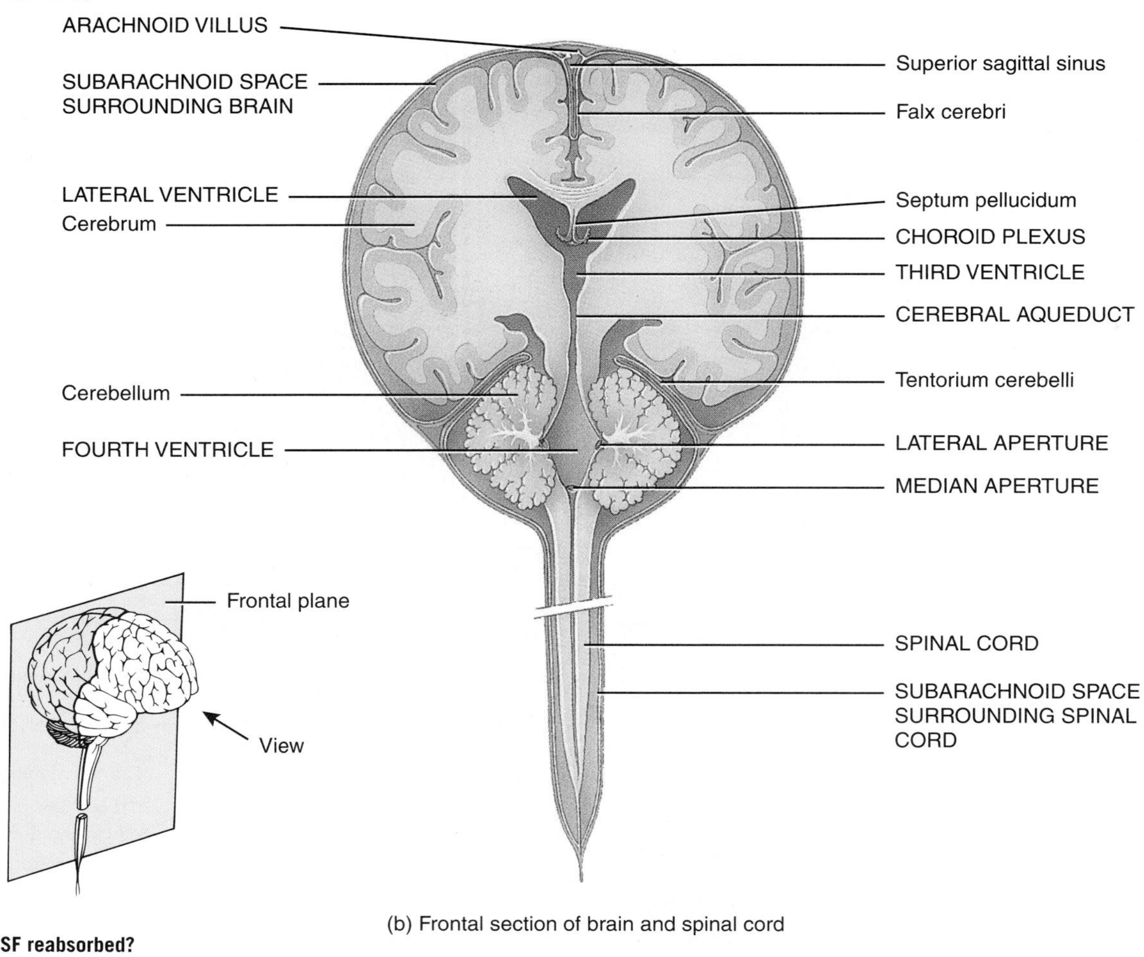

(b) Frontal section of brain and spinal cord

Where is CSF reabsorbed?

the brain and spinal cord. CSF is gradually reabsorbed into the blood through **arachnoid villi,** fingerlike extensions of the arachnoid that project into the dural venous sinuses, especially the **superior sagittal sinus** (Figure 18.4b; also see Figure 18.2). (A sinus is like a vein but has thinner walls.) Normally, CSF is reabsorbed as rapidly as it is formed by the choroid plexuses, at a rate of about 20 mL/hr (480 mL/day). Because the rates of formation and reabsorption are the same, the pressure of CSF normally is constant.

Hydrocephalus

Abnormalities in the brain—tumors, inflammation, or developmental malformations—can interfere with the drainage of CSF from the ventricles into the subarachnoid space. When excess CSF accumulates in the ventricles, the CSF pressure rises. Elevated CSF pressure causes a condition called **hydrocephalus** (hī′-drō-SEF-a-lus; *hydro-* = water; *cephal-* = head). In a baby in which the fontanels have not yet closed, the head bulges due to the increased pressure. If the condition persists, the fluid buildup compresses and damages the delicate nervous tissue. Hydrocephalus is relieved by draining the excess CSF. A neurosurgeon may implant a drain line, called a shunt, into the lateral ventricle to divert CSF into the superior vena cava or abdominal cavity. In adults, hydrocephalus may occur after head injury, meningitis, or subarachnoid hemorrhage. ■

✓ What structures are the sites of CSF production, and where are they located?

✓ What is the difference between the blood–brain barrier and the blood–cerebrospinal fluid barrier?

THE BRAIN STEM

Objective

- Describe the structures and functions of the brain stem.

The brain stem is the part of the brain between the spinal cord and the diencephalon; it consists of the (1) medulla oblongata, (2) pons, and (3) midbrain. Extending through the brain stem is the reticular formation, a netlike region of interspersed gray and white matter.

Medulla Oblongata

The **medulla oblongata** (me-DULL-la ob′-long-GA-ta), or more simply the medulla, is a continuation of the superior part of the spinal cord; it forms the inferior part of the brain stem (Figure 18.5; also see Figure 18.1). The medulla begins at the foramen magnum and extends to the inferior border of the pons, a distance of about 3 cm (1.2 in.). Within the medulla are all ascending (sensory) and descending (motor) tracts that connect the spinal cord with the brain, plus many nuclei (masses of gray matter made up of cell bodies of neurons) that regulate various vital body functions. The medulla also contains nuclei that receive sensory input from or provide motor output to five of the twelve pairs of cranial nerves (see Table 18.2 on page 579).

Figure 18.5 / Medulla oblongata in relation to the rest of the brain stem. (See Tortora, *A Photographic Atlas of the Human Body,* Figure 8.19.)

The brain stem consists of the medulla oblongata, pons, and midbrain.

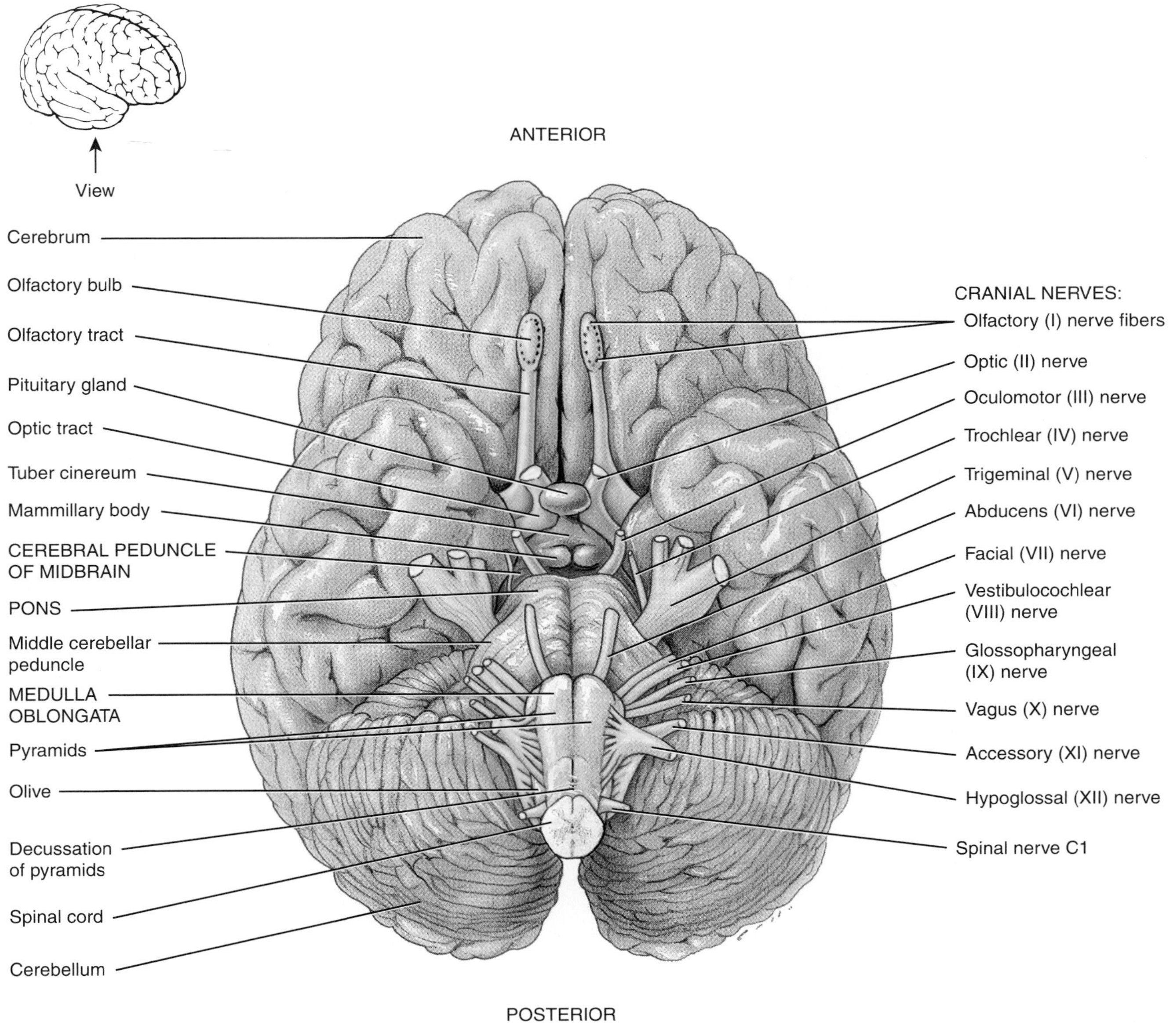

Inferior aspect of brain

What part of the brain stem contains the pyramids? The cerebral peduncles? Literally means "bridge"?

The medulla is organized into several major structural and functional regions. On the anterior aspect of the medulla are two conspicuous external bulges called the **pyramids** (Figures 18.5 and 18.6). The pyramids are formed by the largest motor tracts that pass from the cerebrum to the spinal cord. Just superior to the junction of the medulla with the spinal cord, most of the axons in the left pyramid cross to the right side, and most of the axons in the right pyramid cross over to the left side. This crossing is called the **decussation of pyramids** (dē′-ka-SĀ-shun; *decuss-* = crossing). Thus, neurons in the left cerebral cortex control skeletal muscles on the right side of the body, and neurons in the right cerebral cortex control skeletal muscles on the left side.

Just lateral to each pyramid is an oval-shaped swelling called an **olive** (Figures 18.5 and 18.6). The swelling is caused mostly by the **inferior olivary nucleus** within the medulla. Neurons here relay impulses from proprioceptors, such as muscle spindles, to the cerebellum. Nuclei associated with some somatic sensations (touch, vibration, and proprioception) are located on the posterior aspect of the medulla. These nuclei are the right and left **nucleus gracilis** (gras-I-lis; = slender) and **nucleus cuneatus** (kyū-nē-Ā-tus; = wedge). Many ascending sensory axons form synapses in these nuclei, and postsynaptic neurons then relay the sensory information to the thalamus on the opposite side of the brain (see Figure 19.4a on page 599).

The medulla also contains nuclei that govern several autonomic functions. These nuclei include the **cardiovascular center,** which regulates the rate and force of the heartbeat and the diameter of blood vessels; the **medullary rhythmicity area of the respiratory center** (see Figure 23.14 on page 705), which adjusts the basic rhythm of breathing; and other centers in the medulla that control reflexes for vomiting, coughing, and sneezing.

Finally, the medulla contains nuclei associated with the following five pairs of cranial nerves (see Figure 18.5).

- *Vestibulocochlear (VIII) nerves.* Several nuclei in the medulla receive sensory input from and provide motor output to the cochlea of the internal ear via the cochlear branches of the vestibulocochlear nerves. These nerves convey impulses related to hearing.
- *Glossopharyngeal (IX) nerves.* Nuclei in the medulla relay sensory and motor impulses related to taste, swallowing, and salivation via the glossopharyngeal nerves.
- *Vagus (X) nerves.* Nuclei in the medulla receive sensory impulses from and provide motor impulses to the pharynx and larynx and many thoracic and abdominal viscera via the vagus nerves.
- *Accessory (XI) nerves (cranial portion).* Nuclei in the medulla are the origin for nerve impulses that control swallowing via the cranial portion of the accessory nerves.
- *Hypoglossal (XII) nerves.* Nuclei in the medulla are the origin for nerve impulses that control tongue movements during speech and swallowing via the hypoglossal nerves.

Figure 18.6 / Internal anatomy of the medulla oblongata.

The pyramids of the medulla contain the largest motor tracts that run from the cerebrum to the spinal cord.

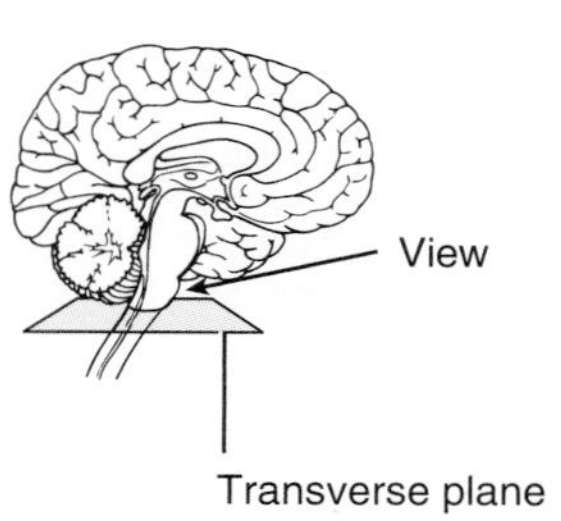

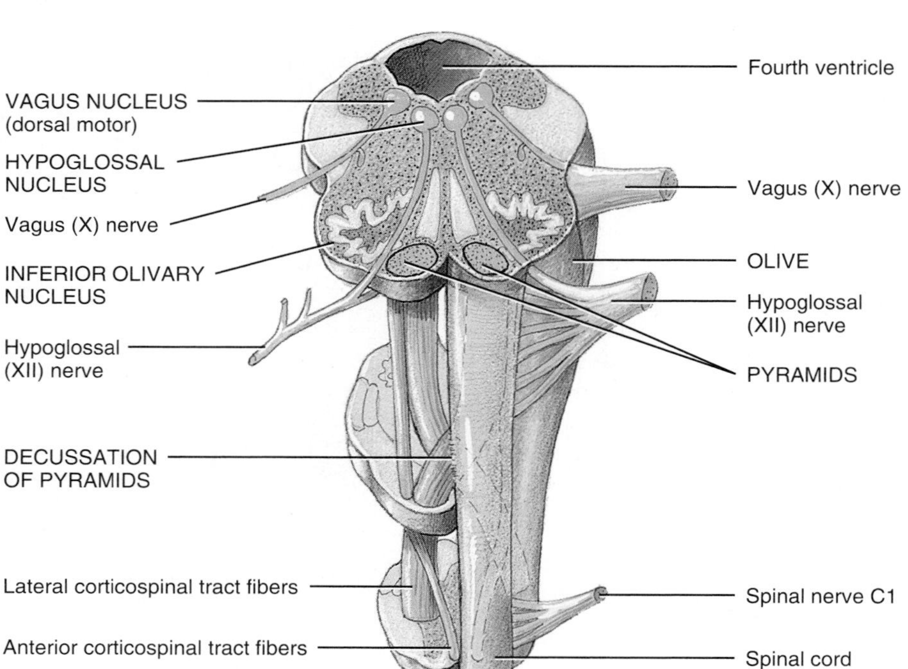

Transverse section and anterior surface of medulla oblongata

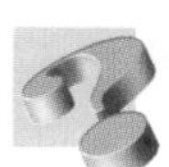

What does decussation mean? What is the functional consequences of decussation of the pyramids?

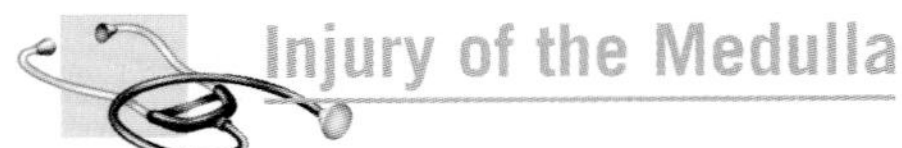

Given the many vital activities controlled by the medulla, it is not surprising that a hard blow to the back of the head or upper neck can be fatal. Damage to the respiratory center is particularly serious and can rapidly lead to death. Symptoms of nonfatal medullary injury may include cranial nerve malfunctions on the same side of the body as the injury, paralysis and loss of sensation on the opposite side of the body, and irregularities in breathing or heart rhythm.

Pons

The anatomical relationship of the **pons** (= bridge) to other parts of the brain can be seen in Figures 18.1 and 18.5. The pons lies directly superior to the medulla and anterior to the cerebellum and is about 2.5 cm (1 in.) long. Like the medulla, the pons consists of both nuclei and tracts. As its name implies, the pons is a bridge that connects parts of the brain with one another. These connections are provided by axons that are organized into tracts. Some of the tracts of the pons connect the cerebral cortex with the cerebellum; others are part of ascending sensory tracts and descending motor tracts.

Important nuclei in the pons are the **pneumotaxic area** (nū-mō-TAK-sik) and the **apneustic area** (ap-NŪ-stik), shown in Figure 23.14 on page 705. Together with the medullary rhythmicity area, the pneumotaxic and apneustic areas help control breathing. The pons also contains nuclei associated with the following four pairs of cranial nerves (see Figure 18.5):

- *Trigeminal (V) nerves.* Nuclei in the pons receive sensory impulses for somatic sensations from the head and face and provide motor impulses that govern chewing via the trigeminal nerves.
- *Abducens (VI) nerves.* Nuclei in the pons provide motor impulses that control eyeball movement via the abducens nerves.
- *Facial (VII) nerves.* Nuclei in the pons receive sensory impulses for taste and provide motor impulses to regulate secretion of saliva and tears and contraction of muscles of facial expression via the facial nerves.
- *Vestibulocochlear (VIII) nerves.* Nuclei in the pons receive sensory impulses from and provide motor impulses to the vestibular apparatus via the vestibular branches of the vestibulocochlear nerves. These nerves convey impulses related to balance and equilibrium.

Midbrain

The **midbrain,** or **mesencephalon,** which extends from the pons to the diencephalon (see Figures 18.1 and 18.5), is about 2.5 cm (1 in.) long. The cerebral aqueduct passes through the midbrain, connecting the third ventricle above with the fourth ventricle below. Like the medulla and the pons, the midbrain contains both tracts (white matter) and nuclei (gray matter).

The anterior part of the midbrain contains a pair of tracts called **cerebral peduncles** (pe-DUNG-kulz; = stalk; see Figure 18.5 and Figure 18.7b). They contain axons of corticospinal, corticopontine, and corticobulbar motor neurons, which conduct nerve impulses from the cerebrum to the spinal cord, medulla, and pons. The cerebral peduncles also contain axons of sensory neurons that extend from the medulla to the thalamus.

The posterior part of the midbrain, called the **tectum** (= roof), contains four rounded elevations, the **corpora quadrigemina** (KOR-po-ra kwad-ri-JEM-in-a; *corpora* = bodies; *quadrigeminus* = four twins) (Figure 18.7a). The two superior elevations, known as the **superior colliculi** (ko-LIK-yū-lī; singular is **colliculus** = small mound), serve as reflex centers that govern movements of the eyes, head, and neck in response to visual and other stimuli. The two inferior elevations, the **inferior colliculi,** are part of the auditory relay from the receptors for hearing in the ear to the sensory cortex of the cerebrum. They are also reflex centers for movements of the head and trunk in response to auditory stimuli.

The midbrain contains several nuclei, including the left and right **substantia nigra** (sub-STAN-shē-a; = substance; NĪ-gra; = black), which are large, darkly pigmented nuclei that control subconscious muscle activities (Figure 18.7b). Also present are the left and right **red nuclei** (Figure 18.7b), whose name derives from their rich blood supply and an iron-containing pigment in their neuronal cell bodies. Fibers from the cerebellum and cerebral cortex form synapses in the red nuclei, which function with the basal ganglia and cerebellum to coordinate muscular movements.

A band of white matter called the **medial lemniscus** (lem-NIS-kus; = ribbon) extends through the medulla, pons, and midbrain. The medial lemniscus contains axons that convey nerve impulses related to sensations of touch, proprioception (joint and muscle position), pressure, and vibration from the cuneate and gracile nuclei in the medulla to the thalamus.

Finally, nuclei in the midbrain are associated with two pairs of cranial nerves (see Figure 18.5):

- *Oculomotor (III) nerves.* Nuclei in the midbrain provide motor impulses that control movements of the eyeball, constriction of the pupil, and changes in shape of the lens via the oculomotor nerves.
- *Trochlear (IV) nerves.* Nuclei in the midbrain provide motor impulses that control movements of the eyeball via the trochlear nerves.

Reticular Formation

The brainstem also contains the **reticular formation** (*ret-* = net), a netlike arrangement of small areas of gray matter interspersed among threads of white matter (Figure 18.7b). The reticular formation, which also extends into the spinal cord and diencephalon, has both sensory and motor functions. The main sensory function is alerting the cerebral cortex to incoming sensory signals. Part of the reticular formation, called the **reticular activating system (RAS),** consists of fibers that project to the cerebral cortex. The RAS is responsible for maintaining con-

Figure 18.7 / Midbrain.

The midbrain connects the pons to the diencephalon.

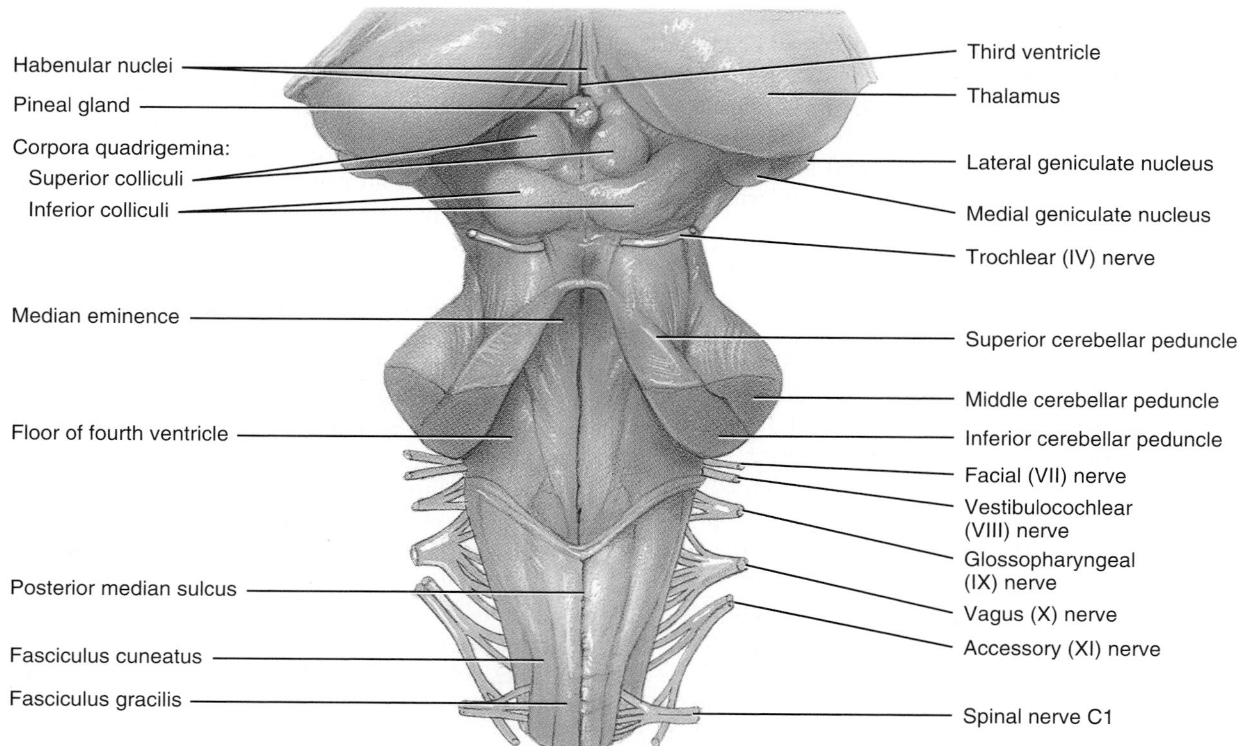

(a) Posterior view of midbrain in relation to brain stem

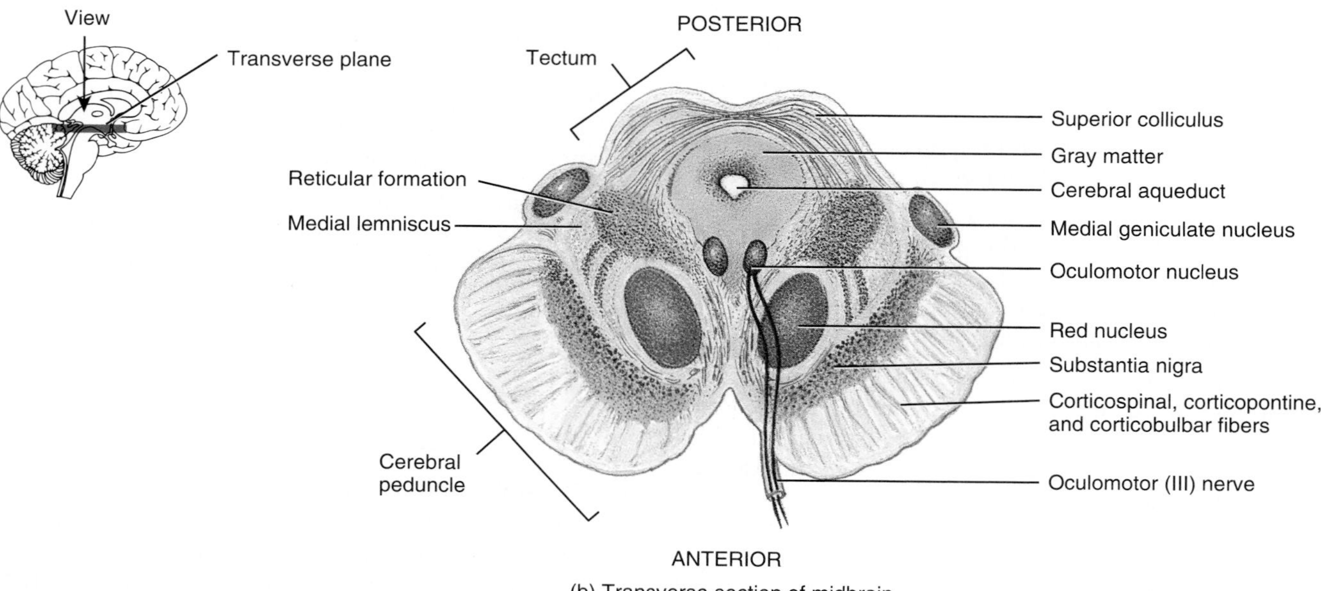

(b) Transverse section of midbrain

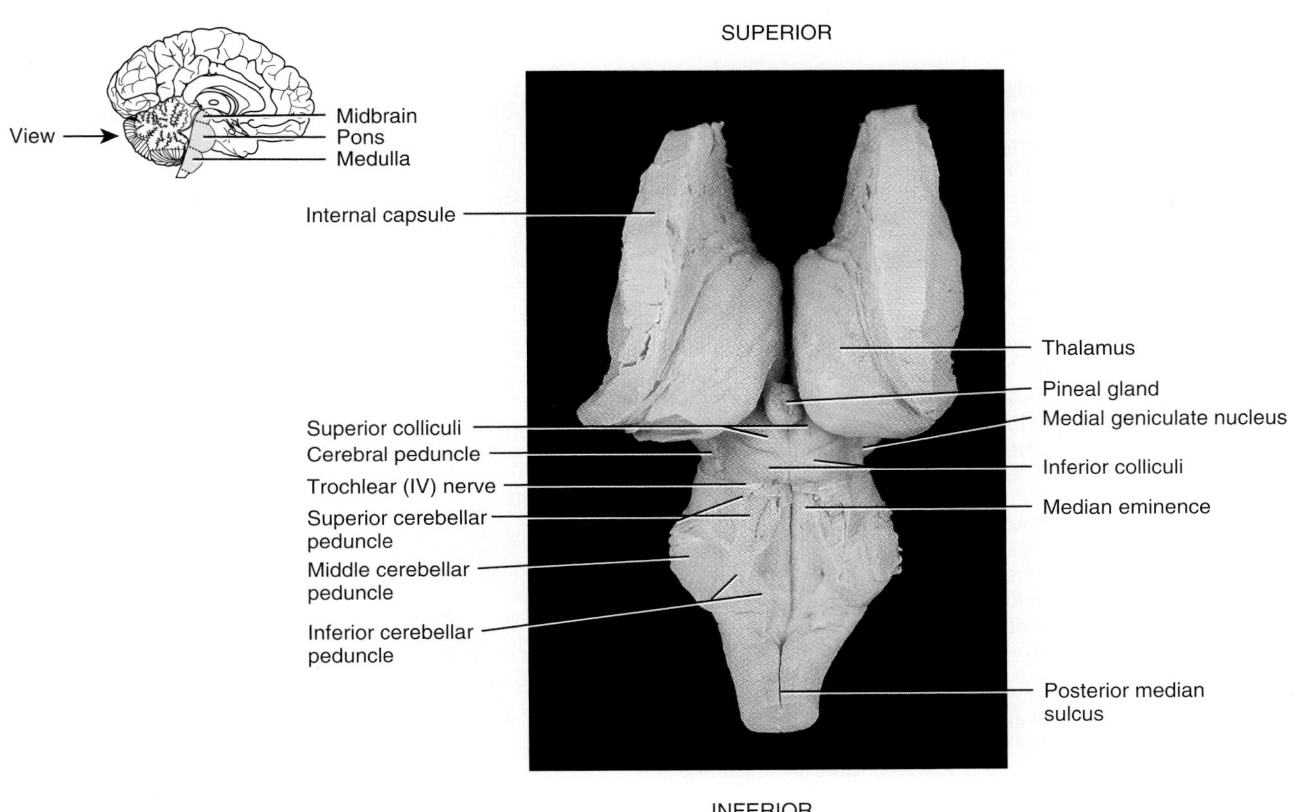

(c) Posterior view of midbrain in relation to brain stem

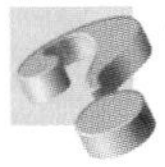

What is the importance of the cerebral peduncles?

sciousness and for awakening from sleep. Incoming impulses from the ears, eyes, and skin are effective stimulators of the RAS. For example, we awaken to the sound of an alarm clock, to a flash of lightning, or to a painful pinch because of RAS activity that arouses the cerebral cortex. The reticular formation's main motor function is to help regulate muscle tone (the slight degree of contraction that characterizes muscles at rest).

The functions of the brain stem are summarized in Table 18.1.

✓ Describe the locations of the medulla, pons, and midbrain relative to one another.
✓ Define decussation of pyramids. Why is it important?
✓ List the body functions that are governed by nuclei in the brain stem.
✓ What are two important functions of the reticular formation?

Table 18.1 Summary of Functions of Principal Parts of the Brain

Part	Function	Part	Function
Brain stem	*Medulla oblongata:* Conveys motor and sensory impulses between other parts of the brain and the spinal cord. Reticular formation (also in pons, midbrain, and diencephalon) functions in consciousness and arousal. Vital centers regulate heartbeat, breathing (together with pons), and blood vessel diameter. Other centers coordinate swallowing, vomiting, coughing, sneezing, and hiccuping. Contains nuclei of origin for cranial nerves VIII, IX, X, XI, and XII.	**Diencephalon**	*Epithalamus:* Consists of pineal gland, which secretes melatonin, and the habenular nuclei. *Thalamus:* Conveys almost all sensory input to the cerebral cortex. Provides crude perception of touch, pressure, pain, and temperature. Includes nuclei involved in movement planning and control.
	Pons: Conveys impulses from one side of the cerebellum to the other and between the medulla and midbrain. Contains nuclei of origin for cranial nerves V, VI, VII, and VIII. Pneumotaxic area and apneustic area, together with the medulla, help control breathing.		*Subthalamus:* Contains the subthalamic nuclei and portions of the red nucleus and the substantia nigra, which are positioned mostly lateral to the midline. These regions communicate with the basal ganglia to help control body movements. *Hypothalamus:* Controls and integrates activities of the autonomic nervous system and pituitary gland. Regulates emotional and behavioral patterns and circadian rhythms. Controls body temperature and regulates eating and drinking behavior. Helps maintain the waking state and establishes patterns of sleep.
	Midbrain: Conveys motor impulses from the cerebral cortex to the pons and sensory impulses from the spinal cord to the thalamus. Superior colliculi coordinate movements of the eyeballs in response to visual and other stimuli, and the inferior colliculi coordinate movements of the head and trunk in response to auditory stimuli. Most of substantia nigra and red nucleus contribute to control of movement. Contains nuclei of origin for cranial nerves III and IV.	**Cerebrum**	Sensory areas interpret sensory impulses, motor areas control muscular movement, and association areas function in emotional and intellectual processes. Basal ganglia coordinate gross, automatic muscle movements and regulate muscle tone. Limbic system functions in emotional aspects of behavior related to survival.
Cerebellum	Compares intended movements with what is actually happening to smooth and coordinate complex, skilled movements. Regulates posture and balance.		

THE CEREBELLUM

Objective

- Describe the structure and functions of the cerebellum.

The **cerebellum,** the second-largest part of the brain, occupies the inferior and posterior aspects of the cranial cavity. It is posterior to the medulla and pons and inferior to the posterior portion of the cerebrum (see Figure 18.1). The cerebellum is separated from the cerebrum by a deep groove known as the **transverse fissure,** and by the **tentorium cerebelli,** which supports the posterior part of the cerebrum (see Figure 18.4b).

In superior or inferior views, the shape of the cerebellum is somewhat like a butterfly. The central constricted area is the **vermis** (= worm), and the lateral "wings" or lobes are the **cerebellar hemispheres** (Figure 18.8a, b). Each hemisphere consists of lobes separated by deep and distinct fissures. The **anterior lobe** and **posterior lobe** govern subconscious movements of skeletal muscles; the **flocculonodular lobe** (*flocculo-* = wool-

Figure 18.8 / Cerebellum. (See Tortora, *A Photographic Atlas of the Human Body,* Figure 8.25.)

The cerebellum coordinates skilled movements and regulates posture and balance.

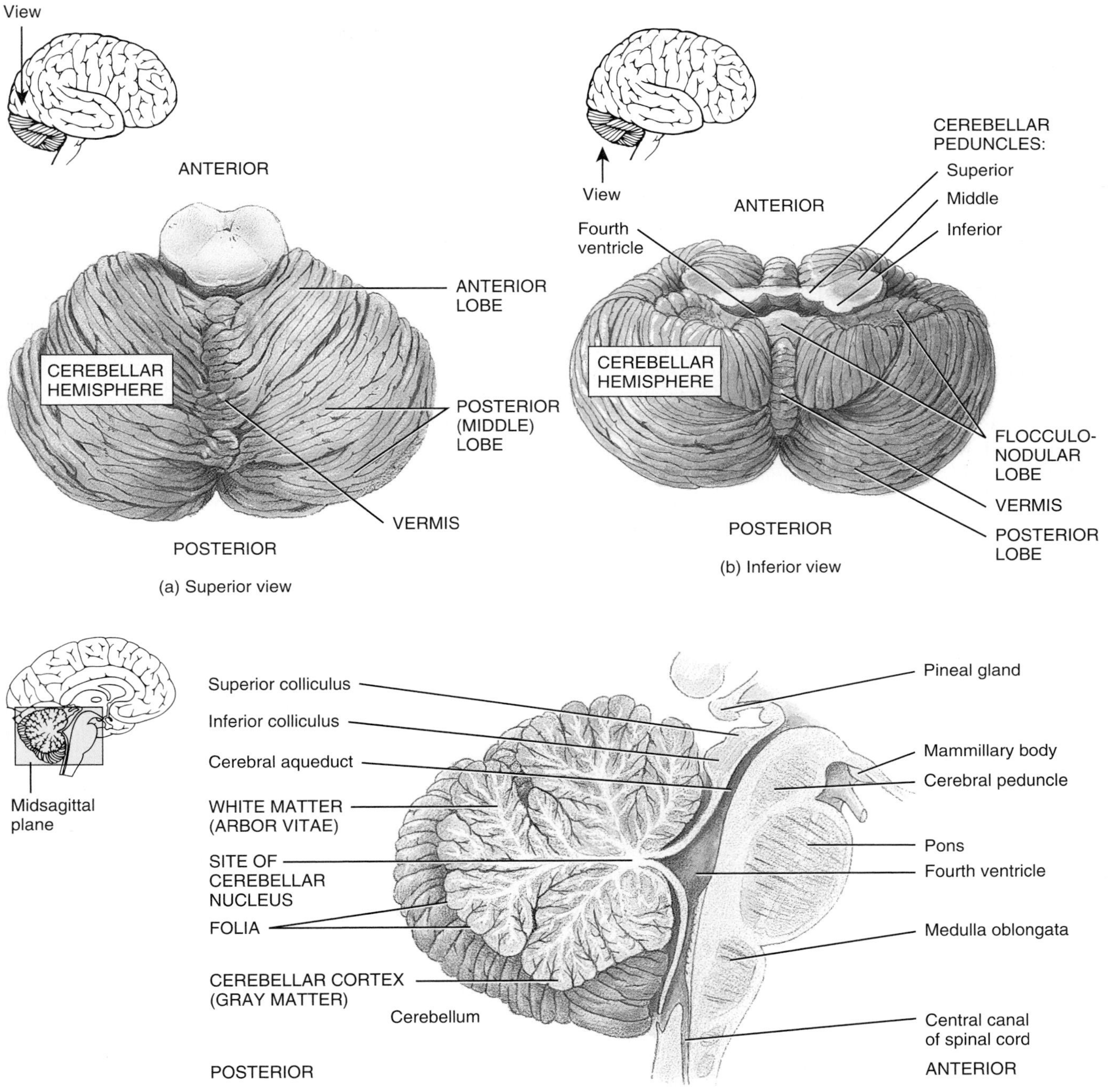

View

ANTERIOR

ANTERIOR LOBE

CEREBELLAR HEMISPHERE

POSTERIOR (MIDDLE) LOBE

VERMIS

POSTERIOR

(a) Superior view

View

CEREBELLAR PEDUNCLES:
Superior
Middle
Inferior

ANTERIOR

Fourth ventricle

CEREBELLAR HEMISPHERE

FLOCCULO-NODULAR LOBE

VERMIS

POSTERIOR

POSTERIOR LOBE

(b) Inferior view

Midsagittal plane

Superior colliculus

Inferior colliculus

Cerebral aqueduct

WHITE MATTER (ARBOR VITAE)

SITE OF CEREBELLAR NUCLEUS

FOLIA

CEREBELLAR CORTEX (GRAY MATTER)

Cerebellum

POSTERIOR

Pineal gland

Mammillary body

Cerebral peduncle

Pons

Fourth ventricle

Medulla oblongata

Central canal of spinal cord

ANTERIOR

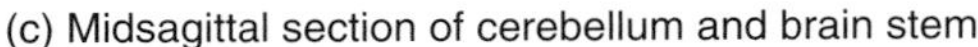

(c) Midsagittal section of cerebellum and brain stem

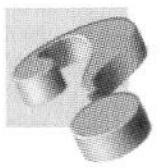

Which fiber tracts carry information into and out of the cerebellum?

like tuft) on the inferior surface is concerned with the sense of equilibrium.

The superficial layer of the cerebellum, called the **cerebellar cortex,** consists of gray matter in a series of slender, parallel ridges called **folia** (= leaves). Deep to the gray matter are tracts called **arbor vitae** (= tree of life) that resemble branches of a tree. Even deeper, within the white matter, are the **cerebellar nuclei,** which give rise to nerve fibers that carry impulses from the cerebellum to other brain centers and to the spinal cord.

The cerebellum is attached to the brain stem by three paired cerebellar peduncles (see Figure 18.7a) which carry information between the cerebellum and other parts of the brain. The **inferior cerebellar peduncles** connect the medulla oblongata to the cerebellum; they contain axons extending from the inferior olivary nucleus of the medulla and the spinal cord to the cerebellum. The **middle cerebellar peduncles** connect the pons to the cerebellum; they contain axons extending from the pons to the cerebellum. The **superior cerebellar peduncles** connect the midbrain to the cerebellum; they contain axons that extend mainly from the cerebellum into the midbrain.

A main function of the cerebellum is to evaluate how well movements initiated by motor areas in the cerebrum are actually being carried out. If the movements initiated by the cerebral motor areas are not being carried out, the cerebellum detects the discrepancies and sends feedback signals to the motor areas to correct the errors and modify the movements. This feedback helps to smooth and coordinate complex sequences of skeletal muscle contractions. Besides coordinating skilled movements, the cerebellum is the main brain region that regulates posture and balance. These aspects of cerebellar function make possible all skilled muscular activities, from catching a baseball to dancing.

The functions of the cerebellum are summarized in Table 18.1 on page 556.

✓ Describe the location and principal parts of the cerebellum.
✓ What are cerebellar peduncles? Explain the function of each.

THE DIENCEPHALON

Objective

- Describe the components and functions of the diencephalon.

The **diencephalon** extends from the brain stem to the cerebrum and surrounds the third ventricle; it includes the thalamus, hypothalamus, epithalamus, and subthalamus.

Thalamus

The **thalamus** (THAL-a-mus; = inner chamber), which measures about 3 cm (1.2 in.) in length and makes up 80% of the diencephalon, consists of paired oval masses of gray matter organized into nuclei with interspersed tracts of white matter (Figure 18.9). Usually, a bridge of gray matter called the **intermediate mass (interthalamic adhesion)** crosses the third ventricle in about 70% of human brains to join the right and left portions of the thalamus.

The thalamus is the principal relay station for most sensory impulses that reach the primary sensory areas of the cerebral cortex from other areas of the cerebrum and from subcortical areas such as the spinal cord, brain stem, and cerebellum. The thalamus allows crude perception of some sensations, such as pain, temperature, and pressure. Precise localization of these sensations depends on the relay of nerve impulses from the thalamus to the cerebral cortex. The thalamus also mediates motor functions by transmitting information from the cerebellum and basal ganglia to the primary motor area of the cerebral cortex. In addition, the thalamus is involved in autonomic activities and the maintenance of consciousness. Connections between the thalamus and cerebral cortex are through the **internal capsule,** a thick band of sensory and motor tracts (see Figure 18.13b). The internal capsule also connects the cerebral cortex to the brain stem and spinal cord.

The gray matter of the right and left sides of the thalamus is divided by a vertical, Y-shaped sheet of white matter called the **internal medullary lamina** (Figure 18.9c). It consists of myelinated fibers that enter and leave the various thalamic nuclei and separates the thalamic nuclei into six major groups (Figure 18.9d, e):

1. **Anterior group nuclei.** Function in emotions, regulation of alertness, and memory.
2. **Medial group nuclei.** Function in emotions, learning, memory, and cognition (awareness and the aquisition of knowledge).
3. **Lateral group nuclei.** Arranged in two tiers called ventral tier nuclei and dorsal tier nuclei. The ventral tier nuclei are as follows. The **ventral anterior nuclei** are involved in motor functions, possibly movement planning. The **ventral lateral nuclei** are concerned with various aspects of motor functions, such as movement planning and control. The **ventral posterior nuclei** relay somatosensory information from the face and body to the cerebral cortex. Included is information related to taste, pain, thermal sensations, crude and discriminative touch, pressure, vibration, and proprioception. The **medial geniculate nuclei,** which are concerned with hearing, and the **lateral geniculate nuclei,** which are concerned with vision, are sometimes included with the ventral tier nuclei.

   The dorsal tier nuclei are the **lateral dorsal nuclei,** which are concerned with emotional expression; the **lateral posterior nuclei,** which are concerned with the integration of sensory information; and **pulvinar nuclei,** which are also involved with the integration of sensory information.
4. **Intralaminar nuclei.** Lie within the internal medullary lamina and assume a role in regulating cortical activities, such as arousal, sensorimotor integration, and pain perception.
5. **Midline nuclei.** Form a thin band adjacent to the third ventricle and have a presumed role in memory and olfaction.

Figure 18.9 / Thalamus. Note the position of the thalamus in (a), the lateral view, and (b), the medial view. Various thalamic nuclei in (d) and (e) are correlated by color to the cortical regions to which they project in (a) and (b). (See Tortora, *A Photographic Atlas of the Human Body,* Figures 8.18 and 18.22.)

The thalamus is the principal relay station for sensory impulses that reach the cerebral cortex from other parts of the brain and the spinal cord.

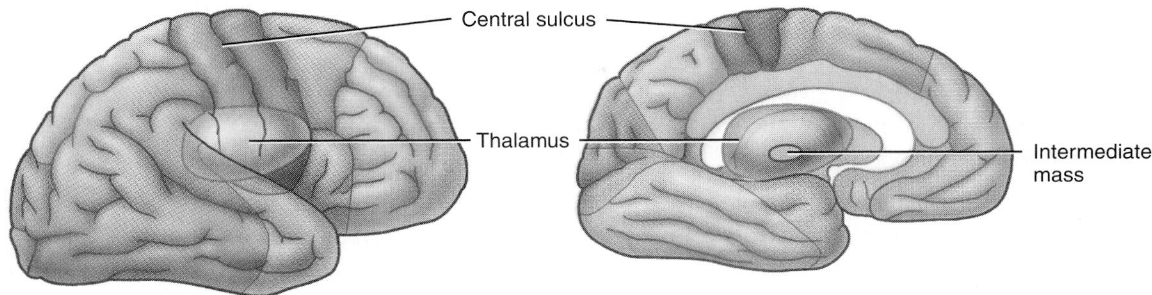

(a) Lateral view of right cerebral hemisphere

(b) Medial view of left cerebral hemisphere

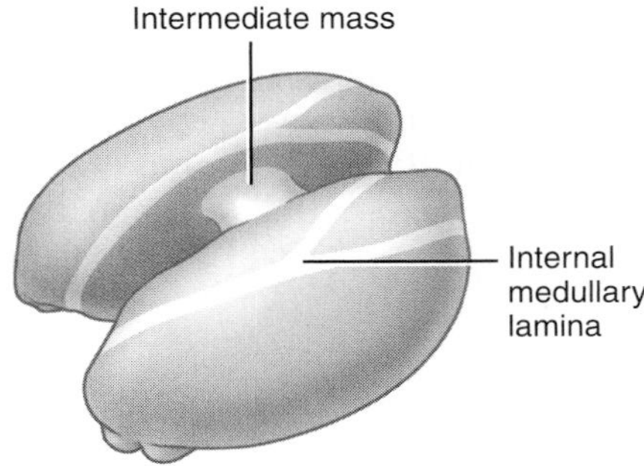

(c) Superolateral view of thalamus showing relationship of intermediate mass and internal medullary lamina

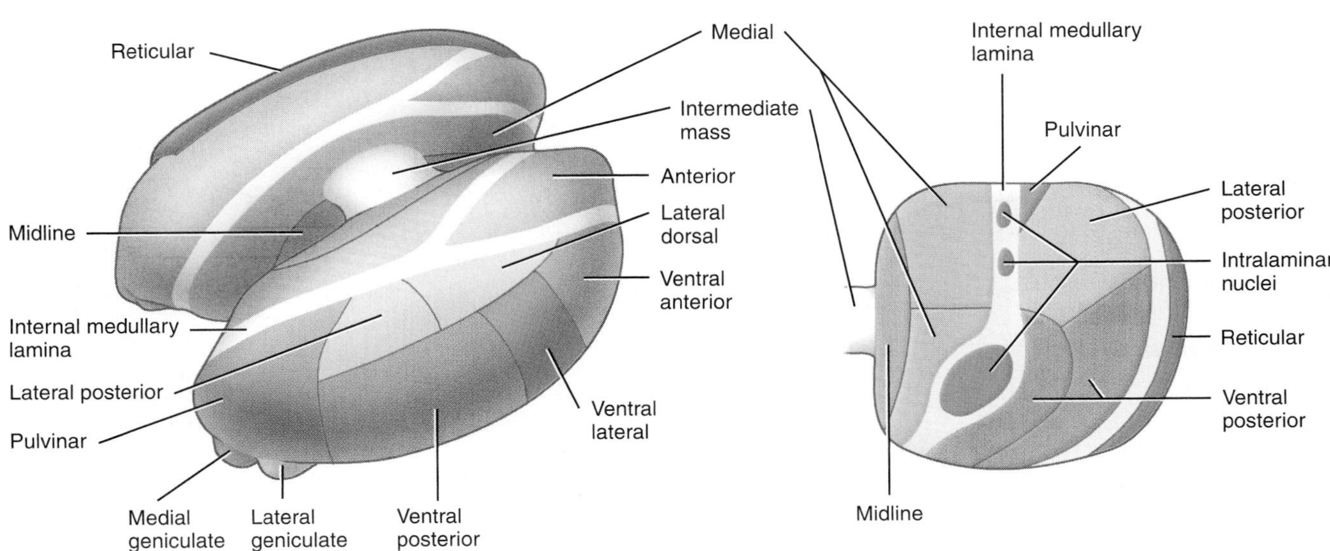

(d) Superolateral view of thalamus showing locations of thalamic nuclei (reticular nucleus is shown on the left side only; all other nuclei are shown on the right side)

(e) Transverse section of right side of thalamus showing locations of thalamic nuclei

What structure connects the right and left sides of the thalamus?

6. **Reticular nuclei.** Surround the lateral aspect of the thalamus and are adjacent to the internal capsule. These thalamic nuclei are the only ones that have an inhibitory effect and do not project to the cerebral cortex. They monitor and integrate the level of activity of thalamic nuclei, thereby indirectly influencing the cerebral cortex.

Hypothalamus

The **hypothalamus** (*hypo-* = under) is a small part of the diencephalon located inferior to the thalamus. It is composed of a dozen or so nuclei in four major regions:

- The *mammillary region* (*mammill-* = nipple-shaped), adjacent to the midbrain, is the most posterior part of the hypothalamus. It includes the mammillary bodies and posterior hypothalamic nucleus (Figure 18.10). The **mammillary bodies** are two, small, rounded projections that serve as relay stations for reflexes related to the sense of smell (see also Figure 18.5).
- The *tuberal region*, the widest part of the hypothalamus, includes the dorsomedial, ventromedial, and arcuate nuclei, plus the stalklike **infundibulum,** which connects the pituitary gland to the hypothalamus (Figure 18.10). The **median eminence** is a slightly raised region that encircles the infundibulum.
- The *supraoptic region* (*supra-* = above; *-optic* = eye) lies superior to the optic chiasm (point of crossing of optic nerves) and contains the paraventricular nucleus, supraoptic nucleus, anterior hypothalamic nucleus, and suprachiasmatic nucleus (Figure 18.10). Axons from the paraventricular and supraoptic nuclei form the hypothalamohypophyseal tract, which extends through the infundibulum to the posterior pituitary.
- The *preoptic region* anterior to the supraoptic region is usually considered part of the hypothalamus because it participates with the hypothalamus in regulating certain autonomic activities. The preoptic region contains the medial and lateral preoptic nuclei (Figure 18.10).

The hypothalamus controls many body activities and is *one of the major regulators of homeostasis.* Sensory impulses related to both somatic and visceral senses arrive at the hypothalamus via afferent pathways, as do impulses from visual, taste, and smell receptors. Other receptors within the hypothalamus itself continually monitor osmotic pressure, certain hormone concentrations, and the temperature of blood. The hypothalamus has several very important connections with the pituitary gland and also produces a variety of hormones, which will be discussed in more detail in Chapter 22. Whereas some functions can be attributed to specific nuclei, others are not so precisely localized. The chief functions of the hypothalamus are as follows:

Figure 18.10 / Hypothalamus. Selected portions of the hypothalamus and a three-dimensional representation of hypothalamic nuclei are shown (after Netter).

The hypothalamus controls many body activities and is an important regulator of homeostasis.

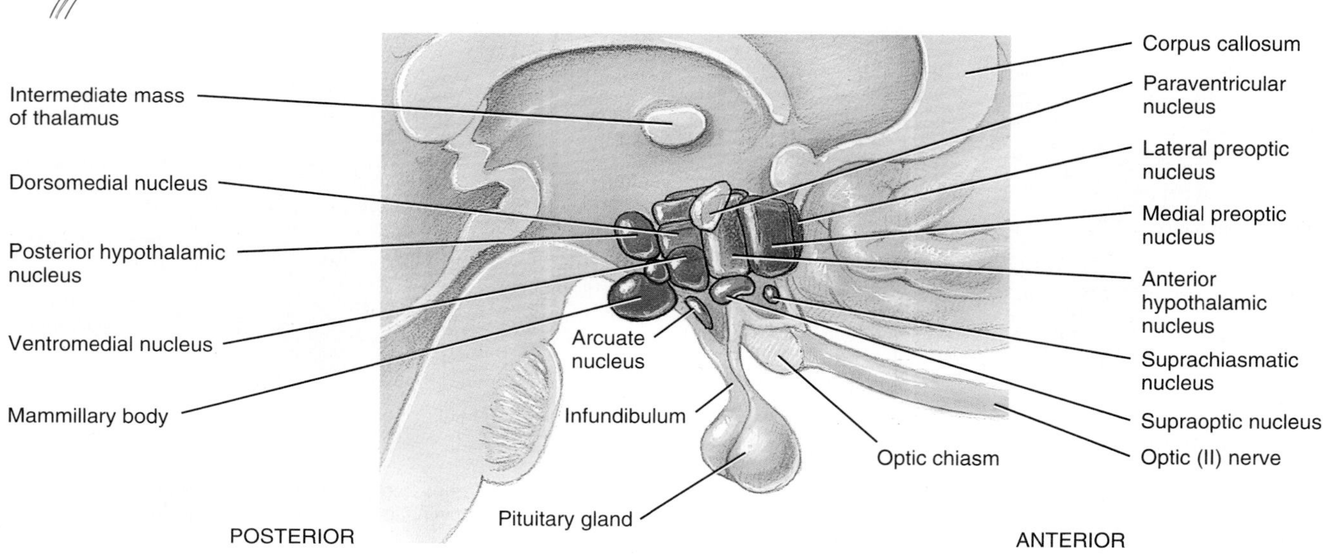

Sagittal section of brain showing hypothalamic nuclei

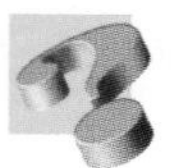

What are the four major regions of the hypothalamus, from posterior to anterior?

1. **Control of the ANS.** The hypothalamus controls and integrates activities of the autonomic nervous system, which regulates contraction of smooth and cardiac muscle and the secretions of many glands. Axons extend from the hypothalamus to sympathetic and parasympathetic nuclei in the brain stem and spinal cord. Through the ANS, the hypothalamus is a major regulator of visceral activities, including regulation of heart rate, movement of food through the gastrointestinal tract, and contraction of the urinary bladder.
2. **Control of the pituitary gland.** The hypothalamus produces several hormones and has two types of important connections with the pituitary gland, an endocrine gland located inferior to the hypothalamus (see Figure 18.1). First, hypothalamic regulating hormones are released into capillary networks in the median eminence. These hormones are carried by the bloodstream directly to the anterior pituitary, where they stimulate or inhibit secretion of pituitary hormones. Second, axons extend from the paraventricular and supraoptic nuclei through the infundibulum into the posterior pituitary. The cell bodies of these neurons make one of two hormones (oxytocin or antidiuretic hormone). Their axons transport the hormones to the posterior pituitary, where they are stored and released.
3. **Regulation of emotional and behavioral patterns.** Together with the limbic system (described shortly), the hypothalamus regulates feelings of rage, aggression, pain, and pleasure, and the behavioral patterns related to sexual arousal.
4. **Regulation of eating and drinking.** The hypothalamus regulates food intake through two centers. The **feeding center** is responsible for hunger sensations; when sufficient food has been ingested, the **satiety center** (sa-TĪ-e-tē; *sati-* = full, satisfied) is stimulated and sends out nerve impulses that inhibit the feeding center. The hypothalamus also contains a **thirst center.** When certain cells in the hypothalamus are stimulated by rising osmotic pressure of the extracellular fluid, they cause the sensation of thirst. The intake of water by drinking restores the osmotic pressure to normal, removing the stimulation and relieving the thirst.
5. **Control of body temperature.** If the temperature of blood flowing through the hypothalamus is above normal, the hypothalamus directs the autonomic nervous system to stimulate activities that promote heat loss. If, however, blood temperature is below normal, the hypothalamus generates impulses that promote heat production and retention.
6. **Regulation of circadian rhythms and states of consciousness.** The suprachiasmatic nucleus establishes patterns of sleep that occur on a circadian (daily) schedule.

Epithalamus

The **epithalamus** (*epi-* = above), a small region superior and posterior to the thalamus, consists of the pineal gland and habenular nuclei. The **pineal gland** (PĪN-ē-al; = pinecone-like) is about the size of a small pea and protrudes from the posterior midline of the third ventricle (see Figure 18.1). Although its physiological role is not completely clear, the pineal gland secretes the hormone **melatonin** and is thus an endocrine gland. In part because more melatonin is liberated during darkness than in light, it is thought to promote sleepiness. Melatonin also appears to contribute to the setting of the body's biological clock. The pineal gland is discussed in more detail in Chapter 22.

The **habenular nuclei** (ha-BEN-yū-lar), shown in Figure 18.7a, are involved in olfaction, especially emotional responses to odors.

Subthalamus

The **subthalamus,** a small area immediately inferior to the thalamus, includes tracts and the paired **subthalamic nuclei,** which connect to motor areas of the cerebrum. Parts of two pairs of midbrain nuclei—the red nucleus and the substantia nigra—also extend into the subthalamus. The subthalamic nuclei, red nuclei, and substantia nigra work together with the basal ganglia, cerebellum, and cerebrum in the control of body movements.

The functions of the four parts of the diencephalon are summarized in Table 18.1 on page 556.

Circumventricular Organs

Parts of the diencephalon, called **circumventricular** (ser′-kum-ven-TRIK-yū-lar) **organs (CVOs)** because they lie in the walls of the third and fourth ventricles, can monitor chemical changes in the blood because they lack the blood–brain barrier. CVOs include part of the hypothalamus, the pineal gland, the pituitary gland, and a few other nearby structures. Functionally, these regions coordinate homeostatic activities of the endocrine and nervous systems, such as the regulation of blood pressure, fluid balance, hunger, and thirst. CVOs are also thought to be the sites of entry into the brain of HIV, the virus that causes AIDS. Once in the brain, HIV may cause dementia (irreversible deterioration of mental state) and other neurological disorders.

✓ Explain why the thalamus is considered a "relay station" in the brain.

✓ Explain why the hypothalamus is considered to be part of both the nervous system and the endocrine system.

THE CEREBRUM

Objective

- Describe the structure and functions of the cerebrum.

The **cerebrum** is supported on the diencephalon and brain stem and forms the bulk of the brain (see Figure 18.1). It is composed of halves called **hemispheres,** each of which consists of an outer layer of gray matter, an internal region of cerebral white matter, and gray matter nuclei within the white matter. The superficial gray matter layer of the cerebrum is called the **cerebral cortex** (*cortex* = rind or bark) (Figure 18.11a). The cortex, which is only 2–4 mm (0.08–0.16 in.) thick, contains billions of neurons.

Figure 18.11 / Cerebrum. Because the insula cannot be seen externally, it has been projected to the surface in (b).

The cerebrum is the "seat of intelligence"; it provides us with the ability to read, write, and speak; to make calculations and compose music; to remember the past and plan for the future; and to create.

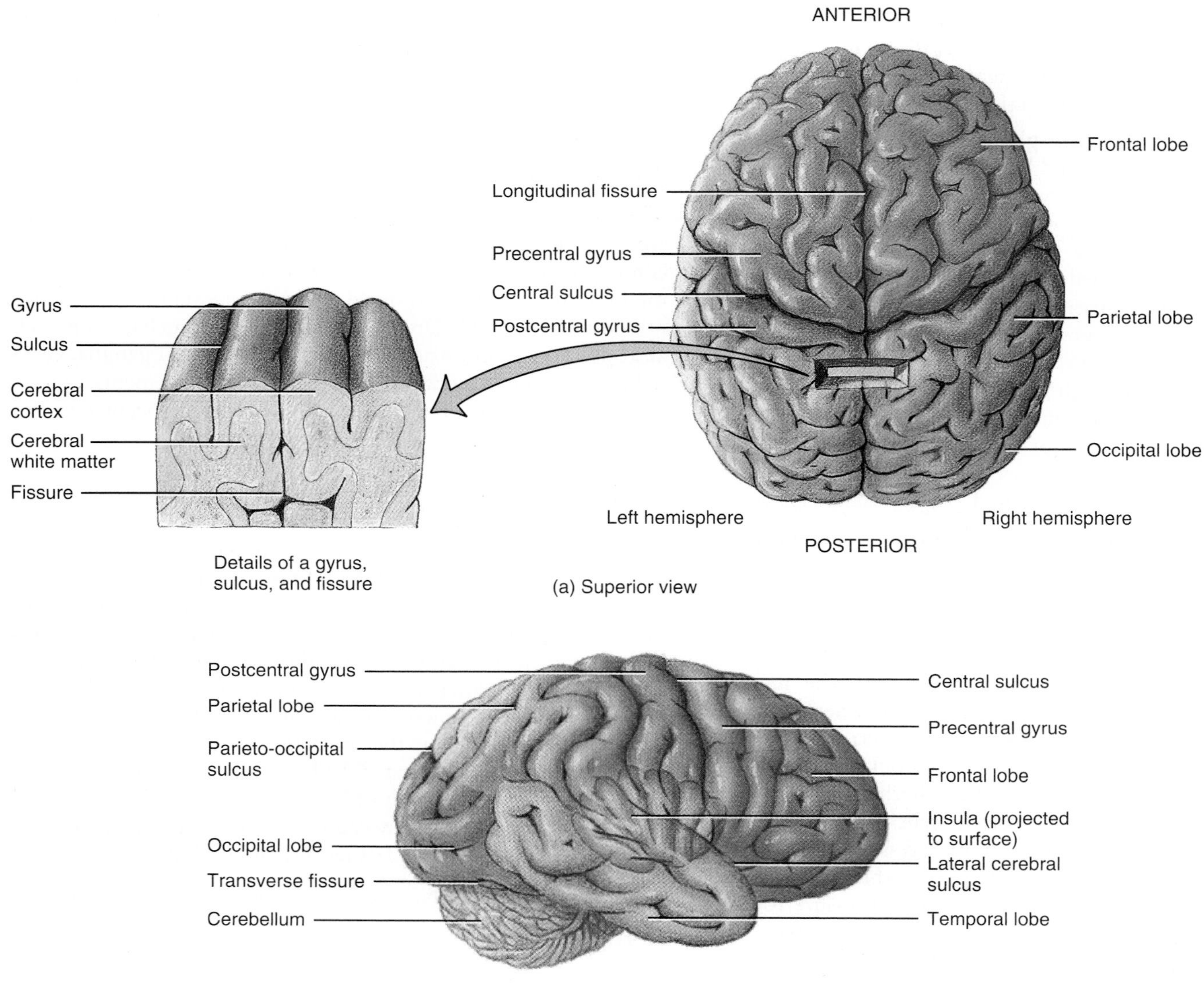

(b) Right lateral view

Deep to the cortex lies the cerebral white matter. The cerebrum is the "seat of intelligence"; it provides us with the ability to read, write, and speak; to make calculations and compose music; and to remember the past, plan for the future, and imagine things that have never existed before.

During embryonic development, when brain size increases rapidly, the gray matter of the cortex enlarges much faster than the deeper white matter. As a result, the cortical region rolls and folds upon itself. The folds are called **gyri** (JĪ-rī; = circles) or **convolutions** (Figure 18.11a, b). The deepest grooves between folds are known as **fissures;** the shallower grooves between folds are termed **sulci** (SUL-sī; = grooves). (The singular terms are gyrus and sulcus.) The most prominent fissure, the **longitudinal fissure,** separates the cerebrum into right and left halves called **cerebral hemispheres.** The hemispheres are connected internally by the **corpus callosum** (kal-LŌ-sum; *corpus* = body; *callosum* = hard), a broad band of white matter containing axons that extend between the hemispheres (see Figure 18.9b).

Lobes of the Cerebrum

Each cerebral hemisphere can be further subdivided into four lobes, named after the bones that cover them: frontal, parietal, temporal, and occipital lobes (Figure 18.11a, b). The **central sulcus** (SUL-kus) separates the **frontal lobe** from the **parietal lobe.** A major gyrus, the **precentral gyrus**—located immedi-

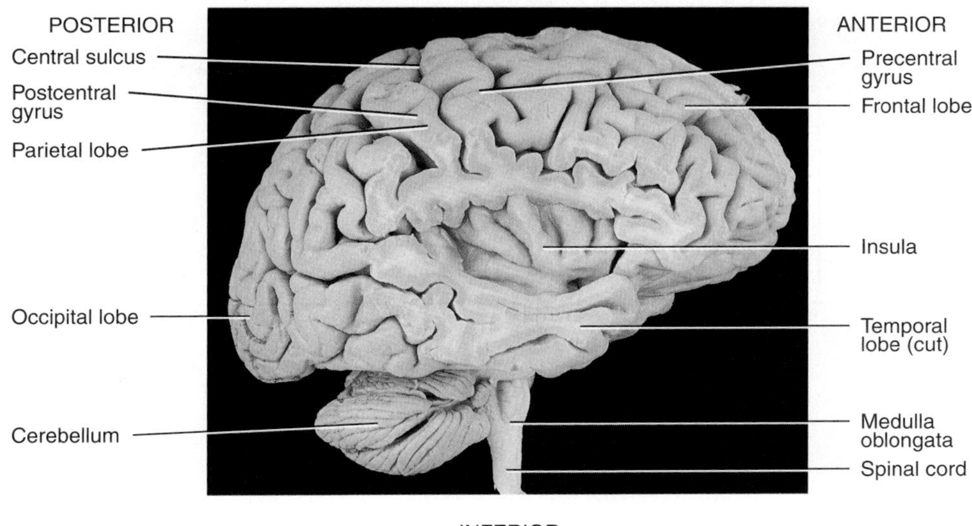

(c) Right lateral view with temporal lobe cut away

During development, which part of the brain—gray matter or white matter—enlarges more rapidly? What are the brain folds, shallow grooves, and deep grooves called?

ately anterior to the central sulcus—contains the primary motor area of the cerebral cortex. Another major gyrus, the **postcentral gyrus,** which is located immediately posterior to the central sulcus, contains the primary somatosensory area of the cerebral cortex. The **lateral cerebral sulcus** separates the **frontal lobe** from the **temporal lobe.** The **parietooccipital sulcus** separates the **parietal lobe** from the **occipital lobe.** A fifth part of the cerebrum, the **insula,** cannot be seen at the surface of the brain because it lies within the lateral cerebral sulcus, deep to the parietal, frontal, and temporal lobes (Figure 18.11b).

Cerebral White Matter

The **cerebral white matter** underlying the cortex consists of myelinated and unmyelinated axons organized into tracts consisting of three principal types of fibers (Figure 18.12):

Figure 18.12 / Organization of fibers into white matter tracts of the left cerebral hemisphere.

Association fibers, commissural fibers, and projection fibers form white matter tracts in the cerebral hemispheres.

Midsagittal plane

View

SUPERIOR

Cerebral cortex

COMMISSURAL AND PROJECTION FIBERS

COMMISSURAL FIBERS:

CORPUS CALLOSUM

ANTERIOR COMMISSURE

ASSOCIATION FIBERS

Septum pellucidum

Mammillary body

POSTERIOR

ANTERIOR

INFERIOR

Medial view of tracts revealed by removing gray matter from a midsagittal section

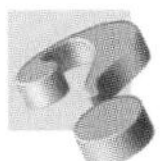

Which fibers carry impulses between gyri of the same hemisphere? Between gyri in opposite hemispheres? Between the cerebrum and thalamus, brain stem, and spinal cord?

14th Century

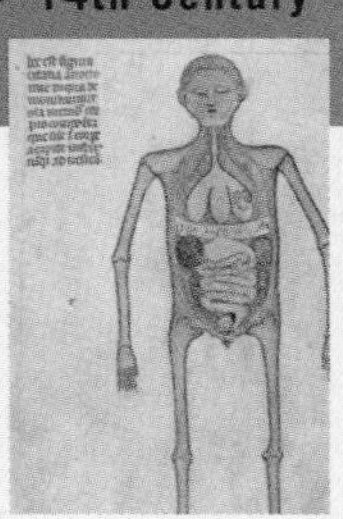

15th Century

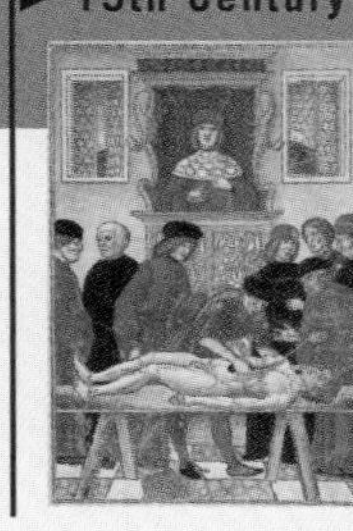

16th Century

17th Century

18th Century

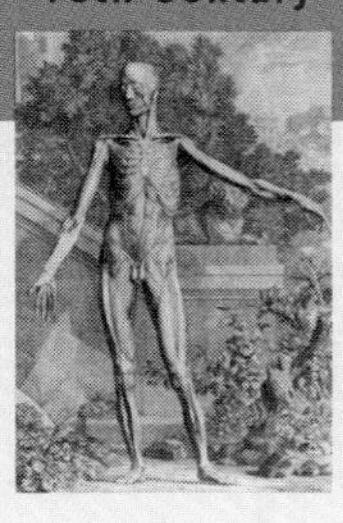

20th Century

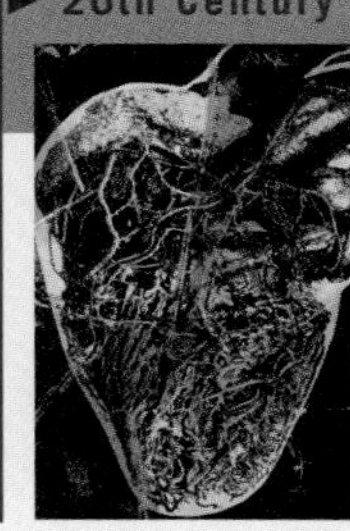

21st Century

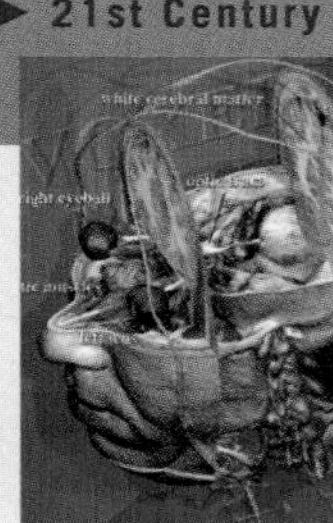

CHANGING IMAGES

The Virtual Couple

In 1989, the National Library of Medicine (NLM) conceived *The Visible Human Project*. This effort to create the most comprehensive image database of human anatomy ever assembled, has been described as "the greatest contribution to anatomy since Vesalius' 1543 publication of *De Humani Corporis Fabrica*." Going beyond traditional images on paper, the NLM intended to create a three-dimensional, high resolution, image database. Anatomy was going digital.

After a competitive proposal process, a $1.4 million dollar contract was awarded to the University of Colorado School of Medicine to generate data for both a male and a female. A 38 year old executed prisoner from Texas, who had donated his body to science, was selected as the initial specimen. From head to toe, data was generated on the intact cadaver using computerized tomography (CT), X-ray, and magnetic resonance imaging (MRI). Cryopreserved (maintained viability of tissues and organs at extremely low temperatures) and sectioned by saw, over 1800, 1-mm thick transverse slices were then photographed via digital photography, and both 35-mm and 70-mm conventional film. It was not until 1994 that The Visible Human Male was complete. One year later The Visible Human Female dataset was released.

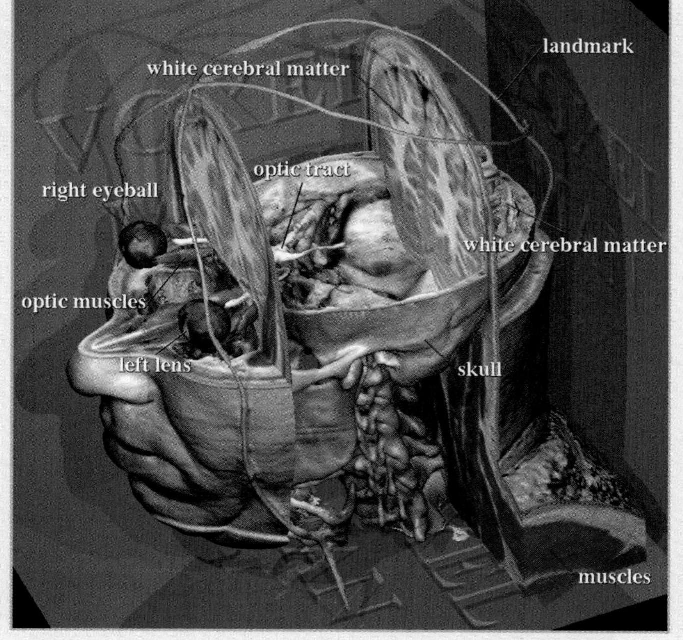

1995 AD

This virtual couple has resulted in a total of 55 gigabytes of data that is now posted on the Internet and available free of charge. Researchers can now download this database and utilize it to reconstruct an image from any desired perspective. This process is now being referred to as reverse engineering. Displayed here is a head image rendered by the University of Hamburg, Germany, Institute of Mathematics and Computer Science in Medicine from Visible Human Male data. Note the unusual perspective and sections that could not have been produced without the use of computers.

As futuristic as the *Visible Human Project* is, it shares much with the history of anatomical studies. As has been the case throughout antiquity, the initial subject was an executed prisoner, and the researchers had concerns about the response to a publicly displayed dismembered human. Today, however, the town square is the Internet.

The NLM is aware of this project's place in history. What will anatomists 50 years from now write about this project? How will they be studying anatomy? Perhaps it is best to not consider the Visible Human Project as futuristic, but as history in the making.

1. **Association fibers** transmit nerve impulses between gyri in the same hemisphere.
2. **Commissural fibers** transmit impulses from the gyri in one cerebral hemisphere to the corresponding gyri in the other cerebral hemisphere. Three important groups of commissural fibers are the **corpus callosum, anterior commissure,** and **posterior commissure.**
3. **Projection fibers** form descending and ascending tracts that establish connections between the cerebrum and thalamus, brain stem, and spinal cord. An example is the **internal capsule,** a thick band of sensory and motor tracts that connect the cerebral cortex with the brain stem and the spinal cord (see Figure 18.13b).

Basal Ganglia

The **basal ganglia** consist of several pairs of nuclei; the two members of each pair are situated in opposite cerebral hemispheres (Figure 18.13). Recall that nuclei are clusters of neuronal cell bodies and consist of gray matter. The largest nucleus in the basal ganglia is the **corpus striatum** (strī-Ā-tum; = striped), which consists of the **caudate nucleus** (*caud-* = tail) and the **lentiform nucleus** (*lentiform* = lentil-shaped). Each lentiform nucleus, in turn, is subdivided into a lateral part called the **putamen** (pu-TĀ-men; = shell) and a medial part called the **globus pallidus** (*globus* = ball; *pallidus* = pale).

Other structures that are functionally linked to (and sometimes considered part of) the basal ganglia are the substantia nigra and red nuclei of the midbrain (see Figure 18.7b) and the subthalamic nuclei of the diencephalon (Figure 18.13b). Axons from the substantia nigra terminate in the caudate nucleus and putamen. The subthalamic nuclei connect with the globus pallidus.

The basal ganglia receive input from and provide output to the cerebral cortex, thalamus, and hypothalamus. In addition, many nerve fibers interconnect the nuclei of the basal ganglia. The caudate nucleus and the putamen control automatic movements of skeletal muscles, such as swinging the arms while walking or laughing in response to a joke. The globus pallidus helps regulate the muscle tone required for specific body movements.

The Limbic System

Encircling the upper part of the brain stem and the corpus callosum is a ring of structures on the inner border of the cerebrum and floor of the diencephalon that constitutes the **limbic system** (*limbic* = border). Among the components of the limbic system are the following structures (Figure 18.14):

1. The **parahippocampal** and **cingulate gyri** (*cingul-* = belt), both gyri of the cerebral hemispheres, plus the **hippocampus** (= seahorse), a portion of the parahippocampal gyrus that extends into the floor of the lateral ventricle, make up the *limbic lobe.*
2. The **dentate gyrus** (*dentate* = toothed) lies between the hippocampus and parahippocampal gyrus.

Figure 18.13 / Basal ganglia. In (a), the basal ganglia have been projected to the surface and are shown in dark green; in (b) they are shown in green and blue. (See Tortora, *A Photographic Atlas of the Human Body,* Figures 8.17 and 8.22.)

The basal ganglia control large, automatic movements of skeletal muscles and muscle tone.

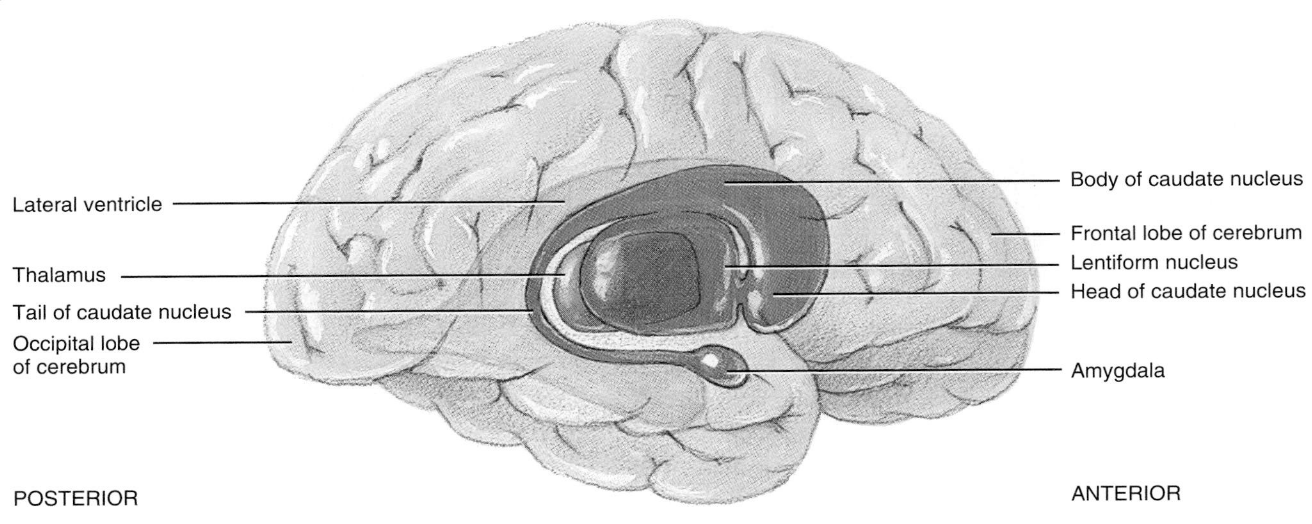

(a) Lateral view of right side of brain

(continues)

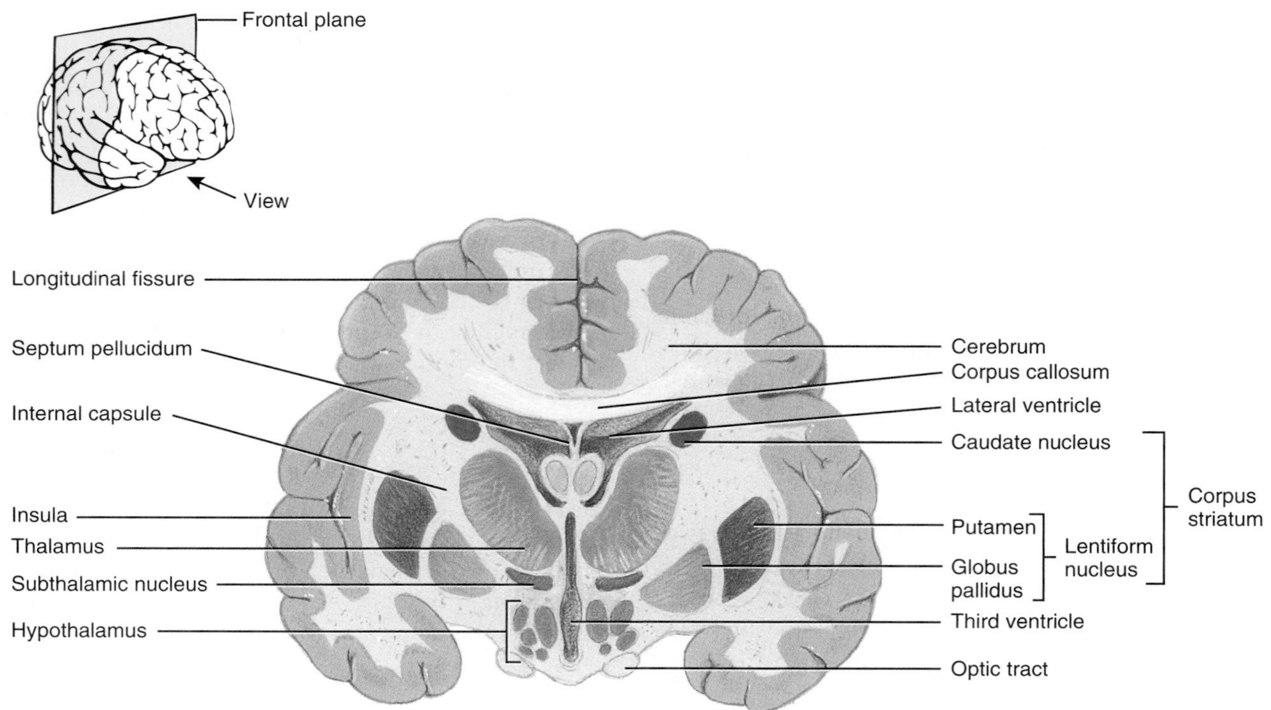

(b) Anterior view of frontal section

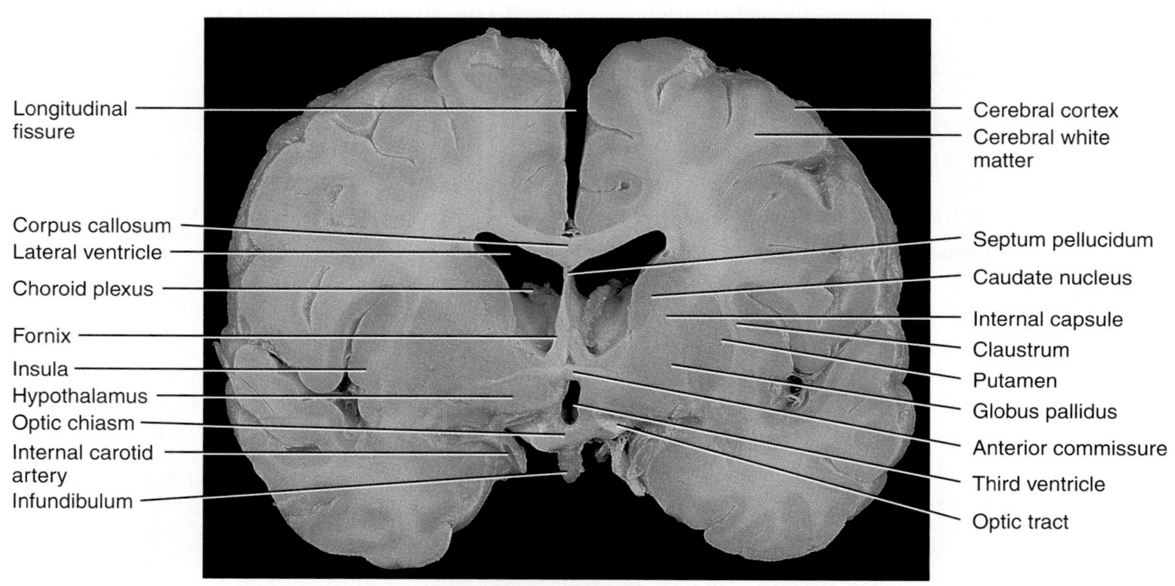

(c) Anterior view of frontal section

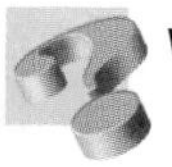

Where are the basal ganglia located relative to the thalamus?

Figure 18.14 / Components of the limbic system and surrounding structures.

The limbic system governs emotional aspects of behavior.

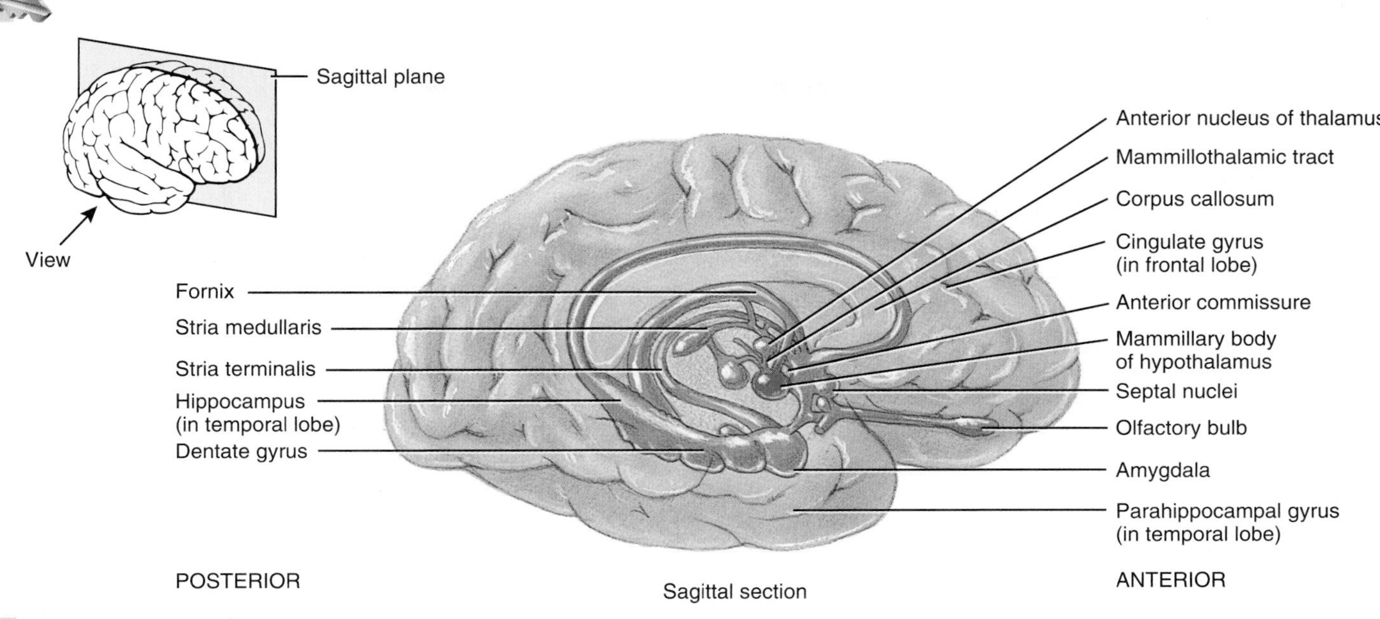

Which part of the limbic system functions with the cerebrum in memory?

3. The **amygdala** (*amygda-* = almond-shaped) is composed of several groups of neurons located at the tail end of the caudate nucleus.
4. The **septal nuclei** are located within the septal area formed by the regions under the corpus callosum and the paraterminal gyrus (a cerebral gyrus).
5. The **mammillary bodies of the hypothalamus** are two round masses close to the midline near the cerebral peduncles.
6. The **anterior nucleus of the thalamus** is located in the floor of the lateral ventricle.
7. The **olfactory bulbs** are flattened bodies of the olfactory pathway that rest on the cribriform plate.
8. The **fornix, stria terminalis, stria medullaris, medial forebrain bundle,** and **mammillothalamic tract** are linked by bundles of interconnecting myelinated axons.

The limbic system is sometimes called the "emotional brain" because it plays a primary role in a range of emotions, including pain, pleasure, docility, affection, and anger. Experiments have shown that when different areas of animals' limbic system are stimulated, the animals' reactions indicate that they are experiencing intense pain or extreme pleasure. Stimulation of other limbic system areas in animals produces tameness and signs of affection. Stimulation of a cat's amygdala or certain nuclei of the hypothalamus produces a behavioral pattern called rage—the cat extends its claws, raises its tail, opens its eyes wide, and hisses and spits.

The hippocampus of the limbic system, in conjunction with other portions of the cerebrum, functions in memory. People with damage to certain limbic system structures forget recent events and cannot commit anything to memory.

The functions of the cerebrum are summarized in Table 18.1 on page 556.

Brain injuries are commonly associated with head trauma and result in part from displacement and distortion of neuronal tissue at the moment of impact. Secondary effects include decreased blood pressure; sustained, elevated intracranial pressure; infections; and respiratory complications. Additional tissue damage occurs when normal blood flow is restored after a period of ischemia (reduced blood flow) because the sudden increase in oxygen level produces large numbers of oxygen free radicals (charged oxygen molecules with an unpaired electron). Brain cells recovering from the effects of a stroke or cardiac arrest also release free radicals. Free radicals cause damage by disrupting cellular DNA and enzymes and by altering plasma membrane permeability.

Various degrees of brain injury are described by specific terms. A **concussion** is an abrupt, but temporary, loss of consciousness (from seconds to hours) following a blow to the head or the sudden stopping of a moving head. It is the most common brain injury. A concussion produces no obvious bruising of the brain. Signs of a concussion are headache, drowsiness, lack of concentration, confusion, or post-traumatic amnesia (memory loss).

A **contusion** is bruising of the brain due to trauma and includes the leakage of blood from microscopic vessels. It is usually associated with a concussion. In a contusion, the pia mater may be torn, allowing blood to enter the subarachnoid space. The area most commonly affected is the frontal lobe. A contusion usually results in an immediate loss of consciousness (generally lasting no longer than 5 minutes), loss of reflexes, transient cessation of respiration, and decreased blood pressure. Vital signs typically stabilize in a few seconds.

A **laceration** is a tear of the brain, usually from a skull fracture or a gunshot wound. A laceration results in rupture of large blood vessels, with bleeding into the brain and subarachnoid space. Consequences include cerebral hematoma (localized pool of blood, usually clotted, that swells against the brain tissue), edema, and increased intracranial pressure. ■

FUNCTIONAL ASPECTS OF THE CEREBRAL CORTEX

Objective

- Describe the locations and functions of the sensory, association, and motor areas of the cerebral cortex.

Specific types of sensory, motor, and integrative signals are processed in certain cerebral regions (Figure 18.15). Generally, **sensory areas** receive and interpret sensory impulses, **motor areas** initiate movements, and **association areas** deal with more complex integrative functions such as memory, emotions, reasoning, will, judgment, personality traits, and intelligence.

Sensory Areas

Sensory impulses arrive mainly in the posterior half of both cerebral hemispheres—that is, in regions posterior to the central

Figure 18.15 / Functional areas of the cerebrum. Broca's speech area is in the left cerebral hemisphere of most people; it is shown here to indicate its relative location. The numbers, still used today, are from K. Brodmann's map of the cerebral cortex, first published in 1909.

Particular areas of the cerebral cortex process sensory, motor, and integrative signals.

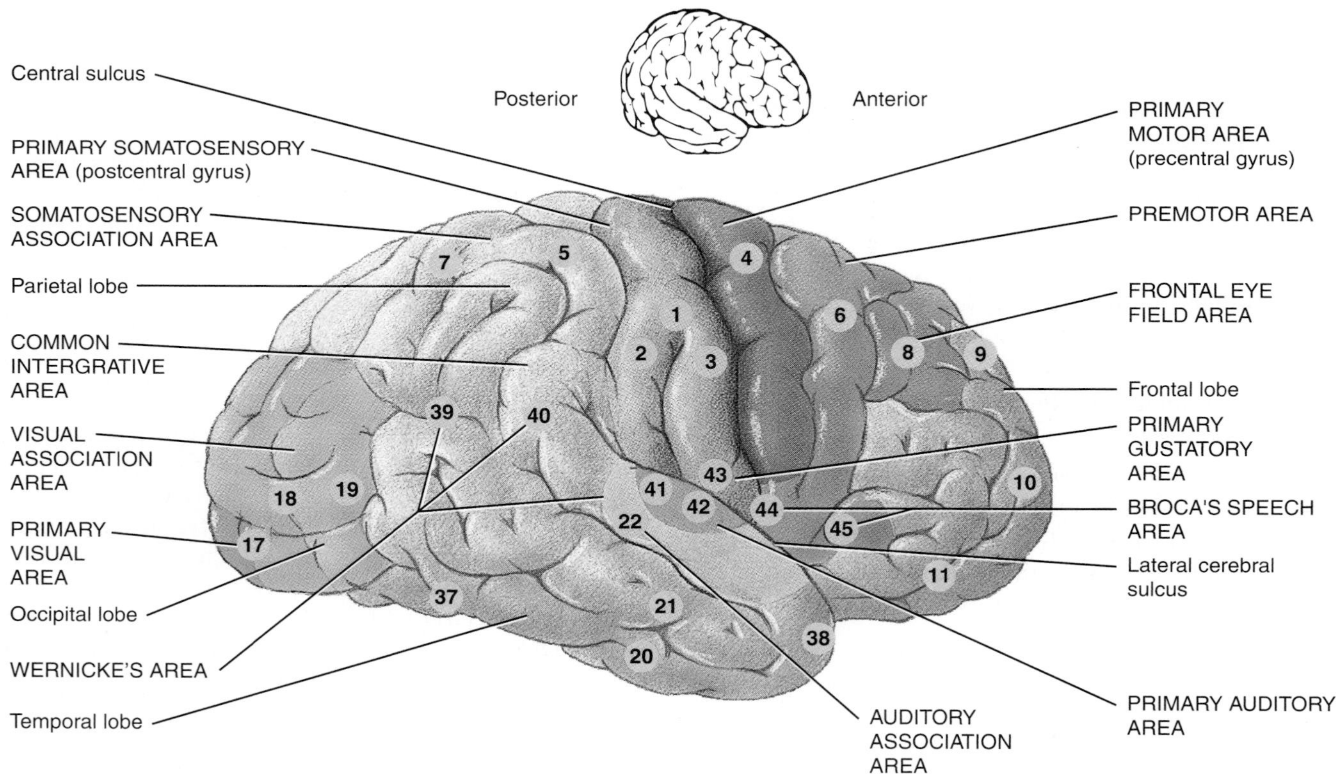

Lateral view of right cerebral hemisphere

What area(s) of the cerebrum integrate(s) interpretation of visual, auditory, and somatic sensations? Translates thoughts into speech? Controls skilled muscular movements? Interprets sensations related to taste? Interprets pitch and rhythm? Interprets shape, color, and movement of objects? Controls voluntary scanning movements of the eyes?

sulci. In the cortex, primary sensory areas have the most direct connections with peripheral sensory receptors. Secondary sensory areas and sensory association areas often are adjacent to the primary areas; they usually receive input from the primary areas and from other diverse regions of the brain.

Secondary sensory areas and sensory association areas integrate sensory experiences to generate meaningful patterns of recognition and awareness. For example, whereas a person with damage in the primary visual area would be blind in at least part of his visual field, a person with damage to a visual association area might see normally yet be unable to recognize a friend.

The following are some important sensory areas (Figure 18.15):

- **Primary Somatosensory Area** (areas 1, 2, and 3) Located directly posterior to the central sulcus of each cerebral hemisphere in the postcentral gyrus of each parietal lobe, the primary somatosensory area extends from the longitudinal fissure on the superior aspect of the cerebrum to the lateral cerebral sulcus.

The primary somatosensory area receives nerve impulses from somatic sensory receptors for touch, proprioception (joint and muscle position), pain, and temperature. Each point within the area receives sensations from a specific part of the body, and the entire body is spatially represented in it. The size of the cortical area receiving impulses from a particular body part depends on the number of receptors present there rather than on the size of the part. For example, a larger portion of the sensory area receives impulses from the lips and fingertips than from the thorax or hip (see Figure 19.5a on page 601). The major function of the primary somatosensory area is to localize exactly the points of the body where sensations originate. Although the thalamus registers sensations in a general way, it cannot distinguish precisely the specific location of stimulation; this capability depends on the primary somatosensory area of the cortex.

- **Primary Visual Area** (area 17) Located on the medial surface of the occipital lobe, it receives impulses that convey visual information. Axons of neurons with cell bodies in the eye form the optic (II) nerves, which terminate in the lateral geniculate nucleus of the thalamus. From the thalamus, neurons project to the primary visual area, carrying information concerning shape, color, and movement of visual stimuli.
- **Primary Auditory Area** (areas 41 and 42) Located in the superior part of the temporal lobe near the lateral cerebral sulcus, it interprets the basic characteristics of sound such as pitch and rhythm.
- **Primary Gustatory Area** (area 43) Located at the base of the postcentral gyrus superior to the lateral cerebral sulcus in the parietal cortex, it receives impulses for taste.
- **Primary Olfactory Area** (area 28) Located in the temporal lobe on the medial aspect (and thus not visible in Figure 18.15), it receives impulses for smell.

Motor Areas

Motor output from the cerebral cortex flows mainly from the anterior part of each hemisphere. Among the most important motor areas are the following (Figure 18.15):

- **Primary Motor Area** (area 4) Located in the precentral gyrus of the frontal lobe, each region in the primary motor area controls voluntary contractions of specific muscles or groups of muscles (see Figure 19.5b on page 601). Electrical stimulation of any point in the primary motor area results in contraction of specific skeletal muscle fibers on the opposite side of the body. As is true for the primary somatosensory area, body parts are not represented in proportion to their size: More cortical area is devoted to those muscles involved in skilled, complex, or delicate movement—as, for example, manipulation of the fingers.
- **Broca's Speech Area** (areas 44 and 45) Speaking and understanding language are complex activities that involve several sensory, association, and motor areas of the cortex. In 97% of the population, these language areas are localized in the *left* hemisphere. The production of speech occurs in Broca's (BRŌ-kaz) speech area, located in one frontal lobe—the *left* frontal lobe in most people—just superior to the lateral cerebral sulcus.

Association Areas

The association areas of the cerebrum consist of some motor and sensory areas, plus large areas on the lateral surfaces of the occipital, parietal, and temporal lobes and on the frontal lobes anterior to the motor areas. Association areas are connected with one another by association tracts and include the following (Figure 18.15):

- **Somatosensory Association Area** (areas 5 and 7) The somatosensory association area is just posterior to and receives input from the primary somatosensory area, as well as from the thalamus and other lower parts of the brain. Its role is to integrate and interpret sensations. This area permits you to determine the exact shape and texture of an object without looking at it, to determine the orientation of one object with respect to another as they are felt, and to sense the relationship of one body part to another. Another role of the somatosensory association area is the storage of memories of past sensory experiences, enabling you to compare current sensations with previous experiences.
- **Visual Association Area** (areas 18 and 19) Located in the occipital lobe, it receives sensory impulses from the primary visual area and the thalamus. It relates present and past visual experiences and is essential for recognizing and evaluating what is seen.
- **Auditory Association Area** (area 22) Located inferior and posterior to the primary auditory area in the temporal cortex, it ascertains whether a sound is speech, music, or noise.
- **Wernicke's (Posterior Language) Area** (area 22, and possibly areas 39 and 40) This area interprets the meaning of

speech by recognizing spoken words; it translates words into thoughts. The regions in the *right* hemisphere that correspond to Broca's and Wernicke's areas in the left hemisphere also contribute to verbal communication by adding tonal inflections and emotional content to spoken language. For example, you can tell if a person is angry or joyful by the tone of voice.

- **Common Integrative Area** (areas 5, 7, 39, and 40) This area, bordered by somatosensory, visual, and auditory association areas, receives nerve impulses from these areas, and from the primary gustatory and primary olfactory areas, the thalamus, and parts of the brain stem. It integrates sensory interpretations from the association areas and impulses from other areas, allowing one thought to be formed on the basis of a variety of sensory inputs. It then transmits signals to other parts of the brain to cause the appropriate response to the interpretation of the sensory signals.
- **Premotor Area** (area 6) Immediately anterior to the primary motor area is a motor association area. Neurons in this area communicate with the primary motor cortex, the sensory association areas in the parietal lobe, the basal ganglia, and the thalamus. The premotor area deals with learned motor activities of a complex and sequential nature. It generates nerve impulses that cause specific groups of muscles to contract in specific sequences. For example, this part of the cortex is active when you write a word. The premotor area controls learned skilled movements and serves as a memory bank for such movements.
- **Frontal Eye Field Area** (area 8) This area in the frontal cortex is sometimes included in the premotor area. It controls voluntary scanning movements of the eyes—reading this sentence, for instance.
- **Language Areas** From Broca's speech area, nerve impulses pass to the premotor regions that control the muscles of the larynx, pharynx, and mouth. The impulses from the premotor area result in specific, coordinated muscle contractions that enable you to speak. Simultaneously, impulses are sent from Broca's speech area to the primary motor area. From here, impulses also control the breathing muscles to regulate the proper flow of air past the vocal cords. The coordinated contractions of your speech and breathing muscles enable you to speak your thoughts.

Table 18.1 on page 556 summarizes the various functions of the cerebrum.

Aphasia

Much of what we know about language areas comes from studies of patients with language or speech disturbances resulting from brain damage. Broca's speech area, the auditory association area, and other language areas are located in the left cerebral hemisphere of most people, regardless of whether they are left-handed or right-handed. Injury to the association or motor speech areas results in **aphasia** (a-FĀ-zē-a; *a-* = without; *-phasia* = speech), an inability to use or comprehend words. Damage to Broca's speech area results in *nonfluent aphasia*, an inability to properly articulate or form words; people with nonfluent aphasia know what they wish to say but cannot speak. Damage to the common integrative area or auditory association area (areas 39 and 22) results in *fluent aphasia*, characterized by faulty understanding of spoken or written words. A person experiencing this type of aphasia may fluently produce strings of words that have no meaning. The underlying deficit may be **word deafness,** an inability to understand spoken words; or **word blindness,** an inability to understand written words; or both.

Hemispheric Lateralization

Although the right and left sides of the brain are fairly symmetrical there are subtle anatomical differences between the two hemispheres. For example, in about two-thirds of the population, the planum temporale, a region of the temporal lobe that includes Wernicke's area, is 50% larger on the left side than on the right side. This asymmetry appears in the human fetus at about 30 weeks of gestation. Moreover, although the two hemispheres share performance of many functions, each hemisphere also specializes in performing certain unique functions (Figure 18.16). This functional asymmetry is termed **hemispheric lateralization.**

The left hemisphere receives sensory signals from and controls the right side of the body, whereas the right hemisphere receives sensory signals from and controls the left side of the body. Beyond these differences, however, in most people the left hemisphere is more important for spoken and written language, numerical and scientific skills, ability to use and understand sign language, and reasoning. Patients with damage in the left hemisphere, for example, often exhibit aphasia. Conversely, the right hemisphere is more important for musical and artistic awareness, spatial and pattern perception, recognition of faces, and emotional content of language, and for generating mental images of sight, sound, touch, taste, and smell to compare relationships among them. Patients with damage in the right hemisphere regions that correspond to Broca's and Wernicke's areas in the left hemisphere speak in a monotonous voice, having lost the ability to impart emotional inflection to what they say. In general, lateralization seems less pronounced in females than in males, both for language (left hemisphere) and for visual and spatial skills (right hemisphere). A possibly related observation is that females tend to have a larger anterior commissure and a larger posterior portion of the corpus callosum than males; both of these white matter tracts provide communication between the two hemispheres.

Brain Waves

At any instant, brain cells are generating millions of nerve impulses (action potentials) in individual neurons. Taken together, these electrical signals are called **brain waves.** Brain waves generated by neurons close to the brain surface, mainly neurons in

Figure 18.16 / Summary of the principal functional differences between the left and right cerebral hemispheres.

Hemispheric lateralization means that each cerebral hemisphere performs unique functions.

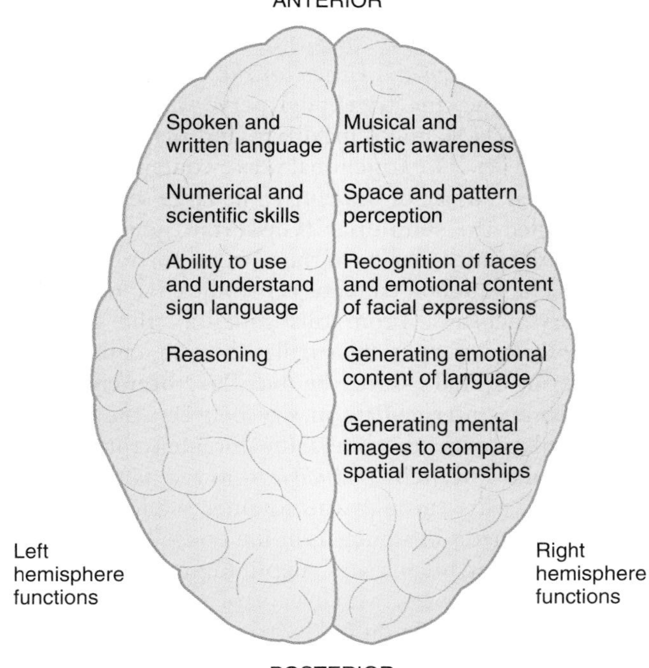

Which two areas of the cerebrum are most important for language?

the cerebral cortex, can be detected by sensors called electrodes placed on the forehead and scalp. A record of such waves is called an **electroencephalogram** (e-lek′-trō-en-SEF-a-lō-gram′) or **EEG.** Electroencephalograms are useful both in studying normal brain functions, such as changes that occur during sleep, and in diagnosing a variety of brain disorders, such as epilepsy, tumors, metabolic abnormalities, sites of trauma, and degenerative diseases.

- ✓ Describe the cortex, convolutions, fissures, and sulci of the cerebrum.
- ✓ List and locate the lobes of the cerebrum. How are they separated from one another? What is the insula?
- ✓ Describe the organization of cerebral white matter. Be sure to indicate the function of each major group of fibers.
- ✓ Name the nuclei that form basal ganglia, list the function of each, and describe the effects of basal ganglia damage.
- ✓ Define the limbic system. List several of its functions.
- ✓ Compare the functions of sensory, motor, and association areas of the cerebral cortex.
- ✓ Describe hemispheric lateralization.
- ✓ What is the diagnostic value of an EEG?

CRANIAL NERVES

Objective

- Identify the cranial nerves by name, number, and type, and give the function of each.

Cranial nerves, like spinal nerves, are part of the peripheral nervous system (PNS). Of the 12 pairs of cranial nerves, ten emerge from the brain stem. The cranial nerves are designated both by roman numerals and names (see Figure 18.5). The roman numerals indicate the order, from anterior to posterior, in which the nerves arise from the brain; the names indicate the nerve's distribution or function.

Cranial nerves emerge from the nose (I), eyes (II), brainstem (III–XII), and spinal cord (a portion of XI). Two cranial nerves (I and II) contain only sensory fibers and thus are called **sensory nerves.** The remainder contain both sensory and motor fibers and are referred to as **mixed nerves,** though a few of them are predominantly motor in function. For example, the oculomotor (III), trochlear (IV), abducens (VI), accessory (XI) and hypoglossal (XII) nerves are referred to as mixed nerves because they contain fibers from proprioceptors, but they are primarily motor, serving to stimulate skeletal muscle contractions.

The cell bodies of sensory neurons are located in ganglia outside the brain, whereas the cell bodies of motor neurons lie in nuclei within the brain. Some cranial nerves include both somatic motor fibers and parasympathetic fibers of the autonomic nervous system.

Although the cranial nerves are mentioned singly in the following description of their type, location, and function, remember that they are paired structures.

Olfactory (I)

The **olfactory (I) nerve** (*olfacere* = to smell) is entirely sensory and conveys nerve impulses related to smell. It arises as bipolar neurons from the olfactory mucosa of the nasal cavity (see Figure 20.1 on page 612). The dendrites and cell bodies of these neurons are generally limited to the mucosa covering the superior nasal conchae and the adjacent nasal septum. Axons from the neurons pass through the cribriform plate of the ethmoid bone and synapse with other olfactory neurons in the **olfactory bulb,** an extension of the brain lying superior to the cribriform plate. The axons of these neurons make up the **olfactory tract.** The fibers from the tract terminate in the primary olfactory area in the temporal lobe of the cerebral cortex.

Optic (II)

The **optic (II) nerve** (*optikos* = vision, eye, or optics) is entirely sensory and conveys nerve impulses related to vision. Signals initiated by rods and cones of the retina are relayed by bipolar neurons to ganglion cells (see Figure 20.7 on page 619). Axons of the ganglion cells, the optic nerve fibers, exit the optic foramen, after which the two optic nerves unite to form the **optic chiasm** (kī-AZM). Within the chiasm, fibers from the medial half of each retina cross to the opposite side; those from the lateral half remain on the same side. From the chiasm, the regrouped fibers pass posteriorly to the **optic tracts.** From the optic tracts the majority of fibers terminate in the lateral geniculate nucleus in the thalamus. They then synapse with neurons that pass to the visual areas of the cerebral cortex (see Figure 20.9 on page 622). Some fibers from the optic chiasm terminate in the superior colliculi of the midbrain. They synapse with neurons whose fibers terminate in the nuclei that convey impulses to the oculomotor (III), trochlear (IV), and abducens (VI) nerves—nerves that control the extrinsic (external) and intrinsic (internal) eye muscles. Through this relay there are widespread motor responses to light stimuli.

Oculomotor (III)

The **oculomotor (III) nerve** (*oculus* = eye; *motor* = mover) is a mixed cranial nerve. It originates from neurons in a nucleus in the ventral portion of the midbrain (Figure 18.17a). It runs anteriorly and divides into superior and inferior divisions, both of which pass through the superior orbital fissure into the orbit. The superior branch is distributed to the superior rectus (an extrinsic eyeball muscle) and the levator palpebrae superioris (the muscle of the upper eyelid). The inferior branch is distributed to the medial rectus, inferior rectus, and inferior oblique muscles—all extrinsic eyeball muscles. These distributions to the levator palpebrae superioris and extrinsic eyeball muscles constitute the motor portion of the oculomotor nerve. Through these distributions are sent nerve impulses that control movements of the eyeball and upper eyelids.

The inferior branch of the oculomotor nerve also provides parasympathetic innervation to the **ciliary ganglion,** a relay center of the autonomic nervous system that connects a nucleus in the midbrain with the intrinsic eyeball muscles. These intrinsic muscles include the ciliary muscle of the eyeball and the sphincter muscle of the iris. Through the ciliary ganglion the oculomotor nerve controls the smooth muscle (ciliary muscle) responsible for adjustment of the lens for near vision and the smooth muscle (sphincter muscle of iris) responsible for constriction of the pupil.

The sensory portion of the oculomotor nerve consists of fibers from proprioceptors in the eyeball muscles supplied by the nerve to the midbrain. These fibers convey nerve impulses related to muscle sense (proprioception).

Trochlear (IV)

The **trochlear** (TRŌK-lē-ar) **(IV) nerve** (*trokhileia* = pulley) is a mixed cranial nerve. It is the smallest of the 12 cranial nerves and is also unique because it is the only cranial nerve to arise from the posterior aspect of the brain stem. The motor portion originates in a nucleus in the midbrain, and axons from the nucleus pass through the superior orbital fissure of the orbit (Figure 18.17b). The motor fibers innervate the superior oblique muscle of the eyeball, another extrinsic eyeball muscle. The trochlear nerve controls movement of the eyeball.

The sensory portion of the trochlear nerve consists of fibers that run from proprioceptors in the superior oblique muscle to a nucleus of the nerve in the midbrain. The sensory portion is responsible for muscle sense, the awareness of the activity of muscles.

Trigeminal (V)

The **trigeminal (V) nerve** (*tri;* = three; *geminus* = twin) is a mixed cranial nerve and the largest of the cranial nerves. As indicated by its name, the trigeminal nerve has three branches: ophthalmic (of-THAL-mik), maxillary, and mandibular (Figure 18.18 on page 574). The trigeminal nerve contains two roots on the ventrolateral surface of the pons. The large sensory root has a swelling called the **semilunar (Gasserian) ganglion** located in a fossa on the inner surface of the petrous portion of the temporal bone. The ganglion contains cell bodies of most of the primary sensory neurons. From this ganglion, the **ophthalmic branch** (*ophthalmos* = eye), the smallest branch, enters the orbit via the superior orbital fissure; the **maxillary branch** (*maxilla* = upper jaw bone), intermediate in size between the ophthalmic and mandibular branches, enters the foramen rotundum; and the **mandibular branch** (*mandible* = lower jaw bone), the largest branch, exits through the foramen ovale. The smaller motor root originates in a nucleus in the pons. The motor fibers join the mandibular branch and supply the muscles of mastication. These motor fibers, which control chewing movements, constitute the motor portion of the trigeminal nerve and nerve to the mylohyoid and anterior belly of the digastric muscles.

The sensory portion of the trigeminal nerve delivers nerve impulses related to touch, pain, and temperature and consists of the ophthalmic, maxillary, and mandibular branches. The ophthalmic branch receives sensory fibers from the skin over the upper eyelid, eyeball, lacrimal glands, upper part of the nasal cavity, side of the nose, forehead, and anterior half of the scalp. The maxillary branch receives sensory fibers from the mucosa of the nose, palate, parts of the pharynx, upper teeth, upper lip, and lower eyelid. The mandibular branch transmits sensory fibers from the anterior two-thirds of the tongue (not taste), cheek and mucosa deep to it, lower teeth, skin over the mandible and side of the head anterior to the ear, and mucosa of the floor of the mouth. Sensory fibers from the three branches of the trigeminal nerve enter the semilunar ganglion and terminate in nuclei in the pons. There are also sensory fibers from proprioceptors in the muscles of mastication.

The inferior alveolar nerve, a branch of the mandibular nerve, supplies all the teeth in one-half of the mandible and is frequently anesthetized in dental procedures. The same procedure will anesthetize the lower lip because the mental nerve is a branch of the inferior alveolar nerve. Because the lingual nerve runs very close to the inferior alveolar nerve near the mental foramen, it is often anesthetized at the same time. For anesthesia to the upper teeth, the superior alveolar nerve endings, which are branches of the maxillary branch, are blocked by inserting the needle beneath the mucous membrane; the anesthetic solution is then infiltrated slowly throughout the area of the roots of the teeth to be treated. ■

Abducens (VI)

The **abducens** (ab-DŪ-sens) **(VI) nerve** (*abducere* = to lead away) is a mixed cranial nerve that originates from a nucleus in the pons (see Figure 18.17c). The motor fibers extend from the nucleus to the lateral rectus muscle of the eyeball, an extrinsic eyeball muscle. Nerve impulses over the fibers bring about abduction of the eyeball (turn it laterally), the reason for the name of the nerve. The sensory fibers run from proprioceptors in the lateral rectus muscle to the pons and mediate muscle sense. The abducens nerve reaches the lateral rectus muscle through the superior orbital fissure of the orbit.

Figure 18.17 / Oculomotor (III), trochlear (IV), and abducens (VI) nerves.

The oculomotor nerve has the widest distribution among extrinsic eye muscles.

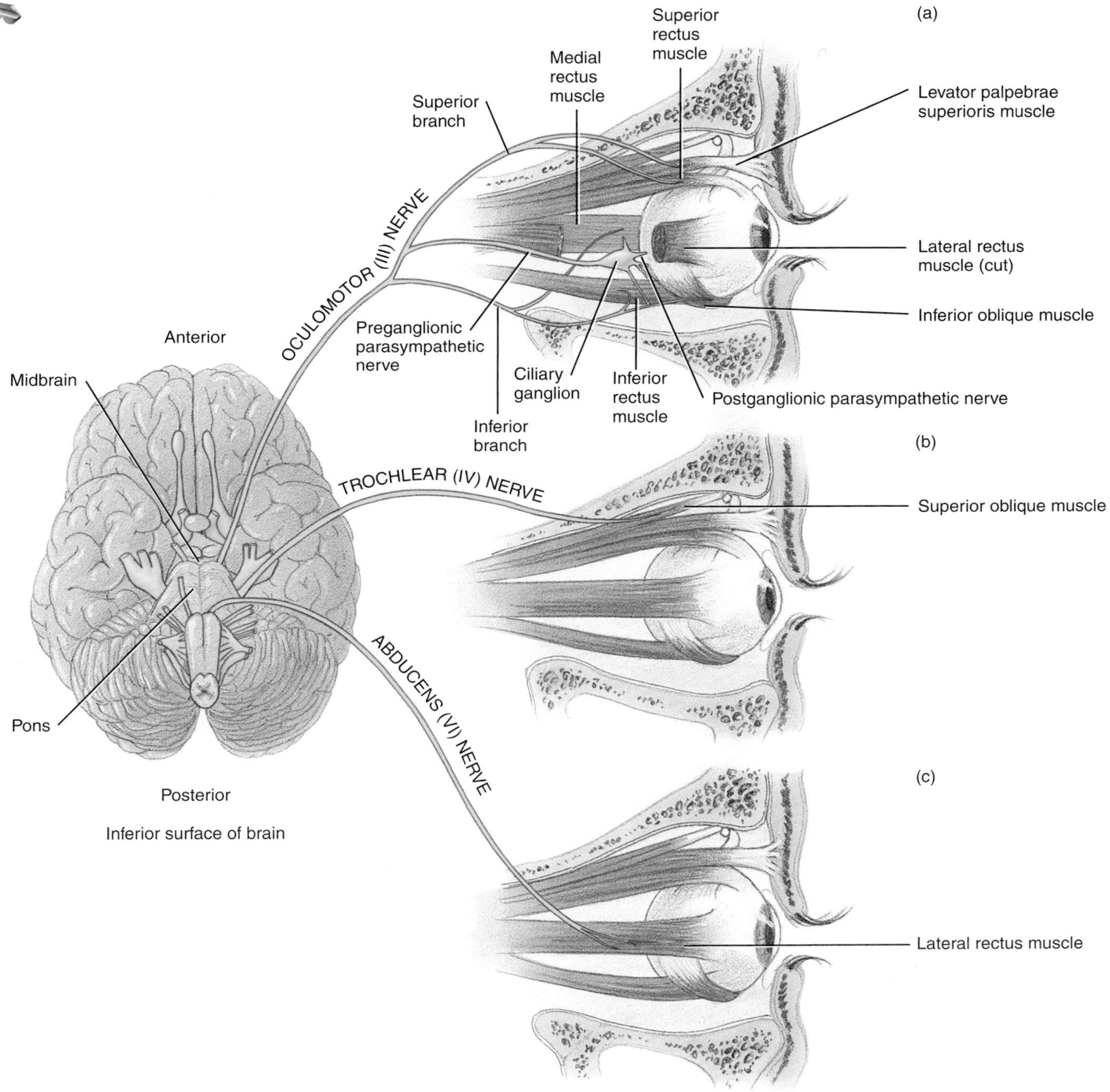

(a)
Superior rectus muscle
Medial rectus muscle
Superior branch
Levator palpebrae superioris muscle
OCULOMOTOR (III) NERVE
Lateral rectus muscle (cut)
Inferior oblique muscle
Preganglionic parasympathetic nerve
Anterior
Midbrain
Ciliary ganglion
Inferior rectus muscle
Postganglionic parasympathetic nerve
Inferior branch
(b)
TROCHLEAR (IV) NERVE
Superior oblique muscle
ABDUCENS (VI) NERVE
Pons
(c)
Posterior
Inferior surface of brain
Lateral rectus muscle

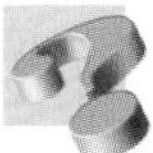

Which branch of the oculomotor nerve is distributed to the superior rectus muscle? Which is the smallest cranial nerve?

Figure 18.18 / Trigeminal (V) nerve.

The three branches of the trigeminal nerve leave the cranium through the superior orbital fissure, foramen rotundum, and foramen ovale.

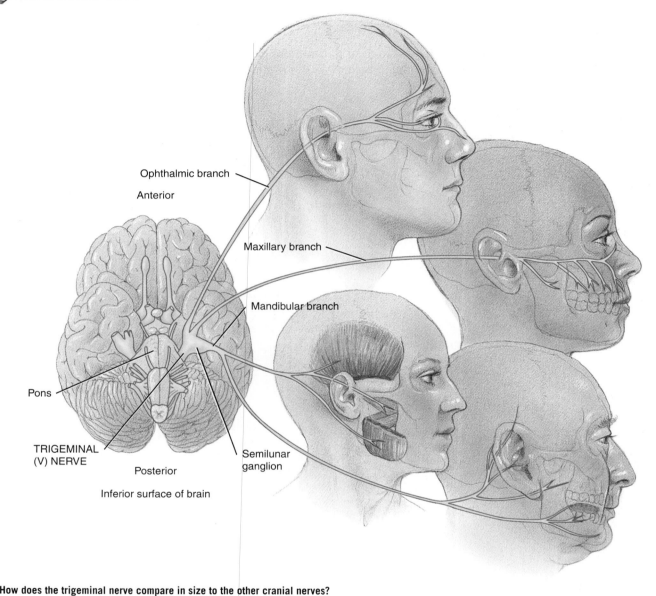

How does the trigeminal nerve compare in size to the other cranial nerves?

Facial (VII)

The **facial (VII) nerve** (*facies* = face) is a mixed cranial nerve. Its motor fibers originate from a nucleus in the pons, enter the petrous portion of the temporal bone, and are distributed to facial, scalp, and neck muscles (Figure 18.19). Nerve impulses along these fibers cause contraction of the muscles of facial expression and posterior belly of the digastric and stylohyoid muscles. Some motor fibers are also distributed via parasympathetics to the lacrimal, nasal, and palatine glands, and to saliva-producing sublingual and submandibular glands.

The sensory fibers extend from the taste buds of the anterior two-thirds of the tongue to the **geniculate ganglion,** a swelling of the facial nerve. From there, the fibers pass to a nucleus in the pons, which sends fibers to the thalamus for relay to the gustatory area of the cerebral cortex, which receives impulses for taste. The sensory portion of the facial nerve also conveys deep general sensations from the face. There are also sensory fibers from proprioceptors in the muscles of the face and scalp.

Vestibulocochlear (VIII)

The **vestibulocochlear** (ves-tib′-yū-lō-KŌK-lē-ar) **(VIII) nerve** (*vestibulum* = vestibule; *kokhlos* = land snail), formerly known as the **acoustic** or **auditory nerve,** is another mainly sensory cranial nerve. It consists of two branches: the cochlear

Figure 18.19 / Facial (VII) nerve.

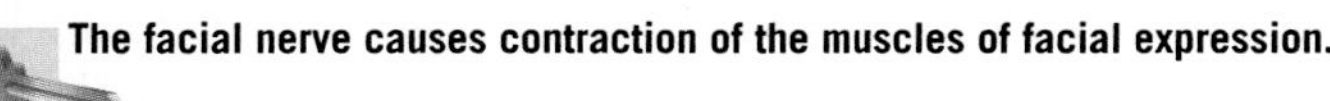

The facial nerve causes contraction of the muscles of facial expression.

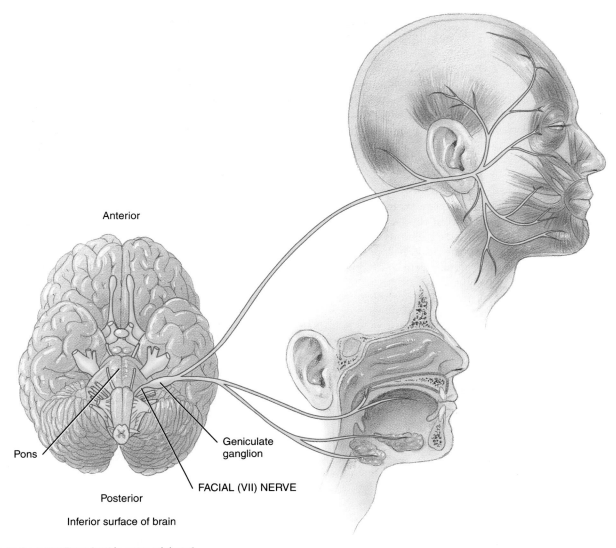

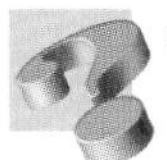

Where do the motor fibers for this nerve originate?

(auditory) branch and the vestibular branch (see Figure 20.13b on page 628). The **cochlear branch,** which conveys nerve impulses associated with hearing, arises in the spiral organ (organ of Corti) in the cochlea of the internal ear. The cell bodies of the cochlear branch are located in the **spiral ganglion** of the cochlea. From there the axons synapse in nuclei in the medulla oblongata. The next neurons synapse in the inferior colliculus and pass on to the medial geniculate nucleus in the thalamus. Ultimately, the fibers synapse with neurons that relay the nerve impulses to the auditory areas of the cerebral cortex.

The **vestibular branch** arises in the semicircular canals, the saccule, and the utricle of the inner ear. Fibers from these structures extend to the **vestibular ganglion,** where the cell bodies are contained (see Figure 20.13b on page 628). The fibers of the neurons with cell bodies in the vestibular ganglion terminate in nuclei in the thalamus. Some fibers also enter the cerebellum. The vestibular branch transmits impulses related to equilibrium. Almost none of the vestibular information reaches the cerebral cortex and thus is not perceived. It is primarily used for reflex responses.

Glossopharyngeal (IX)

The **glossopharyngeal** (glos′ō-fa-RIN-jē-al) **(IX) nerve** (*glossa* = tongue; *pharyngeal* = pharynx) is a mixed cranial nerve. Its motor fibers originate in nuclei in the medulla (Figure

18.20). The nerve exits the skull through the jugular foramen. The motor fibers are distributed to the stylopharyngeus muscle and parotid gland (parasympathetic) to mediate the secretion of saliva. Some of the cell bodies of parasympathetic motor neurons are located in the otic ganglion.

The sensory fibers of the glossopharyngeal nerve supply the pharynx and taste buds of the posterior third of the tongue. Some sensory fibers also originate from receptors in the carotid sinus, which assumes a major role in blood pressure regulation. The sensory fibers terminate in the medulla. There are also sensory fibers from proprioceptors in the muscles innervated by this nerve. Sensory neuron cell bodies are located in the superior and inferior ganglia.

Vagus (X)

The **vagus (X) nerve** (*vagus* = vagrant or wandering) is a mixed cranial nerve that is widely distributed from the head and neck into the thorax and abdomen (Figure 18.21). The nerve derives its name from its wide distribution. In the neck, it is positioned medial and posterior to the internal jugular vein and common carotid artery. Motor fibers of the vagus nerve originate in nuclei of the medulla and terminate in the muscles of the respiratory passageways, lungs, heart, esophagus, stomach, small intestine, most of the large intestine, and gallbladder. Nerve impulses along the motor fibers generate visceral, cardiac, and skeletal muscle movement. Parasympathetic fibers innervate involuntary muscles and glands of the gastrointestinal tract.

Sensory fibers of the vagus nerve, which are more numerous than the motor fibers and carry mostly visceral sensory information, supply essentially the same structures as the motor fibers. They convey nerve impulses for various sensations from the larynx, the viscera, and the ear. The fibers terminate in the medulla and pons. There are also sensory fibers from proprioceptors in the muscles supplied by this nerve.

Accessory (XI)

The **accessory (XI) nerve** (*accessorius* = assisting; formerly the **spinal accessory nerve**) is a mixed cranial nerve. It differs from all other cranial nerves in that it originates from *both* the brain stem and the spinal cord (Figure 18.22 on page 578). The **cranial portion** originates from nuclei in the medulla, passes through the jugular foramen, and supplies the voluntary muscles of the pharynx, larynx, and soft palate that are used in swallowing. The **spinal portion** originates in the anterior gray horn of the first

Figure 18.20 / Glossopharyngeal (IX) nerve.

Sensory fibers of the glossopharyngeal nerve supply the taste buds.

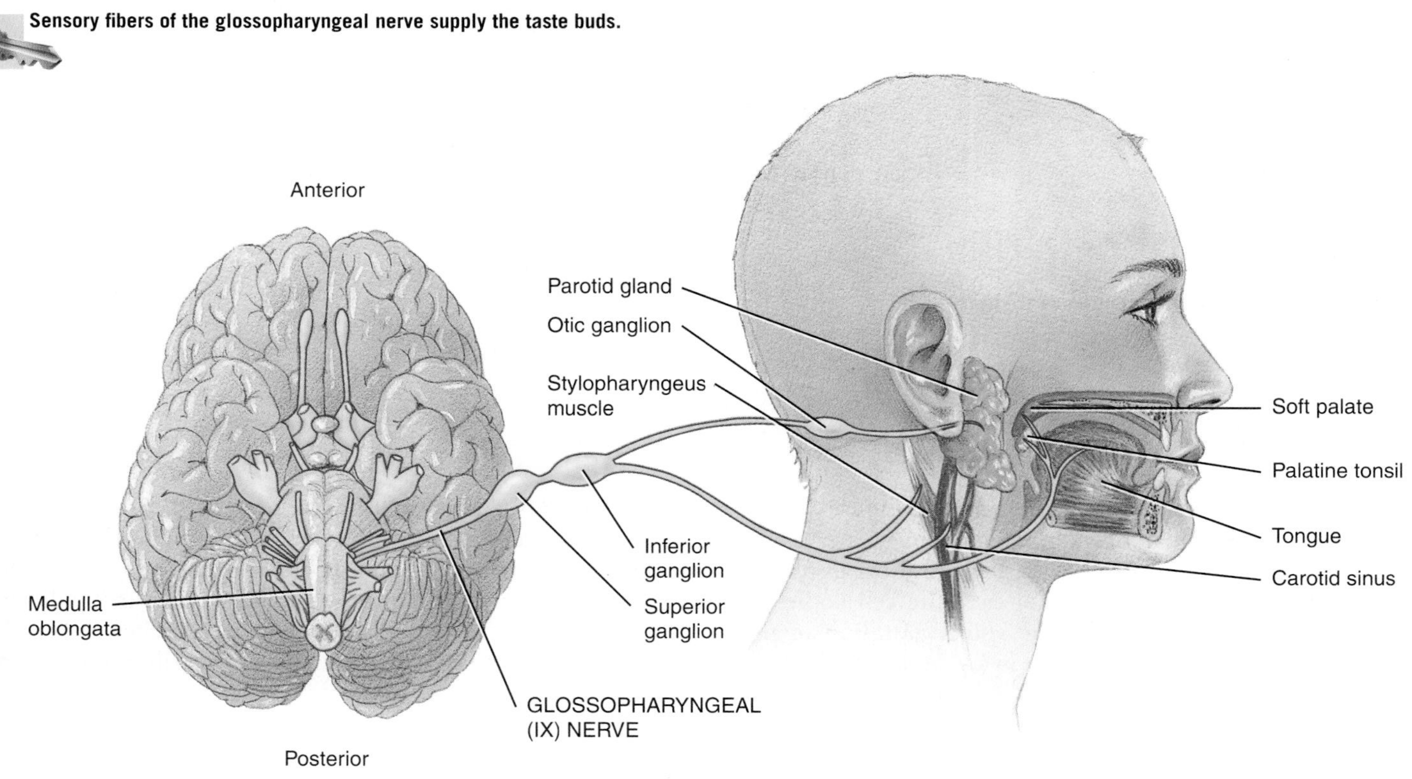

Through which foramen does the glossopharyngeal nerve exit the skull?

Figure 18.21 / Vagus (X) nerve.

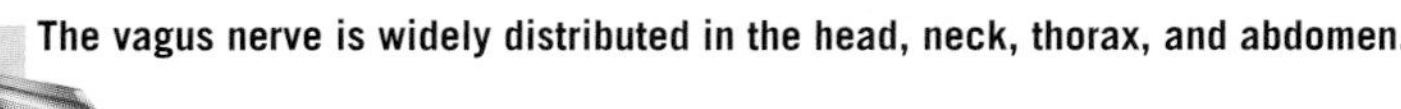

The vagus nerve is widely distributed in the head, neck, thorax, and abdomen.

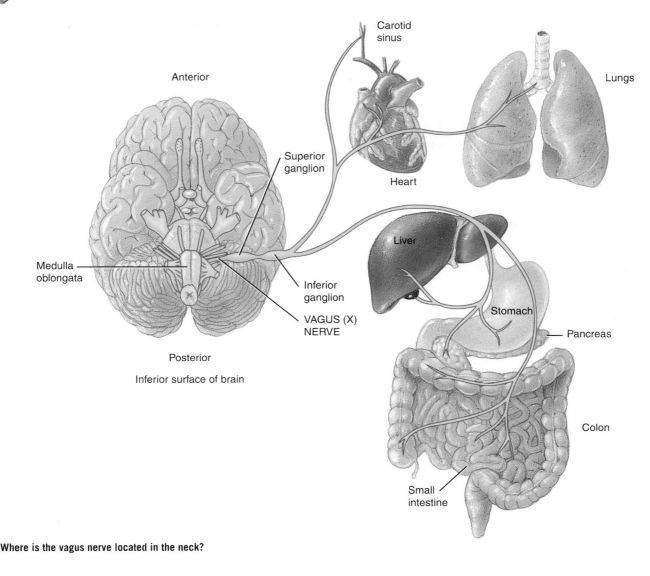

Where is the vagus nerve located in the neck?

five segments of the cervical portion of the spinal cord. The fibers from the segments join, enter the foramen magnum, and exit through the jugular foramen along with the cranial portion.

The spinal portion conveys motor impulses to the sternocleidomastoid and trapezius muscles to coordinate head movements. The sensory fibers originate from proprioceptors in the muscles supplied by its motor neurons and terminate in upper cervical posterior root ganglia.

Hypoglossal (XII)

The **hypoglossal (XII) nerve** (*hypo* = below; *glossa* = tongue) is a mixed cranial nerve. The motor fibers originate in a nucleus in the medulla, pass through the hypoglossal canal, and supply the muscles of the tongue (Figure 18.23). These fibers conduct nerve impulses related to speech and swallowing.

The sensory portion of the hypoglossal nerve consists of fibers originating from proprioceptors in the tongue muscles and terminating in the medulla. The sensory fibers conduct nerve impulses for muscle sense.

A summary of cranial nerves and clinical applications related to their dysfunction is presented in Table 18.2 on page 579.

- ✓ Define a cranial nerve. How are cranial nerves named and numbered?
- ✓ Distinguish between a mixed and a sensory cranial nerve.
- ✓ Devise a test that could reveal damage to each cranial nerve.

Figure 18.22 / Accessory (XI) nerve.

The accessory nerve exits the cranium through the jugular foramen.

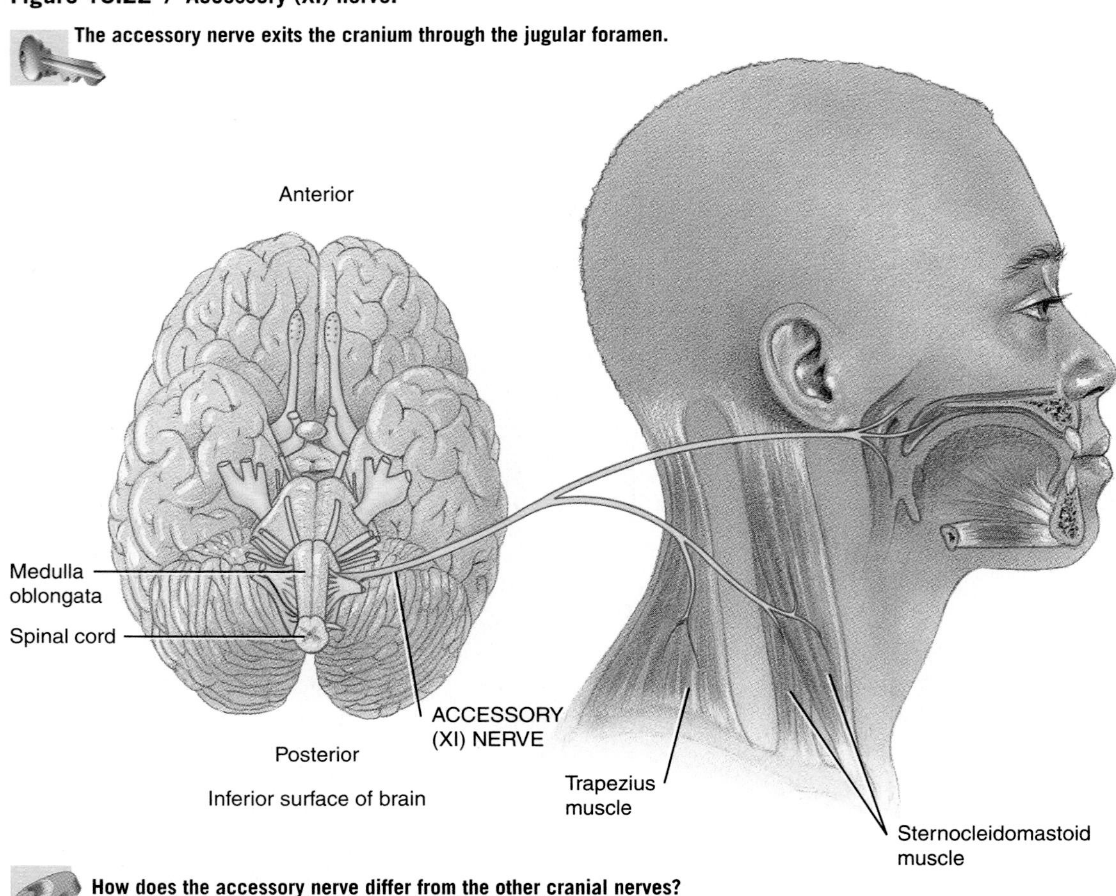

How does the accessory nerve differ from the other cranial nerves?

Figure 18.23 / Hypoglossal (XII) nerve.

The hypoglossal nerve exits the cranium through the hypoglossal canal.

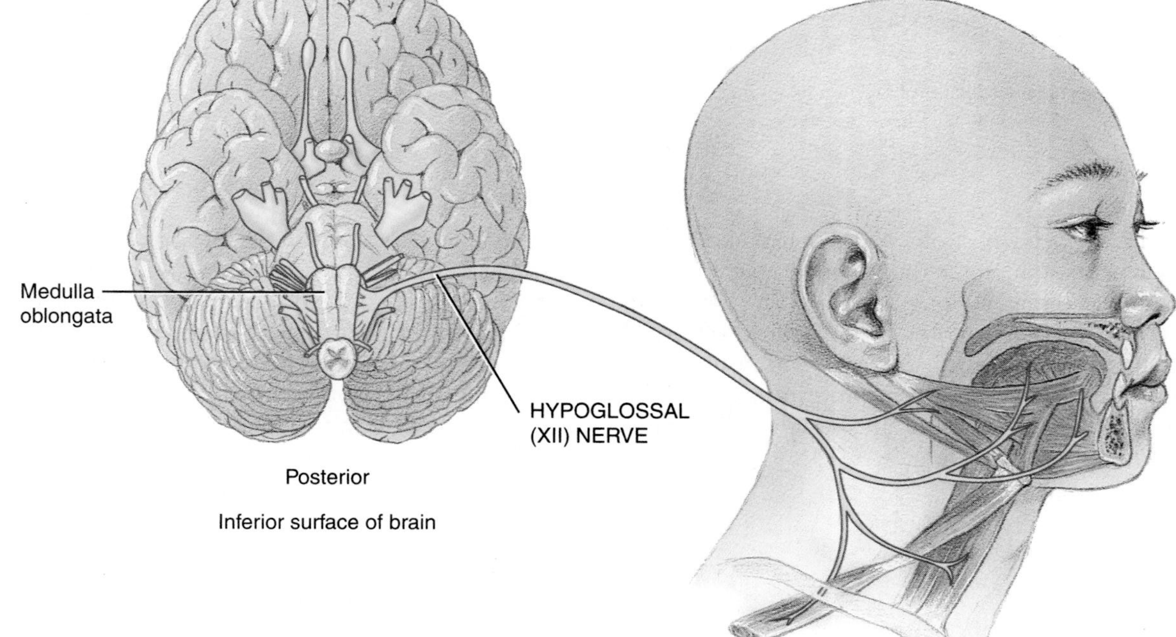

What important motor functions are related to the hypoglossal nerve?

Table 18.2 Summary of Cranial Nerves*

Number and Name	Type and Location	Function and Clinical Application
Cranial Nerve I: **Olfactory** (ol-FAK-tō-rē; *olfact-* = to smell) 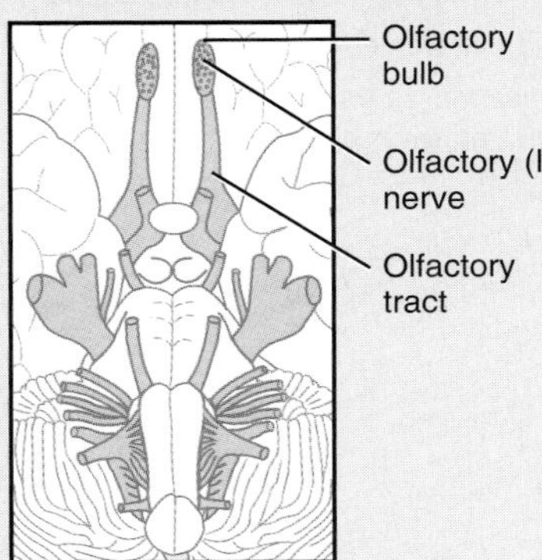	**Sensory** Arises in olfactory mucosa, passes through foramina in the cribriform plate of the ethmoid bone, and ends in the olfactory bulb. The olfactory tract extends via two pathways to olfactory areas of cerebral cortex.	*Function:* Smell. *Clinical application:* Loss of the sense of smell, called *anosmia,* may result from head injuries in which the cribriform plate of the ethmoid bone is fractured and from lesions along the olfactory pathway.
Cranial Nerve II: **Optic** (OP-tik; *opti-* = the eye, vision) 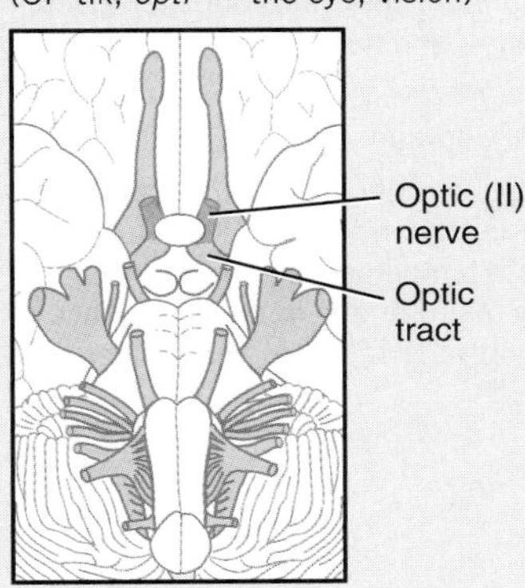	**Sensory** Arises in the retina of the eye, passes through the optic foramen, forms the optic chiasm and then the optic tracts, and terminates in the lateral geniculate nuclei of thalamus. From the thalamus, fibers extend to the primary visual area (area 17) of the cerebral cortex.	*Function:* Vision *Clinical application:* Fractures in the orbit, damage along the visual pathway, and diseases of the nervous system may result in visual field defects and loss of visual acuity. Blindness due to a defect in or loss of one or both eyes is called *anopia.*
Cranial Nerve III: **Oculomotor** (ok′-ū-lō-MŌ-tor; *oculo-* = eye; *-motor* = mover) 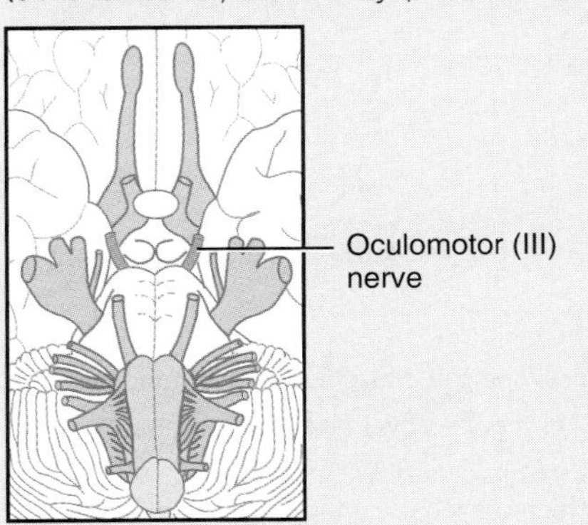	**Mixed (mainly motor)** *Sensory portion:* Consists of fibers from proprioceptors in eyeball muscles that pass through the superior orbital fissure and terminate in the midbrain. *Motor portion:* Originates in the midbrain and passes through the superior orbital fissure. Somatic fibers innervate the levator palpebrae superioris muscles of the upper eyelid and four extrinsic eyeball muscles (superior rectus, medial rectus, inferior rectus, and inferior oblique). Parasympathetic fibers innervate the ciliary muscle of eyeball and the sphincter muscle of the iris.	*Sensory function:* Muscle sense (proprioception). *Motor function:* Movement of eyelid and eyeball, accommodation of lens for near vision, and constriction of pupil. *Clinical application:* Nerve damage causes *strabismus* (a deviation of the eye in which both eyes do not fix on the same object), *ptosis* (drooping) of the upper eyelid, dilation of the pupil, movement of the eyeball downward and outward on the damaged side, loss of accommodation for near vision, and *diplopia* (double vision).

(continues)

*A mnemonic device that can be used to remember the names of the nerves is: "**O**h, **o**h, **o**h, **t**o **t**ouch **a**nd **f**eel **v**ery **g**reen **v**egetables—**AH**!" The boldfaced letters correspond to the initial letter of each pair of cranial nerves.

Table 18.2 Summary of Cranial Nerves (continued)

Number and Name	Type and Location	Function and Clinical Application
Cranial Nerve IV: **Trochlear** (TRŌK-lē-ar; *trochle-* = a pulley) 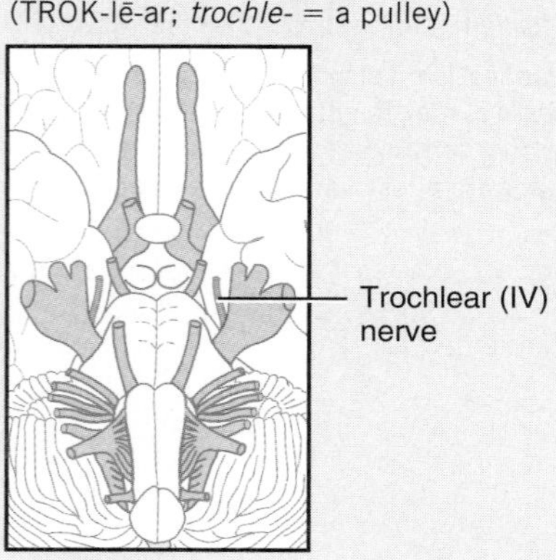	**Mixed (mainly motor)** *Sensory portion:* Consists of fibers from proprioceptors in the superior oblique muscles, which pass through the superior orbital fissure and terminate in the midbrain. *Motor portion:* Originates in the midbrain and passes through the superior orbital fissure. Innervates the superior oblique muscle, an extrinsic eyeball muscle.	*Sensory function:* Muscle sense (proprioception). *Motor function:* Movement of the eyeball. *Clinical application:* In trochlear nerve paralysis, diplopia and strabismus occur.
Cranial Nerve V: **Trigeminal** (trī-JEM-i-nal; = triple, for its three branches) 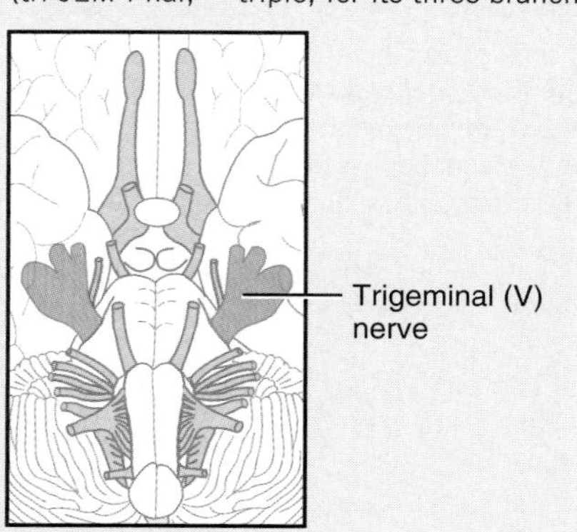	**Mixed** *Sensory portion:* Consists of three branches, all of which end in the pons. (1) The **ophthalamic** (*ophthalm-* = the eye) branch contains fibers from the skin over the upper eyelid, eyeball, lacrimal glands, nasal cavity, side of nose, forehead, and anterior half of scalp that pass through superior orbital fissure. (2) The **maxillary** (*maxilla* = upper jaw bone) branch contains fibers from the mucosa of the nose, palate, parts of the pharynx, upper teeth, upper lip, and lower eyelid that pass through the foramen rotundum. (3) The **mandibular** (*mandibula* = lower jaw bone) branch contains fibers from the anterior two-thirds of the tongue (somatic sensory fibers but not fibers for the special sense of taste), the lower teeth, skin over mandible, cheek and mucosa deep to it, and side of head in front of ear that pass through the foramen ovale. *Motor portion:* Is part of the mandibular branch, which originates in the pons, passes through the foramen ovale, and innervates muscles of mastication (masseter, temporalis, medial pterygoid, lateral pterygoid, anterior belly of digastric, and mylohyoid muscles).	*Sensory function:* Conveys sensations for touch, pain, temperature, and muscle sense (proprioception). *Motor function:* Chewing. *Clinical application:* Injury results in paralysis of the muscles of mastication and a loss of the sensations of touch, temperature, and proprioception. *Neuralgia* (pain) of one or more branches of the trigeminal nerve is called *trigeminal neuralgia (tic douloureux).*
Cranial Nerve VI: **Abducens** (ab-DŪ-senz; *ab-* = away; *-ducens* = to lead) 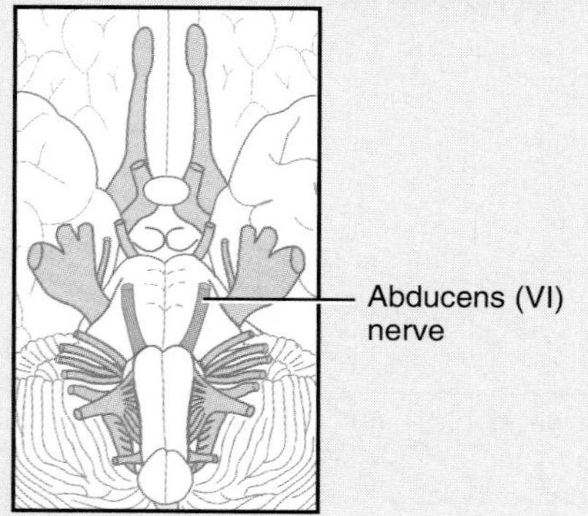	**Mixed (mainly motor)** *Sensory portion:* Consists of fibers from proprioceptors in the lateral rectus muscle that pass through the superior orbital fissure and end in the pons. *Motor portion:* Originates in the pons, passes through the superior orbital fissure, and innervates the lateral rectus muscle, an extrinsic eyeball muscle.	*Sensory function:* Muscle sense (proprioception). *Motor function:* Movement of the eyeball. *Clinical application:* With damage to this nerve, the affected eyeball cannot move laterally beyond the midpoint, and the eye usually is directed medially.

Number and Name	Type and Location	Function and Clinical Application
Cranial Nerve VII: **Facial** (FĀ-shal; = face) 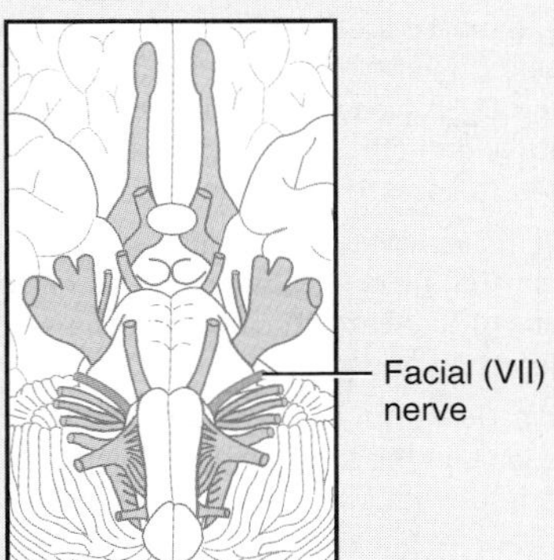	**Mixed** *Sensory portion:* Arises from taste buds on the anterior two-thirds of the tongue, passes through the stylomastoid foramen, and ends in the geniculate ganglion, a nucleus in pons. From there, fibers extend to the thalamus and then to the gustatory areas of cerebral cortex. Also contains fibers from proprioceptors in muscles of the face and scalp. *Motor portion:* Originates in the pons and passes through the stylomastoid foramen. Somatic fibers innervate facial, scalp, and neck muscles. Parasympathetic fibers innervate lacrimal, sublingual, submandibular, nasal, and palatine glands.	*Sensory function:* Muscle sense (proprioception) and taste. *Motor function:* Facial expression and secretion of saliva and tears. *Clinical application:* Injury produces *Bell's palsy* (paralysis of the facial muscles), loss of taste, decreased salivation, and loss of ability to close the eyes, even during sleep.
Cranial Nerve VIII: **Vestibulocochlear** (vest-tib-yū-lō-KŌK-lē-ar; *vestibulo-* = small cavity; *-cochlear* = a spiral, snail-like) 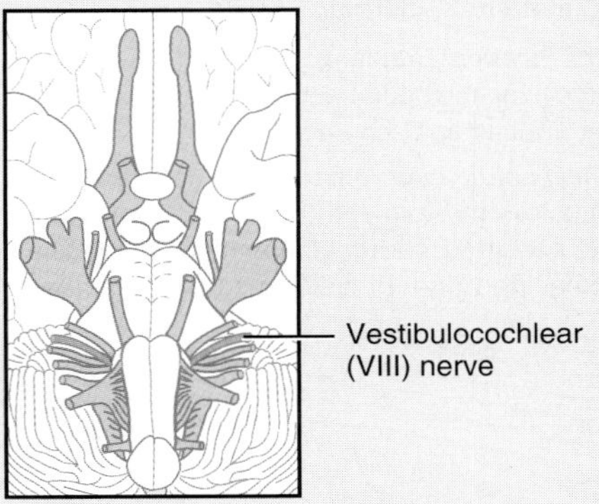	**Mixed (mainly sensory)** *Vestibular branch, sensory portion:* Arises in the semicircular canals, saccule, and utricle and forms the vestibular ganglion. Fibers end in the pons and cerebellum. *Vestibular branch, motor portion:* Fibers innervate hair cells of the semicircular canals, saccule, and utricle. *Cochlear branch, sensory portion:* Arises in the spiral organ (organ of Corti), forms the spiral ganglion, passes through nuclei in the medulla, and ends in the thalamus. Fibers synapse with neurons that relay impulses to the primary auditory area (areas 41 and 42) of the cerebral cortex. *Cochlear branch, motor portion:* Originates in the pons and terminates on hair cells of the spiral organ.	*Vestibular branch sensory function:* Conveys impulses associated with equilibrium. *Vestibular branch motor function:* Adjusts sensitivity of hair cells. *Cochlear branch sensory function:* Conveys impulses for hearing. *Cochlear branch motor function:* May modify function of hair cells by altering their transmission and mechanical response to sound. *Clinical application:* Injury to the vestibular branch may cause *vertigo* (a subjective feeling of rotation), *ataxia* (muscular incoordination), and *nystagmus* (involuntary rapid movement of the eyeball). Injury to the cochlear branch may cause *tinnitus* (ringing in the ears) or deafness.
Cranial Nerve IX: **Glossopharyngeal** (glos'-ō-fa-RIN-jē-al; *glosso-* = tongue; *-pharyngeal* = throat) 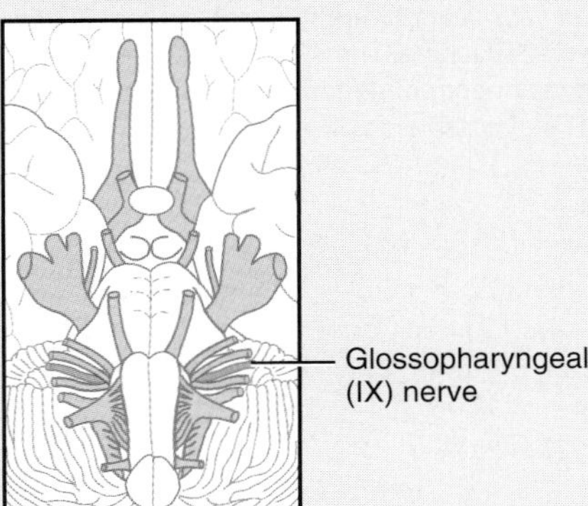	**Mixed** *Sensory portion:* Consists of fibers from taste buds and somatic sensory receptors on posterior one-third of the tongue, from proprioceptors in swallowing muscles supplied by the motor portion, and from stretch receptors in carotid sinus and chemoreceptors in carotid body near the carotid arteries. Fibers pass through the jugular foramen and end in the medulla. *Motor portion:* Originates in the medulla and passes through the jugular foramen. Somatic fibers innervate the stylopharyngeus muscle. Parasympathetic fibers innervate the parotid (salivary) gland.	*Sensory function:* Taste and somatic sensations (touch, pain, temperature) from posterior third of tongue; muscle sense (proprioception) in swallowing muscles; monitoring of blood pressure; monitoring of O_2 and CO_2 in blood for regulation of breathing rate and depth. *Motor function:* Elevates the pharynx during swallowing and speech; stimulates secretion of saliva. *Clinical application:* Injury causes difficulty in swallowing, reduced secretion of saliva, loss of sensation in the throat, and loss of taste sensation.

(continues)

Table 18.2 Summary of Cranial Nerves (continued)

Number and Name	Type and Location	Function and Clinical Application
Cranial Nerve X: **Vagus** (VĀ-gus; *vagus* = vagrant or wandering) 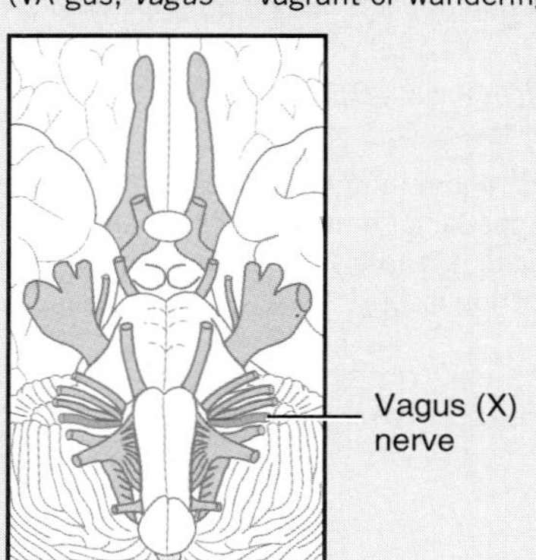	**Mixed** *Sensory portion:* Consists of fibers from small number of taste buds in the epiglottis and pharynx, proprioceptors in muscles of the neck and throat, stretch receptors and chemoreceptors in carotid sinus and carotid body near the carotid arteries, chemoreceptors in aortic body near arch of the aorta, and visceral sensory receptors in most organs of the thoracic and abdominal cavities. Fibers pass through the jugular foramen and end in the medulla and pons. *Motor portion:* Originates in medulla and passes through the jugular foramen. Somatic fibers innervate skeletal muscles in the throat and neck. Parasympathetic fibers innervate smooth muscle in the airways, esophagus, stomach, small intestine, most of large intestine, and gallbladder; cardiac muscle in the heart; and glands of the gastrointestinal (GI) tract.	*Sensory function:* Taste and somatic sensations (touch, pain, temperature) from epiglottis and pharynx; monitoring of blood pressure; monitoring of O_2 and CO_2 in blood for regulation of breathing rate and depth; sensations from visceral organs in thorax and abdomen. *Motor function:* Swallowing, coughing, and voice production; smooth muscle contraction and relaxation in organs of the GI tract; slowing of the heart rate; secretion of digestive fluids. *Clinical application:* Injury interrupts sensations from many organs in the thoracic and abdominal cavities, interferes with swallowing, paralyzes vocal cords, and causes heart rate to increase.
Cranial Nerve XI: **Accessory** (ak-SES-ō-rē; = assisting) 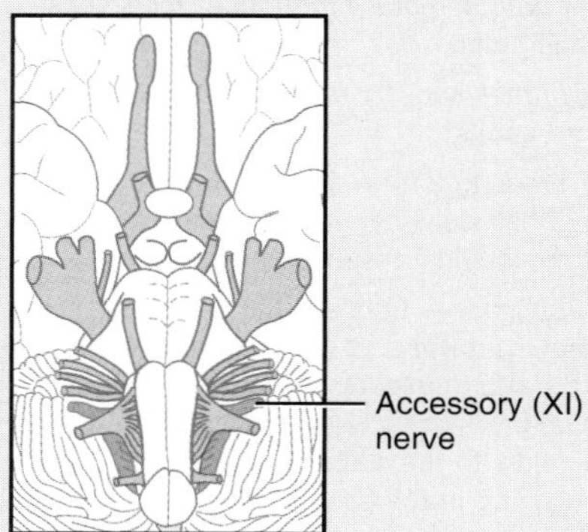	**Mixed (mainly motor)** *Sensory portion:* Consists of fibers from proprioceptors in muscles of the pharynx, larynx, and soft palate that pass through the jugular foramen. *Motor portion:* Consists of a cranial portion and a spinal portion. *Cranial portion* originates in the medulla, passes through the jugular foramen, and supplies muscles of the pharynx, larynx, and soft palate. *Spinal portion* originates in the anterior gray horn of the first five cervical segments of the spinal cord, passes through the jugular foramen, and supplies the sternocleidomastoid and trapezius muscles.	*Sensory function:* Muscle sense (proprioception). *Motor function:* Cranial portion mediates swallowing movements; spinal portion mediates movement of head and shoulders. *Clinical application:* If nerves are damaged, the sternocleidomastoid and trapezius muscles become paralyzed, with resulting inability to raise the shoulders and difficulty in turning the head.
Cranial Nerve XII: **Hypoglossal** (hī-pō-GLOS-al; *hypo-* = below; *-glossal* = tongue) 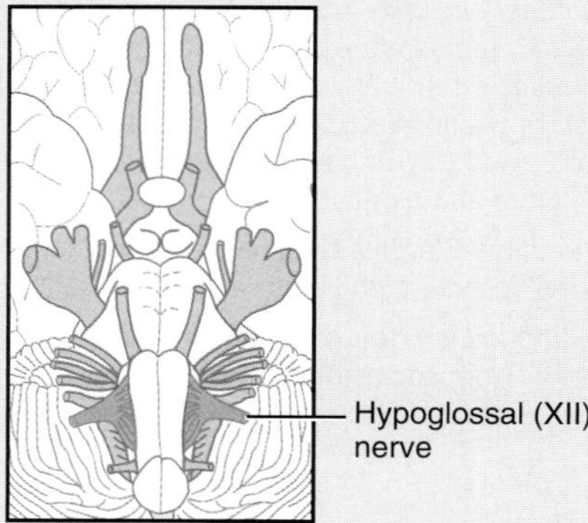	**Mixed (mainly motor)** *Sensory portion:* Consists of fibers from proprioceptors in tongue muscles that pass through the hypoglossal canal and end in the medulla. *Motor portion:* Originates in the medulla, passes through the hypoglossal canal, and supplies muscles of the tongue.	*Sensory function:* Muscle sense (proprioception). *Motor function:* Movement of tongue during speech and swallowing. *Clinical application:* Injury results in difficulty in chewing, speaking, and swallowing. The tongue, when protruded, curls toward the affected side, and the affected side atrophies.

DEVELOPMENTAL ANATOMY OF THE NERVOUS SYSTEM

Objective

- Describe how the parts of the brain develop.

The development of the nervous system begins early in the third week with a thickening of the **ectoderm** called the **neural plate** (Figure 18.24). The plate folds inward and forms a longitudinal groove, the **neural groove.** The raised edges of the neural plate are called **neural folds.** As development continues, the neural folds increase in height and meet to form a tube called the **neural tube.**

Three types of cell layers differentiate from the wall that encloses the neural tube. The outer or **marginal layer cells** develop into the *white matter* of the nervous system; the middle or **mantle layer cells** develop into the *gray matter*; and the inner or **ependymal layer cells** eventually form the *lining of the central canal of the spinal cord and ventricles of the brain.*

The **neural crest** is a mass of tissue between the neural tube and the skin ectoderm (Figure 18.24b). It differentiates and eventually forms the *posterior (dorsal) root ganglia of spinal nerves, spinal nerves, ganglia of cranial nerves, cranial nerves, ganglia of the autonomic nervous system, adrenal medulla,* and *meninges.*

At the end of the fourth week of embryonic development, when the neural tube forms from the neural plate, its anterior part develops into three enlarged areas called **primary vesicles.** These are the **prosencephalon** (prōs′-en-SEF-a-lon; *pros-* = before) or forebrain, **mesencephalon** (mes′ en-SEF-a-lon; *mes-* = middle) or midbrain, and **rhombencephalon** (rom′-

Figure 18.24 / Origin of the nervous system. (a) Dorsal view of an embryo in which the neural folds have partially united, forming the early neural tube. (b) Transverse sections through the embryo showing the formation of the neural tube.

The nervous system begins developing in the third week from a thickening of ectoderm called the neural plate.

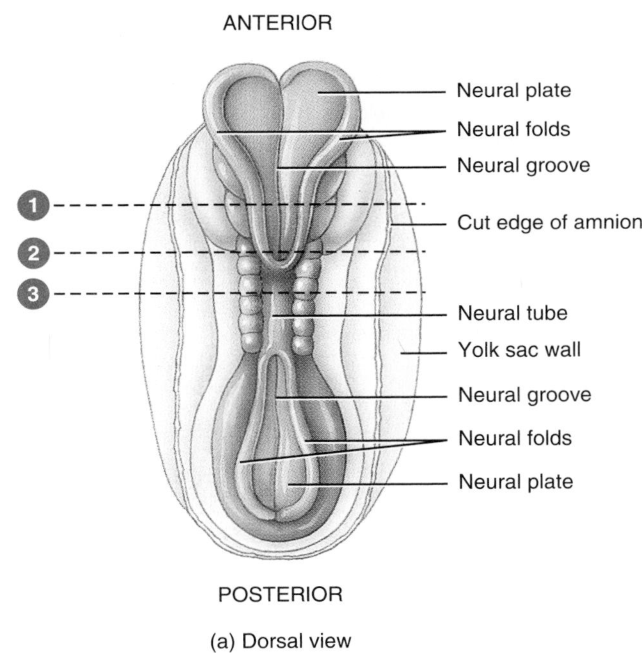

(a) Dorsal view

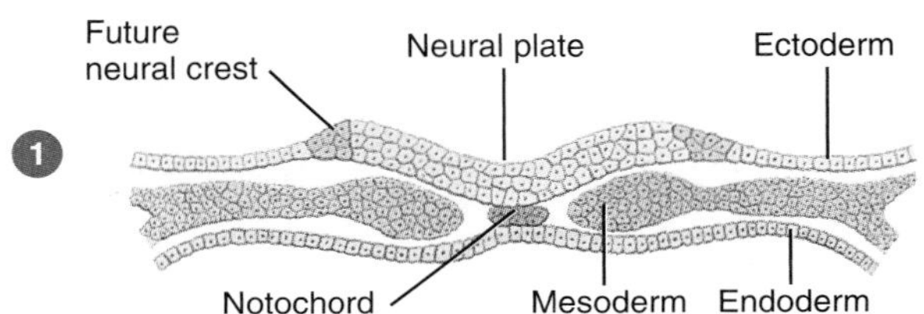

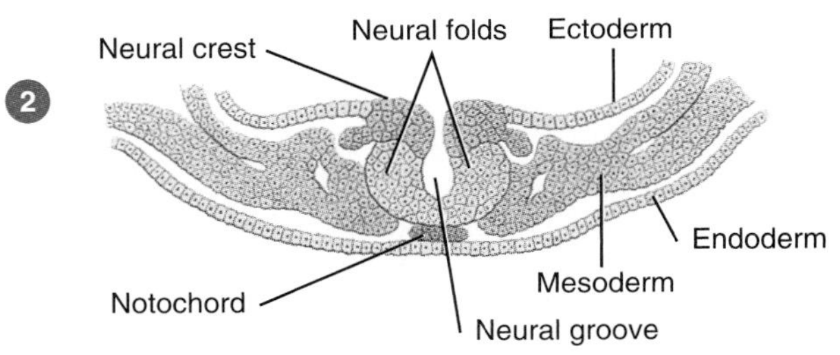

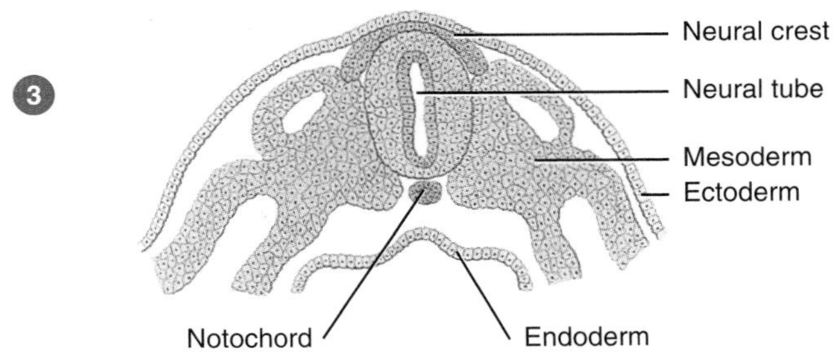

(b) Transverse sections

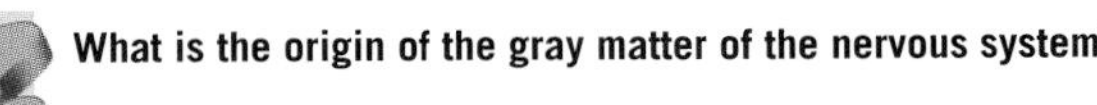
What is the origin of the gray matter of the nervous system?

ben-SEF-a-lon; *rhomb-* = behind) or hindbrain (Figure 18.25a). During the fifth week of development, the prosencephalon develops into two secondary brain vesicles called the **telencephalon** (tel′-en-SEF-a-lon; *tel-* = distant) and the **diencephalon** (dī-en-SEF-a-lon; *di-* = through) (Figure 18.25b). The rhombencephalon also develops into two secondary brain vesicles called the **metencephalon** (met′-en-SEF-a-lon; *met-* = after) and the **myelencephalon** (mī-el-en-SEF-a-lon; *myel-* = marrow). The area of the neural tube inferior to the myelencephalon gives rise to the *spinal cord.*

Including the mesencephalon, then, there are five secondary brain vesicles, which continue to develop as follows:

- The telencephalon develops into the *cerebral hemispheres* and their *basal ganglia* and houses the paired *lateral ventricles.*
- The diencephalon develops into the *epithalamus*, *thalamus*, *subthalamus*, *hypothalamus*, and *pineal gland* and houses the *third ventricle.*
- The mesencephalon develops into the *midbrain* and houses the *cerebral aqueduct.*
- The metencephalon becomes the *pons* and *cerebellum* and houses a portion of the *fourth ventricle.*
- The myelencephalon develops into the *medulla oblongata* and houses a portion of the *fourth ventricle.*

Two neural tube defects—spina bifida (see page 169) and anencephaly (absence of the skull and cerebral hemispheres)—are associated with low levels of folic acid, one of the B vitamins. The incidence of both disorders is greatly reduced when women who may become pregnant take folic acid supplements.

Figure 18.25 / Development of the brain and spinal cord.

The various parts of the brain develop from the primary brain vesicles.

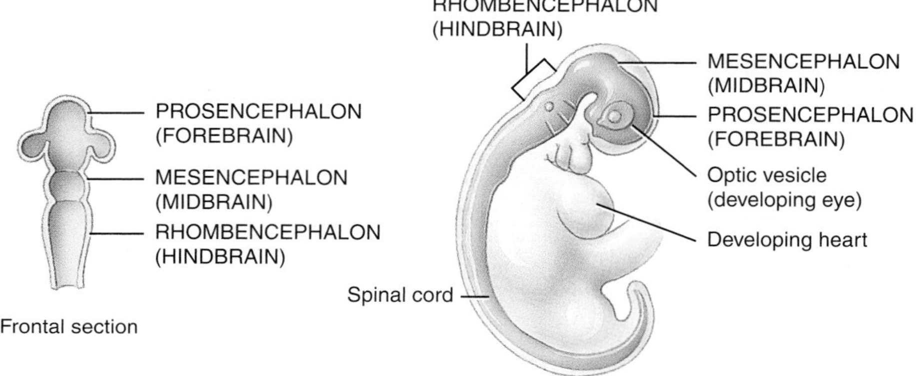

(a) 3–4 week embryo showing primary brain vesicles

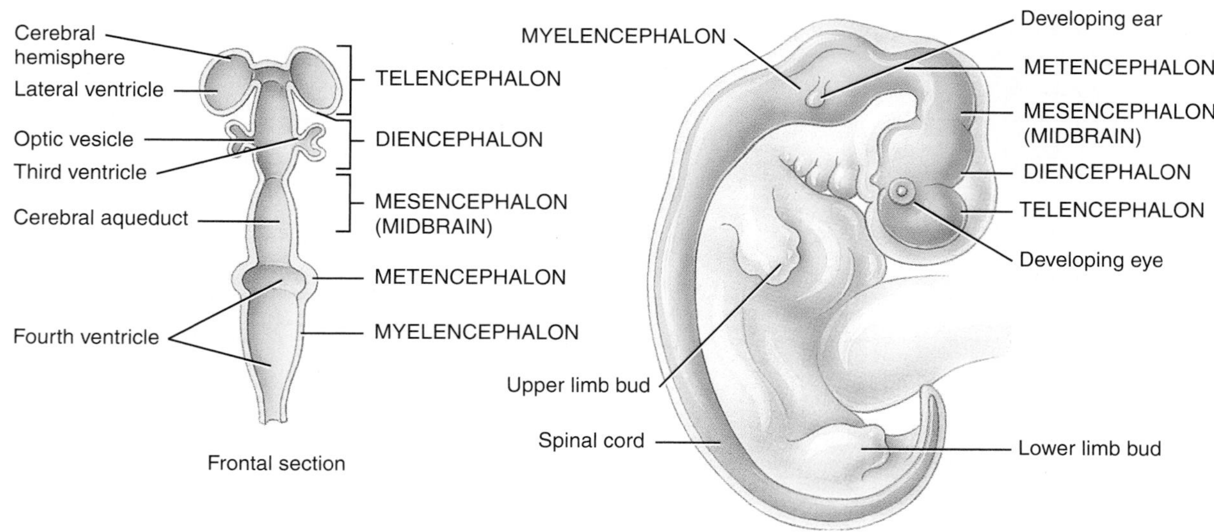

(b) 5–week embryo showing secondary brain vesicles

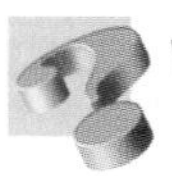

Which primary brain vesicle does not develop into a secondary brain vesicle?

AGING AND THE NERVOUS SYSTEM

Objective

- Describe the effects of aging on the nervous system.

The brain grows rapidly during the first few years of life. Growth is due mainly to an increase in the size of neurons already present, the proliferation and growth of neuroglia, the development of dendritic branches and synaptic contacts, and myelination of the various fiber tracts. From early adulthood onward, brain weight declines. By the time a person reaches 80, the brain weighs about 7% less than it did in the prime of life. Although the number of neurons present does not decrease very much, the number of synaptic contacts declines. Associated with the decrease in brain mass is a decreased capacity for sending nerve impulses to and from the brain; as a result, processing of information diminishes. Conduction velocity decreases, voluntary motor movements slow down, and reflex times increase.

✓ What parts of the brain develop from each primary brain vesicle?

✓ How is brain mass related to age?

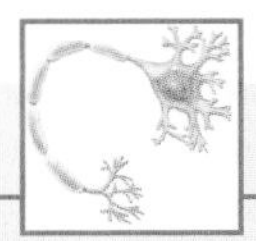

APPLICATIONS TO HEALTH

Cerebrovascular Accident

The most common brain disorder is a **cerebrovascular accident (CVA),** also called a **stroke** or **brain attack.** CVAs affect 500,000 people a year in the United States and represent the third leading cause of death, behind heart attacks and cancer. A CVA is characterized by abrupt onset of persisting neurological symptoms, such as paralysis or loss of sensation, that arise from destruction of brain tissue. Common causes of CVAs are intracerebral hemorrhage (from a blood vessel in the pia mater or brain), emboli (blood clots), and atherosclerosis (formation of cholesterol-containing plaques that block blood flow) of the cerebral arteries.

Among the risk factors implicated in CVAs are high blood pressure, high blood cholesterol, heart disease, narrowed carotid arteries, transient ischemic attacks (TIAs; discussed next), diabetes, smoking, obesity, and excessive alcohol intake.

A clot-dissolving drug called tissue plasminogen activator (t-PA) is now being used to open up blocked blood vessels in the brain. The drug is most effective when administered within three hours of the onset of the CVA, however, and is helpful only for CVAs due to a blood clot. Use of t-PA can decrease the permanent disability associated with CVAs by 50%.

Transient Ischemic Attack

A **transient ischemic attack (TIA)** is an episode of temporary cerebral dysfunction caused by impaired blood flow to the brain. Symptoms include dizziness, weakness, numbness, or paralysis in a limb or in one side of the body; drooping of one side of the face; headache; slurred speech or difficulty understanding speech; and a partial loss of vision or double vision. Sometimes nausea or vomiting also occurs. The onset of symptoms is sudden and reaches maximum intensity almost immediately. A TIA usually persists for 5–10 minutes and only rarely lasts as long as 24 hours; it leaves no persistent neurological deficits. The causes of the impaired blood flow that lead to TIAs are blood clots, atherosclerosis, and certain blood disorders.

It is estimated that about one-third of patients who experience a TIA will have a CVA within 5 years. Therapy for TIAs includes drugs such as aspirin that block the aggregation of blood components involved in clotting (platelets) and anticoagulants; cerebral artery bypass grafting; and carotid endarterectomy (removal of the cholesterol-containing plaques and inner lining of an artery).

Alzheimer Disease

Alzheimer disease (ALTZ-hī-mer) or **AD** is a disabling senile dementia—the loss of reason and ability to care for oneself—that afflicts about 11% of the population over age 65. In the United States, AD afflicts 4 million people and claims over 100,000 lives a year, making it the fourth leading cause of death among the elderly, after heart disease, cancer, and stroke.

Individuals with AD initially have trouble remembering recent events. They then become confused and forgetful, often repeating questions or getting lost while traveling to previously familiar places. Disorientation grows and memories of past events disappear, and episodes of paranoia, hallucination, or violent changes in mood may occur. As their minds continue to deteriorate, they lose their ability to read, write, talk, eat, or walk. The disease ultimately culminates in dementia. A person with AD usually dies of some complication that afflicts bedridden patients, such as pneumonia.

At autopsy, brains of AD victims show three distinct structural abnormalities:

1. *Loss of neurons that liberate acetylcholine.* A major center of neurons that liberate ACh is the nucleus basalis, which is inferior to the globus pallidus. Axons of these neurons project widely throughout the cerebral cortex and limbic system. Their destruction is a hallmark of Alzheimer disease.
2. *Beta-amyloid plaques,* clusters of abnormal proteins deposited outside neurons.
3. *Neurofibrillary tangles,* abnormal bundles of protein filaments inside neurons in affected brain regions.

A variety of risk factors for AD have been identified, although the causes of AD are still not clear. One risk factor for developing AD is a history of head injury. A similar dementia occurs in boxers, probably caused by repeated blows to the head.

Hereditary factors also increase one's AD risk. Three forms (alleles) of a gene on chromosome 19 code for a molecule called apolipoprotein E (apoE) that helps transport cholesterol in the blood. People who have one or two copies of the form that codes for **apolipoprotein E4 (apoE4)** have a much higher risk of developing AD (and an earlier age of onset) when compared with people who have genes for the other forms, apoE2 or apoE3. Although the way in which apoE exerts this influence is not known, some scientists suggest that apoE2 and apoE3 may have a protective effect that apoE4 lacks. In a small number of patients, AD is associated with mutations on chromosome 14 or 21. Genetic flaws do not explain all cases, however. Even in identical twins, one may have AD while the other does not. One study of the blood of a small number of Alzheimer patients revealed defects in certain types of plasma membrane K^+ channels in their platelets. A similar abnormality of neuronal ion channels could cause brain dysfunction. Drugs that inhibit acetylcholinesterase (AChE), the enzyme that inactivates ACh, improve alertness and behavior in about 5% of AD patients. Although more studies are needed, some evidence suggests that vitamin E (an antioxidant), estrogen, ibuprofen, and ginko biloba extract may have slight beneficial effects.

KEY MEDICAL TERMS ASSOCIATED WITH THE BRAIN AND THE CRANIAL NERVES

Agnosia (ag-NŌ-zē-a; *a-* = without; *-gnosia* = knowledge) Inability to recognize the significance of sensory stimuli such as sounds, sights, smells, tastes, and touch.

Apraxia (a-PRAK-sē-a; *-praxia* = coordinated) Inability to carry out purposeful movements in the absence of paralysis.

Delirium (de-LIR-ē-um; = off the track) Also called **acute confusional state (ACS).** A transient disorder of abnormal cognition and disordered attention accompanied by disturbances of the sleep-wake cycle and psychomotor behavior (hyperactivity or hypoactivity of movements and speech).

Dementia (de-MEN-shē-a; *de-* = away from; *-mentia* = mind) A mental disorder that results in permanent or progressive general loss of intellectual abilities, including impairment of memory, judgment, and abstract thinking and changes in personality.

Encephalitis (en′-sef-a-LĪ-tis) An acute inflammation of the brain caused by either a direct attack by any of several viruses or an allergic reaction to any of the many viruses that are normally harmless to the central nervous system. If the virus affects the spinal cord as well, the condition is called **encephalomyelitis.**

Lethargy (LETH-ar-jē) A condition of functional sluggishness.

Nerve block Loss of sensation in a region due to injection of a local anesthetic; an example is local dental anesthesia.

Neuralgia (nū-RAL-jē-a; *neur-* = nerve; *-algia* = pain) Attacks of pain along the entire course or a branch of a peripheral sensory nerve.

Stupor (STŪ-por) Unresponsiveness from which a patient can be aroused only briefly and only by vigorous and repeated stimulation.

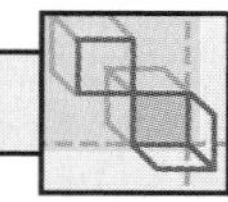

STUDY OUTLINE

Overview of Brain Organization and Blood Supply (p. 545)

1. The principal parts of the brain are the brain stem, cerebellum, diencephalon, and cerebrum.
2. The brain is protected by cranial bones and the cranial meninges.
3. The cranial meninges are continuous with the spinal meninges. From superficial to deep they are the dura mater, arachnoid, and pia mater.
4. Blood flow to the brain is mainly via the internal carotid and vertebral arteries.
5. Any interruption of the oxygen or glucose supply to the brain can result in weakening of, permanent damage to, or death of brain cells.
6. The blood–brain barrier (BBB) causes different substances to move between the blood and the brain tissue at different rates.

Cerebrospinal Fluid Production and Circulation in Ventricles (p. 547)

1. Cerebrospinal fluid (CSF) is formed in the choroid plexuses and circulates through the lateral ventricles, third ventricle, fourth ventricle, subarachnoid space, and central canal. Most of the fluid is absorbed into the blood across the arachnoid villi of the superior sagittal blood sinus.
2. Cerebrospinal fluid provides mechanical protection, chemical protection, and circulation of nutrients.

The Brain Stem (p. 550)

1. The medulla oblongata is continuous with the superior part of the spinal cord and contains both motor and sensory tracts. It contains nuclei that are reflex centers for regulation of heart rate, respiratory rate, vasoconstriction, swallowing, coughing, vomiting, and sneezing. It also contains nuclei associated with cranial nerves VIII (cochlear and vestibular branches) through XII.
2. The pons is superior to the medulla. It connects the spinal cord with the brain and links parts of the brain with one another by way of tracts. It relays nerve impulses related to voluntary skeletal movements from the cerebral cortex to the cerebellum. The pons contains the pneumotaxic and apneustic centers, which help control breathing. It contains nuclei associated with cranial nerves V–VII and the vestibular branch of cranial nerve VIII.
3. The midbrain connects the pons and diencephalon and surrounds the cerebral aqueduct. It conveys motor impulses from the cerebrum to the cerebellum and spinal cord, sends sensory impulses from the spinal cord to the thalamus, and regulates auditory and visual reflexes. It also contains nuclei associated with cranial nerves III and IV.
4. A large part of the brain stem consists of small areas of gray matter and white matter called the reticular formation, which helps maintain consciousness, causes awakening from sleep, and contributes to regulating muscle tone.

The Cerebellum (p. 556)

1. The cerebellum occupies the inferior and posterior aspects of the cranial cavity. It consists of two lateral hemispheres and a medial, constricted vermis.
2. It connects to the brain stem by three pairs of cerebellar peduncles.
3. The cerebellum coordinates contractions of skeletal muscles and maintains normal muscle tone, posture, and balance.

The Diencephalon (p. 558)

1. The diencephalon surrounds the third ventricle and consists of the thalamus, hypothalamus, epithalamus, and subthalamus.
2. The thalamus is superior to the midbrain and contains nuclei that serve as relay stations for all sensory impulses to the cerebral cortex. It also allows crude appreciation of pain, temperature, and pressure and mediates some motor activities.
3. The hypothalamus is inferior to the thalamus. It controls and integrates the autonomic nervous system, connects the nervous and endocrine systems, functions in rage and aggression, controls body temperature, regulates food and fluid intake, and establishes a diurnal sleep pattern.
4. The epithalamus consists of the pineal gland and the habenular nuclei. The pineal gland secretes melatonin, which is thought to promote sleep and to help set the body's biological clock.
5. The subthalamus connects to motor areas of the cerebrum.
6. Circumventricular organs (CVOs) can monitor chemical changes in the blood because they lack the blood–brain barrier.

The Cerebrum (p. 561)

1. The cerebrum is the largest part of the brain. Its cortex contains gyri (convolutions), fissures, and sulci.
2. The cerebrum contains the frontal, parietal, temporal, and occipital lobes.
3. The white matter of the cerebrum is deep to the cortex and consists of myelinated and unmyelinated axons extending to other regions as association, commissural, and projection fibers.
4. The basal ganglia are several groups of nuclei in each cerebral hemisphere. They help control large, automatic movements of skeletal muscles and help regulate muscle tone.
5. The limbic system encircles the upper part of the brain stem and the corpus callosum. It functions in emotional aspects of behavior and memory.

Functional Aspects of the Cerebral Cortex (p. 568)

1. The sensory areas of the cerebral cortex are concerned with the interpretation of sensory impulses. The motor areas are the regions that govern muscular movement. The association areas are concerned with more complex integrative functions.
2. The primary somatosensory area extends from the longitudinal fissure on the superior aspect of the cerebrum to the lateral cerebral sulcus. It receives nerve impulses from somatic sensory receptors for touch, proprioception, pain, and temperature. Each point within the area receives sensations from a specific part of the body.
3. Areas for the special senses include: the primary visual area (area 17), which receives impulses that convey visual information; the primary auditory area (areas 41 and 42), which interprets the basic characteristics of sound such as pitch and rhythm; the primary gustatory area (area 43), which receives impulses for taste; and the primary olfactory area (area 28), which receives impulses for smell.
4. Motor areas include the primary motor area (area 4), which controls voluntary contractions of specific muscles or groups of muscles, and Broca's speech area (areas 44 and 45), which controls production of speech.
5. Association areas of the cerebrum are connected with one another via tracts.
6. The somatosensory association area (areas 5 and 7) permits you to determine the exact shape and texture of an object just by touch, to determine the orientation of one object to another as they are felt, and to sense the relationship of one body part to another.
7. The visual association area (areas 18 and 19) relates present to past visual experiences and is essential for recognizing and evaluating what is seen.
8. The auditory association area (area 22) determines whether a sound is speech, music, or noise.
9. Wernicke's area (area 22 and possibly 39 and 40) interprets the meaning of speech by translating words into thoughts.
10. The common integrative area (areas 5, 7, 39, and 40) integrates sensory interpretations from the association areas and impulses from other areas, allowing one thought to be formed on the basis of a variety of sensory inputs.
11. The premotor area (area 6) generates nerve impulses that cause specific groups of muscles to contract in specific sequences.
12. The frontal eye field area (area 8) controls voluntary scanning movements of the eyes.
13. Table 18.1 on page 556 summarizes the functions of various parts of the brain.
14. Although the brain is fairly symmetrical on its right and left sides, subtle anatomical differences exist between the two hemispheres, and each has unique functions.
15. The left hemisphere receives sensory signals from and controls the right side of the body; it also is more important for language, numerical and scientific skills, and reasoning.
16. The right hemisphere receives sensory signals from and controls the left side of the body; it also is more important for musical and artistic awareness, spatial and pattern perception, recognition of faces, emotional content of language, and generating mental images of sight, sound, touch, taste, and smell.
17. Brain waves generated by the cerebral cortex are recorded from the surface of the head in an electroencephalogram (EEG).
18. The EEG may be used to diagnose epilepsy, infections, and tumors.

Cranial Nerves (p. 571)

1. Twelve pairs of cranial nerves originate from the brain.
2. They are named primarily on the basis of distribution and are numbered I–XII in order of attachment to the brain. Table 18-2 on pages 580–582 summarizes the cranial nerves.

Developmental Anatomy of the Nervous System (p. 583)

1. The development of the nervous system begins with a thickening of a region of the ectoderm called the neural plate.
2. During embryological development, primary brain vesicles are formed and serve as forerunners of various parts of the brain.
3. The telencephalon forms the cerebrum, the diencephalon develops into the thalamus and hypothalamus, the mesencephalon develops into the midbrain, the metencephalon develops into the pons and cerebellum, and the myelencephalon forms the medulla.

Aging and the Nervous System (p. 585)

1. The brain grows rapidly during the first few years of life.
2. Age-related effects involve loss of brain mass and decreased capacity for sending nerve impulses.

SELF-QUIZ QUESTIONS

Choose the one best answer to the following questions:

1. An extension of the dura mater called the ________ separates the cerebrum from the cerebellum. (a) falx cerebelli, (b) tentorium cerebelli, (c) denticulate ligament, (d) falx cerebri, (e) septum pellucidum
2. The meninx that adheres to the surface of the brain and spinal cord and contains blood vessels is the (a) arachnoid, (b) dura mater, (c) pia mater, (d) falx cerebri, (e) denticulate ligament
3. The reason that the motor areas of the right cerebral cortex control voluntary movements on the left side of the body is because (a) the cerebrum contains projection fibers, (b) the cerebellum controls voluntary movements, (c) the medulla oblongata contains decussating pyramids, (d) the pons connects the spinal cord with the brain, (e) the midbrain reroutes all motor impulses.
4. Which of the following is not located in the cerebrum? (a) globus pallidus, (b) corpus striatum, (c) corpus striatum, (d) caudate nucleus (e) lentiform nucleus.
5. Which of the following does not contribute to the formation of a blood-brain barrier? (a) tight junctions between the endothelial cells, (b) the continuous basement membrane of the capillary endothelium (c) a layer of oligodentrocytes (d) processes of astrocytes that contact the capillary walls, (d) none of the above
6. Damage to the occipital lobe of the cerebrum would most likely cause (a) loss of hearing, (b) loss of vision, (c) loss of ability to smell, (d) paralysis, (e) loss of muscle sense (proprioception)

Complete the following.

7. The structures responsible for the production of cerebrospinal fluid are the ________ .
8. The two nuclei of the medulla oblongata that relay sensory information from the spinal cord to the opposite side of the brain are the nucleus gracilis and nucleus ________ .
9. The optic (II) nerve exits from the eye and joins its partner from the other eye at the optic ________ , where some fibers cross to the opposite side of the brain.
10. The caudate and lentiform nuclei, together with several other nuclei, are called ________ . They are islands of gray matter embedded deep within the ________ .
11. White matter fibers that transmit nerve impulses between gyri in the same hemisphere are called ________ fibers.
12. The outer layer of the cerebrum is called the ________ . It is composed of ________ matter, which means that it contains mainly neuron cell bodies. The surface of the cerebrum is a series of tightly packed ridges, called ________ , with shallow grooves between them called ________ .

For the following questions, fill in the blanks as requested:

13. Write "w" for white matter or "g" for gray matter. (a) corpus callosum: ________ , (b) olive: ________ , (c) cerebral cortex: ________ , (d) corpora quadrigemina: ________ , (e) arbor vitae: ________ , (f) caudate and lenticular nuclei: ________ .
14. Write the name of the cranial nerve that fits each descriptions. (a) the only cranial nerve that originates partly from the spinal cord: ________ , (b) eighth (VII) cranial nerve: ________ , (c) is widely distributed into neck, thorax, and abdomen: ________ , (d) senses toothache, pain under a contact lens, wind on the face, (e) the largest cranial nerve; has three parts (ophthalmic, maxillary, and mandibular): ________ , (f) controls contraction of muscle of the iris, causing constriction of the pupil: ________ , (g) innervates muscles of facial expression: ________ , (h) two nerves that contain taste fibers and autonomic fibers to salivary glands: ________ , (i) three purely sensory cranial nerves: ________ .
15. Number the following in the correct sequence, from anterior to posterior. (a) fourth ventricle: ________ , (b) pons and medulla oblongata: ________ , (c) cerebellum ________ .
16. Number the following in the correct sequence, from superior to inferior. (a) thalamus: ________ , (b) hypothalamus: ________ , (c) corpus callosum ________ .
17. Number the following cranial nerves in the order in which they attach to the brain. (a) olfactory: ________ , (b) trochlear: ________ , (c) optic: ________ , (d) trigeminal: ________ , (e) facial: ________ , (f) oculomotor: ________ .
18. Number the following in the correct order for the circulation of CSF, from its production to its reabsorption. (a) arachnoid villi: ________ , (b) median and lateral apertures: ________ , (c) cerebral aqueduct: ________ , (d) lateral ventricle: ________ , (e) third ventricle: ________ , (f) fourth ventricle: ________ , (g) interventricular foramen: ________ , (h) subarachnoid space: ________ , (i) choroid plexuses: ________ , (j) superior sagittal sinus: ________ .

Are the following statements true or false?

19. Match the following.

____	(a) has two main parts, separated by a longitudinal fissure, but joined internally by the corpus callosum	(1) hyopothalamus
____	(b) responsible for coordination of skilled movements and regulation of posture and balance	(2) medulla
____	(c) cranial nerves III-IV attach to this brain part	(3) midbrain
____	(d) cranial nerves V-VIII attach to this brain part	(4) pons
____	(e) cranial nerves VIII-XII attach to this brain part	(5) thalamus
____	(f) regulates food and fluid intake and body temperature	(6) cerebellum
____	(g) all sensations are relayed through here	(7) cerebrum
____	(h) centers for control of heart rate and respiration are located here	
____	(i) constitutes four-fifths of the diencephalon	
____	(j) connects to the pituitary gland via the infundibulum	
____	(k) part of the brain stem; contains tracts that connect the cerebellum, midbrain, and medulla oblongata	
____	(l) surrounds the cerebral aqueduct, and contains nuclei that serve as reflex centers for head and eye movements	

20. The language areas are located in the cerebellar cortex.
21. The limbic system functions in the control of behavior.
22. The three parts of the brain stem are the medulla oblongata, pons, and cerebellum.
23. The cerebral nuclei that control large subconscious movements, such as swinging the arms while walking are the caudate and putamen.

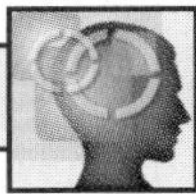

CRITICAL THINKING QUESTIONS

1. An elderly relative suffered a stroke and now has difficulty moving her right arm and also has speech problems. What areas of the brain were damaged by the stroke?
   HINT *What results from decussation of the pyramids of the medulla?*
2. Casey complained to her swim coach "my bathing cap is so tight, it'll squeeze my brains out my ears!" Her coach is majoring in physical therapy and tells her that's anatomically impossible. Explain the coach's position.
   HINT *Put your hands on your head and squeeze. What do you feel?*
3. Bubba "the bull" Bates likes to go to the amateur boxing night at the gym. He made the mistake of challenging a little 175 pound boxer to a match and ended up flat on his back and out cold in 45 seconds. What happened to Bubba's brain (at the boxing match-not before)?
   HINT *He might not remember much of the fight when he wakes up.*
4. Alicia just figured out that her Human Anatomy class actually starts at 9 AM and not at 9: 15 AM, which has been her arrival time since the beginning of the term. One of the other students remarks that Alicia's "gray matter is pretty thin." Should Alicia thank him?
   HINT *Gray matter is something that you DON'T want to lose when you diet.*
5. Dwayne's first trip to the dentist after a 10-year absence resulted in extensive dental work. He received numbing injections of anesthetic in several locations during the session. While having lunch right after the appointment, soup dribbles out of Dwayne's mouth because he has no feeling in his left upper lip, right lower lip, and tip of his tongue. What happened to Dwayne?
   HINT *The dentist wanted to block the transmission of the sensation of pain from the teeth.*

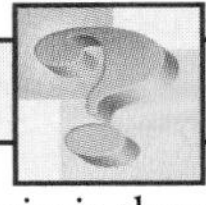

ANSWERS TO FIGURE QUESTIONS

18.1 The largest part of the brain is the cerebrum.

18.2 From superficial to deep, the three cranial meninges are the dura mater, arachnoid, and pia mater.

18.3 The brain stem is anterior to the fourth ventricle, and the cerebellum is posterior to it.

18.4 Cerebrospinal fluid is reabsorbed by the arachnoid villi that project into the dural venous sinuses.

18.5 The medulla oblongata contains the pyramids; the midbrain contains the cerebral peduncles; the pons means "bridge."

18.6 Decussation means crossing to the opposite side. Because the pyramids contain motor tracts that extend from the cortex into the spinal cord and convey impulses for contraction of skeletal muscle, the functional consequence of decussation of the pyramids is that one side of the cerebrum controls muscles on the opposite side of the body.

18.7 The cerebral peduncles are the main connections for tracts running between the superior parts of the brain and the inferior parts of the brain and the spinal cord.

18.8 The cerebellar peduncles carry information into and out of the cerebellum.

18.9 The intermediate mass connects the right and left sides of the thalamus.

18.10 From posterior to anterior, the four major regions of the hypothalamus are the mammillary, tuberal, supraoptic, and preoptic regions.

18.11 The gray matter enlarges more rapidly, in the process producing convolutions or gyri (folds), sulci (shallow grooves), and fissures (deep grooves).

18.12 Association fibers connect gyri of the same hemisphere; commissural fibers connect gyri in opposite hemispheres; projection fibers connect the cerebrum and thalamus, brain stem, and spinal cord.

18.13 The basal ganglia are lateral, superior, and inferior to the thalamus.

18.14 The hippocampus functions in memory.

18.15 Common integrative area; motor speech area; premotor area; gustatory areas; auditory areas; visual areas; frontal eye field area.

18.16 Broca's and Wernicke's areas are important language centers.

18.17 Superior branch; trochlear (IV) nerve.

18.18 It's the largest.

18.19 Pons.

18.20 Jugular foramen.

18.21 Between and behind the internal jugular vein and common carotid artery.

18.22 It originates from both the brain and spinal cord.

18.23 Speech and swallowing.

18.24 Gray matter derives from the mantle layer cells of the neural tube.

18.25 The mesencephalon does not develop into a secondary brain vesicle.

19

GENERAL SENSES AND SENSORY AND MOTOR PATHWAYS

CONTENTS AT A GLANCE

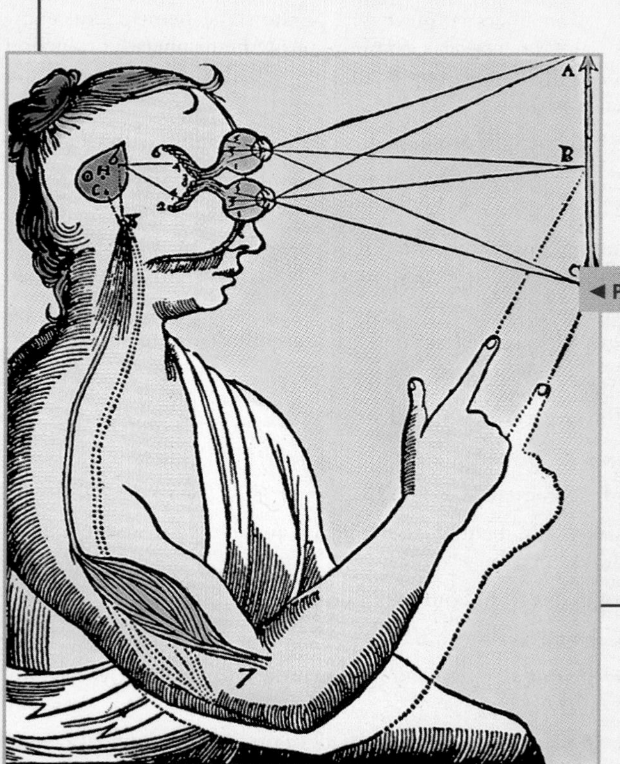

◄ Page 595

Take a good look around you – at this book, your classmates, and your professor. Have you ever wondered how we see an object that is before us? Or how a mental intention is carried out by a physical action?

Page 603 ►

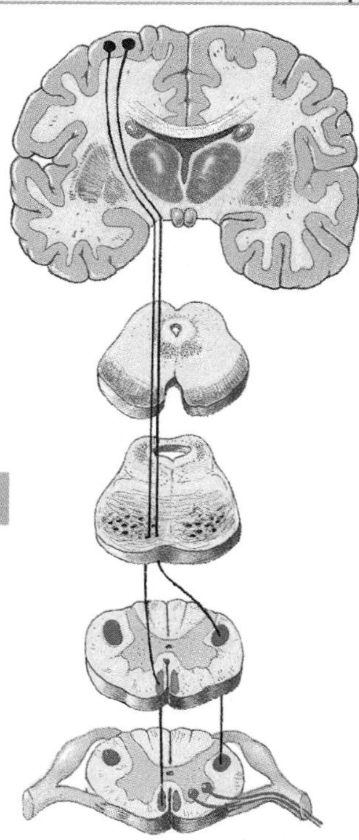

INTRODUCTION

The previous three chapters described the organization of the nervous system. Now we will see how certain parts cooperate to carry out some aspects of its three basic functions: (1) receiving sensory input; (2) integrating, associating, and storing information; and (3) transmitting motor impulses that result in movement or secretion.

In this chapter we will explore the pathways that convey somatic sensory input from the body to the brain and the pathways that carry motor commands for control of movements from the brain to the skeletal muscles. Chapter 20 deals with input from the special senses of smell, taste, vision, hearing, and equilibrium, and Chapter 21 covers output via the autonomic nervous system to smooth muscle, cardiac muscle, and glands.

OVERVIEW OF SENSATIONS

Objectives

- Define a sensation and describe the conditions necessary for a sensation to occur.
- Describe the different ways to classify sensory receptors.

Consider what would happen if you could not feel the pain of a hot pot handle or an inflamed appendix, or if you could not see, hear, smell, taste, or maintain your balance. In short, if you could not sense your environment and make the necessary homeostatic adjustments, you could not survive very well on your own.

Definition of Sensation

Sensation is the conscious or subconscious awareness of external or internal conditions of the body. For a sensation to occur, four conditions must be satisfied:

1. A *stimulus*, or change in the environment, capable of activating certain sensory neurons, must occur.
2. A *sensory receptor* must convert the stimulus to nerve impulses.
3. The nerve impulses must be *conducted* along a neural pathway from the sensory receptor to the brain.
4. A region of the brain must receive and *integrate* the nerve impulses, producing a sensation.

A stimulus that activates a sensory receptor may be in the form of light, heat, pressure, mechanical energy, or chemical energy. A sensory receptor responds to a stimulus by altering its membrane's permeability to small ions. In most types of sensory receptors, the resulting flow of ions across the membrane produces a change that triggers one or more nerve impulses. The impulses are then conducted along the sensory neuron toward the CNS.

Sensory receptors vary in their complexity. The simplest (for example, pain receptors) are *free nerve endings* that have no visible structural specializations. Receptors for other general sensations, such as touch, pressure, and vibration, have *encapsulated nerve endings.* Their dendrites are enclosed in a connective tissue capsule with a distinctive microscopic structure. Still other sensory receptors consist of specialized, separate cells that synapse with sensory neurons.

Characteristics of Sensations

Conscious sensations or **perceptions** are integrated in the cerebral cortex. You seem to see with your eyes, hear with your ears, and feel pain in an injured part of your body. This is because sensory impulses from each part of the body arrive in a specific region of the cerebral cortex, which interprets the sensation as coming from the stimulated sensory receptors.

The distinct quality that makes one sensation different from others is its **modality.** Based on the receptor stimulated, a sensory neuron carries information for one modality only. Neurons relaying impulses for touch, for example, do not also transmit impulses for pain. The specialization of sensory neurons enables nerve impulses from the eyes to be perceived as sight, and those from the ears to be perceived as sounds.

A characteristic of most sensory receptors is **adaptation,** a decrease in sensation during a prolonged stimulus. Adaptation is caused in part by a decrease in the responsiveness of sensory receptors. As a result of adaptation, the perception of a sensation decreases even though the stimulus persists. For example, when you first step into a hot shower, the water may feel very hot, but soon the sensation decreases to one of comfortable warmth even though the stimulus (the high temperature of the water) does not change. Receptors vary in how quickly they adapt.

Classification of Sensations

The senses can be grouped into two classes: general senses and special senses.

1. The **general senses** include both **somatic senses** (*somat-* = of the body) and **visceral senses.** Somatic senses include tactile sensations (touch, pressure, and vibration); thermal sensations (warm and cold); pain sensations; and proprioceptive sensations, which allow perception of both the static (non-moving) positions of limbs and body parts (joint and muscle position sense) and movements of the limbs and head. Visceral senses provide information about conditions within internal organs.
2. The **special senses** include smell, taste, vision, hearing, and equilibrium (balance).

Classification of Sensory Receptors

Sensory receptors are classified on the basis of their location (exteroceptors, interoceptors, and proprioceptors), the type of stimulus that activates them (mechanoreceptors, thermoreceptors, nociceptors, photoreceptors, and chemoreceptors), and their degree of complexity (simple receptors, complex receptors). Table 19.1 describes each of these categories.

✓ Distinguish between sensation and perception.
✓ What is a sensory modality? Adaptation?
✓ What events are needed for a sensation to occur?
✓ Describe the classification of sensory receptors.

Table 19.1 Classification of Sensory Receptors

A. Location

1. **Exteroceptors** (eks′-ter-ō-SEP-tors). Located at or near surface of body; provide information about *external* environment; transmit sensations of hearing, sight, smell, taste, touch, pressure, temperature, and pain.
2. **Interoceptors** (in′-ter-ō-SEP-tors). Located in blood vessels and viscera; provide information about *internal* environment; transmit sensations such as pain, pressure, fatigue, hunger, thirst, and nausea from within the body.
3. **Proprioceptors** (prō′-prē-ō-SEP-tors). Located in muscles, tendons, joints, and the internal ear; provide information about body position and movement; transmit information related to muscle tension, position and tension of joints, and equilibrium (balance).

B. Type of stimulus

1. **Mechanoreceptors.** Detect pressure or stretching; stimuli are related to touch, pressure, proprioception, hearing, equilibrium, and blood pressure.
2. **Thermoreceptors.** Detect changes in temperature.
3. **Nociceptors** (nō′-sē-SEP-tors). Detect pain, usually as a result of physical or chemical damage to tissues.
4. **Photoreceptors.** Detect light in retina of eye.
5. **Chemoreceptors.** Detect taste in mouth; smell in nose; and chemicals such as oxygen, carbon dioxide, water, and glucose in body fluids.

C. Degree of complexity

1. **Simple receptors.** Simple structures and neural pathways that are associated with general senses (touch, pressure, heat, cold, and pain).
2. **Complex receptors.** Complex structures and neural pathways that are associated with special senses (smell, taste, sight, hearing, and equilibrium).

SOMATIC SENSATIONS

Objectives

- Describe the location and function of the receptors for tactile, thermal, and pain sensations.
- Identify the receptors for proprioception and describe their functions.

Somatic sensations arise from stimulation of sensory receptors embedded in the skin or subcutaneous layer; in mucous membranes of the mouth, vagina, and anus; in muscles, tendons, and joints; and in the internal ear. The sensory receptors for somatic sensations are distributed unevenly. Some parts of the body surface are densely populated with receptors, whereas other parts contain only a few. The areas with the highest density of sensory receptors are the tip of the tongue, the lips, and the fingertips. Somatic sensations that result from stimulating the skin surface are called **cutaneous sensations** (kyū-TĀ-nē-us; *cutane-* = skin).

Tactile Sensations

The **tactile sensations** (TAK-tīl; *tact-* = touch) are touch, pressure and vibration, and itch and tickle. Itch and tickle sensations are detected by free nerve endings. All other tactile sensations are detected by a variety of encapsulated mechanoreceptors. Tactile receptors in the skin or subcutaneous layer include corpuscles of touch, hair root plexuses, type I and II cutaneous mechanoreceptors, lamellated corpuscles, and free nerve endings (Figure 19.1).

Touch

Sensations of **touch** generally result from stimulation of tactile receptors in the skin or subcutaneous layer. *Crude touch* is the ability to perceive that something has contacted the skin, even though its exact location, shape, size, or texture cannot be determined. *Fine touch* provides specific information about where the body is touched and the shape, size, and texture of the source of stimulation.

There are two types of rapidly adapting touch receptors. **Corpuscles of touch,** or **Meissner corpuscles** (MĪS-ner), are receptors for fine touch that are located in the dermal papillae of hairless skin. Each corpuscle is an egg-shaped mass of dendrites enclosed by a capsule of connective tissue. These receptors are abundant in the fingertips and palms, eyelids, tip of the tongue, lips, nipples, soles, clitoris, and tip of the penis. **Hair root plexuses** are rapidly adapting touch receptors found in hairy skin; they consist of free nerve endings wrapped around hair follicles. Hair root plexuses detect movements on the skin surface that disturb hairs. For example, a flea landing on a hair causes movement of the hair shaft that stimulates the free nerve endings.

There are also two types of slowly adapting touch receptors. **Type I cutaneous mechanoreceptors,** also known as **Merkel disks,** function in fine touch. These are saucer-shaped, flattened free nerve endings that contact Merkel cells of the stratum basale; they are plentiful in the fingertips, hands, lips, and external genitalia. **Type II cutaneous mechanoreceptors,** or **Ruffini corpuscles,** are elongated, encapsulated receptors located deep in the dermis, and in ligaments and tendons as well. Present in the hands and abundant on the soles, they are most sensitive to stretching that occurs as digits or limbs are moved.

Figure 19.1 / Structure and location of sensory receptors in the skin and subcutaneous layer.

The somatic sensations of touch, pressure, vibration, warmth, cold, and pain arise from sensory receptors in the skin, subcutaneous layer, and mucous membranes.

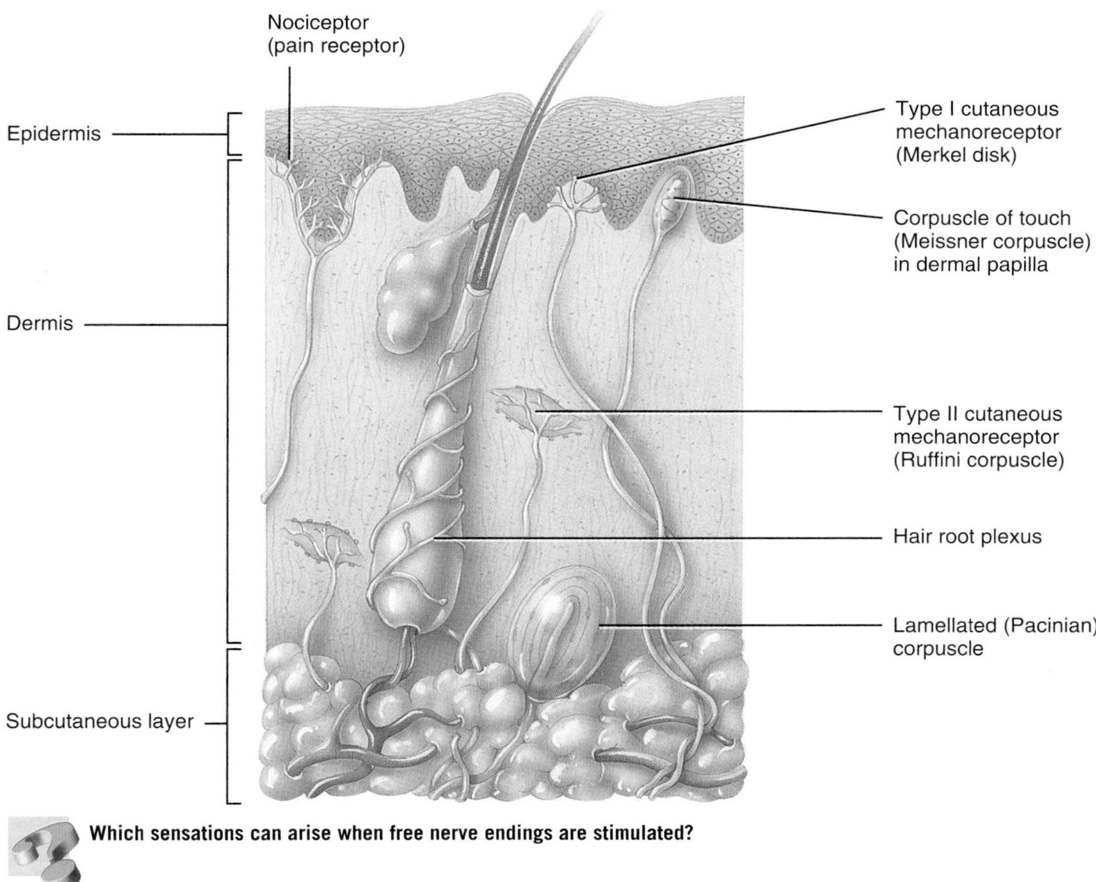

Which sensations can arise when free nerve endings are stimulated?

Pressure and Vibration

Pressure is a sustained sensation that is felt over a larger area than touch. Receptors that contribute to sensations of pressure include corpuscles of touch, type I mechanoreceptors, and lamellated corpuscles. **Lamellated,** or **Pacinian, corpuscles** (pa-SIN-ē-an) are large oval structures composed of a multilayered connective tissue capsule that encloses a nerve ending (Figure 19.1). Like corpuscles of touch, lamellated corpuscles adapt rapidly. They are widely distributed in the body: in the dermis and subcutaneous layer; in tissues that underlie mucous and serous membranes; around joints, tendons, and muscles; in the periosteum; and in the mammary glands, external genitalia, and certain viscera, such as the pancreas and urinary bladder.

Sensations of **vibration** result from rapidly repetitive sensory signals from tactile receptors. The receptors for vibration are the corpuscles of touch and lamellated corpuscles.

Itch and Tickle

The **itch** sensation results from stimulation of free nerve endings by certain chemicals, such as bradykinin, often as a result of a local inflammatory response. Receptors for the **tickle** sensation are thought to be free nerve endings and lamellated corpuscles. This intriguing sensation typically arises only when someone else touches you, not when you touch yourself. The explanation for this puzzle seems to lie in the impulses traveling to and from your cerebellum when you are moving your fingers and tickling yourself that don't occur when someone else is tickling you.

Thermal Sensations

The sensory receptors for **thermal sensations** (sensations of heat and cold) are free nerve endings and consist of two types: cold receptors and warm receptors. **Cold receptors** are located in the stratum basale of the epidermis. Temperatures between 10° and 40°C (50° to 105°F) activate cold receptors. **Warm receptors,** located in the dermis, are activated by temperatures between 32° and 48°C (90° to 118°F). Both cold and warm receptors adapt rapidly at the onset of a stimulus but continue to generate some impulses throughout a prolonged stimulus. Temperatures below 10°C and above 48°C stimulate mainly pain receptors, rather than thermoreceptors, producing painful sensations.

Pain Sensations

The ability to perceive pain is indispensable for a normal life, providing us with information about tissue-damaging stimuli so we can protect ourselves from greater damage. Pain initiates our search for medical assistance, and our description and indication of the location of the pain may help pinpoint the underlying cause of disease.

The sensory receptors for pain, called **nociceptors** (nō′-sē-SEP-tors; *noci-* = harmful), are free nerve endings (Figure 19.1). Pain receptors are found in practically every tissue of the body except the brain, and they respond to several types of stimuli. Excessive stimulation of sensory receptors, excessive stretching of a structure, prolonged muscular contractions, inadequate blood flow to an organ, or the presence of certain chemical substances can all produce the sensation of pain.

During tissue irritation or injury, release of chemicals such as prostaglandins stimulate nociceptors. Nociceptors adapt only slightly or not at all to the presence of these chemicals, which are only slowly removed from the tissues following an injury. This situation explains why pain persists after the initial trauma. If there were adaptation to painful stimuli, irreparable tissue damage could result.

Based on the location of the stimulated receptors, pain may be divided into two types: somatic and visceral. **Somatic pain** that arises from stimulation of receptors in the skin is called **superficial somatic pain,** whereas stimulation of receptors in skeletal muscles, joints, tendons, and fascia causes **deep somatic pain. Visceral pain** results from stimulation of receptors in the viscera.

Recognition of the type and intensity of pain occurs primarily in the cerebral cortex. In most instances of somatic pain, the cortex projects the pain back to the stimulated area. If you burn your finger, you feel the pain in your finger, not in your cortex. In most instances of visceral pain, the sensation is not projected back to the point of stimulation. Rather, the pain is felt in the skin overlying the stimulated organ or in a surface area far from the stimulated organ. This phenomenon is called **referred pain.** It occurs because the area to which the pain is referred and the visceral organ involved are innervated by the same segment of the spinal cord. For example, sensory neurons from the heart as well as from the skin over the heart and left upper limb enter thoracic spinal cord segments T1 to T5. Thus the pain of a heart attack is typically felt in the skin over the heart and along the left arm. Figure 19.2 illustrates skin regions to which visceral pain may be referred.

Patients who have had a limb amputated may experience sensations such as itching, pressure, tingling, or pain as though the limb were still there. This phenomenon is called **phantom limb sensation.** One explanation for phantom sensations is that the cerebral cortex interprets impulses arising in the proximal portions of sensory neurons that previously carried impulses from the limb as coming from the nonexistent (phantom) limb.

Figure 19.2 / Distribution of referred pain. The colored portions of the diagrams indicate skin areas to which visceral pain is referred.

Nociceptors are present in almost every tissue of the body.

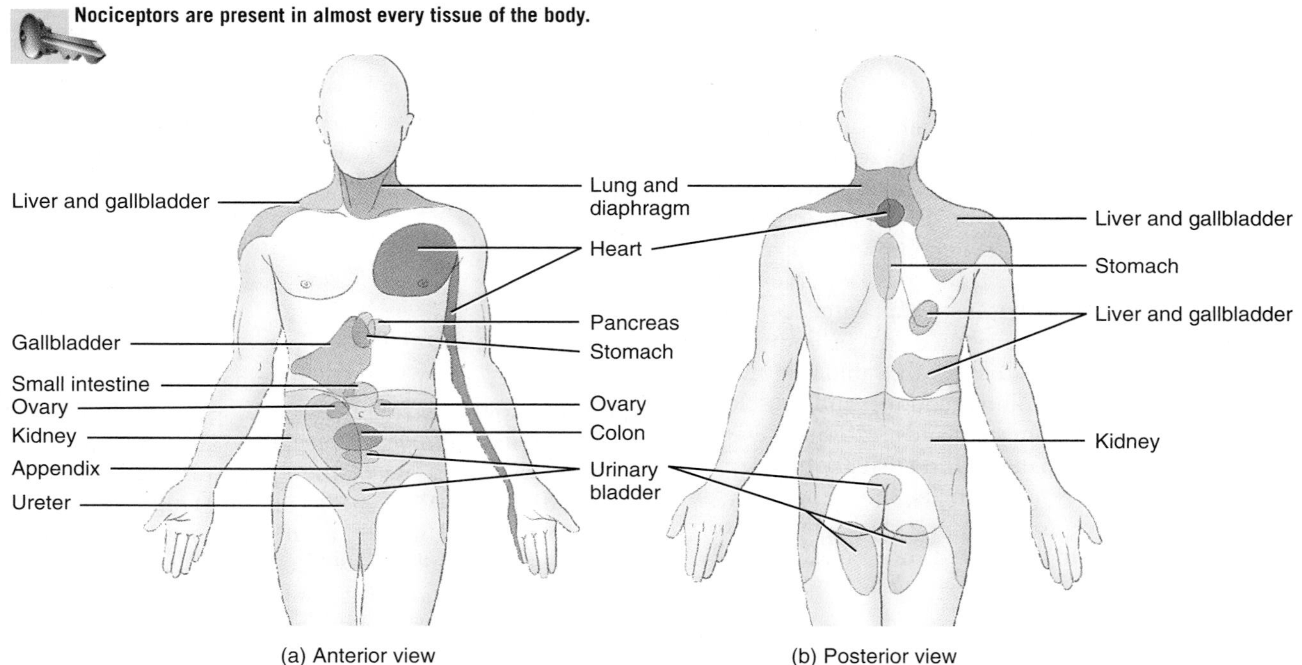

Which visceral organ has the broadest area for referred pain?

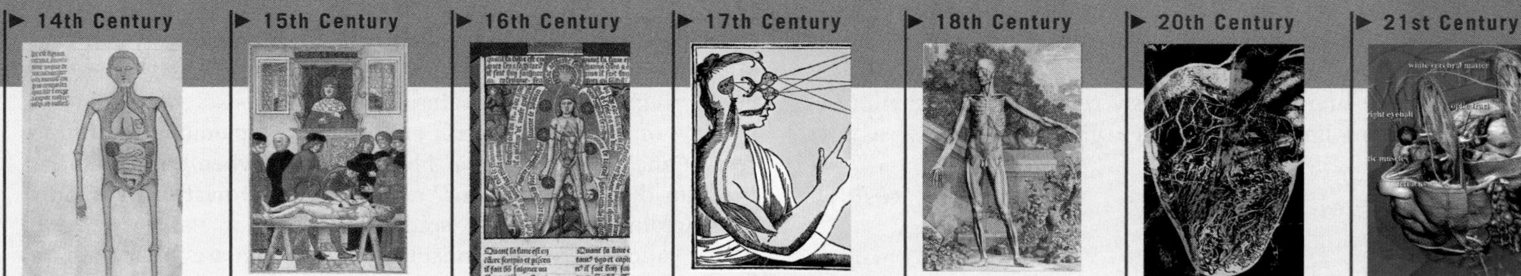

CHANGING IMAGES

Bridging Body and Mind

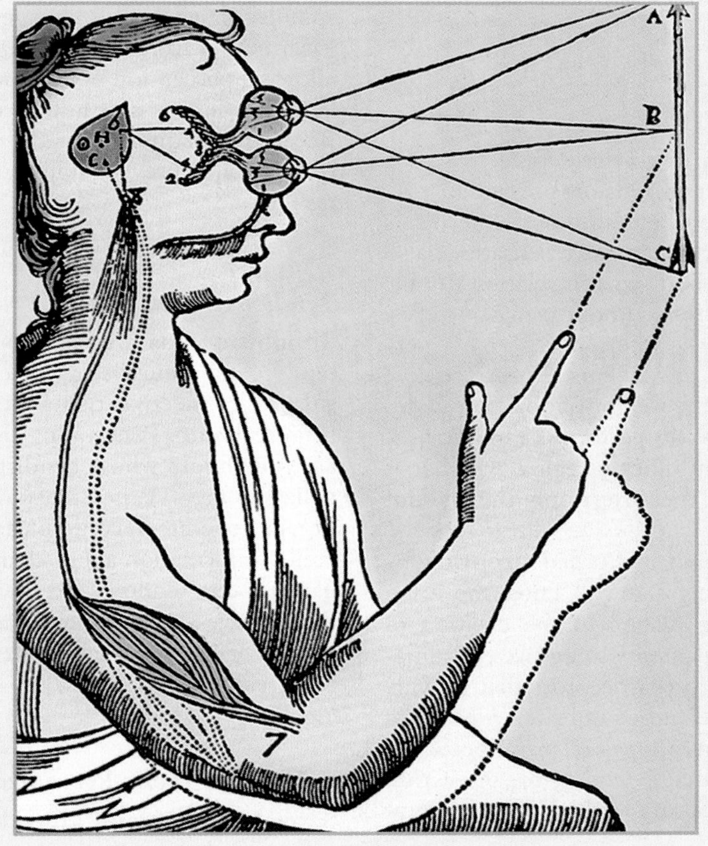

How are we aware of the world around us? How is it that we are capable of looking at an object and then have the brain perceive that object to be there? How are perceived images transformed into physical action, for example, when you decide to raise your hand to catch a ball? These questions of mind and matter, the internal and the external, have puzzled many scientists. Yet it's the writings of René Descartes that are often considered the most compelling historical inquiry into this arena.

Scientist, mathematician, and founder of modern philosophy, Descartes was born in 1596. Author of significant works not only in philosophy, geometry, and astronomy, Descartes explored human structure and function as well. He sought an anatomical connection between a human's physical body and spiritual mind. Descartes pursued an explanation for how touching something hot would register in the mind as a sensation of heat. Such investigations led to him being often considered the founder of reflex theory, a topic considered earlier in Chapter 17.

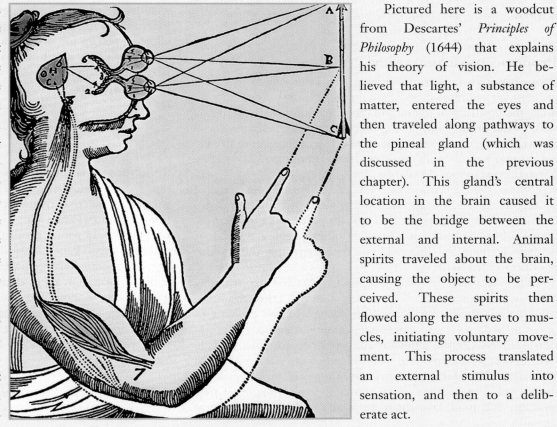

Pictured here is a woodcut from Descartes' *Principles of Philosophy* (1644) that explains his theory of vision. He believed that light, a substance of matter, entered the eyes and then traveled along pathways to the pineal gland (which was discussed in the previous chapter). This gland's central location in the brain caused it to be the bridge between the external and internal. Animal spirits traveled about the brain, causing the object to be perceived. These spirits then flowed along the nerves to muscles, initiating voluntary movement. This process translated an external stimulus into sensation, and then to a deliberate act.

1644 AD

Descartes' elementary diagram suggests anatomical knowledge of the optic nerves, optic chiasm, and optic tracts, which you will study in chapter 20. He seemed to also understand more complicated issues, such as sensory and motor pathways. As you study this chapter about the senses, compare and contrast the present theories of this text with those of Descartes.

Another explanation is that the brain itself contains networks of neurons that generate sensations of body awareness. In this view, neurons in the brain that previously received sensory impulses from the missing limb are still active, giving rise to false sensory perceptions.

Analgesia: Relief from Pain

Some pain sensations are inappropriate; rather than warning of actual or impending damage, they occur out of proportion to minor damage or persist chronically for no obvious reason. In such cases, **analgesia** (*an-* = without; *-algesia* = pain) or pain relief is needed. Analgesic drugs such as aspirin and ibuprofen (for example, Advil) block formation of prostaglandins, which stimulate nociceptors. Local anesthetics, such as Novocaine, provide short-term pain relief by blocking conduction of nerve impulses along first-order pain fibers. Morphine and other opiate drugs alter the quality of pain perception in the brain; pain is still sensed, but it is no longer perceived as being so noxious. ■

Proprioceptive Sensations

Proprioceptive sensations inform us of the degree to which muscles are contracted, the amount of tension present in tendons, the positions of joints, and the orientation of the head relative to the ground and during movement. Such sensations also report the rate of movement of one body part in relation to others, so that we can walk, type, or dress without using our eyes. **Kinesthesia** (kin′-es-THĒ-zē-a; *kin-* = motion; *-esthesia* = perception) is the perception of body movements. Proprioceptive sensations also allow us to estimate the weight of objects and determine the muscular effort necessary to perform a task. For example, when you pick up a bag you quickly realize whether it contains feathers or books, and you then exert only the amount of effort needed to lift it.

The receptors for proprioception are called **proprioceptors,** and they adapt slowly and only slightly. Thus, the brain continually receives nerve impulses related to the position of different body parts and makes adjustments to ensure coordination. Hair cells of the inner ear are proprioceptors that provide information for maintaining balance and equilibrium (described in Chapter 20). Here we discuss three types of proprioceptors: muscle spindles within skeletal muscles, tendon organs within tendons, and joint kinesthetic receptors within synovial joint capsules.

Muscle Spindles

Muscle spindles are specialized groups of muscle fibers interspersed among and oriented parallel to regular skeletal muscle fibers (Figure 19.3a). The ends of the spindles are anchored to endomysium and perimysium. A muscle spindle consists of 3–10 specialized muscle fibers, called **intrafusal muscle fibers,** that are enclosed in a spindle-shaped connective tissue capsule. The central region of each intrafusal fiber contains several nuclei but has few or none of the actin and myosin filaments found in the rest of the fiber. Intrafusal fibers contract when stimulated by medium-diameter neurons called **gamma motor neurons.** Surrounding the muscle spindle are regular skeletal muscle fibers, called **extrafusal muscle fibers,** which are innervated by large-diameter neurons called **alpha motor neurons.** The cell bodies of both types of motor neurons are located in the anterior gray horn of the spinal cord.

The central area of an intrafusal fiber contains two types of slowly adapting sensory fibers. The first is a **type Ia sensory fiber,** which has a large-diameter, rapidly conducting axon. The dendrites of a type Ia sensory fiber spiral around the central area of an intrafusal fiber. The central receptive area of some muscle spindles is also served by a **type II sensory fiber,** whose dendrites are located on either side of the type Ia dendrites. Either sudden or prolonged stretching of the central areas of the intrafusal muscle fibers stimulates the type Ia and type II dendrites. The resulting nerve impulses conduct into the CNS. Muscle spindles thus monitor changes in the length of skeletal muscles. Information from muscle spindles arrives at the cerebral cortex, allowing perception of limb position, and also passes to the cerebellum, where it aids in the coordination of muscle contraction. Because the stimulus for the reflex is stretching of muscles, the reflex helps prevent injury by preventing overstretching of muscles.

Tendon Organs

Tendon organs are proprioceptors found at the junction of a tendon and a muscle. Each tendon organ consists of a thin capsule of connective tissue that encloses a few collagen fibers (Figure 19.3b). Penetrating the capsule are one or more **type Ib sensory fibers** whose dendrites entwine among and around the collagen fibers. When tension is applied to a tendon, the tendon organs generate nerve impulses that conduct into the CNS, providing information about changes in muscle tension. Tendon organs protect tendons and their associated muscles from damage due to excessive tension. Tendon reflexes decrease muscles tension by causing muscle relaxation when muscle force becomes too great.

Joint Kinesthetic Receptors

Several types of **joint kinesthetic receptors** are present within and around the articular capsules of synovial joints. Free nerve endings and type II cutaneous mechanoreceptors (Ruffini corpuscles) in the capsules of joints respond to pressure. Small lamellated (Pacinian) corpuscles in the connective tissue outside articular capsules respond to acceleration and deceleration of joints during movement. Articular ligaments contain receptors similar to tendon organs that adjust reflex inhibition of the adjacent muscles when excessive strain is placed on the joint.

Figure 19.3 / Two types of proprioceptors: a muscle spindle and a tendon organ. (a) In muscle spindles, which monitor changes in skeletal muscle length, type Ia and type II sensory fibers wrap around the central portion of intrafusal muscle fibers. (b) In tendon organs, which monitor the force of muscle contraction, a type Ib sensory fiber is activated by increasing tension on a tendon.

Proprioceptors provide information about movement and the position of the body.

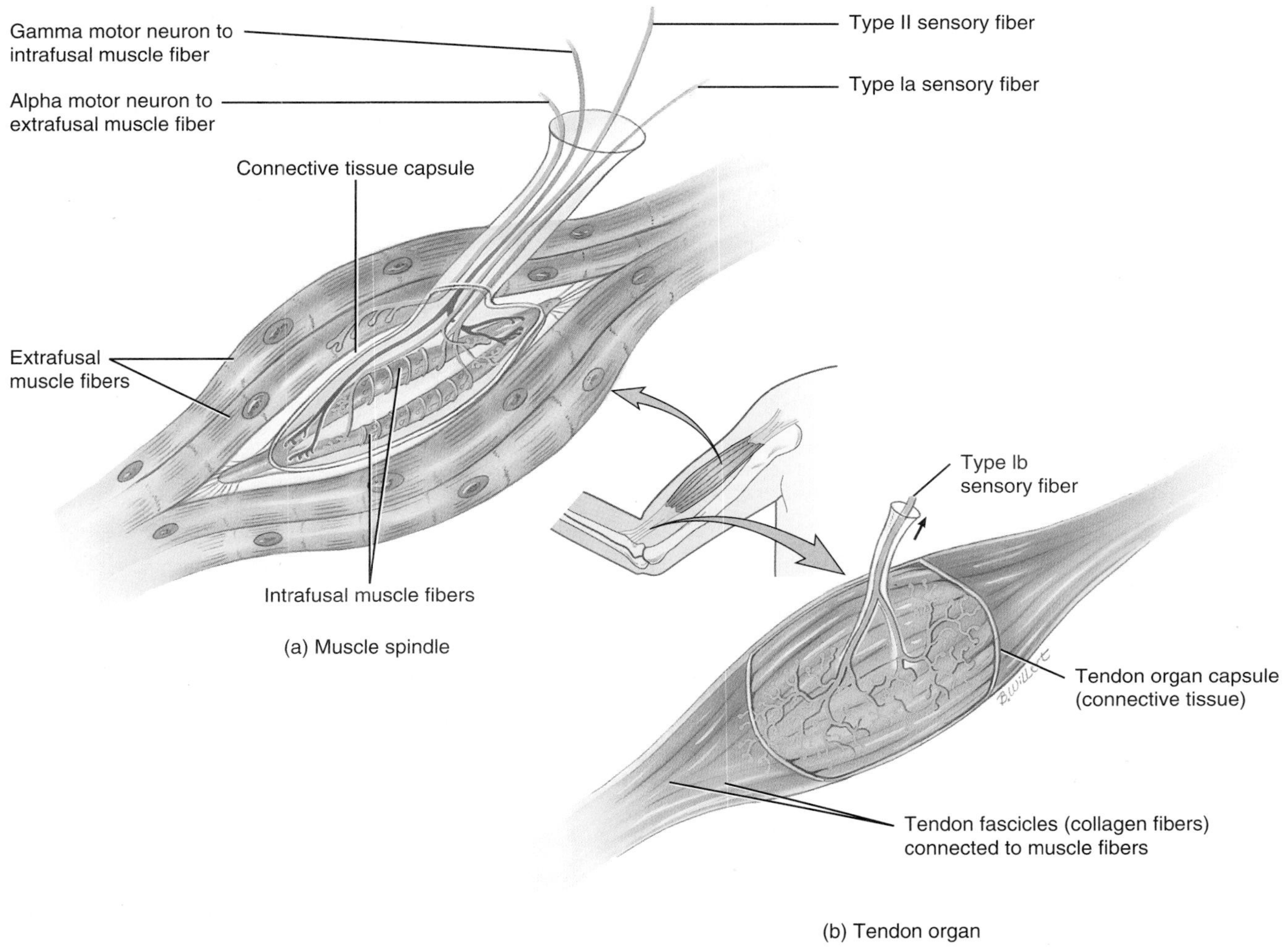

Gamma motor neuron to intrafusal muscle fiber
Alpha motor neuron to extrafusal muscle fiber
Type II sensory fiber
Type Ia sensory fiber
Connective tissue capsule
Extrafusal muscle fibers
Intrafusal muscle fibers
(a) Muscle spindle
Type Ib sensory fiber
Tendon organ capsule (connective tissue)
Tendon fascicles (collagen fibers) connected to muscle fibers
(b) Tendon organ

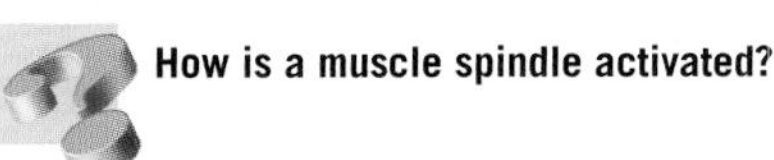

How is a muscle spindle activated?

Table 19.2 summarizes the somatic receptors and the sensations they convey.

✓ Compare and contrast cutaneous and proprioceptive sensations.
✓ Which somatic receptors are encapsulated, and which are not?
✓ List the different kinds of tactile receptors and the function of each.
✓ Contrast somatic, visceral, and phantom pain.
✓ What is referred pain, and how is it useful in diagnosing internal disorders?
✓ What are the functions of the various proprioceptors?

SOMATIC SENSORY PATHWAYS

Objective

- Describe the neuronal components and functions of the posterior column–medial lemniscus, the anterolateral, and the spinocerebellar pathways.

Somatic sensory pathways relay information from somatic receptors to the primary somatosensory area in the cerebral cortex and to the cerebellum. The pathways to the cerebral cortex are composed of thousands of sets of three neurons: a first-order neuron, a second-order neuron, and a third-order neuron.

Table 19.2 Summary of Receptors for Somatic Sensations

Receptor Type	Receptor Structure and Location	Sensations
Tactile Receptors		
Corpuscles of touch (Meissner corpuscles)	Capsule surrounds mass of dendrites in dermal papillae of hairless skin.	Touch, pressure, and slow vibrations.
Hair root plexuses	Free nerve endings wrapped around hair follicles in skin.	Touch.
Type I cutaneous mechanoreceptors (tactile or **Merkel disc)**	Saucer-shaped free nerve endings make contact with Merkel cells in epidermis.	Touch and pressure.
Type II cutaneous mechanoreceptors (Ruffini corpuscles)	Elongated capsule surrounds dendrites deep in dermis and in ligaments and tendons.	Stretching of skin.
Lamellated (Pacinian) corpuscles	Oval, layered capsule surrounds dendrites; present in subcutaneous layer (and sometimes the dermis), submucosal tissues, joints, periosteum, and some viscera.	Pressure, fast vibrations, and tickling.
Itch and tickle receptors	Free nerve endings and lamellated corpuscles in skin and mucous membranes.	Itching and tickling.
Thermoreceptors		
Warm receptors and **cold receptors**	Free nerve endings in skin and mucous membranes of mouth, vagina, and anus.	Warmth or cold.
Pain Receptors		
Nociceptors	Free nerve endings in skin and mucous membranes of mouth, vagina, and anus.	Pain.
Proprioceptors		
Muscle spindles	Dendrites of types Ia and II fibers wrap around central area of encapsulated intrafusal muscle fibers within most skeletal muscles.	Muscle length.
Tendon organs	Capsule encloses collagen fibers and dendrites of type Ib fibers at junction of tendon and muscle.	Muscle tension.
Joint kinesthetic receptors	Lamellated corpuscles, Ruffini corpuscles, tendon organs, and free nerve endings.	Joint position and movement.

1. **First-order neurons** conduct impulses from the somatic receptors into either the brain stem or spinal cord. From the face, mouth, teeth, and eyes, somatic sensory impulses propagate along *cranial nerves* into the brain stem; from the neck, body, and posterior aspect of the head, somatic sensory impulses propagate along *spinal nerves* into the spinal cord.
2. **Second-order neurons** conduct impulses from the spinal cord and brain stem to the thalamus. Axons of second-order neurons decussate (cross over to the opposite side) in the spinal cord or brain stem before ascending to the thalamus. Thus, all somatic sensory information from one side of the body is conveyed to the thalamus on the opposite side.
3. **Third-order neurons** conduct impulses from the thalamus to the primary somatosensory area of the cortex (postcentral gyrus; see Figure 18.15 on page 568), where conscious perception of the sensations results.

Somatic sensory impulses entering the spinal cord ascend to the cerebral cortex by two general pathways: the posterior column–medial lemniscus pathway and the anterolateral (spinothalamic) pathways.

Posterior Column–Medial Lemniscus Pathway to the Cortex

Nerve impulses for conscious proprioception and most tactile sensations ascend to the cortex along a common pathway formed by three-neuron sets (Figure 19.4a). First-order neurons extend from sensory receptors into the spinal cord and up to the medulla oblongata on the same side of the body. The cell bodies of these first-order neurons are in the posterior (dorsal) root ganglia of spinal nerves. In the spinal cord, their axons form the **posterior (dorsal) columns,** which consist of two portions: the **fasciculus gracilis** (fa-SIK-yū-lus gras-I-lis) and the **fasciculus cuneatus** (kyū-nē-Ā-tus). See Table 19.3 on page 600. The axon terminals synapse with second-order neurons whose cell bodies are located in the nucleus gracilis or nucleus cuneatus of the medulla. Impulses from the neck, upper limbs, and upper chest conducted along axons in the fasciculus cuneatus arrive at the

Figure 19.4 / Somatic sensory pathways.

Nerve impulses are conducted along sets of first-order, second-order, and third-order neurons to the primary somatosensory area (postcentral gyrus) of the cerebral cortex.

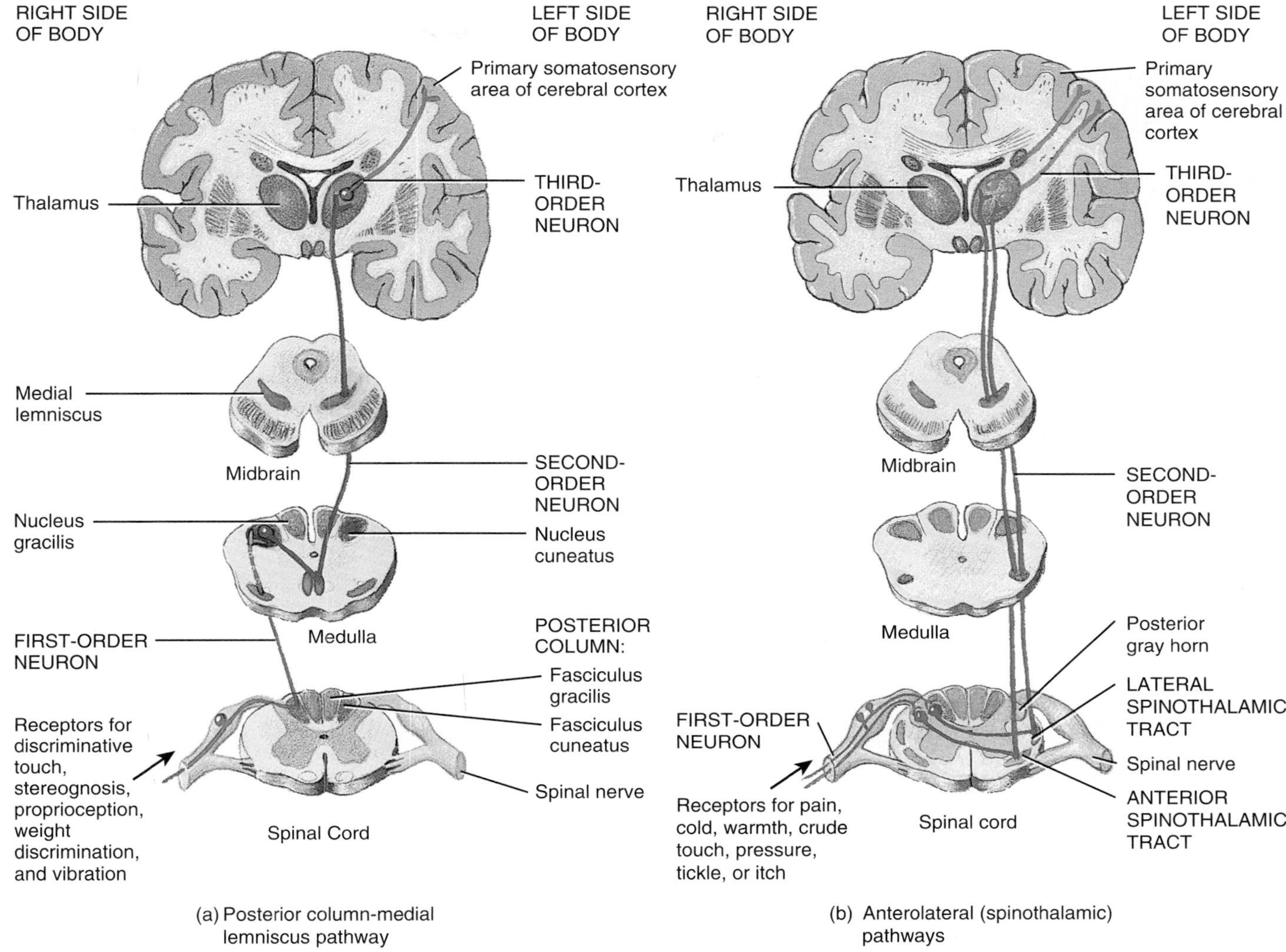

(a) Posterior column-medial lemniscus pathway

(b) Anterolateral (spinothalamic) pathways

What sorts of sensory deficits could be produced by damage to the right lateral spinothalamic tract?

nucleus cuneatus, whereas impulses from the trunk and lower limbs conducted along axons in the fasciculus gracilis arrive at the nucleus gracilis. The axon of the second-order neuron crosses to the opposite side of the medulla and enters the **medial lemniscus,** a projection tract that extends from the medulla to the thalamus. In the thalamus, the axon terminals of second-order neurons synapse with third-order neurons, which project their axons to the primary somatosensory area of the cerebral cortex.

Impulses conducted along the posterior column–medial lemniscus pathway give rise to several highly evolved and refined sensations:

- **Discriminative touch,** the ability to recognize specific information about a touch sensation, such as which point on the body is touched; the shape, size, and texture of the source of stimulation; and two-point discriminations (the ability to distinguish as separate two points of touching that are close together rather than as one).
- **Stereognosis,** the ability to recognize by feel the size, shape, and texture of an object. Examples are reading Braille or identifying a paperclip by feeling it.
- **Proprioception,** the awareness of the precise position of body parts, and **kinesthesia,** the awareness of directions of movement.
- **Weight discrimination,** the ability to assess the weight of an object.
- **Vibratory sensations,** the ability to sense rapidly fluctuating touch.

Table 19.3 Major Somatic Sensory Tracts in the Spinal Cord and Pathways in the Brain

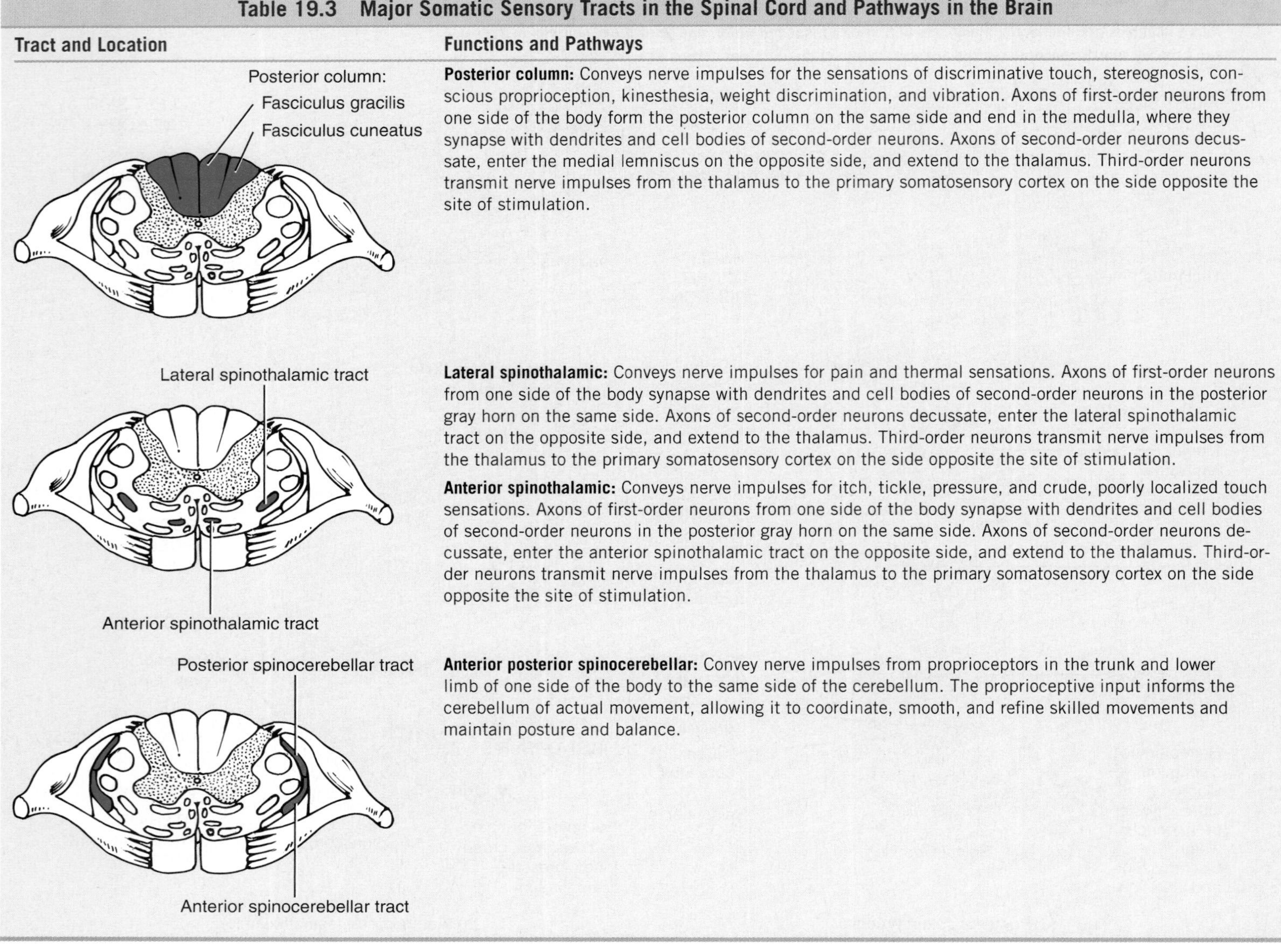

Tract and Location	Functions and Pathways
Posterior column: Fasciculus gracilis; Fasciculus cuneatus	**Posterior column:** Conveys nerve impulses for the sensations of discriminative touch, stereognosis, conscious proprioception, kinesthesia, weight discrimination, and vibration. Axons of first-order neurons from one side of the body form the posterior column on the same side and end in the medulla, where they synapse with dendrites and cell bodies of second-order neurons. Axons of second-order neurons decussate, enter the medial lemniscus on the opposite side, and extend to the thalamus. Third-order neurons transmit nerve impulses from the thalamus to the primary somatosensory cortex on the side opposite the site of stimulation.
Lateral spinothalamic tract; Anterior spinothalamic tract	**Lateral spinothalamic:** Conveys nerve impulses for pain and thermal sensations. Axons of first-order neurons from one side of the body synapse with dendrites and cell bodies of second-order neurons in the posterior gray horn on the same side. Axons of second-order neurons decussate, enter the lateral spinothalamic tract on the opposite side, and extend to the thalamus. Third-order neurons transmit nerve impulses from the thalamus to the primary somatosensory cortex on the side opposite the site of stimulation. **Anterior spinothalamic:** Conveys nerve impulses for itch, tickle, pressure, and crude, poorly localized touch sensations. Axons of first-order neurons from one side of the body synapse with dendrites and cell bodies of second-order neurons in the posterior gray horn on the same side. Axons of second-order neurons decussate, enter the anterior spinothalamic tract on the opposite side, and extend to the thalamus. Third-order neurons transmit nerve impulses from the thalamus to the primary somatosensory cortex on the side opposite the site of stimulation.
Posterior spinocerebellar tract; Anterior spinocerebellar tract	**Anterior posterior spinocerebellar:** Convey nerve impulses from proprioceptors in the trunk and lower limb of one side of the body to the same side of the cerebellum. The proprioceptive input informs the cerebellum of actual movement, allowing it to coordinate, smooth, and refine skilled movements and maintain posture and balance.

Anterolateral Pathways to the Cortex

The **anterolateral,** or **spinothalamic, pathways** (spī-nō-tha-LAM-ik) carry mainly pain and temperature impulses. In addition, they relay the sensations of tickle and itch, as well as some tactile impulses that give rise to a very crude, poorly localized touch or pressure sensation. Like the posterior column–medial lemniscus pathway, the anterolateral pathways are also composed of three-neuron sets (Figure 19.4b). The first-order neuron connects a receptor of the neck, trunk, or limbs with the spinal cord. The cell body of the first-order neuron is in the posterior root ganglion. The axon terminals of the first-order neuron synapse with the second-order neuron, whose cell body is located in the posterior gray horn of the spinal cord. The axon of the second-order neuron crosses to the opposite side of the spinal cord and passes upward to the brain stem in either the **lateral spinothalamic tract** or the **anterior spinothalamic tract.** See Table 19.3. The axon of the second-order neuron ends in the thalamus, where it synapses with the third-order neuron. The axon of the third-order neuron projects to the primary somatosensory area of the cerebral cortex. The lateral spinothalamic tract conveys sensory impulses for pain and temperature, whereas the anterior spinothalamic tract conveys impulses for tickle, itch, crude touch, and pressure.

Mapping of the Somatosensory Cortex

Researchers have mapped out areas of the primary somatosensory cortex (postcentral gyrus) that receive sensory information from different parts of the body. Figure 19.5a shows the location and areas of representation of the somatosensory cortex of the right cerebral hemisphere. The left cerebral hemisphere has a similar somatosensory cortex.

Note that some parts of the body—the lips, face, tongue, and thumb, for example—are represented by large areas in the

Figure 19.5 / (a) Primary somatosensory area (postcentral gyrus) and (b) primary motor area (precentral gyrus) of the right cerebral hemisphere. The left hemisphere has similar representation. (After Penfield and Rasmussen)

Each point on the body surface maps to a specific region in both the primary somatosensory area and the primary motor area.

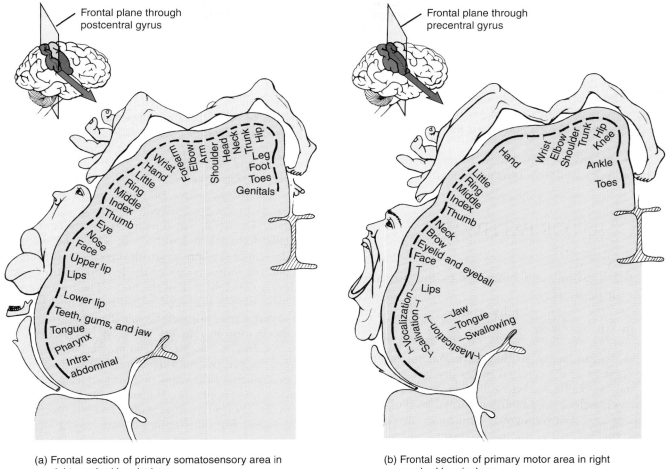

(a) Frontal section of primary somatosensory area in right cerebral hemisphere

(b) Frontal section of primary motor area in right cerebral hemisphere

How do the somatosensory and motor representations compare for the hand, and what does this difference imply?

somatosensory cortex. Other parts of the body, such as the trunk and lower limbs, are represented by much smaller areas. The relative sizes of the areas in the somatosensory cortex are proportional to the number of specialized sensory receptors within the corresponding part of the body. For example, there are many sensory receptors in the skin of the lips but few in the skin of the trunk. The size of the cortical area that represents a body part may expand or shrink somewhat, depending on the quantity of sensory impulses received from that body part. For example, people who learn to read Braille ultimately have a larger representation of the fingertips in the somatosensory cortex.

Somatic Sensory Pathways to the Cerebellum

Two tracts in the spinal cord—the **posterior spinocerebellar tract** (spī-nō-ser-e-BEL-ar) and the **anterior spinocerebellar tract**—are the major routes for proprioceptive impulses to the cerebellum. Although not consciously perceived, sensory impulses conveyed to the cerebellum along these two pathways are critical for posture, balance, and coordination of skilled movements. Two other tracts also convey impulses from proprioceptors of the trunk and upper limbs to the cerebellum: the **cuneocerebellar tract** and the **rostral spinocerebellar tract.**

Table 19.3 summarizes the major somatic sensory tracts in the spinal cord and pathways in the brain.

Syphilis is a common sexually transmitted disease caused by the bacterium *Treponema pallidum.* If the infection is not treated, syphilis progresses through three clinical stages, the third of which—**tertiary syphilis**—typically causes debilitating neurological symptoms. A common outcome is progressive degeneration of the posterior portions of the spinal cord, including the posterior columns, posterior spinocerebellar tracts, and posterior roots. Somatic sensations are lost, and the person's gait becomes uncoordinated because proprioceptive impulses fail to reach the cerebellum. People suffering from tertiary syphilis often watch their feet while walking to maintain their balance. However, vision does not fully compensate for the loss of proprioceptive impulses conducted via the spinocerebellar tracts, and thus body movements are uncoordinated and jerky. ■

- ✓ Describe three differences between the posterior column–medial lemniscus pathway and the anterolateral pathways.
- ✓ How are various parts of the body represented in the somatosensory cortex? Which body parts have the largest representation?
- ✓ What type of sensory information is carried in the spinocerebellar tracts, and what is its usefulness?

SOMATIC MOTOR PATHWAYS

Objectives

- Compare the locations and functions of the direct and indirect motor pathways.
- Explain how the basal ganglia and cerebellum contribute to motor responses.

Control of body movements involves several regions of the brain. Motor portions of the cerebral cortex play the major role in initiating and controlling precise, discrete muscular movements. The basal ganglia help establish a normal level of muscle tone and integrate semivoluntary, automatic movements, whereas the cerebellum assists the motor cortex and basal ganglia by making body movements smooth and coordinated and by helping to maintain normal posture and balance. Two main types of somatic motor pathways extend from the brain to the skeletal muscles: Direct somatic motor pathways extend from the cerebral cortex into the spinal cord and out to skeletal muscles; indirect somatic motor pathways extend from various regions of the brain stem into the spinal cord and out to skeletal muscles.

Before we examine these pathways we consider the role of the motor cortex in voluntary movement.

Mapping of the Motor Cortex

The **primary motor area (precentral gyrus)** of the cerebral cortex is the major control region for initiation of voluntary movements. The adjacent **premotor area** and even the **primary somatosensory area** in the postcentral gyrus also contribute fibers to the descending motor pathways. As is true for somatic sensory representation in the somatosensory area, different muscles are represented unequally in the primary motor area (Figure 19.5b). The degree of representation is proportional to the number of motor units in a particular muscle of the body. For example, the muscles in the thumb, fingers, lips, tongue, and vocal cords have large representations, whereas the trunk has a much smaller representation. By comparing Figures 19.5a and 19.5b, you can see that somatosensory and motor representations are similar but not identical for any given part of the body.

Direct Motor Pathways

Nerve impulses for voluntary movements propagate from the motor cortex to somatic motor neurons that innervate skeletal muscles via the **direct,** or **pyramidal, pathways** (Figure 19.6). The simplest of these pathways consists of sets of two neurons: upper motor neurons and lower motor neurons. About 1 million cell bodies of direct pathway **upper motor neurons (UMNs)** are located in the motor cortex, 60% in the precentral gyrus and 40% in the postcentral gyrus. Their axons descend through the internal capsule of the cerebrum. In the medulla oblongata, the axon bundles form the ventral bulges known as the pyramids. About 90% of the axons of upper motor neurons decussate (cross over) to the contralateral (opposite) side in the medulla oblongata. The 10% that remain on the ipsilateral (same) side eventually decussate at lower levels. Thus, the motor cortex of the right side of the brain controls muscles on the left side of the body, and the motor cortex of the left side of the brain controls muscles on the right side of the body. The upper motor neuron axons terminate in nuclei of cranial nerves in the brain stem or in the anterior gray horn of the spinal cord.

Lower motor neurons (LMNs) extend from the motor nuclei of cranial nerves to skeletal muscles of the face and head, and from the anterior horn of each spinal cord segment to skeletal muscle fibers of the trunk and limbs. Close to their termination point, most upper motor neurons synapse with an interneuron, which, in turn, synapses with a lower motor neuron. Thus, most impulses from the brain are conveyed to interneurons before being received by lower motor neurons. (A few upper motor neurons synapse directly with lower motor neurons). Lower motor neurons are the only neurons that carry information from the CNS to skeletal muscle fibers, and in that role each of them receives and integrates an enormous amount of excitatory and inhibitory input from many presynaptic neurons, both upper motor neurons and interneurons. For this reason, lower motor neurons are also called the **final common pathway.**

The direct pathways convey impulses from the cortex that result in precise, voluntary movements. Three tracts contain axons of upper motor neurons:

1. ***Lateral corticospinal tracts.*** Axons of upper motor neurons that decussate in the medulla form the **lateral corticospinal tracts** (kor′-ti-kō-SPI-nal) in the right and left lateral white columns of the spinal cord (Figure 19.6). Axons of lower motor neurons emerge from all levels of the spinal cord in the anterior roots of spinal nerves and terminate in skeletal muscles. These motor neurons control skilled movements of the limbs, hands, and feet.
2. ***Anterior corticospinal tracts.*** Axons of upper motor neurons that do not decussate in the medulla form the **anterior corticospinal tracts** in the right and left anterior white columns (Figure 19.6). At the spinal cord level where they

Figure 19.6 / Direct motor pathways whereby signals initiated by the primary motor area in the right hemisphere control skeletal muscles on the left side of the body. Spinal cord tracts carrying impulses of direct motor pathways are the lateral corticospinal tract and anterior corticospinal tract.

Direct pathways convey impulses that result in precise, voluntary movements.

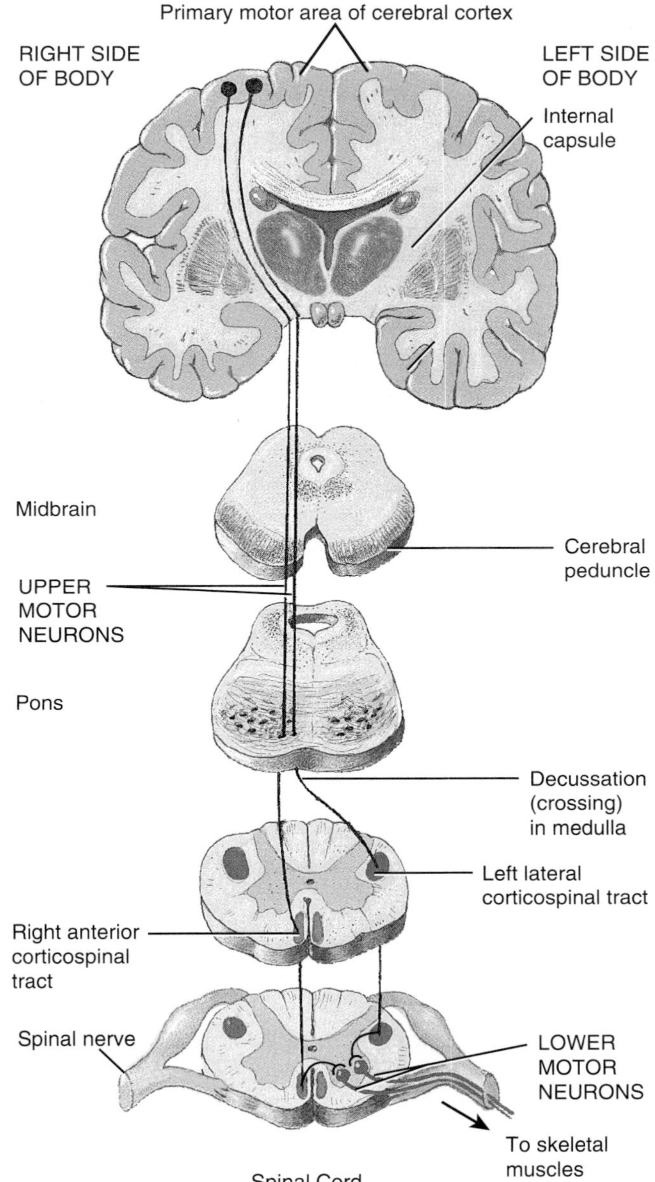

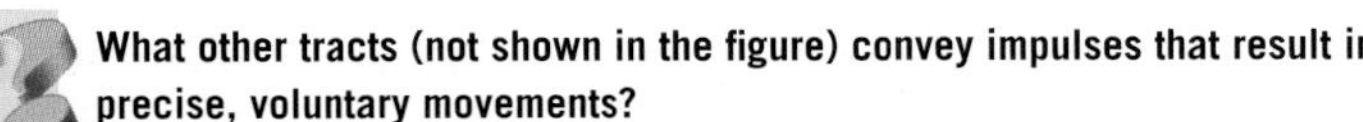
What other tracts (not shown in the figure) convey impulses that result in precise, voluntary movements?

terminate, some of the axons of these upper motor neurons decussate. After crossing over to the opposite side of the cord, they synapse with interneurons or lower motor neurons in the anterior gray horn of the spinal cord. Axons of these lower motor neurons exit the cervical and upper thoracic segments of the cord in the anterior roots of spinal nerves. They terminate in skeletal muscles that control movements of the neck and part of the trunk, thus coordinating movements of the axial skeleton.

3. *Corticobulbar tracts.* Some of the axons of upper motor neurons that conduct impulses for the control of skeletal muscles in the head extend through the internal capsule to the midbrain, where they form the **corticobulbar tracts** (kor′-ti-kō-BUL-bar) in the right and left cerebral peduncles (see Table 19.4). Some of the axons in the corticobulbar tracts have decussated, whereas others have not. The axons terminate in the nuclei of nine pairs of cranial nerves in the pons and medulla oblongata: the oculomotor (III), trochlear (IV), trigeminal (V), abducens (VI), facial (VII), glossopharyngeal (IX), vagus (X), accessory (XI), and hypoglossal (XII). The lower motor neurons of cranial nerves convey impulses that control precise, voluntary movements of the eyes, tongue, and neck, plus chewing, facial expression, and speech.

Table 19.4 summarizes the functions and pathways of the direct motor tracts.

During a neurological exam, assessment of muscle tone, reflexes, and the ability to perform voluntary movements helps pinpoint certain types of motor system dysfunction. Damage or disease of *lower* motor neurons, either of their cell bodies in the anterior horn or of their axons in the anterior root or spinal nerve, produces **flaccid paralysis** of muscles on the same side of the body. There is neither voluntary nor reflex action of the innervated muscle fibers, muscle tone is decreased or lost, and the muscle remains limp or flaccid. Injury or disease of *upper* motor neurons causes **spastic paralysis** of muscles on the opposite side of the body. This condition is characterized by varying degrees of spasticity (increased muscle tone), exaggerated reflexes, and pathological reflexes such as the Babinski sign.

Indirect Motor Pathways

The **indirect,** or **extrapyramidal, pathways** include all somatic motor tracts other than the corticospinal and corticobulbar tracts. Axons of neurons that carry nerve impulses from the indirect pathways descend from various nuclei of the brain stem into five major tracts of the spinal cord and terminate on interneurons or lower motor neurons. These tracts are the **rubrospinal** (RŪ-brō-spī-nal), **tectospinal** (TEK-tō-spī-nal), **vestibulospinal** (ves-TIB-yū-lō-spī-nal), **lateral reticulospinal** (re-TIK-yū-lō-spi-nal), and **medial reticulospinal tracts.**

Table 19.4 Major Somatic Motor Pathways in the Brain and Tracts in the Midbrain and Spinal Cord

Tract and Location	Functions and Pathways
Direct (pyramidal) tracts 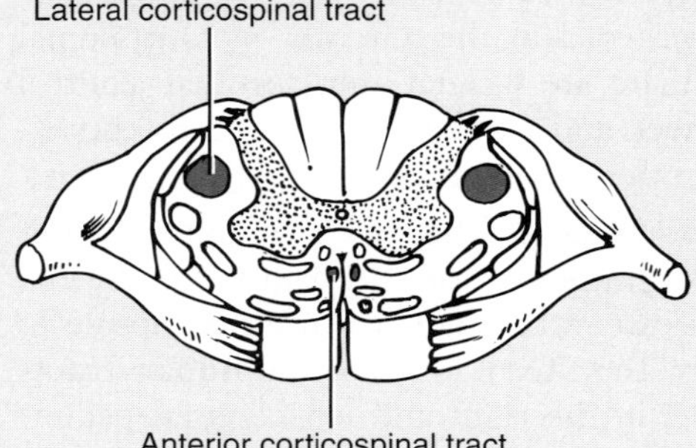	**Lateral corticospinal:** Conveys nerve impulses from the motor cortex to skeletal muscles on opposite side of body for precise, voluntary movements of the limbs, hands, and feet. Axons of upper motor neurons (UMNs) descend from the precentral and postcentral gyri of the cortex into the medulla. Here 90% decussate (cross over to the opposite side) and then enter the contralateral side of the spinal cord to form this tract. At their level of termination, these UMNs end in the anterior gray horn on the same side. They provide input to lower motor neurons, which innervate skeletal muscles. **Anterior corticospinal:** Conveys nerve impulses from the motor cortex to skeletal muscles on opposite side of body for movements of the axial skeleton. Axons of UMNs descend from the cortex into the medulla. Here the 10% that do not decussate enter the spinal cord and form this tract. At their level of termination, these UMNs decussate and end in the anterior gray horn on the opposite side. They provide input to lower motor neurons, which innervate skeletal muscles.
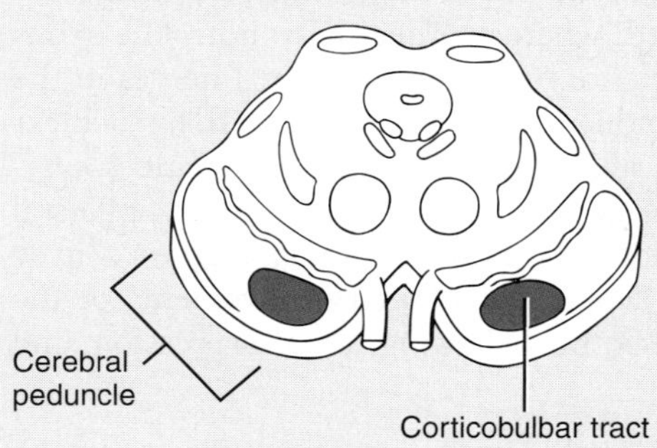	**Corticobulbar:** Conveys nerve impulses from the motor cortex to skeletal muscles of the head and neck to coordinate precise, voluntary movements. Some axons of UMNs descend from the cortex into the brain stem, where some decussate and others do not. They provide input to lower motor neurons in the nuclei of cranial nerves III, IV, V, VI, VII, IX, X, XI, and XII, which control voluntary movements of the eyes, tongue, and neck; chewing; facial expression; and speech.
Indirect (extrapyramidal) tracts 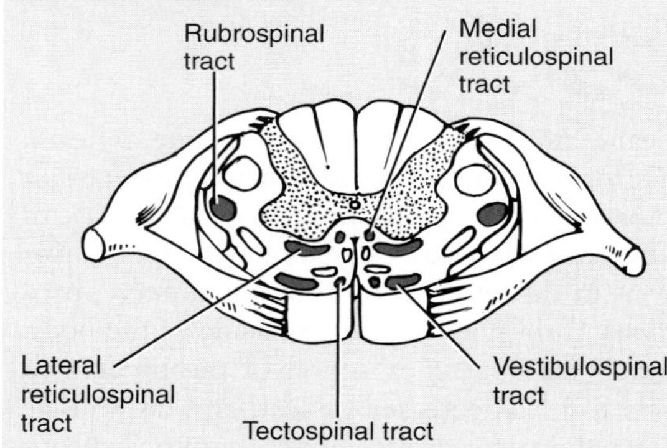	**Rubrospinal:** Conveys nerve impulses from the red nucleus (which receives input from the cortex and cerebellum) to contralateral skeletal muscles that govern precise movements of the distal portions of limbs. **Tectospinal:** Conveys nerve impulses from the superior colliculus to contralateral skeletal muscles that move the head and eyes in response to visual stimuli. **Vestibulospinal:** Conveys nerve impulses from the vestibular nucleus (which receives input about head movements from the vestibular apparatus in the inner ear) to regulate ipsilateral muscle tone for maintaining balance in response to head movements. **Lateral reticulospinal:** Conveys nerve impulses from the reticular formation that facilitate flexor reflexes, inhibit extensor reflexes, and decrease muscle tone in muscles of the axial skeleton and proximal portions of the limbs. **Medial reticulospinal:** Conveys nerve impulses from the reticular formation that facilitate extensor reflexes, inhibit flexor reflexes, and increase muscle tone in muscles of the axial skeleton and proximal portions of the limbs.

1. **Rubrospinal tract.** This tract begins in the red nucleus of the midbrain, which receives impulses from the cerebral cortex and the cerebellum. Like the adjacent lateral corticospinal tract, the rubrospinal tract transmits nerve impulses to the opposite side of the body and governs precise, discrete movements, especially of the distal limbs. An example would be the contractions of finger muscles needed to manipulate scissors while cutting paper.
2. **Tectospinal tract.** This tract begins in the superior colliculus of the midbrain, which receives visual input and governs reflex movements of the head and eyes. Tectospinal fibers transmit impulses to the opposite side of the body to neck muscles that control movements of the head in response to visual stimuli.
3. **Vestibulospinal tract.** This tract begins in the vestibular nucleus of the medulla, which receives impulses from the receptors for balance in the ear (vestibular apparatus). It conveys impulses on the same side of the body that regulate muscle tone in response to movements of the head. This tract therefore plays a major role in balance.
4. **Lateral reticulospinal tract.** This tract originates in the reticular formation of the medulla. Its function is to facilitate flexor reflexes, inhibit extensor reflexes, and decrease muscle tone in muscles of the axial skeleton and proximal limbs.

5. **Medial reticulospinal tract.** This tract originates in the pons. Its function is to facilitate extensor reflexes, inhibit flexor reflexes, and increase muscle tone in muscles of the axial skeleton and proximal limbs.

Table 19.4 summarizes the functions and pathways of the indirect motor tracts.

Roles of the Basal Ganglia

The basal ganglia have many connections with other parts of the brain. Through these connections they help to program habitual or automatic movement sequences such as arm swinging during walking or laughing in response to a joke and to set an appropriate level of muscle tone. The caudate nucleus and the putamen receive input from sensory, motor, and association areas of the cortex and the substantia nigra. Output from the basal ganglia comes mainly from the globus pallidus. (Figure 18.13b on page 566 shows these parts of the basal ganglia.) The globus pallidus, in turn, sends feedback signals to the motor cortex by way of the thalamus. This feedback circuit—from cortex to basal ganglia to thalamus to cortex—appears to function in planning and programming movements. The globus pallidus also sends impulses into the reticular formation that reduce muscle tone. Damage or destruction of certain basal ganglia connections causes a generalized increase in muscle tone that produces abnormal muscle rigidity, as occurs in Parkinson disease (see page 606).

In addition to their motor functions, the basal ganglia are also involved in all aspects of cortical function, including sensory, limbic, cognitive, and linguistic functions.

Roles of the Cerebellum

The cerebellum is active in both learning and performing rapid, coordinated, highly skilled movements such as hitting a golf ball, speaking, and swimming. It also functions to maintain proper posture and equilibrium (balance). Cerebellar function involves four activities:

1. The cerebellum *monitors intentions for movement* by receiving impulses from the motor cortex and basal ganglia via the pontine nuclei in the pons regarding what movements are planned.
2. The cerebellum *monitors actual movement* by receiving input from proprioceptors in joints and muscles that reveals what actually is happening. These nerve impulses travel in the anterior and posterior spinocerebellar tracts. Nerve impulses from the vestibular (equilibrium-sensing) apparatus in the inner ear and from the eyes also enter the cerebellum.
3. The cerebellum *compares the command signals* (intentions for movement) *with sensory information* (actual performance).
4. The cerebellum *sends out corrective signals*, both to nuclei in the brain stem and to the motor cortex via the thalamus.

In summary, the cerebellum receives information from higher brain centers about what the muscles should be doing, and from the peripheral nervous system about what the muscles are doing. If there is a discrepancy between the two, corrective feedback signals are sent from the cerebellum via the thalamus to the cerebrum, where new commands are initiated to decrease the discrepancy and smooth the motion.

Skilled activities such as tennis or volleyball provide good examples of the contribution of the cerebellum to movement. To make a good stop volley or to block a spike, you must bring your racket or arms forward just far enough to make solid contact. How do you stop at exactly the right point? Before you even hit the ball, the cerebellum has sent nerve impulses to the cerebral cortex and basal ganglia informing them where your swing must stop. In response to impulses from the cerebellum, the cortex and basal ganglia transmit motor impulses to opposing body muscles to stop the swing.

- ✓ Which parts of the body have the largest and smallest representation in the motor cortex?
- ✓ Explain why the two main somatic motor pathways are named "direct" and "indirect."
- ✓ Given the normal functions of the basal ganglia and the cerebellum, what differences might you see with disease of the basal ganglia versus damage to the cerebellum?

INTEGRATION OF SENSORY INPUT AND MOTOR OUTPUT

Objective

- Explain how sensory input and motor output are integrated.

Sensory pathways provide the input that keeps the central nervous system informed of changes in the external and internal environment. Output from the CNS is then conveyed through motor pathways, which enable movement and glandular secretions. As sensory information reaches the CNS, it becomes part of a large pool of sensory information. Every input the CNS receives, however, does not necessarily elicit a response. Rather, the incoming information is **integrated;** that is, it is combined with and modified by other information arriving from all other operating sensory receptors.

The integration process occurs not just once but at many places along the pathways of the CNS and at both conscious and subconscious levels. It occurs within the spinal cord, brain stem, cerebellum, basal ganglia, and cerebral cortex. As a result, the output descending along a motor pathway that makes a muscle contract or a gland secrete can be modified and responded to at any of these levels. Motor portions of the cerebral cortex play the major role for initiating and controlling precise movements of muscles. The basal ganglia largely integrate semivoluntary, automatic movements such as walking, swimming, and laughing. The cerebellum assists the motor cortex and basal ganglia by making body movements smooth and coordinated and by contributing significantly to maintaining normal posture and balance.

- ✓ What does integration of sensory input mean?

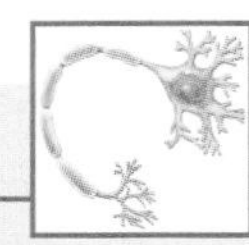

APPLICATIONS TO HEALTH

Spinal Cord Injury

The spinal cord may be damaged by a tumor either within or adjacent to the spinal cord, herniated intervertebral discs, blood clots, penetrating wounds caused by projectile fragments, or other trauma. In many cases of traumatic injury of the spinal cord, the patient has an improved outcome if an anti-inflammatory corticosteroid drug called methylprednisolone is given within 8 hours of the injury.

Depending on the location and extent of spinal cord damage, paralysis may occur. **Monoplegia** (*mono-* = one; *-plegia* = blow or strike) is paralysis of one limb only. **Diplegia** (*di-* = two) is paralysis of both upper limbs or both lower limbs. **Paraplegia** (*para-* = beyond) is paralysis of both lower limbs. **Hemiplegia** (*hemi-* = half) is paralysis of the upper limb, trunk, and lower limb on one side of the body, and **quadriplegia** (*quad-* = four) is paralysis of all four limbs.

Complete transection of the spinal cord means that the cord is severed from one side to the other, thus cutting all sensory and motor tracts. It results in a loss of all sensations and voluntary movement *below* the level of the transection. **Hemisection** is a partial transection of the cord on either the right or left side. Below the level of the hemisection there is a loss of proprioception and discriminative touch sensations on the same side as the injury if the posterior column is cut, paralysis on the same side if the lateral corticospinal tract is cut, and loss of pain, temperature, and crude tactile sensations of the opposite side if the spinothalamic tracts are cut.

Following complete transection, and to varying degrees after hemisection, spinal shock occurs. **Spinal shock** is an immediate response to spinal cord injury characterized by temporary loss of reflex function, called **areflexia** (a′-rē-FLEK-sē-a), below the level of the injury. Signs of acute spinal shock include slow heart rate, low blood pressure, flaccid paralysis of skeletal muscles, loss of somatic sensations, and urinary bladder dysfunction. Spinal shock may begin within 1 hour after injury and may last from several minutes to several months, after which reflex activity gradually returns.

Cerebral Palsy

Cerebral palsy (CP) is a group of motor disorders that cause loss of muscle control and coordination. It is due to damage of the motor areas of the brain during fetal life, birth, or infancy and occurs in about 2 of every 1000 children. One cause is infection of the mother by the German measles (rubella) virus during her first 3 months of pregnancy. Radiation during fetal life, temporary lack of oxygen during birth, and hydrocephalus during infancy may also cause cerebral palsy. Cerebral palsy is not a progressive disease; it does not worsen as time elapses. Once the damage is done, however, it is irreversible.

Parkinson Disease

Parkinson disease (PD) is a progressive disorder of the CNS that typically affects its victims around age 60. The cause is unknown, but toxic environmental factors are suspected, in part because only 5% of PD patients have a family history of the disease. Neurons that extend from the substantia nigra to the basal ganglia, where they release the neurotransmitter dopamine (DA), degenerate in PD. Also in the basal ganglia—in the caudate nucleus—are neurons that liberate the neurotransmitter acetylcholine (ACh). Although the level of ACh does not change as the level of DA declines, the imbalance of neurotransmitter activity—too little DA and too much ACh—is thought to bring about most of the symptoms.

In PD patients, involuntary skeletal muscle contractions often interfere with voluntary movement. For instance, the muscles of the upper limb may alternately contract and relax, causing the hand to shake. This shaking, called **tremor,** is the most common symptom of PD. Also, muscle tone may increase greatly, causing rigidity of the involved body part. Rigidity of the facial muscles gives the face a masklike appearance. The expression is characterized by a wide-eyed, unblinking stare and a slightly open mouth with uncontrolled drooling.

Motor performance is also impaired by **bradykinesia** (*brady-* = slow), in which activities such as shaving, cutting food, and buttoning a blouse take longer and become increasingly more difficult as the disease progresses. Muscular movements are performed not only slowly but with decreasing range of motion, or **hypokinesia** (*hypo-* = under). For example, handwritten letters get smaller, become poorly formed, and eventually become illegible. Often, walking is impaired; steps become shorter and shuffling, and arm swing diminishes. Even speech may be affected.

Treatment of PD is directed toward increasing levels of DA and decreasing levels of ACh. Although people with PD do not manufacture enough dopamine, taking it orally is useless because DA cannot cross the blood–brain barrier. Even though symptoms are partially relieved by a drug developed in the 1960s called levodopa (L-dopa), a precursor of DA, the drug does not slow the progression of the disease. As more and more affected brain cells die, the drug becomes useless. In an alternative approach, a drug called selegiline (deprenyl) inhibits monoamine oxidase, which is one of the enzymes that degrades catecholamine neurotransmitters, including dopamine. This drug slows progression of PD and may be used together with levodopa.

For more than a decade, surgeons have sought to reverse the effects of Parkinson disease by transplanting dopamine-rich fetal nerve tissue into the basal ganglia (usually the putamen) of patients with severe PD. Only a few postsurgical patients have shown any degree of improvement, such as less rigidity and improved quickness of motion. A more recent surgical technique that has produced improvement for some patients is *pallidotomy*, in which a part of the globus pallidus that generates tremors and produces muscle rigidity is destroyed.

KEY MEDICAL TERMS ASSOCIATED WITH GENERAL SENSES AND SENSORY AND MOTOR PATHWAYS

Arousal (a-ROW-zal) Awakening from sleep, a response due to stimulation of the reticular activating system (RAS).

Coma A state of deep unconsciousness from which a person cannot be aroused due to damage to the RAS or other parts of the brain. A comatose patient lies in a sleeplike state with eyes closed. In the lightest stages of coma, brain stem and spinal cord reflexes persist, but in the deepest stages, even reflexes are lost. If respiratory and cardiovascular controls are lost, the patient dies.

Consciousness (KON-shus-nes) A state of wakefulness in which an individual is fully alert, aware, and oriented, partly as a result of feedback between the cerebral cortex and reticular activating system.

Learning The ability to acquire new knowledge or skills through instruction or experience.

Memory The process by which knowledge acquired through learning is retained over time.

Sleep A state of altered consciousness or partial unconsciousness from which an individual can be aroused by many different stimuli.

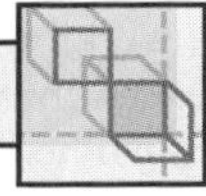

STUDY OUTLINE

Overview of Sensations (p. 591)

1. Sensation is the conscious or subconscious awareness of external and internal conditions of the body.
2. For a sensation to occur, a stimulus must reach a sensory receptor, the stimulus must be converted to a nerve impulse, and the impulse conducted to the brain; finally, the impulse must be integrated by a region of the brain.
3. When stimulated, most sensory receptors ultimately produce one or more nerve impulses.
4. Sensory impulses from each part of the body arrive in specific regions of the cerebral cortex.
5. Modality is the distinct quality that makes one sensation different from others.
6. Adaptation is a decrease in sensation during a prolonged stimulus.
7. Two general classes of senses are general senses, which include somatic senses and visceral senses, and special senses, which include the modalities of smell, taste, vision, hearing, and equilibrium (balance).
8. Table 19.1 on page 592 summarizes the classification of sensory receptors.

Somatic Sensations (p. 592)

1. Somatic sensations include tactile sensations (touch, pressure, vibration, itch, and tickle), thermal sensations (warmth and cold), pain, and proprioception.
2. Receptors for tactile, thermal, and pain sensations are located in the skin, subcutaneous layer, and mucous membranes of the mouth, vagina, and anus.
3. Receptors for proprioceptive sensations (position and movement of body parts) are located in muscles, tendons, joints, and the internal ear.
4. Receptors for touch are (a) hair root plexuses and corpuscles of touch (Meissner corpuscles) and (b) type I cutaneous mechanoreceptors (tactile or Merkel discs). Type II cutaneous mechanoreceptors (Ruffini corpuscles) are sensitive to stretching. Receptors for pressure include corpuscles of touch, type I mechanoreceptors, and lamellated (Pacinian) corpuscles. Receptors for vibration are corpuscles of touch and lamellated corpuscles. Itch receptors are free nerve endings, whereas both free nerve endings and lamellated corpuscles mediate the tickle sensation.
5. Thermoreceptors are free nerve endings. Cold receptors are located in the stratum basale of the epidermis; warm receptors are located in the dermis.
6. Pain receptors (nociceptors) are free nerve endings that are located in nearly every body tissue.
7. Proprioceptors include muscle spindles, tendon organs, joint kinesthetic receptors, and hair cells of the inner ear.
8. Table 19.2 on page 598 summarizes the somatic sensory receptors and the sensations they convey.

Somatic Sensory Pathways (p. 597)

1. Somatic sensory pathways from receptors to the cerebral cortex involve three-neuron sets: first-order, second-order, and third-order neurons.
2. Axon collaterals (branches) of somatic sensory neurons simultaneously carry signals into the cerebellum and the reticular formation of the brain stem.
3. Impulses propagating along the posterior column–medial lemniscus pathway relay discriminative touch, stereognosis, proprioception, weight discrimination, and vibratory sensations.
4. The neural pathway for pain and temperature sensations is the lateral spinothalamic tract.
5. The neural pathway for tickle, itch, crude touch, and pressure sensations is the anterior spinothalamic pathway.
6. Specific regions of the primary somatosensory area (postcentral gyrus) of the cerebral cortex receive information from different parts of the body.
7. The pathways to the cerebellum are the anterior and posterior spinocerebellar tracts, which transmit impulses for subconscious muscle and joint position sense from the trunk and lower limbs.
8. Table 19.3 on page 600 summarizes the major somatic sensory pathways.

Somatic Motor Pathways (p. 602)

1. The primary motor area (precentral gyrus) of the cortex is the major control region for initiation of voluntary movement.
2. Impulses governing voluntary movements propagate from the motor cortex to somatic motor neurons that innervate skeletal muscles via the direct pathways. The simplest pathways consist of upper and lower motor neurons.
3. The direct (pyramidal) pathways include the lateral and anterior corticospinal tracts and corticobulbar tracts.

4. Indirect (extrapyramidal) pathways extend from various regions of the brain stem into the spinal cord and out to skeletal muscles.
5. The basal ganglia program automatic movements, regulate muscle tone, and inhibit other motor neuron circuits.
6. The cerebellum is active in learning and performing rapid, coordinated, highly skilled movements. It also contributes to maintaining balance and posture.
7. Table 19.4 on page 604 summarizes the major somatic motor pathways.

Integration of Sensory Input and Motor Output (p. 605)

1. Sensory input keeps the CNS informed of changes in the environment.
2. Incoming sensory information is integrated at many stations along the CNS at both conscious and subconscious levels.
3. A motor response makes a muscle contract or a gland secrete.

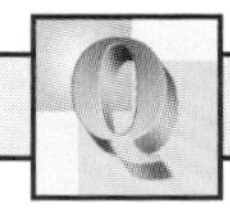

SELF-QUIZ QUESTIONS

Choose the one best answer to these questions.

1. Which of the following is the receptor for pain? (a) chemoreceptor, (b) photoreceptor, (c) nociceptor, (d) thermoreceptor, (e) mechanoreceptor
2. Which of the following sensations is *not* conveyed by the posterior column-medial lemniscus pathway? (a) pain, (b) proprioception, (c) discriminative touch, (d) temperature, (e) kinesthesia
3. The two major routes whereby proprioceptive input reaches the cerebellum are the (a) anterior and posterior spinocerebellar tracts, (b) anterior and lateral spinothalamic tracts, (c) anterior and lateral corticospinal tracts, (d) direct and indirect pathways, (e) fasciculus gracilis and fasciculus cuneatus tracts.
4. Which of the following statements is true for muscle spindles? (a) They are located within and around articular capsules of synovial joints. (b) They are located at the junction of a tendon and a muscle. (c) They respond to acceleration and deceleration of joint movement, and they monitor pressure at joints. (d) They monitor changes in the length of a skeletal muscle.
5. Which of the following statements is true for tendon organs? (a) They are located within and around articular capsules of synovial joints. (b) They are located at the junction of a tendon and a muscle. (c) They respond to acceleration and deceleration of joint movement, and they monitor pressure at joints. (d) They monitor changes in the length of a skeletal muscle.

Complete the following.

6. The five special senses include ________ , ________ , ________ , ________ and ________ .
7. Sensory receptors may be classified by location into three groups: ________ , ________ , and ________ .
8. The receptors for touch, pressure and vibration are all ________-receptors, based on the type of stimulus to which they respond.
9. Upper motor neurons of the indirect pathways begin in the ________ region of the brain and synapse with either lower motor neurons or ________ neurons.
10. The pain felt in the skin of the thorax and left arm during a heart attack is known as ________ pain.
11. Number the following in the correct sequence to indicate the route of nerve impulses along the direct (pyramidal) pathway. (a) anterior gray horn (lower motor neuron): ________ , (b) midbrain and pons: ________ , (c) effector (skeletal muscle): ________ , (d) internal capsule: ________ , (e) lateral corticospinal tract: ________ , (f) medulla oblongata decussation site: ________ , (g) precentral gyrus (upper motor neuron): ________ , (h) anterior root of spinal nerve: ________ .
12. Fill in the blanks with 1 for first-order sensory neurons, 2 for second-order sensory neurons, and 3 for third-order sensory neurons: (a) extend from the thalamus to the postcentral gyrus of the cerebral cortex: ________ , (b) extend from the facial region along cranial nerves into the brain stem: ________ , (c) extend from the spinal cord to the thalamus, or from the brain stem to the thalamus: ________ , (d) extend along spinal nerves into the spinal cord: ________ .
13. The two major routes in the spinal cord by which proprioceptive impulses travel to the cerebellum are the ________ and the ________ tracts.

Are the following statements true or false?

14. Stimulation of pain receptors in skeletal muscles, joints, and tendons results in visceral pain.
15. A sensory pathway, from receptor to the cerebral cortex, involves three neurons.
16. The precentral gyrus of the cerebral cortex is the premotor area.
17. The postcentral gyrus of the cerebral cortex is the primary somatosensory area.
18. Muscle spindles are examples of joint kinesthetic receptors.
19. Match the following.

____	**(a)** located in skeletal muscles	(1) tendon organ
____	**(b)** located within and around articular capsules of synovial joints	(2) muscle spindle
____	**(c)** located at the junction of a tendon and a muscle	(3) joint kinesthetic receptor
____	**(d)** help protect tendons and their associated muscles from damage due to excessive tension	
____	**(e)** respond to acceleration and deceleration of joint movement; monitor pressure at joints	
____	**(f)** monitor changes in the length of a skeletal muscle	

20. Match the following.

____	**(a)** sensitive to movement of hair shaft	(1) corpuscles of touch (Meissner corpuscles)
____	**(b)** egg-shaped masses located in dermal papillae; plentiful in fingertips, palms, and soles; receptors for discriminative touch	(2) tactile (Merkel) discs
____	**(c)** onion-shaped structures sensitive to pressure in skin, membranes, joints and some viscera	(3) lamellated (Pacinian) corpuscles
____	**(d)** dendrites that contact epidermal cells in the stratum basale of skin; function in discriminative touch	(4) hair root plexuses

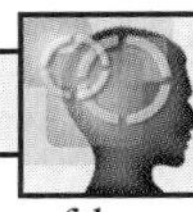

CRITICAL THINKING QUESTIONS

1. When Sau Lan held the cup of hot cocoa in her hands, at first it felt comfortably warm and within minutes, she didn't notice the temperature at all. Absently she took a big gulp and almost choked when she felt the hot cocoa burn her mouth and throat. How did the hot cocoa get hotter?
   HINT *Did the temperature really change or did something change in Sau Lan?*
2. Jenny was laughing uncontrollably. Her brother had her by the leg and was tickling her foot. How does her brain know that her foot is being tickled?
   HINT *If you blocked the pathway for itch, you'd still be able to feel pain and temperature.*
3. Compare the role of the cerebrum and the cerebellum in directing the body's movements while riding a bicycle.
   HINT *Police sobriety checks also test these functions.*
4. Very young children are usually given only spoons (no forks or knives) and cups with lids at the dinner table. Why?
   HINT *This is the same age group that's still in diapers.*
5. Jon used to predict the weather by how much the bunion (abnormally swollen joint on big toe) on his left foot was bothering him. Last year, Jon's left foot was amputated due to complications from diabetes. But sometimes, he still thinks he feels that bunion. Explain Jon's weather toe.
   HINT *His "feelings" aren't affected by the weather anymore.*

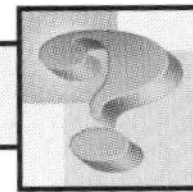

ANSWERS TO FIGURE QUESTIONS

19.1 Pain, thermal sensations, and tickle and itch involve free nerve endings.

19.2 The kidneys have the broadest area for referred pain.

19.3 Muscle spindles are activated when the central areas of their intrafusal fibers are stretched.

19.4 Damage to the right lateral spinothalamic tract could result in loss of pain and thermal sensations on the left side of the body.

19.5 The hand has a larger representation in the motor area than in the somatosensory area, which implies greater precision in the hand's movement control than discriminative ability in its sensation.

19.6 The corticobulbar and rubrospinal tracts (see Table 19.4 on page 604) convey impulses that result in precise, voluntary movements.

20

SPECIAL SENSES

CONTENTS AT A GLANCE

◄ Page 624

How many senses do we have? Reconcile this caricature of the senses with what you read in this chapter, as well what you have already read in the previous chapter.

Page 616 ►

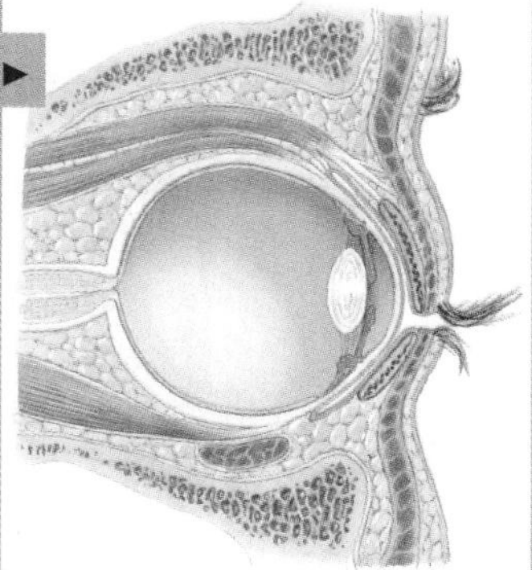

INTRODUCTION

Receptors for the general senses are scattered throughout the body and are relatively simple in structure. They are mostly modified dendrites of sensory neurons. Receptors for the special senses—smell, taste, vision, hearing, and equilibrium—are anatomically distinct from one another and are concentrated in specific locations in the head. They are housed in complex sensory organs such as the eyes and ears. Moreover, cells of receptors for special senses are usually embedded in epithelial tissue within special sense organs. Neural pathways for the special senses are more complex than those for the general senses.

In this chapter we examine the structure and function of the special sense organs, and the pathways involved in conveying information from them to the central nervous system. **Ophthalmology** (of-thal-MOL-ō-jē; *ophthalmo-* = eye; *-ology* = study of) is the science that deals with the eye and its disorders. The other special senses are, in large part, the concern of **otorhinolaryngology** (ō′-tō-rī′-nō-lar-in-GOL-ō-jē; *oto-* = ear; *rhino-* = nose; *laryngo-* = larynx).

OLFACTION: SENSE OF SMELL

Objective

- Describe the olfactory receptors and the neural pathway for olfaction.

Both smell and taste are chemical senses; the sensations arise from the interaction of molecules with smell or taste receptors. Because impulses for smell and taste propagate to the limbic system (and to higher cortical areas as well), certain odors and tastes can evoke strong emotional responses or a flood of memories.

Anatomy of Olfactory Receptors

The nose contains 10–100 million receptors for the sense of smell, or **olfaction** (ol-FAK-shun; *olfact-* = smell). The total area of olfactory epithelium is 5 cm^2 (a little less than 1 $in.^2$). It occupies the superior portion of the nasal cavity, covering the inferior surface of the cribriform plate and extending along the superior nasal concha and upper part of the middle nasal concha (Figure 20.1a). The olfactory epithelium consists of three kinds of cells: olfactory receptor cells, supporting cells, and basal stem cells (Figure 20.1b).

Olfactory receptor cells, which are the first-order neurons of the olfactory pathway, are bipolar neurons whose exposed tip is a knob-shaped dendrite. The sites of olfactory transduction are the **olfactory hairs,** which are cilia that project from the dendrite. Olfactory receptor cells respond to the chemical stimulation of an odorant molecule and ultimately initiate the olfactory response. From the base of each olfactory receptor cell, a single axon projects through the cribriform plate and into the olfactory bulb.

Supporting cells are columnar epithelial cells of the mucous membrane lining the nose. They provide physical support, nourishment, and electrical insulation for the olfactory receptor cells, and they help detoxify chemicals that come in contact with the olfactory epithelium. **Basal cells** lie between the bases of the supporting cells and continually undergo cell division to produce new olfactory receptor cells, which live for only a month or so before being replaced. This process is remarkable because olfactory receptor cells are neurons, and in general, mature neurons are not replaced.

Within the connective tissue that supports the olfactory epithelium are **olfactory (Bowman's) glands,** which produce mucus that is carried to the surface of the epithelium by ducts. The secretion moistens the surface of the olfactory epithelium and dissolves odorants. Supporting cells of the nasal epithelium and olfactory glands are innervated by branches of the facial (VII) nerve, which can be stimulated by certain chemicals. Impulses in these nerves, in turn, stimulate the lacrimal glands in the eyes and nasal mucous glands. The result is tears and a runny nose after inhaling substances such as pepper or the vapors of household ammonia.

The Olfactory Pathway

On each side of the nose, bundles of the slender, unmyelinated axons of olfactory receptor cells extend through about 20 olfactory foramina in the cribriform plate of the ethmoid bone (Figure 20.1b). Collectively, these 40 or so bundles of axons make up the **olfactory (I) nerves.** These nerves terminate in the brain in paired masses of gray matter called the **olfactory bulbs,** which are located inferior to the frontal lobes of the cerebrum and lateral to the crista galli of the ethmoid bone. Within the olfactory bulbs, the axon terminals of olfactory receptor cells—the first-order neurons—form synapses with the dendrites and cell bodies of second-order neurons in the olfactory pathway.

Axons of olfactory bulb neurons extend posteriorly and form the **olfactory tract** (Figure 20.1b). These axons project to the lateral olfactory area, which is located at the inferior and medial surface of the temporal lobe. This cortical region is part of the limbic system and includes some of the amygdala (see Figure 18.14 on page 567). Because many olfactory tract axons terminate in the lateral olfactory area, it is considered the primary olfactory area, where conscious awareness of smells begins. Connections to other limbic system regions and to the hypothalamus probably account for our emotional and memory-evoked responses to odors. Examples include sexual excitement upon smelling a certain perfume, nausea upon smelling a food that once made you violently ill, or an odor-evoked memory of a childhood experience. From the lateral olfactory area, pathways also extend to the frontal lobe, both directly and indirectly via the thalamus. An important region for odor identification and discrimination is the orbitofrontal area, corresponding to

Figure 20.1 / Olfactory epithelium and olfactory receptors. (a) Location of olfactory epithelium in nasal cavity. (b) Anatomy of olfactory receptor cells, which are first-order neurons whose axons extend through the cribriform plate and terminate in the olfactory bulb. (See Tortora, *A Photographic Atlas of the Human Body,* Figure 9.1b.)

The olfactory epithelium consists of olfactory receptor cells, supporting cells, and basal stem cells.

Frontal lobe of cerebrum
Olfactory tract
Olfactory bulb
Cribriform plate of ethmoid bone
Olfactory (I) nerve
Olfactory epithelium
Superior nasal concha

(a) Sagittal view

Olfactory tract
Olfactory bulb neuron
Olfactory bulb
Cribriform plate
Olfactory (I) nerve
Connective tissue
Olfactory gland (produces mucus)
Basal cell
Developing olfactory receptor cell
Olfactory epthelium
Olfactory receptor cell
Supporting cell
Dendrite
Mucus
Olfactory hair
Odorant molecule

(b) Enlarged aspect of olfactory receptors

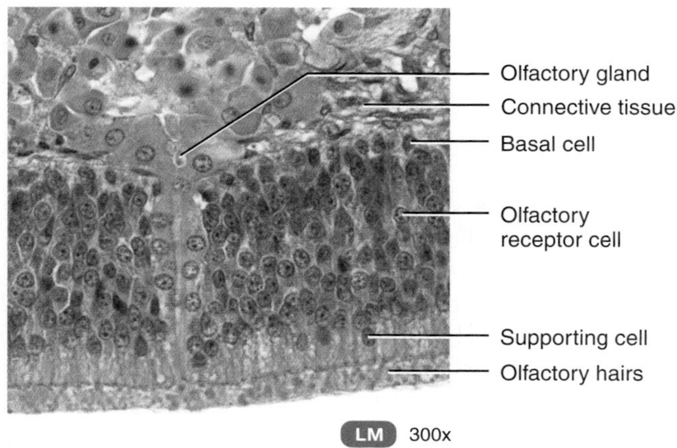

(c) Histology of olfactory epithelium

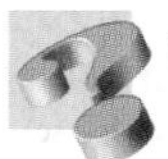

Which part of an olfactory receptor cell detects an odorant molecule?

Brodmann's area 11 (see Figure 18.15 on page 568). People who suffer damage in this area have difficulty identifying different odors. Positron emission tomography (PET) studies suggest some degree of hemispheric lateralization: The orbitofrontal area of the *right* hemisphere exhibits greater activity during olfactory processing.

✓ How do basal stem cells contribute to olfaction?
✓ Describe the pathway from the binding of an odorant molecule to an olfactory hair to the arrival of a nerve impulse in an olfactory bulb.

GUSTATION: SENSE OF TASTE

Objective

- Describe the gustatory receptors and the neural pathway for gustation.

Like olfaction, taste is a chemical sense; to be detected by either sense, stimulating molecules must be dissolved. Taste, or **gustation** (GUS-tā-shun; *gust-* = taste), is much simpler than olfaction in that only four major classes of stimuli can be distinguished: sour, sweet, bitter, and salty. All other "tastes," such as chocolate, pepper, and coffee, are combinations of these four, plus the accompanying olfactory sensations. Odors from food can pass upward from the mouth into the nasal cavity, where they stimulate olfactory receptor cells. Because olfaction is much more sensitive than taste, a given concentration of a food substance may stimulate the olfactory system thousands of times more strongly than it stimulates the gustatory system. When persons with colds or allergies complain that they cannot taste their food, they are reporting blockage of olfaction, not of taste.

Anatomy of Gustatory Receptors

The receptors for sensations of taste are located in the taste buds (Figure 20.2). The nearly 10,000 taste buds of a young adult are mainly on the tongue, but they are also found on the soft palate (posterior portion of roof of mouth), pharynx (throat), and larynx (voice box). The number of taste buds declines with age. Each **taste bud** is an oval body consisting of three kinds of epithelial cells: supporting cells, gustatory receptor cells, and basal cells (Figure 20.2c). The **supporting cells** surround about 50 **gustatory receptor cells.** A single, long microvillus, called a **gustatory hair,** projects from each gustatory receptor cell to the external surface through the **taste pore,** an opening in the taste bud. **Basal cells,** found at the periphery of the taste bud near the connective tissue layer, produce supporting cells, which then develop into gustatory receptor cells with a life span of about 10 days. At their base, the gustatory receptor cells synapse with dendrites to first-order neurons that form the first part of the gustatory pathway. The dendrites of each first-order neuron branch profusely and contact many gustatory receptor cells in several taste buds.

Taste buds are found in elevations on the tongue called **papillae** (pa-PIL-ē), which give the upper surface of the tongue its rough appearance (Figure 20.2a, b). **Circumvallate papillae** (ser-kum-VAL-āt), the largest type, are circular and form an inverted V-shaped row at the posterior portion of the tongue. **Fungiform papillae** (FUN-ji-form) are mushroom-shaped elevations scattered over the entire surface of the tongue. All circumvallate and most fungiform papillae contain taste buds. In addition, the entire surface of the tongue has **filiform papillae** (FIL-i-form), pointed, threadlike structures that rarely contain taste buds. They do, however, contain receptors for touch and increase friction between the tongue and food, making it easier for the tongue to move food in the oral cavity.

The Gustatory Pathway

Three cranial nerves include first-order gustatory fibers from taste buds. The facial (VII) nerve serves the anterior two-thirds of the tongue, the glossopharyngeal (IX) nerve serves the posterior one-third of the tongue, and the vagus (X) nerve serves the throat and epiglottis (cartilage lid over the voice box). From taste buds, impulses propagate along these cranial nerves to the medulla oblongata. From the medulla, some taste fibers project to the limbic system and the hypothalamus, and some project to the thalamus. Taste fibers extend from the thalamus to the primary gustatory area in the parietal lobe of the cerebral cortex (see area 43 in Figure 18.15 on page 568), giving rise to the conscious perception of taste.

✓ How do olfactory receptor cells and gustatoty receptor cells differ in structure and function?
✓ Compare the olfactory and gustatory pathways.

VISION

Objectives

- List and describe the accessory structures of the eye and the structural components of the eyeball and their functions.
- Describe the visual pathway.

More than half the sensory receptors in the human body are located in the eyes, and a large part of the cerebral cortex is devoted to processing visual information. We will examine the accessory structures of the eye, the eyeball, how visual images are formed, the physiology of vision, and the visual pathway.

Accessory Structures of the Eye

The **accessory structures** of the eye are the eyelids, eyelashes, eyebrows, the lacrimal (tearing) apparatus, and extrinsic eye muscles.

Figure 20.2 / The relationship of gustatory receptor cells in taste buds to tongue papillae.
(See Tortora, *A Photographic Atlas of the Human Body,* Figure 9.2.)

Gustatory receptor cells are located in taste buds.

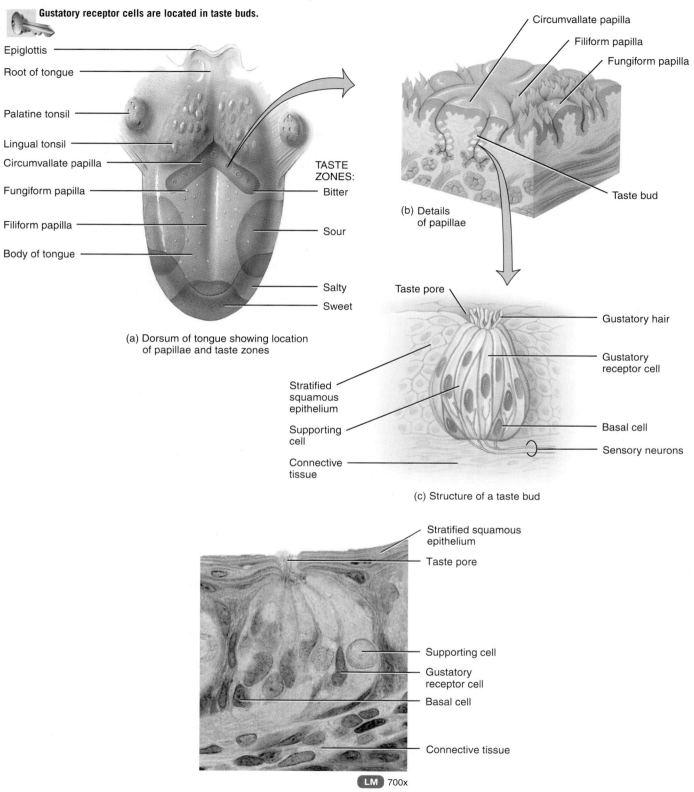

Epiglottis
Root of tongue
Palatine tonsil
Lingual tonsil
Circumvallate papilla
Fungiform papilla
Filiform papilla
Body of tongue
TASTE ZONES:
Bitter
Sour
Salty
Sweet

(a) Dorsum of tongue showing location of papillae and taste zones

Circumvallate papilla
Filiform papilla
Fungiform papilla
Taste bud

(b) Details of papillae

Taste pore
Gustatory hair
Gustatory receptor cell
Stratified squamous epithelium
Supporting cell
Basal cell
Sensory neurons
Connective tissue

(c) Structure of a taste bud

Stratified squamous epithelium
Taste pore
Supporting cell
Gustatory receptor cell
Basal cell
Connective tissue

LM 700x

(d) Histology of a taste bud from a circumvallate papilla

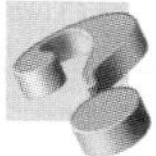

Beginning at the gustatory receptor cells, what structures form the gustatory pathway?

Eyelids

The upper and lower **eyelids,** or **palpebrae** (PAL-pe-brē), shade the eyes during sleep, protect the eyes from excessive light and foreign objects, and spread lubricating secretions over the eyeballs (Figure 20.3; see Figure 11.3 on page 348 also). The upper eyelid is more movable than the lower and contains in its superior region the **levator palpebrae superioris muscle.** The space between the upper and lower eyelids that exposes the eyeball is the **palpebral fissure.** Its angles are known as the **lateral commissure** (KOM-i-shur), which is narrower and closer to the temporal bone, and the **medial commissure,** which is broader and nearer the nasal bone. In the medial commissure is a small, reddish elevation, the **lacrimal caruncle** (KAR-ung-kul), that contains sebaceous (oil) glands and sudoriferous (sweat) glands. The whitish material that sometimes collects in the medial commissure comes from these glands.

From superficial to deep, each eyelid consists of epidermis, dermis, subcutaneous tissue, fibers of the orbicularis oculi muscle, a tarsal plate, tarsal glands, and conjunctiva (Figure 20.3). The **tarsal plate** is a thick fold of connective tissue that gives form and support to the eyelids. Embedded in each tarsal plate is a row of elongated modified sebaceous glands, known as **tarsal** or **Meibomian glands** (mī-BŌ-mē-an), that secrete a fluid that helps keep the eyelids from adhering to each other. Infection of the tarsal glands produces a tumor or cyst on the eyelid called a **chalazion** (ka-LĀ-zē-on). The **conjunctiva** (kon′-junk-TĪ-va) is a thin, protective mucous membrane composed of stratified columnar epithelium with numerous goblet cells that is supported by areolar connective tissue. The **palpebral conjunctiva** lines the inner aspect of the eyelids, and the **bulbar conjunctiva** passes from the eyelids onto the anterior surface of the eyeball lateral to the cornea. Dilation and congestion of the blood vessels of the bulbar conjunctiva due to local irritation or infection are the cause of bloodshot eyes.

Eyelashes and Eyebrows

The **eyelashes,** which project from the border of each eyelid, and the **eyebrows,** which arch transversely above the upper eyelids, help protect the eyeballs from foreign objects, perspiration, and the direct rays of the sun. Sebaceous glands at the base of the hair follicles of the eyelashes, called **sebaceous ciliary glands,** release a lubricating fluid into the follicles. Infection of these glands is called a **sty.**

The Lacrimal Apparatus

The **lacrimal apparatus** (*lacrim-* = tears) is a group of structures that produces and drains **lacrimal fluid,** or **tears.** The **lacrimal glands,** each about the size and shape of an almond, secrete lacrimal fluid, which drains into 6–12 **excretory lacrimal ducts** that empty tears onto the surface of the conjunctiva of the upper lid (Figure 20.3b). From here the tears pass medially over the anterior surface of the eyeball to enter two small openings called **lacrimal puncta.** Tears then pass into two ducts, the **lacrimal canals,** which lead into the **lacrimal sac** and then into the **nasolacrimal duct.** This duct carries the lacrimal fluid into the nasal cavity just inferior to the inferior nasal concha.

Lacrimal fluid is a watery solution containing salts, some mucus, and **lysozyme,** a protective bactericidal enzyme. The fluid protects, cleans, lubricates, and moistens the eyeball. It is spread medially over the surface of the eyeball by the blinking of the eyelids. Each gland produces about 1 ml of lacrimal fluid per day.

Normally, tears are cleared away as fast as they are produced, either by evaporation or by passing into the lacrimal canals and then into the nasal cavity. If an irritating substance contacts the conjunctiva, however, the lacrimal glands are stimulated to oversecrete, and tears accumulate (watery eyes). Lacrimation is a protective mechanism, as the tears dilute and wash away the irritating substance. Watery eyes also occur when an inflammation of the nasal mucosa, such as occurs with a cold, obstructs the nasolacrimal ducts and blocks drainage of tears. Humans are unique in expressing emotions, both happiness and sadness, by **crying.** In response to parasympathetic stimulation, the lacrimal glands produce excessive lacrimal fluid that may spill over the edges of the eyelids and even fill the nasal cavity with fluid.

Extrinsic Eye Muscles

The six extrinsic eye muscles that move each eye receive their innervation from cranial nerves III, IV, or VI. These muscles are the **superior rectus, inferior rectus, lateral rectus, medial rectus, superior oblique,** and **inferior oblique** (see Figure 10.5 on page 269). In general, the motor units in these muscles are small. Some motor neurons serve only two or three muscle fibers—fewer than in any other part of the body except the larynx (voice box)—which permits smooth, precise, and rapid movement of the eyes. As indicated in Exhibit 10.2 on page 268, the extrinsic eye muscles move the eyeball in various directions (laterally, medially, superiorly, or inferiorly). Looking to the right, for example, requires simultaneous contraction of the right lateral rectus and left medial rectus muscles and relaxation of the left lateral rectus and right medial rectus. The oblique muscles preserve rotational stability of the eyeball. Circuits in the brain stem and cerebellum coordinate and synchronize the movements of the eyes.

The surface anatomy of the accessory structures of the eye and the eyeball may be reviewed in Figure 11.3 on page 348.

Anatomy of the Eyeball

The adult **eyeball** measures about 2.5 cm (1 in.) in diameter. Of its total surface area, only the anterior one-sixth is exposed; the remainder is recessed and protected by the orbit, into which it fits.

Anatomically, the wall of the eyeball consists of three layers: fibrous tunic, vascular tunic, and retina.

Fibrous Tunic

The **fibrous tunic,** the superficial coat of the eyeball, is avascular and consists of the anterior cornea and posterior sclera

Figure 20.3 / Accessory structures of the eye.

Accessory structures of the eye are the eyelids, eyelashes, eyebrows, the lacrimal apparatus, and extrinsic eye muscles.

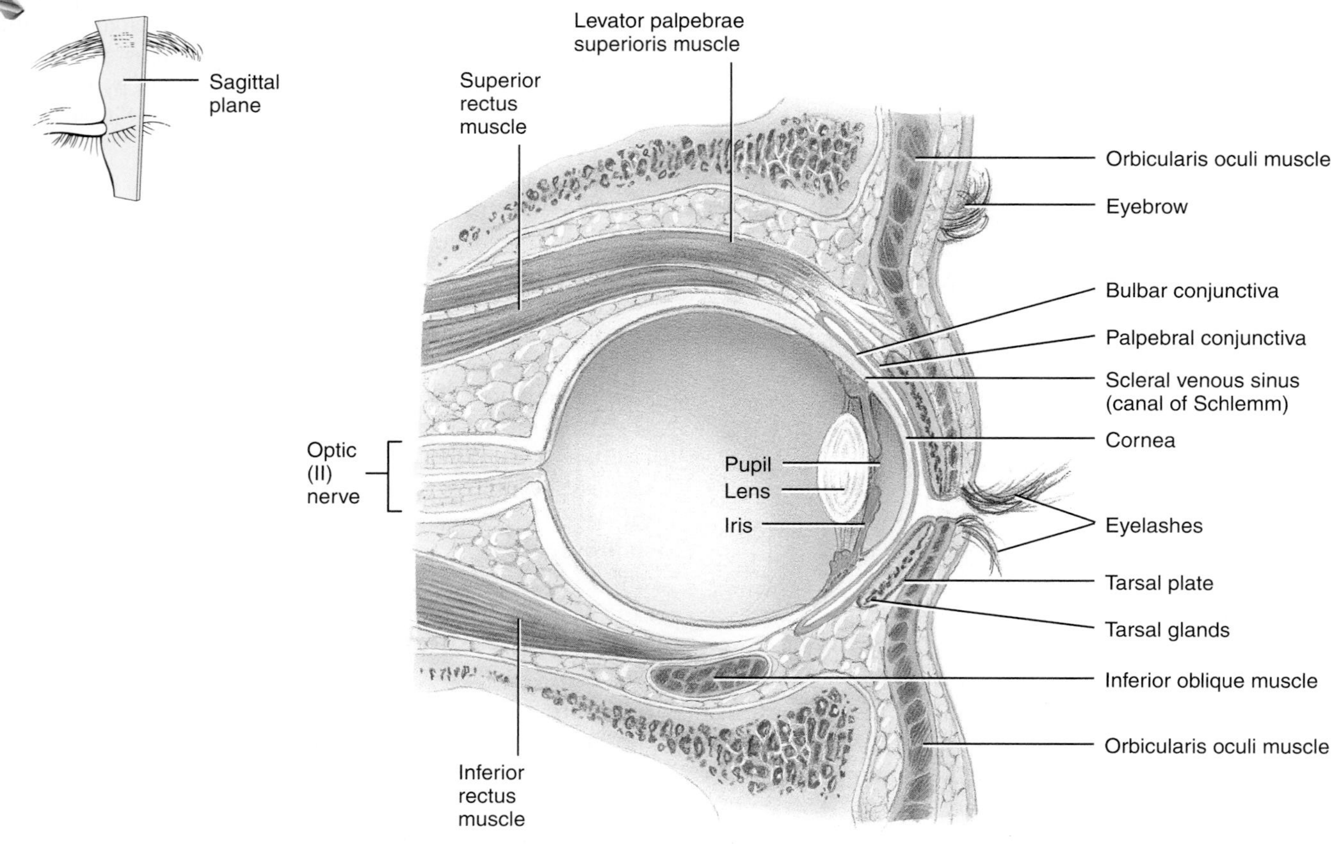

(a) Sagittal section of eye and its accessory structures

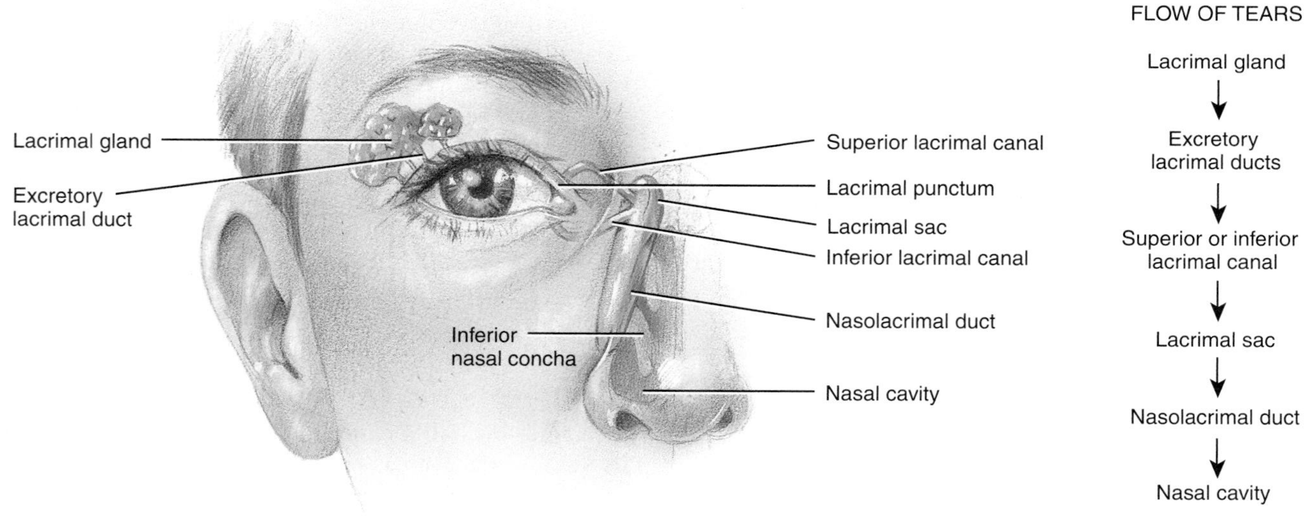

(b) Anterior view of the lacrimal apparatus

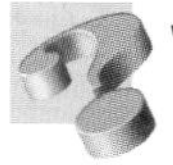

What is lacrimal fluid, and what are its functions?

(Figure 20.4). The **cornea** (KOR-nē-a) is a transparent coat that covers the colored iris. Because it is curved, the cornea helps focus light onto the retina. Its outer surface consists of nonkeratinized stratified squamous epithelium. The middle coat of the cornea consists of collagen fibers and fibroblasts, and the inner surface is simple squamous epithelium. The **sclera** (SKLE-ra; *scler-* = hard), the "white" of the eye, is a layer of dense connective tissue made up mostly of collagen fibers and fibroblasts. The sclera covers all of the eyeball except the cornea; it gives shape to the eyeball, makes it more rigid, and protects its inner parts. At the junction of the sclera and cornea is an opening known as the **scleral venous sinus (canal of Schlemm).**

Corneal Transplants

Corneal transplants are the most common organ transplant operation—and the most successful type of transplant as well. Because the cornea is avascular, blood-borne antibodies that might cause rejection do not enter the transplanted tissue, and rejection rarely occurs. The defective cornea is removed and a donor cornea of similar diameter is sewn in. The shortage of donor corneas has been partially overcome by the development of artificial corneas made of plastic. ■

Figure 20.4 / Gross structure of the eyeball. (See Tortora, *A Photographic Atlas of the Human Body,* Figure 9.3b.)

The wall of the eyeball consists of three layers: the fibrous tunic, the vascular tunic, and the retina.

Transverse plane
Lacrimal sac
Anterior cavity (contains aqueous humor):
Anterior chamber
Posterior chamber
Scleral venous sinus (canal of Schlemm)
Light
Visual axis
Cornea
Pupil
Iris
Lens
Suspensory ligament of lens
Bulbar conjunctiva
Ciliary body:
Ciliary muscle
Ciliary process
Ora serrata
Retina
Choroid
Hyaloid canal
Sclera
Medial rectus muscle
Lateral rectus muscle
Vitreous chamber (contains vitreous body)
MEDIAL
LATERAL
Central retinal artery and vein
Optic (II) nerve
Optic disc (blind spot)
Central fovea

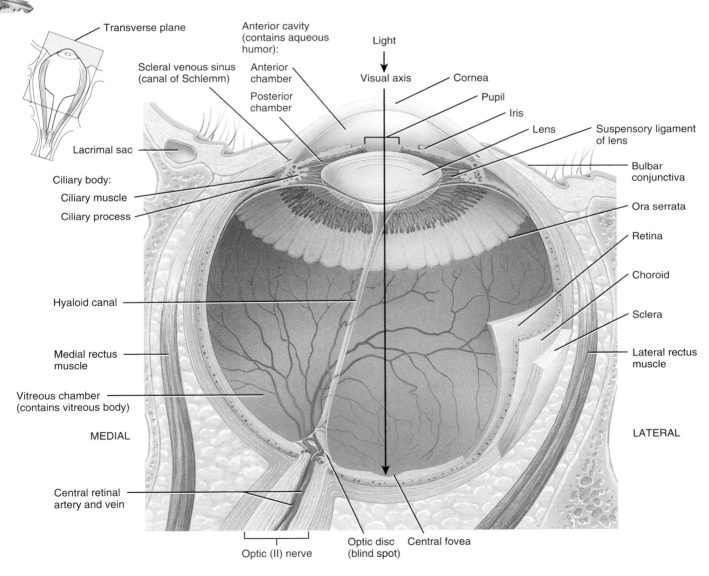

Superior view of transverse section of right eyeball

What are the components of the fibrous tunic and vascular tunic?

Vascular Tunic

The **vascular tunic,** or **uvea** (YŪ-vē-a), is the middle layer of the eyeball and has three parts: choroid, ciliary body, and iris (Figure 20.4). The highly vascularized **choroid** (KŌ-royd), which is the posterior portion of the vascular tunic, lines most of the internal surface of the sclera. It provides nutrients to the posterior surface of the retina.

In the anterior portion of the vascular tunic, the choroid becomes the **ciliary body** (SIL-ē-ar′-ē). It extends from the **ora serrata** (Ō-ra ser-RĀ-ta), the jagged anterior margin of the retina, to a point just posterior to the junction of the sclera and cornea. The ciliary body consists of the ciliary processes and the ciliary muscle. The **ciliary processes** are protrusions or folds on the internal surface of the ciliary body; they contain blood capillaries that secrete aqueous humor, and they also attach to suspensory ligaments, which connect to the lens. The **ciliary muscle** is a circular band of smooth muscle that alters the shape of the lens, adapting it for near or far vision. When the muscle contracts, the ciliary body is pulled anteriorly and inward toward the pupil, the tension in the suspensory ligaments is reduced, and the shape of the lens is altered into a more spherical shape.

The **iris,** the colored portion of the eyeball, is shaped like a flattened donut. It is suspended between the cornea and the lens and is attached at its outer margin to the ciliary processes. It consists of circular and radial smooth muscle fibers. A principal function of the iris is to regulate the amount of light entering the vitreous chamber of the eyeball through the **pupil,** the hole in the center of the iris. Autonomic reflexes regulate pupil diameter in response to light levels (Figure 20.5). When bright light stimulates the eye, parasympathetic neurons stimulate the **circular muscles (constrictor pupillae)** of the iris to contract, causing a decrease in the size of the pupil (constriction). In dim light, sympathetic neurons stimulate the **radial muscles (dilator pupillae)** of the iris to contract, causing an increase in the pupil's size (dilation).

Retina

The third and inner coat of the eyeball, the **retina,** lines the posterior three-quarters of the eyeball and is the beginning of the visual pathway (see Figure 20.4). An ophthalmoscope allows an observer to peer through the pupil, providing a magnified image of the retina and the blood vessels that course across the retina's anterior surface (Figure 20.6). The surface of the retina is the only place in the body where blood vessels can be viewed directly and examined for pathological changes, such as those that occur with hypertension or diabetes mellitus. Several landmarks are visible. The **optic disc** is the site where the optic nerve exits the eyeball. Bundled together with the optic nerve are the **central retinal artery,** a branch of the ophthalmic artery, and the **central retinal vein.** Branches of the central retinal artery fan out to nourish the anterior surface of the retina; the central retinal vein drains blood from the retina through the optic disc.

The retina consists of a pigment epithelium (nonvisual portion) and a neural portion (visual portion). The **pigment epithelium** is a sheet of melanin-containing epithelial cells located between the choroid and the neural portion of the retina. (Some histologists classify it as part of the choroid, not part of the retina.) Melanin in the choroid and in the pigment epithelium absorbs stray light rays, which prevents reflection and scattering of light within the eyeball. As a result, the image cast on the retina by the cornea and lens remains sharp and clear. Albinos lack melanin in all parts of the body, including the eye. They often need to wear sunglasses, even indoors, because even moderately bright light is perceived as bright glare due to light scattering.

The **neural portion** of the retina is a multilayered outgrowth of the brain that extensively processes visual data before transmitting nerve impulses to the thalamus. Three distinct layers of retinal neurons—the **photoreceptor layer,** the **bipolar cell layer,** and the **ganglion cell layer**—are separated by two

Figure 20.5 / Responses of the pupil to light of varying brightness.

Contraction of the circular muscles causes constriction of the pupil; contraction of the radial muscles causes dilation of the pupil.

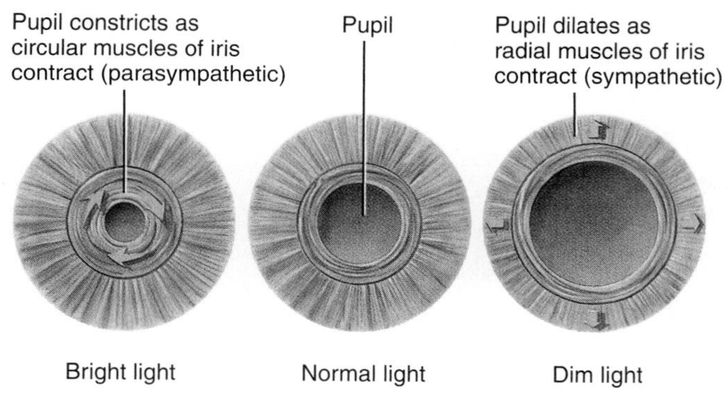

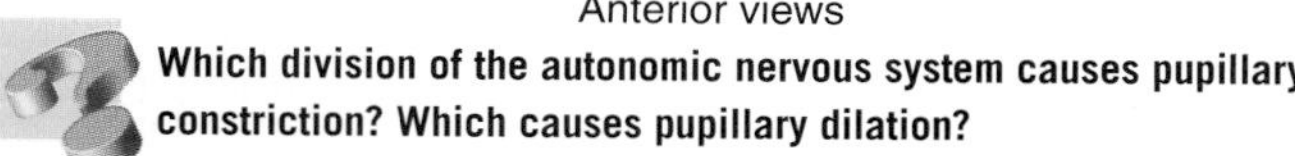

Which division of the autonomic nervous system causes pupillary constriction? Which causes pupillary dilation?

Figure 20.6 / A normal retina, as seen through an ophthalmoscope.

Blood vessels in the retina can be viewed directly and examined for pathological changes.

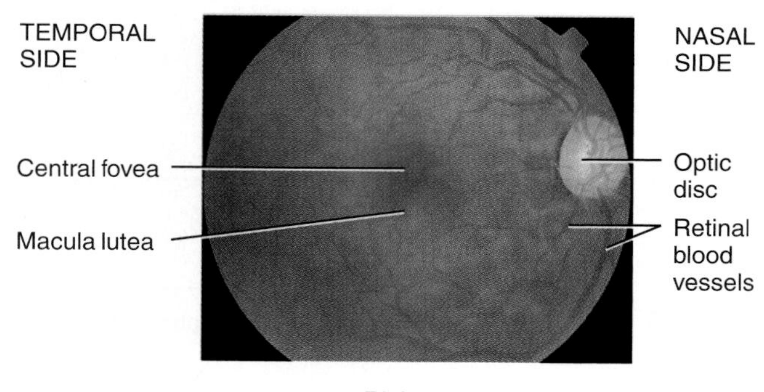

Evidence of what diseases may be seen through an ophthalmoscope?

zones, the outer and inner synaptic layers, where synaptic contacts are made (Figure 20.7). Note that light passes through the ganglion and bipolar cell layers and both synaptic layers before reaching the photoreceptor layer. Two other types of cells present in the retina are called **horizontal cells** and **amacrine cells.** These cells form laterally directed pathways that modify the signals being transmitted along the pathway from photoreceptors to bipolar cells to ganglion cells.

Detached Retina

Detachment of the retina may occur in trauma, such as a blow to the head, or in various eye disorders. The detachment occurs between the neural portion of the retina and the pigment epithelium. Fluid accumulates between these layers, forcing the

Figure 20.7 / Microscopic structure of the retina. The downward blue arrow at left indicates the direction of the signals passing through the neural portion of the retina. Eventually, nerve impulses arise in ganglion cells and propagate along their axons, which make up the optic (II) nerve.

In the retina, visual signals pass from photoreceptors to bipolar cells to ganglion cells.

Path of light through retina

Direction of visual data processing

Pigment epithelium

Photoreceptor layer

Rod

Cone

Outer synaptic layer

Bipolar cell layer

Bipolar cell

Inner synaptic layer

Ganglion cell layer

Ganglion cell

Optic (II) nerve axons

Retinal blood vessel

Nerve impulses propagate along optic (II) nerve axons toward optic disk

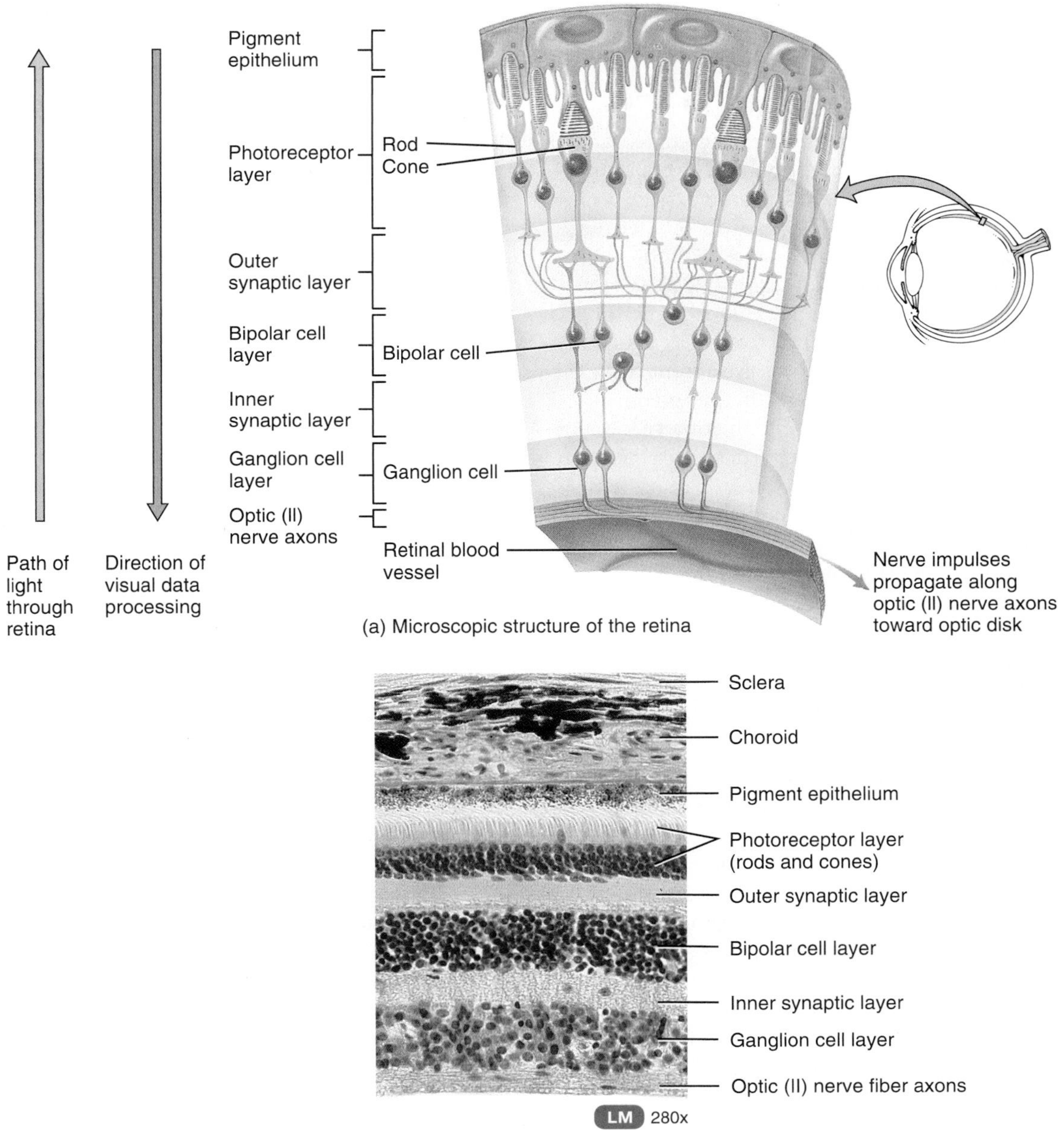

(a) Microscopic structure of the retina

Sclera

Choroid

Pigment epithelium

Photoreceptor layer (rods and cones)

Outer synaptic layer

Bipolar cell layer

Inner synaptic layer

Ganglion cell layer

Optic (II) nerve fiber axons

LM 280x

(b) Histology of a portion of the retina

What are the two types of photoreceptors, and how do their functions differ?

thin, pliable retina to billow outward. The result is distorted vision and blindness in the corresponding field of vision. The retina may be reattached by laser surgery or cryosurgery (which involves the local application of extreme cold). ■

Two types of photoreceptors are specialized to begin the process in which light rays are ultimately converted to nerve impulses: rods and cones. Each retina has about 6 million cones and 120 million rods. **Rods** have a low light threshold, allowing us to see in dim light, such as moonlight. Because they do not provide color vision, in dim light we see only shades of gray. Brighter lights stimulate the **cones,** which have a higher threshold and produce color vision. Most of our visual experiences are mediated by the cone system, the loss of which produces legal blindness. In contrast, a person who loses rod vision mainly has difficulty seeing in dim light and thus should not, for example, drive at night.

The **macula lutea** (MAK-yū-la LŪ-tē-a; *macula* = a small, flat spot; *lute-* = yellowish) is in the exact center of the posterior portion of the retina, at the visual axis of the eye. The **central fovea** (see Figure 20.4), a small depression in the center of the macula lutea, contains only cones. In addition, the layers of bipolar and ganglion cells, which scatter light to some extent, do not cover the cones here; these layers are displaced to the periphery of the fovea. As a result, the central fovea is the area of highest **visual acuity** or **resolution** (sharpness of vision). A main reason that you move your head and eyes while looking at something is to place images of interest on your central fovea—as you do to read each of the words in this sentence! Rods are absent from the central fovea and are more plentiful toward the periphery of the retina. Because rod vision is more sensitive than cone vision, you can see a faint object (such as a dim star) better if you gaze slightly to one side of it rather than looking directly at it.

The principal blood supply of the retina is from the **central retinal artery,** a branch of the ophthalmic artery. The central retinal artery enters the retina at about the middle of the optic disc, an area where the optic (II) nerve exits the eyeball (see Figure 20.4). After passing through the disc, the central retinal artery divides into superior and inferior branches, each of which subdivides into nasal and temporal branches. They fan out to nourish the anterior surface of the retina. The **central retinal vein** drains blood from the retina through the optic disc.

From photoreceptors, information flows to bipolar cells through the outer synaptic layer, and then from bipolar cells through the inner synaptic layer to ganglion cells. The axons of ganglion cells extend posteriorly to the optic disc and exit the eyeball as the optic nerve. The optic disc is also called the **blind spot.** Because it contains no rods or cones, we cannot see an image that strikes the blind spot. Normally, you are not aware of having a blind spot, but you can easily demonstrate its presence. Cover your left eye and gaze directly at the cross below. Then increase or decrease the distance between the book and your eye. At some point the square will disappear as its image falls on the blind spot.

\+ ■

Lens

Lens

Just posterior to the pupil and iris, within the cavity of the eyeball, is the avascular **lens** (see Figure 20.4). The lens is made up of proteins called **crystallins,** which are arranged like the layers of an onion. Normally, the lens is perfectly transparent. It is enclosed by a clear connective tissue capsule and held in position by encircling **suspensory ligaments,** which are attached to the ciliary processes. The lens fine-tunes focusing of light rays onto the retina to facilitate clear vision.

The leading cause of blindness is a loss of transparency of the lens known as a **cataract.** This problem often occurs with aging but may also be caused by injury, exposure to ultraviolet rays, certain medications (such as long-term use of steroids), or complications of other diseases (for example, diabetes). People who smoke also have increased risk of developing cataracts. The lens becomes cloudy or less transparent due to changes in the structure of the lens proteins. Fortunately, sight can usually be restored by surgical removal of the old lens and implantation of a new artificial one. ■

Interior of the Eyeball

The interior of the eyeball is divided by the lens into two cavities: the anterior cavity and vitreous chamber. The **anterior cavity**—the space anterior to the lens—consists of the **anterior chamber,** which lies behind the cornea and in front of the iris, and the **posterior chamber,** which lies behind the iris and in front of the suspensory ligaments and lens (Figure 20.8). The anterior cavity is filled with **aqueous humor** (*aqua* = water), a watery fluid that is continually filtered from blood capillaries in the ciliary processes and that nourishes the lens and cornea. Aqueous humor flows into the posterior chamber, then forward between the iris and the lens, through the pupil, and into the anterior chamber. From the anterior chamber, aqueous humor drains into the scleral venous sinus (canal of Schlemm) and then into the blood. Normally, aqueous humor is completely replaced about every 90 minutes.

The pressure in the eye, called **intraocular pressure,** is produced mainly by the aqueous humor and partly by the vitreous body (described shortly); normally it is about 16 mm of Hg. The intraocular pressure maintains the shape of the eyeball and prevents the eyeball from collapsing.

The second, and larger, cavity of the eyeball is the **vitreous chamber (posterior cavity),** which lies between the lens and the retina. Within the vitreous chamber is the **vitreous body,** a jellylike substance that contributes to intraocular pressure and holds the retina flush against the choroid, so that the retina provides an even surface for the reception of clear images. Unlike the aqueous humor, the vitreous body does not undergo constant replacement. It is formed during embryonic life and is not replaced thereafter. The vitreous body also contains phagocytic cells that remove debris, keeping this part of the eye clear for unobstructed vision. Occasionally, collections of debris may cast a shadow on the retina and create the appearance of specks that

Figure 20.8 / The anterior and posterior chambers of the eye, seen in a section through the anterior portion of the eyeball at the junction of the cornea and sclera.

The lens separates the posterior chamber of the anterior cavity from the vitreous chamber.

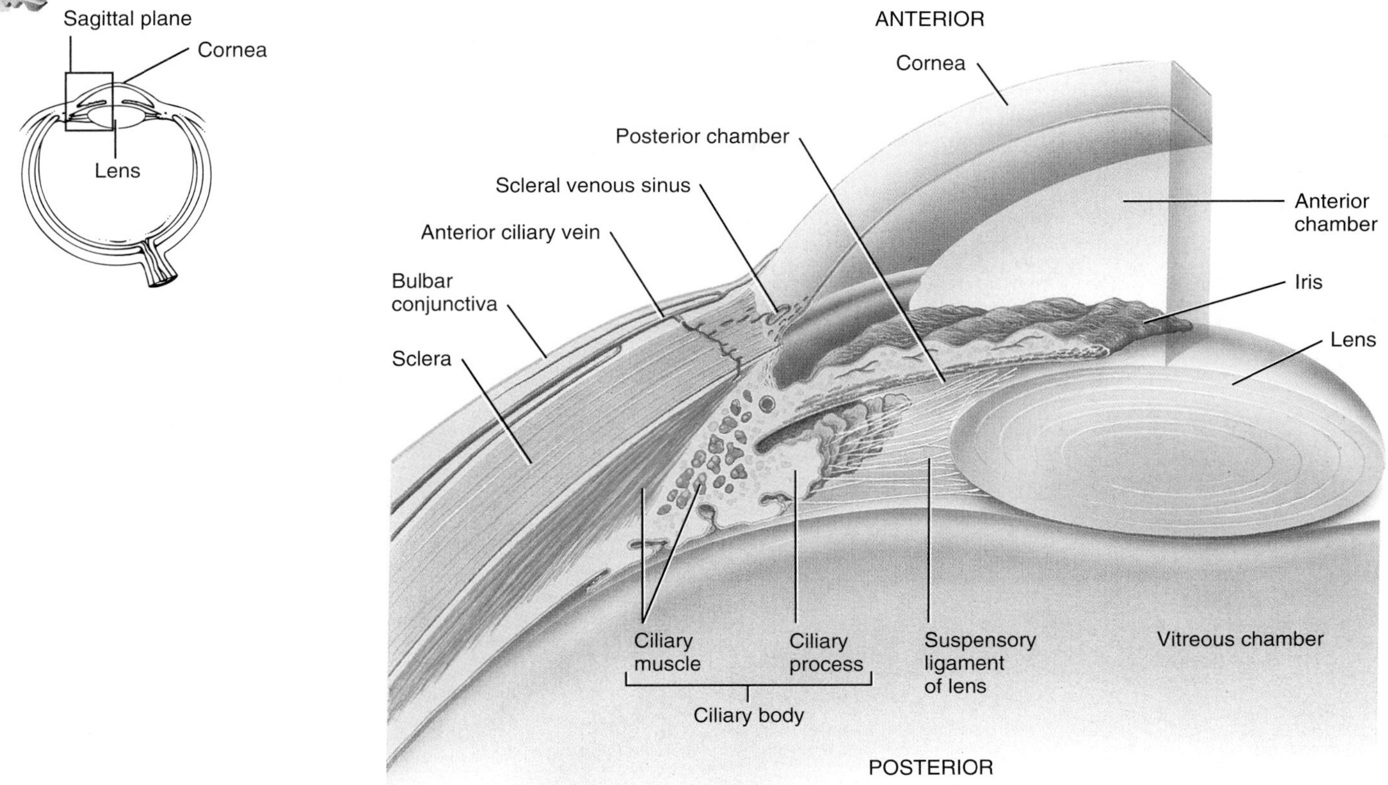

Where is aqueous humor produced, what is its circulation path, and where does it drain from the eyeball?

dart in and out of the field of vision. These "vitreal floaters," which are more common in older individuals, are usually harmless and do not require treatment. The **hyaloid canal** is a narrow channel that runs through the vitreous body from the optic disc to the posterior aspect of the lens. In the fetus, it was occupied by the hyaloid artery.

Table 20.1 summarizes the structures of the eyeball.

The Visual Pathway

Objective

- Describe the processing of visual signals in the retina and the neural pathway for vision.

After considerable processing of visual signals in the retina at synapses among the various types of neurons—the axons of retinal ganglion cells provide output from the retina to the brain. They exit the eyeball via the **optic (II) nerve** (see Figure 20.7).

Processing of Visual Input in the Retina

Within the retina, certain features of visual input are enhanced while other features may be discarded. Input from several cells may either converge upon a smaller number of postsynaptic neurons or diverge to a larger number. On the whole, however, convergence predominates because there are only 1 million ganglion cells receiving input from about 126 million photoreceptor cells.

Chemicals (neurotransmitters) released by rods and cones induce changes in both bipolar cells and horizontal cells that lead to the generation of nerve impulses (see Figure 20.7). Amacrine cells synapse with ganglion cells and transmit information to them. When bipolar or amacrine cells transmit signals to ganglion cells, the ganglion cells initiate nerve impulses.

Pathway in the Brain

The axons of the optic (II) nerve pass through the **optic chiasm** (kī-AZ-m; = a crossover, as in the letter X), a crossing point of the optic nerves (Figure 20.9). Some fibers cross to the opposite side, whereas others remain uncrossed. After passing through the optic chiasm, the fibers, now part of the **optic tract,** enter the brain and terminate in the lateral geniculate nucleus of the thalamus. Here they synapse with neurons whose axons form the **optic radiations,** which project to the primary visual areas in the occipital lobes of the cerebral cortex (area 17 in Figure 18.15 on page 568).

Table 20.1 Structures of the Eyeball

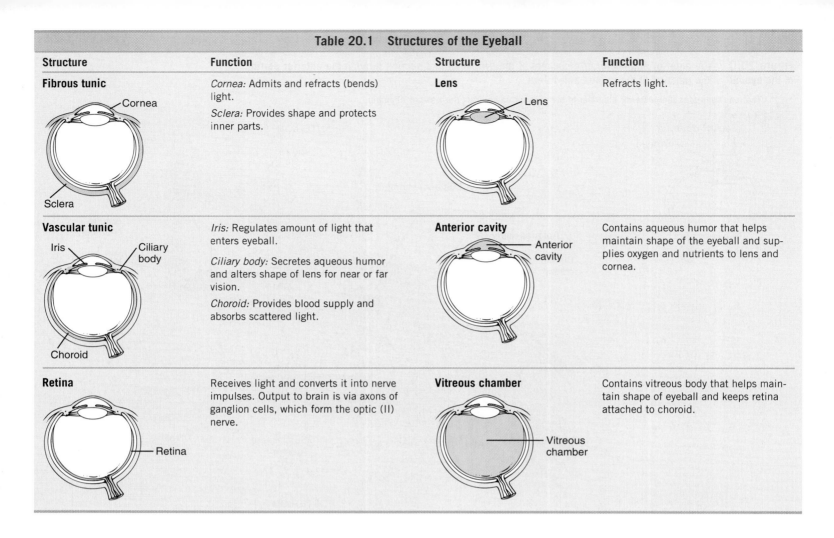

Structure	Function	Structure	Function
Fibrous tunic (Cornea, Sclera)	*Cornea:* Admits and refracts (bends) light. *Sclera:* Provides shape and protects inner parts.	**Lens** (Lens)	Refracts light.
Vascular tunic (Iris, Ciliary body, Choroid)	*Iris:* Regulates amount of light that enters eyeball. *Ciliary body:* Secretes aqueous humor and alters shape of lens for near or far vision. *Choroid:* Provides blood supply and absorbs scattered light.	**Anterior cavity** (Anterior cavity)	Contains aqueous humor that helps maintain shape of the eyeball and supplies oxygen and nutrients to lens and cornea.
Retina (Retina)	Receives light and converts it into nerve impulses. Output to brain is via axons of ganglion cells, which form the optic (II) nerve.	**Vitreous chamber** (Vitreous chamber)	Contains vitreous body that helps maintain shape of eyeball and keeps retina attached to choroid.

Figure 20.9 / The visual pathway. Partial dissection of the brain (a) reveals the optic radiations (axons extending from the thalamus to the occipital lobe).

The optic chiasm is the crossing point of the optic nerves.

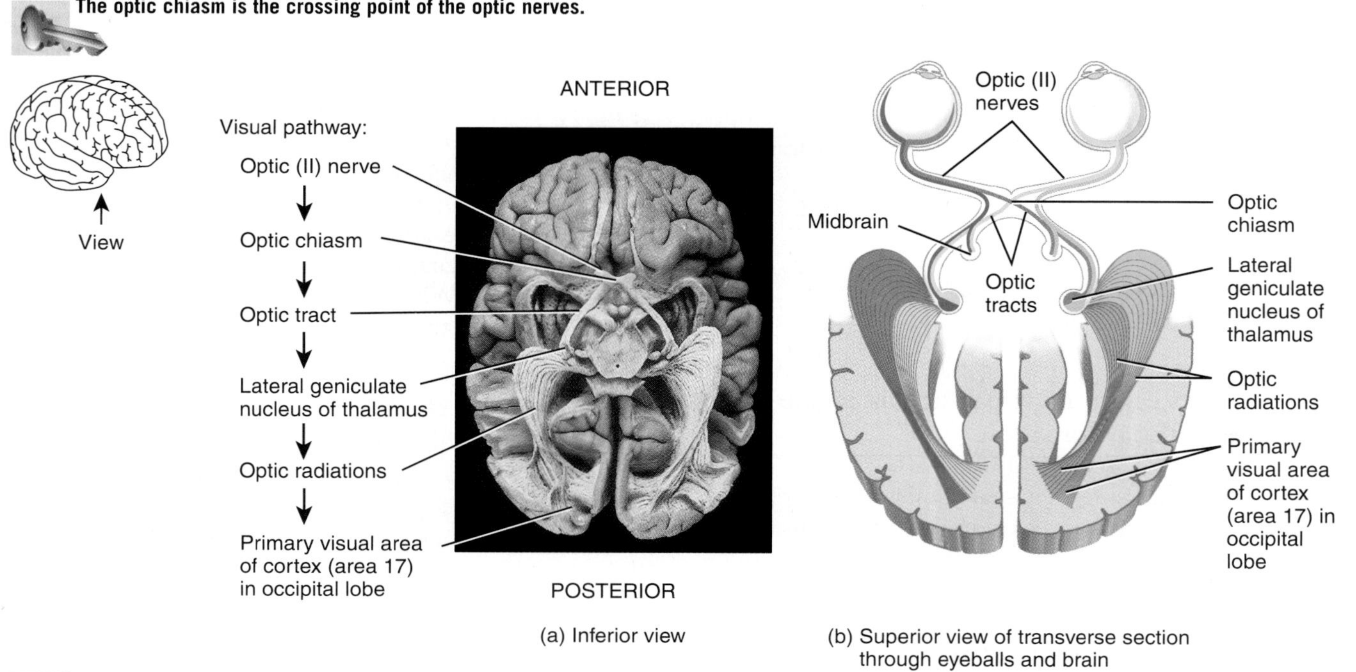

(a) Inferior view

(b) Superior view of transverse section through eyeballs and brain

Where does the optic tract terminate?

✓ Describe the structure and importance of the eyelids, eyelashes, and eyebrows.
✓ What is the function of the lacrimal apparatus?
✓ What are the components of the fibrous tunic, vascular tunic, and retina, and what are their functions?
✓ Describe the histology of the neural portion of the retina.
✓ Describe the pathway for vision.

HEARING AND EQUILIBRIUM

Objectives

- Describe the anatomy of the structures in the three principal regions of the ear.
- List the principal events involved in hearing.
- Identify the receptor organs for equilibrium, and describe how they function.
- Describe the auditory and equilibrium pathways.

Anatomy of the Ear

Besides receptors for sound waves, the ear also contains receptors for equilibrium. The ear is divided into three principal regions: the external ear, which collects sound waves and channels them inward; the middle ear, which conveys sound vibrations to the oval window; and the internal ear, which houses the receptors for hearing and equilibrium.

External (Outer) Ear

The **external (outer) ear** consists of the auricle, external auditory canal, and eardrum (Figure 20.10). The **auricle (pinna)** is a flap of elastic cartilage shaped like the flared end of a trumpet and covered by skin. The rim of the auricle is the **helix;** the inferior portion is the **lobule.** The auricle is attached to the head by ligaments and muscles. The **external auditory canal** (*audit-* = hearing) is a curved tube about 2.5 cm (1 in.) long that lies in the temporal bone and leads from the auricle to the eardrum. The **eardrum,** or **tympanic membrane** (tim-PAN-ik; *tympan-* = a drum), is a thin, semitransparent partition between the ex-

Figure 20.10 / Structure of the ear, illustrated in a frontal section through the right ear and skull.

The ear has three principal regions: the external (outer) ear, the middle ear, and the internal (inner) ear.

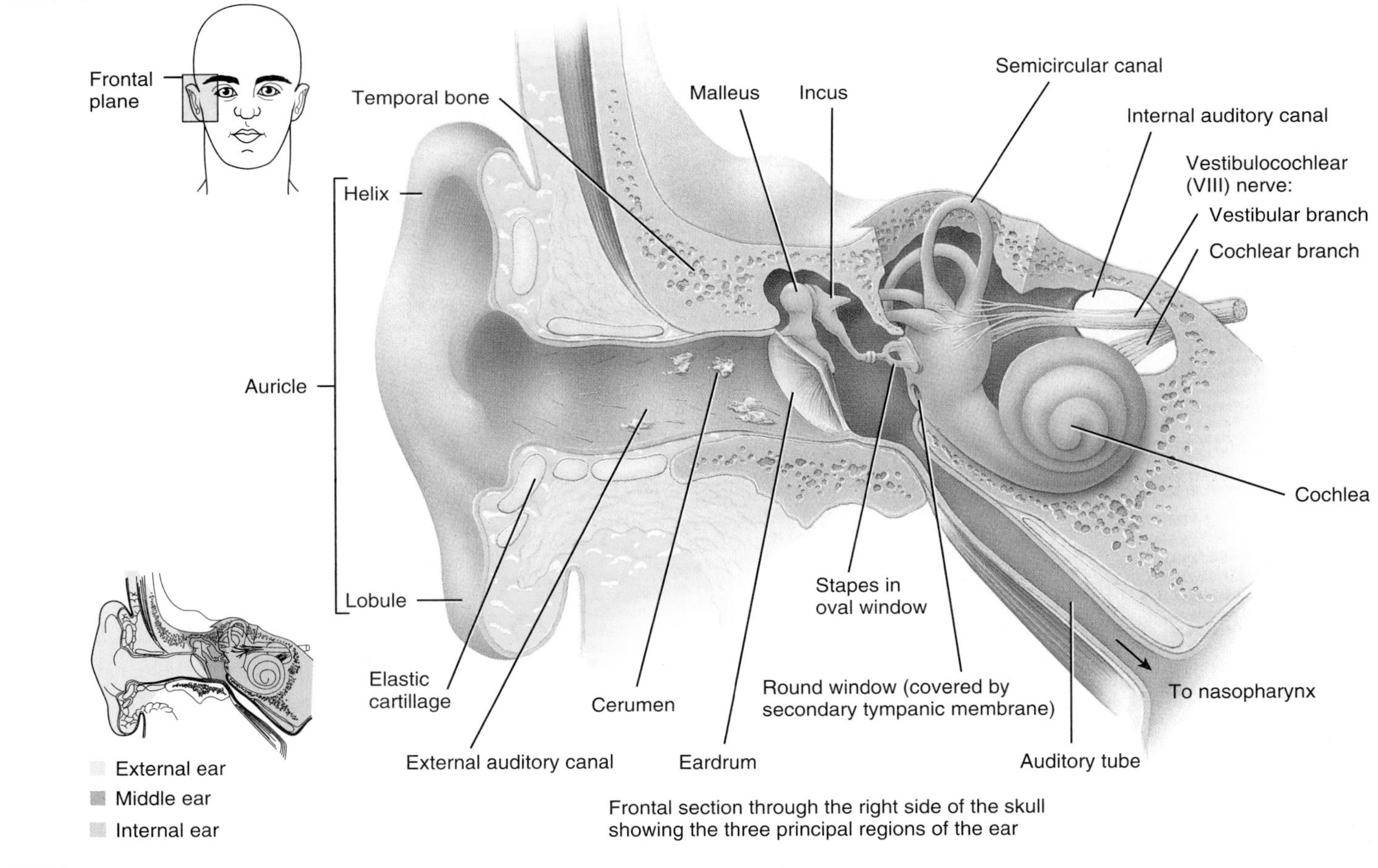

Frontal section through the right side of the skull showing the three principal regions of the ear

To which structure of the external ear does the malleus of the middle ear attach?

14th Century

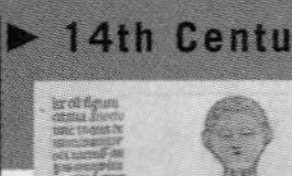

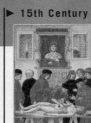

15th Century

16th Century

17th Century

18th Century

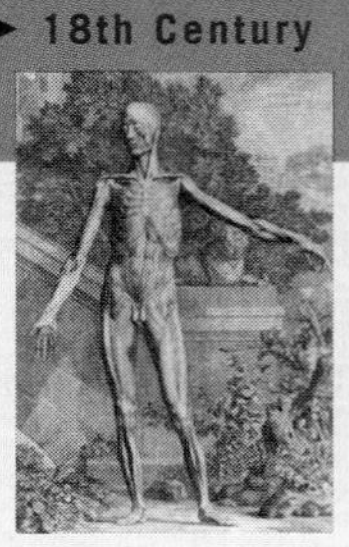

20th Century

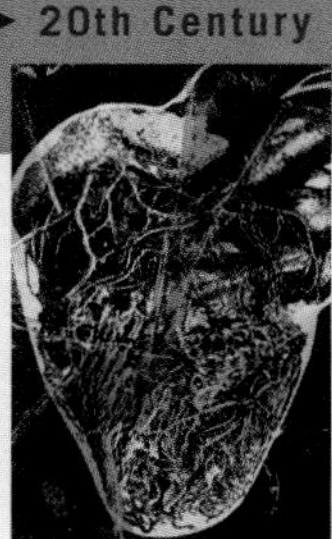

21st Century

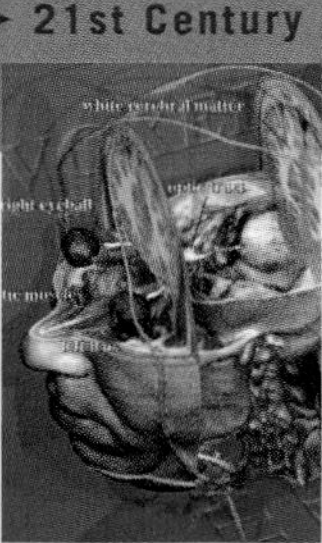

CHANGING IMAGES

Making Sense of the Senses

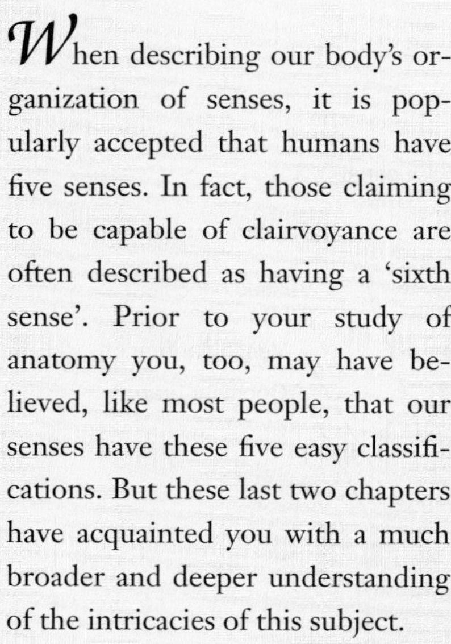

19th Cent.

When describing our body's organization of senses, it is popularly accepted that humans have five senses. In fact, those claiming to be capable of clairvoyance are often described as having a 'sixth sense'. Prior to your study of anatomy you, too, may have believed, like most people, that our senses have these five easy classifications. But these last two chapters have acquainted you with a much broader and deeper understanding of the intricacies of this subject.

In the previous chapter, you learned that what is often simply considered the sensation of touch, is classified as an example of the general or somatic senses. Other cutaneous sensations, as well as the proprioceptive sensations, are evidence of the complexity of touch. In this chapter, the remaining sensations are classified as the special senses. Smell (olfaction) and taste (gustation) are the subject of a brief, initial examination. It is the complexity of vision and hearing that requires a more detailed discussion. So intricate are the auditory sensations that they include static and dynamic equilibrium. Is it appropriate to consider equilibrium a sensation in its own right?

Depicted here are the well-known five senses: sight, smell, taste, touch, and hearing. The French portraitist, Louis-Léopold Boilly, amusingly represented each sensation as a person in this early 19th century caricature. Mr. Sight admires himself in a hand-held looking glass. Miss Smell enjoys the fragrance of her perfume. Mr. Taste is engaged in a gluttonous affair, while Dr. Touch palpates the radial pulse of Miss Smell. Listening closely to the ticking of his timepiece is Mr. Hearing. If you were to include the expanded aspects of the general and special senses that you are now familiar with, how would you do so? Would Ms. Proprioception be a springboard diver in the midst of tumbling maneuver? Would Mr. Equilibrium be riding in a roller coaster or floating within a space ship?

ternal auditory canal and middle ear. The eardrum is covered by epidermis and lined by simple cuboidal epithelium. Between the epithelial layers is connective tissue composed of collagen, elastic fibers, and fibroblasts.

Near the exterior opening, the external auditory canal contains a few hairs and specialized sebaceous (oil) glands called **ceruminous glands** (se-RŪ-mi-nus) that secrete earwax or **cerumen** (se-RŪ-men). The combination of hairs and cerumen helps prevent dust and foreign objects from entering the ear. Cerumen usually dries up and falls out of the ear canal. Some people, however, produce a large amount of cerumen, which can become impacted and can muffle incoming sounds. The treatment for **impacted cerumen** is usually periodic ear irrigation or removal of wax with a blunt instrument by trained medical personnel.

Middle Ear

The **middle ear,** which lies in the temporal bone, is a small, air-filled cavity lined by epithelium (Figure 20.11). It is separated from the external ear by the eardrum and from the internal ear by a thin bony partition that contains two small membrane-covered openings: the oval window and the round window. Extending across the middle ear and attached to it by ligaments are the three smallest bones in the body, the **auditory ossicles** (OS-si-kuls), which are connected by synovial joints. The bones, named for their shapes, are the malleus, incus, and stapes—commonly called the hammer, anvil, and stirrup, respectively. The "handle" of the **malleus** (MAL-ē-us) is attached to the internal surface of the eardrum. The head of the malleus articulates with the body of the incus. The **incus** (ING-kus), the middle bone in the series, articulates with the head of the stapes. The base or footplate of the **stapes** (STĀ-pēz) fits into the **oval window.** Directly below the oval window is another opening, the **round window,** that is enclosed by a membrane called the **secondary tympanic membrane.**

Besides the ligaments, two tiny skeletal muscles also attach to the ossicles. The **tensor tympani muscle** (Figure 20.11), which is innervated by the mandibular branch of the trigeminal (V) nerve, limits movement and increases tension on the eardrum to prevent damage to the inner ear from loud noises. The **stapedius muscle,** which is innervated by the facial (VII) nerve, is the smallest of all skeletal muscles. By dampening large vibrations of the stapes due to loud noises, it protects the oval

Figure 20.11 / The right middle ear containing the auditory ossicles.

Common names for the malleus, incus, and stapes are the hammer, anvil, and stirrup, respectively.

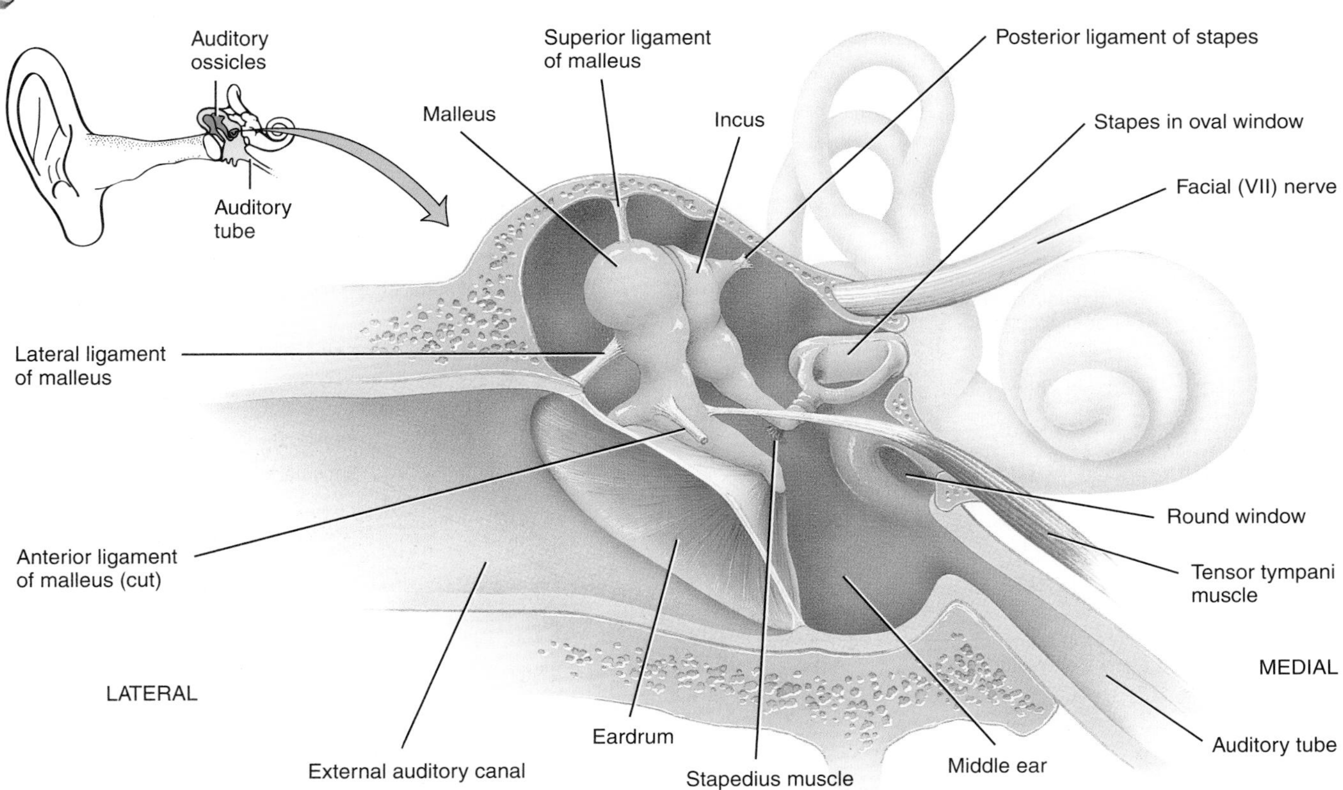

Frontal section showing location of auditory ossicles

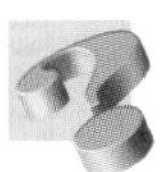
What structures separate the middle ear from the external ear and from the internal ear?

window, but it also decreases the sensitivity of hearing. For this reason, paralysis of the stapedius muscle is associated with **hyperacusia** (abnormally sensitive hearing). Because it takes a fraction of a second for the tensor tympani and stapedius muscles to contract, they can protect the inner ear from prolonged loud noises, but not from brief ones such as a gunshot.

The anterior wall of the middle ear contains an opening that leads directly into the **auditory (Eustachian) tube.** The auditory tube, which consists of both bone and hyaline cartilage, connects the middle ear with the nasopharynx (upper portion of the throat). It is normally closed at its medial (pharyngeal) end; during swallowing and yawning, it opens, allowing air to enter or leave the middle ear until the pressure in the middle ear equals the atmospheric pressure. When the pressures are balanced, the eardrum vibrates freely as sound waves strike it. If the pressure is not equalized, intense pain, hearing impairment, ringing in the ears, and vertigo can develop. The auditory tube is also a route that pathogens can travel from the nose and throat to the middle ear.

Internal (Inner) Ear

The **internal (inner) ear** is also called the **labyrinth** (LAB-i-rinth) because of its complicated series of canals (Figure 20.12). Structurally, it consists of two main divisions: an outer bony labyrinth that encloses an inner membranous labyrinth. The **bony labyrinth,** located in the temporal bone, is a series of cavities divided into three areas: (1) the semicircular canals and (2) the vestibule, both of which contain receptors for equilibrium; and (3) the cochlea, which contains receptors for hearing. The bony labyrinth is lined with periosteum and contains **perilymph.** This fluid, which is chemically similar to cerebrospinal fluid, surrounds the **membranous labyrinth,** a series of sacs and tubes inside the bony labyrinth and having the same general form. The membranous labyrinth is lined by epithelium and contains **endolymph.** The level of K^+ in endolymph is unusually high for an extracellular fluid, and potassium ions play a role in the generation of auditory signals (described shortly).

The **vestibule** (VES-ti-būl) is the oval central portion of the bony labyrinth. In the vestibule, the membranous labyrinth consists of two sacs called the **utricle** (YŪ-tri-kul; = little bag) and the **saccule** (SAK-yūl; = little sac), which are connected by a small duct. Projecting superiorly and posteriorly from the vestibule are the three bony **semicircular canals,** each lying at approximately right angles to the other two. Based on their positions, they are called the anterior, posterior, and lateral semicircular canals. The anterior and posterior semicircular canals

Figure 20.12 / The right internal ear. The outer, cream-colored area is part of the bony labyrinth; the inner, pink-colored area is the membranous labyrinth.

The bony labyrinth contains perilymph, and the membranous labyrinth contains endolymph.

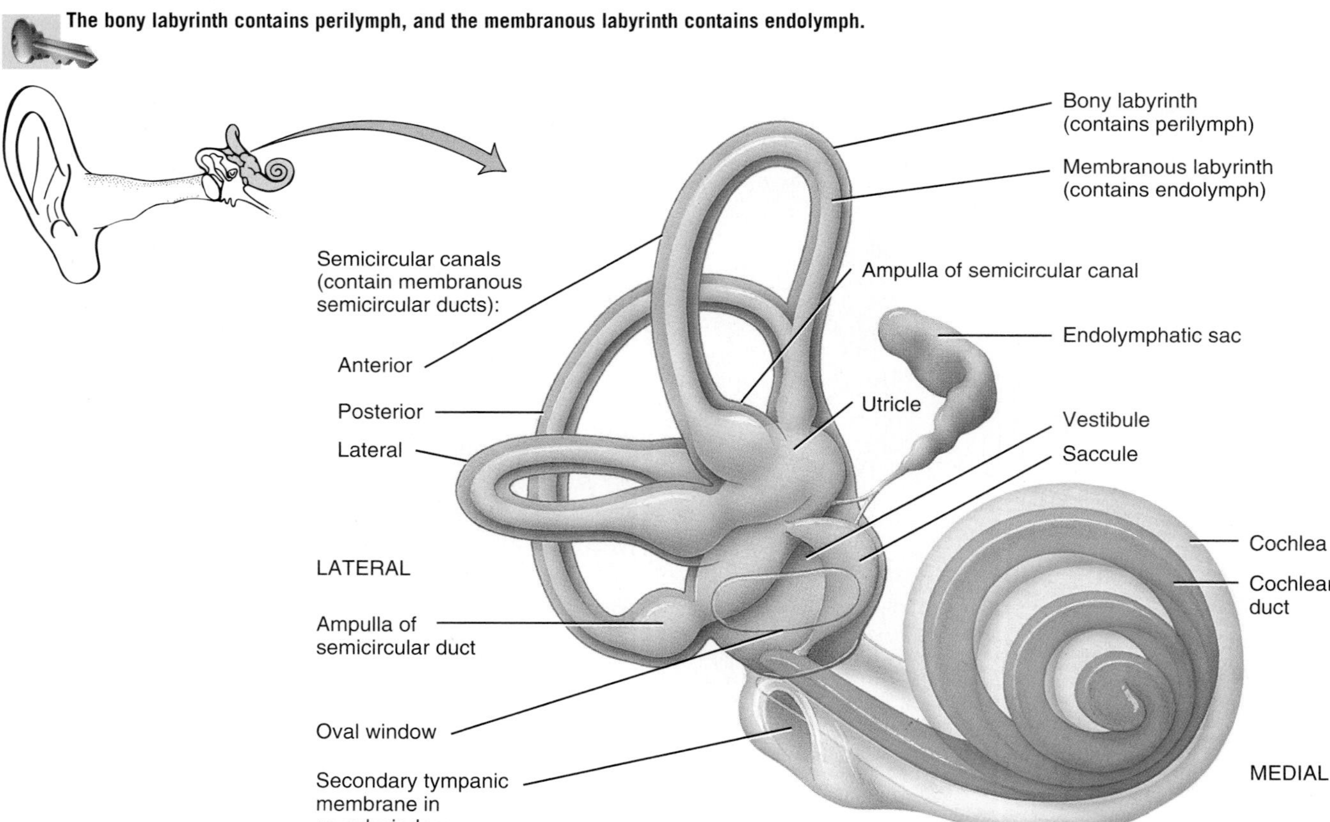

What are the names of the two sacs that lie in the vestibule?

are oriented vertically; the lateral one is oriented horizontally. At one end of each canal is a swollen enlargement called the **ampulla** (am-PŪL-la; = saclike duct). The portions of the membranous labyrinth that lie inside the bony semicircular canals are called the **membranous semicircular ducts.** These structures communicate with the utricle of the vestibule.

The vestibular branch of the vestibulocochlear (VIII) nerve consists of *ampullary, utricular,* and *saccular nerves.* These nerves contain both first-order sensory neurons and motor neurons that synapse with receptors for equilibrium. The first-order sensory neurons carry sensory information from the receptors, and the motor neurons carry feedback signals to the receptors, apparently to modify their sensitivity. Cell bodies of the sensory neurons are located in the **vestibular ganglia** (see Figure 20.13b).

Anterior to the vestibule is the **cochlea** (KŌK-lē-a; = snail-shaped), a bony spiral canal (Figure 20.13a–c) that resembles a snail's shell and makes almost three turns around a central bony core called the **modiolus** (mō-DĪ-ō-lus). Sections through the cochlea show that it is divided into three channels (Figure 20.13c). Together, the partitions that separate the channels are shaped like the letter Y. The stem of the Y is a bony shelf that protrudes into the canal; the wings of the Y are composed mainly of membranous labyrinth. The channel above the bony partition is the **scala vestibuli,** which ends at the oval window; the channel below is the **scala tympani,** which ends at the round window.

The scala vestibuli and scala tympani both contain perilymph and are completely separated, except for an opening at the apex of the cochlea, the **helicotrema** (hel-i-kō-TRĒ-ma; Figure 20.13b). The cochlea adjoins the wall of the vestibule, into which the scala vestibuli opens. The perilymph in the vestibule is continuous with that of the scala vestibuli. The third channel (between the wings of the Y) is the **cochlear duct,** or **scala media.** The **vestibular membrane** separates the cochlear duct from the scala vestibuli, and the **basilar membrane** separates the cochlear duct from the scala tympani.

Resting on the basilar membrane is the **spiral organ,** or **organ of Corti** (see Figure 20.13b, c). The spiral organ is a coiled sheet of epithelial cells, including supporting cells and about 16,000 **hair cells,** which are the receptors for hearing. There are two groups of hair cells: The *inner hair cells* are arranged in a single row and extend the entire length of the cochlea; the *outer hair cells* are arranged in three rows. At the apical tip of each hair cell is a **hair bundle,** consisting of 30–100 *stereocilia* that extend into the endolymph of the cochlear duct. Despite their name,

Figure 20.13 / Semicircular canals, vestibule, and cochlea of the right ear. Note that the cochlea makes nearly three complete turns.

The three channels in the cochlea are the scala vestibuli, the scala tympani, and the cochlear duct.

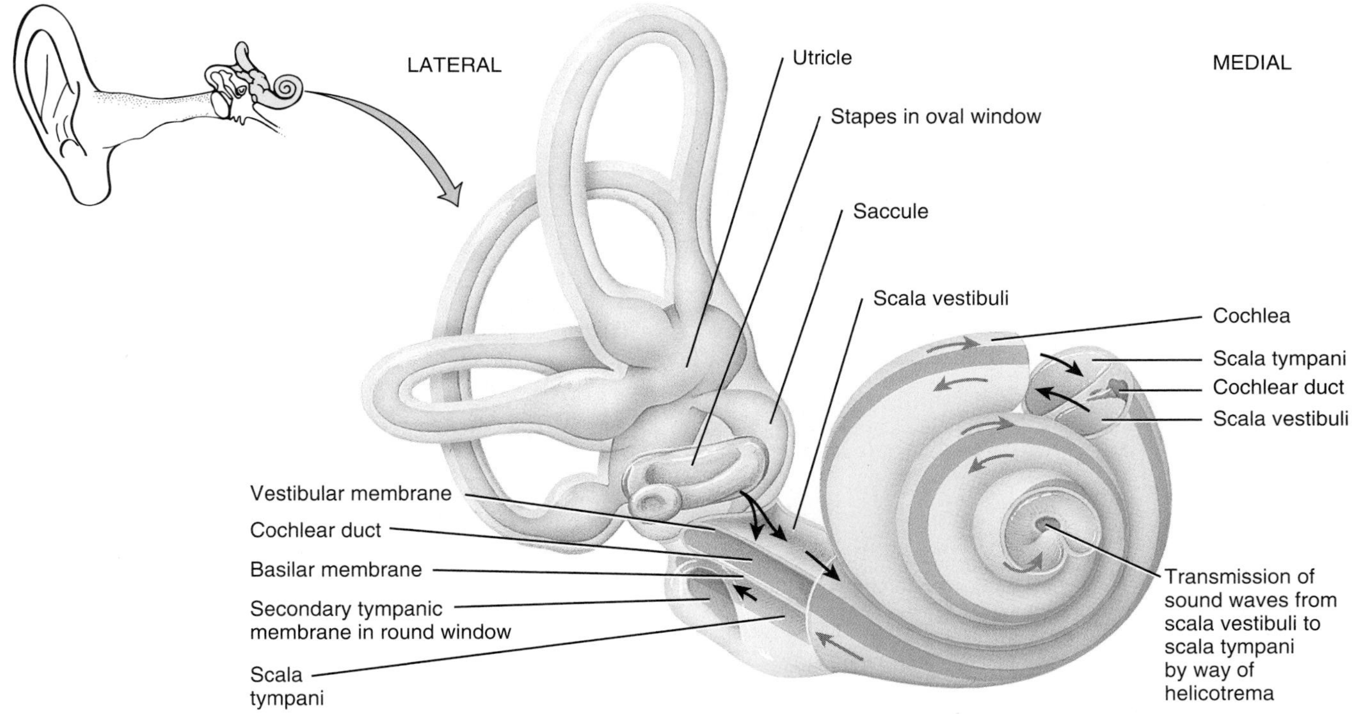

(a) Sections through the cochlea

(continues)

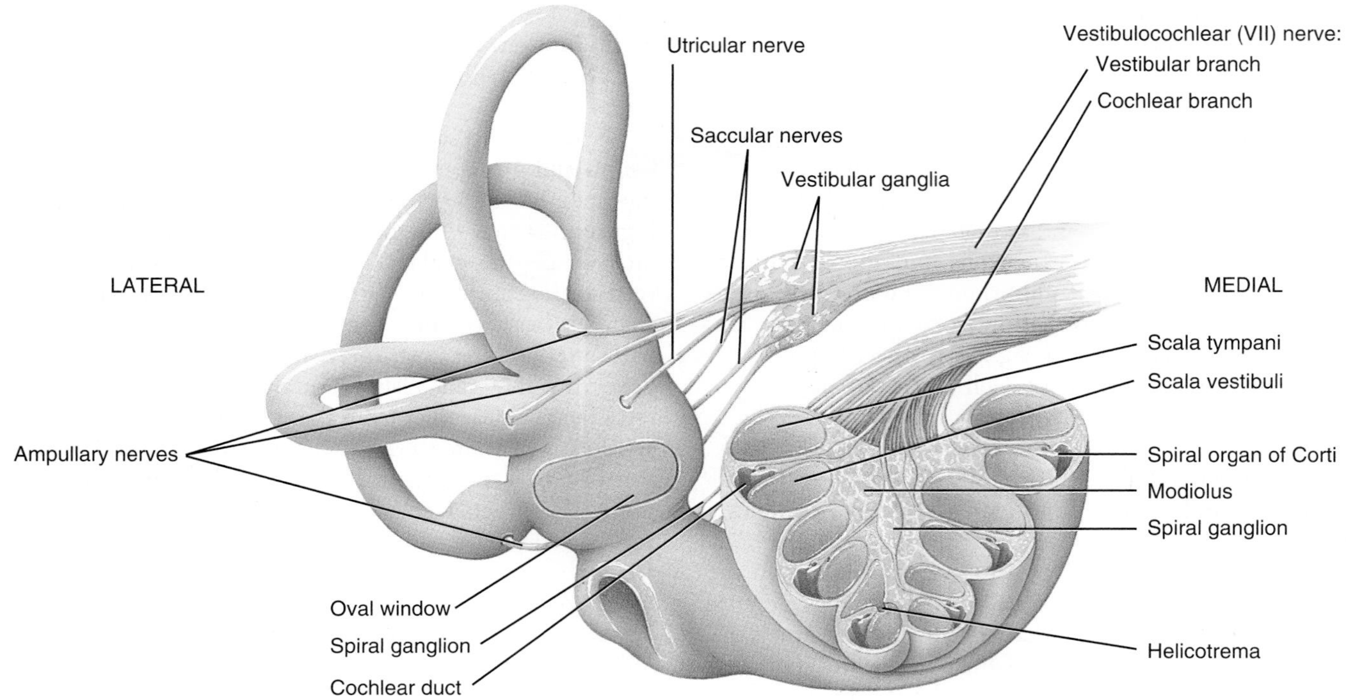

(b) Components of the vestibulocochlear (VIII) nerve

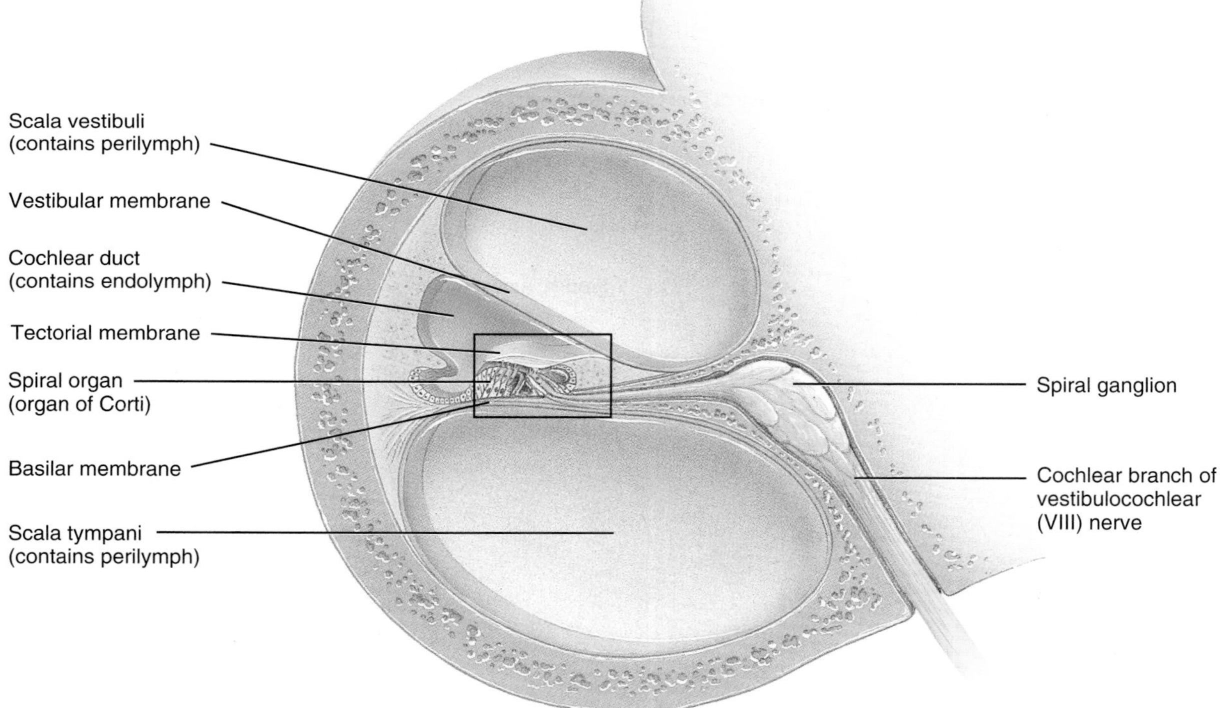

(c) Section through one turn of the cochlea

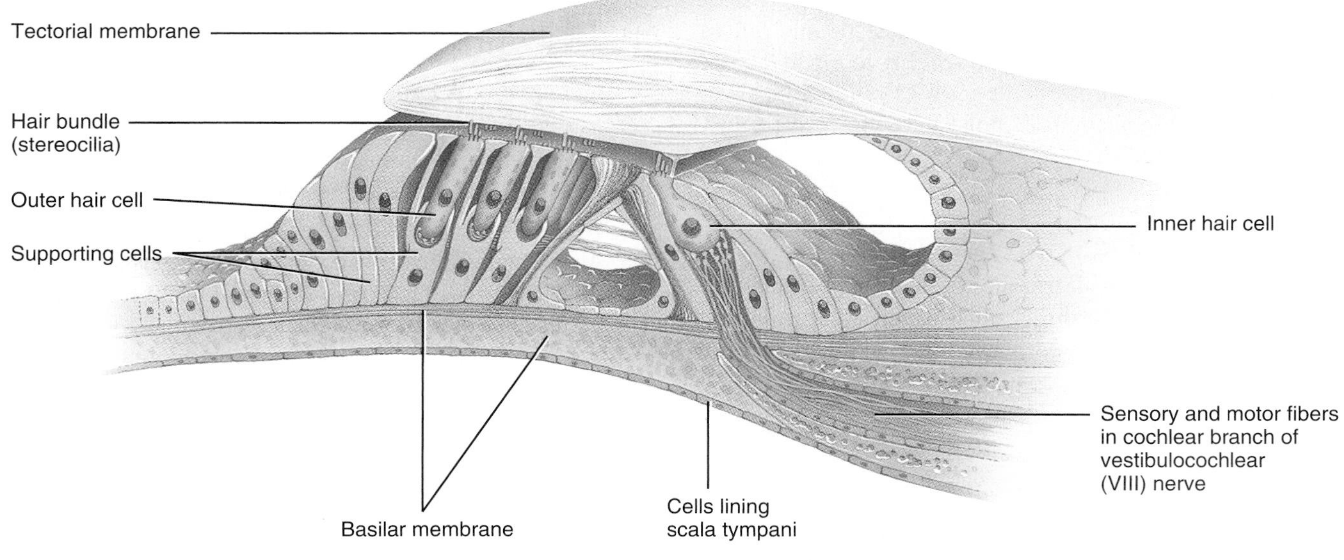

(d) Enlargement of spiral organ (organ of Corti)

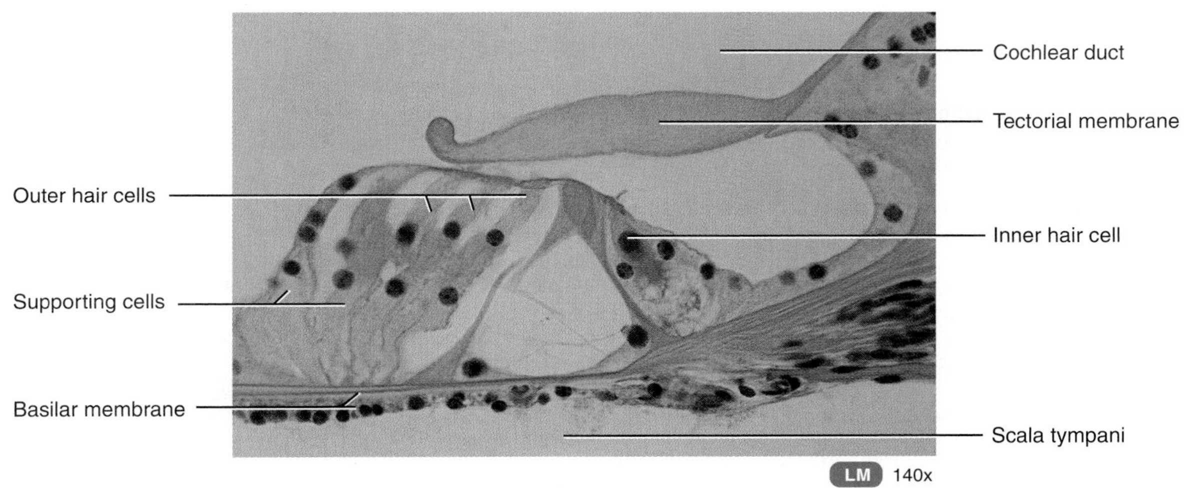

(e) Histology of the spiral organ (organ of Corti)

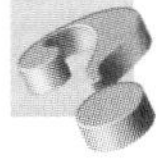

What are the three subdivisions of the bony labyrinth?

stereocilia are actually long, hairlike microvilli arranged in several rows of graded height.

At their basal ends, inner and outer hair cells synapse both with first-order sensory neurons and with motor neurons from the cochlear branch of the vestibulocochlear (VIII) nerve. Cell bodies of the sensory neurons are located in the **spiral ganglion** (Figure 20-13b, c). Although outer hair cells outnumber them by 3 to 1, the inner hair cells synapse with 90–95% of the first-order sensory neurons in the cochlear nerve that relay auditory information to the brain: By contrast, 90% of the motor neurons in the cochlear nerve synapse with outer hair cells. Projecting over and in contact with the hair cells of the spiral organ is the **tectorial membrane** (*tector-* = covering), a flexible gelatinous membrane.

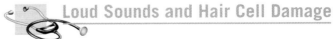

Loud Sounds and Hair Cell Damage

Exposure to loud music, the roar of jet engines, revved-up motorcycles, lawn mowers, and vacuum cleaners damages hair cells of the cochlea. Continued exposure to high-intensity sounds causes **deafness,** a significant or total hearing loss. The louder the sounds, the quicker the loss. Deafness usually begins with loss of sensitivity for high-pitched sounds. Most people fail to notice their progressive hearing loss until destruction is extensive and they begin having difficulty understanding speech. Wearing earplugs while engaging in noisy activities can protect the sensitivity of your ears. ■

Mechanism of Hearing

Sound waves are a series of alternating high- and low-pressure regions traveling in the same direction through some medium (such as air). They originate from a vibrating object in much the same way that ripples arise and travel over the surface of a pond when you toss a stone into it.

The following events are involved in hearing (Figure 20.14):

1. The auricle directs sound waves into the external auditory canal.
2. When sound waves strike the eardrum, the alternating high- and low-pressure of the air cause the eardrum to vibrate back and forth. The distance it moves, which is very small, depends on the intensity and frequency of the sound waves. The eardrum vibrates slowly in response to low-frequency (low-pitched) sounds and rapidly in response to high-frequency (high-pitched) sounds.
3. The central area of the eardrum connects to the malleus, which vibrates along with the eardrum. This vibration is transmitted from the malleus to the incus and then to the stapes.
4. As the stapes moves back and forth, it pushes the membrane of the oval window in and out. The oval window vibrates about 20 times more vigorously than the eardrum because the ossicles efficiently transmit small vibrations spread over a large surface area (the eardrum) into larger vibrations of a smaller surface (the oval window).
5. The movement of the oval window sets up fluid pressure waves in the perilymph of the cochlea. As the oval window bulges inward, it pushes on the perilymph of the scala vestibuli.
6. Pressure waves are transmitted from the scala vestibuli to the scala tympani and eventually to the round window, caus-

Figure 20.14 / Events in the stimulation of auditory receptors in the right ear. The numbers correspond to the events listed in the text. The cochlea has been uncoiled to more easily visualize the transmission of sound waves and their distortion of the vestibular and basilar membranes of the cochlear duct.

The function of hair cells of the spiral organ (organ of Corti) is to ultimately convert a mechanical vibration (stimulus) into an electrical signal (nerve impulse).

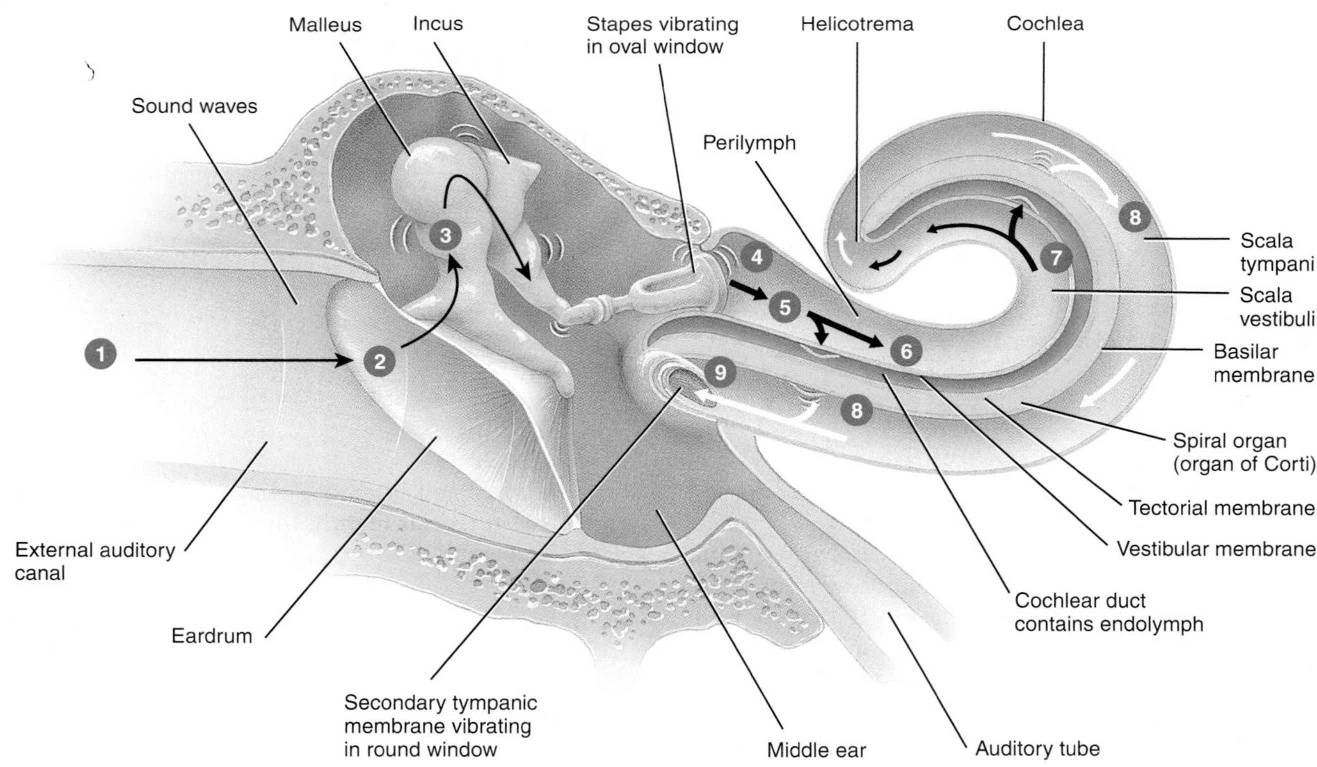

What structure vibrates and sets up pressure waves in the perilymph?

ing it to bulge outward into the middle ear. (See 9 in the illustration)

7 As the pressure waves deform the walls of the scala vestibuli and scala tympani, they also push the vestibular membrane back and forth, creating pressure waves in the endolymph inside the cochlear duct.

8 The pressure waves in the endolymph cause the basilar membrane to vibrate, which moves the hair cells of the spiral organ against the tectorial membrane. Bending of the stereocilia ultimately leads to the generation of nerve impulses in first-order neurons in cochlear nerve fibers.

The Auditory Pathway

First-order sensory neurons in the cochlear branch of each vestibulocochlear (VIII) nerve terminate in the cochlear nuclei of the medulla oblongata on the same side. From there, fibers carrying auditory signals project to the superior olivary nuclei in the pons on both sides. Slight differences in the timing of impulses arriving from the two ears at the olivary nuclei allow us to locate the source of a sound. From both the cochlear and the olivary nuclei, fibers ascend to the inferior colliculus in the midbrain, and then to the medial geniculate body of the thalamus. From the thalamus, auditory signals project to the primary auditory area in the superior temporal gyrus of the cerebral cortex (areas 41 and 42 in Figure 18.15 on page 568). Because many auditory axons decussate (cross over) in the medulla while others remain on the same side, the right and left primary auditory areas receive nerve impulses from both ears.

Cochlear Implants

Cochlear implants translate sounds into electronic signals that can be interpreted by the brain. They are useful for people with deafness caused by a disorder or an injury that has destroyed hair cells in the cochlea. Sound waves are picked up by a tiny microphone and converted into electrical signals by a microprocessor. The signals then travel to electrodes implanted in the cochlea, where they trigger nerve impulses in fibers of the cochlear branch of the vestibulocochlear nerve. These artificially induced nerve impulses propagate over their normal pathways to the brain. The sounds perceived are crude compared to normal hearing, but they provide a sense of rhythm and loudness; information about certain noises, such as those made by telephones and automobiles; and the pitch and cadence of speech. A few patients hear well enough with a cochlear implant to use the telephone. ■

Mechanism of Equilibrium

There are two kinds of **equilibrium** (balance). One kind, called **static equilibrium,** refers to the maintenance of the position of the body (mainly the head) relative to the force of gravity. The second kind, **dynamic equilibrium,** is the maintenance of body position (mainly the head) in response to sudden movements such as rotation, acceleration, and deceleration. Collectively, the receptor organs for equilibrium are called the **vestibular apparatus** (ves-TIB-yū-lar), which includes the saccule, utricle, and membranous semicircular ducts.

Otolithic Organs: Saccule and Utricle

The walls of both the utricle and the saccule contain a small, thickened region called a **macula** (MAK-yū-la; Figure 20.15). The two maculae (plural), which are perpendicular to one another, are the receptors for static equilibrium, and they also contribute to some aspects of dynamic equilibrium. For static equilibrium, they provide sensory information on the position of the head in space and are essential for maintaining appropriate posture and balance. For dynamic equilibrium, they detect linear acceleration and deceleration—for example, the sensations you feel while in an elevator or a car that is speeding up or slowing down.

The two maculae consist of two kinds of cells: **hair cells,** which are the sensory receptors, and **supporting cells.** Hair cells feature **hair bundles** that consist of 70 or more *stereocilia,* which are actually microvilli, plus one *kinocilium,* a conventional cilium anchored firmly to its basal body and extending beyond the longest stereocilia. Scattered among the hair cells are columnar supporting cells that probably secrete the thick, gelatinous, glycoprotein layer, called the **otolithic membrane,** that rests on the hair cells. A layer of dense calcium carbonate crystals called **otoliths** (*oto-* = ear; *-liths* = stones) extends over the entire surface of the otolithic membrane.

Because the otolithic membrane sits on top of the macula, when you tilt your head forward, the otolithic membrane and otoliths are pulled by gravity and slide downhill over the hair cells in the direction of the tilt, bending the hair bundles. The movement of the hair bundles initiates responses that ultimately lead to the generation of nerve impulses. The hair cells synapse with first-order sensory neurons in the vestibular branch of the vestibulocochlear (VIIII) nerve.

Membranous Semicircular Ducts

The three membranous semicircular ducts, together with the saccule and the utricle, function in dynamic equilibrium. The ducts lie at right angles to one another in three planes (Figure 20.16 on page 633): The two vertical ducts are the anterior and posterior membranous semicircular ducts, and the horizontal one is the lateral membranous semicircular duct (see also Figure 20.12). This positioning permits detection of rotational acceleration or deceleration. In the ampulla, the dilated portion of each duct, is a small elevation called the **crista.** Each crista contains a group of **hair cells** and **supporting cells** covered by a mass of gelatinous material called the **cupula** (KŪ-pū-la). When the head moves, the attached membranous semicircular ducts and hair cells move with it. The endolymph, however, is not attached and lags behind due to its inertia. As the moving hair cells drag along the stationary fluid, the hair bundles bend. Bending of the hair bundles produces responses, which lead to nerve impulses that pass along the vestibular branch of the vestibulocochlear (VIII) nerve.

Figure 20.15 / Location and structure of receptors in the maculae of the right ear. Both first-order sensory neurons (blue) and motor neurons (red) synapse with the hair cells.

The movement of stereocilia initiates responses that ultimately lead to the generation of nerve impulses.

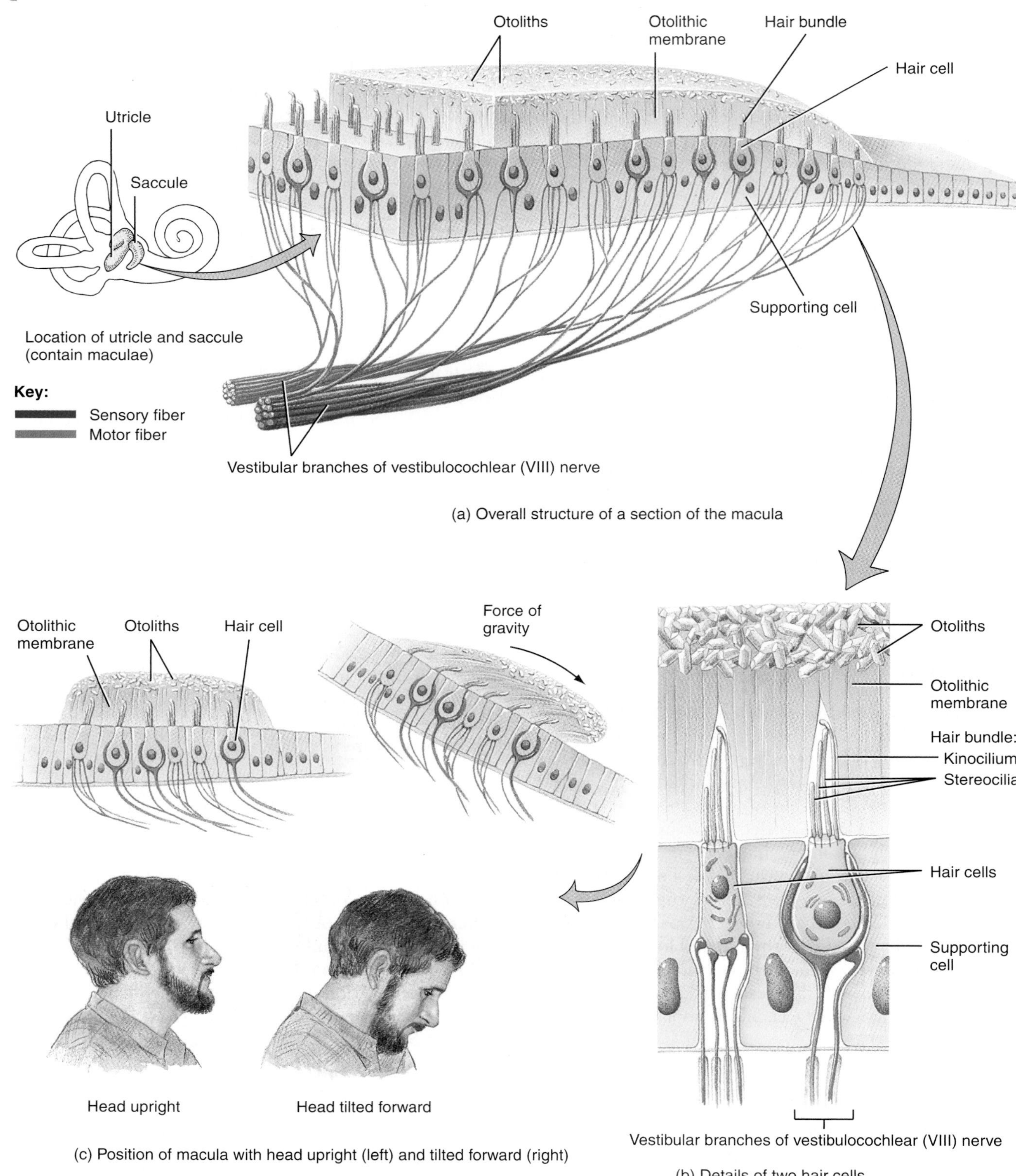

(a) Overall structure of a section of the macula

(b) Details of two hair cells

(c) Position of macula with head upright (left) and tilted forward (right)

With which type of equilibrium are the maculae mainly concerned?

Figure 20.16 / Location and structure of the membranous semicircular ducts of the right ear. Both first-order sensory neurons (blue) and motor neurons (red) synapse with the hair cells. The ampullary nerves are branches of the vestibular division of the vestibulocochlear (VIII) nerve.

The positions of the membranous semicircular ducts permit detection of rotational movements.

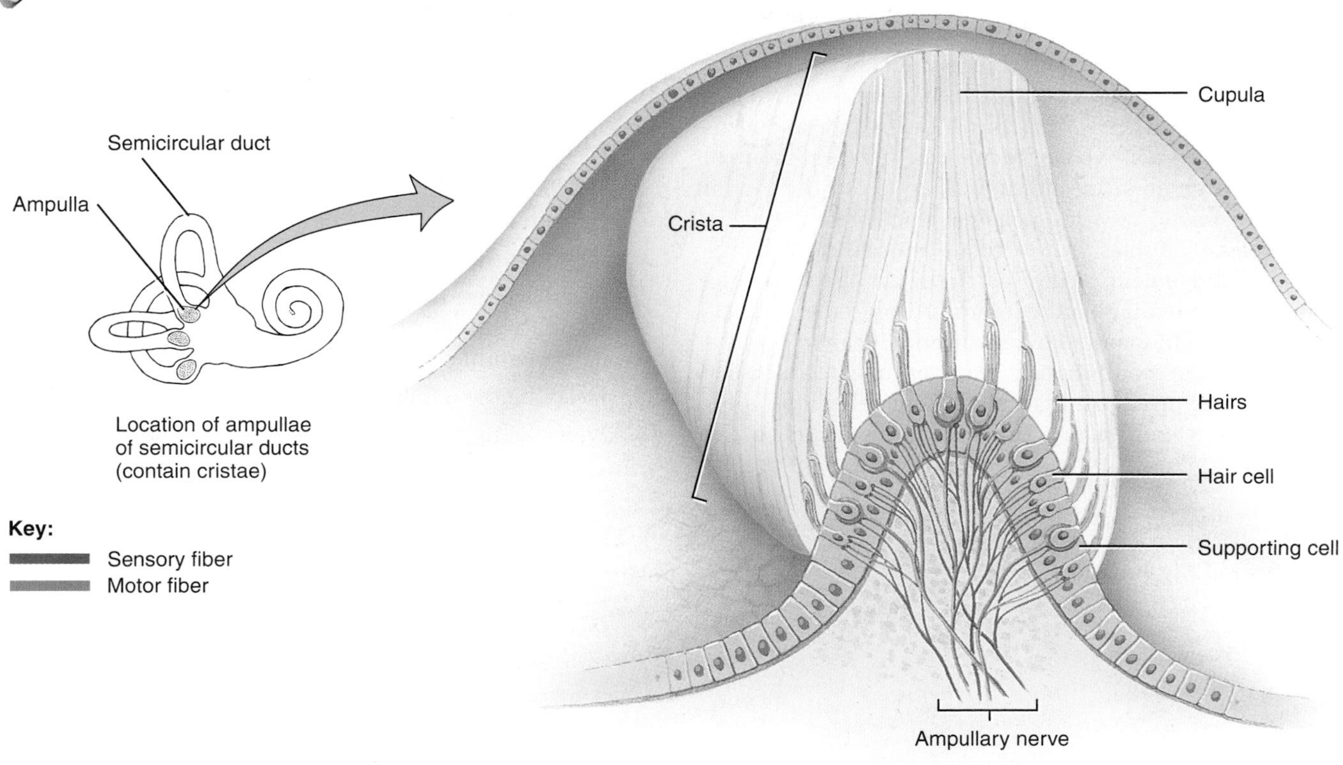

(a) Details of a crista

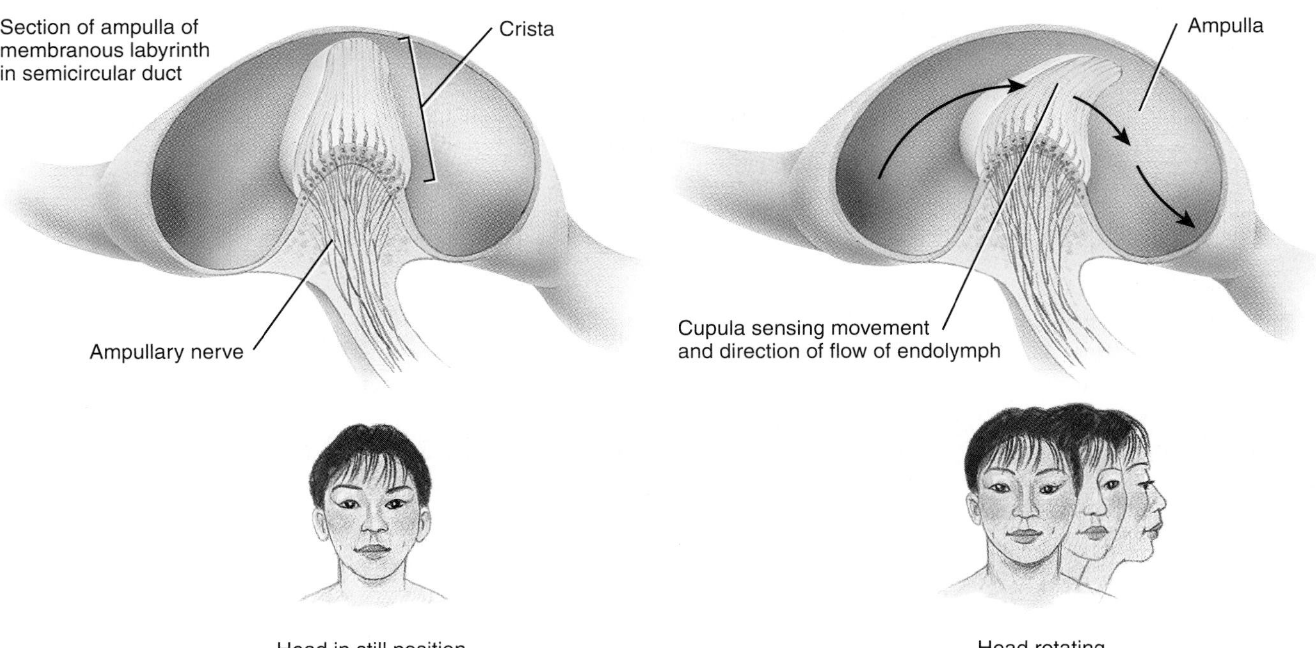

(b) Position of a crista with the head in the still position (left) and when the head rotates (right)

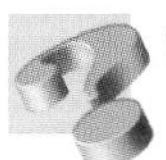

With which type of equilibrium are the membranous semicircular ducts, the utricle, and the saccule associated?

Equilibrium Pathways

Most of the vestibular branch fibers of the vestibulocochlear (VIII) nerve enter the brain stem and terminate in several vestibular nuclei in the medulla and pons. The remaining fibers enter the cerebellum through the inferior cerebellar peduncle (see Figure 18.7c on page 555). Bidirectional pathways connect the vestibular nuclei and cerebellum. Fibers from all the vestibular nuclei extend to the nuclei of cranial nerves that control eye movements—oculomotor (III), trochlear (IV), and abducens (VI)—and to the accessory (XI) nerve nucleus that helps control head and neck movements. In addition, fibers from the lateral vestibular nucleus form the vestibulospinal tract, which conveys impulses to skeletal muscles that regulate muscle tone in response to head movements. Various pathways between the vestibular nuclei, cerebellum, and cerebrum enable the cerebellum to play a key role in maintaining static and dynamic equilibrium. The cerebellum continuously receives updated sensory information from the utricle and saccule. The cerebellum monitors this information and makes corrective adjustments in the motor impulses that originate in the cerebral cortex. Essentially, in response to input from the utricle, saccule, and membranous semicircular ducts, the cerebellum sends continuous nerve impulses to the motor areas of the cerebrum. This feedback allows correction of signals from the motor cortex to specific skeletal muscles to maintain equilibrium.

Table 20.2 summarizes the structures of the ear related to hearing and equilibrium.

- ✓ List the components of the external, middle, and internal ear and their functions.
- ✓ Explain the mechanism of hearing.
- ✓ Describe the auditory pathway.
- ✓ Compare the function of the maculae in maintaining static equilibrium with the role of the cristae in maintaining dynamic equilibrium.
- ✓ Describe the equilibrium pathways.
- ✓ Describe the role of vestibular input to the cerebellum.

Table 20.2 Structures of the Ear

Regions of the Ear and Key Structures	Function
External (outer) ear 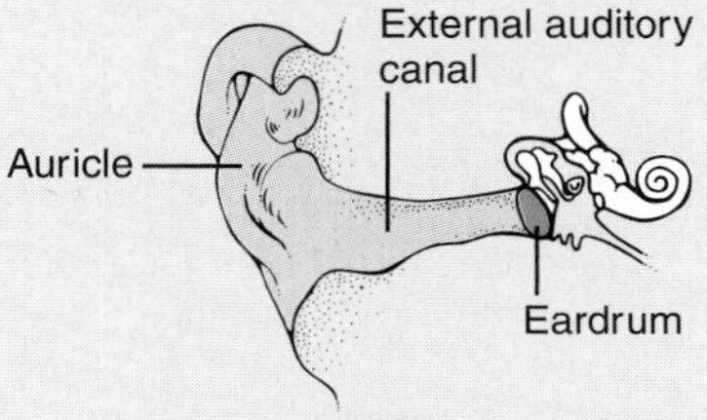 	*Auricle (pinna):* Collects sound waves. *External auditory canal (meatus):* Directs sound waves to eardrum. *Eardrum (tympanic membrane):* Sound waves cause it to vibrate, which, in turn, causes the malleus to vibrate.
Middle ear 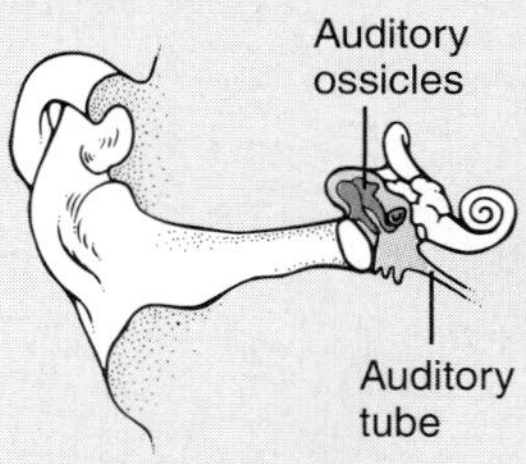 	*Auditory ossicles:* Transmit and amplify vibrations from tympanic membrane to oval window. *Auditory (Eustachian) tube:* Equalizes air pressure on both sides of the tympanic membrane.
Internal (inner) ear 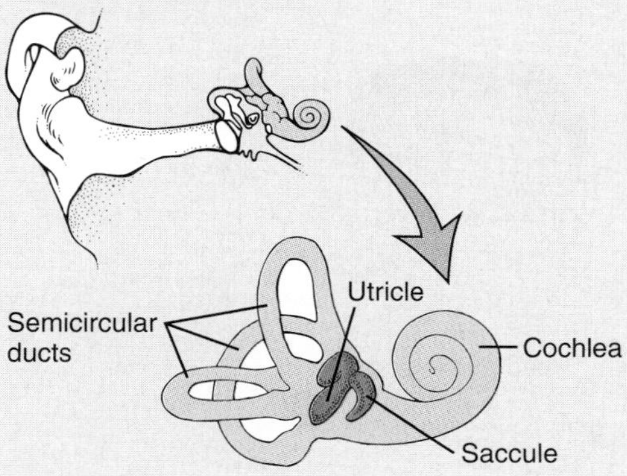 	*Cochlea:* Contains a series of fluids, channels, and membranes that transmit vibrations to the spiral organ (organ of Corti), the organ of hearing; hair cells in the spiral organ ultimately elicit nerve impulses in the cochlear branch of the vestibulocochlear (VIII) nerve. *Semicircular ducts:* Contain cristae, site of hair cells for dynamic equilibrium. *Utricle:* Contains macula, site of hair cells for static and dynamic equilibrium. *Saccule:* Contains macula, site of hair cells for static and dynamic equilibrium.

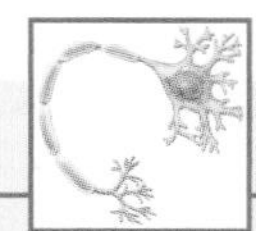

APPLICATIONS TO HEALTH

Glaucoma

Glaucoma is the most common cause of blindness in the United States, afflicting about 2% of the population over age 40. Glaucoma is an abnormally high intraocular pressure due to a buildup of aqueous humor within the anterior chamber. The fluid compresses the lens into the vitreous body and puts pressure on the neurons of the retina. Persistent pressure results in a progression from mild visual impairment to irreversible destruction of neurons of the retina, damage to the optic (II) nerve, and blindness. Glaucoma is painless, and the other eye compensates to a large extent, so a person may experience considerable retinal damage and loss of vision before the condition is diagnosed. Because glaucoma occurs more often with advancing age, regular measurement of intraocular pressure is an increasingly important part of an eye exam as people grow older. Risk factors include race (blacks are more susceptible), increasing age, family history, and past eye injuries and disorders.

Macular Degeneration

Macular degeneration, the leading cause of blindness in individuals over age 75, is a deterioration of the retina in the region of the macula, which is ordinarily the area of most acute vision. Initially, a person may experience blurring and distortion at the center of the visual field. In "dry" macular degeneration, central vision gradually diminishes because the pigment epithelium atrophies and degenerates. There is no effective treatment. "Wet" macular degeneration is due to the growth of new blood vessels that form under the retina and leak plasma or blood. Smokers have a threefold greater risk of developing macular degeneration than nonsmokers. Vision loss can be slowed by using laser surgery to destroy leaking blood vessels.

Deafness

Deafness is significant or total hearing loss. **Sensorineural deafness** is caused by either impairment of hair cells in the cochlea or damage of the cochlear branch of the vestibulocochlear (VIII) nerve. This type of deafness may be caused by atherosclerosis, which reduces blood supply to the ears; by repeated exposure to loud noise, which destroys hair cells of the spiral organ; and by certain drugs such as aspirin and streptomycin. **Conduction deafness** is caused by impairment of the external and middle ear mechanisms for transmitting sounds to the cochlea. Conduction deafness may be caused by otosclerosis, the deposition of new bone around the oval window; by impacted cerumen; by injury to the eardrum; and by aging, which often results in thickening of the eardrum and stiffening of the joints of the auditory ossicles.

Ménière's Disease

Ménière's disease (men′-ē-ĀRZ) results from an increased amount of endolymph that enlarges the membranous labyrinth. Among the symptoms are fluctuating hearing loss (caused by distortion of the basilar membrane of the cochlea) and roaring tinnitus (ringing). Spinning or whirling vertigo is characteristic of Ménière's disease. Almost total destruction of hearing may occur over a period of years.

Otitis Media

Otitis media is an acute infection of the middle ear caused primarily by bacteria and associated with infections of the nose and throat. Symptoms include pain, malaise, fever, and a reddening and outward bulging of the eardrum, which may rupture unless prompt treatment is received (this may involve draining pus from the middle ear). Bacteria from the nasopharynx passing into the auditory tube are the primary cause of all middle ear infections. Children are more susceptible than adults to middle ear infections because their auditory tubes are shorter, wider, and almost horizontal, which decreases drainage.

Motion Sickness

Motion sickness is nausea and vomiting brought on by repetitive angular, linear, or vertical motion. The cause is excessive stimulation of the vestibular apparatus by motion, usually incurred in travel by car, boat, or airplane. Nerve impulses pass from the internal ear to the vomiting center in the medulla. Visual stimuli and emotional factors such as fear or anxiety can also contribute to motion sickness. Susceptible people can take medication (for example, Dramamine) before traveling because prevention is more successful than treatment of symptoms once they have developed.

KEY MEDICAL TERMS ASSOCIATED WITH SPECIAL SENSES

Amblyopia (am′-blē-Ō-pē-a; *ambly-* = dull or dim) Term used to describe the loss of vision in an otherwise normal eye that, because of muscle imbalance, cannot focus in synchrony with the other eye.

Blepharitis (blef-a-RĪ-tis; *blepharo* = eyelid; *itis* = inflammation of) An inflammation of the eyelid.

Conjunctivitis (pinkeye) An inflammation of the conjunctiva; when caused by bacteria such as pneumococci, staphylococci, or *Hemophilus influenzae,* it is very contagious and more common in children. Conjunctivitis may also be caused by irritants, such as dust, smoke, or pollutants in the air, in which case it is not contagious.

Exotropia (ek′-sō-TRŌ-pē-a; *ex* = out; *tropia* = turning) Turning outward of the eyes.

Keratitis (ker′-a-TĪ-tis; *kerat-* = cornea) An inflammation or infection of the cornea.

Mydriasis (mi-DRĪ-a-sis) Dilated pupil.

Myringitis (mir′-in-JĪ-tis; *myringa* = eardrum) An inflammation of the eardrum; also called *tympanitis.*

Nystagmus (nis-TAG-mus; *nystagm-* = nodding or drowsy) A rapid involuntary movement of the eyeballs, possibly caused by a disease of the central nervous system. It is associated with conditions that cause vertigo.

Otalgia (ō-TAL-jē-a; *ot-* = ear; *-algia* = pain) Earache.

Photophobia (fō′-tō-FŌ-bē-a; *photo* = light; *phobia* = fear) Abnormal visual intolerance to light.

Ptosis (TŌ-sis; *ptosis* = fall) Falling or drooping of the eyelid. (This term is also used for the slippage of any organ below its normal position.)

Retinoblastoma (ret′-i-nō-blas-TŌ-ma; *blast* = bud; *oma* = tumor) A tumor arising from immature retinal cells; it accounts for 2% of childhood cancers.

Scotoma (skō-TŌ-ma; = darkness) An area of reduced or lost vision in the visual field.

Strabismus (stra-BIZ-mus) An imbalance in the extrinsic eye muscles that produces a squint.

Tinnitus (ti-NĪ-tus) A ringing, roaring, or clicking in the ears.

Trachoma (tra-KŌ-ma) A serious form of conjunctivitis and the greatest single cause of blindness in the world. It is caused by the bacterium *Chlamydia trachomatis.* The disease produces an excessive growth of subconjunctival tissue and invasion of blood vessels into the cornea, which progresses until the entire cornea is opaque, causing blindness.

Vertigo (VER-ti-gō; = dizziness) A sensation of spinning or movement in which the world seems to revolve or the person seems to revolve in space.

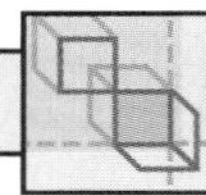

STUDY OUTLINE

Olfaction: Sense of Smell (p. 611)

1. The receptors for olfaction, which are bipolar neurons, are in the nasal epithelium.
2. Axons of olfactory receptor cells form the olfactory (I) nerves, which convey nerve impulses to the olfactory bulb, olfactory tracts, limbic system, and cerebral cortex (temporal and frontal lobes).

Gustation: Sensation of Taste (p. 613)

1. The receptors for gustation, the gustatory receptor cells, are located in taste buds.
2. Before a substance can be tasted, it must be dissolved.
3. Gustatory receptor cells trigger nerve impulses in cranial nerves VII, IX, and X. Taste signals then pass to the medulla oblongata, thalamus, and cerebral cortex (parietal lobe).

Vision (p. 613)

1. Accessory structures of the eyes include the eyebrows, eyelids, eyelashes, lacrimal apparatus, and extrinsic eye muscles.
2. The lacrimal apparatus consists of structures that produce and drain tears.
3. The eye is constructed of three layers: (a) fibrous tunic (sclera and cornea), (b) vascular tunic (choroid, ciliary body, and iris), and (c) retina.
4. The retina consists of pigment epithelium and a neural portion (photoreceptor layer, bipolar cell layer, ganglion cell layer, horizontal cells, and amacrine cells).
5. The anterior cavity contains aqueous humor; the vitreous chamber contains the vitreous body.
6. Chemicals (neurotransmitters) released by rods and cones induce changes in bipolar cells and horizontal cells that ultimately lead to the generation of nerve impulses.
7. Impulses from ganglion cells are conveyed into the optic (II) nerve, through the optic chiasm and optic tract, to the thalamus. From the thalamus, impulses for vision propagate to the cerebral cortex (occipital lobe). Axon collaterals of retinal ganglion cells extend to the midbrain and hypothalamus.

Hearing and Equilibrium (p. 623)

1. The external (outer) ear consists of the auricle, external auditory canal, and eardrum (tympanic membrane).
2. The middle ear consists of the auditory (Eustachian) tube, ossicles, oval window, and round window.
3. The internal (inner) ear consists of the bony labyrinth and membranous labyrinth. The internal ear contains the spiral organ (organ of Corti), the organ of hearing.
4. Sound waves enter the external auditory canal and strike the eardrum causing it to vibrate. The vibrations pass through the ossicles, strike the oval window, set up waves in the perilymph, strike the vestibular membrane and scala tympani, increase pressure in the endolymph, vibrate the basilar membrane, and stimulate hair bundles on the spiral organ (organ of Corti).
5. Bending of stereocilia ultimately leads to generation of nerve impulses in first-order sensory neurons.
6. Sensory fibers in the cochlear branch of the vestibulocochlear (VIII) nerve terminate in the medulla oblongata. Auditory signals then pass to the inferior colliculus, thalamus, and temporal lobes of the cerebral cortex.
7. Static equilibrium is the orientation of the body relative to the pull of gravity. The maculae of the utricle and saccule are the sense organs of static equilibrium.
8. Dynamic equilibrium is the maintenance of body position in response to movement. The cristae in the membranous semicircular ducts are the principal sense organs of dynamic equilibrium.
9. Most vestibular branch fibers of the vestibulocochlear (VIII) nerve enter the brain stem and terminate in the medulla and pons; other fibers enter the cerebellum.

SELF-QUIZ QUESTIONS

Choose the one best answer to these questions.

1. The first synapse in the olfactory pathway occurs in the olfactory (a) glands, (b) nerves, (c) bulbs, (d) cortex, (e) tract.

2. Which of the following sensory cells are also first order sensory neurons? (a) olfactory receptor cells, (b) gustatory receptor cells, (c) hair cells of the spiral organ (organ of Corti), (d) hair cells of the otolithic organs, (e) all of the above.

3. Which of the following is/are *not* part(s) of the vascular tunic of the eyeball? (a) cornea, (b) choroid, (c) ciliary body, (d) iris, (e) all of the above

4. The three layers of retinal neurons, in the order in which they process visual input are the (1) ganglion cell layer, (2) bipolar cell layer, (3) photoreceptor layer.

a. 1,2,3 **b.** 3,2,1 **c.** 2,1,3 **d.** 3,1,2 **e.** 2,3,1

5. The bony labyrinth of the inner ear (a) is lined with ceruminous glands, (b) consists of semicircular canals, vestibule, and tympanic antrum, (c) contains a fluid called perilymph, (d) houses hearing and equilibrium receptors, (e) both c and d.

6. Cones in the eye (a) function best in low light levels, (b) are more plentiful in the peripheral region of the retina, (c) receive sensory impulses from ganglion cells, (d) are concentrated in the central fovea of the macula lutea, (e) form part of the optic disc.

Complete the following.

7. The ________ cells of the olfactory mucosa are responsible for producing new olfactory receptor cells.

8. Taste buds are located on elevated projections of the tongue called ________ . The ________ type is scattered over the entire surface of the tongue.

9. Aqueous humor is produced by the ________ body, and is reabsorbed into the ________ sinus.

10. The three layers of the posterior portion of the eyeball, from superficial to deep are ________ , ________ , and ________ .

11. The mucous membrane that lines the eyelids and covers the anterior surface of the eyeball is called the ________ .

12. The ________ epithelium is a layer of melanin-containing epithelial cells between the choroid and neural portion of the retina.

13. Number the following in the correct sequence for the conduction pathway for the sense of smell. (a) olfactory bulbs: ________ , (b) olfactory receptor cells: ________ , (c) olfactory (I) nerves: ________ , (d) olfactory tract: ________ , (e) cortical region of temporal lobe: ________ .

14. Number the following in the correct sequence for the pathway of sound waves. (a) external auditory canal: ________ , (b) stapes: ________ , (c) malleus: ________ , (d) incus: ________ , (e) oval window: ________ , (f) eardrum (tympanic membrane): ________ .

15. Number the following in correct sequence, from anterior to posterior. (a) anterior chamber: ________ , (b) iris: ________ , (c) cornea: ________ , (d) vitreous chamber: ________ , (e) lens: ________ , (f) optic (II) nerve: ________ , (g) posterior chamber: ________ .

Are the following statements true or false?

16. Dendrites of olfactory receptor cells collectively make up the olfactory (I) nerves.

17. Cones are most important for color vision, whereas rods are most important for seeing shades of gray in dim light.

18. Tears pass medially over the anterior surface of the eyeball.

19. Contraction of the circular muscles of the iris enlarges the pupil of the eye.

20. Match the following terms to their descriptions:

____	**(a)** "white of the eye"	**(1)**	macula lutea
____	**(b)** a clear structure, composed of protein layers arranged like an onion	**(2)**	choroid
____	**(c)** blind spot; area in which there are no cones or rods	**(3)**	cornea
____	**(d)** exact center of the posterior portion of the retina	**(4)**	ciliary muscle
____	**(e)** nonvascular, transparent, fibrous coat; most anterior eye structure	**(5)**	iris
____	**(f)** layer containing neurons; if detached, causes blindness	**(6)**	lens
____	**(g)** dark brown layer; prevents reflection of light rays; also nourishes eyeball because it is vascular	**(7)**	optic disc
____	**(h)** a hole; appears black, like a circular doorway leading into a dark room	**(8)**	ora serrata
____	**(i)** regulates the amount of light entering the eye; colored part of the eye	**(9)**	pupil
____	**(j)** attaches to the lens by means of radially arranged fibers called the suspensory ligaments	**(10)**	retina
____	**(k)** serrated margin of the retina	**(11)**	sclera
____	**(l)** located at the junction of sclera and cornea; drains aqueous humor	**(12)**	scleral venous sinus (canal of Schlemm)

21. Match the following sensory areas of the brain with their locations:

____	**(a)** primary olfactory area	**(1)**	temporal lobe of cerebral cortex
____	**(b)** primary gustatory area	**(2)**	occipital lobe of cerebral cortex
____	**(c)** primary auditory area	**(3)**	parietal lobe of cerebral cortex
____	**(d)** primary visual area		

22. Match the following terms to their descriptions:

____ (a) organs of dynamic equilibrium; located in the ampullae of the semicircular canals

____ (b) eardrum

____ (c) air-filled cavity of temporal bone

____ (d) opening between the middle and inner ear: receives base of stapes

____ (e) oval central portion of the bony labyrinth

____ (f) connects the middle ear to the nasopharynx, allowing for equalization of pressure on both sides of the eardrum

____ (g) organ for hearing

____ (h) ear bones: malleus, incus and stapes

____ (i) receptor for static equilibrium; also contributes to some aspects of dynamic equilibrium; consists of hair cells and supporting cells

____ (j) opening into the middle ear; is enclosed by a membrane called the secondary tympanic membrane

____ (k) contains the spiral organ (organ of Corti)

____ (l) receptor organs for equilibrium; the saccule, utricle, and semicircular canals

(1) macula
(2) round window
(3) oval window
(4) tympanic membrane
(5) cristae
(6) auditory (Eustachian) tube
(7) spiral organ (organ of Corti)
(8) auditory ossicles
(9) middle ear
(10) vestibular apparatus
(11) cochlea
(12) vestibule

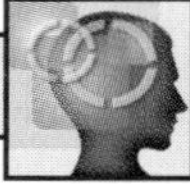

CRITICAL THINKING QUESTIONS

1. Mother held a steaming dish of soup under Aimee's nose. "How can you say it tastes terrible? You haven't even tried it!" Aimee's lips remained clamped tight. Why is Aimee so sure that she won't like the food?
   HINT *How would she know if she'd tried it before?*

2. Deirdre refuses to go to sad movies on a date because whenever she watches a really sad movie, her nose runs. "And that's a horrible way to impress a date!" Explain Deirdre's leaky nose.
   HINT *Her nose isn't the only thing that leaks when she's sad.*

3. Lucas noticed that the colored ring in one of his mother's eyes was a different shape than the other eye. His mother explained that she had gotten poked in the eye with a stick when she was a little girl about his age and that had caused the damage to her eye. What structure of the eye was injured by the stick? What effect would the damage have on her vision?
   HINT *Corrective contact lenses float over this region.*

4. Reuben was on his first cruise and he felt miserable! The ship's doctor mentioned something about messages from his eyes confusing his ears (or the other way around) but Reuben was too sick to listen. He took his medication and went to bed. Explain the cause of Reuben's sea sickness.
   HINT *The ocean wasn't the only fluid moving up and down, up and down, etc.*

5. Why do the eyes in some photographs taken with a flash appear to be red?
   HINT *An albino's eyes will almost always look red in photographs even though the iris color is pale blue.*

ANSWERS TO FIGURE QUESTIONS

20.1 The olfactory hairs detect odorant molecules.

20.2 Gustatory receptor cells → cranial nerves VII, IX, or X → medulla oblongata → either limbic system and hypothalamus, or thalamus → primary gustatory area in the parietal lobe of the cerebral cortex.

20.3 Lacrimal fluid, or tears, is a watery solution containing salts, some mucus, and lysozyme that protects, cleans, lubricates, and moistens the eyeball.

20.4 The fibrous tunic consists of the cornea and sclera; the vascular tunic consists of the choroid, ciliary body, and lens.

20.5 The parasympathetic division of the ANS causes pupillary constriction, whereas the sympathetic division causes pupillary dilation.

20.6 An ophthalmoscopic examination can reveal evidence of hypertension, diabetes mellitus, cataract, and macular degeneration.

20.7 The photoreceptors are rods and cones. Rods provide black-and-white vision in dim light, whereas cones provide high visual acuity and color vision in bright light.

20.8 After its secretion by the ciliary process, aqueous humor flows into the posterior chamber, around the iris, into the anterior chamber, and out of the eyeball through the scleral venous sinus.

20.9 The optic tract terminates in the lateral geniculate nucleus of the thalamus.

20.10 The malleus is attached to the eardrum.

20.11 The eardrum separates the middle ear from the external ear; the oval and round windows separate the middle ear from the internal ear.

20.12 The two sacs in the vestibule are the utricle and saccule.

20.13 The bony labyrinth consists of semicircular canals, vestibule, and cochlea.

20.14 The oval window.

20.15 Maculae mainly operate in static equilibrium.

20.16 The membranous semicircular ducts, utricle, and saccule are mainly associated with dynamic equilibrium.

21

THE AUTONOMIC NERVOUS SYSTEM

CONTENTS AT A GLANCE

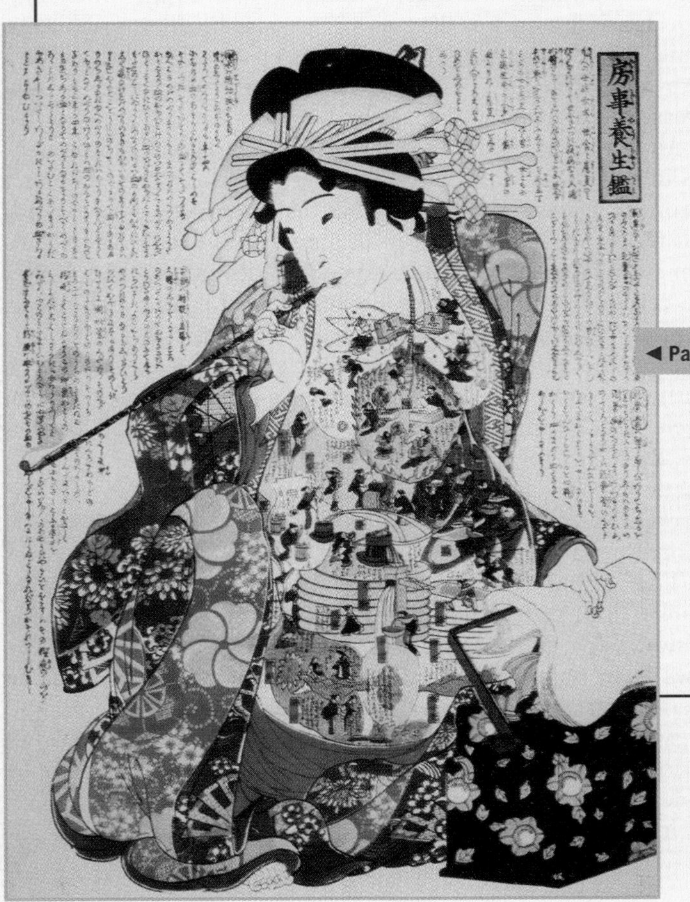

◀ Page 650

Examine this image closely. What do you see? Can you identify any of the organs? Can you guess which processes are being depicted?

Page 644 ▶

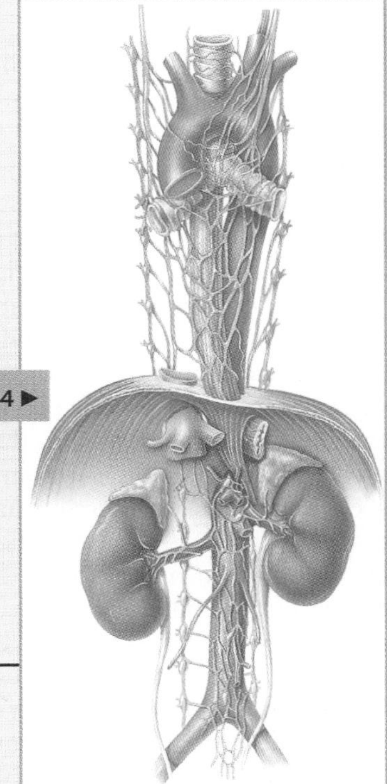

INTRODUCTION

The **autonomic nervous system (ANS)** regulates the activity of smooth muscle, cardiac muscle, and certain glands. The ANS has often been described as a specific *motor output* portion of the peripheral nervous system. To operate, however, the ANS depends on a continual flow of *sensory input* from visceral organs and blood vessels into *integrating centers* in the central nervous system (CNS). Structurally, then, the ANS includes autonomic sensory neurons, integrating centers in the CNS, and autonomic motor neurons.

Functionally, the ANS usually operates without conscious control. The system was originally named *autonomic* because it was thought to function autonomously or in a self-governing manner, without control by the CNS. However, the ANS is regulated by centers in the brain, mainly in the hypothalamus and brain stem, that receive input from the limbic system and other regions of the cerebrum.

In this chapter we compare structural and functional features of the somatic and autonomic nervous systems, then we discuss the anatomy of the motor portion of the ANS and compare the organization and actions of its two major branches, the sympathetic and parasympathetic divisions.

COMPARISON OF SOMATIC AND AUTONOMIC NERVOUS SYSTEMS

Objective

- Compare the structural and functional differences of the somatic and autonomic portions of the nervous system.

The somatic nervous system includes both sensory and motor neurons. The sensory neurons convey input from receptors for the special senses (vision, hearing, taste, smell, and equilibrium) and from receptors for somatic senses (pain, temperature, tactile, and proprioceptive sensations). All these sensations normally are consciously perceived. In turn, somatic motor neurons innervate skeletal muscle—the effector tissue of the somatic nervous system—and produce conscious, voluntary movements. In the somatic nervous system, the effect of a motor neuron is always excitation. When a somatic motor neuron stimulates a skeletal muscle, the muscle contracts. If somatic motor neurons cease to stimulate a muscle, the result is a paralyzed, limp muscle that has no muscle tone. In addition, even though we are generally not conscious of breathing, the muscles that generate respiratory movements are skeletal muscles controlled by somatic motor neurons. If the respiratory motor neurons become inactive, breathing stops. A few skeletal muscles, such as those in the middle ear, cannot be voluntarily controlled.

The main input to the ANS comes from **autonomic sensory neurons.** Mostly, these neurons are associated with interoceptors, such as chemoreceptors that monitor blood CO_2 level and mechanoreceptors that detect the degree of stretch in the walls of organs or blood vessels. These sensory signals are not consciously perceived most of the time, although intense activation of interoceptors may give rise to conscious sensations. Two examples of visceral sensations that are perceived are sensations of pain or nausea from damaged viscera and angina pectoris (chest pain) from inadequate blood flow to the heart. Input that influences the ANS also includes some sensations monitored by somatic sensory and special sensory neurons. For example, pain can produce dramatic changes in some autonomic activities.

Autonomic motor neurons regulate visceral activities by either increasing (exciting) or decreasing (inhibiting) ongoing activities in their effector tissues, which are cardiac muscle, smooth muscle, and glands. Unlike skeletal muscle, these tissues generally function even if their nerve supply is interrupted. The heart continues to beat, for instance, when it is removed for transplantation into another person. Examples of autonomic responses are changes in the diameter of the pupil, dilation and constriction of blood vessels, and adjustment of the rate and force of the heartbeat.

Most autonomic responses cannot be consciously altered or suppressed to any great degree—you probably cannot voluntarily slow your heartbeat to half its normal rate. For this reason, some autonomic responses are the basis for polygraph ("lie detector") tests. Nevertheless, practitioners of yoga or other techniques of meditation may, through long practice, learn how to modulate at least some of their autonomic activities. Input from the general somatic and special senses, acting via the limbic system, also influences responses of autonomic motor neurons. Seeing a bike about to hit you, hearing squealing brakes of a nearby car, or being grabbed by an attacker, for example, would increase the rate and force of your heartbeat.

All autonomic motor pathways consist of two motor neurons in series, one following the other (Figure 21.1a). The first neuron has its cell body in the CNS; its myelinated axon extends from the CNS to an **autonomic ganglion.** (Recall that a ganglion is a collection of neuronal cell bodies outside the CNS.) The cell body of the second neuron is also in that autonomic ganglion; its unmyelinated axon extends directly from the ganglion to the effector (smooth muscle, cardiac muscle, or a gland). In contrast, to cause voluntary contraction of a skeletal muscle fiber, a single myelinated somatic motor neuron extends from the CNS to the effector (Figure 21.1b). Also, whereas all somatic motor neurons release only acetylcholine (ACh) as their neurotransmitter, autonomic motor neurons release either ACh or norepinephrine (NE).

The output (motor) part of the ANS has two principal branches: the **sympathetic division** and the **parasympathetic division.** Most organs have **dual innervation**—they receive impulses from both sympathetic and parasympathetic neurons. In general, nerve impulses from one division stimulate the organ to increase its activity (excitation), whereas impulses from the other division decrease the organ's activity (inhibition). For example, an increased rate of nerve impulses from the sympathetic division increases heart rate, whereas an increased rate of nerve im-

Figure 21.1 / Motor neuron pathways in (a) the autonomic nervous system and (b) the somatic nervous system. Note that autonomic motor neurons release either acetylcholine (ACh) or norepinephrine (NE); somatic motor neurons release ACh.

Stimulation by the autonomic nervous system either excites or inhibits visceral effectors; somatic nervous system stimulation always excites its effectors.

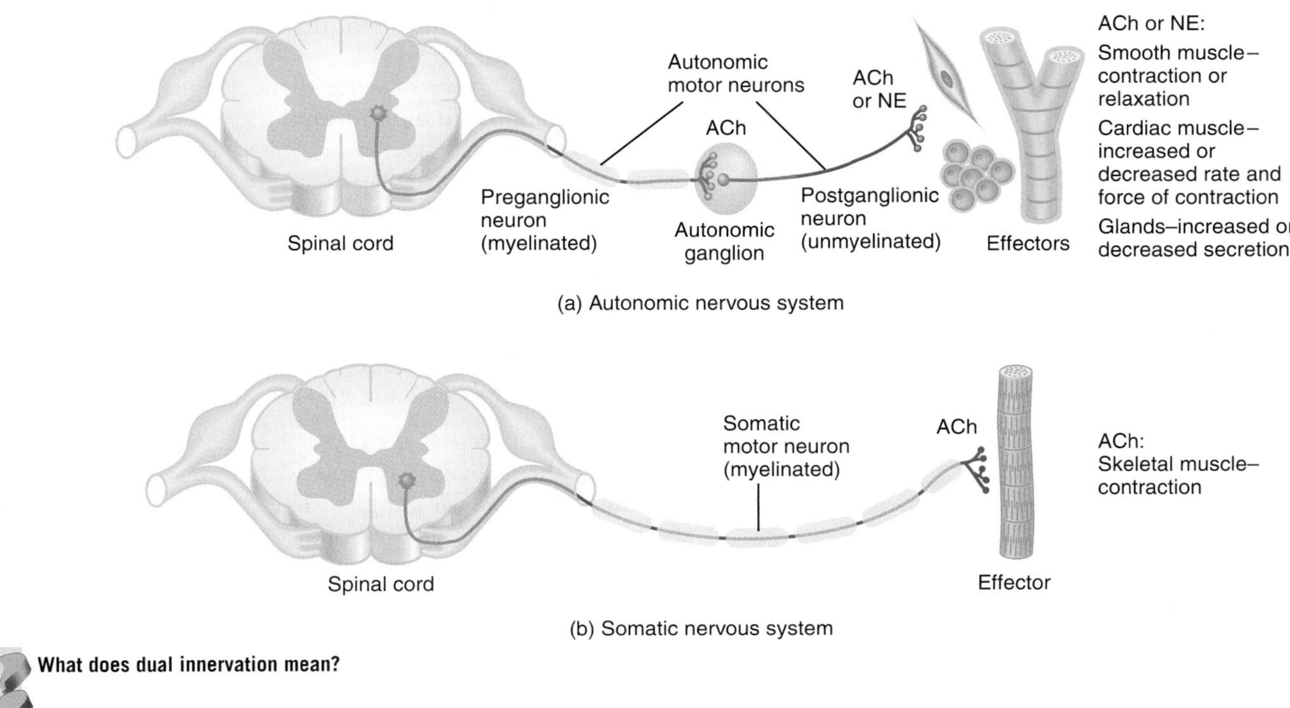

What does dual innervation mean?

pulses from the parasympathetic division decreases heart rate. Table 21.1 summarizes the similarities and differences between the somatic and autonomic nervous systems.

✓ Explain why the autonomic nervous system is so named.

✓ What are the principal input and output components of the autonomic nervous system?

Table 21.1 Summary of Autonomic and Somatic Nervous Systems

	Autonomic Nervous System	Somatic Nervous System
Sensory input	Mainly from interoceptors; some from special senses and somatic senses.	Special senses and somatic senses.
Control of motor output	Involuntary control from limbic system, hypothalamus, brain stem, and spinal cord; limited control from cerebral cortex.	Voluntary control from cerebral cortex, with contributions from basal ganglia, cerebellum, brain stem, and spinal cord.
Motor neuron pathway	Two-neuron pathway: preganglionic neurons extending from CNS synapse with postganglionic neurons in an autonomic ganglion, and postganglionic neurons extending from ganglion synapse with a visceral effector. Also, preganglionic neurons extending from CNS synapse withcells of adrenal medulla.	One-neuron pathway: somatic motor neurons extending from CNS synapse directly with effector.
Neurotransmitters and hormones	Preganglionic axons release acetylcholine (ACh); post-ganglionic axons release ACh (parasympathetic division and sympathetic fibers to sweat glands) or norepinephrine (NE; remainder of sympathetic division); adrenal medulla releases epinephrine and norepinephrine.	All somatic motor neurons release ACh.
Effectors	Smooth muscle, cardiac muscle, glands.	Skeletal muscle.
Responses	Contraction or relaxation of smooth muscle; increased or decreased rate and force of contraction of cardiac muscle; increased or decreased secretions of glands.	Contraction of skeletal muscle.

ANATOMY OF AUTONOMIC MOTOR PATHWAYS

Objectives

- Describe preganglionic and postganglionic neurons of the autonomic nervous system.
- Compare the anatomical components of the sympathetic and parasympathetic divisions of the autonomic nervous system.

Anatomical Components

The first of the two motor neurons in any autonomic motor pathway is called a **preganglionic neuron** (Figure 21.1a). Its cell body is in the brain or spinal cord, and its axon exits the CNS as part of a cranial or spinal nerve. The axon of a preganglionic neuron is a small-diameter, myelinated fiber that extends to an autonomic ganglion, where it synapses with the **postganglionic neuron,** the second neuron in the autonomic motor pathway. Notice that the postganglionic neuron lies entirely outside the CNS; its cell body and dendrites are located in an autonomic ganglion, where it synapses with one or more preganglionic fibers. The axon of a postganglionic neuron is a small-diameter, unmyelinated fiber that terminates in a visceral effector. Thus, preganglionic neurons convey motor impulses from the CNS to autonomic ganglia, and postganglionic neurons relay the impulses from autonomic ganglia to visceral effectors.

Preganglionic Neurons

In the sympathetic division, the preganglionic neurons have their cell bodies in the lateral horns of the gray matter in the 12 thoracic segments and the first two lumbar segments of the spinal cord (Figure 21.2). For this reason, the sympathetic division is also called the **thoracolumbar division** (thō′-ra-kō-LUM-bar), and the axons of the sympathetic preganglionic neurons are known as the **thoracolumbar outflow.**

The cell bodies of the preganglionic neurons of the parasympathetic division are located in the nuclei of the four cranial nerves in the brain stem and in the lateral gray horns of the second through fourth sacral segments of the spinal cord. Hence, the parasympathetic division is also known as the **craniosacral division,** and the axons of the parasympathetic preganglionic neurons are referred to as the **craniosacral outflow.**

Autonomic Ganglia

The autonomic ganglia may be divided into three general groups: Two of the groups are components of the sympathetic division, and one group is a component of the parasympathetic division.

SYMPATHETIC GANGLIA The sympathetic ganglia are the sites of synapses between sympathetic preganglionic and postganglionic neurons. The two groups of sympathetic ganglia are sympathetic trunk ganglia and prevertebral ganglia. **Sympathetic trunk (chain) ganglia** (also called *vertebral chain ganglia* or *paravertebral ganglia*) lie in a vertical row on either side of the vertebral column, extending from the base of the skull to the coccyx (Figure 21.2). Because the sympathetic trunk ganglia are near the spinal cord, most sympathetic preganglionic axons are short. Postganglionic axons from sympathetic trunk ganglia, in general, innervate organs above the diaphragm. Examples of sympathetic trunk ganglia are the **superior, middle,** and **inferior cervical ganglia.**

The second group of sympathetic ganglia, the **prevertebral** *(collateral)* **ganglia,** lie anterior to the spinal column and close to the large abdominal arteries. In general, postganglionic fibers from prevertebral ganglia innervate organs below the diaphragm. Examples of prevertebral ganglia are the **celiac ganglion** (SĒ-lē-ak), on either side of the celiac artery just inferior to the diaphragm; the **superior mesenteric ganglion,** near the beginning of the superior mesenteric artery in the upper abdomen; and the **inferior mesenteric ganglion,** near the beginning of the inferior mesenteric artery in the middle of the abdomen (see Figures 21.2 and 21.3).

PARASYMPATHETIC GANGLIA Preganglionic fibers of the parasympathetic division synapse with postganglionic neurons in **terminal** *(intramural)* **ganglia.** These ganglia are located close to or actually within the wall of a visceral organ. Because the axons of parasympathetic preganglionic neurons extend from the CNS to a terminal ganglion in an innervated organ, they are longer than most of the axons of sympathetic preganglionic neurons. Examples of terminal ganglia include the ciliary ganglion, pterygopalatine ganglion, submandibular ganglion, and otic ganglion (Figure 21.2).

Autonomic Plexuses

In the thorax, abdomen, and pelvis, axons of both sympathetic and parasympathetic neurons form tangled networks called **autonomic plexuses,** many of which lie along major arteries. The autonomic plexuses also may contain sympathetic ganglia and axons of autonomic sensory neurons. Often plexuses are named after the nearest large artery. The major plexuses in the thorax are the **cardiac plexus,** which is at the base of the heart, surrounding the large blood vessels emerging from the heart, and the **pulmonary plexus,** located mostly posterior to each lung (Figure 21.3 on page 644).

The abdomen and pelvis also contain major autonomic plexuses. The **celiac** *(solar)* **plexus** is found at the level of the last thoracic and first lumbar vertebrae. It is the largest autonomic plexus and surrounds the celiac and superior mesenteric arteries. It contains two large celiac ganglia and a dense network of autonomic fibers. Secondary plexuses that arise from the celiac plexus are distributed to the diaphragm, liver, gallbladder, stomach, pancreas, spleen, kidneys, medulla (inner region) of the adrenal gland, testes, and ovaries. The **superior mesenteric plexus,** a secondary plexus of the celiac plexus, contains the superior mesenteric ganglion and supplies the small and large intestine. The **inferior mesenteric plexus,** also a secondary plexus of the celiac plexus, contains the inferior mesenteric ganglion and innervates the large intestine. The **hypogastric plexus** is anterior to the fifth lumbar vertebra and supplies pelvic viscera.

Figure 21.2 / Structure of the sympathetic and parasympathetic divisions of the autonomic nervous system. The parasympathetic division is shown only on one side, and the sympathetic division only on the other side, for diagrammatic purposes; each division actually innervates tissues and organs on both sides of the body.

Sympathetic and parasympathetic stimulation have opposing effects on organs that receive dual innervation.

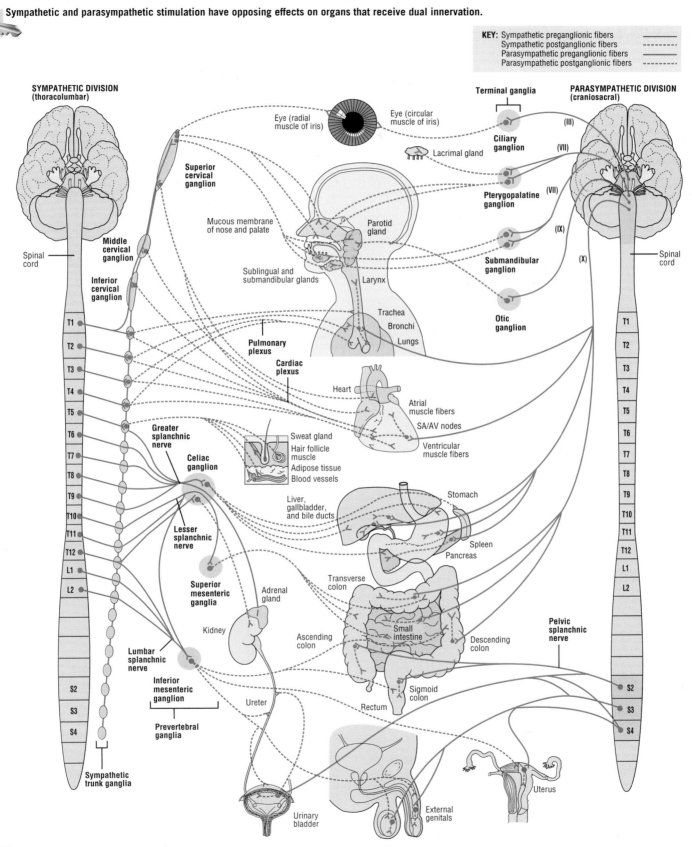

Which division has longer preganglionic fibers? Why?

Figure 21.3 / Autonomic plexuses in the thorax, abdomen, and pelvis.

An autonomic plexus is a network of sympathetic and parasympathetic axons that sometimes also includes autonomic sensory axons and sympathetic ganglia.

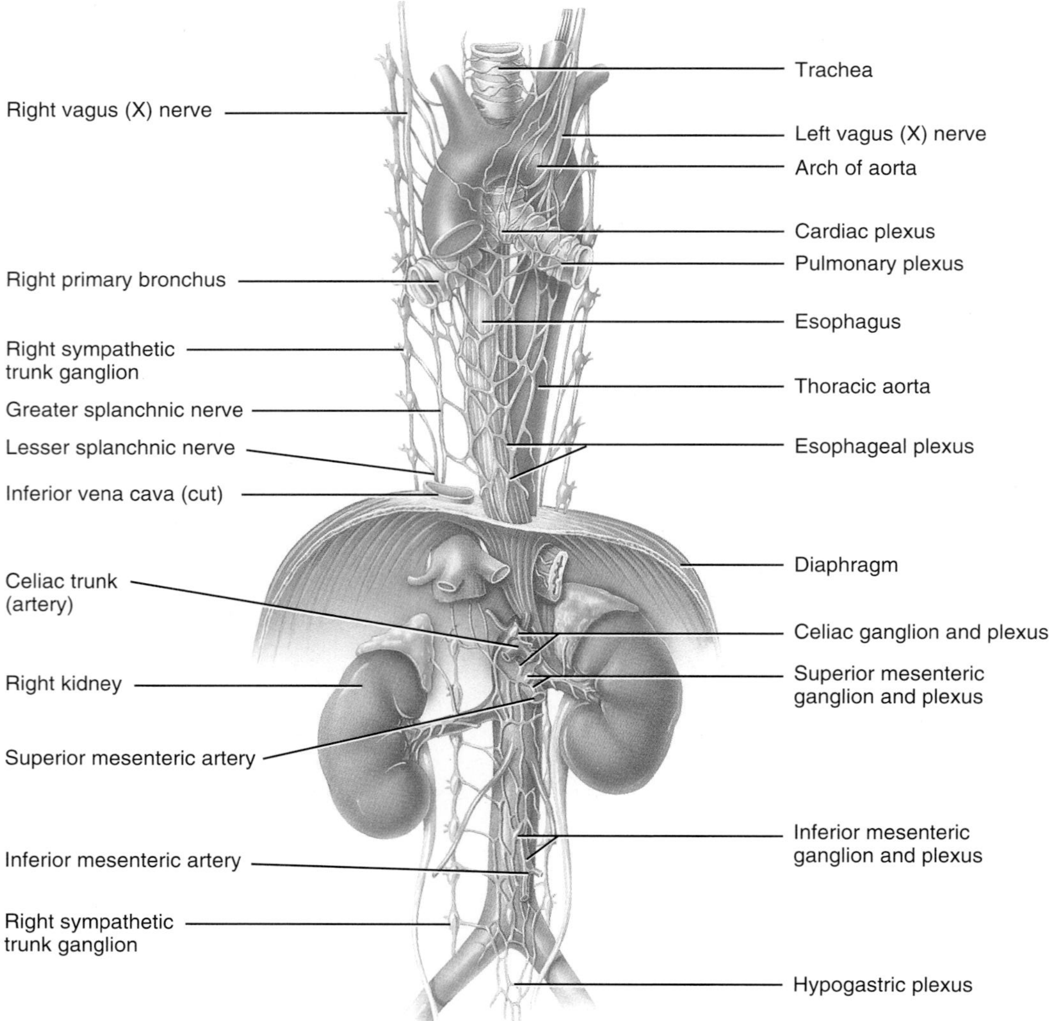

Which is the largest autonomic plexus?

Postganglionic Neurons

Once axons of preganglionic neurons of the sympathetic division pass to sympathetic trunk ganglia, they may connect with postganglionic neurons in one of the following ways (Figure 21.4):

1. An axon may synapse with postganglionic neurons in the ganglion it first reaches.
2. An axon may ascend or descend to a higher or lower ganglion before synapsing with postganglionic neurons.
3. An axon may continue, without synapsing, through the sympathetic trunk ganglion to end at a prevertebral ganglion and synapse with postganglionic neurons there.
4. An axon may extend to and terminate in the adrenal medulla (not shown in the illustration).

A single sympathetic preganglionic fiber has many axon collaterals (branches) and may synapse with 20 or more postganglionic fibers. This pattern of projection is an example of divergence and helps explain why many sympathetic responses affect almost the entire body simultaneously. After exiting their ganglia, the postganglionic fibers typically innervate several visceral effectors (see Figure 21.2).

Axons of preganglionic neurons of the parasympathetic division pass to terminal ganglia near or within a visceral effector (see Figure 21.2). In the ganglion, the presynaptic neuron usually synapses with only four or five postsynaptic neurons, all of

Figure 21.4 / Types of connections between ganglia and postganglionic neurons in the sympathetic division of the autonomic nervous system. Numbers correspond to descriptions in the text. Also illustrated are the gray and white rami communicantes.

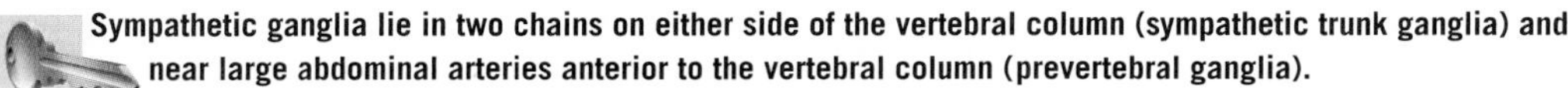

Sympathetic ganglia lie in two chains on either side of the vertebral column (sympathetic trunk ganglia) and near large abdominal arteries anterior to the vertebral column (prevertebral ganglia).

What substance gives the white rami their white appearance?

which supply a single visceral effector. Thus, parasympathetic responses can be localized to a single effector. With this background in mind, we can now examine some specific structural features of the sympathetic and parasympathetic divisions of the ANS.

Structure of the Sympathetic Division

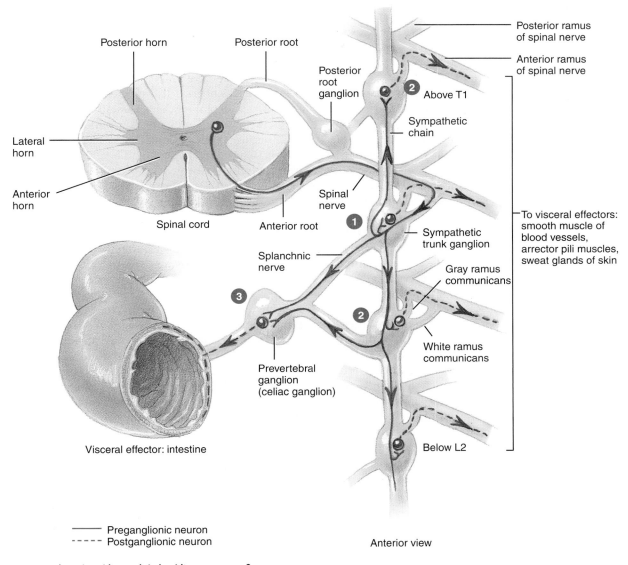

Cell bodies of sympathetic preganglionic neurons are part of the lateral horns of all thoracic segments and of the first two lumbar segments of the spinal cord (see Figure 21.2). The preganglionic axons leave the spinal cord through the anterior root of a spinal nerve along with the somatic motor fibers at the same segmental level. After exiting through the intervertebral foramina, the myelinated preganglionic sympathetic axons enter a short pathway called a **white ramus** before passing to the nearest sympathetic trunk ganglion on the same side (see Figure 21.4).

Collectively, the white rami are called the **white rami communicantes** (kō-myū-ni-KAN-tēz; the singular is **ramus com-**

municans). The "white" in their name indicates that they contain myelinated fibers. Only the thoracic and first two or three lumbar nerves have white rami communicantes. The white rami communicantes connect the anterior ramus of the spinal nerve with the ganglia of the sympathetic trunk. The paired sympathetic trunk ganglia are arranged anterior and lateral to the vertebral column, one on either side. Typically, there are 3 cervical, 11 or 12 thoracic, 4 or 5 lumbar, and 4 or 5 sacral sympathetic trunk ganglia. Although the sympathetic trunk ganglia extend inferiorly from the neck, chest, and abdomen to the coccyx, they receive preganglionic fibers only from the thoracic and lumbar segments of the spinal cord (see Figure 21.2).

The cervical portion of each sympathetic trunk is located in the neck and is subdivided into superior, middle, and inferior ganglia (see Figure 21.2). The **superior cervical ganglion** is anterior to a transverse process of the second cervical vertebra. Postganglionic fibers leaving the superior cervical ganglion serve the head. They are distributed to sweat glands, smooth muscle of the eye, blood vessels of the face, nasal mucosa, and the submandibular, sublingual, and parotid salivary glands. Gray rami communicantes (described shortly) from the ganglion also pass to the upper two to four cervical spinal nerves. The **middle cervical ganglion** lies near the sixth cervical vertebra, and the **inferior cervical ganglion** is located near the first rib, anterior to the transverse processes of the seventh cervical vertebra. Postganglionic fibers from the middle and inferior cervical ganglia innervate the heart.

The thoracic portion of each sympathetic trunk lies anterior to the necks of the corresponding ribs. This portion of the sympathetic trunk receives most of the sympathetic preganglionic fibers. Postganglionic fibers from the thoracic sympathetic trunk innervate the heart, lungs, bronchi, and other thoracic viscera. In the skin, these fibers also innervate sweat glands, blood vessels, and arrector pili muscles of hair follicles.

The lumbar portion of each sympathetic trunk lies lateral to the corresponding lumbar vertebrae. The sacral portion of the sympathetic trunk lies in the pelvic cavity on the medial side of the sacral foramina. Unmyelinated postganglionic fibers from the lumbar and sacral sympathetic trunk ganglia enter a short pathway called a **gray ramus** and then merge with a spinal nerve or join the hypogastric plexus via direct visceral branches. The **gray rami communicantes** are structures containing the postganglionic fibers that connect the ganglia of the sympathetic trunk to spinal nerves (see Figure 21.4). Gray rami communicantes outnumber the white rami because there is a gray ramus leading to each of the 31 pairs of spinal nerves.

As preganglionic fibers extend from a white ramus communicans into the sympathetic trunk, they give off several axon collaterals (branches). These collaterals terminate and synapse in several ways. Some synapse in the first ganglion at the level of entry. Others pass up or down the sympathetic trunk for a variable distance to form the **sympathetic chains,** the fibers on which the ganglia are strung (see Figure 21.4). Many postganglionic fibers rejoin the spinal nerves through gray rami and supply peripheral visceral effectors such as sweat glands, smooth muscle in blood vessels, and arrector pili muscles of hair follicles.

Some preganglionic fibers pass through the sympathetic trunk without terminating in the trunk. Beyond the trunk, they form nerves known as **splanchnic nerves** (SPLANK-nik; see Figure 21.4), which extend to and terminate in the outlying prevertebral ganglia. Splanchnic nerves from the thoracic area terminate in the **celiac ganglion,** where the preganglionic fibers synapse with postganglionic cell bodies. The *greater splanchnic nerve* is formed by preganglionic fibers from thoracic ganglia 5-9 or 10 (T5-T9 or T10), pierces the diaphragm, and enters the celiac ganglion of the celiac plexus. From here, postganglionic fibers extend to the stomach, spleen, liver, kidney, and small intestine (see Figure 21.2). The *lesser splanchnic nerve* is formed by preganglionic fibers from thoracic ganglia 10-11 (T10-T11), pierces the diaphragm, and passes through the celiac plexus to enter the superior mesenteric ganglion of the superior mesenteric plexus. Postganglionic fibers from this ganglion innervate the small intestine and colon. The *lowest splanchnic nerve*, not always present, is formed by preganglionic fibers from thoracic ganglion 12 (T12) or a branch of the lesser splanchnic nerve, passes through the diaphragm, and enters the renal plexus near the kidney. Postganglionic fibers supply kidney arterioles and the ureter.

Preganglionic fibers from lumbar ganglia 1-4 (L1-L4) form the *lumbar splanchnic nerve*, enter the inferior mesenteric plexus, and terminate in the inferior mesenteric ganglion, where they synapse with postganglionic neurons. Axons of postganglionic neurons extend through the hypogastric plexus and supply the distal colon and rectum, urinary bladder, and genital organs. Postganglionic fibers leaving the prevertebral ganglia follow the course of various arteries to abdominal and pelvic visceral effectors.

Sympathetic *preganglionic fibers* also extend to the adrenal medullae. Developmentally, the adrenal medullae and sympathetic ganglia are derived from the same tissue, the neural crest (see Figure 18.24b on page 583). The adrenal medullae are modified sympathetic ganglia, and their cells are similar to sympathetic postganglionic neurons. Rather than extending to another organ, however, these cells release hormones into the blood. Upon stimulation by sympathetic preganglionic neurons, the adrenal medullae release a mixture of catecholamine hormones—about 80% epinephrine, 20% norepinephrine, and a trace amount of dopamine.

In **Horner's syndrome** the sympathetic innervation to one side of the face is lost due to an inherited mutation, an injury, or a disease that affects sympathetic outflow through the superior cervical ganglion. Symptoms occur on the affected side and include ptosis (drooping of the upper eyelid), miosis (constricted pupil), and anhidrosis (lack of sweating).

Structure of the Parasympathetic Division

Cell bodies of parasympathetic preganglionic neurons are found in nuclei in the brain stem and in the lateral horns of the second through fourth sacral segments of the spinal cord (see Figure 21.2). Their axons emerge as part of a cranial nerve or as part of the anterior root of a spinal nerve. The **cranial parasympathetic outflow** consists of preganglionic axons that extend from the brain stem in four cranial nerves. The **sacral parasympathetic outflow** consists of preganglionic axons in anterior roots of the second through fourth sacral nerves. The preganglionic axons of both the cranial and sacral outflows end in terminal ganglia, where they synapse with postganglionic neurons.

The cranial outflow has five components: four pairs of ganglia and the plexuses associated with the vagus (X) nerve. The four pairs of cranial parasympathetic ganglia innervate structures in the head and are located close to the organs they innervate (see Figure 21.2). The **ciliary ganglia** lie lateral to each optic (II) nerve near the posterior aspect of the orbit. Preganglionic axons pass with the oculomotor (III) nerves to the ciliary ganglia. Postganglionic axons from the ganglia innervate smooth muscle fibers in the eyeball. The **pterygopalatine ganglia** (ter′-i-gō-PAL-a-tīn) are lateral to the sphenopalatine foramen, between the sphenoid and palatine bones; they receive preganglionic axons from the facial (VII) nerve and send postganglionic axons to the nasal mucosa, palate, pharynx, and lacrimal glands. The **submandibular ganglia** are found near the ducts of the submandibular salivary glands; they receive preganglionic axons from the facial (VII) nerves and send postganglionic axons to the submandibular and sublingual salivary glands. The **otic ganglia** are situated just inferior to each foramen ovale; they receive preganglionic axons from the glossopharyngeal (IX) nerves and send postganglionic axons to the parotid salivary glands.

Preganglionic axons that leave the brain as part of the vagus (X) nerves carry nearly 80% of the total craniosacral outflow. Vagal fibers extend to many terminal ganglia in the thorax and abdomen. Because the terminal ganglia are close to or in the walls of their visceral effectors, postganglionic parasympathetic axons are very short. As the vagus (X) nerve passes through the thorax, it sends axons to the heart and the airways of the lungs. In the abdomen, it supplies the liver, gallbladder, stomach, pancreas, small intestine, and part of the large intestine.

The sacral parasympathetic outflow consists of preganglionic axons from the anterior roots of the second through fourth sacral nerves (S2-S4). Collectively, they form the *pelvic splanchnic nerves* (see Figure 21.2). These nerves synapse with parasympathetic postganglionic neurons located in terminal ganglia in the walls of the innervated viscera. From the ganglia, parasympathetic postganglionic axons innervate smooth muscle and glands in the walls of the colon, ureters, urinary bladder, and reproductive organs.

The anatomical features of the sympathetic and parasympathetic divisions are compared in Table 21.2.

✓ Explain why the sympathetic division is called the thoracolumbar division even though its ganglia extend from the cervical to the sacral region.
✓ Prepare a list of the organs served by each sympathetic and parasympathetic ganglion.
✓ Describe the locations of sympathetic trunk ganglia, prevertebral ganglia, and terminal ganglia. Which types of autonomic fibers synapse in each type of ganglion?
✓ Explain in anatomical terms why the sympathetic division may produce simultaneous effects throughout the body, whereas parasympathetic effects typically are localized to specific organs.

ANS NEUROTRANSMITTERS AND RECEPTORS

Objective

- Describe the neurotransmitters and receptors involved in autonomic responses.

Based on the neurotransmitter they produce and liberate, autonomic neurons are classified as either cholinergic or adrenergic. The receptors for the neurotransmitters are integral membrane

Table 21.2 Comparative Anatomy of the Sympathetic and Parasympathetic Divisions

	Sympathetic	Parasympathetic
Distribution	Bodywide: skin, sweat glands, arrector pili muscles of hair follicles, adipose tissue, smooth muscle of blood vessels.	Limited mainly to head and to viscera of thorax, abdomen, and pelvis; some blood vessels.
Outflow	Thoracolumbar (T1–L2).	Craniosacral (cranial nerves III, VII, IX, and X; S2–S4).
Associated ganglia	Two types: sympathetic trunk and prevertebral ganglia.	One type: terminal ganglia.
Ganglia locations	Close to CNS and distant from visceral effectors.	Typically near or within wall of visceral effectors.
Fiber length and divergence	Short preganglionic fibers synapse with many long postganglionic neurons that pass to many visceral effectors.	Long preganglionic fibers usually synapse with four to five short postganglionic neurons that pass to a single visceral effector.
Rami communicantes	Both present; white rami communicantes contain myelinated preganglionic fibers, and gray rami communicantes contain unmyelinated postganglionic fibers.	Neither present.

proteins located in the plasma membrane of the postsynaptic neuron or effector cell.

Cholinergic Neurons and Receptors

Cholinergic neurons (kō′-lin-ER-jik) release the neurotransmitter **acetylcholine (ACh).** In the ANS, the cholinergic neurons include (1) all sympathetic and parasympathetic preganglionic neurons, (2) sympathetic postganglionic neurons that innervate most sweat glands, and (3) all parasympathetic postganglionic neurons (Figure 21.5).

ACh is stored in synaptic vesicles in axon terminals and released by exocytosis. It diffuses across the synaptic cleft and binds with specific **cholinergic receptors,** which are integral membrane proteins in the *postsynaptic* plasma membrane. The two types of cholinergic receptors, both of which bind ACh, are nicotinic receptors and muscarinic receptors. **Nicotinic receptors** are present in the plasma membrane of dendrites and cell bodies of both sympathetic and parasympathetic postganglionic neurons (Figure 21.5) and in the motor end plate at the neuromuscular junction. They are so named because nicotine mimics the action of ACh by binding to these receptors. (Nicotine, a

Figure 21.5 / Cholinergic neurons (aqua) and adrenergic neurons (orange) in the sympathetic and parasympathetic divisions. Cholinergic neurons release acetylcholine, whereas adrenergic neurons release norepinephrine. Cholinergic and adrenergic receptors all are integral membrane proteins located in the plasma membrane of a postsynaptic neuron or an effector cell.

Most sympathetic postganglionic neurons are adrenergic; other autonomic neurons are cholinergic.

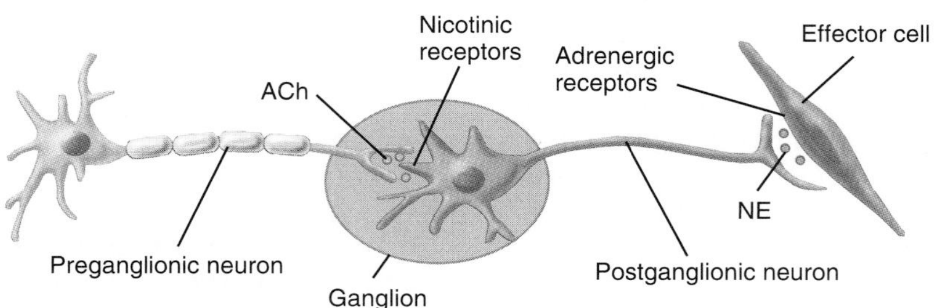

(a) Sympathetic division–innervation to most effector tissues

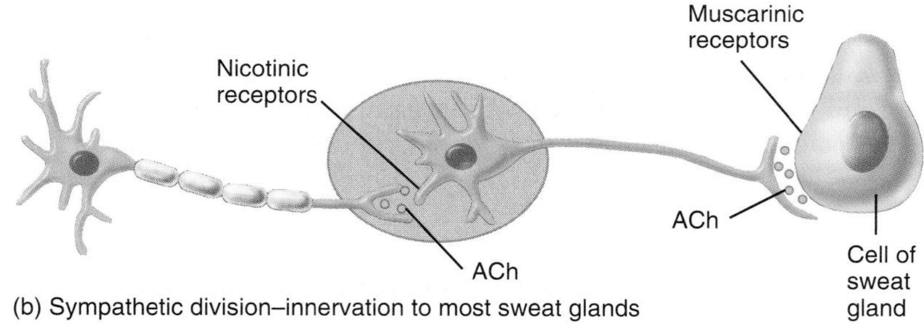

(b) Sympathetic division–innervation to most sweat glands

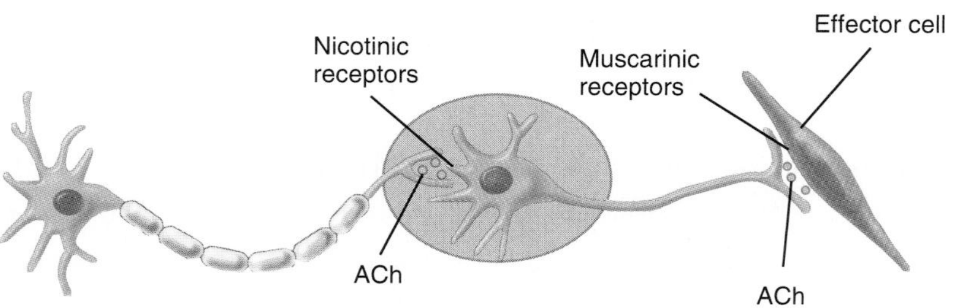

(c) Parasympathetic division

Which neurons are cholinergic and possess nicotinic ACh receptors? What type of receptors for ACh do the effector tissues innervated by these neurons possess?

natural substance in tobacco leaves, is not normally present in nonsmoking humans.) **Muscarinic receptors** are present in the plasma membranes of all effectors (smooth muscle, cardiac muscle, and glands) innervated by parasympathetic postganglionic axons. In addition, most sweat glands receive their innervation from *cholinergic* sympathetic postganglionic fibers and possess muscarinic receptors (Figure 21.5b). These receptors are so named because a mushroom poison called muscarine mimics the actions of ACh by binding to them. Nicotine does not activate muscarinic receptors, and muscarine does not activate nicotinic receptors, but ACh does activate both types of cholinergic receptors.

Activation of nicotinic receptors by ACh causes excitation of the postsynaptic cell—either a postganglionic neuron, an autonomic effector, or a skeletal muscle fiber. Activation of muscarinic receptors by ACh sometimes causes excitation and sometimes causes inhibition, depending on which particular cell bears the muscarinic receptors. The binding of ACh to muscarinic receptors inhibits (relaxes) smooth muscle sphincters in the gastrointestinal tract, for example, but excites smooth muscle fibers in the circular muscles of the iris of the eye, causing them to contract. Because acetylcholine is quickly inactivated by the enzyme **acetylcholinesterase (AChE),** effects triggered by cholinergic fibers are brief.

Adrenergic Neurons and Receptors

In the ANS, **adrenergic neurons** (ad′-ren-ER-jik) release **norepinephrine (NE),** also known as **noradrenalin** (Figure 21.5a). Most sympathetic postganglionic neurons are adrenergic. Like ACh, NE is synthesized and stored in synaptic vesicles and released by exocytosis. The molecules diffuse across the synaptic cleft and bind to specific adrenergic receptors on the postsynaptic membrane, causing either excitation or inhibition of the effector cell.

Adrenergic receptors bind both norepinephrine and epinephrine. Moreover, the receptors are activated by norepinephrine released as a neurotransmitter by sympathetic postganglionic neurons and by epinephrine and norepinephrine released as hormones into the blood by the adrenal medulla. The two main types of adrenergic receptors are **alpha (α) receptors** and **beta (β) receptors,** which are found on visceral effectors innervated by most sympathetic postganglionic axons. These receptors are further classified into subtypes—α_1, α_2, β_1, β_2, and β_3—based on the specific responses they elicit and by their selective binding of drugs that activate or block them. Although there are some exceptions, activation of α_2 and β_1 receptors generally produces excitation, whereas activation of α_2 and β_2 receptors causes inhibition. β_3 receptors are present only on cells of brown adipose tissue, where their activation causes thermogenesis (heat production). Cells of most effectors contain either alpha or beta receptors, although some visceral effector cells contain both. Norepinephrine stimulates alpha receptors more strongly than beta receptors, whereas epinephrine is a potent stimulator of both alpha and beta receptors.

The activity of norepinephrine at a synapse is terminated either when the NE is taken up by the axon that released it or when the NE is enzymatically inactivated by either **catechol-*O*-methyltransferase (COMT)** or **monoamine oxidase (MAO).** Compared to ACh, norepinephrine lingers in the synaptic cleft for a longer period of time. Thus, effects triggered by adrenergic neurons typically are longer lasting than those triggered by cholinergic neurons.

Drugs and Receptor Selectivity

A large variety of drugs and natural products can selectively activate or block specific cholinergic or adrenergic receptors. An **agonist** is a substance that binds to and activates a receptor, in the process mimicking the effect of a natural neurotransmitter or hormone. Phenylephrine, which activates α_1 receptors, is a common ingredient in cold and sinus medications. Because it constricts blood vessels in the nasal mucosa, phenylephrine reduces production of mucus, thus relieving nasal congestion. An **antagonist** is a substance that binds to and blocks a receptor, thereby preventing a natural neurotransmitter or hormone from exerting its effect. For example, atropine, which blocks muscarinic ACh receptors, dilates the pupils, reduces glandular secretions, and relaxes smooth muscle in the gastrointestinal tract. As a result, it is used to dilate the pupils during eye examinations, in the treatment of smooth muscle disorders such as iritis and intestinal hypermotility, and as an antidote for chemical warfare agents that inactivate acetylcholinesterase.

Propranolol (Inderal) often is prescribed for patients with hypertension (high blood pressure). It is a nonselective beta blocker, meaning it binds to all types of beta receptors and prevents their activation by epinephrine and norepinephrine. The desired effects of propranolol are due to its blockade of β_1 receptors—namely, decreased heart rate and force of contraction and a consequent decrease in blood pressure. Undesired effects due to blockade of β_2 receptors may include hypoglycemia (low blood glucose), resulting from decreased glycogen breakdown and decreased gluconeogenesis (the conversion of a noncarbohydrate into glucose in the liver); and mild bronchoconstriction (narrowing of the airways). If these side effects pose a threat to the patient, a selective β_1 blocker such as metoprolol (Lopressor) can be prescribed instead of propranolol. ■

EFFECTS OF THE ANS ON EFFECTORS

As noted earlier, most body organs are innervated by both divisions of the ANS, which typically work in opposition to one another. The balance between sympathetic and parasympathetic activity or "tone" is regulated by the hypothalamus. Typically, the hypothalamus turns up sympathetic tone at the same time it

14th Century

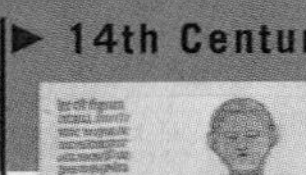

15th Century

16th Century

17th Century

18th Century

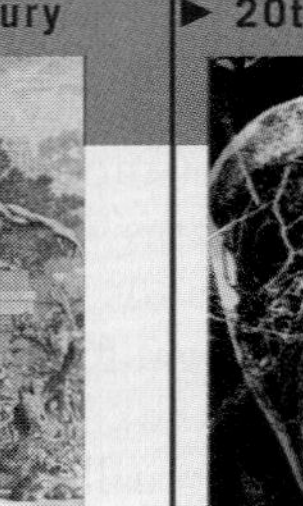

20th Century

21st Century

CHANGING IMAGES

Working the Systems

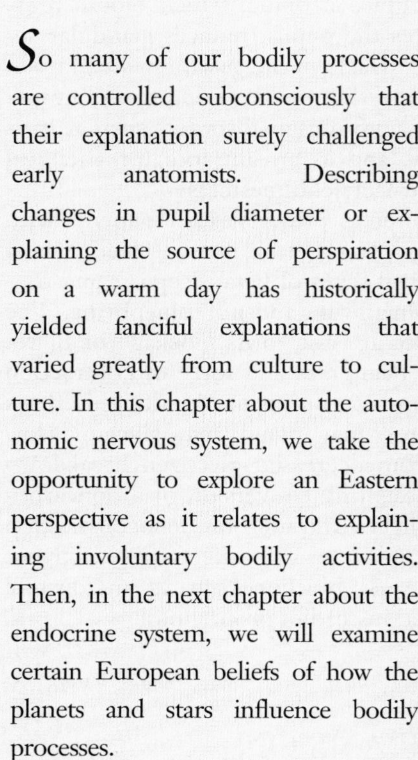

So many of our bodily processes are controlled subconsciously that their explanation surely challenged early anatomists. Describing changes in pupil diameter or explaining the source of perspiration on a warm day has historically yielded fanciful explanations that varied greatly from culture to culture. In this chapter about the autonomic nervous system, we take the opportunity to explore an Eastern perspective as it relates to explaining involuntary bodily activities. Then, in the next chapter about the endocrine system, we will examine certain European beliefs of how the planets and stars influence bodily processes.

19th Cent.

Pictured here is a beautiful image, circa 19th century, of a Japanese woman smoking from a long clay pipe. Her kimono is opened so as to reveal her internal organs. Her autonomic bodily processes are portrayed in a most remarkable manner. While the autonomic neuronal pathways themselves are not depicted, it is their motor effects at various organs that are creatively characterized. Tiny people busily perform such duties as respiration, digestion, and urine formation. Close examination reveals that these laborers operate bellows as they sit atop leaf-like structures that represent the lobes of her lungs. Digestion is regulated by those tending coiled tubes of intestines. Other workers, directing a flow of water through pumps and channels, control urinary activity.

Several other organs and processes are also identifiable. Can you find the trachea and heart? Count the aforementioned lobes of the lungs. Are they correct in number? Look for two sunflower-shaped ovaries budding in a garden. Try to locate the worker who pours menstrual flow into a funnel atop the uterus.

Without doubt Japanese anatomists were intending to be representative, and not literal, in their efforts to describe these complex involuntary activities. Prior to your newfound understanding of such factors as cholinergic and adrenergic neurons, how did you explain these processes?

turns down parasympathetic tone, and vice versa. The two divisions can affect body organs differently because their postganglionic neurons release different neurotransmitters and because the effector organs possess different adrenergic and cholinergic receptors. A few structures receive only sympathetic innervation—sweat glands, arrector pili muscles attached to hair follicles in the skin, the kidneys, most blood vessels, and the adrenal medullae (see Figure 21.2). In these cases there is no opposition from the parasympathetic division. Still, an increase in sympathetic tone has a given effect, whereas a decrease in sympathetic tone produces an opposite effect.

Sympathetic Responses

During physical or emotional stress, the sympathetic division dominates the parasympathetic division. High sympathetic tone favors body functions that can support vigorous physical activity and rapid production of ATP. At the same time, the sympathetic division reduces body functions that favor the storage of energy. Besides physical exertion, a variety of emotions—such as fear, embarrassment, or rage—stimulate the sympathetic division. Visualizing body changes that occur during "E situation" (exercise, emergency, excitement, embarrassment) will help you remember most of the sympathetic responses. Activation of the sympathetic division and release of hormones by the adrenal medullae set in motion a series of physiological responses collectively called the **fight-or-flight response,** which produces the following effects:

1. The pupils of the eyes dilate.
2. Heart rate, force of heart contraction, and blood pressure increase.
3. The airways dilate, allowing faster movement of air into and out of the lungs.
4. The blood vessels that supply nonessential organs such as the kidneys and gastrointestinal tract constrict.
5. Blood vessels that supply organs involved in exercise or fighting off danger—skeletal muscles, cardiac muscle, liver, and adipose tissue—dilate, allowing greater blood flow through these tissues.
6. Liver cells perform glycogenolysis (breakdown of glycogen to glucose), and adipose tissue cells perform lipolysis (breakdown of triglycerides to fatty acids and glycerol).
7. Release of glucose by the liver increases blood glucose level.
8. Processes that are not essential for meeting the stressful situation are inhibited. For example, muscular movements of the gastrointestinal tract and digestive secretions slow down or even stop.

The effects of sympathetic stimulation are longer lasting and more widespread than the effects of parasympathetic stimulation for three reasons:

- Sympathetic postganglionic fibers show great divergence, and as a result, many tissues are activated simultaneously.
- Acetylcholinesterase quickly inactivates acetylcholine, whereas norepinephrine lingers in the synaptic cleft for a longer period of time.
- Epinephrine and norepinephrine secreted into the blood from the adrenal medulla intensify and prolong the responses caused by NE liberated from sympathetic postganglionic axons. These blood-borne hormones circulate throughout the body, affecting all tissues that have alpha and beta receptors. In time, blood-borne NE and epinephrine are inactivated by enzymatic destruction in the liver.

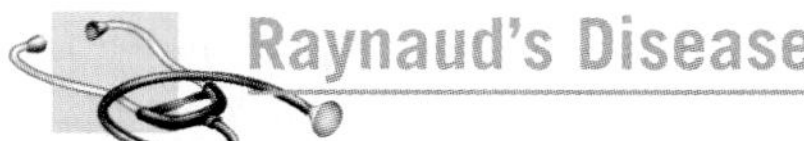

Raynaud's disease is due to excessive sympathetic stimulation of arterioles within the fingers and toes. Its cause is unknown. Because arterioles in the digits vasoconstrict in response to sympathetic stimulation, blood flow is greatly diminished. The digits may even become deprived of blood for minutes to hours and, in extreme cases, may become necrotic from lack of oxygen and nutrients. The disease is most common in young women and is worsened by cold climates.

Parasympathetic Responses

In contrast to the "fight-or-flight" activities of the sympathetic division, the parasympathetic division enhances "rest-and-digest" activities. Parasympathetic responses support body functions that conserve and restore body energy during times of rest and recovery. In the quiet intervals between periods of exercise, parasympathetic impulses to the digestive glands and the smooth muscle of the gastrointestinal tract predominate over sympathetic impulses, allowing energy-supplying food to be digested and absorbed. At the same time, parasympathetic responses reduce body functions that support physical activity.

The acronym "SLUDD" can be helpful in remembering five parasympathetic responses. It stands for salivation (S), lacrimation (L), urination (U), digestion (D), and defecation (D). All of these activities are stimulated mainly by the parasympathetic division. Besides the increasing "SLUDD" responses, other important parasympathetic responses are three "decreases": decreased heart rate, decreased diameter of airways (bronchoconstriction), and decreased diameter (constriction) of the pupils.

Fear typically elicits sympathetic responses, but in the case of "paradoxical fear" there is massive activation of the parasympathetic division. This type of fear occurs when one is backed into a corner, with no escape route or no way to win. It may happen to soldiers in a losing battle, to unprepared students taking an exam, or to athletes before competition. The high level of parasympathetic tone in such situations can cause loss of control over urination or defecation.

Table 21.3 Activities of Sympathetic and Parasympathetic Divisions of the ANS

Visceral Effector	Effect of Sympathetic Stimulation (α or β Adrenergic Receptors, Except as Noted)*	Effect of Parasympathetic Stimulation (Muscarinic ACh Receptors)
Glands		
Adrenal medullae	Secretion of epinephrine and norepinephrine (nicotinic ACh receptors).	No known effect.
Lacrimal (tear)	Slight secretion of tears (α).	Secretion of tears.
Pancreas	Inhibits secretion of digestive enzymes and the hormone insulin (α_2); promotes secretion of the hormone glucagon (β_2).	Secretion of digestive enzymes and the hormone insulin.
Posterior pituitary	Secretion of antidiuretic hormone (ADH) (β_1).	No known effect.
Sweat	Increases sweating in most body regions (muscarinic ACh receptors); sweating on palms of hands and soles of feet (α_1).	No known effect.
Adipose tissue†	Lipolysis (breakdown of triglycerides into fatty acids and glycerol) (β_1); release of fatty acids into blood (β_1 and β_3).	No known effect.
Liver†	Glycogenolysis (conversion of glycogen into glucose); gluconeogenesis (conversion of noncarbohydrates into glucose); decreased bile secretion (α and β_2).	Glycogen synthesis; increased bile secretion.
Kidney, juxtaglomerular cells†	Secretion of renin (β_1).	No known effect.
Cardiac (heart) muscle	Increased heart rate and force of atrial and ventricular contractions (β_1).	Decreased heart rate; decreased force of atrial contraction.
Smooth muscle		
Iris, radial muscle	Contraction → dilation of pupil (α_1).	No known effect.
Iris, circular muscle	No known effect.	Contraction → constriction of pupil.
Ciliary muscle of eye	Relaxation for far vision (β_2).	Contraction for near vision.
Lungs, bronchial muscle	Relaxation → airway dilation (β_2).	Contraction → airway constriction.
Gallbladder and ducts	Relaxation (β_2).	Contraction → increased release of bile into small intestine.

*Subcategories of α and β receptors are listed if known.
†Grouped with glands because adipose cells release substances into the blood.

Table 21.3 summarizes the responses of glands, cardiac muscle, and smooth muscle to stimulation by the sympathetic and parasympathetic divisions of the ANS.

✓ Define cholinergic and adrenergic neurons and receptors.
✓ Give examples of the antagonistic effects of the sympathetic and parasympathetic divisions of the autonomic nervous system.
✓ Describe what happens during the fight-or-flight response.
✓ Why is the parasympathetic division of the ANS called an energy conservation/restoration system?
✓ Describe the sympathetic response in a frightening situation for each of the following body parts: hair follicles, iris of eye, lungs, spleen, adrenal medullae, urinary bladder, stomach, intestines, gallbladder, liver, heart, arterioles of the abdominal viscera, and arterioles of skeletal muscles.

INTEGRATION AND CONTROL OF AUTONOMIC FUNCTIONS

Objectives

- Describe the components of an autonomic reflex.
- Explain the relationship of the hypothalamus to the ANS.

Autonomic Reflexes

Autonomic reflexes are responses that occur when nerve impulses pass over an autonomic reflex arc. These reflexes play a key role in regulating controlled conditions in the body, such as *blood pressure*, by adjusting heart rate, force of ventricular contraction, and blood vessel diameter; *respiration*, by regulating the diameter of bronchial tubes; *digestion*, by adjusting the motility

Visceral Effector	Effect of Sympathetic Stimulation (α or β Adrenergic Receptors, Except as Noted)[a]	Effect of Parasympathetic Stimulation (Muscarinic ACh Receptors)
Smooth muscle (continued)		
Stomach and intestines	Decreased motility and tone (α_1, α_2, β_2); contraction of sphincters (α_1).	Increased motility and tone; relaxation of sphincters.
Spleen	Contraction and discharge of stored blood into general circulation (α_1).	No known effect.
Ureter	Increases motility (α_1)	Increases motility (?).
Urinary bladder	Relaxation of muscular wall (β_2); contraction of sphincter (α_1).	Contraction of muscular wall; relaxation of sphincter.
Uterus	Inhibits contraction in nonpregnant woman (β_2); promotes contraction in pregnant woman (α_1).	Minimal effect.
Sex organs	In males: contraction of smooth muscle of ductus (vas) deferens, seminal vesicle, prostate → ejaculation of semen (α_1).	Vasodilation; erection of clitoris (females) and penis (males).
Hair follicles, arrector pili muscle	Contraction → erection of hairs (α_1).	No known effect.
Vascular smooth muscle		
Salivary gland arterioles	Vasoconstriction, which decreases secretion (β_2).	Vasodilation, which increases K^+ and water secretion.
Gastric gland arterioles	Vasoconstriction, which inhibits secretion (α_1).	Secretion of gastric juice.
Intestinal gland arterioles	Vasoconstriction, which inhibits secretion (α_1).	Secretion of intestinal juice.
Heart arterioles	Relaxation → vasodilation (β_2).	Contraction → vasoconstriction.
Skin and mucosal arterioles	Contraction → vasoconstriction (α_1).	Vasodilation, which may not be physiologically significant.
Skeletal muscle arterioles	Contraction → vasoconstriction (α_1); relaxation → vasodilation (β_2).	No known effect.
Abdominal viscera arterioles	Contraction → vasoconstriction (α_1, β_2).	No known effect.
Brain arterioles	Slight contraction → vasoconstriction (α_1).	No known effect.
Kidney arterioles	Constriction of blood vessels → decreases urine volume (α_1).	No known effect.
Systemic veins	Contraction → constriction (α_1); relaxation → dilation (β_2).	No known effect.

(movement) and muscle tone of the gastrointestinal tract; and *defecation* and *urination*, by regulating the opening and closing of sphincters.

The components of an autonomic reflex arc are as follows:

1. ***Receptor.*** Like the receptor in a somatic reflex arc (see Figure 17.5 on page 524), the receptor in an autonomic reflex arc is the distal end of a sensory neuron, which responds to a stimulus and produces a change that triggers nerve impulses. Autonomic sensory receptors are mostly associated with interoceptors.
2. ***Sensory neuron.*** Conducts nerve impulses from receptors to the CNS.
3. ***Integrating center.*** Interneurons within the CNS relay signals from sensory neurons to motor neurons. The main integrating centers for most autonomic reflexes are located in the hypothalamus and brain stem. Some autonomic reflexes, such as those for urination and defecation, have integrating centers in the spinal cord.
4. ***Motor neurons.*** Nerve impulses triggered by the integrating center propagate out of the CNS along motor neurons to an effector. In an autonomic reflex arc, two motor neurons connect the CNS to an effector: The preganglionic neuron conducts motor impulses from the CNS to an autonomic ganglion, and the postganglionic neuron conducts motor impulses from an autonomic ganglion to an effector (see Figure 21.1).
5. ***Effector.*** In an autonomic reflex arc, the effectors are smooth muscle, cardiac muscle, and glands, and the reflex is called an autonomic reflex.

Autonomic Control by Higher Centers

Normally, we are not aware of muscular contractions of the digestive organs, heartbeat, changes in the diameter of blood vessels, and pupil dilation and constriction because the integrating centers for these autonomic responses are in the spinal cord or the lower regions of the brain. Somatic or autonomic sensory neurons deliver input to these centers, and autonomic motor neurons provide output that adjusts activity in the visceral effector, usually without our conscious perception.

The hypothalamus is the major control and integration center of the ANS. The hypothalamus receives sensory input related to visceral functions, olfaction (smell), and gustation (taste), as well as input related to changes in temperature, osmolarity, and levels of various substances in blood. In addition, it receives input relating to emotions from the limbic system. Output from the hypothalamus influences autonomic centers both in the brain stem (such as the cardiovascular, salivation, swallowing, and vomiting centers) and in the spinal cord (such as the defecation and urination reflex centers in the sacral spinal cord).

Anatomically, the hypothalamus is connected to both the sympathetic and parasympathetic divisions of the ANS by axons of neurons whose dendrites and cell bodies are in various hypothalamic nuclei. The axons form tracts from the hypothalamus to sympathetic and parasympathetic nuclei in the brain stem and spinal cord through relays in the reticular formation. The posterior and lateral portions of the hypothalamus control the sympathetic division. Stimulation of these areas produces an increase in heart rate and force of contraction, a rise in blood pressure due to constriction of blood vessels, an increase in body temperature, dilation of the pupils, and inhibition of the gastrointestinal tract. In contrast, the anterior and medial portions of the hypothalamus control the parasympathetic division. Stimulation of these areas results in a decrease in heart rate, lowering of blood pressure, constriction of the pupils, and increased secretion and motility of the gastrointestinal tract.

Autonomic dysreflexia is an exaggerated response of the sympathetic division of the ANS that occurs in about 85% of individuals with spinal cord injury at or above the level of T6. The condition is seen after recovery from spinal shock (see page 606) and occurs due to interruption of the control of the ANS by higher centers. When certain sensory impulses, such as those resulting from stretching of a full urinary bladder, are unable to ascend the spinal cord, mass stimulation of the sympathetic nerves inferior to the level of injury occurs. Other triggers include stimulation of pain receptors and visceral contractions. Among the effects of increased sympathetic activity is severe vasoconstriction, which elevates blood pressure. In response, the cardiovascular center in the medulla oblongata sends out parasympathetic signals via the vagus (X) nerve that decrease heart rate and dilate blood vessels superior to the level of the injury.

Autonomic dysreflexia is characterized by pounding headache; hypertension; flushed, warm skin with profuse sweating above the injury level; pale, cold, and dry skin below the injury level; and anxiety. It is an emergency condition that requires immediate intervention. If untreated, autonomic dysreflexia can cause seizures, stroke, or heart attack. ■

- ✓ Give three examples of controlled conditions in the body that are kept in homeostatic balance by autonomic reflexes.
- ✓ How does an autonomic reflex arc differ from a somatic reflex arc?

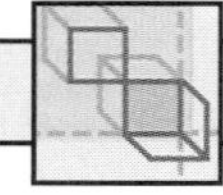

STUDY OUTLINE

Comparison of Somatic and Autonomic Nervous Systems (p. 640)

1. The somatic nervous system operates under conscious control; the ANS usually operates without conscious control.
2. Sensory input for the somatic nervous system is mainly from the special senses and somatic senses; sensory input for the ANS is from interoceptors, in addition to special senses and somatic senses.
3. The axons of somatic motor neurons extend from the CNS and synapse directly with an effector. Autonomic motor pathways consist of two motor neurons in series. The axon of the first motor neuron extends from the CNS and synapses in a ganglion with the second motor neuron; the second neuron synapses with an effector.
4. The output (motor) portion of the ANS has two divisions: sympathetic and parasympathetic. Most body organs receive dual innervation; usually one ANS division causes excitation and the other causes inhibition.
5. Somatic nervous system effectors are skeletal muscles; ANS effectors include cardiac muscle, smooth muscle, and glands.
6. Table 21.1 on page 641 compares the somatic and autonomic nervous systems.

Anatomy of Autonomic Motor Pathways (p. 642)

1. Preganglionic neurons are myelinated; postganglionic neurons are unmyelinated.
2. The cell bodies of sympathetic preganglionic neurons are in the lateral gray horns of the 12 thoracic and the first two or three lumbar segments of the spinal cord; the cell bodies of parasympathetic preganglionic neurons are in four cranial nerve nuclei (III, VII, IX, and X) in the brain stem and lateral gray horns of the second through fourth sacral segments of the spinal cord.
3. Autonomic ganglia are classified as sympathetic trunk ganglia (on both sides of vertebral column), prevertebral ganglia (anterior to vertebral column), and terminal ganglia (near or inside visceral effectors).

4. Sympathetic preganglionic neurons synapse with postganglionic neurons in ganglia of the sympathetic trunk or in prevertebral ganglia; parasympathetic preganglionic neurons synapse with postganglionic neurons in terminal ganglia.
5. Table 21.2 on page 647 compares anatomical features of the sympathetic and parasympathetic divisions.

ANS Neurotransmitters and Receptors (p. 647)

1. Cholinergic neurons release acetylcholine, which binds to nicotinic or muscarinic cholinergic receptors.
2. In the ANS, the cholinergic neurons include all sympathetic and parasympathetic preganglionic neurons, all parasympathetic postganglionic neurons, and sympathetic postganglionic neurons that innervate most sweat glands.
3. In the ANS, adrenergic neurons release norepinephrine. Both epinephrine and norepinephrine bind to alpha and beta adrenergic receptors.
4. Most sympathetic postganglionic neurons are adrenergic.

Effects of the ANS on Effectors (p. 649)

1. The sympathetic division favors body functions that can support vigorous physical activity and rapid production of ATP (fight-or-flight response); the parasympathetic division regulates activities that conserve and restore body energy.
2. The effects of sympathetic stimulation are longer lasting and more widespread than the effects of parasympathetic stimulation.
3. Table 21.3 on pages 652–653 summarizes sympathetic and parasympathetic responses.

Integration and Control of Autonomic Functions (p. 652)

1. An autonomic reflex adjusts the activities of smooth muscle, cardiac muscle, and glands.
2. An autonomic reflex arc consists of a receptor, a sensory neuron, an integrating center, two autonomic motor neurons, and a visceral effector.
3. The hypothalamus is the major control and integration center of the ANS. It is connected to both the sympathetic and the parasympathetic divisions.

SELF-QUIZ QUESTIONS

Choose the one best answer to these questions.

1. Which of the following statements is true for the parasympathetic division of the autonomic nervous system? (a) Ganglia are close to the central nervous system and distant from the effector. (b) It forms the craniosacral outflow. (c) It supplies nerves to blood vessels, sweat glands, and adrenal (suprarenal) glands. (d) It has some of its nerve fibers passing through paravertebral ganglia. (e) All of the above statements are true.

2. Which of the following statements is true? (1) The somatic nervous system and the ANS both include sensory and motor neurons. (2) Motor output of the ANS is controlled by the cerebral cortex and basal ganglia. (3) The effect of a somatic motor neuron is always excitatory; but the effect of an autonomic motor neuron may be either excitatory or inhibitory. (4) Autonomic motor pathways always consist of two motor neurons in series. (5) Autonomic sensory neurons are mostly associated with interoceptors. (6) Somatic sensory receptors are interoceptors.

   a. 1, 2, 3, and 4 **b.** 1, 3, 4, and 5 **c.** 1, 3, 4, and 6 **d.** 2, 3, 4, and 6 **e.** 2, 4, 5, and 6.

3. Which of the following statements about gray rami is false? (a) They carry preganglionic neurons from the anterior rami of spinal nerves to trunk ganglia. (b) They contain only sympathetic nerve fibers. (c) They carry impulses from sympathetic trunk ganglia to spinal nerves. (d) They are located at all levels of the vertebral column. (e) None of the above is false.

4. Which of the following statements about the parasympathetic ganglia is true? (a) They contain adrenergic neurons only. (b) They include the cervical and celiac ganglia. (c) They lie within the spinal cord or brain. (d) They consist of a double chain of structures on either side of the spinal column. (e) None of the above is true.

5. The ANS provides the chief nervous control in which of these activities? (a) following a moving object with the eyes, (b) moving a hand reflexively from a hot object, (c) typing, (d) digesting food, (e) surfing the Internet.

6. Which of the following four statements apply to preganglionic nerve fibers that arise from the thoracic and lumbar parts of the spinal cord? (1) They are also called preganglionic sympathetic fibers. (2) They synapse with postganglionic fibers of the parasympathetic nervous system. (3) They form the posterior roots of the thoracic and lumbar spinal nerves. (4) They are also called preganglionic parasympathetic fibers.

   a. 1 only **b.** 2 only **c.** 3 only **d.** 4 only **e.** 2 and 4

7. Cholinergic fibers are thought to include (a) all preganglionic axons, (b) all postganglionic parasympathetic axons, (c) a few postganglionic sympathetic axons, (d) all axons of somatic motor neurons, (e) all of the above.

8. Which of the following statements would *not* be included in a description of the sympathetic trunks? (a) The trunks lie anterior and lateral to the spinal cord. (b) Each trunk contains 24 ganglia. (c) Each trunk receives preganglionic fibers from the thoracic and sacral regions of the spinal cord. (d) The ganglia of the trunks are called paravertebral ganglia. (e) White rami direct nerve fibers into the trunk; gray rami direct nerve fibers out of the trunk.

9. The axons of neurons lying within the central nervous system and that connect with the ANS are (a) myelinated fibers, (b) postganglionic fibers, (c) preganglionic fibers, (d) cranial nerve fibers, (e) none of the above

Complete the following.

10. The input component of the ANS consists of ________ neurons, and the output component consists of ________ neurons.

11. Organs such as the heart or stomach that receive impulses from both divisions of the ANS are said to have ________ innervation.

12. Adrenergic neurons release the neurotransmitters ________ and ________ .

13. The two types of adrenergic receptors are ________ and receptors.

14. The two types of cholinergic receptors are ________ and ________ .

15. The pterygopalatine, otic, and submandibular ganglia are part of the ________ division of the ANS.

Are the following statements true or false?

16. A major difference between autonomic ganglia and posterior root ganglia is that only autonomic ganglia contain synapses.

17. The ganglia that make up the sympathetic trunks are also called prevertebral or collateral ganglia.

18. Parasympathetic preganglionic neurons originate in the nuclei of the oculomotor (III), facial (VII), glossopharyngeal (IX), or vagus (X) nerves, or in lateral gray horns of the sacral region of the spinal cord.

19. An event occurring during the fight-or-flight response is increased urine production due to blood being shunted to the kidneys.

20. In an autonomic reflex arc, the preganglionic neuron conducts motor impulses from the central nervous system to an autonomic ganglion, and a postganglionic neuron conducts motor impulses from the ganglion to an effector.

21. Match the following plexuses with their descriptions:

____ **(a)** plexus located mostly posterior to each lung

____ **(b)** the largest autonomic plexus; consists of 2 large ganglia and a network of fibers that surround the celiac and superior mesenteric arteries

____ **(c)** plexus located anterior to the fifth lumbar vertebra; supplies pelvic viscera

____ **(d)** plexus containing the inferior mesenteric ganglion; supplies the large intestine

____ **(e)** plexus at the base of the heart; surrounds the large blood vessels emerging from the heart

____ **(f)** plexus containing the superior mesenteric ganglion; supplies the small and large intestine

(1) cardiac plexus
(2) pulmonary plexus
(3) superior mesenteric plexus
(4) inferior mesenteric plexus
(5) celiac (solar) plexus
(6) hypogastric plexus

22. Match the following.

____ **(a)** thoracolumbar outflow

____ **(b)** has long preganglionic fibers and short postganglionic fibers

____ **(c)** has short preganglionic fibers and long postganglionic fibers

____ **(d)** sends some preganglionic fibers through cranial nerves

____ **(e)** has some preganglionic fibers synapsing in sympathetic trunk ganglia

____ **(f)** has more widespread and longer lasting effect in the body

____ **(g)** has some fibers running in gray rami to supply sweat glands, hair follicle muscles, and blood vessels

____ **(h)** has fibers in white rami (connecting spinal nerve with sympathetic trunk ganglia)

____ **(i)** contains fibers that supply viscera with motor impulses

____ **(j)** celiac and superior mesenteric ganglia are sites of postganglionic neuron cell bodies

(1) applies to the sympathetic division of the ANS
(2) applies to the parasympathetic division of the ANS
(3) applies to both divisions of the ANS

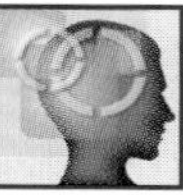

CRITICAL THINKING QUESTIONS

1. Skydiving, hang gliding, and bungee jumping can all give you a great rush (or get you killed). How do these activities cause this rush?
   HINT *A bungee jumper might say she was "pumped with adrenaline" after the jump.*
2. "The quickest way to a man's heart is through his stomach," Sophia's grandmother is fond of saying. Trace the pathway that an impulse would follow from a full stomach to a happy heart.
   HINT *This man will be content to "rest and digest."*
3. After lunch, baby Bobby was looking uncomfortable, squirming, and fretting. Suddenly, a pungent aroma wafted from his diaper and he looked very content. Trace the reflex arc that is responsible for baby Bobby's present state.
   HINT *Baby Bobby's smelly bottom happens automatically (say this three times fast).*
4. The autonomic and the enteric divisions of the nervous system both control the digestive system. How do they compare?
   HINT *Look back at chapter 16.*
5. If you pay attention to the eyes in old paintings, you'll notice that someone who is studying or concentrating has smaller pupils than someone who is relaxed or day dreaming. Why is that?
   HINT *If you're concentrating on something, you are "focused."*

ANSWERS TO FIGURE QUESTIONS

21.1 Dual innervation means that a body organ innervated by the ANS receives both sympathetic and parasympathetic fibers.

21.2 Most parasympathetic preganglionic axons are longer than most sympathetic preganglionic axons because most parasympathetic ganglia are in the walls of visceral organs, whereas most sympathetic ganglia are close to the spinal cord in the sympathetic trunk.

21.3 The largest autonomic plexus is the celiac plexus.

21.4 White rami look white due to the presence of myelin.

21.5 Cholinergic neurons that have nicotinic ACh receptors include sympathetic postganglionic neurons innervating sweat glands and all parasympathetic postganglionic neurons. The effectors innervated by these cholinergic neurons possess muscarinic receptors.

22

THE ENDOCRINE SYSTEM

CONTENTS AT A GLANCE

◀ Page 661

Many people read their horoscopes regularly. Are you one of them? Do you know what sign you were born under? How much do you think the heavens influence our bodies? Is such a belief superstition or science?

Page 660 ▶

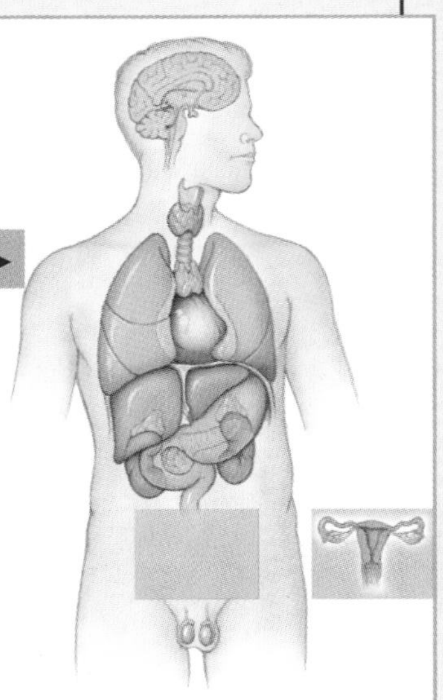

INTRODUCTION

Together, the nervous and endocrine systems coordinate functions of all body systems. The nervous system controls body activities through nerve impulses conducted along axons of neurons. At synapses, nerve impulses trigger the release of mediator molecules called neurotransmitters. In contrast, the glands of the endocrine system release mediator molecules called **hormones** (*hormon* = to excite or get moving) into interstitial fluid and then the bloodstream. The circulating blood delivers the hormones to virtually all the cells of the body. The science that deals with the structure and function of the endocrine glands and the diagnosis and treatment of disorders of the endocrine system is **endocrinology** (en′-dō-kri-NOL-ō-jē; *endo-* = within; *-crin* = to secrete; *-ology* = study of).

The nervous and endocrine systems are coordinated as an interlocking supersystem termed the **neuroendocrine system.** Certain parts of the nervous system stimulate or inhibit the release of hormones, which in turn may promote or inhibit the generation of nerve impulses. The nervous system causes muscles to contract and glands to secrete either more or less of their product. The endocrine system not only helps regulate the activity of smooth muscle, cardiac muscle, and some glands; it affects virtually all other tissues as well. Hormones alter metabolism, regulate growth and development, and influence reproductive processes.

The nervous and endocrine systems respond to a stimulus at different rates. Nerve impulses most often produce an effect within a few milliseconds; some hormones can act within seconds, whereas others can take several hours or more to cause a response. Moreover, the effects of activating the nervous system are generally briefer than the effects produced by the endocrine system. Table 22.1 compares the characteristics of the nervous and endocrine systems.

In this chapter we examine the principal endocrine glands and hormone-producing tissues, and their roles in coordinating body activities.

ENDOCRINE GLANDS DEFINED

Objective

- Distinguish between exocrine glands and endocrine glands.

The body contains two kinds of glands: exocrine glands and endocrine glands. **Exocrine glands** (*exo-* = outside) secrete their products into ducts that carry the secretions into body cavities, into the lumen of an organ, or to the outer surface of the body. Exocrine glands include sudoriferous (sweat), sebaceous (oil), mucous, and digestive glands. **Endocrine glands,** by contrast, secrete their products (hormones) into the interstitial fluid surrounding the secretory cells, rather than into ducts. The secretion then diffuses into capillaries and is carried away by the blood. The endocrine glands of the body, which constitute the **endocrine system,** include the pituitary, thyroid, parathyroid, adrenal, and pineal glands (Figure 22.1). In addition, several organs and tissues of the body contain cells that secrete hormones but are not endocrine glands exclusively. These include the hypothalamus, thymus, pancreas, ovaries, testes, kidneys, stomach, liver, small intestine, skin, heart, adipose tissue, and placenta.

HORMONES

Objective

- Describe how hormones interact with target-cell receptors.

Hormones have powerful effects when present even in very low concentrations. As a rule, most of the 50 or so hormones affect only a few types of cells. The reason that some cells respond to a particular hormone and others do not involves hormone receptors.

Although a given hormone travels throughout the body in the blood, it affects only certain **target cells.** Hormones, like neurotransmitters, influence their target cells by chemically binding to specific protein or glycoprotein **receptors.** Only the target cells for a given hormone have receptors that bind and recognize that hormone. For example, thyroid-stimulating hor-

Table 22.1 Comparison of the Nervous and Endocrine Systems

Characteristic	Nervous System	Endocrine System
Mediator molecules	Neurotransmitters released in response to nerve impulses.	Hormones delivered to tissues throughout the body by the blood.
Cells affected	Muscle cells, gland cells, other neurons.	Virtually all body cells.
Time to onset of action	Typically within milliseconds.	Seconds to hours or days.
Duration of action	Generally briefer.	Generally longer.

Figure 22.1 / Location of many endocrine glands. Also shown are other organs that contain endocrine tissue, and associated structures.

Endocrine glands secrete hormones, which circulating blood delivers to target tissues.

HYPOTHALAMUS
PINEAL GLAND
PITUITARY GLAND
PARATHYROID GLANDS (behind thyroid glands)
THYROID GLAND
Trachea
THYMUS
SKIN
Lung
HEART
LIVER
STOMACH
ADRENAL GLANDS
PANCREAS
KIDNEY
Uterus
SMALL INTESTINE
OVARY
Scrotum
TESTES

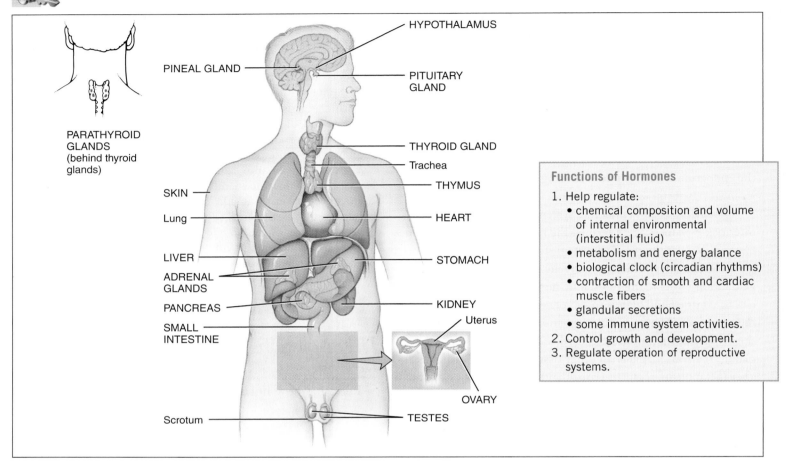

Functions of Hormones

1. Help regulate:
   - chemical composition and volume of internal environmental (interstitial fluid)
   - metabolism and energy balance
   - biological clock (circadian rhythms)
   - contraction of smooth and cardiac muscle fibers
   - glandular secretions
   - some immune system activities.
2. Control growth and development.
3. Regulate operation of reproductive systems.

What is the basic difference between endocrine glands and exocrine glands?

mone (TSH) binds to receptors on cells of the thyroid gland, but it does not bind to cells of the ovaries because ovarian cells do not have TSH receptors.

Receptors, like other cellular proteins, are constantly being synthesized and broken down. Generally, a target cell has 2000–100,000 receptors for a particular hormone. When a hormone is present in excess, the number of target-cell receptors may decrease. This decreases the responsiveness of target cells to the hormone. In contrast, when a hormone (or neurotransmitter) is deficient, the number of receptors may increase. This makes a target tissue more sensitive to a hormone.

Blocking Hormone Receptors

Synthetic hormones that block the receptors for certain naturally occurring hormones are available as drugs. For example, RU486 (mifepristone), which is used to induce abortion, binds to the receptors for progesterone (a female sex hormone) and prevents progesterone from exerting its normal effect. When RU486 is given to a pregnant woman, the uterine conditions needed for nurturing an embryo are not maintained, embryonic development stops, and the embryo is sloughed off along with the uterine lining. This example illustrates an important endocrine principle: If a hormone is prevented from interacting with its receptors, the hormone cannot perform its normal functions. ■

Most hormones are released in short bursts, with little or no secretion between bursts. When increasingly stimulated, an endocrine gland will release its hormone in more frequent bursts, and thus the hormone's overall blood concentration increases; in the absence of stimulation, bursts are minimal or inhibited, and the blood level of the hormone decreases. Regulation of secretion normally maintains homeostasis and prevents overproduction or underproduction of any given hormone.

✓ Distinguish between exocrine and endocrine glands.

✓ Explain the relationship between target cell receptors and hormones.

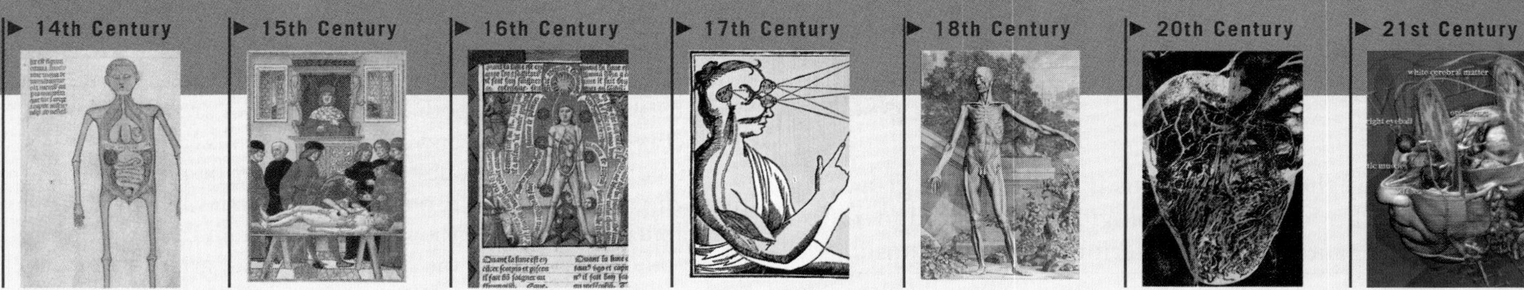

CHANGING IMAGES

Science or Stars?

What is it that controls the menstrual cycle, pregnancy and lactation, growth and development? What about digestion and heart rate? Is it the body itself that commands these processes? Or is it the external world - the heavens above - that governs us? The answer to these questions was not a mystery to the medieval anatomist. Just as a modern day clinician studies basic science in training, the physician of antiquity studied astrology. From the 9th century on, numerous medical texts taught that a physician must know the science of the stars since our bodies change along with them.

Today most scientists would regard it superstitious to consider astral influences when explaining biological phenomena. But centuries ago, science was grounded in a belief system that linked the stars and the outer world with the body, soul, and fate of man. This macrocosm-microcosm principle was almost universally embraced across cultures and throughout history. As you recall from the discussion of René Descartes in Chapter 19, a link between the external and the internal was central to his inquiries. Indeed, the influence of the heavens on health and human function was not deemed superstition; it was considered a vital part of natural science.

1500 AD

Medical texts in medieval and Renaissance Europe often contained illustrations of humans superimposed with astrological icons. Pictured here is an image that is typical of this time and belief. Celestial objects decorate the drawing, with lines indicating the organs they were believed to govern. In this chapter you will explore how it is the endocrine system, not the stars, that regulates many of our body processes. Despite the data that supports this modern day belief, do you believe that the heavens play a role in your body?

HYPOTHALAMUS AND PITUITARY GLAND

Objectives

- Explain why the hypothalamus is an endocrine gland.
- Describe the location, histology, hormones, and functions of the anterior and posterior pituitary glands.

For many years the **pituitary gland** or **hypophysis** (hī-POF-i-sis) was called the "master" endocrine gland because it secretes several hormones that control other endocrine glands. We now know that the pituitary gland itself has a master—the **hypothalamus.** This small region of the brain, inferior to the thalamus, is the major integrating link between the nervous and endocrine systems. It receives input from several other regions of the brain, including the limbic system, cerebral cortex, thalamus, and reticular activating system. It also receives sensory signals from internal organs and from the retina.

Painful, stressful, and emotional experiences all cause changes in hypothalamic activity. In turn, the hypothalamus controls the autonomic nervous system and regulates body temperature, thirst, hunger, sexual behavior, and defensive reactions such as fear and rage. Not only is the hypothalamus an important regulatory center in the nervous system; it is also a crucial endocrine gland. Cells in the hypothalamus synthesize at least nine different hormones, and the pituitary gland secretes seven more. Together, the hormones play important roles in the regulation of virtually all aspects of growth, development, metabolism, and homeostasis.

The pituitary gland is a pea-shaped structure measuring about 1–1.5 cm (0.5 in.) in diameter that lies in the sella turcica of the sphenoid bone and attaches to the hypothalamus by a stalk, the **infundibulum** (= a funnel; Figure 22.2). The pituitary gland has two anatomically and functionally separate portions. The **anterior pituitary gland (anterior lobe)** accounts for about 75% of the total weight of the gland. It develops from an outgrowth of ectoderm called the hypophyseal pouch in the roof of the mouth (see Figure 22.8b). The **posterior pituitary gland (posterior lobe)** also develops from an ectodermal outgrowth, this one called the neurohypophyseal bud (see Figure 22.8b). The posterior pituitary gland contains axons and axon terminals of more than 10,000 neurons whose cell bodies are located in the supraoptic and paraventricular nuclei of the hypothalamus (see Figure 18.10 on page 560). The axon terminals in the posterior pituitary gland are associated with specialized neuroglia called **pituicytes** (pi-TŪ-i-sītz).

A third region called the **pars intermedia,** atrophies during fetal development and ceases to exist as a separate lobe in adults (see Figure 22.8b). However, some of its cells migrate into adjacent parts of the anterior pituitary gland, where they persist.

Anterior Pituitary Gland

The **anterior pituitary gland,** or **adenohypophysis** (ad′-e-nō-hī-POF-i-sis; *adeno-* = gland; *-physis* = organ suspended from the hypothalamus), secretes hormones that regulate a wide range of bodily activities, from growth to reproduction. Release of anterior pituitary gland hormones is stimulated by **releasing hormones** and suppressed by **inhibiting hormones** from the hypothalamus. These hypothalamic hormones are an important link between the nervous and endocrine systems.

Hypothalamic hormones reach the anterior pituitary gland through a portal system. A portal system carries blood between two capillary networks without passing through the heart. Most often, blood passes from the heart through an artery to a capillary to a vein and back to the heart. In the **hypophyseal portal system** (hī′-pō-FIZ-ē-al) blood flows from the median eminence of the hypothalamus into the infundibulum and anterior pituitary gland principally through several **superior hypophyseal arteries** (see Figure 22.2). These arteries are branches of the internal carotid and posterior communicating arteries. The superior hypophyseal arteries form the **primary plexus of the hypophyseal portal system,** a capillary network at the base of the hypothalamus. Near the median eminence and above the optic chiasm are two groups of specialized neurons, called **neurosecretory cells,** that secrete hypothalamic releasing and inhibiting hormones into the primary plexus. These hormones are synthesized in the neuronal cell bodies and packaged inside vesicles, which move into the axon terminals. When nerve impulses reach the axon terminals, they stimulate the vesicles to undergo exocytosis. The hormones then diffuse into the primary plexus of the hypophyseal portal system.

From the primary plexus, blood drains into the **hypophyseal portal veins** that pass down the outside of the infundibulum. In the anterior pituitary gland, the hypophyseal portal veins redivide into another capillary network called the **secondary plexus of the hypophyseal portal system.** This direct route permits hypothalamic hormones to act quickly on anterior pituitary gland cells, before the hormones are diluted or destroyed in the systemic circulation. Hormones secreted by anterior pituitary gland cells pass into the secondary plexus of the hypophyseal portal system and then into the **anterior hypophyseal veins** for distribution to target tissues throughout the body.

The following list describes the seven major hormones secreted by five types of anterior pituitary gland cells:

- **Human growth hormone (hGH),** or **somatotropin** (sō′-ma-tō-TRŌ-pin; *somato-* = body; *-tropin* = change), is secreted by cells called **somatotrophs.** Human growth hormone in turn stimulates several tissues to secrete **insulinlike growth factors,** hormones that stimulate general body growth and regulate various aspects of metabolism.
- **Thyroid-stimulating hormone (TSH),** or **thyrotropin** (thī-rō-TRŌ-pin; *thyro-* = shield), which controls the secretions and other activities of the thyroid gland, is secreted by **thyrotrophs.**
- **Follicle-stimulating hormone (FSH)** and **luteinizing hormone (LH)** (LŪ-tē-in′-īz-ing) are secreted by **gonadotrophs** (*gonado-* = seed). FSH and LH both act on the gonads: They stimulate secretion of estrogens and progesterone and the maturation of oocytes in the ovaries, and they stimulate secretion of testosterone and sperm production in the testes.
- **Prolactin (PRL),** which initiates milk production in the mammary glands, is released by **lactotrophs** (*lacto-* = milk).

Figure 22.2 / Hypothalamus and pituitary gland, and their blood supply. The small figure at right indicates that releasing and inhibiting hormones synthesized by hypothalamic neurons are transported within axons and released from the axon terminals. The hormones diffuse into capillaries of the primary plexus of the hypophyseal portal system and are carried by the hypophyseal portal veins to the secondary plexus of the hypophyseal portal system for distribution to target cells in the anterior pituitary.

Hypothalamic hormones are an important link between the nervous and endocrine systems.

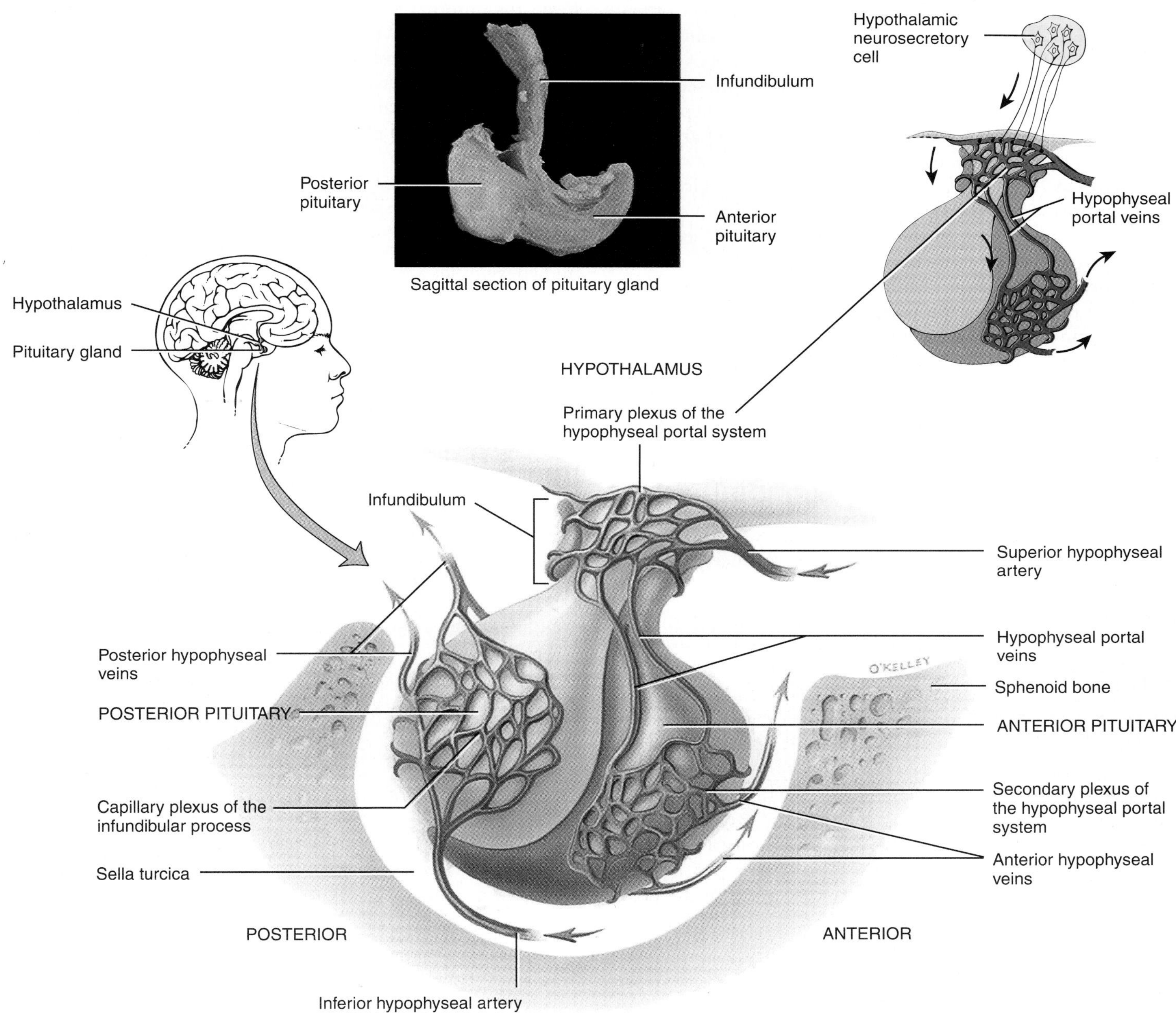

What is the functional importance of the hypophyseal portal veins?

- **Adrenocorticotropic hormone (ACTH),** or **corticotropin** (*cortico-* = rind or bark), which stimulates the adrenal cortex to secrete glucocorticoids, is synthesized by **corticotrophs.** Some corticotrophs, remnants of the pars intermedia, also secrete **melanocyte-stimulating hormone (MSH).**

Hormones that influence another endocrine gland are called **tropic hormones,** or **tropins.** Several of the anterior pituitary hormones are tropins. The two **gonadotropins,** FSH and LH, regulate the functions of the gonads (ovaries and testes). Thyrotropin stimulates the thyroid gland, whereas corticotropin acts on the cortex of the adrenal gland.

The hormones of the anterior pituitary gland are summarized in Table 22.2.

Posterior Pituitary Gland

Although the **posterior pituitary gland,** or **neurohypophysis,** does not *synthesize* hormones, it does *store* and *release* two hormones. As noted earlier, it consists of pituicytes and axon terminals of hypothalamic neurosecretory cells. The cell bodies of the neurosecretory cells are in the paraventricular and supraoptic nuclei of the hypothalamus; their axons form the **hypothalamohypophyseal tract** (hī′-pō-thal′-a-mō-hī-pō-FIZ-ē-al), which begins in the hypothalamus and ends near blood capillaries in the posterior pituitary gland (Figure 22.3). Different neurosecretory cells produce two hormones: **oxytocin (OT;** ok′-sē-TŌ-sin; *oxytoc-* = quick birth) and **antidiuretic hormone (ADH),** also called **vasopressin.**

After their production in the cell bodies of neurosecretory cells, oxytocin and antidiuretic hormone are packed into vesicles and transported to the axon terminals in the posterior pituitary gland. Nerve impulses that propagate along the axon and reach the axon terminals trigger exocytosis of these secretory vesicles.

During and after delivery of a baby, oxytocin has two target tissues: the mother's uterus and breasts. During delivery, oxytocin enhances contraction of smooth muscle cells in the wall of the uterus; after delivery, it stimulates milk ejection ("letdown") from the mammary glands in response to the mechanical stimulus provided by a suckling infant.

An **antidiuretic** is a substance that decreases urine production. ADH causes the kidneys to return more water to the blood, thus decreasing urine volume. In the absence of ADH, urine output increases more than tenfold, from the normal 1–2 liters to about 20 liters a day. ADH also decreases the water lost through sweating and causes constriction of arterioles, which increases blood pressure. This hormone's other name, vasopressin, reflects its effect on blood pressure.

Blood is supplied to the posterior pituitary gland by the **inferior hypophyseal arteries** (see Figure 22.2), which branch from the internal carotid arteries. In the posterior pituitary gland, the inferior hypophyseal arteries drain into the **capillary plexus of the infundibular process,** a capillary network that receives secreted oxytocin and antidiuretic hormone (see Figure 22.2). From this plexus, hormones pass into the **posterior hypophyseal veins** for distribution to target cells in other tissues.

Years before oxytocin was discovered, it was common practice in midwifery to let a first-born twin nurse at the mother's breast to speed the birth of the second child. Now we know why this practice is helpful—it stimulates release of oxytocin. Even after a single birth, nursing promotes expulsion of the placenta (afterbirth) and helps the uterus regain its smaller size. Synthetic OT often is given to induce labor or to increase uterine tone and control hemorrhage just after giving birth.

A summary of posterior pituitary gland hormones is presented in Table 22.3. ■

✓ In what respect is the pituitary gland actually two glands?
✓ Describe the structure and importance of the hypothalamohypophyseal tract.

Table 22.2 Summary of Anterior Pituitary Gland Hormones

Hormone and Target Tissues	Principal Actions
Human growth hormone (hGH) or **somatotropin** (Liver)	Stimulates liver, muscle, cartilage, bone, and other tissues to synthesize and secrete insulinlike growth factors (IGFs); IGFs promote growth of body cells, protein synthesis, tissue repair, lipolysis, and elevation of blood glucose concentration.
Thyroid-stimulating hormone (TSH) or **thyrotropin** (Thyroid gland)	Stimulates synthesis and secretion of thyroid hormones by thyroid gland.
Follicle-stimulating hormone (FSH) (Ovaries, Testes)	In females, initiates development of oocytes and induces ovarian secretion of estrogens. In males, stimulates testes to produce sperm.
Luteinizing hormone (LH) (Ovaries, Testes)	In females, stimulates secretion of estrogens and progesterone, ovulation, and formation of corpus luteum. In males, stimulates interstitial cells in testes to develop and produce testosterone.
Prolactin (PRL) (Mammary glands)	Together with other hormones, promotes milk secretion by the mammary glands.
Adrenocorticotropic hormone (ACTH) or **corticotropin** (Adrenal cortex)	Stimulates secretion of glucocorticoids (mainly cortisol) by adrenal cortex.
Melanocyte-stimulating hormone (MSH) (Skin)	Exact role in humans is unknown but can cause darkening of skin.

Figure 22.3 / Axons of hypothalamic neurosecretory cells form the hypothalamohypophyseal tract, which extends from the paraventricular and supraoptic nuclei to the posterior pituitary. Hormone molecules synthesized in the cell body of a neurosecretory cell are packaged into secretory vesicles that move down to the axon terminals. Nerve impulses trigger exocytosis of the vesicles and release of the hormone.

Oxytocin and antidiuretic hormone are synthesized in the hypothalamus and released into capillaries of the posterior pituitary gland.

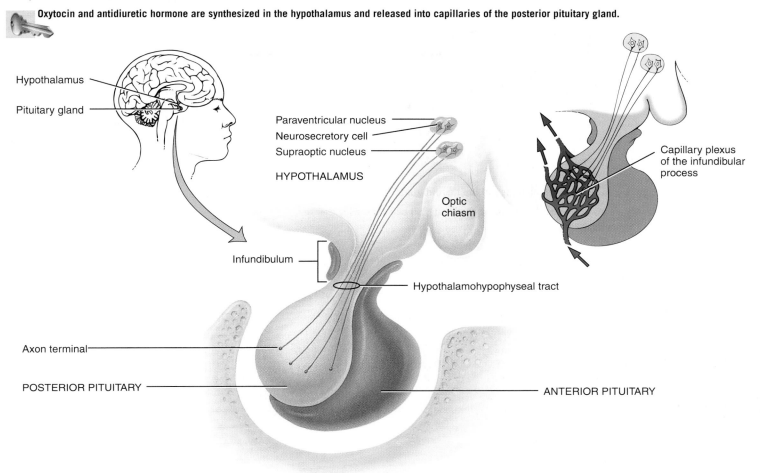

Functionally, how are the hypothalamohypophyseal tract and the hypophyseal portal veins similar? Structurally, how are they different?

Table 22.3 Summary of Posterior Pituitary Gland Hormones

Hormone and Target Tissues	Principal Actions
Oxytocin (OT) Uterus, Mammary glands	Stimulates contraction of smooth muscle cells of uterus during childbirth; stimulates contraction of myoepithelial cells in mammary glands to cause milk ejection.
Antidiuretic hormone (ADH) or **vasopressin** Kidneys, Sudoriferous (sweat) glands, Arterioles	Conserves body water by decreasing urine volume; decreases water loss through perspiration; raises blood pressure by constricting arterioles.

THYROID GLAND

Objective

- Describe the location, histology, hormones, and functions of the thyroid gland.

The butterfly-shaped **thyroid gland** is located just inferior to the larynx (voice box); the right and left **lateral lobes** lie one on either side of the trachea (Figure 22.4a). Connecting the lobes is a mass of tissue called an **isthmus** (IS-mus) that lies anterior to the trachea. A small, pyramidal-shaped lobe sometimes extends upward from the isthmus. The gland usually weighs about 30 g (1 oz) and has a rich blood supply, receiving 80–120 mL of blood per minute.

Microscopic spherical sacs called **thyroid follicles** (Figure 22.4b; also see Figure 22.5c) make up most of the thyroid gland. The wall of each follicle consists primarily of cells called **follicular cells,** most of which extend to the lumen (internal space) of the follicle. When the follicular cells are inactive, their shape is

Figure 22.4 / Location, blood supply, and histology of the thyroid gland.

Thyroid hormones regulate (1) oxygen use and basal metabolic rate, (2) cellular metabolism, and (3) growth and development.

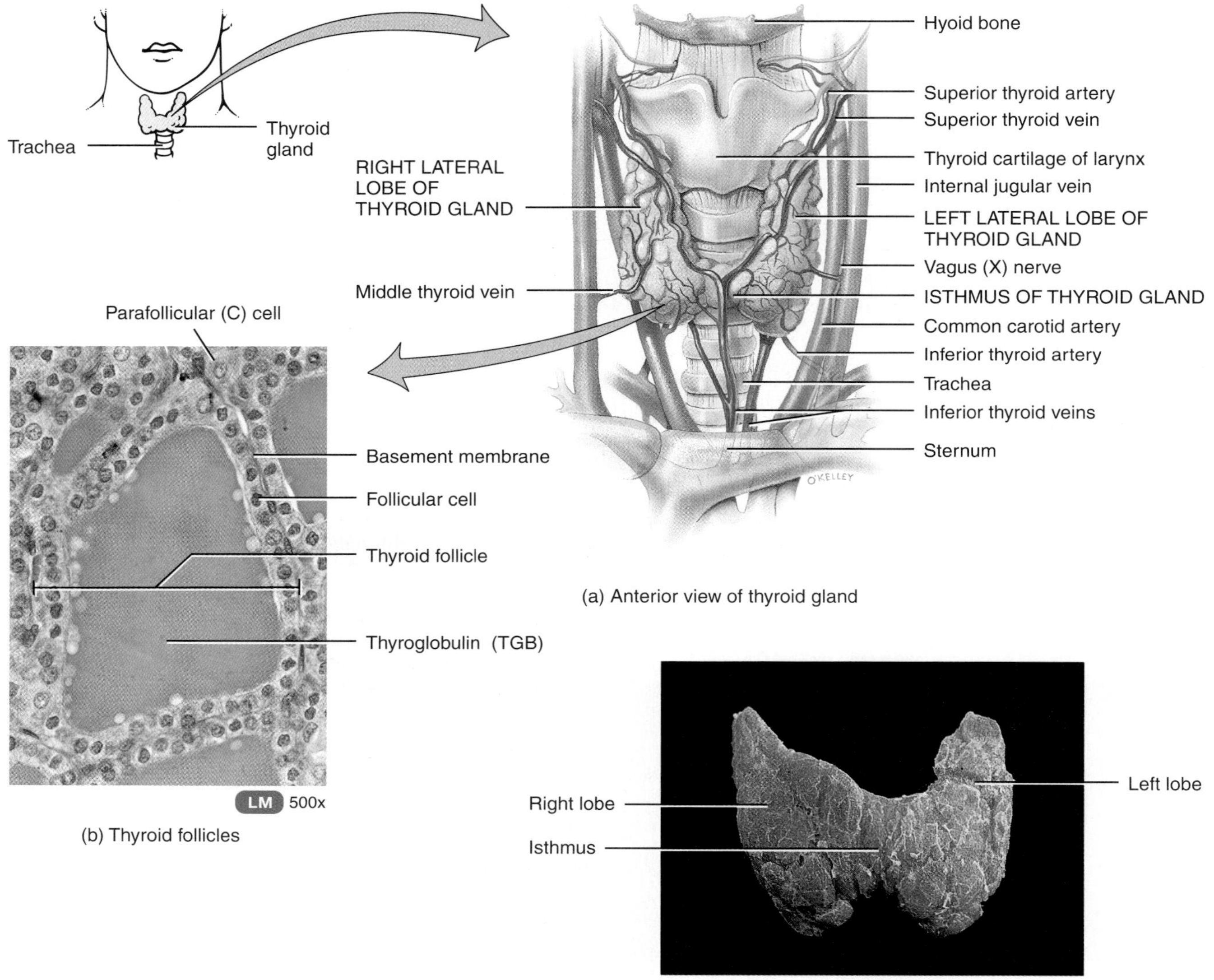

Trachea
Thyroid gland
Hyoid bone
Superior thyroid artery
Superior thyroid vein
Thyroid cartilage of larynx
Internal jugular vein
RIGHT LATERAL LOBE OF THYROID GLAND
LEFT LATERAL LOBE OF THYROID GLAND
Vagus (X) nerve
Middle thyroid vein
ISTHMUS OF THYROID GLAND
Common carotid artery
Inferior thyroid artery
Trachea
Inferior thyroid veins
Sternum

(a) Anterior view of thyroid gland

Parafollicular (C) cell
Basement membrane
Follicular cell
Thyroid follicle
Thyroglobulin (TGB)
LM 500x

(b) Thyroid follicles

Right lobe
Isthmus
Left lobe

(c) Anterior view of thyroid gland

Which cells secrete T_3 and T_4? Which secrete calcitonin? Which of these hormones are also called thyroid hormones?

low cuboidal to squamous, but under the influence of TSH they become cuboidal or low columnar and actively secretory. The follicular cells produce two hormones: **thyroxine** (thī-ROK-sin), which is also called **tetraiodothyronine** (tet-ra-ī-ō-dō-THĪ-rō-nēn), or $\mathbf{T_4}$, because it contains four atoms of iodine, and **triiodothyronine** (trī-ī′-ō-dō-THĪ-rō-nēn), or $\mathbf{T_3}$, which contains three atoms of iodine. T_3 and T_4 are also known as **thyroid hormones.** The thyroid hormones regulate (1) oxygen use and basal metabolic rate, (2) cellular metabolism, and (3) growth and development. A few cells called **parafollicular cells,** or **C cells,** may be embedded within a follicle or lie between follicles. They produce the hormone **calcitonin** (kal-si-TŌ-nin), which helps regulate calcium homeostasis.

The main blood supply of the thyroid gland is from the superior thyroid artery, a branch of the external carotid artery, and the inferior thyroid artery, a branch of the thyrocervical trunk from the subclavian artery. The thyroid gland is drained by the superior and middle thyroid veins, which pass into the internal jugular veins, and the inferior thyroid veins, which join the brachiocephalic veins or internal jugular veins (Figure 22.4a).

The nerve supply of the thyroid gland consists of postganglionic fibers from the superior and middle cervical sympathetic ganglia. Preganglionic fibers from the ganglia are derived from the second through seventh thoracic segments of the spinal cord.

A summary of thyroid gland hormones and their actions is presented in Table 22.4.

✓ Name the hormones produced by follicular and parafollicular cells and describe their actions.

PARATHYROID GLANDS

Objective

- Describe the location, histology, hormone, and functions of the parathyroid glands.

Attached to the posterior surface of the lateral lobes of the thyroid gland are small, round masses of tissue called the **parathyroid glands.** Usually, one superior and one inferior parathyroid gland is attached to each lateral thyroid lobe (Figure 22.5a).

Microscopically, the parathyroids contain two kinds of epithelial cells (Figure 22.5b, c). The more numerous cells, called **principal cells,** probably are the major source of **parathyroid hormone (PTH),** or **parathormone.** The function of the other kind of cell, called an *oxyphil cell,* is not known.

PTH increases the number and activity of osteoclasts (bone-destroying cells). The result is elevated bone resorption, which releases ionic calcium (Ca^{2+}) and phosphates (HPO_4^{2-}) into the blood. PTH also produces two changes in the kidneys: (1) It increases the rate at which the kidneys remove Ca^{2+} and magnesium (Mg^{2+}) from urine that is being formed and returns them to the blood, and (2) it inhibits the reabsorption of HPO_4^{2-} filtered by the kidneys, so that more of it is excreted in urine. More HPO_4^{2-} is lost in the urine than is gained from the bones. Overall, then, PTH decreases blood HPO_4^{2-} level and increases blood Ca^{2+} and Mg^{2+} levels. With respect to blood Ca^{2+} level, PTH and calcitonin are antagonists; that is, they have opposite actions. A third effect of PTH on the kidneys is to promote formation of the hormone **calcitriol,** which is the active form of vitamin D.

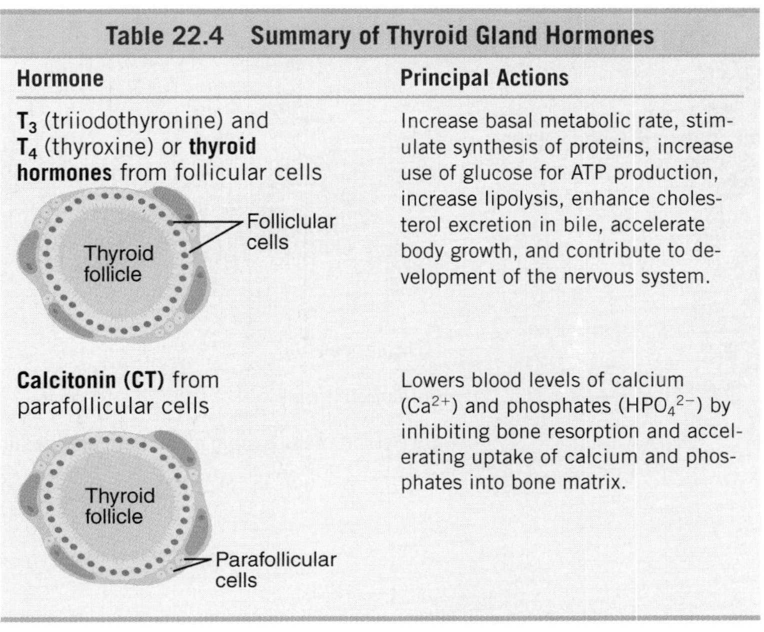

Table 22.4 Summary of Thyroid Gland Hormones

Hormone	Principal Actions
T_3 (triiodothyronine) and T_4 (thyroxine) or **thyroid hormones** from follicular cells	Increase basal metabolic rate, stimulate synthesis of proteins, increase use of glucose for ATP production, increase lipolysis, enhance cholesterol excretion in bile, accelerate body growth, and contribute to development of the nervous system.
Calcitonin (CT) from parafollicular cells	Lowers blood levels of calcium (Ca^{2+}) and phosphates (HPO_4^{2-}) by inhibiting bone resorption and accelerating uptake of calcium and phosphates into bone matrix.

Figure 22.5 / Location, blood supply, and histology of the parathyroid glands.

The parathyroid glands, normally four in number, are embedded in the posterior surface of the thyroid gland.

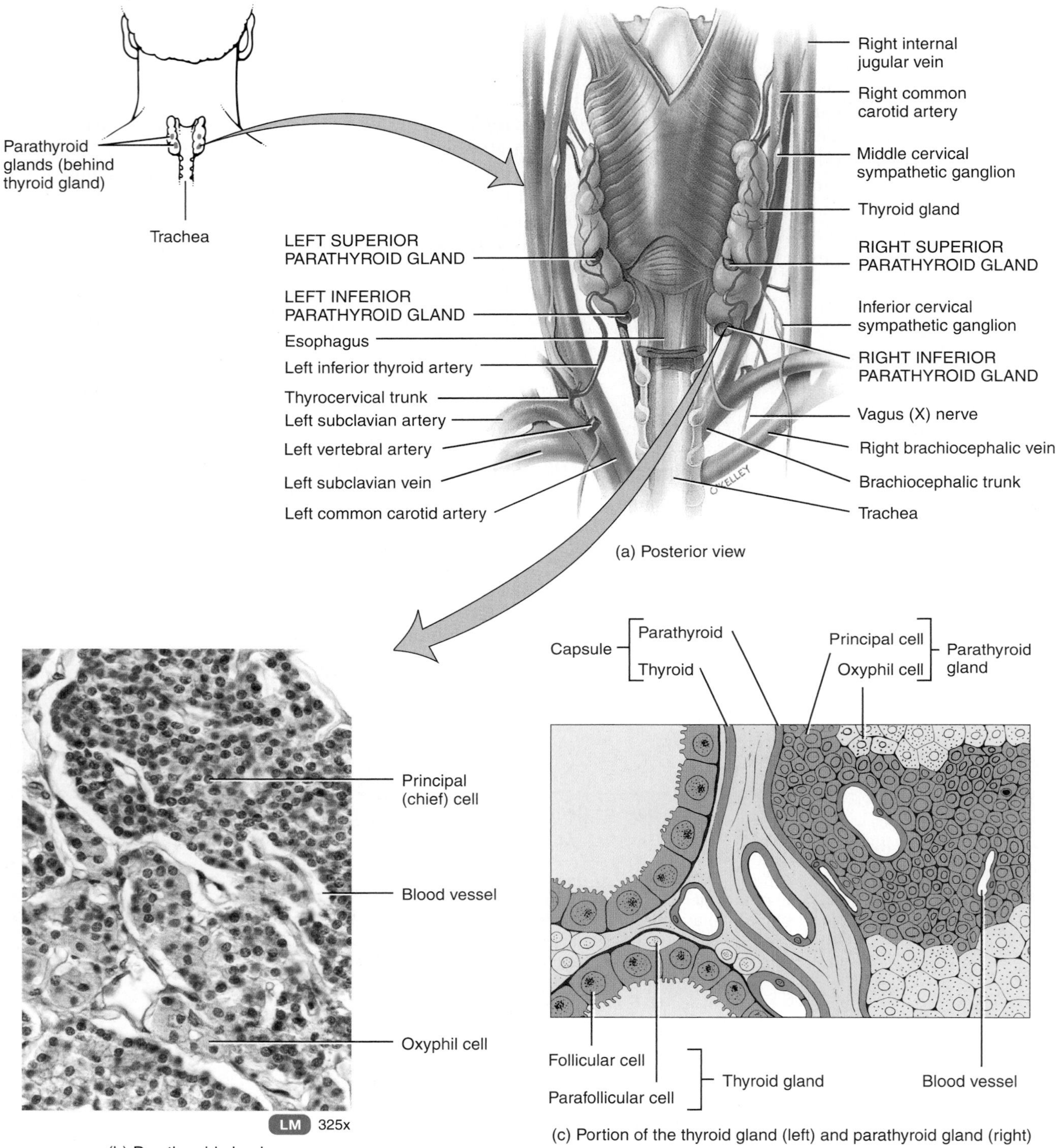

(a) Posterior view

(b) Parathyroid gland

(c) Portion of the thyroid gland (left) and parathyroid gland (right)

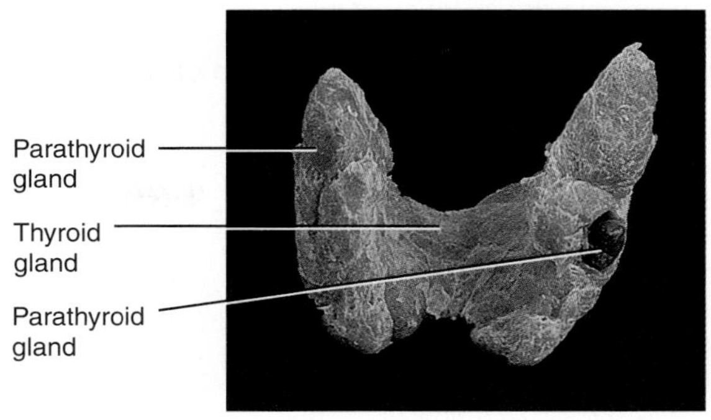

(d) Posterior view of parathyroid glands

What are the secretory products of (1) parafollicular cells of the thyroid gland and (2) principal cells of the parathyroid glands?

Table 22.5 Summary of Parathyroid Gland Hormone

Hormone	Principal Actions
Parathyroid hormone (PTH) from principal cells Principal (chief) cell	Increases blood calcium (Ca^{2+}) and magnesium (Mg^{2+}) levels and decreases blood phosphate (HPO_4^{2-}) level; increases rate of dietary Ca^{2+} and Mg^{2+} absorption; increases bone resorption by osteoclasts; increases Ca^{2+} reabsorption and phosphate excretion by kidneys; and promotes formation of calcitriol (active form of vitamin D).

The parathyroids are abundantly supplied with blood from branches of the superior and inferior thyroid arteries. Blood is drained by the superior, middle, and inferior thyroid veins. The nerve supply of the parathyroids is derived from the thyroid branches of cervical sympathetic ganglia.

A summary of parathyroid hormone and its actions is presented in Table 22.5.

✓ Compare the actions of PTH and calcitonin with respect to blood Ca^{2+} level.

ADRENAL GLANDS

Objective

- Describe the location, histology, hormones, and functions of the adrenal glands.

The paired **adrenal (suprarenal) glands,** one of which lies superior to each kidney (Figure 22.6a), have a flattened pyramidal shape. In an adult, each adrenal gland is 3–5 cm in height, 2–3 cm in width, and a little less than 1 cm thick; it weighs 3.5–5 g, only half its weight at birth. During embryonic development, the adrenal glands differentiate into two structurally and functionally distinct regions: A large, peripherally located **adrenal cortex,** representing 80–90% of the gland by weight, develops from mesoderm; a small, centrally located **adrenal medulla** develops from ectoderm (Figure 22.6b). The adrenal cortex produces steroid hormones that are essential for life. Complete loss of adrenocortical hormones leads to death due to dehydration and electrolyte imbalances in a few days to a week, unless hormone replacement therapy begins promptly. The adrenal medulla produces two catecholamine hormones: norepinephrine and epinephrine. Covering the gland is a connective tissue capsule. The adrenal glands, like the thyroid gland, are highly vascularized.

Adrenal Cortex

The adrenal cortex is subdivided into three zones, each of which secretes different hormones (Figure 22.6d). The outer zone, just deep to the connective tissue capsule, is called the **zona glomerulosa** (*zona* = belt; *glomerul-* = little ball). Its cells, which are closely packed and arranged in spherical clusters and arched columns, secrete hormones called **mineralocorticoids** (min′-er-al-ō-KOR-ti-koyds) because they affect metabolism of the minerals sodium and potassium. The middle zone, or **zona fasciculata** (*fascicul-* = little bundle), is the widest of the three

Figure 22.6 / Location, blood supply, and histology of the adrenal (suprarenal) glands.

The adrenal cortex secretes steroid hormones that are essential for life; the adrenal medulla secretes norepinephrine and epinephrine.

Adrenal glands

Kidney

Inferior phrenic arteries

Right superior suprarenal arteries

RIGHT ADRENAL GLAND

Celiac trunk

LEFT ADRENAL GLAND

Right middle suprarenal artery

Left middle suprarenal artery

Left inferior suprarenal artery

Right inferior suprarenal artery

Left renal artery

Right renal artery

Left renal vein

Right renal vein

O'KELLEY

Superior mesenteric artery

Inferior vena cava

Abdominal aorta

(a) Anterior view

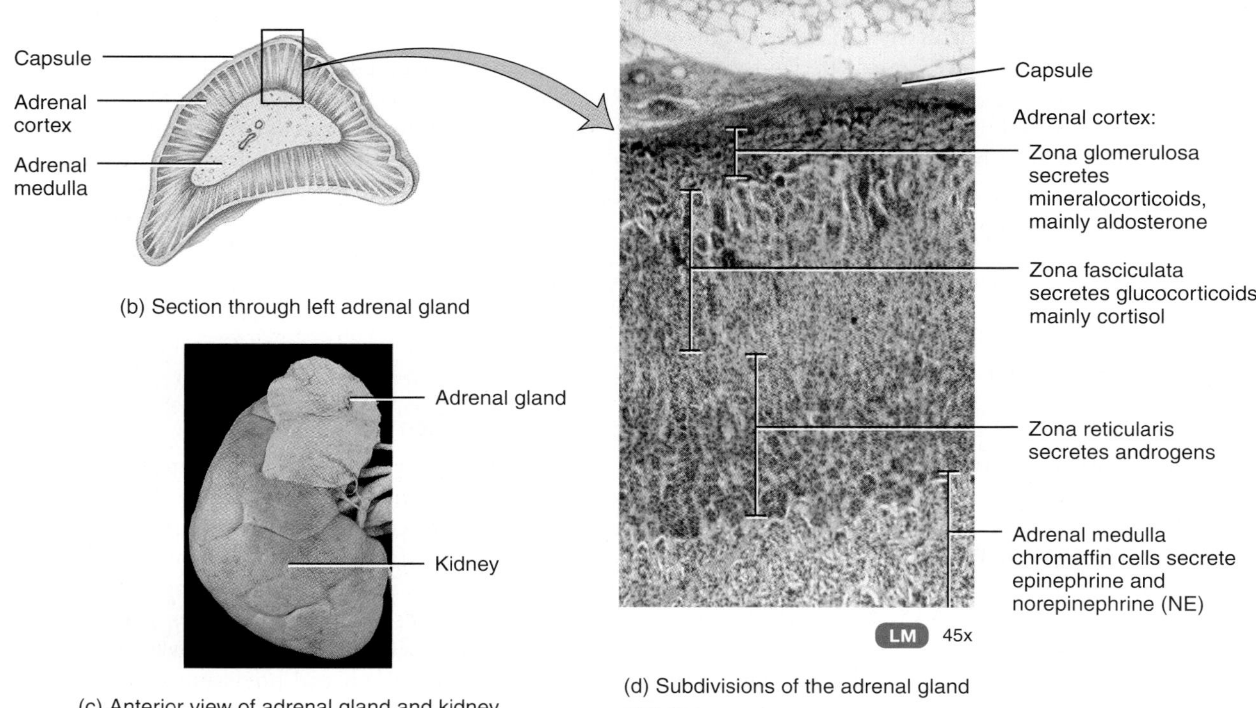

(b) Section through left adrenal gland

(c) Anterior view of adrenal gland and kidney

(d) Subdivisions of the adrenal gland

What is the position of the adrenal glands relative to the kidneys?

zones and consists of cells arranged in long, straight cords. The cells of the zona fasciculata secrete mainly **glucocorticoids** (glū′-kō-KOR-ti-koyds), so named because they affect glucose metabolism. The cells of the inner zone, the **zona reticularis** (*reticul-* = network), are arranged in branching cords. They synthesize small amounts of weak **androgens** (*andro-* = a man), steroid hormones that have masculinizing effects.

Adrenal Medulla

The adrenal medulla consists of hormone-producing cells, called **chromaffin cells** (KRŌ-maf-in; *chrom-* = color; *-affin* = affinity for; Figure 22.6d), which surround large blood vessels. Chromaffin cells receive direct innervation from preganglionic neurons of the sympathetic division of the autonomic nervous system (ANS) and develop from the same embryonic tissue as all other sympathetic postganglionic cells. Thus, they are sympathetic postganglionic cells that are specialized to secrete hormones instead of a neurotransmitter. Because the ANS controls the chromaffin cells directly, hormone release can occur very quickly.

The two principal hormones synthesized by the adrenal medulla are **epinephrine** and **norepinephrine (NE),** also called adrenaline and noradrenaline, respectively. Epinephrine constitutes about 80% of the total secretion of the gland. Both hormones are **sympathomimetic** (sim′-pa-thō-mi-MET-ik)—their effects mimic those brought about by the sympathetic division of the ANS. To a large extent, they are responsible for the fight-or-flight response. Like the glucocorticoids of the adrenal cortex, these hormones help resist stress. Unlike the hormones of the adrenal cortex, however, the medullary hormones are not essential for life.

Table 22.6 Summary of Adrenal Gland Hormones

Hormone	Principal Actions
Adrenal cortical hormones	
Mineralocorticoids (mainly **aldosterone**) from zona glomerulosa cells	Increase blood levels of sodium (Na^+) and water and decrease blood level of potassium (K^+).
Glucocorticoids (mainly **cortisol**) from zona fasciculata cells 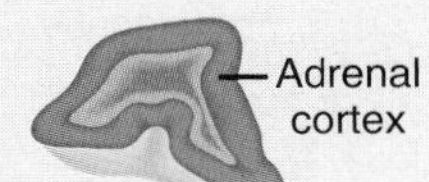	Increase protein breakdown (except in liver), stimulate gluconeogenesis and lipolysis, provide resistance to stress, dampen inflammation, and depress immune responses.
Androgens (mainly **dehydroepiandrosterone** or **DHEA**) from zona reticularis cells	Assist in early growth of axillary and pubic hair in both sexes; in females, contribute to libido and are source of estrogens after menopause.
Adrenal medullary hormones	
Epinephrine and **norepinephrine** from chromaffin cells. 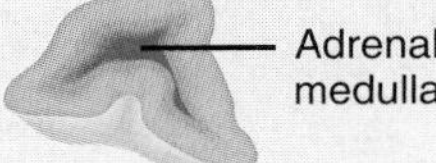	Produce effects that enhance those of the sympathetic division of the autonomic nervous system (ANS) during stress.

The main arteries that supply the adrenal glands are the several superior suprarenal arteries arising from the inferior phrenic artery, the middle suprarenal artery from the aorta, and the inferior suprarenal arteries from the renal arteries. The suprarenal vein of the right adrenal gland drains into the inferior vena cava, whereas the suprarenal vein of the left adrenal gland empties into the left renal vein (see Figure 22.6a).

The principal nerve supply to the adrenal glands is from preganglionic fibers from the thoracic splanchnic nerves, which pass through the celiac and associated sympathetic plexuses. These myelinated fibers end on the secretory cells of the gland found in a region of the medulla.

A summary of adrenal gland hormones and their actions is presented in Table 22.6.

✓ Compare the adrenal cortex and the adrenal medulla with regard to location and histology.

✓ Describe the relationship of the adrenal medulla to the autonomic nervous system.

PANCREAS

Objective

- Describe the location, histology, hormones, and functions of the pancreas.

The **pancreas** (*pan-* = all; *-creas* = flesh) is both an endocrine gland and an exocrine gland. We discuss its endocrine functions here and its exocrine functions in Chapter 24 on the digestive system. The pancreas is a flattened organ that measures about 12.5–15 cm (4.5–6 in.) in length. It is located posterior and slightly inferior to the stomach and consists of a head, a body, and a tail (Figure 22.7a). Roughly 99% of the pancreatic cells are arranged in clusters called **acini** (singular is acinus); these cells produce digestive enzymes, which flow into the gastrointestinal tract through a network of ducts. Scattered among the exocrine acini are 1–2 million tiny clusters of endocrine tissue called **pancreatic islets** or **islets of Langerhans** (LAHNG-er-hanz; Figure 22.7b, c). Abundant capillaries serve both the exocrine and endocrine portions of the pancreas.

Each pancreatic islet includes four types of hormone-secreting cells:

1. **Alpha,** or **A, cells** constitute about 20% of pancreatic islet cells and secrete **glucagon** (GLŪ-ka-gon).
2. **Beta,** or **B, cells** constitute about 70% of pancreatic islet cells and secrete **insulin** (IN-sū-lin).
3. **Delta,** or **D, cells** constitute about 5% of pancreatic islet cells and secrete **somatostatin** (identical to growth hormone inhibiting hormone secreted by the hypothalamus).
4. **F cells** constitute the remainder of pancreatic islet cells and secrete **pancreatic polypeptide.**

The arterial supply of the pancreas is from the superior and inferior pancreaticoduodenal arteries and from the splenic and superior mesenteric arteries. The veins, in general, correspond to

Figure 22.7 / Location, blood supply, and histology of the pancreas.

Pancreatic hormones regulate blood glucose level.

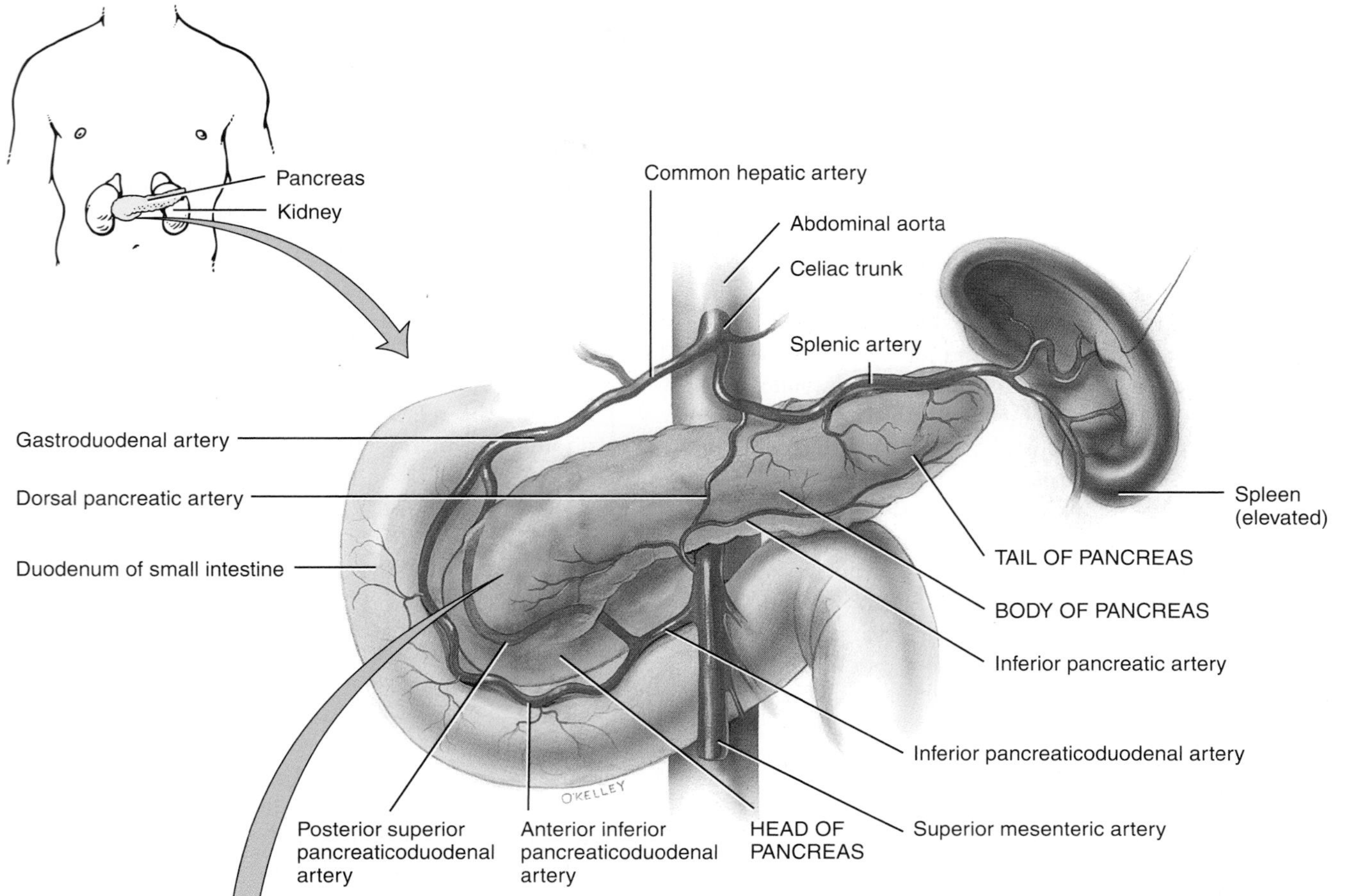

(a) Anterior view

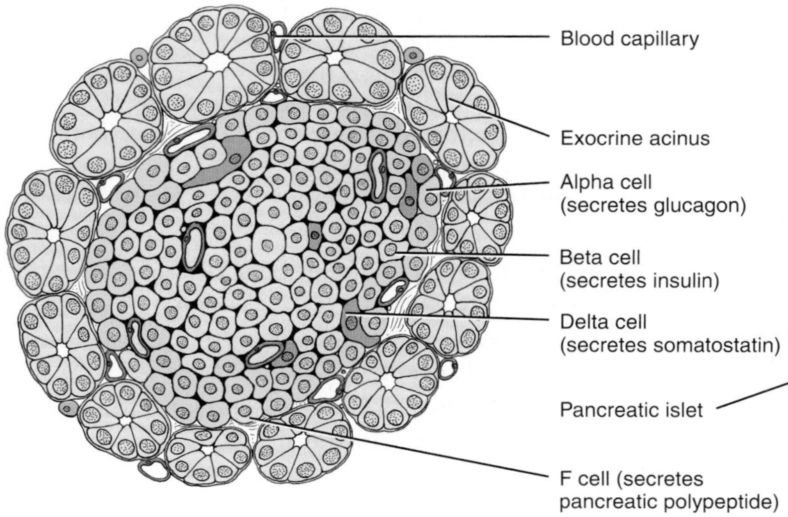

(b) Pancreatic islet and surrounding acini

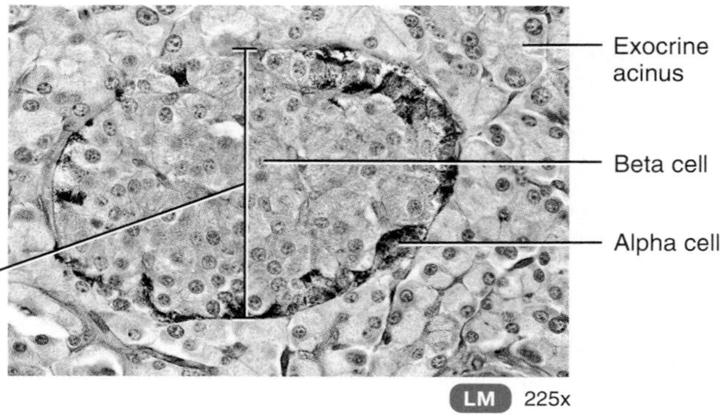

(c) Pancreatic islet and surrounding acini

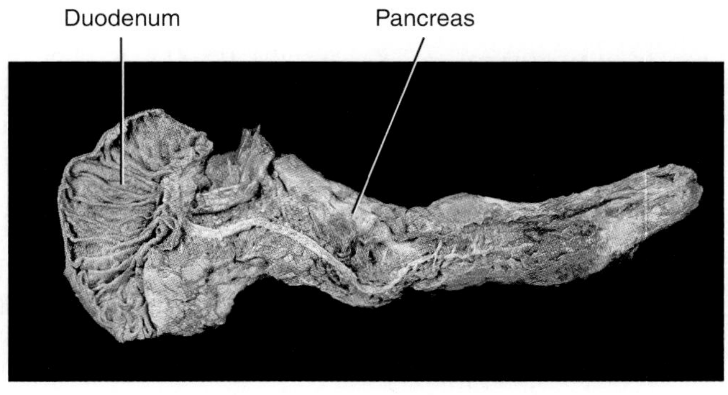

(d) Anterior view of pancreas

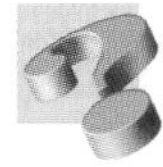

Is the pancreas an exocrine gland or an endocrine gland?

the arteries. Venous blood reaches the hepatic portal vein by means of the splenic and superior mesenteric veins (Figure 22.7a).

The nerves to the pancreas are autonomic nerves derived from the celiac and superior mesenteric plexuses. Included are preganglionic vagal, postganglionic sympathetic, and sensory fibers. Parasympathetic vagal fibers are said to terminate at both acinar (exocrine) and islet (endocrine) cells. Although the innervation is presumed to influence enzyme formation, pancreatic secretion is controlled largely by the hormones secretin and cholecystokinin (CCK) released by the small intestine. The sympathetic fibers that enter the islets (and also end on blood vessels) are vasomotor and are accompanied by sensory fibers, especially for pain.

A summary of pancreatic hormones and their actions is presented in Table 22.7.

✓ Why is the pancreas both an exocrine gland and an endocrine gland?

OVARIES AND TESTES

Objective

- Describe the location, hormones, and functions of the male and female gonads.

The female gonads, called the **ovaries,** are paired oval bodies located in the pelvic cavity. The ovaries produce female sex hormones called **estrogens** and **progesterone.** Along with the gonadotropic hormones of the pituitary gland, the sex hormones regulate the female reproductive cycle, maintain pregnancy, and prepare the mammary glands for lactation. These hormones are also responsible for the development and maintenance of feminine secondary sex characteristics. The ovaries also produce **inhibin,** a protein hormone that inhibits secretion of follicle-stimulating hormone (FSH). During pregnancy, the ovaries and placenta produce a peptide hormone called **relaxin,** which increases the flexibility of the pubic symphysis during pregnancy and

Table 22.7 Summary of Hormones Produced by the Pancreas

Hormone	Principal Actions
Glucagon from alpha or A cells of pancreatic islets 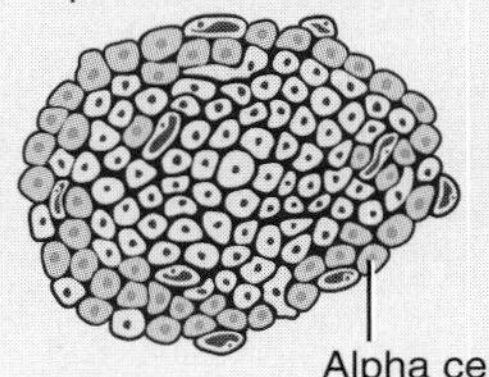	Raises blood glucose level by accelerating breakdown of glycogen into glucose in liver (glycogenolysis), converting other nutrients into glucose in liver (gluconeogenesis), and releasing glucose into the blood.
Insulin from beta or B cells of pancreatic islets 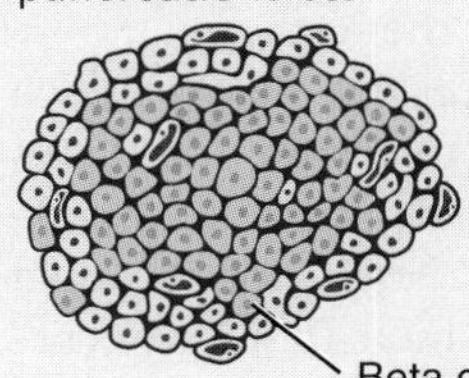	Lowers blood glucose level by accelerating transport of glucose into cells, converting glucose into glycogen (glycogenesis), and decreasing glycogenolysis and gluconeogenesis; also increases lipogenesis and stimulates protein synthesis.
Somatostatin from delta or D cells of pancreatic islets 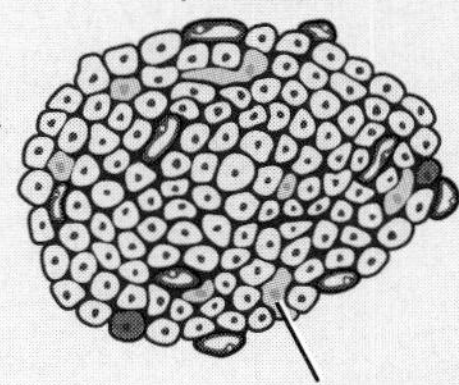	Inhibits secretion of insulin and glucagon and slows absorption of nutrients from the gastrointestinal tract.
Pancreatic polypeptide from F cells of pancreatic islets 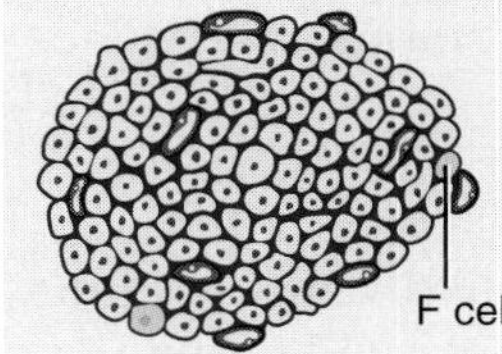	Inhibits somatostatin secretion, gallbladder contraction, and secretion of pancreatic digestive enzymes.

Table 22.8 Summary of Hormones of the Ovaries and Testes

Hormone	Principal Actions
Ovarian hormones	
Estrogens and **progesterone** Ovaries	Together with gonadotropic hormones of the anterior pituitary gland, regulate the female reproductive cycle, maintain pregnancy, prepare the mammary glands for lactation, regulate oogenesis, and promote development and maintenance of feminine secondary sex characteristics.
Relaxin	Increases flexibility of pubic symphysis during pregnancy and helps dilate uterine cervix during labor and delivery.
Inhibin	Inhibits secretion of FSH from anterior pituitary gland.
Testicular hormones	
Testosterone Testes	Stimulates descent of testes before birth, regulates spermatogenesis, and promotes development and maintenance of masculine secondary sex characteristics.
Inhibin	Inhibits secretion of FSH from anterior pituitary gland.

helps dilate the uterine cervix during labor and delivery. These actions help ease the baby's passage by enlarging the birth canal.

The male has two oval gonads, called **testes,** that produce **testosterone,** the primary androgen. Testosterone regulates production of sperm and stimulates the development and maintenance of masculine secondary sex characteristics such as beard growth. The testes also produce inhibin, which inhibits secretion of FSH. The specific roles of gonadotropic hormones and sex hormones are discussed in Chapter 26.

Table 22.8 summarizes the hormones produced by the ovaries and testes and their principal actions.

✓ Explain why the ovaries and testes are endocrine glands.

PINEAL GLAND

Objective

- Describe the location, histology, hormone, and functions of the pineal gland.

The **pineal gland** (PĪN-ē-al; = pinecone shape) is a small endocrine gland attached to the roof of the third ventricle of the brain at the midline (see Figure 22.1). It is part of the epithalamus, positioned between the two superior colliculi, and it weighs 0.1–0.2 g. The gland, which is covered by a capsule formed by the pia mater, consists of masses of neuroglia and secretory cells called **pinealocytes** (pin-ē-AL-ō-sīts). Sympathetic postganglionic fibers from the superior cervical ganglion terminate in the pineal gland.

Although many anatomical features of the pineal gland have been known for years, its physiological role is still unclear. One hormone secreted by the pineal gland is **melatonin.** Melatonin contributes to setting the body's biological clock, which is controlled from the suprachiasmatic nucleus. During sleep, plasma levels of melatonin increase tenfold and then decline to a low level again before awakening. Small doses of melatonin given orally can induce sleep and reset daily rhythms, which might benefit workers whose shifts alternate between daylight and nighttime hours. Melatonin also is a potent antioxidant that may provide some protection against damaging oxygen free radicals. In animals that breed during specific seasons, melatonin inhibits reproductive functions. Whether melatonin influences human reproductive function, however, is still unclear.

The posterior cerebral artery supplies the pineal gland with blood, and the great cerebral vein drains it.

Seasonal affective disorder (SAD) is a type of depression that afflicts some people during the winter months, when day length is short. It is thought to be due, in part, to overproduction of melatonin. Bright light therapy—repeated doses of several hours of exposure to artificial light as bright as sunlight—provides relief for some people. Three to six hours of exposure to bright light also appears to speed recovery from jet lag, the fatigue suffered by travelers who cross several time zones. ■

✓ What is the relationship between melatonin and sleep?

THYMUS

Because of its role in immunity, the details of the structure and functions of the **thymus** are discussed in Chapter 15, which examines the lymphatic system and immunity. At this point, only its hormonal role in immunity will be discussed.

Lymphocytes are one type of white blood cell. These, in turn, are divided into two types, called T cells and B cells, based on their specific roles in immunity. Hormones produced by the thymus, called **thymosin, thymic humoral factor (THF), thymic factor (TF),** and **thymopoietin,** promote the proliferation and maturation of T cells, which destroy foreign substances and microbes. There is also some evidence that thymic hormones may retard the aging process.

✓ What is the role of thymic hormones in immunity?

OTHER ENDOCRINE TISSUES

Objective

- List the hormones secreted by cells in tissues and organs other than endocrine glands, and describe their functions.

Before leaving our discussion of hormones, it should be noted that cells in organs other than those usually classified as endocrine glands also have an endocrine function and secrete hormones. These tissues are summarized in Table 22.9.

✓ List the hormones secreted by the gastrointestinal tract, placenta, kidneys, skin, adipose tissue, and heart, and indicate their functions.

Table 22.9 Summary of Hormones Produced by Organs and Tissues that Contain Endocrine Cells

Hormones	Principal Actions
Gastrointestinal tract	
Gastrin	Promotes secretion of gastric juice and increases motility of the stomach.
Glucose-dependent insulinotropic peptide (GIP)	Stimulates release of insulin by pancreatic beta cells.
Secretin	Stimulates secretion of pancreatic juice and bile.
Cholecystokinin (CCK)	Stimulates secretion of pancreatic juice, regulates release of bile from the gallbladder, and brings about a feeling of fullness after eating.
Placenta	
Human chorionic gonadotropin (hCG)	Stimulates the corpus luteum in the ovary to continue the production of estrogens and progesterone to maintain pregnancy.
Estrogens and **progesterone**	Maintain pregnancy and prepare mammary glands to secrete milk.
Human chorionic somatomammotropin (hCS)	Stimulates the development of the mammary glands for lactation.
Kidneys	
Erythropoietin (EPO)	Increases rate of red blood cell production.
Calcitriol[a] (active form of vitamin D)	Aids in the absorption of dietary calcium and phosphorus.
Heart	
Atrial nutriuretic peptide (ANP)	Decreases blood pressure.
Adipose tissue	
Leptin	Suppresses appetite and may have a permissive effect on activity of GnRH and gonadotropins.

[a]Synthesis begins in the skin, continues in the liver, and ends in the kidneys.

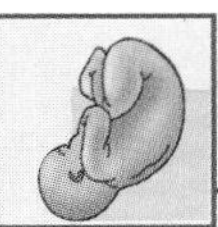

DEVELOPMENTAL ANATOMY OF THE ENDOCRINE SYSTEM

Objective

- Describe the development of endocrine glands.

The development of the endocrine system is not as localized as the development of other systems because endocrine organs develop in widely separated parts of the embryo.

The *pituitary gland (hypophysis)* originates from two different regions of the ectoderm. The *posterior pituitary gland (neurohypophysis)* derives from an outgrowth of ectoderm called the **neurohypophyseal bud,** located on the floor of the hypothalamus (Figure 22.8a). The *infundibulum*, also an outgrowth of the neurohypophyseal bud, connects the posterior pituitary gland to the hypothalamus. The *anterior pituitary gland (adenohypophysis)* is derived from an outgrowth of ectoderm from the roof of the mouth called the **hypophyseal (Rathke's) pouch.** The pouch grows toward the neurohypophyseal bud and eventually loses its connection with the roof of the mouth.

The *thyroid gland* develops as a midventral outgrowth of endoderm, called the **thyroid diverticulum,** from the floor of the pharynx at the level of the second pair of pharyngeal pouches. The outgrowth projects inferiorly and differentiates into the right and left lateral lobes and the isthmus of the gland.

The *parathyroid glands* develop from endoderm as outgrowths from the third and fourth **pharyngeal pouches.**

The adrenal cortex and adrenal medulla have completely different embryological origins. The *adrenal cortex* is derived from intermediate mesoderm from the same region that produces the gonads. The *adrenal medulla* is ectodermal in origin and derives from the **neural crest,** which also gives rise to sympathetic ganglia and other structures of the nervous system (see Figure 18.24b on page 583).

The *pancreas* develops from two outgrowths of endoderm from the part of the **foregut** that later becomes the duodenum (see Figure 24.21 on pages 749–750). The two outgrowths eventually fuse to form the pancreas. The origin of the ovaries and testes is discussed in the section on the reproductive system.

The *pineal gland* arises as an outgrowth between the thalamus and colliculi from ectoderm associated with the **diencephalon** (see Figure 18.25b) on page 584).

The *thymus* arises from endoderm of the third **pharyngeal pouches.**

✓ Compare the origins of the adrenal cortex and adrenal medulla.

Figure 22.8 / Development of the endocrine system.

Glands of the endocrine system develop from all three primary germ layers.

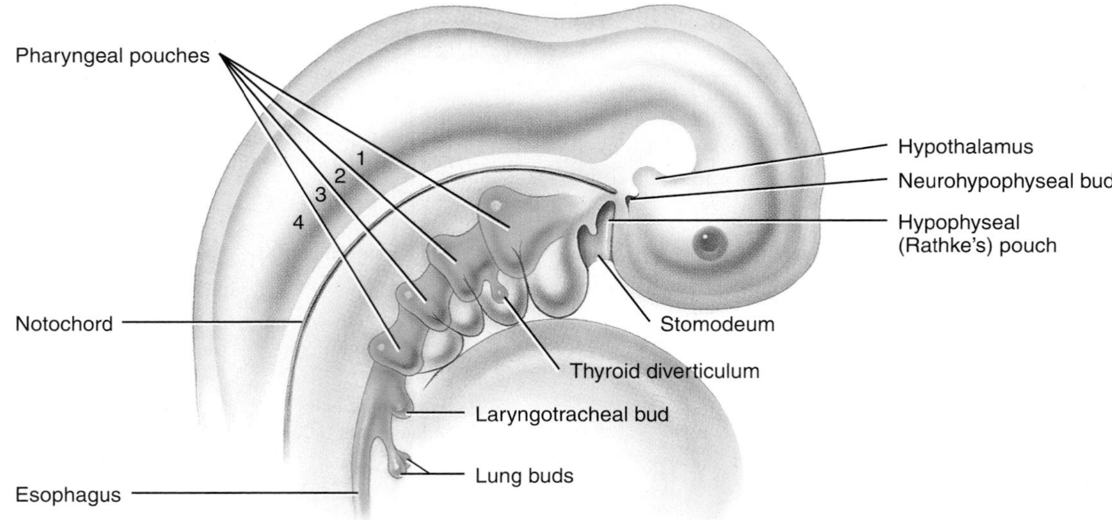

(a) Location of the neurohypophyseal bud, hypophyseal (Rathke's) pouch, and thyroid diverticulum in a 28-day embryo

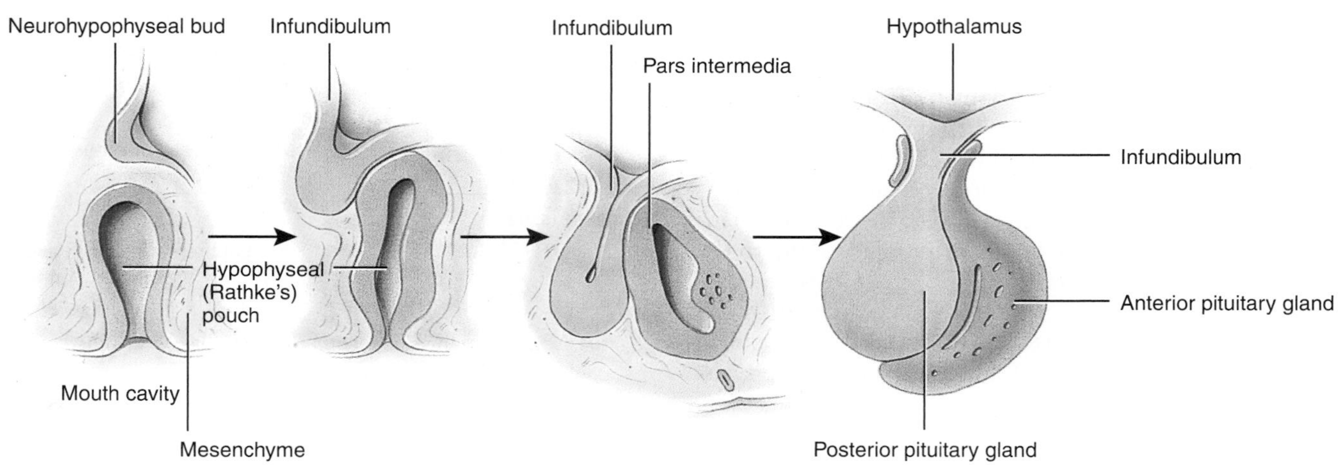

(b) Development of the pituitary gland between 5 and 16 weeks

Which two endocrine glands develop from tissues with two different embryological origins?

AGING AND THE ENDOCRINE SYSTEM

Objective

- Describe the effects of aging on the endocrine system.

Although some endocrine glands shrink as we get older, their performance may or may not be compromised. With respect to the pituitary gland, production of human growth hormone decreases, which is one cause of muscle atrophy as aging proceeds. But production of gonadotropins and thyroid-stimulating hormone actually increases with age. Output of adrenocorticotropic hormone by the pituitary gland is apparently unaffected by age. The thyroid gland often decreases its output of thyroid hormones with age, causing a decrease in metabolic rate, an increase in body fat, and hypothyroidism, which is seen more often in older people.

The thymus is largest in infancy. After puberty its size begins to decrease, and thymic tissue is replaced by adipose and areolar connective tissue. In older adults the thymus has atrophied significantly.

The adrenal glands contain increasingly more fibrous tissue and produce less cortisol and aldosterone with advancing age. However, production of epinephrine and norepinephrine remains normal. The pancreas releases insulin more slowly with age, and receptor sensitivity to glucose declines. As a result, blood glucose levels in older people increase faster and return to normal more slowly than in younger individuals.

The ovaries experience a drastic decrease in size with age, and they no longer respond to gonadotropins. The resultant decreased output of estrogens by the ovaries leads to conditions such as osteoporosis and atherosclerosis. Although testosterone production by the testes decreases with age, the effects are not usually apparent until very old age, and many elderly males can still produce active sperm in normal numbers.

✓ Which hormone is related to muscle atrophy associated with aging?

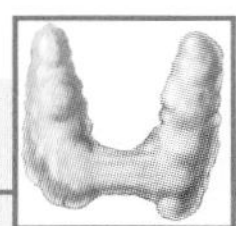

APPLICATIONS TO HEALTH

Most disorders of the endocrine system are due to secretion of too much or too little of a given hormone. However, because hormones cannot function until they bind to their receptors, some endocrine disorders are due to receptor aberrations or defects in second-messenger systems. Because hormones are distributed in the blood to target tissues throughout the body, problems associated with endocrine dysfunction may also be widespread.

Pituitary Gland Disorders

Pituitary Dwarfism, Giantism, and Acromegaly

Hyposecretion of human growth hormone (hGH) during the growth years slows bone growth, and the epiphyseal plates close before normal height is reached. This condition is called **pituitary dwarfism.** Other organs of the body also fail to grow, and a pituitary dwarf is childlike in many physical respects. Treatment requires administration of hGH during childhood, before the epiphyseal plates close.

Hypersecretion of hGH during childhood results in **giantism,** an abnormal increase in the length of long bones. A person with this condition is unusually tall, but has approximately normal body proportions. Hypersecretion of hGH during adulthood causes **acromegaly** (ak′-rō-MEG-a-lē). Further lengthening of the long bones cannot occur because the epiphyseal plates are closed. Instead, the bones of the hands, feet, cheeks, and jaw thicken. Also, the eyelids, lips, tongue, and nose enlarge, and the skin thickens and develops furrows, especially on the forehead and soles (Figure 22.9a).

Diabetes Insipidus

The most common abnormality associated with dysfunction of the posterior pituitary gland is **diabetes insipidus** (dī-a-BĒ-tēs in-SIP-i-dus; *diabetes* = overflow; *insipidus* = tasteless) or **DI.** An inability either to secrete or to respond to antidiuretic hormone (ADH) can cause diabetes insipidus. Neurogenic diabetes insipidus results from hyposecretion of ADH, usually caused by a brain tumor, head trauma, or brain surgery that damages the posterior pituitary gland and the hypothalamic paraventricular and supraoptic nuclei. In nephrogenic diabetes insipidus, the kidneys do not respond to ADH. The ADH receptors may be nonfunctional, or the kidneys may be damaged. A common symptom of both forms of the disorder is excretion of large volumes of urine, with resulting dehydration and thirst. Bedwetting is common in afflicted children. Because so much water is lost in the urine, a person with severe diabetes insipidus may die of dehydration if deprived of water for only a day or so. Treatment of neurogenic diabetes insipidus involves hormone replacement, usually for life. Either subcutaneous injection or nasal spray application of ADH analogs is effective. Treatment of nephrogenic diabetes insipidus is more complex and depends on the nature of the kidney dysfunction. Restriction of salt in the diet and, paradoxically, the use of certain diuretic drugs, are helpful.

Thyroid Gland Disorders

Thyroid gland disorders affect all major body systems and are among the most common endocrine disorders. Hyposecretion of thyroid hormones during fetal life or infancy results in **cretinism** (KRĒ-tin-izm). Cretins exhibit dwarfism because the skeleton fails to grow and mature, and they are severely mentally retarded because the brain fails to develop fully (Figure 22.9b). Because lipid-soluble maternal thyroid hormones cross the placenta and allow normal development, newborn babies usually appear normal, even if they can't produce their own thyroid hormones. Most states require testing of all newborns to ensure adequate thyroid function. If hypothyroidism is detected early, cretinism can be prevented by giving oral thyroid hormone.

Hypothyroidism during the adult years produces **myxedema** (mix-e-DĒ-ma). A hallmark of this disorder is edema (accumulation of interstitial fluid) that causes the facial tissues to swell and look puffy. A person with myxedema has a slow heart rate, low body temperature, sensitivity to cold, dry hair and skin, muscular weakness, general lethargy, and a tendency to gain weight easily. Because the brain has already reached maturity, mental retardation does not occur, but the person's mental functions

Figure 22.9 / Various endocrine disorders.

Disorders of the endocrine system often involve hyposecretion or hypersecretion of hormones.

(a) Acromegaly (excess hGH during adulthood)

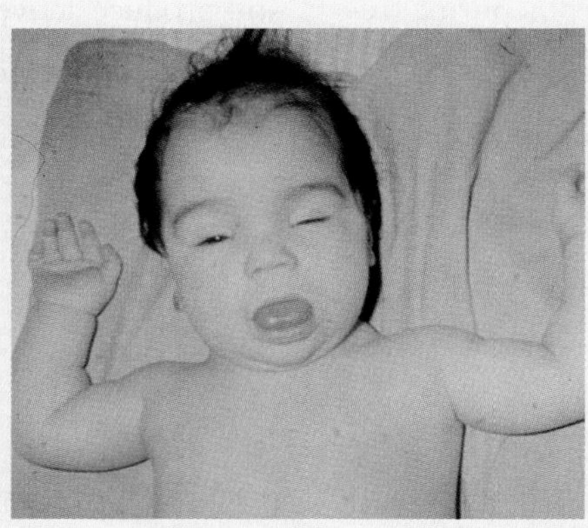

(b) Cretinism (lack of thyroid hormones during fetal life or infancy)

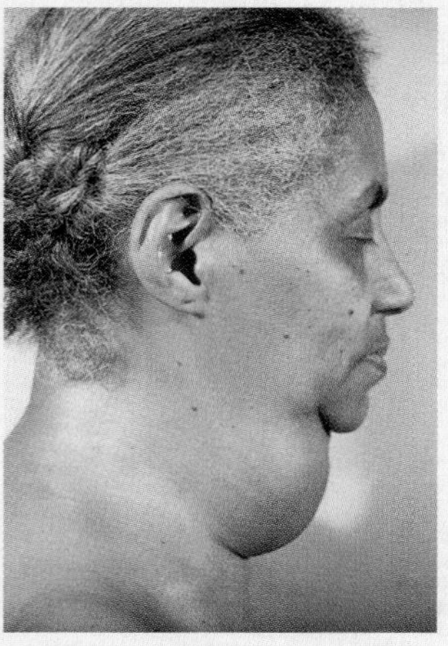

(c) Goiter (enlargement of thyroid gland as in Graves' disease or inadequate iodine intake)

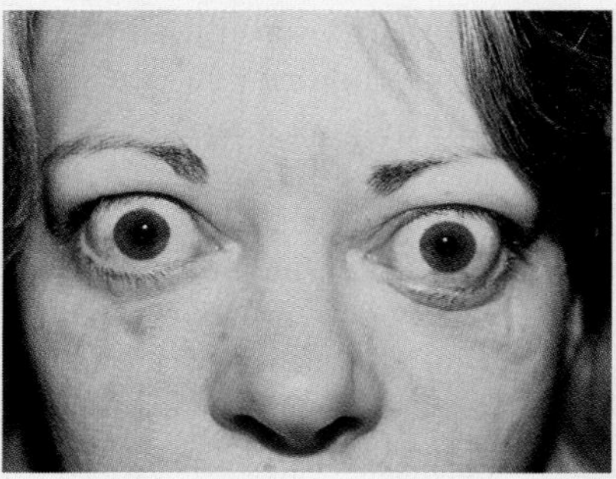

(d) Exophthalmos (excess thyroid hormones, as in Graves' disease)

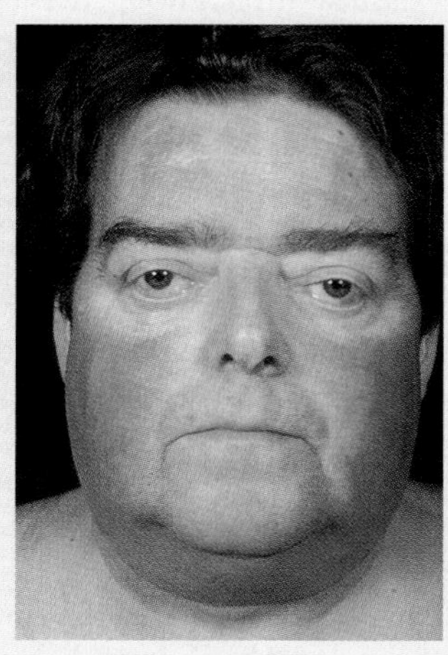

(e) Cushing's syndrome (excess glucocorticoids)

Which of the disorders is due to antibodies that mimic the action of TSH?

may be dulled, such that the person is less alert. Myxedema occurs about five times more often in females than in males. Oral thyroid hormones reduce the symptoms.

The most common form of hyperthyroidism is **Graves' disease,** which is an autoimmune disorder. It occurs seven to ten times more often in females than in males, usually before age 40. Because the person produces antibodies that mimic the action of thyroid-stimulating hormone (TSH), the thyroid gland is continually stimulated to grow and produce thyroid hormones. A primary sign is an enlarged thyroid, which may be two to three times its normal size (Figure 22.9c). Graves' patients often have a peculiar edema behind the eyes, called **exophthalmos** (ek′-sof-THAL-mos), which causes the eyes to protrude (Figure 22.9d). Treatment may include surgically removing part or all of the thyroid gland (thyroidectomy), using radioactive iodine (^{131}I) to selectively destroy thyroid tissue, and using antithyroid drugs to block synthesis of thyroid hormones.

A **goiter** (GOY-ter; *guttur* = throat) is simply an enlarged thyroid gland, and it may be associated with hyperthyroidism, hypothyroidism, or euthyroidism (*eu-* = good), which means normal secretion of thyroid hormone. In some places in the world, dietary iodine intake is inadequate; the resultant low level of thyroid hormone in the blood stimulates secretion of TSH, which causes thyroid gland enlargement.

Parathyroid Gland Disorders

Hypoparathyroidism—too little parathyroid hormone—leads to a deficiency of Ca^{2+}, which causes neurons and muscle fibers to depolarize and produce action potentials spontaneously. This leads to twitches, spasms, and tetany of skeletal muscle. The leading cause of hypoparathyroidism is accidental damage to the parathyroid glands or to their blood supply during thyroidectomy surgery.

Hyperparathyroidism, usually due to a tumor of the parathyroid gland, produces **osteitis fibrosa cystica,** which causes demineralization of bone. The bones become brittle and easily fractured due to excessive loss of Ca^{2+}.

Adrenal Gland Disorders

Disorders of the adrenal cortex involve hyperfunction or hypofunction and result in complex physical, psychological, and metabolic changes that may be life threatening. Disorders of the adrenal medulla involve hyperfunction and are not life threatening.

Cushing's Syndrome

Hypersecretion of cortisol by the adrenal cortex produces **Cushing's syndrome.** Causes include a tumor of the adrenal gland that secretes cortisol, or a tumor elsewhere that secretes adrenocorticotropic hormone (ACTH), which in turn stimulates excessive secretion of cortisol. The condition is characterized by breakdown of muscle proteins and redistribution of body fat, resulting in spindly arms and legs accompanied by a rounded "moon face" (Figure 22.9e), "buffalo hump" on the back, and pendulous (hanging) abdomen. Facial skin is flushed, and the skin covering the abdomen develops stretch marks. The person also bruises easily, and wound healing is poor. The elevated level of cortisol causes hyperglycemia, osteoporosis, weakness, hypertension, increased susceptibility to infection, decreased resistance to stress, and mood swings. People who need long-term, glucocorticoid therapy—for instance, to prevent rejection of a transplanted organ—may develop a cushinoid appearance.

Addison's Disease

Progressive destruction of the adrenal cortex leads to hyposecretion of glucocorticoids and aldosterone and causes **Addison's disease (primary adrenocortical insufficiency).** The majority of cases are thought to be autoimmune disorders in which antibodies either cause adrenal cortex destruction or block binding of ACTH to its receptors. Pathogens, such as the bacterium that causes tuberculosis, also may trigger adrenal cortex destruction. Symptoms, which typically do not appear until 90% of the adrenal cortex has been destroyed, include mental lethargy, anorexia, nausea and vomiting, weight loss, hypoglycemia, and muscular weakness. Loss of aldosterone leads to elevated potassium and decreased sodium in the blood, low blood pressure, dehydration, decreased cardiac output, arrhythmias, and potential cardiac arrest. Excessive skin pigmentation, especially in sun-exposed areas and in mucous membranes, also occurs. Treatment consists of replacing glucocorticoids and mineralocorticoids and increasing sodium in the diet.

Pheochromocytomas

Usually benign tumors of the chromaffin cells of the adrenal medulla, called **pheochromocytomas** (fē-ō-krō′-mō-sī-TŌ-mas; *pheo-* = dusky; *chromo-* = color; *cyto-* = cell), cause hypersecretion of the medullary hormones. Hypersecretion of epinephrine and norepinephrine causes a prolonged version of the fight-or-flight response: rapid heart rate, headache, high blood pressure, high levels of glucose in blood and urine, an elevated basal metabolic rate (BMR), flushed face, nervousness, sweating, and decreased gastrointestinal motility. Treatment is surgical removal of the tumor.

Pancreatic Disorders

Among the pancreatic disorders is the most common endocrine disorder, **diabetes mellitus** (MEL-i-tus; *melli-* = honey sweetened), a chronic disease that affects about 12 million Americans and is the fourth leading cause of death by disease in the United States, primarily because of its widespread cardiovascular effects. Diabetes mellitus is actually a group of disorders caused by an inability to produce or use insulin. The result is an elevation of blood glucose (hyperglycemia) and loss of glucose in the urine (glucosuria). Hallmarks of diabetes mellitus are the three "polys": *polyuria,* excessive urine production due to an inability of the kidneys to reabsorb water; *polydipsia,* excessive thirst; and *polyphagia,* excessive eating.

There are two major types of diabetes mellitus. **Type I diabetes** is caused by an absolute deficiency of insulin. Consequently, Type I diabetes is also called **insulin-dependent diabetes mellitus (IDDM)** because regular injections of insulin are required to prevent death. Most commonly, IDDM develops in people younger than age 20, although the condition persists throughout life. IDDM is an autoimmune disorder in which a person's immune system destroys the pancreatic beta cells. The drug cyclosporine, which suppresses the immune system, shows promise in being able to interrupt the destruction of beta cells.

The cellular metabolism of an untreated type I diabetic is similar to that of a starving person. Because insulin is not present to aid the entry of glucose into body cells, most cells use fatty acids to produce ATP. The accumulation of by-products of fatty acid breakdown—organic acids called ketones (ketone bodies)—cause a form of acidosis called **ketoacidosis,** which lowers the pH of the blood and can result in death. The breakdown of stored triglycerides and proteins also causes weight loss. As lipids are transported by the blood from storage depots to cells, lipid particles are deposited on the walls of blood vessels, leading to atherosclerosis and a multitude of cardiovascular problems, including cerebrovascular insufficiency, ischemic heart disease, peripheral vascular disease, and gangrene. One of the major complications of diabetes is loss of vision due either to cataracts (excessive glucose attaches to lens proteins, causing cloudiness) or to damage to blood vessels of the retina. Severe kidney problems also may result from damage to renal blood vessels.

Type II diabetes, also called **non-insulin-dependent diabetes mellitus (NIDDM),** is much more common than type I, representing more than 90% of all cases. Type II diabetes most often occurs in people who are over 35 and overweight. Clinical symptoms are mild, and the high glucose levels in the blood often can be controlled by diet, exercise, and weight loss. Sometimes, an antidiabetic drug such as *glyburide* (DiaBeta) is used to stimulate secretion of insulin by beta cells of the pancreas. Although some type II diabetics need insulin, many have a sufficient amount (or even a surplus) of insulin in the blood. For these people, diabetes arises not from a shortage of insulin but because target cells become less sensitive to it due to down-regulation of insulin receptors.

Hyperinsulinism most often results when a diabetic injects too much insulin. The principal symptom is **hypoglycemia,** decreased blood glucose level, which occurs because the excess insulin stimulates excessive uptake of glucose by many cells of the body. The resulting hypoglycemia stimulates the secretion of epinephrine, glucagon, and human growth hormone. As a consequence, anxiety, sweating, tremor, increased heart rate, hunger, and weakness occur. When blood glucose falls, brain cells are deprived of the steady supply of glucose they need to function effectively. This condition leads to mental disorientation, convulsions, unconsciousness, and shock and is termed **insulin shock.** Death can occur quickly unless blood glucose is raised.

KEY MEDICAL TERMS ASSOCIATED WITH THE ENDOCRINE SYSTEM

Gynecomastia (gī-ne-kō-MAS-tē-a, *gyneca* = woman; *mast* = breast) Excessive development of male mammary glands. Sometimes a tumor of the adrenal gland may secrete sufficient quantities of estrogen to cause the condition.

Thyroid crisis (storm) A severe state of hyperthyroidism that can be life threatening. It is characterized by high body temperature, rapid heart rate, high blood pressure, gastrointestinal symptoms (abdominal pain, vomiting, diarrhea), agitation, tremors, confusion, seizures, and possibly coma.

Virilism (VIR-il-izm; *vivilis* = masculine) The presence of mature masculine characteristics in females or prepubescent males. In a female, virile characteristics include growth of a beard, development of a much deeper voice, occasionally the development of baldness, development of a masculine distribution of hair on the body and on the pubis, growth of the clitoris such that it may resemble a penis, atrophy of the breasts, infrequent or absent menstruation, and increased muscularity that produces a male-like physique. In prepubescent males the syndrome causes the same characteristics as in females, plus rapid development of the male sexual organs and emergence of sexual desires.

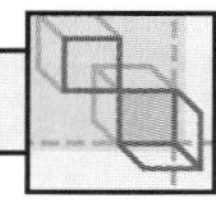

STUDY OUTLINE

Introduction (p. 659)

1. The nervous system controls homeostasis through nerve impulses; the endocrine system uses hormones.
2. The nervous system causes muscles to contract and glands to secrete; the endocrine system affects virtually all body tissues.

Endocrine Glands Defined (p. 659)

1. Exocrine (sudoriferous, sebaceous, and digestive) glands secrete their products through ducts into body cavities or onto body surfaces.
2. Endocrine glands secrete hormones into the blood.
3. The endocrine system consists of endocrine glands and several organs that contain endocrine tissue (see Figure 22.1 on page 659).
4. Hormones regulate the internal environment, metabolism, and energy balance.
5. They also help regulate muscular contraction, glandular secretion, and certain immune responses.
6. Hormones affect growth, development, and reproduction.

Hormones (p. 659)

1. The amount of hormone released is determined by the body's need for the hormone.
2. Cells that respond to the effects of hormones are called target cells.
3. The combination of hormone and receptor activates a chain of events in a target cell that produce the physiological effects of the hormone.

Hypothalamus and Pituitary Gland (p. 662)

1. The hypothalamus is the major integrating link between the nervous and endocrine systems.
2. The hypothalamus and pituitary gland regulate virtually all aspects of growth, development, metabolism, and they also affect other body activities.
3. The pituitary gland is located in the sella turcica and is divided into the anterior pituitary gland (glandular portion), the posterior pituitary gland (nervous portion), and pars intermedia (avascular zone in between).
4. Hormones of the anterior pituitary gland are controlled by releasing or inhibiting hormones produced by the hypothalamus.
5. The blood supply to the anterior pituitary gland is from the superior hypophyseal arteries. It carries releasing and inhibiting hormones from the hypothalamus.
6. Histologically, the anterior pituitary consists of somatotrophs that produce human growth hormone (hGH); lactotrophs that produce prolactin (PRL); corticotrophs that secrete adrenocorticotropic hormone (ACTH) and melanocyte-stimulating hormone (MSH); thyrotrophs that secrete thyroid-stimulating hormone (TSH); and

gonadotrophs that synthesize follicle-stimulating hormone (FSH) and luteinizing hormone (LH).

7. hGH stimulates body growth. TSH regulates thyroid gland activities. FSH and LH both regulate the activities of the ovaries and testes. PRL helps initiate milk secretion. MSH increases skin pigmentation. ACTH regulates the activities of the adrenal cortex.
8. The neural connection between the hypothalamus and posterior pituitary gland is via the supraopticohypophyseal tract.
9. Hormones made by the hypothalamus and stored in the posterior pituitary gland are oxytocin (OT), which stimulates contraction of uterus and ejection of milk, and antidiuretic hormone (ADH), which stimulates water reabsorption by the kidneys and arteriole constriction.

Thyroid Gland (p. 666)

1. The thyroid gland is located inferior to the larynx.
2. Histologically, the thyroid gland consists of thyroid follicles composed of follicular cells, which secrete the thyroid hormones thyroxine (T_4) and triiodothyronine (T_3), and parafollicular cells, which secrete calcitonin (CT).
3. Thyroid hormones regulate the rate of metabolism, growth and development, and the reactivity of the nervous system.
4. Calcitonin (CT) lowers the blood level of calcium.

Parathyroid Glands (p. 667)

1. The parathyroid glands are embedded on the posterior surfaces of the lateral lobes of the thyroid gland.
2. The parathyroids consist of principal and oxyphil cells.
3. Parathyroid hormone (PTH) increases blood calcium level and decreases blood phosphate level.

Adrenal Glands (p. 669)

1. The adrenal glands are located superior to the kidneys. They consist of an outer adrenal cortex and inner adrenal medulla.
2. Histologically, the adrenal cortex is divided into a zona glomerulosa, zona fasciculata, and zona reticularis; the adrenal medulla consists of chromaffin cells and large blood vessels.
3. Cortical secretions are mineralocorticoids, glucocorticoids, and gonadocorticoids.
4. Mineralocorticoids (for example, aldosterone) increase sodium and water reabsorption and decrease potassium reabsorption.
5. Glucocorticoids (for example, cortisol) promote normal protein, glucose, and lipid metabolism, help resist stress, and serve as anti-inflammatory substances.
6. Androgens secreted by the adrenal cortex usually have minimal effects.
7. Medullary secretions are epinephrine and norepinephrine (NE), which produce effects similar to sympathetic responses. They are released during stress.

Pancreas (p. 671)

1. The pancreas is posterior and slightly inferior to the stomach.
2. Histologically, it consists of pancreatic islets, or islets of Langerhans (endocrine cells), and clusters of enzyme-producing cells (acini). The four types of cells in the endocrine portion are alpha, beta, delta, and F cells.
3. Alpha cells secrete glucagon, beta cells secrete insulin, delta cells secrete somatostatin, and F cells secrete pancreatic polypeptide.
4. Glucagon increases blood sugar level.
5. Insulin decreases blood sugar level.

Ovaries and Testes (p. 673)

1. The ovaries are located in the pelvic cavity and produce sex hormones that function in the development and maintenance of female sexual characteristics, the reproductive cycle, pregnancy, lactation, and normal reproductive functions.
2. The testes lie inside the scrotum and produce sex hormones that function in the development and maintenance of male sexual characteristics and normal reproductive functions.

Pineal Gland (p. 674)

1. The pineal gland is attached to the roof of the third ventricle.
2. Histologically, it consists of secretory cells called pinealocytes, neuroglial cells, and scattered postganglionic sympathetic fibers.
3. The pineal gland secretes melatonin, which contributes to setting the body's biological clock (controlled in the suprachiasmatic nucleus). During sleep, plasma levels of melatonin increase tenfold and then decline to a low level again before awakening.

Thymus (p. 674)

1. The thymus secretes several hormones related to immunity.
2. Thymosin, thymic humoral factor (THF), thymic factor (TF), and thymopoietin promote the maturation of T cells.

Other Endocrine Tissues (p. 675)

1. The gastrointestinal tract synthesizes several hormones, including gastrin, gastric inhibitory peptide (GIP), secretin, and cholecystokinin (CCK).
2. The placenta produces human chorionic gonadotropin (hCG), estrogens, progesterone, and human chorionic somatomammotropin (hCS).
3. The kidneys release erythropoietin.
4. The skin begins the synthesis of vitamin D.
5. The atria of the heart produce atrial natriuretic peptide (ANP).
6. Adipose tissue produces leptin.

Developmental Anatomy of the Endocrine System (p. 675)

1. The development of the endocrine system is not as localized as in other systems.
2. The pituitary gland, adrenal medullae, and pineal gland develop from ectoderm; the adrenal cortex develops from mesoderm; and the thyroid gland, parathyroid glands, pancreas, and thymus develop from endoderm.

Aging and the Endocrine System (p. 676)

1. Although endocrine glands shrink with age, their performance may not necessarily be compromised.
2. Whereas the pituitary gland produces less human growth hormone, it produces more gonadotropins and thyroid-stimulating hormone.

SELF-QUIZ QUESTIONS

Choose the one best answer to these questions.

1. A generalized anti-inflammatory effect is most closely associated with (a) glucocorticoids, (b) mineralocorticoids, (c) parathyroid hormone (PTH) (d) insulin, (e) melatonin.

2. A chemical produced by the hypothalamus that causes the anterior pituitary gland to secrete a hormone is called a (a) gonadotropic hormone, (b) tropic hormone, (c) releasing hormone, (d) target hormone, (e) neurotransmitter.

3. A tumor of the beta cells of the pancreatic islets (islets of Langerhans) would probably affect the body's ability to (a) lower blood sugar level, (b) raise blood sugar level, (c) lower blood calcium level, (d) raise blood calcium level, (e) regulate metabolism.

4. Which hormone is involved in milk production? (a) melanocyte-stimulating hormone (MSH), (b) follicle-stimulating hormone (FSH), (c) prolactin (PRL), (d) glucagon, (e) parathyroid hormone (PTH)

5. What the stomach, pancreas, testes, and ovaries have in common is that they (a) are influenced by hormones from the anterior pituitary gland, (b) have tissues that are derived from embryological ectoderm, (c) form hormones that influence secondary sex characteristics, (d) receive their blood supply from the superior mesenteric artery, (e) are considered to be both exocrine and endocrine

6. Which one of these glands is called the "emergency gland" and helps the body meet sudden stress? (a) pituitary gland, (b) pancreas, (c) thyroid gland, (d) thymus, (e) adrenal (suprarenal) glands

Complete the following.

7. Place numbers in the blanks to indicate the correct order the vessels that supply blood to the anterior pituitary gland. (a) hypophyseal portal veins: ________, (b) primary plexus: ________, (c) superior hypophyseal arteries: ________, (d) secondary plexus: ________.

8. Place numbers in the blanks to indicate the correct order of items involved in the route of a hormone. (Note: Not all items apply to this route.) (a) blood capillary: ________, (b) target cells: ________, (c) duct: ________, (d) outer surface of the body or lumen of an organ: ________, (e) interstitial fluid: ________, (f) secretory cell: ________.

9. Thyroid stimulating hormone (TSH) controls the secretion of hormones by follicular cells of the thyroid gland. Therefore TSH is a ________ hormone and the follicular cells are the ________ cells.

10. Glucagon, produced by ________ cells of the pancreatic (islets of Langerhans) causes blood sugar level to ________.

11. A stalk called the ________ attaches the pituitary gland to the hypothalamus. The pituitary gland lies in the ________ of the sphenoid bone.

12. Another term for anterior pituitary gland is ________.

13. The ________ cells of the thyroid gland and the ________ cells of the parathyroid glands secrete hormones that help to regulate the level of calcium in the blood.

Are the following statements true or false?

14. The nervous system produces its effects quickly; the endocrine system can act either quickly or slowly.

15. The secretory portion of the posterior pituitary gland is the axons of neurosecretory cells; the secretory portion of the anterior pituitary gland is glandular epithelium.

16. The middle region of hormone-secreting cells of the adrenal cortex is called the zona reticularis.

17. The gland that assumes the principal role in providing resistance to stress is the adrenal (suprarenal) gland.

18. Chromaffin cells are the principal secreting cells of the pineal gland.

19. The outer region of the adrenal (suprarenal) gland is called the adrenal cortex.

20. Match the following terms with their definitions:

____ (a) adrenal cortex
____ (b) adrenal medulla
____ (c) anterior pituitary gland
____ (d) pineal gland
____ (e) parathyroid glands
____ (f) thymus
____ (g) pancreas

(1) develops from the foregut area that later becomes part of the small intestine
(2) derived from the roof of the stomodeum (mouth) called the hypophyseal (Rathke's) pouch
(3) originates from the neural crest, which also produces sympathetic ganglia
(4) derived from tissue from the same region that forms gonads
(5) arise from the third and fourth pharyngeal pouches (two answers)
(6) arises from the ectoderm of the diencephalon

21. Match the following hormone secreting cells to the hormones they secrete:

____ (a) insulin
____ (b) glucagon
____ (c) calcitonin
____ (d) TSH
____ (e) hGH
____ (f) testosterone
____ (g) ACTH
____ (h) progesterone
____ (i) thyroxine and triiodothyronine
____ (j) FSH and LH

(1) corticotrophs
(2) somatotrophs
(3) thyrotrophs
(4) gonadotrophs
(5) beta cells of pancreatic islets (islets of Langerhans)
(6) alpha cells of pancreatic (islets of Langerhans)
(7) follicular cells of the thyroid gland
(8) parafollicular cells of the thyroid gland
(9) interstitial cells of the testis
(10) corpus luteum of the ovary

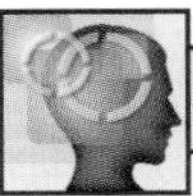

CRITICAL THINKING QUESTIONS

1. You've won a trip to beautiful Tropicanaland, a 12 hour time difference from where you live. Your coworkers gave you a bottle of melatonin, a bottle of melanocyte stimulating hormone, and a very bright flashlight as a bon voyage present. You'll be arriving at 8 P.M. Tropicanaland time, which is 8 A.M. your time. How can you adjust to Tropicanaland time most quickly?
HINT *One of these gifts will be a lot more useful than the others.*

2. As Raj walked into the Human Anatomy class (10 minutes late), he was handed an exam that he had completely forgot about when he decided to stay at the beach for an extra week. Predict how his body will react to this situation.
HINT *The stress he's feeling just wiped out the relaxed mood from the week at the beach.*

3. Beatrice had a large bag of double chunk chocolate chip cookies and a 16 oz cola for lunch- no sandwich 'cause she's on a diet. Explain how her endocrine system will respond to this lunch.
HINT *This meal turns the "food pyramid" into the "sugar cube"!*

4. For several years, military pilots were given radiation treatments in their nasal cavity to reduce sinus problems that interfered with flying. Years later, some of these former pilots began to exhibit problems with their pituitary gland hormones. Can you propose an explanation for this relationship?
HINT *Refer to the structure of the skull in chapter 6.*

5. A patient was found to have markedly elevated blood sugar. Tests showed that his insulin level was actually a bit elevated. How can someone with elevated insulin have elevated blood sugar?
HINT *Hormones can't do their job alone.*

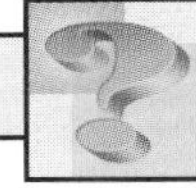

ANSWERS TO FIGURE QUESTIONS

22.1 Secretions of endocrine glands diffuse into interstitial fluid and then into the blood; exocrine secretions flow into ducts that lead into body cavities or to the body surface.

22.2 The hypophyseal portal veins carry blood from the median eminence of the hypothalamus, where hypothalamic releasing and inhibiting hormones are secreted, to the anterior pituitary gland, where these hormones act.

22.3 Functionally, both the hypothalamohypophyseal tract and the hypophyseal portal veins carry hypothalamic hormones to the pituitary gland. Structurally, the tract is composed of axons of neurons that extend from the hypothalamus to the posterior pituitary gland, whereas the portal veins are blood vessels that extend to the anterior pituitary gland.

22.4 Follicular cells secrete T_3 and T_4, also known as thyroid hormones. Parafollicular cells secrete calcitonin.

22.5 Parafollicular cells of the thyroid gland secrete calcitonin; principal cells of the parathyroid gland secrete PTH.

22.6 The adrenal glands are superior to the kidneys in the retroperitoneal space.

22.7 The pancreas is both an endocrine and an exocrine gland.

22.8 The pituitary gland and the adrenal glands both include tissues having two different embryological origins.

22.9 In Graves' disease, antibodies are produced that mimic the action of TSH.

23

THE RESPIRATORY SYSTEM

CONTENTS AT A GLANCE

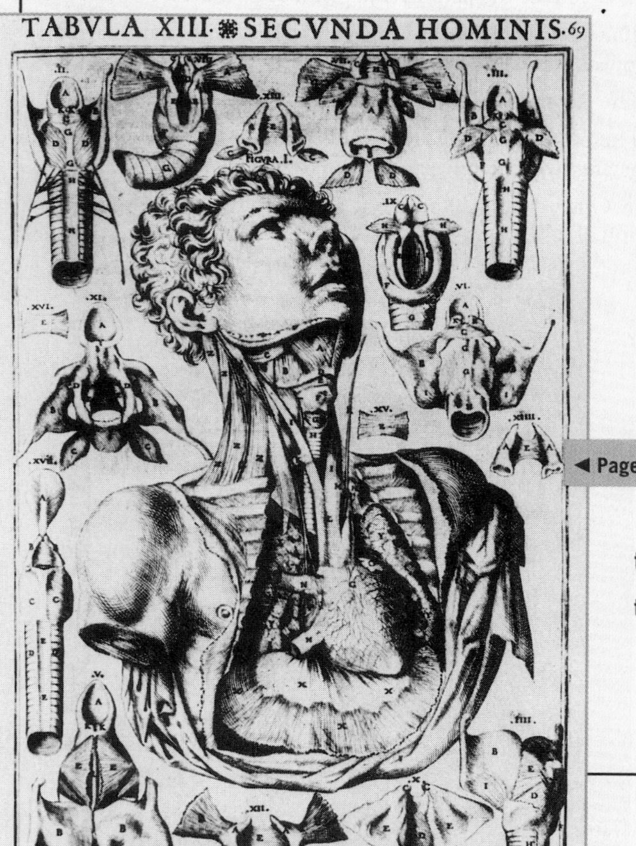

◄ Page 692

Are you able to identify the organs that are depicted in this image?

Page 695 ►

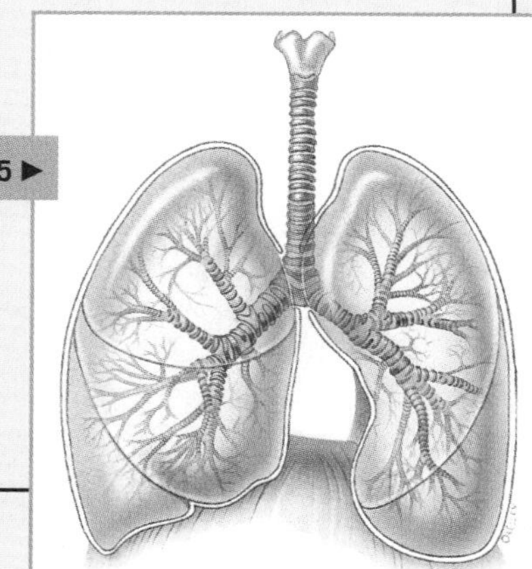

INTRODUCTION

Cells continually use oxygen (O_2) for the metabolic reactions that release energy from nutrient molecules and produce ATP. At the same time, these reactions release carbon dioxide (CO_2). Because an excessive amount of CO_2 produces acidity that can be toxic to cells, the excess CO_2 must be eliminated quickly and efficiently. The two systems that cooperate to supply O_2 and eliminate CO_2 are the cardiovascular and respiratory systems. The respiratory system provides for gas exchange—intake of O_2 and elimination of CO_2—while the blood of the cardiovascular system transports the gases between the lungs and body cells. Failure of either system causes rapid death of cells from oxygen starvation and the buildup of waste products. In addition to functioning in gas exchange, the respiratory system also regulates blood pH, contains receptors for the sense of smell, filters inspired air, produces sounds, and rids the body of some water vapor and heat in exhaled air.

The branch of medicine that deals with the diagnosis and treatment of diseases of the ears, nose, and throat is called **otorhinolaryngology** (ō′-tō-rī′-nō-lar′-in-GOL-ō-jē; *oto-* = ear; *rhino-* = nose; *laryngo-* = voice box; *-ology* = study of). A **pulmonologist** is a specialist in the diagnosis and treatment of diseases of the lungs.

RESPIRATORY SYSTEM ANATOMY

Objectives

- Describe the anatomy and histology of the nose, pharynx, larynx, trachea, bronchi, and lungs.
- Identify the functions of each respiratory system structure.

The **respiratory system** consists of the nose, pharynx (throat), larynx (voice box), trachea (windpipe), bronchi, and lungs (Figure 23.1). Structurally, the respiratory system consists of two portions: (1) the **upper respiratory system,** which includes the nose, pharynx, and associated structures, and (2) the **lower respiratory system,** which includes the larynx, trachea, bronchi, and lungs. Functionally, the respiratory system also consists of two portions: (1) the **conducting portion,** which consists of a series of interconnecting cavities and tubes both outside and within the lungs—the nose, pharynx, larynx, trachea, bronchi, bronchioles, and terminal bronchioles—that filter, warm, and moisten air and conduct it into the lungs, and (2) the **respiratory portion,** which consists of tissues within the lungs where gas exchange occurs—the respiratory bronchioles, alveolar ducts, alveolar sacs, and alveoli, the main sites of gas exchange between air and blood. The volume of the conducting portion in an adult is about 150 mL; that of the respiratory portion is 5–6 liters.

Nose

The **nose** has external and internal portions. The external portion consists of a supporting framework of bone and hyaline cartilage covered with muscle and skin and lined by a mucous membrane. The bony framework of the nose is formed by the frontal bone, nasal bones, and maxillae (Figure 23.2a on page 687). The cartilaginous framework consists of the **septal cartilage,** which forms the anterior portion of the nasal septum; the **lateral nasal cartilages** inferior to the nasal bones; and the **alar cartilages,** which form a portion of the walls of the nostrils. Because it has a framework of pliable hyaline cartilage, the cartilaginous framework of the external nose is somewhat flexible. On the undersurface of the external nose are two openings called the **external nares** (NA-rēz; singular is **naris**), or **nostrils.** The surface anatomy of the nose is shown in Figure 11.5 on page 350. The interior structures of the external portion of the nose have three functions: (1) warming, moistening, and filtering incoming air; (2) detecting olfactory stimuli; and (3) modifying speech vibrations as they pass through the large, hollow resonating chambers.

The internal portion of the nose is a large cavity in the anterior aspect of the skull that lies inferior to the nasal bone and superior to the mouth; it also includes muscle and mucous membrane. Anteriorly, the internal nose merges with the external nose, and posteriorly it communicates with the pharynx through two openings called the **internal nares,** or **choanae** (kō-Ā-nē) (Figure 23.2b). Ducts from the paranasal sinuses (frontal, sphenoidal, maxillary, and ethmoidal sinuses) and the nasolacrimal ducts also open into the internal nose. The lateral walls of the internal nose are formed by the ethmoid, maxillae, lacrimal, palatine, and inferior nasal conchae bones (see Figure 6.14 on page 153); the ethmoid also forms the roof. The floor of the internal nose is formed mostly by the palatine bones and palatine processes of the maxillae, which together constitute the hard palate.

The space inside the internal nose is called the **nasal cavity.** The anterior portion of the nasal cavity, just inside the nostrils, is called the **vestibule** and is surrounded by cartilage; the superior part of the nasal cavity is surrounded by bone. The nasal cavity is divided into right and left sides by a vertical partition called the **nasal septum.** The anterior portion of the septum consists primarily of hyaline cartilage; the remainder is formed by the vomer, perpendicular plate of the ethmoid, maxillae, and palatine bones (see Figure 6.14 on page 153).

When air enters the nostrils, it passes first through the vestibule, which is lined by skin containing coarse hairs that filter out large dust particles. Three shelves formed by projections of the superior, middle, and inferior nasal conchae extend out of each lateral wall of the cavity. The conchae, almost reaching the septum, subdivide each side of the nasal cavity into a series of groovelike passageways—the **superior, middle,** and **inferior meatuses** (mē-Ā-tes-ez; = openings or passages). Mucous membrane lines the cavity and its shelves. The arrangement of conchae and meatuses increases surface area in the internal nose and prevents dehydration by acting as a baffle that traps water droplets during exhalation.

Figure 23.1 / Structures of the respiratory system.

The upper respiratory system includes the nose, pharynx, and associated structures; the lower respiratory system includes the larynx, trachea, bronchi, and lungs.

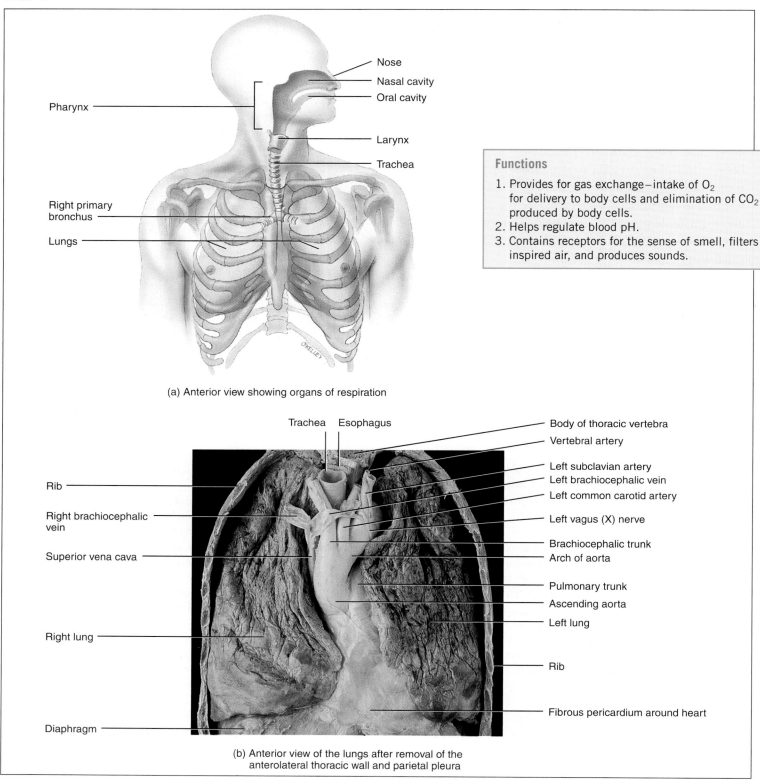

Nose
Nasal cavity
Oral cavity
Pharynx
Larynx
Trachea
Right primary bronchus
Lungs

Functions

1. Provides for gas exchange–intake of O_2 for delivery to body cells and elimination of CO_2 produced by body cells.
2. Helps regulate blood pH.
3. Contains receptors for the sense of smell, filters inspired air, and produces sounds.

O'KELLEY

(a) Anterior view showing organs of respiration

Trachea
Esophagus
Body of thoracic vertebra
Vertebral artery
Left subclavian artery
Left brachiocephalic vein
Left common carotid artery
Left vagus (X) nerve
Brachiocephalic trunk
Arch of aorta
Pulmonary trunk
Ascending aorta
Left lung
Rib
Fibrous pericardium around heart
Rib
Right brachiocephalic vein
Superior vena cava
Right lung
Diaphragm

(b) Anterior view of the lungs after removal of the anterolateral thoracic wall and parietal pleura

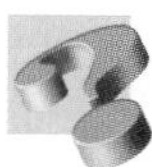

Which structures are part of the conducting portion of the respiratory system?

Figure 23.2 / Respiratory structures in the head and neck. (See Tortora, *A Photographic Atlas of the Human Body,* Figure 11.2.)

As air passes through the nose, it is warmed, filtered, and moistened.

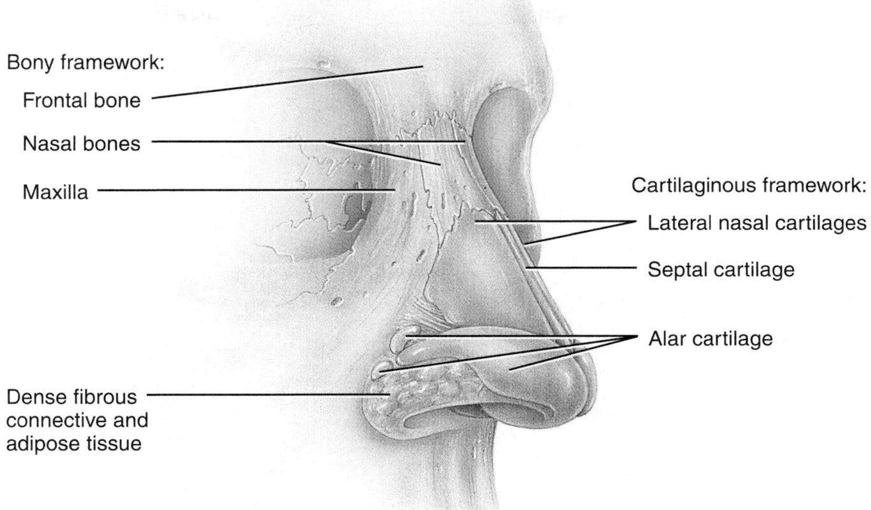

(a) Anterolateral view of external portion of nose showing cartilaginous and bony framework

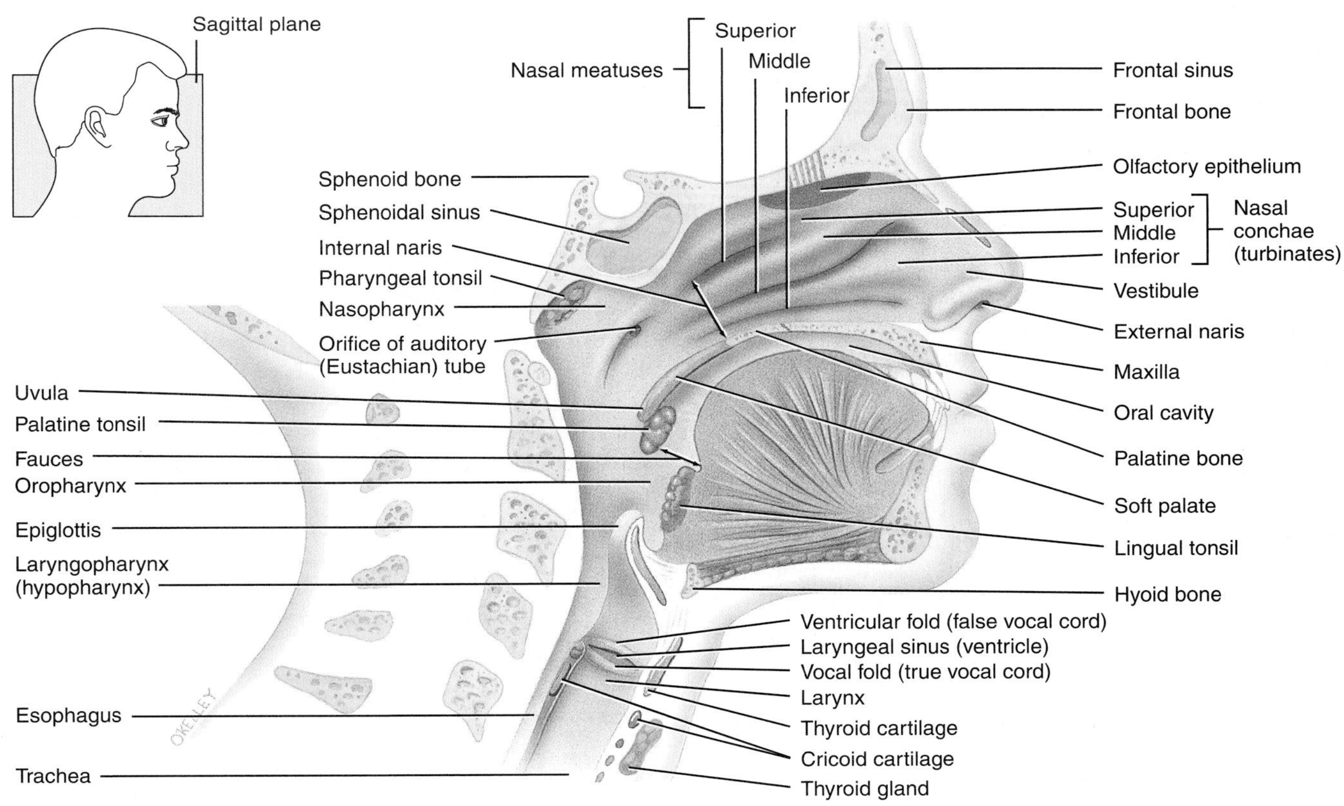

(b) Sagittal section of the left side of the head and neck showing the location of respiratory structures

(continues)

Figure 23.2 (continued)

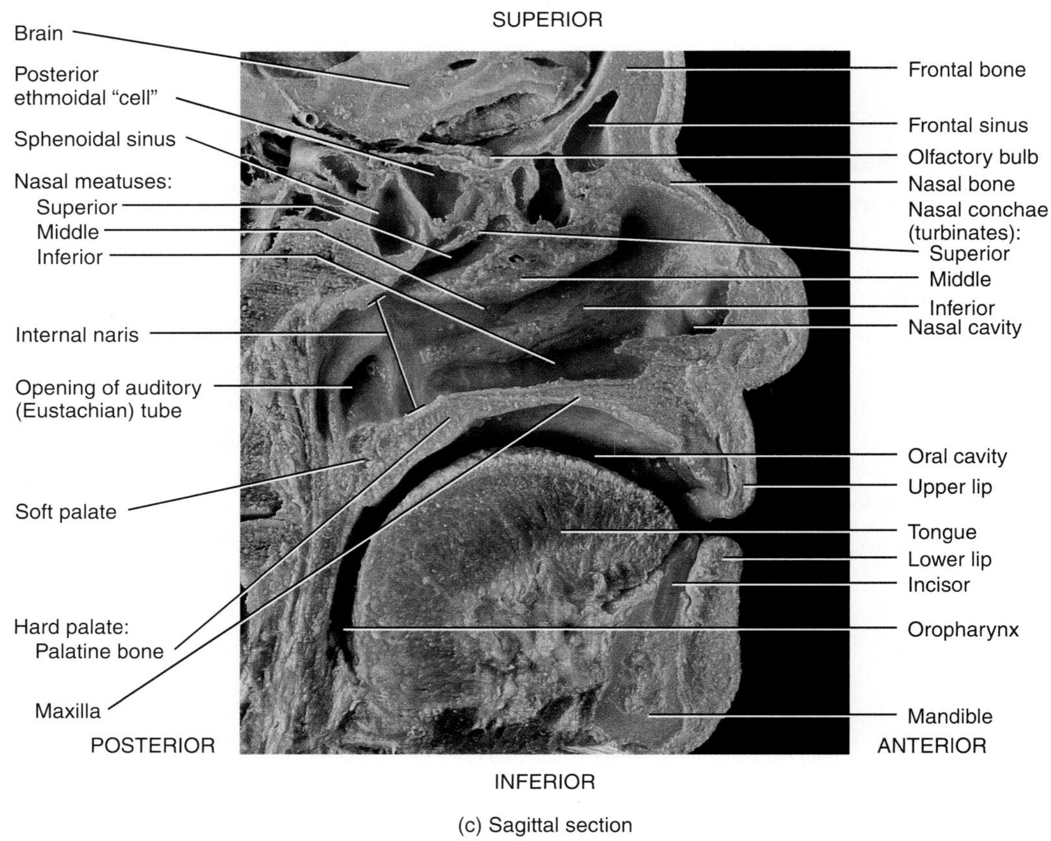

(c) Sagittal section

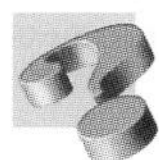

What is the path taken by air molecules into and through the nose?

The olfactory receptors lie in the membrane lining the superior nasal conchae and adjacent septum. This region is called the **olfactory epithelium.** Inferior to the olfactory epithelium, the mucous membrane contains capillaries and pseudostratified ciliated columnar epithelium with many goblet cells. As inspired air whirls around the conchae and meatuses, it is warmed by blood in the capillaries. Mucus secreted by the goblet cells moistens the air and traps dust particles. Drainage from the nasolacrimal ducts and perhaps secretions from the paranasal sinuses also help moisten the air. The cilia move the mucus and trapped dust particles toward the pharynx, at which point they can be swallowed or spit out, thus removing particles from the respiratory tract.

The arterial supply to the nasal cavity is principally from the sphenopalatine branch of the maxillary artery. The remainder is supplied by the ophthalmic artery. The veins of the nasal cavity drain into the sphenopalatine vein, the facial vein, and the ophthalmic vein.

The nerve supply of the nasal cavity consists of olfactory cells in the olfactory epithelium associated with the olfactory (I) nerve and the nerves of general sensation. These nerves are branches of the ophthalmic and maxillary divisions of the trigeminal (V) nerve.

Rhinoplasty (RĪ-nō-plas′-tē; *-plasty* = to mold or to shape), commonly called a "nose job," is a surgical procedure in which the structure of the external nose is altered. Although it is often done for cosmetic reasons, it is sometimes performed to repair a fractured nose or a deviated nasal septum. In the procedure, both a local and general anesthetic are given, and with instruments inserted through the nostrils, the nasal cartilage is reshaped, and the nasal bones are fractured and repositioned, to achieve the desired shape. An internal packing and splint are inserted to keep the nose in the desired position while it heals. ■

Pharynx

The **pharynx** (FAIR-inks), or throat, is a funnel-shaped tube about 13 cm (5 in.) long that starts at the internal nares and extends to the level of the cricoid cartilage, the most inferior cartilage of the larynx (voice box) (Figure 23.3). The pharynx lies just posterior to the nasal and oral cavities, superior to the larynx, and just anterior to the cervical vertebrae. Its wall is composed of skeletal muscles and is lined with a mucous membrane.

Figure 23.3 / Pharynx. (See Tortora, *A Photographic Atlas of the Human Body,* Figure 11.4.)

The three subdivisions of the pharynx are the (1) nasopharynx, (2) oropharynx, and (3) laryngopharynx.

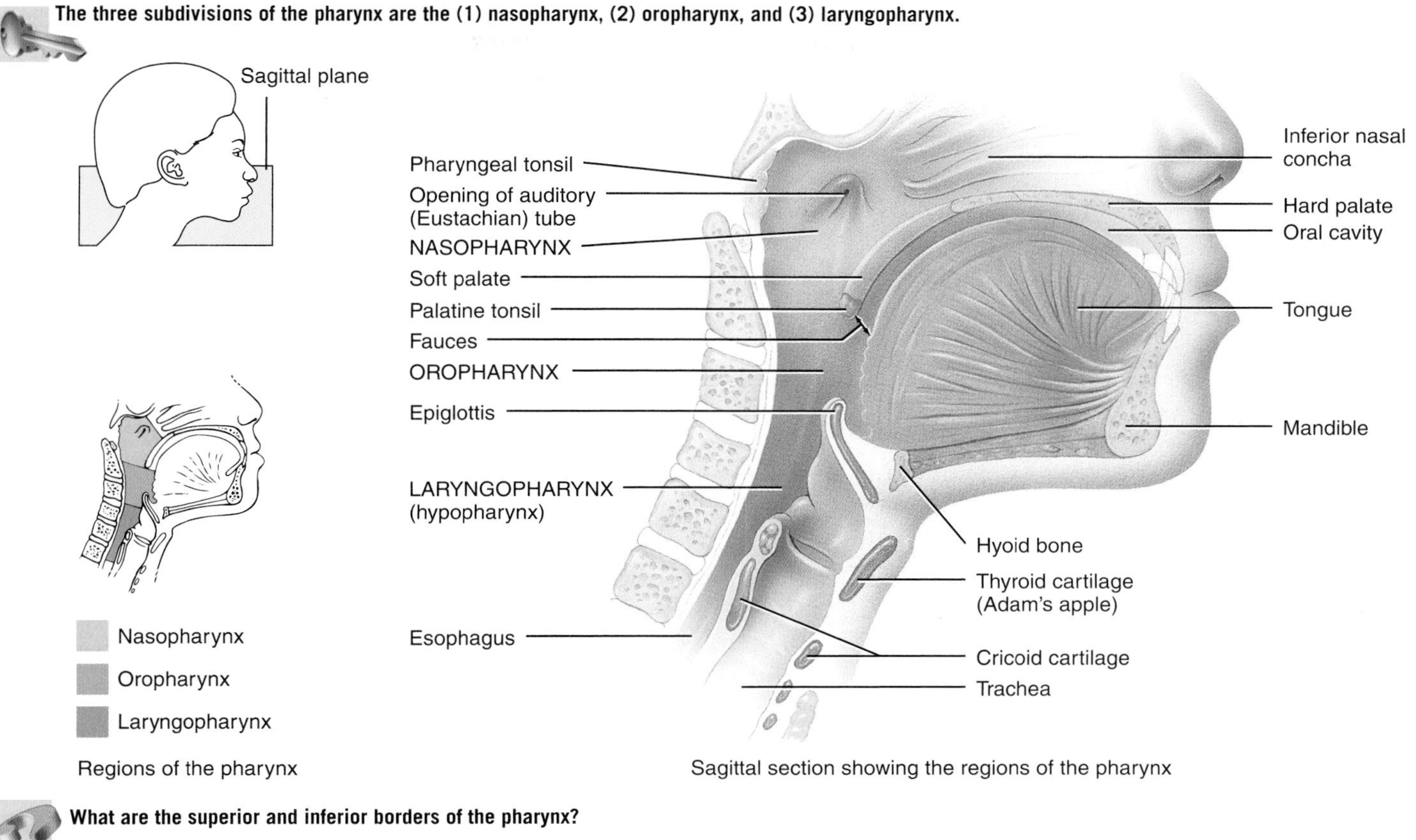

What are the superior and inferior borders of the pharynx?

The pharynx functions as a passageway for air and food, provides a resonating chamber for speech sounds, and houses the tonsils, which participate in immunological reactions against foreign invaders.

The pharynx is divided into three anatomical regions: (1) nasopharynx, (2) oropharynx, (3) laryngopharynx. (See the lower orientation diagram in Figure 23.3.) The muscles of the entire pharynx are arranged in two layers, an outer circular layer and an inner longitudinal layer.

The superior portion of the pharynx, called the **nasopharynx,** lies posterior to the nasal cavity and extends to the plane of the soft palate. There are five openings in its wall: two internal nares, two openings that lead into the auditory (Eustachian) tubes, and the opening into the oropharynx. The posterior wall also contains the **pharyngeal tonsil.** Through the internal nares, the nasopharynx receives air from the nasal cavity and receives packages of dust-laden mucus. The nasopharynx is lined with pseudostratified ciliated columnar epithelium, and the cilia move the mucus down toward the most inferior part of the pharynx. The nasopharynx also exchanges small amounts of air with the auditory (Eustachian) tubes to equalize air pressure between the pharynx and the middle ear.

The intermediate portion of the pharynx, the **oropharynx,** lies posterior to the oral cavity and extends from the soft palate inferiorly to the level of the hyoid bone. It has only one opening, the **fauces** (FAW-sēz; = throat), the opening from the mouth. This portion of the pharynx has both respiratory and digestive functions because it is a common passageway for air, food, and drink. Because the oropharynx is subject to abrasion by food particles, it is lined with nonkeratinized stratified squamous epithelium. Two pairs of tonsils, the **palatine** and **lingual tonsils,** are found in the oropharynx.

The inferior portion of the pharynx, the **laryngopharynx** (la-rin′-gō-FAIR-inks), or **hypopharynx,** begins at the level of the hyoid bone and opens into the esophagus (food tube) posteriorly and the larynx (voice box) anteriorly. Like the oropharynx, the laryngopharynx is both a respiratory as well as a digestive pathway and is lined by nonkeratinized stratified squamous epithelium.

The arterial supply of the pharynx includes the ascending pharyngeal artery, the ascending palatine branch of the facial artery, the descending palatine and pharyngeal branches of the maxillary artery, and the muscular branches of the superior thyroid artery. The veins of the pharynx drain into the pterygoid plexus and the internal jugular veins.

Most of the muscles of the pharynx are innervated by the pharyngeal plexus. This plexus is formed by the pharyngeal branches of the glossopharyngeal (IX), vagus (X), and cranial portion of the accessory (XI) nerves and the superior cervical sympathetic ganglion.

Larynx

The **larynx** (LAIR-inks), or voice box, is a short passageway that connects the laryngopharynx with the trachea. It lies in the midline of the neck anterior to the fourth through sixth cervical vertebrae (C4–C6).

The wall of the larynx is composed of nine pieces of cartilage (Figure 23.4). Three occur singly (thyroid cartilage, epiglottic cartilage, and cricoid cartilage), and three occur in pairs (arytenoid, cuneiform, and corniculate cartilages). Of the paired cartilages, the arytenoid cartilages are the most important

Figure 23.4 / Larynx. (See Tortora, *A Photographic Atlas of the Human Body,* Figures 11.5 and 11.6.)

The larynx is composed of nine pieces of cartilage.

Larynx
Thyroid gland

Epiglottis
Hyoid bone
Thyrohyoid membrane
Corniculate cartilage
Thyroid cartilage (Adam's apple)
Arytenoid cartilage
Cricothyroid ligament
Cricoid cartilage
Cricotracheal ligament
Thyroid gland
Parathyroid glands (4)
Trachea
Tracheal cartilage

(a) Anterior view

(b) Posterior view

Sagittal plane

Epiglottis
Thyrohyoid membrane
Cuneiform cartilage
Corniculate cartilage
Arytenoid cartilage
Cricoid cartilage
Tracheal cartilage
Hyoid bone
Thyrohyoid membrane
Fat body
Ventricular fold (false vocal cord)
Thyroid cartilage
Vocal fold (true vocal cord)
Cricothyroid ligament
Cricotracheal ligament

(c) Sagittal section

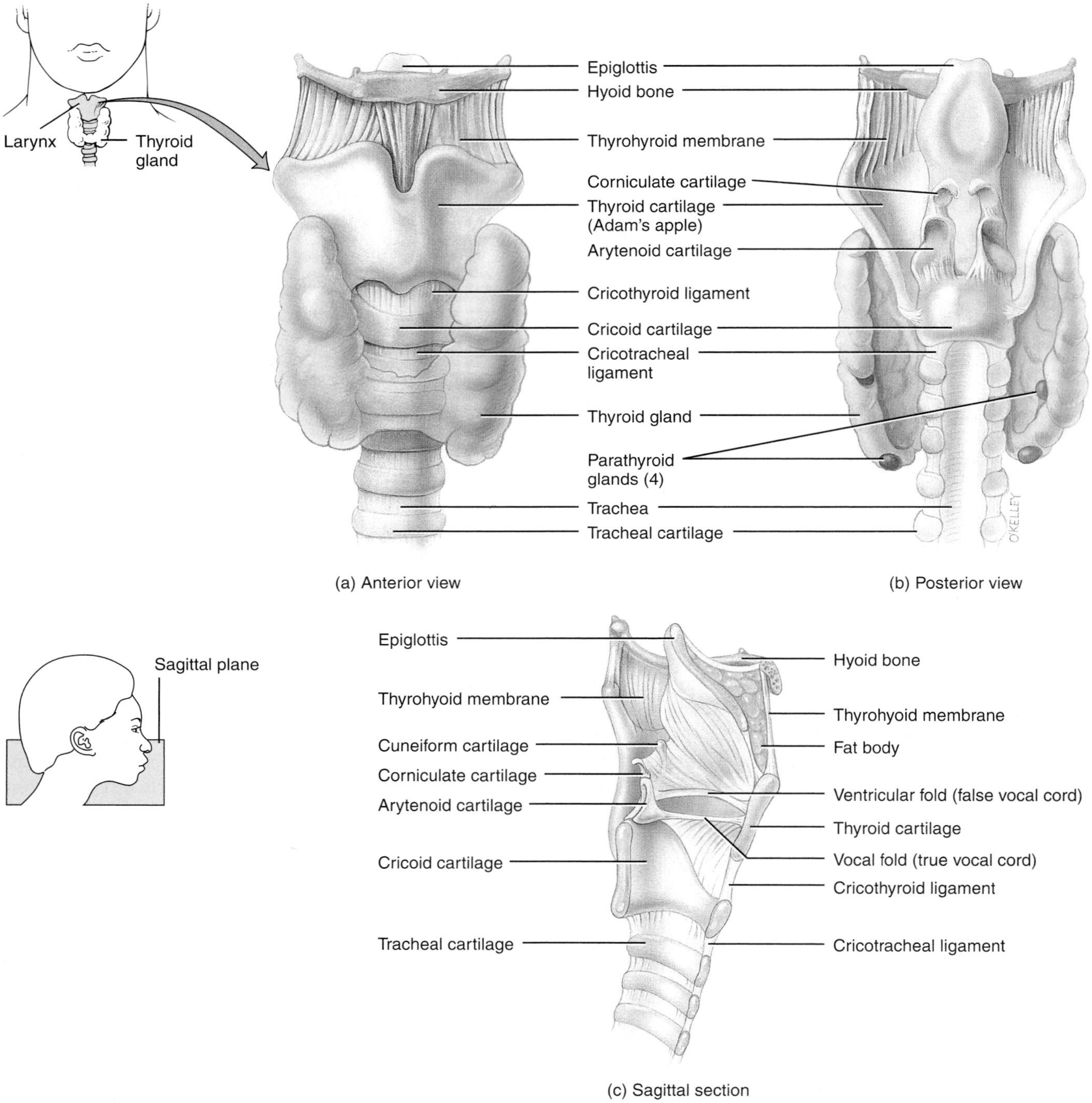

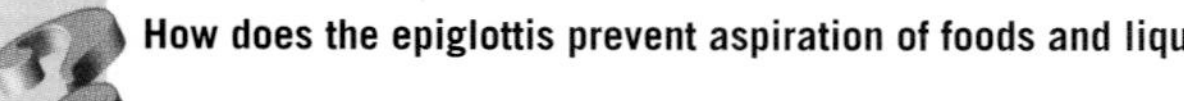

How does the epiglottis prevent aspiration of foods and liquids?

because they influence the positions and tensions of the vocal folds (true vocal cords). Whereas the extrinsic muscles of the larynx connect the cartilages to other structures in the throat, the intrinsic muscles connect the cartilages to each other (see Figure 10.10 on page 281).

The **thyroid cartilage (Adam's apple)** consists of two fused plates of hyaline cartilage that form the anterior wall of the larynx and give it a triangular shape. It is usually larger in males than in females due to the influence of male sex hormones on its growth during puberty. The ligament that connects the thyroid cartilage to the hyoid bone is called the **thyrohyoid membrane.**

The **epiglottis** (*epi-* = over; *glottis* = tongue) is a large, leaf-shaped piece of elastic cartilage that is covered with epithelium (see Figure 23.3). The "stem" of the epiglottis is attached to the anterior rim of the thyroid cartilage, but the "leaf" portion is unattached and free to move up and down like a trap door. During swallowing, the pharynx and larynx rise. Elevation of the pharynx widens it to receive food or drink; elevation of the larynx causes the free edge of the epiglottis to move down and form a lid over the glottis, closing it off. The **glottis** consists of a pair of folds of mucous membrane, the vocal folds (true vocal cords) in the larynx, and the space between them called the **rima glottidis** (RĪ-ma GLOT-ti-dis). The closing of the larynx in this way during swallowing routes liquids and foods into the esophagus and keeps them out of the larynx and airways inferior to it. When small particles of dust, smoke, food, or liquids pass into the larynx, a cough reflex occurs, usually expelling the material.

The **cricoid cartilage** (KRĪ-koyd; = ringlike) is a ring of hyaline cartilage that forms the inferior wall of the larynx. It is attached to the first ring of cartilage of the trachea by the **cricotracheal ligament.** The thyroid cartilage is connected to the cricoid cartilage by the **cricothyroid ligament.** The cricoid cartilage is the landmark for making an emergency airway (a tracheostomy; see page 694).

The paired **arytenoid cartilages** (ar′-i-TĒ-noyd; = ladle-like) are triangular pieces of mostly hyaline cartilage located at the posterior, superior border of the cricoid cartilage. They attach to the vocal folds and intrinsic pharyngeal muscles. Supported by the arytenoid cartilages, the intrinsic pharyngeal muscles contract and thus move the vocal folds.

The paired **corniculate cartilages** (kor-NIK-yū-lāt; = shaped like a small horn), which are horn-shaped pieces of elastic cartilage, are located at the apex of each arytenoid cartilage. The paired **cuneiform cartilages** (kyū-NĒ-i-form; = wedge-shaped), which are club-shaped elastic cartilages anterior to the corniculate cartilages, support the vocal folds and lateral aspects of the epiglottis.

The lining of the larynx superior to the vocal folds is nonkeratinized stratified squamous epithelium. The lining of the larynx inferior to the vocal folds is pseudostratified ciliated columnar epithelium consisting of ciliated columnar cells, goblet cells, and basal cells. Its mucus helps trap dust not removed in the upper passages. Whereas the cilia in the upper respiratory tract move mucus and trapped particles *down* toward the pharynx, the cilia in the lower respiratory tract move them *up* toward the pharynx.

The mucous membrane of the larynx forms two pairs of folds (Figure 23.4c): a superior pair called the **ventricular folds (false vocal cords)** and an inferior pair called simply the **vocal folds (true vocal cords).** The space between the ventricular folds is known as the **rima vestibuli.** The **laryngeal sinus (ventricle)** is a lateral expansion of the middle portion of the laryngeal cavity between the ventricular folds above and the vocal folds below (see Figure 23.2b).

When the ventricular folds are brought together, they function in holding the breath against pressure in the thoracic cavity, such as might occur when a person strains to lift a heavy object. Deep to the mucous membrane of the vocal folds, which is lined by nonkeratinized stratified squamous epithelium, are bands of elastic ligaments stretched between pieces of rigid cartilage like the strings on a guitar. Skeletal muscles of the larynx, called intrinsic muscles, attach to both the rigid cartilage and the vocal folds. When the muscles contract, they pull the elastic ligaments tight and stretch the vocal folds out into the airways so that the rima glottidis is narrowed. If air is directed against the vocal folds, they vibrate and set up sound waves in the column of air in the pharynx, nose, and mouth. The greater the pressure of air, the louder the sound.

When the intrinsic muscles of the larynx contract, they pull on the arytenoid cartilages, which causes them to pivot. Contraction of the posterior cricoarytenoid muscles, for example, moves the vocal folds apart (abduction), thereby opening the rima glottidis (Figure 23.5a on page 693). By contrast, contraction of the lateral cricoarytenoid muscles moves the vocal folds together (adduction), thereby closing the rima glottidis (Figure 23.5b). Other intrinsic muscles can elongate (and place tension on) or shorten (and relax) the vocal folds.

Pitch is controlled by the tension on the vocal folds. If they are pulled taut by the muscles, they vibrate more rapidly, and a higher pitch results. Lower sounds are produced by decreasing the muscular tension on the vocal folds. Due to the influence of androgens (male sex hormones), vocal folds are usually thicker and longer in males than in females, and therefore they vibrate more slowly. Thus, men's voices generally have a lower range of pitch than women's.

Sound originates from the vibration of the vocal folds, but other structures are necessary for converting the sound into recognizable speech. The pharynx, mouth, nasal cavity, and paranasal sinuses all act as resonating chambers that give the voice its human and individual quality. We produce the vowel sounds by constricting and relaxing the muscles in the wall of the pharynx. Muscles of the face, tongue, and lips help us enunciate words.

Whispering is accomplished by closing all but the posterior portion of the rima glottidis. Because the vocal folds do not vibrate during whispering, there is no pitch to this form of speech. However, we can still produce intelligible speech while whispering by changing the shape of the oral cavity as we enunciate. As the size of the oral cavity changes, its resonance qualities change, which imparts a vowel-like pitch to the air as it rushes toward the lips.

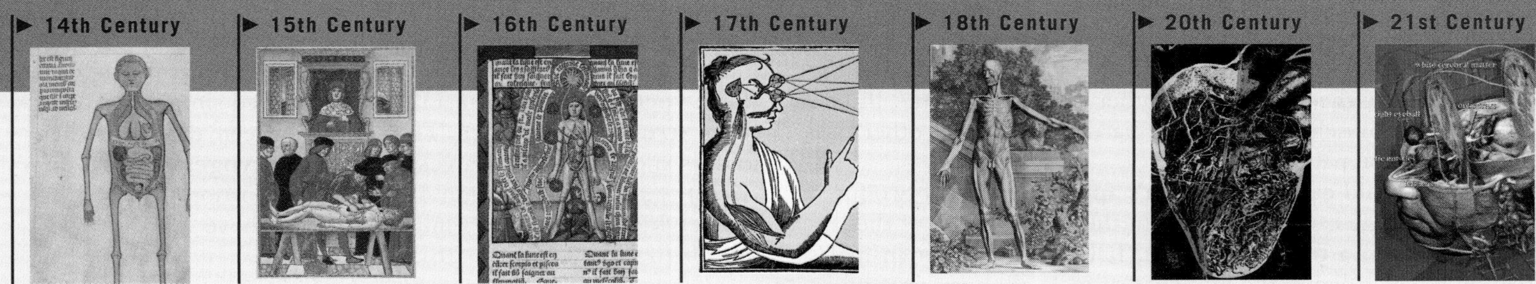

CHANGING IMAGES

Centers of Learning

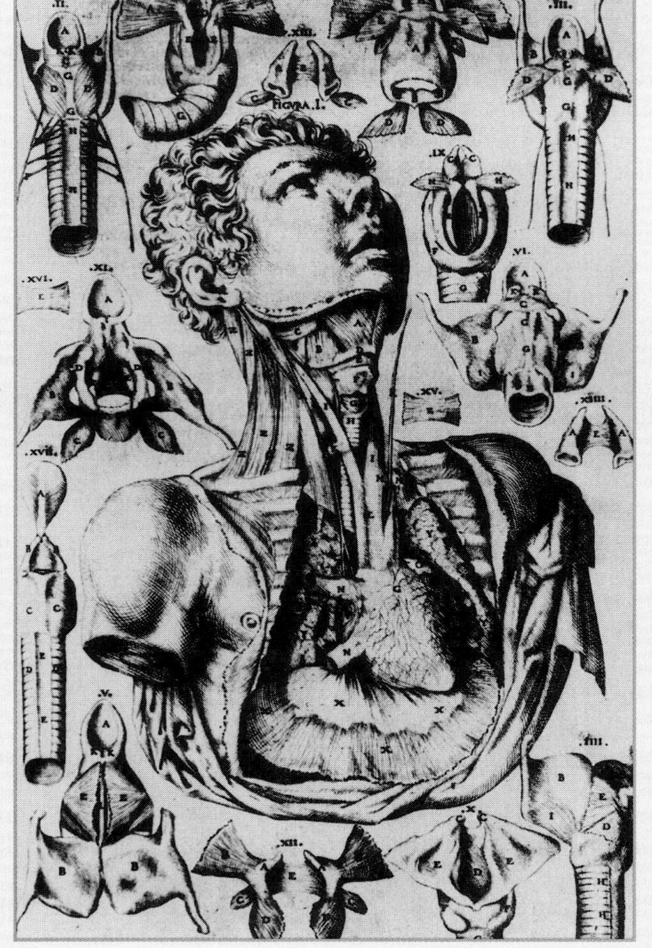

1601 AD

Modern colleges and universities of the West can trace their lineage back to the learning institutions of European Medieval times. Of these, the University of Salerno, Italy, is often considered the oldest of the great European universities. Established in the 10th century, it trained physcians not only from across Europe, but from the near East and Africa as well. Over a period of several centuries, its faculty and students produced numerous anatomical texts, yet they were based primarily on early Greek writings and animal dissections. Human dissection was rare in the early days of Salerno.

By the 12th century, it was the University of Bologna, Italy, that became the European hotbed of anatomical inquiry. The popularity of this university is often attributed to its great enthusiasm for human dissection. Indeed, one of the most famous images of early human dissection, which is reproduced on page 18 in chapter one, was conducted at this institution.

As the centuries passed and the Renaissance commenced, the focal point for not only scientific study, but the humanities as well, shifted to the University of Padua, Italy.

Padua boasted students and faculty of the caliber of Galileo, Vesalius, and Harvey. Another of Padua's significant, but lesser known associates is Giulio Casserio. Professor to Harvey, prolific dissector, and lecturer in anatomy, Casserio specialized in the anatomy of the head and neck as it relates to speech, hearing, and respiration. Pictured here is an illustration from his 1601 achievement: *De Vocis Auditusque Organis Historia Anatomica.* Scattered about the page are depictions of numerous lower respiratory system structures. Can you recognize the thyroid cartilage, epiglottis, and trachea? How about the vocal folds and hyoid bone? Are the other organs related to respiration, such as the heart, diaphragm and lungs, accurately depicted?

Figure 23.5 / Movement of the vocal folds.

The glottis consists of a pair of folds of mucous membrane, the vocal folds in the larynx, and the space between them (the rima glottidis).

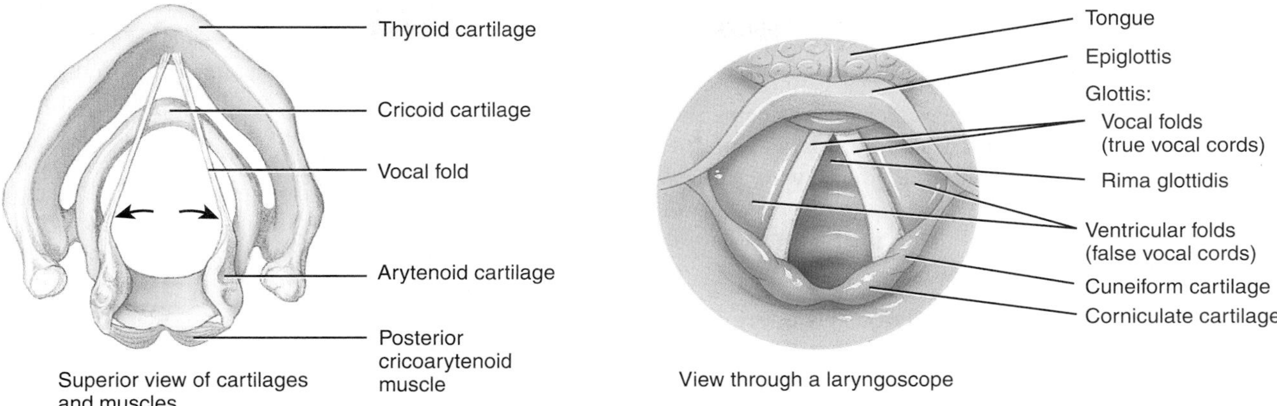

(a) Movement of vocal folds apart (abduction)

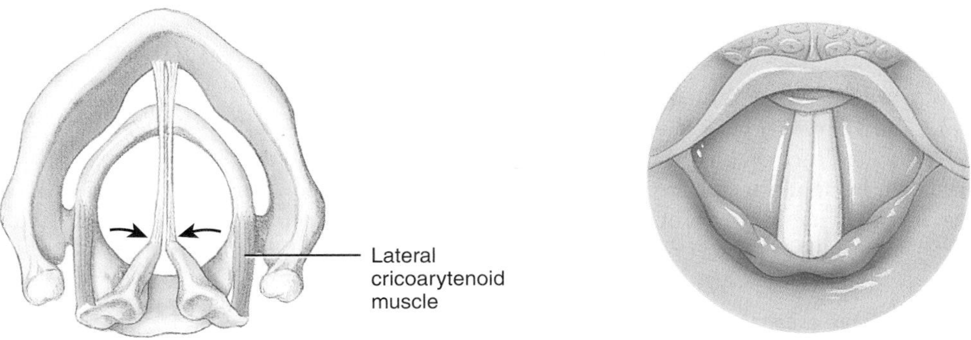

(b) Movement of vocal folds together (adduction)

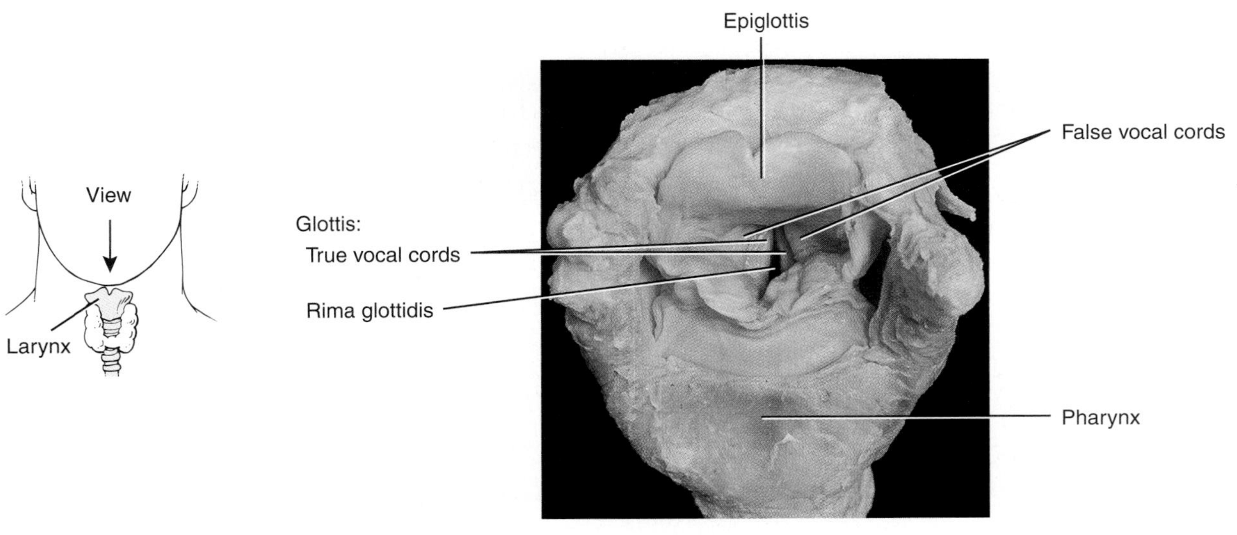

(c) Superior view

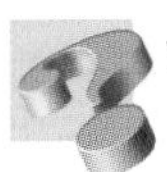

What is the main function of the vocal folds?

Laryngitis and Cancer of the Larynx

Laryngitis is an inflammation of the larynx that is most often caused by a respiratory infection or irritants such as cigarette smoke. Inflammation of the vocal folds causes hoarseness or loss of voice by interfering with the contraction of the folds or by causing them to swell to the point where they cannot vibrate freely. Many long-term smokers acquire a permanent hoarseness from the damage done by chronic inflammation. **Cancer of the larynx** is found almost exclusively in individuals who smoke. The condition is characterized by hoarseness, pain on swallowing, or pain radiating to an ear. Treatment consists of radiation therapy and/or surgery. ■

The arteries of the larynx are the superior and inferior laryngeal arteries. The superior and inferior laryngeal veins accompany the arteries. The superior laryngeal vein empties into the superior thyroid vein, and the inferior vein empties into the inferior thyroid vein.

The nerves of the larynx are the superior and recurrent (inferior) laryngeal branches of the vagus (X) nerves.

Trachea

The **trachea** (TRĀ-kē-a; = sturdy), or windpipe, is a tubular passageway for air that is about 12 cm (5 in.) long and 2.5 cm (1 in.) in diameter. It is located anterior to the esophagus (Figure 23.6) and extends from the larynx to the superior border of the fifth thoracic vertebra (T5), where it divides into right and left primary bronchi (see Figure 23.7).

The layers of the tracheal wall, from deep to superficial, are (1) mucosa, (2) submucosa, (3) hyaline cartilage, and (4) adventitia, which is composed of areolar connective tissue. The tracheal mucosa consists of an epithelial layer of pseudostratified ciliated columnar epithelium and an underlying layer of lamina propria that contains elastic and reticular fibers (see Table 3.1 on page 65). The epithelium consists of ciliated columnar cells and goblet cells that reach the luminal surface, plus basal cells that do not. The epithelium provides the same protection against dust as the membrane lining the nasal cavity and larynx. The submucosa consists of areolar connective tissue with seromucous glands and their ducts. The 16–20 incomplete, horizontal rings of hyaline cartilage resemble the letter C and are stacked one on top of another. They may be felt through the skin inferior to the larynx. The open part of each C-shaped cartilage ring faces the esophagus (Figure 23.6), an arrangement that accommodates slight expansion of the esophagus into the trachea during swallowing. Transverse smooth muscle fibers, called the **trachealis muscle,** and elastic connective tissue stabilize the open ends of the cartilage rings. The solid C-shaped cartilage rings provide a semirigid support so that the tracheal wall does not collapse inward (especially during inspiration) and obstruct the air passageway. The adventitia of the trachea consists of areolar connective tissue that joins the trachea to surrounding tissues.

Tracheostomy and Intubation

Several conditions may block airflow by obstructing the trachea. For example, the rings of cartilage that support the trachea may collapse due to a crushing injury to the chest, inflammation of

Figure 23.6 / Location of the trachea in relation to the esophagus.

The trachea is anterior to the esophagus and extends from the larynx to the superior border of the fifth thoracic vertebra.

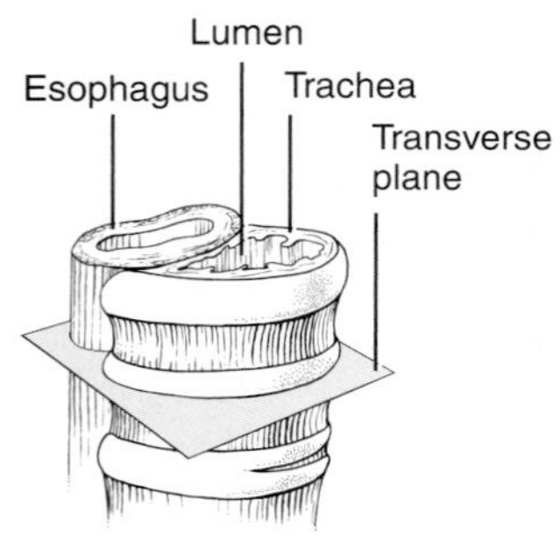

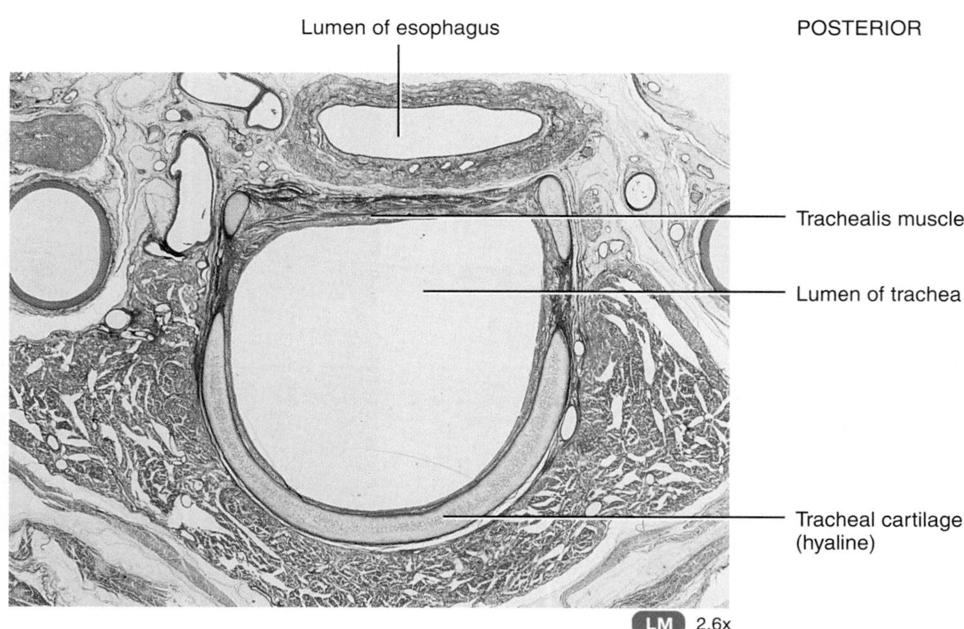

Transverse section of the trachea and esophagus

What is the benefit of not having cartilage between the trachea and the esophagus?

the mucous membrane may cause it to swell so much that the airway closes, or vomit or a foreign object may be aspirated into it. Two methods are used to reestablish airflow past a tracheal obstruction. If the obstruction is superior to the level of the larynx, a **tracheostomy** (trā-kē-OS-tō-mē) may be performed. In this procedure, a skin incision is followed by a short longitudinal incision into the trachea inferior to the cricoid cartilage. The patient can then breathe through a metal or plastic tracheal tube inserted through the incision. The second method is **intubation,** in which a tube is inserted into the mouth or nose and passed inferiorly through the larynx and trachea. The firm wall of the tube pushes aside any flexible obstruction, and the lumen of the tube provides a passageway for air; any mucus clogging the trachea can be suctioned out through the tube. ■

The arteries of the trachea are branches of the inferior thyroid, internal thoracic, and bronchial arteries. The veins of the trachea terminate in the inferior thyroid veins.

The smooth muscle and glands of the trachea are innervated parasympathetically via the vagus (X) nerves directly and by their recurrent laryngeal branches. Sympathetic innervation is through branches from the sympathetic trunk and its ganglia.

Bronchi

At the superior border of the fifth thoracic vertebra, the trachea divides into a **right primary bronchus** (BRON-kus; = windpipe), which goes into the right lung, and a **left primary bronchus,** which goes into the left lung (Figure 23.7). The right primary bronchus is more vertical, shorter, and wider than the left. As a result, an aspirated object is more likely to enter and lodge in the right primary bronchus than the left. Like the trachea, the primary bronchi (BRON-kē) contain incomplete rings of cartilage and are lined by pseudostratified ciliated columnar epithelium.

At the point where the trachea divides into the primary bronchi, there is an internal ridge called the **carina** (ka-RĪ-na; = keel of a boat). It is formed by a posterior and somewhat inferior projection of the last tracheal cartilage. The mucous membrane of the carina is one of the most sensitive areas of the entire larynx and trachea for triggering a cough reflex. Widening

Figure 23.7 / Branching of airways from the trachea: the bronchial tree.

The bronchial tree begins at the trachea and ends at the terminal bronchioles.

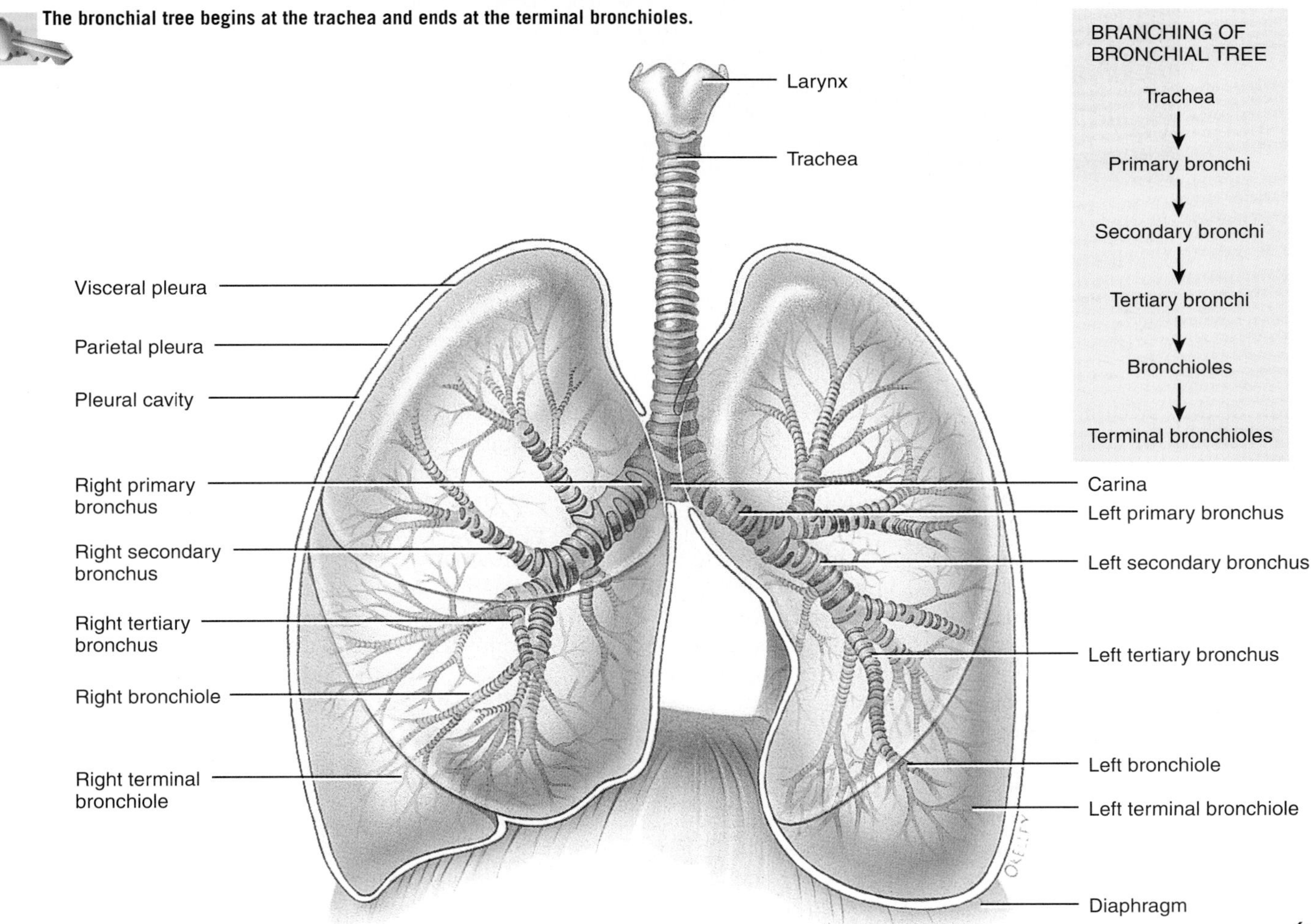

(a) Anterior view

(continues)

Figure 23.7 (continued)

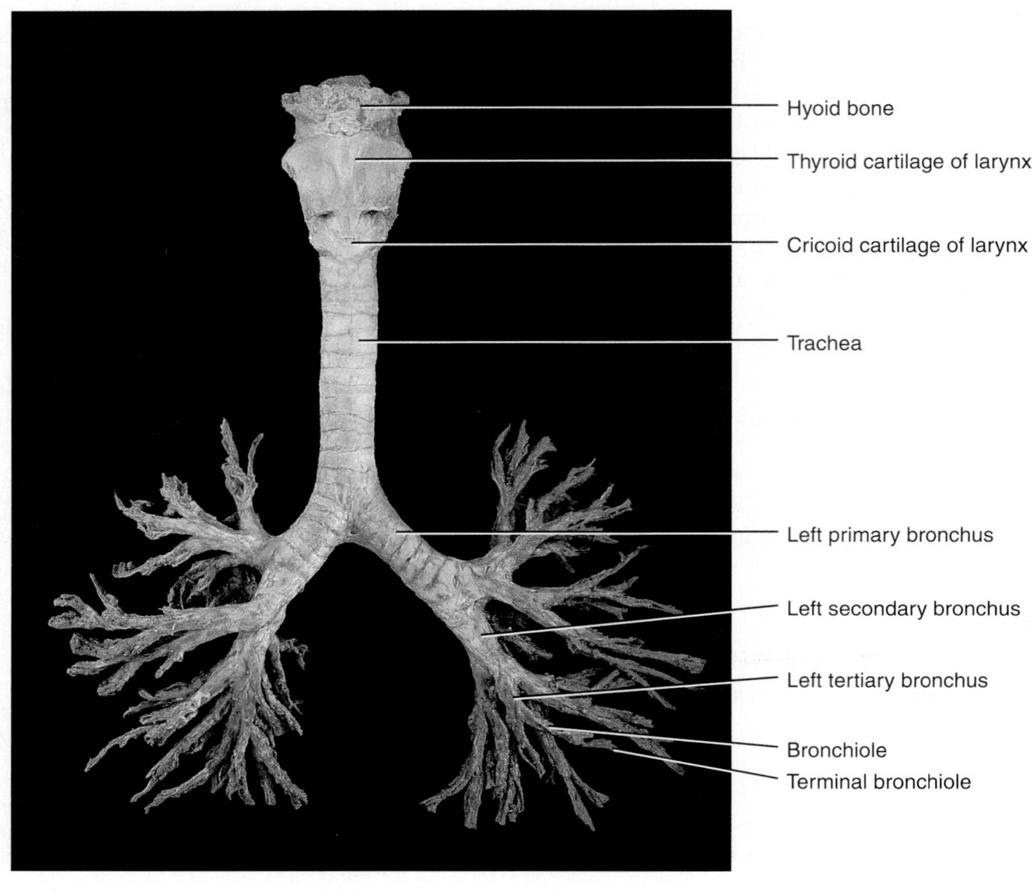

(b) Anterior view

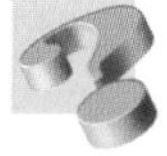

How many lobes and secondary bronchi are present in each lung?

and distortion of the carina is a serious sign because it usually indicates a carcinoma of the lymph nodes around the region where the trachea divides.

On entering the lungs, the primary bronchi divide to form smaller bronchi—the **secondary (lobar) bronchi,** one for each lobe of the lung. (The right lung has three lobes; the left lung has two.) The secondary bronchi continue to branch, forming still smaller bronchi, called **tertiary (segmental) bronchi,** that divide into **bronchioles.** Bronchioles, in turn, branch repeatedly, eventually forming even smaller tubes called **terminal bronchioles.** This extensive branching from the trachea resembles an inverted tree and is commonly referred to as the **bronchial tree.**

As the branching becomes more extensive in the bronchial tree, several structural changes may be noted. First, the mucous membrane in the bronchial tree changes from pseudostratified ciliated columnar epithelium in the primary bronchi, secondary bronchi, and tertiary bronchi to ciliated simple columnar epithelium with some goblet cells in larger bronchioles, to mostly ciliated simple cuboidal epithelium with no goblet cells in smaller bronchioles, to mostly nonciliated simple cuboidal epithelium in terminal bronchioles. (In regions with nonciliated simple cuboidal epithelium, inhaled particles are removed by macrophages.) Second, the incomplete rings of cartilage in primary bronchi are gradually replaced by plates of cartilage that finally disappear. Third, as the amount of cartilage decreases, the amount of smooth muscle increases. Smooth muscle encircles the lumen in spiral bands. Because there is no supporting cartilage, however, muscle spasms can close off the airways; this is what happens during an asthma attack, and it can be a life-threatening situation. During exercise, activity in the sympathetic division of the ANS increases and the adrenal medullae release the hormones epinephrine and norepinephrine, both of which cause relaxation of smooth muscle in the bronchioles, which dilates the airways. The result is improved lung ventilation because air reaches the alveoli more quickly. The parasympathetic division of the ANS and mediators of allergic reactions such as histamine cause contraction of bronchiolar smooth muscle and result in constriction of distal bronchioles.

Many respiratory disorders are treated by means of **nebulization** (neb-yū-li-ZĀ-shun). This procedure consists of administering medication in the form of droplets that are suspended in air into the respiratory tract. The patient inhales the medication as a fine mist. Nebulization therapy can be used with many different types of drugs, such as chemicals that relax the smooth muscle of the airways, chemicals that reduce the thickness of mucus, and antibiotics.

The blood supply to the bronchi is via the left bronchial and right bronchial arteries. The veins that drain the bronchi are the right bronchial vein, which enters the azygos vein, and the left bronchial vein, which empties into the hemiazygos vein or the left superior intercostal vein.

- ✓ What functions do the respiratory and cardiovascular systems have in common?
- ✓ Distinguish between the structural and functional features of the upper and lower respiratory systems.
- ✓ Compare the structure and functions of the external nose and the internal nose.
- ✓ What are the three anatomical regions of the pharynx? List the roles of each in respiration.
- ✓ Explain how the larynx functions in respiration and voice production.
- ✓ Describe the location, structure, and function of the trachea.

LUNGS

Objective

- Describe the anatomy and histology of the lungs.

The **lungs** (= lightweights, because they float) are paired cone-shaped organs lying in the thoracic cavity (Figure 23.8). They are separated from each other by the heart and other structures in the mediastinum, which separates the thoracic cavity into two anatomically distinct chambers. As a result, should trauma cause one lung to collapse, the other may remain expanded. Two layers of serous membrane, collectively called the **pleural**

Figure 23.8 / Relationship of the pleural membranes to the lungs.

The parietal pleura lines the thoracic cavity, whereas the visceral pleura covers the lungs.

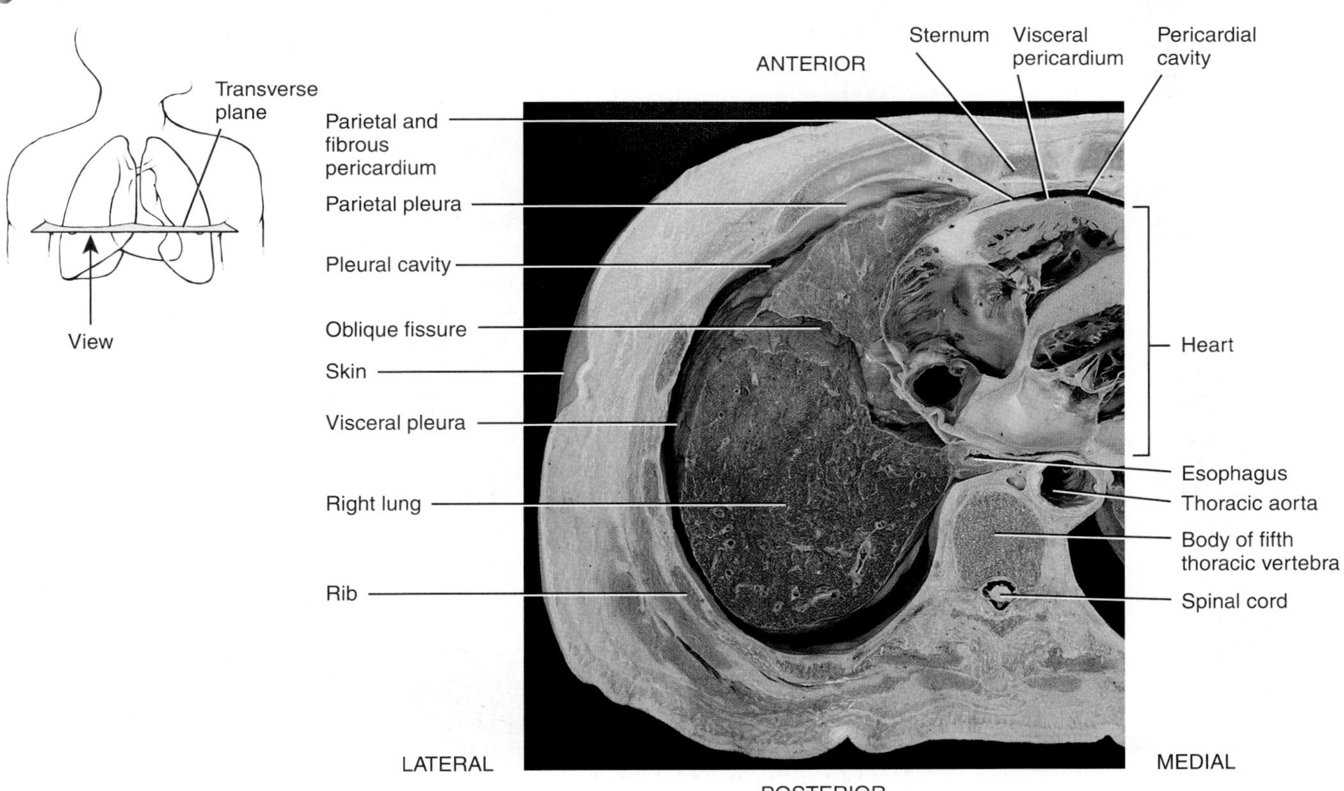

Inferior view of a transverse section through the thoracic cavity showing the pleural cavity and pleural membranes

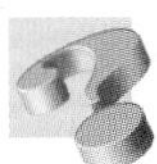

What type of membrane is the pleural membrane?

Figure 23.9 / Surface anatomy of the lungs. (See Tortora, *A Photographic Atlas of the Human Body,* Figure 11.14.)

The subdivisions of the lungs are lobes, bronchopulmonary segments, and lobules.

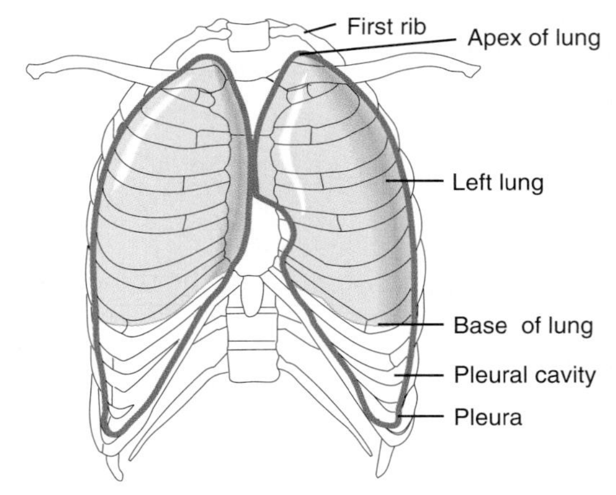

(a) Anterior view of lungs and pleurae in thorax

View (b)
View (c)
Apex
Superior lobe
ANTERIOR
Horizontal fissure
Oblique fissure
Inferior lobe
Cardiac notch
Middle lobe
Base
POSTERIOR
Oblique fissure
Inferior lobe
POSTERIOR

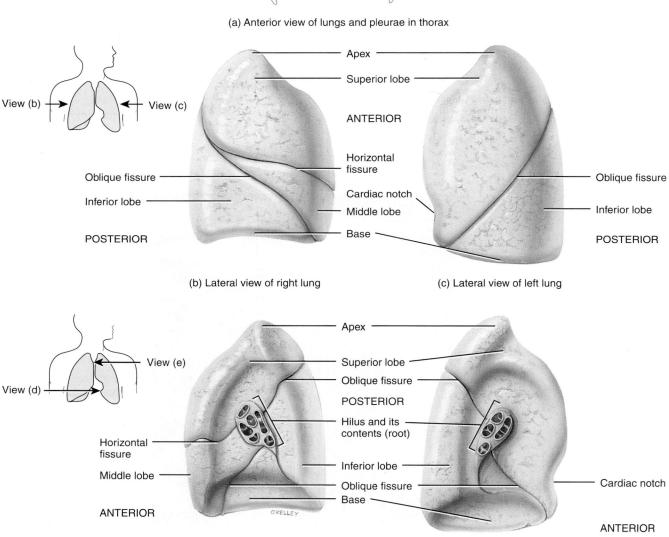

(b) Lateral view of right lung

(c) Lateral view of left lung

View (e)
View (d)
Apex
Superior lobe
Oblique fissure
POSTERIOR
Hilus and its contents (root)
Horizontal fissure
Middle lobe
Inferior lobe
Oblique fissure
Base
Cardiac notch
ANTERIOR
ANTERIOR
O'KELLEY

(d) Medial view of right lung

(e) Medial view of left lung

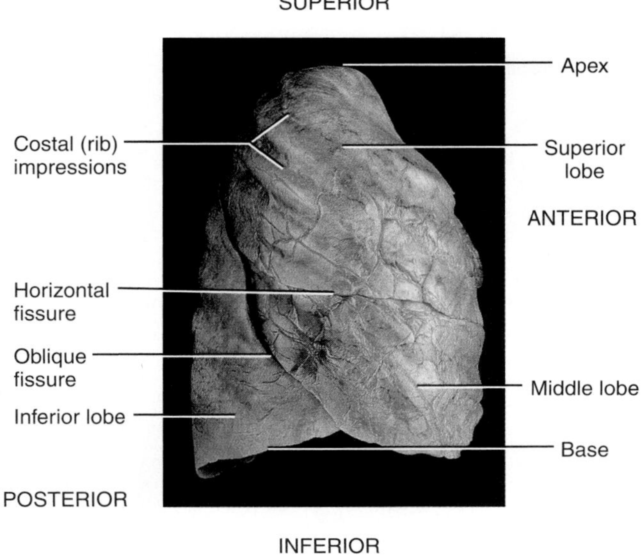

(f) Right lung, lateral view

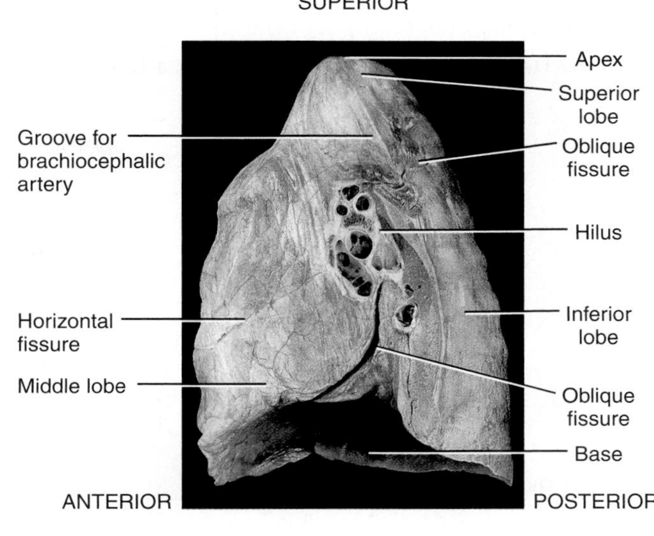

(g) Right lung, medial view

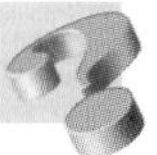

Why are the right and left lungs slightly different in size and shape?

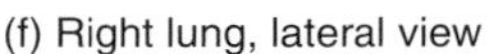

membrane (PLŪR-al; *pleur-* = side), enclose and protect each lung. The superficial layer lines the wall of the thoracic cavity and is called the **parietal pleura;** the deep layer, the **visceral pleura,** covers the lungs themselves. Between the visceral and parietal pleurae is a small space, the **pleural cavity,** which contains a small amount of lubricating fluid secreted by the membranes. This fluid reduces friction between the membranes, allowing them to slide easily over one another during breathing. Pleural fluid also causes the two membranes to adhere to one another just as a film of water causes two glass slides to stick together. Separate pleural cavities surround the left and right lungs. Inflammation of the pleural membrane, called **pleurisy** or **pleuritis,** may in its early stages cause pain due to friction between the parietal and visceral layers of the pleura. If the inflammation persists, excess fluid accumulates in the pleural space, a condition known as **pleural effusion.**

Pneumothorax and Hemothorax

In certain conditions, the pleural cavity may fill with air (**pneumothorax;** *pneumo* = air or breath), blood **(hemothorax),** or pus. Air in the pleural cavity, most commonly introduced in a surgical opening of the chest or as a result of a stab or gunshot wound, may cause the lung to collapse. This condition is called **atelectasis** (at′-e-LEK-ta-sis; *ateles* = incomplete; *ectasis* = expansion). Fluid can be drained from the pleural cavity by inserting a needle, usually posteriorly through the seventh intercostal space. The needle is passed along the superior border of the lower rib to avoid damage to the intercostal nerves and blood vessels. Inferior to the seventh intercostal space there is danger of penetrating the diaphragm.

The lungs extend from the diaphragm to just slightly superior to the clavicles and lie against the ribs anteriorly and posteriorly (Figure 23.9 on pages 698–699). The broad inferior portion of the lung, the **base,** is concave and fits over the convex area of the diaphragm. The narrow superior portion of the lung is the **apex.** The surface of the lung lying against the ribs, the **costal surface,** matches the rounded curvature of the ribs. The **mediastinal (medial) surface** of each lung contains a region, the **hilus,** through which bronchi, pulmonary blood vessels, lymphatic vessels, and nerves enter and exit. These structures are held together by the pleura and connective tissue and constitute the **root** of the lung. Medially, the left lung also contains a concavity, the **cardiac notch,** in which the heart lies. Due to the space occupied by the heart, the left lung is about 10% smaller than the right lung. Although the right lung is thicker and broader, it is also somewhat shorter than the left lung because the diaphragm is higher on the right side, accommodating the liver that lies inferior to it.

The lungs almost fill the thorax (Figure 23.9a). The apex of the lungs lies superior to the medial third of the clavicles and is the only area that can be palpated. The anterior, lateral, and posterior surfaces of the lungs lie against the ribs. The base of the lungs extends from the sixth costal cartilage anteriorly to the spinous process of the tenth thoracic vertebra posteriorly. The pleura extends about 5 cm below the base from the sixth costal cartilage anteriorly to the twelfth rib posteriorly. Thus, the lungs do not completely fill the pleural cavity in this area, and removal of excessive fluid in the pleural cavity can be accom-

Figure 23.10 / Bronchopulmonary segments of the lungs. The bronchial branches are shown in the center of the figure. The bronchopulmonary segments are numbered and named for convenience.

There are ten tertiary (segmental) bronchi in each lung, and each is composed of smaller compartments called lobules.

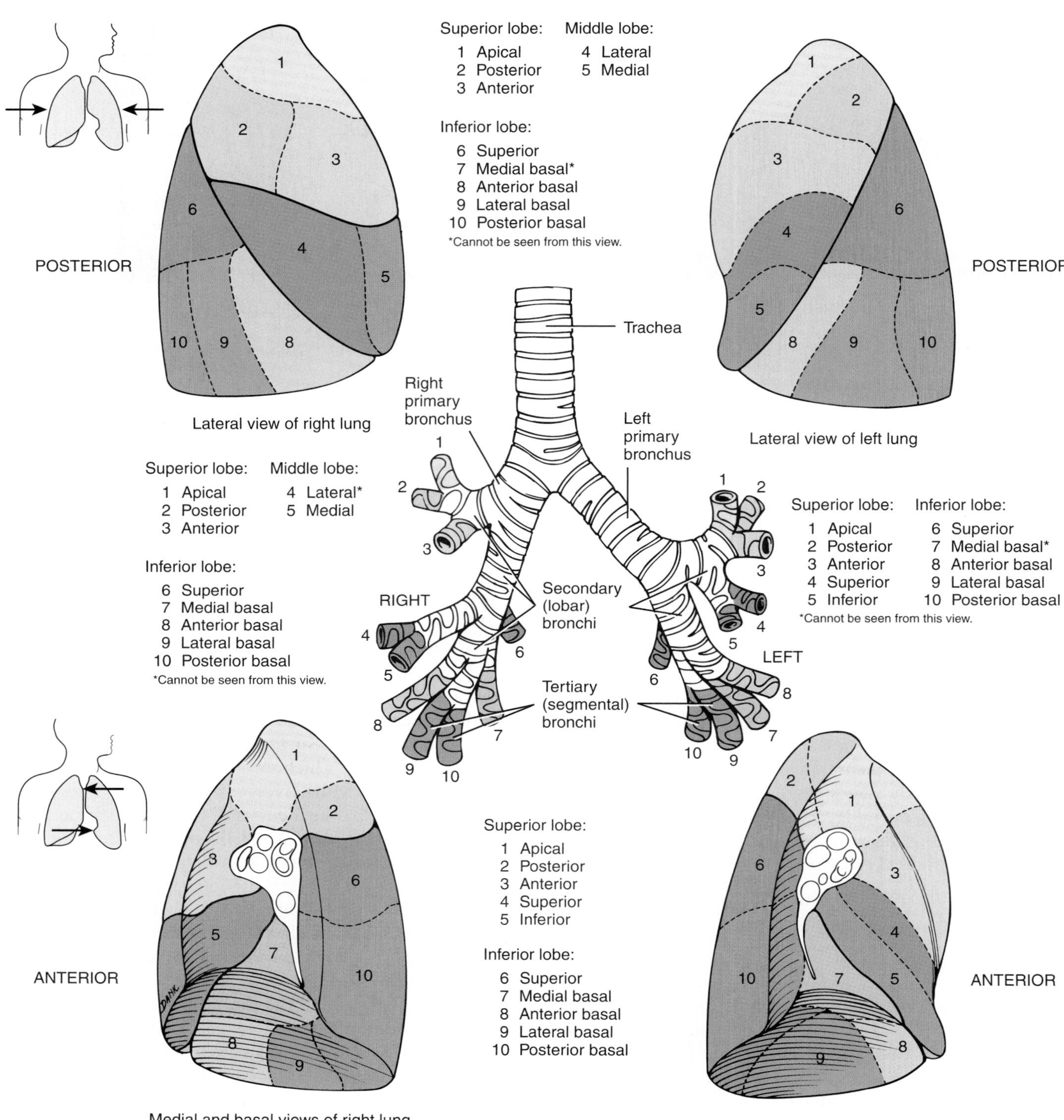

Which bronchi supply a bronchopulmonary segment?

plished without injuring lung tissue by inserting the needle posteriorly through the seventh intercostal space, a procedure termed **thoracentesis** (thor′-a-sen-TĒ-sis; *-centesis* = puncture).

Lobes, Fissures, and Lobules

Each lung is divided into lobes by one or more fissures (Figure 23.9b–e). Both lungs have an **oblique fissure,** which extends inferiorly and anteriorly; the right lung also has a **horizontal fissure.** The oblique fissure in the left lung separates the **superior lobe** from the **inferior lobe.** In the right lung, the superior part of the oblique fissure separates the superior lobe from the inferior lobe, whereas the inferior part of the oblique fissure separates the inferior lobe from the **middle lobe.** The horizontal fissure of the right lung subdivides the superior lobe, thus forming a middle lobe.

Each lobe receives its own secondary (lobar) bronchus. Thus, the right primary bronchus gives rise to three secondary (lobar) bronchi called the **superior, middle,** and **inferior secondary (lobar) bronchi,** whereas the left primary bronchus gives rise to **superior** and **inferior secondary (lobar) bronchi.** Within the substance of the lung, the secondary bronchi give rise to the **tertiary (segmental) bronchi,** which are constant in both origin and distribution—there are ten tertiary bronchi in each lung. The segment of lung tissue that each supplies is called a **bronchopulmonary segment** (Figure 23.10 on page 700). Bronchial and pulmonary disorders (such as tumors or abscesses) that are localized in a bronchopulmonary segment may be surgically removed without seriously disrupting the surrounding lung tissue.

Each bronchopulmonary segment of the lungs has many small compartments called **lobules,** each of which is wrapped in elastic connective tissue and contains a lymphatic vessel, an arteriole, a venule, and a branch from a terminal bronchiole (Figure 23.11a). Terminal bronchioles subdivide into microscopic branches called **respiratory bronchioles** (Figure 23.11b). As the respiratory bronchioles penetrate more deeply into the lungs, the epithelial lining changes from simple cuboidal to simple squamous. Respiratory bronchioles, in turn, subdivide into several (2–11) **alveolar ducts.** The respiratory passages from the trachea to the alveolar ducts contain about 25 orders of branching; that is, branching—from the trachea into primary bronchi (first-order branching) into secondary bronchi (second-order branching) and so on down to the alveolar ducts—occurs about 25 times.

Alveoli

Around the circumference of the alveolar ducts are numerous alveoli and alveolar sacs. An **alveolus** (al-VĒ-ō-lus) is a cup-

Figure 23.11 / Microscopic anatomy of a lobule of the lungs.

Alveolar sacs are two or more alveoli that share a common opening.

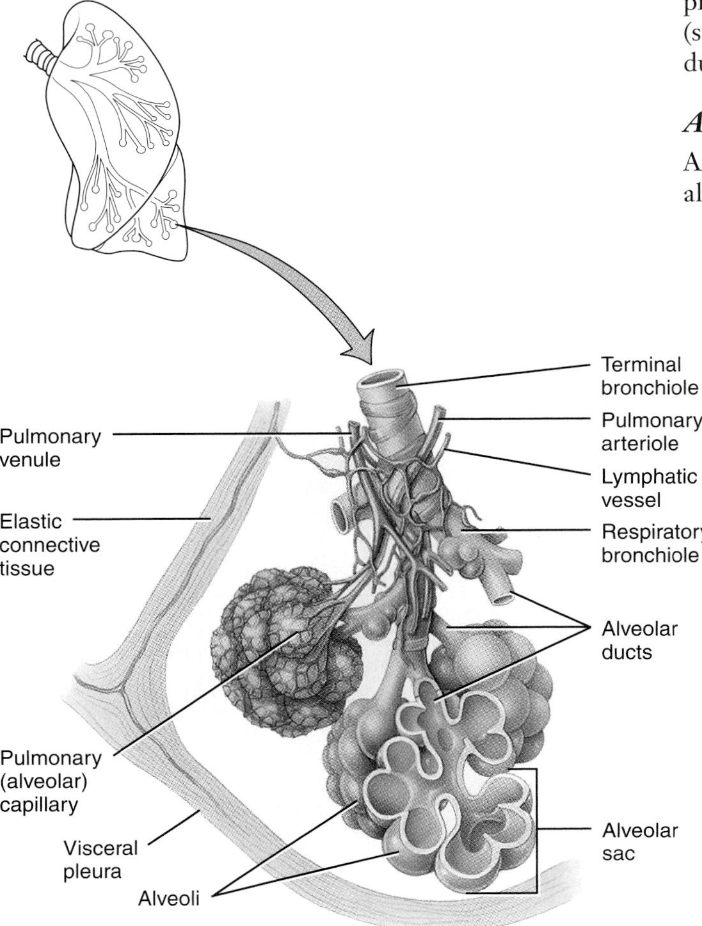

(a) Diagram of a portion of a lobule of the lung

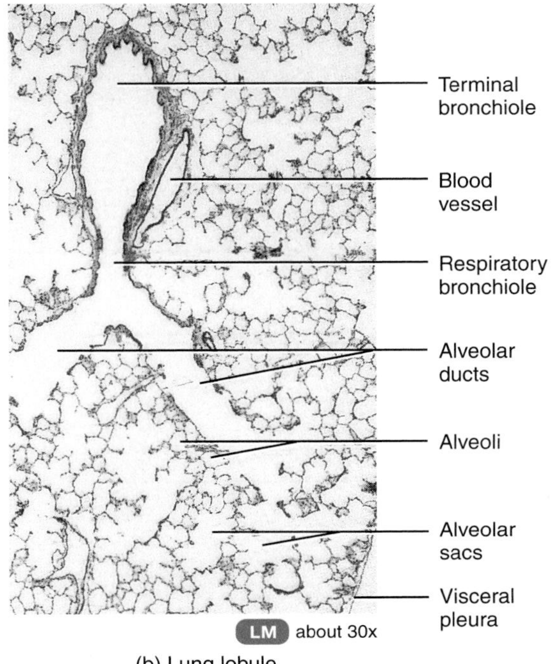

(b) Lung lobule

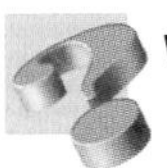

What types of cells make up most of the wall of an alveolus?

shaped outpouching lined by simple squamous epithelium and supported by a thin elastic basement membrane; **alveolar sacs** are two or more alveoli that share a common opening (Figure 23.11a, b). The walls of alveoli consist of two types of alveolar epithelial cells (Figure 23.12). **Type I alveolar cells** are simple squamous epithelial cells that form a mostly continuous lining of the alveolar wall that is interrupted by occasional **type II alveolar cells,** also called **septal cells.** The thin type I alveolar cells are the main sites of gas exchange. Type II alveolar cells, which are rounded or cuboidal epithelial cells whose free surfaces contain microvilli, secrete alveolar fluid, which keeps the surface between the cells and the air moist. Included in the alveolar fluid is **surfactant** (sur-FAK-tant), a complex detergentlike mixture of phospholipids and lipoproteins. Surfactant lowers the surface tension of alveolar fluid, which reduces the tendency of alveoli to collapse. Associated with the alveolar wall are **alveolar macrophages (dust cells),** wandering phagocytes that remove fine dust particles and other debris in the alveolar spaces. Also present are fibroblasts that produce reticular and elastic fibers. Underlying the layer of type I alveolar cells is an elastic basement membrane. Around the alveoli, the lobule's arteriole and venule disperse into a network of blood capillaries that consist of a single layer of endothelial cells and a basement membrane.

The exchange of O_2 and CO_2 between the air spaces in the lungs and the blood takes place by diffusion across alveolar and

Figure 23.12 / Structural components and function of an alveolus.

The exchange of respiratory gases occurs by diffusion across the respiratory membrane.

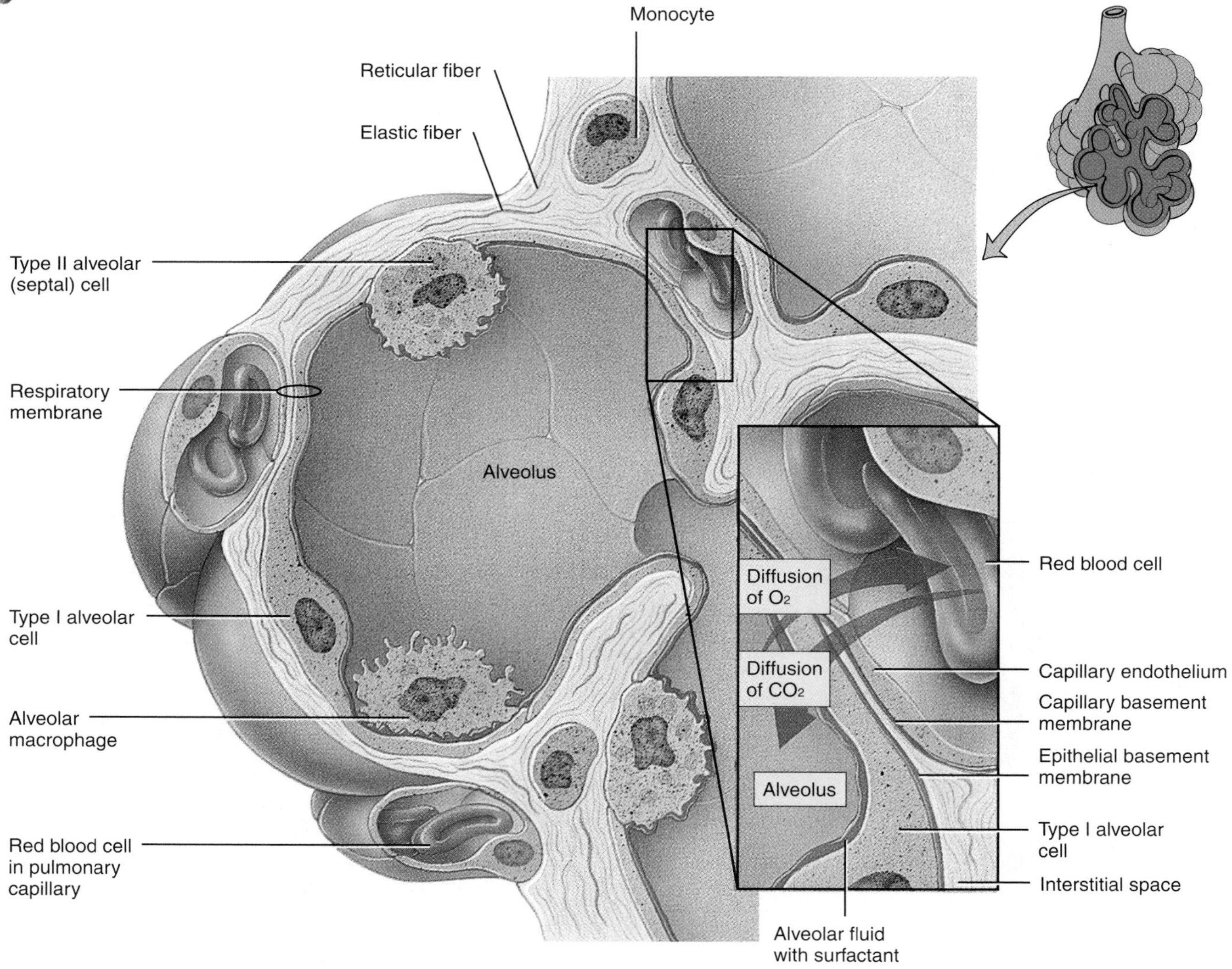

(a) Section through an alveolus showing its cellular components

(b) Details of respiratory membrane

What is the thickness of the respiratory membrane?

capillary walls. The gases diffuse through the **respiratory membrane,** which consists of four layers (Figure 23.12b):

1. A layer of type I and type II alveolar cells and associated alveolar macrophages that constitutes the **alveolar wall.**
2. An **epithelial basement membrane** underlying the alveolar wall.
3. A **capillary basement membrane** that is often fused to the epithelial basement membrane.
4. The **endothelial cells** of the capillary.

Despite having several layers, the respiratory membrane is very thin—only 0.5 μm thick, about one-sixteenth the diameter of a red blood cell. This thinness allows rapid diffusion of gases. Moreover, it has been estimated that the lungs contain 300 million alveoli, providing an immense surface area of 70 m^2 (750 ft^2)—about the size of a handball court—for the exchange of gases.

Blood and Nerve Supply to the Lungs

The lungs receive blood via two sets of arteries: pulmonary arteries and bronchial arteries. Deoxygenated blood passes through the pulmonary trunk, which divides into a left pulmonary artery that enters the left lung and a right pulmonary artery that enters the right lung. Return of the oxygenated blood to the heart occurs by way of the pulmonary veins, which drain into the left atrium (see Figure 14.17 on page 464). A unique feature of pulmonary blood vessels is their constriction in response to localized hypoxia (low O_2 level). In all other body tissues, hypoxia causes dilation of blood vessels, which serves to increase blood flow to a tissue that is not receiving adequate O_2. In the lungs, however, vasoconstriction in response to hypoxia diverts pulmonary blood from poorly ventilated areas to well-ventilated regions of the lungs.

Bronchial arteries, which branch from the aorta, deliver oxygenated blood to the lungs. This blood mainly perfuses the walls of the bronchi and bronchioles. Connections exist between branches of the bronchial arteries and branches of the pulmonary arteries, however, and most blood returns to the heart via pulmonary veins. Some blood, however, drains into bronchial veins, branches of the azygos system, and returns to the heart via the superior vena cava.

The nerve supply of the lungs is derived from the pulmonary plexus, located anterior and posterior to the roots of the lungs. The pulmonary plexus is formed by branches of the vagus (X) nerves and sympathetic trunks. Motor parasympathetic fibers arise from the dorsal nucleus of the vagus (X) nerve, whereas motor sympathetic fibers are postganglionic fibers of the second to fifth thoracic paravertebral ganglia of the sympathetic trunk.

✓ Where are the lungs located? Distinguish the parietal pleura from the visceral pleura.
✓ Define each of the following parts of a lung: base, apex, costal surface, medial surface, hilus, root, cardiac notch, lobe, and lobule.
✓ What is a bronchopulmonary segment?
✓ Describe the histology and function of the respiratory membrane.

MECHANICS OF PULMONARY VENTILATION (BREATHING)

Objectives

- Distinguish among pulmonary ventilation, external respiration, and internal respiration.
- Describe how inspiration and expiration occur.

Respiration is the exchange of gases between the atmosphere, blood, and body cells. It takes place in three basic steps:

1. **Pulmonary ventilation.** The first process, pulmonary (*pulmo* = lung) ventilation, or breathing, is the inspiration (inflow) and expiration (outflow) of air between the atmosphere and the lungs.
2. **External (pulmonary) respiration.** This is the exchange of gases between the air spaces of the lungs and blood in pulmonary capillaries. The blood gains O_2 and loses CO_2.
3. **Internal (tissue) respiration.** The exchange of gases between blood in systemic capillaries and tissue cells is known as internal (tissue) respiration. The blood loses O_2 and gains CO_2.

The flow of air between the atmosphere and lungs occurs for the same reason that blood flows through the body: a pressure gradient (difference) exists. Air moves into the lungs when the pressure inside the lungs is less than the air pressure in the atmosphere. Air moves out of the lungs when the pressure inside the lungs is greater than the pressure in the atmosphere.

Inspiration

Breathing in is called **inspiration (inhalation).** Just before each inspiration, the air pressure inside the lungs equals the pressure of the atmosphere, which is about 760 mm Hg, or 1 atmosphere (atm), at sea level. For air to flow into the lungs, the pressure inside the alveoli of the lungs **(alveolar pressure)** must become lower than the pressure in the atmosphere. This condition is achieved by increasing the volume (size) of the lungs.

For inspiration to occur, the lungs must expand. This increases lung volume and thus decreases the pressure in the lungs below atmospheric pressure. The first step in expanding the alveoli of the lungs during normal quiet breathing involves contraction of the principal inspiratory muscles—the diaphragm and/or external intercostals (Figure 23.13).

The diaphragm, the most important muscle of inspiration, is a dome-shaped skeletal muscle that forms the floor of the thoracic cavity. It is innervated by fibers of the phrenic nerves, which emerge from both sides of the spinal cord at cervical levels 3, 4, and 5. Contraction of the diaphragm causes it to flatten, lowering its dome. This increases the vertical diameter of the thoracic cavity and accounts for the movement of about 75% of the air that enters the lungs during normal quiet inspiration. The distance the diaphragm moves during inspiration ranges from 1 cm (0.4 in.) during normal quiet breathing up to about 10 cm (4 in.) during strenuous exercise. Advanced pregnancy, excessive obesity, or confining abdominal clothing can prevent a

Figure 23.13 / Muscles of inspiration and expiration and their roles in pulmonary ventilation. The pectoralis minor muscle, an accessory inspiratory muscle, is illustrated in Figure 10.13 on page 289.

During deep, labored inspiration, accessory muscles of inspiration (sternocleidomastoids, scalenes, and pectoralis minors) participate.

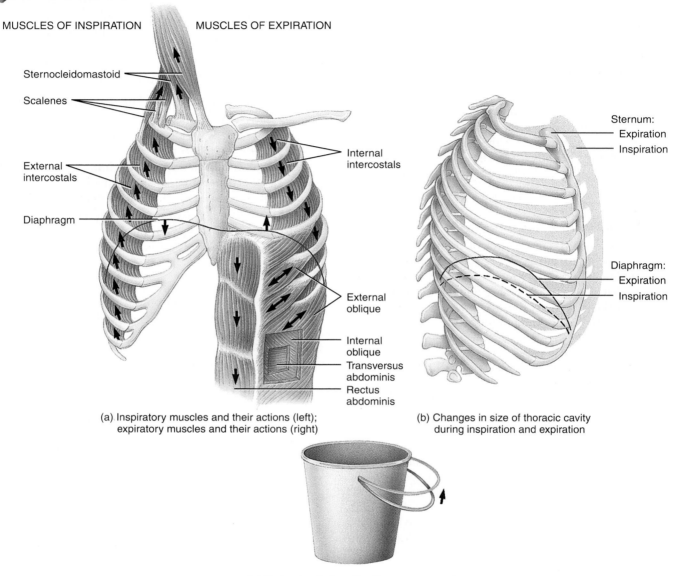

MUSCLES OF INSPIRATION
MUSCLES OF EXPIRATION
Sternocleidomastoid
Scalenes
External intercostals
Diaphragm
Internal intercostals
External oblique
Internal oblique
Transversus abdominis
Rectus abdominis

(a) Inspiratory muscles and their actions (left); expiratory muscles and their actions (right)

Sternum:
Expiration
Inspiration
Diaphragm:
Expiration
Inspiration

(b) Changes in size of thoracic cavity during inspiration and expiration

(c) During inspiration, the ribs move upward and outward like the handle on a bucket

Right now, what muscle is mainly responsible for your breathing?

complete descent of the diaphragm. At the same time the diaphragm contracts, the external intercostals contract. These skeletal muscles run obliquely downward and forward between adjacent ribs, and when these muscles contract, the ribs are pulled superiorly and the sternum is pushed anteriorly. This increases the anteroposterior and lateral diameters of the thoracic cavity.

As the diaphragm and internal intercostals contract and the overall size of the thoracic cavity increases, the walls of the lungs are pulled outward. The parietal and visceral pleurae normally adhere strongly to each other because of the below-atmospheric pressure between them and because of the surface tension created by their moist adjoining surfaces. As the thoracic cavity expands, the parietal pleura lining the cavity is pulled outward in all directions, and the visceral pleura and lungs are pulled along with it.

When the volume of the lungs increases, alveolar pressure decreases from 760 to 758 mm Hg. A pressure gradient is thus established between the atmosphere and the alveoli. Air rushes from the atmosphere into the lungs due to a gas pressure difference, and inspiration takes place. Air continues to move into the lungs as long as the pressure difference exists.

During deep, labored inspiration, accessory muscles of inspiration also participate in increasing the size of the thoracic cavity (Figure 23.13a). These include the sternocleidomastoid, which elevates the sternum; the scalenes, which elevate the superior two ribs; and the pectoralis minor, which elevates the third through fifth ribs.

Expiration

Breathing out, called **expiration (exhalation),** is also achieved by a pressure gradient, but in this case the gradient is reversed: the pressure in the lungs is greater than the pressure of the atmosphere. Normal expiration during quiet breathing depends on two factors: (1) the recoil of elastic fibers that were stretched during inspiration and (2) the inward pull of surface tension due to the film of alveolar fluid.

Expiration starts when the inspiratory muscles relax. As the external intercostals relax, the ribs move inferiorly, and as the diaphragm relaxes, its dome moves superiorly owing to its elasticity. These movements decrease the vertical, anteroposterior, and lateral diameters of the thoracic cavity. Also, surface tension exerts an inward pull and the elastic basement membranes of the alveoli and elastic fibers in bronchioles and alveolar ducts recoil. As a result, lung volume decreases and the alveolar pressure increases to 762 mm Hg. Air then flows from the area of higher pressure in the alveoli to the area of lower pressure in the atmosphere.

During labored breathing and when air movement out of the lungs is impeded, muscles of expiration—abdominal and internal intercostals—contract. Contraction of the abdominal muscles moves the inferior ribs inferiorly and compresses the abdominal viscera, thus forcing the diaphragm superiorly. Contraction of the internal intercostals, which extend inferiorly and posteriorly between adjacent ribs, pulls the ribs inferiorly.

✓ What are the basic differences among pulmonary ventilation, external respiration, and internal respiration?

✓ Compare what happens during quiet versus forceful ventilation.

REGULATION OF RESPIRATION

Objective

- Describe the various factors that regulate the rate and depth of respiration.

Although breathing can be controlled voluntarily for short periods, the nervous system usually controls respirations automatically to meet the body's demand without our conscious concern.

Role of the Respiratory Center

The size of the thorax is altered by the action of the respiratory muscles, which contract and relax as a result of nerve impulses transmitted to them from centers in the brain. The area from which nerve impulses are sent to respiratory muscles consists of clusters of neurons located bilaterally in the medulla oblongata and pons of the brain stem. This area, called the **respiratory center,** consists of a widely dispersed group of neurons functionally divided into three areas: (1) the medullary rhythmicity (rith-MIS-i-tē) area in the medulla oblongata; (2) the pneumotaxic (nū-mō-TAK-sik) area in the pons; and (3) the apneustic (ap-NŪ-stik) area, also in the pons (Figure 23.14).

Medullary Rhythmicity Area

The function of the **medullary rhythmicity area** is to control the basic rhythm of respiration. In the normal resting state, inspiration usually lasts for about 2 seconds and expiration for about 3 seconds. Within the medullary rhythmicity area are both inspiratory and expiratory neurons that constitute inspiratory and expiratory areas, respectively. We first consider the role of the inspiratory neurons in respiration.

The basic rhythm of respiration is determined by nerve impulses generated in the inspiratory area. At the beginning of expiration, the inspiratory area is inactive, but after 3 seconds it automatically becomes active due to impulses generated by autorhythmic neurons. Even when all incoming nerve connections to the inspiratory area are cut or blocked, neurons in this area still rhythmically discharge impulses that result in inspiration. Nerve impulses from the active inspiratory area last for about 2 seconds and reach the diaphragm via the phrenic nerves. When

Figure 23.14 / Locations of areas of the respiratory center.

The respiratory center is composed of neurons in the medullary rhythmicity area in the medulla oblongata plus the pneumotaxic and apneustic areas in the pons.

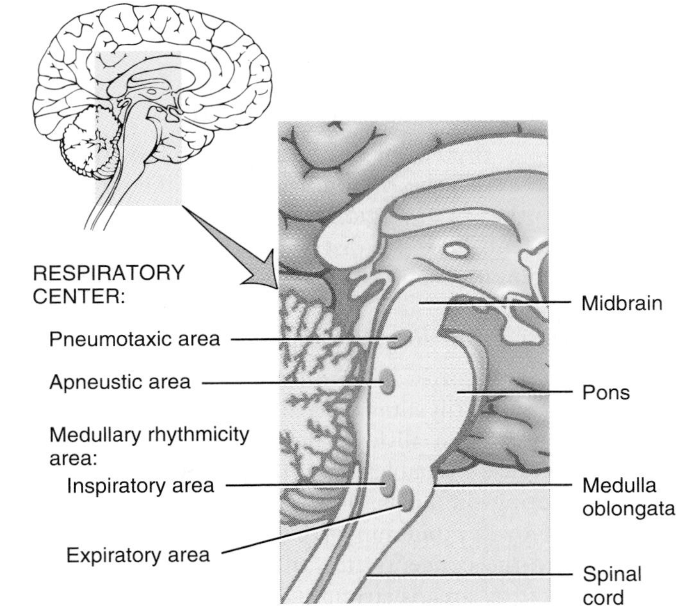

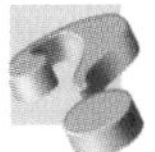

Which area contains neurons that are active and then inactive in a repeating cycle?

the nerve impulses reach the diaphragm, it contracts and inspiration occurs. At the end of 2 seconds, the inspiratory muscles relax for about 3 seconds, and then the cycle repeats.

The neurons of the expiratory area remain inactive during most normal, quiet respirations. During quiet breathing, inspiration is accomplished by active contraction of the diaphragm, and expiration results from passive elastic recoil of the lungs and thoracic wall as the diaphragm relaxes. However, during forceful ventilation, nerve impulses from the inspiratory area activate the expiratory area. Impulses from the expiratory area cause contraction of the internal intercostals and abdominal muscles, which decreases the size of the thoracic cavity and causes forceful expiration.

Pneumotaxic Area

Although the medullary rhythmicity area controls the basic rhythm of respiration, other sites in the brain stem help coordinate the transition between inspiration and expiration. One of these sites is the **pneumotaxic area** (*pneumo-* = air or breath; *-taxic* = arrangement) in the superior portion of the pons (Figure 23.14), which transmits inhibitory impulses to the inspiratory area. The major effect of these nerve impulses is to help turn off the inspiratory area before the lungs become too full of air. In other words, the impulses limit the duration of inspiration and thus facilitate the onset of expiration. When the pneumotaxic area is more active, breathing rate is more rapid.

Apneustic Area

Another part of the brain stem that coordinates the transition between inspiration and expiration is the **apneustic area** in the inferior portions of the pons (Figure 23.14). This area sends stimulatory impulses to the inspiratory area that activate it and prolong inspiration, thus inhibiting expiration. Such stimulation occurs when the pneumotaxic area is inactive; when the pneumotaxic area is active, it overrides the apneustic area.

Regulation of the Respiratory Center

Although the basic rhythm of respiration is set and coordinated by the inspiratory area, the rhythm can be modified in response to inputs from other brain regions and receptors in the peripheral nervous system. Next we discuss several factors that influence the regulation of respiration.

Cortical Influences on Respiration

Because the cerebral cortex has connections with the respiratory center, we can voluntarily alter our pattern of breathing. We can even refuse to breathe at all for a short time. Voluntary control is protective because it enables us to prevent water or irritating gases from entering the lungs. The ability to not breathe, however, is limited by the buildup of CO_2 and H^+ in the blood. When CO_2 and the concentration of H^+ increase to a certain level, the inspiratory area is strongly stimulated, nerve impulses are sent along the phrenic and intercostal nerves to inspiratory muscles, and breathing resumes, whether the person wants it or not. It is impossible for people to kill themselves by holding their breath. Even if a person faints, breathing resumes when consciousness is lost. Nerve impulses from the hypothalamus and limbic system also stimulate the respiratory center, allowing emotional stimuli to alter respirations (as, for example, in crying).

Chemical Regulation of Respiration

Certain chemical stimuli modulate how quickly and deeply we breathe. The respiratory system functions to maintain proper levels of CO_2 and O_2, and the system is, not surprisingly, highly responsive to changes in the blood levels of either. Chemoreceptors in two locations monitor levels of CO_2 and O_2 and provide input to the respiratory center: **Central chemoreceptors** are located in the medulla oblongata (*central* nervous system), whereas **peripheral chemoreceptors** are located in the walls of systemic arteries and relay impulses to the respiratory center over two cranial nerves of the *peripheral* nervous system.

Central chemoreceptors respond to changes in H^+ or CO_2 concentration, or both, in cerebrospinal fluid. Peripheral chemoreceptors are especially sensitive to changes in CO_2 and H^+ in the blood; these receptors are in the **aortic body,** a cluster of chemoreceptors located in the wall of the arch of the aorta, and in the **carotid bodies,** which are oval nodules in the wall of the left and right common carotid arteries where they divide into the internal and external carotid arteries. Sensory fibers from the aortic body join the vagus (X) nerve, whereas those from the carotid bodies join the right and left glossopharyngeal (IX) nerves.

If there is even a slight increase in CO_2, central and peripheral chemoreceptors are stimulated. The chemoreceptors send nerve impulses to the brain that cause the inspiratory area to become highly active, and the rate of respiration increases. This allows the body to expel more CO_2 until the CO_2 is lowered to normal. If arterial CO_2 is lower than normal, the chemoreceptors are not stimulated, and stimulatory impulses are not sent to the inspiratory area. Consequently, the rate of respiration decreases until CO_2 accumulates and the CO_2 rises to normal.

The Inflation Reflex

Located in the walls of bronchi and bronchioles are stretch-sensitive receptors called **baroreceptors** or **stretch receptors.** When these receptors become stretched during overinflation of the lungs, nerve impulses are sent along the vagus (X) nerves to the inspiratory and apneustic areas. In response, the inspiratory area is inhibited, and the apneustic area is inhibited from activating the inspiratory area. As a result, expiration begins. As air leaves the lungs during expiration, the lungs deflate and the stretch receptors are no longer stimulated. Thus, the inspiratory and apneustic areas are no longer inhibited, and a new inspiration begins. Some evidence suggests that this reflex, referred to as the **inflation (Hering–Breuer) reflex,** is mainly a protective mechanism for preventing excessive inflation of the lungs rather than a key component in the normal regulation of respiration.

- ✓ How does the medullary rhythmicity area function in regulating respiration? How are the apneustic and pneumotaxic areas related to the control of respiration?
- ✓ Explain how each of the following modifies respiration: cerebral cortex, inflation reflex, CO_2 and O_2.

EXERCISE AND THE RESPIRATORY SYSTEM

Objective

- Describe the effects of exercise on the respiratory system.

During exercise, the respiratory and cardiovascular systems make adjustments in response to both the intensity and duration of the exercise. The effects of exercise on the heart are discussed on page 406; here we focus on how exercise affects the respiratory system.

The heart pumps the same amount of blood to the lungs as to all the rest of the body. Thus, as cardiac output rises, the rate of blood flow through the lungs also increases. As blood flows more rapidly through the lungs, it picks up more O_2. In addition, the rate at which O_2 diffuses from alveolar air into the blood increases during maximal exercise because blood flows through a larger percentage of the pulmonary capillaries, providing a greater surface area for the diffusion of O_2 into blood.

When muscles contract during exercise, they consume large amounts of O_2 and produce large amounts of CO_2. During vigorous exercise, O_2 consumption and pulmonary ventilation both increase dramatically. At the onset of exercise, an abrupt increase in pulmonary ventilation is followed by a more gradual increase. With moderate exercise, the increase is due mostly to an increase in the depth of ventilation rather than to increased breathing rate. When exercise is more strenuous, the frequency of breathing also increases.

At the end of an exercise session, an abrupt decrease in pulmonary ventilation is followed by a more gradual decline to the resting level. The initial decrease is due mainly to changes in neural factors when movement stops or slows, whereas the more gradual phase reflects the slower return of blood chemistry levels and temperature to the resting state.

Why Smokers Have Lowered Respiratory Efficiency

Smoking may cause a person to become easily "winded" with even moderate exercise because several factors decrease respiratory efficiency in smokers: (1) Nicotine constricts terminal bronchioles, which decreases airflow into and out of the lungs. (2) Carbon monoxide in smoke binds to hemoglobin and reduces its oxygen-carrying capability. (3) Irritants in smoke cause increased mucus secretion by the mucosa of the bronchial tree and swelling of the mucosal lining, both of which impede airflow into and out of the lungs. (4) Irritants in smoke also inhibit the movement of cilia and destroy cilia in the lining of the respiratory system. Thus, excess mucus and foreign debris are not easily removed, which further adds to the difficulty in breathing. (5) With time, smoking leads to destruction of elastic fibers in the lungs and is the prime cause of emphysema (described on page 708). These changes cause collapse of small bronchioles and the trapping of air in alveoli at the end of exhalation; as a result, gas exchange is less efficient. ■

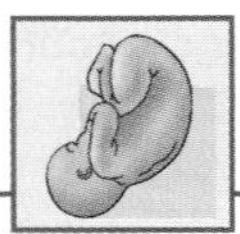

DEVELOPMENTAL ANATOMY OF THE RESPIRATORY SYSTEM

Objective

- Describe the development of the respiratory system.

The development of the mouth and pharynx are discussed on pages 748 and 750; here we consider development of the remainder of the respiratory system. At about 4 weeks of fetal development, the respiratory system begins as an outgrowth of the **endoderm** of the foregut (precursor of some digestive organs) just posterior to the pharynx. This outgrowth is called the **laryngotracheal bud** (see Figure 22.8a on page 676). As the bud grows, it elongates and differentiates into the future epithelial lining of the *larynx* and other structures. Its proximal end maintains a slit-like opening into the pharynx called the *rima glottis.* The middle portion of the bud gives rise to the epithelial lining of the *trachea.* The distal portion divides into two **lung buds,** which grow into the epithelial lining of the *bronchi* and *lungs* (Figure 23.15).

As the lung buds develop, they branch and rebranch and give rise to all the *bronchial tubes.* After the sixth month, the closed terminal portions of the tubes dilate and become the *alve-*

Figure 23.15 / Development of the bronchial tubes and lungs.

The respiratory system develops from endoderm and mesoderm.

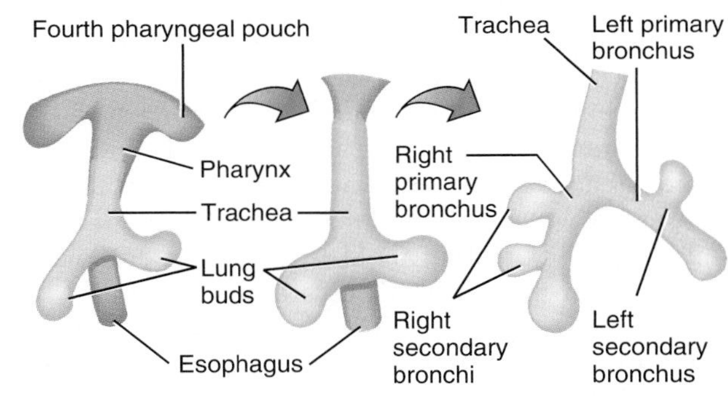

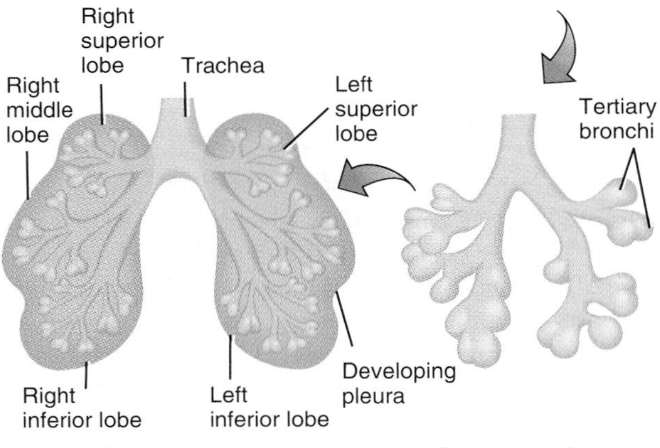

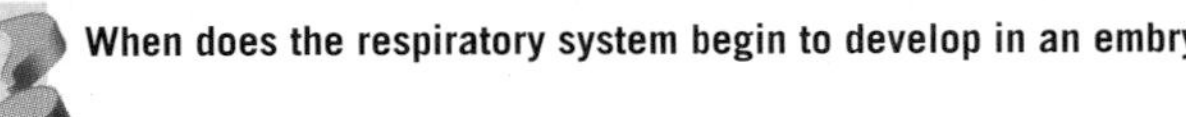

oli of the lungs. The smooth muscle, cartilage, and connective tissues of the bronchial tubes and the pleural sacs of the lungs are contributed by **mesenchymal (mesodermal) cells.** ■

AGING AND THE RESPIRATORY SYSTEM

Objective

- Describe the effects of aging on the respiratory system.

With advancing age, the airways and tissues of the respiratory tract, including the alveoli, become less elastic and more rigid; the chest wall becomes more rigid as well. The result is a decrease in lung capacity. In fact, vital capacity (the maximum amount of air that can be expired after maximal inspiration) can decrease as much as 35% by age 70. Moreover, a decrease in blood levels of O_2, decreased activity of alveolar macrophages, and diminished ciliary action of the epithelium lining the respiratory tract occur. Owing to all these age-related factors, elderly people are more susceptible to pneumonia, bronchitis, emphysema, and other pulmonary disorders.

✓ How does exercise affect the inspiratory area?
✓ What structures develop from the laryngotracheal bud?
✓ What accounts for the decrease in lung capacity with aging?

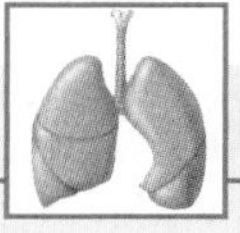

APPLICATIONS TO HEALTH

Asthma

Asthma (AZ-ma; = panting) is a disorder characterized by chronic airway inflammation, airway hypersensitivity to a variety of stimuli, and airway obstruction that is at least partially reversible, either spontaneously or with treatment. It affects 3–5% of the U.S. population and is more common in children than in adults. Airway obstruction may be due to smooth muscle spasms in the walls of smaller bronchi and bronchioles, edema of the mucosa of the airways, increased mucus secretion, and damage to the epithelium of the airway.

Individuals with asthma typically react to low concentrations of agents that do not normally cause symptoms in people without asthma. Sometimes the trigger is an allergen such as pollen, house dust mites, molds, or a particular food. Other common triggers of asthma attacks are emotional upset, aspirin, sulfiting agents (used in wine and beer and to keep greens fresh in salad bars), exercise, and breathing cold air or cigarette smoke.

Symptoms include difficult breathing, coughing, wheezing, chest tightness, tachycardia, fatigue, moist skin, and anxiety. An acute attack is treated by giving an inhaled beta$_2$-adrenergic agonist (albuterol) to help relax smooth muscle in the bronchioles and open up the airways. However, long-term therapy of asthma strives to suppress the underlying inflammation. The anti-inflammatory drugs that are used most often are inhaled corticosteroids (glucocorticoids), cromolyn sodium (Intal), and leukotriene blockers (Accolate).

Chronic Obstructive Pulmonary Disease

Chronic obstructive pulmonary disease (COPD) is a type of respiratory disorder characterized by chronic and recurrent obstruction of airflow, which increases airway resistance. COPD affects about 30 million Americans and is the fourth leading cause of death behind heart disease, cancer, and cerebrovascular disease. The principal types of COPD are emphysema and chronic bronchitis. In most cases, COPD is preventable because its most common cause is cigarette smoking or breathing secondhand smoke. Other causes include air pollution, pulmonary infection, occupational exposure to dusts and gases, and genetic factors. Because men, on average, have more years of exposure to cigarette smoke than women, men are twice as likely as women to suffer from COPD; still, the incidence of COPD in women has risen six-fold in the past 50 years, a reflection of increased smoking among women.

Emphysema

Emphysema (em′-fi-SĒ-ma′; = blown up or full of air) is a disorder characterized by destruction of the walls of the alveoli, producing abnormally large air spaces that remain filled with air during expiration. With less surface area for gas exchange, O_2 diffusion across the damaged respiratory membrane is reduced. Blood O_2 level is somewhat lowered, and any mild exercise that raises the O_2 requirements of the cells leaves the patient breathless. As increasing numbers of alveolar walls are damaged, lung elastic recoil decreases due to loss of elastic fibers, and an increasing amount of air becomes trapped in the lungs at the end of exhalation. Over several years, added inspiratory exertion increases the size of the chest cage, resulting in a "barrel chest."

Emphysema is generally caused by a long-term irritation; cigarette smoke, air pollution, and occupational exposure to industrial dust are the most common irritants. Cigarette smoke not only deactivates a protein that is apparently crucial in preventing emphysema; it also prevents the repair of affected lung tissue.

Treatment consists of cessation of smoking and removing other environmental irritants, exercise training under careful medial supervision, breathing exercises, use of bronchodilators, and oxygen therapy.

Chronic Bronchitis

Chronic bronchitis is a disorder characterized by excessive secretion of bronchial mucus and accompanied by a productive cough (sputum is raised) that lasts for at least three months of the year for two successive years. Cigarette smoking is the leading cause of chronic bronchitis. Inhaled irritants lead to chronic inflammation with an increase in the size and number of mucous glands and goblet cells in the airway epithelium. The thickened and excessive mucus produced narrows the airway and impairs ciliary function. Thus, inhaled pathogens become embedded in airway secretions and multiply rapidly. Besides a productive cough, symptoms of chronic bronchitis are shortness of breath, wheezing, cyanosis, and pulmonary hypertension. Treatment for chronic bronchitis is similar to that for emphysema.

Lung Cancer

In the United States **lung cancer** is the leading cause of cancer death in males and females, accounting for 160,000 deaths annually. At the time of diagnosis, lung cancer is usually well advanced, with distant metastases present in about 55% of patients, and regional lymph node involvement in an additional 25%. Most people with lung cancer die within a year of the initial diagnosis; the overall survival rate is only 10–15%. Cigarette smoke is the most important cause of lung cancer. Roughly 85% of lung cancer cases are related to smoking, and the disease is 10–30 times more common in smokers than nonsmokers. Exposure to secondhand smoke is also associated with lung cancer and heart disease. In the United States, secondhand smoke causes an estimated 4000 deaths a year from lung cancer, and nearly 40,000 deaths a year from heart disease. Other causes of lung cancer are ionizing radiation and inhaled irritants, such as asbestos and radon gas. Emphysema is a common precursor to the development of lung cancer.

The most common type of lung cancer, **bronchogenic carcinoma,** starts in the epithelium of the bronchial tubes. Bronchogenic tumors are named on the basis of where they arise. For example, *adenocarcinomas* develop in peripheral areas of the lungs from bronchial glands and alveolar cells, *squamous cell carcinomas* develop from the epithelium of larger bronchial tubes, and *small (oat) cell carcinomas* develop from epithelial cells in primary bronchi near the hilus of the lungs and tend to involve the mediastinum early on. Depending on the type, bronchogenic tumors may be aggressive, locally invasive, and undergo widespread metastasis. The tumors begin as epithelial lesions that grow to form masses that obstruct the bronchial tubes or invade adjacent lung tissue. Bronchogenic carcinomas metastasize to lymph nodes, the brain, bones, liver, and other organs.

Symptoms of lung cancer are related to the location of the tumor. These may include a chronic cough, spitting blood from the respiratory tract, wheezing, shortness of breath, chest pain, hoarseness, difficulty swallowing, weight loss, anorexia, fatigue, bone pain, confusion, problems with balance, headache, anemia, thrombocytopenia, and jaundice.

Treatment consists of partial or complete surgical removal of a diseased lung (pulmonectomy), radiation therapy, and chemotherapy.

Pneumonia

Pneumonia or **pneumonitis** is an acute infection or inflammation of the alveoli. It is the most common infectious cause of death in the United States, where an estimated 4 million cases occur annually. When certain microbes enter the lungs of susceptible individuals, they release damaging toxins, stimulating inflammation and immune responses that have damaging side effects. The toxins and immune response damage alveoli and bronchial mucous membranes; inflammation and edema cause the alveoli to fill with debris and exudate, interfering with ventilation and gas exchange.

The most common cause of pneumonia is the pneumococcal bacterium *Streptococcus pneumoniae*, but other microbes may also cause pneumonia. Those who are most susceptible to pneumonia are the elderly, infants, immunocompromised individuals (those having AIDS or a malignancy, or those taking immunosuppressive drugs), cigarette smokers, and individuals with an obstructive lung disease. Most cases of pneumonia are preceded by an upper respiratory infection that is frequently viral. Individuals then develop fever, chills, productive or dry cough, malaise, chest pain, and sometimes dyspnea (difficulty breathing), and hemoptysis (spitting blood).

Treatment may involve antibiotics, bronchodilators, oxygen therapy, increased fluid intake, and chest physiotherapy (percussion, vibration, and postural drainage).

Tuberculosis

The bacterium *Mycobacterium tuberculosis* produces an infectious, communicable disease called **tuberculosis (TB)** that most often affects the lungs and the pleurae but may involve other parts of the body. Once the bacteria are inside the lungs, they multiply and cause inflammation, which stimulates neutrophils and macrophages to migrate to the area and engulf the bacteria to prevent their spread. If the immune system is not impaired, the bacteria remain dormant for life, but impaired immunity may enable the bacteria to escape into blood and lymph to infect other organs. In many people, symptoms—fatigue, weight loss, lethargy, anorexia, a low-grade fever, night sweats, cough, dyspnea, chest pain, and hemoptysis—do not develop until the disease is advanced.

During the past several years, the incidence of TB in the United States has risen dramatically. Perhaps the single most important factor related to this increase is the presence of the human immunodeficiency virus (HIV). People infected with HIV are much more likely to develop tuberculosis because their immune systems are impaired. Among the other factors that have contributed to the increased number of cases are homelessness, increased drug abuse, increased immigration from countries with a high prevalence of tuberculosis, increased crowding in housing among the poor, and airborne transmission of tuberculosis in prisons and shelters. In addition, recent outbreaks of tuberculosis involving multi-drug-resistant strains of *Mycobacterium tuberculosis* have occurred because patients fail to complete their antibiotic and other treatment regimens.

Coryza and Influenza

Hundreds of viruses are responsible for **coryza** (ko-RĪ-za), or the **common cold.** A group of viruses called rhinoviruses is responsible for about 40% of all colds in adults. Typical symptoms include sneezing, excessive nasal secretion, dry cough, and congestion. The uncomplicated common cold is not usually accompanied by a fever. Complications include sinusitis, asthma, bronchitis, ear infections, and laryngitis. Recent investigations suggest an association between emotional stress and the common cold: The higher the stress level, the greater the frequency and duration of colds.

Influenza (flu) is also caused by a virus. Its symptoms include chills, fever (usually higher than 101°F = 39°C), headache, and muscular aches. Coldlike symptoms appear as the fever subsides.

Pulmonary Edema

Pulmonary edema is an abnormal accumulation of interstitial fluid in the interstitial spaces and alveoli of the lungs. The edema may arise from increased pulmonary capillary permeability (pulmonary origin) or increased pulmonary capillary pressure (cardiac origin); the latter cause may coincide with congestive heart failure. The most common symptom is dyspnea. Others include wheezing, tachypnea (rapid breathing rate), restlessness, a feeling of suffocation, cyanosis, pallor (paleness), and diaphoresis (excessive perspiration). Treatment consists of administering oxygen, drugs that dilate the bronchioles and lower blood pressure, diuretics to rid the body of excess fluid, and drugs that correct acid–base imbalance; suctioning of airways; and mechanical ventilation.

Cystic Fibrosis

Cystic fibrosis (CF) is an inherited disease of secretory epithelia that affects the airways, liver, pancreas, small intestine, and sweat glands. It is the most common lethal genetic disease in whites: 5% of the population are thought to be genetic carriers. The cause of cystic fibrosis is a genetic mutation affecting a transporter protein that carries chloride ions (Cl^-) across the plasma membranes of many epithelial cells. Because dysfunction of sweat glands causes perspiration to contain excessive sodium chloride (salt), measurement of the excess chloride is one index for diagnosing CF. The mutation also disrupts the normal functioning of several organs by causing ducts within them to become obstructed by thick mucus secretions that do not drain easily from the passageways. Buildup of these secretions leads to inflammation and replacement of injured cells with connective tissue that further blocks the ducts. Clogging and infection of the airways leads to difficulty in breathing and eventual destruction of lung tissue. Lung disease accounts for most deaths from CF. Obstruction of small bile ducts in the liver interferes with digestion and disrupts liver function; clogging of pancreatic ducts prevents digestive enzymes from reaching the small intestine. Because pancreatic juice contains the main fat-digesting enzyme, the person fails to absorb fats or fat-soluble vitamins and thus suffers from vitamin A, D, and K deficiency diseases. With respect to the reproductive systems, blockage of the vas (ductus) deferens leads to infertility in males; the formation of dense mucus plugs in the vagina restricts the entry of sperm into the uterus and can lead to infertility in females.

A child suffering from cystic fibrosis is given pancreatic extract and large doses of vitamins A, D, and K. The recommended diet is high in calories, fats, and proteins, with vitamin supplementation and liberal use of salt.

KEY MEDICAL TERMS ASSOCIATED WITH THE RESPIRATORY SYSTEM

Acute respiratory distress syndrome (ARDS) A form of respiratory failure characterized by excessive leakiness of the respiratory membranes and severe hypoxia. Situations that can cause ARDS include near-drowning, aspiration of acidic gastric juice, drug reactions, inhalation of an irritating gas such as ammonia, allergic reactions, various lung infections such as pneumonia or TB, and pulmonary hypertension. ARDS strikes about 250,000 people a year in the United States, and about 50% of them die despite intensive medical care.

Apnea (AP-nē-a; *a* = without; *pnoia* = air or breath) Absence of ventilatory movements.

Asphyxia (as-FIK-sē-a; *sphyxia* = pulse) Oxygen starvation due to low atmospheric oxygen or interference with ventilation, external respiration, or internal respiration.

Aspiration (as′-pi-RĀ-shun) Inhalation of a foreign substance such as water, food, or a foreign body into the bronchial tree; also, the drawing of a substance in or out by suction.

Bronchiectasis (bron′-kē-EK-ta-sis) A chronic dilation of the bronchi or bronchioles.

Bronchography (bron-KOG-ra-fē) An imaging technique used to visualize the bronchial tree using x-rays. After an opaque contrast medium is inhaled through an intratracheal catheter, radiographs of the chest in various positions are taken, and the developed film, a **bronchogram** (BRON-kō-gram), provides a picture of the bronchial tree.

Bronchoscopy (bron-KOS-kō-pē) Visual examination of the bronchi through a **bronchoscope,** an illuminated, flexible tubular instrument that is passed through the mouth (or nose), larynx, and trachea into the bronchi. The examiner can view the interior of the trachea and bronchi to biopsy a tumor, clear an obstructing object or secretions from an airway, take cultures or smears for microscopic examination, stop bleeding, or deliver drugs.

Cheyne–Stokes respiration (CHĀN STŌKS res′-pi-RĀ-shun) A repeated cycle of irregular breathing that begins with shallow breaths that increase in depth and rapidity and then decrease and cease altogether for 15–20 seconds. Cheyne–Stokes is normal in infants; it is also often seen just before death from pulmonary, cerebral, cardiac, and kidney disease.

Dyspnea (DISP-nē-a; *dys-* = painful, difficult) Painful or labored breathing.

Epistaxis (ep′-i-STAK-sis) Loss of blood from the nose due to trauma, infection, allergy, malignant growths, or bleeding disorders. It can be arrested by cautery with silver nitrate, electrocautery, or firm packing. Also called **nosebleed.**

Heimlich (abdominal thrust) maneuver (HĪM-lik ma-NŪ-ver) First-aid procedure designed to clear the airways of obstructing objects.

It is performed by applying a quick upward thrust between the navel and costal margin that causes sudden elevation of the diaphragm and forceful, rapid expulsion of air in the lungs; this action forces air out the trachea to eject the obstructing object. The Heimlich maneuver is also used to expel water from the lungs of near-drowning victims before resuscitation is begun.

Hemoptysis (hē-MOP-ti-sis; *hemo-* = blood; *-ptysis* = spit) Spitting of blood from the respiratory tract.

Orthopnea (or′-THOP-nē-a; *ortho* = straight) Dyspnea that occurs in the horizontal position. It is an abnormal finding because a normal individual can tolerate the reduction in vital capacity that accompanies the recumbent position.

Rales (RĀLS) Sounds sometimes heard in the lungs that resemble bubbling or rattling. Rales are to the lungs what murmurs are to the heart. Different types are due to the presence of an abnormal type or amount of fluid or mucus within the bronchi or alveoli, or to bronchoconstriction that causes turbulent airflow.

Respirator (RES-pi-rā′-tor) An apparatus fitted to a mask over the nose and mouth, or hooked directly to an endotracheal or tracheotomy tube, that is used to assist or support ventilation or to provide nebulized medication to the air passages.

Respiratory failure A condition in which the respiratory system either cannot supply sufficient O_2 to maintain metabolism or cannot eliminate enough CO_2 to prevent respiratory acidosis (a lower-than-normal pH in extracellular fluid).

Rhinitis (rī-NĪ-tis; *rhin-* = nose) Chronic or acute inflammation of the mucous membrane of the nose.

Sudden infant death syndrome (SIDS) Death of infants between the ages of 1 week and 12 months thought to be due to hypoxia while sleeping in a prone position (on the stomach) and the rebreathing of exhaled air trapped in a depression of the mattress. It is now recommended that normal newborns be placed on their backs for sleeping.

Tachypnea (tak′-ip-NĒ-a; *tachy-* = rapid) Rapid breathing rate.

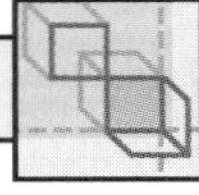

STUDY OUTLINE

Respiratory System Anatomy (p. 685)

1. The respiratory system consists of the nose, pharynx, larynx, trachea, bronchi, and lungs. They act with the cardiovascular system to supply oxygen (O_2) to and remove carbon dioxide (CO_2) from the blood.
2. The external portion of the nose is made of cartilage and skin and is lined with a mucous membrane. Openings to the exterior are the external nares.
3. The internal portion of the nose communicates with the paranasal sinuses and nasopharynx through the internal nares.
4. The nasal cavity is divided by a septum. The anterior portion of the cavity is called the vestibule. The nose warms, moistens, and filters air and functions in olfaction and speech.
5. The pharynx (throat) is a muscular tube lined by a mucous membrane. The anatomic regions of the pharynx are the nasopharynx, oropharynx, and laryngopharynx.
6. The nasopharynx functions in respiration. The oropharynx and laryngopharynx function both in digestion and in respiration.
7. The larynx (voice box) is a passageway that connects the pharynx with the trachea. It contains the thyroid cartilage (Adam's apple); the epiglottis, which prevents food from entering the larynx; the cricoid cartilage, which connects the larynx and trachea; and the paired arytenoid, corniculate, and cuneiform cartilages.
8. The larynx contains vocal folds, which produce sound as they vibrate. Taut folds produce high pitches, and relaxed ones produce low pitches.
9. The trachea (windpipe) extends from the larynx to the primary bronchi. It is composed of C-shaped rings of cartilage and smooth muscle and is lined with pseudostratified ciliated columnar epithelium.
10. The bronchial tree consists of the trachea, primary bronchi, secondary bronchi, tertiary bronchi, bronchioles, and terminal bronchioles. Walls of bronchi contain rings of cartilage; walls of bronchioles contain increasingly smaller plates of cartilage and increasing amounts of smooth muscle.
11. Lungs are paired organs in the thoracic cavity enclosed by the pleural membrane. The parietal pleura is the superficial layer that lines the thoracic cavity; the visceral pleura is the deep layer that covers the lungs.
12. The right lung has three lobes separated by two fissures; the left lung has two lobes separated by one fissure and a depression, the cardiac notch.
13. Secondary bronchi give rise to branches called segmental bronchi, which supply segments of lung tissue called bronchopulmonary segments.
14. Each bronchopulmonary segment consists of lobules, which contain lymphatics, arterioles, venules, terminal bronchioles, respiratory bronchioles, alveolar ducts, alveolar sacs, and alveoli.
15. Alveolar walls consist of type I alveolar cells, type II alveolar cells, and associated alveolar macrophages.
16. Gas exchange occurs across the respiratory membranes.

Mechanics of Pulmonary Ventilation (Breathing) (p. 703)

1. Pulmonary ventilation, or breathing, consists of inspiration and expiration.
2. Inspiration occurs when alveolar pressure falls below atmospheric pressure. Contraction of the diaphragm and external intercostals increases the size of the thorax, thereby decreasing the intrapleural pressure so that the lungs expand. Expansion of the lungs decreases alveolar pressure so that air moves down a pressure gradient from the atmosphere into the lungs.
3. During forceful inspiration, accessory muscles of inspiration (sternocleidomastoids, scalenes, and pectoralis minors) are also used.
4. Expiration occurs when alveolar pressure is higher than atmospheric pressure. Relaxation of the diaphragm and external intercostals results in elastic recoil of the chest wall and lungs, which increases intrapleural pressure; lung volume decreases and alveolar pressure increases, so air moves from the lungs to the atmosphere.
5. Forceful expiration involves contraction of the internal intercostal and abdominal muscles.

Regulation of Respiration (p. 705)

1. The respiratory center consists of a medullary rhythmicity area, a pneumotaxic area, and an apneustic area.
2. The inspiratory area sets the basic rhythm of respiration.
3. The pneumotaxic and apneustic areas coordinate the transition between inspiration and expiration.
4. Respiration may be modified by a number of factors, including cortical influences; the inflation reflex; and chemical stimuli, such as O_2, CO_2, and H^+ levels.

Exercise and the Respiratory System (p. 707)

1. The rate and depth of ventilation change in response to both the intensity and duration of exercise.
2. An increase in pulmonary perfusion and O_2 diffusing capacity occurs during exercise.
3. The abrupt increase in ventilation at the start of exercise is due to neural changes that send excitatory impulses to the inspiratory area in the medulla oblongata. The more gradual increase in ventilation during moderate exercise is due to chemical and physical changes in the bloodstream.

Developmental Anatomy of the Respiratory System (p. 707)

1. The respiratory system begins as an outgrowth of endoderm called the laryngotracheal bud.
2. Smooth muscle, cartilage, and connective tissue of the bronchial tubes and pleural sacs develop from mesoderm.

Aging and the Respiratory System (p. 708)

1. Aging results in decreased vital capacity, decreased blood level of O_2, and diminished alveolar macrophage activity.
2. Elderly people are more susceptible to pneumonia, emphysema, bronchitis, and other pulmonary disorders.

SELF-QUIZ QUESTIONS

Choose the one best answer to these questions.

1. Which of the following is a part of the cartilaginous framework of the nose? (a) arytenoid cartilage, (b) cricoid cartilage, (c) alar cartilage, (d) cartilage of the carina, (e) cuneiform cartilage
2. Which is correct order, from superficial to deep, of the following five structures? (1) parietal pleura, (2) visceral pleura, (3) pleural cavity, (4) lung, (5) wall of thoracic cavity.
   a. 5,1,3,2,4 **b.** 5,3,1,2,4
   c. 4,2,3,1,5 **d.** 5,1,2,4,3
   e. 5,2,3,1,4
3. Which of the following is *not* a part of the nasal septum? (a) ethmoid, (b) nasal conchae (c) hyaline cartilage, (d) vomer, (e) maxillae
4. Which of the following is/are true regarding the pharynx? (1) It serves as a common passage for both the respiratory and digestive systems. (2) Adenoids, or pharyngeal tonsils, are located in the inferior part, the laryngopharynx. (3) The nasal cavity connects with the nasopharynx through two internal nares.
   a. 1 only **b.** 2 only
   c. 3 only **d.** 1 and 3
   e. all of the above
5. Which of the following statements is/are true of the cartilage rings of the trachea? (1) They are made of tough fibrocartilage. (2) They are complete rings of cartilage just like those of the bronchi. (3) They keep the lumen of the trachea open.
   a. 1 only **b.** 2 only
   c. 3 only **d.** all of the above
   e. none of the above
6. Which of the following is *not* a component of the conducting portion of the respiratory system? (a) terminal bronchiole, (b) larynx, (c) bronchus, (d) trachea, (e) respiratory bronchiole
7. Which of the following structures is the smallest in diameter? (a) left primary bronchus, (b) respiratory bronchioles, (c) secondary bronchi, (d) alveolar ducts, (e) right primary bronchus

Complete the following.

8. The broad, inferior portion of the lung that sits on the diaphragm is called the ________ . The ________ surface of each lung contains a hilus. The costal surfaces lie against the ________ .
9. The flap-like structure that prevents aspiration of solid and liquid substances into the trachea is called the ________ .
10. The space between the vocal folds is called the rima ________ , whereas the space between the ventricular folds is called the rima ________ .
11. A section of lung tissue supplied by a tertiary bronchus is called a ________ .
12. Place numbers in the blanks to indicate the route through which air passes as it enters a lobule enroute to alveoli. (a) alveolar ducts: ________ ; (b) respiratory bronchiole: ________ ; (c) terminal bronchiole: ________ .
13. Place numbers in the blanks to indicate the pathway of inspired air. (a) trachea: ________ ; (b) internal nares: ________ ; (c) primary bronchi: ________ ; (d) meatuses: ________ ; (e) larynx: ________ ; (f) tertiary bronchi: ________ ; (g) bronchioles: ________ ; (h) oropharynx: ________ ; (i) external nares ________ ; (j) vestibule: ________ ; (k) nasopharynx: ________ ; (l) secondary bronchi; (m) laryngopharynx: ________ .
14. A comparison of a primary bronchus and a terminal bronchiole would reveal a change in mucous membrane from ________ in the primary bronchus to ________ in the bronchiole.

Are the following statements true or false?

15. Bronchioles have more smooth muscle and less cartilage than bronchi.
16. The pharynx, larynx, and trachea are parts of the upper respiratory system.
17. The opening from the oral cavity into the oropharynx is called the fauces.
18. The cricoid cartilages attach the epiglottis to the glottis.

19. Match the following terms to their descriptions.

____	**(a)** nasal conchae and meatuses are located here	**(1)** nasal cavity
____	**(b)** contains pharyngeal tonsil	**(2)** oropharynx
____	**(c)** palatine tonsils are located here	**(3)** laryngopharynx
____	**(d)** this structure leads directly into the esophagus	**(4)** larynx
____	**(e)** cricoid, epiglottis, and thyroid cartilages are here	**(5)** nasopharynx
____	**(f)** vocal cords here enable voice production	
____	**(g)** both a respiratory and digestive pathway (two answers)	
____	**(h)** fauces opens into this structure	

20. Match the following.

____	**(a)** superior part of nasal cavity	**(1)** pseudostratified ciliated columnar epithelium
____	**(b)** alveoli	**(2)** stratified columnar epithelium
____	**(c)** nasopharynx	**(3)** simple squamous epithelium
____	**(d)** larynx (inferior to vocal folds)	**(4)** nonciliated simple cuboidal epithelium
____	**(e)** primary bronchus	
____	**(f)** oropharynx	
____	**(g)** trachea	
____	**(h)** terminal bronchioles	

21. Match the following terms to their descriptions.

____	**(a)** thicker, broader, and shorter	**(1)** right lung
____	**(b)** has a cardiac notch	**(2)** left lung
____	**(c)** has only two secondary bronchi	
____	**(d)** has a horizontal fissure	
____	**(e)** has a primary bronchus that is shorter, wider, and more vertical	

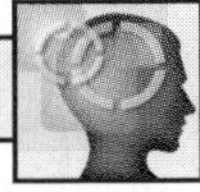

CRITICAL THINKING QUESTIONS

1. Your friend Hedge wants to pierce his nose to go along with the 6 earrings in his ear. He thinks a ring through the center would be awesome but wonders if there's a difference between piercing the center versus the side of the nose. Is there?
   HINT *Feel your entire ear and then your entire nose.*
2. Suzanne just finished listening to one of those cute answering machine messages recorded by a young child. She's still trying to figure out what the kid said in the greeting and wonders if the child was a boy or a girl. Why can you usually tell adult females and males apart by their voices but all little boys and girls sound alike?
   HINT *What happens to boys and girls when they go through puberty?*
3. Dawson was riding his new mountain bike through the woods near his home when he crashed into some brush. He felt a sharp pain in his thorax and when he rolled over, a large stick was impaled in his side. Now he's having trouble breathing. Why?
   HINT *If he removes the stick, he may whistle when he breathes, but the sound won't be coming from his mouth.*
4. Greta was having dinner at the "Fast Bite" diner when the guy next to her, stood up, grabbed his throat, and made gurgling noises. He seemed to be unable to talk or breathe. A waiter grabbed the man around the waist, squeezed, and he was breathing again! How did a belly squeeze save the day?
   HINT *A big chunk of tough roast beef popped out of the man's mouth when he was squeezed.*
5. LaTasha is losing her patience with her little sister, LaTonya. "I'm going to hold my breath 'til I turn blue and die and then you're gonna get it!" screams LaTonya. LaTasha is not too worried. Why not?
   HINT *LaTonya's about to find out that you can't always have your own way.*

ANSWERS TO FIGURE QUESTIONS

22.1 Nose, pharynx, larynx, trachea, bronchi, and bronchioles, except the respiratory bronchioles.

22.2 External nares → vestibule → nasal cavity → internal nares.

22.3 Superior: internal nares; inferior: cricoid cartilage.

22.4 During swallowing, the epiglottis closes over the rima glottidis, the entrance to the trachea.

22.5 Voice production.

22.6 Because the tissues between the esophagus and trachea are soft, the esophagus can bulge into the trachea during swallowing.

22.7 Left lung has two of each; right lung has three of each.

22.8 Serous membrane.

22.9 Because two-thirds of the heart lies to the left of the midline, the left lung contains a cardiac notch to accommodate the position of the heart.

22.10 Tertiary bronchi.

22.11 Type I alveolar cells, type II alveolar cells, and alveolar macrophages.

22.12 It is 0.5μm thick.

22.13 If you are at rest while reading, your diaphragm.

22.14 Medullary inspiratory area.

22.15 At about 4 weeks.

24

THE DIGESTIVE SYSTEM

CONTENTS AT A GLANCE

◄ Page 732

How many body organs can you identify in this illustration?

Page 720 ►

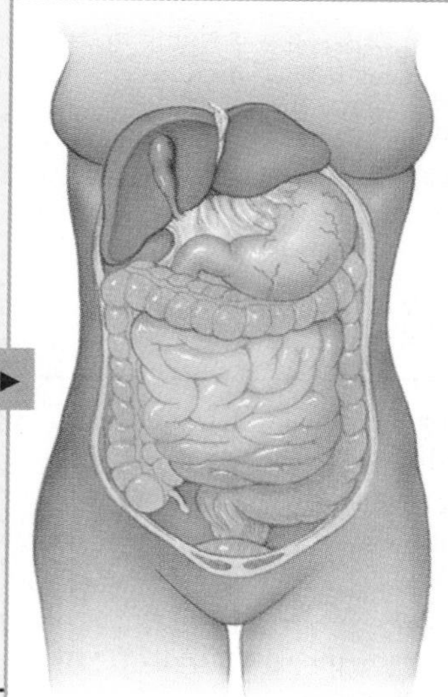

INTRODUCTION

Food contains a variety of nutrients—molecules needed for building new body tissues, repairing damaged tissues, and sustaining needed chemical reactions. Food is also vital for life because it is the source of energy that drives the chemical reactions occurring in every cell. As consumed, however, most food cannot be used as a source of cellular energy. First, it must be broken down into molecules small enough to cross the plasma membranes of cells, a process known as **digestion.** The passage of these smaller molecules through cells into the blood and lymph is termed **absorption.** The organs that collectively perform these functions—the **digestive system**—are the focus of this chapter.

The medical specialty that deals with the structure, function, diagnosis, and treatment of diseases of the stomach and intestines is called **gastroenterology** (gas′-trō-en′-ter-OL-ō-jē; *gastro-* = stomach; *enter-* = intestines; *-ology* = study of). The medical specialty that deals with the diagnosis and treatment of disorders of the rectum and anus is called **proctology** (prok-TOL-ō-jē; *proct-* = rectum).

OVERVIEW OF THE DIGESTIVE SYSTEM

Objectives

- Identify the organs of the digestive system.
- Describe the basic processes performed by the digestive system.

The digestive system (Figure 24.1) is composed of two groups of organs: the gastrointestinal (GI) tract and the accessory digestive organs. The **gastrointestinal (GI) tract,** or **alimentary canal** (*alimentary* = nourishment), is a continuous tube that extends from the mouth to the anus through the ventral body cavity. Organs of the gastrointestinal tract include the mouth, most of the pharynx, esophagus, stomach, small intestine, and large intestine. The length of the GI tract taken from a cadaver is about 9 m (30 ft). In a living person it is shorter because the muscles along the walls of GI tract organs are in a state of tonus (sustained contraction). The **accessory digestive organs** are the teeth, tongue, salivary glands, liver, gallbladder, and pancreas. Teeth aid in the physical breakdown of food, and the tongue assists in chewing and swallowing. The other accessory digestive organs, however, never come into direct contact with food. They produce or store secretions that flow into the GI tract through ducts and aid in the chemical breakdown of food.

The GI tract contains food from the time it is eaten until it is digested and absorbed or eliminated. Muscular contractions in the wall of the GI tract physically break down the food by churning it. The contractions also help to dissolve foods by mixing them with fluids secreted into the tract. Enzymes secreted by accessory structures and cells that line the tract break down the food chemically. Wavelike contractions of the smooth muscle in the wall of the GI tract propel the food along the tract, from the esophagus to the anus.

Overall, the digestive system has six basic functions:

1. **Ingestion.** This process involves taking foods and liquids into the mouth (eating).
2. **Secretion.** Each day, cells within the walls of the GI tract and accessory digestive organs secrete a total of about 7 liters of water, acid, buffers, and enzymes into the lumen of the tract.
3. **Mixing and propulsion.** Alternating contraction and relaxation of smooth muscle in the walls of the GI tract mix food and secretions and propel them toward the anus. This capability of the GI tract to mix and move material along its length is termed **motility.**
4. **Digestion.** Mechanical and chemical processes break down ingested food into small molecules. In **mechanical digestion** the teeth cut and grind food before it is swallowed, and then smooth muscles of the stomach and small intestine churn the food. As a result, food molecules become dissolved and thoroughly mixed with digestive enzymes. In **chemical digestion** the large carbohydrate, lipid, protein, and nucleic acid molecules in food are split into smaller molecules. Digestive enzymes produced by the salivary glands, tongue, stomach, pancreas, and small intestine catalyze these catabolic reactions. A few substances in food can be absorbed without chemical digestion, including amino acids, cholesterol, glucose, vitamins, minerals, and water.
5. **Absorption.** During absorption, the secreted fluids and the small molecules and ions that are products of digestion enter the epithelial cells lining the lumen of the GI tract. The absorbed substances pass into blood or lymph and circulate to cells throughout the body.
6. **Defecation.** Wastes, indigestible substances, bacteria, cells sloughed from the lining of the GI tract, and digested materials that were not absorbed leave the body through the anus in a process called **defecation.** The eliminated material is termed **feces.**

✓ List the GI tract organs and the accessory digestive organs.
✓ Which organs of the digestive system come in contact with food, and what are some of their digestive functions?

LAYERS OF THE GI TRACT

Objective

- Describe the layers that form the wall of the gastrointestinal tract.

The wall of the GI tract, from the esophagus to the anal canal, has the same basic, four-layered arrangement of tissues. The four layers of the tract, from deep to superficial, are the mucosa, submucosa, muscularis, and serosa (Figure 24.2 on page 717).

Figure 24.1 / Organs of the digestive system.

Organs of the gastrointestinal (GI) tract are the mouth, pharynx, esophagus, stomach, small intestine, and large intestine. Accessory digestive organs are the teeth, tongue, salivary glands, liver, gallbladder, and pancreas.

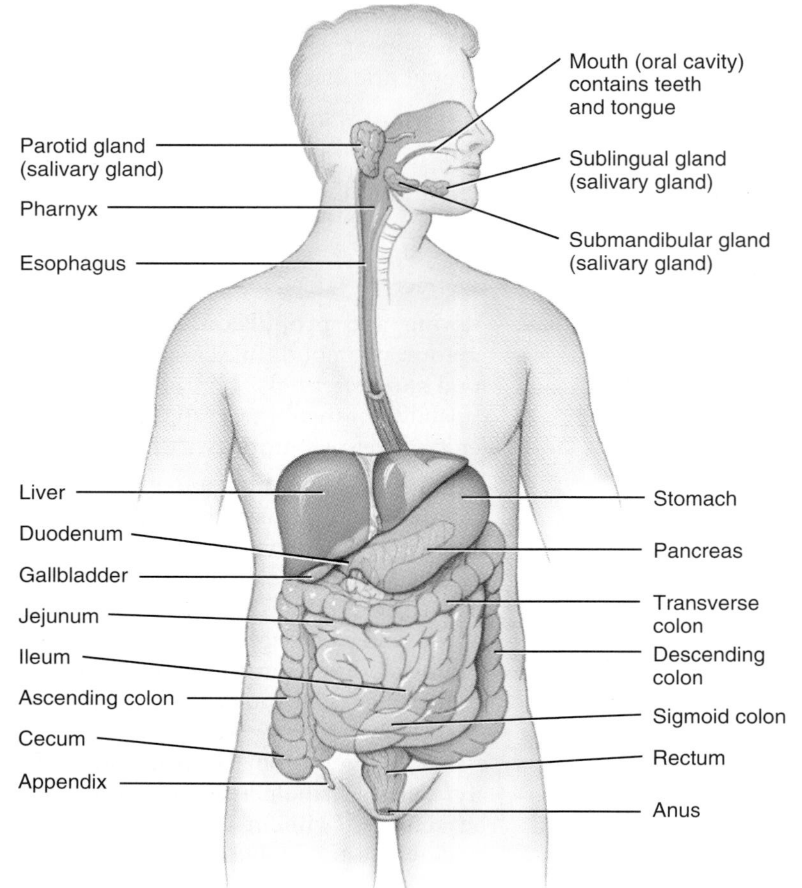

(a) Right lateral view of head and neck and anterior view of trunk

Functions

1. Ingestion. Taking food into the mouth.
2. Secretion. Release of water, acid, buffers, and enzymes into the lumen of the GI tract.
3. Mixing and propulsion. Churning and propulsion of food through the GI tract.
4. Digestion. Mechanical and chemical breakdown of food.
5. Absorption. Passage of digested products from the GI tract into the blood and lymph.
6. Defecation. The elimination of feces from the GI tract.

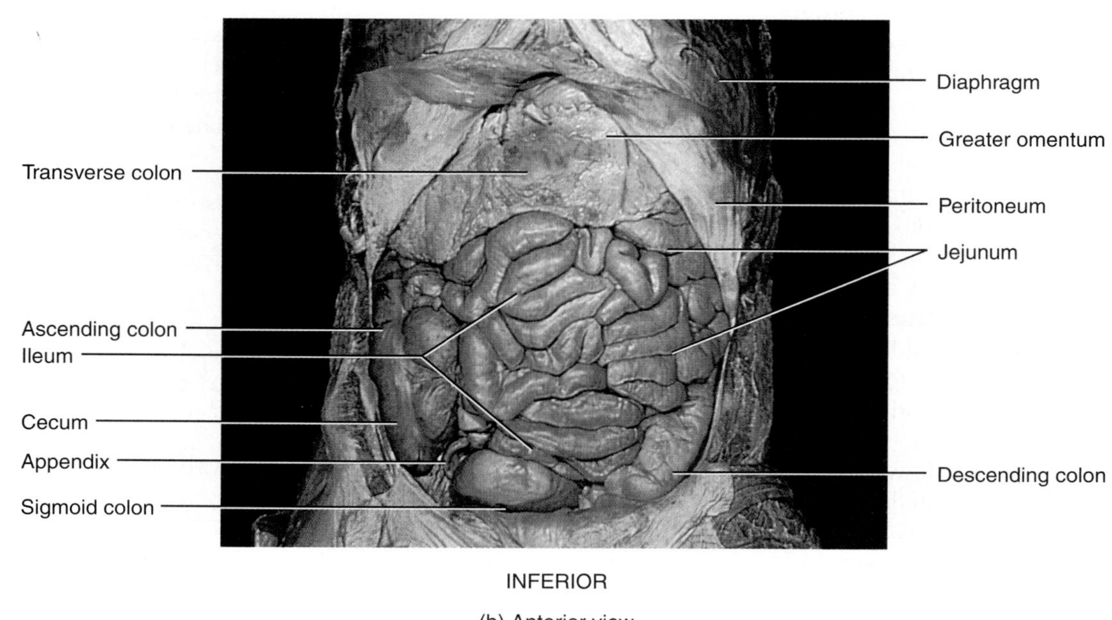

(b) Anterior view

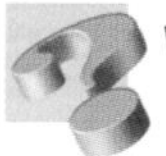

Which structures of the digestive system secrete digestive enzymes?

Figure 24.2 / Three-dimensional depiction of the various layers of the gastrointestinal tract.

The four layers of the GI tract, from deep to superficial, are the mucosa, submucosa, muscularis, and serosa.

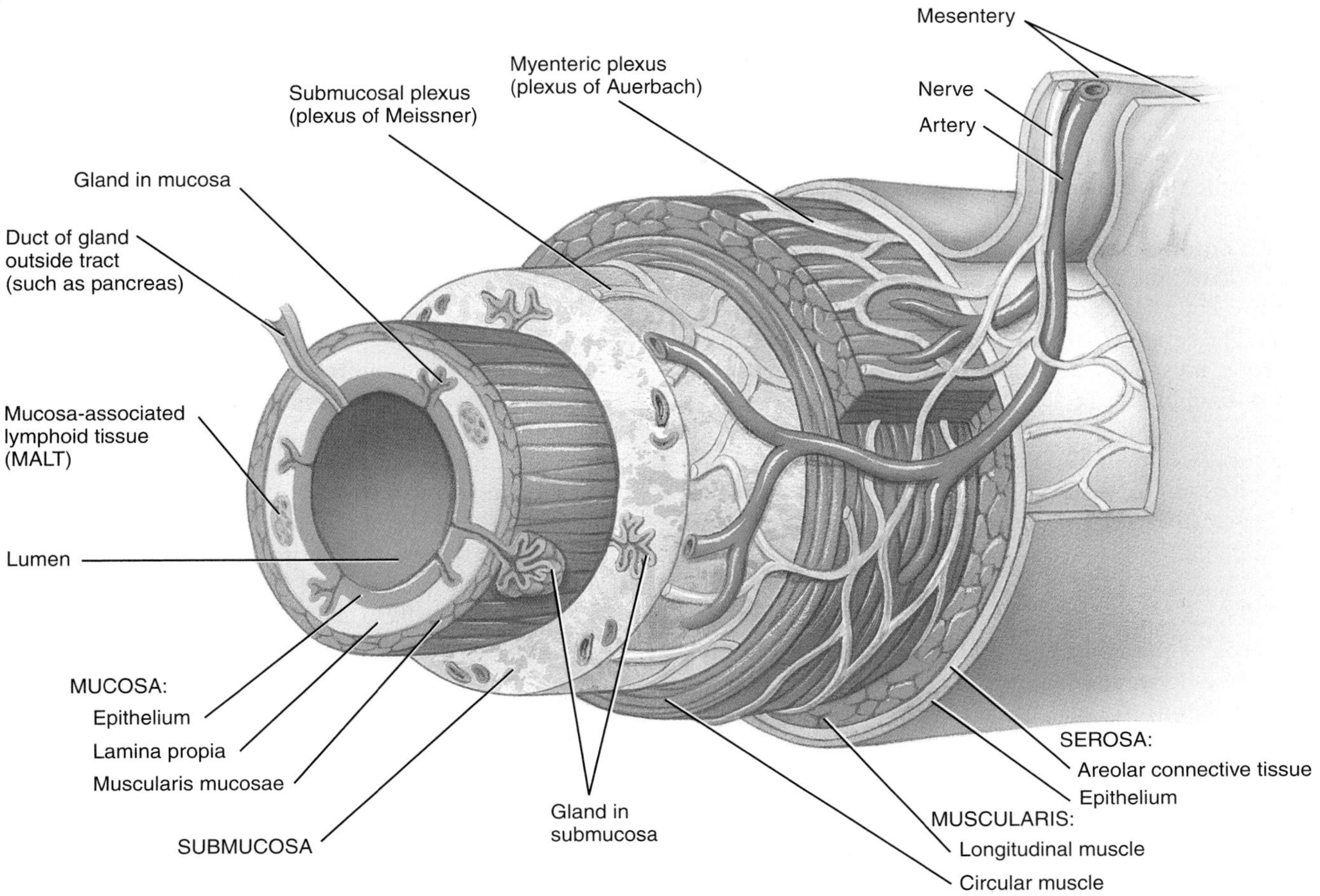

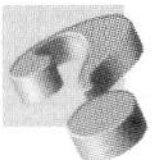

What is the function of the enteric plexuses in the wall of the gastrointestinal tract?

Mucosa

The lumen of the GI tract is lined by a mucous membrane, the **mucosa.** The GI tract mucosa contains three layers: (1) a lining of epithelium in direct contact with the contents of the tract, (2) an underlying layer of areolar connective tissue, and (3) a thin layer of smooth muscle.

1. The **epithelium** in the mouth, pharynx, esophagus, and anal canal is mainly nonkeratinized stratified squamous epithelium that serves a protective function. Simple columnar epithelium, which functions in secretion and absorption, lines the stomach and intestines. Neighboring simple columnar epithelial cells are firmly sealed to each other by tight junctions that restrict leakage between the cells. The rate of renewal of GI tract epithelial cells is rapid: Every 5–7 days they slough off and are replaced by new cells. Located among the absorptive epithelial cells are exocrine cells that secrete mucus and fluid into the lumen of the tract, and several types of endocrine cells, collectively called **enteroendocrine cells,** that secrete hormones into the bloodstream.

2. The **lamina propria** (*lamina* = thin, flat plate) is an areolar connective tissue layer containing many blood and lymphatic vessels that carry the nutrients absorbed into the GI tract to the other tissues of the body. This layer supports the epithelium and binds it to the muscularis mucosae (discussed next). The lamina propria also contains most of the cells of the **mucosa-associated lymphoid tissue (MALT).** These prominent lymphatic nodules contain immune system cells that protect against disease. MALT is present all along the GI tract, especially in the tonsils, small intestine, appendix, and large intestine, and it contains about as many immune cells as are present in all the rest of the body. The lymphocytes and macrophages in MALT mount immune responses against microbes, such as bacteria, that may penetrate the epithelium.

3. A thin layer of smooth muscle fibers called the **muscularis mucosae** throws the mucous membrane of the stomach and small intestine into many small folds that increase the surface area for digestion and absorption. Movements of the muscularis mucosae ensure that all absorptive cells are fully exposed to the contents of the GI tract.

Submucosa

The **submucosa** consists of areolar connective tissue that binds the mucosa to the third layer, the muscularis. It is highly vascular and contains the **submucosal plexus,** or *plexus of Meissner,* a portion of the **enteric nervous system (ENS).** The ENS is the "brain of the gut" and consists of approximately 100 million neurons in two enteric plexuses that extend the entire length of the GI tract. The submucosal plexus contains sensory and motor enteric neurons, plus parasympathetic and sympathetic postganglionic fibers that innervate the mucosa and submucosa. The plexus regulates movements of the mucosa and vasoconstriction of blood vessels. Because it also innervates secretory cells of mucosal glands, it is important in controlling secretions by the GI tract. The submucosa may also contain glands and lymphatic tissue.

Muscularis

The **muscularis** of the mouth, pharynx, and superior and middle parts of the esophagus contains *skeletal muscle* that produces voluntary swallowing. Skeletal muscle also forms the external anal sphincter, which permits voluntary control of defecation. Throughout the rest of the tract, the muscularis consists of *smooth muscle* that is generally found in two sheets: an inner sheet of circular fibers and an outer sheet of longitudinal fibers. Involuntary contractions of the smooth muscles help break down food physically, mix it with digestive secretions, and propel it along the tract. The muscularis also contains the second plexus of the enteric nervous system—the **myenteric plexus** (*my-* = muscle), or *plexus of Auerbach,* which contains enteric neurons, parasympathetic ganglia and postganglionic fibers, and sympathetic postganglionic fibers that innervate the muscularis. This plexus mostly controls GI tract motility, in particular the frequency and strength of contraction of the muscularis.

Serosa

The **serosa** is the superficial layer of those portions of the GI tract that are suspended in the abdominopelvic cavity. It is a serous membrane composed of areolar connective tissue and simple squamous epithelium. As we will see shortly, the esophagus, which passes through the mediastinum, has a superficial layer called the *adventitia* composed of areolar connective tissue. Inferior to the diaphragm, the serosa is also called the **visceral peritoneum;** it forms a portion of the peritoneum, which we examine in detail next.

✓ Where along the GI tract is the muscularis composed of skeletal muscle? Is control of this skeletal muscle voluntary or involuntary?

✓ What two plexuses form the enteric nervous system, and where are they located?

PERITONEUM

Objective

- Describe the peritoneum and its folds.

The **peritoneum** (per′-i-tō-NĒ-um; *peri-* = around) is the largest serous membrane of the body; it consists of a layer of simple squamous epithelium (mesothelium) with an underlying supporting layer of connective tissue. Whereas the **parietal peritoneum** lines the wall of the abdominopelvic cavity, the **visceral peritoneum** covers some of the organs in the cavity and is their serosa (Figure 24.3a). The slim space between the parietal and visceral portions of the peritoneum is called the **peritoneal cavity,** which contains serous fluid. In certain diseases, the peritoneal cavity may become distended by the accumulation of several liters of fluid, a condition called **ascites** (a-SĪ-tēz).

As we will see, some organs lie on the posterior abdominal wall and are covered by peritoneum on their anterior surfaces only. Such organs, including the kidneys and pancreas, are said to be **retroperitoneal** (*retro-* = behind).

Unlike the pericardium and pleurae, which smoothly cover the heart and lungs, the peritoneum contains large folds that weave between the viscera. The folds bind the organs to each other and to the walls of the abdominal cavity and contain blood and lymphatic vessels and nerves that supply the abdominal organs.

The **greater omentum,** the largest peritoneal fold, hangs loosely like a "fatty apron" over the transverse colon and coils of the small intestine (see Figure 24.3a, d). It is a double sheet that folds back upon itself, and thus it is a four-layered structure. From attachments along the stomach and duodenum, the greater omentum extends downward anterior to the small intestine, then turns and extends upward and attaches to the transverse colon. The greater omentum contains considerable adipose tissue and many lymph nodes. It contributes macrophages and antibody-producing plasma cells that help combat an infection of the GI tract and prevent the infection from spreading.

The **falciform ligament** (FAL-si-form; *falc-* = sickle-shaped) attaches the liver to the anterior abdominal wall and diaphragm (Figure 24.3b). The liver is the only digestive organ that is attached to the anterior abdominal wall.

The **lesser omentum** (ō-MENT-um; = fat skin) arises as two folds in the serosa of the stomach and duodenum, and it suspends the stomach and duodenum from the liver (Figure 24.3a, c). It contains some lymph nodes.

Another fold of the peritoneum, called the **mesentery** (MEZ-en-ter′-ē; *mes-* = middle), is an outward fold of the serous coat of the small intestine (see Figure 24.3a, d); the tip of the fold binds the small intestine to the posterior abdominal wall.

A fold of peritoneum called the **mesocolon** (mez′-ō-KŌ-lon) binds the large intestine to the posterior abdominal wall (Figure 24.3a); it also carries blood and lymphatic vessels to the intestines. The mesentery and mesocolon hold the intestines loosely in place, allowing for a great amount of movement as muscular contractions mix and move the luminal contents along the GI tract.

Peritonitis is an acute inflammation of the peritoneum. A common cause of the condition is contamination of the peritoneum by infectious microbes, which can result from accidental or surgical wounds in the abdominal wall, or from perforation or rupture of abdominal organs. If, for example, bacteria gain access to the peritoneal cavity through an intestinal perforation or rupture of the appendix, they can produce an acute, life-threatening form of peritonitis. A less serious (but still painful) form of peritonitis can result from the rubbing together of inflamed peritoneal surfaces.

✓ Describe the locations of the visceral peritoneum and parietal peritoneum.
✓ Describe the attachment sites and functions of the mesentery, mesocolon, falciform ligament, lesser omentum, and greater omentum.

MOUTH

Objectives

- Identify the locations of the salivary glands, and describe the functions of their secretions.
- Describe the structure and functions of the tongue.
- Identify the parts of a typical tooth, and compare deciduous and permanent dentitions.

The **mouth,** also referred to as the **oral** or **buccal cavity** (BUK-al; *bucca* = cheeks), is formed by the cheeks, hard and soft palates, and tongue (Figure 24.4 on page 721). Forming the lateral walls of the oral cavity are the **cheeks**—muscular structures covered externally by skin and internally by nonkeratinized stratified squamous epithelium. The anterior portions of the cheeks end at the lips.

The **lips** or **labia** (= fleshy borders) are fleshy folds surrounding the opening of the mouth. They are covered externally

Figure 24.3 / Relationship of the peritoneal folds to each other and to organs of the digestive system.
The size of the peritoneal cavity in (a) is exaggerated for emphasis.

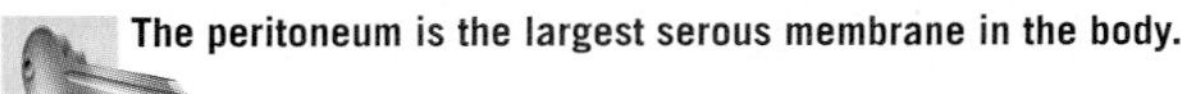

The peritoneum is the largest serous membrane in the body.

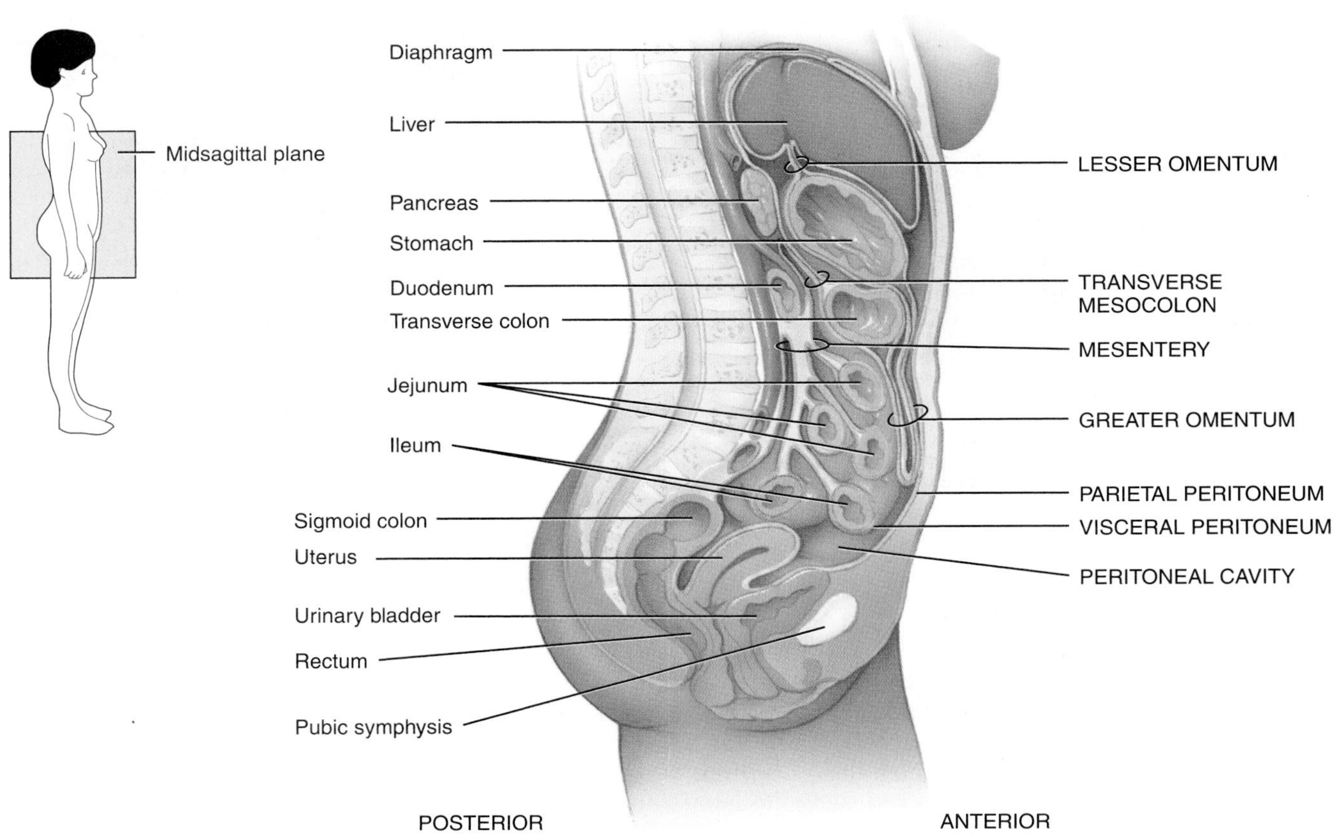

(a) Midsagittal section showing the peritoneal folds

(continues)

Figure 24.3 (continued)

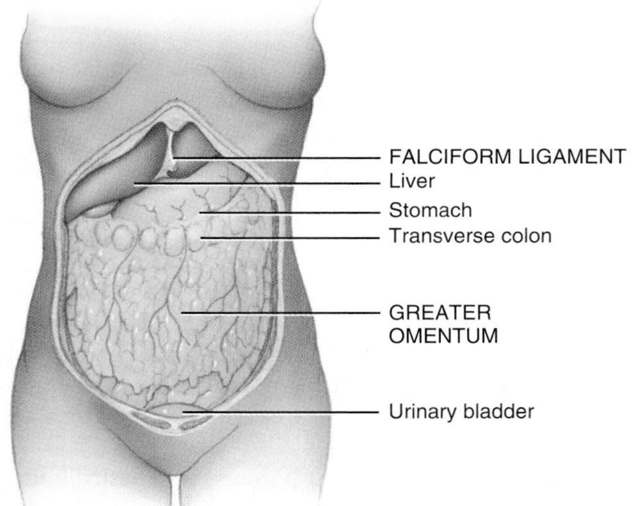

(b) Anterior view

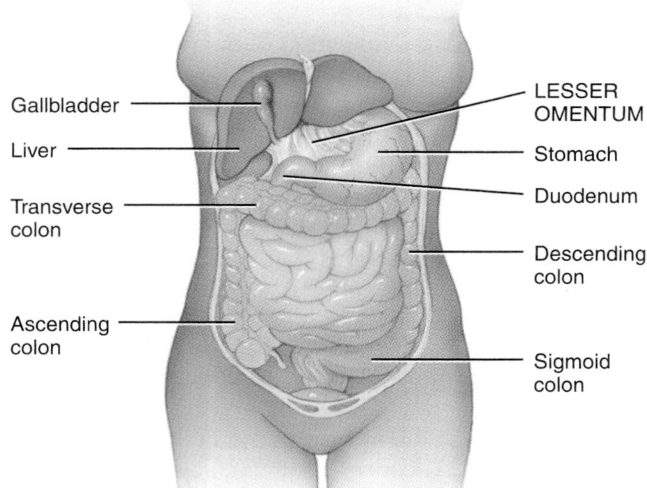

(c) Lesser omentum, anterior view (liver and gallbladder lifted)

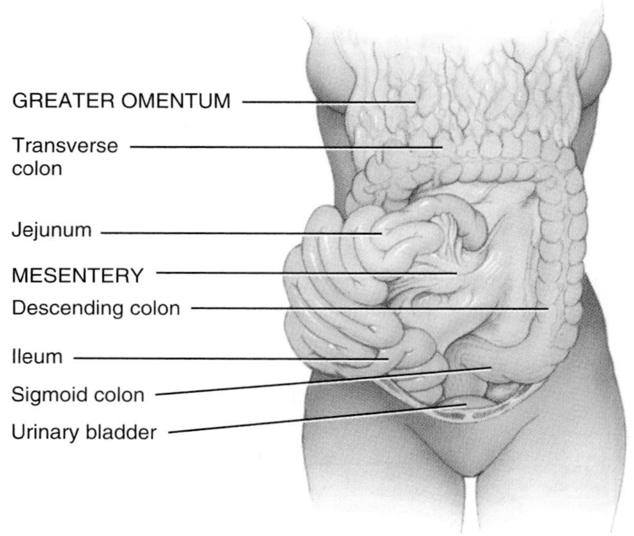

(d) Anterior view (greater omentum lifted and small intestine reflected to right side)

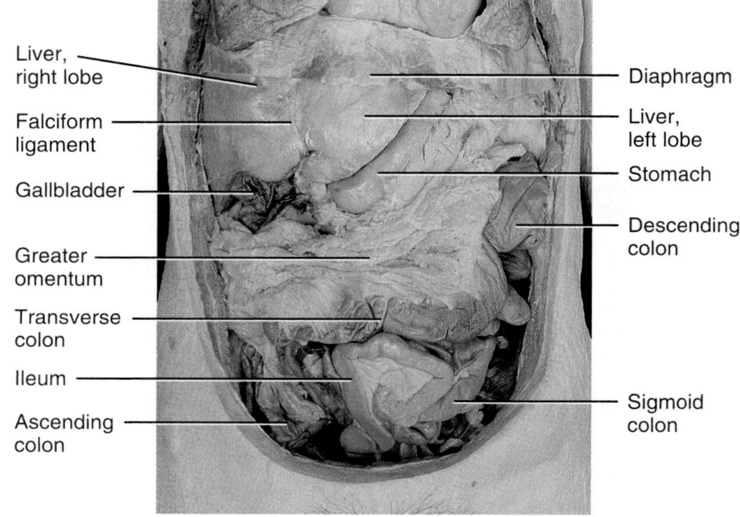

(e) Anterior view

Which peritoneal fold binds the small intestine to the posterior abdominal wall?

by skin and internally by a mucous membrane. There is a transition zone where the two kinds of covering tissue meet. This portion of the lips is nonkeratinized, and the color of the blood in the underlying blood vessels is visible through the transparent surface layer. The inner surface of each lip is attached to its corresponding gum by a midline fold of mucous membrane called the **labial frenulum** (LĀ-bē-al FREN-yū-lum; *frenulum* = small bridle).

The orbicularis oris muscle and connective tissue lie between the skin and the mucous membrane of the oral cavity. During chewing, contraction of the buccinator muscles in the cheeks and orbicularis oris muscle in the lips help keep food between the upper and lower teeth. These muscles also assist in speech.

The **vestibule** (= entrance to a canal) of the oral cavity is a space bounded externally by the cheeks and lips and internally

Figure 24.4 / Structures of the mouth (oral cavity).

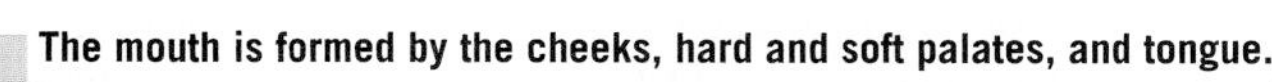

The mouth is formed by the cheeks, hard and soft palates, and tongue.

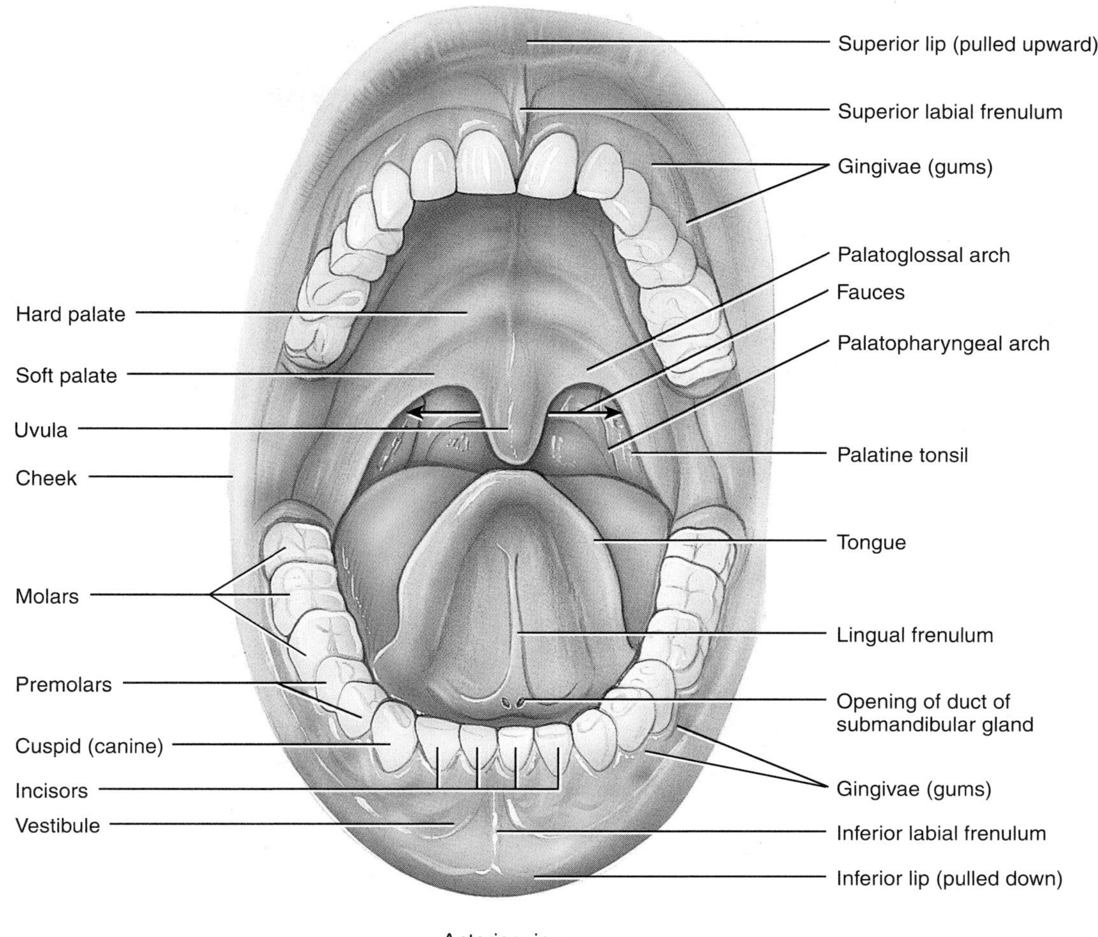

Anterior view

What is the function of the uvula?

by the gums and teeth. The **oral cavity proper** is a space that extends from the gums and teeth to the **fauces** (FAW-sēz; = passages), the opening between the oral cavity and the pharynx or throat.

The **hard palate,** the anterior portion of the roof of the mouth, is formed by the maxillae and palatine bones. It is covered by mucous membrane and forms a bony partition between the oral and nasal cavities. The **soft palate,** which forms the posterior portion of the roof of the mouth, is an arch-shaped muscular partition between the oropharynx and nasopharynx that is lined by mucous membrane.

Hanging from the free border of the soft palate is a conical muscular process called the **uvula** (YŪ-vyū-la; = little grape). During swallowing, the soft palate and uvula are drawn superiorly, closing off the nasopharynx and preventing swallowed foods and liquids from entering the nasal cavity. Lateral to the base of the uvula are two muscular folds that run down the lateral sides of the soft palate: Anteriorly, the **palatoglossal arch** extends to the side of the base of the tongue; posteriorly, the **palatopharyngeal arch** (PAL-a-tō-fa-rin′-jē-al) extends to the side of the pharynx. The palatine tonsils are situated between the arches, and the lingual tonsils are situated at the base of the tongue. At the posterior border of the soft palate, the mouth opens into the oropharynx through the fauces (Figure 24.4).

Salivary Glands

A **salivary gland** releases a secretion called saliva into the oral cavity. Ordinarily, just enough saliva is secreted to keep the mucous membranes of the mouth and pharynx moist and to cleanse the mouth and teeth. When food enters the mouth, however, secretion of saliva increases, and it lubricates, dissolves, and begins the chemical breakdown of the food.

The mucous membrane of the mouth and tongue contains many small salivary glands that open directly, or indirectly via short ducts, to the oral cavity. These glands include *labial, buccal,* and *palatal glands* in the lips, cheeks, and palate, respectively, and *lingual glands* in the tongue, all of which make a small contribution to saliva. However, most saliva is secreted by the **major salivary glands,** which lie beyond the oral mucosa. Their secretions empty into ducts that lead to the oral cavity.

There are three pairs of major salivary glands: the parotid, submandibular, and sublingual glands (Figure 24.5a). The **parotid glands** (*par-* = near; *ot-* = ear) are located inferior and anterior to the ears, between the skin and the masseter muscle. Each secretes saliva into the oral cavity via a **parotid (Stensen's) duct** that pierces the buccinator muscle to open into the vestibule opposite the second maxillary (upper) molar tooth. The **submandibular glands** are found beneath the base of the tongue in the posterior part of the floor of the mouth. Their ducts, the **submandibular (Wharton's) ducts,** run under the mucosa on either side of the midline of the floor of the mouth and enter the oral cavity proper lateral to the lingual frenulum. The **sublingual glands** are superior to the submandibular glands. Their ducts, the **lesser sublingual (Rivinus') ducts,** open into the floor of the mouth in the oral cavity proper.

The parotid gland receives its blood supply from branches of the external carotid artery and is drained by tributaries of the external jugular vein. The submandibular gland is supplied by

Figure 24.5 / Major salivary glands. The light micrograph of the submandibular gland shown in (b) consists mostly of serous acini (serous-fluid–secreting portions of gland) and a few mucous acini (mucus-secreting portions of gland); the parotid glands consist of serous acini only, and the sublingual glands consist of mostly mucous acini and a few serous acini. (See Tortora, *A Photographic Atlas of the Human Body,* Figure 12.6.)

Saliva lubricates and dissolves foods and begins the chemical breakdown of carbohydrates and lipids.

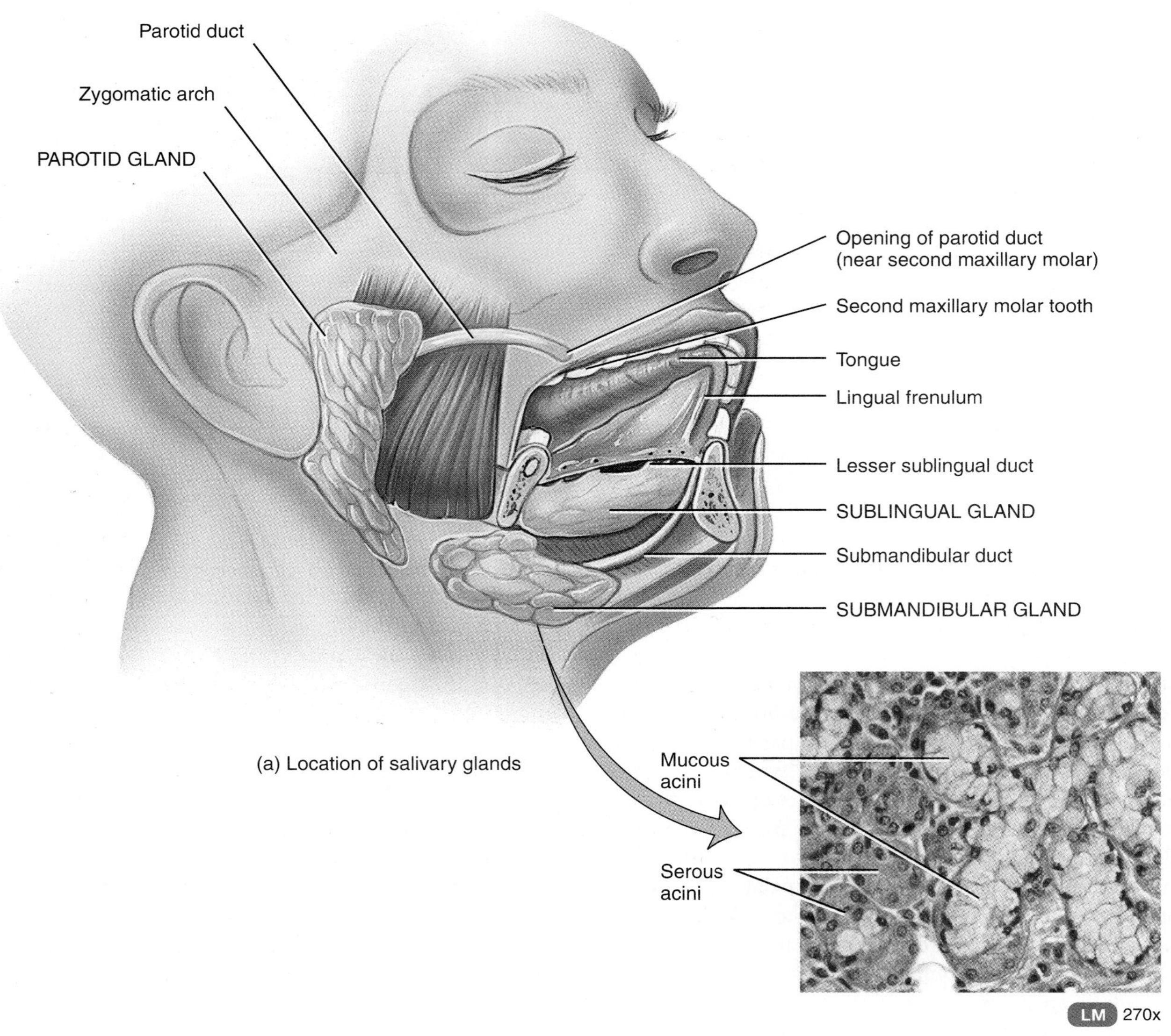

(a) Location of salivary glands

(b) Submandibular gland

The ducts of which salivary glands empty on either side of the lingual frenulum?

branches of the facial artery and drained by tributaries of the facial vein. The sublingual gland is supplied by the sublingual branch of the lingual artery and the submental branch of the facial artery and is drained by tributaries of the sublingual and submental veins.

The salivary glands receive both sympathetic and parasympathetic innervation. The sympathetic fibers form plexuses on the blood vessels that supply the glands and initiate vasoconstriction, which decreases the production of saliva. The parotid gland receives sympathetic fibers from the plexus on the external carotid artery, whereas the submandibular and sublingual glands receive sympathetic fibers that contribute to the sympathetic plexus and accompany the facial artery to the glands. The parasympathetic fibers of the glands produce vasodilation and thus increase the production of saliva.

The fluids secreted by the buccal glands, minor salivary glands, and the three pairs of major salivary glands constitute **saliva.** Amounts of saliva secreted daily vary considerably but range from 1000 to 1500 mL (1 to 1.6 qt). Chemically, saliva is 99.5% water and 0.5% solutes and has a slightly acidic pH (6.35 to 6.85). The solute portion includes mucus, the bacteriolytic enzyme lysozyme, the digestive enzymes salivary amylase and lingual lipase, and traces of salts, proteins, and other organic compounds. **Salivary amylase** initiates the breakdown of starch. **Lingual lipase** is secreted by glands on the dorsum of the tongue. This enzyme, which is active in the stomach, can digest as much as 30% of dietary triglycerides (fats) into simpler fatty acids and monoglycerides.

Secretion of saliva, or **salivation** (sal-i-VĀ-shun), is controlled by the nervous system. Normally, parasympathetic stimulation promotes continuous secretion of a moderate amount of saliva, which keeps the mucous membranes moist and lubricates the movements of the tongue and lips during speech. The saliva is then swallowed and helps moisten the esophagus. Eventually, most components of saliva are reabsorbed, which prevents fluid loss. Sympathetic stimulation dominates during stress, resulting in dryness of the mouth. During dehydration, the salivary glands stop secreting saliva to conserve water; the resulting dryness in the mouth contributes to the sensation of thirst. Drinking will then not only restore the homeostasis of body water but also moisten the mouth.

The touch and taste of food also are potent stimulators of salivary gland secretions. Chemicals in the food stimulate receptors in taste buds on the tongue, and impulses are conveyed from the taste buds to two salivary nuclei in the brain stem (**superior** and **inferior salivatory nuclei**). Returning parasympathetic impulses in fibers of the facial (VII) and glossopharyngeal (IX) nerves stimulate the secretion of saliva. Saliva continues to be heavily secreted for some time after food is swallowed; this flow of saliva washes out the mouth and dilutes and buffers the remnants of irritating chemicals. The smell, sight, sound, or thought of food may also stimulate secretion of saliva.

Although any of the salivary glands may be the target of a nasopharyngeal infection, the mumps virus (myxovirus) typically attacks the parotid glands. **Mumps** is an inflammation and enlargement of the parotid glands accompanied by moderate fever, malaise (general discomfort), and extreme pain in the throat, especially when swallowing sour foods or acidic juices. Swelling occurs on one or both sides of the face, just anterior to the ramus of the mandible. In about 30% of males past puberty, the testes may also become inflamed; sterility rarely occurs because testicular involvement is usually unilateral (one testis only). Since a vaccine became available for mumps in 1967, the incidence of the disease has declined. ■

Tongue

The **tongue** is an accessory digestive organ composed of skeletal muscle covered with mucous membrane. Together with its associated muscles, it forms the floor of the oral cavity. The tongue is divided into symmetrical lateral halves by a median septum that extends its entire length, and it is attached inferiorly to the hyoid bone, styloid process of the temporal bone, and mandible. Each half of the tongue consists of an identical complement of extrinsic and intrinsic muscles.

The **extrinsic muscles** of the tongue, which originate outside the tongue (attach to bones in the area) and insert into connective tissues in the tongue, include the hyoglossus, genioglossus, and styloglossus muscles (see Figure 10.7 on page 275). The extrinsic muscles move the tongue from side to side and in and out to maneuver food for chewing, shape the food into a rounded mass, and force the food to the back of the mouth for swallowing. They also form the floor of the mouth and hold the tongue in position. The **intrinsic muscles** originate in and insert into connective tissue within the tongue and alter the shape and size of the tongue for speech and swallowing. The intrinsic muscles include the longitudinalis superior, longitudinalis inferior, transversus linguae, and verticalis linguae muscles. The **lingual frenulum** (*lingua* = the tongue), a fold of mucous membrane in the midline of the undersurface of the tongue, is attached to the floor of the mouth and aids in limiting the movement of the tongue posteriorly (see Figures 24.4 and 24.5). If a person's lingual frenulum is abnormally short or rigid—a condition called **ankyloglossia** (ang′-kē-lō-GLOSS-ē-a)—then eating and speaking are impaired such that the person is said to be "tongue-tied."

The dorsum (upper surface) and lateral surfaces of the tongue are covered with **papillae** (pa-PIL-ē; = nipple-shaped projections), projections of the lamina propria covered with keratinized epithelium (see Figure 20.2 on page 614). Many papillae contain taste buds, the receptors for gustation (taste). **Fungiform papillae** (FUN-ji-form; = shaped like a mushroom) are mushroomlike elevations distributed among the filiform papillae that are more numerous near the tip of the tongue. They appear as red dots on the surface of the tongue, and most of them contain taste buds. **Circumvallate papillae** (*circum-* = around; *vall-* = wall) are arranged in an inverted V shape on the posterior surface of the tongue; all of them contain taste buds. **Filiform papillae** (FIL-i-form; = threadlike) are whitish, conical projections distributed in parallel rows over the anterior two-thirds of the tongue. Although filiform papillae lack taste buds, they increase friction between the tongue and food, making it easier for the tongue to move food within the oral cavity and they also contain receptors for touch. **Lingual glands** in the lamina propria secrete into saliva both mucus and a watery serous fluid that contains lingual lipase.

Teeth

The **teeth,** or **dentes** (Figure 24.6), are accessory digestive organs located in sockets of the alveolar processes of the mandible and maxillae. The alveolar processes are covered by the **gingivae** (jin-JI-vē), or gums, which extend slightly into each socket to form the gingival sulcus. The sockets are lined by the **periodontal ligament** or **membrane** (*odont-* = tooth), which consists of dense fibrous connective tissue and is attached to the socket walls and the surface of the roots. Thus, it anchors the teeth in position and acts as a shock absorber during chewing.

A typical tooth consists of three principal regions. The **crown** is the visible portion above the level of the gums. Embedded in the socket are one to three **roots.** The **neck** is the constricted junction of the crown and root near the gum line.

Figure 24.6 / A typical tooth and surrounding structures.

Teeth are anchored in sockets of the alveolar processes of the mandible and maxillae.

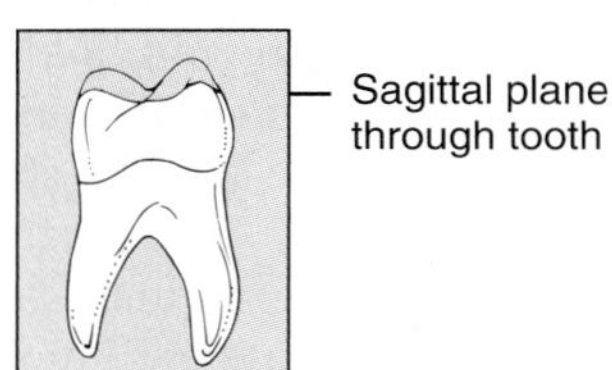

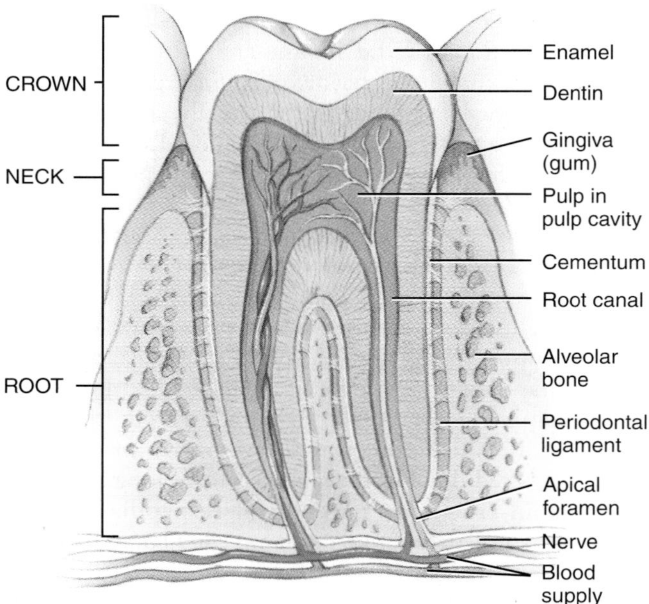

Sagittal section of a mandibular (lower) molar

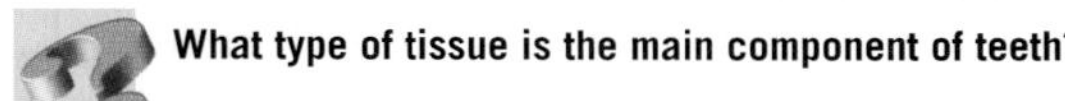
What type of tissue is the main component of teeth?

Teeth are composed primarily of **dentin,** a calcified connective tissue that gives the tooth its basic shape and rigidity. It is harder than bone because of its higher content of calcium salts (70% of dry weight). The dentin encloses a cavity. The enlarged part of the cavity, the **pulp cavity,** lies within the crown and is filled with **pulp,** a connective tissue containing blood vessels, nerves, and lymphatic vessels. Narrow extensions of the pulp cavity, called **root canals,** run through the root of the tooth. Each root canal has an opening at its base, the **apical foramen,** through which blood vessels, lymphatic vessels, and nerves extend.

Root Canal Therapy

Root canal therapy refers to a procedure, accomplished in several phases, in which all traces of pulp tissue are removed from the pulp cavity and root canals of a badly diseased tooth. After a hole is made in the tooth, the root canals are filed out and irrigated to remove bacteria. Then the canals are treated with medication and sealed tightly. The damaged crown is then repaired. ■

The dentin of the crown is covered by **enamel** that consists primarily of calcium phosphate and calcium carbonate. Enamel, the hardest substance in the body and the richest in calcium salts (about 95% of dry weight), protects the tooth from the wear of chewing. It is also a barrier against acids that easily dissolve the dentin. The dentin of the root is covered by **cementum,** another bonelike substance, which attaches the root to the periodontal ligament.

The arteries that supply blood to the teeth are distributed to the pulp cavity and surrounding periodontal ligament. These include the anterior and posterior superior alveolar branches of the maxillary artery and the incisive and dental branches of the inferior alveolar artery.

The teeth receive sensory fibers from branches of the maxillary and mandibular divisions of the trigeminal (V) nerve—the maxillary teeth from branches of the maxillary division and the mandibular teeth from branches of the mandibular division.

The branch of dentistry that is concerned with the prevention, diagnosis, and treatment of diseases that affect the pulp, root, periodontal ligament, and alveolar bone is known as **endodontics** (en′-dō-DON-tiks; *endo-* = within). **Orthodontics** (or′-thō-DON-tiks; *ortho-* = straight) is a branch of dentistry that is concerned with the prevention and correction of abnormally aligned teeth, whereas **periodontics** (per′-ē-ō-DON-tiks) is a branch of dentistry concerned with the treatment of abnormal conditions of the tissues immediately surrounding the teeth.

Humans have two **dentitions,** or sets of teeth: deciduous and permanent. The first of these—the **deciduous teeth** (*deciduu-* = falling out), also called **primary teeth, milk teeth,** or **baby teeth**—begin to erupt at about 6 months of age, and one pair of teeth appears at about each month thereafter, until all 20 are present (Figure 24.7a). The incisors, which are closest to the midline, are chisel-shaped and adapted for cutting into food. They are referred to as either **central** or **lateral incisors** on the

Figure 24.7 / Dentitions and times of eruptions (indicated in parentheses).
(See Tortora, *A Photographic Atlas of the Human Body,* Figure 12.7.)

Deciduous teeth begin to erupt at 6 months of age, and one pair of teeth appears about each month thereafter, until all 20 are present.

Central incisor (8–12 mo.)
Lateral incisor (12–24 mo.)
Cuspid or canine (16–24 mo.)
First molar (12–16 mo.)
Second molar (24–32 mo.)
Upper Teeth
Lower Teeth
Second molar (24–32 mo.)
First molar (12–16 mo.)
Cuspid or canine (16–24 mo.)
Lateral incisor (12–15 mo.)
Central incisor (6–8 mo.)

(a) Deciduous (primary) dentition

Central incisor (7–8 yr.)
Lateral incisor (8–9 yr.)
Cuspid or canine (11–12 yr.)
First premolar or bicuspid (9–10 yr.)
Second premolar or bicuspid (10–12 yr.)
First molar (6–7 yr.)
Second molar (12–13 yr.)
Third molar or wisdom tooth (17–21 yr.)
Upper Teeth
Lower Teeth
Third molar or wisdom tooth (17–21 yr.)
Second molar (11–13 yr.)
First molar (6–7 yr.)
Second premolar or bicuspid (11–12 yr.)
First premolar or bicuspid (9–10 yr.)
Cuspid or canine (9–10 yr.)
Lateral incisor (7–8 yr.)
Central incisor (7–8 yr.)

(b) Permanent (secondary) dentition

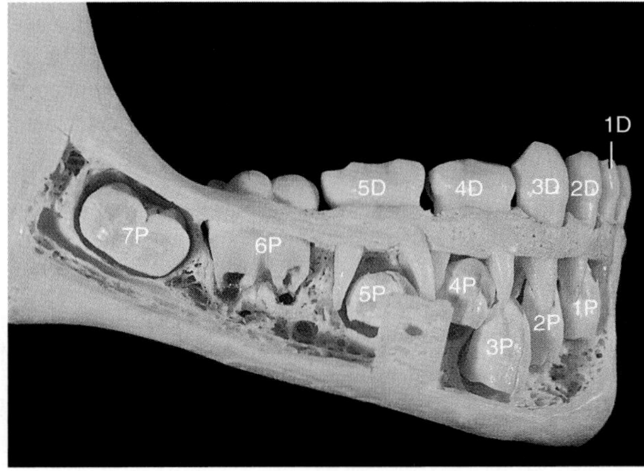

Deciduous teeth:
1D - Central incisor
2D - Lateral incisor
3D - Cuspid (canine)
4D - First molar (bicuspid)
5D - Second molar

Permanent teeth:
1P - Central incisor
2P - Lateral incisor
3P - Cuspid (canine)
4P - First premolar (bicuspid)
5P - Second premolar
6P - First molar
7P - Second molar

(c) Mandible of a six year old child in right lateral view showing erupted deciduous teeth and unerupted permanent teeth (labels provided by Roger P. Santise, D.D.S.)

Which permanent teeth do not replace any deciduous teeth?

basis of their position. Next to the incisors, moving posteriorly, are the **cuspids (canines),** which have a pointed surface called a cusp. Cuspids are used to tear and shred food. Incisors and cuspids have only one root apiece. Posterior to them lie the **first** and **second molars,** which have four cusps. Maxillary (upper) molars have three roots; mandibular (lower) molars have two roots. The molars crush and grind food.

All the deciduous teeth are lost—generally between ages 6 and 12 years—and are replaced by the **permanent (secondary) teeth** (Figure 24.7b). The permanent dentition contains 32 teeth that erupt between age 6 and adulthood. The pattern resembles the deciduous dentition, with the following exceptions. The deciduous molars are replaced by the **first** and **second premolars (bicuspids),** which have two cusps and one root (upper first bicuspids have two roots) and are used for crushing and grinding. The permanent molars, which erupt into the mouth posterior to the bicuspids, do not replace any deciduous teeth and erupt as the jaw grows to accommodate them—the **first molars** at age 6, the **second molars** at age 12, the **third molars (wisdom teeth)** after age 17.

Table 24.1 Surface Orientation of Teeth

Tooth Surface	Description
Labial	Contacting the lips.
Buccal	Contacting or facing the cheeks.
Lingual	Facing the tongue (teeth of the mandible only).
Palatal	Facing the palate (teeth of the maxillae only).
Mesial	Anterior or medial side relative to dental arch.
Distal	Posterior or lateral side relative to dental arch.
Occlusal	The biting surface.

Often the human jaw does not have enough room posterior to the second molars to accommodate the eruption of the third molars. In this case, the third molars remain embedded in the alveolar bone and are said to be "impacted." They often cause pressure and pain and must be removed surgically. In some people, third molars may be dwarfed in size or may not develop at all.

Table 24.1 lists terms used to describe the surface orientation of teeth.

PHARYNX

Through chewing, or **mastication** (mas′ti-KĀ-shun; *masticare* = to chew), the tongue manipulates food, the teeth grind it, and the food is mixed with saliva. As a result, the food is reduced to a soft, flexible mass called a **bolus** (*bolos* = lump) that is easily swallowed. When food is first swallowed, it passes from the mouth into the pharynx.

The **pharynx** (*pharynx* = throat) is a funnel-shaped tube that extends from the internal nares to the esophagus posteriorly and the larynx anteriorly (see Figure 23.3 on page 689). The pharynx is composed of skeletal muscle and lined by mucous membrane. Whereas the nasopharynx functions only in respiration, both the oropharynx and laryngopharynx have digestive and respiratory functions. Swallowing, or **deglutition** (dē-glū-TISH-un), is a mechanism that moves food from the mouth to the stomach. It is helped by saliva and mucus and involves the mouth, pharynx, and esophagus. Food that is swallowed passes from the mouth into the oropharynx and laryngopharynx before passing into the esophagus. Muscular contractions of the oropharynx and laryngopharynx help propel food into the esophagus and then into the stomach.

✓ What structures form the mouth (oral cavity)?
✓ How are the major salivary glands distinguished by location and histologically?
✓ How do the extrinsic and intrinsic muscles of the tongue differ in function?
✓ Contrast the functions of incisors, cuspids, premolars, and molars.
✓ What is a bolus? How is it formed?
✓ Where does the pharynx begin and end?

ESOPHAGUS

Objective

- Describe the location, anatomy, histology, and function of the esophagus.

The **esophagus** (e-SOF-a-gus; = eating gullet) is a collapsible muscular tube that lies posterior to the trachea. It is about 25 cm (10 in.) long. It begins at the inferior end of the laryngopharynx, passes through the mediastinum anterior to the vertebral column, pierces the diaphragm through an opening called the **esophageal hiatus,** and ends in the superior portion of the stomach (see Figure 24.1). Sometimes, a portion of the stomach protrudes above the diaphragm through the esophageal hiatus. This condition is termed **hiatal hernia** (HER-nē-ah).

The arteries of the esophagus are derived from the arteries along its length: inferior thyroid, thoracic aorta, intercostal arteries, phrenic, and left gastric arteries. It is drained by the adjacent veins. Innervation of the esophagus is by recurrent laryngeal nerves, the cervical sympathetic chain, and vagus (X) nerves.

Histology

The **mucosa** of the esophagus consists of nonkeratinized stratified squamous epithelium, lamina propria (areolar connective tissue), and a muscularis mucosae (smooth muscle) (Figure 24.8). Near the stomach, the mucosa of the esophagus also contains mucous glands. The stratified squamous epithelium associated with the lips, mouth, tongue, oropharynx, laryngopharynx, and esophagus affords considerable protection against abrasion and wear-and-tear from food particles that are chewed, mixed with secretions, and swallowed. The **submucosa** contains areolar connective tissue, blood vessels, and mucous glands. The **muscularis** of the superior third of the esophagus is skeletal muscle, the intermediate third is skeletal and smooth muscle, and the inferior third is smooth muscle. The superficial layer is known as the **adventitia** (ad-ven-TISH-a), rather than the serosa, because the areolar connective tissue of this layer is not covered by mesothelium and because the connective tissue merges with the connective tissue of surrounding structures of the mediastinum, through which it passes. The adventitia attaches the esophagus to surrounding structures.

Functions

The esophagus secretes mucus and transports food into the stomach. It does not produce digestive enzymes, and it does not carry on absorption. The passage of food from the laryngopharynx into the esophagus is regulated at the entrance to the esophagus by a sphincter (a circular band or ring of muscle that is normally contracted) called the **upper esophageal sphincter** (e-sof′-a-JĒ-al). It consists of the cricopharyngeus muscle attached to the cricoid cartilage. The elevation of the larynx causes the sphincter to relax, and the bolus enters the esophagus. This sphincter also relaxes during exhalation.

Figure 24.8 / Histology of the esophagus. An enlarged view of nonkeratinized, stratified squamous epithelium is shown in Table 3.1 on page 63. (See Tortora, *A Photographic Atlas of the Human Body,* Figure 12.8.)

The esophagus secretes mucus and transports food to the stomach.

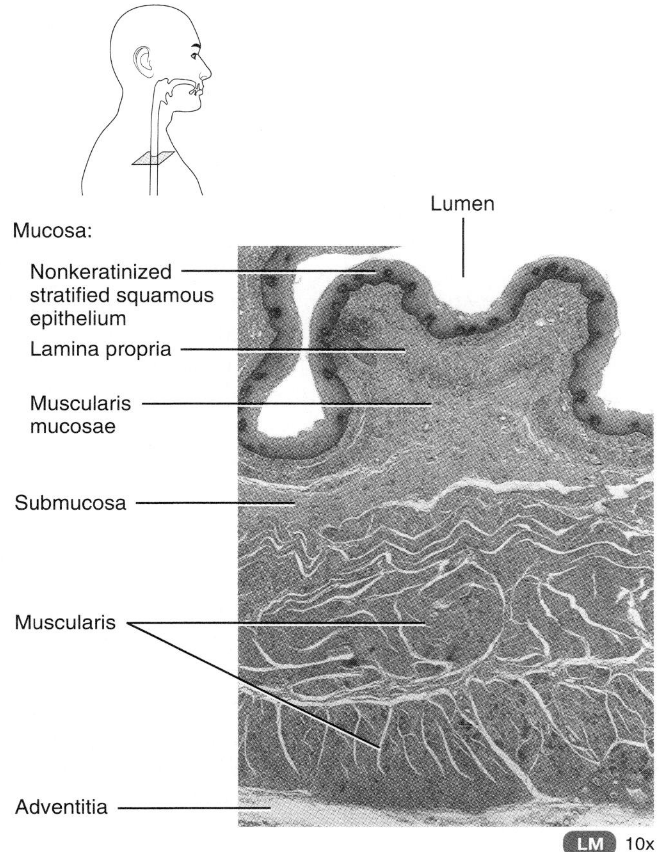

Wall of the esophagus

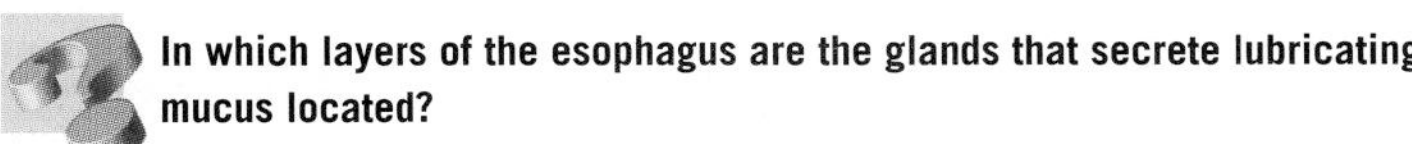

In which layers of the esophagus are the glands that secrete lubricating mucus located?

Food is pushed through the esophagus by a progression of involuntary coordinated contractions and relaxations of the circular and longitudinal layers of the muscularis called **peristalsis** (per′-is-STAL-sis; *stalsis* = constriction) (Figure 24.9). (Peristalsis occurs in other tubular structures, including other portions of the GI tract and the ureters, bile ducts, and uterine tubes; in the esophagus it is controlled by the medulla oblongata.) In the section of the esophagus lying just superior to the bolus, the circular muscle fibers contract, constricting the esophageal wall and squeezing the bolus toward the stomach. Meanwhile, longitudinal fibers inferior to the bolus also contract, which shortens this inferior section and pushes its walls outward so it can receive the bolus. The contractions are repeated in a wave that pushes the food toward the stomach. Mucus secreted by esophageal glands lubricates the bolus and reduces friction.

Figure 24.9 / Peristalsis during deglutition (swallowing).

Peristalsis consists of progressive, wavelike contractions of the muscularis.

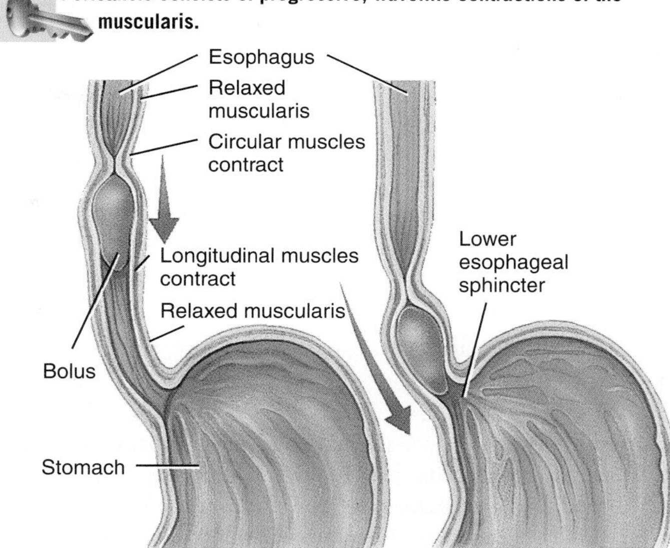

Sagittal section of peristalsis in esophagus

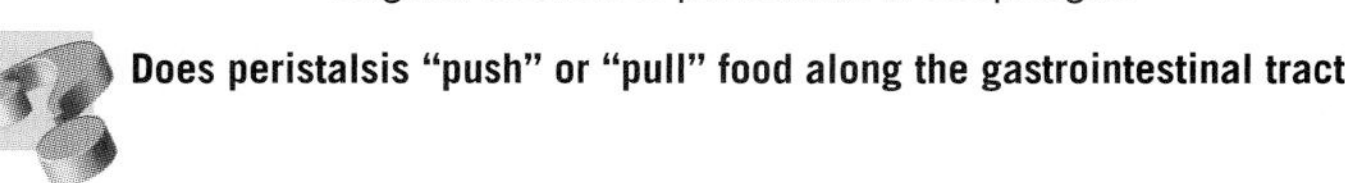

Does peristalsis "push" or "pull" food along the gastrointestinal tract?

Just superior to the level of the diaphragm, the esophagus narrows slightly. This narrowing is a physiological sphincter in the inferior part of the esophagus known as the **lower esophageal (gastroesophageal** or **cardiac) sphincter.** (A *physiological sphincter* is a section of a tubular structure, in this case the esophagus, that functions like a sphincter even though no sphincter muscle is actually present.) The lower esophageal sphincter relaxes during swallowing and thus allows the bolus to pass from the esophagus into the stomach.

Gastroesophageal Reflux Disease

If the lower esophageal sphincter fails to close adequately after food has entered the stomach, the stomach contents can reflux, or back up, into the inferior portion of the esophagus. This condition is known as **gastroesophageal reflux disease (GERD).** Hydrochloric acid (HCl) from the stomach contents can irritate the esophageal wall, resulting in a burning sensation called **heartburn** because it is experienced in a region very near the heart, even though it is unrelated to any cardiac problem. Drinking alcohol and smoking can cause the sphincter to relax, worsening the problem. The symptoms of GERD often can be controlled by avoiding foods that strongly stimulate stomach acid secretion (coffee, chocolate, tomatoes, fatty foods, peppermint, spearmint, and onions). Other acid-reducing strategies include taking over-the-counter histamine-2 (H_2) blockers such as Tagamet HB or Pepcid AC 30–60 minutes before eating, and neutralizing already secreted acid with antacids such as Tums or Maalox. Symptoms are less likely to occur if food is eaten in smaller amounts and if the person does not lie down immediately after a meal. GERD may be associated with cancer of the esophagus.

✓ Describe the location and histology of the esophagus. What is its role in digestion?
✓ Explain the operation of the upper and lower esophageal sphincters.

STOMACH

Objective

- Describe the location, anatomy, histology, and function of the stomach.

The **stomach** is a typically J-shaped enlargement of the GI tract directly inferior to the diaphragm in the epigastric, umbilical, and left hypochondriac regions of the abdomen (see Figure 1.9 on page 16). The stomach connects the esophagus to the duodenum, the first part of the small intestine (Figure 24.10). Because a meal can be eaten much more quickly than the intestines can digest and absorb it, one of the functions of the stomach is to serve as a mixing vat and holding reservoir. At appropriate intervals after food is ingested, the stomach forces a small quantity of material into the first portion of the small intestine. The position and size of the stomach vary continually; the diaphragm pushes it inferiorly with each inspiration and pulls it superiorly with each expiration. Empty, it is about the size of a large sausage, but it is the most distensible portion of the GI tract and can accommodate a large quantity of food. In the stomach, digestion of starch continues, digestion of proteins and triglycerides begins, the semisolid bolus is converted to a liquid, and certain substances are absorbed.

Anatomy

The stomach has four main regions: the cardia, fundus, body, and pylorus (Figure 24.10). The **cardia** (CAR-dē-a) surrounds the superior opening of the stomach. The rounded portion superior and to the left of the cardia is the **fundus** (FUN-dus). Inferior to the fundus is the large central portion of the stomach, called the **body.** The region of the stomach that connects to the duodenum is the **pylorus** (pī-LOR-us; *pyl-* = gate; *-orus* = guard); it has two parts, the **pyloric antrum** (AN-trum; = cave), which connects to the body of the stomach, and the **pyloric canal,** which leads into the duodenum. When the stomach is empty, the mucosa lies in large folds, called **rugae** (RŪ-gē; = wrinkles), that can be seen with the unaided eye. The pylorus communicates with the duodenum of the small intestine via a sphincter called the **pyloric sphincter.** The concave medial border of the stomach is called the **lesser curvature,** and the convex lateral border is called the **greater curvature.**

The arterial supply of the stomach is derived from the celiac trunk. The right and left gastric arteries form an anastomosing arch along the lesser curvature, and the right and left gastroepiploic arteries form a similar arch on the greater curvature. Short gastric arteries supply the fundus. The veins of the same name accompany the arteries and drain, directly or indirectly, into the hepatic portal vein.

The vagus (X) nerves convey parasympathetic fibers to the stomach. These fibers form synapses within the submucosal plexus (plexus of Meissner) in the submucosa and the myenteric plexus (plexus of Auerbach) in the muscularis. The sympathetic nerves arise from the celiac ganglia, and the nerves reach the stomach along the branches of the celiac artery.

Pylorospasm and Pyloric Stenosis

Two abnormalities of the pyloric sphincter can occur in infants. In **pylorospasm** the muscle fibers of the sphincter fail to relax normally, so food does not pass easily from the stomach to the small intestine, the stomach becomes overly full, and the infant vomits often to relieve the pressure. Pylorospasm is treated by drugs that relax the muscle fibers of the sphincter. **Pyloric stenosis** is a narrowing of the pyloric sphincter, and it must be corrected surgically. The hallmark symptom is projectile vomiting—the spraying of liquid vomitus some distance from the infant.

Histology

The stomach wall is composed of the same four basic layers as the rest of the GI tract, with certain modifications (Figure 24.11a on page 730). The surface of the **mucosa** is a layer of simple columnar epithelial cells called **surface mucous cells.** The mucosa contains a **lamina propria** (areolar connective tissue) and a **muscularis mucosae** (smooth muscle). Epithelial cells extend down into the lamina propria, where they form columns of secretory cells called **gastric glands** that line many narrow channels called **gastric pits.** Secretions from several gastric glands flow into each gastric pit and then into the lumen of the stomach.

The gastric glands contain three types of *exocrine gland cells* that secrete their products into the stomach lumen: mucous neck cells, chief cells, and parietal cells. Both mucous surface cells and **mucous neck cells** secrete mucus (Figure 24.11b). **Parietal cells** produce hydrochloric acid and intrinsic factor (needed for absorption of vitamin B_{12}). The **chief (zymogenic) cells** secrete pepsinogen and gastric lipase. The secretions of the mucous, parietal, and chief cells form **gastric juice,** which totals 2000–3000 mL (roughly 2–3 qt.) per day. In addition, gastric glands include a type of enteroendocrine cell, the **G cell,** which is located mainly in the pyloric antrum and secretes the hormone gastrin into the bloodstream. Gastrin stimulates growth of the gastric glands and secretion of large amounts of gastric juice. It also strengthens contraction of the lower esophageal sphincter, increases motility of the stomach, and relaxes the pyloric and ileocecal sphincters (described later).

Three additional layers lie deep to the mucosa. The **submucosa** of the stomach is composed of areolar connective tissue. The **muscularis** has three (rather than two) layers of smooth muscle: an outer longitudinal layer, a middle circular

Figure 24.10 / External and internal anatomy of the stomach. (See Tortora, *A Photographic Atlas of the Human Body,* Figure 12.9.)

The four regions of the stomach are the cardia, fundus, body, and pylorus.

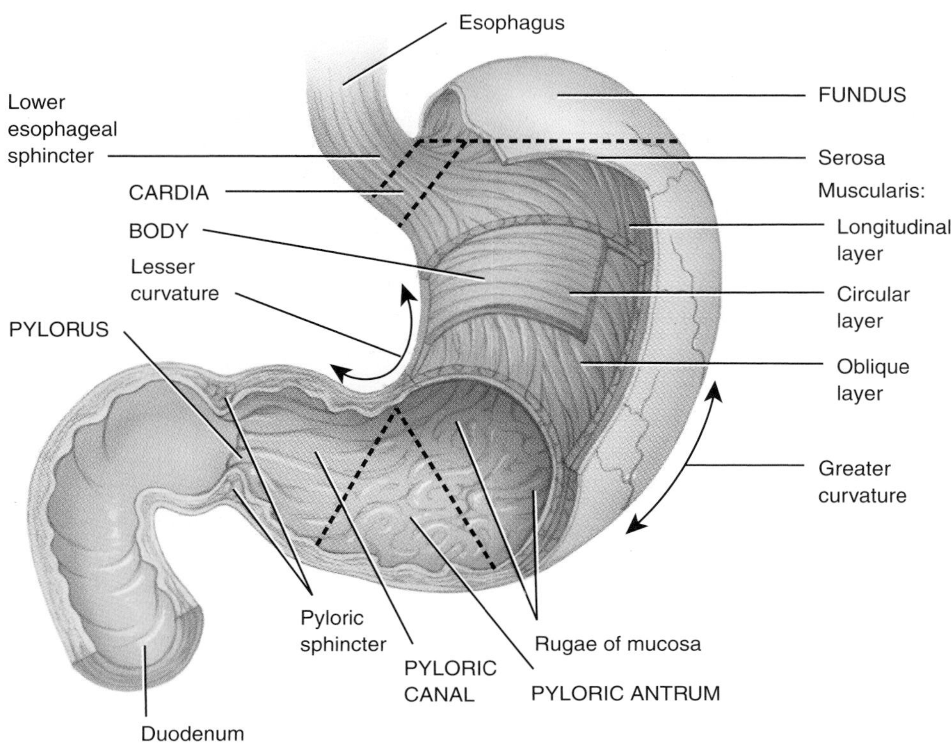

(a) Anterior view of regions of stomach

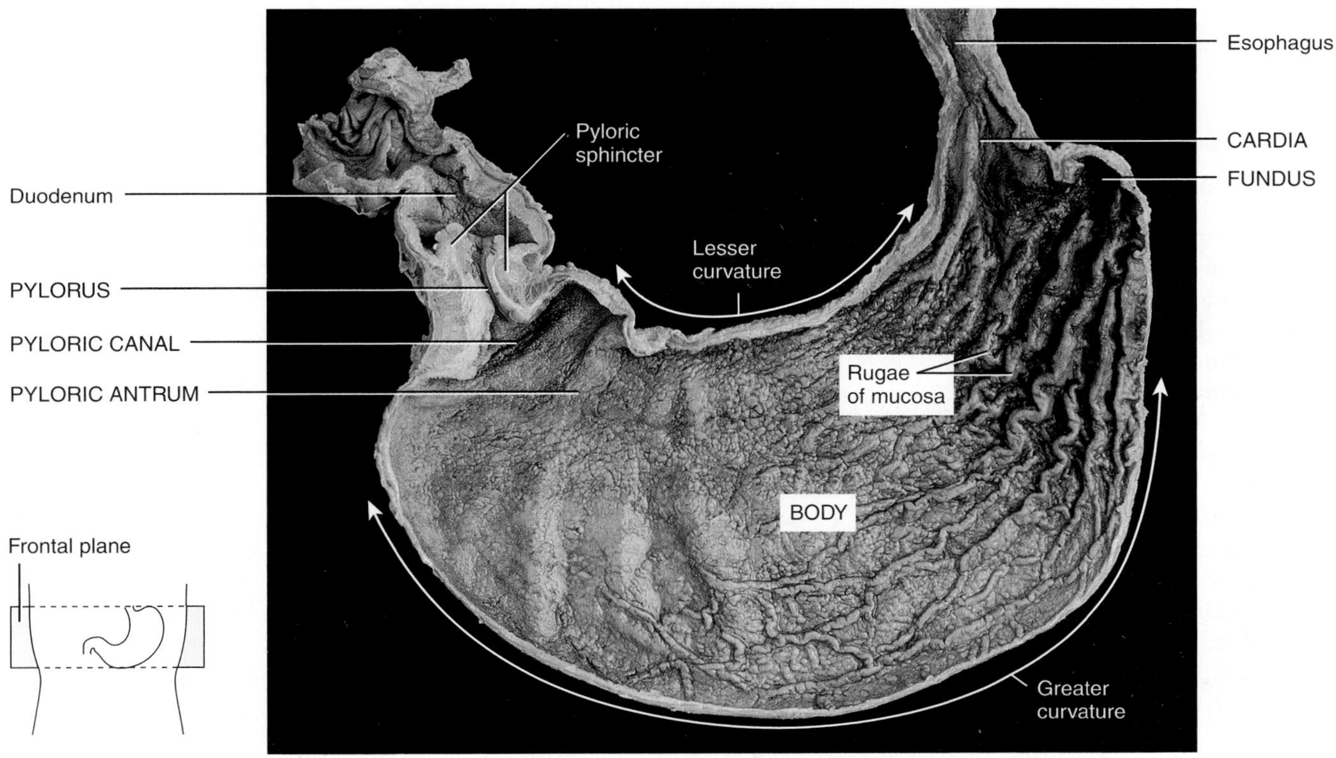

(b) Frontal section of the internal surface

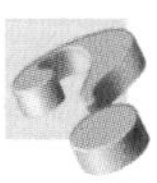

After a very large meal, does your stomach have rugae?

Figure 24.11 / Histology of the stomach.

The muscularis of the stomach has three layers of smooth muscle tissue.

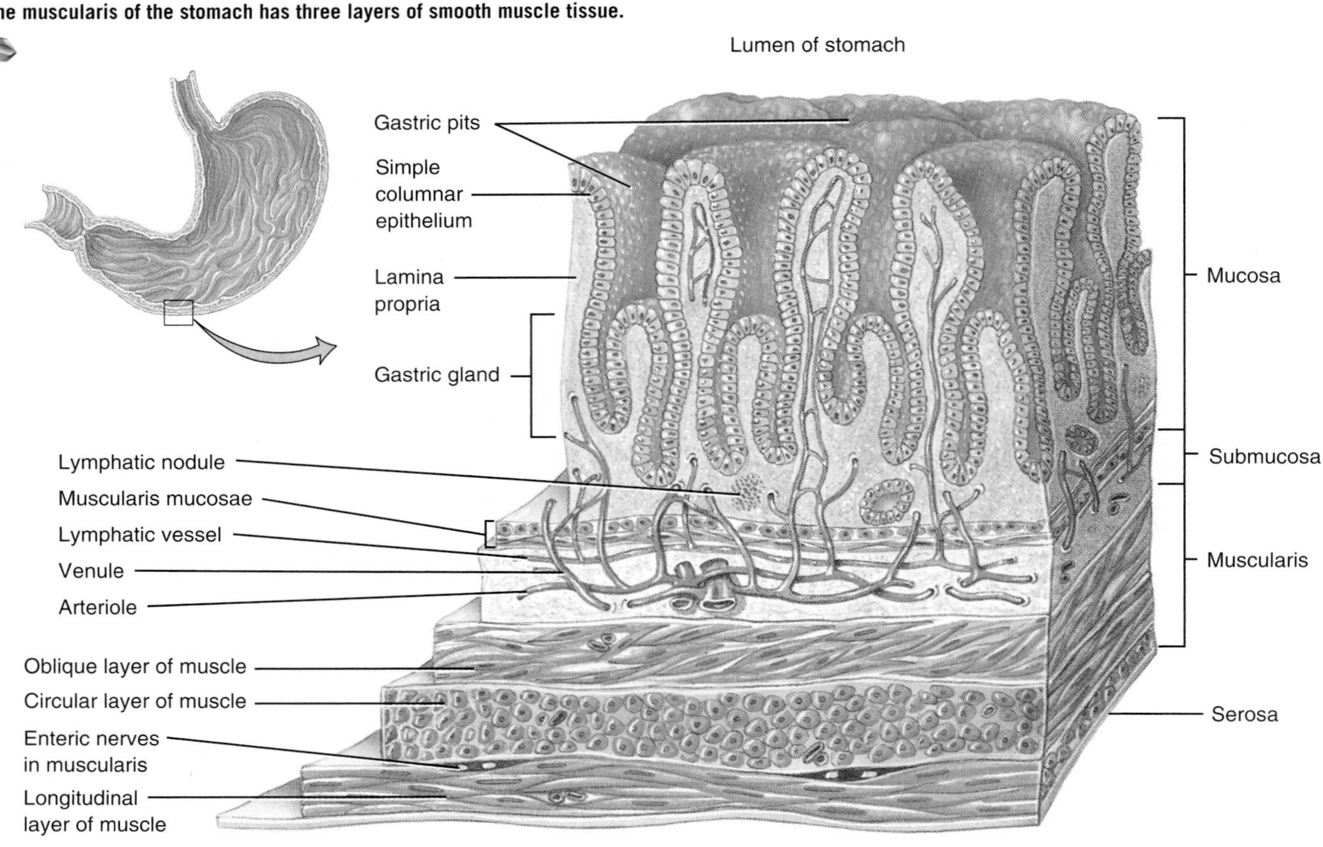

(a) Three dimensional view of layers of the stomach

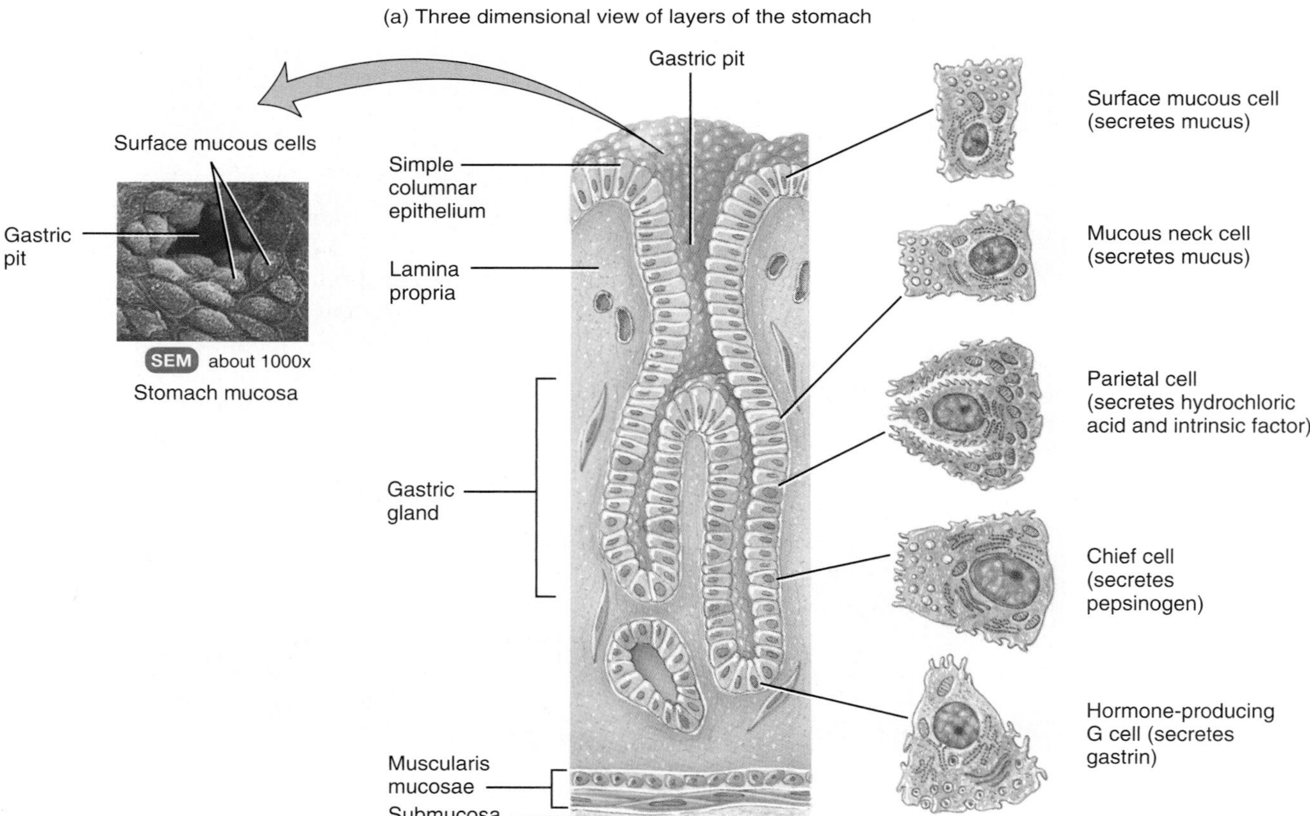

(b) Sectional view of the stomach mucosa showing gastric glands

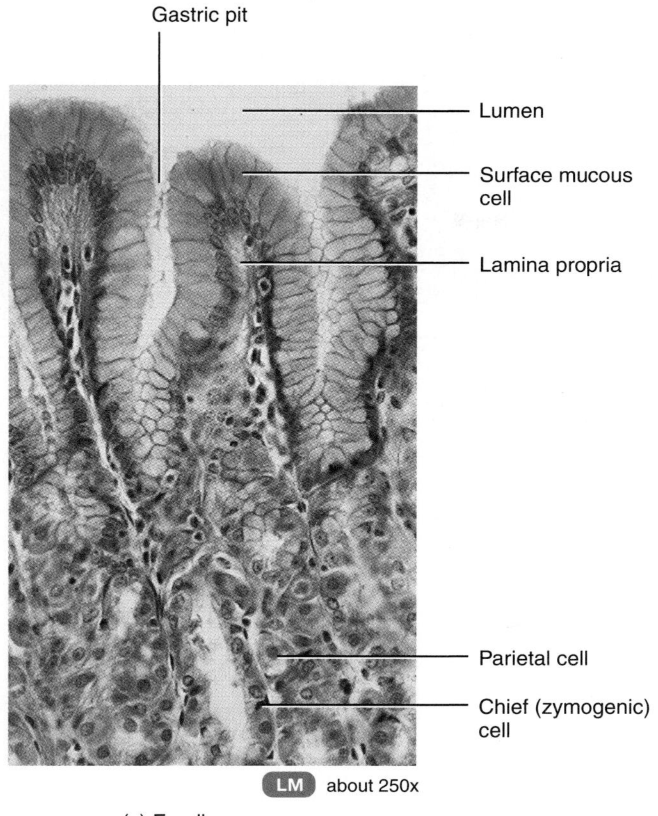

(c) Fundic mucosa

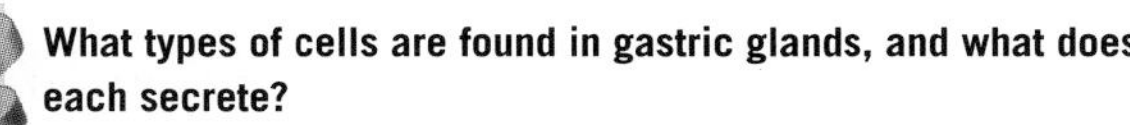
What types of cells are found in gastric glands, and what does each secrete?

layer, and an inner oblique layer. The oblique layer is limited primarily to the body of the stomach. The **serosa** is composed of simple squamous epithelium (mesothelium) and areolar connective tissue and the portion covering the stomach is part of the visceral peritoneum. At the lesser curvature, the visceral peritoneum extends superiorly to the liver as the lesser omentum. At the greater curvature, the visceral peritoneum continues inferiorly as the greater omentum and drapes over the intestines.

Functions

Several minutes after food enters the stomach, gentle, rippling, peristaltic movements called **mixing waves** pass over the stomach every 15–25 seconds. These waves macerate food, mix it with secretions of the gastric glands, and reduce it to a soupy liquid called **chyme** (KĪM; = juice). Few mixing waves are observed in the fundus, which primarily has a storage function. As digestion proceeds in the stomach, more vigorous mixing waves begin at the body of the stomach and intensify as they reach the pylorus. The pyloric sphincter normally remains almost, but not completely, closed; as food reaches the pylorus, each mixing wave forces several milliliters of chyme into the duodenum through the pyloric sphincter. Most of the chyme is forced back into the body of the stomach, where mixing continues. The next wave pushes the chyme forward again and forces a little more into the duodenum. These forward and backward movements of the gastric contents are responsible for most mixing in the stomach.

The enzymatic digestion of proteins begins in the stomach. In the adult, this is achieved mainly through the enzyme **pepsin,** secreted by chief cells (secreted in an inactive form called *pepsinogen*). Pepsin breaks certain peptide bonds between the amino acids making up proteins. Thus a protein chain of many amino acids is broken down into smaller fragments called **peptides.** Pepsin also brings about the clumping and digestion of milk proteins. Another enzyme of the stomach is **gastric lipase.** Gastric lipase splits the short-chain triglycerides (fats) in butterfat molecules found in milk. The enzyme has a limited role in the adult stomach. To digest fats, adults rely almost exclusively on lingual lipase and **pancreatic lipase,** an enzyme secreted by the pancreas into the small intestine.

Within 2–4 hours after eating a meal, the stomach has emptied its contents into the duodenum. Foods rich in carbohydrate spend the least time in the stomach; high-protein foods remain somewhat longer, and emptying is slowest after a fat-laden meal containing large amounts of triglycerides.

Vomiting

Vomiting, or *emesis,* is the forcible expulsion of the contents of the upper GI tract (stomach and sometimes duodenum) through the mouth. The strongest stimuli for vomiting are irritation and distension of the stomach; other stimuli include unpleasant sights, general anesthesia, dizziness, and certain drugs such as morphine and derivatives of digitalis. Nerve impulses are transmitted to the vomiting center in the medulla oblongata, and returning impulses propagate to the upper GI tract organs, diaphragm, and abdominal muscles. Vomiting basically involves squeezing the stomach between the diaphragm and abdominal muscles and expelling the contents through open esophageal sphincters. Prolonged vomiting, especially in infants and elderly people, can be serious because the loss of acidic gastric juice can lead to alkalosis (higher than normal blood pH). ■

The stomach wall is impermeable to the passage of most materials into the blood, so most substances are not absorbed until they reach the small intestine. However, the stomach does participate in the absorption of some water, electrolytes, certain drugs (especially aspirin), and alcohol. The absorption of alcohol by the stomach of females is faster than that in males. The difference is attributed to smaller amounts of the enzyme alcohol dehydrogenase in the stomachs of females. The enzyme breaks down alcohol in the stomach, reducing the amount of alcohol that enters the blood.

- ✓ Describe the location and anatomical features of the stomach.
- ✓ Compare the epithelium of the esophagus with that of the stomach. How is each adapted to the function of the organ?
- ✓ What is the importance of rugae, surface mucous cells, mucous neck cells, chief cells, parietal cells, and G cells in the stomach?
- ✓ Describe mechanical digestion in the stomach.
- ✓ What are the functions of gastric lipase and lingual lipase in the stomach?
- ✓ Describe the role of the stomach in absorption.

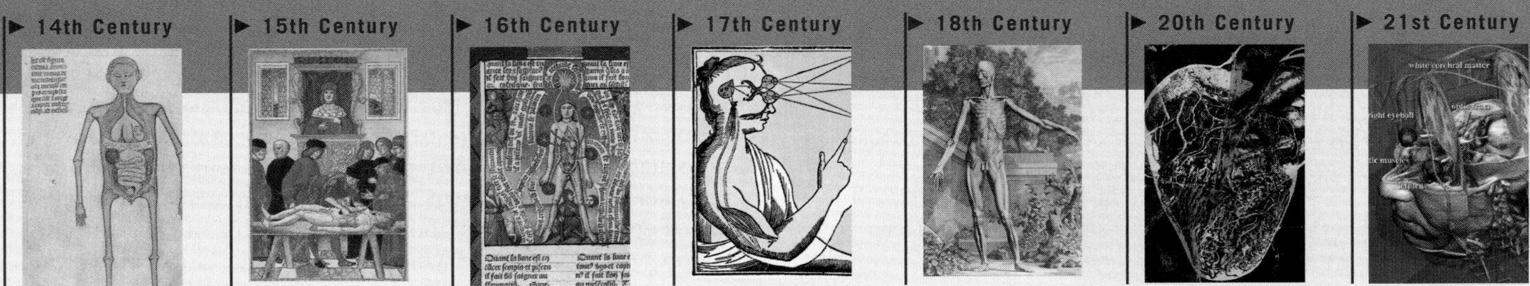

CHANGING IMAGES

Take a Closer Look

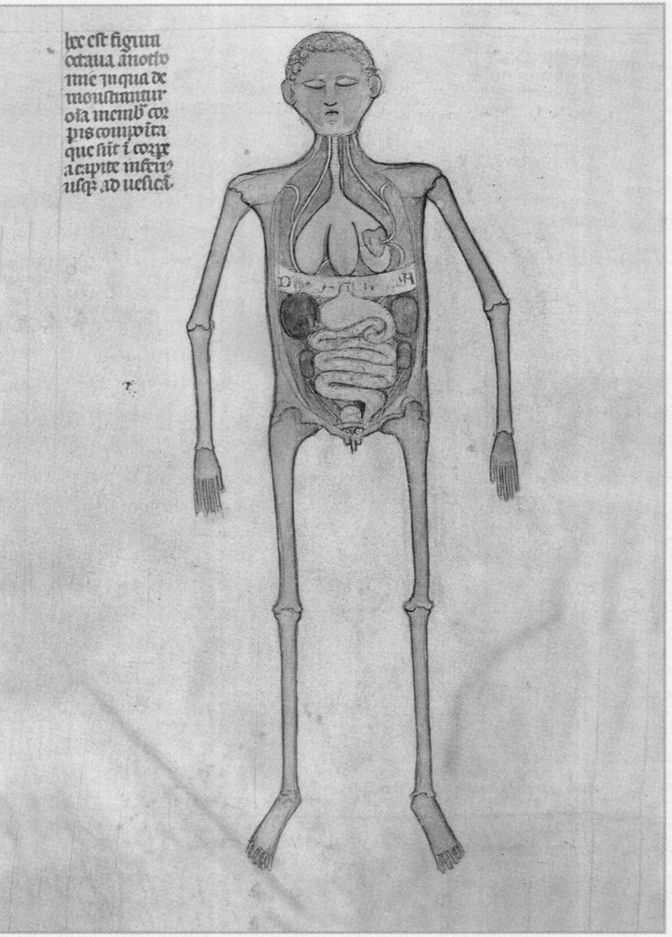

1345 AD

*I*t is often mistakenly believed that the Church in medieval Europe prohibited dissection. The explanation for this myth is most likely a 1299 papal edict by Pope Boniface VIII prohibiting the boiling of human remains. Such practice was customary for Crusaders who died far from their desired place of burial since boiling allowed for easier preservation and transportation of their corpse. Nonetheless, the effect of this proclamation was a frequent interpretation by many church authorities and anatomists that dissection was forbidden. This prohibition resulted in dissection being infrequent in early 14th century Europe.

Surprisingly, dissection experienced somewhat of a revival within a half-century. It was Guido de Vigevano who was at the forefront of this movement in the mid-14th century. His 1345 book, *Liber Notabilium Philippi Septimi*, contains 18 impressive anatomical renderings. His anatomy represented an orderly, sequential method to understanding human form. An approach that is alive to this day.

Pictured here is an illustration from Guido's book. He seemed to understand body cavities as he clearly distinguishes between the thoracic and abdominopelvic cavities. Despite this, many glaring inaccuracies are present. Can you identify any? Note the single three lobed lung and Valentine's Day heart. Having studied the respiratory and cardiovascular systems, you should also be able to discern other misrepresentations of the thorax. Pay attention to the trachea, the vessels arising from the heart, and the orientation of the diaphragm.

In this chapter you are examining the digestive system. As you progress, see if you could determine if the duodenum is appropriately oriented with the stomach? Are the liver and spleen present? Is there a distinction between the small and large intestine? When looking at this figure can you determine any other anatomical errors?

PANCREAS

Objective

- Describe the location, anatomy, histology, and function of the pancreas.

From the stomach, chyme passes into the small intestine. Because chemical digestion in the small intestine depends on activities of the pancreas, liver, and gallbladder, we first consider the activities of these accessory digestive organs and their contributions to digestion in the small intestine.

Anatomy

The **pancreas** (*pan-* = all; *-creas* = flesh), a retroperitoneal gland that is about 12–15 cm (5–6 in.) long and 2.5 cm (1 in.) thick, lies posterior to the greater curvature of the stomach. The pancreas consists of a head, a body, and a tail and is connected to the duodenum usually by two ducts (Figure 24.12). The **head** is the expanded portion of the organ near the curve of the duodenum; superior to and to the left of the head are the central **body** and the tapering **tail.**

Pancreatic secretions pass from the secreting cells into small ducts that ultimately unite to form two larger ducts that convey the secretions into the small intestine. The larger of the two ducts is called the **pancreatic duct (duct of Wirsung).** In most people, the pancreatic duct joins the common bile duct from the liver and gallbladder and enters the duodenum as a common duct called the **hepatopancreatic ampulla (ampulla of Vater).** The ampulla opens on an elevation of the duodenal mucosa, the **major duodenal papilla,** that lies about 10 cm (4 in.) inferior to the pyloric sphincter of the stomach. The smaller of the two ducts, the **accessory duct (duct of Santorini),** leads from the pancreas and empties into the duodenum about 2.5 cm (1 in.) superior to the hepatopancreatic ampulla.

The arterial supply of the pancreas is from the superior and inferior pancreaticoduodenal arteries and from the splenic and superior mesenteric arteries. The veins, in general, correspond to the arteries. Venous blood reaches the hepatic portal vein by means of the splenic and superior mesenteric veins.

Figure 24.12 / Relation of the pancreas to the liver, gallbladder, and duodenum. The inset shows details of the common bile duct and pancreatic duct forming the hepatopancreatic ampulla (ampulla of Vater) and emptying into the duodenum.

Pancreatic enzymes digest starches (polysaccharides), proteins, triglycerides, and nucleic acids.

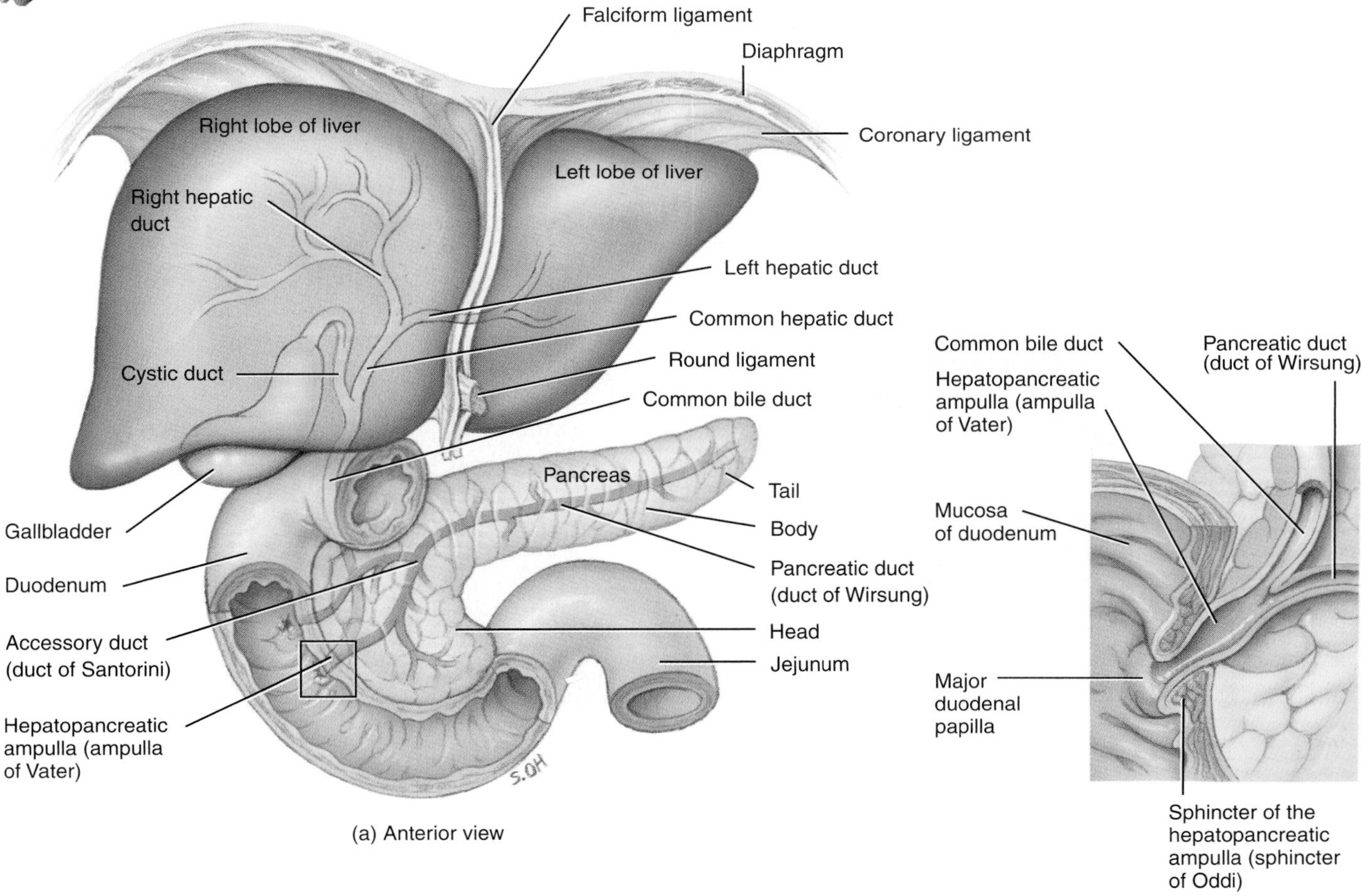

(a) Anterior view

(continues)

The nerves to the pancreas are autonomic nerves derived from the celiac and superior mesenteric plexuses. Included are preganglionic vagal, postganglionic sympathetic, and sensory fibers. Parasympathetic vagal fibers are said to terminate at both acinar (exocrine) and islet (endocrine) cells. Although the innervation is presumed to influence enzyme formation, pancreatic secretion is controlled largely by the hormones secretin and cholecystokinin (CCK) released by the small intestine. The sympathetic fibers enter the islets and also end on blood vessels; these fibers are vasomotor and accompanied by sensory fibers, especially for pain.

Histology

The pancreas is made up of small clusters of glandular epithelial cells, about 99% of which are arranged in clusters called **acini** (AS-i-nē) and constitute the *exocrine* portion of the organ (see Figure 22.7b, c on page 672). The cells within acini secrete a mixture of fluid and digestive enzymes called **pancreatic juice.** The remaining 1% of the cells are organized into clusters called **pancreatic islets (islets of Langerhans),** the *endocrine* portion of the pancreas. These cells secrete the hormones glucagon, insulin, somatostatin, and pancreatic polypeptide. The functions of these hormones are discussed in Chapter 22.

Figure 24.12 (continued)

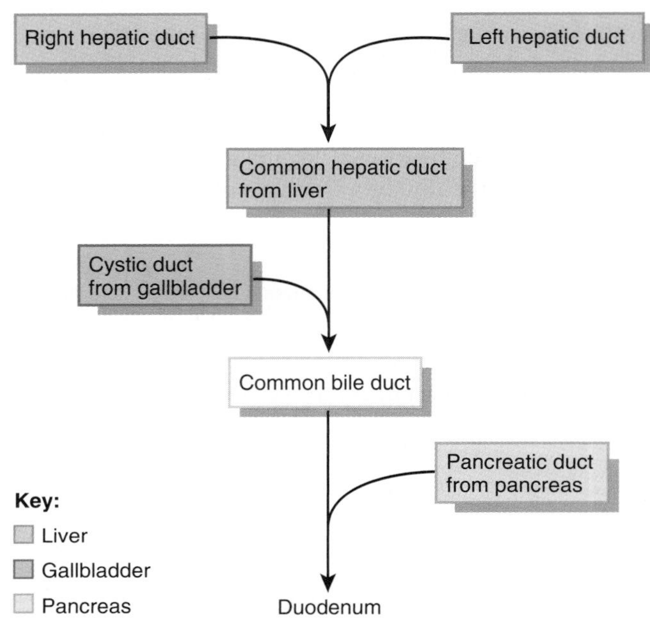

(b) Ducts carrying bile from liver and gallbladder and pancreatic juice from pancreas to the duodenum

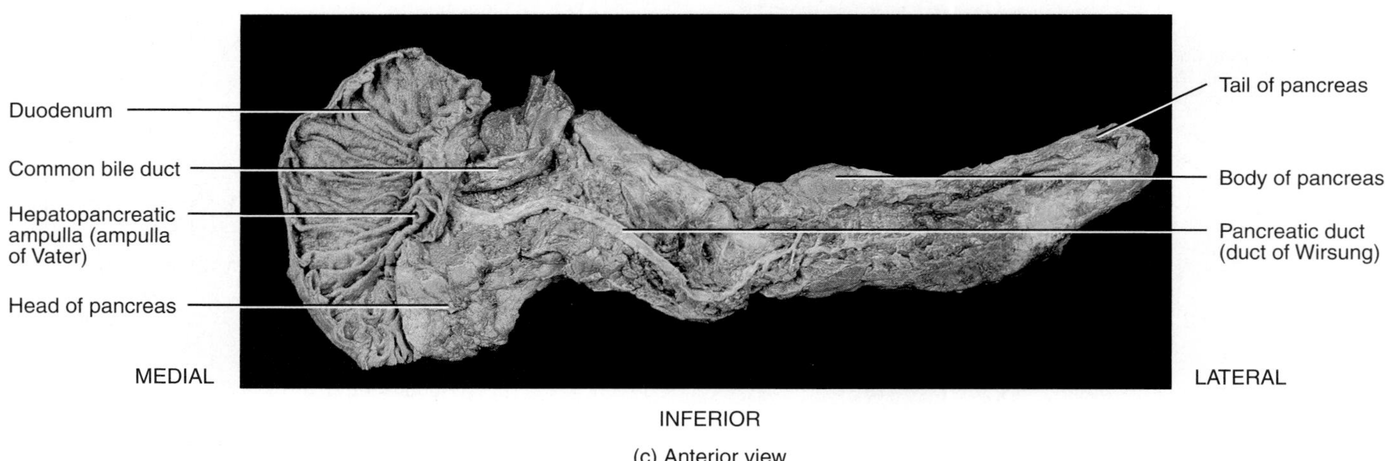

(c) Anterior view

What type of fluid is found in the pancreatic duct? The common bile duct? The hepatopancreatic ampulla?

Functions

Each day the pancreas produces 1200–1500 mL (about 1.2–1.5 qt) of **pancreatic juice,** which is a clear, colorless liquid consisting mostly of water, some salts, sodium bicarbonate, and several enzymes. The sodium bicarbonate gives pancreatic juice a slightly alkaline pH (7.1–8.2) that buffers acidic gastric juice in chyme, stops the action of pepsin from the stomach, and creates the proper pH for the action of digestive enzymes in the small intestine. The enzymes in pancreatic juice include a carbohydrate-digesting enzyme called **pancreatic amylase;** several protein-digesting enzymes called **trypsin** (TRIP-sin), **chymotrypsin** (kī′-mō-TRIP-sin), **carboxypeptidase** (kar-bok′-sē-PEP-ti-dās), and **elastase** (ē-LAS-tās); the principal triglyceride-digesting enzyme in adults, called **pancreatic lipase;** and nucleic acid-digesting enzymes called **ribonuclease** and **deoxyribonuclease.**

Inflammation of the pancreas, as may occur in association with alcohol abuse or chronic gallstones, is called **pancreatitis** (pan′-krē-a-TĪ-tis). In a more severe condition known as **acute pancreatitis,** which is associated with heavy alcohol intake or biliary tract obstruction, the pancreatic cells may release either trypsin instead of trypsinogen or insufficient amounts of trypsin inhibitor, and the trypsin begins to digest the pancreatic cells. Patients with acute pancreatitis usually respond to treatment, but recurrent attacks are the rule. ■

- ✓ Describe the duct system connecting the pancreas to the duodenum.
- ✓ What are pancreatic acini? Contrast their functions with those of the pancreatic islets (islets of Langerhans).
- ✓ Describe the composition and functions of pancreatic juice.

LIVER AND GALLBLADDER

Objective

- Describe the location, anatomy, histology, and functions of the liver and gallbladder.

The **liver** is the heaviest gland of the body, weighing about 1.4 kg (about 3 lb) in an average adult, and after the skin it is the second largest organ of the body. It is inferior to the diaphragm and occupies most of the right hypochondriac and part of the epigastric regions of the abdominopelvic cavity (see Figure 1.9 on page 16).

The **gallbladder** (*gall-* = bile) is a pear-shaped sac that is located in a depression of the posterior surface of the liver. It is 7–10 cm (3–4 in.) long and typically hangs partially below the anterior inferior margin of the liver (Figure 24.12).

Anatomy

The liver is almost completely covered by visceral peritoneum and is completely covered by a dense irregular connective tissue layer that lies deep to the peritoneum. The liver is divided into two principal lobes—a large **right lobe** and a smaller **left lobe**—by the **falciform ligament** (Figures 24.12 and 24.13). Even though the right lobe is considered by many anatomists to include an inferior **quadrate lobe** and a posterior **caudate lobe,** on the basis of internal morphology (primarily the distribution of blood vessels), the quadrate and caudate lobes more appropriately belong to the left lobe. The falciform ligament, a fold of the parietal peritoneum, extends from the undersurface of the diaphragm between the two principal lobes of the liver to the superior surface of the liver, helping to suspend the liver. In the free border of the falciform ligament is the **ligamentum teres (round ligament),** a fibrous cord that is a remnant of the umbilical vein of the fetus; it extends from the liver to the umbilicus. The right and left **coronary ligaments** are narrow reflections of the parietal peritoneum that suspend the liver from the diaphragm.

The parts of the gallbladder are the broad **fundus,** which projects downward beyond the inferior border of the liver; the **body,** the central portion; and the **neck,** the tapered portion. The body and neck project upward.

Histology

The lobes of the liver are made up of many functional units called **lobules** (Figure 24.14 on page 737). A lobule consists of specialized epithelial cells, called **hepatocytes** (*hepat-* = liver; *-cytes* = cells), arranged in irregular, branching, interconnected plates around a **central vein.** Instead of capillaries, the liver has larger, endothelium-lined spaces called **sinusoids,** through which blood passes. Also present in the sinusoids are fixed phagocytes called **stellate reticuloendothelial (Kupffer's) cells,** which destroy worn-out leukocytes and red blood cells, bacteria, and other foreign matter in the venous blood draining from the gastrointestinal tract.

Bile secreted by hepatocytes, enters **bile canaliculi** (kan′-a-LIK-yū-lī; = small canals), which are narrow intercellular canals that empty into small *bile ductules.* The ductules pass bile into *bile ducts* at the periphery of the lobules. The bile ducts merge and eventually form the larger **right** and **left hepatic ducts,** which unite and exit the liver as the **common hepatic duct** (see Figure 24.12). Farther on, the common hepatic duct joins the **cystic duct** (*cystic* = bladder) from the gallbladder to form the **common bile duct.** Bile enters the cystic duct and is temporarily stored in the gallbladder.

The mucosa of the gallbladder consists of simple columnar epithelium arranged in rugae resembling those of the stomach. The wall of the gallbladder lacks a submucosa. The middle, muscular coat consists of smooth muscle fibers; the contraction of these fibers ejects the contents of the gallbladder into the **cystic duct.** The gallbladder's outer coat is the visceral peritoneum. The functions of the gallbladder are to store and concentrate bile (up to tenfold) until it is needed in the small intestine.

Figure 24.13 / External anatomy of the liver. The anterior view is illustrated in Figure 24.12a.

The two principal lobes of the liver, the right and left lobes, are separated by the falciform ligament.

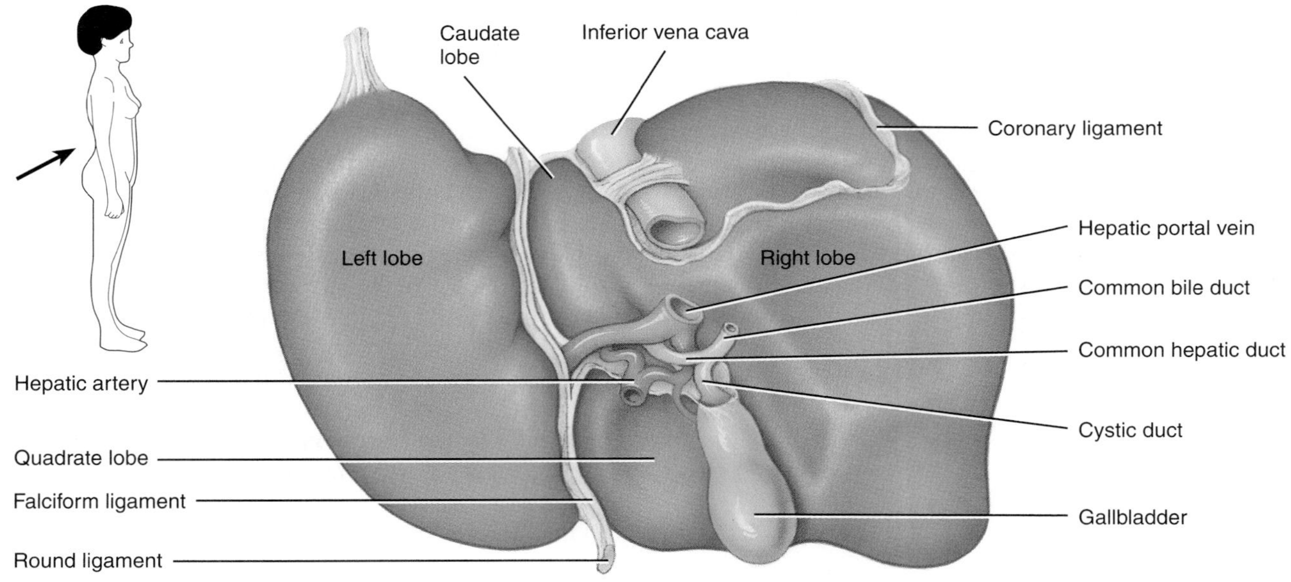

(a) Posteroinferior surface of liver

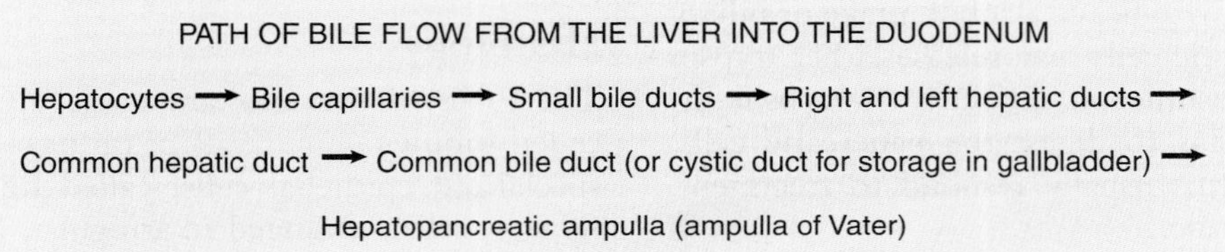

PATH OF BILE FLOW FROM THE LIVER INTO THE DUODENUM

Hepatocytes → Bile capillaries → Small bile ducts → Right and left hepatic ducts →

Common hepatic duct → Common bile duct (or cystic duct for storage in gallbladder) →

Hepatopancreatic ampulla (ampulla of Vater)

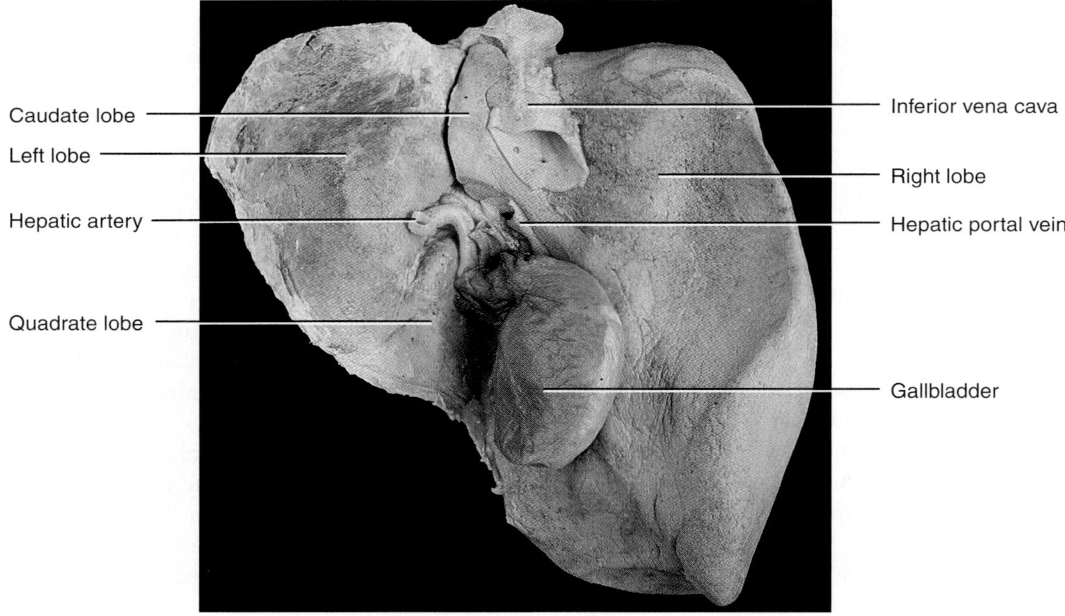

(b) Posteroinferior surface of liver

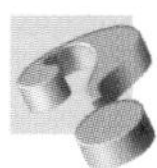

Within which abdominopelvic region (see Figure 1.9a on page 16) could you palpate (feel) most of the liver to decide if it is enlarged?

Figure 24.14 / Histology of a lobule, the functional unit of the liver.

A lobule consists of hepatocytes arranged around a central vein.

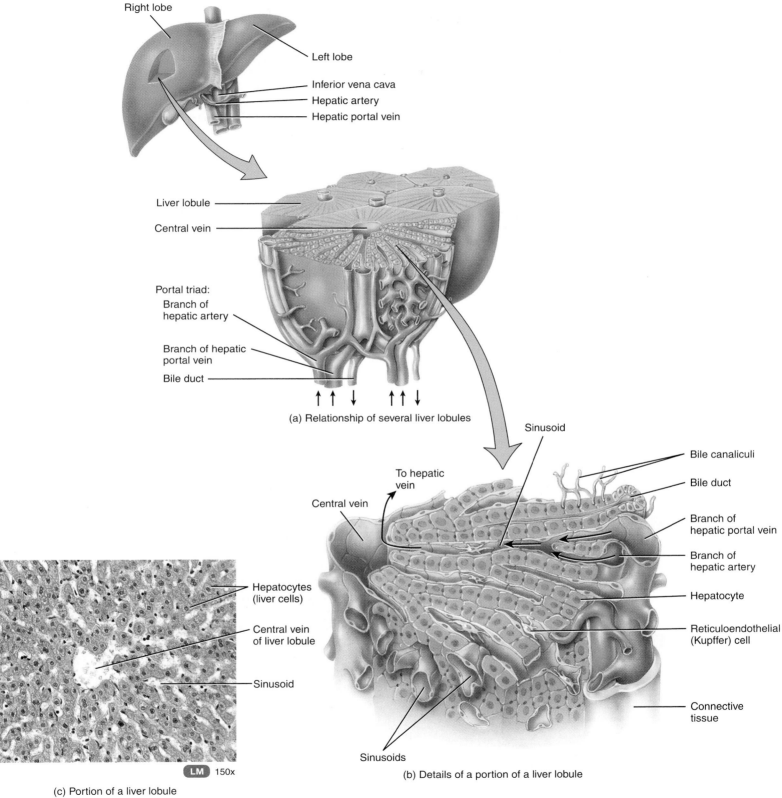

(a) Relationship of several liver lobules

(b) Details of a portion of a liver lobule

(c) Portion of a liver lobule

Which type of liver cell is phagocytic?

Jaundice (*jaune* = yellow) is a yellowish coloration of the sclerae (white of the eyes), skin, and mucous membranes due to a buildup in the body of a yellow compound called bilirubin. Bilirubin, which is formed from the breakdown of the heme pigment in aged red blood cells, is transported to the liver, where it is processed and eventually excreted into bile. The three main categories of jaundice are (1) *prehepatic jaundice*, due to excess production of bilirubin; (2) *hepatic jaundice*, due to congenital liver disease, cirrhosis of the liver, or hepatitis; and (3) *extrahepatic jaundice*, due to blockage of bile drainage by gallstones or cancer of the bowel or the pancreas.

Because the liver of a newborn functions poorly for the first week or so, many babies experience a mild form of jaundice called *neonatal (physiological) jaundice* that disappears as the liver matures. Usually, it is treated by exposing the infant to blue light, which converts bilirubin into substances the kidneys can excrete.

Blood and Nerve Supply

The liver receives blood from two sources (Figure 24.15). From the hepatic artery it obtains oxygenated blood, and from the hepatic portal vein it receives deoxygenated blood containing newly absorbed nutrients, drugs, and possibly microbes and toxins from the gastrointestinal tract. Branches of both the hepatic artery and the hepatic portal vein carry blood into liver sinusoids, where oxygen, most of the nutrients, and certain toxic substances are taken up by the hepatocytes. Products manufactured by the hepatocytes and nutrients needed by other cells are secreted back into the blood, which then drains into the central vein and eventually passes into a hepatic vein. Because blood from the gastrointestinal tract passes through the liver as part of the hepatic portal circulation, the liver is often a site for metastasis of cancer that originates in the GI tract. Branches of the hepatic portal vein, hepatic artery, and bile duct typically accompany each other in their distribution through the liver. Collectively, these three structures are called a **portal triad** (see Figure 24.14a).

The nerve supply to the liver consists of parasympathetic innervation from the vagus (X) nerves and sympathetic innervation from the greater splanchnic nerves through the celiac ganglia.

The gallbladder is supplied by the cystic artery, which usually arises from the right hepatic artery. The cystic veins drain the gallbladder. The nerves to the gallbladder include branches from the celiac plexus and the vagus (X) nerve.

Figure 24.15 / Hepatic blood flow: sources, path through the liver, and return to the heart.

The liver receives oxygen-rich blood via the hepatic artery and nutrient-rich deoxygenated blood via the hepatic portal vein.

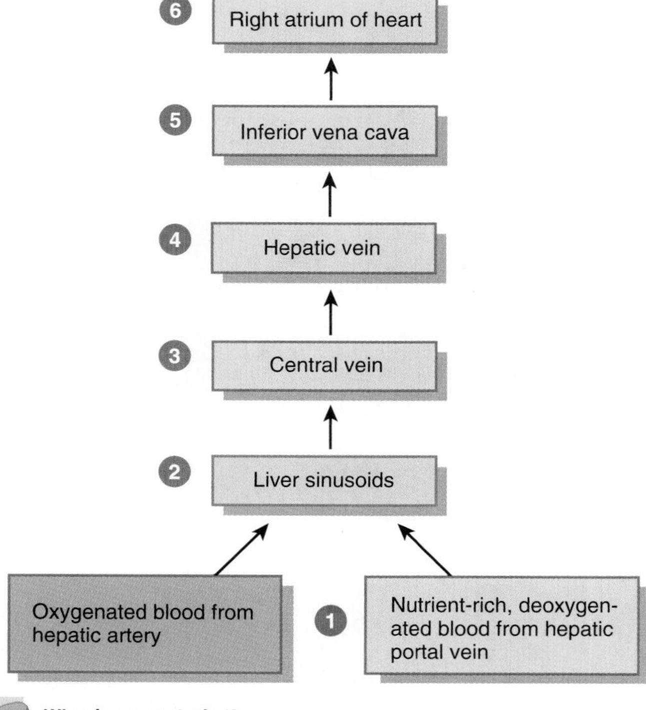

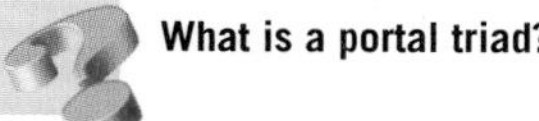
What is a portal triad?

Functions

Hepatocytes continuously secrete 800–1000 mL (about 1 qt) of **bile** per day: Bile is a yellow, brownish, or olive-green liquid and is partially an excretory product and partially a digestive secretion. Bile salts, which are sodium salts and potassium salts of bile acids (mostly cholic acid and chenodeoxycholic acid), play a role in **emulsification,** the breakdown of large lipid globules into a suspension of droplets about 1 μm in diameter, and in the absorption of lipids following their digestion. Between meals, bile flows into the gallbladder for storage because the **sphincter of the hepatopancreatic ampulla** (*sphincter of Oddi;* see Figure 24.12) closes off the entrance to the duodenum. After a meal, several neural and hormonal stimuli promote production and release of bile. Parasympathetic impulses along the vagus (X) nerve fibers can stimulate the liver to increase bile production to more than twice the baseline rate. Fatty acids and amino acids in chyme entering the duodenum stimulate some duodenal enteroendocrine cells to secrete the hormone cholecystokinin (CCK) into the blood. CCK causes contraction of the wall of the gallbladder, which squeezes stored bile out of the gallbladder into the cystic duct and through the common bile duct. CCK also causes relaxation of the sphincter of the hepatopancreatic ampulla, which allows bile to flow into the duodenum.

Besides secreting bile, the liver performs many other vital functions:

- **Carbohydrate metabolism.** The liver is especially important in maintaining a normal blood glucose level. When blood glucose is low, the liver can break down glycogen to glucose and release glucose into the bloodstream. The liver can also convert certain amino acids and lactic acid to glucose, and it can convert other sugars, such as fructose and galactose, into glucose. When blood glucose is high, as occurs just after eating a meal, the liver converts glucose to glycogen and triglycerides for storage.

- **Lipid metabolism.** Hepatocytes store some triglycerides; break down fatty acids to generate ATP; synthesize lipoproteins, which transport fatty acids, triglycerides, and cholesterol to and from body cells; synthesize cholesterol; and use cholesterol to make bile salts.
- **Protein metabolism.** Hepatocytes deaminate (remove the amino group, NH_2, from) amino acids so that the amino acids can be used for ATP production or converted to carbohydrates or fats. The resulting toxic ammonia (NH_3) is then converted into the much less toxic urea, which is excreted in urine. Hepatocytes also synthesize most plasma proteins, such as alpha and beta globulins, albumin, prothrombin, and fibrinogen.
- **Processing of drugs and hormones.** The liver can detoxify substances such as alcohol or excrete drugs such as penicillin, erythromycin, and sulfonamides into bile. It can also chemically alter or excrete thyroid hormones and steroid hormones such as estrogens and aldosterone.
- **Excretion of bilirubin.** As previously noted, bilirubin, derived from the heme of aged red blood cells, is absorbed by the liver from the blood and secreted into bile. Most of the bilirubin in bile is metabolized in the small intestine by bacteria and eliminated in feces.
- **Synthesis of bile salts.** Bile salts are used in the small intestine for the emulsification and absorption of lipids, cholesterol, phospholipids, and lipoproteins.
- **Storage.** In addition to glycogen, the liver is a prime storage site for certain vitamins (A, B_{12}, D, E, and K) and minerals (iron and copper), which are released from the liver when needed elsewhere in the body.
- **Phagocytosis.** The stellate reticuloendothelial (Kupffer's) cells of the liver phagocytize aged red blood cells and white blood cells and some bacteria.
- **Activation of vitamin D.** The skin, liver, and kidneys participate in synthesizing the active form of vitamin D.

If bile contains either insufficient bile salts or lecithin or excessive cholesterol, the cholesterol may crystallize to form **gallstones.** As they grow in size and number, gallstones may cause minimal, intermittent, or complete obstruction to the flow of bile from the gallbladder into the duodenum. Treatment consists of using gallstone-dissolving drugs, lithotripsy (shock-wave therapy), or surgery. For people with recurrent gallstones or for whom drugs or lithotripsy is not indicated, *cholecystectomy*—the removal of the gallbladder and its contents—is necessary. More than half a million cholecystectomies are performed each year in the United States. ■

✓ Draw and label a diagram of a liver lobule.

✓ Describe the pathways of blood flow into, through, and out of the liver.

✓ How are the liver and gallbladder connected to the duodenum?

✓ Describe the functions of the liver and gallbladder.

SMALL INTESTINE

Objective

- Describe the location, anatomy, histology, and function of the small intestine.

The major events of digestion and absorption occur in a long tube called the **small intestine.** Because almost all digestion and absorption of nutrients occur in the small intestine, its structure is specially adapted for this function. Its length alone provides a large surface area for digestion and absorption, and that area is further increased by circular folds, villi, and microvilli. The small intestine begins at the pyloric sphincter of the stomach, coils through the central and inferior part of the abdominal cavity, and eventually opens into the large intestine. It averages 2.5 cm (1 in.) in diameter; its length is about 3 m (10 ft.) in a living person and about 6.5 m (21 ft.) in a cadaver due to the loss of smooth muscle tone after death.

Anatomy

The small intestine is divided into three regions (Figure 24.16). The **duodenum** (dū′-ō-DĒ-num), the shortest region, is retroperitoneal. It starts at the pyloric sphincter of the stomach and extends about 25 cm (10 in.) until it merges with the je-

Figure 24.16 / Regions of the small intestine. See also Figure 24.1b.

Most digestion and absorption occur in the small intestine.

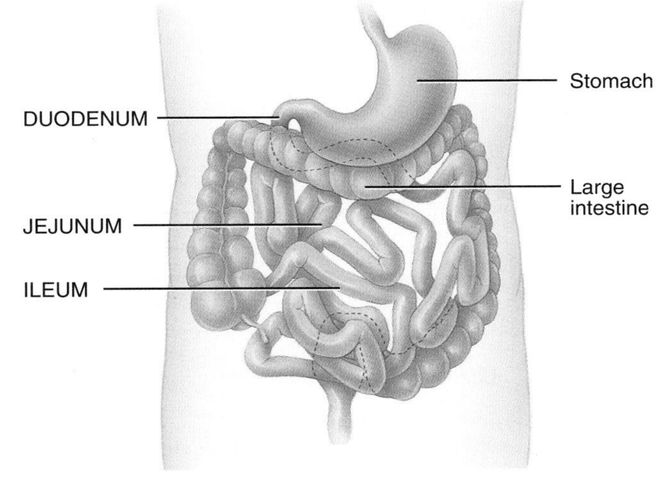

Which portion of the small intestine is the longest?

junum. *Duodenum* means "12"; it is so named because it is about as long as the width of 12 fingers. The **jejunum** (jē-JŪ-num) is about 1 m (3 ft.) long and extends to the ileum. *Jejunum* means "empty," which is how it is found at death. The final and longest region of the small intestine, the **ileum** (IL-ē-um; = twisted), measures about 2 m (6 ft.) and joins the large intestine at the **ileocecal** (il′-ē-ō-SĒ-kal) **sphincter.**

Projections called **circular folds** (or **plicae circulares**) are permanent ridges in the mucosa (see Figure 24.17c) about 10 mm (0.4 in.) high. The circular folds begin near the proximal portion of the duodenum and end at about the midportion of the ileum; some extend all the way around the circumference of the intestine, and others extend only part of the way around. They enhance absorption by increasing surface area and causing the chyme to spiral, rather than move in a straight line, as it passes through the small intestine.

The arterial blood supply of the small intestine is from the superior mesenteric artery and the gastroduodenal artery, coming from the hepatic artery of the celiac trunk. Blood is returned by way of the superior mesenteric vein, which, with the splenic vein, forms the hepatic portal vein.

The nerves to the small intestine are supplied by the superior mesenteric plexus. The branches of the plexus contain postganglionic sympathetic fibers, preganglionic parasympathetic fibers, and sensory fibers. The sensory fibers are both vagal and of spinal nerves. In the wall of the small intestine are two autonomic plexuses: the myenteric plexus between the muscular layers and the submucosal plexus in the submucosa. The nerve fibers are derived chiefly from the sympathetic division of the autonomic nervous system and partly from the vagus (X) nerve.

Histology

Even though the wall of the small intestine is composed of the same four coats that make up most of the GI tract, special features of both the mucosa and the submucosa facilitate the processes of digestion and absorption. The mucosa forms a series of fingerlike **villi** (= tufts of hair), projections that are 0.5–1 mm long (Figure 24.17). The large number of villi (20–40 per square millimeter) vastly increases the surface area of the epithelium available for absorption and digestion and gives the intestinal mucosa a velvety appearance. Each villus (singular form) has a core of lamina propria (areolar connective tissue); embedded in this connective tissue are an arteriole, a venule, a blood capillary network, and a **lacteal** (LAK-tē-al; = milky), which is a lymphatic capillary. Nutrients absorbed by the epithelial cells covering the villus pass through the wall of a capillary or a lacteal to enter blood or lymph, respectively.

Figure 24.17 / Anatomy of the small intestine.

Circular folds, villi, and microvilli increase the surface area of the small intestine for digestion and absorption.

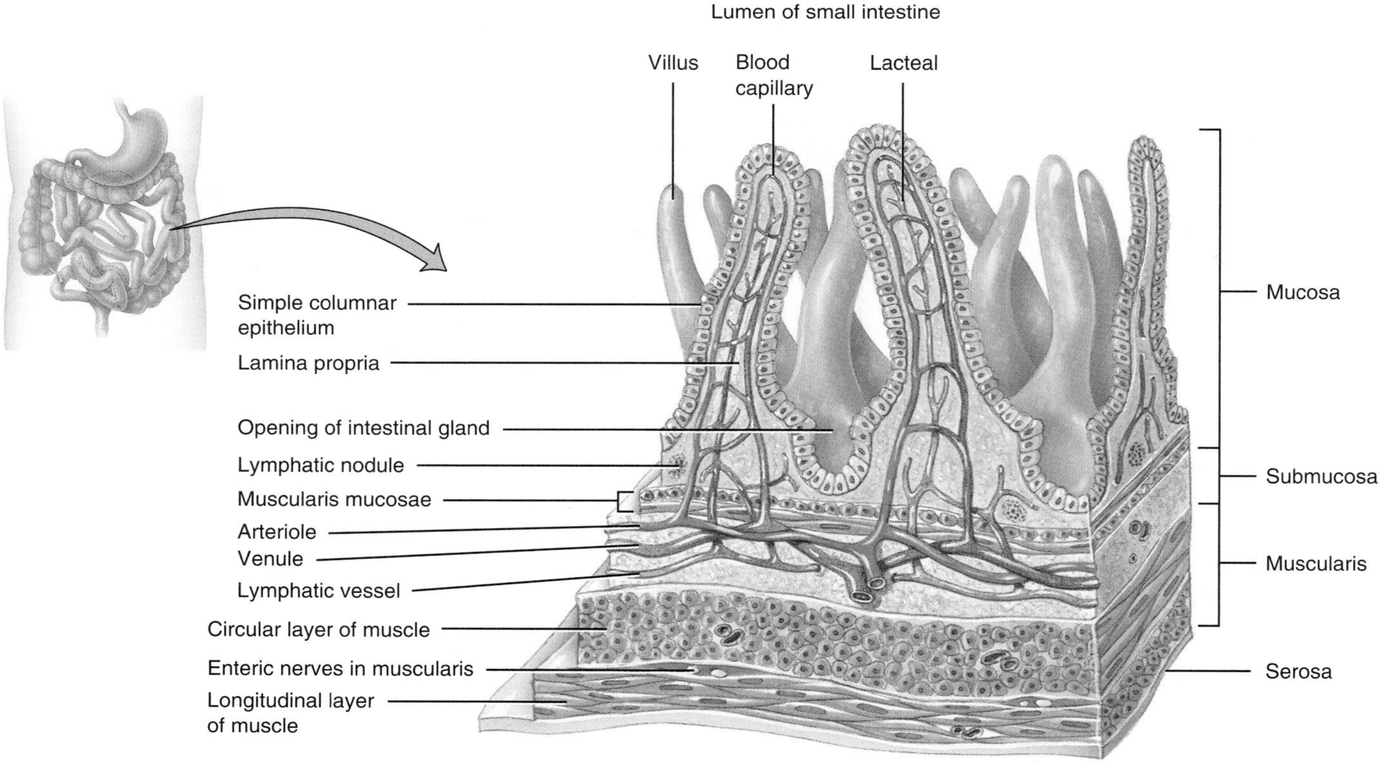

(a) Three-dimensional view of layers of the small intestine showing villi

The epithelium of the mucosa consists of simple columnar epithelium that contains absorptive cells, goblet cells, enteroendocrine cells, and Paneth cells (Figure 24.17b). The apical (free) membrane of absorptive cells features **microvilli** (mī′-krō-VIL-ī; *micro-* = small); each microvillus is a 1 μm-long cylindrical, membrane-covered projection that contains a bundle of 20–30 actin filaments. In a photomicrograph taken through a light microscope, the microvilli are too small to be seen individually; instead they form a fuzzy line, called the **brush border,** extending into the lumen of the small intestine (Figure 24.18b). There are an estimated 200 million microvilli per square millimeter of small intestine. Because the microvilli greatly increase the surface area of the plasma membrane, larger amounts of digested nutrients can diffuse into absorptive cells in a given period of time. The brush border also contains several, brush-border enzymes that have digestive functions (discussed shortly).

The mucosa contains many deep crevices lined with glandular epithelium. Cells lining the crevices form the **intestinal glands (crypts of Lieberkühn)** and secrete intestinal juice. Many of the epithelial cells in the mucosa are goblet cells, which

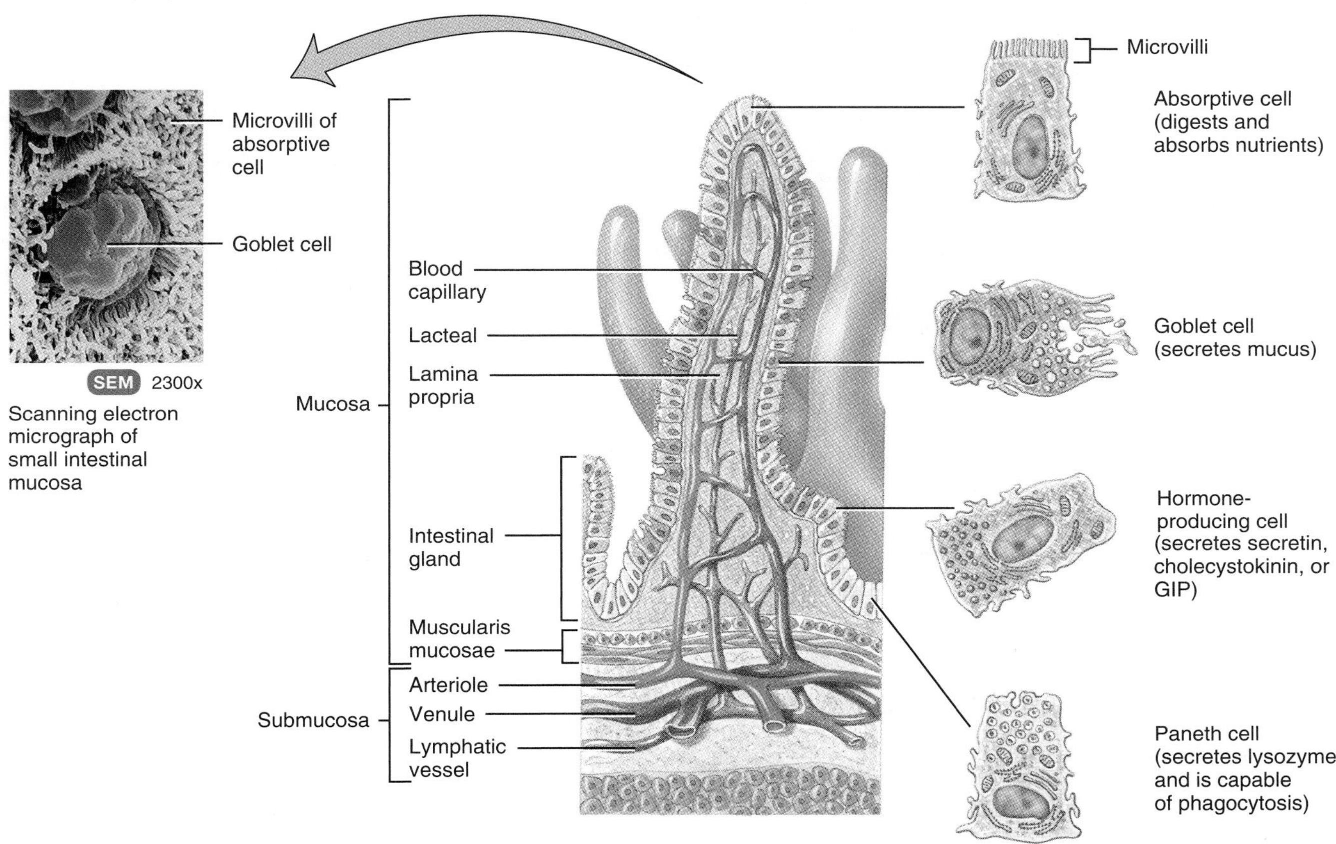

(b) Enlarged villus showing lacteal, capillaries, and intestinal glands

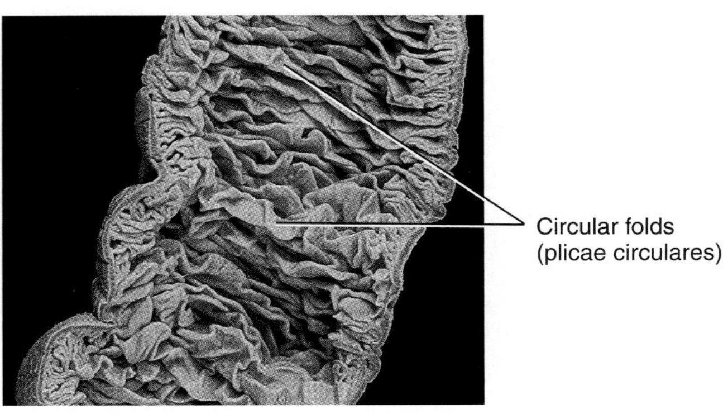

(c) Jejunum cut open to expose the circular folds

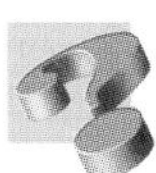

What is the functional significance of the blood capillary network and lacteal in the center of each villus?

Figure 24.18 / Histology of the duodenum.

Microvilli greatly increase the surface area of the small intestine for digestion and absorption.

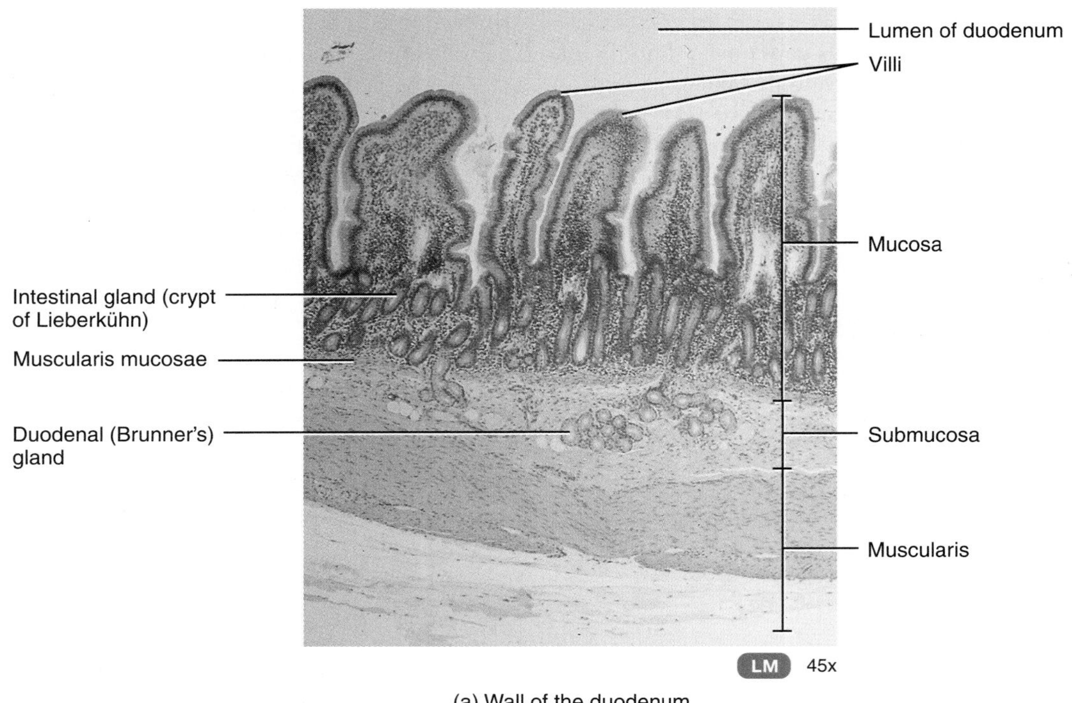

(a) Wall of the duodenum

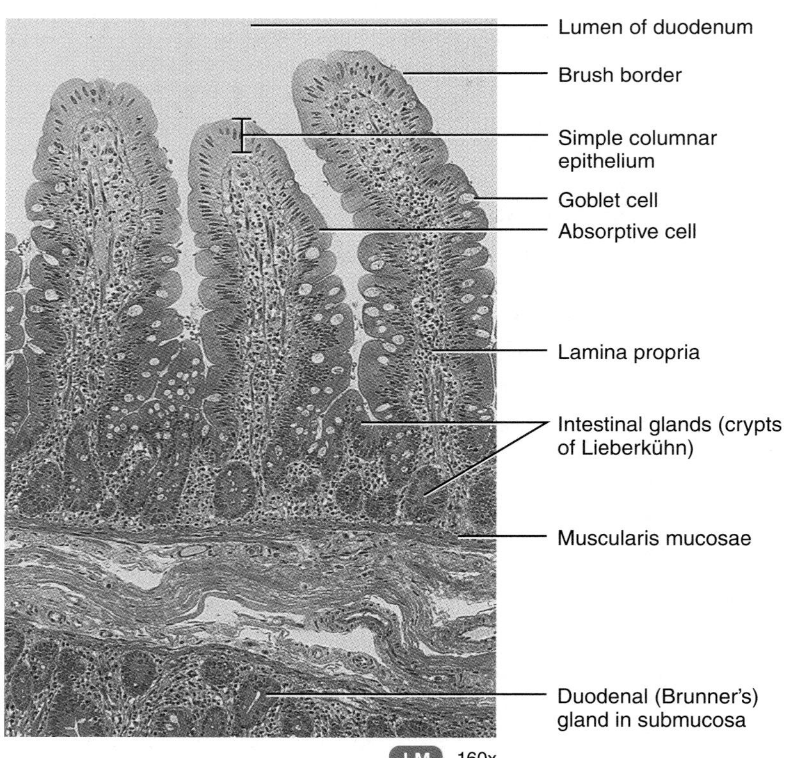

(b) Three villi from the duodenum of the small intestine

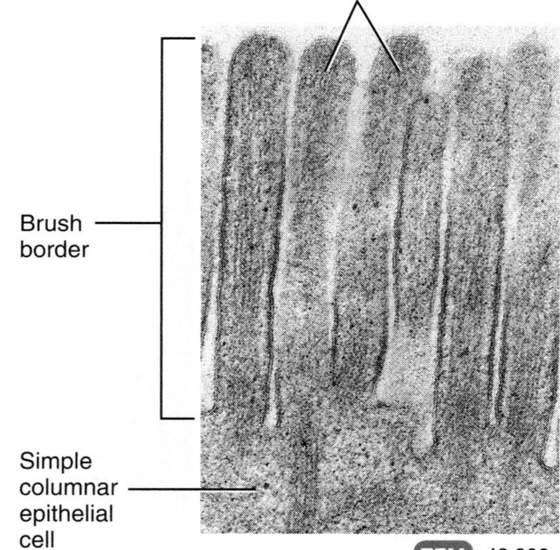

(c) Several microvilli from the duodenum

What is the function of the fluid secreted by duodenal (Brunner's) glands?

secrete mucus. **Paneth cells,** found in the deepest parts of the intestinal glands, secrete the bactericidal enzyme lysozyme, and are also capable of phagocytosis. They may have a role in regulating the microbial population in the intestines. Three types of enteroendocrine cells, also in the deepest part of the intestinal glands, secrete hormones: secretin (by S cells), cholecystokinin (by CCK cells), and glucose-dependent insulinotropic peptide (by K cells). The lamina propria of the small intestine has an abundance of mucosa-associated lymphoid tissue (MALT). **Solitary lymphatic nodules** are most numerous in the distal part of the ileum; groups of lymphatic nodules referred to as **aggregated lymphatic follicles (Peyer's patches)** are also numerous in the ileum. The muscularis mucosae consists of smooth muscle. The submucosa of the duodenum contains **duodenal (Brunner's) glands** (Figure 24.18a), which secrete an alkaline mucus that helps neutralize gastric acid in the chyme.

The **muscularis** of the small intestine consists of two layers of smooth muscle. The outer, thinner layer contains longitudinal fibers; the inner, thicker layer contains circular fibers. Except for a major portion of the duodenum, the serosa (or visceral peritoneum) completely surrounds the small intestine.

Functions

Chyme entering the small intestine contains partially digested carbohydrates, proteins, and lipids (mostly triglycerides). The completion of the digestion of carbohydrates, proteins, and lipids is the result of the collective action of pancreatic juice, bile, and intestinal juice in the small intestine.

Intestinal juice is a clear yellow fluid secreted in amounts of 1 to 2 liters (about 1 to 2 qt) a day. It has a pH of 7.6, which is slightly alkaline, and contains water and mucus. Together, pancreatic and intestinal juice provide a vehicle for the absorption of substances from chyme as they come in contact with the villi.

The absorptive epithelial cells synthesize several digestive enzymes, called **brush-border enzymes,** and insert them in the plasma membrane of the microvilli. Thus, some enzymatic digestion occurs at the surface of the epithelial cells that line the villi, rather than in the lumen exclusively, as occurs in other parts of the GI tract. Among the brush-border enzymes are four carbohydrate-digesting enzymes called **α-dextrinase, maltase, sucrase,** and **lactase;** protein-digesting enzymes called **peptidases** (**aminopeptidase** and **dipeptidase**); and two types of nucleotide-digesting enzymes, **nucleosidases** and **phosphatases.** Also, as cells slough off into the lumen of the small intestine, they break apart and release enzymes that help digest nutrients in the chyme.

The two types of movements of the small intestine—segmentations and a type of peristalsis called migrating motility complexes—are governed mainly by the myenteric plexus. **Segmentations** are a localized, mixing type of contraction that occurs in portions of intestine distended by a large volume of chyme. Segmentations mix chyme with the digestive juices and bring the particles of food into contact with the mucosa for absorption; they do not push the intestinal contents along the tract. A segmentation starts with the contractions of circular muscle fibers in a portion of the small intestine, an action that constricts the intestine into segments. Next, muscle fibers that encircle the middle of each segment also contract, dividing each segment again. Finally, the fibers that first contracted relax, and each small segment unites with an adjoining small segment so that large segments are formed again. As this sequence of events repeats, the chyme sloshes back and forth. Segmentations occur most rapidly in the duodenum, about 12 times per minute, and progressively decrease to about 8 times per minute in the ileum. This movement is similar to alternately squeezing the middle and then the ends of a capped tube of toothpaste.

After most of a meal has been absorbed, which lessens distention of the wall of the small intestine, segmentation stops and peristalsis begins. The type of peristalsis that occurs in the small intestine, termed a **migrating motility complex (MMC),** begins in the lower portion of the stomach and pushes chyme forward along a short stretch of small intestine before dying out. The MMC slowly migrates down the small intestine, reaching the end of the ileum in 90–120 minutes. Then another MMC begins in the stomach. Altogether, chyme remains in the small intestine for 3–5 hours.

All the chemical and mechanical phases of digestion from the mouth through the small intestine are directed toward changing food into forms that can pass through the epithelial cells lining the mucosa into the underlying blood and lymphatic vessels. These forms are monosaccharides (glucose, fructose, and galactose) from carbohydrates; single amino acids, dipeptides, and tripeptides from proteins; fatty acids, glycerol, and monoglycerides from lipids; and pentoses and nitrogenous bases from nucleic acids. Passage of these digested nutrients from the gastrointestinal tract into the blood or lymph is called **absorption.** Absorption occurs by diffusion, facilitated diffusion, osmosis, and active transport.

About 90% of all absorption of nutrients takes place in the small intestine. The other 10% occurs in the stomach and large intestine. Any undigested or unabsorbed material left in the small intestine passes on to the large intestine.

- ✓ What are the regions of the small intestine?
- ✓ In what ways are the mucosa and submucosa of the small intestine adapted for digestion and absorption?
- ✓ Describe the types of movement in the small intestine.
- ✓ Define absorption. In what form are the products of carbohydrate, protein, and lipid digestion absorbed?

LARGE INTESTINE

Objective

- Describe the anatomy, histology, and functions of the large intestine.

The large intestine is the terminal portion of the GI tract and is divided into four principal regions. The overall functions of the large intestine are the completion of absorption, the production of certain vitamins, the formation of feces, and the expulsion of feces from the body.

Anatomy

The **large intestine,** which is about 1.5 m (5 ft) long and 6.5 cm (2.5 in.) in diameter, extends from the ileum to the anus and is attached to the posterior abdominal wall by its **mesocolon,** which is a double layer of peritoneum. Structurally, the four principal regions of the large intestine are the cecum, colon, rectum, and anal canal (Figure 24.19a).

The opening from the ileum into the large intestine is guarded by a fold of mucous membrane called the **ileocecal sphincter,** which allows materials from the small intestine to pass into the large intestine. Hanging inferior to the ileocecal valve is the **cecum,** a blind pouch about 6 cm (2.4 in.) long. Attached to the cecum is a twisted, coiled tube, measuring about 8 cm (3 in.) in length, called the **appendix** or **vermiform appendix** (*vermiform* = worm-shaped; *appendix* = appendage). The mesentery of the appendix, called the **mesoappendix,** attaches the appendix to the inferior part of the mesentery of the ileum.

The open end of the cecum merges with a long tube called the **colon** (= food passage), which is divided into ascending, transverse, descending, and sigmoid portions. Both the ascending and descending colon are retroperitoneal, whereas the transverse and sigmoid colon are not. The **ascending colon** ascends on the right side of the abdomen, reaches the inferior surface of the liver, and turns abruptly to the left to form the **right colic (hepatic) flexure.** The colon continues across the abdomen to the left side as the **transverse colon.** It curves beneath the inferior end of the spleen on the left side as the **left colic (splenic) flexure** and passes inferiorly to the level of the iliac crest as the **descending colon.** The **sigmoid colon** (*sigm-* = S-shaped) begins near the left iliac crest, projects medially to the midline, and terminates as the rectum at about the level of the third sacral vertebra.

The **rectum,** the last 20 cm (8 in.) of the GI tract, lies anterior to the sacrum and coccyx. The terminal 2–3 cm (1 in.) of the rectum is called the **anal canal** (Figure 24.19b). The mucous membrane of the anal canal is arranged in longitudinal folds called **anal columns** that contain a network of arteries and veins. The opening of the anal canal to the exterior, called the **anus,** is guarded by an internal sphincter of smooth muscle (involuntary) and an external sphincter of skeletal muscle (voluntary). Normally the anus is closed except during the elimination of feces.

Inflammation of the appendix, termed **appendicitis,** is preceded by obstruction of the lumen of the appendix by chyme, inflammation, a foreign body, a carcinoma of the cecum, stenosis, or kinking of the organ. It is characterized by high fever, elevated white cell count, and a neutrophil count higher than 75%. The infection that follows may result in edema and ischemia and may progress to gangrene and perforation within 24 to 36 hours. Typically, appendicitis begins with referred pain in the umbilical region of the abdomen, followed by anorexia (loss of appetite for food), nausea, and vomiting. After several hours the pain localizes in the right lower quadrant (RLQ) and is continuous, dull or severe, and intensified by coughing, sneezing, or body movements. Early appendectomy (removal of the appendix) is recommended because it is safer to operate than to risk rupture, peritonitis, and gangrene. ■

The arterial supply of the cecum and colon is derived from branches of the superior mesenteric and inferior mesenteric arteries. The venous return is by way of the superior and inferior mesenteric veins ultimately to the hepatic portal vein and into the liver. The arterial supply of the rectum and anal canal is derived from the superior, middle, and inferior rectal arteries. The rectal veins correspond to the rectal arteries.

The nerves to the large intestine consist of sympathetic, parasympathetic, and sensory components. The sympathetic innervation is derived from the celiac, superior, and inferior mesenteric ganglia and superior and inferior mesenteric plexuses. The fibers reach the viscera by way of the thoracic and lumbar splanchnic nerves. The parasympathetic innervation is derived from the vagus (X) and pelvic splanchnic nerves.

Histology

The wall of the large intestine differs from that of the small intestine in several respects. No villi or permanent circular folds are found in the **mucosa,** which consists of simple columnar epithelium, lamina propria (areolar connective tissue), and muscularis mucosae (smooth muscle) (Figure 24.20 on page 746). The epithelium contains mostly absorptive and goblet cells (Figure 24.20b and c). The absorptive cells function primarily in water absorption, whereas the goblet cells secrete mucus that lubricates the passage of the colonic contents. Both absorptive and goblet cells are located in long, straight, tubular intestinal glands that extend the full thickness of the mucosa. Solitary lymphatic nodules are also found in the mucosa. The **submucosa** of the large intestine is similar to that found in the rest of the GI tract. The **muscularis** consists of an external layer of longitudinal smooth muscle and an internal layer of circular smooth muscle. Unlike other parts of the GI tract, portions of the longitudinal muscles are thickened, forming three conspicuous longitudinal bands called **teniae coli** (TĒ-nē-ē KŌ-lī; *taenia* = flat band), that run most of the length of the large intestine (see Figure 24.19a). The teniae coli are separated by portions of the wall with less or no longitudinal muscle. Tonic contractions of the bands gather the colon into a series of pouches called **haustra** (HAWS-tra; = shaped like pouches), which give the colon a puckered appearance. A single layer of circular smooth muscle lies between the teniae coli. The **serosa** of the large intestine is part of the visceral peritoneum. Small pouches of visceral peritoneum filled with fat are attached to teniae coli and are called **epiploic appendages.**

Functions

The passage of chyme from the ileum into the cecum is regulated by the action of the ileocecal sphincter. Normally, the

Figure 24.19 / Anatomy of the large intestine.

The regions of the large intestine are the cecum, colon, rectum, and anal canal.

Right colic (hepatic) flexure
TRANSVERSE COLON
Left colic (splenic) flexure
DESCENDING COLON
Teniae coli
ASCENDING COLON
Epiploic appendages
Ileum
Mesoappendix
Haustra
Ileocecal sphincter (valve)
CECUM
VERMIFORM APPENDIX
SIGMOID COLON
ANAL CANAL
RECTUM
ANUS

(a) Anterior view of large intestine showing principal subdivisions

Rectum
Anal canal
Internal anal sphincter (involuntary)
External anal sphincter (voluntary)
Anus
Anal column

(b) Frontal section of anal canal

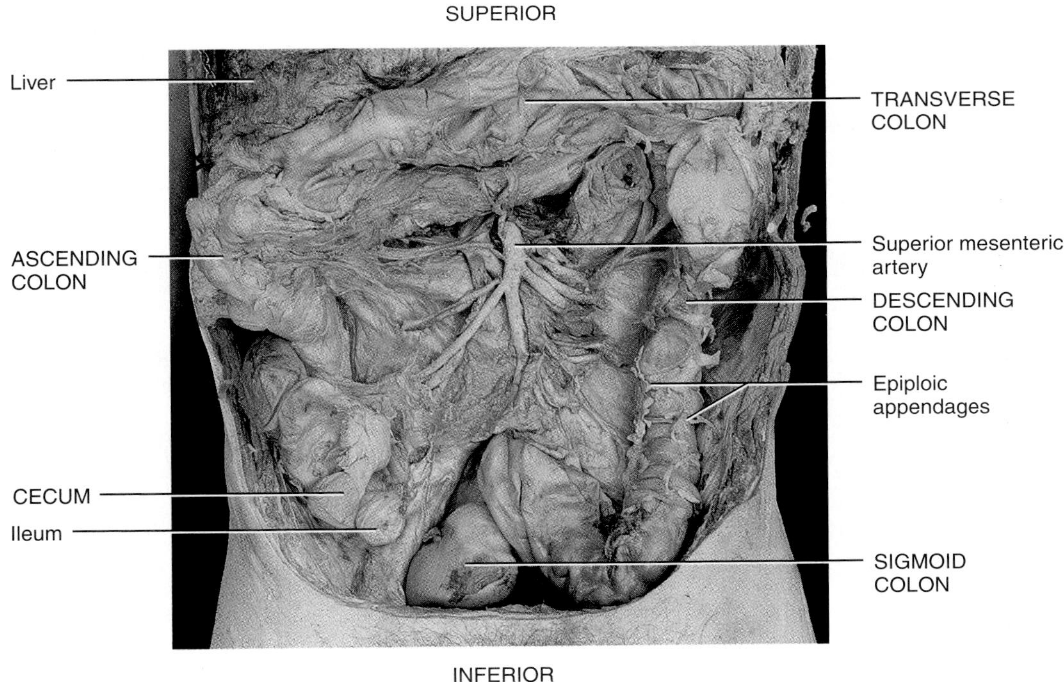

(c) Anterior view

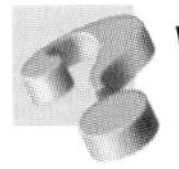

Which portions of the colon are retroperitoneal?

Figure 24.20 / Histology of the large intestine.

Intestinal glands formed by simple columnar epithelial cells and goblet cells extend the full thickness of the mucosa.

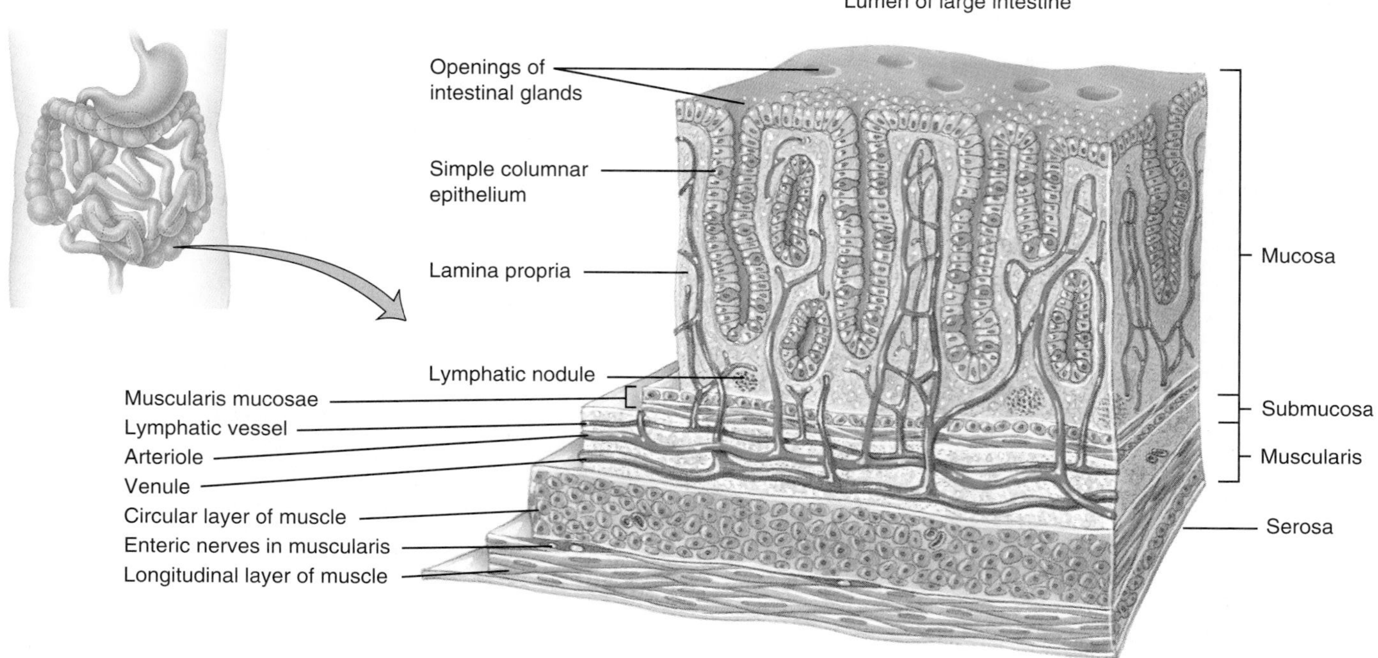

(a) Three-dimensional view of layers of the large intestine

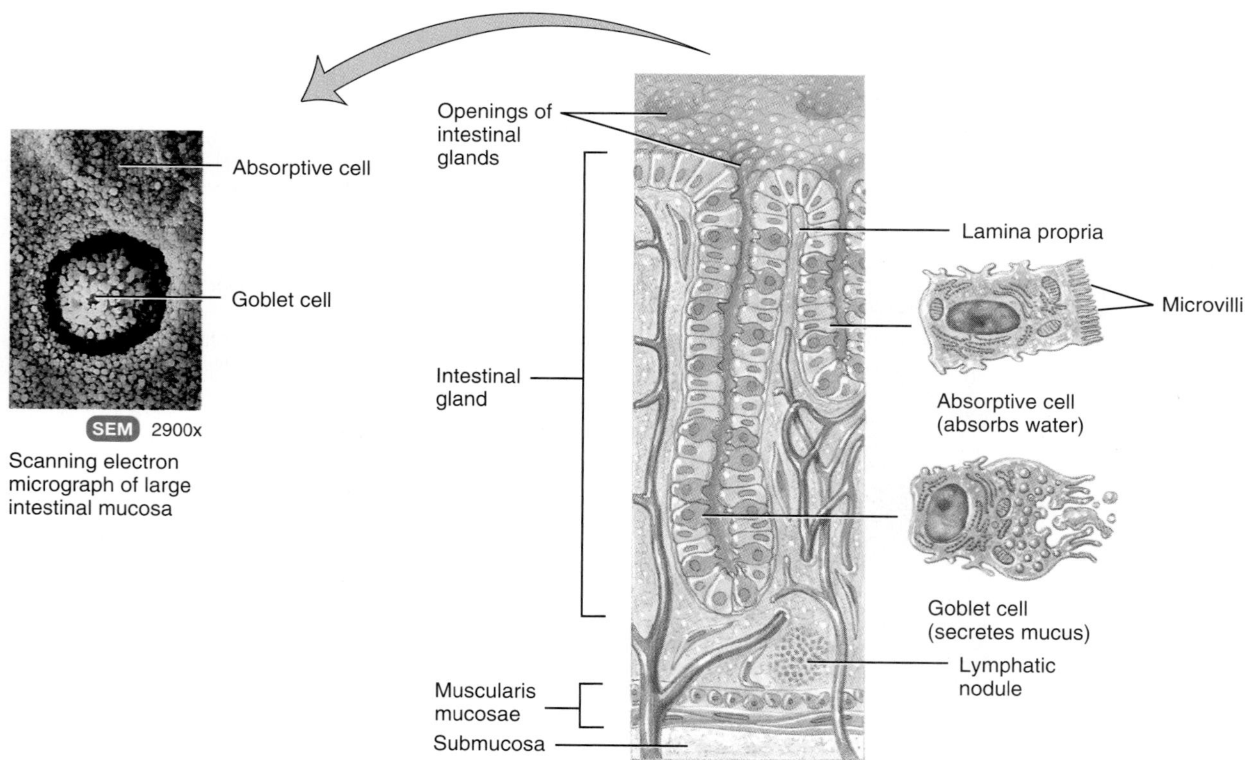

Scanning electron micrograph of large intestinal mucosa

(b) Sectional view of the large intestinal mucosa showing intestinal glands

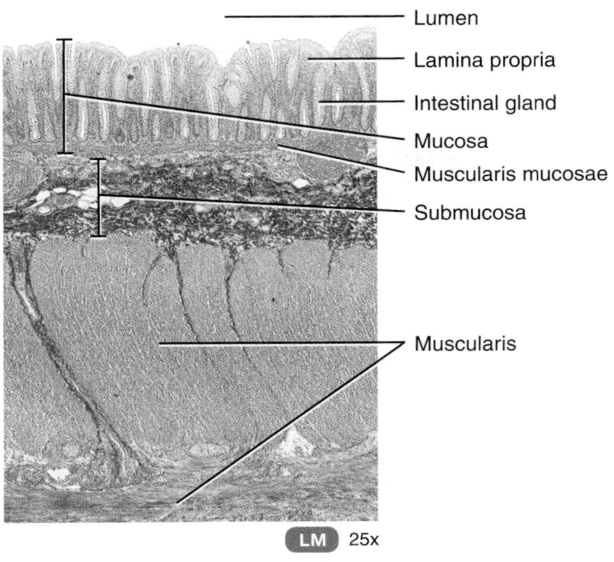

(c) Portion of the wall of the large intestine

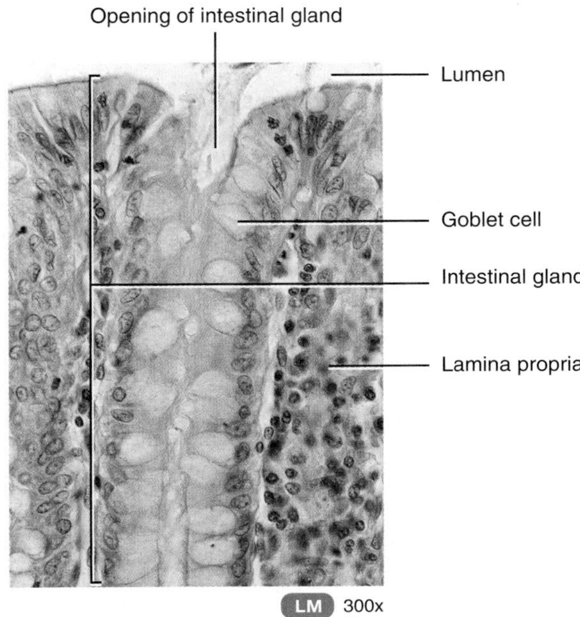

(d) Details of mucosa of large intestine

What is the function of the goblet cells of the large intestine?

valve remains partially closed so that the passage of chyme into the cecum is usually a slow process. Immediately after a meal, ileal peristalsis intensifies, the sphincter relaxes, and chyme is forced from the ileum into the cecum. As food passes through the ileocecal sphincter, it fills the cecum and accumulates in the ascending colon and movements of the colon begin.

One movement characteristic of the large intestine is **haustral churning.** In this process, the haustra remain relaxed and distended while they fill up. When the distension reaches a certain point, the wall contracts and squeezes the contents into the next haustrum. **Peristalsis** also occurs, although at a slower rate (3 to 12 contractions per minute) than in other portions of the tract. A final type of movement is **mass peristalsis,** a strong peristaltic wave that begins at about the middle of the transverse colon and quickly drives the colonic contents into the rectum. Mass peristalsis usually take place three or four times a day, during or immediately after a meal.

The final stage of digestion occurs in the colon through the activity of bacteria that inhabit the lumen. Mucus is secreted by the glands of the large intestine, but no enzymes are secreted. Chyme is prepared for elimination by the action of bacteria, which ferment any remaining carbohydrates and release hydrogen, carbon dioxide, and methane gases. These gases contribute to flatus (gas) in the colon, termed *flatulence* when it is excessive. Bacteria also convert remaining proteins to amino acids and break down the amino acids into simpler substances: indole, skatole, hydrogen sulfide, and fatty acids. Some of the indole and skatole is eliminated in the feces and contributes to their odor; the rest is absorbed and transported to the liver, where these compounds are converted to less toxic compounds and excreted in the urine. Bacteria also decompose bilirubin to simpler pigments, including stercobilin, which give feces their brown color. Several vitamins needed for normal metabolism, including some B vitamins and vitamin K, are bacterial products that are absorbed in the colon.

By the time chyme has remained in the large intestine 3–10 hours, it has become solid or semisolid as a result of water absorption and is now called **feces.** Chemically, feces consist of water, inorganic salts, sloughed-off epithelial cells from the mucosa of the gastrointestinal tract, bacteria, products of bacterial decomposition, unabsorbed digested materials, and indigestible parts of food.

Although 90% of all water absorption occurs in the small intestine, the large intestine absorbs enough to make it an important organ in maintaining the body's water balance. Of the 0.5–1.0 liter of water that enters the large intestine, all but about 100–200 mL is absorbed via osmosis. The large intestine also absorbs electrolytes, including sodium and chloride, and some vitamins.

Occult Blood

The term **occult blood** refers to blood that is hidden; it is not detectable by the human eye. The main diagnostic value of occult blood testing is to screen for colorectal cancer. Two substances frequently examined for occult blood are feces and urine. Several types of products are available for at-home testing for hidden blood in feces. The tests are based on color changes when reagents are added to feces. The presence of occult blood in urine may be detected at home by using dip-and-read reagent strips. ■

Mass peristaltic movements push fecal material from the sigmoid colon into the rectum. The resulting distention of the rectal wall stimulates stretch receptors, which initiates a **defecation reflex** that empties the rectum. The defecation reflex occurs as follows: In response to distention of the rectal wall, the receptors send sensory nerve impulses to the sacral spinal cord. Motor impulses from the cord travel along parasympathetic nerves back to the descending colon, sigmoid colon, rectum, and anus. The resulting contraction of the longitudinal rectal muscles shortens the rectum, thereby increasing the pressure within it. This pressure, along with voluntary contractions of the diaphragm and abdominal muscles, plus parasympathetic stimulation, open the internal sphincter.

The external sphincter is voluntarily controlled. If it is voluntarily relaxed, defecation occurs and the feces are expelled through the anus; if it is voluntarily constricted, defecation can be postponed. Voluntary contractions of the diaphragm and abdominal muscles aid defecation by increasing the pressure within the abdomen, which pushes the walls of the sigmoid colon and rectum inward. If defecation does not occur, the feces back up into the sigmoid colon until the next wave of mass peristalsis again stimulates the stretch receptors, further creating the urge to defecate. In infants, the defecation reflex causes automatic emptying of the rectum because voluntary control of the external anal sphincter has not yet developed.

Diarrhea (dī-a-RĒ-a; *dia-* = through; *rrhea* = flow) is an increase in the frequency, volume, and fluid content of the feces caused by increased motility of and decreased absorption by the intestines. When chyme passes too quickly through the small intestine and feces pass too quickly through the large intestine, there is not enough time for absorption. Frequent diarrhea can result in dehydration and electrolyte imbalances. Excessive motility may be caused by lactose intolerance, stress, and microbes that irritate the gastrointestinal mucosa.

Constipation (kon-sti-PĀ-shun; *con-* = together; *stip-* = to press) refers to infrequent or difficult defecation caused by decreased motility of the intestines. Because the feces remain in the colon for prolonged periods of time, excessive water absorption occurs, and the feces become dry and hard. Constipation may be caused by poor habits (delaying defecation), spasms of the colon, insufficient fiber in the diet, inadequate fluid intake, lack of exercise, emotional stress, and certain drugs. A common treatment is a mild laxative, such as milk of magnesia, which induces defecation. However, many physicians maintain that laxatives are habit-forming, and that adding fiber to the diet, increasing the amount of exercise, and increasing fluid intake are safer ways of controlling this common problem.

Dietary fiber consists of indigestible plant carbohydrates—such as cellulose, lignin, and pectin—found in fruits, vegetables, grains, and beans. **Insoluble fiber,** which does not dissolve in water, includes the woody or structural parts of plants such as the skins of fruits and vegetables and the bran coating around wheat and corn kernels. Insoluble fiber passes through the GI tract largely unchanged and speeds up the passage of material through the tract. **Soluble fiber** dissolves in water and forms a gel, which slows the passage of material through the tract; it is found in abundance in beans, oats, barley, broccoli, prunes, apples, and citrus fruits.

People who choose a fiber-rich diet may reduce their risk of developing obesity, diabetes, atherosclerosis, gallstones, hemorrhoids, diverticulitis, appendicitis, and colorectal cancer. Soluble fiber also may help lower blood cholesterol because the fiber binds bile salts and prevents their reabsorption; as a result, more cholesterol is used to replace the bile salts lost in the feces. ■

A summary of the digestive organs and their functions is presented in Table 24.2.

✓ What are the principal regions of the large intestine?
✓ How does the muscularis of the large intestine differ from that of the rest of the gastrointestinal tract? What are haustra?
✓ Describe the mechanical movements that occur in the large intestine.
✓ Define defecation. How does it occur?
✓ Explain the activities of the large intestine that change its contents into feces.

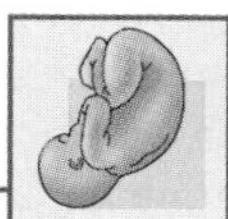

DEVELOPMENTAL ANATOMY OF THE DIGESTIVE SYSTEM

Objective

- Describe the development of the digestive system.

About the 14th day after fertilization, the cells of the endoderm form a cavity referred to as the **primitive gut** (Figure 24.21a). Soon after the mesoderm forms and splits into two layers (somatic and splanchnic), the splanchnic mesoderm associates with the endoderm of the primitive gut; as a result, the primitive gut has a double-layered wall. The **endodermal layer** gives rise to the *epithelial lining* and *glands* of most of the gastrointestinal tract; the **mesodermal layer** produces the *smooth muscle* and *connective tissue* of the tract.

The primitive gut elongates, and during the third week it differentiates into an anterior **foregut,** an intermediate **midgut,** and a posterior **hindgut** (Figure 24.21b). Until the fifth week of development, the midgut opens into the yolk sac; after that time, the yolk sac constricts and detaches from the midgut, and the midgut seals. In the region of the foregut, a depression consisting of ectoderm, the **stomodeum** (stō-mō-DĒ-um), appears (Figure 24.21c). This develops into the *oral cavity.* The **oral membrane** that separates the foregut from the stomodeum ruptures during the fourth week of development, so that the foregut is continuous with the outside of the embryo through the oral cavity. Another depression consisting of ectoderm, the **proc-**

Table 24.2 Summary of Organs of the Digestive System and Their Functions

Organs	Functions
Tongue	Maneuvers food for mastication, shapes food into a bolus, maneuvers food for deglutition, detects sensations for taste, and initiates digestion of triglycerides.
Salivary glands	Saliva softens, moistens, and dissolves foods; cleanses mouth and teeth; initiates the digestion of starch.
Teeth	Cut, tear, and pulverize food to reduce solids to smaller particles for swallowing.
Pancreas	Pancreatic juice buffers acidic gastric juice in chyme, stops the action of pepsin from the stomach, creates the proper pH for digestion in the small intestine, and participates in the digestion of carbohydrates, proteins, triglycerides, and nucleic acids.
Liver	Produces bile, which is required for the emulsification and absorption of lipids in the small intestine.
Gallbladder	Stores and concentrates bile and releases it into the small intestine.
Mouth	See the functions of the tongue, salivary glands, and teeth, all of which are in the mouth. Additionally, the lips and cheeks keep food between the teeth during mastication, and buccal glands lining the mouth produce saliva.
Pharynx	Receives a bolus from the oral cavity and passes it into the esophagus.
Esophagus	Receives a bolus from the pharynx and moves it into the stomach. This requires relaxation of the upper esophageal sphincter and secretion of mucus.
Stomach	Mixing waves combine saliva, food, and gastric juice, which initiates protein digestion, activates pepsin, kills microbes in food, helps absorb vitamin B_{12}, contracts the lower esophageal sphincter, increases stomach motility, relaxes the pyloric sphincter, and moves chyme into the small intestine.
Small intestine	Segmentation mixes chyme with digestive juices; peristalsis propels chyme toward the ileocecal sphincter; digestive secretions from the small intestine, pancreas, and liver complete the digestion of carbohydrates, proteins, lipids, and nucleic acids; circular folds, villi, and microvilli help absorb about 90% of digested nutrients.
Large intestine	Haustral churning, peristalsis, and mass peristalsis drive the colonic contents into the rectum; bacteria produce some B vitamins and vitamin K; absorption of some water, ions, and vitamins occurs; defecation.

Figure 24.21 / Development of the digestive system.

The gastrointestinal tract is derived from endoderm and mesoderm.

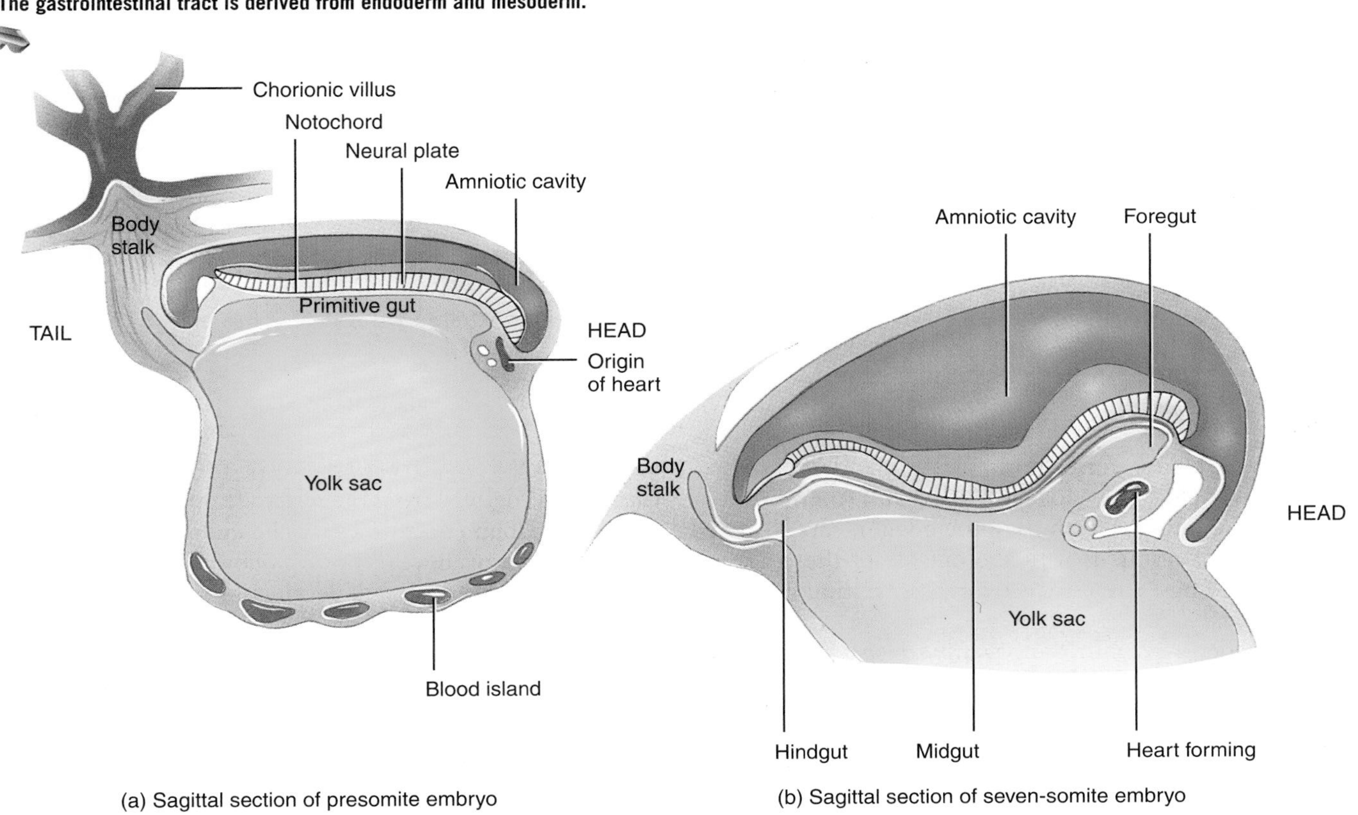

(a) Sagittal section of presomite embryo

(b) Sagittal section of seven-somite embryo

(continues)

Figure 24.21 (continued)

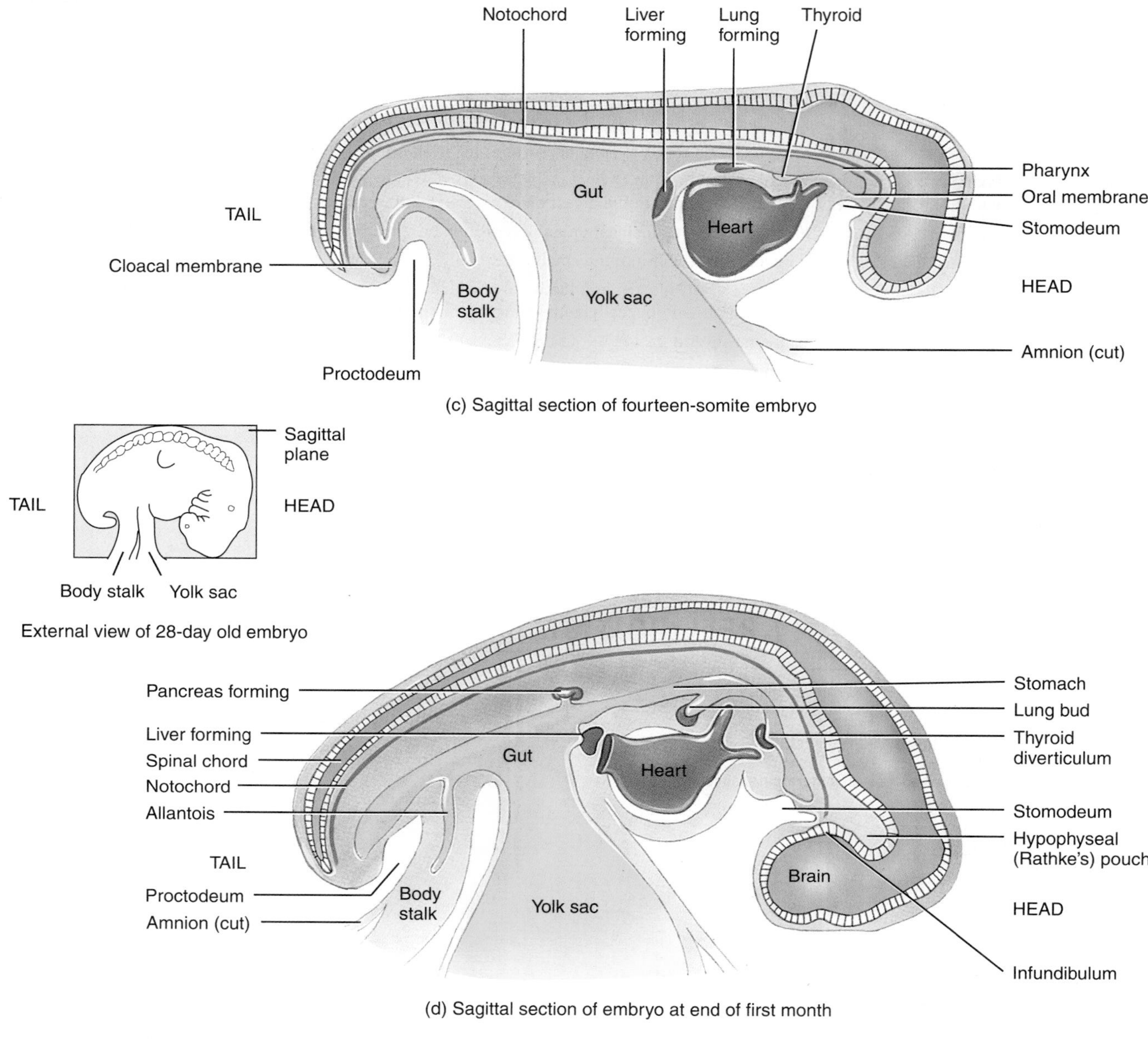

(c) Sagittal section of fourteen-somite embryo

(d) Sagittal section of embryo at end of first month

When does the gastrointestinal tract begin to develop?

todeum (prok-tō-DĒ-um), forms in the hindgut and goes on to develop into the *anus.* The **cloacal** (klō-Ā-kul) **membrane,** which separates the hindgut from the proctodeum, ruptures, so that the hindgut is continuous with the outside of the embryo through the anus. Thus, the GI tract forms a continuous tube from mouth to anus.

The foregut develops into the *pharynx, esophagus, stomach,* and a *portion of the duodenum.* The midgut is transformed into the *remainder of the duodenum,* the *jejunum,* the *ileum,* and *portions of the large intestine* (cecum, appendix, ascending colon, and most of the transverse colon). The hindgut develops into the *remainder of the large intestine,* except for a portion of the anal canal that is derived from the proctodeum.

As development progresses, the endoderm at various places along the foregut develops into hollow buds that grow into the mesoderm. These buds will develop into the *salivary glands, liver, gallbladder,* and *pancreas* (Figure 24.21d). Each of the glands retains a connection with the gastrointestinal tract through ducts. ■

✓ What structures develop from the foregut, midgut, and hindgut?

AGING AND THE DIGESTIVE SYSTEM

Objective

- Describe the effects of aging on the digestive system.

Overall changes of the digestive system associated with aging include decreased secretory mechanisms, decreased motility of the digestive organs, loss of strength and tone of the muscular tissue and its supporting structures, changes in neurosensory feedback regarding enzyme and hormone release, and diminished response to pain and internal sensations. In the upper portion of the GI tract, common changes include reduced sensitivity to mouth irritations and sores, loss of taste, periodontal disease, difficulty in swallowing, hiatal hernia, gastritis, and peptic ulcer disease. Changes that may appear in the small intestine include duodenal ulcers, appendicitis, malabsorption, and maldigestion. Other pathologies that increase in incidence with age are gallbladder problems, jaundice, cirrhosis, and acute pancreatitis. Large intestinal changes such as constipation, hemorrhoids, and diverticular disease may also occur. Cancer of the colon or rectum is quite common.

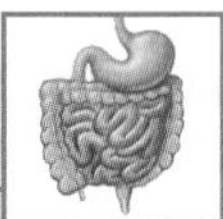

APPLICATIONS TO HEALTH

Dental Caries

Dental caries, or tooth decay, involve a gradual demineralization (softening) of the enamel and dentin. If untreated, microorganisms may invade the pulp, causing inflammation and infection, with subsequent death of the pulp and abscess of the alveolar bone surrounding the root's apex. Such teeth are treated by root canal therapy.

Dental caries begin when bacteria, acting on sugars, produce acids that demineralize the enamel. **Dextran,** a sticky polysaccharide produced from sucrose, causes the bacteria to stick to the teeth. Masses of bacterial cells, dextran, and other debris adhering to teeth constitute **dental plaque.** Saliva cannot reach the tooth surface to buffer the acid because the plaque covers the teeth. Brushing the teeth immediately after eating removes the plaque from flat surfaces before the bacteria can produce acids. Dentists also recommend that the plaque between the teeth be removed every 24 hours with dental floss.

Periodontal Disease

Periodontal disease is a collective term for a variety of conditions characterized by inflammation and degeneration of the gingivae, alveolar bone, periodontal ligament, and cementum. One such condition is called **pyorrhea,** the initial symptoms of which are enlargement and inflammation of the soft tissue and bleeding of the gums. Without treatment, the soft tissue may deteriorate and the alveolar bone may be resorbed, causing loosening of the teeth and recession of the gums. Periodontal diseases are often caused by poor oral hygiene; by local irritants, such as bacteria, impacted food, and cigarette smoke; or by a poor "bite."

Peptic Ulcer Disease

In the United States, 5–10% of the population develops **peptic ulcer disease (PUD).** An **ulcer** is a craterlike lesion in a membrane; ulcers that develop in areas of the GI tract exposed to acidic gastric juice are called **peptic ulcers.** The most common complication of peptic ulcers is bleeding, which can lead to anemia if enough blood is lost. In acute cases, peptic ulcers can lead to shock and death. Three distinct causes of PUD are recognized: (1) the bacterium *Helicobacter pylori;* (2) nonsteroidal anti-inflammatory drugs (NSAIDs) such as aspirin; and (3) hypersecretion of HCl, as occurs in Zollinger–Ellison syndrome, a gastrin-producing tumor usually of the pancreas.

Helicobacter pylori (previously named *Campylobacter pylori*) is the most frequent cause of PUD. The bacterium produces an enzyme called urease, which splits urea into ammonia and carbon dioxide. While shielding the bacterium from the acidity of the stomach, the ammonia also damages the protective mucous layer of the stomach and the underlying gastric cells. *H. pylori* also produces catalase, an enzyme that may protect the microbe from phagocytosis by neutrophils, plus several adhesion proteins that allow the bacterium to attach itself to gastric cells.

Several therapeutic approaches are helpful in the treatment of PUD. Because cigarette smoke, alcohol, caffeine, and NSAIDs can impair mucosal defensive mechanisms, in the process increasing mucosal susceptibility to the damaging effects of HCl, these substances should be avoided. In cases associated with *H. pylori,* treatment with an antibiotic drug often resolves the problem. Oral antacids such as Tums or Maalox can help temporarily by buffering gastric acid. When hypersecretion of HCl is the cause of PUD, H_2 blockers (such as Tagamet) or proton pump inhibitors such as omeprazole (Prilosec), which block secretion of H^+ from parietal cells, may be used.

Diverticulitis

Diverticulosis is the development of diverticula, saclike outpouchings of the wall of the colon in places where the muscularis has become weak. Many people who develop diverticulosis have no symptoms and experience no complications, but about 15% of people with diverticulosis will eventually develop an inflammation known as **diverticulitis.** This condition may be characterized by pain, either constipation or increased frequency of defecation, nausea, vomiting, and low-grade fever. Because diets low in fiber contribute to development of diverticulitis, patients who change to high-fiber diets show marked relief of symptoms. In severe cases, affected portions of the colon may require surgical removal. If diverticula rupture, the release of bacteria into the abdominal cavity can cause peritonitis.

Colorectal Cancer

Colorectal cancer is among the deadliest of malignancies, ranking second to lung cancer in males and third after lung cancer and breast cancer in females. Genetics plays a very important role in that an inherited predisposition contributes to more than half of all cases of colorectal cancer. Intake of alcohol and diets high in animal fat and protein are associated with increased risk of colorectal cancer, whereas dietary fiber, retinoids, calcium, and selenium may be protective. Signs and symptoms of colorectal cancer include diarrhea, constipation, cramping, abdominal pain, and rectal bleeding, either visible or occult. Screening for colorectal cancer includes testing for blood in the feces, digital rectal examination, sigmoidoscopy, colonoscopy, and barium enema. Tumors may be removed endoscopically or surgically.

Hepatitis

Hepatitis is an inflammation of the liver that can be caused by viruses, drugs, and chemicals, including alcohol. Clinically, several types of viral hepatitis are recognized. **Hepatitis A (infectious hepatitis)** is caused by hepatitis A virus and is spread via fecal contamination of objects such as food, clothing, toys, and eating utensils (fecal–oral route). It is generally a mild disease of children and young adults characterized by loss of appetite, malaise, nausea, diarrhea, fever, and chills. Eventually, jaundice appears. This type of hepatitis does not cause lasting liver damage. Most people recover in 4–6 weeks.

Hepatitis B is caused by hepatitis B virus and is spread primarily by sexual contact and contaminated syringes and transfusion equipment. It can also be spread via saliva and tears. Hepatitis B virus can be present for years or even a lifetime, and it can produce cirrhosis and possibly cancer of the liver. Individuals who harbor the active hepatitis B virus are at risk for cirrhosis and also become carriers. Vaccines produced through recombinant DNA technology (for example, Recombivax HB) are available to prevent hepatitis B infection.

Hepatitis C, caused by hepatitis C virus, is clinically similar to hepatitis B and is often spread by blood transfusions. Hepatitis C can cause cirrhosis and possibly liver cancer.

Hepatitis D is caused by hepatitis D virus. It is transmitted like hepatitis B and, in fact, a person must be coinfected with hepatitis B before contracting hepatitis D. Hepatitis D results in severe liver damage and has a higher fatality rate than infection with hepatitis B virus alone.

Hepatitis E is caused by hepatitis E virus and is spread like hepatitis A. Although it does not cause chronic liver disease, hepatitis E virus is responsible for a very high mortality rate in pregnant women.

Obesity

Obesity—body weight more than 20% above some desirable standard due to an excessive accumulation of adipose tissue—affects one-third of the adult population in the United States. (An athlete may be *overweight* due to higher-than-normal amounts of muscle tissue without being obese.) Even moderate obesity is hazardous to health; it is implicated as a risk factor in cardiovascular disease, hypertension, pulmonary disease, non-insulin-dependent diabetes mellitus, arthritis, certain cancers (breast, uterus, and colon), varicose veins, and gallbladder disease. Also, loss of body fat in obese individuals has been shown to elevate HDL cholesterol, the type associated with prevention of cardiovascular disease.

In a few cases, obesity may result from trauma of or tumors in the food-regulating centers in the hypothalamus. In most cases of obesity, no specific cause can be identified. Contributing factors include genetic factors, eating habits taught early in life, overeating to relieve tension, and social customs. Studies indicate that some obese people burn fewer calories during digestion and absorption of a meal. Additionally, obese people who lose weight require about 15% fewer calories to maintain normal body weight than do people who have never been obese. Although leptin suppresses appetite and produces satiety in experimental animals, it is not deficient in most obese people.

Anorexia Nervosa

Anorexia nervosa is a chronic disorder characterized by self-induced weight loss, negative perception of body image, and physiological changes that result from nutritional depletion. Patients with anorexia nervosa have a fixation on weight control and often insist on having a bowel movement every day despite inadequate food intake. They abuse laxatives, which worsens the fluid and electrolyte imbalances and nutrient deficiencies. The disorder is found predominantly in young, single females, and it may be inherited. Abnormal patterns of menstruation, amenorrhea (absence of menstruation), and a lowered basal metabolic rate reflect the depressant effects of starvation. Individuals may become emaciated and may ultimately die of starvation or one of its complications. Also associated with the disorder are osteoporosis, depression, and brain abnormalities coupled with impaired mental performance. Treatment consists of psychotherapy and dietary regulation.

KEY MEDICAL TERMS ASSOCIATED WITH THE DIGESTIVE SYSTEM

Achalasia (ak′-a-LĀ-zē-a; *a-* = without; *chalasis* = relaxation) A condition, caused by malfunction of the myenteric plexus, in which the lower esophageal sphincter fails to relax normally as food approaches. A whole meal may become lodged in the esophagus and enter the stomach very slowly. Distension of the esophagus results in chest pain that is often confused with pain originating from the heart.

Borborygmus (bor′-bō-RIG-mus) A rumbling noise caused by the propulsion of gas through the intestines.

Bulimia (*bu-* = ox; *limia* = hunger) or ***binge–purge syndrome*** A disorder that typically affects young, single, middle-class, white females, characterized by overeating at least twice a week followed by purging by self-induced vomiting, strict dieting or fasting, vigorous exercise, or use of laxatives or diuretics; it occurs in response to fears of being overweight or to stress, depression, and physiological disorders such as hypothalamic tumors.

Canker sore (KANG-ker) Painful ulcer on the mucous membrane of the mouth that affects females more often than males, usually between ages 10 and 40; may be an autoimmune reaction or a food allergy.

Cholecystitis (kō′-lē-sis-TĪ-tis; *chole-* = bile; *cyst-* = bladder; *-itis* = inflammation of) In some cases, an autoimmune inflammation of the gallbladder; other cases are caused by obstruction of the cystic duct by bile stones.

Cirrhosis Distorted or scarred liver as a result of chronic inflammation due to hepatitis, chemicals that destroy hepatocytes, parasites that infect the liver, and alcoholism; the hepatocytes are replaced by fibrous or adipose connective tissue. Symptoms include jaundice, edema in the legs, uncontrolled bleeding, and increased sensitivity to drugs.

Colitis (ko-LĪ-tis) Inflammation of the mucosa of the colon and rectum in which absorption of water and salts is reduced, producing watery, bloody feces and, in severe cases, dehydration and salt depletion. Spasms of the irritated muscularis produce cramps. It is thought to be an autoimmune condition.

Colostomy (kō-LOS-tō-mē; *-stomy* = provide an opening) The diversion of the fecal stream through an opening in the colon, creating a surgical "stoma" (artificial opening) that is affixed to the exterior of the abdominal wall. This opening serves as a substitute anus through which feces are eliminated into a bag worn on the abdomen.

Dysphagia (dis-FĀ-jē-a; *dys-* = abnormal; *phagia* = to eat) Difficulty in swallowing that may be caused by inflammation, paralysis, obstruction, or trauma.

Enteritis (en′-ter-Ī-tis; *enter-* = intestine) An inflammation of the intestine, particularly the small intestine.

Flatus (FLĀ-tus) Air (gas) in the stomach or intestine, usually expelled through the anus. If the gas is expelled through the mouth, it is called **eructation** or **belching** (burping). Flatus may result from gas released during the breakdown of foods in the stomach or from swallowing air or gas-containing substances such as carbonated drinks.

Gastrectomy (gas-TREK-tō-mē; *gastr-* = stomach; *-ectomy* = to cut out) Removal of all or a portion of the stomach.

Gastroscopy (gas-TROS-kō-pē; *-scopy* = to view with a lighted instrument) Endoscopic examination of the stomach in which the examiner can view the interior of the stomach directly to evaluate an ulcer, tumor, inflammation, or source of bleeding.

Heartburn A burning sensation in a region near the heart due to irritation of the mucosa of the esophagus from hydrochloric acid in stomach contents. It is caused by failure of the lower esophageal sphincter to close properly in which the stomach contents enter the inferior esophagus. It is not related to any cardiac problem.

Hemorrhoids (HEM-ō-royds; *hemi* = blood; *rhoia* = flow) Varicosed (enlarged and inflamed) veins in the rectum. Hemorrhoids develop when the veins are put under pressure and become engorged with blood. If the pressure continues, the wall of the vein stretches. Such a distended vessel oozes blood, and bleeding or itching is usually the first sign that a hemorrhoid has developed. Stretching of a vein also favors clot formation, further aggravating swelling and pain. Hemorrhoids may be caused by constipation, which may be brought on by low-fiber diets. Also, repeated straining during defecation forces blood down into the rectal veins, increasing pressure in these veins and possibly causing hemorrhoids. Also called *piles.*

Inflammatory bowel disease (in-FLAM-a-tō′-rē BOW-el) Disorder that exists in two forms: (1) Crohn's disease, an inflammation of the gastrointestinal tract, especially the distal ileum and proximal colon, in which the inflammation may extend from the mucosa through the serosa, and (2) ulcerative colitis, an inflammation of the mucosa of the gastrointestinal tract, usually limited to the large intestine and usually accompanied by rectal bleeding.

Irritable bowel syndrome (IBS) Disease of the entire gastrointestinal tract in which a person reacts to stress by developing symptoms (such as cramping and abdominal pain) associated with alternating patterns of diarrhea and constipation. Excessive amounts of mucus may appear in feces, and other symptoms include flatulence, nausea, and loss of appetite. The condition is also known as **irritable colon** or **spastic colitis.**

Malocclusion (mal′-ō-KLŪ-zhun; *mal-* = bad; *occlusion* = to fit together) Condition in which the surfaces of the maxillary (upper) and mandibular (lower) teeth fit together poorly.

Nausea (NAW-sē-a; *nausia* = seasickness) Discomfort characterized by a loss of appetite and the sensation of impending vomiting. Its causes include local irritation of the gastrointestinal tract, a systemic disease, brain disease or injury, overexertion, or the effects of medication or drug overdosage.

Traveler's diarrhea Infectious disease of the gastrointestinal tract that results in loose, urgent bowel movements, cramping, abdominal pain, malaise, nausea, and occasionally fever and dehydration. It is acquired through ingestion of food or water contaminated with fecal material typically containing bacteria (especially *Escherichia coli*); viruses or protozoan parasites are a less common cause.

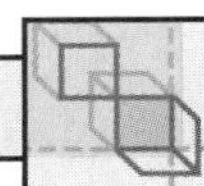

STUDY OUTLINE

Introduction (p. 715)

1. The breaking down of larger food molecules into smaller molecules is called digestion.
2. The passage of these smaller molecules into blood and lymph is termed absorption.

Overview of the Digestive System (p. 715)

1. The organs that collectively perform digestion and absorption constitute the digestive system and are usually composed of two main groups: the gastrointestinal (GI) tract and accessory digestive organs.

2. The GI tract is a continuous tube extending from the mouth to the anus.
3. The accessory digestive organs include the teeth, tongue, salivary glands, liver, gallbladder, and pancreas.
4. Digestion includes six basic processes: ingestion, secretion, mixing and propulsion, mechanical and chemical digestion, absorption, and defecation.
5. Mechanical digestion consists of mastication and movements of the gastrointestinal tract that aid chemical digestion.
6. Chemical digestion is a series of hydrolysis reactions that break down large carbohydrates, lipids, proteins, and nucleic acids in foods into smaller molecules that are usable by body cells.

Layers of the GI Tract (p. 715)

1. The basic arrangement of layers in most of the gastrointestinal tract, from deep to superficial, is the mucosa, submucosa, muscularis, and serosa.
2. Associated with the lamina propria of the mucosa are extensive patches of lymphatic tissue called mucosa-associated lymphoid tissue (MALT).

Peritoneum (p. 718)

1. The peritoneum is the largest serous membrane of the body; it lines the wall of the abdominal cavity and covers some abdominal organs.
2. Folds of the peritoneum include the mesentery, mesocolon, falciform ligament, lesser omentum, and greater omentum.

Mouth (p. 719)

1. The mouth is formed by the cheeks, hard and soft palates, lips, and tongue.
2. The vestibule is the space bounded externally by the cheeks and lips and internally by the teeth and gums.
3. The oral cavity proper extends from the vestibule to the fauces.
4. The tongue, together with its associated muscles, forms the floor of the oral cavity. It is composed of skeletal muscle covered with mucous membrane.
5. The upper surface and sides of the tongue are covered with papillae, some of which contain taste buds.
6. The major portion of saliva is secreted by the salivary glands, which lie outside the mouth and pour their contents into ducts that empty into the oral cavity.
7. There are three pairs of salivary glands: parotid, submandibular (submaxillary), and sublingual glands.
8. Saliva lubricates food and starts the chemical digestion of carbohydrates.
9. Salivation is controlled by the nervous system.
10. The teeth (dentes) project into the mouth and are adapted for mechanical digestion.
11. A typical tooth consists of three principal regions: crown, root, and neck.
12. Teeth are composed primarily of dentin and are covered by enamel, the hardest substance in the body.
13. There are two dentitions: deciduous and permanent.
14. Through mastication, food is mixed with saliva and shaped into a soft, flexible mass called a bolus.
15. Salivary amylase begins the digestion of starches, and lingual lipase acts on triglycerides.

Pharynx (p. 726)

1. Deglutition, or swallowing, moves a bolus from the mouth to the stomach.
2. Muscular contractions of the oropharynx and laryngopharynx propel a bolus into the esophagus.

Esophagus (p. 726)

1. The esophagus is a collapsible, muscular tube that connects the pharynx to the stomach.
2. It passes a bolus into the stomach by peristalsis.
3. It contains an upper and a lower esophageal sphincter.

Stomach (p. 728)

1. The stomach connects the esophagus to the duodenum.
2. The principal anatomic regions of the stomach are the cardia, fundus, body, and pylorus.
3. Adaptations of the stomach for digestion include rugae; glands that produce mucus, hydrochloric acid, pepsin, gastric lipase, and intrinsic factor; and a three-layered muscularis.
4. Mechanical digestion consists of mixing waves.
5. Chemical digestion consists mostly of the conversion of proteins into peptides by pepsin.
6. The stomach wall is impermeable to most substances.
7. Among the substances the stomach can absorb are water, certain ions, drugs, and alcohol.

Pancreas (p. 733)

1. The pancreas consists of a head, a body, and a tail and is connected to the duodenum via the pancreatic duct and accessory duct.
2. Endocrine pancreatic islets (islets of Langerhans) secrete hormones, and exocrine acini secrete pancreatic juice.
3. Pancreatic juice contains enzymes that digest starch (pancreatic amylase), proteins (trypsin, chymotrypsin, carboxypeptidase, and elastase), triglycerides (pancreatic lipase), and nucleic acids (ribonuclease and deoxyribonuclease).

Liver and Gallbladder (p. 735)

1. The liver has left and right lobes; the right lobe includes a quadrate and caudate lobe. The gallbladder is a sac located in a depression on the posterior surface of the liver that stores and concentrates bile.
2. The lobes of the liver are made up of lobules that contain hepatocytes (liver cells), sinusoids, stellate reticuloendothelial (Kupffer's) cells, and a central vein.
3. Hepatocytes produce bile that is carried by a duct system to the gallbladder for concentration and temporary storage. Cholecystokinin (CCK) stimulates ejection of bile into the common bile duct.
4. Bile's contribution to digestion is the emulsification of dietary lipids.
5. The liver also functions in carbohydrate, lipid, and protein metabolism; processing of drugs and hormones; excretion of bilirubin; synthesis of bile salts; storage of vitamins and minerals; phagocytosis; and activation of vitamin D.
6. Bile secretion is regulated by neural and hormonal mechanisms.

Small Intestine (p. 739)

1. The small intestine extends from the pyloric sphincter to the ileocecal sphincter.
2. It is divided into duodenum, jejunum, and ileum.
3. Its glands secrete fluid and mucus, and the circular folds, villi, and microvilli of its wall provide a large surface area for digestion and absorption.
4. Brush-border enzymes digest α-dextrins, maltose, sucrose, lactose,

peptides, and nucleotides at the surface of mucosal epithelial cells.
5. Pancreatic and intestinal brush-border enzymes break down carbohydrates, proteins, and nucleic acids.
6. Mechanical digestion in the small intestine involves segmentation and migrating motility complexes.
7. Absorption is the passage of digested nutrients from the gastrointestinal tract into blood or lymph.
8. The absorbed nutrients include monosaccharides, amino acids, fatty acids, monoglycerides, and pentoses, and nitrogenous bases.

Large Intestine (p. 743)

1. The large intestine extends from the ileocecal sphincter to the anus.
2. Its regions include the cecum, colon, rectum, and anal canal.
3. The mucosa contains many goblet cells, and the muscularis consists of teniae coli and haustra.
4. Mechanical movements of the large intestine include haustral churning, peristalsis, and mass peristalsis.
5. The last stages of chemical digestion occur in the large intestine through bacterial action. Substances are further broken down, and some vitamins are synthesized.
6. The large intestine absorbs water, electrolytes, and vitamins.
7. Feces consist of water, inorganic salts, epithelial cells, bacteria, and undigested foods.
8. The elimination of feces from the rectum is called defecation.
9. Defecation is a reflex action aided by voluntary contractions of the diaphragm and abdominal muscles and relaxation of the external anal sphincter.

Developmental Anatomy of the Digestive System (p. 748)

1. The endoderm of the primitive gut forms the epithelium and glands of most of the gastrointestinal tract.
2. The mesoderm of the primitive gut forms the smooth muscle and connective tissue of the gastrointestinal tract.

Aging and the Digestive System (p. 751)

1. General changes include decreased secretory mechanisms, decreased motility, and loss of tone.
2. Specific changes may include loss of taste, pyorrhea, hernias, peptic ulcer disease, constipation, hemorrhoids, and diverticular diseases.

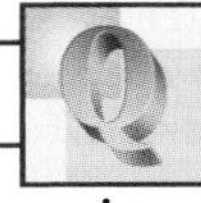

SELF-QUIZ QUESTIONS

Choose the one best answer to these questions.

1. The cells of the gastric glands that produce secretions directly involved in chemical digestion are (a) mucous neck cells, (b) parietal cells, (c) chief cells, (d) G cells, (e) goblet cells.
2. Which anatomical region of the stomach is closest to the esophagus? (a) body, (b) pylorus, (c) fundus, (d) cardia, (e) pyloric sphincter
3. Which of the following four statements is/are characteristic of the large intestine? (1) It is divided into cecum, colon, rectum and anal canal. (2) It contains bacteria that synthesize nutritional factors such as vitamins. (3) It serves as the main absorptive surface for digesting foods. (4) It absorbs much of the water remaining in the waste materials.

   a. 1 only **b.** 2 only **c.** 1 and 2 **d.** 3 and 4 **e.** 1, 2, and 4.
4. Which of the following digestive juices contains enzymes that digest carbohydrates, lipids, and proteins? (a) pancreatic juice, (b) bile, (c) saliva, (d) gastric juice, (e) none of the above (because no one digestive juice contains enzymes for digesting all three classes of foods).
5. Which of the following three statements about the esophagus is/are true? (1) It extends from the pharynx to the stomach. (2) It is approximately 25 cm (9 in.) long in the adult. (3) It is posterior to the trachea, anterior to the vertebral column, and passes through the diaphragm.

   a. 1 only **b.** 2 only **c.** 1 and 2 **d.** 1, 2, and 3 **e.** 1 and 3.
6. The roots of a tooth are covered by a tissue that is harder and denser than bone and is called (a) gingivae, (b) cementum, (c) dentin, (d) periodontal ligament, (e) enamel.
7. Which of the following statements about the parotid glands is/are true? (1) They are the largest of the salivary glands. (2) They are located just anterior to the ear and deep to the mandibular ramus. (3) Their secretions empty into the mouth by a duct that passes through the buccinator muscle.

   a. 1 only **b.** 2 only **c.** 1 and 2 **d.** 1, 2, and 3 **e.** 1 and 3.
8. Which of the following statements is/are true for *every* part of the GI tract from the inferior third of the esophagus to the anus? (a) Enzyme-secreting cells are present in the mucosa. (b) The muscular wall consists of smooth muscle. (c) Lymphoid follicles (Peyer's patches) are present in the submucosa. (d) Epiploic appendages are located on the outer serosa. (e) All of the above are true.

Complete the following.

9. The ________ is a layer of areolar connective tissue in the mucous membrane of the GI tract that contains lymphatic nodules that make up the MALT, an acronym that stands for ________ .
10. Place numbers in the blanks to indicate the correct order for the flow of bile. (a) bile canaliculi: ________ ; (b) common bile duct: ________ ; (c) common hepatic duct: ________ ; (d) right and left hepatic ducts: ________ ; (e) hepatopancreatic ampulla (ampulla of Vater) and duodenum: ________ .
11. A fold of mucous membrane called the ________ attaches the tongue to the floor of the oral cavity.
12. Place an E (for endoderm) or M (for mesoderm) in the blanks to indicate which germ layer gives rise to each of the following structures: (a) epithelial lining and digestive glands of the gastrointestinal tract: ________ ; (b) liver: ________ ; gallbladder, and pancreas: ________ ; (c) muscularis layer and connective tissue of submucosa: ________ .
13. First and second premolars and first, second, and third molars are characteristic of the ________ dentition.
14. The ________ is an organ of the digestive system that assumes a role in phagocytosis, manufacture of plasma proteins, detoxification, and interconversions of nutrients.
15. The duct that conveys bile to and from the gallbladder is the ________ duct.
16. The pouches of the large intestine that become obvious when the taeniae coli contract are called ________ .

17. The two large lobes of the liver are the right and left lobes, and the two smaller lobes are the posterior________ lobe and the inferior ________ lobe.

18. Place numbers in the blanks to arrange the following layers of the GI tract in order, from deepest to most superficial: (a) mucosa: ________ ; (b) muscularis: ________ ; (c) serosa: ________ ; (d) submucosa: ________ .

Are the following statements true or false?

19. The submucosal plexus (plexus of Meissner) is important in controlling motility of the GI tract whereas the myenteric plexus (plexus of Auerbach) controls secretions of the GI tract.

20. The middle portion of the small intestine is the ileum.

21. The concave medial border of the stomach is the lesser curvature.

22. Acini are clusters of cells that make up the exocrine portions of the pancreas.

23. The sinusoids of the liver receive blood from two sources: oxygenated blood from branches of the hepatic artery, and deoxygenated blood from branches of the hepatic portal vein.

24. The ascending colon is located in the left side of the abdominopelvic cavity.

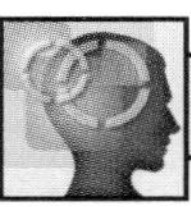

CRITICAL THINKING QUESTIONS

1. If you could leave French fried potatoes in your mouth long enough after chewing them, they would start to taste sweeter and maybe even a little tart. Why?
HINT *What's mixed with the food when you chew?*

2. Shayla enjoys eating fresh lemons. She holds a large lemon wedge with her front teeth and sucks on the juice. Her dentist is concerned that her lemon habit will affect her teeth. Why is he concerned?
HINT *Lemon juice is sometimes used as a cleaner due to its acidity.*

3. The small intestine is the primary site of digestion and absorption in the gastrointestinal tract. What structural modifications are unique to the small intestine and what are their functions?
HINT *Why do people wear pants with an elastic waistband when they're going to the "all-you-can-eat" buffet?*

4. Krystal and her friend were giggling over their milk and fries at her birthday party. In mid-giggle, milk started coming out of Krystal's nose. How did this happen?
HINT *It has something to do with that little thing that hangs from the roof of your mouth and wiggles when you say "ahh."*

5. Ron had just eaten a super size steak sandwich with double fried onions and extra cheese when he felt severe pain in his right side, below the ribs. It wasn't the right spot for heartburn and it was undoubtedly the worst case of indigestion that he'd ever had! The emergency room doctor says it may be time for the gallbladder to come out. What does that have to do with Ron's severe indigestion?
HINT *Bile salts will be needed to emulsify the fats in Ron's fast food feast.*

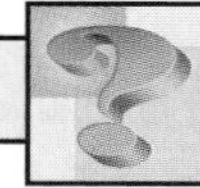

ANSWERS TO FIGURE QUESTIONS

24.1 Digestive enzymes are produced by the salivary glands, tongue, stomach, pancreas, and small intestine.

24.2 The enteric plexuses help regulate secretions and motility of the GI tract.

24.3 Mesentery binds the small intestine to the posterior abdominal wall.

24.4 The uvula helps prevent foods and liquids from entering the nasal cavity during swallowing.

24.5 Submandibular glands.

24.6 The main component of teeth is connective tissue, specifically dentin.

24.7 The first, second, and third molars do not replace any deciduous teeth.

24.8 The esophageal mucosa and submucosa contain mucus-secreting glands.

24.9 Food is pushed along by contraction of smooth muscle behind the bolus and relaxation of smooth muscle in front of it.

24.10 After a large meal, the rugae stretch and disappear as the stomach fills.

24.11 Surface mucous cells and mucous neck cells secrete mucus; chief cells secrete pepsinogen and gastric lipase; parietal cells secrete HCl and intrinsic factor; G cells secrete gastrin.

24.12 The pancreatic duct contains pancreatic juice (fluid and digestive enzymes); the common bile duct contains bile; the hepatopancreatic ampulla contains pancreatic juice and bile.

24.13 The epigastric region.

24.14 The phagocytic cell in the liver is the stellate reticuloendothelial (Kupffer's) cell.

24.15 A portal triad is composed of branches of the hepatic portal vein, hepatic artery, and a bile duct that accompany each other.

24.16 The ileum is the longest part of the small intestine.

24.17 Nutrients being absorbed enter the blood via capillaries or the lymph via lacteals.

24.18 The fluid secreted by duodenal glands—alkaline mucus—neutralizes gastric acid and protects the mucosal lining of the duodenum.

24.19 The ascending and descending portions of the colon are retroperitoneal.

24.20 Goblet cells secrete mucus to lubricate colonic contents.

24.21 The digestive system begins to develop about 14 days after fertilization.

25

THE URINARY SYSTEM

CONTENTS AT A GLANCE

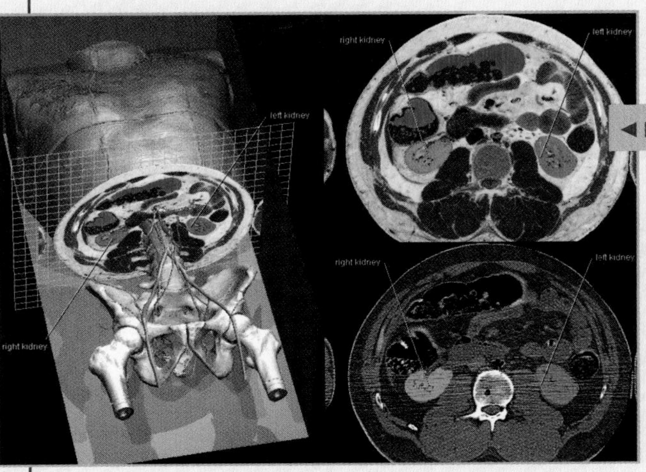

◀ Page 772

When you look at anatomical images that are horizontal slices, are you able to identify relevant structures? Or are you confused by the unusual perspective? Transverse sectional anatomy is not easy to learn. Are you up to the challenge?

Page 759 ▶

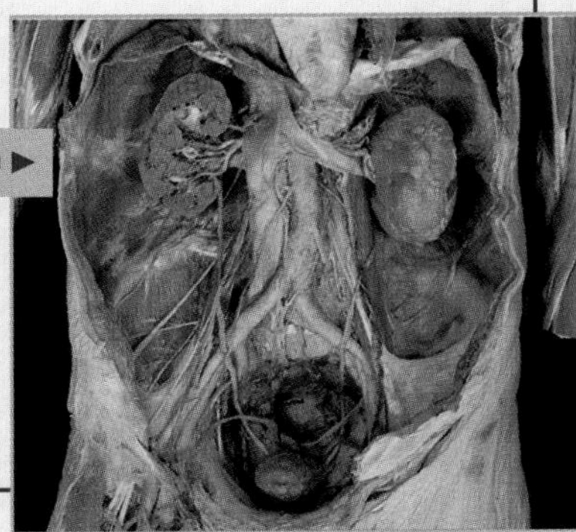

INTRODUCTION

The **urinary system** consists of two kidneys, two ureters, one urinary bladder, and a single urethra (Figure 25.1). The kidneys do the major work of the urinary system, as the other parts of the system are primarily passageways and storage areas. In filtering blood and forming urine, the kidneys perform several functions:

1. **Regulation of blood volume and composition.** The kidneys regulate the composition and volume of the blood and remove wastes from the blood. In the process, urine is formed. They also excrete selected amounts of various wastes including excess H^+, which helps control blood pH.
2. **Regulation of blood pressure.** The kidneys help regulate blood pressure by secreting the enzyme renin, which activates the renin–angiotensin pathway. This results in an increase in blood pressure.
3. **Contribution to metabolism.** The kidneys contribute to metabolism by (1) synthesizing new glucose molecules (gluconeogenesis) during periods of fasting or starvation, (2) secreting erythropoietin, a hormone that stimulates red blood cell production, and (3) participating in synthesis of vitamin D.

Urine is excreted from each kidney through its ureter and is stored in the urinary bladder until it is expelled from the body through the urethra.

Nephrology (nef-ROL-ō-jē; *nephr-* = kidney; *-ology* = study of) is the scientific study of the anatomy, physiology, and pathology of the kidney. The branch of medicine that deals with the male and female urinary systems and the male reproductive system is **urology** (yū-ROL-ō-jē; *uro-* = urine).

ANATOMY AND HISTOLOGY OF THE KIDNEYS

Objectives

- Describe the external and internal gross anatomical features of the kidneys.
- Trace the path of blood flow through the kidneys.
- Describe the structure of renal corpuscles and renal tubules.

The paired **kidneys** are reddish, kidney-bean-shaped organs located just above the waist between the peritoneum and the posterior wall of the abdomen. Because their position is posterior to the peritoneum of the abdominal cavity, they are said to be

Figure 25.1 / Organs of the urinary system, shown in relation to the surrounding structures in a female.
(See Tortora, *A Photographic Atlas of the Human Body,* Figure 13.3.)

Urine formed by the kidneys passes first into the ureters, then to the urinary bladder for storage, and finally through the urethra for elimination from the body.

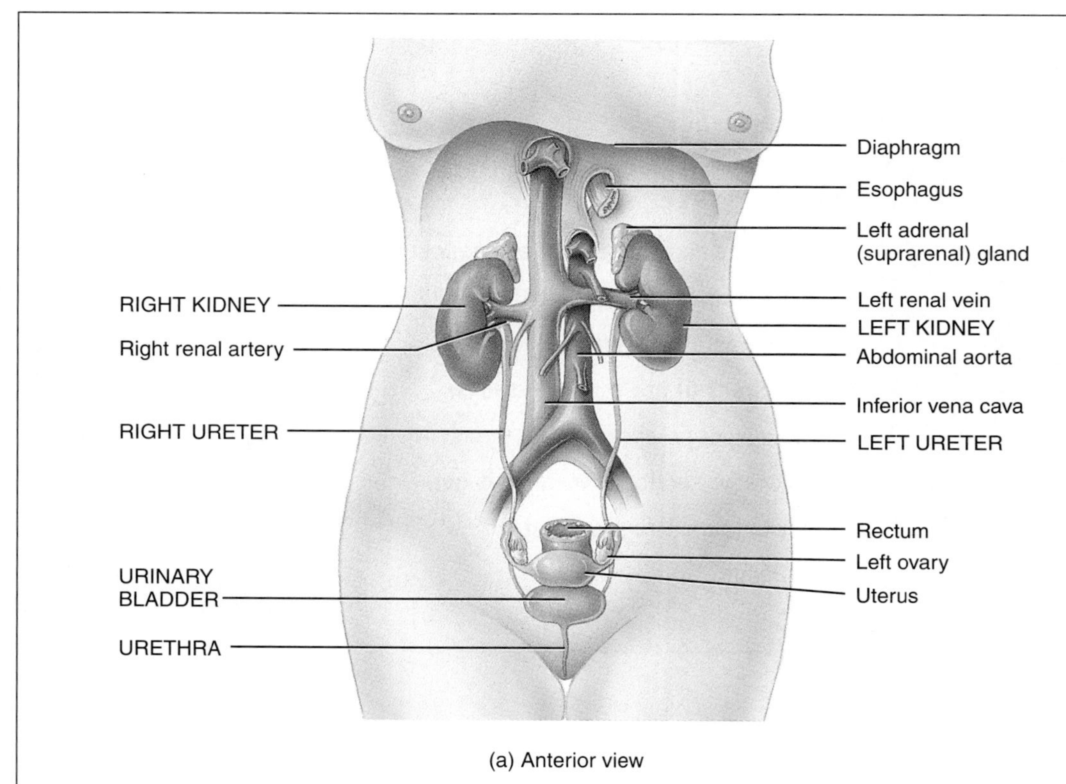

Functions of the Urinary System

1. The kidneys regulate blood volume and composition, help regulate blood pressure, synthesize glucose, release erythropoietin, and participate in vitamin D synthesis.
2. The ureters transport urine from the kidneys to the urinary bladder.
3. The urinary bladder stores urine.
4. The urethra discharges urine from the body.

retroperitoneal (re′-trō-per-i-tō-NĒ-al; *retro-* = behind) organs (Figure 25.2). (Other retroperitoneal structures include the ureters, adrenal glands, ascending colon, and descending colon.) The kidneys are located between the levels of the last thoracic and third lumbar vertebrae, a position where they are partially protected by the eleventh and twelfth pairs of ribs. The right kidney is slightly lower than the left (see Figure 25.1) because the liver occupies considerable space on the right side superior to the kidney.

External Anatomy

A typical kidney in an adult is 10–12 cm (4–5 in.) long, 5–7 cm (2–3 in.) wide, and 3 cm (1 in.) thick—about the size of a bar of bath soap—and has a mass of 135–150 g. The concave medial border of each kidney faces the vertebral column (see Figure 25.1). Near the center of the concave border is a deep vertical fissure called the **renal hilus** (see Figure 25.3) through which the ureter, blood vessels, lymphatic vessels, and nerves leave the kidney.

Three layers of tissue surround each kidney (see Figure 25.2). The deep layer, the **renal capsule** (*ren-* = kidney), is a smooth, transparent, membrane composed of dense irregular connective tissue that is continuous with the outer coat of the ureter. It serves as a barrier against trauma and helps maintain the shape of the kidney. The intermediate layer, the **adipose capsule,** is a mass of fatty tissue surrounding the renal capsule. It also protects the kidney from trauma and holds it firmly in place within the abdominal cavity. The superficial layer, the **renal fascia,** is a thin layer of dense, irregular connective tissue that anchors the kidney to the surrounding structures and to the abdominal wall. On the anterior surface of the kidneys, the renal fascia is deep to the peritoneum.

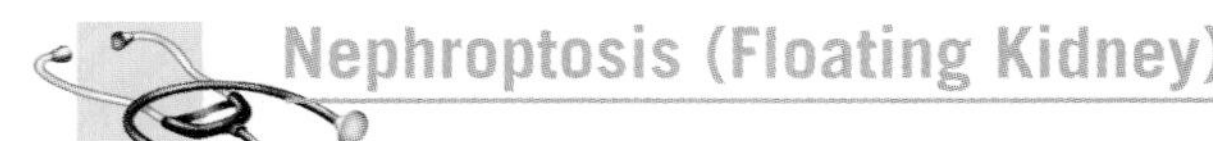

Nephroptosis (nef′-rō-TŌ-sis; *ptosis* = falling), or **floating kidney,** is an inferior displacement (dropping) of the kidney. It occurs when the kidney slips from its normal position because it is not securely held in place by adjacent organs or its covering of fat. People, most often those who are very thin, in whom either the adipose capsule or renal fascia is deficient, may develop nephroptosis. It is dangerous because the ureter may kink and block urine flow. The resulting backup of urine puts pressure on the kidney, which damages the tissue. Twisting of the ureter also causes pain.

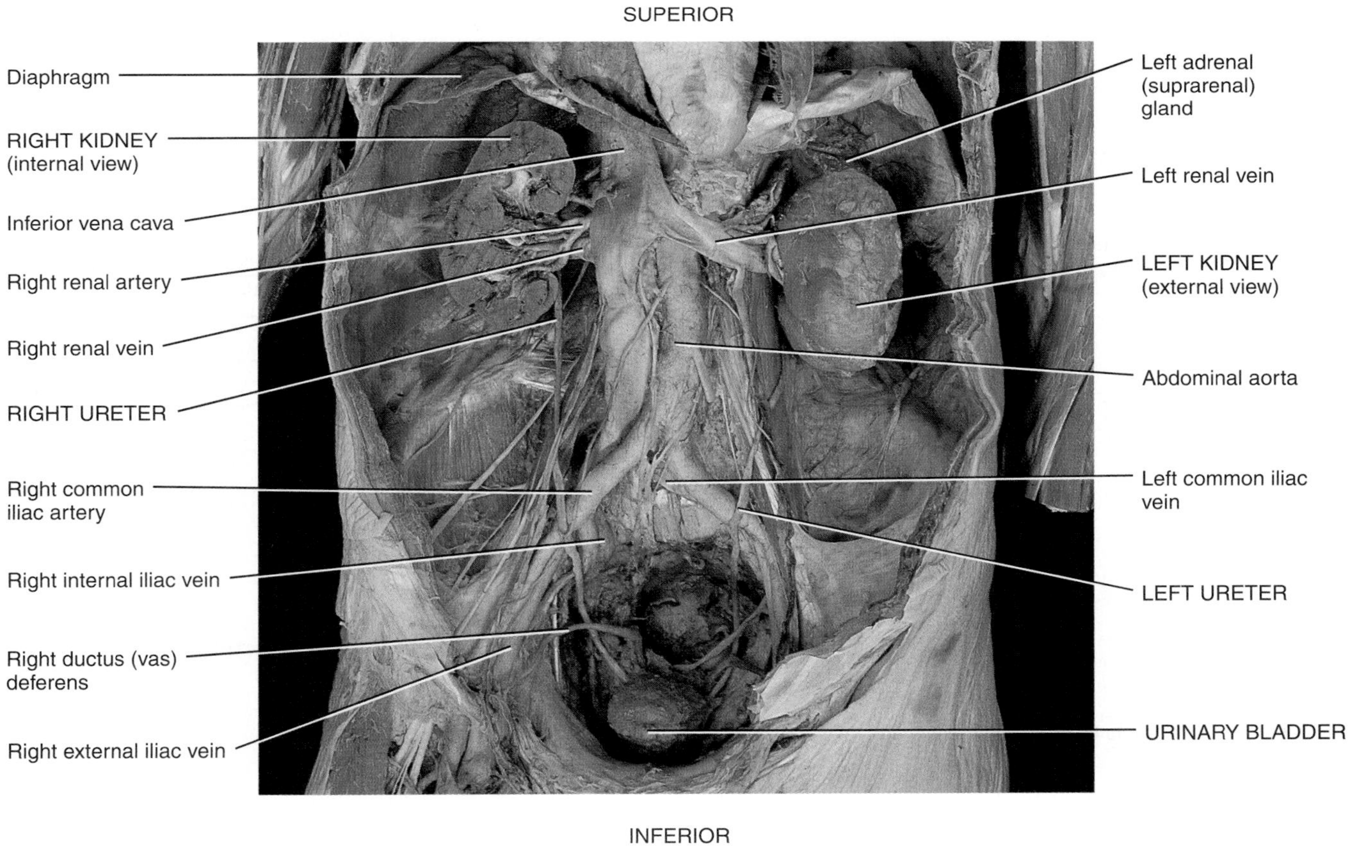

SUPERIOR

Diaphragm

RIGHT KIDNEY (internal view)

Inferior vena cava

Right renal artery

Right renal vein

RIGHT URETER

Right common iliac artery

Right internal iliac vein

Right ductus (vas) deferens

Right external iliac vein

Left adrenal (suprarenal) gland

Left renal vein

LEFT KIDNEY (external view)

Abdominal aorta

Left common iliac vein

LEFT URETER

URINARY BLADDER

INFERIOR

(b) Anterior view

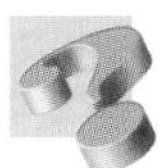

Which organs constitute the urinary system?

Figure 25.2 / Position and coverings of the kidneys.

The kidneys are surrounded by a renal capsule, adipose capsule, and renal fascia.

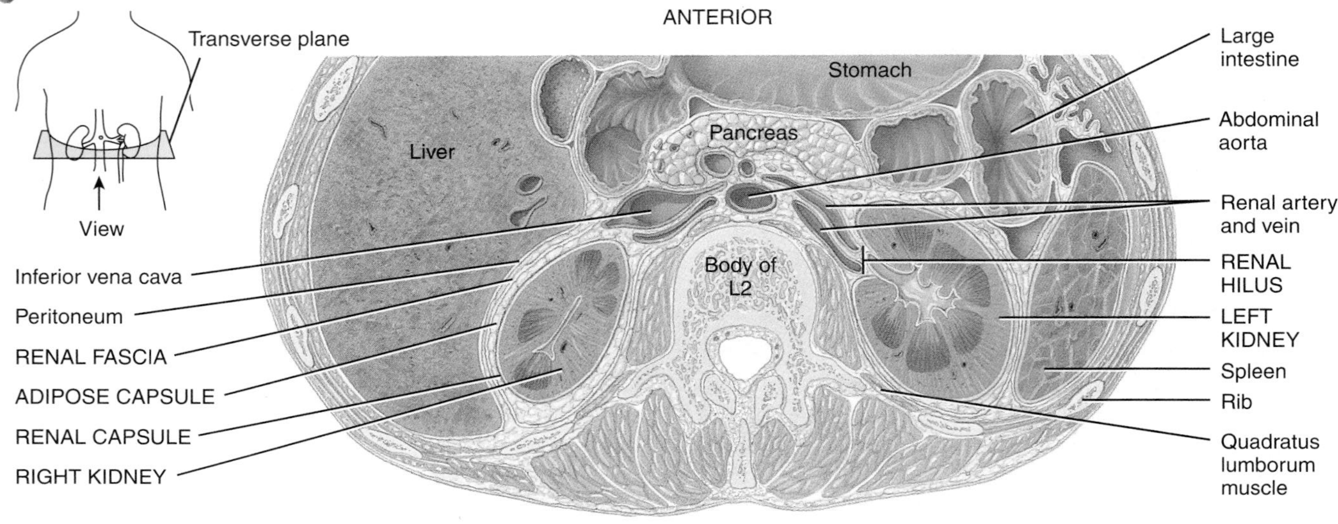

(a) Inferior view of transverse section of abdomen (L2)

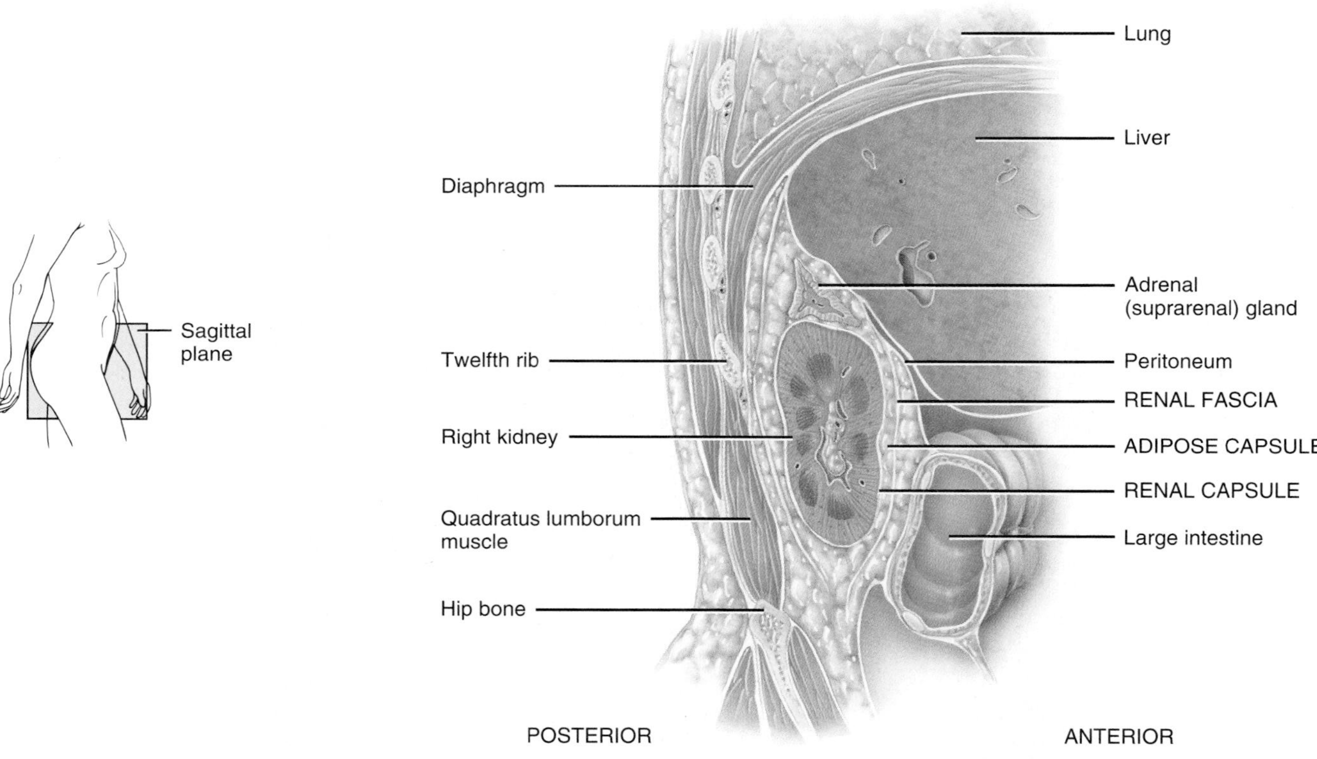

(b) Sagittal section through the right kidney

Why are the kidneys said to be retroperitoneal?

Internal Anatomy

A frontal section through a kidney reveals two distinct regions: a superficial, smooth-textured reddish area called the **renal cortex** (*cortex* = rind or bark) and a deep, reddish-brown region called the **renal medulla** (*medulla* = inner portion) (Figure 25.3). The medulla consists of 8–18 cone-shaped **renal pyramids.** The base (wider end) of each pyramid faces the renal cortex, and its apex (narrower end), called a **renal papilla,** points toward the center

Figure 25.3 / Internal anatomy of the kidneys. (See Tortora, *A Photographic Atlas of the Human Body,* Figure 13.4.)

The two main regions of the kidney parenchyma are the renal cortex and the renal pyramids in the renal medulla.

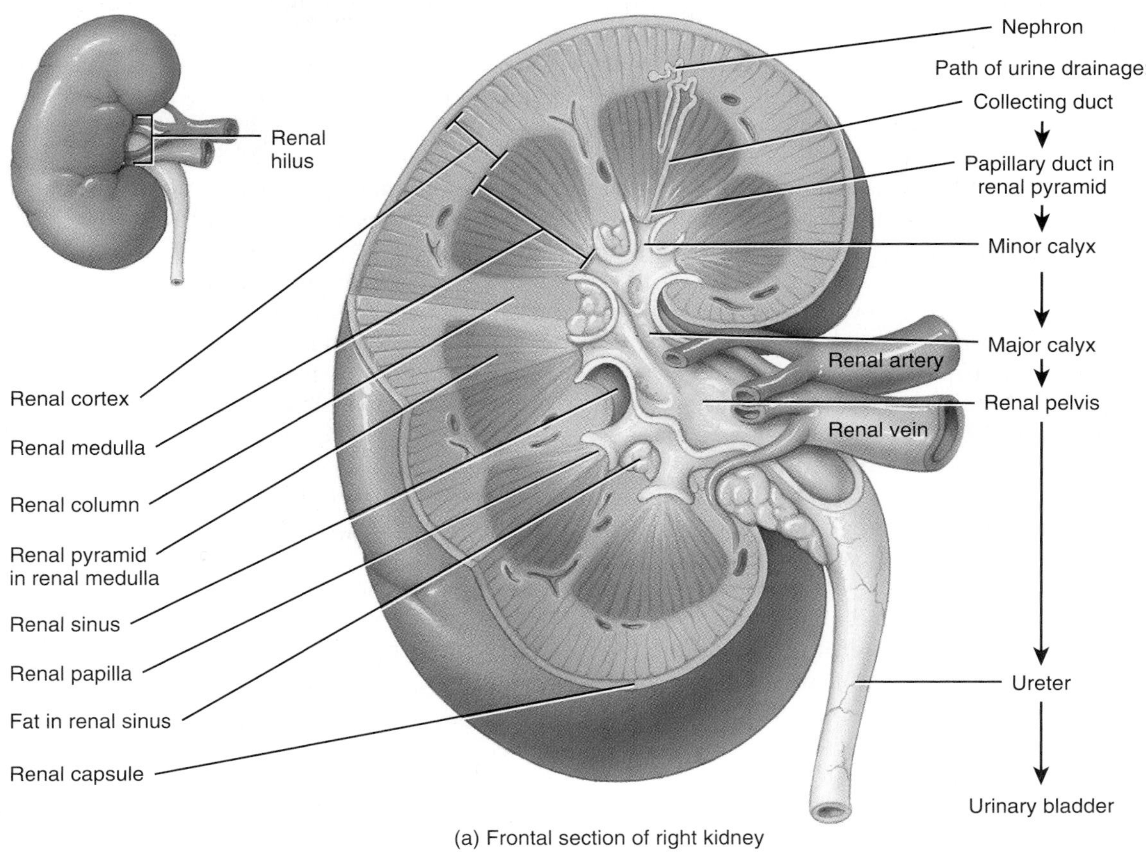

(a) Frontal section of right kidney

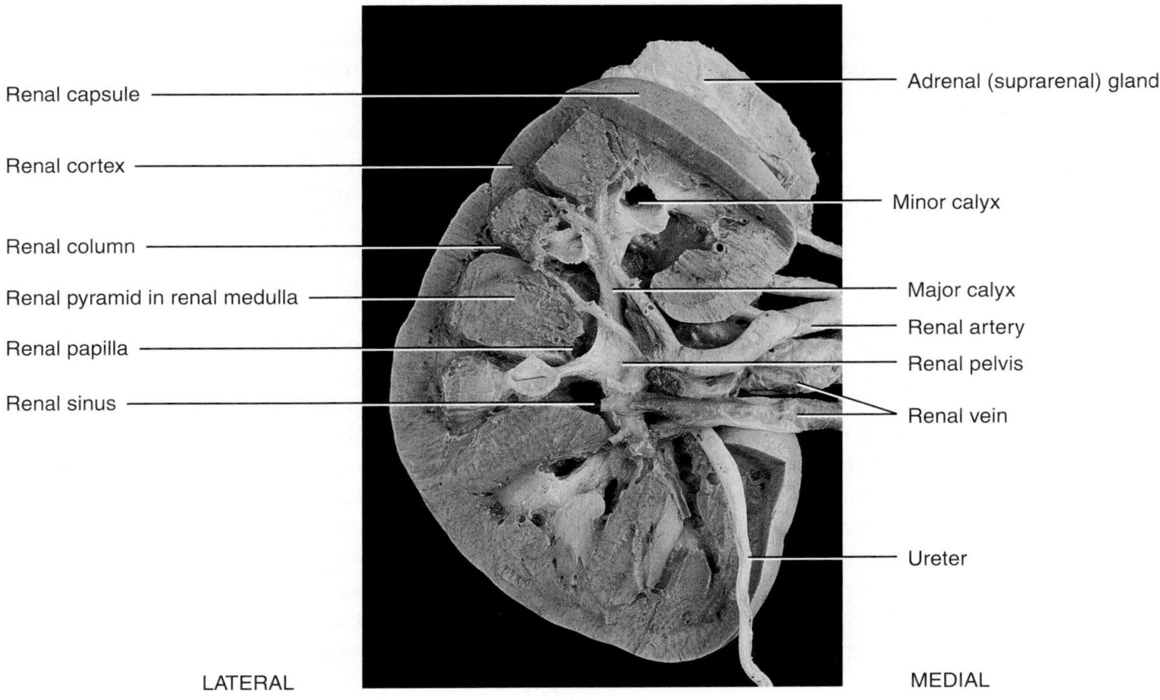

(b) Frontal section of right kidney

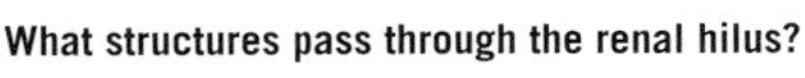

What structures pass through the renal hilus?

of the kidney. The term **renal lobe** is applied to a renal pyramid and its overlying area of renal cortex. The renal cortex is the smooth-textured area extending from the renal capsule to the bases of the renal pyramids and into the spaces between them. It is divided into an outer *cortical zone* and an inner *juxtamedullary zone.* Those portions of the cortex that extend between renal pyramids are called the **renal columns.**

Together, the renal cortex and renal pyramids of the renal medulla constitute the functional portion, or **parenchyma,** of the kidney. Within the parenchyma are the functional units of the kidney—about 1 million microscopic structures called **nephrons** (NEF-rons). Urine formed by the nephrons drains into large **papillary ducts,** which extend through the renal papillae of the pyramids. The papillary ducts drain into cuplike structures called **minor** and **major calyces** (KĀ-li-sēz; = cups; singular is *calyx*). Each kidney has 8–18 minor calyces and 2–3 major calyces. A minor calyx receives urine from the papillary ducts of one renal papilla and delivers it to a major calyx. From the major calyces, urine drains into a single large cavity called the **renal pelvis** (*pelv-* = basin) and then out through the ureter to the urinary bladder.

The hilus expands into a cavity within the kidney called the **renal sinus,** which contains part of the renal pelvis, the calyces, and branches of the renal blood vessels and nerves. Adipose tissue helps stabilize the position of these structures in the renal sinus.

Blood and Nerve Supply

Because the kidneys remove wastes from the blood and regulate its volume and ionic composition, it is not surprising that they are abundantly supplied with blood vessels. Although the kidneys constitute less than 0.5% of total body mass, they receive 20–25% of the resting cardiac output via the right and left **renal arteries** (Figure 25.4). In adults, renal blood flow is about 1200 mL per minute.

Within the kidney, the renal artery divides into several **segmental arteries,** each of which gives off several branches that enter the parenchyma and pass through the renal columns between the renal pyramids as the **interlobar arteries.** At the bases of the renal pyramids, the interlobar arteries arch between the renal medulla and cortex; here they are known as the **arcuate arteries** (*arcuat-* = shaped like a bow). Divisions of the arcuate arteries produce a series of **interlobular arteries,** which enter the renal cortex and give off branches called **afferent arterioles** (*af-* = toward; *-ferrent* = to carry).

Each nephron receives one afferent arteriole, which divides into a tangled, ball-shaped capillary network called the **glomerulus** (glō-MER-yū-lus; = little ball; plural is *glomeruli*). The glomerular capillaries then reunite to form an **efferent arteriole** (*ef-* = out) that drains blood out of the glomerulus. Glomerular capillaries are unique because they are positioned between two arterioles, rather than between an arteriole and a venule. Coordinated vasodilation and vasoconstriction of the afferent and efferent arterioles can produce large changes in both renal blood flow and renal vascular resistance, which in turn affects systemic vascular resistance. Because they are capillary networks, the glomeruli are part of both the cardiovascular and the urinary systems.

The efferent arterioles divide to form a network of capillaries, called the **peritubular capillaries** (*peri-* = around), that surrounds tubular portions of the nephron in the renal cortex. Extending from some efferent arterioles are long loop-shaped capillaries called **vasa recta** (VĀ-sa REK-ta; *vasa* = vessels; *recta* = straight) that supply tubular portions of the nephron in the renal medulla (see Figure 25.5a).

The peritubular capillaries eventually reunite to form **peritubular venules** and then **interlobular veins.** (The interlobular veins also receive blood from the vasa recta.) Then the blood drains through the **arcuate veins** to the **interlobar veins** running between the renal pyramids. Blood leaves the kidney through a single **renal vein** that exits at the renal hilus.

Most renal nerves originate in the **celiac ganglion** and pass through the **renal plexus** into the kidneys along with the renal arteries. All these nerves are part of the sympathetic division of the autonomic nervous system. Most are vasomotor nerves that innervate blood vessels; that is, they regulate the flow of blood through the kidney and renal resistance by altering the diameters of arterioles.

The Nephron

Nephrons are the functional units of the kidneys that engage in three basic processes: filtering blood, returning useful substances to blood so that they are not lost from the body, and removing substances from the blood that are not needed by the body. As a result of these processes, nephrons maintain the homeostasis of blood and urine is produced.

Parts of a Nephron

Each nephron (Figure 25.5 on page 764) consists of two portions: a **renal corpuscle** (KOR-pus-sul; = tiny body), where plasma is filtered, and a **renal tubule** into which the filtered fluid passes. Each renal corpuscle has two components: the **glomerulus** (capillary network) and the **glomerular (Bowman's) capsule,** a double-walled epithelial cup that surrounds the glomerulus. From the glomerular capsule, fluid filtered from the plasma passes into the renal tubule, which has three main sections. In the order that fluid passes through them, the renal tubule consists of a (1) **proximal convoluted tubule,** (2) **loop of Henle (nephron loop),** and (3) **distal convoluted tubule.** *Proximal* denotes the portion of the tubule attached to the glomerular capsule, and *distal* denotes the portion that is farther away. *Convoluted* means the tubule is tightly coiled rather than straight. The renal corpuscle and both convoluted tubules lie within the renal cortex, whereas the loop of Henle extends into the renal medulla, makes a hairpin turn, and then returns to the renal cortex.

The distal convoluted tubules of several nephrons empty into a single **collecting duct.** Collecting ducts then unite and converge until eventually there are only several hundred large **papillary ducts,** which drain into the minor calyces. The collecting ducts and papillary ducts extend from the renal cortex through the renal medulla to the renal pelvis. Although a kidney has about 1 million nephrons, it has a much smaller number of collecting ducts and even fewer papillary ducts.

Figure 25.4 / Blood supply of the kidneys. (See Tortora, *A Photographic Atlas of the Human Body,* Figure 13.6.)

The renal arteries deliver 20–25% of the resting cardiac output to the kidneys.

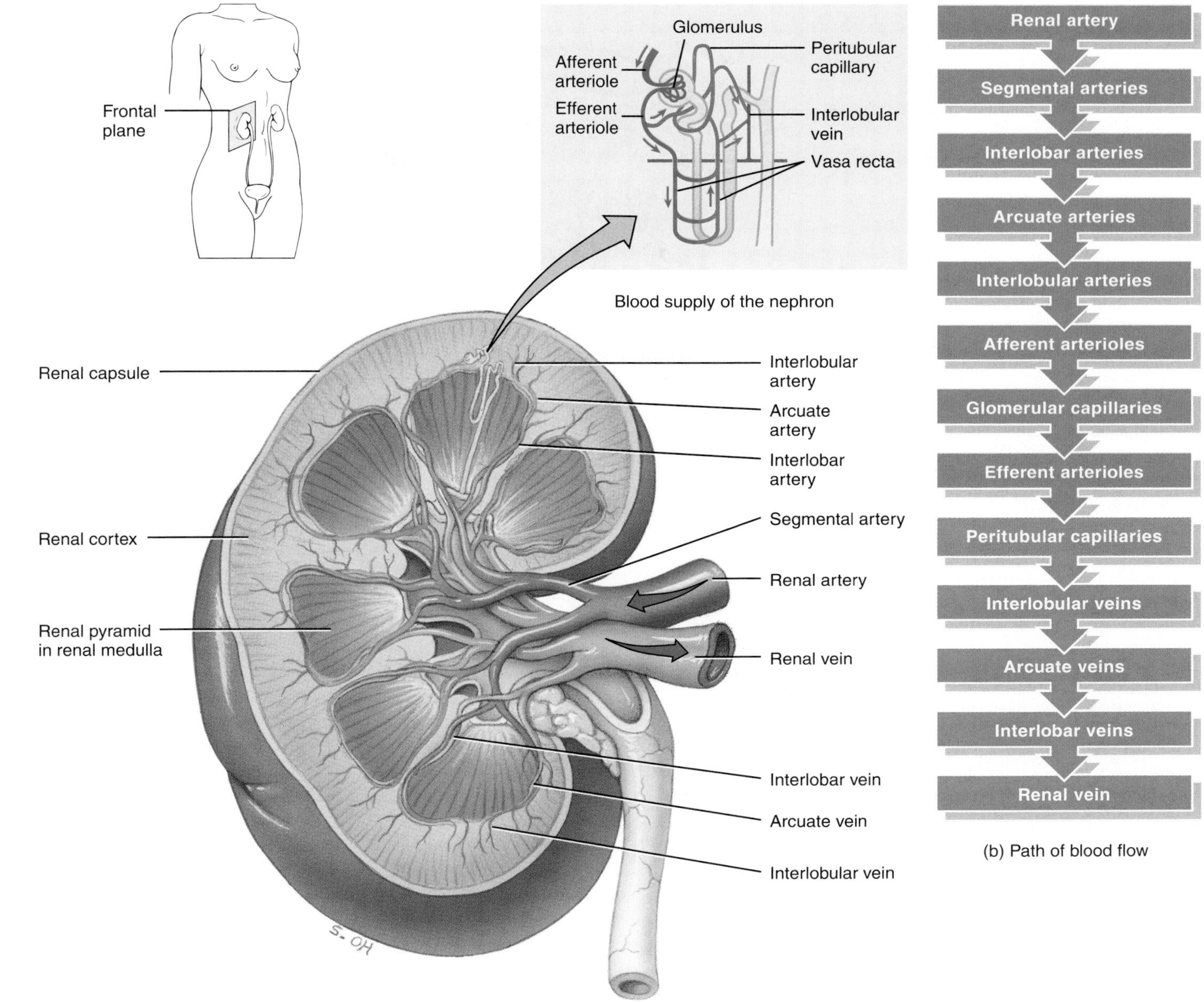

(a) Frontal section of right kidney

(b) Path of blood flow

What volume of blood enters the renal arteries per minute?

In a nephron, the loop of Henle connects the proximal and distal convoluted tubules. The first portion of the loop of Henle dips into the renal medulla, where it is called the **descending limb of the loop of Henle** (Figure 25.5). It then makes that hairpin curve and returns to the renal cortex as the **ascending limb of the loop of Henle.** About 80–85% of the nephrons are termed **cortical nephrons;** their renal corpuscles lie in the outer portion of the renal cortex, and they have *short* loops of Henle that lie mainly in the cortex and penetrate only into the superficial region of the renal medulla (Figure 25.5a). The short loops of Henle receive their blood supply from peritubular capillaries that arise from efferent arterioles. The other 15–20% of the nephrons are called **juxtamedullary nephrons** (*juxta-* = near to); their renal corpuscles are deep in the cortex, close to the medulla, and they have a *long* loop of Henle that extends into the deepest region of the medulla (Figure 25.5b). Long loops of Henle receive their blood supply from peritubular capillaries and vasa recta that arise from efferent arterioles. In addi-

Figure 25.5 / The structure of nephrons (colored gold) and associated blood vessels.
(a) A cortical nephron. (b) A juxtamedullary nephron.

Nephrons are the functional units of the kidneys.

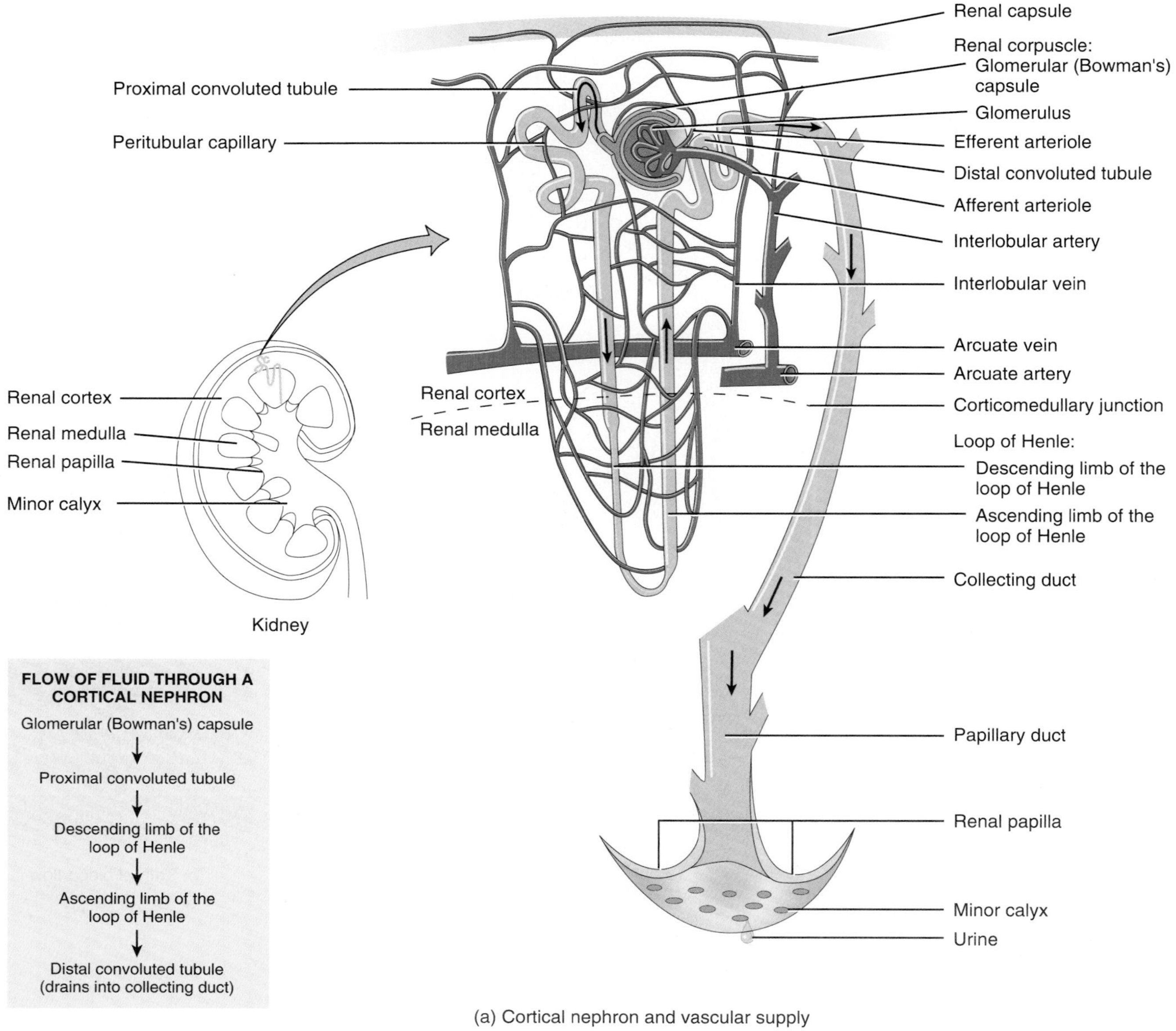

FLOW OF FLUID THROUGH A CORTICAL NEPHRON

Glomerular (Bowman's) capsule
↓
Proximal convoluted tubule
↓
Descending limb of the loop of Henle
↓
Ascending limb of the loop of Henle
↓
Distal convoluted tubule (drains into collecting duct)

(a) Cortical nephron and vascular supply

tion, the ascending limb of the loop of Henle of juxtamedullary nephrons consists of two portions: a **thin ascending limb** followed by a **thick ascending limb** (Figure 25.5b). The lumen of the thin ascending limb is the same as in other areas of the renal tubule; it is only the epithelium that is thinner. Nephrons with long loops of Henle enable the kidneys to excrete very dilute or very concentrated urine.

Histology of the Nephron and Collecting Duct

Beginning at the glomerular capsule, a single layer of epithelial cells forms the entire wall of the glomerular capsule, renal tubule, and ducts. Each part, however, has distinctive histological features that reflect its particular functions. In the order that fluid flows through them, the parts are the glomerular capsule, the renal tubule, and the collecting duct.

GLOMERULAR CAPSULE The glomerular (Bowman's) capsule consists of visceral and parietal layers (Figure 25.6a on page 766). The visceral layer consists of modified simple squamous epithelial cells called **podocytes** (PŌ-dō-cīts; *podo-* = foot; *-cytes* = cells). The many footlike projections of these cells (pedicels) wrap around the single layer of endothelial cells of the glomerular capillaries and form the inner wall of the capsule. The parietal layer of the glomerular capsule consists of simple squamous epithelium and forms the outer wall of the capsule.

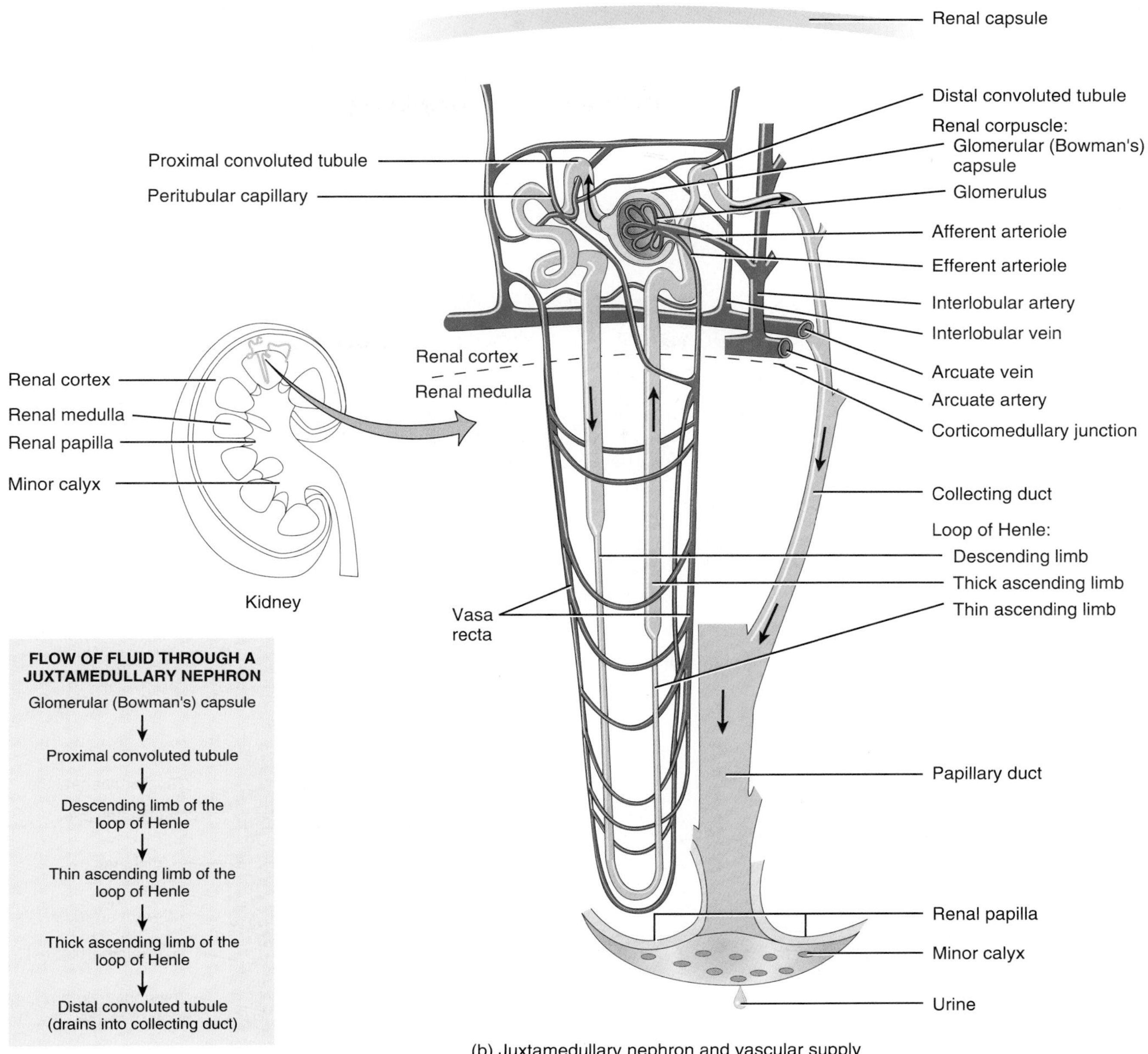

(b) Juxtamedullary nephron and vascular supply

What are the basic differences between cortical and juxtamedullary nephrons?

Fluid filtered from the glomerular capillaries enters the **capsular (Bowman's) space,** the space between the two layers of the glomerular capsule. Think of the relationship between the glomerulus and glomerular capsule in the following way. The glomerulus is a fist punched into a limp balloon (the glomerular capsule) until the fist is covered by two layers of balloon (visceral and parietal layers) with a space in between, the capsular space.

Renal Tubule and Collecting Duct Table 25.1 on page 767 illustrates the histology of the cells that form the renal tubule and collecting duct. In the proximal convoluted tubule, the cells are simple cuboidal epithelial cells and have a prominent brush border of microvilli on their apical surface (surface facing the lumen). These microvilli, like those of the small intestine, increase the surface area for reabsorption and secretion. The descending limb of the loop of Henle and the first part of the ascending limb of the loop of Henle (the thin ascending limb) are simple squamous epithelium. (Recall that cortical, or short-loop, nephrons lack the thin ascending limb.) The thick ascending limb of the loop of Henle is composed of simple cuboidal to low columnar epithelium.

In each nephron, the final portion of the ascending limb of the loop of Henle makes contact with the afferent arteriole serving that renal corpuscle (Figure 25.6a). Because the tubule cells in this region are columnar and crowded together, they are known as the **macula densa** (*macula* = spot; *densa* = dense).

Figure 25.6 / Histology of a renal corpuscle.

A renal corpuscle consists of a glomerular (Bowman's) capsule and a glomerulus.

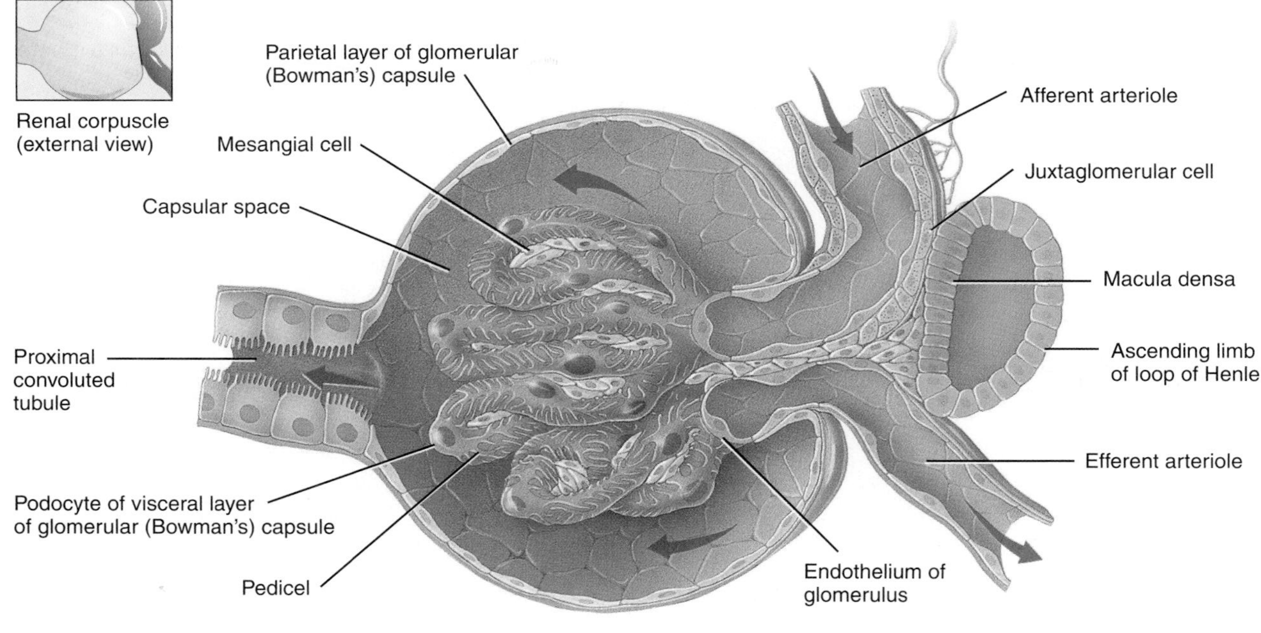

(a) Renal corpuscle (internal view)

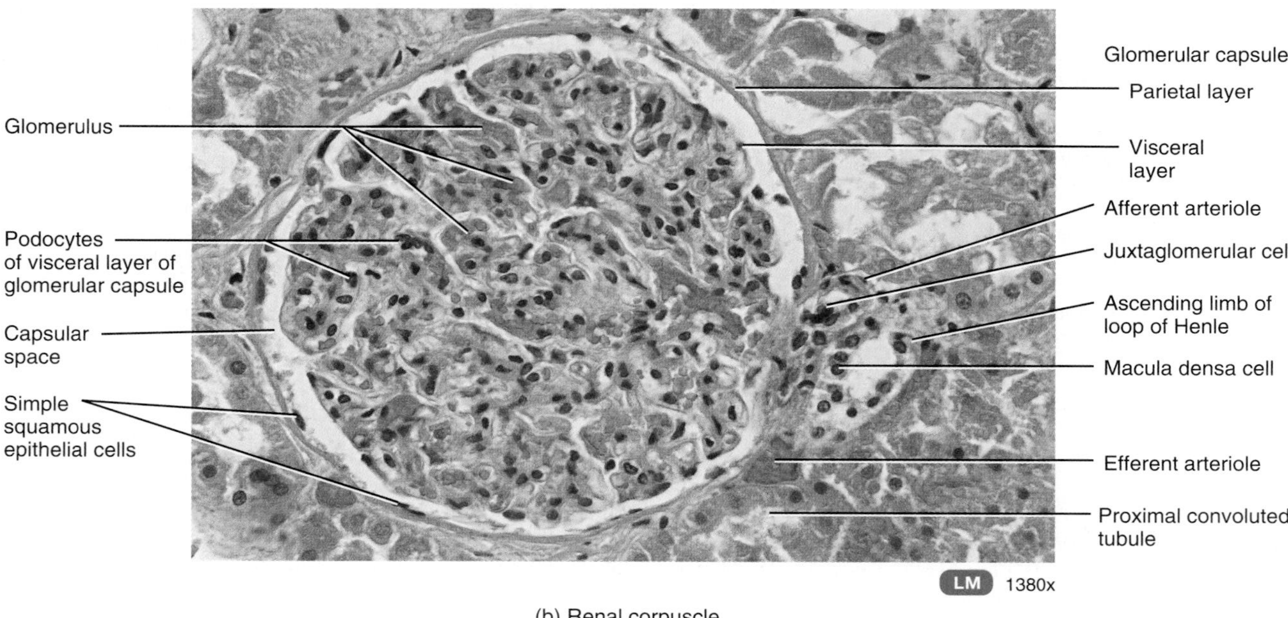

(b) Renal corpuscle

Is the photomicrograph in (b) from a section through the renal cortex or renal medulla? How can you tell?

Alongside the macula densa, the wall of the afferent arteriole (and sometimes the efferent arteriole) contains modified smooth muscle fibers called **juxtaglomerular (JG) cells.** Together with the macula densa, they constitute the **juxtaglomerular apparatus,** or **JGA.** The distal convoluted tubule (DCT) begins a short distance past the macula densa. Even though the cells of the distal convoluted tubule and collecting ducts all are simple cuboidal epithelial cells that have only a few microvilli, beginning in the last portion of the DCT and continuing into the collecting ducts two different types of cells are present. Most are **principal cells,**

Table 25.1 Histological Features of the Renal Tubule and Collecting Duct

Region	Histology	Description
Proximal convoluted tubule (PCT)	Microvilli; Mitochondrion	Simple cuboidal epithelial cells with prominent brush borders of microvilli.
Loop of Henle: descending limb and thin ascending limb		Simple squamous epithelial cells.
Loop of Henle: thick ascending limb		Simple cuboidal to low columnar epithelial cells.
Most of distal convoluted tubule (DCT)		Simple cuboidal epithelial cells.
Final portion of DCT: collecting duct (CD)	Intercalated cell; Principal cell	Simple cuboidal epithelium consisting of principal cells and intercalated cells in last part of DCT and collecting duct.

which have infoldings of the basal membrane; a smaller number are **intercalated cells,** which have microvilli at the apical surface and a large number of mitochondria. Principal cells have receptors for both antidiuretic hormone (ADH) and aldosterone, two hormones that regulate their functions. Intercalated cells play a role in the homeostasis of blood pH. The collecting ducts drain into large papillary ducts, which are lined by simple columnar epithelium.

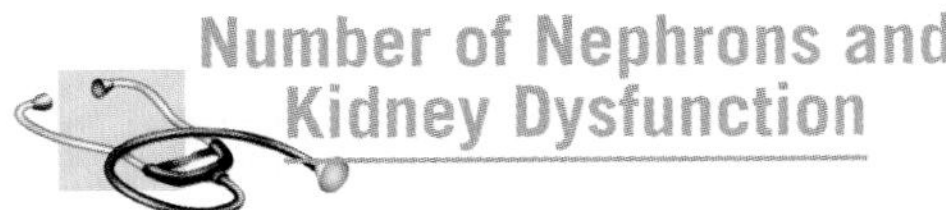

Number of Nephrons and Kidney Dysfunction

The number of nephrons remains constant from birth; any increase in kidney size is due solely to the growth of individual nephrons. If nephrons are injured or become diseased, new ones do not form. Signs of kidney dysfunction do not usually become apparent until function declines to less than 25% of normal because the remaining functional nephrons gradually adapt to handle a larger-than-normal load. Surgical removal of one kidney, for example, stimulates hypertrophy (enlargement) of the remaining kidney, which eventually is able to filter blood at 80% of the rate of two normal kidneys. ■

- ✓ Describe the location of the kidneys. Why are they said to be retroperitoneal?
- ✓ Which branch of the autonomic nervous system innervates renal blood vessels?
- ✓ How do cortical nephrons and juxtamedullary nephrons differ structurally?
- ✓ Describe the histology of the various portions of a nephron and collecting duct.
- ✓ Describe the structure of the juxtaglomerular apparatus (JGA).

FUNCTIONS OF NEPHRONS

Objective

- Identify the three basic tasks performed by nephrons and collecting ducts, and indicate where each task occurs.
- Describe the filtration membrane.

In producing urine, nephrons and collecting ducts perform three basic processes—glomerular filtration, tubular reabsorption, and tubular secretion (Figure 25.7):

1. **Glomerular filtration.** In the first step of urine production, water and most solutes in plasma pass from blood across the wall of glomerular capillaries into the glomerular capsule, which empties into the renal tubule.
2. **Tubular reabsorption.** As filtered fluid flows along the renal tubule and through the collecting duct, most filtered water and many useful solutes are reabsorbed by the tubule cells and returned to the blood as it flows through the peritubular capillaries and vasa recta. Note that *re*absorption refers to the return of substances to the bloodstream, as distinguished from absorption, which means entry of new substances into the body.
3. **Tubular secretion.** As fluid flows along the tubule and through the collecting duct, the tubule and duct cells secrete additional materials, such as wastes, drugs, and excess ions, into the fluid. Tubular secretion *removes* a substance from the blood; in other instances of secretion—for instance, secretion of hormones—a substance often is released into the blood. Solutes in the fluid that drains into the renal pelvis remain in the urine and are excreted.

Glomerular Filtration

The fluid that enters the capsular space is called the **glomerular filtrate.** On average, the daily volume of glomerular filtrate in adults is 150 liters in females and 180 liters in males, a volume that represents about 65 times the entire blood plasma volume. More than 99% of the glomerular filtrate returns to the bloodstream via tubular reabsorption, however, so only 1–2 liters (about 1–2 qt) are excreted as urine.

Together, endothelial cells of glomerular capillaries and podocytes, which completely encircle the capillaries, form a leaky barrier known as the **filtration membrane** or **endothelial–capsular membrane.** This sandwichlike assembly permits filtration of water and small solutes but prevents filtration of most plasma proteins, blood cells, and platelets. Filtered substances move from the bloodstream through three barriers—a glomerular endothelial cell, the basal lamina, and a filtration slit formed by a podocyte (Figure 25.8):

1. Glomerular endothelial cells are quite leaky because they have large **fenestrations** (pores) that are 70–100 nm (0.07–0.1 μm) in diameter. This size permits all solutes in blood plasma to exit glomerular capillaries but prevents filtration of blood cells and platelets. Located among the glomerular capillaries and in the cleft between afferent and efferent arterioles are **mesangial cells** (*mes-* = in the middle; *-angi* = blood vessel), contractile cells that help regulate glomerular filtration (see Figure 26.6a).
2. The **basal lamina,** a layer of acellular material between the endothelium and the podocytes, consists of fibrils in a glycoprotein matrix; it prevents filtration of larger plasma proteins.
3. Extending from each podocyte are thousands of footlike

Figure 25.7 / Relation of a nephron's structure to its three basic functions: glomerular filtration, tubular reabsorption, and tubular secretion. Excreted substances remain in the urine and subsequently leave the body.

Glomerular filtration occurs in the renal corpuscle, whereas tubular reabsorption and tubular secretion occur all along the renal tubule and collecting duct.

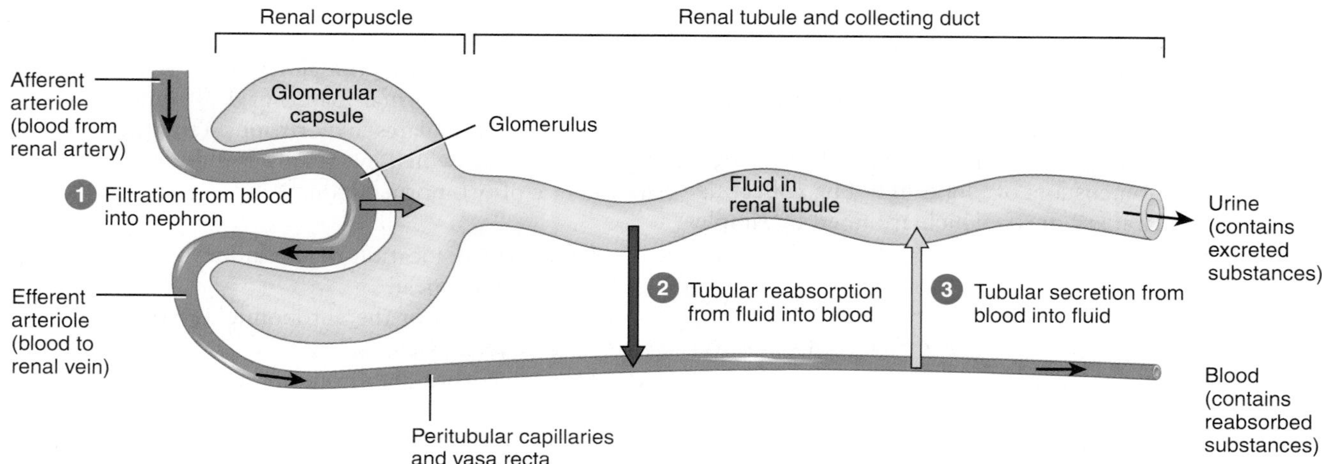

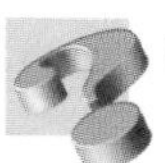

When cells of the renal tubules secrete the drug penicillin, is the drug being added to or removed from the bloodstream?

processes termed **pedicels** (PED-i-sels; = little feet) that wrap around glomerular capillaries. The spaces between pedicels are the **filtration slits.** A thin membrane, the **slit membrane,** extends across each filtration slit; it permits the passage of molecules having a diameter smaller than 6–7 nm (0.006–0.007 μm), including water, glucose, vitamins, amino acids, very small plasma proteins, ammonia, urea, and ions. Because the most plentiful plasma protein—albumin—has a diameter of 7.1 nm, less than 1% of it passes the slit membrane.

Figure 25.8 / The filtration (endothelial–capsular) membrane. The size of the endothelial fenestrations and filtration slits in (a) have been exaggerated for emphasis. Richard K. Kessel and Randy H. Kardon, *Tissues and Organs: A Text-Atlas of Scanning Electron Microscopy.* © 1979 by W. H. Freeman and Company. Reprinted by permission.

During glomerular filtration, water and solutes pass from blood plasma into the capsular space.

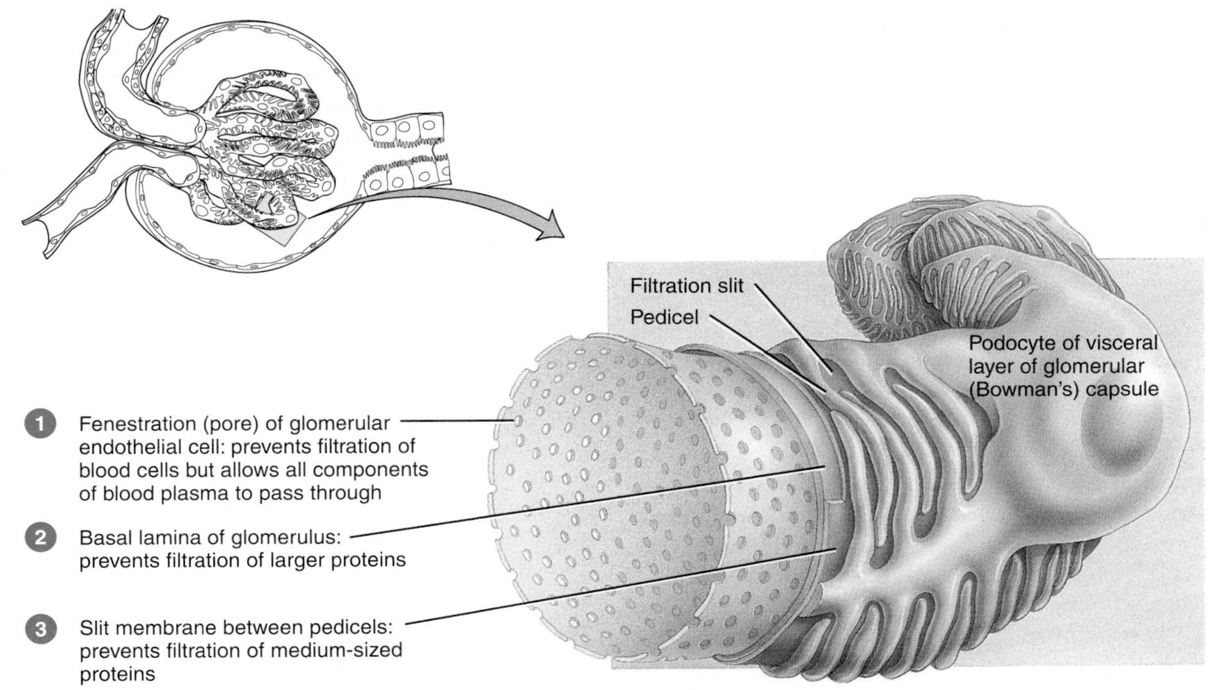

(a) Details of filtration membrane

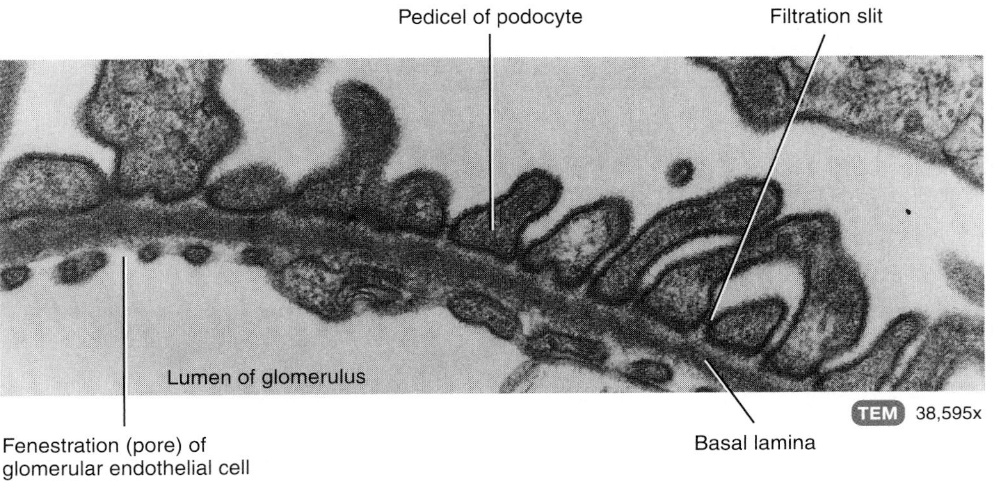

(b) Filtration membrane

Which part of the filtration membrane prevents red blood cells from entering the capsular space?

Tubular Reabsorption

The normal rate of glomerular filtration is so high that the volume of fluid entering the proximal convoluted tubules in half an hour is greater than the total plasma volume. Reabsorption—returning most of the filtered water and many of the filtered solutes to the bloodstream—is the second function of the nephron and collecting duct. Epithelial cells all along the renal tubule and duct carry out reabsorption, but proximal convoluted tubule cells make the largest contribution. Solutes that are reabsorbed by both active and passive processes include glucose, amino acids, urea, and ions such as Na^+ (sodium), K^+ (potassium), Ca^{2+} (calcium), Cl^- (chloride), HCO_3^- (bicarbonate), and HPO_4^{2-} (phosphate). Cells located more distally fine-tune the reabsorption processes to maintain the appropriate concentrations of water and selected ions. Most small proteins and peptides that pass through the filter also are reabsorbed, usually via pinocytosis.

Diuretics are substances that slow renal reabsorption of water and thereby cause *diuresis,* an increased urine flow rate. Naturally occurring diuretics include *caffeine* in coffee, tea, and cola sodas, and *alcohol* in beer, wine, and mixed drinks.

Tubular Secretion

The third function of nephrons and collecting ducts is tubular secretion, the transfer of materials from the blood and tubule cells into tubular fluid. Secreted substances include H^+, K^+, ammonium ions (NH_4^+), creatinine, and certain drugs such as penicillin. Tubular secretion has two important outcomes: The secretion of H^+ helps control blood pH, and the secretion of other substances helps eliminate them from the body.

✓ What are the functions of glomerular filtration, tubular reabsorption, and tubular secretion?

✓ Describe the factors that allow much greater filtration through glomerular capillaries than through capillaries elsewhere in the body.

URINE TRANSPORTATION, STORAGE, AND ELIMINATION

Objective

- Describe the anatomy, histology, and functions of the ureters, urinary bladder, and urethra.

Urine drains through papillary ducts into the minor calyces, which join to become major calyces that unite to form the renal pelvis (see Figure 25.3). From the renal pelvis, urine first drains into the ureters and then into the urinary bladder; urine is then discharged from the body through the single urethra (see Figure 25.1).

Ureters

Each of the two **ureters** (YŪ-re-ters) transports urine from the renal pelvis of one kidney to the urinary bladder. Peristaltic contractions of the muscular walls of the ureters push urine toward the urinary bladder, but hydrostatic pressure and gravity also contribute. Peristaltic waves that pass from the renal pelvis to the urinary bladder vary in frequency from one to five per minute, depending on how fast urine is being formed.

The ureters are 25–30 cm (10–12 in.) long and are thick-walled, narrow tubes that vary in diameter from 1 mm to 10 mm along their course between the renal pelvis and the urinary bladder. Like the kidneys, the ureters are retroperitoneal. At the base of the urinary bladder the ureters curve medially and pass obliquely through the wall of the posterior aspect of the urinary bladder (Figure 25.9).

Even though there is no anatomical valve at the opening of each ureter into the urinary bladder, there is a physiological one that is quite effective. As the urinary bladder fills with urine, pressure within it compresses the oblique openings into the ureters and prevents the backflow of urine. When this physiological valve is not operating properly, it is possible for microbes to travel up the ureters from the urinary bladder to infect one or both kidneys.

Three coats of tissue form the wall of the ureters (Figure 25.10 on page 773). The deepest coat, or **mucosa,** is a mucous membrane with **transitional epithelium** (see Table 3.1E on page 64) and an underlying **lamina propria** of areolar connective tissue with considerable collagen, elastic fibers, and lymphatic tissue. Transitional epithelium is able to stretch—a marked advantage for any organ that must accommodate a variable volume of fluid. Mucus secreted by the mucosa prevents the cells from coming in contact with urine, the solute concentration and pH of which may differ drastically from the cytosol of cells that form the wall of the ureters. Throughout most of the length of the ureters, the intermediate coat, the **muscularis,** is composed of inner longitudinal and outer circular layers of smooth muscle fibers, an arrangement opposite that of the gastrointestinal tract, which contains inner circular and outer longitudinal layers; the muscularis of the distal third of the ureters also contains an outer layer of longitudinal muscle fibers. Thus, the muscularis in the distal third of the ureter is inner longitudinal, middle circular, and outer longitudinal. Peristalsis is the major function of the muscularis. The superficial coat of the ureters is the **adventitia,** a layer of areolar connective tissue containing blood vessels, lymphatic vessels, and nerves that serve the muscularis and mucosa. The adventitia blends in with surrounding connective tissue and anchors the ureters in place.

The arterial supply of the ureters is from the renal, testicular or ovarian, common iliac, and inferior vesical arteries (arising from the internal iliac artery, a trunk with the internal pudendal and superior gluteal arteries, or a branch of the internal pudendal artery). The veins terminate in the corresponding trunks.

Figure 25.9 / Ureters, urinary bladder, and urethra (shown in a female).
(See Tortora, *A Photographic Atlas of the Human Body,* Figures 13.8, 13.10.)

Urine is stored in the urinary bladder before being expelled by micturition.

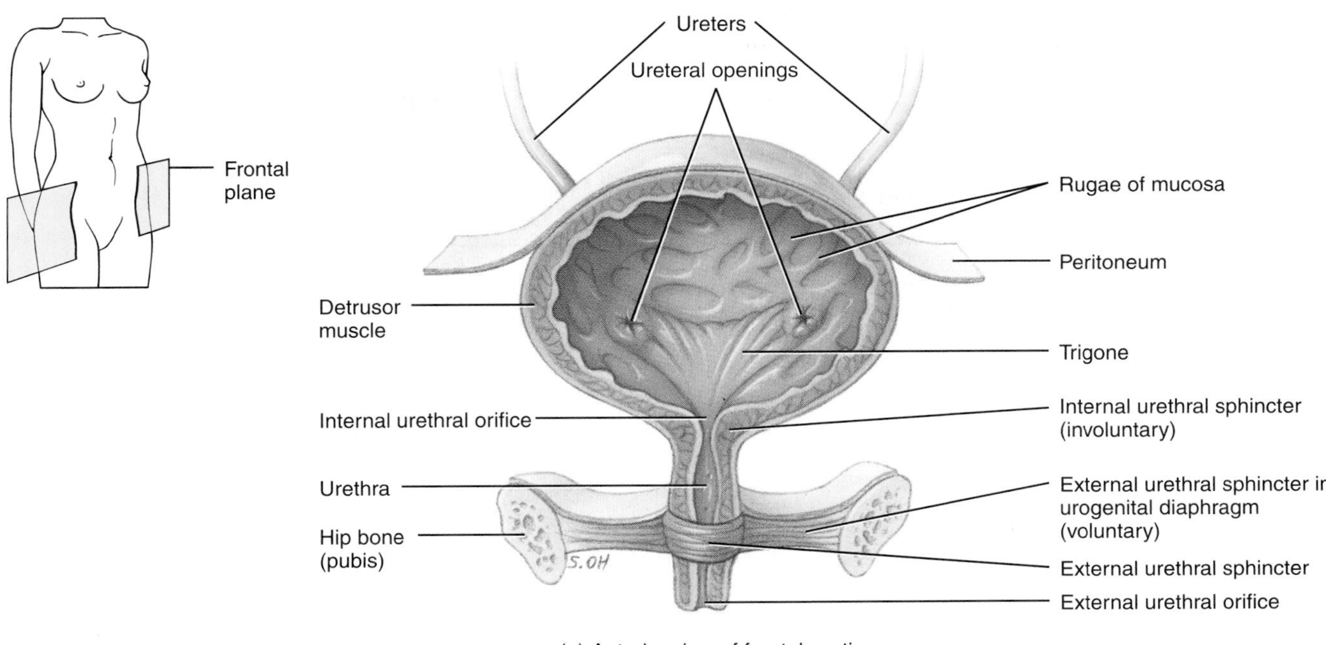

(a) Anterior view of frontal section

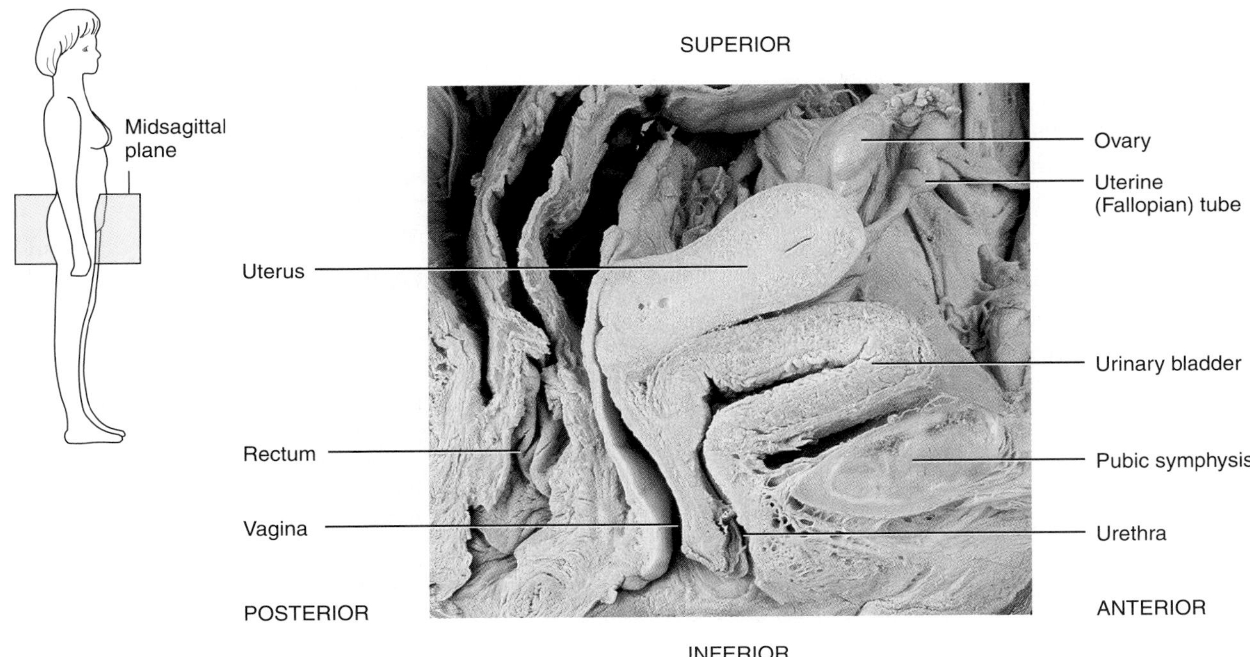

(b) Midsagittal section

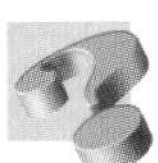

What is a lack of voluntary control over micturition called?

14th Century

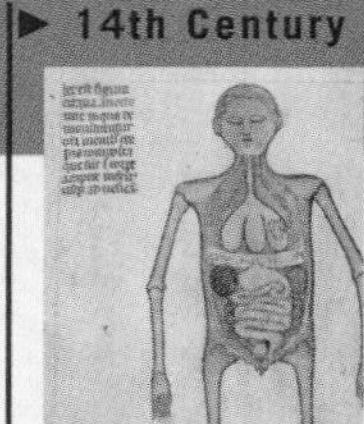

15th Century

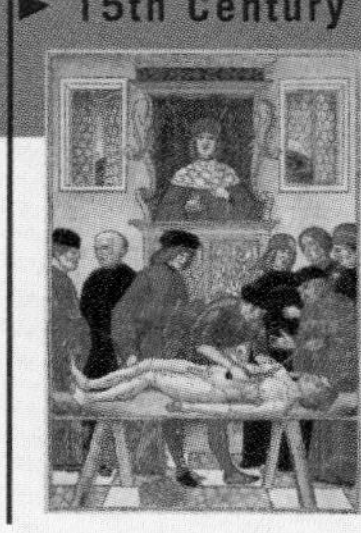

16th Century

17th Century

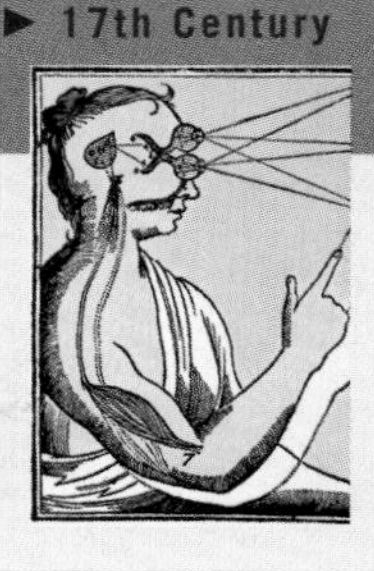

18th Century

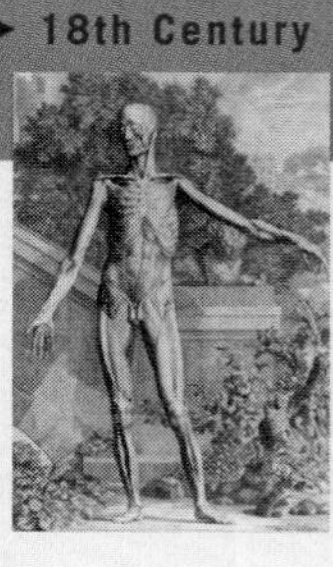

20th Century

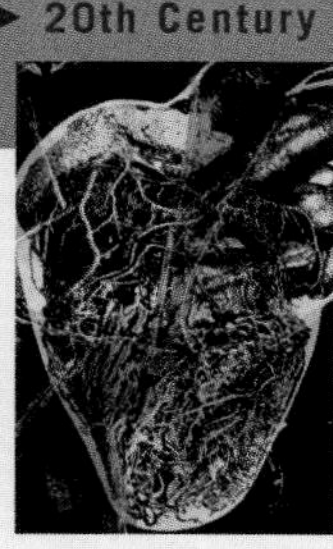

21st Century

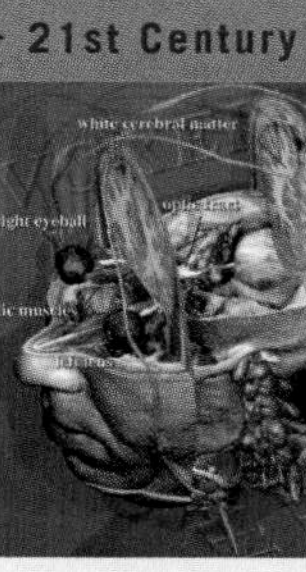

CHANGING IMAGES

A Different Perspective

Spatially visualizing anatomical structures in three dimensions is often difficult for students to master. The vast majority of the images in this text are therefore from an anterior/posterior perspective and the sections are mostly longitudinal—either coronal or sagittal. Perhaps you recall having difficulty understanding the transverse section of the left thigh in Figure 10.23 (page 329) when studying the muscular system. To truly grasp that image, you need to acquire the capacity to envision the three dimensional relationship of a multitude of structures—muscles, bones, nerves, and blood vessels. Such an advanced anatomical skill is required in many clinical settings, as in interpreting radiological images or performing surgery.

Pictured here are transverse sections that were generated from *Visible Human Project* data by the University of Hamburg, Germany, Institute of Mathematics and Computer Science in Medicine. As you may recall, the *Visible Human Project* produced high-resolution, three-dimensional images via transverse sections and an array of imaging techniques of both a male and female cadaver. In the Chapter 18 *Changing Images* essay on page 564, this project was discussed in great detail.

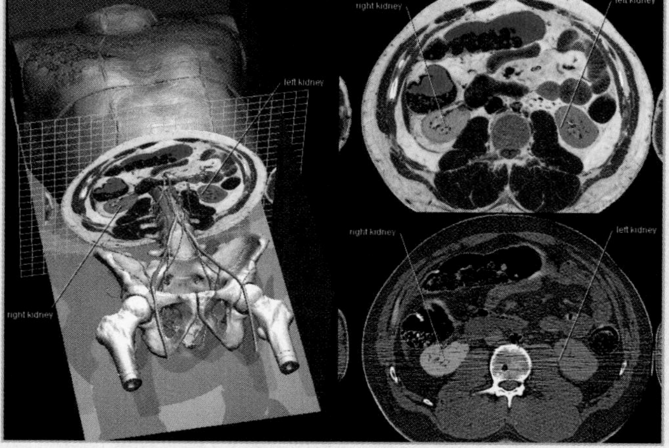

2000 AD

The image shown here correlates the high-resolution photography of the transverse sectional anatomy pictured on the top right, with the computerized tomography (CT) below. To the left you can see where this section resides in the human. Do you notice anything unusual in this perspective? Notice that you are looking at this transverse section from an inferior frame of reference, as if you were standing at the foot of a patient's bed. This is a clinically standard viewpoint. Does this perspective help you understand the spatial relationship of the kidneys, the principal organs of the urinary system, with surrounding structures? How close do the kidneys lie to the vertebrae? Are the muscular layers of the abdominal wall and posterior vertebral region apparent to you? Can you identify the muscles that lie between the kidneys and the vertebrae? Try to identify the pancreas, abdominal aorta, and renal adipose capsule. Use Figure 25.2a on page 760 as a reference. As you near the completion of your anatomical studies, identifying structures from a variety of perspectives is a skill worthy of your efforts. A skill that may be required of you in your profession in the future.

Figure 25.10 / Histology of the ureter.

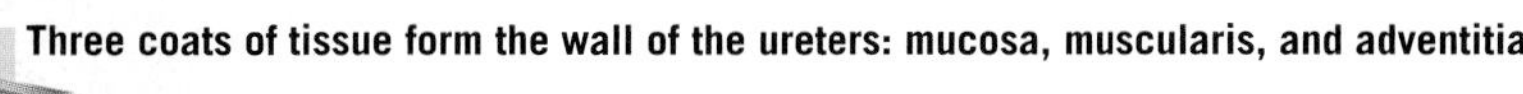

Three coats of tissue form the wall of the ureters: mucosa, muscularis, and adventitia.

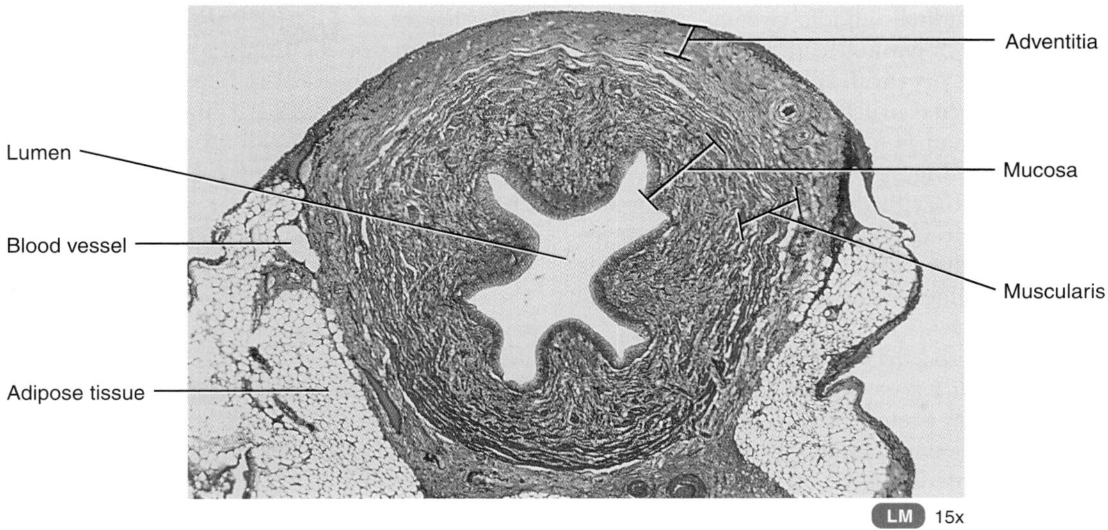

Transverse section of ureter

How does the muscularis of most of the ureters differ from that of the gastrointestinal tract?

The ureters are innervated by the renal plexuses, which are supplied by sympathetic and parasympathetic fibers from the lesser and lowest splanchnic nerves.

Urinary Bladder

The **urinary bladder** is a hollow, distensible muscular organ situated in the pelvic cavity posterior to the pubic symphysis. In males, it is directly anterior to the rectum (see Figure 26.3); in females it is anterior to the vagina and inferior to the uterus (see Figure 26.13 on p. 796). It is held in position by folds of the peritoneum. The shape of the urinary bladder depends on how much urine it contains. Empty, it is collapsed; when slightly distended it becomes spherical; as urine volume increases it becomes pear-shaped and rises into the abdominal cavity. Urinary bladder capacity averages 700–800 mL; it is smaller in females because the uterus occupies the space just superior to the urinary bladder.

In the floor of the urinary bladder is a small triangular area called the **trigone** (TRĪ-gōn; = triangle). The two posterior corners of the trigone contain the two ureteral openings, whereas the opening into the urethra, the **internal urethral orifice,** lies in the anterior corner (see Figure 25.9). Because its mucosa is firmly bound to the muscularis, the trigone has a smooth appearance.

Three coats make up the wall of the urinary bladder (Figure 25.11). The deepest is the **mucosa,** a mucous membrane composed of **transitional epithelium** and an underlying **lamina**

Figure 25.11 / Histology of the urinary bladder.

Discharge of urine from the urinary bladder is a combination of voluntary and involuntary muscular contractions called micturition.

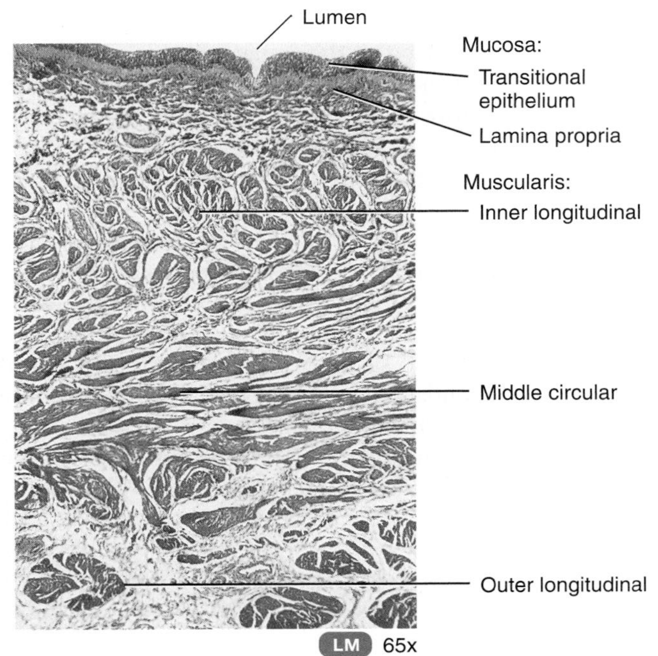

Transverse section of urinary bladder

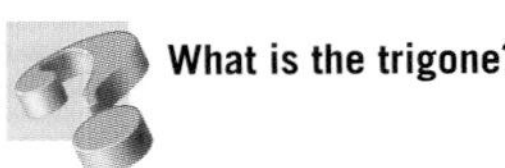

What is the trigone?

propria similar to that of the ureters. Rugae (the folds in the mucosa) are also present. Surrounding the mucosa is the intermediate **muscularis,** also called the **detrusor muscle** (de-TRŪ-ser; = to push down), which consists of three layers of smooth muscle fibers: the inner longitudinal, middle circular, and outer longitudinal layers. Around the opening to the urethra the circular fibers form an **internal urethral sphincter** (see Figure 25.9a); inferior to it is the **external urethral sphincter,** which is composed of skeletal muscle and is a modification of the urogenital diaphragm muscle (see Figure 10.14 on page 291). The most superficial coat of the urinary bladder on the posterior and inferior surfaces is the **adventitia,** a layer of areolar connective tissue that is continuous with that of the ureters. Over the superior surface of the urinary bladder is the **serosa,** a layer of visceral peritoneum.

Discharge of urine from the urinary bladder, called **micturition** (mik′-tū-RISH-un; *mictur-* = urinate), is also known as *urination* or *voiding.* Micturition occurs via a combination of involuntary and voluntary muscle contractions. When the volume of urine in the bladder exceeds 200–400 mL, pressure within the bladder increases considerably, and stretch receptors in its wall transmit nerve impulses into the spinal cord. These impulses propagate to the **micturition center** in sacral spinal cord segments S2 and S3 and trigger a spinal reflex called the **micturition reflex.** In this reflex arc, parasympathetic impulses from the micturition center propagate to the urinary bladder wall and internal urethral sphincter. The nerve impulses cause *contraction* of the detrusor muscle and *relaxation* of the internal urethral sphincter muscle. Simultaneously, the micturition center inhibits somatic motor neurons that innervate skeletal muscle in the external urethral sphincter. Upon contraction of the bladder wall and relaxation of the sphincters, urination takes place. Bladder filling causes a sensation of fullness that initiates a conscious desire to urinate before the micturition reflex actually occurs. Although emptying of the urinary bladder is a reflex, in early childhood we learn to initiate it and stop it voluntarily. Through learned control of the external urethral sphincter muscle and certain muscles of the pelvic floor, the cerebral cortex can initiate micturition or delay its occurrence for a limited period of time.

The arteries of the urinary bladder are the superior vesical (arises from the umbilical artery), the middle vesical (arises from the umbilical artery or a branch of the superior vesical), and the inferior vesical (arises from the internal iliac artery, a trunk with the internal pudendal and superior gluteal arteries, or a branch of the internal pudendal artery). The veins from the urinary bladder pass to the internal iliac trunk.

The nerves are derived partly from the hypogastric sympathetic plexus and partly from the second and third sacral nerves (pelvic splanchnic nerve).

Urethra

The **urethra** is a small tube leading from the internal urethral orifice in the floor of the urinary bladder to the exterior of the body (see Figure 25.1). In both males and females, the urethra is the terminal portion of the urinary system and the passageway for discharging urine from the body; in males it discharges semen as well.

In females, the urethra lies directly posterior to the pubic symphysis, is directed obliquely inferiorly and anteriorly, and has a length of 4 cm (1.5 in.) (see Figure 26.13 on page 796). The opening of the urethra to the exterior, the **external urethral orifice,** is located between the clitoris and the vaginal opening (see Figure 26.22 on page 806). The wall of the female urethra consists of a deep **mucosa** and a superficial **muscularis.** The mucosa is a mucous membrane composed of **epithelium** and **lamina propria** (areolar connective tissue with elastic fibers and a plexus of veins). The muscularis consists of circularly arranged smooth muscle fibers and is continuous with that of the urinary bladder. Near the urinary bladder, the mucosa contains transitional epithelium that is continuous with that of the urinary bladder; near the external urethral orifice the epithelium is nonkeratinized stratified squamous epithelium. Between these areas, the mucosa contains stratified columnar or pseudostratified columnar epithelium.

In males, the urethra also extends from the internal urethral orifice to the exterior, but its length and passage through the body are considerably different than in females (see Figure 26.3 on page 784 and Figure 26.12 on page 795). The male urethra first passes through the prostate gland, then through the urogenital diaphragm, and finally through the penis, a distance of 15–20 cm (6–8 in.).

The male urethra, which also consists of a deep **mucosa** and a superficial **muscularis,** is subdivided into three anatomical regions: (1) The **prostatic urethra** passes through the prostate gland; (2) the **membranous urethra,** the shortest portion, passes through the urogenital diaphragm; and (3) the **spongy urethra,** the longest portion, passes through the penis. The epithelium of the prostatic urethra is continuous with that of the urinary bladder and consists of transitional epithelium that becomes stratified columnar or pseudostratified columnar epithelium more distally. The mucosa of the membranous urethra contains stratified columnar or pseudostratified columnar epithelium. The epithelium of the spongy urethra is stratified columnar or pseudostratified columnar epithelium, except near the external urethral orifice, which is nonkeratinized stratified squamous epithelium. The **lamina propria** of the male urethra is areolar connective tissue with elastic fibers and a plexus of veins.

The muscularis of the prostatic urethra is composed of wisps of mostly circular smooth muscle fibers superficial to the lamina propria; these circular fibers help form the internal urethral sphincter of the urinary bladder. The muscularis of the membranous urethra consists of circularly arranged skeletal muscle fibers of the urogenital diaphragm that help form the external urethral sphincter of the urinary bladder.

Several glands and other structures associated with reproduction deliver their contents into the male urethra. The prostatic urethra contains the openings of (1) ducts that transport secretions from the **prostate gland** and (2) the **seminal vesicles** and **ductus (vas) deferens,** which deliver sperm into the urethra and provide secretions that both neutralize the acidity of the female reproductive tract and contribute to sperm motility and viability. The openings of the ducts of the **bulbourethral (Cowper's) glands** empty into the spongy urethra. They deliver an alkaline substance prior to ejaculation that neutralizes the acidity of the urethra. The glands also secrete mucus, which lubricates the end of the penis during sexual arousal. Throughout the urethra, but especially in the spongy urethra, the openings of the ducts of **urethral (Littré) glands** discharge mucus during sexual arousal and ejaculation.

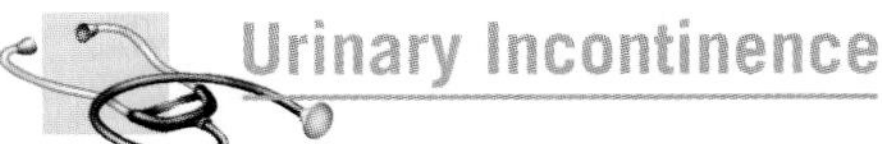

Urinary Incontinence

A lack of voluntary control over micturition is called **urinary incontinence.** In infants and children under 2–3 years old, incontinence is normal because neurons to the external urethral sphincter muscle are not completely developed, and voiding occurs whenever the urinary bladder is sufficiently distended to stimulate the micturition reflex. Urinary incontinence also occurs in adults. Of the more than 10 million U.S. adults who suffer from urinary incontinence, 1–2 million have **stress incontinence.** In this condition, physical stresses that increase abdominal pressure, such as coughing, sneezing, laughing, exercising, pregnancy, or simply walking, cause leakage of urine from the urinary bladder. Other causes of incontinence in adults are injury to the nerves controlling the urinary bladder, loss of bladder flexibility with age, disease or irritation of the bladder or urethra, damage to the external urethral sphincter, and certain drugs. And, compared with nonsmokers, those who smoke have twice the risk of developing incontinence. ■

✓ What forces help propel urine from the renal pelvis to the bladder?
✓ What is micturition? Describe the micturition reflex.
✓ Compare the location, length, and histology of the urethra in males and females.

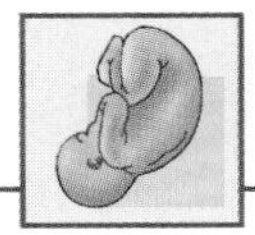

DEVELOPMENTAL ANATOMY OF THE URINARY SYSTEM

Objective

- Describe the development of the urinary system.

Starting in the third week of fetal development, a portion of the mesoderm along the posterior aspect of the embryo, the **intermediate mesoderm,** differentiates into the kidneys. Three pairs of kidneys form within the intermediate mesoderm in successive time periods: pronephros, mesonephros, and metanephros (Figure 25.12). Only the last pair remains as the functional kidneys of the newborn.

Figure 25.12 / Development of the urinary system.

Three pairs of kidneys form within intermediate mesoderm in successive time periods: pronephros, mesonephros, and metanephros.

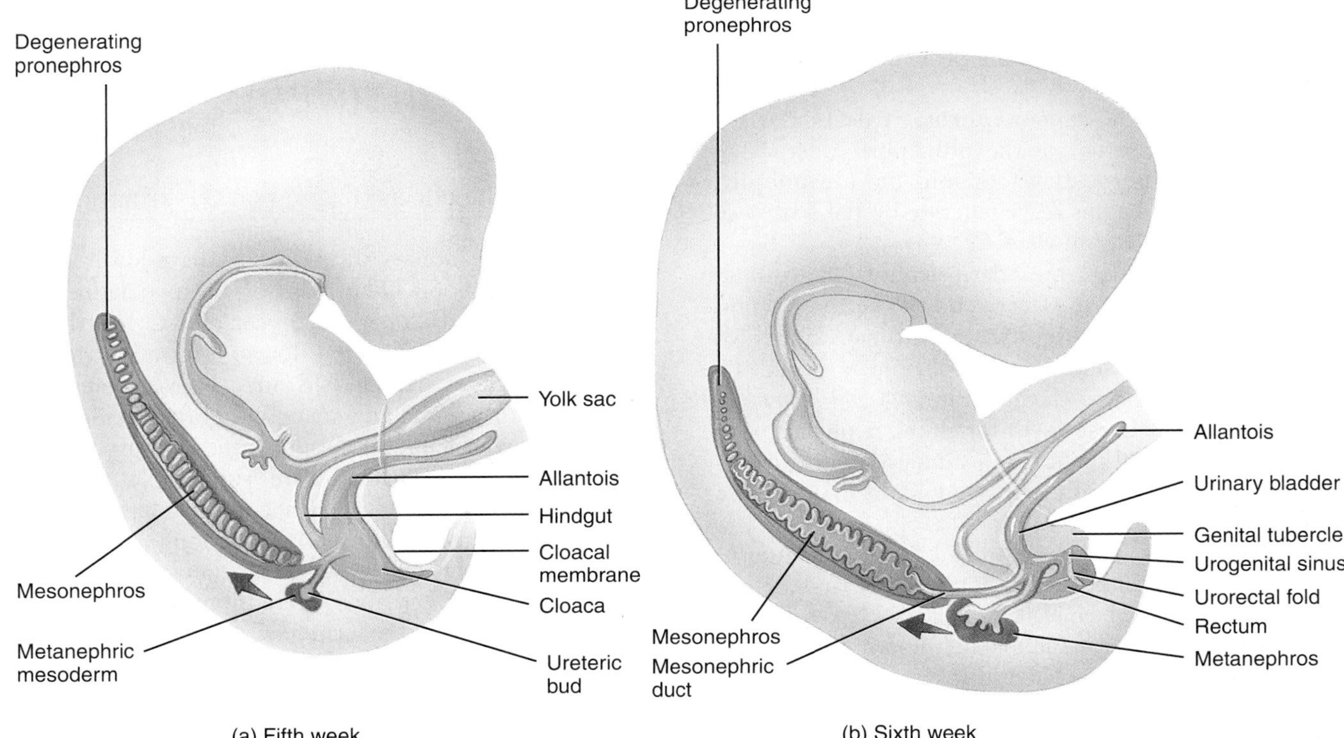

(continues)

Figure 25.12 (continued)

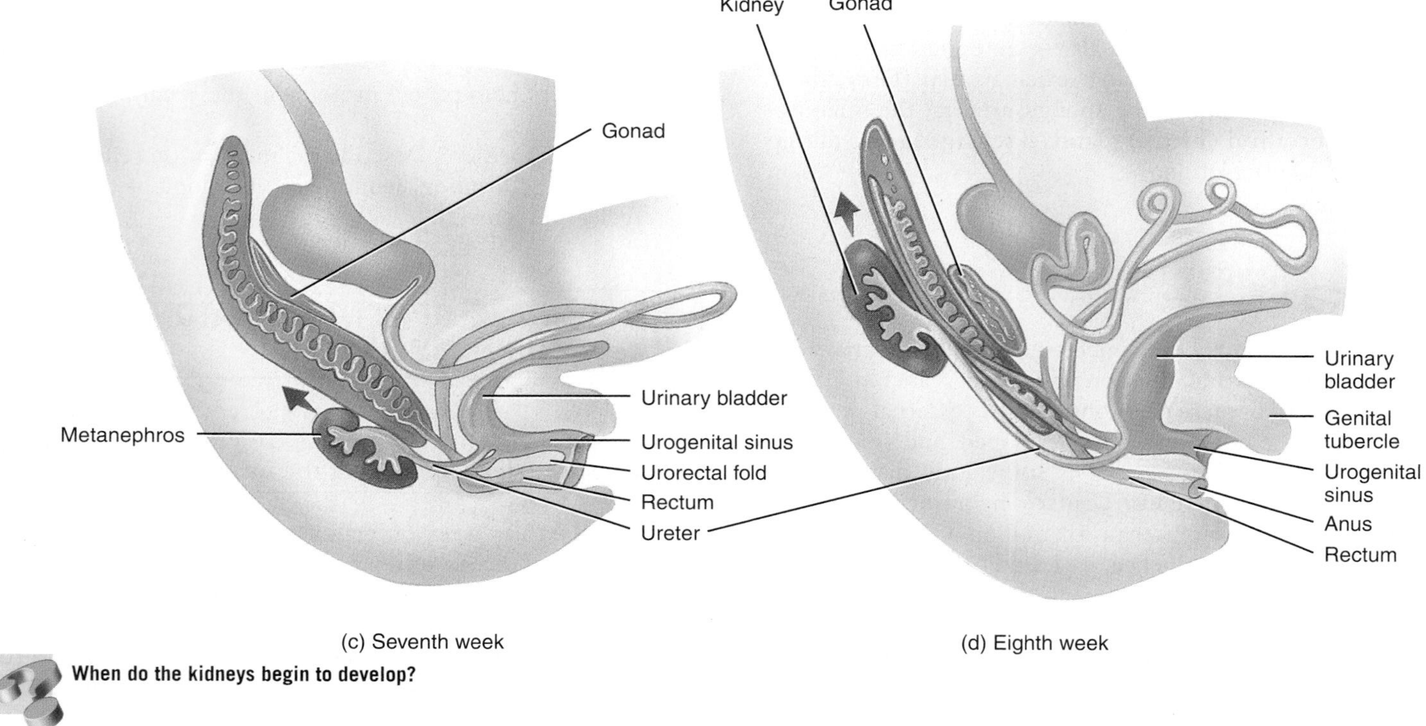

(c) Seventh week

(d) Eighth week

When do the kidneys begin to develop?

The first kidney to form, the **pronephros,** is the most superior of the three and has an associated **pronephric duct.** This duct empties into the **cloaca,** which functions as a common outlet of the urinary, digestive, and reproductive ducts. The pronephros begins to degenerate during the fourth week and is completely gone by the sixth week; the pronephric ducts, however, remain.

The second kidney, the **mesonephros,** replaces the pronephros. The retained portion of the pronephric duct, which connects to the mesonephros, develops into the **mesonephric duct.** The mesonephros begins to degenerate by the sixth week and is almost gone by the eighth week.

At about the fifth week, a mesodermal outgrowth called a **ureteric bud** develops from the distal end of the mesonephric duct near the cloaca. The **metanephros,** or ultimate kidney, develops from the ureteric bud and metanephric mesoderm. The ureteric bud forms the *collecting ducts, calyces, renal pelvis,* and *ureter.* The **metanephric mesoderm** forms the nephrons of the kidneys. By the third month, the fetal kidneys begin excreting urine into the surrounding amniotic fluid; indeed, fetal urine makes up most of the amniotic fluid.

During development, the cloaca divides into a **urogenital sinus,** into which urinary and genital ducts empty, and a *rectum* that discharges into the anal canal. The *urinary bladder* develops from the urogenital sinus. In females, the *urethra* develops as a result of lengthening of the short duct that extends from the urinary bladder to the urogenital sinus. The *vestibule,* into which the urinary and genital ducts empty, is also derived from the urogenital sinus. In males, the urethra is considerably longer and more complicated, but it is also derived from the urogenital sinus. ■

✓ Describe the development of the three pairs of kidneys.

AGING AND THE URINARY SYSTEM

Objective

- Describe the effects of aging on the urinary system.

With aging, the kidneys shrink in size, have lowered blood flow, and filter less blood. The mass of the two kidneys decreases from an average of 260 g in 20-year-olds to less than 200 g by age 80. Likewise, renal blood flow and filtration rate decline by 50% between ages 40 and 70. Kidney diseases that become more common with age include acute and chronic kidney inflammations and renal calculi. Because the sensation of thirst diminishes with age, older individuals are susceptible to dehydration. Urinary tract infections are more common among the elderly, as are polyuria (excessive urine production), nocturia (excessive urination at night), increased frequency of urination, dysuria (painful urination), urinary retention or incontinence, and hematuria (blood in the urine). Cancer of the prostate is the most common malignancy in elderly males (see page 818).

✓ To what extent do kidney mass and filtrate rate decrease with age?

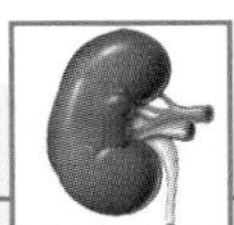

APPLICATIONS TO HEALTH

Renal Calculi

The crystals of salts present in urine occasionally precipitate and solidify into insoluble stones called **renal calculi** (*calculi* = pebbles) or **kidney stones.** They commonly contain crystals of calcium oxalate, uric acid, or calcium phosphate. Conditions leading to calculus formation include the ingestion of excessive calcium, scanty water intake, abnormally alkaline or acidic urine, and overactivity of the parathyroid glands. When a stone lodges in a narrow passage, such as a ureter, the pain can be excruciating. **Shock wave lithotripsy** (LITH-ō-trip′-sē; *litho-* = stone) offers an alternative to surgical removal of kidney stones. A device called a lithotripter delivers brief, high-intensity sound waves through a water-filled cushion. Over a period of 30–60 minutes, 1000 or more hydraulic shock waves pulverize the stone until the fragments are small enough to wash out in the urine.

Urinary Tract Infections

The term **urinary tract infection (UTI)** is used to describe either an infection of a part of the urinary system or the presence of large numbers of microbes in urine. UTIs are more common in females due to their shorter urethra. Symptoms include painful or burning urination, urgent and frequent urination, low back pain, and bed-wetting. UTIs include *urethritis* (inflammation of the urethra), *cystitis* (inflammation of the urinary bladder), and *pyelonephritis* (inflammation of the kidneys). If pyelonephritis becomes chronic, scar tissue forms in the kidneys and severely impairs their function.

Glomerular Diseases

A variety of conditions may damage the kidney glomeruli, either directly or indirectly as a consequence of disease elsewhere in the body. Typically, the filtration membrane sustains damage, and its permeability increases.

Glomerulonephritis is an inflammation of the kidney that involves the glomeruli. One of the most common causes is an allergic reaction to the toxins produced by streptococcal bacteria that have recently infected another part of the body, especially the throat. The glomeruli become so inflamed, swollen, and engorged with blood that the filtration membranes allow blood cells and plasma proteins to enter the filtrate. As a result, the urine contains many erythrocytes (hematuria) and much protein. The glomeruli may be permanently damaged, leading to chronic renal failure.

Nephrotic syndrome is a condition characterized by *proteinuria* (protein in the urine) and *hyperlipidemia* (high blood levels of cholesterol, phospholipids, and triglycerides). The proteinuria is due to an increased permeability of the filtration membrane, which permits proteins, especially albumin, to escape from blood into urine. Loss of albumin results in *hypoalbuminemia* (low blood albumin level) once liver production of albumin fails to meet increased urinary losses. Edema, usually seen around the eyes, ankles, feet, and abdomen, occurs in nephrotic syndrome because loss of albumin from the blood decreases blood colloid osmotic pressure. Nephrotic syndrome is associated with several glomerular diseases of unknown cause, as well as with systemic disorders such as diabetes mellitus, systemic lupus erythematosus (SLE), a variety of cancers, and AIDS.

Renal Failure

Renal failure is a decrease or cessation of glomerular filtration. In **acute renal failure (ARF)** the kidneys abruptly stop working entirely (or almost entirely). The main feature of ARF is the suppression of urine flow, usually characterized either by *oliguria* (*olig-* = scanty; *-uria* = urine production), which is daily urine output less than 250 mL, or by *anuria*, daily urine output less than 50 mL. Causes include low blood volume (for example, due to hemorrhage), decreased cardiac output, damaged renal tubules, kidney stones, the dyes used to visualize blood vessels in angiograms, nonsteroidal antiinflammatory drugs, and some antibiotic drugs. Renal failure causes edema due to salt and water retention; acidosis due to inability of the kidneys to excrete acidic substances; increased levels of urea due to impaired renal excretion of metabolic waste products; elevated potassium levels that can lead to cardiac arrest; anemia, because the kidneys no longer produce enough erythropoietin for adequate red blood cell production; and osteomalacia, because the kidneys are no longer able to convert vitamin D to calcitriol, which is needed for adequate calcium absorption from the small intestine.

Chronic renal failure (CRF) refers to a progressive and usually irreversible decline in glomerular filtration rate (GFR). CRF may result from chronic glomerulonephritis, pyelonephritis, polycystic kidney disease, or traumatic loss of kidney tissue. CRF develops in three stages. In the first stage, *diminished renal reserve*, nephrons are destroyed until about 75% of the functioning nephrons are lost. At this stage, a person may have no signs or symptoms because the remaining nephrons enlarge and take over the function of those that have been lost. Once 75% of the nephrons are lost, the person enters the second stage, called *renal insufficiency*, characterized by a decrease in GFR and increased blood levels of nitrogen-containing wastes and creatinine. Also, the kidneys cannot effectively concentrate or dilute the urine. The final stage, called *end-stage renal failure*, occurs when about 90% of the nephrons have been lost. At this stage, GFR diminishes to 10–15% of normal, oliguria is present, and blood levels of nitrogen-containing wastes and creatinine increase further. People with end-stage renal failure need dialysis therapy and are possible candidates for a kidney transplant operation.

Polycystic Kidney Disease

Polycystic kidney disease (PKD) is one of the most common inherited disorders. In infants it results in death at birth or shortly thereafter; in adults it accounts for 6–12% of kidney transplants. In PKD, the kidney tubules become riddled with hundreds or thousands of cysts (fluid-filled cavities). In addition, inappropriate apoptosis (programmed cell death) of cells in noncystic tubules leads to progressive impairment of renal function and eventually to end-stage renal failure.

People with PKD also may have cysts and apoptosis in the liver, pancreas, spleen, and gonads; increased risk of cerebral aneurysms; heart valve defects; and diverticuli in the colon. Typically, symptoms are not noticed until adulthood, when patients may have back pain, urinary tract infections, blood in the urine, hypertension, and large abdominal masses. Progression to renal failure may be slowed by using drugs to restore normal blood pressure, restricting protein and salt in the diet, and controlling urinary tract infections.

Urinary Bladder Cancer

Each year, nearly 12,000 people die from **urinary bladder cancer,** and its incidence has increased about 36% over the past ten years. It generally strikes people over 50 years of age and is three times more likely to develop in males than females. The disease is typically painless as it develops, but in most cases blood in the urine is a primary sign of the disease. Less often, people experience painful and/or frequent urination.

As long as the disease is identified early and treated promptly, the prognosis is favorable. Fortunately, about 75% of urinary bladder cancers are confined to the epithelium of the urinary bladder and are easily removed by surgery. The lesions tend to be low grade, meaning that they have only a small potential for metastasis.

Urinary bladder cancer is frequently the result of a carcinogen. About half of all cases occur in people who smoke or have at some time smoked cigarettes. The cancer also tends to develop in people who are exposed to chemicals called aromatic amines. Workers in the leather, dye, rubber, and aluminum industries, as well as painters, are often exposed to these chemicals.

KEY MEDICAL TERMS ASSOCIATED WITH THE URINARY SYSTEM

Azotemia (az-ō-TĒ-mē-a; *azot-* = nitrogen; *-emia* = condition of blood) Presence of urea or other nitrogen-containing substances in the blood.

Cystocele (SIS-tō-sēl; *cysto-* = bladder; *-cele* = hernia or rupture) Hernia of the urinary bladder.

Dysuria (dis-YŪ-rē-a; *dys* = painful; *uria* = urine) Painful urination.

Enuresis (en′-yū-RĒ-sis; = to void urine) Involuntary voiding of urine after the age at which voluntary control has typically been attained.

Intravenous pyelogram (in′-tra-VĒ-nus PĪ-e-lō-gram′; *intra-* = within; *veno-* = vein; *pyelo-* = pelvis of kidney; *-gram* = record), or ***IVP*** radiograph (x-ray) of the kidneys after venous injection of a dye.

Nocturnal enuresis (nok-TUR-nal en′-yū-RĒ-sis) Discharge of urine during sleep, resulting in bed-wetting; occurs in about 15% of 5-year-old children and generally resolves spontaneously, afflicting only about 1% of adults. It may have a genetic basis, as bed-wetting occurs more often in identical twins than in fraternal twins and more often in children whose parents or siblings were bed-wetters. Possible causes include smaller-than-normal bladder capacity, failure to awaken in response to a full bladder, and above-normal production of urine at night. Also referred to as **nocturia.**

Polyuria (pol′-ē-YŪ-rē-a; *poly-* = too much) Excessive urine formation.

Stricture (STRIK-chur) Narrowing of the lumen of a canal or hollow organ, as may occur in the ureter, urethra, or any other tubular structure in the body.

Uremia (yū-RĒ-mē-a; *emia* = condition of blood) Toxic levels of urea in the blood resulting from severe malfunction of the kidneys.

Urinary retention A failure to completely or normally void urine; may be due to an obstruction in the urethra or neck of the urinary bladder, to nervous contraction of the urethra, or to lack of urge to urinate. In men, an enlarged prostate may constrict the urethra and cause urinary retention.

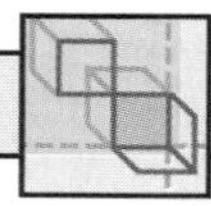

STUDY OUTLINE

Introduction (p. 758)

1. The organs of the urinary system are the kidneys, ureters, urinary bladder, and urethra.
2. After the kidneys filter blood and return most water and many solutes to the bloodstream, the remaining water and solutes constitute urine.
3. The kidneys regulate the ionic composition, osmolarity, volume, and pH of the blood, as well as blood pressure.
4. The kidneys also perform gluconeogenesis, release calcitriol and erythropoietin, and excrete wastes and foreign substances.

Anatomy and Histology of the Kidneys (p. 758)

1. The kidneys are retroperitoneal organs attached to the posterior abdominal wall.
2. Three layers of tissue surround the kidneys: the renal capsule, adipose capsule, and renal fascia.
3. Internally, the kidneys consist of a renal cortex, a renal medulla, renal pyramids, renal papillae, renal columns, calyces, and a renal pelvis.
4. Blood flows into the kidney through the renal artery and successively into segmental, interlobar, arcuate, and interlobular arteries; afferent arterioles; glomerular capillaries; efferent arterioles; peritubular capillaries and vasa recta; and interlobular, arcuate, and interlobar veins before flowing out of the kidney through the renal vein.
5. Vasomotor nerves from the sympathetic division of the autonomic nervous system supply kidney blood vessels; they help regulate bloodflow through the kidney.
6. The nephron is the functional unit of the kidneys. A nephron

consists of a renal corpuscle (glomerulus and glomerular, or Bowman's, capsule) and a renal tubule.

7. A renal tubule consists of a proximal convoluted tubule, a loop of Henle, and a distal convoluted tubule, which drains into a collecting duct (shared by several nephrons). The loop of Henle consists of a descending limb and an ascending limb.
8. A cortical nephron has a short loop that dips only into the superficial region of the renal medulla; a juxtamedullary nephron has a long loop of Henle that stretches through the renal medulla almost to the renal papilla.
9. The wall of the entire glomerular capsule, renal tubule, and ducts consists of a single layer of epithelial cells. The epithelium has distinctive histological features in different parts of the tubule. Table 25.1 on page 767 summarizes the histological features of the renal tubule and collecting duct.
10. The juxtaglomerular apparatus (JGA) consists of the juxtaglomerular cells of an afferent arteriole and the macula densa of the final portion of the ascending limb of the loop of Henle.

Functions of Nephrons (p. 768)

1. Fluid that enters the capsular space is glomerular filtrate.
2. The filtration (endothelial–capsular) membrane consists of the glomerular endothelium, basal lamina, and filtration slits between pedicels of podocytes.
3. Most substances in plasma easily pass through the glomerular filter. However, blood cells and most proteins normally are not filtered.
4. Glomerular filtrate amounts to up to 180 liters of fluid per day. This large amount of fluid is filtered because the filter is porous and thin, the glomerular capillaries are long, and the capillary blood pressure is high.
5. Tubular reabsorption is a selective process that reclaims materials from tubular fluid and returns them to the bloodstream. Reabsorbed substances include water, glucose, amino acids, urea, and ions, such as sodium, chloride, potassium, bicarbonate, and phosphate.
6. Some substances not needed by the body are removed from the blood and discharged into the urine via tubular secretion. Included are ions (K^+, H^+, and NH_4^+), urea, creatinine, and certain drugs.

Urine Transportation, Storage, and Elimination (p. 770)

1. The ureters are retroperitoneal and consist of a mucosa, muscularis, and adventitia. They transport urine from the renal pelvis to the urinary bladder, primarily via peristalsis.
2. The urinary bladder is located in the pelvic cavity posterior to the pubic symphysis; its function is to store urine prior to micturition.
3. The urinary bladder consists of a mucosa with rugae, a muscularis (detrusor muscle), and an adventitia (serosa over the superior surface).
4. The micturition reflex discharges urine from the urinary bladder via parasympathetic impulses that cause contraction of the detrusor muscle and relaxation of the internal urethral sphincter muscle and via inhibition of impulses in somatic motor neurons to the external urethral sphincter.
5. The urethra is a tube leading from the floor of the urinary bladder to the exterior. Its anatomy and histology differ in females and males. In both sexes the urethra functions to discharge urine from the body; in males it discharges semen as well.

Developmental Anatomy of the Urinary System (p. 775)

1. The kidneys develop from intermediate mesoderm.
2. They develop in the following sequence: pronephros, mesonephros, metanephros. Only the metanephros remains and develops into a functional kidney.

Aging and the Urinary System (p. 776)

1. With aging, the kidneys shrink in size, have lowered blood flow, and filter less blood.
2. Common problems related to aging include urinary tract infections, increased frequency of urination, urinary retention or incontinence, and renal calculi.

SELF-QUIZ QUESTIONS

Choose the one best answer to these questions.

1. Which of the following is *not* a function of the kidneys? (a) participation in the formation of the active form of vitamin D, (b) regulation of the volume and composition of the blood, (c) removal of waste from the blood in the form of urine, (d) production of white blood cells, (e) regulation of blood pressure by secretion of renin which activates the renin-angiotensin pathway.

2. Which of the following statements is *not* true? (a) The kidneys are located posterior to the peritoneum (that is, retroperitoneally). (b) The left kidney is usually lower than the right kidney. (c) The hilus is on the concave medial border of the kidney. (d) The cavity of the kidney contains the renal pelvis, which represents the superior expanded portion of the ureter. (e) The kidney exhibits an inner darkened area, the renal medulla, and an outer pale area, the renal cortex.

3. Urine leaving the distal convoluted tubule passes through various structures in which of the following sequences? (a) collecting duct, renal hilus, calyx, ureter, (b) collecting duct, calyx, renal pelvis, ureter, (c) calyx, collecting duct, renal pelvis, ureter, (d) calyx, renal hilus, renal pelvis, ureter, (e) collecting duct, renal hilus, ureter, calyx.

4. The trigone, a landmark in the urinary bladder, is a triangular area bounded by (a) the orifices of the ejaculatory ducts and the urethra, (b) the internal urethral orifice and the inferior border of the detrusor muscle, (c) the ureteral and the internal urethral orifices, (d) the top of the fundus and the ureteral orifices, (e) the major and minor calyces.

5. The notch on the medial surface of the kidney through which blood vessels enter and exit is called the (a) renal medulla, (b) renal column, (c) renal hilus, (d) minor calyx, (e) major calyx.

Complete the following.

6. The apex of a renal pyramid, called the ________ , points toward the interior of the kidney.

7. In the renal corpuscle, the _______ arteriole normally has a smaller diameter than the ________ arteriole.

8. The musculature of the urinary bladder is called the _______ muscle.

9. The three subdivisions of the male urethra, from proximal to distal, are ________ urethra, ________ urethra, and ________ urethra.

10. The three layers of tissue surrounding each kidney, from superficial to deep are named ________ , ________ , and ________ .

11. The conical-shaped structures inside the renal medulla of a kidney are called the ________ .

12. The _____ carry urine from the kidney to the urinary bladder.

13. Place numbers in the blanks to arrange the following vessels in order. (a) arcuate arteries: ________ ; (b) interlobular arteries: ________ ; (c) renal arteries: ________ ; (d) peritubular capillaries and vasa recta: ________ ; (e) glomerular capillaries: ________ ; (f) efferent arteriole: ________ ; (g) afferent arteriole: ________ ; (h) interlobar arteries: ________ ; (i) segmental arteries: ________ .

14. Place numbers in the blanks to arrange the following structures in the correct sequence for the flow of glomerular filtrate. (a) ascending limb of loop of Henle: ________ ; (b) descending limb of loop of Henle: ________ ; (c) collecting duct: ________ ; (d) papillary duct: ________ ; (e) distal convoluted tubule: ________ ; (f) proximal convoluted tubule: ________ .

15. The process of emptying the urinary bladder is called ________ .

Are the following statements true or false?

16. The kidneys are partially protected by the two pairs of floating ribs.

17. Most renal nerves arise from the celiac ganglion and are part of the sympathetic division of the autonomic nervous system.

18. The internal urethral sphincter is composed of voluntary skeletal muscle, whereas the external urethral sphincter is composed of involuntary smooth muscle.

19. Pronephros, mesonephros, and metanephros are the names of the three pairs of kidneys that develop successively in fetal development.

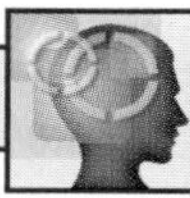

CRITICAL THINKING QUESTIONS

1. Imagine that a new super bug has emerged from a nuclear waste dump. It produces a toxin that blocks renal tubule function but leaves the glomerulus unaffected. Predict the effects of this toxin.
HINT *The kidney does more than just filter the blood.*

2. While traveling down the interstate, little Caitlyn told her father to stop the car NOW! "I'm so full I'm gonna burst!" Her dad thinks she can make it to the next rest stop. What structures of the urinary bladder will help her "hold it?"
HINT *Children often visit the emergency room with broken bones, but rarely with exploding bladders.*

3. Although urinary catheters come in one length only, the number of centimeters that must be inserted to release the urine differs significantly in males and females. Why? Why is volume of urine released from a full bladder different in males and females?
HINT *Why do men and women usually assume different positions when they urinate?*

4. As Juan focused his microscope on his own urine sample during his Human Anatomy lab, he was concerned to see many cells in the field of view; his urine had appeared to be fluid. There's no evidence of blood or infection. What are these cells and where do they come from?
HINT *These cells are shed normally.*

5. Brittany is suffering from anorexia and has lost an excessive amount of weight. Now, along with her eating disorder, she's having problems with urination and lower abdominal pain. What could be causing this new problem?
HINT *Unlike the saying, there really is such a thing as being too thin.*

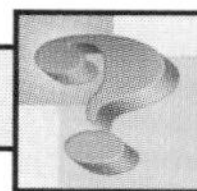

ANSWERS TO FIGURE QUESTIONS

25.1 The kidneys, ureters, urinary bladder, and urethra are the components of the urinary system.

25.2 The kidneys are retroperitoneal because they are posterior to the peritoneum.

25.3 Blood vessels, lymphatic vessels, nerves, and a ureter pass through the renal hilus.

25.4 About 1200 mL of blood enters the renal arteries each minute.

25.5 Cortical nephrons have glomeruli in the superficial renal cortex, and their short loops of Henle penetrate only into the superficial renal medulla; juxtamedullary nephrons have glomeruli deep in the renal cortex, and their long loops of Henle extend through the renal medulla nearly to the renal papilla.

25.6 This section must pass through the renal cortex because there are no renal corpuscles in the renal medulla.

25.7 Secreted penicillin is being removed from the bloodstream.

25.8 Endothelial fenestrations (pores) in glomerular capillaries are too small for red blood cells to pass through.

25.9 Lack of voluntary control over micturition is termed urinary incontinence.

25.10 The muscularis of most of the ureters consists of an inner longitudinal layer and an outer circular layer, an arrangement opposite that of the gastrointestinal tract.

25.11 The trigone is a triangular area in the urinary bladder formed by the ureteral openings (posterior corners) and the internal urethral orifice (anterior corner).

25.12 The kidneys start to develop during the 3rd week of gestation.

26

THE REPRODUCTIVE SYSTEMS

CONTENTS AT A GLANCE

◄ Page 805

Do you recognize this image? Whose work is this? Which human organs are being depicted in this rendering?

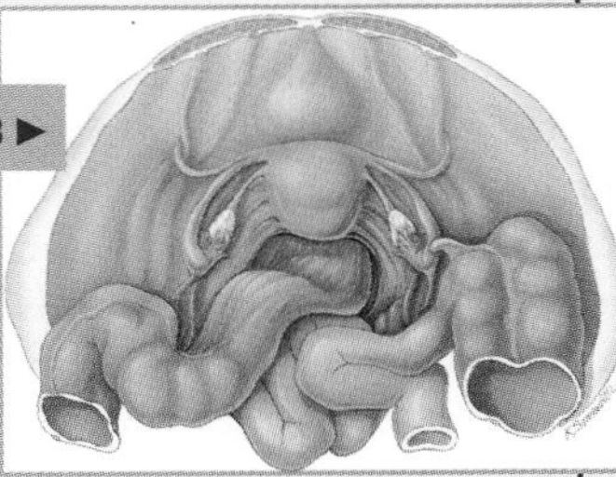

INTRODUCTION

Sexual reproduction is the process by which organisms produce offspring by means of germ cells called **gametes** (GAM-ēts; = spouses). After the male gamete (sperm cell) unites with the female gamete (secondary oocyte)—an event called **fertilization**—the resulting cell contains one set of chromosomes from each parent. Males and females have anatomically distinct reproductive organs that are adapted for producing gametes, facilitating fertilization, and, in females, sustaining the growth of the embryo and fetus.

The male and female reproductive organs can be grouped by function. The **gonads**—testes in males and ovaries in females—produce gametes and secrete sex hormones. Various **ducts** then store and transport the gametes, and **accessory sex glands** produce substances that protect the gametes and facilitate their movement. Finally, **supporting structures,** such as the penis and the uterus, assist the delivery and joining of gametes and, in females, the growth of the fetus during pregnancy.

Gynecology (gī-ne-KOL-ō-jē; *gynec-* = woman; *-ology* = study of) is the specialized branch of medicine concerned with the diagnosis and treatment of diseases of the female reproductive system. As noted in Chapter 25, **urology** (yū-ROL-ō-jē) is the study of the urinary system. Urologists also diagnose and treat diseases and disorders of the male reproductive system.

THE CELL CYCLE IN THE GONADS

Objective

- Describe the process of meiosis.

Before we can discuss the cell cycle in the gonads, we must first understand the distribution of genetic material in cells.

Number of Chromosomes in Somatic Cells and Gametes

In humans, somatic, or body, cells, such as brain, stomach, and kidney cells, contain 23 pairs of chromosomes, or a total of 46 chromosomes; one member of each pair is inherited from each parent. The chromosomes that make up each pair are called **homologous chromosomes** (hō-MOL-ō-gus; *homo-* = same) or **homologs;** they contain similar genes arranged in the same (or almost the same) order. When examined under a light microscope, homologous chromosomes generally look very similar; the exception to this rule is a pair of chromosomes called the **sex chromosomes,** designated X and Y. In females the homologous pair of sex chromosomes consists of two X chromosomes; in males the pair consists of an X and a Y chromosome. The other 22 pairs of chromosomes are called **autosomes.** Because somatic cells contain two sets of chromosomes, they are called **diploid cells** (DIP-loid; *dipl-* = double; *-loid* = form). Geneticists use the symbol n to denote the number of different chromosomes in an organism; in humans, $n = 23$. Diploid cells are $2n$.

In sexual reproduction, each new organism is the result of the union and fusion of two different gametes, one produced by each parent. But if each gamete had the same number of chromosomes as somatic cells, the number of chromosomes would double with each fertilization. Chromosome number does not double at fertilization, however, because the cell cycle in the gonads produces gametes in which the number of chromosomes is reduced by half; gametes contain a single set of chromosomes and thus are **haploid** ($1n$) **cells** (HAP-loyd; *hapl-* = single). Gametes are produced by a special type of cell division called **meiosis** (mī-Ō-sis; *mei-* = lessening; *-osis* = condition of).

Meiosis

Meiosis produces gametes, haploid cells that contain only 23 chromosomes. Meiosis occurs in two successive stages: **meiosis I** and **meiosis II.** During the interphase that precedes meiosis I, the chromosomes replicate in a manner similar to that in the interphase before mitosis in somatic cell division.

Meiosis I

Meiosis I, which begins once chromosomal replication is complete, consists of four phases: prophase I, metaphase I, anaphase I, and telophase I (Figure 26.1a–d). Prophase I is an extended phase in which the chromosomes shorten and thicken, the nuclear envelope and nucleoli disappear, and the mitotic spindle appears. In contrast to prophase of mitosis, the chromosomes become arranged in homologous pairs. In metaphase I, the homologous pairs of chromosomes line up along the metaphase plate of the cell, with homologous chromosomes side by side. (No such pairing of homologous chromosomes occurs during metaphase of mitosis.) The pericentriolar material of a centrosome forms kinetochore microtubules that attach the centromeres to opposite poles of the cell. During anaphase I, the members of each homologous pair separate, with one member of each pair moving to an opposite pole of the cell. The centromeres do not split, and the paired chromatids, held by a centromere, remain together. (During anaphase of mitosis, the centromeres split and the sister chromatids separate.) Telophase I and cytokinesis are similar to telophase and cytokinesis of mitosis. The net effect of meiosis I is that each resulting daughter cell contains the haploid number of chromosomes; each cell contains only one member of each pair of the homologous chromosomes present in the parent cell.

Two events that are not seen in prophase I of mitosis (or in prophase II of meiosis) occur during prophase I of meiosis. First, the two chromatids of each pair of homologous chromosomes pair off, an event called **synapsis.** The resulting four chromatids form a **tetrad.** Second, portions of one chromatid may be exchanged with portions of another; such an exchange is termed **crossing-over** (Figure 26.2). This process, among others, permits an exchange of genes between homologous chromatids so that the resulting daughter cells are genetically unlike each

Figure 26.1 / Meiosis: reproductive cell division. Details of events are discussed in the text.

In reproductive cell division, a single diploid parent cell undergoes meiosis I and meiosis II to produce four haploid gametes that are genetically different from the parent cell.

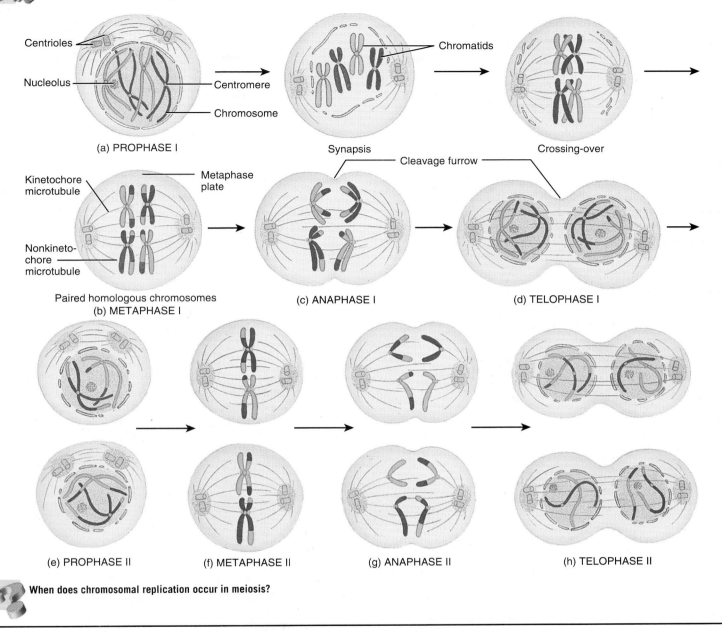

Centrioles

Nucleolus

Centromere

Chromosome

Chromatids

(a) PROPHASE I

Synapsis

Crossing-over

Cleavage furrow

Kinetochore microtubule

Metaphase plate

Nonkinetochore microtubule

Paired homologous chromosomes

(b) METAPHASE I

(c) ANAPHASE I

(d) TELOPHASE I

(e) PROPHASE II

(f) METAPHASE II

(g) ANAPHASE II

(h) TELOPHASE II

When does chromosomal replication occur in meiosis?

Figure 26.2 / Crossing-over within a tetrad during prophase I of meiosis.

Crossing-over permits an exchange of genes between homologous chromosomes.

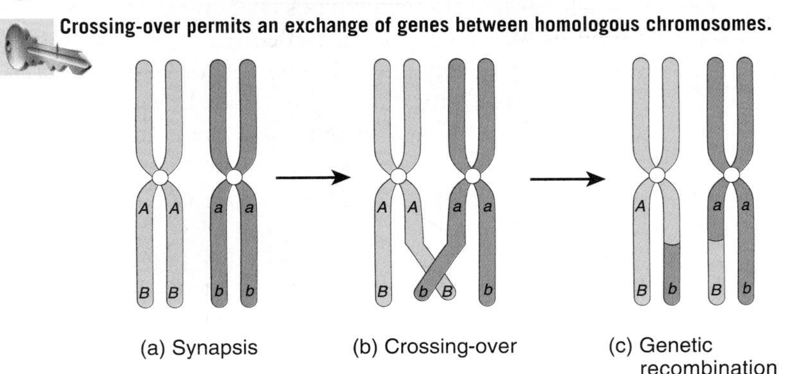

How does crossing-over affect the genetic content of daughter cells?

other and genetically unlike the parent cell that produced them. Crossing-over results in **genetic recombination**—the formation of new combinations of genes—and accounts for part of the great genetic variation among humans and other organisms that form gametes via meiosis.

Meiosis II

The second stage of meiosis, meiosis II, also consists of four phases: prophase II, metaphase II, anaphase II, and telophase II (see Figure 26.1e–h). These phases are similar to those that occur during mitosis; the centromeres split, and the sister chromatids separate and move toward opposite poles of the cell.

In summary, meiosis I begins with a diploid parent cell and ends with two daughter cells, each with the haploid number of chromosomes. During meiosis II, each haploid cell formed during meiosis I divides, and the net result is four genetically different haploid cells.

✓ Distinguish between haploid (n) and diploid ($2n$) cells.
✓ What are homologous chromosomes?
✓ Prepare a table to compare meiosis with mitosis. (HINT: To review mitosis, refer to Figure 2.19 on page 46.)

MALE REPRODUCTIVE SYSTEM

Objective

- Describe the structure and functions of the organs of the male reproductive system.

The organs of the male reproductive system are the testes, a system of ducts (including the ductus deferens, ejaculatory ducts, and urethra), accessory sex glands (seminal vesicles, prostate gland, and bulbourethral gland), and several supporting structures, including the scrotum and the penis (Figure 26.3). The

Figure 26.3 / Male organs of reproduction and surrounding structures.

Reproductive organs are adapted for producing new individuals and passing on genetic material from one generation to the next.

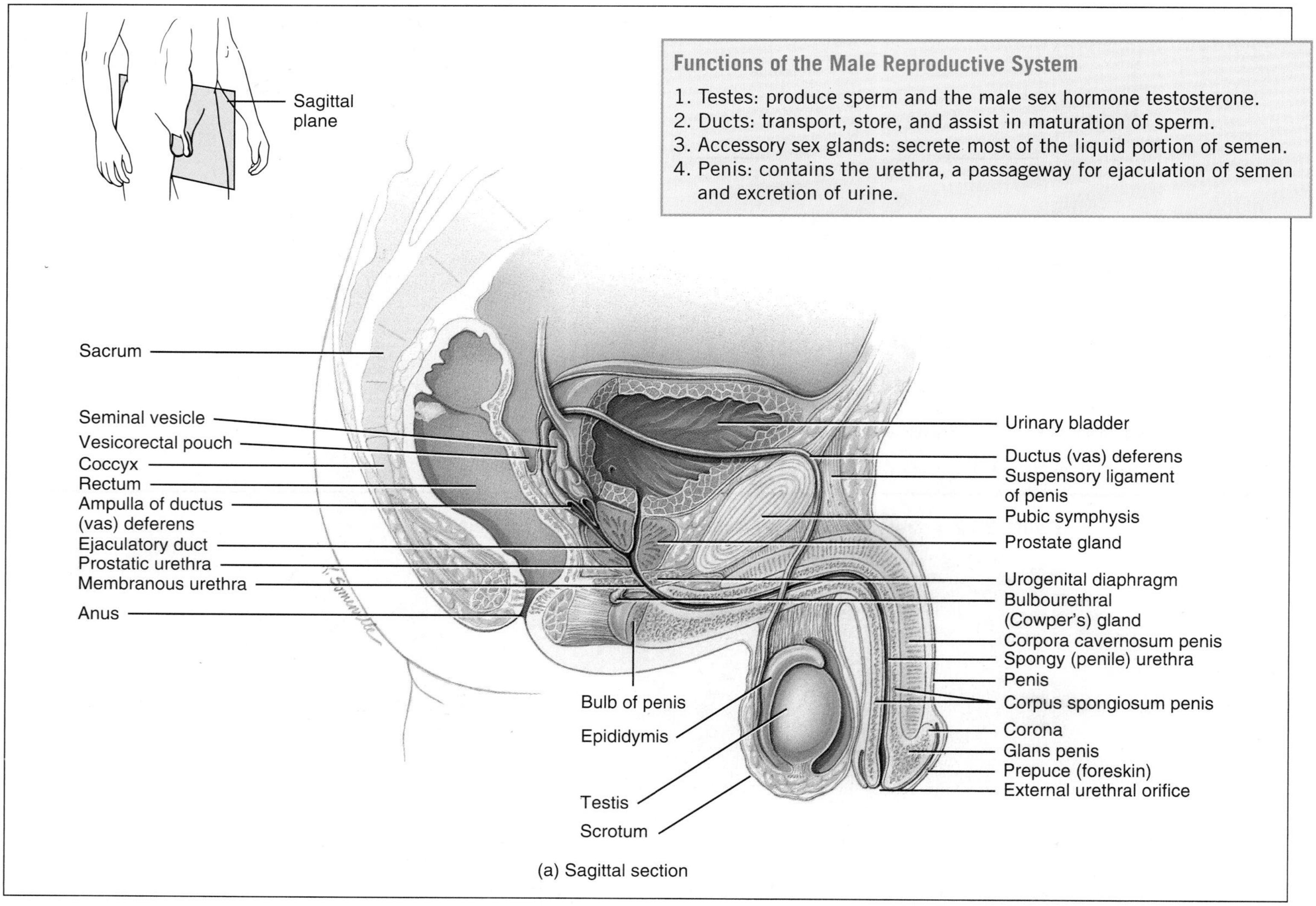

Functions of the Male Reproductive System

1. Testes: produce sperm and the male sex hormone testosterone.
2. Ducts: transport, store, and assist in maturation of sperm.
3. Accessory sex glands: secrete most of the liquid portion of semen.
4. Penis: contains the urethra, a passageway for ejaculation of semen and excretion of urine.

(a) Sagittal section

testes (male gonads) produce sperm and secrete hormones. A system of ducts transports and stores sperm, assists in their maturation, and conveys them to the exterior. Semen contains sperm plus the secretions provided by the accessory sex glands.

Scrotum

The **scrotum** (SKRŌ-tum; = bag), the supporting structure for the testes, is a sac consisting of loose skin and superficial fascia that hangs from the root (attached portion) of the penis (Figure 26.3a). Externally, the scrotum looks like a single pouch of skin separated into lateral portions by a median ridge called the **raphe** (RĀ-fē; = seam); internally, the scrotal septum divides the scrotum into two sacs, each containing a single testis (Figure 26.4). The septum consists of superficial fascia and muscle tissue called the **dartos muscle** (DAR-tōs; = skinned), which consists of bundles of smooth muscle fibers; the dartos muscle is also found in the subcutaneous tissue of the scrotum and is directly continuous with the subcutaneous tissue of the abdominal wall. When it contracts, the dartos muscle causes wrinkling of the skin of the scrotum.

The location of the scrotum and the contraction of its muscle fibers regulate the temperature of the testes. A temperature about 2–3°C below core body temperature, which is required for normal sperm production, is maintained within the scrotum because it is outside the pelvic cavity. The **cremaster muscle** (krē-MAS-ter; = suspender), a small band of skeletal muscle in the spermatic cord that is a continuation of the internal oblique muscle, elevates the testes upon exposure to cold (and during sexual arousal). This action moves the testes closer to the pelvic cavity, where they can absorb body heat. Exposure to warmth reverses the process. The dartos muscle also contracts in response to cold and relaxes in response to warmth.

The blood supply of the scrotum is derived from the internal pudendal branch of the internal iliac artery, the cremasteric branch of the inferior epigastric artery, and the external pudendal artery from the femoral artery. The scrotal veins follow the arteries.

The scrotal nerves are derived from the pudendal nerve, posterior cutaneous nerve of the thigh, and ilioinguinal nerves.

Testes

The **testes** (TES-tēz), or **testicles,** are paired oval glands measuring about 5 cm (2 in.) long and 2.5 cm (1 in.) in diameter (Figure 26.5 on page 787). Each **testis** (singular) weighs 10–15 grams. The testes develop near the kidneys, in the posterior portion of the abdomen, and they usually begin their descent into the scrotum through the inguinal canals (passageways in the anterior abdominal wall; see Figure 26.4) during the latter half of the seventh month of fetal development. A serous membrane called the **tunica vaginalis** (*tunica* = sheath), which is derived

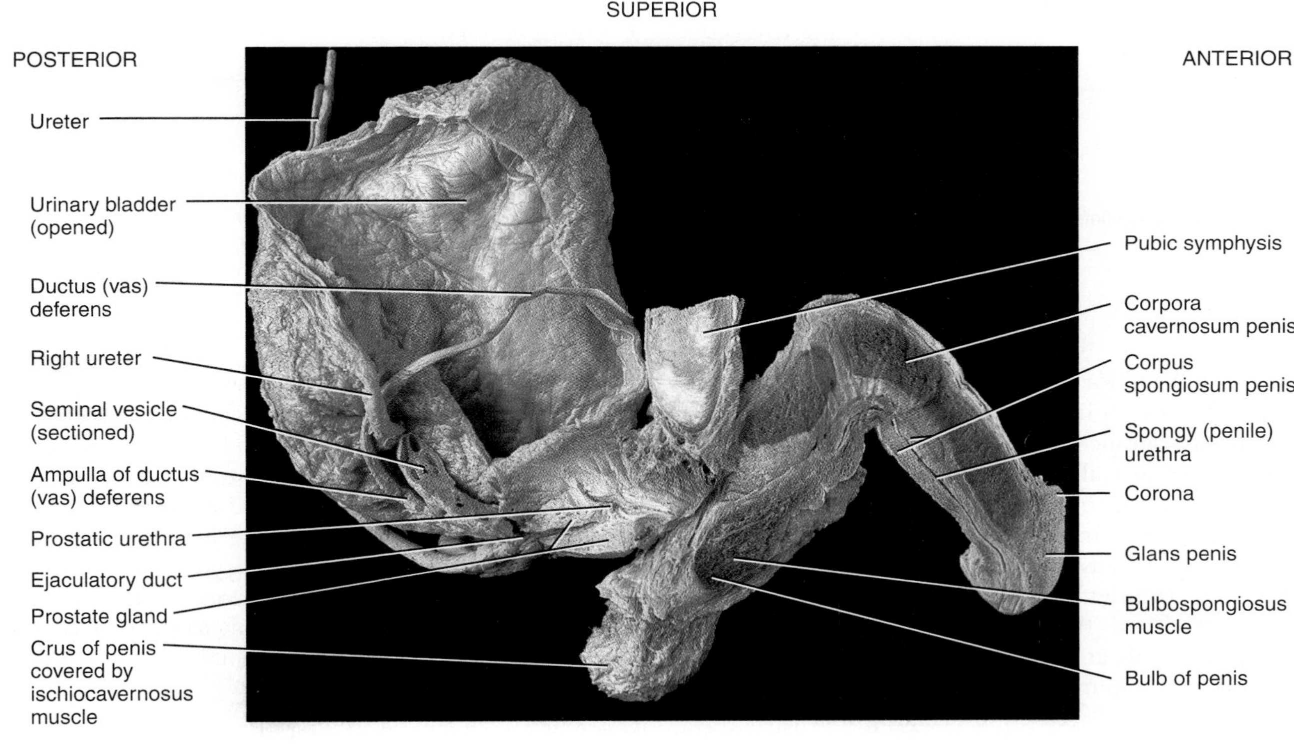

(b) Sagittal dissection

What are the groups of reproductive organs in males, and what are the functions of each group?

Figure 26.4 / The scrotum, the supporting structure for the testes.

The scrotum consists of loose skin and superficial fascia and supports the testes.

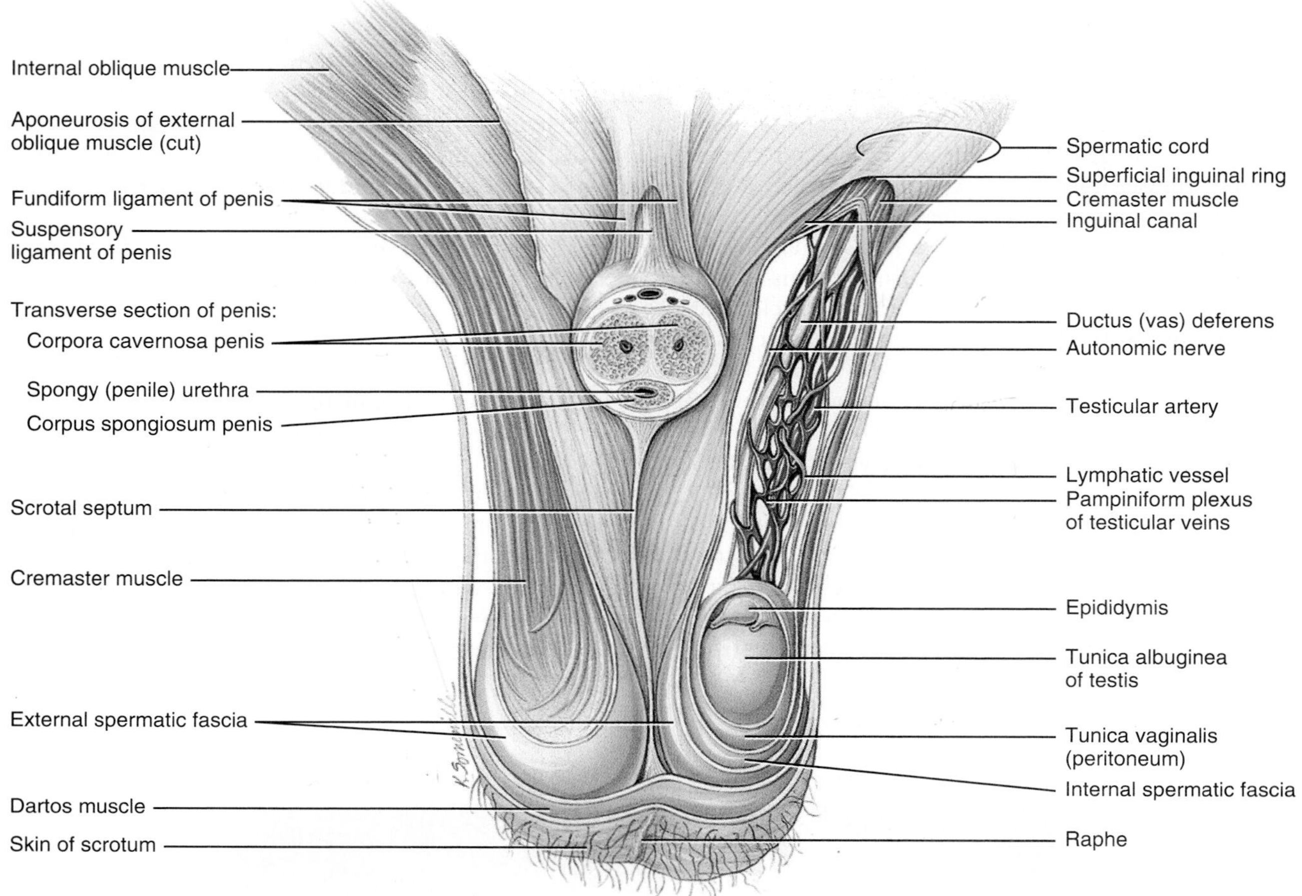

Anterior view of scrotum and testes and transverse section of penis

Which muscles help regulate the temperature of the testes?

from the peritoneum and forms during the descent of the testes, partially covers the testes. Internal to the tunica vaginalis is a dense white fibrous capsule composed of dense irregular connective tissue, the **tunica albuginea** (al′-byū-JIN-ē-a; *albu-* = white); it extends inward, forming septa that divide each testis into a series of internal compartments called **lobules.** Each of the 200–300 lobules contains one to three tightly coiled tubules, the **seminiferous tubules** (*semin-* = seed; *fer-* = to carry), where sperm are produced (Figure 26.6 on page 788).

Spermatogenic cells are cells at any of the sperm-forming stages and are located within seminiferous tubules. Sperm production begins in stem cells called **spermatogonia** (sper′-ma-tō-GŌ-nē-a; *-gonia* = offspring) that line the periphery of the seminiferous tubules (Figure 26.6). These cells develop from **primordial germ cells** (*primordi-* = primitive or early form) that arise from yolk sac endoderm and enter the testes early in development. In the embryonic testes, the primordial germ cells differentiate into spermatogonia, which remain dormant during childhood. At puberty, they begin to undergo mitosis, then meiosis, and finally differentiation to eventually produce sperm. Toward the lumen of the tubule are layers of progressively more mature cells. In order of advancing maturity, these are primary spermatocytes, secondary spermatocytes, spermatids, and sperm. By the time a **sperm cell,** or **spermatozoon** (sper′-ma-tō-ZŌ-on; *-zun* = life) has nearly reached maturity, it is released into the lumen of the seminiferous tubule. (The plural terms are **sperm** and **spermatozoa.**)

Embedded among the spermatogenic cells in the tubules are large **Sertoli cells,** or *sustentacular cells* (sus′-ten-TAK-yū-lar), which extend from the basement membrane to the lumen of the tubule. Just internal to the basement membrane, neighboring Sertoli cells are joined to one another by tight junctions that

Figure 26.5 / Internal and external anatomy of a testis.

The testes are the male gonads, which produce haploid sperm.

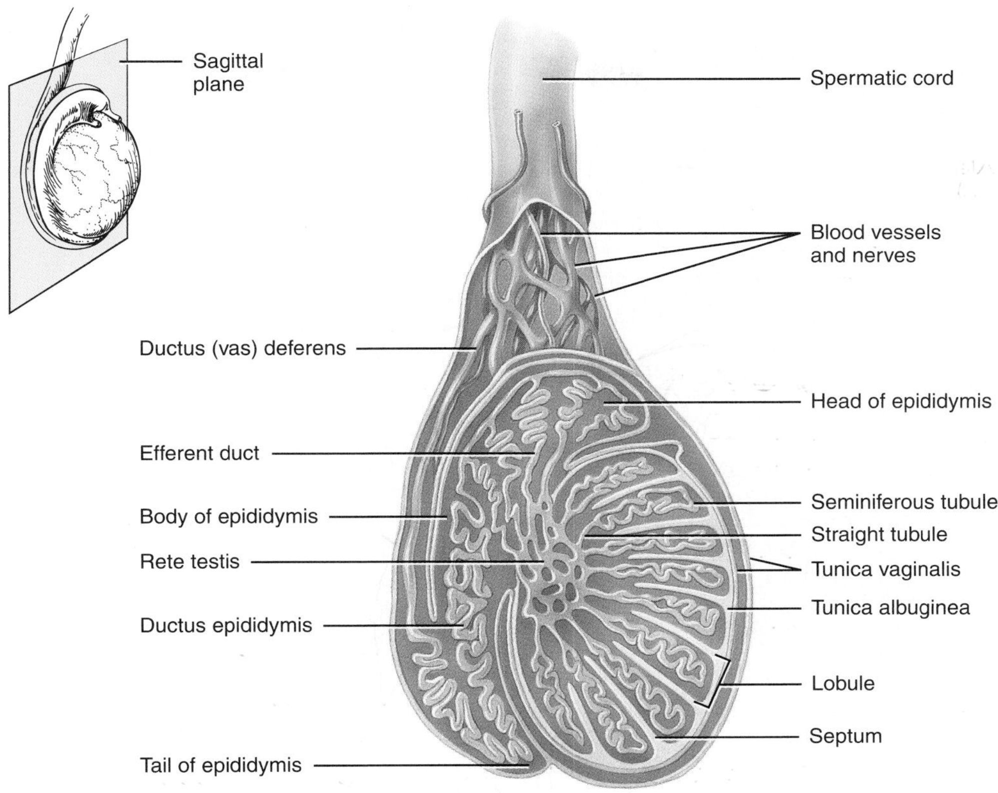

(a) Sagittal section of a testis showing seminiferous tubules

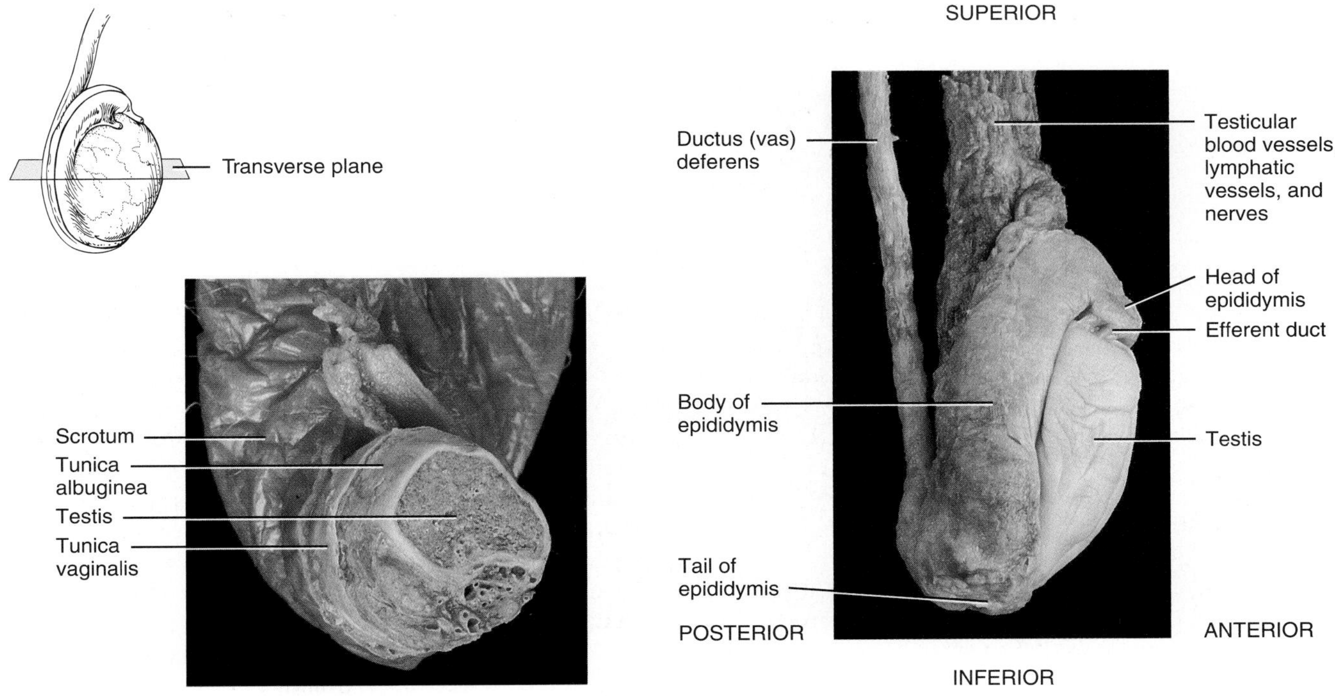

(b) Transverse section

(c) Testis and associated structures (lateral view)

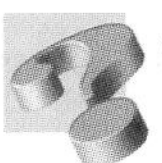

What tissue layers cover and protect the testes?

Figure 26.6 / Microscopic anatomy of the seminiferous tubules and stages of sperm production (spermatogenesis). Arrows in (b) indicate the progression of spermatogenic cells from least mature to most mature. The (n) and ($2n$) refer to haploid and diploid chromosome number, respectively.

Spermatogenesis occurs in the seminiferous tubules of the testes.

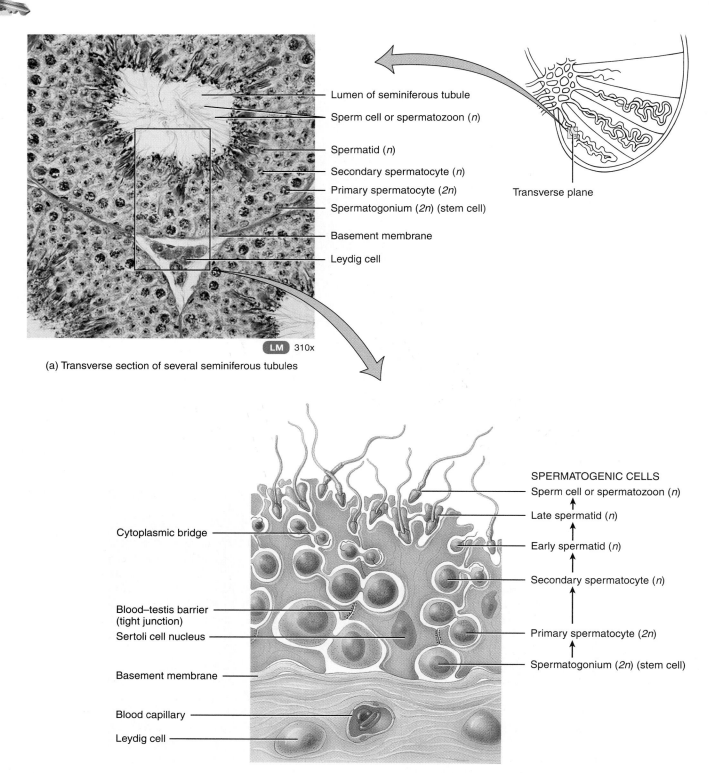

(a) Transverse section of several seminiferous tubules

(b) Transverse section of a portion of a seminiferous tubule

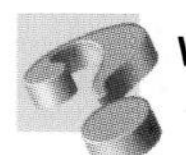

Which cells produce testosterone?

form the **blood–testis barrier.** To reach the developing gametes, substances must first pass through the Sertoli cells. By isolating the spermatogenic cells from the blood, the barrier prevents an immune response against the spermatogenic cell's surface antigens, which are recognized as foreign by the immune system.

Sertoli cells support and protect developing spermatogenic cells; nourish spermatocytes, spermatids, and sperm; phagocytize excess spermatid cytoplasm as development proceeds; and control movements of spermatogenic cells and the release of sperm into the lumen of the seminiferous tubule. They also produce fluid for sperm transport and secrete the hormone inhibin, which decreases the rate of spermatogenesis.

In the spaces between adjacent seminiferous tubules are clusters of cells called **Leydig cells,** or *interstitial endocrinocytes* (Figure 26.6a). These cells secrete testosterone, the most important androgen (male sex hormone).

Cryptorchidism

The condition in which the testes do not descend into the scrotum is called **cryptorchidism** (krip-TOR-ki-dizm; *crypt-* = hidden; *orchid* = testis); it occurs in about 3% of full-term infants and about 30% of premature infants. Untreated bilateral cryptorchidism results in sterility because the cells involved in the initial stages of spermatogenesis are destroyed by the higher temperature of the pelvic cavity. The chance of testicular cancer is 30–50 times greater in cryptorchid testes. The testes of about 80% of boys with cryptorchidism will descend spontaneously during the first year of life. When the testes remain undescended, the condition can be corrected surgically, ideally before 18 months of age. ■

The Process of Spermatogenesis

The process by which the seminiferous tubules of the testes produce haploid sperm is called **spermatogenesis** (sper′-ma-tō-JEN-e-sis; *genesis* = beginning process or production). In humans, this process takes about 65–75 days. Spermatogenesis begins in the spermatogonia, which contain the diploid ($2n$) chromosome number (Figure 26.7). Spermatogonia are *stem cells* because when they undergo mitosis, some of the daughter cells remain near the basement membrane of the seminiferous tubule in an undifferentiated state to serve as a reservoir of cells for future mitosis and subsequent sperm production. The rest of the daughter cells lose contact with the basement membrane, undergo developmental changes, and differentiate into **primary spermatocytes** (SPER-ma-tō-sītz′). Primary spermatocytes, like spermatogonia, are diploid ($2n$); that is, they have 46 chromosomes.

Each primary spermatocyte enlarges before dividing. Then two nuclear divisions occur as part of meiosis (Figure 26.7). In meiosis I, DNA replicates, homologous pairs of chromosomes line up at the metaphase plate, and crossing-over occurs. Then, the meiotic spindle forms and pulls one (duplicated) chromosome of each pair to an opposite pole of the dividing cell. This

Figure 26.7 / Events in spermatogenesis. Diploid cells ($2n$) have 46 chromosomes; haploid cells (n) have 23 chromosomes.

Spermiogenesis involves the maturation of spermatids into sperm.

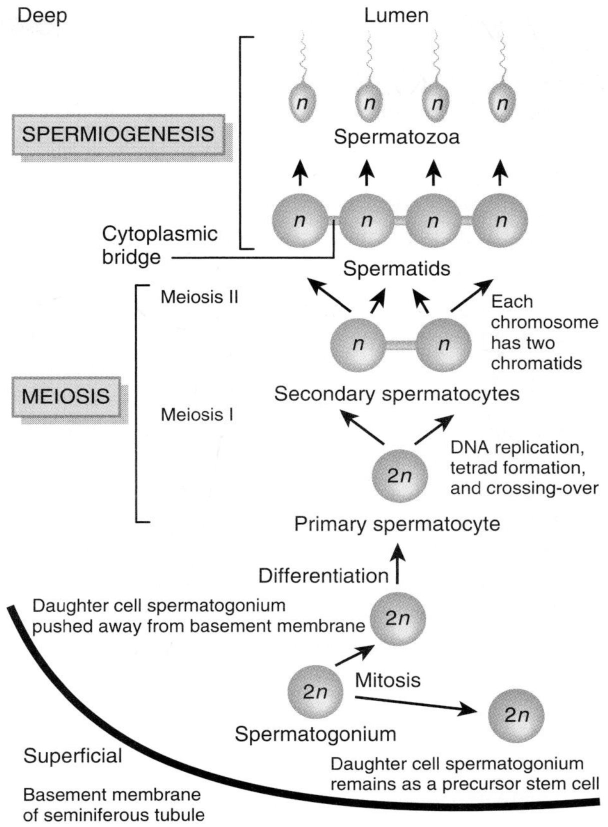

Deep
Lumen
SPERMIOGENESIS
Spermatozoa
Cytoplasmic bridge
Spermatids
Meiosis II
Each chromosome has two chromatids
MEIOSIS
Secondary spermatocytes
Meiosis I
DNA replication, tetrad formation, and crossing-over
Primary spermatocyte
Differentiation
Daughter cell spermatogonium pushed away from basement membrane
Mitosis
Spermatogonium
Superficial
Daughter cell spermatogonium remains as a precursor stem cell
Basement membrane of seminiferous tubule

What is "reduced" during meiosis I?

random assortment of maternally derived chromosomes and paternally derived chromosomes toward opposite poles is another reason for genetic variation among gametes. The cells formed by meiosis I are called **secondary spermatocytes,** and each cell has 23 chromosomes—the haploid number. Each chromosome within a secondary spermatocyte, however, is made up of two chromatids (two copies of the DNA) still attached by a centromere.

In meiosis II, no replication of DNA occurs. The chromosomes line up in single file along the metaphase plate, and the chromatids of each chromosome separate from each other. The cells resulting from meiosis II are called **spermatids;** each spermatid is haploid. A primary spermatocyte therefore produces four spermatids through two rounds of cell division (meiosis I and meiosis II).

A unique and very interesting process occurs during spermatogenesis. As the sperm cells proliferate, they fail to complete cytoplasmic separation (cytokinesis). The four daughter cells remain in contact via cytoplasmic bridges through their entire development (see Figures 26.6b and 26.7). This pattern of development most likely accounts for the synchronized production of sperm in any given area of seminiferous tubule. It may have survival value in that half of the sperm contain an X chromosome and half contain a Y chromosome. The larger X chromosome may carry genes needed for spermatogenesis that are lacking on the smaller Y chromosome.

The final stage of spermatogenesis, called **spermiogenesis** (sper′-mē-ō-JEN-e-sis), is the maturation of spermatids into sperm. Because no cell division occurs in spermiogenesis, each spermatid develops into a single **sperm cell.** The release of a sperm cell from its connection to a Sertoli cell is known as **spermiation.** Sperm then enter the lumen of the seminiferous tubule and flow toward ducts of the testes.

Sperm

Sperm mature at the rate of about 300 million per day and, once ejaculated, most probably do not survive more than 48 hours within the female reproductive tract. A sperm cell consists of structures highly adapted for reaching and penetrating a secondary oocyte: a head, a midpiece, and a tail (Figure 26.8). The **head** contains the nuclear material (DNA) and an **acrosome** (*acro-* = atop), a vesicle that contains hyaluronidase and proteinases, enzymes that aid penetration of the sperm cell into a secondary oocyte. Numerous mitochondria in the **midpiece** carry on the metabolism that provides ATP for locomotion. The **tail,** a typical flagellum, propels the sperm cell along its way.

Figure 26.8 / A sperm cell (spermatozoon).

About 300 million sperm mature each day.

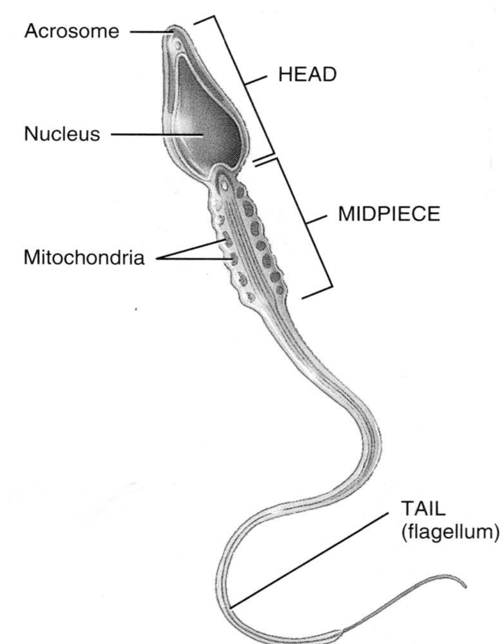

Parts of a sperm cell

What are the functions of each part of a sperm cell?

- ✓ Describe the function of the scrotum in protecting the testes from temperature fluctuations.
- ✓ Describe the internal structure of a testis. Where are sperm cells produced? What are the functions of Sertoli cells and Leydig cells?
- ✓ Describe the principal events of spermatogenesis.
- ✓ Identify the principal parts of a sperm cell, and list the functions of each.

Reproductive System Ducts in Males

Ducts of the Testis

After their release into the lumen of the very convoluted seminiferous tubules, sperm and fluid are propelled toward and enter a series of very short ducts called **straight tubules** by pressure generated by the continual release of sperm and fluid secreted by Sertoli cells. The straight tubules lead to a network of ducts in the testis called the **rete testis** (RĒ-tē; = network) (see Figure 26.5a). From the rete testis, sperm move into a series of coiled **efferent ducts** in the epididymis that empty into a single tube called the **ductus epididymis.**

Epididymis

The **epididymis** (ep′-i-DID-i-mis; *epi-* = over; *-didymis* = testis) is a comma-shaped organ about 4 cm (1.5 in.) long that lies along the posterior border of each testis (see Figure 26.5a). The plural is **epididymides** (ep′-i-did-Ī M-i-dēs). Each epididymis consists mostly of the tightly coiled **ductus epididymis.** The larger, superior portion of the epididymis, the **head,** is where the efferent ducts from the testis join the ductus epididymis. The **body** is the narrow midportion of the epididymis, and the **tail** is the smaller, inferior portion. At its distal end, the tail of the epididymis continues as the ductus (vas) deferens (discussed shortly).

The ductus epididymis is a tightly coiled structure that would measure about 6 m (20 ft) in length if it were straightened out. It is lined with pseudostratified columnar epithelium and encircled by layers of smooth muscle (Figure 26.9). The free surfaces of the columnar cells contain long, branching microvilli called **stereocilia** that increase surface area for the reabsorption of degenerated sperm.

Functionally, the ductus epididymis is the site where sperm motility increases over a 10–14 day period. The ductus epididymis also stores sperm and helps propel them by peristaltic contraction of its smooth muscle into the ductus (vas) deferens. Sperm may remain in storage in the ductus epididymis for a month or more.

Ductus Deferens

Within the tail of the epididymis, the ductus epididymis becomes less convoluted, and its diameter increases; beyond this point, the duct is referred to as the **ductus deferens** or **vas deferens** (see Figure 26.5a). The ductus deferens, which is about 45 cm (18 in.) long, ascends along the posterior border of the epididymis, passes through the inguinal canal (see Figure 26.4), and enters the pelvic cavity; there it loops over the ureter and passes

Figure 26.9 / Histology of the ductus epididymis.

Stereocilia increase the surface area for reabsorption of degenerated sperm.

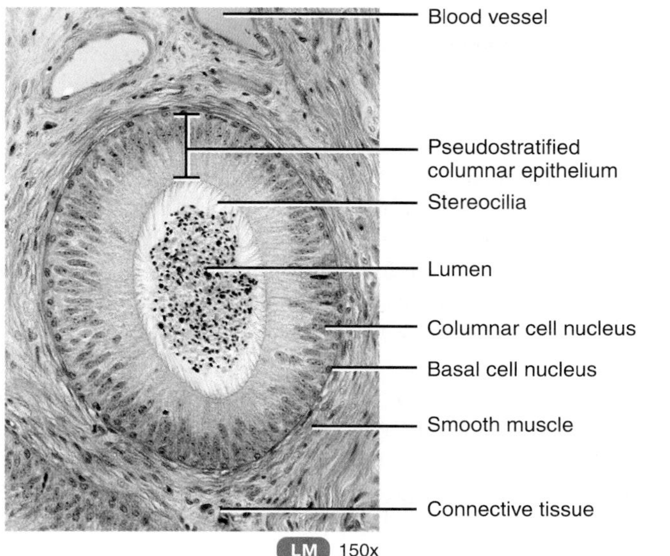

Transverse section of the ductus epididymis

What are the functions of the ductus epididymis?

over the side and down the posterior surface of the urinary bladder (see Figure 26.3a). The dilated terminal portion of the ductus deferens is known as the **ampulla** (am-PŪL-la; = little jar) (see Figure 26.11). The ductus deferens is lined with pseudostratified columnar epithelium and contains a heavy coat of three layers of muscle; the inner and outer layers are longitudinal, and the middle layer is circular (Figure 26.10).

Functionally, the ductus deferens stores sperm; they can remain viable here for up to several months. The ductus deferens also conveys sperm from the epididymis toward the urethra by peristaltic contractions of the muscular coat. Sperm that are not ejaculated are ultimately reabsorbed.

The principal method for sterilization of males is a **vasectomy** (vas-EK-tō-mē; *tome* = incision), in which a portion of each ductus deferens is removed. An incision is made in the posterior side of the scrotum, the ducts are located and cut, each is tied in two places, and the portion between the ties is removed. Although sperm production continues in the testes, sperm can no longer reach the exterior. The sperm degenerate and are destroyed by phagocytosis. Because the blood vessels are not cut, testosterone levels in the blood remain normal, so vasectomy has no effect on sexual desire and performance. If done correctly, it is close to 100% effective. The procedure can be reversed, but the chance of regaining fertility is only 30–40%. ■

Figure 26.10 / Histology of the ductus (vas) deferens.

The ductus (vas) deferens enters the pelvic cavity through the inguinal canal.

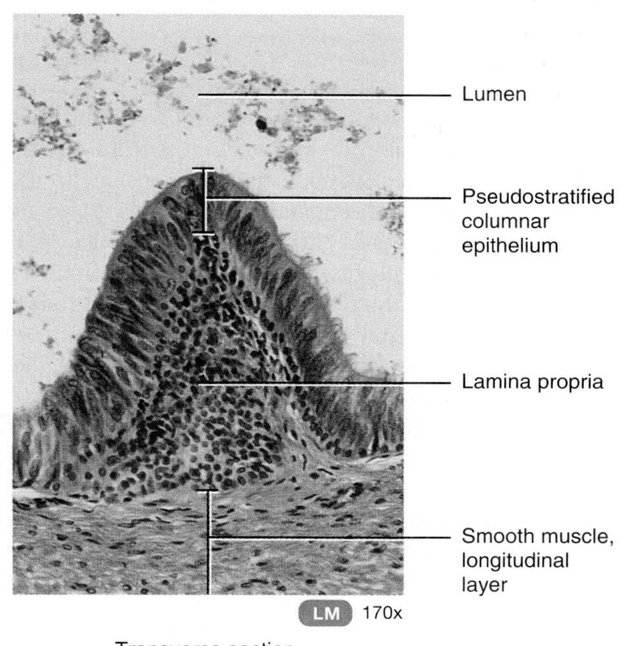

Transverse section

What is the function of the ductus (vas) deferens?

Ejaculatory Ducts

Each **ejaculatory duct** (e-JAK-yū-la-tō′-rē; *ejacul-* = to expel) is about 2 cm (1 in.) long and is formed by the union of the duct from the seminal vesicle and the ampulla of the ductus (vas) deferens (see Figure 26.11). The ejaculatory ducts form just above the base (superior portion) of the prostate gland and pass inferiorly and anteriorly through the prostate gland. They terminate in the prostatic urethra, where they eject sperm and seminal vesicle secretions just before **ejaculation,** the powerful propulsion of semen from the urethra to the exterior. They also transport and eject secretions of the seminal vesicles (described shortly).

Urethra

In males, the **urethra** is the shared terminal duct of the reproductive and urinary systems; it serves as a passageway for both semen and urine. The urethra, which is about 20 cm (8 in.) long, passes through the prostate gland, the urogenital diaphragm, and the penis, and is subdivided into three parts (see Figures 26.3a and 26.11). The **prostatic urethra** is 2–3 cm (1 in.) long and passes through the prostate gland. As this duct continues inferiorly, it passes through the urogenital diaphragm (a muscular partition between the two ischial and pubic rami; see Figure 10.14 on page 291), where it is known as the membranous urethra. The **membranous urethra** is about 1 cm (0.5 in.) in length. As this duct passes through the corpus spongiosum of the penis, it is known as the **spongy (penile) urethra,** which is about 15–20 cm (6–8 in.) long. The spongy urethra ends at the **external urethral orifice.** The histology of the male urethra may be reviewed on page 774.

Spermatic Cord

The **spermatic cord** is a supporting structure of the male reproductive system that ascends out of the scrotum (see Figure 26.4). It consists of the ductus (vas) deferens as it ascends through the scrotum, the testicular artery, autonomic nerves, veins that drain the testes and carry testosterone into the circulation (the pampiniform plexus), lymphatic vessels, and the cremaster muscle. The spermatic cord and ilioinguinal nerve pass through the **inguinal canal** (IN-gwin-al; = groin), an oblique passageway in the anterior abdominal wall just superior and parallel to the medial half of the inguinal ligament. The canal, which is about 4–5 cm (about 2 in.) long, originates at the **deep (abdominal) inguinal ring,** a slitlike opening in the aponeurosis of the transversus abdominis muscle; the canal ends at the **superficial (subcutaneous) inguinal ring** (see Figure 26.4), a somewhat triangular opening in the aponeurosis of the external oblique muscle. In females, the round ligament of the uterus and ilioinguinal nerve pass through the inguinal canal.

Inguinal Hernias

Because the inguinal region is a weak area in the abdominal wall, it is often the site of an **inguinal hernia**—a rupture or separation of a portion of the inguinal area of the abdominal wall resulting in the protrusion of a part of the small intestine. In an *indirect inguinal hernia,* a part of the small intestine protrudes through the deep inguinal ring and enters the scrotum. In a *direct inguinal hernia,* a portion of the small intestine pushes into the posterior wall of the inguinal canal, usually causing a localized bulging in the wall of the canal. Inguinal hernias are much more common in males than in females because the larger inguinal canals in males represent larger weak points in the abdominal wall. ■

- ✓ Which ducts transport sperm within the testes?
- ✓ Describe the location, structure, and functions of the ductus epididymis, ductus (vas) deferens, and ejaculatory duct.
- ✓ Give the locations of the three subdivisions of the male urethra.
- ✓ Trace the course of sperm through the system of ducts from the seminiferous tubules through the urethra.
- ✓ List the structures within the spermatic cord.

Accessory Sex Glands

Whereas the ducts of the male reproductive system store and transport sperm cells, the **accessory sex glands** secrete most of the liquid portion of semen. The accessory sex glands are the seminal vesicles, the prostate gland, and the bulbourethral glands.

Seminal Vesicles

The paired **seminal vesicles** (VES-i-kuls) are convoluted pouchlike structures, about 5 cm (2 in.) in length, lying posterior to and at the base of the urinary bladder anterior to the rectum (Figure 26.11). They secrete an alkaline, viscous fluid that contains fructose (a monosaccharide sugar), prostaglandins, and clotting proteins unlike those found in blood (discussed shortly). The alkaline nature of the fluid helps to neutralize the acidic environment of the male urethra and female reproductive tract that otherwise would inactivate and kill sperm. The fructose is used for ATP production by sperm. Prostaglandins contribute to sperm motility and viability and may also stimulate muscular contractions within the female reproductive tract. Fluid secreted by the seminal vesicles normally constitutes about 60% of the volume of semen.

Prostate Gland

The **prostate gland** (PROS-tāt) is a single, doughnut-shaped gland about the size of a chestnut that lies inferior to the urinary bladder and surrounds the prostatic urethra (Figure 26.11). The prostate secretes a milky, slightly acidic fluid (pH about 6.5) that contains (1) *citric acid,* which can be used by sperm for ATP production via the Krebs cycle; (2) acid phosphatase (the function of which is unknown); and (3) several proteolytic enzymes, such as *prostate-specific antigen (PSA),* pepsinogen, lysozyme, amylase, and hyaluronidase.

Secretions of the prostate gland enter the prostatic urethra through many prostatic ducts. Prostatic secretions make up about 25% of the volume of semen and contribute to sperm motility and viability. The prostate gland slowly increases in size from birth to puberty, and then expands rapidly. The size attained by age 30 remains stable until about age 45, when further enlargement may occur.

Bulbourethral Glands

The paired **bulbourethral glands** (bul′-bō-yū-RĒ-thral), or **Cowper's glands,** each about the size of a pea, lie inferior to the prostate gland on either side of the membranous urethra within the urogenital diaphragm; their ducts open into the spongy urethra (Figure 26.11). During sexual arousal, the bulbourethral glands secrete an alkaline substance that protects the passing sperm by neutralizing acids from urine in the urethra. At the same time, they also secrete mucus that lubricates the end of the penis and the lining of the urethra, thereby decreasing the number of sperm damaged during ejaculation.

Semen

Semen (= seed) is a mixture of sperm and **seminal fluid,** a liquid that consists of the secretions of the seminiferous tubules, seminal vesicles, prostate gland, and bulbourethral glands. The volume of semen in a typical ejaculation is 2.5–5 mL, with a sperm count (concentration) of 50–150 million sperm/mL. A male whose sperm count falls below 20 million/mL is likely to be infertile. The very large number is required because only a tiny fraction ever reach the secondary oocyte.

Figure 26.11 / Locations of several accessory reproductive organs in males. The prostate gland, urethra, and penis have been sectioned to show internal details.

The male urethra has three subdivisions: the prostatic, membranous, and spongy (penile) urethra.

Urinary bladder
Right ductus (vas) deferens
Left ureter
Hip bone (cut)
Prostate gland
Ampulla of ductus (vas) deferens
Seminal vesicle
Prostatic urethra
Seminal vesicle duct
Membranous urethra
Ejaculatory duct
Urogenital diaphragm
Crus of penis
Bulbourethral (Cowper's) gland
Bulb of penis
Corpus spongiosum penis
Corpora cavernosa penis
Spongy (penile) urethra

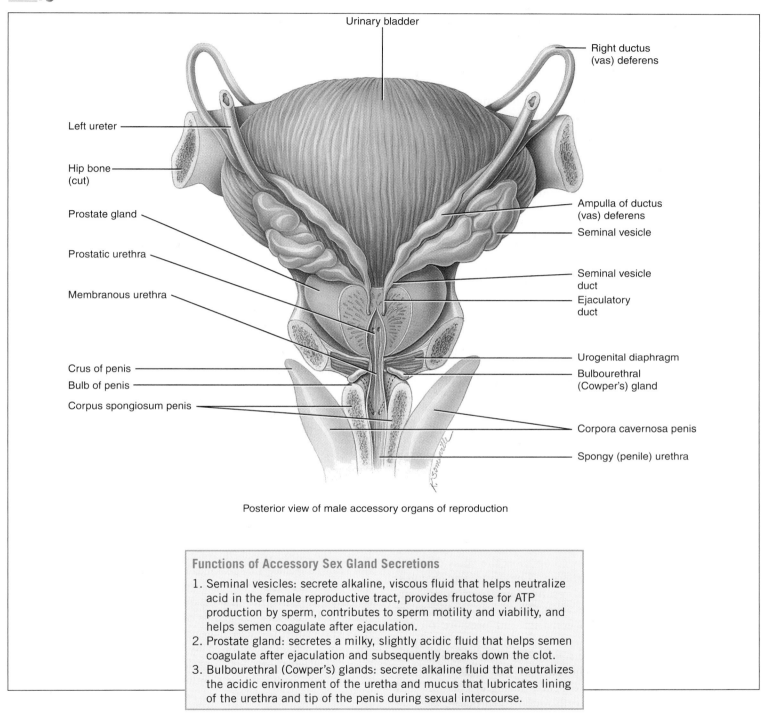

Posterior view of male accessory organs of reproduction

Functions of Accessory Sex Gland Secretions

1. Seminal vesicles: secrete alkaline, viscous fluid that helps neutralize acid in the female reproductive tract, provides fructose for ATP production by sperm, contributes to sperm motility and viability, and helps semen coagulate after ejaculation.
2. Prostate gland: secretes a milky, slightly acidic fluid that helps semen coagulate after ejaculation and subsequently breaks down the clot.
3. Bulbourethral (Cowper's) glands: secrete alkaline fluid that neutralizes the acidic environment of the uretha and mucus that lubricates lining of the urethra and tip of the penis during sexual intercourse.

What accessory sex gland contributes the majority of the seminal fluid?

Despite the slight acidity of prostatic fluid, semen has a slightly alkaline pH of 7.2–7.7 due to the higher pH and large volume of fluid from the seminal vesicles. The prostatic secretion gives semen a milky appearance, whereas fluids from the seminal vesicles and bulbourethral glands give it a sticky consistency. Seminal fluid provides sperm with a transportation medium and nutrients, and it neutralizes the hostile acidic environment of the male's urethra and of the vagina. Semen also contains an antibiotic, *seminalplasmin*, that can destroy certain bacteria. Seminalplasmin may help to control the abundance of bacteria in both semen and the lower female reproductive tract.

Once ejaculated, liquid semen coagulates within 5 minutes due to the presence of clotting proteins from the seminal vesicles. The functional role of semen coagulation is not known, but the proteins involved are different from those that cause blood coagulation. After about 10–20 minutes, semen reliquefies because prostate-specific antigen and other proteolytic enzymes produced by the prostate gland break down the clot. Abnormal or delayed liquefaction of clotted semen may cause complete or partial immobilization of sperm, thereby inhibiting their movement through the cervix of the uterus.

Penis

The **penis** contains the urethra and is a passageway for the ejaculation of semen and the excretion of urine. It is cylindrical in shape and consists of a body, root, and glans penis. The **body** of the penis is composed of three cylindrical masses of tissue, each surrounded by fibrous tissue called the **tunica albuginea** (Figure 26.12b). The paired dorsolateral masses are called the **corpora cavernosa penis** (*corpora* = main bodies; *cavernosa* = hollow) or just simply, the *corpora cavernosa;* the smaller midventral mass, the **corpus spongiosum penis** or just simply, the *corpus spongiosum*, contains the spongy urethra and functions in keeping the spongy urethra open during ejaculation. All three masses are enclosed by fascia and skin and consist of erectile tissue permeated by blood sinuses.

Upon sexual stimulation, which may be visual, tactile, auditory, olfactory, or from the imagination, the arteries supplying the penis dilate, and large quantities of blood enter the blood sinuses. Expansion of these spaces compresses the veins draining the penis, so blood outflow is slowed. These vascular changes, due to a parasympathetic reflex, result in an **erection,** the enlargement and stiffening of the penis. The penis returns to its flaccid state when the arteries constrict and pressure on the veins is relieved.

Ejaculation is a sympathetic reflex. As part of the reflex, the smooth muscle sphincter at the base of the urinary bladder closes. As a result, urine is not expelled during ejaculation, and semen does not enter the urinary bladder. Even before ejaculation occurs, peristaltic contractions in the ampulla of the ductus deferens, seminal vesicles, ejaculatory ducts, and prostate gland propel semen into the penile portion of the urethra (spongy urethra). Typically, this leads to **emission** (ē-MISH-un), the discharge of a small volume of semen before ejaculation. Emission may also occur during sleep (nocturnal emission).

The **root** of the penis is the attached portion (proximal portion) and consists of the **bulb of the penis,** the expanded portion of the base of the corpus spongiosum penis, and the **crura of the penis** (singular is **crus** = resembling a leg), the two separated and tapered portions of the corpora cavernosa penis (Figure 26.12a). The bulb of the penis is attached to the inferior surface of the urogenital diaphragm and enclosed by the bulbospongiosus muscle. Each crus of the penis is attached to the ischial and inferior pubic rami and is surrounded by the ischiocavernosus muscle (see Figure 10.14 on page 291). Contraction of these skeletal muscles aids ejaculation.

The distal end of the corpus spongiosum penis is a slightly enlarged, acorn-shaped region called the **glans penis;** its margin is the **corona.** The distal urethra enlarges within the glans penis and forms a terminal slitlike opening, the **external urethral orifice.** Covering the glans in an uncircumcised penis is the loosely fitting **prepuce** (PRĒ-pyūs), or **foreskin.** The weight of the penis is supported by two ligaments that are continuous with the fascia of the penis: the **fundiform ligament,** which arises from the inferior part of the linea alba, and the **suspensory ligament of the penis,** which arises from the pubic symphysis.

Circumcision (= to cut around) is a surgical procedure in which part or all of the prepuce is removed. It is usually performed just after delivery, 3–4 days after birth, or on the eighth day as part of a Jewish religious rite. Although some health-care professionals can find no medical justification for circumcision, others feel that it has benefits, such as a lower risk of urinary tract infections, protection against penile cancer, and possibly a lower risk for sexually transmitted diseases. ■

The penis has a very rich blood supply from the internal pudendal artery and the femoral artery. The veins drain into corresponding vessels.

The sensory nerves to the penis are branches from the pudendal and ilioinguinal nerves. The corpora have a parasympathetic and a sympathetic supply. Due to parasympathetic stimulation, the blood vessels dilate, increasing the flow of blood into the erectile tissue. The result is that blood is trapped within the penis and erection is maintained. At ejaculation, sympathetic stimulation causes the smooth muscle located in the walls of the ducts and accessory sex glands of the reproductive tract to contract and propel the semen along its course. The musculature of the penis, which is supplied by the pudendal nerve, also contracts at ejaculation. The muscles include the bulbospongiosus muscle, which overlies the bulb of the penis, the ischiocavernosus muscles on either side of the penis, and the superficial transverse perineus muscles on either side of the bulb of the penis (see Figure 10.14 on page 291).

- ✓ Briefly explain the locations and functions of the seminal vesicles, the prostate gland, and the bulbourethral (Cowper's) glands.
- ✓ What is semen? What is its function?
- ✓ How does an erection occur?

Figure 26.12 / Internal structure of the penis. The inset in (b) shows details of the skin and fasciae. (See Tortora, *A Photographic Atlas of the Human Body,* Figure 14.6.)

The penis contains the urethra, a pathway for the ejaculation of semen and the excretion of urine.

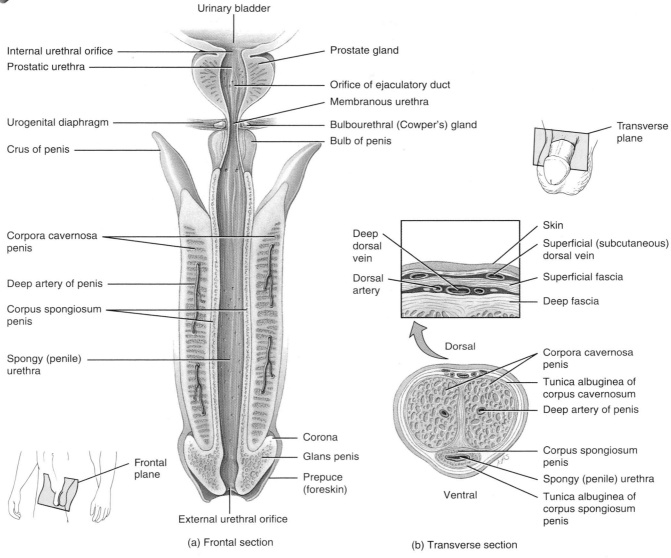

Urinary bladder
Internal urethral orifice
Prostatic urethra
Prostate gland
Orifice of ejaculatory duct
Membranous urethra
Urogenital diaphragm
Bulbourethral (Cowper's) gland
Bulb of penis
Crus of penis
Corpora cavernosa penis
Deep artery of penis
Corpus spongiosum penis
Spongy (penile) urethra
Corona
Glans penis
Prepuce (foreskin)
Frontal plane
External urethral orifice

(a) Frontal section

Transverse plane
Deep dorsal vein
Dorsal artery
Skin
Superficial (subcutaneous) dorsal vein
Superficial fascia
Deep fascia
Dorsal
Corpora cavernosa penis
Tunica albuginea of corpus cavernosum
Deep artery of penis
Corpus spongiosum penis
Spongy (penile) urethra
Ventral
Tunica albuginea of corpus spongiosum penis

(b) Transverse section

Which tissue masses form the erectile tissue in the penis, and why do they become rigid during sexual arousal?

FEMALE REPRODUCTIVE SYSTEM

Objective

- Describe the location, structure, and functions of the organs of the female reproductive system.

The organs of reproduction in females (Figure 26.13) include the ovaries, which produce secondary oocytes and hormones, such as progesterone and estrogens (the female sex hormones), inhibin, and relaxin; the uterine (Fallopian) tubes, or oviducts, which transport secondary oocytes and fertilized ova to the uterus; the uterus, in which embryonic and fetal development occur; the vagina; and external organs that constitute the vulva, or pudendum. The mammary glands also are considered part of the female reproductive system.

Ovaries

The **ovaries** (= egg receptacles) are paired glands that resemble unshelled almonds in size and shape; they are homologous to the testes. (Here *homologous* means that two organs have the same embryonic origin.) The ovaries, one on either side of the uterus, descend to the brim of the superior portion of the pelvic cavity during the third month of development. A series of ligaments holds them in position (Figure 26.14 on page 798). The **broad ligament** of the uterus (see also Figure 26.13b), which is itself part of the parietal peritoneum, attaches to the ovaries by a double-layered fold of peritoneum called the **mesovarium.** The **ovarian ligament** anchors the ovaries to the uterus, and the **suspensory ligament** attaches them to the pelvic wall. Each ovary contains a **hilus,** the point of entrance and exit for blood vessels and nerves and along which the mesovarium is attached.

Histology

Each ovary consists of the following parts (Figure 26.15 on page 799):

- The **germinal epithelium** is a layer of simple epithelium (low cuboidal or squamous) that covers the surface of the ovary and is continuous with the mesothelium that covers the mesovarium. The term *germinal epithelium* is a misnomer because it does not give rise to ova, although at one

Figure 26.13 / Organs of reproduction and surrounding structures in females.

The organs of reproduction in females include the ovaries, uterine (Fallopian) tubes, uterus, vagina, vulva, and mammary glands.

Functions of the Female Reproductive System

1. Ovaries: produce secondary oocytes and hormones, including progesterone and estrogens (female sex hormones), inhibin, and relaxin.
2. Uterine tubes: transport a secondary oocyte to the uterus and normally are the sites where fertilization occurs.
3. Uterus: site of implantation of a fertilized ovum, development of the fetus during pregnancy, and labor.
4. Vagina: receives the penis during sexual intercourse and is a passageway for childbirth
5. Mammary glands: synthesize, secrete, and eject milk for nourishment of the newborn.

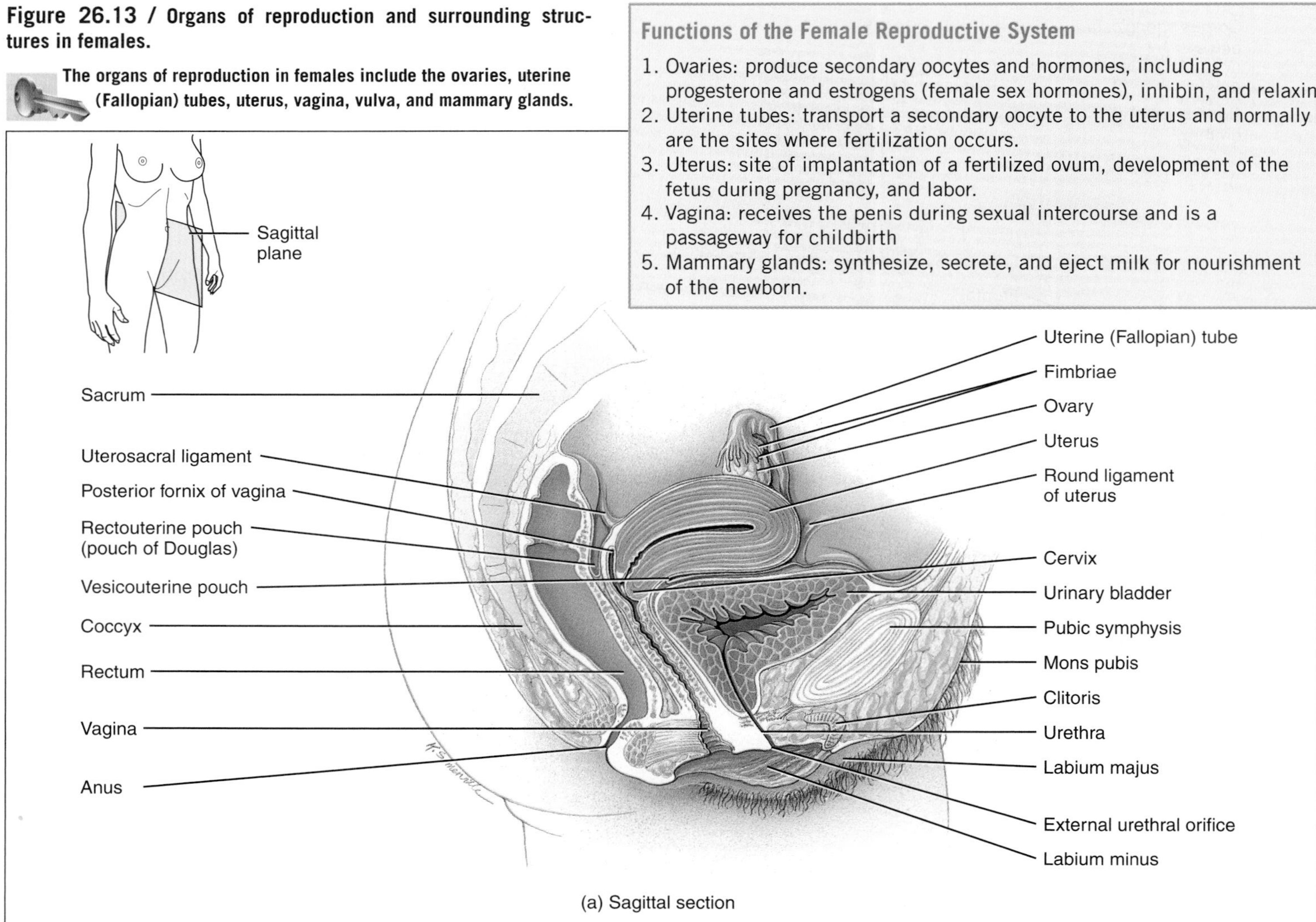

(a) Sagittal section

time people believed that it did. Now we know that the progenitors of ova arise from the endoderm of the yolk sac and migrate to the ovaries during embryonic development.

- The **tunica albuginea** is a whitish capsule of dense, irregular connective tissue immediately deep to the germinal epithelium.
- The **ovarian cortex** is a region just deep to the tunica albuginea that consists of dense connective tissue and contains ovarian follicles (described shortly).
- The **ovarian medulla** is a region deep to the ovarian cortex that consists of loose connective tissue and contains blood vessels, lymphatics, and nerves.
- **Ovarian follicles** (*folliculus* = little bag) lie in the cortex and consist of **oocytes** in various stages of development, plus the surrounding cells. When the surrounding cells form a single layer, they are called **follicular cells;** later in development, when they form several layers, they are referred to as **granulosa cells.** The surrounding cells nourish the developing oocyte and begin to secrete estrogens as the follicle grows larger.
- A **mature (Graafian) follicle** is a large, fluid-filled follicle that soon will rupture and expel a secondary oocyte, a process called **ovulation.**
- A **corpus luteum** (= yellow body) contains the remnants of an ovulated mature follicle. The corpus luteum produces progesterone, estrogens, relaxin, and inhibin until it degenerates and turns into fibrous tissue called a **corpus albicans** (= white body).

The ovarian blood supply is furnished by the ovarian arteries, which anastomose with branches of the uterine arteries. The ovaries are drained by the ovarian veins. On the right side they drain into the inferior vena cava, and on the left side they drain into the renal vein.

Sympathetic and parasympathetic nerve fibers to the ovaries terminate on the blood vessels and enter the substance of the ovaries.

Oogenesis

The formation of gametes in the ovaries is termed **oogenesis** (ō-ō-JEN-e-sis; ū- = egg), and like spermatogenesis it involves

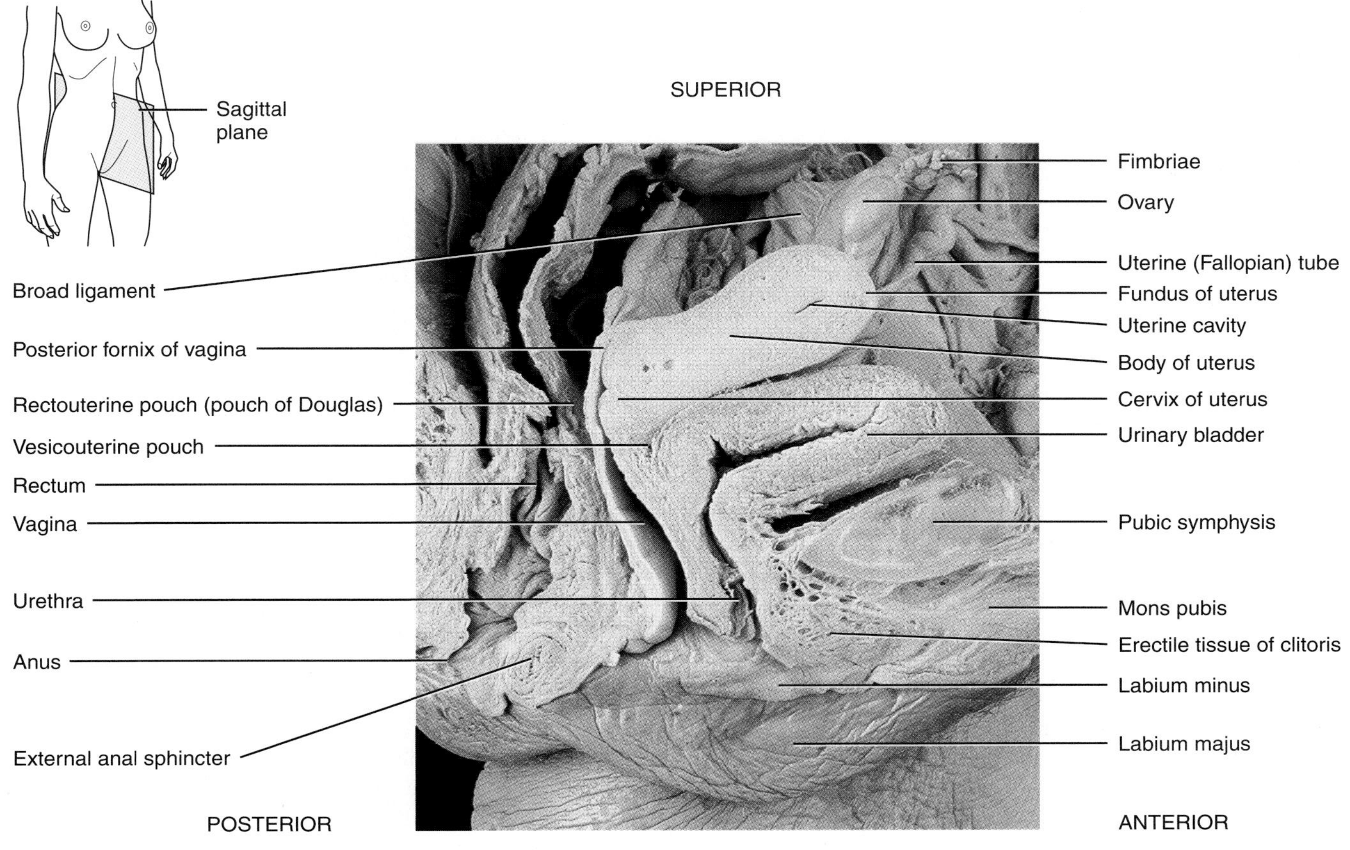

(b) Sagittal section

Which structures in males are homologous to the ovaries, the clitoris, the paraurethral glands, and the greater vestibular glands?

Figure 26.14 / Relative positions of the ovaries, the uterus, and the ligaments that support them.
(See Tortora, *A Photographic Atlas of the Human Body,* Figure 14.9.)

Ligaments holding the ovaries in position are the mesovarium, the ovarian ligament, and the suspensory ligament.

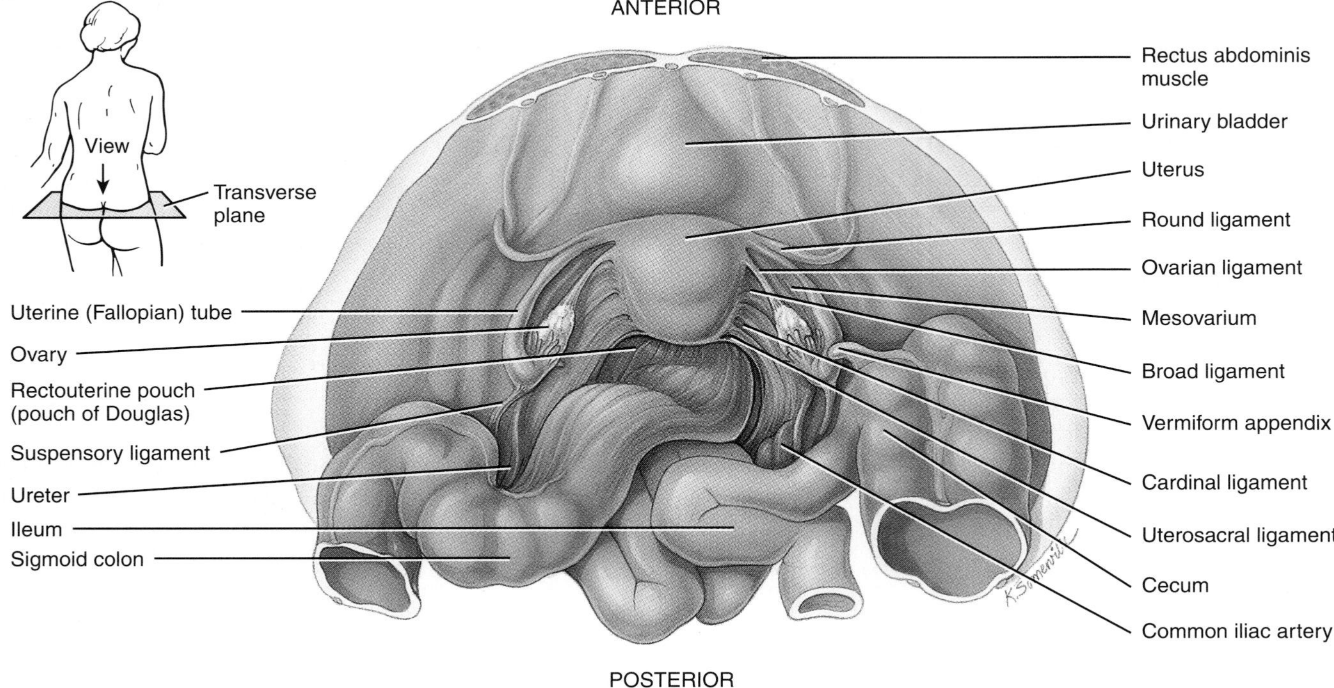

Superior view of transverse section

To which structures do the mesovarium, ovarian ligament, and suspensory ligament anchor the ovary?

meiosis. During early fetal development, primordial (primitive) germ cells migrate from the endoderm of the yolk sac to the ovaries. There, germ cells differentiate within the ovaries into **oogonia** (ō′-o-GŌ-nē-a; singular is **oogonium;** ō′-o-GŌ-nē-um), which are diploid ($2n$) cells that divide mitotically to produce millions of germ cells. Even before birth, most of these germ cells degenerate, a process known as **atresia** (a-TRĒ-zē-a). A few, however, develop into larger cells called **primary oocytes** (Ō-ō-sītz) that enter prophase of meiosis I during fetal development but do not complete that phase until after puberty. At birth, 200,000–2,000,000 oogonia and primary oocytes remain in each ovary. Of these, about 40,000 remain at puberty, and only 400 will mature and ovulate during a woman's reproductive lifetime; the remainder undergo atresia.

Each primary oocyte is surrounded by a single layer of follicular cells, and the entire structure is called a **primordial follicle** (Figure 26.16a). Although the stimulating mechanism is unclear, a few primordial follicles periodically start to grow, even during childhood. They become **primary follicles,** which are surrounded first by one layer of cuboidal-shaped follicular cells and then by six to seven layers of cuboidal and low-columnar cells called **granulosa cells.** As a follicle grows, it forms a clear glycoprotein layer, called the **zona pellucida** (pe-LŪ-si-da), between the primary oocyte and the granulosa cells. The innermost layer of granulosa cells becomes firmly attached to the zona pellucida and is called the **corona radiata** (*corona* = crown; *radiata* = radiation) (Figure 26.16b). The outermost granulosa cells rest on a basement membrane that separates them from the surrounding ovarian stroma; this outer region is called the **theca folliculi.** As a primary follicle continues to grow, the theca differentiates into two layers: (1) the **theca interna,** a vascularized internal layer of secretory cells, and (2) the **theca externa,** an outer layer of connective tissue cells. The granulosa cells begin to secrete follicular fluid, which builds up in a cavity called the antrum in the center of the follicle, which is now termed a **secondary follicle.** During childhood, primordial and developing follicles continue to undergo atresia.

Each month after puberty, the gonadotropic hormones secreted by the anterior pituitary gland stimulate the resumption

Figure 26.15 / Histology of the ovary. The arrows indicate the sequence of developmental stages that occur as part of the maturation of an ovum during the ovarian cycle.

The ovaries are the female gonads; they produce haploid oocytes.

Frontal plane

Primordial follicle

Primary follicle

Secondary follicle

Germinal epithelium

Ovarian cortex

Follicular fluid

Mature (Graafian) follicle

Ovarian medulla

Blood vessels in hilus of ovary

Corpus albicans

Corpus hemorrhagicum (ruptured follicle)

Corona radiata

Ovulation discharges a secondary oocyte

Degenerating corpus luteum

Blood clot

Corpus luteum

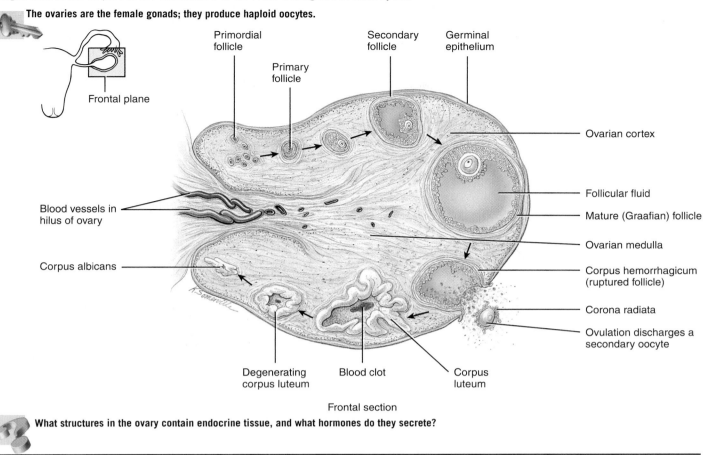

Frontal section

What structures in the ovary contain endocrine tissue, and what hormones do they secrete?

Figure 26.16 / Ovarian follicles. (a) Primordial and primary follicles in the ovarian cortex. (b) A secondary follicle.

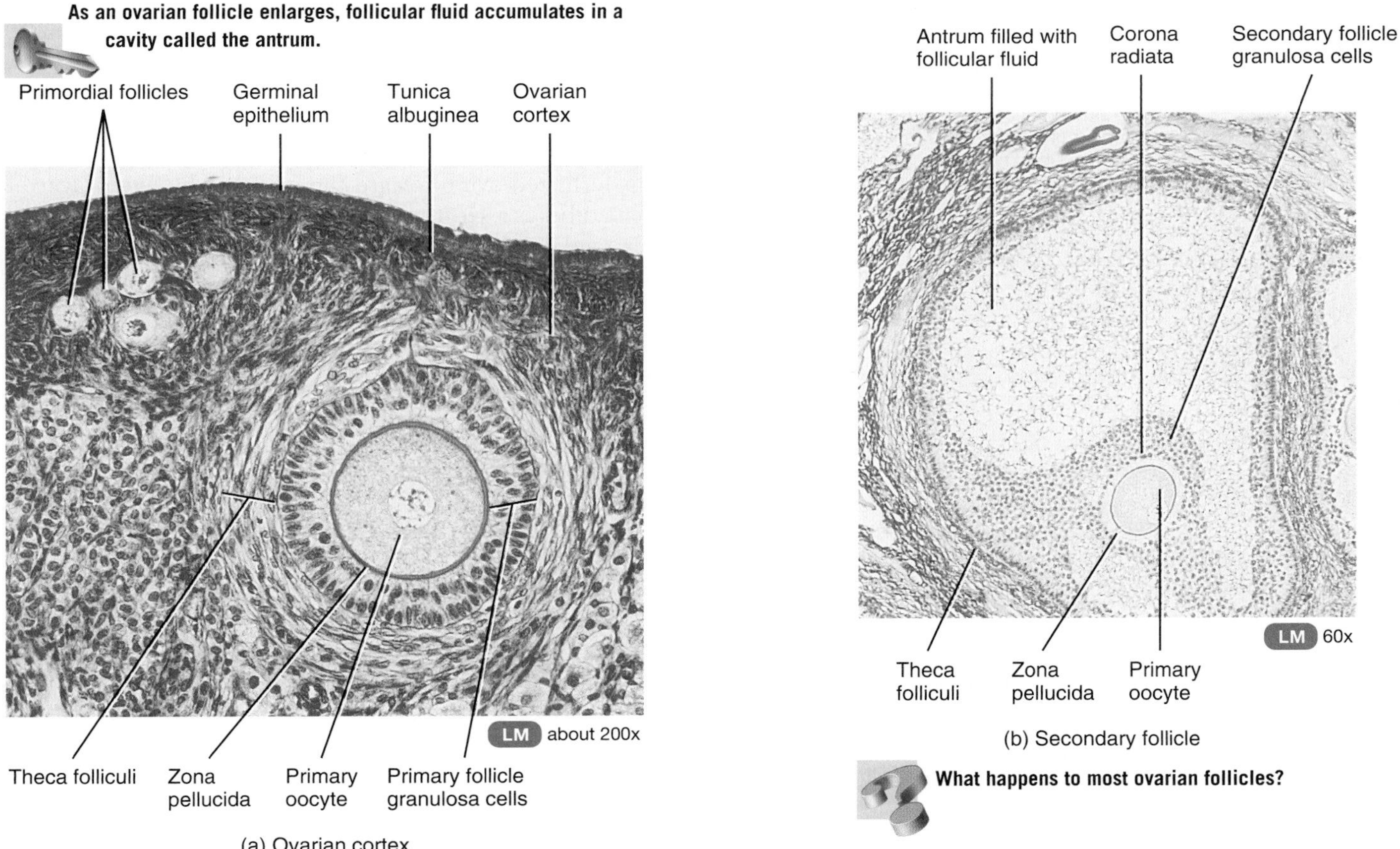

(a) Ovarian cortex

(b) Secondary follicle

What happens to most ovarian follicles?

of oogenesis (Figure 26.17). Meiosis I resumes in several secondary follicles, although only one will eventually reach the maturity needed for ovulation. The diploid primary oocyte completes meiosis I, and two haploid cells of unequal size—both with 23 chromosomes (n) of two chromatids each—are produced. The smaller cell produced by meiosis I, called the **first polar body,** is essentially a packet of discarded nuclear material; the larger cell, known as the **secondary oocyte,** receives most of the cytoplasm. Once a secondary oocyte is formed, it proceeds to the metaphase of meiosis II and then stops. The follicle in which these events are taking place—the **mature (Graafian) follicle**—soon ruptures and releases its secondary oocyte, a process known as **ovulation.**

At ovulation, usually one secondary oocyte (with the first polar body and corona radiata) is expelled into the pelvic cavity. Normally these cells are swept into the uterine tube. If fertilization does not occur, the secondary oocyte degenerates. If sperm are present in the uterine tube and one penetrates the secondary oocyte, however, then meiosis II resumes. The secondary oocyte splits into two haploid (n) cells, again of unequal size. The larger cell is the **ovum,** or mature egg; the smaller one is the **second polar body.** The nuclei of the sperm cell and the ovum then unite, forming a diploid ($2n$) **zygote.** If the first polar body undergoes another division to produce two polar bodies, then the primary oocyte ultimately gives rise to a single haploid (n) ovum and three haploid (n) polar bodies, which all degenerate. Thus, one oogonium gives rise to a single gamete (an ovum), whereas one spermatogonium produces four gametes (sperm).

Figure 26.17 / Oogenesis. Diploid cells ($2n$) have 46 chromosomes; haploid cells (n) have 23 chromosomes.

In an oocyte, meiosis II is completed only if fertilization occurs.

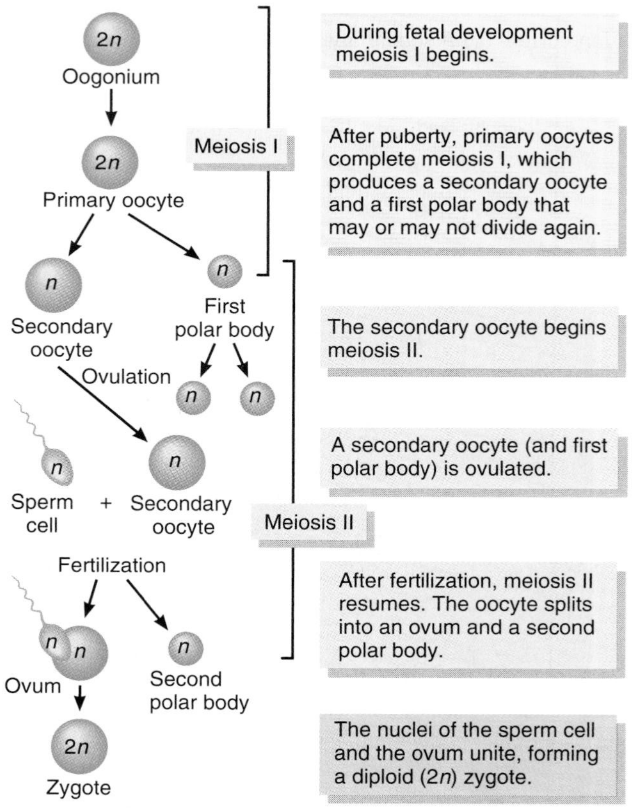

How does the age of a primary oocyte in a female compare with the age of a primary spermatocyte in a male?

✓ How are the ovaries held in position in the pelvic cavity?
✓ Describe the microscopic structure and functions of an ovary.
✓ Describe the principal events of oogenesis.

Uterine Tubes

Females have two **uterine (Fallopian) tubes,** or **oviducts,** that extend laterally from the uterus (Figure 26.18). The tubes, which measure about 10 cm (4 in.) long and lie between the folds of the broad ligaments of the uterus, transport secondary oocytes and fertilized ova from the ovaries to the uterus. The funnel-shaped portion of each tube, called the **infundibulum,** is close to the ovary but is open to the pelvic cavity. It ends in a fringe of fingerlike projections called **fimbriae** (FIM-brē-ē; = fringe), one of which is attached to the lateral end of the ovary. From the infundibulum, the uterine tube extends medially and eventually inferiorly and attaches to the superior lateral angle of the uterus. The **ampulla** of the uterine tube is the widest, longest portion, making up about the lateral two-thirds of its length. The **isthmus** of the uterine tube is the more medial, short, narrow, thick-walled portion that joins the uterus.

Histologically, the uterine tubes are composed of three layers. The internal **mucosa** contains ciliated columnar epithelial cells, which help move the fertilized ovum (or secondary oocyte) along the tube, and secretory cells, which have microvilli and may provide nutrition for the ovum (Figure 26.19 on page 802). The middle layer, the **muscularis,** is composed of an inner, thick, circular ring of smooth muscle and an outer, thin, region of longitudinal smooth muscle. Peristaltic contractions of the muscularis and the ciliary action of the mucosa help move the oocyte or fertilized ovum toward the uterus. The outer layer of the uterine tubes is a serous membrane, the **serosa.**

After ovulation, local currents produced by movements of the fimbriae, which surround the surface of the mature follicle just before ovulation occurs, sweep the secondary oocyte into the uterine tube. A sperm cell usually encounters and fertilizes a secondary oocyte in the ampulla of the uterine tube, although fertilization in the abdominopelvic cavity is not uncommon. Fertilization may occur within the tube at any time up to about 24 hours after ovulation. Some hours after fertilization, the nuclear materials of the haploid ovum and sperm unite; the diploid fertilized ovum is now called a zygote. After several cell divisions, it typically arrives at the uterus about 7 days after ovulation.

The uterine tubes are supplied by branches of the uterine and ovarian arteries (see Figure 26.21). Venous return is via the uterine veins.

The uterine tubes are supplied with sympathetic and parasympathetic nerve fibers from the hypogastric plexus and the pelvic splanchnic nerves. The fibers are distributed to the muscular coat of the tubes and their blood vessels.

Figure 26.18 / Relationship of the uterine (Fallopian) tubes to the ovaries, uterus, and associated structures. In the left side of the drawing the uterine tube and uterus have been sectioned to show internal structures.

After ovulation, a secondary oocyte and its corona radiata move from the pelvic cavity into the infundibulum of the uterine tube. The uterus is the site of menstruation, implantation of a fertilized ovum, development of the fetus, and labor.

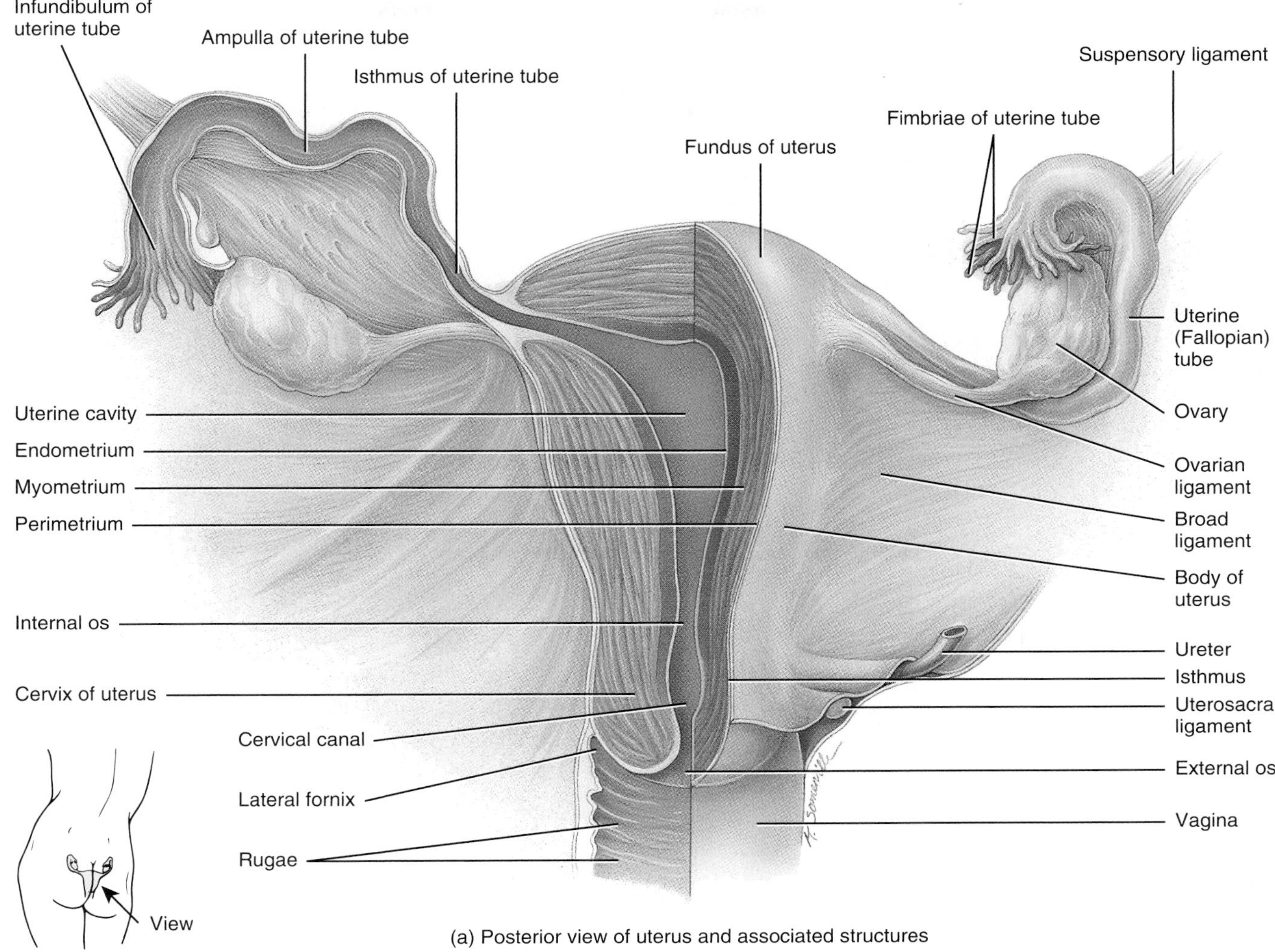

(a) Posterior view of uterus and associated structures

Uterus

The **uterus** (womb) serves as part of the pathway for sperm to reach the uterine tubes (see Figure 26.18). It is also the site of menstruation, implantation of a fertilized ovum, development of the fetus during pregnancy, and labor. Situated between the urinary bladder and the rectum, the uterus is the size and shape of an inverted pear. In females who have never been pregnant, it is about 7.5 cm (3 in.) long, 5 cm (2 in.) wide, and 2.5 cm (1 in.) thick; it is larger in females who have recently been pregnant, and smaller (atrophied) when sex hormone levels are low, as occurs after menopause.

Anatomical subdivisions of the uterus include: (1) a dome-shaped portion superior to the uterine tubes called the **fundus,** (2) a tapering central portion called the **body,** and (3) an inferior narrow portion called the **cervix** that opens into the vagina. Between the body of the uterus and the cervix is the **isthmus** (IS-mus), a constricted region about 1 cm (0.5 in.) long. The interior of the body of the uterus is called the **uterine cavity,** and the interior of the narrow cervix is called the **cervical canal.** The cervical canal opens into the uterine cavity at the **internal os** (*os* = mouthlike opening) and into the vagina at the **external os.**

Normally, the body of the uterus projects anteriorly and superiorly over the urinary bladder in a position called **anteflexion.** The cervix projects inferiorly and posteriorly and enters the anterior wall of the vagina at nearly a right angle (see Figure 26.13). Several ligaments that are either extensions of the parietal peritoneum or fibromuscular cords maintain the position of the uterus (see Figure 26.14). The paired **broad ligaments** are double folds of peritoneum attaching the uterus to either side of the pelvic cavity. The paired **uterosacral ligaments,** also peritoneal extensions, lie on either side of the rectum and connect the uterus to the sacrum. The **cardinal (lateral cervical) liga-**

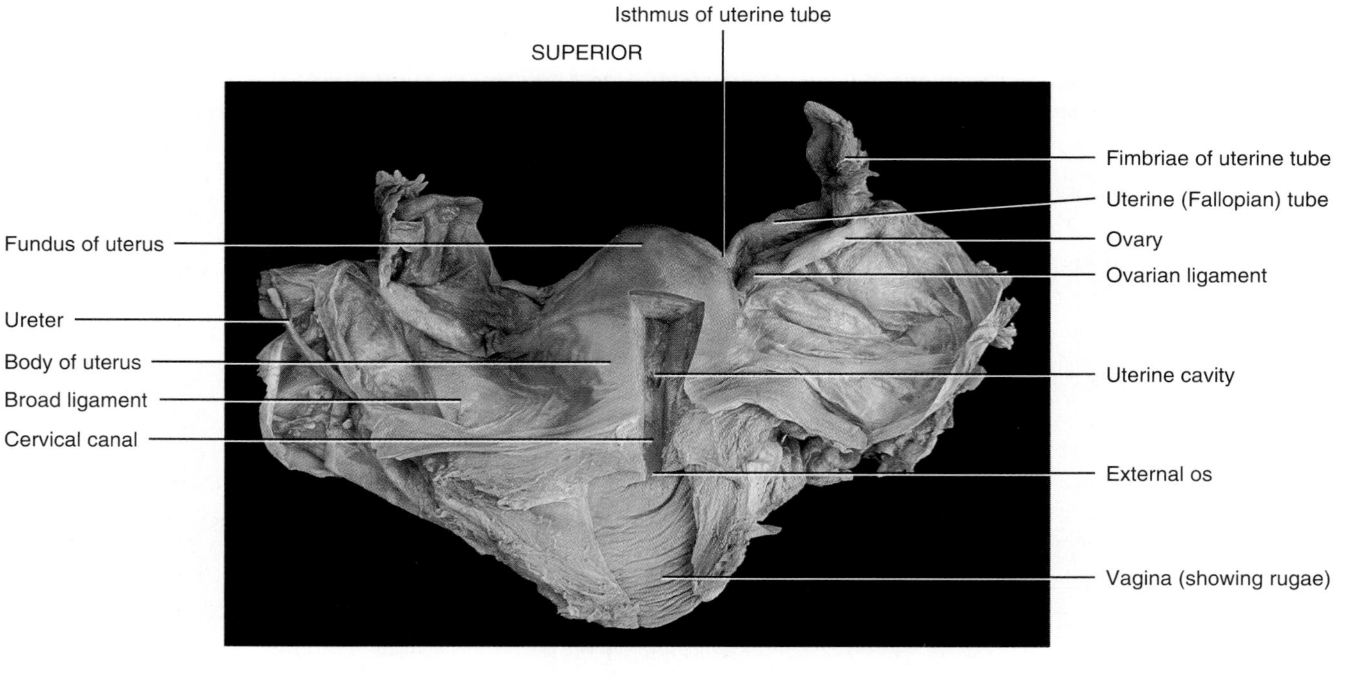

(b) Posterior view of uterus and associated structures

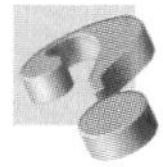

Where does fertilization usually occur?

Figure 26.19 / Mucosal cells of the uterine (Fallopian) tube.

Peristaltic contractions of the muscularis and ciliary action of the mucosa of the uterine tube help move the oocyte or fertilized ovum toward the uterus.

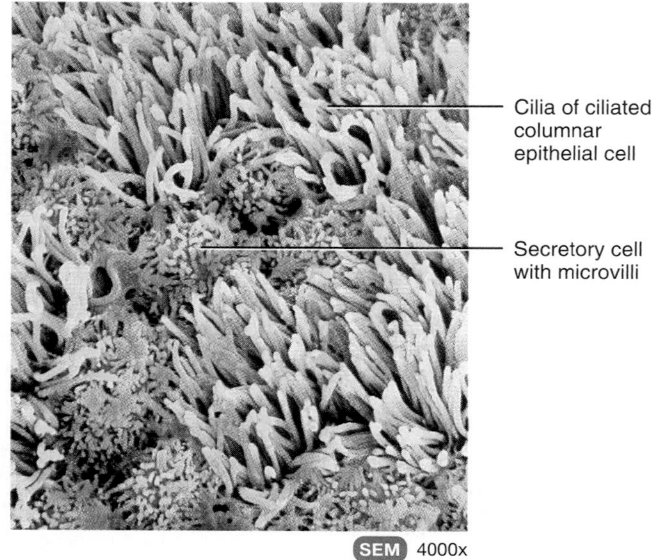

Cilia and secretory cells lining the uterine (Fallopian) tube

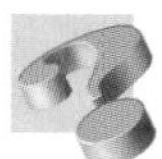

What types of cells line the uterine tubes?

ments extend inferior to the bases of the broad ligaments between the pelvic wall and the cervix and vagina. The **round ligaments** are bands of fibrous connective tissue between the layers of the broad ligament; they extend from a point on the uterus just inferior to the uterine tubes to a portion of the labia majora of the external genitalia. Although the ligaments normally maintain the anteflexed position of the uterus, they also afford the uterine body enough movement such that the uterus may become malpositioned. A posterior tilting of the uterus is called **retroflexion** (*retro-* = backward or behind).

Uterine Prolapse

A condition called **uterine prolapse** (*prolapses* = falling down or downward displacement) may result from weakening of supporting ligaments and pelvic musculature associated with age or disease, traumatic vaginal delivery, chronic straining from coughing or difficult bowel movements, or pelvic tumors. The prolapse may be characterized as *first degree (mild)*, in which the cervix remains within the vagina; *second degree (marked)*, in which the cervix protrudes to the exterior through the vagina; and *third degree (complete)*, in which the entire uterus is outside the vagina. Depending on the degree of prolapse, treatment may involve pelvic exercises, dieting if a patient is overweight, stool softeners to minimize straining during defecation, pessary therapy (placement of a rubber device around the uterine cervix that helps prop up the uterus), and surgery. ■

Histologically, the uterus consists of three layers of tissue: the perimetrium, myometrium, and endometrium (Figure 26.20). The outer layer—the **perimetrium** (*peri-* = around; *-metrium* = uterus) or serosa—is part of the visceral peritoneum; it is composed of simple squamous epithelium and areolar connective tissue. Laterally, it becomes the broad ligament; anteriorly, it covers the urinary bladder and forms a shallow pouch, the **vesicouterine pouch** (ves′-i-kō-YŪ-ter-in; *vesico-* = bladder; see Figure 26.13); posteriorly, it covers the rectum and forms a deep pouch, the **rectouterine pouch** (rek-tō-YŪ-ter-in; *recto-* = rectum) or *pouch of Douglas*—the most inferior point in the pelvic cavity.

The middle layer of the uterus, the **myometrium** (*myo-* = muscle), consists of three layers of smooth muscle fibers and is thickest in the fundus and thinnest in the cervix. The thicker middle layer is circular, whereas the inner and outer layers are longitudinal or oblique. During labor and childbirth, coordinated contractions of the myometrium in response to oxytocin from the posterior pituitary gland help expel the fetus from the uterus.

Figure 26.20 / Histology of the uterus; the superficial perimetrium (serosa) is not shown.

The three layers of the uterus from superficial to deep are the perimetrium (serosa), the myometrium, and the endometrium.

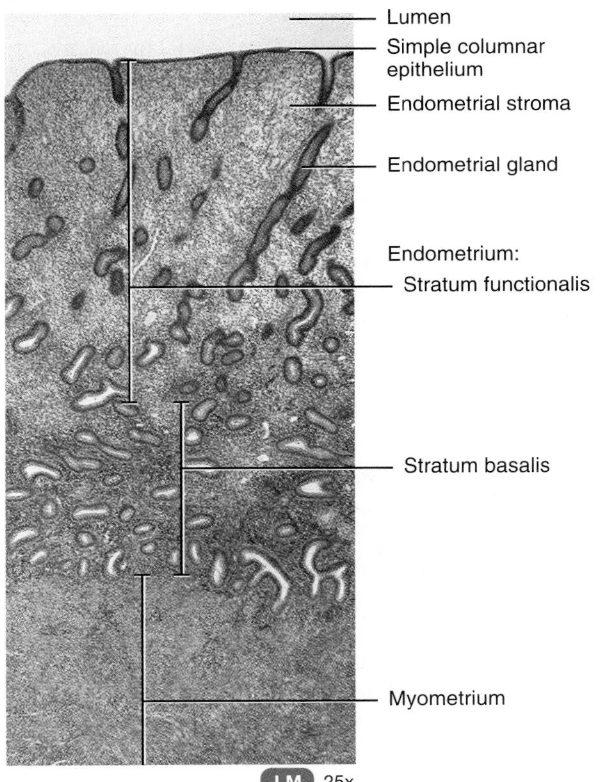

Portion of endometrium and myometrium

What structural features of the endometrium and myometrium contribute to their functions?

The inner layer of the uterus, the **endometrium** (*endo-* = within), is highly vascular and is composed of an innermost layer of simple columnar epithelium (ciliated and secretory cells) that lines the lumen; an underlying endometrial stroma that is a very thick region of lamina propria (areolar connective tissue); and endometrial (uterine) glands that develop as invaginations of the luminal epithelium and extend almost to the myometrium. The endometrium is divided into two layers: The **stratum functionalis** (*functional layer*) lines the uterine cavity and is shed during menstruation; a deeper layer, the **stratum basalis** (*basal layer*), is permanent and gives rise to a new stratum functionalis after each menstruation.

The secretory cells of the mucosa of the cervix produce a secretion called **cervical mucus,** a mixture of water, glycoprotein, serum-type proteins, lipids, enzymes, and inorganic salts. During their reproductive years, females secrete 20–60 mL of cervical mucus per day. Cervical mucus is more hospitable to sperm at or near the time of ovulation because it is then less viscous and more alkaline (pH 8.5); at other times, viscous mucus forms a cervical plug that physically impedes sperm penetration. Cervical mucus supplements the energy needs of sperm, and both the cervix and cervical mucus serve as a sperm reservoir, protect sperm from the hostile environment of the vagina, and protect sperm from phagocytes. They may also play a role in *capacitation*—a functional change that sperm undergo in the female reproductive tract before they are able to fertilize a secondary oocyte.

Blood is supplied to the uterus by branches of the internal iliac artery called **uterine arteries** (Figure 26.21). Branches called **arcuate arteries** (= shaped like a bow) are arranged in a circular fashion in the myometrium and give off **radial arteries** that penetrate deeply into the myometrium. Just before the branches enter the endometrium, they divide into two kinds of arterioles: **straight arterioles** supply the stratum basalis with the materials needed to regenerate the stratum functionalis; **spiral arterioles** supply the stratum functionalis and change markedly during the menstrual cycle. Blood leaving the uterus is drained by the **uterine veins** into the internal iliac veins. The extensive blood supply of the uterus is essential to support regrowth of a new stratum functionalis after menstruation, implantation of a fertilized ovum, and development of the placenta.

Hysterectomy

Hysterectomy (hiss-te-RECT-tō-mē; *hyster-* = uterus), the surgical removal of the uterus, is the most common gynecological operation. It may be indicated in conditions such as endometriosis, pelvic inflammatory disease, recurrent ovarian cysts, excessive uterine bleeding, and cancer of the cervix, uterus, or ovaries. In a partial or subtotal hysterectomy, the body of the uterus is removed but the cervix is left in place; a complete hysterectomy is the removal of the body and cervix of the uterus. A radical hysterectomy includes removal of the body and cervix of the uterus, uterine tubes, possibly the ovaries, superior portion of the vagina, pelvic lymph nodes, and supporting structures, such as ligaments. ■

Figure 26.21 / Blood supply of the uterus. The inset shows histological details of the blood vessels of the endometrium.

Straight arterioles supply the materials needed for regeneration of the stratum functionalis.

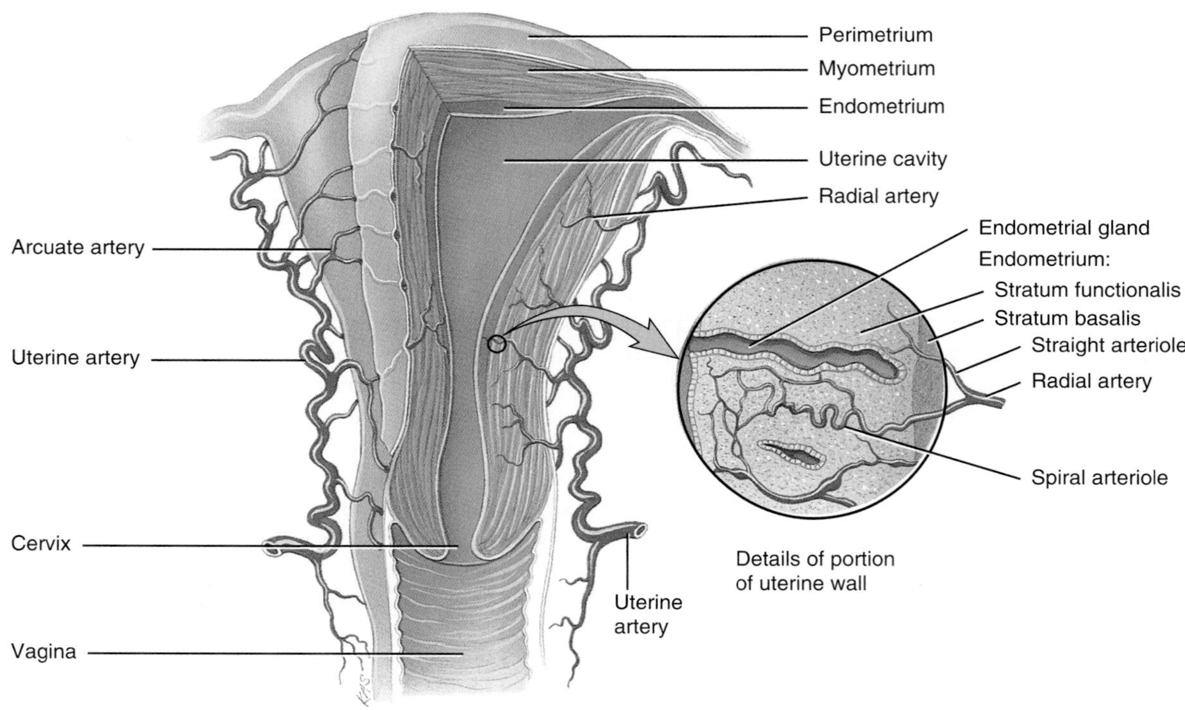

What is the functional significance of the stratum basalis of the endometrium?

- ✓ Where are the uterine tubes located? What is their function?
- ✓ Diagram and label the principal parts of the uterus.
- ✓ Describe the arrangement of ligaments that hold the uterus in its normal position.
- ✓ Describe the histology of the uterus.
- ✓ Discuss the blood supply to the uterus. Why is an abundant blood supply important?

Vagina

The **vagina** (= sheath) serves as a passageway for menstrual flow, for childbirth, and for semen from the penis during sexual intercourse. It is a tubular, 10-cm (4-in.) long fibromuscular organ lined with mucous membrane (see Figures 26.13 and 26.18). Situated between the urinary bladder and the rectum, the vagina is directed superiorly and posteriorly, where it attaches to the uterus. A recess called the **fornix** (= arch or vault) surrounds the vaginal attachment to the cervix.

The **mucosa** of the vagina is continuous with that of the uterus. Histologically, it consists of nonkeratinized stratified squamous epithelium and areolar connective tissue that lies in a series of transverse folds called **rugae.** The mucosa of the vagina contains large stores of glycogen, the decomposition of which produces organic acids. The resulting acidic environment retards microbial growth, but it also is harmful to sperm. Alkaline components of semen, mainly from the seminal vesicles, neutralize the acidity of the vagina and increase viability of the sperm.

The **muscularis** is composed of an outer circular layer and an inner longitudinal layer of smooth muscle that can stretch considerably to accommodate the penis during sexual intercourse and a child during birth.

The **adventitia,** the superficial layer of the vagina, consists of areolar connective tissue; it anchors the vagina to adjacent organs such as the urethra and urinary bladder anteriorly and the rectum and anal canal posteriorly.

At the inferior end of the vaginal opening to the exterior, the **vaginal orifice,** there may be a thin fold of vascularized mucous membrane, called the **hymen** (= membrane), that forms a border around the orifice, partially closing it (see Figure 26.22). Sometimes the hymen completely covers the orifice, a condition called **imperforate hymen** (im-PER-fō-rāt); surgery may be required to open the orifice and permit the discharge of menstrual flow.

14th Century

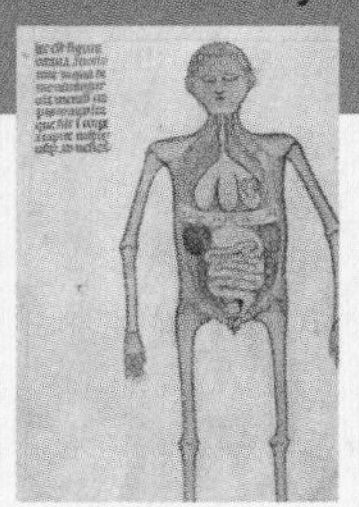

15th Century

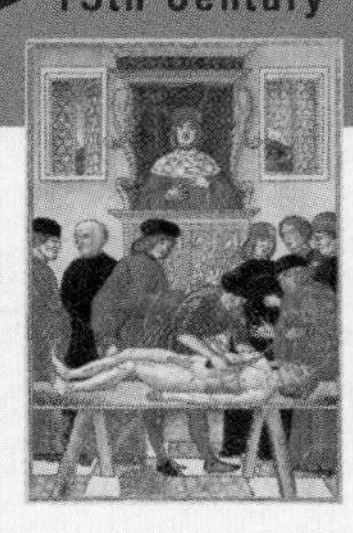

16th Century

17th Century

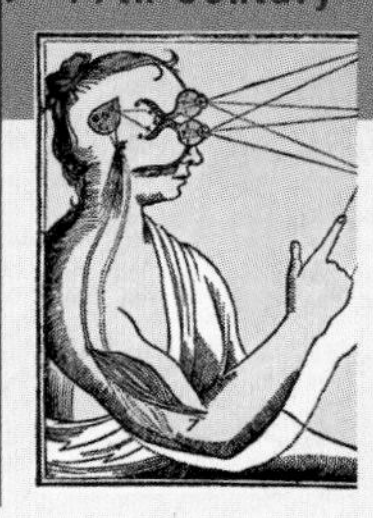

18th Century

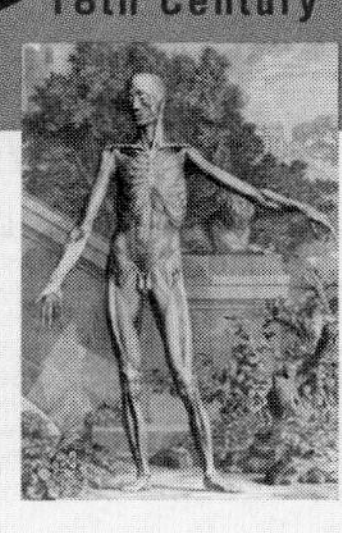

20th Century

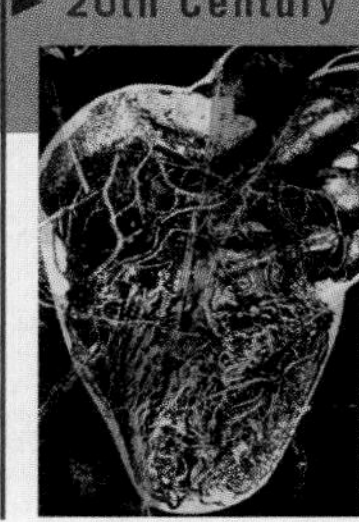

21st Century

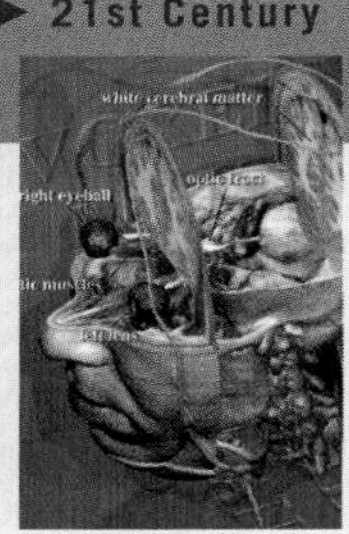

CHANGING IMAGES

How Life Begins

The reproductive system, especially the reproductive capacity of women, has been a focal point of scientific exploration for thousands of years. It was Aristotle who, in the 4th century BC, was the originator of Western embryological thinking with his self-described masterpiece: *Generation of Man.* Although he dissected and experimented on only lower animals and never a human, Aristotle wrote extensively about his beliefs concerning the process of human procreation.

Basing his theories on experiments with the chick egg, Aristotle believed that the human form and psyche resided in the man's semen. It was the mixing of semen with menstrual flow within the uterus that allowed this consolidation of fluids to incubate into a human form. Aristotle wrote that many factors influenced the precision of this process. Indeed, he considered the female to be a "damaged male" as a result of insufficient heat during gestation—a 'defect' that occurred in about half of all pregnancies. Although his thinking may today seem like wild speculation, Aristotle's theories were the basis of embryological thought for nearly two thousand years, influencing the likes of Hippocrates, Galen, and Harvey.

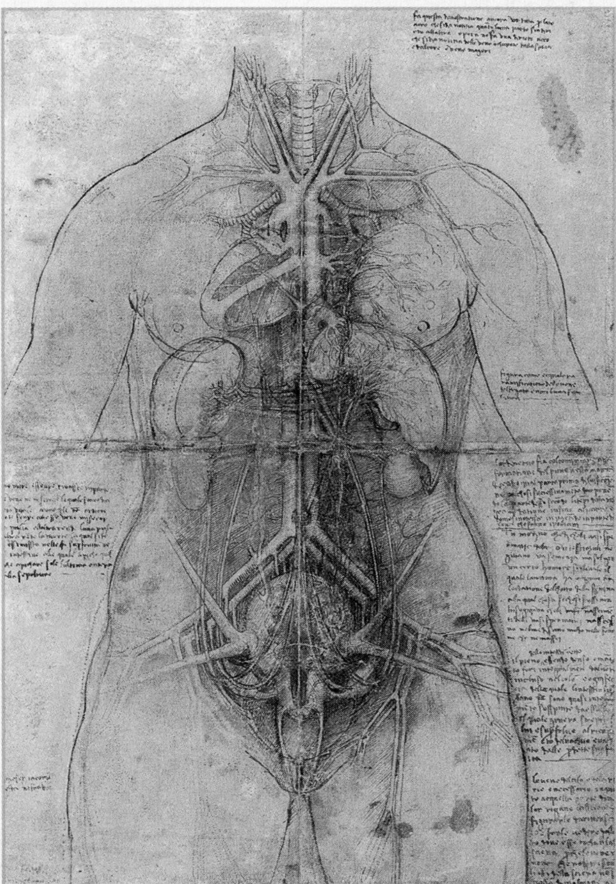

1510 AD

Pictured here is perhaps the most famous reproductive system rendering by Leonardo da Vinci, dated about 1510. In this image, Leonardo, who subscribed to many of Aristotle's beliefs, illustrates the reproductive system of a woman in the early stages of pregnancy. Can you recognize any of the organs? Try to identify the uterus, vagina, and ovaries. Can you consider what da Vinci intended by those large horn-like structures arising from the uterus? Could these be the ovarian and broad ligaments? Examine the organs you studied earlier in this text, such as the heart, liver, and spleen. Are they properly represented? And finally, compare what you learn in this chapter about the uterine and ovarian cycles, to Aristotle's beliefs concerning a woman's contribution to reproduction.

Vulva

The term **vulva** (VUL-va; = to wrap around), or **pudendum** (pyū-DEN-dum), refers to the external genitals of the female (Figure 26.22). The vulva comprises the following components:

- Anterior to the vaginal and urethral openings is the **mons pubis** (MONZ PŪ-bis; *mons* = mountain), an elevation of adipose tissue covered by skin and coarse pubic hair that cushions the pubic symphysis.
- From the mons pubis, two longitudinal folds of skin, the **labia majora** (LĀ-bē-a ma-JŌ-ra; *labia* = lips; *majora* = larger), extend inferiorly and posteriorly. The singular term is *labium majus.* The labia majora are covered by pubic hair and contain an abundance of adipose tissue, sebaceous (oil) glands, and apocrine sudoriferous (sweat) glands. They are homologous to the scrotum.
- Medial to the labia majora are two smaller folds of skin called the **labia minora** (mī-NŌ-ra; *minora* = smaller). The singular term is *labium minus.* Unlike the labia majora, the labia minora are devoid of pubic hair and fat and have few sudoriferous glands, but they do contain many sebaceous glands. The labia minora are homologous to the spongy (penile) urethra.
- The **clitoris** (KLI-to-ris) is a small cylindrical mass of erectile tissue and nerves located at the anterior junction of the labia minora. A layer of skin called the **prepuce** (foreskin) is formed at the point where the labia minora unite and covers the body of the clitoris. The exposed portion of the clitoris is the glans. The clitoris is homologous to the glans penis in males; like the male structure, it is capable of enlargement upon tactile stimulation and has a role in sexual excitement in the female.
- The region between the labia minora is the **vestibule.** Within the vestibule are the hymen (if still present), the vaginal orifice, the external urethral orifice, and the openings of the ducts of several glands. The vestibule is homologous to the membranous urethra of males. The **vaginal orifice,** the opening of the vagina to the exterior, occupies the

Figure 26.22 / Components of the vulva (pudendum). (See Tortora, *A Photographic Atlas of the Human Body,* Figure 14.7.)

The vulva refers to the external genitals of the female.

Mons pubis
Labia majora (spread)
Prepuce of clitoris
Clitoris
Labia minora (spread exposing vestibule)
External urethral orifice
Vaginal orifice (dilated)
Hymen
Anus
Inferior view

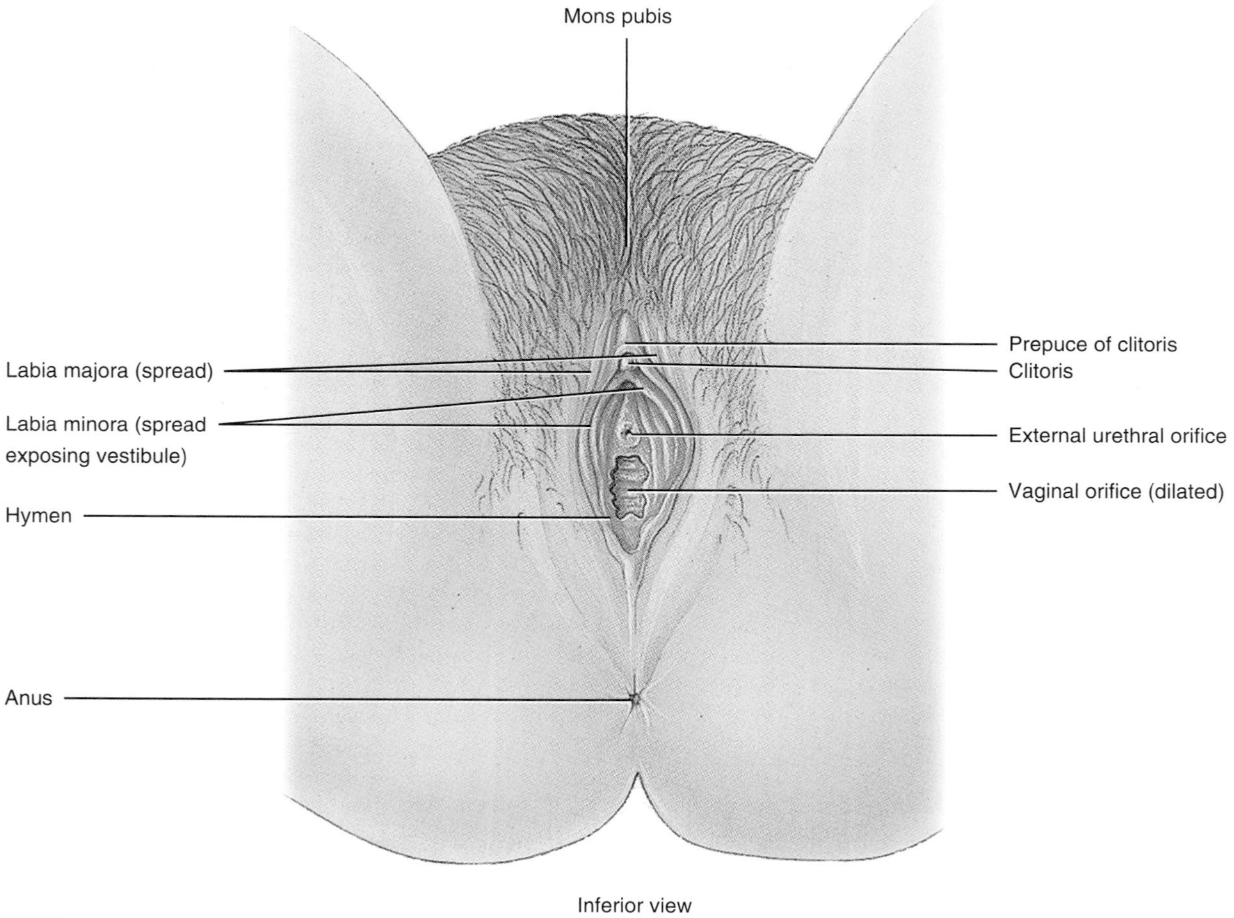

What surface structures are anterior to the vaginal opening? Lateral to it?

greater portion of the vestibule and is bordered by the hymen. Anterior to the vaginal orifice and posterior to the clitoris is the **external urethral orifice,** the opening of the urethra to the exterior. On either side of the external urethral orifice are the openings of the ducts of the **paraurethral** *(Skene's)* **glands,** which are embedded in the wall of the urethra and secrete mucus. The paraurethral glands are homologous to the prostate gland. On either side of the vaginal orifice itself are the **greater vestibular** *(Bartholin's)* **glands** (see Figure 26.23), which open by ducts into a groove between the hymen and labia minora and produce a small quantity of mucus during sexual arousal and intercourse that adds to cervical mucus and provides lubrication. The greater vestibular glands are homologous to the bulbourethral glands in males. Several **lesser vestibular glands** also open into the vestibule.

- The **bulb of the vestibule** (see Figure 26.23) consists of two elongated masses of erectile tissue just deep to the labia on either side of the vaginal orifice. The bulb of the vestibule becomes engorged with blood during sexual arousal, narrowing the vaginal orifice and placing pressure on the penis during intercourse. The bulb of the vestibule is homologous to the corpus spongiosum penis and bulb of the penis in males.

Table 26.1 summarizes the homologous structures of the female and male reproductive systems.

Table 26.1 Summary of Homologous Structures of the Female and Male Reproductive Systems

Female Structures	Male Structures
Ovaries	Testes
Ovum	Sperm cell
Labia majora	Scrotum
Labia minora	Spongy (penile) urethra
Vestibule	Membranous urethra
Bulb of vestibule	Corpus spongiosum penis and bulb of penis
Clitoris	Glans penis
Paraurethral glands	Prostate gland
Greater vestibular glands	Bulbourethral (Cowper's) glands

Perineum

The **perineum** (per′-i-NĒ-um) is the diamond-shaped area medial to the thighs and buttocks of both males and females that contains the external genitals and anus (Figure 26.23). The perineum is bounded anteriorly by the pubic symphysis, laterally by the ischial tuberosities, and posteriorly by the coccyx. A transverse line drawn between the ischial tuberosities divides the perineum into an anterior **urogenital triangle** (yū′-rō-JEN-i-tal) that contains the external genitalia and a posterior anal triangle that contains the anus.

Figure 26.23 / Perineum of a female. (Figure 10.14 on page 291 shows the perineum of a male.)

The perineum is a diamond-shaped area medial to the thighs and buttocks that contains the external genitals and anus.

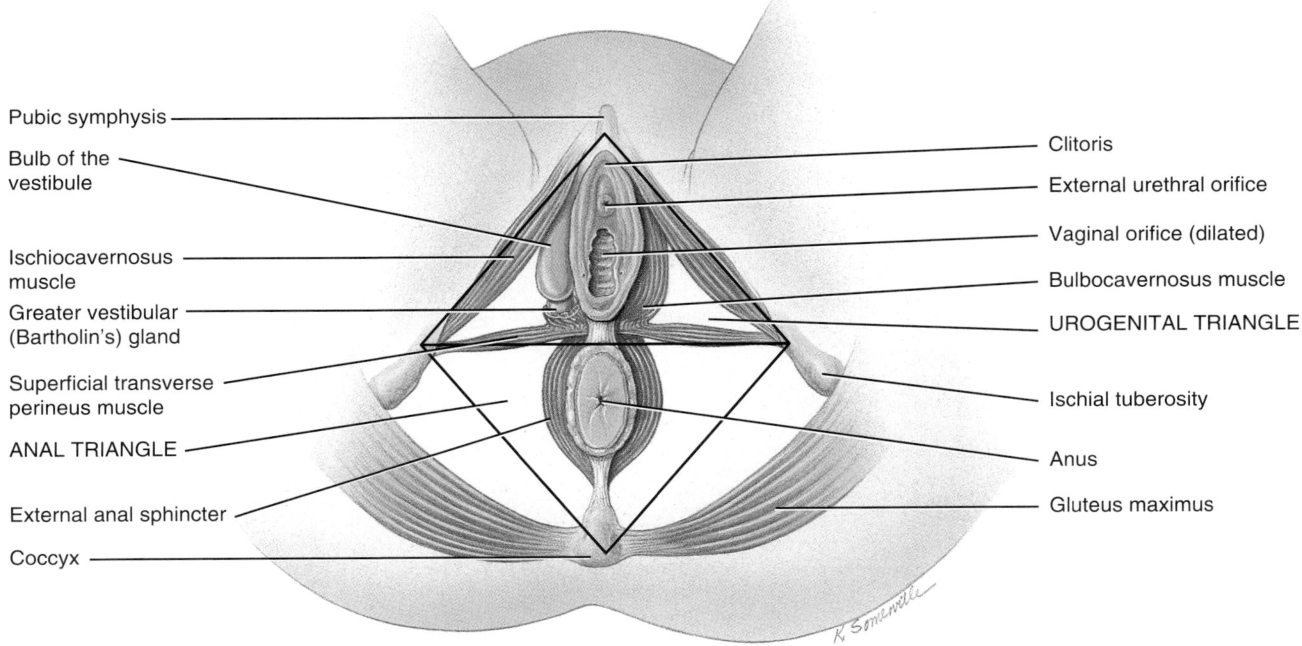

Inferior view

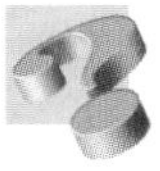
Why is the anterior portion of the perineum called the urogenital triangle?

During childbirth, the emerging fetus stretches the perineal region. To prevent undue stretching and even tearing of this region, a physician sometimes performs an **episiotomy** (e-piz′-ē-OT-ō-mē; *epision* = pubic region; *tome* = incision), a perineal cut made with surgical scissors. This cut enlarges the vaginal opening to make room for the fetus to pass. In effect, a controlled cut is substituted for a jagged, uncontrolled tear. The incision is closed in layers with a continuous suture that is absorbed within a few weeks, so that stitches do not have to be removed.

Mammary Glands

The two **mammary glands** (*mamma* = breast) are modified sudoriferous (sweat) glands that produce milk. They lie over the pectoralis major and serratus anterior muscles and are attached to them by a layer of deep fascia that is dense irregular connective tissue (Figure 26.24).

Each breast has one pigmented projection, the **nipple,** that has a series of closely spaced openings of ducts called **lactiferous ducts,** where milk emerges. The circular pigmented area of skin surrounding the nipple is called the **areola** (a-RĒ-ō-la; = small space); it appears rough because it contains modified sebaceous (oil) glands. Strands of connective tissue called the **suspensory ligaments of the breast (Cooper's ligaments)** run between the skin and deep fascia and support the breast. These ligaments become looser with age or with excessive strain, as occurs in long-term jogging or high-impact aerobics. Wearing a supportive bra slows the appearance of "Cooper's droop."

Internally, the mammary gland consists of 15–20 lobes, or compartments, separated by adipose tissue. The amount of adipose tissue, not the amount of milk produced, determines the size of the breasts. In each lobe are several smaller compartments called **lobules,** composed of grapelike clusters of milksecreting glands termed **alveoli** (= small cavities) embedded in connective tissue (Figure 26.25). Surrounding the alveoli are spindle-shaped cells called **myoepithelial cells,** the contraction of which helps propel milk toward the nipples. When milk is being produced, it passes from the alveoli into a series of **secondary tubules** and then into the **mammary ducts.** Near the nipple, the mammary ducts expand to form sinuses called **lactiferous sinuses** (*lact-* = milk), where some milk may be stored before draining into a lactiferous duct. Each lactiferous duct typically carries milk from one of the lobes to the exterior.

The essential functions of the mammary glands are the synthesis, secretion, and ejection of milk; these functions, called **lactation,** are associated with pregnancy and childbirth. Milk production is stimulated largely by the hormone prolactin, with

Figure 26.24 / Mammary glands.

The mammary glands function in the synthesis, secretion, and ejection of milk (lactation).

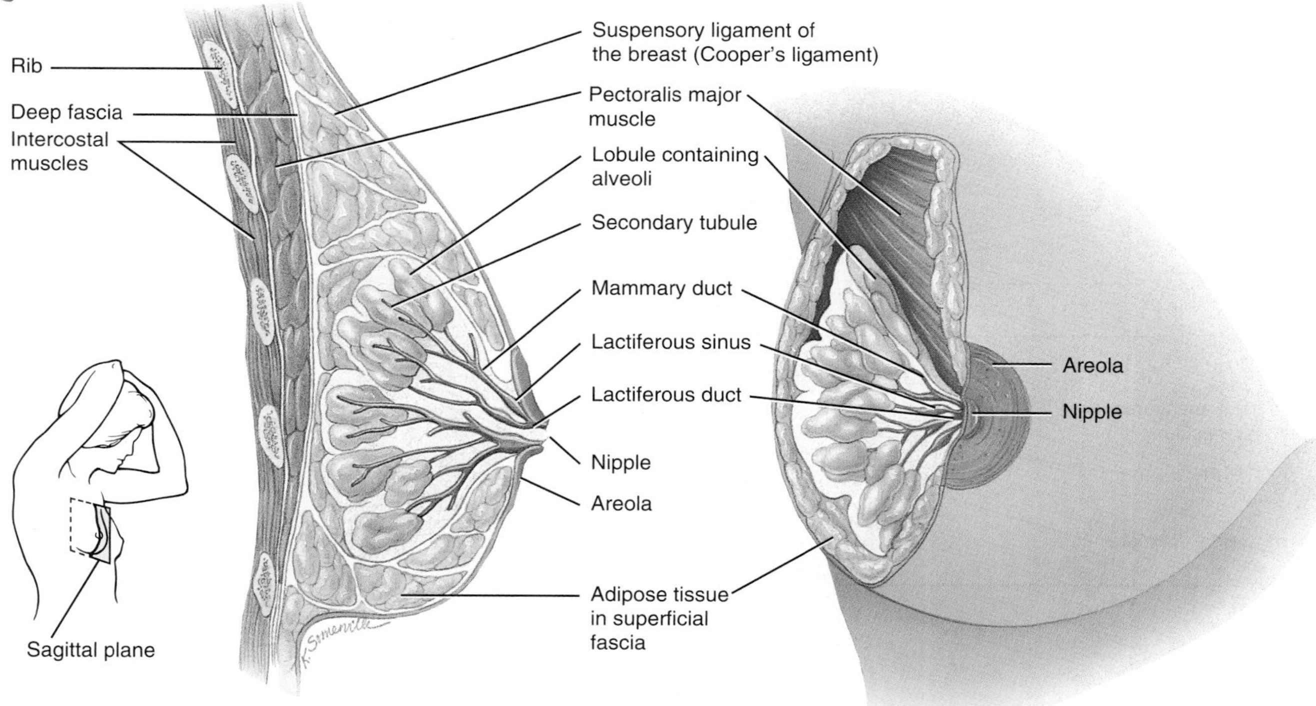

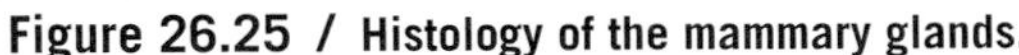

contributions from progesterone and estrogens. The ejection of milk is stimulated by oxytocin, which is released from the posterior pituitary gland in response to the sucking of an infant on the mother's nipple (suckling).

Fibrocystic Disease of the Breasts

The breasts of females are highly susceptible to cysts and tumors. In **fibrocystic disease,** the most common cause of breast lumps in females, one or more cysts (fluid-filled sacs) and thickening of alveoli (clusters of milk-secreting cells) develop. The condition, which occurs mainly in females between the ages of 30 and 50, is probably due to a hormonal imbalance—a relative excess of estrogens or a deficiency of progesterone—in the postovulatory (luteal) phase of the reproductive cycle (discussed shortly). Fibrocystic disease usually causes one or both breasts to become lumpy, swollen, and tender a week or so before menstruation begins.

- ✓ What is the function of the vagina? Describe its histology.
- ✓ List the parts of the vulva, and explain the functions of each.
- ✓ Describe the structure of the mammary glands. How are they supported?
- ✓ Describe the passage of milk from the alveoli of the mammary gland to the nipple.

Figure 26.25 / Histology of the mammary glands.

Contraction of myoepithelial cells helps propel milk toward the nipples.

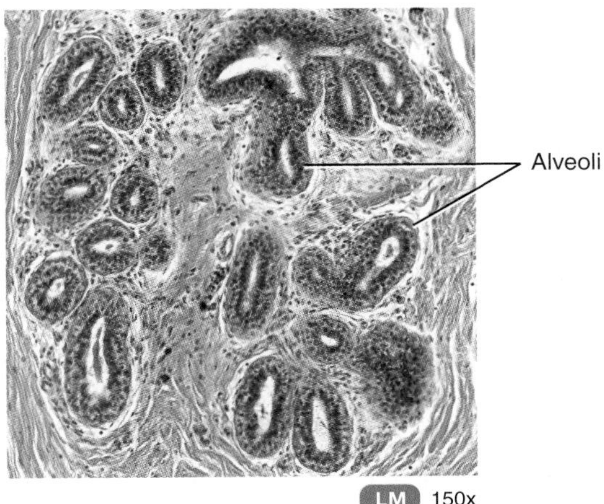

Section of a lactating mammary gland

In which portion of the mammary gland are alveoli located?

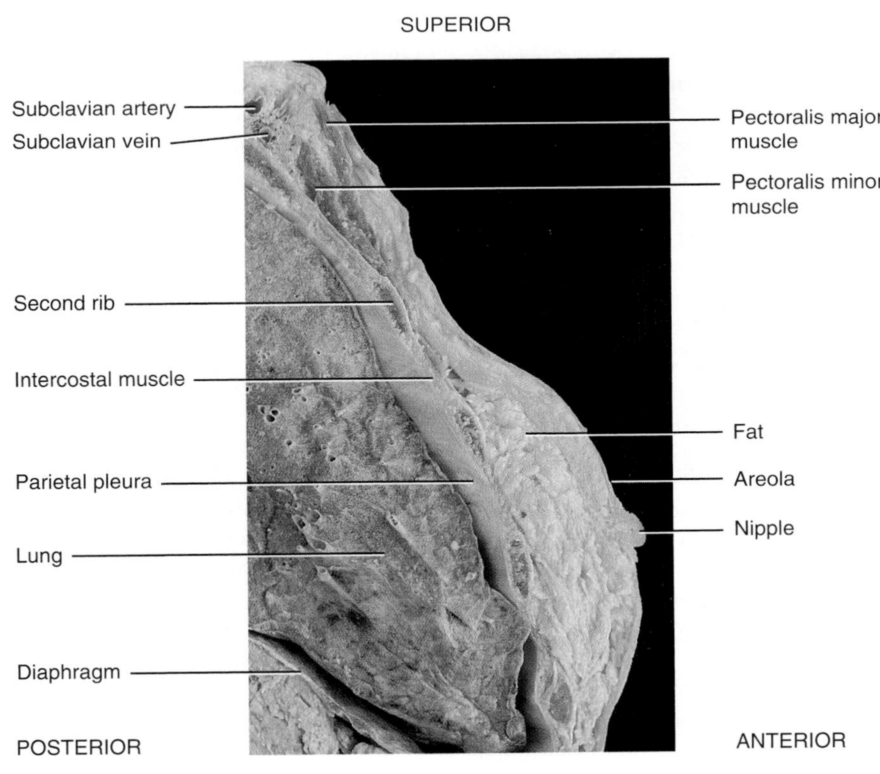

(c) Sagittal section

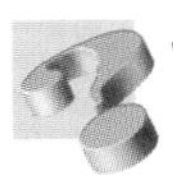

What hormones regulate the synthesis and ejection of milk?

FEMALE REPRODUCTIVE CYCLE

Objective

- Compare the principal events of the ovarian and uterine cycles.

During their reproductive years, nonpregnant females normally experience a cyclical sequence of changes in the ovaries and uterus. Each cycle takes about a month and involves both oogenesis and preparation of the uterus to receive a fertilized ovum. Hormones secreted by the hypothalamus, anterior pituitary gland, and ovaries control the principal events. The **ovarian cycle** is a series of events in the ovaries that occur during and after the maturation of an oocyte. The **uterine (menstrual) cycle** is a concurrent series of changes in the endometrium of the uterus to prepare it for the arrival of a fertilized ovum that will develop in the uterus until birth. If fertilization does not occur, the stratum functionalis of the endometrium is shed. The general term **female reproductive cycle** encompasses the ovarian and uterine cycles, the hormonal changes that regulate them, and the related cyclical changes in the breasts and cervix.

The ovarian and uterine cycles are controlled by **gonadotropin releasing hormone (GnRH)** secreted by the hypothalamus (Figure 26.26). GnRH stimulates the release of **follicle-stimulating hormone (FSH)** and **luteinizing hormone (LH)** from the anterior pituitary gland. FSH, in turn, initiates follicular growth and the secretion of estrogens by the growing follicles. LH stimulates the further development of ovarian follicles and their full secretion of estrogens, brings about ovulation, promotes formation of the corpus luteum, and stimulates the production of estrogens, progesterone, relaxin, and inhibin by the corpus luteum.

The duration of the female reproductive cycle typically is 24–35 days. For this discussion we assume a duration of 28 days, divided into four phases: the menstrual phase, the preovulatory phase, ovulation, and the postovulatory phase (Figure 26.26).

Menstrual Phase

The **menstrual phase** (MEN-strū-al), also called **menstruation** (men′-strū-Ā-shun) or **menses** (= month), lasts for roughly the first 5 days of the cycle. (By convention, the first day of menstruation marks the first day of a new cycle.)

EVENTS IN THE OVARIES During the menstrual phase, 20 or so small secondary follicles, some in each ovary, begin to enlarge. Follicular fluid, secreted by the granulosa cells and oozing from blood capillaries, accumulates in the enlarging antrum while the oocyte remains near the edge of the follicle (see Figure 26.16b).

EVENTS IN THE UTERUS Menstrual flow from the uterus consists of 50–150 mL of blood, tissue fluid, mucus, and epithelial cells derived from the endometrium. This discharge occurs because the declining level of ovarian hormones, especially progesterone, stimulates release of prostaglandins that cause the uterine spiral arterioles to constrict. As a result, the cells they supply become oxygen-deprived and start to die. Eventually, the entire stratum functionalis sloughs off. At this time the endometrium is very thin, about 2–5 mm, because only the stratum basalis remains. The menstrual flow passes from the uterine cavity to the cervix and through the vagina to the exterior.

Preovulatory Phase

The **preovulatory phase,** the second phase of the female reproductive cycle, is the time between menstruation and ovulation. The preovulatory phase of the cycle is more variable in length than the other phases and accounts for most of the difference when cycles are shorter or longer than 28 days. It lasts from days 6 to 13 in a 28-day cycle.

EVENTS IN THE OVARIES Under the influence of FSH, the group of about 20 secondary follicles continues to grow and begins to secrete estrogens and inhibin. By about day 6, one follicle in one ovary has outgrown all the others to become the dominant follicle. Estrogens and inhibin secreted by the dominant follicle decrease the secretion of FSH, which causes the other less well-developed follicles to stop growing and undergo atresia.

The one dominant follicle becomes the **mature (Graafian) follicle,** which continues to enlarge until it is more than 20 mm in diameter and ready for ovulation (see Figure 26.15). This follicle forms a blisterlike bulge on the surface of the ovary. Fraternal (nonidentical) twins may result if two secondary follicles achieve codominance, and both ovulate and are fertilized. During the final maturation process, the dominant follicle continues to increase its estrogen production under the influence of an increasing level of LH. Although estrogens are the primary ovarian hormones before ovulation, small amounts of progesterone are produced by the mature follicle a day or two before ovulation.

With reference to the ovarian cycle, the menstrual and preovulatory phases together are termed the **follicular phase** (fō-LIK-yū-lar) because ovarian follicles are growing and developing.

EVENTS IN THE UTERUS Estrogens liberated into the blood by growing ovarian follicles stimulate the repair of the endometrium; cells of the stratum basalis undergo mitosis and produce a new stratum functionalis. As the endometrium thickens, the short, straight endometrial glands develop, and the arterioles coil and lengthen as they penetrate the stratum functionalis. The thickness of the endometrium approximately doubles, to about 4–10 mm. With reference to the uterine cycle, the preovulatory phase is also termed the proliferative phase because the endometrium is proliferating.

Ovulation

Ovulation, the rupture of the mature (Graafian) follicle with release of the secondary oocyte into the pelvic cavity, usually oc-

Figure 26.26 / The female reproductive cycle. Events in the ovarian and uterine cycles and the release of anterior pituitary gland hormones are correlated with the sequence of the cycle's four phases. In the cycle shown, fertilization and implantation have not occurred.

The length of the female reproductive cycle typically is 24–36 days; the preovulatory phase is more variable in length than the other phases.

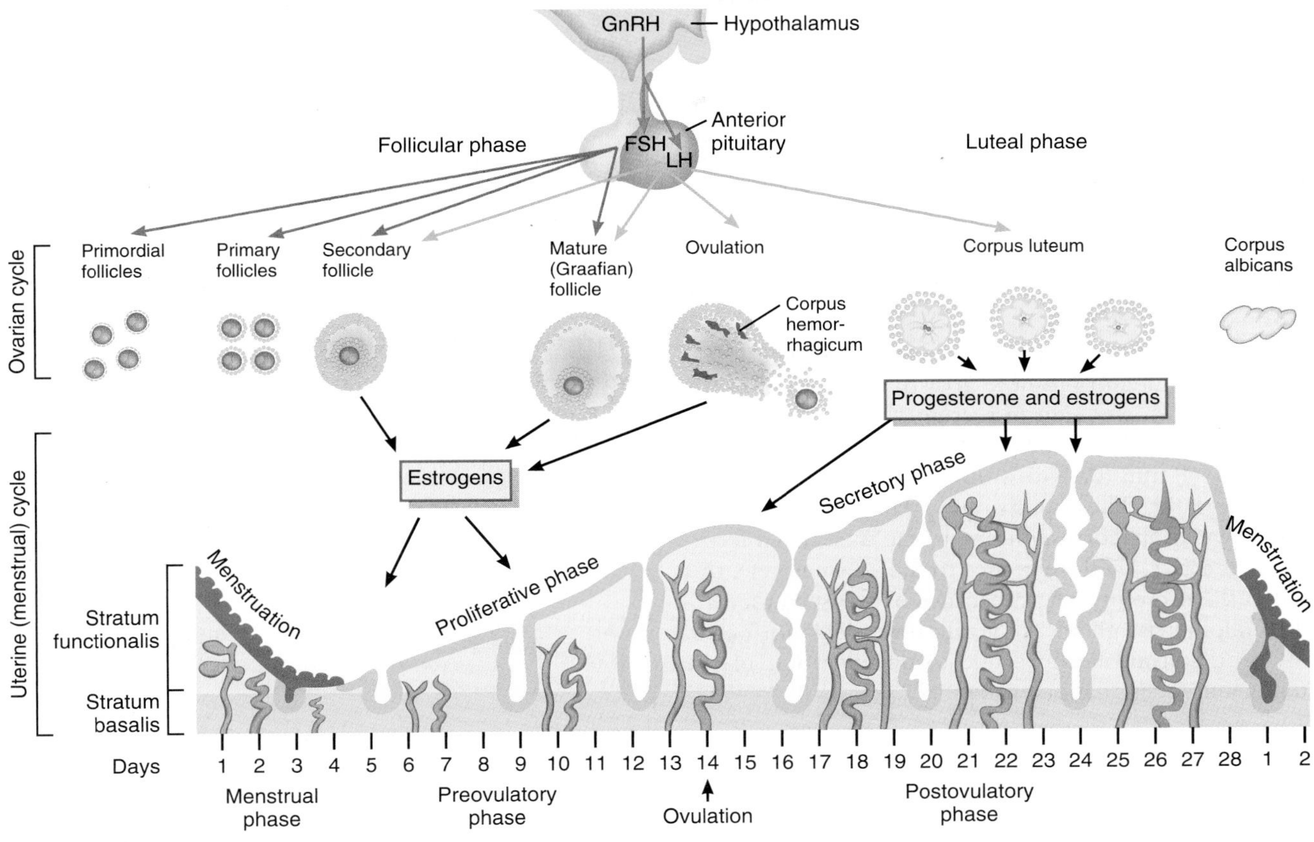

Which hormones are responsible for the proliferative phase of endometrial growth, for ovulation, for growth of the corpus luteum, and for the surge of LH at midcycle?

curs on day 14 in a 28-day cycle. During ovulation, the secondary oocyte remains surrounded by its zona pellucida and corona radiata. It generally takes a total of about 20 days (spanning the last six days of the previous cycle and the first 14 days of the current cycle) for a secondary follicle to develop into a fully mature follicle. During this time the primary oocyte completes meiosis I to become a secondary oocyte, which begins meiosis II but then halts in metaphase. The high levels of estrogens produced during the latter part of the preovulatory phase stimulate the release of LH by the anterior pituitary gland. This LH surge brings about ovulation. An over-the-counter home test may be used to detect the LH surge and can predict ovulation a day in advance.

After ovulation, the mature follicle collapses; once the blood resulting from minor bleeding during rupture of the follicle clots, the follicle becomes the **corpus hemorrhagicum** (*hemo-* = blood; *rrhagic-* = bursting forth) (see Figure 26.15). The clot is absorbed by the remaining follicular cells, which enlarge and form the corpus luteum under the influence of LH. Stimulated by LH, the corpus luteum secretes progesterone, estrogen, relaxin, and inhibin.

Postovulatory Phase

The postovulatory phase of the female reproductive cycle is the most constant in duration: It lasts for 14 days, from days 15 to 28 in a 28-day cycle (see Figure 26.26). It represents the time between ovulation and the onset of the next menses.

EVENTS IN ONE OVARY After ovulation, LH stimulates the remnants of the mature follicle to develop into the corpus luteum, which secretes increasing quantities of progesterone and some estrogens. With reference to the ovarian cycle, this phase is also called the **luteal phase.** Subsequent events in the ovary that ovulated an oocyte depend on whether or not the oocyte becomes fertilized. If the oocyte is not fertilized, the corpus luteum has a lifespan of only 2 weeks; its secretory activity then declines, and it degenerates into a corpus albicans (see Figure 26.15). As the levels of progesterone, estrogens, and inhibin decrease, release of GnRH, FSH, and LH rise due to loss of negative feedback suppression by the ovarian hormones. Then, follicular growth resumes and a new ovarian cycle begins.

If the secondary oocyte is fertilized and begins to divide, the corpus luteum persists past its normal 2-week life span. It is "rescued" from degeneration by **human chorionic gonadotropin** (kō-rē-ON-ik) **(hCG),** a hormone produced by the chorion of the embryo as early as 8–12 days after fertilization; hCG acts like LH in stimulating the secretory activity of the corpus luteum. The presence of hCG in maternal blood or urine is an indicator of pregnancy.

Events in the Uterus Progesterone and estrogens produced by the corpus luteum promote growth and coiling of the endometrial glands (which begin to secrete glycogen), vascularization of the superficial endometrium, thickening of the endometrium to 12–18 mm (0.48–0.72 in), and an increase in the amount of tissue fluid. These preparatory changes peak about one week after ovulation, corresponding to the time that a fertilized ovum might arrive. With reference to the uterine cycle, this phase is called the secretory phase because of the secretory activity of the endometrial glands. If fertilization does not occur, the level of progesterone declines due to degeneration of the corpus luteum, and menstruation ensues.

Menstrual Abnormalities

Amenorrhea (ā-men′-ō-RĒ-a; *a-* = without; *men-* = month; *-rrhea* = a flow) is the absence of menstruation; it may be caused by a hormone imbalance, obesity, extreme weight loss, or very low body fat such as may occur during rigorous athletic training. **Dysmenorrhea** (dis′-men-ō-RĒ-a; *dys-* = difficult or painful) refers to pain associated with menstruation; the term is usually reserved to describe menstrual symptoms that are severe enough to prevent a woman from functioning normally for one or more days each month. Some cases are caused by uterine tumors, ovarian cysts, pelvic inflammatory disease, or intrauterine devices.

Abnormal uterine bleeding includes menstruation of excessive duration or excessive amount, diminished menstrual flow, too frequent menstruation, intermenstrual bleeding, and postmenopausal bleeding. These abnormalities may be caused by disordered hormonal regulation, emotional factors, fibroid tumors of the uterus, and systemic diseases. ■

✓ Describe the function of each of the following hormones in the uterine and ovarian cycles: GnRH, FSH, LH, estrogens, progesterone, and inhibin.

✓ Briefly outline the major events of each phase of the uterine cycle, and correlate them with the events of the ovarian cycle.

BIRTH CONTROL METHODS

Objective

- Compare the various kinds of birth control methods and their effectiveness.

Although no single, ideal method of **birth control** exists, several methods are available, each with advantages and disadvantages. The only method of preventing pregnancy that is 100% reliable is total **abstinence,** the avoidance of sexual intercourse. Other methods discussed here are surgical sterilization, hormonal methods, intrauterine devices, spermatocides, barrier methods, periodic abstinence, coitus interruptus, and induced abortion.

Table 26.2 provides the failure rates for each method.

Surgical Sterilization

Sterilization is a procedure that renders an individual incapable of reproduction. The most common means of sterilization of males is **vasectomy** (vas-EK-tō-mē; *-ectomy* = cut out), which was described on page 791. Sterilization in females most often is achieved by performing a **tubal ligation** (lī-GĀ-shun), in which

Table 26.2 Failure Rates of Several Birth Control Methods

	Failure Rates*	
Method	**Perfect Use†**	**Typical Use**
None	85%	85%
Complete abstinence	0%	0%
Surgical sterilization		
Vasectomy	0.10%	0.15%
Tubal ligation	0.5%	0.5%
Hormonal methods		
Oral contraceptives	0.1%	3%‡
Norplant	0.3%	0.3%
Depo-provera	0.05%	0.05%
Intrauterine device		
Copper T 380A	0.6%	0.8%
Spermatocides	6%	26%‡
Barrier methods		
Male condom	3%	14%‡
Vaginal pouch	5%	21%‡
Diaphragm	6%	20%‡
Periodic abstinence		
Rhythm	9%	25%‡
Sympto-thermal	2%	20%‡
Coitus interruptus	4%	18%‡

*Defined as percentage of women having an unintended pregnancy during the first year of use.
†Failure rate when the method is used correctly and consistently.
‡Includes couples who forgot to use the method.

the uterine tubes are tied closed and then cut. As a result, the secondary oocyte cannot pass into the uterus, and sperm cannot reach the oocyte.

Hormonal Methods

Except for total abstinence or surgical sterilization, hormonal methods are the most effective means of birth control. By using **oral contraceptives** (*"the pill"*) to adjust hormone levels, it is possible to interfere with the production of gametes or implantation of a fertilized ovum in the uterus. The pills used most often contain a higher concentration of a progestin (a substance similar to progesterone) and a lower concentration of estrogens (combination pills). These two hormones decrease the secretion of FSH and LH.

The low levels of FSH and LH usually prevent both follicular development and ovulation; as a result, pregnancy cannot occur because there is no secondary oocyte to fertilize. Even if ovulation does occur, as it does in some cases, oral contraceptives also alter cervical mucus such that it is more hostile to sperm. Oral contraceptives also may be used for **emergency contraception (EC),** the so-called "morning-after pill." When two pills are taken within 72 hours after unprotected intercourse and another two pills are taken 12 hours later, the chance of pregnancy is reduced by 75%.

Other hormonal methods of contraception are Norplant, Depo-provera, and the vaginal ring. **Norplant** consists of six slender hormone-containing capsules that are surgically implanted under the skin of the arm using local anesthesia. They slowly and continually release a progestin, which inhibits ovulation and thickens the cervical mucus. The effects last for 5 years, and Norplant is about as reliable as sterilization. Removing the Norplant capsules restores fertility. **Depo-provera,** which is given as an intramuscular injection once every 3 months, contains progestin that prevents maturation of the ovum and causes changes in the uterine lining that make pregnancy less likely. Expected on the market soon, the **vaginal ring** is a doughnut-shaped ring that fits in the vagina and releases either a progestin alone or a progestin and an estrogen. It is worn for 3 weeks and removed for 1 week to allow menstruation to occur.

The quest for an efficient oral contraceptive for males has been disappointing. The challenge is to find substances that block production of functional sperm without causing erectile dysfunction.

Intrauterine Devices

An **intrauterine device (IUD)** is a small object made of plastic, copper, or stainless steel that is inserted into the cavity of the uterus. IUDs cause changes in the uterine lining that prevent implantation of a fertilized ovum.

Spermatocides

Various foams, creams, jellies, suppositories, and douches that contain sperm-killing agents, or **spermatocides,** make the vagina and cervix unfavorable for sperm survival and are available without prescription. The most widely used spermatocide is nonoxynol-9, which kills sperm by disrupting the plasma membrane. It also inactivates the AIDS virus and decreases the incidence of gonorrhea (described on page 817). A spermatocide is more effective when used together with a diaphragm or a condom.

Barrier Methods

Barrier methods are designed to prevent sperm from gaining access to the uterine cavity and uterine tubes. In addition to preventing pregnancy, barrier methods may also provide some protection against sexually transmitted diseases (STDs) such as AIDS, in contrast to oral contraceptives and IUDs, which confer no such protection. Among the barrier methods are use of a condom, a vaginal pouch, or a diaphragm.

A **condom** is a nonporous, latex covering placed over the penis that prevents deposition of sperm in the female reproductive tract. A **vaginal pouch,** sometimes called a female condom, is made of two flexible rings connected by a polyurethane sheath. One ring lies inside the sheath and is inserted to fit over the cervix; the other ring remains outside the vagina and covers the female external genitals.

A **diaphragm** is a rubber dome-shaped structure that fits over the cervix and is used in conjunction with a spermatocide. It can be inserted up to 6 hours before intercourse. The diaphragm stops most sperm from passing into the cervix; the spermatocide kills most sperm that do get by. Although a diaphragm use does decrease the risk of some STDs, it does not fully protect against HIV infection.

Periodic Abstinence

The **rhythm method** takes advantage of the fact that a secondary oocyte is fertilizable for only 24 hours and is available for only 3–5 days in each reproductive cycle. During this time (3 days before ovulation, the day of ovulation, and 3 days after ovulation) the couple abstains from intercourse. The effectiveness of the rhythm method for birth control is poor in many women due to their irregular cycles.

In the **sympto-thermal method,** couples are instructed to know and understand certain signs of fertility and infertility. The signs of ovulation include increased basal body temperature; the production of clear, stretchy cervical mucus; abundant cervical mucus; and pain associated with ovulation. If a couple abstains from sexual intercourse when the signs of ovulation are present and for 3 days afterward, the chance of pregnancy is decreased. A problem with this method is that fertilization is quite likely if intercourse occurred up to 2 days *before* ovulation.

Coitus Interruptus

Coitus (KŌ-i-tus) **interruptus** is withdrawal of the penis from the vagina just before ejaculation. Failures with this method are due either to failure to withdraw before ejaculation or to preejaculatory emission of sperm-containing fluid from the urethra.

Induced Abortion

Abortion refers to the premature expulsion from the uterus of the products of conception, usually before the 20th week of pregnancy. An abortion may be spontaneous (naturally occurring; also called a miscarriage) or induced (intentionally performed). When birth control methods are not used or fail to prevent an unwanted pregnancy, an **induced abortion** may be performed. Induced abortions may involve vacuum aspiration (suction), infusion of a saline solution, or surgical evacuation (scraping).

Certain drugs, most notably RU 486, can induce a so-called nonsurgical abortion. **RU 486 (mifepristone)** is an antiprogestin; it blocks the action of progesterone. Progesterone prepares the uterine endometrium for implantation and then maintains the uterine lining after implantation. If progesterone levels fall during pregnancy or if the action of the hormone is blocked, menstruation occurs, and the embryo is sloughed off along with the uterine lining. RU 486 can be taken up to 5 weeks after conception.

✓ Explain how oral contraceptives reduce the likelihood of pregnancy.

✓ Why do some methods of birth control protect against sexually transmitted diseases, whereas others do not?

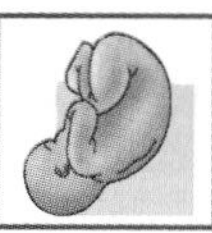

DEVELOPMENTAL ANATOMY OF THE REPRODUCTIVE SYSTEMS

Objective

- Describe the development of the male and female reproductive systems.

The *gonads* develop from the **intermediate mesoderm.** By the sixth week of gestation, they appear as bulges that protrude into the ventral body cavity (Figure 26.27). Adjacent to the gonads are the **mesonephric (Wolffian) ducts,** which eventually develop into structures of the reproductive system in males. A second pair of ducts, the **paramesonephric (Müllerian) ducts,** develop lateral to the mesonephric ducts and eventually form structures of the reproductive system in females. Both sets of ducts empty into the urogenital sinus. An early embryo has the potential to follow either the male or the female developmental pattern because it contains both sets of ducts and primitive gonads that can differentiate into either testes or ovaries.

Cells of male embryos have one X chromosome and one Y chromosome. The male pattern of differentiation is initiated by a Y chromosome "master switch" gene named ***SRY,*** which stands for **S**ex-determining **R**egion of the **Y** chromosome. When *SRY* is activated, primitive Sertoli cells begin to differentiate in the gonadal tissues of male embryos during the seventh week. The developing Sertoli cells secrete a hormone called **Müllerian-inhibiting substance (MIS),** which causes the death of cells within the paramesonephric (Müllerian) ducts. As a result, those cells do not contribute any functional structures to the male reproductive system. Stimulated by human chorionic gonadotropin (hCG), primitive Leydig cells in the gonadal tissue begin to secrete the androgen **testosterone** during the eighth week. Testosterone then stimulates development of the mesonephric duct on each side into the *epididymis, ductus (vas) deferens, ejaculatory duct,* and *seminal vesicle.* The *testes* connect to the mesonephric duct through a series of tubules that eventually become the *seminiferous tubules.* The *prostate* and *bulbourethral glands* are **endodermal** outgrowths of the urethra.

Cells of a female embryo have two X chromosomes and no Y chromosome. Because *SRY* is absent, the gonads develop into *ovaries,* and because MIS is not produced the paramesonephric ducts flourish. The distal ends of the paramesonephric ducts fuse to form the *uterus* and *vagina* whereas the unfused proximal portions become the *uterine (Fallopian) tubes.* The mesonephric ducts degenerate without contributing any functional structures to the female reproductive system because testosterone is absent. The *greater* and *lesser vestibular glands* develop from **endodermal** outgrowths of the vestibule.

The *external genitals* of both male and female embryos also remain undifferentiated until about the eighth week. Before differentiation, all embryos have an elevated midline swelling called the **genital tubercle** (Figure 26.28 on page 816). The tubercle consists of the **urethral groove** (opening into the urogenital sinus), paired **urethral folds,** and paired **labioscrotal swellings.**

In male embryos, some testosterone is converted to a second androgen called **dihydrotestosterone (DHT).** DHT stimulates development of the urethra, prostate gland, and external genitals (scrotum and penis). A portion of the genital tubercle elongates and develops into a penis. Fusion of the urethral folds forms the *spongy (penile) urethra* and leaves an opening to the exterior only at the distal end of the penis, the *external urethral orifice.* The labioscrotal swellings develop into the *scrotum.* In the absence of DHT, the genital tubercle gives rise to the *clitoris* in female embryos. The urethral folds remain open as the *labia minora,* and the labioscrotal swellings become the *labia majora.* The urethral groove becomes the *vestibule.* After birth, androgen levels decline because hCG is no longer present to stimulate secretion of testosterone. ■

✓ Describe the role of hormones in differentiation of the gonads, the mesonephric ducts, the paramesonephric ducts, and the external genitals.

AGING AND THE REPRODUCTIVE SYSTEMS

Objective

- Decribe the effects of aging on the reproductive systems.

During the first decade of life, the reproductive system is in a juvenile state. At about age 10, hormone-directed changes start to occur in both sexes. **Puberty** (PŪ-ber-tē; = a ripe age) is the period when secondary sexual characteristics begin to develop and the potential for sexual reproduction is reached.

In females, the reproductive cycle normally occurs once each month from **menarche** (me-NAR-kē), the first menses, to

Figure 26.27 / Development of the internal reproductive systems.

The gonads develop from intermediate mesoderm.

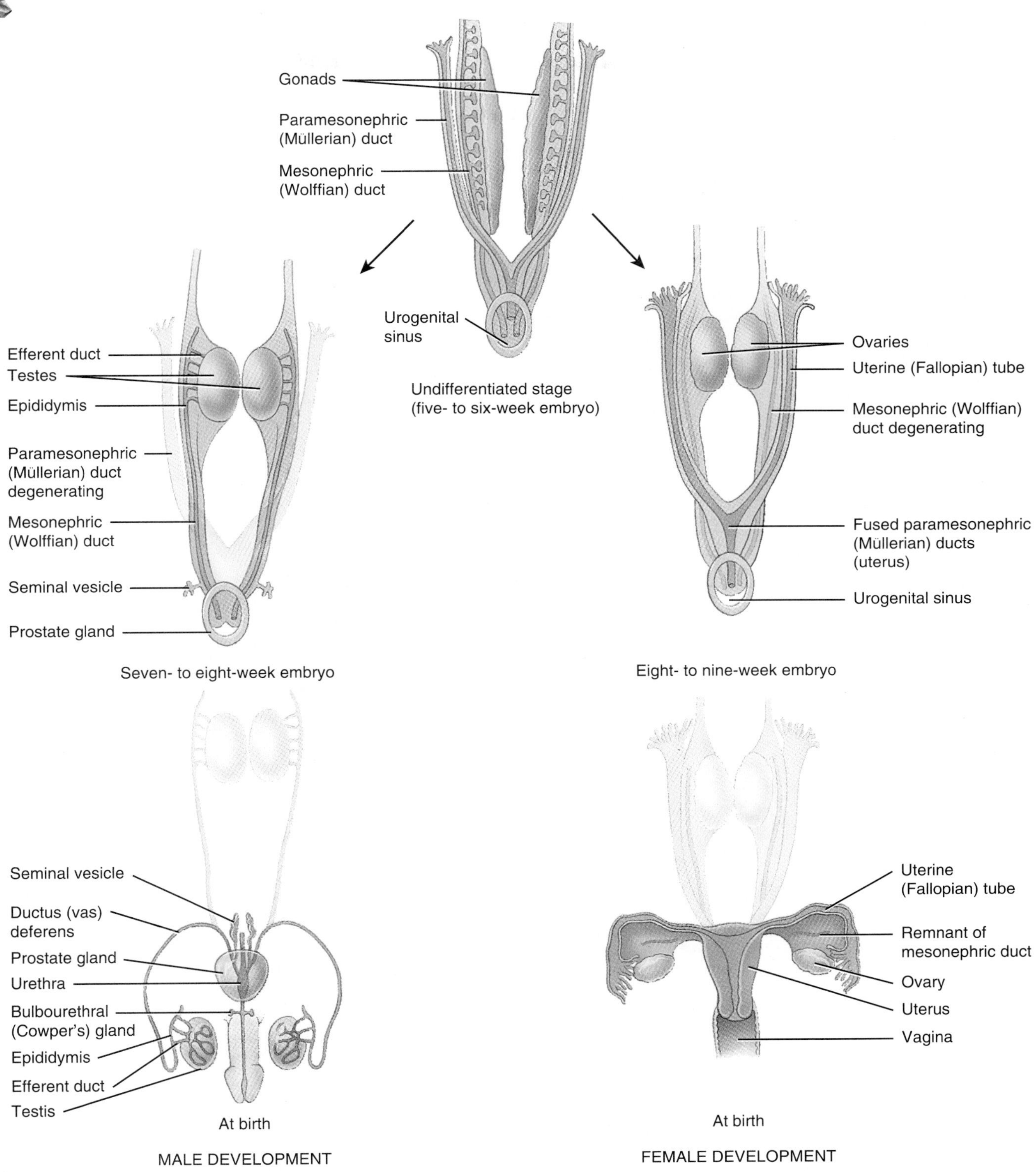

Gonads
Paramesonephric (Müllerian) duct
Mesonephric (Wolffian) duct
Urogenital sinus

Undifferentiated stage (five- to six-week embryo)

Efferent duct
Testes
Epididymis
Paramesonephric (Müllerian) duct degenerating
Mesonephric (Wolffian) duct
Seminal vesicle
Prostate gland

Seven- to eight-week embryo

Ovaries
Uterine (Fallopian) tube
Mesonephric (Wolffian) duct degenerating
Fused paramesonephric (Müllerian) ducts (uterus)
Urogenital sinus

Eight- to nine-week embryo

Seminal vesicle
Ductus (vas) deferens
Prostate gland
Urethra
Bulbourethral (Cowper's) gland
Epididymis
Efferent duct
Testis

At birth

MALE DEVELOPMENT

Uterine (Fallopian) tube
Remnant of mesonephric duct
Ovary
Uterus
Vagina

At birth

FEMALE DEVELOPMENT

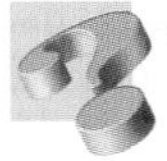

Which gene is responsible for the development of the gonads into testes?

Figure 26.28 / Development of the external genitals.

The external genitals of male and female embryos remain undifferentiated until about the eighth week.

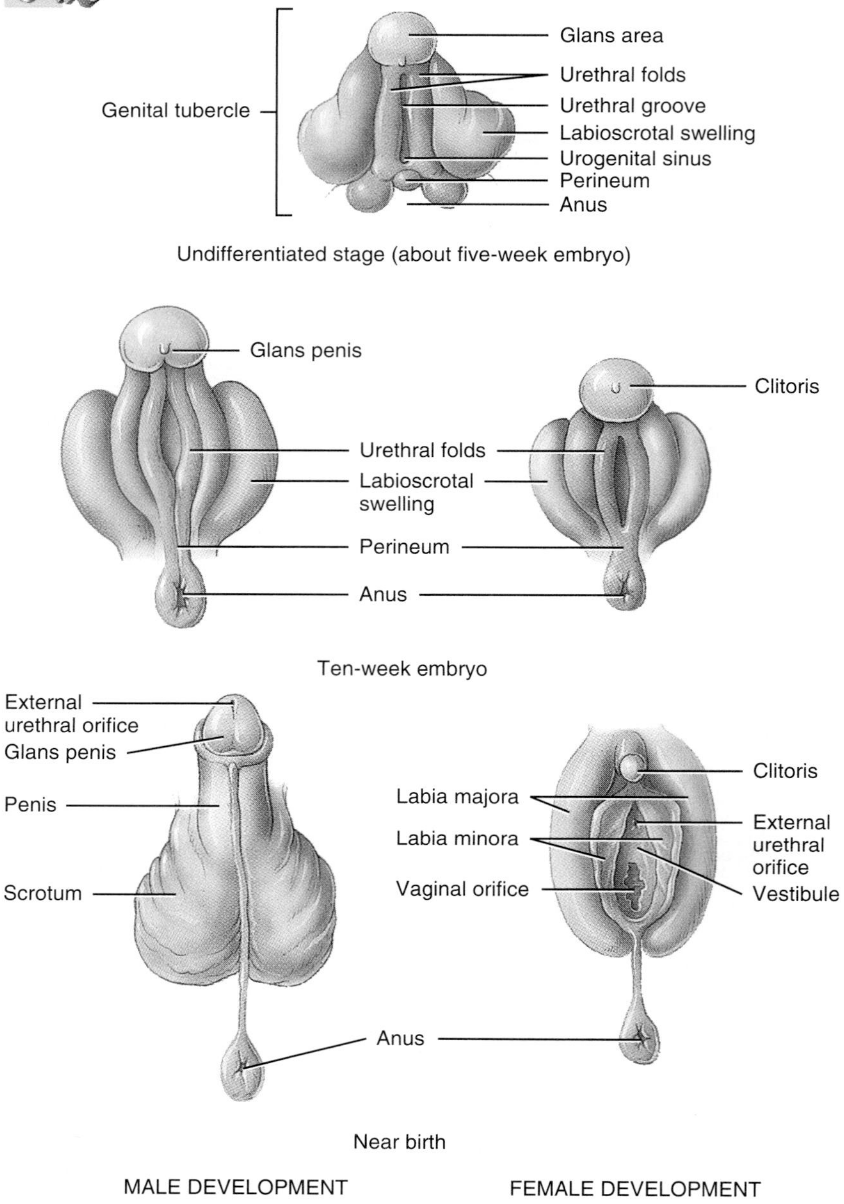

Which hormone is responsible for the differentiation of the external genitals?

menopause, the permanent cessation of menses. Between the ages of 40 and 50 the pool of remaining ovarian follicles becomes exhausted; as a result, the ovaries become less responsive to hormonal stimulation and the production of estrogens declines, despite copious secretion of FSH and LH by the anterior pituitary gland. In addition, changes in the pattern of gonadotropin-releasing hormone (GnRH) release may also contribute to menopausal changes. Some women experience hot flashes and copious sweating, which appear to coincide with increased release of GnRH. Other symptoms of menopause are headache, hair loss, muscular pains, vaginal dryness, insomnia, depression, weight gain, and mood swings. Some atrophy of the ovaries, uterine tubes, uterus, vagina, external genitalia, and breasts occurs in post-menopausal women. Osteoporosis is also a possible occurrence due to the diminished level of estrogens. Sexual desire (libido) does not show a parallel decline; it may be maintained by adrenal sex steroids.

The female reproductive system has a time-limited span of fertility between menarche and menopause. Fertility declines with age, due to the small number of follicles left in the ovaries, less frequent ovulation, and possibly the declining ability of the uterine tubes and uterus to support the young embryo. Uterine cancer peaks at about 65 years of age, but cervical cancer is more common in younger women.

In males, declining reproductive function is much more subtle than in females. Healthy men often retain reproductive capacity into their eighties or nineties. At about age 55 a decline in testosterone synthesis leads to reduced muscle strength, fewer viable sperm, and decreased sexual desire. However, abundant sperm may be present even in old age. Enlargement of the prostate gland is described on page 818.

✓ List some of the changes that occur in males and females at puberty.

✓ Distinguish between menarche and menopause.

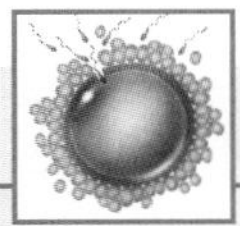

APPLICATIONS TO HEALTH

Sexually Transmitted Diseases

A **sexually transmitted disease (STD)** is one that is spread by sexual contact. In most developed countries of the world, such as those of the European Community, Japan, Australia, and New Zealand, the incidence of STDs has declined markedly during the past 25 years. In the United States, by contrast, STDs have been rising to near epidemic proportions, especially among urban populations. AIDS and hepatitis B, which are sexually transmitted diseases that also may be contracted in other ways, are discussed in Chapters 15 and 24, respectively.

Chlamydia

Chlamydia (kla-MID-ē-a) is a sexually transmitted disease caused by the bacterium *Chlamydia trachomatis* (*chlamy-* = cloak). This unusual bacterium cannot reproduce outside body cells; it "cloaks" itself inside cells, where it divides. At present, chlamydia is the most prevalent sexually transmitted disease, affecting 3–5 million individuals in the United States and causing sterility in more than 20,000 young men and women annually.

In most cases the initial infection is asymptomatic and thus difficult to recognize clinically. In males, urethritis is the principal result, causing a clear discharge, burning on urination, frequent urination, and painful urination. Without treatment, the epididymides may also become inflamed, leading to sterility. In 70% of females with chlamydia, symptoms are absent, but chlamydia is the leading cause of pelvic inflammatory disease. Moreover, the uterine tubes may also become inflamed, which increases the risk of ectopic pregnancy (implantation of a fertilized ovum outside the uterus) and infertility due to the formation of scar tissue in the tubes.

Gonorrhea

Gonorrhea (gon-ō-RĒ-a), or **"the clap,"** is caused by the bacterium *Neisseria gonorrhoeae.* In the United States, 1–2 million new cases of gonorrhea appear each year, most among individuals aged 15–19 years. Discharges from infected mucus membranes are the source of transmission of the bacteria either during sexual contact or during the passage of a newborn through the birth canal. The infection site can be in the mouth and throat after oral-genital contact, in the vagina and penis after genital intercourse, or in the rectum after recto-genital contact.

Males usually experience urethritis with profuse pus drainage and painful urination. The prostate gland and epididymis may also become infected. In females, infection typically occurs in the vagina, often with a discharge of pus. Both infected males and females may harbor the disease without any symptoms, however, until it has progressed to a more advanced stage; about 5–10% of males and 50% of females are asymptomatic. In females, the infection and consequent inflammation can proceed from the vagina into the uterus, uterine tubes, and pelvic cavity. An estimated 50,000–80,000 women in the United States are made infertile by gonorrhea every year as a result of scar tissue formation that closes the uterine tubes. If bacteria in the birth canal are transmitted to the eyes of a newborn, blindness can result. Administration of a 1% silver nitrate solution in the infant's eyes prevents infection.

Syphilis

Syphilis, caused by the bacterium *Treponema pallidum,* is transmitted through sexual contact or exchange of blood, or through the placenta to a fetus. The disease progresses through several stages. During the *primary stage,* the chief sign is a painless open sore, called a **chancre** (SHANG-ker), at the point of contact. The chancre heals within 1 to 5 weeks. From 6 to 24 weeks later, signs and symptoms such as a skin rash, fever, and aches in the joints and muscles usher in the *secondary state,* which is systemic—the infection spreads to all major body systems. When signs of organ degeneration appear, the disease is said to be in the *tertiary stage.* If the nervous system is involved, the tertiary stage is called **neurosyphilis.** As motor areas become extensively damaged, victims may be unable to control urine and bowel movements; eventually they may become bedridden, unable even to feed themselves. Damage to the cerebral cortex produces memory loss and personality changes that range from irritability to hallucinations.

Genital Herpes

Genital herpes is an incurable STD. Type II herpes simplex virus (HSV-2) causes genital infections, producing painful blisters on the prepuce, glans penis, and penile shaft in males and on the vulva or sometimes high up in the vagina in females. The blisters disappear and reappear in most patients, but the virus itself remains in the body. A related virus, type I herpes simplex virus (HSV-1), causes cold sores on the mouth and lips. Infected individuals typically experience recurrences of symptoms several times a year.

Genital Warts

Warts are an infectious disease caused by viruses. Sexual transmission of **genital warts** is common and is caused by the *human papillomavirus (HPV)*. It is estimated that nearly one million persons a year develop genital warts in the United States. Patients with a history of genital warts may be at increased risk for cervical, vaginal, anal, vulval, and penile cancers. There is no cure for genital warts.

Reproductive System Disorders in Males

Testicular Cancer

Testicular cancer is the most common cancer in males between the ages of 20 and 35; it is also one of the most curable cancers in males. More than 95% of testicular cancers arise from spermatogenic cells within the seminiferous tubules. An early sign of testicular cancer is a mass in the testis, often associated with a sensation of testicular heaviness or a dull ache in the lower abdomen; pain usually does not occur. All males should perform regular testicular self-examinations.

Prostate Disorders

Because the prostate surrounds a portion of the urethra, any prostatic infection, enlargement, or tumor in it can obstruct the flow of urine. Acute and chronic infections of the prostate gland are common in postpubescent males, often in association with inflammation of the urethra. In **acute prostatitis,** the prostate gland becomes swollen and tender. **Chronic prostatitis** is one of the most common chronic infections in men of the middle and later years; on examination, the prostate gland feels enlarged, soft, and very tender, and its surface outline is irregular.

An enlarged prostate gland, two to four times the normal size, occurs in approximately one-third of all males over age 60. The condition is called **benign prostatic hyperplasia (BPH)** and is characterized by frequent urination, nocturia (bedwetting), hesitancy in urination, decreased force of urinary stream, postvoiding dribbling, and a sensation of incomplete emptying.

Prostate cancer is the leading cause of death from cancer in men in the United States, having surpassed lung cancer in 1991. Each year it strikes almost 200,000 U.S. men, causing 40,000 deaths. A blood test can measure the level of prostate-specific antigen (PSA) in the blood. The amount of PSA, which is produced only by prostate epithelial cells, increases with enlargement of the prostate gland and may indicate infection, benign enlargement, or prostate cancer. For males over 40, the American Cancer Society recommends annual examination of the prostate gland by a **digital rectal exam,** in which a physician palpates the gland through the rectum with the fingers (digits). Many physicians also recommend an annual PSA test for males over 50. Treatment for prostate cancer may involve surgery, radiation, hormonal therapy, and chemotherapy. Because many prostate cancers grow very slowly, some urologists recommend "watchful waiting" before treating small tumors in men over age 70.

Erectile Dysfunction

Erectile dysfunction (ED), previously termed *impotence*, is the consistent inability of an adult male to ejaculate or to attain or hold an erection long enough for sexual intercourse. Many cases of impotence are caused by insufficient release of nitric oxide, which relaxes the smooth muscle of the penile arteries. The drug Viagra (sildenafil) enhances effects stimulated by nitric oxide in the penis. Other causes of erectile dysfunction include diabetes mellitus, physical abnormalities of the penis, systemic disorders such as syphilis, vascular disturbances (arterial or venous obstructions), neurological disorders, surgery, testosterone deficiency, and drugs (alcohol, antidepressants, antihistamines, antihypertensives, narcotics, nicotine, and tranquilizers). Psychological factors such as anxiety or depression, fear of causing pregnancy, fear of sexually transmitted diseases, religious inhibitions, and emotional immaturity may also cause erectile dysfunction.

Reproductive System Disorders in Females

Premenstrual Syndrome

Premenstrual syndrome (PMS) refers to severe physical and emotional distress that occurs during the postovulatory (luteal) phase of the female reproductive cycle. Signs and symptoms usually increase in severity until the onset of menstruation and then dramatically disappear. The signs and symptoms are highly variable from one woman to another and include edema, weight gain, breast swelling and tenderness, abdominal distension, backache, joint pain, constipation, skin eruptions, fatigue and lethargy, greater need for sleep, depression or anxiety, irritability, mood swings, headache, poor coordination and clumsiness, and cravings for sweet or salty foods. The pathophysiology of PMS is unknown. For some women, getting regular exercise, avoiding caffeine, salt, and alcohol, and eating a diet high in complex carbohydrates and lean proteins can bring considerable relief. Because PMS takes place only after ovulation has occurred, oral contraceptives are an effective treatment for women whose symptoms are incapacitating.

Endometriosis

Endometriosis (en-dō-mē-trē-Ō-sis; *endo-* = within; *metri-* = uterus; *osis* = condition) is characterized by the growth of endometrial tissue outside the uterus. The tissue enters the pelvic cavity via the open uterine tubes and may be found in any of several sites—on the ovaries, the rectouterine pouch, the outer surface of the uterus, the sigmoid colon, pelvic and abdominal lymph nodes, the cervix, the abdominal wall, the kidneys, and the urinary bladder. Endometrial tissue responds to hormonal fluctuations—whether it is inside or outside the uterus—by first proliferating and then breaking down and bleeding, which causes inflammation, pain, scarring, and infertility. Symptoms include premenstrual pain or unusually severe menstrual pain.

Breast Cancer

One in eight U.S. women faces the prospect of **breast cancer.** After lung cancer, it is the second-leading cause of death from cancer in U.S. women; it seldom occurs in men. In females, breast cancer is rarely seen before age 30, and its incidence rises rapidly after menopause. An estimated 5% of the 180,000 cases diagnosed each year in the United States, particularly those that arise in younger women, stem from inherited genetic mutations (changes in the DNA). Researchers have now identified two genes that increase susceptibility to breast cancer: *BRCA1* (*br*east *ca*ncer *1*) and *BRCA2*. Mutation of *BRCA1* also confers a high risk for ovarian cancer. In addition, mutations of the *p53* gene increase the risk of breast cancer in both males and females, and mutations of the androgen receptor gene are associated with the occurrence of breast cancer in some males. Despite the fact that breast cancer generally is not painful until it becomes quite advanced, any lump, no matter how small, should be reported to a physician at once. Early detection—by breast self-examination and mammograms—is the best way to increase the chance of survival.

The most effective technique for detecting tumors less than 1 cm (about 0.5 in.) in diameter is **mammography** (mam-OG-ra-fē; *-graphy* = to record), a type of radiography using very sensitive x-ray film. The image of the breast, called a **mammogram,** is best obtained by compressing the breasts, one at a time, using flat plates. A supplementary procedure for evaluating breast abnormalities is **ultrasound.** Although ultrasound cannot detect tumors smaller than 1 cm in diameter, it can be used to determine whether a lump is a benign, fluid-filled cyst or a solid (and therefore possibly malignant) tumor.

Among the factors that increase the risk of developing breast cancer are (1) a family history of breast cancer, especially in a mother or sister; (2) nulliparity (never having borne a child) or having a first child after age 35; (3) previous cancer in one breast; (4) exposure to ionizing radiation, such as x-rays; (5) excessive alcohol intake; and (6) cigarette smoking.

The American Cancer Society recommends the following steps to help diagnose breast cancer as early as possible:

- All women over 20 should develop the habit of monthly breast self-examination.
- A physician should examine the breasts every 3 years when a woman is between the ages of 20 and 40, and every year after age 40.
- A mammogram should be taken in women between the ages of 35 and 39, to be used later for comparison (baseline mammogram).
- Women with no symptoms should have a mammogram every year or two between ages 40 and 49, and every year after age 50.
- Women of any age with a history of breast cancer, a strong family history of the disease, or other risk factors should consult a physician to determine a schedule for mammography.

Treatment for breast cancer may involve hormone therapy, chemotherapy, radiation therapy, **lumpectomy** (removal of the tumor and the immediate surrounding tissue), a modified or radical mastectomy, or a combination of these approaches. A **radical mastectomy** (*mast-* = breast) involves removal of the affected breast along with the underlying pectoral muscles and the axillary lymph nodes. (Lymph nodes are removed because metastasis of cancerous cells usually occurs through lymphatic or blood vessels.) Radiation treatment and chemotherapy may follow the surgery to ensure the destruction of any stray cancer cells. In some cases of metastatic breast cancer, Herceptin—a monoclonal antibody drug that targets an antigen on the surface of breast-cancer cells—can cause regression of the tumors and retard progression of the disease. Finally, two promising drugs for breast cancer prevention now on the market—Nolvadex (tamoxifen) and Evista (raloxifene)—reduce the incidence of breast cancer.

Ovarian Cancer

Even though ovarian cancer is the sixth most common form of cancer in females, it is the leading cause of death from all gynecological malignancies (excluding breast cancer) because it is difficult to detect before it metastasizes (spreads) beyond the ovaries. Risk factors associated with ovarian cancer include age (usually over age 50); race (whites are at highest risk); family history of ovarian cancer; more than 40 years of active ovulation; nulliparity or first pregnancy after age 30; a high-fat, low-fiber, vitamin A-deficient diet; and prolonged exposure to asbestos and talc. Early ovarian cancer has no symptoms or only mild ones associated with other common problems, such as abdominal discomfort, heartburn, nausea, loss of appetite, bloating, and flatulence. Later-stage signs and symptoms include an enlarged abdomen, abdominal and/or pelvic pain, persistent gastrointestinal disturbances, urinary complications, menstrual irregularities, and heavy menstrual bleeding.

Cervical Cancer

Cervical cancer, carcinoma of the cervix of the uterus, starts with **cervical dysplasia** (dis-PLĀ-sē-a), a change in the shape, growth, and number of cervical cells. If the condition is minimal, the cells may return to normal; if it is severe, it may progress to cancer. In most cases cervical cancer may be detected in its earliest stages by a Pap smear. Some evidence links cervical cancer to the virus that causes genital warts (human papilloma virus). Increased risk is associated with having a large number of sexual partners, having first intercourse at a young age, and smoking cigarettes.

Vulvovaginal Candidiasis

Candida albicans is a yeastlike fungus that commonly grows on mucous membranes of the gastrointestinal and genitourinary tracts. The organism is responsible for **vulvovaginal candidiasis** (vul-vō-VAJ-i-nal can-di-DĪ-a-sis), the most common form of **vaginitis** (vaj′-i′-NĪ-tis), inflammation of the vagina. Candidiasis is characterized by severe itching; a thick, yellow, cheesy discharge; a yeasty odor; and pain. The disorder, experienced at least once by about 75% of females, is usually a result of proliferation of the fungus following antibiotic therapy for another condition. Predisposing conditions include the use of oral contraceptives or cortisone-like medications, pregnancy, and diabetes.

KEY MEDICAL TERMS ASSOCIATED WITH THE REPRODUCTIVE SYSTEMS

Castration (kas-TRĀ-shun; = to prune) Removal, inactivation, or destruction of the gonads; commonly used in reference to removal of the testes only.

Culdoscopy (kul-DOS-kō-pē; *-scopy* = to examine) A procedure in which a culdoscope (endoscope) is inserted through the vagina to view the pelvic cavity.

Endocervical curettage (ku-re-TAZH; *curette* = scraper) A procedure in which the cervix is dilated and the endometrium of the uterus is scraped with a spoon-shaped instrument called a curette; commonly called a D and C (dilation and curettage).

Hermaphroditism (her-MAF-rō-dī-tizm′) Presence of both male and female sex organs in one individual.

Hypospadias (hī′-pō-SPĀ-dē-as; *hypo-* = below) A displaced urethral opening. In males, the displaced opening may be on the underside of the penis, at the penoscrotal junction, between the scrotal folds, or in the perineum; in females, the urethra opens into the vagina.

Leukorrhea (lū′-kō-RĒ-a; *leuko-* = white) A whitish (non-bloody) vaginal discharge that may occur at any age and affects most women at some time.

Oophorectomy (ō′-of-ō-REK-tō-mē; *ūphor-* = bearing eggs) Removal of the ovaries.

Ovarian cyst The most common form of ovarian tumor, in which a fluid-filled follicle or corpus luteum persists and continues growing.

Pelvic inflammatory disease (PID) A collective term for any extensive bacterial infection of the pelvic organs, especially the uterus, uterine tubes, or ovaries, which is characterized by pelvic soreness, lower back pain, abdominal pain, and urethritis. Often the early symptoms of PID occur just after menstruation. As infection spreads, fever may develop, along with painful abscesses of the reproductive organs.

Salpingectomy (sal′-pin-JEK-tō-mē; *salpingo* = tube) Removal of a uterine tube (oviduct).

Smegma (SMEG-ma) The secretion, consisting principally of desquamated epithelial cells, found chiefly around the external genitalia and especially under the foreskin of the male.

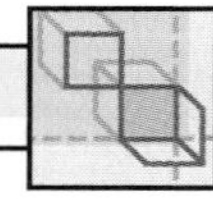

STUDY OUTLINE

The Cell Cycle in the Gonads (p. 782)

1. Reproduction is the process by which new individuals of a species are produced and the genetic material is passed from generation to generation.
2. The organs of reproduction are grouped as gonads (produce gametes), ducts (transport and store gametes), accessory sex glands (produce materials that support gametes), and supporting structures (have various roles in reproduction).
3. Gametes contain the haploid (n) chromosome number, and most somatic cells contain the diploid ($2n$) chromosome number.
4. Meiosis is the process that produces haploid gametes; it consists of two successive nuclear divisions called meiosis I and meiosis II.
5. During meiosis I, homologous chromosomes undergo synapsis (pairing) and crossing-over; the net result is two haploid daughter cells that are genetically unlike each other and unlike the parent cell that produced them.
6. During meiosis II, the two haploid daughter cells divide to form four haploid cells.

Male Reproductive System (p. 784)

1. The male structures of reproduction include the testes, ductus epididymis, ductus (vas) deferens, ejaculatory duct, urethra, seminal vesicles, prostate gland, bulbourethral (Cowper's) glands, and penis.
2. The scrotum is a sac that hangs from the root of the penis and consists of loose skin and superficial fascia; it supports the testes.
3. The temperature of the testes is regulated by contraction of the cremaster muscle and dartos muscle, which either elevates them and brings them closer to the pelvic cavity or relaxes and moves them farther from the pelvic cavity.
4. The testes are paired oval glands (gonads) in the scrotum containing seminiferous tubules, in which sperm cells are made; Sertoli cells (sustentacular cells), which nourish sperm cells and secrete inhibin; and Leydig cells (interstitial endocrinocytes), which produce the male sex hormone testosterone.
5. The testes descend into the scrotum through the inguinal canals during the seventh month of fetal development. Failure of the testes to descend is called cryptorchidism.
6. Secondary oocytes and sperm, both of which are called gametes, are produced in the gonads.
7. Spermatogenesis, which occurs in the testes, is the process whereby immature spermatogonia develop into mature sperm. The spermatogenesis sequence, which includes meiosis I, meiosis II, and spermiogenesis, results in the formation of four haploid sperm (spermatozoa) from each primary spermatocyte.
8. Mature sperm consist of a head, a midpiece, and a tail. Their function is to fertilize a secondary oocyte.
9. The duct system of the testes includes the seminiferous tubules, straight tubules, and rete testis. Sperm flow out of the testes through the efferent ducts.
10. The ductus epididymis is the site of sperm maturation and storage.
11. The ductus (vas) deferens stores sperm and propels them toward the urethra during ejaculation.
12. Each ejaculatory duct, formed by the union of the duct from the seminal vesicle and ductus (vas) deferens, is the passageway for ejection of sperm and secretions of the seminal vesicles into the first portion of the urethra, the prostatic urethra.
13. The urethra in males is subdivided into three portions: the prostatic, membranous, and spongy (penile) urethra.
14. The seminal vesicles secrete an alkaline, viscous fluid that constitutes about 60% of the volume of semen and contributes to sperm viability.
15. The prostate gland secretes a slightly acidic fluid that constitutes about 25% of the volume of semen and contributes to sperm motility.

16. The bulbourethral (Cowper's) glands secrete mucus for lubrication and an alkaline substance that neutralizes acid.
17. Semen is a mixture of sperm and seminal fluid; it provides the fluid in which sperm are transported, supplies nutrients, and neutralizes the acidity of the male urethra and the vagina.
18. The penis consists of a root, a body, and a glans penis.
19. Engorgement of the penile blood sinuses under the influence of sexual excitation is called erection.

Female Reproductive System (p. 796)

1. The female organs of reproduction include the ovaries (gonads), uterine (Fallopian) tubes or oviducts, uterus, vagina, and vulva.
2. The mammary glands are considered part of the reproductive system in females.
3. The ovaries, the female gonads, are located in the superior portion of the pelvic cavity, lateral to the uterus.
4. Ovaries produce secondary oocytes, discharge secondary oocytes (ovulation), and secrete estrogens, progesterone, relaxin, and inhibin.
5. Oogenesis (production of haploid secondary oocytes) begins in the ovaries. The oogenesis sequence includes meiosis I and meiosis II, which goes to completion only after an ovulated secondary oocyte is fertilized by a sperm cell.
6. The uterine (Fallopian) tubes transport secondary oocytes from the ovaries to the uterus and are the normal sites of fertilization. Ciliated cells and peristaltic contractions help move a secondary oocyte or fertilized ovum toward the uterus.
7. The uterus is an organ the size and shape of an inverted pear that functions in menstruation, implantation of a fertilized ovum, development of a fetus during pregnancy, and labor. It also is part of the pathway for sperm to reach the uterine tubes to fertilize a secondary oocyte. Normally, the uterus is held in position by a series of ligaments.
8. Histologically, the layers of the uterus are an outer perimetrium (serosa), a middle myometrium, and an inner endometrium.
9. The vagina is a passageway for sperm and the menstrual flow, the receptacle of the penis during sexual intercourse, and the inferior portion of the birth canal. It is capable of considerable distension.
10. The vulva, a collective term for the external genitals of the female, consists of the mons pubis, labia majora, labia minora, clitoris, vestibule, vaginal and urethral orifices, hymen, bulb of the vestibule, and the paraurethral (Skene's), greater vestibular (Bartholin's), and lesser vestibular glands.
11. The perineum is a diamond-shaped area at the inferior end of the trunk medial to the thighs and buttocks.
12. The mammary glands are modified sweat glands lying superficial to the pectoralis major muscles. Their function is to synthesize, secrete, and eject milk (lactation).
13. Mammary gland development depends on estrogens and progesterone.
14. Milk production is stimulated by prolactin, estrogens, and progesterone; milk ejection is stimulated by oxytocin.

Female Reproductive Cycle (p. 810)

1. The function of the ovarian cycle is to develop a secondary oocyte, whereas that of the uterine (menstrual) cycle is to prepare the endometrium each month to receive a fertilized egg. The female reproductive cycle includes both the ovarian and uterine cycles.
2. The uterine and ovarian cycles are controlled by GnRH from the hypothalamus, which stimulates the release of FSH and LH by the anterior pituitary gland.
3. FSH stimulates development of secondary follicles and initiates secretion of estrogens by the follicles. LH stimulates further development of the follicles, secretion of estrogens by follicular cells, ovulation, formation of the corpus luteum, and the secretion of progesterone and estrogens by the corpus luteum.
4. During the menstrual phase, the stratum functionalis of the endometrium is shed, discharging blood, tissue fluid, mucus, and epithelial cells.
5. During the preovulatory phase, a group of follicles in the ovaries begin to undergo final maturation. One follicle outgrows the others and becomes dominant while the others degenerate. At the same time, endometrial repair occurs in the uterus. Estrogens are the dominant ovarian hormones during the preovulatory phase.
6. Ovulation is the rupture of the dominant mature (Graafian) follicle and the release of a secondary oocyte into the pelvic cavity. It is brought about by a surge of LH. Signs and symptoms of ovulation include increased basal body temperature; clear, stretchy cervical mucus; changes in the uterine cervix; and ovarian pain.
7. During the postovulatory phase, both progesterone and estrogens are secreted in large quantity by the corpus luteum of the ovary, and the uterine endometrium thickens in readiness for implantation.
8. If fertilization and implantation do not occur, the corpus luteum degenerates, and the resulting low level of progesterone allows discharge of the endometrium followed by the initiation of another reproductive cycle.
9. If fertilization and implantation occur, the corpus luteum is maintained by placental hCG, and the corpus luteum and later the placenta secrete progesterone and estrogens to support pregnancy and breast development for lactation.

Birth Control Methods (p. 812)

1. Methods include surgical sterilization (vasectomy, tubal ligation), hormonal methods, intrauterine devices, spermatocides, barrier methods (condom, vaginal pouch, diaphragm), periodic abstinence (rhythm and sympto-thermal methods), coitus interruptus, and induced abortion. See Table 26.2 on page 812 for failure rates for these methods.
2. Contraceptive pills of the combination type contain estrogens and progestins in concentrations that decrease the secretion of FSH and LH and thereby inhibit development of ovarian follicles and ovulation.
3. An abortion is the premature expulsion from the uterus of the products of conception; it may be spontaneous or induced. RU 486 can induce abortion by blocking the action of progesterone.

Developmental Anatomy of the Reproductive Systems (p. 814)

1. The gonads develop from intermediate mesoderm. In the presence of the *SRY* gene, the gonads begin to differentiate into testes during the seventh week. The gonads differentiate into ovaries when the *SRY* gene is absent.
2. In males, testosterone stimulates development of each mesonephric duct into an epididymis, ductus (vas) deferens, ejaculatory duct, and seminal vesicle, and Müllerian-inhibiting substance (MIS) causes the paramesonephric duct cells to die. In females, testosterone and MIS are absent; the paramesonephric ducts develop into the uterine tubes, uterus, and vagina and the mesonephric ducts degenerate.
3. The external genitals develop from the genital tubercle and are

stimulated to develop into typical male structures by the hormone dihydrotestosterone (DHT). The external genitals develop into female structures when DHT is not produced, the normal situation in female embryos.

Aging and the Reproductive Systems (p. 814)

1. Puberty refers to the period of time when secondary sex characteristics begin to develop and the potential for sexual reproduction is reached.
2. In older females, levels of progesterone and estrogens decrease, resulting in changes in menstruation and then menopause; uterine and breast cancer increase in incidence with age.
3. In older males, decreased levels of testosterone are associated with decreased muscle strength, waning sexual desire, and fewer viable sperm; prostate disorders are common.

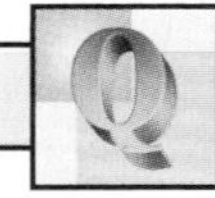

SELF-QUIZ QUESTIONS

Choose the one best answer to these questions.

1. The secondary sex gland(s) that contribute(s) the most volume to the semen is/are the (a) seminal vesicles, (b) bulbourethral (Cowper's) glands, (c) paraurethral (Skene's) glands, (d) greater vestibular (Bartholin's) glands, (e) prostate gland.
2. The efferent ducts empty into the (a) rete testis, (b) epididymis, (c) vas (ductus) deferens, (d) straight tubules, (e) seminiferous tubules.
3. Which of the following statements about ovaries are true? (1) They are female gonads. (2) They produce secondary oocytes and female hormones. (3) They are homologous to the testes. (4) They are attached to the pelvic wall by the suspensory ligament.
   a. 1,3,4 **b.** 2,3,4 **c.** 1,2,4 **d.** 1 and 4
   e. all of the above.
4. Which of the following produce(s) a secretion that helps maintain the motility and viability of sperm? (a) prostate gland, (b) penis, (c) bulbourethral (Cowper's) glands, (d) ejaculatory duct, (e) all of the above.

Complete the following.

5. Place numbers in the blanks to arrange the following structures in the correct sequence for the pathway of sperm. (a) ejaculatory duct: ________ ; (b) testis: ________ ; (c) urethra: ________ ; (d) ductus (vas) deferens: ________ ; (e) epididymis: ________ .
6. Branches of the radial arteries called ________ supply the stratum functionalis of the endometrium.
7. Fill in the blanks with the male homologues for each of the following. (a) labia majora: ________ ; (b) clitoris: ________ ; (c) paraurethral (Skene's) glands: ________ ; (d) greater vestibular (Bartholin's) glands: ________ .
8. The layer of endometrium that is *not* shed during menstruation is the stratum ________ .
9. The thickest layer in the uterine wall is the ________ which is composed of three layers of ________ .
10. If fertilization and implantation do not occur, the corpus luteum degenerates and becomes the ________ .
11. Place numbers in the blanks to arrange the following cells in the correct sequence for the process of sperm production. (a) secondary spermatocyte: ________ ; (b) spermatogonia: ________ ; (c) primary spermatocytes: ________ ; (d) primordial germ cells: ________ ; (e) spermatids: ________ ; (f) spermatozoon: ________ .
12. The ovary is attached to the uterus by the ________ ligament.
13. The rupture of the vesicular ovarian (Graafian) follicle with release of a secondary oocyte into the pelvic cavity is called ________ .
14. Place numbers in the blanks to arrange the following structures in the correct sequence for the pathway of milk in the breasts. (a) lactiferous sinuses: ________ ; (b) alveoli: ________ ; (c) secondary tubules: ________ ; (d) mammary ducts: ________ ; (e) lactiferous ducts: ________ .

Are the following statements true or false?

15. The mass of tissue that surrounds the penile urethra is called the corpus spongiosum.
16. The female urethra is located posterior to the vagina.
17. Contraction of the dartos muscle is responsible for elevation of the testes during sexual arousal or on exposure to cold.
18. The tunica albuginea is a capsule of connective tissue associated with both the ovaries and the testes.
19. The acrosome is the portion of a sperm that contains the genetic material.
20. Ovarian follicles are located in the ovarian cortex.
21. Match the following terms to their descriptions.

____	**(a)** 200–300 of these per testis	**(1)** interstitial endocrinocytes (cells of Leydig)
____	**(b)** tightly coiled tubes (1–3 per lobule); composed of cells that develop into sperm	**(2)** ejaculatory duct
____	**(c)** cells located between developing sperm cells; form the blood-testis barrier and provide nourishment	**(3)** sustentacular (Sertoli) cells
____	**(d)** cells located between seminiferous tubules; secrete testosterone	**(4)** seminiferous tubules
____	**(e)** highly coiled duct in which sperm are stored for maturation	**(5)** ductus (vas) deferens
____	**(f)** ejects sperm and fluid into the urethra just before ejaculation	**(6)** ductus epididymis
____	**(g)** enters the pelvic cavity through the inguinal canal	**(7)** lobule

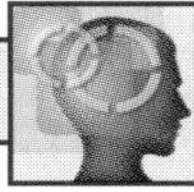

CRITICAL THINKING QUESTIONS

1. Rashim pouted "Mom said she brought my cat Brutus to the vet's to be "fixed" but I think the vet really broke him 'cause now Mom says Brutus can't be making any baby kittens." What really happened to Brutus?
   HINT *Brutus may be careful how he sits for the next few days.*

2. Occasionally, someone feels that he or she has been born the wrong gender and undergoes a "sex-change" or "gender reassignment" process involving hormone treatment and surgery. However, a born male can never truly become a fully biological female or a born female become a fully biological male. Why not?
   HINT *Your gender is determined long before your birth.*

3. Thirty-nine year-old Meg has been advised to have a hysterectomy due to medical problems. She is worried that the procedure will cause menopause. Is this a valid concern?
   HINT *Women often take estrogens to ease the symptoms of menopause.*

4. Once upon a time a handsome primary spermatocyte met a beautiful primary oocyte and they fell happily in live. But, alas, the couple did not live happily ever after. Predict the future outcome of their relationship.
   HINT *They still have some growing up to do.*

5. Phil has promised his wife that he will get a vasectomy after the birth of their next child. However, he's concerned however about possible effects on his virility. How would you respond to Phil's concerns?
   HINT *A vasectomy is used for birth control, not sex control.*

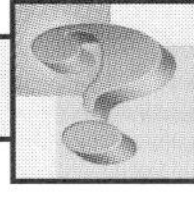

ANSWERS TO FIGURE QUESTIONS

26.1 In meiosis, chromosomal replication occurs during the interphase that precedes meiosis I.

26.2 The result of crossing-over is that daughter cells are genetically unlike each other and genetically unlike the parent cell that produced them.

26.3 The gonads (testes) produce gametes (sperm) and hormones; the ducts transport, store, and receive gametes; and the accessory sex glands secrete materials that support gametes.

26.4 The cremaster and dartos muscles help regulate the temperature of the testes.

26.5 The tunica vaginalis and tunica albuginea are tissue layers that cover and protect the testes.

26.6 Leydig cells secrete testosterone.

26.7 During meiosis I, the number of chromosomes in each cell is reduced by half.

26.8 The sperm head contains DNA and enzymes for penetration of a secondary oocyte; the midpiece contains mitochondria for ATP production; the tail consists of a flagellum that provides propulsion.

26.9 Sperm maturation, sperm storage, and propulsion of sperm into ductus (vas) deferens.

26.10 Stores sperm and conveys sperm toward urethra.

26.11 Seminal vesicles contribute the largest volume to seminal fluid.

26.12 Two corpora cavernosa penis and one corpus spongiosum penis contain blood sinuses that fill with blood that cannot flow out of the penis as quickly as it flows in. The trapped blood engorges and stiffens the tissue, producing an erection. The corpus spongiosum penis keeps the spongy urethra open so that ejaculation can occur.

26.13 The testes are homologous to the ovaries; the glans penis is homologous to the clitoris; the prostate gland is homologous to the paraurethral glands; and the bulbourethral gland is homologous to the greater vestibular glands.

26.14 The mesovarium anchors the ovary to the broad ligament of the uterus and the uterine tube; the ovarian ligament anchors it to the uterus; the suspensory ligament anchors it to the pelvic wall.

26.15 Ovarian follicles secrete estrogens; the corpus luteum secretes progesterone, estrogens, relaxin, and inhibin.

26.16 Most ovarian follicles undergo atresia (degeneration).

26.17 Primary oocytes are present in the ovary at birth, so they are as old as the woman is. In males, primary spermatocytes are continually being formed from stem cells (spermatogonia) and thus are only a few days old.

26.18 Fertilization most often occurs in the ampulla of the uterine tube.

26.19 Ciliated columnar epithelial cells and secretory cells with microvilli line the uterine tubes.

26.20 The endometrium is a highly vascular, secretory epithelium that provides the oxygen and nutrients needed to sustain a fertilized egg; the myometrium is a thick smooth muscle layer that supports the uterine wall during pregnancy and contracts to expel the fetus at birth.

26.21 The stratum basalis of the endometrium provides cells to replace those that shed (the stratum functionalis) during each menstruation.

26.22 Anterior to the vaginal opening are the mons pubis, clitoris, and prepuce. Lateral to the vaginal opening are the labia minora and labia majora.

26.23 The anterior portion of the perineum is called the urogenital triangle because its borders form a triangle that encloses the urethral (uro-) and vaginal (-genital) orifices.

26.24 Prolactin, estrogens, and progesterone regulate the synthesis of milk. Oxytocin regulates the ejection of milk.

26.25 Alveoli are located in lobules.

26.26 The hormones responsible for the proliferative phase of endometrial growth are estrogens; for ovulation, LH; for growth of the corpus luteum, LH; and for the midcycle surge of LH, estrogens.

26.27 The *SRY* gene on the Y chromosome is responsible for the development of the gonads into testes.

26.28 The presence of dihydrotestosterone (DHT) stimulates differentiation of the external genitals in males; its absence allows differentiation of the external genitals in females.

27

DEVELOPMENTAL ANATOMY

CONTENTS AT A GLANCE

▲Page 831

Examine this image and see if you notice anything unusual. In which direction is time being depicted? What is the symbolism of the flowers? Count the number of women and guess why this number was chosen by the artist.

Page 834 ►

INTRODUCTION

Developmental anatomy is the study of the sequence of events from the fertilization of a secondary oocyte to the formation of an adult organism. From fertilization through the eighth week of development, the developing human is an **embryo** (*-bryo* = grow) and this is the period of **embryological development. Embryology** (em-brē-OL-ō-jē) is the study of development from the fertilized egg through the eighth week. **Fetal development** begins at week nine and continues until birth; during this time the developing human is a **fetus** (*feo* = to bring forth).

Obstetrics (ob-STET-riks; *obstetrix* = midwife) is the branch of medicine that deals with the management of pregnancy, labor, and the **neonatal period,** the first 42 days after birth. **Prenatal** (prē-NĀ-tal: *pre* = before; *natus* = born) **development,** is the time from fertilization to birth and includes both embryological and fetal development. It is conveniently divided into three periods of three calendar months each called **trimesters.** Accordingly, obstetricians divide the nine months of pregnancy into three trimesters.

1. The **first trimester** is the period of embryonic and early fetal development. It is the most critical stage of development during which the rudiments of all the major organ systems appear.
2. The **second trimester** is characterized by the almost complete development of organ systems. By the end of this stage, the fetus assumes distinctively human features.
3. The **third trimester** represents a period of rapid fetal growth. During the early stages of this period, most of the organ systems are fully functional.

In this chapter we focus on the developmental sequence from fertilization through implantation, embryonic and fetal development, labor, and birth.

FROM FERTILIZATION TO IMPLANTATION

Objective

- Explain the processes associated with fertilization, morula formation, blastocyst development, and implantation.

Once sperm and a secondary oocyte have developed through meiosis and maturation, and the sperm have been deposited in the vagina, pregnancy can occur. **Pregnancy** is a sequence of events that begins with fertilization, proceeds to implantation, embryonic development, and fetal development, and normally ends with birth about 38 weeks later.

Fertilization

During **fertilization** (fer-til-i-ZĀ-shun; *fertil-* = fruitful) the genetic material from a haploid sperm cell (spermatozoon) and a haploid secondary oocyte merges into a single diploid nucleus. Of the 300–500 million sperm introduced into the vagina, fewer than 1% reach the secondary oocyte. Fertilization normally occurs in the uterine (Fallopian) tube about 12–24 hours after ovulation. Because sperm remain viable in the vagina for about 48 hours and a secondary oocyte is viable for about 24 hours after ovulation, there typically is a 3-day window during which pregnancy is most likely to occur—from 2 days before ovulation to 1 day after ovulation.

The process leading to fertilization begins when peristaltic contractions and the action of cilia transport the oocyte through the uterine tube. Sperm swim up the uterus and into the uterine tubes, propelled by the whiplike movements of their tails (flagella) and possibly guided by chemical attractants released by the oocyte. Additionally, muscular contractions of the uterus, stimulated by prostaglandins in semen, probably aid the movement of sperm toward the uterine tube. Although 100 or so sperm may reach the vicinity of the oocyte within minutes after ejaculation, they *do not become capable* of fertilizing it until several hours later. During this time in the female reproductive tract, sperm undergo **capacitation,** a series of functional changes that cause the sperm's tail to beat even more vigorously and enable its plasma membrane to fuse with the oocyte's plasma membrane. For fertilization to occur, a sperm cell first must penetrate the **corona radiata,** the cloud of granulosa cells that surround the oocyte, and then the **zona pellucida,** the clear glycoprotein layer between the corona radiata and the oocyte's plasma membrane (Figure 27.1a). A glycoprotein in the zona pellucida called ZP3 acts as a sperm receptor; it binds to specific membrane proteins in the sperm head and triggers the **acrosomal reaction,** the release of the contents of the acrosome. The proteolytic acrosomal enzymes digest a path through the zona pellucida as the lashing sperm tail pushes the sperm cell onward. Many sperm bind to ZP3 molecules and undergo acrosomal reactions, but only the first sperm cell to penetrate the entire zona pellucida and reach the oocyte's plasma membrane fuses with the oocyte.

Fusion of a sperm with a secondary oocyte, called **syngamy** (*syn-* = coming together; *-gamy* = marriage), sets in motion events that block **polyspermy,** fertilization by more than one sperm cell. Within 1–3 seconds, the cell membrane of the oocyte depolarizes, which acts as a *fast block to polyspermy*—a depolarized oocyte cannot fuse with another sperm. Depolarization also triggers the intracellular release of calcium ions, which stimulate exocytosis of secretory vesicles from the oocyte. The molecules released by exocytosis inactivate ZP3 and harden the entire zona pellucida, events that constitute the *slow block to polyspermy.*

Once a sperm cell enters a secondary oocyte, the oocyte completes meiosis II. It divides into a larger ovum (mature egg) and a smaller second polar body that fragments and disintegrates. The nucleus in the head of the sperm develops into the **male pronucleus,** and the nucleus of the fertilized ovum develops into the **female pronucleus** (Figure 27.1c). After the pronuclei form, they fuse, producing a single diploid nucleus that contains 23 chromosomes from each pronucleus. Thus, the fusion of the haploid (n) pronuclei restores the diploid number ($2n$) of 46 chromosomes. The fertilized ovum now is called a **zygote** (ZĪ-gōt; *zygosis* = a joining).

Dizygotic (fraternal) twins are produced from the independent release of two secondary oocytes and the subsequent fertilization of each by different sperm. They are the same age and in the uterus at the same time, but they are genetically as dissimilar as any other siblings. Dizygotic twins may or may not be the same sex. Because **monozygotic (identical) twins** develop from a single fertilized ovum, they contain exactly the same genetic material and are always the same sex. Monozygotic twins arise from separation of the developing cells into two embryos, which occurs before 8 days after fertilization 99% of the time. Separations that occur later are likely to produce **conjoined twins,** in which the twins are joined together and share some body structures.

Figure 27.1 / Selected structures and events in fertilization. (a) A sperm cell penetrating the corona radiata and zona pellucida around a secondary oocyte. (b) A sperm cell in contact with a secondary oocyte. (c) Male and female pronuclei.

During fertilization, genetic material from a sperm cell and an oocyte merge to form a single diploid nucleus.

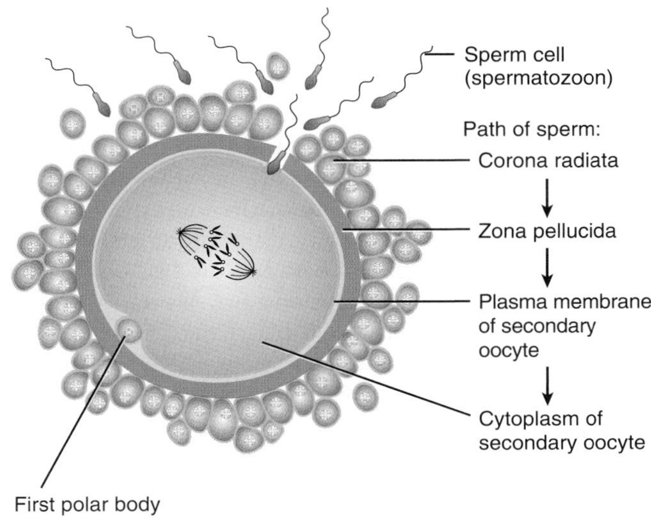

(a) Sperm cell penetrating a secondary oocyte

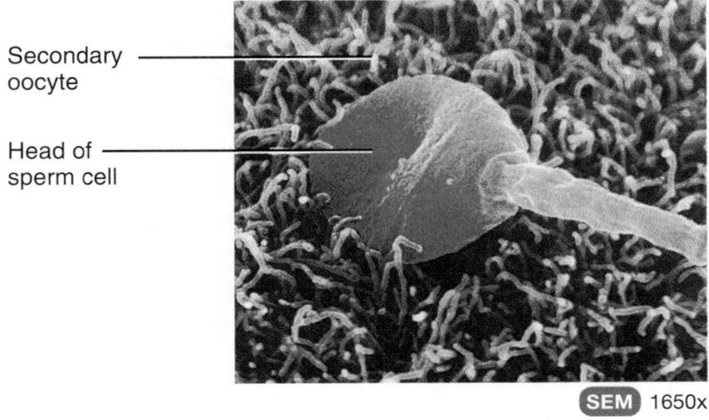

(b) Sperm cell in contact with a secondary oocyte

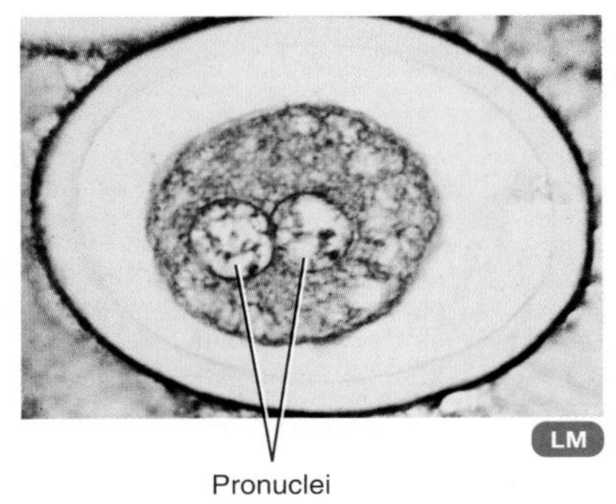

(c) Male and female pronuclei

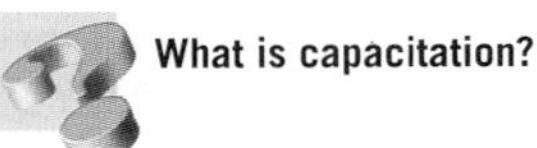

What is capacitation?

Formation of the Morula

After fertilization, the zygote undergoes a series of rapid mitotic cell divisions called **cleavage** (Figure 27.2). The first division begins about 24 hours after fertilization and is completed about 30 hours after fertilization. Each succeeding division takes slightly less time. By the second day after fertilization, the second cleavage is completed and there are four cells (Figure 27.2b); by the end of the third day, there are 16 cells. The progressively smaller cells produced by cleavage are called **blastomeres** (BLAS-tō-mērz; *blasto-* = germ or sprout; *-meres* = parts). Successive cleavages eventually produce the **morula** (MOR-yū-la; *morula* = mulberry), a solid sphere of cells still surrounded by the zona pellucida; the morula is about the same size as the original zygote (Figure 27.2c).

Figure 27.2 / Cleavage and the formation of the morula and blastocyst.

Cleavage refers to the early, rapid mitotic divisions of a zygote.

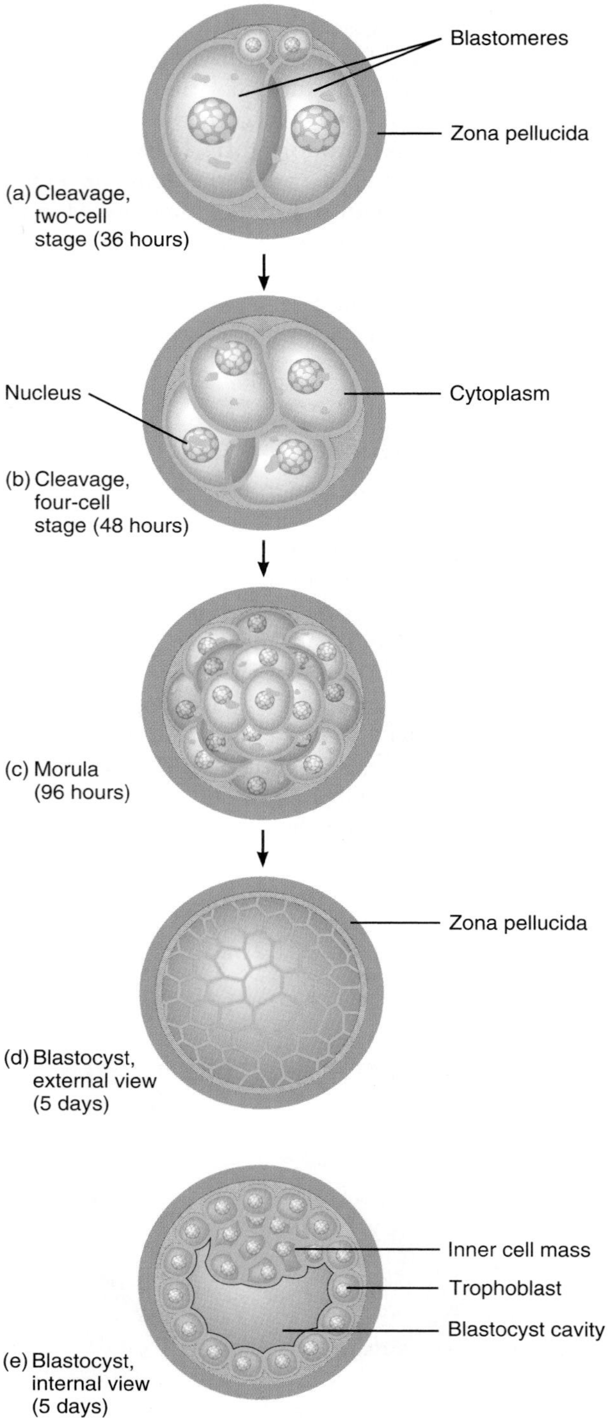

Blastomeres

Zona pellucida

(a) Cleavage, two-cell stage (36 hours)

Nucleus

Cytoplasm

(b) Cleavage, four-cell stage (48 hours)

(c) Morula (96 hours)

Zona pellucida

(d) Blastocyst, external view (5 days)

Inner cell mass

Trophoblast

Blastocyst cavity

(e) Blastocyst, internal view (5 days)

What is the histological difference between a morula and a blastocyst?

Development of the Blastocyst

By the end of the fourth day, the number of cells in the morula increases as it continues to move through the uterine tube toward the uterine cavity. At 4.5–5 days, the dense cluster of cells has developed into a hollow ball of cells that enters the uterine cavity; it is now called a **blastocyst** (*-cyst* = bag) (Figure 27.2d, e).

The blastocyst has an outer covering of cells called the **trophoblast** (TRŌF-ō-blast; *tropho-* = develop or nourish), an **inner cell mass (embryoblast),** and an internal fluid-filled cavity called the **blastocyst cavity.** The trophoblast and part of the inner cell mass ultimately form the membranes composing the fetal portion of the placenta; the rest of the inner cell mass develops into the embryo.

Implantation

The blastocyst remains free within the cavity of the uterus for about 2 days before it attaches to the uterine wall. The endometrium is in its secretory phase, and the blastocyst receives nourishment from the glycogen-rich secretions of the endometrial glands, sometimes called uterine milk. During this time, the zona pellucida disintegrates and the blastocyst enlarges. About 6 days after fertilization the blastocyst attaches to the endometrium, a process called **implantation** (Figure 27.3).

As the blastocyst implants, usually in either the posterior portion of the fundus or the body of the uterus, it is oriented such that the inner cell mass is toward the endometrium (Figure 27.3b). In the region of contact between the blastocyst and endometrium, the trophoblast develops two layers: a **syncytiotrophoblast** (sin-sīt′-ē-ō-TRŌF-ō-blast) that contains no cell boundaries, and a **cytotrophoblast** (sī-tō-TRŌF-ō-blast) between the inner cell mass and syncytiotrophoblast that is composed of distinct cells (Figure 27.3c). These two layers of trophoblast become part of the chorion (one of the fetal membranes) as they undergo further growth (see Figure 27.5b). During implantation, the syncytiotrophoblast secretes enzymes that enable the blastocyst to penetrate the uterine lining by digesting and liquefying the endometrial cells. The endometrial secretions further nourish the burrowing blastocyst for about a week after implantation. Eventually, the blastocyst becomes buried in the endometrium. Another secretion of the trophoblast is human chorionic gonadotropin (hCG), which has actions similar to LH. Human chorionic gonadotropin rescues the corpus luteum from degeneration and sustains its secretion of progesterone and estrogens. These hormones maintain the uterine lining in a secretory state and thereby prevent menstruation.

The principal events associated with fertilization and implantation are summarized in Figure 27.4 on page 830.

Ectopic Pregnancy

Ectopic pregnancy (*ec-* = out of; *-topic* = place) is the development of an embryo or fetus outside the uterine cavity. An ectopic pregnancy usually occurs when passage of the fertilized ovum through the uterine tube is impaired, typically either by

Figure 27.3 / Relation of a blastocyst to the endometrium of the uterus at the time of implantation.

Implantation, the attachment of a blastocyst to the endometrium, occurs about 6 days after fertilization.

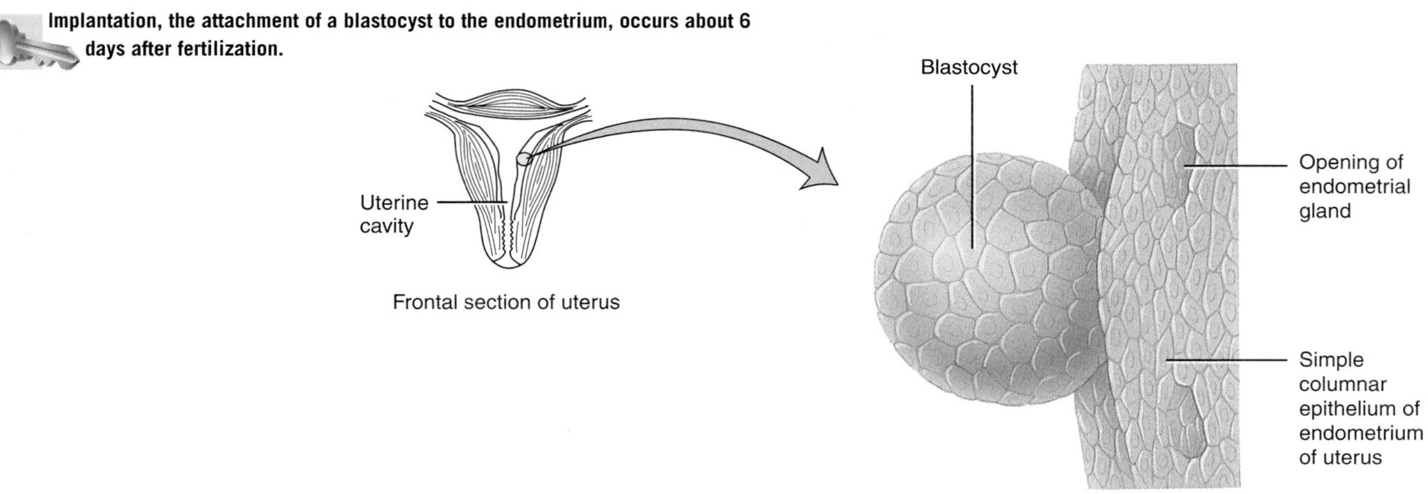

(a) External view, about 6 days after fertilization

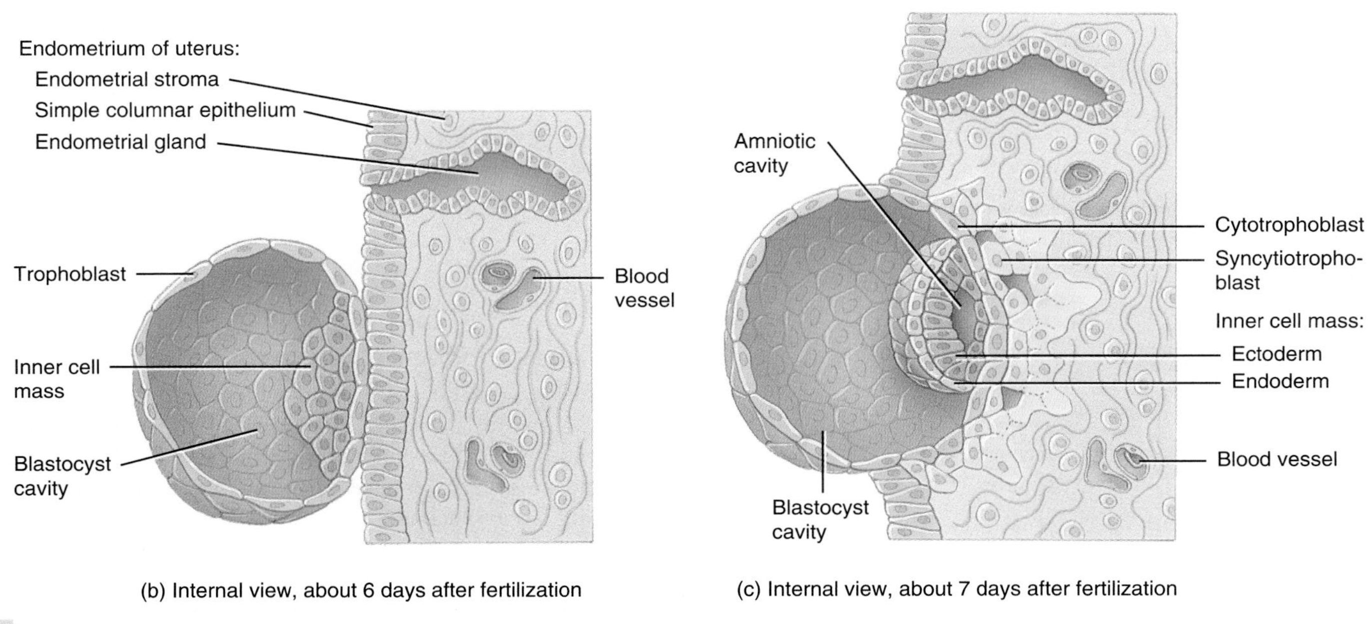

(b) Internal view, about 6 days after fertilization

(c) Internal view, about 7 days after fertilization

How does the blastocyst merge with and burrow into the endometrium?

decreased motility of the uterine tube smooth muscle or abnormal anatomy. Although the most common sites of ectopic pregnancies are the ampullar and infundibular portions of the uterine tube, ectopic pregnancies may also occur in the abdominal cavity or uterine cervix. Compared to nonsmokers, women who smoke and become pregnant are twice as likely to have an ectopic pregnancy because nicotine in cigarette smoke paralyzes the cilia in the lining of the uterine tube (as it does those in the airways). Scars from pelvic inflammatory disease, previous uterine tube surgery, and previous ectopic pregnancy may hinder movement of the fertilized ovum.

The signs and symptoms of ectopic pregnancy include one or two missed menstrual cycles followed by bleeding and acute abdominal and pelvic pain. Unless removed, the developing embryo can rupture the tube, often resulting in death of the mother. ■

- ✓ Define developmental anatomy.
- ✓ Where does fertilization normally occur?
- ✓ How is polyspermy prevented?
- ✓ What is a morula, and how is it formed?
- ✓ Describe the components of a blastocyst.

Figure 27.4 / Summary of events associated with fertilization and implantation.

Fertilization usually occurs in the uterine tube.

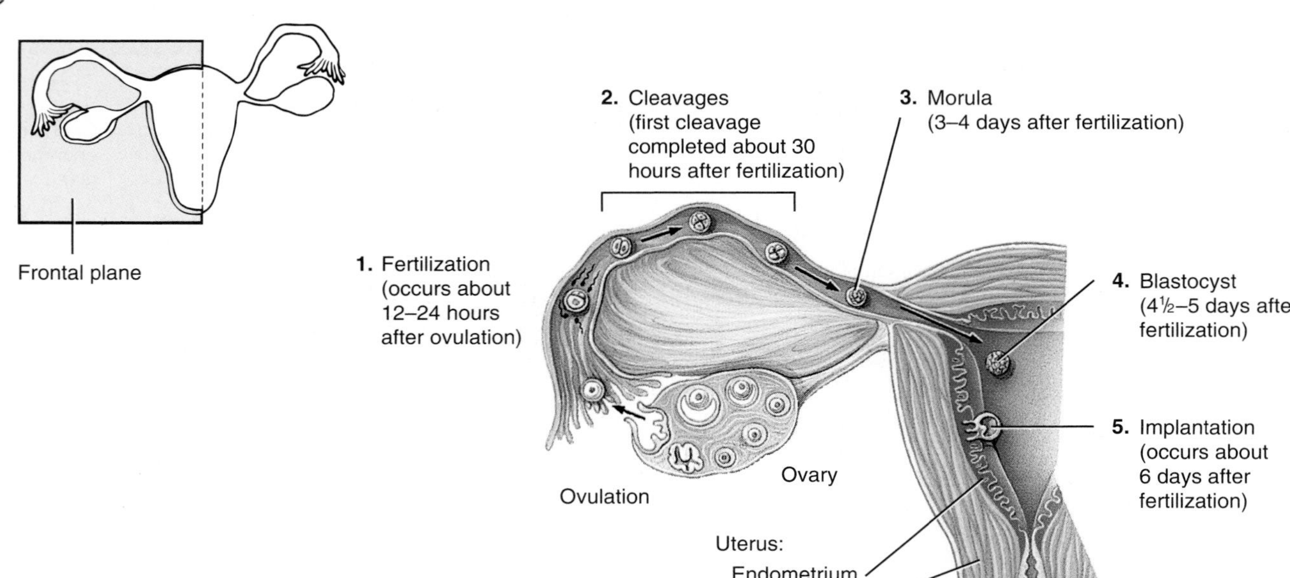

Frontal section through uterus, uterine tube, and ovary

In which phase of the uterine cycle does implantation occur?

EMBRYONIC AND FETAL DEVELOPMENT

Objectives

- Discuss the formation of the primary germ layers and embryonic membranes as the principal events of the embryonic period.
- List representative body structures produced by the primary germ layers.
- Describe the formation of the placenta and umbilical cord.

The time span from fertilization to birth is the **gestation period** (jes-TĀ-shun; *gest-* = to bear). The human gestation period is about 38 weeks, counted from the estimated day of fertilization or 2 weeks after the first day of the last menstruation. By the end of the **embryonic period,** the first two months of development, the rudiments of all the principal adult organs are present, and the embryonic membranes are developed. During the **fetal period,** after the second month, organs established by the primary germ layers grow rapidly, and the fetus takes on a human appearance. By the end of the third month, the placenta, which is the site of exchange of nutrients and wastes between the mother and the fetus, is functioning.

Beginnings of Organ Systems

The first major event of the embryonic period is **gastrulation** (gas′-trū-LĀ-shun), in which the inner cell mass of the blastocyst differentiates into three **primary germ layers:** ectoderm, endoderm, and mesoderm. These germ layers are the major embryonic tissues from which all tissues and organs of the body develop.

Within 8 days after fertilization, the cells of the inner cytotrophoblast proliferate and form the amnion (a fetal membrane) and a space, the **amniotic cavity** (am-nē-OT-ik; *amnio-* = lamb), adjacent to the inner cell mass (Figure 27.5a on page 832). The layer of cells of the inner cell mass that is closer to the amniotic cavity develops into the **ectoderm** (*ecto-* = outside; *derm-* = skin); the layer of cells of the inner cell mass that borders the blastocyst cavity develops into the **endoderm** (*endo-* = inside). As the amniotic cavity forms, the inner cell mass at this stage is called the **embryonic disc** and contains ectodermal and endodermal cells; mesodermal cells are scattered external to the disc.

About the 12th day after fertilization, formation of the primary germ layers and associated structures produces striking changes. The cells of the endodermal layer have been dividing rapidly; groups of them now extend around in a hollow sphere, forming the yolk sac, another fetal membrane (described shortly). Cells of the cytotrophoblast give rise to a loose connective tissue, the **extraembryonic mesoderm** (*meso-* = middle), which completely fills the space between the cytotrophoblast and yolk sac. Soon large spaces

CHANGING IMAGES

Looking at Life Begin

19th Cent.

Throughout history, most anatomical depictions that focused on women were intended to illustrate a woman's reproductive capacity. Existing texts from around the world are replete with illustrations of pregnant uteri and couples engaged in sexual intercourse. Most of these, however, are European in origin.

The image shown here is an intricate Japanese engraving. Made of three wooden panels, as if to symbolize trimesters, this image is circa 19th century. The various phases of fetal development are illustrated as well as some of the changes to a woman's body throughout gestation. Notice that this piece is intended to be read from right to left, as the pregnancy progresses in this direction. Elegantly posed and beautifully portrayed, each of the women holds a different flower blossom. Perhaps this is to represent flourishing life, or possibly the changing seasons of pregnancy. The number of women depicted is also suggestive of how the Japanese, and many other cultures, have historically viewed pregnancy. While many of us today consider a pregnancy to be approximately nine calendar months, it can also be described as being in the order of ten cycles of the moon. As you can see here, ten women are depicted—one for each of the lunar months of pregnancy. Look closely at the growing fetus, which can be seen as if there were a window into her uterus. While the early months show the fetus strangely shaped, it becomes recognizably human in the later months. The fetus has even correctly achieved a head-down position by the conclusion of the pregnancy.

Think about other symbolic aspects of this image. Why is the tenth woman the only one to not have her hair elaborately styled? What is it that she leans upon? Could the less formal hairstyle symbolize her preparation for labor and delivery? Is she reclining upon implements that are to assist with the impending birth of her child? As you analyze this image for anatomical accuracy and metaphor, be sure to not overlook its aesthetic beauty.

Figure 27.5 / Formation of the primary germ layers and associated structures.

The primary germ layers (ectoderm, mesoderm, and endoderm) are the embryonic tissues from which all tissues and organs develop.

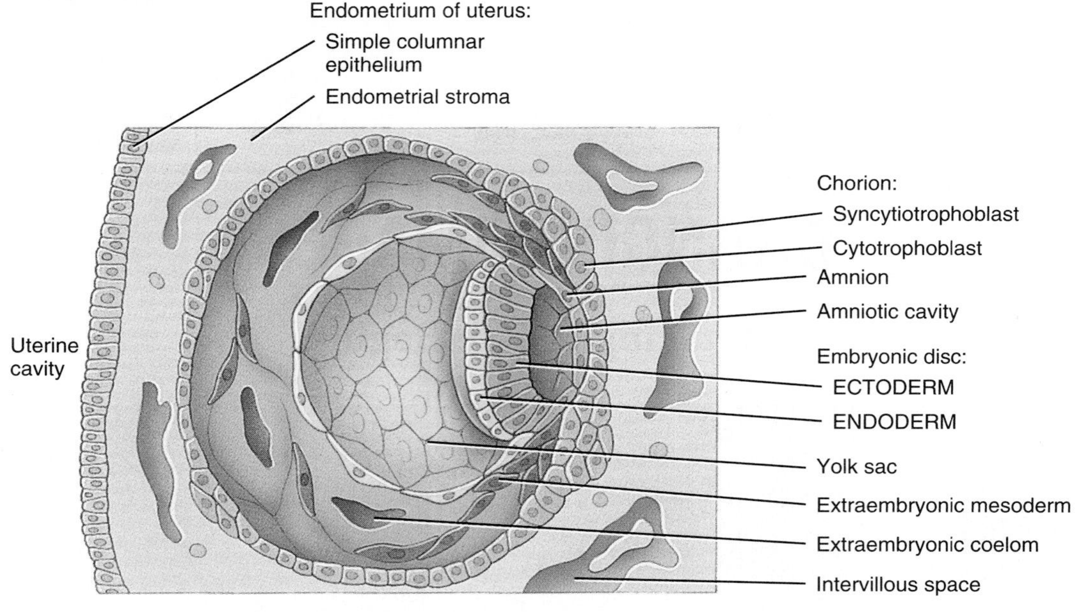

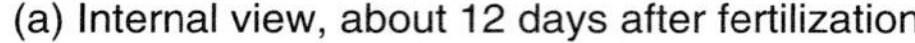

(a) Internal view, about 12 days after fertilization

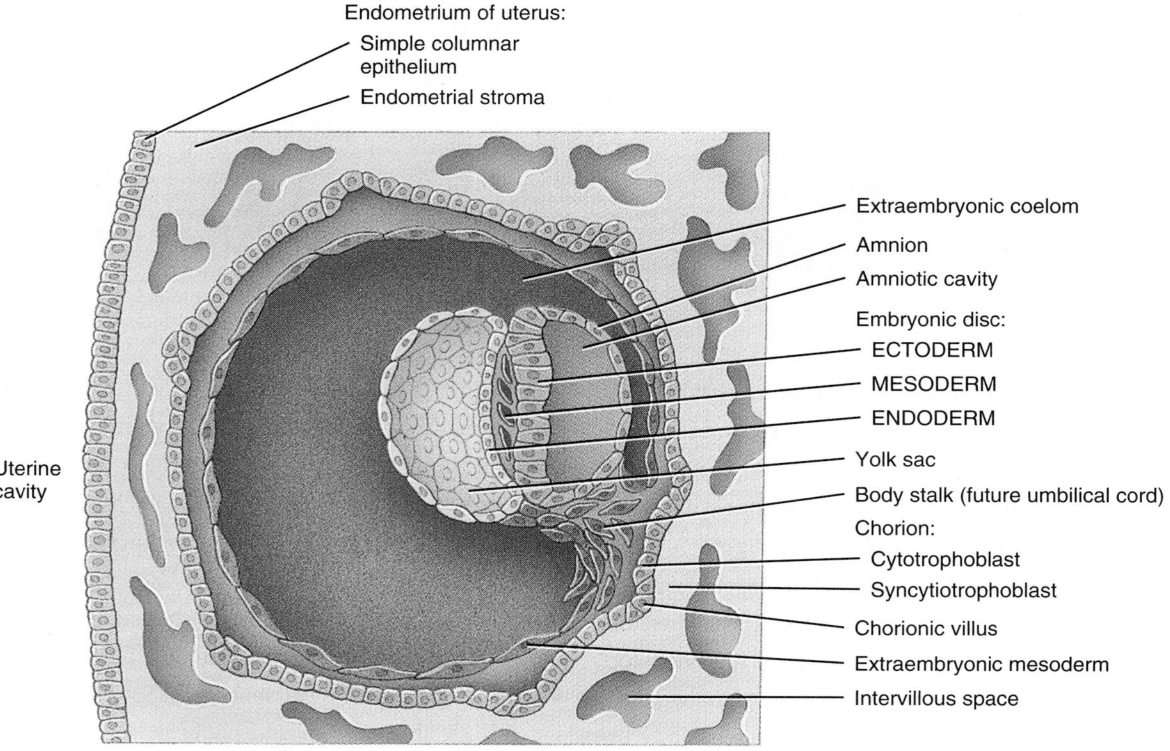

(b) Internal view, about 14 days after fertilization

develop in the extraembryonic mesoderm and come together to form a single, larger cavity called the **extraembryonic coelom** (SĒ-lōm; = cavity). Cavities within the embryonic mesoderm become the future ventral body cavity (Figure 27.5b).

About the 14th day, differentiation of the cells of the embryonic disc produces three distinct layers: the ectoderm, mesoderm, and endoderm (Figure 27.5b). As the embryo develops, the endoderm becomes the epithelial lining of the gastrointesti-

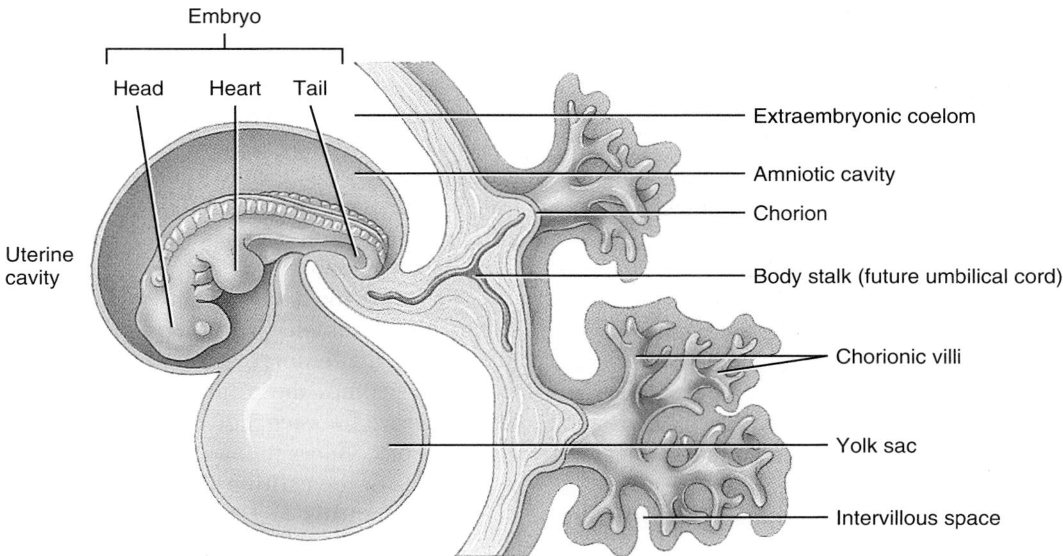

(c) External view, about 25 days after fertilization

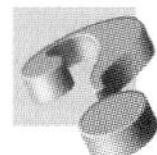

Which cells of the blastocyst give rise to the embryonic disc?

nal and respiratory tracts, and of several other organs. The mesoderm forms muscle, bone and other connective tissues, and the peritoneum. The ectoderm develops into the epidermis of the skin and the nervous system. Table 27.1 provides more details about the fates of these primary germ layers.

Table 27.1 Structures Produced by the Three Primary Germ Layers

Germ Layer	Structures Produced
Endoderm	1. Epithelial lining of gastrointestinal tract (except the oral cavity and anal canal) and the epithelium of its glands.
	2. Epithelial lining of urinary bladder, gallbladder, and liver.
	3. Epithelial lining of pharynx, auditory (Eustachian) tubes, tonsils, larynx, trachea, bronchi, and lungs.
	4. Epithelium of thyroid gland, parathyroid glands, pancreas, and thymus.
	5. Epithelial lining of prostate and bulbourethral (Cowper's) glands, vagina, vestibule, urethra, and associated glands such as the greater (Bartholin's) vestibular and lesser vestibular glands.
Mesoderm	1. All skeletal and cardiac muscle tissue and most smooth muscle tissue.
	2. Cartilage, bone, and other connective tissues.
	3. Blood, red bone marrow, and lymphatic tissue.
	4. Endothelium of blood vessels and lymphatic vessels.
	5. Dermis of skin.
	6. Fibrous tunic and vascular tunic of eye.
	7. Middle ear.
	8. Mesothelium of ventral body cavity.
	9. Epithelium of kidneys, ureters, adrenal cortex, gonads, and genital ducts.
Ectoderm	1. All nervous tissue.
	2. Epidermis of skin.
	3. Hair follicles, arrector pili muscles, nails, and epithelium of skin glands (sebaceous and sudoriferous).
	4. Lens, cornea, and internal eye muscles.
	5. Internal and external ear.
	6. Neuroepithelium of sense organs.
	7. Epithelium of oral cavity, nasal cavity, paranasal sinuses, salivary glands, and anal canal.
	8. Epithelium of pineal gland, pituitary gland, and adrenal medullae.

Formation of Embryonic Membranes

A second major event that occurs during the embryonic period is the formation of the **embryonic membranes.** These membranes lie outside the embryo and protect and nourish the embryo and, later, the fetus. (Recall that the developing embryo becomes a fetus after the second month.) The embryonic membranes are the yolk sac, the amnion, the chorion, and the allantois (Figure 27.6).

In species whose young develop inside a shelled egg (such as birds), the **yolk sac** is the primary source of blood vessels that transport nutrients to the embryo. However, human embryos receive their nutrients from the endometrium; the yolk sac remains small and functions as an early site of blood formation. The yolk sac also contains cells that migrate into the gonads and differentiate into the primitive germ cells (spermatogonia and oogonia).

The **amnion** is a thin, protective membrane that forms by the eighth day after fertilization and initially overlies the embryonic disc. As the embryo grows, the amnion entirely surrounds the embryo, creating a cavity that becomes filled with amniotic fluid. Most amniotic fluid is initially derived from a filtrate of maternal blood; later, the fetus makes daily contributions to the fluid by excreting urine into the amniotic cavity. Amniotic fluid serves as a shock absorber for the fetus, helps regulate fetal body temperature, and prevents adhesions between the skin of the fetus and surrounding tissues. Embryonic cells are sloughed off into amniotic fluid; they can be examined in the procedure called amniocentesis, which is described on page 838. The amnion usually ruptures just before birth; it and its fluid constitutes the "bag of waters."

The **chorion** (KŌR-ē-on) is derived from the trophoblast of the blastocyst and the mesoderm that lines the trophoblast. It surrounds the embryo and, later, the fetus. Eventually the chorion becomes the principal embryonic part of the placenta, the structure for exchange of materials between mother and fetus. It also produces human chorionic gonadotropin (hCG). The inner layer of the chorion eventually fuses with the amnion.

The **allantois** (a-LAN-tō-is; *allant-* = sausage) is a small, vascularized structure that serves as another early site of blood formation. Later its blood vessels form part of the link between mother and fetus.

Placenta and Umbilical Cord

Development of the **placenta** (pla-SEN-ta; = flat cake), which is the site of exchange of nutrients and wastes between the mother and the fetus, occurs during the third month of pregnancy; it is formed by the chorion of the embryo and a portion of the endometrium of the mother. When fully developed, the placenta is shaped like a pancake (see Figure 27.8b). Functionally, the placenta allows oxygen and nutrients to diffuse from maternal blood into fetal blood while carbon dioxide and wastes diffuse from fetal blood into maternal blood.

Figure 27.6 / Embryonic membranes.

Embryonic membranes are outside the embryo; they protect and nourish the embryo and, later, the fetus.

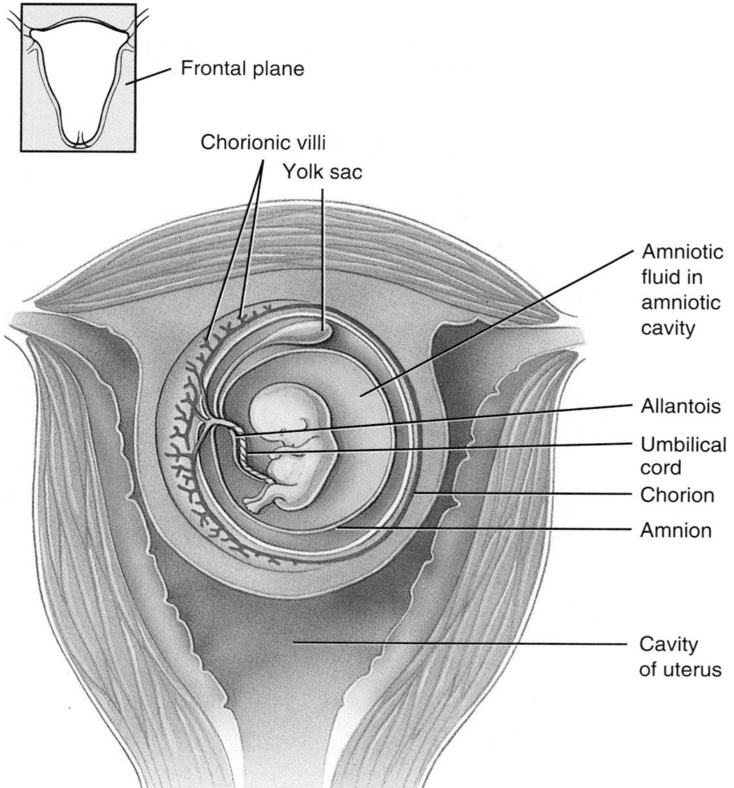

(a) Frontal section through uterus

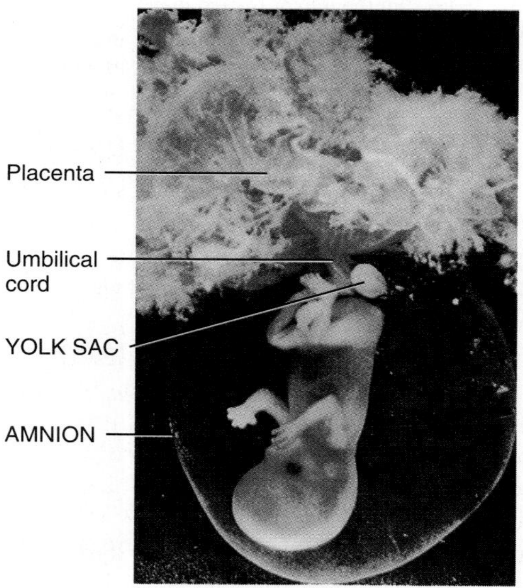

(b) Ten-week fetus

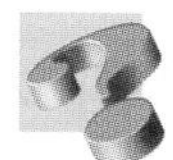

How do the amnion and the chorion differ in function?

The placenta is also a protective barrier because most microorganisms cannot pass through it. However, certain viruses, such as those that cause AIDS, German measles, chickenpox, measles, encephalitis, and poliomyelitis, can cross the placenta. The placenta also stores nutrients such as carbohydrates, proteins, calcium, and iron, which are released into fetal circulation as required, and it produces several hormones that are necessary to maintain pregnancy. Almost all drugs, including alcohol and many other substances that can cause birth defects, pass freely through the placenta.

If implantation occurs, a portion of the endometrium becomes modified and is known as the **decidua** (dē-SID-yū-a; = falling off). The decidua includes all but the stratum basalis layer of the endometrium; it separates from the endometrium after the fetus is delivered just as it does in normal menstruation. Different regions of the decidua, all of which are areas of the stratum functionalis, are named based on their positions relative to the site of the implanted blastocyst (Figure 27.7). The **decidua basalis** is the portion of the endometrium between the chorion and the stratum basalis of the uterus; it becomes the maternal part of the placenta. The **decidua capsularis** is the portion of the endometrium that is located between the embryo and the uterine cavity. The **decidua parietalis** (par-rī-e-TAL-is) is the remaining modified endometrium that lines the noninvolved areas of the rest of the uterus. As the embryo and later the fetus enlarges, the decidua capsularis bulges into the uterine cavity and initially fuses with the decidua parietalis, thereby obliterating the uterine cavity. By about 27 weeks, the decidua capsularis degenerates and disappears.

Connections between mother and developing child are established via the developing placenta and umbilical cord (Figure 27.8a). During embryonic life, fingerlike projections of the chorion, called **chorionic villi** (kō′-rē-ON-ik VIL-ī), grow into the decidua basalis of the endometrium. These projections, which will contain fetal blood vessels of the allantois, continue growing until they are bathed in maternal blood sinuses called **intervillous spaces** (in-ter-VIL-us). The result is that maternal and fetal blood vessels are brought into close proximity. Note, however, that maternal and fetal blood vessels do not join, and the blood they carry does not normally mix. Instead, oxygen and nutrients in the blood of the mother's intervillous spaces diffuse across the cell membranes into the capillaries of the villi while waste products diffuse in the opposite direction. From the capillaries of the villi, nutrients and oxygen enter the fetus through the umbilical vein. Wastes leave the fetus through the umbilical arteries, pass into the capillaries of the villi, and diffuse into the maternal blood. A few materials, such as IgG antibodies, pass from the blood of the mother into the capillaries of the villi.

The **umbilical cord** (um-BIL-i-kul), the vascular connection between mother and fetus, consists of two umbilical arteries that carry deoxygenated fetal blood to the placenta, one umbilical vein that carries oxygenated blood into the fetus, and supporting mucous connective tissue called Wharton's jelly derived from the allantois. The entire umbilical cord is surrounded by a layer of amnion (see Figure 27.8a).

After the birth of the baby, the placenta detaches from the uterus and is termed the **afterbirth.** At this time, the umbilical cord is tied off and then severed, leaving the baby on its own. The small portion (about an inch) of the cord that remains still attached to the infant begins to wither and eventually falls off, usually within 12–15 days after birth. The area where the cord was attached becomes covered by a thin layer of skin, and scar tissue forms. The scar is the **umbilicus (navel).**

Figure 27.7 / Regions of the decidua.

The decidua is a modified portion of the endometrium that develops after implantation.

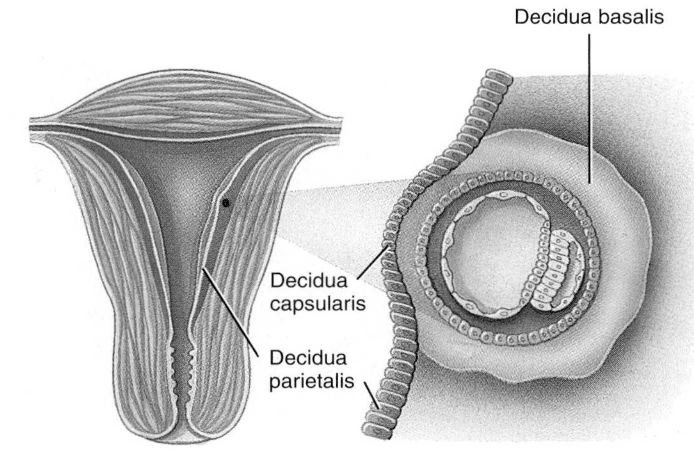

Which part of the decidua helps form the maternal part of the placenta?

Pharmaceutical companies use human placentas as a source of hormones, drugs, and blood; portions of placentas are also used for burn coverage. The placental and umbilical cord veins can also be used in blood vessel grafts, and cord blood can be frozen to provide a future source of pluripotent stem cells.

A summary of changes associated with embryonic and fetal growth is presented in Table 27.2 on page 837.

Throughout the text we have discussed the developmental anatomy of the various body systems in their respective chapters. The following list of these sections is presented here for your review.

Figure 27.8 / Placenta and umbilical cord.

The placenta is formed by the chorion of the embryo and part of the endometrium of the mother.

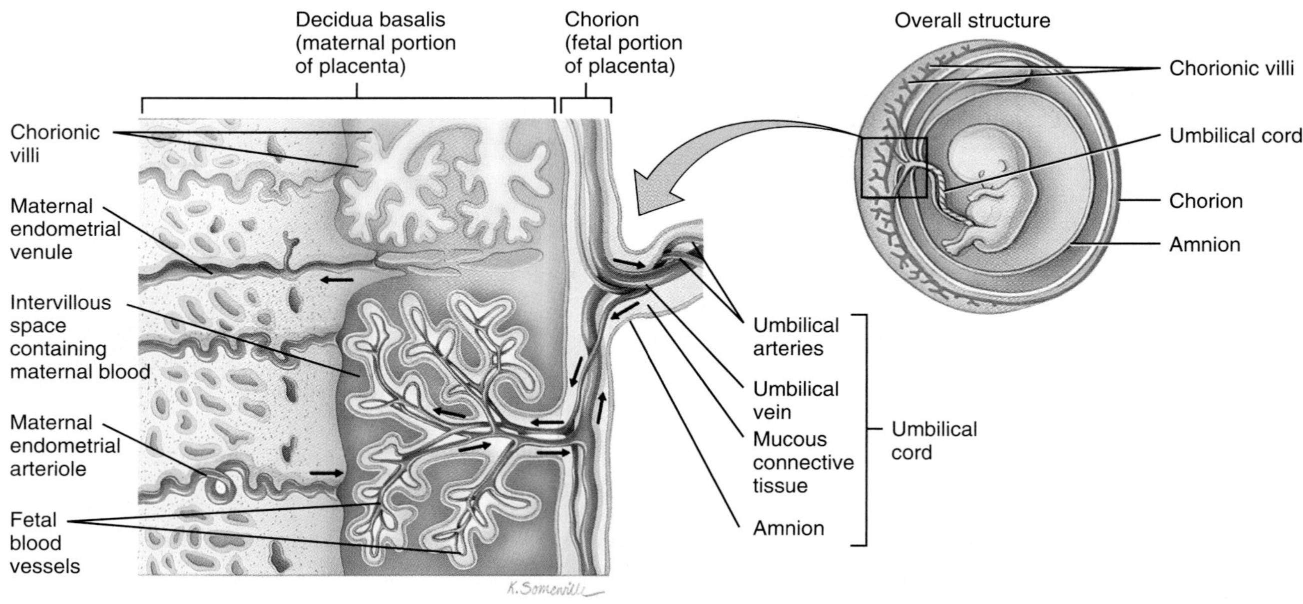

(a) Details of placenta and umbilical cord

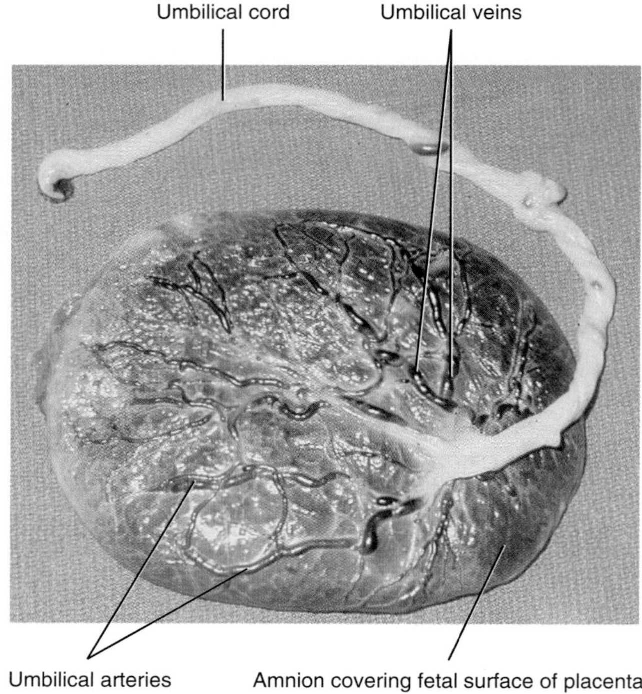

(b) Fetal aspect of placenta

What is the function of the placenta?

Table 27.2 Changes Associated with Embryonic and Fetal Growth

End of Month	Approximate Size and Weight	Representative Changes
1	0.6 cm (3/16 in.)	Eyes, nose, and ears are not yet visible. Vertebral column and vertebral canal form. Small buds that will develop into limbs form. Heart forms and starts beating. Body systems begin to form. The central nervous system appears at the start of the third week.
2	3 cm (1¼ in.) 1 g (1/30 oz)	Eyes are far apart, eyelids fused. Nose is flat. Ossification begins. Limbs become distinct, and digits are well formed. Major blood vessels form. Many internal organs continue to develop.
3	7½ cm (3 in.) 30 g (1 oz)	Eyes are almost fully developed but eyelids are still fused, nose develops a bridge, and external ears are present. Ossification continues. Limbs are fully formed and nails develop. Heartbeat can be detected. Urine starts to form. Fetus begins to move, but it cannot be felt by mother. Body systems continue to develop.
4	18 cm (6½–7 in.) 100 g (4 oz)	Head is large in proportion to rest of body. Face takes on human features, and hair appears on head. Many bones are ossified, and joints begin to form. Rapid development of body systems occurs.
5	25–30 cm (10–12 in.) 200–450 g (½–1 lb)	Head is less disproportionate to rest of body. Fine hair (lanugo) covers body. Brown fat forms and is the site of heat production. Fetal movements are commonly felt by mother (quickening). Rapid development of body systems occurs.
6	27–35 cm (11–14 in.) 550–800 g (1¼–1½ lb)	Head becomes even less disproportionate to rest of body. Eyelids separate and eyelashes form. Substantial weight gain occurs. Skin is wrinkled. Type II alveolar cells begin to produce surfactant.
7	32–42 cm (13–17 in.) 110–1350 g (2½–3 lb)	Head and body are more proportionate. Skin is wrinkled. Seven-month fetus (premature baby) is capable of survival. Fetus assumes an upside-down position. Testes start to descend into scrotum.
8	41–45 cm (16½–18 in.) 2000–2300 g (4½–5 lb)	Subcutaneous fat is deposited. Skin is less wrinkled.
9	50 cm (20 in.) 3200–3400 g (7–7½ lb)	Additional subcutaneous fat accumulates. Lanugo is shed. Nails extend to tips of fingers and maybe even beyond.

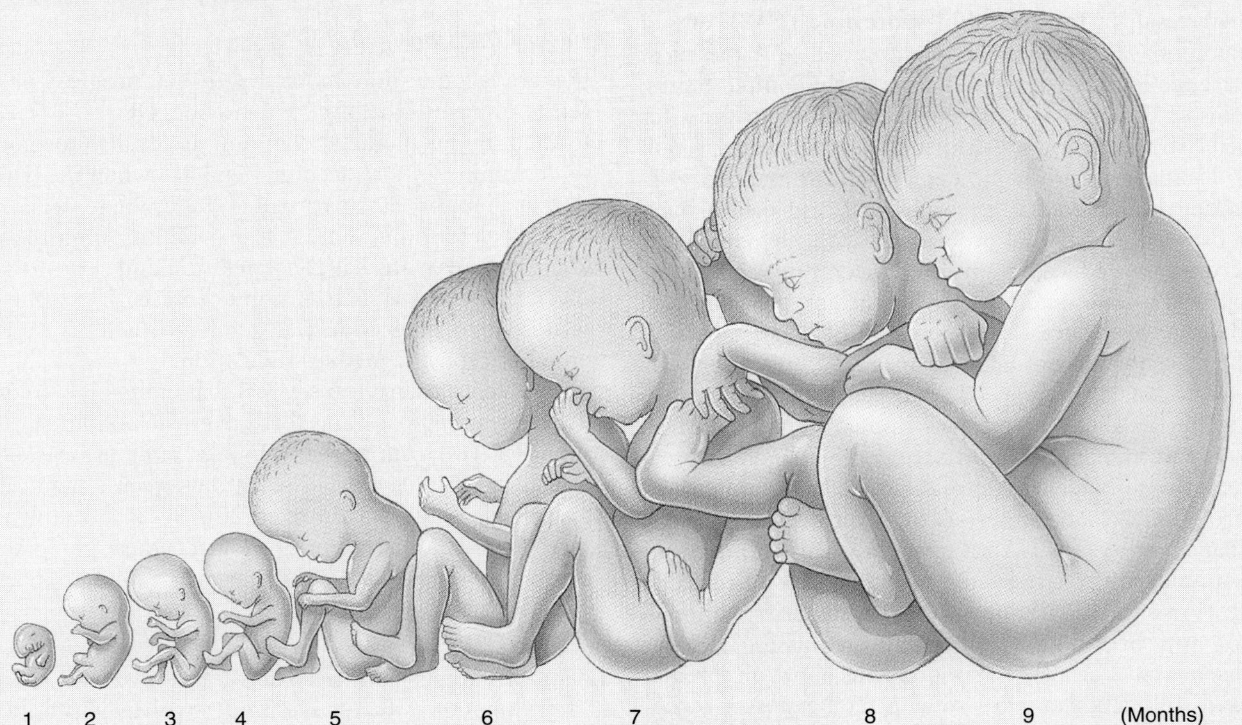

In some cases, part or all of the placenta becomes implanted in the inferior portion of the uterus, near or covering the internal os of the cervix. This condition is called **placenta previa** (PRĒ-vē-a; = before or in front of). Although placenta previa may lead to spontaneous abortion, it also occurs in approximately 1 in 250 live births. It is dangerous to the fetus because it may cause premature birth and intrauterine hypoxia due to maternal bleeding. Maternal mortality is increased due to hemorrhage and infection. The most important symptom is sudden, painless, bright red vaginal bleeding in the third trimester. Cesarean section is the preferred method of delivery in placenta previa. ■

Environmental Influences on Embryonic and Fetal Development

Exposure of a developing embryo or fetus to certain environmental factors can damage the developing organism or even cause death. A **teratogen** (TER-a-tō-jen; *terato-* = monster) is any agent or influence that causes developmental defects in the embryo. In the following sections we briefly discuss several examples.

Chemicals and Drugs

Because the placenta is not an absolute barrier between the maternal and fetal circulations, any drug or chemical that is dangerous to an infant should be considered potentially dangerous to the fetus when given to the mother. Alcohol is by far the number one fetal teratogen. Intrauterine exposure to even a small amount of alcohol may result in **fetal alcohol syndrome (FAS),** one of the most common causes of mental retardation and the most common preventable cause of birth defects in the United States. The symptoms of FAS may include slow growth before and after birth, characteristic facial features (short palpebral fissures, a thin upper lip, and sunken nasal bridge), defective heart and other organs, malformed limbs, genital abnormalities, and central nervous system damage. Behavioral problems, such as hyperactivity, extreme nervousness, reduced ability to concentrate, and an inability to appreciate cause-and-effect relationships, are common. Acetaldehyde, one of the toxic products of alcohol metabolism, may cross the placenta and cause fetal damage.

Other teratogens include pesticides; defoliants (chemicals that cause plants to shed their leaves prematurely); industrial chemicals; some hormones; antibiotics; oral anticoagulants, anticonvulsants, antitumor agents, thyroid drugs, thalidomide, diethylstilbestrol (DES), and numerous other prescription drugs; LSD; marijuana; and cocaine. A pregnant woman who uses cocaine, for example, subjects the fetus to higher risk of retarded growth, attention and orientation problems, hyperirritability, a tendency to stop breathing, malformed or missing organs, strokes, and seizures. The risks of spontaneous abortion, premature birth, and stillbirth also increase with fetal exposure to cocaine.

Cigarette Smoking

Strong evidence implicates cigarette smoking during pregnancy as a cause of low infant birth weight; a strong association between smoking and a higher fetal and infant mortality rate also exists. Women who smoke have a much higher risk of an ectopic pregnancy. Cigarette smoke may be teratogenic and may cause cardiac abnormalities and anencephaly (a developmental defect characterized by the absence of a cerebrum). Maternal smoking also appears to be a significant factor in the development of cleft lip and palate and has tentatively been linked with sudden infant death syndrome (SIDS). Infants nursing from smoking mothers have also been found to have an increased incidence of gastrointestinal disturbances. Even a mother's exposure to secondhand cigarette smoke (breathing air containing tobacco smoke) predisposes her unborn baby to increased incidence of respiratory problems, including bronchitis and pneumonia, during the first year of life.

Irradiation

Ionizing radiation of various kinds is a potent teratogen. Exposure of pregnant mothers to x-rays or radioactive isotopes during the embryo's susceptible period of development may cause microcephaly (small head size relative to the rest of the body), mental retardation, and skeletal malformations. Caution is advised, especially during the first trimester of pregnancy.

Prenatal Diagnostic Tests

Several tests are available to detect genetic disorders and assess fetal well-being. Here we describe fetal ultrasonography, amniocentesis, and chorionic villi sampling (CVS).

Fetal Ultrasonography

If there is a question about the normal progress of a pregnancy, **fetal ultrasonography** (ul′-tra-son-OG-ra-fē) may be performed. By far the most common use of diagnostic ultrasound is to determine a more accurate fetal age when the date of conception is unclear; it is also used to evaluate fetal viability and growth, determine fetal position, identify multiple pregnancies, identify fetal–maternal abnormalities, and serve as an adjunct to special procedures such as amniocentesis. Ultrasound is not used routinely to determine the gender of a fetus; it is performed only for a specific medical indication.

An instrument (transducer) that emits high-frequency sound waves is passed back and forth over the abdomen. The reflected sound waves from the developing fetus are picked up by the transducer and converted to an on-screen image called a **sonogram** (see Table 1.5 on page 20). Because the urinary bladder serves as a landmark during the procedure, the patient needs to drink liquids and not void urine so as to maintain a full bladder.

Amniocentesis

Amniocentesis (am′-nē-ō-sen-TĒ-sis; *amnio-* = amnion; *-centesis* = puncture to remove fluid) involves withdrawing some of the amniotic fluid that bathes the developing fetus and analyzing the fetal cells and dissolved substances. It is used either to test for the presence of certain genetic disorders, such as Down syndrome (DS), spina bifida, hemophilia, Tay–Sachs disease, sickle-cell disease, and certain muscular dystrophies, or to determine fetal maturity and well-being near the time of delivery. To detect suspected genetic abnormalities, the test is usually done at 14–16 weeks of gestation; to assess fetal maturity, it is usually performed after the 35th week of gestation. About 300 chromosomal disorders and over 50 biochemical defects can be detected through amniocentesis. It can also reveal gender, which is important information for the diagnosis of sex-linked disorders, in which an abnormal gene carried by the mother affects her male offspring only (described later in the chapter). If the fetus is female, it will not be afflicted unless the father also carries the defective gene.

During amniocentesis, the position of the fetus and placenta is first identified using ultrasound and palpation. After the skin is prepared with an antiseptic and a local anesthetic is given, a hypodermic needle is inserted through the mother's abdominal wall and uterus into the amniotic cavity, and about 10 mL of fluid are aspirated (Figure 27.9a). The fluid and suspended cells are subjected to microscopic examination and biochemical testing. Elevated levels of alphafetoprotein (AFP) and acetylcholinesterase may indicate failure of the nervous system to develop properly, as occurs in spina bifida or anencephaly (absence of the cerebrum). Chromosome studies, which require growing the cells for 2–4 weeks in a culture medium, may reveal rearranged, missing, or extra chromosomes. Amniocentesis is performed only when a risk for genetic defects is suspected because there is about a 0.5% chance of spontaneous abortion after the procedure.

Chorionic Villi Sampling

In **chorionic villi sampling** (ko-rē-ON-ik VIL-ī) or **CVS,** a catheter is guided through the vagina and cervix of the uterus and then advanced to the chorionic villi under ultrasound guidance (Figure 27.9b). About 30 milligrams of tissue are suctioned out and prepared for chromosomal analysis. Alternatively, the chorionic villi can be sampled by inserting a needle through the abdominal cavity, as performed in amniocentesis.

CVS can identify the same defects as amniocentesis because chorion cells and fetal cells contain the same genome. Moreover, CVS offers several advantages over amniocentesis: It can be performed as early as 8 weeks of gestation, and test results are available in only a few days, permitting an earlier decision on whether or not to continue the pregnancy. In addition, the procedure does not require penetration of the abdomen, uterus, or amniotic cavity by a needle. However, CVS is slightly riskier than amniocentesis; it carries a 1–2% chance of spontaneous abortion after the procedure.

- ✓ Define the embryonic period and the fetal period.
- ✓ Explain the importance of the placenta and umbilical cord to fetal growth.
- ✓ Discuss the principal body changes associated with fetal growth.
- ✓ Describe the following prenatal diagnostic tests: fetal ultrasonography, amniocentesis, and chorionic villi sampling.

Figure 27.9 / Amniocentesis and chorionic villi sampling.

To detect genetic abnormalities, amniocentesis is performed at 14–16 weeks of gestation, whereas chorionic villi sampling may be performed as early as 8 weeks of gestation.

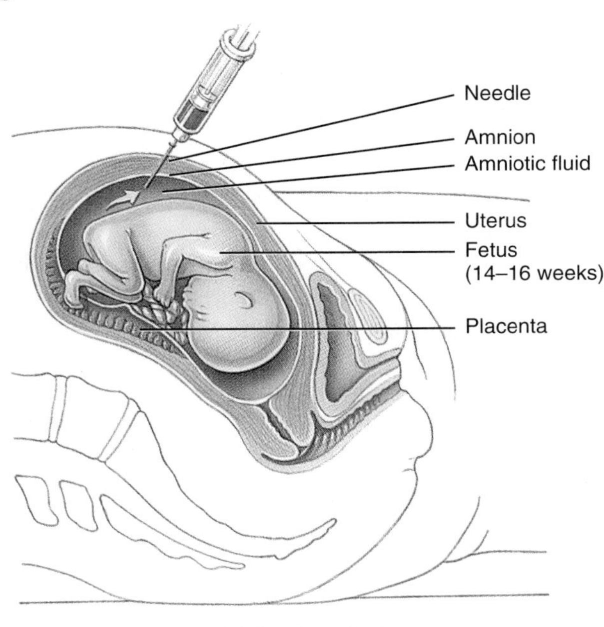

(a) Amniocentesis

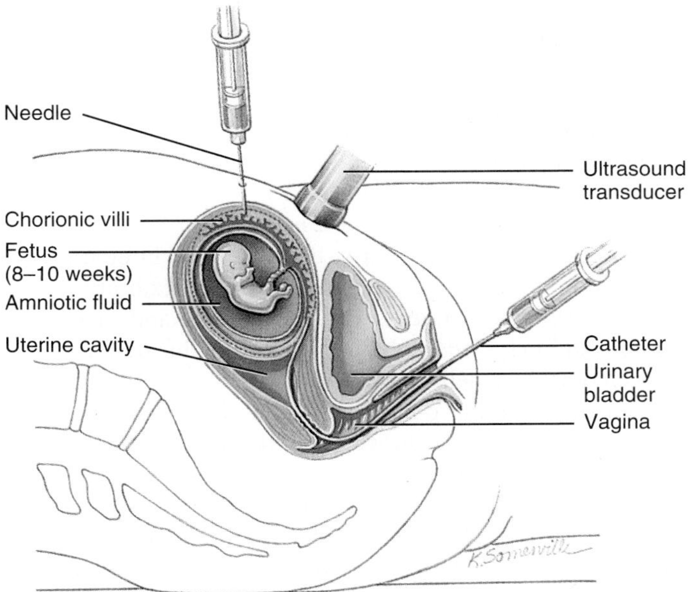

(b) Chorionic villi sampling (CVS)

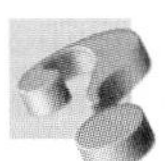
What kinds of information can be provided by amniocentesis?

MATERNAL CHANGES DURING PREGNANCY

Objectives

- Describe the sources and functions of the hormones secreted during pregnancy.
- Describe the hormonal, anatomical, and physiological changes in the mother during pregnancy.

Hormones of Pregnancy

During the first 3–4 months of pregnancy, the corpus luteum continues to secrete **progesterone** and **estrogens,** which maintain the lining of the uterus during pregnancy and prepare the mammary glands to secrete milk. The amounts secreted by the corpus luteum, however, are only slightly more than those produced after ovulation in a normal menstrual cycle. From the third month through the remainder of the pregnancy, the placenta itself provides the high levels of progesterone and estrogens required. As noted previously, the chorion of the placenta secretes **human chorionic gonadotropin (hCG)** into the blood. In turn, hCG stimulates the corpus luteum to continue production of progesterone and estrogens—an activity required to prevent menstruation and for the continued attachment of the embryo and fetus to the lining of the uterus. By the eighth day after fertilization, hCG can be detected in the blood of a pregnant woman. Peak secretion of hCG occurs at about the ninth week of pregnancy; the hCG level decreases sharply during the fourth and fifth months and then levels off until childbirth.

The chorion of the placenta begins to secrete estrogens after the first 3 or 4 weeks of pregnancy and progesterone by the sixth week. These hormones are secreted in increasing quantities until the time of birth. By the fourth month, when the placenta is fully established, the secretion of hCG has been greatly reduced because the secretions of the corpus luteum are no longer essential. Thus, from the third month to the ninth month, the placenta supplies the levels of estrogens and progesterone needed to maintain the pregnancy. A high level of progesterone ensures that the uterine myometrium is relaxed and that the cervix is tightly closed. After delivery, estrogens and progesterone in the blood decrease to normal levels.

Relaxin, a hormone produced first by the corpus luteum of the ovary and later by the placenta, increases the flexibility of the pubic symphysis and ligaments of the sacroiliac and sacrococcygeal joints and helps dilate the uterine cervix during labor. Both of these actions ease delivery of the baby.

A third hormone produced by the chorion of the placenta is **human chorionic somatomammotropin (hCS),** also known as **human placental lactogen (hPL).** The rate of secretion of hCS increases in proportion to placental mass, reaching maximum levels after 32 weeks and remaining relatively constant after that. It is thought to help prepare the mammary glands for lactation, enhance maternal growth by increasing protein synthesis, and regulate certain aspects of metabolism in mother and fetus alike. For example, hCS causes decreased use of glucose by the mother, thus making more available for the fetus. Additionally, hCS promotes the release of fatty acids from adipose tissue, providing an alternative to glucose for the mother's ATP production.

The hormone most recently found to be produced by the placenta is **corticotropin-releasing hormone (CRH),** which in nonpregnant people is secreted only by neurosecretory cells in the hypothalamus. CRH is now thought to be the "clock" that establishes the timing of birth. Secretion of CRH by the placenta begins at about 12 weeks and increases enormously toward the end of pregnancy. Women who have higher levels of CRH earlier in pregnancy are more likely to deliver prematurely, whereas those who have low levels are more likely to deliver after their due date. CRH from the placenta has a second important effect: It increases secretion of cortisol, which is needed for maturation of the fetal lungs and the production of surfactant.

Early pregnancy tests detect the tiny amounts of human chorionic gonadotropin (hCG) in the urine that begin to be excreted about 8 days after fertilization. The test kits can detect pregnancy as early as the first day of a missed menstrual period—that is, at about 14 days after fertilization. Chemicals in the kits produce a color change if a reaction occurs between hCG in the urine and hCG antibodies included in the kit.

Several of the test kits available at pharmacies are as sensitive and accurate as test methods used in many hospitals. Still, false-negative and false-positive results can occur. A false-negative result (the test is negative, even though the woman is pregnant) may be due to testing too soon or to an ectopic pregnancy. A false-positive result (the test is positive, but the woman is not pregnant) may be due to excess protein or blood in the urine or to hCG production due to a rare type of uterine cancer. Thiazide diuretics, hormones, steroids, and thyroid drugs may also affect the outcome of an early pregnancy test. ■

Anatomical and Physiological Changes during Pregnancy

By about the end of the third month of pregnancy, the uterus occupies most of the pelvic cavity; as the fetus continues to grow, the uterus extends higher and higher into the abdominal cavity. Toward the end of a full-term pregnancy, the uterus fills nearly all of the abdominal cavity, reaching above the costal margin nearly to the xiphoid process of the sternum (Figure 27.10). It pushes the maternal intestines, liver, and stomach superiorly, elevates the diaphragm, and widens the thoracic cavity. Pressure on the stomach may force the stomach contents superiorly into the esophagus, resulting in heartburn. In the pelvic cavity, compression of the ureters and urinary bladder occurs.

Besides the anatomical changes associated with pregnancy, pregnancy-induced physiological changes also occur, including weight gain due to the fetus, amniotic fluid, the placenta, uterine enlargement, and increased total body water; increased storage

Figure 27.10 / Normal fetal position at the end of a full-term pregnancy.

The gestation period is the time interval (about 38 weeks) from fertilization to birth.

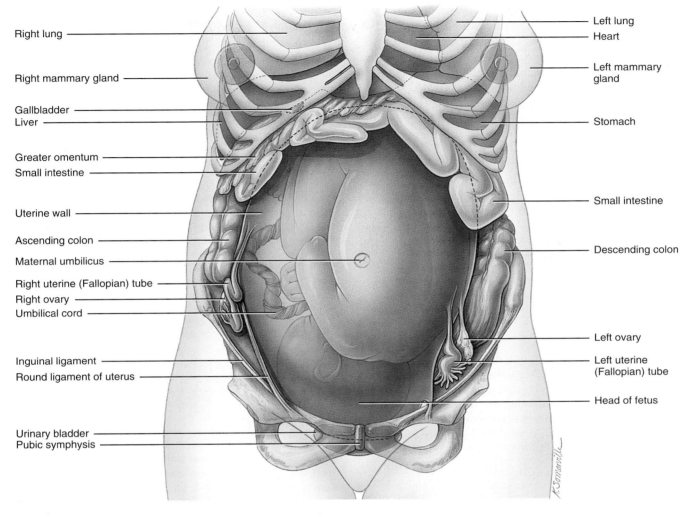

Anterior view of position of organs at end of full-term pregnancy

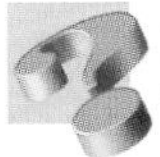

What hormone increases the flexibility of the pubic symphysis and helps dilate the cervix of the uterus to ease delivery of the baby?

of proteins, triglycerides, and minerals; marked breast enlargement in preparation for lactation; and lower back pain due to lordosis (swayback).

Changes in the maternal cardiovascular system include an increase in stroke volume by about 30%; a rise in cardiac output of 20–30% due to increased maternal blood flow to the placenta and increased metabolism; an increase in heart rate of about 10–15%; and an increase in blood volume of 30–50%, mostly during the second half of pregnancy. These increases are necessary to meet the additional demands of the fetus for nutrients and oxygen. When a pregnant woman is lying on her back, the enlarged uterus may compress the aorta, resulting in diminished blood flow to the uterus. Compression of the inferior vena cava also decreases venous return, which leads to edema in the lower limbs and may produce varicose veins. Compression of the renal artery can lead to renal hypertension.

Pulmonary function is also altered during pregnancy to meet the added oxygen demands of the fetus. Tidal volume can increase by 30–40%, expiratory reserve volume can be reduced by up to 40%, functional residual capacity can decline by up to 25%, minute ventilation (the total volume of air inhaled and exhaled each minute) can increase by up to 40%, airway resistance in the bronchial tree can decline by 30–40%, and total body oxygen consumption can increase by about 10–20%. Dyspnea (difficult breathing) also occurs.

With regard to the gastrointestinal tract, pregnant women experience an increase in appetite, but a general decrease in GI tract motility can result in constipation and a delay in gastric emptying time. Nausea, vomiting, and heartburn can also occur.

Pressure on the urinary bladder by the enlarging uterus can produce urinary symptoms, such as increased frequency and urgency of urination, and stress incontinence. An increase in renal plasma flow up to 35% and an increase in glomerular filtration rate up to 40% increase the renal filtering capacity, which allows faster elimination of the extra wastes produced by the fetus.

Changes in the skin during pregnancy are more apparent in some women than others. Included are increased pigmentation around the eyes and cheekbones in a masklike pattern (chloasma), in the areolae of the breasts, and in the linea alba of the lower abdomen (linea nigra). Striae (stretch marks) over the abdomen can occur as the uterus enlarges, and hair loss also increases.

Changes in the reproductive system include edema and increased vascularity of the vulva and increased pliability and vascularity of the vagina. The uterus increases from its nonpregnant mass of 60–80 g to 900–1200 g at term as a result of hyperplasia of muscle fibers in the myometrium in early pregnancy and hypertrophy of muscle fibers during the second and third trimesters.

About 10–15% of all pregnant women in the United States experience **pregnancy-induced hypertension (PIH),** elevated blood pressure associated with pregnancy. The major cause is **preeclampsia** (prē-ē-KLAMP-sē-a), an abnormal condition of pregnancy characterized by sudden hypertension, large amounts of protein in the urine, and generalized edema that typically appears after the 20th week of pregnancy. Other signs and symptoms are generalized edema, blurred vision, and headaches. Preeclampsia might be related to an autoimmune or allergic reaction resulting from the presence of a fetus. When the condition is also associated with convulsions and coma, it is termed **eclampsia.** Other forms of PIH are not associated with protein in the urine. ■

✓ List the hormones involved in pregnancy, and describe the functions of each.

✓ Describe several structural and functional changes that occur in the mother during pregnancy.

EXERCISE AND PREGNANCY

Objective

- Explain the effects of pregnancy on exercise and of exercise on pregnancy.

Only a few changes in early pregnancy affect exercise. A pregnant woman may tire more easily than usual, or morning sickness may interfere with regular exercise. As the pregnancy progresses, weight is gained and posture changes, so more energy is needed to perform activities, and certain maneuvers (sudden stopping, changes in direction, rapid movements) are more difficult to execute. In addition, certain joints, especially the pubic symphysis, become less stable in response to the increased level of the hormone relaxin. As compensation, many mothers-to-be walk with widely spread legs and a shuffling motion.

Although blood shifts from viscera (including the uterus) to the muscles and skin during exercise, there is no evidence of inadequate blood flow to the placenta. The heat generated during exercise may cause dehydration and further increase body temperature. During early pregnancy especially, excessive exercise and heat buildup should be avoided because elevated body temperature has been implicated in neural tube defects. Exercise has no known effect on lactation, provided a woman remains hydrated and wears a bra that provides good support. Moderate physical activity does not endanger the fetuses of healthy women who have a normal pregnancy.

An overall benefit of exercise during pregnancy is a greater sense of well-being, and fewer minor complaints.

LABOR

Objective

- Explain the events associated with the three stages of labor.

Labor is the process by which the fetus is expelled from the uterus through the vagina to the outside. **Parturition** (par′-tū-RISH-un; *parturit-* = childbirth) means giving birth.

The onset of labor is determined by complex interactions of several placental and fetal hormones. Because progesterone inhibits uterine contractions, labor cannot take place until its effects are diminished. Toward the end of gestation, the levels of estrogens in the mother's blood rise sharply, producing changes that overcome the inhibiting effects of progesterone. The rise in estrogens results from increasing secretion by the placenta of corticotropin-releasing hormone, which stimulates the fetal anterior pituitary gland to secrete ACTH (adrenocorticotropic hormone). In turn, ACTH stimulates the fetal adrenal gland to secrete both cortisol and dehydroepiandrosterone (DHEA), the major adrenal androgen. The placenta then converts DHEA into an estrogen. High levels of estrogens cause uterine muscle fibers to display receptors for oxytocin and to form gap junctions with one another. Oxytocin stimulates uterine contractions, and relaxin assists by increasing the flexibility of the pubic symphysis and helping dilate the uterine cervix. Estrogens also stimulate the placenta to release prostaglandins, which induce production of enzymes that digest collagen fibers in the cervix, causing it to soften.

Contractions of the uterine myometrium force the baby's head or body into the cervix, thereby distending (stretching) the cervix. Stretch receptors in the cervix send nerve impulses to neurosecretory cells in the hypothalamus, causing them to release oxytocin into blood capillaries of the posterior pituitary gland. Oxytocin then is carried by the blood to the uterus, where it stimulates the myometrium to contract more forcefully. As the contractions intensify, the baby's body stretches the cervix still more, and the resulting nerve impulses stimulate the secretion of yet more oxytocin. With birth of the infant, the cycle is broken because cervical distention suddenly lessens.

Uterine contractions occur in waves (quite similar to peristaltic waves) that start at the top of the uterus and move downward, eventually expelling the fetus. **True labor** begins when uterine contractions occur at regular intervals, usually producing pain. As the interval between contractions shortens, the contractions intensify. Another symptom of true labor in some women is localization of pain in the back that is intensified by walking. The reliable indicator of true labor is dilation of the cervix and the "show," a discharge of a blood-containing mucus that appears in the cervical canal during labor. In **false labor,** pain is felt in the abdomen at irregular intervals, and it does not intensify and it is not altered significantly by walking. There is no "show" and no cervical dilation.

True labor can be divided into three stages (Figure 27.11):

1. **Stage of dilation.** The time from the onset of labor to the complete dilation of the cervix is the **stage of dilation.** This stage, which typically lasts 6–12 hours, features regular contractions of the uterus, usually a rupturing of the amniotic sac, and complete dilation (to 10 cm) of the cervix. If the amniotic sac does not rupture spontaneously, it is ruptured intentionally.
2. **Stage of expulsion.** The time (10 minutes to several hours) from complete cervical dilation to delivery of the baby is the **stage of expulsion.**
3. **Placental stage.** The time (5–30 minutes or more) after delivery until the placenta or "afterbirth" is expelled by powerful uterine contractions is the **placental stage.** These contractions also constrict blood vessels that were torn during delivery, thereby reducing the likelihood of hemorrhage.

As a rule, labor lasts longer with first babies (typically about 14 hours); for women who have previously given birth, the average duration of labor is about 8 hours—although the time varies enormously among births. Because the fetus may be squeezed through the birth canal (cervix and vagina) for up to several hours, the fetus is stressed during childbirth: The fetal head is compressed, and the fetus undergoes some degree of intermittent hypoxia due to compression of the umbilical cord and the placenta during uterine contractions. In response to this stress, the fetal adrenal medullae secrete very high levels of epinephrine and norepinephrine, the "fight-or-flight" hormones. Much of the protection against the stresses of parturition, as well as preparation of the infant for surviving extrauterine life, are provided by the adrenal medullary hormones. Among other functions, the hormones clear the lungs and alter their physiology in readiness for breathing air, mobilize readily usable nutrients for cellular metabolism, and promote an increased blood flow to the brain and heart.

About 7% of pregnant women have not delivered by 2 weeks after their due date. Such cases carry an increased risk of brain damage to the fetus, and even fetal death, due to inade-

Figure 27.11 / Stages of true labor.

The term parturition refers to birth.

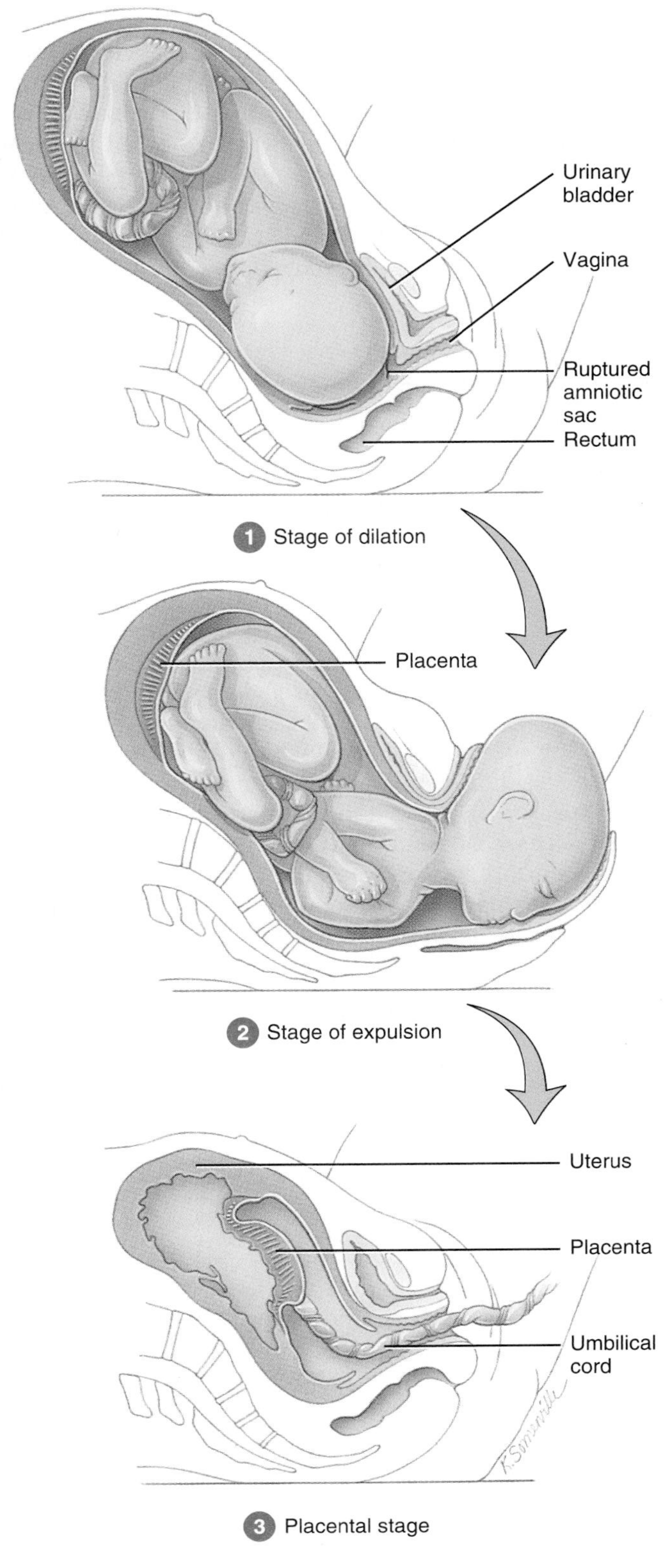

Urinary bladder
Vagina
Ruptured amniotic sac
Rectum
1 Stage of dilation
Placenta
2 Stage of expulsion
Uterus
Placenta
Umbilical cord
3 Placental stage

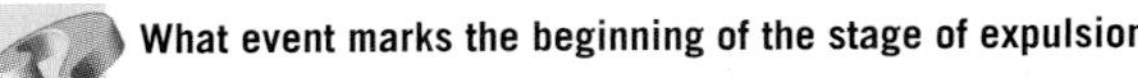
What event marks the beginning of the stage of expulsion?

quate supplies of oxygen and nutrients from an aging placenta. Post-term deliveries may be facilitated by inducing labor (initiated by administration of oxytocin) or by surgical delivery.

Following the delivery of the baby and placenta is a 6-week period during which the maternal reproductive organs and physiology return to the prepregnancy state. This period is called the **puerperium** (pyū′-er-PE-rē-um). Through a process of tissue catabolism, the uterus undergoes a remarkable reduction in size (especially in lactating women), called **involution** (in′-vō-LŪ-shun). The cervix loses its elasticity and regains its prepregnancy firmness. For 2–4 weeks after delivery, women have a uterine discharge called **lochia** (LŌ-kē-a), which consists initially of blood and later of serous fluid derived from the former site of the placenta.

Dystocia and Cesarean Section

Dystocia (dis-TŌ-sē-a; *dys-* = painful or difficult; *toc-* = birth), or difficult labor, may result either from an abnormal position (presentation) of the fetus or a birth canal of inadequate size to permit vaginal birth. In a **breech presentation,** for example, the fetal buttocks or lower limbs, rather than the head, enter the birth canal first; this occurs most often in premature births. If fetal or maternal distress prevents a vaginal birth, the baby may be delivered surgically through an abdominal incision. A low, horizontal cut is made through the abdominal wall and lower portion of the uterus, through which the baby and placenta are removed. Even though it is popularly associated with the birth of Julius Caesar, the procedure for performing this operation is termed a **cesarean section (C-section)** because it was described in Roman Law, *lex cesarea,* 600 years before Julius Caesar was born. Even a history of multiple C-sections need not exclude a pregnant woman from attempting a vaginal delivery. ■

✓ Describe the hormonal changes that induce labor.
✓ Distinguish between false labor and true labor.
✓ Describe what happens during the stage of dilation, the stage of expulsion, and the placental stage of true labor.

ADJUSTMENTS OF THE INFANT AT BIRTH

Objective

- Explain the respiratory and cardiovascular adjustments that occur in an infant at birth.

During pregnancy, the embryo (and later the fetus) is totally dependent on the mother for its existence. The mother supplies the fetus with oxygen and nutrients, eliminates its carbon dioxide and other wastes, protects it against shocks and temperature changes, and provides antibodies that confer protection against certain harmful microbes. At birth, a physiologically mature baby becomes much more self-supporting, and the newborn's body systems must make various adjustments. Next we examine some of the changes that occur in the respiratory and cardiovascular systems.

Respiratory Adjustments

The fetus depends entirely on the mother for obtaining oxygen and eliminating carbon dioxide. The fetal lungs are either collapsed or partially filled with amniotic fluid, which is absorbed at birth. The production of surfactant begins by the end of the sixth month of development, and because the respiratory system is fairly well developed at least 2 months before birth, premature babies delivered at 7 months are able to breathe and cry. After delivery, the baby's supply of oxygen from the mother ceases. Circulation in the baby continues, and as the blood level of carbon dioxide increases, the respiratory center in the medulla oblongata is stimulated, causing the respiratory muscles to contract, and the baby draws its first breath. Because the first inspiration is unusually deep as the lungs contain no air, the baby exhales vigorously and naturally cries. A full-term baby may breathe 45 times a minute for the first 2 weeks after birth. Breathing rate gradually declines until it approaches a normal rate of 12 breaths per minute.

Cardiovascular Adjustments

After the baby's first inspiration, the cardiovascular system must make several adjustments (see Figure 14.18a-b on page 466). Closure of the foramen ovale between the atria of the fetal heart, which occurs at the moment of birth, diverts deoxygenated blood to the lungs for the first time. The foramen ovale is closed by two flaps of septal heart tissue that fold together and permanently fuse. The remnant of the foramen ovale is the fossa ovalis.

Once the lungs begin to function, the ductus arteriosus is shut off by contractions of the muscles in its wall and becomes the ligamentum arteriosum. The muscle contraction is probably mediated by the polypeptide bradykinin, released from the lungs during their initial inflation. The ductus arteriosus generally does not completely and irreversibly close for about 3 months after birth. Incomplete closure results in a condition called **patent ductus arteriosus.**

After the umbilical cord is tied off and severed and blood no longer flows through the umbilical arteries, they fill with connective tissue, and their distal portions become the medial umbilical ligaments. After the umbilical cord is severed, the umbilical vein becomes the ligamentum teres (round ligament) of the liver.

In the fetus, the ductus venosus connects the umbilical vein directly with the inferior vena cava, allowing blood from the placenta to bypass the fetal liver. When the umbilical cord is severed, the ductus venosus collapses, and venous blood from the viscera of the fetus flows into the hepatic portal vein to the liver and then via the hepatic vein to the inferior vena cava. The remnant of the ductus venosus becomes the ligamentum venosum.

At birth, an infant's pulse may range from 120 to 160 beats per minute and may go as high as 180 upon excitation. After birth, oxygen use increases, which stimulates an increase in the rate of red blood cell and hemoglobin production. Moreover, the white blood cell count at birth is very high—sometimes as much as 45,000 cells per microliter (μL)—but the count decreases rapidly by the seventh day.

Premature Infants

Delivery of a physiologically immature baby carries certain risks. A **premature infant** or "preemie" is generally considered to be a baby who weighs less than 2500 g (5.5 lb) at birth. Poor prenatal care, drug abuse, history of a previous premature delivery, and mother's age below 16 or above 35 increase the chance of premature delivery. The body of a premature infant is not yet ready to sustain some critical functions, and thus its survival is uncertain without medical intervention. The major problem after delivery of an infant under 36 weeks of gestation is respiratory distress syndrome (RDS) of the newborn due to insufficient surfactant. RDS can be eased by use of artificial surfactant and a ventilator that delivers oxygen until the lungs can operate on their own. ■

✓ Explain the importance of surfactant at birth and afterward.
✓ What cardiovascular adjustments are made by an infant at birth?

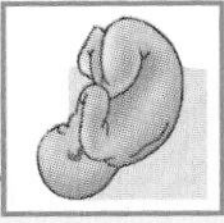

APPLICATIONS TO HEALTH

Infertility

Female infertility, or the inability to conceive, occurs in about 10% of all women of reproductive age in the United States. Female infertility may be caused by ovarian disease, obstruction of the uterine tubes, or conditions in which the uterus is not adequately prepared to receive a fertilized ovum. **Male infertility (sterility)** is an inability to fertilize a secondary oocyte; it does not imply impotence. Male fertility requires production of adequate quantities of viable, normal sperm by the testes, unobstructed transport of sperm though the ducts, and satisfactory deposition in the vagina. The seminiferous tubules of the testes are sensitive to many factors—x-rays, infections, toxins, malnutrition, and higher-than-normal scrotal temperatures—that may cause degenerative changes and produce male sterility.

One cause of infertility in females is inadequate body fat. To begin and maintain a normal reproductive cycle, a female must have a minimum amount of body fat. A moderate deficiency of fat—10–15% below normal weight for height—may delay the onset of menstruation (menarche), inhibit ovulation during the reproductive cycle, or cause amenorrhea, the cessation of menstruation. Both dieting and intensive exercise may reduce body fat below the minimum amount and lead to infertility that is reversible, if weight gain or reduction of intensive exercise or both occurs. Studies of very obese women indicate that they, like very lean ones, experience problems with amenorrhea and infertility. Males also experience reproductive problems in response to undernutrition and weight loss. For example, they produce less prostatic fluid and reduced numbers of sperm that have decreased motility.

Many fertility-expanding techniques now exist for assisting infertile couples to have a baby. The birth of Louise Joy Brown on July 12, 1978, near Manchester, England, was the first recorded case of **in vitro fertilization (IVF)**—fertilization in a laboratory dish. In the IVF procedure, the mother-to-be is given follicle-stimulating hormone (FSH) soon after menstruation, so that several secondary oocytes, rather than the typical single oocyte, will be produced (superovulation). When several follicles have reached the appropriate size, a small incision is made near the umbilicus, and the secondary oocytes are aspirated from the stimulated follicles and transferred to a solution containing sperm, where the oocytes undergo fertilization. Alternatively, an oocyte may be fertilized in vitro by suctioning a sperm or even a spermatid obtained from the testis into a tiny pipette and then injecting it into the oocyte's cytoplasm. This procedure, termed **intracytoplasmic sperm injection (ICSI),** has been used when infertility is due to impairments in sperm motility or to the failure of spermatids to develop into spermatozoa. When the zygote achieved by IVF reaches the 8-cell or 16-cell stage, it is introduced into the uterus for implantation and subsequent growth.

In **embryo transfer,** a man's semen is used to artificially inseminate a fertile secondary oocyte donor. After fertilization in the donor's uterine tube, the morula or blastocyst is transferred from the donor to the infertile woman, who then carries it (and subsequently the fetus) to term. Embryo transfer is indicated for women who are infertile or who do not want to pass on their own genes because they are carriers of a serious genetic disorder.

In **gamete intrafallopian transfer (GIFT)** the goal is to mimic the normal process of conception by uniting sperm and secondary oocyte in the prospective mother's uterine tubes. It is an attempt to bypass conditions in the female reproductive tract that might prevent fertilization, such as high acidity or inappropriate mucus. In this procedure, a woman is given FSH and LH to stimulate the production of several secondary oocytes, which are aspirated from the mature follicles with a laparoscope fitted with a suction device, mixed outside the body, with a solution containing sperm, and then immediately inserted into the uterine tubes.

Down Syndrome

Down syndrome (DS) is a disorder that most often results from nondisjunction of chromosome 21 during meiosis, which causes an extra chromosome to pass to one of the gametes. For this reason, DS is also called trisomy 21. Most of the time the extra chromosome comes from the mother, a not too surprising finding given that all her oocytes began meiosis when she herself was a fetus; they may have been exposed to chromosome-damaging chemicals and radiation for years. Undoubtedly, the kinetochore microtubules responsible for pulling sister chromatids to opposite poles of the cell (see Figure 26.1 on page 783) sustain increasing damage with the passage of time. (Sperm, by contrast, usually are less than 10 weeks old at the time they fertilize a secondary oocyte.) Overall, 1 infant in 800 is born with Down syndrome. However, older women are more likely to have a DS baby. The chance of conceiving a baby with this syndrome, which is less than 1 in 3000 for women under age 30, increases to 1 in 300 in the 35–39 age group and to 1 in 9 at age 48.

Down syndrome is characterized by mental retardation, retarded physical development (short stature and stubby fingers), distinctive facial structures (large tongue, flat profile, broad skull, slanting eyes, and round head), and malformations of the heart, ears, hands, and feet. Sexual maturity is rarely attained.

Fragile X Syndrome

Fragile X syndrome, a recently recognized disorder caused by a defective gene on the X chromosome, is so named because a small portion of the tip of the X chromosome seems susceptible to breakage. It is the leading cause of mental retardation among newborns. Fragile X syndrome affects males more than females and results in learning difficulties, mental retardation, and physical abnormalities such as oversized ears, elongated forehead, enlarged testes, and double jointedness. The syndrome may also be involved in autism in which the individual exhibits extreme withdrawal and refusal to communicate.

Fragile X syndrome can be diagnosed via amniocentesis. Although the syndrome is largely sex-linked, 20–50% of males who inherit the trait are unaffected but can pass it on to their daughters. Although the daughters also are unaffected, their children, male and female alike, may suffer mental retardation. Researchers have suggested that this pattern of inheritance could be explained by maternal imprinting of the gene. According to this theory, the father's fragile X gene becomes chemically changed as it passes through the daughter so that the gene is expressed in the daughter's offspring.

KEY MEDICAL TERMS ASSOCIATED WITH DEVELOPMENTAL ANATOMY

Breech presentation A malpresentation in which the fetal buttocks or lower limbs present into the maternal pelvis; the most common cause is prematurity.

Emesis gravidarum (EM-e-sis gra-VID-ar-um; *emeo* = to vomit; *gravida* = a pregnant woman) Episodes of nausea and possibly vomiting that are most likely to occur in the morning during the early weeks of pregnancy; also called **morning sickness.** Its cause is unknown, but the high levels of human chorionic gonadotropin (hCG) secreted by the placenta, and of progesterone secreted by the ovaries, have been implicated. In some women the severity of these symptoms requires hospitalization for intravenous feeding, and the condition is then known as **hyperemesis gravidarum.**

Karyotype (KAR-ē-ō-tīp; *karyo-* = nucleus) The chromosomal characteristics of an individual presented as a systematic arrangement of pairs of metaphase chromosomes arrayed in descending order of size and according to the position of the centromere; useful in judging whether or not chromosomes are normal in number and structure.

Lethal gene (LĒ-thal jēn; *lethum* = death) A gene that, when expressed, results in death either in the embryonic state or shortly after birth.

Mutation (myū-TĀ-shun; *mutare* = change) A permanent heritable change in a gene that causes it to have a different effect than it had previously.

Puerperal fever (pyū-ER-per-al; *puer* = child, *parere* = to bring forth) An infectious disease of childbirth, also called puerperal sepsis and childbed fever. The disease, which results from an infection originating in the birth canal, affects the endometrium; it may spread to other pelvic structures and lead to septicemia.

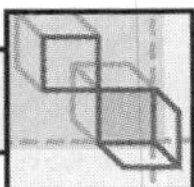

STUDY OUTLINE

From Fertilization to Implantation (p. 826)

1. Pregnancy is a sequence of events that starts with fertilization and normally ends in birth.
2. Its various events are hormonally controlled.
3. Fertilization refers to the penetration of a secondary oocyte by a sperm cell and the subsequent union of their pronuclei to form a zygote.
4. Penetration of the zona pellucida is facilitated by enzymes in the sperm's acrosome.
5. Normally, only one sperm cell fertilizes a secondary oocyte because both fast and slow blocks to polyspermy exist.
6. Early rapid cell division of a zygote is called cleavage, and the cells produced by cleavage are called blastomeres.
7. The solid sphere of cells produced by cleavage is a morula.
8. The morula develops into a blastocyst, a hollow ball of cells differentiated into a trophoblast and an inner cell mass.
9. The attachment of a blastocyst to the endometrium is called implantation; it occurs via enzymatic degradation of the endometrium.

Embryonic and Fetal Development (p. 830)

1. During embryonic growth, the primary germ layers and embryonic membranes are formed.
2. The primary germ layers—ectoderm, mesoderm, and endoderm—form all the tissues of the developing organism. Table 27.1 on page 833 summarizes the structures produced by the three germ layers.
3. The embryonic membranes are the yolk sac, amnion, chorion, and allantois.
4. Fetal and maternal materials are exchanged through the placenta.
5. During the fetal period, organs established by the primary germ layers grow rapidly.
6. The principal changes associated with embryonic and fetal growth are summarized in Table 27.2 on page 837.
7. Teratogens, which are agents that cause physical defects in developing embryos, include chemicals and drugs, alcohol, nicotine, and ionizing radiation.
8. Several prenatal diagnostic tests are used to detect genetic disorders and to assess fetal well-being, including fetal ultrasonography, in which an image of a fetus is displayed on a screen; amniocentesis, the withdrawal and analysis of amniotic fluid and the fetal cells within it; chorionic villi sampling (CVS), which involves withdrawal of chorionic villi tissue for chromosomal analysis. CVS can be done earlier than amniocentesis, and the results are available more quickly, but it is also slightly riskier than amniocentesis.

Maternal Changes during Pregnancy (p. 840)

1. Pregnancy is maintained by human chorionic gonadotropin (hCG), estrogens, and progesterone.
2. Human chorionic somatomammotropin (hCS) assumes a role in breast development, protein anabolism, and glucose and fatty acid catabolism.
3. Relaxin relaxes the pubic symphysis and helps dilate the uterine cervix near the end of pregnancy.
4. Corticotropin-releasing hormone, produced by the placenta, is thought to establish the timing of birth and also stimulates the secretion of cortisol by the fetal adrenal gland.
5. During pregnancy, several anatomical and physiological changes occur in the mother.

Exercise and Pregnancy (p. 842)

1. During pregnancy, some joints become less stable, and certain maneuvers are more difficult to execute.
2. Moderate physical activity does not endanger the fetus in a normal pregnancy.

Labor (p. 842)

1. Labor is the process by which the fetus is expelled from the uterus through the vagina to the outside. True labor involves dilation of the cervix, expulsion of the fetus, and delivery of the placenta.
2. Oxytocin stimulates uterine contractions.

Adjustments of the Infant at Birth (p. 844)

1. The fetus depends on the mother for oxygen and nutrients, the removal of wastes, and protection.
2. Following birth, an infant's respiratory and cardiovascular systems undergo changes in adjusting to becoming self-supporting during postnatal life.

SELF-QUIZ QUESTIONS

Choose the one best answer to the following questions.

1. Which of the following four statements is true? (1) Capacitation occurs in the male reproductive tract. (2) The acrosome produces enzymes that help a sperm cell penetrate a secondary oocyte. (3) A segmentation nucleus contains the diploid (2n) number of chromosomes. (4) The zona pellucida is a gelatinous covering around the acrosome of a sperm cell.
   a. 1, 2 and 3 **b.** 2, 3 and 4 **c.** 2 and 3 **d.** 1 and 3 **e.** All statements are true.
2. How long after fertilization does implantation of an embryo usually occur? (a) 3 weeks, (b) 1 week, (c) 1 day, (d) 3 days, (e) 30 minutes
3. Dizygotic (fraternal) twins result from (a) two secondary oocytes and two sperm, (b) two secondary oocytes and one sperm, (c) one secondary oocyte and two sperm, (d) one secondary oocyte and one sperm, (e) one secondary oocyte and three sperm.
4. Which of the following statements are true of maternal changes during pregnancy? (1) the blood volume increases 30–50%. (2) The heart rate increases 10–15%. (3) There is an increase in motility of the gastrointestinal tract. (4) There may be increased frequency and urgency of urination.
   a. 1, 2 and 4 **b.** 1, 2 and 3 **c.** 2, 3 and 4 **d.** 2 and 3 **e.** 1, 2, 3, and 4.

5. The placenta is formed by the union of the decidua basalis of the endometrium with the (a) yolk sac, (b) amnion, (c) chorion, (d) umbilicus, (e) allantois.

6. Oxygen and nutrients from the maternal blood must pass through which of the following structures before entering the fetal blood? (a) umbilical vein, (b) umbilical artery, (c) decidua capsularis, (d) intervillous spaces, (e) allantois

Complete the following.

7. The 38 weeks from implantation to birth is called the ________ period; the first two months is the ________ period; the remaining portion of the 38 weeks is the ________ period.

8. In a developing embryo, the layer of cells of the inner cell mass near the amniotic cavity develops into ________, whereas the layer of the inner cell mass that borders the blastocele develops into ________ .

9. The inner cell mass of the implanted blastocyst will form the ________ .

10. The embryonic tissues from which all tissues and organs develop are called ________ layers.

11. Place numbers in the blanks to arrange the following events in correct sequence. (a) morula: ________ ; (b) ovum: ________ ; (c) blastocyst: ________ ; (d) zygote: ________ ; (e) fetus: ________ .

12. Place numbers in the blanks to arrange the following events in correct sequence. (a) stage of expulsion; from complete cervical dilation through delivery of the baby: ________ ; (b) time after the delivery of the baby until the placenta ("afterbirth") is expelled; the placental stage: ________ ; (c) time from onset of labor to complete dilation of the cervix; the stage of dilation: ________ .

13. Place numbers in the blanks to arrange the following events in the correct sequence. (a) The blastocyst enters the cavity of the uterus ________ ; (b) The cells of the blastocyst specialize to form the trophoblast and the inner cell mass ________ ; (c) The blastocyst attaches to the endometrium ________ ; (d) The trophoblast develops two specialized layers, the syncytiotrophoblast and the cytotrophoblast ________ ; (e) The zona pellucida disintegrates and the blastocyst enlarges ________ ; (f) Enzymes of the syncytiotrophoblast digest endometrial cells ________ .

Are the following statements true or false?

14. The enzyme acrosin stimulates the movement and migration of sperm in the female reproductive tract.

15. The fusion of the male and female pronuclei is known as syngamy.

16. Development of the placenta occurs during the third month of pregnancy.

17. The primary germ layers which give rise to all tissues and organs are present at the time of implantation.

18. The umbilical cord contains two umbilical arteries that carry oxygenated blood to the fetus, and one umbilical vein that carries fetal blood to the placenta.

19. A high level of maternal corticotropin releasing hormone (CRH) is an indication that a premature delivery is probable.

20. Match the following.

____ **(a)** epithelial lining of all gastrointestinal, respiratory, and genitourinary tracts except near openings to the exterior of the body

____ **(b)** epidermis of skin, epithelial lining of entrances to the body (such as mouth, nose, and anus), hair, nails

____ **(c)** all of the skeletal system (bone, cartilage, and other connective tissues)

____ **(d)** muscle (skeletal, cardiac, and most smooth)

____ **(e)** blood and all blood and lymphatic vessels

____ **(f)** entire nervous system

____ **(g)** thyroid, parathyroid, thymus, and pancreas

(1) develop(s) from ectoderm
(2) develop(s) from endoderm
(3) develop(s) from mesoderm

21. Match the following.

____ **(a)** portion of endometrium that initially separates the embryo from the uterine cavity

____ **(b)** portion of the modified endometrium that lines the region of the uterus not involved with the implanted embryo

____ **(c)** layer of the endometrium between the chorion and the stratum basalis

____ **(d)** fuses with the decidua parietalis, thus obliterating the uterine cavity

____ **(e)** becomes the maternal portion of the placenta

____ **(f)** layer of decidua that disappears by about 27 weeks

(1) decidua basalis
(2) decidua parietalis
(3) decidua capsularis

22. Match the following.

____ **(a)** originally formed from ectoderm, this membrane encloses fluid that acts as a shock absorber for the developing baby

____ **(b)** derived from mesoderm and trophoblast, it becomes the principal embryonic part of the placenta

____ **(c)** serves as an early site of blood cell formation (two answers)

____ **(d)** a small membrane that forms umbilical blood vessels

____ **(e)** the "bag of waters" that ruptures just before birth

____ **(f)** produces human chorionic gonadotropin (hCG)

(1) yolk sac
(2) amnion
(3) chorion
(4) allantois

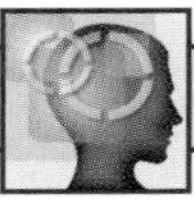

CRITICAL THINKING QUESTIONS

1. Your neighbors from down the street put up a sign to announce the birth of their twins—a girl and a boy. Your next-door neighbor said "Oh, how sweet! I wonder if they're identical?" Without even seeing the twins, what can you tell her?
HINT *Would you expect the parents to dress them in identical outfits?*

2. Some disorders of the nervous system may also exhibit particular skin problems. For example, a person with a type of nervous system tumor may show coffee-colored spots on the skin. How does the structure of the nervous system relate to that of the skin?
HINT *There are 4 main tissue types in adults but only 3 primary germ layers in the embryo.*

3. Josefina, in the last 2 weeks of her first pregnancy, anxiously called her doctor to ask if she should leave for the hospital. She was experiencing irregular "labor" pains which were thankfully eased by walking. She had no other signs to report. The doctor told Josefina to stay home—it wasn't time yet. Why did the doctor tell Josefina to stay home?
HINT *There's a saying in show-biz "another opening, another show."*

4. Tony couldn't help laughing after the last Human Anatomy examination. "Did you see that true and false question about yolk? It must have been a mistake! This isn't poultry 101!" Actually, it was a mistake for Tony to have spent all night watching cable TV. What does yolk have to do with human development?
HINT *The yolk is surrounded by a membrane.*

5. Selma's nurse-midwife has told her to stop smoking and drinking alcoholic beverages during her pregnancy, but Selma is resisting the advice. Selma says "Nothing gets through the placenta. That's what it's for." Is Selma correct?
HINT *How does the baby get its nourishment in utero?*

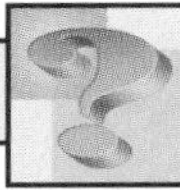

ANSWERS TO FIGURE QUESTIONS

27.1 Capacitation refers to the functional changes in sperm after they have been deposited in the female reproductive tract that enable them to fertilize a secondary oocyte.

27.2 A morula is a solid ball of cells, whereas a blastocyst consists of a rim of cells (trophoblast) surrounding a cavity (blastocyst cavity) and an inner cell mass.

27.3 The blastocyst secretes digestive enzymes that eat away the endometrial lining at the site of implantation.

27.4 Implantation occurs during the secretory phase of the uterine cycle.

27.5 The inner cell mass of the blastocyst gives rise to the embryonic disc.

27.6 The amnion functions as a shock absorber, whereas the chorion forms the placenta.

27.7 The decidua basalis helps form the maternal part of the placenta.

27.8 The placenta functions in exchange of materials between fetus and mother.

27.9 Amniocentesis is used mainly to detect genetic disorders, but it also provides information concerning the maturity (and survivability) of the fetus.

27.10 Relaxin.

27.11 Complete dilation of the cervix marks the onset of the stage of expulsion.

APPENDIX A
MEASUREMENTS

U.S. Customary System			
	Unit	**Relation to Other U.S. Units**	**SI (Metric) Equivalent**
Length	inch	1/12 foot	2.54 centimeters
	foot	12 inches	0.305 meter
	yard	36 inches	9.144 meters
	mile	5,280 feet	1.609 kilometers
Mass	grain	1/1000 pound	64.799 milligrams
	dram	1/16 ounce	1.772 grams
	ounce	16 drams	28.350 grams
	pound	16 ounces	453.6 grams
	ton	2,000 pounds	907.18 kilograms
Volume (Liquid)	ounce	1/16 pint	29.574 milliliters
	pint	16 ounces	0.473 liter
	quart	2 pints	0.946 liter
	gallon	4 quarts	3.785 liters
Volume (Dry)	pint	1/2 quart	0.551 liter
	quart	2 pints	1.101 liters
	peck	8 quarts	8.810 liters
	bushel	4 pecks	35.239 liters

International System (SI)		
BASE UNITS		
Unit	**Quantity**	**Symbol**
meter	length	m
kilogram	mass	kg
second	time	s
liter	volume	L
mole	amount of matter	mol
PREFIXES		
Prefix	**Multiplier**	**Symbol**
tera-	10^{12} = 1,000,000,000,000	T
giga-	10^{9} = 1,000,000,000	G
mega-	10^{6} = 1,000,000	M
kilo-	10^{3} = 1,000	K
hecto-	10^{2} = 100	h
deca-	10^{1} = 10	da
deci-	10^{-1} = 0.1	d
centi-	10^{-2} = 0.01	c
milli-	10^{-3} = 0.001	m
micro-	10^{-6} = 0.000,001	μ
nano-	10^{-9} = 0.000,000,001	n
pico-	10^{-2} = 0.000,000,000,001	p

Apothecary Measures

	Unit	US Equivalent	SI Equivalent
Mass	grain	1 grain	64.8 milligrams
	dram	2.194 drams	3.888 grams
	ounce	1.097 ounces	31.104 grams
	pound	0.823 pound	373.242 grams
Volume	minims	1/60 ounce	0.06 milliliter
	fluid dram	1/8 ounce	3.697 milliliters
	fluid ounce	1 ounce	29.573 milliliters
	pint	1 pint	0.473 liter

U.S. To SI (Metric) Conversion

When you know	Multiply by	To find
inches	2.54	centimeters
feet	30.48	centimeters
yards	0.91	meters
miles	1.61	kilometers
ounces	28.35	grams
pounds	0.45	kilograms
tons	0.91	metric tons
fluid ounces	29.57	milliliters
pints	0.47	liters
quarts	0.95	liters
gallons	3.79	liters

SI (Metric) to U.S. Conversion

When you know	Multiply by	To find
millimeters	0.04	inches
centimeters	0.39	inches
meters	3.28	feet
kilometers	0.62	miles
liters	1.06	quarts
cubic meters	35.32	cubic feet
grams	0.035	ounces
kilograms	2.21	pounds

Temperature Conversion

Fahrenheit (F) to Celsius (C)
°C = (°F − 32) ÷ 1.8
Celsius (C) to Fahrenheit (F)
°F = (°C × 1.8) + 32

APPENDIX B
ANSWERS

ANSWERS TO SELF QUIZ QUESTIONS

Chapter 1

1. (e) 2. (d) 3. (b) 4. (b) 5. (d) 6. (e) 7. epithelia, connective, muscular, nervous 8. pleural, pericardial peritoneal 9. hypochondriac, lumbar, inguinal 10. frontal 11. organs 12. ipsi 13. inferior, superior 14. F 15. T 16. T 17. F 18. T 19. 3(a), 6(b), 5(c), 1(d), 2(e), 8(f), 7(g), 4(h) 20. 1(c), 2(g), 3(a), 4(e), 5(d), 6(h), 7(f), 8(b)

Chapter 2

1. (d) 2. (a) 3. (a) 4. (e) 5. (b) 6. (c) 7. (a) 8. (b) 9. cytosol 10. phagocytosis 11. receptor-mediated endocytosis 12. lipids 13. S 14. selectively permeable 15. sperm 16. T 17. T 18. T 19. 1(j), 2(l), 3(c), 4(k), 5(e), 6(d), 7(h), 8(g), 9(a), 10(i), 11(b), 12(f) 20. 1(a), 2(e), 3(d), 4(b), 5(c)

Chapter 3

1. (b) 2. (d) 3. (e) 4. (d) 5. (b) 6. (e) 7. (a) 8. (b) 9. cardiac 10. exocrine 11. osseous, lamellae 12. plasma 13. serous 14. lamina propria 15. T 16. T 17. F 18. T 19. 1(e), 2(f), 3(h), 4(a), 5(c), 6(g), 7(b), 8(d) 20. 1(d), 2(e), 3(a), 4(h), 5(b), 6(g), 7(c), 8(f), 9(j), 10(i)

Chapter 4

1. (c) 2. (b) 3. (a) 4. (b) 5. (a) 6. (c) 7. ceruminous glands, cerumen 8. dermis, connective 9. sudoriferous 10. T 11. T 12. F 13. T 14. T 15. F 16. 1(b), 2(d), 3(e), 4(a), 5(f), 6(c) 17.1(b), 2(e), 3(a), 4(d), 5(c)

Chapter 5

1. (e), 2. (d) 3. (e) 4. calcium 5. endochondral 6. matrix 7. zone of resting cartilage 8. yellow bone marrow 9. osteoclasts 10. appositional 11. T 12. T 13. F 14. T 15. F 16. T 17. T 18. 1(a), 2(e), 3(c) 4(b), 5(d)

Chapter 6

1. (c) 2. (c) 3. (b) 4. (a) 5. (d) 6. (a) 7. (b) 8. (a) 9. (e) 10. (c) 11. 1(b), 2(j), 3(d), 4(i), 5(h), 6(e), 7(c), 8(f), 9(k), 10(g), 11(l), 12(a) 12. atlas, axis 13. long, short, flat 14. manubrium, body, xiphoid process 15. fontanels 16. T 17. F 18. T 19. F 20. F

Chapter 7

1. (b) 2. (e) 3. (d) 4. (e) 5. (c) 6. (d) 7. (b) 8. (c) 9. (e) 10. (b) 11. subscapular fossa 12. head of the second metacarpal 13. posterior, femur 14. false 15. hip 16. F 17. T 18. T 19. F 20. 1(f), 2(h), 3(a), 4(e), 5(b), 6(i), 7(c), 8(d), 9(g)

Chapter 8

1. (b) 2. (a) 3. (a) 4. (d) 5. (e) 6. (e) 7. (d) 8. amphiarthrotic 9. knee 10. diarthrotic 11. hyaline cartilage (articular cartilage) 12. bursae 13. T 14. T 15. T 16. F 17. F 18. 1(h), 2(d), 3(a), 4(e), 5(b), 6(i), 7(c), 8(f), 9(g) 19. 1(c), 2(e), 3(b), 4(d), 5(a) 20. 1(h), 2(f), 3(d), 4(b), 5(c), 6(a), 7(e), 8(g)

Chapter 9

1. (c) 2. (b) 3. (e) 4. (c) 5. (e) 6. (c) 7. (d) 8. sarcoplasmic reticulum 9. myoglobin, mitochondria 10. T tubules 11. fiber, sarcolemma, sarcoplasm 12. S, U, U, S, S 13. smooth and cardiac 14. T 15. F 16. F 17. T 18. T 19. 1 (b) and (e), 2 (c) and (g), 3 (a), (d), and (f) 20. 1(c), 2(a), 3(e), 4(f), 5(b), 6(d)

Chapter 10

1. (d) 2. (b) 3. (e) 4. (e) 5. (b) 6. (e) 7. (d) 8. prime mover (agonist), synergists 9. parallel, fusiform, pennate, circular, triangular 10. (a) platysma, (b) orbicularis oculi, (c) buccinator, (d) frontalis 11. digastric 12. mastoid process 13. central 14. trapezius, rhomboideus major, rhomboideus minor 15. thenar, hypothenar 16. erector spinae, lumbar 17. flexion 18. anterior, four, tibia, extend 19. gastrocnemius, soleus, plantaris 20.T 21.F 22.F 23. (a) 2 and 3, (b) 3 and 6 (c) 4 and 5 (d) 5 (e) 1 and 6 (f) 3 (g) 6 24. (a)3, (b)2, (c)1, (d)3, (e) 1, (f)2 (g)3 25. (a)3, (b)6, (c)1, (d)11, (e)8, (f)2, (g)10, (h)5, (i)7, (j)9, (k)4

Chapter 11

1. (e) 2. (a) 3. (a) 4. cranial, facial 5. sternocleidomastoid 6. brachial 7. (a) inferior (b) inferior (c) distal (d) superior (e) anterior (f) lateral (g) inferior 8. femoral 9. (a)5, (b)1, (c)2, (d)4, (e)8, (f)3, (g)6, (h)7 10. T 11. T 12. T 13. T 14. T 15. F 16. F 17. (a)5, (b) 1, (c)4, (d)7, (e)3, (f)2, (g)6 18. (a)4, (b)3, (c)7, (d)8, (e)2, (f)1, (g)6, (h)5

Chapter 12

1. (b) 2. (b) 3. (a) 4. (d) 5. (e) 6. (e) 7. (c) 8. (e) 9. (a) 10. (e) 11. (e) 12. (b) 13. differential white blood cell 14. 45%, 55% 15. emigration 16. erythropoiesis 17. albumins, globulins, fibrinogen 18. lysozyme, defensins 19. megakaryocytes 20. T 21. T 22. F 23. (a)5, (b)3, (c)3, (d)1, (e)2, (f)4, (g) 4 and 5, (h) 2, (i)1, (j)4

Chapter 13

1. (a) 2. (e) 3. (a) 4. (d) 5. (d) 6. (a) 7. gap junctions, desmosomes 8. right and left coronary arteries 9. atrioventricular 10. sinoatrial node 11. open, closed 12. meso, first, primitive heart 13. left ventricle 14. inferior left 15. T 16. T 17. F 18. T 19. F 20. T 21. F 22. F 23. (a)3, (b)4, (c)2, (d)1, (e)4, (f)3

Chapter 14

1. (c) 2. (c) 3. (e) 4. (a) 5. (e) 6. (b) 7. (d) 8. (a) 9. common hepatic artery, left gastric artery, splenic artery 10. superior vena cava, inferior vena cava, coronary sinus 11.140, 90 12. inferior vena cava 13. hepatic portal vein 14. anastomosis 15. media 16. vasoconstriction 17. internal jugular 18. pulmonary 19. (a)10, (b)7, (c)8, (d)2, (e)11, (f)1, (g)5, (h)3, (i)9, (j)6, (k)4 20. T 21. F 22. T 23. T 24. (a)5, (b)3, (c)2, (d)5, (e)4, (f)1, (g)2, (h)1, (i)2, (j)1

Chapter 15

1. (b) 2. (c) 3. (d) 4. (e) 5. (a) 6. (e) 7. (e) 8. (d) 9. thymus 10. veins 11. red bone marrow, thymus 12. cisterna chyli 13. meso 14. MALT (mucous-associated lymphatic tissue) 15. F 16. T 17. T 18. T 19. F 20. (a)4, (b)7, (c)1, (d)8, (e)2, (f)6, (g)5, (h) 3

Chapter 16

1. (c) 2. (a) 3. (c) 4. (d) 5. (e) 6. spinal, cranial 7. somatic, autonomic, enteric 8. central 9. epineurium 10. dendrite 11. synaptic vesicles

12. converging 13. (a)3, (b)6, (c)7, (d)8. (e)4, (f)5, (g)1, (h)2 14. (a)3, (b)1, (c)1, (d)2, (e)3, (f)2, (g)1 15. T 16. F 17. T 18. T 19. T 20. F

Chapter 17.

1. (e) 2. (d) 3. (b) 4. (c) 5. (e) 6. medulla oblongata, second lumbar vertebra 7. denticulate ligament 8. dura mater, arachnoid, pia mater 9. endoneurium, perineurium, epineurium 10. anterior, lateral 11. 31, 8, 12, 5, 5, 1 12. F 13. T 14. T 15. F 16. T 17. T 18. (a)1, (b)2, (c)1, (d)5, (e)2, (f)4, (g)3 19. (a)7, (b)5, (c) 6, (d)1, (e)2, (f)4, (g)3

Chapter 18.

1. (b) 2. (c) 3. (c) 4. (b) 5. (c) 6. (b) 7. choroid plexuses 8. cuneatus 9. chiasm 10. basal ganglia, cerebrum 11. association 12. cerebral cortex, gray, gyri, sulci 13. (a)w, (b)g, (c)g, (d)g, (e)w, (f)g 14. (a) accessory, (b) vestibulocochlear, (c) vagus, (d) trigeminal, (e) trigeminal, (f) oculomotor, (g) facial, (h) facial and glossopharyngeal (i) olfactory, optic, and vestibulocochlear. 15. (a)2, (b)1, (c)3 16. (a)2, (b)3, (c)1 17. (a)1, (b)4, (c)2, (d)5, (e)6, (f)3 18. (a)9, (b)7, (c)5, (d)2, (e)4, (f)6, (g)3, (h)8, (i)1, (j)10 19. (a)1, (b)6, (c)3, (d)4, (e)2, (f)1, (g)5, (h)2, (i)5, (j)1, (k)4 20.F 21.T 22.F 23.T

Chapter 19

1. (c) 2. (a) 3. (a) 4. (d) 5. (b) 6.vision, hearing, taste, smell, equilibrium 7. exteroceptors, interoceptors (visceroceptors), proprioceptors 8. mechano 9. brain stem, association 10. referred 11. (a)6, (b)3, (c)8, (d)2, (e)5, (f)4, (g)1, (h)7 12. (a)3, (b)1, (c)2, (d)1 13. posterior spinocerebellar, anterior spinocerebellar 14. F 15. T 16. F 17. T 18. F 19. (a)4, (b)1, (c)3, (d)2 20. (a)4, (b)3, (c)1, (d)2, (e)6, (f)5

Chapter 20

1. (c) 2. (a) 3. (a) 4. (b) 5. (e) 6. (d) 7. basal stem 8. papillae, fungiform 9. ciliary, scleral venous 10. sclera, choroid, retina 11. conjunctiva 12. pigment 13. (a)3, (b)1, (c)2, (d)4, (e)5 14. (a)1, (b)5, (c)3, (d)4, (e)6, (f)2 15. (a)2, (b)3, (c)1, (d)6, (e)5, (f)7, (g)4 16. F 17. T 18. T 19. F 20. (a)11, (b)6, (c)7, (d)1, (e)3, (f)10, (g)2, (h)9, (i)5, (j)4, (k)8, (l)12 21. (a)1, (b)3, (c)1, (d)2 22. (a)5, (b)4, (c)9, (d)3, (e)12, (f)6, (g)7, (h)8, (i)1, (j)2, (k)11, (l)10

Chapter 21

1. (b) 2. (b) 3. (a) 4. (e) 5. (d) 6. (a) 7. (e) 8. (c) 9. (c) 10. general visceral sensory, general visceral motor 11. dual 12. epinephrine, norepinephrine 13. alpha, beta 14. nicotinic, muscarinic 15. parasympathetic 16. T 17. F 18. T 19. F 20. T 21. (a)2, (b)5, (c)6 (d)4, (e)1, (f)3 22. (a)1, (b)2, (c)1, (d)2, (e)1, (f)1, (g)1, (h)1, (i)3, (j)1

Chapter 22

1. (a) 2. (c) 3. (a) 4. (c) 5. (e) 6. (e) 7. (a)3, (b)2, (c)1, (d)4 8. (a)3, (b)4, (e)2, (f)1 9. tropic, target 10. alpha, increase 11. infundibulum, sella turcica 12. adenohypophysis 13. parafollicular or C, principal or chief 14. T 15. T 16. F 17. T 18. F 19. T 20. (a)4, (b)3, (c)2, (d)6, (e)5, (f)5 (g)1 21.(a)5, (b)6, (c)8, (d)3, (e)2, (f)9, (g)1, (h)10, (i)7, (j)4

Chapter 23

1. (c) 2. (a) 3. (b) 4. (d) 5. (c) 6. (e) 7. (d) 8. base, mediastinal or medial, ribs 9. epiglottis 10. glottidis, vestibuli 11. lobule 12. (a)3, (b)2, (c)1 13. (a)9, (b)4, (c)10, (d)3, (e)8, (f)12, (g)13, (h)6, (i)1, (j)2, (k)5, (l)11, (m)7 14. pseudostratified ciliated columnar epithelium with goblet cells, nonciliated simple cuboidal epithelium with no goblet cells 15. T 16. F 17. T 18. F 19. (a)1, (b)5, (c)2, (d)3, (e)4, (f)4, (g)2 and 3 (h)2 20. (a)1, (b)3, (c)1, (d)2, (e)1, (f)2, (g)1, (h)4 21.(a)1, (b)2, (c)2, (d)1, (e)1

Chapter 24

1. (c) 2. (d) 3. (e) 4. (a) 5. (d) 6. (b) 7. (e) 8. (b) 9. lamina propria, mucous-associated lymphoid tissue 10. (a)1, (b)4, (c)3, (d)2, (e)5 11. lingual frenulum 12. (a)E, (b)E, (c)M 13. permanent 14. liver 15. cystic 16. haustra 17. caudate, quadrate 18. (a)1, (b)3, (c)4, (d)2 19. F 20. F 21. T 22. T 23. T 24. F

Chapter 25

1. (d) 2. (b) 3. (b) 4. (c) 5. (c) 6. renal papilla 7. efferent, afferent 8. detrusor 9. prostatic, membranous, spongy (penile) 10. renal fascia, adipose capsule, renal capsule 11. renal pyramids 12. ureters 13. (a)4, (b)5, (c)1, (d)9, (e)7, (f)8, (g)6, (h)3, (i)2 14. (a)3, (b)2, (c)5, (d)6, (e)4, (f)1 15. micturition 16. T 17. T 18. F 19. T

Chapter 26

1. (a) 2. (b) 3. (e) 4. (a) 5. (a) 4, (b)1, (c)5, (d)3, (e)2 6. spiral arterioles 7. scrotum, penis, prostate gland, bulbourethral (Cowper's) glands 8. basalis 9. myometrium, smooth muscle 10. corpus albicans 11. (a)4, (b)2, (c)3, (d)1, (e)5, (f)6 12. ovarian 13. ovulation 14. (a)4, (b)1, (c)2, (d)3, (e)5 15. T 16. F 17. F 18. T 19. T 20. T 21. (a)7, (b)4, (c)3, (d)1, (e) 6, (f)2, (g)5

Chapter 27

1. (c) 2. (b) 3. (a) 4. (a) 5. (c) 6. (d) 7. gestation, embryonic, fetal 8. ectoderm, endoderm 9. embryo 10. primary germ 11. (a)3, (b)1, (c)4, (d)2, (e)5 12. (a)2, (b)3, (c)1 13. (a)2, (b)1, (c)4, (d)5, (e)3, (f)6 14. T 15. F 16. T 17. F 18. F 19. T 20. (a)2, (b)1, (c)3, (d)3, (e)3, (f)1, (g)2 21. (a)3, (b)2, (c)1, (d)3, (e)1, (f)3 22. (a)2, (b)3, (c)1 and 4, (d) 4, (e)2, (f)3

ANSWERS TO CRITICAL THINKING QUESTIONS

Chapter 1

1. The kidneys are located behind the parietal peritoneum. To view the posterior surface of the kidney, a scope would penetrate the posterior body wall but would not enter the peritoneal cavity.
2. Taylor's arm will be extended lateral to the trunk, hand inferior and distal to the elbow, palm facing anteriorly, thumb lateral, and radius lateral to ulna.
3. The alien would have 2 tails, 4 arms, 2 legs, and a mouth where its navel is usually located.
4. A living organism would exhibit some or all of the following: metabolism, responsiveness, movement, growth, differentiation, reproduction.
5. The navel is located on the anterior side of the body, in line with the midsagittal plane, in the umbilical region of the abdominopelvic cavity or at the intersection of the imaginary lines separating the abdominopelvic cavity into quadrants.

Chapter 2

1. The tail is digested by lysosomes in the process of autolysis and the cellular material is reabsorbed. Programmed cell death is called apotosis.
2. Maternal inheritance is due to the DNA found in extranuclear organelles such as the mitochondria. The sperm contributes only nuclear DNA. Mitochondria are inherited only from the mother.
3. Sixty-four (2^6) daughter cells. Start with 1 parent on Sunday, 2^1(2 daughters from 1 division) on Monday, 2^2(4 daughters from 2 divisions) on Tuesday, etc until there are 2^6 cells on Saturday.
4. Arsenic will poison the mitochondrial enzymes involved in cellular respiration. This will halt production of ATP by the mitochondria. Cells with a high metabolic rate such as muscle cells will be particularly affected.
5. Disruption of microtubules would halt cell division. Microtubules are also involved in cell motility, transportation, and movement of cilia, all of which would be affected.

Chapter 3

1. The urinary bladder is lined with stratified transitional epithelium. Cells are connected with tight junctions. An infection damages the epithelial layer allowing urine to leak through to the underlying tissue causing pain and leakage of blood from the underlying the connective tissue.
2. Some tissues found in a chicken leg would include: epithelium in the skin, areolar and adipose tissues in the subcutaneous layer under the skin, skeletal muscle (the meat), blood, hyaline cartilage, and bone.
3. Functions of the mucus membrane include secretion of mucus to prevent dessication and to trap foreign particles. Portions of the nasal cavity will also contain cilia to move mucus along with the trapped particles.
4. The tissue is keratinized stratified squamous epithelium. Epithelium covers all surfaces and is avascular.
5. A thicker layer of compact bone and larger amounts of skeletal muscle to work against the strong gravity. Thicker adipose layer due to the cold. Larger area of simple squamous epithelium in lungs for diffusion of gases and more red blood cells in blood to transport oxygen due to the thin air.

Chapter 4

1. Chronic exposure to UV light causes damage to elastin and collagen promoting wrinkles. The suspicious growth may be skin cancer caused by UV photodamage to skin cells.
2. Exocrine glands include those that produce mucus (protection against drying and traps foreign particles), sweat (controls body temperature), sebum (lubricates skin), cerumen (protective lubricant for ear canal), and many others that produce essential secretions.
3. Felicity's plan is not wise. Synthesis of Vitamin D is dependent on exposure to sunlight. Temperature control is dependent on evaporation of sweat to cool the body
4. The surface layer of the skin is keratinized stratified squamous epithelium. The keratinocytes are dead and filled with the protective protein keratin. Lipids from lamellar granules and sebum also function in water-proofing.
5. Wayne's lumps are calluses--abnormal thickenings of the epidermis caused by the excessive rubbing of his skin by the too tight boots.

Chapter 5

1. Lynda's diet lacks several nutrients necessary for bone health, including essential vitamins (such as A, D), minerals (such as calcium), and proteins. Her lack of exercise will weaken bone strength. Her age and smoking habit may result in a lack of estrogens causing demineralization of bone.
2. Without a skeleton, the skeletal muscles could not provide movement, the internal organs would lack support and protection, mineral storage, (such as calcium, phosphorus) would be inadequate and blood could not be produced in the red bone marrow.
3. At Aunt Edith's age, the production of several hormones necessary for bone remodeling (such as estrogens and human growth hormone) would most be decreased. She may have loss of bone mass, brittleness, and possibly osteoporosis. Increased susceptibility to fractures results in damage to the vertebrae and loss of height.
4. Exercise causes mechanical stress on bones but since there is no gravity in space, the pull of gravity on bones is missing. The lack of stress results in demineralization of bone and weakness.
5. Chantal has stress fractures in her tibia. Aerobic dancing caused repeated, increased stress to the bone resulting in fractures. Chantal's footwear had inadequate cushioning and support.

Chapter 6

1. Fontanels, the soft spots between cranial bones, are fibrous connective tissue membrane-filled spaces. They allow the infant's head to be molded during its passage through the birth canal and allow for brain and skull growth in the infant.
2. Mike does have a vertebral column. The common name may be derived from the presence of the spinous processes which extend posteriorly and form the ridge running down a person's back that is normally associated with the term "spine".
3. Infants are born with a single concave curve. Adults have 4 curves in their vertebral column at the cervical, thoracic, lumbar and sacral regions.
4. An exploded skull is separated at the sutures-the joints between the skull bones.
5. The x-rays were taken to view the paranasal sinuses-the "fuzzy looking holes". John probably had sinusitis due to infected paranasal sinuses.

Chapter 7

1. There are 14 phalanges in each hand: 2 bones in the thumb and 3 in each of the other fingers. Farmer Ramsey has lost 5 phalanges on his left hand so he has 9 remaining on his left and 14 remaining on his right for a total of 23.
2. The talus receives all of the body's weight. Half that weight is normally transmitted to the calcaneus but the weight distribution will be shifted on the hallux by dancing. Problems may appear in the talocrural (ankle) joint and hallux. Rose also has a deviated hallux or bunion that may have been caused by tight ballet shoes. The dancing did strengthen the foot muscles and correct the flat feet.
3. Pelvic and pectoral girdles are necessary to connect the upper and lower limbs to the axial skeleton and are therefore part of the appendicular skeleton.
4. Grandmother Amelia has probably fractured the neck of her femur. This is a common fracture in elderly and is often called a "broken hip".
5. Derrick has patellofemoral syndrome (runner's knee). Running daily in the same direction on a banked track has stressed the downhill knee. The patella is tracking laterally to its proper position resulting in pain after exercise.

Chapter 8

1. Structurally the hip joint is a simple ball-and-socket synovial joint composed of the head of the femur (ball) and the acetabulum of the hip bone (socket). The knee is really three joints: patellofemoral, lateral tibiofemoral, and medial tibiofemoral. It is a combination gliding and hinge joint.
2. Flexion at knee, extension at hip, extension of vertebral column except for hyperextension at neck, abduction at shoulder, flexion of fingers, adduction of thumb, flexion/extension at elbow, and depression of mandible.
3. No more body surfing for Lars. He has a dislocated shoulder. The head of the humerus was displaced from the glenoid cavity causing tearing of the supporting ligaments and tendons (rotator cuff) of the shoulder joint.
4. Taylor's vertebral column has shifted out of alignment. The affected intervertebral joints are the symphysis between the vertebral bodies and synovial (gliding) between the vertebral arches. The sacroiliac joint is synovial (gliding).
5. Articular cartilage, located on the ends of long bones, is hyaline cartilage. The torn cartilage in the knee is a meniscus composed of fibrocartilage. The ear contains elastic cartilage.

Chapter 9

1. The runner's leg muscles will contain a higher amount of slow fibers-rich in myoglobin, mitochondria, and blood. The weight lifter's arm muscles will contain abundant fast fibers- high in glycogen, with less myoglobin and blood capillaries than the slow fibers. The slow fibers appear red while the fast fibers are white in color.
2. Bill's muscles in the casted leg were not being used so they decreased in size due to loss of myofibrils. Bill's leg shows disuse atrophy.
3. Acetylcholine (ACh) is the neurotransmitter used to "bridge the gap" at the neuromuscular junction. If ACh release is blocked, the neuron cannot send a signal to the muscle and the muscle will not contract.
4. Some cardiac muscle cells are autorhythmic and contraction is started intrinsically. Gap junctions connect the cardiac cells, spreading the impulse through the fiber network so it contracts as a unit. Cardiac muscle has a long refractory period so it will exhibit tetanus like skeletal muscle does.
5. The rigid body is in rigor mortis. The lack of ATP prevents the myosin crossbridges from releasing from the actin. The muscle fibers cannot relax and remain fixed in position. After about 24 hours rigor mortis ceases due to decomposition of the tissues.

Chapter 10

1. The orbicularis oris protrudes the lips. The genioglossus protracts the tongue. The buccinator aids in sucking.
2. One likely possibility is the rhomboideus major muscle.
3. Masseter, temporalis, medial pterygoid, and lateral pterygoid
4. The women's favorite is the gluteus maximus. The men's favorite is the biceps brachii.
5. The difficult childbirth damaged the pelvic diaphragm. Most likely the levator ani was damaged by the passage of the large baby through the vagina or an episiotomy may have been performed, damaging the muscle. Muscle damage may result in stress incontinence.

Chapter 11

1. The spine of C7 forms a prominence along the midline at the base of the neck. The spine of T1 forms a second prominence slightly inferior to C7.
2. Palpation is touching or examining using the hands. Palpitation is a forceful beating of the heart which can often be felt by the patient.
3. The ala is the flare of the nostril, the philtrum is the groove in the upper lip, the external naris is the nostril, the helix is the superior and posterior margin of the auricle, the supercilia is the eyebrow, the sclera is the white of the eye.
4. The commonly used site for a blood donation is the median cubital vein which crosses the cubital fossa at an angle in the crock of the arm.
5. The skin superior to the buttocks has a dimple or depression on each side of the spine, due to the attachment of the skin and fascia to bone.

Chapter 12

1. Blood doping is used by some athletes to improve performance but the practice strains the heart. A blood test would reveal polycythemia, an increased number of red blood cells.
2. Basophils appear as bluish-black granular cells in a stained blood smear. An elevated number of basophils may suggest an allergic response such as "hay fever."
3. Males generally have a higher percentage of red blood cells in blood than females. The average hematocrit is 47% for males and 42% for females. More red blood cells mean more hemoglobin so Tony's blood is redder.
4. NO causes vasoconstriction. A treatment for cold feet would increase SNO. SNO would cause vasodilation, increasing blood flow and warming the feet.
5. Raoul's doctor suggested autologous preoperative transfusion (predonation). If Raoul does need a transfusion during surgery it is safer to use his own blood. Possible problems with matching blood types and blood-born disease is eliminated.

Chapter 13

1. The cusps of the valves "catch" the blood like the fabric of the parachute "catches" the air. The cusps are anchored to the papillary muscles of the ventricle by the chordae tendineae, just as the fabric of the parachute is anchored by the lines, so the cusps do not flip open backwards.
2. The anterior ventricular branch is located in the anterior interventricular sulcus and supplies both ventricles. The circumflex branch is located in the coronary sulcus and supplies the left ventricle and the left atrium.
3. The pericardium surrounds the heart. The fibrous pericardium anchors the heart to the diaphragm, sternum, and major blood vessels. The intercalated discs contain desmosomes which hold the cardiac muscle fibers together.
4. Both are correct. Brittany is correct because the two ventricles act as pumps for the pulmonary (right ventricle) and the systemic (left ventricle) circulation. Sergio is correct because the walls and the partition of the atria form a functional network that contracts in unison. The same is true of the structure of the ventricles.
5. The fibrous skeleton of the heart acts as an electrical insulator and blocks the action potential at the atrioventricular boundary. Impulse conduction can occur only through the atrioventricular bundle (bundle of His).

Chapter 14

1. The foramen ovale and ductus arteriosus close to establish the separate pulmonary and systemic circulations. The umbilical artery and veins close since the placenta is no longer functioning. The ductus venosus closes so that the liver is no longer bypassed.
2. A vascular sinus is a type of vein which lacks smooth muscle in its tunica media. The tunica media and externa are replaced by dense connective tissue as in the intracranial sinuses.
3. Varicose veins are caused by faulty valves. The weak valves allow pooling of blood in the veins. Valves are found only in veins and not in arteries.
4. The superior vena cava drains the head, neck, and upper limbs but not the heart. The azygos system can return blood from the lower body to the superior vena cava if the inferior vena cava is obstructed.
5. An anastomosis provides alternate routes for blood flow to an organ, in this case to the brain. The cerebral arterial circle (circle of Willis) is an anastomosis formed by the union of the collateral arteries: the anterior and posterior cerebral arteries, the anterior and posterior communicating arteries, and the internal carotid arteries.

Chapter 15

1. The route is from lymph capillaries to lymphatic vessels to the popliteal nodes to the superficial inguinal nodes to the right lumbar trunk to the cisterna chyli to the thoracic duct to the left subclavian vein.
2. The two palatine tonsils are in the lateral, posterior oral cavity. The two lingual tonsils are at the base of the tongue. The single pharyngeal tonsil is in the posterior nasopharynx. The 5 tonsils function in the immune response to inhaled or ingested foreign invaders.
3. Jason will need a spleenectomy to prevent his bleeding to death from his ruptured spleen. The liver and red bone marrow will take over the spleen's functions following its removal.

4. AIDS is caused by infection with the human immunodeficiency virus (HIV). The virus is transmitted in human blood and body fluids and does not survive long on surfaces outside the body. HIV mainly attacks the T4 lymphocytes. The virus eventually depresses the T4 cell population and the patient acquires opportunistic infections and diseases.
5. The left subclavian trunk drains lymph from the left upper limb. Blockage would cause a build up of lymph and interstitial fluid in the upper limb causes it to swell due to edema.

Chapter 16

1. The jingle that won't stop is similar to a reverberating circuit in that a signal will be repeated over and over again. Reverberating circuits function in breathing, memory, waking, and coordinated muscle activity.
2. Varudhini's neuroglia cell is a Schwann cell (neurolemmocyte). The Schwann cell's membrane is wrapped repeatedly around the neuron's axon to form the myelin sheath.
3. Gray matter appears gray in color due to the absence of myelin. It is composed of cell bodies, dendrites, and unmyelinated axons. White matter's color is due to the presence of myelin.
4. The somatic, afferent, peripheral division would detect sound and smell. The somatic, efferent, peripheral division sends messages to the skeletal muscles for stretching and yawning. Stomach rumbling and salivation are controlled by the autonomic, efferent, peripheral division. The enteric, efferent, peripheral division also controls GI tract organs.
5. In MS, the myelin sheath is gradually destroyed resulting in slower impulse transmission, muscle weakness, and uncoordinated movement. In infants, myelination is not complete. Impulse transmission is slow and movements are uncoordinated.

Chapter 17

1. The senior damaged the cervical region of his spinal cord. Recovery is unlikely.
2. The spinal cord is connected to the periphery by the spinal nerves. The nerves exit through the intervertebral foramina. Spaces are maintained between the vertebrae by the intervertebral discs. The dura mater of the spinal cord fuses with the epineurium. The nerves are connected to the cord by the posterior and anterior roots.
3. The spinal cord is anchored in place by the filum terminale and denticulate ligaments.
4. Helga has shingles. A dormant herpes zoster (chickenpox) virus has reactivated and spread along the sensory neuron from the posterior root ganglion. The infection is probably along spinal nerve T10 as indicated by the line of blisters.
5. Helga's left foot has a normal plantar reflex response. The right foot exhibits the Babinski sign which is normal in infants but indicates damage to the corticospinal tract in adults.

Chapter 18

1. Movement of the right upper limb is controlled by the left hemisphere's primary motor area, located in the post central gyrus. Speech is controlled by Broca's area in the left hemisphere's frontal lobe just superior to the lateral cerebral sulcus.
2. The brain is enclosed by the cranial bones and meninges. The temporal bone houses the middle and inner ear, separating these from the brain.
3. Bubba suffered a concussion and may also have a contusion due to trauma to the brain. The RAS (reticular activating system) was not able to maintain consciousness during this insult to Bubba's brain.
4. The gray matter is located on the outer surface of the cerebrum. Since the cerebrum functions in intelligence, thought, mathematical ability, creativity, and so on, Alicia was being insulted when she was told that her gray matter was thin.
5. The dentist has injected anesthetic into the inferior alveolar nerve, a branch of the mandibular nerve which numbs the lower teeth and the lower lip. The tongue is numbed by blocking the lingual nerve. The upper teeth and lip are numbed by injecting the superior alveolar nerve, a branch of the maxillary branch.

Chapter 19

1. The warm (thermal) receptors in her hands were at first activated by the warmth of the cup, then they adapted to the stimulus as she continued to hold the cup. The high temperature of the hot cocoa stimulated the thermal receptors in her mouth as well as pain receptors.
2. The tickle receptors are free-nerve endings in the foot. The impulse travels along the sensory nerve to the posterior gray horn to synapse with the secondary neuron, crosses to the other side of the cord and runs up the anterior spinothalamic tract to the thalamus. The third-order neuron runs from the thalamus to the "foot" area of the somatosensory area of the cerebral cortex.
3. The primary motor area of the cerebral cortex initiates voluntary movements. The basal ganglia help control the automatic, semivoluntary motions. Sensory input related to bike riding will be received by the brain. The cerebellum and other brain regions will integrate this input to further coordinate the bicycling movements.
4. In an infant or very young child, the myelination of the CNS is not complete. The corticospinal tracts which control fine, voluntary motor movement are not fully myelinated until the child is about 2 years old. An infant would not be able to manipulate a knife or fork safely due to lack of complete motor control.
5. Jon's perception of feeling in his amputated foot is called phantom limb sensation. Impulses from the remaining proximal section of the sensory neuron are perceived by the brain as still coming from the amputated foot

Chapter 20

1. Aimee is probably responding to olfactory stimulation since what is perceived as taste is closely related to smell. The chemical odors from the food dissolve in the mucosal surface of the nasal cavity. The olfactory receptors will be stimulated and the signal will be sent along the olfactory nerves to olfactory bulbs to olfactory tracts to the primary olfactory area of the brain. Connections to the limbic system may be responsible for Aimee's intense reaction.
2. Deirdre's nose is running due to excess production of lacrimal fluid (tears). Tears move medially across the anterior surface of the eye to the lacrimal puncta to the lacrimal canals to the nasolacrimal duct to the nasal cavity to the nostrils.
3. The stick damaged the iris (the colored ring) which is part of the vascular tunic. The iris controls the diameter of the pupil but does not directly affect the focusing of light rays on the retina. If permanent damage was done to the cornea or ciliary muscles, then focusing may be affected.
4. The endolymph in the saccule, utricle, and membranous semicircular ducts moves in response to the up and down motion of the ship on the ocean. The vestibular apparatus is stimulated and sends impulses to the medulla, pons, and cerebellum. The eyes also send feedback to the cerebellum to help maintain balance. Excessive stimulation and conflicting messages (eyes say the ship is still, ears say it's moving) cause motion sickness.
5. Pigment in the iris and choroid are usually sufficient to absorb light entering the eye. A flash photograph captures the picture before the iris has time to constrict in response to the high intensity of light. The picture captures the reflection of the light back through the

pupil (and iris if it is lightly pigmented). The eyes appear red due to the vascular tunic. Albinos have no pigment in their eyes so the "red eye" effect is more pronounced.

Chapter 21

1. The exciting and potentially dangerous activities activate the sympathetic nervous system, resulting in a "fight of flight" reaction. The sympathetic nervous system stimulates the release of epinephrine (adrenalin) and norepinephrine from the adrenal medulla. These hormones prolong the "fight or flight" response.
2. The autonomic sensory neurons detect stretch in the stomach and send impulses to the brain. The hypothalamus sends impulses through the parasympathetic division, along the vagus nerve, to the heart resulting in decreased heart rate and force of contraction.
3. The pressure receptor in the colon is stimulated → autonomic sensory neuron → sacral region of spinal cord → parasympathetic preganglionic neuron → terminal ganglion → postganglionic neuron → smooth muscle in colon/rectum.
4. The autonomic division controls the digestive system and many other organs including the heart, lungs, and eyes. Most organs are controlled by dual innervation with the sympathetic and parasympathetic divisions causing opposite effects. The sympathetic division results in widespread effects over the entire body while the parasympathetic effects are more localized. The enteric system controls only the digestive system and may function independently of the ANS.
5. For parasympathetic: feed and breed or rest and digest or R&R (rest and repose) or defecate and procreate. For sympathetic: Dry mouth and damp hands mean a warm (fast) heart.

Chapter 22

1. Melatonin could be taken to help induce sleep during the evening at Tropicanaland. For the quickest adjustment, exposure to very bright light in the morning would help reset the body's clock. Sunlight would be better for this than a bright flashlight. The MSH may help with getting a tan but won't help reset the body's clock.
2. Raj's sympathetic nervous system will stimulate secretion of the hormones epinephrine and norepinephrine by the adrenal medulla. The hormones are released in response to stress and prolong the fight or flight response.
3. The large amount of sugar in the meal will cause in increase in blood glucose levels. The pancreas will increase secretion of insulin from the beta cells of the pancreatic islets (islets of Langerhans). The insulin will the lower blood glucose level. The secretion of other hormones that influence digestion may also be increased.
4. The pituitary gland is located in the sella turcica in the base of the sphenoid bone. The sphenoid is posterior to the ethmoid bone, which makes up a major portion of the walls and roof of the nasal cavity. Due to its proximity to the nasal cavity, the pituitary gland could have been affected by the radiation.
5. All hormones, including insulin, need an adequate number of functioning receptors to perform their function. The problem in this patient is a lack of functioning receptors, not a problem with insulin level.

Chapter 23

1. The septum is composed of hyaline cartilage covered with mucous membrane. The lateral wall of the nose is skin and muscle, lined with mucous membrane. The upper ear is skin covering elastic cartilage.
2. Due to the influence of androgens, males generally have larger cartilages of the larynx and vocal folds than females. The larger, thicker size of the folds causes them to vibrate more slowly, resulting in a lower pitch of the voice in males.
3. The stick has penetrated the pleural cavity and perhaps the visceral pleura and the lung. Air in the pleural cavity (pneumothorax) has caused the lung to collapse (atelectasis). Only one lung collapsed since the lungs are surrounded by two separate cavities.
4. The man had an object (piece of meat) blocking his airway so that he could not breathe or speak. The waiter performed a Heimlich maneuver. A sharp squeeze to the abdominal wall increases the intrabdominal pressure and pushes up on the diaphragm. This increases the pressure in the lungs and drives out the object.
5. LaTonya's cerebral cortex can allow her to voluntarily hold her breath for a short time. Increasing levels of carbon dioxide and H^+ will stimulate the inspiratory area and normal breathing will resume despite LaTonya's desire to get her sister in trouble.

Chapter 24

1. Saliva contains salivary amylase which will break down the starch of the potatoes into sugar (sweet) and lingual lipase which will break down the fat from the frying into fatty acids (tart) and monoglycerides.
2. The acidity of the lemon's juice could eventually damage the enamel of the front teeth. If the enamel is pitted or dissolved, the dentin underneath would be exposed. Dentin is not as resistant to acid as enamel and would soon be damaged also.
3. Modifications to increase surface area include overall length, circular folds, villi, and microvilli (brush border). The circular folds also enhance absorption by imparting a spiraling motion to the chyme.
4. During swallowing, the soft palate and uvula close off the nasopharynx. If the nervous system sends conflicting signals (such as breathe, swallow, and giggle), the palate may be in the wrong position to block food from entering the nasal cavity.
5. The gallbladder stores and concentrates bile. Gallstones may form from the cholesterol in bile. The gallstones block any of the ducts leading from the gallbladder to the duodenum.

Chapter 25

1. The glomerulus would continue to filter the blood producing filtrate containing water, glucose, ions, and other compounds. The toxin would block the normal reabsorption of 99% of the filtrate by the renal tubules. The infected person would rapidly become dehydrated from lost water and would also lose ions, glucose, and essential vitamins and nutrients. The infected person would most likely soon die.
2. The urinary bladder can stretch considerably due to the presence of transitional epithelium, rugae, and three layers of smooth muscle (detrusor muscle). The internal and external urethral sphincters also function to hold the urine.
3. In females the urethra is about 4 cm long. In males, the urethra is about 15–20 cm long, including its passage through the penis, the urogenital diaphragm, and prostate gland. The female urinary bladder holds less urine than the male bladder due to the presence of the uterus in females.
4. Urine samples contain an abundance of epithelial cells that are shed from the mucosal lining of the urinary system.
5. The excessive loss of weight has caused the loss of adipose tissue around the kidney resulting in nephroptosis. The kidney has dropped out of position twisting the ureter, causing backup of urine into the kidney, blocking flow of urine, and causing pain.

Chapter 26

1. Brutus has been castrated. His testes have been removed so no sperm (and no kittens) will be produced.

2. The presence or absence of the SRY gene on the Y chromosome determines the development of the urogenital structures before birth. The presence of SRY and secretion of testosterone by the fetal testes results in a male. Absence of SRY results in a female. Structures can be altered by surgery and hormones somewhat but the female will still lack receptors and organs necessary for sperm production and the male will lack receptors and organs necessary for the production of ova. The chromosome composition cannot be changed.
3. A hysterectomy will not cause menopause. A hysterectomy is the removal of the uterus, which does not produce hormones. The ovaries are left intact and continue to produce estrogens and progesterone.
4. The number of "daughter cells" following meiosis is different in spermatogenesis versus oogenesis. A primary spermatocyte will undergo meiosis resulting in 4 mature sperm. The primary oocyte will undergo meiosis resulting in one mature ovum and 2 or 3 nonfunctional polar bodies.
5. A vasectomy cuts only the vas (ductus) deferens and leaves the testis untouched. Male secondary sex characteristics and libido (sex drive) are maintained by androgens including testosterone. The secretion of these hormones by the testis is not interrupted by the vasectomy.

Chapter 27

1. Identical (monozygotic) twins develop from the same fertilized ovum or zygote. They are essentially the same person genetically and therefore must be the same gender. These must be fraternal (dizygotic) twins.
2. All tissues and organs form from the embryo's 3 primary germ layers. The ectoderm, the outermost germ layer, develops into the nervous system and the epithelial layer of the skin. This early embryological connection may sometimes result in disorders showing signs in both tissues.
3. Josefina is in false labor. In true labor, the contractions are regular and the pain may localize in the back. The pain may be intensified by walking. True labor is indicated by the "show" of bloody mucus and cervical dilation.
4. The yolk sac provides blood vessels to transport nutrients from the endometrium to the embryo. The germ cells (oogonia and spermatogonia) migrate from the yolk sac.
5. Although the placenta blocks the passage of most microorganisms, many other substances pass though including nutrients, wastes, gases, alcohol, acetaldehyde (a toxic byproduct of alcohol metabolism), antibiotics, and numerous prescription and illegal drugs. Cigarette smoke may be teratogenic and has been associated with low birth weight, SIDS, cleft lip and palate, and other abnormalities.

GLOSSARY

PRONOUNCIATION KEY

1. The most strongly accented syllable appears in capital letters, for example, bilateral (bī-LAT-er-al) and diagnosis (dī-ag-NŌ-sis).
2. If there is a secondary accent, it is noted by a prime (′), for example, physiology (fiz′-ē-OL-ō-jē). Any additional secondary accents are also noted by a prime, for example, decarboxylation (dē′-kar-bok′-si-LĀ-shun).
3. Vowels marked by a line above the letter are pronounced with the long sound as in the following common word:

   ā as in māke ī as in īvy ū as in neuron
   ē as in bē ō as in pōle
4. Vowels not so marked are pronounced with the short sound as in the following words:

   a as in above i as in sip u as in but
   e as in bet o as in not
5. One other phonetic symbol is used to indicate the following sound:

   oy as in oil

A

Abdomen (ab-DŌ-men or AB-do-men) The area between the diaphragm and pelvis.

Abdominal (ab-DOM-i-nal) **cavity** Superior portion of the abdominopelvic cavity that contains the stomach, spleen, liver, gallbladder, pancreas, small intestine, and most of the large intestine.

Abdominopelvic (ab-dom′-i-nō-PEL-vic) **cavity** Inferior component of the ventral body cavity that is subdivided into a superior abdominal cavity and an inferior pelvic cavity.

Abduction (ab-DUK-shun) Movement away from the axis or midline of the body or one of its parts usually in the frontal plane.

Abscess (AB-ses) A localized collection of pus and liquefied tissue in a cavity.

Absorption (ab-SORP-shun) The taking up of liquids by solids or of gases by solids or liquids; intake of fluids or other substances by cells of the skin or mucous membranes; the passage of digested foods from the gastrointestinal tract into blood or lymph.

Accessory duct A duct of the pancreas that empties into the duodenum about 2.5 cm (1 in.) superior to the ampulla of Vater (hepatopancreatic ampulla). Also called the **duct of Santorini** (san′-tō-RE-ne).

Acetabulum (as′-e-TAB-yū-lum) The rounded cavity on the external surface of the hipbone that receives the head of the femur.

Acetylcholine (as′-ē-til-KŌ-lēn) **(ACh)** A neurotransmitter liberated by many peripheral nervous system neurons and some central nervous system neurons. It is excitatory at neuromuscular junctions but inhibitory at some other synapses (slows heart rate).

Achilles tendon *See* **Calcaneal tendon.**

Acini (AS-i-nē) Masses of cells in the pancreas that secrete digestive enzymes.

Acoustic (a-KŪS-tik) Pertaining to sound or the sense of hearing.

Acrosome (AK-rō-sōm) A dense granule in the head of a sperm cell that contains enzymes that facilitate the penetration of a sperm cell into a secondary oocyte.

Actin (AK-tin) The contractile protein that makes up thin myofilaments in muscle fiber (cell).

Action potential A wave of negativity that self-propagates along the outside surface of the membrane of a neuron or muscle fiber (cell); a rapid change in membrane potential that involves a depolarization following a repolarization. Also called a **nerve action potential (nerve impulse)** as it relates to a neuron and a **muscle action potential** as it relates to a muscle fiber (cell).

Active transport The movement of substances, usually ions, across cell membranes, against a concentration gradient, requiring the expenditure of energy (ATP).

Acupuncture (AK-yū-punk′-chur) The insertion of a needle into a tissue for the purpose of drawing fluid or relieving pain. It is also an ancient Chinese practice employed to cure illnesses by inserting needles into specific locations of the skin.

Acute (a-KYŪT) Having rapid onset, severe symptoms, and a short course; not chronic.

Adam's apple *See* **Thyroid cartilage.**

Adduction (ad-DUK-shun) Movement toward the axis or midline of the body or one of its parts usually in the frontal plane.

Adenoids (AD-e-noyds) Inflamed and enlarged pharyngeal tonsils.

Adenosine triphosphate (a-DEN-ō-sēn trī-FOS-fāt) **(ATP)** The universal energy-carrying molecule manufactured in all living cells as a means of capturing and storing energy. It consists of the purine base *adenine* and the five-carbon sugar *ribose*, to which are added, in linear array, three *phosphate* molecules.

Adhesion (ah-HĒ-zhun) Abnormal joining of parts to each other.

Adipocyte (AD-i-pō-sīt) Fat cell, derived from a fibroblast.

Adrenal cortex (a-DRĒ-nal KOR-teks) The outer portion of an adrenal gland, divided into three zones, each of which has a different cellular arrangement and secretes different hormones.

Adrenal (a-DRĒ-nal) **glands** Two glands located superior to each kidney. Also called the **suprarenal** (sū′-pra-RĒ-nal) **glands.**

Adrenal medulla (me-DULL-a) The inner portion of an adrenal gland, consisting of cells that secrete epinephrine and norepinephrine (NE) in response to the stimulation of preganglionic sympathetic neurons.

Adrenergic (ad′-ren-ER-jik) **fiber** A nerve fiber that when stimulated releases norepinephrine (noradrenaline) at a synapse.

Adrenocorticotropic (ad-rē′-nō-kor-ti-kō-TRŌP-ik) **hormone (ACTH)** A hormone produced by the anterior lobe of the pituitary gland that influences the production and secretion of certain hormones of the adrenal cortex.

Adventitia (ad-ven-TISH-ya) The outermost covering of a structure or organ.

Afferent arteriole (AF-er-ant ar-TĒ-rē-ōl) A blood vessel of a kidney that breaks up into the capillary network called a glomerulus; there is one afferent arteriole for each glomerulus.

Agglutination (a-glū′-ti-NĀ-shun) Clumping of microorganisms or blood cells; typically an antigen-antibody reaction.

Agglutinin (a-GLŪ-ti-nin) A specific antibody in blood serum that reacts with a specific agglutinogen and causes the clumping of bacteria, blood cells, or particles. Also called an **isoantibody.**

Agglutinogen (ag′-lū-TIN-ō-gen) A genetically determined antigen located on the surface of red blood cells; basis for the ABO grouping and Rh system of blood classification. Also called an **isoantigen.**

Aggregated lymphatic follicles Aggregated lymph nodules that are most numerous in the ileum. Also called **Peyer's** (PĪ-erz) **patches.**

Agnosia (ag-NŌ-zē-a) A loss of the ability to recognize the meaning of stimuli from the various senses (visual, auditory, touch).

Albumin (al-BYŪ-min) The most abundant (60 percent) and smallest of the plasma proteins, which functions primarily to regulate osmotic pressure of plasma.

Aldosterone (al-do-STĒR-ōn) A mineralocorticoid produced by the adrenal cortex that brings about sodium and water reabsorption and potassium excretion.

Allantois (a-LAN-tō-is) A small, vascularized membrane between the chorion and amnion of the fetus that serves as an early site for blood formation.

Allele (AL-ēl) Genes that control the same inherited trait (such as height or eye color) that are located on the same position (locus) on homologous chromosomes.

Alpha (AL-fa) **cell** A cell in the pancreatic islets (islets of Langerhans) in the pancreas that secretes glucagon.

Alpha receptor Receptor found on visceral effectors innervated by most sympathetic postganglionic axons.

Alveolar-capillary (al-VĒ-ō-lar) **membrane** Structure in the lungs consisting of the alveolar wall and basement membrane and a capillary endothelium and basement membrane through which the diffusion of respiratory gases occurs. Also called the **respiratory membrane.**

Alveolar duct Branch of a respiratory bronchiole around which alveoli and alveolar sacs are arranged.

Alveolar macrophage (MAK-rō-fāj) Cell found in the alveolar walls of the lungs that is highly phagocytic. Also called a **dust cell.**

Alveolar sac A collection or cluster of alveoli that share a common opening.

Alveolus (al-VĒ-ō-lus) A small hollow or cavity; an air sac in the lungs; milk-secreting portion of a mammary gland. *Plural,* **alveoli** (al-VĒ-ō-lī).

Amenorrhea (a-men-ō-RĒ-a) Absence of menstruation.

Amnesia (am-NĒ-zē-a) A lack or loss of memory.

Amniotic (am′-nē-OT-ik) **fluid** Fluid in the amniotic cavity, the space between the developing embryo (or fetus) and amnion; the fluid is initially produced as a filtrate from maternal blood and later from fetal urine.

Amphiarthrosis (am′-fē-ar-THRŌ-sis) A slightly movable articulation midway between a diarthrosis and synarthrosis, in which the articulating bony surfaces are separated by fibrous connective tissue or fibrocartilage to which both are attached.

Ampulla (am-PŪL-la) A saclike dilation of a canal.

Ampulla of Vater *See* **Hepatopancreatic ampulla.**

Anabolism (a-NAB-ō-lizm) Synthetic energy-requiring reactions whereby small molecules are built up into larger ones.

Anal (Ā-nal) **canal** The terminal 2 or 3 cm (1 in.) of the rectum; opens to the exterior through the anus.

Anal column A longitudinal fold in the mucous membrane of the anal canal that contains a network of arteries and veins.

Analgesia (an-al-JĒ-zē-a) Pain relief.

Anal triangle The subdivision of the female or male perineum that contains the anus.

Anaphase (AN-a-fāz) Third stage of mitosis in which the chromatids that have separated at the centromeres move to opposite poles of the cell.

Anastomosis (a-nas-tō-MŌ-sis) An end-to-end union or joining together of blood vessels, lymphatic vessels, or nerves.

Anatomic dead space The volume of air that is inhaled but remains in spaces in the upper respiratory system and does not reach the alveoli to participate in gas exchange; about 150 mL.

Anatomical (an′-a-TOM-i-kal) **position** A position of the body universally used in anatomical descriptions in which the body is erect, facing the observer with the head level and the eyes facing forward, the upper limbs are at the sides, the palms are facing forward, and the feet are on the floor and directed forward.

Anatomy (a-NAT-ō-mē) The structure or study of structure of the body and the relation of its parts to each other.

Androgen (AN-drō-jen) Substance producing or stimulating male characteristics, such as the male hormone testosterone.

Anesthesia (an′-es-THĒ-zē-a) A total or partial loss of feeling or sensation, usually defined with respect to loss of pain sensation; may be general or local.

Ankylosis (ang′-ki-LŌ-sus) Severe or complete loss of movement at a joint.

Anomaly (a-NOM-a-lē) An abnormality that may be a developmental (congenital) defect; a variant from the usual standard.

Anoxia (an-OK-sē-a) Deficiency of oxygen.

Antagonist (an-TAG-ō-nist) A muscle that has an action opposite that of the prime mover (agonist) and yields to the movement of the prime mover.

Anterior (an-TER-ē-or) Nearer to or at the front of the body. Also called **ventral.**

Anterior pituitary (pi-TŪ-i-tar′-ē) **gland** Anterior portion of the pituitary gland. Also called the **adenohypophysis** (ad′-e-nō-hī-POF-i-sis).

Anterior root The structure composed of axons of motor fibers that emerges from the anterior aspect of the spinal cord and extends laterally to join a posterior root, forming a spinal nerve. Also called a **ventral root.**

Antibody (AN-ti-bod′-ē) A protein produced by certain cells in the body in the presence of a specific antigen; the antibody combines with that antigen to neutralize, inhibit, or destroy it. Also called an **immunoglobulin** (im-yoo-nō-GLOB-yoo-lin) or **Ig.**

Antidiuretic hormone (ADH) Hormone produced by neurosecretory cells in the paraventricular and supraoptic nuclei of the hypothalamus that stimulates water reabsorption from kidney cells into the blood and vasoconstriction of arterioles. Also called **vasopressin** (vāz-ō-PRESS-in).

Antigen (AN-ti-jen) Any substance that when introduced into the tissues or blood induces the formation of antibodies and reacts only with those specific antibodies.

Antrum (AN-trum) Any nearly closed cavity or chamber, especially one within a bone, such as a sinus.

Anulus fibrosus (AN-yū-lus fī-BRŌ-sus) A ring of fibrous tissue and fibrocartilage that encircles the pulpy substance (nucleus pulposus) of an intervertebral disc.

Anus (Ā-nus) The distal end and outlet of the rectum.

Aorta (ā-OR-ta) The main systemic trunk of the arterial system of the body; emerges from the left ventricle.

Aortic (ā-OR-tik) **body** Receptor on or near the arch of the aorta that responds to alterations in blood levels of oxygen, carbon dioxide, and hydrogen ions.

Aperture (AP-er-chur) An opening or orifice.

Apex (Ā-peks) The pointed end of a conical structure, such as the apex of the heart.

Apneustic (ap-NŪ-stik) **area** Portion of the respiratory center in the pons that sends stimulatory nerve impulses to the inspiratory area that activate and prolong inspiration and inhibit expiration.

Aponeurosis (ap′-ō-nū-RŌ-sis) A sheetlike tendon joining one muscle with another or with bone.

Apoptosis (a-pōp-TŌ-sis) Orderly, genetically programmed cell death.

Appendage (a-PEN-dij) A structure attached to the body.

Aqueous humor (AK-wē-us HYŪ-mor) The watery fluid, similar in composition to cerebrospinal fluid, that fills the anterior cavity of the eye.

Arachnoid (a-RAK-noyd) The middle of the three coverings (meninges) of the brain or spinal cord.

Arachnoid villus (VIL-us) Berrylike tuft of arachnoid that protrudes into the superior sagittal sinus and through which cerebrospinal fluid is reabsorbed into the bloodstream.

Arbor vitae (AR-bōr VĒ-tē) The treelike appearance of the white matter tracts of the cerebellum when seen in midsagittal section. A series of branching ridges within the cervix of the uterus.

Arch of the aorta (ā-OR-ta) The most superior portion of the aorta, lying between the ascending and descending segments of the aorta.

Areola (a-RĒ-ō-la) Any tiny space in a tissue. The pigmented ring around the nipple of the breast.

Arousal (a-ROW-sal) Awakening from a deep sleep; a response controlled by the thalamic part of the reticular activating system.

Arrector pili (a-REK-tor PI-lē) Smooth muscle attached to hairs; contraction pulls the hairs into a more vertical position, resulting in "goose bumps."

Arrhythmia (a-RITH-mē-a) Irregular heart rhythm. Also called a **dysrhythmia.**

Arteriole (ar-TĒ-rē-ōl) A small, almost microscopic, artery that delivers blood to a capillary.

Artery (AR-ter-ē) A blood vessel that carries blood away from the heart.

Arthritis (ar-THRĪ-tis) Inflammation of a joint.

Arthrology (ar-THROL-ō-jē) The study or description of joints.

Articular (ar-TIK-yū-lar) **capsule** Sleevelike structure around a synovial joint composed of a fibrous capsule and a synovial membrane.

Articular cartilage (KAR-ti-lij) Hyaline cartilage attached to articular bone surfaces.

Articular disc (Fibrocartilage pad between articular surfaces of bones of some synovial joints. Also called a **meniscus** (men-IS-cus).

Articulation (ar-tik′-yū-LĀ-shun) A joint; a point of contact between bones, cartilage and bones, or teeth and bones.

Arytenoid (ar′-i-TĒ-noyd) **cartilages** A pair of small, pyramidal cartilages of the larynx that attach to the vocal folds and intrinsic pharyngeal muscles and can move the vocal folds.

Ascending colon (KŌ-lon) The portion of the large intestine that passes superiorly from the cecum to the inferior edge of the liver where it bends at the right colic (hepatic) flexure to become the transverse colon.

Ascites (as-SĪ-tēz) Serous fluid in the peritoneal cavity.

Aseptic (ā-SEP-tik) Free from any infectious or septic material.

Association area A portion of the cerebral cortex connected by many motor and sensory fibers to other parts of the cortex. The association areas are concerned with motor patterns, memory, concepts of word-hearing and word-seeing, reasoning, will, judgment, and personality traits.

Association neuron (NŪ-ron) A nerve cell lying completely within the central nervous system that carries nerve impulses from sensory neurons to motor neurons. Also called a **connecting neuron.**

Astereognosis (as-ter′-ē-ōg-NŌ-sis) Inability to recognize objects or forms by touch.

Asthenia (as-THĒ-nē-a) Lack or loss of strength.

Astrocyte (AS-trō-sīt) A neuroglial cell having a star shape that supports neurons in the brain and spinal cord and attaches the neurons to blood vessels.

Astresia (a-TRĒ-zē-a) Abnormal closure of a passage, or absence of a normal body opening.

Atrial natriuretic (na′-trē-yū-RET-ik) **peptide (ANP)** Peptide hormone produced by the atria of the heart in response to their stretching that inhibits aldosterone production and thus lowers blood pressure.

Atrioventricular (AV) (ā′-trē-ō-ven-TRIK-yū-lar) **bundle** The portion of the conduction system of the heart that begins at the atrioventricular (AV) node, passes through the cardiac skeleton separating the atria and the ventricles, then runs a short distance down the interventricular septum before splitting into right and left bundle branches. Also called the **bundle of His** (HISS).

Atrioventricular (AV) node The portion of the conduction system of the heart made up of a compact mass of conducting cells located near the orifice of the coronary sinus in the right atrial wall.

Atrioventricular (AV) valve A structure made up of membranous flaps or cusps that allows blood to flow in one direction only, from an atrium into a ventricle.

Atrium (Ā-trē-um) A superior chamber of the heart.

Auditory ossicle (AW-di-tō-rē OS-si-kul) One of the three small bones of the middle ear called the malleus, incus, and stapes.

Auditory tube The tube that connects the middle ear with the nose and nasopharynx region of the throat. Also called the **Eustachian** (yū-STĀ-kē-an) **tube.**

Auscultation (aws-kul-TĀ-shun) Examination by listening to sounds in the body.

Autoimmunity An immunologic response against a person's own tissue antigen.

Autolysis (aw-TOL-i-sis) Spontaneous self-destruction of cells by their own digestive enzymes at death or a pathological process or during normal embryological development.

Autonomic ganglion (aw′-tō-NOM-ik GANG-lē-on) A cluster of sympathetic or parasympathetic cell bodies located outside the central nervous system.

Autonomic nervous system (ANS) Visceral motor neurons, both sympathetic and parasympathetic, that transmit nerve impulses from the central nervous system to smooth muscle, cardiac muscle, and glands; so named because this portion of the nervous system was thought to be self-governing or spontaneous.

Autonomic plexus (PLEK-sus) An extensive network of sympathetic and parasympathetic fibers; the cardiac, celiac, and pelvic plexuses are located in the thorax, abdomen, and pelvis, respectively.

Autophagy (aw-TOF-a-jē) Process by which worn-out organelles are digested within lysosomes.

Autosome (AW-tō-sōm) Any chromosome other than the pair of sex chromosomes.

Axilla (ak-SIL-a) The small hollow beneath the arm where it joins the body at the shoulders. Also called the **armpit.**

Axon (AK-son) The usually single, long process of a nerve cell that carries a nerve impulse away from the cell body.

Axon terminal Terminal branch of an axon and its collateral. Also called a **telodendrium** (tel-ō-DEN-drē-um).

Azygos (AZ-ī-gos) An anatomical structure that is not paired; occurring singly.

B

Ball-and-socket joint A synovial joint in which the rounded surface of one bone moves within a cup-shaped depression or fossa of another bone, as in the shoulder or hip joint. Also called a **spheroid** (SFĒ-roid) **joint.**

Baroreceptor (bar′-ō-re-SEP-tor) Nerve cell capable of responding to changes in blood, air, or fluid pressure. Also called a **pressoreceptor.**

Bartholin's glands *See* **Greater vestibular glands.**

Basal ganglia (GANG-glē-a) Paired clusters of cell bodies that make up the central gray matter in each cerebral hemisphere, including the caudate nucleus, lentiform nucleus, claustrum, and amygdaloid body. Also called **cerebral nuclei** (SER-e-bral NŪ-klē-ī).

Basement membrane Thin, extracellular layer consisting of a basal lamina secreted by epithelial cells and a reticular lamina secreted by connective tissue cells.

Basilar (BAS-i-lar) **membrane** A membrane in the cochlea of the inner ear that separates the cochlear duct from the scala tympani and on which the spiral organ (organ of Corti) rests.

Basophil (BĀ-sō-fil) A type of white blood cell that is characterized by a pale nucleus and large granules that stain readily with basic dyes.

B cell A lymphocyte that develops into an antibody-producing plasma cell or a memory cell.

Belly The abdomen. The gaster or prominent, fleshy part of a skeletal muscle.

Beta (BĀ-ta) **cell** A cell in the pancreatic islets (islets of Langerhans) in the pancreas that secretes insulin.

Beta receptor Receptor found on visceral effectors innervated by most sympathetic postganglionic axons; in general, stimulation of beta receptors leads to inhibition.

Bicuspid (bī-KUS-pid) **valve** Atrioventricular (AV) valve on the left side of the heart. Also called the **mitral valve.**

Bifurcate (bī-FUR-kāt) Having two branches or divisions; forked.

Bilateral (bī-LAT-er-al) Pertaining to two sides of the body.

Bile (BĪL) A secretion of the liver consisting of water, bile salts, bile pigments, cholesterol, lecithin, and several ions; it assumes a role in emulsification of triglycerides prior to their digestion.

Biliary (BIL-ē-er-ē) Relating to bile, the gallbladder, or the bile ducts.

Blastocyst (BLAS-tō-sist) In the development of an embryo, a hollow ball of cells that consists of blastocele (the internal cavity), trophoblast (outer cells), and inner cell mass.

Blastocyst cavity (BLAS-tō-sist) The fluid-filled cavity within the blastocyst.

Blastomere (BLAS-tō-mēr) One of the cells resulting from the cleavage of a fertilized ovum.

Blastula (BLAS-tyū-la) An early stage in the development of a zygote.

Blind spot Area in the retina at the end of the optic (II) nerve in which there are no light receptor cells.

Blood–brain barrier (BBB) A barrier consisting of specialized blood capillaries and astrocytes that prevents the passage of certain substances from the blood to the cerebrospinal fluid and brain.

Blood island Isolated mass and cord of mesenchyme in the mesoderm from which blood vessels develop.

Blood–testis barrier A barrier formed by sustentacular (Sertoli) cells that prevents an immune response against antigens produced by sperm cells and developing cells by isolating the cells from the blood.

Body cavity A space within the body that contains various internal organs.

Bolus (BŌ-lus) A soft, rounded mass, usually food, that is swallowed.

Bony labyrinth (LAB-i-rinth) A series of cavities within the petrous portion of the temporal bone forming the vestibule, cochlea, and semicircular canals of the inner ear.

Bowman's capsule *See* **Glomerular capsule.**

Brachial plexus (BRĀ-kē-al PLEK-sus) A network of nerve fibers of the anterior rami of spinal nerves C5, C6, C7, C8, and T1. The nerves that emerge from the brachial plexus supply the upper limb.

Brain A mass of nervous tissue located in the cranial cavity.

Brain stem The portion of the brain immediately superior to the spinal cord, which is made up of the medulla oblongata, pons, and midbrain.

Broad ligament A double fold of parietal peritoneum attaching the uterus to the side of the pelvic cavity.

Broca's (BRŌ-kaz) **area** Motor area of the brain in the frontal lobe that translates thoughts into speech. Also called the **motor speech area.**

Bronchi (BRONG-kē) Branches of the respiratory passageway including primary bronchi (the two divisions of the trachea), secondary or lobar bronchi (divisions of the primary that are distributed to the lobes of the lung), and tertiary or segmental bronchi (divisions of the secondary that are distributed to bronchopulmonary segments of the lung).

Bronchial tree The trachea, bronchi, and their branching structures.

Bronchiole (BRONG-kē-ōl) Branch of a tertiary bronchus further dividing into terminal bronchioles (distributed to lobules of the lung), which divide into respiratory bronchioles (distributed to alveolar sacs).

Bronchopulmonary (brong′-kō-PUL-mō-ner-ē) **segment** One of the smaller divisions of a lobe of a lung supplied by its own branches of a bronchus.

Bronchus (BRONG-kus) One of the two large branches of the trachea. *Plural,* **bronchi** (BRONG-kē).

Brunner's gland *See* **Duodenal gland.**

Buccal (BUK-al) Pertaining to the cheek or mouth.

Bulb of penis Expanded portion of the base of the corpus spongiosum penis.

Bulbourethral (bul′-bō-yū-RĒ-thral) **gland** One of a pair of glands located inferior to the prostate gland on either side of the urethra that secretes an alkaline fluid into the cavernous urethra. Also called a **Cowper's** (KOW-perz) **gland.**

Bundle branch One of the two branches of the atrioventricular (AV) bundle made up of specialized muscle fibers (cells) that transmit electrical impulses to the ventricles.

Bundle of His *See* **Atrioventricular (AV) bundle.**

Bursa (BUR-sa) A sac or pouch of synovial fluid located at friction points, especially about joints.

Bursitis (bur-SĪ-tis) Inflammation of a bursa.

C

Calcaneal tendon The tendon of the soleus, gastrocnemius, and plantaris muscles at the back of the heel. Also called the **Achilles** (a-KIL-ēz) **tendon.**

Calcification (kal-si-fi-KĀ-shun) Deposition of mineral salts, primarily hydroxyapatite, in a framework formed by collagen fibers in which the tissue hardens. Also called **mineralization.**

Calcitonin (kal-si-TŌ-nin) **(CT)** A hormone produced by the thyroid gland that lowers the calcium and phosphate levels of the blood by inhibiting bone breakdown and accelerating calcium absorption by bones.

Calyx (KĀL-iks) Any cuplike division of the kidney pelvis. *Plural,* **calyces** (KĀ-li-sēz).

Canaliculus (kan′-a-LIK-yū-lus) A small channel or canal, as in bones, where they connect lacunae. *Plural,* **canaliculi** (kan′-a-LIK-yū-lī).

Canal of Schlemm *See* **Scleral venous sinus.**

Capillary (KAP-i-lar′-ē) A microscopic blood vessel located between an arteriole and venule through which materials are exchanged between blood and body cells.

Carcinogen (kar-SIN-ō-jen) Any substance that causes cancer.

Cardiac (KAR-dē-ak) **cycle** A complete heartbeat consisting of systole (contraction) and diastole (relaxation) of both atria plus systole and diastole of both ventricles.

Cardiac muscle An organ specialized for contraction, composed of striated muscle fibers (cells), forming the wall of the heart, and stimulated by an intrinsic conduction system and visceral motor neurons.

Cardiac notch An angular notch in the anterior border of the left lung.

Cardinal ligament A ligament of the uterus, extending laterally from the cervix and vagina as a continuation of the broad ligament.

Cardiology (kar-dē-OL-ō-jē) The study of the heart and diseases associated with it.

Cardiovascular (kar-dē-ō-VAS-kyū-lar) **center** Groups of neurons scattered within the medulla oblongata that regulate heart rate, force of contraction, and blood vessel diameter.

Carina (ka-RĪ-na) A ridge on the inside of the division of the right and left primary bronchi.

Carotene (KAR-o-tēn) Yellow-orange pigment found in the stratum corneum of the epidermis that, together with melanin, accounts for the yellowish coloration of skin.

Carotid (ka-ROT-id) **body** Receptor on or near the carotid sinus that responds to alterations in blood levels of oxygen, carbon dioxide, and hydrogen ions.

Carotid sinus A dilated region of the internal carotid artery immediately superior to the branching of the common carotid artery that contains receptors that monitor blood pressure.

Carpus (KAR-pus) A collective term for the eight bones of the wrist.

Cartilage (KAR-ti-lij) A type of connective tissue consisting of chondrocytes in lacunae embedded in a dense network of collagen and elastic fibers and a matrix of chondroitin sulfate.

Cartilaginous (kar′-ti-LAJ-i-nus) **joint** A joint without a synovial (joint) cavity where the articulating bones are held tightly together by cartilage, allowing little or no movement.

Cast A small mass of hardened material formed within a cavity in the body and then discharged from the body; can originate in different areas and be composed of various materials.

Cauda equina (KAW-da ē-KWĪ-na) A tail-like collection of roots of spinal nerves at the inferior end of the spinal canal.

Cecum (SĒ-kum) A blind pouch at the proximal end of the large intestine to which the ileum is attached.

Celiac plexus (SĒ-lē-ak PLEK-sus) A large mass of ganglia and nerve fibers located at the level of the superior part of the first lumbar vertebra. Also called the **solar plexus.**

Cell The basic structural and functional unit of all organisms; the smallest structure capable of performing all the activities vital to life.

Cell division Process by which a cell reproduces itself that consists of a nuclear division (mitosis) and a cytoplasmic division (cytokinesis); types include somatic and reproductive cell division.

Cellular respiration *See* **Oxidation.**

Cementum (se-MEN-tum) Calcified tissue covering the root of a tooth.

Central canal A circular channel running longitudinally in the center of an osteon (Haversian system) of mature compact bone, containing blood and lymphatic vessels and nerves. Also called a **Haversian** (ha-VĒR-shun) **canal.** A microscopic tube running the length of the spinal cord in the gray commissure.

Central fovea (FŌ-vē-a) A cuplike depression in the center of the macula lutea of the retina, containing cones only; the area of clearest vision.

Central nervous system (CNS) That portion of the nervous system that consists of the brain and spinal cord.

Centrioles (SEN-trē-ōlz) Paired, cylindrical structures within a centrosome, each consisting of a ring of microtubules and arranged at right angles to each other. Assume a role in the formation and regeneration of flagella and cilia.

Centromere (SEN-trō-mēr) The clear, constricted portion of a chromosome where the two chromatids are joined; serves as the point of attachment for the chromosomal microtubules.

Centrosome (SEN-trō-sōm) Organelle that consists of a dense area of cytoplasm (pericentriolar material) that forms microtubules in nondividing cells, the mitotic spindle in dividing cells, and centrioles that assume a role in the formation and regeneration of flagella and cilia.

Cerebellar peduncle (ser-e-BEL-ar pe-DUNG-kul) A bundle of nerve fibers connecting the cerebellum with the brain stem.

Cerebellum (ser-e-BEL-um) The portion of the brain lying posterior to the medulla oblongata and pons, concerned with coordination of movements.

Cerebral aqueduct (SER-ē-bral AK-we-dukt) A channel through the midbrain connecting the third and fourth ventricles and containing cerebrospinal fluid.

Cerebral arterial circle A ring of arteries forming an anastomosis at the base of the brain between the internal carotid and basilar arteries and arteries supplying the brain. Also called the **circle of Willis.**

Cerebral cortex The surface of the cerebral hemispheres, 2–4 mm thick, consisting of six layers of nerve cell bodies (gray matter) in most areas.

Cerebral peduncle (pe-DUNG-kul) One of a pair of nerve fiber bundles located on the ventral surface of the midbrain, conducting nerve impulses between the pons and the cerebral hemispheres.

Cerebrospinal (se-rē′-brō-SPĪ-nal) **fluid (CSF)** A fluid produced in the choroid plexuses and ependymal cells of the ventricles of the brain that circulates in the ventricles and the subarachnoid space around the brain and spinal cord.

Cerebrum (SER-ē-brum) The two hemispheres of the forebrain, making up the largest part of the brain.

Cerumen (se-RŪ-men) Waxlike secretion produced by ceruminous glands in the external auditory meatus (ear canal).

Ceruminous (se-RŪ-mi-nus) **gland** A modified sudoriferous (sweat) gland in the external auditory meatus that secretes cerumen (ear wax).

Cervical ganglion (SER-vi-kul GANG-glē-on) A cluster of nerve cell bodies of postganglionic sympathetic neurons located in the neck, near the vertebral column.

Cervical plexus (PLEX-us) A network of neuron fibers formed by the anterior rami of the first four cervical nerves.

Cervix (SER-viks) Neck; any constricted portion of an organ, such as the inferior cylindrical part of the uterus.

Chemoreceptor (kē′-mō-rē-SEP-tor) Receptor outside the central nervous system on or near the carotid and aortic bodies that detects the presence of chemicals.

Chiasm (kī-AZM) A crossing; especially the crossing of the optic (II) nerve fibers.

Chief cell The secreting cell of a gastric gland that produces pepsinogen, the precursor of the enzyme pepsin, and the enzyme gastric lipase. Also called a **zymogenic** (zī-mō-JEN-ik) **cell.**

Chiropractic (kī-rō-PRAK-tik) A system of treating disease by using one's hands to manipulate body parts, mostly the vertebral column.

Cholesterol (kō-LES-te-rol) Classified as a lipid, the most abundant steroid in animal tissues; located in cell membranes and used for the synthesis of steroid hormones and bile salts.

Cholinergic (kō′-lin-ER-jik) **fiber** A nerve ending that liberates acetylcholine at a synapse.

Chondrocyte (KON-drō-sīt) Cell of mature cartilage.

Chordae tendineae (KOR-dē TEN-din-nē-ē) Tendonlike, fibrous cords that connect the heart valves with the papillary muscles.

Chorion (KŌ-rē-on) The most superficial fetal membrane that becomes the principal embryonic portion of the placenta; serves a protective and nutritive function.

Chorionic villus (kō′-rē-ON-ik VIL-lus) Fingerlike projection of the chorion that grows into the decidua basalis of the endometrium and contains fetal blood vessels.

Chorionic villi sampling (CVS) The removal of a sample of chorionic villus tissue by means of a catheter to analyze the tissue for prenatal genetic defects.

Choroid (KŌ-royd) One of the vascular coats of the eyeball.

Choroid plexus (PLEX-sus) A vascular structure located in the roof of each of the four ventricles of the brain; produces cerebrospinal fluid.

Chromaffin (krō-MAF-in) **cell** Cell that has an affinity for chrome salts, due in part to the presence of the precursors of the neurotransmitter epinephrine; found, among other places, in the adrenal medulla.

Chromatid (KRŌ-ma-tid) One of a pair of identical connected nucleoprotein strands that are joined at the centromere and separate during cell division, each becoming a chromosome of one of the two daughter cells.

Chromatin (KRŌ-ma-tin) The threadlike mass of the genetic material consisting principally of DNA, which is present in the nucleus of a nondividing or interphase cell.

Chromatolysis (krō′-ma-TOL-i-sis) The breakdown of chromatophilic substance (Nissl bodies) into finely granular masses in the cell body of a central or peripheral neuron whose process (axon or dendrite) has been damaged.

Chromatophilic substance Rough endoplasmic reticulum in the cell bodies of neurons that functions in protein synthesis. Also called **Nissl bodies.**

Chromosomal microtubule (mī-krō-TŪB-yūl) Microtubule formed during prophase of mitosis that originates from centromeres, extends from a centromere to a pole of the cell, and assists in chromosomal movement; constitutes a part of the mitotic spindle.

Chromosome (KRŌ-mō-sōm) One of the 46 small, dark-staining bodies that appear in the nucleus of a human diploid ($2n$) cell during cell division.

Chronic (KRON-ik) Long-term or frequently recurring; applied to a disease that is not acute.

Chyle (KĪL) The milky fluid found in the lacteals of the small intestine after digestion.

Chyme (KĪM) The semifluid mixture of partly digested food and digestive secretions found in the stomach and small intestine during digestion of a meal.

Ciliary (SIL-ē-ar′-ē) **body** One of the three portions of the vascular tunic of the eyeball, the others being the choroid and the iris; includes the ciliary muscle and ciliary processes.

Ciliary ganglion (GANG-glē-on) A very small parasympathetic ganglion whose preganglionic fibers come from the oculomotor (III) nerve and whose postganglionic fibers carry nerve impulses to the ciliary muscle and the sphincter muscle of the iris.

Cilium (SIL-ē-um) A hair or hairlike process projecting from a cell that may be used to move the entire cell or to move substances along the surface of the cell.

Circle of Willis *See* **Cerebral arterial circle.**

Circular folds Permanent, deep, transverse folds in the mucosa and submucosa of the small intestine that increase the surface area for absorption. Also called **plicae circulares** (PLĪ-kē SER-kyū-lar-ēs).

Circumduction (ser′-kum-DUK-shun) A movement at a synovial joint in which the distal end of a bone moves in a circle while the proximal end remains relatively stable.

Circumvallate papilla (ser′-kum-VAL-āt pa-PIL-a) One of the circular projections that is arranged in an inverted V-shaped row at the posterior portion of the tongue; the largest of the elevations on the upper surface of the tongue containing taste buds.

Cisterna chyli (sis-TER-na KĪ-lē) The origin of the thoracic duct.

Cleavage The rapid mitotic divisions following the fertilization of a secondary oocyte, resulting in an increased number of progressively smaller cells, called blastomeres, so that the overall size of the zygote remains the same.

Clitoris (KLI-tor-is) An erectile organ of the female located at the anterior junction of the labia minora that is homologous to the male penis.

Clone (KLŌN) A population of cells identical to itself.

Coccyx (KOK-six) The fused bones at the inferior end of the vertebral column.

Cochlea (KŌK-lē-a) The winding, cone-shaped tube forming a portion of the inner ear and containing the spiral organ (organ of Corti).

Cochlear duct The membranous cochlea consisting of a spirally arranged tube enclosed in the bony cochlea and lying along its outer wall. Also called the **scala media** (SCA-la MĒ-dē-a).

Collagen (KOL-a-jen) A protein that is the main organic constituent of connective tissue.

Collateral circulation The alternate route taken by blood through an anastomosis.

Colliculus (ko-LIK-yū-lus) A small elevation.

Colon (KŌ-lon) The division of the large intestine consisting of ascending, transverse, descending, and sigmoid portions.

Colostrum (kō-LOS-trum) A thin, cloudy fluid secreted by the mammary glands a few days prior to or after delivery before true milk is secreted.

Column (KOL-um) Group of white matter tracts in the spinal cord.

Commissure (KOM-i-shūr) The angular junction of the eyelids at either corner of the eyes.

Common bile duct A tube formed by the union of the common hepatic duct and the cystic duct that empties bile into the duodenum at the hepatopancreatic ampulla (ampulla of Vater).

Compact (dense) bone tissue Bone tissue that contains few spaces between osteons (Haversian systems); forms the external portion of all bones and the bulk of the diaphysis (shaft) of long bones.

Concha (KONG-ka) A scroll-like bone found in the skull. *Plural,* **conchae** (KONG-kē). Also called a **turbinate** (TUR-bi-nāt).

Conduction myofiber Muscle fiber (cell) in the subendocardial tissue of the heart specialized for conducting an action potential to the myocardium; part of the conduction system of the heart. Also called a **Purkinje** (pur-KIN-jē) **fiber.**

Conduction system An intrinsic regulating system composed of specialized muscle tissue that generates and distributes electrical impulses that stimulate cardiac muscle fibers (cells) to contract.

Conductivity (kon′-duk-TIV-i-tē) The ability to carry the effect of a stimulus from one part of a cell to another; highly developed in nerve and muscle fibers (cells).

Condyloid (KON-di-loid) **joint** A synovial joint structured so that an oval-shaped condyle of one bone fits into an elliptical cavity of another bone, permitting side-to-side and back-and-forth movements, as at the joint at the wrist between the radius and carpals. Also called an **ellipsoidal** (e-lip-SOY-dal) **joint.**

Cone The light-sensitive receptor in the retina concerned with color vision.

Congenital (kon-JEN-i-tal) Present at the time of birth.

Conjunctiva (kon′-junk-TĪ-va) The delicate membrane covering the eyeballs and lining the eyes.

Connective tissue The most abundant of the four basic tissue types in the body, performing the functions of binding and supporting; consists of relatively few cells in a great deal of intercellular substance.

Consciousness (KON-shus-nes) A state of wakefulness in which an individual is fully alert, aware, and oriented as a result of feedback between the cerebral cortex and reticular activating system.

Contractility (kon′-trak-TIL-i-tē) The capacity of the contractile elements in muscle tissue to shorten and develop tension.

Contralateral (kon′-tra-LAT-er-al) On the opposite side; affecting the opposite side of the body.

Conus medullaris (KŌ-nus med-yū-LAR-is) The tapered portion of the spinal cord inferior to the lumbar enlargement.

Convergence (con-VER-jens) Anatomical arrangement in which the synaptic end bulbs of several presynaptic neurons terminate on one postsynaptic neuron. Medial movement of the two eyeballs so that both are directed toward a near object being viewed to produce a single image.

Cornea (KOR-nē-a) The nonvascular, transparent fibrous coat through which the iris can be seen.

Corona (kō-RŌ-na) Margin of the glans penis.

Coronal (kō-RŌ-nal) **plane** A plane that runs vertical to the ground and divides the body into anterior and posterior portions. Also called **frontal plane.**

Corona radiata Deepest layer of granulosa cells around an oocyte attached to the zona pellucida.

Coronary circulation The pathway followed by the blood from the ascending aorta through the blood vessels supplying the heart and returning to the right atrium. Also called **cardiac circulation.**

Coronary sinus (SĪ-nus) A wide venous channel on the posterior surface of the heart that collects the blood from the coronary circulation and returns it to the right atrium.

Corpora quadrigemina (KOR-por-a kwad-ri-JEM-in-a) Four small elevations (superior and inferior colliculi) on the dorsal region of the midbrain concerned with visual and auditory functions.

Corpus (KOR-pus) The principal part of any organ; any mass or body.

Corpus albicans (AL-bi-kanz) A white fibrous patch in the ovary that forms after the corpus luteum regresses.

Corpus callosum (kal-LŌ-sum) The great commissure of the brain between the cerebral hemispheres.

Corpuscle of touch The sensory receptor for the sensation of touch; found in the dermal papillae, especially in palms and soles. Also called a **Meissner's** (MĪS-nerz) **corpuscle.**

Corpus luteum (LŪ-tē-um) A yellow endocrine gland in the ovary formed when a follicle has discharged its secondary oocyte; secretes estrogens, progesterone, and relaxin.

Corpus striatum (strī-Ā-tum) An area in the interior of each cerebral hemisphere composed of the caudate and lentiform nuclei of the basal ganglia and white matter of the internal capsule, arranged in a striated manner.

Cortex (KOR-teks) An outer layer of an organ. The convoluted layer of gray matter covering each cerebral hemisphere.

Costal cartilage (KOS-tal KAR-ti-lij) Hyaline cartilage that attaches a rib to the sternum.

Cowper's gland *See* **Bulbourethral gland.**

Cranial (KRĀ-nē-al) **cavity** A subdivision of the dorsal body cavity formed by the cranial bones and containing the brain.

Cranial nerve One of 12 pairs of nerves that leave the brain, pass through foramina in the skull, and supply the head, neck, and part of the trunk; each is designated by a Roman numeral and a name.

Craniosacral (krā-nē-ō SĀ-kral) **outflow** The fibers of parasympathetic preganglionic neurons, which have their cell bodies located in nuclei in the brain stem and in the lateral gray matter of the sacral portion of the spinal cord.

Cranium (KRĀ-nē-um) The skeleton of the skull that protects the brain and the organs of sight, hearing, and balance; includes the frontal, parietal, temporal, occipital, sphenoid, and ethmoid bones.

Crista (KRIS-ta) A crest or ridged structure. A small elevation in the ampulla of each semicircular duct that serves as a receptor for dynamic equilibrium.

Crossing-over The exchange of a portion of one chromatid with another in a tetrad during meiosis. It permits an exchange of genes among chromatids and is one factor that results in genetic variation.

Crus (KRUS) **of penis** Separated, tapered portion of the corpora cavernosa penis. *Plural,* **crura** (KROO-ra).

Crypt of Lieberkühn *See* **Intestinal gland.**

Cupula (KUP-yū-la) A mass of gelatinous material covering the hair cells of a crista, a receptor in the ampulla of a semicircular canal stimulated when the head moves.

Cutaneous (kyū-TĀ-nē-us) Pertaining to the skin.

Cystic (SIS-tik) **duct** The duct that transports bile from the gallbladder to the common bile duct.

Cystitis (sis-TĪ-tis) Inflammation of the urinary bladder.

Cytokinesis (sī′-tō-ki-NĒ-sis) Division of the cytoplasm.

Cytology (sī-TOL-ō-jē) The study of cells.

Cytolysis (sī-TOL-i-sis) The destruction of living cells in which the contents leak out.

Cytoplasm (SĪ-tō-plazm) All the cellular contents between the plasma membrane and nucleus.

Cytoskeleton Complex internal structure of cytoplasm consisting of microfilaments, microtubules, and intermediate filaments.

Cytosol (SĪ-tō-sol) The semifluid portion of cytoplasm in which organelles (except the nucleus) are suspended and solutes are dissolved. Also called **intracellular fluid.**

D

Dartos (DAR-tōs) The contractile tissue under the skin of the scrotum.

Debility (dē-BIL-i-tē) Weakness of tonicity in functions or organs of the body.

Decidua (dē-SID-yū-a) That portion of the endometrium of the uterus (all but the deepest layer) that is modified for pregnancy and shed after childbirth.

Deciduous (dē-SID-yū-us) Falling off or being shed seasonally or at a particular stage of development. In the body, referring to the first set of teeth.

Decussation (dē′-ku-SĀ-shun) A crossing-over; usually refers to the crossing of most of the fibers in the large motor tracts to opposite sides in the medullary pyramids.

Deep Away from the surface of the body.

Deep fascia (FASH-ē-a) A sheet of connective tissue wrapped around a muscle to hold it in place.

Deep inguinal (IN-gwi-nal) **ring** A slitlike opening in the aponeurosis of the transversus abdominis muscle that represents the origin of the inguinal canal.

Deep-venous thrombosis (DVT) The presence of a thrombus in a vein, usually a deep vein of the lower limbs.

Defecation (def-e-KĀ-shun) The discharge of feces from the rectum.

Degeneration (dē-jen′-er-Ā-shun) A change from a higher to a lower state; a breakdown in structure.

Deglutition (dē-glū-TISH-un) The act of swallowing.

Delta cell A cell in the pancreatic islets (islets of Langerhans) in the pancreas that secretes somatostatin.

Dendrite (DEN-drīt) A nerve cell process that carries a nerve impulse toward the cell body.

Dendritic (den-DRIT-ik) **cell** One type of antigen-presenting cell with long branchlike projections, for example, Langerhans cells in the epidermis.

Dens (DENZ) Tooth.

Dental caries (KA-rēz) Gradual demineralization of the enamel and dentin of a tooth that may invade the pulp and alveolar bone. Also called **tooth decay.**

Denticulate (den-TIK-yū-lāt) Finely toothed or serrated; characterized by a series of small, pointed projections.

Dentin (DEN-tin) The osseous tissues of a tooth enclosing the pulp cavity.

Dentition (den-TI-shun) The eruption of teeth. The number, shape, and arrangement of teeth.

Deoxyribonucleic (dē-ok′-sē-rī′-bō-nyū-KLĒ-ik) **acid (DNA)** A nucleic acid in the shape of a double helix constructed of nucleotides consisting of one of four nitrogenous bases (adenine, cytosine, guanine, or thymine), deoxyribose, and a phosphate group; encoded in the nucleotides is genetic information.

Depression (dē-PRESS-shun) Movement in which a part of the body moves inferiorly.

Dermal papilla (pa-PILL-a) Fingerlike projection of the papillary region of the dermis that may contain blood capillaries or corpuscles of touch (Meissner's corpuscles).

Dermatology (derm-ma-TOL-ō-jē) The medical specialty dealing with diseases of the skin.

Dermatome (DER-ma-tōm) The cutaneous area developed from one embryonic spinal cord segment and receiving most of its innervation from one spinal nerve.

Dermis (DER-mis) A layer of dense connective tissue lying deep to the epidermis; the true skin or corium.

Descending colon (KŌ-lon) The part of the large intestine descending from the left colic (splenic) flexure to the level of the left iliac crest.

Detrusor (de-TRŪ-ser) **muscle** Muscle in the wall of the urinary bladder.

Developmental anatomy The study of development from the fertilized egg to the adult form. The branch of anatomy called embryology is generally restricted to the study of development from the fertilized egg through the eighth week in utero.

Diagnosis (dī-ag-NŌ-sis) Distinguishing one disease from another or determining the nature of a disease from signs and symptoms by inspection, palpation, laboratory tests, and other means.

Diaphragm (DĪ-a-fram) Any partition that separates one area from another, especially the dome-shaped skeletal muscle between the thoracic and abdominal cavities. Also a dome-shaped structure that fits over the cervix, usually with a spermicide, to prevent conception.

Diaphysis (dī-AF-i-sis) The shaft of a long bone.

Diarthrosis (dī′-ar-THRŌ-sis) Articulation in which opposing bones move freely, as in a hinge joint.

Diastole (dī-AS-tō-lē) In the cardiac cycle, the phase of relaxation or dilation of the heart muscle, especially of the ventricles.

Diencephalon (dī′-en-SEF-a-lon) A part of the brain consisting primarily of the thalamus and the hypothalamus.

Differentiation (dif′-e-ren′-shē-Ā-shun) Acquisition of specific functions different from those of the original general type.

Diffusion (dif-YŪ-zhun) A passive process in which there is a net or greater movement of molecules or ions from a region of high concentration to a region of low concentration until equilibrium is reached.

Dilate (DĪ-lāt) To expand or swell.

Diploid (DIP-loyd) Having the number of chromosomes characteristically found in the somatic cells of an organism. Symbolized $2n$.

Diplopia (di-PLŌ-pē-a) Double vision.

Direct pathways Motor pathways that convey information from the brain down the spinal cord; related to precise, voluntary movements of skeletal muscles. Also called **pyramidal** (pi-RAM-i-dal) **pathways.**

Disease Any change from a state of health.

Distal (DIS-tal) Farther from the attachment of a limb to the trunk or a structure; farther from the point of origin.

Divergence (di-VER-jens) An anatomical arrangement in which the synaptic end bulbs of one presynaptic neuron terminate on several postsynaptic neurons.

Diverticulum (dī-ver-TIK-yū-lum) A sac or pouch in the wall of a canal or organ, especially in the colon.

Dorsal body cavity Cavity near the dorsal surface of the body that consists of a cranial cavity and vertebral canal.

Dorsal ramus (RĀ-mus) A branch of a spinal nerve containing motor and sensory fibers supplying the muscles, skin, and bones of the posterior part of the head, neck, and trunk.

Dorsiflexion (dor′-si-FLEK-shun) Bending the foot in the direction of the dorsum (upper surface).

Duct of Santorini *See* **Accessory duct.**

Duct of Wirsung *See* **Pancreatic duct.**

Ductus arteriosus (DUK-tus ar-tē-rē-Ō-sus) A small vessel connecting the pulmonary trunk with the aorta; found only in the fetus.

Ductus (vas) deferens (DEF-er-ens) The duct that conducts sperm cells from the epididymis to the ejaculatory duct. Also called the **seminal duct.**

Ductus epididymis (ep′-i-DID-i-mis) A tightly coiled tube inside the epididymis, distinguished into a head, body, and tail, in which sperm cells undergo maturation.

Ductus venosus (ve-NŌ-sus) A small vessel in the fetus that helps the circulation bypass the liver.

Duodenal (dū′-ō-DĒ-nal) **gland** Gland in the submucosa of the duodenum that secretes an alkaline mucus to protect the lining of the small intestine from the action of enzymes and to help neutralize the acid in chyme. Also called **Brunner's** (BRUN-erz) **gland.**

Duodenal papilla (pa-PILL-a) An elevation on the duodenal mucosa that receives the hepatopancreatic ampulla (ampulla of Vater).

Duodenum (dū′-ō-DĒ-num) The first 25 cm (10 in.) of the small intestine.

Dura mater (DYŪ-ra MĀ-ter) The outer membrane (meninx) covering the brain and spinal cord.

Dynamic equilibrium (ē-kwi-LIB-rē-um) The maintenance of body position, mainly the head, in response to sudden movements such as rotation.

Dysfunction (dis-FUNK-shun) Absence of complete normal function.

Dystrophia (dis-TRO-fē-a) Progressive weakening of a muscle.

E

Ectoderm The most superficial of the three primary germ layers that gives rise to the nervous system and the epidermis of skin and its derivatives.

Ectopic (ek-TOP-ik) Out of the normal location, as in ectopic pregnancy.

Edema (e-DĒ-ma) An abnormal accumulation of interstitial fluid.

Effector (e-FEK-tor) The organ of the body, either a muscle or a gland, that responds to a motor neuron impulse.

Efferent arteriole (EF-er-ent ar-TĒ-rē-ōl) A vessel of the renal vascular system that transports blood from the glomerulus to the peritubular capillary.

Efferent ducts A series of coiled tubes that transport sperm cells from the rete testis to the epididymis.

Ejaculation (e-jak-yū-LĀ-shun) The reflex ejection or expulsion of semen from the penis.

Ejaculatory (e-JAK-yū-la-tō′-rē) **duct** A tube that transports sperm cells from the ductus (vas) deferens to the prostatic urethra.

Elasticity (e-las-TIS-i-tē) The ability of a tissue to return to its original shape after contraction or extension.

Elevation (el-e-VĀ-shun) Movement in which a part of the body moves superiorly.

Embolism (EM-bō-lizm) Obstruction or closure of a vessel by an embolus.

Embolus (EM-bō-lus) A blood clot, bubble of air, fat from broken bones, mass of bacteria, or other debris or foreign material transported by the blood.

Embryo (EM-brē-ō) The young of any organism in an early stage of development; in humans, the developing organism from fertilization to the end of the eighth week in utero.

Embryology (em′-brē-OL-ō-jē) The study of development from the fertilized egg to the end of the eight week in utero.

Emigration (em′-i-GRĀ-shun) The passage of white blood cells through intact blood vessel walls.

Emission (ē-MISH-un) Propulsion of sperm cells into the urethra in response to peristaltic contractions of the ducts of the testes, epididymides, and ductus (vas) deferens as a result of sympathetic stimulation.

Emphysema (em′-fi-SĒ-ma) A swelling or inflation of air passages due to loss of elasticity in the alveoli.

Emulsification (ē-mul′-si-fi-KĀ-shun) The dispersion of large triglyceride globules to smaller, uniformly distributed particles in the presence of bile.

Enamel (e-NAM-el) The hard, white substance covering the crown of a tooth.

Endocardium (en-dō-KAR-dē-um) The layer of the heart wall, composed of endothelium and smooth muscle, that lines the inside of the heart and covers the valves and tendons that hold the valves open.

Endochondral ossification (en′-dō-KON-dral os′-i-fi-KĀ-shun) The replacement of cartilage by bone. Also called **intracartilaginous** (in′-tra-kar′-ti-LAJ-i-nus) **ossification.**

Endocrine (EN-dō-krin) **gland** A gland that secretes hormones into the blood; a ductless gland.

Endocrinology (en′-dō-kri-NOL-ō-jē) The science concerned with the structure and functions of endocrine glands and the diagnosis and treatment of disorders of the endocrine system.

Endoderm The deepest of the three primary germ layers of the developing embryo that gives rise to the gastrointestinal tract, urinary bladder and urethra, and respiratory tract.

Endodontics (en′-dō-DON-tiks) The branch of dentistry concerned with the prevention, diagnosis, and treatment of diseases that affect the pulp, root, periodontal ligament, and alveolar bone.

Endogenous (en-DOJ-e-nus) Growing from or beginning within the organism.

Endolymph (EN-dō-lymf′) The fluid within the membranous labyrinth of the inner ear.

Endometriosis (en′-dō-MĒ-trē-ō′-sis) The growth of endometrial tissue outside the uterus.

Endometrium (en′-dō-MĒ-trē-um) The mucous membrane lining the uterus.

Endomysium (en′-dō-MĪZ-ē-um) Invagination of the perimysium separating each individual muscle fiber (cell).

Endoneurium (en′-dō-NYŪ-rē-um) Connective tissue wrapping around individual nerve fibers (cells).

Endoplasmic reticulum (en′-dō-PLAZ-mik re-TIK-yū-lum) **(ER)** A network of channels running through the cytoplasm of a cell that serves in intracellular transportation, support, storage, synthesis, and packaging of molecules. Portions of ER where ribosomes are attached to the outer surface are called **rough (granular) ER;** portions that have no ribosomes are called **smooth (agranular) ER.**

End organ of Ruffini *See* **Type II cutaneous mechanoreceptor.**

Endosteum (en-DOS-tē-um) The membrane that lines the medullary cavity of bones, consisting of osteoprogenitor cells and scattered osteoclasts.

Endothelial-capsular (en-dō-THĒ-lē-ar) **membrane** A filtration membrane in a nephron of a kidney consisting of the endothelium and basement membrane of the glomerulus and the epithelium of the visceral layer of the glomerular (Bowman's) capsule.

Endothelium (en′-dō-THĒ-lē-um) The layer of simple squamous epithelium that lines the cavities of the heart, blood vessels, and lymphatic vessels.

Enteroendocrine (en-ter-ō-EN-dō-krin) **cell** A hormone-producing cell in the gastrointestinal mucosa. An example is a G cell in the gastric mucosa that produces the hormone gastrin.

Enzyme (EN-zīm) A substance that affects the speed of chemical changes; an organic catalyst, usually a protein.

Eosinophil (ē′-ō-SIN-ō-fil) A type of white blood cell characterized by granular cytosplasm readily stained by eosin.

Ependymal (e-PEN-de-mal) **cell** Neuroglial cell that lines ventricles of the brain and probably assists in the circulation of cerebrospinal fluid (CSF).

Epicardium (ep′-i-KAR-dē-um) The thin outer layer of the heart wall, composed of serous tissue and mesothelium. Also called the **visceral pericardium.**

Epidemic (ep′-i-DEM-ik) A disease that occurs above the expected level among individuals in a population.

Epidemiology (ep′-i-dē-mē-OL-ō-jē) Medical science concerned with the occurrence and distribution of disease in human populations.

Epidermis (ep-i-DERM-is) The outermost, thinner layer of skin, composed of keratinized stratified squamous epithelium.

Epididymis (ep′-i-DID-i-mis) A comma-shaped organ that lies along the posterior border of the testis and contains the ductus epididymis, in which sperm undergo maturation. *Plural,* **epididymides** (ep′-i-DID-i-mi-dēz).

Epidural (ep′-i-DŪ-ral) **space** A space between the spinal dura mater and the vertebral canal, containing areolar connective tissue and a plexus of veins.

Epiglottis (ep′-i-GLOT-is) A large, leaf-shaped piece of cartilage lying on top of the larynx, with its "stem" attached to the thyroid cartilage and its "leaf" portion unattached and free to move up and down to cover the glottis (vocal folds and rima glottidis).

Epimysium (ep′-i-MĪZ-ē-um) Fibrous connective tissue around muscles.

Epinephrine (ep-ē-NEF-rin) Hormone secreted by the adrenal medulla that produces actions similar to those that result from sympathetic stimulation. Also called **adrenaline** (a-DREN-a-lin).

Epineurium (ep′-i-NYŪ-rē-um) The most superficial covering around the entire nerve.

Epiphyseal (ep′-i-FIZ-ē-al) **line** The remnant of the epiphyseal plate in a long bone.

Epiphyseal plate The cartilaginous plate between the epiphysis and diaphysis that is responsible for the lengthwise growth of long bones.

Epiphysis (ē-PIF-i-sis) The end of a long bone, usually larger in diameter than the shaft (diaphysis).

Epiphysis cerebri (se-RĒ-brē) Pineal gland.

Epithalamus (ep′-i-THAL-a-mus) Part of the diencephalon superior and posterior to the thalamus, comprising the pineal gland and associated structures.

Epithelial (ep′-i-THĒ-lē-al) **tissue** The tissue that forms glands or the superficial part of the skin and lines blood vessels, hollow organs, and passages that lead externally from the body.

Eponychium (ep′-ō-NIK-ē-um) Narrow band of stratum corneum at the proximal border of a nail that extends from the margin of the nail wall. Also called the **cuticle.**

Erection (ē-REK-shun) The enlarged and stiff state of the penis (or clitoris) resulting from the engorgement of the spongy erectile tissue with blood.

Erythema (er′-e-THĒ-ma) Skin redness usually caused by engorgement of the capillaries in the deeper layers of the skin.

Erythrocyte (e-RITH-rō-sīt) Red blood cell.

Erythropoiesis (e-rith′-rō-poy-Ē-sis) The process by which erythrocytes (red blood cells) are formed.

Erythropoietin (e-rith′-rō-POY-ē-tin) A hormone formed from a plasma protein that stimulates erythrocyte (red blood cell) production.

Esophagus (e-SOF-a-gus) A hollow muscular tube connecting the pharynx and the stomach.

Estrogens (ES-tro-jens) Female sex hormones produced by the ovaries concerned with the development and maintenance of female reproductive structures and secondary sex characteristics, fluid and electrolyte balance, and protein anabolism. Examples are β-estradiol, estrone, and estriol.

Etiology (ē′-tē-OL-ō-jē) The study of the causes of disease, including theories of origin and the organisms, if any, involved.

Euphoria (yū-FŌR-ē-a) A subjectively pleasant felling of well-being marked by confidence and assurance.

Eupnea (YŪP-nē-a) Normal quiet breathing.

Eustachian tube *See* **Auditory tube.**

Eversion (ē-VER-zhun) The movement of the sole laterally at the ankle joint.

Exacerbation (eg-zas′-er-BĀ-shun) An increase in the severity of symptoms or of a disease.

Excitability (ek-sīt′-a-BIL-i-tē) The ability of muscle tissue to receive and respond to stimuli; the ability of nerve cells to respond to stimuli and convert them into nerve impulses.

Excretion (eks-KRĒ-shun) The process of eliminating waste products from a cell, tissue, or the entire body; or the products excreted.

Exocrine (EK-sō-krin) **gland** A gland that secretes substances into ducts that empty at covering or lining epithelium or directly onto a free surface.

Exocytosis (ex′-ō-sī-TŌ-sis) A process of discharging cellular products too big to go through the membrane. Particles for export are enclosed by Golgi membranes when they are synthesized. Vesicles pinch off from the Golgi complex and carry the enclosed particles to the interior surface of the cell membrane, where the vesicle membrane and plasma membrane fuse and the contents of the vesicle are discharged.

Exogenous (ex-SOJ-e-nus) Originating outside an organ or part.

Extensibility (ek-sten′-si-BIL-i-tē) The ability of muscle tissue to be stretched when pulled.

Extension (ek-STEN-shun) An increase in the angle between two bones, usually in the sagittal plane; restoring a body part to its anatomical position after flexion.

External Located on or near the surface.

External auditory (AW-di-tōr-ē) **canal** or **meatus** (mē-Ā-tus) A curved tube in the temporal bone that leads to the middle ear.

External ear The outer ear, consisting of the auricle, external auditory canal, and tympanic membrane or eardrum.

External nares (NA-rēz) The external nostrils, or the openings into the nasal cavity on the exterior of the body.

External respiration The exchange of respiratory gases between the lungs and blood. Also called **pulmonary respiration.**

Exteroceptor (eks′-ter-ō-SEP-tor) A receptor adapted for the reception of stimuli from outside the body.

Extracellular fluid (ECF) Fluid outside body cells, such as interstitial fluid and plasma.

Extravasation (eks-trav-a-SĀ-shun) The escape of fluid, especially blood, lymph, or serum, from a vessel into the tissues.

Extrinsic (ek-STRIN-sik) Of external origin.

Exudate (EKS-yoo-dāt) Escaping fluid or semifluid material that oozes from a space that may contain serum, pus, and cellular debris.

F

Falciform ligament (FAL-si-form LIG-a-ment) A sheet of parietal peritoneum between the two principal lobes of the liver. The ligamentum teres, or remnant of the umbilical vein, lies within its fold.

Fallopian tube *See* **Uterine tube.**

Falx cerebelli (FALKS ser′-e-BEL-lē) A small triangular process of the dura mater attached to the occipital bone in the posterior cranial fossa and projecting inward between the two cerebellar hemispheres.

Falx cerebri (SER-e-brē) A fold of the dura mater extending deep into the longitudinal fissure between the two cerebral hemispheres.

Fascia (FASH-ē-a) A fibrous membrane covering, supporting, and separating muscles.

Fascicle (FAS-i-kul) A small bundle or cluster, especially of nerve or muscle fibers (cells). Also called a **fasciculus** (fa-SIK-yū-lus). *Plural,* **fasciculi** (fa-SIK-yū-lī).

Fauces (FAW-sēz) The opening from the mouth into the pharynx.

F-cell A cell in the pancreatic islets (islets of Langerhans) of the pancreas that secretes pancreatic polypeptide.

Feeding (hunger) center A cluster of neurons in the lateral nuclei of the hypothalamus that, when stimulated, brings about feeding.

Female reproductive cycle General term for the ovarian and uterine cycles, the hormonal changes that accompany them, and cyclic changes in the breasts and cervix.

Fertilization (fer′-ti-li-ZĀ-shun) Penetration of a secondary oocyte by a sperm cell and subsequent union of the nuclei of the cells.

Fetal (FĒ-tal) **circulation** The cardiovascular system of the fetus, including the placenta and special blood vessels involved in the exchange of materials between fetus and mother.

Fetus (FĒ-tus) The latter stages of the developing young of an animal; in humans, the developing organism in utero from the beginning of the ninth week to birth.

Fever An elevation in body temperature above its normal temperature of 37°C (98.6° F).

Fibroblast (FĪ-brō-blast) A large, flat cell that forms collagen and elastic fibers and intercellular substance of areolar connective tissue.

Fibrocyte (FĪ-brō-sīt) A mature fibroblast that no longer produces fibers or intercellular substance in connective tissue.

Fibrosis (fī-BRŌ-sis) Abnormal formation of fibrous tissue.

Fibrous (FĪ-brus) **joint** A joint that allows little or no movement, such as a suture and syndesmosis.

Fibrous tunic (TŪ-nik) The outer coat of the eyeball, made up of the posterior sclera and the anterior cornea.

Fight-or-flight response The effects of the stimulation of the sympathetic division of the autonomic nervous system.

Filiform papilla (FIL-i-form pa-PIL-a) One of the conical projections that are distributed in parallel rows over the anterior two-thirds of the tongue and contain no taste buds.

Filtrate (fil-TRĀT) The fluid produced when blood is filtered by the endothelial-capsular membrane.

Filtration (fil-TRĀ-shun) The passage of a liquid through a filter or membrane that acts like a filter.

Filum terminale (FĪ-lum ter-mi-NAL-ē) Nonnervous fibrous tissue of the spinal cord that extends inferiorly from the conus medullaris to the coccyx.

Fimbriae (FIM-brē-ē) Fingerlike structures, especially the lateral ends of the uterine (Fallopian) tubes.

Fissure (FISH-ur) A groove, fold, or slit that may be normal or abnormal.

Fistula (FIS-chū-la) An abnormal passage between two organs or between an organ cavity and the outside.

Fixator A muscle that stabilizes the origin of the prime mover so that the prime mover can act more efficiently.

Fixed macrophage (MAK-rō-fāj) Stationary phagocytic cell found in the liver, lungs, brain, spleen, lymph nodes, subcutaneous tissue, and bone marrow. Also called a **histiocyte** (HISS-tē-ō-sīt).

Flagellum (fla-JEL-um) A hairlike, motile process on the extremity of a sperm cell, bacterium, or protozoan. *Plural,* **flagella** (fla-JEL-a).

Flexion (FLEK-shun) A folding movement in which there is a decrease in the angle between two bones, usually in the sagittal plane.

Follicle (FOL-i-kul) A small secretory sac or cavity.

Follicle-stimulating hormone (FSH) Hormone secreted by the anterior lobe of the pituitary gland that initiates development of ova and stimulates ovaries to secrete estrogens in females and initiates sperm production in males.

Fontanel (fon′-ta-NEL) A membrane-covered spot where bone formation is not yet complete, especially between the cranial bones of an infant's skull.

Foramen (fo-RĀ-men) A passage or opening; a communication between two cavities of an organ or a hole in a bone for passage of vessels or nerves. *Plural,* **foramina** (fo-RĀ-men-a).

Foramen ovale (ō-VAL-ē) An opening in the fetal heart in the septum between the right and left atria. A hole in the greater wing of the sphenoid bone that transmits the mandibular branch of the trigeminal (V) nerve.

Fornix (FOR-niks) An arch or fold; a tract in the brain made up of association fibers, connecting the hippocampus with the mammillary bodies; a recess around the cervix of the uterus where it protrudes into the vagina.

Fossa (FOS-a) A furrow or shallow depression.

Fourth ventricle (VEN-tri-kul) A cavity within the brain lying between the cerebellum and the medulla oblongata and pons.

Frenulum (FREN-yū-lum) A small fold of mucous membrane that connects two parts and limits movement.

Frontal plane A plane at a right angle to a midsagittal plane that divides the body or organs into anterior and posterior portions. Also called a **coronal** (kō-RŌ-nal) **plane.**

Fundus (FUN-dus) The part of a hollow organ farthest from the opening.

Fungiform papilla (FUN-ji-form pa-PIL-a) A mushroomlike elevation on the dorsum of the tongue appearing as a red dot; most contain taste buds.

Furuncle (FYŪR-ung-kul) A boil; painful nodule caused by bacterial infection and inflammation of a hair follicle or sebaceous (oil) gland.

G

Gallbladder A small pouch that stores bile, located inferior to the liver, which is filled with bile and emptied via the cystic duct.

Gamete (GAM-ēt) A male or female reproductive cell; the spermatozoon or ovum.

Ganglion (GANG-glē-on) A group of nerve cell bodies that lie outside the central nervous system. *Plural,* **ganglia** (GANG-glē-a).

Gastric (GAS-trik) **glands** Glands in the mucosa of the stomach that empty in narrow channels called gastric pits. The glands contain chief cells (secrete pepsinogen), parietal cells (secrete hydrochloric acid and intrinsic factor), mucous neck cells (secrete mucus), and G cells (secrete gastrin).

Gastroenterology (gas′-trō-en′-ter-OL-ō-jē) The medical specialty that deals with the structure, function, diagnosis, and treatment of diseases of the stomach and intestines.

Gastrointestinal (gas-trō-in-TES-ti-nal) **(GI) tract** A continuous tube running through the ventral body cavity extending from the mouth to the anus. Also called the **alimentary** (al′-i-MEN-tar-ē) **canal.**

Gastrulation (gas′-trū-LĀ-shun) The various movements of groups of cells that lead to the establishment of the primary germ layers.

Gene (JĒN) Biological unit of heredity; an ultramicroscopic, self-reproducing DNA particle located in a definite position on a particular chromosome.

Genetic engineering The manufacture and manipulation of genetic material.

Genetics The study of heredity.

Genitalia (jen′-i-TĀL-ya) Reproductive organs.

Genome (JĒ-nōm) The complete gene complement of an organism.

Genotype (JĒ-nō-tīp) The total hereditary information carried by an individual; the genetic makeup of an organism.

Geriatrics (jer′-ē-AT-riks) The branch of medicine devoted to the medical problems and care of elderly persons.

Gestation (jes-TĀ-shun) The period of intrauterine development.

Gingivae (jin-JI-vē) Gums. They cover the alveolar processes of the mandible and maxilla and extend slightly into each socket.

Gland Single or group of specialized epithelial cells that secrete substances.

Glans penis (glanz PĒ-nis) The slightly enlarged region at the distal end of the penis.

Gliding joint A synovial joint having articulating surfaces that are usually flat, permitting only side-to-side and back-and-forth movements, as between carpal bones, tarsal bones, and the scapula and clavicle. Also called an **arthrodial** (ar-THRŌ-dē-al) **joint.**

Glomerular (glō-MER-yū-lar) **capsule** A double-walled globe at the proximal end of a nephron that encloses the glomerulus. Also called **Bowman's** (BŌ-manz) **capsule.**

Glomerulus (glō-MER-yū-lus) A rounded mass of nerves or blood vessels, especially the microscopic tuft of capillaries that is surrounded by the glomerular (Bowman's) capsule of each kidney tubule.

Glottis (GLOT-is) The vocal folds (true vocal cords) in the larynx and the space between them (rima glottidis).

Glucagon (GLŪ-ka-gon) A hormone produced by the alpha cells of the pancreas that increases the blood glucose level.

Glucocorticoids (glū-kō-KOR-ti-koyds) A group of hormones of the adrenal cortex.

Glucose (GLŪ-kōs) A six-carbon sugar, $C_6H_{12}O_6$; the major energy source for every cell type in the body. Its metabolism is possible by every known living cell for the production of ATP.

Glycogen (GLĪ-kō-jen) A highly branched polymer of glucose containing thousands of subunits; functions as a compact store of glucose molecules in liver and muscle fibers (cells).

Gnostic (NŌS-tik) **area** Sensory area of the cerebral cortex that receives and integrates sensory input from various parts of the brain so that a common thought can be formed.

Goblet cell A goblet-shaped unicellular gland that secretes mucus. Also called a **mucous cell.**

Golgi (GOL-jē) **complex** An organelle in the cytoplasm of cells consisting of four to eight flattened channels, stacked upon one another, with expanded areas at their ends; functions in packaging secreted proteins, lipid secretion, and carbohydrate synthesis.

Golgi tendon organ *See* **Tendon organ.**

Gomphosis (gom-FŌ-sis) A fibrous joint in which a cone-shaped peg fits into a socket.

Gonad (GŌ-nad) A gland that produces gametes and hormones; the ovary in the female and the testis in the male.

Gonadocorticoids (gō-na-dō-KOR-ti-koydz) Sex hormones (estrogens and androgens) secreted by the adrenal cortex. Also called **adrenal sex hormones.**

Gonadotropic (gō′-nad-ō-TRŌ-pik) **hormone** A hormone that regulates the functions of the gonads.

Gout (GOWT) Hereditary condition associated with excessive uric acid in the blood; the acid crystallizes and deposits in joints, kidneys, and soft tissue.

Graafian follicle *See* **Vesicular ovarian follicle.**

Gray matter Area in the central nervous system and ganglia consisting of nonmyelinated nerve tissue.

Gray ramus communicans (RĀ-mus kō-MYŪ-ni-kans) A short nerve containing postganglionic sympathetic fibers; the cell bodies of the fibers are in a sympathetic chain ganglion, and the nonmyelinated axons run by way of the gray ramus to a spinal nerve and then to the periphery to supply smooth muscle in blood vessels, arrector pili muscles, and sweat glands. *Plural,* **rami communicantes** (RĀ-mē kō-myū-ni-KAN-tēz).

Greater omentum (ō-MEN-tum) A large fold in the serosa of the stomach that hangs down like an apron anterior to the intestines.

Greater vestibular (ves-TIB-yū-lar) **glands** A pair of glands on either side of the vaginal orifice that open by a duct into the space between the hymen and the labia minora. Also called **Bartholin's** (BAR-to-linz) **glands.**

Groin (GROYN) The depression between the thigh and the trunk; the inguinal region.

Gross anatomy The branch of anatomy that deals with structures that can be studied without using a microscope. Also called **macroscopic anatomy.**

Gynecology (gī′-ne-KOL-ō-jē) The branch of medicine dealing with the study and treatment of disorders of the female reproductive system.

Gyrus (JĪ-rus) One of the folds of the cerebral cortex of the brain. *Plural,* **gyri** (JĪ-rī). Also called a **convolution.**

H

Hair follicle (FOL-li-kul) Structure composed of epithelium surrounding the root of a hair from which hair develops.

Hair root plexus (PLEK-sus) A network of dendrites arranged around the root of a hair as free or naked nerve endings that are stimulated when a hair shaft is moved.

Haploid (HAP-loyd) Having half the number of chromosomes characteristically found in the somatic cells of an organism; characteristic of mature gametes. Symbolized *n*.

Hard palate (PAL-at) The anterior portion of the roof of the mouth, formed by the maxillae and palatine bones and lined by mucous membrane.

Haustra (HAWS-tra) The sacculated elevations of the colon.

Haversian canal *See* **Central canal.**

Haversian system *See* **Osteon.**

Heart murmur (MER-mer) An abnormal sound that consists of a flow noise that is heard before the normal lubb-dupp or that may mask normal heart sounds.

Hematology (hē-ma-TOL-ō-jē) The study of blood.

Hematoma (hē-ma-TŌ-ma) A tumor or swelling filled with blood.

Hemoglobin (hē′-mō-GLŌ-bin) **(Hb)** A substance in erythrocytes (red blood cells) consisting of the protein globin and the iron-containing red pigment heme and constituting about 33 percent of the cell volume; involved in the transport of oxygen and carbon dioxide.

Hemopoiesis (hē-mō-poy-Ē-sis) Blood cell production occurring in the red bone marrow of bones. Also called **hematopoiesis** (hem′-a-tō-poy-Ē-sis).

Hemopoietic (hē-mō-poy-E-tic) **stem cell** Red bone marrow cell derived from mesenchyme that gives rise to all formed elements in blood.

Hepatic (he-PAT-ik) **duct** A duct that receives bile from the bile capillaries. Small hepatic ducts merge to form the larger right and left hepatic ducts that unite to leave the liver as the common hepatic duct.

Hepatic portal circulation The flow of blood from the gastrointestinal organs to the liver before returning to the heart.

Hepatopancreatic (hep′-a-tō-pan′-krē-A-tic) **ampulla** A small, raised area in the duodenum where the combined common bile duct and main pancreatic duct empty into the duodenum. Also called the **ampulla of Vater** (VA-ter).

Heterocrine (HET-er-ō-krin) **gland** A gland, such as the pancreas, that is both an exocrine and an endocrine gland.

Hiatus (hī-Ā-tus) An opening; a foramen.

Hilus (HĪ-lus) An area, depression, or pit where blood vessels and nerves enter or leave an organ. Also called a **hilum.**

Hinge joint A synovial joint in which a convex surface of one bone fits into a concave surface of another bone, such as the elbow, knee, ankle, and interphalangeal joints. Also called a **ginglymus** (JIN-gli-mus) **joint.**

Histamine (HISS-ta-mēn) Substance found in many cells, especially mast cells, basophils, and platelets, released when the cells are injured; results in vasodilation, increased permeability of blood vessels, and bronchiole constriction.

Histology (hiss-TOL-ō-jē) Microscopic study of the structure of tissues.

Holocrine (HŌL-ō-krin) **gland** A type of gland in which the entire secreting cell, along with its accumulated secretions, makes up the secretory product of the gland, as in the sebaceous (oil) glands.

Homeostasis (hō′-mē-ō-STĀ-sis) The condition in which the body's internal environment remains relatively constant, within physiological limits.

Homologous (hō-MOL-ō-gus) Correspondence of two organs in structure, position, and origin.

Homologous chromosomes Two chromosomes that belong to a pair. Also called **homologues.**

Hormone (HOR-mōn) Secretion of endocrine tissue that alters the physiological activity of target cells of the body.

Horn Principal area of gray matter in the spinal cord.

Human chorionic gonadotropin (kō-rē-ON-ik gō-nad-ō-TRŌ-pin) **(hCG)** A hormone produced by the developing placenta that maintains the corpus luteum.

Human chorionic somatomammotropin (sō-mat-ō-mam-ō-TRŌ-pin) **(hCS)** A hormone produced by the chorion of the placenta that may stimulate breast tissue for lactation, enhance body growth, and regulate metabolism.

Human growth hormone (hGH) Hormone secreted by the anterior lobe of the pituitary that brings about growth of body tissues, especially skeletal and muscular. Also known as **somatotropin** and **somatotropic hormone (STH).**

Human leukocyte associated (HLA) antigens Surface proteins on white blood cells and other nucleated cells that are unique for each person (except for identical twins) and are used to type tissues and help prevent rejection.

Hyaluronic (hī-a-lū-RON-ik) **acid** A viscous, amorphous extracellular material that binds cells together, lubricates joints, and maintains the shape of the eyeballs.

Hymen (HĪ-men) A thin fold of vascularized mucous membrane at the vaginal orifice.

Hypersecretion (hī-per-se-KRĒ-shun) Overactivity of glands resulting in excessive secretion.

Hypersensitivity (hī-per-sen-si-TI-vi-tē) Overreaction to an allergen that results in pathological changes in tissues. Also called **allergy.**

Hypertonia (hī-per-TŌ-nē-a) Increased muscle tone that is expressed as spasticity or rigidity.

Hypophyseal (hī-pō-FIZ-ē-al) **pouch** An outgrowth of ectoderm from the roof of the stomodeum (mouth) from which the anterior lobe of the pituitary gland develops.

Hypophysis (hī-POF-i-sis) Pituitary gland.

Hypoplasia (hī-pō-PLĀ-zē-a) Defective development of tissue.

Hyposecretion (hī′-pō-se-KRĒ-shun) Underactivity of glands resulting in diminished secretion.

Hypothalamus (hī′-pō-THAL-a-mus) A portion of the diencephalon, lying beneath the thalamus and forming the floor and part of the wall of the third ventricle.

Hypotonic (hī-pō-TON-ik) Having an osmotic pressure lower than that of a solution with which it is compared.

I

Ileocecal (il′-ē-ō-SĒ-kal) **sphincter** A fold of mucous membrane that guards the opening from the ileum into the large intestine. Also called the **ileocecal valve.**

Ileum (IL-ē-um) The terminal portion of the small intestine.

Immunoglobulin (im-yū-nō-GLOB-yū-lin) **(Ig)** An antibody synthesized by plasma cells derived from B lymphocytes in response to the introduction of antigen. Immunoglobulins are divided into five kinds (IgG, IgM, IgA, IgD, IgE) based primarily on the larger protein component present in the immunoglobulin.

Immunology (im′-yū-NOL-ō-jē) The branch of science that deals with the responses of the body when challenged by antigens.

Imperforate (im-PER-fō-rāt) Abnormally closed.

Implantation (im-plan-TĀ-shun) The insertion of a tissue or a part into the body. The attachment of the blastocyst to the lining of the uterus 7–8 days after fertilization.

Inclusion (in-KLŪ-zhun) Temporary structure that contains secretions and storage products of cells.

Indirect pathway Motor pathways that convey information from the brain down the spinal cord; related to automatic movements, coordination of body movements with visual stimuli, skeletal muscle tone and posture, and equilibrium.

Infarction (in-FARK-shun) The presence of a localized area of necrotic tissue, produced by inadequate oxygenation of the tissue.

Infection (in-FEK-shun) Invasion and multiplication of microorganisms in body tissues, which may be inapparent or characterized by cellular injury.

Inferior (in-FĒR-ē-or) Away from the head or toward the lower part of a structure. Also called **caudal** (KAW-dal).

Inferior vena cava (VĒ-na CĀ-va) **(IVC)** Large vein that collects blood from parts of the body inferior to the heart and returns it to the right atrium.

Inflammation (in′-fla-MĀ-shun) Localized, protective response to tissue injury designed to destroy, dilute, or wall off the infecting agent or injured tissue; characterized by redness, pain, heat, swelling, and sometimes loss of function.

Infundibulum (in′-fun-DIB-yū-lum) The stalklike structure that attaches the pituitary gland to the hypothalamus of the brain. The funnel-shaped, open, distal end of the uterine (Fallopian) tube.

Inguinal (IN-gwi-nal) Pertaining to the groin.

Inguinal canal An oblique passageway to the anterior abdominal wall just superior and parallel to the medial half of the inguinal ligament that transmits the spermatic cord and ilioinguinal nerve in the male and the round ligament of the uterus and ilioinguinal nerve in the female.

Inheritance The acquisition of body characteristics and qualities by transmission of genetic information from parents to offspring.

Inhibin A male sex hormone secreted by sustentacular (Sertoli) cells that inhibits FSH release by the anterior pituitary gland and thus spermatogenesis.

Inner cell mass A region of cells of a blastocyst that differentiates into the three primary germ layers—ectoderm, mesoderm, and endoderm—from which all tissues and organs develop; also called an **embryoblast.**

Insertion (in-SER-shun) The manner or place of attachment of a muscle to the bone that it moves.

Inspiration (in-spi-RĀ-shun) The act of drawing air into the lungs.

Insula (IN-sū-la) A triangular area of cerebral cortex that lies deep within the lateral cerebral fissure, under the parietal, frontal, and temporal lobes, and cannot be seen in an external view of the brain. Also called the **island** or **isle of Reil** (RĪL).

Insulin (IN-sū-lin) A hormone produced by the beta cells of the pancreas that decreases the blood glucose level.

Intercalated (in-TER-ka-lāt-ed) **disc** An irregular transverse thickening of sarcolemma that holds cardiac muscle fibers (cells) together and aids in conduction of muscle action potentials.

Intercostal (in′-ter-KOS-tal) **nerve** A nerve supplying a muscle located between the ribs.

Intermediate (in′-ter-MĒ-dē-it) **filament** Protein filament ranging from 8 to 12 nm in diameter that may provide structural reinforcement, hold organelles in place, and give shape to a cell.

Internal capsule A tract of projection fibers connecting various parts of the cerebral cortex and lying between the thalamus and the caudate and lentiform nuclei of the basal ganglia.

Internal ear The inner ear or labyrinth, lying inside the temporal bone, containing the organs of hearing and balance.

Internal nares (NA-rēz) The two openings posterior to the nasal cavities opening into the nasopharynx. Also called the **choanae** (kō-A-nē).

Internal respiration The exchange of respiratory gases between blood and body cells. Also called **tissue respiration.**

Interphase (IN-ter-fāz) The period during its life cycle when a cell is carrying on every life process except division; the stage between two mitotic divisions. Also called **metabolic phase.**

Interstitial cell of Leydig *See* **Interstitial endocrinocyte.**

Interstitial (in′-ter-STISH-al) **endocrinocyte** A cell that is located in the connective tissue between seminiferous tubules in a mature testis that secretes testosterone. Also called an **interstitial cell of Leydig** (LĪ-dig).

Interstitial fluid The portion of extracellular fluid that fills the microscopic spaces between the cells of tissues; the internal environment of the body. Also called **intercellular** or **tissue fluid.**

Interstitial growth Growth from within, as in the growth of cartilage. Also called **endogenous** (en-DOJ-e-nus) **growth.**

Interventricular (in′-ter-ven-TRIK-yū-lar) **foramen** A narrow, oval opening through which the lateral ventricles of the brain communicate with the third ventricle. Also called the **foramen of Monro.**

Intervertebral (in′-ter-VER-te-bral) **disc** A pad of fibrocartilage located between the bodies of two vertebrae.

Intestinal gland Gland that opens onto the surface of the intestinal mucosa and secretes digestive enzymes. Also called a **crypt of Lieberkühn** (LĒ-ber-kyūn).

Intracellular (in′-tra-SEL-yū-lar) **fluid (ICF)** Fluid located within cells.

Intrafusal (in′-tra-FYŪ-zal) **fibers** Three to ten specialized muscle fibers (cells), partially enclosed in a connective tissue capsule that is filled with lymph; the fibers compose muscle spindles.

Intramembranous ossification (in′-tra-MEM-bra-nus os′-i′-fi-KĀ-shun) The method of bone formation in which the bone is formed directly in membranous tissue.

Intraocular (in-tra-OC-yū-lar) **pressure (IOP)** Pressure in the eyeball, produced mainly by aqueous humor.

Intrinsic (in-TRIN-sik) Of internal origin; for example, the intrinsic factor, a mucoprotein formed by the gastric mucosa that is necessary for the absorption of vitamin B_{12}.

Intrinsic factor (IF) A glycoprotein synthesized and secreted by the parietal cells of the gastric mucosa that facilitates vitamin B_{12} absorption in the small intestine.

Invagination (in-vaj′-i-NĀ-shun) The pushing of the wall of a cavity into the cavity itself.

Inversion (in-VER-zhun) The movement of the sole medially at the ankle joint.

Ipsilateral (ip′-si-LAT-er-al) On the same side of the midline, affecting the same side of the body.

Iris The colored portion of the eyeball seen through the cornea that consists of circular and radial smooth muscle; the black hole in the center of the iris is the pupil.

Ischemia (is-KĒ-mē-a) A lack of sufficient blood to a part due to obstruction of circulation.

Island of Reil *See* **Insula.**

Islet of Langerhans *See* **Pancreatic islet.**

Isthmus (IS-mus) A narrow strip of tissue or narrow passage connecting two larger parts.

J

Jejunum (jē-JŪ-num) The middle portion of the small intestine.

Joint kinesthetic (kin′-es-THET-ik) **receptor** A proprioceptive receptor located in a joint, stimulated by joint movement.

Juxtaglomerular (juks-ta-glō-MER-yū-lar) **apparatus (JGA)** Consists of the macula densa (cells of the distal convoluted tubule adjacent to the afferent and efferent arteriole) and juxtaglomerular cells (modified cells of the afferent and sometimes efferent arteriole); secretes renin when blood pressure starts to fall.

K

Keratinocyte (ker-A-tin′-ō-sīt) The most numerous of the epidermal cells that function in the production of keratin.

Kidney (KID-nē) One of the paired reddish organs located in the lumbar region that regulates the composition and volume of blood and produces urine.

Kinesiology (ki-nē′-sē-OL-ō-jē) The study of the movement of body parts.

Kinesthesia (kin-is-THĒ-szē-a) Ability to perceive the extent, direction, or weight of movement; muscle sense.

Kinetochore (ki-NET-ō-kor) Protein complex attached to the outside of a centromere to which kinetochore microtubules attach.

Kupffer's cell *See* **Stellate reticuloendothelial cell.**

Kyphosis (kī-FŌ-sis) An exaggeration of the thoracic curve of the vertebral column, resulting in a "round-shouldered" or hunchback appearance.

L

Labial frenulum (LĀ-bē-al FREN-yū-lum) A medial fold of mucous membrane between the inner surface of the lip and the gums.

Labia majora (LĀ-bē-a ma-JO-ra) Two longitudinal folds of skin extending downward and backward from the mons pubis of the female.

Labia minora (min-OR-a) Two small folds of mucous membrane lying medial to the labia majora of the female.

Labyrinth (LAB-i-rinth) Intricate communicating passageway, especially in the internal ear.

Lacrimal (LAK-ri-mal) **canal** A duct, one on each eyelid, commencing at the punctum at the medial margin of an eyelid and conveying tears medially into the nasolacrimal sac.

Lacrimal gland Secretory cells located at the superior anterolateral portion of each orbit that secrete tears into excretory ducts that open onto the surface of the conjunctiva.

Lacrimal sac The superior expanded portion of the nasolacrimal duct that receives the tears from a lacrimal canal.

Lactation (lak-TĀ-shun) The secretion and ejection of milk by the mammary glands.

Lacteal (LAK-tē-al) One of many intestinal lymphatic vessels in villi that absorb triglycerides from digested food.

Lacuna (la-KŪ-na) A small, hollow space, such as that found in bones in which the osteoblasts lie. *Plural,* **lacunae** (la-KŪ-nē).

Lambdoid (LAM-doyd) **suture** The line of union in the skull between the parietal bones and the occipital bone; sometimes contains sutural bones.

Lamellae (la-MEL-ē) Concentric rings found in compact bone.

Lamellated corpuscle Oval pressure receptor located in subcutaneous tissue and consisting of concentric layers of connective tissue wrapped around a sensory nerve fiber. Also called a **Pacinian** (pa-SIN-ē-an) **corpuscle.**

Lamina (LAM-i-na) A thin, flat layer or membrane, as the flattened part of either side of the arch of a vertebra. *Plural,* **laminae** (LAM-i-nē).

Lamina propria (PRŌ-prē-a) The connective tissue layer of a mucous membrane.

Langerhans (LANG-er-hans) **cell** Epidermal cell that arises from bone marrow that presents antigens to helper T cells, thus activating T cells as part of the immune response.

Lanugo (lan-YŪ-gō) Fine downy hairs that cover the fetus.

Large intestine The portion of the gastrointestinal tract extending from the ileum of the small intestine to the anus, divided structurally into the cecum, colon, rectum, and anal canal.

Laryngopharynx (la-rin′-gō-FAR-inks) The inferior portion of the pharynx, extending downward from the level of the hyoid bone to divide posteriorly into the esophagus and anteriorly into the larynx. Also called the **hypopharynx.**

Laryngotracheal (la-rin′-gō-TRĀ-kē-al) **bud** An outgrowth of endoderm of the foregut from which the respiratory system develops.

Larynx (LAR-inks) The voice box, a short passageway that connects the pharynx with the trachea.

Lateral (LAT-er-al) Farther from the midline of the body or a structure.

Lateral ventricle (VEN-tri-kul) A cavity within a cerebral hemisphere that communicates with the lateral ventricle in the other cerebral hemisphere and with the third ventricle by way of the interventricular foramen.

Lens A transparent organ constructed of proteins (crystallins) lying posterior to the pupil and iris of the eyeball and anterior to the vitreous body.

Lesion (LĒ-zhun) Any localized, abnormal change in tissue formation.

Lesser omentum (ō-MEN-tum) A fold of the peritoneum that extends from the liver to the lesser curvature of the stomach and the commencement of the duodenum.

Lesser vestibular (ves-TIB-yū-lar) **gland** One of the paired mucus-secreting glands that have ducts that open on either side of the urethral orifice in the vestibule of the female.

Leukocyte (LŪ-kō-sīt) A white blood cell.

Libido (li-BĒ-dō) The sexual drive, conscious or subconscious.

Ligament (LIG-a-ment) Dense, regularly arranged connective tissue that attaches bone to bone.

Limbic system A portion of the forebrain, sometimes termed the visceral brain, concerned with various aspects of emotion and behavior, that includes the limbic lobe, denate gyrus, amygdaloid body, septal nuclei, mammillary bodies, anterior thalamic nucleus, olfactory bulbs, and bundles of myelinated axons.

Lingual frenulum (LIN-gwal FREN-yū-lum) A fold of mucous membrane that connects the tongue to the floor of the mouth.

Lingual lipase (LĪ-pās) Digestive enzyme secreted by glands on the dorsum of the tongue that digests triglycerides.

Lipase (LĪ-pās) A fat-splitting enzyme.

Lipid An organic compound composed of carbon, hydrogen, and oxygen that is usually insoluble in water, but soluble in alcohol, ether, and chloroform; examples include triglycerides, phospholipids, steroids, and prostaglandins.

Lipid bilayer Arrangement of phospholipid molecules in two parallel rows in which the hydrophilic "heads" face outward and the hydrophobic "tails" face inward.

Liver Large gland under the diaphragm that occupies most of the right hypochondriac region and part of the epigastric region; functionally, it produces bile salts, heparin, and plasma proteins; converts one nutrient into another; detoxifies substances; stores glycogen, minerals, and vitamins; carries on phagocytosis of blood cells and bacteria; and helps activate vitamin D.

Lobe (lōb) A curved or rounded projection.

Locus coeruleus (LŌ-kus sē-ROO-lē-us) A group of neurons in the brain stem where norepinephrine (NE) is concentrated.

Lordosis (lor-DŌ-sis) An exaggeration of the lumbar curve of the vertebral column.

Lower limb (extremity) The appendage attached at the pelvic (hip) girdle, consisting of the thigh, knee, leg, ankle, foot, and toes.

Lumbar (LUM-bar) Region of the back and side between the ribs and pelvis; loin.

Lumbar plexus (PLEK-sus) A network formed by the anterior branches of spinal nerves L1 through L4.

Lumen (LŪ-men) The space within an artery, vein, duct, or hollow organ.

Lung One of the two main organs of respiration, lying on either side of the heart in the thoracic cavity.

Lunula (LŪ-nyū-la) The moon-shaped white area at the base of a nail.

Luteinizing (LŪ-tē-in′-īz-ing) **hormone (LH)** A hormone secreted by the anterior lobe of the pituitary gland that stimulates ovulation, progesterone secretion by the corpus luteum, and readies the mammary glands for milk secretion in females and stimulates testosterone secretion by the testes in males.

Lymph (LIMF) Fluid confined in lymphatic vessels and flowing through the lymphatic system to be returned to the blood.

Lymphatic (lim-FAT-ik) **capillary** Blind-ended microscopic lymph vessel that begins in spaces between cells and converges with other lymph capillaries to form lymphatic vessels.

Lymphatic tissue A specialized form of reticular tissue that contains large numbers of lymphocytes.

Lymphatic vessel A large vessel that collects lymph from lymph capillaries and converges with other lymphatic vessels to form the thoracic and right lymphatic ducts.

Lymph node An oval or bean-shaped structure located along lymphatic vessels.

Lymphocyte (LIM-fō-sīt) A type of white blood cell, found in lymph nodes, associated with the immune system.

Lysosome (LĪ-sō-sōm) An organelle in the cytoplasm of a cell, enclosed by a single membrane and containing powerful digestive enzymes.

Lysozyme (LĪ-sō-zīm) A bactericidal enzyme found in tears, saliva, and perspiration.

M

Macrophage (MAK-rō-fāj) Phagocytic cell derived from a monocyte. May be fixed or wandering.

Macula (MAK-yū-la) A discolored spot or a colored area. A small, thickened region on the wall of the utricle and saccule that serves as a receptor for static equilibrium.

Macula lutea (LŪ-tē-a) The yellow spot in the center of the retina.

Malaise (ma-LĀYZ) Discomfort, uneasiness, and indisposition, often indicative of infection.

Malignant (ma-LIG-nant) Referring to diseases that tend to become worse and cause death; especially the invasion and spreading of cancer.

Mammary (MAM-ar-ē) **gland** Modified sudoriferous (sweat) gland of the female that secretes milk for the nourishment of the young.

Mammillary (MAM-i-ler-ē) **bodies** Two small rounded bodies posterior to the tuber cinereum that are involved in reflexes related to the sense of smell.

Marrow (MAR-ō) Soft, spongelike material in the cavities of bone. Red bone marrow produces blood cells; yellow bone marrow, formed mainly of fatty tissue, has no blood-producing function.

Mast cell A cell found in areolar connective tissue along blood vessels that produces heparin, a dilator of small blood vessels during inflammation. The name given to a basophil after it has left the bloodstream and entered the tissues.

Matrix (MĀ-trix) Ground substance and fibers external to cells of a connective tissue.

Meatus (mē-Ā-tus) A passage or opening, especially the external portion of a canal.

Mechanoreceptor (me-KAN-ō-rē′-sep-tor) Receptor that detects mechanical deformation of the receptor itself or adjacent cells; stimuli so detected include those related to touch, pressure, vibration, proprioception, hearing, equilibrium, and blood pressure.

Medial (MĒ-dē-al) Nearer the midline of the body or a structure.

Medial lemniscus (lem-NIS-kus) A flat band of myelinated nerve fibers extending through the medulla oblongata, pons, and midbrain and terminating in the thalamus on the same side. Sensory neurons in this tract transmit impulses for proprioception, fine touch, pressure, and vibration sensations.

Median aperture (AP-er-choor) One of the three openings in the roof of the fourth ventricle through which cerebrospinal fluid enters the subarachnoid space of the brain and cord. Also called the **foramen of Magendie.**

Median plane A vertical plane dividing the body into equal right and left sides. Situated in the middle.

Mediastinum (mē′-dē-as-TĪ-num) A broad, median partition, actually a mass of tissue found between the pleurae of the lungs that extends from the sternum to the vertebral column.

Medulla (me-DULL-la) An inner layer of an organ, such as the medulla of the kidneys.

Medulla oblongata (ob′-long-GA-ta) The most inferior part of the brain stem.

Medullary (MED-yū-lar′-ē) **cavity** The space within the diaphysis of a bone that contains yellow bone marrow. Also called the **marrow cavity.**

Medullary rhythmicity (rith-MIS-i-tē) **area** Portion of the respiratory center in the medulla oblongata that controls the basic rhythm of respiration.

Meibomian gland *See* **Tarsal gland.**

Meiosis (mī-Ō-sis) A type of cell division restricted to sex-cell production involving two successive nuclear divisions that result in daughter cells with the haploid (n) number of chromosomes.

Meissner's corpuscle *See* **Corpuscle of touch.**

Melanin (MEL-a-nin) A dark black, brown, or yellow pigment found in some parts of the body such as the skin.

Melanocyte (MEL-a-nō-sīt′) A pigmented cell located between or beneath cells of the deepest layer of the epidermis that synthesizes melanin.

Melanocyte-stimulating hormone (MSH) A hormone secreted by the anterior lobe of the pituitary gland that stimulates the dispersion of melanin granules in melanocytes in amphibians; continued administration produces darkening of skin in humans.

Melatonin (mel-a-TŌN-in) A hormone secreted by the pineal gland that may inhibit reproductive activities.

Membrane A thin, flexible sheet of tissue composed of an epithelial layer and an underlying connective tissue layer, as in an epithelial membrane, or of areolar connective tissue only, as in a synovial membrane.

Membranous labyrinth (mem-BRA-nus LAB-i-rinth) The portion of the labybrinth of the inner ear that is located inside the bony labyrinth and separated from it by the perilymph; made up of the membranous semicircular canals, the saccule and utricle, and the cochlear duct.

Memory The ability to recall thoughts; commonly classified as short-term (activated) and long-term.

Menarche (me-NAR-kē) Beginning of the menstrual function.

Meninges (me-NIN-jēz) Three membranes covering the brain and spinal cord, called the dura mater, arachnoid, and pia mater. *Singular,* **meninx** (MEN-inks).

Menopause (MEN-ō-pawz) The termination of the menstrual cycles.

Menstrual (MEN-strū-al) **cycle** A series of changes in the endometrium of a nonpregnant female that prepares the lining of the uterus to receive a fertilized ovum.

Menstruation (men′-strū-Ā-shun) Periodic discharge of blood, tissue fluid, mucus, and epithelial cells that usually lasts for 5 days; caused by a sudden reduction in estrogens and progestrone. Also called the **menstrual phase** or **menses.**

Merkel (MER-kel) **cell** Cell found in the epidermis of hairless skin that makes contact with a tactile (Merkel) disc that functions in touch.

Merocrine (MER-ō-krin) **gland** A secretory cell that remains intact throughout the process of formation and discharge of the secretory product, as in the salivary and pancreatic glands.

Mesenchyme (MEZ-en-kīm) An embryonic connective tissue from which all other connective tissues arise.

Mesentery (MEZ-en-ter′-ē) A fold of peritoneum attaching the small intestine to the posterior abdominal wall.

Mesocolon (mez′-ō-KŌ-lon) A fold of peritoneum attaching the colon to the posterior abdominal wall.

Mesoderm The middle of the three primary germ layers that gives rise to connective tissues, blood and blood vessels, and muscles.

Mesothelium (mez′-ō-THĒ-lē-um) The layer of simple squamous epithelium that lines serous cavities.

Mesovarium (mez′-ō-VAR-ē-um) A short fold of peritoneum that attaches an ovary to the broad ligament of the uterus.

Metabolism (me-TAB-ō-lizm) The sum of all the biochemical reactions that occur within an organism, including the synthetic (anabolic) reactions and decomposition (catabolic) reactions.

Metacarpus (met′-a-KAR-pus) A collective term for the five bones that make up the palm of the hand.

Metaphase (MET-a-fāz) The second stage of mitosis in which chromatid pairs line up on the equatorial plane of the cell.

Metaphysis (me-TAF-i-sis) Growing portion of a bone.

Metarteriole (met′-ar-TĒ-rē-ōl) A blood vessel that emerges from an arteriole, traverses a capillary network, and empties into a venule.

Metatarsus (met′-a-TAR-sus) A collective term for the five bones located in the foot between the tarsals and the phalanges.

Microfilament (mī-krō-FIL-a-ment) Rodlike cytoplasmic structure about 6 nm in diameter; comprises contractile units in muscle fibers (cells) and provides support, shape, and movement in nonmuscle cells.

Microglia (mī-krō-GLĒ-a) Neuroglial cells that carry on phagocytosis. Also called **brain macrophages** (MAK-rō-fāj-ez).

Microphage (MĪK-rō-fāj) Granular leukocyte that carries on phagocytosis, especially neutrophils and eosinophils.

Microtubule (mī-krō-TŪB-yūl′) Cylindrical cytoplasmic structure, ranging in diameter from 18 to 30 nm, consisting of the protein tubulin; provides support, structure, and transportation.

Microvilli (mī-krō-VIL-ē) Microscopic, fingerlike projections of the cell membranes of small intestinal cells that increase surface area for absorption.

Micturition (mik′-tū-RISH-un) The act of expelling urine from the urinary bladder. Also called **urination** (yoo-ri-NĀ-shun).

Midbrain The part of the brain between the pons and the diencephalon. Also called the **mesencephalon** (mes′-en-SEF-a-lon).

Middle ear A small, epithelial-lined cavity hollowed out of the temporal bone, separated from the external ear by the eardrum and from the internal ear by a thin bony partition containing the oval and round windows; extending across the middle ear are the three auditory ossicles. Also called the **tympanic** (tim-PAN-ik) **cavity.**

Midline An imaginary vertical line that divides the body into equal left and right sides.

Midsagittal plane A vertical plane through the midline of the body that divides the body or organs into equal right and left sides. Also called a **median plane.**

Mineralocorticoids (min′-er-al-ō-KOR-ti-koyds) A group of hormones of the adrenal cortex.

Mitochondrion (mī′-tō-KON-drē-on) A double-membraned organelle that plays a central role in the production of ATP; known as the "powerhouse" of the cell.

Mitosis (mī-TŌ-sis) The orderly division of the nucleus of a cell that ensures that each new daughter nucleus has the same number and kind of chromosomes as the original parent nucleus. The process includes the replication of chromosomes and the distribution of the two sets of chromosomes into two separate and equal nuclei.

Mitotic spindle Collective term for a football-shaped assembly of microtubules (nonkinetochore, kinetochore, and aster) produced by the pericentriolar area of a centrosome and that is responsible for the movement of chromosomes during cell division.

Modality (mō-DAL-i-tē) Any of the specific sensory entities, such as vision, smell, or taste.

Modiolus (mō-DĪ-ō′-lus) The central pillar or column of the cochlea.

Monoclonal antibody (MAb) Antibody produced by in vitro clones of B cells hybridized with cancerous cells.

Monocyte (MON-ō-sīt′) A type of white blood cell characterized by agranular cytoplasm; the largest of the leukocytes.

Mons pubis (MONZ PYŪ-bis) The rounded, fatty prominence over the pubic symphysis, covered by coarse pubic hair.

Morphology (mor-FOL-o-jē) The study of the form and structure of things.

Morula (MOR-yū-la) A solid mass of cells produced by successive cleavages of a fertilized ovum a few days after fertilization.

Motor area The region of the cerebral cortex that governs muscular movement, particularly the precentral gyrus of the frontal lobe.

Motor end plate Portion of the sarcolemma of a muscle fiber (cell) in close approximation with an axon terminal.

Motor neuron (NŪ-ron) A neuron that conveys nerve impulses from the brain and spinal cord to effectors that may be either muscles or glands. Also called an **efferent neuron.**

Motor unit A motor neuron together with the muscle fibers (cells) it stimulates.

Mucous (MYŪ-kus) **cell** A unicellular gland that secretes mucus. Also called a **goblet cell.**

Mucous membrane A membrane that lines a body cavity that opens to the exterior. Also called the **mucosa** (myū-KŌ-sa).

Mucus The thick fluid secretion of mucous glands and mucous membranes.

Muscarinic (mus′-ka-RIN-ik) **receptor** Receptor found on all effectors innervated by parasympathetic postganglionic axons and some effectors innervated by sympathetic postganglionic axons; so named because the actions of acetylcholine (ACh) on such receptors are similar to those produced by muscarine.

Muscle An organ composed of one of three types of muscle tissue (skeletal, cardiac, or visceral), specialized for contraction to produce voluntary or involuntary movement of parts of the body.

Muscle spindle An encapsulated receptor in a skeletal muscle, consisting of specialized muscle fiber (cell) and nerve endings, stimulated by changes in length or tension of muscle fibers; a proprioceptor. Also called a **neuromuscular** (nū-rō-MUS-kyū-lar) **spindle.**

Muscle tissue A tissue specialized to produce motion in response to muscle action potentials by its qualities of contractility, extensibility, elasticity, and excitability. Types include skeletal, cardiac, and smooth.

Muscle tone A sustained, partial contraction of portions of a skeletal muscle in response to activation of stretch receptors.

Muscularis (MUS-kyū-la′-ris) A muscular layer (coat or tunic) of an organ.

Muscularis mucosae (myū-KŌ-sē) A thin layer of smooth muscle fibers (cells) located in the most superficial layer of the mucosa of the gastrointestinal tract, underlying the lamina propria of the mucosa.

Myelin (MĪ-e-lin) **sheath** A white, phospholipid, segmented covering, formed by neurolemmocytes (Schwann cells), around the axons and dendrites of many peripheral neurons.

Myenteric plexus A network of nerve fibers from both autonomic divisions located in the muscularis coat of the small intestine. Also called the **plexus of Auerbach** (OW-er-bak).

Myocardium (mī′-ō-KAR-dē-um) The middle layer of the heart wall, made up of cardiac muscle, comprising the bulk of the heart, and lying between the epicardium and the endocardium.

Myofibril (mī′-ō-FĪ-bril) A threadlike structure, running longitudinally through a muscle fiber (cell) consisting mainly of thick myofilaments (myosin) and thin myofilaments (actin).

Myoglobin (mī-ō-GLŌ-bin) The oxygen-binding, iron-containing conjugated protein complex present in the sarcoplasm of muscle fibers (cells); contributes the red color to muscle.

Myology (mī-OL-ō-jē) The study of muscles.

Myometrium (mī′-ō-MĒ-trē-um) The smooth muscle layer of the uterus.

Myopia (mī-Ō-pē-a) Defect in vision so that objects can be seen distinctly only when very close to the eyes; nearsightedness.

Myosin (MĪ-ō-sin) The contractile protein that makes up the thick myofilaments of muscle fibers (cells).

Myxedema (mix-e-DĒ-ma) Condition caused by hypothyroidism during the adult years characterized by swelling of facial tissues.

N

Nail A hard plate, composed largely of keratin, that develops from the epidermis of the skin to form a protective covering on the dorsal surface of the distal phalanges of the fingers and toes.

Nail matrix (MĀ-triks) The part of the nail beneath the body and root from which the nail is produced.

Nasal (NĀ-zal) **cavity** A mucosa-lined cavity on either side of the nasal septum that opens onto the face at an external naris and into the nasopharynx at an internal naris.

Nasal septum (SEP-tum) A vertical partition composed of bone (perpendicular plate of ethmoid and vomer) and cartilage, covered with amucous membrane, separating the nasal cavity into left and right sides.

Nasolacrimal (nā′-zō-LAK-ri-mal) **duct** A canal that transports the lacrimal secretion (tears) from the nasolacrimal sac into the nose.

Nasopharynx (nā′-zō-FAR-inks) The superior portion of the pharynx, lying posterior to the nose and extending inferiorly to the soft palate.

Nephron (NEF-ron) The functional unit of the kidney.

Nerve A cordlike bundle of nerve fibers (axons and/or dendrites) and their associated connective tissue coursing together outside the central nervous system.

Nerve fiber General term for any process (axon or dendrite) projecting from the cell body of a neuron.

Nerve impulse A wave of negativity (depolarization) that self-propagates along the outside surface of the plasma membrane of a neuron; also called a **nerve action potential.**

Nervous tissue Tissue that initiates and transmits nerve impulses to coordinate body activities.

Neural plate A thickening of ectoderm that forms early in the third week of development and represents the beginning of the development of the nervous system.

Neuroeffector (nū-rō-e-FEK-tor) **junction** Collective term for neuromuscular and neuroglandular junctions.

Neurofibral (nū-rō-FĪ-bral) **node** A space, along a myelinated nerve fiber, between the individual neurolemmocytes (Schwann cells) that form the myelin sheath and the neurolemma. Also called **node of Ranvier** (ron-VĒ-ā).

Neurofibril (nū-rō-FĪ-bril) One of the delicate threads that forms a complicated network in the cytoplasm of the cell body and processes of a neuron.

Neuroglandular (nū-rō-GLAND-yū-lar) **junction** Area of contact between a motor neuron and a gland.

Neuroglia (nū-RŌG-lē-a) Cells of the nervous system that are specialized to perform the functions of connective tissue. The neuroglia of the central nervous system are the astrocytes, oligodendrocytes, microglia, and ependyma; neuroglia of the peripheral nervous system include the neurolemmocytes (Schwann cells) and the ganglion satellite cells. Also called **glial** (GLĒ-al) **cells.**

Neurohypophyseal (nū′-rō-hī-po-FIZ-ē-al) **bud** An outgrowth of ectoderm located on the floor of the hypothalamus that gives rise to the posterior lobe of the pituitary gland.

Neurolemma (nū-rō-LEM-ma) The peripheral, nucleated cytoplasmic layer of the neurolemmocyte (Schwann cell). Also called **sheath of Schwann** (SCHVON).

Neurolemmocyte A neuroglial cell of the peripheral nervous system that forms the myelin sheath and neurolemma of a nerve fiber by wrapping around a nerve fiber in a jelly-roll fashion. Also called a **Schwann** (SCHVON) **cell.**

Neurology (nū-ROL-ō-jē) The branch of science that deals with the normal functioning and disorders of the nervous system.

Neuromuscular (nū-rō-MUS-kyū-lar) **junction** The area of contact between the axon terminal of a motor neuron and a portion of the sarcolemma of a muscle fiber (cell). Also called a **myoneural** (mī-ō-NŪ-ral) **junction.**

Neuron (NŪ-ron) A nerve cell, consisting of a cell body, dendrites, and an axon.

Neurosecretory (nū-rō-SĒC-re-tō-rē) **cell** A cell in a nucleus (paraventricular and supraoptic) in the hypothalamus that produces oxytocin (OT) or antidiuretic hormone (ADH), hormones stored in the posterior lobe of the pituitary gland.

Neurotransmitter One of a variety of molecules synthesized within the nerve axon terminals, released into the synaptic cleft in response to a nerve impulse, and affecting the membrane potential of the postsynaptic neuron. Also called a **transmitter substance.**

Neutrophil (NŪ-trō-fil) A type of white blood cell characterized by granular cytoplasm that stains as readily with acid or basic dyes.

Nicotinic (nik′-ō-TIN-ik) **receptor** Receptor found on both sympathetic and parasympathetic postganglionic neurons so named because the actions of acetylcholine (ACh) in such receptors are similar to those produced by nicotine.

Nipple A pigmented, wrinkled projection on the surface of the mammary gland that is the location of the openings of the lactiferous ducts for milk release.

Nissl bodies *See* **Chromatophilic substance.**

Nociceptor (nō′-sē-SEP-tor) A free (naked) nerve ending that detects pain.

Node of Ranvier *See* **Neurofibral node.**

Norepinephrine (nor′-ep-ē-NEF-rin) **(NE)** A hormone secreted by the adrenal medulla that produces actions similar to those that result from sympathetic stimulation. Also called **noradrenaline** (nor-a-DREN-a-lin).

Notochord (NŌ-tō-cord) A flexible rod of embryonic tissue that lies where the future vertebral column will develop.

Nucleic (nū-KLĒ-ik) **acid** An organic compound that is a long polymer of nucleotides, with each nucleotide containing a pentose sugar, a phosphate group, and one of four possible nitrogenous bases (adenine, cytosine, guanine, and thymine or uracil).

Nucleolus (nū-KLĒ-ō-lus) Nonmembranous spherical body within the nucleus composed of protein, DNA, and RNA that functions in the synthesis and storage of ribosomal RNA.

Nucleosome (NŪ-klē-ō-sōm) Elementary structural subunit of a chromosome consisting of histones and DNA.

Nucleus (NŪ-klē-us) A spherical or oval organelle of a cell that contains the hereditary factors of the cell, called genes. A cluster of unmyelinated nerve cell bodies in the central nervous system. The central portion of an atom made up of protons and neutrons.

Nucleus cuneatus (kyū-nē-Ā-tus) A group of nerve cells in the inferior portion of the medulla oblongata in which fibers of the fasciculus cuneatus terminate.

Nucleus gracilis (gras-I-lis) A group of nerve cells in the inferior portion of the medulla oblongata in which fibers of the fasciculus gracilis terminate.

Nucleus pulposus (pul-PŌ-sus) A soft, pulpy, highly elastic substance in the center of an intervertebral disc, a remnant of the notochord.

O

Oblique (ō-BLĒK) **plane** A plane that passes through the body or an organ at an angle between the transverse plane and either the midsagittal, parasagittal, or frontal plane.

Obstetrics (ob-STET-riks) The specialized branch of medicine that deals with pregnancy, labor, and the period of time immediately following delivery (about 42 days).

Obturator (OB-tyū-rā′-ter) Anything that obstructs or closes a cavity or opening.

Olfactory bulb A mass of gray matter at the termination of an olfactory (I) nerve, lying inferior to the frontal lobe of the cerebrum on either side of the crista galli of the ethmoid bone.

Olfactory receptor cell A bipolar neuron with its cell body lying between supporting cells located in the mucous membrane lining the superior portion of each nasal cavity; converts odor into neural signals.

Olfactory tract A bundle of axons that extends from the olfactory bulb posteriorly to the olfactory portion of the cortex.

Oligodendrocyte (o-lig-ō-DEN-drō-sīt) A neuroglial cell that supports neurons and produces a phospholipid myelin sheath around axons of neurons of the central nervous system.

Oliguria (ol′-i-GYŪ-rē-a) Daily urinary output usually less than 250 ml.

Olive A prominent oval mass on each lateral surface of the superior part of the medulla oblongata.

Oncogene (ONG-kō-jēn) Gene that has the ability to transform a normal cell into a cancerous cell when it is inappropriately activated.

Oncology (ong-KOL-ō-jē) The study of tumors.

Oogenesis (ō′-ō-JEN-e-sis) Formation and development of the ovum.

Ophthalmologist (of′-thal-MOL-ō-jist) A physician who specializes in the diagnosis and treatment of eye disorders with drugs, surgery, and corrective lenses.

Ophthalmology (of′-thal-MOL-ō-jē) The study of the structure, function, and diseases of the eye.

Ophthalmoscopy (of′-thal-MOS-kō-pē) Examination of the interior fundus of the eyeball to detect retinal changes associated with hypertension, diabetes mellitus, atherosclerosis, and increased intracranial pressure.

Optic chiasm (kī-AZM) A crossing point of the optic (II) nerves, anterior to the pituitary gland.

Optic disc A small area of the retina containing openings through which the fibers of the ganglion neurons emerge as the optic (II) nerve. Also called the **blind spot.**

Optician (op-TISH-an) A technician who fits, adjusts, and dispenses corrective lenses on prescription of an ophthalmologist or optometrist.

Optic tract A bundle of axons that transmits nerve impulses from the retina of the eye between the optic chiasm and the thalamus.

Optometrist (op-TOM-e-trist) Specialist with a doctorate degree in optometry who is licensed to examine and test the eyes and treat visual defects by prescribing corrective lenses.

Ora serrata (Ō-ra ser-RĀ-ta) The irregular margin of the retina lying internal and slightly posterior to the junction of the choroid and ciliary body.

Orbit (OR-bit) The bony, pyramid-shaped cavity of the skull that holds the eyeball.

Organ A structure composed of two or more different kinds of tissues with a specific function and usually a recognizable shape.

Organelle (or-gan-EL) A permanent structure within a cell with characteristic morphology that is specialized to serve a specific function in cellular activities.

Orgasm (OR-gazm) Sensory and motor events involved in ejaculation for the male and involuntary contraction of the perineal muscles in the female at the climax of sexual arousal.

Orifice (OR-i-fis) Any aperture or opening.

Origin (OR-i-jin) The place of attachment of a muscle to the more stationary bone, or the end opposite the insertion.

Oropharynx (or′-ō-FAR-inks) The intermediate portion of the pharynx, lying posterior to the mouth and extending from the soft palate inferiorly to the hyoid bone.

Orthopedics (or′-thō-PĒ-diks) The branch of medicine that deals with the preservation and restoration of the skeletal system, articulations, and associated structures.

Osmosis (os-MŌ-sis) The net movement of water molecules through a selectively permeable membrane from an area of high water concentration to an area of lower water concentration until an equilibrium is reached.

Osseous (OS-ē-us) Bony.

Ossicle (OS-si-kul) Small bone, as in the middle ear (malleus, incus, stapes).

Ossification (os′-i-fi-KĀ-shun) Formation of bone. Also called **osteogenesis.**

Osteoblast (OS-tē-ō-blast′) Cell formed from an osteoprogenitor cell that participates in bone formation by secreting some organic components and inorganic salts.

Osteoclast (OS-tē-ō-clast′) A large multinuclear cell that destroys or resorbs bone tissue.

Osteocyte (OS-tē-ō-sīt′) A mature bone cell that maintains the daily activities of bone tissue.

Osteology (os′-tē-OL-ō-jē) The study of bones.

Osteon (OS-tē-on) The basic unit of structure in adult compact bone, consisting of a central (Haversian) canal with its concentrically arranged lamellae, lacunae, osteocytes, and canaliculi. Also called a **Haversian** (ha-VĒR-shun) **system.**

Osteoprogenitor (os′-tē-ō-prō-JEN-i-tor) **cell** Stem cell derived from mesenchyme that has mitotic potential and the ability to differentiate into an osteoblast.

Otolith (Ō-tō-lith) A particle of calcium carbonate embedded in the otolithic membrane that functions in maintaining static equilibrium.

Otolithic (ō-tō-LITH-ik) **membrane** Thick, gelatinous, glycoprotein layer located directly over hair cells of the macula in the saccule and utricle of the inner ear.

Otorhinolaryngology (ō′-tō-rī-nō-lar′-in-GOL-ō-jē) The branch of medicine that deals with the diagnosis and treatment of diseases of the ears, nose, and throat.

Oval window A small opening between the middle ear and inner ear into which the footplate of the stapes fits. Also called the **fenestra vestibuli** (fe-NES-tra ves-TIB-yoo-lē).

Ovarian (ō-VAR-ē-an) **cycle** A monthly series of events in the ovary associated with the maturation of an ovum.

Ovarian follicle (FOL-i-kul) A general name for oocytes (immature ova) in any stage of development, along with their surrounding epithelial cells.

Ovarian ligament (LIG-a-ment) A rounded cord of connective tissue that attaches the ovary to the uterus.

Ovary (Ō-var-ē) Female gonad that produces ova and the hormones estrogen, progesterone, and relaxin.

Ovulation (ō-vyū-LĀ-shun) The rupture of a vesicular ovarian (Graafian) follicle with discharge of a secondary oocyte into the pelvic cavity.

Ovum (Ō-vum) The female reproductive or germ cell; an egg cell.

Oxidation (ok-si-DĀ-shun) The removal of electrons and hydrogen ions (hydrogen atoms) from a molecule or, less commonly, the addition of oxygen to a molecule that results in a decrease in the energy content of the molecule. The oxidation of glucose in the body is also called **cellular respiration.**

Oxyhemoglobin (ok′-sē-HĒ-mō-glō-bin) **(HbO_2)** Hemoglobin combined with oxygen.

Oxyphil cell A cell found in the parathyroid gland that secretes parathyroid hormone (PTH).

Oxytocin (ok′-sē-TŌ-sin) **(OT)** A hormone secreted by neurosecretory cells in the paraventricular and supraoptic nuclei of the hypothalamus that stimulates contraction of the smooth muscle fibers (cells) in the pregnant uterus and contractile cells around the ducts of mammary glands.

P

Pacinian corpuscle *See* **Lamellated corpuscle.**

Palate (PAL-at) The horizontal structure separating the oral and the nasal cavities; the roof of the mouth.

Pancreas (PAN-krē-as) A soft, oblong organ lying along the greater curvature of the stomach and connected by a duct to the duodenum. It is both exocrine (secreting pancreatic juice) and endocrine (secreting insulin, glucagon, somatostatin, and pancreatic polypeptide).

Pancreatic (pan′-krē-AT-ik) **duct** A single, large tube that unites with the common bile duct from the liver and gallbladder and drains pancreatic juice into the duodenum at the hepatopancreatic ampulla (ampulla of Vater). Also called the **duct of Wirsung.**

Pancreatic islet A cluster of endocrine gland cells in the pancreas that secretes insulin, glucagon, somatostatin, and pancreatic polypeptide. Also called an **islet of Langerhans** (LANG-er-hanz).

Paranasal sinus (par′-a-NĀ-zal SĪ-nus) A mucus-lined air cavity in a skull bone that communicates with the nasal cavity. Paranasal sinuses are located in the frontal, maxillary, ethmoid, and sphenoid bones.

Paraplegia (par-a-PLĒ-jē-a) Paralysis of both lower limbs.

Parasagittal plane A vertical plane that does not pass through the midline and that divides the body or organs into *unequal* left and right portions.

Parasympathetic (par′-a-sim-pa-THET-ik) **division** One of the two subdivisions of the autonomic nervous system, having cell bodies of preganglionic neurons in nuclei in the brain stem and in the lateral gray matter of the sacral portion of the spinal cord; primarily concerned with activities that conserve and restore body energy. Also called the **craniosacral** (krā-nē-ō-SĀ-kral) **division.**

Parathyroid (par′-a-THĪ-royd) **gland** One of usually four small endocrine glands embedded on the posterior surfaces of the lateral lobes of the thyroid gland.

Parathyroid hormone (PTH) A hormone secreted by the parathyroid glands that decreases blood phosphate level and increases blood calcium level.

Paraurethral (par′-a-yū-RĒ-thral) **gland** Gland embedded in the wall of the urethra whose duct opens on either side of the urethral orifice and secretes mucus. Also called **Skene's** (SKĒNZ) **gland.**

Parenchyma (par-EN-ki-ma) The functional parts of any organ, as opposed to tissue that forms its stroma or framework.

Parietal (pa-RĪ-e-tal) **cell** The secreting cell of a gastric gland that produces hydrochloric acid and intrinsic factor. Also called an **oxyntic cell.**

Parietal pleura (PLŪ-ra) The outer layer of the serous pleural membrane that encloses and protects the lungs; the layer that is attached to the wall of the pleural cavity.

Parotid (pa-ROT-id) **gland** One of the paired salivary glands located inferior and anterior to the ears connected to the oral cavity via a duct (Stensen's) that opens into the inside of the cheek opposite the maxillary second molar tooth.

Pars intermedia A small avascular zone between the anterior and posterior pituitary glands.

Pathological anatomy The study of structural changes caused by disease.

Pectinate (PEK-ti-nāt) **muscles** Projecting muscle bundles of the anterior atrial walls and the lining of the auricles.

Pediatrician (pē′-dē-a-TRISH-un) A physician who specializes in the care and treatment of children and their illnesses.

Pedicel (PED-i-sel) Footlike structure, as on podocytes of a glomerulus.

Pelvic (PEL-vik) **cavity** Inferior portion of the abdominopelvic cavity that contains the urinary bladder, sigmoid colon, rectum, and internal female and male reproductive structures.

Pelvic splanchnic (SPLANGK-nik) **nerves** Preganglionic parasympathetic fibers from the levels of S2, S3, and S4 that supply the urinary bladder, reproductive organs, and the descending and sigmoid colon and rectum.

Pelvimetry (pel-VIM-e-trē) Measurement of the size of the inlet and outlet of the birth canal.

Pelvis The basinlike structure that is formed by the two hipbones, the sacrum, and the coccyx. The expanded, proximal portion of the ureter, lying within the kidney and into which the major calyces open.

Penis (PĒ-nis) The male copulatory organ, used to introduce sperm cells into the female vagina.

Pepsin Protein-digesting enzyme secreted by chief (zymogenic) cells of the stomach as the inactive form pepsinogen, which is converted to active pepsin by hydrochloric acid.

Perforating canal A minute passageway by means of which blood vessels and nerves from the periosteum penetrate into compact bone. Also called **Volkmann's** (FŌLK-manz) **canal.**

Pericardial (per′-i-KAR-dē-al) **cavity** Small potential space between the visceral and parietal layers of the serous pericardium that contains pericardial fluid.

Pericardium (per′-i-KAR-dē-um) A loose-fitting membrane that encloses the heart, consisting of a superficial fibrous layer and a deep serous layer.

Perichondrium (per′-i-KON-drē-um) The membrane that covers cartilage.

Perikaryon (per′-i-KAR-ē-on) The nerve cell body that contains the nucleus and other organelles. Also called a **soma.**

Perilymph (PER-i-lymf) The fluid contained between the bony and membranous labyrinths of the inner ear.

Perimetrium (per-i-MĒ-trē-um) The serosa of the uterus.

Perimysium (per′-i-MĪZ-ē-um) Invagination of the epimysium that divides muscles into bundles.

Perineum (per′-i-NĒ-um) The pelvic floor; the space between the anus and the scrotum in the male and between the anus and the vulva in the female.

Perineurium (per′-i-NYŪ-rē-um) Connective tissue wrapping around fascicles in a nerve.

Periodontal (per-ē-ō-DON-tal) **ligament** The periosteum lining the alveoli (sockets) for the teeth in the alveolar processes of the mandible and maxillae.

Periosteum (per′-ē-OS-tē-um) The membrane that covers bone and consists of connective tissue, osteoprogenitor cells, and osteoblasts and is essential for bone growth, repair, and nutrition.

Peripheral (pe-RIF-er-al) Located on the outer part or a surface of the body.

Peripheral nervous system (PNS) The part of the nervous system that lies outside the central nervous system—nerves and ganglia.

Peristalsis (per′-i-STAL-sis) Successive muscular contractions along the wall of a hollow muscular structure.

Peritoneum (per′-i-tō-NĒ-um) The largest serous membrane of the body that lines the abdominal cavity and covers the viscera within the cavity.

Peroxisome (pe-ROKS-ī-sōm) Organelle similar in structure to a lysosome that contains enzymes that use molecular oxygen to oxidize various organic compounds. Such reactions produce hydrogen peroxide; abundant in liver cells.

Perspiration Substance produced by sudoriferous (sweat) glands containing water, salts, urea, uric acid, amino acids, ammonia, sugar, lactic acid, and ascorbic acid; helps maintain body temperature and eliminate wastes.

Peyer's patches *See* **Aggregated lymphatic follicles.**

pH A symbol of the measure of the concentration of hydrogen ions (H^+) in a solution. The pH scale extends from 0 to 14, with a value of 7 expressing neutrality, values lower than 7 expressing increasing acidity, and values higher than 7 expressing increasing alkalinity.

Phagocytosis (fag′-ō-sī-TŌ-sis) The process by which cells (phagocytes) ingest particulate matter; especially the ingestion and destruction of microbes, cell debris, and other foreign matter.

Phalanx (FĀ-lanks) The bone of a finger or toe. *Plural,* **phalanges** (fa-LAN-jēz).

Pharmacology (far′-ma-KOL-ō-jē) The science that deals with the effects and uses of drugs in the treatment of disease.

Pharynx (FAR-inks) The throat; a tube that starts at the internal nares and runs partway down the neck where it opens into the esophagus posteriorly and the larynx anteriorly.

Phenotype (FĒ-nō-tīp) The observable expression of genotype; physical characteristics of an organism determined by genetic makeup and influenced by interaction between genes and internal and external environmental factors.

Photopigment A substance that can absorb light and undergo structural changes that can lead to the development of a receptor potential. An example is rhodopsin. Also called **visual pigment.**

Photoreceptor Receptor that detects light on the retina of the eye.

Physiology (fiz′-ē-OL-ō-jē) Science that deals with the functions of an organism or its parts.

Pia mater (PĪ-a MĀ-ter) The inner membrane (meninx) covering the brain and spinal cord.

Pineal (PĪN-ē-al) **gland** The cone-shaped gland located in the roof of the third ventricle. Also called the **epiphysis cerebri** (ē-PIF-i-sis se-RĒ-brē).

Pinealocyte (pin-ē-AL-ō-sīt) Secretory cell of the pineal gland that produces hormones.

Pinna (PIN-na) The projecting part of the external ear composed of elastic cartilage and covered by skin and shaped like the flared end of a trumpet. Also called the **auricle** (OR-i-kul).

Pinocytosis (pi′-nō-sī-TŌ-sis) the process by which cells ingest liquid.

Pituicyte (pi-TŪ-i-sīt) Supporting cell of the posterior lobe of the pituitary gland.

Pituitary (pi-TŪ-i-tar′-ē) **gland** A small endocrine gland lying in the sella turcica of the sphenoid bone and attached to the hypothalamus by the infundibulum; nicknamed the "master gland." Also called the **hypophysis** (hī-POF-i-sis).

Pivot joint A synovial joint in which a rounded, pointed, or conical surface of one bone articulates with a ring formed partly by another bone and partly by a ligament, as in the joint between the atlas and axis and between the proximal ends of the radius and ulna. Also called a **trochoid** (TRŌ-koid) **joint.**

Placenta (pla-SEN-ta) The special structure through which the exchange of materials between fetal and maternal circulations occurs. Also called the **afterbirth.**

Plantar flexion (PLAN-tar FLEX-shun) Bending the foot in the direction of the plantar surface (sole).

Plaque (plak) A cholesterol-containing mass in the tunica media of arteries. A mass of bacterial cells, dextran (polysaccharide), and other debris that adheres to teeth.

Plasma (PLAZ-ma) The extracellular fluid found in blood vessels; blood minus the formed elements.

Plasma cell Cell that produces antibodies and develops from a B cell (lymphocyte).

Plasma (cell) membrane Outer, limiting membrane that separates the cell's internal parts from extracellular fluid and the external environment.

Platelet (PLĀT-let) A fragment of cytoplasm enclosed in a cell membrane and lacking a nucleus; found in the circulating blood; plays a role in blood clotting.

Pleura (PLŪ-ra) The serous membrane that covers the lungs and lines the walls of the chest.

Pleural cavity Small potential space between the visceral and parietal pleurae.

Plexus (PLEX-sus) A network of nerves, veins, or lymphatic vessels.

Plexus of Auerbach *See* **Myenteric plexus.**

Plexus of Meissner *See* **Submucosal plexus.**

Pneumotaxic (nū-mō-TAK-sik) **area** Portion of the respiratory center in the pons that sends inhibitory nerve impulses to the inspiratory area that limit inspiration and facilitate expiration.

Podiatry (pō-DĪ-a-trē) The diagnosis and treatment of foot disorders.

Polar body The smaller cell resulting from the unequal division of cytoplasm during the meiotic divisions of an oocyte. The polar body has no function and is resorbed.

Polyuria (pol′-ē-YŪ-rē-a) An excessive production of urine.

Pons (ponz) The portin of the brain stem that forms a "bridge" between the medulla oblongata and the midbrain, anterior to the cerebellum.

Postcentral gyrus *See* **Primary somatosensory area.**

Posterior (pos-TĒR-ē-or) Nearer to or at the back of the body. Also called **dorsal.**

Posterior column tracts Sensory tracts (fasciculus gracilis and fasciculus cuneatus) that convey information up the spinal cord to the brain related to proprioception, discriminative touch, two-point discrimination, pressure, and vibrations.

Posterior pituitary gland Posterior portion of the pituitary gland. Also called the **neurohypophysis** (nū-rō-hī-POF-i-sis).

Posterior root The structure composed of sensory fibers lying between a spinal nerve and the dorsolateral aspect of the spinal cord. Also called the **dorsal (sensory) root.**

Posterior root ganglion (GANG-glē-on) A group of cell bodies of sensory (afferent) neurons and their supporting cells located along the posterior root of a spinal nerve. Also called a **dorsal (sensory) root ganglion.**

Postganglionic neuron (pōst′-gang-lē-ON-ik NŪ-ron) The second visceral motor neuron in an autonomic pathway, having its cell body and dendrites located in an autonomic ganglion and its unmyelinated axon ending at cardiac muscle, smooth muscle, or a gland.

Postsynaptic (pōst-sin-AP-tik) **neuron** The nerve cell that is activated by the release of a neurotransmitter substance from another neuron and carries nerve impulses away from the synapse.

Pouch of Douglas *See* **Rectouterine pouch.**

Precapillary sphincter (SFINGK-ter) A ring of smooth muscle fibers (cells) at the site of origin of true capillaries that regulate blood flow into true capillaries.

Precentral gyrus *See* **Primary motor area.**

Preganglionic (prē-gang-lē-ON-ik) **neuron** The first visceral motor neuron in an autonomic pathway, with its cell body and dendrites in the brain or spinal cord and its myelinated axon ending at an autonomic ganglion, where it synapses with a postganglionic neuron.

Prepuce (PRĒ-pyūs) The loose-fitting skin covering the glans of the penis and clitoris. Also called the **foreskin.**

Presynaptic (prē-sin-AP-tik) **neuron** A nerve cell that carries nerve impulses toward a synapse.

Prevertebral ganglion (prē-VERT-e-bral GANG-lē-on) A cluster of cell bodies of postganglionic sympathetic neurons anterior to the spinal column and close to large abdominal arteries. Also called a **collateral ganglion.**

Primary germ layer One of three layers of embryonic tissue, called ectoderm, mesoderm, and endoderm, that give rise to all tissues and organs of the organism.

Primary motor area A region of the cerebral cortex in the precentral gyrus of the frontal lobe of the cerebrum that controls specific muscles or groups of muscles.

Primary somatosensory area A region of the cerebral cortex posterior to the central sulcus in the postcentral gyrus of the parietal lobe of the cerebrum that localizes exactly the points of the body where somatic sensations originate.

Prime mover The muscle directly responsible for producing the desired motion. Also called an **agonist** (AG-ō-nist).

Primitive gut Embryonic structure composed of endoderm and mesoderm that gives rise to most of the gastrointestinal tract.

Primordial (prī-MŌR-dē-al) Existing first; especially primordial egg cells in the ovary.

Principal cell Cell found in the parathyroid glands that secretes parathyroid hormone (PTH). Also called a **chief cell.**

Proctology (prok-TOL-ō-jē) The branch of medicine that treats the rectum and its disorders.

Progesterone (prō-JES-te-rōn) A female sex hormone produced by the ovaries that helps prepare the endometrium for implantation of a fertilized ovum and the mammary glands for milk secretion.

Prolactin (prō-LAK-tin) **(PRL)** A hormone secreted by the anterior lobe of the pituitary gland that initiates and maintains milk secretion by the mammary glands.

Prolapse (PRŌ-laps) A dropping or falling down of an organ, especially the uterus or rectum.

Proliferation (pro-lif′-er-Ā-shun) Rapid and repeated reproduction of new parts, especially cells.

Pronation (prō-NĀ-shun) A movement of the forearm in which the palm of the hand is turned posteriorly or inferiorly.

Prophase (PRŌ-fāz) The first stage of mitosis during which chromatid pairs are formed and aggregate around the equatorial plane region of the cell.

Proprioceptor (prō′-prē-ō-SEP-tor) A receptor located in muscles, tendons, or joints that provides information about body position and movements.

Prostaglandin (pros′-ta-GLAN-din) **(PG)** A membrane-associated lipid composed of 20-carbon fatty acids with 5 carbon atoms joined to form a cyclopentane ring; synthesized in small quantities and basically mimics the activities of hormones.

Prostate (PROS-tāt) **gland** A doughnut-shaped gland inferior to the urinary bladder that surrounds the superior portion of the male urethra and secretes a slightly acid solution that contributes to sperm motility and viability.

Protein An organic compound consisting of carbon, hydrogen, oxygen, nitrogen, and sometimes sulfur and phosphorus, and made up of amino acids linked by peptide bonds.

Proto-oncogene (prō′-tō-ONG-kō-jēn) Gene responsible for some aspect of normal growth and development; it may transform into an oncogene, a gene capable of causing cancer.

Protraction (prō-TRAK-shun) The movement of the mandible or shoulder girdle forward on a plane parallel with the ground.

Proximal (PROK-si-mal) Nearer the attachment of a limb to the trunk or a structure; nearer to the point of origin.

Pseudopods (SŪ-dō-pods) Temporary, protruding projections of cytoplasm.

Pterygopalatine ganglion (ter′-i-gō-PAL-a-tīn GANG-glē-on) A cluster of cell bodies of parasympathetic postganglionic neurons ending at the lacrimal and nasal glands.

Pubic (PYŪ-bik) **symphysis** A slightly movable cartilaginous joint between the anterior surfaces of the hipbones.

Pudendum (pū-DEN-dum) A collective designation for the external genitalia of the female.

Pulmonary (PUL-mo-ner′-ē) Concerning or affected by the lungs.

Pulmonary circulation The flow of deoxygenated blood from the right ventricle to the lungs and the return of oxygenated blood from the lungs to the left atrium.

Pulmonary ventilation The inflow (inspiration) and outflow (expiration) of air occurring between the atmosphere and the lungs. Also called **breathing.**

Pulp cavity A cavity within the crown and neck of a tooth, filled with pulp, a connective tissue containing blood vessels, nerves, and lymphatic vessels.

Pulse (PULS) The rhythmic expansion and elastic recoil of an artery with each contraction of the left ventricle.

Pupil The hole in the center of the iris, the area through which light enters the posterior cavity of the eyeball.

Purkinje fiber *See* **Conduction myofiber.**

Pus The liquid product of inflammation containing leukocytes or their remains and debris of dead cells.

Pyloric (pī-LOR-ik) **sphincter** A thickened ring of smooth muscle through which the pylorus of the stomach communicates with the duodenum. Also called the **pyloric valve.**

Pyogenesis (pī′-ō-JEN-e-sis) Formation of pus.

Pyorrhea (pī-ō-RĒ-a) A discharge or flow of pus, especially in the alveoli (sockets) and the tissues of the gums.

Pyramid (PIR-a-mid) A pointed or cone-shaped structure; one of two roughly triangular structures on the ventral side of the medulla oblongata composed of the largest motor tracts that run from the cerebral cortex to the spinal cord; a triangular-shaped structure in the renal medulla composed of the straight segments of renal tubules.

Pyuria (pī-PYŪ-rē-a) The presence of leukocytes and other components of pus in urine.

Q

Quadrant (KWOD-rant) One of four parts.

Quadriplegia (kwod′-ri-PLĒ-jē-a) Paralysis of the two upper and two lower extremities.

R

Radiographic (rā′-dē-ō-GRAF-ik) **anatomy** Diagnostic branch of anatomy that includes the use of X rays.

Rami communicantes (RĀ-mē ko-myū-ni-KAN-tēz) Branches of a spinal nerve. *Singular,* **ramus communicans** (RĀ-mus ko-MYŪ-ni-kans).

Rathke's pouch *See* **Hypophyseal pouch.**

Receptor A specialized cell or a nerve cell terminal modified to respond to some specific sensory modality, such as touch, pressure, cold, light, or sound, and convert it to a nerve impulse by way of a generator or receptor potential. A specific molecule or arrangement of molecules organized to accept only molecules with a complementary shape.

Receptor-mediated endocytosis A highly selective process in which cells take up large molecules or particles. In the process, successive compartments called endocytic vesicles and endosomes form. The molecules or particles are eventually broken down by enzymes in lysosomes.

Recombinant DNA Synthetic DNA, formed by joining a fragment of DNA from one source to a portion of DNA from another.

Rectouterine pouch A pocket formed by the parietal peritoneum as it moves posteriorly from the surface of the uterus and is reflected onto the rectum; the most inferior point in the pelvic cavity. Also called the **pouch** or **cul de sac of Douglas.**

Rectum (REK-tum) The last 20 cm (7 in.) of the gastrointestinal tract, from the sigmoid colon to the anus.

Recumbent (re-KUM-bent) Lying down.

Red nucleus A cluster of cell bodies in the midbrain, occupying a large portion of the tectum and sending fibers into the rubroreticular and rubrospinal tracts.

Red pulp That portion of the spleen that consists of venous sinuses filled with blood and cords of splenic tissue called splenic (Billroth's) cords.

Reflex Fast response to a change (stimulus) in the internal or external environment that attempts to restore body activities within normal limits; passes over a reflex arc.

Reflex arc The most basic conduction pathway through the nervous system, connecting a receptor and an effector and consisting of a receptor, a sensory neuron, a center in the central nervous system for a synapse, a motor neuron, and an effector.

Regeneration (rē-jen′-er-Ā-shun) The natural renewal of a structure.

Regimen (REJ-i-men) A strictly regulated scheme of diet, exercise, or activity designed to achieve certain ends.

Regional anatomy The division of anatomy dealing with a specific region of the body, such as the head, neck, chest, or abdomen.

Relapse (RĒ-laps) The return of a disease weeks or months after its apparent cessation.

Relaxin A female hormone produced by the ovaries that relaxes the pubic symphysis and helps dilate the uterine cervix to facilitate delivery.

Remodeling Replacement of old bone by new bone tissue.

Renal corpuscle (KOR-pus′-l) A glomerular (Bowman's) capsule and its enclosed glomerulus.

Renal pelvis A cavity in the center of the kidney formed by the expanded, proximal portion of the ureter, lying within the kidney, and into which the major calyces open.

Renal pyramid A triangular structure in the renal medulla composed of the straight segments of renal tubules.

Reproduction (rē′-prō-DUK-shun) Either the formation of new cells for growth, repair, or replacement, or the production of a new individual.

Reproductive cell division Type of cell division in which sperm and egg cells are produced; consists of meiosis and cytokinesis.

Respiration (res-pi-RĀ-shun) Overall exchange of gases between the atmosphere, blood, and body cells consisting of pulmonary ventilation, external respiration, and internal respiration.

Respiratory center Neurons in the reticular formation of the brain stem that regulate the rate of respiration.

Rete (RĒ-tē) **testis** The network of ducts in the testes.

Reticular (re-TIK-yū-lar) **activating system (RAS)** An extensive network of branched nerve cells running through the core of the brain stem. When these cells are activated, a generalized alert or arousal behavior results.

Reticular formation A network of small groups of nerve cells scattered among bundles of fibers beginning in the medulla oblongata as a continuation of the spinal cord and extending upward through the central part of the brain stem.

Reticulocyte (re-TIK-yū-lō-sīt) An immature red blood cell.

Reticulum (re-TIK-yū-lum) A network.

Retina (RET-i-na) The deep coat of the eyeball, lying only in the posterior portion of the eye and consisting of nervous tissue and a pigmented layer composed of epithelial cells lying in contact with the choroid. Also called the **nervous tunic** (TOO-nik).

Retraction (rē-TRAK-shun) The movement of a protracted part of the body posteriorly on a plane parallel to the ground, as in pulling the lower jaw back in line with the upper jaw.

Retroflexion (re-trō-FLEK-shun) A malposition of the uterus in which it is tilted posteriorly.

Retrograde degeneration (RE-trō-grād dē-jen-er-Ā-shun) Changes that occur in the proximal portion of a damaged axon only as far as the first neurofibral node (node of Ranvier); similar to changes that occur during Wallerian degeneration.

Retroperitoneal (re′-trō-per-i-tō-NĒ-al) External to the peritoneal lining of the abdominal cavity.

Rh factor An inherited agglutinogen (antigen) on the surface of red blood cells.

Rhinology (rī-NOL-ō-jē) The study of the nose and its disorders.

Ribonucleic (rī′-bō-nyū-KLĒ-ik) **acid (RNA)** A single-stranded nucleic acid constructed of nucleotides consisting of one of four possible nitrogenous bases (adenine, cytosine, guanine, or uracil), ribose, and a phosphate group; three types are messenger RNA (mRNA), transfer RNA (tRNA), and ribosomal RNA (rRNA), each of which cooperates with DNA for protein synthesis.

Ribosome (RĪ-bō-sōm) An organelle in the cytoplasm of cells, composed of ribosomal RNA and ribosomal proteins, that synthesizes proteins; nicknamed the "protein factory."

Right lymphatic (lim-FAT-ik) **duct** A vessel of the lymphatic system that drains lymph from the upper right side of the body and empties it into the right subclavian vein.

Rigidity (ri-JID-i-tē) Hypertonia characterized by increased muscle tone, but reflexes are not affected.

Rima glottidis (RĪ-ma GLOT-ti-dis) Space between the vocal folds (true vocal cords).

Rod A visual receptor in the retina of the eye that is specialized for vision in dim light.

Root canal A narrow extension of the pulp cavity lying within the root of a tooth.

Root of penis Attached portion of penis that consists of the bulb and crura.

Rotation (rō-TĀ-shun) Moving a bone around its own axis, with no other movement.

Round ligament (LIG-a-ment) A band of fibrous connective tissue enclosed between the folds of the broad ligament of the uterus, emerging from a point on the uterus just inferior to the uterine (Fallopian) tube, extending laterally along the pelvic wall, and penetrating the abdominal wall through the deep inguinal ring to end in the labia majora.

Round window A small opening between the middle and inner ear, directly inferior to the oval window, covered by the secondary tympanic membrane. Also called the **fenestra cochlea** (fe-NES-tra KŌK-lē-a).

Rugae (RŪ-jē) Large folds in the mucosa of an empty hollow organ, such as the stomach and vagina.

S

Saccule (SAK-yūl) The inferior and smaller of the two chambers in the membranous labyrinth inside the vestibule of the inner ear containing a receptor organ for static equilibrium.

Sacral hiatus (hī-Ā-tus) Inferior entrance to the vertebral canal formed when the laminae of the fifth sacral vertebra (and sometimes fourth) fail to meet.

Sacral plexus (PLEK-sus) A network formed by the anterior branches of spinal nerves L4 through S3.

Sacral promontory (PROM-on-tor′-ē) The superior surface of the body of the first sacral vertebra that projects anteriorly into the pelvic cavity; a line from the sacral promontory to the superior border of the pubic symphysis divides the abdominal and pelvic cavities.

Saddle joint A synovial joint in which the articular surface of one bone is saddle-shaped and the articular surface of the other bone fits into the saddle as a sitting rider would, as in the joint between the trapezium and the metacarpal of the thumb. Also called a **sellaris** (sel-LA-ris) **joint.**

Sagittal (SAJ-i-tal) **plane** A vertical plane that divides the body or organs into left and right portions. Such a plane may be **midsagittal (median),** in which the divisions are equal, or **parasagittal,** in which the divisions are unequal.

Saliva (sa-LĪ-va) A clear, alkaline, somewhat viscous secretion produced by the three pairs of salivary glands; contains various salts, mucin, lysozyme, and salivary amylase.

Salivary amylase (SAL-i-ver-ē AM-i-lās) An enzyme in saliva that initiates the chemical breakdown of starch, mostly in the mouth.

Salivary gland One of three pairs of glands that lie external to the mouth and pour their secretory roduct (called saliva) into ducts that empty into the oral cavity; the parotid, submandibular, and sublingual glands.

Sarcolemma (sar′-kō-LEM-ma) The cell membrane of a muscle fiber (cell), especially of a skeletal muscle fiber.

Sarcoma (sar-KŌ-ma) A connective tissue tumor, often highly malignant.

Sarcomere (SAR-kō-mēr) A contractile unit in a striated muscle fiber (cell) extending from one Z disc to the next Z disc.

Sarcoplasm (SAR-kō-plazm) The cytoplasm of a muscle fiber (cell).

Sarcoplasmic reticulum (sar′-kō-PLAZ-mik re-TIK-yū-lum) A network of saccules and tubes surrounding myofibrils of a muscle fiber (cell), comparable to endoplasmic reticulum; functions to reabsorb calcium ions during relaxation and to release them to cause contraction.

Satiety center A collection of nerve cells located in the ventromedial nuclei of the hypothalamus that, when stimulated, brings about the cessation of eating.

Scala tympani (SKA-la TIM-pan-ē) The inferior spiral-shaped channel of the bony cochlea, filled with perilymph.

Scala vestibuli (ves-TIB-yū-lē) The superior spiral-shaped channel of the bony cochlea, filled with perilymph.

Schwann cell *See* **Neurolemmocyte.**

Sclera (SKLE-ra) The white coat of fibrous tissue that forms the superficial protective covering over the eyeball except in the most anterior portion; the posterior portion of the fibrous tunic.

Scleral venous sinus A circular venous sinus located at the junction of the sclera and the cornea through which aqueous humor drains from the anterior chamber of the eyeball into the blood. Also called the **canal of Schlemm** (SHLEM).

Sclerosis (skle-RŌ-sis) A hardening with loss of elasticity of tissues.

Scoliosis (skō′-lē-Ō-sis) An abnormal curvature from the normal vertical line of the backbone.

Scrotum (SKRŌ-tum) A skin-covered pouch that contains the testes and their accessory structures.

Sebaceous (se-BĀ-shus) **gland** An exocrine gland in the dermis of the skin, almost always associated with a hair follicle, that secretes sebum. Also called an **oil gland.**

Sebum (SĒ-bum) Secretion of sebaceous (oil) glands.

Secondary sex characteristic A feature characteristic of the male or female body that develops at puberty under the stimulation of sex hormones but is not directly involved in sexual reproduction, such as distribution of body hair, voice pitch, body shape, and muscle development.

Secretion (se-KRĒ-shun) Production and release from a gland cell of a fluid, especially a functionally useful product as opposed to a waste product.

Selective permeability (per′-mē-a-BIL-i-tē) The property of a membrane by which it permits the passage of certain substances but restricts the passage of others.

Sella turcica (SEL-a TUR-si-ka) A depression on the superior surface of the sphenoid bone that houses the pituitary gland.

Semen (SĒ-men) A fluid discharged at ejaculation by a male that consists of a mixture of sperm cells and the secretions of the seminal vesicles, prostate gland, and bulbourethral (Cowper's) glands.

Semicircular canals Three bony channels (anterior, posterior, lateral), filled with perilymph, in which lie the membranous semicircular canals filled with endolymph. They contain receptors for equilibrium.

Semicricular ducts The membranous semicircular canals filled with endolymph and floating in the perilymph of the bony semicircular canals. They contain cristae that are concerned with dynamic equilibrium.

Semilunar (sem′-ē-LŪ-nar) **valve** A valve guarding the entrance into the aorta or the pulmonary trunk from a ventricle of the heart.

Seminal vesicle (SEM-i-nal VES-i-kul) One of a pair of convoluted, pouchlike structures, lying posterior and inferior to the urinary bladder and anterior to the rectum, that secrete a component of semen into the ejaculatory ducts.

Seminiferous tubule (sem′-i-NI-fer-us TŪ-byūl) A tightly coiled duct, located in a lobule of the testis, where spermatozoa are produced.

Sensory area A region of the cerebral cortex concerned with the interpretation of sensory impulses.

Sensory neuron (NŪ-ron) A neuron that carries nerve impulses toward the central nervous system. Also called an **afferent neuron.**

Septal defect An opening in the septum (interatrial or interventricular) between the left and right sides of the heart.

Septum (SEP-tum) A wall dividing two cavities.

Serosa (ser-Ō-sa) Any serous membrane. The external layer of an organ formed by a serous membrane. The membrane that lines the pleural, pericardial, and peritoneal cavities.

Serous (SIR-us) **membrane** A membrane that lines a body cavity that does not open to the exterior. Also called the **serosa** (se-RŌ-sa).

Serum Plasma minus its clotting proteins.

Sesamoid (SES-a-moyd) **bones** Small bones usually found in tendons.

Sex chromosomes The twenty-third pair of chromosomes, designated X and Y, which determines the genetic sex of an individual; in males, the pair is XY; in females, XX.

Sheath of Schwann *See* **Neurolemma.**

Sigmoid colon (SIG-moyd KŌ-lon) The S-shaped portion of the large intestine that begins at the level of the left iliac crest, projects inward to the midline, and terminates at the rectum at about the level of the third sacral vertebra.

Sign Any objective evidence of disease that can be observed or measured such as a lesion, swelling, or fever.

Sinoatrial (si-nō-Ā-trēal) **SA) node** A compact mass of cardiac muscle fibers (cells) specialized for conduction, located in the right atrium inferior to the opening of the superior vena cava. Also called the **sinuatrial node** or **pacemaker.**

Sinus (SĪ-nus) A hollow in a bone (paranasal sinus) or other tissue; a channel for blood (vascular sinus): any cavity having a narrow opening.

Sinusoid (SĪN-yū-soyd) A microscopic space or passage for blood in certain organs such as the liver or spleen.

Skeletal muscle An organ specialized for contraction, composed of striated muscle fibers (cells), supported by connective tissue, attached to a bone by a tendon or an aponeurosis, and stimulated by somatic motor neurons.

Skene's gland *See* **Paraurethral gland.**

Skull The skeleton of the head consisting of the cranial and facial bones.

Sliding-filament mechanism The most commonly accepted explanation for muscle contraction in which actin and myosin myofilaments move into interdigitation with each other, decreasing the length of the sarcomeres.

Small intestine A long tube of the gastrointestinal tract that begins at the pyloric sphincter of the stomach, coils through the central and inferior part of the abdominal cavity, and ends at the large intestine; divided into three segments: duodenum, jejunum, and ileum.

Smooth muscle An organ specialized for contraction, composed of smooth muscle fibers (cells), located in the walls of hollow internal structures, and innervated by a visceral motor neuron.

Soft palate (PAL-at) The posterior portion of the roof of the mouth, extending posteriorly form the palatine bones and ending at the uvula. It is a muscular partition lined with mucous membrane.

Somatic cell division Type of cell division in which a single starting cell (parent cell) duplicates itself to produce two identical cells (daughter cells); consists of mitosis and cytokinesis.

Somatic (sō-MAT-ik) **nervous system (SNS)** The portion of the peripheral nervous system made up of the somatic motor fibers that run between the central nervous system and the skeletal muscles and skin.

Somite (SŌ-mīt) Block of mesodermal cells in a developing embryo that is distinguished into a myotome (which forms most of the skeletal muscles), dermatome (which forms connective tissues), and sclerotome (which forms the vertebrae).

Spermatic (sper-MAT-ik) **cord** A supporting structure of the male reproductive system, extending from a testis to the deep inguinal ring, that includes the ductus (vas) deferens, arteries. veins, lymphatic vessels, nerves, cremaster muscle, and connective tissue.

Spermatogenesis (sper′-ma-tō-JEN-e-sis) The formation and development of sperm cells in the seminiferous tubules of the testes.

Spermatozoon (sper,′-ma-tō-ZŌ-on) A mature sperm cell.

Sperm cell *See* **Spermatozoon.**

Spermiogenesis (sper′-Mē-ō-JEN-e-sis) The maturation of spermatids into sperm cells.

Spincter (SFINGK-ter) A circular muscle constricting an orifice.

Sphincter of Oddi *See* **Sphincter of the hepatopancreatic ampulla.**

Sphincter of the hepatopancreatic ampulla A circular muscle at the opening of the common bile and main pancreatic ducts in the duodenum. Also called the **sphincter of Oddi** (OD-ē).

Spinal (SPĪ-nal) **cord** A mass of nerve tissue located in the vertebral canal from which 31 pairs of spinal nerves originate.

Spinal nerve One of the 31 pairs of nerves that originate on the spinal cord from posterior and anterior roots.

Spinothalamic (spi-nō-THAL-a-mik) **tracts** Sensory tracts that convey information up the spinal cord to the brain related to pain, temperature, crude touch, and deep pressure.

Spinous (SPĪ-nus) **process** A sharp or thornlike process or projection, Also called a **spine.** A sharp ridge running diagonally across the posterior surface of the scapula.

Spiral organ The organ of hearing, consisting of supporting cells and hair cells that rest on the basilar membrane and extend into the endolymph of the cochlear duct. Also called the **organ of Corti** (KOR-tē).

Splanchnic (SPLANK-nik) Pertaining to the viscera.

Spleen (SPLĒN) Large mass of lymphatic tissue between the fundus of the stomach and the diaphragm that functions in phagocytosis, production of lymphocytes, and blood storage.

Spongy (cancellous) bone tissue Bone tissue that consists of an irregular lattice work of thin columns of bone called trabeculae; spaces between trabeculae of some bones are filled with red bone marrow; found inside short, flat, and irregular bones and in the epiphyses (ends) of long bones.

Sputum (SPŪ-tum) Substance ejected from the mouth containing saliva and mucus.

Squamous (SKWĀ-mus) Scalelike.

Stasis (STĀ-sis) Stagnation or halt of normal flow of fluids, as blood, urine, or of the intestinal fluid.

Static equilibrium (ē-kwi-LIB-rē-um) The maintenance of posture in response to changes in the orientation of the body, mainly the head, relative to the ground.

Stellate reticuloendothelial (STEL-āt re-tik′-yū-lō-en′-dō-THĒ-lē-al) **cell** Phagocytic cell that lines a sinusoid of the liver. Also called a **Kupffer's** (KŪP-ferz) **cell.**

Stereocilia (ste′-rē-ō-SIL-ē-a) Groups of extremely long, slender, nonmotile microvilli projecting from epithelial cells lining the epididymis.

Stomach The J-shaped enlargement of the gastrointestinal tract directly under the diaphragm in the epigastric, umbilical, and left hypochondriac regions of the abdomen, between the esophagus and small intestine.

Straight tubule (TŪ-byūl) A duct in a testis leading from a convoluted seminiferous tubule to the rete testis.

Stratum (STRĀ-tum) A layer.

Stratum basalis (STRĀ-tum ba-SAL-is) The superficial layer of the endometrium, next to the myometrium, that is maintained during menstruation and gestation and produces a new functionalis following menstruation or parturition.

Stratum functionalis (funk′-shun-AL-is) The deep layer of the endometrium, the layer next to the uterine cavity, that is shed during menstruation and that forms the maternal portion of the placenta during gestation.

Stretch receptor Receptor in the walls of bronchi, bronchioles, and lungs that sends impulses to the respiratory center that prevents overinflation of the lungs.

Stroma (STRŌ-ma) The tissue that forms the ground substance, foundation, or framework of an organ, as opposed to its functional parts.

Subarachnoid (sub′-a-RAK-noyd) **space** A space between the arachnoid and the pia mater that surrounds the brain and spinal cord through which cerebrospinal fluid circulates.

Subcutaneous (sub′-kyū-TĀ-nē-us) Beneath the skin. Also called **hypodermic** (hī-pō-DER-mik).

Subcutaneous layer A continuous sheet of areolar connective tissue and adipose tissue between the dermis of the skin and the deep fascia of the muscles. Also called the **superficial fascia** (FASH-ē-a) and hypodermis.

Subdural (sub-DŪ-ral) **space** A space between the dura mater and the arachnoid of the brain and spinal cord that contains a small amount of fluid.

Sublingual (sub-LING-gwal) **gland** One of a pair of salivary glands situated in the floor of the mouth deep to the mucous membrane and to the side of the lingual frenulum, with a duct (Rivinus's) that opens into the floor of the mouth.

Submandibular (sub′-man-DIB-yū-lar) **gland** One of a pair of salivary glands inferior to the base of the tongue under the mucous membrane in the posterior part of the floor of the mouth, posterior to the sublingual glands, with a duct (Wharton's) situated to the side of the lingual frenulum. Also called the **submaxillary** (sub′-MAK-si-ler-ē) **gland.**

Submucosa (sub-myū-KŌ-sa) A layer of connective tissue located deep to a mucous membrane, as in the gastrointestinal tract or the urinary bladder; the submucosa connects the mucosa to the muscularis layer.

Submucosal plexus A network of autonomic nerve fibers located in the superficial portion of the submucous layer of the small intestine. Also called the **plexus of Meissner** (MĪS-ner).

Subserous fascia (sub-SE-rus FASH-ē-a) A layer of connective tissue internal to the deep fascia, lying between the deep fascia and the serous membrane that lines the body cavities.

Subthalamus (sub-THAL-a-mus) Part of the diencephalon inferior to the thalamus; the substantia nigra and red nucleus extend from the midbrain into the subthalamus.

Sudoriferous (soo′-dor-IF-er-us) **gland** An eccrine or merocrine exocrine gland in the dermis or subcutaneous layer that produces perspiration. Also called a **sweat gland.**

Sulcus (SUL-kus) A groove or depression between parts, especially between the convolutions of the brain. *Plural,* **sulci** (SUL-sē).

Superficial (soo′-per-FISH-al) Located on or near the surface of the body.

Superficial fascia (FASH-ē-a) A continuous sheet of fibrous connective tissue between the dermis of the skin and the deep fascia of the muscles. Also called **subcutaneous** (sub′kyū-TĀ-nē-us) **layer.**

Superficial inguinal (IN-gwi-nal) **ring** A triangular opening in the aponeurosis of the external oblique muscle that represents the termination of the inguinal canal.

Superior (sū-PĒR-ē-or) Toward the head or upper part of a structure. Also called **cephalic** (SEF-a-lic) or **cranial.**

Superior vena cava (VĒ-na CĀ-va) **(SVC)** Large vein that collects blood from parts of the body superior to the heart and returns it to the right atrium.

Supination (sū-pī-NĀ-shun) A movement of the forearm in which the palm is turned anteriorly or superiorly.

Suppuration (sup′-yū-RĀ-shun) Pus formation and discharge.

Supraopticohypophyseal (sū′pra-op′tik-ō-hī-po-FIZ-ē-al) **tract** A bundle of nerve processes made up of fibers that have their cell bodies in the hypothalamus but release their neurosecretions in the posterior pituitary gland.

Surface anatomy The study of the structures that can be identified from the outside of the body by visualization and palpation.

Surfactant (sur-FAK-tant) A phospholipid substance produced by the lungs that decreases surface tension.

Susceptibility (sus-sep′-ti-BIL-i-tē) Lack of resistance of a body to the deleterious or other effects of an agent such as pathogenic microorganisms.

Suspensory ligament (sus-PEN-so-rē LIG-a-ment) A fold of peritoneum extending laterally from the surface of the ovary to the pelvic wall.

Sustentacular (sus′-ten-TAK-yū-lar) **cell** A supporting cell of seminiferous tubules that produces secretions for supplying nutrients to sperm cells and the hormone inhibin. Also called a **Sertoli** (ser-TŌ-lē) **cell.**

Sutural (SŪ-cher-al) **bone** A small bone located within a suture between certain cranial bones.

Suture (SŪ-cher) An immovable fibrous joint in the skull where bone surfaces are closely united.

Sympathetic (sim′-pa-THET-ik) **division** One of the two subdivisions of the autonomic nervous system, having cell bodies of preganglionic neurons in the lateral gray columns of the thoracic segment and first two or three lumbar segments of the spinal cord; it is primarily concerned with processes that involve the expenditure of energy. Also called the **thoracolumbar** (thō′-ra-kō-LUM-bar) **division.**

Sympathetic trunk ganglion (GANG-lē-on) A cluster of cell bodies of postganglionic sympathetic neurons lateral to the vertebral column, close to the body of a vertebra. These ganglia extend inferiorly through the neck, thorax, and abdomen to the coccyx on both sides of the vertebral column and are connected to one another to form a chain on each side of the vertebral column. Also called **lateral,** or **sympathetic, chain** or **vertebral chain ganglia.**

Symphysis (SIM-fi-sis) A line of union. A slightly movable cartilaginous joint such as the pubic symphysis between the anterior surface of the hipbones.

Symptom (SIMP-tum) A subjective change in body function not apparent to an observer, such as fever or nausea, that indicates the presence of a disease or disorder of the body.

Synapsis (SIN-aps) The functional junction between two neurons or between a neuron and an effector, such as a muscle or gland; may be electrical or chemical.

Synapsis (sin-AP-sis) The pairing of homologous chromosomes during prophase I of meiosis.

Synaptic (sin-AP-tik) **cleft** The narrow gap that separates the axon terminal of one nerve cell from another nerve cell or muscle fiber (cell) and across which a neurotransmitter diffuses to affect the postsynaptic cell.

Synaptic end bulb Expanded distal end of an axon terminal that contains synaptic vesicles. Also called a **synaptic knob** or **end foot.**

Synaptic gutter Invaginated portion of a sarcolemma under an axon terminal. Also called a **synpatic trough** (TROF).

Synaptic vesicle Membrane-enclosed sac in a synaptic end bulb that stores neurotransmitters.

Synarthrosis (sin′-ar-THRŌ-sis) An immovable joint.

Synchondrosis (sin′-kon-DRŌ-sis) A cartilaginous joint in which the connecting material is hyaline cartilage.

Syndesmosis (sin′-dez-MŌ-sis) A fibrous joint in which articulating bones are united by dense fibrous tissue.

Syndrome (SIN-drōm) A group of signs and symptoms that occur together in a pattern that is characteristic of a particular disease or abnormal condition.

Synergist (SIN-er-jist) A muscle that assists the prime mover by preventing unwanted movements in intermediate joints or otherwise aiding the movement of the prime mover.

Synostosis (sin′-os-TŌ-sis) A joint in which the dense fibrous connective tissue that unites bones at a suture has been replaced by bone, resulting in a complete fusion across the suture line.

Synovial (si-NŌ-vē-al) **cavity** The space between the articulating bones of a synovial (diarthrotic) joint, filled with synovial fluid. Also called a **joint cavity.**

Synovial fluid Secretion of synovial membranes that lubricates joints and nourishes articular cartilage.

Synovial joint A fully movable or diarthrotic joint in which a synovial (joint) cavity is present between the two articulating bones.

Synovial membrane The deeper of the two layers of the articular capsule of a synovial joint, composed of areolar connective tissue that secretes synovial fluid into the synovial (joint) cavity.

System An association or organs that have a common function.

Systemic anatomy The study of particular systems of the body, such as the skeletal, muscular, nervous, cardiovascular, or urinary systems.

Systemic circulation The routes through which oxygenated blood flows from the left ventricle through the aorta to all the organs of the body and deoxygenated blood returns to the right atrium.

Systole (SIS-tō-lē) In the cardiac cycle, the phase of contraction of the heart muscle, especially of the ventricles.

Systolic (sis-TO-lik) **blood pressure** The force exerted by blood on arterial walls during ventricular contraction; the highest pressure measured in the large arteries, about 120 mm Hg under normal conditions for a young, adult male.

T

Tachycardia (tak′-i-KAR-dē-a) A rapid heartbeat or pulse rate.

Tactile disc Modified epidermal cell in the stratum basale of hairless skin that functions as a cutaneous receptor for discrimination touch. Also called **Merkel's** (MER-kelz) **disc.**

Taenia coli (TĒ-nē-a KŌ-lī) One of three flat bands of thickened, longitudinal muscles running the length of the large intestine.

Target cell A cell whose activity is affected by a particular hormone.

Tarsal gland Sebaceous (oil) gland that opens on the edge of each eyelid. Also called a **Meibomian** (Mī-BŌ-mē-an) **gland.**

Tarsal plate A thin, elongated sheet of connective tissue, one in each eyelid, giving the eyelid form and support. The aponeurosis of the levator palpebrae superioris is attached to the tarsal plate of the superior eyelid.

Tarsus (TAR-sus) A collective term for the seven bones of the ankle.

T cell A lymphocyte that becomes mature in the thymus gland and differentiates into one of several kinds of cells that function in cell-mediated immunity.

Tectorial (tek-TŌ-rē-al) **membrane** A gelatinous membrane projecting over and in contact with the hair cells of the spiral organ (organ of Corti) in the cochlear duct.

Telophase (TEL-ō-fāz) The final stage of mitosis in which the daughter nuclei become established.

Temporomandibular joint (TMJ) syndrome A disorder of the temporomandibular joint (TMJ) characterized by dull pain around the ear, tenderness of jaw muscles, a clicking or popping noise when opening or closing the mouth, limited or abnormal opening of the mouth, headache, tooth sensitivity, and abnormal wearing of the teeth.

Tendon (TEN-don) A white fibrous cord of dense, regularly arranged connective tissue that attaches muscle to bone.

Tendon organ A proprioceptive receptor sensitive to changes in muscle tension and force of contraction, found chiefly near the junction of tendons and muscles. Also called a **Golgi** (GOL-jē) **tendon organ.**

Tentorium cerebelli (ten-TŌ-rē-um ser′-e-BEL-ē) A transverse shelf of dura mater that forms a partition between the occipital lobe of the cerebral hemispheres and the cerebellum and that covers the cerebellum.

Terminal ganglion (TER-mi-nal GANG-glē-on) A cluster of cell bodies of postganglionic parasympathetic neurons either lying very close to the visceral effectors or located within the walls of the visceral effectors supplied by the postganglionic fibers.

Testis (TES-tis) Male gonad that produces sperm and the hormones testosterone and inhibin. Also called a **testicle.**

Testosterone (tes-TOS-te-rōn) A male sex hormone (androgen) secreted by interstitial endocrinocytes (cells of Leydig) of a mature testis; controls the growth and development of male sex organs, secondary sex characteristics, sperm cells, and body growth.

Tetralogy of Fallot (tet-RAL-ō-jē of fal-Ō) A combination of four congenital heart defects: (1) constricted pulmonary semilunar valve, (2) interventricular septal opening, (3) emergence of the aorta from both ventricles instead of from the left only, and (4) enlarged right ventricle.

Thalamus (THAL-a-mus) A large, oval structure located superior to the midbrain, consisting of two masses of gray matter covered by a thin layer of white matter.

Thermoreceptor (THER-mō-rē-sep-tor) Receptor that detects changes in temperature.

Third ventricle (VEN-tri-kul) A slitlike cavity between the right and left halves of the thalamus and between the lateral ventricles.

Thirst center A cluster of neurons in the hypothalamus that is sensitive to the osmotic pressure of extracellular fluid and brings about the sensation of thirst.

Thoracic (thō-RAS-ik) **cavity** Superior component of the ventral body cavity that contains two pleural cavities, the mediastinum, and the pericardial cavity.

Thoracic duct A lymphatic vessel that begins as a dilation called the cisterna chyli, receives lymph from the left side of the head, neck, and chest, the left upper limb, and the entire body below the ribs, and empties into the left subclavian vein. Also called the **left lymphatic** (lim-FAT-ik) **duct.**

Thoracolumbar (thō′-ra-kō-LUM-bar) **outflow** The fibers of the sympathetic preganglionic neurons, which have their cell bodies in the lateral gray columns of the thoracic segment and first two or three lumbar segments of the spinal cord.

Thorax (THŌ-raks) The chest.

Thrombus A clot formed in an unbroken blood vessel, usually a vein.

Thymus (THĪ-mus) A bilobed organ, located in the superior mediastinum posterior to the sternum and between the lungs, that plays a role in the immune mechanism of the body.

Thyroid cartilage (THĪ-royd KAR-ti-lij) The largest single cartilage of the larynx, consisting of two fused plates that form the anterior wall of the larynx. Also called the **Adam's apple.**

Thyroid colloid (KOL-loyd) A complex in thyroid follicles consisting of thyroglobulin and stored thyroid hormones.

Thyroid follicle (FOL-i-kul) Spherical sac that forms the parenchyma of the thyroid gland and consists of follicular cells that produce thyroxine (T_4) and triiodothyronine (T_3) and parafollicular cells that produce calcitonin (CT).

Thyroid gland An endocrine gland with right and left lateral lobes on either side of the trachea connected by an isthmus located anterior to the trachea just inferior to the cricoid cartilage.

Thyroid-stimulating hormone (TSH) A hormone secreted by the anterior lobe of the pituitary gland that stimulates the synthesis and secretion of hormones produced by the thyroid gland.

Thyroxine (thī-ROK-sēn) **(T_4)** A hormone secreted by the thyroid gland that regulates organic metabolism, growth and development, and the activity of the nervous system.

Tissue A group of similar cells and their intercellular substance joined together to perform a specific system.

Tissue macrophage system General term that refers to wandering and fixed macrophages.

Tissue rejection Phenomenon by which the body recognizes the protein (HLA antigens) in transplanted tissues or organs as foreign and produces antibodies against them.

Tonsil (TON-sil) A multiple aggregation of large lymphatic nodules embedded in mucous membrane.

Topical (TOP-i-kal) Applied to the surface rather than ingested or injected.

Trabecula (tra-BEK-yū-la) Irregular latticework of thin columns of spongy bone. Fibrous cord of connective tissue serving as supporting fiber by forming a septum extending into an organ from its wall or capsule. *Plural,* **trabeculae** (tra-BEK-yū-lē).

Trabeculae carneae (tra-BEK-yū-lē KAR-nē-ē) Ridges and folds of the myocardium in the ventricles.

Trachea (TRĀ-kē-a) Tubular air passageway extending from the larynx to the fifth thoracic vertebra. Also called the **windpipe.**

Tract A bundle of nerve fibers in the central nervous system.

Transverse colon (trans-VERS KŌ-lon) The portion of the large intestine extending across the abdomen from the right colic (hepatic) flexure to the left colic (splenic) flexure.

Transverse fissure (FISH-er) The deep cleft that separates the cerebrum from the cerebellum.

Transverse plane A plane that runs parallel to the ground and divides the body or organs into superior and inferior portions. Also called a **horizontal plane.**

Transverse tubules (TŪ-byūls) **(T tubules)** Minute, cylindrical invaginations of the muscle fiber (cell) membrane that carry the muscle action potentials deep into the muscle fiber.

Trauma (TRAW-ma) An injury, either a physical wound or psychic disorder, caused by an external agent or force, such as a physical blow or emotional shock; the agent or force that causes the injury.

Triad (TRĪ-ad) A complex of three units in a muscle fiber (cell) composed of a transverse tubule and the segments of sarcoplasmic reticulum on both sides of it.

Tricuspid (trī-KUS-pid) **valve** Atrioventricular (AV) valve on the right side of the heart.

Trigone (TRĪ-gon) A triangular area at the base of the urinary bladder.

Triiodothyronine (trī-ī-ōd-ō-THĪ-rō-nēn) **(T_3)** A hormone produced by the thyroid gland that regulates organic metabolism, growth and development, and the activity of the nervous system.

Trophoblast (TRŌF-ō-blast) The superficial covering of cells of the blastocyst.

Tropic (TRŌ-pik) **hormone** A hormone whose target is another endocrine gland.

Trunk The part of the body to which the upper and lower limbs are attached.

Tunica albuginea (TŪ-ni-ka al′-byū-JIN-ē-a) A dense layer of white fibrous tissue covering a testis or deep to the surface of an ovary.

Tunica externa (eks-TER-na) The superficial coat of an artery or vein, composed mostly of elastic and collagen fibers. Also called the **adventitia.**

Tunica interna (in-TER-na) The deep coat of an artery or vein, consisting of a lining of endothelium, basement membrane, and internal elastic lamina. Also called the **tunica intima** (IN-ti-ma).

Tunica media (MĒ-dē-a) The middle coat of an artery or vein, composed of smooth muscle and elastic fibers.

Tympanic antrum (tim-PAN-ik AN-trum) An air space in the posterior wall of the middle ear that leads into the mastoid air cells or sinus.

Tympanic (tim-PAN-ik) **membrane** A thin, semitransparent partition of fibrous connective tissue between the external auditory meatus and the middle ear. Also called the **eardrum.**

Type II cutaneous mechanoreceptor A receptor embedded deeply in the dermis and deeper tissues that detects heavy and continuous touch sensations. Also called an **end organ of Ruffini.**

U

Umbilical (um-BIL-i-kal) **cord** The long, ropelike structure, containing the umbilical arteries and vein that connect the fetus to the placenta.

Umbilicus (um-BIL-i-kus or um-bil-Ī-kus) A small scar on the abdomen that marks the former attachment of the umbilical cord to the fetus. Also called the **navel.**

Upper limb (extremity) The appendage attached at the shoulder girdle, consisting of the arm, forearm, wrist, hand, and fingers.

Uremia (yū-RĒ-mē-a) Accumulation of toxic levels of urea and other nitrogenous waste products in the blood, usually resulting form severe kidney malfunction.

Ureter (YŪ-re-ter) One of two tubes that connect the kidney with the urinary bladder.

Urethra (yū-RĒ-thra) The duct from the urinary bladder to the exterior of the body that conveys urine in females and urine and semen in males.

Urinary (YŪ-ri-ner-ē) **bladder** A hollow, muscular organ situated in the pelvic cavity posterior to the pubic symphysis.

Urine The fluid produced by the kidneys that contains wastes or excess materials and is excreted from the body through the urethra.

Urogenital (yū′-rō-JEN-i-tal) **triangle** The region of the pelvic floor inferior to the public symphysis, bounded by the pubic symphysis and the ischial tuberosities and containing the external genitalia.

Urology (yū-ROL-ō-jē) The specialized branch of medicine that deals with the structure, function, and diseases of the male and female urinary systems and the male reproductive system.

Uterine (YŪ-ter-in) **tube** Duct that transports ova from the ovary to the uterus. Also called the **Fallopian** (fal-LŌ-pē-an) **tube** or **oviduct.**

Uterosacral ligament (yū′-ter-ō-SĀ-kral LIG-a-ment) A fibrous band of tissue extending from the cervix of the uterus laterally to attach to the sacrum.

Uterovesical (yū′-ter-ō-VES-i-kal) **pouch** A shallow pouch formed by the reflection of the peritoneum form the anterior surface of the urterus, at the junction of the cervix and the body, to the posterior surface of the urinary bladder.

Uterus (YŪ-te-rus) The hollow, muscular organ in females that is the site of menstruation, implantation, development of the fetus, and labor. Also called the **womb.**

Utricle (YŪ-tri-kul) The larger of the two divisions of the membranous labyrinth located inside the vestibule of the inner ear, containing a receptor organ for static equilibrium.

Uvea (YŪ-vē-a) The three structures that together make up the vascular tunic of the eye.

Uvula (YŪ-vyū-la) A soft, fleshy mass, especially the V-shaped pendant part, descending from the soft palate.

V

Vacuole (VAK-yū-ōl) Membrane-bound organelle that, in animal cells, frequently functions in temporary storage or transportation.

Vagina (va-JĪ-na) A muscular, tubular organ that leads form the uterus to the vestibule, situated between the urinary bladder and the rectum of the female.

Valvular stenosis (VAL-vyū-lar ste-NŌ-sis) A narrowing of a heart valve, usually the bicuspid (mitral) valve.

Variocele (VAR-i-kō-sēl) A twisted vein; especially, the accumulation of blood in the veins of the spermatic cord.

Varicose (VAR-i-kōs) pertaining to an unnatural swelling, as in the case of a varicose vein.

Vas A vessel or duct.

Vasa recta (VĀ-sa REK-ta) Extensions of the efferent arteriole of a juxtaglomerular nephron that run alongside the loop of the nephron (Henle) in the medullary region.

Vascular (VAS-kyū-lar) Pertaining to or containing many blood vessels.

Vascular spasm Contraction of the smooth muscle in the wall of a damaged blood vessel to prevent blood loss.

Vascular tunic (TŪ-nik) The middle layer of the eyeball, composed of the choroid, ciliary body, and iris. Also called the **uvea** (YŪ-vē-a).

Vascular (venous) sinus A vein with a thin endothelial wall that lacks a tunica media and externa and is supported by surrounding tissue.

Vasectomy (va-SEK-tō-mē) A means of sterilization of males in which a portion of each ductus (vas) deferens is removed.

Vasoconstriction (vāz-ō-kon-STRIK-shun) A decease in the size of the lumen of a blood vessel caused by contraction of the smooth muscle in the wall of the vessel.

Vasodilation (vās′-sō-DĪ-lā-shun) An increase in the size of the lumen of a blood vessel caused by relaxation of the smooth muscle in the wall of the vessels.

Vasomotor (vā-sō-MŌ-tor) **center** A cluster of neurons in the medulla oblongata that controls the diameter of blood vessels, especially arteries.

Vein A blood vessel that conveys blood from tissues back to the heart.

Vena cava (VĒ-na KĀ-va) One of two large veins that open into the right atrium, returning to the heart all of the deoxygenated blood from the systemic circulation except from the coronary circulation.

Venereal (ve-NĒ-rē-al) **disease (VD)** *See* **Sexually transmitted disease.**

Ventral (VEN-tral) Pertaining to the anterior or front side of the body; opposite of dorsal.

Ventral body cavity Cavity near the ventral aspect of the body that contains viscera and consists of a superior thoracic cavity and an inferior abdominopelvic cavity.

Ventral ramus (RĀ-mus) The anterior branch of a spinal nerve, containing sensory and motor fibers to the muscles and skin of the anterior surface of the head, neck, trunk, and the limbs.

Ventricle (VEN-tri-kul) A cavity in the brain or an inferior chamber of the heart.

Ventricular fibrillation (ven-TRIK-yū-lar fib-ri-LĀ-shun) Asynchronous ventricular contractions that result in cardiovascular failure.

Venule (VEN-yūl) A small vein that collects blood from capillaries and delivers it to a vein.

Vermiform appendix (VER-mi-form a-PEN-diks) A twisted, coiled tube attached to the cecum.

Vermilion (ver-MIL-yon) The area of the mouth where the skin on the outside meets the mucous membrane on the inside.

Vermis (VER-mis) The central constricted area of the cerebellum that separates the two cerebellar hemispheres.

Vertebral (VER-te-bral) **canal** A cavity within the vertebral column formed by the vertebral foramina of all kthe vertebrae and containing the spinal cord. Also called the **spinal canal.**

Veretebral column The 26 vertebrae; encloses and protects the spinal cord and serves as a point of attachment for the ribs and back muscles.

Vesicle (VES-i-kul) A small bladder or sac containing liquid.

Vesicular ovarian follicle A relatively large, fluid-filled follicle that will rupture and release a secondary oocyte, a process called ovulation. Also called a **Graafian** (GRAF-ē-an) **follicle.**

Vestibular (ves-TIB-yū-lar) **apparatus** Collective term for the organs for equilibrium, which includes the saccule, utricle, and semicircular ducts.

Vestibular (ves-TIB-yū-lar) **membrane** the membrane that separates the cochlear duct form the scala vestibuli.

Vestibule (VES-ti-byūl) A small space or cavity at the beginning of a canal, especially the inner ear, larynx, mouth, nose, and vagina.

Villus (VIL-lus) A projection of the intestinal mucosal cells containing connective tissue, blood vessels, and a lymphatic vessel; functions in the absorption of the end products of digestion. *Plural,* **villi** (VIL-ī).

Viscera (VIS-er-a) The organs inside the ventral body cavity. *Singular,* **viscus** (VIS-kus).

Visceral effector (e-FEK-tor) Cardiac muscle, smooth muscle, and glandular epithelium.

Visceral muscle An organ specialized for contraction, composed of smooth muscle fibers (cells), located in the walls of hollow internal structures, and stimulated by visceral motor neurons.

Visceral pleura (PLŪ-ra) The deep layer of the serous membrane that covers the lungs.

Visceroceptor (vis′-er-ō-SEP-tor) Receptor that provides information about the body's internal environment.

Vitreous (VIT-rē-us) **body** A soft, jellylike substance that fills the vitreous chambers of the eyeball, lying between the lens and the retina.

Vocal folds Pair of mucous membrane folds below the ventricular folds that function in voice production. Also called **true vocal cords.**

Volkmann's canal *See* **Perforating canal**

Vulva (VUL-va) Collective designation for the external genitalia of the female. Also called the **pudendum** (pū-DEN-dum).

W

Wallerian (wal-LE-rē-an) **degeneration** Degeneration of the portion of the axon and myelin sheath of a neuron distal to the site of injury.

Wandering macrophage (MAK-rō-fāj) Phagocytic cell that develops from a monocyte, leaves the blood, and migrates to infected tissues.

White matter Aggregations or bundles of myelinated axons located in the brain and spinal cord.

White pulp The portion of the spleen composed of lymphatic tissue, mostly lymphocytes, arranged around central arteries; in some areas of white pulp, lymphocytes are thickened into lymphatic nodules called splenic nodules (Malpighian corpuscles).

White ramus communicans (RĀ-mus ko-MYŪ-ni-kans) The portion of a preganglionic sympathetic nerve fiber that branches away from the anterior ramus of a spinal nerve to enter the nearest sympathetic trunk ganglion.

X

Xiphoid (ZĪ-foyd) Sword-shaped. The most inferior portion of the sternum.

Y

Yolk sac An extraembryonic membrane that connects with the midgut during early embryonic development, but is nonfunctional in humans.

Z

Zona fasciculata (ZŌ-na fa-sik′-yū-LA-ta) The middle zone of the adrenal cortex that consists of cells arranged in long, straight cords and that secretes glucocorticoid hormones.

Zona glomerulosa (glo-mer′-yū-LŌ-sa) The external zone of the adrenal cortex, directly under the connective tissue covering, that consists of cells arranged in arched loops or round balls and that secretes mineralocorticoid hormones.

Zona pellucida (pe-LŪ-si-da) Clear glycoprotein layer between a secondary oocyte and granulosa cells to the corona radiata that surrounds a secondary oocyte.

Zona reticularis (ret-ik′-yū-LAR-is) The internal zone of the adrenal cortex, consisting of cords of branching cells that secrete sex hormones, chiefly androgens.

Zygote (ZĪ-gōt) The single cell resulting from the union of a male and female gamete: the fertilized ovum.

CREDITS

PHOTO CREDITS

Chapter 1 Pages 1 (left) & 18: Courtesy Yale University Library. Page 1 (right): Courtesy Technicare Corporation. Page 3: John Wilson White. Page 9 (top): Stephen A. Kieffer and E. Robert Heitzman, *An Atlas of Cross-Sectional Anatomy.* Harper & Row, Publishers, New York, 1979. Page 9 (center): Lester V. Bergman/Project Masters, Inc. Page 9 (bottom): Martin Rotker. Pages 13 & 14: Courtesy Mark Nielsen. Page 19 (top left): Biophoto Associates/Photo Researchers. Page 19 (top right): Courtesy Bayer Corporation. Page 19 (bottom left): Simon Fraser/Photo Researchers. Page 19 (bottom right): Courtesy Andrew Joseph Tortora and Damaris Soler. Page 21 (left): Courtesy Technicare Corporation. Page 21 (right): Howard Sochurek/Medical Images, Inc.

Chapter 2 Pages 25 (top) & 48 (top): John Reader/Photo Researchers. Pages 25 (bottom) & 48 (bottom): Biophoto Associates/Photo Researchers. Page 31: Courtesy Abbott Laboratories. Page 34: Courtesy Kent McDonald. Page 36: D. W. Fawcett/Photo Researchers. Page 37: Biophoto Associates/Photo Researchers. Page 39: Courtesy Daniel S. Friend, Harvard Medical School. Page 40: D.W. Fawcett/Photo Researchers. Page 41: CNRI/Photo Researchers. Page 46: Courtesy Courtesy Michael H. Ross.

Chapter 3 Pages 55 & 80: Kazuyoshi Nomachi/Photo Researchers. Page 60: Biophoto Associates/Photo Researchers. Pages 61, 62, 63 (bottom), 64, 65, 66 (bottom), 72, 73, 74, 77, 78, 79 & 82: Courtesy Michael H. Ross. Page 63 (top): Biophoto Associates/Photo Researchers. Page 66 (top): Lester V. Bergman/The Bergman Collection. Page 75: Courtesy Andrew J. Kuntzman. Page 76 (top): Ed Reschke. Page 76 (bottom): Biophoto Associates/Photo Researchers. Page 83: R. Calentine/Visuals Unlimited. Page 84: ©Ed Reschke.

Chapter 4 Pages 89 & 95: Sylvain Grandadam/Photo Researchers. Page 92: ©L.V. Bergman/Bergman Collection. Page 97: Science Photo Library/Photo Researchers.

Chapter 5 Pages 109 & 127: National Library of Medicine/Photo Researchers. Page 111: Courtesy Mark Nielsen. Page 119 (top): The Bergman Collection. Page 119 (bottom): Biophoto Associates/Photo Researchers. Page 126: P. Motta, Dept. of Anatomy, University La Sapienza, Rome/Photo Researchers.

Chapter 6 Pages 131 & 156: From *Andreas Versalius of Brussels*, Dover Publications. Pages 137, 139, 143, 145, 159, 161 & 166: Courtesy Mark Nielsen. Page 150: Omikron/Science Photo Library/Photo Researchers.

Chapter 7 Pages 172 & 178: The Royal Collection ©HM Queen Elizabeth II. Pages 181 & 192: Courtesy Mark Nielsen.

Chapter 8 Pages 196 & 201: ©Werner Forman Archive, Philip Goldman Collection, London/Art Resource. Pages 200, 220 & 223: Courtesy Mark Nielsen. Pages 205, 206, 207, 208 & 209: John Wilson White.

Chapter 9 Pages 229 & 237: ©Art Resource. Page 235: Courtesy Denah Appelt and Clara Franzini-Armstrong. Page 239: Courtesy Fujita. Page 242: Biophoto Associates/Photo Researchers.

Chapter 10 Pages 253 & 262: Bernhard Siegfried Albinus, *Ecorché* 1747, Bibl. Nationale, Paris/Art Resource. Pages 267, 270, 287, 302, 305, 310, 311, 326, 334 & 335: Courtesy Mark Nielsen.

Chapter 11 Pages 344 & 354: Michelangelo Buonarroti, *Creation of Adam.* Sistine Chapel, Vatican Palace, Vatican State/Art Resource. Pages 346-349, 351, 353, 356- 359, 361-366: John Wiley & Sons. Page 350: Courtesy Lynne Marie Barghesi.

Chapter 12 Page 369 (bottom right) & 373: Lennart Nilsson, *Our Body Victorious*, Boehringer Ingelheim International GmbH. Page 369 & 377: Courtesy National Library of Medicine. Page 378: John D. Cunningham/Visuals Unlimited. Page 382: Lewin/Royal Free Hospital/Photo Researchers.

Chapter 13 Pages 387 & 391: Andy Warhol, *Human Heart*, 1979, The Andy Warhol Foundation, Inc./Art Resource. Page 389: John Wiley & Sons. Pages 393, 305, 397 & 400: Courtesy Mark Nielsen. Page 408 (left): ©Vu/Cabisco/Visuals Unlimited. Page 408 (right): W. Ober/Visuals Unlimited.

Chapter 14 Pages 414 & 422: Omikron/Photo Researchers. Page 416 (center): Courtesy Dennis Strete. Page 416 (bottom): Courtesy Michael H. Ross. Pages 419, 426, 432, 439, 443, 448, 452 & 455: Courtesy Mark Nielsen.

Chapter 15 Pages 473 & 479: Erich Lessing/Art Resource. Page 480: Courtesy Michael H. Ross. Page 481 (center): Leroy, Biocosmos/Photo Researchers. Page 481 (bottom left): Lester Bergman & Associates. Page 481 (bottom right): Courtesy Mark Nielsen. Page 483 (top): Courtesy Mark Nielsen. Page 483 (bottom): Courtesy Michael H. Ross.

Chapter 16 Pages 499 & 502: Michelangelo Merisi Caravaggio, *Boy Bitten by a Lizard*, Private Collection, London/Art Resource. Page 505: Ed Reschke. Page 507: Dennis Kunkel/Phototake.

Chapter 17 Pages 515 & 525: National Library of Medicine/Photo Researchers. Pages 517, 519, 521, 533 & 536: Courtesy Mark Nielsen. Page 526: From Richard Kessel and Randy Kardon, *Tissues and Organs: A Text-Atlas of Scanning Electron Microscopy.* Copyright ©1979 by W. H. Freeman and Company. Reprinted by permission.

Chapter 18 Page 544 (left) & 564: Courtesy Institute of Mathematics and Computer Science in Medicine, University of Hamburg, Germany. Page 544 (right) & 566: N. Gluhbegovic and T.H. Williams, *The Human Brain: A Photographic Guide*, Harper and Row, Publishers, Inc. Hagerstown, MD, 1980. Pages 546 & 563 (top): Courtesy Mark Nielsen. Page 555: Courtesy Mark Nielsen. Page 563 (bottom): N. Gluhbegovic and T.H. Williams, *The Human Brain: A Photographic Guide*, Harper and Row, Publishers, Inc. Hagerstown, MD, 1980.

Chapter 19 Pages 590 & 595: Dr. Jeremy Burgess/Photo Researchers.

Chapter 20 Pages 610 & 624: George Bernard/Photo Researchers. Pages 612, 614, 618 & 619: Courtesy Michael H. Ross. Page 622: N. Gluhbegovic and T.H. Williams, *The Human Brain: A Photographic Guide*, Harper and Row, Publishers, Inc. Hagerstown, MD, 1980. Page 629: John D. Cunningham/Visuals Unlimited.

Chapter 21 Pages 639 & 649: Courtesy National Library of Medicine.

Chapter 22 Pages 658 & 661: Giraudon/Art Resource. Pages 663, 666 (left), 668 & 672: Courtesy Michael H. Ross. Pages 666 (right), 669, 670 (left): Courtesy Mark Nielsen. Page 670 (right): Andrew J. Kuntzman. Page 673: Courtesy Jim Sheets, University of Alabama. Page 678 (top and center, top right and bottom left): Lester V. Bergman/Project Masters, Inc. Page 678 (top right): Martin Rotker/Phototake. Page 678 (bottom right): Biophoto Associates/Photo Researchers.

Chapter 23 Pages 684 & 691: Courtesy National Library of Medicine. Pages 686, 688, 693, 696, 697 & 699: Courtesy Mark Nielsen. Page 694: John Cunningham/Visuals Unlimited. Page 701: Biophoto Associates/Photo Researchers.

Chapter 24 Pages 714 & 732: Giraurdon/Art Resource. Page 716: From *A Stereoscopic Atlas of Human Anatomy* by David L. Bassett. Pages 720, 729, 734, 736, 741 (bottom) & 745: Courtesy Mark Nielsen. Pages 722, 727, 737, 742 (bottom right) & 747: Courtesy Michael H. Ross. Page 725: Courtesy D. Gentry Steele. Page 730: Hessler/Vu/Visuals Unlimited. Page 731: Ed Reschke. Page 741 (top): P. Motta and A. Familiari, Univeristy of La Sapinza, Rom/SPL/Photo Researchers. Page 742 (bottom left): Willis/Biological Photo Service. Page 742 (top): Fred E. Hossler/Visuals Unlimited. Page 746: P. Motta, University La Sapienza, Rome/SPL/Photo Researchers.

Chapter 25 Pages 757 (left) & 772: Courtesy Karl Heinz Hoehne, Inst. of Math. and Computer Science in Medicine (IMDM), University of Hamburg. Pages 757 (right), 759, 761 & 771: Courtesy Mark Nielsen. Page 766: Courtesy Dennis Strete. Page 769: © R.G. Kessel and R. H. Kardon, *Tissues and Organs: A Text-Atlas of Scanning Electron Microscopy*, W. Freeman. All rights reserved. Page 773: Biophoto Associates/Photo Researchers.

Chapter 26 Pages 781 & 800: The Royal Collection Picture Library © HM Queen Elizabeth II. Pages 785, 787, 797, 803 (top) & 809 (bottom): Courtesy Mark Nielsen. Page 788: Ed Reschke. Page 791 (left) and 809 (top): Courtesy Michael H. Ross. Page 791 (right): Michler/Photo Researchers. Page 799: Biophoto Associates/Photo Researchers. Page 803 (bottom): P. Motta, University La Sapienza, Rome/Photo Researchers. Page 804: Andrew J. Kuntzman.

Chapter 27 Pages 825 (center) & 831: Courtesy National Library of Medicine. Page 825 (bottom right) & 834: From Robert Rugh, Landrum B. Shettles, with Richard Einhorn: *Conception to Birth: The Drama of Life's Beginnings*. By permission of Harper & Row, Publishers, Inc. Page 827 (left): David Phillips/Photo Researchers. Page 827 (right): CABISCO/Phototake. Page 836: Siu, Biomedical Comm./ Custom Medical Stock Photo.

INDEX

B

C

P

T

COMBINING FORMS, WORD ROOTS, PREFIXES, AND SUFFIXES

Many medical terms are "compound" words; that is, they are made up of word roots with prefixes or suffixes. For example, *leukocyte* (white blood cell) is a combination of *leuko*, the combining form for the word root meaning "white," and *cyt*, the word root meaning "cell." Learning the medical meanings of the fundamental word parts will enable you to analyze many long, complicated terms.

The following list includes some of the most commonly used combining forms, word roots, prefixes, and suffixes used in medical terms, and an example for each.

COMBINING FORMS AND WORD ROOTS

Acous- hear Acoustics
Acr-, Acro- extremity Acromegaly
Aden- gland Adenoma
Angi- vessel Angiocardiography
Anthr-, Anthro- joint Arthritis
Aut-, Auto- self Autolysis

Bio- life, living Biopsy
Blast- germ, bud Blastocyte
Brachi- arm Brachial plexus
Bucc- cheek Buccal

Capit- head Decapitate
Carcin- cancer Carcinogen
Cardi-, Cardia-, Cardio- heart Cardiogram
Cephal- head Hydrocephalus
Cerebro- brain Cerobrospinal fluid
Chondr-, Chond- cartilage Chondrocyte
Cost- rib Intercostal
Crani- skull Cranial cavity

Derm- skin Dermatology
Dur- hard Dura mater

Enter-, Entero- intestine Gastroenterology
Erythr-, Erythro- red Erythrocyte

Gastr-, Gastro- stomach Gastarointestinal tract
Gloss-, Glosso- tongue Hypoglossal
Glyco- sugar Glycogen
Gyn- female, woman Gynecology

Hem-, Hemato- blood Hematoma
Hepar-, Hepat- liver Hepatic duct
Hist-, Histo- tissue Histology
Hyster- uterus Hysterectomy

Labi- lip Labia majora
Lip- fat Lipid
Lumb- lower back, loin Lumbar

Macul- spot, blotch Macula
Mamm- breast Mammary gland
Meningo- membrane Meninges
Morpho- form, shape Morphologly
Myelo- marrow, spinal cord Myeloblast
Myo- muscle Myoma

Nephr-, Nephro- kidney Nephrosis
Neuro- nerve Neuroeffector

Oculo- eye Ocular
Onco- mass, tumor Oncology
Oo- egg Oocyte
Osteo- bone Osteocyte
Oto- ear Otosclerosis

Phag-, Phago- to eat Phagocytosis
Phleb- vein Phlebitis
Pneumo- lung, air Pneumothorax
Pod- foot Podiatry
Procto- anus, rectum Proctology
Pyo- pus Pyogenesis

Ren- kidney Renal
Rhin- nose Rhinology

Scler-, Sclero- hard Atherosclerosis
Stasis- stand still Homeostasis
Sten-, Steno- narrow Stenosis

Therm- heat Thermoreceptor
Thromb- clot, lump Thrombus

Vas-, Vaso- vessel, duct Vasoconstriction
Viscer-, Viscero- organ Visceroreceptor

Zoo- animal Zoology

PREFIXES

A-, An- without, lack of, deficient Antresia; anesthesia
Ab- away from, from Abduction
Ad- to, near, toward Adduction
Andro- male, man Androgen
Ante- before Antepartum
Anti- against Anticoagulant

Basi- base, foundation Basal ganglia
Bi- two, double, twice Bifurcate

Cata- down, lower, under Catabolism
Circum- around Circumduction
Co-, Con- with, together Congenital
Contra- against, opposite Contralateral
Crypt- hidden, concealed Cryptorchidism

De- down, from Deciduous
Di-, Diplo- two Diploid
Dis- away from Disarticulate
Dys- painful, difficult Dysfunction

E-, Ec-, Ex- out from, out of Excretion
Ecto-, Exo- outer, outside Exocrine
Em-, En- in, on Empyema
End-, Endo- inside Endocardium
Epi- upon, on, above Epidermis
Eu- well Eupnea
Extra- outside, beyond, in addition to Extracellular

Gen- originate, produce, form Genetics

Hemi- half Hemiplegia
Heter-, Hetero- other, different Heterogeneous
Homeo-, Homo- unchanging, the same, steady Heomestasis
Hyper- beyond, extensive Hypersecretion
Hypo- under, below, deficient Hypoplasia

In-, Im- in, inside, not Incontinent
Inter- among, between Interphase
Intra- within, inside Interacellular

Later- side Lateral

Macro- large, great Macrophage
Mal- bad, abnormal Malignant
Medi-, Meso- middle Medial
Megalo- great, large Magakaryocyte
Meta- after, beyond Metacarpus
Micro- small Microfilament
Mono- one Monocyte

Neo- new Neonatal

Oglio- small, deficent Oliguria
Ortho- straight, normal Orthopedics

Para- near, beyond, apart from, beside Paranasal sinus
Peri- around Pericardium
Poly- much, many Polycythemia
Post- after, beyond Postnatal
Pre-, Pro- before, in front of Pronation
Pseud- Pseudo- false Pseudopods

Retro- backward, located behind Retroflexion

Semi- half Semicircular
Sub- under, beneath, below Submucosa
Super- above, beyond Superficial
Supra- above, over Supraopticohypophyseal
Sym-, Syn- with, together, jointed Symphysis; Syndrome

Tachy- rapid Tachycardia
Trans- across, through, beyond Transverse colon
Tri- three Trigone

SUFFIXES

-able capable of, having ability to Viable
-ac, -al pertaining to Cardiac cycle
-algia painful condition Myalgia
-asis, -asia, esis, -osis condition or state of Hemostasis

-ferent carry Efferent ducts

-gen agent that produces or originates Pathogen
-genic produced from, producing Carcinogenic

-ia state, condition Hypertonia
-ician person associated with Pediatrician
-itis inflammation Neuritis

-logy, -ology the study or science of Neurology
-lyso, -lysis solution, dissolve, loosening Hemolysis

-malacia softening Osteomalacia
-megaly enlarged Cardiomegaly

-oma tumor Hematoma

-penia deficiency Thrombocytopenia
-poiesis production Hematopoiesis

-tomy cutting into, incision into Appendectomy
-trophy state relating to nutrition or growth Atrophy

-uria urine Polyuria